General Chemistry

Table of Atomic Masses*

Name	Symbol	Atomic number	Atomic mass
actinium	Ac	89	[227.0277]
aluminum	Al	13	26.9815386(8)
americium	Am	95	[243.0614]
antimony	Sb	51	121.760(1)
argon	Ar	18	39.948(1)
arsenic	As	33	74.92160(2)
astatine	At	85	[209.9871]
barium	Ba	56	137.327(7)
berkelium	Bk	97	[247.0703]
beryllium	Be	4	9.012182(3)
bismuth	Bi	83	208.98040(1)
bohrium	Bh	107	[264.12]
boron	B	5	10.811(7)
bromine	Br	35	79.904(1)
cadmium	Cd	48	112.411(8)
calcium	Ca	20	40.078(4)
californium	Cf	98	[251.0796]
carbon	C	6	12.0107(8)
cerium	Ce	58	140.116(1)
cesium	Cs	55	132.9054519(2)
chlorine	Cl	17	35.453(2)
chromium	Cr	24	51.9961(6)
cobalt	Co	27	58.933195(5)
copper	Cu	29	63.546(3)
curium	Cm	96	[247.0704]
darmstadtium	Ds	110	[271]
dubnium	Db	105	[262.1141]
dysprosium	Dy	66	162.500(1)
einsteinium	Es	99	[252.0830]
erbium	Er	68	167.259(3)
europium	Eu	63	151.964(1)
fermium	Fm	100	[257.0951]
fluorine	F	9	18.9984032(5)
francium	Fr	87	[223.0197]
gadolinium	Gd	64	157.25(3)
gallium	Ga	31	69.723(1)
germanium	Ge	32	72.61(1)
gold	Au	79	196.966569(4)
hafnium	Hf	72	178.49(2)
hassium	Hs	108	[277]
helium	He	2	4.002602(2)
holmium	Ho	67	164.93032(2)
hydrogen	H	1	1.00794(7)
indium	In	49	114.818(3)
iodine	I	53	126.90447(3)
iridium	Ir	77	192.217(3)
iron	Fe	26	55.845(2)
krypton	Kr	36	83.798(2)
lanthanum	La	57	138.90547(7)
lawrencium	Lr	103	[262.1097]
lead	Pb	82	207.2(1)
lithium	Li	3	6.941(2)
lutetium	Lu	71	174.967(1)
magnesium	Mg	12	24.3050(6)
manganese	Mn	25	54.938045(5)
meitnerium	Mt	109	[268.1388]
mendelevium	Md	101	[258.0984]
mercury	Hg	80	200.59(2)
molybdenum	Mo	42	95.96(2)
neodymium	Nd	60	144.242(3)
neon	Ne	10	20.1797(6)
neptunium	Np	93	[237.0482]
nickel	Ni	28	58.6934(2)
niobium	Nb	41	92.90638(2)
nitrogen	N	7	14.0067(2)
nobelium	No	102	[259.1010]
osmium	Os	76	190.23(3)
oxygen	O	8	15.9994(3)
palladium	Pd	46	106.42(1)
phosphorus	P	15	30.973762(2)
platinum	Pt	78	195.084(9)
plutonium	Pu	94	[244.0642]
polonium	Po	84	[208.9824]
potassium	K	19	39.0983(1)
praseodymium	Pr	59	140.90765(2)
promethium	Pm	61	[144.9127]
protactinium	Pa	91	231.03588(2)
radium	Ra	88	[226.0254]
radon	Rn	86	[222.0176]
rhenium	Re	75	186.207(1)
rhodium	Rh	45	102.90550(2)
roentgenium	Rg	111	[272.1535]
rubidium	Rb	37	85.4678(3)
ruthenium	Ru	44	101.07(2)
rutherfordium	Rf	104	[261.1088]
samarium	Sm	62	150.36(2)
scandium	Sc	21	44.955912(6)
seaborgium	Sg	106	[266.1219]
selenium	Se	34	78.96(3)
silicon	Si	14	28.0855(3)
silver	Ag	47	107.8682(2)
sodium	Na	11	22.98976928(2)
strontium	Sr	38	87.62(1)
sulfur	S	16	32.065(5)
tantalum	Ta	73	180.94788(2)
technetium	Tc	43	[97.9072]
tellurium	Te	52	127.60(3)
terbium	Tb	65	158.92535(2)
thallium	Tl	81	204.3833(2)
thorium	Th	90	232.03806(2)
thulium	Tm	69	168.93421(2)
tin	Sn	50	118.710(7)
titanium	Ti	22	47.867(1)
tungsten	W	74	183.84(1)
ununbium	Uub	112	[285]
ununhexium	Uuh	116	[289]
ununquadium	Uuq	114	[289]
uranium	U	92	238.02891(3)
vanadium	V	23	50.9415(1)
xenon	Xe	54	131.293(6)
ytterbium	Yb	70	173.054(3)
yttrium	Y	39	88.90585(2)
zinc	Zn	30	65.38(4)
zirconium	Zr	40	91.224(2)

*Numbers in brackets are the atomic mass or mass number of the longest lived isotope. Numbers in parentheses indicate the uncertainty in the last significant digit. Data from the *CRC Handbook of Chemistry and Physics,* 87th ed., 2006–2007.

ABOUT THE COVER

Chinese is one of the few languages to have its own symbol for each element. The cover shows the calligraphic symbols of nine elements drawn in flowing-grass style. The calligrapher, Wang Zhaozheng, was born in 1928 and served as personal secretary to both Chiang Kai-shek and Deng Xiaoping. His works have been published in numerous collections and are on display around the world. He is currently an emeritus professor at Southwest Jiaotong University in China.

General Chemistry

FOURTH EDITION

Donald A. McQuarrie
University of California, Davis

Peter A. Rock
University of California, Davis

Ethan B. Gallogly
Santa Monica College

Illustrations by
George Kelvin and Laurel Muller

University Science Books
www.uscibooks.com

University Science Books
20 Edgehill Road
Mill Valley, CA 94941
www.uscibooks.com

Produced by Wilsted & Taylor Publishing Services
Project Manager: Jennifer Uhlich
Developmental Editor: John Murdzek
Copy Editor: Jennifer McClain
Editorial Assistance: Nancy Evans, Andrew Joron, Antonia Angress
Illustrations: George Kelvin and Laurel Muller
Book and Cover Design: Yvonne Tsang
Composition: Yvonne Tsang, Laurel Muller, Hassan Herz, Lucy Brank, Jody Hanson
Proofreading: Mervin Hanson
Printing and Binding: Transcontinental

This book is printed on acid-free paper.

Library of Congress Cataloging-in-Publication Data
McQuarrie, Donald A. (Donald Allan)
General chemistry / Donald A. McQuarrie, Peter A. Rock. — 4th ed. / Ethan B. Gallogly.
p. cm.
Includes index.
ISBN 978-1-891389-60-3 (alk. paper)
1. Chemistry—Textbooks. I. Rock, Peter A., 1939– II. Gallogly, Ethan B., 1965– III. Title.
QD31.3.M356 2010
540—dc22 2010004450

Printed in Canada
10 9 8 7 6 5 4 3 2

This book is dedicated to the memory of
Peter A. Rock, 1939–2006, and
Donald A. McQuarrie, 1937–2009.

CONTENTS IN BRIEF

CONTENTS

Interchapters (online at www.McQuarrieGeneralChemistry.com)

PREFACE

The last edition of McQuarrie and Rock came out in 1991. Over the years it has been gratifying to be told by so many people how much they regretted not seeing a fourth edition. It is a great pleasure to have the opportunity to present this new edition, especially with the new perspective of having been away from it for almost twenty years.

Unlike many subsequent textbook editions, we have made a number of significant changes. Perhaps the most significant of these is that we are using the "atoms first" approach, which has made such inroads into the general chemistry curriculum since the third edition. After an introductory chapter on "Chemistry and the Scientific Method," we go on to discuss elements, compounds, and chemical nomenclature along with a brief introduction to atoms, molecules, and the nuclear model of the atom. In Chapter 3 we emphasize the periodic properties of the elements by way of a few selected chemical reactions of the various groups of elements. Having introduced the periodic table, probably the most important topic in general chemistry, we then have six chapters where we use the quantum theory to present the underlying explanation of the periodic properties of the elements. The first of these six chapters, Chapter 4, discusses atomic spectra and the concept of the quantization of energy levels. Then in Chapter 5 we discuss multielectron atoms and show the connection between the electron configurations of multielectron atoms and chemical periodicity. Chapter 6 discusses ionic bonds, the simplest type of bonding. After a rather thorough discussion of Lewis formulas in Chapter 7, we go on to use Lewis formulas to predict molecular geometries using VSEPR theory in Chapter 8. This introduces the students to a great variety of molecules and compounds and gives more practice writing Lewis formulas. In Chapter 9, the last of the six consecutive chapters on quantum theory and atomic and molecular structure, we present a fairly detailed introduction to covalent bonding, using simple molecular orbital theory for diatomic molecules and hybrid orbitals to describe the bonding in polyatomic molecules.

Finally, in Chapter 10, we embark on a fairly conventional sequence of chapters on chemical reactivity, chemical calculations, the properties of gases, thermochemistry, liquids and solids, solutions, chemical kinetics, chemical equilibria, acids and bases, thermodynamics, oxidation-reduction reactions, electrochemistry, and transition metals.

We have noticed that many general chemistry texts do not make a clear distinction between a chemical reaction, which is an actual physical process that takes place in the laboratory, and the chemical equation that we use to express this reaction. How we choose to write a chemical equation to describe a chemical reaction is somewhat arbitrary in the sense that the balancing (stoichiometric) coefficients are arbitrary. We can write an equation with one set of balancing coefficients or any multiple of them. Thus we can express the reaction of hydrogen and oxygen as

$$2\,H_2(g) + O_2(g) \rightarrow 2\,H_2O(l)$$

or as

$$H_2(g) + \frac{1}{2}O_2(g) \rightarrow H_2O(l)$$

if we want to emphasize the combustion of one mole of hydrogen. Thus, for example, the value of the enthalpy of combustion of the reaction is –237.1 kilojoules per mole in the first case and –118.5 kilojoules per mole in the second case, where mole refers to a mole of the reaction as described by the equation as written. This also strongly points out that the balancing coefficients are *relative* quantities and consequently are unitless. Both of these statements are in accord with just about every physical chemistry book and should be adhered to.

Another important feature, one that we used in previous editions but was not always appreciated, is that equilibrium constants as we define them in the introductory chapters on equilibria have units. There is no way to get around this. They are looking right at you when you define an equilibrium constant in terms of concentrations, K_c, or pressures, K_p. You can appeal to some sort of standard state of unit concentration or unit pressure to make the units mysteriously disappear, but certainly such an arbitrary convention is not justified at this point. Furthermore, when doing equilibrium calculations, the resulting equilibrium concentrations should come out in terms of concentration or pressure, which they don't if K_c or K_p are taken to be unitless. The reason for suppressing the units in equilibrium constants is the anticipation of using the thermodynamic equation

$$\Delta G^\circ_{rxn} = -RT \ln K$$

Clearly K cannot have units in order to take its logarithm. It is important to realize in this case that K is *not* the same as K_c or K_p. It is the *thermodynamic equilibrium constant* that is defined by

$$K = K_c/Q^\circ_c \qquad \text{or} \qquad K = K_p/Q^\circ_p$$

where Q°_c is the standard reaction quotient which has a numerical value of unity with units of molarity and Q°_p is the similar quantity for pressure. Now, and only now, is K unitless. A formal introduction of the concept of the thermodynamic equilibrium constant is not just another way of saying the same thing as in the earlier chapters, but is an entirely new equilibrium constant. All this is in accord with the 1982 recommendation of the International Union of Pure and Applied Chemistry (IUPAC).

We have usually adhered to the IUPAC recommendations, but could not bring ourselves to do it in the case of pressure units. IUPAC recommends the use of the SI units of bars and Pascals, but atmospheres are so ingrained in the chemistry curriculum that it is difficult to not use them. Consequently we use both bars and atmospheres throughout the text and require the students to be bilingual in this regard. Along the same lines, we have eschewed the use of the term STP, which is woefully ambiguous. The IUPAC definition of STP is the conditions at one bar and 0°C, whereas the older, fully ingrained definition still permeating chemistry texts is the conditions at one atmosphere and 0°C. An informal survey of many high school chemistry teachers shows that the venerable fact that one mole of an ideal gas occupies 22.414 liters at one atmosphere and

0°C is still in great use, whereas under the IUPAC recommendations one mole of an ideal gas occupies 22.711 liters at one bar and 0°C.

One final innovation is the use of what we call Interchapters for the introduction of descriptive chemistry. These Interchapters are available at www.McQuarrieGeneralChemistry.com. Every general chemistry author knows all too well that how to present descriptive chemistry is a nagging problem, as numerous articles over the years in the *Journal of Chemical Education* attest. In the third edition, for example, we included two full chapters on "The Chemistry of the Main Group Elements." Unfortunately, many instructors simply do not have the time, or perhaps even the inclination, to cover these chapters because these chapters typically come toward the end of the text. We have elected to present descriptive chemistry in a number of short online segments (about ten pages) that can be covered readily or assigned as reading; references to relevant interchapters are given throughout the book. For example, some of the interchapters are called "Hydrogen and Oxygen," "The Alkali Metals," "Nitrogen," "Saturated Hydrocarbons," "Unsaturated Hydrocarbons," "Aromatic Hydrocarbons," "The Main-Group Metals," and so on. It seems particularly worthwhile that the students be introduced to an *elementary* discussion of organic chemistry at an early stage so that organic molecules can be used as examples. Although we have avoided references to the plethora of websites out there because of their volatility, we do strongly recommend the *Journal of Chemical Education* website called Periodic Table Live!, which you can link to at www.McQuarrieGeneralChemistry.com. When you click on an element in the periodic table in this website, you get a list of its chemical and physical properties and even photos and videos of a number of its reactions. Students should be encouraged to refer to this website frequently.

Donald A. McQuarrie

PREFACE

When Don McQuarrie, Peter Rock, and I agreed to collaborate on this fourth edition of their classic chemistry text, we shared a common vision of what the new text should offer.

We decided to begin with atomic theory and then to discuss chemical bonding and molecules before introducing reaction classes and other chemical properties. Because chemical reaction classes and chemical properties flow naturally from chemical bonding and structure, we believed this sequence would lead students to a more comprehensive understanding of these complex topics. For example, we use Lewis formulas to show why acetic acid is acidic, sodium hydroxide is basic, and methanol is neutral, despite all three having what appear to be hydroxyl (OH) groups. Such a presentation is nearly impossible if reactions are taught before structure, but follows naturally from an "atoms first" approach.

Another important change from earlier editions was in reformatting the chapter frontispieces in the form of profiles of prominent scientists in the style of Don's other books. We hope these brief biographies of great pioneers in the sciences will serve as role models to inspire students considering careers in the field, and that these will prove interesting and valuable to instructors and students alike.

We also wanted to integrate many concepts from organic, polymer, biological, and descriptive chemistry as supplements to the main chapters. The second edition had inserted this material as short "interchapters" interspersed throughout the text. In this edition we bring back, and expand upon, the interchapters but make them available via the Internet and cross-reference them in the printed version. This enables the instructor to pick and choose among them and permits us to include additional interchapters as the need arises. We also hope that others will submit short interchapters for general chemistry that we can collect on our website for public distribution. Moreover, making these supplements available electronically, rather than incorporating them in the text, reduces the physical size and the cost of the text to students.

In developing this edition we worked closely with Sapling Learning to provide an optional electronic homework system to accompany the text. This provides students with instantaneous feedback on assignments so that students can improve their understanding of the chemical principles of each chapter. This system will improve the use of class time and includes short practice exercises to aid in the mastery of concepts.

Additionally, there are a number of other innovations in this edition, such as the use of IUPAC conventions throughout, careful attention to significant figures in all problems and illustrative examples, and utilization of the *CRC Handbook of Chemistry and Physics* as the source of most data.

For readers familiar with earlier editions, the current version divides both the chapter on quantum theory and that on kinetics into two chapters each to allow for more examples and applications. For instance, we now include a section on enzyme kinetics. We also changed how we present nuclear chemistry in this edition. Rather than including a chapter that focuses mostly on nuclear physics, we chose instead to add a new interchapter on the chemical applications of radioisotopes, while integrating the material on first-order nuclear decay and radio-dating into the chapters on kinetics. Finally, we added several new interchapters on current topics of interest, such as "The World Supply of Energy."

Little could I have imagined when I began work on this text that both of the original authors would pass away during the course of this project. We lost Peter shortly after commencing this revision. Fortunately, Don survived to see the book's manuscript through to completion. I am most grateful to Carole McQuarrie, Don's wife and a chemist in her own right, whose help was as invaluable as Don's, and who carried on heroically collaborating on the work after his passing.

Working with Don on this project has expanded my professional growth and knowledge more than any other endeavor I have previously undertaken. I have no way to repay the wisdom and experience that he so generously shared with me during the course of our collaboration. I only hope that this book, the fruit of our efforts, will be of as much benefit to the students who use it as it has been to me in helping to create it.

This book could not have been completed without the assistance of a great many people and institutions. First and foremost, I wish to thank Don McQuarrie and Peter Rock, whose brilliance as scientists and chemical educators live on within these pages.

I would also like to thank our publisher, Bruce Armbruster of University Science Books, who made this fourth edition possible; my department and colleagues at Santa Monica College, who provided me with time to work on this as well as valued pedagogical advice; Mervin Hanson, for his herculean effort in reworking all the problem solutions and for his invaluable suggestions with the text; Nate Lewis, for his contributions to the Energy interchapter; Miriam Bennet, Lisa Dysleski, Harry B. Gray, Hal Harris, Mark L. Kearley, Joseph Kushick, Robert Lamoreaux, Jacob Morris, and Alan Van Orden, who all helped contribute to this work; George Kelvin and Laurel Muller, for their excellent artwork throughout; Wang Zhaozheng, for the cover art; Jane Ellis, for her hours of help in securing photos and rights and her efforts in marketing; Jennifer Uhlich and the staff of Wilsted & Taylor Publishing Services for their excellent layout. I also wish to thank Kate Liba, who first taught me the art of writing, and William M. Jackson, my Ph.D. advisor at the University of California, Davis, from whom I learned the art of research. I also thank my family for their patience and forbearance during the years in which I labored on this text.

This edition is dedicated to the memory of Don McQuarrie and Peter Rock.

Ethan B. Gallogly

NOTE TO THE INSTRUCTOR

The coverage of topics in this book has been arranged to facilitate a smooth transition to the atoms-first teaching approach. We were very conscious of the difficulty some institutions face in altering published course descriptions. Therefore, while arranging the chapters in this text, we intentionally kept the major topics presented in the first and second halves of the book unaltered.

Some concerns naturally arise over the coordination of lectures and laboratories as a result of the fact that the mole concept and chemical calculations are not introduced until Chapter 11. In fact, a review of most first-semester chemistry laboratory experiments reveals that roughly half of these do not require stoichiometric calculations. In most cases a simple rearrangement of existing laboratory experiments is all that is required to coordinate with an atoms-first approach.

We have provided a sample course syllabus and laboratory schedule on our website at www.mcquarriegeneralchemistry.com/sample_syllabus.pdf that shows a hypothetical pairing of laboratories and lectures for our text. Here we present experiments on measurements, separations, periodic properties, spectroscopy, and chemical structure prior to experiments requiring stoichiometric calculations. In this way student experiments develop from fundamental principles to chemical calculations in a manner that mirrors the text. The only experiment where moles and molarity might be deemed essential is the spectroscopic determination of concentration. However, even this can be performed by presenting concentration in units of milligrams per milliliter rather than molarity.

For instructors whose laboratory program requires the early introduction of moles and molarity, the material in Sections 11-1 and 11-2 on the mole concept and Sections 12-1 and 12-2 on solutions and molarity may easily be introduced at the end of Chapter 1 as a continuation of the discussion of chemical measurement. This minor rearrangement may be accomplished without loss of continuity in order to support the use of these units in the laboratory.

Thus, although the atoms-first approach may require some minor adjustments to the course syllabus and laboratory schedule, often a simple rearrangement of existing experiments will prove sufficient to enjoy the pedagogical advantages of what we believe to be a more logical approach to teaching chemistry.

For those who wish to go beyond the printed text, this book also features a complimentary selection of online Interchapters. These explore supplemental topics such as Sources of Energy, Polymers, Nuclear Chemistry, Descriptive Chemistry, Biochemistry, and Organic Chemistry. Because these are online and easily accessible, they are flexible and easy to assign, while keeping the printed book slim and affordable. Moreover, they contain some of the most beautiful color images in the text! Not to be missed, the Interchapters are online at: www.McQuarrieGeneralChemistry.com.

We welcome your comments, questions, and feedback regarding this new edition of *General Chemistry*!

Ethan B. Gallogly
gallogly@uscibooks.com

General Chemistry

Antoine-Laurent Lavoisier (1743–1794) was born in Paris to a wealthy family that later entered the nobility. He obtained a Bachelor of Law degree from the University of Paris, but despite his father's wishes for him to pursue a career in law, he chose to follow one in science. He was elected a member of the French Academy of Sciences at the early age of 27. In 1771, he married **Marie-Anne Pierrette Paulze** (1758–1836), the daughter of a parliamentary lawyer and financier. Soon after their marriage, she learned chemistry on her own in order to assist her husband in his work and became more than just an assistant. She was fluent in several languages and was able to translate chemical publications in other languages into French, so that they were current in the research being carried out in other countries. She also studied painting with the well-known artist, Jacques Louis David, who painted the portrait of the couple shown above. Her exacting attention to detail in her illustrations of the equipment used, especially in his famous book, *Elementary Treatise on Chemistry*, allowed their results to be reproduced and verified. Using the most sensitive balances available, Lavoisier showed that the masses of the reactants and the products of a chemical reaction are the same, and thereby discovered the law of conservation of mass, which was to place chemistry on a firm quantitative basis. He also was the first to show that combustion is a reaction with oxygen, and later confirmed that water was not an element but rather was composed of hydrogen and oxygen. Because of his financial connection with a much hated tax-collecting firm, Lavoisier was denounced, arrested, and guillotined in 1794 by supporters of the French Revolution. Jean-Paul Marat, a key figure in the Revolution, had developed a hatred of him because Lavoisier had shown Marat to be a poor chemist and denied his admission to the French Academy of Sciences. In spite of being impoverished after his death, Lavoisier's wife saw that all of their manuscripts were published and distributed. Lavoisier is generally considered to be the father of modern chemistry, but more accurately the Lavoisiers together should be regarded as the parents of modern chemistry because of Marie-Anne Lavoisier's invaluable collaboration in their achievements.

1. Chemistry and the Scientific Method

You and about one million other students around the world are about to begin your first college course in chemistry. Most of you do not plan to become professional chemists; probably only about one in a hundred of you will graduate with a bachelor's degree in chemistry. Whatever your chosen field of study, however, there is a good chance that you will need a knowledge of elementary chemistry.

Chemists use the scientific method to describe the immense variety of the world's substances, from a grain of sand to the components of the human body. As you will see, they can do this because chemistry is a quantitative science, based on experimental measurements and scientific calculations. You must therefore begin with a clear understanding of the methods scientists use to measure and calculate physical quantities. This chapter gives you these foundations.

1-1. Why Should You Study Chemistry?

Chemistry is the study of the properties of substances and how they react with one another. Chemical substances and chemical reactions pervade all aspects of the world around us. The new substances formed in reactions have properties different from those of the substances that reacted with one another, properties that chemists can predict and put to use. Hundreds of materials that we use every day, directly and indirectly, are products of chemical research (Figure 1.1).

The examples of useful products of chemical reactions are limitless. The development of fertilizers, one of the major focuses of the chemical industry, has profoundly affected agricultural production. Equally important is the pharmaceutical industry. Who among us has not taken an antibiotic to cure an infection or used a drug to alleviate the pain associated with dental work, an accident, or surgery? Modern medicine, which rests firmly upon chemistry, has increased our life expectancy by about 18 years since the 1920s. It is hard to believe that, little over a century ago, many people died from simple infections.

Figure 1.1 A modern chemical research laboratory.

Perhaps the chemical products most familiar to us are plastics. About 50% of industrial chemists are involved with the development and production of plastics. The United States alone produces over 50 million metric tons (110 billion pounds) of plastics a year, some 5 billion kilograms (11 billion pounds) of which are synthetic fibers used in bed sheets, clothing, backpacks, shoes, and other woven materials. This corresponds to about 160 kg (350 lb) of plastics and 16 kg (35 lb) of synthetic fibers per person living in the United States per year. Names such as nylon, polyethylene, Formica, Saran, Teflon, Hollofil, Gore-Tex, polyester, Nalgene, PVC, and silicone are familiar to us in our homes, our clothing, and the activities of our daily life. Chemistry also underlies the products that make our daily life possible—computer chips, paper, fuels, cement, liquid crystal displays, detergents, magnetic storage media, refrigerants, batteries, scents, flavorings, preservatives, paint, ceramics, solar cells, and cosmetics, to name only a few. In addition, metals such as steel, lightweight alloys of titanium and aluminum, and materials made from carbon fibers make possible modern ships, automobiles, aircraft, and satellites.

Chemistry is also needed for a study and understanding of our environment. Unfortunately, a great many people today have a fear of chemicals, owing in part to the legacy of various pesticides such as DDT, chemical contamination of waterways, and air pollution. However, an understanding of these problems and their solutions also comes from the study of the chemistry involved. Biodegradable packing materials, hydrogen fuel cells, recyclable carpeting, and non-ozone-depleting refrigerants are just some of the new environmentally friendly "green" substances being developed by today's chemists.

It is remarkable that all chemicals are built up from only about 100 different basic units, called atoms. Atomic theory pictures substances as atoms, or groups of atoms, joined together into units called molecules and ions. You will start by exploring atomic theory, then go on to study chemical bonding and chemical reactions, and then learn to do calculations involving chemical reactions. You will learn to make predictions about what reactions take place, under what conditions they take place, and how quickly they take place; what substances are produced in these reactions; and what the structure, properties, and behavior of these substances will be. You will learn the chemistry behind many of the materials and processes we have already mentioned. We are confident that you will find your study of chemistry both interesting and enjoyable.

1-2. Chemistry Is an Experimental Science

Chemistry is an experimental science based on the scientific method. The essence of the **scientific method** is the use of carefully controlled experiments to answer scientific questions (Figure 1.2).

To use the scientific method, we must first define our goal; that is, we must first formulate the question we wish to answer. After defining our goal, our next step is to collect information or data about the subject under consideration. The data we collect will be of two sorts: **qualitative data,** consisting of descriptive observations, and **quantitative data,** consisting of numbers obtained by measurement. If we gather enough data about our subject, we will be able to form a hypothesis. A **hypothesis** is a proposition put forth as the possible explanation for, or prediction of, an observation or a phenomenon. If hypotheses are supported by a sufficient number of experimental observations obtained under a wide variety of conditions, they evolve into scientific theories.

To test a hypothesis, we perform experiments. If the experiments support the hypothesis, then we perform further experiments to see whether our results are reproducible under a variety of experimental conditions. After many experiments, a pattern may emerge in the form of a constant relationship among phenomena under the same conditions. A concise statement of this relationship is called a **law of nature** or a **scientific law.** A law summarizes the relationship but does not explain it.

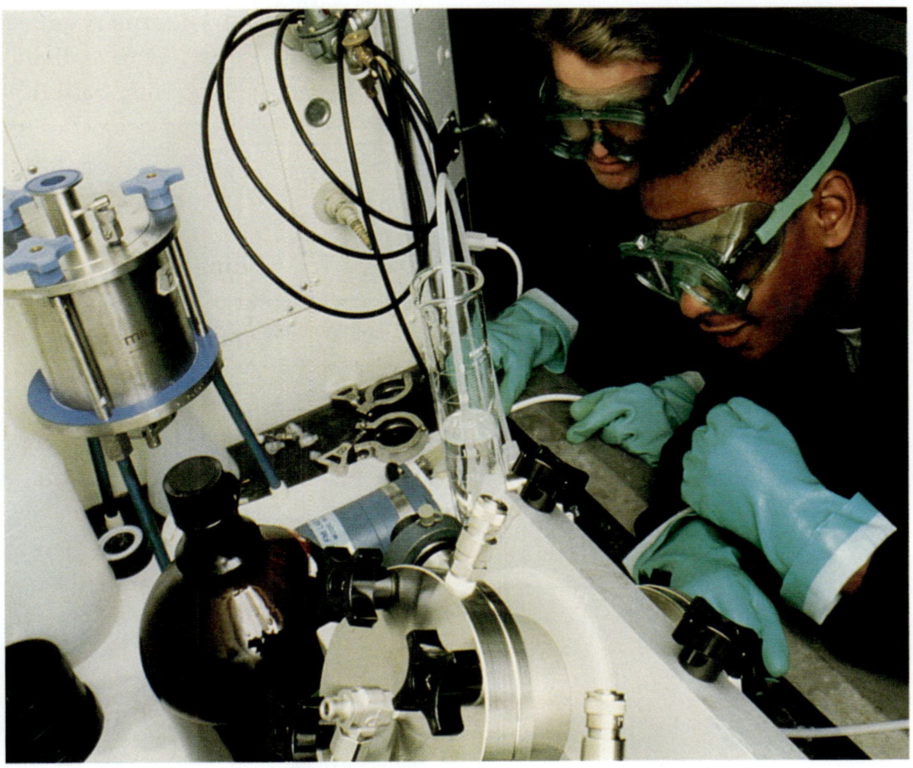

Figure 1.2 Chemistry is based firmly on the results of experiments. Carefully planned experiments are an endless source of fascination, excitement, and challenge.

Once a law has been formulated, scientists try to develop a **theory,** or a unifying principle, that explains the law based on the experimental observations. Eventually, the theory also will be tested and perhaps modified or rejected as a result of further experimentation. Theories are ever evolving as new experiments are carried out. The scientific method underlies chemistry. When we study atomic theory in Chapter 2, we will see how the results of a large body of experiments led to the discovery of several important laws, which were in turn explained by a unifying atomic theory of matter.

It is important to realize that no theory can ever be proved correct by experiment. Experimental results can provide supporting data for a theory, but no matter how many experiments yield results consistent with a theory, the possibility always remains that additional experiments will demonstrate a flaw in the theory. This is the primary reason why experiments should be designed to disprove a hypothesis or theory rather than simply to provide additional support for the theory. The role of experiments, hypotheses, laws, and theories in the scientific method is outlined in Figure 1.3.

As Figure 1.3 shows, scientific theories are subject to ongoing revision. For example, the theory that the sun goes around the earth was replaced by one in which the earth orbits the sun, and later by one in which each orbits their combined center of mass. Still later this theory was replaced by one involving both space and time, as proposed by Albert Einstein. Most theories in use have known limitations. An imperfect theory is often useful, however, even though we cannot have complete confidence in its theoretical predictions. For example, a theory that correctly predicts the result, say, 90% of the time is quite useful. Because scientific theories produce a unification of ideas, imperfect theories generally are not abandoned until a better theory is developed.

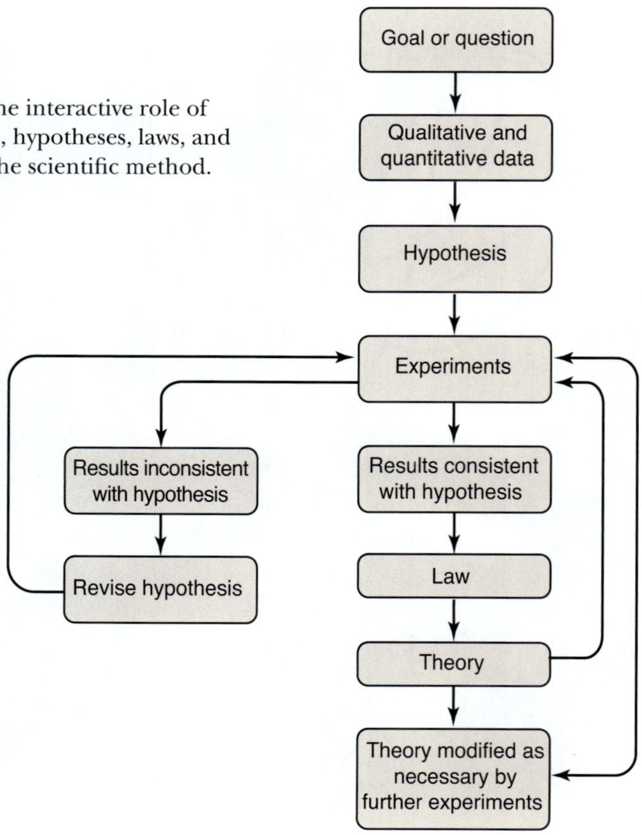

Figure 1.3 The interactive role of experiments, hypotheses, laws, and theories in the scientific method.

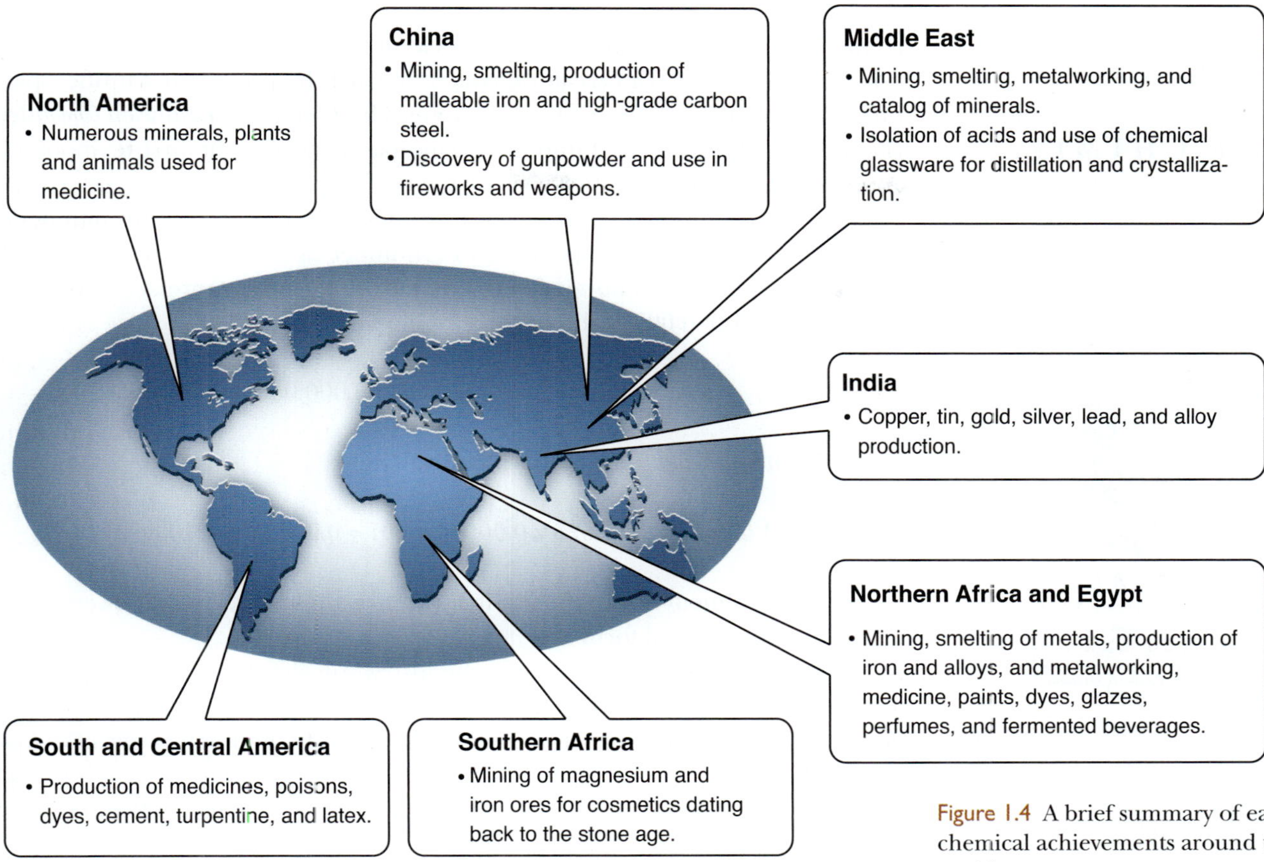

Figure 1.4 A brief summary of early chemical achievements around the world.

1-3. Modern Chemistry Is Based on Quantitative Measurements

Although early peoples around the world practiced various forms of rudimentary chemistry and made many important technological discoveries (Figure 1.4), chemistry didn't begin to develop as a modern science until the eighteenth century. Modern physical sciences are based firmly on **quantitative measurements,** measurements in which the result is expressed as a number. For example, the determinations that the mass of 1.00 cubic centimeter (cm^3) of gold is 19.3 grams and that 1.25 grams of calcium react with 1.00 gram of sulfur express the results of quantitative measurements. Compare these determinations with **qualitative observations,** where we note general characteristics, such as color, odor, taste, and the tendency to undergo chemical change in the presence of other substances. An example of a qualitative statement is that lead is much denser than aluminum. As we shall see later in this chapter, the corresponding quantitative statement is that the mass of 1.00 cm^3 of lead is 11.3 grams, whereas the mass of 1.00 cm^3 of aluminum is 2.70 grams.

Our modern word *chemistry* and the Arabic word *alchemy* derive from the Greek word *chemeia,* which refers to metalworking and transmutation, the belief that base metals could be converted into gold. *Chemeia* probably derives from the word *Khem,* the name for ancient Egypt, in honor of their techniques in metalworking and early theories of transmutation. However, some scholars believe that *chemeia* is of Chinese origin, derived from the words *kim mi* in southern dialect, meaning "the secret of gold."

The French scientist Antoine Lavoisier (Frontispiece) was the first chemist to fully appreciate the importance of carrying out quantitative chemical measurements in the modern sense. Lavoisier designed special balances that were more accurate than any devised before, and he used these balances to discover the **law of conservation of mass:** in a chemical reaction, the total mass of the reacting substances is equal to the total mass of the products formed. In other words, by careful quantitative measurements Lavoisier was able to show that mass is conserved in chemical reactions. Lavoisier's influence on the develop-

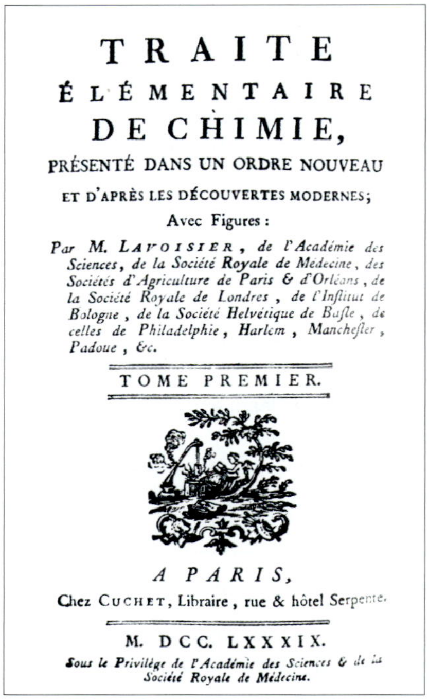

TRAITE
ÉLÉMENTAIRE
DE CHIMIE,
PRÉSENTÉ DANS UN ORDRE NOUVEAU
ET D'APRÈS LES DÉCOUVERTES MODERNES;
Avec Figures :
Par M. LAVOISIER, de l'Académie des Sciences, de la Société Royale de Médecine, des Sociétés d'Agriculture de Paris & d'Orléans, de la Société Royale de Londres, de l'Institut de Bologne, de la Société Helvétique de Basle, de celles de Philadelphie, Harlem, Manchester, Padoue, &c.
TOME PREMIER.
A PARIS,
Chez CUCHET, Libraire, rue & hôtel Serpente.
M. DCC. LXXXIX.
Sous le Privilége de l'Académie des Sciences & de la Société Royale de Médecine.

Figure 1.5 The title page to Lavoisier's textbook of chemistry.

ment of chemistry as a modern science cannot be overstated. In 1789 he published his *Elementary Treatise on Chemistry,* in which he presented a unified picture of the chemical knowledge of the time. The *Elementary Treatise on Chemistry* (Figure 1.5) was translated into many languages and was the first textbook of chemistry based on quantitative experiments.

Throughout this text we use **scientific notation** to represent many numbers. Scientific notation is the expression of a number as multiplied by a power of 10. For example, a large number such as 6 000 000 may be expressed as 6×10^6, because $10^6 = 1\ 000\ 000$ and

$$6 \times 10^6 = 6 \times 1\ 000\ 000 = 6\ 000\ 000$$

Similarly, the number 1.626×10^{-9} is equivalent to

$$1.626 \times 10^{-9} = 1.626 \times 0.000\ 000\ 001 = 0.000\ 000\ 001\ 626$$

It is more convenient to express numbers such as these in terms of powers of 10 rather than in decimal form.

To be successful in your study of chemistry, you should be proficient in the use and mathematical manipulation of numbers in scientific notation. A more detailed review of working with numbers in scientific notation is given in Appendix A.

It is good practice always to include a zero before a leading decimal point so that the decimal point does not get overlooked. For example, you should write 0.345 instead of just .345. You can see here that the zero nicely alerts you to the presence of the following decimal point.

EXAMPLE 1-1: Express the numbers (a) 24 000 and (b) 0.000 000 572 using scientific notation.

Solution: (a) The number 24 000 has four powers of 10 following the first digit:

24 000
1 2 3 4

Thus, the number may be expressed as

$$2.4 \times 10\ 000 = 2.4 \times 10^4$$

in scientific notation.

(b) Counting backwards from the 5 in the number 0.000 000 572, we see that there are seven powers of 10 between the 5 and the decimal place.

0.000 000 572
7 6 5 4 3 2 1

Because the decimal place is to the left of the 5, this number is expressed in scientific notation as 5.72×10^{-7}.

PRACTICE PROBLEM 1-1: Perform the addition: $2.26 \times 10^{-5} + 1.7201 \times 10^{-3}$. State your answer in scientific notation.

Answer: 1.7427×10^{-3}

Figure 1.6 Comparison of a meter stick with a yardstick (the meter stick is about 10% longer than the yardstick); a liter with a quart (the volume of a liter is about 6% larger than the volume of a quart); and a kilogram with a pound (the mass of a kilogram is about 2.2 times larger than the mass of a pound).

1-4. The Metric System of Units and Standards Is Used in Scientific Work

With every number that represents a measurement, the units of that measurement must be indicated. If we measure the thickness of a wire and find it to be 1.35 millimeters (mm), then we express the result as 1.35 mm. To say that the thickness of the wire is 1.35 would be meaningless.

The preferred system of units used in scientific work is the **metric system.** Several sets of units make up the metric system, but nowadays we express all measurements in terms of just one set of metric units called **SI units** (for *S*ystème *I*nternational). Some basic SI units are given in Table 1.1.

The basic SI unit of **length** is the **meter** (m). Prior to the advent of SI units, the meter was defined as the length of a special platinum rod maintained in a repository in France. However, this standard for the meter is not precise enough for modern scientific work of the highest accuracy, primarily because of the variation in the length of the "meter rod" with temperature. The SI definition of the meter is now given in terms of the speed of light in a vacuum (see Appendix B), a fundamental constant of nature that is neither dependent on temperature nor subject to mechanical damage or loss, as is a platinum rod. In more familiar terms, a meter is equivalent to 1.094 yards, or to 39.37 inches. Thus, a meter stick is 3.37 inches (about 10%) longer than a yardstick (36 inches) (Figure 1.6).

TABLE 1.1 Basic SI units

What is measured	Unit of measurement	Symbol
length	meter	m
mass	kilogram	kg
temperature	kelvin	K
time	second	s
amount of substance	mole	mol

Figure 1.7 Road sign showing distances in miles and kilometers.

The SI system of units uses a series of prefixes to indicate factors-of-10 multiples and fractions of SI units. For example, 1000 meters is called a **kilometer** (km), which is equivalent to 0.621 miles and is the unit used to express distances between towns and cities on maps and road signs in most countries throughout the world (Figure 1.7). A **centimeter** (cm) is one one-hundredth of a meter—there are 2.54 cm in one inch. Common SI-unit prefixes used in chemistry are given in Table 1.2 (a more complete list is given in Appendix B). Some of these prefixes, kilo- (kilogram and kilometer), centi- (centimeter), milli- (milliliter and millimeter), mega- (megabytes and megatons), and giga- (gigabytes), are in everyday use. We shall see later that the two prefixes pico- and atto- are commonly used to describe molecular sizes and molecular energies.

EXAMPLE 1-2: Using the prefixes given in Table 1.2, explain what is meant by (a) a microsecond; (b) a milligram; and (c) 100 picometers.

Solution: (a) Table 1.2 shows that the prefix micro- means 10^{-6}, so a microsecond (μs) is 10^{-6} seconds or one one-millionth (1/1 000 000) of one second. Events that take place in microseconds are common in scientific experiments. (b) The prefix milli- means 10^{-3}, so a milligram is 10^{-3} grams or one one-thousandth (1/1000) of a gram. (c) The prefix pico- means 10^{-12}, so a picometer is 10^{-12} meters or one-millionth of one-millionth of a meter (10^{-12} is equal to $10^{-6} \times 10^{-6}$ and 10^{-6} is one one-millionth). Thus, 100 picometers, or 100 pm, is equal to 100×10^{-12} meters or 1.00×10^{-10} meters. We shall see that the picometer is a convenient unit of length when discussing the sizes of atoms and molecules.

PRACTICE PROBLEM 1-2: What is meant by (a) 400 nm and (b) 20 ps?

Answer: (a) 4.00×10^{-7} m; (b) 2.0×10^{-11} s

TABLE 1.2 Common prefixes for SI units

Prefix	Symbol	Multiple	Example*
giga-	G	10^9 or 1 000 000 000	1 gigajoule, 1 GJ = 1×10^9 joules
mega-	M	10^6 or 1 000 000	1 megajoule 1 MJ = 1×10^6 joules
kilo-	k	10^3 or 1000	1 kilometer, 1 km = 1×10^3 meters
centi-	c	10^{-2} or 1/100	1 centimeter, 1 cm = 1×10^{-2} meters
milli-	m	10^{-3} or 1/1000	1 milliliter, 1 mL = 1×10^{-3} liters
micro-	μ	10^{-6} or 1/1 000 000	1 microsecond, 1 μs = 1×10^{-6} seconds
nano-	n	10^{-9} or 1/1 000 000 000	1 nanometer, 1 nm = 1×10^{-9} meters
pico-	p	10^{-12} or 1/1 000 000 000 000	1 picometer, 1 pm = 1×10^{-12} meters
femto-	f	10^{-15} or 1/1 000 000 000 000 000	1 femtosecond, 1fs = 1×10^{-15} seconds
atto-	a	10^{-18} or 1/1 000 000 000 000 000 000	1 attojoule, 1 aJ = 1×10^{-18} joules

*We define liters and joules later in this chapter.

Measures of **volume,** denoted by the symbol V, are derived from the basic SI unit for length, which is the meter. A cubic meter (1 m^3) is the volume of a cube that is one meter on each edge. However, a more convenient measure of volume for laboratory work is the liter. A **liter** is equal to the volume of a cube that is 10 centimeters (or one-tenth of a meter) on each edge (Figure 1.8). The volume of a cube is equal to the cube of the length of an edge of the cube; thus,

$$1\text{ L} = (10\text{ cm})^3 = 1000\text{ cm}^3$$

A **milliliter,** 1 mL, is one one-thousandth (1/1000) of a liter; in other words, there are 1000 mL in one liter. Because 1000 mL and 1000 cm^3 are both equal to one liter, we conclude that 1 mL = 1 cm^3; that is, a milliliter (1 mL) and a **cubic centimeter** (1 cm^3, sometimes also abbreviated "cc") are equal to each other. One liter is equal to 1.057 liquid U.S. quarts (see the inside of the back cover); thus, the volume of a liter is 5.7% larger than the volume of a quart (Figure 1.6). A U.S. gallon contains exactly 4 U.S. quarts and corresponds to 3.785 liters.

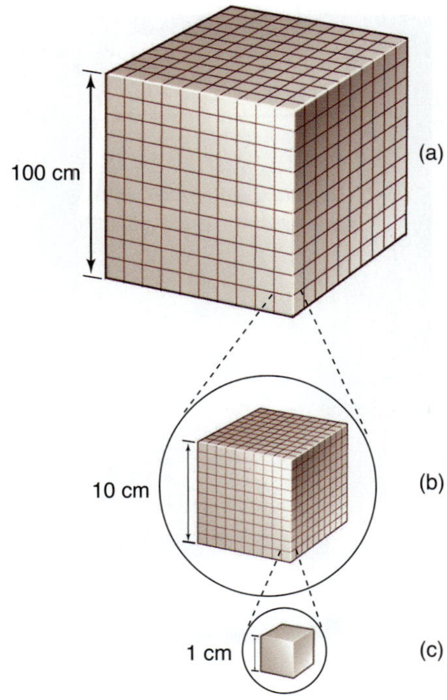

Figure 1.8 (a) Each edge of a cubic meter (1 m^3) is 100 cm in length. A cubic meter contains 1000 liters. (b) Each edge of a one-liter cube is 10 cm in length. A liter contains 1000 milliliters. (c) Each edge of a one-milliliter cube is 1 cm in length. A milliliter is therefore equivalent in volume to a cubic centimeter (1 mL = 1 cm^3).

The SI unit of mass is the **kilogram** (kg). The mass of a cylinder of a platinum-iridium alloy kept by the International Bureau of Weights and Measures in Sèvres, France (near Paris), represents the standard kilogram, which is the only basic SI unit still defined by an artifact. One kilogram is equal to 1000 **grams** (1 kg = 1000 g).

The mass of an object is determined in the laboratory by balancing its weight against the weight of a reference set of masses (Figure 1.9). The mass values of the set of reference masses are fixed by comparison with the standard kilogram.

Because a mass is determined by balancing its weight against reference masses, the terms *mass* and *weight* are often used interchangeably (for example, "the sample weighs 28 grams"). However, strictly speaking, these terms are not the same. The **mass** of an object characterizes the object's inertia, or resistance to being moved, and is an intrinsic property of the object. The **weight** of an object is equal to the force of attraction of the object to a large body, such as the earth or the moon. An object on the moon weighs about one-sixth as much as it does on the earth, but the mass of the object is the same in both places. To avoid such ambiguities, we generally use the term *mass* rather than the term *weight* throughout this book. A 1-kilogram mass weighs 2.205 pounds on earth; therefore, one pound is equivalent to 453.6 grams (Figure 1.6).

A modern analytical balance (Figure 1.9, top) has the standard masses enclosed by the balance housing. The masses are controlled by an internal set of movable levers. The basic principle of operation is the same as that for the beam balance (Figure 1.9, middle), except that the balance point is detected optically with a light beam rather than visually with the naked eye. An electronic or digital balance (Figure 1.9, bottom) uses a pressure-sensitive crystal to measure mass. The electronic balances found in most general chemistry labs are precise to only about ±1 mg. For more sensitive measurements an analytical balance must be used.

Temperature is a property that constitutes a quantitative measure of the relative tendency of heat to escape from an object. The higher the temperature of an object, the greater is the tendency of heat to escape from the object. When we say that water is "hot" to the touch, we mean that heat flows readily from the water to our fingers, which are at a lower temperature than the water. When the water is "cold," the flow of heat is from our fingers to the water, which is at a lower temperature than our fingers. Numerical **temperature scales** are established by assigning temperatures to two reference systems. For example, a temperature scale can be established by assigning a temperature of exactly zero degrees to the freezing point of water and exactly 100 degrees to the boiling point of water at one atmosphere of pressure, the pressure of air at sea level on a clear day. We shall study pressure in more detail in Chapter 13.

A **thermometer** is a device used to measure temperature. A thermometer contains a substance whose properties change in a reproducible way with changes in temperature. For example, the property might be the volume of a certain liquid, such as mercury. Because the volume of an enclosed sample of mercury increases with increasing temperature, it can be used to measure temperature. For instance, using the temperature scale proposed above, we mark on the glass rod the position of the mercury column when the thermometer is in contact with an ice-water mixture; we label this position 0.0. Then we mark the position of the mercury column as 100.0, when the thermometer is in contact with boiling water at one atmosphere of pressure. The temperature scale is then determined by marking off the thermometer scale linearly between the two calibration points.

We discuss temperature in more detail in Chapter 13. It is sufficient for our purposes at this stage to know that there are three different (but interrelated) temperature scales in common use (Figure 1.10): the Celsius temperature scale, the Fahrenheit temperature scale, and the Kelvin temperature scale.

The most fundamental temperature scale is the **Kelvin temperature scale**, and it defines the SI unit for temperature, which is called the **kelvin** (K). The lowest possible temperature of the Kelvin scale is zero kelvin (0 K), which we shall learn later is the lowest temperature that any substance can have. Note that the degree sign is omitted, and all temperatures on the Kelvin scale are positive. The **Celsius temperature scale** (°C, denoted by t and once called the centigrade temperature scale) is related to the Kelvin temperature scale (denoted by T) by the equation

$$T\text{ (in K)} = t\text{ (in °C)} + 273.15 \tag{1.1}$$

Thus, a Kelvin temperature of 373.15 K corresponds to 373.15 – 273.15 = 100.00°C (see Figure 1.10). One degree on the Celsius temperature scale corresponds to the same temperature interval as one degree on the Kelvin temperature scale. The two scales differ only in their zero points.

The United States is one of the very few countries in the world to use the **Fahrenheit temperature scale** (°F). On the Fahrenheit temperature scale, the temperature of an ice-water mixture is set as 32°F and that of boiling water at one atmosphere of pressure is set at 212°F. The Celsius temperature scale is related to the Fahrenheit temperature scale by the equation

$$t\text{ (in °C)} = (5/9)\ [t\text{ (in °F)} - 32.0] \tag{1.2}$$

Note that t (in °C) = 0°C when t (in °F) = 32°F and that t (in °C) = 100°C when t (in °F) = 212°F. Using Equation 1.2, we see that a Fahrenheit temperature of 98.6°F ("normal" body temperature) corresponds to a Celsius temperature of

$$t = (5/9)\ (98.6 - 32.0)\text{°C} = 37.0\text{°C}$$

and a Kelvin temperature of

$$T = (273.15 + 37.0)\text{ K} = 310.2\text{ K}$$

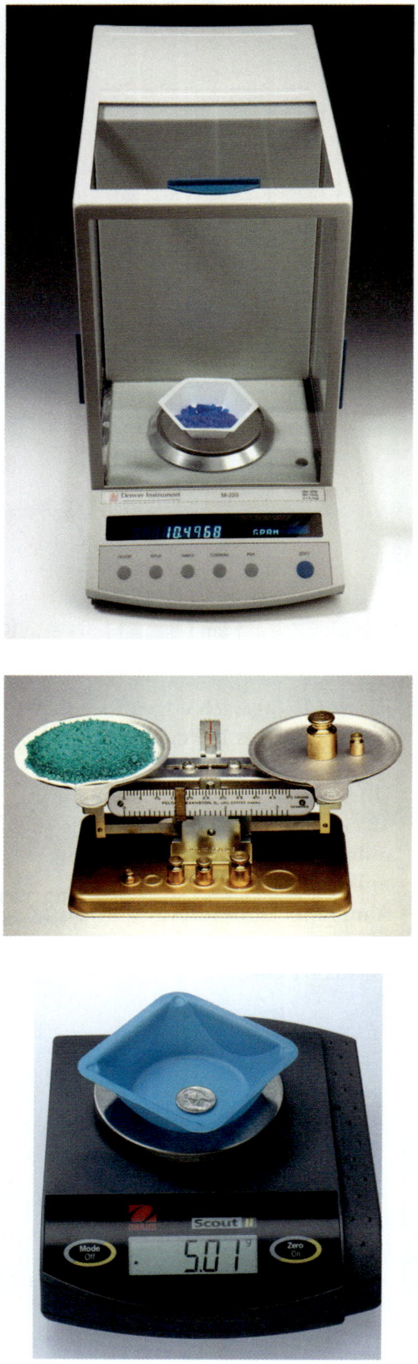

Figure 1.9 (*top*) An automatic analytical balance, (*middle*) a laboratory beam balance, and (*bottom*) an electronic balance; are all used to determine mass. The beam balance requires the placement of standard masses on one of the pans to achieve a mass balance.

EXAMPLE 1-3: Derive Equation 1.2 by considering the information in Figure 1.10.

Solution: We first note that there are exactly 100°C between the boiling and freezing points of water on the Celsius scale. Comparison with the Fahrenheit scale shows that there are 212°F − 32°F = 180°F between the same two points on the Fahrenheit scale. Thus, a degree Fahrenheit is five-ninths (100°C/180°F = 5°C/9°F) of a degree Celsius. We also note the 0°C corresponds to 32°F; thus, there is a (5/9) × 32-degree shift in the zero point on going from the Fahrenheit to the Celsius scale. Combination of these two results yields

$$t\text{ (in °C)} = (5/9)\ t\text{ (in °F)} - (5/9)\ (32.0\text{ °F})$$

or

$$t\text{ (in °C)} = (5/9)\ [t\text{ (in °F)} - 32.0]$$

PRACTICE PROBLEM 1-3: (a) Convert −90°F to degrees Celsius and to kelvin. (b) Find the one value of the temperature at which the temperatures on the Celsius and Fahrenheit scales coincide numerically.

Answer: (a) −68°C, 205 K; (b) −40°C = −40°F

Density is an example of a property expressed in **compound units** because it involves both the unit of mass and the unit of volume. **Density** is defined as the mass per unit volume of a substance, density = mass/volume, or in symbols

$$d = \frac{m}{V} \tag{1.3}$$

To find the density of a material, we simply determine the volume of a known mass of the material and then use Equation 1.3.

The concept of density allows us to compare the masses of equal volumes of materials. The old joke that asks, "Which is heavier, a pound of lead or a pound of feathers?" plays upon our intuition regarding density. Often, when we say that a material is heavy or light, we really mean that its density is high or low, respectively.

Equation 1.3 indicates that the **dimensions** of density are mass per unit volume, which can be expressed in a variety of units. If we express the mass in grams and the volume in cubic centimeters, then the units of density are grams per cubic centimeter. For example, the density of ice is 0.92 g/cm^3, where the slash denotes "per." In other words, 0.92 grams of ice occupies a volume of 1 cm^3 (the "unit volume" in this case).

From algebra, we know that $1/a^n = a^{-n}$, where n is an exponent. Thus, $1/cm^3 = cm^{-3}$. Therefore, we also can express density as $g{\cdot}cm^{-3}$ instead of g/cm^3. The use of centered dots in compound units is an SI convention (Appendix B) and is used to avoid ambiguities. For example, m·s denotes meter-second, whereas ms (without the dot) denotes millisecond.

We also often use a subscript to represent a variable at a specific time or

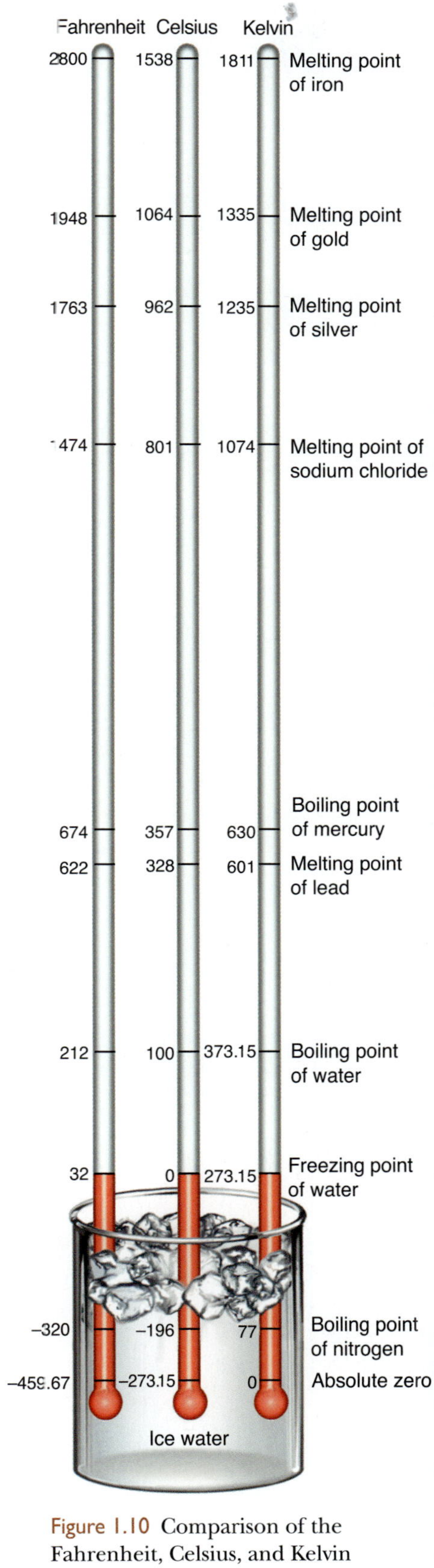

Figure 1.10 Comparison of the Fahrenheit, Celsius, and Kelvin temperature scales.

condition. For example, v_{15} might represent the velocity after 15 seconds, T_a might represent the temperature of substance a, and so forth. We generally use a subscript zero or i to represent the initial or starting conditions. For example, t_0 represents the initial time. Similarly, the subscript f is often used to represent the final state of a system. Throughout this text we define new symbols as we encounter them.

EXAMPLE 1-4: Calculate the density of gold, given that, at 20°C, 5.00 cm^3 of gold has a mass of 96.5 grams.

Solution: Using Equation 1.3, we find that

$$d = \frac{m}{V} = \frac{96.5\ \text{g}}{5.00\ \text{cm}^3} = 19.3\ \text{g·cm}^{-3}$$

As always, answers must be given with the accompanying units, in this case g·cm^{-3}. An answer, "the density is equal to 19.3" has no meaning because it is unclear whether the measurement refers to 19.3 g·cm^{-3}, 19.3 g·L^{-1}, 19.3 lb·ft^{-3} or some other combination of units expressing density.

PRACTICE PROBLEM 1-4: Mercury is the only metal that is a liquid at 25°C. Given that 1.667 mL of mercury has a mass of 22.55 grams at 25°C, calculate the density of mercury in units of g·mL^{-1} and g·cm^{-3}.

Answer: 13.53 g·mL^{-1} = 13.53 g·cm^{-3}

The density of a substance depends on the temperature. Because the volume of a given mass of most substances increases as its temperature increases, the density of most substances decreases as the temperature increases. For example, because a 10.0-gram sample of mercury occupies a volume of 0.735 mL at 0°C and 0.749 mL at 100°C, the densities of mercury at these two temperatures are

$$d = \frac{10.0\ \text{g}}{0.735\ \text{mL}} = 13.6\ \text{g·mL}^{-1} \quad (0°\text{C})$$

and

$$d = \frac{10.0\ \text{g}}{0.749\ \text{mL}} = 13.4\ \text{g·mL}^{-1} \quad (100°\text{C})$$

Both density and temperature are examples of the **intensive properties** of a substance, which are properties whose values are independent of the amount of a substance. For example, the density of gold at 20°C is 19.3 g·cm^{-3}, whether the gold sample consists of 5.00 g or 5.00 kg. In contrast to intensive properties, **extensive properties** are directly proportional to the amount of a substance. Thus, mass and volume are both extensive properties. If we double the amount of a substance, then the mass and the volume also double, but the density remains the same (at a given temperature).

1-5. The SI Unit of Energy Is a Joule

Up to this point we have introduced the SI or metric units for length, volume, mass, and temperature. Energy plays a central role in chemistry, so in this section we shall present a brief introduction to the concept of energy along with its units.

Energy can be defined as the ability to cause a change in a physical system such as the compression of a cylinder, the increase in temperature of a substance, or the movement of a mass. At this point it is useful to introduce two forms of energy: the energy associated with a moving object, called **kinetic energy**; and the energy of an object due to its location relative to a specific reference point, called **potential energy**.

The kinetic energy of an object is defined as

$$E_k = \frac{1}{2}mv^2 \tag{1.4}$$

where E_k is the kinetic energy, m is the mass of the object, and v is its velocity. In SI units, mass is given in kilograms and velocity in meters per second, yielding $kg{\cdot}m^2{\cdot}s^{-2}$ as the units of energy. The SI unit of energy is the **joule (J)**, where $1\ J = 1\ kg{\cdot}m^2{\cdot}s^{-2}$ (see Appendix B). Joules are named for James Prescott Joule (Chapter 14 Frontispiece) whose research led to the law of conservation of energy, which we discuss below. One joule is approximately the energy needed to lift a 100-gram mass (about the mass of an orange) to a height of one meter. Table 1.3 lists some other examples for you to get a better idea of the magnitude

TABLE 1.3 Comparative measures of energy

Example	Approximate energy
Energy from the sun that hits the earth in one day	1.5×10^{23} J
Approximate world energy consumption for 2005	4.8×10^{20} J
Largest nuclear device ever tested	2.5×10^{17} J
Approximate energy of the atomic bomb dropped on Hiroshima	8.4×10^{13} J
One kiloton of TNT	4.2×10^{12} J
Nutritional energy needed by a person per day	7×10^{6} J
One Calorie (nutritional)	4184 J
One British Thermal Unit (BTU)	1055 J
Energy to lift 100 grams to a height of one meter	1 J
Energy of an X-ray photon	10^{-14} J
Energy of an electron in a hydrogen atom	2×10^{-18} J
Energy of a typical chemical bond	2×10^{-19} J
Energy of a photon of visible light	10^{-19} J
Average kinetic energy of a molecule at room temperature	4×10^{-21} J

of the joule. One thing to notice about Table 1.3 is that atomic and molecular energies are about 10^{-18} joules. This unit is called an **attojoule** (**aJ**) and is typical of atomic and molecular energies.

EXAMPLE 1-5: Calculate the kinetic energy of a 5.0-gram bullet moving at a speed of 515 $m\cdot s^{-1}$.

Solution: Using Equation 1.4,

$$E_k = (\tfrac{1}{2})(0.0050\ \text{kg})(515\ \text{m}\cdot\text{s}^{-1})^2 = 6.6 \times 10^2\ \text{kg}\cdot\text{m}^2\cdot\text{s}^{-2}$$
$$= 6.6 \times 10^2\ \text{J} = 0.66\ \text{kJ}$$

PRACTICE PROBLEM 1-5: Calculate the kinetic energy of a fully loaded Boeing 747 (322 000 kg) moving at a speed of 950 kph (kilometers per hour).

Answer: 1.12×10^{10} J

Potential energy is the energy stored in an object because of its position relative to some reference point. For example, a 142-gram baseball at a height of 30 meters has an energy relative to the ground because it will speed up if dropped. For an object located above the ground, the potential energy, which we shall denote by E_p, is directly proportional to the mass, m, and to the height, h, of the mass above the ground. Thus, we have

$$E_p = mgh \tag{1.5}$$

where g is a proportionality constant, called the **gravitational acceleration constant**, whose value is determined experimentally to be 9.81 $m\cdot s^{-2}$ on earth. If a baseball (0.142 kg) is 15 meters above the ground, then we calculate its potential energy using Equation 1.5 to be

$$E_p = mgh = (0.142\ \text{kg})(9.81\ \text{m}\cdot\text{s}^{-2})(15\ \text{m})$$
$$= 21\ \text{kg}\cdot\text{m}^2\cdot\text{s}^{-2} = 21\ \text{J}$$

One of the most fundamental laws of science is the **law of conservation of energy**: during any process, energy is neither created nor destroyed. Energy can be converted from one form to another or transferred from one system to another, but the total amount of energy never changes.

The law of conservation of energy says that the sum of the kinetic energy and potential energy of an object in flight is a constant. That is,

$$E_{\text{total}} = E_k + E_p = \text{constant}$$

Therefore, an object moving upwards will decelerate as kinetic energy is changed into potential energy. An object at the apex of its upward trajectory will have only potential energy and no kinetic energy. As the object falls back down, it exchanges its potential energy for kinetic energy, reaching its maximum energy of motion just before hitting the ground (Figure 1.11).

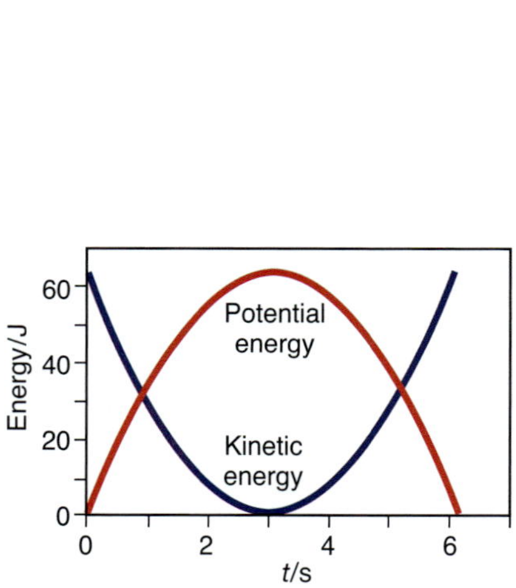

Figure 1.11 The kinetic energy and potential energy of an object in flight.

EXAMPLE 1-6: Calculate the maximum height that a bullet of mass 25 grams will reach if it is shot straight up with an initial speed (v_0) of 450 m·s^{-1}. (Ignore the effects of air resistance.)

Solution: All the energy of the bullet on exit from the gun is kinetic energy, so we have (Equation 1.4)

$$\begin{aligned} E_k\ (\text{initial}) &= \tfrac{1}{2}mv_0^2 = \left(\tfrac{1}{2}\right)(0.025\ \text{kg})(450\ \text{m}\cdot\text{s}^{-1})^2 \\ &= 2.5\times10^3\ \text{kg}\cdot\text{m}^2\cdot\text{s}^{-2} \\ &= 2.5\times10^3\ \text{J} \\ &= 2.5\ \text{kJ} \end{aligned}$$

and the total energy is

$$\begin{aligned} E_{\text{total}}\ (\text{initial}) &= E_k\ (\text{initial}) + E_p\ (\text{initial}) \\ &= 2.5\ \text{kJ} + 0\ \text{kJ} \\ &= 2.5\ \text{kJ} \end{aligned}$$

At the apex of the bullet's flight, the kinetic energy will be zero (because its speed is zero) and so all its energy will be potential energy. Thus, we have

$$\begin{aligned} E_{\text{total}}\ (\text{apex}) &= E_k\ (\text{apex}) + E_p\ (\text{apex}) \\ &= 0\ \text{kJ} + 2.5\ \text{kJ} \\ &= 2.5\ \text{kJ} \end{aligned}$$

We use Equation 1.5 to solve for the maximum height

$$E_p(\text{apex}) = 2.5\ \text{kJ} = 2.5\times10^3\ \text{J} = mgh(\text{apex})$$

Thus, we have for h(apex)

$$\begin{aligned} h(\text{apex}) = \frac{E_p(\text{apex})}{gm} &= \frac{2.5\times10^3\ \text{J}}{(9.81\ \text{m}\cdot\text{s}^{-2})(0.025\ \text{kg})} \\ &= \frac{2.5\times10^3\ \text{kg}\cdot\text{m}^2\cdot\text{s}^{-2}}{(9.81\ \text{m}\cdot\text{s}^{-2})(0.025\ \text{kg})} \\ &= 1.0\times10^4\ \text{m} \end{aligned}$$

PRACTICE PROBLEM 1-6: Electricity is generated at hydroelectric facilities by converting the potential energy of water in a lake to kinetic energy as the water falls down large pipes that are connected to turbines (Figure 1.12). The turbines convert the kinetic energy of the falling water into electricity. Given that the water falls 200 meters, calculate the kinetic energy available at the turbine input connection per kilogram of water.

Answer: 2×10^3 J or 2 kJ

Figure 1.12 Hoover Dam at Lake Mead. The potential energy of the water in Lake Mead is converted to electrical energy when it drops from near the lake surface to the bottom of the dam. The falling water drives the turbines that are within the dam and generates electricity.

The rate at which energy is produced or utilized is called **power.** The SI unit of power is a **watt** (W), which is defined as exactly one joule per second; that is, $1\ \text{W} = 1\ \text{J}\cdot\text{s}^{-1}$. For example, a 40-watt incandescent light bulb emits energy in the form of light (20%) and heat (80%) at a rate of $40\ \text{J}\cdot\text{s}^{-1}$. Home energy consumption is often expressed in units of **kilowatt-hours.** A kilowatt-hour (kW·h) is the energy used by a one-kilowatt device operating for one hour. Thus,

$$\begin{aligned}1\ \text{kilowatt-hour} &= \left(\frac{1\ \text{kJ}}{\text{s}}\right)(1\text{h}) \\ &= \left(\frac{1\ \text{kJ}}{\text{s}}\right)(1\text{h})\left(\frac{3600\ \text{s}}{\text{h}}\right) \\ &= 3600\ \text{kJ}\end{aligned}$$

EXAMPLE 1-7: The electrical energy requirements of an average U.S. home are about 30 kW·h per day. How many joules of electrical energy does a typical U.S. home use per day? If electrical energy costs \$0.10 per kW·h, how much does it cost to provide electrical energy to an average U.S. home for a typical 30-day month?

Solution:

$$\begin{aligned}30\ \text{kW}\cdot\text{h} &= \left(\frac{30\ \text{kJ}}{\text{s}}\right)(1\text{h}) \\ &= \left(\frac{30 \times 10^3\ \text{kJ}}{\text{s}}\right)(1\text{h})\left(\frac{3600\ \text{s}}{\text{h}}\right) \\ &= 1 \times 10^8\ \text{J}\end{aligned}$$

A typical U.S. home uses about 1×10^8 J of electrical energy per day. At a rate of \$0.10 per kW·h, the cost per 30-day month is

$$\text{Cost} = (30\text{ days})\left(\frac{30\text{ kW}\cdot\text{h}}{\text{day}}\right)\left(\frac{\$0.10}{\text{kW}\cdot\text{h}}\right) = \$90$$

PRACTICE PROBLEM 1-7: Modern electric clothes dryers use about 7.5 kW when in the heating mode. Assuming it takes 55 minutes to dry a large load of clothes, calculate the cost of the electricity used to dry a load of clothes (use the energy cost of \$0.10 per kW·h from the previous problem).

Answer: \$0.69

1-6. Percentage Error Can Be Used to Measure Accuracy

The counting of objects is the only type of experiment that can be carried out with complete accuracy, that is, without any inherent error. Let's consider the problem of determining how many coins there are in a jar. We can determine the exact number of coins simply by counting them. Now suppose that there are 1542 coins in the jar. The number 1542 is exact; there is no uncertainty associated with it. It is a different matter, however, when we wish to determine the mass of 1542 coins with a balance (Figure 1.9). Suppose that the balance we use is capable of measuring the mass of an object to the nearest one-tenth of a gram and that we use it to determine the mass of the 1542 coins as 4776.2 ± 0.1 grams. Because our balance was capable of measuring only to the nearest one-tenth of a gram, we know that the 0.2 in 4776.2 is not an exact number. It could actually be any number between 0.1 and 0.3; for example, the mass could be 4776.13, 4776.262, or any number between 4776.1 and 4776.3. The ±0.1 indicates the uncertainty in the last digit in the result 4776.2 grams.

Now suppose we had a more sensitive balance capable of measuring the mass of the coins to the nearest hundredth of a gram. Our result might be 4776.23 ± 0.01 grams, meaning any number in the range 4776.22 grams to 4776.24 grams. We would need a still more sensitive balance to determine the mass of the coins to the nearest milligram (±0.001 g). Thus, unlike with counting, we can never determine with absolute certainty what the mass of our 1542 coins is.

In scientific work we need to distinguish between the accuracy of a result and the precision of a result. **Accuracy** refers to how close our result is to the actual value. To clarify the distinction between precision and accuracy, let's return to our measurement of the mass of the 1542 coins. The result 4776.2 ± 0.1 grams for the mass of the coins may differ significantly from the actual mass because perhaps the balance was improperly calibrated, or we misread the result displayed by the balance, or we failed to put all 1542 coins back in the jar, or we forgot to correct for the mass of the jar.

The **precision** of a result conveys both how well repeated measurements of a quantity give results that agree with one another and how sensitive a measuring instrument was used (Figure 1.13). But high precision is no guarantee of high accuracy, because the same error source may be present in each and every measurement. The difference between accuracy and precision in experimental data is illustrated in Figure 1.14.

One measure of the accuracy of an experiment is the **percentage error**. Percentage error is defined as the difference between the average value of the

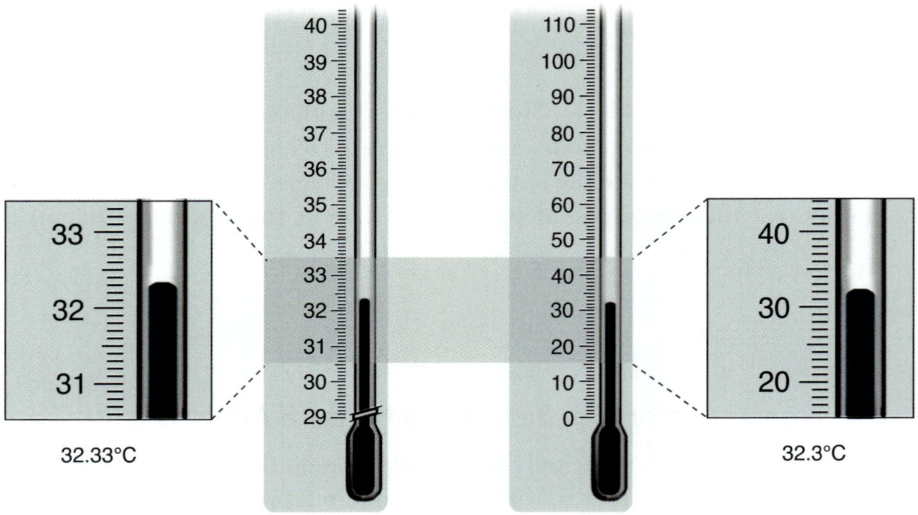

Figure 1.13 Precise measurements are essential to obtain reproducible experimental results. A measurement of 32.33°C is more precise than a measurement of 32.3°C.

experimental results and the accepted or true value, divided by the true value, multiplied by 100. In the form of an equation we have

$$\text{percentage error} = \frac{\text{average value} - \text{true value}}{\text{true value}} \times 100 \qquad (1.6)$$

The *CRC Handbook of Chemistry and Physics* has been a standard source of chemical and physical data for almost a century.

The true value in Equation 1.6 is a value that represents the most accurate currently known value for the quantity measured to the best knowledge of the experimenter. For example, in first-year chemistry experiments we typically take a value published in a textbook or reference handbook such as the *CRC Handbook of Chemistry and Physics* (Figure 1.15) to be a true value. However, the values in handbooks are revised when scientists perform better and more refined experiments; so what is taken as a true value for a particular measurement may change with time.

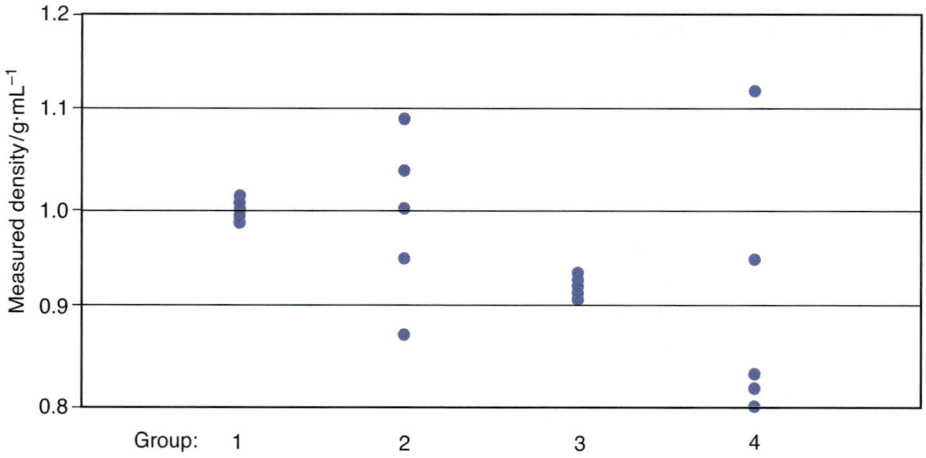

Figure 1.14 Five experimental measurements of the density of water collected at 25°C and one atmosphere of pressure by four groups of students, where the true value is accepted to be 1.00 g·mL^{-1}. The data collected by group 1 are both accurate and precise. The data collected by group 2 are accurate because the average of the five data points is close to the true value; however, these data are less precise than those collected by group 1, because the data points are much more scattered about the true value. The data collected by group 3 shows good precision, but poor accuracy compared with the previous groups' data. The data collected by group 4 are neither accurate nor precise in comparison to the other groups' data.

EXAMPLE 1-8: A student measures the mass and volume of a block of copper metal at 20°C and collects the following data:

Trial	Mass	Volume
1	5.051 g	0.571 cm^3
2	5.052 g	0.577 cm^3
3	5.055 g	0.575 cm^3

(a) Use these data to determine the average density of the block. (b) Using a reference source such as the *CRC Handbook of Chemistry and Physics* (available at most libraries) or an online source, look up the density of copper at 20°C and determine the percentage error in the average experimental density.

Solution: (a) The density is given by Equation 1.3 as the mass divided by the volume for the block. Thus, the density for each of the three trials is as follows:

$$\text{Trial 1} \quad d = \frac{5.051\text{ g}}{0.571\text{ cm}^3} = 8.85\text{ g}\cdot\text{cm}^{-3}$$

$$\text{Trial 2} \quad d = \frac{5.052\text{ g}}{0.577\text{ cm}^3} = 8.76\text{ g}\cdot\text{cm}^{-3}$$

$$\text{Trial 3} \quad d = \frac{5.055\text{ g}}{0.575\text{ cm}^3} = 8.79\text{ g}\cdot\text{cm}^{-3}$$

The average of these results is simply the sum of the densities divided by the number of trials (3), so that

$$\text{average density} = \frac{(8.85 + 8.76 + 8.79)\text{ g}\cdot\text{cm}^{-3}}{3} = 8.80\text{ g}\cdot\text{cm}^{-3}$$

(b) The true value for the density of copper metal is given in the *CRC Handbook of Chemistry and Physics* as 8.96 $g\cdot cm^{-3}$ at 20°C. Using Equation 1.6 to determine the percentage error, we find

$$\text{percentage error} = \frac{\text{average value} - \text{true value}}{\text{true value}} \times 100$$

$$= \frac{8.80\text{ g}\cdot\text{cm}^{-3} - 8.96\text{ g}\cdot\text{cm}^{-3}}{8.96\text{ g}\cdot\text{cm}^{-3}} \times 100 = -1.8\%$$

Note that the negative sign in −1.8% tells us that our answer is 1.8% *below* the true value. Often only the absolute values of percentage errors are reported. Thus, the percentage error for this experiment could be reported as either −1.8% or simply 1.8%, depending on the convention used. Throughout this text we will include the signs when reporting percentage errors.

Figure 1.15 The *CRC Handbook of Chemistry and Physics* is a standard reference book that contains data agreed upon by a number of scientists for a wide variety of important measurements. Most school libraries contain a copy of this handbook and some have a subscription to the online edition as well. Most of the data found in the tables of this textbook are derived from this reference.

PRACTICE PROBLEM 1-8: A chemist is testing a new microscope designed to measure the diameter of ultrathin wires. In a test of the device, the following results for a wire with a known diameter of 2.00×10^{-7} meters were obtained:

Trial	Diameter
1	0.203 μm
2	0.209 μm
3	0.199 μm

Determine the percentage error for the average of these measurements.

Answer: 2%

1-7. The Precision of a Measured Quantity Is Indicated by the Number of Significant Figures

It is important in reporting scientific data to give both the value and the uncertainty (or precision) of the value. For example, if we walked into a pub and observed a darts player hit the bull's-eye of a target, we would have no information as to whether it was a "lucky shot" or the player was an international darts champion or somewhere in between. Only after observing a series of throws could we determine how good a darts player she is. For this same reason, experimental results are generally not considered valid without a number of repeated trials and a stated experimental precision or uncertainty.

Denoting the uncertainty of a measured quantity by the ± notation is desirable in describing scientific results, but such a notation can often be cumbersome. In this text we indicate the precision of measured quantities by the number of **significant figures** used to express a result.

All digits in a numerical result are significant if only the last digit has some uncertainty. A result expressed as 4776.2 grams means that there is an uncertainty of at least one unit in the last digit and that there are five significant figures in the result. Zeros are considered to be significant figures if they are not included just to locate the decimal point; thus, both zeros in 1001.2 are significant figures. Zeros are not considered to be significant figures if their presence serves only to position the decimal point, as in 0.001 25, which has only three significant figures. In certain cases, it is not clear just how many significant figures are implied. Consider the number 100. With the number presented as 100, we might mean that the value of the number is exactly 100. But the two zeros might not be significant and we might mean that the value is approximately 100—say, 100±10. The number of significant figures in such a case is uncertain. Usually, the number of significant figures can be deduced from the statement of the problem. It is the writer's obligation to indicate clearly the number of significant figures by appropriate use of decimal points and zeros. Thus, 100.0 has four significant figures and 35×10^3 has two significant figures.

A common convention is that 100 is interpreted as having one significant figure (the two zeros are taken to represent the placeholders), and 100. (with a decimal place after the last zero) is interpreted as having three significant figures. However, this convention is by no means a standard practiced by all authors, and it still leaves no way to express 100±10 except by the use of scientific notation, as 1.0×10^2. Thus, it is preferable to express numbers such as 100 in scientific notation to avoid any ambiguity in the number of significant figures implied.

The rules for determining the significant figures in a measured quantity can be summarized as follows:

1. All nonzero digits and zeros between nonzero digits are significant figures. For example, 4023 mL has four significant figures.
2. Zeros used solely to position the decimal point are not significant figures. For example, 0.000 <u>206</u> L has three significant figures (underlined). The zeros to the left of the 2 are not significant figures.
3. If a numerical result ends in one or more zeros to the right of the decimal point, then those zeros are significant figures. For example, 2.200 grams has four significant figures.
4. If a numerical result ends in zeros that are not to the right of a decimal point, then those zeros may or may not be significant figures. In such cases we must deduce the number of significant figures from the statement of the problem. For example, the statement "350 000 spectators lined the parade route" involves a number that probably has only two significant figures at best; we infer this because it is obvious that no one actually counted the spectators.
5. A useful rule of thumb to use for determining whether or not zeros are significant figures is that they are not significant if they disappear when scientific notation is used. For example, in the number 0.0197 ($= 1.97 \times 10^{-2}$) the zeros are not significant figures; in the number 0.01090 ($= 1.090 \times 10^{-2}$) the first two zeros are not significant and the second two are.
6. Numbers that can be counted exactly, such as 5 trials in an experiment or 60 carbon atoms in the formula C_{60}, and defined conversion factors, such as 1 meter = 100 centimeters, are considered **exact numbers**. For example, if you rode to class with 3 other people, there is no uncertainty to the number; it is not possible that there were actually 3.17 other people in the vehicle. Exact numbers have no limit to their precision and do not obey the rules given for significant figures (or alternatively, may be treated as having an infinite number of significant figures).

EXAMPLE 1-9: State the number of significant figures in each of the following numbers: (a) 0.0312; (b) 0.031 20; (c) 312 pages; (d) 3.1200×10^5 grams

Solution:

(a) 0.0<u>312</u> has three significant figures (underlined).

(b) 0.0<u>31</u> <u>20</u> has four significant figures (underlined).

(c) 312 pages is an exact (countable) number.

(d) <u>3.1200</u> $\times 10^5$ grams has five significant figures (underlined); grams is a measured (not counted) unit.

PRACTICE PROBLEM 1-9: Determine the number of significant figures in the following numbers: (a) The human population of the United States is about 301 000 000; (b) 30 006; (c) 0.002 9060; (d) 12 helium atoms.

Answer: (a) only 3; (b) 5; (c) 5; (d) exact

1-8. Calculated Numerical Results Should Show the Correct Number of Significant Figures

In scientific calculations we must express the final numerical result with the correct number of significant figures; otherwise, an incorrect impression of the precision of the results is conveyed.

In multiplication and division, the calculated result should not be expressed to more significant figures than the factor in the calculation with the least number of significant figures. For example, if we perform the multiplication 8.3145 × 298.2 on a hand calculator, the following result comes up on the calculator display: 2479.3839. However, not all the figures in this result are significant. The correct result is 2479 because the factor 298.2 has only four significant figures; thus, the result cannot have more than four significant figures. The extra figures are not significant and should be discarded.

EXAMPLE 1-10: Determine the result to the correct number of significant figures:

$$y = \frac{2.90 \times 0.082\,05 \times 298}{0.93}$$

Solution: Using a hand calculator, we obtain

$$y = 76.244\,742$$

The factor 0.93 has the least number of significant figures—only two. Thus, the calculated result is valid to only two significant figures. The correct result is $y = 76$.

PRACTICE PROBLEM 1-10: Determine the result of the following calculation to the correct number of significant figures:

$$y = \frac{8.314 \times 298.15}{96\,485.3}$$

Answer: 0.02569

In addition or subtraction, the calculated result should have no more figures after the decimal point than the least number of figures after the decimal point in any of the numbers that are being added or subtracted. Consider the sum

$$\begin{array}{r} 6.939 \\ +\ 1.007\,07 \\ \hline 7.946\,07 \end{array} = 7.946$$

The last two digits in 7.946 07 are not significant, because we know the value of the first number in the addition, 6.939, to only three digits beyond the decimal point. Thus, the result cannot be accurate to more than three digits past the decimal point. Therefore, our result expressed to the correct number of significant figures is 7.946. Here we have drawn a vertical red line to emphasize the cutoff for the last significant figure based on the number with the least number of significant figures after the decimal point.

Now consider the subtraction

$$y = 1.750 \times 10^6 - 2.50 \times 10^4$$

Here we must express each number to the same power of 10 and then perform the subtraction to determine the number of significant figures in the result. Thus, we begin by rewriting 2.50×10^4 as 0.0250×10^6 and then perform the subtraction.

$$\begin{array}{r} 1.750 \times 10^6 \\ -\ 0.0250 \times 10^6 \\ \hline 1.725 \times 10^6 \end{array}$$

The correct answer is $y = 1.725 \times 10^6$ to four significant figures. (We could just as well have solved this problem by converting 1.750×10^6 to a power of 10^4.)

In discarding nonsignificant figures, we use the following convention: if the digit following the last digit retained is a 5, 6, 7, 8, or 9, then the preceding digit should be increased by 1; otherwise (i.e., for 0, 1, 2, 3, and 4), the preceding digit should be left unchanged. Thus, rounding off the following numbers to three significant figures, we obtain 27.35 → 27.4, 27.34 → 27.3, and 27.348 → 27.3. Note that in half of the cases (0, 1, 2, 3, 4) we discard the nonsignificant digit, and in the other half (5, 6, 7, 8, 9) we increase the preceding digit by 1 when we discard the nonsignificant digits.

When combining mathematical operations, such as a calculation with a multiplication step followed by a subtraction step, the number of significant figures must be calculated separately for each step in the same order that the operations are performed.

EXAMPLE 1-11: Determine the result of the following calculation to the correct number of significant figures:

$$y = 2796.8 - 2795$$

Solution: Subtraction yields 1.8, which we must round off to 2, because the second number (2795) has no digits to the right of the decimal point; therefore, the correct result should have no digits to the right of the decimal point and have only one significant figure.

PRACTICE PROBLEM 1-11: Determine the result of the following calculation to the correct number of significant figures:

$$y = \frac{7.2960}{8.9000} - 132.0$$

Answer: −131.2

EXAMPLE 1-12: A student measures the density of an aluminum object and obtains the following results:

Trial	Density
1	3.1 g·cm^{-3}
2	3.0 g·cm^{-3}
3	2.7 g·cm^{-3}
4	3.3 g·cm^{-3}

Calculate the average density of the object to the correct number of significant figures.

Solution: The average density is given by

$$\text{average density} = \frac{(3.1 + 3.0 + 2.7 + 3.3)\ \text{g}\cdot\text{cm}^{-3}}{4}$$

$$= \frac{12.1\ \text{g}\cdot\text{cm}^{-3}}{4} = 3.03\ \text{g}\cdot\text{cm}^{-3}$$

The rule for addition and subtraction of significant figures is different from that for multiplication and division. You should keep this in mind when calculating averages or determining percentage errors.

This calculation involves an addition and a division step. Following the rules for addition and subtraction of significant figures, the addition step yields the value 12.1 g·cm^{-3}. Because the number of trials (4) is an exact number, we ignore the rules for significant figures in the division by 4, so our answer has three significant figures. Note that in this case the average experimental density has more significant figures than any of the individual measurements. This is not unusual because an average is generally more precise than any of the individual measurements from which it results. Thus, even for experiments with relatively large inherent measurement errors, one method of increasing the precision is to make many repeated measurements and average the results.

PRACTICE PROBLEM 1-12: A reference book lists the density of aluminum as 2.70 g·cm^{-3}. Calculate the percentage error in the average density of the aluminum object determined in Example 1-12 to the correct number of significant figures.

Answer: 12% (two significant figures)

In general, to avoid compounding rounding errors, it is best to perform all mathematical calculations to extra significant figures first and then trim the final answer to the correct number of significant figures, based on the operation in each step. For example, consider the operation

$$y = (1.0 + 0.46)^3$$

The addition in the first step yields $1.0 + 0.46 = 1.46$, which is equal to 1.5 to the correct number of significant figures. Cubing this number yields

$$y = (1.5)^3 = 3.375 = 3.4$$

to two significant figures (because cubing a number is the same as multiplying it by itself three times, we apply the rules for multiplication to powers). However, if we perform the same calculation on a handheld calculator and then trim the result to two significant figures, we get

$$y = (1.0 + 0.46)^3 = (1.46)^3 = 3.112\ 136 = 3.1$$

Thus, rounding in each step yields an overall error greater than our assumed precision in the last significant digit. That is,

$$3.4 \pm 0.1 \neq 3.1 \pm 0.1$$

In general, the more steps in an operation, the more such rounding errors can compound. Thus, to avoid rounding errors, it's best to perform all mathematical operations to extra significant figures first and then trim the final answer to the correct number of significant figures, based on the operations performed.

In carrying out calculations involving several steps, the final answer depends upon how many digits you keep in the intermediate steps, which is somewhat arbitrary. Although it is important to report answers to the correct number of significant figures, answers that differ a little in the last significant digit are *not* significantly different because of the uncertainty associated with this last digit. For example, a result reported to four significant figures that reads 0.3456 or 0.3457 is not significantly different.

1-9. Dimensional Analysis Is Used to Simplify Many Types of Chemical Calculations

Dimensional analysis is a particularly useful method for calculations that involve quantities with units. The basic idea involved in **dimensional analysis** is to treat the units of the various quantities involved in the calculations as quantities that follow the rules of algebra. The calculation is set up in such a way that the undesired units cancel out and the numerical answer is obtained in the desired units. If the answer obtained does not have the desired units, then we know that the procedure used to obtain the result is incorrect.

Let's consider some specific examples of the use of dimensional analysis in calculations. Only numbers that have the same units can be added or subtracted. If we add 2.12 cm and 4.73 cm, we obtain 6.85 cm. If we wish to add 76.4 cm to 1.19 m, we must first convert 76.4 cm to meters or 1.19 m to centimeters. We convert from one unit to another by using a **unit conversion factor**. Suppose we want to convert meters to centimeters. In Appendix B we find that

$$1\text{ m} = 100\text{ cm} \tag{1.7}$$

Equation 1.7 is a definition and therefore exact; there is no limit to the number of significant figures on either side of the equation. If we divide both sides of Equation 1.7 by 1 m, we get

$$1 = \frac{100\ \text{cm}}{1\ \text{m}} \tag{1.8}$$

Equation 1.8 is called a unit conversion factor, because we can use it to convert a number in meters to a number in centimeters by multiplying the number in meters by the unit conversion factor in Equation 1.8. A unit conversion factor, as shown in Equation 1.8, is equal to unity; thus, we can multiply any quantity by a unit conversion factor without changing its intrinsic value. If we multiply 1.19 meters by the unit conversion factor in Equation 1.8, then we obtain

$$(1.19\ \cancel{\text{m}})\left(\frac{100\ \text{cm}}{1\ \cancel{\text{m}}}\right) = 119\ \text{cm}$$

Notice that the unit of meter cancels, giving the final result in centimeters.

To convert 76.4 centimeters to meters, we use the reciprocal of Equation 1.8:

$$(76.4\ \cancel{\text{cm}})\left(\frac{1\ \text{m}}{100\ \cancel{\text{cm}}}\right) = 0.764\ \text{m}$$

Notice in this case that the unit of centimeter cancels out, giving the final result in meters. From these results, we see that the sum of 76.4 cm and 1.19 meters is

$$76.4\ \text{cm} + (1.19\ \text{m})\left(\frac{100\ \text{cm}}{1\ \text{m}}\right) = 195\ \text{cm}$$

or

$$(76.4\ \text{cm})\left(\frac{1\ \text{m}}{100\ \text{cm}}\right) + 1.19\ \text{m} = 1.95\ \text{m}$$

Notice that both results are given to three significant figures.

As another example of converting from one set of units to another, let's convert 55 miles per hour (conventionally abbreviated mph) to kilometers per hour. From the inside of the back cover, we find that

$$1\ \text{mile} = 1.609\ \text{km} \tag{1.9}$$

Dividing both sides of Equation 1.9 by 1 mile yields the unit conversion factor

$$1 = \frac{1.609\ \text{km}}{1\ \text{mile}} \tag{1.10}$$

Equation 1.10 is the unit conversion factor that is used to convert a speed given in miles per hour to a speed in kilometers per hour. Thus,

$$\left(\frac{55\ \text{miles}}{1\ \text{hour}}\right)\left(\frac{1.609\ \text{km}}{1\ \text{mile}}\right) = 89\ \text{km}\cdot\text{h}^{-1}$$

Note that the use of the proper units for each quantity provides an internal check on the correctness of the calculation. We must multiply 55 mph by 1.609 km/1 mile to obtain the result in the desired units of $\text{km}\cdot\text{h}^{-1}$. Also note that if we had used the conversion 1 km = 0.6214 miles from the inside of the back cover, then the unit conversion would be

$$\left(\frac{55\ \text{miles}}{1\ \text{hour}}\right)\left(\frac{1\ \text{km}}{0.6214\ \text{miles}}\right) = 89\ \text{km}\cdot\text{h}^{-1}$$

and the unit of miles once again cancels out.

EXAMPLE 1-13: Given that the unit conversion factor for the conversion of U.S. quarts to milliliters is 946.3 mL per quart (qt) and 1 L = 1000 mL, compute the number of liters in 1.00 U.S. gallon (gal) of gasoline. (There are 4 quarts in a U.S. gallon.)

Solution: Before starting, it is a good idea to create a plan of how we will perform the given conversion. Our objective is to convert gallons to liters. We can illustrate this process using an arrow to represent the plan as follows:

$$\text{gallons} \xrightarrow{?} \text{liters}$$

Because we are given conversions from gallons to quarts, we can write

$$\text{gallons} \xrightarrow{4\ \text{qt/gal}} \text{quarts} \xrightarrow{?} \text{liters}$$

as the first step in our plan. Continuing in this way, we outline a plan for the overall required conversion:

$$\text{gallons} \xrightarrow{4\ \text{qt/gal}} \text{quarts} \xrightarrow{946.3\ \text{mL/qt}} \text{milliliters} \xrightarrow{1\ \text{L}/1000\ \text{mL}} \text{liters}$$

Then using the given conversion factors and making sure the units cancel (as shown in red), we find number of liters of gasoline in 1.00 gallon of gasoline is

$$(1.00\ \text{gal})\left(\frac{4\ \text{qt}}{1\ \text{gal}}\right)\left(\frac{946.3\ \text{mL}}{1\ \text{qt}}\right)\left(\frac{1\ \text{L}}{1000\ \text{mL}}\right) = 3.79\ \text{L}$$

After some practice you will begin to "see" the plan in your head and will no longer need to write it out stepwise when working problems of this nature, but you should still always write out all the conversion factors and their units when performing a calculation. The canceling of unit conversion factors is an excellent double-check that you have performed all your calculations correctly.

PRACTICE PROBLEM 1-13: Suppose you are driving in France and the next town is 72 km away. How far away is it in miles? If the elevation of the town is 1260 meters, what is its elevation in feet?

Answer: 45 miles, 4130 ft

EXAMPLE 1-14: In most countries meat is sold in the market by the kilogram. Suppose the price of a certain cut of beef is 1400 pesos per kilogram and the exchange rate is 124 pesos to the U.S. dollar. What is the cost of the meat in dollars per pound (lb)?

Solution: The cost of the meat per pound obtained by using dimensional analysis is

$$\left(\frac{1400\ \cancel{\text{peso}}}{1\ \cancel{\text{kg}}}\right)\left(\frac{1\ \text{dollar}}{124\ \cancel{\text{peso}}}\right)\left(\frac{1\ \cancel{\text{kg}}}{2.20\ \text{lb}}\right) = 5.13\ \text{dollar}\cdot\text{lb}^{-1}$$

PRACTICE PROBLEM 1-14: A certain hard disk advertises that its average write-speed is 45 MBps (megabytes per second). Use the conversion factors in Appendix B to determine how long it would take to save 25 digital photos with a size of 1.2 GB (gigabytes) each to the hard disk (assuming that the write-speed of the hard disk is the rate-limiting factor in transferring the data).

Answer: 11 min

Figure 1.16 Gold ingots weighing one kilogram. Each ingot is stamped with its weight, purity, and reference number.

Now let's consider an example with several unit conversions involving quantities in compound units. It has been estimated that all the gold that has ever been mined would occupy a cube 19 meters on a side (Figure 1.16). Given that the density of gold is 19.3 $g\cdot cm^{-3}$, let's calculate the mass of all this gold. The volume of a cube 19 meters on a side is

$$\text{volume} = (19\ \text{m})^3 = 6859\ \text{m}^3 \tag{1.11}$$

Although the volume calculated in Equation 1.11 is good to only two significant figures (because the value 19 meters is good to only two significant figures), we shall carry extra significant figures through the calculation and then round off the final result to two significant figures. (As discussed before, this procedure minimizes accumulation of round-off errors in the calculation.) The mass of the gold is obtained by multiplying the density by the volume; but before doing this, we must convert cubic meters to cubic centimeters because the density is given in $g\cdot cm^{-3}$. From Appendix B, we find that

$$1\ \text{m} = 100\ \text{cm}$$

By cubing both sides of this expression, we obtain

$$(1\ \text{m})^3 = (100\ \text{cm})^3 = 1 \times 10^6\ \text{cm}^3$$

so the unit conversion factor is

$$1 = \frac{10^6\ \text{cm}^3}{1\ \text{m}^3}$$

Thus, the volume in cubic centimeters is equal to

$$\text{volume} = (6859\ \text{m}^3)\left(\frac{10^6\ \text{cm}^3}{1\ \text{m}^3}\right) = 6.859 \times 10^9\ \text{cm}^3$$

If we multiply the volume of the gold by the density of gold, then we obtain the mass of the gold:

$$\begin{aligned}\text{mass} &= \text{volume} \times \text{density}\\ &= (6.859 \times 10^9\ \text{cm}^3)\ (19.3\ \text{g}\cdot\text{cm}^{-3}) = 1.3 \times 10^{11}\ \text{g}\end{aligned}$$

The result is rounded off to two significant figures, because, as we mentioned before, the side of the cube (19 meters) is given to only two significant figures. In obtaining this result, we have used the fact that $(\text{cm}^3)(\text{cm}^{-3}) = 1$.

Let's see what this mass of gold would be worth at \$850 per troy ounce (troy oz), the price of gold in February 2009. (Gold is sold by the troy ounce, which is about 10% heavier than the avoirdupois ounce, the unit used for foods.) There are 31.1 grams in 1 troy ounce, so the unit conversion factor is

$$1 = \frac{1\ \text{troy oz}}{31.1\ \text{g}}$$

The mass of the gold in troy ounces is

$$\begin{aligned}\text{mass} &= (1.3 \times 10^{11}\ \text{g})\left(\frac{1\ \text{troy oz}}{31.1\ \text{g}}\right)\\ &= 4.2 \times 10^9\ \text{troy oz}\end{aligned}$$

At \$850 per troy ounce, the value in U.S. dollars of all the gold ever mined is

$$\begin{aligned}\text{value} &= (4.2 \times 10^9\ \text{troy oz})\left(\frac{\$850}{1\ \text{troy oz}}\right)\\ &= \$3.6 \times 10^{12} = 3.6\ \text{trillion U.S. dollars}\end{aligned}$$

Scientific calculations involve ways of handling numbers and units that may be new to you. If, however, you carefully set up the necessary conversion factors and make certain that the appropriate units cancel to give the units required for the answer—that is, if you use the dimensional analysis approach—then making unit conversions becomes a straightforward procedure.

1-10. The Guggenheim Notation Is Used to Label Table Headings and the Axes of Graphs

In presenting tables of quantities with units, it is convenient to use a column heading to specify the units rather than writing out the units next to each numerical entry. The least ambiguous way to do this is to write the name or symbol of the quantity followed by a slash and the symbol for the units. For example, the heading "Distance/m" indicates that the units of the numerical entries are distances expressed in meters. This notation is called the **Guggenheim notation** after E. A. Guggenheim, the British chemist who proposed its use.

To see how Guggenheim notation works, let's suppose that we wish to tabulate a number of masses, such as 1.604 grams, 2.763 grams, and 3.006 grams. We use the Guggenheim notation, "Mass/g," in the heading and list the masses as numbers without units, as shown in Table 1.4, column (a). Now suppose that later we wish to retrieve the values with their units. The heading indicates that the numbers in the column are masses divided by grams, so we write, for example, mass/g = 1.604. We can multiply both sides of the equation by g to obtain mass = 1.604 grams. Note that the heading is treated as an algebraic quantity and that we retrieve the data through an algebraic process.

TABLE 1.4 Tabulated data with headings in the Guggenheim notation

(a) Mass/g	(b) Mass/10^{-4} g
1.604	1.29
2.763	3.58
3.006	7.16

The Guggenheim notation is particularly convenient when the values to be tabulated are expressed in scientific notation. Suppose that we wish to tabulate the masses 1.29×10^{-4} grams, 3.58×10^{-4} grams, and 7.16×10^{-4} grams. In this case we can use the heading "Mass/10^{-4} g" to simplify these tabulated data, as shown in Table 1.4, column (b). To retrieve the data from the table, we write, for example, mass/10^{-4} g = 1.29, from which we get mass = 1.29×10^{-4} grams. Notice that the Guggenheim notation enables us to list tabular entries as unitless numbers.

EXAMPLE 1-15: Consider the following tabulated data:

Time/10^{-5} s	Speed/10^5 m·s^{-1}
1.00	3.061
1.50	4.153
2.00	6.302
2.50	8.999

Retrieve the actual data—the values and their units for time and speed, respectively—as four data pairs.

Solution: To find the actual times, we use, for example, time/10^{-5} s = 1.00, from which we obtain time = 1.00×10^{-5} s. The corresponding speed is given by speed/10^5 m·s^{-1} = 3.061, or speed = 3.061×10^5 m·s^{-1}. The other data pairs are 1.50×10^{-5} s, 4.153×10^5 m·s^{-1}; 2.00×10^{-5} s, 6.302×10^5 m·s^{-1}; and 2.50×10^{-5} s, 8.999×10^5 m·s^{-1}.

PRACTICE PROBLEM 1-15: The SI unit for the quantity of electrical charge is a coulomb (C). Use the Guggenheim notation to tabulate the following data: 7.05×10^{-15} C, 3.24×10^{-15} C, and 9.86×10^{-16} C.

Answer:

Charge/10^{-15} C
7.05
3.24
0.986

Looking through this book, you will see that the axes of graphs in figures are labeled like the column headings of tabulated data and that the numbers on the axes are unitless. The Guggenheim notation is especially useful for the graphical presentation of data because only dimensionless quantities can be graphed, and the retrieval of quantities with appropriate units from the graph is unambiguous when the Guggenheim notation is used on the graph axes.

EXAMPLE 1-16: Use the Guggenheim notation to label the graph axes and plot the following data:

v/m·s^{-1}	t/s
0	0
16	1.0
64	2.0
144	3.0
256	4.0

Solution: On a suitable piece of graph paper (Figure 1.17), we mark off the vertical axis (v/m·s^{-1}) and the horizontal axis (t/s). The data pairs are the coordinates of the various points on the graph and thus are used to position the points on the graph. Once the points are located, we draw a smooth curve through the points.

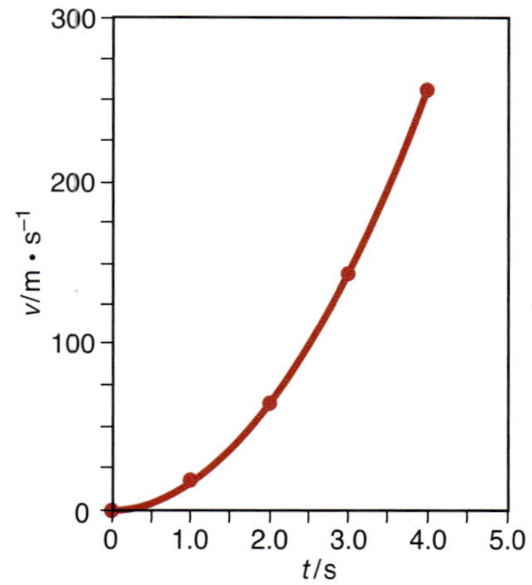

Figure 1.17 Plot of the data in Example 1-16 using the Guggenheim notation to label the axes.

PRACTICE PROBLEM 1-16: Replot the data given in Example 1-16 in the form v versus t^2. Label the axes according to the Guggenheim notation. What does your result tell you about the mathematical dependence of v on t?

Answer: The plot is a straight line. Therefore v is proportional to t^2.

If you encounter difficulty with Example 1-16 or Practice Problem 1-16, you should study Appendix A, which contains a review of mathematics.

The International Union of Pure and Applied Chemistry (IUPAC) is an international organization that proposes the units, symbols, conventions, and nomenclature to be used by scientists throughout the world. We shall refer to IUPAC proposals a number of times in later chapters.

The Guggenheim notation has been adopted by the International Union of Pure and Applied Chemistry (IUPAC), and is the officially recommended notation for labeling columns of data in tables and axes of graphs. Although older, less convenient, and often ambiguous, alternative conventions are still in use, they are being phased out in the scientific literature.

This section concludes our discussion of units and various aspects of scientific calculations, including dimensional analysis and significant figures. The next chapter is devoted to developing an understanding of the nature of elements and chemical compounds within the framework of Dalton's atomic theory.

SUMMARY

Chemistry is an experimental science based on the scientific method. Scientific questions are answered by carrying out appropriate experiments. Scientific laws are concise summaries of large numbers of experimental observations. Scientific theories are designed to provide explanations for the laws and observations, and are subject to ongoing refinement based on new observations and results.

The internationally sanctioned units for scientific measurements are the SI units. The basic SI unit for length is the meter, the basic SI unit for mass is the kilogram, and the basic SI unit for temperature is the kelvin. Compound units are combinations of basic SI units.

The SI unit of energy is the joule. Two forms of energy that we discussed in this chapter are kinetic energy, which is due to the motion of an object, and potential energy, which is due to its position. Although kinetic, potential, and other forms of energy may be exchanged, the total energy must always be conserved.

Accuracy and precision can be assessed using percentage error and significant figures, respectively. To carry out scientific calculations, you need to understand significant figures, units of measurement, and unit conversion factors. There are rules for determining the correct number of significant figures in a calculated result. Units must always be included with numbers that represent the results of measurements; otherwise, the numbers are meaningless. Similar quantities must be converted to the same type of units before they can be used together in calculations. Unit conversions are made with unit conversion factors. The general procedure for setting up a scientific calculation in such a way that all the units but those desired in the final result cancel out is called dimensional analysis.

Guggenheim notation is used to label table entries and graph axes so that we can retrieve without ambiguity the entries in a table or a graph and their appropriate units.

TERMS YOU SHOULD KNOW

scientific method *3*
qualitative data *3*
quantitative data *3*
hypothesis *3*
law of nature or scientific law *3*
theory *4*
quantitative measurement *5*
qualitative observation *5*
law of conservation of mass *5*
scientific notation *6*
metric system *7*
SI units *7*
length, l *7*
meter, m *7*
kilometer, km *8*
centimeter, cm *8*
giga-, G *8*
mega-, M *8*
kilo-, k *8*
centi-, c *8*
milli-, m *8*
micro-, μ *8*
nano-, n *8*
pico-, p *8*
femto-, f *8*
atto-, a *8*
volume, V *9*
liter, L *9*
milliliter, mL *9*
cubic centimeter, cm^3 *9*
kilogram, kg *9*
gram, g *9*
mass, m *9*
weight *9*
temperature *9*
temperature scale *9*
thermometer *10*
Kelvin temperature, T *10*
kelvin, K *10*
Celsius temperature, t *10*
Fahrenheit temperature, t_F *10*
compound units *11*
density, d *11*
dimensions *11*
intensive property *12*
extensive property *12*
kinetic energy, E_k *13*
potential energy, E_p *13*
joule, J *13*
attojoule, aJ *14*
gravitational acceleration constant, g *14*
law of conservation of energy *14*
power *16*
watt, W *16*
kilowatt-hour, kW·h *16*
accuracy *17*
precision *17*
percentage error *17*
true value *18*
significant figures *20*
exact number *21*
dimensional analysis *25*
unit conversion factor *25*
Guggenheim notation *30*

EQUATIONS YOU SHOULD KNOW HOW TO USE

$T\text{ (in K)} = t\text{ (in °C)} + 273.15$ (1.1) (relation between Kelvin and Celsius temperature scales)

$t\text{ (in °C)} = (5/9)\ [t\text{ (in °F)} - 32.0]$ (1.2) (relation between Celsius and Fahrenheit temperature scales)

$d = \frac{m}{V}$ (1.3) (definition of density)

$E_k = \frac{1}{2}mv^2$ (1.4) (kinetic energy)

$E_p = mgh$ (1.5) (potential energy of an object as a function of its height relative to the ground)

$\text{percentage error} = \frac{\text{average value} - \text{true value}}{\text{true value}} \times 100$ (1.6) (percentage error)

PROBLEMS

THE SCIENTIFIC METHOD

1-1. What is the difference between a hypothesis, a law, and a theory?

1-2. An experiment is performed that disproves a long-standing theory. According to the scientific method, how should the scientists involved proceed?

1-3. In the context of the scientific method, comment on the statement, "The theory of evolution is a fact."

1-4. In the context of the scientific method, comment on the statement, "No two snowflakes are alike."

SCIENTIFIC NOTATION

1-5. Express the following numbers in scientific notation:

(a) 0.000 000 0277

(b) 0.001 82

(c) 123 000 000

(d) 1254

1-6. Convert the following numbers expressed in scientific notation into decimal form

(a) 2.998×10^{8}

(b) $5.485\ 80 \times 10^{-4}$

(c) 5.292×10^{-11}

(d) 1.55×10^{15}

1-7. What is 1.206×10^{-23} expressed to the power of 10^{-25}?

1-8. What is 5.560×10^{5} expressed to the power of 10^{3}?

UNITS

1-9. What SI prefixes and symbols correspond to the following multipliers?

(a) 10^{-2} (b) 10^{-6}

(c) 10^{-3} (d) 10^{9}

(e) 10^{6} (f) 10^{-18}

1-10. What multipliers correspond to the following SI prefixes?

(a) pico (b) giga

(c) nano (d) kilo

(e) atto (f) femto

1-11. Arrange the following quantities in order of increasing length:

(a) 100 nm (b) 1.0 km

(c) 1.0×10^{3} cm (d) 100 pm

(e) 1.00×10^{3} nm (f) 1000 m

1-12. Arrange the following quantities in order of increasing volume:

(a) 10 L (b) 100 mL

(c) 0.10 ML (d) 1.0×10^{3} μL

(e) 20 cL (f) 1.0×10^{4} nL

1-13. The volume of a sphere is given by $V = (4/3)\pi r^{3}$, where r is the radius. Compute the volume of a sphere with a radius of 100.0 pm. State your answer in units of m^{3}.

1-14. The volume of a cube is given by $V = l^{3}$, where l is the length of an edge of the cube. Compute the volume of a cube with $l = 200.0$ pm. State your answer in units of m^{3}.

1-15. A 20.4-gram mass of a substance has a volume of 1.50 cm^{3}. Compute the density of the substance in $g\cdot cm^{-3}$.

1-16. The density of gold is 19.3 $g\cdot cm^{-3}$. Compute the volume in cm^{3} of 31.1 grams (1 troy ounce) of gold.

1-17. Helium has a normal boiling point of 4.22 K. Determine this temperature in degrees Celsius and Fahrenheit.

1-18. The title of a classic Ray Bradbury novel is *Fahrenheit 451*. What would the title of this novel be in degrees Celsius? In kelvin?

CONSERVATION OF ENERGY

1-19. What is the maximum height achieved if a 0.500-kg mass is thrown straight upward with an initial speed of 55 $m\cdot s^{-1}$? Ignore the effect of air resistance.

1-20. If a rock climber accidentally drops a 56-gram piton from a height of 375 meters, what would its speed be before striking the ground? Ignore the effects of air resistance.

1-21. A pole vaulter with a mass of 75 kg is running at a speed of 9.2 $m \cdot s^{-1}$ at the takeoff point. What height can this person clear if all this kinetic energy is directed vertically?

1-22. A 1200-kg car traveling at 89 $km \cdot h^{-1}$ (55 mph) runs out of gas at the bottom of a hill. Neglecting air resistance and the rolling resistance due to friction, calculate the height of the highest hill that the car can get over.

1-23. A typical U.S. home uses 40 kW·h of energy per day (remember that a watt is a joule per second). If one gallon of gasoline can generate 121 MJ of energy, how many gallons of gasoline does it take to power a typical U.S. home for a day? Which consumes more energy, your home or your car?

1-24. If electricity costs $0.10 per kW·h (remember that a watt is a joule per second) in the United States, what is the cost of operating a 300-watt halogen floor lamp five hours per day for a year? Why do you think halogen lamps are gradually being replaced by fluorescent lamps for use in the home?

PERCENTAGE ERROR

1-25. The melting point of a substance is found to be 128.7°C. The correct answer according to a reference handbook is 129.5°C. Calculate the percentage error in these results.

1-26. A physics student measures the speed of sound in air as 352 $m \cdot s^{-1}$. A reference source lists the speed as being 344 $m \cdot s^{-1}$. Calculate the percentage error in the student's experimental result.

1-27. The label on a package of vitamins claims that each pill contains 150 mg of vitamin C. In quantitative measurements of the vitamin C content, five random samplings of the product are found to have a vitamin C content of: 153.2 mg, 151.1 mg, 152.0 mg, 146.9 mg, and 149.8 mg. What is the percentage error in the average vitamin C content of the pills, using the label as the presumed true value?

1-28. A student is measuring the density of liquid mercury at 35°C and obtains the following results:

Trial	Density/$g \cdot cm^{-3}$
1	13.56
2	13.62
3	13.59

Using the *CRC Handbook of Chemistry and Physics* (which may be found in most libraries), another standard reference source, or the Internet, look up the density of mercury metal at 35°C and calculate the percentage error in this experiment. Be sure to cite your reference.

SIGNIFICANT FIGURES

1-29. Determine the number of significant figures in each of the following:

(a) 0.0390

(b) 6.022×10^{23}

(c) 3.652×10^{-5}

(d) 1 300 000 000 which is the 2007 population of China

(e) a hexagon has 6 sides

1-30. Determine the number of significant figures in each of the following:

(a) 578

(b) 0.000 578

(c) There are 1000 m in 1 km.

(d) $\pi \approx 3.141\,592\,65$

(e) 93 000 000 miles (the distance from the earth to the sun)

1-31. Calculate z to the correct number of significant figures in each part:

(a) $656.29 - 654 = z$

(b) $(27.5)^3 = z$

(c) $51/18.02 = z$

(d) $1.187 \times 10^{-3} - 9.5 \times 10^{-4} = z$

1-32. Calculate z to the correct number of significant figures in each part:

(a) $213.3642 + 17.54 + 32\,978 = z$

(b) $373.26 - 119 = z$

(c) $(6.626\,075 \times 10^{-34})(2.997\,925 \times 10^{8}) / (1.380\,66 \times 10^{-23}) = z$

(d) $(9.109\,390 \times 10^{-31} + 1.672\,62 \times 10^{-27} - 1.674\,93 \times 10^{-27})(2.997\,925 \times 10^{8})^2 = z$

1-33. The area of a circle is πr^2, where r is the radius. Calculate the area of a circle 20.55 cm in diameter to the correct number of significant figures. Express your answer in units of cm^2.

1-34. The edge length of a cubic crystal is 133 pm. Calculate the volume of the crystal to the correct number of significant figures. Express your answer in units of pm^3.

UNIT CONVERSIONS AND DIMENSIONAL ANALYSIS

1-35. Perform the following SI unit conversions:

(a) 21.5 cm to m

(b) 7.56×10^{-6} m to μm

(c) 1.2×10^{-7} mg to pg

(d) 15 000 000 kilobytes to gigabytes

(e) 6.67 mm^3 to cm^3

1-36. Perform the following SI unit conversions:

(a) 1.259×10^3 J to kJ

(b) 2.18×10^{-18} J to aJ

(c) 5.5 MJ to kJ

(d) 7.5×10^{-3} ns to fs

(e) 2.0 m^3 to mL

1-37. Use the information from the inside of the back cover to make the following unit conversions, expressing your results to the correct number of significant figures:

(a) 1.00 liters to quarts

(b) 186 000 miles per second to meters per second

(c) 8.3145 $J \cdot K^{-1} \cdot mol^{-1}$ to $cal \cdot K^{-1} \cdot mol^{-1}$

1-38. Use the information from the inside of the back cover to make the following unit conversions, expressing your results to the correct number of significant figures:

(a) 325 feet to meters

(b) 1.54 angstroms (Å) to picometers and to nanometers

(c) 175 pounds to kilograms

1-39. How many minutes does it take for light to travel from the sun to the earth, given that the sun is about 93 million miles from the earth?

1-40. In older U.S. cars, total cylinder volume is expressed in cubic inches. Compute the total cylinder volume in liters of a 454-cubic-inch engine.

1-41. An office heater puts out 2440 BTUs (British thermal units) of energy in the form of heat per hour. Given that 1 BTU = 1.055 kJ (kilojoules), how many megajoules of energy in the form of heat can be produced per year by the heater (assume continuous operation)?

1-42. A particular MP3 encoder can convert 75 kilobytes of music per second from a typical music CD. How many minutes will it take to convert 12.8 megabytes of music from a CD?

1-43. A carpet sells for $28.99 a square yard. What is the price of the carpet per square meter? How much will it cost to carpet an area of 1475 square feet?

1-44. Using the conversion factors 100 cm = 1 m, 1 cm^3 = 1 mL, and 1 L = 1000 mL, determine the number of (a) milliliters and (b) liters in a cubic meter.

1-45. A certain process can plate out 80.6 μm^3 of gold onto a metallic surface per second. How long will it take to create a 0.010-μm-thick coating on a surface area of 0.30 cm^2 using this process?

1-46. A 200-mL Erlenmeyer flask weighs 215.8 grams when empty and 297.1 grams when filled with an unknown alcohol to the 125-mL mark. Determine the density of the alcohol in $g \cdot mL^{-1}$.

GUGGENHEIM NOTATION

1-47. In an experiment to determine the relationship between the temperature of a gas and its volume, the following data were determined: at 0.0°C, the volume of the gas was 1.00 liters; at 100°C, the volume was 1.37 liters; at 200°C, the volume was 1.73 liters; and at 300°C, the volume was 2.10 liters. Make a table of these data using the Guggenheim notation.

1-48. In an experiment to determine the acceleration of a car, the following data were determined: at 0 seconds, the car was at rest; at 2.0 seconds, the distance traveled was 51 feet; at 4.0 seconds, the distance traveled was 204 feet; and at 6.0 seconds, the distance traveled was 459 feet. Make a table of these data using the Guggenheim notation.

1-49. Plot the following data on a piece of graph paper, using the Guggenheim notation to label the axes:

$[N_2O_5]/10^{-2}$ M	t/min
1.24	0
0.62	23
0.31	46
0.16	69
0.080	92

1-50. Consider the following tabulated data:

Height/ft	Time/s
0.87	6.09
1.40	11.65
1.99	18.11
4.24	30.41

Convert the height to centimeters and construct a new table.

ADDITIONAL PROBLEMS

1-51. Which of the following are *extensive* properties?

(a) mass (b) volume

(c) boiling point (d) density

1-52. The following statements (numbered here for convenience) are excepted from a description of the element lead in a reference handbook:

i. The molecular mass of lead is 207.2.

ii. Its elemental symbol is Pb.

iii. It is a bluish-white metal of bright luster.

iv. Lead has a density of 11.3 g·cm^{-3}.

v. It is very soft, highly malleable, ductile, and a poor conductor of electricity.

vi. Lead has a melting point of 327.462°C and a boiling point of 1749°C.

Classify each of these statements as qualitative or quantitative descriptions of lead.

1-53. What is the difference between precision and accuracy?

1-54. What information does the number of significant figures convey about a measurement?

1-55. What information does the percentage error convey about a measurement?

1-56. Can a value ever have 100% precision?

1-57. A general rule of thumb when performing scientific calculations is to use at least one more significant figure in all known constants than are present in one's experimentally measured data. Why is this?

1-58. A joke reads, "A couple is visiting a museum when the tour guide tells them that a particular moon rock is one million and six years old. The couple is amazed and asks how the age was determined so precisely. The guide says, 'I don't know how they do it, but when I started working here six years ago the rock was labeled as being one million years old, so now that makes it one million and six years old.'" Using the concept of significant figures, explain the error made by the guide.

1-59. Several groups of students are attempting to determine the density of a lead weight by various methods. The data collected by each of the groups are indicated below.

	Measured density/g·cm^{-3}		
Group	Trial 1	Trial 2	Trial 3
1	12.7	11.2	10.3
2	11.5	11.4	11.4
3	10.9	11.3	11.1

A handbook lists the density of lead as 11.3 g·cm^{-3}. (a) For each of the three group's data, calculate the average experimental density of the lead weight determined by the group and the percentage error in the group's average density to the correct number of significant figures. (b) Which of the group's data are the most accurate? (c) Which are the most precise?

1-60. (a) Using *all* of the density data given in the previous problem, calculate the entire class' average experimental density for the lead weight and the percentage error in the class' average density to the correct number of significant figures. (b) Comment on the precision and accuracy of the overall class data compared to those of the individual groups calculated in the previous problem.

1-61. Estimate the volume of liquid in this graduated cylinder to the correct number of significant figures.

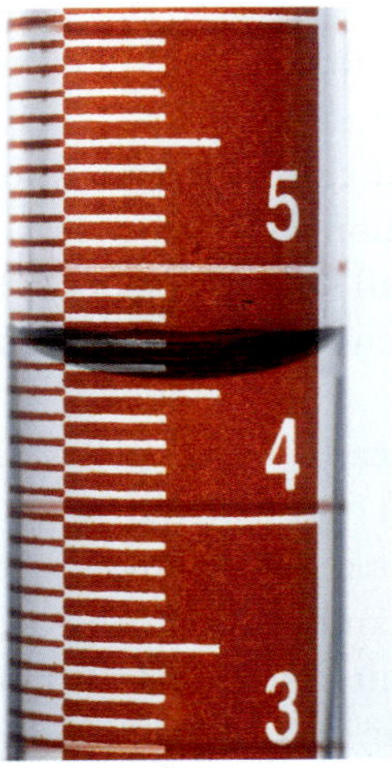

1-62. Estimate the temperature shown on this thermometer to the correct number of significant figures.

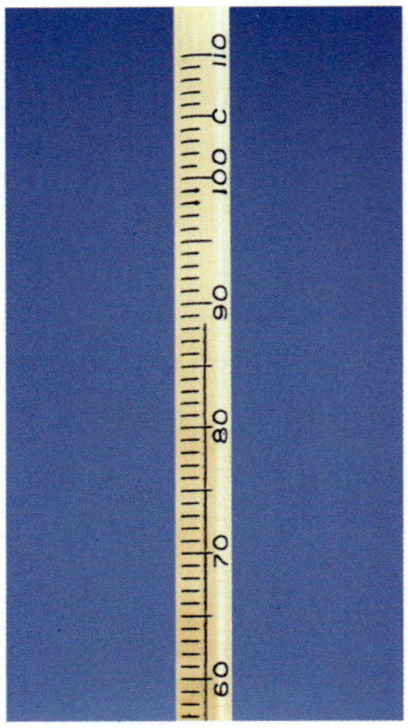

1-63. Sulfuric acid sold for laboratory use consists of 96.7% by mass sulfuric acid; the rest is water. The density of the solution is 1.845 g·cm^{-3}. Compute the number of kilograms and pounds of sulfuric acid in a 2.20-liter bottle of laboratory sulfuric acid.

1-64. Calculate the volume of 55 grams of benzene, given that the density of benzene is 0.879 g·cm^{-3}.

1-65. According to the official major league baseball rules, a regulation baseball "shall weigh not less than five nor more than 5¼ ounces avoirdupois and measure not less than nine nor more than 9¼ inches in circumference." Compute the allowed range in density of an official major league baseball in ounces per cubic inch.

1-66. The new Freedom Tower being built in Manhattan is planned to be 1776 feet tall and will become the world's tallest building. How much energy is required to transport a 180-pound man to the top of the tower on one of the building's elevators? (Assume the only energy expended is in lifting the man.)

1-67. If electricity costs $0.10 per kW·h in the United States, what is the annual cost savings in electricity associated with exchanging a 100-watt incandescent lightbulb in a lamppost used six hours per day for an energy-efficient 25-watt fluorescent bulb that produces the same amount of light? A typical incandescent bulb costs $0.89 and lasts for about a year; a typical energy-efficient fluorescent bulb costs about $3.49 and lasts for about 3 years. Is the additional cost of the fluorescent bulb justified?

1-68. Today, the best commercially available solar panels are about 18% efficient at converting the energy from sunlight into electricity. A typical home in the United States uses about 40 kW·h of electricity a day, and U.S. electricity costs currently average about $0.10 per kW·h. (a) Assuming 8.0 hours of useful sunlight per day with an incident solar energy of 0.5 kW·m^{-2}, calculate the minimum panel area needed to provide for a typical home's electrical needs. Will a solar panel of this size fit on the roof of a typical single-family home? (b) If it costs $15 000 to install a solar panel of this size, how long will it take to pay for itself? Assume U.S. energy prices remain constant over this time.

1-69. *Horsepower* is a unit of power; one horsepower (1 hp) is equivalent to 745 watts. The world record for climbing a 25-feet rope was set by Garvin Smith in Los Angeles in 1947, when he climbed the 25 feet in 4.7 seconds. Assuming that Mr. Smith has a mass of 65 kg, calculate the average power (in both watts and horsepower) that he generated during his climb.

1-70. Certainly one of the most famous equations in science is $E = mc^2$, which comes from Einstein's theory of relativity. Einstein was the first to show that mass can be converted into energy, and that energy can be converted into mass. This equation relates the amount of energy produced when the mass of a system decreases, or conversely, the amount of energy that is required to increase the mass of a system. In either case, the equation relates the change in energy associated with a change in mass, and in the notation of this chapter, Einstein's mass-energy relation is written as

$$\Delta E = c^2 \Delta m$$

where ΔE is the change in energy, c is the speed of light, and Δm is the change in mass. Calculate the amount of energy produced when a 1.0-gram mass is completely converted into energy.

1-71. The speed of sound at sea level is about 770 mph. Calculate the time it takes for sound to travel 1.00 miles at sea level and compare that value with the time it takes light to travel the same distance.

1-72. The density of air at sea level and 98°F is about 1.20 $g \cdot L^{-1}$. The volume of the adult human lungs in the expanded state is about 6.0 liters. Given that air is 20% oxygen by mass, compute the number of grams of oxygen in your lungs.

1-73. The density of pure gold is 19.3 $g \cdot cm^{-3}$ at 20°C. A quantity of what appears to be gold has a mass of 465 grams and a volume of 26.5 milliliters. Is the substance likely to be pure gold?

1-74. A simple laboratory gas burner (Bunsen burner) has a flame temperature of about 1200°C. Calculate the corresponding Fahrenheit and Kelvin temperatures.

1-75. Derive the following relationship between the Kelvin and Fahrenheit temperature scales:

$$T \text{ (in K)} = 255.37 + \frac{5}{9} t \text{ (in °F)}$$

What is the Fahrenheit temperature that corresponds to absolute zero on the Kelvin temperature scale?

1-76. A container that can hold 6780 grams of mercury can hold only 797 grams of carbon tetrachloride. Given that the density of mercury is 13.6 $g \cdot mL^{-1}$, calculate the density of carbon tetrachloride.

1-77. There are exactly 640 acres in a square mile. Determine the number of square meters in an acre.

1-78. A hectare is an old metric unit corresponding to a square of 100 meters on a side. Using the result of the previous problem, calculate how many acres there are in a hectare.

1-79. Aerogel is the substance with the lowest known density of any solid. Aerogel has some remarkable properties, for example, it is the best known insulator of any solid. NASA recently used aerogel to capture dust particles from the tail of a comet, returning these particles safely to earth in its Stardust mission. Determine the mass in kilograms of a cubic meter of aerogel, which has a density of 1.5 $mg \cdot cm^{-3}$.

1-80. Aluminum has a density of 2.70 $g \cdot cm^{-3}$. If a sheet of aluminum foil measuring 20.5 centimeters in length and 15.2 centimeters in width has a mass of 1.683 grams, what is the thickness of the foil in millimeters?

1-81. Copper metal is highly ductile, meaning it can be drawn into thin wires. How many meters of gauge-10 (0.1019-inch diameter) cylindrical copper wire can be made from a 10.0-kilogram block of copper with a density of 8.96 $g \cdot cm^{-3}$? How many meters of gauge-24 wire (0.0201-inch diameter) can be made from the same block?

1-82. Devise an experiment to determine the total volume of your own body and your density.

1-83. Given the air in a building has a density of 1.184×10^{-3} $g \cdot cm^{-3}$, how much does the air weigh (in pounds) in a room measuring 30.0 feet and 7.0 inches wide by 41.0 feet long by 9.00 feet high?

1-84. When you follow the manufacturer's specifications, 1.0 gallon of a latex, gloss enamel paint supposedly covers 350 square feet of a sealed surface. What is the average thickness (in millimeters) of a coat of this paint?

1-85. (*) The outer atmosphere of Jupiter contains frozen clouds of ammonia gas. Suppose that a Jovian being creates a temperature scale based on the freezing point of ammonia as 0°J (degrees Jove) and boiling point of ammonia as 100°J, which correspond to –77.7°C and –33.4°C, respectively. The being divides its temperature scale into 100 evenly spaced degrees. (a) Derive a formula for the conversion factor between °J and °C. (b) How does the magnitude of one °J compare with the magnitude of one °C? (c) What are the normal freezing and boiling points of water in °J?

1-86. (*) The U.S. penny was originally minted entirely of copper. However, after the year 1857 the U.S. penny was made of a variety of metals because of the high cost of copper. Since 1982 pennies have been made of zinc plated with a thin layer of copper. The modern penny weighs 2.500 grams, has a diameter of 19.05 millimeters, and an average thickness of 1.224 millimeters. Given that the density of copper and zinc are 8.96 $g \cdot cm^{-3}$ and 7.13 $g \cdot cm^{-3}$, respectively, determine the percentage by mass of copper in a modern penny.

Marie Curie, born Maria Sklodowska (1867–1934), right, was born in Warsaw (now Poland, then part of the Russian Empire) to schoolteachers. She worked as a governess to support her older sister's study of medicine in Paris. In 1891, she joined her sister to study at the University of Paris. She received her doctorate of science degree there in 1903, the first woman in France to do so. While there she met and married Pierre Curie, a member of the University's physics faculty. Together they studied radioactive materials and isolated the radioactive elements radium and polonium (which they named) in their dilapidated laboratory. In 1903, she and Pierre were awarded the Nobel Prize in Physics along with Henri Becquerel, the discoverer of radioactivity. In 1911, she was awarded the Nobel Prize in Chemistry for her discovery and isolation of radium and polonium. In spite of wide scientific acclaim, she was not elected to the French Academy of Sciences because of the prevailing prejudice against women. She died from aplastic anemia, almost certainly due to exposure to radiation.

Irene Joliet-Curie (1897–1956), left, was born in Paris to Marie and Pierre Curie. She had an unconventional early education, but eventually entered the Sorbonne. Her education was interrupted by World War I, during which she and her mother ran 20 mobile field hospitals equipped with primitive X-ray equipment. After the war, she entered the Radium Institute, founded by her parents in Paris, and completed her doctorate in 1925. While a student there, she instructed Frederic Joliet in techniques for research involving radioactive material. They married and together continued the study of atomic nuclei. The two created radioactive nitrogen, phosphorus, and silicon from nonradioactive isotopes. In 1935, they were awarded the Nobel Prize in Chemistry for their discovery of artificial radioactivity. A laboratory accident involving polonium led to Irene's developing a fatal case of leukemia. Marie and Irene Curie are the only mother-daughter recipients of the Nobel Prize.

2. Atoms and Molecules

The millions of chemicals known today are made up of little more than 100 elementary components—the chemical elements. In the early 1800s, an English schoolteacher named John Dalton drew on the idea of the elements and their chemical combinations to propose the atomic theory. Dalton began with the idea of structureless, solid spheres, which he called atoms. In Dalton's atomic theory, all substances consist either of atoms or molecules (which are groups of atoms joined together). Dalton's theory gives a simple picture of chemical reactions and explains countless chemical observations. Building on Dalton's work, experiments in the last years of the nineteenth century and the early years of the twentieth century led to a dramatically new view of the atom—the nuclear model.

2-1. Elements Are the Simplest Substances

Almost all the millions of different chemicals known today can be broken down into simpler substances. Any substance that cannot be broken down into simpler substances is called an **element.** A substance that can be broken down into two or more elements is called a **compound.** Before the early 1800s, many substances were classified incorrectly as elements because methods to break them down had not yet been developed, but these errors were gradually rectified over the years. Although our definition of an element is a satisfactory working definition at this stage, we shall soon learn the modern definition of an element: *An element is a substance that consists only of atoms with the same nuclear charge.*

There are currently almost 120 known chemical elements. It turns out that 99.99% of all known substances are made from only about 40 elements, which makes the other 80 fairly rare. Table 2.1 lists the most common elements found in the earth's crust, oceans, and atmosphere (but not the core, which is thought to be mostly iron). Note that only 10 elements make up over 99% of the total mass. Oxygen and silicon are the most common elements on earth; they are the major constituents of sand, soil, and rocks. Oxygen also occurs as a free element in the atmosphere and in combination with hydrogen in water. Table 2.2 lists the most common elements found in the human body. Again, only 10 elements constitute over 99% of the total mass of the human body. Because about 70% of the mass of the human body is water, much of this mass is oxygen and hy-

TABLE 2.1 Elemental composition of the earth's surface, which includes the crust, oceans, and atmosphere

Element	Percent by mass
oxygen	49.1
silicon	26.1
aluminum	7.5
iron	4.7
calcium	3.4
sodium	2.6
potassium	2.4
magnesium	1.9
hydrogen	0.88
titanium	0.58
chlorine	0.19
carbon	0.09
all others	0.56

TABLE 2.2 Elemental composition of the human body

Element	Percent by mass
oxygen	61
carbon	23
hydrogen	10
nitrogen	2.6
calcium	1.4
phosphorus	1.1
sulfur	0.20
potassium	0.20
sodium	0.14
chlorine	0.12
other trace elements	0.24

drogen. Outside of planet Earth, evidence from the study of the spectra of stars and nebulae suggest that one element, hydrogen, accounts for more than 90% of the atoms and about 75% of the observable mass of the universe. The other elements are thought to be the by-products of nuclear processes occurring in stars.

The elements can be divided into two broad classes: **metals** and **nonmetals.** We are all familiar with the properties of solid metals. They have a characteristic luster, can be cast into various shapes, and are usually good conductors of electricity and heat. In addition, they are **malleable,** a term that means they can be rolled or hammered into sheets, and **ductile,** a term that means they can be drawn into wires.

About three-fourths of the elements are metals. All the metals except mercury are solids at room temperature (about 20°C). Mercury is a shiny, silver-colored liquid at room temperature (Figure 2.1). Mercury used to be called quicksilver because of its silvery luster and the tendency of drops of mercury to roll rapidly on nonlevel surfaces.

Table 2.3 lists some common metals and their chemical symbols, and Figure 2.2 shows some of these metals. **Chemical symbols** are abbreviations used to designate the elements. They usually consist of the first one or two letters in the name of the element, but some chemical symbols do not seem to correspond at all to the elements' names. These symbols are derived from the Latin names of the elements (Table 2.4). You should memorize the chemical symbols of the more common elements because we shall be using them throughout this book. The origins of the names of the elements are discussed in Interchapter A.

WWW

See Interchapter A at www.McQuarrieGeneralChemistry.com.

Unlike metals, nonmetals vary greatly in their appearance. Over half of the nonmetals are gases at room temperature; the others are solids, except for bromine, which is a red-brown, corrosive liquid (Figure 2.1). In contrast to metals,

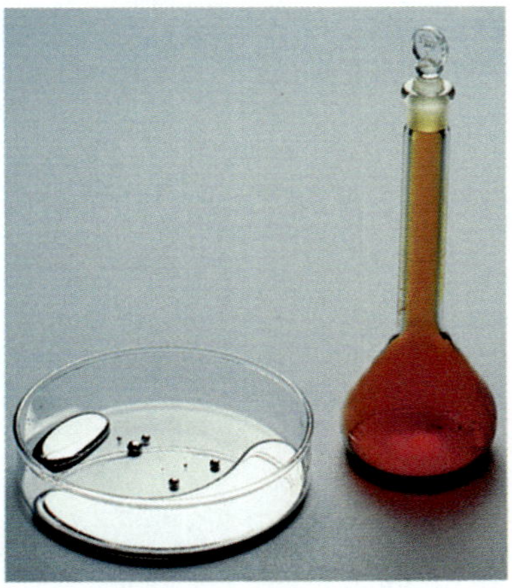

Figure 2.1 Mercury and bromine are the only elements that are liquids at room temperature (20°C). The red-brown vapor above the liquid bromine is bromine gas.

TABLE 2.3 Common metals and their chemical symbols

Element	Symbol	Element	Symbol
aluminum	Al	mercury	Hg
barium	Ba	nickel	Ni
cadmium	Cd	platinum	Pt
calcium	Ca	potassium	K
chromium	Cr	silver	Ag
cobalt	Co	sodium	Na
copper	Cu	strontium	Sr
gold	Au	tin	Sn
iron	Fe	titanium	Ti
lead	Pb	tungsten	W
lithium	Li	uranium	U
magnesium	Mg	zinc	Zn
manganese	Mn		

Figure 2.2 Common metals. Clockwise starting with the cylinder at top center are titanium, nickel, copper, aluminum, iron, and zinc.

Figure 2.3 Common nonmetals. Top row (*left to right*): arsenic, iodine, and selenium; bottom row (*left to right*): sulfur, carbon, boron, and phosphorus.

nonmetals are poor conductors of electricity and heat, they cannot be rolled into sheets or drawn into wires, and they do not have a characteristic luster. Table 2.5 lists several common nonmetals, their chemical symbols, and their appearances and Figure 2.3 shows some of these common nonmetals. Note that several of the symbols of the nonmetallic elements in Table 2.5 have a subscript 2. This number indicates that these elements—hydrogen (H_2), nitrogen (N_2), oxygen (O_2), fluorine (F_2), chlorine (Cl_2), bromine (Br_2), and iodine (I_2)—exist in nature as a unit consisting of two **atoms** that are joined together. A unit consisting of two or more atoms that are joined together is called a **molecule.** A molecule consisting of just two atoms is called a **diatomic molecule.** Scale models of diatomic molecules are shown in Figure 2.4.

When we refer to any of the naturally occurring diatomic elements by name, we shall assume the diatomic formulas listed here unless otherwise specified. For example, *oxygen* refers to the diatomic formula O_2, but an *oxygen atom* is O. It so happens that only the natural diatomic elements have names ending with *-ine* or *-gen;* no other element's name has these endings.

TABLE 2.4 Elements whose symbol corresponds to the Latin name

Element	Symbol	Latin name
antimony	Sb	stibnum
copper	Cu	cuprum
gold	Au	aurum
iron	Fe	ferrum
lead	Pb	plumbum
mercury	Hg	hydrargyrum
potassium	K	kalium
silver	Ag	argentum
sodium	Na	natrium
tin	Sn	stannum

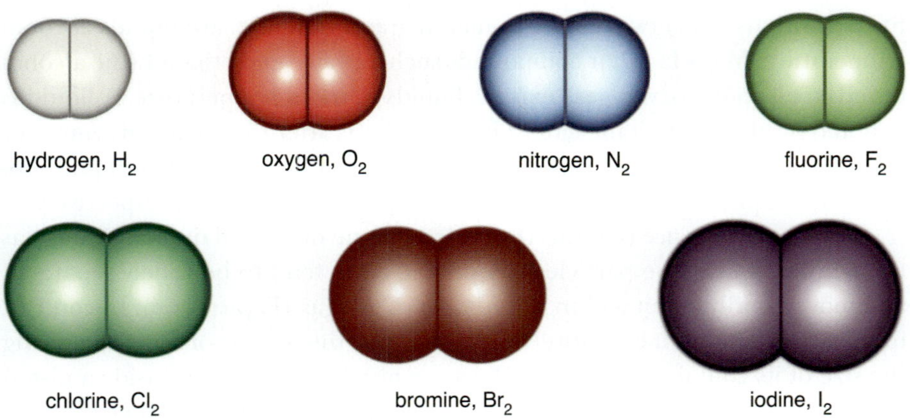

Figure 2.4 Scale models of molecules of hydrogen, oxygen, nitrogen, fluorine, chlorine, bromine, and iodine. These substances exist as diatomic molecules in their natural states but are still classified as elements because their molecules consist of identical atoms.

WWW

See the Interchapters at www.McQuarrieGeneralChemistry.com.

Both the metals and nonmetals can be further subdivided into various subclasses and groups based on their physical and chemical properties. These will be treated in detail in Chapter 3 and the Interchapters.

2-2. The States of Matter Include Solids, Liquids, and Gases

In nature, pure elements and compounds are found in various physical states. These **states of matter** include solids, liquids, and gases. Although you probably already have some familiarity with these various states of matter, we shall briefly define each of these states in turn.

We have already seen that all the metallic elements (except mercury) are solids at room temperature. Many compounds you are familiar with, such as table salt, limestone, and quartz, also exist as solids at room temperature. A **solid** can be characterized as having a fixed volume and a fixed shape because the particles that make up a solid are held together in a rigid, well-defined lattice (Figure 2.5a). Later on we shall learn about the forces between the constituent particles of a solid that hold them in fixed lattices and learn to characterize various kinds of different solids. Not all solids are hard or rigid. For example, gold is quite soft and malleable, a property that allows it to be worked into jewelry or thin sheets known as gold leaf.

TABLE 2.5 Some common nonmetals and their appearances at room temperature

Element	Symbol*	Appearance
Gases		
hydrogen	H_2	colorless
helium	He	colorless
nitrogen	N_2	colorless
oxygen	O_2	colorless
fluorine	F_2	pale yellow
neon	Ne	colorless
chlorine	Cl_2	green-yellow
argon	Ar	colorless
krypton	Kr	colorless
xenon	Xe	colorless
Liquids		
bromine	Br_2	red-brown
Solids		
carbon	C	black (in the form of coal or graphite)
phosphorus	P	pale yellow or red
sulfur	S	lemon yellow
iodine	I_2	violet-black

*A subscript 2 tells us that, at room temperature, the element exists as a diatomic molecule.

A solid is denoted by placing an italic letter *s* in parenthesis after the chemical symbol or formula of the solid element or molecule. For example, at room temperature iron and table salt are solids, so we write their chemical formulas as Fe(*s*) and NaCl(*s*) to indicate the states of these substances. We shall do this throughout the book to help you remember the state of various substances at room temperature. (We shall learn how to write the chemical formulas of compounds later in this text; for now these are simply given by way of illustration.)

A **liquid** is a substance that has a definite volume, but not a specific shape. When poured into a container, a liquid fills up the container to the extent of its volume and conforms to the shape of the container. In doing so, it forms a surface within the container. Like a solid, the particles in a liquid are held together by the forces between them. However, in contrast to a solid, the particles in a liquid are not fixed in their positions but are free to move about within the volume of the liquid (Figure 2.5b). Even though a liquid has a fixed volume, it does not have a fixed shape because its constituent particles are not held in fixed positions. Only two elements, mercury and bromine, are liquids at room temperature. Many familiar compounds such as water and ethanol (the alcohol found in alcoholic beverages) exist as liquids at room temperature. A liquid is denoted by placing an italic letter *l* in parenthesis after the chemical symbol or formula of the liquid; for example, the formulas of liquid mercury, liquid bromine, and liquid water are written as Hg(*l*), Br_2(*l*) and H_2O(*l*), respectively.

We shall include the phases of compounds at room temperature. You are certainly not expected to know these (few chemists do). If you ever want to find out the phase of a compound, simply enter the name of the compound into an Internet search engine.

A **gas** is a substance that fills the entire volume of its container and thus has no definite shape. The particles comprising a gas tend to be widely separated and move rapidly about within the volume of the gas (Figure 2.5c). In contrast to solids and liquids, it is relatively easy to change the volume of a gas by altering the size of its container. For example, a gas may be compressed inside a piston such as a bicycle pump. We shall study gases in more detail in Chapter 13. A gas is denoted by placing an italic letter *g* in parenthesis after the chemical symbol or formula of the gaseous element or compound. For example, nitrogen, oxygen, and carbon dioxide are all gases at room temperature and normal at-

Figure 2.5 (a) A solid, like the one illustrated here, is rigid and does not assume the shape of its container. From a microscopic perspective we can picture a solid as an ordered lattice or a network of atoms or molecules. (b) A liquid fills its container to a definite volume and forms a surface, such as the one illustrated here. From a microscopic view, in a liquid the particles are held together by forces between the particles, but are still free to move about randomly throughout the volume of the liquid. (c) A gas fills the entire container. From a microscopic view, a gas consists of rapidly moving particles that are free to move about the full volume of the container. However, most of the volume of a gas is empty space.

mospheric pressure, so under these conditions we write their formulas as $N_2(g)$, $O_2(g)$, and $CO_2(g)$, respectively.

2-3. A Mixture Can Be Separated by Taking Advantage of the Different Physical Properties of Its Components

Most substances in nature occur as **mixtures**, in which the component substances exist together without combining chemically. A familiar example of a mixture is air, which consists of 78% nitrogen and 21% oxygen with small amounts of argon, water vapor, and carbon dioxide. Another example is a mixture of salt and pepper. Not only most naturally occurring substances but also many laboratory preparations consist of mixtures.

When determining physical properties such as density or melting point or the chemical properties of an element or a compound, chemists must be certain that the substance is reasonably pure, otherwise the results have very limited applicability in that they are restricted to the particular impure sample that was studied. Thus, it is often necessary for chemists to separate the various pure components from a mixture.

Let's consider the problem of separating a mixture composed of sugar, sand, iron filings, and gold dust into its four components (Figure 2.6). The first thing to recognize about the mixture is that it is **heterogeneous**; that is, it is not uniform from point to point. The heterogeneity of the mixture can be seen clearly with the aid of a microscope (Figure 2.6c). We could separate the four components of the mixture by using tweezers, a microscope, and a lot of time and patience; however, we can achieve much more rapid separations with other methods. We can separate the iron filings from the mixture by using a

(a) (b) (c) (d)

Figure 2.6 (a) A mixture of sugar, sand, iron filings, and gold dust. The components of the mixture cannot be determined by casual inspection. (b) The pure, separated components of the mixture. (c) A microscopic view of the mixture. Note that the mixture is heterogeneous (i.e., not uniform from point to point) and that each of the four components is clearly distinguishable. (d) A magnet can be used to separate iron filings from the mixture. The iron filings are attracted by the magnet, but the other three components are not.

magnet (Figure 2.6d), which attracts the magnetic iron particles but has no effect on the other three components. The same technique is used on a much larger scale in waste recycling to separate substances that are strongly attracted to a magnetic (e.g., iron, steel, nickel) from those that are not (e.g., aluminum, glass, paper, and plastics).

After the iron has been removed from our mixture, the sugar can be separated from the remaining components by adding water. We call this process **dissolution**. Only the sugar dissolves in the water to form a solution of sugar in water.

A **solution** is a **homogeneous** (i.e., uniform from point to point) mixture of two or more components. The components of a solution do not have to be a solid and a liquid; there are many types of solutions. For example, a mixture of various nonreactive gases is a solution. However, the most common type of solution is a solid dissolved in a liquid. The solid that is dissolved is called the **solute,** and the liquid in which it is dissolved is called the **solvent.** The terms solvent and solute are merely terms of convenience, because all the components of a solution are uniformly dispersed throughout the solution (Figure 2.7). The most common solvent that we shall encounter in chemistry is water, $H_2O(l)$. When a substance is dissolved in water, we say that it forms an **aqueous solution** and denote its state using the symbol (aq). The sugar dissolved in water that

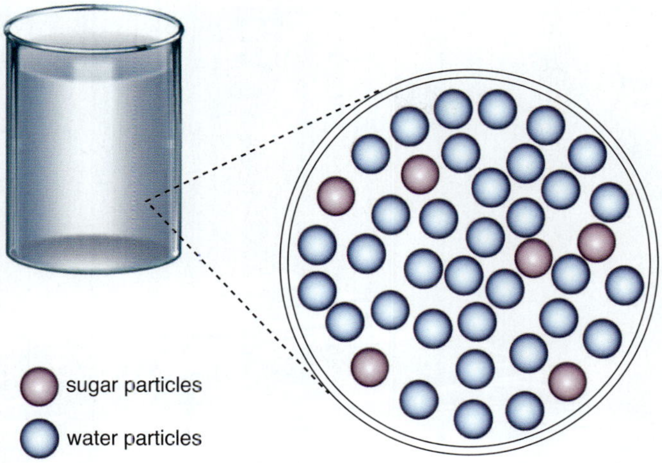

Figure 2.7 From a microscopic perspective a solution is a homogeneous mixture of a solvent and a solute, in this case water and sugar, respectively.

we have been discussing is an example of an aqueous solution. We denote the chemical formula of the sugar in the solution as $C_{12}H_{22}O_{11}(aq)$ to indicate that it is dissolved in water.

Let's return to our example. The formation of the sugar-water solution leaves the sand and gold particles at the bottom of the container. The heterogeneous mixture of gold, sand, and sugar-water solution can then be separated by **filtration** (Figure 2.8). The sugar-water solution passes readily through the small pores in the filter paper, but the solid gold and sand particles are too large to pass through and are trapped on the paper. The sugar can be recovered from the sugar-water solution by **evaporating** the water, a process that leaves the recrystallized sugar in the container. Salts are separated from seawater and brines on a commercial scale by the evaporation of the water from the brines (Figure 2.9).

Figure 2.8 Filtration can be used to separate a liquid from a solid. The liquid passes through filter paper, but the solid particles are too large to do so. Filter paper is available in a wide range of pore sizes, down to pores small enough (2.5×10^{-8} m) to remove bacteria (the smallest bacteria are about 1×10^{-7} m in diameter). Micropore filters can be used in place of pasteurization to produce bacteria-free liquids such as canned draft beer and bottled water.

The sand and gold dust can be separated by panning or by sluice-box techniques, which rely on the differences in density of the two solids to achieve a separation. In simple panning, water is added to the mixture of sand and gold and the slurry is swirled in a shallow, saucer-shaped metal pan. The dense ($19.3\ g\cdot cm^{-3}$) gold particles collect near the center of the pan, whereas the less dense sand particles (2 to $3\ g\cdot cm^{-3}$) swirl out of the pan. In the sluice-box technique, running water is passed over an agitated sand-gold mixture; the less dense sand particles, which rise higher in the water than the gold, are swept away in the stream of water.

When fine gold particles are firmly attached to sand particles, the gold can be separated by shaking the mixture with liquid mercury, in which the gold dissolves but the sand does not. The sand, which floats on the mercury, is removed.

We can separate the resulting solution of gold in mercury by taking advantage of the different physical properties of each component of the mixture. Because mercury has a relatively low boiling point (357°C), while that of gold is quite high (2856°C), the solution of gold in mercury can be separated by **distillation**, in which the mercury boils away and the solid gold remains behind. The mercury vapors are cooled and thereby converted back to liquid in a condenser. We call this process **condensation**. A simple but typical distillation apparatus is

Figure 2.9 Sodium and potassium compounds such as NaCl(s) and KCl(s) are obtained commercially by evaporation of brines. This photo shows a solar evaporation pond. The blue color is due to a dye that is mixed with the solution to enhance heat absorption and hence speed up evaporation.

shown in Figure 2.10. Mercury distillation is usually carried out in an iron flask, and the mercury is collected and reused to extract more gold. Another example of distillation is the extraction of fresh water from seawater; the dissolved salts remain behind as solids in the distillation flask after the water is boiled away.

The simple distillation apparatus shown in Figure 2.10 is suitable for the separation of a liquid from a solution when a solid is dissolved in the liquid. The liquid is the only component that vaporizes or, in other words, is the only **volatile** (i.e., easily vaporized) component. The idea is that the liquid is boiled away, leaving the solid behind. If a solution contains two or more volatile components, however (e.g., ethanol and water), then the components can be separated by taking advantage of differences in boiling point. The separation of a

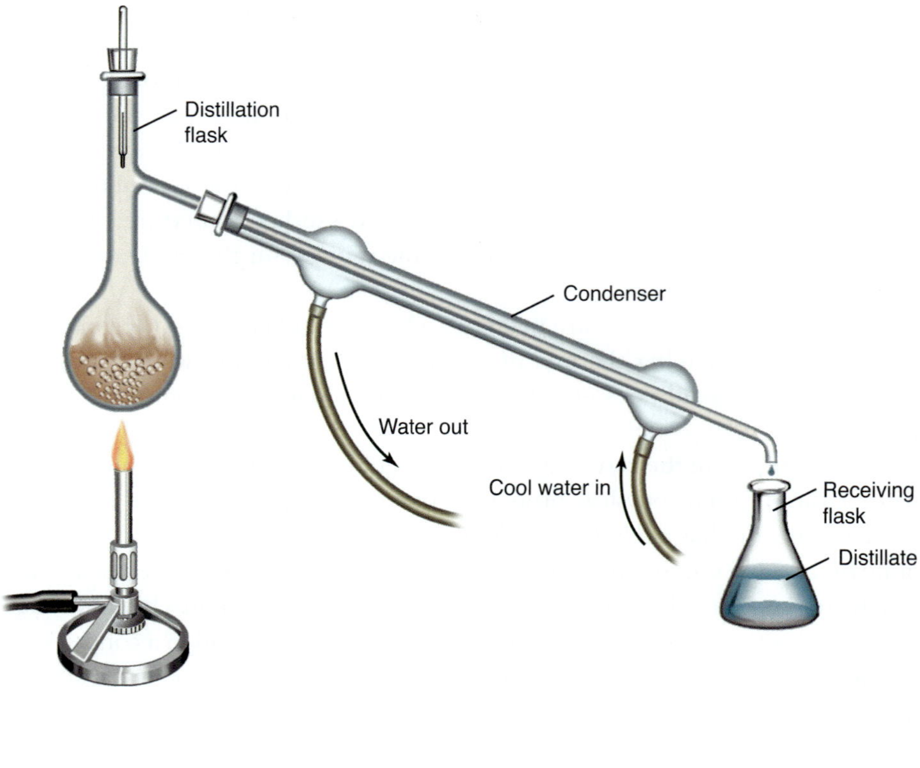

Figure 2.10 A simple distillation apparatus can be used to separate a solid from a liquid in which the solid is dissolved. The solution in the distillation flask is heated and the liquid is vaporized. The vapors rise in the distillation flask and pass into the condenser (the long, horizontal tube with the two hoses connected to it). The condenser is surrounded by a water jacket through which cooling water circulates. The vapor cools and condenses as it flows down the condenser tube. It is collected in the receiving flask. The solid component of the solution remains behind in the distillation flask.

solution with two or more volatile components is achieved by a process called **fractional distillation** (see Chapter 16).

2-4. The Law of Constant Composition States That the Relative Amount of Each Element in a Compound Is Always the Same

The quantitative approach pioneered by Lavoisier was used in the chemical analysis of compounds. The quantitative chemical analysis of a great many compounds led to the **law of constant composition:** The relative amount of each element in a particular compound is always the same, regardless of the source of the compound or how the compound was prepared.

For example, if calcium metal is heated with sulfur in the absence of water and oxygen, the compound called calcium sulfide, CaS(*s*), which is used in fluorescent paints, is formed (Figure 2.11). We can specify the relative amounts of calcium and sulfur in calcium sulfide as the **mass percentage** of each element. The mass percentages of calcium and sulfur in calcium sulfide are given by

$$\begin{matrix}\text{mass percentage of calcium} \\ \text{in calcium sulfide}\end{matrix} = \frac{\text{mass of calcium}}{\text{mass of calcium sulfide}} \times 100$$

where the factor of 100 is necessary to convert the ratio of masses to a percentage. Similarly,

$$\begin{matrix}\text{mass percentage of sulfur} \\ \text{in calcium sulfide}\end{matrix} = \frac{\text{mass of sulfur}}{\text{mass of calcium sulfide}} \times 100$$

Suppose we analyze 1.630 grams of calcium sulfide and find that it consists of 0.906 grams of calcium and 0.724 grams of sulfur. Then the mass percentages of calcium and sulfur in calcium sulfide are

$$\begin{matrix}\text{mass percentage of calcium} \\ \text{in calcium sulfide}\end{matrix} = \frac{\text{mass of calcium}}{\text{mass of calcium sulfide}} \times 100$$

$$= \frac{0.906\text{ g}}{1.630\text{ g}} \times 100 = 55.6\%$$

$$\begin{matrix}\text{mass percentage of sulfur} \\ \text{in calcium sulfide}\end{matrix} = \frac{\text{mass of sulfur}}{\text{mass of calcium sulfide}} \times 100$$

$$= \frac{0.724\text{ g}}{1.630\text{ g}} \times 100 = 44.4\%$$

Because calcium and sulfur are the only two elements present in calcium sulfide, the sum of the mass percentages of calcium and sulfur must add up to 100% (55.6% + 44.4% = 100.0%).

The law of constant composition says that the mass percentage of calcium in pure calcium sulfide is always 55.6%. It does not matter whether the calcium sulfide is prepared by heating a large amount of calcium with a small amount of sulfur or by heating a small amount of calcium with a large amount of sulfur. Similarly, the mass percentage of sulfur in calcium sulfide is always 44.4%. Any

Figure 2.11 The reaction between the elements calcium and sulfur yields calcium sulfide, CaS(*s*). The calcium sulfide is being formed at high temperature and is emitting light.

excess of calcium or of sulfur simply does not react to form calcium sulfide. If calcium is in excess, then, in addition to the reaction product calcium sulfide, we have unreacted calcium metal remaining in the reaction vessel. If sulfur is in excess, then, in addition to the reaction product, unreacted sulfur remains.

EXAMPLE 2-1: Suppose we analyze 2.83 grams of a compound of lead and sulfur and find that it consists of 2.45 grams of lead and 0.380 grams of sulfur. Calculate the mass percentages of lead and sulfur in the compound, which is called lead sulfide, $PbS(s)$.

Solution: The mass percentage of lead in lead sulfide is

$$\begin{aligned}\text{mass percentage of lead in lead sulfide} &= \frac{\text{mass of lead}}{\text{mass of lead sulfide}} \times 100 \\ &= \frac{2.45\text{ g}}{2.83\text{ g}} \times 100 = 86.6\%\end{aligned}$$

The mass percentage of sulfur in lead sulfide is

$$\begin{aligned}\text{mass percentage of sulfur in lead sulfide} &= \frac{\text{mass of sulfur}}{\text{mass of lead sulfide}} \times 100 \\ &= \frac{0.380\text{ g}}{2.83\text{ g}} \times 100 = 13.4\%\end{aligned}$$

The law of constant composition assures us that the mass percentage of lead in lead sulfide is independent of the source of the lead sulfide. The principal natural source of lead sulfide is an ore known as *galena* (Figure 2.12).

Figure 2.12 Metal sulfides are valuable ores of metals. Shown here is galena, $PbS(s)$, the principal ore of lead.

PRACTICE PROBLEM 2-1: A 5.650-gram sample of a compound containing the elements potassium, nitrogen, and oxygen was found to contain 38.67% K and 13.86% N. Calculate the number of grams of each element in the sample. (*Hint*: Recall that the sum of the mass percentages of all the elements in a compound must total 100%.)

Answer: 2.185 grams of K, 0.7831 grams of N, and 2.682 grams of O.

Quantitative chemical analysis also led to the discovery of the **law of multiple proportions**. This law says that when a given element, call it X, combines with another element, call it Y, to form two different compounds, then X combines in such a way that for a fixed mass of X, the ratio of the masses of Y in the two compounds consists of small whole numbers. This law is best shown by example. Carbon combines with oxygen to form two different compounds. Experiments show that 1.33 grams of oxygen always combine with one gram of carbon to form the first compound and that 2.66 grams of oxygen always combine with one gram of carbon to form the second compound:

carbon + oxygen:	Mass of carbon/grams	Mass of oxygen/grams
first compound	1.00	1.33
second compound	1.00	2.66

The ratio of the mass of oxygen in the second compound to that of the first is 2.66/1.33 = 2/1, a ratio of small whole numbers. These data can be readily understood if we assume that the first and second compounds have the chemical formulas CO and CO_2, respectively, because for a fixed amount of carbon, CO_2 contains twice the mass of oxygen as CO.

In another experiment it was found that sulfur and oxygen combine to form two compounds with the following data for the masses of oxygen that combined with one gram of sulfur in each of the two compounds:

sulfur + oxygen:	Mass of sulfur/grams	Mass of oxygen/grams
first compound	1.00	1.00
second compound	1.00	1.50

The ratio of the mass of oxygen in the second compound to that of the first is 1.50/1.00 = 3/2, another small whole number ratio. These data can be understood if we know that the two compounds have the chemical formulas SO_2 and SO_3, respectively. Later in this text we shall learn how to deduce chemical formulas from such data.

The law of constant composition and the law of multiple proportions were some of the observations that led to the atomic theory of the elements.

Figure 2.13 John Dalton (1766–1844) was born in England to a poor Quaker family. He began teaching at age twelve in a Quaker academy, while continuing his own studies in mathematics and natural philosophy. His attempts to earn his living as a public lecturer were not successful because by all accounts he was a dreadful speaker. Dalton's first scientific paper was on self-diagnosed color blindness, still known as "Daltonism." His main scientific interest was meteorology, and he kept a daily diary of his weather observations up to the day he died. He presented his atomic theory in lectures and publications using wooden spheres to represent the various elements, which helped the theory to become widely accepted by the scientific community. He was awarded honorary degrees by Oxford University in 1832, and Edinburgh University in 1834, although he was not able to obtain a position at any university because he was a Quaker.

2-5. Dalton's Atomic Theory Explains the Law of Constant Composition

By the end of the eighteenth century, scientists had analyzed many compounds and had amassed a large amount of experimental data. But they lacked a theory that could bring all these data into a single framework. In 1803 John Dalton (Figure 2.13), an English elementary school teacher, proposed an **atomic theory**. His theory provided a simple and beautiful explanation of both the law of constant composition and the law of conservation of mass. We can express the postulates of Dalton's atomic theory in modern terms as follows:

1. Matter is composed of small, indivisible particles called atoms.
2. The atoms of a given element all have the same mass and are identical in all respects, including chemical behavior.
3. The atoms of different elements differ in mass and in chemical behavior.
4. Chemical compounds are composed of two or more atoms of different elements joined together. The particle that results when two or more atoms join together is called a molecule.
5. In a chemical reaction, the atoms involved are rearranged, separated, or recombined to form new substances. No atoms are created or destroyed, and the atoms themselves are not changed.

As we shall see, some of these postulates were later modified, but the main features of Dalton's atomic theory still are accepted today.

The law of conservation of mass in chemical reactions follows directly from Dalton's postulate that atoms are neither created nor destroyed in chemical reactions; rather, they are simply rearranged to form new substances. The law of constant composition follows from Dalton's postulate that atoms are indivisible and that compounds are formed by joining together different types of atoms. That is, compounds have constant composition because they contain fixed ratios of the different types of atoms. For example, suppose it is found that calcium sulfide is formed when calcium and sulfur are combined in a one-to-one ratio. In such a case, the ratio of calcium atoms to sulfur atoms is one-to-one, no matter how the sample is prepared. The law of multiple proportions follows from the fact that some elements can combine to form more than one compound, such as in the case of CO and CO_2. Although Dalton's application of his theory was marred by several incorrect guesses about the relative numbers of atoms in compounds (Figure 2.14), these errors were eventually resolved. Meanwhile, the atomic theory gained wide recognition and is now universally accepted.

Figure 2.14 Dalton's symbols for chemical elements. Some of these "elements" are now known to be compounds, not elements (e.g., lime, soda, and potash). Some of Dalton's atomic masses were in error because of incorrect assumptions regarding the relative numbers of atoms in compounds.

Dalton's atomic theory enables us to set up a scale of relative atomic masses. Consider calcium sulfide, which we know consists of 55.6% calcium and 44.4% sulfur by mass. Suppose there is one calcium atom for each sulfur atom in calcium sulfide. Because we know that the mass of a calcium atom relative to that of a sulfur atom must be the same as the mass percentages in calcium sulfide, we know that the ratio of the mass of a calcium atom to that of a sulfur atom is

$$\frac{\text{mass of a calcium atom}}{\text{mass of a sulfur atom}} = \frac{55.6}{44.4} = 1.25$$

or

$$\text{mass of a calcium atom} = 1.25 \times (\text{mass of a sulfur atom})$$

Thus, even though we cannot easily determine the mass of any individual atom, we can use the quantitative results of chemical analyses to determine the *relative* masses of atoms. Of course, we have based our result for calcium and sulfur on the assumption that there is one atom of calcium for each atom of sulfur in calcium sulfide.

Let's consider another compound, hydrogen chloride. Quantitative chemical analysis shows that the mass percentages of hydrogen and chlorine in hydrogen chloride are 2.76% and 97.24%, respectively. Once again, assuming (correctly, it turns out) that one atom of hydrogen is combined with one atom of chlorine, we find that

$$\frac{\text{mass of a chlorine atom}}{\text{mass of a hydrogen atom}} = \frac{97.24}{2.76} = 35.2$$

or

$$\text{mass of a chlorine atom} = 35.2 \times (\text{mass of a hydrogen atom})$$

By continuing in this manner with other compounds, it is possible to build up a table of relative atomic masses. We define a quantity called the **atomic mass**

ratio (usually referred to simply as the **atomic mass**) as the ratio of the mass of a given atom to the mass of some particular reference atom. At one time the mass of hydrogen, the lightest atom, was arbitrarily given the value of exactly one and used as the reference to which all other atomic masses were compared. As we point out in Chapter 11, however, a form of carbon is now used as the standard. Thus, today the atomic mass of hydrogen is 1.008 instead of exactly one. The presently accepted atomic masses of the elements are given on the inside front cover of the text.

Because "atomic masses" are actually ratios of masses, they have no units. Nevertheless, it is sometimes useful to assign to atomic masses a unit called the **atomic mass unit**. The atomic mass unit was once denoted by **amu**, but the symbol **u** is now the symbol recommended by the International Union of Pure and Applied Chemistry (IUPAC). In biochemistry, however, the unit dalton, with the symbol Da, is often used. Thus, we can say that the atomic mass of carbon is 12.01 or 12.01 u or 12.01 Da; all three statements are correct. Although we shall refer to atomic mass ratios for elements as simply atomic masses, it is important to recognize that atomic masses are actually relative, dimensionless quantities. The particular values assigned to atomic masses (but not their ratios) depend on the reference atom chosen to set up the scale.

EXAMPLE 2-2: Suppose the atomic mass of hydrogen were set at exactly one. Refer to the table of atomic masses on the inside front cover to calculate the atomic mass of carbon on the H = 1 scale.

Solution: Because the ratio of the masses of two atoms is independent of the value of the mass chosen for the reference element, we have in the present system

$$\frac{\text{mass of C}}{\text{mass of H}} = \frac{12.0107}{1.00794} = 11.9161$$

Thus, for the revised (H = 1 exactly) system, we would have

$$\frac{\text{mass of C}}{\text{mass of H}} = \frac{\text{mass of C}}{1} = 11.9161$$

and the atomic mass of carbon on the H = 1 scale would be 11.9161 rather than 12.0107.

PRACTICE PROBLEM 2-2: Prior to the adoption of the present carbon-based scale of atomic mass, the atomic mass of oxygen was set equal to exactly 16 and oxygen was used as the standard. Refer to the table of atomic masses given on the inside front cover to calculate the atomic mass of carbon to five significant figures on the O = 16 scale.

Answer: 12.011. The difference between this value and the currently used value is not significant to five significant figures.

Elements are substances that consist of only one kind of atom (Fe, C, H_2). *Molecules* are units consisting of two or more atoms (Cl_2, HCl, CO_2). *Compounds* are substances that consist of molecules that contain atoms of different elements (H_2O, NaCl, $CaCO_3$).

2-6. Molecules Are Groups of Atoms Joined Together

Dalton postulated in the original version of his atomic theory that an element is a substance consisting of identical atoms and that a compound is a substance consisting of molecules containing different kinds of atoms. Although Dalton did not realize it at the time, some of the elements occur naturally as molecules containing more than one of the same kind of atoms. As we already noted in Table 2.5, the elements hydrogen, nitrogen, oxygen, fluorine, chlorine, bromine, and iodine exist as diatomic molecules of the same kind of atoms (see Figure 2.4). Consequently, these substances are classified as elements. Compounds, on the other hand, are made up of molecules containing different kinds of atoms. Examples of the chemical formulas for molecules of chemical compounds are as follows:

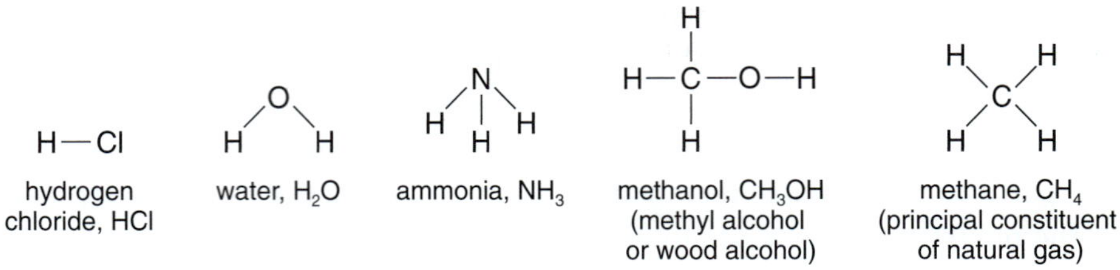

These structural formulas indicate how the atoms are joined together in the molecules. The lines represent bonds between connected atoms. We shall learn how to write such formulas in Chapter 7. Scale models of these molecules are shown in Figure 2.15.

Dalton's atomic theory provides a microscopic view of chemical reactions. Recall that Dalton proposed that, in a chemical reaction, atoms in the reactant molecules are separated and then rearranged into product molecules. According to this view, the chemical reaction between hydrogen and oxygen to form water may be represented by the following rearrangement:

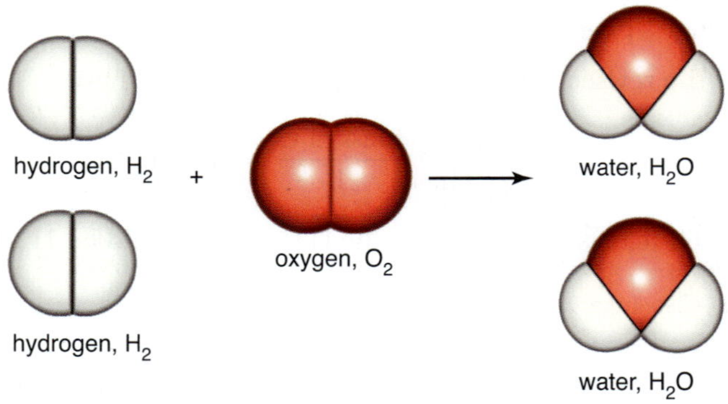

Note that completely different molecules, and hence completely different substances, are formed in a chemical reaction. Hydrogen and oxygen are gases, whereas water is a liquid at room temperature and normal atmospheric pressure.

As another example, consider the burning of carbon in oxygen to form carbon dioxide, which may be expressed as follows:

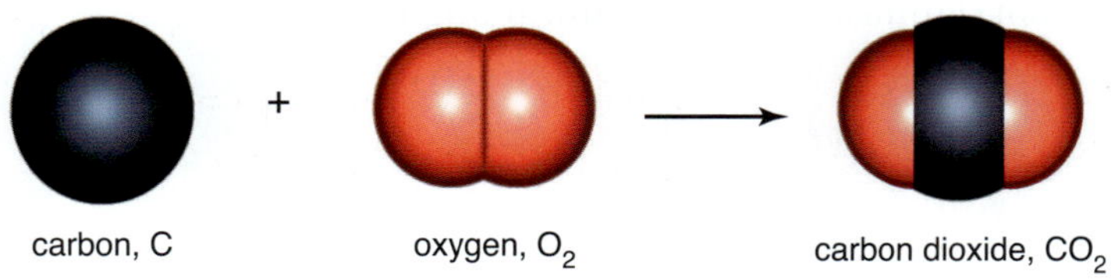

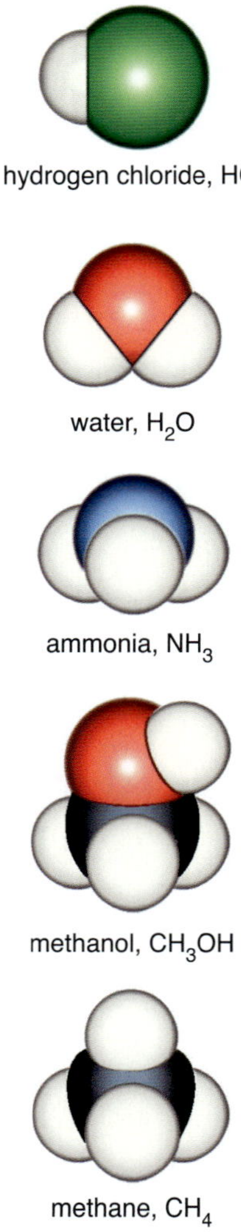

Figure 2.15 Scale models of molecules of hydrogen chloride, water, ammonia, methanol, and methane.

Once again, a completely new substance is formed. Carbon is a black solid; the product, carbon dioxide, is a colorless gas under the conditions given above.

As a final example, consider the reaction between steam (hot gaseous water) and red-hot carbon to form hydrogen and carbon monoxide, which may be represented:

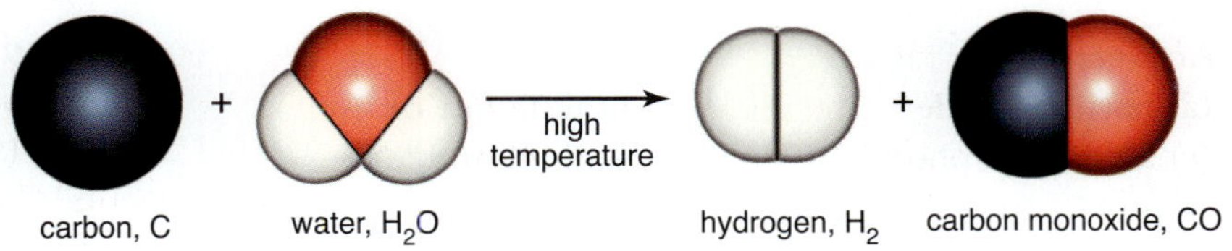

In each of the three representations of the reactions above, the numbers of each kind of atom do not change. Atoms are neither created nor destroyed in chemical reactions; they are simply rearranged into new molecules, in accordance with the conservation of atoms and of mass in chemical reactions.

2-7. Compounds Are Named by an Orderly System of Chemical Nomenclature

The system for the assignment of names to compounds is called **chemical nomenclature.** Throughout this text as we encounter new classes of compounds, we shall learn the IUPAC (International Union of Pure and Applied Chemistry) rules for naming these compounds (and sometimes the older classical names as well). These rules were developed to promote a single worldwide standard for chemical nomenclature and allow for fast and easy searching of literature and databases for information regarding chemical compounds. Knowing these rules will be of invaluable assistance to you throughout your study of science. For easy reference, a summary of the rules for naming the different classes of compounds presented in this text is given in Appendix C.

In this chapter we discuss only the system of naming compounds consisting of two elements, that is, **binary compounds.** When the two elements that make up a binary compound are a metal and a nonmetal that combine in only one fixed ratio, we name the compound by first writing the name of the metal and then that of the nonmetal, with the ending of the name of the nonmetal changed to *-ide*. For example, we saw that the name of the compound formed between calcium (a metal, Table 2.3) and sulfur (a nonmetal, Table 2.5) is calcium sulf*ide*. Because calcium sulfide consists of one atom of calcium for each atom of sulfur, we write the **chemical formula** of calcium sulfide as CaS(*s*); in other words, we simply join the chemical symbols of the two elements. In a different case, calcium combines with two atoms of chlorine (a nonmetal) to form calcium chloride; thus, the formula of calcium chloride is $CaCl_2(s)$. Note that the number of atoms is indicated by a subscript. The subscript 2 in $CaCl_2(s)$ means that there are two chlorine atoms per calcium atom in calcium chloride. Table 2.6 lists the *-ide* nomenclature for some common nonmetals.

TABLE 2.6 The *-ide* nomenclature of the nonmetals

Element	*-ide* nomenclature
arsenic	arsenide
bromine	bromide
carbon	carbide
chlorine	chloride
fluorine	fluoride
hydrogen	hydride
iodine	iodide
nitrogen	nitride
oxygen	oxide
phosphorus	phosphide
selenium	selenide
sulfur	sulfide

EXAMPLE 2-3: Name the following binary compounds:
(a) $K_2O(s)$ (b) $AlBr_3(s)$ (c) CdSe(*s*) (d) $MgH_2(s)$

Solution: Using Table 2.6 for the correct *-ide* nomenclature, we have

(a) potassium oxide (b) aluminum bromide
(c) cadmium selenide (d) magnesium hydride

PRACTICE PROBLEM 2-3: Name the following binary compounds:
(a) $BaI_2(s)$ (b) $Li_3N(s)$ (c) $AlP(s)$ (d) $Na_2S(s)$

Answer:
(a) barium iodide (b) lithium nitride
(c) aluminum phosphide (d) sodium sulfide

Many binary compounds involve combinations of two nonmetals (Table 2.5). Because more than one binary compound may result from the combination of the same two nonmetallic elements, we distinguish the various possibilities by means of Greek numerical prefixes (Table 2.7). For example,

$CO(g)$ carbon monoxide $CO_2(g)$ carbon dioxide

Some other examples are

$SO_2(g)$ sulfur dioxide $SO_3(g)$ sulfur trioxide
$SF_4(g)$ sulfur tetrafluoride $SF_6(g)$ sulfur hexafluoride
$PCl_3(l)$ phosphorus trichloride $PCl_5(s)$ phosphorus pentachloride

Ball-and-stick models of these compounds are shown in Figure 2.16.

TABLE 2.7 Greek prefixes used to indicate the number of atoms of a given type in a molecule

Number	Greek prefix*	Example
1	*mono-*	carbon monoxide, CO
2	*di-*	carbon dioxide, CO_2
3	*tri-*	sulfur trioxide, SO_3
4	*tetra-*	carbon tetrachloride, CCl_4
5	*penta-*	phosphorus pentachloride, PCl_5
6	*hexa-*	sulfur hexafluoride, SF_6
7	*hepta-*	*(examples of compounds using prefixes greater than six will be given later in the text)*
8	*octa-*	
9	*nona-*	
10	*deca-*	

*The final *a* or *o* is dropped from the prefix when it is combined with a name beginning with a vowel.

When naming binary compounds the prefix *mono-* is not used for naming the first element and is generally dropped from the second; notable exceptions are carbon monoxide and sometimes nitrogen monoxide. For example,

$NO(g)$	nitrogen oxide or nitrogen monoxide
$CO(g)$	carbon monoxide

In addition, as illustrated in the previous examples (and noted in the footnote to Table 2.7), the final *a* or *o* is dropped from the prefix when it is combined with a name beginning with a vowel. For example, penta + chloride is written as pentachloride; but penta + iodide is changed to pentiodide (the *a* in penta is dropped). Likewise, mono + hydride is written as monohydride; but mono + oxide is changed to monoxide. This is not the case with the *di-* and *tri-* prefixes; carbon dioxide and boron triiodide are both correct.

Hydrogen is another important exception: it can act as either a metal or nonmetal. When hydrogen is listed first in a binary formula, it is generally treated as a metal and named accordingly; for example, $H_2S(g)$, hydrogen sulfide. When hydrogen is listed at the end of a binary formula, it is generally treated as a nonmetal; for example, $NaH(s)$, sodium hydride, and $AsH_3(s)$, arsenic trihydride.

The names of some additional compounds containing hydrogen that you should know are **water**, $H_2O(l)$, **ammonia**, $NH_3(g)$, and **methane**, $CH_4(g)$.

EXAMPLE 2-4: Name the following binary compounds:
(a) $BrF_5(l)$ (b) $XeF_4(s)$ (c) $NH_3(g)$ (d) $N_2O_4(g)$ (e) $HBr(g)$

Solution: Because these compounds involve two nonmetallic elements, we must denote the relative numbers of the two types of atoms in the name. (a) Bromine is written first in the formula; thus we name the compound bromine pentafluoride (the prefix *mono-* is usually omitted on the first named element); (b) xenon tetrafluoride; (c) ammonia; (d) dinitrogen tetroxide; (e) hydrogen bromide.

PRACTICE PROBLEM 2-4: Name the following compounds:
(a) $N_2O(g)$ (b) $NO(g)$ (c) $N_2O_3(l)$ (d) $N_2O_5(g)$ (e) $NO_2(g)$

Answer:
(a) dinitrogen oxide, (b) nitrogen oxide (or nitrogen monoxide), (c) dinitrogen trioxide, (d) dinitrogen pentoxide, (e) nitrogen dioxide.

The compound dinitrogen oxide, $N_2O(g)$, (common name, nitrous oxide) was the first known general anesthetic (laughing gas) and is still sometimes used in dentistry. It is also used as a propellant for canned whipped cream and shaving cream. Except for $N_2O_3(l)$, all the nitrogen oxides are gases at room temperature and normal pressure.

At this point you should understand how to name binary compounds when you are given the formula. In Chapter 6, we shall learn how to write a correct chemical formula from the name of a compound.

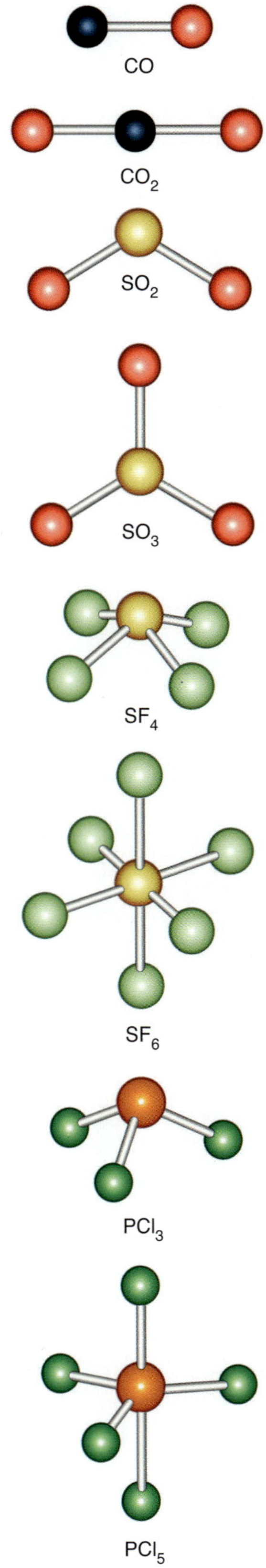

Figure 2.16 Structures of several binary nonmetallic compounds.

2-8. Molecular Mass Is the Sum of the Atomic Masses of the Atoms in a Molecule

Now that we can distinguish different compounds by their chemical formulas, we can make our explanation of the law of constant composition still clearer and more useful by introducing the idea of molecular mass. The sum of the atomic masses of the atoms in a molecule is called the **molecular mass** of the substance. For example, a water molecule, H_2O, consists of two atoms of hydrogen and one atom of oxygen. Using the table of atomic masses given on the inside front cover, we see that the molecular mass of water is

$$\begin{aligned}\text{molecular mass of } H_2O &= 2\ (\text{atomic mass of H}) + (\text{atomic mass of O})\\ &= 2\ (1.008) + (16.00) = 18.02\end{aligned}$$

to four significant figures. Similarly, using the table of atomic masses given on the inside front cover, we see that the molecular mass of dinitrogen pentoxide, N_2O_5, is

$$\begin{aligned}\text{molecular mass of } N_2O_5 &= 2\ (\text{atomic mass of N}) + 5\ (\text{atomic mass of O})\\ &= 2\ (14.01) + 5\ (16.00) = 108.02\end{aligned}$$

The following example shows how to use atomic and molecular masses to calculate the mass percentage composition of compounds.

EXAMPLE 2-5: Using the fact that the atomic mass of lead is 207.2 and that of sulfur is 32.07, calculate the mass percentages of lead and sulfur in the compound lead sulfide, PbS(*s*).

Solution: As the formula PbS(*s*) indicates, lead sulfide consists of one atom of lead for each atom of sulfur. The molecular mass of lead sulfide is

$$\begin{aligned}\text{molecular mass of lead sulfide} &= \underset{\text{of lead}}{\text{atomic mass}} + \underset{\text{of sulfur}}{\text{atomic mass}}\\ &= 207.2 + 32.07 = 239.3\end{aligned}$$

The mass percentages of lead and sulfur in lead sulfide are

$$\begin{aligned}\text{mass percentage of lead} &= \frac{\text{atomic mass of lead}}{\text{molecular mass of lead sulfide}} \times 100\\ &= \frac{207.2}{239.3} \times 100 = 86.59\%\end{aligned}$$

$$\begin{aligned}\text{mass percentage of sulfur} &= \frac{\text{atomic mass of sulfur}}{\text{molecular mass of lead sulfide}} \times 100\\ &= \frac{32.07}{239.3} \times 100 = 13.40\%\end{aligned}$$

Note that this result is the same as that calculated in Example 2-1. The table of atomic masses must be consistent with experimental values of mass percentages. The two mass percentages in this example do not add up to exactly 100% because of a slight round-off error.

PRACTICE PROBLEM 2-5: Calculate the mass percentages of bromine and fluorine in $BrF_5(l)$.

Answer: 45.687% Br and 54.313% F

One of the great advantages of Dalton's atomic theory was that he was able to use it to devise a table of atomic masses that could then be used in chemical calculations like those in Example 2-5. What Dalton did not know, however, is that not all atoms of a given element have the same atomic mass. This discovery, which was made in the twentieth century, required a new model of the atom.

2-9. Most of the Mass of an Atom Is Concentrated in Its Nucleus

For most of the nineteenth century, atoms were considered to be indivisible, stable particles, as proposed by Dalton. Toward the end of the century, however, new experiments indicated that an atom is composed of even smaller **subatomic particles**.

One of the first experiments on subatomic particles was carried out by the English physicist J. J. Thomson in 1897 (Figure 2.17). Some years earlier, it had been discovered that an electric discharge (glowing current) flows between metallic electrodes that are sealed in a partially evacuated glass tube, as shown in

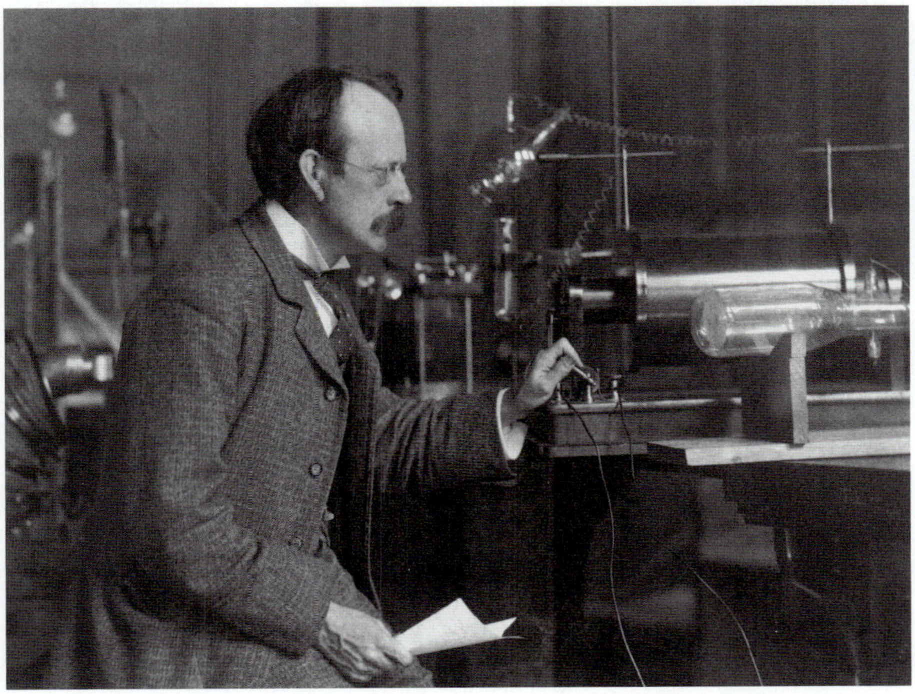

Figure 2.17 English scientist Sir Joseph John Thomson (1856–1940). Thomson's work on cathode rays led to the discovery of the electron in 1897. His later work with positive ion beams led to a method of separating atoms and molecules by mass and the discovery of neon. Thomson was awarded the Nobel Prize in Physics in 1906 and was knighted in 1908. He is buried in Westminster Abbey.

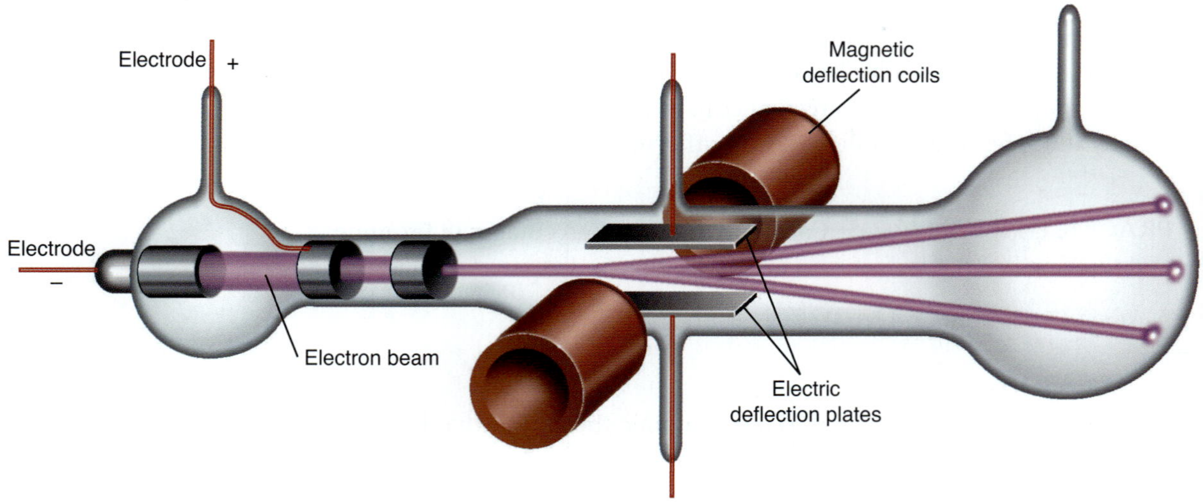

Figure 2.18 Schematic of a discharge tube. When a voltage is applied across electrodes that are sealed in a partially evacuated glass tube, the space between the electrodes glows.

Figure 2.18. These glowing discharges were called **cathode rays**. Scientists knew that these rays were not caused by atoms or heavier particles because the mass of the two plates in the apparatus remained constant. Much debate among physicists ensued over the nature of the rays. Using an apparatus of the type depicted in Figure 2.18, Thomson deflected the rays with electric and magnetic fields and showed that they were actually streams of identical, negatively charged particles. He correctly reasoned that these particles, which are now called **electrons,** are constituents of atoms. The electron was the first subatomic particle to be discovered.

Thomson's discharge tube experiment along with later experiments showed that the mass of an electron is 9.109×10^{-31} kg and that its charge is -1.602×10^{-19} C, where C is the symbol for **coulomb**, the SI unit of charge. It is often convenient to express charges as multiples of 1.602×10^{-19} C. Thus the charge on an electron is −1 in this convention. The mass of an electron is only 1/1800 the mass of a hydrogen atom, confirming Thomson's assumption that an electron is a subatomic particle.

Thomson's discharge tube was the forerunner of the **cathode ray tube** (CRT) used widely in television and video monitors before the advent of the liquid crystal display or LCD screen (see Section 15-12).

If an atom contains electrons, which are negatively charged particles, then it also must contain positively charged particles because atoms are electrically neutral. The total amount of negative charge in a neutral atom must be balanced by an equal amount of positive charge. The question is: How are the positively charged particles and electrons arranged within an atom?

About the same time that Thomson discovered the electron, the French scientist Antoine-Henri Becquerel discovered **radioactivity,** the process by which certain atoms spontaneously break apart. Becquerel showed that uranium atoms are **radioactive.** Shortly after his discovery, Marie and Pierre Curie (Frontispiece), working in Paris, discovered other radioactive elements such as radium (so named because it emits rays) and polonium (named for Poland, Marie Curie's native country). Then in the early 1900s, the New Zealand–born physicist Ernest Rutherford (Figure 2.19) began to study radioactivity. He discovered that the radiation emitted by radioactive substances consists of three types, which are now called **α-particles** (alpha particles), **β-particles** (beta par-

TABLE 2.8 Properties of the three radioactive emissions discovered by Rutherford

Original name	Modern name	Mass*	Charge†
α-ray	α-particle	4.00	+2
β-ray	β-particle (electron)	5.49×10^{-4}	−1
γ-ray	γ-ray	0	0

*In atomic mass units.

†Relative to the charge on a proton, which is defined as +1. The actual charge on a proton is 1.602×10^{-19} coulomb. The charges given here are, in effect, in units of the charge on the proton.

Figure 2.19 Sir Ernest Rutherford (1871–1937) was born in a rural community in New Zealand. In 1895 he was awarded a scholarship to attend Cambridge University. There Rutherford studied under Professor J. J. Thomson. In 1896 after the discovery of X-rays, Rutherford began to study ionizing radiation. During his long and distinguished career, he characterized various forms of radioactivity, naming the α- and β-particles, the γ-ray, and coining the term half-life to describe the rate of nuclear decay. His work on α-particle scattering led to the discovery of the nucleus, the nuclear model for the atom, and the discovery of the proton (which he also named). He was awarded the Nobel Prize in Chemistry in 1908.

ticles), and **γ-rays** (gamma rays). Experiments by Rutherford and others showed that α-particles have a charge equal in magnitude to that of two electrons but of opposite (i.e., positive) sign and a mass equal to the mass of a helium atom (4.00 u); that β-particles are simply electrons that result from radioactive disintegrations; and that γ-rays are electromagnetic radiation similar to X-rays (see Chapter 4). Table 2.8 summarizes the properties of these three common emissions from radioactive substances.

Rutherford became intrigued with the idea of using the newly discovered α-particles as subatomic projectiles. At the time Rutherford was working with Hans Geiger (who later developed the Geiger counter). In a now-famous experiment, Ernest Marsden, a 20-year-old undergraduate student who was being trained by Geiger, took a piece of gold and formed it into an extremely thin foil (gold is very malleable). He then directed a beam of α-particles at the gold foil and observed the paths of the α-particles by watching for flashes as they struck a fluorescent screen surrounding the foil (Figure 2.20). Because most scientists believed at the time that the positive charge in atoms was spread uni-

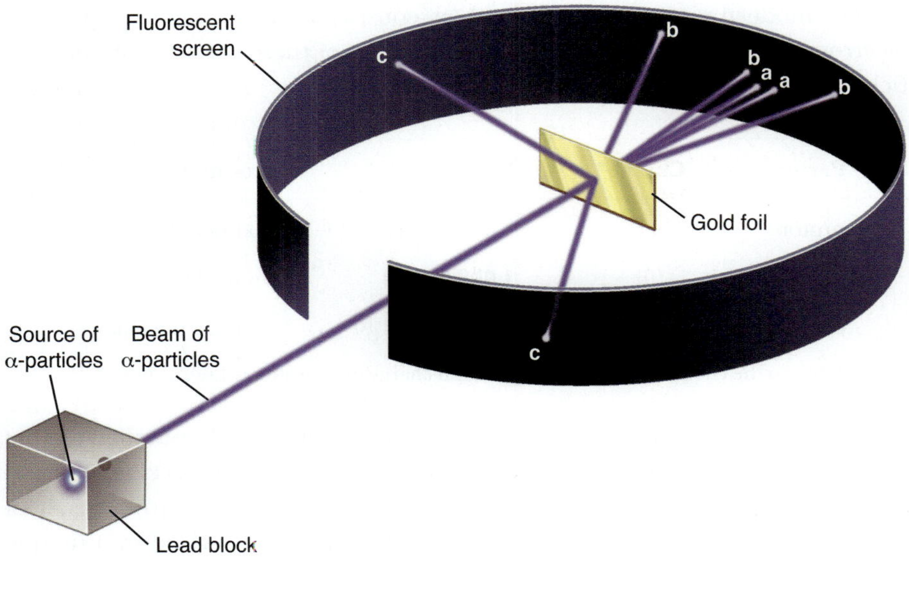

Figure 2.20 In 1911, Rutherford, Geiger, and Marsden set up an experiment in which a thin gold foil was bombarded with α-particles. Most of the particles passed through the foil (pathway a). Some were deflected only slightly (pathway b) when they passed near a gold nucleus in the foil, and a few were deflected backward (pathway c) when they collided with a nucleus.

formly throughout the atom, Rutherford and Marsden expected the tiny fast-moving alpha particles to tear through the foil. They assumed that the trajectories of the α-particles would be altered only slightly (at most by two degrees).

Contrary to expectations, while most of the α-particles passed straight through the foil, a few "bounced back," being deflected through large angles (pathway c in Figure 2.20). These results astounded Rutherford as he was to relate in a later lecture, "It was quite the most incredible event that ever happened to me in my life. It was almost as incredible as if you fired a 15-inch shell at a piece of tissue paper and it came back and hit you!"

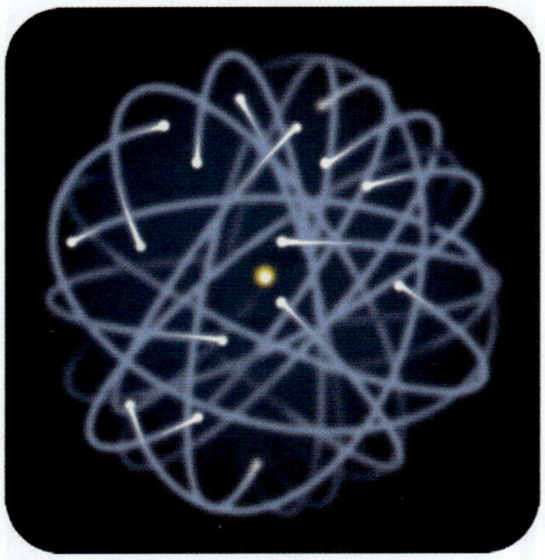

Figure 2.21 Rutherford's nuclear model of the atom. The nucleus is very small and located at the center. The electrons are located in the space around the nucleus. In fact, as we shall see in Chapter 4, electrons do *not* travel around a nucleus in well-defined orbits, as depicted here.

Rutherford interpreted this astonishing result to mean that all the positive charge and the bulk of the mass of an atom are concentrated in a very small volume at the center, which he called the **nucleus** (Figure 2.21). He named this the **nuclear model of the atom**.

By counting the number of α-particles deflected in various directions, Rutherford was able to show that the radius of a gold nucleus is about 1/100 000 of the radius of the atom. If the radius of the nucleus were the size of a golf ball, then the radius of the atom would be about 3000 meters (10 000 feet). Thus, the nucleus represents a very small fraction of the volume of an atom, as Figure 2.21 implies.

The positively charged particles found in the atomic nucleus are called **protons.** These subatomic particles have a positive charge equal in magnitude to that of an electron but opposite in sign. The mass of a proton is almost the same as the mass of a hydrogen atom, about 1800 times the mass of an electron.

The electrons in an atom are located throughout the space surrounding the nucleus (Figure 2.21). We shall discuss how the electrons are arranged in an atom in Chapters 4 and 5.

2-10. Atoms Consist of Protons, Neutrons, and Electrons

Our picture of the atom is not yet complete. Experiments suggested that the mass of a nucleus cannot be attributed to the protons alone. Scientists hypothesized in the 1920s, and in 1932 it was experimentally verified, that there is another type of particle in the nucleus. This particle has a slightly greater mass than a proton and is called a **neutron** because it is electrically neutral.

The modern picture of an atom, then, consists of three types of particles—electrons, protons, and neutrons. The properties of these three subatomic particles are shown below.

Particle	Charge*	Mass/u	Located
proton	+1	1.007 276 47	in nucleus
neutron	0	1.008 664 90	in nucleus
electron	−1	$5.485\ 7990 \times 10^{-4}$	outside nucleus

*Relative to the charge on a proton. The actual charge on a proton is 1.602×10^{-19} C.

The number of protons in an atom is called the **atomic number** of that atom and is denoted by Z. In a **neutral atom,** the number of electrons is equal to the number of protons. The differences between elements are a result of the different atomic numbers, and each element is characterized by a unique atomic number. In other words, no two elements have the same atomic number. For example, hydrogen has an atomic number of 1 (1 proton in the nucleus), helium has an atomic number of 2 (2 protons in the nucleus), and uranium has an atomic number of 92 (92 protons in the nucleus). The table of the elements given on the inside front cover of this book lists the atomic numbers of all the known elements. The total number of protons and neutrons in an atom is called the **mass number** of that atom and is denoted by A.

EXAMPLE 2-6: Use the mass data given in the table above to calculate the percentage of the mass of a hydrogen atom that is located in the nucleus. Assume that the hydrogen nucleus consists of a single proton.

Solution: The mass percentage in the nucleus of a hydrogen atom is given by the ratio of the mass of a proton to the mass of a proton plus the mass of an electron multiplied by 100 to convert the result to a percentage:

$$\frac{1.007\,276\,47}{1.007\,276\,47 + 0.000\,548\,579\,90} \times 100 = 99.945\,5679\%$$

A hydrogen atom has a lower mass percentage in its nucleus than any other atom.

PRACTICE PROBLEM 2-6: Given that the diameter of a nucleus is roughly 1×10^{-5} times that of an atom, calculate the percentage by volume of an atom that is occupied by the nucleus. (*Hint:* Recall that the volume of a sphere is given by $V = 4/3\, \pi r^3$.)

Answer: Roughly 1×10^{-13} % of the volume is occupied by the nucleus. An even smaller percentage is occupied by the electrons; the remainder is empty space.

2-11. Most Elements Occur in Nature as Mixtures of Isotopes

Nuclei are made up of protons and neutrons, each of which has a mass of approximately 1 u; therefore, you might expect atomic masses to be approximately equal to whole numbers. Although many atomic masses are approximately whole numbers (for example, the atomic mass of oxygen is 16.00 and the atomic mass of fluorine is 19.00), many others are not. Chlorine ($Z = 17$) has an atomic mass of 35.45, magnesium ($Z = 12$) has an atomic mass of 24.31, and copper ($Z = 29$) has an atomic mass of 63.55. The explanation for these variations lies in the fact that many elements consist of two or more **isotopes,** which are atoms of one element that contain the same number of protons but different numbers of neutrons. Recall that it is the number of *protons* (the atomic number) that characterizes a particular element, but nuclei of the same element may have different numbers of neutrons. For example, the most common isotope of the simplest element, hydrogen, contains one proton and one electron, but no neutrons. Another less common isotope of hydrogen contains one proton, one neutron, and one electron. These two hydrogen isotopes both undergo the same chemical reactions. The heavier isotope is called heavy hydrogen—or, more commonly, **deuterium**—and is often denoted by the special symbol D. Water that is made from deuterium is called **heavy water** and is usually denoted by $D_2O(l)$.

An isotope is specified by its atomic number (Z) and its mass number (A). The notation used to designate isotopes is the chemical symbol of the element written with its atomic number as a left subscript and its mass number as a left superscript:

mass number → $^{A}_{Z}X$ ← chemical symbol
atomic number ↗

For example, an ordinary hydrogen atom is denoted $^{1}_{1}H$ and a deuterium atom is denoted $^{2}_{1}H$ (or sometimes $^{2}_{1}D$). The number of neutrons, N, in an atom is equal to

$$N = A - Z \tag{2.1}$$

Because all isotopes of a given element have the same number of protons, the atomic number (Z) and the chemical symbol (X) are redundant. Consequently, the atomic number is often omitted. For example, the isotope of carbon used in radiocarbon dating (containing six protons and eight neutrons) may be written as $^{14}_{6}C$, ^{14}C, or carbon-14.

EXAMPLE 2-7: Fill in the blanks.

Symbol	Atomic number	Number of neutrons	Mass number
(a)	22		48
(b)		110	184
(c) $^{?}_{?}Co$			60

Solution:

(a) The number of neutrons is the mass number minus the atomic number (Equation 2.1), or $48 - 22 = 26$ neutrons. The element with atomic number 22 is titanium (see the inside front cover), so the symbol of this isotope is $^{48}_{22}Ti$. It is called titanium-48.

(b) The atomic number equals the number of protons, which is the mass number minus the number of neutrons, or $184 - 110 = 74$. The element with atomic number 74 is tungsten, so the symbol is $^{184}_{74}W$. It is called tungsten-184.

(c) According to the symbol, the element is cobalt, whose atomic number is 27. The symbol of the particular isotope is $^{60}_{27}Co$, and the isotope has $60 - 27 = 33$ neutrons. Cobalt-60 is used as a γ-radiation source for the treatment of cancer in some radiation therapies.

PRACTICE PROBLEM 2-7: The radioactive isotope phosphorus-32 is used extensively in biochemistry and medicine to monitor chemical reactions. Give the number of protons, neutrons, and electrons in a neutral phosphorous-32 atom.

Answer: 15, 17, 15, respectively

Although one of the postulates of Dalton's atomic theory was that all the atoms of a given element have the same mass, we now see that this is not usually so. Isotopes of the same element have different masses, but all atoms of a given isotope have the same mass. Several common natural isotopes and their corresponding masses are given in Table 2.9. Note that the isotopic mass of carbon-12 is exactly 12. The modern atomic mass scale is based on this convention. All atomic masses are given relative to the mass of the carbon-12 isotope, which is defined by international convention to be *exactly* 12.

Table 2.9 also shows two isotopes for helium: helium-3 and helium-4. The atomic number of helium is 2; in other words, a helium nucleus has two protons and a nuclear charge of +2. A helium-4 nucleus has a charge of +2 and an

TABLE 2.9 Naturally occurring isotopes of some common elements*

Element	Isotope	Isotopic mass/u	Natural abundance/%	Protons	Neutrons	Mass number
hydrogen	$^{1}_{1}H$	1.007 825 0321	99.9850	1	0	1
(deuterium)	$^{2}_{1}H$	2.014 101 7780	0.0115	1	1	2
(tritium)	$^{3}_{1}H$	3.016 049 2675	trace	1	2	3
helium	$^{3}_{2}He$	3.016 029 3097	0.000137	2	1	3
	$^{4}_{2}He$	4.002 603 2497	99.999863	2	2	4
carbon	$^{12}_{6}C$	12 (Exact)	98.93	6	6	12
	$^{13}_{6}C$	13.003 354 8378	1.07	6	7	13
	$^{14}_{6}C$	14.003 2420	trace	6	8	14
oxygen	$^{16}_{8}O$	15.994 914 6221	99.757	8	8	16
	$^{17}_{8}O$	16.999 131 50	0.038	8	9	17
	$^{18}_{8}O$	17.999 1604	0.205	8	10	18
fluorine	$^{19}_{9}F$	18.998 403 20	100	9	10	19
magnesium	$^{24}_{12}Mg$	23.985 041 90	78.99	12	12	24
	$^{25}_{12}Mg$	24.985 837 02	10.00	12	13	25
	$^{26}_{12}Mg$	25.982 593 04	11.01	12	14	26
chlorine	$^{35}_{17}Cl$	34.968 852 71	75.78	17	18	35
	$^{37}_{17}Cl$	36.965 902 60	24.22	17	20	37

*Data like these are available for all the naturally occurring elements in reference sources such as the *CRC Handbook* (Figure 1.15).

atomic mass of 4.00, the same as an alpha particle (Table 2.8). In fact, we now know that an α-particle is simply the nucleus of a helium-4 isotope.

As Table 2.9 implies, many elements occur in nature as mixtures of isotopes. The naturally occurring percentages of the isotopes of a particular element are referred to as the **natural abundances** of that element. Naturally occurring chlorine consists of two isotopes: 75.78% chlorine-35 and 24.22% chlorine-37. These percentages are nearly independent of the natural source of the chlorine. In other words, chlorine obtained from, say, salt deposits in Africa or Australia or North America has nearly the same isotopic composition as that given in Table 2.9.

The mass and percentage of each isotope of an element can be determined by using an instrument called a **mass spectrometer** (Figure 2.22). When a gas consisting of atoms is bombarded with electrons from an external source, the bombarding electrons knock electrons out of the neutral atoms in the gas. A neutral atom contains an equal number of positively charged protons (Z) and negatively charged electrons. An atom or molecule that gains or loses one or more electrons becomes charged and is called an **ion**. These gas-phase ions are accelerated by an electric field and pass through slits to form a narrow, well-focused beam (Figure 2.23). The beam of ions then passes through an electric or magnetic field that deflects each ion by an amount proportional to its mass. Thus, the original beam of ions splits into several separate beams, one for each isotope of the gas. The intensities of the separated ion beams, which can be determined experimentally, are a direct measure of the number of ions in each beam. In this manner, we can determine not only the mass of each isotope of any element (by the amount of deflection of each beam) but also the percentage of each isotope (by the intensity of each beam). Also note that the mass spectrometer separates the mixture of isotopes into isotopically pure components; therefore, we can use it to prepare isotopically pure samples of elements.

The atomic mass of chlorine is the sum of the masses of each isotope, each multiplied by its natural abundance. This is called a **weighted average.** Using the isotopic masses and natural abundances of chlorine given in Table 2.9, we obtain

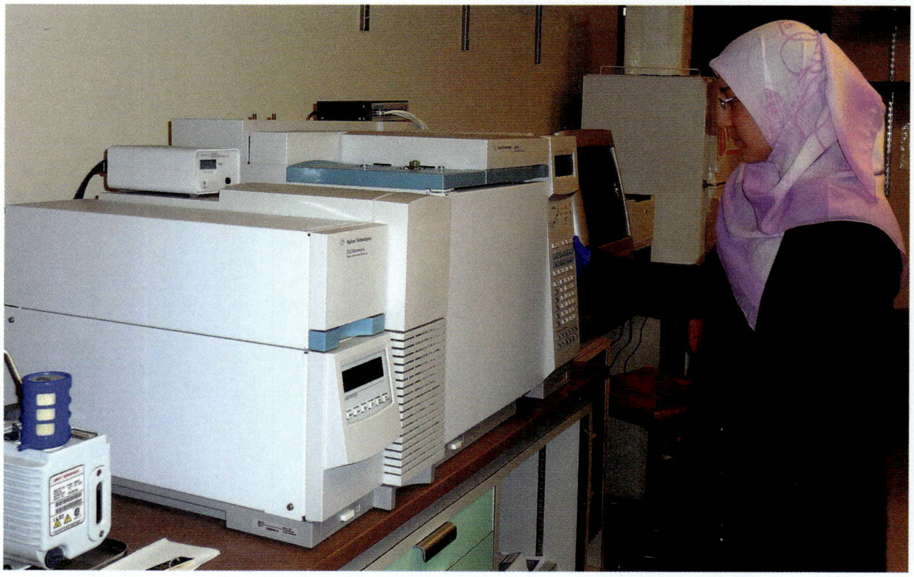

Figure 2.22 A mass spectrometer like the one shown here can be used to separate the isotopes of an element. In addition to determining the mass of isotopes, mass spectrometers are also a powerful tool for chemical analysis. The mass spectrometer shown here is used by students to check for traces of cocaine on U.S. currency.

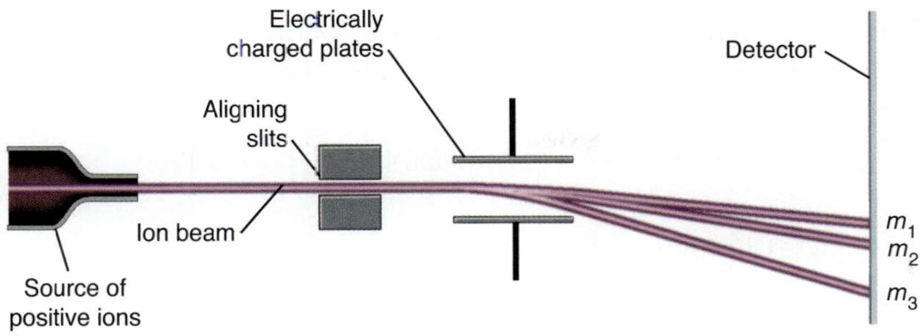

Figure 2.23 Schematic diagram of a mass spectrometer. A gas is bombarded with electrons that knock other electrons out of the gas atoms or molecules and produce positively charged ions. These ions are then accelerated by an electric field and form a narrow beam as they pass through the aligning slits. The beam then passes through an electric or magnetic field. The ions of different masses (m) are deflected to different extents, and the beam is split into several separate beams. The particles of different masses strike a detector, such as a photographic plate, at different places, and the amount of exposure at various places is proportional to the number of particles having a particular mass. Here, $m_1 > m_2 > m_3$. (Modern mass spectrometers replace the photographic plate with an electronic detector positioned at a fixed angle and scan the electric field strength needed to deflect the ion beam enough to reach the detector.)

$$
\begin{aligned}
{}^{35}_{17}\text{Cl:}\quad (34.968\,852\,71)\left(\frac{75.78}{100}\right) &= 26.50 \\
{}^{37}_{17}\text{Cl:}\quad (36.965\,902\,60)\left(\frac{24.22}{100}\right) &= 8.953 \\
\text{Average atomic mass} &= 35.45
\end{aligned}
$$

This value is the atomic mass of chlorine given on the inside front cover of the book. Note the number of significant figures obtained from the separate multiplication and addition steps involved in this calculation. The factors (75.78/100) and (24.22/100) must be included in order to take into account the relative natural abundance of each isotope. Thus, tables of atomic masses of the elements contain the relative average masses of atoms. The average mass is related to the masses of the individual isotopes of the element in the manner just illustrated for chlorine.

EXAMPLE 2-8: Naturally occurring chromium is a mixture of four isotopes with the following isotopic masses and natural abundances:

Mass number	Isotopic mass	Natural abundance/%
50	49.946 0496	4.345
52	51.940 5119	83.789
53	52.940 6538	9.501
54	53.938 8849	2.365

Calculate the average atomic mass of chromium.

Solution: The average atomic mass is the sum of the masses of the four isotopes each weighted by their respective abundances:

$$\begin{array}{lll} \text{Chromium-50} & (49.946\,0496)\left(\dfrac{4.345}{100}\right) & = 2.170 \\ \text{Chromium-52} & (51.940\,5119)\left(\dfrac{83.789}{100}\right) & = 43.520 \\ \text{Chromium-53} & (52.940\,6538)\left(\dfrac{9.501}{100}\right) & = 5.030 \\ \text{Chromium-54} & (53.938\,8849)\left(\dfrac{2.365}{100}\right) & = 1.276 \\ \hline & \text{Average atomic mass} & = 51.996 \end{array}$$

PRACTICE PROBLEM 2-8: Naturally occurring lithium is composed of two isotopes, lithium-6 (6.015 1223) and lithium-7 (7.016 0040). Given that the atomic mass of lithium is 6.941, compute the natural abundances of lithium-6 and lithium-7. (*Hint:* If we denote the percentage of lithium-6 in naturally occurring lithium by x, then the percentage of lithium-7 is $100 - x$, and the calculation proceeds like that in Example 2-8, except that now we know the atomic mass of naturally occurring lithium and we seek the mass percentages of the two isotopes.)

Answer: 7.5% lithium-6 and 92.5% lithium-7. Note the number of significant figures in each answer.

Small variations in natural abundances of the isotopic compositions of the elements limit the precision with which atomic masses can be specified, and so the masses of individual isotopes are known to much greater precision than the atomic masses given in the periodic table.

2-12. Ions Are Charged Particles

As we saw in the previous section, an atom or molecule with a charge due to the loss or gain of one or more electrons is called an ion. Positively charged ions are called **cations**; and negatively charged ions are called **anions**. We will encounter ions throughout our study of chemistry, so we introduce a notation for them here. An atom that has lost one electron has a net charge of +1; an atom that has lost two electrons has a charge of +2; an atom that has gained an electron has a charge of −1; and so on. We denote an ion by the chemical symbol of the element with a right-hand superscript to indicate its charge:

K^+	singly charged potassium ion
Mg^{2+}	doubly charged magnesium ion
Cl^-	singly charged chloride ion
S^{2-}	doubly charged sulfide ion

A neutral potassium atom has 19 protons and 19 electrons and a neutral chlorine atom has 17 protons and 17 electrons. A K^+ ion has 18 electrons ($19 - 1 = 18$), and a Cl^- ion also has 18 electrons ($17 + 1 = 18$). Species that contain the same number of electrons are said to be **isoelectronic.**

Because electrons are negatively charged, a charge of +3 means that the element has *lost* three of its electrons, not that it has three electrons or has gained three.

The names of cations are simply the element name plus the word ion or cation. Anions are given an *-ide* ending (like the second-named element in binary compounds) plus the word ion or anion. For example,

Ca^{2+} calcium ion or calcium cation

H^- hydride ion or hydride anion

We should mention that when writing out the name of an ion, we do not always indicate that the ion is "singly charged" or "doubly charged" when its charge is obvious from the context.

EXAMPLE 2-9: How many electrons are there in a Mg^{2+} cation and a S^{2-} anion?

Solution: From the table on the inside of the front cover, we see that the atomic number of magnesium is 12. A Mg^{2+} ion is a magnesium atom that has lost two electrons, so a Mg^{2+} ion has 10 electrons. The atomic number of sulfur is 16. The S^{2-} ion has two more electrons than a neutral sulfur atom, so a S^{2-} ion has 18 electrons.

PRACTICE PROBLEM 2-9: Give an example of a cation that is isoelectronic with the oxide ion, O^{2-}.

Answer: Na^+, Mg^{2+}, Al^{3+}.

EXAMPLE 2-10: Write the symbol for the element that has 8 protons, 10 neutrons, and 10 electrons.

Solution: ${}^{18}_{8}O^{2-}$. (Oxygen-18 is often used by climatologists to estimate past global temperatures in ice-covered regions. The warmer the temperature at which snow forms, the greater the oxygen-18 content and vice versa. Thus, scientists drilling down through layers in the polar ice caps can determine both how much precipitation fell in a given year and the average annual temperature. These data can be used to analyze long-term trends such as global climate change.)

PRACTICE PROBLEM 2-10: How many protons, neutrons, and electrons does each of the following species have?

(a) ${}^{28}_{14}Si^{4-}$ (b) ${}^{186}W^{5+}$ (c) ${}^{235}U$ (d) ${}^{58}_{26}Fe^{3+}$

Answer: (a) Silicon has 14 protons ($Z = 14$); because it has a mass number of 28, there are (28 − 14) = 14 neutrons. Because the charge is −4, there must be 4 extra electrons for a total of (14 + 4) = 18 electrons. (b) 74 protons. Although the atomic number is not given, all tungsten atoms have 74 protons. There are 112 neutrons and 69 electrons in $^{186}W^{5+}$. (c) 92 protons, 143 neutrons, and 92 electrons (no charge is given so we assume the atom is neutral). (d) 26 protons, 32 neutrons, and 23 electrons.

SUMMARY

The Lavoisiers' work led directly to the discovery of the law of constant composition and to Dalton's atomic theory. Dalton was able to use the atomic theory to determine the relative masses of atoms and molecules and to use these values in interpreting the results of chemical analyses. According to the atomic theory, the atoms in reactant molecules are separated and rearranged into product molecules in a chemical reaction. Because atoms are neither created nor destroyed in chemical reactions, chemical reactions obey the law of conservation of mass.

Elements are substances that consist of only one kind of atom. There are almost 120 known elements, about three-quarters of which are metals. Elements combine chemically to form compounds, whose constituent particles are called molecules, which are groups of atoms joined together. Chemists represent elements by chemical symbols and compounds by chemical formulas. The system of naming compounds is called chemical nomenclature.

Chemicals can exist as solids, liquids, or gases. We denote these states of matter by attaching the symbols (s), (l), or (g) to the chemical formula of a substance. Many of the substances found in nature exist as mixtures. Separation techniques such as filtration, evaporation, and distillation are used in chemistry to obtain pure substances from mixtures.

Protons and neutrons are found in the nucleus of an atom, the small center containing all the positive charge and essentially all the mass of the atom. The number of protons in an atom is the atomic number (Z) of that atom. The total number of protons and neutrons in an atom is the mass number (A) of that atom.

Each element is characterized by its atomic number. Nuclei with the same number of protons but different numbers of neutrons are called isotopes. Most elements occur naturally as mixtures of isotopes, and atomic masses are weighted averages of the isotopic masses. Because isotopes of an element have the same atomic number, they are chemically identical and they undergo the same chemical reactions.

When an atom or molecule gains or loses electrons, it becomes charged and is called an ion. Positively charged ions are called cations; negatively charged ions, anions. Cations have a positive charge that is equal in magnitude to the number of electrons lost and anions have a negative charge that is equal in magnitude to the number of electrons gained. Two species with the same number of electrons are said to be isoelectronic.

In this chapter we have established the fundamental concepts of atoms and molecules. These concepts are the foundations on which we shall build our study of chemistry.

TERMS YOU SHOULD KNOW

element *41*
compound *41*
metal *42*
nonmetal *42*
malleable *42*
ductile *42*
chemical symbol *42*
atom *43*
molecule *43*
diatomic molecule *43*
states of matter *44*
solid (*s*) *44*
liquid (*l*) *44*
gas (*g*) *44*
mixture *45*
heterogeneous *45*
dissolution *46*
solution *46*
homogeneous *46*
solute *46*
solvent *46*
aqueous solution (*aq*) *46*
filtration *47*
evaporation *47*
distillation *47*
condensation *47*
volatile *48*
fractional distillation *49*
law of constant composition *49*
mass percentage *49*
law of multiple proportions *50*
atomic theory *51*
atomic mass ratio *52–53*
atomic mass *53*
atomic mass unit, u *53*
IUPAC (International Union of Pure and Applied Chemistry) *53*
chemical nomenclature *55*
binary compound *55*
chemical formula *55*
water, $H_2O(l)$ *57*
ammonia, $NH_3(g)$ *57*
methane, $CH_4(g)$ *57*
molecular mass, u *58*
subatomic particle *59*
cathode rays *60*
electron *60*
coulomb, C *60*
cathode ray tube (CRT) *60*
radioactivity *60*
radioactive *60*
α-particle *60*
β-particle *60*
γ-rays *61*
nucleus *62*
nuclear model of the atom *62*
proton *62*
neutron *62*
atomic number, *Z* *62*
neutral atom *62*
mass number, *A* *63*
isotope *63*
deuterium *64*
heavy water *64*
natural abundance *66*
mass spectrometer *66*
ion *66*
weighted average *66*
cation *68*
anion *68*
isoelectronic *69*

EQUATION YOU SHOULD KNOW HOW TO USE

$N = A - Z$ (2.1) (relation between the number of neutrons *N*, the mass number *A*, and the atomic number *Z*)

PROBLEMS

CHEMICAL SYMBOLS

2-1. Give the chemical symbols for the following elements:

(a) selenium (b) indium
(c) manganese (d) silver
(e) mercury (f) krypton
(g) palladium (h) thallium
(i) uranium (j) tungsten

2-2. Give the chemical symbols for the following elements:

(a) tin (b) gold
(c) zirconium (d) bismuth
(e) copper (f) rubidium
(g) bromine (h) neon
(i) antimony (j) arsenic

2-3. Name the elements with the following chemical symbols:

(a) Ge	(b) Sc	(c) Ir	(d) Cs
(e) Sr	(f) Am	(g) Mo	(h) S
(i) Pu	(j) Xe		

2-4. Name the elements with the following chemical symbols:

(a) Pt	(b) Te	(c) Pb	(d) Ta
(e) Ba	(f) Ti	(g) Re	(h) La
(i) Eu	(j) Pr		

SEPARATIONS

2-5. Explain how you could separate iron filings from aluminum powder.

2-6. Explain how you could separate table salt, NaCl(*s*), from sand.

2-7. What is the role of the condenser in distillation?

2-8. What is meant when a liquid is said to be volatile?

2-9. Describe how evaporation can be used to separate certain components in a solution.

2-10. What is distillation used for?

2-11. What are the contrasting physical properties of gold and sand on which panning of gold depends?

2-12. What is meant by a heterogeneous mixture?

MASS PERCENTAGES IN COMPOUNDS

2-13. A 1.659-gram sample of a compound of sodium and oxygen contains 0.978 grams of sodium and 0.681 grams of oxygen. Calculate the mass percentages of sodium and oxygen in the compound.

2-14. The compound lanthanum oxide is used in the production of optical glass and fluorescent phosphors. An 8.29-gram sample is found to contain 7.08 grams of lanthanum and 1.21 grams of oxygen. Calculate the mass percentages of lanthanum and oxygen in lanthanum oxide.

2-15. A 1.28-gram sample of copper is heated with sulfur to produce 1.60 grams of a copper sulfide compound. Calculate the mass percentages of copper and sulfur in the compound.

2-16. Stannous fluoride, which is an active ingredient in some toothpastes and helps to prevent cavities, contains tin and fluorine. A 1.793-gram sample was found to contain 1.358 grams of tin. Calculate the mass percentages of tin and fluorine in stannous fluoride.

2-17. Potassium cyanide is used in extracting gold and silver from their ores. A 12.63-milligrams sample is found to contain 7.58 milligrams of potassium, 2.33 milligrams of carbon, and 2.72 milligrams of nitrogen. Calculate the mass percentages of potassium, carbon, and nitrogen in potassium cyanide.

2-18. Ethanol, the alcohol in alcoholic beverages, is a compound of carbon, hydrogen, and oxygen. A 3.70-gram sample of ethanol contains 1.93 grams of carbon and 0.49 grams of hydrogen. Calculate the mass percentages of carbon, hydrogen, and oxygen in ethanol.

NOMENCLATURE

2-19. Name the following binary compounds:

(a) Li_2S	(b) BaO
(c) Mg_3P_2	(d) CsBr

2-20. Name the following binary compounds:

(a) BaF_2	(b) Mg_3N_2
(c) CsCl	(d) CaS

2-21. Name the following binary compounds:

(a) SiC	(b) GaP
(c) Al_2O_3	(d) $BeCl_2$

2-22. Name the following binary compounds:

(a) MgF_2	(b) AlN
(c) MgSe	(d) Li_3P

2-23. Name the following pairs of compounds:

(a) ClF_3 and ClF_5	(b) SF_4 and SF_6
(c) KrF_2 and KrF_4	(d) BrO and BrO_2

2-24. Name the following pairs of compounds:

(a) $SbCl_3$ and $SbCl_5$	(b) ICl_3 and ICl_5
(c) SeO_2 and SeO_3	(d) CS and CS_2

2-25. Name the following compounds:

(a) NO_2 (b) NH_3

(c) ZnS (d) K_2O

2-26. Name the following compounds:

(a) BaH_2 (b) Li_2S

(c) BeO (d) CH_4

MOLECULAR MASSES

2-27. Calculate the molecular mass for each of the following oxides to five significant figures:

(a) TiO_2 (white pigment)

(b) Fe_2O_3 (rust)

(c) V_2O_5 (a catalyst)

(d) P_4O_{10} (dehydrating agent)

2-28. Calculate the molecular mass for each of the following ores to five significant figures:

(a) $CaWO_4$ (scheelite, an ore of tungsten)

(b) Fe_3O_4 (magnetite)

(c) Na_3AlF_6 (cryolite)

(d) $Be_3Al_2Si_6O_{18}$ (beryl)

(e) Zn_2SiO_4 (willemite)

2-29. Calculate the molecular mass for each of the following halogen compounds to five significant figures:

(a) BrN_3 (explosive)

(b) $NaIO_3$ (antiseptic)

(c) CCl_2F_2 (former refrigerant)

(d) $C_{14}H_9Cl_6$ (DDT)

2-30. Calculate the molecular mass for each of the following vitamins to five significant figures:

(a) $C_{20}H_{30}O$ (vitamin A)

(b) $C_{12}H_{17}ClN_4OS$ (vitamin B_1, thiamine)

(c) $C_{17}H_{20}N_4O_6$ (vitamin B_2, riboflavin)

(d) $C_{56}H_{88}O_2$ (vitamin D_1)

(e) $C_6H_8O_6$ (vitamin C, ascorbic acid)

2-31. Calculate the molecular mass of the compounds represented by the following chemical structures to five significant figures (lines represent bonds between adjacent atoms):

(a)

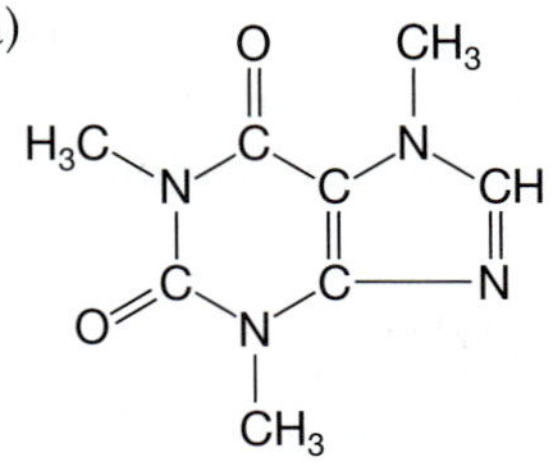

caffeine

(b)

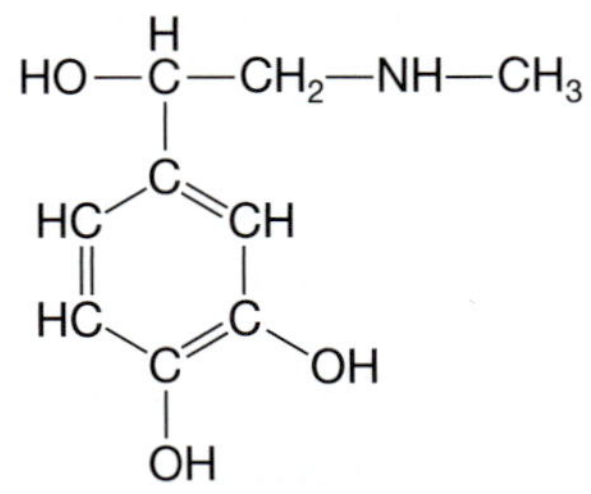

epinephrine

2-32. Calculate the molecular mass of the compounds represented by the following chemical structures to five significant figures (lines represent bonds between adjacent atoms):

(a)

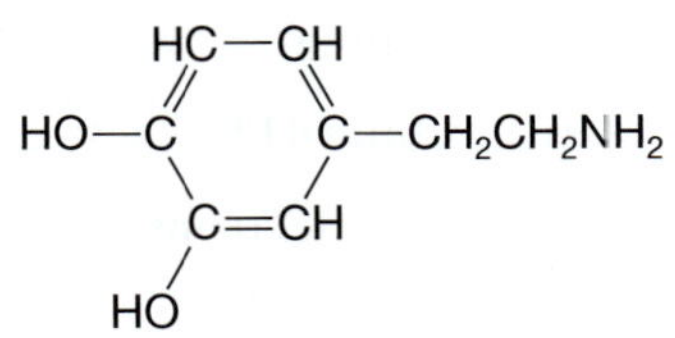

dopamine

(b)

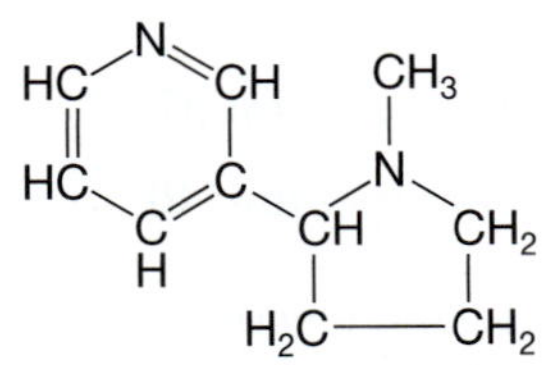

nicotine

MASS PERCENTAGES AND ATOMIC MASSES

2-33. Use the atomic masses given on the inside front cover of the text to calculate the mass percentages of chlorine and fluorine in chlorine trifluoride.

2-34. Use the atomic masses given on the inside front cover of the text to calculate the mass percentages of nitrogen and oxygen in dinitrogen oxide.

2-35. Ordinary table sugar, or sucrose, has the chemical formula $C_{12}H_{22}O_{11}$. Calculate the mass percentages of carbon, hydrogen, and oxygen in sucrose.

2-36. A key compound in the production of aluminum metal is cryolite, $Na_3AlF_6(s)$. Calculate the mass percentages of sodium, aluminum, and fluorine in this compound.

2-37. Calculate the number of grams of xenon in 2.000 grams of the compound xenon tetrafluoride.

2-38. Calculate the number of grams of sulfur in 5.585 grams of the compound sulfur trioxide.

2-39. The label of a fertilizer states that the fertilizer contains 18.0% P_4O_{10} by mass. What is the mass percentage of phosphorus in the fertilizer?

2-40. A sample of an impure ore contains 42.7% Cr_2O_3 by mass. What is the mass percentage of chromium in the ore?

PROTONS, NEUTRONS, AND ELECTRONS

2-41. The following isotopes are used widely in medicine or industry:

(a) iodine-131 (b) cobalt-60

(c) potassium-43 (d) indium-113

How many protons, neutrons, and electrons are there in a neutral atom of each of these isotopes?

2-42. The following isotopes do not occur naturally but are produced in nuclear reactors:

(a) phosphorus-30 (b) technetium-97

(c) iron-55 (d) americium-240

How many protons, neutrons, and electrons are there in a neutral atom of each of these isotopes?

2-43. Fill in the blanks in the following table:

Symbol	Atomic number	Number of neutrons	Mass number
$^{14}_{6}\text{C}$			
$^{?}_{?}\text{Am}$			241
	53		123
		10	18

2-44. Fill in the blanks in the following table:

Symbol	Atomic number	Number of neutrons	Mass number
$^{?}_{?}\text{Ca}$			48
	40		90
		78	131
$^{?}_{?}\text{Mo}$		57	

2-45. Fill in the blanks in the following table:

Symbol	Atomic number	Number of neutrons	Mass number
	31	36	
		8	15
	27		58
$^{?}_{?}\text{Xe}$			133

2-46. Fill in the blanks in the following table:

Symbol	Atomic number	Number of neutrons	Mass number
$^{39}_{19}\text{K}$			
$^{?}_{?}\text{Fe}$			56
	36		84
		70	120

ISOTOPIC COMPOSITION

Pay special attention to significant figures when working these problems.

2-47. Naturally occurring oxygen consists of three isotopes, with the isotopic masses and abundances given in Table 2.9. Calculate the average atomic mass of naturally occurring oxygen.

2-48. Naturally occurring magnesium consists of three isotopes, with the isotopic masses and abundances given in Table 2.9. Calculate the average atomic mass of naturally occurring magnesium.

2-49. Naturally occurring neon is a mixture of three isotopes with the following isotopic masses and abundances:

Isotope	Isotopic mass	Abundance/%
^{20}Ne	19.992 440 1759	90.48
^{21}Ne	20.993 846 74	0.27
^{22}Ne	21.991 385 51	9.25

Calculate the average atomic mass of naturally occurring neon.

2-50. Naturally occurring silicon consists of three isotopes with the following isotopic masses and abundances:

Isotope	Isotopic mass	Abundance/%
^{28}Si	27.976 926 5327	92.2297
^{29}Si	28.976 494 72	4.6832
^{30}Si	29.973 770 22	3.0872

Calculate the average atomic mass of naturally occurring silicon.

2-51. Copper metal has a wide variety of uses, including copper pipes, coins, and wires. Naturally occurring copper consists of two isotopes, copper-63 and copper-65. Using the average atomic mass of copper listed in the periodic table inside the cover of this book, fill in the missing data in the table below. (*Hint:* Because copper consists of only two isotopes, what can you conclude about the sum of their abundances?)

Isotope	Isotopic mass	Abundance/%
^{63}Cu	62.929 6011	69.17
^{65}Cu		

2-52. Gallium is a metal with a wide variety of uses, including computer memory chips, light-emitting diodes, and lasers. Radioactive isotopes of gallium are used to image the human body and locate tumors. Naturally occurring gallium consists of two isotopes. Using the average atomic mass of gallium listed in the periodic table inside the cover of this book, fill in the missing data in the table below. (*Hint:* Because gallium consists of only two isotopes, what can you conclude about the sum of their abundances?)

Isotope	Isotopic mass	Abundance/%
$^{?}Ga$		60.108
^{71}Ga	70.924 7050	

2-53. Naturally occurring bromine consists of two isotopes, ^{79}Br and ^{81}Br, with isotopic masses of 78.9183 and 80.9163, respectively. Given that the observed atomic mass of bromine is 79.904, calculate the percentages of ^{79}Br and ^{81}Br in naturally occurring bromine.

2-54. Naturally occurring boron consists of two isotopes with the isotopic masses 10.013 and 11.009. The observed atomic mass of boron is 10.811. Calculate the abundance of each isotope.

2-55. Nitrogen has two naturally occurring isotopes, ^{14}N and ^{15}N, with isotopic masses of 14.0031 and 15.0001, respectively. The average atomic mass of nitrogen is 14.0067. Use these data to compute the percentage of ^{15}N in naturally occurring nitrogen.

2-56. Naturally occurring europium consists of two isotopes, ^{151}Eu and ^{153}Eu, with isotopic masses of 150.9199 and 152.9212, respectively. Given that the average atomic mass of europium is 151.964, calculate the percent abundance of each isotope.

IONS

2-57. How many electrons are there in the following ions?

(a) Cs^{+} (b) I^{-}
(c) Se^{2-} (d) N^{3-}

2-58. How many electrons are there in the following ions?

(a) Br^{-} (b) P^{3-}
(c) Ag^{+} (d) K^{+}

2-59. Determine the number of electrons in the following ions:

(a) Ba^{2+} (b) S^{2-}
(c) Ga^{3+} (d) Ti^{4+}

2-60. Determine the number of electrons in the following ions:

(a) Te^{2-} (b) La^{3+}
(c) Be^{2+} (d) Ge^{4+}

2-61. Give three ions that are isoelectronic with each of the following:

(a) K^+ (b) Kr

(c) N^{3-} (d) I^-

2-62. Give three ions that are isoelectronic with each of the following:

(a) F^- (b) Se^{2-}

(c) Ba^{2+} (d) La^{3+}

2-63. Name each of the following ions. Use the designation "cation" or "anion" when writing the names.

(a) O^{2-} (b) H^+

(c) Na^+ (d) F^-

2-64. Name each of the following ions. Use the designation "cation" or "anion" when writing the names.

(a) S^{2-} (b) Al^{3+}

(c) H^- (d) V^{3+}

2-65. Use the atomic masses given in the inside front cover to compute molecular masses of the following ions to five significant figures:

(a) OH^- (b) H_3O^+

(c) AlF_6^{3-} (d) PCl_4^+

2-66. Use the atomic masses given in the inside front cover to compute molecular masses of the following ions to five significant figures:

(a) NH_4^+ (b) HO_2^-

(c) $AgCl_2^-$ (d) PCl_6^-

ADDITIONAL PROBLEMS

2-67. Write the name of five elements that have symbols due to Latin names.

2-68. Which elements have names similar to those of planets?

2-69. Describe what is meant by a subatomic particle. What are the three subatomic particles discussed in this chapter? List several properties of each of these three subatomic particles, such as their mass, charge, and where they are found in an atom.

2-70. What is an isotope? Do isotopes of the same element have the same chemical properties?

2-71. Although water, ammonia, and methane are never called by their systematic (or IUPAC) names, what would they be called using this system?

2-72. (*) Heavy water has the formula D_2O, and semideuterated water has the formula DHO, where D stands for deuterium. Using the data from Table 2.9, estimate the percentages of D_2O and DHO present in natural water.

2-73. Rank the following fertilizers in decreasing order of mass percentage of nitrogen:

(a) $NH_4NO_3(s)$ (b) $NH_3(g)$

(c) $(NH_4)_2SO_4(s)$ (d) $(NH_4)_2HPO_4(s)$

(e) $(NH_4)H_2PO_4(s)$ (f) $KNO_3(s)$

2-74. A certain protein was found to contain 0.168% cobalt by mass. Determine the minimum molecular mass of the protein.

2-75. A sample of rutile, an ore of titanium consisting principally of $TiO_2(s)$, was found to be 65.2% $TiO_2(s)$ by mass, with the remainder being sand impurities. What is the minimum number of metric tons of the ore that must be processed to obtain 10.0 metric tons of titanium?

2-76. Tungsten is the metal used in the filaments of incandescent lightbulbs. Naturally occurring tungsten consists of five isotopes with the following isotopic masses and abundances:

Isotope	Isotopic mass	Abundance/%
^{180}W	179.946 706	0.12
^{182}W	181.948 206	26.50
^{183}W	182.950 2245	14.31
^{184}W	183.950 9326	30.64
^{186}W	185.954 362	28.43

Calculate the average atomic mass of naturally occurring tungsten.

2-77. An isotope of iodine used to treat hyperactive thyroids is iodine-131. It forms the iodide ion I^-.

(a) How many protons are there in a nucleus of I and in a nucleus of I^-?

(b) How many neutrons are there in a nucleus of I and in a nucleus of I^-?

(c) How many electrons are there in an iodine atom and in an iodide ion?

2-78. In one compound of nitrogen and oxygen, 0.615 grams of nitrogen combines with 0.703 grams of oxygen. In another 1.27 grams of nitrogen combines with 2.90 grams of oxygen. Show how these data illustrate the law of multiple proportions.

2-79. Sulfur forms two oxides. In one of them, 1.87 grams of sulfur combines with 1.87 grams of oxygen, and in the other 3.94 grams of sulfur combines with 5.91 grams of oxygen. Show how these data illustrate the law of multiple proportions.

2-80. Suppose we decide to establish an atomic mass scale by setting the atomic mass of ^{12}C exactly equal to one. What would be the atomic masses of naturally occurring hydrogen and oxygen on this scale?

2-81.(*) A 40.0-milligrams sample of the compound X_4O_{10} contains 22.5 milligrams of oxygen atoms. What is the atomic mass of element X?

2-82.(*) At one time, chemists thought that the formula of a binary compound was the simplest formula possible. For example, the chemical formula of water was thought to be HO and that of ammonia was thought to be NH. Given that 0.832 grams of oxygen combines with 0.104 grams of hydrogen, calculate the atomic mass of oxygen based upon HO as the formula for water. Similarly, given that 0.403 grams of nitrogen combines with 0.0864 grams of hydrogen, calculate the atomic mass of nitrogen based upon NH as the formula for ammonia.

Dmitri Ivanovich Mendeleev (1834–1907), was born in Tobolsk, Siberia, to early settlers of Siberia. He started his studies at the St. Petersburg Pedagogical Institute and, after studying for several years at various laboratories in Europe, returned to St. Petersburg to earn his doctorate. In 1863, at the age of 33, he was appointed as a professor of chemistry at the University of St. Petersburg. In 1869, he published the first version of his periodic table in his textbook *Principles of Chemistry*. In addition to his great achievement with the periodic table, he was involved in the early development of the extensive oil fields at Baku, helping to create the first Russian oil refinery, was a founding member of the Russian Chemical Society, developed and patented the standard formula for Russian vodka, and is credited with bringing the metric system to Russia.

Mendeleev had a prickly personality and often quarreled with his first wife, Feozva Leshceva. In 1882, he divorced her in order to marry his niece's best friend, the young Anna Ivanova Popova, with whom he stayed happily married. In his later years, Mendeleev grew notorious for his lack of concern over his personal appearance. It is said that he cut his hair and beard only once a year and even refused to trim them for the Czar.

Throughout his life, Mendeleev worked to abolish social inequity and became an outspoken critic of the Russian government. Because of his participation in student protests, he was forced to resign from the university in 1890. His last lecture was broken up by police fearing riots.

3. The Periodic Table and Chemical Periodicity

Lavoisier and Dalton opened the way to a systematic exploration of pure substances and their properties. With the new tools of atomic mass, atomic number, and quantitative measurement, scientists could probe the vast variety of chemical behavior. Their studies, and our studies, are made easier because many elements share common characteristics. Instead of looking separately at each element and its compounds, we can separate the elements into groups and study the reactions of each group. In addition, when we arrange the elements in order of increasing atomic number, we find a repetitive order in their chemical properties. This arrangement, known as the periodic table, will be the basis for understanding the chemistry of the elements throughout this book and in your future studies. We begin this chapter by looking at some simple chemical reactions and how we express these using a notation called a chemical equation. In later chapters, as we learn more about the properties of atoms and molecules, we shall see why certain chemicals react to form new compounds and why others do not.

3-1. New Substances Are Formed in Chemical Reactions

Let's begin with some of the simplest chemical reactions, those in which a metal reacts directly with a nonmetal. For example, consider the reaction between sodium and chlorine. Sodium is a very reactive metal. It reacts spontaneously with the oxygen and water vapor in the air and is usually stored under kerosene, an oily liquid that is unreactive in the absence of sparks or flames. Chlorine, a nonmetal, is a greenish-yellow, highly reactive, toxic gas that attacks many metals. When sodium metal is dropped into a container of chlorine gas, there is a vigorous, spontaneous reaction. The product of the reaction is a white, crystalline solid—sodium chloride—which is ordinary table salt. We can represent the reaction of sodium with chlorine as

$$\underset{\text{very reactive metal}}{\text{sodium metal}} + \underset{\text{very reactive nonmetal}}{\text{chlorine gas}} \rightarrow \underset{\substack{\text{nonreactive compound}\\ \text{(ordinary table salt)}}}{\text{sodium chloride}} \quad (3.1)$$

Figure 3.1 When sodium (a very reactive metal) reacts with chlorine gas (a very reactive nonmetal), the product is sodium chloride (ordinary table salt). Note that the product is an entirely new substance and is chemically and physically different from both reactants.

Equation 3.1 illustrates the fact that the chemical properties of a product of a chemical reaction need not bear any resemblance to the chemical properties of the reactants (Figure 3.1). Entirely new substances are formed in chemical reactions.

Another example of the differences between reactants and products is found in the reaction between hydrogen and oxygen to form water. Both hydrogen and oxygen are colorless, odorless gases. Together they form an explosive mixture that is set off easily by a spark or a flame. The reaction between hydrogen and oxygen can be described by the equation

$$\underset{\text{colorless gas}}{\text{hydrogen gas}} + \underset{\text{colorless gas}}{\text{oxygen gas}} \rightarrow \underset{\text{colorless liquid}}{\text{water}} \qquad (3.2)$$

The properties of water are radically different from those of either hydrogen or oxygen. Because hydrogen is the least dense gas, being about 15 times less dense than air, balloons and dirigibles used to be filled with hydrogen. This practice was discontinued after 1937, when the hydrogen-filled German dirigible *Hindenburg* exploded at Lakehurst, New Jersey (Figure 3.2). Modern blimps and weather balloons are filled with helium gas, which is nonreactive and about seven times less dense than air.

It is cumbersome to write out the full names of the reactants and products of chemical reactions as we have been doing. Consequently, chemists have devised a shorthand way of describing the chemical changes that occur in chemical reactions. A chemical reaction is represented by a chemical equation that denotes the reactants and the products in terms of their chemical formulas. The **reactants** are the substances that react with each other, and the **products** are the substances formed in the reaction. A left-to-right arrow is used to indicate

Figure 3.2 The explosion of the hydrogen gas-filled German dirigible *Hindenburg* during landing at Lakehurst, New Jersey.

that the reactants are converted to products. For example, the reaction between sodium metal and chlorine gas to form sodium chloride is expressed by

$$\mathrm{Na}(s) + \mathrm{Cl_2}(g) \rightarrow \mathrm{NaCl}(s) \quad \text{(not balanced)} \tag{3.3}$$

The plus sign on the left-hand side of this equation means "reacts with." The arrow separates the reactants from the products and means "to yield" or "to produce." The arrow also shows the direction in which the reaction proceeds. Recall that the symbol (s) after a chemical formula tells us that the substance is a solid and the symbol (g) tells us that the substance is a gas. All substances in elemental form are represented simply by the symbol of the element (e.g., Na) in chemical equations. The subscript 2 following the element chlorine (Cl_2) in Equation 3.3 indicates that there are two atoms of chlorine in its molecular formula—recall from Chapter 2 that chlorine is one of the naturally occurring diatomic elements. At this point in your study the formulas of all the reactants and products will be given; in later chapters you will learn to use chemical principles to predict the products of some of the more common chemical reactions yourself.

The representation of a chemical reaction by the chemical formulas of the reactants and products separated by an arrow, as we have done above, is called a **chemical equation**. Equation 3.3 still omits an essential feature of chemical equations; that is, the number of chlorine atoms is not the same on both the left (reactant) and the right (product) sides. A systematic procedure for balancing simple chemical equations is presented in the next section. This will be expanded to include more complex equations in later chapters.

3-2. A Chemical Equation Must Be Balanced

Dalton's atomic theory showed that Lavoisier's law of conservation of mass in chemical reactions is a direct result of the conservation of each type of atom involved in the chemical reaction. Although new substances are formed in chemical reactions as a result of new arrangements of the atoms, *the individual atoms of various types are neither created nor destroyed in a chemical reaction.* Thus, the number of atoms of each element remains the same in a chemical reaction. Consequentially, a complete chemical equation must always be balanced; that is, it must have the same number of each type of atom on both sides. Note that Equation 3.3 contains two atoms of chlorine on the left but only one atom of chlorine on the right. We can balance Equation 3.3 with respect to the chlorine atoms by placing a 2 in front of the NaCl(s) on the right-hand side of the equation:

$$\mathrm{Na}(s) + \mathrm{Cl_2}(g) \rightarrow 2\,\mathrm{NaCl}(s) \quad \text{(not balanced)}$$

Now there are two sodium atoms on the right but only one on the left. If we place a 2 in front of the Na(s) on the left, then we obtain

$$2\,\mathrm{Na}(s) + \mathrm{Cl_2}(g) \rightarrow 2\,\mathrm{NaCl}(s) \quad \text{(balanced)} \tag{3.4}$$

Equation 3.4 is a **balanced chemical equation** for the reaction of sodium with chlorine; both sides contain the same number of each kind of atom.

We balance chemical equations by placing the appropriate numbers, called

balancing coefficients, in front of the chemical formulas. The chemical formulas of the reactants and products themselves are fixed. They cannot be altered. If a subscript is changed, the formula is incorrect. We cannot balance Equation 3.3 by changing NaCl to $NaCl_2$. The chemical formula of sodium chloride is NaCl, not $NaCl_2$. In fact, no such compound as $NaCl_2$ exists.

Note that it is conventional to omit the number 1 when writing chemical formulas and equations. It would be unconventional to express Equation 3.4 as

$$2\,Na_1(s) + 1\,Cl_2(g) \rightarrow 2\,Na_1Cl_1(s) \qquad \text{(unconventional)}$$

Let's consider now the reaction of hydrogen with oxygen. Recall that hydrogen and oxygen exist as diatomic molecules. We first write Equation 3.2 in terms of the chemical formulas of the reactants and products, that is,

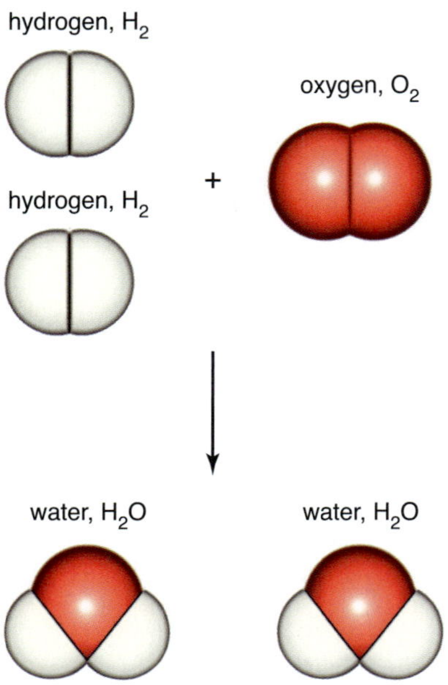

Figure 3.3 A representation using molecular models of the reaction between hydrogen and oxygen to form water. A chemical reaction is a rearrangement of the atoms in the reactant molecules to form product molecules. The numbers of each kind of atom are the same both before and after the reaction.

$$H_2(g) + O_2(g) \rightarrow H_2O(l) \qquad \text{(not balanced)}$$

where the (l) after the formula of water means that it is a liquid. In this equation, there are two oxygen atoms on the left in $O_2(g)$ and one oxygen atom on the right in $H_2O(l)$. If we place a 2 in front of the $H_2O(l)$, then the equation is balanced with respect to oxygen atoms:

$$H_2(g) + O_2(g) \rightarrow 2\,H_2O(l) \qquad \text{(not balanced)}$$

Now there are four hydrogen atoms on the right (2×2) and only two on the left. We balance the hydrogen atoms by placing a 2 in front of the $H_2(g)$:

$$2\,H_2(g) + O_2(g) \rightarrow 2\,H_2O(l) \qquad \text{(balanced)}$$

This gives us the balanced equation for the reaction of hydrogen with oxygen (Figure 3.3). The balanced chemical equation can be expressed in words as follows: two molecules of hydrogen react with one molecule of oxygen to yield two molecules of water. Once again, note that a chemical equation is balanced by placing coefficients in front of the formulas of the reactants and products; the formulas themselves are not altered.

The method that we used to balance the chemical equations in the two examples above is called the method of **balancing by inspection**. The essence of this method is to adjust the coefficients of the chemical formulas on each side of the arrow such that the number of atoms of every element involved is exactly the same on both sides of the equation. The procedure is summarized as follows:

1. Write down the chemical formulas of the reactants and products. Separate the reactants from the products by an arrow and the individual reactants and products using plus (+) signs.
2. Look for elements that appear only once on each side of the equation, including any subscripts, and adjust the balancing coefficients so that the number of atoms of the element on both sides is the same.
3. Inspect the remaining elements in turn and balance them as needed.
4. Perform a final check of each element in your chemical equation to be sure that every element is balanced.

5. Add the symbols (s) for solids, (l) for liquids, (g) for gases, or (aq) for aqueous (or water-based) solutions after each chemical formula, if the states of the reactants and products are known.

WWW

Links to many interesting chemically relevant websites can be found at www.McQuarrieGeneralChemistry.com. There is a link to an excellent website called Periodic Table Live! that you should explore. If you click on any element in the periodic table that is displayed, it gives you the chemical and physical properties of the element, including still photos and sometimes videos of reactions. We shall encourage you to go to this website many times throughout the book.

EXAMPLE 3-1: Sodium metal reacts vigorously with water. The reactants and products of the reaction are

$$\underset{\text{sodium metal}}{Na(s)} + \underset{\text{water}}{H_2O(l)} \rightarrow \underset{\text{sodium hydroxide}}{NaOH(aq)} + \underset{\text{hydrogen gas}}{H_2(g)}$$

Balance this chemical equation.

Solution: Step 1 has been done for us. All the reactants and products have been written out with the appropriate plus signs and arrows.

$$Na + H_2O \rightarrow NaOH + H_2 \qquad \text{(not balanced)} \qquad (3.5)$$

Step 2 requires us to look for elements that appear only once on each side of the equation and balance these first. Examining Equation 3.5 we see that sodium appears once on both sides in Na and NaOH. Because both contain only one atom of sodium, no adjustment to the coefficients is necessary. However, should we need to change either of these coefficients later, we must also change the other to maintain this equality. Likewise oxygen appears once on both sides of the equation in H_2O and NaOH, but because there is only one oxygen atom on each side, no adjustment of the coefficients involving oxygen atoms is necessary.

Step 3 requires us to adjust the coefficients of the remaining elements, in this case hydrogen. Because there are two hydrogen atoms on the left side of the equation but three on the right, we need to adjust the coefficients. We need more H atoms on the left, so we place a 2 in front of H_2O, a change that in turn requires a 2 in front of NaOH to maintain the equality of the oxygen atoms:

$$Na + 2H_2O \rightarrow 2NaOH + H_2 \qquad \text{(not balanced)}$$

The 2 in front of NaOH in turn requires a 2 in front of Na:

$$2Na + 2H_2O \rightarrow 2NaOH + H_2 \qquad \text{(balanced)}$$

Step 4 is to make a final check of the number of atoms on each side of the equation; this can be tabulated as follows:

Atom	Number on left	Number on right	Equal?
hydrogen	4	4	√ yes
oxygen	2	2	√ yes
sodium	2	2	√ yes

The table shows that the number of each atom on the left and right sides of the equation are equal, so the equation is now balanced.

Step 5 is to write out the final equation including the states of the compounds, if known (see note in margin on page 83). Because these were given initially, our final balanced chemical equation is

$$2\,Na(s) + 2\,H_2O(l) \rightarrow 2\,NaOH(aq) + H_2(g) \qquad \text{(final balanced equation including states)}$$

When dried, the sodium hydroxide formed in the reaction between sodium and water is a white, translucent solid used in the manufacture of paper and soaps and in petroleum refining. It is extremely corrosive to the skin and other tissues and is often called caustic soda or lye. A paste of sodium hydroxide and water is used in several commercially available oven cleaners.

PRACTICE PROBLEM 3-1: Black phosphorus, $P(s)$, reacts with excess oxygen gas, $O_2(g)$, to form the oxide $P_4O_{10}(s)$. Write a balanced chemical equation for this reaction.

Answer: $4\,P(s) + 5\,O_2(g) \rightarrow P_4O_{10}(s)$

BALANCING CHEMICAL EQUATIONS ALGEBRAICALLY. There is another way to balance chemical equations that might appeal to you if you like algebra. Consider the (unbalanced) equation in Example 3-1. Let's write it as

$$a\,Na + b\,H_2O \rightarrow c\,NaOH + d\,H_2$$

where a through d are to be determined. If we balance the sodium atoms, then we must have

$$a = c \qquad \text{(Na atoms balance)}$$

The hydrogen and oxygen atoms give us

$$2b = c + 2d \qquad \text{(H atoms balance)}$$

$$b = c \qquad \text{(O atoms balance)}$$

There are three equations and four unknowns. The balancing coefficients are relative quantities, so we may set any one of them equal to whatever we wish. Let's set $a = 1$. Then, the first equation gives $c = 1$, the third gives $b = 1$, and the second gives $2 = 1 + 2d$, or $d = \frac{1}{2}$. Thus, the balanced equation is

$$Na(s) + H_2O(l) \rightarrow NaOH(aq) + \tfrac{1}{2}H_2(g)$$

or

$$2\,Na(s) + 2\,H_2O(l) \rightarrow 2\,NaOH(aq) + H_2(g)$$

This method will always work, although the set of algebraic equations for the balancing coefficients can be fairly large. We simply present this method for you as an alternative to the method of balancing by inspection if you like to work with algebraic equations.

With a little practice, you can become proficient in balancing certain types of chemical equations by inspection. Problems 3-1 through 3-8 at the end of this chapter will give you the practice you need.

We should emphasize here that a chemical equation is a representation of a chemical reaction. It is somewhat arbitrary how we choose to write a chemical equation. Sometimes you will see chemical equations like Equation 3.4,

$$2\,Na(s) + Cl_2(g) \rightarrow 2\,NaCl(s)$$

written as

$$Na(s) + \tfrac{1}{2}Cl_2(g) \rightarrow NaCl(s) \tag{3.6}$$

Of course, we do not mean to imply by this equation that one atom of sodium reacts with half a molecule of chlorine to give one unit of sodium chloride. The balancing coefficients lend themselves to many interpretations. For example, we could interpret the balancing coefficients to be dozens if we wished. In this case we would read Equation 3.6 as one dozen sodium atoms reacts with half a dozen chlorine molecules to give one dozen sodium chloride units. In Chapter 11 we're going to learn about a convenient unit called a mole (you may or may not remember about moles from high school chemistry). Using this unit, we could interpret Equation 3.6 as one mole of sodium reacts with one-half mole of chlorine to give one mole of sodium chloride. All these interpretations are equally valid. You may use whichever one is most convenient for you at the time. Because the interpretation of balancing coefficients is so arbitrary, they do not have units. Strictly speaking, they simply represent the *relative* values of the quantities (be they atoms, or molecules, or whatever) in the equation.

All these considerations emphasize that a chemical equation is a *representation* of a chemical reaction. Sometimes chemists refer to a chemical equation such as Equations 3.4 or 3.6 as a chemical reaction, but this is somewhat careless. A chemical reaction is a chemical event that occurs in nature, and a chemical equation is simply a convenient way for chemists to represent that reaction.

3-3. Elements Can Be Grouped According to Their Chemical Properties

By the 1860s, more than 60 elements had been discovered, and many chemists had begun to notice patterns in the chemical properties of certain elements. For example, consider the three metals lithium, Li(s), sodium, Na(s), and potassium, K(s). All three of these metals are less dense than water (Figure 3.4), are soft enough to be cut with a knife (Figure 3.5), have fairly low melting points (below 200°C), and are very reactive. In fact, they all react spontaneously with oxygen and water. Just as sodium reacts vigorously with chlorine, so do lithium and potassium, as described by the chemical equations

$$2\,Li(s) + Cl_2(g) \rightarrow 2\,LiCl(s)$$
$$2\,Na(s) + Cl_2(g) \rightarrow 2\,NaCl(s)$$
$$2\,K(s) + Cl_2(g) \rightarrow 2\,KCl(s)$$

Figure 3.4 The alkali metals lithium, sodium, and potassium are less dense than water. Lithium floats on oil, which floats on water.

Figure 3.5 The alkali metals are soft. Sodium can be cut with a pocketknife.

Figure 3.6 Reactions of the alkali metal potassium and the alkaline-earth metal calcium with water. (*top*) Potassium reacts violently with water to produce potassium hydroxide and hydrogen gas. The flame is a result of the explosive reaction of the evolved hydrogen gas with oxygen in the air. The yellow sparks are molten pieces of potassium reacting with oxygen in the air. The molten potassium was blown out of the reaction vessel by the explosion. (*bottom*) Calcium reacts slowly with water to produce hydrogen gas (bubbles) and calcium hydroxide.

The product in all three cases is a white, unreactive, crystalline, ionic solid (i.e., a solid composed of ions) that dissolves readily in water, and melts at high temperatures.

Example 3-1 shows the chemical equation for the reaction of sodium with water to produce sodium hydroxide, $NaOH(s)$, and hydrogen. Lithium and potassium (Figure 3.6, top) undergo similar reactions, as described by the equations

$$2\,Li(s) + 2\,H_2O(l) \rightarrow 2\,LiOH(aq) + H_2(g)$$
$$2\,K(s) + 2\,H_2O(l) \rightarrow 2\,KOH(aq) + H_2(g)$$

Just like sodium hydroxide, lithium hydroxide, $LiOH(s)$, and potassium hydroxide, $KOH(s)$, are water-soluble, corrosive, white, translucent solids.

Lithium, sodium, and potassium have similar chemical properties and can be considered as a group. Because the hydroxides of these metals are **alkaline**, that is, they form corrosive aqueous solutions that feel slippery (like soap solutions) and react with acids (see Section 10-3), they are called **alkali metals**. The term **alkali** comes from an Arabic word meaning "the ash" and refers to the fact that sodium and potassium were discovered in the ashes left from the combustion of plants.

The members of other groups of elements also share similar chemical properties. For example, magnesium, $Mg(s)$, calcium, $Ca(s)$, strontium, $Sr(s)$, and barium, $Ba(s)$, have many chemical properties in common. As a group, these metals are called the **alkaline-earth metals** because their compounds often occur in alkaline soil deposits. When heated, these metals all burn brightly in oxygen to form white, crystalline, ionic oxides. The reaction of magnesium with oxygen is shown in Figure 3.7. These reactions may be described in terms of equations as

$$2\,Mg(s) + O_2(g) \rightarrow 2\,MgO(s)$$
$$2\,Ca(s) + O_2(g) \rightarrow 2\,CaO(s)$$

$$2\,Sr(s) + O_2(g) \rightarrow 2\,SrO(s)$$
$$2\,Ba(s) + O_2(g) \rightarrow 2\,BaO(s)$$

Calcium (Figure 3.6, bottom), strontium, and barium react slowly with cold water to yield a metal hydroxide and hydrogen gas, as represented by the chemical equations

$$Ca(s) + 2\,H_2O(l) \rightarrow Ca(OH)_2(s) + H_2(g)$$
$$Sr(s) + 2\,H_2O(l) \rightarrow Sr(OH)_2(s) + H_2(g)$$
$$Ba(s) + 2\,H_2O(l) \rightarrow Ba(OH)_2(s) + H_2(g)$$

Magnesium undergoes a similar reaction at high temperatures. We see, then, that these four metals have similar chemical properties and can be placed into a group, just as lithium, sodium, and potassium can.

Figure 3.7 Magnesium ribbon burns rapidly in oxygen to form magnesium oxide, a white solid. Magnesium is used in flares, fuses, and pyrotechnics.

EXAMPLE 3-2: Given that magnesium reacts with sulfur to form magnesium sulfide, $MgS(s)$, write a balanced chemical equation for the reaction between calcium and sulfur.

Solution: The chemical equation for the reaction of $Mg(s)$ with $S(s)$ is

$$Mg(s) + S(s) \rightarrow \underset{\text{magnesium sulfide}}{MgS(s)}$$

Therefore, by analogy, we predict

$$Ca(s) + S(s) \rightarrow \underset{\text{calcium sulfide}}{CaS(s)}$$

which is correct.

PRACTICE PROBLEM 3-2: Given that calcium reacts with chlorine to form $CaCl_2(s)$, predict the product of the reaction between strontium and chlorine and write a balanced chemical equation for the reaction.

Answer: $Sr(s) + Cl_2(g) \rightarrow SrCl_2(s)$

Figure 3.8 *Left to right:* Chlorine, $Cl_2(g)$, bromine, $Br_2(l)$, and iodine, $I_2(s)$, members of the halogen family. Fluorine is too reactive to be stored in a flask.

Another group of elements having similar chemical properties consists of the nonmetals fluorine, $F_2(g)$, chlorine, $Cl_2(g)$, bromine, $Br_2(l)$, and iodine, $I_2(s)$, (Figure 3.8). As we noted in Chapter 2, these elements exist as diatomic molecules. All four elements are very reactive; that is, they react with most metals and nonmetals. As a group, they are called **halogens,** a Greek word meaning "salt formers." The halogens react with the alkali metals to give white, crystalline, ionic solids called **halides.** The specific names of the halides are fluoride, chloride, bromide, and iodide. For example, in the case of sodium we have

$$2\,Na(s) + F_2(g) \rightarrow 2\,NaF(s) \qquad \text{sodium fluoride}$$
$$2\,Na(s) + Cl_2(g) \rightarrow 2\,NaCl(s) \qquad \text{sodium chloride}$$
$$2\,Na(s) + Br_2(l) \rightarrow 2\,NaBr(s) \qquad \text{sodium bromide}$$
$$2\,Na(s) + I_2(s) \rightarrow 2\,NaI(s) \qquad \text{sodium iodide}$$

The halogens react with the alkaline earth metals to yield the salts $MF_2(s)$, $MCl_2(s)$, $MBr_2(s)$, and $MI_2(s)$, where M is any alkaline earth metal.

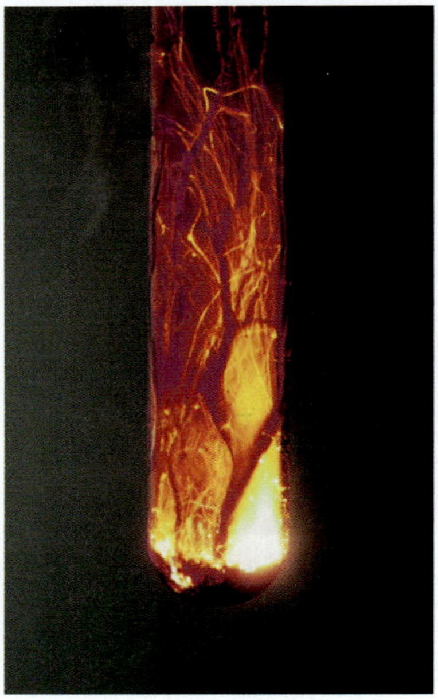

Figure 3.9 The reaction of aluminum metal with bromine to form aluminum bromide, $AlBr_3(s)$.

EXAMPLE 3-3: Given that bromine reacts with aluminum to produce aluminum bromide, $AlBr_3(s)$ (Figure 3.9), predict the product of the reaction between fluorine and aluminum and write a balanced chemical equation for the reaction.

Solution: We assume that the halogen fluorine reacts with aluminum in a manner analogous to the halogen bromine, thus we have

$$Al(s) + F_2(g) \rightarrow AlF_3(s) \qquad \text{(not balanced)}$$

Balancing this equation yields

$$2Al(s) + 3F_2(g) \rightarrow 2AlF_3(s)$$

PRACTICE PROBLEM 3-3: Chlorine forms the compounds $CCl_4(l)$ and $NCl_3(l)$. Predict the formulas of the analogous iodine compounds.

Answer: $CI_4(s)$ and $NI_3(s)$

The key question at this stage is: How can we explain why elements fall into groups characterized by similar chemical properties? The answer to this question leads us to still more significant patterns of reactivity among the elements.

3-4. The Elements Show a Periodic Pattern When Listed in Order of Increasing Atomic Number

Throughout the history of chemistry many attempts have been made to find patterns and to classify the elements. After Dalton proposed his atomic theory, the concept of atomic mass and the experimental determination of atomic masses took on increasing importance. In the middle of the nineteenth century, many chemists discovered various trends related to the masses of the elements, but most of these early classification systems had significant shortcomings. However, in 1869 a table put forth by the Russian chemist Dmitri Mendeleev (Frontispiece) began to gain acceptance and gradually evolved into our modern periodic table. A brief history of the development of the periodic table is presented in Interchapter B.

See Interchapter B at www.McQuarrieGeneralChemistry.com.

Mendeleev arranged the elements in order of increasing atomic mass and was able to show that the chemical properties of the elements exhibit repetitive patterns in chemical behavior. To illustrate the idea of Mendeleev's observation, let's start with the element lithium and arrange the succeeding elements in order of increasing atomic mass, as shown in Table 3.1. If we examine Table 3.1 carefully, then we see that the chemical properties of the elements show a remarkably repetitive, or periodic, pattern. The variations in the properties of the elements increasing in atomic number from lithium to neon are repeated in the properties of the elements from sodium to argon. The repeating pattern

TABLE 3.1 The chemical properties of 16 elements, listed in order of increasing atomic mass

Atomic mass	Element	Symbol	Properties	Formula of element	Formula of halide*
6.9	lithium	Li	very reactive metal	Li	LiX
9.0	beryllium	Be	reactive metal	Be	BeX_2
10.8	boron	B	semimetal†	B	BX_3
12.0	carbon	C	nonmetallic solid	C	CX_4
14.0	nitrogen	N	nonmetallic diatomic gas	N_2	NX_3
16.0	oxygen	O	nonmetallic, moderately reactive diatomic gas	O_2	OX_2
19.0	fluorine	F	very reactive diatomic gas	F_2	FX
20.2	neon	Ne	very unreactive monatomic gas	Ne	none
23.0	sodium	Na	very reactive metal	Na	NaX
24.3	magnesium	Mg	reactive metal	Mg	MgX_2
27.0	aluminum	Al	metal	Al	AlX_3
28.1	silicon	Si	semimetal†	Si	SiX_4
31.0	phosphorus	P	nonmetallic solid	P	PX_3
32.1	sulfur	S	nonmetallic solid	S	SX_2
35.5	chlorine	Cl	very reactive diatomic gas	Cl_2	ClX
39.9	argon	Ar	very unreactive monatomic gas	Ar	none

*X stands for F, Cl, Br, or I.

†Boron and silicon are called semimetals because they have properties that are intermediate between those of the metals and those of the nonmetals.

(or periodicity) is seen more clearly if we arrange the elements horizontally in two rows, or **periods**.

First row (or period) of elements:

Li	Be	B	C	N	O	F	Ne
lithium	beryllium	boron	carbon	nitrogen	oxygen	fluorine	neon

Second row (or period) of elements:

Na	Mg	Al	Si	P	S	Cl	Ar
sodium	magnesium	aluminum	silicon	phosphorus	sulfur	chlorine	argon

Sodium is placed below lithium to start a new period (row) because the chemical properties of sodium are similar to those of lithium. This placement of sodium leads, by continuation of the listing by atomic mass order, to the placement of Mg below Be and Cl below F (Cl_2 and F_2 have similar chemical behaviors). Mendeleev did not know about the existence of the noble gases (see Section 3-5). He then began a new period by placing K in the same group as Na and Li, because K is similar to Na in chemical behavior. He continued, listing the elements in

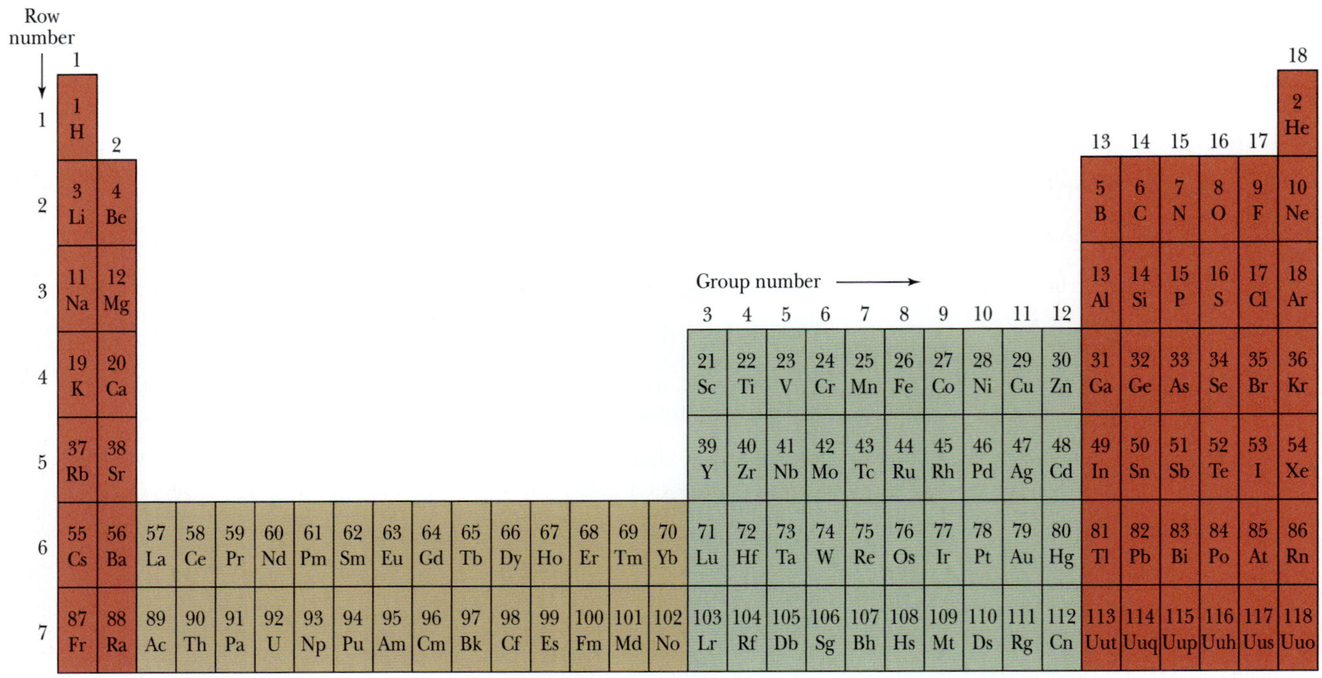

Figure 3.10 A modern version of the periodic table of the elements. In this version the elements are ordered according to atomic number rather than atomic mass. The chemical properties of the elements show a periodic pattern, that is, elements in a given column have similar chemical properties. Rows (or periods) are numbered from 1 to 7 along the left, groups (or families) from 1 to 18 along the top. As we shall see, the elements shaded in yellow are generally set off as an insert to the table.

order of increasing atomic mass, making certain in each case that the elements in the same group (column) had similar chemical properties.

Mendeleev's genius was not in the arrangement of elements in order of increasing atomic mass. His genius lay in the **periodic arrangement** of the elements by atomic mass and in his realization that apparent gaps in the periodic arrangement, that is, in the **periodicity,** must correspond to missing elements. He used the concept of periodicity to predict many of the chemical and physical properties of these, at that time unknown, elements.

Figure 3.10 presents a modern version of the **periodic table of the elements**. This version is more complicated than the version with which we began because it contains more elements than just the 16 listed in Table 3.1, and many more elements than were known to Mendeleev. In the modern periodic table in Figure 3.10, the elements are arranged in order of increasing atomic number instead of increasing atomic mass. With a few exceptions, the order is the same in both cases. The idea of atomic number was not developed until the early 1900s, about 40 years after Mendeleev's first periodic table was published.

WWW

The chemistry of selected groups or families of the periodic table is discussed in the Interchapters of the text. These include the alkali metals (Interchapter D), main group metals (Interchapter I), halogens (Interchapter Q), and noble gases (Interchapter K). These can be found at www.McQuarrieGeneralChemistry.com.

3-5. Elements Assigned to the Same Column in the Periodic Table Have Similar Chemical Properties

Notice in Figure 3.10 that lithium, sodium, and potassium occur in the leftmost column of the periodic table. All the elements in that column have similar chemical properties. Although we have not discussed rubidium, Rb(*s*), cesium, Cs(*s*), or francium, Fr(*s*), the fact that these elements occur in the same column as lithium, sodium, and potassium suggests that they undergo similar chemical reactions. Francium is a radioactive element not found in nature, but rubidium and cesium are light, soft, very reactive metals (Figure 3.11). Rubidium and cesium react explosively with the halogens, water, hydrogen, oxygen, and many other substances. By analogy with the chemical equations describing reactions of sodium that we discussed earlier, we predict that, for example,

$$2\,\mathrm{Rb}(s) + \mathrm{Cl_2}(g) \rightarrow 2\,\mathrm{RbCl}(s)$$
$$2\,\mathrm{Cs}(s) + 2\,\mathrm{H_2O}(l) \rightarrow 2\,\mathrm{CsOH}(s) + \mathrm{H_2}(g)$$

Other reactions of rubidium and cesium also are similar to those of lithium, sodium, and potassium.

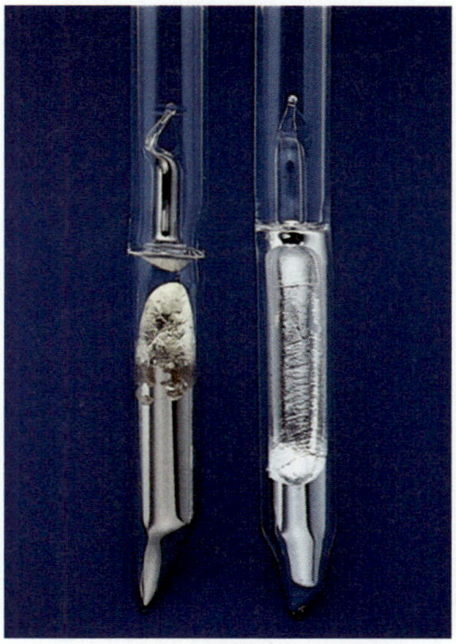

Figure 3.11 Cesium (golden color) and rubidium (silver color) are stored in vacuum-sealed ampoules to prevent their reacting with air.

EXAMPLE 3-4: Predict the product of the reaction between beryllium and oxygen.

Solution: Because beryllium, like calcium, is an alkaline-earth metal, we predict that $\mathrm{BeO}(s)$ is the reaction product, analogous to $\mathrm{CaO}(s)$. The balanced chemical equation for the reaction is

$$2\,\mathrm{Be}(s) + \mathrm{O_2}(g) \rightarrow 2\,\mathrm{BeO}(s)$$

PRACTICE PROBLEM 3-4: Write a balanced chemical equation to describe the reaction between rubidium and water.

Answer: $2\,\mathrm{Rb}(s) + 2\,\mathrm{H_2O}(l) \rightarrow 2\,\mathrm{RbOH}(aq) + \mathrm{H_2}(g)$

Elements assigned to the same column of the periodic table are said to belong to the same **group** or **family.** The leftmost column in the periodic table is labeled 1, so the elements in that column are referred to as the **Group 1 metals** (Interchapter D). As noted previously, the Group 1 metals are also called the alkali metals. The metals in the column labeled 2 are called the **Group 2 metals,** or the alkaline earth metals. Every element in Groups 1 and 2 is a reactive metal and undergoes chemical reactions similar to those of the other members of its respective group. At this stage we will regard the group numbers as identification devices. Later we will see that the group numbers have further significance in terms of the arrangement of the electrons in the atoms.

Elements in the same column in the rest of the periodic table also share similar chemical properties. This similarity is particularly strong in Groups 1–2 and 13–18, the **main-group elements** (Interchapter I). The halogens, which we know behave similarly, occur in Group 17 (Interchapter Q). The rightmost column of the periodic table (Group 18) contains the **noble gases**, which are characterized primarily by their relative lack of chemical reactivity (Figure 3.12). Prior to 1962, they often were called the **inert gases** because no compounds of these gases were known. In 1962, however, xenon was shown to form compounds with fluorine and oxygen, the most reactive nonmetals. Krypton fluorides are also known, but no stable compounds of helium, neon, or argon are yet known. The noble-gas elements are the least reactive group of elements in the periodic table (Interchapter K).

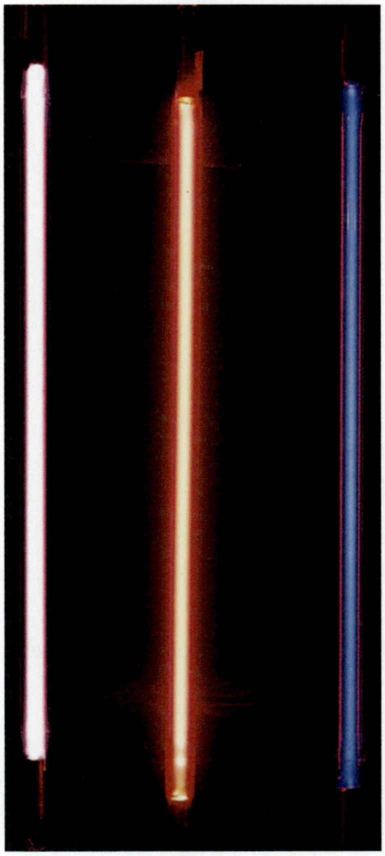

Figure 3.12 When an electric discharge is passed through a noble gas, light of a characteristic color is emitted. The tubes in the photo contain helium (*left*), neon (*center*), and argon (*right*).

Because the elements in a given group have similar chemical properties, the simple compounds of the elements from that group usually have similar chemical formulas. This relation can be seen in Table 3.2, which shows the known binary hydrides of the main-group elements, shown in red in Figure 3.10.

TABLE 3.2 The binary hydrides of the main-group elements

LiH	BeH_2	BH_3	CH_4	NH_3	H_2O	HF
NaH	MgH_2	AlH_3	SiH_4	PH_3	H_2S	HCl
KH	CaH_2	GaH_3	GeH_4	AsH_3	H_2Se	HBr
RbH	SrH_2	InH_3	SnH_4	SbH_3	H_2Te	HI
CsH	BaH_2	TlH_3	PbH_4	BiH_3		

EXAMPLE 3-5: Phosphorus is a nonmetallic solid that occurs in white, red, and black forms. White phosphorus, $P_4(s)$, spontaneously bursts into flame in the presence of oxygen to produce the oxide $P_4O_6(s)$. Use the periodic table to predict the reaction between yellow arsenic, $As_4(s)$, and oxygen.

Solution: Arsenic occurs in the same group (Group 15) as phosphorus, so we predict, by analogy with phosphorus, that $As_4(s)$ reacts with oxygen to produce $As_4O_6(s)$. The balanced chemical equation for the reaction is

$$As_4(s) + 3\,O_2(g) \rightarrow As_4O_6(s)$$

Arsenic compounds are well known to be poisonous, the lethal dose of $As_4O_6(s)$ being about 0.1 grams for an average adult. However, very small amounts of arsenic compounds may be beneficial or even necessary. The human body normally contains about 10–20 mg of arsenic, mostly bound to and excreted with food products.

PRACTICE PROBLEM 3-5: Predict the chemical formulas of the chlorides of the third-row elements of Groups 1, 2, and 13–16.

Answer: NaCl, $MgCl_2$, $AlCl_3$, $SiCl_4$, PCl_3, SCl_2

www

See Interchapter B at www.McQuarrieGeneralChemistry.com.

TABLE 3.3 Comparison of Mendeleev's predictions and actual experimental values for the properties of gallium

Property	Predicted	Observed
atomic mass	69	69.7
density/$g\cdot cm^{-3}$	6.0	5.9
melting point	low	30°C
boiling point	high	2400°C
formula of oxide	M_2O_3	Ga_2O_3

As we have noted, Mendeleev left gaps in his periodic table to accommodate his predictions of the existence of undiscovered elements. For example, in 1869 the element following zinc in order of atomic mass was arsenic. Yet Mendeleev's concept of periodicity led him to place arsenic in Group 15 rather than in Group 13 or 14 because its chemical behavior was similar to that of phosphorus. He then boldly proposed that there were two, as yet undiscovered, elements to fill the gaps this left between zinc and arsenic. Mendeleev accurately predicted many of the properties of these elements prior to their discovery (see Interchapter B). Table 3.3 compares Mendeleev's 1869 predictions with the actual properties of the element gallium (atomic number 31), which was not discovered until 1875. Gallium has an unusually low melting point for a metal (30°C) and melts when held in the hand (Figure 3.13).

The periodic table is the most useful concept in chemistry. Almost every general chemistry classroom and laboratory in the world has a periodic table hanging on the wall. You may have noticed that the periodic table hanging in your classroom or laboratory is slightly different from the one given in Figure 3.10.

A more common version of the periodic table is shown in Figure 3.14. The major difference between Figure 3.10 and Figure 3.14 involves the treatment of elements 57 through 71—collectively called the **lanthanide series**—and elements 89 through 103—called the **actinide series**. These elements have been removed from the normal numerical sequence of elements and placed at the bottom of the table. This change produces a more compact (short-form) periodic table. Some versions of the periodic table place lanthanum (La) and actinium (Ac) in the table and place elements 58 through 71 (cerium through lutetium) and elements 90 through 103 (thorium through lawrencium) below. In addition, many versions have different headings for the columns. Older versions of the periodic table label the six rightmost groups as 3A through 8A and the transition metals as 3B, 4B, and so on. These differences are not important and are no cause for concern.

The periodic table contains all the known chemical elements and shows the periodic relationships among them. As we have noted, elements in the same

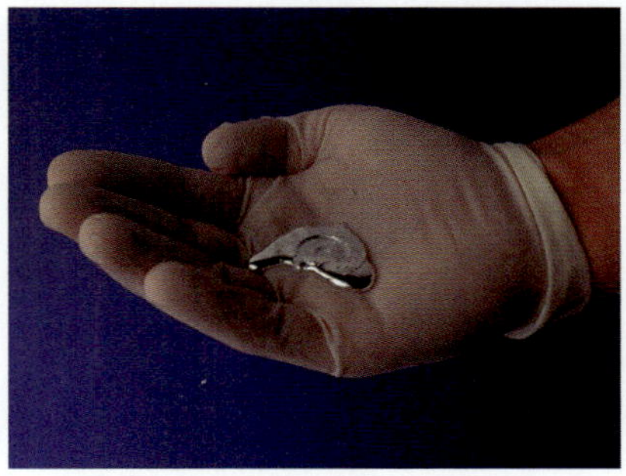

Figure 3.13 Gallium metal has a melting point of 30°C and therefore melts when held in the hand (human body temperature is about 37°C).

1	2	3	4	5	6	7	8	9	10	11	12	13	14	15	16	17	18
1 **H** 1.008																	2 **He** 4.003
3 **Li** 6.941	4 **Be** 9.012											5 **B** 10.81	6 **C** 12.01	7 **N** 14.01	8 **O** 16.00	9 **F** 19.00	10 **Ne** 20.18
11 **Na** 22.99	12 **Mg** 24.31											13 **Al** 26.98	14 **Si** 28.09	15 **P** 30.97	16 **S** 32.07	17 **Cl** 35.45	18 **Ar** 39.95
19 **K** 39.10	20 **Ca** 40.08	21 **Sc** 44.96	22 **Ti** 47.87	23 **V** 50.94	24 **Cr** 52.00	25 **Mn** 54.94	26 **Fe** 55.85	27 **Co** 58.93	28 **Ni** 58.69	29 **Cu** 63.55	30 **Zn** 65.41	31 **Ga** 69.72	32 **Ge** 72.64	33 **As** 74.92	34 **Se** 78.96	35 **Br** 79.90	36 **Kr** 83.80
37 **Rb** 85.47	38 **Sr** 87.62	39 **Y** 88.91	40 **Zr** 91.22	41 **Nb** 92.91	42 **Mo** 95.94	43 **Tc** (98)	44 **Ru** 101.1	45 **Rh** 102.9	46 **Pd** 106.4	47 **Ag** 107.9	48 **Cd** 112.4	49 **In** 114.8	50 **Sn** 118.7	51 **Sb** 121.8	52 **Te** 127.6	53 **I** 126.9	54 **Xe** 131.3
55 **Cs** 132.9	56 **Ba** 137.3	57–71	72 **Hf** 178.5	73 **Ta** 180.9	74 **W** 183.8	75 **Re** 186.2	76 **Os** 190.2	77 **Ir** 192.2	78 **Pt** 195.1	79 **Au** 197.0	80 **Hg** 200.6	81 **Tl** 204.4	82 **Pb** 207.2	83 **Bi** 209.0	84 **Po** (209)	85 **At** (210)	86 **Rn** (222)
87 **Fr** (223)	88 **Ra** (226)	89–103	104 **Rf** (261)	105 **Db** (262)	106 **Sg** (266)	107 **Bh** (264)	108 **Hs** (277)	109 **Mt** (268)	110 **Ds** (281)	111 **Rg** (272)	112 **Cn** (285)	113 **Uut**	114 **Uuq** (289)	115 **Uup**	116 **Uuh** (289)	117 **Uus**	118 **Uuo**

Lanthanide series

57 **La** 138.9	58 **Ce** 140.1	59 **Pr** 140.9	60 **Nd** 144.2	61 **Pm** (145)	62 **Sm** 150.4	63 **Eu** 152.0	64 **Gd** 157.3	65 **Tb** 158.9	66 **Dy** 162.5	67 **Ho** 164.9	68 **Er** 167.3	69 **Tm** 168.9	70 **Yb** 173.0	71 **Lu** 175.0

Actinide series

89 **Ac** (227)	90 **Th** 232.0	91 **Pa** 231.0	92 **U** 238.0	93 **Np** (237)	94 **Pu** (244)	95 **Am** (243)	96 **Cm** (247)	97 **Bk** (247)	98 **Cf** (251)	99 **Es** (252)	100 **Fm** (257)	101 **Md** (258)	102 **No** (259)	103 **Lr** (262)

Figure 3.14 A short-form version of the periodic table. In this version, currently recommended by IUPAC, the lanthanide series (elements 57 through 71) and the actinide series (elements 89 through 103) have been placed at the bottom of the table for two reasons. First, the separation leads to a more compact table. Second, and more important, all the elements in each of these two series have exceptionally similar chemical properties. The number under the symbol for each element is the atomic mass. An atomic mass in parentheses indicates that all the isotopes of that element are radioactive; the atomic mass given is that for the longest-lived isotope.

Other versions of the periodic table have the elements cerium (58) through lutetium (71) and thorium (90) through lawrencium (103) placed below. See, however, the paper by William B. Jensen in the *Journal of Chemical Education* **59**, 634 (1982).

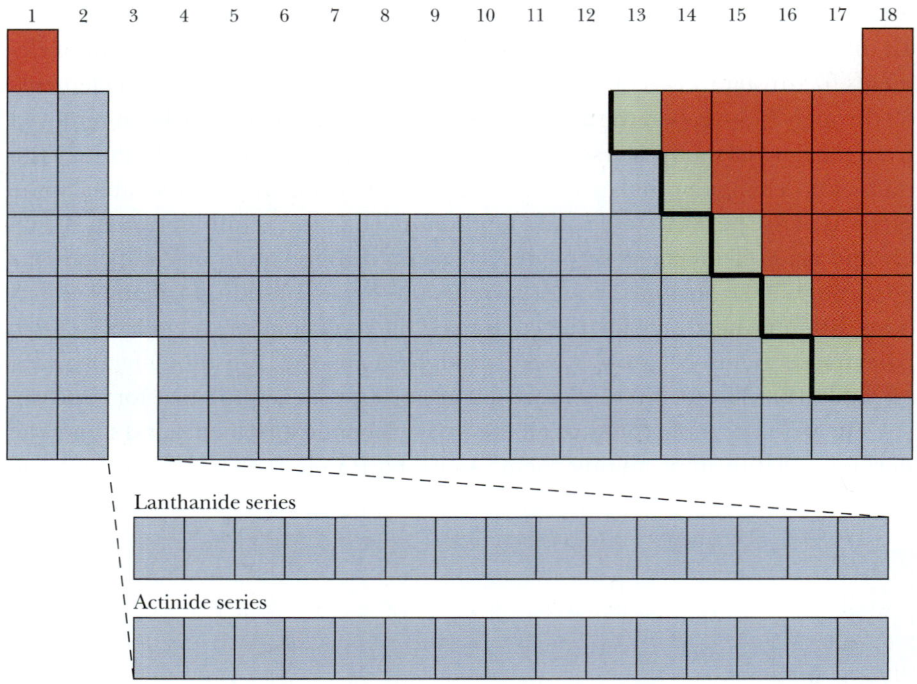

Figure 3.15 The position of metals (*light blue*) and nonmetals (*light red*) in the periodic table. The nonmetals appear only at the far right of the table, and the metals appear at the left. The elements along the steplike border between metals and nonmetals are the semimetals (*light green*).

column are said to belong to the same group or family, and the horizontal rows in the periodic table often are called periods. Your progress in learning basic chemistry will be aided greatly by an understanding and an appreciation of the periodic table.

3-6. Elements Are Arranged as Main-Group Elements, Transition Metals, and Inner Transition Metals

Metals and nonmetals occur in separate regions of the periodic table, as shown in Figure 3.15. The nonmetals are on the right side and are separated from the metals by a zigzag line. As might be expected, the elements on the border between metals and nonmetals (light green elements in Figure 3.15) have properties that are intermediate between those of metals and those of nonmetals. Such elements, called **semimetals** (or **metalloids**), are brittle, semilustrous solids (Figure 3.16). Because semimetals do not conduct electricity and heat as well as metals do—but better than nonmetals do—they are called **semiconductors**. The semimetals silicon and germanium are widely used in the manufacture of semiconducting devices, transistors, and integrated circuits.

Figure 3.16 The semimetals (or metalloids) are brittle solids. Two semimetals, boron (*top*) and silicon (*bottom*), are shown here.

Contrasting properties of the metals, the semimetals, and the nonmetals are given in Table 3.4. The metallic character of the elements, in the sense of their reactivity with nonmetals, increases as we move down a column in the periodic table and as we move from right to left in a row of the periodic table (Figure 3.17). Thus, francium is the most metallic element and fluorine is the most nonmetallic element. The noble gases (Group 18) are nonmetals, but they are so unreactive that with the exception of xenon, and to a lesser extent krypton, they do not form compounds. The transition from metallic to semimetallic to nonmetallic behavior as we move from left to right across the periodic table is not sharp; rather, there is a gradual change in properties of the main-group elements from distinctly metallic (Group 1) to distinctly nonmetallic (Group 17).

TABLE 3.4 Comparison of physical properties of metals, semimetals, and nonmetals

Metals	Semimetals	Nonmetals
high electrical and thermal conductance	intermediate electrical and thermal conductance	insulators
electrical resistance increases with increasing temperature	electrical resistance decreases with increasing temperature	resistance insensitive to temperature
malleable and ductile	brittle	not malleable, not ductile
nonvolatile and high-melting oxides, halides, and hydrides	volatile and low-melting halides and hydrides	volatile and low-melting oxides, halides, and hydrides

The ten elements in each row of Groups 3–12 are called **transition metals**, because they span the region of the periodic table where the transition from metallic to nonmetallic behavior of the main-group elements occurs. Unlike the main-group elements, the transition metals within any row have similar chemical properties. Many of the transition metals are probably familiar to you (Figure 3.18). Iron, nickel, chromium, copper, tungsten, and titanium are widely used in alloys for structural materials and play a key role in the world's technology. The precious metals—gold, platinum, and silver—are used as hard currency, in jewelry, and in high-quality electronic circuits. The transition metals vary greatly in abundance. Iron and titanium are plentiful, whereas rhenium (Re) and hafnium (Hf) are rare.

A more detailed discussion of the chemistry of the transition metals is given in Chapter 26.

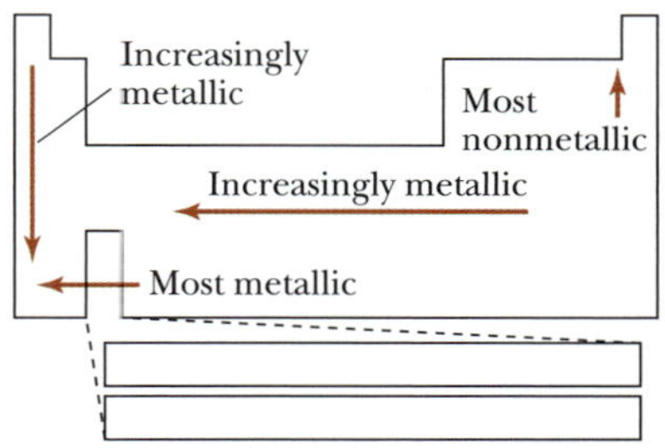

Figure 3.17 Trends in the metallic character of the elements.

The properties of the transition metals vary from group to group; yet they all are characterized by high densities and high melting points. In addition, unlike the compounds of the Group 1 and Group 2 metals, many compounds of the transition metals are colored. The metals with the greatest densities—iridium, Ir(*s*), 22.65 g·cm^{-3}, and osmium, Os(*s*), 22.61 g·cm^{-3}—and the highest melting point—tungsten (W), 3410°C—are transition metals.

The elements in the lanthanide series and the actinide series—the two series that begin with lanthanum (Z = 57) and actinium (Z = 89) in Figure

Figure 3.18 The first-row transition series metals. *Top row* (*left to right*): Sc, Ti, V, Cr, and Mn. *Bottom row:* Fe, Co, Ni, Cu, and Zn.

3.14—together are called the **inner transition metals**. The elements in each of these two series have remarkably similar chemical properties. The lanthanides are also called the **rare-earth elements** because they were once thought to occur only in very small quantities. The actinides are radioactive elements, most of which do not occur in nature but are produced in nuclear reactions. Figure 3.19 indicates the positions in the periodic table of the three main classes of elements and Figure 3.20 indicates the positions of the major groups, or families, of elements.

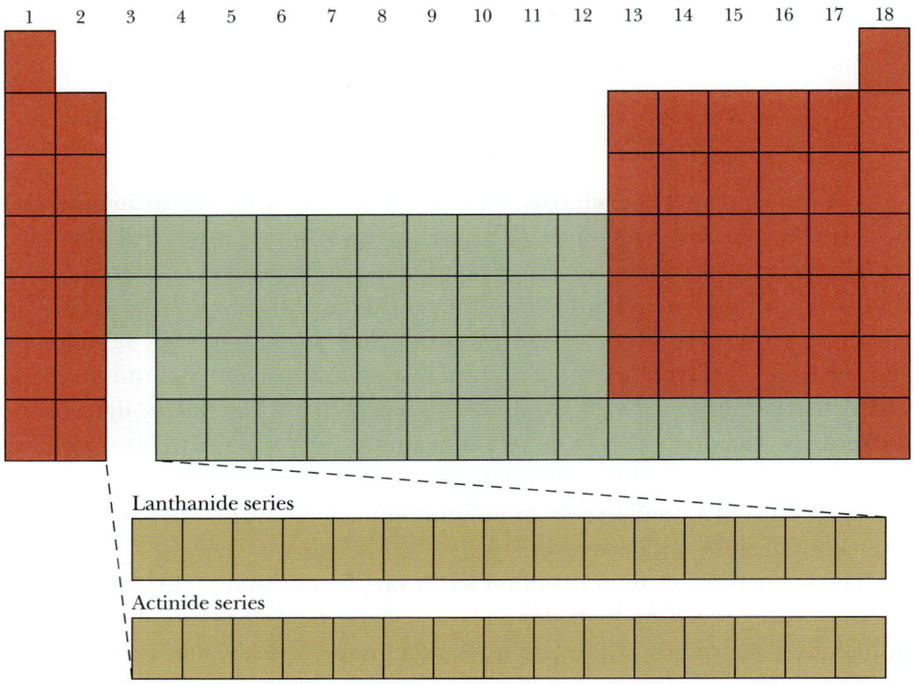

Figure 3.19 The usual form of the periodic table, showing the main-group elements (*red*), the transition metals (*green*), and the inner transition metals (*yellow*).

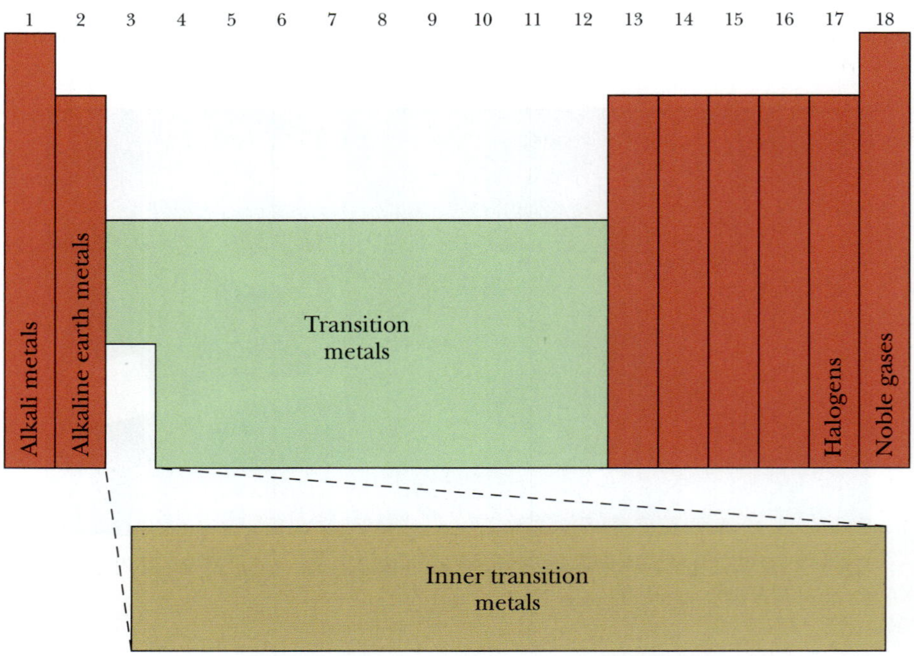

Figure 3.20 A schematic of the periodic table indicating the major groups (families) of elements.

EXAMPLE 3-6: Use the periodic table to classify each of the following elements as either a main-group element, a transition metal, or an inner transition metal:

Sb Sg Sc Se Th

If the element is a main-group element, then indicate its group number and state whether it is a metal, a nonmetal, or a semimetal.

Solution:

Symbol	Name	Classification
Sb	antimony	a Group 15 (main-group) semimetal
Sg	seaborgium	a Group 6 transition metal
Sc	scandium	a Group 3 transition metal
Se	selenium	a Group 16 (main-group) nonmetal
Th	thorium	an actinide (inner transition metal)

PRACTICE PROBLEM 3-6: Classify each of the following elements:

Fr Am Ge Kr Pb

Answer: Francium (Fr) is a Group 1 (main-group) metal; americium (Am) is an actinide (an inner transition metal); germanium (Ge) is a Group 14 (main-group) semimetal; krypton (Kr) is a Group 18 (main-group) non-metal noble gas; lead (Pb) is a Group 14 (main-group) metal.

3-7. Periodic Trends Contain Some Irregularities

Even though the periodic table is our most important guide to chemistry, it would be overly optimistic to expect that the great diversity of the chemical reactions of all the elements could be summarized or condensed into a single diagram. To begin with, hydrogen is unusual because it does not fit nicely into any group. It usually is placed in Group 1 with the alkali metals and occasionally in Group 17 with the halogens. Some versions of the periodic table place hydrogen in both groups; others list it separately from the table. Hydrogen is not a metal like the Group 1 metals; yet it forms many compounds whose formulas are similar to those of the Group 1 metal compounds. For example, we have

$HCl(g)$ hydrogen chloride and $NaCl(s)$ sodium chloride

$H_2S(g)$ hydrogen sulfide and $Na_2S(s)$ sodium sulfide

However, sodium chloride and sodium sulfide are white, crystalline solids, whereas hydrogen chloride and hydrogen sulfide are suffocating, toxic gases.

Hydrogen is a diatomic gas like the halogens and forms many compounds with formulas similar to those of halogen compounds:

$NaH(s)$ sodium hydride and $NaCl(s)$ sodium chloride

$NH_3(g)$ ammonia and $NCl_3(l)$ nitrogen trichloride

Figure 3.21 When antimony, $Sb(s)$, powder is placed in an atmosphere of chlorine, a vigorous reaction involving the formation of $SbCl_3(s)$ takes place.

This analogy is superficial, however. Even though the formulas may be similar, the chemical and physical properties of these pairs of compounds are very different. For example, $NaH(s)$ reacts vigorously with water to produce hydrogen gas and sodium hydroxide, whereas $NaCl(s)$ simply dissolves in water.

Perhaps most important is the difference in the properties of the first member of a group from those of the other members. Even in Group 1, which is a relatively homogeneous group, we find that lithium differs in a number of ways from the other members of the group. For example, most salts of sodium, potassium, rubidium, and cesium are soluble in water, but many lithium salts are only sparingly soluble. Lithium reacts directly with nitrogen at room temperature, but the other Group 1 metals react with nitrogen only at temperatures above 500°C. The chemical properties of beryllium are different from those of magnesium; and boron (a semimetal) is quite different from aluminum. The Group 15 elements phosphorus, arsenic, antimony, and bismuth react directly with chlorine (Figure 3.21), according to the chemical equations

$$2\,P(s) + 3\,Cl_2(g) \rightarrow 2\,PCl_3(l)$$

$$2\,As(s) + 3\,Cl_2(g) \rightarrow 2\,AsCl_3(l)$$

$$2\,Sb(s) + 3\,Cl_2(g) \rightarrow 2\,SbCl_3(s)$$

$$2\,Bi(s) + 3\,Cl_2(g) \rightarrow 2\,BiCl_3(s)$$

but nitrogen does not. The most striking example of this difference, however, occurs in Group 14, where the chemistry of carbon is significantly different from that of the other Group 14 elements. In fact, the chemistry of carbon is so diverse that an entire subfield of chemistry, called organic chemistry, is devoted to it (see Interchapters F, G, and H). Carbon atoms bonded together form the backbones of innumerable, extremely complex molecules that can each involve thousands of atoms. In chemistry we refer to compounds containing carbon as **organic compounds**. The chemistry of the other Group 14 elements is not as diverse as that of carbon.

See Interchapters F, G, and H at www.McQuarrieGeneralChemistry.com.

Surprising similarities exist in reactivity between the first element in a group and the second element in the following group, for example, between lithium and magnesium, beryllium and aluminum, and boron and silicon. These similarities are referred to as **diagonal relationships**, because of the relative locations of these pairs of elements in the periodic table (Figure 3.22). As we shall see in Chapter 5, the diagonal relationships are due in part to a similarity in the sizes of the ions in each of the diagonal pairs of elements (e.g., Li^+ and Mg^{2+} are similar in size, as are Be^{2+} and Al^{3+}).

1 H							2 He
3 Li	4 Be	5 B	6 C	7 N	8 O	9 F	10 Ne
11 Na	12 Mg	13 Al	14 Si	15 P	16 S	17 Cl	18 Ar

Figure 3.22 Diagonal relationships among the elements in the periodic table. Note that the transition metals are not shown in the figure. The arrows connect pairs of elements in different groups that have some similarities in chemical properties.

Although Mendeleev's organization of the table was based purely on chemical knowledge, as we shall see in the next two chapters, the periodic trends he observed are based on a fundamental repeating pattern in the electronic structure of the atoms.

SUMMARY

Chemical reactions are described in terms of chemical formulas as balanced chemical equations. The numbers of each kind of atom are the same on both sides of a balanced chemical equation (law of conservation of mass), even though the atoms are rearranged so that new substances are formed in the chemical reaction.

When the elements are ordered according to atomic number, there is a repetitive, or periodic, pattern of chemical properties. These periodic patterns of chemical behavior are displayed in the periodic table of the elements, which lists all the known chemical elements. Elements are denoted as main-group elements, transition metals, or inner transition metals. Most elements are metals, which appear on the left side of the periodic table. The nonmetals appear on the right side of the periodic table, and the semimetals appear on the border between metals and nonmetals.

TERMS YOU SHOULD KNOW

reactant *80*
product *80*
chemical equation *81*
balanced chemical equation *81*
balancing coefficient *82*
balancing by inspection *82*
alkaline *86*
alkali metal *86*
alkaline-earth metal *86*
halogen *87*
halide *87*
period *89*
periodic arrangement *90*
periodicity *90*
periodic table of the elements *90*
group (family) *91*
Group 1 metal *91*
Group 2 metal *91*
main-group element *91*
noble gas (inert gas) *91*
lanthanide series *93*
actinide series *93*
semimetal *94*
metalloid *94*
semiconductor *94*
transition metal *95*
inner transition metal *96*
rare-earth element *96*
organic compounds *98*
diagonal relationship *98*

PROBLEMS

BALANCING EQUATIONS

3-1. Balance the following chemical equations:

(a) $P(s) + Br_2(l) \rightarrow PBr_3(l)$

(b) $H_2O_2(l) \rightarrow H_2O(l) + O_2(g)$

(c) $CoO(s) + O_2(g) \rightarrow Co_2O_3(s)$

(d) $PCl_5(s) + H_2O(l) \rightarrow H_3PO_4(l) + HCl(g)$

3-2. Balance the following chemical equations:

(a) $KHF_2(s) \rightarrow KF(s) + H_2(g) + F_2(g)$

(b) $C_3H_8(g) + O_2(g) \rightarrow CO_2(g) + H_2O(l)$

(c) $P_4O_{10}(s) + H_2O(l) \rightarrow H_3PO_4(l)$

(d) $N_2H_4(g) \rightarrow NH_3(g) + N_2(g)$

3-3. Balance the following chemical equations:

(a) $CaH_2(s) + H_2O(l) \rightarrow Ca(OH)_2(aq) + H_2(g)$

(b) $CaCO_3(s) + HCl(aq) \rightarrow CaCl_2(aq) + CO_2(g) + H_2O(l)$

(c) $C_6H_{12}O_2(aq) + O_2(g) \rightarrow CO_2(g) + H_2O(l)$

(d) $Li(s) + CO_2(g) + H_2O(l) \rightarrow LiHCO_3(s) + H_2(g)$

3-4. Balance the following chemical equations:

(a) $H_2SO_4(aq) + KOH(aq) \rightarrow K_2SO_4(aq) + H_2O(l)$

(b) $Li_3N(s) + H_2O(l) \rightarrow LiOH(aq) + NH_3(g)$

(c) $Al_4C_3(s) + HCl(aq) \rightarrow AlCl_3(aq) + CH_4(g)$

(d) $ZnS(s) + HBr(aq) \rightarrow ZnBr_2(aq) + H_2S(g)$

3-5. Balance the following chemical equations and name the reactants and products in each case:

(a) $NaH(s) + H_2O(l) \rightarrow NaOH(aq) + H_2(g)$

(b) $SO_2(g) + O_2(g) \rightarrow SO_3(g)$

(c) $H_2S(g) + LiOH(aq) \rightarrow Li_2S(aq) + H_2O(l)$

(d) $ZnO(s) + CO(g) \rightarrow Zn(s) + CO_2(g)$

3-6. Balance the following chemical equations and name the reactants and products in each case:

(a) $PCl_3(g) + Cl_2(g) \rightarrow PCl_5(s)$

(b) $Sb(s) + Cl_2(g) \rightarrow SbCl_3(s)$

(c) $GaBr_3(s) + Cl_2(g) \rightarrow GaCl_3(s) + Br_2(l)$

(d) $Mg_3N_2(s) + HCl(g) \rightarrow MgCl_2(s) + NH_3(g)$

3-7. Working by analogy to the reactions given in this chapter, complete and balance the following chemical equations and name the product(s) formed in each case.

(a) $Na(s) + I_2(s) \rightarrow$

(b) $Sr(s) + H_2(g) \rightarrow$

(c) $Ba(s) + Cl_2(g) \rightarrow$

(d) $Mg(s) + O_2(g) \rightarrow$

3-8. Working by analogy to the reactions given in this chapter, complete and balance the following chemical equations and name the product(s) formed in each case.

(a) $Sr(s) + S(s) \rightarrow$

(b) $K(s) + H_2O(l) \rightarrow$

(c) $Ca(s) + H_2O(l) \rightarrow$

(d) $Al(s) + Cl_2(g) \rightarrow$

PERIODIC TABLE

3-9. Astatine is a radioactive halogen that concentrates in the thyroid gland. Predict from its position in the periodic table the following properties of astatine:

(a) physical state at 25°C (solid, liquid, or gas)

(b) formula of sodium salt

(c) color of sodium salt

(d) formula of gaseous astatine

(e) color of solid astatine

3-10. Radon is a radioactive noble gas that has been used as a tracer in detecting gas leaks. Radon occurs naturally as a decay product of uranium and is sometimes concentrated in the basements of homes, where it constitutes a health hazard because of its carcinogenic character. Predict from its position in the periodic table the following properties of radon:

(a) color

(b) odor

(c) formula of gaseous radon

(d) reaction with water

3-11. By referring to the periodic table, classify each of the following as a main-group element, a transition metal, or an inner transition metal. If a main-group or transition metal element, indicate which group and whether the element is a metal, a nonmetal, or a semimetal.

Tl Eu Xe Hf Ru Am B

3-12. By referring to the periodic table, classify each of the following as a main-group element, a transition metal, or an inner transition metal. If a main-group or transition metal element, indicate which group and whether the element is a metal, a nonmetal, or a semimetal.

Se As Mo Rn Ta Bi In

3-13. What properties do the elements in Group 1 (the alkali metals) have in common? How do they react with bromine, water, and oxygen? Which reacts the most vigorously with these? Which element in the group exhibits slightly different chemistry from the others?

3-14. What properties do the elements in Group 17 (the halogens) have in common? How do they react with potassium, strontium, and aluminum? Which reacts the most vigorously with these? Which is the only halogen that forms known compounds with xenon and krypton?

3-15. What properties do the elements in Group 18 (the noble gases) have in common? How do they react with other elements? Which of these elements is radioactive in nature? Which of these elements is used in modern balloons? What is the advantage of using this element instead of hydrogen? What is the disadvantage?

3-16. What properties do the elements in Group 2 (the alkali earths) have in common? How do they react with iodine, water, and oxygen? Which reacts the most vigorously with these? Which element in the group is the most metallic in character?

3-17. Radium is a brilliant white radioactive metal that was discovered by Pierre and Marie Curie. It was isolated from the mineral pitchblende from North Bohemia, and it took seven tons of pitchblende to recover only one gram of radium. By analogy with the reactions presented in this chapter, predict the reaction of radium with

(a) oxygen (b) iodine

(c) chlorine (d) sulfur

3-18. Francium is a radioactive metal that does not occur in nature. Predict the reaction of francium with

(a) bromine (b) hydrogen

(c) water (d) sulfur

3-19. Which of the following elements are main-group elements?

(a) tin (b) antimony (c) iron

(d) titanium (e) argon

3-20. Which of the following metals are transition metals?

(a) francium (b) palladium (c) silver

(d) lead (e) zinc

3-21. Identify the following elements as metals, semimetals, or nonmetals:

(a) Si (b) Sn (c) Sc (d) Sb (e) Sm

3-22. Identify the following elements as metals, semimetals, or nonmetals:

(a) Ge (b) Ga (c) Gd (d) Al (e) At

ADDITIONAL PROBLEMS

3-23. What does the position of an element in the periodic table tell you about its chemical properties? Give several examples.

3-24. One of the groups of the periodic table is commonly referred to as the "coinage metals" (not an IUPAC-approved name). What is the number of this group?

3-25. Elements in the modern periodic table are arranged by the number of protons (or atomic number). However, Mendeleev was unaware of the existence of protons in 1869 when he published his table and instead arranged his elements based on atomic mass. In several cases Mendeleev was forced to break strict mass order to keep chemically similar families together. Which elements in the modern table do not follow strict mass order?

3-26. The discovery of argon (the first noble gas to be isolated) was initially troubling for Mendeleev, who argued that it probably was not a new element at all but rather a newly discovered form of nitrogen (for example, N_3, similar to ozone, O_3). Why was it difficult for Mendeleev to accept argon as a new element?

3-27. A radioactive isotope of strontium is produced in nuclear explosions and is one of the dangerous components of nuclear fallout. Using the periodic table, suggest a reason for its impact on living organisms.

3-28. For each of the following equations, write the chemical formulas and balance the equation. Try to indicate all the physical states (solid, liquid, gas) as well.

(a) potassium + water → potassium hydroxide + hydrogen

(b) potassium hydride + water → potassium hydroxide + hydrogen

(c) silicon dioxide + carbon → silicon carbide + carbon monoxide

(d) silicon dioxide + hydrogen fluoride → silicon tetrafluoride + water

(e) phosphorus + chlorine → phosphorus trichloride

3-29. Without using any references, list as many elements as you can from memory and classify each as a metal or a nonmetal. Check your results and score yourself as follows:

more than 95	hall-of-famer
80 to 95	major leaguer
60 to 79	triple A player
40 to 59	semipro player
fewer than 40	little leaguer

3-30. When lithium nitride, $Li_3N(s)$, is treated with

water, $NH_3(g)$ is produced. Predict the formula of the gas produced when sodium phosphide, $Na_3P(s)$, is treated with water.

3-31. According to the Star Trek "Fifth Interstellar Geophysical Conference Standard" version of the periodic table, there are 140 known elements. Element 117, "topaline," is reported to have been discovered on Ganymede in 2021. To which group in the periodic table does element 117 belong? Predict the formula of the compound that presumably would result from the reaction between calcium and topaline, and name the compound accordingly. Using the Internet, look up element 117. Has it yet been isolated? If so, when was it first discovered and what is it currently called?

3-32. As a group, most of the alkali metal compounds dissolve readily in water. Based on the text discussions, which of the following compounds do you think would dissolve the least in water?

(a) $NaF(s)$ (b) $LiF(s)$ (c) $KF(s)$

3-33. Which of the following equations represents the reaction that occurs least vigorously?

(a) $Be(s) + 2HNO_3(aq) \rightarrow Be(NO_3)_2(aq) + H_2(g)$

(b) $Ca(s) + 2HNO_3(aq) \rightarrow Ca(NO_3)_2(aq) + H_2(g)$

(c) $Ba(s) + 2HNO_3(aq) \rightarrow Ba(NO_3)_2(aq) + H_2(g)$

3-34. Balance the following chemical equations:

(a) $F_2(g) + Al_2O_3(s) \rightarrow AlF_3(s) + O_2(g)$

(b) $NH_3(g) + O_2(g) \rightarrow NO(g) + H_2O(l)$

(c) $C_6H_6(l) + O_2(g) \rightarrow CO_2(g) + H_2O(l)$

(d) $H_2SO_4(aq) + Al_2(CO_3)_3(s) \rightarrow Al_2(SO_4)_3(aq) + CO_2(g) + H_2O(l)$

3-35. Balance the following chemical equations and name the reactants and products in each case.

(a) $Al(s) + Cl_2(g) \rightarrow AlCl_3(s)$

(b) $Al(s) + O_2(g) \rightarrow Al_2O_3(s)$

(c) $\underset{\text{phosgene}}{COCl_2(g)} + Na(s) \rightarrow NaCl(s) + CO(g)$

(d) $Be(s) + O_2(g) \rightarrow BeO(s)$

(e) $K(s) + S(l) \rightarrow K_2S(s)$

3-36. For each of the following reactions, write the chemical formulas of the reactants and product and balance the equations:

(a) sodium + sulfur $\rightarrow$ sodium sulfide

(b) calcium + bromine $\rightarrow$ calcium bromide

(c) barium + oxygen $\rightarrow$ barium oxide

(d) sulfur dioxide + oxygen $\rightarrow$ sulfur trioxide

(e) magnesium + nitrogen $\rightarrow$ magnesium nitride

3-37. Astatine is a radioactive metal; its longest-lived isotope has a half-life of only 8.1 hours. Although astatine is thought to occur naturally along with uranium and thorium isotopes, less than one ounce is thought to be present in the entire earth's crust. It is still unknown whether astatine forms diatomic molecules like the other halogens. Predict the reaction of diatomic astatine with each of the following elements and name the compounds formed:

(a) cesium (b) hydrogen (c) calcium

(d) bromine (e) aluminum

3-38. Using the *CRC Handbook of Chemistry and Physics* (available in most libraries) or an online reference, find the following information for the element antimony:

(a) Melting point

(b) Color

(c) Density

(d) Uses

3-39. Using the *CRC Handbook of Chemistry and Physics* (available in most libraries) or an online reference, find the following information for the element tellurium:

(a) Melting point

(b) Color

(c) Density

(d) Uses

3-40. The first metal to be prepared from its ore was copper, perhaps as early as 6000 BC in the Middle East. The ore, which contained copper oxide, was heated with charcoal, which was prepared by the incomplete burning of wood and is mainly elemental carbon, $C(s)$. Later, iron and tin were prepared in the same way. Write the chemical equations for the preparation of these metals. Assume the ores to be $CuO(s)$, $SnO_2(s)$, and $Fe_2O_3(s)$.

3-41. Balance the chemical equation

$$NaOH(aq) + NaNO_2(aq) + Al(s) + H_2O(l) \rightarrow NH_3(aq) + NaAlO_2(aq)$$

3-42. Balance the chemical equation

$$HCl(aq) + NaI(aq) + NaClO(aq) \rightarrow I_2(s) + NaCl(aq) + H_2O(l)$$

3-43. (*) Balance the equation

$$FeCl_2(aq) + KMnO_4(aq) + HCl(aq) \rightarrow FeCl_3(aq) + MnCl_2(aq) + H_2O(l) + KCl(aq)$$

3-44. (*) Balance the equation

$$HCl(aq) + K_2CrO_7(aq) + C_2H_5OH(aq) \rightarrow CrCl_3(aq) + CO_2(g) + KCl(aq) + H_2O(l)$$

3-45. (*) Balance the equation

$$HCl(aq) + As_2O_3(s) + NaNO_3(aq) + H_2O(l) \rightarrow NO(g) + H_3AsO_4(aq) + NaCl(aq)$$

3-46. (**) (*Only for the strong of heart*) Balance the chemical equation

$$As_2S_5(s) + NaNO_3(aq) + HCl(aq) \rightarrow H_3AsO_4(aq) + NaHSO_4(aq) + NO_2(g) + H_2O(l) + NaCl(aq)$$

3-47. The metals in the lanthanide series, the rare earth elements, have very similar chemical properties. However, they have very different magnetic properties. Which lanthanide metal has the following properties?

Its alloys are the strongest magnets; they are used for earrings for non-pierced ears.

Its alloys change shape in magnetic fields.

It is used as a contrast agent for MRI scans.

It concentrates the magnetic field in MRI machines when added to another magnet.

3-48. All the elements in the actinide series are radioactive.

Which are naturally occurring?

Which is the most abundant and how abundant is it?

Which metal is used in many smoke detectors?

Which is used in neutron activation analysis because it is a strong neutron emitter?

Niels Bohr (1885–1962), left, was born in Copenhagen, Denmark. His father was a distinguished physiologist. In 1911, Bohr received his Ph.D. in physics from the University of Copenhagen. He then spent a year with J. J. Thomson and Ernest Rutherford in England, where he formulated his theory of the hydrogen atom and its atomic spectrum. In 1913, he returned to the University of Copenhagen. In 1920, he was named director of the Institute of Theoretical Physics at the University, which was supported largely by the Carlsberg Brewery. The institute was an international center for theoretical physics during the 1920s and 30s, when quantum theory was being developed. Almost every scientist who was active in the development of quantum theory worked at Bohr's institute at one time or another. In 1943, because the Germans planned to arrest him to force him to work on their atomic bomb project, Bohr and his family fled to England under great secrecy and spent the remaining war years in the United States, where he participated in the Manhattan Project at Los Alamos. After World War II, Bohr worked energetically for peaceful uses of atomic energy. He organized the first Atoms for Peace Conference in 1955 and received the first Atoms for Peace prize in 1957. Bohr was awarded the Nobel Prize for Physics in 1922 "for his investigation of the structure of atoms and of the radiation emanating from them."

"The Father of Quantum Theory" Max Planck (1858–1948), right, was born in Kiel, Germany (then Prussia), where his father was a law professor. Planck was a slow, focused, methodical thinker. He was often surprised that others could pursue several lines of intellectual work simultaneously. He began his studies at the University of Munich and received his Ph.D. in theoretical physics in 1879 for his dissertation on the second law of thermodynamics. He joined the faculty of the University of Kiel in 1885, and in 1888, was appointed director of the Institute of Theoretical Physics, which was formed for him, at the University of Berlin. Planck was president of the Kaiser Wilhelm Society, later renamed the Max Planck Society, from 1930 until 1937, when he was forced to retire by the Nazi government. Planck's personal life was clouded by tragedy. His two daughters died in childbirth, one son died in World War I, and another son was executed in World War II for his part in the attempt to assassinate Hitler in 1944. Planck was awarded the Nobel Prize in Physics in 1918 "in recognition of services he rendered to the advancement of physics by his discovery of energy quanta."

4. Early Quantum Theory

Why do some atoms join together to form molecules and others do not? Why do the chemical and physical properties of the elements vary, and what accounts for the periodic repetition of these properties, as seen in the periodic table? To answer these and other fundamental questions in chemistry, we must learn how electrons are arranged around the nucleus of an atom. Recall that our picture of an atom is a small, massive central nucleus containing all the positive charge of the atom and essentially all the mass with electrons distributed about the nucleus in some sort of diffuse manner. The distribution of the electrons is described by quantum theory.

Quantum theory, as we shall use it here, evolved slowly in two phases over a period of time from about 1900 to 1930. The first phase, which occurred from 1900 to 1925, was stimulated by a number of experiments that could not be explained by the theoretical models of the time. Attempts to explain these experiments led to a very different view of the behavior of matter and light than had existed before. In this chapter we shall describe some of these experiments and then discuss the attempts to explain them. We shall see that particles can display wavelike behavior and that waves can display particle-like behavior and that the energies of electrons in an atom are restricted to certain discrete values that are characteristic of the atom. We shall then utilize this information to explain the spectra associated with atomic systems. In the next chapter, we shall discuss the development of quantum theory from 1925 onward, which is used today to understand the structure of atoms and molecules.

4-1. First Ionization Energy Is One of Many Periodic Properties of the Elements

The periodic table offers a great deal of insight into the **electronic structure,** or **electron arrangements,** of atoms. For example, in Chapter 3 we learned that, in general, elements in the same column in the periodic table are similar chemically. So we might expect that their outermost, and hence most chemically important, electrons have similar arrangements.

A direct indication of the arrangement of electrons about a nucleus is given by the ionization energies of the atom or ion. The **ionization energy** of an atom or an ion is the minimum energy required to remove an electron completely from the gaseous atom or ion. This energy can be determined experimentally. The **first ionization energy,** I_1, of an atom is the minimum energy required to remove an electron from a neutral gaseous atom, A, to produce a positively charged gaseous ion, A^+, and an electron, e^-:

$$A(g) \rightarrow A^+(g) + e^-(g) \qquad \text{(first ionization energy, } I_1\text{)}$$

The **second ionization energy,** I_2, is the minimum energy required to remove an electron from a gaseous A^+ ion to produce a gaseous A^{2+} ion:

$$A^+(g) \rightarrow A^{2+}(g) + e^-(g) \qquad \text{(second ionization energy, } I_2\text{)}$$

Depending on the total number of electrons in the atom, we can go on to define and measure the third (I_3), fourth (I_4), and successive ionization energies. We expect each successive ionization energy to be greater than the preceding one because, in removing successive electrons from an ion, we must overcome an increasingly greater electrical attraction between the positively charged ion and the electron that is being removed. Thus, we expect and find that $I_1 < I_2 < I_3 < I_4$, and so forth, for any given gaseous atom.

When we plot the first ionization energies of the elements against atomic number, we find that there is a periodic pattern in these data (Figure 4.1). Note that the noble gases have relatively large first ionization energies; in other words, it is relatively difficult to remove electrons from noble-gas atoms. This observation suggests that the electronic structures of the noble-gas atoms are more stable than those of the elements that precede and follow them in the periodic table. Furthermore, the alkali metals have relatively low ionization energies (Figure 4.1), in accord with their extremely reactive nature. Thus, we see that ionization energies as well as chemical properties display a periodic character because both depend on electronic structure. Figure 4.2 illustrates the trends in the first ionization energies of the elements in the periodic table.

Figure 4.1 First ionization energy (I_1) plotted against atomic number shows the periodic nature of the properties of the elements. The symbols for the noble gases are in red and the symbols for the alkali metals are in blue.

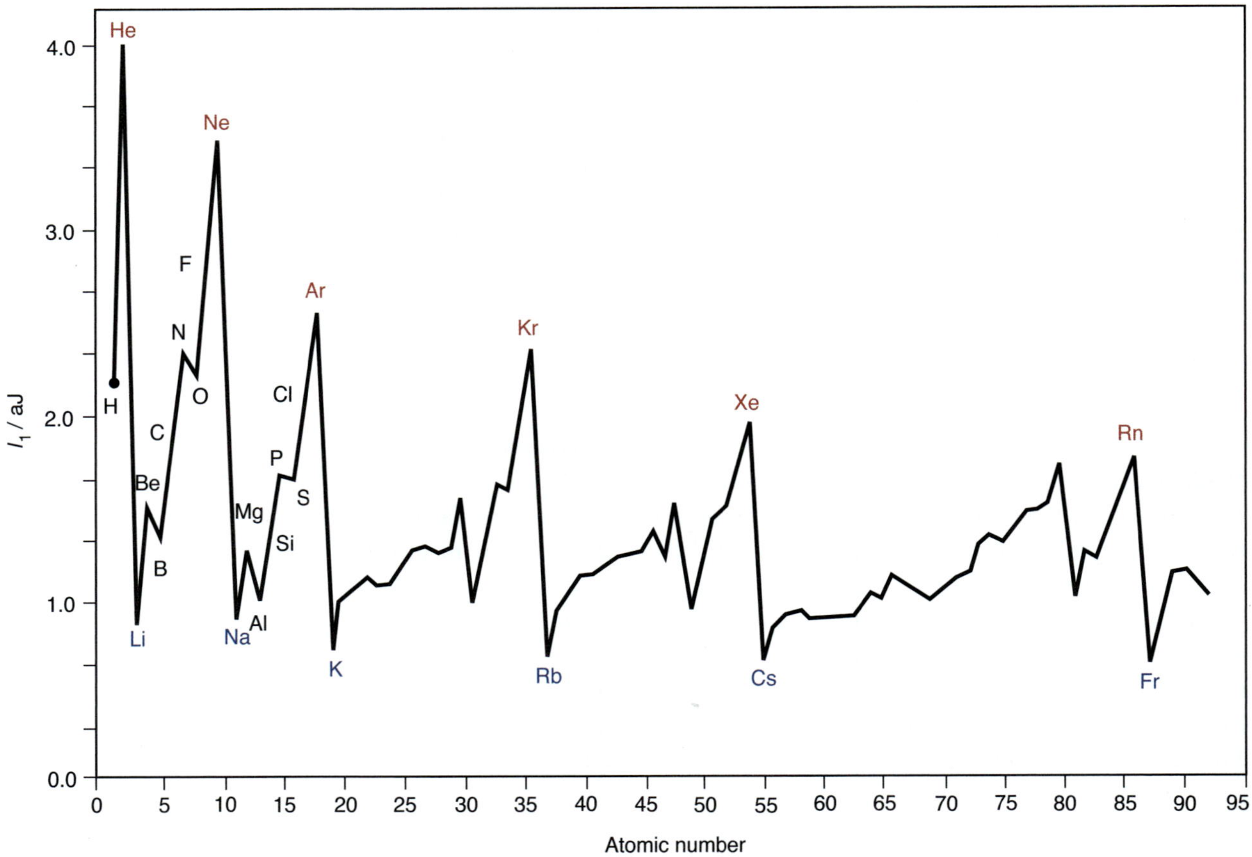

4-2. The Values of Successive Ionization Energies of Atoms Suggest a Shell Structure

We can gain more insight into electronic structure by looking at not just the first ionization energies but successive ionization energies. Table 4.1 lists values of I_1 through I_{11} for the atoms of hydrogen through sodium. Note that the energies in Table 4.1 are in units of attojoules, or 10^{-18} J. As we mentioned in Section 1-5, attojoules are the natural units for atomic and molecular energies.

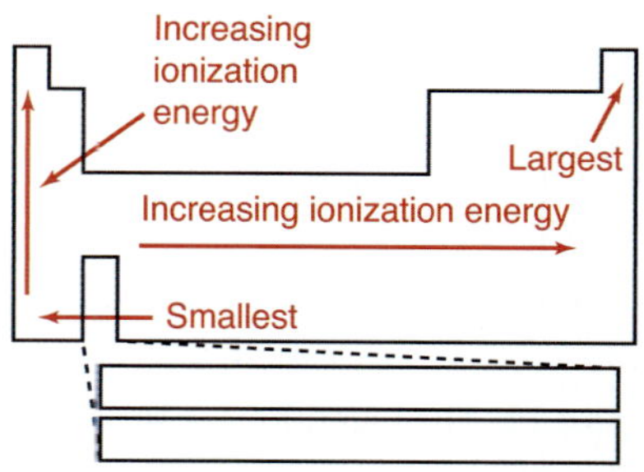

Figure 4.2 The trend in first ionization energy in the periodic table. Ionization energy increases as we go from left to right across each row and as we go up each column.

EXAMPLE 4-1: Referring to the ionization energies listed in Table 4.1, determine the energy required to form a single $Be^{2+}(g)$ ion from a neutral beryllium atom in the gas phase, $Be(g)$.

Solution: To convert $Be(g)$ to $Be^{2+}(g)$ requires removing two electrons. The relevant equations are

$$\text{Step 1: } Be(g) \rightarrow Be^{+}(g) + e^{-}(g) \qquad I_1 = 1.49 \text{ aJ}$$

$$\text{Step 2: } Be^{+}(g) \rightarrow Be^{2+}(g) + e^{-}(g) \qquad I_2 = 2.92 \text{ aJ}$$

The total energy required is $I_1 + I_2 = 4.41$ aJ.

PRACTICE PROBLEM 4-1: Use the data in Table 4.1 to calculate the energy needed to convert a neutral boron atom in the gas phase into a $B^{3+}(g)$ ion.

Answer: 11.45 aJ.

TABLE 4.1 Successive ionization energies of the elements hydrogen through sodium*

		Ionization energy/aJ										
Z	Element	I_1	I_2	I_3	I_4	I_5	I_6	I_7	I_8	I_9	I_{10}	I_{11}
1	H	2.18										
2	He	3.94	8.72									
3	Li	0.86	12.1	19.6								
4	Be	1.49	2.92	24.7	34.9							
5	B	1.33	4.04	6.08	41.5	54.5						
6	C	1.81	3.90	7.67	10.3	62.8	78.5					
7	N	2.32	4.75	7.60	12.4	15.7	88.4	107				
8	O	2.17	5.63	8.80	12.4	18.2	22.7	118	140			
9	F	2.78	5.60	10.0	14.0	18.3	25.2	29.7	153	177		
10	Ne	3.45	6.56	10.2	15.6	20.2	25.3	33.2	38.3	192	218	
11	Na	0.83	7.57	11.5	15.8	22.2	27.6	33.4	42.3	48.0	238	264

*The red lines separate regions of relatively low and relatively high ionization energies.

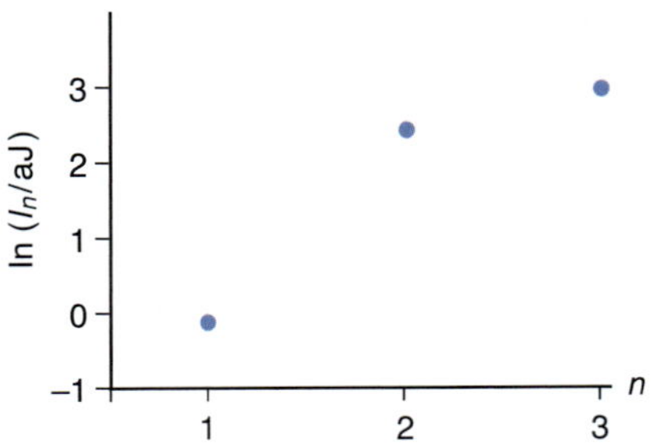

Figure 4.3 The logarithms of the three successive ionization energies (ln I_n/aJ) of a lithium atom versus the number of electrons removed (n). This graph suggests that the electrons in a lithium atom are arranged in two shells: an inner shell consisting of two electrons and an outer shell consisting of one electron. (Logarithms of the ionization energies are plotted rather than the ionization energies themselves in order to compress the vertical scale.)

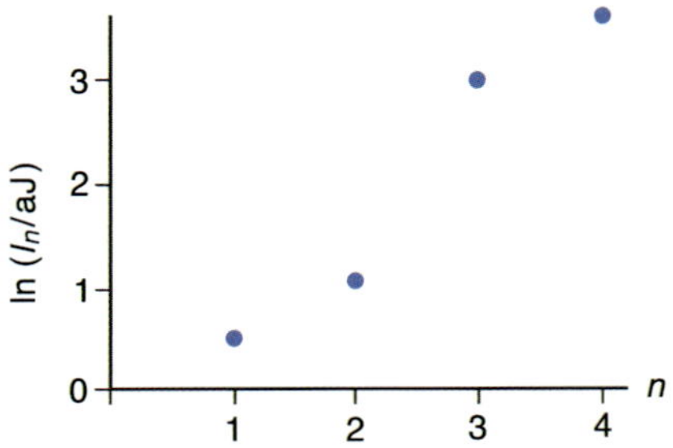

Figure 4.4 The logarithms of the four successive ionization energies (ln I_n/aJ) of a beryllium atom versus the number of electrons removed (n). This graph suggests that the electrons in a beryllium atom are arranged in two shells: an inner shell consisting of two electrons and an outer shell consisting of two electrons.

We can use the data in Table 4.1 to learn a lot about how electrons arrange themselves in atoms. Let's look at a helium atom. The first ionization energy of a helium atom (3.94 aJ) is much greater than that of a hydrogen atom (2.18 aJ) or a lithium atom (0.86 aJ), an observation that suggests the extraordinary stability of a helium atom. The second ionization energy of a helium atom, in other words, the energy required to remove the second electron, is even higher (8.72 aJ), being more than twice as large as the first. Realize, however, that here we are removing an electron from a positively charged He^+ ion, so we should expect I_2 to be greater than I_1 because of the electrical attraction between the positively charged ion and the negatively charged electron we are removing.

Figure 4.3 shows a plot of the energy required to remove the nth electron, I_n, against n for a lithium atom. (We actually plot the logarithm of the energy, ln (I_n/aJ), against n to compress the range of the vertical axis and produce a tidier graph. We divide the values of I_n by 1 aJ here so that I_n/aJ is unitless; you can take logarithms of only unitless quantities.) Note that this figure suggests that the first electron is much easier to remove than the second one. From the values given in Table 4.1, we see that the difference between the first and second ionization energies for a lithium atom ($I_2 - I_1$) is 11.2 aJ. This value is quite a bit larger than we might expect on the basis that we are simply removing the second electron from a $Li^+(g)$ ion. For example, the difference between the first and second ionization energies for a helium atom is only 4.78 aJ.

Let's make a similar graph for the Group 2 metal beryllium (Figure 4.4). In this case, it appears that the first two electrons are relatively easily removed, leaving a core of two electrons that are much more tightly bound. The large energy difference between the two sets of electrons suggests that the four electrons in a beryllium atom are arranged such that two of them are more easily detached, and hence more chemically active, whereas the other two constitute a relatively stable, helium-like inner core. We call these two sets of electrons **shells**. Thus, Figure 4.4 suggests that the electrons in a beryllium atom are arranged in two distinct shells.

EXAMPLE 4-2: Use the data in Table 4.1 for a carbon atom to plot ln(I_n/aJ) versus the number of electrons removed, n, as done in Figure 4.3 and Figure 4.4.

Solution: We make the following table:

n	I_n/aJ	ln(I_n/aJ)
1	1.81	0.593
2	3.90	1.36
3	7.67	2.04
4	10.3	2.33
5	62.8	4.14
6	78.5	4.36

The plot of $\ln(I_n/\text{aJ})$ versus n for a carbon atom gives

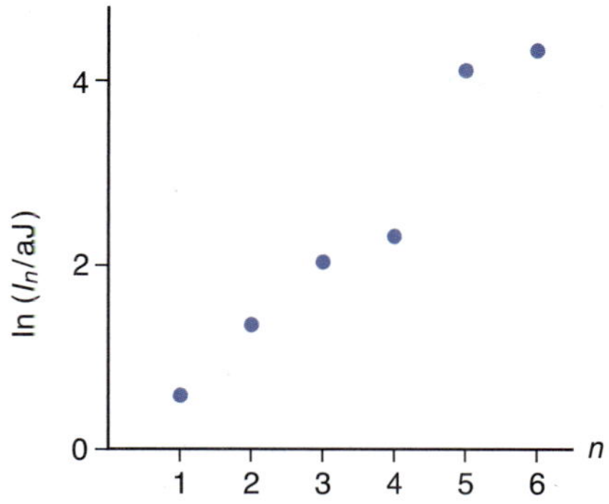

Notice that the first four electrons are much easier to remove from a carbon atom than the fifth and sixth electrons. This suggests a carbon atom has two shells; an inner shell containing two electrons and an outer shell containing four electrons.

PRACTICE PROBLEM 4-2: Use the data in Table 4.1 for fluorine to construct a plot of $\ln(I_n/\text{aJ})$ against n, and describe your results in terms of the apparent number of shells.

Answer: The electronic structure of a fluorine atom is arranged in two shells: an inner helium-like shell consisting of two electrons and an outer shell consisting of seven electrons. Note that fluorine is in the seventh position in the second row of the periodic table.

If we continue making these kinds of plots for each of the elements in the second row of the periodic table through neon (Figure 4.5), we see that the electrons in these elements fall into two sets: an inner helium-like core of two relatively tightly bound electrons and an outer shell consisting of one through eight electrons for the elements lithium through neon. For each of these plots the number of electrons in the outer shell corresponds to the position of the element in the periodic table.

The atoms of the elements sodium through argon in the third row of the periodic table show a pattern similar to those of the second row, but for sodium through argon it appears that the inner-core electronic structure is more like that of a neon atom than that of a helium atom. For example, Figure 4.6 suggests that the electronic structure of a sodium atom is a neonlike core with a relatively loosely bound outer electron. Recall from Chapter 3 that sodium is a highly reactive element in the alkali metal family of the periodic table, whereas neon is an unreactive element in the noble-gas family. It turns out that elements like neon that have a completely full outer electron shell are particularly stable and unreactive. As we shall see, each of the elements in the noble-gas family of the periodic table has a completely full outer electron shell, thus explaining their relative inertness compared with other elements. (Notice again the relatively large first ionization energies of the noble-gas elements in Figure 4.1.) If we were to make a similar plot for atoms of each of the elements in the third row

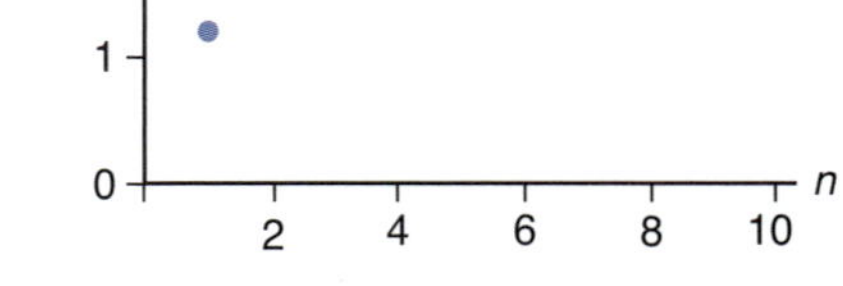

Figure 4.5 The logarithms of the ten successive ionization energies ($\ln I_n/\text{aJ}$) of a neon atom versus the number of electrons removed (n). This graph suggests that the electrons in a neon atom are arranged in two shells: an inner shell consisting of two electrons and an outer shell consisting of eight electrons.

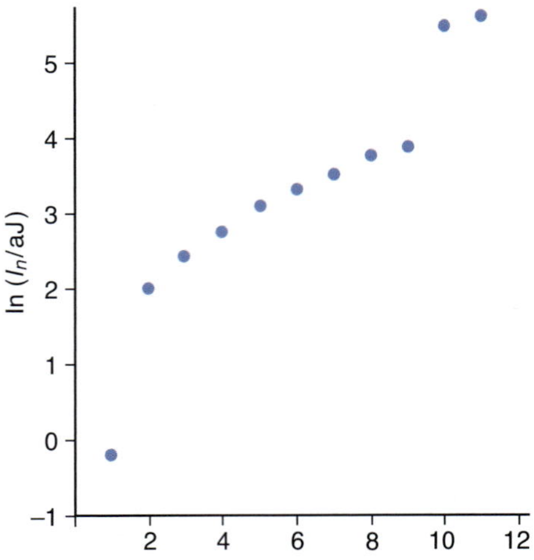

Figure 4.6 The logarithms of the 11 successive ionization energies ($\ln I_n/\text{aJ}$) of a sodium atom versus the number of electrons removed (n).

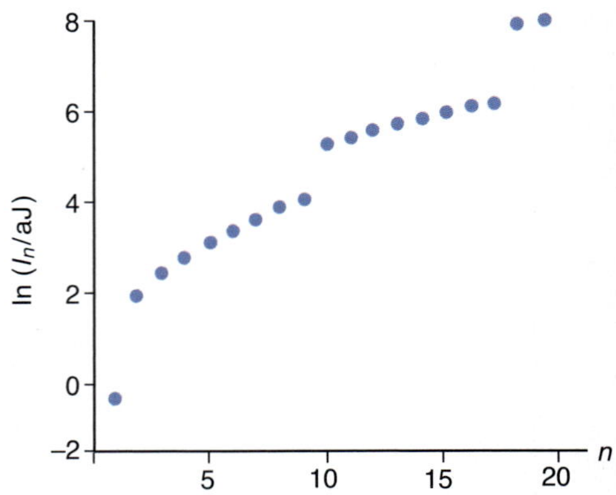

Figure 4.7 The logarithms of the 19 successive ionization energies ($\ln I_n/\text{aJ}$) of a potassium atom versus the number of electrons removed (n).

of the periodic table, we would see that the electronic structure of the atoms consists of an inner shell of two electrons and a second shell of eight electrons (constituting a neonlike core), with an outer shell consisting of the number of electrons corresponding to the position of the element in the periodic table.

Just as the chemical properties of potassium are similar to those of lithium and sodium, a plot of $\ln(I_n/\text{aJ})$ against n for a potassium atom is similar to the plots for a lithium atom and a sodium atom (Figure 4.7). In the case of a potassium atom, the electronic structure is a noble-gas (argon) core with one relatively loosely bound electron. Thus, the electronic structure of a potassium atom consists of four shells: an inner shell containing two electrons, a second shell containing eight electrons, a third shell containing eight electrons, and a fourth and outer shell containing just one electron, which is an argon-like core with one outer electron.

We can summarize these ideas by presenting the atoms as shown in Table 4.2. In the second column, we indicate the noble-gas-like core by using the symbol for the gas enclosed in brackets; the outer-shell electrons are indicated by the dots. These electrons in the outermost occupied shell of an atom are also called **valence electrons.** Thus, for example, we represent beryllium by a helium-like inner core with two valence electrons outside this core; the placement of the dots is arbitrary. As we shall see, the valence electrons play the dominant role in chemical bonding.

The third column in Table 4.2 is an abbreviated version of the second column. Only the valence electrons are indicated, and the appropriate noble-gas core is understood. The number of valence electrons increases from one to eight as we go from alkali metal to noble gas across a row of main-group elements in the periodic table. This pattern repeats itself from row to row. It is the valence electrons that play the key role in determining the chemical properties of the elements; thus, fluorine and chlorine, for example, are chemically similar because they both have seven valence electrons. The representation of valence electrons as dots around the symbol for an element is called a **Lewis electron-dot formula;** it was introduced in 1916 by G. N. Lewis, one of the great-

TABLE 4.2 A simple representation of the first 18 elements, indicating their noble-gas-like inner core and their outer electrons

Symbol	Inner-core representation	Lewis electron-dot formula
H	H	H·
He	[He]	He:
Li	[He]·	Li·
Be	[He]:	Be:
B	$[\dot{\text{H}}\text{e}]:$	$\dot{\text{B}}:$
C	$\cdot[\dot{\text{H}}\text{e}]:$	$\cdot\dot{\text{C}}:$
N	$\cdot[\underset{\cdot}{\dot{\text{H}}}\text{e}]:$	$\cdot\underset{\cdot}{\dot{\text{N}}}:$
O	$\cdot[\underset{\cdot\cdot}{\ddot{\text{H}}}\text{e}]\cdot$	$\cdot\underset{\cdot\cdot}{\ddot{\text{O}}}\cdot$
F	$:[\underset{\cdot}{\ddot{\text{H}}}\text{e}]:$	$:\underset{\cdot}{\ddot{\text{F}}}:$
Ne	[Ne]	$:\underset{\cdot\cdot}{\ddot{\text{Ne}}}:$
Na	[Ne]·	Na·
Mg	[Ne]:	Mg:
Al	$[\dot{\text{N}}\text{e}]:$	$\dot{\text{A}}\text{l}:$
Si	$\cdot[\dot{\text{N}}\text{e}]:$	$\cdot\dot{\text{S}}\text{i}:$
P	$\cdot[\underset{\cdot}{\ddot{\text{N}}}\text{e}]\cdot$	$\cdot\underset{\cdot}{\ddot{\text{P}}}\cdot$
S	$\cdot[\underset{\cdot\cdot}{\ddot{\text{N}}}\text{e}]\cdot$	$\cdot\underset{\cdot\cdot}{\ddot{\text{S}}}\cdot$
Cl	$:[\underset{\cdot}{\ddot{\text{N}}}\text{e}]:$	$:\underset{\cdot}{\ddot{\text{C}}}\text{l}:$
Ar	[Ar]	$:\underset{\cdot\cdot}{\ddot{\text{Ar}}}:$
K	[Ar]·	K·
Ca	[Ar]:	Ca:

est American chemists (Chapter 7 Frontispiece). Lewis electron-dot formulas show only the valence electrons, that is, the chemically important electrons.

To picture more clearly the electronic structure of atoms, we must first briefly consider electromagnetic radiation. Knowledge of how electrons and radiation interact dramatically increases our understanding of atoms and the nature of matter.

4-3. The Electromagnetic Spectrum Is Characterized by Radiation of Different Wavelengths

Radio waves, microwaves, infrared light, visible light, ultraviolet light, X-rays, and γ-rays (gamma rays) are all forms of electromagnetic radiation (Figure 4.8). For many years scientists disagreed over whether electromagnetic radia-

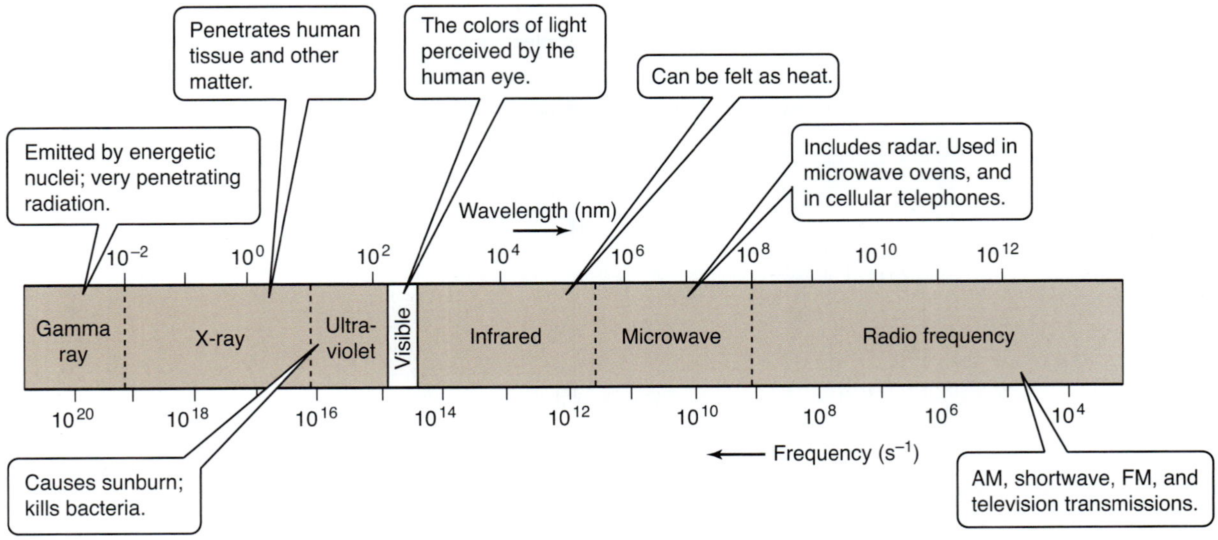

Figure 4.8 The regions of the electromagnetic spectrum. Wavelengths are given in nanometers (1 nm = 10^{-9} m); frequencies are given in cycles per second (s^{-1}), which are the same as hertz (Hz).

tion exists as beams of particles or as continuous waves. Many experiments supported one viewpoint or the other; but toward the end of the nineteenth century, most evidence favored a wave picture of electromagnetic radiation.

In the 1860s the Scottish physicist James Clerk Maxwell developed the **electromagnetic theory of radiation.** This theory states that all forms of radiation are propagated through space as oscillating electric and magnetic fields (Figure 4.9). The electric and magnetic fields, which oscillate together but at right angles to each other, arise from vibrating electric charges in a material, such as a quartz crystal in a radio transmitter or a hot tungsten filament in an incandescent light bulb.

Figure 4.10 depicts a simplified version of electromagnetic waves. The distance between successive crests or troughs is called the **wavelength** and is denoted by the Greek letter lambda, λ. If we picture each wave as moving across the page, then the number of crests that pass a given point per second is called the **frequency** and is denoted by the Greek letter nu, ν. Although the wavelength is the distance per cycle and the frequency is the number of cycles per second, the term *cycle* is understood; so we omit it when the units of λ and ν are given. Thus, for example, the SI units of wavelength are meters and the units of frequency are reciprocal seconds (1/s or s^{-1}). The product of the wavelength and the fre-

Figure 4.9 The oscillating electric and magnetic fields in electromagnetic radiation. The electric and magnetic fields oscillate together as sine waves at right angles to one another.

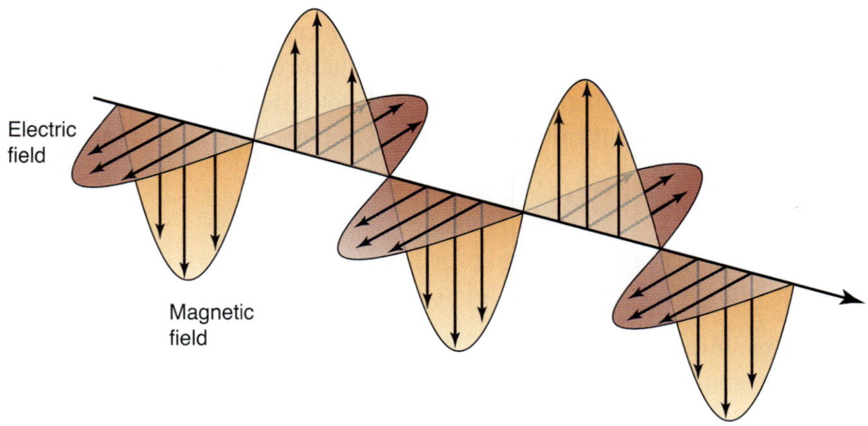

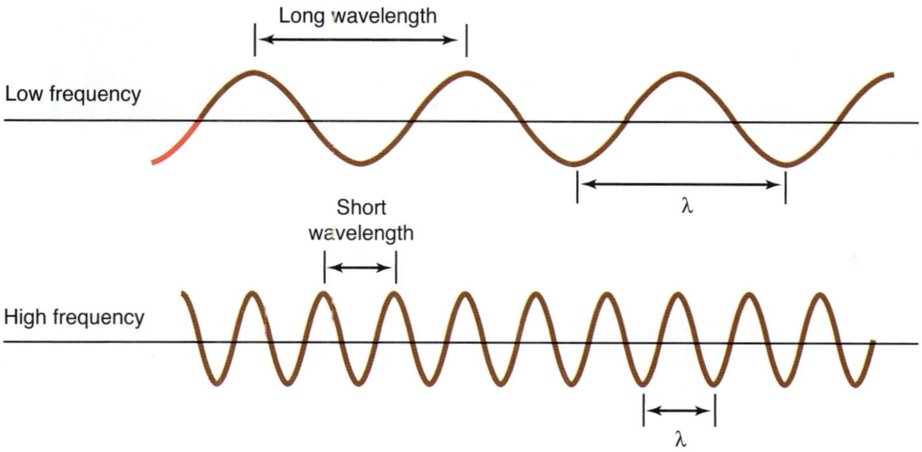

Figure 4.10 Two wave forms, one with a wavelength (λ) three times larger than the other. If both waves moved across the page with the same speed, three crests of the bottom wave would pass by a given point for every crest of the top wave. Thus, the frequency of the bottom wave is three times greater than the frequency of the top wave.

quency, $\lambda\nu$, is the speed at which the wave travels. You may know the speed of light as 186 000 miles per second, which corresponds to $2.9979 \times 10^8 \text{ m·s}^{-1}$. All forms of electromagnetic radiation travel at a speed of $2.9979 \times 10^8 \text{ m·s}^{-1}$, which usually is called simply the **speed of light.** If we denote the speed of light by c, then we can write

$$\lambda\nu = c \tag{4.1}$$

The various forms of electromagnetic radiation differ only in their frequency or their wavelength. For example, 100 MHz (megahertz) is in the middle of the frequency range for the FM radio dial. The unit **hertz** is the same as the unit reciprocal second ($1 \text{ Hz} = 1 \text{ s}^{-1}$), so 100 MHz is equal to $100 \times 10^6 \text{ s}^{-1}$. The wavelength of a 100-MHz signal is

$$\lambda = \frac{c}{\nu} = \frac{2.9979 \times 10^8 \text{ m·s}^{-1}}{100 \times 10^6 \text{ s}^{-1}} = 3.00 \text{ m}$$

EXAMPLE 4-3: Lasers are often used to study chemical reactions (Figure 4.11). A $CO_2(g)$ laser produces an intense beam of radiation whose wavelength is 10.6 μm. Calculate the frequency of this radiation.

Solution: Because $\lambda = 10.6 \times 10^{-6}$ m, we have, using Equation 4.1,

$$\nu = \frac{c}{\lambda} = \frac{2.9979 \times 10^8 \text{ m·s}^{-1}}{10.6 \times 10^{-6} \text{ m}} = 2.83 \times 10^{13} \text{ s}^{-1} = 2.83 \times 10^{13} \text{ Hz}$$

PRACTICE PROBLEM 4-3: Calculate the wavelength of electromagnetic radiation with a frequency of 1.50×10^{18} Hz and identify the type of radiation (Figure 4.8).

Answer: 2.00×10^{-10} m = 0.200 nm; X-ray.

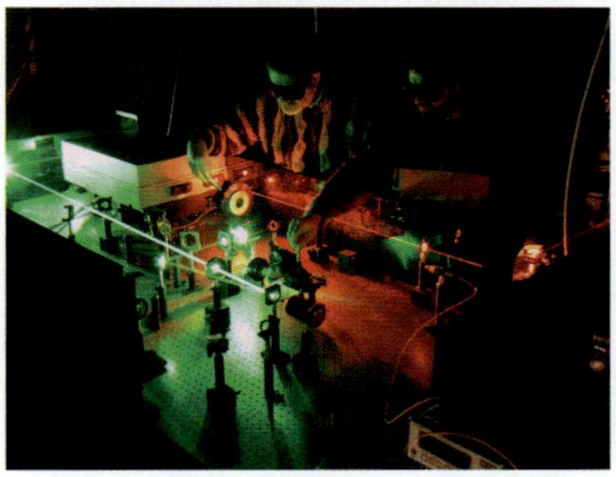

Figure 4.11 Lasers are used to study a wide variety of chemical reactions, ranging from the simplest (e.g., $D + H_2 \rightarrow HD + H$) to the most complex (e.g., combustion reactions). Lasers provide incredible details regarding the energy transfer processes involved in chemical reactions. Lasers are also used in the bar readers at supermarkets, in CD and DVD players, in surgery, in the manufacture of precision mechanical parts, and in many other applications. The development of modern quantum theory led to the invention of the laser.

We see from these calculations that the range of wavelengths and frequencies of electromagnetic radiation, called the **electromagnetic spectrum,** is enormous (Figure 4.8).

Figure 4.12 The visible spectrum runs from violet to red. The sequence of colors in the visible spectrum is remembered readily using the mnemonic name **ROY G BIV** (**R**ed-**O**range-**Y**ellow-**G**reen-**B**lue-**I**ndigo-**V**iolet), where the sequence of colors is from right to left in the figure.

4-4. The Emission Spectra of Atoms Consist of Series of Lines

When we pass white light through a prism, we see that the white light separates into many colors. We see the same effect in a rainbow, where white light from the sun passes through water droplets in the atmosphere and is separated into its component colors. The wavelengths of the radiation in white light range from about 400 nm to about 750 nm. The human eye is sensitive to this region of the electromagnetic spectrum, and we call it the **visible region** (Figure 4.12). The short-wavelength end of the visible region (400 nm) is violet, and the long-wavelength end (750 nm) is red.

The spectrum of white light has no gaps in it; we call it a **continuous spectrum,** meaning that radiation is emitted at all wavelengths in the region. Yet, if we use a prism to examine radiation emitted from a chemical sample placed in a flame or a gas through which an electric spark is passed (such as that from sodium lamps used as street lights or from hydrogen gas lamps), then we find that the radiation emitted is not continuous but consists of separate lines (Figure 4.13). This observation suggests that the radiation is emitted at only certain (discrete) wavelengths. We call this type of spectrum a **line spectrum.** A line spectrum is characteristic of the particular sample used. If the sample consists of individual atoms, then we call the emitted spectrum an **atomic emission spectrum.** The simplest atomic emission spectrum is that of the hydrogen atom. The visible portion of the emission spectrum of hydrogen is shown in Figure 4.14a and other regions in Figure 4.14b.

Figure 4.13 An emission spectrum is generated by passing the light emitted from an excited sample (in a discharge tube, flame, or other source) through a prism or diffraction grating to separate the light according to wavelength. The image is then recorded on a film or with an electronic detector. An emission spectrum appears as a series of bright lines against a dark background.

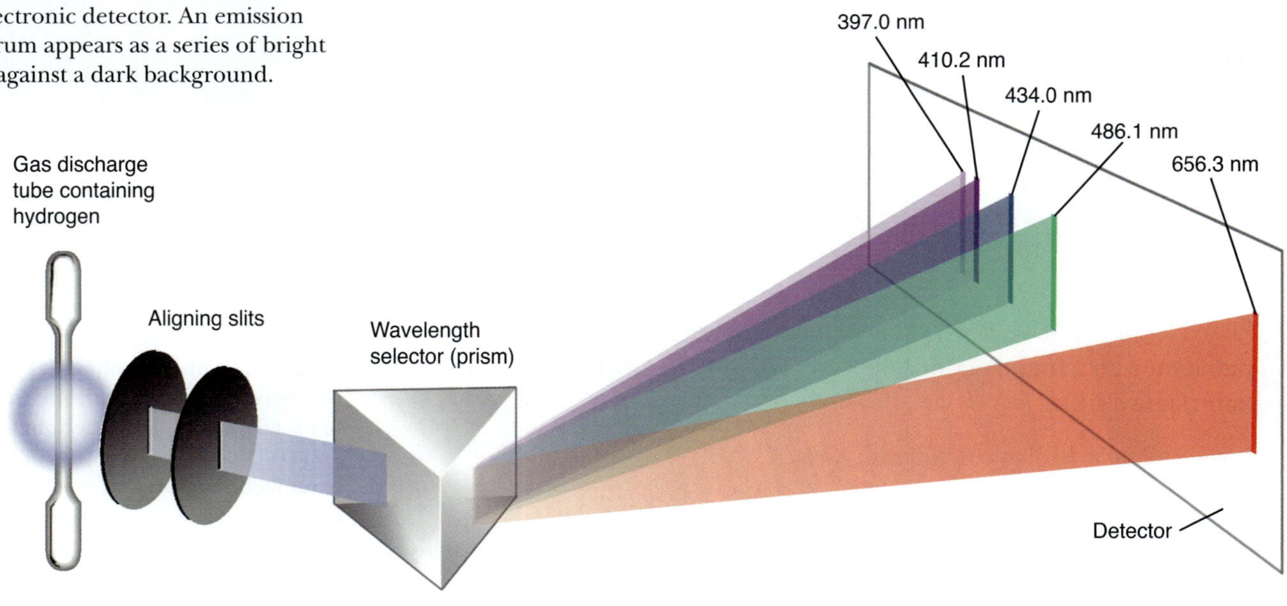

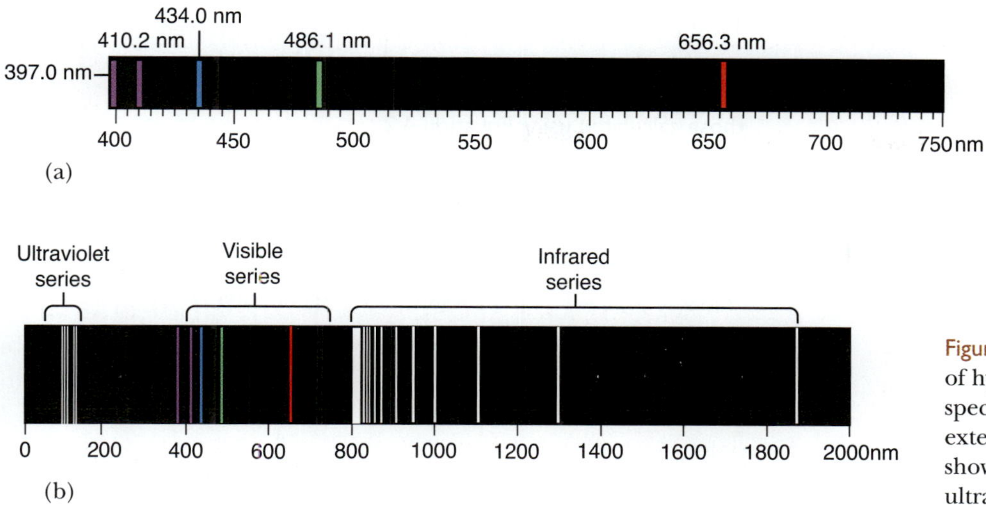

Figure 4.14 The atomic spectrum of hydrogen. (a) The visible line spectrum of hydrogen. (b) The extended line spectrum of hydrogen showing the infrared, visible, and ultraviolet regions of the spectrum.

For many years scientists tried to find some pattern in the wavelengths or frequencies corresponding to the lines in the atomic emission spectrum of hydrogen. Finally, in 1885, Johann Balmer, an amateur Swiss scientist, showed that a plot of the reciprocal of the wavelength ($1/\lambda$) corresponding to the lines shown in Figure 4.14a versus $1/n^2$ (where n is an integer and $n = 3$ is the red line, $n = 4$ is the green line, and so on) is a straight line (Figure 4.15). Balmer's discovery was extended by Johannes Rydberg, a Swedish physicist, who gave the following empirical equation for the wavelengths of lines in the visible spectrum of hydrogen:

$$\frac{1}{\lambda} = (1.097 \times 10^7 \text{ m}^{-1})\left(\frac{1}{4} - \frac{1}{n^2}\right) \qquad n = 3, 4, 5, \ldots \qquad (4.2)$$

Equation 4.2 is known as the **Rydberg-Balmer equation** and the constant 1.097×10^7 m^{-1} is known as the **Rydberg constant.** The Rydberg-Balmer equation accurately predicts the wavelengths of the lines in the visible spectrum of hydrogen (Table 4.3) and is the equation for the line plotted in Figure 4.15.

The atomic emission spectra of all elements consist of series of lines similar to the Balmer series for hydrogen (Figure 4.16). However, the number of lines

Figure 4.15 Plot of $1/\lambda$ versus $1/n^2$ for the lines in the visible spectrum of atomic hydrogen.

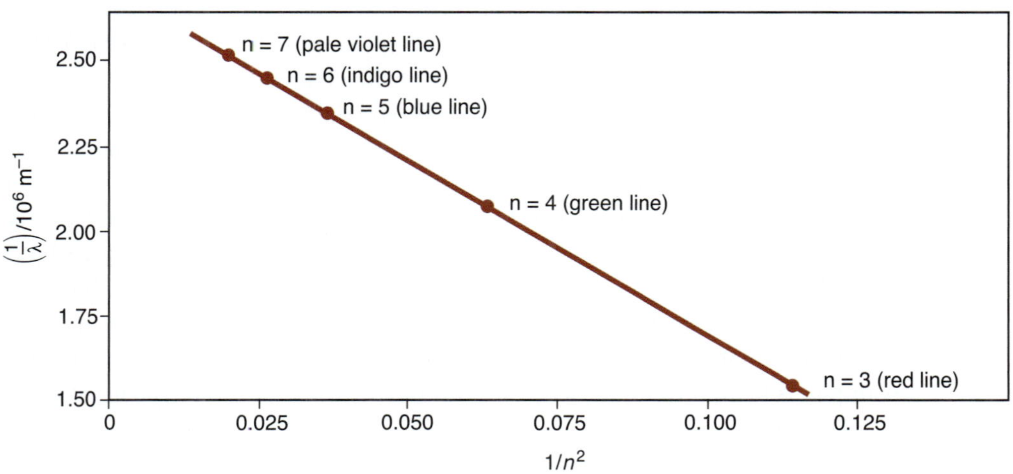

TABLE 4.3 Comparison of the experimentally observed wavelengths corresponding to the lines in the visible region of the hydrogen atomic emission spectrum and those calculated using Equation 4.2

Line color	n	Observed λ/nm	Calculated λ/nm
red	3	656.3	656.3
green	4	486.1	486.2
blue	5	434.0	434.1
indigo	6	410.2	410.2
violet	7	397.0	397.0

WWW

The National Institute of Standards (NIST) maintains a number of free online databases. The experimentally observed wavelengths of the emission lines of most elements can be found there. A link to the NIST databases may be found at: www.McQuarrieGeneralChemistry.com

and wavelengths of the lines are different for every element. Thus, the line spectrum associated with an element serves as a fingerprint for that element.

The study of the spectral lines, or, more generally, the study of the interaction of electromagnetic radiation and atoms, is called **atomic spectroscopy.** We can identify each type of atom present in a sample by comparing the observed line spectrum with those in a handbook or a database of atomic spectra. Atomic spectroscopy is a standard tool in analytical chemistry, the part of chemistry involved with chemical analysis. Atomic spectroscopy is used in archaeology, art preservation, astronomy, criminal investigations, environmental science, medicine, and many other fields.

4-5. Electromagnetic Radiation Can Be Viewed as a Beam of Photons

If light has a continuous spectrum, then how can the discrete spectral lines in atomic spectra be explained? Atomic spectra were only the first of many experimental results that were impossible to interpret within the framework of the physics of the late nineteenth century, a theoretical framework now called **classical physics.** The theoretical analysis of these experiments stimulated some radically new ideas about the nature of matter and energy that have had profound effects in science. These ideas led to a new general theory of atomic structure that is in excellent agreement with our picture of the electronic structure of atoms and with the periodic properties of the elements.

The first person to break with the ideas of classical physics was the German physicist, Max Planck (Frontispiece), in his study of **blackbody radiation**, the emission of light from a heated body. All solid objects at temperatures greater than zero kelvin emit thermal radiation. The term blackbody is used to distinguish the object as an ideal radiator. The radiation from such an object depends only on its temperature and not its specific composition. It happens that a good emitter of radiation is also a good absorber. An ideal emitter will absorb all radiation incident upon it and will thus appear black. As a metal, for example, is heated to increasingly high temperatures, the color of the glowing metal changes continuously from red to white; this behavior is the origin of the terms *red hot* and *white hot.* Classical physics predicted that radiation from

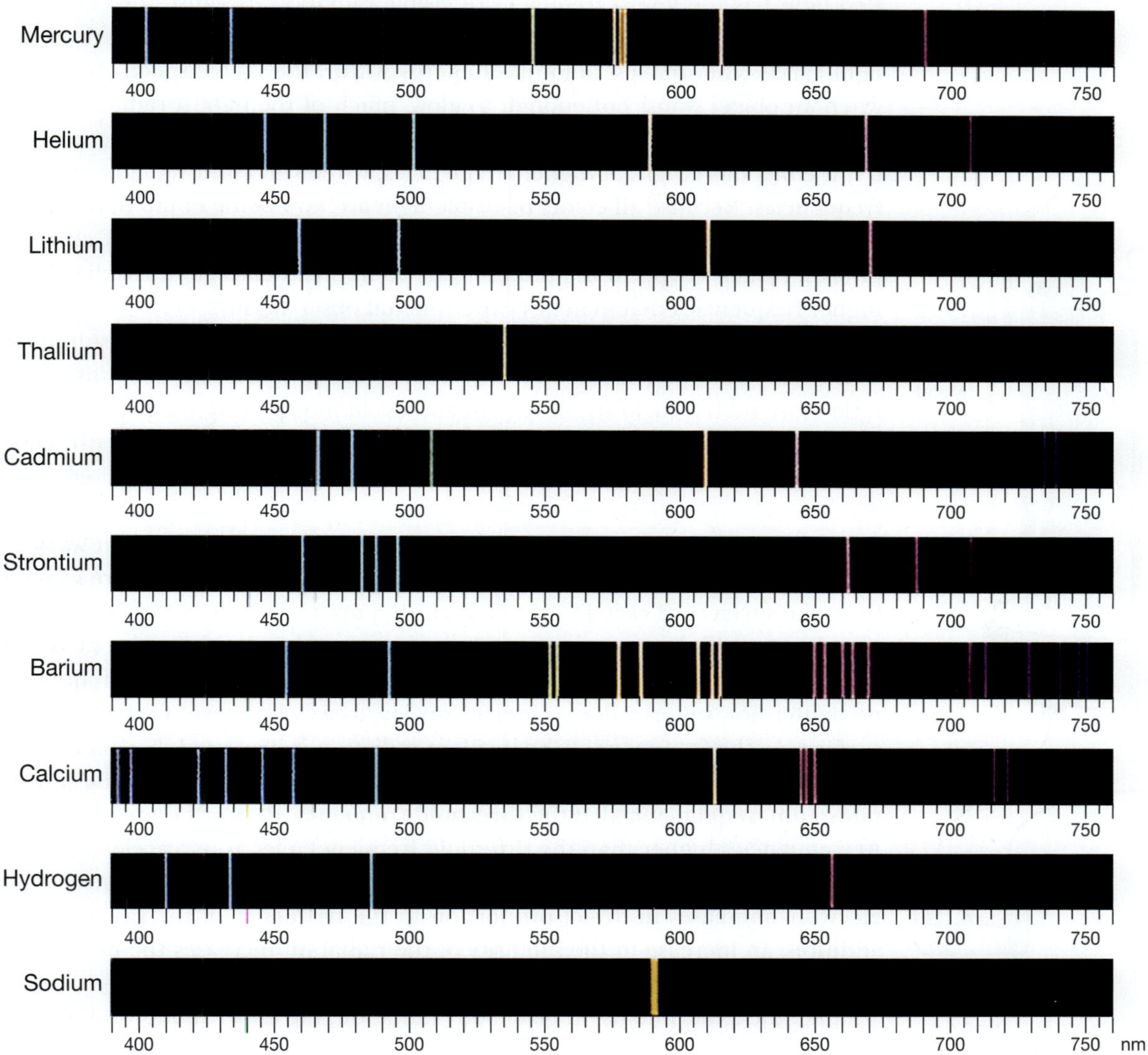

Figure 4.16 Atomic spectra. The visible emission spectra of atomic mercury, helium, lithium, thallium, cadmium, strontium, barium, calcium, hydrogen, and sodium.

a heated body would be produced at all frequencies, with increasing intensity at greater frequencies. Thus, a hot object, such as an incandescent light bulb filament, would give off not only visible light, but also significant quantities of ultraviolet radiation, x-rays, and so forth, in contradiction to what is actually observed. With revolutionary insight, Planck proposed that the radiation is emitted in discrete packets, called **quanta** (singular, **quantum**), and that the energy associated with a quantum of radiation is proportional to the frequency of the emitted radiation. In terms of an equation, Planck proposed that,

$$E = h\nu \tag{4.3}$$

where E is the energy of a quantum of radiation, ν is the frequency of the radiation, and h is a proportionality constant, now called the **Planck constant**. Planck was able to reproduce all the data on blackbody radiation by assigning h the value of 6.626×10^{-34} J·s.

Planck's hypothesis explains why a metal object gives off red light when it first begins to glow and then glows white hot as it reaches higher temperatures.

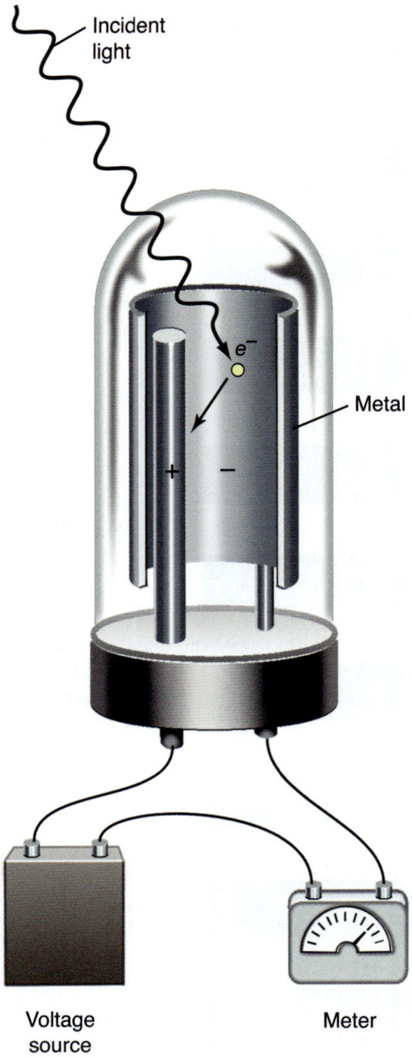

Figure 4.17 An apparatus for studying the photoelectric effect. Light of a certain frequency falls on a clean metal surface, ejecting electrons that are attracted toward the positive electrode. The number of electrons ejected is registered by a detecting meter.

Red light has the lowest frequency of visible radiation. Because the energy of a quantum of radiation is proportional to the frequency of the radiation, red light carries the smallest amount of energy within the visible spectrum. Thus, when an object is just hot enough to glow, much of the light it radiates corresponds to the lowest energy form of visible radiation. As the object becomes hotter, it emits more light at higher frequencies, while still emitting light at lower frequencies. Because all colors of visible light are now being emitted, the glowing object appears white. Moreover, in contrast to the prediction of classical physics, Planck's hypothesis predicts a fall-off in the intensity of the radiation at higher frequencies that matches experimental observations.

Planck's theory of blackbody radiation, with its unconventional assumption that a heated body emits radiation only in small discrete packets, was considered little more than a curiosity, despite its success in interpreting the experimental data on blackbody radiation. Most scientists believed that in time a more satisfactory, which is to say classical, theory of blackbody radiation would emerge. This was not to be, however.

Another phenomenon that classical physics was unable to explain was the **photoelectric effect.** In the 1880s, it was discovered that electrons are ejected from the surfaces of certain metals when they are exposed to ultraviolet radiation. Figure 4.17 shows an apparatus for measuring the photoelectric effect and Figure 4.18 shows some typical experimental data. Radiation of low frequency, no matter how intense, does not eject any electrons whatsoever from the metal surface. As the frequency of the incident radiation is increased, however, a frequency is reached beyond which electrons are now ejected. The value of this minimum frequency, called the **threshold frequency**, depends upon the metal. At frequencies higher than the threshold frequency, electrons are ejected from the surface of the metal. At these frequencies, the kinetic energy of the ejected electrons increases in direct proportion to the frequency of the radiation. In addition, an increase in the intensity of the radiation increases the number of electrons emitted but does not affect their energies.

Like the experiments on blackbody radiation, the experiments on the photoelectric effect defied theoretical explanation for many years. Not until Albert Einstein (Figure 4.19), using the ideas behind Equation 4.3, interpreted the experimental results were these observations understood. Einstein's physical picture was beautifully simple. He pictured electromagnetic radiation as a beam of particles that he called **photons**. Each of the photons in the beam is a little packet of energy with the value $E = h\nu$, just as in Equation 4.3. The intensity of the radiation is proportional to the number of photons in the beam. If we denote the threshold frequency (below which no electrons can be ejected from the metallic surface), by ν_0, then the minimum energy required to eject an electron is $h\nu_0$. The energy in excess of the energy required to eject an electron is given by $E = h\nu - h\nu_0$. This excess energy goes into the kinetic energy, E_k, of the ejected electrons, and so we can write the equation

$$\begin{aligned} E_k &= 0 && (\text{when } \nu < \nu_0) \\ &= h\nu - h\nu_0 && (\text{when } \nu > \nu_0) \end{aligned} \tag{4.4}$$

When E_k in Equation 4.4 is plotted against ν, you obtain exactly the same type of curve as shown in Figure 4.18. The slope of the straight line is equal to h, and when Einstein used Equation 4.4 to fit experimental data, he had to use the very same value of h that Planck had used to fit the experimental data for blackbody radiation. Furthermore, because the intensity of the radiation is proportional to the number of photons in the beam, the number of electrons ejected is proportional to the intensity of the radiation, in accord with the experimental observation.

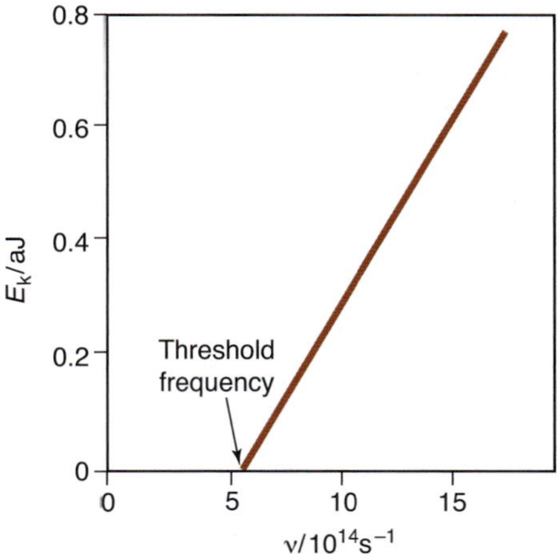

Figure 4.18 When a metallic surface is exposed to ultraviolet radiation of frequency ν, electrons are ejected from the surface. The kinetic energy, E_k, of these electrons is directly proportional to ν, as shown here for sodium. The frequency below which no electrons are ejected is called the threshold frequency and is equal to 5.51×10^{14} Hz for sodium metal.

EXAMPLE 4-4: Calculate the kinetic energy of the electrons ejected from the surface of copper by radiation of wavelength 210.0 nm. Take the threshold frequency of copper to be $\nu_0 = 1.076 \times 10^{15}$ Hz.

Solution: We first calculate the frequency of the incident radiation according to Equation 4.1.

$$\nu = \frac{c}{\lambda} = \frac{2.9979 \times 10^8 \text{ m·s}^{-1}}{210.0 \times 10^{-9} \text{ m}} = 1.428 \times 10^{15} \text{ s}^{-1}$$

Using Equation 4.4, we have

$$E_k = h\nu - h\nu_0 = h(\nu - \nu_0) = (6.626 \times 10^{-34} \text{ J·s})(1.428 \times 10^{15} \text{ s}^{-1} - 1.076 \times 10^{15} \text{ s}^{-1})$$
$$= 2.33 \times 10^{-19} \text{ J} = 0.233 \text{ aJ}$$

PRACTICE PROBLEM 4-4: Selecting any two points on the experimentally determined line plotted in Figure 4.18, estimate the value of h from the slope of the straight line.

Solution: Choosing, for instance, the points 0.6 aJ and 0.2 aJ on the vertical axis, we can estimate the corresponding points as 14×10^{14} s^{-1} and 8×10^{14} s^{-1}, respectively, on the horizontal axis to obtain a slope of about 6.7×10^{-34} J·s. For a review of working with the equation of a straight line, see Appendix A6.

We can also express the energy of a photon in terms of wavelength. Combining Equations 4.1 and 4.3 yields

$$E = \frac{hc}{\lambda} \tag{4.5}$$

Because the energy of a photon is inversely proportional to its wavelength, the shorter its wavelength, the larger its energy. For example, photons of blue light have a shorter wavelength and therefore a larger energy than photons of red light. Energetic photons such as X-rays have very short wavelengths and, consequentially, very high frequencies (Equation 4.1).

Figure 4.19 Albert Einstein (1879–1955). It is interesting to note that Einstein received the 1921 Nobel Prize in Physics for his work on the photoelectric effect and not for his work on the theory of relativity, as many people think.

EXAMPLE 4-5: Calculate the wavelength of a photon that has an energy equal to the ionization energy of one hydrogen atom.

Solution: According to Table 4.1, the ionization energy of atomic hydrogen is 2.18 aJ or 2.18×10^{-18} J. We use Equation 4.5 to calculate the wavelength. Solving Equation 4.5 for λ gives

$$\lambda = \frac{hc}{E} = \frac{(6.626 \times 10^{-34} \text{ J·s})(2.9979 \times 10^8 \text{ m·s}^{-1})}{2.18 \times 10^{-18} \text{ J}}$$
$$= 9.11 \times 10^{-8} \text{ m} = 91.1 \text{ nm}$$

PRACTICE PROBLEM 4-5: The red line in the visible region of the emission spectrum of hydrogen occurs at 656.3 nm. Calculate the energy and the frequency of these photons.

Answer: 0.3027 aJ; $\nu = 4.568 \times 10^{14}\ \text{s}^{-1}$

In spite of the enormous success of Einstein's treatment of the photoelectric effect, the idea of energy existing in discrete little packets, that is, the **quantization of energy**, contradicted classical physics and was resisted by most scientists at that time. Nevertheless, in the two very different experiments, blackbody radiation and the photoelectric effect, which had previously defied explanation, the very same constant, h, arose. Surely, this was no coincidence. These early theories were the first steps in the development of **quantum theory** that invoked the idea of quantization of energy, or of energy existing in little discrete packets. Planck, who first recognized this in 1900, is called the father of quantum theory. In the next sections, we shall see that all phenomena that occur on the atomic and molecular level, including the atomic spectra that we discussed earlier, are described by quantum theory.

Figure 4.20 Born in 1892 to a French aristocratic family, Duke Louis Victor de Broglie studied history as an undergraduate. His interest turned to science as a result of working with his older brother Maurice, who built his own private X-ray research laboratory in his family's Paris mansion. His interest in wave properties matured after his assignment to radio communications in World War I, during which he was stationed at the top of the Eiffel Tower. De Broglie received a doctoral degree in physics from the University of Paris in 1924. After reading de Broglie's doctoral thesis on the wavelike properties of matter, Einstein praised the work, saying, "He has lifted one corner of the great veil!" De Broglie was awarded the Nobel Prize in Physics in 1929 "for his discovery of the wave nature of electrons."

4-6. De Broglie Was the First to Propose That Matter Has Wavelike Properties

Scientists have always had difficulty describing the nature of light. In many experiments light exhibits a definite wavelike character, but in others it seems to behave as a stream of little particles (photons). Because light appears sometimes to be wavelike and sometimes to be particle-like, we talk of the **wave-particle duality** of light. In 1924, Louis de Broglie (Figure 4.20), a French physicist, proposed an unusual idea in his doctoral thesis: if light, which has wave properties, also can display particle-like properties under certain conditions, then matter, which certainly appears to be particle-like, might also display wavelike properties under certain conditions. This proposal is rather strange at first sight, but it does suggest a nice symmetry in nature. Certainly, if light can appear to be particle-like at times, why shouldn't matter appear to be wavelike at times?

Using the theory of relativity, De Broglie put this idea into a quantitative scheme by proposing that both light and matter obey the equation

$$\lambda = \frac{h}{p} \tag{4.6}$$

where λ is the **de Broglie wavelength,** h is the Planck constant, and p is the momentum of the particle. For a particle that is moving at a speed not too near the speed of light, the **momentum,** p, is given by mv, where m is the mass of the particle at rest and v is its speed. Momentum is proportional to both the mass of an object and its speed. For example, a heavy truck has a greater momentum than a light car when both are moving at the same speed. Using the relation $p = mv$, Equation 4.6 also can be written as

$$\lambda = \frac{h}{mv} \tag{4.7}$$

EXAMPLE 4-6: Calculate the de Broglie wavelength of an electron traveling at 1.00% of the speed of light.

Solution: The mass of an electron is 9.1094×10^{-31} kg (see the inside of the back cover). Its speed is

$$v = (0.0100)(2.9979 \times 10^{8} \text{ m·s}^{-1}) = 3.00 \times 10^{6} \text{ m·s}^{-1}$$

so its momentum is

$$mv = (9.1094 \times 10^{-31} \text{ kg})(3.00 \times 10^{6} \text{ m·s}^{-1}) = 2.73 \times 10^{-24} \text{ kg·m·s}^{-1}$$

The de Broglie wavelength of this electron is

$$\lambda = \frac{h}{mv} = \frac{6.626 \times 10^{-34} \text{ J·s}}{2.73 \times 10^{-24} \text{ kg·m·s}^{-1}}$$
$$= 2.43 \times 10^{-10} \text{ m} = 0.243 \text{ nm}$$

We have used the fact that $1 \text{ J} = 1 \text{ kg·m}^2\text{·s}^{-2}$. By referring to Figure 4.8, we see that the wavelength of the electron in this example corresponds to the wavelength of X-rays.

PRACTICE PROBLEM 4-6: Calculate the de Broglie wavelength in meters of a golf ball (45.9 grams) traveling at 120 miles per hour. Compare your result to the diameter of a golf ball.

Answer: 2.69×10^{-34} m; the value of λ is totally negligible relative to the dimensions of a golf ball.

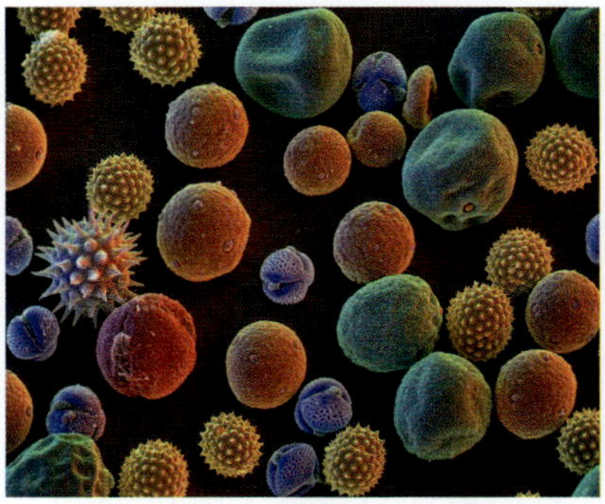

Figure 4.21 An electron micrograph of pollen particles collected by filtering a pollen-laden air sample. The pollen grains range in size from about 15 μm to 40 μm.

Although Practice Problem 4-6 shows that Equation 4.7 is of trivial consequence for a macroscopic object like a golf ball, Example 4-6 predicts that electrons can act like X-rays. This effect is demonstrated in the **electron microscope**, in which the wave nature of electrons is used to determine the size and shapes of objects, much as light is used in a light microscope (Figure 4.21).

4-7. Electrons Exhibit Both Particle-Like and Wavelike Properties

Substantiation of de Broglie's hypothesis was provided by the work of the British scientist, G. P. Thomson, and the American scientist, C. Davisson, who shortly after the publication of de Broglie's thesis independently discovered **electron diffraction** in 1926 and 1927, respectively. The two later shared the 1937 Nobel Prize in Physics for this work.

Diffraction is inherently a wave property. For instance, when a beam of X-rays is directed at a thin foil of a crystalline substance, the beam is scattered in a definite manner characteristic of the atomic structure of the crystalline substance. This phenomenon, called **X-ray diffraction,** happens because the length of the interatomic spacings in the crystal is close to the wavelength of the X-rays. The X-ray diffraction pattern obtained by irradiating a thin piece of aluminum foil with X-rays is shown in Figure 4.22 (*top*). It is known in phys-

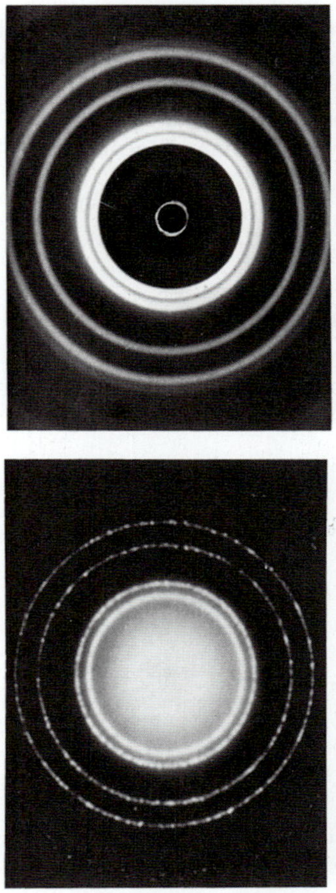

Figure 4.22 Bombarding aluminum foil with X-rays and with electrons produces characteristic diffraction patterns. (*top*) The X-ray diffraction pattern. (*bottom*) The electron diffraction pattern. The similarity of the patterns shows that electrons can behave like X-rays and display wavelike properties.

Figure 4.23 A pictorial metaphor of the concept of wave-particle duality. If you stare at the illustration, we see either a young woman or an old witch. You never see both simultaneously.

ics that a diffraction pattern like the one shown in Figure 4.22 (*top*) can be obtained only if X-rays act as waves. However, a beam of electrons produces a similar pattern—called an electron diffraction pattern (Figure 4.22 *bottom*). The similarity of the two patterns demonstrates that indeed electrons can exhibit wavelike behavior.

The contemporary view of nature is that electromagnetic radiation and matter can exhibit either wavelike or particle-like properties, depending on the experiment. For example, electromagnetic radiation appears to be wavelike when diffraction experiments are performed, but it appears to be particle-like when photoelectric effects are studied. Electrons, on the other hand, appear to be particles when they are studied in an apparatus like the one that J. J. Thomson used to discover them (Figure 2.18), but they appear to act like waves when directed at a suitable target in that they give rise to a diffraction pattern (Figure 4.22 *bottom*). This wave-particle duality appears to be an inherent property of matter that manifests itself whenever the mass of the particles involved is very small ($\lambda = h/mv$), as is the case for electrons and atoms.

An interesting visual metaphor for the concept of wave-particle duality is presented in Figure 4.23. What we see in this picture depends, in part, on how we look at it. We see either one image or the other, but not both simultaneously. This same property is involved in the wave-particle duality concept. In effect, the electron has two faces—a wave face and a particle face. The two electron faces are mutually exclusive in that they are never seen simultaneously in a single experiment. We see either the wave face or the particle face, but not both together. In general, which face we see depends on the type of experiment we are performing.

Another way to look at this seeming paradox is to view the problem as one of misclassification. The wave-particle duality can be interpreted as the result of our attempt to classify physical phenomena as either wavelike or particle-like based on our macroscopic experiences, excluding the possibility that some things don't fit neatly into either box. This is analogous to claiming the platypus is a strange animal because it is neither strictly a mammal nor a reptile. In fact the platypus is not strange; it's just a platypus. It's the artificial line drawn by us between mammals and reptiles that makes it seem strange. In the same manner, one can view the apparent physical contradictions that arise in quantum theory as something requiring us to adjust our paradigms of what things really are.

It is interesting to note that it was J. J. Thomson who first showed, in 1895, that the electron is a subatomic particle; and it was G. P. Thomson who, along with others, first showed experimentally, in 1926, that the electron could act as a wave. The two Thomsons were father and son. The father received a Nobel Prize in 1906 for showing that the electron is a particle, and the son received a Nobel Prize in 1937 for showing that it is a wave.

4-8. The Energy of the Electron in a Hydrogen Atom Is Quantized

In 1913, a young Danish physicist named Niels Bohr (Frontispiece) formulated a description of the hydrogen atom that successfully explained the observed atomic spectrum of hydrogen. A key postulate of the Bohr theory is that the electron in a hydrogen atom is restricted to only certain circular orbits about the nucleus. Ten years later, when de Broglie proposed the wave nature of elec-

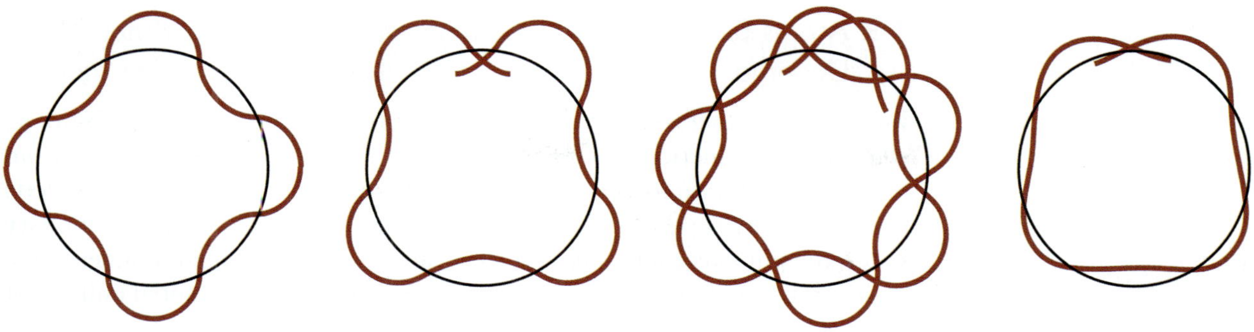

Figure 4.24 An illustration of matching and mismatching de Broglie waves bound in Bohr orbits. If the wavelengths of the de Broglie waves are such that an integral number of them fit around the circle, then they match after a complete revolution, as shown in (a), where $4\lambda = 2\pi r$. If the waves do not match after a complete revolution, cancellation will result and the waves will progressively disappear, as seen in (b) through (d). Only those electron energies that correspond to matching de Broglie waves are allowed.

trons, it became clear why a stable electron orbit results only when the de Broglie wave associated with the bound electron has a suitable wavelength. Figure 4.24 shows that if the de Broglie waves of the electron do not match after a complete revolution, then the waves cancel and the waves will progressively disappear. The orbits will be stable only if the wave pattern matches with each revolution. This will only occur if some whole number of complete wavelengths, n, fit around the circumference of the orbit. But we know that for an orbit of radius r, the circumference is $2\pi r$, so we have the **quantum condition**

$$2\pi r = n\lambda; \qquad n = 1, 2, 3, \ldots \tag{4.8}$$

Using what is essentially this condition, together with the force balance condition between the orbiting electron and the proton, Bohr showed that the only possible energies of the electron in these orbits are given by

$$E_n = \frac{-2.1799 \times 10^{-18}\,\text{J}}{n^2} = \frac{-2.1799\,\text{aJ}}{n^2} \qquad n = 1, 2, 3, \ldots \tag{4.9}$$

Note that n is restricted to integer values. When $n = 1$, $E_1 = -2.18$ aJ (to three significant figures); when $n = 2$, $E_2 = -0.545$ aJ; and so on. We say that the energy is **quantized** because E_n can take on only discrete values. The E_n values given by Equation 4.9 correspond to the allowed **energy states** of electrons in a hydrogen atom. A plot of these energy states as a function of n is given by the horizontal lines or levels shown in Figure 4.25. Note that the spacing between subsequent levels gets closer as the value of n in Equation 4.9 increases.

Because of the minus sign, all the energies given by Equation 4.9 are negative. The lowest energy is E_1; and $E_1 < E_2 < E_3$, and so on. By convention the state of zero energy occurs when $n = \infty$, that is, $E_\infty = 0$. In this state of zero energy, the proton and electron are so far apart that they do not attract each other at all (that is, the atom is ionized), so we take their interaction energy to be zero. At closer distances, the proton and electron attract each other because they have opposite charges. A negative energy state is more stable than a state of zero energy.

Allowed energy states are called **stationary states** in quantum theory. The stationary state of lowest energy is called the **ground state,** and the allowed states of higher energies are called **excited states.** In the hydrogen atom, the state with $n = 1$ is the ground state; the state with $n = 2$ is the **first excited state;** the state with $n = 3$ is the **second excited state;** and so on up to $n = \infty$, where the atom is ionized (the electron is no longer bound to the atom).

4-9. Atoms Emit or Absorb Electromagnetic Radiation When They Undergo Transitions from One Stationary State to Another

Bohr assumed that when an atom is in a stationary state, it does not absorb or emit electromagnetic radiation. When an atom undergoes a transition from one stationary state to another, however, it emits or absorbs electromagnetic radiation. Because the Bohr model proposed a limited number of stationary states, the photons emitted or absorbed can have only a limited number of energies and, therefore, a limited number of frequencies. This limited number of frequencies results in a line spectrum.

Consider a hydrogen atom that undergoes a transition from the $n = 4$ state to the $n = 2$ state (Figure 4.25). In this case, the electron goes from a higher energy state to a lower energy state and a photon is released. The initial state here is a hydrogen atom in the $n = 4$ state and the final state is the $n = 2$ state and a photon. Conservation of energy gives us

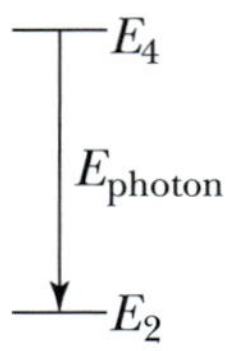

$$E_4 = E_2 + E_{\text{photon}}$$

Solving this equation for the energy of the photon and using Equation 4.9 for E_4 and E_2 gives

$$\begin{aligned} E_{\text{photon}} = E_4 - E_2 &= \left(\frac{-2.1799 \text{ aJ}}{4^2}\right) - \left(\frac{-2.1799 \text{ aJ}}{2^2}\right) \\ &= (2.1799 \text{ aJ})\left(\frac{1}{2^2} - \frac{1}{4^2}\right) \\ &= 0.40873 \text{ aJ} = 4.0873 \times 10^{-19} \text{ J} \end{aligned}$$

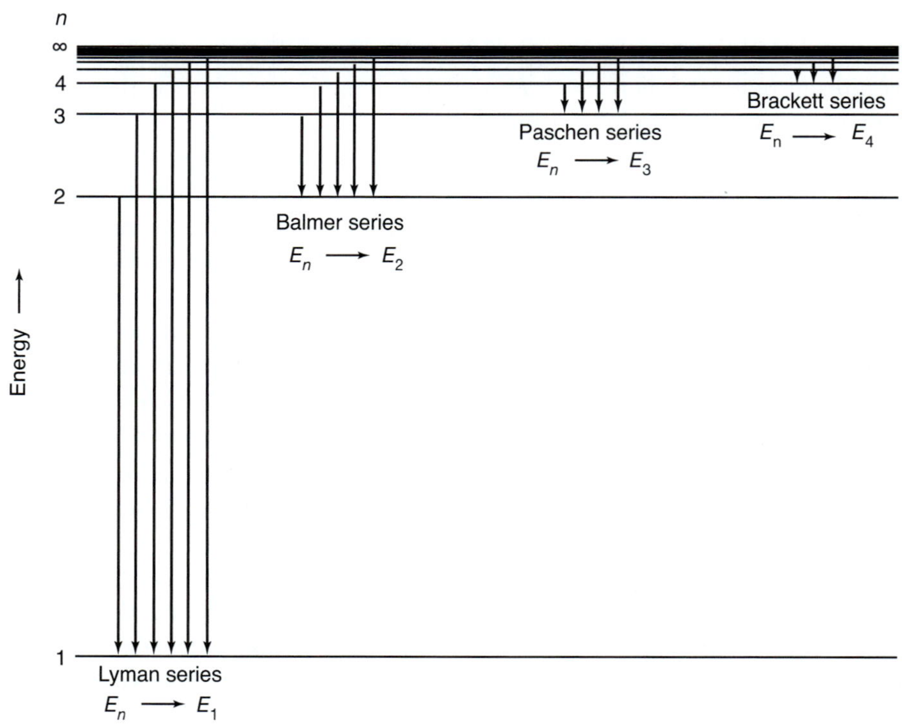

Figure 4.25 Transitions from higher energy states to lower energy states for a hydrogen atom. Each transition is accompanied by the emission of a photon. The various series of lines are called the Lyman series ($n > 1$ to $n = 1$); the Balmer series ($n > 2$ to $n = 2$); the Paschen series ($n > 3$ to $n = 3$); and the Brackett series ($n > 4$ to $n = 4$). The series are named after their discoverers.

Using the relation, $E = h\nu$, we find that the frequency of the light emitted during the transition is

$$\nu = \frac{E_{\text{photon}}}{h} = \frac{4.0873 \times 10^{-19}\,\text{J}}{6.626 \times 10^{-34}\,\text{J·s}} = 6.169 \times 10^{14}\,\text{s}^{-1}$$

The wavelength corresponding to this frequency is given by Equation 4.1:

$$\lambda = \frac{c}{\nu} = \frac{2.9979 \times 10^{8}\,\text{m·s}^{-1}}{6.169 \times 10^{14}\,\text{s}^{-1}} = 4.860 \times 10^{-7}\,\text{m} = 486.0\,\text{nm}$$

in nice agreement with the experimental value given in Table 4.3.

We can derive a general formula for the wavelength associated with the transition from some arbitrary initial excited state, n_i, to some arbitrary final state, n_f, where the energy of the initial state is greater than that of the final state, or $E_i > E_f$. Applying the law of conservation of energy to the transition, we see that

$$E_i = E_f + E_{\text{photon}}$$

Solving for E_{photon} and substituting Equation 4.9 into this expression yields

$$E_{\text{photon}} = E_i - E_f = \left(\frac{-2.1799\,\text{aJ}}{n_i^2}\right) - \left(\frac{-2.1799\,\text{aJ}}{n_f^2}\right)$$

or

$$E_{\text{photon}} = (2.1799\,\text{aJ})\left(\frac{1}{n_f^2} - \frac{1}{n_i^2}\right) \qquad (n_i > n_f) \qquad (4.10)$$

If we substitute Equation 4.5 ($E = hc/\lambda$) into Equation 4.10 and solve for $1/\lambda$, then we obtain

$$\frac{1}{\lambda} = \frac{E}{hc} = \left[\frac{2.1799 \times 10^{-18}\,\text{J}}{(6.626 \times 10^{-34}\,\text{J·s})(2.9979 \times 10^{8}\,\text{m·s}^{-1})}\right]\left(\frac{1}{n_f^2} - \frac{1}{n_i^2}\right)$$

$$= (1.097 \times 10^{7}\,\text{m}^{-1})\left(\frac{1}{n_f^2} - \frac{1}{n_i^2}\right) \qquad (4.11)$$

Equation 4.11 predicts that there is a series of lines in the hydrogen atom emission spectrum that corresponds to the transitions from states n_i to n_f (where $n_i > n_f$). Some of these transitions are shown in Figure 4.25.

If we let $n_f = 2$ in Equation 4.11, we obtain the Rydberg-Balmer equation (Equation 4.2). Thus, we find that the Rydberg-Balmer equation is the equation that predicts the series of lines from states n_i (= 3, 4, 5, . . .) to the state $n_f = 2$ in the emission spectrum of hydrogen. This series of emission lines is called the **Balmer series** (Figure 4.26). The theoretical derivation of the Rydberg-Balmer equation was one of the great successes of Bohr's theory of the hydrogen atom.

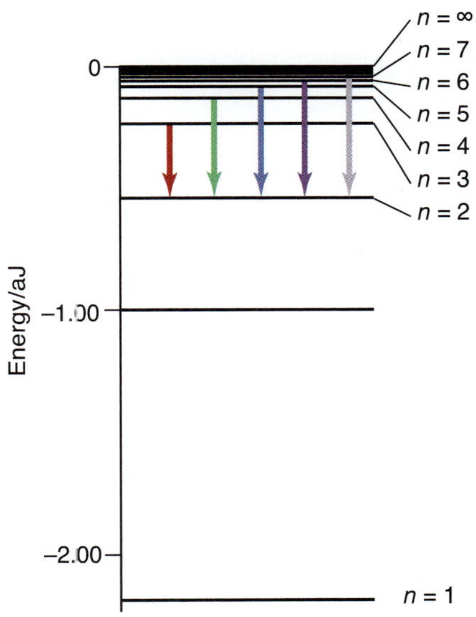

Figure 4.26 The Balmer series. The arrows representing the transitions giving rise to the lines in the visible region of the spectrum are highlighted with the colors corresponding to the wavelengths of the spectral lines.

EXAMPLE 4-7: Calculate the frequency and wavelength of the $n = 3$ to $n = 1$ atomic transition in the hydrogen atom.

Solution: The energy of the photon emitted is given by Equation 4.10 as

$$E_{\text{photon}} = (2.1799 \text{ aJ})\left(\frac{1}{1^2} - \frac{1}{3^2}\right) = 1.9377 \text{ aJ} = 1.9377 \times 10^{-18} \text{ J}$$

Now using Equation 4.3, $E = h\nu$, we find that the frequency of this photon is

$$\nu = \frac{E}{h} = \frac{1.9377 \times 10^{-18} \text{ J}}{6.626 \times 10^{-34} \text{ J}\cdot\text{s}} = 2.924 \times 10^{15} \text{ s}^{-1}$$

Similarly, using the relation $E = hc/\lambda$, we find that the wavelength of the emitted photon is

$$\lambda = \frac{hc}{E} = \frac{(6.626 \times 10^{-34} \text{ J}\cdot\text{s})(2.9979 \times 10^{8} \text{ m}\cdot\text{s}^{-1})}{1.9377 \times 10^{-18} \text{ J}\cdot\text{s}} = 1.025 \times 10^{-7} \text{ m} = 102.5 \text{ nm}$$

The series of lines that results from transition between states $n_i > 1$ and the ground state ($n_f = 1$) in the hydrogen atom occurs in the ultraviolet region of the spectrum and is called the **Lyman series** (Figure 4.25). The values obtained for the frequencies of these lines using Equations 4.10 and 4.3 are in excellent agreement with those obtained from experiment.

PRACTICE PROBLEM 4-7: Use Equation 4.11 to calculate the value of the wavelength for the $n = 5$ to $n = 3$ transition for a hydrogen atom. In what region of the spectrum does this line appear?

Answer: 1282 nm; infrared.

EXAMPLE 4-8: The frequency of the indigo line in the visible emission spectrum of the hydrogen atom is 7.308×10^{14} s^{-1}. If the line originates from the $n = 6$ level, what is the level of the final state of the electron in this transition?

Solution: We must first solve Equation 4.3 for the energy of this transition:

$$E_{\text{photon}} = h\nu = (6.626 \times 10^{-34} \text{ J}\cdot\text{s})(7.308 \times 10^{14} \text{ s}^{-1}) = 4.842 \times 10^{-19} \text{ J} = 0.4842 \text{ aJ}$$

Substituting this result into Equation 4.10, we obtain

$$0.4842 \text{ aJ} = (2.1799 \text{ aJ})\left(\frac{1}{n_f^2} - \frac{1}{6^2}\right)$$

Solving this expression for the value of n_f yields

$$\frac{1}{n_f^2} = \frac{0.4842 \text{ aJ}}{2.1799 \text{ aJ}} + \frac{1}{6^2}$$

$$\frac{1}{n_f^2} = 0.2221 + 0.0278$$

$$n = \sqrt{4.000} = 2$$

This is one of the lines in the Balmer series (Figure 4.26). Transitions from $n = 3, 4, 5, 6$, and 7 to $n = 2$ in the Balmer series result in the five lines seen in the visible spectrum of hydrogen.

PRACTICE PROBLEM 4-8: The wavelength of the red emission line in the Balmer series is 656.3 nm. Calculate the level, n, from which the electron originates. What is this excited state called?

Answer: $n = 3$, in agreement with the values found in Table 4.3. The state $n = 3$ is called the *second excited state* because it is two levels above the ground state of hydrogen ($n = 1$).

The emission spectra that we have studied so far are obtained when a spark is discharged through the gas. The discharge is a pulse of energy that dissociates molecules into atoms and promotes the atoms into excited states. As the atoms return to their ground state, the electrons fall down through the allowed energy states, and the photons emitted produce the observed atomic emission spectrum (Figure 4.13). As we have noted, atomic emission spectra can be used to identify elements; we apply this principle in the **flame tests** that we use to identify the alkali metals in samples. Each alkali metal has a characteristic flame color (Figure 4.27).

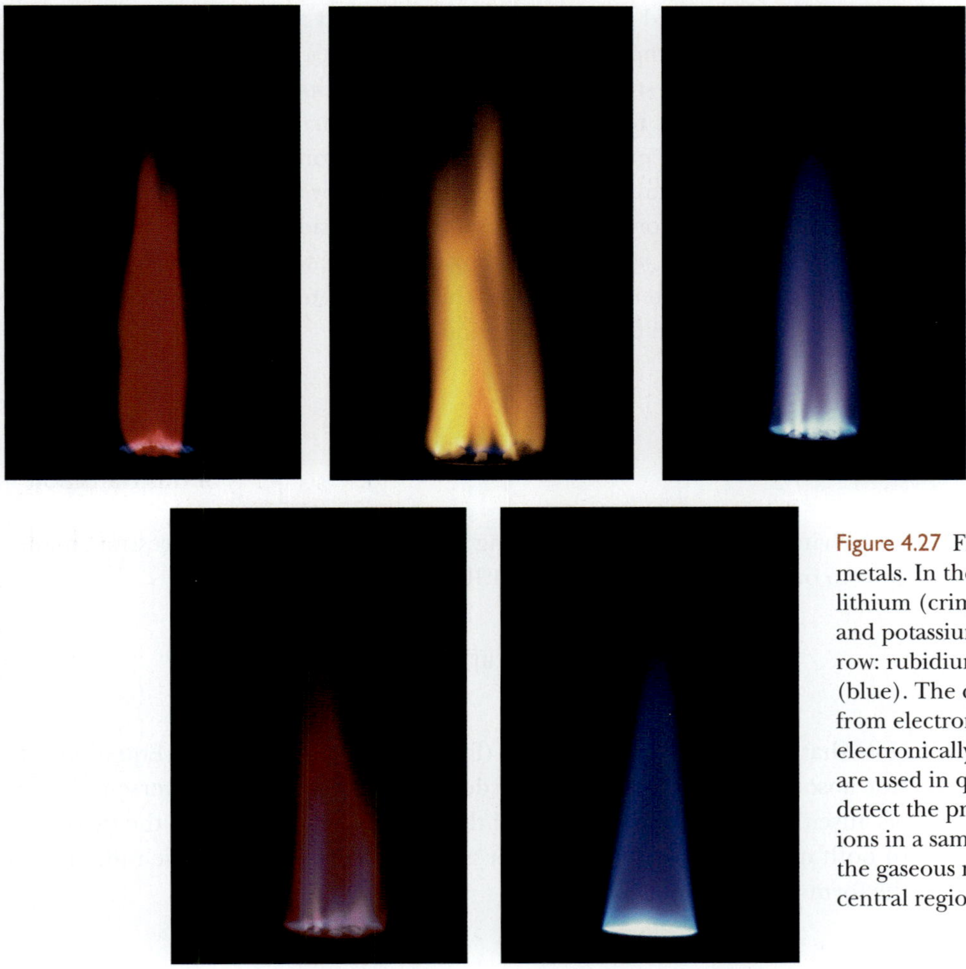

Figure 4.27 Flames of the Group 1 metals. In the top row, from the left: lithium (crimson), sodium (yellow), and potassium (violet); in the second row: rubidium (red) and cesium (blue). The colors, which arise from electronic transitions in the electronically excited metal atoms, are used in qualitative analysis to detect the presence of alkali metal ions in a sample. Ions are reduced to the gaseous metal atoms in the lower central region of the flame.

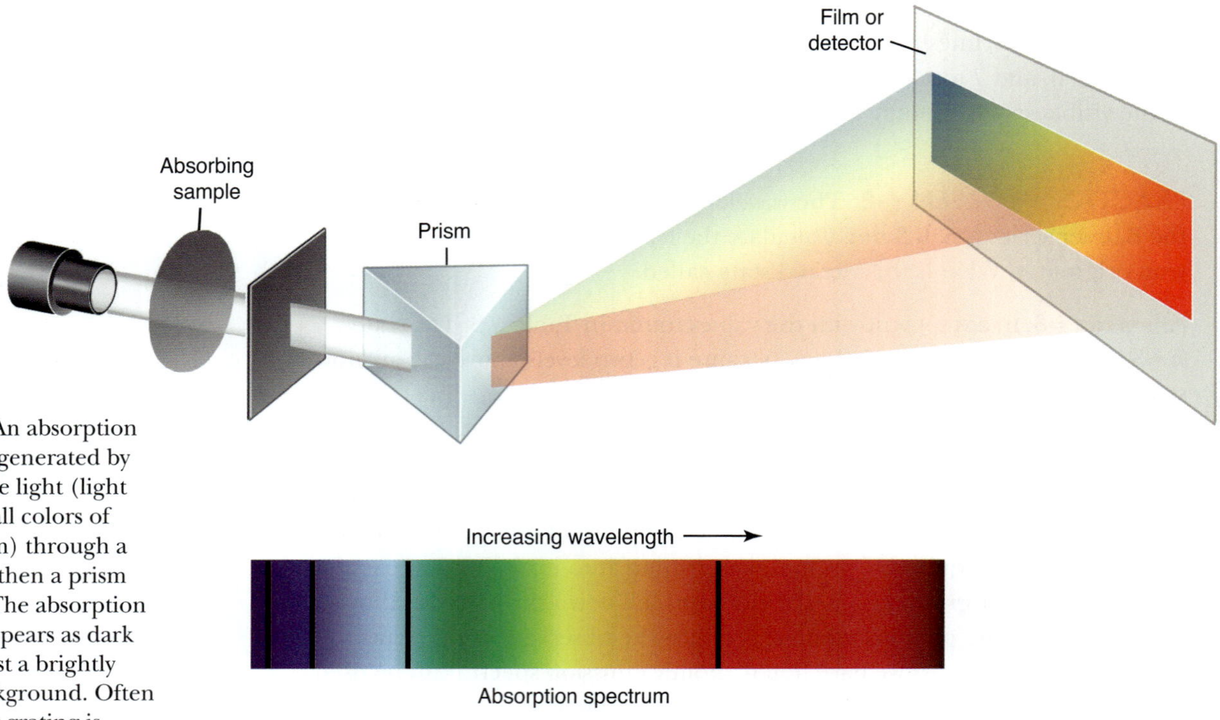

Figure 4.28 An absorption spectrum is generated by passing white light (light containing all colors of the spectrum) through a sample and then a prism or grating. The absorption spectrum appears as dark bands against a brightly colored background. Often the prism or grating is placed before the sample to enable the measurement of the absorption at selected wavelengths.

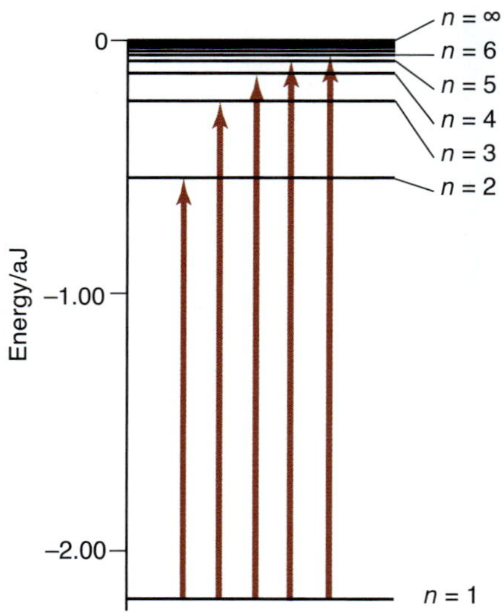

Figure 4.29 A hydrogen atom in its ground electronic state ($n = 1$) can absorb electromagnetic radiation. When this occurs, the electron is promoted to an excited ($n > 1$) state.

An **absorption spectrum** is observed experimentally when a sample is irradiated with electromagnetic radiation (Figure 4.28). For the hydrogen atom, this can be interpreted in terms of the Bohr theory as arising from transitions in which electrons jump from lower energy states to higher energy states. The spectrometer used to study such transitions is equipped with a wavelength selector that enables us to sweep through the electromagnetic radiation range of interest. The incident radiation promotes electrons to excited energy states. Only radiation of certain frequencies is absorbed by the sample.

To see how an absorption spectrum arises, consider a transition from some arbitrary initial state, n_i, to some arbitrary final state, n_f, where the energy of the final state is greater than that of the initial state, or $E_f > E_i$ (Figure 4.29). Applying conservation of energy, we find

$$E_f = E_i + E_{\text{photon}}$$

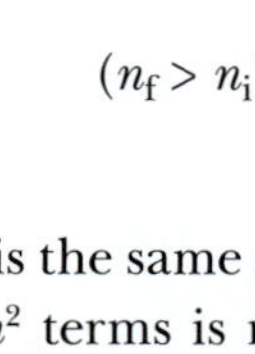

Solving for E_{photon} and substituting Equation 4.9 into the expression, analogous to our derivation of Equation 4.10, yields

$$E_{\text{photon}} = (2.1799\text{ aJ})\left(\frac{1}{n_i^2} - \frac{1}{n_f^2}\right) \qquad (n_f > n_i) \tag{4.12}$$

Note that the form of Equation 4.10 (for emission) is the same as Equation 4.12 (for absorption), except that the order of the $1/n^2$ terms is reversed. This is because the initial and final states of the electron are reversed in the two cases. In both cases, the energy emitted or absorbed is electromagnetic radiation in the form of a photon.

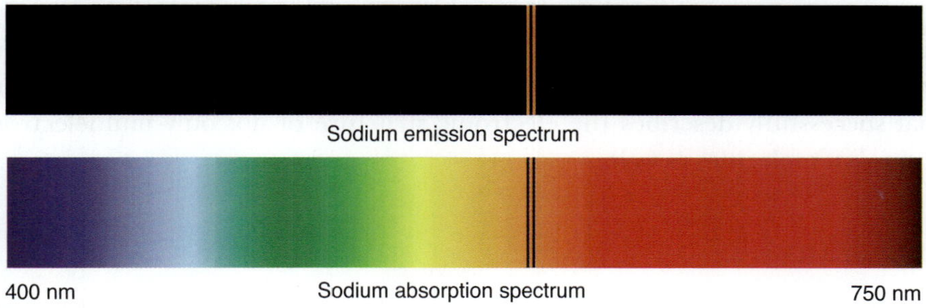

Figure 4.30 The strongest transitions in the emission and absorption spectra of sodium are at 589.0 nm and 589.6 nm in the yellow region of the visible spectra.

Whereas an emission spectrum appears as a series of colored lines against a dark background, an absorption spectrum is a series of dark bands at the frequencies where light is absorbed against a bright background. Figure 4.30 shows an emission and absorption spectrum for sodium. Sodium lamps are often used in streetlights, giving rise to their bright yellow color.

Equation 4.12 can be used to predict the absorption spectra of a hydrogen atom. The calculations are analogous to those found in the previous examples. Figure 4.29 shows the transitions that correspond to the $n = 1$ to $n > 1$ series in the absorption spectrum of atomic hydrogen. Emission and absorption spectroscopy are used extensively in chemical analysis.

It is important to note that only frequencies of light where the energy corresponds to the energy difference between any two states in the atom are readily absorbed or emitted (Figure 4.31). Were this not the case, the spectrum would not consist of a series of lines, but rather would be continuous.

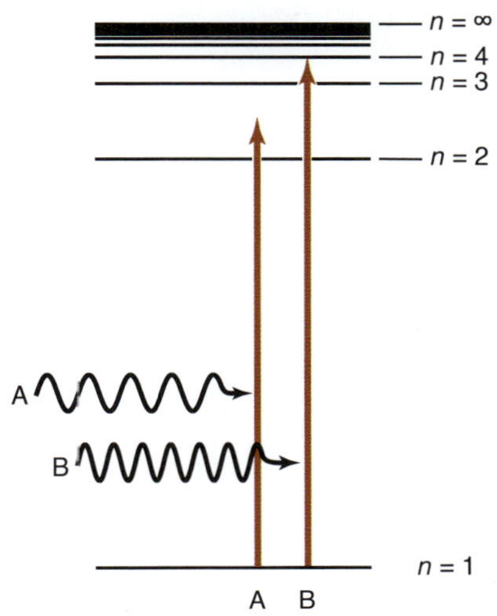

Figure 4.31 Two photons (waves) incident on a hydrogen atom and their relative energies (arrows). Photon B can be absorbed because its energy corresponds to the difference in energy between the $n = 1$ and $n = 4$ states. Photon A cannot be absorbed because its energy does not correspond to the difference in energy between any of the states. Even though photon A has more than enough energy to promote an electron from the $n = 1$ to $n = 2$ state, it cannot dissipate the energy in excess of that needed for the transition.

EXAMPLE 4-9: Calculate the ionization energy of a hydrogen atom.

Solution: The ionization energy, as noted in Section 4-1, is the energy required to completely remove the outermost electron from the ground state of the atom. Therefore, the ionization of a hydrogen atom in the ground state corresponds to a transition from the $n = 1$ state to the $n = \infty$ state. Because the atom is absorbing energy, the energy required is given by Equation 4.12

$$E_{\text{ionization}} = (2.1799\ \text{aJ})\left(\frac{1}{1^2} - \frac{1}{\infty^2}\right) = (2.1799\ \text{aJ})(1 - 0) = 2.18\ \text{aJ}$$

where we have used the fact that $1/\infty^2 = 0$ and rounded our result to three significant figures. This result is in excellent agreement with the value given in Table 4.1.

PRACTICE PROBLEM 4-9: Compute the maximum wavelength of a photon capable of ionizing a hydrogen atom in its ground state.

Answer: 91.1 nm.

The Bohr theory of the hydrogen atom was tremendously successful, but it could not be extended to atoms having more than one electron, despite the efforts of a great many people. It turned out that an entirely new approach to electronic structure was needed, and this was developed independently by Werner Heisenberg and Erwin Schrödinger in 1925. We shall discuss this approach in the next chapter, and in particular, we shall discuss the Schrödinger equation that successfully describes the electronic structure of not only multielectron atoms but molecules as well.

SUMMARY

First ionization energy is a periodic property (Figure 4.1). The values of successive ionization energies suggest that electrons in atoms are arranged in shells (Figures 4.3 through 4.7), as depicted in Lewis electron-dot formulas given in Table 4.2.

In 1900 Planck initiated quantum theory by postulating that electromagnetic radiation is emitted from heated bodies only in quanta, or little packets, with energies given by $E = h\nu$. Five years later Einstein used the same ideas to describe the photoelectric effect. He showed that light, or electromagnetic radiation in general, can be viewed as a stream of particles called photons, whose energies are given by Planck's formula, $E = h\nu$. Later de Broglie argued that if light can display particle-like properties in the form of photons, then particles should display wavelike properties. The wavelength associated with a moving particle is given by $\lambda = h/mv$. De Broglie's postulate was substantiated experimentally by electron diffraction, where a crystal irradiated with electrons gives the same type of diffraction pattern as with X-rays.

In 1911 Neils Bohr formulated the first theory of the electronic structure of a hydrogen atom. He proposed that the electron in a hydrogen atom is restricted to only certain circular orbits, with the nucleus at the center. The electron in each allowed orbit has a different energy that results in the energy of the electron being quantized. The radiation emitted from an excited chemical sample consists of a number of discrete lines, called a line spectrum. The emission spectrum generated by an excited atom is called an atomic emission spectrum. Atoms absorb radiation of only certain wavelengths producing an atomic absorption spectrum. The emission spectrum of a hydrogen atom consists of several series of lines. The hydrogen emission lines that appear in the visible portion of the spectrum are called the Balmer series. The wavelengths of the lines in the Balmer series are given by the Rydberg-Balmer equation. In 1911 Bohr developed a model of the hydrogen atom that accounted beautifully for all of the lines in its atomic emission spectrum.

TERMS YOU SHOULD KNOW

electronic structure *105*
electron arrangement *105*
ionization energy, I *105*
first ionization energy, I_1 *105*
second ionization energy, I_2 *106*
shell *108*
valence electron *110*
Lewis electron-dot formula *110*
electromagnetic theory of radiation *112*
wavelength, λ *112*
frequency, ν *112*
speed of light, c (2.9979×10^8 m·s^{-1}) *113*
hertz (Hz) *113*
electromagnetic spectrum *113*
visible region *114*
continuous spectrum *114*
line spectrum *114*
atomic emission spectrum *114*
Rydberg-Balmer equation *115*
Rydberg constant (1.097×10^7 m^{-1}) *115*
atomic spectroscopy *116*
classical physics *116*
blackbody radiation *116*
quantum *117*
Planck constant, h (6.626×10^{-34} J·s) *117*
photoelectric effect *118*

threshold frequency, ν_0 *118*
photon *118*
quantization of energy *120*
quantum theory *120*
wave-particle duality *120*
de Broglie wavelength *120*
momentum, p *120*
electron microscope *121*
electron diffraction *121*
X-ray diffraction *121*
quantum condition *123*
quantized *123*
energy state *123*
stationary state *123*
ground state *123*
excited state *123*
first excited state *123*
second excited state *123*
Balmer series *125*
Lyman series *126*
flame test *127*
absorption spectrum *128*

EQUATIONS YOU SHOULD KNOW HOW TO USE

$$\lambda\nu = c$$ (4.1) (relation between wavelength and frequency)

$$\frac{1}{\lambda} = (1.097 \times 10^7\ \mathrm{m^{-1}})\left(\frac{1}{4} - \frac{1}{n^2}\right),\ n = 3, 4, 5, \ldots$$ (4.2) (Rydberg-Balmer equation; for wavelengths of visible lines in the hydrogen atom spectrum)

$$E = h\nu$$ (4.3) (energy of a photon in terms of frequency)

$$E_k = h\nu - h\nu_0 \quad (\text{when } \nu > \nu_0)$$ (4.4) (photoelectric effect)

$$E = \frac{hc}{\lambda}$$ (4.5) (energy of a photon in terms of wavelength)

$$\lambda = \frac{h}{mv}$$ (4.7) (de Broglie wavelength)

$$E_n = \frac{-2.1799\ \mathrm{aJ}}{n^2} \qquad n = 1, 2, 3, \ldots$$ (4.9) (energies of the electron in a hydrogen atom)

$$E_{\text{photon}} = (2.1799\ \mathrm{aJ})\left(\frac{1}{n_f^2} - \frac{1}{n_i^2}\right) \qquad (n_i > n_f)$$ (4.10) (energy of a photon emitted by a hydrogen atom in transition from n_i to n_f, where $E_i > E_f$)

$$E_{\text{photon}} = (2.1799\ \mathrm{aJ})\left(\frac{1}{n_i^2} - \frac{1}{n_f^2}\right) \qquad (n_f > n_i)$$ (4.12) (energy of a photon absorbed by a hydrogen atom in transition from n_i to n_f, where $E_f > E_i$)

PROBLEMS

IONIZATION ENERGIES

4-1. Arrange the following species in order of decreasing first ionization energy:

He Be Kr Ne

4-2. Arrange the following species in order of decreasing first ionization energy:

Ca Mg Ba Sr

4-3. From the data in Table 4.1, calculate the energy required to form a single Be^{2+} ion from a neutral beryllium atom in the gas phase.

4-4. From the data in Table 4.1, calculate the energy required to form a single B^{3+} ion from a neutral boron atom in the gas phase.

4-5. Using the data in Table 4.1, calculate the energy required to convert 6.022×10^{23} neutral lithium atoms in the gas phase into $Li^+(g)$ ions.

4-6. Using the data in Table 4.1, calculate the energy required to convert 6.022×10^{23} neutral helium atoms in the gas phase into $He^{2+}(g)$ ions.

4-7. Use the data in Table 4.1 to plot the logarithms of the ionization energies of a boron atom versus the number of electrons removed. What does the plot suggest about the electronic structure of boron?

4-8. Use the data in Table 4.1 to plot the logarithms of the ionization energies of a beryllium atom versus the number of electrons removed. Compare your plot to Figure 4.4.

LEWIS ELECTRON-DOT FORMULAS

4-9. Write Lewis electron-dot formulas for all the alkali metal atoms and for all the halogen atoms. What is the similarity in all the alkali metal atom formulas and in all the halogen atom formulas?

4-10. Write the Lewis electron-dot formulas for the Group 16 elements. Comment on the similarities in the valence-electron configurations.

4-11. Write the Lewis electron-dot formula for

Ar S S^{2-} Al^{3+} Cl^-

4-12. Write the Lewis electron-dot formula for

B^{3+} N^{3-} F^- O^{2-} Na^+

ELECTROMAGNETIC RADIATION

4-13. Compare photons of ultraviolet and gamma radiation. Which has (a) the longer wavelength, (b) the greater frequency, (c) the greater energy?

4-14. Compare yellow and green light from the visible spectrum. Which has (a) the longer wavelength, (b) the greater frequency, (c) the greater energy?

4-15. A helium-neon laser produces light of wavelength 633 nm. What is the frequency of this light?

4-16. The radiation given off by a sodium lamp, which is used in streetlights and darkroom safelights, has a wavelength of 589 nm. What is the frequency of this radiation?

4-17. A campus radio station broadcasts at 89.9 MHz (megahertz) on the FM dial. What is the wavelength of this transmission in meters?

4-18. The Ka Wide-Band radar guns most commonly used by law enforcement agencies to measure a driver's speed operate in a frequency range of 34.2 to 35.2 GHz (gigahertz). What is the wavelength range of this radiation?

4-19. The first ionization energy of a potassium atom is 0.696 aJ. What is the wavelength of light that is just sufficient to ionize a potassium atom?

4-20. The first ionization energy of an argon atom is 2.52 aJ. Do X-rays with a wavelength of 80 nm have sufficient energy to ionize argon?

PHOTONS

4-21. The human eye is able to detect as little as 2.35×10^{-18} J of green light of wavelength 510 nm. Calculate the minimum number of photons of green light that can be detected by the human eye.

4-22. One Einstein is a unit used in spectroscopy that is defined as 6.022×10^{23} photons. Calculate the energy of one Einstein of X-ray photons of wavelength 210 pm.

4-23. A carbon dioxide laser produces radiation of wavelength 10.6 μm. Calculate the energy of one photon produced by this laser. If the laser produces about one joule of energy per pulse, how many photons are produced per pulse?

4-24. Argon-fluoride excimer lasers are now commonly used in corrective eye surgery. The short, intense pulses of light generated by these lasers are able to cleanly ablate microscopic layers of corneal tissue, effectively reshaping the lens with little or no trauma to the rest of the eye. Calculate the energy delivered to the cornea per laser pulse, if one pulse of light from an argon-fluoride excimer laser contains 2.5×10^{16} photons with a wavelength of 193 nm.

4-25. Using the data in Table 4.1, calculate the frequency and wavelength of photons capable of just ionizing 1.0×10^{16} sodium atoms in the gas phase. Assuming an ionization efficiency of 16%, how many such photons are required to ionize the whole sample?

4-26. A dental X-ray machine produces X-rays with a wavelength of 1.54 angstroms, where one angstrom (symbol Å) is equal to 10^{-10} meters and is a unit commonly used in spectroscopy. Calculate the energy of 6.022×10^{23} of these X-ray photons in joules.

PHOTOELECTRIC EFFECT

4-27. The threshold frequency, ν_0, of gold metal is 1.2×10^{15} s^{-1}. Will ultraviolet radiation of wavelength 200 nm eject electrons from the surface of metallic gold?

4-28. The photocells that are used in automatic door openers are an application of the photoelectric effect. A beam of light strikes a metal surface, from which electrons are emitted, producing an electric current. When the beam of light is blocked by a person walking through the beam, the electric circuit is broken, thereby opening the door. If the source of light is a solid-state gallium-arsenide diode laser that emits light at a wavelength of 840 nm, would copper be a satisfactory metal to use in the photocell? The threshold frequency of copper is 1.01×10^{15} Hz.

4-29. Given that the threshold frequency of cesium metal is 4.38×10^{14} Hz, calculate the kinetic energy of an electron ejected from the surface of cesium metal when it is irradiated with light of wavelength 400.0 nm.

4-30. The threshold frequency of a metal can be determined from measurements of the speed of the ejected electrons. Electrons were ejected from a metal with a speed of 5.00×10^5 m·s^{-1} when irradiated by light having a wavelength of 390.0 nm. Determine the threshold frequency of this metal.

DE BROGLIE WAVELENGTH

4-31. Calculate the de Broglie wavelength of a proton traveling at a speed of 1.00×10^5 m·s^{-1}. The mass of a proton is 1.67×10^{-27} kg.

4-32. A certain rifle bullet has a mass of 5.00 grams. Calculate the de Broglie wavelength of the bullet traveling at 1250 miles per hour.

4-33. Calculate the de Broglie wavelength of a hydrogen molecule traveling at a speed of 2.0×10^3 m·s^{-1}. Take the mass of a hydrogen molecule to be 3.35×10^{-27} kg.

4-34. The de Broglie wavelength of electrons used in a particular electron microscope is 96.0 pm. What is the speed of one of these electrons?

4-35. Estimate the de Broglie wavelength of a 75-kg person walking at a pace of 3 kilometers per hour. Comment on your answer.

4-36. Assuming that the smallest measurable wavelength in an experiment is 0.10 fm (femtometers), what is the maximum mass of an object traveling at 100 m·s^{-1} for which the de Broglie wavelength is observable?

HYDROGEN ATOMIC SPECTRUM

4-37. A hydrogen atom absorbs a photon and is excited to the $n = 3$ state. What is this state called? If the electron relaxes from this state down to the $n = 2$ state, will the wavelength of light emitted be greater or less than it would be if the electron were to relax directly back to the $n = 1$ state?

4-38. When an electron in a hydrogen atom is in the $n = 3$ state, is it on average closer to or farther from the nucleus than in the ground state? Does it take more or less energy to ionize a hydrogen atom in the $n = 3$ state than one in which the electron is in the ground state? Why is this?

4-39. How much energy is required for an electron in a hydrogen atom to make a transition from the $n = 2$ state to the $n = 3$ state? What is the wavelength of a photon having this energy?

4-40. What is the frequency of the emitted photon when an electron in a hydrogen atom returns to the ground state from the second excited state? To what series does this line belong? In what region of the spectrum is this line observed?

4-41. Calculate the frequency of the first five lines in the Lyman series of hydrogen (Figure 4.25). In what region(s) of the spectrum are these lines observed?

4-42. Calculate the wavelength of the first five lines in the Paschen series of hydrogen (Figure 4.25). In what region(s) of the spectrum are these lines observed?

4-43. A line in the Brackett series of hydrogen (Figure 4.25) has a wavelength of 2626 nm. From what state did the electron originate? In what region of the spectrum is this line observed?

4-44. Wavenumbers are a unit of frequency commonly used in spectroscopy. Wavenumbers are the reciprocal of wavelength and are typically given in units of inverse centimeters (cm^{-1}). A line in the Paschen series of hydrogen (Figure 4.25) has a frequency of 9140 cm^{-1}. From what state did the electron originate?

4-45. Compute the ionization energy of a hydrogen atom in its first excited state.

4-46. A ground state hydrogen atom absorbs a photon of light having a wavelength of 97.2 nm. It then gives off a photon having a wavelength of 486 nm. What is the final state of the hydrogen atom?

4-47. The first four energy levels in arbitrary units of energy of an unknown element are listed below:

n	1	2	3	4	...	∞
E	−20	−9	−4	−3	...	0

A gaseous sample of the element is bombarded by photons with energies in these units of 20, 16, 15, 11, 9, and 1. Which of these photons could potentially be absorbed by an electron in the $n = 1$ state of the element? In each case indicate the transition that would occur.

4-48. Based on the data from the previous problem, if an electron in the element is excited to the $n = 3$ state and then allowed to relax back down to the ground state, what possible transitions could occur? What are the energies corresponding to these possible transitions?

4-49. The energy levels of one-electron ions, such as He^+ and Li^{2+}, are given by the equation

$$E_n = (-2.1799 \text{ aJ})\frac{Z^2}{n^2}; \qquad n = 1, 2, 3, \ldots$$

where Z is the atomic number. Compare the measured ionization energies (Table 4.1) for He^+, Li^{2+}, and Be^{3+} ions with the values calculated from this equation.

4-50. The series in the He^+ spectrum that corresponds to the set of transitions where the electron falls from a higher level to the $n = 4$ state is called the Pickering series, an important series in solar astronomy. Using the equation given in the previous problem, calculate the wavelengths of the first five lines of this series. In what region(s) of the spectrum do they occur? Why do you think this series is important for solar astronomy?

ADDITIONAL PROBLEMS

4-51. Standard household electricity is between 50 and 60 cycles per second in most countries. Is this a wavelength or a frequency?

4-52. What does the term *quantum* in quantum theory refer to?

4-53. Summarize the difference between atomic emission and atomic absorption spectroscopy.

4-54. A student has a solution that she believes may contain cesium ions. Based on the reading of this chapter, what is one test that she might perform to suggest the presence or absence of cesium in her sample?

4-55. What evidence suggests that the electron is particle-like? What evidence suggests that the electron is wavelike?

4-56. Why isn't the wave-particle duality observed for large everyday objects, such as a baseball approaching home plate?

4-57. What would atomic emission spectra look like if the energy levels in atoms were not quantized?

4-58. Explain the physical significance of the state characterized by $n = \infty$ and $E = 0$ for a hydrogen atom.

4-59. Why are the energies of the bound states ($n = 1$ to $n < \infty$) of the hydrogen atom negative?

4-60. A space probe identifies a new element in a sample collected from an asteroid. Successive ionization energies in attojoules per atom for the new element are shown below.

I_1	I_2	I_3	I_4	I_5	I_6	I_7
0.507	1.017	4.108	5.074	6.147	7.903	8.294

To what family of the periodic table does this new ele ment probably belong?

4-61. A metal foil has a threshold frequency of 5.45×10^{14} Hz. Which of the colors of visible light (red, orange, yellow, green, blue, indigo, or violet) have enough energy to eject electrons from this metal? Which do not?

4-62. Data for the photoelectric effect of silver are given in the following table:

Frequency of incident radiation/10^{15} s^{-1}	**Kinetic energy of ejected electrons/10^{-19} J**
2.00	5.90
2.50	9.21
3.00	12.52
3.50	15.84
4.00	19.15

Using these data, find the experimentally determined value of Planck's constant and the threshold frequency for silver.

4-63. In order to resolve an object in the electron microscope, the wavelength of the electrons must be close to the diameter of the object. What kinetic energy must the electrons have in order to resolve a DNA molecule, which is 2.00 nm in diameter? Take the mass of an electron to be 9.11×10^{-31} kg.

4-64. An excited hydrogen atom emits a photon with a frequency of 1.141×10^{14} Hz to reach the $n = 4$ state. From what state did the electron originate?

4-65. In a process called *multiphoton absorption* requiring a very intense light source, two or more photons, the sum of whose energy is equal to the gap between two energy levels in an atom, may be simultaneously absorbed. In a laser absorption experiment, a ground state hydrogen atom undergoes multiphoton absorption and is promoted to an excited state. If the wavelengths of the two simultaneously absorbed photons are both 194.5 nm, to what state is the hydrogen atom promoted?

4-66. Make a graph of frequency as a function of $1/n^2$ for the lines in the Lyman series of atomic hydrogen (Problem 4-41). What is the slope of this line equal to?

4-67. Recall from Chapter 1 that a watt is a unit of energy per unit time, and one watt (W) is equal to one joule per second ($J \cdot s^{-1}$). A 100-W incandescent lightbulb produces about 4% of its energy as visible light. Assuming that the light has an average wavelength of 510 nm, calculate how many such photons are emitted per second by a 100-W incandescent lightbulb.

4-68. At noon on a clear day, sunlight reaches the earth's surface at Madison, Wisconsin, with an average power of approximately 1.0 $kJ \cdot s^{-1} \cdot m^{-2}$. If the sunlight consists of photons with an average wavelength of 510 nm, how many photons strike a 1.0-cm^2 area per second?

4-69. Hydrogen atoms are excited by a laser to the $n = 4$ state and then allowed to emit. What is the maximum number of distinct spectral lines (lines of different wavelengths) that can be observed from this system? Calculate the wavelength of each of these lines and sketch the observed spectrum.

4-70. (*) For a particle moving in a circular orbit, the quantity mvr (mass × velocity × radius of the orbit) is a fundamental quantity called the angular momentum of the particle. Show that Equation 4.8 is equivalent to the condition that the angular momentum of the electron in a hydrogen atom must be an integral multiple of $h/2\pi$ (i.e., that the angular momentum is quantized).

Erwin Schrödinger (1887–1961) was born in Vienna, Austria. He received his Ph.D. in theoretical physics in 1910 from the University of Vienna. He then held a number of positions in Germany and in 1927 succeeded Max Planck at the University of Berlin. Schrödinger left Berlin in 1933 because of his opposition to Hitler and Nazi policies and eventually moved to the University of Graz in Austria in 1936. After the invasion of Austria by Germany, he was forcibly removed from his professorship. He then moved to the Institute of Advanced Studies, which was created for him, at the University College, Dublin, Ireland. He remained there for 17 years and then retired to his native Austria. Schrödinger shared the Nobel Prize for Physics with Paul Dirac in 1933 for the "discovery of new productive forms of atomic theory." Oddly enough, Schrödinger never accepted the probabilistic interpretation of wave functions, and he, along with Einstein, remained skeptical of this interpretation of quantum theory their entire lives. Schrödinger preferred to work alone, and so no school developed around him, as it did for several other developers of quantum theory. His pioneering influential book, *What Is Life?*, stimulated a number of physicists to work on biological problems.

His personal life was at best unconventional, as he apparently had numerous extramarital affairs. He formulated what we now call the Schrödinger equation during a romantic vacation in the Swiss Alps with one of his paramours, whose identity has never been discovered. These stories have been engagingly related by Walter Moore in his biography, *Schrödinger* (Cambridge University Press, 1989).

5. Quantum Theory and Atomic Structure

In the previous chapter, we discussed the early development of quantum theory. We saw that the Bohr theory provided a spectacular explanation of the hydrogen atomic spectrum, but unfortunately could not be extended to atoms with more than one electron. In this chapter we shall discuss a radical departure from the Bohr theory due to Heisenberg and Schrödinger that is applicable to multielectron atoms and molecules. The central result of this theory is an equation called the Schrödinger equation that, as we shall see, can be used to describe the arrangement of the electrons in atoms and molecules and that leads to a complete correlation between atomic structure and the periodic table. This theory ultimately explains most of the chemical properties of the elements as well as the structure and shapes of molecules. Quantum theory is essential to your understanding of what is to come, and this chapter is a pivotal one in your general chemistry course.

5-1. The Schrödinger Equation Is the Central Equation of Quantum Theory

As successful as the Bohr theory was in explaining the spectrum of a hydrogen atom, it could not explain the spectrum of any atom with more than one electron. Furthermore, later work showed that Bohr's description of a well-defined orbit and momentum of the electron in a hydrogen atom was inconsistent with a fundamental principle of nature called the **Heisenberg uncertainty principle**. In the mid-1920s, a young German physicist, Werner Heisenberg (Figure 5.1), showed that it is not possible to measure accurately both the position and the momentum (mv, mass times velocity) of a particle simultaneously. This uncertainty is not due to poor measurement or poor experimental technique, but is a fundamental property of the act of measurement itself.

Let's consider a measurement of the position of an electron. If we wish to locate the electron to within a distance Δx (where Δx denotes the uncertainty in the position x) along the x axis, then we must use electromagnetic radiation with a wavelength at least as small as Δx. In other words, for the electron to be "seen," a photon must interact or collide with it in some way; otherwise the photon will just pass right by the electron and it will appear transparent. According

Figure 5.1 Werner Heisenberg (1901–1976) was a leader in the development of quantum theory in the 1920s and formulated the uncertainty principle that bears his name. Heisenberg was awarded the 1932 Nobel Prize for Physics. However, his role in Nazi Germany is somewhat clouded. While he was in charge of the German project to build an atomic bomb, it is still unclear whether he actually helped or hindered the effort. This prompted author David Cassidy to title his biography of Heisenberg *Uncertainty* (W.H. Freeman, 1993).

to Equation 4.6, the photon has a momentum $p = h / \lambda$, and during the collision some of this momentum will be transferred to the electron. The act of locating the electron, therefore, leads to a change in its momentum.

Heisenberg's analysis showed that it is not possible to determine exactly how much of the momentum of the incoming photon is transferred to the electron when the photon bounces off the electron. So the mere attempt to locate an electron to within a distance Δx causes an uncertainty in its momentum. If we assume that the minimum uncertainty in the position of the electron Δx is on the order of λ, the wavelength associated with the photon, and the uncertainty in the momentum Δp is on the order of h / λ, then we find that $(\Delta x)(\Delta p)$ is approximately greater than or equal to h. A more detailed analysis finds

$$(\Delta x)(\Delta p) \geq \frac{h}{4\pi} \tag{5.1}$$

where h is the Planck constant. Equation 5.1 expresses the Heisenberg uncertainty principle, which imposes a fundamental limitation on the accuracy of determining *simultaneously* the position and the momentum of a particle like the electron due to the act of measurement itself.

Just as in the case of the de Broglie wavelength of a moving particle, the Heisenberg uncertainty principle is of no practical consequence for everyday objects such as airplanes and golf balls. In the macroscopic world, the forces involved in the observation of an object are so small that they have virtually no effect on the position or the momentum of the object. On the other hand, in the world of atomic and subatomic particles, the object being measured and the force acting on it (via the momentum of the photon) are of similar orders of magnitude, and the very act of observation changes the position and momentum of the object. Therefore, Heisenberg's principle has significant consequences when applied to atomic and subatomic particles. The following Example and Practice Problem illustrate the effects of observation on the object being observed.

EXAMPLE 5-1: Suppose that we wish to locate the position of an electron to within 5×10^{-11} m, which is a few percent of the size of an atom. Estimate the uncertainty in the velocity of the electron according to the Heisenberg uncertainty principle. If the electron is moving at a speed of $5.0 \times 10^{6}\ \text{m}\cdot\text{s}^{-1}$, what fraction of this speed does the uncertainty represent?

Solution: We use Equation 5.1 to estimate the uncertainty Δp:

$$
\begin{aligned}
(\Delta p) &\geq \frac{h}{4\pi(\Delta x)} \\
&= \frac{6.626 \times 10^{-34}\ \text{J}\cdot\text{s}}{4\pi(5 \times 10^{-11}\ \text{m})} \\
&= 1 \times 10^{-24}\ \text{kg}\cdot\text{m}\cdot\text{s}^{-1}
\end{aligned}
$$

where we have used the fact that $1\ \text{J} = 1\ \text{kg}\cdot\text{m}^2\cdot\text{s}^{-2}$. Because the mass of an electron is 9.11×10^{-31} kg (inside back cover), the uncertainty in its velocity is given by

$$
\begin{aligned}
\Delta v &\geq \frac{\Delta p}{m} \\
&= \frac{1 \times 10^{-24}\ \text{kg}\cdot\text{m}\cdot\text{s}^{-1}}{9.11 \times 10^{-31}\ \text{kg}} \\
&= 1 \times 10^{6}\ \text{m}\cdot\text{s}^{-1}
\end{aligned}
$$

This result represents an extremely large uncertainty in the velocity, being about 0.4% of the speed of light, or about 20% of the stated speed of the electron. Also note these are the minimum uncertainties; the actual uncertainties may be even greater than this. These figures represent the best measurements possible under the given conditions.

PRACTICE PROBLEM 5-1: Consider a golf ball with a mass of 45.9 grams traveling at 200 kilometers per hour. If an experiment is designed to measure the position of the golf ball at some instant of time with a precision of 1 mm, then what will be the uncertainty in the speed of the golf ball? What percentage of the speed of the golf ball does this uncertainty represent? Compare your results to those obtained in Example 5-1.

Answer: $1 \times 10^{-30}\ \text{m}\cdot\text{s}^{-1}$; $2 \times 10^{-30}\%$. Thus the uncertainty principle is of little consequence for macroscopic objects.

In assuming that the electron in the hydrogen atom is restricted to discrete, precisely defined orbits, the Bohr theory is in conflict with the Heisenberg uncertainty principle in that it attempts to provide a too-detailed picture of the motion of the electron. In 1926, the Austrian physicist Erwin Schrödinger (Frontispiece) presented what has become one of the most famous equations in science: the **Schrödinger equation,** the central equation of quantum theory. This equation is consistent with both the wave nature of particles and the Heisenberg uncertainty principle. Furthermore, unlike the Bohr theory, it predicts the properties of multielectron atoms and molecules. Although the

Schrödinger equation is too complicated to present here, we must discuss some of its consequences.

When we solve the Schrödinger equation for a hydrogen atom, we find that the energy of the electron is restricted to a discrete set of values that are the same as those predicted by the Bohr theory. That is, the energy of the electron is quantized and is restricted to the values given by Equation 4.9. So, the Schrödinger (or quantum) theory and the Bohr theory give the same results for the atomic spectrum of hydrogen.

The Bohr theory and quantum theory differ completely, however, in their descriptions of the location of the electron about the nucleus. Instead of restricting the electron to certain, precisely defined orbits, the Schrödinger wave equation provides one or more functions, called **wave functions**, associated with each allowed energy. Wave functions, customarily denoted by the Greek letter psi, ψ, are functions of the position of the electron. We emphasize this dependence by writing $\psi = \psi(x, y, z)$, where x, y, and z are the coordinates used to denote the position of the electron about the nucleus. The square of the wave function, $\psi^2(x, y, z)$, has the following physical interpretation: the value of the square of the wave function is a **probability density,** in the sense that $\psi^2\, \Delta V$ is the probability that the electron will be found in a small volume element ΔV surrounding the point (x, y, z). This is a remarkable statement. It says that we cannot locate the electron position precisely; we can only assign a *probability* that the electron is in a certain region of space.

Being three-dimensional quantities, the wave functions for the electron in a hydrogen atom, called hydrogen **atomic orbitals**, depend upon three quantum numbers that are commonly denoted by n, l, and m_l. We shall discuss each of these quantum numbers in turn. (Orbitals are just a special type of wave function, and we shall often use the two terms interchangeably.)

The quantum number n is called the **principal quantum number.** It alone determines the energy of the electron in a hydrogen atom (as we shall see in Section 5-5, this is not the case for multielectron atoms). We have already seen that n can take on the values $n = 1, 2, 3$, and so on. These values correlate to the shells we discussed in Section 4-2. For example, orbitals sharing the quantum number $n = 1$ may be referred to as being in the first shell, $n = 2$ as in the second shell, and so forth.

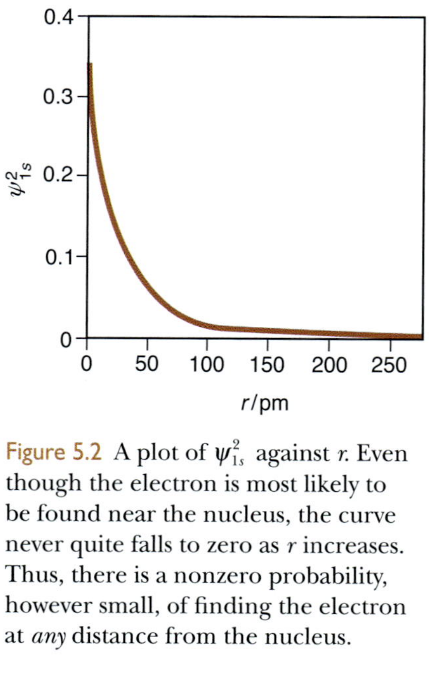

Figure 5.2 A plot of ψ^2_{1s} against r. Even though the electron is most likely to be found near the nucleus, the curve never quite falls to zero as r increases. Thus, there is a nonzero probability, however small, of finding the electron at *any* distance from the nucleus.

When $n = 1$, the energy is the lowest allowed value, which corresponds to the ground state of the hydrogen atom. The wave function that describes the ground state of the hydrogen atom depends only on the distance of the electron from the proton and can be written $\psi(r)$, where r is the distance of the electron from the proton. For reasons that we shall soon see, the ground state wave function of the hydrogen atom is denoted by ψ_{1s} rather than by just ψ_1, and this orbital is called the hydrogen 1s orbital. The probability density, ψ^2_{1s}, is plotted in Figure 5.2. Note that probability density falls off rapidly with distance. Because ψ^2_{1s} depends on only the magnitude of r and not on the direction of r in space, ψ^2_{1s} is said to be **spherically symmetric.**

Several other representations of atomic orbitals give a more informative picture than a simple plot of the square of the wave function versus distance from the nucleus. For example, we can represent the hydrogen 1s orbital by the stippled diagram shown in Figure 5.3a. The number of dots in a volume element ΔV is proportional to the probability of finding the electron in that volume. The

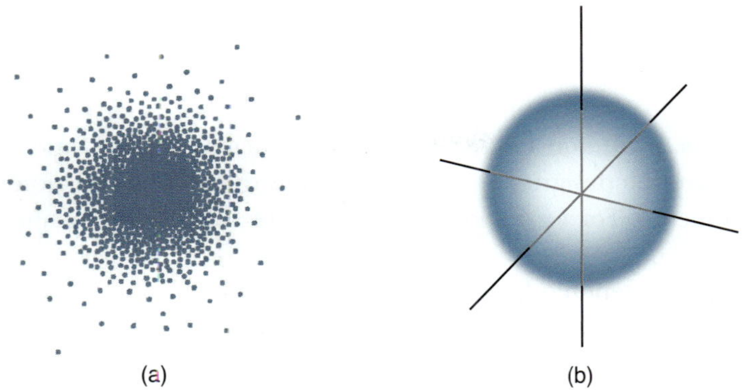

Figure 5.3 Two different representations of a hydrogen 1*s* orbital. (a) The density of the stippled dots in any small region is proportional to the probability of finding the electron in that region. (b) The sphere encloses a volume in which there is a 99% probability of finding the electron. Recall that the 1*s* orbital is spherically symmetric and note that (a) represents a cross section through a sphere.

relation between a plot of ψ_{1s}^2 versus r and the stippled representation of Figure 5.3 is shown in Figure 5.4. Once again, notice that the likelihood of finding an electron decreases as we move away from the nucleus.

Another representation of the 1*s* orbital shows the volume within which the electron has a certain chance of being found. The sphere in Figure 5.3b represents the volume within which there is a 99% chance of finding the electron. The representation in Figure 5.3b has the advantage of portraying clearly the three-dimensional shape of the orbital. Note in Figure 5.2 that ψ_{1s}^2 never becomes identically equal to zero, no matter how large r becomes.

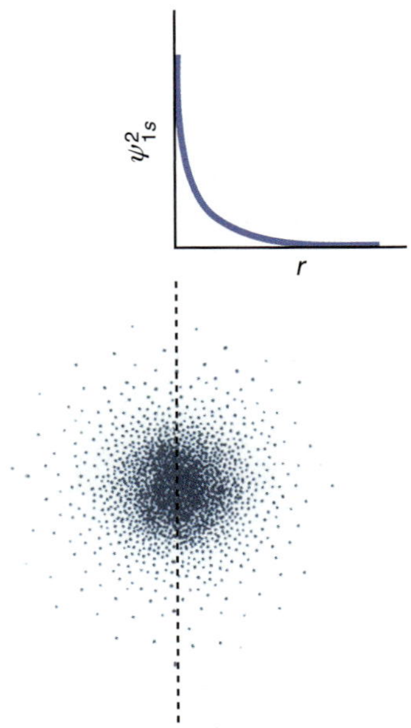

Figure 5.4 The relation between a plot of ψ_{1s}^2 versus r and the stippled representation of a 1*s* orbital. Both show that the probability of finding the electron around some point decreases rapidly with distance from the nucleus.

5-2. The Shape of an Orbital Depends on the Value of the Azimuthal Quantum Number

The principal quantum number, n, specifies the effective size, or the extent, of an orbital. The quantum number l specifies the shape of an orbital. Orbitals with different values of l have different shapes. This second quantum number is called the **azimuthal quantum number,** although we could just as well call it the shape quantum number because, as we shall see, the value of l determines the shape of an orbital. A direct result of solving the Schrödinger equation is that l is restricted to the values 0, 1, . . . , $n-1$. The allowed values of l for a given value of n are summarized in the following table.

n	l
1	0
2	0, 1
3	0, 1, 2
4	0, 1, 2, 3
·	·
·	·
·	·
n	0, 1, 2, 3, . . . , $n-1$

Note that for each value of n, l is an integer that ranges from 0 to $n-1$. For historical reasons, the values of l are designated by letters:

l	0	1	2	3	4	...
designation	*s*	*p*	*d*	*f*	*g*	...

The letters **s**, **p**, **d**, and **f** stand for *s*harp, *p*rincipal, *d*iffuse, and *f*undamental, which are the designations of the series in the atomic emission spectra of the alkali metals. For $l = 4$ and greater, the letters follow in alphabetical order after *f*.

Orbitals are denoted by first writing the numerical value of n (1, 2, 3, . . .) and then following this by the letter designation for the value of l (*s, p, d, f*, . . .). For example, an orbital for which $n = 1$ and $l = 0$ is called a 1*s* orbital, as we have already seen. An orbital for which $n = 3$ and $l = 2$ is called a 3*d* orbital. Table 5.1 lists the orbitals for $n = 1$ through $n = 4$.

TABLE 5.1 The designation of orbitals by letters

n	l	Designation
1	0	1*s*
2	0	2*s*
	1	2*p*
3	0	3*s*
	1	3*p*
	2	3*d*
4	0	4*s*
	1	4*p*
	2	4*d*
	3	4*f*

EXAMPLE 5-2: Why is there no 2*d* orbital listed in Table 5.1?

Solution: When $n = 2$, l can have only the values 0 or 1, because $n - 1 = 1$ when $n = 2$. Thus, when $n = 2$, the maximum possible value of l is 1. Because a *d* orbital has $l = 2$, there is no such orbital as a 2*d* orbital.

PRACTICE PROBLEM 5-2: Why is there no 3*f* orbital listed in Table 5.1?

Answer: A 3*f* orbital would require $n = 3$ and $l = 3$, a state that is not allowed. The maximum value of l is 2 when $n = 3$.

In a hydrogen atom, an electron occupying a 1*s* orbital has an energy that is obtained from Equation 4.9 by setting $n = 1$. When $n = 2$, l can be 0 or 1, so we have two possibilities, a 2*s* ($n = 2$, $l = 0$) or a 2*p* ($n = 2$, $l = 1$) orbital. Both of these orbitals have a principal quantum number n of 2, so an electron in a hydrogen atom described by either of these orbitals has an energy E_2 given by Equation 4.9. In quantum theory, two or more orbitals that have the same energy are said to be **degenerate**. The 2*s* and 2*p* orbitals have different shapes, however, because they are associated with different values of l.

All *s* orbitals are spherically symmetric. In Figure 5.5, ψ^2_{2s} is plotted versus r. The radius of a sphere that encloses a 99% probability of finding the electron in a 2*s* orbital is about 600 pm; the corresponding radius for a 1*s* orbital is about 200 pm (Figure 5.2). A 2*s* orbital has a higher electron density farther from the nucleus than a 1*s* orbital does. Figure 5.5 also shows that the electron probability density is zero over a spherical surface whose radius is about 100 pm. All orbitals except the 1*s* orbital have a surface (or surfaces) on which the electron probability density is zero. Such surfaces are called **nodal surfaces.** Figure 5.5 also illustrates the relation between a plot of ψ^2_{2s} versus r and a stippled diagram representing the 2*s* orbital probability density. The surface of 99% probability for a 2*s* orbital looks the same as that shown in Figure 5.3b for a 1*s* orbital, only larger.

The surface of 99% probability for a 3*s* orbital looks the same as that of a 2*s* orbital, only larger, having a radius of about 1300 pm. Figure 5.6 shows the relationship between a plot of ψ^2_{3s} versus r and a stippled representation of a 3*s* orbital. This plot shows that a 3*s* orbital has two spherical nodal surfaces.

We also have a 2*p* orbital to consider when $n = 2$. The most obvious feature

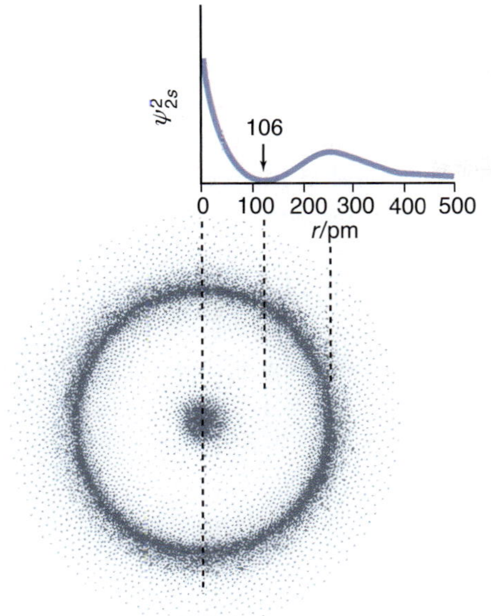

Figure 5.5 The relationship between a plot of ψ^2_{2s} versus r and a stippled representation of a 2*s* orbital. This representation shows that a 2*s* orbital has one (spherical) nodal surface, where the electron density is zero. Remember that *s* orbitals are spherically symmetric and that this is a cross section of a three-dimensional diagram.

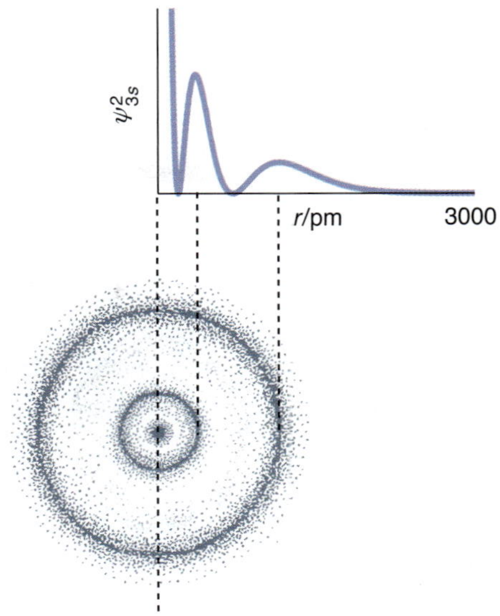

Figure 5.6 The relationship between a plot of ψ^2_{3s} versus r and a stippled representation of a 3*s* orbital. This representation shows that a 3*s* orbital has two (spherical) nodal surfaces, where the electron density is zero. Again, this is a cross section of a three-dimensional diagram.

of a 2*p* orbital is that it is *not* spherically symmetric. The representation shown in Figure 5.7 shows the three-dimensional shape of a 2*p* orbital. When viewed along its long axis, a 2*p* orbital appears to be circular. We say that a 2*p* orbital is **cylindrically symmetric** about its long axis (the *z* axis in Figure 5.7). Note that the *xy* plane that bisects the 2*p* orbital shown in Figure 5.7 is a nodal surface; the 2*p* orbital vanishes everywhere on that surface. Just as all *s* orbitals are spherically symmetric, all *p* orbitals are cylindrically symmetric about their long axis. The most important property of *p* orbitals for our purposes is that they are directed along an axis, as shown in Figure 5.7. Figure 5.8 shows a stippled diagram of a 2*p* orbital. The stippled diagram representation illustrates the probability density of a 2*p* electron.

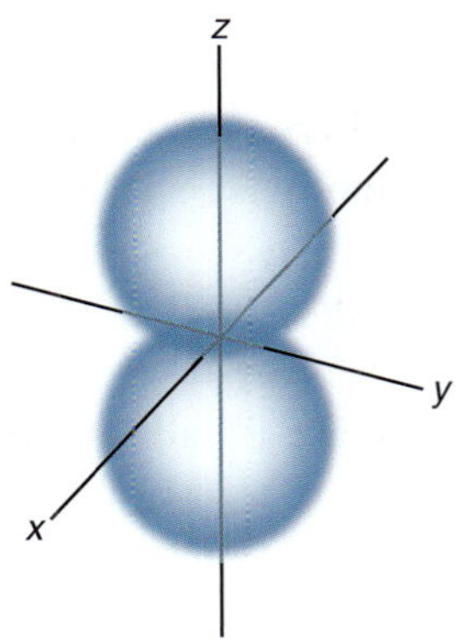

Figure 5.7 The surfaces that enclose a 99% probability of finding a 2*p* electron within them. This representation is a somewhat simplified version of the actual shape, but the important point is that the orbital is directed along an axis. The orbital is said to be cylindrically symmetric along this axis.

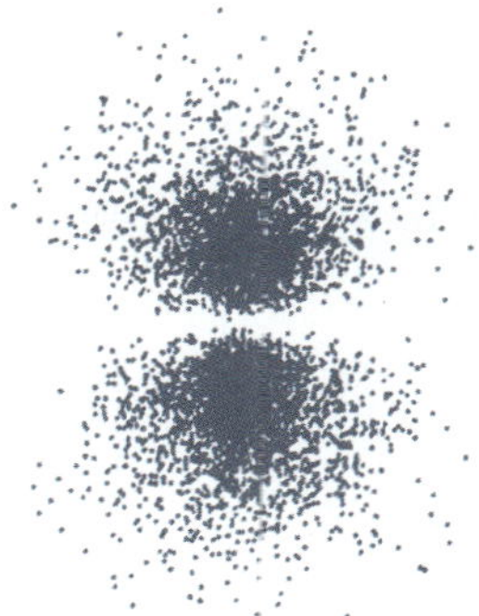

Figure 5.8 A stippled diagram representation of a cross section through the 2*p* orbital shown in Figure 5.7. In this case the diagram represents the probability density of a 2*p* electron and so the shapes in Figures 5.7 and 5.8 are a little different. Both illustrations clearly show the directional character of a 2*p* orbital, however.

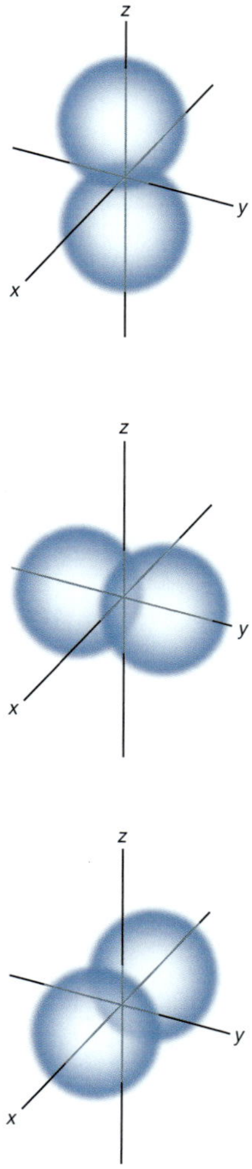

Figure 5.9 The three $2p$ orbitals. They have the same shape but different spatial orientations because they have the same azimuthal quantum number ($l = 1$), but different magnetic quantum numbers. Recall that the shape of an orbital depends on the value of l and that its orientation depends on the value of m_l. The three p orbitals have different m_l values; are directed along the x, y, and z axes; and are designated by p_x, p_y, and p_z, respectively.

5-3. The Spatial Orientation of an Orbital Depends on the Value of the Magnetic Quantum Number

The third quantum number, m_l, called the **magnetic quantum number**, determines the spatial orientation of an orbital. It turns out that the magnetic quantum number can assume integer values between l and $-l$. Namely,

$$l, (l-1), (l-2), \ldots, 0, -1, -2, \ldots, -l$$

or $m_l = 0, \pm1, \pm2, \ldots, \pm l$. The allowed values of m_l depend on the value of l according to the following table:

l	m_l
0	0
1	+1, 0, −1
2	+2, +1, 0, −1, −2
3	+3, +2, +1, 0, −1, −2, −3

For an s orbital, $l = 0$, so the only value that m_l can have is 0. For a p orbital, $l = 1$, so m_l can have the values +1, 0, −1. Table 5.2 summarizes the allowed values of l and m_l for $n = 1$ through $n = 4$.

For a given value of n and l, each unique value of m_l describes a different orbital. Table 5.2 shows that there is only one s orbital for each value of n, three p orbitals for $n \geq 2$, five d orbitals for $n \geq 3$, and seven f orbitals for $n \geq 4$. For example, for a given shell (or value of n) each of the three p orbitals differs by the value of the magnetic quantum number, m_l (= +1, 0, −1). All three orbitals have the same shape because they all have the same value of l (= 1), but they have different orientations in space because they all have different values of m_l. The three $2p$ orbitals are shown in Figure 5.9. One $2p$ orbital is directed along the z axis, as depicted in Figure 5.7. The other two have the same shape as the one directed along the z axis but are directed along the x axis and the y axis, respectively. The p orbitals are designated by p_x, p_y, and p_z, with the subscripts indicating the axis along which the orbital is directed. Note that because an atom may have any orientation in space and because the energies of these three orbitals are degenerate, the specific designations (x, y, or z) assigned to the orbitals or axes are arbitrary.

EXAMPLE 5-3: Without referring to Table 5.2, list all the values of l and m_l that are allowed for $n = 3$.

Solution: When $n = 3$, l can have the values 0, 1, and 2. Thus, for $l = 0$ we have a $3s$ orbital (for the $3s$ orbital, $l = 0$, so m_l must also equal 0). For $l = 1$ we have a set of three $3p$ orbitals (for the $3p$ orbitals, $l = 1$, so m_l can be +1, 0, and −1); and for $l = 2$ we have a set of five $3d$ orbitals (for the $3d$ orbitals, $l = 2$, so m_l can be +2, +1, 0, −1, and −2).

TABLE 5.2 The allowed values of l and m_l for n = 1 through n = 4

n	l	m_l	Orbital	Number of orbitals
1	0	0	1*s*	1
2	0	0	2*s*	1
	1	1, 0, −1	2*p*	3
3	0	0	3*s*	1
	1	1, 0, −1	3*p*	3
	2	2, 1, 0, −1, −2	3*d*	5
4	0	0	4*s*	1
	1	1, 0, −1	4*p*	3
	2	2, 1, 0, −1, −2	4*d*	5
	3	3, 2, 1, 0, −1, −2, −3	4*f*	7

PRACTICE PROBLEM 5-3: Extend Table 5.2 to include the n = 5 case.

Answer:

n	l	m_l	Orbital	Number of orbitals
5	0	0	5*s*	1
	1	1, 0, −1	5*p*	3
	2	2, 1, 0, −1, −2	5*d*	5
	3	3, 2, 1, 0, −1, −2, −3	5*f*	7
	4	4, 3, 2, 1, 0, −1, −2, −3, −4	5*g*	9

As shown in Table 5.2, there are five *d* orbitals with m_l values of +2, +1, 0, −1, and −2. These are given the designations d_{xy}, d_{yz}, d_{xz}, $d_{x^2-y^2}$, and d_{z^2}. The five *3d* orbitals are shown in Figure 5.10. Each of these orbitals has two nodal surfaces. All the nodal surfaces are planar except for d_{z^2}, which has a unique shape defined by two conically shaped nodal surfaces. The *d* orbitals are chemically significant for the transition metal elements and in the bonding of some nonmetals as well. We shall discuss the chemistry of the *d* orbitals in Chapter 26.

5-4. An Electron Has an Intrinsic Spin

For each orbital, the Schrödinger equation yields three quantum numbers: n, l, and m_l. When first presented, this equation explained a great deal of experimental data, but some scattered observations still did not fit into the picture. For example, close examination of some atomic spectral lines shows that they actually consist of two closely spaced lines, such as the two lines at 589.0 nm and 589.6 nm in the atomic spectrum of sodium (Figure 4.30). Even though a fine detail, the splitting of spectral lines was perplexing. In 1926, the German physi-

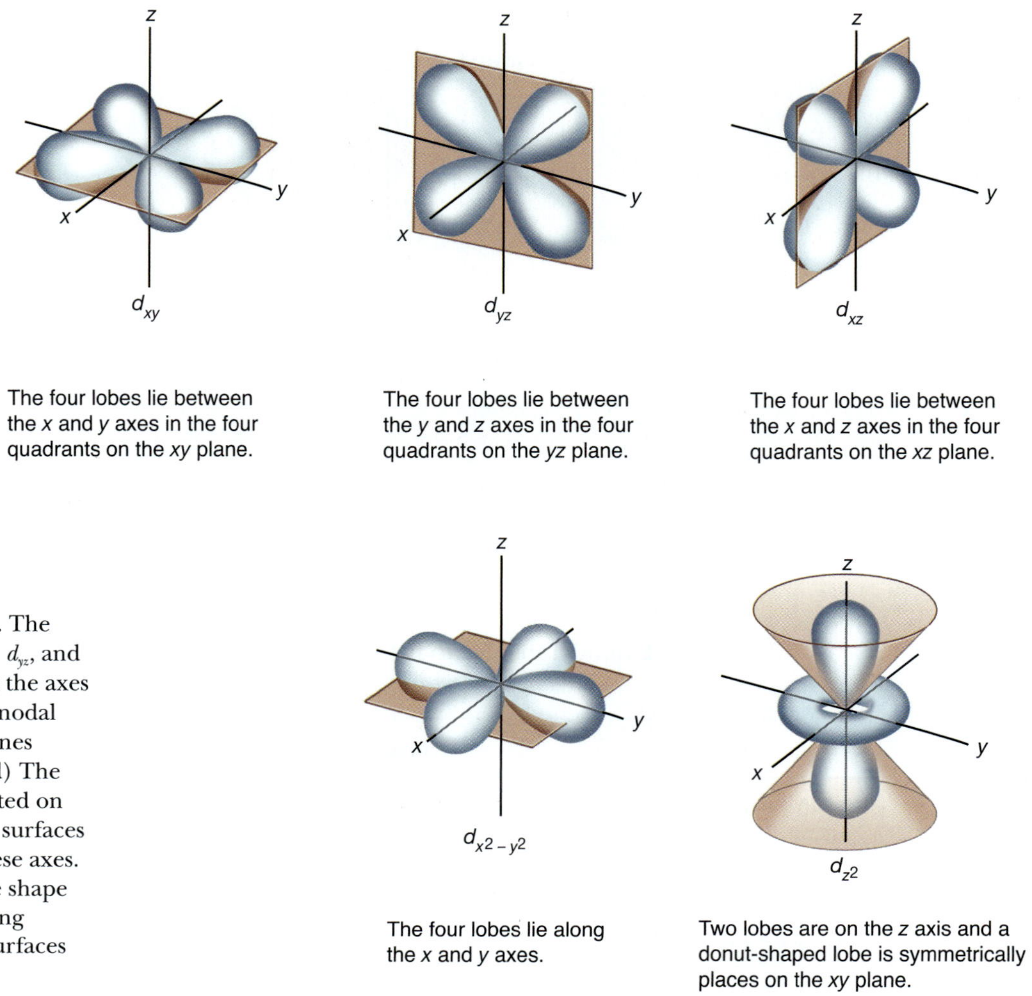

Figure 5.10 The five *3d* orbitals. The first three *d* orbitals (a) d_{xy}, (b) d_{yz}, and (c) d_{xz} all have lobes that bisect the axes for which they are named and nodal surfaces directed along the planes perpendicular to these axes. (d) The $d_{x^2-y^2}$ orbital has four lobes located on the *x* and *y* axes and two nodal surfaces along the planes that bisect these axes. (e) The d_{z^2} orbital has a unique shape consisting of two lobes and a ring defined by two conical nodal surfaces oriented along the *z* axis.

cist Wolfgang Pauli (Figure 5.13) argued that this splitting could be explained if the electron exists in two different states. Shortly after this, two Dutch scientists, George Uhlenbeck and Samuel Goudsmit, identified these two different states with a property called the **intrinsic electron spin,** much as if an electron could literally spin like a top in one of two directions about its axis. Furthermore, a rotating charge creates a magnetic field, so that in a sense electrons act as little magnets, as illustrated below. It should be emphasized that the concept of spin is a classical analogy; the electron as described by quantum theory does not physically spin. Rather, the two states described by the term *spin* arise due to relativistic effects not included in Schrödinger's original wave equation.

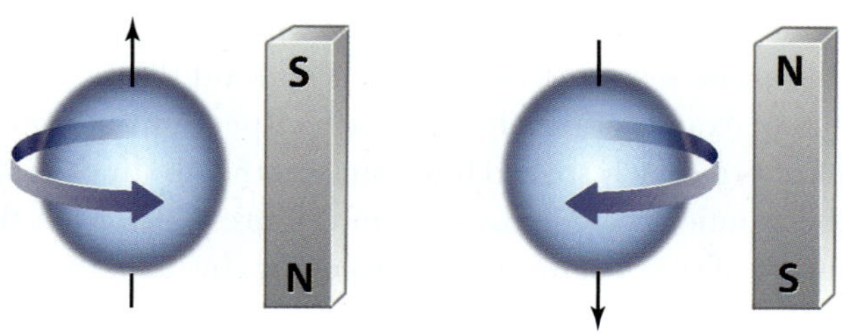

THE STERN-GERLACH EXPERIMENT. It is interesting to note that in 1922, four years before the advent of the Schrödinger equation, two German scientists, Otto Stern and Walther Gerlach, were able to separate a beam of gaseous silver atoms into two distinct beams of equal intensity by passing the beam through an inhomogeneous magnetic field. As we shall see in Section 5-11, silver atoms behave magnetically as though they have one electron. The beam splitting is a consequence of the two different possibilities for the spin of this one electron. The electrons act like tiny magnets and are pulled into the large magnetic field if their spin magnetic moment is aligned with the field ($m_s = +\frac{1}{2}$) or are repelled by the field if their spin magnetic moment is opposed to the field ($m_s = -\frac{1}{2}$).

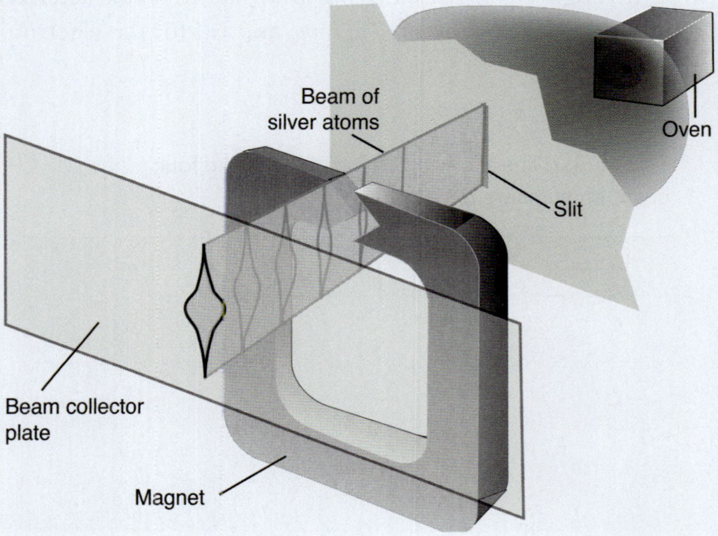

Because half the silver atoms have an electron with $m_s = +\frac{1}{2}$ and half have one with $m_s = -\frac{1}{2}$, the two resulting beams have the same intensity. Below is shown a 1922 postcard from Walther Gerlach to Niels Bohr showing the results of his and Stern's experiment. The left side shows the pattern of the beam of silver atoms without a magnetic field, and the right side shows the pattern with the inhomogeneous magnetic field.

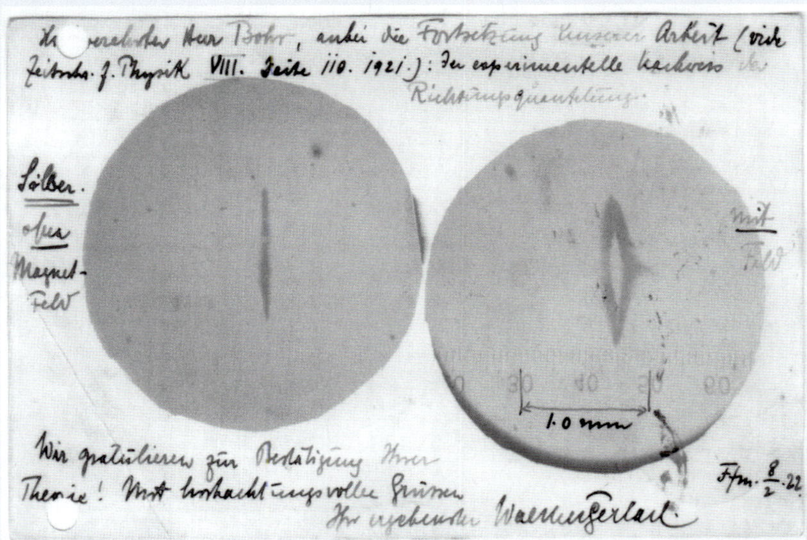

The intrinsic spin of an electron introduces a fourth quantum number, called the **spin quantum number,** denoted by m_s. It designates the spin state of the electron and takes on one of two possible values: $+\frac{1}{2}$ or $-\frac{1}{2}$.

With the introduction of the spin quantum number, it takes a total of four quantum numbers to specify the state of the electron in a hydrogen atom. These quantum numbers are

$$n = 1, 2, 3, \ldots$$
$$l = 0, 1, 2, \ldots, n-1$$
$$m_l = l, (l-1), \ldots, 0, -1, \ldots, -l$$
$$m_s = +\tfrac{1}{2} \text{ or } -\tfrac{1}{2}$$

Table 5.3 summarizes the allowed combinations of the four quantum numbers for $n = 1$ through $n = 3$. An electron in an atom is characterized by the values of the four quantum numbers (n, l, m_l, and m_s) for the electron.

TABLE 5.3 The allowed combinations of the four quantum numbers for n = 1 through n = 3

n	l	m_l	m_s
1	0	0	$+\frac{1}{2}$ or $-\frac{1}{2}$
2	0	0	$+\frac{1}{2}$ or $-\frac{1}{2}$
	1	+1	$+\frac{1}{2}$ or $-\frac{1}{2}$
		0	$+\frac{1}{2}$ or $-\frac{1}{2}$
		−1	$+\frac{1}{2}$ or $-\frac{1}{2}$
3	0	0	$+\frac{1}{2}$ or $-\frac{1}{2}$
	1	+1	$+\frac{1}{2}$ or $-\frac{1}{2}$
		0	$+\frac{1}{2}$ or $-\frac{1}{2}$
		−1	$+\frac{1}{2}$ or $-\frac{1}{2}$
	2	+2	$+\frac{1}{2}$ or $-\frac{1}{2}$
		+1	$+\frac{1}{2}$ or $-\frac{1}{2}$
		0	$+\frac{1}{2}$ or $-\frac{1}{2}$
		−1	$+\frac{1}{2}$ or $-\frac{1}{2}$
		−2	$+\frac{1}{2}$ or $-\frac{1}{2}$

EXAMPLE 5-4: Without reference to Table 5.2 or Table 5.3, deduce the possible sets of the four quantum numbers (n, l, m_l, and m_s) for an electron in an atom when $n = 2$.

Solution: When $n = 2$, l can be 0 or 1. Let's consider the case $l = 0$ first. If $l = 0$, then $m_l = 0$. The spin quantum number can have the value $+\frac{1}{2}$ or $-\frac{1}{2}$, regardless of the values of the other three quantum numbers. We thus have two possible sets of quantum numbers:

n	l	m_l	m_s
2	0	0	$+\frac{1}{2}$
2	0	0	$-\frac{1}{2}$

Now consider the case $n = 2$ and $l = 1$. When $l = 1$, m_l can be +1, 0, or −1 and for each value of m_l, the value of m_s can be $m_s = +\frac{1}{2}$ or $-\frac{1}{2}$, so we get the following six possible sets of quantum numbers:

n	l	m_l	m_s
2	1	+1	$+\frac{1}{2}$
2	1	+1	$-\frac{1}{2}$
2	1	0	$+\frac{1}{2}$
2	1	0	$-\frac{1}{2}$
2	1	−1	$+\frac{1}{2}$
2	1	−1	$-\frac{1}{2}$

Thus, there are eight possible sets of the four quantum numbers when $n = 2$. Two of these sets have $l = 0$ and correspond to two electrons with opposite spins in a 2s orbital, and six of these sets have $l = 1$ and correspond to two electrons with opposite spins in each of the three 2p orbitals ($2p_x$, $2p_y$, and $2p_z$).

PRACTICE PROBLEM 5-4: Which of the following sets of four quantum numbers (n, l, m_l, m_s) are *not* allowed for an electron in an atom?

$(4, 2, 2, +\frac{1}{2})$ $(4, 1, 0, -\frac{1}{2})$ $(4, 2, 3, +\frac{1}{2})$

Answer: $(4, 2, 3, +\frac{1}{2})$ is not allowed because when $l = 2$, the maximum allowed value of m_l is 2.

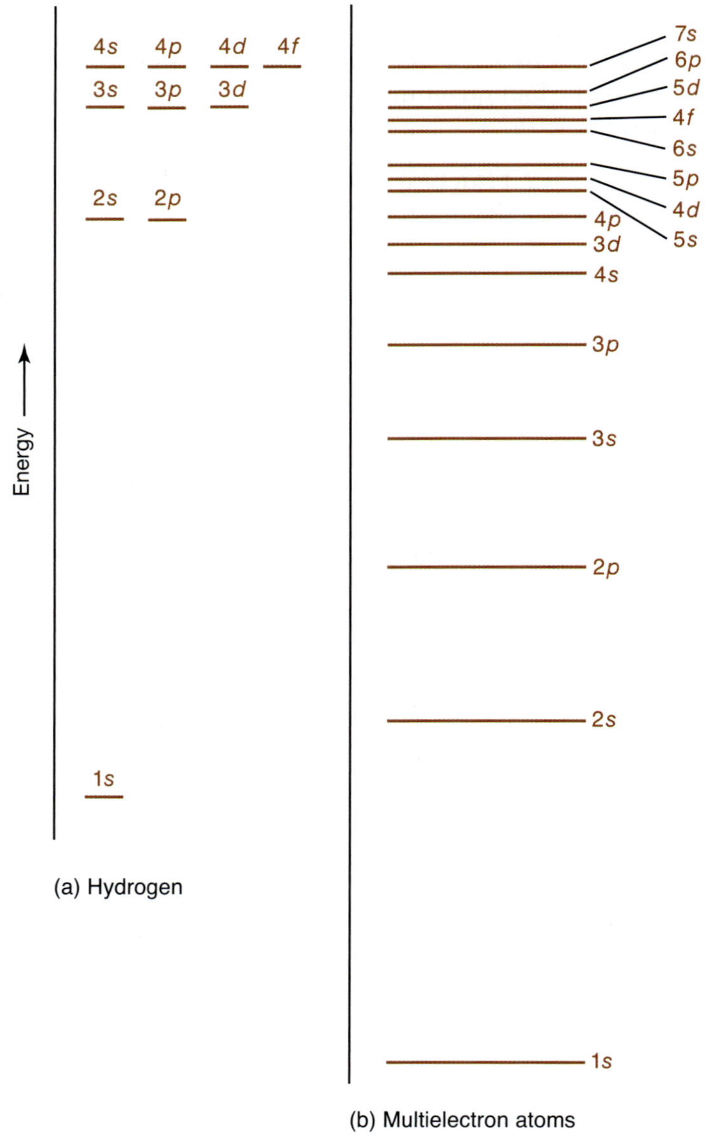

Figure 5.11 The relative energies of atomic orbitals. (a) For hydrogen, the energy depends on only the principal quantum number; thus, orbitals with the same value of n have the same energy. (b) For atoms containing more than one electron, the orbital energies depend on both the principal quantum number n and the azimuthal quantum number l. Thus, orbitals with the same value of n but different values of l have different energies. Although there are exceptions, the relative energy levels shown here apply to the majority of multielectron atoms we shall discuss.

5-5. The Energy States of Atoms with Two or More Electrons Depend on the Values of Both n and l

The hydrogen atom wave functions serve as the prototype wave functions for all other atoms. Equation 4.9 indicates that the energy of an electron in a hydrogen atom depends only on the principal quantum number, n, and not on the other quantum numbers l, m_l, and m_s. Consequently, orbitals having the same value of n, such as the 3s, 3p, and 3d orbitals, in a hydrogen atom are degenerate (have the same energy), as shown in Figure 5.11a. This is *not* the case, however, for atoms with more than one electron. In multielectron atoms there are not only electron-nucleus interactions but also electron-electron interactions. Because of these electron-electron interactions, for multielectron atoms the relationship between energy and quantum numbers cannot be described using Equation 4.9. The electronic energies of multielectron atoms depend in a complicated way on the azimuthal quantum number, l, as well as on the principal quantum number, n. Thus, for example, the 2s and 2p orbitals for atoms other than hydrogen have different energies. The ordering of the orbital energies, shown in Figure 5.11b, is $1s < 2s < 2p < 3s < 3p < 4s < 3d < \ldots$. As n increases, the dependence of the energy on l becomes so pronounced that the energy of the $4s$ orbital is less than that of the $3d$ orbital. As in the case of a hydrogen atom, the orbital energies bunch together as n increases, so this type of reversal becomes even more pronounced at higher energies. Fortunately, there is a simple mnemonic device that helps us remember the order of the orbitals in most atoms (Figure 5.12).

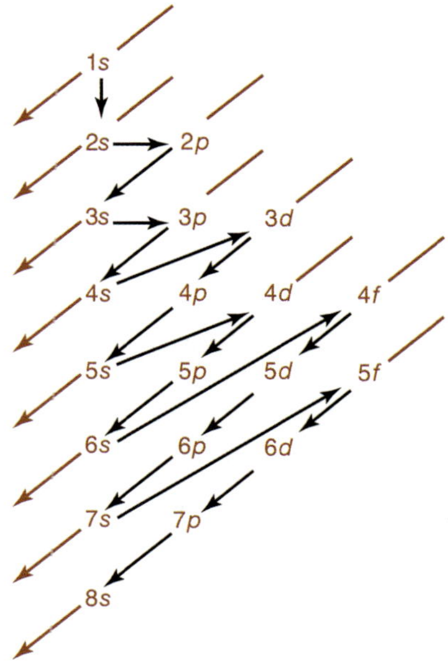

Figure 5.12 Mnemonic device for remembering the order of the orbital energies of neutral atoms containing more than one electron. The correct order of the orbital energies of most atoms beyond hydrogen is obtained by going down a diagonal line as far as possible and then jumping to the top of the next diagonal line.

5-6. The Pauli Exclusion Principle States That No Two Electrons in the Same Atom Can Have the Same Set of Four Quantum Numbers

Before we can correlate electronic structure with the periodic table, we must learn how to assign the electrons to the various orbitals. It was Wolfgang Pauli (Figure 5.13) who in 1926 first determined how to make this assignment. He proposed that no two electrons in the same atom can have the same set of four quantum numbers. This idea is now called the **Pauli exclusion principle.**

Table 5.4 lists the allowed sets of four quantum numbers (n, l, m_l, m_s) for electrons in atoms with $n = 1$ to $n = 4$. Only two combinations are allowed for $n = 1$: $(1, 0, 0, +\frac{1}{2})$ and $(1, 0, 0, -\frac{1}{2})$. Both combinations have $n = 1$ and $l = 0$, so they correspond to two electrons in a $1s$ orbital. The two electrons differ only in their spin quantum numbers. We can represent this pictorially by a line with two vertical arrows:

$$\underset{1s}{\underline{\uparrow\downarrow}}$$

The line represents the orbital, and the two arrows represent the two electrons with different spin quantum numbers. The arrow pointing upward represents an electron with $m_s = +\frac{1}{2}$, and the arrow pointing downward represents one with $m_s = -\frac{1}{2}$. This pictorial representation is so ingrained that chemists often use the terms **spin up** and **spin down** to refer to electrons with $m_s = +\frac{1}{2}$ and $m_s = -\frac{1}{2}$, respectively. When two electrons occupy an orbital, they are said to have their electron spins **paired.** A single electron in an orbital has its electron spin **unpaired** and is said to be an **unpaired electron.** According to the Pauli exclusion principle, the spin quantum numbers of the electrons in a given orbital cannot be the same; if they were, the electrons would have the same set of four quantum numbers. Thus, the representations $\underline{\uparrow\uparrow}$ and $\underline{\downarrow\downarrow}$ are not allowed; they are forbidden.

The $n = 1$ level is complete with two electrons because there are only two possible sets of four quantum numbers with $n = 1$. When $n = 2$, there are two possible values of l, namely, 0 and 1. The $l = 0$ value corresponds to a $2s$ orbital, which can hold two electrons of opposite spins. The $l = 1$ value corresponds to three $2p$ orbitals ($m_l = +1, 0, -1$), each of which can hold two electrons of opposite spins, giving a total of six electrons in the three $2p$ orbitals. The $n = 2$ level, then, can hold a total of eight electrons (two in the $2s$ orbital and six in the three $2p$ orbitals). No more than two electrons can occupy any one orbital:

$$\underset{2s}{\underline{\uparrow\downarrow}} \quad \underbrace{\underline{\uparrow\downarrow}\,\underline{\uparrow\downarrow}\,\underline{\uparrow\downarrow}}_{2p}$$

Figure 5.13 Wolfgang Pauli (1900–1958). Born in Vienna, Pauli mastered Einstein's papers on relativity while in high school and wrote a highly acclaimed monograph on the theory of relativity when he was only 20 years old. He received his Ph.D. from the University of Munich at the age of 21. Pauli won the 1945 Nobel Prize in Physics "for the discovery of the exclusion principle" that bears his name.

TABLE 5.4 The occupation of orbitals according to the Pauli exclusion principle

n	l	m_l	m_s
1 (first shell) (2 electrons)	0 (s subshell) (2 electrons)	0	$+\frac{1}{2}$ or $-\frac{1}{2}$
2 (second shell) (8 electrons)	0 (s subshell) (2 electrons)	0	$+\frac{1}{2}$ or $-\frac{1}{2}$
	1 (p subshell) (6 electrons)	+1	$+\frac{1}{2}$ or $-\frac{1}{2}$
		0	$+\frac{1}{2}$ or $-\frac{1}{2}$
		−1	$+\frac{1}{2}$ or $-\frac{1}{2}$
3 (third shell) (18 electrons)	0 (s subshell) (2 electrons)	0	$+\frac{1}{2}$ or $-\frac{1}{2}$
	1 (p subshell) (6 electrons)	+1	$+\frac{1}{2}$ or $-\frac{1}{2}$
		0	$+\frac{1}{2}$ or $-\frac{1}{2}$
		−1	$+\frac{1}{2}$ or $-\frac{1}{2}$
	2 (d subshell) (10 electrons)	+2	$+\frac{1}{2}$ or $-\frac{1}{2}$
		+1	$+\frac{1}{2}$ or $-\frac{1}{2}$
		0	$+\frac{1}{2}$ or $-\frac{1}{2}$
		−1	$+\frac{1}{2}$ or $-\frac{1}{2}$
		−2	$+\frac{1}{2}$ or $-\frac{1}{2}$
4 (fourth shell) (32 electrons)	0 (s subshell) (2 electrons)	0	$+\frac{1}{2}$ or $-\frac{1}{2}$
	1 (p subshell) (6 electrons)	+1	$+\frac{1}{2}$ or $-\frac{1}{2}$
		0	$+\frac{1}{2}$ or $-\frac{1}{2}$
		−1	$+\frac{1}{2}$ or $-\frac{1}{2}$
	2 (d subshell) (10 electrons)	+2	$+\frac{1}{2}$ or $-\frac{1}{2}$
		+1	$+\frac{1}{2}$ or $-\frac{1}{2}$
		0	$+\frac{1}{2}$ or $-\frac{1}{2}$
		−1	$+\frac{1}{2}$ or $-\frac{1}{2}$
		−2	$+\frac{1}{2}$ or $-\frac{1}{2}$
	3 (f subshell) (14 electrons)	+3	$+\frac{1}{2}$ or $-\frac{1}{2}$
		+2	$+\frac{1}{2}$ or $-\frac{1}{2}$
		+1	$+\frac{1}{2}$ or $-\frac{1}{2}$
		0	$+\frac{1}{2}$ or $-\frac{1}{2}$
		−1	$+\frac{1}{2}$ or $-\frac{1}{2}$
		−2	$+\frac{1}{2}$ or $-\frac{1}{2}$
		−3	$+\frac{1}{2}$ or $-\frac{1}{2}$

As we have already mentioned, the levels designated by n are called shells. The groups of orbitals designated by different l values within these shells are called **subshells.** For $n = 2$, there are two subshells: the s subshell, which consists of one orbital that can contain a maximum of two electrons, and the p subshell that consists of three orbitals and can contain a maximum of six electrons (Table 5.4).

For $n = 3$, we have $3s$, $3p$, and $3d$ subshells. The only new feature here is the d subshell. Because each d subshell contains five d orbitals and each orbital can contain only two electrons with opposite spins, the d subshell can contain up to 10 electrons. Thus, as Table 5.4 shows, the $n = 3$ level, or third shell, can contain up to 18 (= 2 + 6 + 10) electrons. The only new feature for $n = 4$ is the f subshell. Because there are seven f orbitals and each one can contain only two electrons with opposite spins, the f subshell can contain up to 14 electrons, giving a total capacity of 32 (= 2 + 6 + 10 + 14) electrons for the $n = 4$ level.

5-7. Electron Configurations Designate the Occupancy of Electrons in Atomic Orbitals

We are now ready to use Table 5.4 to interpret some of the principal features of the periodic table in terms of electronic structure. Consider first a helium atom with its two electrons. The lowest energy state of a helium atom is achieved by placing both electrons in the $1s$ orbital because this orbital has the lowest energy. Thus, we can represent the **ground electronic state** (the allowed electronic state of lowest energy) in a helium atom by $\underline{\uparrow\downarrow}$ or by $1s^2$. The latter notation is standard. The $1s$ means that we are considering a $1s$ orbital, and the superscript 2 denotes the two electrons in the $1s$ orbital. It is understood that the electrons have different spin quantum numbers, or opposite spins. If we are depicting five electrons in the $3p$ orbitals, we write $3p^5$. The arrangement of electrons in the orbitals is called the **electron configuration** of the atom. Thus, we say that the electron configuration of the ground state of a helium atom is $1s^2$.

Let's go on now and consider the case of a lithium atom with its three electrons. It is not possible to place three electrons in a $1s$ orbital without violating the Pauli exclusion principle because two of the electrons would have the same set of four quantum numbers. The $1s$ orbital is completely filled by two of the electrons, so the third electron must be assigned to the next available orbital, the $2s$ orbital. The electron in the $2s$ orbital can have $m_s = +\frac{1}{2}$ or $-\frac{1}{2}$, so we can represent a lithium atom by

$$\underset{1s}{\underline{\uparrow\downarrow}}\ \underset{2s}{\underline{\uparrow}} \quad \text{or} \quad \underset{1s}{\underline{\uparrow\downarrow}}\ \underset{2s}{\underline{\downarrow}}$$

The direction of the arrow in the $2s$ orbital is not important here, but it is customary to use the spin-up picture on the left. The standard notation is $1s^22s^1$.

In Section 4-2 we used the experimental values of the ionization energies for a lithium atom given in Table 4.1 to argue that lithium can be represented as a helium core with one outer electron. In Table 4.2, we represented a lithium atom by the electron-dot formula [He]• or Li•, which shows one valence electron. Now we see that this same conclusion follows naturally from quantum theory.

The ground state of a beryllium atom ($Z = 4$) is obtained by placing the fourth electron in the $2s$ orbital such that the two electrons there have opposite spins. Pictorially, we have

$$\text{beryllium}\quad \underset{1s}{\underline{\uparrow\downarrow}}\ \ \underset{2s}{\underline{\uparrow\downarrow}}$$

and the standard notation for this ground state electron configuration is $1s^2 2s^2$.

In a boron atom ($Z = 5$) both the $1s$ and $2s$ orbitals are filled, so we must use the $2p$ orbitals. Thus, we have for a boron atom

$$\text{boron}\quad \underset{1s}{\underline{\uparrow\downarrow}}\ \ \underset{2s}{\underline{\uparrow\downarrow}}\ \ \underset{2p}{\underline{\ \uparrow\ }\ \underline{\ \ \ \ }\ \underline{\ \ \ \ }}$$

In the absence of any external electric or magnetic fields, the three $2p$ orbitals are degenerate, so it does not matter into which of the three $2p$ orbitals we place the electron; however, it is customary to represent it pictorially as being in the first available orbital, as we have done here. The ground state electron configuration of a boron atom is written as $1s^2 2s^2 2p^1$.

EXAMPLE 5-5: The ground state electron configuration of ions can be described by the same notation that we have discussed for atoms. What is the ground state electron configuration of a B^+ ion?

Solution: A neutral boron atom has five electrons ($Z = 5$), and a B^+ ion has one fewer electron than a neutral boron atom; therefore, a B^+ ion has four electrons. The ground electronic state is obtained by placing two of these electrons in the $1s$ orbital and two in the $2s$ orbital:

$$B^+\quad \underset{1s}{\underline{\uparrow\downarrow}}\ \ \underset{2s}{\underline{\uparrow\downarrow}}$$

or $1s^2 2s^2$.

Recall from Chapter 2 that two species which have the same number of electrons, and hence the same ground-state electron configuration, are said to be isoelectronic.

PRACTICE PROBLEM 5-5: Give the ground state electron configuration for a F^- ion. What neutral atom is a F^- ion isoelectronic with?

Answer: A fluoride ion has 10 electrons and its ground state electron configuration is $1s^2 2s^2 2p^6$. It is isoelectronic with neon.

5-8. Hund's Rule Is Used to Predict Ground State Electron Configurations

A carbon atom ($Z = 6$) in its ground electronic state has two of its electrons in $2p$ orbitals. We have three distinct choices for the placement of the two $2p$ electrons. The three configurations that obey the Pauli exclusion principle are as follows:

$$(1)\quad \underset{1s}{\underline{\uparrow\downarrow}}\ \ \underset{2s}{\underline{\uparrow\downarrow}}\ \ \underset{2p}{\underline{\uparrow\downarrow}\ \underline{\ \ \ \ }\ \underline{\ \ \ \ }}$$

$$(2)\quad \underset{1s}{\underline{\uparrow\downarrow}}\ \ \underset{2s}{\underline{\uparrow\downarrow}}\ \ \underset{2p}{\underline{\ \uparrow\ }\ \underline{\ \uparrow\ }\ \underline{\ \ \ \ }}$$

$$(3)\quad \underset{1s}{\underline{\uparrow\downarrow}}\ \ \underset{2s}{\underline{\uparrow\downarrow}}\ \ \underset{2p}{\underline{\ \uparrow\ }\ \underline{\ \downarrow\ }\ \underline{\ \ \ \ }}$$

There are small differences in the energies of these three configurations. In configuration (1), both electrons occupy the same *p* orbital and hence are restricted, on the average, to the same region in space. In the other two cases, the two electrons occupy different *p* orbitals, so they are, on average, in different regions of space. Because electrons have the same charge and so repel each other, the placement of the two electrons into different *p* orbitals, and hence different regions of space, minimizes the repulsion between the electrons. Thus, we conclude that configurations (2) and (3) have lower energies and are favored over configuration (1). It has been determined experimentally that the configuration in which the two *p* electrons are placed in different *p* orbitals with **parallel spins** leads to the lowest energy, or ground state, configuration. Therefore, the ground state electron configuration of a carbon atom is

$$\text{carbon} \quad \underset{1s}{\underline{\uparrow\downarrow}} \;\; \underset{2s}{\underline{\uparrow\downarrow}} \;\; \underline{\uparrow}\;\underset{2p}{\underline{\uparrow}}\;\underline{}$$

The standard notation is $1s^22s^22p_x^12p_y^1$. This notation often is condensed to $1s^22s^22p^2$. In both cases the reader is assumed to know that the two $2p$ electrons are unpaired and have parallel spins in the ground state. The choice of the *x* and *y* axes for the two occupied *p* orbitals is arbitrary because the atom may have any orientation in space and all three $2p$ orbitals are degenerate. We could just as well have written $1s^22s^22p_x^12p_z^1$ or $1s^22s^22p_y^12p_z^1$.

The arguments given above can be generalized to give what is called **Hund's rule**, named after its originator, the German scientist Friedrich Hund: For any set of orbitals of the same energy, that is, for any subshell, the ground state electron configuration is obtained by placing the electrons in different orbitals of this set with parallel spins. No orbital in the subshell contains two electrons until each orbital contains one electron. Using Hund's rule, we write for a nitrogen atom ($Z = 7$)

$$\text{nitrogen} \quad \underset{1s}{\underline{\uparrow\downarrow}} \;\; \underset{2s}{\underline{\uparrow\downarrow}} \;\; \underline{\uparrow}\;\underset{2p}{\underline{\uparrow}}\;\underline{\uparrow}$$

The standard notation is $1s^22s^22p_x^12p_y^12p_z^1$ or in condensed notation, $1s^22s^22p^3$. Again, the three $2p$ electrons are understood to have parallel spins in the ground state.

For an oxygen atom ($Z = 8$), we begin to pair up the *p* electrons and obtain

$$\text{oxygen} \quad \underset{1s}{\underline{\uparrow\downarrow}} \;\; \underset{2s}{\underline{\uparrow\downarrow}} \;\; \underline{\uparrow\downarrow}\;\underset{2p}{\underline{\uparrow}}\;\underline{\uparrow}$$

or $1s^22s^22p_x^22p_y^12p_z^1$, or simply $1s^22s^22p^4$. Once again, it does not matter into which $2p$ orbital we place the paired electrons. The electron configurations $1s^22s^22p_x^12p_y^22p_z^1$ and $1s^22s^22p_x^12p_y^12p_z^2$ are equivalent to each other and to $1s^22s^22p_x^22p_y^12p_z^1$.

EXAMPLE 5-6: What is the ground state electron configuration of an O^+ ion?

Solution: A neutral oxygen atom has eight electrons, and so an O^+ ion has seven electrons (for oxygen, $Z = 8$; for O^+ we have $8 - 1 = 7$ electrons). Four of the electrons are in the $1s$ and $2s$ orbitals. The other three are in the $2p$

orbitals. According to Hund's rule, the three $2p$ electrons are in different $2p$ orbitals and all have the same spin. The ground state electron configuration is $1s^2 2s^2 2p_x^1 2p_y^1 2p_z^1$, or simply $1s^2 2s^2 2p^3$.

PRACTICE PROBLEM 5-6: What is the ground state electron configuration of an O^{2-} ion?

Answer: An oxide ion has 10 electrons and so its ground state electron configuration is $1s^2 2s^2 2p^6$.

TABLE 5.5 Ground state electron configurations of the first 10 elements

Element	Ground state electron configuration
hydrogen	$1s^1$
helium	$1s^2$
lithium	$1s^2 2s^1$
beryllium	$1s^2 2s^2$
boron	$1s^2 2s^2 2p^1$
carbon	$1s^2 2s^2 2p^2$
nitrogen	$1s^2 2s^2 2p^3$
oxygen	$1s^2 2s^2 2p^4$
fluorine	$1s^2 2s^2 2p^5$
neon	$1s^2 2s^2 2p^6$

The ground state electron configurations of the first ten elements are shown in Table 5.5. Note that a helium atom has a filled $n = 1$ shell and that a neon atom has a filled $n = 2$ shell. The ground state electron configuration is obtained by filling up the atomic orbitals of lowest energy in accord with the Pauli exclusion principle and Hund's rule.

5-9. When an Atom Absorbs Electromagnetic Radiation, Electrons Are Promoted to Excited States

We saw in Section 4-9 that an atom can absorb electromagnetic radiation. In this process an electron is promoted to an orbital of higher energy, and the atom is said to be in an excited state. For example, a lithium atom absorbs electromagnetic radiation of wavelength 671 nm and undergoes the electronic transition

$$\underset{\text{ground state}}{\mathrm{Li}(1s^2 2s^1)} + \underset{\text{photon}}{h\nu} \rightarrow \underset{\text{excited state}}{\mathrm{Li}^*(1s^2 2p^1)}$$

where $h\nu$ represents the energy of the absorbed photon. We see that the electron in the $2s$ orbital is promoted to a $2p$ orbital in the process (Figure 5.14). The resulting lithium atom is in an excited state (as denoted by the asterisk), and its excited state electron configuration is $1s^2 2p^1$. The first excited state is obtained by promoting the electron of highest energy in the ground state to the next available orbital. For our purposes in this chapter, we are interested primarily in ground electronic states, but we should realize that the ground state is just the lowest of a set of allowed atomic energy states.

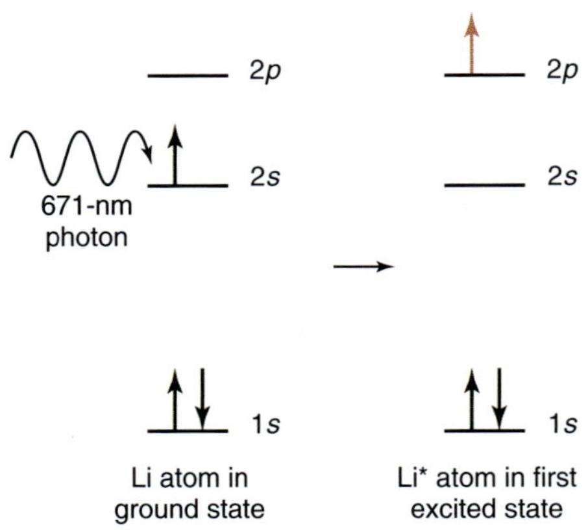

Figure 5.14 A lithium atom in the ground state absorbs a 671-nm photon, promoting it to the first excited state. The difference in energy between the $2s$ and $2p$ orbitals in lithium must equal the energy of the photon absorbed, 0.296 aJ, as determined from its wavelength using the equation $E = hc/\lambda$.

EXAMPLE 5-7: What is the electron configuration of the first excited state of a neon atom?

Solution: The ground state electron configuration of a neon atom is $1s^2 2s^2 2p^6$. The electron of highest energy is any one of the $2p$ electrons. The next available orbital is the $3s$ orbital, so

$$\mathrm{Ne}^* \text{ (first excited state): } 1s^2 2s^2 2p^5 3s^1$$

PRACTICE PROBLEM 5-7: What is the electronic configuration of the first excited state of an O^{2-} ion?

Answer: First excited state is $1s^2 2s^2 2p^5 3s^1$.

The differences in energy between the various states of an atom yield the atomic absorption and emission spectra discussed in Section 4-4. Whereas Bohr's equation is valid only for single-electron atoms, Schrödinger's equation may be solved using a computer to find the energies of both the ground and excited state orbitals of multielectron elements. The differences in energy between these orbitals predict the atomic spectrum of that element.

TABLE 5.6 Ground state electron configurations of third-row elements

Element	Ground state electron configuration	Abbreviated form of ground state electron configuration
sodium	$1s^22s^22p^63s^1$	$[Ne]3s^1$
magnesium	$1s^22s^22p^63s^2$	$[Ne]3s^2$
aluminum	$1s^22s^22p^63s^23p^1$	$[Ne]3s^23p^1$
silicon	$1s^22s^22p^63s^23p^2$	$[Ne]3s^23p^2$
phosphorus	$1s^22s^22p^63s^23p^3$	$[Ne]3s^23p^3$
sulfur	$1s^22s^22p^63s^23p^4$	$[Ne]3s^23p^4$
chlorine	$1s^22s^22p^63s^23p^5$	$[Ne]3s^23p^5$
argon	$1s^22s^22p^63s^23p^6$	$[Ne]3s^23p^6$ or [Ar]

5-10. Elements in the Same Column of the Periodic Table Have Similar Valence-Electron Configurations

According to either Figure 5.11b or Figure 5.12, after neon we use the 3*s* and 3*p* orbitals to obtain the electron configurations of the atoms in the next row of the periodic table. In this series of elements we are filling up the 3*s* and 3*p* orbitals outside a neon inner-shell or core structure. It is common practice to use the abbreviated form of the electron configurations shown in the right-hand column of Table 5.6. Notice how nicely the ground state electron configurations of the elements correlate with the Lewis electron-dot formulas (Table 4.2 and Table 5.7). In each case, the number of dots displayed in the Lewis electron-

TABLE 5.7 Comparison of Lewis electron-dot formulas and ground state electron configurations

Element	Lewis electron-dot formula	Ground state electron configuration
carbon	·Ċ·	$[He]2s^22p^2$
fluorine	:F̈: (seven dots)	$[He]2s^22p^5$
neon	:N̈e: (eight dots)	$[He]2s^22p^6$ or [Ne]
sodium	Na·	$[Ne]3s^1$
chlorine	:C̈l: (seven dots)	$[Ne]3s^23p^5$

dot formula is the same as the total number of electrons in the outer (valence) shell, as indicated in the electron configuration.

Recall from Section 4-2 that electrons in the outermost occupied shell (highest *n*-value shell with an electron in it) of a neutral atom or a monatomic ion of a main-group element are called **valence electrons**. A main-group atomic cation for which the outermost occupied shell is completely filled (e.g., Na^+ with a neon configuration) has no valence electrons. A main-group atomic anion for which the outermost *ns* and *np* subshells are filled completely (e.g., F^- with a neon configuration) has eight (ns^2np^6 and $2 + 6 = 8$) valence electrons.

EXAMPLE 5-8: How many valence electrons are there in (a) an O^{2-} ion and (b) a Ne^+ ion?

Solution: (a) The ground state electron configuration of an O^{2-} ion is $1s^22s^22p^6$, so there are eight valence electrons in an O^{2-} ion. The Lewis electron-dot formula of an O^{2-} ion is $[:\ddot{\underset{..}{O}}:]^{2-}$.
(b) The ground state electron configuration of a Ne^+ ion is $1s^22s^22p^5$, so there are seven valence electrons in a Ne^+ ion. The Lewis electron-dot formula of a Ne^+ ion is $[:\ddot{\underset{..}{Ne}}\cdot]^{+}$.

PRACTICE PROBLEM 5-8: How many valence electrons are there in (a) a Ne atom, (b) an Al^{3+} ion, (c) a Mg^{2+} ion, (d) a P atom, and (e) a Cl^- ion?

Answer: (a) 8; (b) 0; (c) 0; (d) 5; (e) 8

If we compare the electron configurations of a sodium atom through an argon atom (Table 5.6) with those of a lithium atom through a neon atom (Table 5.5), we see why these two series of elements have a periodic correlation in chemical properties, as described in Chapter 3. Their valence electron configurations range from ns^1 to ns^2np^6 ($n = 2$ and $n = 3$, respectively) in the same manner. That is, elements in the same column of the periodic table such as fluorine and chlorine have valence electron configurations of the same form:

$$\text{fluorine: [He]}2s^22p^5 \qquad \text{chlorine: [Ne]}3s^23p^5$$

As we shall see when we discuss bonding, the outermost or valence electrons determine how an atom behaves in a chemical reaction.

Figure 5.11b shows that the 4*s* orbital follows the 3*p* orbital. Thus, the electron configurations of the next two elements after argon are

$$\text{potassium: [Ar]}4s^1 \qquad \text{calcium: [Ar]}4s^2$$

where [Ar] denotes the ground state electron configuration of an argon atom. If we consider the ground state electron configurations of a lithium, sodium, and potassium atom, we see why they fall naturally into the same column of the periodic table. Each has an ns^1 configuration outside a noble-gas configuration, that is,

1	2	3	4	5	6	7	8	9	10	11	12	13	14	15	16	17	18
ns^1																	ns^2np^6
1 **H** $1s^1$	ns^2											ns^2np^1	ns^2np^2	ns^2np^3	ns^2np^4	ns^2np^5	2 **He** $1s^2$
3 **Li** $2s^1$	4 **Be** $2s^2$											5 **B** $2s^22p^1$	6 **C** $2s^22p^2$	7 **N** $2s^22p^3$	8 **O** $2s^22p^4$	9 **F** $2s^22p^5$	10 **Ne** $2s^22p^6$
11 **Na** $3s^1$	12 **Mg** $3s^2$											13 **Al** $3s^23p^1$	14 **Si** $3s^23p^2$	15 **P** $3s^23p^3$	16 **S** $3s^23p^4$	17 **Cl** $3s^23p^5$	18 **Ar** $3s^23p^6$
19 **K** $4s^1$	20 **Ca** $4s^2$	21 **Sc** $4s^23d^1$	22 **Ti** $4s^23d^2$	23 **V** $4s^23d^3$	24 **Cr** $4s^13d^5$	25 **Mn** $4s^23d^5$	26 **Fe** $4s^23d^6$	27 **Co** $4s^23d^7$	28 **Ni** $4s^23d^8$	29 **Cu** $4s^13d^{10}$	30 **Zn** $4s^23d^{10}$	31 **Ga** $4s^24p^1$	32 **Ge** $4s^24p^2$	33 **As** $4s^24p^3$	34 **Se** $4s^24p^4$	35 **Br** $4s^24p^5$	36 **Kr** $4s^24p^6$
37 **Rb** $5s^1$	38 **Sr** $5s^2$	39 **Y** $5s^24d^1$	40 **Zr** $5s^24d^2$	41 **Nb** $5s^14d^4$	42 **Mo** $5s^14d^5$	43 **Tc** $5s^24d^5$	44 **Ru** $5s^14d^7$	45 **Rh** $5s^14d^8$	46 **Pd** $4d^{10}$	47 **Ag** $5s^14d^{10}$	48 **Cd** $5s^24d^{10}$	49 **In** $5s^25p^1$	50 **Sn** $5s^25p^2$	51 **Sb** $5s^25p^3$	52 **Te** $5s^25p^4$	53 **I** $5s^25p^5$	54 **Xe** $5s^25p^6$
55 **Cs** $6s^1$	56 **Ba** $6s^2$	57–71	72 **Hf** $6s^25d^2$	73 **Ta** $6s^25d^3$	74 **W** $6s^25d^4$	75 **Re** $6s^25d^5$	76 **Os** $6s^25d^6$	77 **Ir** $6s^25d^7$	78 **Pt** $6s^15d^9$	79 **Au** $6s^15d^{10}$	80 **Hg** $6s^25d^{10}$	81 **Tl** $6s^26p^1$	82 **Pb** $6s^26p^2$	83 **Bi** $6s^26p^3$	84 **Po** $6s^26p^4$	85 **At** $6s^26p^5$	86 **Rn** $6s^26p^6$
87 **Fr** $7s^1$	88 **Ra** $7s^2$	89–103	104 **Rf** $7s^26d^2$	105 **Db** $7s^26d^3$	106 **Sg** $7s^26d^4$	107 **Bh** $7s^26d^5$	108 **Hs** $7s^26d^6$	109 **Mt** $7s^26d^7$	110 **Ds** $7s^16d^9$	111 **Rg** $7s^16d^{10}$	112 **Cn** $7s^2 6d^{10}$	113 **Uut**	114 **Uuq**	115 **Uup**	116 **Uuh**	117 **Uus**	118 **Uuo**

Lanthanide series

57 **La** $6s^25d^1$	58 **Ce** $6s^25d^14f^1$	59 **Pr** $6s^24f^3$	60 **Nd** $6s^24f^4$	61 **Pm** $6s^24f^5$	62 **Sm** $6s^24f^6$	63 **Eu** $6s^24f^7$	64 **Gd** $6s^25d^14f^7$	65 **Tb** $6s^24f^9$	66 **Dy** $6s^24f^{10}$	67 **Ho** $6s^24f^{11}$	68 **Er** $6s^24f^{12}$	69 **Tm** $6s^24f^{13}$	70 **Yb** $6s^24f^{14}$	71 **Lu** $6s^25d^14f^{14}$

Actinide series

89 **Ac** $7s^26d^1$	90 **Th** $7s^26d^2$	91 **Pa** $7s^26d^15f^2$	92 **U** $7s^26d^15f^3$	93 **Np** $7s^26d^15f^4$	94 **Pu** $7s^25f^6$	95 **Am** $7s^25f^7$	96 **Cm** $7s^26d^15f^7$	97 **Bk** $7s^25f^9$	98 **Cf** $7s^25f^{10}$	99 **Es** $7s^25f^{11}$	100 **Fm** $7s^25f^{12}$	101 **Md** $7s^25f^{13}$	102 **No** $7s^25f^{14}$	103 **Lr** $7s^26d^15f^{14}$

Figure 5.15 A periodic table showing the ground state electron configurations of the outer electrons of the elements. The general valence-electron configurations of the main-group elements are given above each group. Thus, the alkali metals have the valence electron configuration ns^1, the alkaline-earth metals ns^2, and so on.

lithium: $[\text{He}]2s^1$ sodium: $[\text{Ne}]3s^1$ potassium: $[\text{Ar}]4s^1$

Also, the principal quantum number of the outer *s* orbital coincides with the number of the row of the periodic table (Figure 5.15). Each row starts off with an alkali metal, whose electron configuration is [noble gas]ns^1. For example, cesium, which follows xenon and begins the sixth row of the table, has the electron configuration

cesium: $[\text{Xe}]6s^1$

The same type of observation can be used to explain why the alkaline-earth metals undergo similar chemical reactions. The electron configuration of an alkaline-earth metal is [noble gas]ns^2 (Table 5.8).

One of the most amazing consequences of quantum theory is that Schrödinger's mathematical formulation not only predicts the energy level transi-

TABLE 5.8 Ground state electron configurations of the alkaline-earth metals

Element	Ground state electron configuration
beryllium	$[\text{He}]2s^2$
magnesium	$[\text{Ne}]3s^2$
calcium	$[\text{Ar}]4s^2$
strontium	$[\text{Kr}]5s^2$
barium	$[\text{Xe}]6s^2$
radium	$[\text{Rn}]7s^2$

tions that correlate with atomic spectra but also predicts the periodic similarity of the elements and provides a theoretical basis for the entire periodic table.

TABLE 5.9 Ground state electron configurations of the $3d$ transition metals. Exceptions to the normal filling order are shown in red.

Element	Ground state electron configuration
scandium	$[Ar]4s^2 3d^1$
titanium	$[Ar]4s^2 3d^2$
vanadium	$[Ar]4s^2 3d^3$
chromium	$[Ar]4s^1 3d^5$
manganese	$[Ar]4s^2 3d^5$
iron	$[Ar]4s^2 3d^6$
cobalt	$[Ar]4s^2 3d^7$
nickel	$[Ar]4s^2 3d^8$
copper	$[Ar]4s^1 3d^{10}$
zinc	$[Ar]4s^2 3d^{10}$

5-11. The Chemistry of Transition Metal Elements Depends upon Their *d* Orbital Electrons

Once we reach calcium ($Z = 20$), the $4s$ orbital is completely filled. Figure 5.11b shows that the next available orbitals are the five $3d$ orbitals. Each of these orbitals can be occupied by two electrons of opposite spin, giving a maximum of 10 electrons in all. Note that this number corresponds perfectly with the 10 transition metals that occur between calcium and gallium in the periodic table. Thus, in the first set of transition metals, we see the sequential filling of the five $3d$ orbitals. Because of this, the first set of transition metals is called the **3*d* transition metal series**. You may think that the ground state electron configurations of these 10 elements go smoothly from [Ar] $4s^2 3d^1$ to [Ar] $4s^2 3d^{10}$, but this is not so. The actual ground state electron configurations of the $3d$ transition metals are as shown in Table 5.9. We see that chromium and copper have only one $4s$ electron. In each case an electron has been taken from the $4s$ orbital in order to either half-fill or completely fill all the $3d$ orbitals. This filling pattern results because an extra stability is realized by the valence electron configurations

$$\text{Cr: } \underset{4s}{\underline{\uparrow}} \quad \underset{3d}{\underline{\uparrow}\ \underline{\uparrow}\ \underline{\uparrow}\ \underline{\uparrow}\ \underline{\uparrow}} \qquad \text{and} \qquad \text{Cu: } \underset{4s}{\underline{\uparrow}} \quad \underset{3d}{\underline{\uparrow\downarrow}\ \underline{\uparrow\downarrow}\ \underline{\uparrow\downarrow}\ \underline{\uparrow\downarrow}\ \underline{\uparrow\downarrow}}$$

relative to the *incorrect* $4s^2 3d^4$ and $4s^2 3d^9$ ground state valence electron configurations for the neutral gaseous atoms of these elements. It so happens that the energies of the electrons in the $4s$ and $3d$ orbitals are very similar (Figure 5.11b), and deviations from the regular filling order of these two orbitals—as well as for the $5s$ and $4d$ and $6s$ and $5d$ orbitals—are found, especially when such deviations result in a half or completely filled d subshell (Figure 5.15). If you look carefully at Figure 5.15, you'll see that even more irregularities show up in the filling order of the $4d$ series—at niobium ($5s^1 4d^4$), molybdenum ($5s^1 4d^5$), ruthenium ($5s^1 4d^7$), rhodium ($5s^1 4d^8$), palladium ($4d^{10}$), and silver ($5s^1 4d^{10}$)—than in the $3d$ series. The unpaired $5s$ electron in silver explains the magnetic properties first observed by Stern and Gerlach, as described in the sidebox in Section 5-4. In the $5d$ series there are only three irregularities in the filling order.

According to Figure 5.11b, after the $3d$ orbitals are filled, the next available orbitals are the $4p$ orbitals, which fill up as shown in Table 5.10. For these six elements, the $4p$ orbitals are sequentially filled, and these elements fall naturally into the fourth row of the periodic table under the sequence of elements boron through neon and aluminum through argon, which fill the $2p$ and $3p$ orbitals, respectively (Figure 5.15). When writing electron configurations, we generally list orbitals in order of increasing energy (filling order). For example, we write $5s$ before $4d$ in the electronic configuration of a silver atom, $[Kr]5s^1 4d^{10}$.

TABLE 5.10 Ground state electron configurations of the p-block fourth-row elements

Element	Ground state electron configuration
gallium	$[Ar]4s^2 3d^{10} 4p^1$
germanium	$[Ar]4s^2 3d^{10} 4p^2$
arsenic	$[Ar]4s^2 3d^{10} 4p^3$
selenium	$[Ar]4s^2 3d^{10} 4p^4$
bromine	$[Ar]4s^2 3d^{10} 4p^5$
krypton	$[Ar]4s^2 3d^{10} 4p^6$ or [Kr]

A krypton atom, like all the noble-gas atoms, has a completely filled set of p orbitals where the principal quantum number corresponds to the row in which it is located in the periodic table. Figure 5.11b shows that the $5s$ orbital follows the $4p$ orbital, so we are back to the left-hand column of the periodic table with the alkali metal rubidium followed by the alkaline-earth metal strontium. The atoms of these two metals have the ground state electron configurations $[Kr]5s^1$

and [Kr]$5s^2$, respectively. The next available orbitals are the $4d$ orbitals, which lead to the **$4d$ transition metal series,** yttrium through cadmium. The ground state electron configurations of the outer electrons of the atoms of these 10 metals (Figure 5.15) show irregularities like those found in the $3d$ transition metal series. After a cadmium atom, [Kr]$5s^2 4d^{10}$, the $5p$ orbitals are filled to give the six elements indium through the noble gas xenon, which has the ground state electron configuration [Kr]$5s^2 4d^{10} 5p^6$ or simply [Xe]. As before, the completion of a set of p orbitals leads to a noble gas located in the far right-hand column of the periodic table. The two reactive metals cesium and barium follow xenon by filling the $6s$ orbital to give atomic ground state electron configurations [Xe]$6s^1$ and [Xe]$6s^2$, respectively.

After filling the $6s$ orbital, we begin to fill the seven $4f$ orbitals. Because each of these seven orbitals can hold two electrons of opposite spin, we expect that the atoms of the next 14 elements should involve the filling of the $4f$ orbitals. The elements lanthanum ($Z = 57$) through ytterbium ($Z = 70$) constitute what is called the **lanthanide series** because the series begins with the element lanthanum in the periodic table. Figure 5.15 shows that, except for a few irregularities like those found for the d transition metal series, the atoms of the lanthanides involve a sequential filling of the seven $4f$ orbitals. The chemistry of these elements is so similar that for many years it proved quite difficult to separate them from the naturally occurring mixtures. However, separations are now achieved using chromatographic and other methods.

There are seven f orbitals, so you might expect a total of 14 elements to occupy the lanthanide and actinide series. However, the current IUPAC version of the periodic table (Figure 5.15) places 15 members in each series because of both chemical and electronic structure considerations. In our discussion of orbital filling we take lutetium and lawrencium as beginning the $5d$ and $6d$ transition metal series, but we could just as well have chosen lanthanum and actinium. Such differences are of no concern to us here.

If we consider that the lanthanides differ only in the number of electrons in the $4f$ subshells, with the $6s$ and $5p$ subshells already filled, the reason for their chemical similarity becomes clear. According to quantum theory, the average distance of an electron from a nucleus depends on both the principal quantum number n and the azimuthal quantum number l. Although the average distance of an electron from the nucleus increases with n, it increases less as l increases than as n increases. For this reason, the average distance of $4f$ ($n = 4$, $l = 3$) electrons from the nucleus is *less* than that of $6s$ ($n = 6$, $l = 0$) or $5p$ ($n = 5$, $l = 1$) electrons. The $4f$ electron density, then, is concentrated toward the interior of the atom, so $4f$ electrons have little effect on the chemical activity of the atom, which is dominated by the outer s and p electrons. For this reason, the lanthanides are also called **inner transition metals.** The outer electron configuration, which plays a principal role in determining chemical activity, is the same for all the lanthanides ($5p^6 6s^2$) and accounts for their similar chemical properties.

Following the lanthanides is a third transition metal series (the $5d$ transition metal series) consisting of the elements lutetium ($Z = 71$) through mercury ($Z = 80$). This series, in which the $5d$ orbitals are filled, is followed by the six elements thallium ($Z = 81$) through radon ($Z = 86$). Radon, a radioactive noble gas with the ground state electron configuration [Xe]$6s^2 4f^{14} 5d^{10} 6p^6$, or [Rn], finishes the sixth row of the table.

The next two elements, the radioactive metals francium, [Rn]$7s^1$, and radium, [Rn]$7s^2$, are followed by another inner transition metal series in which the $5f$ orbitals are filled. This series begins with actinium ($Z = 89$) and ends with nobelium ($Z = 102$), and is called the **actinide series.** All the elements in this series are radioactive. In fact, with the exception of trace quantities of plutonium, the elements beyond uranium ($Z = 92$), called the **transuranium elements,** have not been found in nature. They are synthesized in nuclear reactors (see Interchapter O).

WWW

See Interchapter O at www.McQuarrieGeneralChemistry.com.

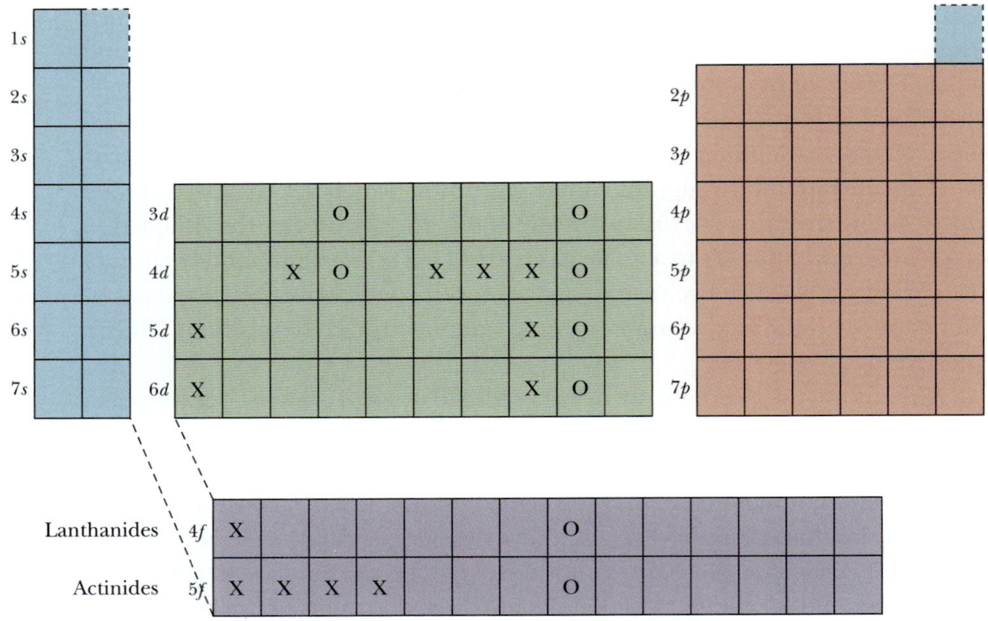

Figure 5.16 An expanded periodic table indicating which orbitals are filling up with electrons as we move through various regions of the table. The colors designate the *s*-block (*blue*), *p*-block (*orange*), *d*-block (*green*), and *f*-block (*purple*) elements. X's mark exceptions to the normal filling order; O's mark exceptions with half- or completely filled subshells. (The elements Lu and Lr have been placed in the *d*-block; see Figure 3.10.) Strong, Judith A. *The Journal of Chemical Education* **1986**, *63*, 834.

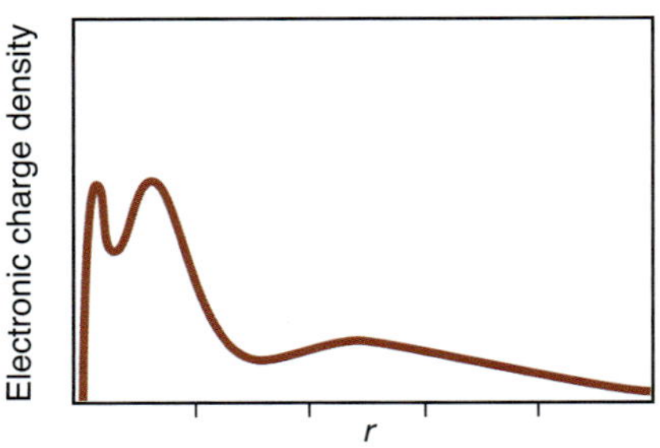

Figure 5.17 The distribution of the electronic charge density versus the distance from the nucleus for an argon atom can be obtained by solving the Schrödinger equation with a computer. Note that there appear to be three shells. Two of these are well defined and close to the nucleus. The third, outermost shell is more diffuse.

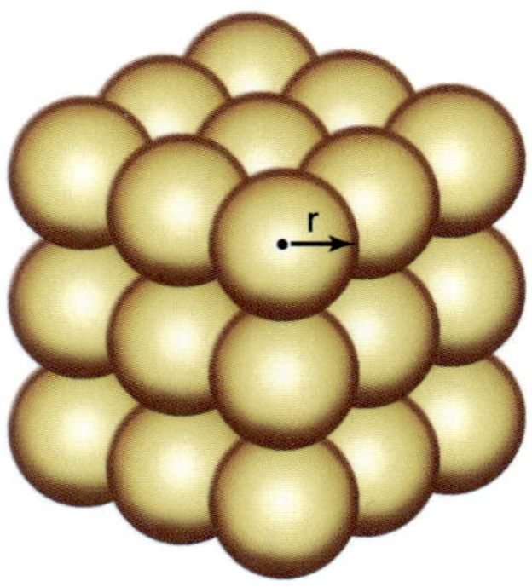

Figure 5.18 A simple cubic arrangement of atoms in a crystal.

Figure 5.16 shows an expanded periodic table summarizing which orbitals are filled as we move through the various regions of the table. These regions are referred to as the ***s*-block elements** (Groups 1 and 2), the ***p*-block elements** (Groups 13 through 18), the ***d*-block elements** (the transition metals, Groups 3 through 12), and the ***f*-block elements** (the inner transition metals).

5-12. Atomic Radii and Ionization Energies Are Periodic Properties

As can be seen from Figures 5.4, 5.5, and 5.6, the probability of finding an electron at some distance, r, from the nucleus decreases with increasing r. Even though the decrease occurs fairly rapidly with increasing r, it never actually becomes zero, even at very large distances from the nucleus. As a consequence, it is not possible to define unambiguously an outer "edge" of an atom, that is, a distance beyond which there is zero probability of finding an electron associated with the nucleus. In other words, an atom has no sharp boundary. Although the Schrödinger equation is complicated for multielectron atoms, it can be solved with a computer. The results of such a calculation for an argon atom are sketched in Figure 5.17. We can clearly discern three shells: the inner two shells, the first shell and the second shell, are relatively well defined; the third, outermost shell is much more diffuse.

Even though atoms do not have well-defined edges, we can propose practical definitions for **atomic radii** based on models. For example, the atoms in a crystal of an element are arranged in ordered arrays. A simple version of such an ordered array is shown in Figure 5.18, in which the atoms are arranged in a simple cubic array. If we propose that one-half of the distance between adjacent nuclei in a simple cubic array constitutes an effective atomic radius, then we can determine atomic radii. Real crystals usually exist in more complicated geometric patterns than a simple cubic pattern, but effective atomic radii can still be deduced. Atomic radii obtained in this manner are called **crystallographic**

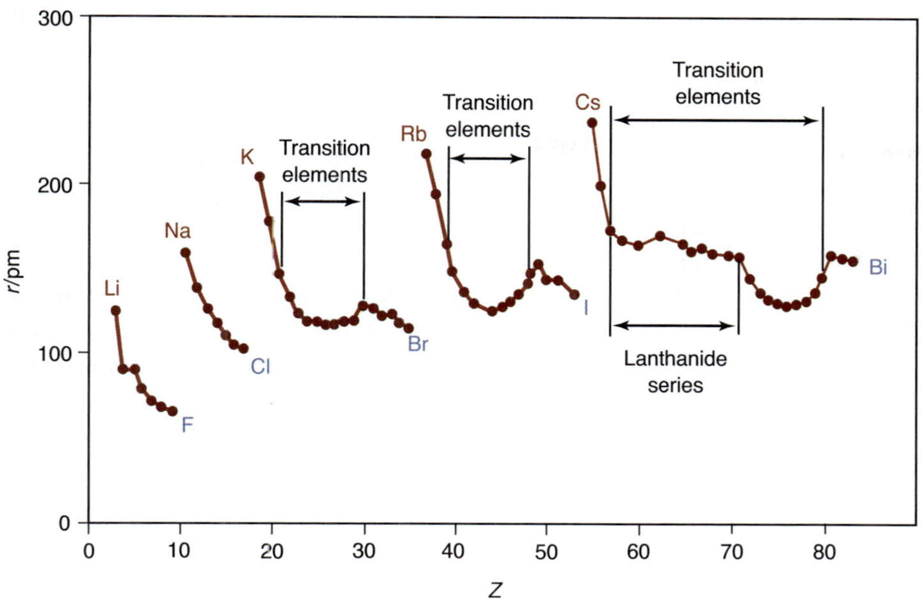

Figure 5.19 Crystallographic radii of the elements versus atomic number. Note that atomic radius is a periodic property.

radii. The crystallographic radii of the elements are plotted against atomic number in Figure 5.19; the resulting patterns indicate the periodic dependence of crystallographic radii on atomic number.

The crystallographic radii of the atoms of the elements lithium through fluorine decrease uniformly as we look from left to right across the periodic table. As the nuclear charge increases, the nucleus attracts the electrons more strongly. This same trend is seen in Figure 5.19 for the other rows of the periodic table. Atomic radii of the main-group elements usually decrease as we look from left to right in a row across the periodic table, as a consequence of the steady increase in nuclear charge within the row.

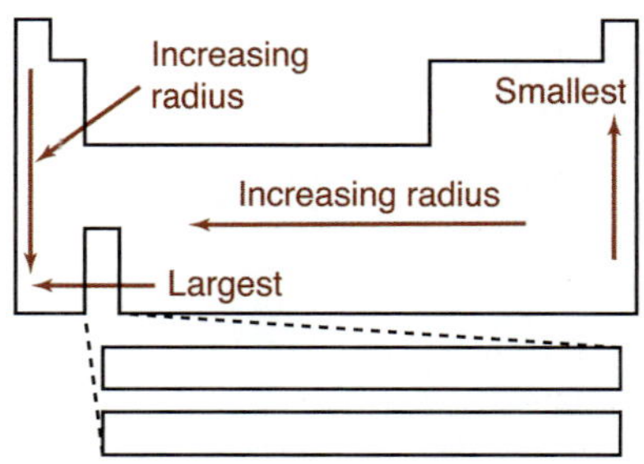

Figure 5.20 The trend of atomic radii in the periodic table.

The crystallographic radii of the atoms of the alkali metal group also increase as we look down the periodic table within the group. Although the nuclear charge increases, the outermost electrons begin new shells, and, as shown in Figure 5.20, this effect outweighs the increased nuclear attraction. Similar behavior is found for other groups in the periodic table.

The reasoning we have just used to explain the variation of atomic radii in the periodic table also can be used to explain variations in first ionization energies (Figure 4.1). The atomic radius increases as we go down a group in the periodic table. The farther the electron is from the nucleus, the less the nuclear attraction, so the more easily the electron is removed. Therefore, first ionization energies are seen to decrease as we look down the periodic table within a group. Similarly, the decrease in atomic size as we move from left to right across a row of the periodic table due to the increase in nuclear charge is also reflected in the corresponding increase in first ionization energies. Trends in atomic radii and ionization energies are thus seen to follow directly from quantum theory.

SUMMARY

The Bohr theory of the hydrogen atom could not be used to explain the atomic spectra of multielectron atoms and also was inconsistent with the Heisenberg uncertainty principle. In 1925 Schrödinger proposed the central equation of quantum theory. One consequence of the Schrödinger equation is that the electrons in atoms and molecules can have only certain discrete, or quantized, energies. In addition, Schrödinger showed that an electron in an atom or molecule must be described by a wave function, or orbital, that is obtained by solving the Schrödinger equation. The square of a wave function gives the probability density associated with finding the electron in some region of space. The hydrogen atom orbitals serve as the prototype wave functions for all other atoms. The orbitals are specified by three quantum numbers: n, the principal quantum number; l, the azimuthal quantum number; and m_l, the magnetic quantum number. Orbitals with $l = 0$ are called s orbitals; orbitals with $l = 1$ are called p orbitals; orbitals with $l = 2$ are called d orbitals; and orbitals with $l = 3$ are called f orbitals. For various values of n, there is one s orbital, three p orbitals (for $n \geq 2$), five d orbitals (for $n \geq 3$), and seven f orbitals (for $n \geq 4$) (see Table 5.4).

To explain certain fine details in atomic spectra, a fourth quantum number is required. This is the spin quantum number, m_s, that specifies the intrinsic spin of an electron, a quantum property. The spin quantum number can have the value of $+\frac{1}{2}$ or $-\frac{1}{2}$.

The energy states of the hydrogen atom depend only on the principal quantum number, n; for atoms with more than one electron (multielectron atoms), the energy states depend on both n and the azimuthal quantum number, l. According to the Pauli exclusion principle, no two electrons in an atom can have the same set of four quantum numbers (n, l, m_l, m_s). Using this principle and the order of the energy states given in Figures 5.11b and 5.12, together with Hund's rule, we are able to write ground state electron configurations and correlate these with the periodic table. For main-group elements, the number of valence electrons is equal to the number of electrons beyond the noble-gas core. (The electronic configuration of the transition metals is discussed more thoroughly in the next two chapters.) Electron configurations enable us to understand the trends of chemical reactivity, atomic radii, and ionization energies within the periodic table.

TERMS YOU SHOULD KNOW

Heisenberg uncertainty principle *137*
Schrödinger equation *139*
wave function, ψ *140*
probability density *140*
atomic orbital *140*
principal quantum number, n *140*
spherically symmetric *140*
azimuthal quantum number, l *141*
s orbital *142*
p orbital *142*
d orbital *142*
f orbital *142*
degenerate *142*
nodal surface *142*
cylindrically symmetric *143*
magnetic quantum number, m_l *144*
intrinsic electron spin *146*
spin quantum number, m_s *148*
Pauli exclusion principle *151*
spin up *151*
spin down *151*
paired electrons *151*
unpaired electrons *151*
subshell *153*
ground electronic state *153*
electron configuration *153*
parallel spins *155*
Hund's rule *155*
valence electrons *158*
$3d$ transition metal series *160*
$4d$ transition metal series *161*
lanthanide series *161*
inner transition metals *161*
actinide series *161*
transuranium elements *161*
s-block elements *162*
p-block elements *162*
d-block elements *162*
f-block elements *162*
atomic radius *162*
crystallographic radius *162–163*

EQUATIONS YOU SHOULD KNOW HOW TO USE

$$(\Delta x)(\Delta p) \geq \frac{h}{4\pi} \qquad \text{(5.1) (Heisenberg uncertainty principle)}$$

$$\left.\begin{aligned} &n = 1, 2, 3 \ldots \\ &l = 0, 1, 2, \ldots, n-1 \\ &m_l = l, (l-1), (l-2), \ldots, 0, -1, -2, \ldots, -l \\ &m_s = +\tfrac{1}{2} \text{ or } -\tfrac{1}{2} \end{aligned}\right\} \text{quantum numbers}$$

PROBLEMS

HEISENBERG UNCERTAINTY PRINCIPLE

5-1. Estimate the uncertainty in the position of a nitrogen molecule that is traveling with a speed of 500 ± 2 m·s^{-1}. Take the mass of a nitrogen molecule to be 4.65×10^{-23} grams.

5-2. What is the greatest precision with which the speed of an alpha particle may be measured if its position is known to ±1 nm? Take the mass of an alpha particle to be 6.65×10^{-24} grams.

5-3. What is the uncertainty in the velocity of an electron whose position is known to within 2×10^{-8} meters? If the electron is moving at a speed of 5.0×10^5 m·s^{-1}, what fraction of this speed does the uncertainty represent?

5-4. Consider a 2750-pound automobile clocked by law enforcement radar at a speed of 85.5 miles per hour. If the position of the car is known to within 5.0 feet at the time of the measurement, what is the uncertainty in the velocity of the car? Could the driver of the car reasonably evade a speeding ticket by invoking the Heisenberg uncertainty principle?

QUANTUM NUMBERS AND ORBITALS

5-5. Indicate which of the following atomic orbital designations are impossible for a hydrogen atom:

(a) 7*s* (b) 1*p* (c) 5*d* (d) 2*d* (e) 4*f*

5-6. Without referring to the text, sketch the 1*s*, 2*s*, 2*p*, and 3*d* orbitals. Be sure to indicate their orientations with respect to the *x*-*y*-*z* axes and sketch any nodal surfaces. Label each of the orbitals you have sketched (for example, $2p_x$). Compare your drawings to those in the text.

5-7. Give the corresponding atomic orbital designations (that is, 1*s*, 3*p*, and so on) for electrons with the following sets of quantum numbers:

	n	l	m_l	m_s
(a)	4	1	0	$-\frac{1}{2}$
(b)	3	2	0	$+\frac{1}{2}$
(c)	4	2	-1	$-\frac{1}{2}$
(d)	2	0	0	$-\frac{1}{2}$

5-8. Give the corresponding atomic orbital designations for electrons with the following sets of quantum numbers:

	n	l	m_l	m_s
(a)	3	1	-1	$+\frac{1}{2}$
(b)	5	0	0	$+\frac{1}{2}$
(c)	2	1	0	$+\frac{1}{2}$
(d)	4	3	-2	$+\frac{1}{2}$

5-9. If $l = 2$, what can you deduce about n? If $m_l = +3$, what can you say about l?

5-10. Indicate which of the following sets of quantum numbers are allowed (that is, possible) for an electron in an atom:

	n	l	m_l	m_s
(a)	2	1	0	$+\frac{1}{2}$
(b)	3	0	$+1$	$-\frac{1}{2}$
(c)	3	2	-2	$-\frac{1}{2}$
(d)	1	1	0	$+\frac{1}{2}$
(e)	2	1	0	0

ORBITALS AND ELECTRONS

5-11. Give all the possible sets of four quantum numbers for an electron in a $5d$ orbital.

5-12. Give all the possible sets of four quantum numbers for an electron in a $4f$ orbital.

5-13. Without referring to the text, deduce the maximum number of electrons that can occupy an s orbital, a subshell of p orbitals, a subshell of d orbitals, and a subshell of f orbitals.

5-14. Without referring to the text, deduce the maximum number of electrons that can occupy the $n = 1$ shell, the $n = 2$ shell, the $n = 3$ shell, and the $n = 4$ shell.

5-15. Explain why there are 10 members of each d-block series.

5-16. Explain why there are 14 members of each f-block series.

ELECTRON CONFIGURATIONS OF ATOMS

5-17. Indicate which of the following electron configurations are ruled out by the Pauli exclusion principle:

(a) $1s^22s^22p^7$ (b) $1s^22s^22p^63s^3$

(c) $1s^22s^22p^63s^23p^64s^23d^{12}$ (d) $1s^22s^22p^63s^23p^6$

5-18. Explain why the following ground state electron configurations are not possible:

(a) $1s^22s^32p^3$ (b) $1s^22s^22p^33s^6$

(c) $1s^22s^22p^73s^23p^8$ (d) $1s^22s^22p^63s^23p^14s^23d^{14}$

5-19. Write the corresponding electron configuration for each of the following pictorial representations. Name the element, assuming that the configuration describes a neutral atom.

(a)

↑↓	↑↓	↑↓ ↑↓ ↑↓	↑↓	↑ ↑ —
$1s$	$2s$	$2p$	$3s$	$3p$

(b)

↑↓	↑↓	↑↓ ↑↓ ↑↓	↑↓	↑↓ ↑↓ ↑↓
$1s$	$2s$	$2p$	$3s$	$3p$

↑	↑ ↑ ↑ ↑ ↑
$4s$	$3d$

(c)

↑↓	↑↓	↑↓ ↑↓ ↑↓	↑↓	↑↓ ↑↓ ↑↓
$1s$	$2s$	$2p$	$3s$	$3p$

↑↓	↑↓ ↑↓ ↑↓ ↑↓ ↑↓	↑ ↑ —
$4s$	$3d$	$4p$

(d)

↑↓	↑↓	↑↓ ↑↓ ↑↓	↑↓	↑↓ ↑↓ ↑↓
$1s$	$2s$	$2p$	$3s$	$3p$

↑↓	↑↓ ↑↓ ↑↓ ↑↓ ↑↓	↑↓ ↑↓ ↑
$4s$	$3d$	$4p$

(e)

↑↓	↑↓	↑ — —
$1s$	$2s$	$2p$

5-20. Write the corresponding electron configuration for each of the following pictorial representations. Name the element that each represents, assuming neutral atoms.

(a)

↑↓	↑↓	↑↓ ↑↓ ↑↓	↑↓	↑ — —
$1s$	$2s$	$2p$	$3s$	$3p$

(b)

↑↓	↑↓	↑↓ ↑↓ ↑↓	↑↓	↑↓ ↑↓ ↑↓
$1s$	$2s$	$2p$	$3s$	$3p$

↑↓	↑ ↑ ↑ — —
$4s$	$3d$

(c)

↑↓	↑↓	↑↓ ↑↓ ↑
$1s$	$2s$	$2p$

(d)

↑↓	↑↓	↑↓ ↑↓ ↑↓	↑↓	↑↓ ↑↓ ↑↓
$1s$	$2s$	$2p$	$3s$	$3p$

↑↓	↑↓ ↑↓ ↑↓ ↑↓ ↑↓	↑ — —
$4s$	$3d$	$4p$

(e)

↑↓	↑↓	↑↓ ↑ ↑
$1s$	$2s$	$2p$

5-21. Using the noble-gas shorthand notation, write the ground state electron configurations for the following elements:

(a) Ti (b) K (c) Fe (d) As (e) F

5-22. Using the noble-gas shorthand notation, write the ground state electron configurations for the following elements:

(a) Si (b) Ni (c) Se (d) Cd (e) Mg

5-23. Using the noble-gas shorthand notation, write the ground state electron configurations for the following neutral atoms:

(a) Br (b) Hg (c) Lr (d) Cu (e) Rn

5-24. Using the noble-gas shorthand notation, write the ground state electron configurations for the following neutral atoms:

(a) Ba (b) Ag (c) Gd (d) Pd (e) Sn

5-25. Write out the complete pictorial representation of the electron configurations beyond the noble-gas core, using up and down arrows to represent each of the electron spins for the atoms Si and Fe.

5-26. Write out the complete pictorial representation of the electron configurations beyond the noble-gas core, using up and down arrows to represent each of the electron spins for the atoms S and Cr.

ELECTRON CONFIGURATIONS AND THE PERIODIC TABLE

5-27. Using only the periodic table, write the ground state electron configuration of the following elements using the noble-gas shorthand notation:

(a) Ca (b) Br (c) B (d) Zn (e) W

5-28. Using only the periodic table, write the ground state electron configuration of the following elements using the noble-gas shorthand notation:

(a) Y (b) Po (c) Co (d) Es (e) Pb

5-29. Give two examples of

(a) an atom with a half-filled subshell

(b) an atom with a completed outer shell

(c) an atom with its outer electrons occupying a half-filled subshell and a filled subshell

5-30. Determine the element of lowest atomic number whose ground state contains

(a) an *f* electron (b) three *d* electrons

(c) a complete *d* subshell (d) ten *p* electrons

5-31. How many unpaired electrons are there in the ground state of each of the following atoms?

(a) Ge (b) Se (c) V (d) Fe

5-32. How many unpaired electrons are there in the ground state of each of the following atoms?

(a) Al (b) Cr (c) S (d) Hg

GROUND STATE ELECTRON CONFIGURATIONS OF IONS

5-33. Nonmetals add electrons under certain conditions in order to attain a noble-gas electron configuration. How many electrons must be gained in this process by the following elements? What noble-gas electron configuration is attained in each case?

(a) H (b) O (c) C (d) S

5-34. Metals lose electrons to attain a noble-gas electron configuration. How many electrons are lost by the following elements when they attain such a configuration? What is the corresponding noble-gas-like inner core of the ion in each case?

(a) Ca (b) Li (c) Na (d) Mg

5-35. Use Hund's rule to write the ground state electron configuration for the following ions:

(a) P^{3-} (b) Br^{-} (c) Se^{2-} (d) Ba^{2+}

What do these electron configurations have in common?

5-36. Use Hund's rule to write ground state electron configurations for the following ions:

(a) O^{+} (b) C^{-} (c) F^{+} (d) O^{2+}

In each case indicate the neutral atom that is isoelectronic with the given ion.

5-37. How many unpaired electrons are there in the ground state of each of the following ions?

(a) Cl^{-} (b) O^{+} (c) Al^{3+} (d) Xe^{+} (e) K^{+}

5-38. How many unpaired electrons are there in the ground state of each of the following ions?

(a) O^{2-} (b) Ca^{+} (c) He^{+} (d) Pb^{2+} (e) N^{3-}

5-39. For each of the ions listed in Problem 5-37, indicate the neutral atom that is isoelectronic with the given ion.

5-40. For each of the ions listed in Problem 5-38, indicate the neutral atom that is isoelectronic with the given ion.

5-41. Describe the following processes in terms of the electron configurations of the species involved:

(a) $O(g) + 2e^- \rightarrow O^{2-}(g)$

(b) $Ca(g) + Sr^{2+}(g) \rightarrow Sr(g) + Ca^{2+}(g)$

5-42. Describe the following processes in terms of the electron configurations of the species involved:

(a) $I(g) + e^- \rightarrow I^-(g)$

(b) $K(g) + F(g) \rightarrow K^+(g) + F^-(g)$

EXCITED-STATE ELECTRON CONFIGURATIONS

5-43. Write the electron configuration for the first excited state of

(a) Be^{2+} (b) He^+ (c) F^- (d) Li

5-44. Which of the following electron configurations of neutral atoms represents an excited state? Identify the atom in each case.

(a) $1s^22s^22p^53s^1$ (b) $1s^22s^22p^53s^2$

(c) $1s^12s^1$ (d) $1s^22s^22p^63s^23p^64s^1$

VALENCE ELECTRONS

5-45. Determine the number of valence electrons in each of the following main-group elements or ions and draw the corresponding Lewis dot formula:

(a) C (b) N (c) O (d) Br^- (e) Mg^{2+}

5-46. Determine the number of valence electrons in each of the following main-group elements or ions and draw the corresponding Lewis dot formula:

(a) He (b) N^{3-} (c) F (d) Na (e) K^+

ATOMIC RADII

5-47. Determine the member of each of the following pairs of atoms that has the larger radius (do not use any references except the periodic table):

(a) N and P (b) P and S

(c) S and Ar (d) Ar and Kr

5-48. Determine the member of each of the following pairs of atoms that has the larger radius (do not use any references except the periodic table):

(a) O and F (b) Kr and Xe

(c) F and Cl (d) Mg and Ca

5-49. Without using any references except the periodic table, arrange the following sets of atoms in order of increasing atomic radius:

(a) Kr, He, Ar, Ne (b) K, Na, Rb, Li (c) Be, Ne, F, N

5-50. Without using any references except the periodic table, arrange the members of the following groups in order of increasing size:

(a) Li, Na, Cs, Rb (b) Al, Na, Mg, P (c) Ca, Ba, Sr, Mg

5-51. The radii of lithium and its ions are shown below.

Species	Radius/pm
Li	135
Li^+	60
Li^{2+}	18

Explain why the radii decrease from Li to Li^{2+}.

5-52. Arrange the following gaseous ions in order of increasing size:

Mg^{2+} Na^+ N^{3-} O^{2-} F^-

Explain your rationale.

ADDITIONAL PROBLEMS

5-53. In Bohr's model of the hydrogen atom, what does the value of the quantum number designated by n represent? In the quantum theory described by the Schrödinger equation, what do the values of the quantum numbers designated by n, l, m_l, and m_s represent? What are these quantum numbers called?

5-54. What is the physical meaning of the Δx and Δp terms in the Heisenberg uncertainty formula, $(\Delta x)(\Delta p) \geq h/4\pi$? If we know the momentum of an electron with 100% certainty, what then is the uncertainty in its position?

5-55. Suppose that the electrons in atoms did not have to obey the Pauli exclusion principle. What would the ground state electronic configuration of a sodium atom be under these conditions?

5-56. Without using any references except the periodic table, arrange the following sets of atoms in order of increasing ionization energy:

(a) B, O, Ne, F (b) Te, I, Sn, Xe

(c) K, Ca, Rb, Cs (d) Ar, Na, S, Al

5-57. Why is the m_s quantum number called the "spin" quantum number? Does an electron in an atom literally spin?

5-58. What is meant by the term "valence electrons"?

5-59. On which quantum numbers does the energy of a hydrogen atom depend? On which quantum numbers does the energy of a multielectron atom depend? Why are these different?

5-60. Without counting the total number of electrons, determine the neutral atom whose ground state electron configuration is

(a) $1s^22s^22p^63s^23p^64s^23d^8$

(b) $1s^22s^22p^63s^23p^64s^23d^{10}4p^65s^14d^{10}$

(c) $1s^22s^22p^63s^23p^4$

(d) $1s^22s^22p^63s^23p^64s^23d^{10}4p^65s^24d^{10}5p^66s^24f^{14}5d^{10}6p^2$

5-61. Without looking at a periodic table, deduce the atomic numbers of the other elements that are in the same family as the element with atomic number (a) 16 and (b) 11.

5-62. Name each of the atoms with the following ground state electron configuration for its valence electrons:

(a) $3s^23p^1$ (b) $2s^22p^4$ (c) $4s^23d^{10}$ (d) $4s^23d^{10}4p^6$

5-63. For elements of atomic number (a) 15, (b) 26, and (c) 32 in their ground states, answer the following questions without reference to the text or to a periodic table:

(i) How many d electrons in each element?

(ii) How many electrons having quantum number $l = 1$ in each element?

(iii) How many unpaired electrons in each element?

5-64. Which of the following electron configurations of neutral atoms represent excited states? Identify the atom in each case.

(a) [Ar] $4s^23d^3$ (b) $2s^2$ (c) $1s^22s^22p^63s^23p^63d^2$

(d) [Kr] $5s^14d^5$ (e) [Xe] $6s^24f^1$

5-65. Arrange the following species into groups of isoelectronic species:

F^- Sc^{3+} Be^{2+} Rb^+ O^{2-} Na^+ Ti^{4+}

Ar B^{3+} He Se^{2-} Y^{3+}

5-66. Determine the number of unpaired electrons in the ground state of the following species:

(a) F^+ (b) Sn^{2+} (c) Bi^{3+} (d) Ar^+

In each case indicate the neutral atom that is isoelectronic with the given ion.

5-67. Show what the periodic table would look like if the filling order of atomic orbitals were regular; that is, if the order were $1s < 2s < 2p < 3s < 3p < 3d < 4s$ and so on.

5-68. How would the ground state electron configurations of the elements in the second row of the periodic table differ if the $2s$ and $2p$ orbitals had the same energy, as they do for a hydrogen atom?

5-69. The order of the orbitals given in Figure 5.12b can be deduced by the following argument. The energy of an orbital increases with the sum $n + l$. For orbitals with the same value of $n + l$, those with the smaller value of n have lower energies. This observation is also known as Hund's rule (there are several Hund's rules). Show that this rule is consistent with the order given in Figure 5.11b.

5-70. (*) Although in general the ionization energy of an atom increases from left to right as we move across a period, there are some anomalies in the trend. For example, referring to Figure 4.1, we observe that there are small dips in the ionization energy trend in the second period between beryllium and boron and again between nitrogen and oxygen. Using the concepts we have outlined in this chapter, see if you can explain why the ionization energy of boron is less than that of beryllium and why that of oxygen is less than that of nitrogen.

Svante Arrhenius (1859–1927) was born in Wijk near Uppsala, Sweden. He showed early talent in mathematics, but decided to study chemistry at the University of Uppsala. He received his doctorate in 1884 from the University of Uppsala for his dissertation on the theory that electrolyte solutions contain ions that are formed when a salt dissolves in water and that the ions are the conductors of electricity. His thesis work was quite controversial at the time and was not accepted; in fact, he barely obtained his doctorate. He then received a traveling scholarship, which allowed him to spend five years studying in Europe, where his theory was enthusiastically accepted by the scientific community. In spite of the conservative scientific attitude in Sweden, he returned there, but he could not immediately obtain a university position. Swedish scientists did not accept his ionic theory and he became a teacher at the Technical High School in Stockholm. In 1891, he became a lecturer at Stockholm University. Two years later, his position was elevated to professor at the University of Stockholm after he underwent an oral examination by a hostile committee. He was awarded the 1903 Nobel Prize in Chemistry "for his electrolytic theory of dissociations." In 1904, Arrhenius became the first director of the newly created Nobel Institute for Physical Research in Stockholm. In 1896, he studied the effect of carbon dioxide on the temperature of the earth's atmosphere, now known as the greenhouse effect. In addition to his research interests, Arrhenius was also an accomplished writer and published a number of popular books.

6. Ionic Bonds and Compounds

In Chapter 5 we showed how the electronic structure of atoms underlies the structure of the periodic table. Therefore, it seems reasonable to suppose that an understanding of atomic structure should help us to understand the chemical bonding that occurs between atoms in molecules. For example, we can use the electron configurations of the sodium atom and the chlorine atom to understand why the chemical formula of sodium chloride is NaCl, instead of $NaCl_2$ or Na_2Cl. We also can understand why sodium chloride is an **ionic compound**, capable of conducting an electric current when it is dissolved in water or melted. Similarly, in the next few chapters we shall study **covalent compounds** and learn why carbon and hydrogen combine to form the stable molecule methane with a formula CH_4 instead of CH or CH_2, and why nitrogen is a diatomic gas with a formula N_2 at room temperature. All these observations relate to the bonding that occurs between atoms. The general properties of ionic and covalent compounds differ greatly, as can be seen from the trends listed in Table 6.1. In this and the next few chapters we shall develop an understanding of these observations in terms of chemical bonds.

6-1. The Electrostatic Force That Binds Oppositely Charged Ions Together Is Called an Ionic Bond

As our first step toward understanding ionic bonding, we discuss an important experimental property of ionic compounds in aqueous solution. When most ionic compounds dissolve in water, the crystals break up into mobile ions rather than neutral molecules. For example, an aqueous solution of sodium chloride consists of $Na^+(aq)$ and $Cl^-(aq)$ ions that are dispersed and free to move throughout the water. If electrodes (for example, strips of an inert metal such as platinum) are connected to the poles of a battery and dipped into a solution containing ions, then the positive ions are attracted to the negative electrode and the negative ions are attracted to the positive electrode (Figure 6.1). The movement of the ions toward the respective electrodes constitutes an electric current through the solution.

In contrast, covalent compounds yield neutral molecules when they dissolve in water and consequentially are poor conductors of an electric current because

TABLE 6.1 General comparison of ionic and covalent compounds

Properties	Ionic compounds	Covalent compounds
molecular structure	tend to form extended crystal lattices of alternating ions	tend to exist as individual molecules with shared electrons
in solutions or liquid phase	form ions, good conductors of electricity	do not form ions, poor conductors of electricity
melting point	high (all are solids at 25°C)	varies, typically low

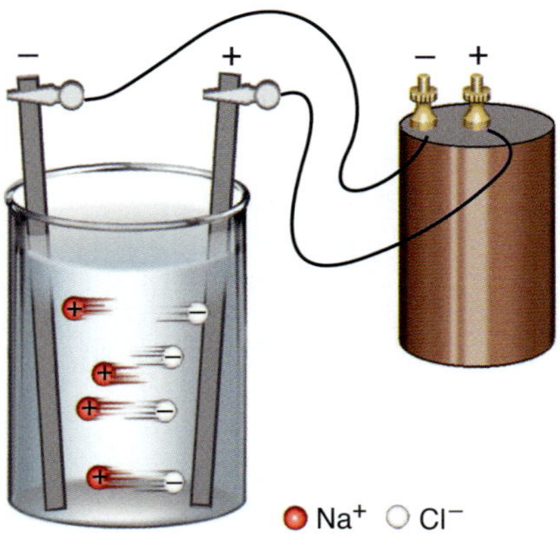

Figure 6.1 An aqueous solution of NaCl(s) conducts an electric current. An electric voltage is applied by dipping metal strips (electrodes) attached to the poles of a battery into the solution. Like the poles of a battery, one of the electrodes is positive and the other is negative. The positively charged sodium ions are attracted to the negative electrode, and the negatively charged chloride ions are attracted to the positive electrode. Thus, the $Na^+(aq)$ ions migrate to the left in the figure and the $Cl^-(aq)$ ions migrate to the right. The migration of the ions constitutes an electric current through the solution.

no charge carriers like those in solutions of ionic compounds are present. For example, an aqueous solution of sucrose (table sugar), $C_{12}H_{22}O_{11}(aq)$, contains neutral sucrose molecules, and so it does not conduct an electric current (Figure 6.2).

The key to explaining observations such as these lies in understanding the difference between ionic and covalent bonding. To understand the bonding in ionic compounds, let's first consider the reaction between a sodium atom and a chlorine atom. The ground state electron configurations of the sodium and chlorine atoms are

$$\text{Na: [Ne]}3s^1 \qquad \text{Cl: [Ne]}3s^23p^5$$

The electron configuration of a sodium atom consists of a neonlike inner core with a $3s$ valence electron outside the core. If the sodium atom loses this $3s$ electron, then the resultant species is a sodium ion, with an electron configuration like that of the noble gas neon. We can describe the ionization process by the equation

$$\text{Na([Ne]}3s^1) \rightarrow \text{Na}^+\text{([Ne])} + \text{e}^-$$

Recall from Chapter 4 that once a sodium atom loses its $3s$ electron, the resultant neonlike electron configuration is relatively stable to further ionization.

If a chlorine atom accepts an electron, then the resultant species is a chloride ion, with an electron configuration like that of the noble gas argon. We can describe the ionization process by the equation

$$\text{Cl([Ne]}3s^23p^5) + \text{e}^- \rightarrow \text{Cl}^-\text{([Ar])}$$

Thus, we see that both a sodium atom and a chlorine atom can simultaneously achieve noble-gas electron configurations through the transfer of an electron from the sodium atom to the chlorine atom. We can describe the electron transfer by the equation

$$\text{Na([Ne]}3s^1) + \text{Cl([Ne]}3s^23p^5) \rightarrow \text{Na}^+\text{([Ne])} + \text{Cl}^-\text{([Ar])}$$

or, in terms of Lewis electron-dot formulas,

$$\text{Na}\cdot + \cdot\ddot{\underset{\cdot\cdot}{\text{Cl}}}\text{:} \longrightarrow \underbrace{\text{Na}^+ + \text{:}\ddot{\underset{\cdot\cdot}{\text{Cl}}}\text{:}^-}_{\text{Na}^+\text{Cl}^-}$$

Figure 6.2 Comparison of the currents (as measured using ammeters) through equal concentrations of sodium chloride, NaCl(*aq*), and sucrose, $C_{12}H_{22}O_{11}(aq)$. The solution of the strong electrolyte, NaCl(*aq*), is a much better conductor of electricity than the solution of the nonelectrolyte, $C_{12}H_{22}O_{11}(aq)$.

The sodium ion and the chloride ion have opposite charges, so they attract each other. This **electrostatic force** binds the ions together and is called an **ionic bond.**

We have seen that noble-gas electron configurations are relatively stable to the gain or loss of additional electrons. Because both sodium ions and chloride ions achieve a noble-gas electron configuration, the above reaction occurs easily, and there is no tendency for additional electron transfer. Because a sodium atom readily loses one and only one electron, whereas a chlorine atom readily gains one and only one electron, when a sodium atom reacts with a chlorine atom, the transfer of an electron from the sodium atom to the chlorine atom results in one sodium ion and one chloride ion. The compound, sodium chloride, like all chemical compounds, is electrically neutral. Therefore, the chemical formula of sodium chloride must be NaCl and not $NaCl_2$, Na_2Cl, or anything other than NaCl. From this we see that ionic compounds are composed of ions and must be electrically neutral.

Like sodium, the other metallic elements in Group 1 typically lose one electron in reactions to become +1 ions and achieve the electron arrangement of the preceding noble gas.

EXAMPLE 6-1: Predict the charge on a calcium ion.

Solution: Calcium belongs to Group 2. Because a calcium atom has two more electrons than an argon atom, the electron configuration of a calcium atom is Ca: [Ar]$4s^2$. If it loses two electrons, then the Ca^{2+} ion achieves the relatively stable argonlike noble-gas electron configuration. When a calcium atom loses two electrons, the charge of the resulting calcium ion is +2. We also can conclude that the other Group 2 elements also form stable +2 ions.

PRACTICE PROBLEM 6-1: Predict the charge on an aluminum ion.

Answer: Al^{3+}

Let's consider the elements in Groups 16 and 17. As with fluorine, each of the halogens comes right before a noble gas in the periodic table, so we predict that all the halogen atoms can gain one electron to form halide ions with a charge of −1. Oxygen occurs two positions before neon in the periodic table, so an oxygen atom has two fewer electrons than a neon atom. An oxygen atom can achieve a neonlike electron arrangement by gaining two electrons to become an oxide ion, which is written as O^{2-}.

EXAMPLE 6-2: Predict the charge on a sulfide ion.

Solution: Like oxygen, sulfur ($Z = 16$) is a Group 16 element with an electron configuration of $[Ne]3s^23p^4$. Sulfur can achieve an argonlike electron arrangement by gaining two electrons. It thereby forms the sulfide ion, S^{2-}.

PRACTICE PROBLEM 6-2: Predict the charge on a selenide ion, a nitride ion, and a phosphide ion.

Answer: Se^{2-}, N^{3-}, and P^{3-}

The reaction between sodium and chlorine is an example of a reaction between a reactive metal and a reactive nonmetal that produces ions with noble-gas electron configurations. We shall see repeatedly that a noble-gas electron configuration is a particularly stable electron arrangement and that there is a strong tendency for this configuration to occur. This is especially true for the elements in the first two rows of the periodic table. Figure 6.3 shows some of the atoms commonly encountered that lose or gain electrons to achieve a noble-gas electron configuration. All the ions in Figure 6.3 form a noble-gas electron configuration of ns^2np^6, except for Li^+ and Be^{2+}, which form a heliumlike electron configuration of $1s^2$ when they form ions. The tendency of main-group elements to form stable noble-gas-like ions with the configuration ns^2np^6 is often referred to as the **octet rule** because there are eight (= 2 + 6) electrons in the outer shell.

Metallic elements lose electrons to become positively charged ions called **cations** and nonmetallic elements gain electrons to become negatively charged ions called **anions**. We can use the charge of these ions to predict the formulas of neutral ionic compounds.

EXAMPLE 6-3: Write the electron transfer equation for the formation of the ionic compound $MgBr_2(s)$ from its neutral elements, showing the electronic configuration of each of the species involved.

Solution: A magnesium atom loses two electrons to form the noble-gas-like Mg^{2+} ion, according to the equation

$$Mg([Ne]3s^2) \rightarrow Mg^{2+}([Ne]) + 2e^-$$

Likewise, we add one electron to each of two bromine atoms to form two Br^- ions with noble-gas-like configurations

$$2\,Br([Ar]4s^24p^5) + 2e^- \rightarrow 2\,Br^-([Kr])$$

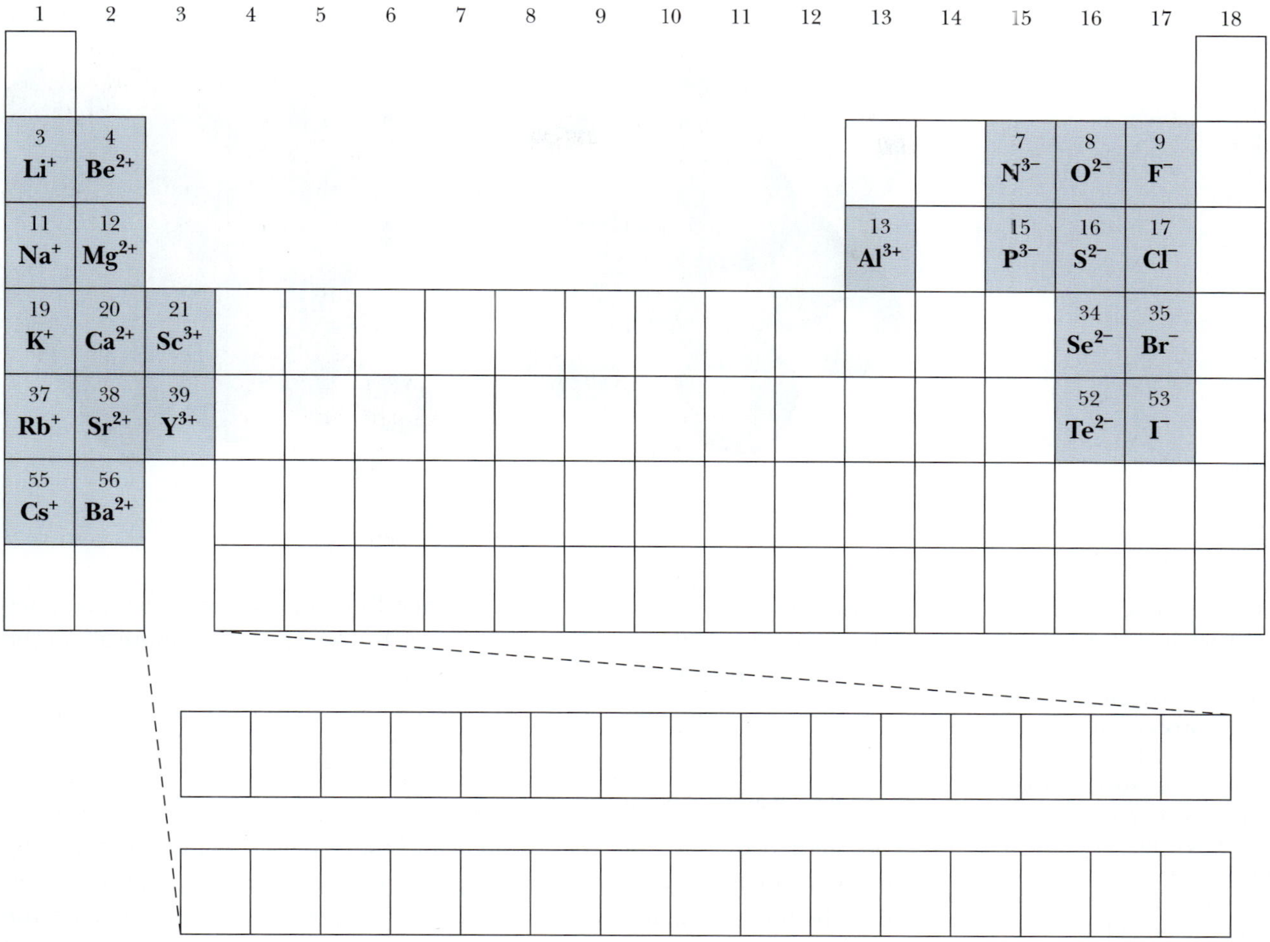

Figure 6.3 Commonly encountered ions with a noble-gas outer electron configuration.

Therefore, the corresponding electron transfer equation is

$$\mathrm{Mg([Ne]3}s^2) + 2\ \mathrm{Br([Ar]4}s^2 4p^5) \rightarrow \mathrm{Mg^{2+}([Ne])} + 2\ \mathrm{Br^-([Kr])}$$

PRACTICE PROBLEM 6-3: Write the electron transfer equation for the formation of $K_2O(s)$ from its neutral elements using Lewis electron-dot formulas.

Answer: $2\ \mathrm{K}\cdot + \cdot\ddot{\underset{\cdot\cdot}{\mathrm{O}}}\cdot \longrightarrow \mathrm{K}^+ :\ddot{\underset{\cdot\cdot}{\mathrm{O}}}:^{2-} \mathrm{K}^+$

One major difference between ionic and covalent compounds (Table 6.1) is that ionic compounds tend to form extended crystal lattices of alternating ions, while covalent compounds tend to exist as individual molecules, even when they are in the solid state (Figure 6.4). Although NaCl is the simplest formula that describes sodium chloride, it is in fact rare in nature to find individual sodium chloride molecules. For ionic compounds, the chemical formula is really just the simplest formula that describes the ratio of the number of cations to the number of anions in the extended crystal (Figure 6.4a). In contrast, solids

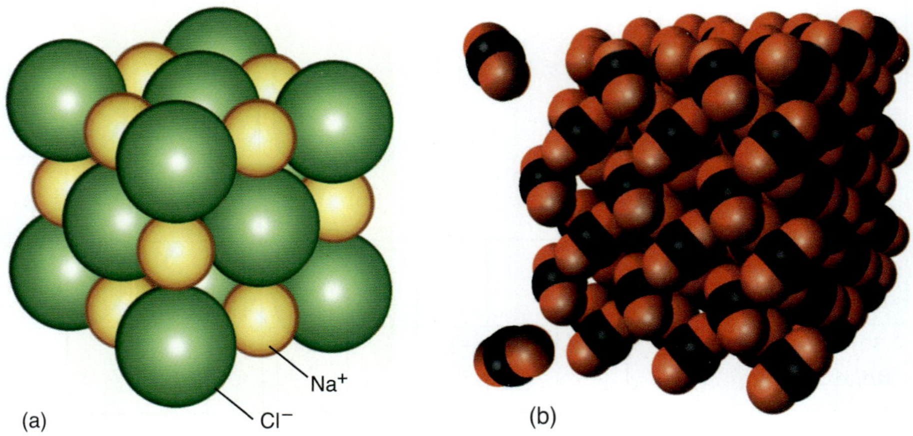

Figure 6.4 (a) A sodium chloride, $NaCl(s)$, crystal. Ionic compounds tend to form extended crystal lattices of alternatingly charged ions. Because of the strong forces created by such networks of alternating ions, all ionic substances are solids at room temperature. (b) Dry ice is a solid composed of individual carbon dioxide molecules, $CO_2(s)$. As we shall see in the next chapter, carbon dioxide is a covalent compound. Covalent compounds can be solids, liquids, or gases at room temperature. At normal atmospheric pressure dry ice sublimes (changes directly from a solid to a gas) at −78.5°C. Notice that even in the solid phase, the carbon dioxide molecules are individual units and do not form ions. (We shall study the forces that hold covalent solids together in Chapter 15.) (Images are not drawn to the same scale.)

formed from covalent compounds are generally comprised of individual molecules rather than a network of ions (Figure 6.4b). We shall learn more about the differences between ionic and covalent compounds in the upcoming chapters.

6-2. Chemical Formulas of Ionic Compounds Are Based on Ionic Charge

The method of deducing the formulas of binary ionic compounds involving ions with noble-gas electron arrangements is straightforward. First we determine the positive or negative ionic charge of an element. We then balance the total positive and negative charges by using appropriate numbers of cations and anions. The ionic charges of some commonly encountered elements are shown in Figure 6.3. The ionic charges in Figure 6.3 are the charges on the ions that have a noble-gas electron arrangement. Notice the correspondence between the ionic charge of the main-group elements and their position in the periodic table.

A correct chemical formula is obtained by combining the atomic ions such that the total positive and negative charges are equal. For example, the correct formula of strontium fluoride is $SrF_2(s)$ because a strontium ion has an ionic charge of +2 and a fluoride ion has an ionic charge of −1. It requires two fluoride ions to balance the +2 ionic charge of one strontium ion. The chemical formulas of some other ionic compounds are as follows:

$LiI(s)$ lithium iodide	$Sr_3N_2(s)$ strontium nitride
$AlF_3(s)$ aluminum fluoride	$CaO(s)$ calcium oxide

EXAMPLE 6-4: Use Figure 6.3 to write the chemical formulas of (a) sodium sulfide, (b) aluminum oxide, and (c) aluminum nitride.

Solution: (a) According to Figure 6.3, the ionic charge of a sodium cation is +1 and the ionic charge of a sulfide anion is −2. We must combine the

sodium and sulfide ions into sodium sulfide in such a way that the total positive charge is equal to the total negative charge in the formula unit. Thus, we combine two sodium ions and one sulfide ion to obtain the formula $Na_2S(s)$ for sodium sulfide.
(b) The ionic charge of an aluminum cation is +3 and that of an oxide anion is −2. We must combine two aluminum ions [total charge on two Al^{3+} ions is $2 \times (+3) = +6$] with three oxide ions [total charge on three O^{2-} ions is $3 \times (-2) = -6$] in order to obtain a neutral formula unit for aluminum oxide. Thus, $Al_2O_3(s)$ is the formula of aluminum oxide.
(c) The ionic charge of an aluminum cation is +3 and that of a nitride anion is −3, so we simply balance the charge in the compound by taking one of each ion, yielding a formula of $AlN(s)$.

PRACTICE PROBLEM 6-4: (a) Write the complete ground state electron configurations of Mg^{2+} and N^{3-} and predict the formula of magnesium nitride. (b) Is magnesium nitride a solid, liquid, or gas at room temperature?

Answer: (a) $Mg^{2+}(1s^22s^22p^6)$, $N^{3-}(1s^22s^22p^6)$, $Mg_3N_2(s)$. (b) As noted in Table 6.1, all pure ionic compounds are solids at room temperature.

6-3. The Common Ionic Charges of Transition Metal Ions Can Be Understood in Terms of Electron Configurations

There are many metal ions not listed in Figure 6.3. For example, let's consider silver ($Z = 47$), which has the ground state electron configuration Ag: $[Kr]5s^14d^{10}$. The silver atom would have to lose eleven electrons or gain seven electrons to achieve a noble-gas electron configuration. A review of the data in Table 4.1 for other atoms suggests that the first of these alternatives would require an enormous amount of energy. The energy required for the addition of seven electrons is also prohibitively large. Each successive electron would have to overcome a larger and larger repulsion as the negative charge on the ion increased. Consequently, atomic ions with a charge greater than three are rare, but not unknown.

Although a silver atom cannot reasonably be expected to achieve a noble-gas configuration, its outer electron configuration would be $4s^24p^64d^{10}$ if it lost its $5s$ electron. This configuration, with 18 electrons in the outer shell, is a relatively stable electron configuration and is sometimes called an **18-outer electron configuration.** The unusual stability of the $ns^2np^6nd^{10}$ outer electron configuration is often referred to simply as the **18-electron rule.** Silver forms an 18-electron rule ion, Ag^+, by the loss of one electron, as described by

$$Ag([Kr]5s^14d^{10}) \rightarrow Ag^+([Kr]4d^{10}) + e^-$$

or

$$Ag([Kr]5s^14d^{10}) \rightarrow Ag^+(1s^22s^22p^63s^23p^63d^{10}4s^24p^64d^{10}) + e^-$$

EXAMPLE 6-5: Use the 18-electron rule to predict the electron configuration and the charge of a zinc ion.

Solution: A zinc atom has a $1s^22s^22p^63s^23p^64s^23d^{10}$ ground state electron configuration. A zinc atom can achieve the 18-outer electron configuration $3s^23p^63d^{10}$ by losing its two $4s$ electrons. Thus, we predict that the electron configuration of a zinc ion is

$$Zn^{2+}(1s^22s^22p^63s^23p^63d^{10}) \quad \text{or} \quad Zn^{2+}([Ar]3d^{10})$$

and its charge is +2, which is correct. Note that a Zn^{2+} ion has a completely filled $n = 3$ shell.

PRACTICE PROBLEM 6-5: Use the 18-electron rule to predict a possible charge on an indium ion.

Answer: In^{3+}

Some of the other metals that form 18-electron ions are shown in Figure 6.5. Note that these metals occur near the ends of the d transition series and that the charge on the ions increases by one unit as we look left to right along a row of the periodic table.

Another outer-electron configuration that is often found in ions is illustrated by the element thallium ($Z = 81$). The electron configuration of a thallium atom is $[Xe]6s^25d^{10}6p^1$. Loss of the $6p$ electron yields the electron configuration $[Xe]\ 6s^25d^{10}$ for a Tl^+ ion. Although a Tl^+ ion does not have a noble-gas octet-rule electron configuration or an 18-electron-rule configuration, it does have all its subshells completely filled, and this type of electron configuration is also relatively stable. Other elements that behave like thallium are shown in Figure 6.6. Note that these elements are in Groups 13, 14, and 15. The

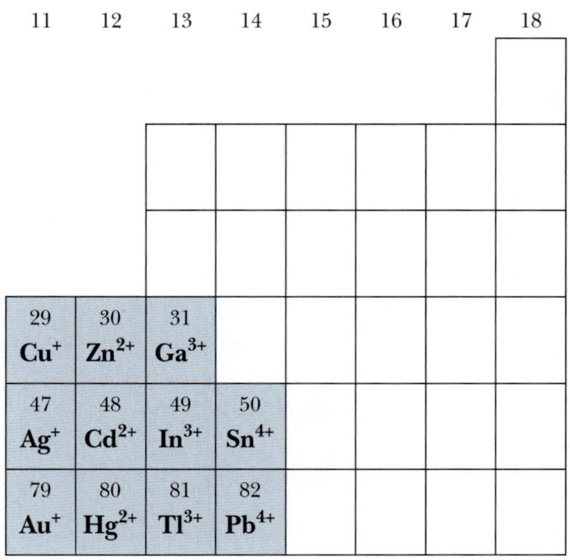

Figure 6.5 Metal ions with an 18-outer electron configuration, $ns^2np^6nd^{10}$.

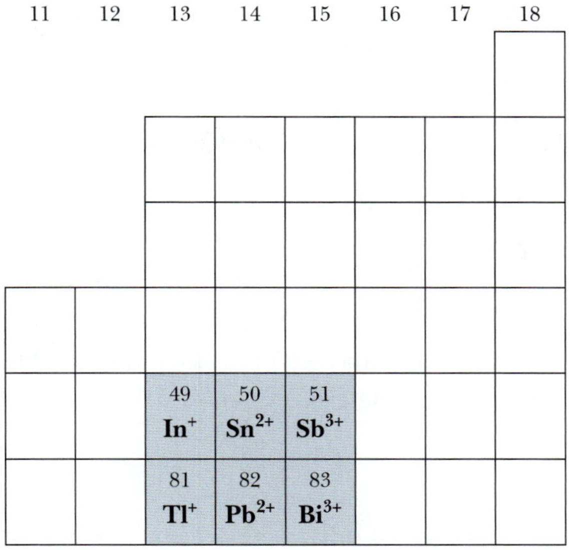

Figure 6.6 Ions with the outer electron configuration [noble gas]$nd^{10}(n+1)s^2$.

singly charged ions in Figure 6.6 can lose two more electrons to achieve an 18-electron-rule configuration. For example, if Tl^{+} loses its two 6*s* electrons, then the resulting configuration for Tl^{3+} is [Xe]$5d^{10}$, which appears in Figure 6.5. Thus, we see that thallium and indium have two common ionic charges, +1 and +3; tin and lead also have two common ionic charges, +2 and +4. The common ionic charges of some selected metals found in typical ionic compounds are given in Table 6.2.

TABLE 6.2 Common ionic charges of selected metals in ionic compounds*

Metals with one common ionic charge	
Group 1 metals:	all +1; e.g., Na^{+}
Group 2 metals:	all +2; e.g., Mg^{2+}
Ag^{-}	Ni^{2+}
Cd^{2+}	Sc^{3+}
Zn^{2+}	Al^{3+}
Metals with two common ionic charges	
Au^{+} and Au^{3+}	Co^{2+} and Co^{3+}
Cu^{+} and Cu^{2+}	Fe^{2+} and Fe^{3+}
$Hg_2^{2+(\dagger)}$ and Hg^{2+}	Tl^{+} and Tl^{3+}
Pb^{2+} and Pb^{4+}	Sb^{3+} and Sb^{5+}
Sn^{2+} and Sn^{4+}	Ti^{3+} and Ti^{4+}
Metals with three common ionic charges	
Cr^{2+}, Cr^{3+}, Cr^{6+}	Mn^{2+}, Mn^{4+}, Mn^{7+}

*Only the most commonly observed ionic charges have been listed; many of these metals have additional less commonly observed ionic charges as well.

†Mercury(I) ion is dimeric (Hg_2^{2+}); that is, it involves a molecular ion composed of two Hg(I) ions bonded together.

6-4. The Ionic Charge of Transition Metal Ions with More Than One Common Ionic Charge Is Indicated by a Roman Numeral

Some of the main-group metals and many of the transition metals can form ions with more than one ionic charge. Thus, it is necessary to indicate the ionic charge of these metal ions when naming compounds containing such a metal. For those metals with more than one ionic charge, the ionic charge of the metal in the compound is indicated by a Roman numeral in parentheses following the name of the metal; this convention gives us the systematic or IUPAC name of the compound. For example, according to Table 6.2, iron ions have two possible charges, +2 and +3. The two chlorides of iron are shown below.

$FeCl_2(s)$ iron(II) chloride $\quad$ $FeCl_3(s)$ iron(III) chloride

EXAMPLE 6-6: Name the compounds: (a) $AuCl_3(s)$ (b) $Fe_2O_3(s)$

Solution: (a) The charge on a chloride ion is −1, so the total charge on the three chloride ions is −3. Thus, the charge on the gold ion in $AuCl_3$ is +3. The systematic name of $AuCl_3(s)$ is gold(III) chloride.

(b) Taking the charge on an oxide ion as −2, we determine that the total charge on the three oxide ions in Fe_2O_3 is −6. This charge must be balanced by the total charge on the two iron ions. Thus, the charge on each iron ion Fe_2O_3 is +3. The systematic name of $Fe_2O_3(s)$ is iron(III) oxide, a common component of rust.

PRACTICE PROBLEM 6-6: (a) Give the systematic names of $Hg_2Cl_2(s)$ and $HgI_2(s)$. (b) Give the chemical formulas for the compounds thallium(III) sulfide and lead(IV) oxide.

Answer: (a) mercury(I) chloride, mercury(II) iodide (b) $Tl_2S_3(s)$, $PbO_2(s)$ Note that, as indicated in the footnote to Table 6.2, dimeric mercury, Hg_2^{2+}, is called mercury(I).

When a metal or transition metal has only one common ionic charge, we do not include the charge in the name. For example, $AgCl(s)$ is written as silver chloride and *not* silver(I) chloride because the only common ionic charge of silver is +1. The common ionic charges of all the elements are given in the periodic table on the inside front cover of this book. It should be noted that while we have given the common, or most typical, ionic charges of the elements found in chemical compounds, other ionic charges are also possible.

An older nomenclature, which is still in use, changes the ending of the metals' names according to their ionic charge, rather than using Roman numerals. In the older nomenclature system, an *-ous* ending is used to indicate the lower common ionic charge, and an *-ic* ending to indicate the higher common ionic charge. Those ions with only one common ionic charge (like zinc and silver) are given no special ending. Because some ions like cobalt can be +2 and +3, while others like titanium are +3 and +4, this system requires that you memorize the common ionic charges in order to name a compound and so is less convenient than the modern IUPAC system of nomenclature. Some examples of older names that follow these rules are shown below.

Ion	Systematic name	Older name	Ion	Systematic name	Older name
Co^{2+}	cobalt(II)	cobaltous	Ti^{3+}	titanium(III)	titanous
Co^{3+}	cobalt(III)	cobaltic	Ti^{4+}	titanium(IV)	titanic
Tl^{+}	thallium(II)	thallous	Zn^{2+}	zinc	zinc
Tl^{3+}	thallium(III)	thallic			

In addition to those metals that follow these rules, many others were named using their Latin roots in the older system. A list of such compounds is given in Table 6.3. For example, iron(III) bromide is called ferric bromide in the older nomenclature. A compound that became famous as an ingredient in the first fluoride-containing toothpastes is stannous fluoride, $SnF_2(s)$, the modern name for which is tin(II) fluoride. For compounds with more than two common ionic charges, the naming is inconsistent. Often the name of the element was used with a Greek prefix indicating the number of anions. For example,

TABLE 6.3 Metals with special older names and other exceptions (italicized)

Metal	Ionic charge	Symbols	Modern systematic (IUPAC) names	Old names
copper	+1	Cu^{+}	copper(I)	*cuprous*
	+2	Cu^{2+}	copper(II)	*cupric*
gold	+1	Au^{+}	gold(I)	*aurous*
	+3	Au^{3+}	gold(III)	*auric*
iron	+2	Fe^{2+}	iron(II)	*ferrous*
	+3	Fe^{3+}	iron(III)	*ferric*
lead	+2	Pb^{2+}	lead(II)	*plumbous*
	+4	Pb^{4+}	lead(IV)	*plumbic*
mercury	+1	Hg_2^{2+}	mercury(I)*	mercurous
	+2	Hg^{2+}	mercury (II)	mercuric
tin	+2	Sn^{2+}	tin(II)	*stannous*
	+4	Sn^{4+}	tin(IV)	*stannic*

*Mercury(I) ion is dimeric (Hg_2^{2+}). Although the symbol Hg comes from the Latin word *hydrargyrum*, meaning liquid silver, this name is not used in the older nomenclature.

chromium trioxide is the older name for what we now call chromium(VI) oxide, $CrO_3(s)$. In this book, in most cases we shall use the modern IUPAC (systematic) names of compounds or ions. However, you should still have some familiarity with the older names in case you should encounter them elsewhere, especially in the older literature. The use of Roman numerals to denote the charge on the metal ion eliminates the need to specify the number of anions. Thus, iron(II) dichloride is redundant because iron(II) denotes Fe^{2+}, which in turn requires two chloride ions per formula unit to balance the charge on Fe^{2+}. For this reason, the systematic name of $FeCl_2(s)$ is simply iron(II) chloride.

6-5. The Filling Order of Most Transition Metal Ions Is Regular

As we have seen in the previous sections, the ground state electron configurations of transition metal ions do not follow the same filling order as neutral atoms. Nevertheless, in most cases the ground state electron configurations of transition metal ions are relatively easy to deduce. We learned in Chapter 5 that the $3d$ orbitals are filled after the $4s$ orbital in neutral atoms. This order of filling occurs because the order of energy levels for most *neutral atoms* is that shown in Figures 5.11b and 5.12. However, when electrons are lost from a neutral atom, the charge on the ion alters the order of the orbital energies such that in most transition metal ions, the energy of the $3d$ orbitals is less than that of the $4s$ orbital. A similar situation occurs for the $4d$ and $5s$ orbitals and the $5d$ and $6s$ orbitals. Therefore, the filling order of the orbitals for the majority of transition metal ions is regular, in the sense that the order of orbital energies is

$$1s < 2s < 2p < 3s < 3p < 3d < 4s < 4p < 4d < 4f < \ldots$$

Thus, although the ground state electron configuration of a nickel atom is $[Ar]4s^23d^8$, that for a singly charged nickel ion, Ni^+, is $[Ar]3d^9$ (as confirmed by spectroscopic evidence) and is not $[Ar]4s^23d^7$ or $[Ar]4s^13d^8$. Similarly, the ground state electron configuration of a doubly-charged Ni^{2+} ion is $[Ar]3d^8$ and not $[Ar]4s^23d^6$ (Figure 6.7).

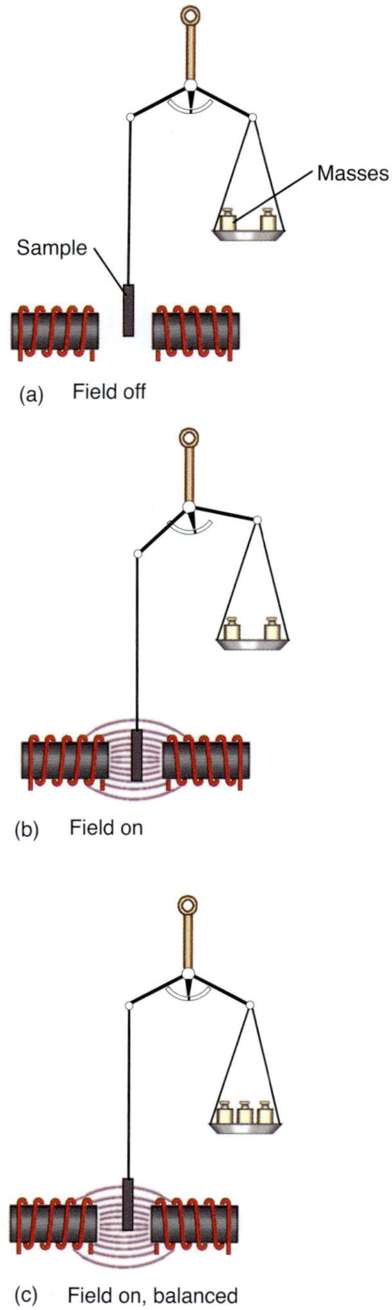

Figure 6.7 One method for determining the electronic configuration of transition metal ions is to use a magnetic field. Because unpaired electrons act like tiny magnets, the number of unpaired electrons in a transition metal ion can be determined by weighing the sample in a magnetic field. If the sample contains unpaired electrons, the magnetic attractive force on the sample makes it appear heavier. The number of unpaired electrons in the sample and the electron configuration can be calculated from the apparent mass gain when the field is on. Masses are added to the balance pan until balance is restored with the field on.

EXAMPLE 6-7: Predict the ground state electron configuration of Fe^{3+}.

Solution: The ground state electron configuration of an iron atom ($Z = 26$) is

$$[Ar]4s^23d^6$$

Although the $4s$ orbital has a lower energy than the $3d$ orbitals have in most neutral atoms, this situation is not true for the majority of ions. In Fe^{3+} the $3d$ orbitals are lower in energy than the $4s$ orbital. Therefore, the ground state electron configuration of Fe^{3+}, which has three fewer electrons than a neutral iron atom, is

$$[Ar]3d^5$$

PRACTICE PROBLEM 6-7: Predict the ground state electron configuration of a Pd^{2+} ion.

Answer: Pd^{2+}: $[Kr]4d^8$

EXAMPLE 6-8: Referring only to the periodic table, predict the ground state electron configurations of the ions Cu^+ and Cu^{2+}.

Solution: The atomic number of copper is 29, so a Cu^+ ion has 28 electrons, and a Cu^{2+} ion has 27 electrons. The order of filling the orbitals of the transition-metal ions is regular, so the electron configurations are as follows:

$$Cu^+\text{: } 1s^22s^22p^63s^23p^63d^{10} \quad \text{or} \quad [Ar]3d^{10}$$

$$Cu^{2+}\text{: } 1s^22s^22p^63s^23p^63d^9 \quad \text{or} \quad [Ar]3d^9$$

The Cu(II) ionic charge is the most common ionic charge of copper ions.

PRACTICE PROBLEM 6-8: Predict the ground state electron configurations of the Cr(II), Cr(III), and Cr(VI) ions.

Answers: Cr(II): $[Ar]3d^4$; Cr(III): $[Ar]3d^3$; Cr(VI): $[Ar]$

Note that the common ionic charges of the transition metals are all positive, so there is no need to consider the case of transition metal anions here.

6-6. Cations Are Smaller and Anions Are Larger Than Their Neutral Parent Atoms

Because atoms and ions are different species, we should expect atomic radii and **ionic radii** to have different values. For example, the average distance from the nucleus of the 3*s* electron in a sodium atom is greater than that of the 1*s*, 2*s*, and 2*p* electrons because the 3*s* electron is in the $n = 3$ shell. When a sodium atom loses its 3*s* electron, only the $n = 1$ and $n = 2$ shells are occupied, so a Na^+ ion is smaller than a sodium atom. In addition, the excess positive charge draws the remaining electrons toward the nucleus, causing the electron distribution to contract. Positive atomic ions are always smaller than their corresponding neutral atoms for this reason. As discussed in Section 5-12, we shall use the crystallographic radii of ions taken from X-ray data on ionic solids to estimate the size of these ions.

The relative sizes of the alkali metal atoms and ions are shown in the first column of Figure 6.8. The Group 2 metals lose two outer *s* electrons in becoming M^{2+} ions. The excess positive charge of +2 contracts the remaining electron shells even more than in the case of the Group 1 metals, as you can see by comparing the sizes in the first and second columns of Figure 6.8.

The atoms of nonmetals gain electrons in becoming ions. The addition of an extra electron increases the electron-electron repulsion and causes the electron distribution to expand. Negative ions are always larger than their corresponding neutral atoms for this reason. For example, the relative atomic and ionic sizes of the halogen atoms and halide ions are shown in the column headed by 17 in Figure 6.8. Numerical values of the crystallographic radii of many ions are given in Table 6.4.

TABLE 6.4 Crystallographic radii of ions in picometers (1 pm = 10^{-12} m)

Ion	Radius	Ion	Radius	Ion	Radius	Ion	Radius	Ion	Radius	Ion	Radius	Ion	Radius
Cations:								**Anions:**					
Ag^+	115	Ba^{2+}	135	Al^{3+}	54	Ce^{4+}	87	Br^-	196	O^{2-}	140	N^{3-}	171
Cs^+	167	Ca^{2+}	100	B^{3+}	23	Ti^{4+}	61	Cl^-	181	S^{2-}	184	P^{3-}	212
Cu^+	77	Cd^{2+}	95	Cr^{3+}	62	U^{4+}	89	F^-	133	Se^{2-}	198		
K^+	138	Co^{2+}	65	Fe^{3+}	55	Zr^{4+}	84	H^-	154	Te^{2-}	221		
Li^+	76	Cu^{2+}	73	Ga^{3+}	62			I^-	220				
Na^+	102	Fe^{2+}	61	In^{3+}	80								
Rb^+	152	Mg^{2+}	72	La^{3+}	103								
Tl^+	150	Ni^{2+}	69	Tl^{3+}	89								
		Sr^{2+}	118	Y^{3+}	90								
		Zn^{2+}	74										

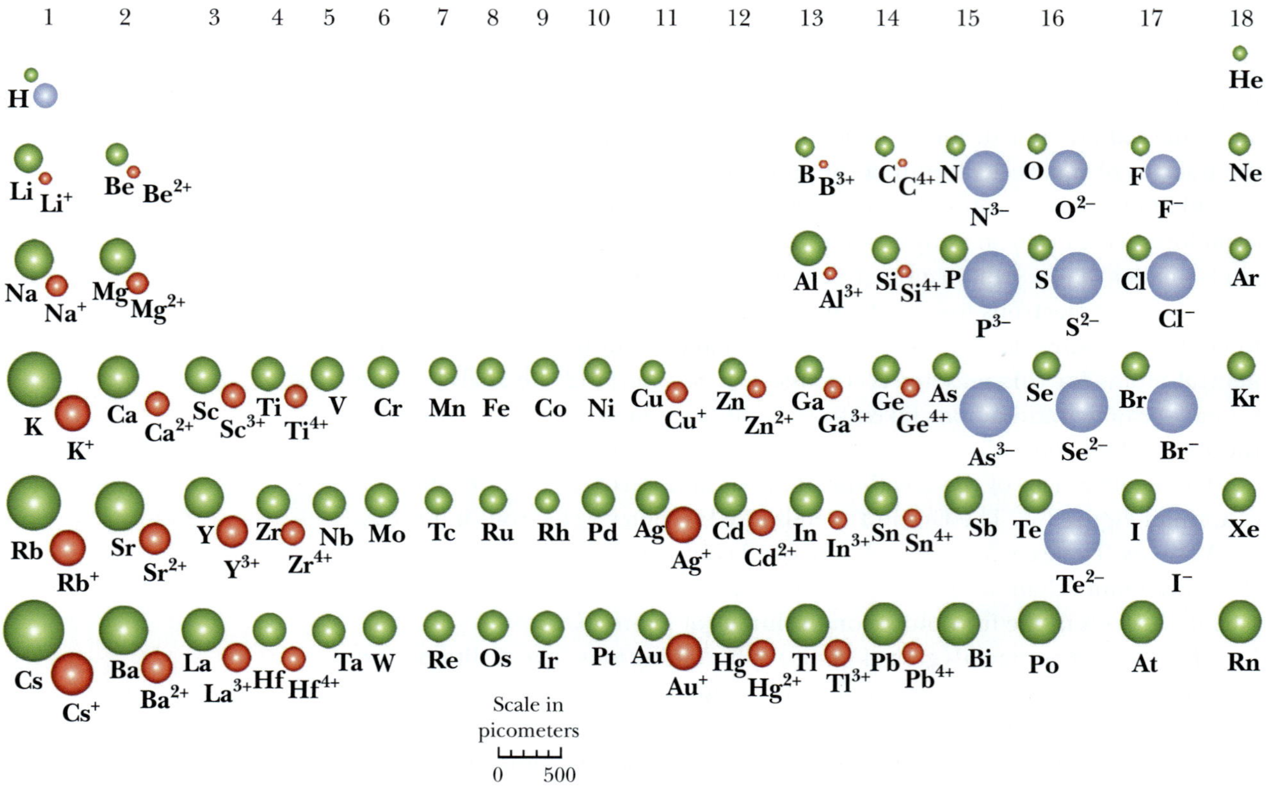

Figure 6.8 The relative sizes of atoms and ions. Neutral atoms are shown in green, cations in red, and anions in blue. Cations are smaller and anions are larger than their respective neutral parent atoms.

EXAMPLE 6-9: Without reference to Table 6.4 or to Figure 6.8, predict which is the larger ion, K^+ or Cl^-.

Solution: The ground state electron configuration of both K^+ and Cl^- is $1s^2 2s^2 2p^6 3s^2 3p^6$. Both ions have the same number of electrons, 18 (that is, they are **isoelectronic**), but a potassium atom has a nuclear charge of +19 and a chlorine atom has a nuclear charge of only +17. The excess positive charge of a K^+ ion contracts the electron distribution, and the excess negative charge of a Cl^- ion leads to an enlargement of the electron distribution, so we predict that a Cl^- ion is larger than a K^+ ion. The radius of a K^+ ion is 138 pm and that of a Cl^- ion is 181 pm.

PRACTICE PROBLEM 6-9: Without reference to Table 6.4 or to Figure 6.8, arrange the following ions in order of increasing size:

Mg^{2+} Na^+ I^- Br^- Al^{3+}

Answer: (smallest) $Al^{3+} < Mg^{2+} < Na^+ < Br^- < I^-$ (largest)

6-7. Coulomb's Law Is Used to Calculate the Energy of an Ion Pair

Up to now our discussion of ionic bonds has been qualitative. We now show by calculations that when an ionic bond is formed, the energy of the ionic product is lower than that of the atomic reactants. Let's consider the reaction described by the equation

$$Na(g) + Cl(g) \rightarrow Na^{+}Cl^{-}(g) \qquad (6.1)$$

Because the energy of the $Na^{+}Cl^{-}(g)$ product is less than that of the atomic reactants, energy is released in this process and a stable bond results.

The net energy change for the reaction can be calculated conveniently by imagining the reaction to take place in three separate steps and then calculating the sum of the energy changes for each of these steps:

1. The electron is removed from the sodium atom (ionization). The energy required to ionize a sodium atom is 0.824 aJ.
2. The electron removed from the sodium atom is added to the chlorine atom. Energy is released in the process, and this energy is called the **electron affinity** of chlorine. (We discuss the idea of electron affinity below.) The electron affinity of atomic chlorine is −0.580 aJ.
3. The sodium ion and chloride ion are brought together as shown in Figure 6.9. From Table 6.4 we see that the radius of a sodium ion is 102 pm and that of a chloride ion is 181 pm. Thus, the centers of the two ions are 181 + 102 = 283 pm apart when the two ions are just touching, assuming that the ions behave like hard spheres. We refer to the distance between the centers of an ionically bonded ion pair as the **equilibrium ion-pair separation distance,** denoted by d_{eq}.

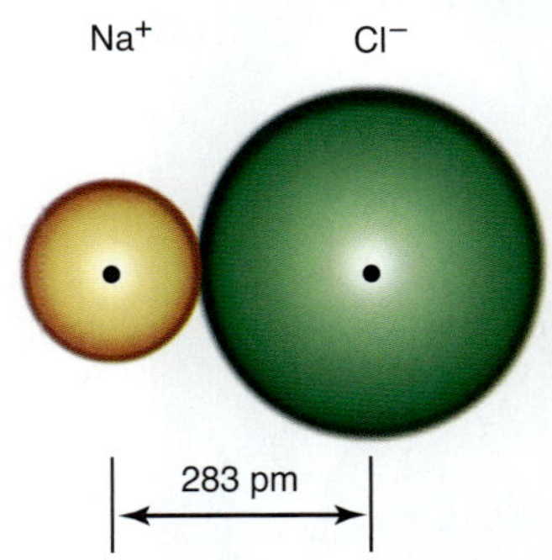

Figure 6.9 A solid-sphere representation of the ion pair $Na^{+}Cl^{-}(g)$. According to Table 6.4, the sodium ion can be represented as a hard sphere of radius 102 pm and the chloride ion can be represented as a hard sphere of radius 181 pm. Because the two ions have opposite charges, they draw together until they touch, with a distance between their centers of 102 pm + 181 pm = 283 pm. They are bound together at this distance in an ionic bond. We call this value the equilibrium ion-pair separation distance, d_{eq}.

Step 1 involves ionization energy, which we discussed in Section 4-1. We now need to discuss the energy associated with steps 2 and 3. Consider the process of adding an electron to a gaseous atom:

$$A(g) + e^{-} \rightarrow A^{-}(g)$$

The energy associated with this process is called the **first electron affinity,** EA_1, of the atom $A(g)$. For example, for a chlorine atom we have

$$Cl(g) + e^{-} \rightarrow Cl^{-}(g) \qquad EA_1 = -0.580 \text{ aJ}$$

The value of EA_1 in this case is negative because energy is released in the process.

Just as we define successive ionization energies, we define successive electron affinities. For example, the first two electron affinities for an oxygen atom are

$$O(g) + e^{-} \rightarrow O^{-}(g) \qquad EA_1 = -0.234 \text{ aJ}$$
$$O^{-}(g) + e^{-} \rightarrow O^{2-}(g) \qquad EA_2 = +1.30 \text{ aJ}$$

TABLE 6.5 Electron affinities of the atoms of some reactive nonmetals

Atom	EA/aJ
H	−0.12
F	−0.545
Cl	−0.580
Br	−0.540
I	−0.490
O	−0.234
	+1.30 (EA_2)
S	−0.332
	+0.980 (EA_2)

Figure 6.10 Charles-Augustin de Coulomb (1736–1806) was born in Angouleme, France into a well-to-do family. He studied mathematics early in his education but due to a reversal of his family's fortune, he entered a military school where he was trained in engineering. He worked on military projects in France and on the island of Martinique in the Caribbean. He returned to supervise various engineering projects in France, and during this time, he discovered the relationship between electric charges and the distance between them, now known as Coulomb's law. As a result of the importance of his experimental work, he was stationed permanently in Paris doing research. He worked on the committee revising the standards of weights and measure, which led to the metric system.

Notice that the value of the **second electron affinity,** EA_2, is positive; it requires energy to overcome the repulsion between the negatively charged ion $O^-(g)$ and the electron. The most important electron affinities for our purposes are those of the reactive nonmetals (Table 6.5).

So far then, for NaCl we can write for steps 1 and 2

$$\text{step 1: } Na(g) \rightarrow Na^+(g) + e^- \qquad I_1 = 0.824 \text{ aJ}$$

$$\text{step 2: } Cl(g) + e^- \rightarrow Cl^-(g) \qquad EA_1 = -0.580 \text{ aJ}$$

If we add these two equations, we find that

$$\text{step 1 + step 2: } Na(g) + Cl(g) \rightarrow Na^+(g) + Cl^-(g) \qquad E_{1+2} = I_1 + EA_1 = 0.244 \text{ aJ}$$

Each of the species in this process is in the gas phase. The products $Na^+(g)$ and $Cl^-(g)$ are so far apart that they are effectively isolated entities. We now must calculate the energy change involved in bringing the two widely separated ions (where their energy of interaction is taken to be zero) to their equilibrium ion-pair separation distance of 283 pm (Figure 6.9). To calculate this energy, we use Coulomb's law.

Coulomb's law (Figure 6.10) states that the energy of interaction of two ions is directly proportional to the product of their electrical charges and is inversely proportional to the distance between their centers. Thus, for the energy of interaction we have

$$E_{\text{coulomb}} = k\frac{Q_1 Q_2}{d} \tag{6.2}$$

where Q_1 and Q_2 are the charges of the two ions, d is the distance between the centers of the two ions, and k is a proportionality constant. The value of the proportionality constant k depends on the units of the charges and of the distance d. If the charges are measured in units of the charge on a proton (+1 for Na^+, −1 for Cl^-, and so on), d is measured in picometers (1 pm = 10^{-12} m), and E_{coulomb} is expressed in attojoules, then E_{coulomb} is given by

$$E_{\text{coulomb}} = (231 \text{ aJ}\cdot\text{pm})\frac{Q_1 Q_2}{d} \tag{6.3}$$

This relationship is illustrated in Figure 6.11. We see from Equation 6.3 that as d, the ion-pair distance, gets very large, the electrostatic energy of interaction goes to zero. If the charges of the ions have the same sign, then E_{coulomb} in Equation 6.3 is positive. If the ions are oppositely charged, then E_{coulomb} is negative. Remember that like charges repel and opposite charges attract so that only ions with opposite charges can form stable ionic bonds. For a Na^+Cl^- ion pair, the charge on the sodium ion, Q_1, is +1, the charge on the chloride anion, Q_2, is −1, and the distance between the ions, d, is 283 pm (= 102 pm + 181 pm). Thus,

$$E_{\text{coulomb}} = (231 \text{ aJ}\cdot\text{pm})\frac{(+1)(-1)}{283 \text{ pm}} = -0.816 \text{ aJ} = -8.16 \times 10^{-19} \text{ J}$$

The minus sign means that the ions attract each other and that energy is released when the ions are brought together. Thus, the energy of the pair of ions at 283 pm is less than it is when the ions are very far from each other. The quantity −0.816 aJ is the amount of energy released by the formation of one Na^+Cl^- (*g*) ion pair. We can express the Na^+Cl^-(*g*) ion-pair formation as a chemical equation:

$$\text{step 3: } Na^+(g) + Cl^-(g) \underset{d = 283\text{ pm}}{\rightarrow} Na^+Cl^-(g) \qquad E_3 = -0.816\text{ aJ}$$

Adding the chemical equation for step 3 to the sum of steps 1 and 2 and the corresponding changes in energy gives the total energy change for Equation 6.1.

$$Na(g) + Cl(g) \underset{d = 283\text{ pm}}{\rightarrow} Na^+Cl^-(g)$$

$$E_{rxn} = I_1 + EA_1 + E_3 = E_{1+2} + E_3$$
$$= 0.244\text{ aJ} - 0.816\text{ aJ} = -0.572\text{ aJ} = -5.72 \times 10^{-19}\text{ J}$$

The fact that energy is released in the process means that the energy of the Na^+Cl^-(*g*) ion pair is lower than that of the two separated gaseous atoms. Because the ion pair has a lower energy than two separated atoms, it is stable with respect to the atoms. The overall process is illustrated in Figure 6.12.

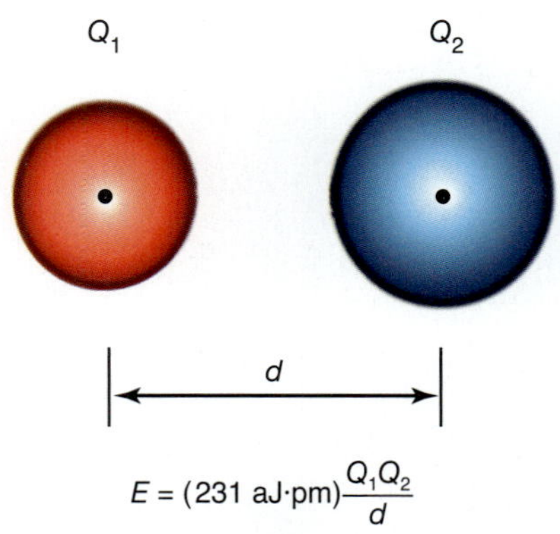

Figure 6.11 Two ions separated by a distance *d*. The charges of the ions are Q_1 and Q_2. The energy of interaction of two ions is given by Coulomb's law (Equation 6.3).

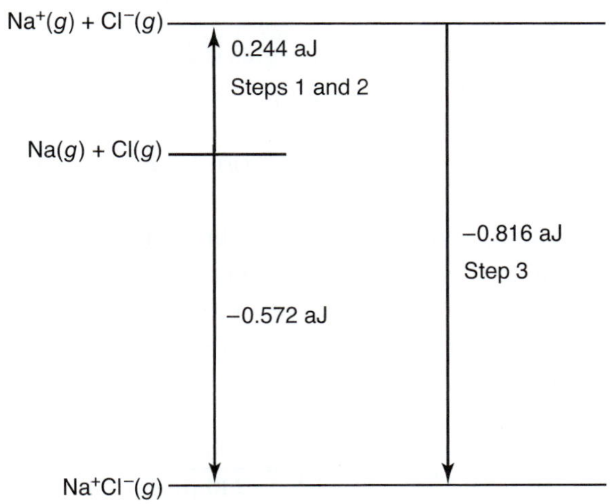

Figure 6.12 Steps used to calculate the energy released in the process $Na(g) + Cl(g) \rightarrow Na^+Cl^-(g)$. First the atoms are converted to ions (steps 1 and 2), and then the two ions are brought together to a distance equal to the sum of their crystallographic radii (step 3). These first and second steps use the ionization energy of sodium and the electron affinity of chlorine. The third step uses Coulomb's law to calculate the energy involved in bringing the two isolated ions together.

EXAMPLE 6-10: Calculate the energy released in the reaction described by the equation

$$Ca(g) + O(g) \rightarrow CaO(g)$$

given that the first and second ionization energies of $Ca(g)$ are 0.980 aJ and 1.90 aJ, respectively.

Solution: We can describe this reaction using the following three steps: (1) the ionization of $Ca(g)$; (2) the addition of two electrons to $O(g)$; and (3) the bringing together of $Ca^{2+}(g)$ and $O^{2-}(g)$ to a distance equal to their equilibrium ion-pair separation.

1. From the first and second ionization energies of $Ca(g)$, we write

$$Ca(g) \rightarrow Ca^{+}(g) + e^{-} \qquad I_1 = +0.980 \text{ aJ}$$

$$Ca^{+}(g) \rightarrow Ca^{2+}(g) + e^{-} \qquad I_2 = +1.90 \text{ aJ}$$

The sum of these ionization energies is $E_1 = I_1 + I_2 = 2.88$ aJ.

2. From Table 6.5 we see that the first electron affinity of $O(g)$ is −0.234 aJ and its second electron affinity is +1.30 aJ,

$$O(g) + e^{-} \rightarrow O^{-}(g) \qquad EA_1 = -0.234 \text{ aJ}$$

$$O^{-}(g) + e^{-} \rightarrow O^{2-}(g) \qquad EA_2 = +1.30 \text{ aJ}$$

The sum of these electron affinity energies is $E_2 = EA_1 + EA_2 = 1.07$ aJ. If we add the results of steps 1 and 2, we get

$$Ca(g) + O(g) \rightarrow Ca^{2+}(g) + O^{2-}(g) \qquad E_{1+2} = +3.95 \text{ aJ}$$

3. We now calculate the energy involved in bringing $Ca^{2+}(g)$ and $O^{2-}(g)$ to a distance equal to their equilibrium ion-pair separation. According to Table 6.4, the crystallographic radius of a Ca^{2+} ion is 100 pm and that of an O^{2-} ion is 140 pm. Thus, their equilibrium ion-pair distance is 240 pm. We now use Equation 6.3:

$$\begin{aligned} E_{\text{coulomb}} &= (231 \text{ aJ}\cdot\text{pm})\frac{(+2)(-2)}{240 \text{ pm}} \\ &= -3.85 \text{ aJ} = -3.85 \times 10^{-18} \text{ J} \end{aligned}$$

The minus sign indicates that energy is released in the process. We can express this result in the form of an equation:

$$Ca^{2+}(g) + O^{2-}(g) \xrightarrow[d=240 \text{ pm}]{} Ca^{2+}O^{2-}(g) \qquad E_3 = -3.85 \text{ aJ}$$

If we combine this result with the net result of steps 1 and 2, then for the equation

$$Ca(g) + O(g) \rightarrow Ca^{2+}O^{2-}(g)$$

we get

$$E_{rxn} = E_{1+2} + E_3 = 3.95 \text{ aJ} - 3.85 \text{ aJ} = 0.10 \text{ aJ}$$

Although this value is positive, CaO forms a stable ionic solid; we have not yet considered the energy involved in forming a crystal lattice.

The rules for the subtraction of significant figures leaves only two significant figures in this case.

Notice that the coulombic energy released from the formation of a CaO(*g*) ion pair (−3.85 aJ) from the separated ions in step 3 is about four times greater than that released from the formation of an NaCl(*g*) ion pair (−0.816 aJ) due to the doubly charged calcium and oxygen ions.

PRACTICE PROBLEM 6-10: Calculate the value of E_{rxn} for the reaction described by the equation

$$\text{Cs}(g) + \text{Cl}(g) \rightarrow \text{Cs}^+\text{Cl}^-(g)$$

given that the first ionization energy of Cs(*g*) is 0.624 aJ.

Answer: $E_{rxn} = -0.619$ aJ

Purely ionic chemical bonds are the simplest type of chemical bonds. They are the result of an electrostatic attraction between oppositely charged ions. If we know the ionic charges involved and their equilibrium ion pair distance, then we can use Equation 6.3 to calculate the energy released when an ionic bond is formed. This energy is equal and opposite to the amount of energy that must be supplied to break the ionic bond when separating the ions.

The reactions that we have discussed so far have been simplified in the sense that we have discussed reactions between only gaseous atoms to form gaseous ion pairs. At room temperature, sodium chloride exists as a solid consisting of sodium ions and chloride ions, as depicted in Figure 6.13. If you look carefully at the two representations in Figure 6.13 (especially the one in b), then you'll see that each sodium ion is surrounded by a set of six (nearest neighbor) chloride ions, then a set of 12 (next nearest neighbor) sodium ions, and so on. Using Coulomb's law for all these interactions, it is possible to calculate the energy of

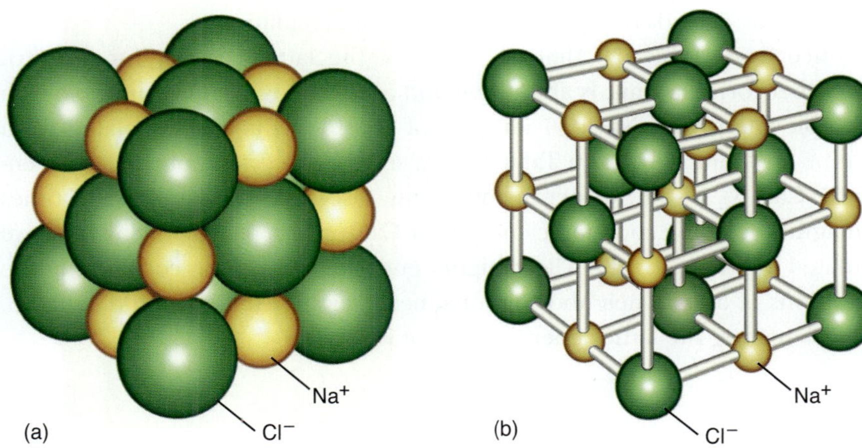

Figure 6.13 The crystalline structure of NaCl(*s*). Each Na^+ ion is surrounded by six Cl^- ions, and each Cl^- ion is in turn surrounded by six Na^+ ions. (a) This representation shows the ions as spheres drawn to scale and illustrates the packing that occurs in the crystal. (b) A more open illustration of the NaCl(*s*) crystal structure.

TABLE 6.6 Calculated and experimental lattice energies of some selected ionic compounds

Compound	Calculated lattice energies/ aJ·formula unit^{-1}	Measured lattice energies/ aJ·formula unit^{-1}
NaF	–1.51	–1.54
NaCl	–1.28	–1.31
NaBr	–1.22	–1.25
KF	–1.34	–1.38
KCl	–1.16	–1.20
KBr	–1.11	–1.15
CaF_2	–4.38	–4.40
$CaCl_2$	–3.77	–3.77
Na_2O	–4.12	–4.11
K_2O	–3.72	–3.71
CaO	–5.67	–5.65

the entire crystal, and such calculations have been done by chemists for many different types of crystals. The resultant energy is called the **lattice energy** of the crystal. In calculating lattice energies, we picture a crystal as made up of small hard (impenetrable) spherical particles situated at lattice positions, as depicted in Figure 6.13a. In actuality, the particles are not completely impenetrable and wiggle, or vibrate, a little bit about their equilibrium positions, but our hard-sphere model is perfectly adequate for such calculations. One of the reasons why all ionic compounds are solids at room temperature is because of the large amount of energy that must be supplied to break up these extended crystal lattices.

Table 6.6 lists the calculated values of the lattice energy per formula unit for a number of ionic solids. You can see that the agreement between the calculated and experimental values is quite good, which lends credence to the hard-sphere model of ionic crystals that we have used for our calculations.

Notice from Table 6.6 that the value of the lattice energy for NaCl(*s*), –1.28 aJ per formula unit, is about one and a half times the magnitude of the coulombic energy per NaCl(*g*) ion pair of –0.816 aJ that we calculated from Equation 6.3. Similarly, from Table 6.6 we see that the value of the lattice energy for CaO(*s*) is –5.67 aJ per formula unit, also about one and a half times the magnitude of the coulombic energy per CaO(*g*) ion pair of –3.85 aJ that we calculated in Example 6.10. These higher energies result from considering the interactions of all the neighboring ions, next neighboring ions, and so forth, when calculating the lattice energy of an ionic crystal.

SUMMARY

An ionic bond is the electrostatic force that binds oppositely charged ions together. An ionic compound is formed between two reactants if one of them has a relatively low ionization energy and the other has a relatively high electron affinity. This event occurs when a reactive metal reacts with a reactive nonmetal. When it occurs, one or more electrons are transferred completely from the metal atom to the nonmetal atom, a process resulting in an ionic bond.

Many ions are stable because they achieve a noble-gas electron configuration, although there are stable ions that have other types of electron configurations as well, such as the 18-outer electron configuration. For most main-group elements we can deduce the most stable ionic charge from the atom's location in the periodic table. Ions are different from atoms, so ionic radii are not the same as atomic radii. The relative sizes of ions and their trends in the periodic table can be understood in terms of electron configurations.

Most transition metals and some main-group metals can have multiple ionic charges. When naming compounds containing a metal that has more than one common ionic charge, we include the ionic charge of the metal after its name in the compound using Roman numerals in parentheses.

A quantitative discussion of the energetics of the formation of ionic compounds from their respective elements in the gas phase involves the concepts of ionization energy, electron affinity, ion-pairing energy, and lattice energy.

TERMS YOU SHOULD KNOW

ionic compound *171*
covalent compound *171*
electrostatic force *173*
ionic bond *173*
octet rule *174*
cation *174*
anion *174*
18-outer electron configuration *177*
18-electron rule *177*
ionic radii *183*
isoelectronic *184*
electron affinity *185*
equilibrium ion-pair separation distance, d_{eq} *185*
first electron affinity, EA_1 *185*
second electron affinity, EA_2 *186*
Coulomb's law *186*
lattice energy *190*

AN EQUATION YOU SHOULD KNOW HOW TO USE

$$E_{coulomb} = (231\ \text{aJ}\cdot\text{pm})\frac{Q_1Q_2}{d} \qquad (6.3)$$

(Coulomb's law; d is the separation of the centers of the ions in picometers and Q_1 and Q_2 are the charges on the ions in units of the charge on a proton.)

PROBLEMS

NOMENCLATURE AND CHEMICAL FORMULAS

6-1. Indicate which of the following ions have a noble-gas electron arrangement and, for such ions, identify the corresponding noble gas:

(a) Cs^+ (b) Ga^{3+} (c) S^{2-}
(d) P^{3-} (e) Al^{3+} (f) O^-

6-2. Indicate which of the following ions have a noble-gas electron arrangement and, for such ions, identify the corresponding noble gas:

(a) Ba^{2+} (b) Ca^+ (c) Br^-
(d) N^{3-} (e) Se^{2-} (f) Fe^{2+}

6-3. Determine the ionic charge on each atom in the following compounds, and name the compound:

(a) MgS (b) AlP

(c) BaF_2 (d) Ga_2O_3

6-4. Determine the ionic charge on each atom in the following compounds, and name the compound:

(a) Li_2O (b) CaS

(c) Mg_3N_2 (d) Al_2S_3

6-5. Write the chemical formula and systematic name for the binary compound formed from each of the following pairs of ions:

(a) Fe^{3+} and O^{2-} (b) Cd^{2+} and S^{2-} (c) Ni^{2+} and Br^-

(d) Ru^{3+} and F^- (e) Tl^+ and S^{2-}

6-6. Write the chemical formula and systematic name for the binary compound formed from each of the following pairs of ions:

(a) Ga^{3+} and S^{2-} (b) Fe^{3+} and Se^{2-} (c) Pb^{4+} and O^{2-}

(d) Ba^{2+} and At^- (e) Zn^{2+} and N^{3-}

6-7. Write the chemical formula for

(a) gallium selenide (b) aluminum phosphide

(c) potassium iodide (d) strontium fluoride

6-8. Write the chemical formula for

(a) aluminum sulfide (b) sodium oxide

(c) barium fluoride (d) lithium hydride

6-9. Write the chemical formula for

(a) lithium nitride (b) gallium telluride

(c) barium nitride (d) magnesium bromide

6-10. Write the chemical formula for

(a) cesium oxide (b) sodium selenide

(c) lithium sulfide (d) calcium iodide

6-11. Write the systematic name of

(a) $CuI(s)$ (b) $Hg_2Br_2(s)$ (c) $Fe_2O_3(s)$

(d) $CoF_2(s)$ (e) $FeO(s)$

6-12. Write the systematic name of

(a) $SnO_2(s)$ (b) $FeF_3(s)$ (c) $PbO_2(s)$

(d) $CoN(s)$ (e) $HgSe(s)$

6-13. Write the systematic name of

(a) $AgBr(s)$ (b) $PbCl_2(s)$ (c) $ZnO(s)$

(d) $CuF_2(s)$ (e) $Ba_3N_2(s)$

6-14. Write the systematic name of

(a) $NaH(s)$ (b) $SnI_2(s)$ (c) $Au_2S(s)$

(d) $CdS(s)$ (e) $K_2O(s)$

6-15. Write the chemical formula for

(a) cobalt(III) phosphide (b) titanium(IV) chloride

(c) manganese(IV) oxide (d) chromium(VI) oxide

(e) vanadium(V) oxide

6-16. Write the chemical formula for

(a) ruthenium(III) sulfide (b) scandium fluoride

(c) osmium(VIII) oxide (d) manganese(II) sulfide

(e) platinum(IV) chloride

6-17. Write the chemical formula for

(a) yttrium sulfide (b) tungsten(VI) bromide

(c) magnesium telluride (d) rubidium nitride

6-18. Write the chemical formula for

(a) thallium(III) chloride (b) cadmium iodide

(c) zinc arsenide (d) aluminum bromide

6-19. For each of the compounds in Problem 6-11, give the older names for the compounds if different from the systematic name.

6-20. For each of the compounds in Problem 6-12, give the older names for the compounds if different from the systematic name.

6-21. Give the systematic name and chemical formula for each of the following compounds named using the older nomenclature:

(a) ferric bromide (b) stannic nitride

(c) ferrous bromide (d) aurous sulfide

6-22. Give the systematic name and chemical formula for each of the following compounds named using the older nomenclature:

(a) plumbic fluoride (b) mercurous chloride

(c) plumbous sulfide (d) mercuric oxide

6-23. For each of the following equations, write the chemical formulas and balance the equation:

(a) sodium + sulfur → sodium sulfide

(b) calcium + bromine → calcium bromide

(c) barium + oxygen → barium oxide

(d) sulfur dioxide + oxygen → sulfur trioxide

6-24. For each of the following equations, write the chemical formulas and balance the equation:

(a) carbon monoxide + oxygen → carbon dioxide

(b) cesium + bromine → cesium bromide

(c) nitrogen monoxide + oxygen → nitrogen dioxide

(d) ammonia + oxygen → nitrogen monoxide + water

ELECTRON CONFIGURATIONS AND CHEMICAL REACTIONS

6-25. Describe the following equations in terms of the electron configurations of the various species involved:

(a) $Ca(g) + 2F(g) \rightarrow CaF_2(g)$

(b) $Sr(g) + 2Br(g) \rightarrow SrBr_2(g)$

(c) $2Al(g) + 3O(g) \rightarrow Al_2O_3(g)$

6-26. Use electron configurations to describe the formation of the following ionic compounds from the atoms:

(a) $GaF_3(g)$ (b) $AgCl(g)$ (c) $Li_3N(g)$

6-27. Describe the following equations in terms of Lewis electron-dot formulas for the various species involved:

(a) $3Li(g) + N(g) \rightarrow Li_3N(g)$

(b) $Na(g) + H(g) \rightarrow NaH(g)$

(c) $Al(g) + 3I(g) \rightarrow AlI_3(g)$

6-28. Predict the products of the following reactions from a consideration of the Lewis electron-dot formulas of the reactants and the achievement of noble-gas electron configurations in the product ions:

(a) calcium and nitrogen (as N)

(b) aluminum and chlorine (as Cl)

(c) lithium and oxygen (as O)

ELECTRON CONFIGURATIONS OF IONS

6-29. Predict the ground state electron configuration of

(a) Cr^{2+} (b) Cu^{2+} (c) Co^{3+} (d) Mn^{2+}

6-30. Predict the ground state electron configuration of

(a) Ru^{2+} (b) W^{3+} (c) Pd^{2+} (d) Ti^{+}

6-31. Which *d* transition metal ions with a +2 charge have

(a) six *d* electrons (b) ten *d* electrons

(c) one *d* electron (d) five *d* electrons

Do you see a connection between the number of *d* electrons in the +2 ion and the position of the ion in its transition metal series?

6-32. How many *d* electrons are there in

(a) Fe^{2+} (b) Zn^{2+} (c) V^{2+} (d) Ni^{2+}

Can you see a pattern between the number of *d* electrons and the position of these ions in the first transition metal series?

6-33. Using only a periodic table, predict the ground state 18-outer electron configuration and the charge of a

(a) cadmium ion (b) indium(III) ion

(c) zinc ion (d) thallium(III) ion

6-34. Using only a periodic table, predict the ground state 18-outer electron configuration and the charge of a

(a) copper(I) ion (b) gallium ion

(c) mercury(II) ion (d) gold(I) ion

6-35. Determine which of the following salts are composed of isoelectronic cations and anions:

(a) LiF (b) NaF (c) KBr

(d) KCl (e) BaI_2 (f) AlF_3

6-36. Determine which of the following salts are composed of isoelectronic cations and anions:

(a) NaCl (b) RbBr (c) $SrCl_2$

(d) $SrBr_2$ (e) MgF_2 (f) KI

IONIC RADII

6-37. Without referring to Figure 6.8 or Table 6.4, predict which of the following pairs has the larger relative size:

(a) Li and Li^{+} (b) Cu^{+} and Cu^{2+}

(c) Cl and Cl^{-} (d) O^{2-} and O^{2+}

6-38. Without referring to Figure 6.8 or Table 6.4, predict which of the following pairs has the larger relative size:

(a) H^+ and H (b) Fe^{2+} and Fe^{3+}

(c) S and S^{2-} (d) O^- and O^{2-}

6-39. The following pairs of ions are isoelectronic. Predict which is the larger ion in each pair:

(a) K^+ and Cl^- (b) Ag^+ and Cd^{2+}

(c) Cu^+ and Zn^{2+} (d) F^- and O^{2-}

6-40. The following pairs of ions are isoelectronic. Predict which is the larger ion in each pair:

(a) Ca^{2+} and Cl^- (b) Au^+ and Hg^{2+}

(c) Mn^{2+} and Cr^+ (d) P^{3-} and S^{2-}

6-41. List the following ions in order of increasing radius:

Na^+ O^{2-} Mg^{2+} F^- Al^{3+}

6-42. List the following ions in order of increasing radius:

Y^{3+} Br^- Se^{2-} Rb^+ Mo^{6+}

IONIZATION ENERGIES AND ELECTRON AFFINITIES

6-43. List the following atoms in order of the ease with which they gain an electron to form an anion:

Br I H Cl

6-44. List the following atoms in order of the ease with which they lose electron(s) to form cations:

B K Na He H

6-45. Use the electron affinity data for Cl, Br, and I in Table 6.5, together with your knowledge of periodic trends, to estimate the electron affinity of astatine.

6-46. Explain why the magnitude of the first electron affinity of a chlorine atom is greater than that of a sulfur atom.

6-47. Using Table 4.1 and Table 6.5, calculate E_{rxn} for the reaction equations below.

(a) $Li(g) + Br(g) \rightarrow Li^+(g) + Br^-(g)$

(b) $I^-(g) + Cl(g) \rightarrow I(g) + Cl^-(g)$

(c) $2Na(g) + S(g) \rightarrow 2Na^+(g) + S^{2-}(g)$

6-48. Using Table 4.1 and Table 6.5, calculate E_{rxn} for the reaction equations below. Take the first two ionization energies of a magnesium atom to be 1.23 aJ and 2.41 aJ.

(a) $Na(g) + H(g) \rightarrow Na^+(g) + H^-(g)$

(b) $Mg(g) + O(g) \rightarrow Mg^{2+}(g) + O^{2-}(g)$

(c) $Mg(g) + 2Br(g) \rightarrow Mg^{2+}(g) + 2Br^-(g)$

CALCULATIONS INVOLVING COULOMB'S LAW

6-49. Use Coulomb's law to calculate the energy of a zinc ion and an oxide ion that are at their equilibrium ion-pair separation distance.

6-50. Use Coulomb's law to calculate the energy of a sodium ion and a fluoride ion that are at their equilibrium ion-pair separation distance.

6-51. Calculate the value of the energy released (in aJ per ion pair) in the reaction described by the equation

$$K(g) + Br(g) \rightarrow K^+Br^-(g)$$

The first ionization energy of a potassium atom is 0.696 aJ.

6-52. Calculate the value of the energy change in aJ per ion pair for the reaction described by the equation

$$Mg(g) + O(g) \rightarrow Mg^{2+}O^{2-}(g)$$

Take the first two ionization energies of a magnesium atom to be 1.23 aJ and 2.41 aJ.

6-53. Calculate the value of the energy change in aJ per ion pair for the reaction described by the equation

$$Na(g) + H(g) \rightarrow Na^+H^-(g)$$

6-54. Calculate the value of E_{rxn} for the following equation:

$$Zn(g) + S(g) \rightarrow Zn^{2+}S^{2-}(g)$$

The ionization energy for the process

$$Zn(g) \rightarrow Zn^{2+}(g) + 2e^- \quad E = 4.38 \text{ aJ}$$

6-55. Construct a diagram like that shown in Figure 6.12 for $Li^+F^-(g)$ (see Table 4.1, Table 6.4, and Table 6.5 for the necessary data).

6-56. Construct a diagram like that shown in Figure 6.12 for $Na^+H^-(g)$ (see Table 4.1, Table 6.4, and Table 6.5 for the necessary data).

ADDITIONAL PROBLEMS

6-57. What are the forces that hold an ionic bond together?

6-58. Calcium ions exist in a +2 ionic charge in ionic compounds, but fluorine ions exist in a −1 ionic charge. Explain how you can predict this using only the periodic table.

6-59. Your lab partner believes the chemical formula of an unknown salt your group has been asked to identify is $Al_2O_7(s)$. Explain why you have serious doubts about this answer.

6-60. Explain why cations are smaller and anions are larger than their neutral parent atoms.

6-61. List three ions that are isoelectronic with F^-.

6-62. Predict the charge on (a) a lutetium ion; (b) a lawrencium ion.

6-63. Tin, antimony, and tellurium can each form negatively charged ions. Predict the ionic charge of each of the anions of these metals.

6-64. Explain the errors in each of the following systematic names, and write the correct systematic name and chemical formula for each of these compounds.

(a) zinc(II) chloride (b) sodium monohydride

(c) stannous(II) nitride (d) cadmium dichloride

6-65. A transition metal ion with x outer-shell d electrons is said to be a d^x ion.

(a) Which +3 transition metal ions are d^4 ions?

(b) Which +3 transition metal ions are d^6 ions?

(c) Which +1 transition metal ions are d^3 ions?

6-66. Referring to the definition of a d^x ion given in the previous problem, give three examples of

(a) +2 transition metal ions that are d^{10}

(b) +4 transition metal ions that are d^0

6-67. How many unpaired electrons are there in

(a) an Fe^{2+} ion (b) an Fe^{3+} ion (c) a Cu^+ ion

6-68. The ionic radius of K^+ is 138 pm, while the ionic radius of Cu^+ is 77 pm. Explain why the radius of Cu^+ is smaller than that of K^+ even though a Cu^+ ion has ten more electrons than a K^+ ion.

6-69. Why does the energy evolved in the process

$$Na(g) + X(g) \rightarrow Na^+X^-(g)$$

decrease in the order (X = Cl) < (X = Br) < (X = I)?

6-70. Which of the following anions would you predict to be the smallest: phosphide, sulfide, or chloride?

6-71. Which of the following cations would you predict to be the largest: Mn^+, Fe^{2+}, or Co^{3+}?

6-72. Predict which compound in each of the following pairs will have the greatest magnitude of the coulombic energy per ion pair. Explain your reasoning.

(a) LiCl and LiBr (b) NaF and MgO

(c) LiF and CaO

6-73. Suggest an explanation for the fact that the Group 13 elements have a common +3 ionic charge.

6-74. Suggest why fluoride salts generally have higher melting points than other salts of the same metal.

6-75. Suggest why the second electron affinity (EA_2) of oxygen is much more positive than the second electron affinity (EA_2) of sulfur.

6-76. (*) Use Coulomb's law to compute the ionic bond energy in aJ per formula unit of the linear $Cl^-Ba^{2+}Cl^-(g)$ ionic species.

6-77. (*) Lead is known to form highly colored oxides, some of which have been used as natural pigments (coloring agents) since at least the second century. Some of these oxides include: PbO (yellow), PbO_2 (brown), and Pb_2O_3 (reddish-yellow). Given that the only known ionic charges of lead are Pb(II) and Pb(IV), how can you explain the existence of the Pb_2O_3 salt?

6-78. (*) Although each ion in a sodium chloride crystal has six nearest neighbors (Figure 6.13), the value of the lattice energy for NaCl(s) listed in Table 6.6 of 1.28 aJ per formula unit is only about one and a half times the magnitude of the coulombic energy per NaCl ion pair of 0.816 aJ that we calculated from Equation 6.3 (see Section 6-7). Why do you think the value of the energy for NaCl(s) isn't about six times that of the coulombic energy per ion pair?

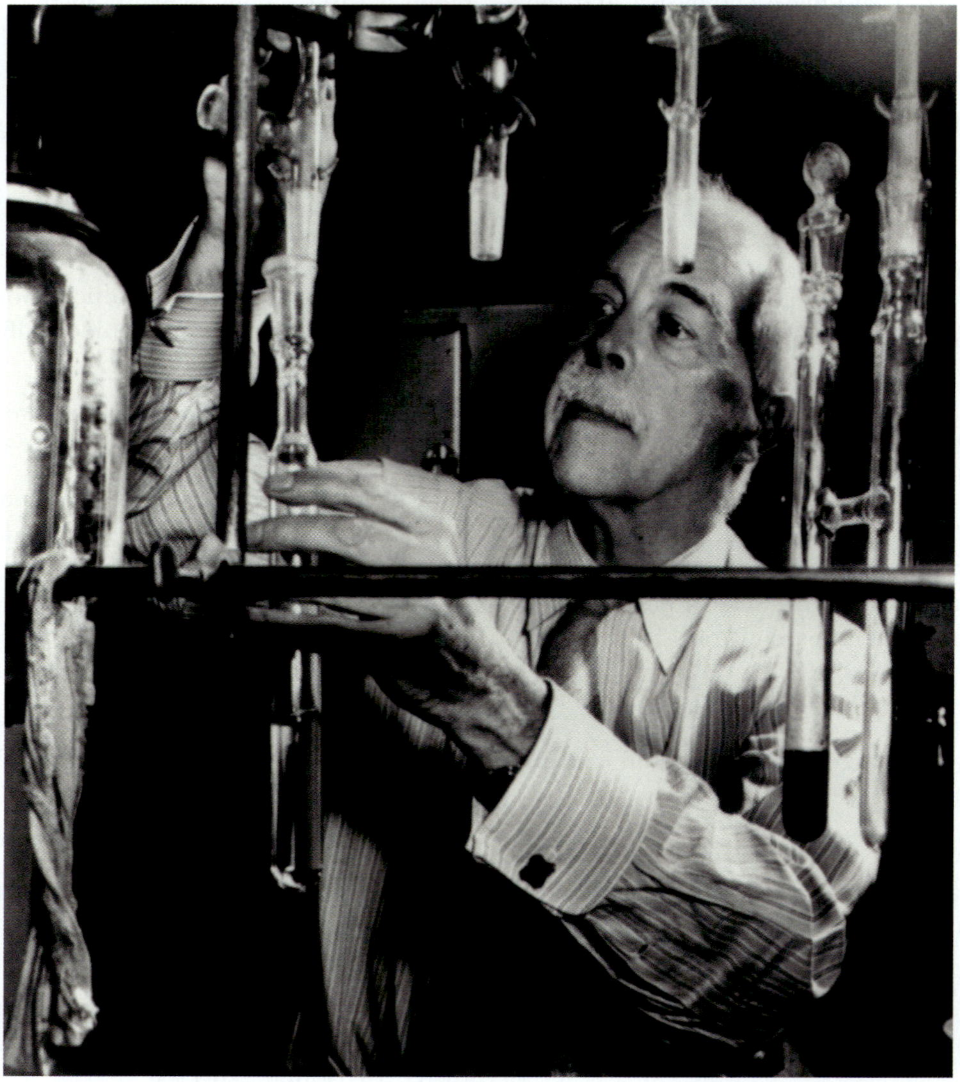

Gilbert Newton Lewis (1875–1946) was born in West Newton, Massachusetts. He was home schooled until his teens, having learned to read at age three. His own children were also home schooled. At age 14, he entered the University of Nebraska and three years later transferred to Harvard University, where he obtained his Ph.D. in 1899. After spending a year studying in Germany, he returned to Harvard as an instructor. Lewis left there in 1904 to become Superintendent of Weights and Measures in the Philippines, and a year later took a position at the Massachusetts Institute of Technology. There he became interested in atomic structure, leading to his development of electron-dot formulas. In 1912, he accepted the position of Dean of the College of Chemistry at the University of California at Berkeley. At Berkeley, Lewis held weekly workshops and discussions, where often the whole department would brainstorm new ideas and results. The department he ran became greater than the sum of its individuals, producing many outstanding chemists and not a few Nobel Prize winners. Lewis remained at Berkeley for the rest of his life, suffering a fatal heart attack while working in his laboratory. Lewis made many important contributions in chemistry. In the 1920s, he introduced Lewis formulas and described a covalent bond (which he named) as a shared pair of electrons. Lewis was one of America's outstanding chemists. The chemistry department Lewis developed at Berkeley remains one of the finest in the world. Lewis is one of the most outstanding chemists never to win a Nobel Prize.

7. Lewis Formulas

In the last chapter, we discussed how metals react with nonmetals to form ionic compounds. Outer-shell electrons are transferred completely from one atom to another, a process resulting in electrostatic attraction—the ionic bond. In 1916, the American chemist G. N. Lewis (Frontispiece) postulated another kind of chemical bond in which two atoms share a pair of electrons—a covalent bond. Lewis published his results nearly a decade before the birth of the quantum theory that was to give his idea of electron pairs, or covalent bonding, a firm theoretical basis.

Covalent bonds make possible an enormous class of compounds. The compounds are poor conductors of an electric current, and many are gases or liquids at room temperature (see Table 6.1). In this chapter, we investigate covalent bonding by studying the method of writing molecular formulas that Lewis introduced. We also learn how to characterize bonds that are intermediate between covalent and ionic bonds. Although a full understanding of covalent bonding depends on quantum theory, which we consider in Chapter 9, Lewis formulas remain one of the most useful concepts in chemistry.

7-1. A Covalent Bond Can Be Described as a Pair of Electrons Shared by Two Atoms

Consider the chlorine molecule, Cl_2. The Lewis electron-dot formula for a chlorine atom is

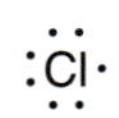

Recall that a Lewis electron-dot formula for an atom shows only the **valence electrons** (generally the outer-shell electrons) and that the number of valence electrons in a main-group element is determined by its position in the periodic table. The chlorine atom is one electron short of having eight electrons in its outer shell and achieving an argonlike electron configuration. One chlorine atom in Cl_2 could get this electron from the other chlorine atom, but then that chlorine atom would be two electrons short of a full octet, making it less stable. In a sense there is a stalemate with respect to electron transfer. Both atoms have the same driving force to gain an electron. From a more quantitative point of view, although a chlorine atom has a large electron affinity (−0.580 aJ), it has a much higher ionization energy (2.09 aJ) and does not lose an electron easily.

Ionic bonds result in binary compounds only when one atomic reactant is a metal with a relatively low ionization energy and the other is a nonmetal with a relatively high electron affinity.

Although we have ruled out the formation of an ionic bond in Cl_2, there is a way for the two chlorine atoms to achieve an argonlike electron configuration *simultaneously*. If the two chlorine atoms *share* a pair of electrons between them, then the resulting distribution of valence electrons can be pictured as

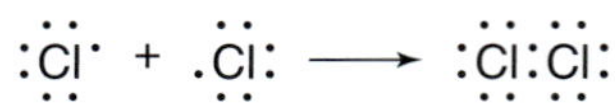

Notice that *each* chlorine atom has eight electrons in its outer shell:

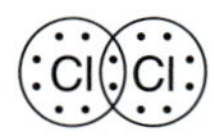

By sharing a pair of electrons, each chlorine atom is able to achieve the stable argonlike outer electron configuration of eight electrons. According to Lewis's picture, the shared electron pair is responsible for holding the two chlorine atoms together as a chlorine molecule. The bond formed between two atoms by a shared electron pair is called a **covalent bond.**

The electron-dot formula we have depicted for a Cl_2 molecule is called a **Lewis formula.** It is conventional to indicate the electron-pair bond as a line joining the two atoms and the other electrons as pairs of dots surrounding the atoms:

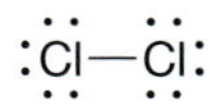

In general, the halogens have the Lewis formula :X–X:, where X is a F, Cl, Br, or I atom. The pairs of electrons that are not shared between the chlorine atoms are called **lone electron pairs,** or simply **lone pairs.** A Lewis formula depicts a covalent bond as a pair of electrons shared between two atoms.

When $Cl_2(g)$ is solidified (its freezing point is –101°C), it forms a **molecular crystal** (Figure 7.1). In contrast to the extended lattice of alternating ions in an ionic crystal, the constituent particles of a molecular crystal are individual molecules. The low melting point of most molecular crystals, in this case chlorine, indicates that the attraction between the molecules is weak relative to the attraction between ions in an ionic crystal. Because chlorine molecules are neutral, there is no net electrostatic attraction between them in the crystal. The interactions of neutral molecules are discussed in Chapter 15.

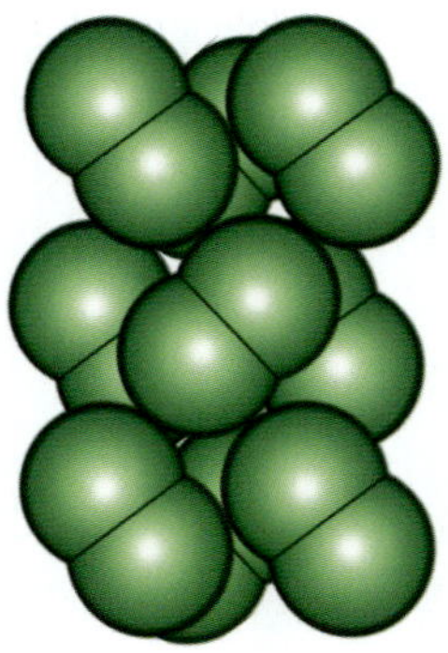

Figure 7.1 The regular arrangement of the chlorine molecules in crystalline $Cl_2(s)$. This pattern is repeated throughout the crystal. Chlorine molecules are neutral, so they do not attract one another as strongly as neighboring ions in an ionic lattice. Consequently, molecular crystals like $Cl_2(s)$ usually have lower melting points than ionic crystals. The melting point of $Cl_2(s)$ is –101°C. For comparison, the melting point of $NaCl(s)$ is 800°C.

Figure 7.2, which shows molecular models of the halogen molecules, can be used to help define **bond length.** The drawings suggest that the nuclei of the two halogen atoms, like the ions in ionic bonds, are held at a certain equilibrium separation distance. We refer to this internuclear distance as the bond length. Table 7.1 lists the bond lengths of the halogens. Note that the bond lengths of the diatomic halogen molecules increase as atomic number increases.

7-2. We Always Try to Satisfy the Octet Rule When Writing Lewis Formulas

As we saw in the previous section, each chlorine atom in a chlorine molecule has eight electrons in its outer shell. Lewis generalized this result into what we call the **octet rule:** each element forms covalent bonds such that eight electrons occupy its outer shell. It was Lewis's genius that he noted this generalization rationalized the bonding in the overwhelming majority of chemical compounds. When writing Lewis formulas, we always try to satisfy the octet rule. The octet rule has its origin in the special stability of the noble-gas electron configuration and was formulated by Lewis well before quantum theory was developed. Thus, for example, carbon, nitrogen, oxygen, and fluorine atoms achieve a neonlike electron configuration when they are surrounded by eight valence electrons. We shall show that, although there are exceptions to the octet rule, it is still quite useful because of the large number of compounds that do obey it. We do not violate the octet rule in writing Lewis formulas unless there is a good reason to do so. The following example illustrates how to apply the octet rule.

When fluorine gas is bubbled through an aqueous solution of sodium hydroxide, NaOH(*aq*), the pale yellow gas oxygen difluoride, $OF_2(g)$, is formed. We can deduce the Lewis formula for an OF_2 molecule by first writing the Lewis electron-dot formulas for the atoms:

We wish to join these three atoms such that each has eight electrons in its outer shell. Pictorially, we wish to join these atoms such that each one can be written with eight valence electrons surrounding the nucleus. By bringing the fluorine atoms in toward the oxygen atom, we see that the electron-dot formula

allows each of the three atoms to be surrounded simultaneously by eight electrons. Thus, we conclude that a satisfactory Lewis formula for an OF_2 molecule in which each atom satisfies the octet rule is

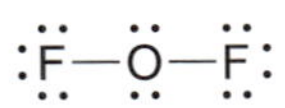

As a final check of this formula, note that there are 20 valence electrons indicated in the Lewis formula for the molecule and that there is a total of 20 valence electrons (seven from each of the two fluorine atoms and six from the oxygen atom) in the Lewis electron-dot formulas for the individual atoms.

The Lewis formula for oxygen difluoride depicts the two fluorine atoms attached to a central oxygen atom. One great utility of Lewis formulas is that they suggest which atoms are actually bonded to each other in a molecule.

We can write Lewis formulas in a systematic manner by using the following four-step procedure:

1. Arrange the symbols of the atoms that are bonded together in the molecule next to one another. For oxygen difluoride we would write

   F O F

   Although it may seem like a difficult step for you to decide how to arrange the atoms at this stage, you will become more confident with experience. Later in this chapter we show why the O–F–F structure is unacceptable. At this stage, if there is only one atom of a particular element, it is a good first try to assume that this atom is the central atom (as in OF_2) and that the

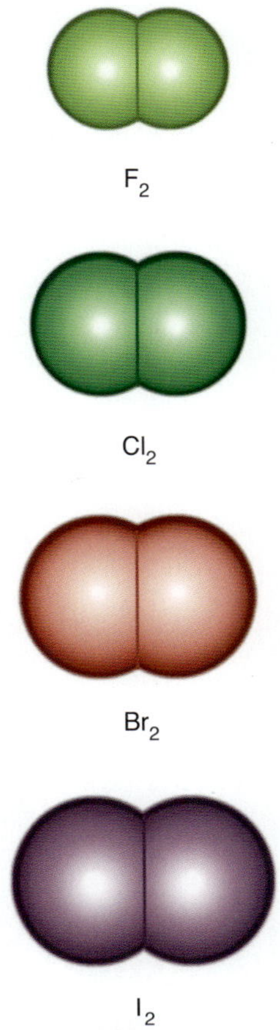

Figure 7.2 Space-filling models of the halogen molecules, drawn to scale to indicate the relative sizes of the atoms. Note that the halogens become larger as we go down the group in the periodic table.

TABLE 7.1 Bond lengths of the halogen molecules

Molecule	Bond length/pm
F_2	141
Cl_2	199
Br_2	228
I_2	267

other atoms are bonded to it. Sometimes, however, the correct arrangement can be found only by trial and error.

2. Compute the total number of valence electrons in the molecule by adding the number of valence electrons for all the atoms in the molecule. If the species is an ion rather than a molecule, then you must take the charge of the ion into account by adding electrons if it is a negative ion or subtracting electrons if it is a positive ion. The table that follows provides several examples of this rule.

Species	Valence electrons from individual atoms			Charge adjustment		Total valence electrons
C_2H_4	2C		4H			
	(2×4)	+	(4×1)	none	=	12
NH_2^-	N		2H			
	(1×5)	+	(2×1)	+1	=	8
NH_4^+	N		4H			
	(1×5)	+	(4×1)	−1	=	8

3. Represent a two-electron covalent bond by placing a line between the atoms that are assumed to be bonded to each other. For oxygen difluoride, we have

F—O—F

4. Arrange the remaining valence electrons as lone pairs about each atom so that the octet rule is satisfied for each one:

:F̤̈—Ö̤—F̤̈:

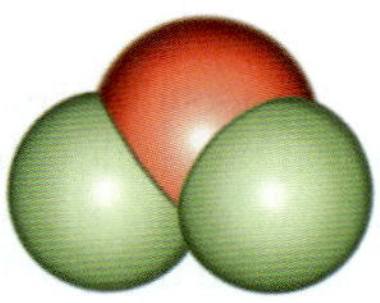

Molecular model of oxygen difluoride, OF_2

The use of this procedure is illustrated in the following Examples.

EXAMPLE 7-1: Write the Lewis formula for a carbon tetrachloride molecule, CCl_4.

Solution: Because carbon is the unique atom in this molecule, we shall assume that it is the central atom and that each chlorine atom is covalently bonded to it:

```
     Cl
Cl   C   Cl
     Cl
```

Molecular model of carbon tetrachloride, CCl_4

The total number of valence electrons is $(1 \times 4) + (4 \times 7) = 32$. We use eight of these electrons to form carbon-chlorine bonds, an arrangement that satisfies the octet rule about the carbon atom. We place the remaining 24 valence electrons as lone pairs on the chlorine atoms to satisfy the octet rule on the chlorine atoms. The Lewis formula is

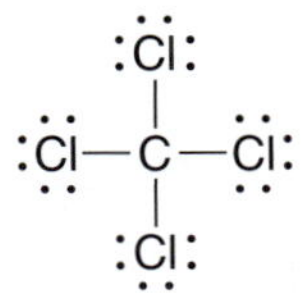

PRACTICE PROBLEM 7-1: Silicon tetrachloride, $SiCl_4(l)$, is a colorless, fuming liquid that forms a dense and persistent cloud when exposed to moist air. Because of this property, silicon tetrachloride is used to produce smoke screens. The equation for the reaction is

$$SiCl_4(l) + 2\,H_2O(l) \rightarrow SiO_2(s) + 4\,HCl(g)$$

The white cloud consists of highly dispersed particles of silicon dioxide, $SiO_2(s)$. Write the Lewis formula of a silicon tetrachloride molecule, which is a covalent compound.

Answer:

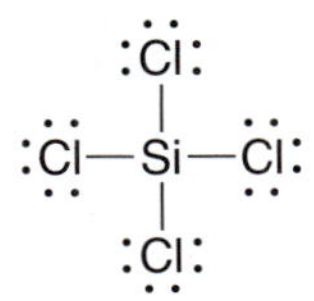

EXAMPLE 7-2: Write the Lewis formula of a nitrogen trifluoride molecule, NF_3.

Solution: Because nitrogen is the unique atom in this molecule, we shall assume that it is the central atom and that each fluorine atom is covalently bonded to it:

F N F

F

The total number of valence electrons is $(1 \times 5) + (3 \times 7) = 26$. We use six of these valence electrons to form nitrogen-fluorine bonds. We now place valence electrons as lone pairs on each fluorine atom (accounting for 18 of the 20 remaining valence electrons) and the remaining two valence electrons as a lone pair on the nitrogen atom. The completed Lewis formula is

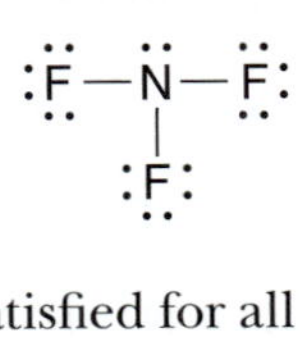

Notice that the octet rule is satisfied for all four atoms.

PRACTICE PROBLEM 7-2: Phosphorus pentachloride, $PCl_5(s)$, is a solid at room temperature and consists of ion pairs of the type $[PCl_4]^+[PCl_6]^-$. Write the Lewis formula of the PCl_4^+ ion. (We shall examine the PCl_6^- anion in a later problem.)

Answer:

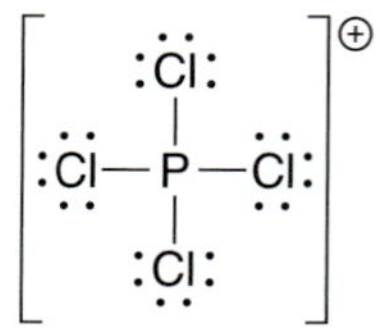

(We enclose the charge in a circle for clarity; conventionally the charge is shown without a circle.)

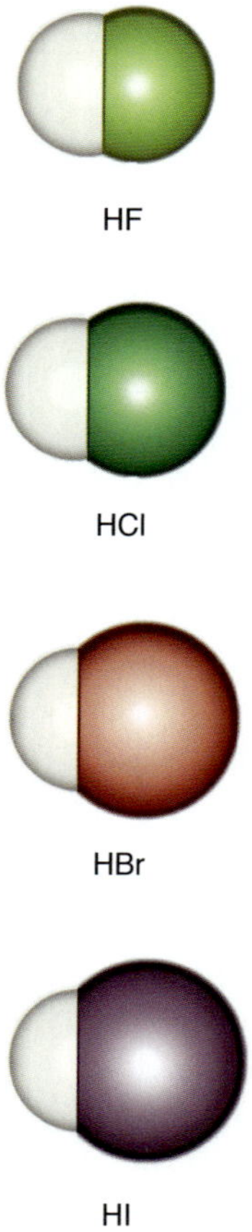

Figure 7.3 Space-filling models of the hydrogen halide molecules.

TABLE 7.2 Bond lengths of the hydrogen halides

Compound	Bond length/pm
HF	92
HCl	128
HBr	141
HI	161

7-3. Hydrogen Atoms Are Almost Always Terminal Atoms in Lewis Formulas

We have said that there are exceptions to the octet rule. One important exception is the hydrogen atom. The noble gas closest to hydrogen in the periodic table is helium. We might expect, then, that hydrogen needs only two electrons in order to attain a noble-gas electron configuration. For example, let's consider a hydrogen molecule, H_2. The electron-dot formula for a hydrogen atom is

H·

Each hydrogen atom can be surrounded by two electrons if the two hydrogen atoms in the H_2 molecule share electrons:

H· + ·H ⟶ H:H or H—H

In this way each hydrogen atom in the molecule achieves a heliumlike electron configuration.

The Lewis formulas for the hydrogen halides are obtained directly from the electron-dot formulas of the individual atoms. If we let X be F, Cl, Br, or I, we can write

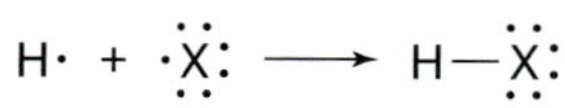

The hydrogen atom has two electrons surrounding it, and the halogen atom has eight. Molecular models of the hydrogen halides are shown in Figure 7.3. The bond lengths of the hydrogen halides are given in Table 7.2.

EXAMPLE 7-3: Write the Lewis formula for an ammonium ion, NH_4^+.

Solution: We first arrange the atoms as

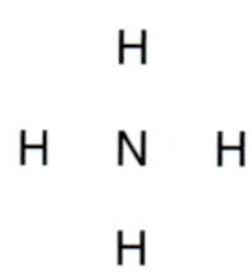

There is a total of $5 + (4 \times 1) - 1 = 8$ valence electrons. We use all eight of these electrons to form the hydrogen-nitrogen bonds:

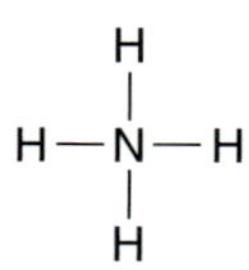

We indicate that this species has a charge of +1 by writing

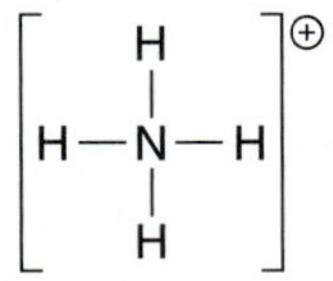

Note that the nitrogen atom has eight electrons around it and each hydrogen atom has two.

PRACTICE PROBLEM 7-3: Phosphine, $PH_3(g)$, is a colorless, toxic gas with a garliclike odor. It is used as a doping agent for semiconductor production and also as a fumigant. Write the Lewis formula of a phosphine molecule.

Answer:

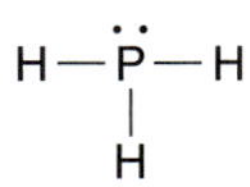

Because a hydrogen atom completes its valence shell with a total of two electrons, hydrogen atoms almost always form a covalent bond to only one other atom and so almost always are terminal atoms in Lewis formulas.

EXAMPLE 7-4: Write the Lewis formula for a chloroform molecule, $CHCl_3$.

Solution: The chloroform molecule has three different types of atoms. The first step is to decide how to arrange these atoms in the Lewis formula. Hydrogen is almost always a terminal atom in a Lewis formula. Of the remaining four atoms (one C and three Cl atoms), the carbon atom is unique; therefore, we guess that it is the central atom. We have the postulated arrangement:

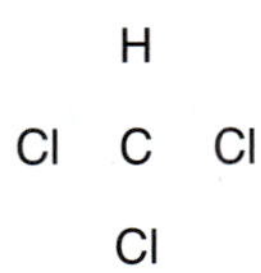

There is a total of $4 + (3 \times 7) + 1 = 26$ valence electrons to be accommodated in the Lewis formula. If we use eight of these electrons to form four bonds,

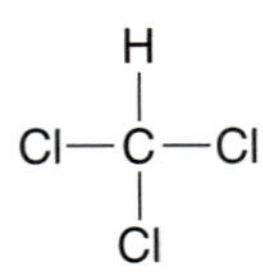

then the remaining 18 electrons can be accommodated as nine lone pairs (three lone pairs on each of the three chlorine atoms). Thus, the Lewis formula of a $CHCl_3$ molecule is

Figure 7.4 A space-filling model of a trichloromethane molecule, better known as chloroform, $CHCl_3$. Chloroform was once used as an anesthetic and replaced ether for a time until it was found to cause fatal cardiac arrhythmia in some patients. Chloroform has been misrepresented in films and TV dramas as a harmless "knock-out" gas, quickly putting a villain's victim under using just a few drops on a handkerchief. In fact, the amount needed to do this would probably be lethal.

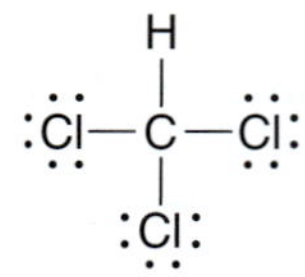

Note that the octet rule is satisfied for the carbon atom and for each chlorine atom, whereas hydrogen has a closed shell of two electrons. A three-dimensional molecular model (or space-filling model) of chloroform is shown in Figure 7.4.

PRACTICE PROBLEM 7-4: Sodium metal reacts with ammonia to produce sodium amide, $NaNH_2(s)$, and hydrogen gas. Write the Lewis formula of an amide anion, NH_2^-.

Answer:

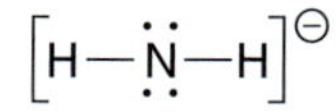

Up to now we have considered only molecules for which the arrangement of atoms in the Lewis formula is based on placing the unique atom (other than hydrogen, which is generally a terminal atom) in the central position. However, consider the problem of writing a Lewis formula for a hydrazine molecule, N_2H_4. For this molecule, there is no unique atom to place in the central position, so we assume that the two nitrogen atoms must be bonded to each other, with the four hydrogen atoms in terminal positions. Thus, we write

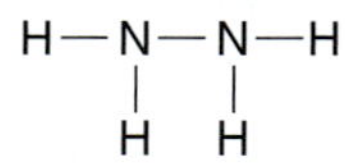

This molecule has $(2 \times 5) + (4 \times 1) = 14$ valence electrons. The five bonds require a total of ten valence electrons. The remaining four valence electrons are placed as lone pairs on the nitrogen atoms in order to complete the octets on the nitrogen atoms and simultaneously accommodate the 14 valence electrons.

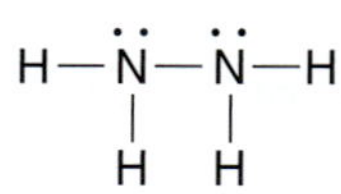

Molecular model of hydrazine, N_2H_4

Figure 7.5 A space-filling model of a hydrazine molecule, N_2H_4. Hydrazine is used as a rocket fuel and as a propellant on the space shuttle orbiter.

A molecular model of a hydrazine molecule is shown in Figure 7.5.

EXAMPLE 7-5: Write the Lewis formula for a methanol molecule, CH_3OH, commonly called methyl alcohol.

Solution: Because we expect the hydrogen atoms to be terminal, the carbon and oxygen atoms must be bonded to each other:

C—O

We must now position the four hydrogen atoms. An oxygen atom has six valence electrons; thus, it usually completes its octet by forming two bonds, whereas a carbon atom has only four valence electrons, and so usually completes its octet by forming four bonds. Thus, we write

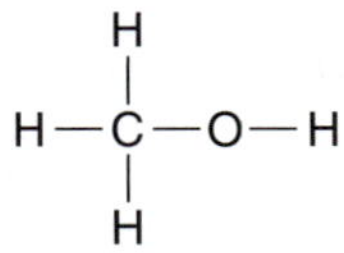

The CH_3OH molecule contains $4 + 6 + (4 \times 1) = 14$ valence electrons. The five bonds require a total of 10 valence electrons. The remaining four valence electrons are accommodated in two lone pairs on the oxygen atom; in fact, these electrons are needed to complete the octet on the oxygen atom. Thus, the completed Lewis formula of a methanol molecule is

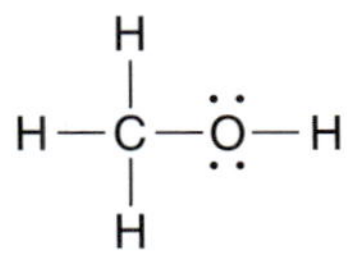

A space-filling model of a methanol molecule is shown in Figure 7.6.

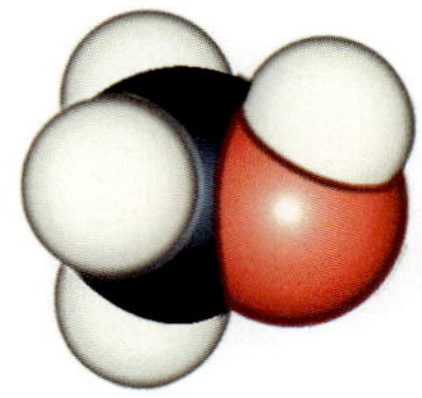

Figure 7.6 A space-filling model of a methanol molecule, CH_3OH. Methanol is sometimes called "wood alcohol" because it can be obtained from the distillation of wood. Methanol is poisonous and can be fatal or cause blindness if ingested.

PRACTICE PROBLEM 7-5: Methane, CH_4, is the simplest hydrocarbon, that is, the simplest compound that consists of only carbon and hydrogen (Figure 7.7). The next simplest hydrocarbon is ethane, $C_2H_6(g)$ (Figure 7.8). Write the Lewis formulas of a methane molecule and an ethane molecule.

Answer:

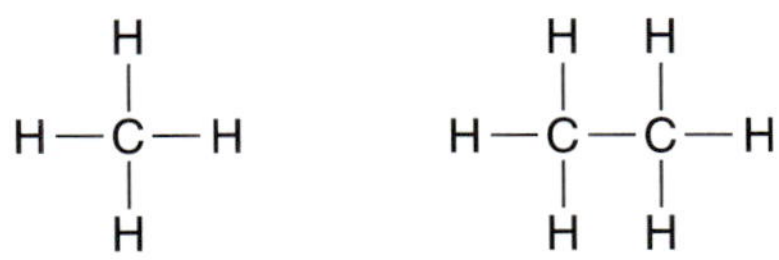

methane ethane

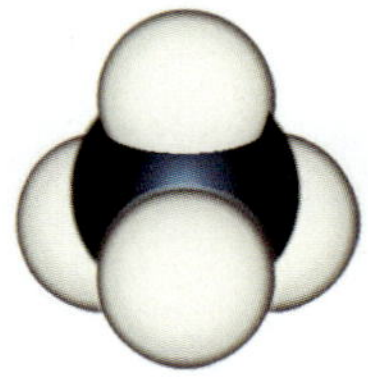

Figure 7.7 A space-filling model of a methane molecule, CH_4. Methane is the main component in natural gas and is commonly used for heating and cooking. Although methane is colorless and odorless, natural gas is often mixed with small quantities of highly pungent sulfur-containing compounds to aid in the detection of gas leaks in the home. Methane is also a greenhouse gas and one of the factors in global climate change.

Compounds based on the chemistry of carbon, such as those in the preceding Example and Practice Problem, are called **organic compounds**. We discuss the chemistry of various organic compounds in Interchapters F, G, and H.

Figure 7.8 A space-filling model of an ethane molecule, C_2H_6. Ethane is a colorless, odorless gaseous hydrocarbon produced by the refining of petroleum. Like most hydrocarbons, ethane gas is highly flammable.

7-4. Formal Charges Can Be Assigned to Atoms in Lewis Formulas

It is often possible when writing Lewis formulas to come up with more than one arrangement of atoms, bonds, and lone pairs that satisfy the octet rule. The question then arises whether one of these is better than the others in describing the bonding in the species being considered. As an aid to answering this question, we assign a charge to each atom in a molecule or ion. Such charges are called **formal charges** because they are assigned by a set of arbitrary rules and do not necessarily represent the actual charges on the atoms. To assign formal

See Interchapters F, G, and H at www.McQuarrieGeneralChemistry.com

charges, we *assume* that each pair of shared electrons is shared *equally* between the two atoms and assign one of these electrons to each atom. Lone-pair electrons are assigned to the atom on which they are located. The formal charge is the assigned net charge on the atom. The formal charge on an atom in a Lewis formula is calculated by using the following equation:

$$\begin{pmatrix}\text{formal charge}\\ \text{on an atom in}\\ \text{a Lewis formula}\end{pmatrix} = \begin{pmatrix}\text{total number of}\\ \text{valence electrons}\\ \text{in the free atom}\end{pmatrix} - \begin{pmatrix}\text{total number}\\ \text{of lone-pair}\\ \text{electrons}\end{pmatrix} - \frac{1}{2}\begin{pmatrix}\text{total number}\\ \text{of shared}\\ \text{electrons}\end{pmatrix} \tag{7.1}$$

Consider the ammonium ion NH_4^+:

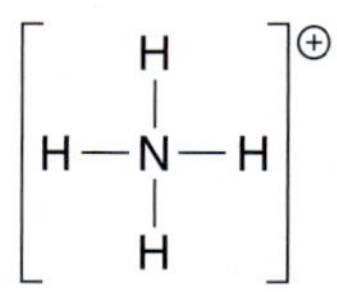

A hydrogen atom has one valence electron, and there are no lone-pair electrons in NH_4^+. Each hydrogen atom shares two electrons, so the formal charge assigned to each hydrogen atom is given by Equation 7.1 as

$$\text{formal charge on each H atom in } NH_4^+ = 1 - 0 - \tfrac{1}{2}\,(2) = 0$$

A nitrogen atom has five valence electrons. The nitrogen atom in NH_4^+ shares eight electrons, so the formal charge on N in $NH_4^+ = 5 - 0 - \frac{1}{2}\,(8) = +1$. Thus, NH_4^+ is often written as

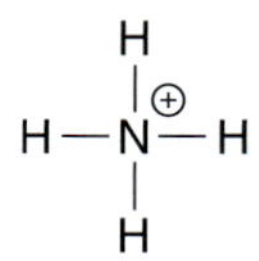

where the ⊕ on N denotes a formal charge of +1 on the nitrogen atom (conventionally, we only indicate non-zero formal charges). Notice that the sum of the formal charges on the various atoms is equal to the net charge on the molecular ion.

EXAMPLE 7-6: Assign formal charges to each of the atoms in the Lewis formula of the hydronium ion, H_3O^+.

Solution: The Lewis formula of a hydronium ion is

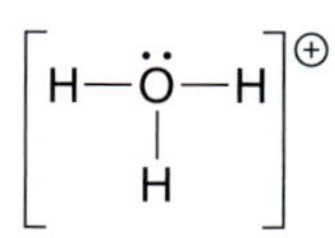

The formal charges on the oxygen atom and the hydrogen atoms in a hydronium ion are calculated using Equation 7.1:

$$\text{formal charge on the O atom} = 6 - 2 - \tfrac{1}{2}\,(6) = +1$$

$$\text{formal charge on each H atom} = 1 - 0 - \tfrac{1}{2}\,(2) = 0$$

Thus, the Lewis formula for H_3O^+, with the formal charge indicated is

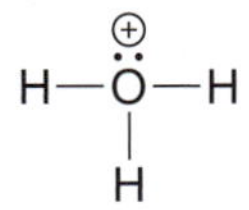

Once again, notice that the sum of the formal charges on the various atoms is equal to the net charge on the species, which is +1 for a H_3O^+ ion.

PRACTICE PROBLEM 7-6: Write the Lewis formula of a tetrafluoroborate ion, BF_4^-, and assign formal charges to each atom.

Answer:

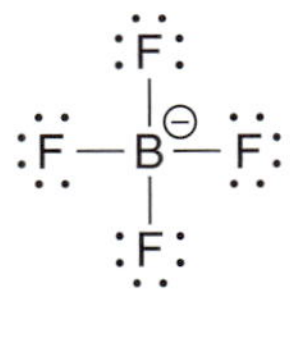

We discussed the oxygen difluoride molecule, OF_2, in Section 7-2 and wrote its Lewis formula as

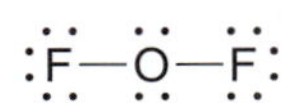

I

The OF_2 molecule was used as an example of the principle that it is a good first try to place the unique atom in the center. The formal charge on each atom in formula I is 0. There is another Lewis formula for an OF_2 molecule, however, that satisfies the octet rule:

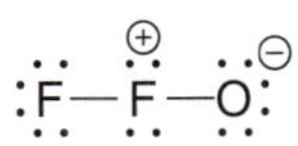

II

These two Lewis formulas predict an entirely different bonding pattern in OF_2, and a totally different structure. The first predicts that the oxygen atom is in the center of the molecule and that there are two oxygen–fluorine bonds. The second predicts that one of the fluorine atoms is in the center and that there is one fluorine–fluorine bond and one oxygen–fluorine bond.

We can use formal charges to select one of these Lewis formulas for an OF_2 molecule over the other. Although formal charges do *not* represent the actual charges on the atoms in a molecule, it is sometimes convenient to consider them as if they were real. For example, consider formula II. Although it satisfies the octet rule, there is a separation of formal charges in the formula as written. There is no separation of formal charges in formula I, however. We predict (correctly), then, that the actual structure of an OF_2 molecule is represented by formula I, with the oxygen atom in the center. Usually, the Lewis formula with the lower formal charges or the least separated formal charges represents the preferred (lowest energy) Lewis formula.

EXAMPLE 7-7: Use formal charges to determine which of the two Lewis formulas better represents the structure of a hydroxylamine molecule, NH_3O.

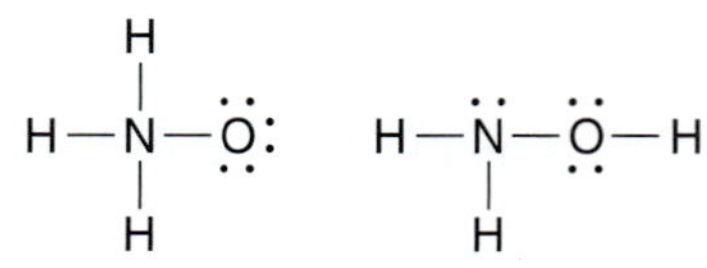

I II

Solution: The formal charges of the nitrogen and oxygen atoms in formula I are

$$\text{formal charge on the N atom} = 5 - 0 - \tfrac{1}{2}(8) = +1$$

$$\text{formal charge on the O atom} = 6 - 6 - \tfrac{1}{2}(2) = -1$$

and those in formula II are

$$\text{formal charge on the N atom} = 5 - 2 - \tfrac{1}{2}(6) = 0$$

$$\text{formal charge on the O atom} = 6 - 4 - \tfrac{1}{2}(4) = 0$$

If we write the Lewis formulas with the calculated formal charges, then we have

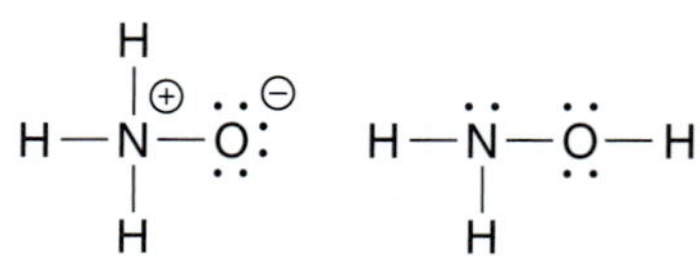

I II

Because formula II has zero formal charges everywhere, we predict (correctly) that formula II represents the actual structure of a hydroxylamine molecule. The chemical formula of hydroxylamine is usually written as NH_2OH to reflect its Lewis formula.

Molecular model of hydroxylamine, NH_2OH

PRACTICE PROBLEM 7-7: Use formal charges to determine which of the two Lewis formulas best represents the structure of a hydrogen peroxide molecule, H_2O_2.

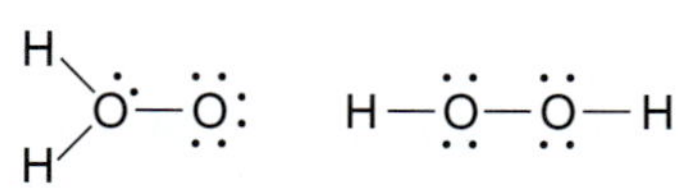

I II

Answer: Formula II has a formal charge of 0 on each atom and is the preferred Lewis formula.

7-5. It Is Not Always Possible to Satisfy the Octet Rule Using Only Single Bonds

In all the molecules that we have discussed so far, exactly the correct number of valence electrons remained after step 3 (page 200) to be used in step 4. In this section we consider cases in which there are not enough electrons to satisfy the octet rule for each atom using only single bonds. The ethene molecule, C_2H_4, commonly known as ethylene, serves as a good example (Figure 7.9). Using the same reasoning that we did for hydrazine, we arrange the atoms as

H C C H
H H

An ethylene molecule has a total of $(2 \times 4) + (4 \times 1) = 12$ valence electrons. We use ten of them to join the atoms:

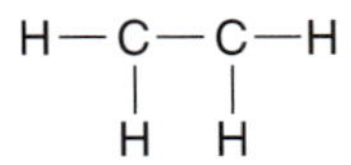

If we use only single bonds, it is not possible to satisfy the octet rule for each carbon atom with only the two remaining valence electrons. We are short two electrons. When this situation occurs, we add one more bond for each two electrons that we are short. In the case of ethylene, we add another bond between the carbon atoms to get

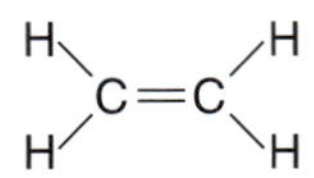

Notice that now the octet rule is satisfied for each carbon atom.

When two atoms are joined by two pairs of electrons, we say that there is a **double bond** between the atoms. A double bond between two atoms is shorter and stronger than a single bond between the same two atoms. The carbon–carbon double bond in an C_2H_4 molecule is indeed shorter and stronger than, for example, the carbon–carbon single bond in an ethane molecule, $H_3C–CH_3$ (Figure 7.8). Table 7.3 gives typical bond lengths and bond energies for various single and double bonds.

It is also possible to have a **triple bond,** as we now show for a N_2 molecule. There are ten valence electrons in a N_2 molecule. If we start with one bond and then try to satisfy the octet rule for each nitrogen atom, we find that we are four electrons short. For example,

$$\ddot{\underset{..}{N}}—\ddot{\underset{..}{N}} \quad \text{(violates the octet rule)}$$

Thus, we add two more bonds (one for each two electrons we are short) to obtain

$$N{\equiv}N$$

The four remaining valence electrons are now added according to step 4 to obtain

$$:N{\equiv}N:$$

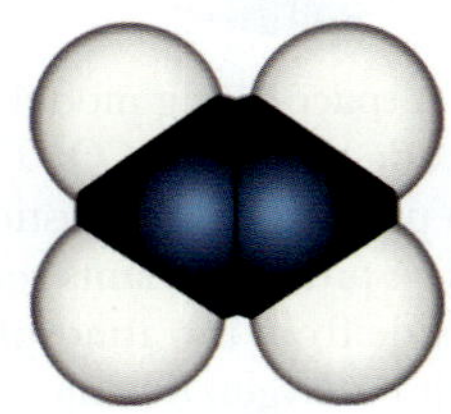

Figure 7.9 A space-filling model of an ethene molecule, C_2H_4, better known as ethylene. Ethylene is a colorless, odorless gas that is widely used in the production of plastics, such as polyethylene (see Interchapter S). It is also a plant hormone used in commercial fruit ripening where fruit is picked green for easy transport and then artificially ripened using ethylene gas before sale.

See Interchapter S at www.McQuarrieGeneralChemistry.com.

TABLE 7.3 Average bond lengths and bond energies of single, double, and triple bonds

Bond	Average bond length/pm	Average bond energy/aJ
C—O	142	0.581
C=O	121	1.21
C—C	153	0.581
C=C	134	1.02
C≡C	120	1.35
N—N	145	0.266
N=N	118	0.698
N≡N	113	1.58

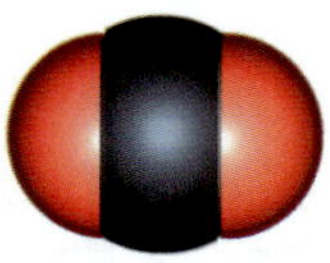

Figure 7.10 A space-filling model of a carbon dioxide molecule, CO_2. Carbon dioxide is a product of combustion and animal respiration. Plants remove carbon dioxide from the atmosphere, converting it to oxygen during photosynthesis. Carbon dioxide has extensive uses from dry ice, $CO_2(s)$, to fire extinguishers, to the carbonation of beverages. It is also used as a solvent in the decaffeination of tea and coffee and as an inert gas in propellants. Carbon dioxide is also a potent greenhouse gas (Interchapter L). The atmosphere of the planet Venus is believed to consist mostly of carbon dioxide gas.

www

See Interchapter L at www.McQuarrieGeneralChemistry.com.

EXAMPLE 7-8: Write a Lewis formula for a CO_2 molecule (Figure 7.10).

Solution: We arrange the atoms with the carbon atom in the center:

O C O

There is a total of $(1 \times 4) + (2 \times 6) = 16$ valence electrons. If we add one bond between each oxygen atom and the carbon atom and try to satisfy the octet rule for each atom, we find that we are four electrons short. For example,

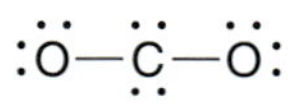 (violates the octet rule about the oxygen atoms)

Thus, we go back to step 3 and add two more bonds:

O=C=O

Now we use step 4 and arrange the remaining eight valence electrons as lone pairs to satisfy the octet rule for each atom:

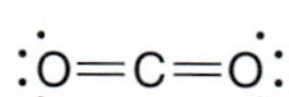

The Lewis formula of a CO_2 molecule shows two carbon–oxygen double bonds. Incidentally, we could also have made one single bond and one triple bond instead of two double bonds. This would, however, produce a Lewis formula with a formal charge separation.

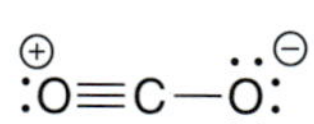

The Lewis formula with two double bonds gives no charge separation; therefore, it is the preferred one.

PRACTICE PROBLEM 7-8: Methanal, $H_2CO(g)$, better known by its common name, formaldehyde, is a gas with a pungent, characteristic odor. An aqueous solution of formaldehyde is called formalin and is sometimes used to preserve biological specimens. Formaldehyde is used extensively in the production of certain plastics (e.g., bakelite, melamine). (a) Write the Lewis formula of a formaldehyde molecule. (b) If one of the hydrogen atoms in a formaldehyde molecule is replaced by a methyl group $—CH_3$, the compound is called ethanal (better known by its common name, acetaldehyde), a colorless liquid with a pungent, fruity odor. Write the Lewis formula of this molecule.

Answer:

(a) 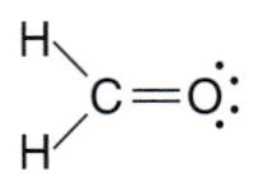

(b) 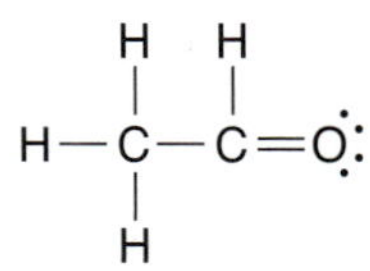

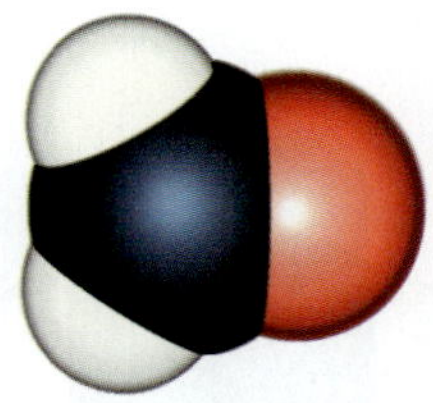

Molecular model of formaldehyde, H_2CO

Molecular model of acetaldehyde, CH_3CHO

EXAMPLE 7-9: Write the Lewis formula for a hydrogen cyanide molecule, HCN.

Solution: Either the carbon atom or the nitrogen atom might be the central atom in this case. When in doubt, it's a good guess to arrange the atoms as the formula is written (why else write the formula that way?), but let's try both possible arrangements:

H C N or H N C

Use four of the ten valence electrons to write

H—C—N or H—N—C

We are four electrons short of satisfying the octet rule for both the carbon atom and the nitrogen atom, so we add two more bonds to each possible Lewis formula. The hydrogen atom already has two electrons around it, so in this case we must form a triple bond between the carbon atom and the nitrogen atom:

H—C≡N or H—N≡C

The remaining two valence electrons are placed on the nitrogen atom as a lone pair in one case and on the carbon atom as a lone pair in the other case, so that both the carbon atom and the nitrogen atom satisfy the octet rule. The two possible Lewis formulas are

H—C≡N: or H—N≡C:

We now add formal charges to each possibility to write

$$\mathrm{H{-}C{\equiv}N{:}} \quad \text{or} \quad \mathrm{H{-}\overset{\oplus}{N}{\equiv}\overset{\ominus}{C}{:}}$$

The case of placing the nitrogen atom in the center results in a molecule with a separation of formal charges. Therefore, our choice of putting the carbon atom in the center (in accordance with the formula as it was written) is the preferred one.

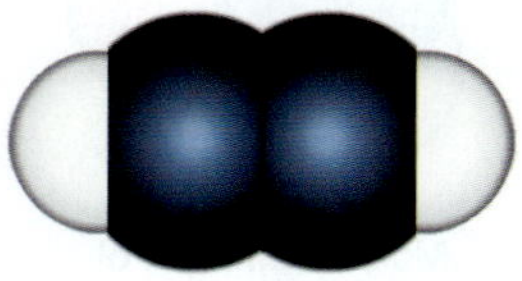

Figure 7.11 A space-filling model of an ethyne molecule, C_2H_2, better known as acetylene. Acetylene is the fuel in an oxyacetylene torch, but it finds even greater use as a raw material in the plastics industry (see Interchapter S).

See Interchapter S at www.McQuarrieGeneralChemistry.com.

PRACTICE PROBLEM 7-9: Ethyne, $C_2H_2(g)$, better known by its common name, acetylene, is a colorless gas that burns in oxygen to give a high flame temperature (Figure 7.11). Write the Lewis formula of an acetylene molecule.

Answer:

$$H{-}C{\equiv}C{-}H$$

7-6. A Resonance Hybrid Is a Superposition of Lewis Formulas

For many molecules and ions, we can write two or more equally satisfactory Lewis formulas. For example, let's consider a nitrite ion, NO_2^-. One Lewis formula for a NO_2^- ion is

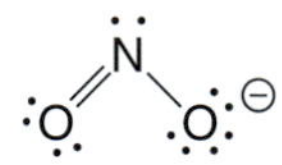

One of the oxygen atoms, the right-hand one as written, has a formal charge of −1, having three lone pairs and one bond. Another equally acceptable Lewis formula for a NO_2^- ion is

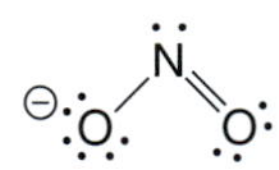

In this case the negative formal charge is on the other oxygen atom. Both of these Lewis formulas satisfy the octet rule. When it is possible to write two or more satisfactory Lewis formulas *without altering the positions of the nuclei,* the actual formula is viewed as an average, or as a superposition, of the individual formulas. Each of the individual Lewis formulas is said to be a **resonance form,** and the use of multiple Lewis formulas is called **resonance.** We indicate resonance forms by means of a two-headed arrow, as in the following notation:

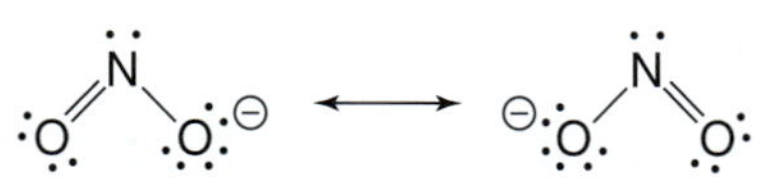

Neither of the individual Lewis formulas taken separately accurately reflects the actual bonding. Two Lewis formulas taken together are necessary to describe the bonding in a NO_2^- ion.

There is no generally accepted way to represent resonance pictorially, but one way is to write the formula of a NO_2^- ion as

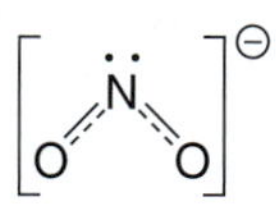

where the two dashed lines taken together represent a pair of bonding electrons that spread over the two bonds. Such a superimposed formula is called a **resonance hybrid** because it is a hybrid of the various resonance forms. Each of the nitrogen–oxygen bonds in a NO_2^- ion can be thought of as an *average* of a single bond and a double bond. The superimposed Lewis formulas suggest that the two nitrogen–oxygen bonds in a NO_2^- ion are equivalent. This idea is in accord with experimental observation: the two bonds do have exactly the same length, 113 pm. Notice that using only one Lewis formula suggests that the two nitrogen–oxygen bonds are not equivalent, one being a single bond and the other being a double bond.

The resonance hybrid for a NO_2^- ion suggests that the –1 charge of the ion is shared equally by the two oxygen atoms, rather than being completely on either one of the two oxygen atoms as suggested by either of the two resonance forms. In such a case, we say that the charge is **delocalized.** Resonance hybrids with delocalized charges have lower energies than their (hypothetical) individual resonance forms. This difference in energy is called **resonance energy.**

Another example of the need to use resonance forms to obtain a satisfactory picture of the distribution of electronic charge occurs in a nitrate ion, NO_3^-. Three equally satisfactory Lewis formulas for a NO_3^- ion are shown below:

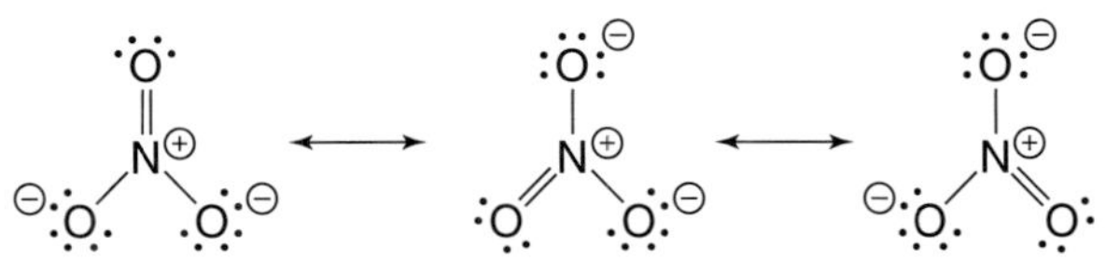

Because each of these formulas is equally satisfactory, the actual structure is viewed as a superposition, or an average, of the three formulas and can be represented pictorially as the resonance hybrid

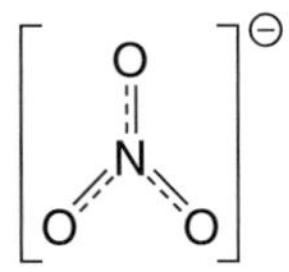

where the three dashed lines taken together represent a pair of bonding electrons spread over the three bonds. In this case each of the nitrogen-oxygen bonds is an average of a double bond and two single bonds (from the three averaged resonance formulas). As the superimposed representation suggests, the three nitrogen–oxygen bonds are equivalent, in agreement with experimental observation that each nitrogen–oxygen bond is 122 pm in length. Furthermore, there are no chemical reactions that can be used to distinguish one oxygen atom from another in a nitrate ion, another observation suggesting that all three oxygen atoms are bonded equivalently to the nitrogen atom.

The need for resonance forms arises from the fact that the bonds represented by Lewis formulas involve equally shared electron-pairs. If the species involves a bond intermediate between a single and a double bond, then we need to write two or more Lewis formulas to describe the bonding in the molecule. Resonance is in no sense a real phenomenon (the molecule does not "alternate" between the different representations). It is just a device that enables us to give a more realistic picture of the electron distribution in a species when we use Lewis formulas.

The following Example is particularly important because it involves several of the concepts that we have discussed in this chapter.

EXAMPLE 7-10: Sulfur dioxide, $SO_2(g)$, is well known for its unpleasant suffocating odor, which smells like a struck match. It is often used in sparing quantities as a preservative for wines and dried fruits. Write Lewis formulas for the two resonance forms of sulfur dioxide. Indicate formal charges and discuss the bonding in this molecule.

Solution: We arrange the atoms as

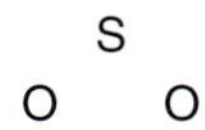

Here we have drawn a slight angle between the atoms. Although in Lewis formalism we could have just as well drawn SO_2 as a linear molecule. As we shall see in the next chapter, there is in fact about a 120-degree angle between the three atoms, as we have drawn them here.

Sulfur dioxide has 18 valence electrons. The two resonance forms are as follows:

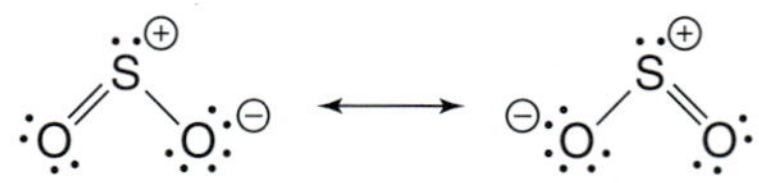

The formal charges indicated are calculated by using Equation 7.1:

$$\text{formal charge on the S atom in } SO_2 = 6 - 2 - \tfrac{1}{2}\,(6) = +1$$

$$\text{formal charge on the singly bonded O atom in } SO_2 = 6 - 6 - \tfrac{1}{2}\,(2) = -1$$

$$\text{formal charge on the doubly bonded O atom in } SO_2 = 6 - 4 - \tfrac{1}{2}\,(4) = 0$$

These two Lewis formulas constitute equivalent resonance forms, so the actual formula is an average of the two, which can be represented by the resonance hybrid

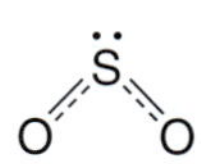

This formula suggests that the two sulfur–oxygen bonds in a SO_2 molecule are equivalent, and that the two sulfur–oxygen bond lengths in a SO_2 molecule are equal in length. This prediction is in agreement with experimental results.

PRACTICE PROBLEM 7-10: Sodium carbonate, $NaCO_3(s)$, is an ionic solid that is used in the manufacture of glass. Write Lewis formulas for the resonance forms of the carbonate ion, CO_3^{2-}. Indicate formal charges and discuss the bonding in this ion.

Answer:

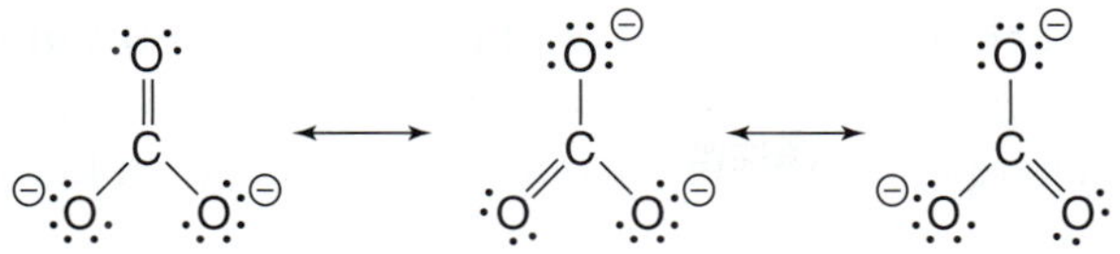

The three carbon–oxygen bonds are equivalent.

An important example of resonance and its consequences is provided by a benzene molecule, C_6H_6. Benzene is a clear, colorless, highly flammable liquid with a characteristic odor (Figure 7.12). It is obtained from petroleum and coal tar and has many chemical uses. A benzene molecule has two principal resonance forms:

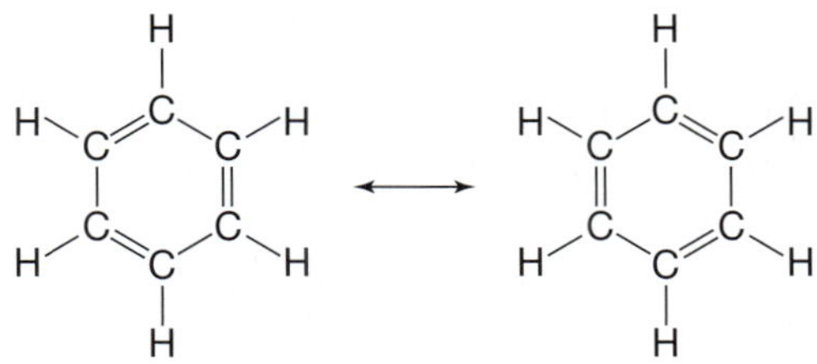

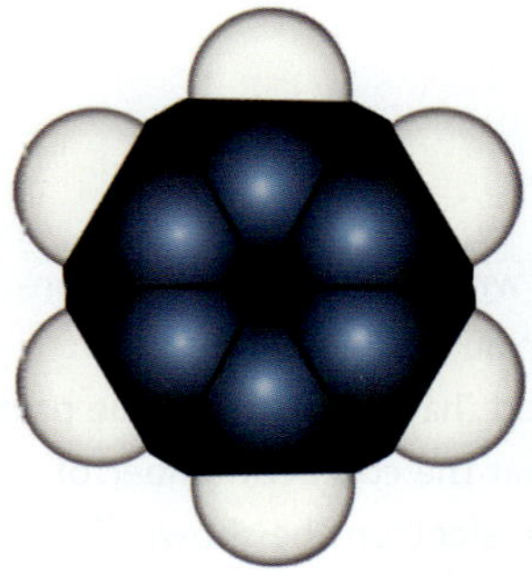

Figure 7.12 A space-filling model of a benzene molecule, C_6H_6. Benzene was commonly used as a solvent in paints, thinners, and glues until it was discovered to be a carcinogen. Although still an important industrial chemical (see Appendix H), its use is now highly regulated.

We depict the superposition or average of these two Lewis formulas by the resonance hybrid

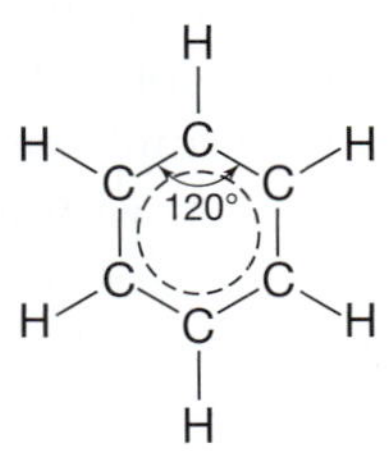

This formula predicts that all the carbon–carbon bonds in a benzene molecule are equivalent, as observed experimentally, and that each is the average of a single bond and a double bond. The carbon–carbon bond distance in a benzene molecule is 140 pm, which is intermediate between the usual carbon–carbon single-bond (153 pm) and double-bond (134 pm) distances. Benzene is a beautifully symmetric molecule. It is a **planar molecule** (all the atoms lie in the same plane), with the ring of carbon atoms constituting a perfect hexagon with 120° interior carbon–carbon bond angles. A commonly used representation of a benzene molecule is simply a regular hexagon with a circle drawn within it:

Each vertex represents a carbon atom to which is attached a hydrogen atom. This formula emphasizes that the six carbon–carbon bonds and the six hydrogen–carbon bonds in a benzene molecule are equivalent. The benzene ring is part of the chemical formula of a great many organic compounds. Ben-

See Interchapter H at www.McQuarrieGeneralChemistry.com.

zene behaves chemically as a substance with no double bonds and is a relatively unreactive molecule (see Interchapter H). The unusual stability of a benzene molecule is ascribed to what chemists call **resonance stabilization:** the energy of the actual molecule, represented by a superposition of Lewis formulas, is lower than the energy of any of its (hypothetical) individual Lewis formulas.

7-7. A Species with One or More Unpaired Electrons Is Called a Free Radical

As useful as the octet rule is, in some cases it cannot be satisfied. In particular, the octet rule cannot be satisfied by each atom in a species with an odd total number of electrons. For example, consider a nitrogen oxide molecule, NO, which has 5 + 6 = 11 valence electrons. The Lewis electron-dot formulas of a nitrogen and an oxygen atom are

When writing the Lewis electron-dot formulas for atoms it is not required that the electrons be paired, only that the correct number of valence electrons be shown. Thus, for example, we can write the Lewis electron-dot formula for an oxygen atom as 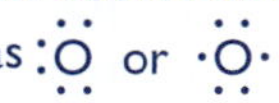

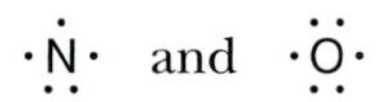

If we try to write a Lewis formula for a NO molecule, we find that it is not possible to satisfy the octet rule. The best that we can do is to write

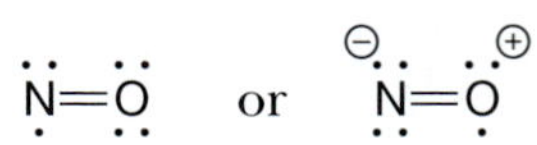

The difficulty here is that the total number of valence electrons is an odd number (11), so it is impossible to pair up all the electrons as we have been doing. A species that has one or more unpaired electrons is called a **free radical.** Because of the unpaired electron(s), free radicals are usually very reactive.

Another example of a free radical is provided by chlorine dioxide, $ClO_2(g)$, a yellow to reddish-yellow gas with an unpleasant odor similar to that of chlorine; chlorine dioxide reacts explosively with many substances. The chlorine atom has seven valence electrons, and each oxygen atom has six valence electrons. Therefore, a ClO_2 molecule has an odd number (19) of valence electrons. The two resonance forms with the lowest formal charge separation are

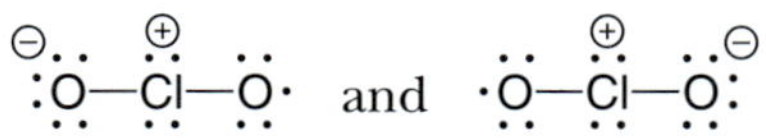

The ClO_2 free radical is viewed as a hybrid of the resonance forms, and the two chlorine–oxygen bonds in a ClO_2 molecule are equivalent.

The molecules NO and ClO_2 are free radicals. They have an odd number of electrons, so they cannot satisfy the octet rule. Other compounds, called **electron-deficient compounds,** have an even number of outer electrons but do not have enough electrons to form octets about each atom. Compounds of beryllium and boron serve as good examples of electron-deficient compounds. Consider the beryllium hydride molecule, BeH_2. The electron-dot formulas for the beryllium atom and the hydrogen atom are

H· and ·Be·

A Lewis formula for a BeH_2 molecule is

H—Be—H

The beryllium atom is four electrons short of satisfying the octet rule. Electron-deficient molecules, like free radicals, generally are highly reactive.

NO, AN AMAZING MOLECULE. Nitrogen oxide, NO(*g*), is a simple molecule that consists of just one oxygen atom and one nitrogen atom, but it has some amazing properties. It is both a toxic pollutant and an important chemical signal in biological systems. The NO molecule is a free radical, having one unpaired electron that makes it a highly reactive species. It occurs as an atmospheric pollutant because it is produced by the reaction between $O_2(g)$ and $N_2(g)$ at the high temperatures that exist in the internal combustion engine. Its presence in the atmosphere leads to the production of a number of toxic substances, such as $NO_2(g)$ and nitric acid. One function of catalytic converters is to convert NO(*g*) back to $O_2(g)$ and $N_2(g)$.

Nitrogen oxide is produced naturally in various organs in the body. Although NO(*g*) has a lifetime of a few seconds, it readily diffuses across cell membranes. One important function of NO(*g*) is to signal smooth muscles in blood vessels to relax, resulting in vasodilation and increased blood flow. Nitroglycerin, which is used to treat heart conditions, acts by producing NO(*g*) in the body, thereby relaxing the muscles in the blood vessels. Similarly, the drugs Viagra, Levitra, and Cialis all work by producing NO(*g*) and relaxing blood vessels. Nitric oxide is an important chemical signal for a variety of other biological systems, such as the nervous system and the immune system. The discoverers of the role of NO(*g*) as a biological messenger, Robert F. Furchgott, Fedid Murad, and Louis J. Ignarro, were awarded the Nobel Price in Medicine in 1998.

EXAMPLE 7-11: Suggest a Lewis formula for a boron trifluoride molecule, BF_3, that satisfies the octet rule. Suggest a reason why the electron-deficient formula is preferred.

Solution: Boron trifluoride is an electron-deficient molecule. Each of the three fluorine atoms has seven valence electrons and the boron atom has three, for a total of 24 valence electrons. A Lewis formula for a BF_3 molecule using 12 electron pairs is

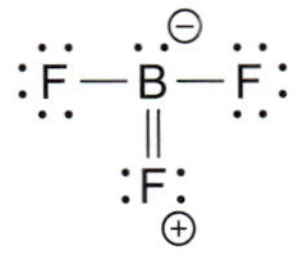

The formal charge separation in this case can be used to decide that this Lewis formula is less favorable than the formula

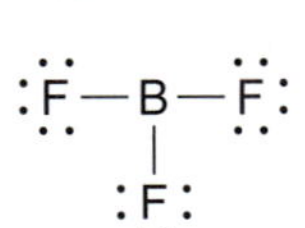

for which the formal charges are all 0.

PRACTICE PROBLEM 7-11: Nitrogen dioxide, $NO_2(g)$, is one of the substances responsible for the brownish haze in photochemical smog found over many large cities. Write the Lewis formula for a NO_2 molecule.

Answer: Two resonance forms are

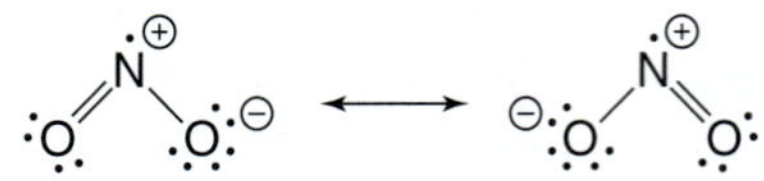

As noted, electron-deficient compounds are usually highly reactive species. For example, the electron-deficient molecule BF_3 readily reacts with an NH_3 molecule to form H_3NBF_3 according to

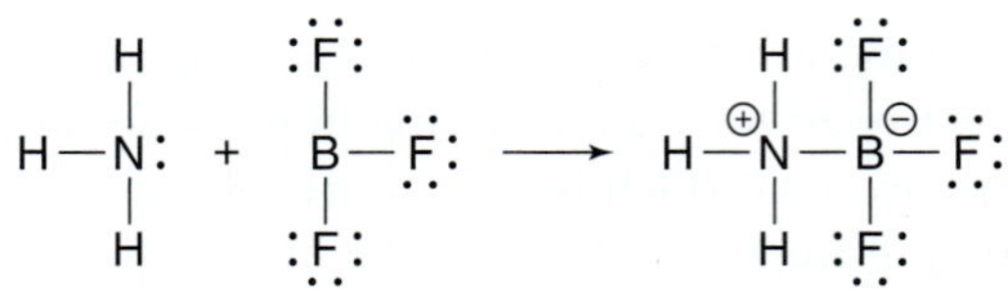

The lone electron pair in NH_3 can be shared between the nitrogen atom and the boron atom so that the octet rule is now satisfied for each atom.

7-8. Atoms of Elements Below the Second Row in the Periodic Table Can Expand Their Valence Shells

There is some debate in the chemical literature about whether Lewis formulas using expanded valence shells accurately portray the electron density in molecules; see L. Suidan, et al., *Journal of Chemical Education* **72**, 583 (1995); G.H. Purser, *Journal of Chemical Education* **78**, 981 (2001).

We have not yet considered the case in which there are more valence electrons than are needed to satisfy the octet rule on each atom. Such a case can happen when one of the elements in the compound lies below the second-row elements carbon, nitrogen, oxygen, and fluorine in the periodic table so that the principal quantum number of the valence electrons is $n > 2$. Such elements will usually be the central atom. In this case, we assign the "extra" electrons as lone pairs to that element, which we say has an **expanded valence shell.**

As an example, let's write the Lewis formula for a sulfur tetrafluoride molecule, SF_4. First we arrange the atoms as

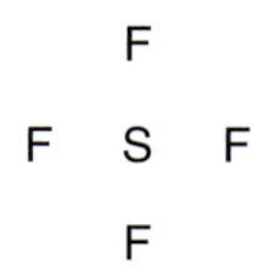

Of the $6 + (4 \times 7) = 34$ valence electrons, we use eight electrons to form four sulfur-fluorine bonds. We can satisfy the octet rule for each atom by using only 24 of the remaining valence electrons:

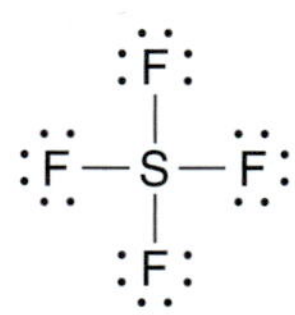 (two valence electrons unassigned)

Two valence electrons are still to be accounted for. Because sulfur lies in the third row of the periodic table, we add these as a lone pair to the sulfur atom. The completed Lewis formula is

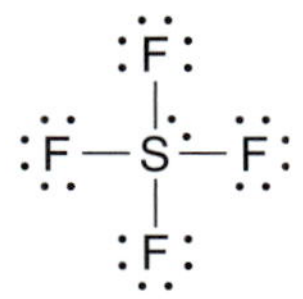

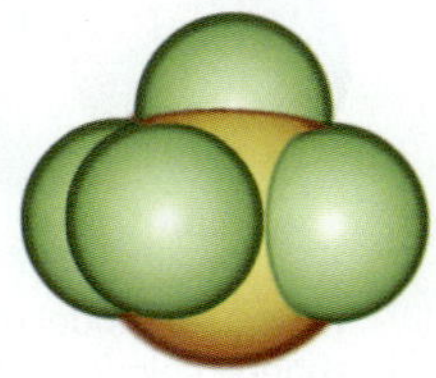

Molecular model of sulfur tetrafluoride, SF_4

The exact position of the lone pair of electrons on the sulfur atom is not important; for example, we could just as well have placed the pair to the upper left of the sulfur atom. Notice that the formal charges are 0 on all the atoms in the Lewis formula.

In a sulfur tetrafluoride molecule, SF_4, sulfur expands its valence shell by using its $3d$ orbitals. It is not possible for the atoms of elements in the second row of the periodic table to expand their valence shells beyond eight electrons because second-row elements complete the $n = 2$ shell, which does not contain d orbitals, when they satisfy the octet rule. Second-row elements would have to use orbitals in the $n = 3$ shell to accommodate more electrons, but the energies of the orbitals in the $n = 3$ shell are much higher than those of the orbitals in the $n = 2$ shell. Thus, although $SF_4(g)$ has been synthesized, $OF_4(g)$ has never been observed.

EXAMPLE 7-12: Xenon difluoride, $XeF_2(s)$, was one of the first noble-gas-containing compounds to be prepared (see Interchapter K). Write the Lewis formula for the XeF_2 molecule.

See Interchapter K at www.McQuarrieGeneralChemistry.com.

Solution: We arrange the atoms as

F Xe F

Of the $8 + (2 \times 7) = 22$ valence electrons, four are used to form the two xenon-fluorine bonds. We can use 12 of the remaining 18 valence electrons to satisfy the octet rule on each fluorine atom:

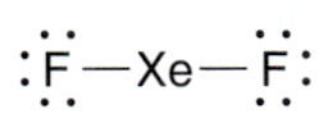 (six valence electrons unassigned)

The remaining six valence electrons are placed on the xenon atom as three lone pairs. The Lewis formula is

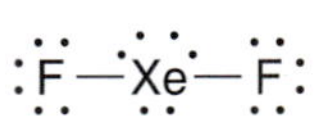

which has 0 formal charges on all the atoms.

PRACTICE PROBLEM 7-12: Phosphoryl chloride, $POCl_3(l)$, is a colorless, clear, strongly fuming liquid with a pungent odor. It is used as a chlorinating agent, especially to replace oxygen atoms with chlorine atoms in

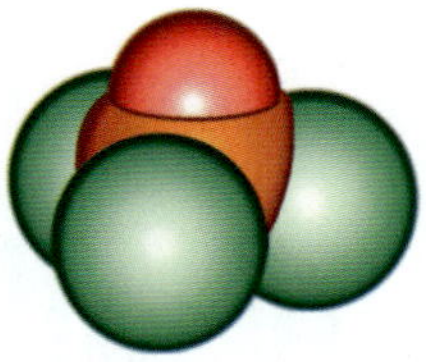

Molecular model of phosphoryl chloride, $POCl_3$

organic compounds. Write a Lewis formula for a phosphoryl chloride molecule that has no formal charges.

Answer:

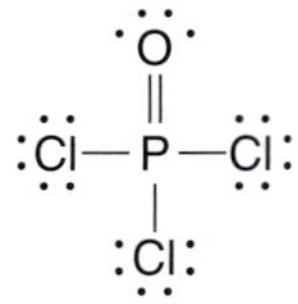

Other examples of species in which using step 4 leads to more than eight electrons around the central atom are shown as follows.

Molecular model of xenon tetrafluoride, XeF_4

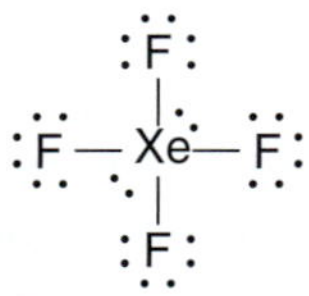

xenon tetrafluoride, XeF_4

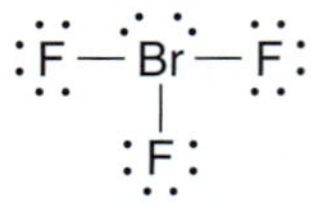

bromine trifluoride, BrF_3

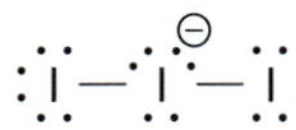

triiodide ion, I_3^-

Because the atoms of elements below the second row of the periodic table can accommodate more than eight electrons in their valence shells, they are able to bond to more than four atoms. Some examples are the following molecules

Molecular model of bromine trifluoride, BrF_3

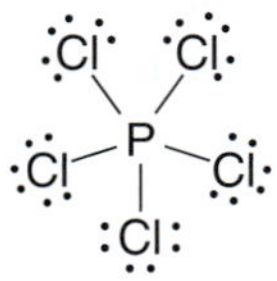

phosphorous pentachloride, PCl_5

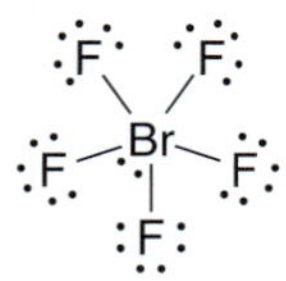

bromine pentafluoride, BrF_5

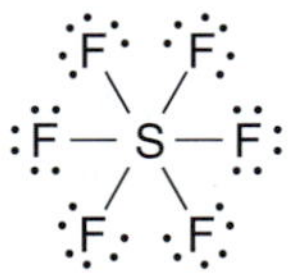

sulfur hexafluoride, SF_6

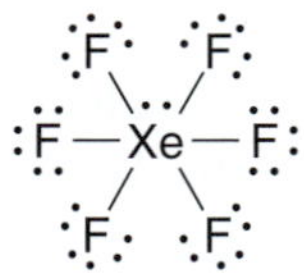

xenon hexafluoride, XeF_6

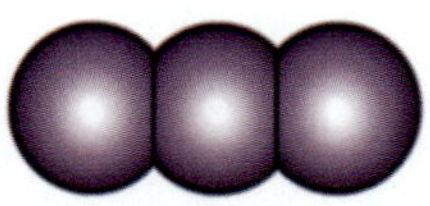

Molecular model of a triiodide ion, I_3^-

EXAMPLE 7-13: Tellurium forms the pentafluorotellurate ion, TeF_5^-, when $KF(s)$ and $TeO_2(s)$ are dissolved in $HF(aq)$. Write the Lewis formula of a TeF_5^- ion.

Solution: A tellurium atom has six valence electrons, and each fluorine atom has seven valence electrons. The negative charge on the ion gives a total of 42 valence electrons. Of these 42 valence electrons, ten are used to form bonds between the five fluorine atoms and the central tellurium atom, 30 more are used as lone pairs on the fluorine atoms to satisfy the octet rule, and the remaining two valence electrons constitute a lone pair on the tellurium atom. Thus, the Lewis formula of a TeF_5^- ion is

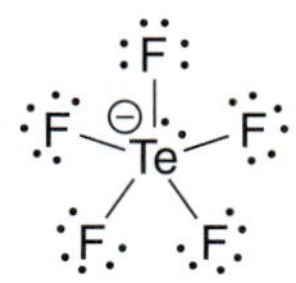

The formal charge on the tellurium atom is given by Equation 7.1:

$$\text{formal charge on the Te atom} = 6 - 2 - \tfrac{1}{2}\,(10) = -1$$

PRACTICE PROBLEM 7-13: In Practice Problem 7-2 we saw that solid phosphorus pentachloride consists of ion pairs of the type $[PCl_4]^+[PCl_6]^-$ and wrote the Lewis formula of the PCl_4^+ cation. Now write the Lewis formula of the PCl_6^- anion.

Answer:

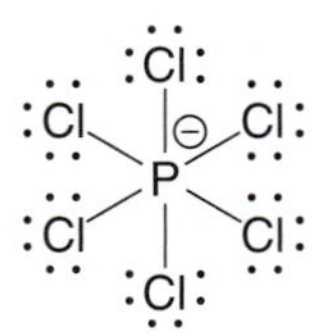

The fact that atoms of elements below the second row of the periodic table can expand their valence shells leads to additional resonance formulas for many of the compounds involving those atoms. For example, consider a sulfuryl chloride molecule, SO_2Cl_2. According to the rules that we have presented, the Lewis formula for a SO_2Cl_2 molecule is

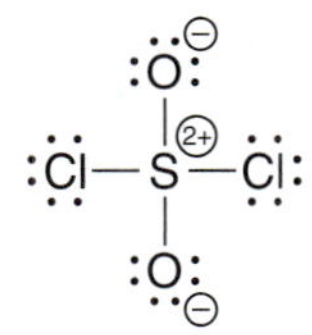

This Lewis formula, however, displays a large formal charge separation that can be reduced by writing additional Lewis formulas, recognizing the ability of sulfur to expand its valence shell:

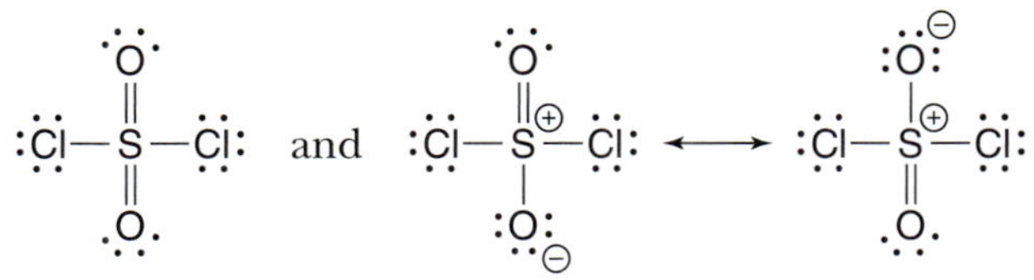

All four of these Lewis formulas are resonance forms of a SO_2Cl_2 molecule, so we have

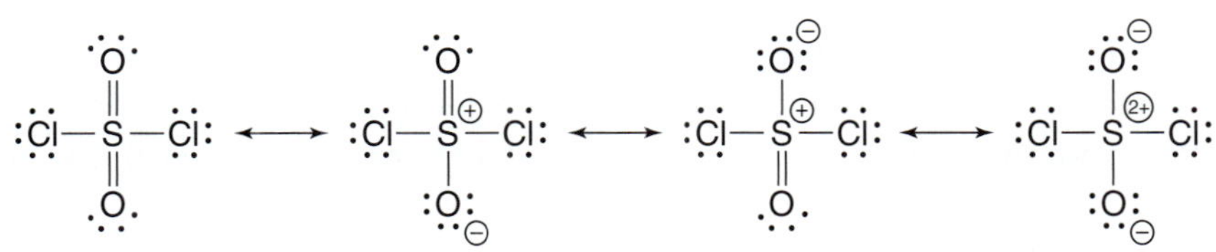

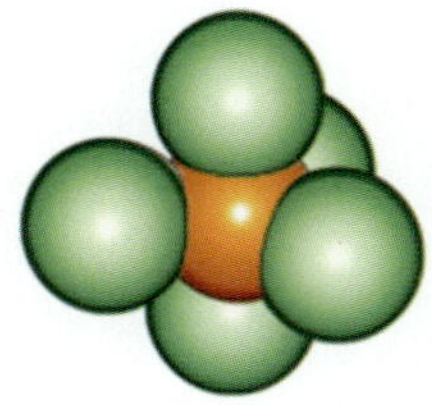

Molecular model of phosphorus pentachloride, PCl_5

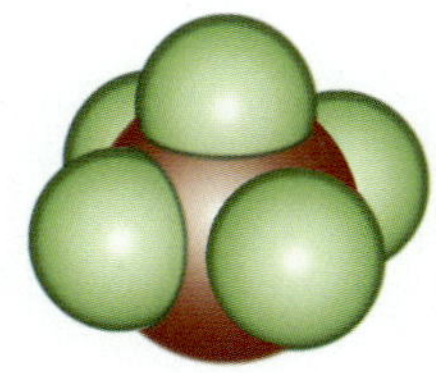

Molecular model of bromine pentafluoride, BrF_5

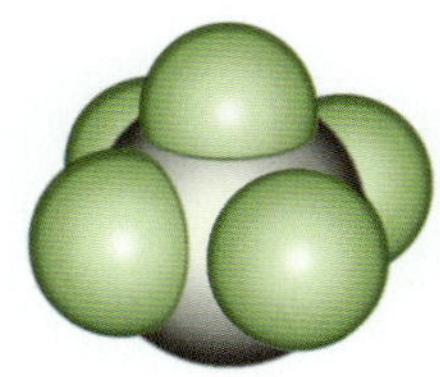

Molecular model of a pentafluorotellurate ion, TeF_5^-

Molecular model of sulfuryl chloride, SO_2Cl_2

EXAMPLE 7-14: Write Lewis formulas for the various resonance forms of a sulfur trioxide molecule, SO_3, including those in which the sulfur atom has an expanded valence shell. Indicate formal charges and discuss the bonding in a SO_3 molecule. *Hint:* There are seven resonance forms.

Solution: A sulfur atom has six valence electrons and each oxygen atom has six valence electrons for a total of 24 valence electrons. The various resonance forms of a SO_3 molecule are

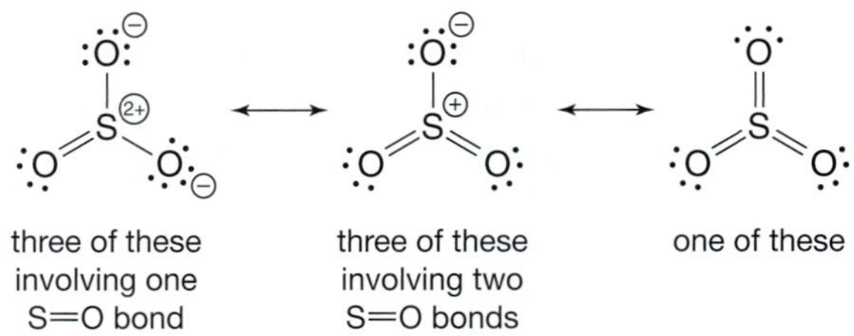

A superposition of all seven resonance formulas suggests (correctly) that the three sulfur–oxygen bonds in a SO_3 molecule are equivalent.

PRACTICE PROBLEM 7-14: Write Lewis formulas for the various resonance forms of a phosphate ion, PO_4^{3-}, including those in which the phosphate atom has an expanded valence shell. Indicate formal charges and discuss the bonding in a phosphate ion.

Answer:

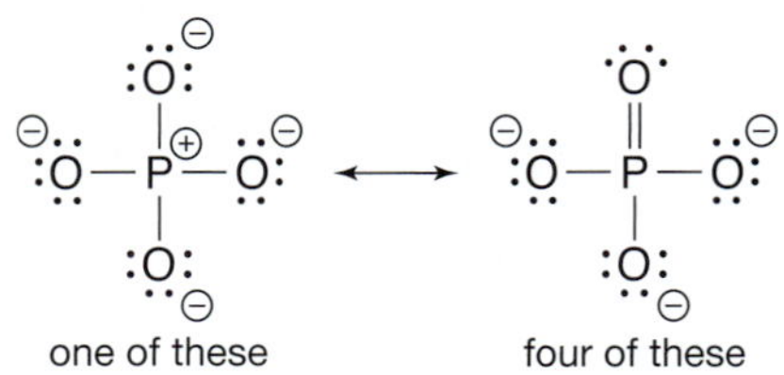

All phosphorus–oxygen bonds are equivalent. Nevertheless, we cannot overemphasize the fact that the resonance hybrid is the only structure that is observed experimentally and that the individual resonance forms are simply hypothetical constructs that allow us to deduce the complete resonance hybrid.

7-9. Electronegativity Is a Periodic Property

Although we have discussed ionic bonds and covalent bonds in this and the previous chapter as distinct cases, most chemical bonds are neither purely ionic nor purely covalent, but rather are intermediate between the two. The bond in a hydrogen chloride molecule, HCl, serves as a good example of this point.

When we introduced the concept of formal charge, we *arbitrarily* assigned

1 H 2.1																	2 He —
3 Li 0.98	4 Be 1.57											5 B 2.04	6 C 2.55	7 N 3.04	8 O 3.44	9 F 3.98	10 Ne —
11 Na 0.93	12 Mg 1.31											13 Al 1.61	14 Si 1.90	15 P 2.19	16 S 2.58	17 Cl 3.16	18 Ar —
19 K 0.82	20 Ca 1.00	21 Sc 1.36	22 Ti 1.54	23 V 1.63	24 Cr 1.66	25 Mn 1.55	26 Fe 1.83	27 Co 1.88	28 Ni 1.91	29 Cu 1.90	30 Zn 1.65	31 Ga 1.81	32 Ge 2.01	33 As 2.18	34 Se 2.55	35 Br 2.96	36 Kr —
37 Rb 0.82	38 Sr 0.95	39 Y 1.22	40 Zr 1.33	41 Nb 1.6	42 Mo 2.16	43 Tc 1.9	44 Ru 2.2	45 Rh 2.28	46 Pd 2.20	47 Ag 1.93	48 Cd 1.69	49 In 1.78	50 Sn 1.96	51 Sb 2.05	52 Te 2.1	53 I 2.66	54 Xe —
55 Cs 0.79	56 Ba 0.89	57–71 1.1–1.2	72 Hf 1.3	73 Ta 1.5	74 W 2.36	75 Re 1.9	76 Os 2.2	77 Ir 2.20	78 Pt 2.28	79 Au 2.54	80 Hg 2.00	81 Tl 2.04	82 Pb 2.33	83 Bi 2.02	84 Po 2.0	85 At 2.2	86 Rn —
87 Fr 0.7	88 Ra 0.9	89+ 1.1–1.3															

Figure 7.13 Electronegativities of the elements as calculated by Linus Pauling. Note that the electronegativities of the elements in the second and third rows increase from left to right, and that they increase from bottom to top in a given column and that the noble gases all have a zero electronegativity, as shown in Figure 7.15.

one of the electrons in the covalent bond to each atom. The formal charges of the H atom and the Cl atom in a HCl molecule are 0. We emphasize here that this procedure is formal and arbitrary, however useful. When we assign formal charges we are tacitly assuming that the electrons in the covalent bond are shared equally by the hydrogen atom and the chlorine atom. We know, however, that different isolated atoms have different ionization energies and different electron affinities. It seems reasonable to assume that different atoms will also attract electrons differently when they are covalently bonded.

Electronegativity is a measure of the tendency with which an atom in a molecule attracts the electrons in its covalent bonds with other atoms. The larger the electronegativity of an atom, the greater the attraction of the atom for the electrons in its covalent bonds. As we shall see, electronegativity differences can also be used to predict the charge distribution in a bond and to select or reject a Lewis formula. However, electronegativity is a derived quantity in that it is not directly measurable. Over the years various electronegativity scales have been proposed. The electronegativity scale most commonly used today was proposed in the 1930s by the American chemist Linus Pauling (Chapter 9 Frontispiece). The Pauling electronegativities range from 0 (least electronegative) to 4 (most electronegative) for each of the elements in the periodic table (Figure 7.13). Note that the noble gases are not assigned a value because they typically do not form compounds with other elements.

In Figure 7.14, the Pauling electronegativities are plotted against atomic number. Figure 7.14 shows clearly that electronegativity is a periodic property. We see that electronegativities increase from left to right across the short (second and third) rows of the periodic table, as the elements become increasingly nonmetallic. Furthermore, electronegativities decrease in going down a column (Figure 7.13) because the nuclear attraction of the outer electrons decreases as the size of the atom increases. From the values in Figure 7.13 we see

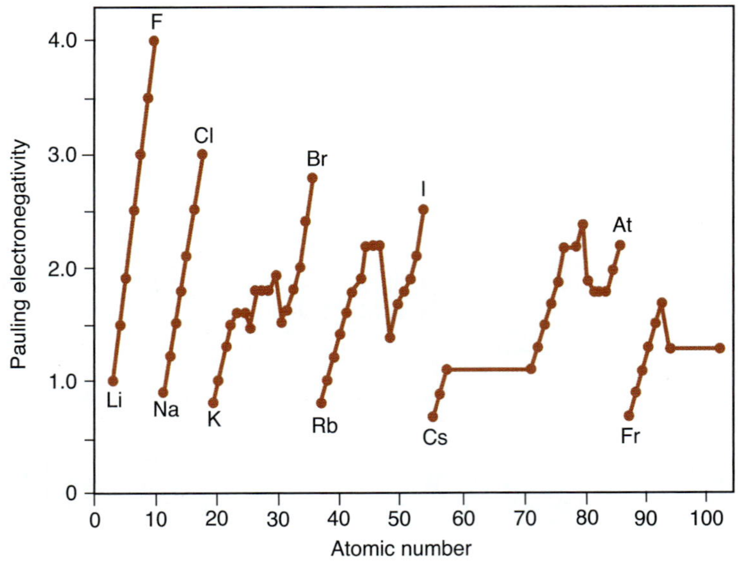

Figure 7.14 Pauling electronegativities plotted against atomic number.

that fluorine is the most electronegative atom and that the least electronegative atoms are cesium and francium. The order for the generally most useful electronegativities is as follows:

F	>	O	>	Cl	>	N	>	S	>	C	>	P	>	H
3.98		3.44		3.16		3.04		2.58		2.55		2.19		2.1

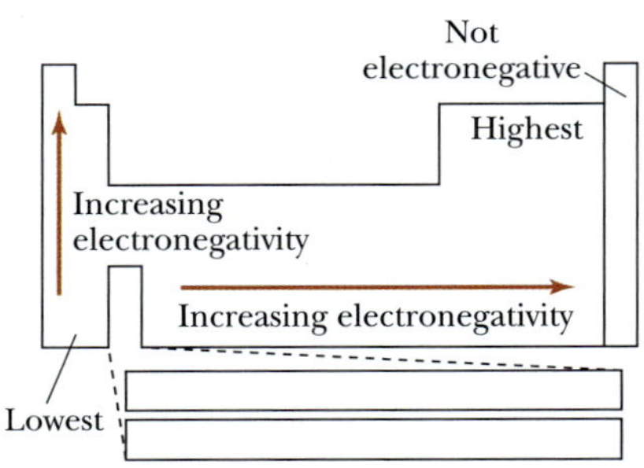

Figure 7.15 The trends of electronegativities in the periodic table.

Because electronegativity is a derived quantity based on an arbitrary scale, only differences in electronegativities are meaningful. It is significant that the electronegativity difference between fluorine and hydrogen is 1.9, but the electronegativity of fluorine is twice that of hydrogen only because of the scale we are using. A comparison of Figures 7.13 and Figure 7.15 shows the inverse relationship between metallic character and electronegativity.

7-10. We Can Use Electronegativity Differences to Predict the Polarity of Chemical Bonds

It is the difference in electronegativities of the two atoms in a covalent bond that determines how the electrons in the bond are shared. If the electronegativities are nearly the same, or less than about 0.4, then the electrons in the bond are essentially shared equally and the bond is called a **pure covalent bond,** or a **nonpolar bond.** Equal sharing of bonding electrons occurs in homonuclear diatomic molecules. If the electronegativities of the two atoms differ by about 0.4 or more, then electrons in the bond are not shared equally and the bond is said to be a **polar covalent bond.** The extreme case of a polar bond occurs when the difference in electronegativities is large, say, greater than about 2.0. For such a case, the electron pair is completely on the more electronegative atom, and the bond is a **pure ionic bond.**

Bond character:	Covalent	Polar covalent	Ionic
Electronegativity difference:	≈0 – 0.3	≈0.4 – 2.0	≈2.1 – 4.0

A polar covalent bond can be illustrated by a HCl molecule. According to Figure 7.13, the electronegativity of a hydrogen atom is 2.1 and that of a chlorine atom is 3.16, and the difference between them is 1.1. Thus, the electrons in the bond are not shared equally. Because the electronegativity of a chlorine atom is greater than that of a hydrogen atom, the chlorine atom attracts the electron pair more strongly than the hydrogen atom does. The bonding electrons are shifted a little toward the chlorine atom, so it acquires a *partial* negative charge, thereby leaving the hydrogen atom with a partial positive charge; that is, the bond is polar. We indicate **partial charges** by the lowercase Greek letter delta, δ, and we write

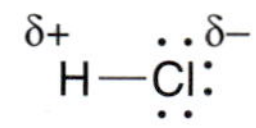

It is important to understand that δ+ and δ– represents only the fraction of an electronic charge that results from the unequal sharing of the electrons in the covalent bond. The numerical value of δ is of no importance to us at this stage. It denotes only that the hydrogen atom is slightly positively charged and that the chlorine atom is slightly negatively charged. From a quantum theoretical point of view, δ represents the fact that the two electrons in a covalent bond are more likely to be found near the chlorine atom than near the hydrogen atom. We say that the bond in a HCl molecule has **partial ionic character.**

EXAMPLE 7-15: Describe the charge distribution in a molecule of the interhalogen compound chlorine fluoride, ClF.

Solution: According to Figure 7.13, the electronegativity of a F atom is 3.98 and that of a Cl atom is 3.16, so the chlorine-fluorine bond is a polar covalent bond. The electron pair is somewhat more likely to be found near the fluorine atom than near the chlorine atom. Thus, the fluorine atom has a slightly negative charge, δ–, and the chlorine atom has a slightly positive charge, δ+. We can represent this polar bond by writing its Lewis formula

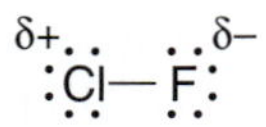

PRACTICE PROBLEM 7-15: Describe the charge distribution in a water molecule, which has a bent shape.

Answer:

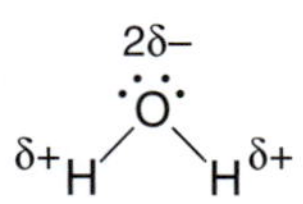

EXAMPLE 7-16: Using the electronegativities given in Figure 7.13, predict whether the bonds in calcium oxide and mercury(II) chloride are covalent, polar covalent, or ionic.

Solution: The difference in the electronegativities of a calcium atom and an oxygen atom is 3.44 – 1.00 = 2.44. This difference is larger than 2.0, so we predict that the bonding in CaO is ionic. The difference in the electronegativities of a mercury atom and a chlorine atom is 3.16 – 2.00 = 1.16, so we predict that the bonding in $HgCl_2$ is polar covalent (or covalent with ionic character).

PRACTICE PROBLEM 7-16: Using the electronegativities given in Figure 7.13, discuss the bonding in a silane molecule, SiH_4, and a $BeCl_2$ molecule.

Answer: The bonding in a silane molecule is covalent with a small degree of ionic character and the bonding in a $BeCl_2$ molecule is polar covalent with a fair degree of ionic character.

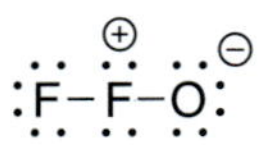
oxygen difluoride

Electronegativity can also be used to help write (or reject) Lewis formulas. For example, in Section 7-4 we rejected the Lewis formula shown in the margin for an OF_2 molecule because of the separation of formal charges in the molecule as written. In addition to this, a formal charge of +1 on a very electronegative atom (recall that fluorine is the most electronegative atom) is not chemically reasonable because a fluorine atom tends to gain electrons, not lose them. Notice also that in Practice Problem 7-14 the formulas

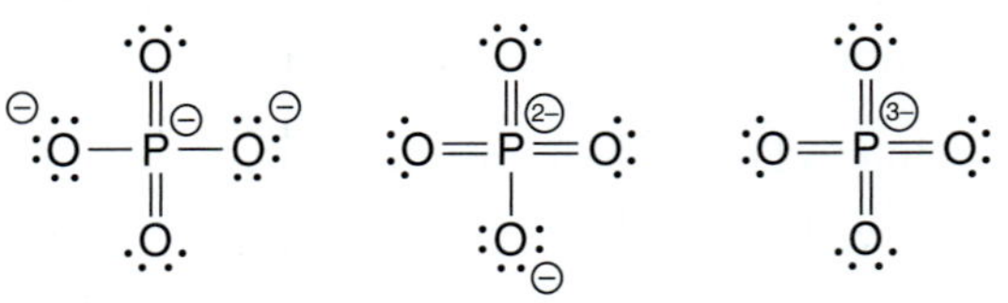

and their equivalents were not considered as resonance forms of the phosphate ion. We did not consider these formulas because in each case there is a negative formal charge on the less electronegative phosphorous atom. In general, if there is a choice, negative formal charges should be placed on the more electronegative elements and positive formal charges on the less electronegative elements.

7-11. Polyatomic Molecules with Polar Bonds May Be Polar or Nonpolar

One measure of the polarity of a diatomic molecule is its **dipole moment.** The dipole moment customarily is represented as an arrow pointing along the bond from δ– to δ+. For a HCl molecule and a ClF molecule, for example, we have

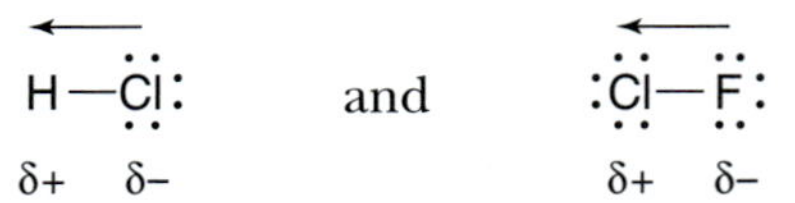

This notation indicates the direction of the dipole moment. The magnitude of the dipole moment is the absolute value of the product of the length of the bond and the net charge on either atom. Dipole moments can be measured

experimentally, and the dipole moments of the hydrogen halides are given in Table 7.4. Note the dependence of the dipole moment on the electronegativity difference. The larger the difference in electronegativity, the larger the dipole moment.

A quantity that has magnitude and direction is called a **vector**. Dipole moments are vector quantities. To understand the properties of a vector quantity, consider a familiar example: the force with which you pull on something. A force must be described by both the direction in which it is applied and its magnitude. Like forces, vectors cancel if applied with the same magnitude in exactly opposite directions. In a stalemated tug of war, both teams are pulling with the same magnitude of force but in opposing directions. The net result is an effective cancellation. This idea can be illustrated pictorially as follows:

 or  = no net force

Even if they are equal in magnitude, if the forces are not applied in opposing directions, then there is a resultant net force:

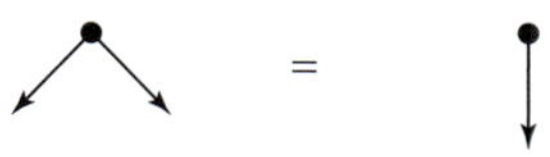

two forces on an object = net force on the object

It will not be necessary for us to be able to calculate the magnitude of the net force. We need only see pictorially the direction of the net force.

In a **polyatomic molecule** (a molecule with more than two atoms), the polarity of each bond, like the dipole moment of a diatomic molecule, can be represented by an arrow and is a quantity that has both magnitude and direction. Bond polarities, then, must have the same vector properties as forces.

As we shall see in the next chapter, CO_2 is a **linear molecule**, that is, all three atoms lie in a straight line (Figure 7.16). In a carbon dioxide molecule, each carbon–oxygen bond is polar because an oxygen atom is more electronegative than a carbon atom, but the bond polarities point in opposite directions because CO_2 is a linear molecule.

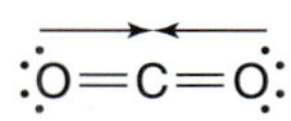

Unlike physics books and physical chemistry books, many general chemistry books still define the direction of the dipole moment to be from the positive charge to the negative charge of a molecule. We define its direction to be from the negative to the positive charge because the International Union of Pure and Applied Chemistry (IUPAC), which is the official organization that monitors the international exchange of scientific information in the field of chemistry, says "When a dipole is composed of two charges Q and −Q separated by a distance *r*, the direction of the dipole is taken to be from the negative to the positive charge. The opposite convention is sometimes used but is to be discouraged."

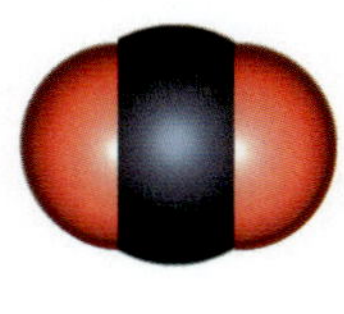

CO_2

Figure 7.16 A space-filling model of a carbon dioxide molecule. CO_2 is a linear molecule. The O–C–O bond angle is 180°.

TABLE 7.4 The dipole moments of the gaseous hydrogen halides

Molecule	Electronegativity difference	Dipole moment/10^{-30} C·m*
HF	1.9	6.36
HCl	1.1	3.43
HBr	0.9	2.63
HI	0.6	1.27

*The units of dipole moment are charge × distance, or coulombs × meters (C·m) in SI units.

The bond polarities cancel exactly, so the CO_2 molecule has no net dipole moment. A molecule with no net dipole moment is a **nonpolar molecule.**

Conversely, if we know the dipole moment experimentally, then we can apply our representation to learn about the direction of the bonds. For example, because a CO_2 molecule has no net dipole moment, the molecule must be linear.

Let's consider a H_2O molecule as another example. The oxygen–hydrogen bonds are polar because an oxygen atom is more electronegative than a hydrogen atom; we can illustrate this as

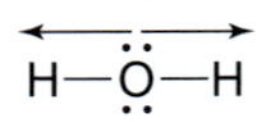

However, this diagram implies that water is a nonpolar molecule. In fact, a H_2O molecule has a large dipole moment, so it is a **polar molecule.** The discrepancy is due to our assumption that H_2O is a linear molecule. We will see in the next chapter that the water molecule is bent. The H–O–H bond angle is not 180°; rather it is 104.5° (Figure 7.17). We can illustrate the bent structure by

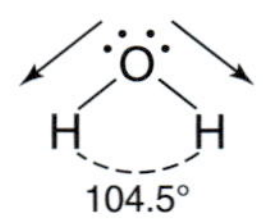

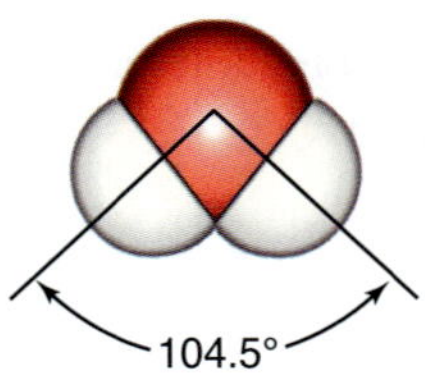

H_2O

Figure 7.17 Model of a water molecule. H_2O is a bent molecule. The H–O–H bond angle is 104.5°.

Thus, the net dipole moment in a H_2O molecule has the orientation

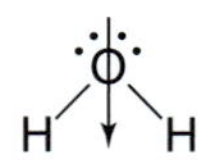

There is an important lesson for us here. The Lewis formulas that we have learned to write in this chapter suggest only which atoms are bonded to which in the molecule, not the spatial arrangement of the atoms. Although Lewis formulas are very useful in this regard, they are *not* meant to indicate the three-dimensional arrangement of the atoms in a molecule. In the next chapter we learn some simple, useful rules for predicting the shapes of molecules by using Lewis formulas.

SUMMARY

A Lewis formula shows the arrangement of the valence electrons of the constituent atoms in a molecule and suggests how these atoms are bonded to each other. A covalent bond is represented as a pair of electrons shared between two atoms. The octet rule states that each element in a species forms covalent bonds that will give it eight electrons in its outer shell. The octet rule is especially useful for compounds involving carbon, nitrogen, oxygen, and fluorine atoms. Two atoms may form a single bond, a double bond, or a triple bond between them in order to satisfy the octet rule.

There are a number of exceptions to the octet rule. Hydrogen is almost always a terminal atom, requiring only two electrons. Species with an odd number of electrons do not satisfy the octet rule and form free radicals. Some electron-deficient species have an insufficient number of electrons to satisfy the octet rule. Elements in the third and lower rows of the periodic table may expand their valence shell by using *d* orbitals to accommodate more than eight electrons in their outer shell.

Another aid for writing Lewis formulas is the assignment of formal charges to the various atoms in a Lewis formula. The preferred Lewis formula has the lowest formal charge or the least separated formal charges. If there is a choice, negative formal charges should be placed on the more electronegative elements and positive formal charges on the less electronegative elements in a molecule.

When it is possible to write two or more Lewis formulas for a molecule without altering the positions of the atoms, each individual formula is said to be a contributing form to a resonance hybrid. The actual bonding is best represented by an average, or a superposition, of the individual contributing forms, called a resonance hybrid.

Most chemical bonds are neither purely ionic nor purely covalent. A good measure of the ionic character in a bond is determined from the electronegativity difference between the atoms making up the bond. When the electronegativities of the two atoms joined by a covalent bond are different, the bond is said to be polar.

Lewis formulas indicate only the bonding within a molecule, *not* the shape of a molecule.

TERMS YOU SHOULD KNOW

valence electron *197*
covalent bond *198*
Lewis formula *198*
lone electron pair *198*
lone pair *198*
molecular crystal *198*
bond length *198*
octet rule *198*
organic compound *205*
formal charge *205*
double bond *209*
triple bond *209*
resonance form *212*
resonance *212*
resonance hybrid *213*
delocalized bond *213*
resonance energy *213*
planar molecule *215*
resonance stabilization *216*
free radical *216*
electron-deficient compound *216*
expanded valence shell *218*
electronegativity *223*
pure covalent bond *224*
nonpolar bond *224*
polar covalent bond *224*
pure ionic bond *224*
partial charge, δ *225*
partial ionic character *225*
dipole moment (C·m) *226*
vector *227*
polyatomic molecule *227*
linear molecule *227*
nonpolar molecule *228*
polar molecule *228*

AN EQUATION YOU SHOULD KNOW HOW TO USE

$$\begin{pmatrix}\text{formal charge}\\ \text{on an atom in}\\ \text{a Lewis formula}\end{pmatrix} = \begin{pmatrix}\text{total number of}\\ \text{valence electrons}\\ \text{in the free atom}\end{pmatrix} - \begin{pmatrix}\text{total number}\\ \text{of lone-pair}\\ \text{electrons}\end{pmatrix} - \frac{1}{2}\begin{pmatrix}\text{total number}\\ \text{of shared}\\ \text{electrons}\end{pmatrix} \quad (7.1) \text{ (calculation of formal charge)}$$

PROBLEMS

COMPOUNDS INVOLVING SINGLE BONDS

7-1. Pure hydrogen peroxide, $H_2O_2(l)$, is a colorless liquid that is caustic to the skin, but a 3% aqueous solution is a mild bleaching agent. Write the Lewis formula for a H_2O_2 molecule.

7-2. Tetrafluorohydrazine, $N_2F_4(l)$, is a colorless liquid that is used as rocket fuel. Write the Lewis formula for a tetrafluorohydrazine molecule.

7-3. Write the Lewis formula for the molecules

(a) SCl_2 (b) $GeCl_4$ (c) $AsBr_3$ (d) PH_3

7-4. Write the Lewis formula for the molecules

(a) PBr_3 (b) SiF_4 (c) NI_3 (d) H_2Se

7-5. Write the Lewis formula for the molecules

(a) methane, CH_4

(b) fluoromethane, CH_3F

(c) aminomethane, CH_3NH_2

7-6. Write the Lewis formula for the molecules

(a) methyl mercaptan, CH_3SH

(b) dimethyl ether, CH_3OCH_3

(c) trimethyl amine, $N(CH_3)_3$

MULTIPLE BONDS

7-7. Write the Lewis formula for the molecules

(a) acetylene, C_2H_2 (b) diazine, N_2H_2

(c) phosgene, $COCl_2$ (d) fluorine cyanide, FCN

7-8. Write the Lewis formula for the molecules

(a) nitrous acid, HNO_2

(b) silicon dioxide, SiO_2

(c) propylene, CH_3CHCH_2

(d) perchloroethylene, CCl_2CCl_2

7-9. Vinyl chloride is an important industrial chemical used in the manufacture of polyvinyl chloride. Its chemical formula is C_2H_3Cl. Write the Lewis formula for a vinyl chloride molecule.

7-10. Azides of heavy metals explode when struck sharply and are used in detonation caps. Write the Lewis formula for the azide ion, N_3^-.

7-11. Formic acid (systematic name: methanoic acid) is a colorless liquid with a penetrating odor. It is the irritating ingredient in the bite of ants. Its chemical formula is HCOOH. Write the Lewis formula for a formic acid molecule. (*Hint:* Each of the atoms in the correct formula has a zero formal charge.)

7-12. Acetone (systematic name: 2-propanone) is an organic compound widely used in the chemical industry as a solvent, for example, in paints and varnishes. You may be familiar with its sweet odor because it is used as a fingernail polish remover. Its chemical formula is CH_3COCH_3. Write the Lewis formula for an acetone molecule.

FORMAL CHARGE

7-13. Use formal charge considerations to rule out the Lewis formula for a NF_3 molecule in which the nitrogen atom and the three fluorine atoms are connected in a row.

7-14. Use formal charge considerations to predict the arrangement of the atoms in a NOCl molecule.

7-15. Laughing gas, an anesthetic and a propellant in whipped-cream-dispensing cans, has the chemical formula N_2O. Use Lewis formulas and formal charge considerations to predict which structure, NNO or NON, is the more likely.

7-16. Use formal charge considerations to rule out the Lewis formula for a NO_2^- ion in which the arrangement of the atoms is O–O–N.

RESONANCE

7-17. Write Lewis formulas for the resonance forms of the formate ion, $HCOO^-$. Indicate formal charges and discuss the bonding of this ion.

7-18. Write Lewis formulas for the resonance forms of the acetate ion, CH_3COO^-. Indicate formal charges and discuss the bonding in this ion.

7-19. Write Lewis formulas for the resonance forms of the hydrogen carbonate (bicarbonate) ion, HCO_3^-. Indicate formal charges and discuss the bonding in this ion.

7-20. Write Lewis formulas for the resonance forms of an ozone molecule, O_3. Indicate formal charges and discuss the bonding in this molecule.

7-21. Write the Lewis formula for the following benzene derivatives:

(a) chlorobenzene, C_6H_5Cl

(b) aminobenzene, $C_6H_5NH_2$

(c) benzoic acid, C_6H_5COOH

(d) phenol, C_6H_5OH

7-22. Naphthalene, which gives mothballs their characteristic odor, has the formula $C_{10}H_8$. Given that its structure is two benzene rings fused together along one carbon-carbon bond, write its Lewis formula.

FREE RADICALS

7-23. Which of the following species contain an odd number of electrons?

(a) NO_2 (b) CO (c) O_3^- (d) O_2^-

Write a Lewis formula for each of these species.

7-24. Which of the following species contain an odd number of electrons?

(a) BrO_3 (b) SO_3 (c) HNO (d) HO_2

Write a Lewis formula for each of these species.

7-25. Nitrosamines are carcinogens that are found in tobacco smoke. They can also be formed in the body from the nitrites and nitrates used to preserve processed meats, especially bacon and sausage. The simplest nitrosamine molecule is methylnitrosamine, H_3CNNO. Write the Lewis formula for this molecule. Is it a free radical?

7-26. Many free radicals combine to form molecules that do not contain any unpaired electrons. The driving force for the radical-radical combination reaction is the formation of a new electron-pair bond. Write Lewis formulas for the reactant and product species in the following chemical equations:

(a) $CH_3(g) + CH_3(g) \rightarrow H_3CCH_3(g)$

(b) $N(g) + NO(g) \rightarrow NNO(g)$

(c) $2\ OH(g) \rightarrow H_2O_2(g)$

EXPANDED OCTETS

7-27. Write the Lewis formula for the ions

(a) PCl_6^- (b) I_3^- (c) SiF_6^- (d) $IO_2F_2^-$

7-28. Write a Lewis formula for each of the following compounds of xenon:

Compound	Form at 25°C	Melting point/°C
XeF_2	colorless crystals	129
XeF_4	colorless crystals	117
XeF_6	colorless crystals	50
$XeOF_4$	colorless liquid	−46
XeO_2F_2	colorless crystals	31

7-29. Write the Lewis formulas of the interhalogen compounds IF_3 and IF_5.

7-30. Write the Lewis formulas of the interhalogen ions ICl_4^-, IF_4^+, and IF_2^-.

7-31. Write Lewis formulas for a sulfur tetrafluoride, SF_4, and sulfur hexafluoride, SF_6, molecule.

7-32. Write Lewis formulas for a sulfinyl fluoride, SOF_2, and sulfonyl fluoride, SO_2F_2, molecule.

ELECTRONEGATIVITY AND DIPOLE MOMENTS

7-33. Using only a periodic table as a guide, arrange the following atoms in order of decreasing electronegativity:

Na S Si F O

7-34. Using only a periodic table as a guide, arrange the following atoms in order of decreasing electronegativity:

In Se Sb Cl S

7-35. Arrange the following groups of molecules in order of increasing dipole moment:

(a) HCl HF HI HBr (linear molecules)

(b) PH_3 NH_3 AsH_3 (tripod-shaped molecules)

(c) Cl_2O F_2O H_2O (bent molecules)

7-36. Arrange the following groups of molecules in order of increasing dipole moment:

(a) ClF_3 BrF_3 IF_3 (T-shaped molecules)

(b) H_2O H_2S H_2Te H_2Se (bent molecules)

(c) O_3 SO_2 H_2S (bent molecules)

7-37. Describe the charge distribution in the following molecules:

(a) nitrogen trifluoride, NF_3

(b) oxygen difluoride, OF_2

(c) oxygen dibromide, OBr_2

7-38. Describe the charge distribution in the following molecules:

(a) hydrogen fluoride, HF

(b) phosphine, PH_3

(c) hydrogen sulfide, H_2S

ADDITIONAL PROBLEMS

7-39. A solution of sugar in water is a poor conductor of electricity, whereas a solution of table salt in water is a good conductor of electricity. What does this tell you about the bonding in the two substances?

7-40. What is a chemical bond? What kinds of chemical bonds are there? How do we distinguish between each type of chemical bond?

7-41. The octet rule states that atoms and ions with eight electrons in their valence shell are more stable than those with fewer than eight electrons. Does this rule hold for all atoms in the periodic table?

7-42. What is meant by "delocalized charge"?

7-43. Explain why phosphorous pentafluoride molecules, PF_5, exist in nature, whereas nitrogen pentafluoride molecules, NF_5, do not—even though the elements phosphorous and nitrogen are in the same family in the periodic table.

7-44. Write the Lewis formula, show the formal charges, and discuss the bonding for

(a) the sulfate ion, SO_4^{2-}

(b) the phosphate ion, PO_4^{3-}

(c) the acetate ion, $CH_3CO_2^-$

7-45. The halogens form a number of interhalogen compounds. For example, chlorine pentafluoride, $ClF_5(g)$, can be prepared according to the following chemical equation:

$$KCl(s) + 3F_2(g) \rightarrow \underset{\text{colorless}}{ClF_5(g)} + KF(s)$$

The halogen fluorides are very reactive, combining explosively with water, for example. Write the Lewis formula for each of the following halogen fluoride species:

(a) ClF_5 (b) IF_3 (c) IF_7 (d) IF_4^+

7-46. Write the Lewis formula for

(a) the tetrafluoroammonium ion, NF_4^+

(b) the tetrafluorochlorinium ion, ClF_4^+

(c) the phosphonium ion, PH_4^+

(d) the hexafluoroarsenate ion, AsF_6^-

(e) the tetrafluorobromate ion, BrF_4^-

7-47. Write the Lewis formula for each of the following oxychlorine species:

(a) perchlorate ion, ClO_4^-

(b) chlorine oxide, ClO

(c) chlorate ion, ClO_3^-

(d) chlorine dioxide, ClO_2

(e) hypochlorite ion, ClO^-

7-48. Write the Lewis formula for each of the following acid molecules:

(a) $HClO_3$ (b) HNO_2

(c) HIO_4 (d) $HBrO_2$

7-49. Write the Lewis formula for each of the following oxyacids of sulfur:

(a) sulfuric acid, H_2SO_4

(b) thiosulfuric acid, $H_2S_2O_3$

(c) disulfuric acid, $H_2S_2O_7$ (has a S–O–S bond)

(d) dithionic acid, $H_2S_2O_6$ (has a S–S bond)

(e) peroxydisulfuric acid, $H_2S_2O_8$ (has an O–O bond)

7-50. Sulfur forms a number of fluorides. Some of them are

(a) SF_2 (b) SF_4 (c) S_2F_{10} (d) S_2F_2

Write a Lewis formula for each of these fluoride molecules.

7-51. Phosphorus forms a number of oxohalides, X_3PO, in which X may be a F, Cl, or Br atom. The most common of these, phosphoryl chloride, is obtained as follows:

$$2\,PCl_3(g) + O_2(g) \rightarrow 2\,Cl_3PO(g)$$

Write the Lewis formula for a phosphoryl halide molecule.

7-52. (*) In the P_4O_6 molecule each phosphorus atom is bonded to three oxygen atoms and each oxygen atom is bonded to two phosphorus atoms. Write a Lewis formula for the P_4O_6 molecule.

7-53. Indicate whether or not molecules of the following species have a triple bond.

(a) CH_2CHCN (b) HOOCCOOH

(c) C_2^{2-} (d) CH_3CHCH_2

7-54. Use Lewis formulas and formal charge considerations to suggest that the structure of an N_2F_2 molecule is FNNF rather than FFNN or NFFN.

7-55. Use Lewis formulas and formal charge considerations to suggest that the structure of a hydrocyanic acid molecule is HCN rather than HNC.

7-56. An alcohol is an organic compound containing an OH group. Write the Lewis formula for molecules of the following alcohols (common names are given in parentheses):

(a) ethanol (ethyl alcohol), CH_3CH_2OH

(b) 1-propanol (*n*-propyl alcohol), $CH_3CH_2CH_2OH$

(c) 2-propanol (isopropyl alcohol), $(CH_3)_2CHOH$

7-57. Write all the possible Lewis formulas for a molecule with the chemical formula $C_2H_4Cl_2$.

7-58. Write the Lewis formula for a bromine chloride molecule and indicate its dipole moment.

7-59. A hydroxide ion, OH^-, results when a proton is removed from a H_2O molecule. The ion that results when a proton is removed from a H_2O_2 molecule is called a hydroperoxide ion, HO_2^-. Write the Lewis formula for a hydroperoxide ion. Name the following ionic compounds: $NaHO_2$ and $Ba(HO_2)_2$.

7-60. A methoxide ion, CH_3O^-, results when a proton is removed from a methanol molecule, CH_3OH. Write the Lewis formula for a CH_3O^- ion. Name the compounds $KOCH_3$ and $Al(OCH_3)_3$.

7-61. Write the Lewis formula for each of the following species. Give resonance forms where appropriate and indicate formal charges.

(a) CS_3^{2-} (b) $C_2O_4^{2-}$ (c) NCS^-

7-62. Write the Lewis formula for each of the following nitrogen oxide molecules. Indicate formal charge and resonance forms.

Formula	Name	Form at 25°C, 1 atm
N_2O	dinitrogen oxide	colorless gas
NO	nitrogen oxide	colorless gas
N_2O_3	dinitrogen trioxide	dark blue gas
NO_2	nitrogen dioxide	brown gas
N_2O_4	dinitrogen tetroxide	colorless gas
N_2O_5	dinitrogen pentoxide	white solid

7-63. Solid sulfur consists of eight-membered rings of sulfur atoms. Write the Lewis formula for a S_8 molecule.

7-64. Write Lewis formulas for the resonance forms of the thiosulfate ion, $S_2O_3^{2-}$. Indicate formal charges and discuss the bonding in this ion.

7-65. (*) White phosphorous has the chemical formula $P_4(s)$. A P_4 molecule has 20 valence electrons. Draw a Lewis formula for a white phosphorous molecule in which none of the atoms violates the octet rule and the formal charge on each atom is zero.

Ronald J. Gillespie (1924–), left, was born in London, England. He was educated at the University of London, obtaining his doctorate in 1957. In the same year, he and Sir Ronald Nyholm put forth the VSEPR theory of molecular geometry, now an essential part of all general chemistry courses. He is an Emeritus Professor of Chemistry at McMaster University, Hamilton, Ontario, where he has taught inorganic chemistry and general chemistry for over thirty years. He is also a Fellow of the Royal Society of London and the Royal Society of Canada. Professor Gillespie has an international reputation for chemical education, having received the Manufacturing Chemists College Chemistry Teacher Award, the Union Carbide Award of the Chemical Institute of Canada for Chemical Education, and the McMaster Students Union Award for Excellence in Teaching and four honorary degrees.

Sir Ronald Sydney Nyholm (1917–1971), right, was born in the mining town of Broken Hill, New South Wales, Australia. He was educated at the University of Sydney and received his doctorate at University College in London in 1950. He then took a post as an associate professor of inorganic chemistry at the New South Wales University of Technology, becoming President of the Royal Society of New South Wales in 1954. In 1955, he returned to London as a professor of chemistry and head of the department at University College in London. There he and Gillespie developed the VSEPR model in 1957. In the following year, he was elected a Fellow of the Royal Society. Professor Nyholm was an internationalist, fascinated by both language and culture, and mentored research students from over 30 different countries. In 1967, he was made a Knight Bachelor for his scientific achievements and especially for his service to the teaching of science. Tragically, his distinguished career was cut short when he died from injuries sustained in an automobile accident.

8. Prediction of Molecular Geometries

The shapes of molecules play a major role in determining a wide variety of chemical properties, including chemical reactivity, odor, taste, and drug action. In this chapter we devise a set of simple, systematic rules that allow us to predict the shapes of thousands of molecules. These rules are based on the Lewis formulas that we developed in Chapter 7 and are collectively called the valence-shell electron-pair repulsion (VSEPR) theory. In spite of its rather imposing name, VSEPR theory is easy to understand, easy to apply, and remarkably reliable. We should perhaps emphasize here that we use these rules to predict the shapes of *isolated* molecules, such as molecules in the gas phase. The shapes of molecules even in the solid phase, however, typically differ little from the predictions that we shall make here.

8-1. Lewis Formulas Do Not Give Us the Shapes of Molecules

Lewis formulas show the bonding relationships among atoms in a molecule. Although a Lewis formula shows the connectivity between atoms, it does not show us the shape or geometrical arrangement of the nuclei in the molecule. Consider the molecule dichloromethane, CH_2Cl_2. One Lewis formula for dichloromethane is

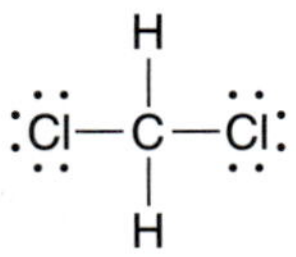

I

If we infer from the Lewis formula that dichloromethane is flat, or **planar,** then we must conclude that the Lewis formula

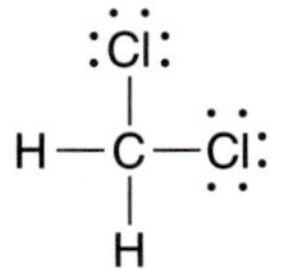

II

represents a different geometry for dichloromethane. In formula I the two chlorine atoms lie 180° apart, whereas in formula II they lie 90° apart. Molecules that have the same chemical formula (CH_2Cl_2, in this case) and atom-to-atom bonding but different spatial arrangements of the atoms are called **stereoisomers**. Stereoisomers are different molecular species and therefore have different chemical and physical properties.

Two stereoisomers of a dichloromethane molecule have never been observed; there is only one kind of dichloromethane molecule. This finding suggests that the four bonds are oriented such that there is only one distinct way of bonding the two hydrogen atoms and the two chlorine atoms to the central carbon atom. Therefore, our assumption that dichloromethane is a planar molecule is incorrect.

A geometric arrangement that shows why there are no geometric isomers of a dichloromethane molecule was reported independently in 1874 by the Dutch chemist Jacobus H. van't Hoff and the French chemist Joseph Le Bel. They proposed that the four bonds about a central carbon atom in a molecule such as methane, CH_4, or dichloromethane, CH_2Cl_2, are directed toward the vertices of a **tetrahedron** (Figure 8.1). A regular tetrahedron is a four-sided figure that has four equivalent vertices and four identical faces, each of which is an equilateral triangle (Figure 8.2). A molecule whose three-dimensional form is that of a tetrahedron is described as being **tetrahedral** in shape (we change the ending to *al* to form the adjective). Cardboard models of tetrahedral molecules originally made by J. H. van't Hoff are shown in Figure 8.3.

We shall see some examples of compounds that do form stereoisomers in Section 8-10.

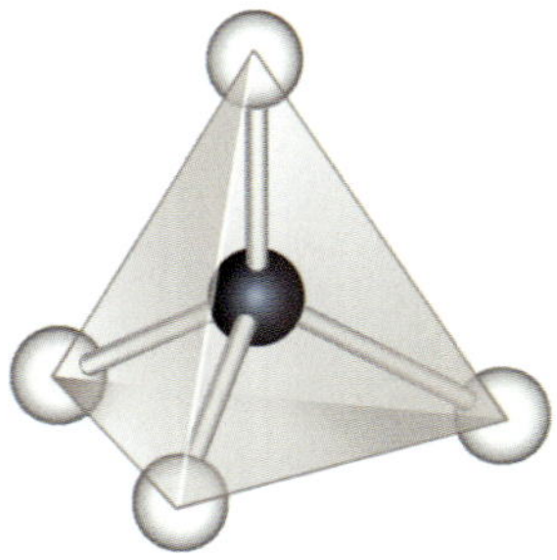

(a)

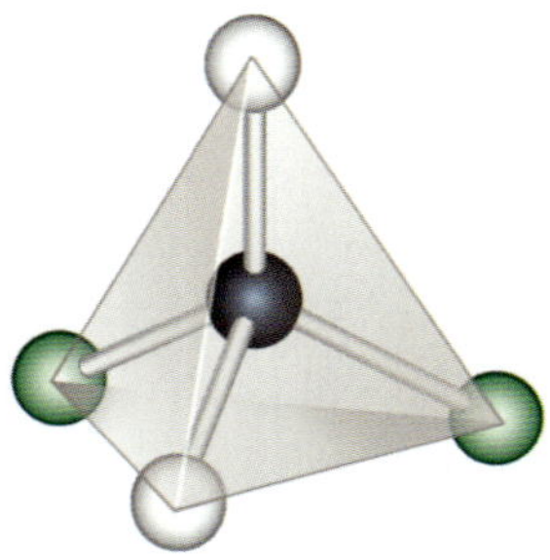

(b)

Figure 8.1 (a) A ball-and-stick model of a methane, CH_4, molecule. Each carbon–hydrogen bond in a CH_4 molecule points toward the vertex of a regular tetrahedron. The positions of all four hydrogen atoms are equivalent by symmetry. All the H–C–H bond angles are the same, 109.5°. (b) A ball-and-stick model of a dichloromethane, CH_2Cl_2, molecule. Note that it makes no difference at which two vertices we place the two chlorine atoms; in each case, exactly the same molecule results.

8-2. All Four Vertices of a Regular Tetrahedron Are Equivalent

You can verify from Figure 8.1b or by constructing a molecular model that the four vertices of a tetrahedron are equivalent and there is only one way of bonding two hydrogen atoms and two chlorine atoms directly to a central carbon atom. The tetrahedral model for a CH_2Cl_2 molecule is thus in accord with the experimental fact that dichloromethane has no isomers.

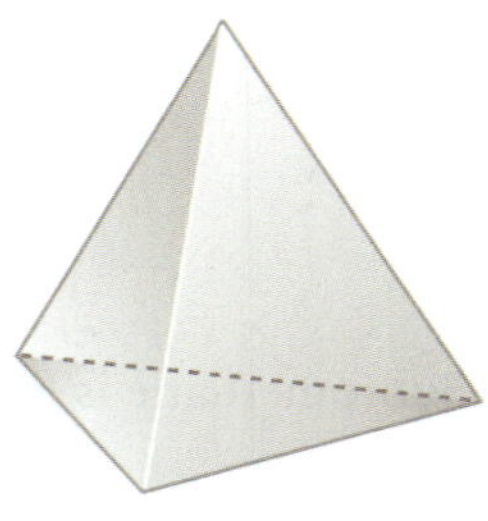

(a)

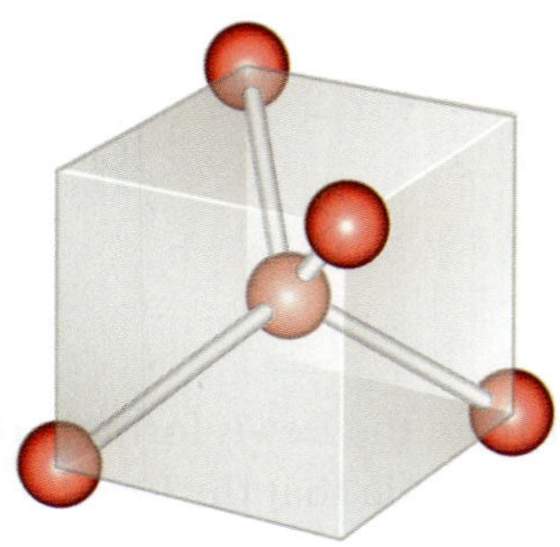

(b)

Figure 8.2 (a) A regular tetrahedron is a symmetric body consisting of four equivalent vertices and four equivalent faces. Each face is an equilateral triangle. A tetrahedron differs from the more familiar square pyramid, which has a square base and four triangular sides. (b) A tetrahedron also can be viewed as derived from a cube by placing atoms at four of the eight vertices as shown, and then placing an atom in the center of the cube. From this diagram you can derive the tetrahedral angle of 109.5° (see Problem 8-80).

Figure 8.3 J. H. van't Hoff (Chapter 16 Frontispiece), the Dutch chemist who first proposed the tetrahedral geometry of methane and related compounds, built these cardboard models to illustrate molecular shapes. In 1901 van't Hoff received the first Nobel Prize awarded in chemistry.

Molecular models of the type shown in Figure 8.4 are called **space-filling molecular models.** Such models give fairly accurate representations of the angles between bonds and of the relative sizes of the atoms in molecules. A less realistic molecular model, but one in which the geometry is usually easier to see, is the **ball-and-stick molecular model** used in Figure 8.1.

In a tetrahedral molecule like methane, CH_4, all the H–C–H bond angles are equal to 109.5°, which is called the **tetrahedral bond angle.** The tetrahedral bond angle of 109.5° is a direct consequence of the geometrical properties of a regular tetrahedron. It is the angle between any two vertices of the tetrahedron and a point located exactly in the center of the tetrahedron (Figure 8.5).

A carbon atom that is bonded to four other atoms is called a **tetravalent** carbon atom. The hypothesis of van't Hoff and Le Bel that the bonds of a tetravalent carbon atom are tetrahedrally oriented was the beginning of what is called **structural chemistry,** the area of chemistry in which the shapes and sizes of molecules are studied. Many experimental methods have been developed to determine molecular geometries. Most of the methods involve the interaction of electromagnetic radiation or electrons with molecules. Using such methods, we can measure bond lengths and bond angles in molecules and thereby determine molecular geometries, such as the tetrahedral structure of a CH_2Cl_2 molecule. It turns out that molecules occur in a fascinating variety of shapes. We noted in Chapter 7 that CO_2 is a linear molecule and that H_2O is a bent molecule. We have also noted that CH_4 is an example of a tetrahedral molecule. Some examples of other molecular geometries are shown in Figure 8.6.

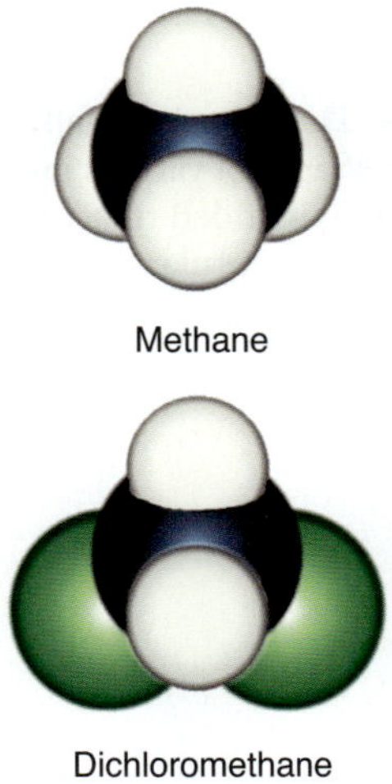

Figure 8.4 Space-filling molecular models of methane and dichloromethane.

8-3. Valence-Shell Electron-Pair Repulsion Theory Is Used to Predict the Shapes of Molecules

A simple theory proposed in 1957 by Ronald J. Gillespie and Sir Ronald S. Nyholm (Frontispiece) enables us to predict the shapes of molecules such as those shown in Figure 8.6. The prediction method of the theory is based on the total number of bonds and lone electron pairs in the valence shell of the central atom in a molecule. The key postulate of the theory is that the shape of a molecule is determined by minimizing the mutual repulsion of the electron pairs

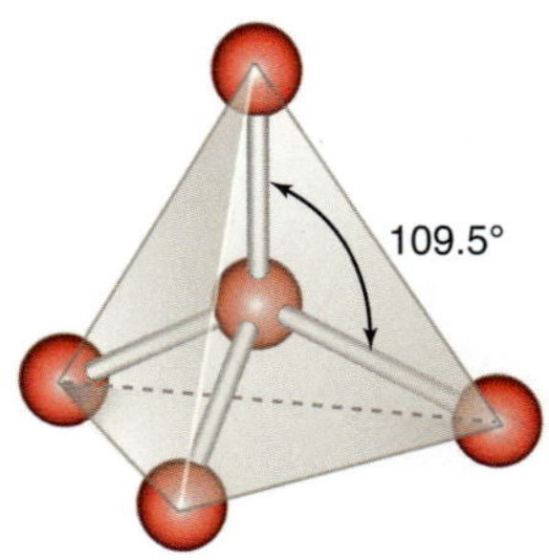

Figure 8.5 A tetrahedron showing the six equivalent tetrahedral bond angles.

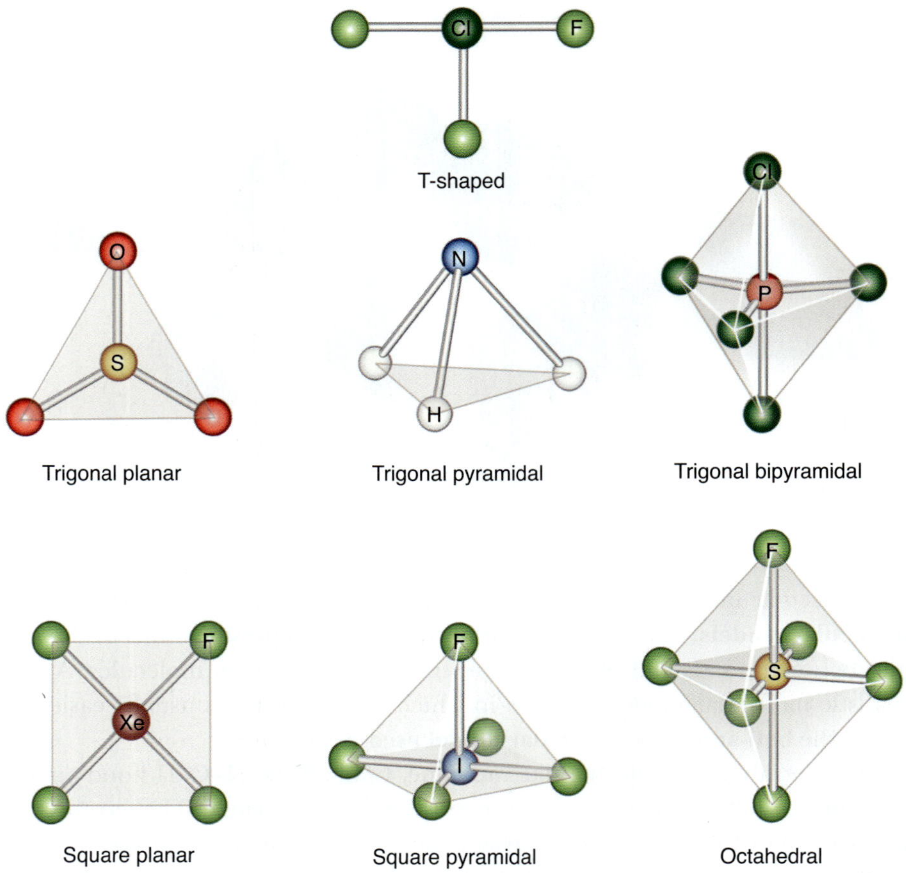

Figure 8.6 The experimentally observed shapes of various molecules. The shading is added to help us visualize the molecular shape.

in the valence shell of the central atom. For this reason, the theory is called the **valence-shell electron-pair repulsion theory,** or the **VSEPR theory**.

Consider a beryllium chloride molecule, $BeCl_2$, an electron-deficient compound. The Lewis formula for a $BeCl_2$ molecule is

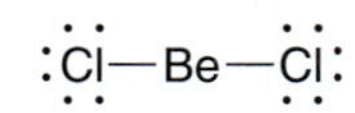

Although the central beryllium atom has no lone electron pairs, it does have two covalent bonds and thus has two bonding electron pairs in its valence shell. These valence-shell electron pairs repel each other and can minimize their mutual repulsion by being as far apart as possible. If we visualize the central beryllium atom as a sphere and the two valence-shell electron pairs (the two covalent bonds) as being on the surface of the sphere, then the two bonds minimize their mutual repulsion by being at opposite poles of the sphere. Thus, the two bonds are on opposite sides of the central beryllium atom, and the Cl–Be–Cl bond angle is 180°. The shape of a molecule is characterized by the positions of the atomic nuclei in the molecule, so we say that $BeCl_2$ is a **linear** molecule. This prediction is in accord with experimental studies of $BeCl_2$ molecules in the gas phase. The positioning of the two valence-shell electron pairs on opposite sides of the central atom is shown in Figure 8.7a. Note that we do not consider the valence electrons making up the lone pairs on the two chlorine atoms when determining the shape of a $BeCl_2$ molecule; we only consider the valence elec-

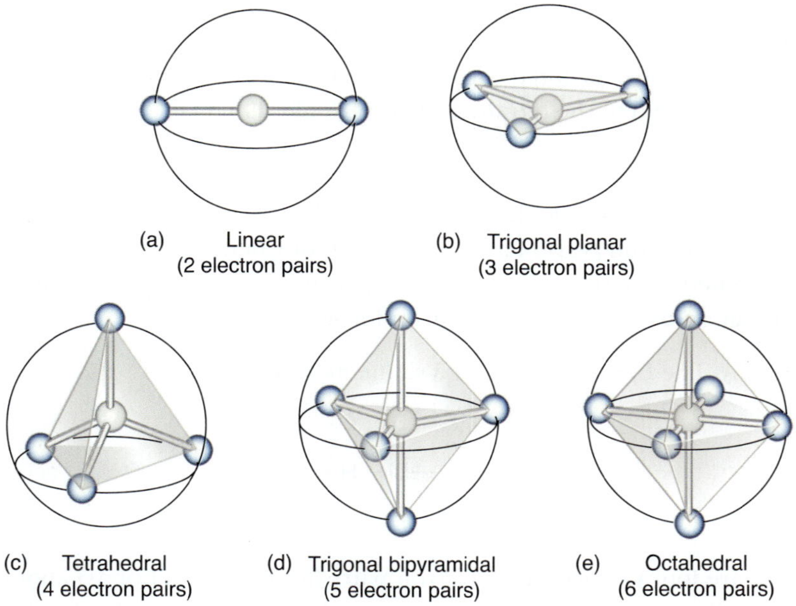

Figure 8.7 Sets of electron pairs (blue spheres) arranged on the surfaces of spheres so as to minimize the mutual repulsion between them. (a) Two electron pairs lie at opposite poles of a sphere. (b) Three electron pairs lie on an equator at the vertices of an equilateral triangle. (c) Four electron pairs lie at the vertices of a regular tetrahedron. (d) Five electron pairs are arranged such that two lie at poles and the other three lie on the equator at the vertices of an equilateral triangle. (e) Six electron pairs lie at the vertices of a regular octahedron. The black arcs indicate the geometrical arrangement of the electron pairs—they do *not* indicate chemical bonds.

trons directly attached to the central atom. We should point out that beryllium chloride is a solid at room temperature (20°C), but the high-temperature vapor consists of individual beryllium chloride molecules with a linear geometry.

Consider now a molecule with three electron pairs in the valence shell of the central atom. An example is the electron-deficient compound boron trifluoride, $BF_3(g)$, which acts as a catalyst for a number of reactions and is a gas at room temperature. The Lewis formula for a BF_3 molecule is

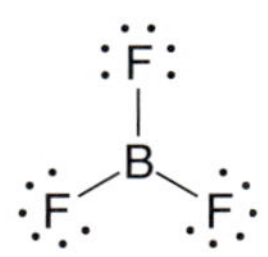

The three valence-shell electron pairs (that is, the three covalent bonds) surrounding the boron atom can minimize their mutual repulsion by maximizing their mutual separation. This response results in a **trigonal planar** arrangement of the three electron pairs (Figure 8.7b). We predict, then, that BF_3 is a planar, symmetric molecule with F–B–F bond angles equal to 120° (360°/3 = 120°). This prediction is in agreement with the experimentally determined structure of BF_3 molecules in the gas phase.

8-4. The Number of Valence-Shell Electron Pairs Determines the Shape of a Molecule

Methane, CH_4, is an example of a molecule with four covalent bonds surrounding a central atom (margin). The four electron pairs minimize their mutual repulsion by pointing toward the vertices of a regular tetrahedron (Figures 8.1a and 8.7c). Any angular movement of one of the four electron pairs in a tetrahedral array will bring the moved electron pair closer to two of the other electron pairs and thus increase the electron-electron repulsion energy. We see,

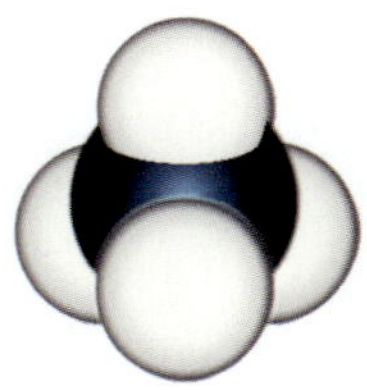

Methane, CH_4

then, that the tetrahedral geometry of a methane molecule is a result of the minimization of the mutual repulsion of the four electron pairs making up its four covalent bonds. As noted earlier, all the H–C–H bond angles in a methane molecule are equal to the tetrahedral bond angle, 109.5°.

EXAMPLE 8-1: The element silicon lies below carbon in Group 14 of the periodic table and is also tetravalent (has four valence electrons). Predict the geometry of a silane molecule, SiH_4. Silane is a gas that is used in the preparation of extremely pure silicon for semiconductors.

Solution: The Lewis formula for a silane molecule is

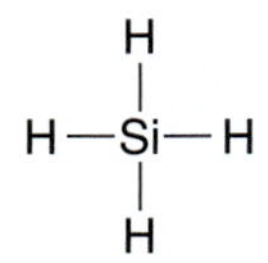

There are four valence-shell electron pairs (four covalent bonds) about the central silicon atom, so we predict (correctly) that a silane molecule is tetrahedral and that the H–Si–H bond angles are 109.5°.

PRACTICE PROBLEM 8-1: Predict the geometry of a chloroform molecule, $CHCl_3$, which was once widely used as an anesthetic.

Answer: tetrahedral

The following Example shows that VSEPR theory can be applied to molecular ions.

EXAMPLE 8-2: Predict the geometry of the an ammonium ion, NH_4^+.

Solution: The Lewis formula for an NH_4^+ ion is

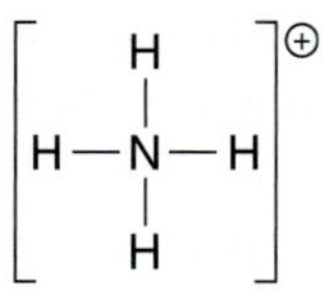

Because the valence shell of the nitrogen atom in NH_4^+ contains a total of four electron pairs (four covalent bonds), we predict that an ammonium ion is tetrahedral, which is the observed structure of an NH_4^+ ion.

PRACTICE PROBLEM 8-2: Predict the geometry of a tetrafluoroborate ion, BF_4^-.

Answer: tetrahedral

Many molecules have five electron pairs in the form of five covalent bonds in the valence shell of the central atom. An example is a phosphorus pentachloride molecule, PCl_5, whose Lewis formula is

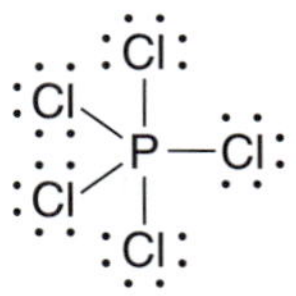

The arrangement of the five electron pairs in the valence shell of the phosphorus atom that minimizes their mutual electron-pair repulsions is a **trigonal bipyramid** (Figure 8.7d and Figure 8.8). Notice that the three vertices on the equator of the sphere in Figure 8.7d form an equilateral triangle and that the vertices at the polar axis lie above and below the center of the equilateral triangle. The five vertices of a trigonal bipyramid are *not* equivalent. The three vertices lying on the equator in Figure 8.7d and Figure 8.8 are equivalent and are called **equatorial vertices;** the two vertices lying at the polar axes are equivalent and are called **axial vertices.** The axial vertices are not geometrically equivalent to the equatorial vertices; and, as we will show, this nonequivalence has important structural consequences. Some other examples of trigonal bipyramidal molecules are antimony pentachloride, $SbCl_5$, and arsenic pentafluoride, AsF_5.

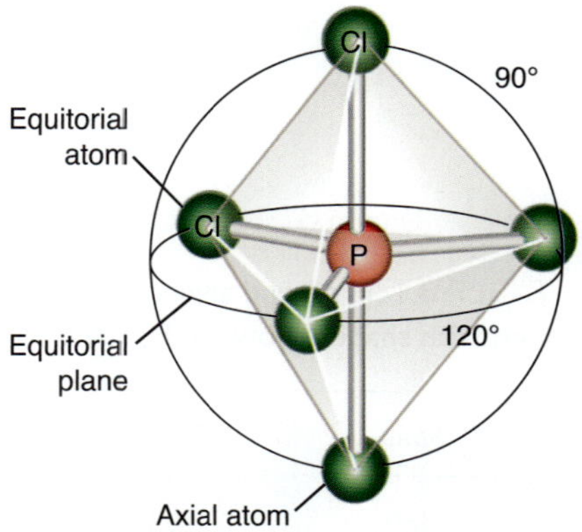

Figure 8.8 The shape of a gas-phase phosphorus pentachloride molecule. The equatorial Cl–P–Cl angles are 120° (360°/3 = 120°), and the axial-to-equatorial angles are 90°.

As an example of a molecule with six electron pairs in the valence shell of the central atom, consider the molecule sulfur hexafluoride, SF_6, whose Lewis formula is

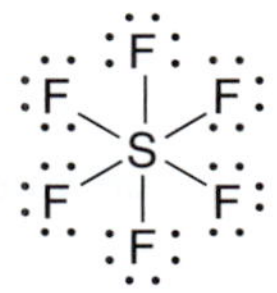

As the Lewis formula shows, there are six electron pairs, which also constitute six covalent bonds in this case, in the valence shell of the central sulfur atom. These six electron pairs mutually repel one another. The mutual repulsion of the six pairs is minimized if the six electron pairs point toward the vertices of a regular **octahedron** (Figure 8.9). This geometric figure has six vertices and eight faces. All eight faces are identical equilateral triangles. An important property of a regular octahedron is that all six vertices are equivalent. We see, then, that a SF_6 molecule has octahedral symmetry and that the six fluorine atoms are geometrically equivalent (Figure 8.7e). There is no way, by either chemical or physical methods, to distinguish among the six sulfur-fluorine bonds in a SF_6 molecule. All the F–S–F bond angles that involve adjacent fluorine atoms in a SF_6 molecule are 90°.

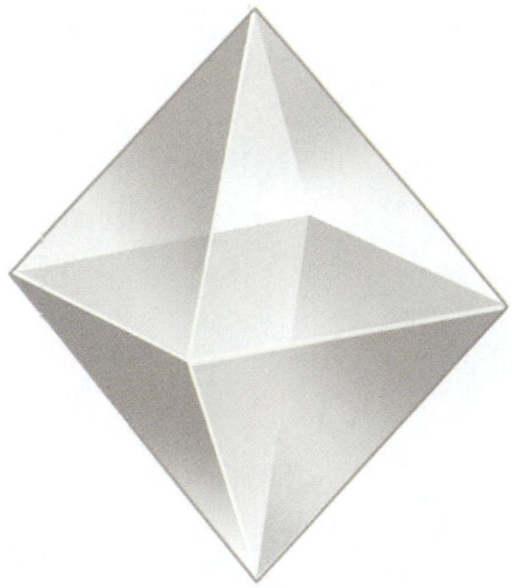

Figure 8.9 A regular octahedron is a symmetric body consisting of six equivalent vertices and eight identical faces that are equilateral triangles.

EXAMPLE 8-3: Predict the shape of a hexachlorophosphate ion, PCl_6^-.

Solution: The Lewis formula for a PCl_6^- ion is

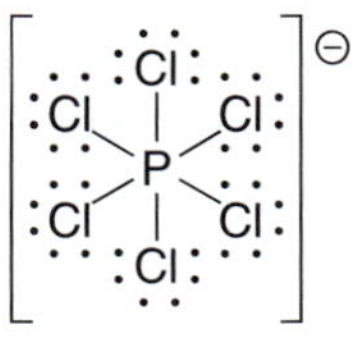

The six covalent bonds are directed toward the vertices of a regular octahedron, and we predict (correctly) that a PCl_6^- ion is octahedral.

PRACTICE PROBLEM 8-3: Predict the geometry and the F–Al–F bond angles involving adjacent fluorine atoms in an AlF_6^{3-} ion.

Answer: An AlF_6^{3-} ion is octahedral with F–Al–F angles involving adjacent fluorine atoms of 90°.

TABLE 8.1 The bond angles associated with shapes shown in Figure 8.7

Shape	Structure
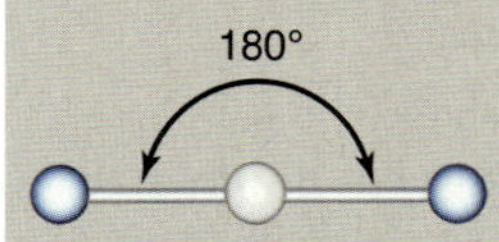	linear
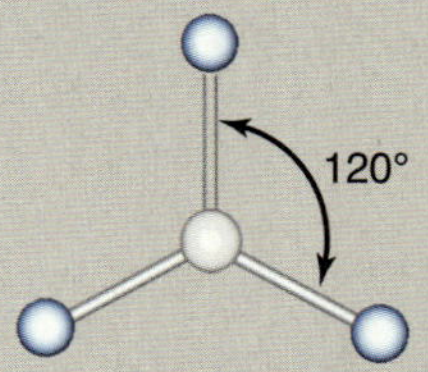	trigonal planar
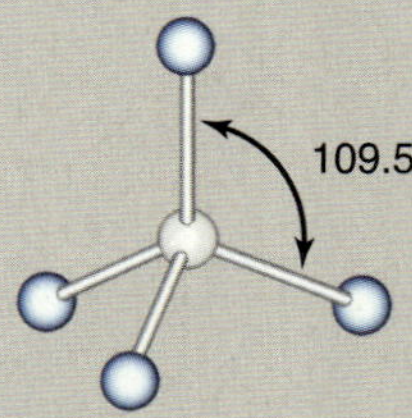	tetrahedral
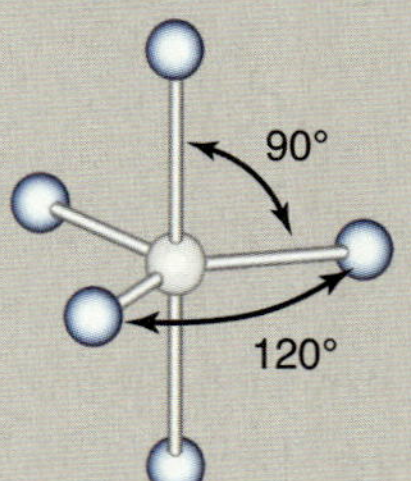	trigonal bipyramidal
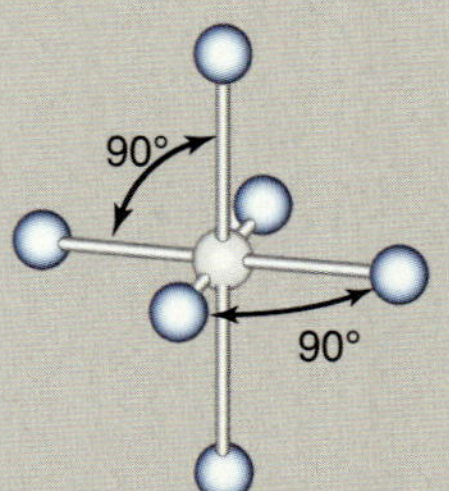	octahedral

Table 8.1 shows the bond angles associated with the molecular shapes we have discussed thus far.

8-5. Lone Electron Pairs in the Valence Shell Affect the Shapes of Molecules

In each case we have discussed so far, all the electron pairs in the valence shell of the central atom have been in covalent bonds. Now let's consider cases in which there are lone pairs of electrons as well as covalent bonds in the valence shell of the central atom. As an example, consider an ammonia, NH_3, molecule. The Lewis formula for an NH_3 molecule is

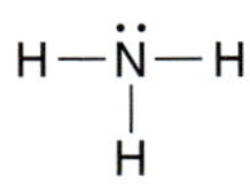

There are four electron pairs in the valence shell of the nitrogen atom. Three of them are in covalent bonds, and one is a lone pair. These four valence-shell electron pairs mutually repel one another and, therefore, are directed toward the corners of a tetrahedron (Figure 8.10b). The three hydrogen atoms form an equilateral triangle, and the nitrogen atom sits directly above the center of the plane of the triangle. Such a structure is called a triangular pyramid or **trigonal pyramid** and an ammonia molecule is described as being **trigonal pyramidal** in shape. An ammonia molecule is shaped like a tripod, with the three N–H bonds forming the legs of the tripod. A space-filling molecular model of NH_3 is shown in Figure 8.11. It is important to keep in mind that what we mean by the shape of a molecule, or the **molecular shape**, is defined by the positions of the nuclei in the molecule because it is only the positions of these nuclei that can be located in most methods used to experimentally determine molecular structures. The lone-pair electrons, on the other hand, are relatively diffuse and are generally not located in structural determinations.

If the four electron pairs in an NH_3 molecule pointed to the corners of a regular tetrahedron, then the H–N–H bond angles would be 109.5°. The four electron pairs in this case are not equivalent, however. Three of them occur in covalent bonds, and the fourth is a lone pair. Thus, we might expect some distortion from a regular tetrahedral geometry. Electron pairs in covalent bonds are shared between two atoms and are localized between them. A lone pair of electrons, on the other hand, is associated only with the central atom and therefore is not as localized as the pair of electrons in a covalent bond. Thus, a lone pair of electrons is more spread out, or bulkier, than a bonding pair of

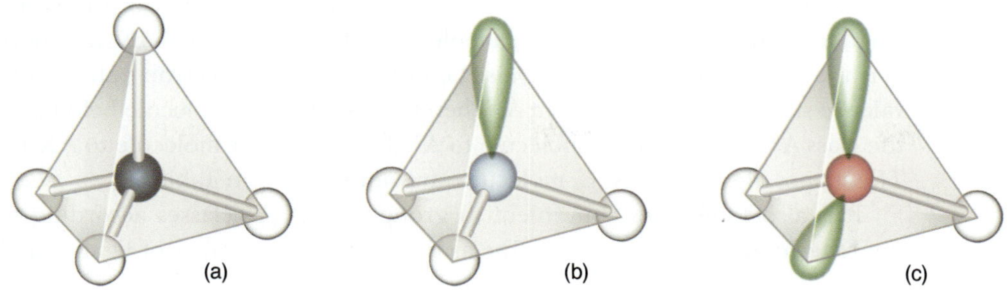

Figure 8.10 The role of bond pairs and lone pairs of electrons in determining molecular geometry. (a) CH_4, (b) NH_3, (c) H_2O.

electrons; consequently, a lone pair of electrons takes up more room around a central atom than a covalent bond does. This arrangement means that the repulsion between a lone pair of electrons and the electron pair in a covalent bond is greater than the repulsion between the electron pairs in two adjacent covalent bonds. This effect causes the H–N–H bond angles in an NH_3 molecule to decrease slightly from the 109.5° regular tetrahedral angle to 107.3° (Figure 8.12).

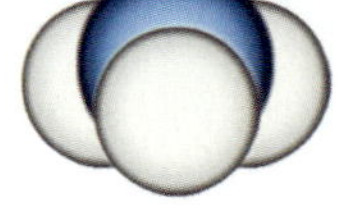

Figure 8.11 A space-filling molecular model of an ammonia molecule, NH_3.

We cannot predict quantitatively from VSEPR theory that the H–N–H bond angles in an NH_3 molecule are 107.3°, but we can predict that they are somewhat less than the ideal tetrahedral angle of 109.5°. We say that VSEPR is a qualitative rather than a quantitative theory. However, as a rough rule of thumb, applicable whenever C, N, or O are central atoms in a tetrahedral arrangement, each lone pair decreases the remaining bond angles involving adjacent atoms bonded to the central atom by about 2°.

The example of an NH_3 molecule shows that it is the *total* number of electron pairs in the valence shell of the central atom that determines the geometry of the nuclei in a molecule. As a second example, consider a water molecule. The Lewis formula for a H_2O molecule is

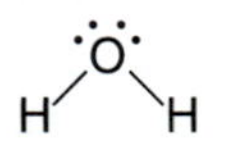

The four valence-shell electron pairs on the oxygen atom in a water molecule are directed toward the vertices of a tetrahedron (Figure 8.10c), so we see that H_2O is a **bent** or V-shaped molecule.

The two lone electron pairs take up more room around the oxygen atom than the two electron pairs in the covalent bonds, so we expect the repulsion between the lone electron pairs to be greater than either that between a lone electron pair and a covalent bond or that between two covalent bonds. We predict, then, that the H–O–H bond angle in a H_2O molecule is less than the regular tetrahedral angle of 109.5° and that it is even smaller than the H–N–H bond angle (107.3°) in an NH_3 molecule. The experimentally measured bond angle in a H_2O molecule is 104.5°.

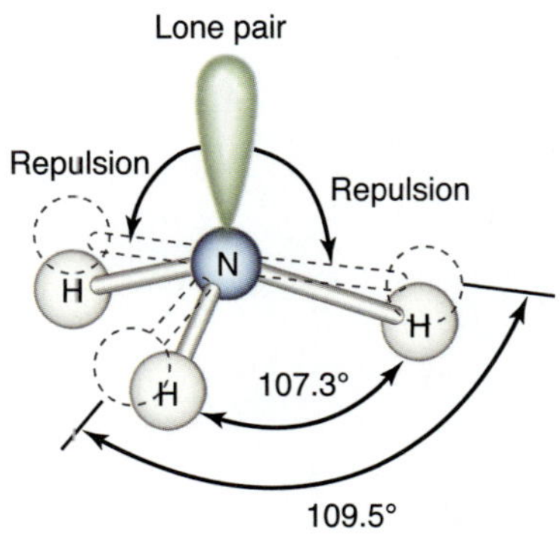

Figure 8.12 A lone pair of electrons is more spread out, or bulkier, than a bonding pair of electrons, causing the H–N–H bond angles in an NH_3 molecule to decrease slightly from the 109.5° regular tetrahedral angle to 107.3°.

Each of the molecules CH_4, NH_3, and H_2O has four electron pairs in the valence shell of the central atom. The four electron pairs are directed roughly toward the corners of a tetrahedron, as shown in Figure 8.10. The shape of the molecule is described by giving the positions of the nuclei. Thus, a CH_4 molecule is tetrahedral; an NH_3 molecule is trigonal pyramidal; and a H_2O molecule is bent.

We can classify molecules by introducing the following terminology. Let A represent a central atom, X an atom bonded to the central atom, and E a lone pair of electrons on the central atom. We call an atom bonded to a central atom a **ligand.** Thus, the central atom in a molecule can be classified as $\mathbf{AX_mE_n}$, where m is the number of ligands and n is the number of lone electron pairs in the valence shell of the central atom A. Therefore, a methane molecule belongs to the class AX_4; an ammonia molecule to AX_3E; and a water molecule to AX_2E_2. The shapes of various AX_mE_n molecules are summarized in Table 8.2.

It is also sometimes convenient to group the AX_mE_n classes according to their **steric number** which we define as the sum of the number of ligands and lone pairs attached to the central atom,

$$\begin{aligned}\text{steric number} &= \begin{pmatrix}\text{number of atoms}\\ \text{bonded to central atom}\end{pmatrix} + \begin{pmatrix}\text{number of lone pairs}\\ \text{on central atom}\end{pmatrix}\\ &= m + n \qquad (8.1)\end{aligned}$$

The sum of $m + n$ for the molecules CH_4, NH_3, and H_2O is four, and so these molecules have a steric number of four. Various classes of molecules arranged by steric number and the number of lone pairs, E_n, are given in Figure 8.13.

EXAMPLE 8-4: Predict the shape and the H–O–H bond angles for a hydronium ion, H_3O^+.

Solution: The Lewis formula for a H_3O^+ ion is

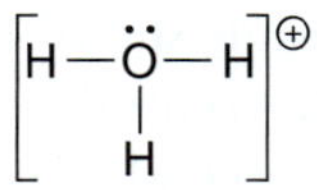

There are three hydrogen ligands and one lone pair attached to the central oxygen atom. Thus, it belongs to the class AX_3E. We see from Table 8.2 that it has a trigonal pyramidal shape. The H–O–H bond angles for a H_3O^+ ion will be somewhat less than that of an ideal tetrahedron and so we predict that the angle is <109.5°.

PRACTICE PROBLEM 8-4: Predict the shape and the F–S–F bond angle for the sulfur difluoride, SF_2, molecule.

Answer: AX_2E_2, bent, <109.5°

8-6. VSEPR Theory Is Applicable to Molecules That Contain Multiple Bonds

In predicting the shapes of molecules by VSEPR theory, we count a double or triple bond as one group of electrons connecting the ligand X to the central atom A. We use this rule because all the bonding electrons between two atoms must be shared between those same two atoms. For example, the Lewis formula for a carbon dioxide molecule is

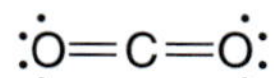

TABLE 8.2 Molecular shapes. The white spheres represent the central atom, the red spheres represent ligands, and the green lobes represent lone pairs.

Molecular class	Ideal shape	Examples	Molecular class	Ideal shape	Examples
AX_2	 linear	CO_2, HCN, $BeCl_2$	AX_4E	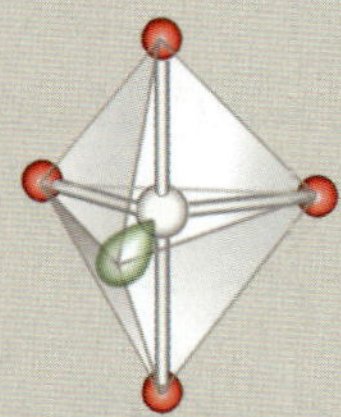 seesaw-shaped	SF_4, XeO_2F_2, IF_4^+, $IO_2F_2^-$
AX_3	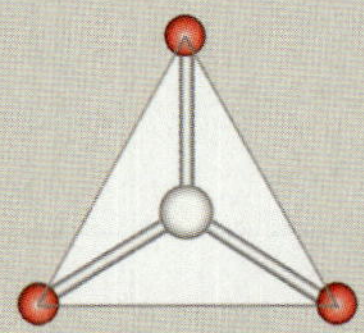 trigonal planar	SO_3, BF_3, NO_3^-, CO_3^{2-}	AX_3E_2	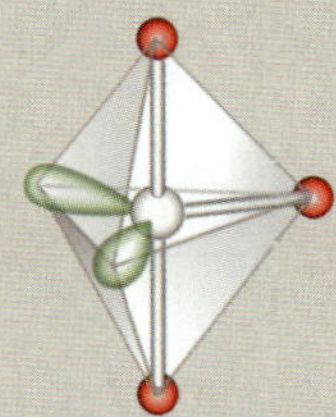 T-shaped	ClF_3, BrF_3
AX_2E	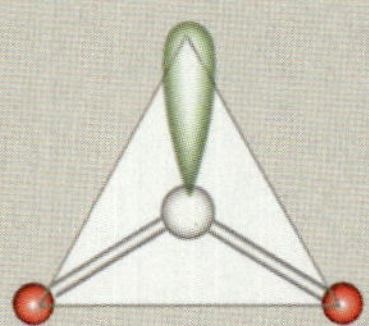 bent	SO_2, O_3, PbX_2, SnX_2 (where X is a halogen)	AX_2E_3	 linear	XeF_2, I_3^-, IF_2^-
AX_4	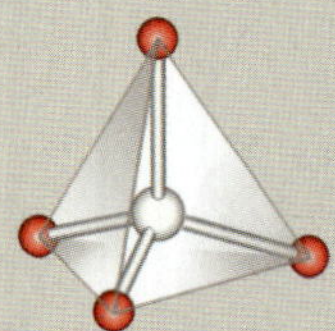 tetrahedral	SiH_4, CH_4, SO_4^{2-}, ClO_4^-, PO_4^{3-}, XeO_4	AX_6	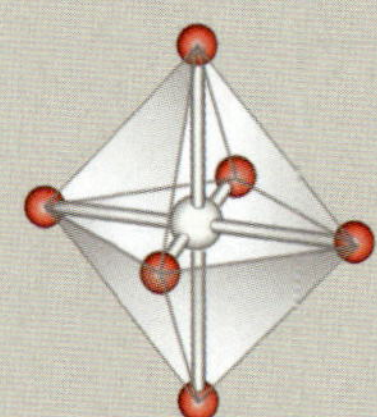 octahedral	SF_6, IOF_5
AX_3E	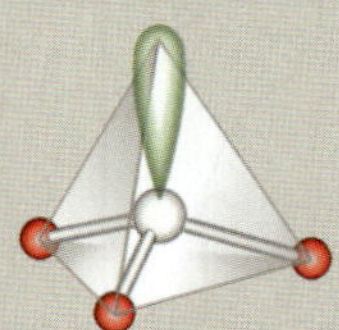 trigonal pyramidal	NH_3, PF_3, $AsCl_3$, ClO_3^-, H_3O^+, XeO_3	AX_5E	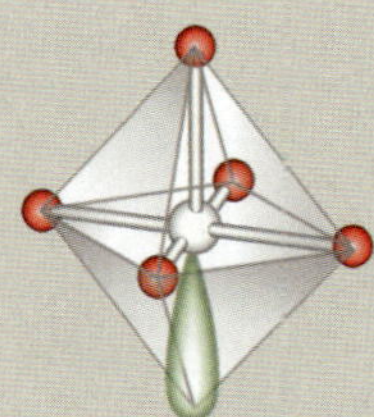 square pyramidal	IF_5, TeF_5^-, $XeOF_4$
AX_2E_2	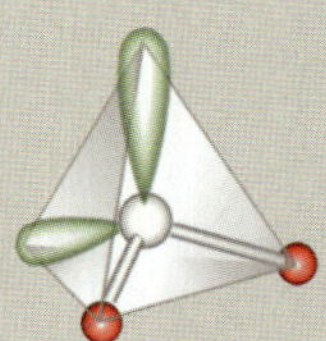 bent	H_2O, OF_2, SF_2	AX_4E_2	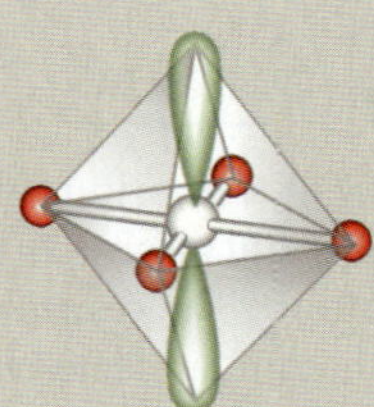 square planar	XeF_4, ICl_4^-
AX_5	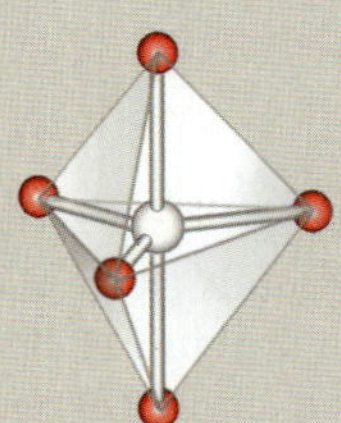 trigonal bipryamidal	PCl_5, AsF_5, SOF_4			

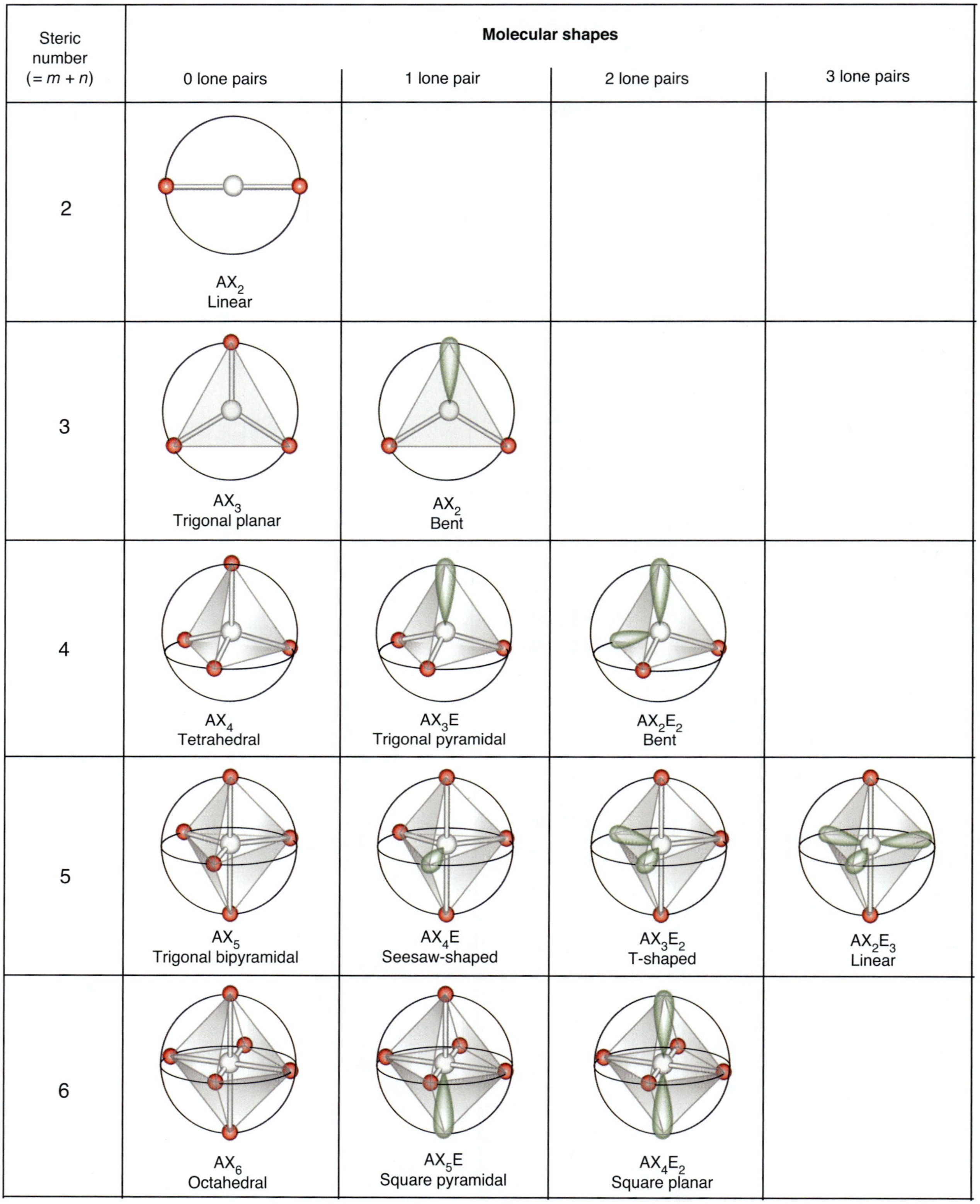

Figure 8.13 A summary of the various observed molecular shapes that result when m ligands (X) and n lone electron pairs (E) surround a central atom (A) to form an AX_mE_n molecule. The steric number is a sum of the number of ligands and lone pairs surrounding the central atom (steric number = $m + n$). The white sphere at the center of each sphere represents the central atom, and the red spheres on the surfaces of the spheres represent the attached ligands. The lone electron pairs, shown as green lobes, are at the vertices.

Thus, a CO_2 molecule is classified as an AX_2 molecule and so we predict that it has a linear geometry. Another example of an AX_2 molecule that contains a multiple bond is a hydrogen cyanide molecule, HCN, the Lewis formula for which is

H—C≡N:

There are two groups of electrons around the central carbon atom (one single bond and one triple bond), so we predict that HCN is a linear molecule.

EXAMPLE 8-5: Predict the shape of a methanal, H_2CO, molecule, better known by its common name, formaldehyde.

Solution: The Lewis formula for a H_2CO molecule is

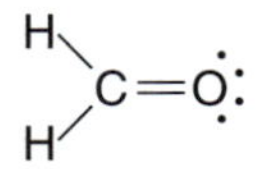

We treat the double bond as one group of electrons and classify H_2CO as an AX_3 molecule. We predict (correctly) that H_2CO is a trigonal planar molecule; in other words, all four atoms lie in a single plane. Figure 8.14 shows a space-filling model of a formaldehyde molecule.

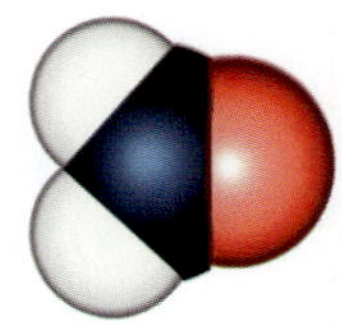

Figure 8.14 A space-filling molecular model of a formaldehyde molecule, H_2CO.

PRACTICE PROBLEM 8-5: Predict the shape of a carbon disulfide molecule, CS_2. Carbon disulfide is a volatile liquid that readily dissolves sulfur.

Answer: linear

Although we have treated a double or triple bond as a single group of electrons, when a molecule is placed in an AX_mE_n class, we must recognize that a multiple bond involves more than two electrons and thus is larger (bulkier) than a single covalent bond. Consequently, multiple bonds repel single bonds more strongly than single bonds repel other single bonds. In terms of their spatial requirements, multiple bonds act like lone electron pairs. Because of this effect, we predict that the H–C–H bond angle in a H_2CO molecule (Example 8-5) is slightly less than 120° and that the H–C–O bond angles are slightly larger than 120°. The actual experimental values are 116° and 122°, respectively:

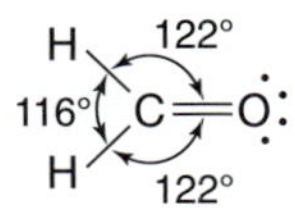

EXAMPLE 8-6: Phosgene is a colorless, highly toxic gas. When diluted with air, it has an odor resembling newly mown hay. Thionyl chloride is used as a chlorinating agent in organic synthesis and as a solvent in some

lithium batteries. Compare the shapes of the phosgene, $COCl_2$, and the thionyl chloride molecules, $SOCl_2$.

Solution: The Lewis formulas for these two molecules are

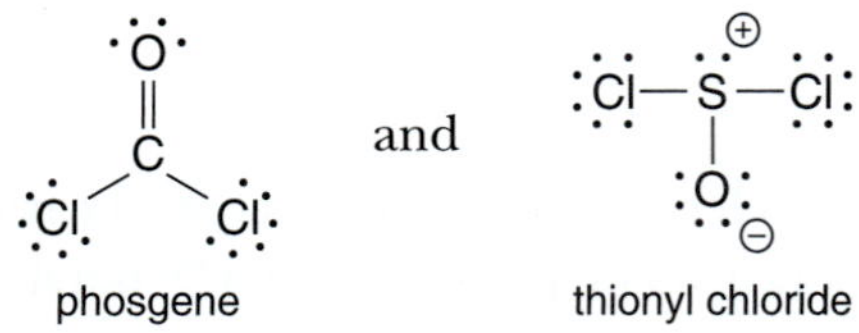

A phosgene molecule belongs to the class AX_3 and therefore is trigonal planar, like a formaldehyde molecule. The lone pair on the sulfur atom in a thionyl chloride molecule puts this molecule in the AX_3E class. Thus, a thionyl chloride molecule is trigonal pyramidal, with the two chlorine atoms and the oxygen atom lying in a plane and the sulfur atom lying above the plane.

Notice that even though the chemical formulas for a phosgene molecule and a thionyl chloride molecule are similar, the shapes of the two molecules are different because of the lone electron pair on the central sulfur atom in the thionyl chloride molecule.

PRACTICE PROBLEM 8-6: Predict the shapes of the chlorite and chlorate ions, ClO_2^-, and ClO_3^-, respectively.

Answer: ClO_2^-, bent; ClO_3^-, trigonal prymidal

In Example 8-6 we ignored the resonance form

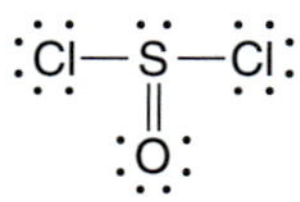

for a $SOCl_2$ molecule. Because different resonance forms of a species always have the same number of *groups* of electrons around the central atom, VSEPR theory applies to molecules that are described by a resonance hybrid just as well as it does to molecules that can be represented by just one Lewis formula, as illustrated in the following example.

EXAMPLE 8-7: Predict the shape of the carbonate ion, CO_3^{2-}.

Solution: There are three equivalent resonance forms of a CO_3^{2-} ion:

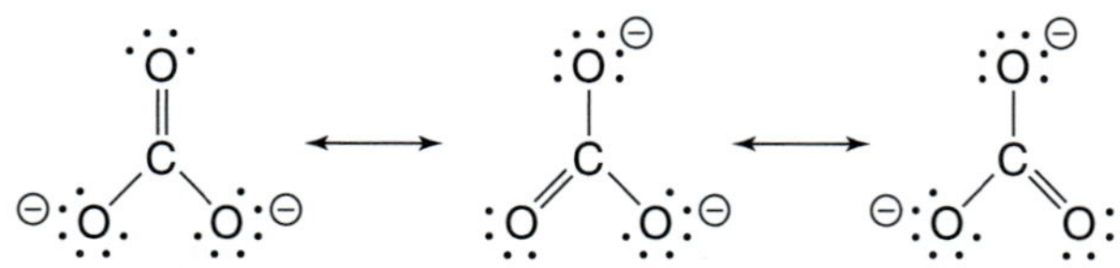

The three equivalent resonance forms tell us that all three C–O bonds in a CO_3^{2-} ion are equivalent. In each resonance form there are three groups of electrons and no lone pairs around the central carbon atom. Thus, a CO_3^{2-} ion is of the class AX_3. We predict that a carbonate ion is trigonal planar with O–C–O bond angles of 120°. This prediction is in agreement with experimental results. The equivalence of the three C–O bonds in a CO_3^{2-} ion makes the three O–C–O bond angles equivalent (120°).

PRACTICE PROBLEM 8-7: What is the molecular class, shape, and O–S–O bond angle predicted by VSEPR theory for the sulfur dioxide, SO_2, molecule?

Answer: AX_2E, bent, <120°

8-7. Lone-Pair Electrons Occupy the Equatorial Vertices of a Trigonal Bipyramid

Recall that the five vertices of a trigonal bipyramid are not equivalent to one another (Figure 8.7d and Figure 8.8). They form a set of three equivalent equatorial vertices and two equivalent axial vertices. Consequently, in considering the class of molecules designated by AX_4E, for example, we have two nonequivalent choices for the position of the lone pair. We can place it at an equatorial vertex or at an axial vertex.

Figure 8.15a shows that an equatorial pair of electrons has only two nearest neighbors at 90°, and Figure 8.15b shows that an axial electron pair has three nearest neighbors at 90°. The two other neighbors in Figure 8.15a lie at 120° from the position occupied by the lone electron pair and thus are sufficiently far away that their interaction with the lone pair is much less than those at 90°. Consequently, the electron repulsions due to a lone electron pair are minimized by placing the lone pair at an equatorial vertex rather than at an axial vertex. Thus, molecules that belong to the classes AX_4E, AX_3E_2, and AX_2E_3 have the lone electron pairs at the equatorial positions, as shown in Figures 8.15a and 8.16. Because the shape of a molecule is defined by the positions of the atomic

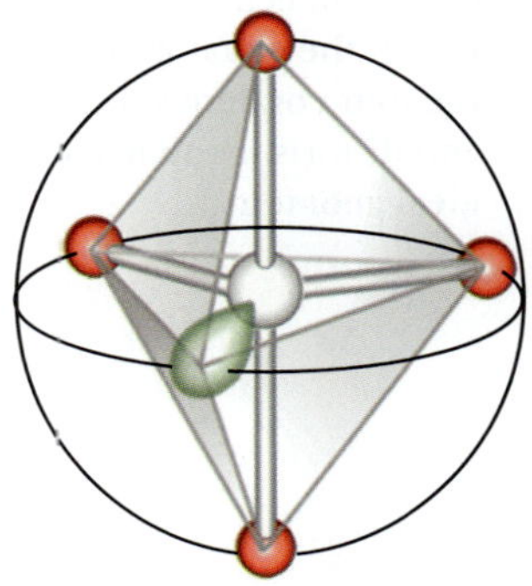

(a) Equatorial lone pair, AX_4E
Two neighbor atoms at 90°

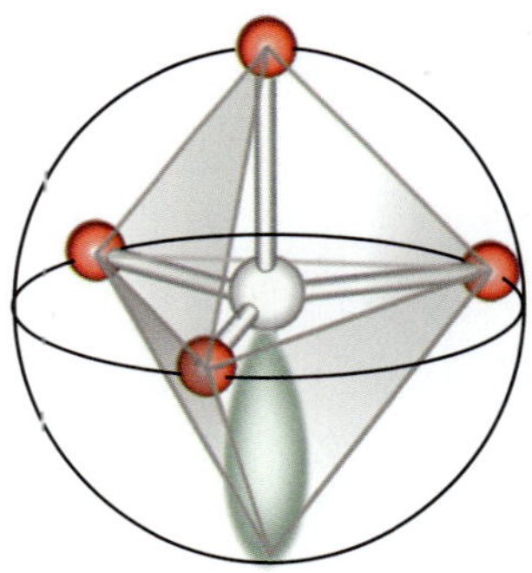

(b) Axial lone pair, AX_4E
Three neighbor atoms at 90°

Figure 8.15 Because the five vertices of a trigonal bipyramid fall into two distinct classes, there are two nonequivalent positions available for the lone electron pair in a molecule of the class AX_4E. (a) A lone pair at an equatorial position has only two nearest neighbors at 90°. (b) A lone pair at an axial position has three nearest neighbors at 90°. Consequently, the repulsion due to a lone electron pair is minimized by placing it at an equatorial vertex, and molecules with the general formula AX_4E are shaped like a seesaw, as shown in (a), where the two legs of the seesaw are formed from the equatorial X atoms.

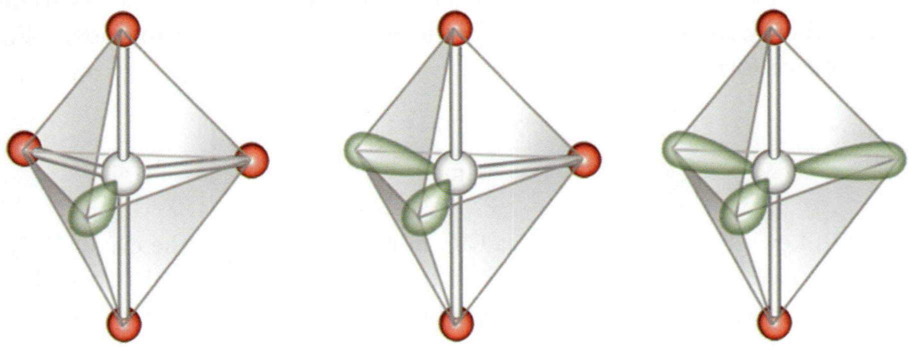

(a) AX_4E Seesaw

(b) AX_3E_2 T-shaped

(c) AX_2E_3 Linear

Figure 8.16 The shapes of molecules that belong to the classes (a) AX_4E, (b) AX_3E_2, and (c) AX_2E_3. The lone pairs occupy the equatorial positions in each case.

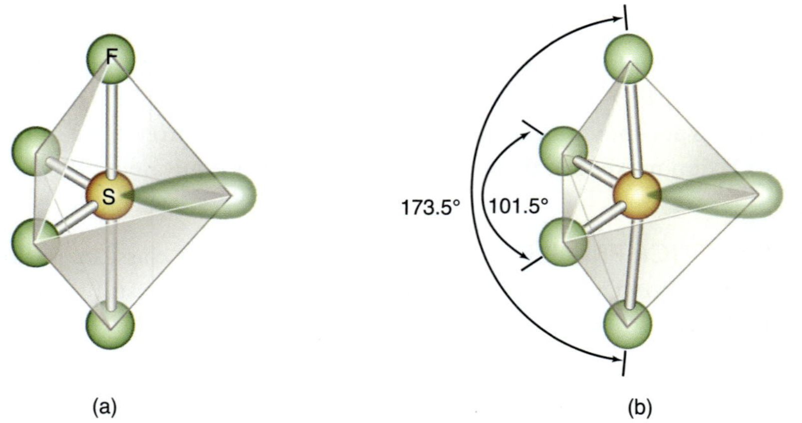

Figure 8.17 The geometry of a sulfur tetrafluoride molecule, which belongs to the class AX_4E. (a) The ideal shape of the molecule. (b) The lone electron pair at the equatorial position repels the four covalent sulfur-fluorine bonds and distorts the molecule away from ideal geometry.

nuclei, we see from Figure 8.16 that an AX_4E molecule is shaped like a **seesaw,** an AX_3E_2 molecule is **T-shaped,** and an AX_2E_3 molecule is linear.

The Lewis formula for a sulfur tetrafluoride molecule,

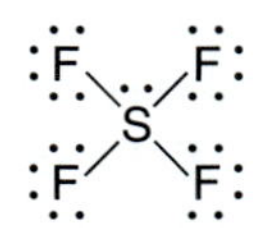

shows that this molecule belongs to the class AX_4E. The lone pair is placed at one of the equatorial positions of a trigonal bipyramid, so the ideal shape of a SF_4 molecule is the seesaw shown in Figure 8.17a. This ideal shape predicts an axial to axial F–S–F bond angle of 180° and an equatorial to equatorial F–S–F bond angle of 120°. The lone pair at the equatorial position causes a small distortion from ideal behavior, however; and the actual shape of a SF_4 molecule is that shown in Figure 8.17b. The experimentally observed bond angles in $SF_4(g)$, which are indicated in Figure 8.17b, are in accord with the rule that lone pairs take up more space than single covalent bonds.

There are a number of compounds formed between halogen atoms in which a less electronegative central halogen atom is bonded to more electronegative halogen atoms. Most of the **interhalogen compounds** are listed in Table 8.3. The molecules of all known interhalogen compounds obey the predictions of VSEPR theory. As an example, consider the molecule chlorine trifluoride, ClF_3, whose Lewis formula is

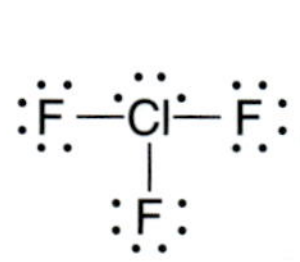

This Lewis formula shows that a ClF_3 molecule belongs to the class AX_3E_2. The ClF_3 molecule is T-shaped, as shown in Figure 8.18a. The ideal T-shape has axial to equatorial F–Cl–F bond angles of exactly 90°. However, the two equatorial lone pairs cause small distortions, so these bond angles are somewhat less than 90°, as shown in Figure 8.18b.

TABLE 8.3 Interhalogen compounds

AXE_3*	AX_3E_2	AX_5E
IF	IF_3	IF_5
BrF	BrF_3	BrF_5
ClF	ClF_3	ClF_5
ICl		
BrCl		
IBr		

*Because they have only two nuclei, all AXE_n molecules are linear.

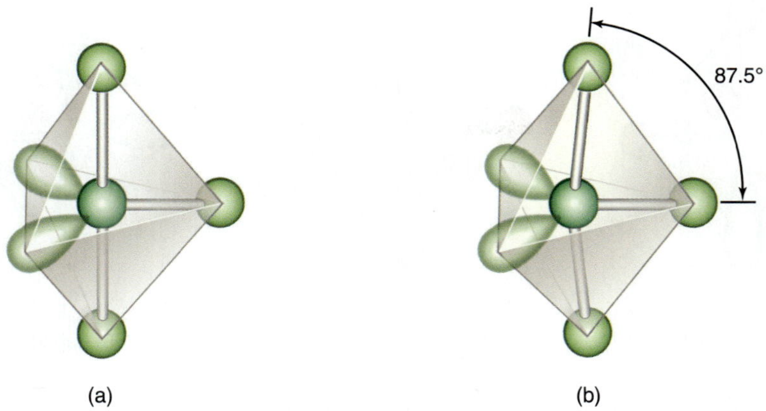

Figure 8.18 The geometry of a chlorine trifluoride molecule, ClF_3, which belongs to the class AX_3E_2. (a) The ideal shape of the molecule. (b) The two lone electron pairs at the equatorial positions repel the chlorine-fluorine bonds and distort the molecule away from ideal geometry.

EXAMPLE 8-8: Although elemental iodine, $I_2(s)$, is not very soluble in water, it is very soluble in aqueous solutions of potassium iodide. The increased solubility is due to the formation of a triiodide ion, $I_3^-(aq)$. The equation for the reaction between iodine and iodide is

$$I_2(aq) + I^-(aq) \rightarrow I_3^-(aq)$$

Predict the geometry of a triiodide ion.

Solution: The Lewis formula for an I_3^- ion is

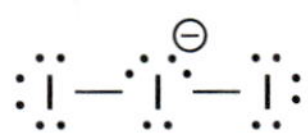

This Lewis formula shows that an I_3^- ion belongs to the class AX_2E_3. Figure 8.16c indicates that the three lone pairs occupy equatorial positions in a trigonal bipyramid; thus, I_3^- is a linear ion (Figure 8.19).

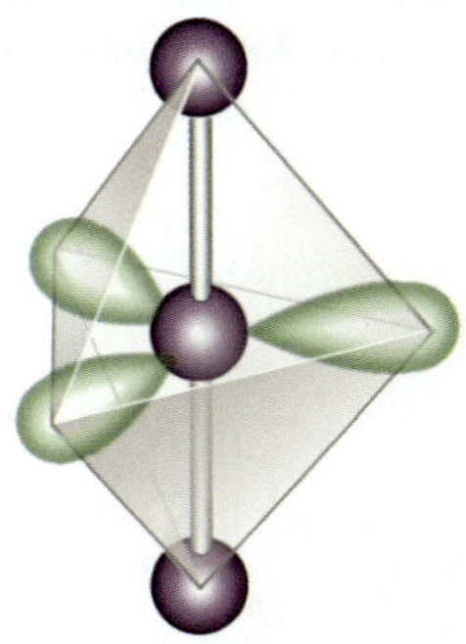

Figure 8.19 The ion I_3^- belongs to the class AX_2E_3. The three lone pairs occupy the equatorial vertices of a trigonal bipyramid. The two iodine ligands occupy the axial positions, so I_3^- is a linear ion.

PRACTICE PROBLEM 8-8: Predict the shape of a xenon difluoride molecule.

Answer: Linear

One of the impressive successes of VSEPR theory is its correct prediction of the structures of noble-gas compounds. Some noble-gas compounds that have been synthesized are the xenon fluorides: $XeF_2(s)$, $XeF_4(s)$; xenon oxyfluorides: $XeOF_4(l)$, $XeO_2F_2(s)$; xenon oxides: $XeO_3(s)$; and krypton fluorides: $KrF_2(s)$, $KrF_4(s)$; see interchapter K.

See Interchapter K at www.McQuarrieGeneralChemistry.com.

8-8. Two Lone Electron Pairs Occupy Opposite Vertices of an Octahedron

The octahedral group, that is, compounds with a steric number of six, involves species of the type AX_6, AX_5E, AX_4E_2, and AX_3E_3. The corresponding structures are shown in Figure 8.20. Because all six vertices of a regular octahedron

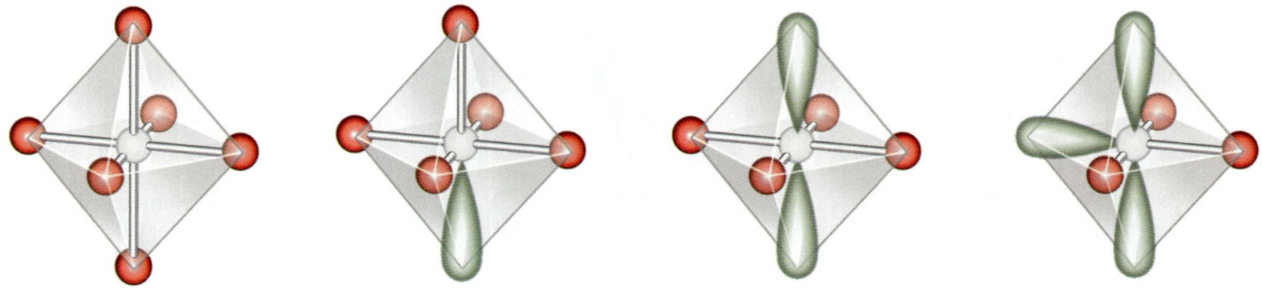

(a) AX_6 Octahedral

(b) AX_5E Square pyramidal

(c) AX_4E_2 Square planar

(d) AX_3E_3 T-shaped

Figure 8.20 The ideal shapes associated with the classes (a) AX_6, (b) AX_5E, (c) AX_4E_2, and (d) AX_3E_3. In (c) and (d), two of the lone electron pairs occupy opposite vertices because this placement minimizes the relatively strong lone-pair–lone-pair electron repulsions.

are equivalent, so are all six possible positions of the lone pair in an AX_5E molecule. To minimize the lone-pair–lone-pair repulsion in an AX_4E_2 molecule, however, the two lone pairs must be placed at opposite vertices, as shown in Figure 8.20c. No AX_3E_3 molecules are known, but Figure 8.20d predicts that they would be T-shaped.

An example of an AX_6 molecule is sulfur hexafluoride, SF_6, which has the predicted octahedral shape (Figure 8.20a). The Lewis formula for the interhalogen molecule bromine pentafluoride, BrF_5, is

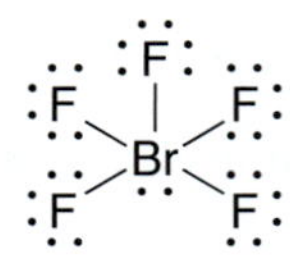

The Lewis formula shows that a BrF_5 molecule belongs to the class AX_5E. According to Figure 8.20b, we predict that a BrF_5 molecule has a **square pyramidal** shape, as shown in Figure 8.21. The adjacent F–Br–F bond angles in a BrF_5 molecule are slightly less than the ideal angle of 90°, because of the lone pair sitting at one vertex. The following example shows that an AX_4E_2 molecule has a **square planar** geometry.

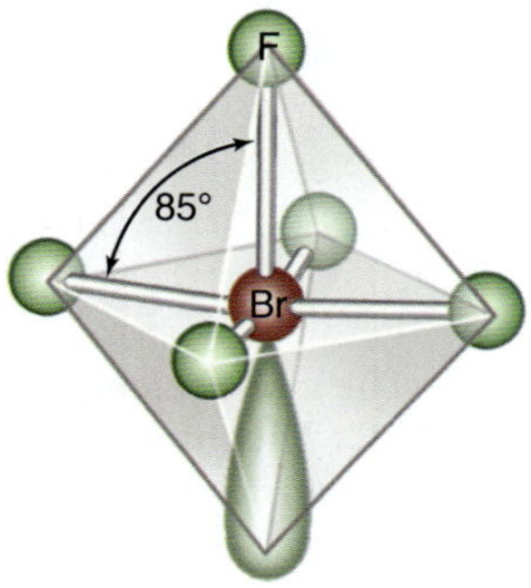

Figure 8.21 The shape of the interhalogen molecule bromine pentafluoride, BrF_5. The lone electron pair repels the bromine-fluorine bonds, causing the bromine atom to lie slightly below the plane formed by four of the fluorine atoms. The BrF_5 molecule has a shape somewhat like an opened umbrella.

EXAMPLE 8-9: Xenon tetrafluoride is prepared by heating Xe(g) and F_2(g) at 400°C at high pressure in a nickel container. The equation is

$$Xe(g) + 2\,F_2(g) \rightarrow XeF_4(s)$$

Predict the shape of a XeF_4 molecule.

Solution: The Lewis formula for a XeF_4 molecule is

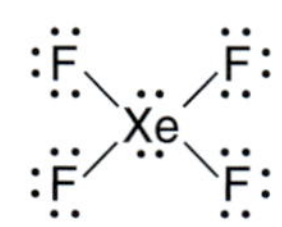

This molecule belongs to the class AX_4E_2; so according to Figure 8.20c, we predict that XeF_4 is a square planar molecule. This shape is indeed that observed, as shown in Figure 8.22 (see also Interchapter K).

PRACTICE PROBLEM 8-9: When $XeF_6(s)$ is dissolved in water and the solution is carefully evaporated, the dangerously explosive compound $XeO_3(s)$ is produced. Predict the structure of a XeO_3 molecule.

Answer: trigonal pyramidal

8-9. Molecular Geometry Determines Whether or Not a Molecule Has a Net Dipole Moment

We learned in Section 7-11 that although both bonds in a carbon dioxide molecule are polar, the molecule itself does not have a dipole moment. Because CO_2 is a linear molecule, the two polar bonds, which can be treated as vector quantities, oppose each other exactly. Similar reasoning applies to a square planar molecule in which all four ligands are the same, such as xenon tetrafluoride. Although the four bonds in a XeF_4 molecule are polar, the square planar geometry of the molecule

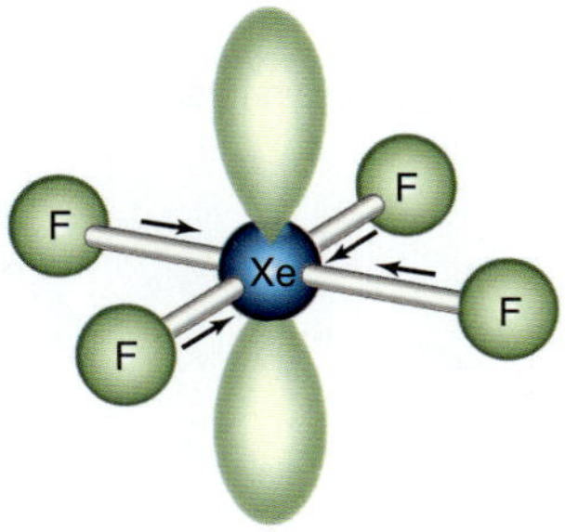

yields no net dipole moment. These two examples show that even though a molecule has polar bonds, the symmetry of the molecule may cause it to have no net dipole moment. In order for a molecule to have a net dipole moment, it must meet two criteria: (1) it must have polar bonds (i.e., there must be an electronegativity difference between two or more of the atoms in the molecule), and (2) the polar bonds must be asymmetric (i.e., their dipole moments must not cancel). Molecules with no net dipole moment are said to be **nonpolar molecules.** Some other examples of nonpolar molecules are

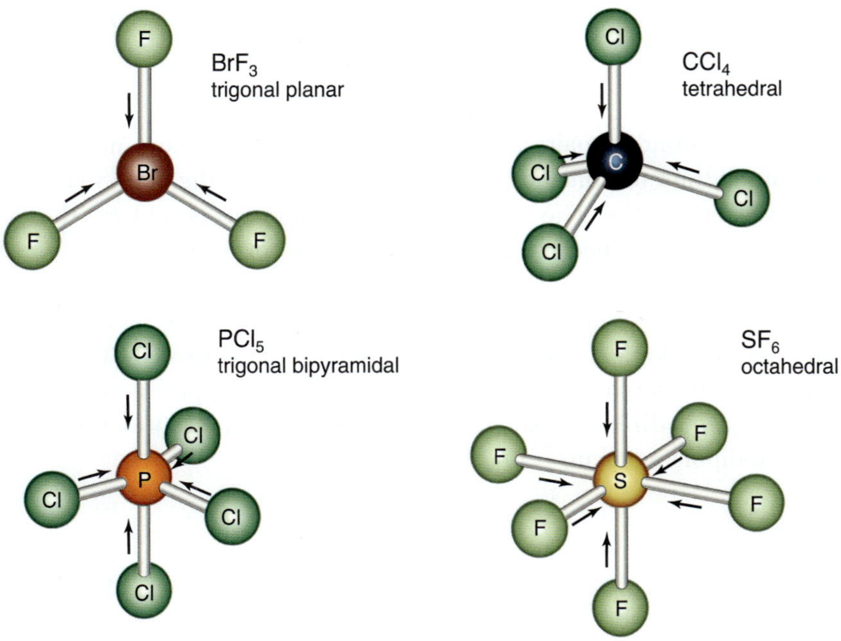

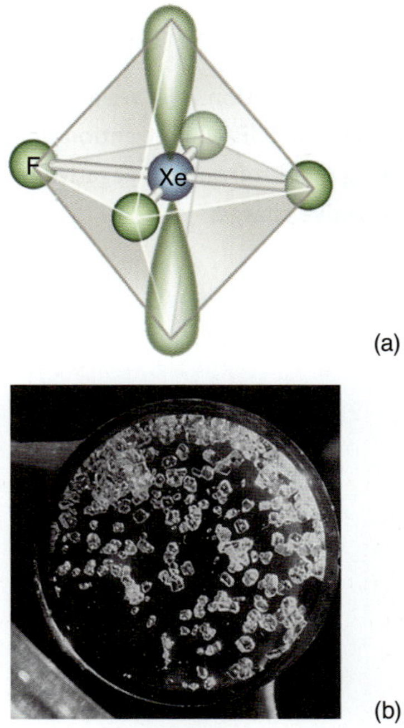

Figure 8.22 (a) The geometry of a xenon tetrafluoride molecule, XeF_4, which belongs to the class AX_4E_2. The two lone electron pairs occupy opposite vertices of the octahedron, so XeF_4 is a planar molecule. (b) Crystals of xenon tetrafluoride, one of the first noble-gas compounds synthesized.

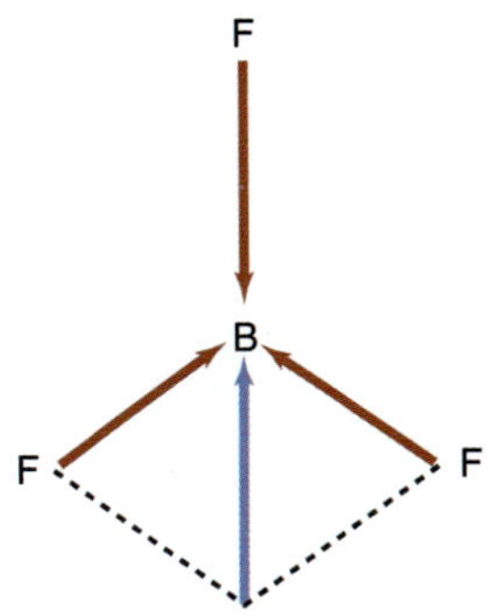

Figure 8.23 Cancellation of bond moments in a BF_3 molecule. The blue arrow is the resultant moment of the two inward-pointing B–F bond moments, and it is exactly equal in magnitude and opposite in direction to the downward-pointing B–F bond moment. The resultant moment of two bond moments is obtained as the diagonal of the parallelogram formed from the two bond moments and their corresponding parallels, as shown in the figure.

As noted in Section 7-11, we are following the IUPAC convention of drawing dipoles from the negative to the positive charge.

The net dipole moments in these examples may not be immediately evident, but a study of the structures shown in Table 8.2 shows that in each case the geometrical orientations of the polar bonds produces a zero net dipole moment (see, for example, Figure 8.23).

Examples of **polar molecules** (those possessing a net dipole moment) are

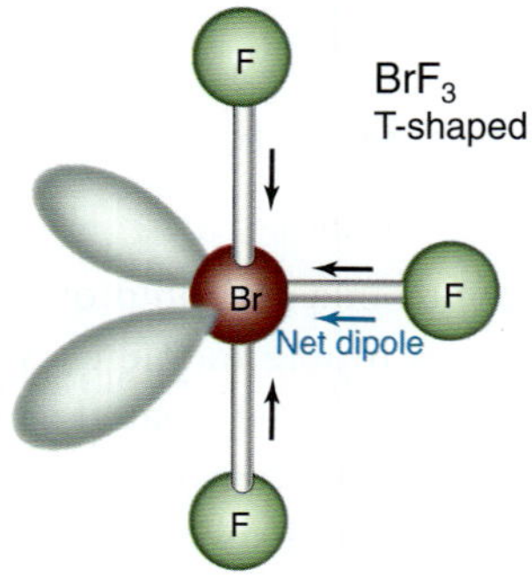

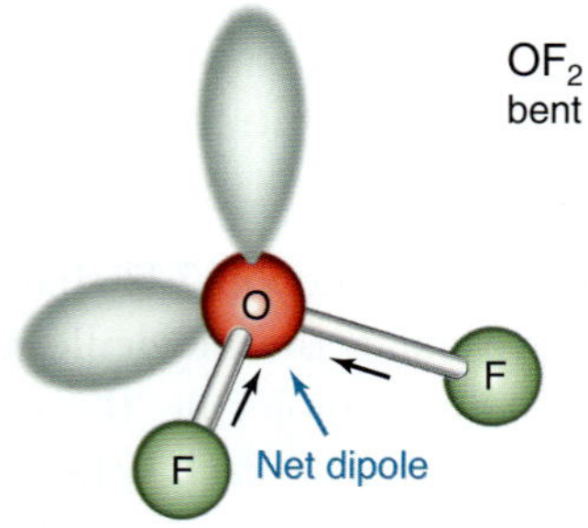

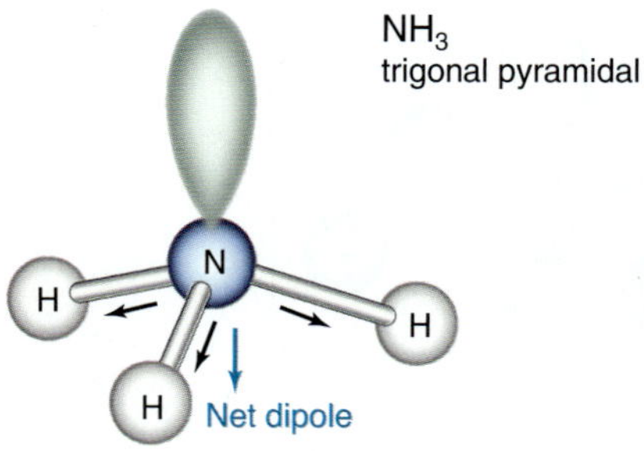

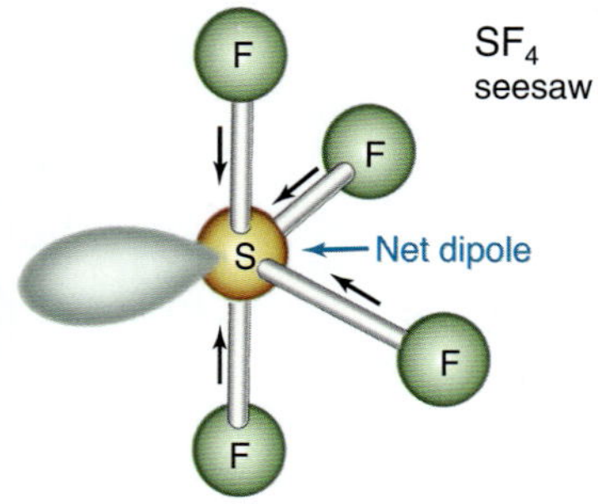

EXAMPLE 8-10: Predict whether the molecule dichloromethane, CH_2Cl_2, has a net dipole moment.

Solution: The Lewis formula for a CH_2Cl_2 molecule is

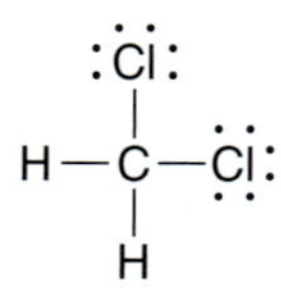

There are four bonding pairs and no lone pairs of electrons around the carbon atom; thus, we conclude from VSEPR theory that the molecular geometry is tetrahedral. The C–Cl bond dipole moments point from the Cl atom to the C atom because the electronegativity of the Cl atom is greater than that of a C atom. The C–H bond dipole moments point from the C atom to the H atom, because the electronegativity of a C atom is greater than that for a H atom. Thus, the resultant of the two C–Cl bond dipole moments and the resultant of the two C–H bond dipole moments point in the same direction, as indicated by the blue arrow representing the overall dipole moment.

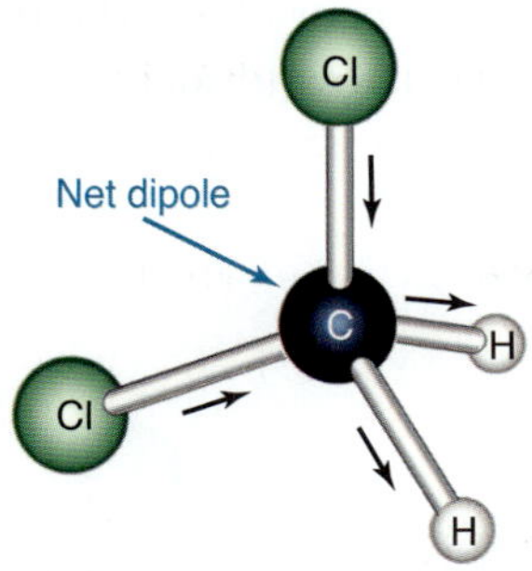

Thus, we predict (correctly) that a dichloromethane molecule is polar.

PRACTICE PROBLEM 8-10: Determine which, if any, of the following molecules are polar:

(a) CH_3Cl (b) BrF_3 (c) BrF_5 (d) AsF_5

Answer: CH_3Cl, BrF_3, and BrF_5 are polar molecules.

8-10. Isomers Play a Major Role in Determining Odor, Taste, and Drug Action

It is often possible to draw two or more satisfactory Lewis formulas for a given chemical formula by changing the connectivity of the atoms. For example, the alcohols commonly called propanol and isopropanol (systematic names 1-propanol and 2-propanol, respectively) both have the chemical formula $C_3H_8O(l)$ (Interchapter P). However, as we can see from their Lewis formulas,

See Interchapter P at www.McQuarrieGeneralChemistry.com.

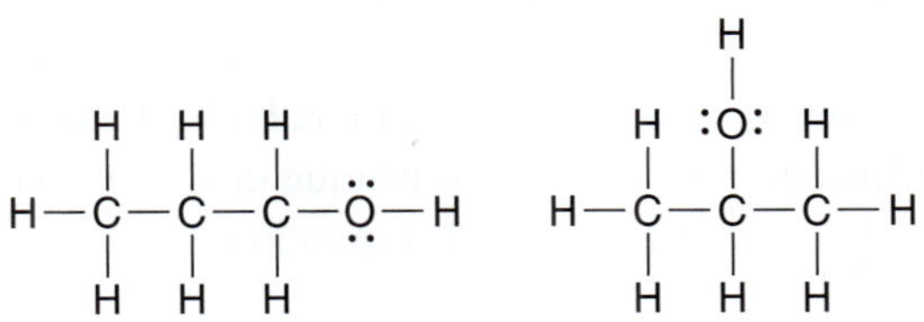

propanol isopropanol

the bonding in each molecule is different. These molecules are examples of **structural isomers**. Structural isomers are molecules that share the same chemical formula but have different bonding or connectivity between the atoms, as indicated by their respective Lewis formulas. Structural isomers are chemically distinct species with their own unique physical and chemical properties.

In addition to structural isomers, it is also possible to form chemically distinct isomers by changing the spatial arrangement of the atoms without altering the atom-to-atom connectivity in a molecule. Recall from Section 8-1 that such isomers are called **stereoisomers**. Consider the case of a SF_4Cl_2 molecule. Its Lewis formula is

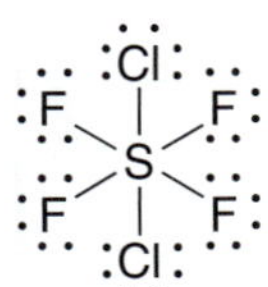

and the shape predicted by VSEPR theory is octahedral. However, in this case we see that it is possible to draw two nonequivalent three-dimensional representations of a SF_4Cl_2 molecule:

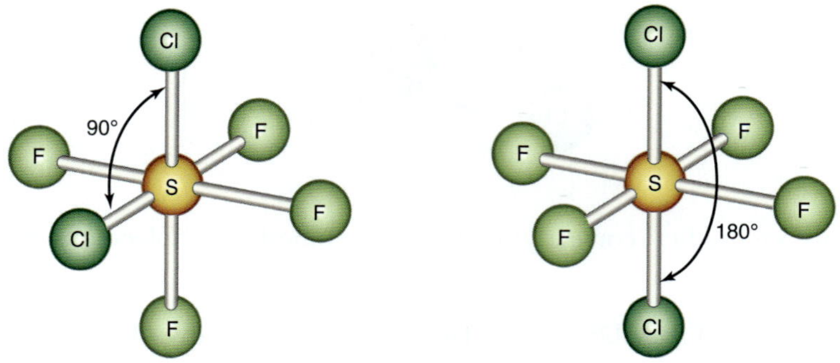

These two representations are a class of stereoisomers known as **geometrical isomers** because the geometric arrangement of the atoms differs about a center. These are *not* structural isomers because the atom-to-atom connectivity in each case is the same—both have the same six ligands bonded to a central sulfur atom.

Optical isomers are another class of stereoisomers that play an important role in biological chemistry. Optical isomers are nonsuperimposable isomers that are mirror images of each other. Your two hands serve as a good example of what a nonsuperimposable mirror image is. Your right hand is a mirror image of your left hand, and the two cannot be superimposed (Figure 8.24). A molecule can exist in optically isomeric forms if the mirror image of the molecule cannot be superimposed on itself.

Optical isomers occur when we have a central atom bonded to four *different* atoms (or groups of atoms) arranged in a tetrahedral geometry. Consider, for example, the molecule CHFBrI, where carbon is the central atom. Because this molecule has a tetrahedral geometry and each of the atoms bonded to the central atom is different, it can form two unique optical isomers (Figure 8.25). We can represent these two isomers as shown below.

Building models can help you to visualize why these two isomers are nonsuperimposable.

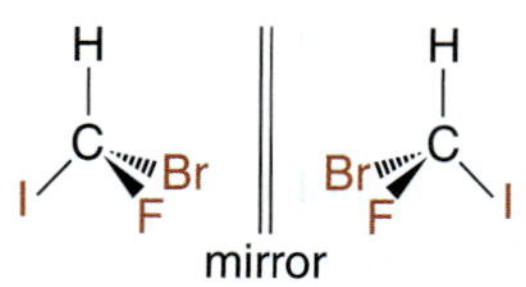

(lone electron pairs not shown for clarity)

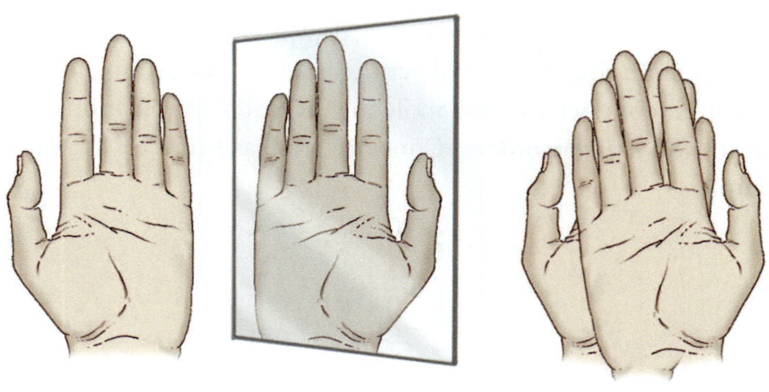

Figure 8.24 Mirror images may be superimposable or nonsuperimposable. Your two hands are an excellent example of nonsuperimposable mirror images. Your right hand is not superimposable on your left hand.

Here the dashed wedge-shaped bonds indicate that the bromine atoms lie below the page and the solid, wedge-shaped bonds indicate that the fluorine atoms lie above the page. These isomers are mirror images of each other and cannot be superimposed, just as a right hand cannot be superimposed onto a left hand. Molecules that exhibit optical isomerism are said to be **chiral**.

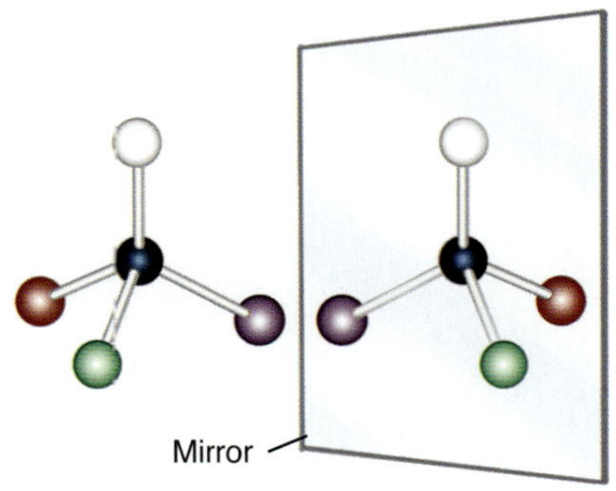

Figure 8.25 The tetrahedral CHFBrI molecule forms two unique optical isomers because its mirror image cannot be superimposed onto itself.

Many biologically significant molecules are chiral. Alanine is an amino acid (one of about twenty molecules from which the proteins are formed, see Interchapter T). The Lewis formula for an alanine molecule is

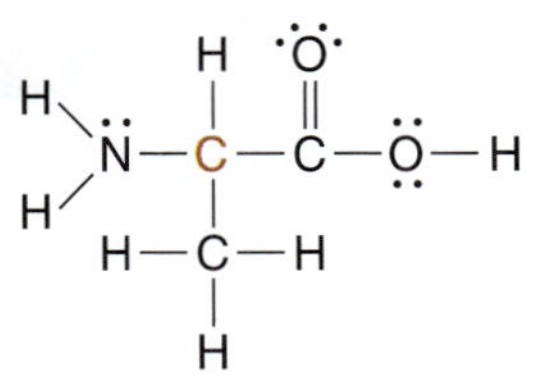

alanine

See Interchapter T at www.McQuarrieGeneralChemistry.com.

We can make the chirality clearer by reducing each of the groups of atoms around the central carbon atom (red) to its shortened chemical formula:

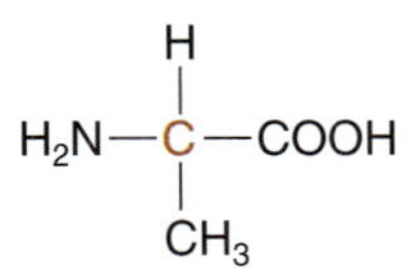

Applying VSEPR theory to the central carbon atom (red) and treating each of the surrounding groups of atoms (–H, $-NH_2$, –COOH, and $-CH_3$) as single entities, we see that alanine has a tetrahedral geometry around this carbon center. Because each of the four ligands attached to this carbon atom is unique, alanine is chiral.

Using wedge-shaped bonds, the two optical isomers of alanine may be represented as

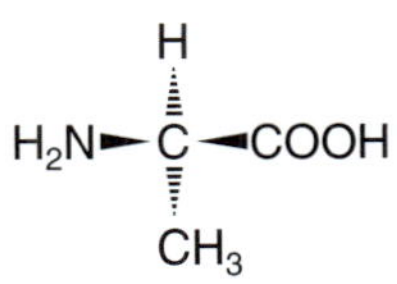

D-alanine

and

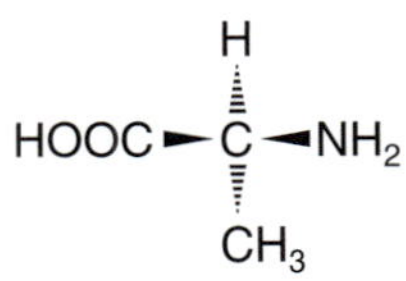

L-alanine

where the prefixes D and L are conventionally used to distinguish the two forms. Space-filling models of D and L isomers of an alanine molecule are shown in Figure 8.26.

EXAMPLE 8-11: Indicate whether the molecules (a) CF_2Br_2 and (b) SiHFBrCl (the first atom listed is the central atom) can exhibit optical isomerism.

Solution: (a) As predicted by VSEPR theory, the molecule CF_2Br_2 is tetrahedral.

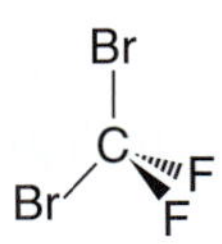

(lone electron pairs not shown for clarity)

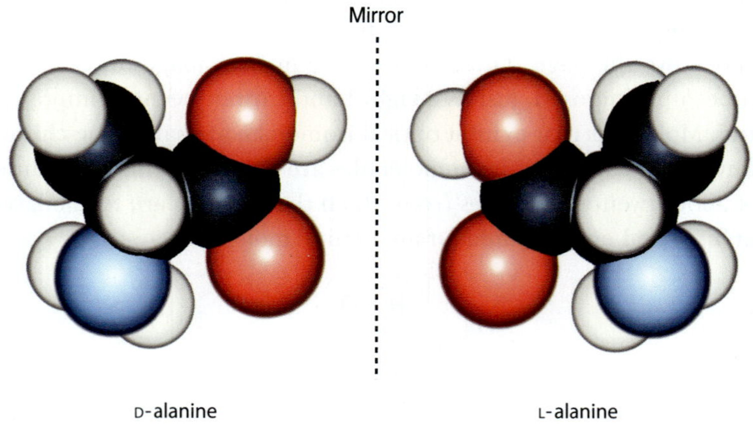

Figure 8.26 Space-filling models of D and L isomers of an alanine molecule.

Because a CF_2Br_2 molecule does not have four different kinds of atoms bonded to the central carbon atom, it cannot form optical isomers.
(b) As predicted by VSEPR theory, the molecule SiHFBrCl is tetrahedral:

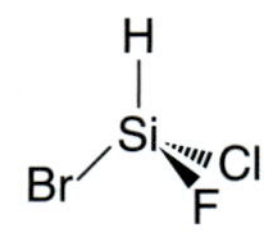

Because each of the atoms bonded to the tetrahedral silicon center is different, the molecule exhibits optical isomerism. The two isomers are shown below.

PRACTICE PROBLEM 8-11: The simplest amino acid is glycine. Its Lewis formula is

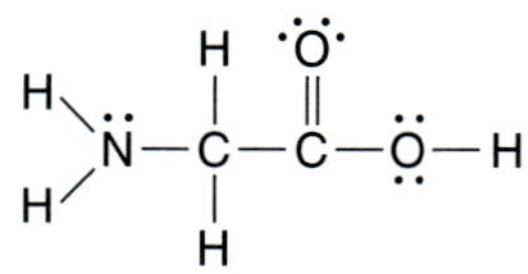

Is glycine a chiral molecule?

Answer: No. It has only three different types of attached groups to the tetrahedral carbon atom (–H, $-NH_2$ and –COOH).

For the most part, optical isomers display the same chemical properties. Their biochemical properties, however, are often radically different. The biological receptors in our body are highly **stereospecific**, that is, they can distinguish between the D and L forms of various chiral compounds. It turns out that only the L isomers of the amino acids occur in most biological systems. This has great significance in the chemistry of biologically active compounds. For example, the antibiotic penicillin attacks proteins containing D-alanine that are found in the cell walls of bacteria, but does not attack the L-isomer found in humans. As a result penicillin kills bacteria but not people. Fragrances and flavors are another example where the stereochemistry of the molecules plays an important role. The molecule D-carvone is the principal component of the oil from caraway seeds and smells like rye, whereas its mirror image, L-carvone, is one of the principal components of spearmint oil and smells like spearmint.

As we have seen in this chapter, the three-dimensional shape of molecules enables us to see why some molecules are polar and helps us to understand the physical properties of various chemicals and biologically active substances. The shape of molecules also explains why certain chemical reactions proceed and others do not (as we shall see in Chapter 18). The VSEPR theory is simple and useful, and it correctly predicts the shapes of thousands of molecules and ions. As we mentioned earlier, VSEPR is a qualitative theory, in that it does not in all cases predict precise numerical values of bond angles. Also, it cannot be used to predict bond lengths. Although the theory cannot predict that the H–N–H bond angles in an NH_3 molecule are 107.3°, it can predict that they are slightly less than the ideal tetrahedral angle of 109.5°. However, VSEPR theory fails to predict correctly the bond angle in a H_2Se molecule, which is approximately 90°, or the H–P–H bond angles in a PH_3 molecule, which also are approximately 90°. Although such exceptions are rare, it is a good idea to keep them in mind when applying VSEPR theory.

SUMMARY

The valence-shell electron-pair repulsion (VSEPR) theory is used to predict the shapes of molecules and ions in which there is a central atom bonded to ligands. Molecular shapes are defined by the arrangements of the nuclei. However, VSEPR theory takes into account all the valence-shell electrons, including lone pairs, to predict the geometrical arrangement of the ligands around the central atom. The theory is based on the premise that the valence-shell electron pairs around a central atom arrange themselves to minimize their mutual repulsion. The procedure for using VSEPR theory to predict molecular shapes can be summarized as follows:

1. Use the Lewis formula to determine the class, AX_mE_n, to which the molecule or ion belongs, where A represents a central atom, X represents a ligand atom, and E represents a lone pair of electrons.
2. Given the class AX_mE_n to which the molecule or ion belongs, use Table 8.2 to predict its shape, along with electronegativity differences to predict its polarity.

Lone pairs of electrons and multiple bonds have a larger spatial requirement than bonded pairs and produce small distortions of the molecular geometry from regular geometric shapes.

A polar bond results when the two atoms in the bond have different electronegativities. The molecular geometry of a molecule with polar bonds determines whether or not the molecule has a net dipole movement.

Chemicals that share the same formula but that have different connectivity are called structural isomers. Chemicals that have the same connectivity but can have different spatial arrangements of their ligands are called geometric isomers. Compounds that are chiral form two nonsuperimposable mirror images and are called optical isomers. Both geometric isomers and optical isomers are collectively known as stereoisomers. Stereoisomers are significant in the chemistry of many biochemical processes.

TERMS YOU SHOULD KNOW

planar *235*
stereoisomers *236*
tetrahedron, tetrahedral *236*
space-filling molecular model *237*
ball-and-stick molecular model *237*
tetrahedral bond angle (109.5°) *237*
tetravalent *237*
structural chemistry *237*
valence-shell electron-pair repulsion (VSEPR) theory *238*
linear *238*
trigonal planar *239*
trigonal bipyramid, trigonal bipyramidal *241*
equatorial vertex *241*
axial vertex *241*
octahedron, octahedral *241*
trigonal pyramid, trigonal pyramidal *242*
molecular shape *242*
bent *243*
ligand *244*
AX_mE_n *244*
steric number *244*
seesaw *250*
T-shaped *250*
interhalogen compound *250*
square pyramid, square pyramidal *252*
square planar *252*
nonpolar molecule *253*
polar molecule *254*
structural isomer *255*
stereoisomer *255*
geometric isomers *256*
optical isomers *256*
chiral *257*
stereospecific *258*

AN EQUATION YOU SHOULD KNOW HOW TO USE

$$\text{steric number} = \begin{pmatrix}\text{number of atoms}\\ \text{bonded to central atom}\end{pmatrix} + \begin{pmatrix}\text{number of lone pairs}\\ \text{on central atom}\end{pmatrix} \quad (8.1)$$

$$= m + n$$

(the number of ligands and lone pairs attached to a central atom)

PROBLEMS

MOLECULES AND IONS INVOLVING ONLY SINGLE BONDS

8-1. Which of the following molecules have bond angles of 90°?

(a) TeF_6 (b) $AsBr_5$ (c) GaI_3 (d) XeF_4

8-2. Which of the following species have bond angles of 90°?

(a) NH_4^+ (b) PF_5 (c) AlF_6^{3-} (d) $SiCl_4$

8-3. Which of the following species have 120° bond angles?

(a) ClF_3 (b) $SbBr_6^-$ (c) $SbCl_5$ (d) $InCl_3$

8-4. Which of the following species have 180° bond angles?

(a) SeF_6 (b) BrF_2^- (c) SCl_2 (d) $SiBr_4$

8-5. Which of the following triatomic molecules are linear? Which are bent?

(a) TeF_2 (b) $SnBr_2$ (c) KrF_2 (d) OF_2

8-6. Which of the following triatomic ions are linear? Which are bent?

(a) NH_2^- (b) PF_2^+ (c) IF_2^+ (d) Br_3^-

8-7. Which of the following molecules are tetrahedral?

(a) XeF_4 (b) XeO_2F_2 (c) NF_3O (d) SeF_4

8-8. Which of the following molecules are trigonal pyramidal?

(a) NH_2Cl (b) ClF_3 (c) PF_3 (d) BF_3

8-9. Give the molecular class and steric number and name the geometry that describes the shape of each of the following molecules:

(a) SF_4 (b) SF_6 (c) $XeOF_4$ (d) IOF_5

8-10. Give the molecular class and steric number and name the geometry that describes the shape of each of the following ions:

(a) IBr_4^- (b) PCl_4^+ (c) BF_4^- (d) IF_4^+

8-11. Give the molecular class, shape, and bond angles (90°, 109.5°, <120°, etc.) for each of the following halides:

(a) OF_2 (b) KrF_4 (c) CF_4 (d) $GeCl_4$

8-12. Give the molecular class, shape, and bond angles (90°, 109.5°, <120°, etc.) for each of the following fluorides:

(a) SeF_6 (b) GeF_4 (c) BrF_3 (d) IF_5

8-13. Give the molecular class, shape, and bond angles for each of the following species:

(a) TeF_6 (b) $SbCl_5$ (c) ICl_4^- (d) $InBr_3$

8-14. Give the molecular class, shape, and bond angles for each of the following ions:

(a) BH_4^- (b) SiF_6^{2-} (c) SiF_3^+ (d) $SnCl_6^{2-}$

8-15. Give the molecular class, shape, and bond angles for each of the following ions:

(a) AlH_4^- (b) SbF_6^- (c) BrF_4^- (d) $AsCl_4^+$

8-16. Give the molecular class, shape, and bond angles for each of the following molecules:

(a) $GeCl_4$ (b) $SbCl_3$ (c) $TeCl_6$ (d) SCl_4

8-17. VSEPR theory has been successful in predicting the geometry of interhalogen molecules and ions. Predict the shapes of the interhalogen molecules given in Table 8.3.

8-18. Predict the shapes and bond angles of the following iodofluorine ions:

(a) IF_2^+ (b) IF_6^+ (c) IF_4^+ (d) IF_4^-

MOLECULES OR IONS THAT MAY INVOLVE MULTIPLE BONDS

8-19. Predict the shape and bond angles of each of the following molecules:

(a) $SeOCl_2$ (b) SO_2Cl_2

(c) SOF_4 (d) ClO_3F

8-20. Predict the shape and bond angles of each of the following species:

(a) $XeOF_4$ (b) IOF_5

(c) $PO_2F_2^-$ (d) PO_3F^{2-}

8-21. Predict the shape and bond angles of each of the following species:

(a) CCl_2O (b) NSF_3 (c) N_3^- (d) SbOCl

8-22. Predict the shape and bond angles of each of the following species:

(a) $IO_2F_2^-$ (b) ClO_2^- (c) NOCl (d) NO_2Cl

8-23. Predict the shapes and bond angles of the following ions:

(a) BrO_2^- (b) TeF_5^- (c) SO_3Cl^- (d) SF_3^+

8-24. Predict the shapes and bond angles of the following molecules:

(a) NF_3O (b) GeO_2 (c) $AsOCl_3$ (d) XeO_2

8-25. Predict the shape and bond angles of each of the following molecules:

(a) XeO_2F_4 (b) IO_2F_3 (c) IO_2F (d) IO_3F

8-26. Predict the shape and bond angles of each of the following ions:

(a) TlF_4^- (b) IO_2^- (c) CS_3^{2-} (d) BrO_3^-

8-27. The species NO_2^+ and NO_2^- have O–N–O bond angles of 180° and 115°, respectively. Use VSEPR theory to explain the difference in bond angles.

8-28. Compare the shapes and bond angles of the oxynitrogen ions

(a) NO_2^- (b) NO_3^- (c) NO_2^+ (d) NO_4^{3-}

MOLECULES WITH MORE THAN ONE CENTRAL ATOM

8-29. Describe the bonding around the oxygen atom and the two carbon atoms (the central atoms) in dimethyl ether and make a three-dimensional sketch of the molecule. The Lewis formula of dimethyl ether is

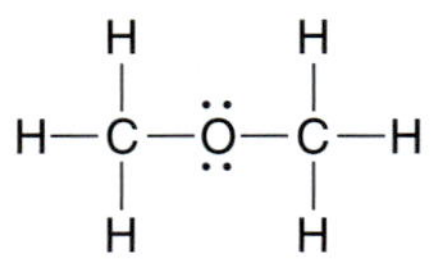

8-30. Describe the bonding around the carbon atom and nitrogen atom (the central atoms) in cyanamide and make a three-dimensional sketch of the molecule. The Lewis formula of cyanamide is

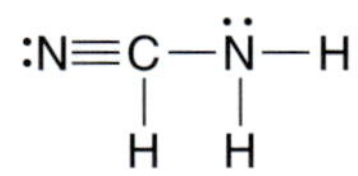

8-31. Draw the Lewis formula and describe the bonding around each of the central atoms in an acetone molecule, $CH_3(CO)CH_3$ (systematic name: 2-propanone), the active compound in most nail polish removers. Make a three-dimensional sketch of the molecule.

8-32. Draw the Lewis formula and describe the bonding around each of the central atoms in an acetic acid molecule, CH_3COOH, the acidic component of vinegar. Make a three-dimensional sketch of the molecule.

MOLECULAR SHAPES AND DIPOLE MOMENTS

8-33. Describe the bond polarities in the following molecules and predict which ones are polar.

(a) NF_3 (b) OF_2 (c) OBr_2 (d) BH_3

8-34. Describe the bond polarities in the following molecules and predict which ones are polar.

(a) CCl_4 (b) PCl_3 (c) ClF_3 (d) CF_2Cl_2

8-35. For each of the following molecules, predict the shape and bond angles. Indicate which ones are polar.

(a) XeF_2 (b) AsF_5 (c) $TeCl_4$ (d) Cl_2O

8-36. For each of the following molecules, predict the shape and bond angles. Indicate which ones are polar.

(a) $GeCl_4$ (b) SCl_2 (c) PoF_6 (d) BrF_3

8-37. For each of the following molecules, predict the shape and bond angles. Indicate which ones are polar.

(a) $GaCl_3$ (b) $TeCl_2$ (c) TeF_4 (d) $SbCl_5$

8-38. For each of the following molecules, predict the shape and bond angles. Indicate which ones are polar.

(a) TeF_6 (b) ClF_5 (c) $SiCl_4$ (d) $SeCl_2$

8-39. Predict which of the following molecules are polar:

(a) CF_4 (b) AsF_3 (c) CH_2Cl_2 (d) IOF_5

8-40. Predict which of the following molecules are polar:

(a) $TeBr_4$ (b) BCl_3 (c) SF_5Br (d) SOF_4

ISOMERS

8-41. Draw structural formulas showing the bonding for the three possible structural isomers of pentane, C_5H_{12}.

8-42. Draw structural formulas showing the bonding for the five possible structural isomers of hexane, C_6H_{14}.

8-43. Describe the possible geometric isomers of a trigonal bipyramidal molecule whose formula is

(a) AX_4Y (b) AX_3Y_2 (c) AX_2Y_3

8-44. Describe the possible geometric isomers of an octahedral molecule whose formula is

(a) AX_5Y (b) AX_4Y_2 (c) AX_3Y_3

8-45. Describe the possible geometric isomers of $C_2H_4Cl_2$.

8-46. Describe the possible geometric isomers of $N_2H_2F_2$.

8-47. Indicate which of the following compounds (written in shorthand notation to indicate the bonding) can exist as optical isomers:

(a) CH_3Cl

(b) 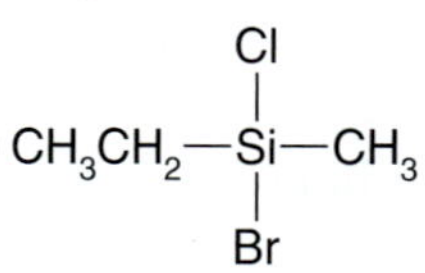

(c) 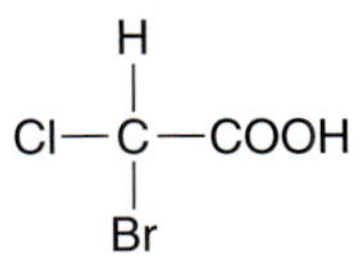

(d) 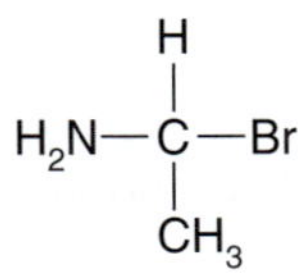

8-48. Indicate which of the following compounds (written in shorthand notation) can exist as optical isomers:

(a) 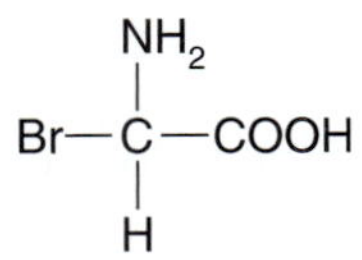

(b) CH_2Cl_2

(c) $(CH_3)_2SiCl_2$

(d) 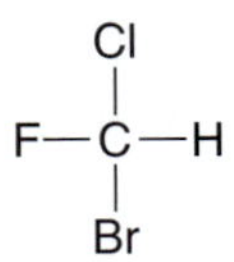

ADDITIONAL PROBLEMS

8-49. Explain why a Lewis formula alone is insufficient to predict the actual shape of a molecule.

8-50. Use the concepts from this chapter to explain why a tetrahedral geometry is preferred over a square planar geometry for a molecule such as methane, CH_4, with an AX_4 configuration.

8-51. In the trigonal bipyramidal structure we place lone pairs in the equatorial plane first. Use the concepts of lone-pair–lone-pair repulsion and lone-pair–bonding-pair repulsion from this chapter to show why a T-shaped molecular geometry is preferred over other possibilities for molecules with an AX_3E_2 configuration.

8-52. List two requirements that a molecule must meet in order to be polar.

8-53. Explain the difference between a structural isomer and a stereoisomer.

8-54. Give one example of each of the following:

(a) bent molecule

(b) bent ion

(c) tetrahedral ion

(d) octahedral molecule

8-55. Give one example of each of the following:

(a) trigonal planar molecule

(b) trigonal pyramidal molecule

(c) T-shaped molecule

(d) octahedral ion

8-56. VSEPR theory has been successful in predicting the molecular geometry of noble-gas compounds. Predict the molecular geometry of each of the following xenon species:

(a) XeO_3 (b) XeO_4 (c) XeO_2F_2 (d) XeO_6^{4-}

8-57. Predict the geometry of each of the following phosphorus-containing species:

(a) POF_3 (b) $POCl$ (c) PH_2^- (d) PCl_4^+

8-58. Compare the shapes of the following oxysulfur ions:

(a) sulfoxylate ion, SO_2^{2-}

(b) sulfite ion, SO_3^{2-}

(c) sulfate ion, SO_4^{2-}

8-59. Compare the shapes of the following oxychlorine ions:

(a) chlorite, ClO_2^-

(b) chlorate, ClO_3^-

(c) perchlorate, ClO_4^-

8-60. In Section 8-4 we saw that the shape of a gaseous PCl_5 molecule is trigonal bipyramidal. However, solid PCl_5, unlike gaseous PCl_5, is composed of PCl_4^+ cations and PCl_6^- anions. There are no PCl_5 molecules in $PCl_5(s)$. Determine the shape of the PCl_4^+ and PCl_6^- ions.

8-61. Arrange the following groups of molecules in order of increasing dipole moment:

(a) HCl HF HI HBr

(b) PH_3 NH_3 AsH_3

(c) ClF_3 BrF_3 IF_3

(d) H_2O H_2S H_2Te H_2Se

8-62. Which of the following molecules is trigonal pyramidal?

(a) SOF_2 (b) ClF_3 (c) NO_2Cl (d) BF_3

8-63. Which of the following fluorides has 90° bond angles?

(a) XeF_4 (b) CF_4 (c) SF_2 (d) XeF_2

8-64. Which of the following molecules is linear?

(a) $BeCl_2$ (b) O_3 (c) OCl_2 (d) NOF

8-65. Which of the following species is polar?

(a) CO_2 (b) BF_3 (c) I_3^- (d) ClF_3

8-66. Predict the shapes of the following bromofluoride ions:

(a) BrF_2^- (b) BrF_4^- (c) BrF_2^+ (d) BrF_4^+

8-67. Predict the shapes of the following oxyfluoro compounds of sulfur:

(a) SOF_4 (b) SOF_2 (c) SO_2F_2

8-68. From the accompanying list, select the appropriate description(s) of the bond angles that occur in each of the following fluorides:

(a) SeF_6
(b) GeF_4
(c) BrF_3
(d) IF_5

(1) exactly 90°
(2) slightly less than 90°
(3) exactly 109.5°
(4) slightly less than 109.5°
(5) exactly 120°
(6) slightly less than 120°
(7) slightly greater than than 120°
(8) exactly 180°
(9) slightly less than 180°

8.69. From the accompanying list, select the appropriate description(s) of the bond angles that occur in each of the following molecules:

(a) $GeCl_4$
(b) $SbCl_3$
(c) TeF_6
(d) SF_4

(1) exactly 90°
(2) slightly less than 90°
(3) exactly 109.5°
(4) slightly less than 109.5°
(5) exactly 120°
(6) slightly less than 120°
(7) slightly greater than than 120°
(8) exactly 180°
(9) slightly less than 180°

8.70. Indicate which, if any, of the listed bond angles occur in the following species:

(a) TeF_6
(b) $SbCl_5$
(c) ICl_4^-
(d) $InBr_3$

(1) 90°
(2) 109.5°
(3) 120°

8.71. Indicate which, if any, of the listed bond angles occur in the following species:

(a) BF_4^-
(b) SiF_6^{2-}
(c) SiF_3^+
(d) $SnCl_6^{2-}$

(1) 90°
(2) 109.5°
(3) 120°

8-72. Give one example of each of the following:

(a) bent molecule
(b) bent ion
(c) tetrahedral ion
(d) octahedral ion

8-73. Give one example of each of the following:

(a) trigonal planar molecule
(b) trigonal pyramidal molecule
(c) T-shaped molecule
(d) octahedral ion

8-74. Draw the Lewis formula and use VSEPR theory to predict the bonding around each of the central atoms in the dimethyl sulfide, CH_3SCH_3, molecule. Make a three-dimensional sketch of the dimethyl sulfide molecule. Is dimethyl sulfide a polar or nonpolar molecule?

8-75. Describe the possible isomers of $C_2H_2F_2Cl_2$.

8-76. Describe the possible stereoisomers of the following compounds (if any) where H, X, Y, and Z represent different ligands around a central atom, A.

(a) a tetrahedral molecule AX_3Y
(b) a square planar molecule AX_2Y_2
(c) a tetrahedral molecule AHXYZ
(d) a square planar molecule AX_3Y

8-77. Square planar molecules are never chiral. Use models or drawings to show why this is the case.

8-78. The amino acid serine has the Lewis formula

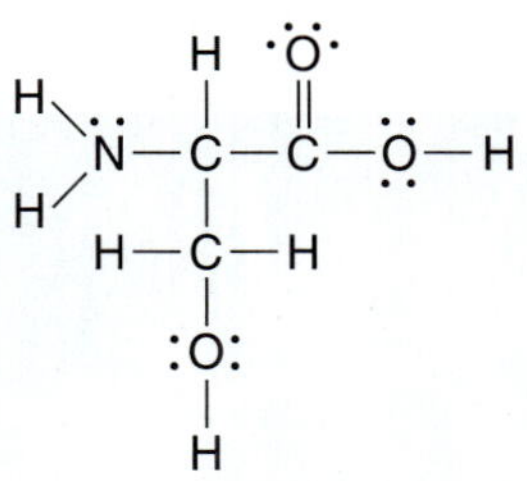

Like alanine, serine is also chiral. Locate the chiral center in serine and make a sketch similar to those of the D and L isomers of alanine on page 261, using wedged bonds to show the three-dimensional location of the various groups.

8-79. (*) Natural crystals of tartaric acid will rotate polarized light in a specific direction, while synthetically made crystals do not. In 1849 Louis Pasteur deduced that this "optical activity" was due to the existence of two distinct isomeric forms of tartaric acid, leading to the discovery of chirality in molecules. The formula of tartaric acid is

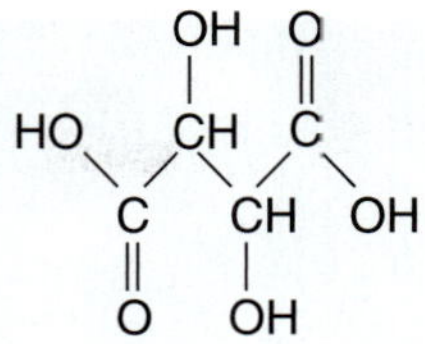

Using wedged bonds, sketch the two optical isomers of a tartaric acid molecule.

8-80. (*) Use Figure 8.2b to show that the tetrahedral angle in a regular tetrahedron is 109.5°. *Hint*: If we let the edge of the cube be of length a, then the diagonal on the face of the cube has a length $\sqrt{2}a$ by the Pythagorean theorem. From this information, you can determine the distance in terms of a from the center of the cube to a vertex from which, in turn, you can determine the tetrahedral angle.

Linus Pauling (1901–1994) was born in Portland, Oregon. He attended Oregon State University, where he also taught various chemistry classes as an undergraduate student. His future wife was a student in his chemistry class for home economics majors. He received his Ph.D. in chemistry in 1925 from the California Institute of Technology for his dissertation on X-ray crystallography of organic compounds and the structure of crystals. After spending a year studying at the University of Munich, he joined the faculty at the California Institute of Technology, where he remained for almost 40 years. Pauling was a pioneer in the application of quantum theory to chemistry. His book *The Nature of the Chemical Bond* (1939) is one of the most influential chemistry texts of the twentieth century. In the 1930s, he became interested in biological molecules and developed a structural theory of protein molecules—work that led to elucidating that sickle cell anemia is caused by a faulty structure of hemoglobin. In the early 1950s, he proposed the alpha helix as the basic structure of proteins. Pauling was awarded the Nobel Prize in Chemistry in 1954 "for his research into the nature of the chemical bond and its application to the elucidation of the structure of complex structures." During the 1950s, Pauling was in the forefront of the fight against nuclear testing, for which he was awarded the Nobel Peace Prize in 1963. From the early 1980s until his death, he was embroiled in the controversy of advocating the use of vitamin C as protection against the common cold and some serious maladies such as cancer. Pauling is the only person to have been awarded two unshared Nobel Prizes.

9. Covalent Bonding

As we have seen, the sharing of electrons between two atoms results in a covalent bond. We used this information to draw Lewis formulas, and along with VSEPR theory, to infer molecular shapes. However, we did not quantitatively explain why such bonds form in the first place. To answer this, we need a more fundamental theory of covalent bonding. We showed in Chapter 5 how quantum theory describes the electrons in atoms in terms of atomic orbitals. Now we shall use quantum theory to show how molecular orbitals may be formed by combining atomic orbitals on different atoms. Recall from Chapter 5 that an orbital is just a special type of wave function. We shall use the two terms interchangeably throughout this chapter. In the first few sections, we introduce molecular orbital theory and use it to describe the bonding in simple homonuclear diatomic molecules (in which the two atoms are the same) and heteronuclear diatomic molecules (in which the two atoms are different). We then go on to discuss polyatomic molecules. To describe the bonding in polyatomic molecules, we shall use special orbitals called hybrid orbitals that are combinations of atomic orbitals on the same atom and are constructed with the geometry of the molecule in mind. A nice feature of using hybrid orbitals to describe the bonding in polyatomic molecules is that the bonding is closely related to the Lewis formulas that we learned in Chapter 7. Following this chapter, we shall use our knowledge of molecular shape, structure, and orbitals to understand the physical properties of molecules and chemical reactivity.

9-1. A Molecular Orbital Is a Combination of Atomic Orbitals on Different Atoms

The simplest neutral molecule is diatomic hydrogen, H_2, which has only two electrons. The Schrödinger equation (Section 5-1) that describes the motion of the electrons in H_2 can be solved to a high degree of accuracy using a computer. The results are valuable because they are similar to the results for more complicated molecules. Let's look at the approach of quantum theory in more detail.

As a first step in setting up the Schrödinger equation for a H_2 molecule, the two nuclei are fixed at some given separation. Then the two electrons are included, and the equation is solved to give the wave functions and energies that describe the two electrons. The wave function that corresponds to the lowest energy, the **ground state wave function,** can be used to compute contour diagrams, much like the maps used to show peaks and valleys in hilly terrain. These diagrams show the distribution of electron density around the two nuclei.

Figure 9.1 shows contour diagrams of the ground state electron density of a H_2 molecule as a function of the separation of the nuclei of the two hydrogen atoms. We see that at large separations the two atoms hardly interact, so the electron density is just that of two electrons, each in a 1*s* orbital about each of the two hydrogen atoms. As the separation decreases, however, the two 1*s* orbitals overlap and combine into one orbital that is distributed around both nuclei. Such an orbital is called a **molecular orbital** because it extends over both nuclei in the molecule. Throughout this chapter, we shall build molecular orbitals by combining atomic orbitals on different atoms. When electrons occupy these orbitals, the buildup of electron density between the nuclei results in a covalent bond. Note how the detailed quantum theoretical results shown in Figure 9.1 correspond to the Lewis formula; both approaches picture a covalent bond as the sharing of an electron pair between two nuclei.

The lower part of Figure 9.1 shows the energies that correspond to the electron densities. Notice that interaction energies are negative for any distances at which the atoms attract each other. These negative values mean that the energy of a H–H bond is less than that of two separated hydrogen atoms, which in turn means that a H_2 molecule is more stable than two isolated hydrogen atoms under these conditions. The graph shows that, for a H_2 molecule, the interaction energy has a minimum at the internuclear separation R = 74 pm. This value of R is the predicted length of a H_2 bond and is in excellent agreement with the experimental value.

9-2. The Hydrogen Molecular Ion H_2^+ Is the Simplest Diatomic Species

In this section we discuss a theory of bonding, called **molecular orbital theory**, that gives us insight into why, for example, two hydrogen atoms join to form a stable molecule, whereas two helium atoms do not. Although molecular orbital theory can be applied to all molecules, for simplicity we shall begin by considering only **homonuclear diatomic molecules** (diatomic molecules in which both nuclei are the same).

Recall that we describe the electronic structure of atoms in terms of atomic orbitals, which are based on the set of orbitals that were given for a hydrogen atom. Because a hydrogen atom has only one electron, its atomic orbitals are relatively simple to calculate from the Schrödinger equation and serve as approximate orbitals for more complicated atoms. A one-electron system that ap-

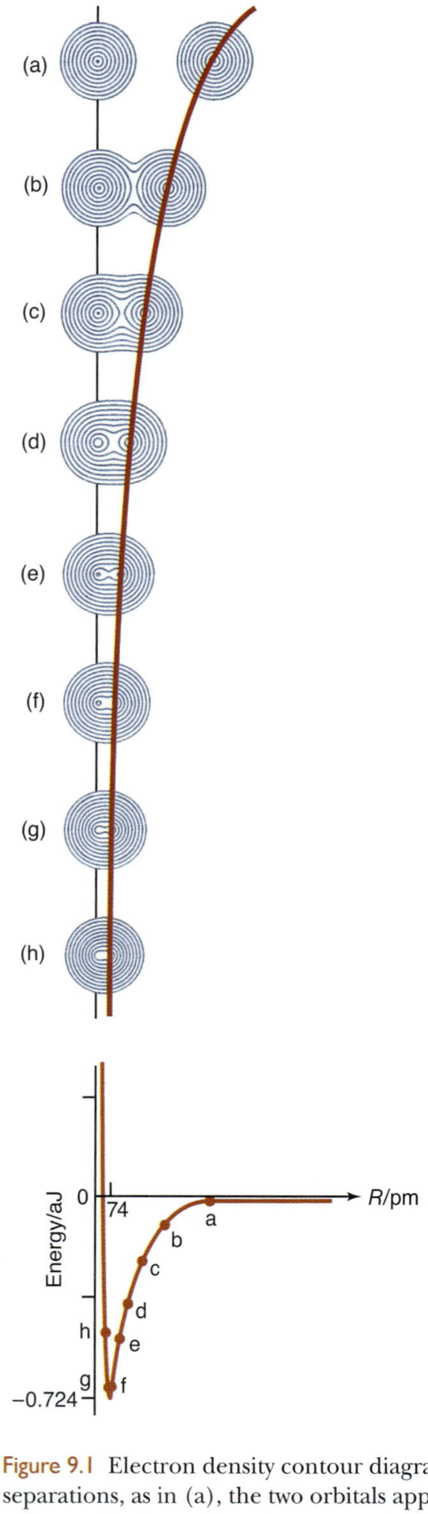

Figure 9.1 Electron density contour diagrams of two hydrogen atoms as a function of their separation (upper part). At large separations, as in (a), the two orbitals appear simply as those of two separate atoms. As the atoms come together, the two separate atomic orbitals combine into one molecular orbital encompassing both nuclei, as in (b) through (h). The lower part of the figure shows the energy of two hydrogen atoms as a function of their separation R. The labels (a) through (h) correspond to those in the upper part of the figure. At large distances, the two hydrogen atoms do not interact, so their interaction energy is zero. As the two atoms come together, they attract each other, and so their interaction energy becomes negative. When they are less than 74 pm apart, the interaction energy increases and they repel each other. The bond length of a H_2 molecule is the distance at which the energy is a minimum, that is, 74 pm. The energy at this distance is –0.724 aJ, which is the energy required to dissociate the H_2 molecule into two separate hydrogen atoms.

plies to homonuclear diatomic molecules is the **hydrogen molecular ion**, H_2^+, which consists of two protons and one electron. The H_2^+ ion can be produced experimentally by using $H_2(g)$ in a discharge tube (see Figure 2.18). It is a stable species, with a bond length of 106 pm and a bond energy of 0.423 aJ.

The Schrödinger equation for a H_2^+ ion, like that for a hydrogen atom, is relatively easy to solve, and we obtain a set of wave functions, or orbitals, and their corresponding energies. As noted earlier, these orbitals extend over both nuclei in H_2^+ and therefore are called molecular orbitals. In Chapter 5 we discussed the shapes of the various hydrogen atomic orbitals and then used them to build up the electronic structures of more complicated atoms. In the same way, we now use the various H_2^+ molecular orbitals to build up the electronic structures of more complicated diatomic molecules.

Figure 9.2 shows the shapes of the first few molecular orbitals of H_2^+. Each

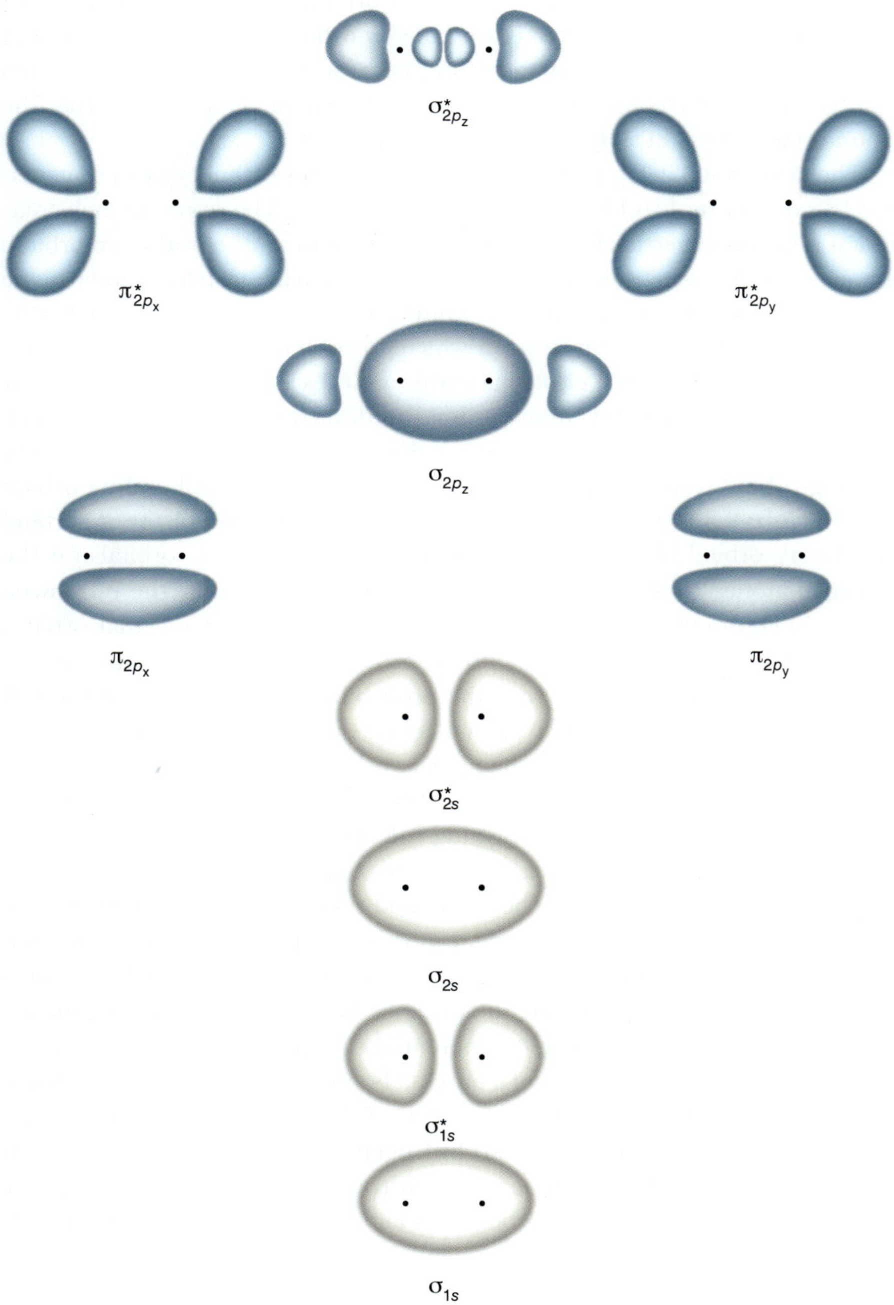

Figure 9.2 The three-dimensional surfaces that depict the shapes but not the relative sizes of the first few H_2^+ molecular orbitals. The orbitals are listed in order of increasing energy. The nuclei are shown as heavy black dots. Some molecular orbitals have nodal planes between the nuclei. The two molecular orbitals designated by π_{2p_x} and π_{2p_y} have the same energy and the two designated by $\pi^*_{2p_x}$ and $\pi^*_{2p_y}$ also have the same energy.

shape represents the three-dimensional surface that encloses a certain probability of finding the electron within the volume enclosed by that surface. We shall explain the notation that we use to designate these orbitals (σ_{1s}, $\pi^*_{2p_y}$, for example) as we go along.

Quantum chemists (chemists who apply quantum theory to molecules) have discovered that all the orbitals in Figure 9.2 can be generated by the addition or subtraction of orbitals on neighboring atoms. For example, the lowest H_2^+ molecular orbital shown in Figure 9.2 can be generated by adding 1*s* orbitals on the two hydrogen atoms, much like that depicted in Figure 9.1. The resultant concentration of electronic charge between the nuclei tends to draw the nuclei together and is responsible for the bond in the ground electronic state of a H_2^+ molecular ion. Such a molecular orbital is called a **bonding orbital**. This orbital has a circular cross section along the **internuclear axis**, the axis or line drawn between the nuclei of the two bonding atoms. Because atomic *s* orbitals have a circular cross section (Figure 5.3), molecular orbitals that have a circular cross section along the internuclear axis are called **σ orbitals** (sigma, σ, is the Greek letter corresponding to *s*). Furthermore, because this orbital is constructed from two hydrogen 1*s* orbitals, it is called a σ_{1s} orbital. Thus, we see that the ground state orbital of a H_2^+ ion is a σ_{1s} orbital.

The next orbital in Figure 9.2, the one of second-lowest energy, can be generated by subtracting 1*s* orbitals on the hydrogen atoms. The resultant molecular orbital is also symmetric about the internuclear axis, and so is also a σ orbital. However, it is different from the ground state σ_{1s} orbital in that it has a nodal plane midway between the nuclei and perpendicular to the internuclear axis. (Recall from Chapter 5 that surfaces on which an orbital has a value of zero are called nodal surfaces.) The electrons in this orbital are concentrated on the far sides of the two nuclei and therefore tend to draw the nuclei away from each other. Such an orbital is called an **antibonding orbital**. To distinguish this antibonding orbital from the bonding orbital, we denote the antibonding orbital by σ^*_{1s}. Figure 9.3 is a pictorial description of the formation of the σ_{1s} orbital and the σ^*_{1s} orbital by the addition and the subtraction of 1*s* orbitals on the hydrogen atoms. Note that the orbitals in Figure 9.3 look like the two lowest orbitals in Figure 9.2, justifying the process of constructing molecular orbitals by the addition and subtraction of atomic orbitals. Recall from Chapter 5 that the square of a wave function gives the probability of where the electron will be found, and so the buildup of the molecular orbital between the nuclei corresponds to the electron density being accumulated there, which results in a chemical bond. When the 1*s* orbitals are subtracted, however, they cancel each other midway between the nuclei, giving the results for the molecular orbital shown in the upper drawing in Figure 9.3. This cancellation leads to a region in which the electron is unlikely to be found and thus a nodal plane between the nuclei. The resulting molecular orbital in this case is an antibonding orbital. Figure 9.4 shows the potential energies associated with these orbitals as a function of the **internuclear distance**, the distance between the two atomic nuclei. As in Figure 9.1, the minimum in the potential energy curve of the σ_{1s} orbital in Figure 9.4 depicts the formation of a bond (a bonding orbital), and the absence of a minimum in the potential energy curve of the σ^*_{1s} orbital (an antibonding orbital) reflects the fact that the two hydrogen atoms in this case repel each other. Finally, Figure 9.5 is a schematic illustration of an energy diagram for the formation of the σ_{1s} and σ^*_{1s} orbitals. As Figure 9.5 implies, when 1*s* orbitals

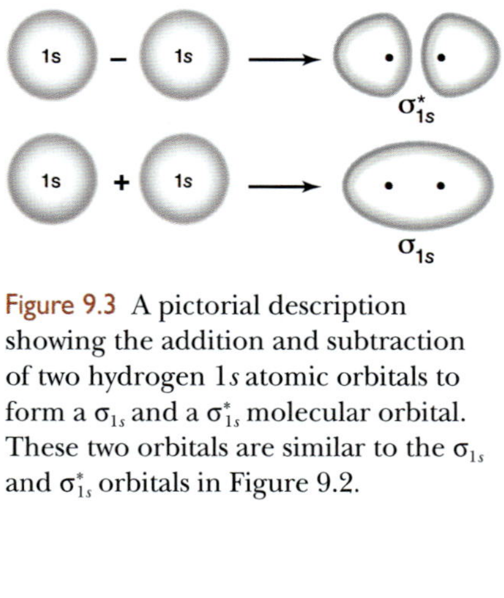

Figure 9.3 A pictorial description showing the addition and subtraction of two hydrogen 1*s* atomic orbitals to form a σ_{1s} and a σ^*_{1s} molecular orbital. These two orbitals are similar to the σ_{1s} and σ^*_{1s} orbitals in Figure 9.2.

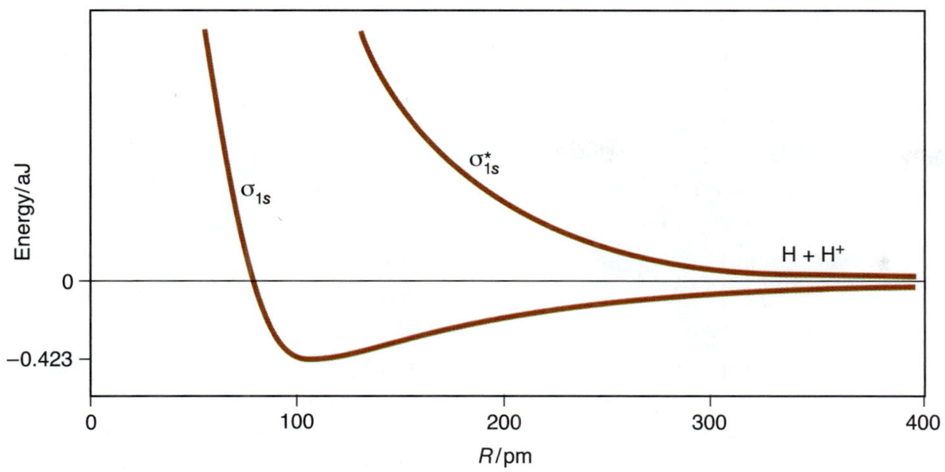

Figure 9.4 Potential energy of the σ_{1s} bonding molecular orbital and σ^*_{1s} antibonding molecular orbital for H_2^+ as a function of the internuclear separation between the two hydrogen nuclei. The minimum in the lower curve represents the equilibrium nuclear separation or average bond length of the molecular ion (106 pm). This energy minimum depicts the formation of a stable bond. In contrast, the energy of the σ^*_{1s} orbital (upper curve) is always above that of the separated atoms, making this an antibonding orbital.

are combined (added or subtracted), a bonding orbital of lower energy and an antibonding orbital of higher energy are formed.

The next two molecular orbitals in Figure 9.2 are formed from a combination of the atomic 2*s* orbitals. Even though a ground state hydrogen atom does not have electrons in these orbitals, they still are used for excited states and we can use their combinations mathematically. Just as in the case of combining 1*s* orbitals, we can add and subtract 2*s* orbitals to form a bonding orbital and an antibonding orbital. These molecular orbitals also have cylindrical symmetry and thus are σ orbitals. In fact, they look much like σ_{1s} orbitals, except they are larger in extent. Because they are formed from 2*s* orbitals, they are designated by σ_{2s} and σ^*_{2s}.

We can also combine 2*p* atomic orbitals on the hydrogen atoms. Recall that there are three 2*p* orbitals, directed along three perpendicular axes. As Figure 9.6 indicates, we can combine one pair of 2*p* orbitals along the internuclear axis (which we shall arbitrarily call the *z* axis), while the other 2*p* orbitals (oriented along the *x* and *y* axes) are combined perpendicular to the internuclear axis. Let's consider the combination along the internuclear axis (the *z* axis) first. Figure 9.7 focuses on this combination. In this figure we have labeled one of the lobes of each 2*p* orbital with a positive sign (+) and the other with a negative sign (−) to indicate the regions in which the wave function has positive and negative values. In Figure 9.7a, the $2p_z$ orbitals are combined in such a way that their positive lobes overlap, resulting in a molecular orbital in which the electronic charge is concentrated between the two nuclei, thus forming a bonding orbital. It is a σ orbital because it is cylindrically symmetric about the internuclear axis, and we denote it by σ_{2p_z} to convey that it is formed from the two $2p_z$ orbitals. Figure 9.7b shows that we can also combine two $2p_z$ orbitals along the internuclear

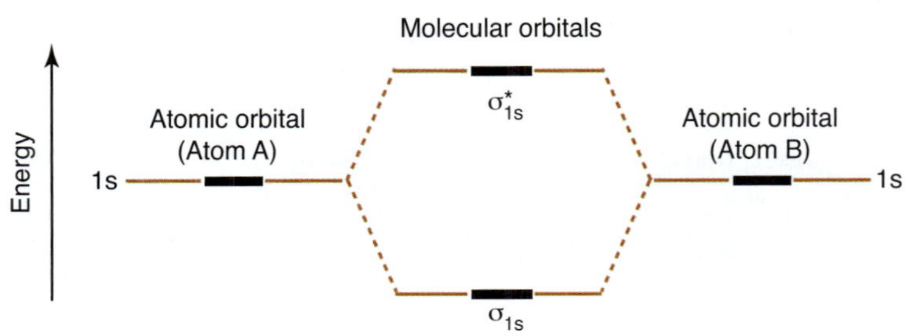

Figure 9.5 A schematic energy diagram showing the resultant energies when 1*s* orbitals are added and subtracted to give a σ_{1s} and σ^*_{1s} molecular orbital.

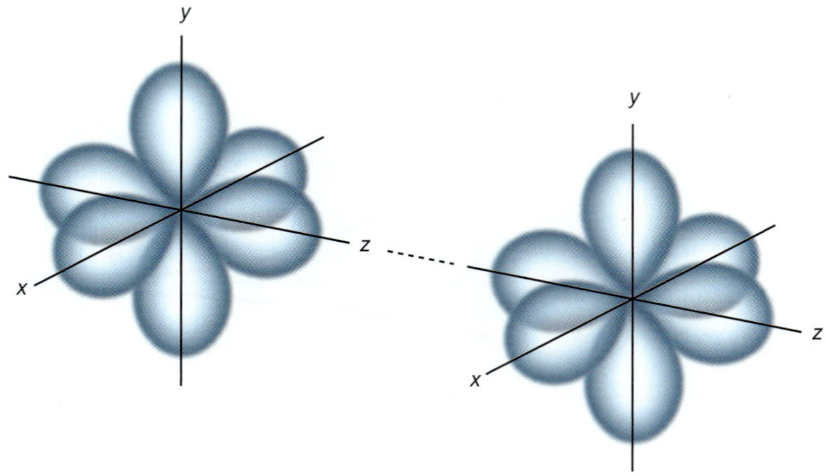

Figure 9.6 The three $2p$ atomic orbitals are directed along three perpendicular axes. These orbitals can be combined in two ways: the p orbitals that are directed along the internuclear axis (what we shall call the z axis) can be combined in one way (see Figure 9.7) and those directed perpendicular to the internuclear axis (the x and y axes) can be combined in another way (see Figure 9.8).

axis such that the positive lobe on one orbital combines with the negative lobe on the other. This results in a molecular orbital with a nodal plane between the two nuclei and hence an antibonding orbital. We distinguish this orbital from the bonding σ_{2p_z} orbital by designating it as $\sigma^*_{2p_z}$.

Now let's consider combining the $2p$ orbitals perpendicular to the internuclear axis in Figure 9.6, and those directed along the y axis in particular. Figure 9.8 shows that these orbitals can be combined in two ways. In Figure 9.8a, they are combined such that their positive lobes overlap and their negative lobes overlap. As the figure shows, the resulting molecular orbital has a cross section along the internuclear axis that is similar to the p orbitals, and so is designated a **π orbital** (the Greek letter π corresponds to the letter p). Furthermore, the electron density is concentrated between the two nuclei and hence is a bonding orbital, which we denote by π_{2p_y}. Figure 9.8b shows that we can combine the $2p_y$ orbitals such that the positive lobe on one orbital overlaps the negative lobe on the other. This also leads to a π orbital, but one with a nodal plane between the nuclei and hence an antibonding orbital. We designate this molecular orbital by $\pi^*_{2p_y}$. We have considered only the $2p$ orbitals directed along the y axis in Figure 9.6. A combination of the $2p$ orbitals directed along the x axis in Figure

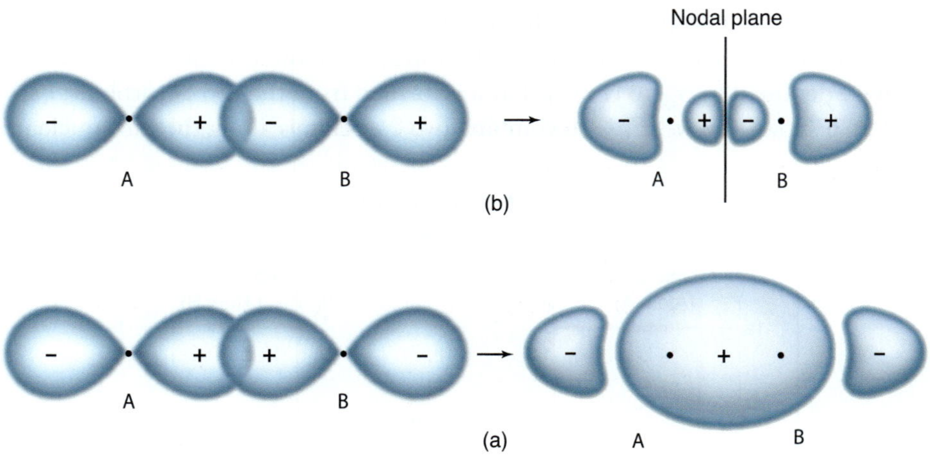

Figure 9.7 The combination of $2p$ orbitals that lie along the internuclear (z) axis. One lobe of a p orbital has a positive sign and the other a negative sign, as indicated in the figure. In (a), the $2p_z$ orbitals are combined such that the lobes having the same sign overlap, resulting in a bonding σ_{2p_z} orbital. In (b), they are combined such that lobes of different signs overlap, resulting in an antibonding $\sigma^*_{2p_z}$ orbital.

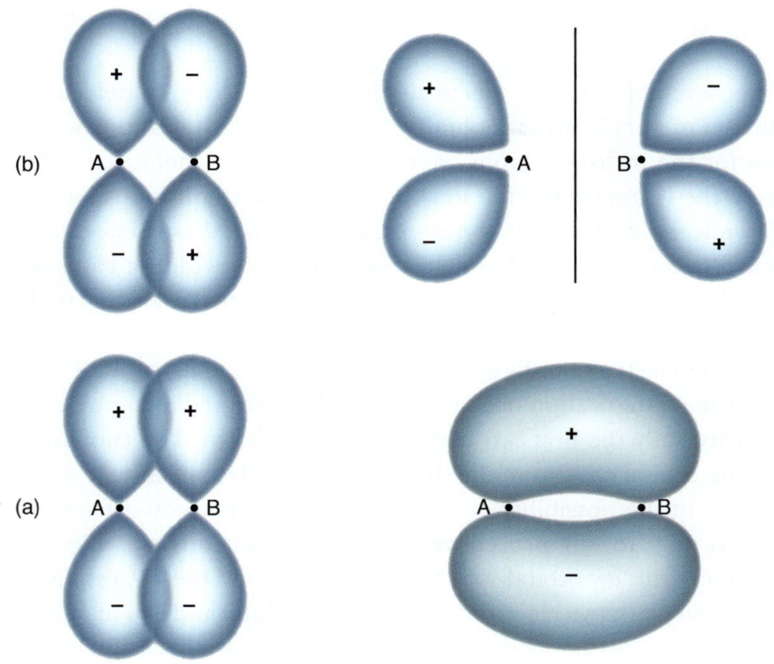

Figure 9.8 The combination of $2p$ orbitals that lie perpendicular to the internuclear axis. In (a), the p_y orbitals are combined such that lobes of the same sign overlap, resulting in a bonding π_{2p_y} orbital. In (b), the p_y orbitals are combined such that lobes of opposite sign overlap, resulting in an antibonding $\pi^*_{2p_y}$ orbital. There is a similar picture for the p_x orbitals.

9.6 yields a pair of molecular orbitals that are essentially the same as those obtained from the $2p$ orbitals directed along the y axis, only differing in orientation by 90 degrees. Thus, the $2p$ orbitals directed along the x and y axes in Figure 9.6 lead to a pair of π_{2p} orbitals (π_{2p_x} and π_{2p_y}) having the same energy and a pair of π^*_{2p} orbitals ($\pi^*_{2p_x}$ and $\pi^*_{2p_y}$) having the same energy, as shown in Figure 9.2. Each pair of molecular orbitals is, therefore, doubly degenerate. (Recall from Chapter 5 that two or more orbitals that have the same energy are said to be degenerate.) In the case of a H_2^+ ion, the energy of the doubly degenerate π_{2p} orbitals lies below that of the σ_{2p_z} molecular orbital, which in turn lies below that of the doubly degenerate π^*_{2p} molecular orbitals, and these lie below the $\sigma^*_{2p_z}$ molecular orbital as shown in Figure 9.2.

Just as a hydrogen atom has atomic orbitals of ever-increasing energy, a H_2^+ ion has molecular orbitals of higher energy than those presented here as well. However, we require only those shown in Figure 9.2 for our subsequent discussion of diatomic molecules from the second row of the periodic table. Again, keep in mind that all the molecular orbitals shown in Figure 9.2 come in bonding-antibonding pairs. Thus, we have both a σ_{1s} and a σ^*_{1s} orbital, a σ_{2s} and a σ^*_{2s} orbital, and so on.

Before we leave this section, we should comment on one thing. We have taken combinations of two $1s$ orbitals, two $2s$ orbitals, and various $2p$ orbitals to construct molecular orbitals. We never took a combination of a $1s$ and a $2s$ orbital, for example. The reason for this is that only orbitals of similar energies combine effectively. There is nothing wrong with taking combinations of orbitals of different energies; they just don't contribute significantly to the final result. Similarly, we did not consider combinations of orbitals such as $2p_x$ and $2p_z$ because they also do not combine effectively.

9-3. The Strength of a Covalent Bond Is Predicted by Its Bond Order

Just as we were able to write electron configurations for multielectron atoms using the set of atomic orbitals obtained from the solution of the Schrödinger equation for the hydrogen atom, we can use the set of molecular orbitals derived from the H_2^+ molecular ion to write electron configurations for multielectron diatomic molecules. As in the atomic case, each of these molecular orbitals is occupied by a maximum of two electrons, in accord with the Pauli exclusion principle. The hydrogen molecule, H_2, has two electrons. According to the Pauli exclusion principle, we place two electrons of opposite spins in the σ_{1s} orbital and write the electron configuration of a H_2 molecule as $(\sigma_{1s})^2$. The ground state electron configuration of a H_2 molecule is illustrated in Figure 9.9, where, for simplicity, only the first two energy levels are shown. The two electrons in the bonding orbital constitute the single bond of a H_2 molecule.

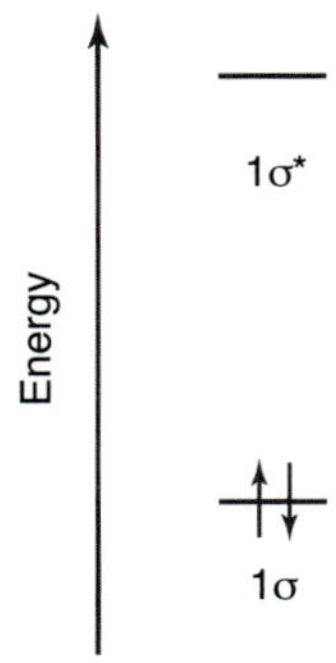

Figure 9.9 A pictorial representation of the electron configuration of a H_2 molecule. The two electrons occupy the molecular orbital corresponding to the lowest energy (σ_{1s}) and have opposite spins in accord with the Pauli exclusion principle.

What about the possibility of a He_2 molecule, which would have four electrons? According to our procedure, two of the electrons of a He_2 molecule should occupy the σ_{1s} orbital and two should occupy the σ^*_{1s} orbital (Figure 9.10). Thus, we place two electrons in a bonding orbital and two in an antibonding orbital. Electrons in a bonding orbital tend to draw the nuclei together, whereas electrons in an antibonding orbital tend to draw the nuclei away from each other. The two effects cancel, and so we predict there should be no net bonding, which is the case under normal conditions (Table 9.1).

To make our theory more quantitative, we can define a quantity called **bond order** by the equation

$$\text{bond order} = \frac{\begin{pmatrix}\text{number of}\\ \text{electrons in}\\ \text{bonding orbitals}\end{pmatrix} - \begin{pmatrix}\text{number of}\\ \text{electrons in}\\ \text{antibonding orbitals}\end{pmatrix}}{2} \tag{9.1}$$

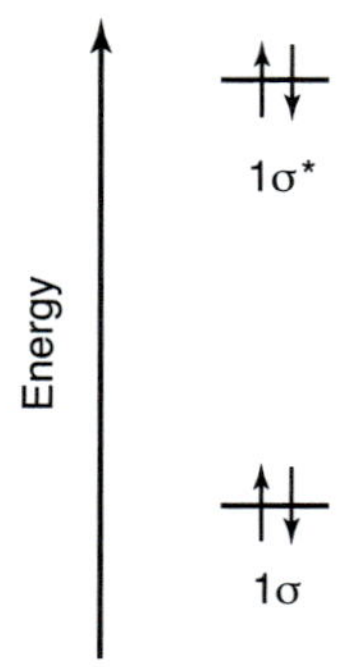

Figure 9.10 A pictorial representation of the electron configuration of a He_2 molecule. There are two electrons in a bonding orbital (σ_{1s}) and two in an antibonding orbital (σ^*_{1s}), so a He_2 molecule has no net bonding. The molecule He_2 is not observed under normal conditions.

A bond order of ½ indicates a one-electron bond (one-half of an electron pair); a bond order of 1 indicates a single bond (one pair of electrons); a bond order of 2 indicates a double bond (two pairs of electrons); and so on. Table 9.1 summarizes the properties of the molecular species, H_2^+, H_2, He_2^+, and He_2. The bond order of 0 for a He_2 molecule indicates that two helium atoms do not form a stable covalent bond under normal conditions. Note from Table 9.1 that bond lengths decrease and bond energies increase with increasing bond order. Thus, bond order is also a qualitative measure of bond strength.

TABLE 9.1 Molecular properties of H_2^+, H_2, He_2^+, and He_2

Species	Number of electrons	Ground state configuration	Bond order	Bond length/pm	Bond energy/aJ
H_2^+	1	$(\sigma_{1s})^1$	½	106	0.423
H_2	2	$(\sigma_{1s})^2$	1	74	0.724
He_2^+	3	$(\sigma_{1s})^2(\sigma^*_{1s})^1$	½	108	0.400
He_2	4	$(\sigma_{1s})^2(\sigma^*_{1s})^2$	0	not observed	not observed

9-4. Molecular Orbital Theory Predicts the Electron Configurations of Diatomic Molecules

When we wrote electron configurations for multielectron atoms in Chapter 5, we used the ordering of the energies of the atomic orbitals shown in Figures 5.11 and 5.12. Recall that the order of the energies of the orbitals for multielectron atoms does not follow the order of the energies of the hydrogen atomic orbitals. For instance, the energy of the 4*s* orbital is less than that of the 3*d* orbitals for multielectron atoms. A similar thing happens when we use the H_2^+ molecular orbitals that we have constructed from the combination of hydrogen atomic orbitals. The ordering of the molecular orbitals shown in Figure 9.2 can be used for the homonuclear diatomic molecules H_2 through N_2, that is, for $Z = 1$ through $Z = 7$ (recall from Chapter 2 that Z is the atomic number or number of protons in an atom). However, for $Z > 7$, the σ_{2p} and π_{2p} orbitals interchange energies so that the energy of the σ_{2p} orbital is less than that of the π_{2p} orbitals. The ordering of the energies of the molecular orbitals that we use to write electron configuration of the second-row homonuclear diatomic molecules Li_2 through Ne_2 is shown in Figure 9.11. Observe that the order of the σ_{2p} and π_{2p} orbitals changes in going from N_2 to O_2 molecules. Figure 9.11 is, in a sense, the homonuclear diatomic molecule analog of Figures 5.11 and 5.12 for multielectron atoms. We have already discussed the homonuclear molecules H_2 through He_2, so now we'll use Figure 9.11 to write electron configurations for the homonuclear diatomic molecules Li_2 through Ne_2 and use these electron configurations to discuss the bonding in these molecules.

Lithium vapor contains diatomic lithium molecules, Li_2. A lithium atom has three electrons, so a Li_2 molecule has a total of six electrons. In the ground state of a Li_2 molecule, the six electrons occupy the lowest molecular orbitals shown in Figure 9.11, in accord with the Pauli exclusion principle. The ground state electron configuration of a Li_2 molecule is $(\sigma_{1s})^2(\sigma^*_{1s})^2(\sigma_{2s})^2$. There are four bonding electrons and two antibonding electrons, so the bond order is 1

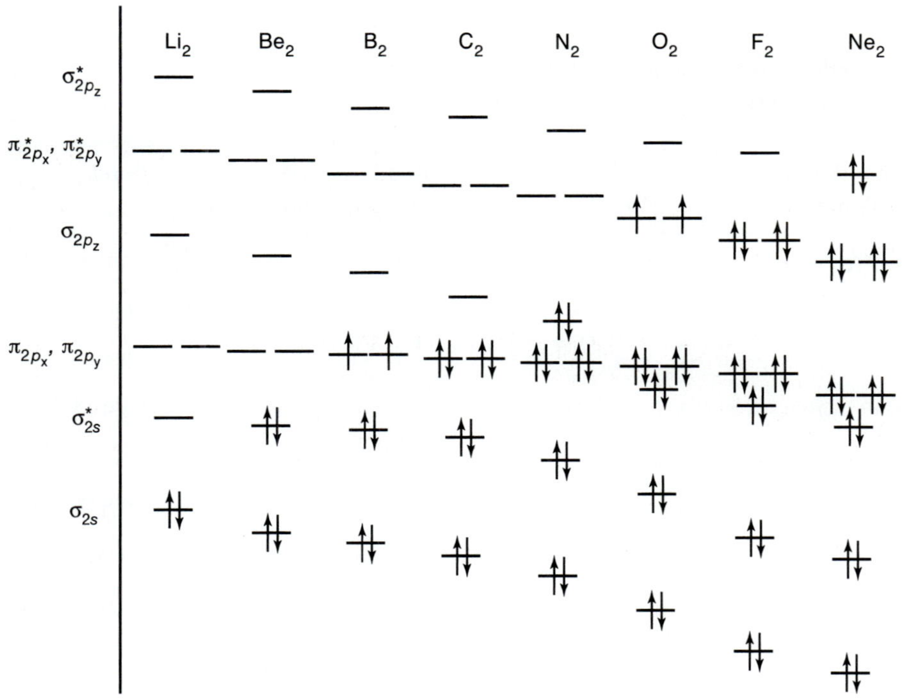

Figure 9.11 The relative energies (not to scale) of the molecular orbitals for the homonuclear diatomic molecules Li_2 through Ne_2. Notice that for O_2 through Ne_2 the energy of the σ_{2p_z} orbital is below that of the π_{2p} orbitals.

(Equation 9.1). Thus, we predict (correctly) that a Li_2 molecule is more stable than two separated lithium atoms. Table 9.2 shows that a Li_2 molecule has a bond length of 267 pm and a bond energy of 0.174 aJ.

We write π_{2p} and π^*_{2p} for convenience instead of π_{2p_x}, π_{2p_y} and $\pi^*_{2p_x}$, $\pi^*_{2p_y}$. We use the full notation only when we need to denote specific electron configurations.

EXAMPLE 9-1: Use Figure 9.11 to write the ground-state electron configuration of a N_2 molecule. Calculate the bond order of a N_2 molecule and compare your result with its Lewis formula.

Solution: There are 14 electrons in a N_2 molecule. Using Figure 9.11, we see that its ground state electron configuration is $(\sigma_{1s})^2(\sigma^*_{1s})^2(\sigma_{2s})^2(\sigma^*_{2s})^2(\pi_{2p})^4(\sigma_{2p})^2$. According to Equation 9.1, the bond order in a N_2 molecule is

$$\text{bond order} = \frac{10-4}{2} = 3$$

The Lewis formula for a N_2 molecule, :N≡N:, is thus in agreement with molecular orbital theory. The triple bond in a N_2 molecule accounts for its short bond length (110 pm) and its unusually large bond energy (1.57 aJ). The bond in a N_2 molecule is one of the strongest known bonds.

PRACTICE PROBLEM 9-1: Use molecular orbital theory to explain why neon does not form a stable diatomic molecule under normal conditions.

Answer: Using Figure 9.13, we see that the ground state electron configuration of a Ne_2 molecule is $(\sigma_{1s})^2(\sigma^*_{1s})^2(\sigma_{2s})^2(\sigma^*_{2s})^2(\sigma_{2p})^2(\pi_{2p})^4(\pi^*_{2p})^4(\sigma^*_{2p})^2$, giving a bond order of (10 – 10)/2 = 0. Like a He_2 molecule, a Ne_2 molecule does not exist under normal conditions.

One of the most impressive aspects of molecular orbital theory is its ability to predict that oxygen molecules are **paramagnetic**. This property means that

TABLE 9.2 Properties of the homonuclear diatomic molecules of the second-row elements

Species	Ground state configuration	Bond order	Bond length/pm	Bond energy/aJ
Li_2	$(\sigma_{1s})^2(\sigma^*_{1s})^2(\sigma_{2s})^2$	1	267	0.174
Be_2	$(\sigma_{1s})^2(\sigma^*_{1s})^2(\sigma_{2s})^2(\sigma^*_{2s})^2$	0	245	≈0.01
B_2	$(\sigma_{1s})^2(\sigma^*_{1s})^2(\sigma_{2s})^2(\sigma^*_{2s})^2(\pi_{2p_x})^1(\pi_{2p_y})^1$	1	159	0.493
C_2	$(\sigma_{1s})^2(\sigma^*_{1s})^2(\sigma_{2s})^2(\sigma^*_{2s})^2(\pi_{2p})^4$	2	124	1.01
N_2	$(\sigma_{1s})^2(\sigma^*_{1s})^2(\sigma_{2s})^2(\sigma^*_{2s})^2(\pi_{2p})^4(\sigma_{2p})^2$	3	110	1.57
O_2	$(\sigma_{1s})^2(\sigma^*_{1s})^2(\sigma_{2s})^2(\sigma^*_{2s})^2(\sigma_{2p})^2(\pi_{2p})^4(\pi^*_{2p_x})^1(\pi^*_{2p_y})^1$	2	121	0.827
F_2	$(\sigma_{1s})^2(\sigma^*_{1s})^2(\sigma_{2s})^2(\sigma^*_{2s})^2(\sigma_{2p})^2(\pi_{2p})^4(\pi^*_{2p})^4$	1	141	0.264
Ne_2	$(\sigma_{1s})^2(\sigma^*_{1s})^2(\sigma_{2s})^2(\sigma^*_{2s})^2(\sigma_{2p})^2(\pi_{2p})^4(\pi^*_{2p})^4(\sigma^*_{2p})^2$	0	not observed	not observed

oxygen is weakly attracted to a region between the poles of a magnet (Figure 9.12). Most substances are **diamagnetic**, meaning that they are slightly repelled by a magnetic field. Let's see how the paramagnetism of $O_2(g)$ is related to its electron structure.

Each oxygen atom has eight electrons; thus, an O_2 molecule has a total of 16 electrons. When the 16 electrons are placed according to the molecular orbital diagram given in Figure 9.11, the last two go into the π^*_{2p} orbitals. As in the atomic case, we apply Hund's rule (Section 5.8). Because the two π^*_{2p} orbitals have the same energy, we place one electron in each π^*_{2p} orbital such that the two electrons have unpaired spins, as shown in Figure 9.11. The ground state electron configuration of an O_2 molecule is $(\sigma_{1s})^2(\sigma^*_{1s})^2(\sigma_{2s})^2(\sigma^*_{2s})^2(\sigma_{2p})^2(\pi_{2p})^4(\pi^*_{2px})^1(\pi^*_{2py})^1$. According to Hund's rule, each π^*_{2p} orbital is occupied by one electron and the spins are unpaired. Therefore, an oxygen molecule has a net electron spin and so acts like a tiny magnet. Thus, $O_2(g)$ is attracted into a region between the poles of a magnet (Figure 9.12).

The amount of oxygen in air can be monitored by measuring its paramagnetism. Because oxygen is the only major component of air that is paramagnetic, the measured paramagnetism of air is directly proportional to the amount of oxygen present. Linus Pauling (Frontispiece) developed a method using the paramagnetism of oxygen to monitor oxygen levels in submarines and airplanes in World War II. A similar method is still used by physicians to monitor the oxygen content in blood during anesthesia.

The Lewis formula of an O_2 molecule does not account for the paramagnetism of $O_2(g)$. According to the octet rule, we should write the Lewis formula of an O_2 molecule as :Ö=Ö: , but this formula implies incorrectly that all the electrons are paired. The oxygen molecule is an exception to the utility of Lewis formulas, whereas the more fundamental molecular orbital theory is able to account successfully for the distribution of the electrons in molecular oxygen.

Table 9.2 gives the ground state electron configurations of the homonuclear diatomic molecules Li_2 through Ne_2.

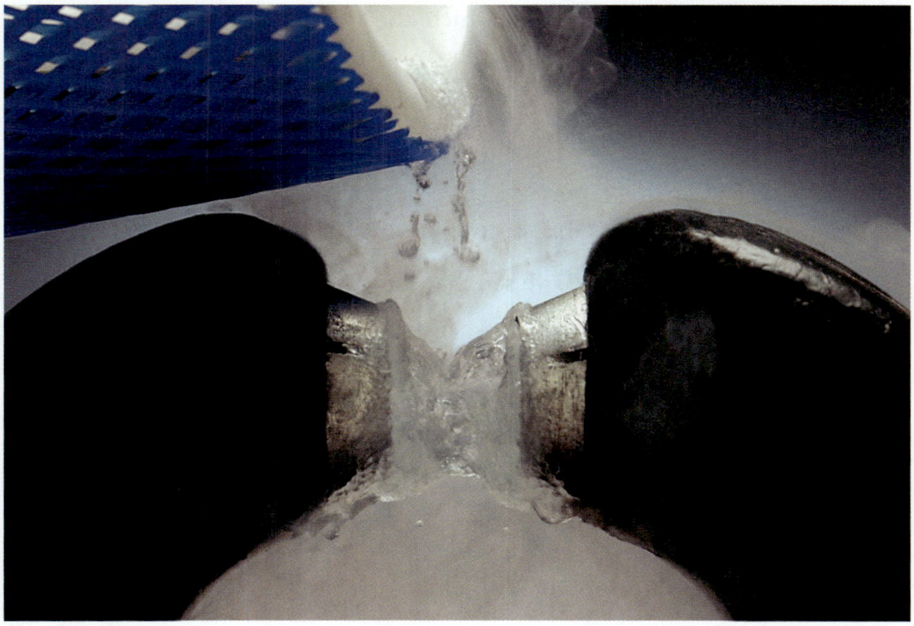

Figure 9.12 Liquid oxygen is attracted to the magnetic field between the poles of a magnet because oxygen is paramagnetic. The attraction of paramagnetic materials to a magnetic field is much weaker than that of ferromagnetic materials (such as iron), and so a strong magnet is used to illustrate this effect.

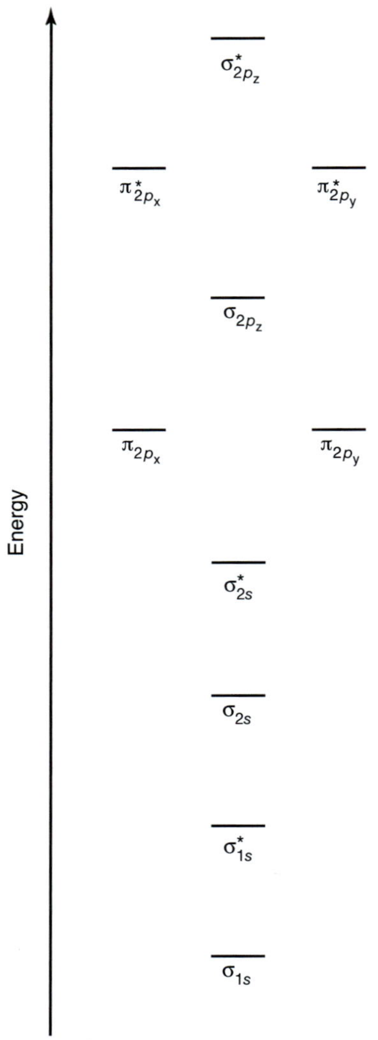

Figure 9.13 The molecular-orbital energy levels of a heteronuclear diatomic molecule for which the atomic numbers of the two atoms in the molecule differ by only one or two units. The order of the orbitals is the same as that of the molecules H_2 through N_2 in Figure 9.11.

EXAMPLE 9-2: Use Figure 9.11 to determine which species has the greater bond length, F_2 or F_2^-.

Solution: The ground state electron configurations of the two species are

$$F_2\text{: } (\sigma_{1s})^2(\sigma^*_{1s})^2(\sigma_{2s})^2(\sigma^*_{2s})^2\,(\sigma_{2p})^2\,(\pi_{2p})^4(\pi^*_{2p})^4$$

$$F_2^-\text{: } (\sigma_{1s})^2(\sigma^*_{1s})^2(\sigma_{2s})^2(\sigma^*_{2s})^2\,(\sigma_{2p})^2\,(\pi_{2p})^4(\pi^*_{2p})^4(\sigma^*_{2p})^1$$

The bond orders are

$$\text{bond order } F_2 = \frac{10-8}{2} = 1$$

$$\text{bond order } F_2^- = \frac{10-9}{2} = \frac{1}{2}$$

Thus, we predict that F_2^- has a longer bond length than F_2.

PRACTICE PROBLEM 9-2: A ground state F_2 molecule may be promoted into an excited state using a laser. Calculate the bond order of an excited F_2 molecule with the electron configuration $(\sigma_{1s})^2(\sigma^*_{1s})^2(\sigma_{2s})^2(\sigma^*_{2s})^2\,(\sigma_{2p})^1\,(\pi_{2p})^4$ $(\pi^*_{2p})^4\,(\sigma^*_{2p})^1$. Comment on the stability of this excited state.

Answer: The bond order of this excited state is 0, making it unstable. In most cases the excited molecule will simply reemit the absorbed light, returning to its ground state. However, a small fraction of these unstable molecules will dissociate (or fall apart). Because a single laser pulse can contain over 10^{25} photons, laser light may be used to dissociate large numbers of molecules into their component atoms. This process is called **photodissociation**.

Molecular orbital theory can also be applied to **heteronuclear diatomic molecules** (diatomic molecules in which the two atoms are different). The energy-level scheme in Figure 9.13 can be used if the atomic numbers of the two atoms in the molecule differ by only one or two atomic numbers.

EXAMPLE 9-3: Using Figure 9.13, predict which of the following species you would expect to have the shortest bond length, CO^+, CO, or CO^-?

Solution: Using Figure 9.13, we see that the ground state electron configurations of these three species are

$$CO^+\text{: } (\sigma_{1s})^2(\sigma^*_{1s})^2(\sigma_{2s})^2(\sigma^*_{2s})^2(\pi_{2p})^4(\sigma_{2p})^1$$

$$CO\text{: } (\sigma_{1s})^2(\sigma^*_{1s})^2(\sigma_{2s})^2(\sigma^*_{2s})^2(\pi_{2p})^4(\sigma_{2p})^2$$

$$CO^-\text{: } (\sigma_{1s})^2(\sigma^*_{1s})^2(\sigma_{2s})^2(\sigma^*_{2s})^2(\pi_{2p})^4(\sigma_{2p})^2\,(\pi^*_{2p})^1$$

with bond orders of 2, 3, and 2, respectively. Thus, we predict that a CO molecule has the shortest bond, which is correct.

PRACTICE PROBLEM 9-3: Using Figure 9.13, predict which of the following species has the largest bond energy, CN^+, CN, or CN^-?

Answer: CN^-

9-5. The Bonding in Polyatomic Molecules Can Be Described Using Localized Bonds

Molecular orbital theory not only applies to diatomic molecules, it can be applied to **polyatomic molecules** as well. In doing so we construct molecular orbitals this time by combining atomic orbitals from all the atoms in the molecule. Then, using the corresponding energy-level diagram, we place electrons into the molecular orbitals in accord with the Pauli exclusion principle. Because molecular orbitals are combinations of atomic orbitals from all the atoms in a polyatomic molecule, they are often spread over the entire molecule. Rather than use molecular orbital theory as we have described it for diatomic molecules, however, we shall use a simplified approach which recognizes that many chemical bonds have properties such as bond lengths, bond angles, and bond energies that are fairly constant from molecule to molecule. For example, the carbon-hydrogen bond lengths in many molecules are about 110 pm and their bond energies are a little greater than 0.7 aJ. A large amount of experimental data such as these suggests that the bonding in most polyatomic molecules can be analyzed in terms of orbitals that are localized between pairs of bonded atoms.

Consider the methane molecule, CH_4, in which each hydrogen atom is joined to the central carbon atom by a covalent bond. As Figure 9.14 suggests, the bonding electrons, and hence the orbitals that describe them, are localized along the line joining each hydrogen atom to the central carbon atom. These electrons are said to occupy **localized bond orbitals**, and each pair of electrons that occupies a localized bond orbital is said to constitute a **localized covalent bond**. Note the similarity between this bonding picture in a CH_4 molecule and its Lewis formula.

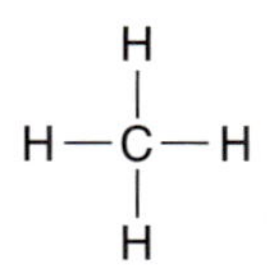

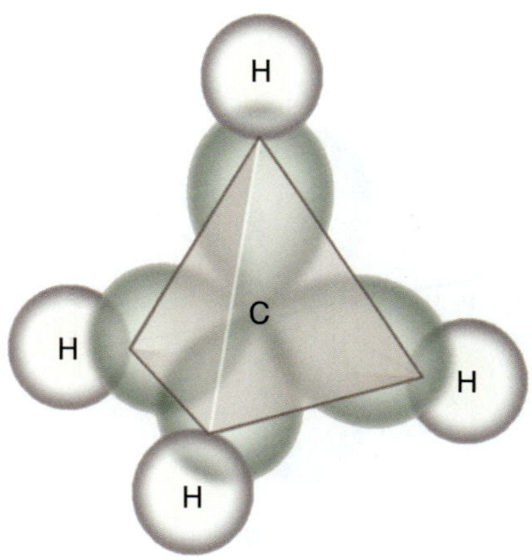

Figure 9.14 The bond orbitals in a methane molecule can be pictured as four carbon-hydrogen bond orbitals, directed toward the vertices of a tetrahedron. A localized bond orbital that is occupied by two electrons of opposite spin constitutes a covalent bond localized between two atoms.

The simplest neutral polyatomic molecule is BeH_2. Beryllium hydride is an electron-deficient compound. Its Lewis formula, H–Be–H, does not satisfy the octet rule. According to VSEPR theory, beryllium hydride is a symmetric linear molecule; the two Be–H bonds are 180° apart and are equivalent. Therefore, we must form two equivalent bond orbitals that are localized along the molecular H–Be–H axis.

In order to describe the bonding in a BeH_2 molecule and many other polyatomic molecules, we introduce the idea of a hybrid orbital. We define a **hybrid orbital** as a combination of atomic orbitals *of the same atom*. In the case of BeH_2, we consider combinations of the valence atomic orbitals on a beryllium atom, namely, the 2*s* and the 2*p* orbitals (Figure 9.15). Because a BeH_2 molecule has two equivalent bonds 180° apart, we wish to construct two hybrid orbitals of the same symmetry. It turns out that it is possible to combine the $2s$ and *one* of the $2p$ orbitals to produce such orbitals by adding and subtracting the two atomic orbitals as we did when forming molecular orbitals.

Figure 9.16 illustrates the formation of two hybrid orbitals from a $2s$ and a $2p$ orbital. As we see from Figure 9.16, there is a buildup of the hybrid orbital where the values of the combining $2s$ and $2p$ orbitals have the same signs and partial cancellation where they have opposite signs. Figure 9.16 shows that one of our hybrid orbitals is directed toward the right (as drawn) and that the other has the same shape but is directed toward the left (as drawn).

These two hybrid orbitals on the beryllium atom are called ***sp* orbitals** because they are formed from the $2s$ orbital and one of the $2p$ orbitals. The *sp* orbitals have two important features: (1) each one provides a large region to combine with a hydrogen $1s$ orbital (the positive region in Figure 9.16), and (2) they are 180° apart. The two empty $2p$ orbitals that are not used to form the *sp* hybrid orbitals are perpendicular to each other and perpendicular to the line formed by the *sp* hybrid orbitals (Figure 9.17).

We now form two covalent bond orbitals by combining each *sp* hybrid orbital with a hydrogen $1s$ atomic orbital. As shown in Figure 9.18, these are σ-bond orbitals. Each is localized between the beryllium atom and a hydrogen atom, and they are 180° apart. Recall from Section 9.2 that the H_2^+ molecular orbitals occur as bonding and antibonding pairs. Similarly, when each beryllium *sp* orbital overlaps with a hydrogen $1s$ orbital, a σ-bonding orbital and a σ*-antibonding orbital results. The two σ*-antibonding orbitals have much higher energies than the σ-bonding orbitals, so they are not used in the bonding of a BeH_2 molecule in the ground state. Thus, they are not indicated in Figure 9.18. If we consider only molecules in their ground states, we can ignore the higher-

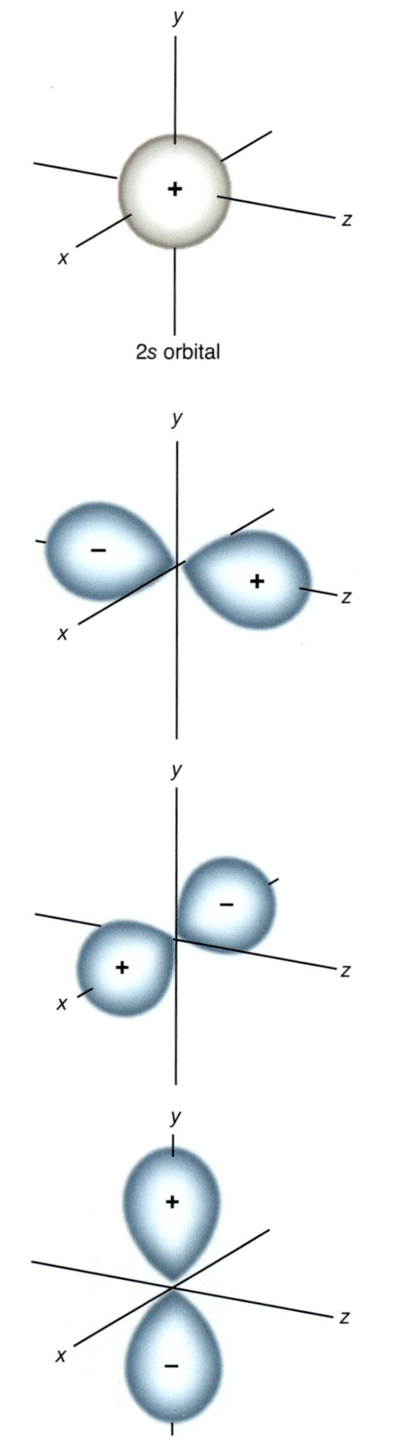

Figure 9.15 The $2s$ and $2p$ orbitals, showing the regions in which they have positive values and regions in which they have negative values. The $2s$ orbital has a positive value in its outer region, where it combines with other orbitals. In contrast, the $2p$ orbitals have a positive value in one lobe and a negative value in the other.

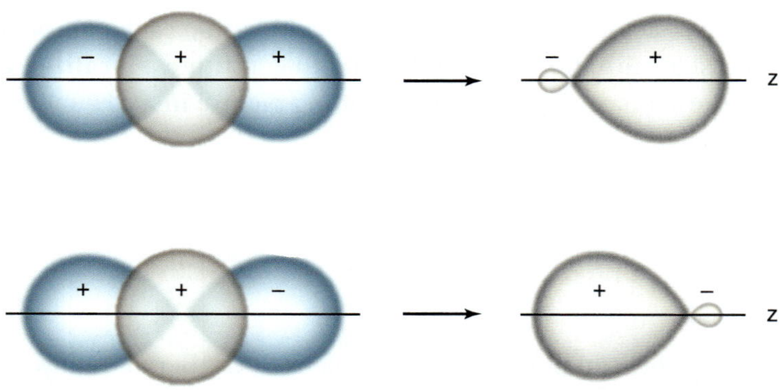

Figure 9.16 The formation of two *sp* hybrid orbitals from a $2s$ and a $2p$ atomic orbital oriented along the bonding axis. The two resulting *sp* orbitals are 180° apart. Addition and subtraction of the $2s$ and $2p$ atomic orbitals may be illustrated by reversing the signs on the $2p$ orbital, as shown here. The $2s$ and $2p$ orbitals reinforce each other in regions where they have values with the same sign and partially cancel each other in regions where they have values of opposite sign. Consequently, each *sp* orbital consists of a large lobe of positive value and a small lobe of negative value. For simplicity, we often shall omit the little lobes of negative value.

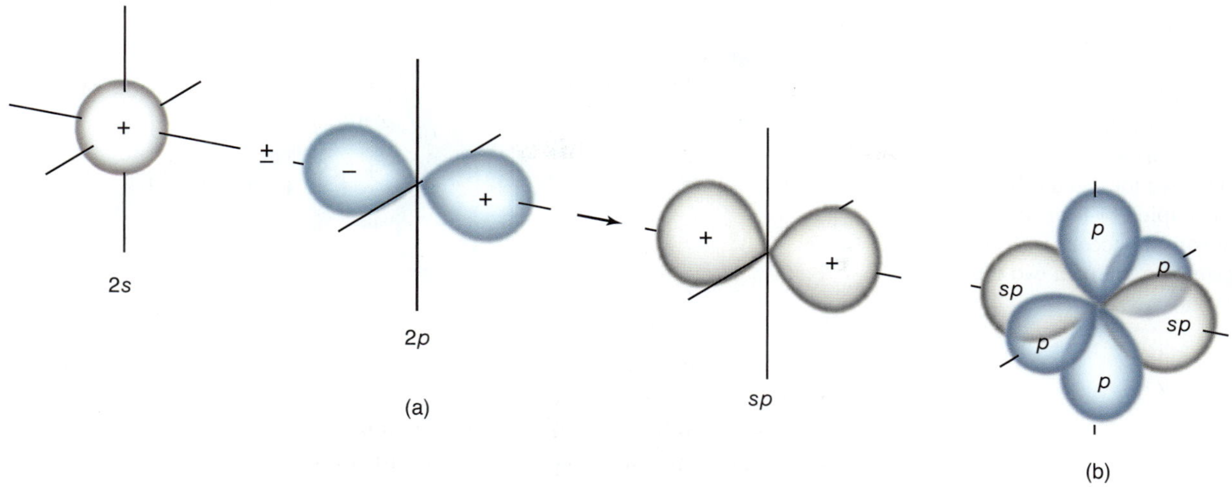

Figure 9.17 The formation of sp hybrid orbitals results from combining the $2s$ orbital and one $2p$ orbital on a single atom. The two sp orbitals are equivalent and are 180° apart. In (a), for simplicity, only the $2p$ orbital that is combined with the $2s$ orbital is shown. In (b), all the orbitals are shown. The two $2p$ orbitals that are not combined with the $2s$ orbital are perpendicular to each other and to the line formed by the sp hybrid orbitals. The little lobes of negative sign cannot be seen in this figure.

energy antibonding orbitals that result when orbitals on different atoms are combined in this way.

When a σ orbital is occupied by two electrons of opposite spins, the result is a **σ bond**. A BeH_2 molecule has 2 + (2 × 1) = 4 valence electrons. We can describe the bonding in a BeH_2 molecule by placing these four electrons into two $Be(sp) + H(1s)$ σ bonds. Each σ-bond orbital is occupied by two electrons, so that the bond order in each σ bond in a BeH_2 molecule is 1.

EXAMPLE 9-4: Describe the localized bond orbitals in a BeF_2 molecule, which is a linear molecule.

Solution: Because BeF_2 is a linear molecule (VSEPR theory), it is appropriate to use sp hybrid orbitals on the beryllium atom to form localized bonding orbitals with the fluorine atoms. Thus, the central beryllium atom in a BeF_2 molecule can be depicted as in Figure 9.19.

The ground state electron configuration of a fluorine atom is $1s^22s^22p_x^2 2p_y^22p_z^1$. The only orbital on the fluorine atom that can be occupied by another electron is the $2p_z$ orbital. The $2p_z$ orbitals on each fluorine atom can combine with an sp orbital on the beryllium atom to form two localized σ-bond orbitals, as shown in Figure 9.19. The beryllium sp orbitals and the fluorine $2p_z$ orbitals are collinear to maximize their overlap. Each bond orbital is occupied by two electrons of opposite spin, thereby forming two localized $Be(sp) + F(2p_z)$ σ bonds of bond order 1.

Figure 9.18 The formation of the two equivalent localized σ-bond orbitals in a beryllium hydride, BeH_2, molecule. Each bond orbital is formed by the combination of a beryllium sp hybrid orbital and a hydrogen $1s$ atomic orbital. There are four valence electrons in a BeH_2 molecule: two from the beryllium atom and one from each of the two hydrogen atoms. The four valence electrons occupy the two localized σ-bond orbitals, forming the two localized beryllium-hydrogen bonds in a BeH_2 molecule.

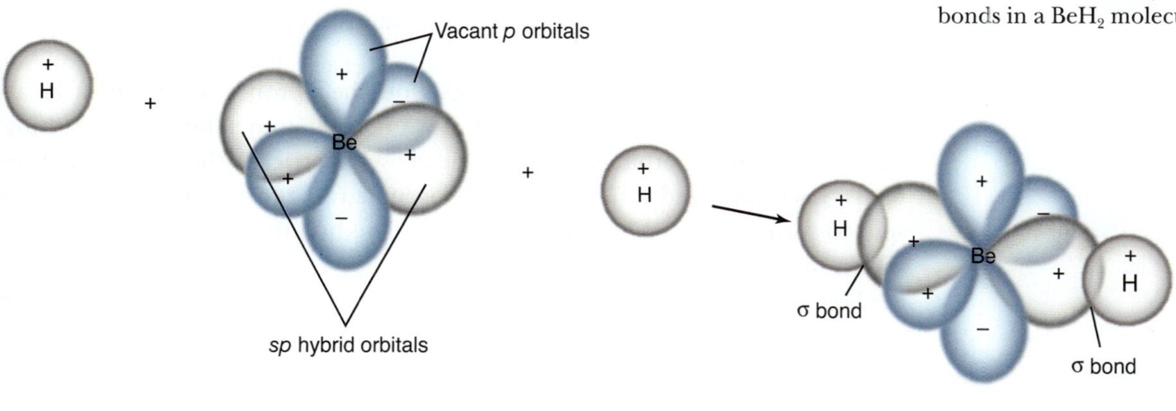

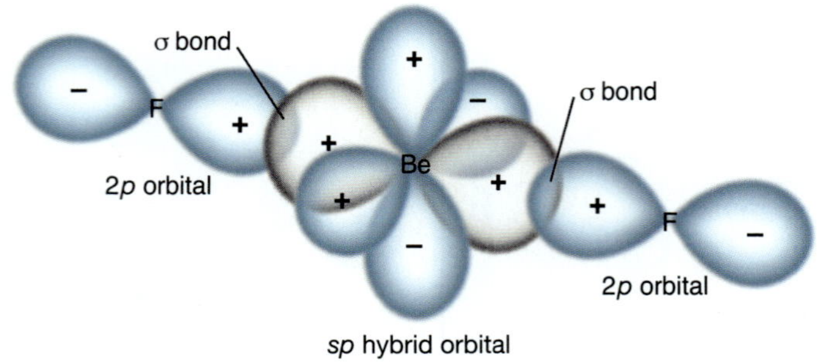

Figure 9.19 An illustration of the bonding in a beryllium fluoride, BeF_2, molecule. Each of the beryllium–fluorine bond orbitals is formed by the combination of a beryllium *sp* orbital and a fluorine $2p$ orbital. The two localized σ-bond orbitals are occupied by four of the valence electrons, in accord with the Pauli exclusion principle, to form two σ bonds.

PRACTICE PROBLEM 9-4: Although zinc chloride is an ionic solid at room temperature, it exists as molecules in the vapor phase. Describe the localized bond orbitals in a covalently bonded $ZnCl_2$ molecule. (Take the number of valence electrons in the zinc atom to be two. We shall learn the rules for assigning the number of valence electrons for transition metal elements in covalent molecules in Chapter 26.)

Answer: The Lewis formula of a $ZnCl_2$ molecule is :Ċ̤l–Zn–Ċ̤l: . According to VSEPR theory, $ZnCl_2$ is a linear molecule. Consequently, it is appropriate to use *sp* hybrid orbitals on the zinc atom to describe the bonding of a $ZnCl_2$ molecule as two $\text{Zn}(sp) + \text{Cl}(2p_z)$ σ bonds of bond order 1.

9-6. sp^2 Hybrid Orbitals Have Trigonal Planar Symmetry

An example of a molecule with three equivalent covalent bonds is BF_3. Boron trifluoride is an electron-deficient compound whose Lewis formula is

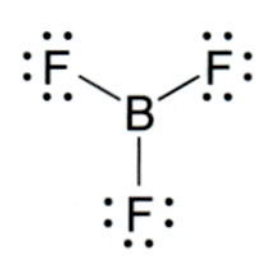

As we saw in Chapter 8, a BF_3 molecule has a trigonal planar geometry, and each F–B–F bond angle is 120°. In order to describe the bonding in a BF_3 molecule using localized bonds, we must construct three equivalent hybrid orbitals on the boron atom that lie in a plane and are directed 120° apart. The valence-shell orbitals on a boron atom are $2s$, $2p_x$, $2p_y$, and $2p_z$. It turns out that if we form a combination of the $2s$ orbital and *two* of the $2p$ orbitals, we get three equivalent hybrid orbitals that lie in a plane 120° apart (Figure 9.20). Because the hybrid orbitals are formed from the $2s$ and two of the $2p$ orbitals, they are called ***sp*² orbitals**.

Each of the three sp^2 hybrid orbitals on the boron atom forms a localized bond orbital with the $2p$ orbital of a fluorine atom. The three resulting $\text{B}(sp^2) + \text{F}(2p)$ bond orbitals are cylindrically symmetric, so they are σ-bond orbitals. Each localized σ-bond orbital is occupied by two electrons of opposite spin to form a σ bond of bond order of 1. The bonding in a BF_3 molecule is shown in Figure 9.21.

From these examples, we see that when we combine a $2s$ orbital with one $2p$

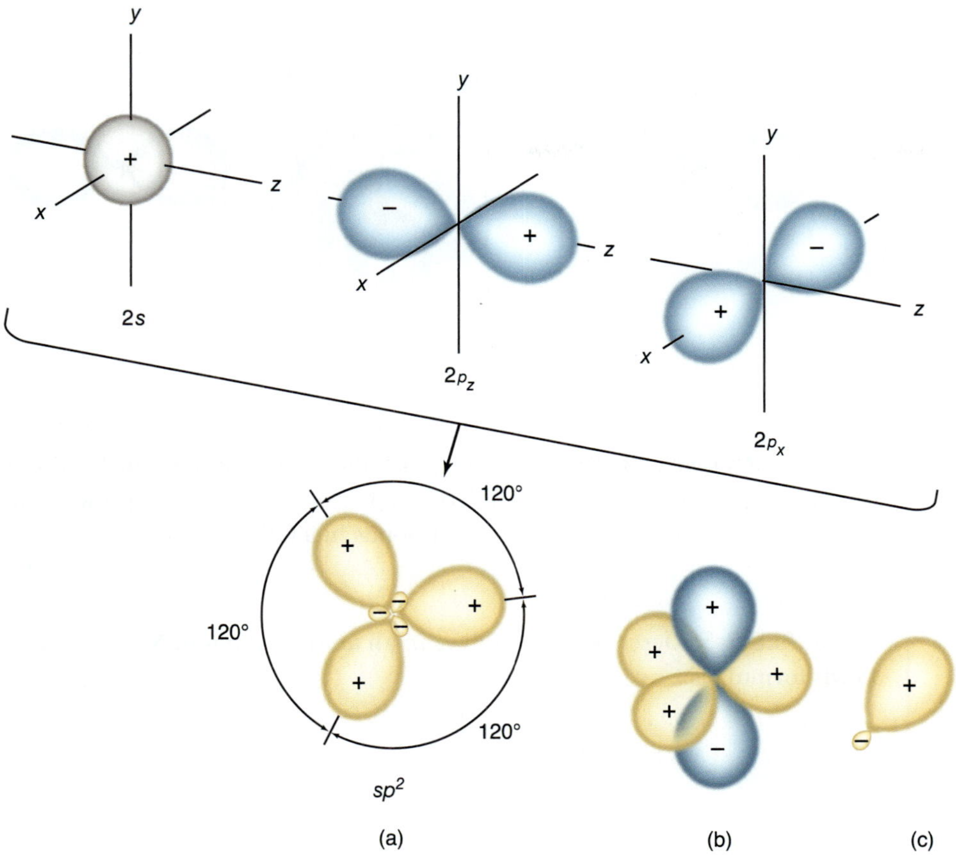

Figure 9.20 The formation of sp^2 hybrid orbitals by combining the $2s$ orbital and two $2p$ orbitals on a single atom. (a) The three sp^2 orbitals formed are equivalent, lie in a plane, and are 120° apart. For simplicity, only the two $2p$ orbitals that are combined with the $2s$ orbital are shown. (b) The $2p$ orbital that is not combined with the $2s$ orbital is perpendicular to the plane formed by the three sp^2 orbitals. The relatively small negative value regions of the sp^2 hybrid orbitals are omitted. (c) A single complete sp^2 hybrid orbital.

orbital on the same atom, we get two *sp* hybrid orbitals; and when we combine a $2s$ orbital with two $2p$ orbitals on the same atom, we get three sp^2 hybrid orbitals. These two results are an example of the **principle of conservation of orbitals**: If we combine atomic orbitals *on the same atom* to form hybrid orbitals, then the number of resulting hybrid orbitals is equal to the number of atomic orbitals combined.

9-7. sp^3 Hybrid Orbitals Point Toward the Vertices of a Tetrahedron

As predicted by VSEPR theory, methane, CH_4, is a tetrahedral molecule, and its four carbon-hydrogen bonds are equivalent. Thus, in order to describe the bonding in a methane molecule using localized bonds, we must construct four equivalent bond orbitals on the central carbon atom. If we combine the $2s$ orbital and all three $2p$ orbitals on the carbon atom, we get four equivalent hybrid orbitals, each pointing to a vertex of a tetrahedron (Figure 9.22). Because the four equivalent hybrid orbitals result from combining the $2s$ and all three $2p$ orbitals on the carbon atom, they are called $\boldsymbol{sp^3}$ **orbitals**.

The four equivalent localized σ-bond orbitals in a CH_4 molecule are formed by combining each sp^3 orbital with a hydrogen $1s$ orbital (Figure 9.23). (We are ignoring the four σ^*-antibonding orbitals that are also formed.) There are

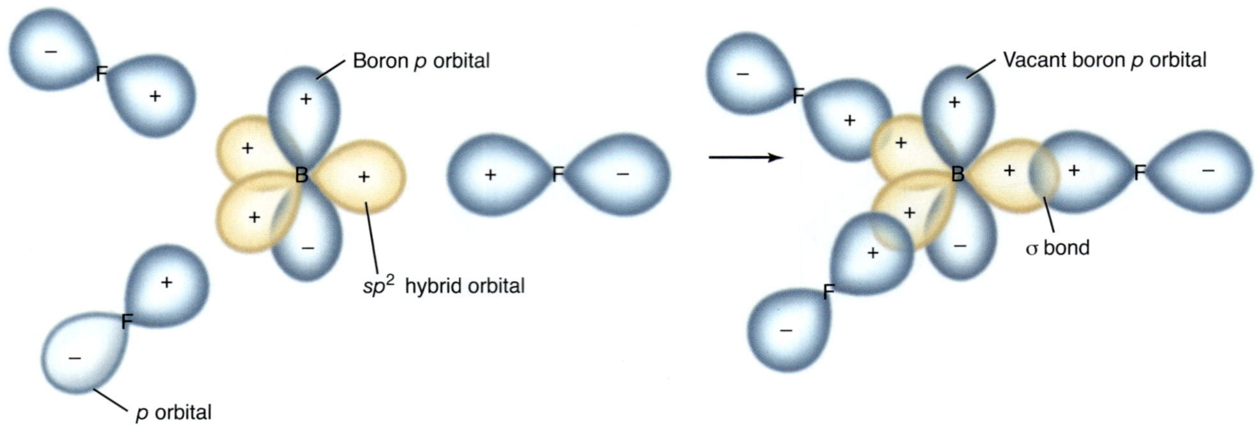

Figure 9.21 A schematic illustration of the bonding in a boron fluoride, BF_3, molecule. Each of the three boron–fluorine σ-bond orbitals is formed by the combination of a boron sp^2 orbital and a fluorine $2p$ orbital. The three localized boron–fluorine bond orbitals $B(sp^2) + F(2p)$ are occupied by six of the molecule's valence electrons and constitute the three covalent boron-fluorine σ bonds of bond order 1.

$4 + (4 \times 1) = 8$ valence electrons in a CH_4 molecule. Each of the four localized $C(sp^3) + H(1s)$ σ-bond orbitals is occupied by a pair of electrons of opposite spin, accounting for the four localized covalent bonds in a CH_4 molecule.

EXAMPLE 9-5: Describe the bonding in an ammonium ion, NH_4^+, whose Lewis formula is

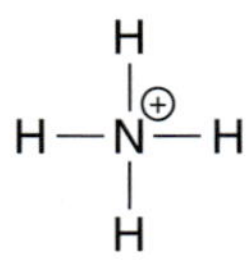

Solution: We learned in Chapter 8 that an NH_4^+ ion has a tetrahedral geometry. Thus, we wish to form four localized bond orbitals that point toward the vertices of a tetrahedron. To do so we form sp^3 hybrid orbitals on the nitrogen atom by combining its $2s$ orbital and all three of its $2p$ orbitals.

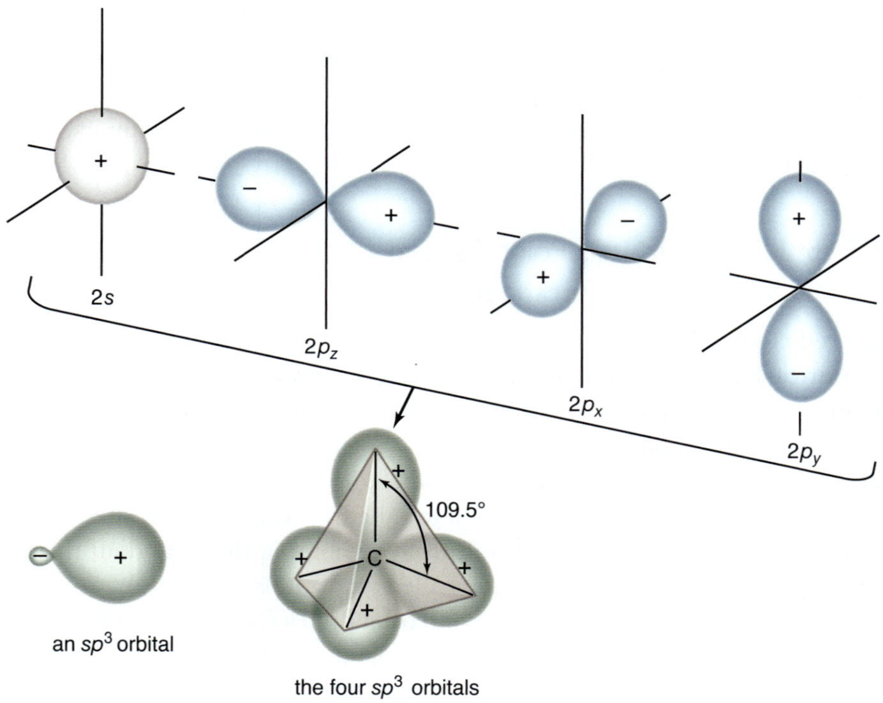

Figure 9.22 The $2s$ and three $2p$ orbitals on an atom can be combined to form four sp^3 hybrid orbitals that are all equivalent and point toward the vertices of a tetrahedron. The angle between sp^3 orbitals is the tetrahedral bond angle, 109.5°. The small negative value regions of the sp^3 orbitals are omitted from the tetrahedral figure to simplify the figure. A single sp^3 orbital is shown to the left of the set of four sp^3 orbitals.

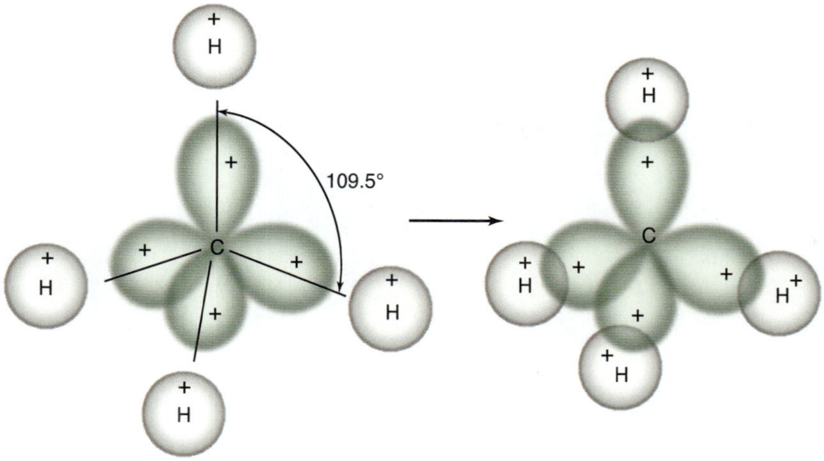

Figure 9.23 Four equivalent localized σ-bond orbitals in a methane, CH_4, molecule are formed by combining each of the four carbon sp^3 orbitals with a hydrogen 1*s* orbital. There are eight valence electrons in a CH_4 molecule (four from the carbon atom and one from each hydrogen atom). Each of the four localized bond orbitals is occupied by a pair of electrons of opposite spin, accounting for the four localized carbon–hydrogen σ bonds in a methane molecule.

The resulting sp^3 hybrid orbitals are similar to the carbon atom sp^3 orbitals shown in Figure 9.22. We next form four equivalent localized bond orbitals by combining each sp^3 orbital on the nitrogen atom with a hydrogen 1*s* orbital. There are $5 + (4 \times 1) - 1 = 8$ valence electrons in an NH_4^+ ion (because the overall charge on the ammonium ion is +1, we have one less electron than in the neutral species). Two valence electrons of opposite spin occupy each of the four σ-bond orbitals, thereby forming the four $N(sp^3) + H(1s)$ covalent bonds in an NH_4^+ ion. The bonding and the shape of an ammonium ion are similar to that shown for a methane molecule in Figure 9.23.

PRACTICE PROBLEM 9-5: Use localized bond orbitals to describe the bonding in a tetrafluoroborate ion, BF_4^-.

Answer: A BF_4^- ion has tetrahedral geometry, so it is appropriate to form sp^3 hybrid orbitals on the boron atom and bond these to each of the half-filled 2*p* orbitals on the fluorine atoms. Thus, we combine each of the sp^3 orbitals on the boron atom with a 2*p* orbital on the fluorine atom to form four localized $B(sp^3) + F(2p)$ σ-bond orbitals. The eight valence electrons in a BF_4^- ion occupy these four σ-bond orbitals in accord with the Pauli exclusion principle.

We can also use sp^3 orbitals to describe the bonding in molecules that have no single central atom. An example is an ethane molecule, C_2H_6, whose Lewis formula is

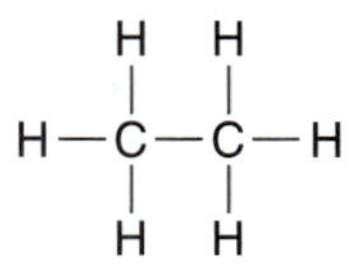

Figure 9.24 shows ball-and-stick and space-filling models of ethane. The disposition of the bonds about each carbon atom is tetrahedral, so it is appropri-

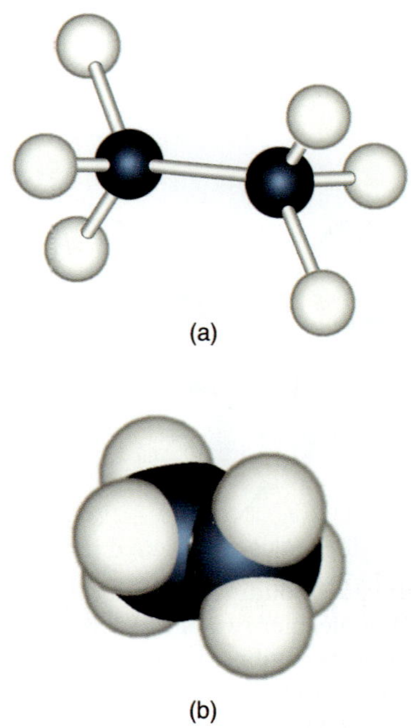

Figure 9.24 Molecular models of an ethane, C_2H_6, molecule. (a) Ball-and-stick model. (b) Space-filling model. Notice that the bonds about each carbon atom are tetrahedrally oriented.

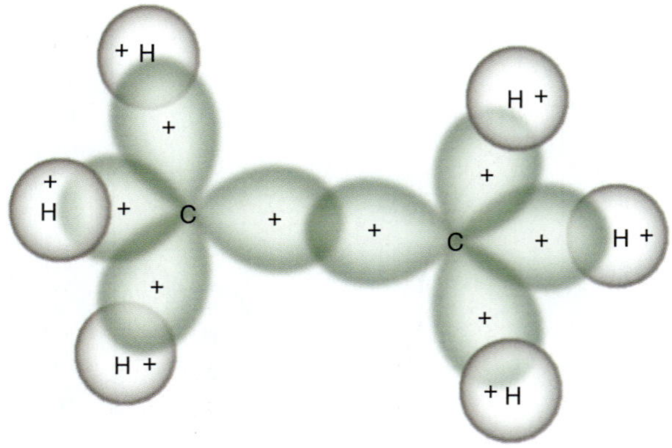

Figure 9.25 The six carbon-hydrogen σ-bond orbitals in an ethane molecule result from the combination of sp^3 orbitals on the carbon atoms and $1s$ orbitals on the hydrogen atoms. The carbon–carbon σ-bond orbital results from the combination of two sp^3 orbitals, one from each carbon atom. There are $(2 \times 4) + (6 \times 1) = 14$ valence electrons in an ethane molecule. Each of the seven σ bond orbitals is occupied by two valence electrons of opposite spin, accounting for the seven σ bonds in an ethane molecule.

ate to use sp^3 hybrid orbitals on the carbon atoms to describe the bonding. The carbon–carbon bond orbital in an ethane molecule is formed by the combination of two sp^3 orbitals, one from each carbon atom. The six carbon–hydrogen bond orbitals in an ethane molecule result from the overlap of the three remaining sp^3 orbitals on each carbon atom with a hydrogen $1s$ atomic orbital. There are seven σ-bond orbitals in an ethane molecule and $(2 \times 4) + (6 \times 1) = 14$ valence electrons. The 14 valence electrons occupy the seven σ-bond orbitals in an ethane molecule such that each bonding orbital has two electrons of opposite spin. The resulting bonding in an ethane molecule is shown in Figure 9.25.

We should emphasize at this point that hybrid orbitals are "after-the-fact" constructions. The geometry associated with a set of hybrid orbitals does *not* determine the geometry of the molecule; rather, the geometry of the molecule determines which type of hybrid orbitals are appropriate to describe the bonding in that molecule. For example, a methane molecule is not tetrahedral because the valence orbitals of the carbon atom are sp^3. The molecule is tetrahedral because that shape (as predicted by VSEPR theory) gives a methane molecule its lowest-possible energy.

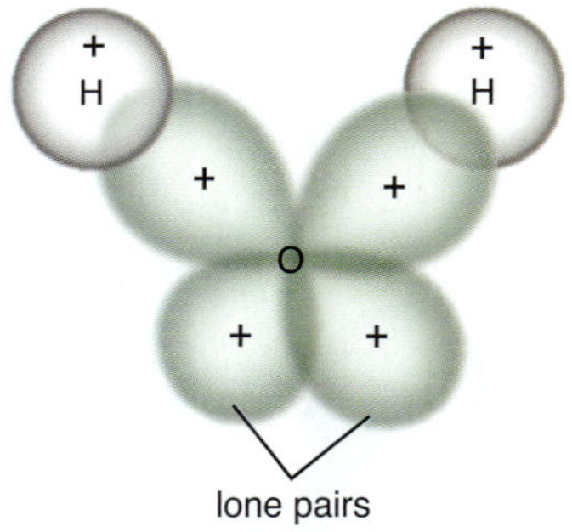

Figure 9.26 Bonding in a water, H_2O, molecule. Two of the oxygen sp^3 orbitals combine with hydrogen $1s$ orbitals to form two equivalent localized σ-bond orbitals. Of the eight valence electrons, four occupy the two bonding orbitals and four occupy the two nonbonded sp^3 orbitals on the oxygen atom. The latter are lone electron pairs.

9-8. sp^3 Orbitals Can Describe Molecules with Four Electron Pairs About the Central Atom

None of the molecules that we have considered up to now has had a lone pair of electrons. Let's now look at a water molecule, which has two lone pairs of electrons:

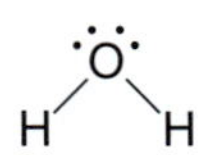

The oxygen atom in a water molecule is surrounded by four pairs of electrons: two pairs in covalent bonds and two lone pairs. We expect from VSEPR theory that the four pairs of electrons will be tetrahedrally disposed. Thus, it is appropriate to use sp^3 hybrid orbitals on the oxygen atom to describe the bonding in a H_2O molecule. Each hydrogen $1s$ orbital combines with one of the sp^3 hybrid orbitals on the oxygen atom to produce an oxygen–hydrogen σ-bond orbital (Figure 9.26). Next, we account for the eight valence electrons in a H_2O

molecule. Four of the valence electrons occupy the two $O(sp^3) + H(1s)$ σ-bond orbitals. The other four occupy the two $O(sp^3)$ **nonbonded orbitals**, constituting the two lone electron pairs on the oxygen atom.

On the basis of this bonding, we predict that the H–O–H bond angle in a water molecule will be 109.5°. This prediction differs from the experimental value of 104.5° because the four orbitals surrounding the oxygen atom are not used in the same way. Two are used to form bonds with the hydrogen atoms, and two are used for the lone-pair electrons. Recall from our discussion of VSEPR theory that the H–O–H bond angle in a H_2O molecule will be somewhat less than the tetrahedral value of 109.5° because the lone electron pairs repel the two hydrogen–oxygen bonds.

EXAMPLE 9-6: Use hybrid orbitals to describe the bonding in an ammonia molecule, NH_3. The Lewis formula for NH_3 is

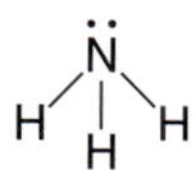

Solution: An ammonia molecule has three covalent bonds and one lone pair of electrons. We know from VSEPR theory that the four electron pairs in the valence shell of the nitrogen atom point toward the vertices of a tetrahedron. Therefore, it is appropriate to use sp^3 hybrid orbitals on the nitrogen atom. Three of these sp^3 orbitals form localized bond orbitals by combining with the hydrogen 1*s* orbitals. Thus, we can describe the bonding in an ammonia molecule in terms of three localized σ-bond orbitals and one nonbonded lone pair in an sp^3 orbital on the nitrogen atom (Figure 9.27).

There are eight valence electrons in an NH_3 molecule. Six of them occupy the three localized $N(sp^3) + H(1s)$ σ-bond orbitals and two occupy the nonbonded $N(sp^3)$ orbital. The use of sp^3 orbitals implies that the H–N–H bond angles are 109.5°. The four valence orbitals in an NH_3 molecule are not used equivalently, however, because one describes a lone pair. Thus, we should expect to find small deviations from a regular tetrahedral shape; and, in fact, the observed H–N–H bond angles in an NH_3 molecule are 107.3°.

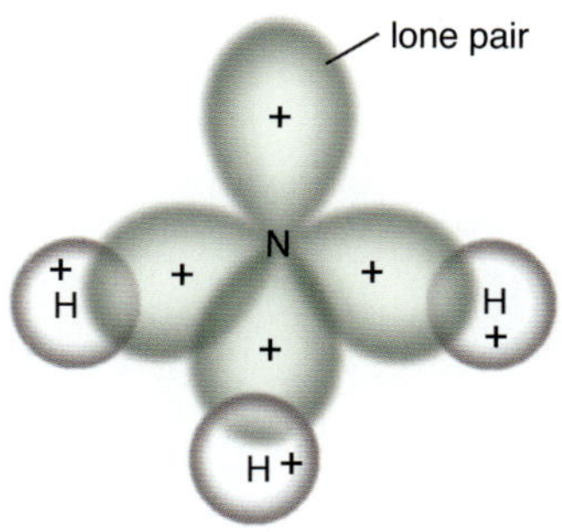

Figure 9.27 The use of sp^3 hybrid orbitals on the nitrogen atom describes the bonding in an ammonia, NH_3, molecule. Three of the nitrogen sp^3 orbitals combine with hydrogen 1*s* orbitals to form three equivalent localized σ-bond orbitals. The fourth nitrogen sp^3 orbital is a nonbonded orbital and is occupied by the lone pair of electrons in ammonia.

PRACTICE PROBLEM 9-6: Use localized bond orbitals to describe the bonding in an amide ion, NH_2^-.

Answer: An NH_2^- ion is isoelectronic with a H_2O molecule, and the bonding in an NH_2^- ion is similar to that in a H_2O molecule.

The sp^3 orbitals are also appropriate to describe the bonding of the oxygen atom in alcohols, which are organic compounds involving an –OH group bonded to a carbon atom. The simplest alcohol is

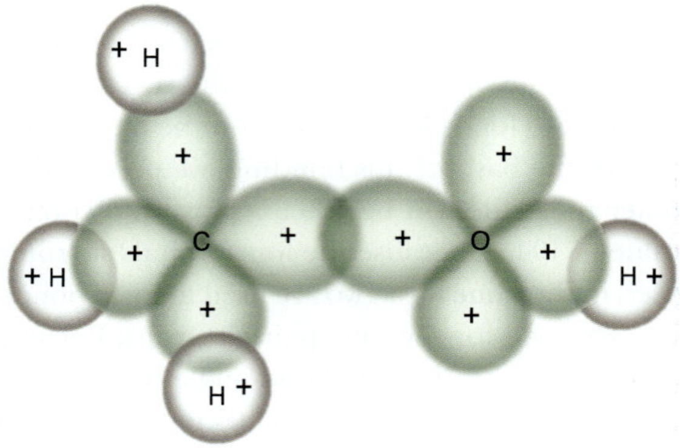

Figure 9.28 A schematic representation of the bond orbitals in a methanol, CH_3OH, molecule. We use sp^3 orbitals on both the carbon atom and the oxygen atom.

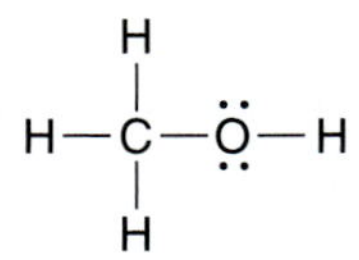

methanol, CH_3OH

The bonding in a methanol molecule is illustrated in Figure 9.28. Both the carbon atom and the oxygen atom are surrounded by four pairs of electrons, which according to VSEPR theory, we expect to be tetrahedrally disposed. Therefore, it is appropriate to use sp^3 hybrid orbitals on both the carbon atom and the oxygen atom to describe the bonding in a methanol molecule. The carbon–oxygen bond orbital results from the combination an sp^3 orbital from the carbon atom and an sp^3 orbital from the oxygen atom. The carbon–hydrogen and oxygen–hydrogen bonds result from the combination of sp^3 and hydrogen 1*s* orbitals.

There are a total of five σ-bond orbitals in a CH_3OH molecule. Ten of $4 + (4 \times 1) + 6 = 14$ valence electrons in a CH_3OH molecule occupy these five σ-bond orbitals, and the other four valence electrons occupy the two remaining nonbonded sp^3 orbitals as lone pairs on the oxygen atom.

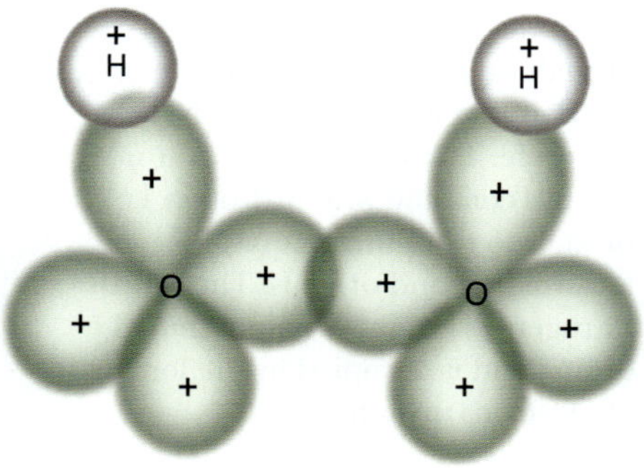

Figure 9.29 A schematic representation of the bonding in a hydrogen peroxide, H_2O_2, molecule. The oxygen–oxygen σ-bond orbital is formed by the combination of an sp^3 orbital from each oxygen atom; each hydrogen–oxygen σ-bond orbital is formed by the combination of an oxygen sp^3 orbital and a hydrogen 1*s* orbital.

EXAMPLE 9-7: Describe the bonding in a hydrogen peroxide molecule, H_2O_2. Its Lewis formula is $\mathrm{H{-}\ddot{\underset{\cdot\cdot}{O}}{-}\ddot{\underset{\cdot\cdot}{O}}{-}H}$.

Solution: Each oxygen atom has four pairs of electrons, which we predict on the basis of VSEPR theory will be tetrahedrally disposed. Therefore, it is appropriate to use sp^3 hybrid orbitals on the oxygen atoms. The oxygen–oxygen σ-bond orbital is formed by the combination of an sp^3 orbital on each oxygen atom. Each hydrogen–oxygen σ-bond orbital is formed by the combination of an oxygen sp^3 orbital and a hydrogen 1*s* orbital.

There are $(2 \times 1) + (2 \times 6) = 14$ valence electrons in a H_2O_2 molecule. Six of the valence electrons occupy the three σ-bond orbitals, forming the three σ bonds in the Lewis formula for a H_2O_2 molecule. The other eight valence electrons occupy the four sp^3 orbitals, two on each oxygen atom, forming the two lone electron pairs shown on each oxygen atom in the Lewis formula. Figure 9.29 illustrates the bonding in a H_2O_2 molecule.

PRACTICE PROBLEM 9-7: Use localized bond orbitals to describe the bonding in a hydrazine, N_2H_4, molecule.

Answer: The N–H bonds are N(sp^3) + H(1s) localized σ-bond orbitals. The N–N bond is a N(sp^3) + N(sp^3) localized σ-bond orbital, and the lone pairs occupy the remaining two nonbonded N(sp^3) orbitals.

9-9. Hybrid Atomic Orbitals Can Involve *d* Orbitals

In Chapter 8 we learned about molecules that are trigonal bipyramidal (for example, phosphorus pentachloride, PCl_5) and octahedral (for example, sulfur hexafluoride, SF_6). The central atom in each of these molecules has an expanded valence shell, and one way of describing the bonding in such molecules is to include *d* orbitals in the construction of hybrid orbitals.

A combination of a 3*s* orbital, three 3*p* orbitals, and one 3*d* orbital gives five hybrid atomic orbitals that have trigonal bipyramidal symmetry (Figure 9.30). These five ***sp*³*d* orbitals** are interesting because they are not equivalent to one another. In fact, there are two sets of orbitals: a set of three equivalent equatorial orbitals and a set of two equivalent axial orbitals. This result is consistent with the experimental fact that the five chlorine atoms in a PCl_5 molecule are not equivalent (Section 8.4). The five phosphorus–chlorine σ-bond orbitals are formed by combining each phosphorus sp^3d hybrid orbital with a half-filled chlorine 3*p* orbital. Ten of the valence electrons (five from the phosphorus atom and one from each of the chlorine atoms) occupy the five localized bond orbitals (two electrons in each orbital) to form the five localized covalent bonds.

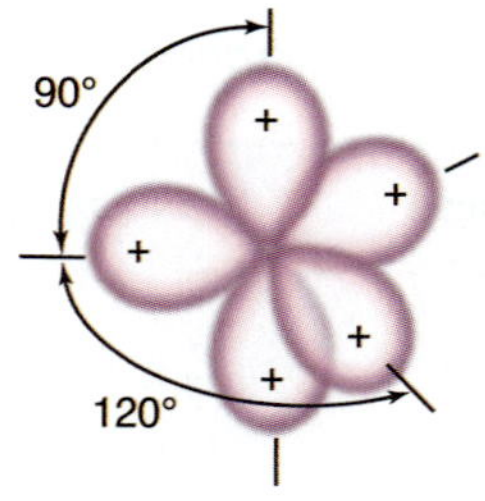

Figure 9.30 The five sp^3d hybrid orbitals. The five orbitals shown here are constructed from the 3*s* orbital, the three 3*p* orbitals, and one 3*d* orbital. They point to the vertices of a trigonal bipyramid.

To describe the bonding in the octahedral SF_6 molecule, we need six equivalent hybrid orbitals on the sulfur atom that point toward the vertices of an octahedron. This arrangement can be achieved by combining the 3*s* orbital, the three 3*p* orbitals, and two of the 3*d* orbitals on the sulfur atom. The resulting six ***sp*³*d*² orbitals** point toward the vertices of a regular octahedron (Figure 9.31). The six sulfur–fluorine σ-bond orbitals in a SF_6 molecule are formed by combining each sulfur sp^3d^2 orbital with a fluorine 2*p* orbital. Twelve of the valence electrons (six from the sulfur atom and one from each of the fluorine atoms) occupy the six localized bond orbitals to form the six localized covalent bonds.

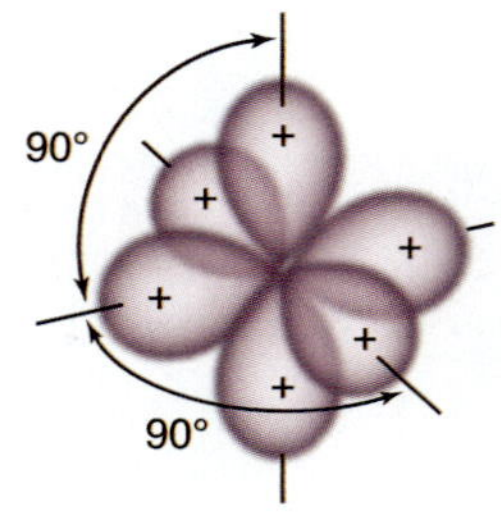

Figure 9.31 The six sp^3d^2 hybrid orbitals. The six orbitals shown here are constructed from the 3*s* orbital, the three 3*p* orbitals, and two 3*d* orbitals. They point to the vertices of a regular octahedron.

For the PCl_5 and SF_6 molecules, we use 3*s*, 3*p*, and some of the 3*d* orbitals on the central atom to form hybrid atomic orbitals. As we pointed out earlier, quantum theory tells us that only orbitals of similar energy combine effectively. In other words, we can combine the 3*s*, 3*p*, and 3*d* orbitals because they have similar energies. The combination of 3*d* orbitals with 2*s* and 2*p* orbitals does not produce hybrid orbitals that are effective in forming bonds because the 3*d* orbitals are much higher in energy than the 2*s* or 2*p* orbitals are. This restriction explains why only elements in the third and higher rows of the periodic table can expand their valence shells, as we saw in Chapter 7. For example, for atoms of elements such as phosphorus and sulfur, which have valence electrons that occupy the 3*s* and 3*p* orbitals, we can use their 3*d* orbitals to expand their valence shells, but for atoms of second-row elements such as carbon and nitrogen, which have valence electrons that occupy the 2*s* and 2*p* orbitals, we cannot use 3*d* orbitals to form hybrid orbitals.

Table 9.3 summarizes the hybrid atomic orbitals that we have introduced.

TABLE 9.3 Properties of hybrid orbitals

Hybrid	Number	Orbital geometry	Orbital angle	Examples	Hybridization required
sp	2	linear	180°	BeH_2, BeF_2	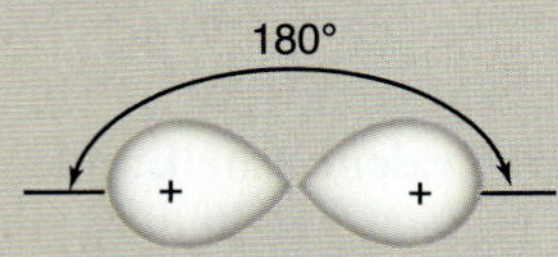
sp^2	3	trigonal planar	120°	BF_3	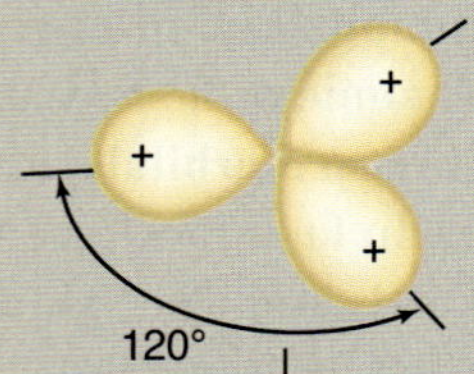
sp^3	4	tetrahedral	109.5°	CH_4, BF_4^-, NH_4^+	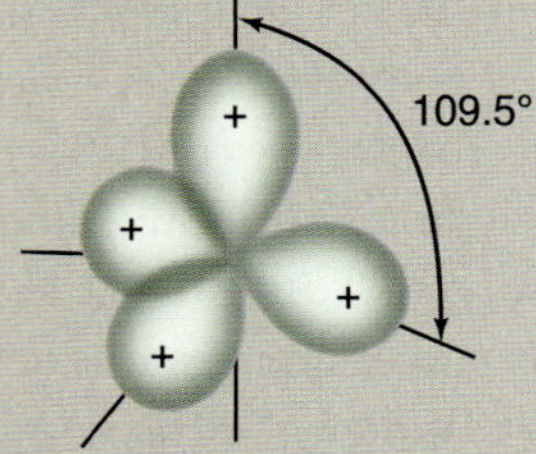
sp^3*d*	5	trigonal bipyramidal	90°, 120°	PCl_5	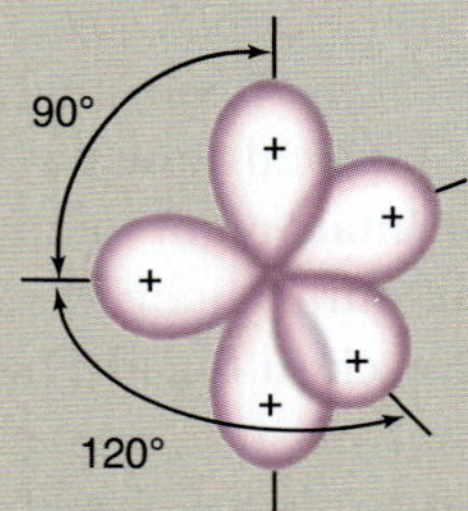
sp^3*d*^{2}	6	octahedral	90°	SF_6, AlF_6^{3-}	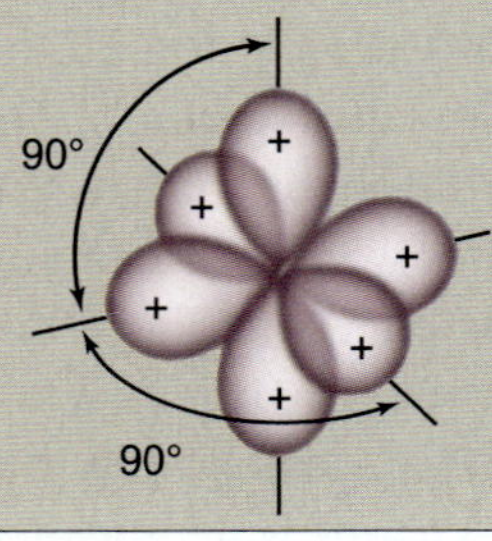

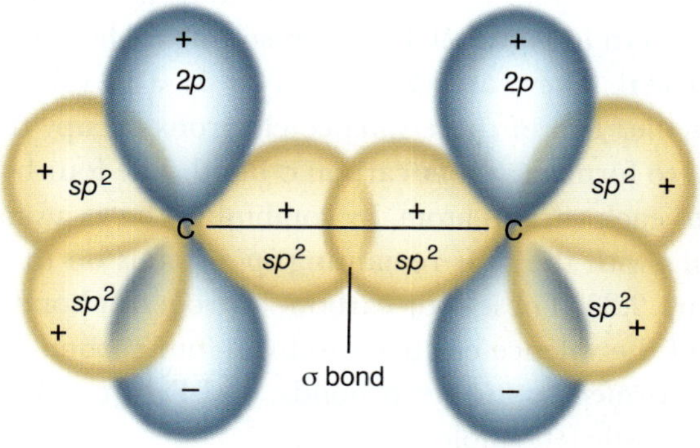

Figure 9.32 Two carbon atoms in an ethylene molecule are joined by the combination of an sp^2 orbital from each. The resulting bond orbital is cylindrically symmetric around the carbon–carbon axis and therefore is a σ-bond orbital. The carbon–carbon σ-bond orbital constitutes part of the double bond in an ethylene molecule.

We again note that the hybridization of orbitals is predicted by the classification of the molecule according to VSEPR theory and not the other way around. It is also significant that in each case the number of resulting hybrid orbitals is equal to the number of atomic orbitals used to construct them, in accord with the principle of the conservation of orbitals.

9-10. A Double Bond Can Be Represented by a σ Bond and a π Bond

All the molecules that we have discussed so far in the sections on hybrid orbitals have only single bonds. One of the simplest molecules in which there is a double bond is ethene, C_2H_4, which is generally referred to by its common name, ethylene. Its Lewis formula is

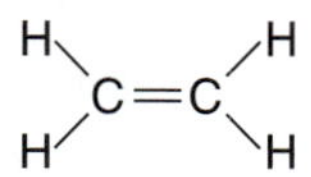

The geometry of an ethylene molecule is quite different from that of an ethane molecule (Section 9.7). All six atoms in an ethylene molecule lie in one plane, and each carbon atom is bonded to three other atoms. The geometry about each carbon atom is trigonal planar, and we saw in Section 9.6 that sp^2 hybrid orbitals are appropriate for describing this geometry. Therefore, we shall describe the bonding in ethylene using sp^2 orbitals on each carbon atom.

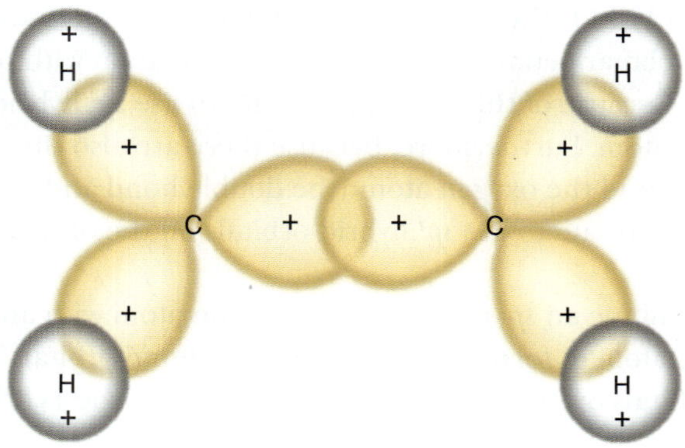

Figure 9.33 The σ-bond framework in an ethylene, $CH_2{=}CH_2$, molecule. The carbon–carbon σ-bond orbital results from the combination of two sp^2 orbitals, one from each carbon atom. The four carbon–hydrogen σ-bond orbitals result from the combination of carbon sp^2 hybrid orbitals and hydrogen 1*s* atomic orbitals. The remaining 2*p* atomic orbitals on the carbon atoms are not shown but are perpendicular to the page.

The first step is to join the two carbon atoms by combining an sp^2 orbital from each, as shown in Figure 9.32. The resulting carbon–carbon bond orbital is a σ-bond orbital. There are $(2 \times 4) + (4 \times 1) = 12$ valence electrons in an ethylene molecule. Two of these valence electrons occupy the carbon–carbon σ-bond orbital to form a carbon–carbon σ bond. The four hydrogen atoms are bonded, two to each carbon atom, by combining the 1*s* atomic orbitals on the hydrogen atoms with the four remaining sp^2 orbitals on the carbon atoms, as shown in Figure 9.33. These four carbon–hydrogen σ-bond orbitals are occupied by eight of the valence electrons to form four carbon–hydrogen σ bonds. All five bonds formed so far are σ bonds, and Figure 9.33 shows the **σ-bond framework** in an ethylene molecule.

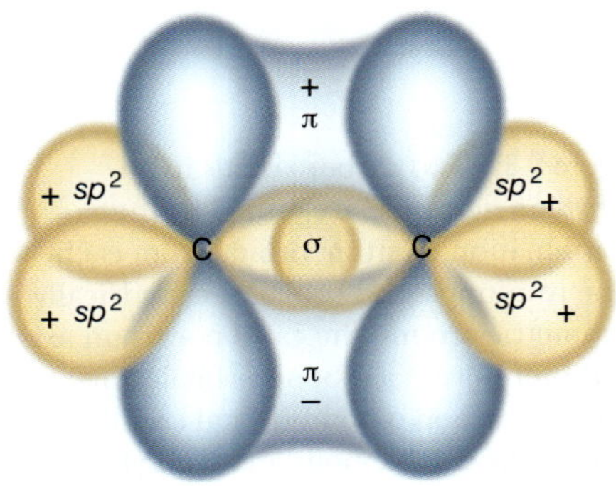

Figure 9.34 A double bond in an ethylene molecule consists of a σ bond and a π bond. The σ bond results from the combination of two sp^2 orbitals, one from each carbon atom. The π bond results from the combination of two 2*p* orbitals, one from each carbon atom. The π orbital maintains the σ-bond framework in a planar shape and prevents rotation about the double bond.

Recall that there is a 2*p* atomic orbital on each carbon atom perpendicular to each H–C–H plane (Figure 9.32). If the two ends of the molecule are oriented such that these two 2*p* orbitals are parallel, then their overlap is maximized and the π orbital illustrated in Figure 9.34 results. This π orbital is occupied by the remaining two valence electrons to form a **π bond**. Note that the positive and negative lobes of a π bond are formed from the positive and negative lobes of the two *p* orbitals from which it is formed. Consequently, the two lobes taken together represent *one* π bond. The double bond in an ethylene molecule is described by the σ bond *and* the π bond, as shown in Figure 9.34. Thus, we have four $C(sp^2) + H(1s)$ σ bonds, one $C(sp^2) + C(sp^2)$ σ bond, and one $C(2p) + C(2p)$ π bond. The carbon–carbon σ and π bonds are *collectively* referred to as the carbon-carbon double bond. For the double bond, the bond order due to the σ bond is 1 and that due to the π bond is 1, for a total bond order of 2. A σ bond and a π bond do not have the same energy, so a double bond, although much stronger than a single bond, is not twice as strong as a single bond. Carbon–carbon single-bond energies are about 0.6 aJ, whereas carbon–carbon double-bond energies are about 1 aJ. As Figure 9.34 shows, the overlap of the two 2*p* orbitals that form the π orbital locks the molecule into a planar configuration.

EXAMPLE 9-8: Describe the bonding in the methanal molecule, generally referred to by its common name, formaldehyde, which has the Lewis formula

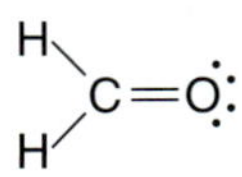

Solution: From VSEPR theory, we conclude that a formaldehyde molecule has a trigonal planar geometry around the carbon atom with bond angles of about 120°. Because of this, it is appropriate to use sp^2 hybrid orbitals on the carbon atom. Furthermore, because there are also three groups of electrons around the oxygen atom (the double bond and two lone pairs), it is appropriate to use sp^2 hybrid orbitals on the oxygen atom as well.

First we combine an sp^2 orbital on the carbon atom with an sp^2 orbital on the oxygen atom to form a carbon–oxygen σ-bond orbital. The remaining two sp^2 orbitals on the carbon atom combine with the 1*s* orbitals on the

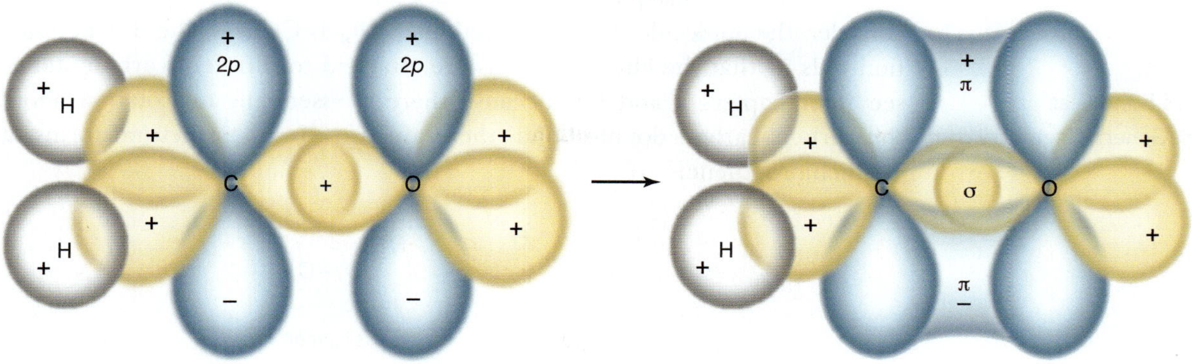

Figure 9.35 The bonding in a formaldehyde, H_2CO, molecule. (a) The σ-bond framework, showing the $2p$ orbitals that are perpendicular to the plane formed by the four atoms. These two $2p$ orbitals combine to form a π-bond orbital, which is occupied by two of the valence electrons. (b) The carbon–oxygen double bond consists of one σ bond and one π bond.

hydrogen atoms to form the two carbon–hydrogen σ-bond orbitals. There are $4+6+(2\times 1)=12$ valence electrons in a formaldehyde molecule. Two of these valence electrons occupy the carbon–oxygen σ-bond orbital to form a carbon–oxygen σ bond, and four of the valence electrons occupy the two carbon–hydrogen σ-bond orbitals to form two carbon–hydrogen σ bonds. The remaining two sp^2 orbitals on the oxygen atom are occupied by four of the valence electrons, constituting the two lone pairs on the oxygen atom.

The remaining $2p$ orbital on the carbon atom and the remaining $2p$ orbital on the oxygen atom, both of which are perpendicular to the plane of the molecule, are now combined to form a carbon–oxygen π orbital that is occupied by the two remaining valence electrons. Thus, the carbon–oxygen double bond is composed of a σ bond and a π bond and has a bond order of two. The bonding in a formaldehyde molecule is shown in Figure 9.35.

PRACTICE PROBLEM 9-8: Use localized bond orbitals to describe the bonding in a phosgene molecule, Cl_2CO.

Answer: As with a formaldehyde molecule, it is appropriate to use sp^2 hybrid orbitals on both the carbon atom and the oxygen atom in a phosgene molecule. The resulting bonding can be summarized by a $C(sp^2)+O(sp^2)$ σ bond, a $C(2p)+O(2p)$ π bond, two $Cl(3p)+C(sp^2)$ σ bonds, and two lone pairs occupying the two $O(sp^2)$ nonbonded orbitals.

9-11. There Is Limited Rotation About a Double Bond

The double bond in an ethene molecule consists of a σ bond and a π bond. The π bond locks the molecule into a planar shape (Figure 9.34). A significant amount of energy is required to break a π bond, so essentially no rotation occurs about a double bond at room temperature.

See Interchapters F and G at www.McQuarrieGeneralChemistry.com.

To see the consequences of the lack of rotation about double bonds, consider the molecule 1,2-dichloroethene, ClCH=CHCl. (The 1,2 designation tells us that the chlorine atoms are attached to different carbon atoms; see Interchapters F and G). Because there is essentially no rotation about the carbon–carbon double bond, there are two distinct structural forms of 1,2-dichloroethene:

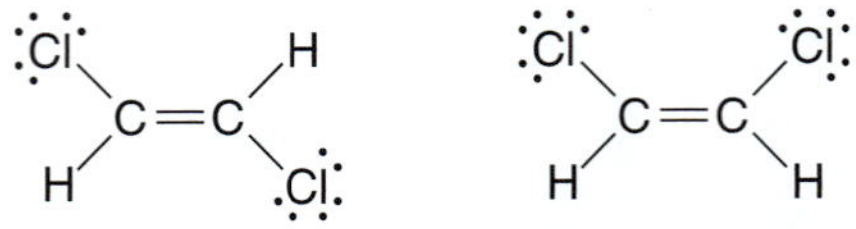

trans isomer *cis* isomer

The first of these is called *trans*-1,2-dichloroethene because the chlorine atoms lie across the double bond from each other (*trans* means "across"). The other is called *cis*-1,2-dichloroethene because the chlorine atoms lie on the same side of the double bond (*cis* means "on the same side").

As we saw in the previous chapter, molecules with the same atom-to-atom bonding but different spatial arrangements are called stereoisomers. The particular type of stereoisomerism that is displayed by 1,2-dichloroethene is called ***cis-trans* isomerism**. Stereoisomers, and *cis-trans* isomers in particular, have different physical properties. Although both the *trans* and *cis* isomers of 1,2-dichloroethene have polar bonds, the *trans* isomer has no net dipole moment, but the *cis* isomer does. The boiling point of the *trans* isomer of 1,2-dichloroethene is 48°C and that of the *cis* isomer is 60°C. We shall show in Chapter 15 why the *cis* isomer, with its net dipole moment, has a higher boiling point than the nonpolar *trans* isomer.

An example of the importance of *cis-trans* isomerism occurs in the chemistry of vision. Although we have stated that no rotation is allowed about double bonds in the ground state of a molecule, *cis* and *trans* isomers can interconvert if the molecule is supplied with sufficient energy in the form of heat or light:

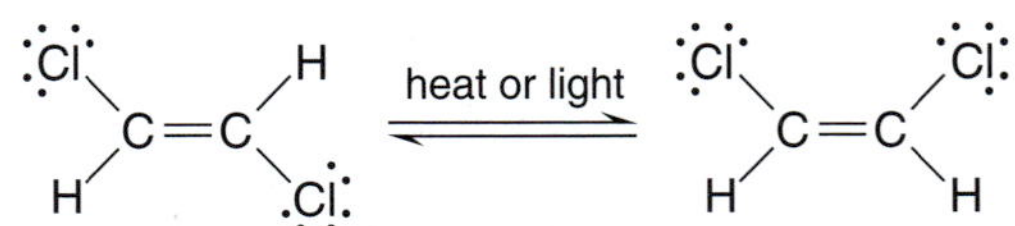

It was determined in the 1950s that the chemistry of vision involves *cis* to *trans* isomerization. The retina of the eye contains a substance called rhodopsin that consists of a molecule called 11-*cis*-retinal combined with a protein called

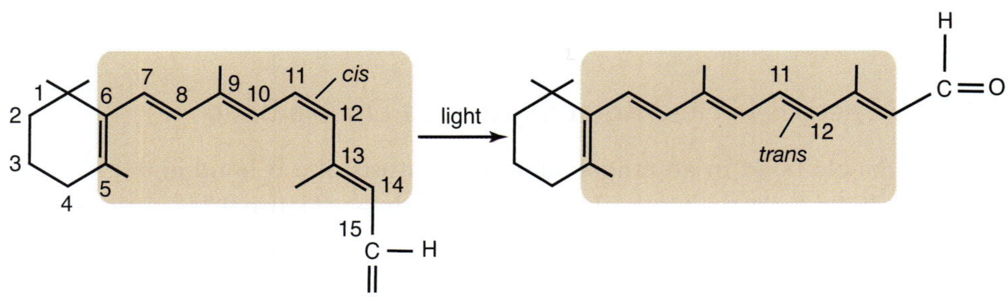

11-*cis*-retinal 11-*trans*-retinal

Figure 9.36 When a photon of light strikes the retina of the eye, the molecule 11-*cis*-retinal is converted to the *trans* isomer, 11-*trans*-retinal. The reaction involves the breaking of the π bond between the 11th and 12th carbon atoms, a rotation about the remaining σ bond, and then re-forming the π bond to lock the molecule into its *trans* form.

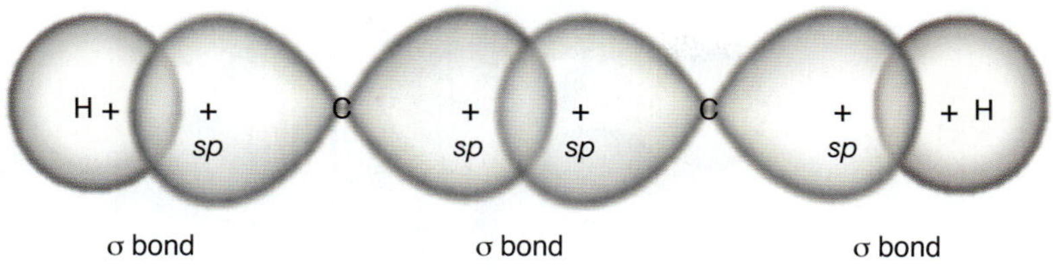

Figure 9.37 The σ-bond framework of an acetylene, HC≡CH, molecule. The carbon–carbon σ-bond orbital results from combining two *sp* orbitals, one from each carbon atom. Each of the two carbon–hydrogen bond orbitals results from combining a carbon *sp* orbital and a hydrogen 1*s* orbital.

opsin. When an 11-*cis*-retinal molecule is struck by a photon of visible light, it isomerizes at the *cis* double bond to give 11-*trans*-retinal (Figure 9.36).

The numbers in Figure 9.36 refer to carbon atoms that are not specifically shown; recall the shorthand representation of benzene in Chapter 7. The shaded area in these formulas shown in Figure 9.36 represents a planar region in the molecule. The shapes of the *cis* and *trans* isomers are significantly different, and the light-induced change in shape triggers a response in the optic nerve cells that is transmitted to the brain and perceived as vision. The vision response occurs through a sequence of processes that has been investigated quite thoroughly. The primary event, however, is the conversion of the *cis* to the *trans* isomer of retinal.

9-12. A Triple Bond Can Be Represented by One σ Bond and Two π Bonds

Let's next consider a molecule that contains a triple bond. A good example is an ethyne molecule, C_2H_2, which is generally referred to by its common name, acetylene. The Lewis formula for an acetylene molecule is H−C≡C−H. An acetylene molecule is linear, with each carbon atom bonded to only two other atoms. We saw in Section 9.5 that *sp* hybrid orbitals are appropriate to describe the bonding of an atom that forms two bonds separated by 180° (Figure 9.17).

We can build a σ-bond framework for an acetylene molecule in two steps: We first form a carbon-carbon σ-bond orbital by combining two *sp* orbitals, one from each carbon atom. We then form the two carbon–hydrogen σ-bond orbitals by combining a hydrogen 1*s* orbital with the *sp* orbital on each carbon atom.

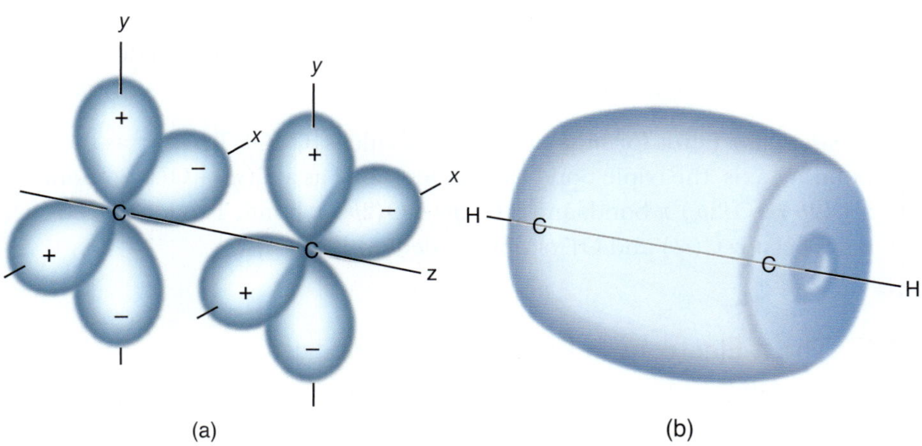

Figure 9.38 (a) The 2*p* orbitals on the carbon atoms in an acetylene molecule. If we take *z* to be the bonding axis, then the 2*p* orbitals that are directed along the *x* axis combine to form one π-bond orbital, and the 2*p* orbitals directed along the *y* axis combine to form another π-bond orbital. (b) The two π bonds in acetylene constitute a barrel-shaped distribution of electron density in the bond region.

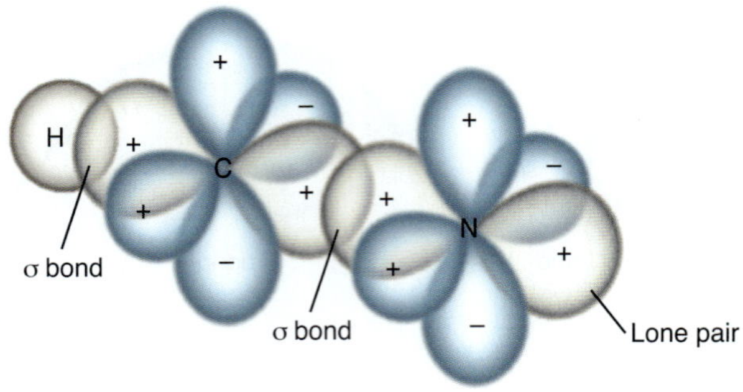

Figure 9.39 The σ-bond framework of a hydrogen cyanide, HCN, molecule.

These three σ-bond orbitals are occupied by six of the (2 × 4) + (2 × 1) = 10 valence electrons in an acetylene molecule and form the σ-bond framework of an acetylene molecule (Figure 9.37).

The remaining carbon 2*p* orbitals are perpendicular to the H–C–C–H bonding axis in an acetylene molecule, as shown in Figure 9.38a. These orbitals can combine to produce two π-bond orbitals. These two π-bond orbitals are occupied by the four remaining valence electrons to form two π bonds. The total bond order for the carbon–carbon triple bond is 3, due to the two electrons in the σ bond and the four electrons that occupy the two π bonds. The triple bond consists of one σ bond and two π bonds (Figure 9.38b).

EXAMPLE 9-9: Compare the bonding in an acetylene molecule with that in a hydrogen cyanide molecule, HCN.

Solution: The Lewis formula for a HCN molecule is H−C≡N∶. Because VSEPR theory predicts that the molecule is linear, it is appropriate to use *sp* orbitals on both the carbon atom and the nitrogen atom in HCN. The σ-bond framework of a HCN molecule is shown in Figure 9.39; it is similar to that of an C_2H_2 molecule (Figures 9.37 and 9.38). The 2*p* orbitals of the carbon and nitrogen atoms combine to form the two π-bond orbitals. There are four bond orbitals in HCN: two are σ-bond orbitals and two are π-bond orbitals. There are ten valence electrons in a HCN molecule: eight occupy the four bond orbitals, and two occupy the nitrogen *sp* nonbonded orbital and constitute a lone electron pair on the nitrogen atom.

PRACTICE PROBLEM 9-9: Use localized bonding orbitals to describe the bonding in a carbon monoxide molecule, CO.

Answer: The Lewis formula of a CO molecule is ∶C≡O∶. Taking *z* as the bonding axis, the triple bond can be described as a C(*sp*) + O(*sp*) σ bond, a $C(2p_x) + O(2p_x)$ π bond, and a $C(2p_y) + O(2p_y)$ π bond. Two lone electron pairs occupy C(*sp*) and O(*sp*) nonbonded orbitals.

9-13. The π Electrons in Benzene Are Delocalized

In many molecules and ions, there are orbitals that extend over more than two adjacent atoms. One of the most important examples is a benzene molecule,

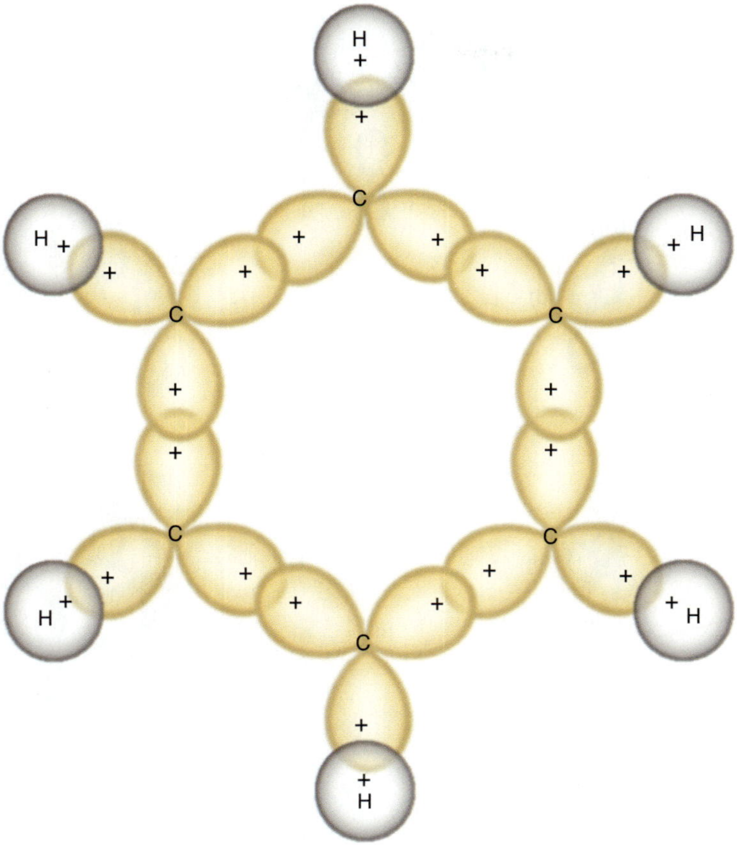

Figure 9.40 The σ-bond framework in a benzene, C_6H_6, molecule. Each carbon-carbon σ-bond orbital results from the combination of sp^2 orbitals, and each carbon–hydrogen σ-bond orbital results from the combination of a carbon sp^2 orbital and a hydrogen 1*s* orbital. All twelve atoms lie in a single plane, so benzene is a planar molecule. The six carbon atoms form a regular hexagon. Not shown are the six 2*p* orbitals, one on each carbon atom, that are perpendicular to the hexagonal plane. (See Figure 9.41.)

C_6H_6. Recall that the Lewis formula of benzene (Section 7-6) involves two resonance forms. We expressed the superposition of these two resonance forms as

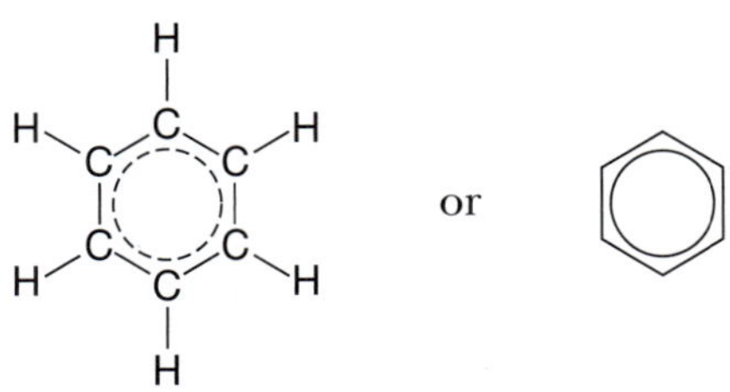

We can describe the bonding in a benzene molecule in terms of σ bonds and π bonds. Benzene is a planar molecule in the shape of a regular hexagon. The angles in a regular hexagon are 120°, so the three bonds surrounding each carbon atom lie in a plane at an angle of 120°. Thus, it is appropriate to use sp^2 hybrid orbitals on the carbon atoms to describe the bonding in a benzene molecule. This depiction leads directly to the σ-bond framework shown in Figure 9.40. Note that there are 12 σ-bond orbitals.

Each carbon atom also has a 2*p* orbital that is perpendicular to the hexagonal plane. These six 2*p* orbitals combine to give a total of six π molecular orbitals (conservation of orbitals). The six combinations of the six 2*p* orbitals perpendicular to the benzene ring and the resulting corresponding molecular orbitals are shown in Figure 9.41. The order of the sets of orbitals in Figure 9.41 corresponds to their relative energies, with the lowest energy occurring at the

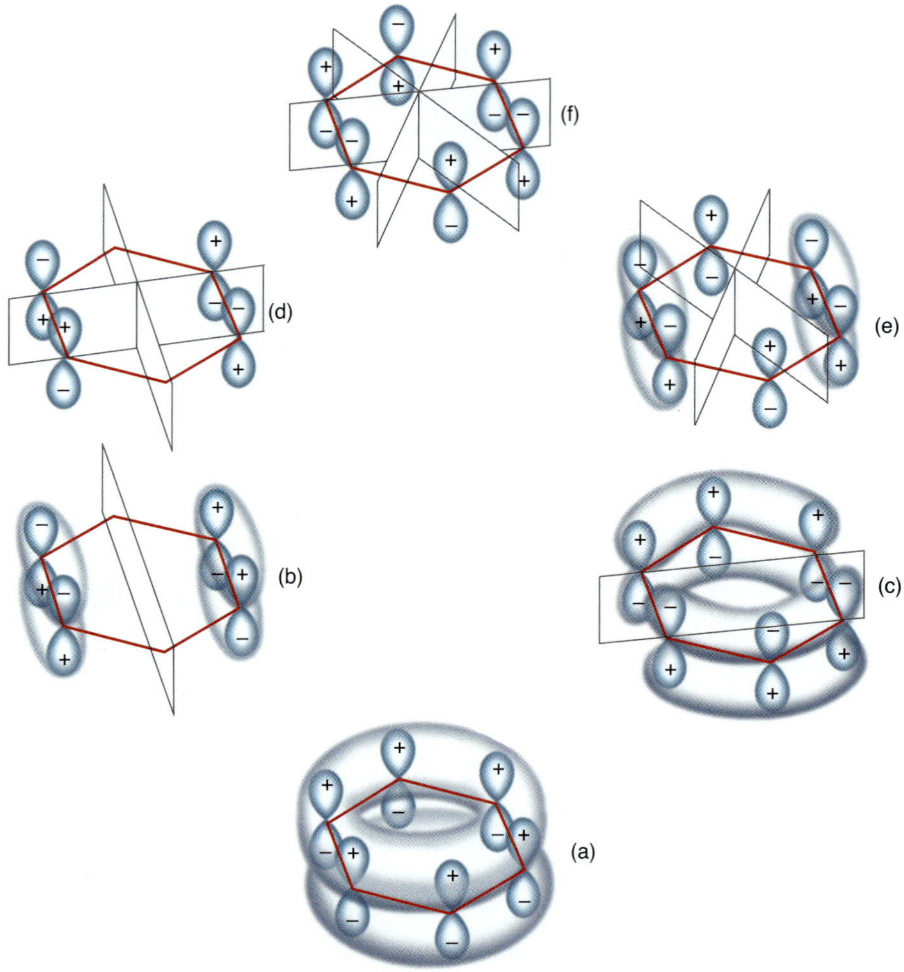

Figure 9.41 The combination of the six $2p$ orbitals perpendicular to the hexagonal plane of the σ-bond framework of a benzene molecule. Each picture illustrates how the positive lobes and negative lobes of the $2p$ orbitals perpendicular to the planar σ-bond framework are oriented with respect to each other. The transparent boxes show the location of the nodal planes. The six resultant molecular orbitals of a benzene molecule are ordered according to their energies. The three π orbitals of lowest energy are bonding orbitals, and the other three are antibonding orbitals. As you can see, the energy of the orbitals increases as the number of nodal planes increases.

bottom. In Figure 9.41, the signs of the $2p$ orbitals illustrate how they are oriented when they are combined. In the orbital of lowest energy (Figure 9.41a), all six $2p$ orbitals are aligned, which means that the resultant molecular orbital consists of a sum of all six $2p$ orbitals. Observe that the molecular orbital looks like one ring above the plane of the molecule and one below it. This molecular orbital is a bonding orbital. Because it is spread over several nuclei it is said to be **delocalized**, and the electrons that occupy this orbital are said to be delocalized as well. The next set of orbitals (Figure 9.41b and c) is doubly degenerate: in other words, the two molecular orbitals have the same energy. It turns out that one of the molecular orbitals actually consists of a combination of only four $2p$ orbitals. In both cases, however, the $2p$ orbitals are combined in such a way that there is one nodal plane in the molecular orbital that is perpendicular to the plane of the molecule, as you can see in Figure 9.41. Although higher in energy than the first molecular orbital because of the nodal plane, these two doubly degenerate molecular orbitals are bonding orbitals. They are also delocalized because one ranges over four nuclei and the other over six nuclei. The third set of orbitals (Figure 9.41d and e) shows that the $2p$ orbitals are combined such that the resultant molecular orbital has two nodal planes that are perpendicular to the plane of the molecule. These doubly degenerate molecular orbitals turn out to be antibonding orbitals. Finally, the molecular orbital of highest energy (Figure 9.41f) is formed by combining the six $2p$ orbitals in an alternating fashion. The resultant molecular orbital has three nodal planes that are perpendicular to the plane of the molecule and is also an antibonding orbital.

A benzene molecule serves as a nice illustration of the energy of a wave function and its number of nodes. A result from quantum theory says that the more nodes that a wave function has, the greater its energy. We see from Figure 9.41 that the lowest-energy orbital of a benzene molecule has no nodes perpendicular to its plane; the next two (doubly degenerate) orbitals each have one such node; the next two (doubly degenerate) orbitals each have two such nodes; and the highest-energy orbital has three such nodes.

A benzene molecule has $(6 \times 4) + (6 \times 1) = 30$ valence electrons. In its ground electronic state, 24 of these electrons occupy the 12 σ-bond orbitals and the remaining six occupy the three π orbitals of lowest energy in Figure 9.41. All six of these π electrons are in delocalized molecular orbitals, which gives benzene an extraordinary degree of stability. Delocalization in our quantum theoretical description is what resonance in the Lewis formula description attempts to show. The use of delocalized π molecular orbitals provides a more satisfying and clearer picture than do the Lewis formula resonance forms.

SUMMARY

In Chapter 5 we discussed the atomic orbitals of the simplest atom, hydrogen, and then used them to write electron configurations for other atoms. In this chapter, we presented the molecular orbitals of the simplest diatomic species, H_2^+, and then used them to discuss the bonding in other diatomic molecules. The bonding properties of diatomic molecules depend on the number of electrons in bonding and antibonding orbitals. Molecular orbital theory correctly predicts that the diatomic molecules He_2 and Ne_2 do not exist under normal conditions and that $O_2(g)$ is paramagnetic.

In our treatment of polyatomic molecules, we use localized bond orbitals. Once the geometry of a molecule is known, we can choose a hybrid orbital to describe the localized bonding. Table 9.3 summarizes the associated geometries of various hybrid atomic orbitals.

Bond orbitals for single bonds are cylindrically symmetric about the bond axis and are called σ-bond orbitals. A σ-bond orbital occupied by two electrons of opposite spin constitutes a σ bond.

Double bonds can be represented by one σ bond and one π bond. The π bond is formed by the combination of *p* orbitals on adjacent atoms. The atoms bonded directly to double-bonded atoms all lie in one plane. Because a molecule in its ground state cannot rotate about a double bond, *cis* and *trans* isomers can result. Triple bonds can be represented by one σ bond and two π bonds.

In some molecules, such as benzene, the π-bond orbitals are spread more or less uniformly over many atoms and are said to be delocalized. Delocalization of electrons confers an extra degree of stability on a molecule or ion, which accounts for the relative stability of benzene.

In describing molecules such as benzene, we use localized bonds to treat σ bonds and molecular orbital theory for π bonds. We use this mixed approach because of its close connection with Lewis formulas and because in a great many molecules, the σ bonds are localized and the π bonds are delocalized. As we shall see later, this approach will be useful when studying chemical reactivity.

TERMS YOU SHOULD KNOW

ground state wave function *268*
molecular orbital *268*
molecular orbital theory *268*
homonuclear diatomic molecule *268*
hydrogen molecular ion, H_2^+ *269*
bonding orbital *270*
internuclear axis *270*
σ orbital *270*
antibonding orbital *270*
σ^* orbital *270*
internuclear distance *270*
π orbital *272*
π^* orbital *272*
bond order *274*
paramagnetic *276*
diamagnetic *277*
photodissociation *278*
heteronuclear diatomic molecule *278*
polyatomic molecule *279*
localized bond orbital *279*
localized covalent bond *279*
hybrid orbital *279*
sp (hybrid) orbital *280*
σ bond *281*
sp^2 (hybrid) orbital *282*
principle of conservation of orbitals *283*
sp^3 (hybrid) orbital *283*
nonbonded orbital *287*
sp^3d (hybrid) orbital *289*
sp^3d^2 (hybrid) orbital *289*
σ-bond framework *292*
π bond *292*
cis-trans isomerism *294*
delocalized orbital *298*
delocalized electron *298*

AN EQUATION YOU SHOULD KNOW HOW TO USE

$$\text{bond order} = \frac{\begin{pmatrix}\text{number of}\\ \text{electrons in}\\ \text{bonding orbitals}\end{pmatrix} - \begin{pmatrix}\text{number of}\\ \text{electrons in}\\ \text{antibonding orbitals}\end{pmatrix}}{2} \qquad (9.1) \quad \text{(definition of bond order)}$$

PROBLEMS

DIATOMIC MOLECULES

9-1. Use molecular orbital theory to explain why a diatomic beryllium molecule does not exist under normal conditions.

9-2. Use molecular orbital theory to calculate the bond order of a diatomic boron molecule.

9-3. Use molecular orbital theory to predict whether or not a diatomic carbon molecule is paramagnetic.

9-4. Use molecular orbital theory to predict whether or not a diatomic boron molecule is paramagnetic.

9-5. Use molecular orbital theory to explain why the bond energy of a N_2 molecule is greater than that of a N_2^+ ion.

9-6. Use molecular orbital theory to explain why the bond energy of an O_2 molecule is less than that of an O_2^+ ion.

9-7. Use molecular orbital theory to predict the relative bond energies and bond lengths of a F_2 molecule and a F_2^+ ion.

9-8. Use molecular orbital theory to predict the relative bond energies and bond lengths of a diatomic carbon molecule, C_2, and the acetylide ion, C_2^+.

9-9. Use molecular orbital theory and Figure 9.13 to determine the ground state electron configurations and bond orders of NF, NF^+, and NF^-. Which of these species do you predict to be paramagnetic?

9-10. Use molecular orbital theory and Figure 9.13 to determine the ground state electron configurations and bond orders of NO, NO^+, and NO^-. Which of these species do you predict to have the longest bond length?

9-11. Write the ground state electron configurations and determine the bond orders of the following ions:

(a) O_2^- (b) C_2^+ (c) Be_2^+ (d) N_2^+

9-12. Write the ground state electron configurations and determine the bond orders of the following ions:

(a) C_2^- (b) B_2^+ (c) He_2^+ (d) F_2^{2-}

9-13. For each of the following molecular electron configurations, decide whether it describes a ground electronic state or an excited electronic state.

a. $(\sigma_{1s})^2(\sigma^*_{1s})^2(\sigma^*_{2s})^1$

b. $(\sigma_{1s})^2(\sigma^*_{1s})^2(\sigma_{2s})^2(\sigma^*_{2s})^1$

c. $(\sigma_{1s})^2(\sigma^*_{1s})^2(\sigma_{2s})^2(\sigma^*_{2s})^2(\pi_{2p})^3(\sigma_{2p})^1$

d. $(\sigma_{1s})^2(\sigma^*_{1s})^2(\sigma_{2s})^2(\sigma^*_{2s})^2(\pi_{2p})^4(\sigma_{2p})^2$

9-14. Which of the following excited state configurations are unstable? That is, which of the following configurations could result in the dissociation of an excited molecule?

a. $(\sigma_{1s})^2(\sigma^*_{1s})^2(\pi_{2p})^1$

b. $(\sigma_{1s})^2(\sigma^*_{1s})^2(\sigma_{2s})^1(\sigma^*_{2s})^1$

c. $(\sigma_{1s})^2(\sigma^*_{1s})^2(\sigma_{2s})^2(\sigma^*_{2s})^2(\pi_{2p})^4(\pi^*_{2p})^1$

d. $(\sigma_{1s})^2(\sigma^*_{1s})^2(\sigma_{2s})^2(\sigma^*_{2s})^2(\pi_{2p})^4(\pi^*_{2p})^4(\sigma^*_{2p})^1$

9-15. An excited state of the C_2 molecule has the electron configuration $(\sigma_{1s})^2(\sigma^*_{1s})^2(\sigma_{2s})^2(\sigma^*_{2s})^1(\pi_{2p})^4(\sigma_{2p})^1$. Would you expect the bond length in this excited state to be longer or shorter than that in the ground state?

9-16. In some cases the removal of an electron from a species can result in a stronger net bonding (e.g., an O_2^+ ion versus an O_2 molecule). Give an example in which the addition of an electron to a species produces a stronger net bonding.

9-17. The energy-level diagram in Figure 9.11 can be continued to higher energies. The next few orbitals in order of increasing energy are σ_{3s}, σ^*_{3s}, π_{3p}, σ_{3p}, π^*_{3p} and σ^*_{3p}; in other words, the σ_{2s} to σ^*_{2p} pattern is repeated. Use this extended energy-level diagram to predict whether or not a Mg_2 molecule is stable.

9-18. Use the extended energy-level diagram given in the previous problem to determine whether a P_2 molecule is paramagnetic or diamagnetic. Calculate the bond order of a P_2 molecule and compare this with its Lewis formula. Although phosphorous can exist as a stable diatomic species in the gas phase, at room temperature pure phosphorous forms tetrahedral P_4 molecules, called white phosphorous. Several other forms (or allotropes) of phosphorus exist as well (see Interchapter N).

POLYATOMIC MOLECULES

9-19. How many valence electrons are there in a BH_3 molecule? Describe the bonding in a BH_3 molecule in terms of hybrid orbitals.

9-20. How many valence electrons are there in a $HgCl_2$ molecule? Describe the covalent bonding in a $HgCl_2$ molecule in terms of hybrid orbitals. (Take the number of valence electrons in the mercury atom to be two. We shall learn the rules for assigning the number of valence electrons for transition metal elements in covalent molecules in Chapter 26.)

9-21. How many valence electrons are there in a CF_4 molecule? Use hybrid orbitals to describe the bonding in a CF_4 molecule.

9-22. How many valence electrons are there in a SF_2 molecule? Use hybrid orbitals to describe the bonding in a SF_2 molecule.

9-23. The hydronium ion, H_3O^+, has a trigonal pyramidal geometry with H–O–H bond angles of 110°. Describe the bonding in the H_3O^+ ion using localized bond orbitals.

9-24. Use localized bond orbitals to describe the bonding in a NF_3 molecule.

9-25. Use localized bond orbitals to describe the bonding in the tetrahydroborate ion, BH_4^-.

9-26. Use localized bond orbitals to describe the bonding in a methyl cation, CH_3^+.

9-27. How many valence electrons are there in a chloromethane molecule, CH_3Cl? Describe the bonding in a CH_3Cl molecule.

9-28. How many valence electrons are there in a chloroform molecule, $HCCl_3$? Describe the bonding in a chloroform molecule.

9-29. How many valence electrons are there in an IF_4^+ ion? Use hybrid orbitals to describe the bonding in an IF_4^+ ion.

9-30. How many valence electrons are there in a XeF_2 molecule? Use hybrid orbitals to describe the bonding in a XeF_2 molecule.

9-31. How many valence electrons are there in a PCl_6^- ion? Use hybrid orbitals to describe the bonding in a PCl_6^- ion.

9-32. How many valence electrons are there in a TeF_5^- ion? Use hybrid orbitals to describe the bonding in a TeF_5^- ion.

MOLECULES WITH NO UNIQUE CENTRAL ATOM

9-33. Use hybrid orbitals to describe the bonding in a hydroxylamine molecule, $HONH_2$.

9-34. A class of organic compounds called alcohols may be viewed as derived from HOH (a water molecule) by replacing one of the hydrogen atoms by an alkyl group, which is a hydrocarbon group, such as $-CH_3$ (methyl) or $-CH_2CH_3$ (ethyl). A simple alcohol is ethyl alcohol, or ethanol, whose Lewis formula is

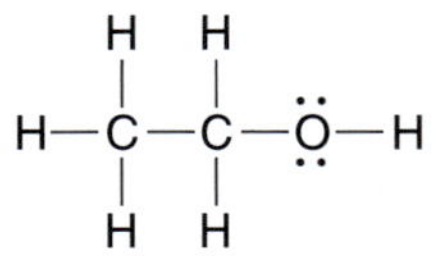

Describe the bonding and shape around both carbon atoms and the oxygen atom in an ethanol molecule.

9-35. A class of organic compounds called amines may be viewed as derived from NH_3 (an ammonia molecule) with one or more hydrogen atoms replaced by alkyl groups (see previous problem). Examples of amines are

CH_3NH_2	$(CH_3)_2NH$	$(CH_3)_3N$
methylamine	dimethylamine	trimethylamine

Describe the bonding and the shape of a methylamine molecule.

9-36. Describe the bonding and shape of dimethylamine (see previous problem). How many σ bonds are there? How many lone electron pairs?

9-37. If both hydrogen atoms in HOH (a water molecule) are replaced by alkyl groups (see Problem 9-34), the result is an ether, ROR′, where R and R′ are alkyl groups that may or may not be different. The simplest ether is dimethyl ether:

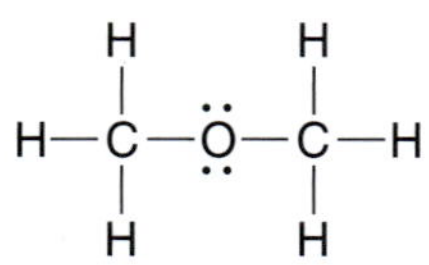

dimethyl ether

Describe the bonding around both carbon atoms and the oxygen atom in a dimethyl ether molecule.

9-38. Describe the bonding in an ethyl methyl ether molecule, whose Lewis formula is

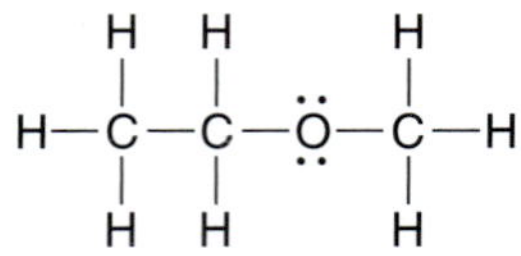

MULTIPLE BONDS

9-39. How many σ bonds and π bonds are there in each of the following molecules?

(a) ClHC=CHCl

(b) $H_2C=CHCH=CHCH_3$

(c) CH_3COOH

(d)

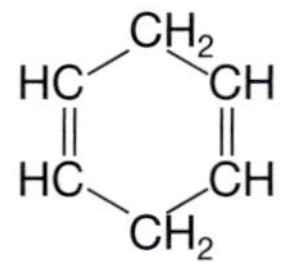

9-40. How many σ bonds and π bonds are there in each of the following molecules?

(a) $H_2C=CCl_2$

(b) HOOC–COOH

(c) FHC=C=CHF

(d)

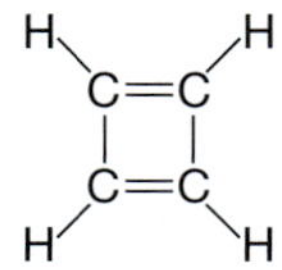

9-41. How many σ bonds are there in a 1-butyne (ethylacetylene) molecule, $CH{\equiv}CCH_2CH_3$? How many π bonds? How many valence electrons? Calculate the bond orders of each of the three carbon-carbon bonds in the molecule.

9-42. How many σ bonds and π bonds are there in a methyl cyanide molecule, CH_3CN? How many valence electrons? Calculate the bond orders of the C–C and C–N bonds.

9-43. Describe the bonding in a carbon monoxide molecule, CO, in terms of localized bond orbitals. Compare the localized bonding description of CO and that obtained from molecular orbital theory.

9-44. Describe the bonding in the acetylide ion, C_2^{2-}, in terms of localized bond orbitals. Compare the localized bonding description of C_2^{2-} and that obtained from molecular orbital theory.

GEOMETRIC ISOMERS

9-45. Draw the *cis* and *trans* isomers of 2-butene, $CH_3CHCHCH_3$. Label each isomer.

9-46. Draw the *cis* and *trans* isomers of 1,2-dichloro-1,2-dibromoethene, BrClC=CBrCl. Label each isomer.

9-47. Which of the species shown in Problem 9-39 can exhibit *cis-trans* isomerism? In each case where isomerism is possible, sketch the possible isomers and label them *cis* or *trans*.

9-48. Which of the species shown in Problem 9-40 can exhibit *cis-trans* isomerism? In each case where isomerism is possible, sketch the possible isomers and label them *cis* or *trans*.

DELOCALIZED BONDS

9-49. Naphthalene is a white, crystalline solid with an odor characteristic of mothballs. Write the complete Lewis formula for and describe the bonding in a naphthalene molecule, $C_{10}H_8$,

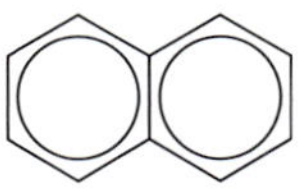

How many σ bonds and π bonds are there in a naphthalene molecule? How many valence electrons occupy σ-bond orbitals and how many occupy π-bond orbitals?

9-50. Anthracene is a yellow, crystalline solid found in coal tar. Write the complete Lewis formula for and describe the bonding in an anthracene molecule, $C_{14}H_{10}$,

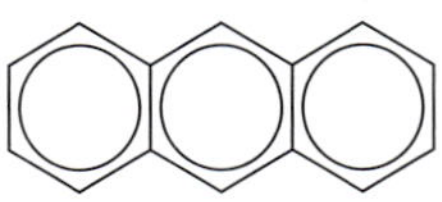

How many σ bonds and π bonds are there in an anthracene molecule? How many valence electrons occupy σ-bond orbitals and how many occupy π-bond orbitals?

ADDITIONAL PROBLEMS

9-51. Discuss the differences between the molecular orbital and the localized bond descriptions of a molecule. What can each theory tell us about a molecule? What are the advantages and disadvantages of each description?

9-52. What is the difference between a bonding and an antibonding molecular orbital?

9-53. Discuss the difference between a σ-bond orbital and a π-bond orbital.

9-54. How does bond order relate to bond length and bond strength?

9-55. What is the meaning of a "zero" bond order?

9-56. Explain why it is possible to form *cis-trans* isomers around a carbon–carbon double bond, but not around a carbon–carbon single or triple bond.

9-57. What is meant by a "delocalized" electron? How are delocalized electrons described using Lewis formulas? How are they described using molecular orbital theory?

9-58. In our localized bond description of σ-bonding orbitals in molecules, we generally chose to ignore the presence of antibonding orbitals. Why are we able to do this?

9-59. Which of the following is the least stable energetically: a H_2^+ molecular ion in the ground electronic state, $(\sigma_{1s})^1$; a H_2^+ molecular ion in the first excited (antibonding) state, $(\sigma_{1s}^*)^1$; or a separated H atom and H^+ ion?

9-60. Use Figure 9.13 to determine the bond order in a boron oxide molecule, BO.

9-61. Use molecular orbital theory to determine the relative bond lengths and bond energies of a CO molecule and a CO^+ ion.

9-62. Use molecular orbital theory to determine the ground state electron configuration of the cyanide ion, CN^-. What other species do you know that have the same bond order as that of a CN^- ion?

9-63. Write the ground state electron configuration and determine the bond order of a Na_2 molecule (see Problem 9-17).

9-64. An excited state of the N_2 molecule has the electron configuration $(\sigma_{1s})^2(\sigma_{1s}^*)^2(\sigma_{2s})^2(\sigma_{2s}^*)^2(\pi_{2p})^4(\sigma_{2p})^1(\pi_{2p}^*)^1$. Compare the bond length in this excited state to the bond length in the ground state of a N_2 molecule.

9-65. One method of experimentally determining whether a species is paramagnetic is to weigh it in an instrument called a magnetic susceptibility balance. This is a balance with a strong electromagnet placed next to the sample holder. If the species is paramagnetic, the mass reading of the balance will increase when the field is switched on (Figure 6.7). Which of the following species are paramagnetic? Which do you predict will have the strongest mass shift on a magnetic susceptibility balance? Why?

(a) CO (b) F_2 (c) O_2 (d) NO

9-66. If $F_2(g)$ molecules are excited by a laser to the $(\sigma_{1s})^2(\sigma^*_{1s})^2(\sigma_{2s})^2(\sigma^*_{2s})^2(\sigma_{2p})^1(\pi_{2p})^4(\pi^*_{2p})^4(\sigma^*_{2p})^1$ state, what are some of the ways that these excited molecules can release this additional electronic energy?

9-67. Aldehydes are organic compounds that have the general Lewis formula

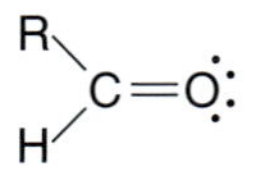

where R is either a hydrogen atom (giving formaldehyde) or an alkyl group such as $-CH_3$ (methyl) or $-CH_2CH_3$ (ethyl). Describe the bonding in an acetaldehyde molecule, CH_3CHO.

9-68. Ketones are organic compounds with the general Lewis formula

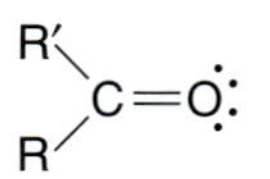

where R and R′ are alkyl groups (see previous problem) that may or may not be different. The simplest ketone is 2-propanone, commonly known as acetone, $(CH_3)_2CO$, one of the most important solvents. Describe the bonding and shape of an acetone molecule.

9-69. Describe the bonding and shape of a propyne (methyl acetylene) molecule, $CH_3C{\equiv}CH$. How many σ and π bonds are there? How many valence electrons?

9-70. Describe the bonding and shape of an ethyl cyanide molecule, CH_3CH_2CN. How many σ and π bonds are there? How many valence electrons?

9-71. Determine the type of hybrid orbitals that would be appropriate to use for the valence orbitals on the carbon atoms in the following molecules:

(a) C_2F_4 (b) CS_2

9-72. How many σ bonds and π bonds are there in the following molecules?

(a) CH_3CHCH_2 (b) CH_3CHO

(c) CH_3CN (d) CH_3OCH_3

9-73. Explain why the bonding in an ethene (ethylene) molecule, C_2H_4, requires that all the atoms in the molecule lie in the same plane.

9-74. There are three different possible isomers of a dibromoethene molecule, $C_2H_2Br_2$. One of them has no net dipole moment, but the other two do. Write Lewis formulas for each of these isomers and indicate the class of isomers they belong to. Indicate the net dipole moment of each of these isomers.

9-75. Caffeine, the active ingredient in coffee, tea, and some carbonated beverages, has the structural formula

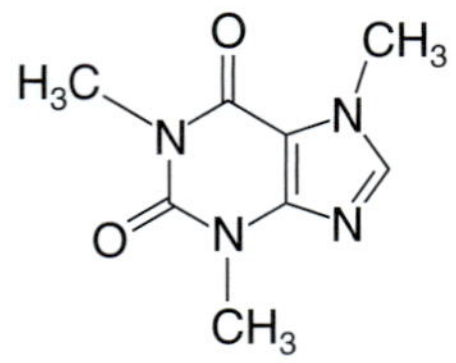

Draw the complete Lewis formula for caffeine (including all lone electron pairs) and indicate the hybridization of each of the atoms in the molecule.

9-76. The molecule cyanazine, once used as a herbicide, has now been phased out by the EPA because of its significant human toxicity. The structural formula of cyanazine is

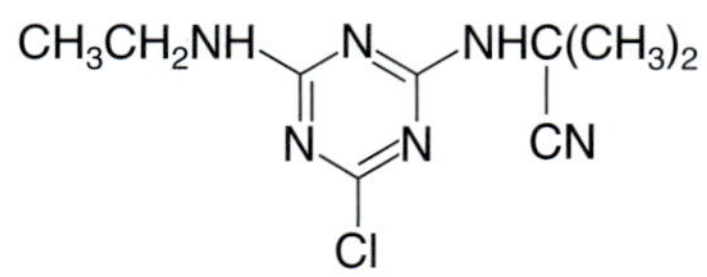

Draw the complete Lewis formula for cyanazine (including all lone electron pairs) and indicate the hybridization of each of the atoms in the molecule.

9-77. A ground state BO molecule may be promoted into an excited state using a laser. Using Figure 9.13, calculate the bond order of an excited BO molecule in its excited state of lowest energy. Comment on the stability of this excited state.

9-78. (*) The π energy-level diagram for a trigonal planar species such as a nitrate ion, NO_3^-, is

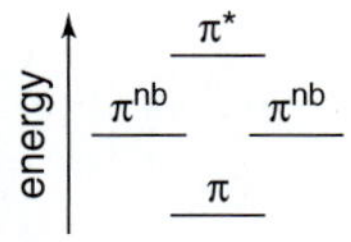

where π^{nb} designates a nonbonding π orbital. Use this π energy-level diagram to describe the bonding in a NO_3^- ion. What is the bond order of each nitrogen-oxygen bond in a NO_3^- ion? Can you rationalize your result with Lewis formulas for a NO_3^- ion?

9-79. (*) Use the π energy-level diagram shown in the previous problem to describe the bonding in a carbonate ion, CO_3^{2-}. Calculate the total bond order. Can you rationalize your results with the Lewis formulas for a CO_3^{2-} ion?

9-80. (*) Although the CO_2 molecule has double bonds on either side of the central carbon atom, as described by its Lewis formula,

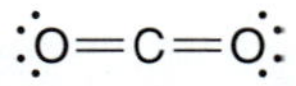

it does not exhibit delocalization of its electrons (resonance) across the whole molecule. Using the concepts from this chapter, explain why the electrons in each of the two double bonds in a CO_2 molecule are localized.

9-81. (*) Explain why the first ionization energy of a hydrogen molecule (2.47 aJ) is greater than that of a hydrogen atom (2.18 aJ); whereas the first ionization energy of an oxygen molecule (1.93 aJ) is less than that of an oxygen atom (2.18 aJ). Based on these observations, would you predict the first ionization energy of a nitrogen molecule to be greater or less than that of a nitrogen atom?

Percy Julian (1899–1975) was born in Montgomery, Alabama. The segregated school system at that time usually did not go beyond the 8th grade for African-Americans, but due to his parents' efforts, Julian was able to continue his education through high school. He was then accepted by DePauw University in Indiana as a probationary student, and in spite of the inadequacy of his early education, he graduated Phi Beta Kappa and valedictorian in 1920. With the encouragement of one of his chemistry professors, Julian went on to graduate school at Harvard University. After attaining his master's degree, he was unable to obtain a teaching assistantship in order to support himself for his doctorate degree because of the racial prejudices of the time. Nevertheless, he received a Rockefeller Foundation fellowship to study at the University of Vienna, where he received his Ph.D. in 1931. Although he enjoyed the liberal social environment and the freedom from racial prejudices in Vienna, Julian returned to the United States to accept a position at Howard University, where he met his future wife, Anna Johnson, the first African-American woman to earn a Ph.D. in sociology. He became embroiled in university politics at Howard and was forced to leave in 1932. His former professor offered him a position in organic chemistry at DePauw University. There his eloquent synthesis of a steroid used to treat glaucoma derived from an inexpensive extract of Calabar beans earned him an international reputation. He later joined the Glidden Company as director of research of the Soya Products Division, where he continued his research on steroids and earned 109 patents. In 1953 he founded Julian Laboratories, a chemical company that employed more African-American chemists than any other facility at the time and made him one of the wealthiest black businessmen in America. In addition to almost 20 honorary degrees, Julian was elected to the National Academy of Sciences in 1973 in recognition of his scientific achievements. His life was the subject of a NOVA program entitled *A Forgotten Genius*, which appeared on PBS.

10. Chemical Reactivity

Having laid the foundations, we are finally able to begin what is at the heart of chemistry, the study of chemical reactions. In the first few chapters, we learned about chemical elements and saw how the elements can be classified based on reactivity into a periodic arrangement, known as the periodic table of the elements. We then learned about modern quantum theory and how the electrons are arranged about the nucleus of an atom. We used these concepts to understand chemical bonding and why certain chemical elements bond together to form molecules.

Now in this chapter, based on our understanding of atoms, molecules, ions, and bonding, we shall begin to learn how chemicals react with one another. In the following chapters, we shall learn how to carry out chemical calculations and how chemists can predict the amount of products formed from a chemical reaction. But first we must learn a little more about chemical reactions and how chemists classify the various types of reactions.

One convenient classification scheme for chemical reactions has four categories: (1) combination reactions, (2) decomposition reactions, (3) single-replacement reactions, and (4) double-replacement reactions. Although not all chemical reactions fall into these four categories, many do, and it is helpful when first learning about chemical reactions to try to classify them into these types. Discussion of reaction types leads naturally to chemical nomenclature, to relative activities of metals, and to an introduction to the chemistry of acids and bases.

Finally, we present another classification system for chemical reactions. All chemical reactions can also be assigned to one of two classes: reactions in which electrons are transferred from one chemical species to another and reactions in which electrons are not transferred. Reactions in which electrons are transferred from one species to another are called oxidation-reduction ("redox") reactions or electron-transfer reactions.

Here we present only a qualitative introduction to these various reaction classes. Many important reaction types, such as acid-base reactions, precipitation reactions, and oxidation-reduction reactions, are further developed on a quantitative basis in later chapters.

10-1. A Combination Reaction Is the Reaction of Two Substances to Form a Single Product

One of the simplest classes of reactions is a **combination reaction**. In a combination reaction the reactants combine to form a single product. This type of reaction occurs between either two elements, an element and a compound, or two compounds. A combination reaction is easily recognized because it involves two (or occasionally three) reactants and only a single product.

Figure 10.1 Sulfur burns in oxygen with a blue flame. The production of sulfur dioxide, $SO_2(g)$, by burning sulfur is a key reaction in the manufacture of sulfuric acid. Sulfur dioxide is also produced when sulfur-containing coal or petroleum is burned.

Aluminum is resistant to extensive air oxidation because the initial oxidation of the aluminum metal by oxygen in air produces a tough, tightly adherent layer of aluminum oxide that is impervious to oxygen. This layer protects the underlying metal from further oxidation. In contrast, iron is oxidized (corroded) extensively by oxygen in air because the product iron(III) oxide is nonadherent. Thus, iron can be corroded completely by air oxidation. The oxidation of aluminum and iron by oxygen are examples of combination reactions involving a metal and a nonmetal, as described by the following chemical equations:

$$4\,\text{Al}(s) + 3\,\text{O}_2(g) \rightarrow 2\,\text{Al}_2\text{O}_3(s)$$
$$4\,\text{Fe}(s) + 3\,\text{O}_2(g) \rightarrow 2\,\text{Fe}_2\text{O}_3(s)$$

When the reactants in a combination reaction are a metal and a nonmetal, the product is often an ionic compound. Because ionic compounds almost always have high melting points [for example, NaCl(s) melts at 800°C and CaO(s) melts at 2850°C], we shall assume that the ionic reaction products are formed as solids, unless the reaction temperature is very high or a solvent in which the ionic compound is soluble is present. As we saw in Chapter 6, the high melting points of ionic compounds are a consequence of the strong electrostatic forces between oppositely charged ions in the solid.

Many combination reactions occur between two nonmetals. When two nonmetals react, the product is a covalently bonded compound. For example, carbon burns in oxygen to give carbon dioxide, as described by the equation

$$\text{C}(s) + \text{O}_2(g) \rightarrow \text{CO}_2(g) \qquad (10.1)$$

Sulfur burns in oxygen with a blue flame (Figure 10.1) to form sulfur dioxide, which is a toxic, colorless gas with a sharp, irritating odor.

$$\text{S}(s) + \text{O}_2(g) \rightarrow \text{SO}_2(g) \qquad (10.2)$$

Figure 10.2 Although steel wool does not burn in air, it burns vigorously in pure oxygen.

Carbon dioxide and sulfur dioxide are covalently bonded molecules. Covalent compounds are generally much more **volatile** than ionic compounds; that is, they are more readily vaporized than ionic solids and consequently tend to have lower melting points and boiling points. Thus, the covalent compounds $CO_2(g)$ and $SO_2(g)$ are gases at 25°C, whereas, as noted, the ionic compounds NaCl(s) and CaO(s) are solids even at high temperatures.

The reactions described in Equations 10.1 and 10.2 are also examples of **combustion reactions** (Figure 10.2). A combustion reaction is one in which a substance (or fuel) is burned in oxygen or some other oxidizer (we shall discuss oxidation reactions in Section 10-11).

The combination reactions presented so far are combinations of two elements. We can also have combination reactions between compounds. Our first example is the reaction between sodium oxide $Na_2O(s)$ and carbon dioxide to form the ionic compound sodium carbonate $Na_2CO_3(s)$, which we represent as

$$Na_2O(s) + CO_2(g) \rightarrow \underset{\text{sodium carbonate}}{Na_2CO_3(s)}$$

This combination reaction can be used to remove carbon dioxide from air and to measure the amount of carbon dioxide present in a gaseous sample by weighing the solid before and after addition of the gaseous sample.

When the gases ammonia and hydrogen chloride are mixed (Figure 10.3), they combine to form the ionic compound ammonium chloride as described by the equation

$$NH_3(g) + HCl(g) \rightarrow \underset{\text{ammonium chloride}}{NH_4Cl(s)}$$

Another example of a combination reaction involving two compounds is the reaction of sulfur trioxide, $SO_3(g)$, with magnesium oxide, $MgO(s)$, to form the ionic compound magnesium sulfate, $MgSO_4(s)$, according to

$$SO_3(g) + MgO(s) \rightarrow \underset{\text{magnesium sulfate}}{MgSO_4(s)}$$

This combination reaction can be used to remove sulfur trioxide from air.

Figure 10.3 Ammonia and hydrogen chloride are colorless gases. They react in a combination reaction to produce the solid white compound ammonium chloride $NH_4Cl(s)$. The white cloud in the picture consists of small ammonium chloride particles formed where the ammonia vapor from one bottle comes into contact with the hydrogen chloride vapor from the other.

10-2. Polyatomic Ions Remain Intact in Aqueous Solution

When a covalent compound, such as ethanol, $CH_3CH_2OH(l)$, is dissolved in water to form a solution, the ethanol and water molecules become more or less evenly distributed throughout the solution (Figure 10.4). In contrast, when an ionic compound, such as sodium chloride, $NaCl(s)$, dissolves in water, the species dissociates into positive and negative ions that tend to be surrounded by shells of water molecules in the solution (Figure 10.5). It turns out that these shells of solvated ions have a lower energy than the ions in the crystal lattice, thus accounting for the solubility of sodium chloride in water.

We can represent the process of dissolution of an ionic solid by a chemical equation, such as

$$NaCl(s) \xrightarrow[H_2O(l)]{} Na^+(aq) + Cl^-(aq) \qquad (10.3)$$

The placement of $H_2O(l)$ below the reaction arrow indicates that water is acting as a solvent for the products; that is, water is the liquid in which the sodium chloride dissolves. The symbols $Na^+(aq)$ and $Cl^-(aq)$ indicate that sodium chloride forms ions in aqueous solution. It is perfectly acceptable to write the formula $NaCl(aq)$ to indicate that sodium chloride has been dissolved in water, but the dissolution equation given above shows a clearer picture of what is actually occurring in the solution. As we shall see in the upcoming sections, many important chemical reactions occur between the various ions present in aqueous solutions.

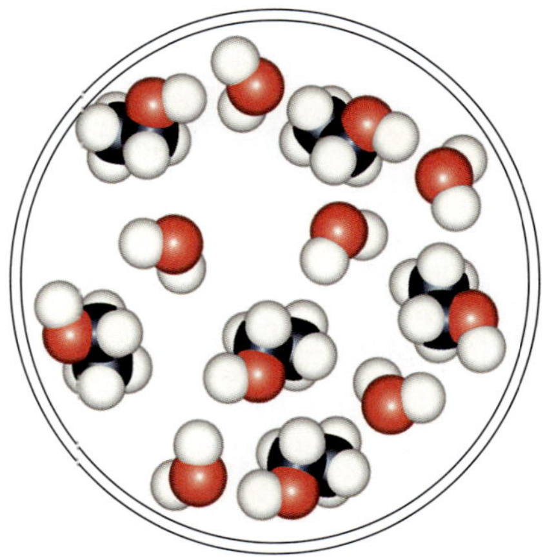

Figure 10.4 When a covalent compound such as ethanol, $CH_3CH_2OH(l)$, is dissolved in water, it tends to be dispersed throughout the solution.

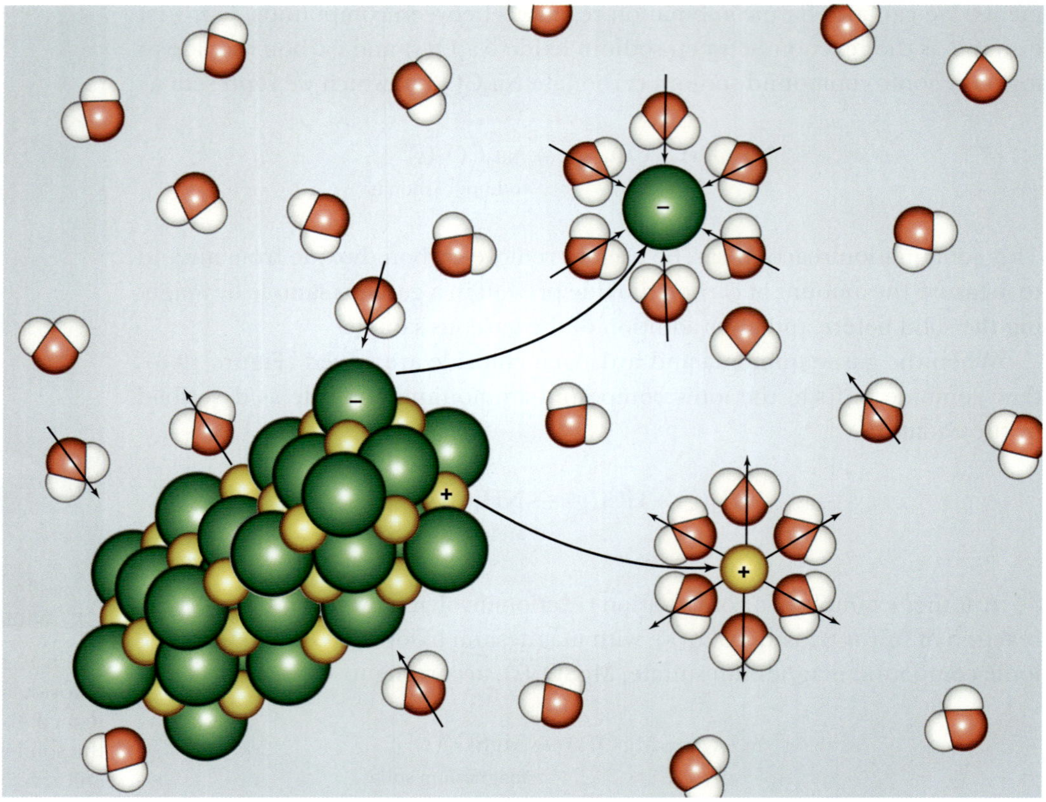

Figure 10.5 When an ionic compound such as sodium chloride, NaCl(*s*), is dissolved in water, the water molecules tend to form shells that solvate the resultant ions. Recall that water is a polar molecule. Notice that the negative ends of the water molecules in these shells are oriented toward the positive sodium ions and that the positive ends of the water molecules are oriented toward the negative chloride anions.

Sodium chloride is an example of a simple binary compound: one consisting of just two elements, a metal and a nonmetal. Let's now consider ammonium chloride, $NH_4Cl(s)$. Solid ammonium chloride is highly soluble in water. On the basis of the dissolution reaction of sodium chloride (Equation 10.3), you might predict (*incorrectly*) that when ammonium chloride is dissolved in water it breaks up into ions of its component elements according to the equation

$$NH_4Cl \xrightarrow[H_2O(l)]{} N^{3-}(aq) + 4H^+(aq) + Cl^-(aq) \quad (\textit{incorrect})$$

but this is not the case. It turns out that when $NH_4Cl(s)$ is dissolved in water, essentially only two ions per formula unit are produced:

$$\underset{\text{ammonium chloride}}{NH_4Cl(s)} \xrightarrow[H_2O(l)]{} \underset{\text{ammonium ion}}{NH_4^+(aq)} + \underset{\text{chloride ion}}{Cl^-(aq)}$$

What we call an ammonium ion, NH_4^+, tends to remain intact as a **polyatomic ion** in aqueous solution. As we shall see, a polyatomic ion is a charged species consisting of two or more atoms held together by covalent bonds. Polyatomic ions are constituents of many ionic substances.

Let's look at another example. When magnesium sulfate, $MgSO_4(s)$, is dissolved in water, only two ions per $MgSO_4$ formula unit are produced as described by

$$\underset{\text{magnesium sulfate}}{MgSO_4(s)} \xrightarrow[H_2O(l)]{} \underset{\text{magnesium ion}}{Mg^{2+}(aq)} + \underset{\text{sulfate ion}}{SO_4^{2-}(aq)}$$

The SO_4^{2-} ion remains intact as a polyatomic ion in aqueous solution (Figure 10.6).

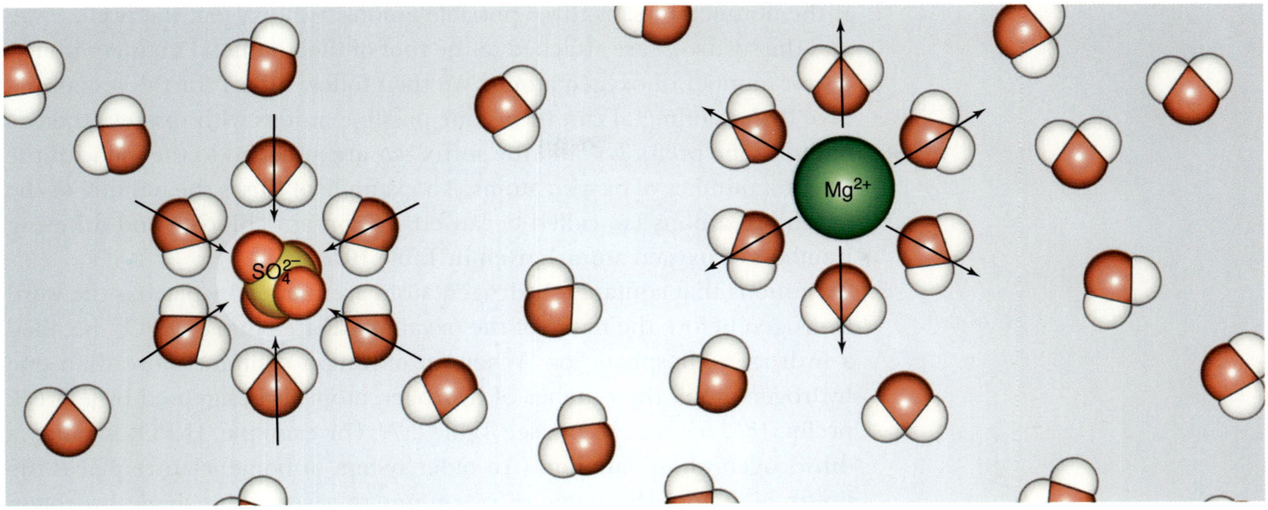

Figure 10.6 $MgSO_4(s)$ dissolves in water to form $Mg^{2+}(aq)$ and $SO_4^{2-}(aq)$ ions.

As we have already noted, polyatomic ions like NH_4^+ and SO_4^{2-} are charged species held together by covalent bonds. We can illustrate this fact by using the rules we learned in Chapter 7 to write the Lewis formulas of these ions. The Lewis formula for an NH_4^+ ion is

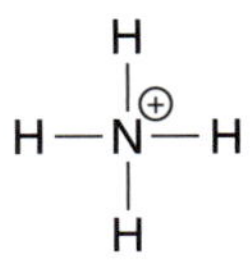

and that for a SO_4^{2-} ion is

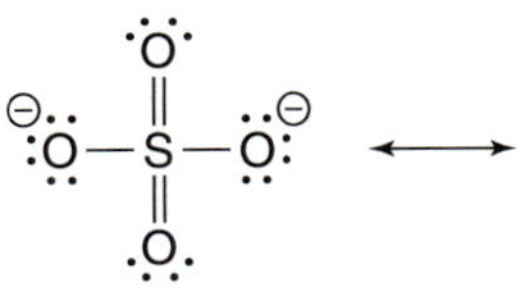

The polyatomic ion SO_4^{2-} is called a sulfate ion; similarly, the name of $MgSO_4(s)$ is magnesium sulfate.

There are many polyatomic ions that are important in chemistry. Table 10.1 lists some of the most common of these. The table is grouped by charge and arranged so that compounds with similar chemical formulas and names are grouped together. Many of the polyatomic ions have irregular names, generally because they were named prior to our modern systematic nomenclature. However, in the case of the **oxyanions**, anions involving nonmetals combined with oxygen, we can often apply the following systematic rules:

1. If a nonmetal forms two possible anions with oxygen, the one with the fewer number of oxygen atoms has the suffix *-ite*, and the greater number of oxygen atoms has the suffix *-ate* attached to the root of the nonmetal's name. For example,

   NO_2^- is nitr*ite*
   NO_3^- is nitr*ate*

2. If the nonmetal forms three possible anions with oxygen, the prefix *hypo-* and the suffix *-ite* are attached to the root of the nonmetal's name with the fewest number of oxygen atoms. We then follow Rule 1 for the remaining two. If the nonmetal can form four possible anions with oxygen, then in addition the prefix *per-* and the suffix *-ate* are attached to the ion with the greatest number of oxygen atoms. An example of this is the naming of the oxychloro anions (so called because they contain chlorine and differing numbers of oxygen atoms) given in Table 10.2.
3. Oxyanions that contain a hydrogen atom are named by placing the word hydrogen before the name of the oxyanion. For example, HPO_4^{2-} is called a hydrogen phosphate ion. When an oxyanion contains more than one hydrogen atom, the number of hydrogen atoms is designated by a Greek prefix, (*di-*, *tri-*, and so on, see Table 2.7). For example, $H_2PO_4^-$ is called a dihydrogen phosphate ion. (An older system of nomenclature places the prefix *bi-* before the name of oxyanions that contain a single hydrogen atom. For example, a hydrogen carbonate ion, HCO_3^-, is sometimes called a bicarbonate ion. These older names are given in parentheses in Table 10.1.)

Compounds containing the ions in Table 10.1 are named according to the rules for naming binary compounds. For example, NaOH(*s*) is called sodium hydroxide, KCN(*s*) is called potassium cyanide, and NH_4Cl(*s*) is called ammonium chloride.

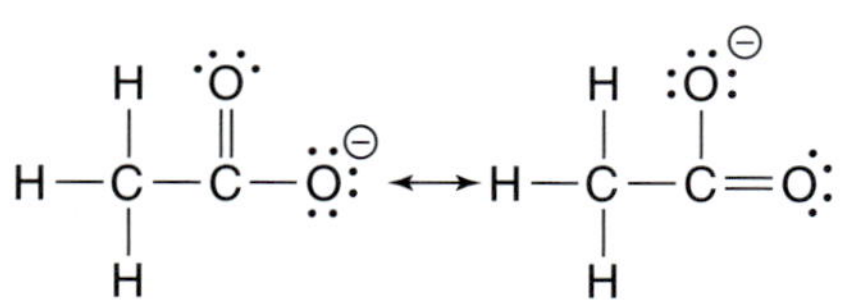

Although the acetate ion is often written compactly as $C_2H_3O_2^-$, it is better written as CH_3COO^- which more closely resembles its Lewis formula.

TABLE 10.1 Common polyatomic ions*

OH^-	hydroxide	O_2^{2-}	peroxide
CN^-	cyanide	CO_3^{2-}	carbonate
SCN^-	thiocyanate	SO_3^{2-}	sulfite
HCO_3^-	hydrogen carbonate (bicarbonate)	SO_4^{2-}	sulfate
HSO_3^-	hydrogen sulfite (bisulfite)	$S_2O_3^{2-}$	thiosulfate
HSO_4^-	hydrogen sulfate (bisulfate)	$C_2O_4^{2-}$	oxalate
$C_2H_3O_2^-$	acetate (also written as CH_3COO^-)	CrO_4^{2-}	chromate
NO_2^-	nitrite	$Cr_2O_7^{2-}$	dichromate
NO_3^-	nitrate		
MnO_4^-	permanganate	PO_3^{3-}	phosphite
ClO^-	hypochlorite	PO_4^{3-}	phosphate
ClO_2^-	chlorite		
ClO_3^-	chlorate	NH_4^+	ammonium
ClO_4^-	perchlorate	Hg_2^{2+}	mercury(I)

*Often encountered common names are given in parentheses.

TABLE 10.2 Nomenclature of oxychloro anions

Formula for oxyanion	Number of oxygen atoms	Prefix used	Suffix used	Name of oxyanion
ClO^-	1	*hypo-*	*-ite*	hypochlorite
ClO_2^-	2		*-ite*	chlorite
ClO_3^-	3		*-ate*	chlorate
ClO_4^-	4	*per-*	*-ate*	perchlorate

EXAMPLE 10-1: Name the following ionic compounds:
(a) $KMnO_4(s)$ (b) $Co(NO_2)_2(s)$ (c) $CrPO_4(s)$ (d) $NaHSO_3(s)$

Solution: (a) From Table 10.1, we note that MnO_4^- is called the permanganate ion, so $KMnO_4(s)$ is called potassium permanganate. Aqueous solutions of potassium permangate are a beautiful purple color (Figure 10.7).
(b) The NO_2^- ion is called the nitrite ion, so $Co(NO_2)_2(s)$ is called cobalt(II) nitrite (see Table 6.2). (c) The PO_4^{3-} ion is called the phosphate ion, so $CrPO_4(s)$ is called chromium(III) phosphate (see Table 6.2).
(d) The HSO_3^- ion is an oxyanion formed by the addition of a hydrogen atom to a sulfite, SO_3^{2-}, ion, and so it is called a hydrogen sulfite ion. Therefore, $NaHSO_3(s)$ is called sodium hydrogen sulfite. Its older name is sodium bisulfite.

Figure 10.7 Potassium permanganate, $KMnO_4(s)$, dissolving in water.

PRACTICE PROBLEM 10-1: Name the following ionic compounds:
(a) $NH_4CH_3OO(s)$ (b) $PbCrO_4(s)$ (c) $K_2Cr_2O_7(s)$ (d) $NaH_2PO_3(s)$

Answer: (a) ammonium acetate (b) lead(II) chromate
(c) potassium dichromate (d) sodium dihydrogen phosphite

EXAMPLE 10-2: Write the chemical formulas for (a) sodium thiosulfate, (b) copper(II) perchlorate, and (c) calcium hydroxide.

Solution: (a) Sodium thiosulfate involves the ions Na^+ and $S_2O_3^{2-}$. Because the $S_2O_3^{2-}$ ion has an ionic charge of −2, it requires two Na^+ ions for each $S_2O_3^{2-}$ anion, and so $Na_2S_2O_3(s)$ is the formula of sodium thiosulfate.
(b) Copper(II) perchlorate requires two ClO_4^- anions for each Cu^{2+} ion, so the formula of the compound is $Cu(ClO_4)_2(s)$. Note that the entire perchlorate anion, which exists as a unit, is enclosed in parentheses and the subscript 2 lies outside the parentheses. One formula unit of $Cu(ClO_4)_2(s)$ contains one copper atom, two chlorine atoms, and eight oxygen atoms.
(c) Calcium hydroxide involves the ions Ca^{2+} and OH^-; thus, it has the formula $Ca(OH)_2(s)$. Once again, note the use of parentheses.

PRACTICE PROBLEM 10-2: Write the chemical formula for
(a) sodium phosphate (b) mercury(I) nitrate (c) nickel(II) sulfite

Answer: (a) $Na_3PO_4(s)$ (b) $Hg_2(NO_3)_2(s)$ (c) $NiSO_3(s)$

As we have already seen, when compounds containing polyatomic ions are dissolved in water, the polyatomic ions usually persist in aqueous solution as covalently bonded units.

EXAMPLE 10-3: Write the chemical equations for the dissolution reactions that occur when the following ionic compounds are dissolved in water: (a) $(NH_4)_2S_2O_3(s)$ (b) $Hg_2(NO_3)_2(s)$ (c) $K_2Cr_2O_7(s)$

Solution: We solve this problem by using our knowledge of the formulas for polyatomic ions (Table 10.1).

(a) $(NH_4)_2S_2O_3(s) \xrightarrow[H_2O(l)]{} 2\,NH_4^+(aq) + S_2O_3^{2-}(aq)$

(b) $Hg_2(NO_3)_2(s) \xrightarrow[H_2O(l)]{} Hg_2^{2+}(aq) + 2\,NO_3^-(aq)$

(c) $K_2Cr_2O_7(s) \xrightarrow[H_2O(l)]{} 2\,K^+(aq) + Cr_2O_7^{2-}(aq)$

PRACTICE PROBLEM 10-3: Write the chemical formulas of the ions produced when the following ionic compounds are dissolved in water:
(a) $NH_4CH_3OO(s)$ (b) $Hg(ClO_4)_2(s)$ (c) $Ba(OH)_2(s)$

Answer: (a) $NH_4^+(aq)$ and $CH_3COO^-(aq)$ (b) $Hg^{2+}(aq)$ and $2\,ClO_4^-(aq)$
(c) $Ba^{2+}(aq)$ and $2\,OH^-(aq)$

10-3. Some Metal Oxides Yield Bases and Some Hydrogen-Containing Compounds Yield Acids When Combined with Water

An important class of combination reactions are those that involve the combination of metal oxides or of nonmetal oxides with water. For example, if we combine solid sodium oxide with an excess of water, then the water acts both as a reactant and as a solvent for the products according to

$$Na_2O(s) + H_2O(l) \rightarrow \underset{\text{sodium hydroxide}}{2\,NaOH(aq)} \tag{10.4}$$

Similarly, barium oxide reacts with an excess of water according to

$$BaO(s) + H_2O(l) \rightarrow \underset{\text{barium hydroxide}}{Ba(OH)_2(aq)} \tag{10.5}$$

In aqueous solution $NaOH(aq)$ exists as the ions $Na^+(aq)$ and $OH^-(aq)$, and $Ba(OH)_2(aq)$ exists as the ions $Ba^{2+}(aq)$ and $OH^-(aq)$. Thus, the above two equations often are written in the more explicit form

$$Na_2O(s) + H_2O(l) \rightarrow 2\,Na^+(aq) + 2\,OH^-(aq)$$

and

$$BaO(s) + H_2O(l) \rightarrow Ba^{2+}(aq) + 2\,OH^-(aq)$$

A compound that yields **hydroxide ions** (OH^-) when dissolved in water is called a **base.** Both sodium hydroxide and barium hydroxide are bases.

Many metal oxides do not react with water and so do not form basic solutions. An example of such a nonreactive metal oxide is aluminum oxide, $Al_2O_3(s)$. This compound gives the surface of aluminum doors and window frames their dull appearance and is **inert** (unreactive) in water. Thus, we write

$$Al_2O_3(s) + H_2O(l) \rightarrow \text{no reaction}$$

The only metal oxides that react with water are the Group 1 metal oxides and some of the Group 2 metal oxides (Figure 10.8).

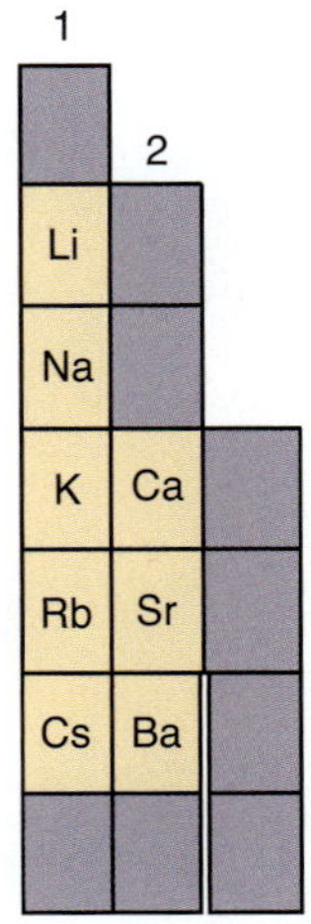

Figure 10.8 Metals whose oxides react with water, forming basic solutions.

EXAMPLE 10-4: Write a balanced chemical equation for the reaction of potassium oxide, $K_2O(s)$, with water.

Solution: Potassium is a Group 1 metal, so its oxide reacts with water much as sodium oxide does. Using Equation 10.4 for the reaction of $Na_2O(s)$ with water as a guide, we write

$$K_2O(s) + H_2O(l) \rightarrow 2\,K^+(aq) + 2\,OH^-(aq)$$

Because hydroxide ions are produced when $K_2O(s)$ reacts with water, $KOH(aq)$ is a base.

PRACTICE PROBLEM 10-4: Which one of the following metal oxides can react with water to form a basic solution?
(a) $Fe_2O_3(s)$ (b) $ZnO(s)$ (c) $SrO(s)$ (d) $TiO_2(s)$
Write a balanced chemical equation for the reaction that occurs.

Answer: $SrO(s)$ is the only metal oxide containing a Group 1 or Group 2 metal.

$$SrO(s) + H_2O(l) \rightarrow Sr^{2+}(aq) + 2\,OH^-(aq)$$

Many nonmetal oxides yield acids when combined with water. A simple definition of an **acid** is a compound that yields **hydrogen ions,** $H^+(aq)$, when dissolved in water. Experimental evidence suggests that $H^+(aq)$ ions in aqueous solution exist in several forms, such as $H_3O^+(aq)$, $H_5O_2^+(aq)$, and $H_9O_4^+(aq)$. However, the species $H_3O^+(aq)$, called the **hydronium ion,** is the dominant species. Using Lewis formulas, we can write the equation for the formation of $H_3O^+(aq)$ as

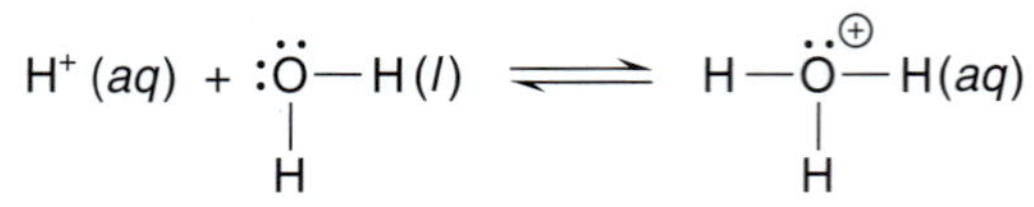

We can see here that a hydronium ion is a combination of a hydrogen ion and a water molecule. For simplicity, we sometimes denote a hydrogen ion in water by $H^+(aq)$ instead of $H_3O^+(aq)$.

It is not possible, in most cases, to decide whether a compound is an acid simply from the presence of hydrogen atoms in its chemical formula. The fact that a compound contains hydrogen atoms does not guarantee that it will yield $H^+(aq)$ ions in water. For example, hydrogen, $H_2(g)$, methane, $CH_4(g)$, and ethanol, $CH_3CH_2OH(l)$, are not acids. We shall learn the criteria for determining if a compound is an acid in Chapter 20; at this point we shall only introduce some of the more common acids and their reactions.

Some common polyatomic acids and their corresponding anions are given in Table 10.3. Notice that in each of these cases an acid is formed when a hydrogen cation is combined with a polyatomic anion from Table 10.1. Also note that in the case of acetic acid only the hydrogen atom in the –COOH group is acidic; the three hydrogen atoms in the $-CH_3$ group are not acidic. The equation that describes the acidic nature of acetic acid is

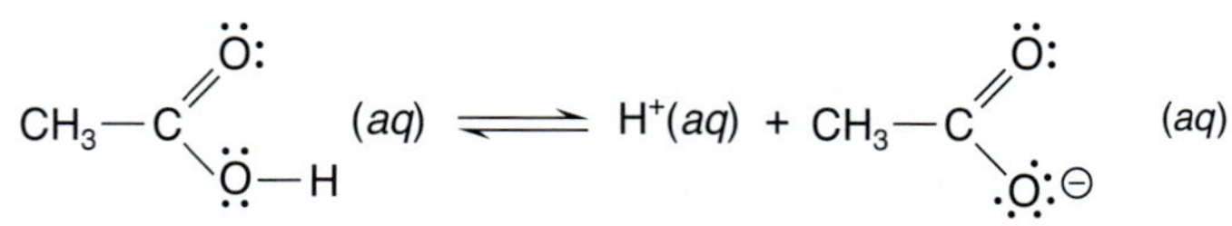

The hydrogen atoms in the chemical formula of acids that yield $H^+(aq)$ ions when the acid is dissolved in water are called **acidic hydrogen atoms** or **acidic protons.** Recall from Chapter 2 that a neutral hydrogen atom contains one proton and one electron, so that a $H^+(aq)$ ion is simply a proton. Nitric acid, $HNO_3(aq)$, is said to be a **monoprotic acid** because it has one acidic proton; sulfuric acid, $H_2SO_4(aq)$, is said to be a **diprotic acid** because it has two acidic protons.

$$H_2SO_4(aq) \rightarrow 2\,H^+(aq) + SO_4^{2-}(aq)$$

Phosphoric acid, $H_3PO_4(aq)$ is said to be a **triprotic acid** because it has three acidic protons. Acids that have more than one acidic proton are also referred to collectively as **polyprotic acids**.

EXAMPLE 10-5: Write a chemical equation showing the dissolution of nitric acid into ions in water.

Solution: According to Table 10.3, the chemical formula of nitric acid is HNO_3 and the formula of its corresponding anion is NO_3^-. Consequently, we write

$$HNO_3(aq) \rightarrow H^+(aq) + NO_3^-(aq)$$

TABLE 10.3 Common polyatomic acids and their anions

Acid	Formula	Anion	Formula
acetic acid*	$HC_2H_3O_2$ (CH_3COOH)	acetate	$C_2H_3O_2^-$ (CH_3COO^-)
carbonic acid	H_2CO_3	carbonate	CO_3^{2-}
nitric acid	HNO_3	nitrate	NO_3^-
perchloric acid	$HClO_4$	perchlorate	ClO_4^-
phosphoric acid	H_3PO_4	phosphate	PO_4^{3-}
sulfuric acid	H_2SO_4	sulfate	SO_4^{2-}

*Formulas in parentheses are commonly written alternative forms of acetic acid and acetate. The systematic name of acetic acid is ethanoic acid; however, the name acetic acid is still preferred in general usage.

PRACTICE PROBLEM 10-5: Oxalic acid, which is formed from the oxalate polyatomic anion (Table 10.1), can be extracted from the roots and leaves of plants such as rhubarb and tea. Predict whether oxalic acid is a monoprotic, diprotic, or triprotic acid.

Answer: diprotic

The acids listed in Table 10.3 are called **oxyacids** because their chemical formulas contain oxygen atoms. The nomenclature of the oxyacids is also based on the name of the anion from which the acid is derived and can be summarized by the following two rules:

A summary of all the nomenclature rules is given in Appendix C.

1. If the name of the anion ends in *-ite*, then the corresponding acid name ends in *-ous acid*.
2. If the name of the anion ends in *-ate*, then the corresponding acid name ends in *-ic acid*.

The following Example illustrates the application of these rules.

EXAMPLE 10-6: Name the following oxyacids:
(a) $HNO_2(aq)$ (b) $H_2C_2O_4(aq)$ (c) $HClO_4(aq)$

Solution: (a) $HNO_2(aq)$ is formed from a proton and a nitrite anion. The name nitrite ends in *-ite*, so the acid is named nitrous acid. (b) $H_2C_2O_4(aq)$ is formed from two protons and an oxalate anion. Oxalate ends in *-ate*, so the acid is named oxalic acid. (c) $HClO_4(aq)$ is formed from a proton and a perchlorate anion, so the acid is named perchloric acid.

PRACTICE PROBLEM 10-6: The nomenclature of the various oxychloro acids formed from the oxychloro anions (Table 10.2) is given in Table 10.4.

Bromine and iodine (but not fluorine) form oxyanions and oxyacids that have formulas similar to those of chlorine, and so are named accordingly. For example, $IO_4^-(aq)$ is a periodate anion and $HIO_4(aq)$ is periodic acid. Similarly, $BrO^-(aq)$ is a hypobromite anion and $HBrO(aq)$ is hypobromous acid. Name the following oxyanions and oxyacids:

(a) $IO^-(aq)$ (b) $HIO_2(aq)$ (c) $BrO_2^-(aq)$

Answer: (a) hypoiodite ion (b) iodous acid (c) bromite ion

There is another group of acids called **binary acids** that consist of only two elements, one of which must be hydrogen. The most important binary acid is hydrochloric acid, $HCl(aq)$, which is obtained by dissolving hydrogen chloride gas in water:

$$HCl(g) \xrightarrow[H_2O(l)]{} \underset{\text{hydrochloric acid}}{H^+(aq) + Cl^-(aq)}$$

The formulas and names of four common binary acids are given in Table 10.5. When naming binary acids, we simply add the prefix *hydro-* to the name of the anion and the suffix *-ic* plus the word *acid.* For example, the binary acid $H_2S(aq)$ is named *hydrosulfuric acid.* Note that an acidic substance must be dissolved in a solution to be an acid. For example, $HCl(g)$ is hydrogen chloride, whereas $HCl(aq)$ is hydrochloric acid.

In addition to the polyatomic acids (Table 10.3) and binary acids (Table 10.5), soluble organic substances with a Lewis formula that contains the functional group –COOH are also acidic. Such substances are called **organic acids** (Interchapter R). The two simplest organic acids (also known as **carboxylic acids**) are formic acid and acetic acid.

See Interchapter R at www.McQuarrieGeneralChemistry.com

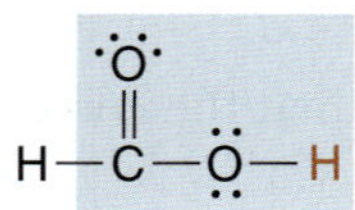

formic acid (found in ant and bee stings)

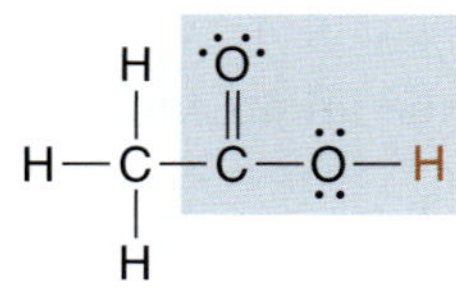

acetic acid (found in vinegar)

TABLE 10.4 Nomenclature of oxychloro acids

Formula for oxyanion	Number of oxygen atoms	Prefix used	Suffix used	Name of oxyanion
$HClO$	1	*hypo-*	*-ous*	hypochlorous acid
$HClO_2$	2		*-ous*	chlorous acid
$HClO_3$	3		*-ic*	chloric acid
$HClO_4$	4	*per-*	*-ic*	perchloric acid

TABLE 10.5 Common binary acids

Acid	Formula	Anion	Corresponding gas
hydrobromic acid	$HBr(aq)$	$Br^-(aq)$	hydrogen bromide, $HBr(g)$
hydrochloric acid	$HCl(aq)$	$Cl^-(aq)$	hydrogen chloride, $HCl(g)$
hydroiodic acid	$HI(aq)$	$I^-(aq)$	hydrogen iodide, $HI(g)$
hydrosulfuric acid	$H_2S(aq)$	$S^{2-}(aq)$	hydrogen sulfide, $H_2S(g)$

The acidic functional groups are shaded and the acidic protons are shown in red. The chemical formula of acetic acid is often written as $CH_3COOH(aq)$, rather than as $HC_2H_3O_2(aq)$, to emphasize the presence of the –COOH group (Table 10.3). This is actually the preferable way to write it.

We should point out that formic acid and acetic acid are actually the older or common names of what are now systematically named methanoic acid and ethanoic acid, respectively. However, because the common names of these substances are still the ones preferred in general usage, we use the older names here. (We discuss the systematic naming of carboxylic acids in Interchapter R.)

We can use the Lewis formulas that we developed in Chapter 7 to understand the acidity of carboxylic acids. Let's look, for example, at acetic acid. When acetic acid dissociates into ions, the acetate anion formed may be described by the two resonance formulas

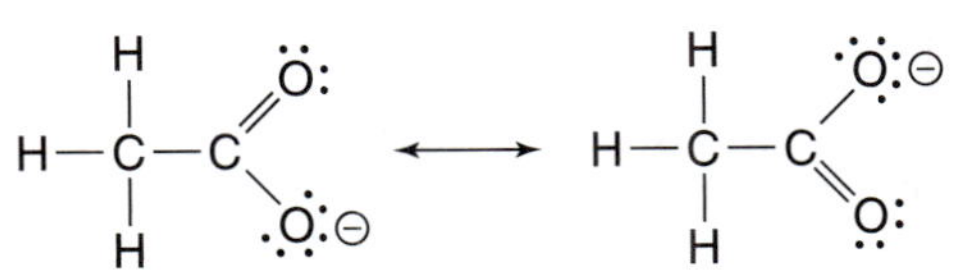

or by the resonance hybrid

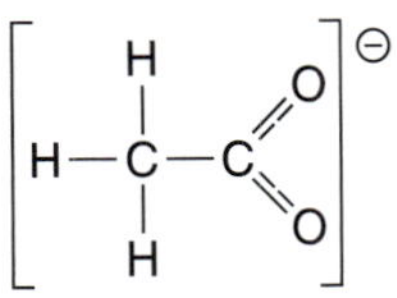

The Lewis formula of the hybrid shows that the negative charge is distributed equally between the two oxygen atoms. This delocalization of the negative charge over the two oxygen atoms confers an added degree of stability to the acetate anion. Recall from Chapters 7 and 9 that resonance structures are particularly stable. Therefore, we can explain the acidity of organic acids (acids containing a –COOH group) by the fact that these acids leave behind a resonance-stabilized anion when they dissociate in solution.

Because acids and bases are so important in chemistry, we have listed a few of their properties in Table 10.6. (We shall discuss acids and bases in greater depth in Chapters 20 and 21.) As the table shows, an **acidic solution** tastes sour. Vinegar tastes sour because it is a dilute solution of acetic acid; lemon juice tastes sour because it contains citric acid; and rhubarb tastes sour because it

TABLE 10.6 Properties of acids and bases

Acids	Bases
solutions taste sour (don't taste any, but recall the taste of vinegar or lemon juice)	solutions taste bitter and feel slippery to the touch (don't taste any, but recall the taste of soap)
produce hydrogen ions, $H^+(aq)$, when dissolved in water	produce hydroxide ions, $OH^-(aq)$, when dissolved in water
neutralize bases to produce salts and water	neutralize acids to produce salts and water
solutions turn blue litmus paper red	solutions turn red litmus paper blue
react with many metals to produce hydrogen gas	

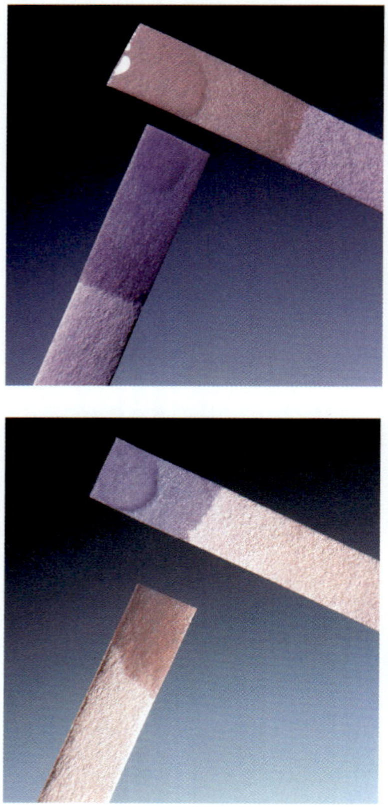

Figure 10.9 Litmus paper is used as a simple test for acidity or basicity. (*top*) Blue litmus paper turns red when placed in an acidic solution, and remains blue when placed in water or a basic solution. (*bottom*) Red litmus paper turns blue when placed in a basic solution, and remains red when placed in water or an acidic solution.

contains oxalic acid. A **basic solution** tastes bitter and feels slippery; an example of a basic solution is soapy water. An interesting and important property of acidic and basic solutions is their effect on certain dyes and vegetable matter. Litmus, which is a vegetable substance obtained from lichens, is red in acidic solutions and blue in basic solutions. Paper impregnated with litmus, called **litmus paper,** provides us with the means to perform a quick test to see whether a solution is acidic or basic (Figure 10.9). Because many acids and bases are poisonous, we *never* test for acidity or basicity by tasting chemicals.

10-4. In a Decomposition Reaction, a Substance Is Broken Down into Two or More Less Complex Substances

A **decomposition reaction** is the opposite of a combination reaction because decomposition involves the breaking apart of a substance into simpler substances. Such a reaction is easy to recognize because there is only one reactant and more than one product. For example, when heated, many metal oxides decompose by giving off oxygen gas. When mercury(II) oxide is heated (Figure 10.10), it decomposes into its constituent elements, mercury and oxygen according to

$$2\,HgO(s) \xrightarrow{\text{high } T} 2\,Hg(g) + O_2(g)$$

Figure 10.10 When mercury(II) oxide is heated, it decomposes into elemental mercury and oxygen gas. The red compound shown here is mercury(II) oxide. Liquid mercury has condensed on the walls of the test tube and the evolved oxygen gas exits at the tube opening.

The designation "high T" (T for temperature) over the arrow indicates that the reaction must be run at a high temperature. (In some cases we specify the actual temperature, as in $\xrightarrow{500°\text{C}}$.) The mercury(II) oxide decomposition reaction is the reaction that was used by the English chemist Joseph Priestley in 1774 in the first preparation and identification of oxygen as an element.

Many metal carbonates, such as calcium carbonate, $CaCO_3(s)$, decompose to yield a metal oxide and carbon dioxide gas upon heating. For example, the high temperature decomposition of $CaCO_3(s)$ is described by the equation

$$CaCO_3(s) \xrightarrow{\text{high } T} CaO(s) + CO_2(g)$$

Calcium carbonate, $CaCO_3(s)$, occurs in nature as the principal constituent of limestone, seashells, and eggshells (Figure 10.11); it is also the major ingredient in blackboard chalk.

Many metal sulfites and sulfates undergo decomposition reactions, similar to those of metal carbonates, to form metal oxides and sulfur oxides:

$$CaSO_3(s) \xrightarrow{\text{high } T} CaO(s) + SO_2(g)$$

$$MgSO_4(s) \xrightarrow{\text{high } T} MgO(s) + SO_3(g)$$

Figure 10.11 Calcium carbonate is the principal constituent of eggshells. (*top*) A magnified eggshell and the underlying membrane. (*bottom*) Further enlargement of the surface of the shell itself shows its crystalline structure. An eggshell is about 0.6 mm thick.

EXAMPLE 10-7: When potassium chlorate is heated, oxygen gas is evolved and solid potassium chloride formed. This decomposition reaction can be used to generate small quantities of oxygen in the laboratory. Write a balanced chemical equation for the process.

Solution: The chemical formulas of the reactant and products are

$$KClO_3(s) \xrightarrow{\text{high } T} KCl(s) + O_2(g) \quad \textit{not balanced}$$

We can balance the equation with respect to oxygen by placing a 2 in front of $KClO_3(s)$ ($2 \times 3 = 6$ oxygen atoms) and a 3 in front of $O_2(g)$ ($3 \times 2 = 6$ oxygen atoms):

$$2\,KClO_3(s) \xrightarrow{\text{high } T} KCl(s) + 3\,O_2(g) \quad \textit{not balanced}$$

We can finish balancing the equation by placing a 2 in front of $KCl(s)$ to obtain

$$2\,KClO_3(s) \xrightarrow{\text{high } T} 2\,KCl(s) + 3\,O_2(g) \quad \textit{balanced}$$

This reaction is called a thermal decomposition reaction because the decomposition is brought about by heating.

PRACTICE PROBLEM 10-7: When white crystals of sodium azide, $NaN_3(s)$, are heated in a test tube under a vacuum, the solid melts to a clear liquid, which on further heating decomposes with the evolution of a colorless unreactive gas and the deposition of a metallic mirrorlike deposit on the inside of the tube (Figure 10.12). Write a balanced chemical equation for the decomposition reaction.

Answer: $2\,NaN_3(s) \rightarrow 2\,Na(s) + 3\,N_2(g)$

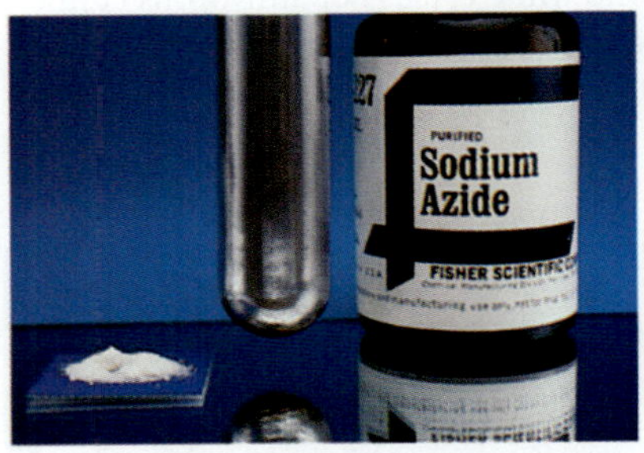

Figure 10.12 When sodium azide, $NaN_3(s)$, is heated in a vacuum, it decomposes into nitrogen gas and sodium metal vapor. The sodium metal vapor condenses on the wall of the tube to form a sodium mirror.

Sodium azide is the primary ingredient in the explosive charge of automobile safety air bags (Figure 10.13). In contrast to the relatively safe decomposition reaction of sodium azide, many other metal azides decompose violently when heated or when subjected to even mild mechanical shock. Lead(II) azide, $Pb(N_3)_2(s)$, and mercury(II) azide, $Hg(N_3)_2(s)$, are compounds that are used in blasting caps to help detonate other more-difficult-to-detonate explosives such as dynamite.

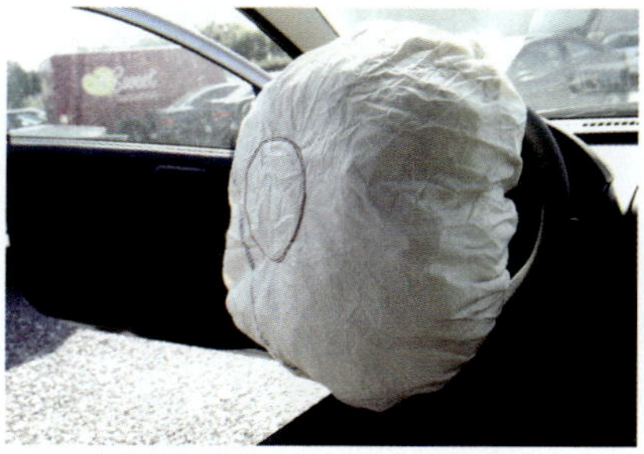

Figure 10.13 An automobile safety air bag undergoes an explosive expansion (driven by nitrogen gas evolution from sodium azide) to provide crash protection for the occupant.

Figure 10.14 Bright blue copper sulfate pentahydrate crystals can be converted into white crystals of anhydrous copper sulfate by heating gently in a test tube to drive off the waters of hydration.

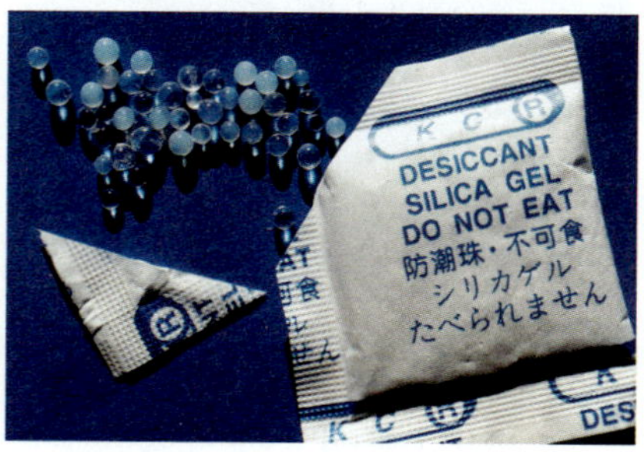

Figure 10.15 Silica gel is a common dessicant used in packaging and dessicators to keep hygroscopic compounds dry.

10-5. Hydrates Result from a Combination Reaction Between Water and an Anhydrous Salt

Some salts can combine with water to form new compounds that incorporate water molecules into their molecular structure in specific whole-number ratios. Such water-containing compounds are called **hydrates**. For example, when white copper sulfate crystals are combined with water, the reaction produces a bright blue solid. The chemical equation describing this reaction is

$$\underset{\text{copper sulfate (white)}}{CuSO_4(s)} + 5\,H_2O(l) \rightarrow \underset{\text{copper sulfate pentahydrate (bright blue)}}{CuSO_4{\cdot}5\,H_2O(s)}$$

The bright blue crystals that form are a hydrate of copper sulfate known as copper sulfate pentahydrate. Unlike ionic and covalent bonds, the water molecules in a hydrate (**the waters of hydration**) are loosely bound to the salt and can generally be driven off by gentle heating. For example, the five waters of hydration in copper sulfate pentahydrate can be removed by heating the hydrate gently to re-form the salt (Figure 10.14). When the waters of hydration are removed, the resulting salt is called an **anhydrous salt**; anhydrous means "without water."

$$\underset{\text{hydrate}}{CuSO_4\cdot 5\,H_2O(s)} \xrightarrow{\text{heat}} \underset{\text{anhydrous salt}}{CuSO_4(s)} + \underset{\text{water}}{5\,H_2O(l)}$$

Because the water molecules are not strongly bound to the salt, we write the formula of hydrates by writing the chemical formula of the anhydrous salt followed by a dot and then the number of waters of hydration in the chemical formula. For example, the building material gypsum, used in drywall and plaster of Paris, is a hydrate that contains calcium sulfate and two waters of hydration. The formula of gypsum is written as $CaSO_4{\cdot}2\,H_2O(s)$. To name a hydrate, we write the name of the anhydrous salt followed by the appropriate Greek prefix (Table 2.7) to indicate the number of waters of hydration and then add the word *hydrate.* For example, the proper chemical name of gypsum is calcium sulfate dihydrate.

The anhydrous salts of many hydrates are highly **hygroscopic**, that is, they strongly absorb water, even from the atmosphere. Because of this, these salts can be used as **desiccants**, or drying agents, for other substances (Figure 10.15). Some examples of common hydrates are given in Table 10.7.

EXAMPLE 10-8: Write the chemical formula of lithium chromate dihydrate.

Solution: Referring to Table 10.1, we see a chromate anion has the formula CrO_4^{2-}. Because a lithium cation has an ionic charge of +1, the formula of anhydrous lithium chromate is $Li_2CrO_4(s)$. The prefix *di-* before the word hydrate tells us that the formula contains two waters of hydration, and so we write the chemical formula of the hydrate as $Li_2CrO_4{\cdot}2\,H_2O(s)$.

TABLE 10.7 Names and formulas of some common hydrates

Formula of hydrate	Chemical name	Common name	Use
$CaSO_4 \cdot 2H_2O$	calcium sulfate dihydrate	gypsum	plaster of Paris, drywall
$CuSO_4 \cdot 5H_2O$	copper(II) sulfate pentahydrate	blue vitriol	algae control in pools, copper plating, dyes, and fireworks
$MgSO_4 \cdot 7H_2O$	magnesium sulfate heptahydrate	Epsom salt	medicinal bath salt, agriculture
$KAl(SO_4)_2 \cdot 12H_2O$	potassium aluminum sulfate dodecahydrate	alum	used in dyes, pickling, canning of foods, and papermaking
$Na_2CO_3 \cdot 10H_2O$	sodium carbonate decahydrate	washing soda	detergent
$Na_2B_4O_7 \cdot 10H_2O$	sodium tetraborate decahydrate	borax	detergent, soap, disinfectant, pesticide
$Na_2S_2O_3 \cdot 5H_2O$	sodium thiosulfate pentahydrate	hypo	silver based photographic fixing agent

PRACTICE PROBLEM 10-8: Name the compound $NaC_2H_3O_2 \cdot 3H_2O(s)$ (which may also be written as $NaCH_3COO \cdot 3H_2O(s)$ or $NaOOCCH_3 \cdot 3H_2O(s)$ to emphasize its structure).

Answer: sodium acetate trihydrate

10-6. In a Single-Replacement Reaction, One Element in a Compound Is Replaced by Another

Titanium is used to make lightweight, high-strength alloys for airplanes, boats, bicycles, spacecraft, and missiles. Titanium metal is prepared by reacting titanium tetrachloride with molten magnesium as described by

$$2\,Mg(l) + TiCl_4(g) \rightarrow 2\,MgCl_2(s) + Ti(s)$$

We can see that the magnesium takes the place of the titanium in the chloride salt. A reaction in which an element in a compound is replaced by another element is called a **single-replacement reaction** or a **substitution reaction.**

An important type of single-replacement reaction involves the reaction between a metal, such as iron, and a dilute solution of an acid, such as sulfuric acid. As the reaction takes place, bubbles of hydrogen gas appear at the surface of the iron (Figure 10.16). The equation for the single-replacement reaction that occurs is

$$Fe(s) + H_2SO_4(aq) \rightarrow FeSO_4(aq) + H_2(g)$$

In this case an iron atom replaces the two hydrogen atoms in sulfuric acid. We call the metals that react with acids in this way **active metals**. A key property of an acid is its ability to attack active metals and produce hydrogen gas.

Figure 10.16 The reaction between iron metal and a dilute aqueous solution of sulfuric acid. The bubbles generated at the surface of the iron nail are hydrogen gas. The iron replaces the hydrogen atoms in $H_2SO_4(aq)$ and thereby enters the solution as $Fe^{2+}(aq)$.

EXAMPLE 10-9: Magnesium is an active metal. Write a balanced chemical equation that describes the reaction of magnesium metal with hydrobromic acid.

Solution: Because magnesium, $Mg(s)$, is an active metal, it reacts with hydrobromic acid, $HBr(aq)$, to produce hydrogen gas, $H_2(g)$, and bromide ions, $Br^-(aq)$. Because magnesium is a Group 2 metal, it forms doubly charged $Mg^{2+}(aq)$ ions when it replaces hydrogen ions from the solution. These magnesium ions combine with the bromide ions to form magnesium bromide, $MgBr_2(aq)$, which remains in solution (we shall learn to predict the solubility of ionic compounds in Section 10-9). The balanced chemical equation for the single-replacement reaction that occurs is

$$Mg(s) + 2\,HBr(aq) \rightarrow MgBr_2(aq) + H_2(g)$$

PRACTICE PROBLEM 10-9: Write a balanced chemical equation for the single-replacement reaction of aluminum metal with sulfuric acid.

Answer: $2\,Al(s) + 3\,H_2SO_4(aq) \rightarrow Al_2(SO_4)_3(aq) + 3\,H_2(g)$

All reactions between metals and aqueous solutions of acids that evolve hydrogen gas may be pictured as the replacement of the acidic hydrogen atoms in the acid by the metal.

10-7. Metals Can Be Ordered in Terms of Relative Activity Based on Single-Replacement Reactions

Silver nitrate dissolves in water to form a colorless transparent solution. When we place a copper wire in a $AgNO_3(aq)$ solution, the solution gradually becomes blue and crystals of metallic silver form on the copper wire, drop off, and sink to the bottom of the solution (Figure 10.17). The equation for this reaction is

$$\underset{\text{colorless}}{Cu(s) + 2\,AgNO_3(aq)} \rightarrow \underset{\text{blue}}{Cu(NO_3)_2(aq)} + 2\,Ag(s)$$

The resulting solution is blue because copper(II) nitrate dissolved in water forms a blue solution. On the basis of this reaction, we conclude that copper is a more active metal than silver because copper metal displaces silver ions from solution.

When we carry out a similar reaction with zinc metal and a copper(II) nitrate solution, we find that the zinc metal displaces the copper ions from solution (Figure 10.18):

$$Zn(s) + Cu(NO_3)_2(aq) \rightarrow Cu(s) + Zn(NO_3)_2(aq)$$

We conclude from this reaction that zinc is a more active metal than copper.

On the basis of these two reactions we can arrange zinc, copper, and silver metals in order of their relative activities:

Zn
Cu ↑ increasing activity
Ag

Figure 10.17 (*top*) A silver nitrate solution, $AgNO_3(aq)$, is colorless. (*bottom*) When a copper wire is inserted, the solution slowly turns blue as a result of the formation of aqueous copper nitrate, $Cu(NO_3)_2(aq)$. Note the silver metal crystals on the bottom of the beaker.

TABLE 10.8 Activity series for some common metals

	Metal	Reactivity
Increasing activity →	Li	react directly with cold water and vigorously with dilute acids to produce $H_2(g)$
	K	
	Ba	
	Ca	
	Na	
	Mg	react with steam or hot water and acids to produce $H_2(g)$
	Al	
	Mn	
	Zn	
	Cr	
	Fe	
	Co	react with acids to produce $H_2(g)$
	Ni	
	Sn	
	Pb	
	Cu	do not react with water or acids to produce $H_2(g)$
	Hg	
	Ag	

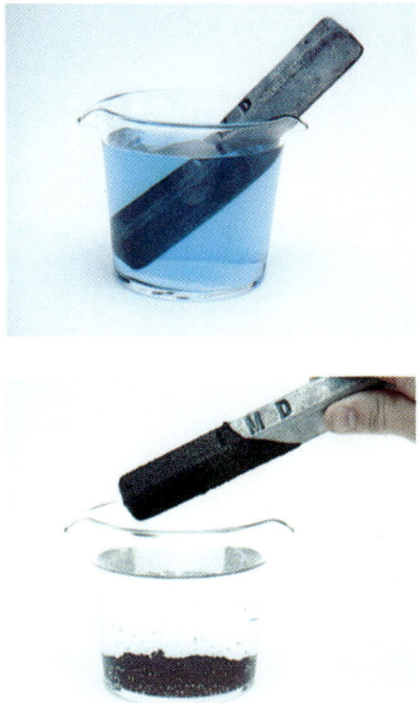

Figure 10.18 (*top*) A zinc rod is placed in a copper nitrate (blue) solution. (*bottom*) Zinc has replaced the copper to form a colorless zinc nitrate solution $Zn(NO_3)_2(aq)$. Elemental copper has coated the rod and has fallen to the bottom of the flask.

By performing additional experiments similar to those just described, we can determine the position of other metals in an **activity series** of the metals (Table 10.8). In general, a metal of greater activity will displace an aqueous metal ion of a less active metal.

EXAMPLE 10-10: Use Table 10.8 to predict whether or not a reaction occurs in the following cases. If a reaction does occur, complete and balance the chemical equation.

(a) $Zn(s) + HgCl_2(aq) \rightarrow$

(b) $Zn(s) + Ca(ClO_4)_2(aq) \rightarrow$

Solution: (a) Zinc lies above mercury in the activity series, therefore zinc displaces mercury from its chloride compound according to

$$Zn(s) + HgCl_2(aq) \rightarrow ZnCl_2(aq) + Hg(l)$$

(b) Zinc lies below calcium in the activity series, therefore zinc does not displace calcium from the perchlorate compound, and no reaction occurs.

$$Zn(s) + Ca(ClO_4)_2(aq) \rightarrow \text{no reaction}$$

PRACTICE PROBLEM 10-10: (a) Place gold in the activity series for metals given in Table 10.8. (*Hint:* Gold is prized not only for its luster but also for its resistance to corrosion.) (b) Although hydrogen gas is not a metal, it is generally included in the activity series of metals because it can be displaced from acids (or sometimes even water) by an active metal. Where does hydrogen belong in Table 10.8?

Answer: (a) Gold should be placed below silver at the bottom of the activity series. Unlike the other metals, which oxidize or tarnish easily, gold is a precious metal prized for its nonreactivity, as manifested by its ability to remain lustrous for millennia. (b) Hydrogen should be placed between copper and lead. Referring to Table 10.8 we see that the metals above lead can displace hydrogen from acids or even water in the case of the most active metals; whereas copper cannot displace hydrogen from water or acids. Therefore, hydrogen is more active than copper but less active than lead.

Although the single-replacement reactions that we have discussed so far involve the replacement of one metal by another, or hydrogen by a metal, there are many other types of single-replacement reactions. One particularly important type is the reaction between a metal oxide and carbon, such as those described by the following chemical equations:

$$3\,C(s) + 2\,Fe_2O_3(s) \rightarrow 4\,Fe(l) + 3\,CO_2(g)$$
$$C(s) + 2\,ZnO(s) \rightarrow 2\,Zn(l) + CO_2(g)$$

Reactions such as these are used in **smelting**, the production of metals from their ores, which are often either metal oxides or compounds such as sulfides that are readily convertible to oxides.

EXAMPLE 10-11: When metal sulfides are roasted in air, they are converted to metal oxides and sulfur dioxide gas. Write a balanced chemical equation for conversion of zinc sulfide to zinc oxide by reaction with oxygen gas.

Solution:

$$2\,ZnS(s) + 3\,O_2(g) \xrightarrow{\text{high } T} 2\,ZnO(s) + 2\,SO_2(g)$$

In this case oxygen is replacing sulfur, rather than one metal replacing another.

PRACTICE PROBLEM 10-11: Propose a method for the removal of the sulfur dioxide gas from the products of the reaction in the previous Example. (*Hint*: Use a combination reaction.)

Answer: $MgO(s) + SO_2(g) \rightarrow MgSO_3(s)$

10-8. The Relative Activity of the Halogens Is $F_2 > Cl_2 > Br_2 > I_2$

The reactions of the nonmetals are too diversified to summarize in a single table similar to Table 10.8. However, the relative activities of the halogens can be readily established by means of single-replacement reactions. For example, if bromine is added to an aqueous solution of sodium iodide, free iodine is produced according to

$$Br_2(l) + 2\,NaI(aq) \rightarrow 2\,NaBr(aq) + I_2(s)$$

This result implies that bromine is more active than iodine. The iodine produced appears as a solid because $I_2(s)$ is only slightly soluble in water.

Similarly, if chlorine gas is bubbled into an aqueous solution of sodium bromide, as shown in Figure 10.19, then free bromine is produced:

$$Cl_2(g) + 2\,NaBr(aq) \rightarrow 2\,NaCl(aq) + Br_2(aq)$$

This result implies that chlorine is more active than bromine. Finally, fluorine not only is the most active halogen but also is the most active element. Fluorine readily displaces chlorine from chlorides; for example,

$$F_2(g) + 2\,KCl(s) \rightarrow 2\,KF(s) + Cl_2(g)$$

Thus, the halogens may be ranked in order of activity as

$$F_2 > Cl_2 > Br_2 > I_2$$

Generally, the activities of nonmetals (except for the noble gases) increase as we go up within a group in the periodic table. This situation is in sharp contrast to the activities of metals, which generally increase as we go down within a group (Figure 10.20).

Figure 10.19 When chlorine gas, $Cl_2(g)$, is bubbled into a colorless aqueous sodium bromide, $NaBr(aq)$, solution, the chlorine replaces the bromide ions to form sodium chloride, $NaCl(aq)$, and a reddish-brown solution of bromine in water, $Br_2(aq)$.

EXAMPLE 10-12: Predict whether or not a reaction occurs when liquid bromine, $Br_2(l)$, is mixed with an aqueous solution of calcium iodide, $CaI_2(aq)$. If a reaction does occur, give the balanced chemical equation.

Solution: Bromine is more active than iodine, so it will replace the iodide ions in calcium iodide. Thus, we predict

$$Br_2(l) + CaI_2(aq) \rightarrow CaBr_2(aq) + I_2(s)$$

This prediction is correct.

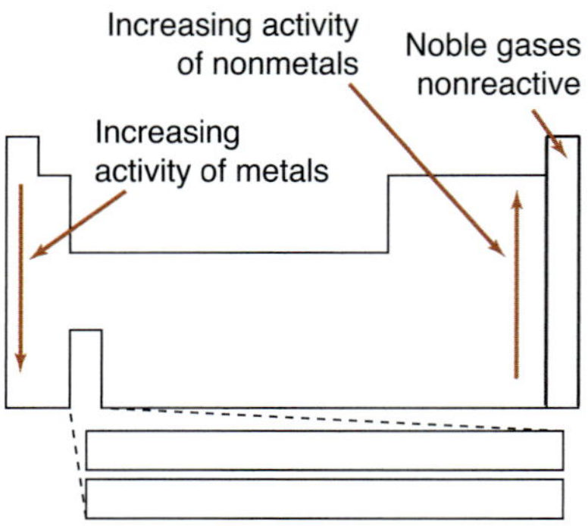

Figure 10.20 Activity trends of metals and nonmetals.

PRACTICE PROBLEM 10-12: Elemental bromine is obtained on a commercial scale from bromide-containing brines by bubbling a chlorine-gas-plus-air mixture through the brine. The volatile bromine produced is swept out of the reaction mixture by the air current. Write a balanced chemical equation for the reaction that occurs between the bromide ion and chlorine in this process.

Answer: $2\,Br^-(aq) + Cl_2(g) \rightarrow 2\,Cl^-(aq) + Br_2(g)$

10-9. In a Double-Replacement Reaction, the Cations and Anions of Two Ionic Compounds Exchange to Form New Compounds

A simple and impressive double-replacement reaction is the reaction between an aqueous solution of sodium chloride, $NaCl(aq)$, and an aqueous solution of silver nitrate, $AgNO_3(aq)$. Both solutions are clear; yet when they are mixed, a white precipitate forms immediately (Figure 10.21). A **precipitate** is an insoluble product of a reaction that occurs in solution. The chemical reaction between sodium chloride and silver nitrate can be described by the equation

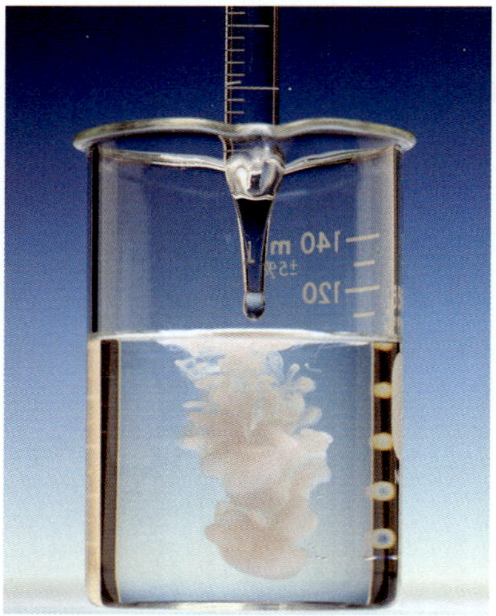

Figure 10.21 The reaction of sodium chloride, $NaCl(aq)$, with silver nitrate, $AgNO_3(aq)$, yields the white precipitate, silver chloride, $AgCl(s)$.

$$NaCl(aq) + AgNO_3(aq) \rightarrow NaNO_3(aq) + AgCl(s)$$

The white precipitate that forms is silver chloride, $AgCl(s)$, an insoluble compound. The reaction is called a **double-replacement reaction** because we visualize it as an exchange of anions between the two cations, $Na^+(aq)$ and $Ag^+(aq)$.

A double-replacement reaction involving the formation of a precipitate is called a **precipitation reaction.** It is instructive to analyze the above equation for the silver chloride precipitation reaction in terms of the ions involved. Sodium chloride and silver nitrate are both soluble ionic compounds. Therefore, an aqueous solution of sodium chloride consists of $Na^+(aq)$ and $Cl^-(aq)$ ions. Similarly, an aqueous solution of silver nitrate consists of $Ag^+(aq)$ and $NO_3^-(aq)$ ions. At the very instant the sodium chloride and silver nitrate solutions are mixed, these four kinds of ions exist in one solution. As we show in later chapters, the ions in a solution constantly move around and collide with water molecules and with one another. If a $Na^+(aq)$ ion collides with a $Cl^-(aq)$ ion, they simply drift apart because sodium chloride is soluble in water. Similarly, a collision between a $Ag^+(aq)$ ion and a $NO_3^-(aq)$ ion is of no consequence because silver nitrate is also soluble in water. However, when a $Ag^+(aq)$ ion collides with a $Cl^-(aq)$ ion, silver chloride is formed. Because silver chloride is insoluble in water, it precipitates out of solution as $AgCl(s)$. We say that the **driving force** of the chemical reaction between sodium chloride and silver nitrate is the formation of the silver chloride precipitate. In other words, the formation of an insoluble precipitate drives a precipitation reaction toward the product side of the reaction.

In order to appreciate precipitation reactions, we must have some idea of which compounds are soluble and which are insoluble. While it is not possible to predict the solubility of all compounds, it is possible to state a few rules of thumb that can be used to predict whether a compound is likely to be soluble or insoluble in water (Table 10.9).

TABLE 10.9 Solubility Rules (applied in the order given)

1. Most alkali metal salts and ammonium salts are soluble.
2. Most nitrates, acetates, and perchlorates are soluble.
3. Most silver, lead, and mercury(I) salts are insoluble.
4. Most chlorides, bromides, and iodides are soluble.
5. Most carbonates, chromates, sulfides, oxides, phosphates, and hydroxides are insoluble, except for hydroxides of Ba^{2+}, Ca^{2+}, and Sr^{2+}, which are slightly soluble.
6. Most sulfates are soluble, except for calcium sulfate and barium sulfate, which are insoluble.

The solubility rules are also given in Appendix F.

The solubility rules must be applied in the order given. For example, sodium sulfide, $Na_2S(s)$ is soluble because Rule 1 states that sodium salts are soluble. We ignore Rule 5 concerning sulfides because Rule 1 takes precedence in this case.

EXAMPLE 10-13: Predict the solubility of the following compounds in water: (a) $(NH_4)_2SO_4(s)$ (b) $CaCO_3(s)$ (c) $Al_2O_3(s)$ (d) $Pb(NO_3)_2(s)$

Solution: Applying the solubility rules in the order given, we find that (a) ammonium sulfate is soluble (Rule 1); (b) calcium carbonate is insoluble (Rule 5); (c) aluminum oxide is insoluble (Rule 5); (d) lead(II) nitrate is soluble (Rule 2).

PRACTICE PROBLEM 10-13: Predict the solubility of the following compounds in water: (a) $K_2SO_4(s)$ (b) $BaSO_4(s)$ (c) $Ga(NO_3)_2(s)$ (d) $(NH_4)_3PO_4(s)$

Answer: (a) soluble; (b) insoluble; (c) soluble; (d) soluble.

The solubility rules are extremely useful for predicting the products of various chemical reactions in aqueous solution.

EXAMPLE 10-14: Use the solubility rules to predict the products of the following three reactions. In each case write the balanced chemical equation for the reaction that occurs or "no reaction" if no precipitates form.

(a) $BaCl_2(aq) + Na_2SO_4(aq) \rightarrow$

(b) $LiOH(aq) + Pb(NO_3)_2(aq) \rightarrow$

(c) $NaCH_3COO(aq) + CaBr_2(aq) \rightarrow$

Solution: (a) Because the reactants are both aqueous species, we have $Ba^{2+}(aq)$, $Cl^-(aq)$, $Na^+(aq)$, and $SO_4^{2-}(aq)$ ions in solution. Combining the positive ions with the negative anions to form two new compounds, we have the reaction described by the equation

$$BaCl_2(aq) + Na_2SO_4(aq) \rightarrow 2\,NaCl + BaSO_4$$

Applying the solubility rules to each of the proposed products in turn, we find that sodium chloride is soluble (Rule 1), but that barium sulfate is insoluble (Rule 6). Thus, we write

$$BaCl_2(aq) + Na_2SO_4(aq) \rightarrow 2\,NaCl(aq) + BaSO_4(s)$$

(b) Combining the positive and negative ions formed in solution from the reactants, we obtain $LiNO_3$ and $Pb(OH)_2$ as possible double-replacement products. Applying the solubility rules to each of these, we find that lithium nitrate is soluble, but lead(II) hydroxide is insoluble and thus forms a precipitate. Therefore, the balanced chemical equation is

$$2\,LiOH(aq) + Pb(NO_3)_2(aq) \rightarrow 2\,LiNO_3(aq) + Pb(OH)_2(s)$$

(c) Combining the positive and negative ions formed in solution from the reactants, we obtain $Ca(CH_3COO)_2$ and NaBr as possible double-replacement products. Applying the solubility rules to each of these compounds, we find that both calcium acetate and sodium bromide are soluble. Because no precipitate is formed, there is no driving force for this reaction, and so we predict that no reaction occurs. In other words, because calcium acetate and sodium bromide will produce the *same ions* in solution that the reactants sodium acetate and calcium bromide produce, no net chemical change occurs. Thus, we write

$$NaCH_3COO(aq) + CaBr_2(aq) \rightarrow \text{no reaction}$$

PRACTICE PROBLEM 10-14: Use the solubility rules to predict the products of the following reactions. In each case write the balanced chemical equation for the reaction that occurs, or write "no reaction" if no reaction occurs.

(a) $Hg_2(NO_3)_2(aq) + NaBr(aq) \rightarrow$

(b) $NH_4Cl(aq) + KClO_4(aq) \rightarrow$

(c) $Na_2S(aq) + Cd(NO_3)_2(aq) \rightarrow$

Answer: (a) $Hg_2(NO_3)_2(aq) + 2\,NaBr(aq) \rightarrow Hg_2Br_2(s) + 2\,NaNO_3(aq)$

(b) no reaction

(c) $Na_2S(aq) + Cd(NO_3)_2(aq) \rightarrow 2\,NaNO_3(aq) + CdS(s)$

It is convenient to use **net ionic equations** to describe double-replacement reactions that occur in solution. In the reaction between sodium chloride and silver nitrate described by the equation

$$NaCl(aq) + AgNO_3(aq) \rightarrow NaNO_3(aq) + AgCl(s)$$

the ions $Na^+(aq)$ and $NO_3^-(aq)$ do not participate directly in the precipitation reaction that is the formation of solid silver chloride from the ions $Ag^+(aq)$ and $Cl^-(aq)$. Because the $Na^+(aq)$ and $NO_3^-(aq)$ ions are present initially in the reaction solution and because they remain unchanged in the product solution, we say that these ions are **spectator ions.** In other words, these ions are not involved directly in the formation of the silver chloride precipitate.

If we write the chemical equation in terms of all the ions involved, that is, as the **complete ionic equation**,

$$Na^+(aq) + Cl^-(aq) + Ag^+(aq) + NO_3^-(aq) \rightarrow Na^+(aq) + NO_3^-(aq) + AgCl(s)$$

then we see that the spectator ions, $Na^+(aq)$ and $NO_3^-(aq)$, appear on both sides of the equation. Silver chloride is insoluble in water and so it is not broken up into ions in this equation. Because they appear on both sides of the complete ionic equation, the spectator ions can be canceled:

$$\cancel{Na^+}(aq) + Cl^-(aq) + Ag^+(aq) + \cancel{NO_3^-}(aq) \rightarrow \cancel{Na^+}(aq) + \cancel{NO_3^-}(aq) + AgCl(s)$$

This gives the net ionic equation

$$Ag^+(aq) + Cl^-(aq) \rightarrow AgCl(s) \qquad (10.6)$$

The net ionic equation describes the essence of the reaction, namely, the formation of solid silver chloride from the $Ag^+(aq)$ and $Cl^-(aq)$ ions in solution. Spectator ions do not appear in the net ionic equation.

The use of net ionic equations helps us focus on the key species in a reaction that occurs in solution. For example, the net ionic equation corresponding to the double-replacement reaction between potassium chloride, $KCl(aq)$, and silver perchlorate, $AgClO_4(aq)$, is the same as that for the reaction between sodium chloride, $NaCl(aq)$, and silver nitrate, $AgNO_3(aq)$, described in Equation 10.6. In this case, the complete ionic equation is

$$K^+(aq) + Cl^-(aq) + Ag^+(aq) + ClO_4^-(aq) \rightarrow K^+(aq) + ClO_4^-(aq) + AgCl(s)$$

for which the corresponding net ionic equation is

$$Ag^+(aq) + Cl^-(aq) \rightarrow AgCl(s)$$

EXAMPLE 10-15: Write the net ionic equation for the reaction between cadmium nitrate, $Cd(NO_3)_2(aq)$, and sodium sulfide, $Na_2S(aq)$, which forms an orange-colored precipitate.

Solution: Using the solubility rules, we find that the resulting double-replacement reaction produces a cadmium sulfide precipitate. The complete ionic equation is

$$Cd^{2+}(aq) + 2\,NO_3^-(aq) + 2\,Na^+(aq) + S^{2-}(aq) \rightarrow 2\,Na^+(aq) + 2\,NO_3^-(aq) + CdS(s)$$

We obtain the net ionic equation by canceling the $Na^+(aq)$ and $NO_3^-(aq)$ spectator ions on both sides of the complete ionic equation:

$$Cd^{2+}(aq) + S^{2-}(aq) \rightarrow \underset{\text{orange}}{CdS(s)}$$

This reaction is shown in Figure 10.22.

Figure 10.22 When colorless aqueous solutions of cadmium nitrate, $Cd(NO_3)_2(aq)$, and sodium sulfide, $Na_2S(aq)$, are mixed, an orange precipitate of cadmium sulfide, $CdS(s)$, forms immediately.

PRACTICE PROBLEM 10-15: Write the net ionic equation for the reaction between an aqueous solution of silver perchlorate and an aqueous solution of sodium chromate.

Answer: $2\,Ag^{+}(aq) + CrO_4^{2-}(aq) \rightarrow Ag_2CrO_4(s)$

10-10. An Acid-Base Reaction Is an Example of a Double-Replacement Reaction

We have seen that the driving force for double-replacement reactions can be the formation of a precipitate. Another type of driving force for a double-replacement reaction is the formation of a covalent compound from ionic reactants. The most important example of this type of reaction is that between an acid and a base, such as that between hydrochloric acid and sodium hydroxide as described by the chemical equation

$$HCl(aq) + NaOH(aq) \rightarrow H_2O(l) + NaCl(aq)$$

The complete ionic equation is

$$H^{+}(aq) + Cl^{-}(aq) + Na^{+}(aq) + OH^{-}(aq) \rightarrow H_2O(l) + Na^{+}(aq) + Cl^{-}(aq)$$

and the net ionic equation is

$$H^{+}(aq) + OH^{-}(aq) \rightarrow H_2O(l)$$

Because water exists almost exclusively as covalently bonded molecules, the $H^{+}(aq)$ and $OH^{-}(aq)$ are depleted from the reaction mixture by the formation of liquid water. Note the similarity between the two driving forces for double-replacement reactions: the formation of a precipitate and the formation of a covalent compound. In each case reactant ions are depleted from the reaction mixture.

The reaction between hydrochloric acid, $HCl(aq)$, and sodium hydroxide, $NaOH(aq)$, is amazing. We probably all know that acids like hydrochloric acid are corrosive. They react with many metals, and their concentrated solutions damage flesh, causing painful burns and blisters. Less familiar, perhaps, are the chemical properties of bases. Bases like sodium hydroxide are also very corrosive, causing painful burns and blisters on the skin, just as hydrochloric acid does. Thus, the reaction between hydrochloric acid and sodium hydroxide is one between two reactive and hazardous substances, but the products are sodium chloride and water, a harmless aqueous solution of table salt. We say that the acid and the base have **neutralized** each other. The chemical reaction between an acid and a base is called a **neutralization reaction.** The ionic compound that is formed along with the water in a neutralization reaction is called a **salt.** For example,

$$\underset{\text{acid}}{H_2SO_4(aq)} + \underset{\text{base}}{2\,KOH(aq)} \rightarrow \underset{\text{salt}}{K_2SO_4(aq)} + \underset{\text{water}}{2\,H_2O(l)}$$

In this case the salt that is formed, potassium sulfate, $K_2SO_4(aq)$, is water-soluble and exists in solution primarily in the form of its constituent ions, $2K^+(aq)$ and $SO_4^{2-}(aq)$.

The essential chemical characteristic of an acid is its production of $H^+(aq)$ ions in aqueous solutions, and the essential chemical characteristic of a base is its production of $OH^-(aq)$ ions in aqueous solutions (Table 10.6). When an acid and a base neutralize each other, the $H^+(aq)$ ions and $OH^-(aq)$ ions react to produce $H_2O(l)$; and in so doing they nullify the acidic and basic character of the separate solutions:

$$\underset{\text{acid}}{H^+(aq)} + \underset{\text{base}}{OH^-(aq)} \rightarrow \underset{\text{water}}{H_2O(l)}$$

EXAMPLE 10-16: Complete and balance the equation for the following neutralization reaction and name the salt formed. Write the corresponding net ionic equation for the reaction.

$$HNO_3(aq) + Ba(OH)_2(aq) \rightarrow$$

Solution: This reaction is between an acid and a base and so the products will be water and a salt. The acid is nitric acid and the base is barium hydroxide, so the salt formed is barium nitrate (Table 10.1), which we see from the solubility rules (Table 10.9) is a soluble salt:

$$\underset{\substack{\text{nitric acid}\\\text{(acid)}}}{2HNO_3(aq)} + \underset{\substack{\text{barium hydroxide}\\\text{(base)}}}{Ba(OH)_2(aq)} \rightarrow \underset{\substack{\text{barium nitrate}\\\text{(salt)}}}{Ba(NO_3)_2(aq)} + \underset{\text{water}}{2H_2O(l)}$$

Rewriting the equation in terms of the ions involved gives

$$2H^+(aq) + 2NO_3^-(aq) + Ba^{2+}(aq) + 2OH^-(aq) \rightarrow Ba^{2+}(aq) + 2NO_3^-(aq) + 2H_2O(l)$$

Cancellation of the spectator ions yields the net ionic equation

$$2H^+(aq) + 2OH^-(aq) \rightarrow 2H_2O(l)$$

Dividing both sides of this equation by two yields

$$H^+(aq) + OH^-(aq) \rightarrow H_2O(l)$$

which is the same net ionic equation found above for the neutralization reaction between hydrochloric acid and sodium hydroxide.

PRACTICE PROBLEM 10-16: Write a balanced chemical equation to describe the reaction between aqueous sulfuric acid and solid lithium oxide. Write the corresponding net ionic equation for the reaction.

Answer: $H_2SO_4(aq) + Li_2O(s) \rightarrow Li_2SO_4(aq) + H_2O(l)$

$$2H^+(aq) + Li_2O(s) \rightarrow 2Li^+(aq) + H_2O(l)$$

We have shown that double-replacement reactions are driven by the formation of a precipitate or the formation of a neutral covalent compound, such as water, from ions in the solution. A third type of driving force for a double-replacement reaction is a variation of the latter type, namely, the formation of a gaseous covalent product.

An example is provided by the reaction of calcium carbonate (limestone), $CaCO_3(s)$, with a dilute solution of hydrochloric acid as described by the chemical equation

$$CaCO_3(s) + 2\,HCl(aq) \rightarrow CaCl_2(aq) + H_2CO_3(aq)$$

The corresponding net ionic equation is

$$CaCO_3(s) + 2\,H^+(aq) \rightarrow Ca^{2+}(aq) + H_2CO_3(aq)$$

The carbonic acid, $H_2CO_3(aq)$, formed in this double-replacement reaction is unstable and decomposes to form carbon dioxide and water:

$$H_2CO_3(aq) \rightarrow H_2O(l) + CO_2(g)$$

Thus, the overall reaction equation may be written as

$$CaCO_3(s) + 2\,HCl(aq) \rightarrow CaCl_2(aq) + H_2O(l) + CO_2(g)$$

where it is understood that this is a double-replacement reaction because the water and carbon dioxide gas come from the subsequent decomposition of carbonic acid.

The carbon dioxide gas formed is not very soluble in water and so escapes from the solution. The evolution of carbon dioxide gas from the reaction mixture drives the reaction toward the product side. The evolution of an odorless gas by limestone when treated with dilute acids is used as a simple geological field test for this mineral (Figure 10.23).

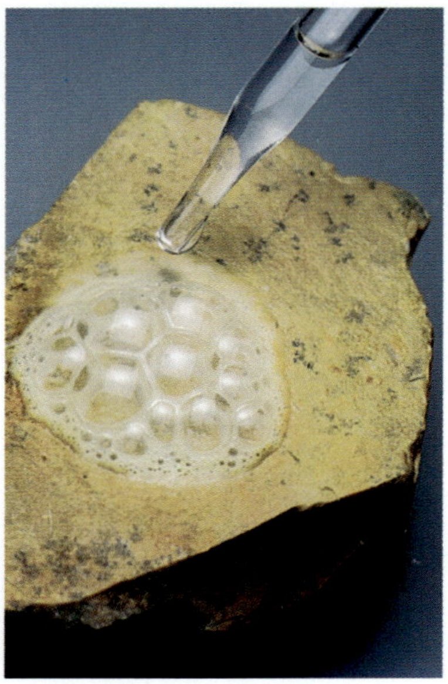

Figure 10.23 A field test for limestone uses hydrochloric acid. The $HCl(aq)$ deposited on the surface of the limestone reacts to produce bubbles of carbon dioxide, an odorless gas.

EXAMPLE 10-17: Treatment of zinc sulfide with a dilute aqueous hydrochloric acid solution yields a toxic gas that smells like rotten eggs. Write a balanced chemical equation describing the reaction and give the corresponding net ionic equation.

Solution: The gas produced is hydrogen sulfide, $H_2S(g)$, and the balanced equation for the reaction is

$$ZnS(s) + 2\,HCl(aq) \rightarrow ZnCl_2(aq) + H_2S(g)$$

The corresponding net ionic equation is

$$ZnS(s) + 2\,H^+(aq) \rightarrow Zn^{2+}(aq) + H_2S(g)$$

PRACTICE PROBLEM 10-17: When sodium hydrogen carbonate (household bicarbonate of soda or baking soda) is mixed with vinegar (a 5% aqueous solution of acetic acid), a gas is evolved (Figure 10.24). Write a balanced chemical equation describing this reaction.

Answer: $NaHCO_3(s) + CH_3COOH(aq) \rightarrow NaCH_3COO(aq) + CO_2(g) + H_2O(l)$

Figure 10.24 The reaction between sodium bicarbonate and vinegar.

10-11. Oxidation-Reduction Reactions Involve the Transfer of Electrons Between Species

In addition to the classification scheme for chemical reactions described in the previous sections, all chemical reactions can also be assigned to one of two classes: reactions in which electrons are transferred from one chemical species (i.e., an element, a compound, or an ion) to another and reactions in which electrons are not transferred. Reactions in which electrons are transferred from one species to another are called **oxidation-reduction ("redox") reactions** or **electron-transfer reactions**. The simplest example of an oxidation-reduction reaction is the reaction between a metal and a nonmetal. For example, the reaction of sodium metal with sulfur produces the ionic compound sodium sulfide (Figure 10.25), as described by the equation

$$2\,Na(s) + S(s) \rightarrow Na_2S(s)$$

We can depict this equation in terms of Lewis electron-dot formulas by writing

$$\underset{[Ne]3s^1}{2\,Na\cdot} \; + \; \underset{[Ne]3s^23p^4}{\cdot\ddot{\underset{\cdot\cdot}{S}}\cdot} \longrightarrow \underset{[Ne]}{2\,Na^+} \; + \; \underset{[Ne]3s^23p^6 \text{ or } [Ar]}{:\ddot{\underset{\cdot\cdot}{S}}:^{2-}}$$

The ionic products in this equation both have noble-gas electron configurations. The stability of these ions accounts for the driving force of this reaction; we go from relatively reactive reactants to relatively stable products.

We see here that two sodium atoms react with one sulfur atom. The two sodium atoms give up two electrons and the sulfur atom acquires the two electrons.

As this reaction shows, the electron transfer between species in a chemical equation must be balanced; that is, in any oxidation-reduction reaction the total number of electrons lost always equals the total number of electrons gained.

When an atom in a reaction gives up electrons to another species, we say the atom is **oxidized.** The term **oxidation** denotes a loss of electrons. When an atom in a reaction accepts electrons from another species, we say the atom is **reduced.** The term **reduction** denotes a gain of electrons.

Figure 10.25 The alkali metal sodium reacts with the nonmetal sulfur to produce sodium sulfide.

It turns out that many of the reactions we have already studied are also oxidation-reduction reactions. For example, the combination reaction between iron metal and oxygen gas described by the equation

$$4\,Fe(s) + 3\,O_2(g) \rightarrow 2\,Fe_2O_3(s) \qquad (10.7)$$

is also an oxidation-reduction reaction. A free element such as iron metal or oxygen gas has no ionic charge. As we saw in Chapter 6, the addition of two electrons to a neutral oxygen atom gives it a noble-gas electron configuration, so oxygen atoms often have a −2 ionic charge in ionic compounds. Taking the ionic charge of each oxygen atom in $Fe_2O_3(s)$ as −2, we see that in order to maintain electronic neutrality, the ionic charge of each iron atom in the compound must be +3, because $2(+3) + 3(-2) = 0$.

$$\overset{(0)}{4\,Fe(s)} + \overset{(0)}{3\,O_2(g)} \rightarrow 2\,\overset{(+3)}{Fe_2}\overset{(-2)}{O_3}(s)$$

Because the ionic charge of each of the four iron atoms in Equation 10.7 changes from 0 to +3, we say that the iron atoms are oxidized in the reaction. Similarly, because the ionic charge of each of the six oxygen atoms changes from 0 to −2, we say that the oxygen atoms are reduced in the reaction. Therefore, we see that this reaction is an oxidation-reduction reaction. Notice also that the total number of electrons lost by the iron atoms in this reaction is equal to the total number of electrons gained by the oxygen atoms. That is, four iron atoms go from an ionic charge of 0 to +3 for a total loss of $4 \times 3 = 12$ electrons and six oxygen atoms go from an ionic charge of 0 to −2 for a total gain of $6 \times 2 = 12$ electrons.

We show the changes in ionic charges in the chemical equation. Recall that $Fe_2O_3(s)$ is named iron(III) oxide, where the Roman numeral three indicates the ionic charge of each iron atom in the compound. Iron(III) oxide, $Fe_2O_3(s)$, and its hydrates are the principal components of rust. In fact, the formation of rust is often referred to as the oxidation of iron.

EXAMPLE 10-18: Show that the single-replacement reaction between copper metal and silver nitrate described by the chemical equation

$$Cu(s) + 2\,AgNO_3(aq) \rightarrow Cu(NO_3)_2(aq) + 2\,Ag(s)$$

is an oxidation-reduction reaction. Which atom is oxidized and which atom is reduced in this reaction?

Solution: The ionic charge on copper metal, $Cu(s)$, is 0. Because the nitrate polyatomic anion, $NO_3^-(aq)$, has an ionic charge of −1 (Table 10.1), we conclude that the copper ion in $Cu(NO_3)_2(aq)$ must have an ionic charge of +2 (recall from Chapter 6 that copper ions commonly have a +1 or +2 ionic charge in compounds). Therefore, the ionic charge of the copper atom changes from 0 to +2 in this reaction.

The ionic charge of the silver ion in silver nitrate, $AgNO_3(aq)$, is +1 (recall from Chapter 6 that the ionic charge of silver is +1 in most compounds), and the ionic charge of silver metal, $Ag(s)$, is 0. Therefore, the ionic charge of each of the two silver atoms changes from +1 to 0 in this reaction.

$$\overset{(0)}{\underset{|}{Cu}}(s) + 2\,\overset{(+1)}{\underset{|}{Ag}}NO_3(aq) \rightarrow \overset{(+2)}{\underset{|}{Cu}}(NO_3)_2(aq) + 2\,\overset{(0)}{\underset{|}{Ag}}(s)$$

Because this reaction involves the transfer of electrons from one species to another, it is an oxidation-reduction reaction. In this reaction, the copper atom gives up two electrons and each of the two silver ions gains an electron. Thus, the copper atom is oxidized and the silver ion is reduced. Notice that both the elements involved and the electrons transferred in this reaction equation are balanced, as they must be.

PRACTICE PROBLEM 10-18: Determine which of the reactions described by the chemical equations below are oxidation-reduction reactions:

(a) $2\,Fe(s) + 3\,Cl_2(g) \rightarrow 2\,FeCl_3(s)$

(b) $2\,AgNO_3(aq) + Na_2S(aq) \rightarrow Ag_2S(s) + 2\,NaNO_3(aq)$

(c) $Zn(s) + HgCl_2(aq) \rightarrow ZnCl_2(aq) + Hg(l)$

Answer: (a) and (c) are oxidation-reduction reactions.

In an oxidation-reduction reaction, the substance that supplies electrons, or the **electron donor**, is called the **reducing agent.** The substance that gains electrons, or the **electron acceptor**, is called the **oxidizing agent.** Thus, in the reaction described by the equation

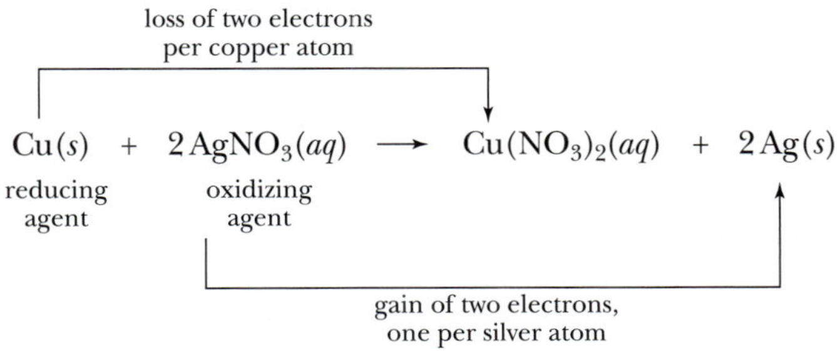

copper is the reducing agent because it is the substance that supplies electrons for the reduction of the silver ions in silver nitrate; and silver nitrate is the oxidizing agent because it is the substance that accepts electrons from the oxidation of the copper. Table 10.10 summarizes the properties of an oxidizing agent and a reducing agent.

TABLE 10.10 Summary of oxidation-reduction reactions

The reducing agent:
contains the atom that is oxidized
contains the atom whose ionic charge increases
is the electron donor
The oxidizing agent:
contains the atom that is reduced
contains the atom whose ionic charge decreases
is the electron acceptor

See Interchapter U at www.McQuarrieGeneralChemistry.com.

EXAMPLE 10-19: Silver button batteries, such as those used to power wristwatches (Interchapter U), get their energy from the electron-transfer reaction between silver oxide and zinc metal as described by the chemical equation

$$Zn(s) + Ag_2O(s) \rightarrow ZnO(s) + 2Ag(s)$$

Identify the oxidizing and reducing agents in this equation.

Solution: The ionic charge of the zinc atom changes from 0 in zinc metal to +2 in zinc oxide. Similarly, the ionic charge of the silver atom changes from +1 in silver oxide to 0 in silver metal. Thus, in the chemical equation, one zinc atom is oxidized ($0 \rightarrow +2$) and two silver atoms are reduced ($+1 \rightarrow 0$), so $Zn(s)$ is the reducing agent and $Ag_2O(s)$ is the oxidizing agent. There is a transfer of two electrons in this equation. Although only the silver atom is being reduced, we say that the reactant containing the silver atom, $Ag_2O(s)$, is the oxidizing agent.

PRACTICE PROBLEM 10-19: Identify the oxidizing and reducing agents in the equation

$$2Al(s) + Mn_2O_3(s) \rightarrow 2Mn(s) + Al_2O_3(s)$$

Answer: Oxidizing agent, $Mn_2O_3(s)$; reducing agent, $Al(s)$

Figure 10.26 The sulfur deposits that occur around hot springs usually result from the oxidation-reduction reaction between sulfur dioxide and hydrogen sulfide.

Oxidation-reduction reactions are one of the most important classes of chemical reactions. Many important chemical processes involve oxidation-reduction reactions (Figure 10.26). We shall devote an entire chapter to oxidation-reduction reactions later on (Chapter 24).

One of the most difficult challenges that a beginning student of chemistry faces is predicting the products of a chemical reaction when only the reactants are given. This question is often difficult even for a trained chemist. The classification of reaction types presented in this chapter is a helpful start in this regard, but one still requires additional chemical experience to develop confidence. As you see more chemical reactions throughout this book, think about each one and try to classify it according to the schemes developed in this chapter.

SUMMARY

Many chemical reactions can be classified as being one of four types:

1. Combination reaction: a reaction of two substances to form a single product.
2. Decomposition reaction: a reaction in which a substance breaks down into two or more simpler substances.
3. Single-replacement reaction: a reaction involving the substitution of one element in a compound by another (also called a substitution reaction).

4. Double-replacement reaction: a reaction in which the cations of two ionic compounds exchange anionic partners. Double-replacement reactions may be divided into the following additional categories: reactions in which a precipitate is formed (the products of which may be predicted using the solubility rules), reactions in which a covalent compound is formed, such as acid-base reactions that produce water, and reactions that produce a gas.

Polyatomic ions are covalently bonded ions that remain intact in water. Table 10.1 lists some chemically important examples of polyatomic ions.

The solubility rules (Table 10.9) predict if an ionic compound is soluble in water. These must be applied in order and may be used to predict which products will precipitate out of solution during a reaction.

Acids are substances that produce hydrogen ions, $H^+(aq)$, when dissolved in water; bases are substances that produce hydroxide ions, $OH^-(aq)$, when dissolved in water. Three classes of acids are binary acids, oxyacids, and organic acids. Acids and bases cause certain dyes to change color. For example, litmus turns red in acidic solutions and blue in basic solutions. An acid and a base react with each other to produce water and a salt; such reactions are called neutralization reactions. The naming of acids follows a simple set of rules.

Hydrates are formed by the combination of an anhydrous salt with water. The formation of hydrates is reversible and the waters of hydration may be driven off by heating. The formula of a hydrate is written with a dot between the anhydrous salt and the number of waters of hydration, and the name is written as the name of the salt plus the word hydrate using the appropriate Greek prefix.

Data from single-replacement reactions can be used to order the metals in terms of relative activity. The resulting activity series of the metals (Table 10.8) is used to predict whether or not other single-replacement reactions involving metals occur. The nonmetals are too varied to allow a correspondingly simple activity series, but the relative activities of the halogens are $F_2 > Cl_2 > Br_2 > I_2$.

Reactions involving ions can be written in a compact form by using net ionic equations. We obtain net ionic equations by canceling the spectator ions from both sides of the complete ionic equation.

All chemical reactions can be classified either as reactions that involve the transfer of electrons between reactants, called oxidation-reduction or redox reactions, or as reactions that do not involve electron transfer. In an oxidation-reduction reaction, electrons are transferred from the reducing agent, which contains the atom that is oxidized, to the oxidizing agent, which contains the atom that is reduced. Oxidation denotes a loss of electrons and reduction denotes a gain of electrons. The number of electrons lost by the reducing agent must equal the number of electrons gained by the oxidizing agent.

TERMS YOU SHOULD KNOW

combination reaction *308*
volatile *308*
combustion reaction *308*
polyatomic ion *310*
oxyanion *311*
hydroxide ion, $OH^-(aq)$ *315*
base *315*
inert *315*
acid *315*
hydrogen ion, $H^+(aq)$ *315*
hydronium ion, $H_3O^+(aq)$ *315*
acidic hydrogen atom *316*
acidic proton *316*
monoprotic, diprotic, and triprotic acids *316*

PROBLEMS

CHEMICAL NOMENCLATURE AND POLYATOMIC ANIONS

10-1. Name the following ionic compounds:

(a) $Ca(CN)_2(s)$ (b) $AgClO_4(s)$

(c) $KMnO_4(s)$ (d) $SrCrO_4(s)$

10-2. Name the following ionic compounds:

(a) $NaCH_3COO(s)$ (b) $Ca(ClO_3)_2(s)$

(c) $(NH_4)_2CO_3(s)$ (d) $Ba(NO_3)_2(s)$

10-3. Name the following ionic compounds, which are used as fertilizers:

(a) $(NH_4)_2SO_4(s)$ (b) $(NH_4)_3PO_4(s)$

(c) $Ca_3(PO_4)_2(s)$ (d) $K_3PO_4(s)$

10-4. Name the following ionic compounds, which are used in silver-based photography:

(a) $(NH_4)_2S_2O_3(s)$ (fixer) (b) $Na_2SO_3(s)$ (preservative)

(c) $K_2CO_3(s)$ (activator) (d) $Na_2S_2O_3(s)$ (fixer)

10-5. Write the systematic name for

(a) $Hg_2Cl_2(s)$ (b) $Cr(NO_3)_3(s)$

(c) $CoBr_2(s)$ (d) $CuCO_3(s)$

10-6. Write the systematic name for

(a) $CrSO_4(s)$ (b) $Co(CN)_2(s)$

(c) $Sn(NO_3)_2(s)$ (d) $Cu_2CO_3(s)$

10-7. Write the chemical formula for

(a) sodium thiosulfate

(b) potassium hydrogen carbonate

(c) sodium hypochlorite

(d) calcium sulfite

10-8. Write the chemical formula for

(a) acetic acid (b) chloric acid

(c) carbonic acid (d) perchloric acid

10-9. Write the chemical formula for

(a) sodium sulfite (b) potassium phosphate

(c) silver sulfate (d) ammonium nitrate

10-10. Write the chemical formula for

(a) sodium perchlorate

(b) potassium permanganate

(c) calcium sulfite

(d) lithium cyanide

10-11. Write the chemical formula for

(a) chromium(III) oxide

(b) tin(II) hydroxide

(c) copper(II) acetate

(d) cobalt(III) sulfate

10-12. Write the chemical formula for

(a) mercury(I) acetate

(b) mercury(II) cyanide

(c) iron(II) perchlorate

(d) chromium(II) sulfite

10-13. Write the chemical formula for

(a) hydrosulfuric acid

(b) aluminum oxide

(c) potassium dichromate

(d) nickel(II) acetate

10-14. Write the formulas of the following common household chemicals:

(a) sodium hypochlorite (the active ingredient bleach)

(b) hydrogen peroxide (used medicinally in dilute solutions)

(c) potassium hydroxide (found in some drain cleaners)

(d) acetic acid (the active ingredient in vinegar)

ACIDS AND BASES

10-15. Decide, on the basis of your personal experience and the information in Table 10.6, which of the following solutions are acidic and which are basic:

(a) carbonated soft drinks

(b) apple cider

(c) an antacid dissolved in water

(d) tomatoes

(e) hand soap

10-16. Decide, on the basis of your personal experience and the information in Table 10.6, which of the following solutions are acidic and which are basic:

(a) laundry detergent in water

(b) orange juice

(c) jam

(d) bicarbonate of soda (baking soda) in water

(e) household ammonia

10-17. Which of the following compounds are acidic in aqueous solution?

(a) $HCl(g)$ (b) $NaOH(s)$

(c) $HClO(aq)$ (d) $CH_3COOH(aq)$

10-18. Which of the following compounds are basic in aqueous solution?

(a) $LiOH(s)$ (b) $Na_2O(s)$

(c) $PbO(s)$ (d) $HNO_3(aq)$

OXYACIDS AND OXYANIONS

10-19. Classify each of the following acids as an oxyacid, binary acid, or organic acid.

(a) $C_6H_5COOH(aq)$ (b) $HF(aq)$

(c) $HC(COOH)_3(aq)$ (d) $HClO_4(aq)$

10-20. Classify each of the following acids as an oxyacid, binary acid, or organic acid.

(a) $CH_3CH_2COOH(aq)$ (b) $HNO_3(aq)$

(c) $H_2C_2O_4(aq)$ (d) $H_3PO_4(aq)$

10-21. Name the following oxyacids:

(a) $H_2SO_3(aq)$ (b) $HBrO_3(aq)$

(c) $H_3PO_2(aq)$ (d) $HIO_4(aq)$

10-22. Name the following oxyacids:

(a) $HNO_2(aq)$ (b) $H_2SO_2(aq)$

(c) $HClO_2(aq)$ (d) $HIO_3(aq)$

10-23. Name the following salts containing oxyanions:

(a) $KBrO(s)$ (b) $CaHPO_3(s)$

(c) $Pb(ClO_2)_2(s)$ (d) $Ni(ClO_4)_2(s)$

10-24. Name the following salts containing oxyanions:

(a) $Cu(ClO)_2(s)$ (b) $Sc(IO_3)_3(s)$

(c) $Fe(BrO_3)_3(s)$ (d) $Ru(IO_4)_3(s)$

10-25. Draw Lewis formulas for the following oxyanions, indicating the formal charge of each atom:

(a) nitrate ion (b) nitrite ion

(c) sulfate ion (d) sulfite ion

10-26. Draw Lewis formulas for the following oxyanions, indicating the formal charge of each atom:

(a) perchlorate ion (b) carbonate ion

(c) phosphate ion (d) acetate ion

10-27. Using the Lewis formulas you drew in Problem 10-25, name and draw the Lewis formulas of the corresponding acids formed from these anions.

10-28. Using the Lewis formulas you drew in Problem 10-26, name and draw the Lewis formulas of the corresponding acids formed from these anions.

HYDRATES

10-29. Name the following hydrates:

(a) $ZnSO_4 \cdot H_2O(s)$ (b) $BaCl_2 \cdot 2H_2O(s)$

(c) $NiSO_4 \cdot 6H_2O(s)$ (d) $Na_2S \cdot 9H_2O(s)$

10-30. Name the following hydrates:

(a) $Ba(OH)_2 \cdot 8H_2O(s)$ (b) $PbCl_2 \cdot 2H_2O(s)$

(c) $LiOH \cdot H_2O(s)$ (d) $Li_2CrO_4 \cdot 2H_2O(s)$

10-31. Write the chemical formula for the following hydrates:

(a) iridium(III) bromide tetrahydrate

(b) tin(IV) chloride pentahydrate

(c) zinc nitrate hexahydrate

(d) sodium carbonate monohydrate

10-32. Write the chemical formula for the following hydrates:

(a) oxalic acid dihydrate

(b) aluminum sulfate octahydrate

(c) neodymium(III) iodide nonahydrate

(d) disodium hydrogen phosphate heptahydrate

CLASSIFICATION OF REACTIONS AND PREDICTION OF REACTION PRODUCTS

10-33. Classify each of the reactions described by the following chemical equations as combination, decomposition, single replacement, or double replacement.

(a) $CaCO_3(s) \rightarrow CaO(s) + CO_2(g)$

(b) $NH_3(g) + HCl(g) \rightarrow NH_4Cl(s)$

(c) $2AgBr(s) + Cl_2(g) \rightarrow 2AgCl(s) + Br_2(l)$

(d) $Ag_2SO_4(s) + 2NaI(aq) \rightarrow 2AgI(s) + Na_2SO_4(aq)$

10-34. Classify each of the reactions described by the following chemical equations as combination, decomposition, single replacement, or double replacement.

(a) $2KClO_3(s) \rightarrow 2KCl(s) + 3O_2(g)$

(b) $V_2O_5(s) + 5Ca(l) \rightarrow 2V(l) + 5CaO(s)$

(c) $2NaCl(s) + H_2SO_4(l) \rightarrow 2HCl(g) + Na_2SO_4(s)$

(d) $Fe(s) + 2HBr(aq) \rightarrow FeBr_2(aq) + H_2(g)$

10-35. Classify each of the reactions described by the following chemical equations as combination, decomposition, single replacement, or double replacement. If the equation is not balanced, balance it.

(a) $BaCO_3(s) \rightarrow BaO(s) + CO_2(g)$

(b) $Fe(s) + O_2(g) \rightarrow Fe_2O_3(s)$

(c) $Al(s) + Mn_2O_3(s) \rightarrow Mn(s) + Al_2O_3(s)$

(d) $AgNO_3(aq) + H_2SO_4(aq) \rightarrow Ag_2SO_4(s) + HNO_3(aq)$

10-36. Classify each of the reactions described by the following chemical equations as combination, decomposition, single replacement, or double replacement. If the equation is not balanced, balance it.

(a) $NaClO_3(s) \rightarrow NaCl(s) + O_2(g)$

(b) $CaO(s) + SO_3(g) \rightarrow CaSO_4(s)$

(c) $H_2(g) + AgCl(s) \xrightarrow[H_2O(l)]{} Ag(s) + HCl(aq)$

(d) $Hg_2(NO_3)_2(aq) + CH_3COOH(aq) \rightarrow Hg_2(CH_3COO)_2(s) + HNO_3(aq)$

10-37. Complete and balance the equations for the following combination reactions:

(a) $Mg(s) + N_2(g) \rightarrow$

(b) $H_2(g) + S(s) \rightarrow$

(c) $MgO(s) + SO_2(g) \rightarrow$

10-38. Complete and balance the equations for the following combination reactions:

(a) $Li(s) + O_2(g) \rightarrow$

(b) $MgO(s) + CO_2(g) \rightarrow$

(c) $H_2(g) + O_2(g) \rightarrow$

(d) $N_2(g) + H_2(g) \rightarrow$

10-39. Complete and balance the equations for the following single-replacement reactions:

(a) $Zn(s) + HBr(aq) \rightarrow$

(b) $Al(s) + Fe_2O_3(s) \rightarrow$

(c) $Pb(s) + Cu(NO_3)_2(aq) \rightarrow$

(d) $Br_2(l) + NaI(aq) \rightarrow$

10-40. Complete and balance the equations for the following single-replacement reactions:

(a) $Ba(s) + H_2O(g) \rightarrow$

(b) $Fe(s) + H_2SO_4(aq) \rightarrow$

(c) $Ca(s) + HBr(aq) \rightarrow$

(d) $Pb(s) + HCl(aq) \rightarrow$

NET IONIC EQUATIONS

10-41. Write the net ionic equation corresponding to

(a) $Na_2S(aq) + 2HCl(aq) \rightarrow 2NaCl(aq) + H_2S(g)$

(b) $PbCl_2(aq) + Na_2S(aq) \rightarrow 2NaCl(aq) + PbS(s)$

(c) $H_2SO_4(aq) + 2KOH(aq) \rightarrow K_2SO_4(aq) + 2H_2O(l)$

(d) $Na_2O(s) + 2HCl(aq) \rightarrow 2NaCl(aq) + H_2O(l)$

(e) $NH_3(g) + HCl(aq) \rightarrow NH_4Cl(aq)$

10-42. Write the net ionic equation corresponding to

(a) $HClO_3(aq) + KOH(aq) \rightarrow KClO_3(aq) + H_2O(l)$

(b) $Pb(NO_3)_2(aq) + Na_2CO_3(aq) \rightarrow 2NaNO_3(aq) + PbCO_3(s)$

(c) $2AgClO_4(aq) + (NH_4)_2SO_4(aq) \rightarrow 2NH_4ClO_4(aq) + Ag_2SO_4(s)$

(d) $K_2S(aq) + Zn(NO_3)_2(aq) \rightarrow 2KNO_3(aq) + ZnS(s)$

10-43. Balance each equation and write the corresponding net ionic equation.

(a) $Fe(NO_3)_3(aq) + NaOH(aq) \rightarrow Fe(OH)_3(s) + NaNO_3(aq)$

(b) $Zn(ClO_4)_2(aq) + K_2S(aq) \rightarrow ZnS(s) + KClO_4(aq)$

(c) $Pb(NO_3)_2(aq) + KOH(aq) \rightarrow Pb(OH)_2(s) + KNO_3(aq)$

(d) $Zn(NO_3)_2(aq) + Na_2CO_3(aq) \rightarrow ZnCO_3(s) + NaNO_3(aq)$

10-44. Balance each equation and write the corresponding net ionic equation.

(a) $AgNO_3(aq) + Na_2S(aq) \rightarrow Ag_2S(s) + NaNO_3(aq)$

(b) $H_2SO_4(aq) + Pb(NO_3)_2(aq) \rightarrow PbSO_4(s) + HNO_3(aq)$

(c) $Hg(NO_3)_2(aq) + NaI(aq) \rightarrow HgI_2(s) + NaNO_3(aq)$

(d) $CdCl_2(aq) + AgClO_4(aq) \rightarrow AgCl(s) + Cd(ClO_4)_2(aq)$

SOLUBILITY RULES

10-45. Use the solubility rules to predict whether the following compounds are soluble or insoluble in water:

(a) $AgI(s)$ (b) $Pb(ClO_4)_2(s)$

(c) $NH_4Br(s)$ (d) $K_2SO_4(s)$

(e) $SrCO_3(s)$

10-46. Use the solubility rules to predict whether the following compounds are soluble or insoluble in water:

(a) $Al_2O_3(s)$ (b) $CuCl_2(s)$

(c) $KNO_3(s)$ (d) $Hg_2Br_2(s)$

(e) $PbCl_2(s)$

10-47. Use the solubility rules to predict whether the following barium salts are soluble or insoluble in water:

(a) $BaCO_3(s)$ (b) $Ba(ClO_4)_2(s)$

(c) $BaCl_2(s)$ (d) $BaS(s)$

(e) $BaSO_4(s)$

10-48. Use the solubility rules to predict whether the following silver salts are soluble or insoluble in water:

(a) $AgBr(s)$ (b) $AgNO_3(s)$

(c) $Ag_2S(s)$ (d) $AgClO_4(s)$

(e) $Ag_2CO_3(s)$

10-49. Use the solubility rules to predict whether the following compounds are soluble or insoluble in water. If a compound is soluble, write the net ionic equation for the dissolution of the compound into its component ions in water.

(a) potassium chromate

(b) ammonium nitrate

(c) calcium carbonate

(d) sodium hydroxide

10-50. Use the solubility rules to predict whether the following compounds are soluble or insoluble in water. If a compound is soluble, write the net ionic equation for the dissolution of the compound into its component ions in water.

(a) iron(III) bromide

(b) calcium sulfate

(c) ammonium carbonate

(d) potassium sulfide

DOUBLE-REPLACEMENT REACTIONS

10-51. Use the solubility rules to predict the products of the following reactions. In each case, complete and balance the equation and write the corresponding net ionic equation. If no precipitate forms, then write "no reaction."

(a) $CuCl_2(aq) + Na_2S(aq) \rightarrow$

(b) $MgBr_2(aq) + K_2CO_3(aq) \rightarrow$

(c) $BaCl_2(aq) + K_2SO_4(aq) \rightarrow$

(d) $Hg_2(NO_3)_2(aq) + KCl(aq) \rightarrow$

10-52. Use the solubility rules to predict the products of the following reactions. In each case complete and balance the equation and write the net ionic equation. If no precipitate forms, then write "no reaction."

(a) $H_2SO_4(aq) + Ca(ClO_4)_2(aq) \rightarrow$

(b) $AgNO_3(aq) + NaClO_4(aq) \rightarrow$

(c) $Hg_2(NO_3)_2(aq) + NaC_7H_5O_2(aq) \rightarrow$

(d) $Pb(CH_3COO)_2(aq) + KBr(aq) \rightarrow$

10-53. Predict the products and write balanced chemical equations and net ionic equations for the following gas-forming reactions:

(a) $H_2SO_4(aq) + NaHCO_3(s) \rightarrow$

(b) $HNO_3(aq) + CaS(s) \rightarrow$

(c) $HCl(aq) + Na_2SO_3(aq) \rightarrow$ (*Hint*: Analogous to the reaction of acids with the carbonate ion.)

10-54. Predict the products and write balanced chemical equations and net ionic equations for the following gas-forming reactions:

(a) $NH_4NO_3(aq) + NaOH(aq) \rightarrow$ (*Hint:* Water is one of the products.)

(b) $HNO_3(aq) + BaCO_3(s) \rightarrow$

(c) $H_2O_2(aq) \rightarrow$ (*Hint*: Water is one of the products of this decomposition reaction.)

10-55. Complete and balance the equation for each of the following acid-base reactions, name the reaction products, and write the corresponding net ionic equation:

(a) $HClO_3(aq) + Ba(OH)_2(aq) \rightarrow$

(b) $CH_3COOH(aq) + KOH(aq) \rightarrow$

(c) $HI(aq) + Mg(OH)_2(s) \rightarrow$

(d) $H_2SO_4(aq) + RbOH(aq) \rightarrow$

10-56. Complete and balance the equation for each of the following acid-base reactions, name the reaction products, and write the corresponding net ionic equation:

(a) $HClO_4(aq) + Ca(OH)_2(aq) \rightarrow$

(b) $HCl(aq) + CaCO_3(s) \rightarrow$

(c) $HNO_3(aq) + Al_2O_3(s) \rightarrow$

(d) $H_2SO_4(aq) + Cu(OH)_2(s) \rightarrow$

10-57. Predict the products of the following reactions and write the balanced chemical equation. If no reaction occurs, then write "no reaction."

(a) An aqueous solution of ammonium sulfate is mixed with an aqueous solution of barium chloride.

(b) An aqueous solution of sodium sulfate is mixed with an aqueous solution of ammonium nitrate.

(c) Hydrochloric acid is added to an aqueous solution of lead(II) nitrate.

(d) An aqueous solution of perchloric acid is added to an aqueous solution of potassium hydroxide.

10-58. Predict the products of the following reactions and write the balanced chemical equation. If no reaction occurs, then write "no reaction."

(a) An aqueous solution of potassium chromate is mixed with an aqueous solution of lead(II) nitrate.

(b) Hydrochloric acid is added to an aqueous solution of sodium sulfide.

(c) An aqueous solution of barium hydroxide is mixed with an aqueous solution of zinc sulfate.

(d) Solid calcium oxide is added to an aqueous solution of nitric acid.

10-59. Classify each of the reactions in Problem 10-57 as a precipitation, acid-base, or gas-forming reaction.

10-60. Classify each of the reactions in Problem 10-58 as a precipitation, acid-base, or gas-forming reaction.

OXIDATION-REDUCTION REACTIONS

10-61. Indicate which element is oxidized and which is reduced in the reactions described by the following chemical equations:

(a) $Ca(s) + Cl_2(g) \rightarrow CaCl_2(s)$

(b) $4Al(s) + 3O_2(g) \rightarrow 2Al_2O_3(s)$

(c) $2Rb(s) + Br_2(l) \rightarrow 2RbBr(s)$

(d) $2Na(s) + S(s) \rightarrow Na_2S(s)$

10-62. Indicate which species is the oxidizing agent and which is the reducing agent in the reactions described by the following chemical equations:

(a) $2Li(s) + Se(s) \rightarrow Li_2Se(s)$

(b) $2Sc(s) + 3I_2(s) \rightarrow 2ScI_3(s)$

(c) $4Ga(s) + P_4(s) \rightarrow 4GaP(s)$

(d) $2K(s) + F_2(g) \rightarrow 2KF(s)$

10-63. For each oxidation-reduction reaction equation given in Problem 10-61, indicate how many electrons are transferred in the formation of one formula unit of product.

10-64. For each oxidation-reduction reaction equation given in Problem 10-62, indicate how many electrons are transferred in the formation of one formula unit of product.

10-65. In a hydrogen fuel cell, hydrogen gas and oxygen gas are combined on a catalyst to form water, generating electricity. Write the balanced chemical equation describing this reaction and show that this is an oxidation-reduction reaction. Identify the oxidizing agent and reducing agent and determine the number of electrons transferred in the balanced chemical equation.

10-66. The combustion of methane, $CH_4(g)$, is used to power natural gas vehicles. Write the chemical equation describing this reaction and show that this is an oxidation-reduction reaction. Identify the oxidizing agent and reducing agent and determine the number of electrons transferred in the balanced chemical equation.

ADDITIONAL PROBLEMS

10-67. List two tests discussed in this chapter that could be used to determine if a solution is acidic. Could these tests also be used to determine if a solution is basic?

10-68. Explain the difference between a single replacement reaction and a double replacement reaction.

10-69. The reaction described by the chemical equation

$$2H_2(g) + O_2(g) \rightarrow 2H_2O(l)$$

falls into several of the reaction classes we have discussed in this chapter. List each of the classes that can be used to describe this reaction. In each case explain why the reaction fits into the particular class.

10-70. What two products are formed when a hydrocarbon (a substance containing only the elements hydrogen and carbon) undergoes complete combustion in oxygen gas?

10-71. Sodium hydroxide, $NaOH(s)$, methanol, $CH_3OH(l)$, and acetic acid, $CH_3COOH(aq)$, are highly water-soluble compounds that contain "OH" groups in their formulas. Use your knowledge of bonding and Lewis formulas to explain why sodium hydroxide forms highly basic solutions, solutions of methanol are essentially neutral, and acetic acid readily loses a hydrogen ion to form acidic solutions.

10-72. For each of the following reactions, write the chemical formulas and balance the equations:

(a) sodium + hydrogen → sodium hydride

(b) aluminum + sulfur → aluminum sulfide

(c) steam + carbon → carbon monoxide + hydrogen

(d) phosphorus trichloride + chlorine → phosphorus pentachloride

10-73. You are presented in the lab with an unknown metal that is either silver or nickel. What chemical test could you perform to determine the identity of the metal?

10-74. Red ants contain an appreciable amount of formic acid, $\underline{H}CHO_2$, where only the underlined proton is acidic. Ants sprayed with window cleaner containing ammonia die quickly. Write the neutralization reaction between formic acid and ammonia.

10-75. Vinegar, which is an aqueous solution of acetic acid, is used to remove deposits of calcium carbonate (lime scale) from automatic coffee makers. Write the equation describing the reaction between acetic acid and calcium carbonate.

10-76. Write the balanced chemical equation for the following reactions. If no reaction occurs, write "no reaction."

(a) Chlorine gas is bubbled through an aqueous solution of sodium iodide.

(b) Liquid bromine is added to an aqueous solution of sodium iodide.

(c) Solid iodine is added to an aqueous solution of potassium chloride.

(d) Liquid bromine is added to an aqueous solution of sodium fluoride.

10-77. Write the balanced chemical equation for the following reactions:

(a) Solid potassium chlorate decomposes to yield solid potassium chloride and oxygen gas.

(b) Potassium metal reacts with water with the evolution of hydrogen gas.

(c) Excess hydrogen gas is injected into a flask containing a small amount of liquid bromine. Upon heating, the bromine liquid disappears and a colorless gas is formed.

10-78. Write the balanced chemical equation for the following reactions:

(a) Zinc sulfide dissolves in aqueous hydrochloric acid with the emission of hydrogen sulfide gas.

(b) Lead(IV) oxide decomposes to yield lead(II) oxide and a colorless gas.

(c) An aqueous solution containing calcium chloride is added to dilute aqueous phosphoric acid, and a white precipitate forms.

10-79. Write the balanced chemical equation for the following reactions:

(a) The combustion of methane gas, $CH_4(g)$, in oxygen yields carbon dioxide and water.

(b) When nickel metal is placed into a blue solution containing copper(II) nitrate, the solution slowly changes to green and a metallic precipitate forms.

(c) When calcium metal is placed in water, it produces bubbles until the metal is dissolved, leaving a highly alkaline (basic) solution.

10-80. Write the balanced chemical equation for the following reactions:

(a) When table sugar, $C_{12}H_{22}O_{11}(s)$, is heated on a stove, it decomposes, forming a black solid and water.

(b) When chlorine gas is bubbled through a clear solution of sodium bromide, the solution turns light brown.

(c) When lithium oxide is added to water, it forms a basic solution.

10-81. Write the balanced chemical equation for the following reactions:

(a) When sulfuric acid is placed on chalk (calcium carbonate), the chalk bubbles, forming a gas.

(b) When tin metal is placed in a solution of hydrochloric acid, bubbles form and the metal is slowly dissolved.

(c) Magnesium ribbon burns in air with a bright blue flame, leaving behind a white powdery solid residue.

10-82. Write the balanced chemical equation for the following reactions:

(a) When washing soda (sodium carbonate decahydrate) is dried in an oven, the water is driven off, leaving behind an anhydrous salt.

(b) When lead(II) nitrate is added to an aqueous solution containing sodium sulfate, a milky white precipitate forms.

(c) When an iron rod is placed into an aqueous solution containing lead(II) nitrate, the iron is oxidized to iron(III) and the lead is reduced.

10-83. Fire extinguishers are rated as class A, B, C, or D depending on the types of fire that they are effective in extinguishing:

Class A extinguishers are for normal combustibles such as wood and paper.

Class B extinguishers are for flammable liquids such as oil and gasoline.

Class C extinguishers are for electrical fires.

Class D extinguishers are for specific flammable metals.

(a) Water is often used in class A extinguishers. Why is water an effective fire extinguisher for normal combustibles?

(b) Carbon dioxide is rated as a class B-C extinguisher (carbon dioxide is not effective against class A fires because the material usually reignites after the CO_2 gas dissipates). Why would you *not* want to use water to extinguish fires of each of these two classes?

(c) Sodium is listed as a flammable metal and requires a special class D extinguisher (typically, sodium chloride is used to extinguish sodium metal fires). Why would you *never* want to use water to extinguish a fire involving sodium metal?

10-84. A student spills a beaker containing concentrated hydrochloric acid on the laboratory bench. Would it be better to clean the spill with water, a $NaOH(aq)$ solution, or powdered calcium carbonate? Explain why.

10-85. The active ingredient in most drain cleaners is concentrated $NaOH(aq)$. Although both acids and bases are effective at dissolving hair and other items that might clog a drain, why are drain cleaners exclusively made from bases? (Concentrated acids and bases are both hazardous substances.)

10-86. The metal the ancient Greeks produced from lead-containing ores was an alloy of lead and silver. They were able to separate the two metals by melting the alloy and blowing air over the molten metal. Write the equation for the reaction that takes place.

10-87. Lead production became important in Roman times because of its use in making pipes to carry water to the famous Roman baths. The manufacture of lead had been developed by the Greeks as a byproduct of the silver mines outside Athens. Lead occurs in silver ores as the sulfide called galena, $PbS(s)$. The silver occurs as the oxide. The ore is first roasted, that is, heated in air to convert the lead to an oxide. The ore is then smelted, that is, heated with charcoal to obtain the pure metal. Write the chemical equations for the reactions that occur in the roasting and smelting of the lead-silver ore.

10-88. (*) Mercury has been known since early times, although we do not know when it was first discovered. Probably because it is a liquid metal, mercury has been thought to have mystical properties throughout history. It occurs in the ore cinnabar as the sulfide $HgS(s)$. The metal was prepared by oxidation of the ore by heating in air, then decomposition of the oxide by continued heating, followed by condensation of the mercury vapor. Write the equations for these reactions.

10-89. The first metal known to be prepared from its ore was copper, perhaps as early as 6000 BC in the Middle East. The ore, which contained copper(II) oxide, was heated with charcoal, which was prepared by the incomplete burning of wood, and is mainly elemental carbon. Later, iron and tin were prepared in the same way. Bronze, an alloy of copper and tin, was made by mixing copper-containing ore with tin ore and heating the mixture in the presence of charcoal. Write the chemical equations for the preparation of the metals $Cu(s)$, $Sn(s)$, and $Fe(s)$ from the ores $CuO(s)$, $SnO_2(s)$, and $Fe_2O_3(s)$, respectively.

10-90. (*) An unknown aqueous solution may contain one or more of the following ions: $NH_4^+(aq)$, $Pb^{2+}(aq)$, $Ca^{2+}(aq)$, and $Fe^{3+}(aq)$. Devise a procedure (it may help to make a flowchart) by which you determine with certainty which ion(s) are present in the solution.

10-91. (*) Nitrogen forms oxyacids with formulas of HNO_2 and HNO_3, whereas phosphorous forms oxyacids with formulas of H_3PO_3 and H_3PO_4 and arsenic forms oxyacids with formulas of H_3AsO_3 and H_3AsO_4. Explain why nitrogen, which is in the same group as phosphorous and arsenic, does not form oxyacids analogous to those of the other two elements.

10-92. (*) Police breathalyzers measure the percent alcohol (ethanol) in a person's breath. A person is asked to breathe into the device, and a fraction of this breath sample is bubbled through a solution of potassium dichromate. If ethanol is present, it will turn the yellowish-orange solution of potassium dichromate into a green solution of chromium(III) sulfate according to the unbalanced reaction

$$CH_3CH_2OH(aq) + \underset{\text{(yellowish-orange)}}{K_2Cr_2O_7(aq)} + H_2SO_4(aq) \rightarrow$$

$$CH_3COOH(aq) + \underset{\text{(green)}}{Cr_2(SO_4)_3(aq)} + K_2SO_4(aq) + H_2O(l)$$

A spectrometer is then used to correlate the absorption of green light to the concentration of alcohol in the person's breath. Balance this equation.

Justus von Liebig (1803–1873) was born in Darmstadt, Germany, to a poor merchant family. He developed an interest in chemistry working in his father's shop, where he experimented with fulminates, a class of highly unstable and explosive compounds. This led him to be expelled from middle school after he caused an explosion with some chemicals he brought to school. After spending six months as an apprentice to a pharmacist and reading books on chemistry borrowed from the open library of Duke Ludwig, he became an assistant to Professor Karl Wilhelm Kastner, who saw that Liebig received a formal education and an honorary doctoral degree. Liebig went on to learn analytical techniques in Paris that led to an appointment as a professor at the University of Giessen at just 21 years of age. At Giessen, he stressed the importance of a laboratory component to a chemical education and under state patronage built the first modern teaching laboratory, complete with fume hoods, from an old army barracks. As his reputation grew, hundreds of chemistry students came to study under him, thus disseminating his new experimental approach to teaching. Among his great achievements, Liebig perfected the technique for rapid elemental analysis using combustion analysis that is still in use today; he was the founder of agricultural chemistry; and his work in chemical analysis laid the foundation upon which modern organic chemistry is built. Because of his irascible nature and hot temper, Liebig alienated many scientific friends and made many enemies. He had few interests outside of chemistry. It is an ironic twist of history that Justus von Liebig, a great chemistry teacher to whom many Nobel Laureates trace their scientific genealogy, was both expelled from middle school and never completed his own doctoral dissertation.

11. Chemical Calculations

In the last chapter, we learned to treat chemical reactions qualitatively. We saw how atoms, molecules, and ions can react with one another to form new compounds, and we discussed how chemists classify many common types of reactions. In this chapter we shall learn how to treat chemical reactions quantitatively. We shall see how we can use chemical calculations to predict the amount of a compound that can be prepared from a given initial quantity of reactants. We also learn to predict how much reactant should be used to obtain a specified amount of product. The basis for all such chemical calculations is the fundamental concept of a mole (one of the most important ideas in chemistry) and the balanced chemical equation that describes a reaction.

11-1. The Quantity of a Substance That is Equal to Its Formula Mass in Grams Is Called a Mole

We learned in Chapter 2 that the atomic mass of an element is a relative quantity; it is the mass of one atom of the element relative to the mass of one atom of carbon-12, which by convention has a mass of exactly 12 atomic mass units, or 12 u. Consider the following table of four elements:

Element	Atomic mass/u
helium, He	4
carbon, C	12
titanium, Ti	48
molybdenum, Mo	96

This table shows that

- One carbon atom has a mass three times that of one helium atom.
- One titanium atom has a mass four times that of one carbon atom and 12 times that of one helium atom.
- One molybdenum atom has a mass twice that of one titanium atom, eight times that of one carbon atom, and 24 times that of one helium atom.

It is important to realize that we have not deduced the absolute mass of any one atom; at this point we can determine only relative masses based on our arbitrarily defined scale of carbon-12 equals 12 atomic mass units (12 u).

Consider 12 grams of carbon, 48 grams of titanium, and 96 grams of molybdenum (Figure 11.1). One titanium atom has a mass four times that of one carbon atom; therefore, 48 grams of titanium atoms must contain the same number of atoms as 12 grams of carbon. Similarly, one molybdenum atom has twice the mass of one titanium atom, so 96 grams of molybdenum must contain the same number of atoms as 48 grams of titanium. We conclude that 12 grams of carbon, 48 grams of titanium, and 96 grams of molybdenum all contain the same number of atoms. If we continue this line of reasoning, we find that exactly the same number of atoms is contained in that quantity of an atomic element whose mass in grams is numerically equal to the element's atomic mass. Thus, we find from the atomic masses given on the periodic table on the inside front cover that 10.8 grams of boron, 23.0 grams of sodium, 63.6 grams of copper, and 200.6 grams of mercury all contain the same number of atoms.

All the substances we have considered here so far are **atomic substances,** that is, substances composed of only one type of atom. Now consider the following **molecular substances:**

Substance	Molecular mass/u
methane, CH_4	$12 + (4 \times 1) = 16$
oxygen, O_2	$2 \times 16 = 32$
ozone, O_3	$3 \times 16 = 48$

Figure 11.1 12 grams of carbon contains the same number of atoms as 48 grams of titanium and 96 grams of molybdenum.

Like atomic masses, molecular masses are relative masses. A molecule of oxygen, O_2, has a mass of 32 u, twice that of a molecule of methane, 16 u. A molecule of ozone has a mass of 48 u, three times that of a molecule of methane. Using the same reasoning we used for atomic substances, we conclude that 16 grams of methane, 32 grams of oxygen, and 48 grams of ozone must all contain the same number of *molecules*. In addition, because the atomic mass of titanium is equal to the molecular mass of ozone, the number of *atoms* in 48 grams of titanium must equal the number of *molecules* in 48 grams of ozone.

We can eliminate the necessity of using the two separate terms atomic mass and molecular mass by using the single term **formula mass** to cover both. Likewise, a **formula unit** can refer to an atom, a molecule, or an ion. We now can extend our previous statement to say that identical numbers of formula units are contained in those quantities of different substances whose masses in grams are numerically equal to their respective formula masses. Thus, 4 grams of helium, 12 grams of carbon, 16 grams of methane, and 32 grams of oxygen all contain the same number of formula units. The formula units are atoms in the cases of helium and carbon, and molecules in the cases of methane and oxygen.

To aid in chemical calculations, chemists use a unit called a **mole**. *The quantity of a substance whose mass in grams is numerically equal to the formula mass of the substance is called a mole.* The abbreviation for mole is **mol** and is often used whenever a specific quantity of moles is written, for example, 3.12 mol $CH_4(g)$. The **molar mass** of a compound is the number of grams needed to make up one mole of the compound. For example, methane, $CH_4(g)$, has a formula mass

Figure 11.2 *(Clockwise from far left)* Molar quantities of sulfur (32.1 grams), sucrose (table sugar) (342.3 grams), copper sulfate pentahydrate (249.7 grams), sodium chloride (58.4 grams), copper (63.6 grams), and mercury(II) oxide (216.6 grams). One mole of a substance is that quantity containing the number of grams numerically equal to its formula mass.

of 16.04, and so its molar mass is 16.04 $g\cdot mol^{-1}$. Notice that formula mass is a unitless quantity, whereas molar mass has units of grams per mole. Figure 11.2 shows one mole of each of six common substances.

As you perform chemical calculations, you will often need to convert between mass and moles. For example, the formula mass of $CH_4(g)$ is 16.04, so there are 16.04 grams per mole of methane. This fact yields two unit conversion factors:

$$\left(\frac{1\text{ mol CH}_4}{16.04\text{ g CH}_4}\right) = 1 \quad \text{or} \quad \left(\frac{16.04\text{ g CH}_4}{1\text{ mol CH}_4}\right) = 1$$

The factor with mass in grams in the denominator may be used to convert from grams to moles of $CH_4(g)$, and its inverse with moles in the denominator to convert from moles of $CH_4(g)$ to mass in grams.

For example, suppose you needed to know the number of moles in 50.0 grams of methane. Using the unit conversion factor with the mass in grams in the denominator, we find

$$\text{moles of CH}_4 = \underbrace{(50.0\text{ g CH}_4)}_{\substack{\text{3 significant}\\ \text{figures}}}\underbrace{\left(\frac{1\text{ mol CH}_4}{16.04\text{ g CH}_4}\right)}_{\substack{\text{4 significant}\\ \text{figures (1 is exact)}}} = \underbrace{3.12\text{ mol CH}_4}_{\substack{\text{3 significant}\\ \text{figures}}}$$

Notice that the result is expressed to three significant figures and assigned the units moles of $CH_4(g)$, as required.

You can also calculate the mass of a certain number of moles of a substance. For example, let's calculate the mass of 2.16 moles of sodium chloride, NaCl(*s*).

The formula mass of NaCl(s) is 58.44. The mass of NaCl(s) in 2.16 moles is

$$\text{mass of NaCl} = \underbrace{(2.16\text{ mol NaCl})}_{\substack{\text{3 significant}\\ \text{figures}}}\underbrace{\left(\frac{58.44\text{ g NaCl}}{1\text{ mol NaCl}}\right)}_{\substack{\text{4 significant}\\ \text{figures (1 is exact)}}} = \underbrace{126\text{ g NaCl}}_{\substack{\text{3 significant}\\ \text{figures}}}$$

Notice that we express the final result to three significant figures and that the units assigned to the result are grams of NaCl(s).

When using a conversion factor, it is generally best if the numbers used in the conversion factor (if not exact) have *at least one more significant figure* than the data being converted; otherwise the conversion factor, not the actual data, will limit the number of significant figures in the result. It is important to remember that defined conversion factors such as 1 m ≡ 100 cm are always exact numbers and never limit the number of significant figures in the result.

EXAMPLE 11-1: Calculate the number of moles in (a) 28.0 grams of water (about 1 oz) and (b) 324 mg of aspirin, $C_9H_8O_4(s)$ (324 mg is the mass of aspirin in one 5-grain aspirin tablet).

Solution: (a) The formula mass of $H_2O(l)$ is 18.02. Consequently, the number of moles of water in 28.0 grams is

$$\text{moles of H}_2\text{O} = (28.0\text{ g H}_2\text{O})\left(\frac{1\text{ mol H}_2\text{O}}{18.02\text{ g H}_2\text{O}}\right) = 1.55\text{ mol H}_2\text{O}$$

(b) The chemical formula of aspirin is $C_9H_8O_4(s)$, so its formula mass is $(9 \times 12.01) + (8 \times 1.008) + (4 \times 16.00) = 180.2$. (Recall that the atomic mass of naturally occurring carbon is 12.01.) The number of moles of aspirin in 324 milligrams is

$$\text{moles of C}_9\text{H}_8\text{O}_4 = (324\text{ mg C}_9\text{H}_8\text{O}_4)\left(\frac{1\text{ g}}{1000\text{ mg}}\right)\left(\frac{1\text{ mol C}_9\text{H}_8\text{O}_4}{180.2\text{ g C}_9\text{H}_8\text{O}_4}\right)$$

$$= 1.80 \times 10^{-3}\text{ mol C}_9\text{H}_8\text{O}_4$$

Notice that we have to convert milligrams to grams before dividing by 180.2 grams.

PRACTICE PROBLEM 11-1: Some data for annual world chemical production are given in Appendix H. (a) Calculate the number of moles that are produced annually in the United States for each of the top five chemicals listed by mass produced: sulfuric acid, $H_2SO_4(l)$, nitrogen, $N_2(g)$, ethylene, $C_2H_4(g)$, oxygen, $O_2(g)$, and hydrogen, $H_2(g)$. (b) Which one has the largest annual production on a molar basis? (one metric ton = 1000 kg)

Answer: (a) 3.83×10^{11} mol $H_2SO_4(l)$, 9.52×10^{11} mol $N_2(g)$, 9.16×10^{11} mol $C_2H_4(g)$, 6.11×10^{11} mol $O_2(g)$, 8.78×10^{12} mol $H_2(g)$; (b) $H_2(g)$

Example 11-1 illustrates an important point. In order to calculate the number of moles in a given mass of a chemical compound, it is necessary to know the chemical formula of the compound. A mole of any compound is defined only in terms of its chemical formula. If a substance (such as coal or wood) cannot be represented by a single chemical formula, then we can give only the mass of the substance.

11-2. One Mole of Any Substance Contains Avogadro's Number of Formula Units

It has been determined experimentally that one mole of any substance contains 6.022×10^{23} formula units (to four significant figures). This number is called **Avogadro's number** after the Italian scientist Amedeo Avogadro, who was one of the earliest scientists to distinguish between atoms and molecules (Chapter 13 Frontispiece). We say not only that one mole of any substance contains Avogadro's number of formula units but also that one mole is that mass of a substance containing Avogadro's number of formula units, or "elementary entities." For example, the atomic mass of the pure isotope carbon-12 is taken to be exactly 12, so 12.00 grams of carbon-12 contains 6.022×10^{23} atoms. Likewise, the molecular mass of water is 18.02, so 18.02 grams of water contains 6.022×10^{23} molecules.

A mole is simply a designation for Avogadro's number of "things" such as atoms and molecules, just as a dozen of eggs is a designation for twelve eggs. It is often helpful to think of one mole as a "counting unit" representing Avogadro's number of things, just as one dozen is a counting unit representing twelve things. But instead of the number 12 implied by the term dozen, the number of things in a mole is 6.022×10^{23}. A mole may be more intimidating because of the huge magnitude of Avogadro's number, but it is really the same concept as a dozen. A mole of eggs would be 6.022×10^{23} eggs, but it is not a practical measure of the number of eggs. A mole of atoms or molecules, on the other hand, is a practical measure of the number of atoms or molecules in a substance because of their small size. A few examples of things that we count in chemistry using moles are given in Table 11.1.

We now have an alternative definition for a mole: *One mole is the mass of a substance containing Avogadro's number of formula units.* For example, referring to

TABLE 11.1 Some things for which we use the counting unit "mole"

Counting unit	Number of things	Examples of things counted in moles	Mass of one mole
1 mole	6.022×10^{23}	atoms, such as aluminum, Al	26.98 g
		molecules, such as water, H_2O	18.02 g
		ions, such as Na^+	22.99 g
		elementary particles, such as electrons, e^-	0.5486 mg

Table 11.1, one mole of aluminum atoms may be expressed as either 6.022×10^{23} aluminum atoms or 26.98 grams of aluminum. The formula mass of a substance is the mass in grams of one mole or 6.022×10^{23} formula units. That is,

$$1 \text{ mol Al} = 6.022 \times 10^{23} \text{ Al atoms} = 26.98 \text{ g Al}$$

Avogadro's number is an enormous number. If we were to express Avogadro's number without using scientific notation, we would have 602 200 000 000 000 000 000 000. In order to appreciate the magnitude of Avogadro's number another way, let's compute how many years it would take to spend Avogadro's number of dollars at a rate of one million dollars per second. Because there are 3.15×10^7 seconds in one year, the number of years required to spend 6.022×10^{23} dollars is

$$\begin{aligned}\text{number of years} &= (6.022 \times 10^{23} \text{ dollars})\left(\frac{1 \text{ s}}{10^6 \text{ dollars}}\right)\left(\frac{1 \text{ year}}{3.15 \times 10^7 \text{ s}}\right) \\ &= 1.91 \times 10^{10} \text{ years}\end{aligned}$$

or 19.1 billion years (1 billion = 10^9). This interval is over four times longer than the estimated age of the earth (4.6 billion years) and is somewhat larger than the estimated age of the universe (14 billion years). This calculation illustrates just how large Avogadro's number is and, consequently, how small atoms and molecules are. Look again at the samples in Figure 11.2. Each of these contains 6.022×10^{23} formula units of the indicated substance.

We can use Avogadro's number to calculate the mass of a single atom or molecule, as illustrated by the following Example.

EXAMPLE 11-2: Using Avogadro's number, calculate the mass of one nitrogen molecule.

Solution: Recall that nitrogen occurs as a diatomic molecule. The formula of molecular nitrogen is N_2, so its formula mass, or molecular mass, is 28.02. Thus, there are 28.02 grams of nitrogen in one mole. Using the fact that one mole of any substance contains 6.022×10^{23} formula units, the mass of one nitrogen molecule is

$$\begin{aligned}\left(\begin{array}{c}\text{mass of one} \\ \text{nitrogen molecule}\end{array}\right) &= \left(\frac{28.02 \text{ g N}_2}{1 \text{ mol N}_2}\right)\left(\frac{1 \text{ mol}}{6.022 \times 10^{23} \text{ molecules}}\right) \\ &= 4.653 \times 10^{-23} \text{ g} \cdot \text{molecule}^{-1}\end{aligned}$$

PRACTICE PROBLEM 11-2: In Chapter 13 when we study gases, we shall use the masses of molecules in kilograms. Calculate the mass of a carbon dioxide, $CO_2(g)$, molecule and of a sulfur hexafluoride, $SF_6(g)$, molecule in kilograms.

Answer: CO_2, 7.308×10^{-26} kg; SF_6, 2.425×10^{-25} kg

Avogadro's number also can be used to calculate the number of atoms or molecules in a given mass of a substance. The next Example illustrates this type of calculation.

EXAMPLE 11-3: Calculate how many methane molecules and how many hydrogen and carbon atoms there are in a picogram of methane, $CH_4(g)$.

Solution: The formula mass of methane, $CH_4(g)$, is $12.01 + (4 \times 1.008) = 16.04$, so one picogram of $CH_4(g)$ consists of

$$\text{molecules of } CH_4 = (1.00 \text{ pg } CH_4)\left(\frac{1 \times 10^{-12} \text{ g}}{1 \text{ pg}}\right)\left(\frac{1 \text{ mol } CH_4}{16.04 \text{ g } CH_4}\right)\left(\frac{6.022 \times 10^{23} \text{ molecules } CH_4}{1 \text{ mol } CH_4}\right)$$

$$= 3.75 \times 10^{10} \text{ } CH_4 \text{ molecules}$$

Each molecule of methane contains one carbon atom and four hydrogen atoms, so

$$\text{number of C atoms} = (3.75 \times 10^{10} \text{ } CH_4 \text{ molecules})\left(\frac{1 \text{ C atom}}{1 \text{ } CH_4 \text{ molecule}}\right)$$

$$= 3.75 \times 10^{10} \text{ C atoms}$$

$$\text{number of H atoms} = (3.75 \times 10^{10} \text{ } CH_4 \text{ molecules})\left(\frac{4 \text{ H atoms}}{1 \text{ } CH_4 \text{ molecule}}\right)$$

$$= 1.50 \times 10^{11} \text{ H atoms}$$

Realize that a picogram is a millionth of a millionth of a gram, a quantity so small that were all the molecules condensed into a liquid you would still need a high-quality microscope to see it. Nevertheless, there are well over a thousand million methane molecules in the sample. This illustrates just how small atoms and molecules are.

PRACTICE PROBLEM 11-3: Some inkjet printers produce picoliter-sized drops. How many water molecules are there in one picoliter of water? How many hydrogen and oxygen atoms does this correspond to? Take the density of water to be 1.00 g·mL^{-1}.

Answer: 3.34×10^{13} water molecules, 6.68×10^{13} hydrogen atoms, 3.34×10^{13} oxygen atoms.

Table 11.2 summarizes the relationships between molar quantities.

We conclude this section with the official SI definition of a mole: "The mole is the amount of substance of a system which contains as many elementary entities as there are atoms in exactly 0.012 kilograms of carbon-12. When the mole is used, the elementary entities must be specified and may be atoms, molecules, ions, electrons, other particles, or specified groups of such particles." Note that because the atomic mass of carbon-12 is exactly 12 by definition, a mole of carbon-12 contains exactly 12 grams (= 0.012 kg) of carbon. This SI definition of a mole is equivalent to the other definitions given in this section.

TABLE 11.2 Representative relationships between molar quantities

Substance	Formula	Formula mass	Molar mass/ $g{\cdot}mol^{-1}$	Number of particles in one mole	Number of moles
atomic chlorine	Cl	35.45	35.45	6.022×10^{23} chlorine atoms	1 mole of Cl atoms
chlorine gas	Cl_2	70.90	70.90	6.022×10^{23} chlorine molecules	1 mole of Cl_2 molecules
				12.044×10^{23} chlorine atoms	2 moles of Cl atoms
water	H_2O	18.02	18.02	6.022×10^{23} water molecules	1 mole of H_2O molecules
				12.044×10^{23} hydrogen atoms	2 moles of H atoms
				6.022×10^{23} oxygen atoms	1 mole of O atoms
sodium chloride	NaCl	58.44	58.44	6.022×10^{23} NaCl formula units	1 mole of NaCl formula units
				6.022×10^{23} sodium ions	1 mole of Na^+ ions
				6.022×10^{23} chloride ions	1 mole of Cl^- ions
barium fluoride	BaF_2	175.3	175.3	6.022×10^{23} BaF_2 formula units	1 mole of BaF_2 formula units
				6.022×10^{23} barium ions	1 mole of Ba^{2+} ions
				12.044×10^{23} fluoride ions	2 moles of F^- ions
nitrate ion	NO_3^-	62.01	62.01	6.022×10^{23} nitrate ions	1 mole of NO_3^- ions
				6.022×10^{23} nitrogen atoms	1 mole of N atoms
				18.066×10^{23} oxygen atoms	3 moles of O atoms

The concept of a mole is one of the most important concepts, if not the most important, in all of chemistry. We shall be using moles in most of the following chapters of the book. It is important that you understand what a mole is and how to use it to do chemical calculations involving moles. If you are not comfortable with the mole concept at this point, be sure to do as many of the problems from the end of this chapter as needed in order to become confident in your use of the concept of a mole.

11-3. Simplest Formulas Can Be Determined by Chemical Analysis

We introduced Avogadro's number and the concept of a mole in the previous two sections. We now introduce another fundamental concept in our study of chemistry, that of **stoichiometry** (stoi´ke om´i tre). Stoichiometry is the calculation of the quantities of elements or compounds involved in chemical reactions. The word stoichiometry is derived from the Greek words *stoicheio,* meaning "simplest components or parts," and *metrein,* meaning "to measure." The concept of a mole is central to carrying out stoichiometric calculations.

For example, we can use the concept of a mole to determine the simplest chemical formula of a substance. Zinc oxide is found by chemical analysis to be 80.3 percent (by mass) zinc and 19.7 percent (by mass) oxygen. When working with mass percentages in chemical calculations, it is convenient to consider a 100-gram sample so that the mass percentages can be converted easily to grams. For example, a 100-gram sample of zinc oxide contains 80.3 grams of zinc and 19.7 grams of oxygen. We can write this schematically as

$$80.3 \text{ g Zn} \approx 19.7 \text{ g O}$$

where the symbol $\approx$ means "is **stoichiometrically equivalent to**" or, in this case, "combines with." If we divide 80.3 grams of zinc by the atomic mass of zinc (65.38), we find that

$$\text{moles of Zn} = (80.3 \text{ g Zn})\left(\frac{1 \text{ mol Zn}}{65.41 \text{ g Zn}}\right) = 1.23 \text{ mol Zn}$$

and if we divide 19.7 grams of oxygen by the atomic mass of oxygen (16.00), we obtain

$$\text{moles of O} = (19.7 \text{ g O})\left(\frac{1 \text{ mol O}}{16.00 \text{ g O}}\right) = 1.23 \text{ mol O}$$

Thus, we have

$$1.23 \text{ mol Zn} \approx 1.23 \text{ mol O}$$

or, dividing both sides of this expression by 1.23,

$$1.00 \text{ mol Zn} \approx 1.00 \text{ mol O}$$

Because 1.00 mole corresponds to Avogadro's number of atoms, we can divide both sides by Avogadro's number and get

$$1.00 \text{ atom of Zn} \approx 1.00 \text{ atom of O}$$

This expression says that one atom of zinc combines with one atom of oxygen and that the **simplest formula** of zinc oxide is therefore ZnO. We call ZnO the simplest chemical formula of zinc oxide because chemical analysis provides us with only the *ratios* of atoms in a chemical formula and not the actual number of atoms. Mass percentages alone cannot be used to distinguish among ZnO, Zn_2O_2, Zn_3O_3, or any other multiple of the formula ZnO. For this reason, simplest formulas are often called **empirical formulas.** The following Example illustrates another calculation of a simplest, or empirical, formula.

EXAMPLE 11-4: Chemical analysis shows that 2-propanol, commonly called isopropyl alcohol, is 60.0% carbon, 13.4% hydrogen, and 26.6% oxygen by mass. Determine the empirical formula of isopropyl alcohol. Given that the molecular mass of the compound is 60.09, determine its molecular formula.

Solution: As usual, we take a 100-gram sample and write

$$60.0 \text{ g C} \Bumpeq 13.4 \text{ g H} \Bumpeq 26.6 \text{ g O}$$

We divide each value by the corresponding atomic mass and get

$$\text{moles of C} = (60.0 \text{ g C})\left(\frac{1 \text{ mol C}}{12.01 \text{ g C}}\right) = 5.00 \text{ mol C}$$

$$\text{moles of H} = (13.4 \text{ g H})\left(\frac{1 \text{ mol H}}{1.008 \text{ g H}}\right) = 13.3 \text{ mol H}$$

$$\text{moles of O} = (26.6 \text{ g O})\left(\frac{1 \text{ mol O}}{16.00 \text{ g O}}\right) = 1.66 \text{ mol O}$$

When determining empirical formulas we use the masses of atoms and not those of molecules in our calculations because we want to determine the number of atoms in the compound.

or

$$5.00 \text{ mol C} \Bumpeq 13.3 \text{ mol H} \Bumpeq 1.66 \text{ mol O}$$

To find a simple, whole-number relationship for these values, we divide through by the smallest value (1.66) and get

$$3.00 \text{ mol C} \Bumpeq 8.00 \text{ mol H} \Bumpeq 1.00 \text{ mol O}$$

Dividing through by Avogadro's number, we find that the empirical formula of an isopropyl alcohol molecule consists of three carbon atoms, eight hydrogen atoms, and one oxygen atom, or that the empirical formula is C_3H_8O. The formula mass of C_3H_8O is 60.09, and so we see in this case that the empirical formula and the molecular formula are the same.

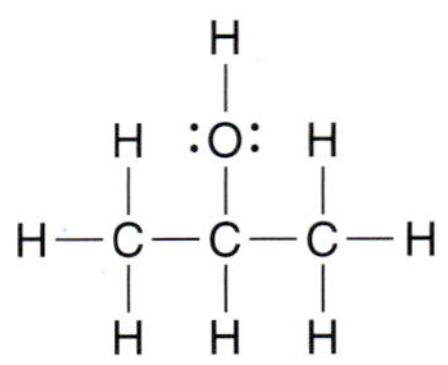

The Lewis formula of 2-propanol or isopropyl alcohol, the active component in rubbing alcohol solutions.

PRACTICE PROBLEM 11-4: Ethanol, or ethyl alcohol, is 52.1% carbon, 13.2% hydrogen, and 34.7% oxygen. Determine the empirical formula of ethanol.

Answer: C_2H_6O (see margin)

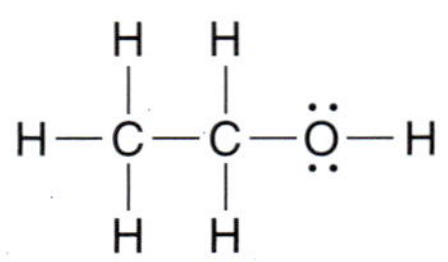

The Lewis formula of ethanol, the alcohol present in alcoholic beverages. Ethanol is sometimes referred by its common name, ethyl alcohol. We discuss the chemistry of alcohols in Interchapter P.

In contrast to the previous Examples, if we already know the chemical formula, then we can determine the mass percentage of an element in the formula. To do so, we multiply the number of atoms in the chemical formula by the atomic mass of the atom, divide by the formula mass of the compound, and then multiply by 100 to get a percentage. For example, the chemical formula of aluminum oxide is Al_2O_3 (formula mass = 101.96). Thus, the mass percentages of aluminum and oxygen in aluminum oxide are given by

$$\begin{pmatrix}\text{mass percentage}\\ \text{of aluminum in}\\ \text{aluminum oxide}\end{pmatrix} = \left(\frac{2 \times \text{atomic mass of aluminum}}{\text{formula mass of aluminum oxide}}\right) \times 100$$

$$= \left(\frac{2 \times 26.98}{101.96}\right) \times 100 = 52.92\%$$

WWW See Interchapter P at www.McQuarrieGeneralChemistry.com.

and

$$\begin{pmatrix}\text{mass percentage}\\ \text{of oxygen in}\\ \text{aluminum oxide}\end{pmatrix} = \left(\frac{3 \times \text{atomic mass of oxygen}}{\text{formula mass of aluminum oxide}}\right) \times 100$$

$$= \left(\frac{3 \times 16.00}{101.96}\right) \times 100 = 47.08\%$$

Notice that the total is equal to 100.00%, as it should.

The next Example illustrates an experimental procedure for determining empirical formulas.

EXAMPLE 11-5: A 0.450-gram sample of magnesium, Mg(s), is reacted completely with nitrogen, $N_2(g)$, to produce 0.623 grams of magnesium nitride (Figure 11.3). Use these data to determine the empirical formula of magnesium nitride.

Solution: We can represent this combination reaction schematically as follows:

0.450 g magnesium metal	+	excess nitrogen gas	→	0.623 g $Mg_?N_?$ magnesium nitride (formula unknown)

The 0.623 grams of magnesium nitride formed contains 0.450 grams of magnesium, so the mass of nitrogen in the product is

$$\text{mass of N atoms in product} = 0.623\text{ g} - 0.450\text{ g} = 0.173\text{ g N}$$

We can convert 0.450 grams of magnesium to moles of magnesium by dividing by its atomic mass (24.31):

$$\text{moles of Mg} = (0.450\text{ g Mg})\left(\frac{1\text{ mol Mg}}{24.31\text{ g Mg}}\right) = 0.0185\text{ mol Mg}$$

Similarly, by dividing the mass of the nitrogen atoms by the atomic mass of nitrogen (14.01), we obtain

$$\text{moles of N} = (0.173\text{ g N})\left(\frac{1\text{ mol N}}{14.01\text{ g N}}\right) = 0.0123\text{ mol N}$$

Thus, we have the relation

$$0.0185\text{ mol Mg} \approx 0.0123\text{ mol N}$$

When we divide both quantities by the smaller number (0.0123), we obtain

$$1.50\text{ mol Mg} \approx 1.00\text{ mol N}$$

Because we are seeking the simplest whole-number relationship, we multiply both sides by 2 to get

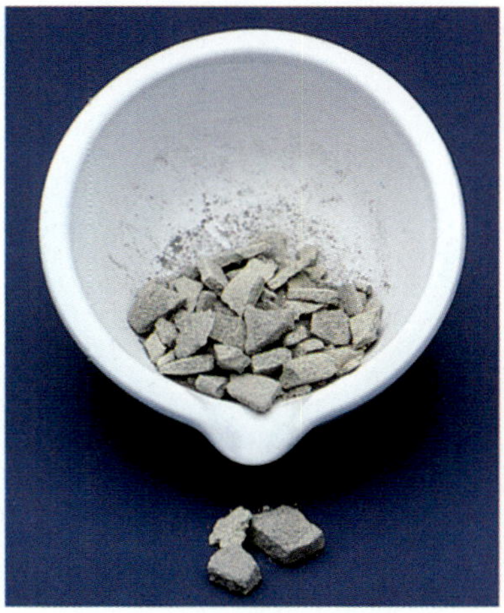

Figure 11.3 Magnesium nitride is produced when magnesium is burned in an atmosphere of nitrogen.

$$3.00 \text{ mol Mg} \Leftrightarrow 2.00 \text{ mol N}$$

We see that 3.00 moles of magnesium atoms combine with 2.00 moles of nitrogen atoms. Thus, the empirical formula for magnesium nitride is Mg_3N_2.

PRACTICE PROBLEM 11-5: A 2.18-gram sample of scandium is burned in oxygen to produce 3.34 grams of scandium oxide. Use these data to determine the empirical formula of scandium oxide.

Answer: Sc_2O_3

11-4. Empirical Formulas Can Be Used to Determine an Unknown Atomic Mass

If we know the empirical formula of a compound, then we can determine the atomic mass of one of the elements in the compound if the atomic masses of the other elements in the compound are known. This determination is a standard experiment in many general chemistry laboratory courses.

EXAMPLE 11-6: The empirical formula of a metal oxide is MO, where M stands for the chemical symbol of the metal. A weighed quantity of metal, 0.490 grams, is burned in oxygen, and the metal oxide produced is found to have a mass of 0.813 grams. Given that the atomic mass of oxygen is 16.00, determine the atomic mass of the metal.

Solution: The mass of oxygen in the metal oxide is

$$\text{mass of O in oxide} = 0.813 \text{ g MO} - 0.490 \text{ g M} = 0.323 \text{ g O}$$

The number of moles of O is

$$\text{moles of O} = (0.323 \text{ g O})\left(\frac{1 \text{ mol O}}{16.00 \text{ g O}}\right) = 0.0202 \text{ mol O}$$

The empirical formula MO tells us that 0.0202 moles of M are combined with 0.0202 moles of O. Because we started with 0.490 grams of M, we have

$$0.490 \text{ g M} \Leftrightarrow 0.0202 \text{ mol M}$$

The atomic mass of the metal can be obtained if we determine how many grams of the metal correspond to 1.00 mole. To determine this, we divide both sides of the stoichiometric correspondence by 0.0202 to get

$$24.3 \text{ g M} \Leftrightarrow 1.00 \text{ mol M}$$

In other words, the atomic mass of the metal is 24.3. By consulting a table of atomic masses, we see that the metal is magnesium.

PRACTICE PROBLEM 11-6: The empirical formula of a metal oxide is M_2O_3. A 3.058-gram sample of the metal is burned in oxygen, and the M_2O_3 produced is found to have a mass of 4.111 grams. Given that the atomic mass of oxygen is 16.00, determine the atomic mass of the metal. What is the metal?

Answer: 69.70; gallium

You can see from these examples that it is possible to determine the empirical formula of a compound if the atomic masses are known, and that it is possible to determine the atomic mass of an element if the empirical formula of one of its compounds and the atomic masses of the other elements that make up the compound are known. We are faced with a dilemma here. Atomic masses can be determined if empirical formulas are known (Example 11-6), but atomic masses must be known to determine empirical formulas (Examples 11-4 and 11-5). This was a serious problem in the early 1800s, shortly after Dalton formulated his atomic theory. At the time, it was necessary to guess the empirical formula of a compound—even though an incorrect guess led to an incorrect atomic mass. In fact, many of the atomic masses originally proposed by Dalton were in error for exactly this reason. We show in Chapter 13 that it was the quantitative study of gases and of reactions between gases that provided the key for resolving the difficulty in determining reliable values of atomic masses. Today mass spectrometry (Section 2-11) allows for the *direct* determination of atomic masses, so ambiguities no longer exist in the table of atomic masses.

11-5. An Empirical Formula Along With the Molecular Mass Determines the Molecular Formula

Suppose that the chemical analysis of a compound gives 85.7% carbon and 14.3% hydrogen by mass. We then have

$$85.7\text{ g C} \Leftrightarrow 14.3\text{ g H}$$
$$7.14\text{ mol C} \Leftrightarrow 14.2\text{ mol H}$$
$$1\text{ mol C} \Leftrightarrow 2\text{ mol H}$$

and conclude that the empirical formula is CH_2. However, the actual formula, the **molecular formula**, might be C_2H_4, C_3H_6, or, generally, C_nH_{2n} for any whole number n. The chemical analysis gives us only ratios of the number of atoms. If we know the molecular mass from another experiment, however, we can determine the molecular formula unambiguously. For example, suppose we know that the molecular mass of our compound, as determined from other experiments, is 42. By listing the various possible formulas, all of which have the empirical formula CH_2,

Formula	Formula mass
CH_2	14
C_2H_4	28
C_3H_6	42
C_4H_8	56

we see that C_3H_6 is the molecular formula that has a formula mass of 42. The molecular formula of the compound is therefore C_3H_6. You can understand now why the empirical formula deduced from chemical analysis is also called the simplest formula. It must be supplemented by molecular mass data to determine the molecular formula.

Note that the molecular formula in the above example is "three times" the empirical formula. This factor of three can be obtained directly by using the following relationship:

$$\text{“factor”} = \frac{\text{molecular mass}}{\text{empirical formula mass}}$$
$$= \frac{42}{14} = 3$$
$$CH_2 \times 3 = C_3H_6$$

See Interchapter F at www.McQuarrieGeneralChemistry.com.

EXAMPLE 11-7: There are many compounds, called **hydrocarbons**, that consist of only carbon and hydrogen (see Interchapter F). Gasoline typically is a mixture of over 100 different hydrocarbons—the exact formulation varies widely depending on both the crude oil and the refinery process used. Chemical analysis of one of the common constituents of gasoline yields 92.30% carbon and 7.70% hydrogen by mass. (a) Determine the simplest formula of this compound. (b) Given that its molecular mass as determined by mass spectrometry is 78, determine its molecular formula.

Solution: (a) The determination of the simplest formula can be summarized by

$$92.30\text{ g C} \approx 7.70\text{ g H}$$
$$7.685\text{ mol C} \approx 7.64\text{ mol H}$$
$$1\text{ mol C} \approx 1\text{ mol H}$$

The simplest formula is CH, which has a formula mass of 13.

(b) The molecular mass is 78 and the empirical formula mass is 13, so the molecular formula must be 78/13 = 6 times the empirical formula. Thus, the molecular formula is C_6H_6, which as we have already seen is the molecule benzene (margin).

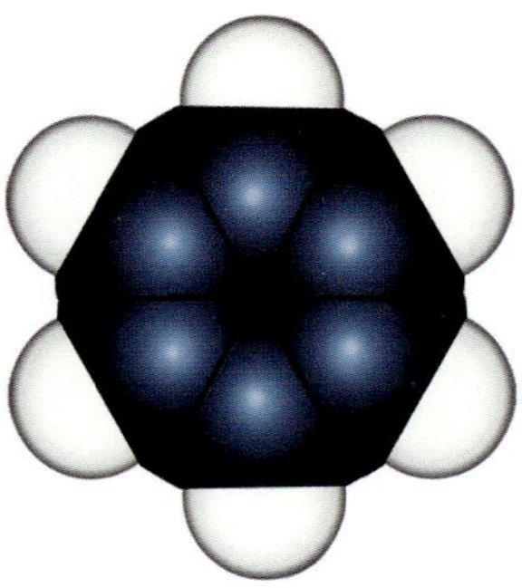

Molecular model of benzene, C_6H_6

PRACTICE PROBLEM 11-7: The United States produced approximately twelve million metric tons of 1,2-dichloroethane in 2005 (listed under its common name ethylene dichloride, in Appendix H). Chemical analysis shows that 1,2-dichloroethane is 24.27% carbon, 4.075% hydrogen, and 71.65% chlorine by mass. Given that the molecular mass of 1,2-dichloroethane is 98.95, determine its molecular formula.

Answer: $C_2H_4Cl_2$

It is important to realize that the molecular formula of a compound does not tell us in what arrangement the atoms are bonded together in the molecules of the compound. For example, the two compounds ethanol (the alcohol found in alcoholic beverages) and dimethyl ether (once used as an anesthetic) both have the molecular formula C_2H_6O, but their molecular structures are very different, as shown in the margin.

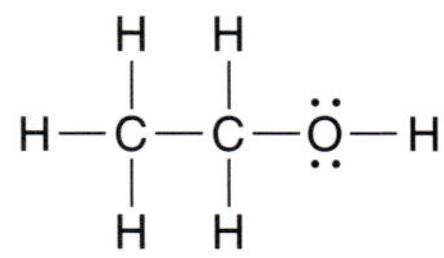

ethanol

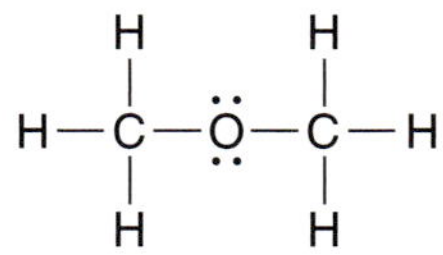

dimethyl ether

Chemists use a variety of techniques for determining the structures of molecules. Many of these involve the use of spectroscopy.

11-6. The Percentage Composition of Many Compounds Can Be Determined by Combustion Analysis

Combustion analysis, originally developed by Lavoisier (Chapter 1 Frontispiece) is one of the oldest analytical techniques in chemistry and is used for determining the elemental composition of many molecules (Figure 11.4). This important

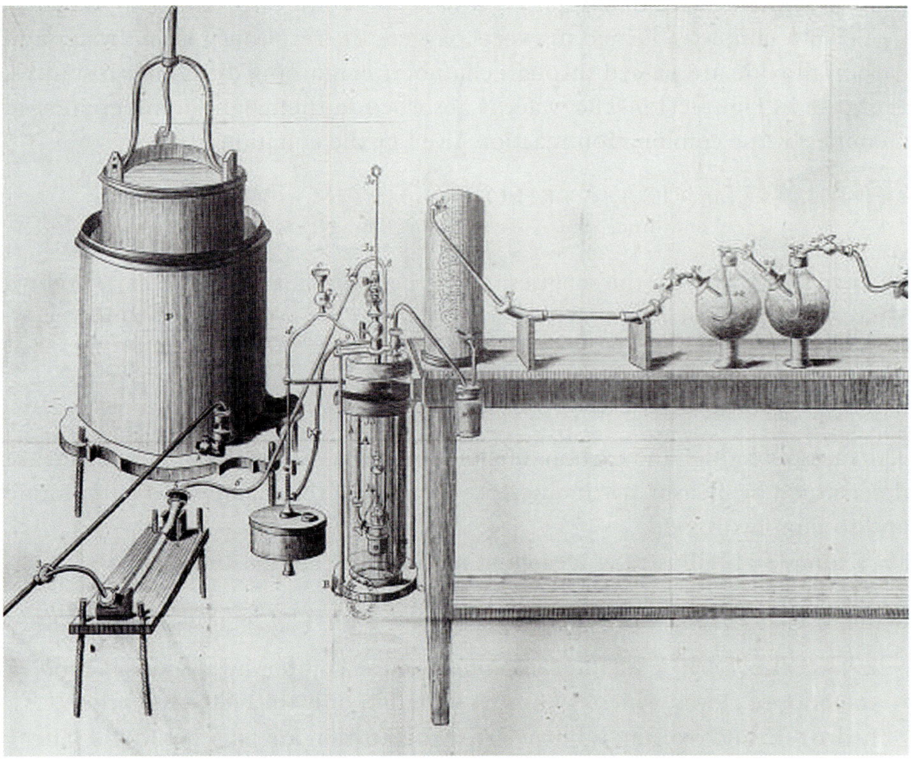

Figure 11.4 A combustion analysis device designed by Lavoisier (Chapter 1 Frontispiece) in 1789 for the analysis of oils. The drawing was etched by his wife, Marie-Anne Lavoisier. It is uncertain whether Lavoisier ever actually constructed this device, a complex precursor to the much simpler technique perfected by Liebig (Frontispiece).

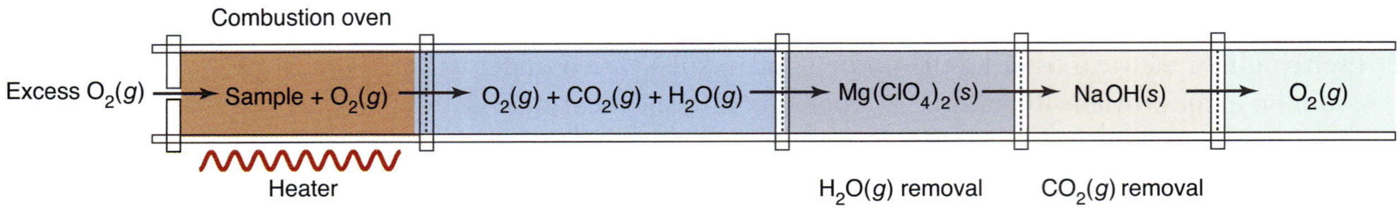

Figure 11.5 A schematic illustration of the removal of $H_2O(l)$ and $CO_2(g)$ from the combustion gases. Anhydrous magnesium perchlorate $Mg(ClO_4)_2(s)$ removes the water vapor according to

$$Mg(ClO_4)_2(s) + 6H_2O(g) \rightarrow Mg(ClO_4)_2{\cdot}6H_2O(s)$$

Then sodium hydroxide removes the $CO_2(g)$:

$$NaOH(s) + CO_2(g) \rightarrow NaHCO_3(s)$$

Excess $O_2(g)$ is used as a carrier gas to "sweep" all the $H_2O(g)$ and $CO_2(g)$ from the combustion zone (left-hand compartment).

technique was perfected into a simple method for routine chemical analysis by the pioneering German chemist Justus von Liebig (Frontispiece).

Many organic compounds consist of only carbon and hydrogen atoms, or of carbon, hydrogen, and oxygen atoms. When these compounds are burned completely in an excess of $O_2(g)$, all the carbon in the original sample ends up in $CO_2(g)$ and all the hydrogen ends up in $H_2O(g)$. These facts are the basis of the determination of the percentage composition of such compounds by **combustion analysis**, the analysis of the products obtained when an organic compound is burned in an excess of oxygen.

After a sample is burned in excess oxygen, the resulting gaseous water and carbon dioxide are passed through chambers containing different substances, as shown in Figure 11.5. The water is absorbed in the magnesium perchlorate chamber by the combination reaction given by the equation

$$\underset{\text{anhydrous}}{Mg(ClO_4)_2(s)} + 6H_2O(g) \rightarrow \underset{\text{hydrated}}{Mg(ClO_4)_2{\cdot}6H_2O(s)}$$

After passing through the magnesium perchlorate chamber, the carbon dioxide reacts with the sodium hydroxide in the next chamber according to

$$NaOH(s) + CO_2(g) \rightarrow NaHCO_3(s)$$

The masses of water and carbon dioxide formed in the combustion reaction are determined by measuring the increase in mass of the magnesium perchlorate and sodium hydroxide.

Example 11-8 illustrates a calculation involving combustion analysis.

EXAMPLE 11-8: A 1.250-gram sample of the compound responsible for the odor of cloves, which is known to contain only the elements carbon, hydrogen, and oxygen, is burned in a combustion analysis apparatus. The mass of carbon dioxide produced is 3.350 grams, and the mass of water produced is 0.823 grams. (a) Determine the mass percentage of carbon, hydrogen, and oxygen in the sample. (b) Determine the molecular formula of the compound, given that its molecular mass is 164.

Solution: (a) We begin by writing a general outline of the combustion of our compound:

$$C_?H_?O_? + O_2(g)\ \text{(excess)} \longrightarrow CO_2(g) + H_2O(g)\quad (\textit{unbalanced})$$

From this we see that all the carbon in the compound ends up in the carbon dioxide produced and that all the hydrogen in the compound ends up in the water produced (this assumes complete combustion, which in an excess of pure oxygen gas is generally the case).

Thus, the mass of carbon in the original sample is equal to the mass of carbon in the carbon dioxide formed; and the mass of hydrogen in the original sample is equal to the mass of hydrogen in the water formed.

$$\text{mass of C} = \begin{pmatrix}\text{mass of}\\ CO_2\\ \text{formed}\end{pmatrix}\begin{pmatrix}\text{fraction of the}\\ \text{mass of } CO_2\\ \text{due to C}\end{pmatrix}$$

$$= \begin{pmatrix}\text{mass of}\\ CO_2\\ \text{formed}\end{pmatrix}\left(\frac{\text{atomic mass of C}}{\text{formula mass of } CO_2}\right)$$

$$= (3.350\text{ g } CO_2)\left(\frac{12.01\text{ g C}}{44.01\text{ g } CO_2}\right) = 0.9142\text{ g C}$$

$$\text{mass of H} = \begin{pmatrix}\text{mass of}\\ H_2O\\ \text{formed}\end{pmatrix}\begin{pmatrix}\text{fraction of the}\\ \text{mass of } H_2O\\ \text{due to H}\end{pmatrix}$$

$$= \begin{pmatrix}\text{mass of}\\ H_2O\\ \text{formed}\end{pmatrix}\left(\frac{2\times\text{atomic mass of H}}{\text{formula mass of } H_2O}\right)$$

$$= (0.823\text{ g } H_2O)\left(\frac{2\times 1.008\text{ g H}}{18.02\text{ g } H_2O}\right) = 0.0921\text{ g H}$$

Now using the mass of the original compound, we can find the mass percentages of carbon and hydrogen in the compound:

$$\text{mass percentage of C} = \left(\frac{\text{mass of C}}{\text{mass of sample}}\right)\times 100$$

$$= \left(\frac{0.9142\text{ g}}{1.250\text{ g}}\right)\times 100 = 73.14\%$$

$$\text{mass percentage of H} = \left(\frac{\text{mass of H}}{\text{mass of sample}}\right)\times 100$$

$$= \left(\frac{0.0921\text{ g}}{1.250\text{ g}}\right)\times 100 = 7.37\%$$

However, we cannot use this method for determining the mass of oxygen in our compound because the compound was burned in *excess* oxygen. Nevertheless, because the only remaining unknown mass percentage in our formula is oxygen, we can determine the mass percentage of oxygen by difference:

$$\text{mass percentage of O} = 100\% - \begin{pmatrix}\text{mass percentage}\\ \text{of C}\end{pmatrix} - \begin{pmatrix}\text{mass percentage}\\ \text{of H}\end{pmatrix}$$

$$= 100\% - 73.14\% - 7.37\% = 19.49\%$$

Thus, we find that the mass percentages of carbon, hydrogen, and oxygen in our compound are 73.14% C, 7.37% H, and 19.49% O.

(b) To determine the molecular formula, first we calculate the empirical formula from the mass percentages, as described in Section 11-3. As usual, we consider a 100-gram sample and write

$$73.14\text{ g C} \Leftrightarrow 7.37\text{ g H} \Leftrightarrow 19.49\text{ g O}$$

$$6.090\text{ mol C} \Leftrightarrow 7.31\text{ mol H} \Leftrightarrow 1.218\text{ mol O}$$

$$5.00\text{ mol C} \Leftrightarrow 6.00\text{ mol H} \Leftrightarrow 1.00\text{ mol O}$$

The empirical formula is therefore C_5H_6O. The molecular mass is known to be 164 and the empirical formula mass is 82, so the molecular formula must be $164/82 = 2$ times the empirical formula. Thus, the molecular formula is $C_{10}H_{12}O_2$.

PRACTICE PROBLEM 11-8: A 2.475-gram sample of vitamin C is burned in a combustion analysis apparatus. The mass of $CO_2(g)$ produced is found to be 3.710 grams and the mass of $H_2O(l)$ produced is found to be 1.013 grams. Assuming that the chemical formula of vitamin C consists of carbon, hydrogen, and oxygen, and has a molecular mass of 176, determine the empirical formula and molecular formula of vitamin C.

Answer: $C_3H_4O_3$; $C_6H_8O_6$

11-7. The Coefficients in Chemical Equations Can Be Interpreted as Numbers of Moles

A subject of great practical importance in chemistry is the determination of what quantity of product can be obtained from a given quantity of reactants. For example, the reaction between hydrogen and nitrogen to produce ammonia, $NH_3(g)$, can be described by the equation

$$3\,H_2(g) + N_2(g) \rightarrow 2\,NH_3(g) \tag{11.1}$$

where the coefficients in the equation are called **balancing coefficients** or **stoichiometric coefficients**. We might wish to know how much $NH_3(g)$ is produced when 10.0 grams of $N_2(g)$ reacts with excess $H_2(g)$. Recall from Section 3-2 that we can interpret the balancing coefficients in a chemical equation in a number of ways. To determine how much of one substance can be obtained from another, we interpret the equation in terms of moles. Thus, we interpret Equation 11.1 as

$$3 \text{ mol } H_2(g) + 1 \text{ mol } N_2(g) \rightarrow 2 \text{ mol } NH_3(g)$$

This result is important. It tells us that the stoichiometric or balancing coefficients are the relative numbers of moles of each substance in a balanced chemical equation.

We can also interpret the hydrogen-nitrogen reaction in terms of masses. If we convert moles to masses by multiplying by the appropriate molar masses, then we get

$$6.05 \text{ g } H_2(g) + 28.02 \text{ g } N_2(g) \rightarrow 34.07 \text{ g } NH_3(g)$$

Note that the total mass is the same on the two sides of the equation, in accord with the law of conservation of mass. Table 11.3 summarizes the various interpretations of the equation for the reaction of hydrogen with nitrogen to produce ammonia as well as that of the reaction of sodium with chlorine to produce sodium chloride.

We are now ready to calculate how much ammonia is produced when a given quantity of nitrogen or hydrogen is used. Let's calculate how many moles of $NH_3(g)$ can be produced from 10.0 moles of $N_2(g)$, assuming that an excess amount of $H_2(g)$ is available. According to Equation 11.1, two moles of $NH_3(g)$ are produced from each mole of $N_2(g)$. We can express this relationship as a **stoichiometric unit conversion factor**,

$$1 \text{ mol } N_2 \approx 2 \text{ mol } NH_3$$

TABLE 11.3 The various interpretations of two chemical equations

Interpretation	**$3H_2$**	+	**N_2**	→	**$2NH_3$**
molecular:	3 molecules	+	1 molecule	→	2 molecules
molar:	3 moles	+	1 mole	→	2 moles
mass:	6.05 grams	+	28.02 grams	→	34.07 grams
Interpretation	**2Na**	+	**Cl_2**	→	**2NaCl**
molecular:	2 atoms	+	1 molecule	→	2 ion pairs or 2 formula units
molar:	2 moles	+	1 mole	→	2 moles
mass:	45.98 grams	+	70.90 grams	→	116.88 grams

which can also be expressed as

$$\frac{2\text{ mol NH}_3}{1\text{ mol N}_2} = 1 \qquad \text{or} \qquad \frac{1\text{ mol N}_2}{2\text{ mol NH}_3} = 1$$

When working with stoichiometric unit conversion factors, we choose the form suitable to the conversion of the given units. In this case, because we are converting from moles of nitrogen to moles of ammonia, we use the conversion factor with moles of ammonia in the numerator and moles of nitrogen in the denominator. Thus, we find that 10.0 moles of $N_2(g)$ yields

$$\text{moles of NH}_3 = (10.0\text{ mol N}_2)\left(\frac{2\text{ mol NH}_3}{1\text{ mol N}_2}\right) = 20.0\text{ mol NH}_3$$

We can also calculate the number of grams of $NH_3(g)$ produced by using the fact that one mole of $NH_3(g)$ corresponds to 17.03 grams of $NH_3(g)$:

$$\text{mass of NH}_3\text{ produced} = (20.0\text{ mol NH}_3)\left(\frac{17.03\text{ g NH}_3}{1\text{ mol NH}_3}\right)$$
$$= 341\text{ g NH}_3$$

Thus, we see that ratios of the balancing coefficients in a chemical equation are also unit conversion factors that allow us to convert from the number of moles of one substance into the number of moles of any other species consumed or produced in that reaction. The following Examples and Practice Problems illustrate the use of stoichiometric coefficients as conversion factors in chemical equations.

EXAMPLE 11-9: How many grams of $NH_3(g)$ can be produced from 8.50 grams of $H_2(g)$, assuming that an excess amount of $N_2(g)$ is available? What is the minimum mass of $N_2(g)$ required?

Solution: The equation for the reaction is $3H_2(g) + N_2(g) \rightarrow 2NH_3(g)$. The number of moles of $H_2(g)$ corresponding to 8.50 grams is

$$\text{moles of H}_2 = (8.50\text{ g H}_2)\left(\frac{1\text{ mol H}_2}{2.016\text{ g H}_2}\right) = 4.22\text{ mol H}_2$$

The number of moles of $NH_3(g)$ is obtained by using the stoichiometric unit conversion factor between $NH_3(g)$ and $H_2(g)$,

$$\frac{2\text{ mol NH}_3}{3\text{ mol H}_2} = 1$$

which we obtain directly from the balanced chemical equation. Therefore,

$$\text{moles of NH}_3 = (4.22 \text{ mol H}_2)\left(\frac{2 \text{ mol NH}_3}{3 \text{ mol H}_2}\right) = 2.81 \text{ mol NH}_3$$

The mass of $NH_3(g)$ is given by

$$\text{mass of NH}_3 = (2.81 \text{ mol NH}_3)\left(\frac{17.03 \text{ g NH}_3}{1 \text{ mol NH}_3}\right) = 47.9 \text{ g NH}_3$$

To calculate the minimum mass of $N_2(g)$ required to react completely with 8.50 grams, or 4.22 moles of $H_2(g)$, we first calculate the number of moles of $N_2(g)$ required:

$$\text{moles of N}_2 = (4.22 \text{ mol H}_2)\left(\frac{1 \text{ mol N}_2}{3 \text{ mol H}_2}\right) = 1.41 \text{ mol N}_2$$

As always, the stoichiometric unit conversion factor is obtained from the balanced chemical equation. The minimum mass of $N_2(g)$ required is given by

$$\text{mass of N}_2 = (1.41 \text{ mol N}_2)\left(\frac{28.02 \text{ g N}_2}{1 \text{ mol N}_2}\right) = 39.5 \text{ g N}_2$$

PRACTICE PROBLEM 11-9: A frequently used method for preparing oxygen in the laboratory is by the thermal decomposition of potassium chlorate, $KClO_3(s)$ (Figure 11.6). This reaction is described by the unbalanced equation

$$\text{KClO}_3(s) \rightarrow \text{KCl}(s) + \text{O}_2(g) \quad (\textit{unbalanced})$$

(a) Balance this equation. (b) How many moles of $O_2(g)$ can be prepared from 0.50 moles of $KClO_3(s)$? (c) How many grams of $O_2(g)$ can be prepared from 30.6 grams of $KClO_3(s)$?

Answer: (a) $2\,KClO_3(s) \rightarrow 2\,KCl(s) + 3\,O_2(g)$; (b) 0.75 moles; (c) 12.0 grams

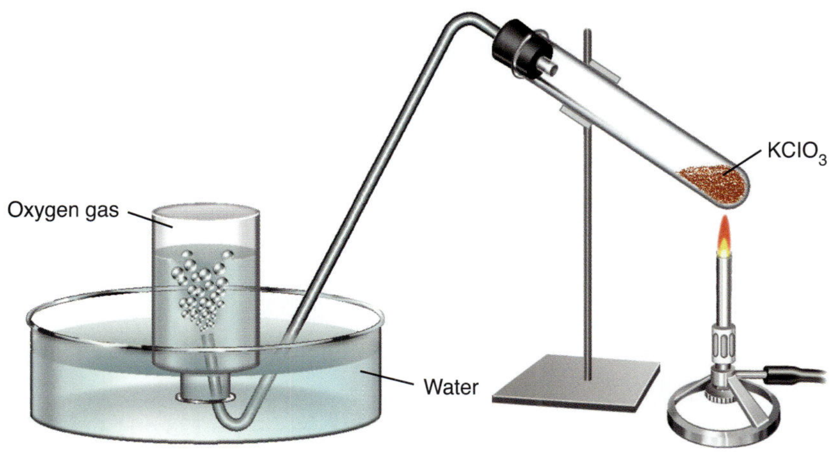

Figure 11.6 A typical experimental setup for the production of small amounts of oxygen by gently heating potassium chlorate, $KClO_3(s)$. Because it is only slightly soluble in water, the oxygen is collected by the displacement of water from an inverted bottle.

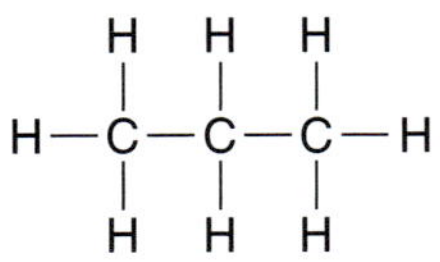

The Lewis formula of propane, $C_3H_8(g)$. Its formula is sometimes written as $CH_3CH_2CH_3(g)$ to show the structure. Propane is stored in metal tanks and often used as a fuel in areas not serviced by natural gas pipelines.

EXAMPLE 11-10: Propane, $C_3H_8(g)$, a common fuel, burns in oxygen according to the equation

$$C_3H_8(g) + 5\,O_2(g) \rightarrow 3\,CO_2(g) + 4\,H_2O(l)$$

(a) How many grams of $O_2(g)$ are required to burn 75.0 grams of $C_3H_8(g)$?
(b) How many grams of $H_2O(l)$ and $CO_2(g)$ are then produced?

Solution: (a) The chemical equation states that five moles of $O_2(g)$ are required to burn one mole of $C_3H_8(g)$. The molecular mass of $C_3H_8(g)$ is 44.10, so 75.0 grams of $C_3H_8(g)$ corresponds to

$$\text{moles of } C_3H_8 = (75.0 \text{ g } C_3H_8)\left(\frac{1 \text{ mol } C_3H_8}{44.09 \text{ g } C_3H_8}\right) = 1.70 \text{ mol } C_3H_8$$

The number of moles of $O_2(g)$ required is

$$\text{moles of } O_2 = (1.70 \text{ mol } C_3H_8)\left(\frac{5 \text{ mol } O_2}{1 \text{ mol } C_3H_8}\right) = 8.50 \text{ mol } O_2$$

To find out how many grams of $O_2(g)$ this is, we multiply the number of moles by the molar mass of $O_2(g)$:

$$\text{mass of } O_2 = (8.50 \text{ mol } O_2)\left(\frac{32.00 \text{ g } O_2}{1 \text{ mol } O_2}\right) = 272 \text{ g } O_2$$

Thus, we see that 272 grams of $O_2(g)$ are required to burn 75.0 grams of $C_3H_8(g)$. (Remember whenever possible to use at least one more significant figure in your constants than are present in your data.)

(b) According to the chemical equation, three moles of $CO_2(g)$ and four moles of $H_2O(l)$ are produced for each mole of $C_3H_8(g)$ burned. Therefore, the number of moles of $CO_2(g)$ produced is

$$\text{moles of } CO_2 = (1.70 \text{ mol } C_3H_8)\left(\frac{3 \text{ mol } CO_2}{1 \text{ mol } C_3H_8}\right) = 5.10 \text{ mol } CO_2$$

The number of moles of $H_2O(l)$ produced is

$$\text{moles of } H_2O = (1.70 \text{ mol } C_3H_8)\left(\frac{4 \text{ mol } H_2O}{1 \text{ mol } C_3H_8}\right) = 6.80 \text{ mol } H_2O$$

The moles of $CO_2(g)$ and $H_2O(l)$ are converted to the mass in grams by multiplying by the respective molar masses:

$$\text{mass of } CO_2 = (5.10 \text{ mol } CO_2)\left(\frac{44.01 \text{ g } CO_2}{1 \text{ mol } CO_2}\right) = 224 \text{ g } CO_2$$

$$\text{mass of } H_2O = (6.80 \text{ mol } H_2O)\left(\frac{18.02 \text{ g } H_2O}{1 \text{ mol } H_2O}\right) = 123 \text{ g } H_2O$$

These are the quantities of $CO_2(g)$ and $H_2O(l)$ produced when 75.0 grams of propane are burned.

We can summarize the results of this example by

$$\begin{array}{cccc} C_3H_8(g) + & 5\,O_2(g) \rightarrow & 3\,CO_2(g) + & 4\,H_2O(l) \\ 75.0\text{ g} & 272\text{ g} & 224\text{ g} & 123\text{ g} \end{array}$$

Notice that the total mass on each side of the chemical reaction is the same, as it must be according to the principle of conservation of mass.

PRACTICE PROBLEM 11-10: Deuterated ammonia, $ND_3(g)$, can be prepared by reacting lithium nitride, $Li_3N(s)$, with heavy water, $D_2O(l)$, according to

$$Li_3N(s) + 3\,D_2O(l) \rightarrow 3\,LiOD(s) + ND_3(g)$$

How many milligrams of heavy water are required to produce 7.15 milligrams of $ND_3(g)$? Given that the density of heavy water is $1.106\text{ g}\cdot\text{mL}^{-1}$ at 25°C, how many milliliters of heavy water are required? Take the atomic mass of deuterium, D, to be 2.014.

Answer: 21.4 mg; 0.0194 mL

In Example 11-9, we represented the reaction between $H_2(g)$ and $N_2(g)$ to form $NH_3(g)$ by Equation 11.1. We could also have represented it by

$$\tfrac{3}{2}H_2(g) + \tfrac{1}{2}N_2(g) \rightarrow NH_3(g) \qquad (11.2)$$

Before leaving this section, we shall show that the result we obtained in Example 11-9 does not depend upon which equation we use to represent the reaction. Starting again with 8.50 grams of $H_2(g)$, or 4.22 moles of $H_2(g)$, the number of moles of $NH_3(g)$ produced may be determined using the stoichiometric unit conversion factors from Equation 11.2, namely,

$$\frac{1\text{ mol }NH_3}{\frac{3}{2}\text{ mol }H_2} = 1$$

Therefore,

$$\text{moles of }NH_3(g) = (4.22\text{ mol }H_2)\left(\frac{1\text{ mol }NH_3}{\frac{3}{2}\text{ mol }H_2}\right) = 2.81\text{ mol }NH_3$$

which is exactly the same result as we obtained in Example 11-9 using Equation 11.1. You can see that this is so because

$$\frac{1\text{ mol }NH_3}{\frac{3}{2}\text{ mol }H_2} = \frac{2\text{ mol }NH_3}{3\text{ mol }H_2}$$

In other words, the stoichiometric unit conversion factors are identical because they consist of *ratios* of stoichiometric coefficients. Of course, this must be so for physical reasons; it would be ridiculous to obtain different amounts of products depending upon how we (arbitrarily) choose to represent the reaction by a chemical equation.

11-8. Calculations Involving Chemical Reactions Are Carried Out in Terms of Moles

For calculations involving chemical reactions and mass, the procedure is first to convert mass to moles, then convert moles of one substance to moles of another by using the balancing coefficients in the chemical equation, and then convert moles into mass. An understanding of the flowchart in Figure 11.7 will allow you to do most calculations involving chemical equations and mass. The following calculation further illustrates the use of Figure 11.7.

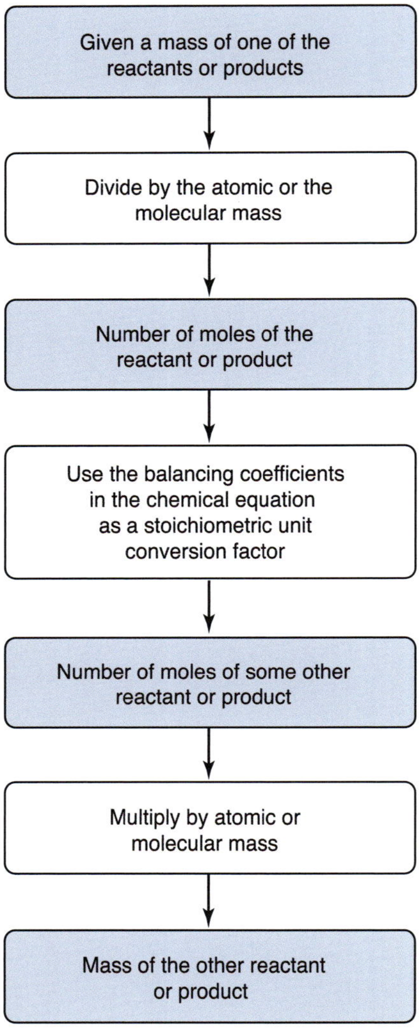

Figure 11.7 Flow diagram of the procedure for calculating the mass or the number of moles from chemical equations. The essence of the method is to realize that we convert from moles of one substance to moles of another substance in a chemical equation by using ratios of the stoichiometric coefficients.

Small quantities of bromine can be prepared in the laboratory by heating manganese(IV) oxide, concentrated sulfuric acid, and potassium bromide under a fume hood. The equation describing this reaction is

$$MnO_2(s) + 2H_2SO_4(aq) + 2KBr(aq) \rightarrow MnSO_4(aq) + K_2SO_4(aq) + 2H_2O(l) + Br_2(l)$$

Let's calculate how many grams of $MnO_2(s)$ and $KBr(aq)$ are required to produce 225 grams of $Br_2(l)$ in an excess of $H_2SO_4(aq)$. From the balanced chemical equation, we see that

$$1\text{ mol } Br_2 \Leftrightarrow 1\text{ mol } MnO_2 \quad \text{and} \quad 1\text{ mol } Br_2 \Leftrightarrow 2\text{ mol KBr}$$

The number of moles of $Br_2(l)$ corresponding to 225 grams is

$$\text{moles of } Br_2 = (225\text{ g } Br_2)\left(\frac{1\text{ mol } Br_2}{159.8\text{ g } Br_2}\right) = 1.41\text{ mol}$$

To convert from moles of $Br_2(l)$ to moles of $MnO_2(s)$, we use the stoichiometric coefficients from the balanced equation as outlined in Figure 11.7:

$$\text{moles of } MnO_2 \text{ required} = (1.41\text{ mol } Br_2)\left(\frac{1\text{ mol } MnO_2}{1\text{ mol } Br_2}\right) = 1.41\text{ mol}$$

and

$$\text{moles of KBr required} = (1.41\text{ mol } Br_2)\left(\frac{2\text{ mol KBr}}{1\text{ mol } Br_2}\right) = 2.82\text{ mol}$$

Therefore, the number of grams of each reactant required is

$$\text{mass of } MnO_2 \text{ required} = (1.41\text{ mol } MnO_2)\left(\frac{86.94\text{ g } MnO_2}{1\text{ mol } MnO_2}\right) = 123\text{ g}$$

and

$$\text{mass of KBr required} = (2.82 \text{ mol KBr})\left(\frac{119.0 \text{ g KBr}}{1 \text{ mol KBr}}\right) = 336 \text{ g}$$

The next two Examples further illustrate the calculation of quantities involved in chemical reactions.

EXAMPLE 11-11: Diiodine pentoxide, $I_2O_5(s)$, is a reagent used for the quantitative determination of carbon monoxide. The equation for the reaction is

$$I_2O_5(s) + 5\,CO(g) \rightarrow I_2(s) + 5\,CO_2(g)$$

A gas sample containing carbon monoxide is collected from the exhaust of an engine. If 0.098 grams of $I_2(s)$ are produced from the reaction of the $CO(g)$ in the gas sample with excess $I_2O_5(s)$, then how many grams of $CO(g)$ are present in the sample?

Solution: We see from the chemical equation that five moles of $CO(g)$ produce one mole of $I_2(s)$, or that

$$1 \text{ mol } I_2 \approx 5 \text{ mol CO}$$

The number of moles of $I_2(s)$ corresponding to 0.098 grams is

We carry the calculations through to one more significant figure than the final answer to avoid rounding errors.

$$\text{moles of } I_2 = (0.098 \text{ g } I_2)\left(\frac{1 \text{ mol } I_2}{253.8 \text{ } I_2}\right) = 3.86 \times 10^{-4} \text{ mol } I_2$$

The number of moles of $CO(g)$ is

$$\text{moles of } CO(g) = (3.86 \times 10^{-4} \text{ mol } I_2)\left(\frac{5 \text{ mol CO}}{1 \text{ mol } I_2}\right) = 0.00193 \text{ mol CO}$$

and the mass of $CO(g)$ is given by

$$\text{mass of CO} = (0.00193 \text{ mol CO})\left(\frac{28.01 \text{ g CO}}{1 \text{ mol CO}}\right) = 0.054 \text{ g CO}$$

PRACTICE PROBLEM 11-11: Finely divided sulfur ignites spontaneously in fluorine gas to produce sulfur hexafluoride, as described by the chemical equation

$$S(s) + 3\,F_2(g) \rightarrow SF_6(g)$$

How many grams of sulfur hexafluoride, $SF_6(g)$, can be produced from 5.00 grams of sulfur? How many grams of fluorine gas are required to react with the 5.00 grams of sulfur?

Answer: 22.8 grams of $SF_6(g)$; 17.8 grams of $F_2(g)$

EXAMPLE 11-12: Phosphorus reacts directly with sodium metal to produce sodium phosphide, as described by the chemical equation

$$3\,\mathrm{Na}(s) + \mathrm{P}(s) \rightarrow \mathrm{Na_3P}(s)$$

How many grams of sodium phosphide, $Na_3P(s)$, can be produced from 10.0 grams of sodium metal?

Solution: We see from the chemical equation that one mole of $Na_3P(s)$ is produced from three moles of $Na(s)$, or that

$$1\text{ mol Na}_3\text{P} \approx 3\text{ mol Na}$$

The number of moles of $Na(s)$ is given by

$$\text{moles of Na} = (10.0\text{ g Na})\left(\frac{1\text{ mol Na}}{22.99\text{ g Na}}\right) = 0.435\text{ mol}$$

The number of moles of $Na_3P(g)$ produced is given by

$$\text{moles of Na}_3\text{P} = (0.435\text{ mol Na})\left(\frac{1\text{ mol Na}_3\text{P}}{3\text{ mol Na}}\right) = 0.145\text{ mol}$$

and the number of grams of Na_3P produced is given by

$$\text{mass of Na}_3\text{P} = (0.145\text{ mol Na}_3\text{P})\left(\frac{99.94\text{ g Na}_3\text{P}}{1\text{ mol Na}_3\text{P}}\right) = 14.5\text{ g Na}_3\text{P}$$

This calculation can also be done in one operation:

$$\text{mass of Na}_3\text{P} = (10.0\text{ g Na})\left(\frac{1\text{ mol Na}}{22.99\text{ g Na}}\right)\left(\frac{1\text{ mol Na}_3\text{P}}{3\text{ mol Na}}\right)\left(\frac{99.94\text{ g Na}_3\text{P}}{1\text{ mol Na}_3\text{P}}\right)$$
$$= 14.5\text{ g Na}_3\text{P}$$

You should try to become comfortable doing a calculation like this in one step.

PRACTICE PROBLEM 11-12: Phosphorus triiodide can be prepared by the direct combination of phosphorus and iodine:

$$2\,\mathrm{P}(s) + 3\,\mathrm{I_2}(s) \rightarrow 2\,\mathrm{PI_3}(s)$$

(a) How many grams of $PI_3(s)$ can be prepared from 1.25 grams of $P(s)$? (b) How many grams of $I_2(s)$ are required to react with the 1.25 grams of $P(s)$?

Answer: (a) 16.6 grams $PI_3(s)$; (b) 15.4 grams $I_2(s)$

11-9. It Is Not Always Necessary to Know the Chemical Equation to Carry Out Stoichiometric Calculations

In the previous section we calculated the quantities involved in chemical reactions by using the chemical equation. Sometimes it is not necessary to know the complete chemical equation in order to carry out such calculations. For example, sulfuric acid, $H_2SO_4(l)$, is the most widely used and important industrial chemical. Let's calculate how much sulfuric acid can be produced from 1.00 metric ton of sulfur (Figure 11.8). (A **metric ton** is equal to 1000 kg or 2200 pounds.) If all the sulfur ends up in sulfuric acid, then we can write that one mole of $H_2SO_4(l)$ is produced from one mole of $S(s)$, or

Figure 11.8 Elemental sulfur is mined in huge quantities. Most of the sulfur mined is used in the manufacture of sulfuric acid.

$$1 \text{ mol S} \approx 1 \text{ mol H}_2\text{SO}_4$$

because one molecule of H_2SO_4 contains one atom of sulfur. One metric ton of $S(s)$ corresponds to

$$\text{moles of S} = (1.00 \text{ metric ton S})\left(\frac{10^3 \text{ kg}}{1 \text{ metric ton}}\right)\left(\frac{10^3 \text{ g}}{1 \text{ kg}}\right)\left(\frac{1 \text{ mol S}}{32.065 \text{ g S}}\right)$$
$$= 3.12 \times 10^4 \text{ mol}$$

Using the above stoichiometric equality, we see that

$$3.12 \times 10^4 \text{ mol S} \approx 3.12 \times 10^4 \text{ mol H}_2\text{SO}_4$$

The molecular mass of H_2SO_4 is 98.08, so the quantity of H_2SO_4 produced is

$$\text{mass of H}_2\text{SO}_4 = (3.12 \times 10^4 \text{ mol H}_2\text{SO}_4)\left(\frac{98.08 \text{ g H}_2\text{SO}_4}{1 \text{ mol H}_2\text{SO}_4}\right)\left(\frac{1 \text{ kg}}{10^3 \text{ g}}\right)\left(\frac{1 \text{ metric ton}}{10^3 \text{ kg}}\right)$$
$$= 3.06 \text{ metric tons}$$

Thus, we are able to calculate the quantity of sulfuric acid produced without knowing the chemical equations involved, provided all the sulfur ends up in sulfuric acid as we assumed. The following two Examples involve similar calculations.

EXAMPLE 11-13: Calculate the *maximum* number of grams of $Ca_3(PO_4)_2(s)$ that can be obtained from 10.0 grams of $P_4O_{10}(s)$, assuming that all the phosphorus from $P_4O_{10}(s)$ ends up in the $Ca_3(PO_4)_2(s)$.

Solution: Because all the phosphorus from $P_4O_{10}(s)$ ends up in the $Ca_3(PO_4)_2(s)$, we have the correspondence

$$1 \text{ mol P}_4\text{O}_{10} \approx 2 \text{ mol Ca}_3(\text{PO}_4)_2$$

Note that $P_4O_{10}(s)$ has four phosphorus atoms per formula unit, whereas $Ca_3(PO_4)_2(s)$ has two phosphorus atoms per formula unit. Ten grams of $P_4O_{10}(s)$ corresponds to

$$\text{mass of } P_4O_{10} = (10.0 \text{ g } P_4O_{10})\left(\frac{1 \text{ mol } P_4O_{10}}{283.9 \text{ g } P_4O_{10}}\right)$$

$$= 0.0352 \text{ mol}$$

The number of moles of $Ca_3(PO_4)_2(s)$ is given by

$$\text{moles of } Ca_3(PO_4)_2 = (0.0352 \text{ mol } P_4O_{10})\left(\frac{2 \text{ mol } Ca_3(PO_4)_2}{1 \text{ mol } P_4O_{10}}\right)$$

$$= 0.0704 \text{ mol } Ca_3(PO_4)_2$$

and the number of grams of $Ca_3(PO_4)_2(s)$ is given by

$$\text{mass of } Ca_3(PO_4)_2 = (0.0704 \text{ mol } Ca_3(PO_4)_2)\left(\frac{310.2 \text{ g } Ca_3(PO_4)_2}{1 \text{ mol } Ca_3(PO_4)_2}\right)$$

$$= 21.8 \text{ g } Ca_3(PO_4)_2$$

Figure 11.9 Chalcopyrite, $CuFeS_2(s)$ (*left*), is one of the principal ores of copper. Malachite, $CuCO_3{\cdot}Cu(OH)_2(s)$ (*middle*), and chalcocite, $Cu_2S(s)$ (*right*), are two other important copper ores.

PRACTICE PROBLEM 11-13: The most important ore of copper is chalcopyrite, $CuFeS_2(s)$ (Figure 11.9). The extraction of copper metal from the ore is a complex process, partly because the ore is impure and contains much sand and clay. What is the *maximum* amount of copper that can be obtained from one metric ton (1000 kg) of an ore that is 30% $CuFeS_2(s)$?

Answer: 0.104 metric tons

We can apply the techniques that we have learned in this chapter to mixtures, as the next Example shows.

EXAMPLE 11-14: A mixture of $NaCl(s)$ and $KCl(s)$ weighing 1.250 grams is dissolved in water. When $AgNO_3(s)$ is added to the solution, a precipitate of $AgCl(s)$ that weighs 2.500 grams is obtained. Assuming that all the chloride in the mixture is precipitated as $AgCl(s)$, calculate the mass percentages of sodium chloride and of potassium chloride in the mixture.

Solution: Because all the chloride in the mixture was precipitated as $AgCl(s)$, we have

$$\begin{pmatrix}\text{number of}\\ \text{chloride ions}\\ \text{in NaCl}\end{pmatrix} + \begin{pmatrix}\text{number of}\\ \text{chloride ions}\\ \text{in KCl}\end{pmatrix} = \begin{pmatrix}\text{number of}\\ \text{chloride ions}\\ \text{in AgCl}\end{pmatrix}$$

or in terms of moles,

$$\text{moles of NaCl} + \text{moles of KCl} = \text{moles of AgCl}$$

If we let x be the number of grams of $NaCl(s)$ in the mixture, then $1.250 - x$ is the number of grams of $KCl(s)$. The number of moles of $NaCl(s)$, $KCl(s)$, and $AgCl(s)$ are given by

$$\text{moles of NaCl} = (x\text{ g NaCl})\left(\frac{1\text{ mol NaCl}}{58.44\text{ g NaCl}}\right)$$

$$\text{moles of KCl} = [(1.250 - x)\text{ g KCl}]\left(\frac{1\text{ mol KCl}}{74.55\text{ g KCl}}\right)$$

$$\text{moles of AgCl} = (2.500\text{ g AgCl})\left(\frac{1\text{ mol AgCl}}{143.3\text{ g AgCl}}\right) = 0.01744\text{ mol}$$

Using the relation

$$\text{mol NaCl}(s) + \text{mol KCl}(s) = \text{mol AgCl}(s)$$

gives

$$(x\text{ g NaCl})\left(\frac{1\text{ mol NaCl}}{58.44\text{ g NaCl}}\right) + [(1.250 - x)\text{ g KCl}]\left(\frac{1\text{ mol KCl}}{74.55\text{ g KCl}}\right) = 0.01744\text{ mol AgCl}$$

from which we have

$$0.01711x + 0.01677 - 0.01341x = 0.01744$$

Collecting like terms gives

$$3.70 \times 10^{-3}\, x = 6.7 \times 10^{-4}$$

Solving for x, we obtain that $x = 0.18$ grams of NaCl(s). Thus, the mass percentages of NaCl(s) and KCl(s) in the mixture are

$$\text{mass \% NaCl} = \left(\frac{0.18\text{ g}}{1.250\text{ g}}\right) \times 100 = 14\%$$

and

$$\text{mass \% KCl} = 100\% - 14\% = 86\%\text{ KCl}(s)$$

PRACTICE PROBLEM 11-14: A mixture of NaCl(s) and $BaCl_2(s)$ weighing 2.86 grams is dissolved in water. When $AgNO_3(s)$ is added to the solution, a precipitate of AgCl(s) that weighs 4.81 grams is obtained. Calculate the mass percentages of NaCl(s) and of $BaCl_2(s)$ in the sample.

Answer: 29% NaCl(s) and 71% $BaCl_2(s)$

The significant figures have been determined stepwise using the rules for multiplication and division or addition and subtraction for each operation.

11-10. When Two or More Substances React, the Mass of the Product Is Determined by the Limiting Reactant

If you look back over the Examples in this chapter, you will notice that in no case did we start out with the masses of more than one of the given reactants, or if we did, we assumed that one of them was in excess. In other words, we have always assumed that there was sufficient material present to react with all that reactant for which the quantity was given. Let's now consider an example in which we do have the quantities of two reactants given. Cadmium sulfide, CdS(*s*), which is used in light meters, solar cells, and other light-sensitive devices, can be made by the direct combination of the two elements according to

$$\text{Cd}(s) + \text{S}(s) \rightarrow \text{CdS}(s) \tag{11.3}$$

How much CdS(*s*) is produced if we start out with 2.00 grams of cadmium and 2.00 grams of sulfur? As in all stoichiometric calculations, we first determine the number of moles of each reactant:

$$\text{moles of Cd} = (2.00 \text{ g Cd})\left(\frac{1 \text{ mol Cd}}{112.4 \text{ g Cd}}\right) = 0.0178 \text{ mol Cd}$$

and

$$\text{moles of S} = (2.00 \text{ g S})\left(\frac{1 \text{ mol S}}{32.065 \text{ g S}}\right) = 0.0624 \text{ mol S}$$

We know from inspection of the balanced chemical equation that one mole of cadmium requires one mole of sulfur, so the 0.0178 moles of cadmium require 0.0178 moles of sulfur. Thus, there is excess sulfur; only 0.0178 moles of sulfur react and (0.0624 – 0.0178) moles = 0.0446 moles of sulfur remain after the reaction is completed. The cadmium reacts completely, and the number of moles of cadmium consumed determines how much CdS(*s*) is produced. The reactant that is consumed completely and thereby limits the amount of product formed is called the **limiting reactant**, and any other reactants are called **excess reactants**. In this example, cadmium is the limiting reactant and sulfur is in excess. Because the limiting reactant is completely consumed while the excess reactant is not, the initial mass of the limiting reactant must be used to calculate how much product is formed.

In Equation 11.3, 0.0178 moles of cadmium react with 0.0178 moles of sulfur to produce 0.0178 moles of CdS(*s*). The mass of CdS(*s*) produced is

$$\text{mass of CdS} = (0.0178 \text{ mol CdS})\left(\frac{144.5 \text{ g CdS}}{1 \text{ mol CdS}}\right) = 2.57 \text{ g CdS}(s)$$

The sulfur is in excess by 0.0446 moles. The corresponding mass of the excess sulfur is given by

$$\text{mass of excess S} = (0.0446 \text{ mol S})\left(\frac{32.065 \text{ g S}}{1 \text{ mol S}}\right) = 1.43 \text{ g S}(s)$$

Notice that before the reaction there are 2.00 grams of cadmium and 2.00 grams of sulfur, or 4.00 grams of reactants. After the reaction, there are 2.57 grams of cadmium sulfide and 1.43 grams of sulfur, or a total of 4.00 grams, as required by the law of conservation of mass.

When the masses of two or more reactants are given in a problem, we must check to see which, if any of them, is a limiting reactant. If one of the reactants is limiting, then it is the one to be used in the calculation of the mass of product obtained.

EXAMPLE 11-15: A mixture is prepared from 25.0 grams of aluminum and 58.0 grams of iron(III) oxide, $Fe_2O_3(s)$. When the mixture is ignited, the reaction that occurs is described by the equation

$$Fe_2O_3(s) + 2\,Al(s) \rightarrow Al_2O_3(s) + 2\,Fe(l)$$

(a) How much iron is produced in the reaction? (b) Which reactant is in excess and how many grams of this reactant remain? (c) How much $Al_2O_3(s)$ is produced?

Solution: (a) Because the masses of both reactants are given, we must check to see which, if either, is a limiting reactant. The number of moles of the two reactants, $Al(s)$ and $Fe_2O_3(s)$, is given by

$$\text{moles of Al} = (25.0\text{ g Al})\left(\frac{1\text{ mol Al}}{26.98\text{ g Al}}\right) = 0.927\text{ mol Al}$$

$$\text{moles of Fe}_2\text{O}_3 = (58.0\text{ g Fe}_2\text{O}_3)\left(\frac{1\text{ mol Fe}_2\text{O}_3}{159.7\text{ g Fe}_2\text{O}_3}\right) = 0.363\text{ mol Fe}_2\text{O}_3$$

According to the chemical equation, 0.363 moles of $Fe_2O_3(s)$ require

$$\text{moles of Al} = (0.363\text{ mol Fe}_2\text{O}_3)\left(\frac{2\text{ mol Al}}{1\text{ mol Fe}_2\text{O}_3}\right) = 0.726\text{ mol Al}$$

and so we see that aluminum is in excess by (0.927 moles – 0.726 moles) = 0.201 moles. Thus, $Fe_2O_3(s)$ is the limiting reactant and the one that we should use to calculate the amount of iron produced. This is given by

$$\text{mass of Fe}(l)\text{ produced} = (0.363\text{ mol Fe}_2\text{O}_3)\left(\frac{2\text{ mol Fe}}{1\text{ mol Fe}_2\text{O}_3}\right)\left(\frac{55.85\text{ g Fe}}{1\text{ mol Fe}}\right)$$

$$= 40.6\text{ g Fe}$$

(b) As we found above, $Al(s)$ is in excess. The number of grams of $Al(s)$ in excess is given by

$$\text{mass of aluminum in excess} = (0.201\text{ mol Al})\left(\frac{26.98\text{ g Al}}{1\text{ mol Al}}\right) = 5.42\text{ g Al}$$

(c) The amount of $Al_2O_3(s)$ produced is given by

$$\text{mass of Al}_2\text{O}_3(s)\text{ produced} = (0.363\text{ mol Fe}_2\text{O}_3)\left(\frac{1\text{ mol Al}_2\text{O}_3}{1\text{ mol Fe}_2\text{O}_3}\right)\left(\frac{102.0\text{ g Al}_2\text{O}_3}{1\text{ mol Al}_2\text{O}_3}\right)$$

$$= 37.0\text{ g Al}_2\text{O}_3$$

Although we started out with more grams of $Fe_2O_3(s)$ than aluminum, $Fe_2O_3(s)$ was the limiting reactant in this Example. Once again, we have conservation of mass, as we must. The total mass of the reactants is

Figure 11.10 Thermite reaction. A spectacular example of a single-replacement reaction is the reaction between powdered aluminum metal and iron(III) oxide described by the equation

$$2\,Al(s) + Fe_2O_3(s) \rightarrow 2\,Fe(l) + Al_2O_3(s)$$

Once this reaction is initiated by a heat source such as a burning magnesium ribbon, it proceeds vigorously, producing so much heat that the iron is formed as a liquid.

25.0 grams + 58.0 grams = 83.0 grams and after the reaction is completed we have 40.6 grams of $Fe(l)$, 37.0 grams of $Al_2O_3(s)$, and 5.4 grams of excess $Al(s)$, or a total mass of 83.0 grams.

The reaction of aluminum metal with a metal oxide is called a **thermite reaction** and has numerous applications (Figure 11.10). Thermite reactions were once used to weld railroad tracks and are used in thermite grenades, which are employed by the military to destroy heavy equipment. In a thermite reaction, the reaction temperature can exceed 3500°C.

PRACTICE PROBLEM 11-15: Calcium sulfide, which is used in luminous paints and as a depilatory, can be made by heating calcium sulfate with charcoal at a high temperature. The unbalanced equation is

$$CaSO_4(s) + C(s) \rightarrow CaS(s) + CO(g) \quad (\textit{unbalanced})$$

How many grams of $CaS(s)$ can be prepared from 125 grams each of $CaSO_4(s)$ and $C(s)$?

Answer: 66.2 g

There are many instances in which it is important to add reactants in stoichiometric proportions so as not to have any reactants left over. The propulsion system of satellites and space vehicles serves as a good example. Satellite thrusters, the Lunar Lander rocket engines, the Space Shuttle Orbiter, and the spacecraft that carried the Mars Exploration Rovers (Figure 11.11) were all powered by a reaction similar to the one described by

$$\underset{\text{dinitrogen tetroxide}}{N_2O_4(l)} + \underset{\text{hydrazine}}{2\,N_2H_4(l)} \rightarrow 3\,N_2(g) + 4\,H_2O(g)$$

Dinitrogen tetroxide and hydrazine react explosively when brought into contact. These two reactants are typically kept in separate tanks and pumped through pipes into the rocket engines, where they react. The gases produced (water is a gas at the exhaust temperatures of the rocket engines) exit through the exhaust chamber of the engine to propel or maneuver the spacecraft. The cost of carrying materials into space is enormous, and the two fuels must be combined in the correct proportions. It would be wasteful to carry any excess reactant.

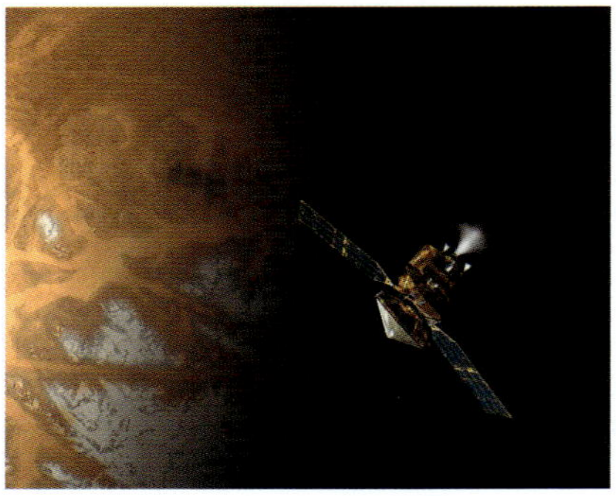

Figure 11.11 The Mars Reconnaissance Orbiter (MRO) was positioned over the surface of Mars using rocket engines powered by a stoichiometric mixture of hydrazine, $N_2H_4(l)$, and dinitrogen tetroxide, $N_2O_4(l)$.

11-11. For Many Chemical Reactions the Amount of the Desired Product Obtained Is Less Than the Theoretical Amount

In all the Examples considered up to now, we have assumed that the amount of products produced can be calculated from the complete reaction of the limiting reactant. However, it frequently happens that less than this **theoretical yield** of product is obtained because (1) the reaction may fail to go to completion; (2) there may be side reactions that give rise to undesired products; (3) some of the desired product may not be readily recoverable or may be lost in the purification process; or (4) the original reactants may be impure or contaminated.

A QUICK METHOD FOR FINDING LIMITING REACTANTS. A quick method for determining which species, if any, is a limiting reactant is to divide the number of moles of each reactant by its stoichiometric coefficient in the balanced chemical equation. The reactant with the least number of moles divided by its stoichiometric coefficient is a limiting reactant. For instance, in Example 11-15 we had to determine whether 25.0 grams of $Al(s)$ or 58.0 grams of $Fe_2O_3(s)$ was a limiting reactant in the thermite reaction. We then found that these masses corresponded to 0.363 moles of $Fe_2O_3(s)$ and 0.927 moles of $Al(s)$. Let's use the stoichiometric coefficients of the two reactants in the balanced chemical equation,

$$Fe_2O_3(s) + 2\,Al(s) \rightarrow Al_2O_3(s) + 2\,Fe(l)$$

to determine which, if either, of these is a limiting reactant. Dividing the number of moles of each of these species by its stoichiometric coefficient gives

$$\frac{0.927 \text{ mol Al}}{2} = 0.464 \text{ mol}$$

and

$$\frac{0.363 \text{ mol } Fe_2O_3}{1} = 0.363 \text{ mol}$$

Because the number of moles of $Fe_2O_3(s)$ divided by it stoichiometric coefficient is less than the number of moles of $Al(s)$ divided by its stoichiometric coefficient, we conclude that $Fe_2O_3(s)$ is a limiting reactant. This method is particularly useful for working problems involving more than two reactants.

In these instances, the mass of the product that actually is obtained is called the **actual yield**, and the efficiency of conversion of reactants into recovered products can be expressed as the **percentage yield** (% yield). The percentage yield is defined as

$$\% \text{ yield} = \left(\frac{\text{actual yield}}{\text{theoretical yield}}\right) \times 100 \tag{11.4}$$

The industrial production of methanol, $CH_3OH(l)$, from the high-pressure reaction

$$CO(g) + 2\,H_2(g) \rightarrow CH_3OH(l)$$

serves to illustrate the difference between the theoretical yield and the actual yield of a reaction. For a variety of reasons, this reaction does not give a 100% yield. Suppose that 5.12 metric tons of $CH_3OH(l)$ is obtained from 1.00 metric ton of $H_2(g)$ reacting with an excess of $CO(g)$. Let's calculate the percentage yield of $CH_3OH(l)$. 1 metric ton equals 1×10^6 grams, and so the theoretical yield is

$$\text{theoretical yield} = (1 \text{ metric ton } H_2)\left(\frac{10^6 \text{ g}}{1 \text{ metric ton}}\right)\left(\frac{1 \text{ mol } H_2}{2.016 \text{ g } H_2}\right)$$
$$\times\left(\frac{1 \text{ mol } CH_3OH}{2 \text{ mol } H_2}\right)\left(\frac{32.03 \text{ g } CH_3OH}{1 \text{ mol } CH_3OH}\right)$$
$$= 7.94 \times 10^6 \text{ g } CH_3OH = 7.94 \text{ metric tons } CH_3OH$$

The percentage yield is given by Equation 11.4:

$$\% \text{ yield} = \left(\frac{\text{actual yield}}{\text{theoretical yield}}\right) \times 100 = \left(\frac{5.12 \text{ metric tons}}{7.94 \text{ metric tons}}\right) \times 100 = 64.5\%$$

EXAMPLE 11-16: A 0.473-gram sample of phosphorus, $P_4(s)$, is reacted with an excess of chlorine, $Cl_2(g)$. 2.12 grams of phosphorus pentachloride, $PCl_5(s)$, is collected. The equation for the reaction is

$$P_4(s) + 10\,Cl_2(g) \rightarrow 4\,PCl_5(s)$$

(a) What is the percentage yield of $PCl_5(s)$? (b) Given this yield, how much phosphorous is required to produce 5.00 grams of $PCl_5(s)$ if chlorine is in excess?

Solution: (a) The theoretical yield of $PCl_5(s)$ is

$$\text{theoretical yield} = (0.473 \text{ g } P_4)\left(\frac{1 \text{ mol } P_4}{123.9 \text{ g } P_4}\right)\left(\frac{4 \text{ mol } PCl_5}{1 \text{ mol } P_4}\right)\left(\frac{208.2 \text{ g } PCl_5}{1 \text{ mol } PCl_5}\right)$$
$$= 3.18 \text{ g } PCl_5$$

The percentage yield is

$$\% \text{ yield} = \left(\frac{\text{actual yield}}{\text{theoretical yield}}\right) \times 100 = \left(\frac{2.12 \text{ g}}{3.18 \text{ g}}\right) \times 100 = 66.7\%$$

(b) The amount of phosphorus required to produce 5.00 grams of $PCl_5(s)$ with a 66.7% reaction yield is

$$\text{mass of } P_4 = (5.00 \text{ g } PCl_5)\left(\frac{1 \text{ mol } PCl_5}{208.2 \text{ g } PCl_5}\right)\left(\frac{1 \text{ mol } P_4}{4 \text{ mol } PCl_5}\right)\left(\frac{123.9 \text{ g } P_4}{1 \text{ mol } P_4}\right)\left(\frac{100\%}{66.7\%}\right)$$
$$= 1.12 \text{ g}$$

PRACTICE PROBLEM 11-16: Tin(IV) chloride can be made by heating tin in an atmosphere of excess dry chlorine:

$$Sn(s) + 2\,Cl_2(g) \rightarrow SnCl_4(l)$$

If the percentage yield of this process is 64.3%, how many grams of tin are required to produce 0.106 grams of $SnCl_4(l)$?

Answer: 0.0751 grams

Most reactions take place in solution, so in the next chapter we shall discuss how to express quantitatively the concentrations of solutions and how to calculate reaction quantities involving reactions that take place in solutions.

SUMMARY

The quantity of a substance that is numerically equal to its formula mass in grams is called a mole of that substance. In order to calculate the number of moles in a given mass of a substance, it is necessary to know its chemical formula. Another definition of a mole is the mass of a substance that contains Avogadro's number (6.022×10^{23}) of formula units or elementary entities. The mole can be described as a counting unit, similar to a dozen. The mole concept is central to our study of chemistry.

Stoichiometric calculations are based on the concept of a mole. We use the idea of a mole to determine chemical formulas from chemical analysis. One of the oldest techniques of chemical analysis is combustion analysis, from which we can determine the empirical formula of a compound. If we know the molecular mass from another experiment, we can then determine the molecular formula as well. The molecular formula is usually insufficient for determining chemical structure, which is generally determined using various spectroscopic methods.

The mole concept is also central to chemical calculations. By interpreting the balancing coefficients in chemical equations in terms of moles, we can use chemical equations to calculate quantities of substances involved in chemical reactions. For example, we can calculate how much product can be obtained from a given amount of reactant, or how much reactant to use to obtain a given amount of product. Finally, percentage yields are a measure of how much product is actually formed in a given experiment.

TERMS YOU SHOULD KNOW

atomic substances *350*
molecular substances *350*
formula mass *350*
formula unit *350*
mole (mol) *350*
molar mass *350*
Avogadro's number *353*
stoichiometry *356*
stoichiometrically equivalent to ($\approx$) *357*
simplest formula *357*
empirical formula *357*
molecular formula *361*
hydrocarbon *362*
combustion analysis *364*
balancing coefficient *366*
stoichiometric coefficient *366*
stoichiometric unit conversion factor *367*
metric ton *375*
limiting reactant *378*
excess reactant *378*
thermite reaction *380*
theoretical yield *380*
actual yield *381*
percentage yield (% yield) *381*

AN EQUATION YOU SHOULD KNOW HOW TO USE

$$\% \text{ yield} = \left(\frac{\text{actual yield}}{\text{theoretical yield}}\right) \times 100 \qquad (11.4) \text{ (calculation of percentage yield)}$$

PROBLEMS

MOLES AND MASS PERCENT

11-1. Calculate the number of moles in

(a) 52.0 grams of calcium carbonate

(b) 250.0 milliliters of ethanol, $CH_3CH_2OH(l)$, with a density of 0.76 $g{\cdot}mL^{-1}$

(c) 28.1 grams of carbon dioxide gas

(d) 5.55×10^{22} molecules of sulfur hexafluoride gas

11-2. Calculate the mass in grams of

(a) 3.00 moles of $Hg(l)$

(b) 1.872×10^{24} molecules of iron(III) hydroxide

(c) 1.0 mole of ^{18}O atoms

(d) 2.0 moles of nitrogen gas

11-3. Naproxen sodium, $C_{14}H_{13}O_3Na(s)$, the active ingredient in the pain reliever Aleve®, is a nonsteroidal anti-inflammatory drug. Calculate the mass percentage of each of the elements in naproxen sodium to four significant figures and show that the sum of these mass percentages totals one hundred percent.

11-4. The chemical N,N'-diethyl-*m*-toluamide, commonly known as DEET, was developed by the U.S. Army following World War II as an insect repellent for use in jungle warfare. DEET works by blocking insect receptors used to locate hosts. The chemical formula of DEET is $C_{12}H_{17}NO$. Determine the mass percentage of each of the elements in DEET to four significant figures and show that the sum of these mass percentages totals one hundred percent.

11-5. Calculate the mass percentage of each element in the following organic compounds to four significant figures.

(a) methane, $CH_4(g)$

(b) ethanol, $CH_3CH_2OH(l)$

(c) acetic acid, $CH_3COOH(aq)$

(d) benzene, $C_6H_6(l)$

11-6. Calculate the mass percentage of each element in each of the following compounds to four significant figures.

(a) methanol, $CH_3OH(l)$

(b) water, $H_2O(l)$

(c) hydrogen peroxide, $H_2O_2(l)$

(d) Epsom salt, $MgSO_4{\cdot}7H_2O(s)$

11-7. Determine the number of grams of

(a) carbon in 58.5 grams of calcium carbonate

(b) lead in 10.0 grams of lead(IV) oxide

(c) oxygen in 5.0 grams of copper(II) sulfate pentahydrate

(d) hydrogen in 100.0 grams of hydrogen sulfide

11-8. Determine the number of grams of

(a) oxygen in 13.0 grams of calcium acetate

(b) fluorine in 0.22 grams of xenon tetrafluoride

(c) water in 25.0 grams of barium chloride dihydrate

(d) carbon in 2.00 grams of hexane, C_6H_{14}

SIMPLEST OR EMPIRICAL FORMULAS

11-9. Calcium carbide produces acetylene when water is added to it. The acetylene evolved is burned to provide the light source on spelunkers' helmets. Chemical analysis shows that calcium carbide is 62.5% (by mass) calcium and 37.5% (by mass) carbon. Determine the empirical formula of calcium carbide.

11-10. Rust occurs when iron metal reacts with the oxygen in the air. Chemical analysis shows that dry rust is 69.9% iron and 30.1% oxygen by mass. Determine the empirical formula of rust.

11-11. A 2.46-gram sample of copper metal is reacted completely with chlorine gas to produce 5.22 grams of copper chloride. Determine the empirical formula of this chloride.

11-12. A 3.78-gram sample of iron metal is reacted with sulfur to produce 5.95 grams of iron sulfide. Determine the empirical formula of this compound.

11-13. A 28.1-gram sample of cobalt metal was reacted completely with excess chlorine gas. The mass of the compound formed was 61.9 grams. Determine its empirical formula.

11-14. A 5.00-gram sample of aluminum metal is burned in an oxygen atmosphere to produce 9.45 grams of aluminum oxide. Use these data to determine the empirical formula of aluminum oxide.

11-15. Given the following mass percentages of the elements in certain compounds, determine the empirical formulas in each case and name the compound.

(a) 46.45% Li 53.55% O

(b) 59.78% Li 40.22% N

(c) 14.17% Li 85.83% N

(d) 36.11% Ca 63.89% Cl

11-16. Given the following mass percentages of the elements in certain compounds, determine the empirical formula in each case and name the compound.

(a) 71.89% Tl 28.11% Br

(b) 74.51% Pb 25.49% Cl

(c) 82.24% N 17.76% H

(d) 72.24% Mg 27.76% N

DETERMINATION OF ATOMIC MASS

11-17. A 1.443-gram sample of metal is reacted with excess oxygen to yield 1.683 grams of the oxide $M_2O_3(s)$. Calculate the atomic mass of the element M and identify the metal.

11-18. An element forms a chloride whose formula is XCl_4 that is known to consist of 74.8% chlorine by mass. Calculate the atomic mass of X and identify it.

11-19. A sample of a compound with the formula $MCl_2 \cdot 2H_2O$ has a mass of 0.642 grams. When the compound is heated to remove the waters of hydration, 0.0949 grams of water are collected. What element is M?

11-20. The formula of an acid is only partially known as HXO_3. The mass of 0.0133 moles of this acid is 1.123 grams. Find the atomic mass of X and identify the element represented by X.

MOLECULAR FORMULAS

11-21. The chemical 2-propanone, commonly known as acetone, is an important chemical solvent; a familiar home use is as a nail polish remover. Chemical analysis shows that acetone is 62.0% carbon, 10.4% hydrogen, and 27.5% oxygen by mass. Determine the empirical formula of acetone. In a separate experiment, the molecular mass is found to be 58.1. What is the molecular formula of acetone?

11-22. Glucose, one of the main sources of energy used by living organisms, has a molecular mass of 180.2. Chemical analysis shows that glucose is 40.0% carbon, 6.71% hydrogen, and 53.3% oxygen by mass. Determine its molecular formula.

11-23. A class of compounds called sodium metaphosphates was used as additives to detergents to improve cleaning ability. One of them has a molecular mass of 612. Chemical analysis shows that this sodium metaphosphate consists of 22.5% sodium, 30.4% phosphorus, and 47.1% oxygen by mass. Determine the molecular formula of this compound.

11-24. A hemoglobin sample is found to be 0.373% iron by mass. Given that there are four iron atoms per hemoglobin molecule, determine the molecular mass of hemoglobin.

COMBUSTION ANALYSIS

11-25. Combustion analysis of a 1.000-gram sample of a compound known to contain only carbon, hydrogen, and oxygen produces 1.500 grams of $CO_2(g)$ and 0.409 grams of $H_2O(g)$. Determine the empirical formula of the compound.

11-26. Combustion analysis of a 1.000-gram sample of a compound known to contain only carbon, hydrogen, and iron produces 2.367 grams of $CO_2(g)$ and 0.4835 grams of $H_2O(g)$. Determine the empirical formula of the compound.

11-27. Diethyl ether, often called simply ether, is a common solvent that contains carbon, hydrogen, and oxygen. A 1.23-gram sample was burned under controlled conditions to produce 2.92 grams of $CO_2(g)$ and 1.49 grams of $H_2O(g)$. Determine the empirical formula of diethyl ether.

11-28. Butylated hydroxytoluene, BHT, a food preservative, contains carbon, hydrogen, and oxygen. A sample of 15.42 milligrams of BHT was burned in a stream of oxygen and yielded 46.20 milligrams of $CO_2(g)$ and 15.13 milligrams of $H_2O(g)$. Calculate the empirical formula of BHT.

11-29. Pyridine is recovered from coke-oven gases and is used extensively in the chemical industry, in particular, in the synthesis of vitamins and drugs. Pyridine contains carbon, hydrogen, and nitrogen. A 0.5460-gram sample was burned to produce 1.518 grams of $CO_2(g)$ and 0.311 grams of $H_2O(g)$. Determine the empirical formula of pyridine.

11-30. Prior to their phaseout in the 1980s, chemicals containing lead were commonly added to gasoline as anti-knocking agents. A 5.83-gram sample of one such additive containing lead, carbon, and hydrogen was burned in an apparatus like that shown in Figure 11.5, and 6.34 grams of $CO_2(g)$ and 3.26 grams of $H_2O(g)$ were produced. Determine the empirical formula of the additive.

CALCULATIONS INVOLVING CHEMICAL REACTIONS

11-31. The combustion of propane may be described by the chemical equation

$$C_3H_8(g) + 5\,O_2(g) \rightarrow 3\,CO_2(g) + 4\,H_2O(g)$$

How many grams of oxygen are required to burn completely 10.0 grams of propane?

11-32. Iodine is prepared both in the laboratory and commercially by adding $Cl_2(g)$ to an aqueous solution containing sodium iodide according to

$$2\,NaI(aq) + Cl_2(g) \rightarrow I_2(s) + 2\,NaCl(aq)$$

How many grams of sodium iodide must be used to produce 50.0 grams of iodine?

11-33. Small quantities of chlorine can be prepared in the laboratory by the reaction described by the equation

$$MnO_2(s) + 4\,HCl(aq) \rightarrow MnCl_2(aq) + Cl_2(g) + 2\,H_2O(l)$$

How many grams of chlorine can be prepared from 100.0 grams of manganese(II) oxide?

11-34. Small quantities of oxygen can be prepared in the laboratory by heating potassium chlorate, $KClO_3(s)$. The equation for the reaction is

$$2\,KClO_3(s) \rightarrow 2\,KCl(s) + 3\,O_2(g)$$

Calculate how many grams of $O_2(g)$ can be produced from heating 10.0 grams of $KClO_3(s)$.

11-35. Lithium nitride reacts with water to produce ammonia and lithium hydroxide according to the equation

$$Li_3N(s) + 3\,H_2O(l) \rightarrow NH_3(g) + 3\,LiOH(aq)$$

Heavy water is water with the isotope deuterium in place of ordinary hydrogen, and its formula is D_2O. The above reaction can be used to produce heavy ammonia, $ND_3(g)$, according to the equation

$$Li_3N(s) + 3\,D_2O(l) \rightarrow ND_3(g) + 3\,LiOD(aq)$$

Calculate how many grams of heavy water are required to produce 200.0 milligrams of $ND_3(g)$. The atomic mass of deuterium is 2.014.

11-36. A common natural source of phosphorus is phosphate rock, an ore found in extensive deposits in areas that were originally ocean floor. The formula of one type of phosphate rock is $Ca_{10}(OH)_2(PO_4)_6$. Phosphate rock is converted to phosphoric acid by the reaction described by the equation

$$Ca_{10}(OH)_2(PO_4)_6(s) + 10\,H_2SO_4(l) \rightarrow 6\,H_3PO_4(l) + 10\,CaSO_4(s) + 2\,H_2O(l)$$

Calculate how many metric tons of phosphoric acid can be produced from 100 metric tons of phosphate rock (1 metric ton = 1000 kilograms).

11-37. Zinc is produced from its principal ore, sphalerite, $ZnS(s)$, by the two-step process described by

$$(1)\ 2\,ZnS(s) + 3\,O_2(g) \rightarrow 2\,ZnO(s) + 2\,SO_2(g)$$

$$(2)\ ZnO(s) + C(s) \rightarrow Zn(s) + CO(g)$$

How many kilograms of zinc can be produced from 2.00×10^5 kilograms of $ZnS(g)$?

11-38. Titanium is produced from its principal ore, rutile, $TiO_2(s)$, by the two-step process described by

$$(1)\ TiO_2(s) + 2\,Cl_2(g) + 2\,C(s) \rightarrow TiCl_4(g) + 2\,CO(g)$$

$$(2)\ TiCl_4(g) + 2\,Mg(s) \rightarrow Ti(s) + 2\,MgCl_2(s)$$

How many kilograms of titanium can be produced from 4.10×10^3 kilograms of $TiO_2(s)$?

CALCULATIONS WITHOUT THE CHEMICAL EQUATION

11-39. Hydrazine, $N_2H_4(l)$, is produced commercially by the reaction of ammonia, $NH_3(g)$, with sodium hypochlorite, $NaOCl(s)$ (the Raschig synthesis). Assuming that all the nitrogen in ammonia ends up in hydrazine, calculate how many metric tons of hydrazine can be produced from 10.0 metric tons of $NH_3(g)$.

11-40. Sulfur is obtained in large quantities from the hydrogen sulfide in so-called sour natural gas resulting from the removal of sulfur from petroleum. How many metric tons of sulfur can be obtained from 10.0 metric tons of hydrogen sulfide?

11-41. How many metric tons of aluminum can be obtained from 100.0 metric tons of bauxite, $Al_2O_3 \cdot 2H_2O(s)$, the principal ore of aluminum?

11-42. Sodium reacts with anhydrous ammonia to produce sodium amide, $NaNH_2(s)$. How many grams of sodium amide can be produced from 10.0 grams of sodium?

MIXTURES

11-43. Table salt, $NaCl(s)$, and sugar, $C_{12}H_{22}O_{11}(s)$, are accidentally mixed. A 5.00-gram sample is burned, and 2.20 grams of $CO_2(g)$ are produced. What is the mass percentage of the table salt in the mixture?

11-44. A mixture of $KCl(s)$ and $KClO_3(s)$ having a mass of 18.17 grams is heated to convert the $KClO_3(s)$ to $KCl(s)$ according to the chemical equation

$$2\,KClO_3(s) \rightarrow 2\,KCl(s) + 3\,O_2(g)$$

If the mixture has a mass of 12.62 grams after heating, what was the mass percentage of $KClO_3(s)$ in the original mixture?

11-45. A 4.07-gram sample of zinc and magnesium is reacted completely with hydrochloric acid. If 0.204 grams of hydrogen gas are produced by the reaction, calculate the mass percentage composition of the sample.

11-46. A 3.00-gram mixture of $K_2SO_4(s)$ and $MnSO_4(s)$ is dissolved in water and all the sulfate is precipitated as $BaSO_4(s)$ by the addition of $Ba(NO_3)_2(aq)$. If the $BaSO_4(s)$ precipitate has a mass of 4.37 grams, calculate the mass percentages of $K_2SO_4(s)$ and $MnSO_4(s)$ in the sample.

11-47. A 6.76-gram mixture of $CaCO_3(s)$ and $MgCO_3(s)$ is heated to drive off $CO_2(g)$. If the $CaO(s)$-$MgO(s)$ mixture that results has a mass of 3.38 grams, calculate the mass percentages of $CaCO_3(s)$ and $MgCO_3(s)$ in the original mixture.

11-48. A 9.87-gram sample of an alloy of aluminum and magnesium is completely reacted with hydrochloric acid and yields 0.998 grams of hydrogen gas. Calculate the percentage by mass of each metal in the alloy.

LIMITING REACTANT

11-49. Potassium nitrate is widely used as a fertilizer because it provides two essential elements, potassium and nitrogen. It is made by mixing potassium chloride and nitric acid in the presence of oxygen according to the equation

$$4\,KCl(aq) + 4\,HNO_3(aq) + O_2(g) \rightarrow 4\,KNO_3(aq) + 2\,Cl_2(g) + 2\,H_2O(l)$$

How many kilograms of potassium nitrate will be produced from a solution containing 50.0 kilograms of potassium chloride and one containing 50.0 kilograms of nitric acid? An important by-product is chlorine. How many kilograms of chlorine gas will be produced?

11-50. Phosphorus forms a compound similar to ammonia. The compound has the chemical formula PH_3 and is called phosphine. It can be prepared by the reaction described by the equation

$$P_4(s) + 3\,NaOH(aq) + 3\,H_2O(l) \rightarrow PH_3(g) + 3\,NaH_2PO_2(aq)$$

If 20.0 grams of phosphorus and a solution containing 50.0 grams of NaOH are reacted with $H_2O(l)$ in excess, how many grams of phosphine will be obtained?

11-51. Sodium hydroxide reacts with sulfuric acid according to the equation

$$2NaOH(aq) + H_2SO_4(aq) \rightarrow Na_2SO_4(aq) + 2\,H_2O(l)$$

Suppose that a solution containing 60.0 grams of sodium hydroxide is added to one containing 20.0 grams of sulfuric acid. How many grams of sodium sulfate will be produced?

11-52. Bromine can be prepared by adding chlorine to an aqueous solution of sodium bromide. The reaction equation is

$$2\,NaBr(aq) + Cl_2(g) \rightarrow Br_2(l) + 2\,NaCl(aq)$$

How many grams of bromine are formed if a solution containing 25.0 grams of aqueous $NaBr(s)$ and 25.0 grams of $Cl_2(g)$ are reacted?

11-53. Cryolite, $Na_3AlF_6(s)$, an ore used in the production of aluminum, can be synthesized by the reaction described by the equation

$$Al_2O_3(s) + NaOH(l) + HF(g) \xrightarrow{High\ T} Na_3AlF_6(s) + H_2O(g) \quad (unbalanced)$$

(a) Balance this equation. (b) If 10.0 kilograms of $Al_2O_3(s)$, 50.00 kilograms of NaOH(l), and 50.0 kilograms of HF(g) react completely, how many kilograms of cryolite will be produced? (c) Which reactants will be in excess and how many kilograms of each of these reactants will remain?

11-54. When calcium carbonate is added to hydrochloric acid, calcium chloride, carbon dioxide, and water are produced. (a) Write the balanced chemical equation for this reaction. (b) How many grams of calcium chloride will be produced when 25.0 grams of calcium carbonate are combined with 15.0 grams of hydrochloric acid? (c) Which reactant is in excess and how many grams of this reactant will remain after the reaction is complete?

11-55. A solution containing 12.5 grams of silver nitrate is added to a solution containing 67.2 grams of sodium sulfate. (a) Write the balanced chemical equation for this reaction. (b) Determine how many grams of solid silver sulfate are formed after the reaction comes to completion. (c) How many grams of the reactant in excess will remain after the reaction?

11-56. A solution containing 17.5 grams of cadmium chloride is mixed with a solution containing 35.5 grams of silver perchlorate. (a) Write the balanced chemical equation for this reaction. (b) Determine how many grams of silver chloride are produced after the reaction comes to completion. (c) How many grams of the reactant in excess will remain after the reaction?

11-57. When 18.0 grams of iron(III) nitrate are mixed with a solution containing 50.0 grams of sodium hydroxide, a dark brown precipitate forms. Determine the mass of the precipitate formed if the reaction goes to completion. How many grams of the reactant in excess will remain after the reaction?

11-58. If a solution containing 25.0 grams of mercury(II) nitrate is allowed to react completely with a solution containing 15.0 grams of sodium bromide, how many grams of solid precipitate will be formed? How many grams of the reactant in excess will remain after the reaction?

PERCENTAGE YIELD

11-59. Titanium dioxide is converted to titanium tetrachloride by reaction with chlorine gas and carbon according to the equation

$$TiO_2(s) + 2\,Cl_2(g) + 2\,C(s) \rightarrow TiCl_4(g) + 2\,CO(g)$$

Suppose 50.0 grams of $TiO_2(s)$ are reacted with excess $Cl_2(g)$ and C(s), and 55.0 grams of $TiCl_4(g)$ are obtained. Calculate the percentage yield of $TiCl_4(g)$.

11-60. Antimony is produced by the reaction of antimony(III) oxide, $Sb_4O_6(s)$, with carbon according to the equation

$$Sb_4O_6(s) + 6\,C(s) \rightarrow 4\,Sb(s) + 6\,CO(g)$$

Given that 60.0 grams of $Sb_4O_6(s)$ are reacted with excess C(s), and 49.0 grams of Sb(s) are obtained, calculate the percentage yield of antimony.

11-61. Ethyl propionate, $CH_3CH_2COOCH_2CH_3(l)$, is obtained from the reaction between propionic acid, $CH_3CH_2COOH(aq)$, and ethanol, using sulfuric acid as a catalyst. The reaction may be described by the equation

$$CH_3CH_2OH(aq) + CH_3CH_2COOH(aq) \xrightarrow{H_2SO_4} CH_3CH_2COOCH_2CH_3(aq) + H_2O(l)$$

Ethyl propionate has a pineapple-like odor and is used as a flavoring agent in fruit syrup. In an experiment 349 grams of ethyl propionate were obtained from 255 grams of ethanol, with propionic acid in excess. Calculate the percentage yield of this reaction.

11-62. Ethanol, $CH_3CH_2OH(l)$, is produced commercially from the reaction of water with ethene, $C_2H_4(g)$, commonly known as ethylene. Ethylene is obtained from petroleum and is one of the top produced industrial chemicals in the United States (Appendix H). It is the basis for the synthesis of a variety of important chemicals and polymers. The reaction equation for the synthesis of ethanol is

$$C_2H_4(g) + H_2O(l) \xrightarrow{H_2SO_4} CH_3CH_2OH(l)$$

Given that 13.5 kilograms of ethanol were produced from 10.0 kilograms of ethylene, calculate the percentage yield in this synthesis.

11-63. When ammonia gas is reacted with copper(II) oxide, nitrogen gas, water vapor, and copper metal are produced. (a) Write the balanced equation for this reaction. (b) How many kilograms of copper can be produced from the reaction of 12.0 kilograms of copper(II) oxide with 12.0 kilograms of ammonia if the reaction is found to have a 45% yield?

11-64. The addition of iron to sulfuric acid is found to produce iron(II) sulfate and hydrogen gas with a 95% yield. (a) Write a balanced equation for this reaction. (b) How many kilograms of iron are required to generate 1.00 kilograms of hydrogen gas?

ADDITIONAL PROBLEMS

11-65. (a) What is the difference between an empirical formula and a molecular formula? (b) Is it possible for the empirical and molecular formulas of a compound to be the same?

11-66. When analyzing the results of combustion analysis, we use the mass percentage of carbon in the $CO_2(g)$ produced to find the mass of carbon from our original sample and the mass percentage of hydrogen in the $H_2O(g)$ produced to find the mass of hydrogen from our original sample. Why can't the same procedure be used to determine the mass of oxygen present in the original sample?

11-67. When left exposed to air and water, iron metal will rust, forming an oxide. If a bar of pure iron is allowed to rust and then both the rust and remaining iron are weighed, will the total mass be greater than, less than, or equal to the mass of the original iron bar? If all the rust is scraped off, will the mass of the bar now be greater than, less than, or equal to the original mass of the bar? Justify your answers.

11-68. A student working in the laboratory finds that her percentage yield in a synthesis is only 55%. What factors might have resulted in this yield?

11-69. Why is it not possible for a reaction to have a percentage yield of the desired product greater than 100% (what law is this in violation of)? What is the most likely explanation for a percentage yield measured in the laboratory that is in excess of 100%?

11-70. Determine the mass of ammonium nitrate, $NH_4NO_3(s)$, that has the same number of nitrogen atoms as 2.0 liters of liquid nitrogen, $N_2(l)$. Take the density of liquid nitrogen to be 0.808 g·mL^{-1}.

11-71. Determine the number of grams of each element in 5.00 grams of copper(II) sulfate pentahydrate.

11-72. The following reaction equation describes one of the side reactions in the manufacture of rayon from wood pulp:

$$3\,CS_2(g) + 6\,NaOH(aq) \rightarrow 2\,Na_2CS_3(aq) + Na_2CO_3(aq) + 3\,H_2O(l)$$

How many grams of each product are formed when 1.00 kilogram of each reactant is used?

11-73. Chlorine is produced industrially by the electrolysis of brine (a solution of naturally occurring salts and consists mainly of sodium chloride). The reaction is described by the equation

$$2\,NaCl(aq) + 2\,H_2O(l) \xrightarrow{\text{electrolysis}} 2\,NaOH(aq) + Cl_2(g) + H_2(g)$$

The other products, sodium hydroxide and hydrogen, are also valuable commercial compounds. How many kilograms of each product can be obtained from the electrolysis of 1.00 kilogram of salt that is 95% sodium chloride by mass?

11-74. Lithium is the only Group 1 metal that yields the normal oxide, $Li_2O(s)$, when it is burned in excess oxygen. The other alkali metals react with excess oxygen as shown below.

(a) $2\,Na(s) + O_2(g) \rightarrow \underset{\text{sodium peroxide}}{Na_2O_2(s)}$

(b) $K(s) + O_2(g) \rightarrow \underset{\text{potassium superoxide}}{KO_2(s)}$

(c) $Rb(s) + O_2(g) \rightarrow \underset{\text{rubidium superoxide}}{RbO_2(s)}$

(d) $Cs(s) + O_2(g) \rightarrow \underset{\text{cesium superoxide}}{CsO_2(s)}$

Calculate how much product is formed when 0.600 grams of each alkali metal are burned in excess oxygen.

11-75. The arsenic in an ore sample was converted into water-soluble sodium arsenate, $Na_3AsO_4(aq)$. From the solution of $Na_3AsO_4(aq)$, insoluble silver arsenate, $Ag_3AsO_4(s)$, was precipitated and weighed. A 5.00-gram ore sample produced 3.09 grams of silver arsenate. Calculate the mass percentage of arsenic in the ore sample.

11-76. Glucose is used as an energy source by the human body. The overall reaction in the body is described by the equation

$$C_6H_{12}O_6(aq) + 6\,O_2(g) \rightarrow 6\,CO_2(g) + 6\,H_2O(l)$$

Calculate the number of grams of oxygen required to convert 28.0 grams of glucose to $CO_2(g)$ and $H_2O(l)$. Also compute the number of grams of $CO_2(g)$ produced.

11-77. A hydrated form of copper(II) sulfate, $CuSO_4 \cdot nH_2O(s)$, is heated to drive off all the waters of hydration. If we start with 9.40 grams of hydrated salt and have 5.25 grams of anhydrous $CuSO_4(s)$ after heating, find the number of water molecules, n, associated with each $CuSO_4$ formula unit.

11-78. A 12.42-gram sample of a mixture of $NaCl(s)$ and $CaCl_2(s)$ is dissolved in water and all the chloride is precipitated as $AgCl(s)$. If the $AgCl(s)$ precipitate has a mass of 31.70 grams, calculate the mass percentage of each component in the original mixture.

11-79. An ore is analyzed for its lead content as follows. A sample is dissolved in water; then sodium sulfate is added to precipitate the lead as lead(II) sulfate, $PbSO_4(s)$. The net ionic equation for the reaction is

$$Pb^{2+}(aq) + SO_4^{2-}(aq) \rightarrow PbSO_4(s)$$

It was found that 13.73 grams of lead(II) sulfate were precipitated from a sample of ore having a mass of 53.92 grams. How many grams of lead are there in the sample? What is the mass percentage of lead in the ore?

11-80. Nickel sulfate forms a blue-green hydrate with the formula $NiSO_4 \cdot nH_2O(s)$. If this hydrate is heated to a high enough temperature, $H_2O(g)$ can be driven off, leaving the greenish-yellow anhydrous salt $NiSO_4(s)$. A 12.060-gram sample of the hydrate was heated to 300°C. The resulting $NiSO_4(s)$ had a mass of 7.101 grams. Calculate the value of n in $NiSO_4 \cdot nH_2O(s)$.

11-81. Tin(IV) chloride can be produced by heating metallic tin in a stream of dry chlorine gas. If the process has a yield of 84%, how many grams of tin should be heated in excess chlorine to produce 200 grams of tin(IV) chloride?

11-82. Boron and hydrogen form a number of compounds called boranes (see Interchapter I). A certain borane is found to contain 85.63% boron by mass, the rest being hydrogen. What is the empirical formula of the borane?

11-83. A scientist is analyzing a white crystalline compound with a pungent odor similar to mothballs. Combustion analysis of 5.28 milligrams of the sample yields 18.13 milligrams of $CO_2(g)$ and 2.97 milligrams of $H_2O(l)$. Analysis of the compound in a mass spectrometer finds it to have a formula mass of 128. What is the molecular formula of the compound?

11-84. A mixture of $Na_2SO_4(s)$ and $NaHSO_4(s)$ having a mass of 2.606 grams is dissolved in water and excess $Ba(OH)_2(s)$ is added, precipitating $BaSO_4(s)$ according to

$$Na_2SO_4(s) + Ba(OH)_2(aq) \rightarrow BaSO_4(s) + 2\,NaOH(aq)$$

$$NaHSO_4(s) + Ba(OH)_2(aq) \rightarrow BaSO_4(s) + NaOH(aq) + H_2O(l)$$

The precipitate, $BaSO_4(s)$, has a mass of 4.688 grams. Calculate the mass percentages of $Na_2SO_4(s)$ and $NaHSO_4(s)$ in the mixture.

11-85. An oxide of molybdenum with the chemical formula $Mo_2O_3(s)$ is converted completely to another oxide. The oxide $Mo_2O_3(s)$ had a mass of 12.64 grams and the new oxide has a mass of 13.48 grams. Determine the empirical formula of the new oxide.

11-86. Nitric acid, $HNO_3(aq)$, is made commercially from ammonia by the Ostwald process, which was developed by the German chemist Wilhelm Ostwald. The process consists of three steps:

(1) $4\,NH_3(g) + 5\,O_2(g) \rightarrow 4\,NO(g) + 6\,H_2O(g)$

(2) $2\,NO(g) + O_2(g) \rightarrow 2\,NO_2(g)$

(3) $3\,NO_2(g) + H_2O(l) \rightarrow 2\,HNO_3(aq) + NO(g)$

How many kilograms of nitric acid can be produced from 6.40×10^4 kilograms of ammonia?

11-87. Antimony is usually found in nature as the mineral stibnite, $Sb_2S_3(s)$. Pure antimony can be obtained by first converting the sulfide to an oxide and then heating the oxide with coke (carbon). The reaction equations are

(1) $2\,Sb_2S_3(s) + 9\,O_2(g) \rightarrow Sb_4O_6(s) + 6\,SO_2(g)$

(2) $Sb_4O_6(s) + 6\,C(s) \rightarrow 4\,Sb(s) + 6\,CO(g)$

How many kilograms of antimony are obtained from 0.50 kilograms of stibnite?

11-88. A police forensics lab is analyzing a sample of white powder found at a crime scene to determine if it is cocaine. Elemental analysis of the powder shows that it is 67.31% carbon, 6.978% hydrogen, 4.618% nitrogen, and 21.10% oxygen by mass. The chemical formula of cocaine is $C_{17}H_{21}NO_4$. From this evidence, can the investigators conclude that the white powder is cocaine?

11-89. Chromium metal can be obtained by the thermite reaction between powdered aluminum metal and chromium(III) oxide. The reaction gives off so much heat that the chromium metal formed is molten. The other product is aluminum oxide. (a) Write a balanced chemical equation for this reaction. (b) How many grams of chromium can be produced if 500.0 grams of chromium(III) oxide are reacted with 100.0 grams of aluminum?

11-90. A website promoting the use of alternative energy vehicles and hybrid technologies claims that, "A typical automobile in the USA uses about 500 gallons of gasoline per year, producing about 5 tons of carbon dioxide." Does this statement make sense? Make your own estimate assuming that the primary ingredient in gasoline is octane, $C_8H_{18}(l)$, which has a density of 0.7 $g \cdot mL^{-1}$.

11-91. The concept of determining which reactant is limiting and which is in excess is akin to determining the number of sandwiches that can be made from a set number of ingredients. Assuming that a cheese sandwich consists of 2 slices of bread and 3 slices of cheese, determine the number of cheese sandwiches that can be prepared from a loaf of 24 slices of bread and a package of 40 slices of cheese. Which of the two ingredients limits the number of sandwiches that can be made? What quantity of the ingredient in excess remains?

11-92. (*) A 30.450-milligram sample of a chemical known to contain only carbon, hydrogen, oxygen, and sulfur is put into a combustion analysis apparatus, yielding 54.246 milligrams of carbon dioxide and 22.206 milligrams of water. In another experiment, 23.725 milligrams of the compound is reacted with excess oxygen to produce 10.255 milligrams of sulfur dioxide. What is the empirical formula of the compound?

11-93. (*) Two unidentified elements X and Y can combine to form three different compounds:

Compound	Grams of X	Grams of Y
1	25.2	28.8
2	72.8	41.6
3	22.4	64.0

If compound 1 is found to be a binary compound with the formula XY, find the simplest formulas of compounds 2 and 3.

11-94. (*) In 1773 Ben Franklin wrote in a letter about calming the waves on Clapham pond using a small quantity of oil,

> "*...the oil, though not more than a teaspoonful, produced an instant calm over a space several yards square which spread amazingly and extended itself gradually till it reached the lee side, making all that quarter of the pond, perhaps half an acre, as smooth as a looking glass.*"

Assuming Franklin used castor oil, which has a formula mass of about 180 and a density of about 0.96 $g \cdot mL^{-1}$, and that the oil forms a one-molecule-thick monolayer on the pond, *estimate* Avogadro's number using Franklin's data. Why must we assume the oil forms a monolayer? What other assumptions must we make?

Joel Hildebrand (1881–1983) was born in Camden, New Jersey, where he attended a Quaker school and in his spare time read science books from his grandfather's library. His high-school principal, recognizing that Hildebrand knew more chemistry than he did, gave him a copy of his Harvard chemistry textbook and the keys to the lab. Through a series of experiments, Hildebrand systematically determined that many of the assertions in the text were wrong. This had a profound effect on him and sparked his interest in performing research. He attended the University of Pennsylvania, earning a bachelor's degree in 1903 and a Ph.D. in chemistry in 1906. He then traveled to Europe, attending lectures at various universities, and then returned to the University of Pennsylvania in 1907. In 1913, he was offered a position by G. N. Lewis at the University of California at Berkeley, where he worked for the next 65 years, publishing some 280 research papers (more than half after his "retirement"). One of his most famous achievements was solving the problem of "the bends" in diving—where dissolved nitrogen would bubble out of the blood during decompression—by replacing nitrogen with the less-soluble helium in breathing mixtures. Hildebrand packed lecture halls with students attending his lively lectures and demonstrations. He served as president of the American Chemical Society in 1955, and in 1962 was awarded its highest scientific honor, the Priestly Medal. Hildebrand was also an active outdoors enthusiast. In 1921, he taught himself to ski and managed the United States ski team at the 1936 Winter Olympics. He wrote books on camping and mountain climbing and from 1937 to 1940 was president of the Sierra Club. Joel Hildebrand continued working long past his official retirement and died at the age of 101. His wife, Emily, lived to be 103.

12. Chemical Calculations for Solutions

Most chemical and biological processes take place in solution, particularly in aqueous solution. In this chapter we discuss how to carry out a number of calculations for chemical reactions that take place in solution. First we discuss what a solution is, then we discuss ways to calculate the concentrations of the components of a solution. Concentration can be expressed in a variety of units, but by far the most important unit of concentration in chemistry is molarity. We shall use molarity to carry out stoichiometric calculations for reactions where one or more of the reactants or products occurs in solution. Finally, we briefly discuss titration experiments involving acid-base neutralization reactions.

12-1. A Solution Is a Homogeneous Mixture of Two or More Substances

As we have seen in Section 2-3, a solution is a mixture that is homogeneous at the molecular level. From a molecular point of view, the species in a solution are uniformly dispersed among one another (Figure 12.1). The **components** of a solution are the pure substances that are mixed to form the solution. The components do not have to be a solid and a liquid; there are many types of solutions (Table 12.1).

The most common type of solution is a solid dissolved in a liquid. The solid that is dissolved is called the **solute**, and the liquid in which it is dissolved is called the **solvent**. The terms solvent and solute are merely terms of convenience because all the components of a solution are uniformly dispersed throughout the solution. When $NaCl(s)$ is dissolved in water, we say that $NaCl(s)$ is the solute and $H_2O(l)$ is the solvent. The process of dissolving $NaCl(s)$ in water is represented by the equation

$$NaCl(s) \xrightarrow[H_2O(l)]{} Na^+(aq) + Cl^-(aq)$$

where $H_2O(l)$ under the arrow tells us that water is the solvent. The species $Na^+(aq)$ and $Cl^-(aq)$ represent a sodium ion and a chloride ion in an aqueous solution. As Figure 12.2 illustrates, these ions are solvated by water molecules; that is, they are surrounded by a loosely bound shell of water molecules.

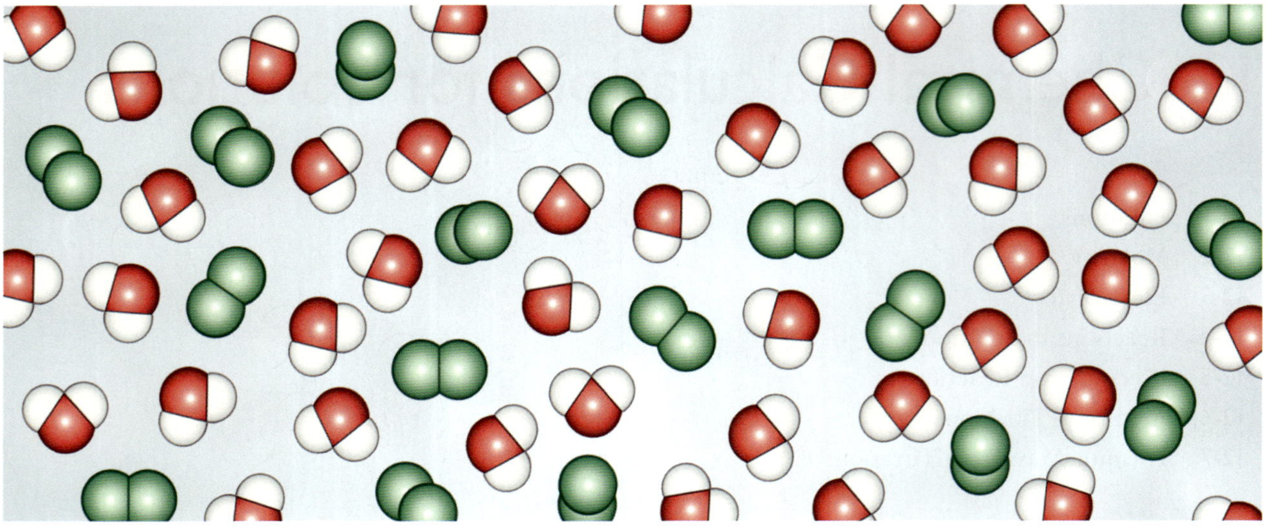

Figure 12.1 A solution is a homogeneous mixture at the molecular level. The species in a solution are uniformly dispersed among one another.

When a small quantity of sodium chloride is added to a beaker of water, the sodium chloride dissolves completely, leaving no crystals at the bottom of the beaker. As more and more sodium chloride is added, we reach a point where no more sodium chloride can dissolve, so that any further sodium chloride crystals that we add simply remain at the bottom of the beaker. Such a solution is called a **saturated solution**, and the maximum quantity of solute dissolved is called the **solubility** of that solute. Solubility can be expressed in a variety of units, but quite commonly it is expressed as grams of solute per 100 grams of solvent. For example, we can say that the solubility of NaCl(*s*) in water at 20°C is about 36 grams per 100 grams of $H_2O(l)$.

It is important to realize that the solubility of a substance is the maximum quantity that can be dissolved in a saturated solution at a particular temperature. The solubility of NaCl(*s*) at 20°C is about 36 grams per 100 grams of $H_2O(l)$. If we add 50 grams of NaCl(*s*) to 100 grams of $H_2O(l)$ at 20°C, then 36 grams dissolve and 14 grams are left as undissolved NaCl(*s*). The solution is

TABLE 12.1 Types and examples of solutions

State of component 1	State of component 2	State of resulting solution	Examples
gas	gas	gas	air; vaporized gasoline-air mixture in the combustion chambers of a car
gas	liquid	liquid	oxygen in water; carbon dioxide in carbonated beverages
gas	solid	solid	hydrogen in palladium and platinum
liquid	liquid	liquid	water and alcohol
liquid	solid	solid	mercury in gold or silver
solid	liquid	liquid	sodium chloride in water
solid	solid	solid	metal alloys

saturated. If we add 25 grams of NaCl(s) to 100 grams of $H_2O(l)$, then all the NaCl(s) dissolves to form what is called an **unsaturated solution**; that is, a solution in which we can dissolve more of the solute.

In most cases the solubility of a substance depends on temperature. The effect of temperature on the solubility of several salts in water is shown in Figure 12.3. Almost all substances become more soluble in water as the temperature increases. For example, potassium nitrate is about five times more soluble in water at 40°C than it is at 0°C.

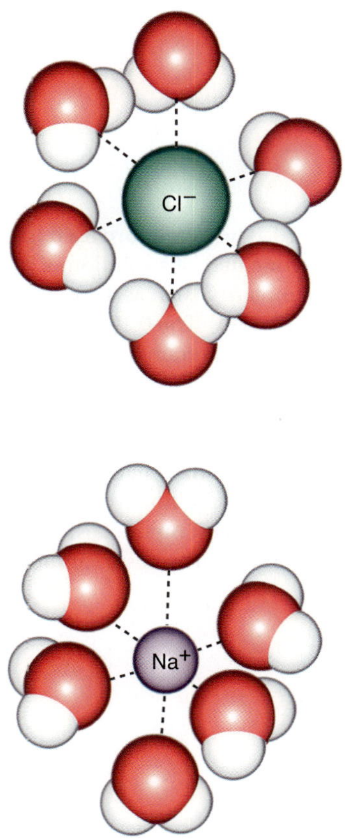

Figure 12.2 Ions in aqueous solutions are surrounded by a loosely bound shell of water molecules. Such ions are said to be solvated.

12-2. Molarity Is the Most Common Unit of Concentration

The **concentration** of solute in a solution describes the quantity of solute dissolved in a given quantity of solvent or a given quantity of solution. A common method of expressing the concentration of a solute is **molarity,** which is denoted by the symbol M. Molarity is defined as the number of moles of solute per liter of solution:

$$\text{molarity} = \frac{\text{moles of solute}}{\text{liters of solution}} \tag{12.1}$$

Equation 12.1 can be expressed symbolically as

$$M = \frac{n}{V} \tag{12.2}$$

where M represents the molarity of the solution, n is the number of moles of solute dissolved in the solution, and V is the total volume of the solution in liters. To see how to use Equation 12.2, let's calculate the molarity of a solution prepared by dissolving 62.3 grams of sucrose, $C_{12}H_{22}O_{11}(s)$, in enough water to form 0.500 liters of solution. The formula mass of sucrose is 342.3, so 62.3 grams corresponds to

$$\text{moles of sucrose} = (62.3 \text{ g sucrose})\left(\frac{1 \text{ mol sucrose}}{342.3 \text{ g sucrose}}\right) = 0.182 \text{ mol}$$

The molarity of the solution is given by

$$M = \frac{n}{V} = \frac{0.182 \text{ mol}}{0.500 \text{ L}} = 0.364 \text{ mol·L}^{-1} = 0.364 \text{ M}$$

We say that the concentration of sucrose in the solution is 0.364 **molar**, which we write as 0.364 M. The unit of molarity is written as M.

The definition of molarity involves the total volume of the solution, not just the volume of the solvent. Suppose we wish to prepare one liter of a 0.100-M aqueous solution of potassium dichromate, $K_2Cr_2O_7(aq)$. We would prepare the solution by weighing out 0.100 moles (29.4 grams) of $K_2Cr_2O_7(s)$, dissolving it in less than one liter of water, say, about 500 mL, and then adding water while stirring until the final volume of the solution is precisely one liter. As shown in Figure 12.4, we use a **volumetric flask**, which is a piece of glassware used to prepare precise volumes. It would be incorrect to add 0.100 moles of $K_2Cr_2O_7(s)$

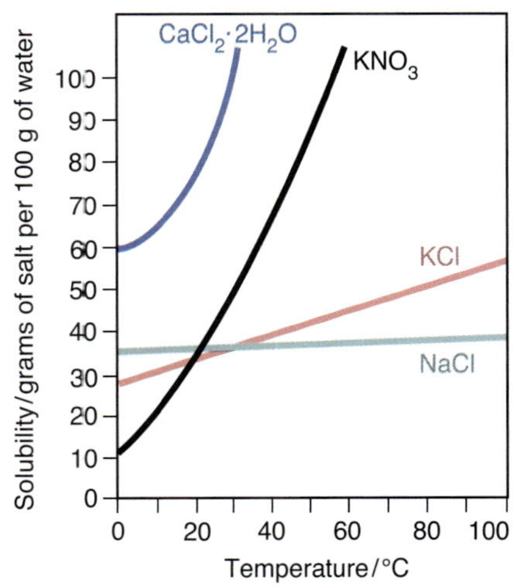

Figure 12.3 The solubility of most salts increases with increasing temperature.

(a) (b) (c)

Figure 12.4 The procedure used to prepare one liter of a solution of a specific molarity, such as 0.100 M $K_2Cr_2O_7(aq)$. (a) The 0.100 moles of $K_2Cr_2O_7(s)$ (29.4 grams) are weighed out and (b) added to a one-liter volumetric flask that is only partially filled with water. (c) The $K_2Cr_2O_7(s)$ is dissolved, and then more water is added to bring the final volume up to the one-liter mark on the flask. The solution is swirled to ensure uniform mixing.

to one liter of water; the final volume of such a solution is not precisely one liter because the added $K_2Cr_2O_7(s)$ changes the volume from 1.00 liter to 1.02 liters. The following example illustrates the procedure for making up a solution of a specified molarity.

THE PRECISION OF VOLUMETRIC GLASSWARE. The exact precision of volumetric glassware varies with the size and quality of the glassware used. The precision is rated using the International Organization for Standardization (ISO) system where "Class A" volumetric glassware (the sort found in most general chemistry laboratories) has the lowest precision, with subsequent letters designating glassware of increasing precision. Class A volumetric glassware has a precision that is at least 0.1% of the volume being measured. Thus, a Class A 250-mL volumetric flask has a precision of at least ± 0.25 mL, or about four significant figures at the stated temperature. For good analytical work, it is critical that the precision of the glassware be greater than that of the desired solution. Throughout this text we do not specify significant figures for volumetric measurements; rather, we assume that the glassware used has a precision greater than that of the desired solutions.

EXAMPLE 12-1: Potassium bromide, $KBr(s)$, is used by veterinarians to treat epilepsy in dogs. Explain how you would prepare 250 mL of a 0.600-M aqueous $KBr(aq)$ solution.

Solution: From Equation 12.2 and the specified volume and concentration, we can calculate the number of moles of KBr(*s*) required. Equation 12.2 can be written as

$$n = MV \tag{12.3}$$

so

$$\text{moles of KBr} = (0.600\ \text{M})(250\ \text{mL})\left(\frac{1\ \text{L}}{1000\ \text{mL}}\right) = 0.150\ \text{mol}$$

We can convert moles to grams by multiplying by the formula mass of KBr(*s*):

$$\text{mass of KBr} = (0.150\ \text{mol KBr})\left(\frac{119.0\ \text{g KBr}}{1\ \text{mol KBr}}\right) = 17.9\ \text{g}$$

To prepare the solution, we add 17.9 grams of KBr(*s*) to a 250-mL volumetric flask that is partially filled with distilled water. We swirl the flask until the salt is dissolved and then dilute the solution to the 250-mL mark on the flask and swirl it again to assure uniformity. We would not add the KBr(*s*) to 250 mL of water, because the volume of the resulting solution would not necessarily be 250 mL.

PRACTICE PROBLEM 12-1: Ammonium selenate, $(NH_4)_2SeO_4(s)$, is used as a mothproofing agent. Describe how you would prepare 0.500 L of a 0.155-M aqueous solution of ammonium selenate.

Answer: Dissolve 13.9 grams of ammonium selenate in less than 500 mL of water and then dilute to 0.500 L using a volumetric flask.

Occasionally, the concentration of a solution is given as the mass percentage of the solute. For example, commercial sulfuric acid is sold as a solution that is 96.7% H_2SO_4 and 3.3% water by mass. If you know the density of such a solution, you can calculate its molarity. The density of the sulfuric acid solution is 1.84 $g{\cdot}mL^{-1}$ at 20°C. The mass of H_2SO_4 in one liter of solution is given by

$$\left(\begin{matrix}\text{mass of H}_2\text{SO}_4\\ \text{per liter of solution}\end{matrix}\right) = \left(\frac{1000\ \text{mL}}{1\ \text{L}}\right)\left(\frac{1.84\ \text{g solution}}{1\ \text{mL}}\right)\left(\frac{96.7\ \text{g H}_2\text{SO}_4}{100\ \text{g solution}}\right)$$
$$= 1780\ \text{g H}_2\text{SO}_4\ \text{per liter of solution}$$

and the number of moles of $H_2SO_4(aq)$ per liter of solution—or, in other words, the molarity—is given by

$$\text{molarity of H}_2\text{SO}_4(aq) = \left(\frac{1780\ \text{g H}_2\text{SO}_4}{1\ \text{L solution}}\right)\left(\frac{1\ \text{mol H}_2\text{SO}_4}{98.08\ \text{g H}_2\text{SO}_4}\right) = 18.1\ \text{M}$$

EXAMPLE 12-2: Ammonia is sold as an aqueous solution that is 28% NH_3 by mass and has a density of 0.90 g·mL^{-1} at 20°C. Calculate the molarity of this solution.

Solution: The mass of NH_3 in one liter of solution is

$$\left(\begin{array}{c}\text{mass of NH}_3\\ \text{per liter of solution}\end{array}\right) = \left(\frac{1000\text{ mL}}{1\text{ L}}\right)\left(\frac{0.90\text{ g solution}}{1\text{ mL solution}}\right)\left(\frac{28\text{ g NH}_3}{100\text{ g solution}}\right)$$

$$= 250\text{ g NH}_3\text{ per liter of solution}$$

The molarity is given by

$$\text{molarity of NH}_3 = \left(\frac{250\text{ g NH}_3}{1\text{ L solution}}\right)\left(\frac{1\text{ mol NH}_3}{17.03\text{ g NH}_3}\right) = 15\text{ M}$$

PRACTICE PROBLEM 12-2: A concentrated sodium hydroxide solution is 50.0% NaOH by mass and has a density of 1.525 g·mL^{-1} at 20°C. Calculate the molarity of the solution.

Answer: 19.1 M

It is often necessary in laboratory work to prepare a more dilute solution from a more concentrated stock solution, as we might do with the concentrated sodium hydroxide solution above. In such cases, a certain volume of a solution of known molarity is diluted with a certain volume of pure solvent to produce the final solution with the desired molarity. The key point to recognize in carrying out such **dilution** calculations is that the number of moles of solute does not change on dilution with solvent (Figure 12.5). Thus, from Equation 12.3, we have

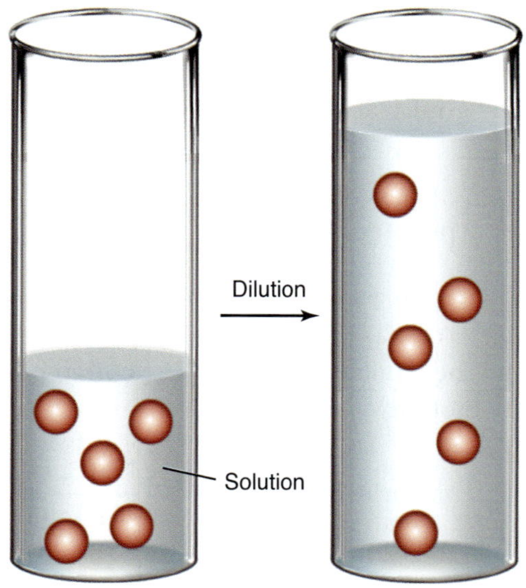

Figure 12.5 When we dilute a solution, the volume of solvent is increased but the number of moles of solute (spheres) stays the same.

$$\text{number of moles of solute before dilution} = n_1 = M_1V_1$$

and

$$\text{number of moles of solute after dilution} = n_2 = M_2V_2$$

But $n_1 = n_2$, and so we have

$$M_1V_1 = M_2V_2 \qquad \text{(dilution)} \qquad (12.4)$$

The following Example illustrates a dilution calculation.

EXAMPLE 12-3: Compute the volume of the 19.1-M concentrated NaOH(*aq*) solution in Practice Problem 12-2 required to produce 500 mL of 3.0-M NaOH(*aq*).

Solution: From Equation 12.4, we have

$$M_1V_1 = M_2V_2$$

$$(19.1\ \text{mol}\cdot\text{L}^{-1})\ (V_1) = (3.0\ \text{mol}\cdot\text{L}^{-1})\ (0.500\ \text{L})$$

Thus,

$$V_1 = \frac{(0.500\ \text{L})(3.0\ \text{mol}\cdot\text{L}^{-1})}{19.1\ \text{mol}\cdot\text{L}^{-1}} = 0.079\ \text{L}$$

or $V_1 = 79$ mL. To make the 3.0 M NaOH(*aq*) solution, we add 79 mL of 19.1 M NaOH(*aq*) to a 500-mL volumetric flask that is about half-filled with water, swirl the solution, and dilute with water to the 500-mL mark on the flask. Finally, we swirl again to make the new solution homogeneous.

PRACTICE PROBLEM 12-3: Commercial nitric acid, HNO_3(*aq*), is a 15.9-M aqueous solution. How would you prepare one liter of 6.00 M HNO_3(*aq*) solution from commercial nitric acid?

Answer: Dilute 377 mL of the 15.9 M HNO_3(*aq*) to one liter using a volumetric flask.

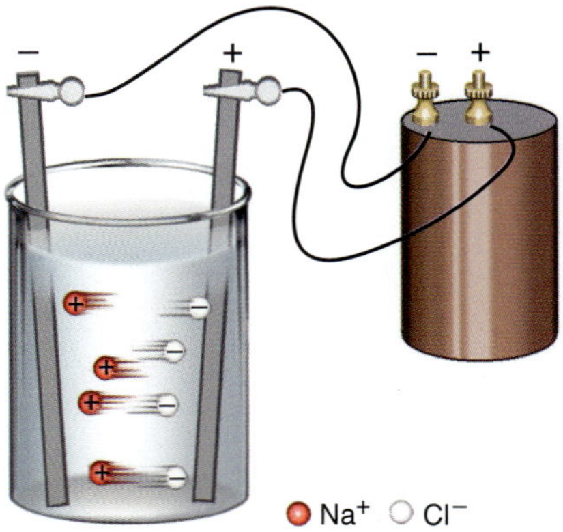

Figure 12.6 An aqueous solution of NaCl(*aq*) conducts an electric current. An electric voltage is applied by dipping metal strips (electrodes) attached to poles of a battery into the solution. Like the poles of a battery, one of the electrodes is positive and the other is negative. The positively charged sodium ions are attracted to the negative electrode, and the negatively charged chloride ions are attracted to the positive electrode. Thus, the Na^+(*aq*) ions migrate to the left in the figure and the Cl^-(*aq*) ions migrate to the right. The migration of the ions constitutes an electric current through the solution.

12-3. Solutions That Contain Ions Conduct an Electric Current

As we have already seen in Section 6-1, when ionic substances dissolve, the crystals separate into ions that are free to move about in solution and thus can conduct an electric current. In contrast, covalent compounds tend to yield neutral molecules when they dissolve in water and consequentially are poor conductors of an electric current. Ionic substances like sodium chloride, NaCl(*s*), or calcium chloride, $CaCl_2$(*s*), whose aqueous solutions conduct an electric current, are called **electrolytes** (Figure 12.6). Covalent substances like sucrose (table sugar), $C_{12}H_{22}O_{11}$(*s*), whose aqueous solutions do not conduct an electric current, are called **nonelectrolytes** (Figure 12.7).

Not all solutions of electrolytes conduct an electric current to the same

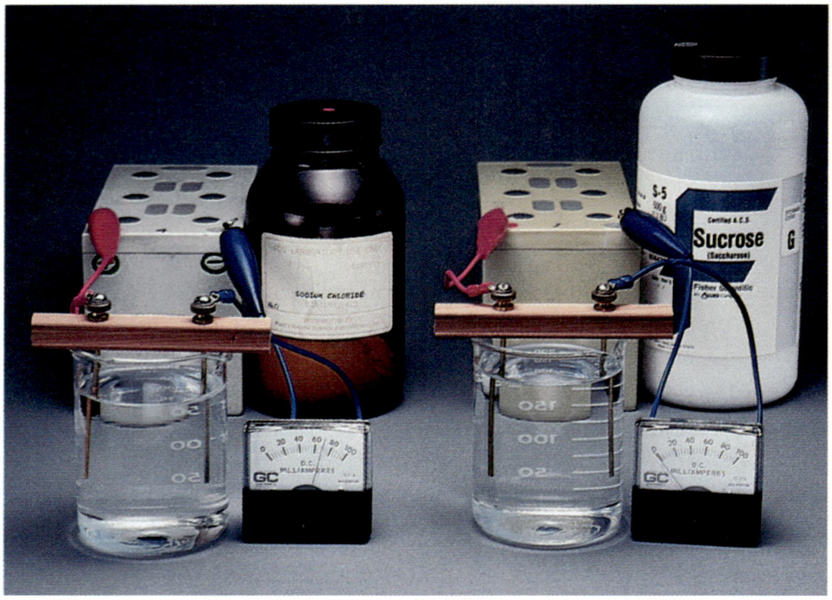

Figure 12.7 Comparison of the currents (as measured using ammeters) through equal concentrations of aqueous sodium chloride and sucrose. Note that the solution of a strong electrolyte (NaCl) is a much better conductor of electricity than the solution of a nonelectrolyte (sucrose).

extent. For example, although a 0.10 M $HgCl_2(aq)$ solution does conduct an electric current, it is a much poorer conductor of electricity than is a 0.10 M $CaCl_2(aq)$ solution. For this reason, we call calcium chloride a **strong electrolyte** and mercury(II) chloride a **weak electrolyte**. When a strong electrolyte such as calcium chloride dissolves in water, essentially all the calcium chloride formula units **dissociate** (break up) in solution into free ions, which are available to conduct an electric current. However, when a weak electrolyte such as mercury(II) chloride dissolves in water, only a small fraction of the mercury(II) chloride formula units dissociate into ions; most exist as molecular mercury(II) chloride units. Because a $HgCl_2(aq)$ solution contains far fewer ions to conduct a current than does a $CaCl_2(aq)$ solution of the same concentration, a $HgCl_2(aq)$ solution is a poorer conductor than is a $CaCl_2(aq)$ solution at the same concentration.

The following simple rules can be used to predict whether a substance is a strong electrolyte, a weak electrolyte, or a nonelectrolyte.

1. The acids $HCl(aq)$, $HBr(aq)$, $HI(aq)$, $HNO_3(aq)$, $H_2SO_4(aq)$, and $HClO_4(aq)$ are strong electrolytes. Most other acids are weak electrolytes. In other words, if an acid is not on this short list of strong electrolytes, then it is a weak electrolyte.
2. The soluble hydroxides of the Group 1 and 2 metals are strong electrolytes. Most other bases, and particularly ammonia, are weak electrolytes.
3. Most soluble salts (Table 10.9) are strong electrolytes in aqueous solution.
4. The halides and cyanides of the "heavy metals" (i.e., those with high atomic numbers), for example, mercury and lead, are often weak electrolytes.
5. Most organic compounds, that is, compounds that consist of carbon, hydrogen, and possibly other atoms, are nonelectrolytes. Notable exceptions are organic acids and bases, which are usually weak electrolytes.

EXAMPLE 12-4: Classify each of the following compounds as either a strong electrolyte, a weak electrolyte, or a nonelectrolyte in aqueous solution: (a) $NaNO_3(aq)$; (b) $C_2H_5OH(aq)$ (ethanol); (c) $Ba(OH)_2(aq)$; (d) $AuCl_3(aq)$.

Solution: (a) Sodium nitrate is a water-soluble salt (see solubility rules, Table 10.9)and a strong electrolyte. (b) Ethanol is an organic compound and a nonelectrolyte. (c) Barium hydroxide is a water-soluble Group 2 hydroxide and a strong electrolyte. (d) Gold(III) chloride is a heavy-metal halide; thus, we predict that it is a weak electrolyte in aqueous solution. This prediction is correct.

PRACTICE PROBLEM 12-4: Classify each of the following compounds either as a strong electrolyte, a weak electrolyte, or a nonelectrolyte in aqueous solution: (a) potassium chlorate; (b) acetone, $(CH_3)_2CO(aq)$; (c) sulfurous acid; (d) mercury(II) cyanide.

Answer: (a) $KClO_3(aq)$ (strong electrolyte); (b) $(CH_3)_2CO(aq)$ (nonelectrolyte); (c) $H_2SO_3(aq)$ (weak electrolyte); (d) $Hg(CN)_2(aq)$ (weak electrolyte)

Recall that a weak electrolyte differs from a strong electrolyte in the extent to which the formula units of the compound dissociate into ions when it dissolves. For example, when sufficient $CaCl_2(s)$ is dissolved in water to form, say, a 0.10-M solution, essentially all the calcium chloride in the solution exists as the ions $Ca^{2+}(aq)$ and $Cl^-(aq)$ (Figure 12.8). When $HgCl_2(s)$ dissolves in water, most of it exists as undissociated $HgCl_2(aq)$ units, with only traces of the ions $HgCl^+(aq)$, $Hg^{2+}(aq)$, and $Cl^-(aq)$. For this situation, we can write the chemical equations

$$CaCl_2(s) \xrightarrow[H_2O(l)]{100\%} Ca^{2+}(aq) + 2\,Cl^-(aq)$$

$$HgCl_2(s) \xrightarrow[H_2O(l)]{} \begin{cases} \xrightarrow{99.8\%} HgCl_2(aq) \\ \xrightarrow{0.18\%} HgCl^+(aq) + Cl^-(aq) \\ \xrightarrow{0.02\%} Hg^{2+}(aq) + 2\,Cl^-(aq) \end{cases}$$

where the percentages refer to the extent of dissociation in the 0.10-M solution.

The extent to which a compound dissociates into ions in solution is called the **degree of dissociation**. The degree of dissociation of a dissolved compound is obtained by determining the **electrical conductance** of the solution. At a particular concentration of salt, the greater the degree of dissociation, the more ions there will be in the solution, so the electrical conductance will be greater due to the greater number of ions. We use the **molar conductance** to com-

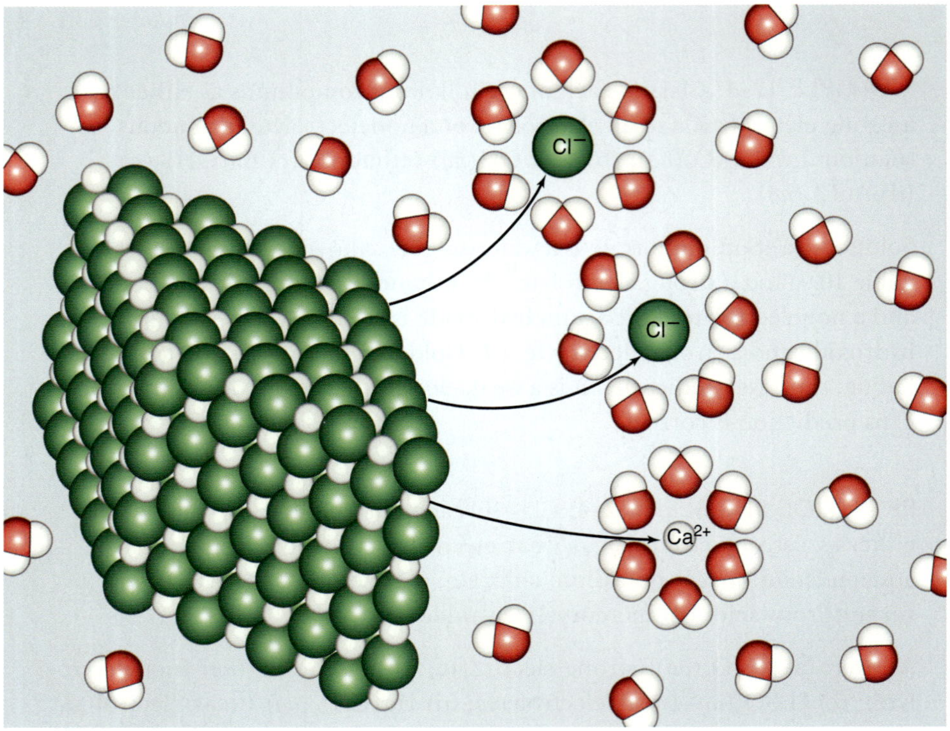

Figure 12.8 When calcium chloride, $CaCl_2(s)$, a strong electrolyte, is dissolved in water, each formula unit dissociates into one $Ca^{2+}(aq)$ ion and two $Cl^-(aq)$ ions, according to its chemical formula.

pare the conductivities of salts on a per mole basis—the molar conductance is the electrical conductivity of the solution per mole of the dissolved compound. Weak electrolytes have much lower molar conductances than strong electrolytes (Table 12.2).

For strong electrolytes, the concentration of the ions in solution depends upon the chemical formula of the undissociated compound. For example, a solution that is 0.100 M in $CaCl_2(aq)$ is 0.100 M in $Ca^{2+}(aq)$ ions and 0.200 M in $Cl^-(aq)$ ions because

$$1\text{ mol } CaCl_2 \Leftrightarrow 1\text{ mol } Ca^{2+}(aq) \Leftrightarrow 2\text{ mol } Cl^-(aq)$$

Each mole of calcium chloride salt dissolved in solution dissociates into one mole of calcium ions and two moles of chloride ions, according to its chemical formula.

When expressing the concentrations of specific ions in units of molarity, we enclose the formula for the ion in square brackets. Thus, we write the concentrations of calcium and chloride ions as $[Ca^{2+}] = 0.100$ M and $[Cl^-] = 0.200$ M.

EXAMPLE 12-5: Aluminum nitrate is a strong electrolyte. (a) Calculate the concentration of aluminum ions and nitrate ions in a 0.300 M $Al(NO_3)_3(aq)$ solution. (b) What is the total number of moles of ions present in 125 mL of a 0.300 M $Al(NO_3)_3(aq)$ solution?

Solution: (a) The chemical formula of aluminum nitrate is $Al(NO_3)_3$. Thus, for each mole of aluminum nitrate in solution, we have one mole of aluminum ions and three moles of nitrate ions. This may be written as a dissociation equation in water,

$$Al(NO_3)_3(s) \xrightarrow[H_2O(l)]{100\%} Al^{3+}(aq) + 3\,NO_3^-(aq)$$

Thus, we find

$$[Al^{3+}] = \left(\frac{0.300 \text{ mol } Al(NO_3)_3}{1 \text{ L}}\right)\left(\frac{1 \text{ mol } Al^{3+}}{1 \text{ mol } Al(NO_3)_3}\right) = 0.300 \text{ M}$$

$$[NO_3^-] = \left(\frac{0.300 \text{ mol } Al(NO_3)_3}{1 \text{ L}}\right)\left(\frac{3 \text{ mol } NO_3^-}{1 \text{ mol } Al(NO_3)_3}\right) = 0.900 \text{ M}$$

(b) The number of moles of each ion is given by Equation 12.3 as

$$\text{moles of } Al^{3+}(aq) = MV = (0.300 \text{ mol·L}^{-1})\ (0.125 \text{ L}) = 0.0375 \text{ mol}$$

$$\text{moles of } NO_3^-(aq) = MV = (0.900 \text{ mol·L}^{-1})\ (0.125 \text{ L}) = 0.113 \text{ mol}$$

Thus, the total number of moles of ions in the solution is 0.0375 moles + 0.113 moles = 0.151 moles.

PRACTICE PROBLEM 12-5: Copper(II) chloride is a strong electrolyte. What is the total number of moles of ions present in 35.0 mL of a 0.250-M copper(II) chloride aqueous solution?

Answer: 0.0263 moles

TABLE 12.2 Molar conductances of strong and weak electrolytes (25°C, 0.10-M aqueous solutions)

Compound	Molar conductance/ $ohm^{-1}·cm^2·mol^{-1}$
Strong electrolytes	
$HCl(aq)$	391
$KCl(aq)$	129
$NaOH(aq)$	221
$AgNO_3(aq)$	109
$BaCl_2(aq)$	210
$NaCH_3COO(aq)$ (sodium acetate)	73
Weak electrolytes	
$CH_3COOH(aq)$ (acetic acid)	5.2
$NH_3(aq)$	3.5
$HgCl_2(aq)$	2

12-4. Molarity Is Used in Stoichiometric Calculations for Reactions That Take Place in Solution

The concept of molarity allows us to extend the types of calculations we discussed in Chapter 11 to reactions that take place in solution. For example, a standard laboratory preparation of small quantities of bromine involves the reaction described by the chemical equation

$$MnO_2(s) + 4\,HBr(aq) \rightarrow MnBr_2(aq) + Br_2(l) + 2\,H_2O(l)$$

What volume of 8.84 M HBr(*aq*) solution would be required to react completely with 3.62 grams of $MnO_2(s)$?

From the chemical equation, we see that one mole of $MnO_2(s)$ requires four moles of HBr(*aq*) to react completely, or that

$$1 \text{ mol } MnO_2 \approx 4 \text{ mol HBr}$$

The number of moles of $MnO_2(s)$ is given by

$$\text{moles of } MnO_2 = (3.62 \text{ g } MnO_2)\left(\frac{1 \text{ mol } MnO_2}{86.94 \text{ g } MnO_2}\right) = 0.0416 \text{ mol}$$

and the corresponding number of moles of HBr(*aq*) required is

$$\text{moles of HBr} = (0.0416 \text{ mol } MnO_2)\left(\frac{4 \text{ mol HBr}}{1 \text{ mol } MnO_2}\right) = 0.166 \text{ mol}$$

When working with molarity in chemical calculations it is often helpful to rewrite the unit M as a unit conversion factor between liters and moles. For example, the concentration of the 8.84 M HBr(*aq*) solution may be expressed as either

$$\frac{8.84 \text{ mol HBr}}{1 \text{ L}} = 1 \quad \text{or} \quad \frac{1 \text{ L}}{8.84 \text{ mol HBr}} = 1$$

Making use of the second expression above, the volume of the solution is given by

$$\text{volume of solution} = (0.166 \text{ mol HBr})\left(\frac{1 \text{ L}}{8.84 \text{ mol HBr}}\right)\left(\frac{1000 \text{ mL}}{1 \text{ L}}\right) = 18.8 \text{ mL}$$

We could have also determined the volume required by solving Equation 12.3 for V to obtain

$$V = \frac{n}{M} = \frac{0.166 \text{ mol}}{8.84 \text{ mol·L}^{-1}} = 0.0188 \text{ L} = 18.8 \text{ mL}$$

The following Example illustrates a calculation involving a reaction between a solution and a solid.

EXAMPLE 12-6: Zinc reacts with hydrochloric acid, $HCl(aq)$ (Figure 12.9), according to the equation

$$Zn(s) + 2\,HCl(aq) \rightarrow ZnCl_2(aq) + H_2(g)$$

Calculate how many grams of zinc react with 50.0 mL of 6.00 M $HCl(aq)$.

Solution: The equation for the reaction indicates that one mole of $Zn(s)$ reacts with two moles of $HCl(aq)$. First we determine the number of moles of $HCl(aq)$ in 50.0 mL of a 6.00 M $HCl(aq)$ solution:

$$\text{moles of HCl} = (50.0\text{ mL})\left(\frac{1\text{ L}}{1000\text{ mL}}\right)\left(\frac{6.00\text{ mol}}{1\text{ L}}\right) = 0.300\text{ mol}$$

The number of grams of zinc that react is given by

$$\text{mass of Zn} = (0.300\text{ mol HCl})\left(\frac{1\text{ mol Zn}}{2\text{ mol HCl}}\right)\left(\frac{65.41\text{ g Zn}}{1\text{ mol Zn}}\right) = 9.81\text{ g Zn}$$

In essence the molarity of the solution and the stoichiometric coefficients from the balanced equation are conversion factors that allow us to convert the volume of $HCl(aq)$ given to the mass of zinc reacted. This practice follows the same format outlined in Figure 11.7.

Figure 12.9 Zinc metal reacts with an aqueous solution of hydrochloric acid. The bubbles are hydrogen gas escaping from the solution.

PRACTICE PROBLEM 12-6: Aluminum reacts with a moderately concentrated sodium hydroxide solution (Figure 12.10) according to:

$$2\,Al(s) + 2\,NaOH(aq) + 6\,H_2O(l) \rightarrow 2\,NaAl(OH)_4(aq) + 3\,H_2(g)$$

How many grams of aluminum will react with 30.0 mL of 6.00 M $NaOH(aq)$ solution?

Answer: 4.86 grams

Figure 12.10 Drāno® consists of a mixture of pieces of aluminum and $NaOH(s)$. When Drāno® is added to water, the aluminum reacts with the $NaOH(aq)$ to produce hydrogen gas.

12-5. Molarity Can Be Used to Calculate Quantities in Precipitation Reactions

The calculations that we have done so far involve the reaction of a solid with a solution. We can do similar calculations for a reaction between two solutions. We saw in Section 10-9 that precipitate formation is often the driving force in double-replacement reactions. For example, when solutions of $Hg_2(NO_3)_2(aq)$ and $KI(aq)$ are mixed, a yellow precipitate of $Hg_2I_2(s)$ is formed (Figure 12.11). The equation that describes this reaction is

$$Hg_2(NO_3)_2(aq) + 2\,KI(aq) \rightarrow 2\,KNO_3(aq) + Hg_2I_2(s)$$

and the net ionic equation is

$$Hg_2^{2+}(aq) + 2\,I^-(aq) \rightarrow Hg_2I_2(s)$$

Remember that mercury(I) exists in aqueous solution as Hg_2^{2+} (*aq*) ions.

Figure 12.11 A mercury(I) iodide, $Hg_2I_2(s)$, precipitate is produced by the addition of $Hg_2(NO_3)_2(aq)$ to $KI(aq)$.

(Recall that we can predict the formation of the insoluble mercury(I) iodide precipitate, $Hg_2I_2(s)$, by applying the solubility rules that we learned in Chapter 10.)

Let's calculate the volume of 0.400 M $KI(aq)$ required to react completely with 35.0 mL of 0.250 M $Hg_2(NO_3)_2(aq)$. The chemical equation indicates that two moles of $KI(aq)$ react completely with one mole of $Hg_2(NO_3)_2(aq)$, or

$$1 \text{ mol } Hg_2(NO_3)_2 \approx 2 \text{ mol KI}$$

The number of moles of $Hg_2(NO_3)_2(aq)$ in 35.0 mL of 0.250 M $Hg_2(NO_3)_2(aq)$ solution is given by

$$\text{moles of } Hg_2(NO_3)_2 = MV = (35.0 \text{ mL})\left(\frac{1 \text{ L}}{1000 \text{ mL}}\right)\left(\frac{0.250 \text{ mol}}{1 \text{ L}}\right) = 8.75 \times 10^{-3} \text{ mol}$$

The number of moles of $KI(aq)$ required is given by

$$\text{moles of KI} = (8.75 \times 10^{-3} \text{ mol } Hg_2(NO_3)_2)\left(\frac{2 \text{ mol KI}}{1 \text{ mol } Hg_2(NO_3)_2}\right)$$

$$= 1.75 \times 10^{-2} \text{ mol}$$

The corresponding volume of 0.400 M $KI(aq)$ solution is given by

$$\text{volume} = \frac{n}{M} = (1.75 \times 10^{-2} \text{ mol KI})\left(\frac{1 \text{ L}}{0.400 \text{ mol KI}}\right) = 0.0438 \text{ L} = 43.8 \text{ mL}$$

EXAMPLE 12-7: Nickel(II) sulfide, $NiS(s)$, which is insoluble in water, can be formed according to the equation

$$NiCl_2(aq) + K_2S(aq) \rightarrow 2\,KCl(aq) + NiS(s)$$

The net ionic equation is

$$Ni^{2+}(aq) + S^{2-}(aq) \rightarrow NiS(s)$$

Calculate the volume of 0.655 M $K_2S(aq)$ that is required to precipitate all the nickel from 42.5 mL of 0.165 M $NiCl_2(aq)$.

Solution: The number of moles of $NiCl_2(aq)$ in the solution is given by

$$\text{moles of } NiCl_2 = MV = (42.5 \text{ mL } NiCl_2)\left(\frac{1 \text{ L}}{1000 \text{ mL}}\right)\left(\frac{0.165 \text{ mol } NiCl_2}{1 \text{ L}}\right)$$

$$= 7.01 \times 10^{-3} \text{ mol}$$

The volume of $K_2S(aq)$ required is given by

$$\text{volume of } K_2S = (7.01 \times 10^{-3} \text{ mol } NiCl_2)\left(\frac{1 \text{ mol } K_2S}{1 \text{ mol } NiCl_2}\right)\left(\frac{1 \text{ L}}{0.655 \text{ mol } K_2S}\right)$$

$$= 0.0107 \text{ L} = 10.7 \text{ mL}$$

PRACTICE PROBLEM 12-7: Barium sulfate, $BaSO_4(s)$, is insoluble in water and can be formed according to the net ionic equation

$$Ba^{2+}(aq) + SO_4^{2-}(aq) \rightarrow BaSO_4(s)$$

Calculate the volume of 0.500 M $K_2SO_4(aq)$ solution that is required to precipitate all the barium from 25.0 mL of 0.350 M $Ba(NO_3)_2(aq)$.

Answer: 17.5 mL

When mixing two solutions, we must always check to see whether one of the reactants is a limiting reactant. For example, suppose that 50.0 mL of 0.150 M $AgNO_3(aq)$ and 50.0 mL of 0.200 M $Na_2CrO_4(aq)$ are mixed (Figure 12.12); how many grams of silver chromate, $Ag_2CrO_4(s)$, will be produced? The equation for the reaction is

Figure 12.12 A silver chromate, $AgCrO_4(s)$, precipitate is produced by the addition of $AgNO_3(aq)$ to $Na_2CrO_4(aq)$.

$$2\,AgNO_3(aq) + Na_2CrO_4(aq) \rightarrow 2\,NaNO_3(aq) + Ag_2CrO_4(s)$$

We first must determine which, if either, of these reactants is a limiting reactant. The number of moles of each reactant is given by

$$\text{moles of } AgNO_3 = MV = (50.0\text{ mL})\left(\frac{1\text{ L}}{1000\text{ mL}}\right)\left(\frac{0.150\text{ mol } AgNO_3}{1\text{ L}}\right)$$
$$= 7.50 \times 10^{-3}\text{ mol}$$

and

$$\text{moles of } Na_2CrO_4 = MV = (50.0\text{ mL})\left(\frac{1\text{ L}}{1000\text{ mL}}\right)\left(\frac{0.200\text{ mol } Na_2CrO_4}{1\text{ L}}\right)$$
$$= 10.0 \times 10^{-3}\text{ mol}$$

According to the chemical equation, it requires two moles of $AgNO_3(aq)$ for every mole of $Na_2CrO_4(aq)$. Because there is insufficient $AgNO_3(aq)$ to react with the 10.0×10^{-3} moles of $Na_2CrO_4(aq)$, we see that $AgNO_3(aq)$ is the limiting reactant and that $Na_2CrO_4(aq)$ is in excess. The mass of $Ag_2CrO_4(s)$ that will be precipitated is given by

$$\text{mass of } Ag_2CrO_4 = (7.50 \times 10^{-3}\text{ mol } AgNO_3)\left(\frac{1\text{ mol } Ag_2CrO_4}{2\text{ mol } AgNO_3}\right)\left(\frac{331.7\text{ g } Ag_2CrO_4}{1\text{ mol } Ag_2CrO_4}\right)$$
$$= 1.24\text{ g } Ag_2CrO_4$$

EXAMPLE 12-8: Calculate how many grams of lead(II) oxalate, $PbC_2O_4(s)$, can be precipitated by mixing 25.0 mL of 2.00 M $Pb(NO_3)_2(aq)$ with 40.0 mL of 1.50 M $K_2C_2O_4(aq)$. The equation for the reaction is

$$Pb(NO_3)_2(aq) + K_2C_2O_4(aq) \rightarrow 2\,KNO_3(aq) + PbC_2O_4(s)$$

and the net ionic equation is

$$Pb^{2+}(aq) + C_2O_4^{2-}(aq) \rightarrow PbC_2O_4(s)$$

Solution: Because the quantities of both reactants are given, we must determine which, if either, of these reactants is a limiting reactant. The number of moles of each reactant is given by

$$\text{moles of } Pb(NO_3)_2 = MV = (25.0 \text{ mL})\left(\frac{1 \text{ L}}{1000 \text{ mL}}\right)\left(\frac{2.00 \text{ mol } Pb(NO_3)_2}{1 \text{ L}}\right)$$

$$= 5.00 \times 10^{-2} \text{ mol}$$

$$\text{moles of } K_2C_2O_4 = MV = (40.0 \text{ mL})\left(\frac{1 \text{ L}}{1000 \text{ mL}}\right)\left(\frac{1.50 \text{ mol } K_2C_2O_4}{1 \text{ L}}\right)$$

$$= 6.00 \times 10^{-2} \text{ mol}$$

According to the chemical equation, the reactants react on a one-to-one mole basis, so we see that $K_2C_2O_4(aq)$ is in excess and that $Pb(NO_3)_2(aq)$ is a limiting reactant. The mass of $PbC_2O_4(s)$ that will be precipitated is given by

$$\text{mass of } PbC_2O_4 = (5.00 \times 10^{-2} \text{ mol } Pb(NO_3)_2)\left(\frac{1 \text{ mol } PbC_2O_4}{1 \text{ mol } Pb(NO_3)_2}\right)\left(\frac{295.2 \text{ g } PbC_2O_4}{1 \text{ mol } PbC_2O_4}\right)$$

$$= 14.8 \text{ g } PbC_2O_4$$

PRACTICE PROBLEM 12-8: When a solution of cadmium(II) nitrate, $Cd(NO_3)_2(aq)$, and a solution of sodium sulfide, $Na_2S(aq)$, are mixed, a yellow-orange precipitate forms (Figure 12.13). (a) Write a balanced chemical equation describing this reaction. (b) Use the solubility rules that we learned in Chapter 10 to determine the identity of the yellow-orange precipitate. (c) Calculate how many grams of precipitate form when 25.0 mL of 0.100 M $Cd(NO_3)_2(aq)$ is mixed with 20.0 mL of 0.150 M $Na_2S(aq)$.

Answer: (a) $Cd(NO_3)_2(aq) + Na_2S(aq) \rightarrow 2NaNO_3(aq) + CdS(s)$; (b) $CdS(s)$ is the precipitate; (c) 0.361 grams

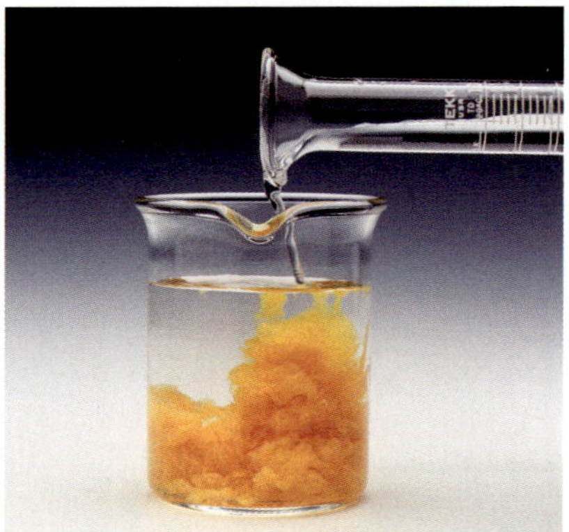

Figure 12.13 When colorless solutions of cadmium(II) nitrate, $Cd(NO_3)_2(aq)$, and sodium sulfide, $Na_2S(aq)$, are mixed, a yellow-orange precipitate forms.

12-6. The Concentration of an Acid or a Base Can Be Determined by Titration

We showed in Chapter 10 that a double-replacement reaction may be driven by the formation of a covalent compound from ionic reactants. The most important example is a **neutralization reaction** between an acid and a base.

Suppose we have a basic solution whose concentration is not known. We measure out a certain volume and then slowly add an acidic solution of known concentration until the base is completely neutralized. (We shall learn in Chapter 21 that the point at which a neutralization reaction is just completed can be signaled by substances called indicators that change colors when acids or bases are neutralized.) Such a process is called a **titration** and can be carried out with the apparatus shown in Figure 12.14. Knowledge of the volume and concentration of the acidic solution that is required to neutralize the base and the reaction stoichiometry are sufficient to determine the concentration of the basic solution. As an example, suppose we find that it requires 27.25 mL of 0.150 M HCl(*aq*) solution to neutralize 30.00 mL of a NaOH(*aq*) solution. The chemical equation that describes the reaction between HCl(*aq*) and NaOH(*aq*) is

$$\mathrm{HCl}(aq) + \mathrm{NaOH}(aq) \rightarrow \mathrm{NaCl}(aq) + \mathrm{H_2O}(l)$$

and the net ionic equation is

$$\mathrm{H^+}(aq) + \mathrm{OH^-}(aq) \rightarrow \mathrm{H_2O}(l)$$

This equation indicates that one mole of HCl(*aq*) is required to neutralize one mole of NaOH(*aq*). Therefore, the number of moles of NaOH(*aq*) that are neutralized by the HCl(*aq*) in the titration is given by

$$\text{moles of NaOH} = (27.25\text{ mL HCl})\left(\frac{1\text{ L}}{1000\text{ mL}}\right)\left(\frac{0.150\text{ mol HCl}}{1\text{ L}}\right)\left(\frac{1\text{ mol NaOH}}{1\text{ mol HCl}}\right)$$

$$= 4.09 \times 10^{-3}\text{ mol}$$

There are, therefore, 4.09×10^{-3} moles of NaOH(*aq*) in the 30.00 mL of the NaOH(*aq*) solution, so the concentration of NaOH(*aq*) in the solution is given by

$$M = \frac{n}{V} = \frac{4.09 \times 10^{-3}\text{ mol}}{30.00 \times 10^{-3}\text{ L}} = 0.136\text{ M}$$

You may have noticed in many of the calculations that we have done in this chapter that a factor of one liter per 1000 mL, or 10^{-3} L·mL^{-1}, occurs throughout the calculations. This factor arises because the volumes in chemical experiments are usually expressed in milliliters; so when we use the equation $n = MV$ to calculate the number of moles, the stated volume in milliliters must be multiplied by 10^{-3} L·mL^{-1} to convert milliliters to liters. We can avoid these factors of 10^{-3} L·mL^{-1} by using units of **millimoles** (mmol) instead of moles and thus expressing our concentrations as millimoles per milliliter rather than as moles per liter. The key relation is

$$\left(\frac{\text{mmol}}{\text{mL}}\right)\left(\frac{10^{-3}\text{ mol·mmol}^{-1}}{10^{-3}\text{ L·mL}^{-1}}\right) = \left(\frac{\text{mol}}{\text{L}}\right) = \text{M} \qquad (12.5)$$

Thus, we see that units of mmol·mL^{-1} are equivalent to mol·L^{-1} or M.

Let's redo the earlier titration calculation of the neutralization of 30.00 mL of a NaOH(*aq*) solution of unknown concentration by 27.25 mL of 0.150 M HCl(*aq*). Expressing the molar concentrations as mmol·mL^{-1} rather than as mol·L^{-1}, we write

$$\text{millimoles of NaOH} = (27.25\text{ mL HCl})\left(\frac{0.150\text{ mmol HCl}}{1\text{ mL}}\right)\left(\frac{1\text{ mmol NaOH}}{1\text{ mmol HCl}}\right)$$

$$= 4.09\text{ mmol}$$

Figure 12.14 (*top*) Setup of a titration experiment. The tall piece of equipment is a **buret**, a precision-made piece of glassware used to measure and deliver precise volumes of a solution. (*bottom*) When reading a buret, the volume is read from the bottom of the **meniscus** (the curve typically formed by the liquid due to adherence of the solution to the glass) viewed at eye level. The 50-mL burets found in most general chemistry laboratories have a precision of about ±0.02 mL. By taking the difference between the final and initial volumes, the amount of liquid dispensed from a buret may be measured with a high degree of precision.

$$M = \frac{n}{V} = \frac{4.09 \text{ mmol NaOH}}{30.00 \text{ mL}} = 0.136 \text{ M NaOH}$$

Notice that the use of millimoles instead of moles has eliminated the factor of 10^{-3} L·mL^{-1}, thus simplifying the calculation. The following Example illustrates the calculation of the concentration of a solution from titration data using millimoles rather than moles.

EXAMPLE 12-9: In a titration experiment, it is found that 37.60 mL of 0.210 M NaOH(aq) is required to neutralize 25.05 mL of $H_2SO_4(aq)$. Calculate the concentration of the $H_2SO_4(aq)$ solution.

Solution: The equation for the reaction

$$H_2SO_4(aq) + 2\,NaOH(aq) \rightarrow Na_2SO_4(aq) + 2\,H_2O(l)$$

indicates that it requires two moles of NaOH(aq) to neutralize one mole of $H_2SO_4(aq)$. This is equivalent to saying that it requires two millimoles of NaOH(aq) to neutralize one millimole of $H_2SO_4(aq)$. Consequently, we have

$$\text{millimoles of } H_2SO_4 = (37.60 \text{ mL NaOH})\left(\frac{0.210 \text{ mmol NaOH}}{1 \text{ mL}}\right)\left(\frac{1 \text{ mmol } H_2SO_4}{2 \text{ mmol NaOH}}\right)$$

$$= 3.95 \text{ mmol}$$

The concentration of the $H_2SO_4(aq)$ solution is

$$M = \frac{n}{V} = \frac{3.95 \text{ mmol}}{25.05 \text{ mL}} = 0.158 \text{ M}$$

PRACTICE PROBLEM 12-9: In a titration experiment, it was found that 40.05 mL of 0.1065 M KOH(aq) are required to titrate 32.10 mL of an aqueous solution of oxalic acid, $H_2C_2O_4(aq)$. Calculate the concentration of the oxalic acid solution. The equation for the reaction is

$$H_2C_2O_4(aq) + 2\,KOH(aq) + K_2C_2O_4(aq) + 2\,H_2O(l)$$

Answer: 0.06644 M

Oxalic acid is a dicarboxylic acid with a Lewis formula of

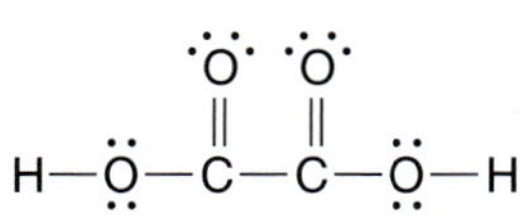

The chemistry of carboxylic acids is discussed in Interchapter R, at www.McQuarrieGeneralChemistry.com

12-7. The Formula Mass of an Unknown Acid Can Be Determined from Titration Data

The following Example illustrates how we can determine the formula mass of an acid from titration data.

EXAMPLE 12-10: A 2.50-gram sample of an unknown acid is dissolved in water to make 100.0 mL of solution and is neutralized with 0.400 M NaOH(*aq*). The volume of NaOH(*aq*) solution required to neutralize the acid is 84.25 mL. Assuming that the acid has only one acidic proton per formula unit, calculate its formula mass.

Solution: Because the acid has one acidic proton, we can write the equation for the neutralization reaction as

$$\text{NaOH}(aq) + \text{HA}(aq) \rightarrow \text{NaA}(aq) + \text{H}_2\text{O}(l)$$

where A represents the formula of the anion in the unknown acid. The number of millimoles of NaOH(*aq*) required to neutralize the acid is given by

$$\text{millimoles of NaOH} = MV = (84.25\text{ mL NaOH})\left(\frac{0.400\text{ mmol NaOH}}{1\text{ mL NaOH}}\right)$$

$$= 33.7\text{ mmol}$$

From the chemical equation, we see that

$$1\text{ mmol NaOH} \approx 1\text{ mmol acid}$$

so the amount of acid in the solution is 33.7 mmol. Because this is equivalent to the 2.50 grams of unknown acid we started with, we write

$$2.50\text{ g acid} \approx 33.7\text{ mmol acid} = 3.37 \times 10^{-2}\text{ mol acid}$$

Dividing by 3.37×10^{-2} gives

$$74.2\text{ g acid} \approx 1.00\text{ mol acid}$$

Thus, the formula mass of the unknown acid is 74.2.

PRACTICE PROBLEM 12-10: A 2.50-gram sample of malonic acid is dissolved to make 100 mL of solution. The solution is neutralized with 28.5 mL of 1.684 M KOH(*aq*). Given that the formula mass of malonic acid is 104.1, determine how many acidic protons it has per formula unit.

Answer: two

The following Example and Practice Problem incorporate many of the concepts that we have learned so far.

EXAMPLE 12-11: Calculate the concentrations of all the ions in solution following the reaction of 25.0 mL of 0.200 M $CaCl_2$(*aq*) with 25.0 mL of 0.300 M $AgNO_3$(*aq*).

Solution: Applying the solubility rules (Table 10.9), we find that the reaction produces an insoluble silver chloride, AgCl(*s*), precipitate. The balanced chemical equation for this reaction is

$$\text{CaCl}_2(aq) + 2\,\text{AgNO}_3(aq) \rightarrow \text{Ca(NO}_3)_2(aq) + 2\,\text{AgCl}(s)$$

We must now consider if either of the reactants is a limiting reactant. The number of millimoles of each reactant is given by

$$\text{mmol of CaCl}_2 = (25.0\text{ mL CaCl}_2)\left(\frac{0.200\text{ mmol CaCl}_2}{1\text{ mL}}\right) = 5.00\text{ mmol}$$

and

$$\text{mmol of AgNO}_3 = (25.0\text{ mL AgNO}_3)\left(\frac{0.300\text{ mmol AgNO}_3}{1\text{ mL}}\right) = 7.50\text{ mmol}$$

Interpreting the coefficients of the balanced chemical equation as millimoles, we see that it requires two millimoles of $AgNO_3(aq)$ for one millimole of $CaCl_2(aq)$. Therefore, it takes 10.0 mmol of $AgNO_3(aq)$ to react completely with the 5.00 mmol of $CaCl_2(aq)$. Because there is insufficient $AgNO_3(aq)$ to react with the 5.00 millimoles of $CaCl_2(aq)$, we see that $AgNO_3(aq)$ is the limiting reactant and $CaCl_2(aq)$ is in excess. Therefore, the number of millimoles of $AgCl(s)$ formed is given by

$$\text{mmol of AgCl}(s) = (7.50\text{ mmol AgNO}_3)\left(\frac{2\text{ mmol AgCl}}{2\text{ moles AgNO}_3}\right) = 7.50\text{ mmol}$$

At the end of the reaction, both calcium ions and chloride ions will be present in solution (because they are in excess). Nitrate ions will also be present because, according to the solubility rules, calcium nitrate is soluble. However, there will be essentially no silver ions remaining in solution because silver nitrate is the limiting reactant, and essentially all the silver ions will have precipitated out as $AgCl(s)$.

The number of millimoles of $Ca^{2+}(aq)$ and $NO_3^-(aq)$ will be the same as their initial values because neither appears in the net ionic equation, and so both remain in solution as spectator ions. However, the *concentration* of each of these ions has changed because the total volume of solution has increased. Thus, we start by finding the initial number of millimoles of each of these species.

$$\text{mmol of Ca}^{2+} = (25.0\text{ mL CaCl}_2)\left(\frac{0.200\text{ mmol CaCl}_2}{1\text{ mL}}\right)\left(\frac{1\text{ mmol Ca}^{2+}}{1\text{ mmol CaCl}_2}\right) = 5.00\text{ mmol}$$

$$\text{mmol of NO}_3^- = (25.0\text{ mL AgNO}_3)\left(\frac{0.300\text{ mmol AgNO}_3}{1\text{ mL}}\right)\left(\frac{1\text{ mmol NO}_3^-}{1\text{ mmol AgNO}_3}\right) = 7.50\text{ mmol}$$

Because we mixed two solutions with a volume of 25.0 mL each, the total volume of the resulting solution is 50.0 mL, and the concentrations of $Ca^{2+}(aq)$ and $NO_3^-(aq)$ are given by

$$[Ca^{2+}] = \frac{5.00 \text{ mmol } Ca^{2+}}{50.0 \text{ mL}} = 0.100 \text{ M}$$

$$[NO_3^-] = \frac{7.50 \text{ mmol } NO_3^-}{50.0 \text{ mL}} = 0.150 \text{ M}$$

To calculate the concentration of chloride ions remaining, we must first determine the number of millimoles of chloride ions in excess. We have already determined that we began the reaction with 5.00 mmol of $CaCl_2(aq)$. Because there are two chloride ions per $CaCl_2$ formula unit, we have

$$\text{mmol of } Cl^- = (5.00 \text{ mmol } CaCl_2)\left(\frac{2 \text{ mmol } Cl^-}{1 \text{ mmol } CaCl_2}\right) = 10.00 \text{ mmol}$$

We have already calculated that 7.50 millimoles of $AgCl(s)$ are formed in the reaction. Because there is one chloride ion per AgCl formula unit, there are

$$\begin{pmatrix}\text{mmol } Cl^-(aq)\\ \text{in excess}\end{pmatrix} = \begin{pmatrix}10.0 \text{ mmol } Cl^-(aq)\\ \text{from } CaCl_2\end{pmatrix} - \begin{pmatrix}7.50 \text{ mmol } Cl^-(aq)\\ \text{precipitated as } AgCl(s)\end{pmatrix} = 2.5 \text{ mmol}$$

Dividing this by the new total volume, we find that the concentration of chloride ions following the reaction is

$$[Cl^-] = \frac{2.5 \text{ mmol}}{50.0 \text{ mL}} = 0.050 \text{ M}$$

PRACTICE PROBLEM 12-11: Dilute acids can react with calcium carbonate to form carbon dioxide gas (Figure 12.15). Calculate the number of millimoles of carbon dioxide gas produced and the concentration of all the ions remaining in solution following the reaction of 25 mL of 3.0 M $HCl(aq)$ with 1.5 grams of $CaCO_3(s)$. (Assume that the volume of the solution does not increase appreciably with the addition of the calcium carbonate.)

Answer: 15 mmol of $CO_2(g)$ are produced; $[Ca^{2+}] = 0.60$ M; $[Cl^-] = 3.0$ M; $[H^+] = 1.8$ M

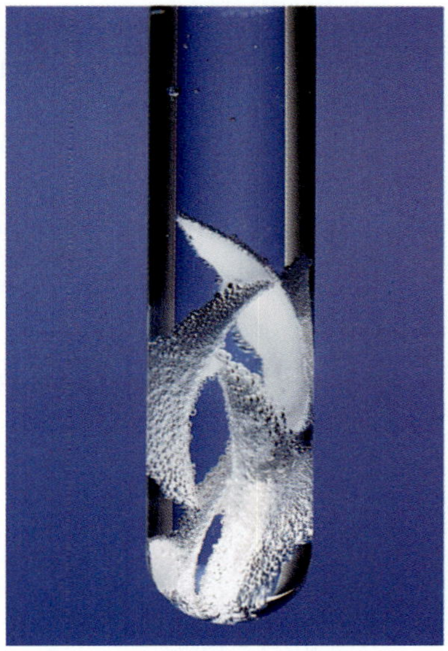

Figure 12.15 Many metal carbonates react with dilute acids to produce carbon dioxide, $CO_2(g)$, as one of the products. Here we see an eggshell [principally calcium carbonate, $CaCO_3(s)$] in dilute hydrochloric acid, $HCl(aq)$. The bubbles on the surfaces of the eggshell fragments are $CO_2(g)$, which is not very soluble in water.

SUMMARY

Chemical calculations involving solutions are based on the concept of the mole. For this reason, one of the most common and important measures of concentration of a solution is molarity. The molarity is defined as the number of moles of solute in exactly one liter of solution. Soluble salts that conduct electricity are known as electrolytes. The strength of an electrolyte depends upon the degree of dissociation into ions. For strong electrolytes, the concentrations of ions in solution depend on the chemical formula of the substance dissolved.

Many reactions take place in solution, and stoichiometric calculations involving reactants or products are carried out readily if the concentrations of the solutions are expressed in units of molarity. A common type of reaction between solutions is a double-replacement reaction in which a precipitate is formed. Another important type of reaction that occurs in solution is an acid-base neutralization reaction, best exemplified by a titration experiment. A titration can be used to determine the concentration of a solution of an acid or a base. The formula mass of an unknown acid can also be determined by titration.

TERMS YOU SHOULD KNOW

component *393*
solute *393*
solvent *393*
saturated solution *394*
solubility, grams of solute per 100 g of solvent *394*
unsaturated solution *395*
concentration *395*
molarity, moles of solute per 1 L of solution, *M* *395*
molar, M *395*
volumetric flask *395*
dilution *398*
electrolytes *399*
nonelectrolytes *399*
strong electrolyte *400*
weak electrolyte *400*
dissociate *400*
degree of dissociation *401*
electrical conductance *401*
molar conductance *401*
neutralization reaction *408*
titration *409*
buret *409*
meniscus *409*
millimoles (mmol) *409*

EQUATIONS YOU SHOULD KNOW HOW TO USE

$M = \frac{n}{V}$ (12.2) definition of molarity

$n = MV$ (12.3) number of moles of solute in a volume of solution (volume expressed in liters)

$M_1V_1 = M_2V_2$ (12.4) used for dilution calculations

PROBLEMS

PREPARATION OF SOLUTIONS

12-1. Calculate the molarity of a saturated solution of sodium hydrogen carbonate (baking soda), $NaHCO_3(aq)$ that contains 69.0 grams in 1.00 liter of solution.

12-2. Sodium hydroxide is extremely soluble in water. A saturated solution contains 572 grams of $NaOH(s)$ per liter of solution. Calculate the molarity of a saturated $NaOH(aq)$ solution.

12-3. A saturated solution of calcium hydroxide, $Ca(OH)_2(aq)$, contains 0.185 grams per 100 milliliters of solution. Calculate the molarity of a saturated calcium hydroxide solution.

12-4. A cup of coffee may contain as much as 300 milligrams of caffeine, $C_8H_{10}N_4O_2(s)$. Calculate the molarity of caffeine in one cup of coffee (4 cups = 0.946 liters).

12-5. Calculate the number of moles of solute in

(a) 25.46 milliliters of a 0.1255 M $K_2Cr_2O_7(aq)$ solution

(b) 50 μL of a 0.020 M $C_6H_{12}O_6(aq)$ solution

12-6. Calculate the number of moles of solute in

(a) 50.0 μL of a 0.200 M $NaCl(aq)$ solution

(b) 2.00 milliliters of a 2.00-mM $H_2SO_4(aq)$ solution

12-7. How many milliliters of 18.0 M $H_2SO_4(aq)$ are required to prepare 500 milliliters of 0.30 M $H_2SO_4(aq)$?

12-8. How many milliliters of 12.0 M $HCl(aq)$ are required to prepare 250 milliliters of 1.0 M $HCl(aq)$?

12-9. Explain how you would prepare 500 milliliters of a 0.250-M aqueous solution of sucrose, $C_{12}H_{22}O_{11}(aq)$. This solution is used frequently in biological experiments.

12-10. How would you prepare 50.0 milliliters of a 0.200 M $CuSO_4(aq)$ solution, starting with solid $CuSO_4{\cdot}5H_2O(s)$?

12-11. A stock solution of perchloric acid is 70.5% by mass $HClO_4$ and 29.5% by mass water. The density of the solution is 1.67 $g{\cdot}mL^{-1}$ at 20°C. Calculate the molarity of the solution.

12-12. Concentrated phosphoric acid is sold as a solution of 85% phosphoric acid and 15% water by mass. Given that its molarity is 15 M, calculate the density of concentrated phosphoric acid.

CONCENTRATION OF IONS

12-13. Classify each of the following as strong electrolytes, weak electrolytes, or nonelectrolytes in aqueous solution: HI, NH_4Br, CH_3COOH (acetic acid), KOH, CH_3OH (methanol).

12-14. Classify each of the following household chemicals as strong electrolytes, weak electrolytes, or nonelectrolytes in aqueous solution: sugar ($C_{12}H_{22}O_{11}$), table salt (NaCl), baking soda ($NaHCO_3$), ammonia (NH_3), nail polish remover (acetone, CH_3COCH_3).

12-15. Calcium hydroxide is a strong electrolyte. Determine the concentration of each of the individual ions in a 0.30 M $Ca(OH)_2(aq)$ solution.

12-16. Nickel(III) chloride is a strong electrolyte. Determine the concentration of each of the individual ions in a 0.050 M $NiCl_3(aq)$ solution.

REACTIONS IN SOLUTION

12-17. When silicon is heated with an aqueous solution of sodium hydroxide, sodium silicate and hydrogen gas are formed according to

$$Si(s) + 2NaOH(aq) + H_2O(l) \rightarrow Na_2SiO_3(aq) + 2H_2(g)$$

How many milliliters of 6.00 M $NaOH(aq)$ are required to react with 12.5 grams of silicon? How many grams of hydrogen gas will be produced?

12-18. Hydrogen peroxide can be prepared by the reaction of barium peroxide with sulfuric acid according to the equation

$$BaO_2(s) + H_2SO_4(aq) \rightarrow BaSO_4(s) + H_2O_2(aq)$$

How many milliliters of 3.75 M $H_2SO_4(aq)$ are required to react completely with 17.6 grams of $BaO_2(s)$?

12-19. When 15.0 grams of copper(II) oxide are reacted with 31.4 milliliters of 6.00 M $H_2SO_4(aq)$, what is the concentration of the resulting copper(II) sulfate solution? The equation for the reaction is

$$CuO(s) + H_2SO_4(aq) \rightarrow CuSO_4(aq) + H_2O(l)$$

12-20. Cobalt(II) chloride can be prepared by the reaction of cobalt(II) carbonate with aqueous hydrochloric acid according to the equation

$$CoCO_3(s) + 2HCl(aq) \rightarrow CoCl_2(aq) + H_2O(l) + CO_2(g)$$

The cobalt(II) chloride crystallizes from solution as a red-colored hexahydrate, $CoCl_2{\cdot}6H_2O(s)$. How many grams of $CoCl_2{\cdot}6H_2O(s)$ can be prepared by reacting 100.0 milliliters of 0.375 M $HCl(aq)$ with 2.17 grams of $CoCO_3(s)$?

12-21. Elemental iodine can be prepared by the reaction of manganese(IV) oxide with potassium iodide in the presence of sulfuric acid according to the equation

$$MnO_2(s) + 2KI(s) + 2H_2SO_4(aq) \rightarrow MnSO_4(aq) + K_2SO_4(aq) + 2H_2O(l) + I_2(s)$$

How many grams of iodine can be prepared by reacting 250.0 milliliters of 1.75 M $H_2SO_4(aq)$ with 20.0 grams of $MnO_2(s)$ and 75.0 grams of $KI(s)$?

12-22. Silane, $SiH_4(g)$, can be prepared by the reaction of magnesium silicide with aqueous hydrochloric acid according to the equation

$$Mg_2Si(s) + 4HCl(aq) \rightarrow 2MgCl_2(aq) + SiH_4(g)$$

How many grams of silane can be prepared by reacting 1.09 grams of $Mg_2Si(s)$ and 50.0 milliliters of 1.25 M $HCl(aq)$?

12-23. Zinc reacts with hydrochloric acid according to the reaction equation

$$Zn(s) + 2HCl(aq) \rightarrow ZnCl_2(aq) + H_2(g)$$

How many milliliters of 2.00 M $HCl(aq)$ are required to react with 2.55 grams of $Zn(s)$?

12-24. Bromine is obtained commercially from natural brines from wells in Michigan and Arkansas by the reaction described by the equation

$$Cl_2(g) + 2NaBr(aq) \rightarrow 2NaCl(aq) + Br_2(l)$$

If the concentration of $NaBr(aq)$ is 4.00×10^{-3} M, how many grams of bromine can be obtained per cubic meter of brine? How many grams of chlorine are required?

12-25. Sodium hypochlorite, $NaClO(s)$, is used as a bleaching agent in many commercial bleaches. Sodium hypochlorite can be prepared by the reaction described by the equation

$$Cl_2(g) + 2NaOH(aq) \rightarrow NaClO(aq) + NaCl(aq) + H_2O(l)$$

How many grams of $Cl_2(g)$ are required to react with 5.00 liters of 6.00 M $NaOH(aq)$?

12-26. Silver chloride dissolves in an aqueous solution of ammonia according to

$$AgCl(s) + 2NH_3(aq) \rightarrow [Ag(NH_3)_2]^+(aq) + Cl^-(aq)$$

How many liters of a 0.100 M $NH_3(aq)$ solution are required to dissolve 0.231 grams of $AgCl(s)$?

12-27. How many grams of metallic copper will be formed if 30.0 grams of zinc are allowed to react with 375 milliliters of 0.165 M $CuSO_4(aq)$? What will be the concentration of all the ions in solution following the reaction?

12-28. How many grams of silver iodide will precipitate if 175 milliliters of 0.850 M $AgNO_3(aq)$ are added to 125 milliliters of 0.765 M $CaI_2(aq)$? What will be the concentration of all the ions in solution following the reaction?

CALCULATIONS INVOLVING ACID-BASE REACTIONS

12-29. By titration it is found that 27.5 milliliters of 0.155 M $NaOH(aq)$ are required to neutralize 25.0 milliliters of $HCl(aq)$. Calculate the concentration of the hydrochloric acid solution.

12-30. By titration it is found that 24.6 milliliters of 0.300 M $H_2SO_4(aq)$ are required to neutralize 20.0 milliliters of $NaOH(aq)$. Calculate the concentration of the NaOH solution.

12-31. (a) What volume of a 0.108 M $HNO_3(aq)$ solution is required to neutralize 15.0 μL of 0.010 M $Ca(OH)_2(aq)$? (b) What volume of a 0.300 M $H_2SO_4(aq)$ solution is required to neutralize 25.0 milliliters of 0.200 M $NaOH(aq)$?

12-32. Commercial antacid tablets contain a base, often an insoluble metal hydroxide, that reacts with stomach acid. Two such bases are $Mg(OH)_2(s)$ and $Al(OH)_3(s)$. Given that stomach acid is about 0.10 M $HCl(aq)$, calculate the number of milliliters of stomach acid that can be neutralized by 500 milligrams of each of these bases.

12-33. A 40.0-gram sample of $KOH(s)$ is dissolved in water to a final volume of 0.200 liters, and the resulting solution is added to 2.00 liters of 0.125 M $HCl(aq)$. Calculate the molarity of $KCl(aq)$ in the resulting solution.

12-34. A 500-milliliters sample of 0.200 M $NaOH(aq)$ is added to 200 milliliters of 0.100 M $HBr(aq)$. Calculate the molarities of the each of the species present after the reaction occurs.

12-35. A 0.365-gram sample of a mixture of $NaOH(s)$ and $NaCl(s)$ requires 31.7 milliliters of 0.150 M $HCl(aq)$ to react with all the $NaOH(s)$. What is the mass percentage of the $NaOH(s)$ in the mixture?

12-36. To test the purity of $NaOH(s)$ after its manufacture, 0.400 grams are dissolved in enough water to make 100.0 milliliters of solution. The solution is titrated with 0.100 M $HCl(aq)$ to determine the concentration of $NaOH(aq)$. It is found that 25.0 milliliters of $NaOH(aq)$ is neutralized by 23.2 milliliters of $HCl(aq)$. Calculate the purity of the solid $NaOH(s)$. What assumption do you have to make?

12-37. A 1.00-gram sample of an unknown acid is dissolved to make 100.0 milliliters of solution and neutralized with 0.250 M $NaOH(aq)$. The volume of

$NaOH(aq)$ required to neutralize the acid was 86.9 milliliters. Assume that the acid has two acidic protons per molecule and compute the formula mass of the acid.

12-38. A 1.00-gram sample of an unknown acid is dissolved in water to make 100.0 milliliters of solution and is neutralized with 0.250 M $NaOH(aq)$. The volume of $NaOH(aq)$ required to neutralize the acid was 66.6 milliliters. Assume that the acid has only one acidic proton per formula unit. Compute the formula mass of the acid.

ADDITIONAL PROBLEMS

12-39. A student observes that a solution of washing powder dissolved in water is a strong conductor of electricity; a solution of window cleaner weakly conducts an electrical current; and a solution of rubbing alcohol does not conduct an electrical current. What does this tell you about the properties of these three substances?

12-40. How might you use setups like the ones shown in Figures 12.6 and 12.7 to distinguish between a solution of pure water, one containing acetic acid, and one containing potassium chloride?

12-41. Water is a poor conductor of electricity. Given this, why is it dangerous to shower or bathe while using electrical appliances?

12-42. Construct a flow diagram similar to the one given in Figure 11.7 for determining the mass of another reactant or product starting from the concentration and volume of a reactant of a reaction taking place in an aqueous solution.

12-43. Calculate the molarity of scandium nitrate and of $Sc^{3+}(aq)$ ions and $NO_3^-(aq)$ ions in a solution containing 2.86 grams of scandium nitrate dissolved in 250.0 milliliters of solution.

12-44. A stock solution of hydrazine, $N_2H_4(aq)$, is 95.0% by mass hydrazine and 5.0% by mass water, and its density is 1.011 $g\cdot mL^{-1}$ at 20°C. Calculate the molarity of the solution.

12-45. What molarity of a $CaCl_2(aq)$ solution should you use if you want the molarity of the $Cl^-(aq)$ ions in the solution to be 0.100 M?

12-46. A saturated hydrochloric acid solution has a density of 1.20 $g\cdot mL^{-1}$ and is 40% by mass HCl. What is the molarity of a saturated $HCl(aq)$ solution?

12-47. Chromium(III) chloride forms a hexahydrate, $CrCl_3\cdot 6H_2O(s)$. How many grams of $CrCl_3\cdot 6H_2O(s)$ are needed to prepare 250 milliliters of an aqueous solution that is 0.500 M in $Cr^{3+}(aq)$?

12-48. Calcium chloride forms a hexahydrate, $CaCl_2\cdot 6H_2O(s)$. How many grams of $CaCl_2\cdot 6H_2O(s)$ are required to make 500 milliliters of a solution having the same concentration of $Cl^-(aq)$ as one that was prepared by dissolving 75.6 grams of $NaCl(s)$ in enough water to make exactly one liter of solution?

12-49. How many milliliters of commercial phosphoric acid (14.6 M) are required to prepare one liter of 0.650 M $H_3PO_4(aq)$?

12-50. How many milliliters of 6.00 M $HNO_3(aq)$ are required to prepare 50.0 milliliters of 0.50 M $HNO_3(aq)$?

12-51. Describe how you would prepare 500 milliliters of 0.050 M sodium dihydrogen phosphate, $NaH_2PO_4(aq)$, starting with a 1.00-M solution.

12-52. Describe how you would prepare 250 milliliters of 0.12-M sulfuric acid from a stock solution labeled 8.0-M sulfuric acid.

12-53. A solution is prepared by dissolving 5.85 grams of sodium chloride in water and diluting to 50.0 milliliters in a volumetric flask. A 10.00-milliliter sample of this solution is then added to a 100-milliliter volumetric flask to which water is then added to the mark. Calculate the molarity of the final solution.

12-54. The unit ppm (parts per million) is used for very dilute concentrations and represents one gram of solute per 10^6 grams of solvent, the equivalent of one milligram of solute per kilogram of solvent. The EPA currently limits the concentration of sodium fluoride in water (either natural or added for fluoridation) to 4.0 ppm. What is this concentration in units of molarity?

12-55. The unit ppb (parts per billion) is used for extremely dilute concentrations and represents one gram of solute per 10^9 grams of solvent, or one microgram of solute per kilogram of solvent. Suppose that 3.5 grams of phenol red indicator (chemical formula: $C_{19}H_{14}O_5S$) are placed in an Olympic-sized swimming pool to determine the pH. (a) If the pool measures 50.0 meters long by 25.0 meters wide with an average depth of 2.75 meters, calculate the concentration of the indicator in parts per billion after it becomes thoroughly mixed with the volume of the pool. (b) What is its concentration in units of molarity?

12-56. How many grams of a $CaCl_2(aq)$ solution that is 14% $CaCl_2$ by mass contains 3.25 grams of $CaCl_2(s)$?

12-57. What volume of a 6.00 M $K_2SO_4(aq)$ solution contains 8.60 grams of $K_2SO_4(s)$?

12-58. How many grams of $Al_2(SO_4)_3{\cdot}18H_2O(s)$ are required to prepare 500 milliliters of 3.00 M $Al_2(SO_4)_3(aq)$?

12-59. A 7.55-gram sample of an aqueous sulfuric acid solution is diluted by adding it to water and then excess $BaCl_2(aq)$ is added to the solution. If the resulting $BaSO_4(s)$ precipitate has a mass of 6.11 grams, what is the mass percentage of H_2SO_4 in the original solution?

12-60. A sample of $KOH(aq)$ was found to contain 8.75% water by mass. How many grams of this material are required to make 500 milliliters of a 6.00 M $KOH(aq)$ solution?

12-61. How many grams of $CuSO_4{\cdot}5H_2O(s)$ are required to produce 500 milliliters of an aqueous solution that contains 250-milligrams $Cu^{2+}(aq)$ per liter?

12-62. What volume of 1.250 M $HCl(aq)$ is required to react with 11.78 grams of $Na_2CO_3(s)$? The equation for the reaction is

$$Na_2CO_3(s) + 2HCl(aq) \rightarrow 2NaCl(aq) + H_2O(l) + CO_2(g)$$

12-63. What volume of 0.865 M $CaCl_2(aq)$ is required to precipitate all the $Ag^+(aq)$ from 35.0 milliliters of 0.500 M $AgNO_3(aq)$? How many grams of $AgCl(s)$ will precipitate?

12-64. A 6.076-gram sample of succinic acid, $H_6C_4O_4(s)$, is titrated with 3.965 M $NaOH(aq)$. It requires 25.95 milliliters to react completely with the succinic acid. Determine the number of acidic hydrogen atoms in succinic acid.

12-65. When 250.0 milliliters of an aqueous solution of hydrochloric acid is reacted with an excess of zinc, 0.955 grams of hydrogen is evolved. What is the molarity of the original $HCl(aq)$ solution?

12-66. The amount of nitrogen in an organic substance can be determined by an analytical method called the Kjeldahl method, in which all the nitrogen in the organic substance is converted to ammonia. The ammonia, which is a weak base, can be neutralized with hydrochloric acid, as described by the equation

$$NH_3(aq) + HCl(aq) \rightarrow NH_4Cl(aq)$$

If it requires 27.5 milliliters of 0.150 M $HCl(aq)$ to neutralize all the $NH_3(g)$ from a 2.25-gram sample of organic material, calculate the mass percentage of nitrogen in the sample.

12-67. Quicklime is a commercial product consisting primarily of calcium oxide. If it requires 37.05 milliliters of 2.565 M $HCl(aq)$ to react with all the $CaO(s)$ in a 2.710-gram sample of quicklime, calculate the mass percentage of $CaO(s)$ in the sample. Assume that none of the impurities react with the $HCl(aq)$.

12-68. How many grams of $BaSO_4(s)$ are precipitated when 20.0 milliliters of 0.450 M $BaCl_2(aq)$ are mixed with 36.0 milliliters of a 0.250 M $K_2SO_4(aq)$ solution?

12-69. How many grams of lead(II) iodate, $Pb(IO_3)_2(s)$, are precipitated when 25.0 milliliters of 3.00 M $Pb(NO_3)_2(aq)$ are mixed with 30.0 milliliters of a 5.00 M $KIO_3(aq)$ solution?

12-70. How many grams of $ZnS(s)$ are precipitated when 30.0 milliliters of 1.76 M $Zn(NO_3)_2(aq)$ are mixed with 30.0 milliliters of a 2.18 M $Na_2S(aq)$ solution?

12-71. How many grams of silver oxalate, $Ag_2C_2O_4(s)$, are precipitated when 50.0 milliliters of 0.751 M $AgClO_4(aq)$ are mixed with 50.0 milliliters of a 0.400 M $K_2C_2O_4(aq)$ solution?

12-72. Suppose we have a solution of lead nitrate, $Pb(NO_3)_2(aq)$. A solution of $NaCl(aq)$ is added slowly until no further precipitation of $PbCl_2(s)$ occurs. The $PbCl_2(s)$ precipitate is collected by filtration, dried, and weighed. A total of 12.79 grams of $PbCl_2(s)$ is obtained from 200.0 milliliters of the original solution. Calculate the molarity of the $Pb(NO_3)_2(aq)$ solution.

12-73. Suppose we have a solution of lead acetate, $Pb(C_2H_3O_2)_2(aq)$. We add a solution of $KI(aq)$ until no further precipitation of $PbI_2(s)$ occurs. The $PbI_2(s)$ precipitate is collected by filtration, dried, and weighed. A total of 22.6 grams of $PbI_2(s)$ is obtained from 250.0 milliliters of the original solution. Calculate the molarity of the $Pb(C_2H_3O_2)_2(aq)$ solution.

12-74. A certain pesticide contains arsenic in the form of $As_4O_6(s)$. The arsenic in a 11.75-gram sample of the pesticide was converted to arsenate, $AsO_4^{3-}(aq)$, that was precipitated completely as $Ag_3AsO_4(s)$ by the addition of 37.5 milliliters of 0.655 M $AgNO_3(aq)$. Calculate the mass percentage of $As_4O_6(s)$ in the sample.

12-75. The most important ore of lead is galena that is principally $PbS(s)$. The ore may be analyzed for its lead content by treating the ore with nitric acid and then precipitating the $Pb^{2+}(aq)$ with potassium chromate, $K_2CrO_4(aq)$, to give $PbCrO_4(s)$. A 6.053-gram sample of galena yielded 7.248 grams of $PbCrO_4(s)$ precipitate when treated in the above manner. Calculate the mass percentage of $PbS(s)$ in the ore.

12-76. Ammonia can be titrated with hypobromite according to

$$2\,NH_3(aq) + 3\,OBr^-(aq) \rightarrow N_2(g) + 3\,Br^-(aq) + 3\,H_2O(l)$$

Calculate the molarity of a hypobromite solution if 10.00 milliliters of the $OBr^-(aq)$ solution requires 3.52 milligrams of ammonia for a complete reaction.

12-77. Fuels can be analyzed for their sulfur content by burning them in oxygen and passing the resulting exhaust gases through a dilute solution of hydrogen peroxide in which the sulfur oxides, $SO_2(g)$ and $SO_3(g)$, are converted to sulfuric acid. Calculate the mass percentage of sulfur in a 5.63-gram sample of fuel that required 21.35 milliliters of 0.1006 M $NaOH(aq)$ to neutralize completely the sulfuric acid that was formed when the exhaust gases were passed through $H_2O_2(aq)$.

12-78. Aspirin, $HC_9H_7O_4(s)$, is a monoprotic acid (one acidic proton per formula unit) that can be analyzed by titration with a strong base. Calculate the mass percentage of aspirin in a 1.00-gram tablet that requires 23.52 milliliters of 0.1500 M $NaOH(aq)$ for complete neutralization.

12-79. A Rolaids® antacid tablet contains about 0.33 grams of $NaAl(OH)_2CO_3(s)$. Calculate how much stomach acid, assumed to be 0.14 M $HCl(aq)$, can be neutralized by one Rolaids® tablet. The equation for the neutralization reaction is

$$NaAl(OH)_2CO_3(s) + 4\,HCl(aq) \rightarrow NaCl(aq) + AlCl_3(aq) + 3\,H_2O(l) + CO_2(g)$$

12-80. Ethanol is produced by the action of certain yeasts on sugars such as glucose:

$$\underset{\text{glucose}}{C_6H_{12}O_6(aq)} \rightarrow 2\,\underset{\text{ethanol}}{CH_3CH_2OH(aq)} + 2\,CO_2(g)$$

Wine is made by adding yeast to grape juice. What concentration of glucose in grams per liter must the grape juice contain to produce wine that is 11% ethanol by volume? (Take the density of ethanol to be 0.79 g·mL^{-1}.)

12-81. A 0.450-gram sample of impure $CaCO_3(s)$ is dissolved in 50.0 milliliters of 0.150 M $HCl(aq)$. The equation for the reaction is

$$CaCO_3(s) + 2\,HCl(aq) \rightarrow CaCl_2(aq) + H_2O(l) + CO_2(g)$$

The excess $HCl(aq)$ is titrated by 8.75 milliliters of 0.125 M $NaOH(aq)$. Calculate the mass percentage of $CaCO_3(s)$ in the sample.

12-82. The equation that describes the reaction used in some police breath analyzer devices is

$$3\,CH_3CH_2OH(aq) + 2\,Cr_2O_7^{2-}(aq) + 16\,H^+(aq) \rightarrow 3\,HC_2H_3O_2(aq) + 4\,Cr^{3+}(aq) + 11\,H_2O(l)$$

How many grams of $HC_2H_3O_2$ are produced by the reaction of 55.0 milliliters of 0.560 M $Cr_2O_7^{2-}$ with 100.0 milliliters of 0.963 M $CH_3CH_2OH(aq)$?

12-83. The amount of $I_3^-(aq)$ in a solution can be determined by titration with a solution containing a known concentration of $S_2O_3^{2-}$ (thiosulfate ion). The determination is based on the net ionic equation

$$2\,S_2O_3^{2-}(aq) + I_3^-(aq) \rightarrow S_4O_6^{2-}(aq) + 3\,I^-(aq)$$

Given that it requires 36.4 milliliters of 0.330 M $Na_2S_2O_3(aq)$ to titrate the $I_3^-(aq)$ in a 15.0-milliliters sample, calculate the molarity of $I_3^-(aq)$ in the solution.

12-84. How many milliliters of 0.475 M $KMnO_4(aq)$ are required to react with 50.0 milliliters of 0.336 M $K_2C_2O_4(aq)$? The equation for the reaction is

$$2\,KMnO_4(aq) + 5\,K_2C_2O_4(aq) + 16\,HCl(aq) \rightarrow 2\,MnCl_2(aq) + 12\,KCl(aq) + 10\,CO_2(g) + 8\,H_2O(l)$$

12-85. (*) A 0.3146-gram sample of a mixture of $NaCl(s)$ and $KBr(s)$ was dissolved in water. The resulting solution required 37.60 milliliters of 0.08765 M $AgNO_3(aq)$ to precipitate the $Cl^-(aq)$ and $Br^-(aq)$ as $AgCl(s)$ and $AgBr(s)$. Calculate the mass percentages of $NaCl(s)$ and $KBr(s)$ in the mixture.

Lorenzo Romano Amedeo Carlo Avogadro, Conte di Quaregna e di Cerreto, left, (1776–1856) was born in Turin, Italy, to a noble family. Avogadro was a brilliant student, graduating with an advanced degree in ecclesiastical law at the age of 20. He practiced law for a few years, but then decided to teach physics and mathematics at a high school in Verceli, where his family owned property. In 1820 he became professor of physics at the University of Turin. Shortly after, his participation in a failed plot against the king of Sardinia caused him to lose his teaching post. Eventually, the political climate changed and in 1833 he was able to return to the university where he taught for another 20 years. It was in 1811 when he was teaching high school in Verceli that he published what we now call "Avogadro's hypothesis" that says that equal volumes of gases at the same temperature and pressure contain the same number of molecules. Unfortunately, other chemists at that time had no appreciation of the difference between atoms and molecules, and could not accept the possibility that elements such as nitrogen and oxygen could consist of diatomic molecules. Avogadro's work was largely ignored for almost 50 years, until Cannizzaro convincingly showed that it leads to a consistent set of atomic masses.

Stanislao Cannizzaro, right, (1826–1910) was born in Palermo, Sicily. He enrolled as a medical student at his local university at the age of 15, but soon turned to chemistry. After a few years he went on to the University of Pisa. His native Sicily was in open rebellion against the current government, and he returned to serve as an artillery officer with the insurgents. When the rebellion was crushed a few years later, he escaped to Paris, where he carried out important research in organic chemistry. Upon receiving his doctorate in chemistry, he held professorships at a number of universities, and in 1861 accepted a chair of chemistry at Palermo. He spent 10 years there, after which he became chair of chemistry at the University of Rome, where he stayed until one year before his death in 1910. Cannizzaro's greatest contribution to chemistry was an 1858 memoir in which he used Avogadro's hypothesis to deduce a consistent set of atomic and molecular masses. Prior to this, the state of atomic and molecular masses was in total confusion. For example, because of the lack of a consistent set of atomic masses, there were 18 different formulas for acetic acid in use! This situation came to a head in 1860 when over one hundred of the most prominent chemists of the world convened a conference at Karlsrule, Germany, to discuss the muddled state of atomic masses. After many heated exchanges, Cannizzaro won over the majority of the participants, and shortly afterward chemistry had the first accurate version of the table of atomic masses that we use today. Many people consider this work of Cannizzaro to be one of the greatest contributions to chemistry.

13. Properties of Gases

In this chapter, we examine the properties of gases. Many chemical reactions involve gases as reactants or products or both, so we must determine how the properties of gases depend on conditions such as temperature, pressure, volume, and number of moles. We show how gases respond to changes in pressure and temperature and then discuss how the pressure, temperature, and volume of a gas are related to one another. After presenting a number of experimental observations concerning gases, we discuss the kinetic theory of gases, which gives us insight into the molecular nature of the gaseous state.

13-1. Most of the Volume of a Gas Is Empty Space

Before we discuss the nature of gases, we must first consider the three physical states of matter: solid, liquid, and gas. Recall from Section 2-2 that a solid has a fixed volume and shape; a liquid has a fixed volume but assumes the shape of the container into which it is poured; and a gas has neither a fixed volume nor a fixed shape, but rather expands to occupy the entire volume of any closed container into which it is placed. We shall examine each of these states of matter in more detail here.

The molecular picture of a crystalline solid is that of an ordered array, called a lattice, of particles (atoms, molecules, or ions), as shown in Figure 13.1. The

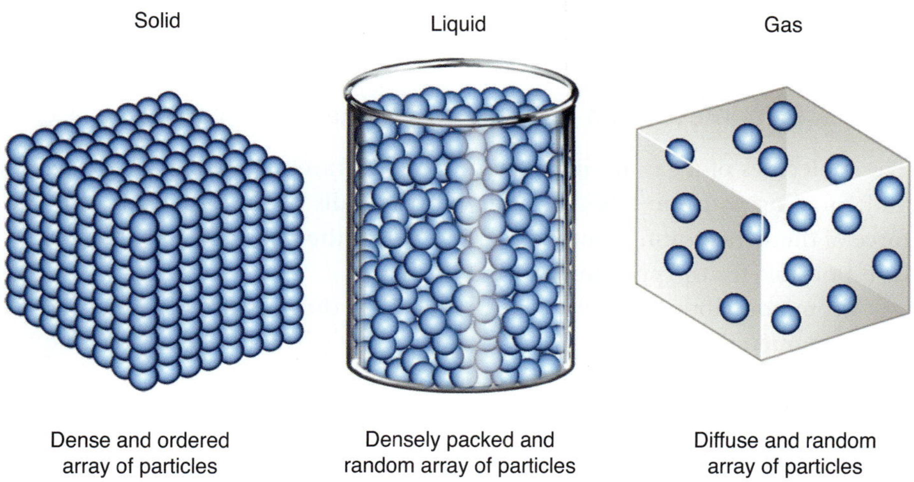

Figure 13.1 Molecular views of the particles in (*left*) a solid, (*center*) a liquid, and (*right*) a gas.

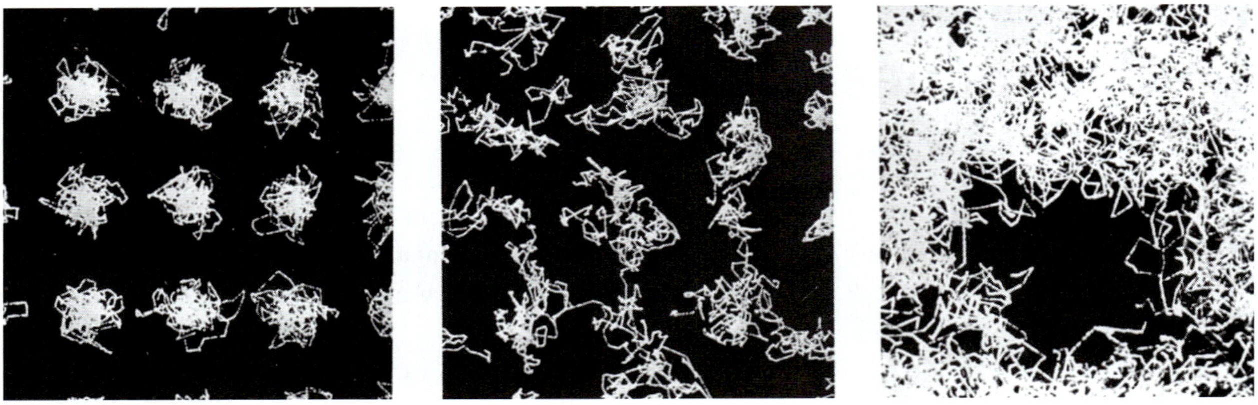

(a) (b) (c)

Figure 13.2 Computer-calculated paths of particles. (a) Motion of atoms in an atomic crystal (note that the atoms move only about fixed positions). (b) A crystal in the process of melting (note the breakdown of the ordered array). (c) A liquid and its vapor (the dark area represents a gas bubble surrounded by particles whose motions characterize a liquid).

individual particles vibrate just a little around fixed lattice positions but are not free to move about (Figure 13.2a). The restrictions on the motion of the particles of a solid are reflected by the fixed volume and shape that characterize a solid.

A molecular view of a liquid (Figure 13.1) shows the particles in continuous contact with one another but free to move about throughout the liquid. There is no orderly, fixed arrangement of particles in a liquid as there is in a solid. When a solid melts and becomes a liquid, the lattice breaks down and the constituent particles are no longer held in fixed positions (Figure 13.2b). The fact that the densities of the solid phase and the liquid phase of any substance do not differ greatly from each other indicates that the distance between particles is similar in the two phases. Furthermore, the solid and liquid phases of a substance have similar, small **compressibilities**, meaning that their volumes do not change appreciably with increasing pressure. The similar compressibilities of the solid and liquid phases of a substance are further evidence that the particles in the two phases have similar separations.

When a given mass of liquid is **vaporized** (converted to a gas), there is a huge increase in its volume. For example, one mole of liquid water occupies 17.3 mL at 100°C, whereas one mole of water vapor occupies over 30 000 mL under similar conditions. Upon vaporization, the molecules of a substance become widely separated, as indicated in Figure 13.1. The picture of a gas as a substance with widely separated particles accounts nicely for the relative ease with which gases can be compressed. The particles take up only a small fraction of the total space occupied by a gas; most of the volume of a gas is empty space. As we note later in this chapter, the volume of a gas decreases markedly as the pressure increases; in other words, gases have large compressibilities.

13-2. A Manometer Can Be Used to Measure the Pressure of a Gas

The molecules of a gas are in constant motion, traveling about at high speeds and colliding with one another and with the walls of the container. It is the force of these incessant, numerous collisions with the walls of the container that is responsible for the **pressure** exerted by a gas.

A common laboratory setup used to measure the pressure exerted by a gas is a **manometer**, which is a glass U-shaped tube partially filled with a liquid (Figure 13.3). Mercury is commonly used as the liquid in a manometer because it has a high density and is fairly unreactive. Figure 13.3 illustrates the measure-

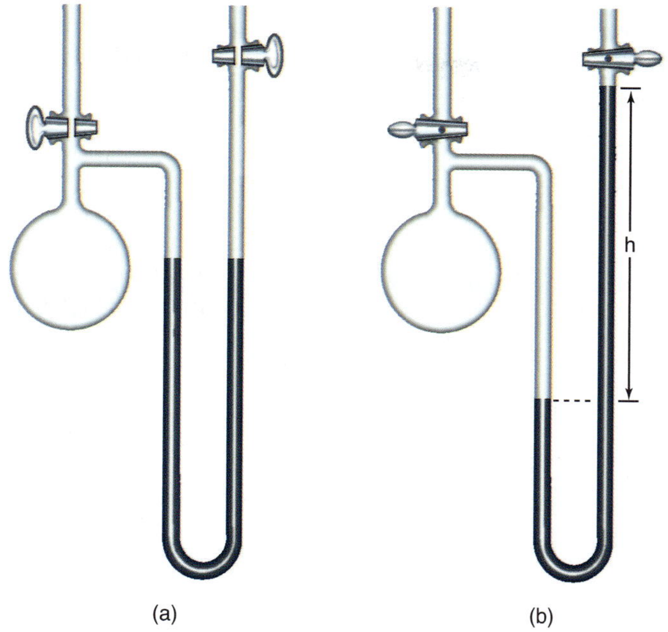

Figure 13.3 A mercury manometer. (a) Both stopcocks are open to the atmosphere, so both columns are exposed to atmospheric pressure. Both columns are at the same height because the pressure is the same on both surfaces. (b) The two stopcocks are closed, and the air in the right-hand column has been evacuated so that the pressure on the top of the right-hand column of mercury is essentially zero. The heights of the columns are no longer the same. The difference in heights, *h*, is a direct measure of the pressure of the gas in the flask.

ment of gas pressure using a manometer. In Figure 13.3b the right-hand side of the tube contains a vacuum so that the height *h* of the column of mercury supported by the gas in the flask is directly proportional to the pressure of the gas. Because of this direct proportionality, it is convenient to express pressure in terms of the height of a column of mercury that the gas will support. This height is usually measured in millimeters, so pressure is expressed in terms of millimeters of mercury (mm Hg). The pressure unit mm Hg is called a **torr**, after the Italian scientist Evangelista Torricelli. Torricelli invented the **barometer**, which is used to measure atmospheric pressure (Figure 13.4). Thus, we say, for example, that the pressure of a gas is 600 Torr.

Although mercury is most often used as the liquid in a manometer, other liquids can be used. The height of the column of liquid that can be supported by a gas is inversely proportional to the density of the liquid; that is, the less dense the liquid, the taller the column will be.

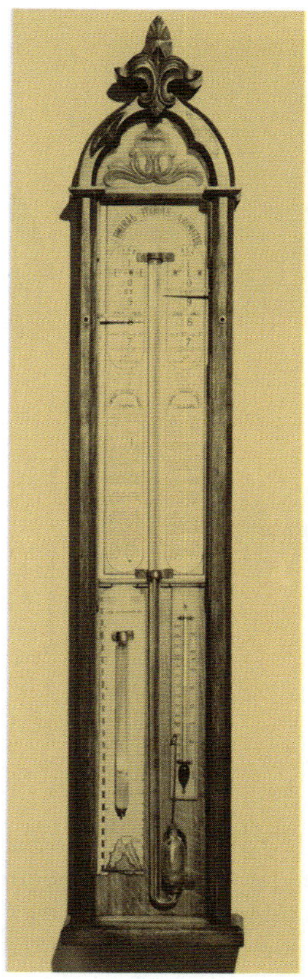

Figure 13.4 The pressure exerted by the atmosphere can support a column of mercury that is about 760 mm high, as seen in the central tube of the barometer shown here. This barometer is at the National Maritime Museum in Greenwich, England.

EXAMPLE 13-1: Calculate the height of a column of water that will be supported by a pressure of 755 Torr. Take the density of mercury to be 13.6 $g \cdot mL^{-1}$ and that of water to be 1.00 $g \cdot mL^{-1}$.

Solution: A pressure of 755 Torr corresponds to a column of mercury that is 755 mm high. Because mercury is 13.6 times more dense than water, the column of water supported will be 13.6 times higher than that of mercury, or

$$\begin{aligned}\text{height of water column} &= 13.6 \times \text{height of mercury column}\\ &= (13.6)(755\text{ mm}) = 1.03 \times 10^4\text{ mm}\\ &= 10.3\text{ m (or 33.9 ft)}\end{aligned}$$

For a liquid to be useful as a manometric fluid, typical heights supported must be large enough to be accurately measured, but small enough to avoid the necessity for holes in the ceiling. Mercury is the only liquid that is dense enough to avoid the latter problem for measurements of atmospheric pressure at room temperature. Various organic liquids, such as certain silicone oils and di-*n*-butyl phthalate, $C_6H_4(COOC_4H_9)_2(l)$, are occasionally used for measurements of low pressures.

PRACTICE PROBLEM 13-1: Di-*n*-butyl phthalate, $C_6H_4(COOC_4H_9)_2(l)$, is an oily, unreactive liquid with a density of 1.046 g·mL^{-1} at 20°C. Calculate the height of a column of di-*n*-butyl phthalate that will be supported by a pressure of 2.00 Torr.

Answer: 26.0 mm

In addition to manometers and barometers, a variety of other meters and gauges are also used to measure pressure. Some of these are shown in Figure 13.5.

Figure 13.5 A few of the various meters and gauges used to measure pressure. (*top*) A tire pressure gauge uses a slider to measure pressures above 760 Torr. (*second from top*) A mechanical pressure gauge uses a compressible material to measure pressures up to about 10^6 Torr. (*second from bottom*) A thermocouple is an electronic meter that uses the relationship between gas temperature and pressure to measure pressures between 1 Torr and 10^{-3} Torr. (*bottom*) An ion gauge uses an ion current created by a filament to measure vacuum pressures below 10^{-4} Torr.

13-3. The SI Unit of Pressure Is the Pascal

Although we do not "feel" it, the atmosphere surrounding the earth is a gas that exerts a pressure (Figure 13.6). The manometer pictured in Figure 13.3 can be used to demonstrate this pressure. If the flask is open to the atmosphere and the air in the right-hand side is evacuated, a column of mercury will be supported by the atmospheric (barometric) pressure. The height of the mercury column depends on elevation above sea level, temperature, and climatic conditions, but at sea level on a clear day it is about 760 mm as shown in Figure 13.4.

Several units are used to express pressure. One **standard atmosphere** (atm) is defined as a pressure of 760 Torr. Although not an SI unit, it is still common to express pressure in terms of atmospheres. Strictly speaking, torr and atmosphere are not units of pressure but rather are quantities that are directly proportional to pressure. Pressure is defined as a *force per unit area:*

$$P = \frac{F}{A} \tag{13.1}$$

The SI unit of pressure is the **pascal** (Pa), which is defined as one newton per square meter. (A newton is the SI unit of force: 1 N = 1 kg·m·s^{-2}. A newton is roughly equal to the gravitational force exerted on an apple at the earth's surface.) Thus, we have

$$1\text{ Pa} = \frac{1\text{ N}}{\text{m}^2} = \frac{1\text{ kg·m·s}^{-2}}{\text{m}^2} = 1\text{ kg·m}^{-1}\text{·s}^{-2} = 1\text{ J·m}^{-3} \tag{13.2}$$

where 1 J = 1 kg·m^2·s^{-2}.

Although the pascal is the official SI unit of pressure, it is not a convenient size for dealing with pressures of gases under atmospheric conditions. It turns

Figure 13.6 (*left*) In 1654 Otto van Guericke, the inventor of the vacuum pump, famously demonstrated before the emperor's court in Vienna (and later in Berlin) that due to the pressure of the atmosphere, two teams of eight horses were unable to pull apart two evacuated hollow hemispheres 35.5 cm (14 inches) in diameter. (*right*) The two hemispheres on display in the Deutsches Museum, Munich.

out that one atmosphere is about 10 000 pascals, so a more convenient SI unit of pressure is the **bar** that is defined to be exactly 10 000 pascals, or 100 kilopascals. The numerical value of a bar and an atmosphere are almost equal to each other; the actual relation between them is

$$1 \text{ atm} = 1.01325 \text{ bar}$$

In spite of the fact that the use of pascals and bars has been recommended by the International Union of Pure and Applied Chemistry (IUPAC) for decades, most chemists and chemistry textbook authors still cling to the use of torr and atmospheres as units of pressure. These units are being replaced by pascals and bars, but at a glacial pace, although meteorologists frequently express atmospheric pressure in millibars and geologists deal with gigapascal pressures when discussing the formation of mineral deposits. Because of all this, we must be bilingual and be able to use both sets of units fluently. Table 13.1 summarizes the relations between the various units that we shall use for pressure.

TABLE 13.1 Various units for expressing pressure

SI units		Older units	
1 pascal (Pa) = 1 $kg{\cdot}m^{-1}{\cdot}s^{-2}$ = 1 $J{\cdot}m^{-3}$ (standard pressure = 100 kPa or 1 bar)		1 "standard atmosphere" or atm (defined as equal to 760 Torr)	
100 kPa	= 1×10^5 Pa	1 atm	= 1.01325×10^5 Pa
	= 1 bar		= 1.01325 bar
	= 0.9869 atm		= 101.325 kPa
	= 750.1 Torr		= 760 Torr
	= 750.1 mm Hg		= 760 mm Hg
	= 14.5 $lb{\cdot}in^{-2}$ (psi)		= 14.7 $lb{\cdot}in^{-2}$ (psi)

Figure 13.7 Robert Boyle (1627–1691) was born in Ireland. His father, the first Earl of Cork, was one of the wealthiest men in Britain. Boyle was educated at Eton, but finished his education under a private tutor. His father sent him on a tour of Europe to complete his education, where he learned French, Latin, Italian, rhetoric, theology, and mathematics. After his father's death, he moved to his father's manor in Stalbridge, England, setting up a laboratory there. Boyle took regular trips to London to discuss science at gatherings of "The Invisible College." This informal meeting of scientists later become the Royal Society of London, the oldest continuous scientific society in the world. Boyle's goal was to apply mathematical analysis to chemistry, believing that "nature was a complex system governed by a small number of mathematical laws." In his most famous experiments, he demonstrated the relationship between the pressure and the volume of a gas—known today as Boyle's law.

EXAMPLE 13-2: The diamond anvil is a device in which small amounts of material may be squeezed between the faces of two diamonds to pressures of up to 360 gigapascals (GPa). This is similar to the pressure found in the earth's core and allows for both chemical and geological experimentation on small samples under conditions of extreme pressure. How many atmospheres are 360 GPa equivalent to?

Solution: We refer to Table 13.1 and see that 1 atm is equivalent to 1.013×10^5 Pa. Using the fact that the prefix *giga-* stands for 10^9 (see the inside of the back cover), we write

$$(360\text{ GPa})\left(\frac{1\times10^9\text{ Pa}}{1\text{ GPa}}\right)\left(\frac{1\text{ atm}}{1.013\times10^5\text{ Pa}}\right)=3.55\times10^6\text{ atm}$$

or about three and a half million atmospheres!

PRACTICE PROBLEM 13-2: In meteorology, pressures are expressed in units of millibars (mbar). Using Table 13.1, convert a pressure 985 mbar to torr and to atmospheres.

Answer: 739 Torr; 0.972 atm

13-4. The Volume of a Gas Is Inversely Proportional to Its Pressure and Directly Proportional to Its Kelvin Temperature

The first systematic study of the behavior of gases under different applied pressures was carried out in the 1660s by Robert Boyle (Figure 13.7). Boyle was able to show that, at constant temperature, the volume of a given sample of gas is inversely proportional to the pressure:

$$V \propto \frac{1}{P}$$

We can also write this relationship as

$$V=\frac{c}{P}\qquad\text{(constant temperature)}\qquad(13.3)$$

where c denotes a proportionality constant whose value in each case is determined by the amount and the temperature of the gas. The relationship between pressure and volume expressed in Equation 13.3 is known as **Boyle's law**. Equation 13.3 is plotted in Figure 13.8. Note that the greater the pressure on a gas, the smaller is the volume at constant temperature. If we double the pressure on a gas, then its volume decreases by a factor of two.

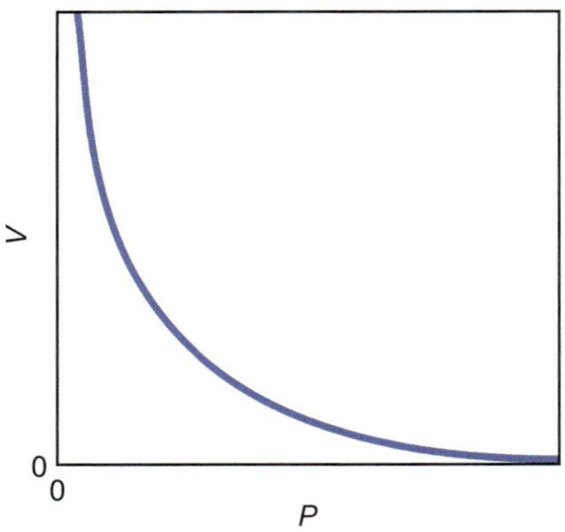

Figure 13.8 Boyle's law states that the volume of a gas at constant temperature is inversely proportional to its pressure.

Our molecular picture of a gas provides us with a nice interpretation of Boyle's law. The pressure that a gas exerts on the walls of its container is due to the incessant collisions of the molecules with the walls. If we decrease the volume, then the molecules collide more frequently with the walls, and hence

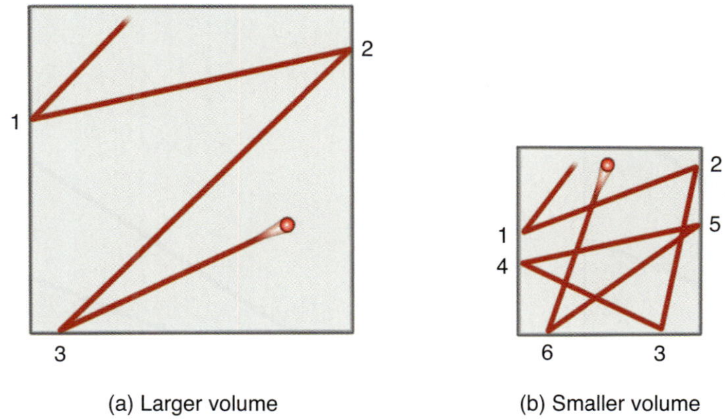

Figure 13.9 A molecular interpretation of Boyle's law. When the volume of a gas is decreased, the gas molecules collide with the walls more frequently, resulting in an increase in the pressure. The total path lengths traveled by the gas particles in diagrams (a) and (b) are identical, but the gas particle in the smaller volume (b) undergoes more collisions with the walls of its chamber than that in the larger volume (a), thus exerting a greater pressure on the walls.

exert a larger pressure (Figure 13.9). Thus, we see that a decrease in volume leads to an increase in pressure, in accord with Boyle's law.

Jacques Charles (Figure 13.10), a French scientist and adventurer, showed that there is a linear relationship between the volume of a gas and its temperature at constant pressure. This relationship is known as **Charles's law**. Charles's law also has a nice interpretation in terms of our molecular picture of a gas. We shall see in Section 13-10 that the speeds of the molecules increase with increasing temperature. Therefore, as the temperature of the gas increases, the speeds of the molecules increase and the collisions with the walls are more energetic, and so exert a higher pressure. If the container is not rigid, this in turn leads to a larger volume of the gas. Thus, we see that an increase in temperature leads to an increase in volume, in accord with Charles's law.

Typical experimental data illustrating Charles's law are plotted in Figure 13.11. Notice in Figure 13.11 that all three sets of experimental data extrapolate to the same point. Careful experimental measurements show that this point corresponds to a temperature of −273.15°C. The lines in Figure 13.11 suggest that if we add 273.15 to the temperatures expressed on the Celsius scale, then all three of the lines in Figure 13.11 will meet at the origin, as shown in Figure 13.12. This new temperature scale that we have proposed—adding 273.15 to the Celsius temperatures—turns out to be a fundamental temperature scale. We call it the **absolute** or the **Kelvin temperature scale**, after the British scientist Lord Kelvin

Figure 13.10 French physicist and inventor Jacques Alexandre César Charles (1746–1823). Although known for his formulation of Charles's law, he was also the inventor of the hydrogen balloon. In 1783, shortly after the Montgolfier brothers demonstrated their hot-air balloon for the French court, Charles and Ainé Roberts took off from Paris in a hydrogen-filled balloon before a crowd of 400 000 people. The balloon (pictured here) covered 43 kilometers before landing. Charles brought onboard a barometer and thermometer, making this both the first manned hydrogen balloon flight and the first scientific flight.

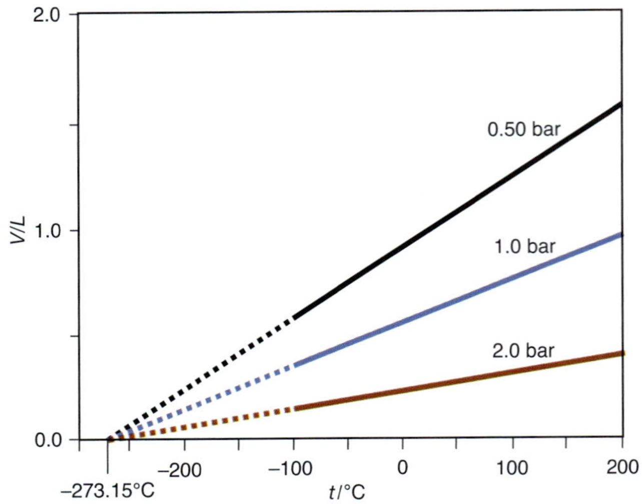

Figure 13.11 The volume (V) of 0.580 grams of air plotted as a function of temperature (t) at three different pressures. Note that all three curves extrapolate to $V = 0$ at −273.15°C. These plots suggest that we can define a temperature scale that reflects the basic relationship of volume and temperature by adding 273.15 to the Celsius scale.

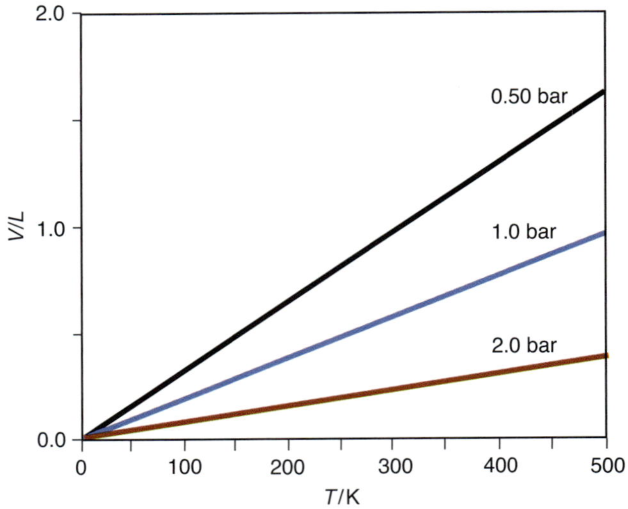

Figure 13.12 The volume of 0.580 grams of air plotted as a function of the Kelvin temperature at three different pressures. All three curves intersect at the origin (compare with Figure 13.11).

who first proposed it. The unit of this temperature scale is the **kelvin** (denoted by K), which we introduced in Chapter 1 when we discussed SI units. Using T to denote a Kelvin temperature and t to denote a Celsius temperature, we can express the relation between the two temperature scales as

$$T/\mathrm{K} = t/{}^\circ\mathrm{C} + 273.15 \tag{13.4}$$

We have expressed Equation 13.4 in Guggenheim notation, which we discussed in Section 1-10.

We call the units of T "kelvins," not "degrees kelvin." The Kelvin temperature scale is fundamental because zero kelvin is the lowest possible temperature. We refer to 0 K as **absolute zero**. By setting $T = 0\,\mathrm{K}$ in Equation 13.4, we see that the lowest temperature on the Celsius scale is −273.15°C. Figure 1.10 compares the temperature values of a number of fixed points on the various temperature scales.

The Kelvin temperature scale allows us to deduce a simple mathematical relation between volume and temperature. Recall from algebra that the equation of a straight line may be expressed as

$$y = mx + b \tag{13.5}$$

where m and b are constants. The quantity b is the **intercept** of the straight line with the y axis (the vertical axis), and m is the **slope** of the line and is a measure of the line's steepness (see Appendix A). All three curves in Figure 13.12 intercept the vertical axis at the origin, so the value of the intercept (b) is zero. Replacing y by V and x by T in Equation 13.5, we have

$$V = mT \qquad \text{(constant pressure)} \tag{13.6}$$

which is the algebraic equation for the lines plotted in Figure 13.12. The value of the slope (m) depends on the pressure and the amount of gas.

Equation 13.6 says that the volume of a fixed quantity of a gas at a fixed pressure is directly proportional to the Kelvin temperature of the gas. Equation 13.6 is an expression of Charles's law, and we can use it to calculate how the volume of a gas changes if we change its temperature, keeping the amount of gas and its pressure fixed. Suppose that a sample of helium gas in a flexible container such as a balloon (Figure 13.13) occupies a volume of 1.25 liters at 10.0°C and that we cool the helium by placing it in liquid nitrogen at −196°C. What will be the new volume of the helium gas, assuming that the surrounding pressure remains constant? We can use Equation 13.6 to write

$$\frac{V_i}{T_i} = m \quad \text{and} \quad \frac{V_f}{T_f} = m$$

where the subscripts i and f stand for "initial" and "final," respectively. From these two equations, we can write

$$\frac{V_i}{T_i} = \frac{V_f}{T_f}$$

Solving for V_f gives

$$V_f = V_i\left(\frac{T_f}{T_i}\right) \tag{13.7}$$

Equation 13.7 is a useful expression of Charles's law.

Recall from Chapter 1 that the rule for the addition and subtraction of significant figures is different from that for multiplication and division, (see Section 1-8).

We must always use absolute (Kelvin) temperatures in such calculations. So, for the helium sample, we have

$$T_i = 10.0\text{°C} + 273.15 = 283.2\text{ K}$$

$$T_f = -196\text{°C} + 273.15 = 77\text{ K}$$

Substituting these values and $V_i = 1.25$ L into Equation 13.7 gives

$$V_f = (1.25\text{ L})\left(\frac{77\text{ K}}{283.2\text{ K}}\right) = 0.34\text{ L}$$

(a) (b) (c) (d)

Figure 13.13 Liquid nitrogen is poured over an inflated balloon (a). As the temperature decreases, the volume of the gas inside the balloon decreases, (b) and (c). As the gas in the balloon warms up, its volume increases, thus inflating the balloon again (d).

Notice that the volume decreases because the temperature decreases (Figure 13.13). The following Example illustrates the use of Charles's law to devise a gas thermometer.

EXAMPLE 13-3: A simple gas thermometer can be made by trapping a small sample of a gas beneath a drop of mercury in a glass capillary tube that is sealed at one end and open at the other (Figure 13.14). Suppose that in such a thermometer, the gas occupies a volume of 0.180 mL at 0°C. The thermometer is then immersed in a liquid, and the final volume of the gas is 0.232 mL. What is the temperature of the liquid?

Solution: The sample of gas trapped in the tube is a fixed quantity; as long as the pressure on the gas remains the same, we can use Charles's law. We must always remember to use absolute temperatures in Charles's law:

$$T_i = 0°\text{C} + 273.15 = 273 \text{ K}$$

We are seeking T_f, so we solve Equation 13.7 for T_f to get

$$T_f = T_i\left(\frac{V_f}{V_i}\right) = (273 \text{ K})\left(\frac{0.232 \text{ mL}}{0.180 \text{ mL}}\right) = 352 \text{ K}$$

The corresponding temperature in degrees Celsius is

$$t_f = 352 \text{ K} - 273 = 79°\text{C}$$

PRACTICE PROBLEM 13-3: A sample of $Cl_2(g)$ occupies 600.0 mL at 0°C. What volume will the same quantity of chlorine gas occupy at 250.0°C, assuming constant pressure?

Answer: 1149 mL

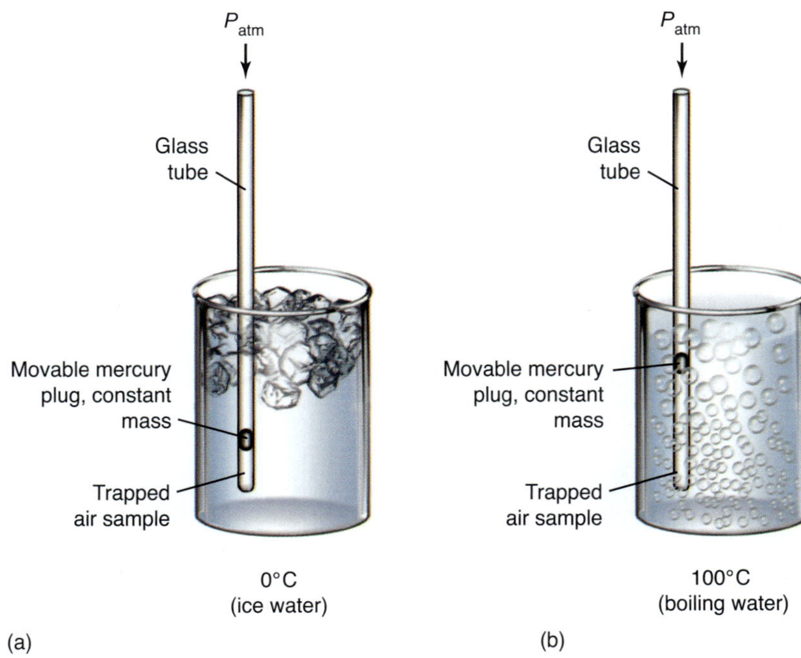

Figure 13.14 A gas thermometer. A sample of air is trapped by a drop of mercury in a capillary tube that is sealed at the bottom. According to Charles's law, the volume of the air is directly proportional to the Kelvin temperature. The atmosphere maintains a constant pressure on the air trapped below the mercury, which moves up or down to a position at which the pressure of the trapped air equals the atmospheric pressure plus that exerted by the mercury. As the temperature increases, as in (b), the drop of mercury rises because the volume of the trapped air increases.

13-5. Equal Volumes of Gases at the Same Pressure and Temperature Contain Equal Numbers of Molecules

Early experiments led to the discovery of a remarkable property of gaseous reactions. It was observed by the French chemist Joseph Louis Gay-Lussac (Figure 13.15) in the early 1800s that if all volumes are measured at the same pressure and temperature, then the volumes in which gases combine in chemical reactions are related to each other by small whole numbers. For example,

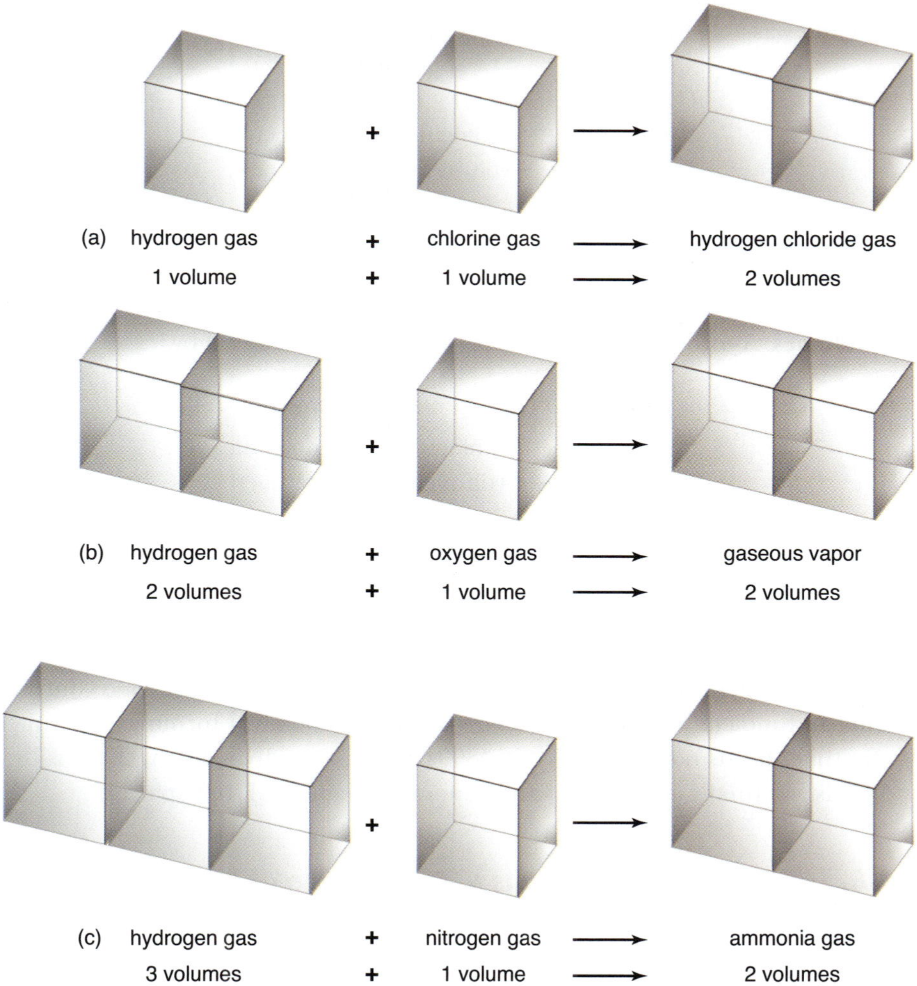

In each of these cases the relative volumes of reactants and products are in the proportion of simple whole numbers. This observation is known as **Gay-Lussac's law of combining volumes** and was one of the earliest indications of the existence of atoms and molecules. The interpretation of Gay-Lussac's law by the Italian chemist Amedeo Avogadro (Frontispiece) in 1811 led to the realization that many of the common gaseous elements, such as hydrogen, oxygen, nitrogen, and chlorine, occur naturally as diatomic molecules (H_2, O_2, N_2, and Cl_2) rather than as single atoms. This was such an important result in the development of chemistry that we shall review Avogadro's line of reasoning to arrive at his conclusion.

Following Gay-Lussac's observations, Avogadro postulated that equal volumes of gases at the same pressure and temperature contain equal numbers of molecules. This statement was known as **Avogadro's hypothesis** at the time, but now it is accepted as a law. Consider the reaction between hydrogen and chlorine to form hydrogen chloride. Experimentally, it is observed that *one volume* of hydrogen reacts with *one volume* of chlorine to produce *two volumes* of hydrogen chloride. According to Avogadro's reasoning, this means that *one molecule* of hydrogen reacts with *one molecule* of chlorine to produce *two molecules* of hydrogen chloride (Figure 13.16). If one molecule of hydrogen can form two molecules of hydrogen chloride, then a hydrogen molecule must consist of two atoms (at least) of hydrogen. Avogadro pictured both hydrogen and chlorine as diatomic gases and was able to represent the reaction between them as

$$\begin{array}{llll} H_2(g) & + \; Cl_2(g) & \rightarrow & 2\,HCl(g) \\ 1 \text{ molecule} & + \; 1 \text{ molecule} & \rightarrow & 2 \text{ molecules} \\ 1 \text{ volume} & + \; 1 \text{ volume} & \rightarrow & 2 \text{ volumes} \end{array}$$

Figure 13.15 The French scientist Joseph Louis Gay-Lussac (1778–1850). Gay-Lussac performed detailed quantitative experiments on gases. He is best known for his law of combining volumes in which he demonstrated that during reactions, gas volumes combine in simple whole-number ratios. Gay-Lussac independently discovered Charles's law and, like Charles, made several ascents in hydrogen balloons to collect data on the atmosphere.

Prior to Avogadro's explanation, it was difficult to see how one volume of hydrogen could produce two volumes of hydrogen chloride. If hydrogen occurred simply as atoms, there would be no way to explain Gay-Lussac's law of combining volumes. Another example of Avogadro's law is illustrated by

$$\begin{array}{llll} 2\,H_2(g) & + \; O_2(g) & \rightarrow & 2\,H_2O(g) \\ 2 \text{ molecules} & + \; 1 \text{ molecule} & \rightarrow & 2 \text{ molecules} \\ 2 \text{ volumes} & + \; 1 \text{ volume} & \rightarrow & 2 \text{ volumes} \end{array}$$

The work of Gay-Lussac and Avogadro's subsequent interpretation represent a great turning point in the history of chemistry—allowing chemists to quantitatively discuss chemical reactions in the unifying language of atoms and molecules. In spite of the beautiful simplicity of Avogadro's explanation of these reactions, his work was largely ignored and chemists continued to confuse atoms and molecules and to use many incorrect chemical formulas for quite some time. It was not until the mid-1800s, after Avogadro's death, that his hypothesis was finally appreciated and generally accepted (Frontispiece).

13-6. The Ideal-Gas Equation Is a Combination of Boyle's, Charles's, and Avogadro's Laws

Avogadro postulated that equal volumes of gases at the same pressure and temperature contain the same number of molecules. This statement implies that equal volumes of gases at the same pressure and temperature contain equal numbers of moles, n. Thus, we can write Avogadro's law as

$$V \propto n \qquad \text{(fixed } P \text{ and } T\text{)}$$

Boyle's law and Charles's law are, respectively,

$$V \propto \frac{1}{P} \qquad \text{(fixed } T \text{ and } n\text{)}$$

$$V \propto T \qquad \text{(fixed } P \text{ and } n\text{)}$$

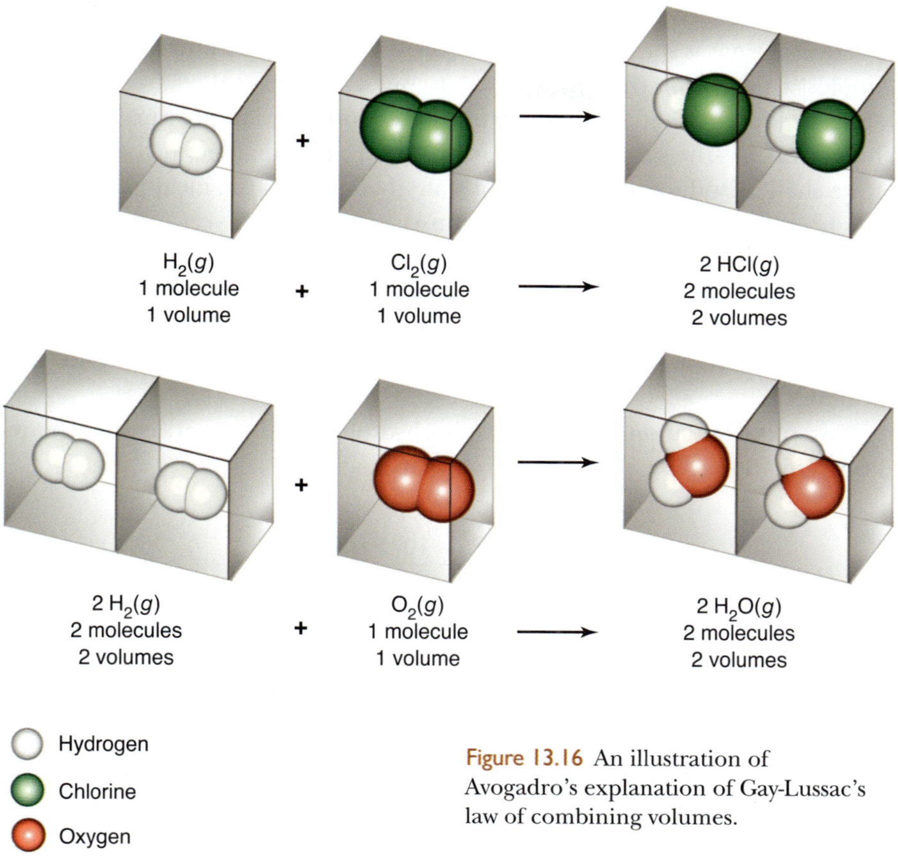

Figure 13.16 An illustration of Avogadro's explanation of Gay-Lussac's law of combining volumes.

We can combine these three proportionality statements for volume, V, into one by writing

$$V \propto \frac{nT}{P}$$

Note how the three individual statements for V are all included in the combined statement. For example, if P and n are fixed, then only T can vary; and we see that $V \propto T$, which is Charles's law. If P and T are fixed, then only n can vary; and we have Avogadro's law, $V \propto n$. Finally, if T and n are fixed, then we have Boyle's law, $V \propto 1/P$.

We can convert the combined proportionality statement for V to an equation by introducing a proportionality constant, R, and writing the equation as

$$PV = nRT \tag{13.8}$$

This expression is called the **ideal-gas law** or **ideal-gas equation**. It is based on Boyle's law, Charles's law, and Avogadro's law. Boyle's law and Charles's law are valid only at low pressures (less than a few bars or atmospheres, say), so Equation 13.8 is likewise valid only at low pressures. It turns out, however, that most gases do not deviate from Equation 13.8 by more than a few percent at pressures up to tens of bars or atmospheres, so Equation 13.8 is quite useful. Gases that satisfy the ideal-gas equation are said to behave ideally, or to be **ideal gases**.

The ideal-gas equation is an example of what we call an **equation of state**. The state of a gas is specified by its pressure, volume, temperature, and number of moles; an equation that relates these to one another is called an equation of state. The ideal-gas equation, which is accurate only for low pressures and high temperatures, is the simplest equation of state. In the final section of this chapter, we shall introduce an equation of state that extends the ideal-gas equation to higher pressures and lower temperatures.

Before we can use Equation 13.8, we must determine the value of R, the **molar gas constant**. It has been determined experimentally that one mole of an ideal gas at 0°C and one atm occupies 0.0224141 m^3 or 22.4141 liters. This volume, shown in Figure 13.17, is the **molar volume** of an ideal gas at 0°C and one atm.

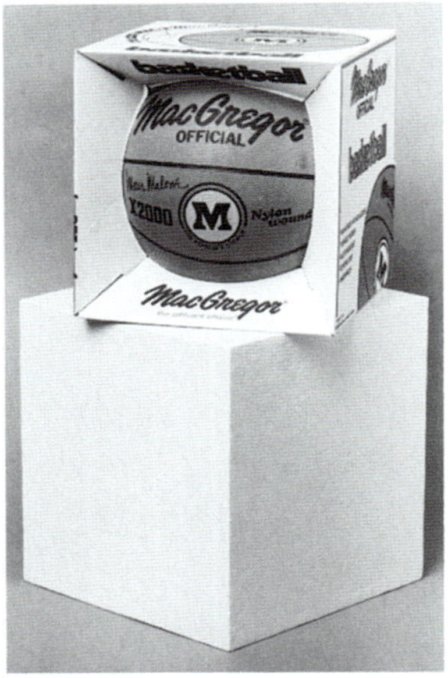

Figure 13.17 The volume of one mole of an ideal gas at 0°C and one atm. A volume of 22.4141 L is represented by this cube whose edges measure 28.2 cm. A basketball is shown for comparison.

If we solve Equation 13.8 for R and substitute this information into the resulting equation using SI units, we find that

$$R = \frac{PV}{nT} = \frac{(1\text{ atm})\left(\dfrac{1.01325 \times 10^5\text{ Pa}}{1\text{ atm}}\right)(0.0224141\text{ m}^3)}{(1\text{ mol})(273.15\text{ K})}$$

$$= 8.3145\text{ Pa}\cdot\text{m}^3\cdot\text{mol}^{-1}\cdot\text{K}^{-1} = 8.3145\text{ J}\cdot\text{mol}^{-1}\cdot\text{K}^{-1} \qquad (13.9)$$

where we have made use of Equation 13.2 ($1\text{ Pa} = 1\text{ J}\cdot\text{m}^{-3}$) to convert from $\text{Pa}\cdot\text{m}^3$ to joules and the fact that one atm is 1.01325×10^5 Pa (Table 13.1).

As we have seen, when working with SI units, P is expressed in units of pascals (Pa), V in cubic meters (m^3), n in moles, and T in kelvins. In contrast, when working with liters and atmospheres, P must be expressed in standard atmospheres, V in liters, n in moles, and T in kelvins. In this case the value of R is given by

$$R = \frac{(1\text{ atm})(22.4141\text{ L})}{(1.000\text{ mol})(273.15\text{ K})} = 0.082058\text{ L}\cdot\text{atm}\cdot\text{mol}^{-1}\cdot\text{K}^{-1} \qquad (13.10)$$

The values of the molar gas constant R in the different unit systems commonly used in chemistry are listed in Table 13.2. Which value of R to use depends upon the units of the pressure and the volume. The following examples illustrate the uses of various values of R.

TABLE 13.2 The value of the molar gas constant R in four sets of units

P	V	n	T	R
Pa	m^3	mol	K	$8.314472\text{ J}\cdot\text{mol}^{-1}\cdot\text{K}^{-1}$
bar	L	mol	K	$0.08314472\text{ L}\cdot\text{bar}\cdot\text{mol}^{-1}\cdot\text{K}^{-1}$
atm	L	mol	K	$0.0820575\text{ L}\cdot\text{atm}\cdot\text{mol}^{-1}\cdot\text{K}^{-1}$
Torr	L	mol	K	$62.3637\text{ L}\cdot\text{Torr}\cdot\text{mol}^{-1}\cdot\text{K}^{-1}$

EXAMPLE 13-4: A spherical balloon is inflated with helium to a diameter of 30.0 meters. If the pressure is 740.0 Torr and the temperature is 27.0°C, what is the mass of helium in the balloon?

Solution: The volume of the (spherical) balloon is

$$V = \frac{4}{3}\pi r^3 = \frac{4}{3}\pi\left(\frac{30.0\text{ m}}{2}\right)^3 = 1.41\times 10^4\text{ m}^3$$

We know the value of three variables—V, P, T—and we wish to calculate the fourth—n. If we use the ideal-gas equation with $R = 0.08206\text{ L·atm·mol}^{-1}\text{·K}^{-1}$ (taken to one more significant figure than the data), then we must express V in liters, P in atmospheres, and T in kelvins:

$$V = (1.41\times 10^4\text{ m}^3)\left(\frac{100\text{ cm}}{1\text{ m}}\right)^3\left(\frac{1\text{ mL}}{1\text{ cm}^3}\right)\left(\frac{1\text{ L}}{1000\text{ mL}}\right) = 1.41\times 10^7\text{ L}$$

$$P = (740.0\text{ Torr})\left(\frac{1\text{ atm}}{760\text{ Torr}}\right) = 0.9737\text{ atm}$$

$$T = 27.0°\text{C} + 273.15 = 300.2\text{ K}$$

Solving Equation 13.8 for n gives

$$n = \frac{PV}{RT} = \frac{(0.9737\text{ atm})(1.41\times 10^7\text{ L})}{(0.08206\text{ L·atm·mol}^{-1}\text{·K}^{-1})(300.2\text{ K})}$$
$$= 5.57\times 10^5\text{ mol}$$

The number of grams of helium is given by

$$\text{mass of helium} = (5.57\times 10^5\text{ mol He})\left(\frac{4.003\text{ g He}}{1\text{ mol He}}\right)$$
$$= 2.23\times 10^6\text{ g} = 2.23\text{ metric tons}$$

We also could have solved this problem using $R = 8.3145\text{ Pa·m}^3\text{·mol}^{-1}\text{·K}^{-1}$ and SI units. In this case we leave V in cubic meters and convert P to pascals:

$$P = (740.0\text{ Torr})\left(\frac{1.01325\times 10^5\text{ Pa}}{760\text{ Torr}}\right) = 9.866\times 10^4\text{ Pa}$$

Making use of Equation 13.8 we find

$$n = \frac{PV}{RT} = \frac{(9.866\times 10^4\text{ Pa})(1.41\times 10^4\text{ m}^3)}{(8.3145\text{ Pa·m}^3\text{·mol}^{-1}\text{·K}^{-1})(300.2\text{ K})} = 5.57\times 10^5\text{ mol}$$

You might try solving Example 13-4 again using the value $R = 62.3637\ \text{L}\cdot\text{Torr}\cdot\text{mol}^{-1}\cdot\text{K}^{-1}$.

which is the same answer we obtained using units of L·atm. In general we choose the units of greatest convenience (those requiring the fewest conversions) although, as we have just seen, any set of units may be used.

PRACTICE PROBLEM 13-4: The pressure in a 10.0-liter gas cylinder containing $N_2(g)$ is 4.15 atm at 20.0°C. What is the mass of nitrogen in the cylinder?

Answer: 48.3 g

EXAMPLE 13-5: Ammonia, $NH_3(g)$, can be produced by heating ammonium chloride, $NH_4Cl(s)$, with calcium hydroxide, $Ca(OH)_2(s)$, according to the chemical equation

$$2\,NH_4Cl(s) + Ca(OH)_2(s) \rightarrow CaCl_2(s) + 2\,H_2O(l) + 2\,NH_3(g)$$

If all the $NH_3(g)$ produced by reacting 2.50 grams of $NH_4Cl(s)$ with excess $Ca(OH)_2(s)$ occupies a 500.0-mL container at 15°C, what will the pressure of the $NH_3(g)$ be in atmospheres?

Solution: The number of moles of $NH_3(g)$ generated is given by

$$\text{moles of NH}_3 = (2.50\ \text{g NH}_4\text{Cl})\left(\frac{1\ \text{mol NH}_4\text{Cl}}{53.49\ \text{g NH}_4\text{Cl}}\right)\left(\frac{2\ \text{mol NH}_3}{2\ \text{mol NH}_4\text{Cl}}\right)$$

$$= 4.67 \times 10^{-2}\ \text{mol}$$

The pressure is calculated from Equation 13.8. Because we want to express our answer in terms of atmospheres, we use $R = 0.08206\ \text{L}\cdot\text{atm}\cdot\text{mol}^{-1}\cdot\text{K}^{-1}$ and write

$$P = \frac{nRT}{V} = \frac{(4.67 \times 10^{-2}\ \text{mol})(0.08206\ \text{L}\cdot\text{atm}\cdot\text{mol}^{-1}\cdot\text{K}^{-1})(288\ \text{K})}{0.5000\ \text{L}}$$

$$= 2.21\ \text{atm}$$

PRACTICE PROBLEM 13-5: If all the $O_2(g)$ generated by the thermal decomposition of 1.34 grams of potassium chlorate, $KClO_3(s)$, occupies 250.0 mL at 20.0°C, what will the pressure be in bars? The equation for the reaction is $2\,KClO_3(s) \rightarrow 2\,KCl(s) + 3\,O_2(g)$

Answer: 1.60 bar

EXAMPLE 13-6: How many milliliters of $H_2(g)$ will be generated by reacting 0.914 grams of $Zn(s)$ with 50.0 mL of 0.650 M $HCl(aq)$ at one bar and 0.00°C?

Solution: The equation for the reaction is

$$\mathrm{Zn}(s) + 2\,\mathrm{HCl}(aq) \rightarrow \mathrm{ZnCl_2}(aq) + \mathrm{H_2}(g)$$

We are given quantities of both reactants, so we first must see which, if either, of them is a limiting reactant. The number of moles of each reactant is given by

$$\text{moles of Zn} = (0.914\ \text{g Zn})\left(\frac{1\ \text{mol Zn}}{65.41\ \text{g Zn}}\right) = 0.0140\ \text{mol Zn}$$

$$\text{moles of HCl} = (50.0\ \text{mL})\left(\frac{1\ \text{L}}{1000\ \text{mL}}\right)\left(\frac{0.650\ \text{mol HCl}}{1\ \text{L}}\right) = 0.0325\ \text{mol HCl}$$

According to the chemical equation for the reaction, 0.0140 moles of $\mathrm{Zn}(s)$ require 0.0280 moles of $\mathrm{HCl}(aq)$, so we see that $\mathrm{HCl}(aq)$ is in excess and that $\mathrm{Zn}(s)$ is the limiting reactant. The number of moles of $\mathrm{H_2}(g)$ that will be produced from this quantity of zinc is given by

$$\text{moles of H}_2\text{ produced} = (0.0140\ \text{mol Zn})\left(\frac{1\ \text{mol H}_2}{1\ \text{mol Zn}}\right) = 0.0140\ \text{mol H}_2$$

$$V = \frac{nRT}{P} = \frac{(0.0140\ \text{mol})(0.08314\ \text{L·bar·mol}^{-1}\text{·K}^{-1})(273.15\ \text{K})}{1\ \text{bar}}$$

$$= 0.318\ \text{L} = 318\ \text{mL}$$

where we have used the value $R = 0.08314\ \text{L·bar·mol}^{-1}\text{·K}^{-1}$ from Table 13.2.

You may have encountered the term STP, which stands for standard temperature and pressure, if you have studied chemistry before. We shall intentionally eschew this term here because it has become ambiguous. It used to stand for the conditions 0°C and one atmosphere, but in 1982 the International Union of Pure and Applied Chemistry redefined it be the conditions at 0°C and one bar. Just as with the IUPAC suggestions for the units of pressure, the chemical community has been slow to fully adopt this new definition, and some authors use it and some don't. If and when you do come across this term in other sources, you must be sure to ascertain which definition is being used. It is really the obligation of the author to specify what he or she means by STP.

PRACTICE PROBLEM 13-6: Powdered charcoal, $\mathrm{C}(s)$, ignites spontaneously in $\mathrm{F_2}(g)$ at room temperature and forms carbon tetrafluoride, $\mathrm{CF_4}(g)$. What volume of $\mathrm{CF_4}(g)$ at 20.0°C and 760.0 Torr will be produced by reacting 250.0 mL of $\mathrm{F_2}(g)$ at 20.0°C and 1200.0 Torr with 26.2 mg of charcoal?

Answer: 52.4 mL

In Examples 13-4 through 13-6 we were given three quantities and had to calculate a fourth. Another application of the ideal-gas equation involves changes from one set of conditions to another.

EXAMPLE 13-7: One mole of $\mathrm{O_2}(g)$ occupies 22.7 L at 0°C and 1.00 bar. What volume does it occupy at 175°C and 4.00 bar?

Solution: In this problem we are given V at one set of conditions (T and P) and asked to calculate V under another set of conditions (that is, different T and P). Because R is a constant and because n is a constant in this problem, we can write the ideal-gas equation as

$$\frac{PV}{T} = nR = \text{constant}$$

This equation says that the ratio PV/T remains constant, so we can write

$$\frac{P_i V_i}{T_i} = \frac{P_f V_f}{T_f} \tag{13.11}$$

where the subscripts i and f denote initial and final, respectively. Recall that we wish to calculate a final volume, so we solve Equation 13.11 for V_f.

$$V_f = V_i \left(\frac{P_i}{P_f}\right)\left(\frac{T_f}{T_i}\right)$$

If we substitute the given quantities into this equation, we obtain

$$V_f = (22.7 \text{ L})\left(\frac{1.00 \text{ bar}}{4.00 \text{ bar}}\right)\left(\frac{448 \text{ K}}{273 \text{ K}}\right) = 9.31 \text{ L}$$

In this case, the increase in pressure (from 1.00 bar to 4.00 bar) leads to a decrease of the volume of the gas, whereas an increase in temperature leads to an increase in the volume. The pressure increases by a factor of 4.00, whereas the temperature increases by a factor of only 448/273 = 1.64; thus, the *net* effect is a decrease in the volume of the gas.

PRACTICE PROBLEM 13-7: A 2-liter cylinder contains a gas at a pressure of 250 kilopascals at 0°C. If the cylinder cannot withstand pressures in excess of 500 kilopascals, will the cylinder burst if it is heated to 400°C?

Answer: Yes, the final pressure would be 620 kPa.

In Example 13-7 we multiplied the initial volume (22.7 L) by a pressure ratio and a temperature ratio. The pressure increased from 1.00 bar to 4.00 bar, and the pressure ratio used was 1.00/4.00; this multiplication yields a smaller volume, as you would expect. Similarly, the temperature increased from 273 K to 448 K, and the temperature ratio used was 448/273; this multiplication yields a larger volume. A "commonsense" method of solving this problem is to write

$$V_f = V_i \times \text{pressure ratio} \times \text{temperature ratio}$$

and to decide by simple reasoning whether each ratio to be used is greater or less than unity.

EXAMPLE 13-8: Calculate the volume of $CO_2(g)$ produced from the complete combustion of 0.520 L of acetylene gas, $C_2H_2(g)$, at one bar and 0.00°C.

Solution: The balanced chemical equation for this reaction is

$$2\,C_2H_2(g) + 5\,O_2(g) \rightarrow 4\,CO_2(g) + 2\,H_2O(l)$$

Because the temperature and pressure are held fixed in this problem, the ideal-gas equation, $PV = nRT$, says that the number of moles of gas is proportional to the volume of gas. Therefore, we can read the above chemical equation as two liters of $C_2H_2(g)$ produce four liters of $CO_2(g)$. The volume of $CO_2(g)$ produced from 0.520 L of $C_2H_2(g)$ is therefore given by

$$\text{volume of } CO_2 = (0.520\text{ L } C_2H_2)\left(\frac{4\text{ L } CO_2}{2\text{ L } C_2H_2}\right) = 1.04\text{ L}$$

We can only use this sort of conversion between two *gaseous* species in a reaction. We could not, for example, determine the volume of liquid water produced in this way. The result of this problem is simply an affirmation of Gay-Lussac's law.

PRACTICE PROBLEM 13-8: Use the proportionality between pressure and number of moles at constant volume and temperature to determine the pressure of $CO_2(g)$ produced when propane, $CH_3CH_2CH_3(g)$, undergoes complete combustion with a stoichiometric quantity of $O_2(g)$ at 200°C at a pressure of 0.20 bar.

Answer: 0.60 bar

13-7. The Ideal-Gas Equation Can Be Used to Calculate the Molecular Masses of Gases

One of the most important applications of the ideal-gas equation involves the determination of the molecular mass of a gas. Let's see how we know that chlorine is a diatomic species. Suppose we find that a 0.286-gram sample of chlorine gas occupies 0.250 L at 300.0 Torr and 25°C. We are given V, P, and T, so we can use the ideal-gas law, Equation 13.8, and the appropriate value of R from Table 13.2 to calculate the number of moles, n:

$$n = \frac{PV}{RT} = \frac{(300.0\text{ Torr})(0.250\text{ L})}{(62.36\text{ L·Torr·mol}^{-1}\text{·K}^{-1})(298\text{ K})} = 4.036 \times 10^{-3}\text{ mol}$$

Thus, 0.286 grams of chlorine gas corresponds to 4.036×10^{-3} mol:

$$0.286\text{ g} \simeq 4.036 \times 10^{-3}\text{ mol}$$

Dividing both sides by 4.036×10^{-3} we obtain

$$70.9\text{ g} \simeq 1.00\text{ mol}$$

Thus, we find that the molecular mass of chlorine is 70.9. This result implies that chlorine is a diatomic gas with a molecular formula of $Cl_2(g)$ because the atomic mass of chlorine is 35.45.

EXAMPLE 13-9: The density of dry air at 1.00 atm and 20.0°C is 1.205 g·L^{-1}. Calculate the effective or average molecular mass of air; in other words, calculate the molecular mass of air as if it were a pure gas.

Solution: The density of the air in moles per liter is given by

$$\frac{n}{V} = \frac{P}{RT} = \frac{1.00 \text{ atm}}{(0.08206 \text{ L·atm·mol}^{-1}\text{·K}^{-1})(293.2 \text{ K})} = 0.0416 \text{ mol·L}^{-1}$$

Taking a one-liter sample, we see that

$$0.0416 \text{ mol} \approx 1.205 \text{ g}$$

Dividing through by 0.0416, we obtain

$$1 \text{ mol} \approx 29.0 \text{ g}$$

so the effective molecular mass of dry air is 29.0.

PRACTICE PROBLEM 13-9: Meteorites contain small quantities of trapped argon, which originates from radioactive processes. The density of a gas extracted from a particular meteorite is found to be 1.67 kg·m^{-3} at 15.0°C and 100.0 kPa. Calculate the molecular mass of the gas from this meteorite sample and determine if it is consistent with the gas being argon.

Answer: 40.0; yes.

We can combine a calculation in which we determine the molecular mass with a determination of the empirical (simplest) formula of the compound from chemical analysis to determine the molecular formula.

EXAMPLE 13-10: Chemical analysis shows that the gas acetylene is 92.3% carbon and 7.70% hydrogen by mass. It has a density of 1.602 kg·m^{-3} at 20.0°C and 150.0 kPa. Use these data to show that the molecular formula of acetylene is C_2H_2.

Solution: The determination of the empirical formula from chemical analysis is explained in Section 11-3. Following the procedure given there, we write

$$92.3 \text{ g C} \approx 7.70 \text{ g H}$$

Dividing the left side by 12.01 g·mol^{-1} of C and the right by 1.008 g·mol^{-1} of H gives

$$7.69 \text{ mol C} \approx 7.64 \text{ mol H}$$

Dividing both sides by 7.64 and rounding yields

$$1 \text{ mol C} \approx 1 \text{ mol H}$$

so the empirical formula of acetylene is CH.

We use the density data to determine the molecular mass of acetylene. Solving Equation 13.8 for n/V and using the appropriate SI units, we have

$$\frac{n}{V} = \frac{P}{RT} = \frac{1.500 \times 10^5 \text{ Pa}}{(8.3145 \text{ Pa·m}^3\text{·mol}^{-1}\text{·K}^{-1})(293.2 \text{ K})} = 61.53 \text{ mol·m}^3$$

Taking a one cubic meter sample, we see that

$$61.53 \text{ mol} \approx 1.602 \text{ kg}$$

Dividing through by 61.53, we obtain

$$1 \text{ mol} \approx 0.02604 \text{ kg}$$

or

$$1 \text{ mol} \approx 26.04 \text{ g}$$

so the molecular mass of acetylene is 26.04. Its empirical formula is CH and the empirical formula mass is 13.02, so the molecular formula of acetylene must be C_2H_2.

PRACTICE PROBLEM 13-10: Chemical analysis shows that the gas propene is 85.6% carbon and 14.4% hydrogen. Given that it has a density of 1.55 g·L^{-1} at 40.0°C and 720.0 Torr, determine the molecular formula of propene.

Answer: $C_3H_6(g)$

We can also use the ideal-gas equation to calculate the densities of gases in grams per liter. If we solve Equation 13.8 for n/V, we get

$$\frac{n}{V} = \frac{P}{RT}$$

The ratio n/V is equal to the molar density of the gas, which can be expressed in units of moles per liter. We can convert from moles per liter to grams per liter by multiplying both sides of the equation by the number of grams per mole, or the **molar mass**, M. (Recall that molar masses are numerically equal to formula masses but have the units of g·mol^{-1}. For example, the formula mass of $O_2(g)$ is 32.0 and its molar mass is 32.0 g·mol^{-1}. If we denote the density in grams per liter by the symbol ρ (the Greek letter rho), we can write

$$\rho = \frac{Mn}{V} = \frac{MP}{RT} \tag{13.12}$$

Equation 13.12 shows that the density of a gas is directly proportional to its pressure and inversely proportional to its temperature. In other words, gas density increases as pressure increases and as temperature decreases.

EXAMPLE 13-11: The density of a nonreactive colorless gas found in an undersea vent was carefully determined to be 1.250 g·L^{-1} at 0.00°C, and 1.000 atm. Determine the molar mass of the gas and identity of the gas.

Solution: Solving Equation 13.12 for molar mass, we obtain

$$M = \rho\left(\frac{RT}{P}\right) = \left(\frac{m}{V}\right)\left(\frac{RT}{P}\right) \tag{13.13}$$

where m is the mass of a gas occupying a specific volume. This equation allows for a direct determination of the molar mass of a gas from its density at a given temperature and pressure. Substituting in the density of the unknown gas and the temperature and pressure, we find

$$M = \left(\frac{1.250\ \text{g}}{1\ \text{L}}\right)\left(\frac{(0.082058\ \text{L·atm·mol}^{-1}\text{·K}^{-1})(273.15\ \text{K})}{1.000\ \text{atm}}\right)$$
$$= 28.02\ \text{g·mol}^{-1}$$

The gas is nitrogen, $N_2(g)$.

PRACTICE PROBLEM 13-11: Calculate the density of $NO_2(g)$ at 0°C and 1.00 bar.

Answer: 2.03 g·L^{-1}

Nitrogen dioxide is more dense than air under the same conditions, so it can be poured from one container to another, as shown in Figure 13.18.

Figure 13.18 Like liquids, gases have flow properties and can be poured from one container to another if they are denser than air. This photo shows $NO_2(g)$ being poured.

You should realize that the ideal-gas law, Equation 13.8, is independent of the identity of the gas; that is, it does not contain a term for molar mass or any other factor specific to the gas in question. It is because of this that Avogadro's law is valid. That is, equal volumes of gases, at the same temperature and pressure, contain equal numbers of molecules (or moles) because $n = PV/RT$ for *any* gas that obeys this relationship. In contrast, the density of a gas depends on its molar mass, Equation 13.12. Thus, gas densities are *not* independent of the identity of the gas and so allow for a direct determination of molar mass. In the 1860s Stanislao Cannizzaro (Frontispiece) used measurements of gas densities together with Avogadro's law to experimentally compile the first consistent table of atomic masses.

13-8. The Total Pressure of a Mixture of Ideal Gases Is the Sum of the Partial Pressures of All the Gases in the Mixture

Up to this point we have not considered explicitly mixtures of gases, and yet mixtures of gases are of great importance. For example, air is a mixture of about 78% nitrogen, 21% oxygen, and 1% argon by volume with lesser amounts of other gases, such as carbon dioxide. Many chemical processes involve gaseous mixtures (Figure 13.19). For example, the commercial production of ammonia involves the reaction described by the equation

$$3\,H_2(g) + N_2(g) \xrightarrow[500^\circ C]{300\text{ bar}} 2\,NH_3(g)$$

Thus, the reaction vessel contains a mixture of $N_2(g)$, $H_2(g)$, and $NH_3(g)$.

In a mixture of ideal gases, each gas exerts a pressure as if it were present alone in the container. For a mixture of two ideal gases, we have

$$P_{\text{total}} = P_1 + P_2 \tag{13.14}$$

The pressure exerted by each gas is called its **partial pressure**, and Equation 13.14 is known as **Dalton's law of partial pressures**. Each of the gases in Equation 13.14 obeys the ideal-gas equation, so

$$P_1 = \frac{n_1RT}{V} \qquad P_2 = \frac{n_2RT}{V} \tag{13.15}$$

Notice that the volume occupied by each gas is V because each gas in a mixture occupies the entire container. If the partial pressures P_1 and P_2 are substituted into Equation 13.14, then for our two-gas mixture, we have

Figure 13.19 Chemists can manipulate gases and mixtures of gases at various pressures using an apparatus called a vacuum rack. This photo shows a vacuum rack in a research laboratory at the Scripps Institution in San Diego.

$$P_{\text{total}} = \frac{n_1RT}{V} + \frac{n_2RT}{V} = (n_1 + n_2)\frac{RT}{V} = n_{\text{total}}\frac{RT}{V} \tag{13.16}$$

The total pressure exerted by a mixture of gases is determined by the total number of moles of gas in the mixture.

If we divide each of the Equations 13.15 by Equation 13.16, we obtain

$$\frac{P_1}{P_{\text{total}}} = \frac{n_1}{n_1 + n_2} \qquad \frac{P_2}{P_{\text{total}}} = \frac{n_2}{n_1 + n_1} \tag{13.17}$$

We now write

$$x_1 = \frac{n_1}{n_1 + n_2} = \frac{n_1}{n_{\text{total}}} \qquad x_2 = \frac{n_2}{n_1 + n_2} = \frac{n_2}{n_{\text{total}}} \tag{13.18}$$

where x_1 is the mole fraction of gas 1 in the mixture and x_2 is the mole fraction of gas 2. Notice that a **mole fraction** is unitless (it is a fraction) and that

$$x_1 + x_2 = 1 \tag{13.19}$$

Equation 13.17 can be written in terms of mole fractions as

$$P_1 = x_1P_{\text{total}} \qquad P_2 = x_2P_{\text{total}} \tag{13.20}$$

which expresses the partial pressure of each species in terms of its mole fraction and the total pressure.

If we multiply both the numerators and the denominators in Equations 13.18 by Avogadro's number, then x_1 and x_2 will be expressed as molecular fractions,

$$x_1 = \frac{N_1}{N_{\text{total}}} \qquad x_2 = \frac{N_2}{N_{\text{total}}} \tag{13.21}$$

where N_1 and N_2 are the number of molecules of gases 1 and 2, respectively. Thus, x_1, for example, is the fraction of molecules in a mixture that are gas 1 molecules. With this in mind, we can give a simple molecular interpretation of Dalton's law of partial pressures. The pressure exerted by a gas or a gaseous mixture is due to the constant collisions of the molecules of the gas with the walls of the container. According to the ideal-gas law, at a fixed temperature and volume, the pressure exerted by a gas is proportional only to the number of molecules of the gas. Thus, in a mixture of gases, the partial pressure of each gas is the total pressure multiplied by the fraction of molecules of each gas.

Equations 13.14 through 13.21 may be expressed for a mixture of N gases by a more general form of Equation 13.14, namely, $P_{\text{total}} = P_1 + P_2 + \ldots + P_N$, which yields

$$P_{\text{i}} = \frac{n_{\text{i}}RT}{V} = x_{\text{i}}P_{\text{total}}, \quad x_{\text{i}} = \frac{n_{\text{i}}}{n_{\text{total}}}, \quad x_1 + x_2 + \cdots + x_N = 1$$

where i represents the ith species in the gas.

EXAMPLE 13-12: A 0.428-gram mixture of gases contained in a vessel at 1.75 atm is found to be 15.6% $N_2(g)$, 46.0% $N_2O(g)$, and 38.4% $CO_2(g)$ by mass. What is the partial pressure of each gas in the mixture?

Solution: The masses of each of the three gases in the mixture are given by

$$\text{mass of } N_2 = (15.6\%)(0.428 \text{ g}) = 0.0668 \text{ g } N_2$$

$$\text{mass of } N_2O = (46.0\%)(0.428 \text{ g}) = 0.197 \text{ g } N_2O$$

$$\text{mass of } CO_2 = (38.4\%)(0.428 \text{ g}) = 0.164 \text{ g } CO_2$$

and the numbers of moles of each gas are given by

$$\text{moles of } N_2 = (0.0668 \text{ g } N_2)\left(\frac{1 \text{ mol } N_2}{28.01 \text{ g } N_2}\right) = 2.38 \times 10^{-3} \text{ mol}$$

$$\text{moles of } N_2O = (0.197 \text{ g } N_2O)\left(\frac{1 \text{ mol } N_2O}{44.01 \text{ g } N_2O}\right) = 4.48 \times 10^{-3} \text{ mol}$$

$$\text{moles of } CO_2 = (0.164 \text{ g } CO_2)\left(\frac{1 \text{ mol } CO_2}{44.01 \text{ g } CO_2}\right) = 3.73 \times 10^{-3} \text{ mol}$$

The total number of moles is 10.59×10^{-3} mol and the various mole fractions are given by

$$x_{N_2} = \frac{2.38 \times 10^{-3} \text{ mol}}{10.59 \times 10^{-3} \text{ mol}} = 0.225$$

$$x_{N_2O} = \frac{4.48 \times 10^{-3} \text{ mol}}{10.59 \times 10^{-3} \text{ mol}} = 0.423$$

$$x_{CO_2} = \frac{3.73 \times 10^{-3} \text{ mol}}{10.59 \times 10^{-3} \text{ mol}} = 0.352$$

The partial pressures are

$$P_{N_2} = x_{N_2}P_{\text{total}} = (0.225)(1.75 \text{ atm}) = 0.394 \text{ atm}$$

$$P_{N_2O} = x_{N_2O}P_{\text{total}} = (0.423)(1.75 \text{ atm}) = 0.740 \text{ atm}$$

$$P_{CO_2} = x_{CO_2}P_{\text{total}} = (0.352)(1.75 \text{ atm}) = 0.616 \text{ atm}$$

PRACTICE PROBLEM 13-12: Consider two flasks connected by a stopcock (Figure 13.20). One flask has a volume of 500.0 mL and contains $N_2(g)$ at a pressure of 700.0 Torr. The other flask has a volume of 400.0 mL and contains $O_2(g)$ at a pressure of 950.0 Torr. If the stopcock is opened so that the two gases mix completely, calculate the partial pressures of $N_2(g)$ and $O_2(g)$ and the total pressure of the resultant mixture.

Answer: $P_{O_2} = 422.2$ Torr; $P_{N_2} = 388.9$ Torr; $P_{\text{total}} = 811.1$ Torr

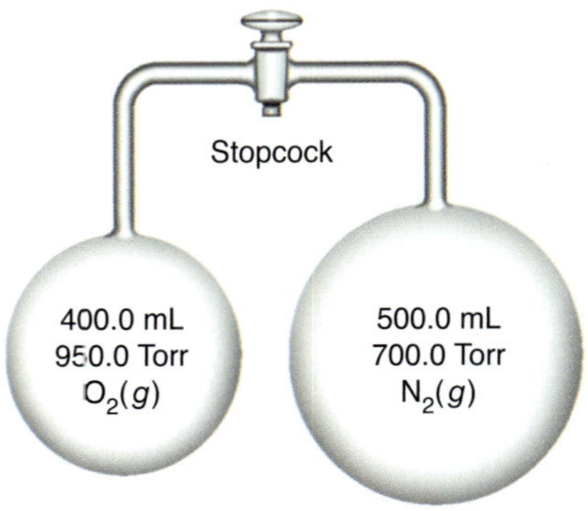

Figure 13.20 Two flasks connected by a stopcock. One flask has a volume of 400.0 mL and the other has a volume of 500.0 mL.

EXAMPLE 13-13: Suppose that we have a mixture of $N_2(g)$ and $O_2(g)$ of unknown composition whose total pressure is 385 Torr. If all the $O_2(g)$ is removed from the mixture by reaction with phosphorus, which does not react directly with nitrogen, then the new pressure is 251 Torr. Calculate the mole fraction of each gas in the original mixture.

Solution: The pressure before the removal of the $O_2(g)$ is

$$P_{\text{total}} = 385 \text{ Torr} = P_{N_2} + P_{O_2}$$

After removal of the $O_2(g)$

$$P_{N_2} = 251 \text{ Torr}$$

Therefore, the pressure due to $O_2(g)$ initially is

$$P_{O_2} = P_{\text{total}} - P_{N_2} = 385 \text{ Torr} - 251 \text{ Torr} = 134 \text{ Torr}$$

Now let x_{N_2} be the mole fraction of $N_2(g)$ in the mixture and $x_{O_2} = 1 - x_{N_2}$ be the mole fraction of the $O_2(g)$. Using Equation 13.20, the mole fractions of $N_2(g)$ and $O_2(g)$ in the mixture are given by

$$x_{N_2} = \frac{P_{N_2}}{P_{\text{total}}} = \frac{251 \text{ Torr}}{385 \text{ Torr}} = 0.652 \qquad x_{O_2} = \frac{P_{O_2}}{P_{\text{total}}} = \frac{134 \text{ Torr}}{385 \text{ Torr}} = 0.348$$

Note that $x_{N_2} + x_{O_2} = 1$, as it must.

PRACTICE PROBLEM 13-13: A mixture of $N_2(g)$ and $H_2(g)$ has mole fractions of 0.40 and 0.60, respectively. Determine the density of the mixture at one bar and 0.00°C.

Answer: $0.54 \text{ g} \cdot \text{L}^{-1}$

Practical applications of Dalton's law of partial pressures often arise in the laboratory. A standard method for determining the quantity of a water-insoluble gas evolved in a chemical reaction is diagramed in Figure 13.21. The gas displaces the water from an inverted container that is initially filled with water. When the reaction is completed, the container is raised or lowered until the water levels inside and outside are the same. When the two levels are the same, the pressure inside the container is equal to the atmospheric pressure. The pressure inside the container, however, is not due just to the gas collected; there is also water vapor present. Thus, the pressure inside the container is

$$P_{\text{total}} = P_{\text{gas}} + P_{H_2O} = P_{\text{atmosphere}} \tag{13.22}$$

The vapor pressure of $H_2O(l)$, P_{H_2O}, depends only on the temperature; Table 15.7 gives the vapor pressure of water at various temperatures. We shall study the pressure due to water vapor in more detail in Chapter 15.

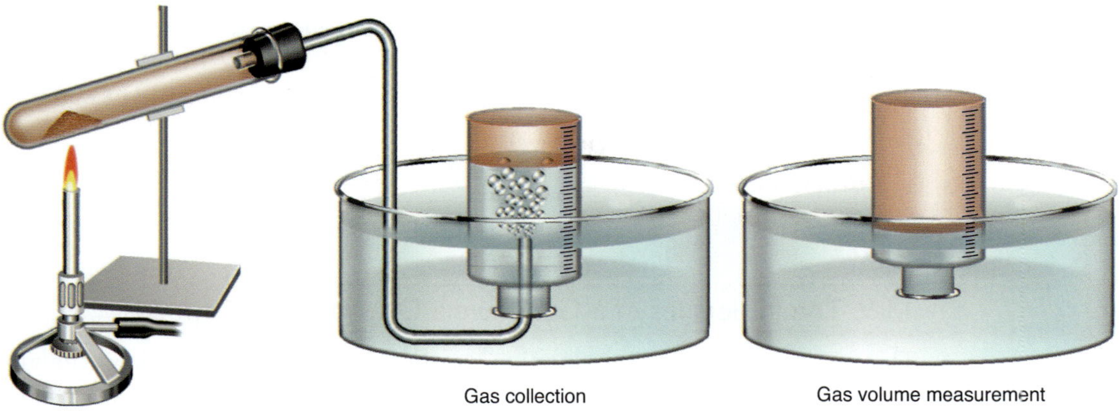

Figure 13.21 (a) The collection of a gas over water. (b) When the water levels inside and outside the container are equal, the pressure inside the container and the atmospheric pressure must be equal.

EXAMPLE 13-14: The reaction described by the chemical equation

$$2\,KClO_3(s) \rightarrow 2\,KCl(s) + 3\,O_2(g)$$

represents a common laboratory procedure for producing small quantities of pure oxygen. A 0.250-liter flask is filled with oxygen that has been collected over water at an atmospheric pressure of 729 Torr (Figure 13.21). The temperature of the water and the gas is 14°C. Calculate the molar volume of $O_2(g)$ at 0°C and 760 Torr. The vapor pressure of $H_2O(l)$ at 14°C is 12.0 Torr.

Solution: To calculate the molar volume of $O_2(g)$, we must first determine its partial pressure. The atmospheric pressure is 729 Torr, and the vapor pressure of $H_2O(l)$ at 14°C is 12.0 Torr; therefore,

$$P_{O_2} = P_{\text{total}} - P_{H_2O} = (729 - 12.0)\text{ Torr} = 717\text{ Torr}$$

From this value, we calculate the number of moles of $O_2(g)$ produced using the ideal-gas equation:

$$n = \frac{PV}{RT} = \frac{(717\text{ Torr})(0.250\text{ L})}{(62.36\text{ L·Torr·mol}^{-1}\text{·K}^{-1})(287\text{ K})} = 0.0100\text{ mol}$$

To find the volume of $O_2(g)$ at 0°C and 760 Torr, we use the form of Equation 13.11 we derived earlier:

$$V_f = V_i\left(\frac{P_i}{P_f}\right)\left(\frac{T_f}{T_i}\right) = (0.250\text{ L})\left(\frac{717\text{ Torr}}{760\text{ Torr}}\right)\left(\frac{273\text{ K}}{287\text{ K}}\right) = 0.224\text{ L}$$

Thus, the molar volume of $O_2(g)$ at 0°C and 760 Torr is

$$\text{molar volume of } O_2 = \frac{0.224\text{ L}}{0.0100\text{ mol}} = 22.4\text{ L·mol}^{-1}$$

PRACTICE PROBLEM 13-14: The hydrogen generated in a reaction is collected over water and occupies 425 mL at 18°C and 742 Torr. What is the volume of the dry $H_2(g)$ at 0°C and 760 Torr? The vapor pressure of water at 18°C is 15.5 Torr.

Answer: 381 mL

13-9. The Molecules of a Gas Have a Distribution of Speeds

The fact that the ideal-gas equation can be used for *all* gases at low enough pressures suggests that this law reflects the fundamental nature of gases. As we have seen, the molecules in a gas are widely separated and in constant motion and exert a pressure as they collide with the walls of the container. By applying the laws of physics to the motion of the molecules, it is possible to calculate the pressure exerted by the molecules as a result of their collisions with the walls of the container. Because this approach focuses on the motion of the molecules, it is called the **kinetic theory of gases**.

Recall from Chapter 1 that a body in motion has energy by virtue of the fact that it is in motion. The energy associated with the motion of a body, called **kinetic energy** (E_k), is given by Equation 1.4 as

$$E_k = \frac{1}{2}mv^2$$

where m is the mass of the body and v is its velocity (or speed). If m is expressed in kilograms and v in meters per second, then E_k has the units of joules (J). Recall that $1\text{ J} = 1\text{ kg·m}^2\text{·s}^{-2}$ (see Appendix B).

We shall see in Example 13-15 that the typical speed of a nitrogen molecule at room temperature is about 500 m·s^{-1}, or about 1100 miles per hour (mph). Let's calculate the kinetic energy of a mole of nitrogen molecules if each molecule has a speed of 500 m·s^{-1}. The kinetic energy per mole of molecules, each traveling with a speed, v, is given by

$$E_k = \frac{1}{2}M_{kg}v^2 \tag{13.23}$$

where M_{kg} is the molar mass in kilograms per mole. The molar mass must be expressed in units of kilograms per mole (kg·mol^{-1}) so that the units of E_k will be joules per mole ($\text{J·mol}^{-1} = \text{kg·m}^2\text{·s}^{-2}\text{·mol}^{-1}$). Thus, the kinetic energy per mole of nitrogen molecules is given by

$$E_k = \frac{1}{2}(28.0\text{ g·mol}^{-1})\left(\frac{1\text{ kg}}{1000\text{ g}}\right)(500\text{ m·s}^{-1})^2$$

$$= 3500\text{ J·mol}^{-1} = 3.5\text{ kJ·mol}^{-1}$$

if each nitrogen molecule has a speed of 500 m·s^{-1}. Dividing by Avogadro's number, we see that the kinetic energy in joules per nitrogen molecule at room temperature is about 6×10^{-21} J, or about 0.006 aJ.

Although, for simplicity, we have just implied that all the molecules in a gas travel at the same speed; in fact, they do not. Figure 13.22 illustrates an experimental apparatus that can be used to determine the distribution of the speeds of the molecules in a gas, and Figure 13.23 shows the results of such an

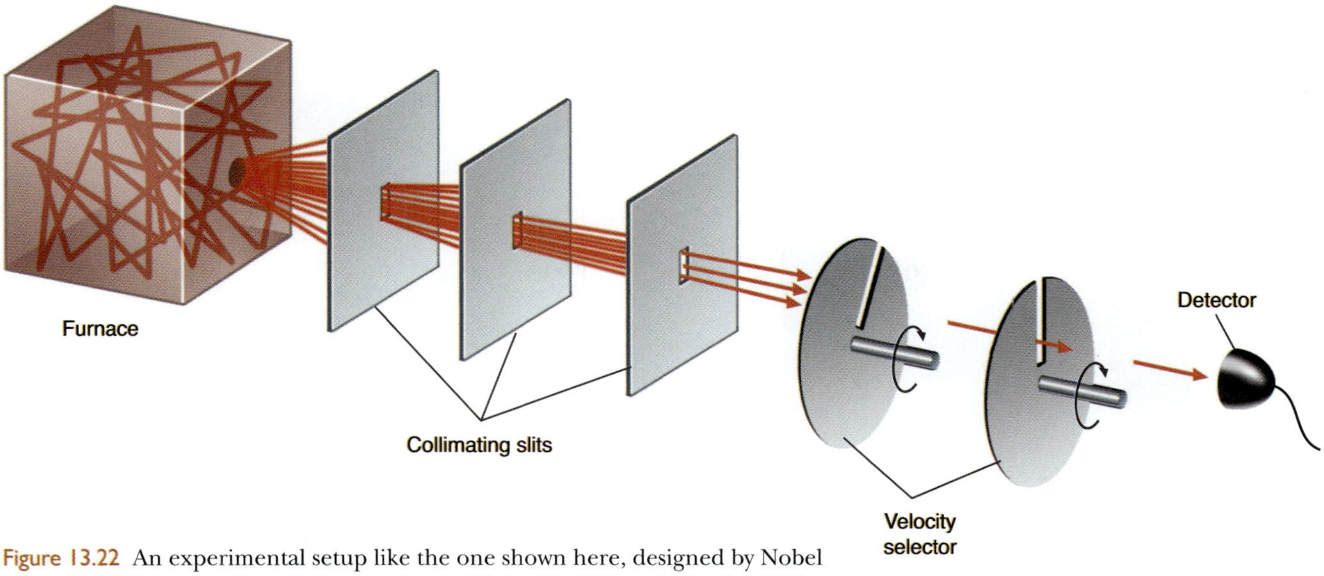

Figure 13.22 An experimental setup like the one shown here, designed by Nobel laureate Polykarp Kusch and his coworkers at Columbia University in the 1950s, can be used to measure the speed distribution of the molecules in a gas. A beam of high-temperature gaseous atoms or molecules emerges from a small hole in a furnace and is passed through a series of collimating slits. The beam is then passed through a velocity selector consisting of a series of rotating disks with small slits set at fixed angles from each other. Only those particles traveling at the right speed to pass through each of the successive slits can exit the selector. Particles moving at different speeds can be selected by varying the rate of rotation of the disks. A detector counts the number of particles exiting the apparatus at a given speed. By counting the number of atoms exiting the apparatus as a function of speed, one can measure the velocity distribution of a gas at a given temperature.

experiment. In Figure 13.23 we plot the fraction of molecules that have a speed, v, versus v at two different temperatures. Both curves start at the origin, rise to a maximum, and then fall off to zero as the speed increases. Notice that more molecules travel at higher speeds at higher temperatures. The distribution of molecular speeds in a gas is called a **Maxwell-Boltzmann distribution**, after James Clerk Maxwell (Figure 13.25) and Ludwig Boltzmann (Figure 13.26),

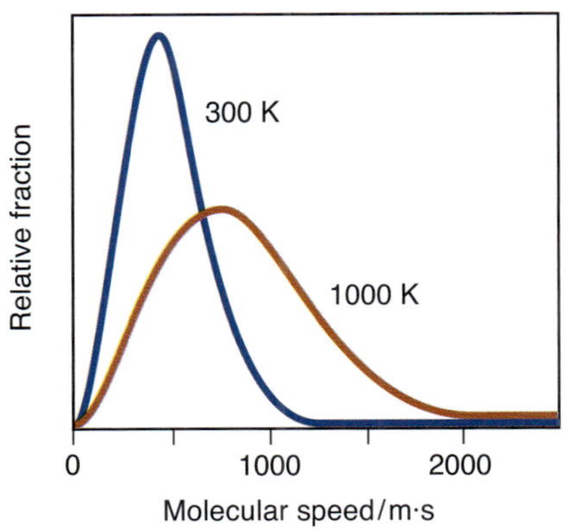

Figure 13.23 The distribution of speeds for nitrogen molecules at 300 K and 1000 K. The distribution is represented by the fraction of nitrogen molecules that have speed v plotted against that speed. Note, for example, that the fraction of molecules with a speed of 1000 m·s^{-1} is greater at 1000 K than at 300 K.

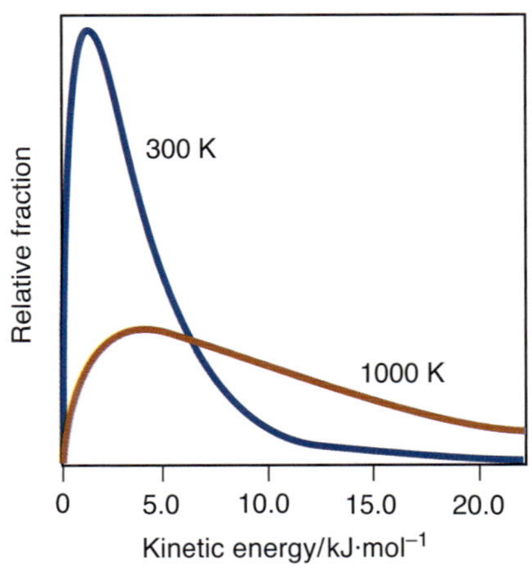

Figure 13.24 The distribution of kinetic energies for nitrogen molecules at 300 K and 1000 K. The distribution is represented by the fraction of nitrogen molecules that have kinetic energy E_k plotted against that kinetic energy.

Figure 13.25 James Clerk Maxwell (1831–1879), born in Edinburgh, Scotland, made significant contributions to many fields of science. He and Boltzmann applied the methods of probability to describe the motions of molecules in gases and showed that their velocities follow the Maxwell-Boltzmann distribution. His greatest work was his theory of electricity and magnetism, that is summarized by four equations called Maxwell's equations, known by all students of physics. Using these equations, he calculated that the speed of propagation of an electromagnetic wave is the same as the speed of light, thus showing that light is a form of electromagnetic radiation. In 1871, he became the first Cavendish Professor of Physics at Cambridge University, now one of the most prestigious chairs in all of science.

Figure 13.26 Ludwig Boltzmann (1844–1906), born in Vienna, Austria, received his doctorate from the University of Vienna, where he did experimental work on gases and radiation. He became known for his theoretical work on thermodynamics and the kinetic theory of gases. He was an early proponent of atomic theory, and his work involved a study of the atomic theory of matter. The distribution of the speeds and the energies of gas molecules is now called the Maxwell-Boltzmann distribution. At the time, the atomic nature of matter was not generally accepted and Boltzmann's work was criticized by a number of eminent scientists.

two scientists who developed the kinetic theory of gases in the latter half of the nineteenth century.

Because the molecules in a gas travel at different speeds, they also have different kinetic energies. Thus, there is a distribution of molecular kinetic energies in a gas. This distribution is shown in Figure 13.24, where the fraction of molecules with kinetic energy E_k is plotted against E_k. Note that the curves are somewhat similar to those in Figure 13.23, in the sense that they rise through a maximum and then fall off to zero as v increases.

13-10. The Kinetic Theory of Gases Allows Us to Calculate the Root-Mean-Square Speed of a Molecule

We are now ready to discuss the kinetic theory of gases. The postulates of the kinetic theory of gases are as follows:

1. The molecules in a gas are incessantly in motion. They collide randomly with one another and with the walls of the container.
2. All collisions between the gas particles or the walls of the chamber are **elastic**. That is, no energy is lost due to heat or friction during collisions.

3. The average distance between the molecules in a gas is much larger than the size of the molecules. In other words, the gas consists mostly of empty space.
4. Any interactions between the molecules in a gas are negligible. We assume that the molecules in a gas neither repel nor attract one another.
5. The mean (or average) kinetic energy of the molecules in a gas is proportional to the Kelvin temperature of the gas.

This last postulate involves the ***mean* kinetic energy** because there is a distribution of molecular kinetic energies. It is common notation to denote the mean of a quantity by placing a bar over the symbol for that quantity, so, using Equation 13.23, we can write the mean kinetic energy per mole as

$$\overline{E_k} = \tfrac{1}{2}M_{kg}\overline{v^2} \qquad (13.24)$$

Our fifth postulate of the kinetic theory of gases states that

$$\overline{E_k} = \tfrac{1}{2}M_{kg}\overline{v^2} = cT \qquad (13.25)$$

where c is a proportionality constant.

The first four postulates of the kinetic theory of gases describe the model of a gas that we have already introduced, and we shall see the importance of the final postulate shortly. From these postulates, we can calculate the properties of gases in terms of molecular quantities. For example, we can calculate an expression for the pressure of a gas in terms of the speeds of its constituent molecules. A detailed analysis of the collisions between molecules in a gas shows that the proportionality constant in Equation 13.25 is $c = 3R/2$, where R is the molar gas constant. Therefore, Equation 13.25 becomes

$$\overline{E_k} = \tfrac{3}{2}RT \qquad (13.26)$$

Equation 13.26 states that the temperature of an ideal gas is directly related to its mean kinetic energy, which is the idea behind the fifth postulate of the kinetic theory of gases. It tells us the meaning of temperature, an *observed* quantity, in terms of the mean value of a *molecular* property, the kinetic energy of the gas molecules.

Because $\overline{E_k}$ is expressed in units of joule·mole^{-1}, we must express R in units of joule·mole^{-1}·kelvin^{-1} when we use Equation 13.26. Thus, we use the value $R = 8.3145$ J·mol^{-1}·K^{-1} when calculating the mean kinetic energy of gases. We can use Equations 13.25 and 13.26 to calculate an important measure of the speed at which the molecules in a gas travel. Equating these two expressions for mean kinetic energy yields

$$\tfrac{1}{2}M_{kg}\overline{v^2} = \tfrac{3}{2}RT \qquad (13.27)$$

We first solve Equation 13.27 for the mean value of v^2,

$$\overline{v^2} = \frac{3RT}{M_{kg}}$$

and then take the square root of both sides (recall that raising a function to the ½ power is equivalent to taking its square root):

$$\left(\overline{v^2}\right)^{1/2} = \left(\frac{3RT}{M_{\text{kg}}}\right)^{1/2} \tag{13.28}$$

The quantity $\left(\overline{v^2}\right)^{1/2}$ has the units of speed ($\text{m}\cdot\text{s}^{-1}$). It is called the **root-mean-square (rms) speed** because it is the square *root* of the *mean* (or average) of the *square* of the molecular speeds. We will denote the root-mean-square speed by v_{rms}, so Equation 13.28 can be written as

$$v_{\text{rms}} = \left(\frac{3RT}{M_{\text{kg}}}\right)^{1/2} \tag{13.29}$$

The root-mean-square speed is a good measure of the average molecular speed in a gas.

Strictly speaking, the average molecular speed in a gas is given by

$$\bar{v} = \left(\frac{8RT}{\pi M_{\text{kg}}}\right)^{1/2}$$

However, because $v_{\text{rms}}/\bar{v} = 1.085$ this difference is of no concern to us here.

EXAMPLE 13-15: Calculate the root-mean-square speed of a nitrogen molecule at 20°C.

Solution: The molar mass of $N_2(g)$ in kilograms per mole is

$$M_{\text{kg}} = (28.0\ \text{g}\cdot\text{mol}^{-1})\left(\frac{1\ \text{kg}}{1000\ \text{g}}\right) = 0.0280\ \text{kg}\cdot\text{mol}^{-1}$$

Thus, using Equation 13.29, we calculate

$$v_{\text{rms}} = \left(\frac{3RT}{M_{\text{kg}}}\right)^{1/2} = \left[\frac{(3)(8.3145\ \text{J}\cdot\text{mol}^{-1}\cdot\text{K}^{-1})(293\ \text{K})}{0.0280\ \text{kg}\cdot\text{mol}^{-1}}\right]^{1/2}$$

$$= (2.61\times10^{5}\ \text{J}\cdot\text{kg}^{-1})^{1/2} = (2.61\times10^{5}\ \text{m}^{2}\cdot\text{s}^{-2})^{1/2} = 511\ \text{m}\cdot\text{s}^{-1}$$

Note that

$$511\ \text{m}\cdot\text{s}^{-1}\left(\frac{1\ \text{km}}{1000\ \text{m}}\right)\left(\frac{3600\ \text{s}}{1\ \text{hour}}\right) = 1840\ \text{kph (kilometers per hour)}$$

This is about 1140 mph (miles per hour), which is comparable to the muzzle velocity of a high-speed rifle bullet.

Values of v_{rms} for several gases are given in Table 13.3. Note that v_{rms} decreases with increasing molar mass at constant temperature, as is required by Equation 13.29.

PRACTICE PROBLEM 13-15: The speed of sound in an ideal diatomic gas is given by

$$v_{\text{sound}} = \left(\frac{7RT}{5M_{\text{kg}}}\right)^{1/2} \tag{13.30}$$

Calculate the speed of sound in $N_2(g)$ at 20°C. Compare this value to the speed of sound in air at 20°C and one bar (see sidebox).

Answer: 349 m·s^{-1}; this is comparable to the speed of sound in air at 20°C and one bar because air is mostly $N_2(g)$.

TABLE 13.3 Values of v_{rms} for four gases at 20°C and 1000°C

		$v_{rms}/\text{m·s}^{-1}$	
Gas	**Molar mass/kg·mol^{-1}**	**t = 20°C**	**t = 1000°C**
H_2	0.0020	1900	4000
N_2	0.0280	510	1060
O_2	0.0320	480	1000
CO_2	0.0440	410	850

THE SPEED OF SOUND A sound wave is a pressure disturbance that travels through a substance. The speed with which a sound wave travels through a gas depends on the speeds of the molecules in the gas through which it travels. It can be shown from the kinetic theory of gases that the speed of sound through a gas is about 0.7 v_{rms}. The speed of sound in air at 20°C and one bar is about 340 m·s^{-1} or 1200 kph (770 mph). When something travels at supersonic speeds, it moves faster than the molecules in the media, effectively creating a vacuum in its wake. The sound of the air rushing back in creates the "sonic boom" made by supersonic aircraft, missiles, and spacecraft on reentry (Figure 13.27). The first man-made object to travel faster than sound was the whip, the tip of which can move at supersonic speeds, creating its characteristic "cracking" sound.

Figure 13.27 When an aircraft first breaks the sound barrier, the sudden change in pressure can condense the water vapor in the atmosphere, forming a vapor cloud in the wake of the plane.

13-11. We Can Use Effusion to Determine the Formula Mass of a Gas

We can use Equation 13.28 to derive a formula for the relative rates at which gases leak from a container through one or more small holes, a process called **effusion** (Figure 13.28). For two gases at the same pressure and temperature, the rate of effusion is directly proportional to the root-mean-square speed of the molecules. We let v_{rms_A} and v_{rms_B} be the root-mean-square speeds of two gases A and B, and we use Equation 13.29 to write

$$v_{rms_A} = \left(\frac{3RT}{M_{kg_A}}\right)^{1/2} \quad \text{and} \quad v_{rms_B} = \left(\frac{3RT}{M_{kg_B}}\right)^{1/2}$$

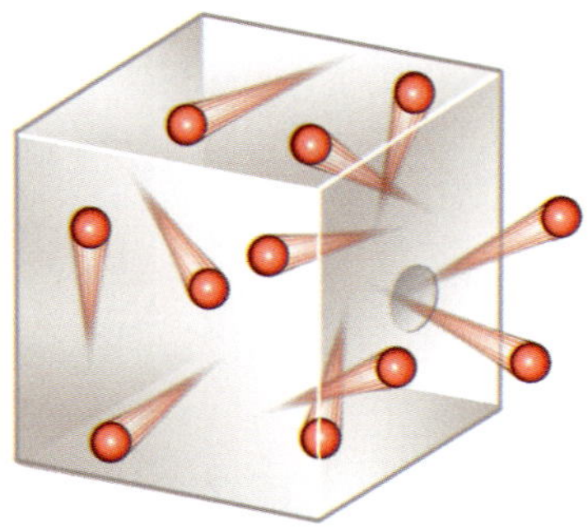

Figure 13.28 The rate at which a gas effuses through a small hole in a container is proportional to the root-mean-square speed of the gas molecules. On average, the faster the gas molecules are moving the sooner they encounter the hole and can escape from the container.

The temperature does not have a subscript because both gases are at the same temperature. If we divide $v_{\mathrm{rms_A}}$ by $v_{\mathrm{rms_B}}$, we obtain

$$\frac{v_{\mathrm{rms_A}}}{v_{\mathrm{rms_B}}} = \left(\frac{M_{\mathrm{kg_B}}}{M_{\mathrm{kg_A}}}\right)^{1/2} = \left(\frac{M_\mathrm{B}}{M_\mathrm{A}}\right)^{1/2}$$

where M_A and M_B are the molar masses of gases A and B. The rate of effusion is directly proportional to v_rms, so

$$\frac{\mathrm{Rate_A}}{\mathrm{Rate_B}} = \left(\frac{M_\mathrm{B}}{M_\mathrm{A}}\right)^{1/2} \tag{13.31}$$

This relation was observed experimentally by the British scientist Thomas Graham in the 1840s and is called **Graham's law of effusion**. For Graham's law to hold, the gas molecules must pass directly through the holes of the container without colliding with any other molecules. For this to occur, the diameters of the holes must be small enough and the pressure of the gas must be low enough that the molecules pass through the holes individually, that is, without colliding with any other gas molecules.

EXAMPLE 13-16: A porous container is filled with fluorine, $F_2(g)$. It is observed that the pressure within the container decreased by 20% in 6.0 hours. If the same container is filled with He(g) under the same conditions, how long would it take for the pressure to decrease by 20%?

Solution: We must first recognize that rate and time are inversely proportional to each other; the shorter the time for a certain amount of effusion to occur, the greater is the rate of effusion. Thus, Equation 13.31 can be written in the form

$$\frac{\mathrm{Rate_A}}{\mathrm{Rate_B}} = \frac{\mathrm{time_B}}{\mathrm{time_A}} = \left(\frac{M_\mathrm{B}}{M_\mathrm{A}}\right)^{1/2} \tag{13.32}$$

If we let A denote $F_2(g)$ and B denote He(g), then Equation 13.32 becomes

$$\mathrm{time_{He}} = (\mathrm{time_{F_2}})\left(\frac{M_\mathrm{He}}{M_{\mathrm{F_2}}}\right)^{1/2} = (6.0\ \mathrm{h})\left(\frac{4.003}{38.00}\right)^{1/2} = 2.0\ \mathrm{h}$$

Helium takes less time to escape because, being lighter, the atoms travel at a higher mean speed than fluorine molecules do at a given temperature.

PRACTICE PROBLEM 13-16: A porous container is filled with equal amounts of He(g) and an unknown noble gas. The He(g) escaped 5.73 times faster than the unknown gas did. What is the identity of the unknown noble gas?

Answer: Xe(g)

Notice from Practice Problem 13-16 that we can use Graham's law of effusion to determine the formula mass of an unknown gas.

13-12. The Average Distance a Molecule Travels Between Collisions Is Called the Mean Free Path

Although most of the molecules in a gas at one bar and 20°C travel with speeds of hundreds of meters per second, they do not travel any appreciable distances that rapidly. We all have observed that it may take several minutes for an odor to spread through a draft-free room. The explanation for this observation lies in the fact that the molecules in a gas undergo many collisions, so their actual path is a chaotic, zigzag path like that shown in Figure 13.29. Between collisions, gas molecules travel with speeds of hundreds of meters per second, but their net progress is quite slow. The average distance traveled between collisions is called the **mean free path** (l).

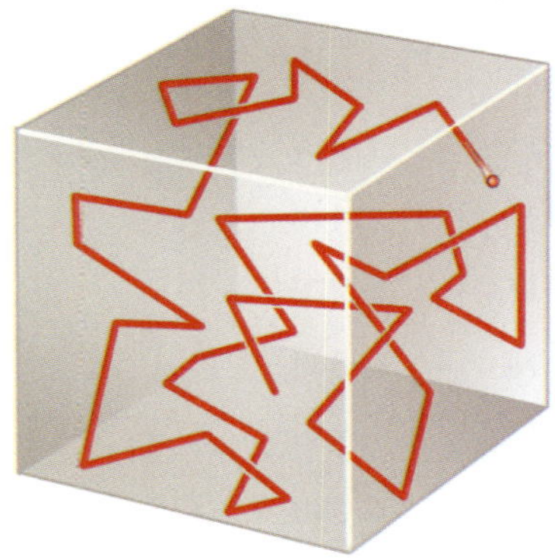

Figure 13.29 A typical path followed by a gas molecule. The molecule travels in a straight line until it collides with another molecule, at which point its direction is changed in an almost random manner. At 20°C and one bar, a molecule undergoes about 10^{10} collisions per second.

We can estimate the mean free path by reasoning that the greater the density of molecules, the more often a molecule will collide; so the mean free path will be shorter. Therefore, the mean free path should be inversely proportional to the number density of molecules, N/V, where N is the number of molecules in a given volume, V. Furthermore, the greater the cross-sectional area of a molecule, the greater a target it will present for other molecules; and again the mean free path will be shorter. If a molecule has a diameter equal to d, then its cross-sectional area will be approximately πr^2 (the area of a circle) or $\pi(d/2)^2$. Therefore, the mean free path should be inversely proportional to the *square* of the diameter of the molecule. If we let l be the mean free path, then we write

$$l \propto \left(\frac{1}{N/V}\right)\left(\frac{1}{d^2}\right) \quad \text{or} \quad l = \frac{c}{d^2(N/V)} \tag{13.33}$$

where c is a proportionality constant. It turns out that the value of this constant is $1/(\pi\sqrt{2})$. Using this result, and the fact that $N = n \cdot N_A$ (where N_A is Avogadro's number), and $n/V = P/RT$ from the ideal-gas equation, Equation 13.33 may be rewritten as

$$l = \frac{RT}{(\pi\sqrt{2})\, d^2 N_A P} \tag{13.34}$$

Molecular diameters are obviously very small and are commonly expressed in units of picometers (pm), 1 pm = 10^{-12} meters. Table 13.4 lists the molecular diameters of some selected molecules.

EXAMPLE 13-17: Use Equation 13.34 and the data in Table 13.4 to calculate the mean free path in $N_2(g)$ at 1.00 bar and 20°C. How many molecular diameters does a typical nitrogen molecule travel between collisions?

Solution: The values used in Equation 13.34 should all be in SI units. From Table 13.4 we find that the molecular diameter of N_2 is 380 pm, which is equivalent to 3.8×10^{-10} meters. A pressure of 1.00 bar is equal

TABLE 13.4 Atomic or molecular diameters for various gases

Gas	Diameter/pm
He	210
Ne	250
Ar	370
Kr	410
Xe	490
H_2	270
N_2	380
O_2	360
Cl_2	540
CH_4	410

to 1.00×10^5 Pa, or $1.00 \times 10^5\ \text{J·m}^{-3}$ in SI units (Table 13.1). Using these values, a temperature of 293 K, and $R = 8.3145\ \text{J·mol}^{-1}\text{·K}^{-1}$, we find that the mean free path of a nitrogen molecule under these conditions is

$$l = \frac{RT}{(\pi\sqrt{2})\,d^2 N_A P}$$

$$= \frac{(8.3145\ \text{J·mol}^{-1}\text{·K}^{-1})(293\ \text{K})}{(\pi\sqrt{2})(3.8 \times 10^{-10}\ \text{m})^2(6.022 \times 10^{23}\ \text{mol}^{-1})(1.00 \times 10^5\ \text{J·m}^{-3})}$$

$$= 6.3 \times 10^{-8}\ \text{m}$$

This is equivalent to a distance of 6.3×10^4 pm. Because the molecular diameter of a nitrogen molecule is 380 pm, we find that

$$\frac{l}{d} = \frac{6.3 \times 10^4\ \text{pm}}{380\ \text{pm}} = 170$$

This means that at 1.00 bar and 20°C, a nitrogen molecule travels an average distance of almost 200 molecular diameters between collisions. Indeed the volume of a gas is essentially all empty space!

PRACTICE PROBLEM 13-17: Calculate the mean free path of a hydrogen molecule at 0°C at the low pressure of 10^{-5} Torr, which is a typical pressure used in experiments carried out under vacuum.

Answer: 9 meters

We can also derive a formula for the **collision frequency** of a molecule, that is, the number of collisions that a molecule undergoes per second. If we denote the collision frequency by z, then

$$z = \frac{\text{collisions}}{\text{second}}$$

$$= \frac{\text{distance traveled per second}}{\text{distance traveled per collision}}$$

$$= \frac{v_{\text{rms}}}{l} \tag{13.35}$$

Let's use this result to calculate the collision frequency of a nitrogen molecule at 1.00 bar and 20°C. From Example 13-15 we see that $v_{\text{rms}} = 511\ \text{m·s}^{-1}$, and from Example 13-17 we found that $l = 6.3 \times 10^{-8}$ m under the same conditions. For $N_2(g)$ at 20°C and 1.00 bar, we get

$$z = \frac{v_{\text{rms}}}{l} = \frac{511\ \text{m·s}^{-1}}{6.3 \times 10^{-8}\ \text{m·collision}^{-1}} = 8.1 \times 10^9\ \text{collisions·s}^{-1}$$

Thus, we see that one nitrogen molecule undergoes about eight billion collisions per second at 20°C and one bar!

13-13. The van der Waals Equation Accounts for Deviations from Gas Ideality

The ideal-gas equation is valid for all gases at sufficiently low pressures. As the pressure on a given quantity of gas is increased, however, deviations from the ideal-gas equation appear. These deviations can be displayed graphically by plotting PV/RT as a function of pressure, as shown in Figure 13.30. For one mole of an ideal gas, PV/RT is equal to unity for any value of P, so deviations from ideal-gas behavior occur as deviations of the ratio PV/RT from unity. The extent of **deviation from ideality** at a given pressure depends on the temperature and the nature of the gas. The closer the gas is to the point at which it liquefies, the larger the deviation from ideal behavior will be. There are many equations that correct the ideal-gas law for these deviations from ideality. We shall examine the simplest and most famous of them here.

The ideal-gas equation is based upon the physical assumptions that (1) the sizes of the molecules of the gas are essentially zero and (2) that they do not attract each other. Both of these assumptions are incorrect: molecules do have a certain size and they do attract each other. We can modify the ideal-gas equation to take these two factors into account. We'll first subtract the actual volume of the molecules of the gas from V to write $V - nb$, where b is a constant whose value depends upon the gas and is a measure of the size of the molecules. The quantity $V - nb$ can be interpreted as the actual volume that is free for the molecules to move about in (Figure 13.31). Substituting $V - nb$ for V in the ideal-gas equation gives us

$$P(V - nb) = nRT \tag{13.36}$$

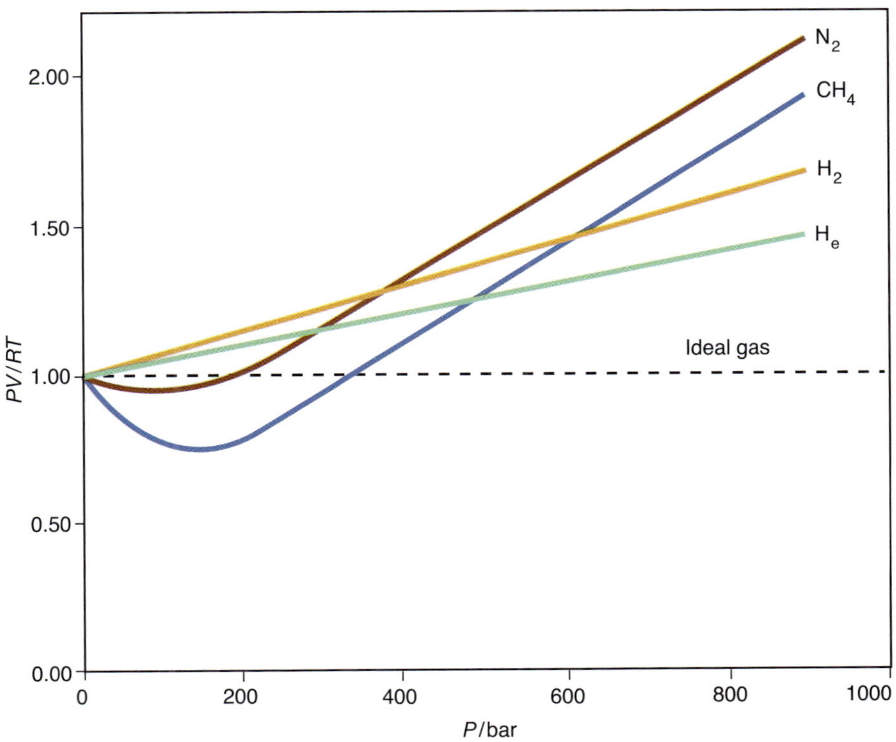

Figure 13.30 A plot of (PV/RT) versus P for one mole of helium, hydrogen, nitrogen, and methane at 300 K. This figure shows that the ideal-gas equation, for which $PV/RT = 1$, is not valid at high pressures.

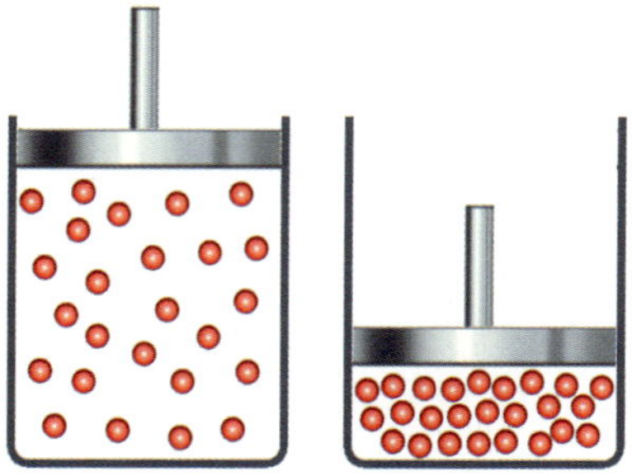

Figure 13.31 At high pressures, the volume of the molecules of a gas is no longer negligible relative to the volume of the container.

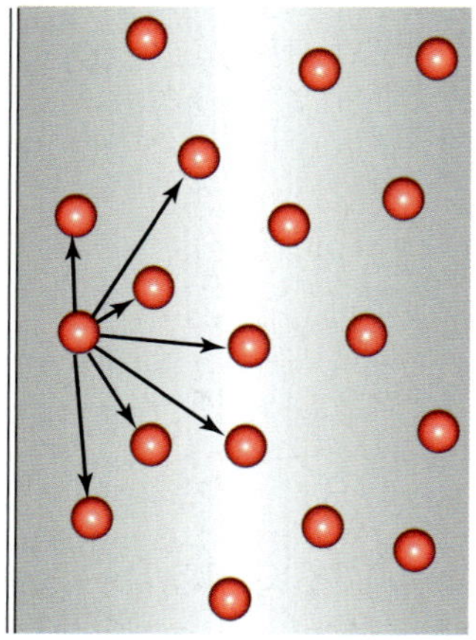

Figure 13.32 An illustration of the fact that a gas molecule near a wall of its container experiences a net inward drag due to its attraction by the other molecules. Consequentially, it strikes the wall with less force than if the gas were ideal.

This equation recognizes the finite size of the molecules, but does not take into account the attraction between the molecules. Because of the attractive forces between the molecules, the actual pressure of the gas is less than that given by the ideal-gas law (Figure 13.32). Because the attraction is between pairs of molecules, this factor is proportional to the square of the density of the gas. If we let the proportionality constant be a, then we can write

$$P_{\text{actual}} = P_{\text{ideal}} - a\left(\frac{n}{V}\right)^2 \tag{13.37}$$

Substituting P_{ideal} from Equation 13.37 for the pressure, P, given by the ideal-gas law in Equation 13.36, we obtain

$$\left(P + a\frac{n^2}{V^2}\right)(V - nb) = nRT \tag{13.38}$$

Equation 13.38 is called the **van der Waals equation**, after the Dutch scientist and Nobel laureate Johannes van der Waals (Figure 13.33), who was the first to recognize the importance of molecular size and molecular attractions on the equation of state that describes a gas. The two constants a and b, whose values depend upon the particular gas, are called **van der Waals constants** (Table 13.5).

Let's use the van der Waals equation to calculate the pressure exerted at 0°C by 1.00 mole of methane, $CH_4(g)$, occupying a 0.250-liter container. From Table 13.5 we find that $a = 2.303\ \text{L}^2\cdot\text{bar}\cdot\text{mol}^{-2}$ and $b = 0.0431\ \text{L}\cdot\text{mol}^{-1}$ for methane. If we divide Equation 13.38 by $V - nb$ and solve for P, then we obtain

$$P = \frac{nRT}{V - nb} - \frac{n^2a}{V^2} \tag{13.39}$$

Substituting $n = 1.00$ mol, $R = 0.083145\ \text{L}\cdot\text{bar}\cdot\text{mol}^{-1}\cdot\text{K}^{-1}$, $T = 273.15$ K, $V = 0.250$ L, and the values of a and b into Equation 13.39, we obtain

TABLE 13.5 The van der Waals constants of some gases

Gas	Formula	$a/\text{L}^2\cdot\text{bar}\cdot\text{mol}^{-2}$	$b/\text{L}\cdot\text{mol}^{-1}$
ammonia	NH_3	4.3044	0.037847
carbon dioxide	CO_2	3.6551	0.042816
methane	CH_4	2.3026	0.043067
neon	Ne	0.2167	0.017383
nitrogen	N_2	1.3661	0.038577
oxygen	O_2	1.3820	0.031860
propane	C_3H_8	9.3919	0.090494

$$P = \left[\frac{(1.00 \text{ mol})(0.083145 \text{ L·bar·mol}^{-1}\text{·K}^{-1})(273.15 \text{ K})}{0.250 \text{ L} - (1.00 \text{ mol})(0.0431 \text{ L·mol}^{-1})}\right] - \left[\frac{(1.00 \text{ mol})^2(2.303 \text{ L}^2\text{·bar·mol}^{-2})}{(0.250 \text{ L})^2}\right]$$

$$= 73 \text{ bar}$$

By comparison, the ideal-gas equation predicts that

$$P = \frac{nRT}{V} = \frac{(1.00 \text{ mol})(0.083145 \text{ L·bar·mol}^{-1}\text{·K}^{-1})(273.15 \text{ K})}{0.250 \text{ L}}$$

$$= 90.8 \text{ bar}$$

The prediction of the van der Waals equation is in much better agreement with the experimental value of 75.3 bar than the ideal-gas equation is.

Figure 13.33 Johannes Diderik van der Waals (1837–1923) was born in Leiden, the Netherlands. He did his doctoral thesis on the properties of gases and, soon after, published what is now known as the van der Waals equation. In 1876 he became the first professor of physics at the University of Amsterdam, which under his influence became a major center of theoretical and experimental research on fluids. Van der Waals received the 1910 Nobel Prize in Physics for his work on the equation of state for gases and liquids.

SUMMARY

In a gas, the particles are widely separated and travel throughout the entire volume of their container in a chaotic manner, colliding with one another and with the walls of the container. The pressure exerted by a gas is due to the incessant collisions of the gas molecules on the walls of the container. At constant temperature, the volume and the pressure of a gas are related by Boyle's law, which says that volume and pressure are inversely related. The relation between the volume of a gas and its temperature is given by Charles's law, which serves also to define the absolute or Kelvin temperature scale.

The experimental study of the combining volumes of reacting gases led to Gay-Lussac's law of combining volumes. Gay-Lussac's law in turn led to Avogadro's law, which states that equal volumes of gases at the same pressure and temperature contain equal numbers of molecules. Boyle's, Charles's, and Avogadro's laws can be combined into the ideal-gas law, which relates the pressure, volume, temperature, and number of moles of a gas.

All the experimental gas laws can be explained by the kinetic theory of gases. One of the central tenets of the kinetic theory of gases is that the mean kinetic energy of a gas is directly proportional to its absolute temperature. The kinetic theory provides equations that can be used to calculate the root-mean-square speed of a molecule, its mean free path, collision frequency, relative rate of effusion, and other molecular quantities.

The ideal-gas equation is valid for all gases at sufficiently low densities and sufficiently high temperatures. As the pressure on a given quantity of gas is increased, however, deviations from the ideal-gas equation are observed. The van der Waals equation, which takes into account that the molecules of a gas attract one another and have a finite size, is valid at higher densities and lower temperatures than is the ideal-gas equation.

TERMS YOU SHOULD KNOW

compressibility *422*
vaporization *422*
pressure *422*
manometer *422*
Torr (mm Hg) *423*
barometer *423*
standard atmosphere (atm) *424*
pascal (Pa) *424*
bar *425*
Boyle's law *426*
Charles's law *427*
absolute (Kelvin) temperature scale *427*
kelvin *428*
absolute zero *428*
intercept *428*
slope *428*
Gay-Lussac's law of combining volumes *431*
Avogadro's hypothesis *432*
ideal-gas equation (ideal-gas law) *433*
ideal gas *433*
equation of state *434*
molar gas constant, R *434*
molar volume *434*
molar mass *441*
partial pressure *443*
Dalton's law of partial pressures *443*
mole fraction, x_i *444*
kinetic theory of gases *448*
kinetic energy, E_k *448*
Maxwell-Boltzmann distribution *449*
elastic *450*
mean kinetic energy, $\overline{E_k}$ *451*
root-mean-square speed, v_{rms} *452*
speed of sound *453*
effusion *453*
Graham's law of effusion *454*
mean free path, l *455*
collision frequency, z *456*
deviation from ideality *457*
van der Waals equation *458*
van der Waals constants *458*

EQUATIONS YOU SHOULD KNOW HOW TO USE

$P = \dfrac{F}{A}$	(13.1) definition of pressure
$V = \dfrac{c}{P}$	(13.3) Boyle's law
$T/\mathrm{K} = t/{}^\circ\mathrm{C} + 273.15$	(13.4) definition of Kelvin temperature
$V = mT$	(13.6) Charles's law
$PV = nRT$	(13.8) ideal-gas equation
$\dfrac{P_i V_i}{T_i} = \dfrac{P_f V_f}{T_f}$	(13.11) alternative form of ideal-gas law
$\rho = \dfrac{MP}{RT}$	(13.12) density of an ideal gas
$P_{total} = P_1 + P_2 + \ldots + P_N$	(13.14) Dalton's law of partial pressure
$x_i = \dfrac{n_i}{n_{total}}$	(13.18) mole fraction
$x_1 + x_2 + \ldots + x_N = 1$	(13.19) sum of mole fractions
$P_i = x_i \mathrm{P}_{total}$	(13.20) partial pressure of a gas
$\overline{E_k} = \frac{1}{2} M_{kg} \overline{v^2}$	(13.24) mean kinetic energy per mole of gas
$\overline{E_k} = \frac{3}{2} RT$	(13.26) relationship between mean kinetic energy per mole of an ideal gas and temperature

$$v_{rms} = \left(\frac{3RT}{M_{kg}}\right)^{1/2}$$ (13.29) root-mean-square speed of gas molecules

$$\frac{\text{Rate}_A}{\text{Rate}_B} = \frac{\text{time}_B}{\text{time}_A} = \left(\frac{M_B}{M_A}\right)^{1/2}$$ (13.32) Graham's law of effusion

$$l = \frac{RT}{(\pi\sqrt{2})\,d^2 N_A P}$$ (13.34) mean free path of gas molecules

$$z = \frac{v_{rms}}{l}$$ (13.35) collision frequency of gas molecules

$$\left(P + a\frac{n^2}{V^2}\right)(V - nb) = nRT$$ (13.38) van der Waals equation

PROBLEMS

PRESSURE AND BOYLE'S LAW

13-1. Carry out the following unit conversions:

(a) The atmospheric pressure at the surface of Venus is about 75 atm. Convert 75 atm to torr and to bars.

(b) The atmospheric pressure in Mexico City is about 580 Torr. Convert this pressure to atmospheres and to millibars.

(c) The pressure of carbon dioxide in a gas cylinder is 5.2 atm. Convert this pressure to pascals and to kilopascals.

(d) The pressure of a sample of nitrogen gas is 920 Torr. Convert this pressure to pascals and to atmospheres.

13-2. In meteorology pressures are expressed in units of millibars (mbar). Convert 985 mbar to torr and atmospheres.

13-3. A gas occupies a volume of 0.120 m^3 at a pressure of 12 kPa; what will the volume of the gas be at a pressure of 25 kPa and the same temperature?

13-4. Nitrogen is often injected into cans and bottles to keep the contents from oxidizing. If the nitrogen injected into a particular bottle occupies a volume of 10.0 milliliters at a pressure of 1.2 bar, what volume will it occupy at a pressure of 0.95 bar at the same temperature?

13-5. A gas bubble has a volume of 0.650 milliliters at the bottom of a lake, where the pressure is 3.46 atm. What is the volume of the bubble at the surface of the lake, where the pressure is 1.00 atm? What are the diameters of the bubble at the two depths? Assume that the temperature is constant and that the bubble is spherical.

13-6. Suppose we wish to inflate a weather balloon with helium. The balloon should have a volume of 100 m^3 when inflated to a pressure of 0.10 bar. If we use 50.0-liter cylinders of compressed helium gas at a pressure of 100 bar, how many cylinders do we need? Assume that the temperature remains constant.

TEMPERATURE AND CHARLES'S LAW

13-7. Convert the following temperatures to the Kelvin scale:

(a) –183°C (the melting point of oxygen)

(b) 6000°C (temperature at the surface of the sun)

(c) –269°C (the boiling point of helium)

(d) 800°C (the melting point of sodium chloride)

13-8. A balloon has a volume of 1.2 liters on a warm 32°C day. If the same balloon is placed in the freezer and cooled to –18°C, what volume will it occupy? (Assume constant pressure.)

13-9. A piston containing 25.0 dm^3 of gas at 45°C is expanded to a new volume of 55.0 dm^3. To what temperature must the gas be heated to maintain the same pressure as before the expansion?

13-10. Suppose that the gas in a gas thermometer occupies 12.6 milliliters at 20.0°C. The thermometer is immersed in a container of solid carbon dioxide chips (dry ice); the gas then occupies 8.4 milliliters. What is the temperature of the dry ice?

GAY-LUSSAC'S LAW

13-11. A scientist finds that two volumes of an unknown gas X combine with three volumes of another gas Y to form one volume of a new gas. Assuming that each of the reactants are monatomic gases, write a balanced chemical equation, including the formula of the new gas in terms of X and Y.

13-12. A scientist finds that one volume of an unknown gas X combines with two volumes of another gas Y to form two volumes of a new gas with a formula XY_2. Are any of the gases diatomic? Write a balanced equation for this reaction.

13-13. Methane burns according to the equation

$$CH_4(g) + 2\,O_2(g) \rightarrow CO_2(g) + 2\,H_2O(g)$$

What volume of air, which is 21% oxygen by volume, is required to burn 5.0 liters of methane when both are the same temperature and pressure?

13-14. Hydrogen and oxygen react violently with each other once the reaction is initiated. For example, a spark can set off the reaction and cause the mixture to explode. What volume of oxygen gas will react with 0.55 liters of hydrogen gas if both are at 300°C and one bar? What volume of water vapor will be produced at 300°C and one bar?

IDEAL-GAS LAW

13-15. Calculate the volume that 0.65 moles of $NH_3(g)$ occupies at 37°C and 575 Torr.

13-16. Calculate the number of grams of propane, $C_3H_8(g)$, in a 50.0-liter container at a pressure of 7.50 atm and a temperature of 25°C.

13-17. Calculate the number of $Cl_2(g)$ molecules in a volume of 5.00 milliliters at 40.0°C and 2.15×10^4 Pa.

13-18. Calculate the pressure in pascals that is exerted by 6.15 milligrams of $CO_2(g)$ occupying 2.10 milliliters at 75°C.

13-19. Calculate the Celsius temperature at which 0.0100 grams of $CH_4(g)$ occupies a volume of 250.0 cm^3 with a pressure of 8.27 kPa.

13-20. A sample of radon gas occupies 7.12 μL at 22°C and 8.72×10^4 Pa. What is the mass of the radon gas?

13-21. The ozone molecules in the stratosphere absorb much of the ultraviolet radiation from the sun. The temperature of the stratosphere is −23°C, and the pressure due to the ozone is 1.4×10^{-7} atm. Calculate the number of ozone molecules present in 1.0 m^3 in the stratosphere.

13-22. A low pressure of 1.0×10^{-3} Torr is readily obtained in the laboratory by means of a mechanical vacuum pump. Calculate the number of molecules in 1.00 milliliter of gas at this pressure and 20°C.

GAS DENSITY

13-23. Calculate the density of water in the gas phase at 100.0°C and 1.00 atm. Compare this value with the density of liquid water at 100.0°C and 1.00 atm, which is 0.958 $g \cdot mL^{-1}$.

13-24. Calculate the density of the chlorofluorocarbon, $CF_2Cl_2(g)$, at 0°C and 1.00 atm.

13-25. A certain gaseous hydrocarbon was determined to be 82.66% carbon and 17.34% hydrogen by mass. A 6.09-gram sample of the gas occupied 2.48 liters at 1.00 atm and 15°C. Determine the molecular formula of the hydrocarbon.

13-26. Upon chemical analysis, a gaseous hydrocarbon is found to contain 88.82% carbon and 11.18% hydrogen by mass. A 62.6-milligram sample of the gas occupies 34.9 milliliters at 772 Torr and 100.0°C. Determine the molecular formula of the hydrocarbon.

13-27. The chemical xylene is used as a solvent and in the synthesis of polyester fibers. Chemical analysis shows that xylene is 90.50% carbon and 9.50% hydrogen by mass. A 2.334-gram sample of xylene was vaporized in a sealed 500.0-milliliters container at 100.0°C, producing a pressure of 1.346 atm. Use these data to determine the molecular formula of xylene.

13-28. Ethylene is a gas produced in petroleum cracking and is used to synthesize a variety of important chemicals, such as polyethylene and polyvinylchloride. Chemical analysis shows that ethylene is 85.60%

carbon and 14.40% hydrogen by mass. It has a density of 0.9588 g·L^{-1} at 25°C and 635 Torr. Use these data to determine the molecular formula of ethylene.

13-29. The Dumas method was one of the first reliable techniques for determining the molar mass of a volatile liquid. In this method a liquid is boiled inside an open glass bulb so that its vapor completely fills the bulb, displacing any air present. The vapor inside the bulb is then cooled and the condensate weighed. From this mass, the boiling point of the liquid, and the atmospheric pressure, one can determine the molar mass of the liquid. A chemist using this technique finds that a pure volatile liquid with a boiling point of 80.1°C fills a 500.0-cm^3 bulb at an atmospheric pressure of 755.3 Torr. If the weighed condensate has a mass of 1.335 grams, what is the molecular mass of the liquid? If the empirical formula of the unknown liquid is CH, what is the molecular formula?

13-30. A chemist using the Dumas method (see previous Problem) finds that a pure volatile liquid with a boiling point of 10.8°C fills a 250.0-cm^3 bulb at an atmospheric pressure of 0.998 bar. If the weighed condensate has a mass of 635 milligrams, what is the molecular mass of the liquid? Chemical analysis shows that when 1.200 grams of the liquid is combusted, 2.637 grams of $CO_2(g)$ and 1.439 grams of $H_2O(l)$ are formed. Determine the molecular formula of the unknown liquid. Assume that the unknown liquid is composed of only the elements carbon, hydrogen, and oxygen.

PARTIAL PRESSURES

13-31. A gaseous mixture consisting of 0.513 grams of $H_2(g)$ and 16.1 grams of $N_2(g)$ occupies 10.0 liters at 20.0°C. Calculate the partial pressures of $H_2(g)$ and $N_2(g)$ in the mixture in units of atmospheres.

13-32. A gaseous mixture contains 400.0 Torr of $H_2(g)$, 355.1 Torr of $N_2(g)$, and 75.2 Torr of $Ar(g)$. Calculate the mole fraction of each of these gases.

13-33. A 0.622-gram mixture of gases contained in a vessel at 1450 Torr is found to be 22.2% argon, 68.5% helium, and 9.3% fluorine by mass. What is the partial pressure of each of the gases in the mixture?

13-34. In Example 13-9 we experimentally determined 29.0 to be the effective molecular mass of dry air. Assuming that air consists only of nitrogen and oxygen, use this value to determine the mass percentages of oxygen and nitrogen in the air. How well does this agree with the actual percentages? What does this result suggest about the actual composition of air?

13-35. A mixture of $O_2(g)$ and $N_2(g)$ is reacted with white phosphorus, $P_4(s)$, which removes the oxygen. If the volume of the mixture decreases from 50.0 milliliters to 35.0 milliliters, calculate the partial pressures of $O_2(g)$ and $N_2(g)$ in the mixture. Assume that the total pressure remains constant at 745 Torr.

13-36. A gaseous mixture inside a rigid steel vessel contains 75% $CO_2(g)$ and 25% $H_2O(g)$ by volume at 175°C and 225 kPa. The mixture is then cooled to 0°C, thereby condensing the water vapor. What is the pressure of the $CO_2(g)$ at 0°C? (Assume that there is no water vapor present after condensation.)

13-37. A 2.0-liter sample of $H_2(g)$ at 1.00 atm, an 8.0-liter sample of $N_2(g)$ at 3.00 atm, and a 4.0-liter sample of $Kr(g)$ at 0.50 atm are transferred to a 10.0-liter container at the same temperature. Calculate the partial pressure of each gas and the total final pressure in the new container.

13-38. Two glass bulbs are connected by a valve. One bulb has a volume of 650.0 milliliters and is occupied by $N_2(g)$ at 825 Torr. The other has a volume of 500.0 milliliters and is occupied by $O_2(g)$ at 732 Torr. The valve is opened and the two gases mix. Calculate the total pressure and the partial pressures of $N_2(g)$ and $O_2(g)$ in the resulting mixture.

IDEAL-GAS LAW AND CHEMICAL REACTIONS

13-39. Cellular respiration occurs according to the overall equation

$$\underset{\text{glucose}}{C_6H_{12}O_6(s)} + 6\,O_2(g) \rightarrow 6\,CO_2(g) + 6\,H_2O(l)$$

Calculate the volume of $CO_2(g)$ produced at 37°C (body temperature) and 1.00 bar when 1.00 gram of glucose is metabolized.

13-40. Chlorine gas is produced by the electrolysis of an aqueous solution of sodium chloride:

$$2\,NaCl(aq) + 2\,H_2O(l) \xrightarrow{\text{electrolysis}} 2\,NaOH(aq) + H_2(g) + Cl_2(g)$$

If the $H_2(g)$ and $Cl_2(g)$ from this reaction are collected separately at 10.0 atm and 25°C, what volume of each can be obtained from 2.50 kilograms of dissolved sodium chloride?

13-41. Chlorine can be prepared in the laboratory by the reaction of manganese dioxide with hydrochloric acid, $HCl(aq)$, as described by the chemical equation

$$MnO_2(s) + 4\,HCl(aq) \rightarrow MnCl_2(aq) + 2\,H_2O(l) + Cl_2(g)$$

How much $MnO_2(s)$ should be added to excess $HCl(aq)$ to obtain 255 milliliters of $Cl_2(g)$ at 25°C and 755 Torr?

13-42. Acetylene, $C_2H_2(g)$, is prepared by the reaction of calcium carbide, $CaC_2(g)$, with water, as described by the balanced chemical equation

$$CaC_2(s) + 2\,H_2O(l) \rightarrow Ca(OH)_2(s) + C_2H_2(g)$$

What volume of $C_2H_2(g)$ can be obtained from 100.0 grams of $CaC_2(s)$ and 100.0 grams of water at 0°C and 1.00 atm? What volume results when the temperature is 125°C and the pressure is 1.00 atm?

13-43. Lithium metal reacts with $N_2(g)$ at room temperature (20°C) according to the equation

$$6\,Li(s) + N_2(g) \rightarrow 2\,Li_3N(s)$$

A sample of $Li(s)$ was placed under a $N_2(g)$ atmosphere in a sealed 1.00-liter container at a pressure of 1.23 atm. One hour later the pressure dropped to 0.92 atm. Calculate the number of grams of $N_2(g)$ that reacted with the $Li(s)$. Assuming that all the $Li(s)$ reacted, calculate the mass of $Li(s)$ originally present.

13-44. About 50% of U.S. and most Canadian sulfur is produced by the Claus process, in which sulfur is obtained from the $H_2S(g)$ that occurs in natural gas deposits or is produced when sulfur is removed from petroleum. The reactions are described by the equations

(1) $2\,H_2S(g) + 3\,O_2(g) \rightarrow 2\,SO_2(g) + 2\,H_2O(g)$

(2) $SO_2(g) + 2\,H_2S(g) \rightarrow 3\,S(l) + 2\,H_2O(g)$

How many metric tons of sulfur can be produced from 2.00 million liters of $H_2S(g)$ at 6.00 bar and 200.0°C?

13-45. Nitroglycerin decomposes according to the equation

$$4\,C_3H_5(NO_3)_3(s) \rightarrow 12\,CO_2(g) + 10\,H_2O(l) + 6\,N_2(g) + O_2(g)$$

What is the total volume of the gases produced when collected at 1.00 bar and 25°C from the decomposition of 10.0 grams of nitroglycerin? What pressure is produced if the reaction is confined to a volume of 0.500 liters at 25°C? Assume that you can use the ideal-gas equation. Neglect any pressure due to water vapor.

13-46. Explosions occur when a substance decomposes rapidly with the production of a large volume of gases. When detonated, TNT (trinitrotoluene), decomposes according to the equation

$$2\,C_7H_5(NO_2)_3(s) \rightarrow 2\,C(s) + 12\,CO(g) + 5\,H_2(g) + 3\,N_2(g)$$

What is the total volume of gases produced from 1.00 kilogram of TNT at 0°C and 1.00 atm? What pressure is produced if the reaction is confined to a 50-liter container at 500°C? Assume that you can use the ideal-gas equation.

13-47. Sulfur dioxide can combine with oxygen to form sulfur trioxide according to the equation

$$2\,SO_2(g) + O_2(g) \rightarrow 2\,SO_3(g)$$

A researcher studying the reaction under various conditions introduces 725.0 Pa of sulfur dioxide into a rigid stainless steel reaction chamber maintained at a constant temperature. She then introduces 500.0 Pa of oxygen. Assuming that all the sulfur dioxide is converted to sulfur trioxide, calculate the final total pressure inside the chamber.

13-48. A rigid stainless steel chamber contains 125 Torr of methane and excess oxygen at 200.0°C. A spark is ignited inside the chamber, completely combusting the methane. What is the change in total pressure within the chamber following the reaction? Assume a constant temperature throughout the process.

MOLECULAR SPEEDS

13-49. Calculate the root-mean-square speed, v_{rms}, of a fluorine molecule at 25°C.

13-50. Calculate the root-mean-square speed, v_{rms}, for $N_2O(g)$ at 20°C, 200°C, and 2000°C.

13-51. If the temperature of a gas is doubled, how much is the root-mean-square speed of the molecules comprising the gas increased?

13-52. Consider a mixture of hydrogen and iodine gas. Calculate the ratio of the root-mean-square speeds of $H_2(g)$ and $I_2(g)$ gas molecules in the reaction mixture.

GRAHAM'S LAW OF EFFUSION

13-53. Two identical balloons are filled, one with helium and one with nitrogen, at the same temperature and pressure. If the nitrogen leaks out from its balloon at the rate of 75 $mL \cdot h^{-1}$, what will be the rate of leakage from the helium-filled balloon?

13-54. Two identical porous containers are filled, one with hydrogen and one with carbon dioxide at the same temperature and pressure. After one day, 1.50 milliliters of carbon dioxide have leaked out of its container. How much hydrogen has leaked out in one day?

13-55. It takes 145 seconds for 1.00 milliliter of $N_2(g)$ to effuse from a certain porous container. Given that it takes 230 seconds for 1.00 milliliter of an unknown gas to effuse under the same temperature and pressure, calculate the molecular mass of the unknown gas.

13-56. Suppose that it takes 175.0 seconds for 1.00 milliliter of $N_2(g)$ to effuse from a porous container under a certain temperature and pressure and that it takes 200.0 seconds for 1.00 milliliter of a CO-$CO_2(g)$ mixture to effuse under the same conditions. What is the volume percentage of $CO(g)$ in the mixture?

MEAN FREE PATH

13-57. Interstellar space has an average temperature of about 10 K and an average density of hydrogen atoms of about one hydrogen atom per cubic meter. Calculate the mean free path of hydrogen atoms in interstellar space. Take d = 100 pm for a hydrogen atom.

13-58. Calculate the pressures in atmospheres at which the mean free path of a hydrogen molecule will be 1.00 μm, 1.00 mm, and 1.00 m at 20.0°C.

13-59. Calculate the number of collisions per second that one hydrogen molecule undergoes at 20°C and 1.0 bar.

13-60. Calculate the number of collisions per second that one molecule of nitrogen undergoes at 20°C and a pressure of 1.0×10^{-3} Torr (the pressure achieved by most mechanical vacuum pumps).

VAN DER WAALS EQUATION

13-61. Explain what factors not accounted for by the ideal-gas equation are corrected for by the van der Waals constants *a* and *b*.

13-62. Show that in the limit where the volume of a gas is large compared to the number of particles occupying that volume, Equation 13.38, the van der Waals equation, becomes identical to Equation 13.8, the ideal-gas equation.

13-63. Use the van der Waals equation to calculate the pressure in bars exerted by 24.5 grams of $NH_3(g)$ confined to a 2.15-liter container at 300 K. Compare your answer with the pressure calculated using the ideal-gas equation.

13-64. Use the van der Waals equation to calculate the pressure in bars exerted by 45 grams of propane, $C_3H_8(g)$, confined to a 2.2-liter container at 300.0°C. Compare your answer with the pressure calculated using the ideal-gas equation.

ADDITIONAL PROBLEMS

13-65. Under normal conditions, why is it relatively easy to compress a gas, but much more difficult to compress a liquid?

13-66. Why do most barometers use liquid mercury, a toxic metal, rather than, say, water or alcohol, as a medium for measuring atmospheric pressure?

13-67. A child dips a straw into a glass of milk and covers the top end with his finger. Why doesn't the milk spill out of the straw when he lifts the straw out of the glass? Why, when he removes his finger, does the milk now spill out of the straw?

13-68. In the United States, atmospheric pressure is often reported in "inches of mercury." Calculate the pressure of 1.00 bar in inches of mercury.

13-69. At a certain temperature and pressure, the density of $CO_2(g)$ was determined to be $1.7192\ g\cdot L^{-1}$ and the density of $O_2(g)$ to be $1.2500\ g\cdot L^{-1}$. Using these data and the known atomic mass of oxygen (15.9994), calculate the atomic mass of carbon to five significant figures.

13-70. Gallium metal can be used as a manometer fluid at high temperatures because of its wide liquid range (30 to 2400°C). Compute the height of a column of liquid gallium in a gallium manometer when the temperature is 850°C and the pressure is 1300 Torr. Take the density of liquid gallium to be $6.0\ g\cdot mL^{-1}$.

13-71. It takes 0.3625 grams of $N_2(g)$ to fill a glass container at 298.2 K and 0.0100 bar pressure. It takes 0.9175 grams of an unknown homonuclear diatomic gas to fill the same container under the same conditions. What is this gas?

13-72. Given below are pressure-volume data for a sample of 0.28 grams of $N_2(g)$ at 25°C. Verify Boyle's law for these data. Plot the data so that a straight line is obtained.

P/atm	0.26	0.41	0.83	1.20	2.10	2.63	3.14
V/L	0.938	0.595	0.294	0.203	0.116	0.093	0.078

13-73. A container of carbon dioxide has a volume of 50.0 liters. The carbon dioxide was originally at a pressure of 10.0 atm at 25°C, but after the tank had been used for a month at 25°C, the pressure dropped to 4.7 atm. Calculate the number of grams of carbon dioxide used during this period of time.

13-74. A website claims that it requires 10 000 gallons of air to burn one gallon of gasoline. Using octane, $C_8H_{18}(l)$, as the chemical formula of gasoline and the fact that air is 21% oxygen by volume, calculate the volume of air at 0°C and 1.00 bar that is required to burn a gallon of gasoline. Is the information on the website correct? Take the density of octane to be $0.70\ g\cdot mL^{-1}$.

13-75. Lactic acid is produced by the muscles when insufficient oxygen is available and is responsible for muscle cramps during vigorous exercising. It also provides the acidity found in dairy products. Chemical analysis shows that lactic acid is 39.99% carbon, 6.73% hydrogen, and 53.28% oxygen by mass. A 0.3338-gram sample of lactic acid was vaporized in a sealed 300.0-milliliters container at 150.0°C, producing a pressure of 326 Torr. Use these data to determine the molecular formula of lactic acid.

13-76. Calculate the volume of 0.200 M NaOH required to prepare 150 milliliters of $H_2(g)$ at 10.0°C and 750 Torr from the reaction described by the equation

$$2\,Al(s) + 2\,NaOH(aq) + 2\,H_2O(l) \rightarrow 2\,NaAlO_2(aq) + 3\,H_2(g)$$

13-77. Calculate the temperature at which a carbon dioxide molecule would have the same root-mean-square speed as a neon atom at 125°C.

13-78. Ammonia gas reacts with hydrogen chloride gas as described by the equation

$$NH_3(g) + HCl(g) \rightarrow NH_4Cl(s)$$

Suppose 5.0 grams of $NH_3(g)$ are reacted with 10.0 grams of $HCl(g)$ in a 1.00-liter vessel at 75°C. Compute the final pressure (in bar) of gas in the vessel.

13-79. Compare the mass of oxygen that you utilize to the mass of solid food that you consume each day. Assume that you breathe in 0.5 liters of air with each breath, that your respiratory rate is 14 breaths per minute, that you utilize about 25% of the oxygen inhaled, and that you eat 1 kilogram of solid food daily.

13-80. Using the data given in Table 13.5, predict which compound of those listed shows the largest deviation from ideal-gas behavior at 1000 bar.

13-81. How many grams of zinc must be added to a 100-milliliters solution of 6.00 M $HCl(aq)$ to produce 275 milliliters of hydrogen gas at 20.0°C and 0.989 bar?

13-82. Sodium peroxide is used in self-contained breathing devices to absorb exhaled carbon dioxide and simultaneously produce oxygen. The equation for the reaction is

$$2\,Na_2O_2(s) + 2\,CO_2(g) \rightarrow 2\,Na_2CO_3(s) + O_2(g)$$

How many liters of $CO_2(g)$, measured at 0°C and one atm, can be absorbed by one kilogram of $Na_2O_2(s)$? How many liters of $O_2(g)$ will be produced under the same conditions?

13-83. Nitrous oxide, $N_2O(g)$, can be prepared by gently heating equimolar quantities of potassium nitrate and ammonium chloride:

$$NH_4Cl(s) + KNO_3(s) \rightarrow KCl(s) + 2\,H_2O(l) + N_2O(g)$$

How many grams of $NH_4Cl(s)$ and $KNO_3(s)$ are required to produce 825 milliliters of $N_2O(g)$ at 25°C and 748 Torr?

13-84. Calculate the volume of $Cl_2(g)$ produced at 815 Torr and 15°C if 6.75 grams of $KMnO_4(s)$ are added to 300.0 milliliters of 0.1150 M $HCl(aq)$. The equation for the reaction is

$$2\,KMnO_4(s) + 16\,HCl(aq) \rightarrow 2\,MnCl_2(aq) + 2\,KCl(aq) + 5\,Cl_2(g) + 8\,H_2O(l)$$

13-85. The pressure in a ceramic vessel that contained nitrogen gas dropped from 1850 Torr to 915 Torr in 30.0 minutes. When the same vessel was filled with another gas, the pressure dropped from 1850 Torr to 915 Torr in 54.3 minutes. Calculate the molecular mass of the second gas, assuming that the gases effuse from the container.

13-86. On a hot, humid day the partial pressure of water vapor in the atmosphere is typically 30 to 40 Torr. Suppose that the partial pressure of water vapor is 35 Torr and that the temperature is 35°C. If all the water vapor in a room that measures 3.0 meters by 5.0 meters by 6.0 meters were condensed, how many milliliters of $H_2O(l)$ would be obtained?

13-87. (*) A mixture of zinc and aluminum with a total mass of 5.62 grams reacts completely with hydrochloric acid, liberating 2.67 liters of hydrogen gas at 23°C and 773 Torr. Calculate the mass percentage of zinc in the mixture.

13-88. (*) Both sodium hydride, $NaH(s)$, and calcium hydride, $CaH_2(s)$, react with water to produce $H_2(g)$ and the respective hydroxides. Suppose that a 3.75-gram sample of a mixture of $NaH(s)$ and $CaH_2(s)$ is added to water and the evolved $H_2(g)$ is collected. The $H_2(g)$ occupies a volume of 4.12 liters at 742 Torr and 17°C. Calculate the mass percentages of $NaH(s)$ and $CaH_2(s)$ in the mixture.

13-89. Each of the hemispheres shown in Figure 13.6 has a diameter of 14 inches. Estimate the effective force in pounds required to pull the two hemispheres apart when evacuated at one atmosphere of external pressure. Use the information on the inside back cover of the text for the necessary unit conversions.

13-90. While working after school in his high school chemistry laboratory, Joel Hildebrand (Chapter 12 Frontispiece) showed that the formula for a certain oxide of nitrogen published in a college chemistry textbook as N_2O_2 was wrong. He did so by demonstrating that two volumes of the oxide in question combined with one volume of oxygen to form one volume of the brown gas, dinitrogen tetroxide. Use Gay-Lussac's law of combining volumes to identify the correct formula for this oxide of nitrogen.

13-91. (*) The minimum velocity necessary for objects to leave the earth's gravitational field and escape into space is $11.2\ km\cdot s^{-1}$. Calculate v_{rms} for He at 2500°C in units of kilometers per second. Assuming that this is the average temperature of the helium in the earth's upper atmosphere, can any of this helium escape into space?

13-92. (*) A 0.428-gram sample of a mixture of $KCl(s)$ and $KClO_3(s)$ is heated; 80.7 milliliters of $O_2(g)$ is collected over water at 18°C and 756 Torr. Calculate the mass percentage of $KClO_3(s)$ in the mixture. (P_{H_2O} = 15.5 Torr at 18°C.)

13-93. (*) A mixture of neon and argon has a density of $1.64\ g\cdot L^{-1}$ at 0°C and 800.0 Torr. Compute the ratio of the number of moles of neon to the number of moles of argon in the mixture.

13-94. The speed of sound in an ideal monatomic gas is given by

$$v_{sound} = \left(\frac{5RT}{3M_{kg}}\right)^{1/2}$$

Derive an equation for the ratio v_{rms}/v_{sound}. Calculate v_{rms} for an argon atom at 20°C and compare your result to the speed of sound in argon.

James Prescott Joule (1818–1889) was born in Salford, near Manchester, England, to a wealthy brewer. He and his elder brother were tutored at home by John Dalton, who was then in his 70s. Due to his father's wealth, Joule never engaged in any profession nor did he manage the family brewery. He conducted his pioneering experiments in laboratories that he built in his home or in his father's brewery at his own expense. From 1837 to 1847, he carried out a series of experiments that led to the general law of the conservation of energy. Joule announced his measurements in a public lecture at St. Ann's Church in Manchester, England, because his earlier reports had been rejected by the British Association. He later had his lecture published in the *Manchester Courier*, a newspaper for which his brother wrote musical critiques. In 1847, he presented his results to the British Association meeting in Oxford, where the importance of his work was finally appreciated. Due to his lack of mathematical training, Joule did not contribute further to the new and expanding field of thermodynamics. Joule was elected to the Royal Society in 1850. His wife died young, leaving Joule to raise his two children by himself. Later in life he suffered severe financial losses, and in 1878 friends obtained a pension for him from the government. Unfortunately, his mental powers had begun to decline and he died after a long illness. The SI unit of energy is named in his honor.

14. Thermochemistry

In this chapter we begin our discussion of thermodynamics, which is that part of chemistry that deals with the energy changes involved in chemical reactions and changes in the physical states of substances. **Thermochemistry** is the branch of thermodynamics that deals with the evolution or absorption of energy as heat in chemical processes. Most chemical reactions are accompanied by the evolution or the absorption of energy as heat. Energy evolved in the burning of petroleum, natural gas, and coal supplies over 85% of the energy used annually in the United States. All living organisms store energy in the form of certain chemicals. Rockets, missiles, and explosives all derive their power from the energy released in chemical reactions (Figure 14.1). In this chapter we develop an understanding of the energy changes involved in chemical reactions.

Figure 14.1 Blastoff of a space shuttle. The energy released by the three main engines is equivalent to the output of 23 Hoover Dams. The two solid rocket boosters on the sides of the main fuel tank consume 5 metric tons of fuel per second and generate a combined thrust equivalent to 44 million horsepower, which is equivalent to the horsepower of 15 000 six-axle diesel locomotives or 400 000 subcompact cars.

14-1. The Transfer of Energy Between a Reaction System and Its Surroundings Occurs as Work or Heat

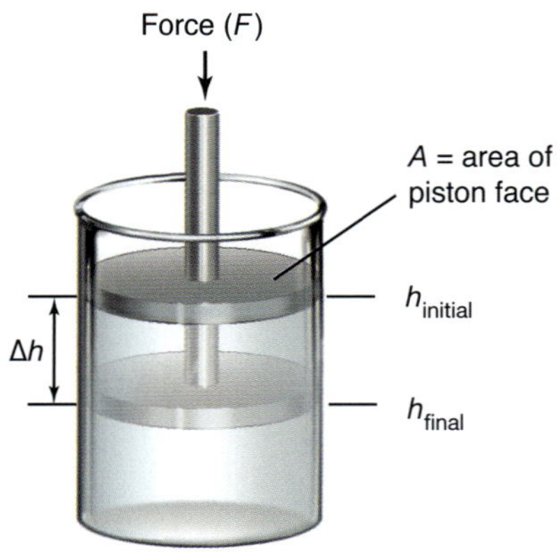

Figure 14.2 A system enclosed within a cylinder equipped with a piston. A force F is applied to a piston, thereby causing a displacement $\Delta h = h_{final} - h_{initial}$ of the piston. The area of the piston face is A.

Chemical reactions almost always involve a transfer of energy between the reaction **system** and its **surroundings**. The reaction system includes all the chemicals involved in the reaction. The surroundings consist of everything else, such as containers, water baths, and the atmosphere.

There are two ways in which energy can be transferred between a reaction system and its surroundings: either as work or as heat. **Thermodynamics** (*thermo* = heat, *dynamics* = change) is the study of these energy transfers. The transfer of energy as **work** involves the action of a force through a distance. Suppose our system consists of the contents of a gas inside a cylinder equipped with a piston at some initial height h, as shown in Figure 14.2. The application of a constant force, F, to the piston causes a displacement of the system by an amount Δh. Recall that the Greek letter delta, Δ, indicates a change in a quantity. The value of Δh is determined by subtracting the *initial value* of the quantity h from the *final value*. Thus, $\Delta h = h_{final} - h_{initial}$. The work done on a system is given by the magnitude of the force times the displacement (Δh) caused by the force. Thus, we can express the work, w, done on the system by

$$w = -F\Delta h \tag{14.1}$$

The minus sign occurs in Equation 14.1 because we shall use the convention that the work done *on* a system is a positive quantity. Observe that when the piston in Figure 14.2 is pushed inward, $\Delta h = h_{final} - h_{initial}$ has a negative value because $h_{final} < h_{initial}$ (compression). The minus sign in Equation 14.1 makes the right-hand side of the equation a positive value, as required for work done upon compressing a system (the system being compressed gains energy from the surroundings). Conversely, the work done *by* the system during expansion is a negative quantity because energy is lost by the system to the surroundings when the system does work. This concept is easy to grasp if you remember that *when the work is positive the system gains energy, and when the work is negative the system loses energy.*

We learned in the previous chapter that pressure is defined as force per unit area, or $P = F/A$, so that the force on the cylinder is given by $F = PA$. Substituting this expression into Equation 14.1 gives

$$w = -PA\Delta h \tag{14.2}$$

During compression, work is done on the system by the surroundings. The system gains energy and the sign of the work is positive ($w > 0$). During expansion, work is done on the surroundings by the system. The system loses energy and the sign of the work is negative ($w < 0$).

Because the volume of a cylinder is the area of the base times the height, that is, $V = Ah$, the change in volume of the system when the piston is displaced by Δh is $\Delta V = A\Delta h$. Thus, Equation 14.2 can be rewritten as

$$w = -P\Delta V = -P(V_f - V_i) \tag{14.3}$$

where V_f is the final volume and V_i is the initial volume. Equation 14.3 gives the work done on a system that is compressed or expanded at a constant pressure P. Equation 14.3 is especially useful in thermochemistry because many reactions take place in vessels open to the atmosphere, and the atmosphere exerts a constant pressure during the course of the reaction. Upon compression, $V_f < V_i$, and so the work done on a system as a result of compression is a positive quantity. Conversely, after expansion, $V_f > V_i$ and so the work done on a system as a result of its expansion is a negative quantity.

In SI units, work is measured in joules ($1\text{ J} = 1\text{ kg}\cdot\text{m}^2\cdot\text{s}^{-2}$), pressure in pascals ($1\text{ Pa} = 1\text{ J}\cdot\text{m}^{-3}$) and volume in cubic meters (m^3). However, if the pressure is expressed in atmospheres and the volume is expressed in liters, as is commonly done, then Equation 14.3 requires that the units of work be atmospheres × liters (usually written as **liter-atmosphere,** L·atm). The conversion factor between the energy units of joule and liter-atmosphere is

$$1\text{ L}\cdot\text{atm} = 101.325\text{ J} \qquad (14.4)$$

A simple way to derive and to remember this conversion factor is to write the molar gas constant, R, in both sets of units (see the inside of the back cover), $R = 8.3145\text{ J}\cdot\text{mol}^{-1}\cdot\text{K}^{-1}$ and $R = 0.082058\text{ L}\cdot\text{atm}\cdot\text{mol}^{-1}\cdot\text{K}^{-1}$. Dividing these two expressions for R by $0.082058\text{ L}\cdot\text{atm}\cdot\text{mol}^{-1}\cdot\text{K}^{-1}$ yields the relationship given in Equation 14.4.

EXAMPLE 14-1: How much work is required to compress a gas from a volume of 5.0 L to a volume of 2.5 L by exerting a constant pressure of 1.5 atm?

Solution: We use Equation 14.3:

$$w = -P\Delta V = -P(V_f - V_i) = -(1.5\text{ atm})(2.5\text{ L} - 5.0\text{ L}) = 3.8\text{ L}\cdot\text{atm}$$

$$= (3.8\text{ L}\cdot\text{atm})\left(\frac{101.325\text{ J}}{1\text{ L}\cdot\text{atm}}\right) = 380\text{ J}$$

The work is positive, meaning that we have done work *on* the system; or, in other words, energy was transferred from the surroundings to the gas (system). The energy of the system is greater after it is compressed.

PRACTICE PROBLEM 14-1: How much work is done when a gas expands from a volume of 8.50 L to 12.00 L against a constant pressure of 2.00 atm?

Answer: −709 J. The work is negative in this case; the energy of the gas decreases upon expansion.

In contrast to the transfer of energy as work, the transfer of energy as **heat** does not require the application of a force. Energy transfer as heat occurs whenever there is a temperature difference between a system and its surroundings. Energy flows spontaneously as heat from a region of higher temperature to a region of lower temperature. We denote the energy transferred as heat by the symbol q.

The sign convention for heat (q) is the same as that for work (w). That is, when energy as heat is lost by the system to the surroundings, the sign of q is negative ($q < 0$). We call a reaction that releases energy as heat into the surroundings an **exothermic reaction** (*exo* = out). Conversely, when energy as heat is transferred from the surroundings into the system, the sign of q is positive ($q > 0$) and we call the reaction an **endothermic reaction** (*endo* = in).

You should realize that a system does not contain heat or work. Rather, heat and work are ways in which energy is transferred, as the following analogy illustrates. If a container is left outside in the rain, the question, "How much rain is in the container?" is answered correctly as, "There is no rain in the container, only water." Rain is the means by which water was transferred from the clouds to the container.

In thermodynamics, the energy of a system is denoted by the symbol U. We denote the **energy change** of a system by ΔU where $\Delta U = U_f - U_i$. In Chapter 1 we discussed the law of conservation of energy, which states that during any process, energy is neither created nor destroyed. Energy can only be converted from one form to another or transferred from one system to another. As we noted earlier, the energy of a system can change as a result of a transfer of energy as heat (q) or as work (w). Application of the law of conservation of energy to a reaction system yields

$$\Delta U = q + w \tag{14.5}$$

where q is the energy transferred as heat to or from the system and w is the energy transferred as work to or from the system. Equation 14.5, which constitutes a mathematical expression of the **first law of thermodynamics**, is simply an energy balance equation and is equivalent to the law of conservation of energy. It says that in going from some particular initial energy state to some particular final energy state, the energy transferred as work and the energy transferred as heat must always add up to give the same value as ΔU for the system. Because ΔU is equal to $U_{final} - U_{initial}$, the value of ΔU depends *only* on the initial and final states of the system. The value of ΔU does not depend on how the system goes from the initial to the final state. In thermodynamics, functions that depend only on the state of a system and not on how that state is achieved are called **state functions**.

Work and heat are not state functions, but **energy transfer functions.** Their values depend on how the energy transfer is carried out or the conditions under which the energy is transferred. Even though the individual values of q and w depend upon how the process is carried out, Equation 14.5 says that their sum does not. Their sum is a state function.

We shall denote state functions by uppercase letters (for example, P for pressure, V for volume, T for temperature, and U for energy) and energy transfer functions by lowercase letters (for example, w for work and q for heat).

If we apply the first law of thermodynamics to a chemical reaction, then Equation 14.5 becomes

$$\Delta U_{rxn} = q + w \tag{14.6}$$

where the subscript rxn stands for *reaction*. The value of ΔU_{rxn} is the **energy change of the reaction.**

The sign convention for the change in energy (ΔU) is the same as that for the work (w) and the heat (q). When $\Delta U < 0$, energy flows from the system into the surroundings, and when $\Delta U > 0$, energy is transferred from the surroundings into the system. Figure 14.3 illustrates the relationships between the system, the surroundings, and the work, heat, and energy transferred.

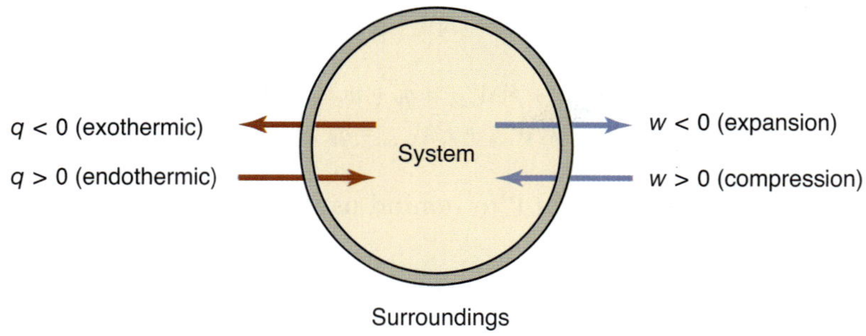

Figure 14.3 By convention, when energy in the form of work is transferred from the system to the surroundings the sign of the work is negative ($w < 0$), and when it is transferred from the surroundings to the system the sign is positive ($w > 0$). Similarly, when energy in the form of heat is transferred from the system to the surroundings the sign of the heat is negative ($q < 0$), and when it is transferred from the surroundings to the system the sign is positive ($q > 0$). Because $\Delta U = q + w$, when the net flow of energy is from the system to the surroundings $\Delta U < 0$, and when the net flow of energy is from the surroundings to the system $\Delta U > 0$. The sign convention of all three quantities (q, w, and ΔU) is the same.

14-2. Enthalpy Is a State Function

As we have already noted, the energy transferred as heat is not a state function. This creates a problem because if the energy evolved as heat by a reaction depends upon the conditions under which the reaction is run, then it is possible for different laboratories to obtain different measurements of this quantity for the same reaction, depending on how that reaction is carried out. In this section we shall introduce a new state function, called enthalpy, that does not depend on how a reaction is carried out.

Many chemical reactions occur at constant pressure (for example, open to the atmosphere). For reactions that occur at constant pressure, it is convenient to introduce a thermodynamic state function H, called **enthalpy,** for the reaction system, defined by the equation

$$H = U + PV \tag{14.7}$$

Application of Equation 14.7 to a chemical reaction yields

$$\Delta H_{\text{rxn}} = H_{\text{f}} - H_{\text{i}} = \Delta U_{\text{rxn}} + \Delta(PV)_{\text{rxn}} \tag{14.8}$$

where ΔH_{rxn} is the **enthalpy change of the reaction**.

We can express the $\Delta(PV)_{\text{rxn}}$ term in this equation as

$$\Delta(PV)_{\text{rxn}} = P_{\text{f}}V_{\text{f}} - P_{\text{i}}V_{\text{i}}$$

in terms of the final and initial values of the pressure and the volume. If the reaction occurs at constant pressure, then $P_{\text{f}} = P_{\text{i}} = P$, and so

$$\Delta(PV)_{\text{rxn}} = P(V_{\text{f}} - V_{\text{i}}) = P\Delta V_{\text{rxn}} \tag{14.9}$$

Substituting Equation 14.9 into Equation 14.8, we have

$$\Delta H_{\text{rxn}} = \Delta U_{\text{rxn}} + P\Delta V_{\text{rxn}} \qquad \text{(constant pressure)} \tag{14.10}$$

From this equation we see that the difference between ΔU_{rxn} and ΔH_{rxn} is just the energy that is required for the system to expand or contract against a constant external pressure. The difference between ΔU_{rxn} and ΔH_{rxn} is usually small, as we shall show in the next section.

Equation 14.10 takes on a simple form if we subsitute Equation 14.6 ($\Delta U = q + w$) and Equation 14.3 ($w = -P\Delta V_{rxn}$) into Equation 14.10 to obtain

$$\begin{aligned}\Delta H_{rxn} &= \Delta U_{rxn} + P\Delta V_{rxn} = q_P + w_P + P\Delta V_{rxn} \\ &= q_P - P\Delta V_{rxn} + P\Delta V_{rxn} = q_P\end{aligned} \tag{14.11}$$

where we have used a subscript P to remind us that Equation 14.11 applies at constant pressure.

From Equation 14.11 we find that when we specify conditions of constant pressure, $q_P = \Delta H_{rxn}$, and so the transfer of energy as heat at constant pressure is a state function. Thus, we see that the enthalpy change for a reaction has the useful property that

$$\Delta H_{rxn} = q_P = \begin{pmatrix}\text{energy evolved or absorbed} \\ \text{as heat at constant pressure}\end{pmatrix} \tag{14.12}$$

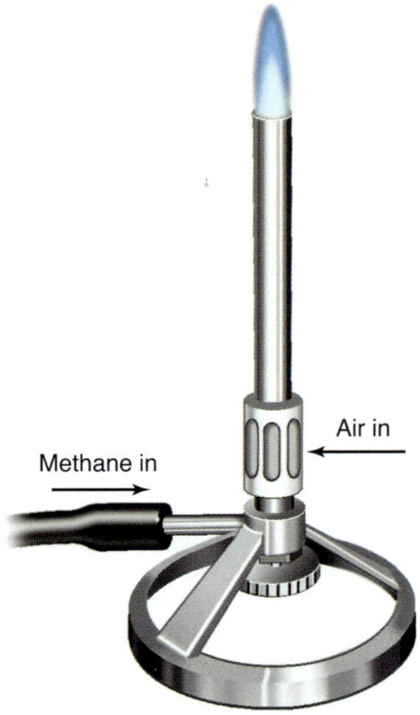

Figure 14.4 Combustion of methane, $CH_4(g)$, in air using a Bunsen burner. Because the combustion reaction takes place open to the atmosphere, it is a constant-pressure reaction.

Because enthalpy is a state function, we can tabulate values for the change in enthalpy of various reactions regardless of how a particular experiment is carried out. As we shall see in the upcoming sections, these values have a wide range of chemical applications.

In Chapter 10 we found that **fuels** are reactants that are oxidized to provide energy as heat. For example, the equation for the combustion (air-oxidation) of methane, $CH_4(g)$, can be written as

$$\underset{\text{fuel}}{CH_4(g)} + \underset{\text{oxidizer}}{2\,O_2(g)} \rightarrow CO_2(g) + 2\,H_2O(l) \quad (\textit{exothermic})$$

Because this reaction releases energy as heat, it is an exothermic reaction (Figure 14.4). Most combustion reactions are highly exothermic.

Not all chemical reactions are exothermic. The cold packs used in sports medicine contain separate compartments of chemicals. When the pack is folded, a plastic barrier breaks, allowing the contents to mix. These react, cooling the pack rapidly. One pair of chemicals used in such packs is solid ammonium nitrate, $NH_4NO_3(s)$, and water. When these are mixed, the resulting dissolution reaction is highly endothermic.

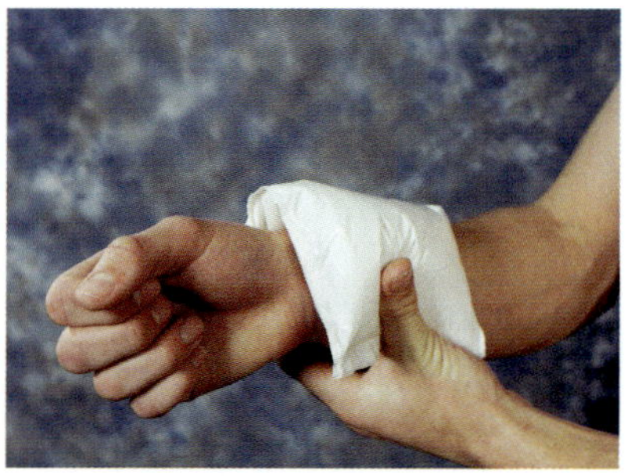

Figure 14.5 A commercial cold pack. Chemical cold and hot packs are used in sports medicine to rapidly respond to injuries without the need for ice or hot water.

$$NH_4NO_3(s) \xrightarrow[H_2O(l)]{} NH_4^+(aq) + NO_3^-(aq) \quad (\textit{endothermic})$$

Sports packs using this dissolution reaction cool from room temperature to below freezing in less than a minute. When combined with a gel, such packs can stay cool for up to 30 minutes (Figure 14.5).

Exothermic and endothermic reactions are illustrated schematically in Figure 14.6. Figure 14.6a represents an exothermic reaction; the enthalpy of the reactants is greater than the enthalpy of the products, so energy is evolved as heat as the reaction proceeds. Figure 14.6b represents an endothermic reaction; the enthalpy of the reactants is less than the enthalpy of the products, so energy in the form of heat must be supplied in order to drive the reaction up the enthalpy "hill."

The enthalpy change for a chemical reaction is equal to the total enthalpy of the products minus the total enthalpy of the reactants:

$$\Delta H_{rxn} = H_{prod} - H_{react} \tag{14.13}$$

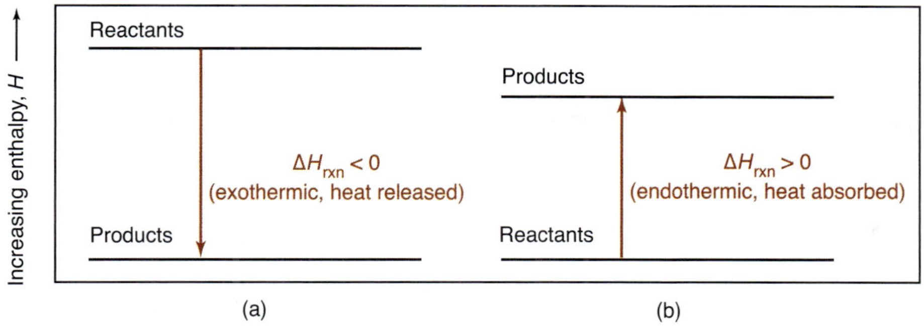

Figure 14.6 An enthalpy diagram for (a) an exothermic reaction and (b) an endothermic reaction. (a) In an exothermic reaction, the enthalpy of the products is less than that of the reactants. The enthalpy difference is released as energy in the form of heat during the reaction at constant pressure. (b) In an endothermic reaction, the enthalpy of the products is greater than that of the reactants. The enthalpy difference must be supplied as energy in the form of heat in order for the reaction to occur at constant pressure.

For an exothermic reaction, H_{prod} is less than H_{react}, so $\Delta H_{\text{rxn}} < 0$. For an endothermic reaction, H_{prod} is greater than H_{react}, so $\Delta H_{\text{rxn}} > 0$.

When specifying a value for the enthalpy change of a particular chemical reaction that takes place at a pressure of one bar, we use the notation $\Delta H^{\circ}_{\text{rxn}}$. The superscript degree sign tells us that the enthalpy change is the **standard enthalpy change** for the reaction. The standard enthalpy change refers to pure reactants at a pressure of one bar and to pure products at a pressure of one bar. These states are called the **standard states** for substances. The standard state for a solute is a one-molar (1 M) solution of the solute.

When stating the numerical value of the standard enthalpy change of a reaction, we must include both the value of $\Delta H^{\circ}_{\text{rxn}}$ and the corresponding chemical equation. For example, the chemical equation describing the standard enthalpy change for the reaction of $H_2(g)$ and $O_2(g)$ can be written as

$$\underset{1\text{ bar}}{2H_2(g)} + \underset{1\text{ bar}}{O_2(g)} \rightarrow \underset{1\text{ bar}}{2H_2O(l)} \tag{14.14}$$

for which we find experimentally that $\Delta H^{\circ}_{\text{rxn}} = -571.6\text{ kJ}\cdot\text{mol}^{-1}$ at 25°C. The negative value of $\Delta H^{\circ}_{\text{rxn}}$ tells us that the reaction is exothermic. The subscript rxn means that $\Delta H^{\circ}_{\text{rxn}}$ is expressed in units of kilojoules per mole *for the chemical equation as written*; in other words, the value of $\Delta H^{\circ}_{\text{rxn}}$ stated is the enthalpy change for the reaction of *two moles* of $H_2(g)$ with *one mole* of $O_2(g)$ at 25°C and a pressure of one bar. If instead we had chosen to represent the reaction by the equation

$$\underset{1\text{ bar}}{H_2(g)} + \underset{1\text{ bar}}{\tfrac{1}{2}O_2(g)} \rightarrow \underset{1\text{ bar}}{H_2O(l)} \tag{14.15}$$

then the corresponding value of $\Delta H^{\circ}_{\text{rxn}}$ would be $-285.8\text{ kJ}\cdot\text{mol}^{-1}$ at 25°C and a constant pressure of one bar, meaning that this is the value of $\Delta H^{\circ}_{\text{rxn}}$ when *one mole* of $H_2(g)$ reacts with *half a mole* of $O_2(g)$ under these conditions. We see then that both the chemical equation and the corresponding value of $\Delta H^{\circ}_{\text{rxn}}$ must be given. Simply stating that $\Delta H^{\circ}_{\text{rxn}} = -571.6$ kJ for the reaction of $H_2(g)$ with $O_2(g)$ is ambiguous.

Enthalpies of reactions may be used to predict the amount of energy transferred as heat during a given process, as the following Example illustrates.

Figure 14.7 The thermite reaction between powdered aluminum metal and iron(III) oxide described by

$$2\,Al(s) + Fe_2O_3(s) \rightarrow 2\,Fe(l) + Al_2O_3(s)$$
$$\Delta H^\circ_{rxn} = -851.5\ kJ\cdot mol^{-1}$$

is one of the most exothermic reactions known. Once this reaction is initiated by a heat source such as a burning magnesium ribbon, it proceeds vigorously, producing so much heat that the iron is formed as a liquid.

EXAMPLE 14-2: Calculate the energy released as heat when 30.0 grams of iron(III) oxide, $Fe_2O_3(s)$, is combined with 15.0 grams of $Al(s)$ in the thermite reaction (Figure 14.7) at 25°C and a constant pressure of one bar. The equation for the reaction is

$$Fe_2O_3(s) + 2\,Al(s) \rightarrow Al_2O_3(s) + 2\,Fe(s) \qquad \Delta H^\circ_{rxn} = -851.5\ kJ\cdot mol^{-1}$$

Solution: Because the masses of both reactants are given, we must check to see if either is a limiting reactant. The number of moles of each reactant is given by

$$\text{moles of Al} = (15.0\ g\ Al)\left(\frac{1\ mol\ Al}{26.98\ g\ Al}\right) = 0.556\ mol\ Al$$

$$\text{moles of } Fe_2O_3 = (60.0\ g\ Fe_2O_3)\left(\frac{1\ mol\ Fe_2O_3}{159.7\ g\ Fe_2O_3}\right) = 0.376\ mol\ Fe_2O_3$$

Because 0.556 moles of $Al(s)$ requires (0.556/2 =) 0.278 moles of $Fe_2O_3(s)$ to react, we see that aluminum is the limiting reactant and that $Fe_2O_3(s)$ is in excess by (0.376 − 0.278 =) 0.098 moles.

From the stoichiometric coefficients of the balanced chemical equation, we see that −851.5 kJ of energy is produced as heat for every two moles of aluminum consumed. Therefore, the energy liberated as heat by the combustion of 15.0 grams of aluminum at one bar is

$$q = (15.0\ g\ Al)\left(\frac{1\ mol\ Al}{26.98\ g\ Al}\right)\left(\frac{-851.5\ kJ}{2\ mol\ Al}\right) = -237\ kJ$$

This is about −16 kJ per gram of aluminum and is one of the most exothermic reactions known.

PRACTICE PROBLEM 14-2: Calculate the amount of energy released as heat by the formation of 1.00 mL of water from a stoichiometric mixture of $H_2(g)$ and $O_2(g)$ at 25°C and a constant pressure of one bar. Calculate the amount of energy released as heat by this reaction per mole and per gram of $H_2(g)$. Take the density of water to be 1.00 grams per milliliter.

$$2\,H_2(g) + O_2(g) \rightarrow 2\,H_2O(l) \qquad \Delta H^\circ_{rxn} = -572\ kJ\cdot mol^{-1}$$

Answer: −15.9 kJ; −286 $kJ\cdot mol^{-1}$; −142 $kJ\cdot g^{-1}$

You may have noticed that we specified the temperature (25°C) when we gave the values of ΔH°_{rxn} for the above reactions. We did this because the value of ΔH°_{rxn} depends on the temperature at which a reaction occurs. The variation in ΔH°_{rxn} with temperature usually is not great, but strictly speaking, we should specify the reaction temperature. In this book we shall ignore the variation in ΔH°_{rxn} with temperature; this approximation is satisfactory if the temperature change is not large.

14-3. The Difference Between the Values of ΔH°_{rxn} and ΔU°_{rxn} Is Usually Small

The relationship between ΔH°_{rxn} and ΔU°_{rxn} is given by Equation 14.10.

$$\Delta H^\circ_{rxn} = \Delta U^\circ_{rxn} + P\Delta V^\circ_{rxn} \quad \text{(constant pressure)} \tag{14.16}$$

with P equal to exactly one bar (recall that the superscript degree sign implies that these quantities are for a pressure of one bar). This equation shows that the difference between the values of ΔH°_{rxn} and ΔU°_{rxn} is determined by the value of $P\Delta V^\circ_{rxn}$ at a pressure of one bar. Notice again that the subscript rxn on ΔV°_{rxn} means that this term in Equation 14.16 is interpreted as the change in the molar volume of gas (with units of volume per mole) for the corresponding chemical equation as written. Because the molar volume of a gas at one bar is about 1700 times greater than the molar volume of a liquid or a solid, only the volume terms corresponding to gaseous products or gaseous reactants will make an appreciable contribution to the value of ΔV°_{rxn}. For example, for the reaction described by

$$2\,H_2(g) + O_2(g) \rightarrow 2\,H_2O(l) \qquad \Delta H^\circ_{rxn} = -571.6\ \text{kJ}\cdot\text{mol}^{-1}$$

we have

$$\Delta V^\circ_{rxn} = V^\circ_{prod} - V^\circ_{react} = 2\,V^\circ_{H_2O(l)} - (2\,V^\circ_{H_2(g)} + V^\circ_{O_2(g)}) \approx -2\,V^\circ_{H_2(g)} - V^\circ_{O_2(g)}$$

or

$$P\Delta V^\circ_{rxn} \approx -2\,PV^\circ_{H_2(g)} - PV^\circ_{O_2(g)}$$

Because $V^\circ_{H_2(g)}$ and $V^\circ_{O_2(g)}$ represent molar volumes of these gases, from the ideal-gas equation, $PV = nRT$, applied to exactly one mole ($n = 1$) of each gas, we obtain

$$PV^\circ_{H_2(g)} = RT \qquad \text{and} \qquad PV^\circ_{O_2(g)} = RT$$

Thus, for this reaction at 25°C, we have

$$P\Delta V^\circ_{rxn} = -2\,PV^\circ_{H_2(g)} - PV^\circ_{O_2(g)} = -2\,RT - RT = -3\,RT$$

As another example, consider the reaction for the decomposition of potassium chlorate, $KClO_3(s)$, to generate $O_2(g)$ according to

$$2\,KClO_3(s) \rightarrow 2\,KCl(s) + 3\,O_2(g)$$

In this case,

$$\Delta V^\circ_{rxn} = 2\,V^\circ_{KCl(s)} + 3\,V^\circ_{O_2(g)} - 2\,V^\circ_{KClO_3(s)} \approx 3\,V^\circ_{O_2(g)}$$

and

$$P\Delta V^\circ_{rxn} = 3\,V^\circ_{O_2(g)} = 3\,RT$$

If we define the change in the stoichiometric coefficients of the gaseous species, $\Delta\nu_{gas}$, for a given chemical equation, as

$$\Delta\nu_{gas} = \begin{pmatrix}\text{sum of the stoichiometric} \\ \text{coefficients of gaseous} \\ \text{products}\end{pmatrix} - \begin{pmatrix}\text{sum of the stoichiometric} \\ \text{coefficients of gaseous} \\ \text{reactants}\end{pmatrix} \qquad (14.17)$$

then, the ideal-gas equation may be written as

$$P\Delta V^\circ_{rxn} = \Delta\nu_{gas}RT \qquad (14.18)$$

Note that the stoichiometric coefficients of only gaseous species are included in Equations 14.17 and 14.18 because the molar volumes of solids and liquids are negligible compared to the molar volumes of the gaseous species.

EXAMPLE 14-3: Calculate the value of ΔU°_{rxn} at 25°C for the reaction described by

$$2\,H_2(g) + O_2(g) \rightarrow 2\,H_2O(l) \qquad \Delta H^\circ_{rxn} = -571.6\text{ kJ}\cdot\text{mol}^{-1}$$

Solution: In this case, water is a liquid and so

$$\Delta\nu_{gas} = 0 - (\nu_{H_2} + \nu_{O_2}) = 0 - (2 + 1) = -3$$

Thus at 25°C,

$$P\Delta V^\circ_{rxn} = \Delta\nu_{gas}RT = -3\,RT = (-3)(8.3145\text{ J}\cdot\text{mol}^{-1}\cdot\text{K}^{-1})(298\text{ K})$$
$$= -7.43\text{ kJ}\cdot\text{mol}^{-1}$$

and, therefore, ΔH°_{rxn} and ΔU°_{rxn} differ by about 7 kJ·mol^{-1}. We now use Equation 14.10 to calculate the value of ΔU°_{rxn}:

$$\Delta U^\circ_{rxn} = \Delta H^\circ_{rxn} - P\Delta V^\circ_{rxn} = -571.6\text{ kJ}\cdot\text{mol}^{-1} - (-7.43\text{ kJ}\cdot\text{mol}^{-1})$$
$$= -564.2\text{ kJ}\cdot\text{mol}^{-1}$$

In most cases the difference between the value of ΔH°_{rxn} and ΔU°_{rxn} is small.

Hence, in this case, the values of ΔU°_{rxn} and ΔH°_{rxn} differ by only about 1%; this will usually be the case.

PRACTICE PROBLEM 14-3: Calculate the value of ΔU°_{rxn} at 25°C for the reaction described by

$$H_2(g) + \tfrac{1}{2}\,O_2(g) \rightarrow H_2O(l) \qquad \Delta H^\circ_{rxn} = -285.8\text{ kJ}\cdot\text{mol}^{-1}$$

and compare your result with the one obtained in Example 14-3.

Answer: The answer is one-half that obtained in Example 14-3 because the chemical equation that we use here is one half that in Example 14-3.

14-4. Enthalpy Changes for Chemical Equations Are Additive

A useful property of ΔH°_{rxn} values for chemical equations is their additivity. This property follows directly from the fact that enthalpy is a state function. If we add two chemical equations to obtain a third chemical equation, then the value of ΔH°_{rxn} for the resulting equation is equal to the sum of the values of ΔH°_{rxn} for the two equations that are added together. The additivity of ΔH°_{rxn} values is best illustrated by example. Consider the two chemical equations

(1) $\quad Sn(s) + Cl_2(g) \rightarrow SnCl_2(s) \qquad \Delta H^\circ_{rxn}(1) = -325.1\ \text{kJ}\cdot\text{mol}^{-1}$

(2) $\quad SnCl_2(s) + Cl_2(g) \rightarrow SnCl_4(l) \qquad \Delta H^\circ_{rxn}(2) = -186.2\ \text{kJ}\cdot\text{mol}^{-1}$

If we add these two chemical equations as if they were algebraic equations, then we get

$$Sn(s) + SnCl_2(s) + 2\,Cl_2(g) \rightarrow SnCl_2(s) + SnCl_4(l)$$

If we cancel $SnCl_2(s)$ from both sides, we get

(3) $\quad Sn(s) + 2\,Cl_2(g) \rightarrow SnCl_4(l)$

The additive property of ΔH°_{rxn} values tells us that ΔH°_{rxn} for equation (3) is simply

$$\begin{aligned}\Delta H^\circ_{rxn}(3) &= \Delta H^\circ_{rxn}(1+2) = \Delta H^\circ_{rxn}(1) + \Delta H^\circ_{rxn}(2) \\ &= -325.1\ \text{kJ}\cdot\text{mol}^{-1} + (-186.2\ \text{kJ}\cdot\text{mol}^{-1}) \\ &= -511.3\ \text{kJ}\cdot\text{mol}^{-1}\end{aligned} \tag{14.19}$$

In effect we can imagine equations (1) and (2) as representing a two-step process with the same initial and final steps as equation (3). The total enthalpy change for the two equations together must, therefore, be the same as if the reaction proceeded in a single step (Figure 14.8).

The additivity property of ΔH°_{rxn} values is known as **Hess's law.** If two or more chemical equations are added together, then the value of ΔH°_{rxn} for the resulting equation is equal to the sum of the ΔH°_{rxn} values for the separate equations.

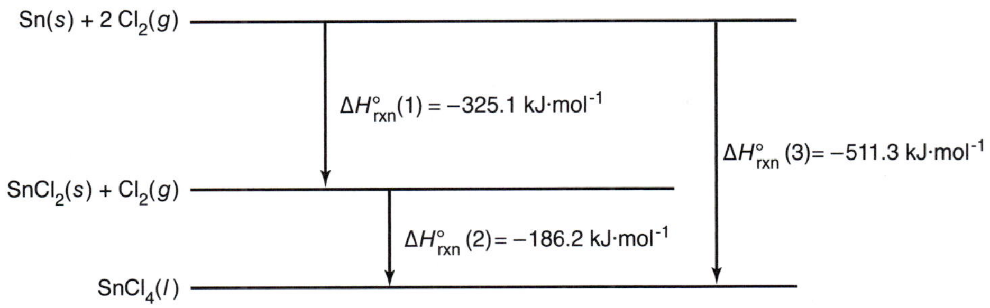

Figure 14.8 A schematic illustration of the application of Hess's law to the reaction described by $Sn(s) + 2\,Cl_2(g) \rightarrow SnCl_4(l)$. $\Delta H^\circ_{rxn}(3) = \Delta H^\circ_{rxn}(1) + \Delta H^\circ_{rxn}(2)$.

Suppose now that we add a chemical equation to itself, for example,

(1)	$SO_2(g) \rightarrow S(s) + O_2(g)$	$\Delta H^\circ_{rxn}(1)$
(2)	$SO_2(g) \rightarrow S(s) + O_2(g)$	$\Delta H^\circ_{rxn}(2) = \Delta H^\circ_{rxn}(1)$
(3)	$2\,SO_2(g) \rightarrow 2\,S(s) + 2\,O_2(g)$	$\Delta H^\circ_{rxn}(3)$

In this case

$$\Delta H^\circ_{rxn}(3) = \Delta H^\circ_{rxn}(1) + \Delta H^\circ_{rxn}(1) = 2\Delta H^\circ_{rxn}(1) \qquad (14.20)$$

Notice that adding a chemical equation to itself is equivalent to multiplying both sides of the chemical equation by two, that is, $2\,[SO_2(g) \rightarrow S(s) + O_2(g)]$ becomes $2\,SO_2(g) \rightarrow 2\,S(s) + 2\,O_2(g)$. Equation 14.20 can be generalized to cover multiplication of a chemical equation by any numerical factor n. For example, $n\,[SO_2(g) \rightarrow S(s) + O_2(g)]$ becomes $n\,SO_2(g) \rightarrow n\,S(s) + n\,O_2(g)$. Multiplication of the chemical equation by n is equivalent to writing the equation out n times and adding the n equations together, so the value of ΔH°_{rxn} for the resulting equation is

$$\Delta H^\circ_{rxn} = n\Delta H^\circ_{rxn}(1) \qquad (14.21)$$

Now let's consider the following combination of chemical equations:

(1)	$SO_2(g) \rightarrow S(s) + O_2(g)$	$\Delta H^\circ_{rxn}(1)$
(2)	$S(s) + O_2(g) \rightarrow SO_2(g)$	$\Delta H^\circ_{rxn}(2)$

The value of ΔH°_{rxn} for the sum of these two equations is

$$\Delta H^\circ_{rxn}(1+2) = \Delta H^\circ_{rxn}(1) + \Delta H^\circ_{rxn}(2)$$

But the addition of these two equations yields no net reactants and no net products or, in other words, no net chemical change whatsoever. Because there is no net change, the value of $\Delta H^\circ_{rxn}(1{+}2)$ must be zero and we conclude that $\Delta H^\circ_{rxn}(2) = -\Delta H^\circ_{rxn}(1)$. Because equation (2) above is simply the reverse of equation (1), we conclude from Hess's law that

$$\Delta H^\circ_{rxn}(\text{reverse}) = -\Delta H^\circ_{rxn}(\text{forward}) \qquad (14.22)$$

Equation 14.22 is easy to apply. If we reverse a chemical equation, then the reactants become the products and the products become the reactants, and so the sign of ΔH°_{rxn} changes. That is, if energy is released as heat in one direction, then it is absorbed as heat in the other direction. Equation 14.22 also says that $\Delta H = 0$ for a **cyclic process**, in other words, one in which the initial state is the same as the final state. We can express this result as an equation by writing

$$\Delta H_{cyclic} = 0 \qquad (14.23)$$

EXAMPLE 14-4: Given the following ΔH°_{rxn} values,

(1) $SO_2(g) \rightarrow S(s) + O_2(g)$ $\Delta H^\circ_{rxn}(1) = +296.8\ kJ\cdot mol^{-1}$

(2) $2S(s) + 3O_2(g) \rightarrow 2SO_3(g)$ $\Delta H^\circ_{rxn}(2) = -791.4\ kJ\cdot mol^{-1}$

calculate the value of ΔH°_{rxn} for the equation

(3) $2SO_2(g) + O_2(g) \rightarrow 2SO_3(g)$

Solution: To obtain equation (3) from equations (1) and (2), it is first necessary to multiply equation (1) by two because equation (3) involves two moles of $SO_2(g)$ as a reactant. For the equation

(4) $2SO_2(g) \rightarrow 2S(s)+2O_2(g)$

we have from Equation 14.21

$$\Delta H^\circ_{rxn}(4) = 2\Delta H^\circ_{rxn}(1) = (2)(296.8\ kJ\cdot mol^{-1}) = 593.6\ kJ\cdot mol^{-1}$$

Addition of equations (2) and (4) yields

$$2S(s) + 3O_2(g) + 2SO_2(g) \rightarrow 2S(s) + 2O_2(g) + 2SO_3(g)$$

If we cancel $2S(s)$ and $2O_2(g)$ from both sides, then we get equation (3):

$$2SO_2(g) + O_2(g) \rightarrow 2SO_3(g)$$

The corresponding value of ΔH°_{rxn} for this equation is

$$\begin{aligned}\Delta H^\circ_{rxn}(3) &= \Delta H^\circ_{rxn}(2) + \Delta H^\circ_{rxn}(4) = -791.4\ kJ\cdot mol^{-1} + 593.6\ kJ\cdot mol^{-1}\\ &= -197.8\ kJ\cdot mol^{-1}\end{aligned}$$

The conversion of $SO_2(g)$ to $SO_3(g)$ is an exothermic reaction. This reaction occurs in the catalytic converters of automobiles that run on gasoline containing traces of sulfur compounds (Figure 14.9). This reaction also occurs in the manufacture of sulfuric acid, one of the most important industrial chemicals worldwide (see Appendix H).

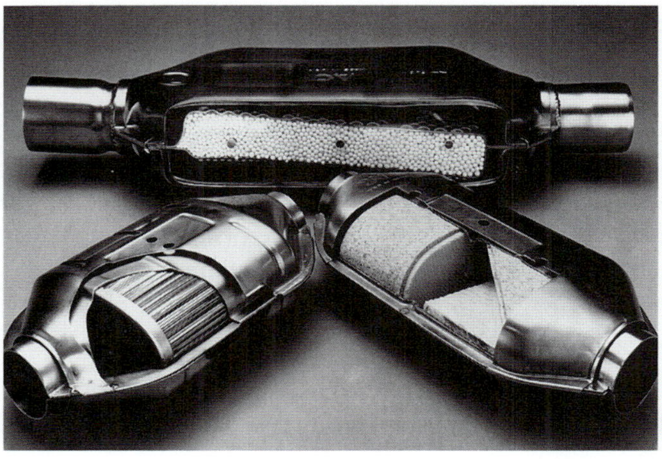

Figure 14.9 An automobile catalytic converter, which converts $SO_2(g)$ to $SO_3(g)$, facilitates the reduction of $NO(g)$ and $NO_2(g)$ to $N_2(g)$, and the oxidation of unburned or partially oxidized hydrocarbons to $CO_2(g)$ and $H_2O(l)$. Inside stainless steel shells, catalytic converters carry metal catalysts on ceramic pellets, ceramic blocks, or sheaves of a special alloy.

PRACTICE PROBLEM 14-4: Given the following values of ΔH°_{rxn},

(1) $Cu(s) + \frac{1}{2}O_2(g) \rightarrow CuO(s) \quad \Delta H^\circ_{rxn}(1) = -157.3\ kJ\cdot mol^{-1}$

(2) $NO_2(g) \rightarrow \frac{1}{2}N_2(g) + O_2(g) \quad \Delta H^\circ_{rxn}(2) = -33.2\ kJ\cdot mol^{-1}$

calculate the value of ΔH°_{rxn} for the equation

(3) $4Cu(s) + 2NO_2(g) \rightarrow 4CuO(s) + N_2(g)$

Answer: $-695.6\ kJ\cdot mol^{-1}$

EXAMPLE 14-5: Use the following ΔH°_{rxn} data

(1) $2P(s) + 3Cl_2(g) \rightarrow 2PCl_3(l) \qquad \Delta H^\circ_{rxn}(1) = -639.4\ kJ\cdot mol^{-1}$

(2) $2P(s) + 5Cl_2(g) \rightarrow 2PCl_5(s) \qquad \Delta H^\circ_{rxn}(2) = -887.0\ kJ\cdot mol^{-1}$

to calculate the value of ΔH°_{rxn} for the equation

(3) $PCl_3(l) + Cl_2(g) \rightarrow PCl_5(s)$

Solution: In this case we reverse equation (1)

$$2PCl_3(l) \rightarrow 2P(s) + 3Cl_2(g) \qquad \Delta H^\circ_{rxn} = -\Delta H^\circ_{rxn}(1)$$

and add this to equation (2), canceling $3Cl_2(g)$ and $2P(s)$, to obtain:

(4) $2PCl_3(l) + 2Cl_2(g) \rightarrow 2PCl_5(s)$

Thus, from Hess's law, we have

$$\Delta H^\circ_{rxn}(4) = \Delta H^\circ_{rxn}(2) - \Delta H^\circ_{rxn}(1)$$
$$= -887.0\ kJ\cdot mol^{-1} + 639.4\ kJ\cdot mol^{-1} = -247.6\ kJ\cdot mol^{-1}$$

We now multiply equation (4) by $\frac{1}{2}$ to obtain equation (3). Using Equation 14.21, we obtain

$$\Delta H^\circ_{rxn}(3) = \tfrac{1}{2}\Delta H^\circ_{rxn}(4) = \tfrac{1}{2}(-247.6\ kJ\cdot mol^{-1}) = -123.8\ kJ\cdot mol^{-1}$$

$CH_3CH_2CH_2CH_3$

n-butane

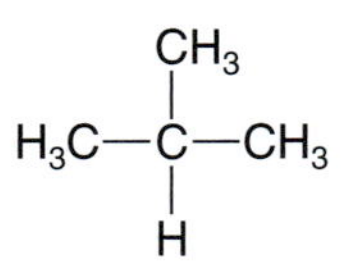

isobutane

PRACTICE PROBLEM 14-5: The hydrocarbon butane, $C_4H_{10}(g)$, exists as two structurally different isomers called *n*-butane and isobutane (see margin). The standard molar enthalpies of combustion at 25°C of *n*-butane and isobutane are $-2877\ kJ\cdot mol^{-1}$ and $-2869\ kJ\cdot mol^{-1}$, respectively. Calculate the value of ΔH°_{rxn} for the conversion of *n*-butane to isobutane according to

$$n\text{-}C_4H_{10}(g) \rightarrow i\text{-}C_4H_{10}(g)$$

(*n*-butane) (isobutane)

Hint: First write the respective combustion reaction equations for each isomer.

Answer: $-8\ kJ\cdot mol^{-1}$

Hess's law illustrates the utility of state functions. Many times it is difficult or impractical to measure the energy evolved as heat by a particular reaction. By applying Hess's law, we are often able to circumvent such difficulties by imagining a pathway from reactants to products where the enthalpies of reaction for the individual steps are known—even if the reaction itself does not actually follow this mechanism.

A nice example of such an application of Hess's law is given by the reaction described by

$$\text{(1)} \qquad C(s) + \tfrac{1}{2}O(g) \rightarrow CO(g)$$

It is not possible to determine the value of ΔH°_{rxn} for this equation directly because a large quantity of $CO_2(g)$ will also be formed when $C(s)$ is burned in $O_2(g)$ according to

$$\text{(2)} \qquad C(s) + O_2(g) \rightarrow CO_2(g) \qquad \Delta H^{\circ}_{rxn}(2) = -393.5\ \text{kJ}\cdot\text{mol}^{-1}$$

In fact, all the $C(s)$ will be converted to $CO_2(g)$ unless there is a deficiency in $O_2(g)$, in which case there will be a mixture of $CO(g)$ and $CO_2(g)$. Thus, it is easy to determine the value of $\Delta H^{\circ}_{rxn}(2)$ by assuring an excess of $O_2(g)$. Furthermore, $CO(g)$ burns readily in $O_2(g)$ to form $CO_2(g)$ according to

$$\text{(3)} \qquad CO(g) + \tfrac{1}{2}O_2(g) \rightarrow CO_2(g) \qquad \Delta H^{\circ}_{rxn}(3) = -283.0\ \text{kJ}\cdot\text{mol}^{-1}$$

and it is easy to determine the value of $\Delta H^{\circ}_{rxn}(3)$.

We can use the values of $\Delta H^{\circ}_{rxn}(2)$ and $\Delta H^{\circ}_{rxn}(3)$ to determine the value of $\Delta H^{\circ}_{rxn}(1)$. If we reverse equation (3) and add it to equation (2), canceling the $\frac{1}{2}O_2(g)$ and $CO_2(g)$ from both sides, then we obtain equation (1). The corresponding value of ΔH°_{rxn} for equation (1) is given by

$$\begin{aligned}\Delta H^{\circ}_{rxn}(1) &= \Delta H^{\circ}_{rxn}(2) - \Delta H^{\circ}_{rxn}(3)\\ &= -393.5\ \text{kJ}\cdot\text{mol}^{-1} - (-283.0\ \text{kJ}\cdot\text{mol}^{-1}) = -110.5\ \text{kJ}\cdot\text{mol}^{-1}\end{aligned}$$

Even though we can't measure the value of $\Delta H^{\circ}_{rxn}(1)$ directly, we can use Hess's law to obtain it from a two-step process for which we can measure the values of ΔH°_{rxn} experimentally (Figure 14.10).

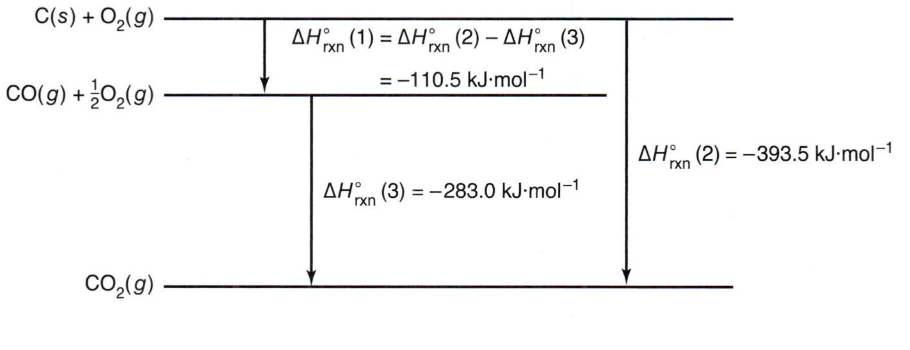

Figure 14.10 A schematic illustration of the application of Hess's law to the reaction described by $C(s) + \frac{1}{2}O_2(g) \rightarrow CO(g)$ where $\Delta H^{\circ}_{rxn}(1) = \Delta H^{\circ}_{rxn}(2) - \Delta H^{\circ}_{rxn}(3)$

TABLE 14.1 Rules for Hess's law calculations involving chemical equations

Operation	Result
Addition of two or more chemical equations	$\Delta H^\circ_{rxn}(1+2) = \Delta H^\circ_{rxn}(1) + \Delta H^\circ_{rxn}(2)$
Multiplication of a chemical equation by a factor of n	$\Delta H^\circ_{rxn} = n\Delta H^\circ_{rxn}(1)$
Reversing a chemical equation	$\Delta H^\circ_{rxn}(\text{reverse}) = -\Delta H^\circ_{rxn}(\text{forward})$

The utility of Hess's law is that it enables us to calculate a ΔH°_{rxn} value for a chemical equation from ΔH°_{rxn} values for related equations. A summary of the rules for Hess's law is given in Table 14.1.

14-5. Enthalpies of Reactions Can Be Calculated from Tabulated Molar Enthalpies of Formation

The value of ΔH°_{rxn} for the reaction of $C(s)$ with $O_2(g)$ is

$$C(s) + O_2(g) \rightarrow CO_2(g) \qquad \Delta H^\circ_{rxn} = \Delta H^\circ_f[CO_2(g)] = -393.5 \text{ kJ·mol}^{-1}$$

We refer to this value of ΔH°_{rxn} as the **standard molar enthalpy of formation** for $CO_2(g)$ and denote it by the symbol $\Delta H^\circ_f[CO_2(g)]$. The standard molar enthalpy of formation is defined as the enthalpy change for the equation in which one mole of a compound is formed in its one-bar standard state from the most stable form of its constituent elements in their one-bar standard states at the temperature of interest.

The subscript f on ΔH°_f stands for *formation* from the elements and also indicates that the value is for one mole. A $\Delta H^\circ_f[CO_2(g)]$ value of -393.5 kJ·mol^{-1} tells us that one mole of $CO_2(g)$ lies 393.5 kJ "downhill" on the enthalpy scale relative to its constituent elements (Figure 14.11a).

The standard molar enthalpy of formation of water from its elements is equal to the ΔH°_{rxn} value for the reaction in which one mole of $H_2O(l)$ is formed from its elements at one bar (Figure 14.11b):

$$H_2(g) + \tfrac{1}{2}O_2(g) \rightarrow H_2O(l) \qquad \Delta H^\circ_{rxn} = \Delta H^\circ_f[H_2O(l)] = -285.8 \text{ kJ·mol}^{-1}$$

Most compounds cannot be formed directly from their elements. For example, an attempt to make the hydrocarbon acetylene, $C_2H_2(g)$, by reaction of $C(s)$ with $H_2(g)$, as described by

$$2C(s) + H_2(g) \rightarrow C_2H_2(g)$$

yields not just $C_2H_2(g)$ but a complex mixture of various hydrocarbons such as $C_2H_4(g)$ and $C_2H_6(g)$, among others. Nevertheless, we can determine the value of ΔH°_f for acetylene by using Hess's law, together with the available ΔH°_{rxn} data on combustion reactions. All three species in the above chemical equation burn in oxygen according to

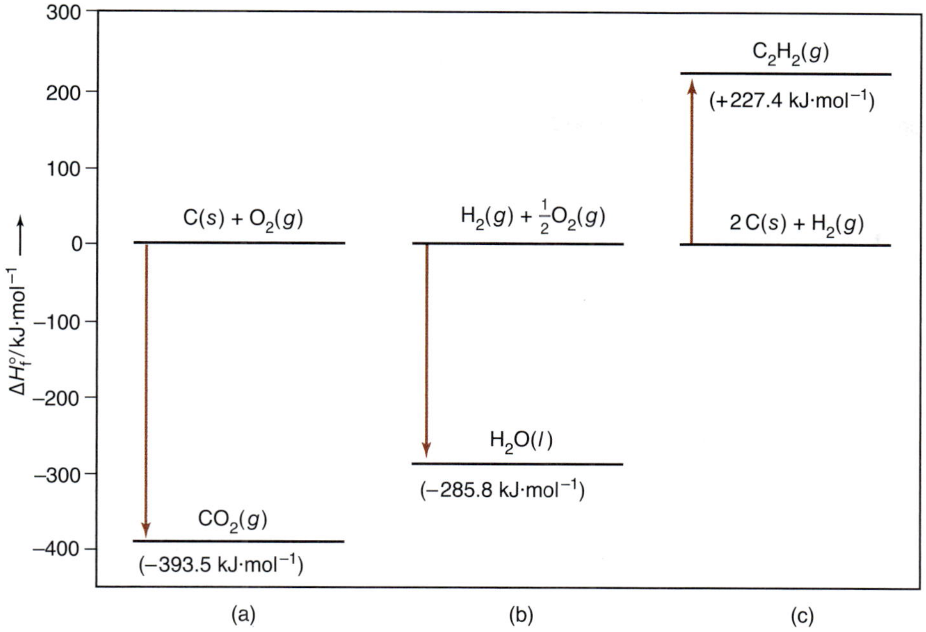

Figure 14.11 Enthalpy changes involved in the formation of $CO_2(g)$, $H_2O(l)$, and $C_2H_2(g)$ from their elements at 25°C, 1 bar. Note that (a) $CO_2(g)$ lies 393.5 kJ·mol^{-1} (on the enthalpy scale) below its constituent elements; (b) $H_2O(l)$ lies 285.8 kJ·mol^{-1} below its constituent elements; and (c) $C_2H_2(g)$ lies 227.4 kJ·mol^{-1} above its constituent elements.

$$\text{(1)} \quad C(s) + O_2(g) \rightarrow CO_2(g) \qquad \Delta H^\circ_{rxn}(1) = -393.5 \text{ kJ·mol}^{-1}$$

$$\text{(2)} \quad H_2(g) + \tfrac{1}{2}O_2(g) \rightarrow H_2O(l) \qquad \Delta H^\circ_{rxn}(2) = -285.8 \text{ kJ·mol}^{-1}$$

$$\text{(3)} \quad C_2H_2(g) + \tfrac{5}{2}O_2(g) \rightarrow 2CO_2(g) + H_2O(l) \quad \Delta H^\circ_{rxn}(3) = -1300.2 \text{ kJ·mol}^{-1}$$

If we multiply equation (1) by two, reverse equation (3), and add the results to equation (2), we obtain

$$\text{(4)} \qquad 2C(s) + H_2(g) \rightarrow C_2H_2(g)$$

From Hess's law, we know that the value of ΔH°_{rxn} for equation (4) is

$$\begin{aligned}\Delta H^\circ_{rxn}(4) &= 2\Delta H^\circ_{rxn}(1) + \Delta H^\circ_{rxn}(2) - \Delta H^\circ_{rxn}(3)\\ &= (2)(-393.5 \text{ kJ·mol}^{-1}) + (-285.8 \text{ kJ·mol}^{-1}) - (-1300.2 \text{ kJ·mol}^{-1})\\ &= +227.4 \text{ kJ·mol}^{-1}\end{aligned}$$

Because equation (4) represents the formation of one mole of $C_2H_2(g)$ from its elements, we conclude that $\Delta H^\circ_f[C_2H_2(g)] = +227.4$ kJ·mol^{-1} (Figure 14.11c). Thus, we see that it is possible to obtain values of ΔH°_f even if the compound cannot be formed directly from its elements.

EXAMPLE 14-6: Given that the standard enthalpy of combustion at 25°C for $C(s)$, $H_2(g)$, and $CH_4(g)$ are −393.5 kJ·mol^{-1}, −285.8 kJ·mol^{-1}, and −890.2 kJ·mol^{-1}, respectively, calculate the standard molar enthalpy of formation of methane, $CH_4(g)$, at 25°C.

Solution: The chemical equations for the three combustion reactions are

(1) $C(s) + O_2(g) \rightarrow CO_2(g)$ $\Delta H^\circ_{rxn}(1) = -393.5\ kJ\cdot mol^{-1}$

(2) $H_2(g) + \frac{1}{2}O_2(g) \rightarrow H_2O(l)$ $\Delta H^\circ_{rxn}(2) = -285.8\ kJ\cdot mol^{-1}$

(3) $CH_4(g) + 2O_2(g) \rightarrow CO_2(g) + 2H_2O(l)$ $\Delta H^\circ_{rxn}(3) = -890.2\ kJ\cdot mol^{-1}$

If we reverse equation (3), multiply equation (2) by two, and add the results to equation (1), then we obtain the equation for the formation of $CH_4(g)$ from its elements

(4) $C(s) + 2H_2(g) \rightarrow CH_4(g)$ $\Delta H^\circ_{rxn}(4) = \Delta H^\circ_f[CH_4(g)]$

for which

$$\begin{aligned}\Delta H^\circ_{rxn}(4) &= \Delta H^\circ_{rxn}(1) + 2\Delta H^\circ_{rxn}(2) - \Delta H^\circ_{rxn}(3)\\ &= (-393.5\ kJ\cdot mol^{-1}) + (2)(-285.8\ kJ\cdot mol^{-1}) - (-890.2\ kJ\cdot mol^{-1})\\ &= -74.9\ kJ\cdot mol^{-1}\end{aligned}$$

Because equation (4) represents the formation of one mole of $CH_4(g)$ directly from its elements, we have $\Delta H^\circ_f[CH_4(g)] = -74.9\ kJ\cdot mol^{-1}$.

PRACTICE PROBLEM 14-6: Diborane, $B_2H_6(g)$, cannot be made directly from boron and hydrogen, but its standard molar enthalpy of formation can be determined from the following ΔH°_{rxn} data for combustion reactions, which we write as

(1) $4B(s) + 3O_2(g) \rightarrow 2B_2O_3(s)$ $\Delta H^\circ_{rxn}(1) = -2547.0\ kJ\cdot mol^{-1}$

(2) $2H_2(g) + O_2(g) \rightarrow 2H_2O(l)$ $\Delta H^\circ_{rxn}(2) = -571.6\ kJ\cdot mol^{-1}$

(3) $B_2H_6(g) + 3O_2(g) \rightarrow B_2O_3(s) + 3H_2O(l)$ $\Delta H^\circ_{rxn}(3) = -2167.3\ kJ\cdot mol^{-1}$

Calculate the value of $\Delta H^\circ_f[B_2H_6(g)]$ from these data.

Answer: $+36.4\ kJ\cdot mol^{-1}$

TABLE 14.2 The states of representative elemental forms for which we take $\Delta H^\circ_f = 0$ at 25°C

Element	Formula
hydrogen	$H_2(g)$
oxygen	$O_2(g)$
nitrogen	$N_2(g)$
chlorine	$Cl_2(g)$
fluorine	$F_2(g)$
bromine	$Br_2(l)$
mercury	$Hg(l)$
sodium	$Na(s)$
magnesium	$Mg(s)$
carbon (graphite)	$C(s)$
sulfur (rhombic)	$S(s)$
iron	$Fe(s)$

As suggested by Figure 14.11, we can set up a table of ΔH°_f values for compounds by setting the ΔH°_f values for the most stable form of the elements equal to zero. That is, for each element in its most stable physical state at one bar at the temperature of interest (usually, but not always, 25°C), we set ΔH°_f equal to zero, as illustrated below (see also Table 14.2):

$$\Delta H^\circ_f[O_2(g)] = 0 \qquad \Delta H^\circ_f[Br_2(l)] = 0 \qquad \Delta H^\circ_f[C(\text{graphite}, s)] = 0$$

Thus, standard molar enthalpies of formation of compounds are given relative to the elements in their normal physical states at one bar.

Realize that our choice of defining the enthalpy of formation for elements in their normal states at one bar as equal to zero is arbitrary and is merely a choice of convenience. This definition can be compared to our convention of using sea level as the zero point for determining elevation. For example, the height of

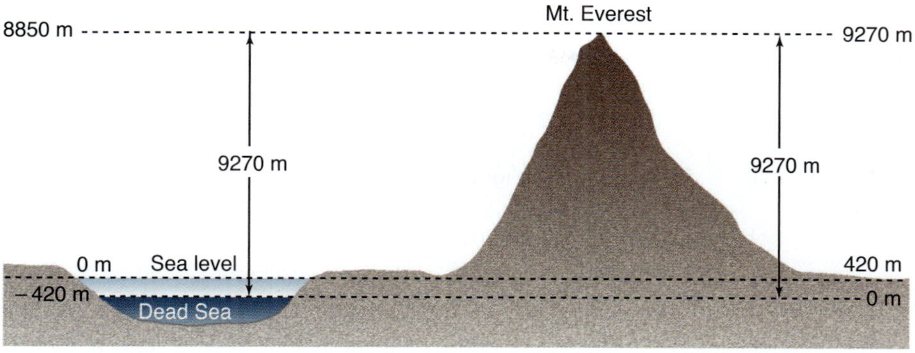

Figure 14.12 It makes no difference whether we define sea level (*left-hand scale*) or the Dead Sea (*right-hand scale*) as our zero point for measuring elevation; in both cases the *difference* in elevation between the level of the Dead Sea and the top of Mt. Everest is the same.

Mt. Everest (the tallest mountain on earth), is 8850 meters above sea level. The lowest point on the earth's surface is the Dead Sea, which lies −420 meters below sea level. The difference in elevation between these two points is 9270 meters. If instead of sea level we had chosen to take the Dead Sea as our arbitrary zero point for determining elevation, then the height of the Dead Sea would be zero meters, and that of Mt. Everest would be 9270 meters *above the Dead Sea.* But notice that the difference in elevation between the two points remains the same, 9270 meters (Figure 14.12). When studying chemical reactions, what we are interested in is the *change* (or difference) in enthalpy between the products and the reactants. This quantity is independent of what we select as our reference state, and so we simply choose the one of greatest convenience, which for our purposes is pure elements in their normal physical state at one bar.

Table 14.3 lists values of ΔH_f° at 25°C for a number of substances. If you look at Table 14.3, you will see that $\Delta H_f^\circ[C(\text{diamond})] = +1.897\ \text{kJ·mol}^{-1}$, $\Delta H_f^\circ[Br_2(g)] = +30.9\ \text{kJ·mol}^{-1}$, and $\Delta H_f^\circ[I_2(g)] = +62.4\ \text{kJ·mol}^{-1}$. The ΔH_f° values for these forms of the elements are not equal to zero because C(diamond), $Br_2(g)$, and $I_2(g)$ are not the most stable physical states of these elements at 25°C and one bar. The most stable physical states of these elements at 25°C are C(graphite), $Br_2(l)$, and $I_2(s)$.

Once we have tabulated standard molar enthalpies of formation, we can then use Hess's law to calculate the standard enthalpy change of any chemical equation for which we know the ΔH_f° values of all the products and reactants. For example, let's consider one of the reactions that is used in the production of tin from its principal ore, cassiterite, $SnO_2(s)$, which we write as

(1) $SnO_2(s) + C(s) \rightarrow Sn(s) + CO_2(g)$

We can write this equation as the sum of the two equations,

(2) $C(s) + O_2(g) \rightarrow CO_2(g)$

(3) $SnO_2(s) \rightarrow Sn(s) + O_2(g)$

Notice that ΔH_{rxn}° for equation (2) is equal to $\Delta H_f^\circ[CO_2(g)]$. Moreover, ΔH_{rxn}° for equation (3) is equal to $-\Delta H_f^\circ[SnO_2(s)]$ because it is the equation for the reverse of the formation of $SnO_2(s)$ directly from its elements. If we add these two equations along with their corresponding values of ΔH_{rxn}°, we obtain

$$SnO_2(s) + C(s) \rightarrow Sn(s) + CO_2(g) \qquad \Delta H_{rxn}^\circ = \Delta H_f^\circ[CO_2(g)] - \Delta H_f^\circ[SnO_2(s)]$$

TABLE 14.3 Standard enthalpies of formation, ΔH_f°, for various substances at 25°C

Substance	Formula	$\Delta H_f^\circ/kJ \cdot mol^{-1}$
aluminum oxide	$Al_2O_3(s)$	-1675.7
ammonia	$NH_3(g)$	-45.9
benzene	$C_6H_6(l)$	+49.1
benzoic acid	$C_6H_5COOH(s)$	-385.2
bromine vapor	$Br_2(g)$	+30.9
butane	$C_4H_{10}(g)$	-125.7
calcium carbonate	$CaCO_3(s)$	-1207.6
carbon (diamond)	$C(s)$	+1.897
carbon (graphite)	$C(s)$	0
carbon (buckminster fullerene)	$C_{60}(s)$	+2327.0
carbon dioxide	$CO_2(g)$	-393.5
carbon monoxide	$CO(g)$	-110.5
carbon tetrachloride	$CCl_4(l)$ $CCl_4(g)$	-128.2 -95.7
chromium(III) oxide	$Cr_2O_3(s)$	-1139.7
cyclohexane	$C_6H_{12}(l)$	-156.4
ethane	$C_2H_6(g)$	-84.0
ethanol (ethyl alcohol)	$CH_3CH_2OH(l)$	-277.6
ethene (ethylene)	$C_2H_4(g)$	+52.4
ethyne (acetylene)	$C_2H_2(g)$	+227.4
freon-12 (dichloro difluoromethane)	$CF_2Cl_2(g)$	-477.4
glucose	$C_6H_{12}O_6(s)$	-1273.3
hexane	$C_6H_{14}(l)$	-198.7
hydrazine	$N_2H_4(l)$ $N_2H_4(g)$	+50.6 +95.4
hydrogen bromide	$HBr(g)$	-36.3
hydrogen chloride	$HCl(g)$	-92.3
hydrogen fluoride	$HF(g)$	-273.3
hydrogen iodide	$HI(g)$	+26.5
hydrogen peroxide	$H_2O_2(l)$	-187.8
iodine vapor	$I_2(g)$	+62.4
magnesium carbonate	$MgCO_3(s)$	-1095.8
magnesium oxide	$MgO(s)$	-601.6
magnesium sulfide	$MgS(s)$	-346.0
methane	$CH_4(g)$	-74.6
methanol (methyl alcohol)	$CH_3OH(l)$ $CH_3OH(g)$	-239.2 -201.0
methyl chloride	$CH_3Cl(g)$	-81.9
nitrogen dioxide	$NO_2(g)$	+33.2
nitrogen oxide	$NO(g)$	+91.3
dinitrogen tetroxide	$N_2O_4(g)$ $N_2O_4(l)$	+11.1 -19.5
octane	$C_8H_{18}(l)$	-250.1
pentane	$C_5H_{12}(l)$	-173.5
propane	$C_3H_8(g)$	-103.8
sodium carbonate	$Na_2CO_3(s)$	-1130.7
sodium oxide	$Na_2O(s)$	-414.2
sucrose	$C_{12}H_{22}O_{11}(s)$	-2226.1
sulfur dioxide	$SO_2(g)$	-296.8
sulfur trioxide	$SO_3(g)$	-395.7
tin(IV) oxide	$SnO_2(s)$	-577.6
water	$H_2O(l)$ $H_2O(g)$	-285.8 -241.8

Data from *CRC Handbook of Chemistry and Physics*, 86th Ed., Ed. David R. Lide, CRC Press, 2005–2006. (More thermodynamic data are given in Appendix D.)

Because, by our convention for ΔH_f° values of elements, we have $\Delta H_f^\circ[C(s)] = 0$ and $\Delta H_f^\circ[Sn(s)] = 0$, we can write ΔH_{rxn}° for equation (1) as

$$\Delta H_{rxn}^\circ = \{\Delta H_f^\circ[Sn(s)] + \Delta H_f^\circ[CO_2(g)]\} - \{\Delta H_f^\circ[SnO_2(s)] + \Delta H_f^\circ[C(s)]\}$$

This equation has the form

$$\Delta H_{rxn}^\circ = \Delta H_f^\circ[\text{products}] - \Delta H_f^\circ[\text{reactants}] \qquad (14.24)$$

Using the values of $\Delta H_f^\circ[CO_2(g)]$ and $\Delta H_f^\circ[SnO_2(s)]$ given in Table 14.3, we have

$$\begin{aligned}\Delta H_{rxn}^\circ &= (0\ \text{kJ}\cdot\text{mol}^{-1}) + (-393.5\ \text{kJ}\cdot\text{mol}^{-1}) - \{(-577.6\ \text{kJ}\cdot\text{mol}^{-1}) + (0\ \text{kJ}\cdot\text{mol}^{-1})\}\\ &= +184.1\ \text{kJ}\cdot\text{mol}^{-1}\end{aligned}$$

EXAMPLE 14-7: Express ΔH_{rxn}° for the equation

(1) $NH_3(g) + 3\,Cl_2(g) \rightarrow NCl_3(l) + 3\,HCl(g)$

in terms of the standard molar enthalpies of formation of the reactants and the products.

Solution: The equations for the formation of one mole of the reactant and product compounds directly from their elements are

(2) $\frac{1}{2}N_2(g) + \frac{3}{2}H_2(g) \rightarrow NH_3(g) \qquad \Delta H_{rxn}^\circ = \Delta H_f^\circ[NH_3(g)]$

(3) $\frac{1}{2}N_2(g) + \frac{3}{2}Cl_2(g) \rightarrow NCl_3(l) \qquad \Delta H_{rxn}^\circ = \Delta H_f^\circ[NCl_3(l)]$

(4) $\frac{1}{2}H_2(g) + \frac{1}{2}Cl_2(g) \rightarrow HCl(g) \qquad \Delta H_{rxn}^\circ = \Delta H_f^\circ[HCl(g)]$

Equation (1) can be obtained by combining equations (2), (3), and (4) as follows:

$$\text{equation (1)} = \text{equation (3)} + 3 \times \text{equation (4)} - \text{equation (2)}$$

so

$$\Delta H_{rxn}^\circ(1) = \Delta H_{rxn}^\circ(3) + 3\Delta H_{rxn}^\circ(4) - \Delta H_{rxn}^\circ(2)$$

or

$$\Delta H_{rxn}^\circ(1) = \Delta H_f^\circ[NCl_3(l)] + 3\Delta H_f^\circ[HCl(g)] - \Delta H_f^\circ[NH_3(g)]$$

Notice that this result is in the form of Equation 14.24 and that $\Delta H_f^\circ[Cl_2(g)]$ does not appear explicitly, because it is equal to zero by definition.

PRACTICE PROBLEM 14-7: Express ΔH_{rxn}° for the equation

$$CS_2(l) + 3\,O_2(g) \rightarrow CO_2(g) + 2\,SO_2(g)$$

in terms of the standard molar enthalpies of formation of the reactants and products.

Answer: $\Delta H_{rxn}^\circ = \Delta H_f^\circ[CO_2(g)] + 2\Delta H_f^\circ[SO_2(g)] - \Delta H_f^\circ[CS_2(l)]$

Figure 14.13 A Sterno burner in operation. Sterno is a mixture of ethanol and methanol in a gelling agent, which makes the fuel a solidlike material.

EXAMPLE 14-8: Use the ΔH_f° data in Table 14.3 to calculate the value of ΔH_{rxn}° for the combustion of ethanol, $CH_3CH_2OH(l)$, at 25°C, described by the chemical equation

$$CH_3CH_2OH(l) + 3O_2(g) \rightarrow 2CO_2(g) + 3H_2O(l)$$

Solution: Referring to Tables 14.2 and 14.3, we find that $\Delta H_f^\circ[CO_2(g)] = -393.5\ kJ\cdot mol^{-1}$; $\Delta H_f^\circ[H_2O(l)] = -285.8\ kJ\cdot mol^{-1}$; $\Delta H_f^\circ[O_2(g)] = 0$; and $\Delta H_f^\circ[CH_3CH_2OH(l)] = -277.6\ kJ\cdot mol^{-1}$. Application of Equation 14.24 yields

$$\begin{aligned}\Delta H_{rxn}^\circ &= \{2\Delta H_f^\circ[CO_2(g)] + 3\Delta H_f^\circ[H_2O(l)]\} - \{\Delta H_f^\circ[CH_3CH_2OH(l)] + 3\Delta H_f^\circ[O_2(g)]\}\\ &= \{2(-393.5\ kJ\cdot mol^{-1}) + 3(-285.8\ kJ\cdot mol^{-1})\} - \\ &\qquad \{(-277.6\ kJ\cdot mol^{-1}) + 3(0\ kJ\cdot mol^{-1})\}\\ &= -1366.8\ kJ\cdot mol^{-1}\end{aligned}$$

The ethanol combustion reaction is highly exothermic and is used extensively in alcohol burners of various types to keep food warm in chafing dishes (Figure 14.13). Ethanol is also used as a fuel and as an additive to gasoline mixtures.

PRACTICE PROBLEM 14-8: Use the data in Appendix D to calculate the standard enthalpy of combustion per gram for propane, $C_3H_8(g)$, at 25°C.

Answer: $\Delta H_{rxn}^\circ = -50.34\ kJ\cdot g^{-1}$

Although we have illustrated the validity of Equation 14.24 for only a few reactions, it is true in general.

FOOD IS FUEL The food we eat constitutes the fuel needed to maintain body temperature, accomplish numerous other physiological functions, and provide the energy we need. The common unit for the energy content of food is the **calorie (cal)**. A calorie was formerly defined as the amount of energy in the form of heat required to raise the temperature of one gram of water from 14.5°C to 15.5°C (one degree Celsius). In the SI unit system, a calorie is defined as exactly 4.184 J. The calorie used to be the unit of energy used in thermochemistry, but it has been replaced by the joule. The popular term "calorie" used by nutritionists, physicians, and the popular press is actually a kilocalorie and is sometimes written Calorie (abbreviated Cal) with a capital C. This is the term that appears on nutritional labels throughout the United States. Thus, 4184 J = 1000 cal = 1 kcal = 1 nutritional Calorie. About 100 kJ per kilogram of body weight per day is required to keep the body functioning at a minimal level.

If we consume more food than we require for our normal activity level, then the excess that is not eliminated is stored in the body as fat. One gram of body fat contains about 39 kJ of stored energy. Carbohydrates and proteins have less than one-half the energy content of fats per gram. A one-hour brisk walk over average terrain consumes only about 700 kJ of stored energy. The message is that calories (or better, kilojoules) count.

Both ethanol (Example 14-8) and propane (Practice Problem 14-8) are used as fuels. Table 14.4 lists values of the enthalpy of combustion per gram for some common fuels and the grams of carbon produced in the form of $CO_2(g)$ per kilojoule (see Interchapter L). The molar enthalpy of combustion data in Table 14.4 shows that the most energy-rich fuel on a per mass basis is hydrogen, which has an energy content per gram of well over twice that of the next-best fuel. Because of its unusually high energy content per gram, and because the fuel must be lifted as part of the space vehicle, liquid hydrogen was used in the first stage of the Apollo spaceships that traveled to the moon and is the fuel carried in the Space Shuttle's external tank that is used to help lift it into orbit. The main disadvantages of liquid hydrogen as a fuel are that $H_2(l)$ can be maintained as a liquid only at very low temperatures (about 20 K at 1 bar) and that hydrogen readily forms explosive mixtures with air. The second and third stages of the Apollo spaceships were powered by the reaction between kerosene and liquid oxygen (LOX), both of which were stored on the spaceship.

See Interchapter L at www.McQuarrieGeneralChemistry.com.

TABLE 14.4 Standard enthalpies of combustion of selected fuels and number of grams of carbon produced in the form of $CO_2(g)$ per kilojoule

Fuel	ΔH°_{rxn}/kJ·g^{-1}	carbon/g·kJ^{-1}
hydrogen, $H_2(g)$	−141.8	0.0
methane, $CH_4(g)$	−55.5	13.5
propane, $C_3H_8(g)$	−50.3	16.2
methanol, $CH_3OH(l)$	−22.7	16.5
ethanol, $CH_3CH_2OH(l)$	−29.7	17.6
fuel oil	−40.9	21.3
gasoline	−46.5	17.6
kerosene	−46.4	18.5
coal, anthracite	−34.6	27.3
wood, oak	−18.9	25.3
charcoal, wood	−34.7	26.8

The power used to propel the Apollo Lunar Lander spaceships to and from the surface of the moon and to maneuver satellites and the various missions to Mars (once in space) is the energy released by the reaction of *1,1*-dimethylhydrazine, $H_2NN(CH_3)_2(l)$, with dinitrogen tetroxide, $N_2O_4(l)$:

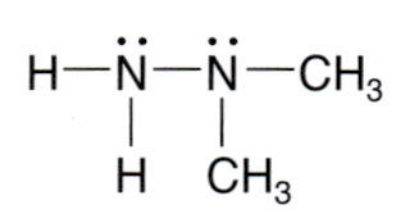

1,1-dimethylhydrazine

$$H_2NN(CH_3)_2(l) + 2\,N_2O_4(l) \rightarrow 3\,N_2(g) + 2\,CO_2(g) + 4\,H_2O(g)$$

$$\Delta H^\circ_{rxn} = -29 \text{ kJ per gram of fuel}$$

Figure 14.14 Underwater firing of a Polaris A-3 test missile from a nuclear submarine.

This reaction is especially suitable for a space vehicle because the reaction starts spontaneously on mixing. No battery or spark plugs, with associated electrical circuitry, are required.

The principal solid fuel used in the Space Shuttle booster rockets and a number of missiles (Figure 14.14) consists of a mixture of 70% ammonium perchlorate, 18% aluminum metal powder, and 12% binder. Ammonium perchlorate is a self-contained solid fuel—the fuel NH_4^+ and the oxidizer ClO_4^- are together in the solid. The overall equation for the $NH_4ClO_4(s)$ decomposition reaction is

$$4\,NH_4ClO_4(s) \rightarrow 2\,N_2(g) + 4\,HCl(g) + 6\,H_2O(g) + 5\,O_2(g)$$

Note that the $NH_4ClO_4(s)$ is oxygen rich; that is, $O_2(g)$ is a reaction product. To utilize this available oxygen and thereby to provide more rocket thrust, aluminum powder and a binder are added. The aluminum is oxidized to aluminum oxide:

$$4\,Al(s) + 3\,O_2(g) \rightarrow 2\,Al_2O_3(s)$$

and the binder is oxidized to $CO_2(g)$ and $H_2O(g)$. The aluminum powder also promotes a more rapid and even decomposition of the $NH_4ClO_4(s)$.

14-6. The Value of ΔH°_{rxn} Is Determined Primarily by the Difference in the Molar Bond Enthalpies of the Reactant and Product Molecules

Now that we know more about the role of ΔH°_{rxn} in reactions, let's see what molecular properties of reactants and products give rise to the observed values of ΔH°_{rxn}.

The enthalpy change for the equation

$$H_2O(g) \rightarrow O(g) + 2\,H(g)$$

is $\Delta H^\circ_{rxn} = +925$ kJ. We know that there are two oxygen-hydrogen bonds in a water molecule (H–O–H). The positive value of ΔH°_{rxn} indicates that an input of 925 kJ of energy is required to break the two moles of oxygen-hydrogen bonds in one mole of water molecules. Thus, the average **molar bond enthalpy**, H_{bond}(oxygen-hydrogen) in a water molecule is equal to one-half the total energy required to break the two moles of oxygen-hydrogen bonds:

$$H_{bond}(\text{O–H}) = \tfrac{1}{2}(+925\ \text{kJ·mol}^{-1}) = +463\ \text{kJ·mol}^{-1}$$

Strictly speaking, this is the average O–H molar bond enthalpy of a water molecule. The values of the oxygen-hydrogen molar bond enthalpies in a variety of compounds are found to have about the same value as the O–H molar bond enthalpy in a water molecule. For example, the standard enthalpy change for the reaction described by

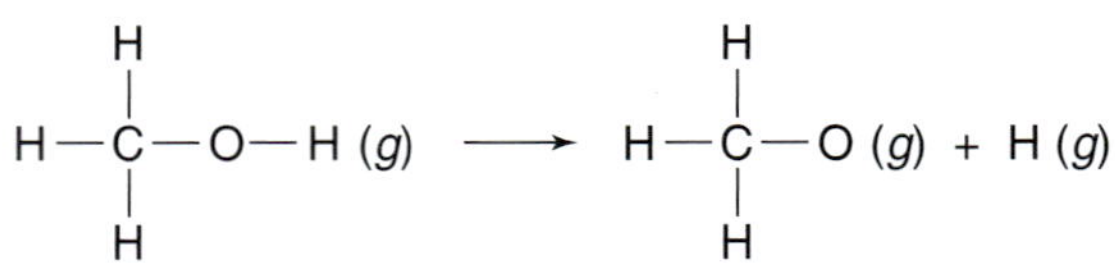

in which a single oxygen-hydrogen bond is broken, is $\Delta H^\circ_{rxn} = +464$ kJ·mol^{-1}, which is very similar to that of the O–H bond in a water molecule.

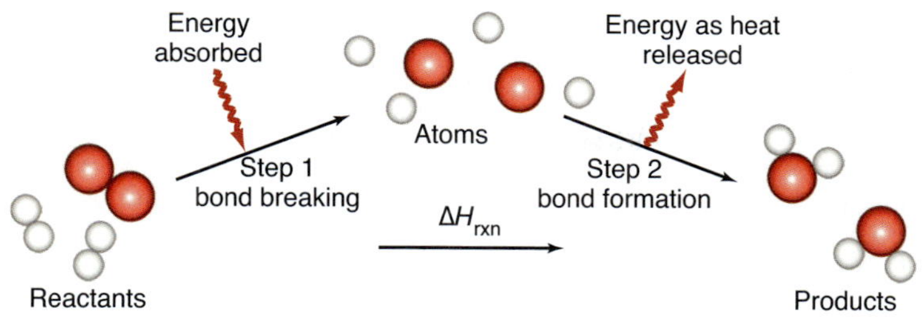

Figure 14.15 The enthalpy change of a reaction is given approximately by the difference between the energy as heat required to break all the chemical bonds in the reactant molecules and the energy released as heat on the formation of all the chemical bonds in the product molecules.

In general, we can picture a chemical reaction as taking place in two steps (Figure 14.15). In Step 1 the reactant molecules are broken down into their constituent atoms, and in Step 2 the atoms are rejoined to form the product molecules. Step 1 requires an input of energy to break all the bonds in the reactant molecules, whereas Step 2 evolves energy as the bonds in the product molecules are formed. The total enthalpy change for the equation is

$$\Delta H^\circ_{rxn} \approx \begin{pmatrix} \text{energy required} \\ \text{to break all the} \\ \text{bonds in the} \\ \text{reactant molecules} \end{pmatrix} - \begin{pmatrix} \text{energy evolved as heat} \\ \text{upon the formation} \\ \text{of all the bonds in the} \\ \text{product molecules} \end{pmatrix} \quad (14.25)$$

Equation 14.25 is simply an accounting of the energy involved in the breaking and the formation of the bonds of the species involved in the reaction. The energy required to break all the bonds in the reactant molecules is equal to H_{bond} [reactants], the sum of the molar bond enthalpies for all the reactant bonds. The energy evolved as heat upon the formation of all the bonds in the product molecules is equal to $-H_{bond}$ [products], the sum of the molar bond enthalpies for all the product bonds in the chemical equation for the reaction. The negative sign in $-H_{bond}$ [products] occurs because the formation of the bonds is the reverse of the breaking of the bonds. Thus, we can write Equation 14.25 as

$$\Delta H^\circ_{rxn} \approx H_{bond}\,[\text{reactants}] - H_{bond}\,[\text{products}] \quad (14.26)$$

The reason for the "approximately equals to sign" here is that molar bond enthalpies are *average* values and so values of ΔH°_{rxn} calculated based on average molar bond enthalpies are only approximations to actual experimental values. Notice the similarity of Equation 14.26 to Equation 14.24, but with the enthalpies of the product molecules and reactant moecules reversed; this arises from the fact that molar bond enthalpies are defined as the enthalpy required to break one mole of a given bond, whereas molar enthalpies of formation are defined as the enthalpy released upon the formation of one mole of bonds.

We see from either Equation 14.25 or 14.26 that if more energy is released on formation of the product bonds than is required to break the reactant bonds, then the value of ΔH°_{rxn} is negative (energy released; exothermic). If less energy is released on the formation of the product bonds than is required to break the reactant bonds, then the value of ΔH°_{rxn} is positive (energy consumed; endothermic). Thus, the value of ΔH°_{rxn} is determined primarily by the difference in the molar bond enthalpies of the reactant and product molecules. Values of average molar bond enthalpies for a variety of chemical bonds are given in Table 14.5.

TABLE 14.5 Average molar bond enthalpies

Bond	Molar bond enthalpy, $H_{bond}/kJ\cdot mol^{-1}$	Bond	Molar bond enthalpy, $H_{bond}/kJ\cdot mol^{-1}$
O–H	464	C≡N	890
O–O	142	N–H	390
C–O	351	N–N	159
O=O	502	N=N	418
C=O	730	N≡N	945
C–C	347	F–F	155
C=C	615	Cl–Cl	243
C≡C	811	Br–Br	192
C–H	414	H–H	435
C–F	439	H–F	565
C–Cl	331	H–Cl	431
C–Br	276	H–Br	368
C–N	293	H–S	364
C=N	615		

In most cases we can use Hess's law and tables of standard enthalpies of formation to determine the enthalpy of a reaction as we did in Section 14-5. However, in some cases we need to estimate the enthalpy of a reaction for which no experimental data are available, or we simply want a "quick estimate" of the enthalpy of reaction without searching for molar enthalpy of formation values in the literature. In such cases, molar bond enthalpies can yield an adequate approximation of the actual enthalpies of reaction.

Let's use the molar bond enthalpies in Table 14.5 to estimate the value of ΔH°_{rxn} for the reaction described by the equation

$$CH_4(g) + Cl_2(g) \rightarrow CH_3Cl(g) + HCl(g)$$

and compare this value to the actual enthalpy of reaction as calculated using Hess's law and the data in Table 14.3. The bonding in these molecules can be represented by

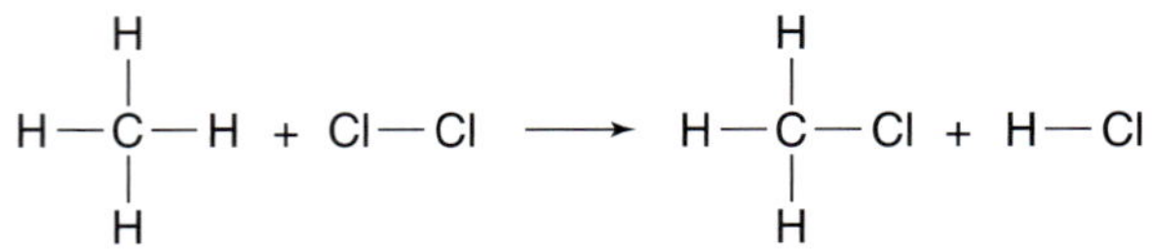

This representation makes it clear that the reaction involves *breaking* one C–H bond and one Cl–Cl bond and *forming* one C–Cl bond and one H–Cl bond. The value of H_{bond} [reactants] is

$$H_{bond}\,[\text{reactants}] = H_{bond}\,(\text{C–H}) + H_{bond}\,(\text{Cl–Cl})$$
$$= 414\ kJ\cdot mol^{-1} + 243\ kJ\cdot mol^{-1} = 657\ kJ\cdot mol^{-1}$$

and the value of H_{bond} [products] is

$$H_{bond}\,[\text{products}] = H_{bond}\,(\text{C–Cl}) + H_{bond}\,(\text{H–Cl})$$
$$= 331\ \text{kJ·mol}^{-1} + 431\ \text{kJ·mol}^{-1} = 762\ \text{kJ·mol}^{-1}$$

The standard enthalpy of reaction is given by Equation 14.26 as

$$\Delta H^{\circ}_{rxn} \approx H_{bond}\,[\text{reactants}] - H_{bond}\,[\text{products}]$$
$$= 657\ \text{kJ·mol}^{-1} - 762\ \text{kJ·mol}^{-1} = -105\ \text{kJ·mol}^{-1}$$

Using Equation 14.24 and the experimentally determined standard molar enthalpies of formation data from Table 14.3, we find that

$$\Delta H^{\circ}_{rxn} = \{\Delta H^{\circ}_{f}[CH_3Cl(g)] + \Delta H^{\circ}_{f}[HCl(g)]\} - \{\Delta H^{\circ}_{f}[CH_4(g)] + \Delta H^{\circ}_{f}[Cl_2(g)]\}$$
$$\approx \{(-81.9\ \text{kJ·mol}^{-1}) + (-92.3\ \text{kJ·mol}^{-1})\} - \{(-74.6\ \text{kJ·mol}^{-1}) + (0\ \text{kJ·mol}^{-1})\}$$
$$= -99.6\ \text{kJ·mol}^{-1}$$

Thus, there is only a 5% difference between the approximate value of -105 kJ·mol^{-1} determined using average molar bond enthalpies and the value of -99.6 kJ·mol^{-1} determined using the standard molar enthalpies of formation.

EXAMPLE 14-9: Chemical reactions can be used to produce flames for heating. The nonnuclear reaction with the highest attainable flame temperature (approximately 6000°C, about the surface temperature of the sun) is that between hydrogen and fluorine:

$$H_2(g) + F_2(g) \rightarrow 2\,HF(g)$$

Use the molar bond enthalpies in Table 14.5 to estimate the value of ΔH°_{rxn} for this equation. Compare this to the value calculated using the molar enthalpies of formation in Table 14.3.

Solution: The reaction involves the breaking of one hydrogen–hydrogen bond and one fluorine–fluorine bond and the formation of two hydrogen–fluorine bonds. Thus,

$$\Delta H^{\circ}_{rxn} \approx H_{bond}(\text{H–H}) + H_{bond}\,(\text{F–F}) - 2\,H_{bond}\,(\text{H–F})$$
$$= (435\ \text{kJ·mol}^{-1}) + (155\ \text{kJ·mol}^{-1}) - (2)(565\ \text{KJ·mol}^{-1})$$
$$= -\,540\ \text{kJ·mol}^{-1}$$

Because this reaction is the formation reaction for two moles of $HF(g)$ from its constituent elements, the experimentally determined value of ΔH°_{rxn} is twice the molar enthalpy of formation of $HF(g)$ listed in Table 14.3, or -546.6 kJ·mol^{-1}, a value with less than a 2% difference from that which we found using average molar bond enthalpies.

PRACTICE PROBLEM 14-9: Hydrazine, $N_2H_4(l)$, and its derivatives are used as rocket fuels. Use the molar bond enthalpies in Table 14.5 to estimate the molar enthalpy of formation of $N_2H_4(g)$. Calculate the percentage difference between this value and the tabulated value of $\Delta H^{\circ}_{f}[N_2H_4(g)]$.

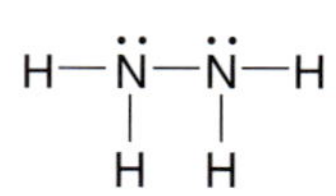

hydrazine

Answer: $\Delta H^{\circ}_{f} \approx 96$ kJ·mol^{-1}; 0.6%

Although the value of ΔH°_{rxn} is determined primarily by the difference in bond enthalpies of the reactant molecules and the product molecules, the attractive forces *between* chemical species in the liquid and in solid phases can also make significant contributions to the value of ΔH°_{rxn}. As an example, consider the vaporization of liquid water at 25°C, $H_2O(l) \rightarrow H_2O(g)$. Using Table 14.3, we calculate that $\Delta H^\circ_{rxn} = +44.0$ kJ for the vaporization of one mole of $H_2O(l)$. No internal oxygen–hydrogen bonds are broken in this process. In the vaporization of water, it is the attractive forces *between* the water molecules that must be overcome. (We shall study the nature of these forces in the next chapter.) For this reason, the calculation of ΔH°_{rxn} values from molar bond enthalpies alone is restricted to gas-phase reactions. Again, keep in mind that the values in Table 14.5 are *average* values and that values of ΔH°_{rxn} determined using molar bond enthalpies are only approximations to the actual experimentally determined values.

14-7. Heat Capacity Measures the Ability of a Substance to Take Up Energy as Heat

The ΔH°_f values given in Table 14.3 are derived from experimentally determined enthalpies of reaction. We shall see how to measure enthalpies of reaction in the next section, but first we must introduce a quantity called heat capacity.

The **heat capacity** of a sample of a substance is defined as the energy in the form of heat required to raise the temperature of the sample by one degree Celsius, or equivalently, by one kelvin (see sidebox). If the substance is heated at constant pressure, then the heat capacity is denoted by c_P, where the subscript P denotes constant pressure. We can write an equation for the definition of the heat capacity at constant pressure:

$$c_P = \frac{q_P}{\Delta T} \tag{14.27}$$

Because $q_P = \Delta H$ (Equation 14.12), Equation 14.27 can also be written as

$$c_P = \frac{\Delta H}{\Delta T} \tag{14.28}$$

THE CHANGE IN TEMPERATURE IN DEGREES CELSIUS IS EQUIVALENT TO THE CHANGE IN KELVIN

$$\Delta T(K) = \Delta t(°C)$$

To show the validity of this equation, realize that the Kelvin and Celsius scales differ by a constant amount because

$$T/K = t/°C + 273.15$$

Therefore, if we let $T_{initial} = t_{initial} + 273.15$ and $T_{final} = t_{final} + 273.15$, then

$$T_{final} - T_{initial} = (t_{final} + 273.15) - (t_{initial} + 273.15) = t_{final} - t_{initial}$$

or

$$\Delta T = \Delta t$$

Thus,

$$q_P = \Delta H = c_P \Delta T \tag{14.29}$$

where q_P is the energy added as heat at constant pressure and ΔT is the increase in temperature of the substance arising from the energy input. All that is needed to determine the heat capacity of a substance is to add a known quantity of energy as heat and then measure the resulting increase in temperature at some fixed pressure. The heat capacity of a substance is always positive. The following Example illustrates the use of Equation 14.27.

EXAMPLE 14-10: When 4.219 kJ of energy are added as heat to 36.0 grams of water at one bar, the temperature of the water increases from 10.00°C to 38.05°C. Calculate the heat capacity of the 36.0 grams of $H_2O(l)$.

Solution: We use Equation 14.27. The value of ΔT is equal to 38.05°C − 10.00°C = 28.05°C. Because a kelvin and a Celsius degree are the same size (see sidebox), we have

$$c_P = \frac{q_P}{\Delta T} = \frac{4219\ \text{J}}{28.05\ \text{K}} = 150.4\ \text{J}\cdot\text{K}^{-1}$$

for the heat capacity of 36.0 grams of $H_2O(l)$.

PRACTICE PROBLEM 14-10: The heat capacity of 18.0 grams of ice is 37.7 J·K^{-1}. Calculate the final temperature of 18.0 grams of ice, initially at −20.0°C, that results when 200.0 joules of heat are absorbed by the ice at constant pressure.

Answer: −14.7°C

We denote the heat capacity per mole, or the **molar heat capacity,** of any substance by C_P. (A capital C is used to denote molar heat capacity.) The heat capacity per mole of water can be computed from the value of 150.4 J·K^{-1} for 36.0 grams of water calculated in Example 14-10. Because one mole of water has a mass of 18.016 grams, the value of C_P for water is

$$C_P = \frac{150.4\ \text{J}\cdot\text{K}^{-1}}{(36.0\ \text{g})\left(\dfrac{1\ \text{mol}}{18.016\ \text{g}}\right)} = 75.3\ \text{J}\cdot\text{mol}^{-1}\cdot\text{K}^{-1}$$

Notice from this expression that $C_P = c_p/n$. Thus, using molar heat capacities, Equation 14.29 may be also be written as

$$q_P = \Delta H = nC_P \Delta T \tag{14.30}$$

where n is the number of moles of the substance gaining or losing energy as heat. The values of the molar heat capacities for a variety of substances are given in Table 14.6.

TABLE 14.6 Molar heat capacities at constant pressure for various substances at 25°C*

Name	Formula	C_P/J·mol^{-1}·K^{-1}	Name	Formula	C_P/J·mol^{-1}·K^{-1}
aluminum	$Al(s)$	24.4	lithium	$Li(s)$	24.8
ammonia	$NH_3(g)$	35.1	mercury	$Hg(g)$	20.8
argon	$Ar(g)$	20.8		$Hg(l)$	28.0
carbon dioxide	$CO_2(g)$	37.1	methane	$CH_4(g)$	35.7
carbon monoxide	$CO(g)$	29.1	neon	$Ne(g)$	20.8
copper	$Cu(s)$	24.4	nitrogen	$N_2(g)$	29.1
ethane	$C_2H_6(g)$	52.5	oxygen	$O_2(g)$	29.4
ethene (ethylene)	$C_2H_4(g)$	42.0	silver	$Ag(s)$	25.4
ethyne (acetylene)	$C_2H_2(g)$	44.0	sodium	$Na(s)$	28.2
gold	$Au(s)$	25.4	tin	$Sn(s)$	27.0
helium	$He(g)$	20.8	water	$H_2O(s)$	37.7
hydrogen	$H_2(g)$	28.8		$H_2O(l)$	75.3
iron	$Fe(s)$	25.1		$H_2O(g)$	33.6

*Data from *CRC Handbook of Chemistry and Physics,* 86th Ed., Ed. David R. Lide, CRC Press, 2005–2006.

EXAMPLE 14-11: Calculate the energy as heat required to raise the temperature of 150 kg of water (40 gallons, the volume of a typical home water heater in the U.S.) from 18°C to 60°C, assuming no loss of energy to the surroundings.

Solution: The energy in the form of heat required can be calculated from Equation 14.30:

$$\begin{aligned} q_P &= nC_P\Delta T \\ &= (150\text{ kg})\left(\frac{1000\text{ g}}{1\text{ kg}}\right)\left(\frac{1\text{ mol}}{18.016\text{ g}}\right)(75.3\text{ J}\cdot\text{mol}^{-1}\cdot\text{K}^{-1})(333-291)\text{K} \\ &= 2.6\times10^{7}\text{ J} = 26\text{ MJ} \end{aligned}$$

where 1 MJ = 1 megajoule = 1×10^6 J. If natural gas provides the heating energy at a cost of about $1.50 per 100 MJ, it costs about 40 cents to heat 40 gallons of water from 18°C to 60°C, assuming that all the energy transferred as heat goes into the water. In practice, only about half the energy transferred as heat is absorbed by the water; the remainder is lost to the surroundings, so the actual cost is about 80 cents.

PRACTICE PROBLEM 14-11: Consider a swimming pool 25 feet long by 12 feet wide with an average depth of 6 feet. Take the molar heat capacity of water as 75.3 J·K^{-1}·mol^{-1} and the density of water as 1.0 g·cm^{-3}. Calculate the number of kilojoules of energy in the form of heat required to raise the temperature of the pool water by 10°C, assuming no energy is lost as heat to the surroundings. Given that the cost of natural gas is \$1.55 per therm (1 therm = 10^5 Btu; 1 Btu = 1.055 kJ) and assuming that 25% of the energy is lost as heat to the surroundings during the heating cycle, calculate the cost of raising the pool temperature by 10°C.

Answer: 2×10^6 kJ; \$40

(Note that the cost of *maintaining* a heated swimming pool at a temperature greater than its surroundings is much higher.)

The heat capacity per gram of a substance is called the **specific heat** and is denoted by c_{sp}. Specific heats are especially useful in dealing with heat capacities of substances for which we cannot write a chemical formula, such as glass, steel, wood, or cement. To obtain the specific heat of a substance, we determine its heat capacity (Equation 14.27) and divide the result by the mass in grams, that is,

$$c_{sp} = \frac{c_P}{m} = \frac{q_P}{m\Delta T} \tag{14.31}$$

EXAMPLE 14-12: Calculate the specific heat of a sample of glass, given that the input of 150 joules of energy as heat to a 50.0-gram sample causes a temperature increase of 3.57°C.

Solution: Application of Equation 14.31 yields for the specific heat

$$c_{sp} = \frac{q_P}{m\Delta T} = \frac{150\,\text{J}}{(50.0\text{ g})(3.57\text{ K})} = 0.84\,\text{J}\cdot\text{g}^{-1}\cdot\text{K}^{-1}$$

The specific heat of glass of $0.84\,\text{J}\cdot\text{g}^{-1}\cdot\text{K}^{-1}$ is also approximately equal to the specific heat of rocks, cement, or dirt.

PRACTICE PROBLEM 14-12: The specific heat of a certain steel is $0.46\,\text{J}\cdot\text{g}^{-1}\cdot\text{K}^{-1}$ and the specific heat of a "dry" wood is $1.8\,\text{J}\cdot\text{g}^{-1}\cdot\text{K}^{-1}$. Suppose we add a given amount of energy as heat to 10.0-gram samples of the steel and wood. (a) Which sample will undergo the largest temperature increase? (b) Calculate the ratio of the temperature increase for the two samples.

Answer: (a) steel; (b) 3.9 = (steel/wood)

The relatively high specific heat and low thermal conductivity of wood compared with metal is why wooden handles are often used as grips on steel frying pans.

If we combine two nonreactive substances at different temperatures, then the higher temperature substance will transfer energy to the lower temperature sub-

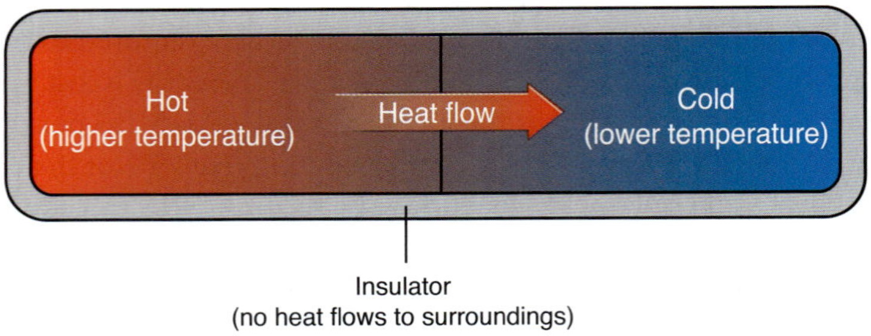

Figure 14.16 Energy in the form of heat flows spontaneously from an object at a higher temperature to one at a lower temperature until thermal equilibrium is attained.

stance until the combined system attains a uniform temperature (Figure 14.16). The value of the final temperature lies somewhere between the two initial temperatures. If no energy is transferred as heat to or from the surroundings, then all the energy lost as heat by the high-temperature substance is gained by the low-temperature substance; that is,

$$q_h(\text{lost}) + q_l(\text{gained}) = 0 \tag{14.32}$$

where the subscripts h and l denote the high- and low-temperature substances, respectively. Recall that $q_P = \Delta H = c_P \Delta T$ (Equation 14.29), so we can rewrite Equation 14.32 as

$$c_{P_h}(T_f - T_h) + c_{P_l}(T_f - T_l) = 0 \tag{14.33}$$

where T_f is the final temperature of the two substances.

Observe that T_f must be less than T_h, so that the first term is negative, and T_f must be greater than T_l, so that the second term is positive. In addition, the value of T_f will be closer numerically to the initial temperature of the system with the greater heat capacity. The higher the heat capacity of a system, the smaller the temperature change that the system undergoes when it loses or gains a given amount of energy as heat. The following Example illustrates the application of Equation 14.33.

EXAMPLE 14-13: A 50.0-gram piece of copper metal at 80.0°C is placed in 100.0 mL (100.0 g) of water at 10.0°C. Assuming no energy is lost to the surroundings as heat, what will be the final temperature of the copper-water system? Take the molar heat capacities of Cu(*s*) and $H_2O(l)$ as $24.4\ \text{J}\cdot\text{mol}^{-1}\cdot{}^{\circ}\text{C}^{-1}$ and $75.3\ \text{J}\cdot\text{mol}^{-1}\cdot{}^{\circ}\text{C}^{-1}$, respectively.

Solution: When the hot copper metal is placed in the cool water, the copper will cool down and the water will warm up until they both come to the same final temperature. Let this (unknown) final temperature be T_f. The heat capacities of the copper and the water samples are

$$c_P[\mathrm{Cu}(s)] = (50.0\text{ g})\left(\frac{1\text{ mol}}{63.55\text{ g}}\right)(24.4\text{ J}\cdot\text{mol}^{-1}\cdot{}^\circ\text{C}^{-1}) = 19.2\text{ J}\cdot{}^\circ\text{C}^{-1}$$

$$c_P[\mathrm{H_2O}(l)] = (100.0\text{ g})\left(\frac{1\text{ mol}}{18.016\text{ g}}\right)(75.3\text{ J}\cdot\text{mol}^{-1}\cdot{}^\circ\text{C}^{-1}) = 418\text{ J}\cdot{}^\circ\text{C}^{-1}$$

Application of Equation 14.33, with Cu(s) as the high-temperature sample and $H_2O(l)$ as the low-temperature sample, yields

$$\begin{aligned}0 &= c_{P_h}(T_f - T_h) + c_{P_l}(T_f - T_l)\\ &= (19.2\text{ J}\cdot{}^\circ\text{C}^{-1})(T_f - 80.0^\circ\text{C}) + (418\text{ J}\cdot{}^\circ\text{C}^{-1})(T_f - 10.0^\circ\text{C})\end{aligned}$$

or, collecting like terms,

$$0 = [19.2\text{ J}\cdot{}^\circ\text{C}^{-1} + 418\text{ J}\cdot{}^\circ\text{C}^{-1}]T_f - [(19.2)(80.0)\text{J} + (418)(10.0)\text{J}]$$

from which we obtain

$$T_f = \frac{(19.2)(80.0)\text{J} + (418)(10.0)\text{J}}{19.2\text{ J}\cdot{}^\circ\text{C}^{-1} + 418\text{ J}\cdot{}^\circ\text{C}^{-1}} = \frac{5716\text{ J}}{437\text{ J}\cdot{}^\circ\text{C}^{-1}} = 13.1^\circ\text{C}$$

The temperature of the copper decreases by about 66.9°C (= 80.0°C − 13.1°C), whereas the temperature of the water increases by only 3.1°C (= 13.1°C − 10.0°C) because the heat capacity of the water is over twenty times greater than that of the copper.

PRACTICE PROBLEM 14-13: Calculate the final temperature if a 500.0-gram piece of steel at 1200°C is dropped into 100.0 kg of water at 15°C. Assume no energy is lost to the surroundings as heat. Take the specific heat of steel as $0.46\text{ J}\cdot{}^\circ\text{C}^{-1}\cdot\text{g}^{-1}$.

Answer: 16°C

14-8. A Calorimeter Is a Device Used to Measure the Amount of Energy Evolved or Absorbed as Heat in a Reaction

The value of ΔH for a chemical reaction can be measured in a device called a **calorimeter.** A simple calorimeter, consisting of a **Dewar flask** ("thermos bottle") equipped with a high-precision thermometer, is shown in Figure 14.17. A calorimeter works on the principle that the total energy is always conserved. For a chemical reaction occurring at fixed pressure, the value of ΔH is equal to the energy evolved or absorbed as heat by the reaction. Consider an exothermic reaction run in a Dewar flask. The energy evolved as heat by the reaction cannot escape from the flask, so it is absorbed by the calorimeter contents (reaction mixture, thermometer, stirrer, and so on). The absorption of the energy evolved as heat by the calorimeter contents leads to an increase in the temperature of the calorimeter contents. Because all the energy evolved as heat by the reaction is absorbed by the calorimeter contents, we can write

$$\Delta H = -\Delta H_{cal} \tag{14.34}$$

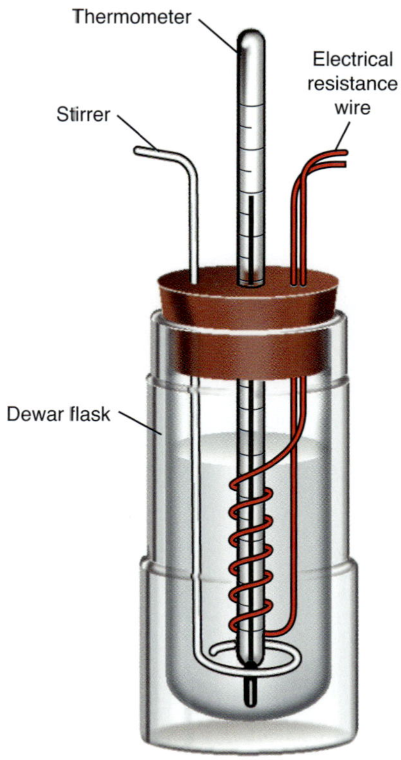

Figure 14.17 A simple calorimeter, consisting of a Dewar flask and its cover (which prevents a significant transfer of energy as heat to or from the surroundings); a high-precision thermometer (which gives the temperature to within ±0.001 K); a simple ring-type stirrer; and an electrical resistance heater. One reactant is placed in the Dewar flask and then the other reactant, at the same temperature, is added. As the reaction mixture is stirred, the change in temperature is measured. The resistance heater is used to measure the heat capacity of the calorimeter.

where the subscript cal stands for calorimeter. Using Equation 14.29, we can write ΔH_{cal} as

$$\Delta H_{cal} = c_{P,cal}\Delta T \quad (14.35)$$

where ΔT is the observed temperature change. Substitution of Equation 14.35 into Equation 14.34 yields

$$\Delta H = -c_{P,cal}\Delta T \quad (14.36)$$

Recall that one definition of standard states is that all species in solution have a concentration of one molar (1 M).

Equation 14.36 tells us that if we run a chemical reaction in a calorimeter with a known heat capacity ($c_{P,cal}$) and determine the temperature change at standard conditions, then the value of ΔH° can be determined. The value of $c_{P,cal}$ can be found by electrical resistance heating, for example (Figure 14.17). Using the observed value of ΔH°, we can then calculate the value of ΔH°_{rxn}, as shown below in Example 14.14. The value of ΔH°_{rxn} for reactions involving solutions depends on concentration, but this effect is usually not large. We shall ignore the effect of concentration on ΔH°_{rxn} values and assume that $\Delta H^\circ_{rxn} \approx \Delta H^\circ$.

EXAMPLE 14-14: A 0.500-L sample of 0.200 M NaCl(aq) is added to 0.500 L of 0.200 M $AgNO_3(aq)$ in a calorimeter with a known total heat capacity equal to $4.60 \times 10^3\ J\cdot K^{-1}$ at a constant pressure of one bar. The observed temperature change is +1.423 K. Use these data to determine the value of ΔH°_{rxn} for the equation

$$AgNO_3(aq) + NaCl(aq) \rightarrow AgCl(s) + NaNO_3(aq)$$

Solution: Stoichiometrically equivalent quantities of NaCl(aq) and $AgNO_3(aq)$ are added to each other, and so there is no limiting reactant in this case. The observed increase in temperature arises from the formation of the precipitate AgCl(s). The number of moles of AgCl(s) formed in the reaction is given by

$$\begin{aligned}\text{moles of AgCl}(s)\text{ formed} &= \text{moles of AgNO}_3(aq)\text{ or NaCl}(aq)\\ &= (0.500\ L)(0.200\ mol\cdot L^{-1}) = 0.100\ mol\end{aligned}$$

The value of ΔH for the formation of 0.100 moles of AgCl(s) is

$$\Delta H = -c_{P,cal}\Delta T = -(4.60 \times 10^3\ J\cdot K^{-1})(1.423\ K) = -6550\ J$$

This is the value of ΔH for the formation of 0.100 moles; thus, we have

$$\Delta H^\circ_{rxn} = \frac{-6550\ J}{0.100\ mol} = -65.5 \times 10^3\ J\cdot mol^{-1} = -65.5\ kJ\cdot mol^{-1}$$

for the value of ΔH°_{rxn} for the formation of one mole of AgCl(s).

PRACTICE PROBLEM 14-14: When 2.50 grams of ammonium chloride are dissolved in 100.0 mL of water inside an insulated Dewar flask with a total heat capacity of 418 $J\cdot K^{-1}$ at a constant pressure of one bar, a temperature decrease of 1.65°C is observed. Calculate the value of ΔH°_{rxn} for the dissolution of ammonium chloride in units of $kJ\cdot mol^{-1}$. Is this an exothermic or endothermic process? Assume that no energy is lost as heat through the Dewar flask during the reaction.

Answer: +14.8 $kJ\cdot mol^{-1}$; endothermic

14-9. The Energy of Reaction Can Be Measured in a Bomb Calorimeter

Let's go back to the equation:

$$\Delta U_{rxn} = q + w \tag{14.37}$$

If a reaction occurs in a rigid, closed container, then there is no change in volume ($\Delta V = 0$), and so no energy is transferred as work ($w = -P\Delta V = 0$). In this case, Equation 14.37 becomes

$$\Delta U_{rxn} = q_V = \begin{pmatrix}\text{energy evolved or absorbed}\\ \text{as heat at constant volume}\end{pmatrix} \tag{14.38}$$

The subscript V on q emphasizes that q_V is the energy evolved or absorbed as heat when the volume of the reaction system is constant. Because we have specified the manner in which the reaction takes place, $q_V = \Delta U_{rxn}$ is a state function. Equation 14.38 is the constant volume analog of Equation 14.12.

Let's think about the difference between a reaction run at constant volume and one run at constant pressure. In the case of a reaction run at constant pressure, any gas that is produced has to expend energy as work to push back the surrounding atmosphere, whereas one run at constant volume does no work. Because $\Delta U_{rxn} = q + w$ is a state function, the sum of q and w must be the same for both reactions. Therefore, q, the energy evolved or absorbed as heat in a reaction run at constant pressure (where $w \neq 0$), is not the same as that for the reaction run at constant volume (where $w = 0$).

The value of ΔU for the combustion of a substance can be determined in a **bomb calorimeter** like that shown in Figure 14.18. A bomb calorimeter has a fixed total volume; thus, the reaction takes place at constant volume, and so Equation 14.38 applies. A known mass of the substance whose energy of combustion is to be determined is loaded into the bomb calorimeter along with an ignition wire. The calorimeter is then pressurized with excess oxygen gas at about 30 bar. The combustion reaction is initiated by passing a short burst of high-voltage current through the ignition wire. The value of ΔU of combustion is determined by measuring the temperature increase of the calorimeter and of the water in which it is immersed. From the known heat capacity of the calorimeter assembly and the observed value of ΔT, the value of ΔU can be computed in a manner analogous to that described for the Dewar-flask calorimeter.

Although a bomb calorimeter gives us a value of ΔU rather than ΔH, we can usually assume that $\Delta H \approx \Delta U$, as shown in the next Example.

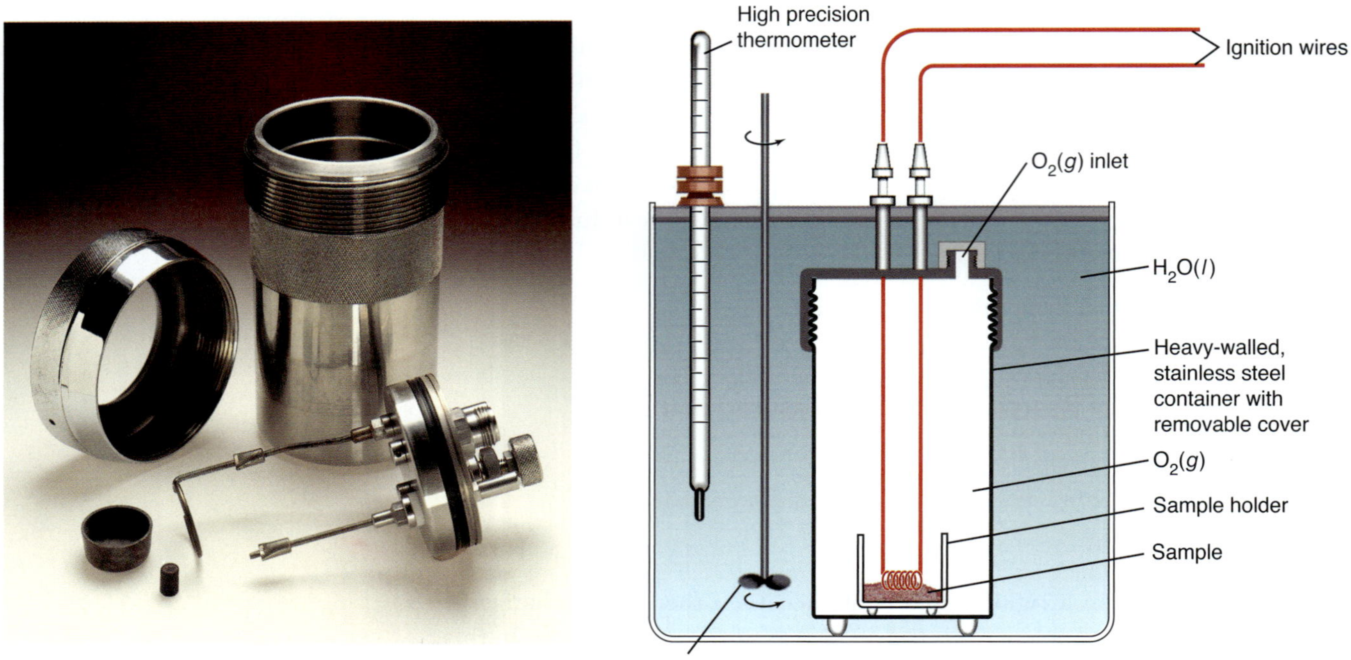

(a) (b)

Figure 14.18 (a) A bomb calorimeter and (b) its cross section. The inner container ("bomb"), within which the combustion reaction occurs, is placed in the outer container filled with water, as shown here. The temperature change of the surrounding water bath is measured using a high precision thermometer. The water is stirred to ensure a uniform distribution of heat.

EXAMPLE 14-15: A 1.000-gram sample of octane, $C_8H_{18}(l)$, is burned in a calorimeter like that shown in Figure 14.18, and the observed temperature increase is 1.679 K. The total heat capacity of the calorimeter is $c_{V,cal} = 28.46\ kJ\cdot K^{-1}$ (where we write $c_{V,cal}$ to emphasize that the value of the heat capacity of the calorimeter is that at constant volume in this case).
(a) Calculate the energy of combustion per gram and per mole of $C_8H_{18}(l)$.
(b) Using the molar enthalpy of formation values listed in Table 14.3, show that $\Delta H_{rxn} \approx \Delta U_{rxn}$ for this reaction at 25°C.

Solution: (a) The equation for the combustion reaction is

$$2\,C_8H_{18}(l) + 25\,O_2(g) \rightarrow 16\,CO_2(g) + 18\,H_2O(l)$$

The energy of combustion is given by

$$\Delta U = -c_{V,cal}\Delta T$$

which is analogous to Equation 14.36 under conditions of constant volume. Using this equation, we find that the value of ΔU for 1.000 grams of $C_8H_{18}(l)$ is

$$\Delta U_{rxn}\,[\text{per }1.000\text{ g }C_8H_{18}(l)] = -(28.46\ kJ\cdot K^{-1})(1.679\ K\cdot g^{-1})$$
$$= -47.78\ kJ\cdot g^{-1}$$

because the observed ΔT is associated with the combustion of 1.000 grams of octane. The value of ΔU_{rxn} per mole of octane (molar mass = 114.22 $g\cdot mol^{-1}$) is

$$\Delta U_{rxn}\,[\text{per mol }C_8H_{18}(l)] = (-47.78\ kJ\cdot g^{-1})(114.22\ g\cdot mol^{-1})$$
$$= -5457\ kJ\cdot mol^{-1}$$

or in terms of the overall reaction equation we write,

$$C_8H_{18}(l) + \tfrac{25}{2}O_2(g) \rightarrow 8\,CO_2(g) + 9\,H_2O(l) \quad \Delta U_{rxn} = -5457 \text{ kJ}\cdot\text{mol}^{-1}$$

Here we balanced the chemical equation using fractional coefficients rather than whole number coefficients because we are interested in the standard enthalpy of combustion *per mole of octane.*

(b) Using the values listed in Table 14.3 and applying Equation 14.24, we find that

$$\begin{aligned}\Delta H_{rxn} &= \{8\Delta H^\circ_f[CO_2(g)] + 9\Delta H^\circ_f[H_2O(l)]\} - \{\Delta H^\circ_f[C_8H_{18}(l)] + \tfrac{25}{2}\Delta H^\circ_f[O_2(g)]\}\\ &= \{8(-393.5 \text{ kJ}\cdot\text{mol}^{-1}) + 9(-285.8 \text{ kJ}\cdot\text{mol}^{-1})\} - \{(-250.1 \text{ kJ}\cdot\text{mol}^{-1})\}\\ &= -5470.1 \text{ kJ}\cdot\text{mol}^{-1}\end{aligned}$$

Hence, ΔH_{rxn} and ΔU_{rxn} differ by two-tenths of a percent, showing that $\Delta H_{rxn} \approx \Delta U_{rxn}$ is a good approximation in this case.

PRACTICE PROBLEM 14-15: A 2.218-gram sample of benzoic acid, $C_6H_5COOH(s)$, is burned in a bomb calorimeter whose total heat capacity is $18.94 \text{ kJ}\cdot\text{K}^{-1}$. (a) If the observed temperature increase is 3.094°C, calculate the standard energy of combustion per gram and per mole of benzoic acid. (b) Using the molar enthalpy of formation values listed in Table 14.3, calculate ΔH_{rxn} at 25°C and a pressure of one bar. (c) Is $\Delta H_{rxn} \approx \Delta U_{rxn}$ a good approximation in this case?

Answer: (a) $-26.42 \text{ kJ}\cdot\text{g}^{-1}$, $-3226 \text{ kJ}\cdot\text{mol}^{-1}$; (b) $-3226.7 \text{ kJ}\cdot\text{mol}^{-1}$; (c) yes

We intentionally did not write ΔU°_{rxn} in Example 14-15 and Practice Problem 14-15 because bomb calorimetry experiments are usually carried out under fairly high pressures of oxygen to assure that combustion is complete. Even so, the value of ΔU_{rxn} that is obtained is close to the standard (one bar) value, ΔU°_{rxn}.

14-10. The Magnitudes of Heat Capacities Have a Molecular Interpretation

If you look at the values of the constant-pressure molar heat capacities given in Table 14.6, you will see that all the monatomic gases—He(g), Ne(g), Ar(g), Hg(g)—have a C_P value of $20.8 \text{ J}\cdot\text{mol}^{-1}\cdot\text{K}^{-1}$ at 25°C. Being monatomic, these species can increase their energy only by traveling at a faster speed. Polyatomic molecules, on the other hand, can take up energy in other ways. In addition to moving along in a straight line (called **translational motion**), polyatomic molecules can rotate and vibrate, as shown in Figure 14.19. Energy can be taken up by these **rotational and vibrational motions**, in addition to the translational motion, so the polyatomic gases in Table 14.6—for instance, $H_2(g)$, $NH_3(g)$, and $CH_4(g)$—all have values of C_P that are greater than $20.8 \text{ J}\cdot\text{K}^{-1}\cdot\text{mol}^{-1}$. Each polyatomic gaseous molecule rotates and vibrates to a different extent at different temperatures, so we can only say that $C_P > 20.8 \text{ J}\cdot\text{mol}^{-1}\cdot\text{K}^{-1}$ in each case. However, generally speaking, the larger the molecule, the greater the number of ways in which it can rotate and vibrate, and therefore, the greater its heat capacity.

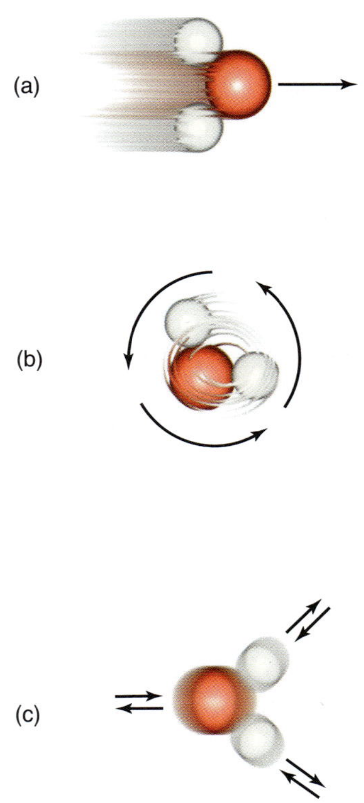

Figure 14.19 Molecular motion may be translational, rotational, or vibrational. (a) Translation is the movement of an entire molecule through space. (b) Rotation is the spinning of a molecule in space. (c) Vibration is the back-and-forth movement of nuclei about fixed relative positions within the molecule.

We can derive the value of C_P for monatomic gases ($20.8\ \mathrm{J\cdot mol^{-1}\cdot K^{-1}}$) by using Equation 13.26 from the kinetic theory of gases, which says that the mean kinetic energy per mole of a gas is $\overline{E_k} = \frac{3}{2}RT$. Excluding electronic energy, kinetic energy is the only energy available to a monatomic gas, so we can associate $\overline{E_k}$ with the thermodynamic energy U. Thus, we can write

$$\Delta U = U_f - U_i = \tfrac{3}{2}RT_f - \tfrac{3}{2}RT_i = \tfrac{3}{2}R(T_f - T_i) = \tfrac{3}{2}R\Delta T \tag{14.39}$$

If we substitute $\Delta U = \frac{3}{2}R\Delta T$ into Equation 14.8, $\Delta H = \Delta U + \Delta(PV)$, we obtain

$$\Delta H = \tfrac{3}{2}R\Delta T + \Delta(PV) \tag{14.40}$$

But $PV = RT$ for one mole of an ideal gas, and so $\Delta(PV) = \Delta(RT) = R\Delta T$. If we substitute this result into Equation 14.40 for ΔH, we obtain

$$\Delta H = \tfrac{3}{2}R\Delta T + R\Delta T = \tfrac{5}{2}R\Delta T \tag{14.41}$$

Using this result in Equation 14.28 (where $n = 1$) gives

$$C_P = \frac{\Delta H}{\Delta T} = \frac{\frac{5}{2}R\Delta T}{\Delta T} = \tfrac{5}{2}R \tag{14.42}$$

Finally, substituting $R = 8.3145\ \mathrm{J\cdot mol^{-1}\cdot K^{-1}}$ into this equation gives $C_P = 20.78\ \mathrm{J\cdot mol^{-1}\cdot K^{-1}}$, in agreement with the values listed in Table 14.6 for the monatomic gases. Thus, the kinetic theory of gases actually enables us to calculate the correct experimental values of the molar heat capacities of monatomic gases.

Another correspondence in Table 14.6 can be found by looking at the values of C_P for the solid metals: Al(*s*), Au(*s*), Cu(*s*), Fe(*s*), Ag(*s*), and Li(*s*), for example. Notice that in each case the value of C_P is approximately equal to $3R = 3(8.3145\ \mathrm{J\cdot mol^{-1}\cdot K^{-1}}) = 24.9\ \mathrm{J\cdot mol^{-1}\cdot K^{-1}}$. This observation, which is easy to use but has a theoretical explanation too involved to give here, was observed experimentally in the early 1800s by two French chemists, Pierre Dulong and Alexis Petit, and is known as the **rule of Dulong and Petit.** This rule played an important role in the early investigations of atomic masses and the chemical formulas of compounds.

One final observation to be made regarding Table 14.6 is the relatively large heat capacity of liquid water, which is twice as large as the heat capacity of ice. Because of its large heat capacity, a given quantity of water is able to absorb a relatively large quantity of energy as heat, so water is an excellent coolant. We shall see in Chapter 15 that water molecules have rather strong intermolecular interactions in the liquid state and that the relatively large amount of energy that is absorbed as heat goes into breaking up these interactions as well as increasing the temperature.

SUMMARY

A chemical reaction involves a transfer of energy between the reaction system and its surroundings. The transfer of energy can be as work or heat. Work involves the action of a force through a distance. When the work is positive, a system gains energy, and when the work is negative, the system loses energy. Reactions that give off energy as heat are called exothermic reactions, and reactions that take in energy as heat are called endothermic reactions. The total energy involved is always conserved; that is, energy is neither created nor destroyed, but simply transferred from one system to another, a principle known as the first law of thermodynamics. The study of energy transfers is called thermodynamics; and thermochemistry is the branch of thermodynamics that deals with the energy evolved or absorbed as heat in chemical reactions.

For a reaction run at constant pressure, the energy evolved or absorbed as heat, q_P, is equal to the enthalpy change; thus, $\Delta H_{rxn} = q_P$. The state function property of the enthalpy leads to Hess's law that states that the enthalpy changes for chemical equations are additive. For a constant-volume process, the energy evolved or absorbed as heat, q_V, is equal to the change in energy for the reaction; thus, $\Delta U_{rxn} = q_V$.

The standard molar enthalpy of formation of a compound from its elements, ΔH_f°, is determined by assigning a value of zero to the standard enthalpies of formation of the most stable state of each element at a pressure of one bar. A table of ΔH_f° values can be used to compute ΔH_{rxn}° values.

The molar bond enthalpy is defined as the enthalpy required to break one mole of a given bond. A table of average molar bond enthalpies can be used to compute approximate ΔH_{rxn}° values.

The heat capacity of a substance is a measure of its capacity to take up energy as heat. The higher the heat capacity of a substance, the smaller the resulting temperature increase for a given amount of energy in the form of heat added.

Combustion reactions are those in which a fuel is burned (usually in oxygen) to provide energy in the form of heat. The amount of energy absorbed or evolved as heat by a chemical reaction can be measured using a calorimeter.

TERMS YOU SHOULD KNOW

thermochemistry *469*
system *470*
surroundings *470*
thermodynamics *470*
work, w *470*
liter-atmosphere, L·atm *471*
heat, q *471*
exothermic reaction *471*
endothermic reaction *472*
energy change, ΔU *472*
first law of thermodynamics *472*
state function *472*
energy transfer function *472*
energy change of a reaction, ΔU_{rxn} *472*
enthalpy, H *473*
enthalpy change of a reaction, ΔH_{rxn} *473*
fuel *474*
standard enthalpy change, ΔH_{rxn}° *475*
standard state *475*
Hess's law *479*
cyclic process *480*
standard molar enthalpy of formation, ΔH_f° *484*
calorie, cal *490*
molar bond enthalpy *492*
heat capacity, c_P *496*
molar heat capacity, C_P *497*
specific heat, c_{sp} *499*
calorimeter *501*
Dewar flask *501*
bomb calorimeter *503*
translational motion *505*
rotational motion *505*
vibrational motion *505*
rule of Dulong and Petit *506*

EQUATIONS YOU SHOULD KNOW HOW TO USE

$w = -P\Delta V$ (constant pressure)	(14.3)	(expansion or compression work at constant P)
$\Delta U_{rxn} = q + w$	(14.6)	(first law of thermodynamics)
$\Delta H_{rxn} = \Delta U_{rxn} + P\Delta V_{rxn}$ (constant pressure)	(14.10)	(relation between ΔH_{rxn} and ΔU_{rxn} at constant pressure)
$\Delta H_{rxn} = q_P$ (constant pressure)	(14.12)	(energy transferred as heat at constant pressure)
$\Delta H_{rxn} = H_{prod} - H_{react}$	(14.13)	(definition of ΔH_{rxn})
$P\Delta V^{\circ}_{rxn} = \Delta \nu_{gas} RT$	(14.18)	(a form of the ideal-gas equation)
$\Delta H^{\circ}_{rxn}(1 + 2) = \Delta H^{\circ}_{rxn}(1) + \Delta H^{\circ}_{rxn}(2)$	(14.19)	(Hess's law)
ΔH°_{rxn} (reverse) $= -\Delta H^{\circ}_{rxn}$ (forward)	(14.22)	(a form of Hess's law)
$\Delta H_{cyclic} = 0$	(14.23)	(the enthalpy change for a cyclic process is equal to zero; enthalpy is a state function)
$\Delta H^{\circ}_{rxn} = \Delta H^{\circ}_{f}[\text{products}] - \Delta H^{\circ}_{f}[\text{reactants}]$	(14.24)	(calculation of ΔH°_{rxn} values from tabulated ΔH°_{f} values of the products and the reactants in a chemical equation)
$\Delta H^{\circ}_{rxn} \approx \Delta H^{\circ}_{bond}[\text{reactants}] - \Delta H^{\circ}_{bond}[\text{products}]$	(14.26)	(approximation of ΔH°_{rxn} for gas phase reactions using average molar bond enthalpies)
$c_P = \dfrac{q_P}{\Delta T} = \dfrac{\Delta H}{\Delta T}$	(14.27 and 14.28)	(definition of heat capacity)
$C_P = \dfrac{q_P}{n\Delta T}$	(14.30)	(definition of molar heat capacity)
$c_{sp} = \dfrac{q_P}{m\Delta T}$	(14.31)	(definition of specific heat)
$c_{P_h}(T_f - T_h) + c_{P_l}(T_f - T_l) = 0$	(14.33)	(calculation of final temperature when two substances at different temperatures are allowed to transfer energy as heat)
$\Delta H = -c_{P,cal}\Delta T$	(14.36)	(calorimetric determination of ΔH)
$\Delta U_{rxn} = q_V$ (constant volume)	(14.38)	(energy transferred as heat at constant volume)

PROBLEMS

HEAT AND ENERGY

14-1. How much work (in joules) does a system do if its volume increases from 10 liters to 25 liters against a constant pressure of 3.5 atm?

14-2. How much work (in joules) must be done on a system to decrease its volume from 10.0 liters to 2.0 liters by exerting a constant pressure of 4.0 atm?

14-3. A gas-forming reaction produces 0.75 m^3 of gas against a constant pressure of 110.0 kPa. Calculate the work done by the gas in joules.

14-4. Calculate the work done in joules when a mechanical compressor exerting a constant pressure of 350.0 kPa compresses an air sample from a volume of 500.0 cm^3 to a volume of 250.0 cm^3.

14-5. Calculate the work done when 1.0 milliliter of 6.0 M $HCl(aq)$ is placed on a block of limestone, $CaCO_3(s)$, and allowed to react completely at 1.00 bar and 25°C. Assume the volume of the liquids and solids participating in the reaction are negligible.

14-6. Calculate the work done when 2.0 liters of methane gas, $CH_4(g)$, undergoes combustion in excess oxygen at 0°C and 1.00 bar. Assume the volume of water formed is negligible.

14-7. When 30.0 grams of methane, $CH_4(g)$, burn in oxygen, −1503 kJ of energy are evolved as heat. Calculate the amount of energy (in kilojoules) evolved as heat when 1.00 mole of methane burns.

14-8. When 2.46 grams of barium reacts with chlorine, −15.4 kJ of energy is evolved as heat. Calculate the energy evolved as heat (in kilojoules) when 1.00 mole of barium chloride is formed from barium and chlorine.

14-9. When 1.280 grams of carbon react with sulfur to give carbon disulfide, $CS_2(l)$, 9.52 kJ of energy are absorbed as heat. Calculate the energy absorbed as heat (in kilojoules) when 1.00 mole of carbon disulfide is formed from carbon and sulfur.

14-10. When 0.165 grams of magnesium are burned in oxygen, −4.08 kJ of energy are evolved as heat. Calculate the energy evolved as heat (in kilojoules) when 1.00 mole of magnesium oxide is formed from magnesium and oxygen.

14-11. The oxidation of copper(I) oxide, $Cu_2O(s)$, to copper(II) oxide, $CuO(s)$, is an exothermic process,

$$2\,Cu_2O(s) + O_2(g) \rightarrow 4\,CuO(s) \qquad \Delta H^\circ_{rxn} = -292.0\ \text{kJ}\cdot\text{mol}^{-1}$$

Calculate the energy released as heat when 25.0 grams of $Cu_2O(s)$ undergo oxidation at constant pressure.

14-12. The oxidation of iron is highly exothermic and is sometimes used to generate heat in commercial heat packs. The reaction that takes place may be represented as

$$4\,Fe(s) + 3\,O_2(g) \rightarrow 2\,Fe_2O_3(s) \qquad \Delta H^\circ_{rxn} = -1648\ \text{kJ}\cdot\text{mol}^{-1}$$

Calculate the energy released as heat when 10.0 grams of iron undergo oxidation at constant pressure.

14-13. Calculate the value of ΔU°_{rxn} at 25°C for the reaction equation

$$Cu(s) + Cl_2(g) \rightarrow CuCl_2(s) \qquad \Delta H^\circ_{rxn} = -220.1\ \text{kJ}\cdot\text{mol}^{-1}$$

14-14. Calculate the value of ΔU°_{rxn} at 25°C for the reaction equation

$$4\,NH_3(g) + 5\,O_2(g) \rightarrow 4\,NO(g) + 6\,H_2O(l) \qquad \Delta H^\circ_{rxn} = -1166\ \text{kJ}\cdot\text{mol}^{-1}$$

14-15. For the reaction given in Problem 14-11, calculate the work and energy change, ΔU_{rxn}, when 25.0 grams of $Cu_2O(s)$ undergo oxidation at a constant pressure of 1.00 bar and a constant temperature of 25°C.

14-16. For the reaction given in Problem 14-12, calculate the work and energy change, ΔU_{rxn}, when 10.0 grams of iron metal undergo oxidation at a constant pressure of 1.00 bar and a constant temperature of 25°C.

HESS'S LAW

14-17. The value of ΔH°_{rxn} for the reaction described by

$$2\,ZnS(s) + O_2(g) \rightarrow 2\,ZnO(s) + 2\,S(s)$$

is -289.0 kJ·mol^{-1}. Calculate the value of ΔH°_{rxn} for the equation

$$ZnO(s) + S(s) \rightarrow ZnS(s) + \tfrac{1}{2}O_2(g)$$

14-18. The value of ΔH°_{rxn} for the equation

$$CaO(s) + H_2O(l) \rightarrow Ca(OH)_2(s)$$

is -64.50 kJ·mol^{-1}. Calculate the number of kilojoules of energy in the form of heat required to convert 1.00 grams of $Ca(OH)_2(s)$ to $CaO(s)$.

14-19. Given that

$$CH_3CH_2OH(l) + 3\,O_2(g) \rightarrow 2\,CO_2(g) + 3\,H_2O(g) \qquad \Delta H^\circ_{rxn} = -1234.8\ \text{kJ·mol}^{-1}$$

$$CH_3OCH_3(l) + 3\,O_2(g) \rightarrow 2\,CO_2(g) + 3\,H_2O(g) \qquad \Delta H^\circ_{rxn} = -1309.1\ \text{kJ·mol}^{-1}$$

calculate the value of ΔH°_{rxn} for the equation

$$CH_3CH_2OH(l) \rightarrow CH_3OCH_3(l)$$

14-20. Use the values of ΔH°_{rxn} given for the equations

$$Cu(s) + Cl_2(g) \rightarrow CuCl_2(s) \qquad \Delta H^\circ_{rxn} = -220.1\ \text{kJ·mol}^{-1}$$

$$2\,Cu(s) + Cl_2(g) \rightarrow 2\,CuCl(s) \qquad \Delta H^\circ_{rxn} = -137.2\ \text{kJ·mol}^{-1}$$

to calculate the value of ΔH°_{rxn} for the equation

$$CuCl_2(s) + Cu(s) \rightarrow 2\,CuCl(s)$$

14-21. Use the values of ΔH°_{rxn} given for the equations

$$2\,Fe(s) + \tfrac{3}{2}O_2(g) \rightarrow Fe_2O_3(s) \qquad \Delta H^\circ_{rxn} = -824.2\ \text{kJ·mol}^{-1}$$

$$3\,Fe(s) + 2\,O_2(g) \rightarrow Fe_3O_4(s) \qquad \Delta H^\circ_{rxn} = -1118.4\ \text{kJ·mol}^{-1}$$

to calculate the value of ΔH°_{rxn} for the equation

$$3\,Fe_2O_3(s) \rightarrow 2\,Fe_3O_4(s) + \tfrac{1}{2}O_2(g)$$

14-22. Given that

$$H_2(g) + F_2(g) \rightarrow 2\,HF(g) \qquad \Delta H^\circ_{rxn} = -546.6\ \text{kJ·mol}^{-1}$$

$$2\,H_2(g) + O_2(g) \rightarrow 2\,H_2O(l) \qquad \Delta H^\circ_{rxn} = -571.6\ \text{kJ·mol}^{-1}$$

calculate the value of ΔH°_{rxn} for

$$2\,F_2(g) + 2\,H_2O(l) \rightarrow 4\,HF(g) + O_2(g)$$

14-23. The standard molar enthalpies of combustion at 25°C for sucrose, glucose, and fructose are given below:

Compound	**ΔH°_{rxn}/kJ·mol^{-1}**
$C_{12}H_{22}O_{11}(s)$, sucrose	−5639.7
$C_6H_{12}O_6(s)$, glucose	−2802.5
$C_6H_{12}O_6(s)$, fructose	−2810.2

Use these combustion data along with Hess's law to calculate the value of ΔH°_{rxn} at 25°C for the equation

$$\underset{\text{sucrose}}{C_{12}H_{22}O_{11}(s)} + H_2O(l) \rightarrow \underset{\text{glucose}}{C_6H_{12}O_6(s)} + \underset{\text{fructose}}{C_6H_{12}O_6(s)}$$

14-24. The standard molar enthalpies of combustion of the isomers *m*-xylene and *p*-xylene, both with formulas of $(CH_3)_2C_6H_4$, are −4550.1 kJ·mol^{-1} and −4551.1 kJ·mol^{-1}, respectively. Use these data, together with Hess's law, to calculate the value of ΔH°_{rxn} for the reaction described by

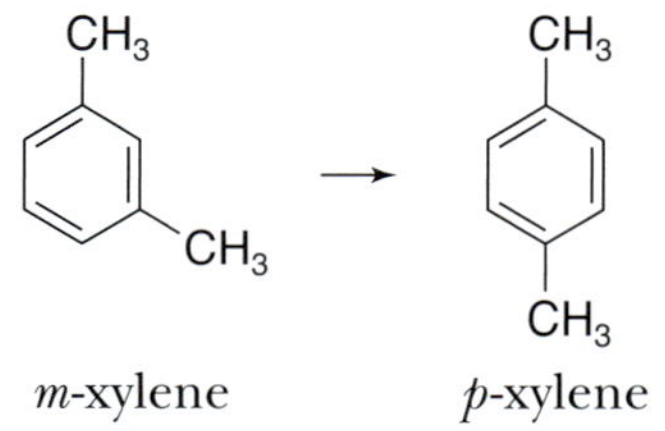

14-25. Given that

$$4\,NH_3(g) + 5\,O_2(g) \rightarrow 4\,NO(g) + 6\,H_2O(l) \qquad \Delta H^\circ_{rxn} = -1166\ \text{kJ·mol}^{-1}$$

$$4\,NH_3(g) + 3\,O_2(g) \rightarrow 2\,N_2(g) + 6\,H_2O(l) \qquad \Delta H^\circ_{rxn} = -1531\ \text{kJ·mol}^{-1}$$

calculate the value of ΔH°_{rxn} for the equation

$$N_2(g) + O_2(g) \rightarrow 2\,NO(g)$$

14-26. Given that

$$\mathrm{Xe}(g) + \mathrm{F_2}(g) \rightarrow \mathrm{XeF_2}(s) \qquad \Delta H^\circ_{\mathrm{rxn}} = -123\ \mathrm{kJ{\cdot}mol^{-1}}$$

$$\mathrm{Xe}(g) + 2\,\mathrm{F_2}(g) \rightarrow \mathrm{XeF_4}(s) \qquad \Delta H^\circ_{\mathrm{rxn}} = -262\ \mathrm{kJ{\cdot}mol^{-1}}$$

calculate the value of $\Delta H^\circ_{\mathrm{rxn}}$ for the equation

$$\mathrm{XeF_2}(s) + \mathrm{F_2}(g) \rightarrow \mathrm{XeF_4}(s)$$

MOLAR ENTHALPIES OF FORMATION

14-27. Given that $\Delta H^\circ_{\mathrm{f}} = 142.7\ \mathrm{kJ{\cdot}mol^{-1}}$ for $\mathrm{O_3}(g)$ and $\Delta H^\circ_{\mathrm{f}} = 249.2\ \mathrm{kJ{\cdot}mol^{-1}}$ for $\mathrm{O}(g)$, calculate the value of $\Delta H^\circ_{\mathrm{rxn}}$ for the equation

$$\mathrm{O_2}(g) + \mathrm{O}(g) \rightarrow \mathrm{O_3}(g)$$

This is one of the reactions that produce ozone in the atmosphere.

14-28. The $\Delta H^\circ_{\mathrm{f}}$ values for $\mathrm{Cu_2O}(s)$ and $\mathrm{CuO}(s)$ are $-168.6\ \mathrm{kJ{\cdot}mol^{-1}}$ and $-157.3\ \mathrm{kJ{\cdot}mol^{-1}}$, respectively. Calculate the value of $\Delta H^\circ_{\mathrm{rxn}}$ for the equation

$$\mathrm{CuO}(s) + \mathrm{Cu}(s) \rightarrow \mathrm{Cu_2O}(s)$$

14-29. Use the $\Delta H^\circ_{\mathrm{f}}$ data in Table 14.3 to calculate the values of $\Delta H^\circ_{\mathrm{rxn}}$ for the following equations:

(a) $\mathrm{N_2H_4}(l) + \mathrm{O_2}(g) \rightarrow \mathrm{N_2}(g) + 2\,\mathrm{H_2O}(g)$

(b) $\mathrm{C_2H_4}(g) + \mathrm{H_2O}(l) \rightarrow \mathrm{CH_3CH_2OH}(l)$

(c) $\mathrm{CH_4}(g) + 4\,\mathrm{Cl_2}(g) \rightarrow \mathrm{CCl_4}(l) + 4\,\mathrm{HCl}(g)$

In each case state whether the reaction is endothermic or exothermic.

14-30. Use the $\Delta H^\circ_{\mathrm{f}}$ data in Table 14.3 to calculate the values of $\Delta H^\circ_{\mathrm{rxn}}$ for the following equations:

(a) $2\,\mathrm{H_2O_2}(l) \rightarrow 2\,\mathrm{H_2O}(l) + \mathrm{O_2}(g)$

(b) $\mathrm{MgO}(s) + \mathrm{CO_2}(g) \rightarrow \mathrm{MgCO_3}(s)$

(c) $4\,\mathrm{NH_3}(g) + 5\,\mathrm{O_2}(g) \rightarrow 4\,\mathrm{NO}(g) + 6\,\mathrm{H_2O}(g)$

In each case state whether the reaction is endothermic or exothermic.

14-31. Use the $\Delta H^\circ_{\mathrm{f}}$ data in Table 14.3 to calculate the values of $\Delta H^\circ_{\mathrm{rxn}}$ for the combustion reactions described by the following equations:

(a) $\mathrm{CH_3CH_2OH}(l) + 3\,\mathrm{O_2}(g) \rightarrow 2\,\mathrm{CO_2}(g) + 3\,\mathrm{H_2O}(l)$

(b) $\mathrm{C_2H_6}(g) + \frac{7}{2}\mathrm{O_2}(g) \rightarrow 2\,\mathrm{CO_2}(g) + 3\,\mathrm{H_2O}(l)$

Compare the standard enthalpy of combustion per gram of the fuels $\mathrm{CH_3CH_2OH}(l)$ and $\mathrm{C_2H_6}(g)$.

14-32. Use the $\Delta H^\circ_{\mathrm{f}}$ data in Table 14.3 to calculate the values of $\Delta H^\circ_{\mathrm{rxn}}$ for the combustion reactions described by the following equations:

(a) $\mathrm{CH_3OH}(l) + \frac{3}{2}\mathrm{O_2}(g) \rightarrow \mathrm{CO_2}(g) + 2\,\mathrm{H_2O}(l)$

(b) $\mathrm{N_2H_4}(l) + \mathrm{O_2}(g) \rightarrow \mathrm{N_2}(g) + 2\,\mathrm{H_2O}(l)$

Compare the standard enthalpy of combustion per gram of the fuels $\mathrm{CH_3OH}(l)$ and $\mathrm{N_2H_4}(l)$.

14-33. Given that $\Delta H^\circ_{\mathrm{rxn}} = -2810.2\ \mathrm{kJ{\cdot}mol^{-1}}$ for the combustion of fructose, $\mathrm{C_6H_{12}O_6}(s)$, as described by the equation

$$\mathrm{C_6H_{12}O_6}(s) + 6\,\mathrm{O_2}(g) \rightarrow 6\,\mathrm{CO_2}(g) + 6\,\mathrm{H_2O}(l)$$

Use the $\Delta H^\circ_{\mathrm{f}}$ data in Table 14.3 together with the given $\Delta H^\circ_{\mathrm{rxn}}$ value to calculate the value of $\Delta H^\circ_{\mathrm{f}}$ for fructose.

14-34. Given that $\Delta H^\circ_{\mathrm{rxn}} = -5639.7\ \mathrm{kJ{\cdot}mol^{-1}}$ for the combustion of sucrose, $\mathrm{C_{12}H_{22}O_{11}}(s)$, as described by the equation

$$\mathrm{C_{12}H_{22}O_{11}}(s) + 12\,\mathrm{O_2}(g) \rightarrow 12\,\mathrm{CO_2}(g) + 11\,\mathrm{H_2O}(l)$$

Use the $\Delta H^\circ_{\mathrm{f}}$ data in Table 14.3 together with the given $\Delta H^\circ_{\mathrm{rxn}}$ value to calculate the value of $\Delta H^\circ_{\mathrm{f}}$ for sucrose.

14-35. Using Table 14.3, calculate the values of $\Delta H^\circ_{\mathrm{rxn}}$ for

(a) $\mathrm{H_2}(g) + \mathrm{F_2}(g) \rightarrow 2\,\mathrm{HF}(g)$

(b) $2\,\mathrm{CO}(g) + \mathrm{O_2}(g) \rightarrow 2\,\mathrm{CO_2}(g)$

(c) $3\,\mathrm{H_2}(g) + \mathrm{N_2}(g) \rightarrow 2\,\mathrm{NH_3}(g)$

(d) $2\,\mathrm{NO}(g) + \mathrm{O_2}(g) \rightarrow 2\,\mathrm{NO_2}(g)$

State whether each reaction is endothermic or exothermic.

14-36. Using Table 14.3, calculate the energy in the form of heat required to vaporize

(a) 1.00 mole of water at 25°C

(b) 1.00 mole of $\mathrm{CCl_4}(l)$ at 25°C

14-37. Using the data in Table 14.3, calculate the molar enthalpy change for the reaction described by

$$2\,\mathrm{SO_2}(g) + \mathrm{O_2}(g) \rightarrow 2\,\mathrm{SO_3}(g)$$

How much energy in the form of heat is produced when a volume of 1.50 liters of $\mathrm{SO_2}(g)$ is converted to 1.50 liters of $\mathrm{SO_3}(g)$ according to this process at a constant pressure and temperature of 1.00 bar and 25.0°C? Assume ideal-gas behavior.

14-38. The roasting of calcium may be used to produce calcium carbonate according to the reaction described by

$$2\,Ca(s) + 4\,O_2(g) + 3\,C(s) \rightarrow CO_2(g) + 2\,CaCO_3(s)$$

Use the data in Table 14.3 to calculate the value of ΔH°_{rxn} in kilojoules per mole and calculate the energy released as heat by the production of 500.0 kg of calcium carbonate using this process at constant pressure.

MOLAR BOND ENTHALPIES

14-39. The enthalpy change for the equation

$$ClF_3(g) \rightarrow Cl(g) + 3\,F(g)$$

is 523 kJ·mol^{-1}. Calculate the average molar bond enthalpy of a chlorine-fluorine bond in a ClF_3 molecule. The bonding in a ClF_3 molecule is

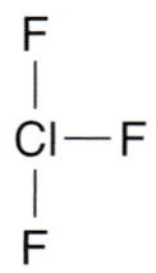

14-40. The enthalpy change for the equation

$$OF_2(g) \rightarrow O(g) + 2\,F(g)$$

is 384 kJ·mol^{-1}. Calculate the average molar bond enthalpy of an oxygen-fluorine bond in an OF_2 molecule. The bonding in an OF_2 molecule is

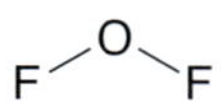

14-41. Use the molar bond enthalpy data in Table 14.5 to estimate the value of ΔH°_{rxn} for the equation

$$CCl_4(g) + 2\,F_2(g) \rightarrow CF_4(g) + 2\,Cl_2(g)$$

The bonding in a CCl_4 and that in a CF_4 molecule are

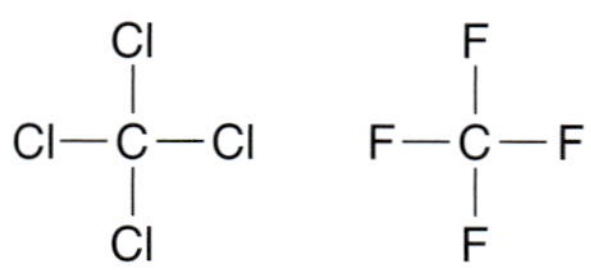

14-42. Use the molar bond enthalpy data in Table 14.5 to estimate the value of ΔH°_{rxn} for the equation

$$CH_3OH(g) + F_2(g) \rightarrow FCH_2OH(g) + HF(g)$$

The bonding in a CH_3OH molecule and that in a FCH_2OH molecule are

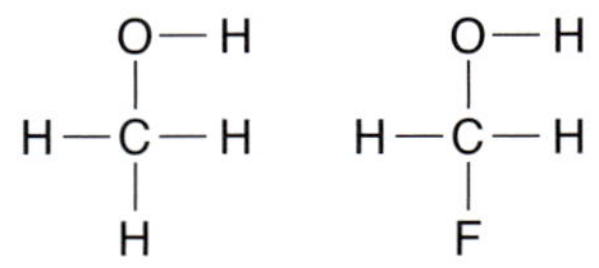

14-43. The formation of water from oxygen and hydrogen involves the reaction described by the equation

$$2\,H_2(g) + O_2(g) \rightarrow 2\,H_2O(g)$$

Use the molar bond enthalpies given in Table 14.5 and the ΔH°_f value for $H_2O(g)$ given in Table 14.3 to calculate the molar bond enthalpy of the oxygen–oxygen double bond in an O_2 molecule. Compare your result to the value listed in Table 14.5.

14-44. The formation of ammonia from hydrogen and nitrogen involves the reaction described by the equation

$$N_2(g) + 3\,H_2(g) \rightarrow 2\,NH_3(g)$$

Use the molar bond enthalpies given in Table 14.5 and the ΔH°_f value for $NH_3(g)$ given in Table 14.3 to calculate the molar bond enthalpy of the nitrogen–nitrogen triple bond in a N_2 molecule. Compare your result to the value listed in Table 14.5.

14-45. Given that

$$\Delta H^\circ_f[H(g)] = 218.0\ \text{kJ·mol}^{-1}$$
$$\Delta H^\circ_f[C(g)] = 716.7\ \text{kJ·mol}^{-1}$$
$$\Delta H^\circ_f[CH_4(g)] = -74.6\ \text{kJ·mol}^{-1}$$

calculate the average molar bond enthalpy of a carbon–hydrogen bond in a CH_4 molecule. Why is there a slight difference between this value and the molar bond enthalpy of a carbon–hydrogen bond listed in Table 14.5?

14-46. Given that

$$\Delta H^\circ_f[Cl(g)] = 121.3\ \text{kJ·mol}^{-1}$$
$$\Delta H^\circ_f[C(g)] = 716.7\ \text{kJ·mol}^{-1}$$
$$\Delta H^\circ_f[CCl_4(g)] = -95.7\ \text{kJ·mol}^{-1}$$

calculate the average molar bond enthalpy of a carbon–chlorine bond in a CCl_4 molecule. Why is there a slight difference between this value and the molar bond enthalpy of a carbon–chlorine bond listed in Table 14.5?

HEAT CAPACITY

14-47. When 1105 joules of energy as heat are added to 36.5 grams of ethanol, $CH_3CH_2OH(l)$, the temperature increases by 12.3°C. Calculate the molar heat capacity of $CH_3CH_2OH(l)$.

14-48. When 285 joules of energy as heat are added to 33.6 grams of hexane, $C_6H_{14}(l)$, a component of gasoline, the temperature rises from 25.00°C to 28.74°C. Calculate the molar heat capacity of $C_6H_{14}(l)$.

14-49. A 10.0-kilogram sample of liquid water is used to cool an engine. Calculate the energy in the form of heat removed (in joules) from the engine when the temperature of the water is raised from 20.0°C to 90.0°C. Take $C_P = 75.3\ J\cdot K^{-1}\cdot mol^{-1}$ for $H_2O(l)$.

14-50. Liquid sodium is being considered as an engine coolant. How many grams of liquid sodium are needed to absorb 1.00 MJ of energy in the form of heat if the temperature of the sodium is not to increase by more than 10°C? Take $C_P = 30.8\ J\cdot K^{-1}\cdot mol^{-1}$ for $Na(l)$ at 500 K.

14-51. A 25.0-gram sample of copper at 90.0°C is placed in 100.0 grams of water at 20.0°C. The copper and water quickly come to the same temperature by the process of heat transfer from copper to water. Calculate the final temperature of the water. Take the molar heat capacity of copper to be $24.4\ J\cdot K^{-1}\cdot mol^{-1}$ and that of $H_2O(l)$ to be $75.3\ J\cdot K^{-1}\cdot mol^{-1}$.

14-52. If a 50.0-gram piece of copper is heated to 100.0°C and then put into a vessel containing 250.0 milliliters of water at 1.5°C, what will be the final temperature of the water? Take the molar heat capacity of copper to be $24.4\ J\cdot K^{-1}\cdot mol^{-1}$ and that of $H_2O(l)$ to be $75.3\ J\cdot K^{-1}\cdot mol^{-1}$.

14-53. A 1.00-kilogram block of aluminum metal ($C_P = 24.4\ J\cdot K^{-1}\cdot mol^{-1}$) at 500.0°C is placed in contact with a 1.00-kg block of copper ($C_P = 24.4\ J\cdot K^{-1}\cdot mol^{-1}$) at 10.0°C. What will be the final temperature of the two blocks? Assume that no energy is lost to the surroundings as heat.

14-54. A 50.0-gram sample of a metal alloy at 25.0°C is placed in 100.0 grams of water at 55.0°C. The final temperature of the water and metal is 48.1°C. Calculate the specific heat of the alloy.

CALORIMETRY

14-55. A 100.0-mL sample of 0.200-M aqueous hydrochloric acid is added to 100.0 mL of 0.200-M aqueous ammonia in a calorimeter with a total heat capacity of $480\ J\cdot K^{-1}$. The temperature increase is 2.34 K. Calculate the value of ΔH°_{rxn} for the equation

$$HCl(aq) + NH_3(aq) \rightarrow NH_4Cl(aq)$$

which describes the reaction that occurs when the two solutions are mixed.

14-56. A 0.0500-L sample of 0.500-M barium nitrate is added to 0.0500 L of 0.500-M magnesium sulfate in a calorimeter with a total heat capacity of $455\ J\cdot K^{-1}$. The observed increase in temperature is 1.43 K. Calculate the value of ΔH°_{rxn} for the equation

$$Ba(NO_3)_2(aq) + MgSO_4(aq) \rightarrow BaSO_4(s) + Mg(NO_3)_2(aq)$$

which describes the reaction that occurs when the two solutions are mixed.

14-57. A 2.50-gram sample of powdered zinc is added to 100.0 mL of a 2.0-M aqueous solution of hydrochloric acid in a calorimeter with a total heat capacity of $481\ J\cdot K^{-1}$. The observed increase in temperature is 12.2 K at a constant pressure of one bar. Using these data, calculate the value of ΔH°_{rxn} for the equation,

$$Zn(s) + 2HCl(aq) \rightarrow ZnCl_2(aq) + H_2(g)$$

which describes the reaction that occurs when the two substances are mixed.

14-58. Calcium hydroxide, $Ca(OH)_2(s)$, is prepared by adding calcium oxide (lime), $CaO(s)$, to water. It is important to know how much energy is evolved as heat in order to provide for adequate cooling. A 10.0-gram sample of $CaO(s)$ is added to 1.00 liter of water in a calorimeter with a total heat capacity of $4.37\ kJ\cdot K^{-1}$. The observed increase in temperature is 2.70 K. Calculate the value of ΔH°_{rxn} for the production of 1.00 mole of $Ca(OH)_2(s)$. The equation describing the reaction is

$$CaO(s) + H_2O(l) \rightarrow Ca(OH)_2(s)$$

14-59. A 5.00-gram sample of potassium chloride is dissolved in 1.00 liter of water in a calorimeter that has a total heat capacity of $4.51\ kJ\cdot K^{-1}$. The temperature decreases by 0.256 K. Calculate the molar heat of solution of potassium chloride.

14-60. A 5.00-gram sample of pure nitric acid is dissolved in 1.00 liter of water in a calorimeter that has a total heat capacity of 5.16 $kJ \cdot K^{-1}$. The temperature increases by 0.511 K. Calculate the molar heat of solution of nitric acid in water.

BOMB CALORIMETERS

14-61. Propane is often used as a home fuel in areas where natural gas is not available. When 3.00 grams of propane, $C_3H_8(g)$, are burned in excess oxygen in a bomb calorimeter that has a total heat capacity of 32.7 $kJ \cdot K^{-1}$, the temperature of the calorimeter increases by 4.25 K. Calculate the standard enthalpy of combustion in kilojoules per gram and kilojoules per mole of propane. Assume that $\Delta H_{rxn} \approx \Delta U_{rxn}$.

14-62. Fructose, $C_6H_{12}O_6(s)$, is a sugar found in fruits and a source of energy for the body. The combustion of fructose takes place according to the equation

$$C_6H_{12}O_6(s) + 6O_2(g) \rightarrow 6CO_2(g) + 6H_2O(l)$$

When 5.00 grams of fructose are burned in excess oxygen in a bomb calorimeter with a heat capacity of 29.7 $kJ \cdot K^{-1}$, the temperature of the calorimeter increases by 2.635 K. Calculate the standard enthalpy of combustion per gram and per mole of fructose. Assume that $\Delta H_{rxn} \approx \Delta U_{rxn}$.

14-63. Based on your answer to Problem 14-61, how many grams of propane, $C_3H_8(g)$, are required to warm all the water in a 100-liter (25-gallon) storage tank from 20°C to 55°C? Assume that the heater has an 85% efficiency.

14-64. Based on your answer to Problem 14-62, determine how much energy (in kilojoules) is released as heat when 15.0 grams of fructose, $C_6H_{12}O_6(s)$, is converted to $CO_2(g)$ and $H_2O(l)$ in the body? How many nutritional Calories does this correspond to?

14-65. When 2.50 grams of oxalic acid, $H_2C_2O_4(s)$, are burned in a bomb calorimeter that has a total heat capacity of 8.75 $kJ \cdot K^{-1}$, the temperature increases by 0.780 K. Calculate the standard molar enthalpy of combustion of oxalic acid. Assume $\Delta H_{rxn} \approx \Delta U_{rxn}$. Using this result and the ΔH_f° values for $CO_2(g)$ and $H_2O(l)$ given in Table 14.3, calculate the molar enthalpy of formation of oxalic acid at 25°C.

14-66. When 2.620 grams of lactic acid, $C_3H_6O_3(s)$, are burned in a bomb calorimeter that has a total heat capacity of 21.70 $kJ \cdot K^{-1}$, the temperature of the calorimeter increases by 1.800 K. Calculate the standard molar enthalpy of combustion of lactic acid. Assume $\Delta H_{rxn} \approx \Delta U_{rxn}$. Using this result and the ΔH_f° values for $CO_2(g)$ and $H_2O(l)$ given in Table 14.3, calculate the value of ΔH_f° for lactic acid. Lactic acid is produced in muscle when there is a shortage of oxygen, such as during vigorous exercise. A buildup of lactic acid is responsible for muscle cramps.

ADDITIONAL PROBLEMS

14-67. Water on a hot stove begins to boil, giving off steam. Assuming that the boiling water and steam comprise the system, describe this process in terms of the energy transferred as heat and work.

14-68. In an endothermic reaction, is the total enthalpy of the products formed greater or less than that of the reactants?

14-69. A student is arguing with his father that because energy is a state function, it takes the same amount of energy, and thus the same quantity of gasoline, to accelerate a car from 0 to 60 miles per hour in 10 seconds as it does to accelerate it in 30 seconds. What is the flaw in the student's argument?

14-70. Why must we use a calorimeter to measure the heat evolved during a reaction? Why can't we simply use a thermometer?

14-71. Explain why chemists collecting thermodynamic data create tables of enthalpy changes for chemical reactions rather than listing the energy transferred in the form of heat and work for chemical reactions.

14-72. The dissolution of ammonium chloride in water is endothermic. Would you predict ammonium chloride to be more soluble in warm or cool water?

14-73. For each of the following reaction equations at constant pressure, indicate whether the system does work on the surroundings, the surroundings does work on the system, or essentially no work is performed.

(a) $2H_2O_2(g) \rightarrow O_2(g) + 2H_2O(g)$

(b) $CH_4(g) + 2O_2(g) \rightarrow CO_2(g) + 2H_2O(g)$

(c) $NH_3(g) + HCl(g) \rightarrow NH_4Cl(s)$

(d) $2HCl(aq) + CaCO_3(s) \rightarrow H_2O(l) + CO_2(g) + CaCl_2(aq)$

14-74. According to a U.S. Nutrition Facts label on a package of potato chips, a 10.5-ounce package provides 130 nutritional Calories per serving. One serving is listed by the manufacturer as being 1.125 ounces. What is the total nutritional Calorie content of the package? In most other countries food labels list the total energy content of the package in kilojoules. How would the energy content of this package be labeled in other countries?

14-75. One proposal for an "effortless" method of losing weight is to drink large amounts of cold water. The body must provide energy in the form of heat in order to bring the temperature of the cold water to body temperature, 37°C. This is provided by the burning of stored carbohydrates or fat. (a) How much energy in the form of heat must the body provide to warm 1.0 liter of cold water at 4°C (a typical refrigerator temperature) to 37°C? (b) Calculate how many grams of body fat must be burned to provide this energy. Assume that the burning of fat produces about 39 kilojoules per gram of fat. (c) How many liters of 4°C water must be consumed to lose 1.0 kilogram (2.2 lbs) of body fat?

14-76. Glucose is used as fuel in the body according to the reaction

$$C_6H_{12}O_6(aq) + 6\,O_2(g) \rightarrow 6\,CO_2(g) + 6\,H_2O(l) \qquad \Delta H_{rxn} = -2802.5\ \text{kJ}\cdot\text{mol}^{-1}$$

During a fever, body temperature rises to about 39°C. How many grams of glucose must be burned to raise the temperature of the body from a normal temperature of 36.8°C to 39.0°C for an 82-kilogram person? Assume that all the energy released as heat from the combustion of glucose is used to heat the body. Assume also that the heat capacity per gram (specific heat) of the body is that of water.

14-77. Bicycle riding at 13 miles per hour (a moderate pace) consumes 2000 kilojoules of energy per hour for a 150-pound person. How many miles must this person ride in order to lose one pound of body fat if fat provides 39 kilojoules of energy per gram?

14-78. A 2.0-gram sample of dry ice, $CO_2(s)$, is placed inside a balloon at 1.00 bar and 25°C. The dry ice sublimes, becoming a gas and filling the balloon. Calculate the work done by the system. Assume that $CO_2(g)$ behaves as an ideal gas and that the initial volume of the solid is negligible.

14-79. Using the data in Table 14.3 calculate the standard enthalpy of combustion per gram of octane, $C_8H_{18}(l)$ and that of ethanol, $CH_3CH_2OH(l)$, at 1.00 bar and 25°C. Which fuel provides more energy per gram?

14-80. The French chemists Pierre L. Dulong and Alexis T. Petit noted in 1819 that the molar heat capacity of many solids at ordinary temperatures is proportional to the number of atoms per formula unit of the solid. They quantified their observations in what is known as Dulong and Petit's rule that says that the molar heat capacity of a solid can be expressed as

$$C_P \approx N \times 25\ \text{J}\cdot\text{K}^{-1}\cdot\text{mol}^{-1}$$

where N is the number of atoms per formula unit. The observed heat capacity per gram of a compound containing thallium and chlorine is 0.208 $\text{J}\cdot\text{K}^{-1}\cdot\text{g}^{-1}$. Use Dulong and Petit's rule to determine the formula of the compound.

14-81. The heat capacity per gram of an oxide of rubidium is 0.64 $\text{J}\cdot\text{K}^{-1}\cdot\text{g}^{-1}$. Use Dulong and Petit's rule (see the previous Problem) to determine the formula of the compound.

14-82. Calculate the standard enthalpy of combustion per gram of $C_2H_2(g)$, $C_2H_4(g)$, and $C_2H_6(g)$ at 25°C. Is there a trend?

14-83. For each of the following reaction equations, calculate the value of ΔU°_{rxn} at 25°C and 1.00 bar and compare the result to the value of ΔH°_{rxn}.

Reaction equation	$\Delta H^\circ_{rxn}/\text{kJ}\cdot\text{mol}^{-1}$
(a) $2\,H_2(g) + O_2(g) \rightarrow 2\,H_2O(l)$	−571.6
(b) $Sn(s) + 2\,Cl_2(g) \rightarrow SnCl_4(l)$	−511.3
(c) $H_2(g) + Cl_2(g) \rightarrow 2\,HCl(g)$	−184.6

14-84. Calculate the value of ΔH°_{rxn} for the reaction described by the equation

$$3\,Cu(s) + 8\,HNO_3(aq) \rightarrow 3\,Cu(NO_3)_2(aq) + 2\,NO(g) + 4\,H_2O(l)$$

Use the data in Table 14.3 and the molar enthalpy of formation data for the ions listed below:

Ion	$\Delta H^\circ_f/\text{kJ}\cdot\text{mol}^{-1}$
$H^+(aq)$	0
$Cu^{2+}(aq)$	64.39 $\text{kJ}\cdot\text{mol}^{-1}$
$NO_3^-(aq)$	−207.35 $\text{kJ}\cdot\text{mol}^{-1}$

14-85. The Apollo Lunar Module was powered by a reaction similar to the following:

$$2\,N_2H_4(l) + N_2O_4(l) \rightarrow 3\,N_2(g) + 4\,H_2O(g)$$

hydrazine dinitrogen tetroxide

Using the values of ΔH_f° given in Table 14.3, calculate the value of ΔH_{rxn}° for this reaction.

14-86. When a solution of hydrochloric acid, $HCl(aq)$, is neutralized by a solution of sodium hydroxide, $NaOH(aq)$, the standard molar enthalpy of reaction is −55.7 kJ per mole of each of the reactants. Calculate the energy in the form of heat released when 100.0 mL of 0.100 M $HCl(aq)$ is neutralized by 100.0 mL of 0.100 M $NaOH(aq)$. If the reaction is performed inside an insulated Dewar flask at constant pressure, what will the temperature change of the solution be? Assume no energy is transferred to the surroundings as heat and that the heat capacity and density of the mixture is the same as that of pure water.

14-87. Ammonium nitrate is an explosive at high temperatures and can decompose according to the equation

$$2\,NH_4NO_3(s) \rightarrow 2\,N_2(g) + 4\,H_2O(g) + O_2(g)$$

A 1.00-gram sample of $NH_4NO_3(s)$ is detonated in a bomb calorimeter with a total heat capacity of 4.92 $kJ\cdot K^{-1}$. The temperature increase is 0.300 K. Calculate the energy evolved as heat (in kilojoules) for the decomposition of 1.00 kilogram of ammonium nitrate under these conditions.

14-88. A researcher studying the nutritional value of a new candy places a 5.00-gram sample of the candy inside a bomb calorimeter and combusts it in excess oxygen. The observed temperature increase is 2.89°C. If the heat capacity of the calorimeter is 38.70 $kJ\cdot K^{-1}$, how many nutritional Calories are there per gram of the candy? Assume $\Delta H_{rxn}^\circ \approx \Delta U_{rxn}^\circ$.

14-89. A certain high-performance airplane consumes 290 liters of aviation fuel per minute. If the density of the fuel is 0.72 $g\cdot mL^{-1}$ and the standard enthalpy of combustion of the fuel is −50.0 $kJ\cdot g^{-1}$, calculate the maximum power (in units of kilowatts = kilojoules per second) that can be produced by this aircraft.

14-90. Use the data in Table 14.3 to calculate the energy in kilojoules required to make a 32-mile round-trip commute in a vehicle that has a fuel efficiency of 25 miles per gallon. Assume the energy from burning fuel may be approximated as the standard enthalpy of combustion of octane, $C_8H_{18}(l)$, and take the density of octane as 0.7 $g\cdot mL^{-1}$. If one barrel of oil can provide 6000 MJ of energy from combustion, calculate how many barrels of oil are consumed in making this commute five days a week for a year.

14-91. (*) Show that the combustion of coal produces more $CO_2(g)$ per unit of energy than any other fossil fuel—about twice that produced by burning natural gas. Take the formula of coal to be $C(s)$ and that of natural gas to be $CH_4(g)$. Use the chemical equations describing the combustion of several fossil fuels to show why this is the case.

14-92. The human body converts the molecule adenosine diphosphate, ADP, into adenosine triphosphate, ATP, as a means of storing and transferring energy in the body. The chemical equation for the reaction is

$$ADP(aq) + H_2PO_4^-(aq) \rightarrow ATP(aq)$$

for which the enthalpy of reaction is ΔH_{rxn}° = 30.5 $kJ\cdot mol^{-1}$. Use the information provided in Problem 14-76 to calculate the number of molecules of ATP that can be generated from the combustion of one molecule of glucose within the body. Assume that the process is 100% efficient.

14-93. One of the most famous equations in science is $E = mc^2$, which comes from Einstein's theory of relativity. Einstein was the first to show that mass can be converted into energy, and that energy can be converted into mass. This equation relates the amount of energy produced when the mass of a system decreases, or conversely, the amount of energy that is required to increase the mass of a system. In either case, the equation relates the change in energy associated with a change in mass, and in the notation of this chapter, Einstein's mass-energy relation is written as

$$\Delta U = c^2\Delta m$$

where ΔU is the change in energy and Δm is the change in mass. Calculate the amount of energy produced when a 1.00 gram mass is converted into energy. Compare this result to the magnitude of the energy changes in chemical reactions that are from about −400 $kJ\cdot mol^{-1}$ to 600 $kJ\cdot mol^{-1}$.

14-94. Nitrogen oxide gas can be converted to nitrogen gas by using carbon monoxide. Write the chemical equation for the reaction and calculate the value of ΔH°_{rxn} for the equation.

14-95. Ethanol, $CH_3CH_2OH(l)$, can be formed in a combination reaction between ethylene gas, $C_2H_4(g)$, and water vapor. Write the chemical equation for the reaction and calculate the value of ΔH°_{rxn} for the equation.

14-96. In this problem, we shall show that work is a path function. Let's start with an ideal gas enclosed in a one-liter container at a pressure of 10.0 bar. Calculate the work done by the gas if it expands against a constant pressure of 1.0 bar to a final volume of 10.0 liters, as shown in (a) below. Now let's carry out this process in two steps. In the first step, let the gas expand against a constant pressure of 5.0 bar to a volume of 2.0 liters. In the second step, let the gas expand against a constant pressure of 1.0 bar to a final volume of 10.0 liters, as shown in (b) below. Note that the final state in the one-step expansion and in the two-step expansion is the same (10.0 liters at 1.0 bar), and so ΔU, which is a state function, is the same for both processes. Now calculate the work done by the gas in the two-step expansion and compare your result to the one that you obtained for the one-step expansion.

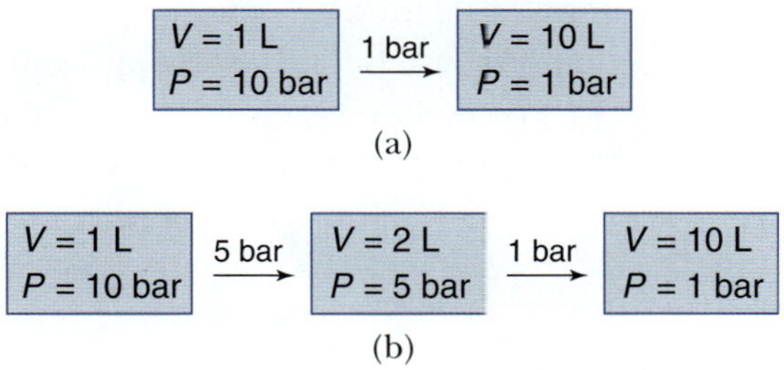

A schematic illustration of (a) the one-step expansion and (b) the two-step expansion of an ideal gas.

Dorothy Crowfoot Hodgkin (1910–1994) was born in Cairo, Egypt, where her father was serving with the Egyptian Ministry of Education. She entered Oxford at a time when few women studied science subjects and received her doctorate in 1934 from Cambridge University for her dissertation on X-ray crystallography of large molecules. While a graduate student, she recorded the first X-ray diffraction pattern from a protein (pepsin) crystal. She later determined the three-dimensional structure of a number of biologically important molecules such as cholesterol, penicillin, vitamin B12, and zinc insulin (which contains almost 800 atoms). In 1965, Hodgkin was awarded the Order of Merit, the highest civilian honor in the United Kingdom, becoming the second woman after Florence Nightingale to be thus honored. In 1970, she became the chancellor of Bristol University, where she established the Hodgkin Scholarship and Hodgkin House, in honor of her husband, for students from third-world countries. She was actively involved in the campaign for peace and disarmament and in the 1970s served as president of the Pugwash Conference on Science and World Affairs (an international organization that brings together scholars and public figures to work toward reducing the danger of armed conflict and to seek solutions to global security threats). In 1964, Hodgkin was awarded the Nobel Prize in Chemistry "for her determination by X-ray techniques of the structure of important biological substances."

15. Liquids and Solids

Your study of chemistry has taken you from a fundamental understanding of how atoms combine to form compounds to calculations involving chemical equations. Now we begin to build on our picture of molecular structure in order to predict chemical changes and the properties of matter. In this chapter, we discuss how the behavior of liquids and solids depends on the attraction between molecules.

If molecules did not attract one another, then gases would not condense into liquids. Without an orderly arrangement of atoms in a solid, a diamond would not retain its hardness and the "lead" in pencils its softness. In this chapter we shall discuss these and other properties of liquids and solids in terms of intermolecular forces. We shall examine the processes of melting, freezing, and vaporization and then introduce phase diagrams that give the temperatures and pressures at which a substance occurs as a solid, a liquid, or a gas. We then discuss how various types of crystal structures are described and how these contrast with other solids. Finally, we briefly discuss liquid crystals and colloids.

15-1. The Molecules in Solids and Liquids Are Close to One Another

We learned in Chapter 13 that the molecules in a gas are in constant motion, traveling distances more than a hundred molecular diameters between collisions at one bar. In addition to traveling in straight lines (**translational motion**), the molecules rotate and vibrate freely as they travel between collisions (Figure 15.1). Because the average distance between the molecules is so large, gases have low densities and can be readily compressed. In addition, the volume of a gas is always the same as the volume of its container.

By contrast, the molecules of a solid are close together and restricted to fixed positions in space. The molecules vibrate about these fixed positions, but they usually do not rotate or move easily from site to site. The molecules in a liquid are also close together, but they are not restricted to fixed positions; they rotate and move throughout the fluid. The molecules in a liquid are essentially in constant interaction with one another, and the distance that a molecule travels between collisions is usually less than a molecular diameter. The

TABLE 15.1 Characteristics of solids, liquids, and gases

Phase	Translation	Rotation	Vibration	Average distance between particles
solid	none	none or hindered	about fixed positions in space	less than one molecular diameter
liquid	hindered	hindered	free	less than one molecular diameter
gas	free	free	free	about 10 molecular diameters at one atm

translational and rotational motions of molecules in a liquid are hindered by frequent collisions between the molecules. The molecular distinctions between a solid, a liquid, and a gas are summarized in Table 15.1.

Because the particles in solids are fixed in space and are not able to move from site to site, a solid body has a fixed shape, and the volume and shape of a solid are independent of its container. Although the molecules in a liquid are essentially in continual contact with one another, they are able to move past one another and move throughout the liquid. Thus, unlike a solid body, a liquid can be poured from one container to another. Furthermore, a liquid has a fixed volume because the molecules are constantly attracting one another. But, unlike a solid, a liquid assumes the shape of its container because its constituent molecules are not held in fixed positions in space.

Liquids and solids have densities that are about a thousand times greater than the densities of gases at one atm. The densities of gases are often expressed in units of grams per liter, whereas the densities of liquids and solids are expressed in units of grams per milliliter or grams per cubic centimeter. The relative densities of gases, liquids, and solids imply that the distances between molecules in liquids and solids are much smaller than they are in gases.

The molar volume of a substance is a direct indication of the average separation between the molecules. The molar volume of the liquid phase of a compound is approximately equal to the molar volume of the solid phase of the same compound. Yet the molar volume of a gas is much larger than that of a liquid or a solid at the same pressure and temperature. The molar volumes for solid, liquid, and gaseous water at 0°C and 100.0°C at one atm are shown in Table 15.2. Note that the molar volume of water increases by a factor of 1630

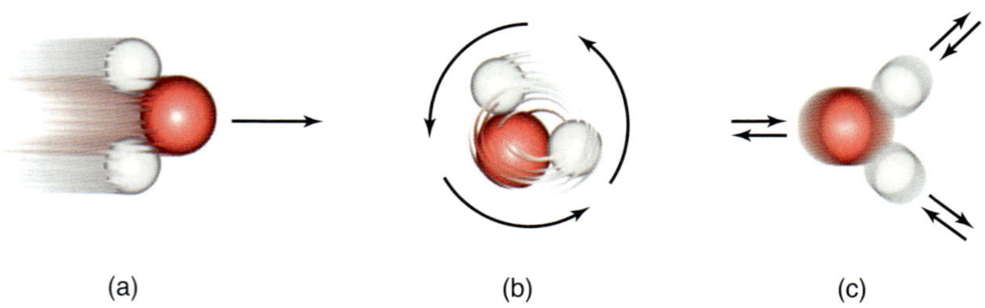

Figure 15.1 Molecular motion may be translational, rotational, or vibrational. (a) Translation is movement of the entire molecule through space. (b) Rotation is the spinning of a molecule in space. (c) Vibration is the back-and-forth movement of nuclei about their equilibrium positions.

TABLE 15.2 Molar volume of solid, liquid, and gaseous water at 0°C and 100°C and one atm (1.0135 bar)

Phase of H_2O	Molar volume at 0°C/mL·mol^{-1}	Molar volume at 100°C/mL·mol^{-1}
solid	19.8	—
liquid	18.0	18.8
gas	—	30600

(30600 mL/18.8 mL) upon vaporization at 100.0°C. The average separation between the molecules on going from the liquid to the gas phase at one atm increases over 10-fold.

15-2. The Processes of Melting and Boiling Appear as Horizontal Lines on a Heating Curve

Let's consider an experiment in which a pure substance is converted from a solid to a liquid to a gas by the application of heat at a constant rate at one atm. Figure 15.2 shows how the temperature of the substance—water, in this case—varies with time when it is heated at a constant rate. Such a plot is called a **heating curve.** Initially, the water is at a temperature of −10°C, so it is in the form of ice. As energy in the form of heat is added, the temperature of the ice

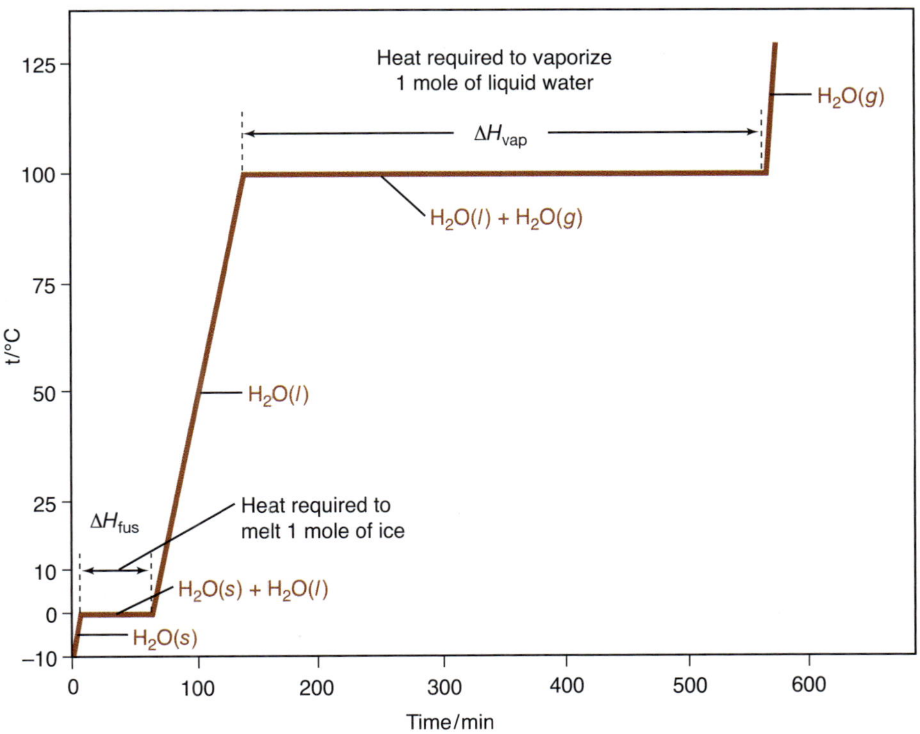

Figure 15.2 The heating curve for one mole of water at one atm starting with ice at −10°C. The energy is added as heat at a constant rate of 100 J·min^{-1}. The most noteworthy features of the heating curve are the horizontal portions, which represent the molar enthalpy of fusion, H_{fus}, and the molar enthalpy of vaporization, H_{vap}. You can see that the molar enthalpy of vaporization is much larger than the molar enthalpy of fusion.

increases until 0°C, the melting point of ice, is reached. At 0°C, the temperature remains constant for 60 min even though the ice is being heated at a constant rate of 100 J·min^{-1}. The energy as heat being added at 0°C melts the ice to liquid water. The temperature of the ice-water mixture remains at 0°C until all the ice is melted. The energy absorbed as heat that is required to melt one mole of any substance is called the **molar enthalpy of fusion** and is denoted by ΔH_{fus}. The experimental data plotted in Figure 15.2 show that 60.1 minutes are required to melt one mole of ice when heat is added at the rate of 100.0 J·min^{-1}; so the molar enthalpy of fusion for ice is given by

$$\Delta H_{\text{fus}} = (60.1\ \text{min·mol}^{-1})(100.0\ \text{J·min}^{-1}) = 6010\ \text{J·mol}^{-1} = 6.01\ \text{kJ·mol}^{-1}$$

After all the ice is melted, and not until then, the temperature increases from 0°C. The temperature continues to increase until 100.0°C, the boiling point of water, is reached. At 100.0°C, the temperature remains constant for 406.5 minutes even though heat is being added at a rate of 100.0 J·min^{-1}. The energy being absorbed at 100.0°C vaporizes the liquid water to water vapor. The temperature of the liquid-vapor mixture remains at 100.0°C until all the water is vaporized. The energy absorbed as heat that is required to vaporize one mole of any substance is called the **molar enthalpy of vaporization** and is denoted by ΔH_{vap}. According to Figure 15.2, the value of ΔH_{vap} for water is

$$\Delta H_{\text{vap}} = (406.5\ \text{min·mol}^{-1})(100.0\ \text{J·min}^{-1}) = 40\,650\ \text{J·mol}^{-1} = 40.65\ \text{kJ·mol}^{-1}$$

Once all the water is vaporized, the temperature increases from 100.0°C, as shown in Figure 15.2.

The horizontal portions of the heating curve of a pure substance represent the enthalpy of fusion and the enthalpy of vaporization. The other regions represent pure phases being heated at a constant rate. We learned in Chapter 14 that it requires energy to raise the temperature of a substance. The heat absorbed in raising the temperature of a substance from T_1 to T_2 at constant pressure without a change in phase is given by

$$q_{\text{P}} = nC_{\text{P}}(T_2 - T_1) = nC_{\text{P}}\Delta T \tag{15.1}$$

(see Equation 14.30), where C_{P} is the molar heat capacity at constant pressure and n is the number of moles. Recall that the heat capacity is the measure of the ability of a substance to take up energy as heat. Different rates of temperature increase for $H_2O(s)$, $H_2O(l)$, and $H_2O(g)$ appear in Figure 15.2 (steep segments of the heating curve) because

$$C_{\text{P}}[H_2O(l)] > C_{\text{P}}[H_2O(s)] > C_{\text{P}}[H_2O(g)]$$

15-3. Energy Is Required to Melt a Solid and to Vaporize a Liquid

The melting, or fusion, of one mole of ice can be represented by

$$H_2O(s) \rightarrow H_2O(l) \qquad \Delta H_{\text{fus}} = 6.01\ \text{kJ·mol}^{-1}$$

or, in general, we write

$$X(s) \rightarrow X(l) \qquad \Delta H_{fus}$$

where X stands for any element or compound. The energy as heat that is required to melt n moles of a substance is given by

$$q_{fus} = n\,\Delta H_{fus} \tag{15.2}$$

The enthalpy of fusion is necessarily positive because it requires energy to break up the crystal lattice. Recall that a positive value of ΔH means that energy as heat is absorbed in a process.

The vaporization of one mole of water can be represented by

$$H_2O(l) \rightarrow H_2O(g) \qquad \Delta H_{vap} = 40.65\ \text{kJ·mol}^{-1}$$

or, in general,

$$X(l) \rightarrow X(g) \qquad \Delta H_{vap}$$

The energy that is absorbed as heat and is required to vaporize n moles of a substance is given by

$$q_{vap} = n\,\Delta H_{vap} \tag{15.3}$$

The value of ΔH_{vap} is always positive because energy is required to separate the molecules in a liquid from one another. Gas phase molecules are so far apart from one another that they interact only very weakly relative to liquid phase molecules. Essentially, all the energy put in as enthalpy of vaporization is required to separate the molecules of the liquid from one another. Because the temperature does not change during vaporization, the average kinetic energy of the molecules does not change. Table 15.3 gives the molar enthalpies of fusion and vaporization of several substances.

EXAMPLE 15-1: How long would it take to convert 50.0 grams of ice at 0°C to water at 100.0°C if the heating rate is 250.0 J·min^{-1}? The molar heat capacity of $H_2O(l)$ is 75.3 J·mol^{-1}·K^{-1}.

Solution: A mass of 50.0 grams of water corresponds to 2.77 moles. First we calculate how long it takes to melt the ice at 0°C. The value of ΔH_{fus} for ice given in Table 15.3 is 6.01 kJ·mol^{-1}. Then we calculate how long it takes to heat the water from 0°C to 100.0°C. The energy required to melt 2.77 moles of ice is given by Equation 15.2:

$$q_{fus} = n\Delta H_{fus} = (2.77\ \text{mol})\ (6.01\ \text{kJ·mol}^{-1}) = 16.6\ \text{kJ}$$

The time it would take to add this much energy is given by q_{fus} divided by the rate of heating, or

TABLE 15.3 Normal melting points, boiling points, and molar enthalpies of vaporization and fusion at these temperatures and one atm (1.01325 bar) pressure

Compound	Chemical formula	Melting point/K	Boiling point/K	ΔH_{fus} /kJ·mol^{-1}	ΔH_{vap} /kJ·mol^{-1}
ammonia	NH_3	195.42	239.82	5.66	23.33
argon	Ar	83.79	87.30	1.18	6.43
bromine	Br_2	266.0	332.0	10.57	29.96
carbon dioxide	CO_2	216.59	(195) sublimes	9.02	(25.2) sublimes
chlorine	Cl_2	171.7	239.11	6.40	20.41
chloromethane	CH_3Cl	175.5	249.06	6.43	21.40
dichloromethane	CH_2Cl_2	176.0	313	4.60	28.06
helium	He	—	4.22	0.014	0.08
hydrogen	H_2	13.81	20.28	0.12	0.90
hydrogen bromide	HBr	186.4	206.77	2.41	17.6
hydrogen chloride	HCl	158.98	188	2.00	16.15
hydrogen sulfide	H_2S	187.7	213.60	2.38	18.67
iodine	I_2	386.9	457.6	15.52	41.57
krypton	Kr	115.77	119.93	1.64	9.08
lithium bromide	LiBr	825	1583	17.6	148.1
mercury	Hg	234.32	629.88	2.29	59.11
methanal (formaldehyde)	H_2CO	181	252	—	24.5
methane	CH_4	90.68	111.67	0.94	8.19
neon	Ne	24.54	27.07	0.328	1.71
nitrogen	N_2	63	77.36	0.71	5.57
oxygen	O_2	54.36	90.20	0.44	6.82
sulfur dioxide	SO_2	198	263.10	7.4	24.94
water	H_2O	273.15	373.15	6.01	40.65
xenon	Xe	161.36	165.04	2.27	12.57

$$\text{time} = \frac{16.6\ \text{kJ}}{250.0\ \text{J}\cdot\text{min}^{-1}} = \frac{16.6\times10^3\ \text{J}}{250.0\ \text{J}\cdot\text{min}^{-1}} = 66.4\ \text{min}$$

The energy required to heat the water from 0°C to 100.0°C is (Equation 15.1)

$$q_P = nC_P\Delta T = (2.77\ \text{mol})(75.3\ \text{J}\cdot\text{mol}^{-1}\cdot\text{K}^{-1})(100.0\ \text{K}) = 2.086\times10^4\ \text{J}$$

The time it takes to add this much energy as heat is given by

$$\text{time} = \frac{2.086\times10^4\ \text{J}}{250.0\ \text{J}\cdot\text{min}^{-1}} = 83.4\ \text{min}$$

The total time it would take to convert 50.0 grams of ice at 0°C to water at 100.0°C at this rate of heating is

$$\text{time} = 66.4\ \text{min} + 83.4\ \text{min} = 149.8\ \text{min}$$

PRACTICE PROBLEM 15-1: How long would it take to convert 100.0 grams of solid sodium at 20.0°C to sodium vapor at 1000.0°C if the heating rate at a pressure of one atm is 8.0 $\text{kJ}\cdot\text{min}^{-1}$? The melting point of sodium is 97.8°C; its boiling point is 883°C; its molar enthalpy of fusion is 2.60 $\text{kJ}\cdot\text{mol}^{-1}$; its molar enthalpy of vaporization is 97.4 $\text{kJ}\cdot\text{mol}^{-1}$; and the heat capacities of solid, liquid, and gaseous sodium are 28.2 $\text{J}\cdot\text{mol}^{-1}\cdot\text{K}^{-1}$, 30.8 $\text{J}\cdot\text{mol}^{-1}\cdot\text{K}^{-1}$, and 20.8 $\text{J}\cdot\text{mol}^{-1}\cdot\text{K}^{-1}$, respectively.

Answer: 70 minutes

EXAMPLE 15-2: Compute the energy released as heat at a pressure of one atm when 28 grams of liquid water at 18°C is converted to ice at 0°C. (An ice cube contains about 28 grams, or one ounce, of water.) The molar heat capacity of $H_2O(l)$ is $C_P = 75.3\ \text{J}\cdot\text{mol}^{-1}\cdot\text{K}^{-1}$, and $\Delta H_{fus} = 6.01\ \text{kJ}\cdot\text{mol}^{-1}$ for ice.

Solution: The overall process must be broken down into two steps. We must first bring the $H_2O(l)$ from 18°C to 0°C (the freezing point of water) and then consider the process $H_2O(l) \rightarrow H_2O(s)$ at 0°C:

$$\underset{\text{at }18^\circ\text{C}}{28\ \text{g}\ H_2O(l)} \xrightarrow{\text{step 1}} \underset{\text{at }0^\circ\text{C}}{28\ \text{g}\ H_2O(l)} \xrightarrow{\text{step 2}} \underset{\text{at }0^\circ\text{C}}{28\ \text{g}\ H_2O(s)}$$

Twenty-eight grams of water corresponds to 1.55 moles. From Equation 15.1, we have

We will use three significant figures for the number of moles of water to avoid later round-off errors.

$$\begin{aligned}q_P &= nC_P(T_2 - T_1)\\ &= (1.55\text{ mol})(75.3\text{ J·mol}^{-1}\text{·K}^{-1})(273\text{K} - 291\text{K})\\ &= -2100\text{ J} = -2.1\text{ kJ}\end{aligned}$$

The negative sign for q_P reflects the fact that energy must be removed to lower the temperature of the water. For step 2,

$$q_P = n(-\Delta H_{fus}) = (1.55\text{ mol})(-6.01\text{ kJ·mol}^{-1}) = -9.3\text{ kJ}$$

where the minus sign in front of ΔH_{fus} arises because freezing is the reverse of melting, that is, $\Delta H_{freezing} = -\Delta H_{fus}$. The total amount of energy that must be *removed* as heat from the 28 grams of water is (–2.1 kJ) + (–9.3 kJ) = –11.8 kJ.

PRACTICE PROBLEM 15-2: Using the data in Table 15.3 and Table 14.6, compute the energy released as heat at one atm when 25.0 grams of liquid mercury at 300.0°C is converted to solid mercury at –60.0°C. Take the heat capacity of Hg(*s*) to be 28.3 J·mol⁻¹·K⁻¹.

Answer: –1.54 kJ

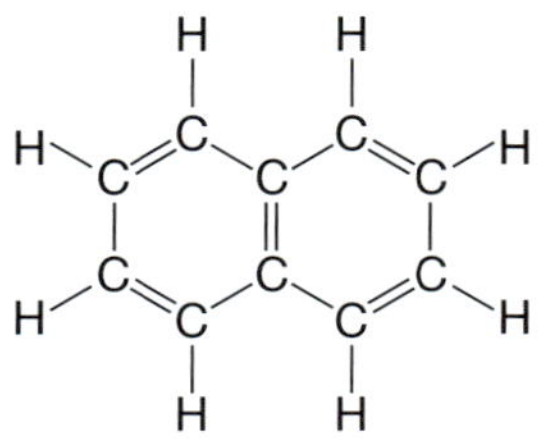

naphthalene

It is possible for a solid to be converted directly to a gas without passing through the liquid phase. This process is called **sublimation.** The **molar enthalpy of sublimation,** ΔH_{sub}, is the energy absorbed as heat when one mole of a solid is sublimed at constant pressure. Essentially all the energy put in as heat in the sublimation process is used to separate the molecules in the solid from one another. The larger the value of ΔH_{sub}, the stronger the intermolecular attractions in the solid are.

The best-known example of sublimation is the conversion of dry ice, $CO_2(s)$, to carbon dioxide gas according to

$$CO_2(s) \rightarrow CO_2(g) \qquad \Delta H_{sub} = 25.2\text{ kJ·mol}^{-1}$$

The name "dry ice" is used because the $CO_2(s)$ does not become a liquid at one atmosphere. Dry ice at one atmosphere has a temperature of –78°C and is widely used as a one-time, low-temperature refrigerant. The sublimation of 44.0 grams (one mole) of dry ice requires 25.2 kJ of heat.

Ice, $H_2O(s)$, sublimes at temperatures below its melting point (0°C). The equation for the sublimation of ice is

$$H_2O(s) \rightarrow H_2O(g) \qquad \Delta H_{sub} = 46.7\text{ kJ·mol}^{-1}$$

Figure 15.3 Naphthalene, $C_{10}H_8(s)$, is one of a number of substances that readily sublime at room temperature. Here we see crystals of naphthalene condensing directly from vapor to crystals on the surface of a cold tube.

Snow often sublimes, and so does the ice in the freezer compartment of your refrigerator. Perhaps you've noticed that ice cubes left in the freezer get smaller as time passes. Another substance that sublimes at temperatures below its melting point is naphthalene, $C_{10}H_8(s)$ (margin), a principal component of one type of mothball (Figure 15.3).

15-4. Van Der Waals Forces Are Attractive Forces Between Molecules

In the process of vaporization or sublimation, the molecules of the liquid or solid, which are in contact with one another, become separated and widely dispersed. The value of ΔH_{vap} or ΔH_{sub} reflects how strongly the molecules attract one another in the liquid or solid phase. The more strongly the molecules attract one another, the greater the value of ΔH_{vap} or ΔH_{sub}.

The simplest force at the atomic or molecular level is that between ions. We studied this force in Chapter 6 when we discussed ionic compounds. Ions with opposite charges attract each other; ions with like charges repel each other. The electrical force between ions is relatively strong; it requires a relatively large amount of energy to separate ions of opposite charge. Therefore, the molar enthalpies of vaporization of ionic compounds are much larger than those of most nonionic compounds. Molar enthalpies of vaporization of ionic compounds are typically at least 100 $kJ \cdot mol^{-1}$. The boiling points of ionic compounds also are higher than those of nonionic compounds. This is the reason that all ionic compounds are solids at room temperature.

Most of the compounds listed in Table 15.3 are covalent compounds. In Chapter 7 we learned that some molecules have dipole moments; in other words, they are polar. An example of a polar molecule is methanal, $CH_2O(g)$, better known by its common name, formaldehyde, whose Lewis formula is

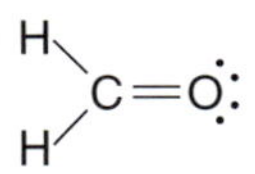

The bond dipole moments in a formaldehyde molecule can be illustrated by

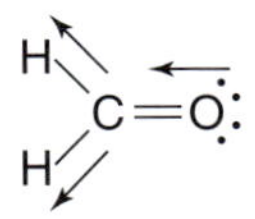

Recall that we define the direction of a dipole moment to be from the negative charge to the positive charge, in agreement with the IUPAC convention.

because an oxygen atom is more electronegative than a carbon atom, which in turn is more electronegative than a hydrogen atom. The net dipole moment is illustrated by

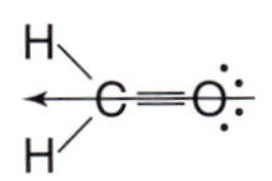

Thus, even though a formaldehyde molecule is electrically neutral overall, it has a positively charged end and a negatively charged end and therefore is a polar molecule.

EXAMPLE 15-3: Recall Section 8-9 where we discussed how to use the symmetry of a molecule to predict whether or not it has a net dipole moment. Use VSEPR theory to predict which of the following molecules has a net dipole moment:

(a) SF_6 (b) SF_4 (c) PF_5

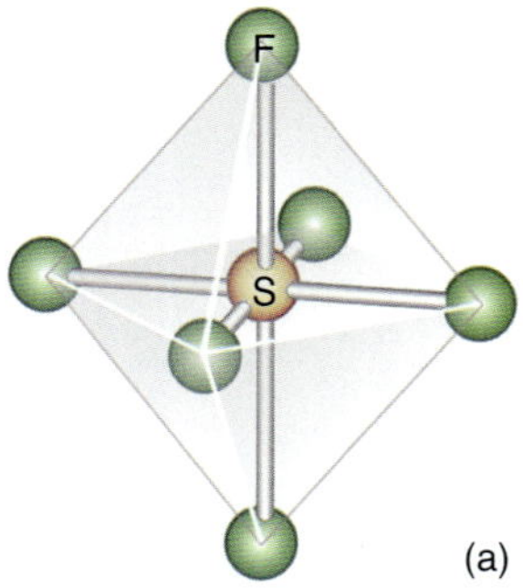

(a)

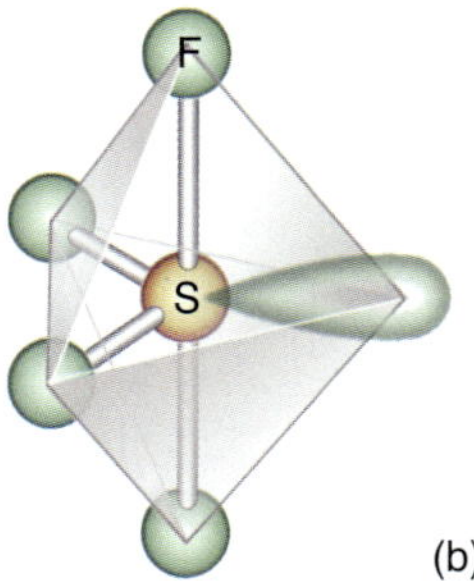

(b)

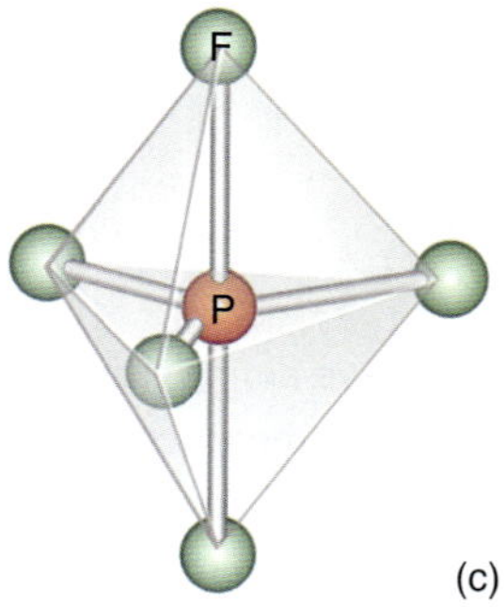

(c)

Solution: See margin drawings. (a) Sulfur hexafluoride, SF_6, is an AX_6 molecule; so it is octahedral and has no net dipole moment. (b) Sulfur tetrafluoride, SF_4, is an AX_4E molecule; so it is seesaw-shaped and has a net dipole moment. (c) Phosphorus pentafluoride, PF_5, is an AX_5 molecule; so it is trigonal bipyramidal and has no net dipole moment.

PRACTICE PROBLEM 15-3: Which of the following molecules are polar?
(a) C_2Cl_4 (b) PCl_3 (c) SO_2 (d) CH_3Cl (e) CCl_4

Answer: (b), (c), and (d) are polar.

Polar molecules attract one another (Figure 15.4), and this attraction is called **dipole-dipole attraction.** The charge separation in polar molecules is considerably smaller than the full electronic charges on ions, so dipole-dipole forces are weaker than ion-ion forces. Molar enthalpies of vaporization for typical polar compounds are around 10–20 $kJ\cdot mol^{-1}$. We must emphasize here that the interaction energies *between* neutral molecules are much weaker than the covalent bonds *within* molecules, which have energies of hundreds of kilojoules per mole. The forces between neutral molecules are collectively referred to as **intermolecular forces**; the prefix *inter* means *between*. The forces within a molecule (i.e., chemical bonds) are called **intramolecular forces**; the prefix *intra* means *within*.

A particularly important dipole-dipole attraction occurs when one or more hydrogen atoms in a molecule are directly bonded to a highly electronegative atom, as in the case of water and ammonia. Let's consider the molar enthalpies of vaporization of water, ammonia, and methane:

Molecule:	H_2O	NH_3	CH_4
ΔH_{vap} /$kJ\cdot mol^{-1}$	40.7	23.4	8.2

These three compounds have approximately the same molecular mass (18, 17, and 16, respectively), but the amount of energy required to separate the molecules of these three liquids are very different.

Figure 15.4 Although polar molecules are electrically neutral overall, they attract one another by a dipole-dipole force. The molecules orient themselves as shown because the positive end of one attracts the negative end of another. The dipoles are said to be oriented head-to-tail.

Water molecules in liquid water attract one another through the electrostatic interaction between a hydrogen atom and the oxygen atom on different molecules:

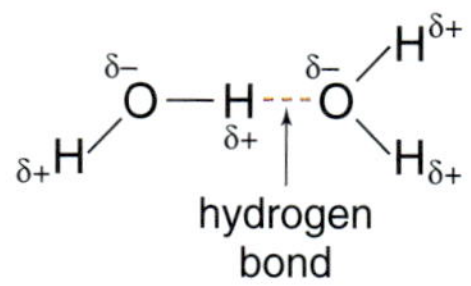

The electrostatic attraction that occurs between molecules in which a hydrogen atom is bonded to a highly electronegative atom, such as an oxygen, nitrogen, or fluorine atom, is called a **hydrogen bond.** Because a hydrogen atom is so small, the charge on it is highly concentrated, so it strongly attracts electronegative atoms in neighboring molecules.

Hydrogen bonds are a particularly strong form of dipole-dipole attraction yielding molar enthalpies of vaporization around 20–40 kJ·mol^{-1}. The pattern shown in Figure 15.5 extends throughout liquid water and gives water its large value of ΔH_{vap} (40.7 kJ·mol^{-1}) and its high boiling point. We shall see that hydrogen bonds give water many special properties.

Hydrogen bonding greatly affects the structure of ice (Figure 15.6), which is described as an open structure because of the significant amount of empty space between the molecules. The open structure of ice is a direct consequence of the fact that each hydrogen atom is hydrogen-bonded to the oxygen atom of an adjacent molecule. Note the tetrahedral arrangement of the oxygen atoms in ice. Every oxygen atom sits in the center of a tetrahedron formed by four other oxygen atoms.

The structure of liquid water is less open than the structure of ice because, when ice melts, the total number of hydrogen bonds decreases. Unlike most other substances, the density of water *increases* upon going from solid to liquid because of a partial breakdown of the hydrogen-bonded structure (that's why ice floats on water). The extent of hydrogen bonding in liquid water is only about 80%, whereas in ice nearly 100% of the oxygen atoms are hydrogen-bonded. The extent of hydrogen bonding in water decreases as the temperature increases.

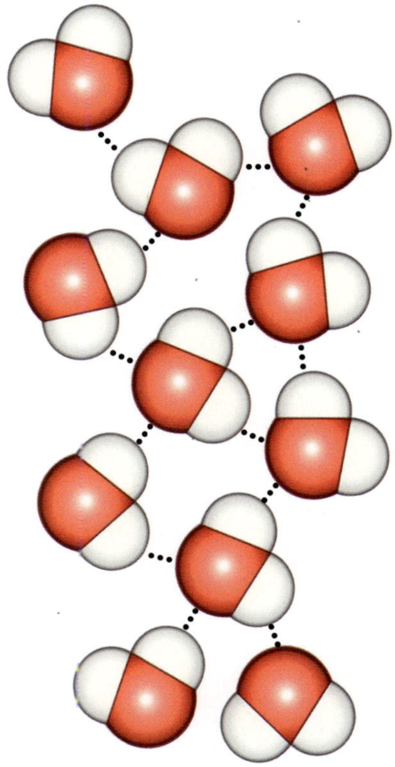

Figure 15.5 There are many hydrogen bonds in liquid water; each oxygen atom can form two hydrogen bonds because each oxygen atom has two lone pairs of electrons. Thus, each water molecule has the ability to form four hydrogen bonds. At 25°C about 80% of the hydrogen atoms in water are hydrogen bonded.

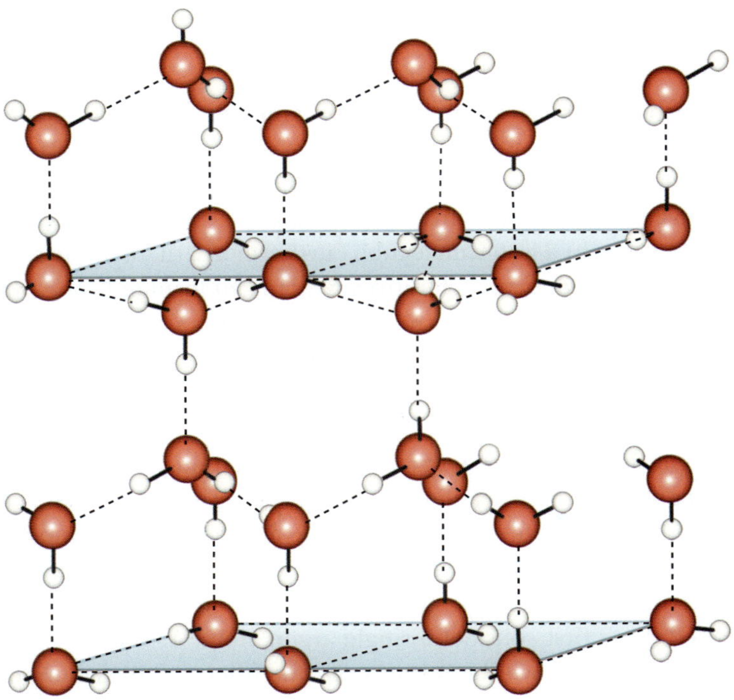

Figure 15.6 The crystalline structure of ice. Each water molecule can form four hydrogen bonds. Each oxygen atom is located in the center of a tetrahedron formed by four other oxygen atoms. The entire structure is held together by hydrogen bonds.

HYDROGEN BONDING HAS IMPORTANT CONSEQUENCES FOR LIFE Water contains two hydrogen atoms covalently bound to a central oxygen atom. We know that oxygen falls into Group 16 in the periodic table. If we compare the boiling points of other substances from this group with the formula H_2X, we find an interesting trend:

Molecule	H_2Te	H_2Se	H_2S
Boiling point/°C	−2.0	−41.25	−59.55

By extrapolation water, $H_2O(l)$, should have a boiling point around −90°C (these data are plotted in Figure 15.10). The remarkable difference between this and the actual boiling point of water of 100.0°C is due to hydrogen bonding. Were it not for hydrogen bonding, most of the water on earth would be gaseous and life as we know it could not exist. Moreover, unlike most other substances, the density of water increases upon going from a solid to a liquid because of a partial breakdown of its hydrogen-bonded structure. Because of this ice floats on water, whereas for most other substances the solid form is the more dense and sinks when placed into its liquid phase. The fact that ice floats on water causes the surfaces of lakes and ponds to freeze before the bottom in winter. This surface ice forms an insulating layer that allows fish and plants to survive below. Were this not the case, most rivers and ponds would freeze solid from the bottom up, altering aquatic life and the food chain. These amazing properties of water that arise from hydrogen bonding are what makes finding water one of the criteria used when searching for life on other planets.

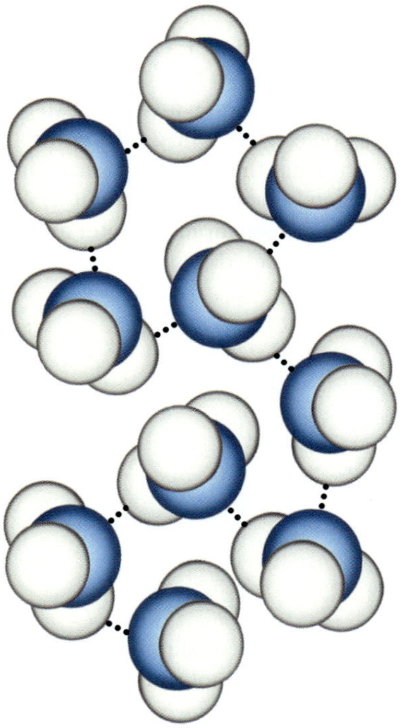

Figure 15.7 There are fewer hydrogen bonds in $NH_3(l)$ than in $H_2O(l)$ because an NH_3 molecule has only one lone electron pair, so each nitrogen atom can form only one hydrogen bond.

Hydrogen bonding also occurs in liquid ammonia, $NH_3(l)$, but the individual hydrogen bonds are weaker than those in water. This is because nitrogen is less electronegative than oxygen and the fractional charges on the nitrogen and hydrogen atoms in an ammonia molecule are less than those on the oxygen and hydrogen atoms in a water molecule. Furthermore, there are fewer hydrogen bonds in $NH_3(l)$ because each nitrogen atom can form only one hydrogen bond (Figure 15.7). Consequently, the value of ΔH_{vap} for ammonia is only about half that for water. Methane is a nonpolar molecule because it is tetrahedral and the four bond moments cancel. Thus, of $H_2O(l)$, $NH_3(l)$, and $CH_4(l)$; $CH_4(l)$ has the lowest value of ΔH_{vap}.

We must consider one more attractive force. Even though methane is nonpolar, it liquefies at 91 K and has a molar enthalpy of vaporization of 8.2 $kJ\cdot mol^{-1}$; therefore, methane molecules must attract one another. Even the noble gases, which consist of single, spherical atoms, can be liquefied. How neutral, nonpolar molecules attract one another was not understood until quantum theory was developed. Let's first consider a single argon atom. As the electrons move about in the atom, there will be instants of time when the distribution of the total charge will not be perfectly symmetric. For example, perhaps there will be more electrons on one side of the atom than on the other side, as shown in Figure 15.8. Over a period of time, these little fluctuations in the electronic charge distribution will average out, thereby yielding a perfectly symmetric distribution. But at any instant of time, there will be an instantaneous *asymmetry* in the electronic charge distribution, resulting in an instantaneous dipole moment (Figure 15.8). Now let's consider two argon atoms separated by a small distance. The motion of the electrons in one atom influences the motion of the electrons in the other atom in such a way that the instantaneous dipole moments line up head-to-tail (Figure 15.9). This synchronized motion of the electrons leads to an effective dipole-dipole attraction that accounts for the attractive force between atoms, nonpolar molecules, and a sizeable fraction of the attractive forces between large molecules as well. This force was first explained by the German physicist Fritz London and is now called a **London force**.

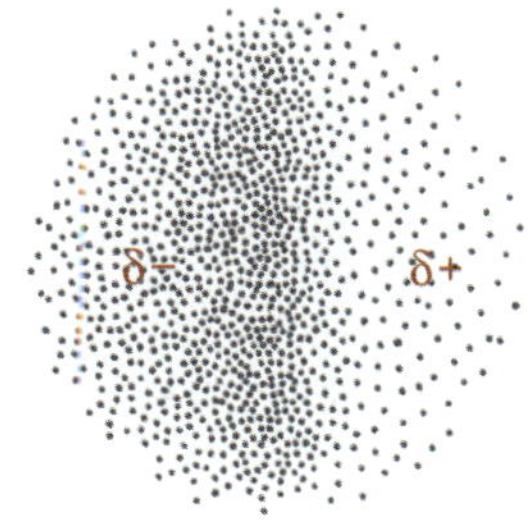

Figure 15.8 If it were possible to take an instantaneous view of an atom, it might look like this drawing. The instantaneous position of the electrons leads to an instantaneous dipole moment. The negative charge is due to a greater-than-average electronic charge density, and the positive charge is due to a less-than-average electronic charge density. As the electrons move around, the dipole moment points in all directions at different times and averages to zero.

Because London forces are due to the motion of electrons, their strength depends on the number of electrons. The more electrons there are in the two interacting molecules, the stronger their attraction for each other. Therefore, for purely nonpolar substances, which attract each other only by way of London forces, we expect the value of ΔH_{vap} to increase with the number of electrons, or even with the size of the molecules. The data in Table 15.4 support this prediction. Notice that, within each group, the value of ΔH_{vap} increases with the number of electrons.

Figure 15.10 shows plots of the boiling points of the noble gases and the hydrides of the nonmetallic elements. The hydrogen-bonded compounds $H_2O(l)$, $NH_3(l)$, and $HF(l)$ have unusually high boiling points. Except for the hydrogen-bonded compounds, there is a general increase of boiling point with increasing molecular mass. This increase in boiling point is due to the increase in London forces, which increases with the number of electrons in a molecule. In Figure 15.10, we see that the boiling points of the noble gases (plotted in purple) and the hydrides of Group 14 (plotted in green) are lower than those of the other molecules. This is because the noble gases and Group 14 hydrides are nonpo-

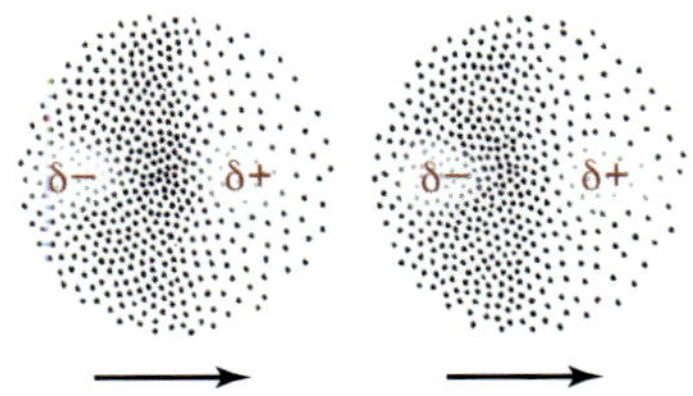

Figure 15.9 When two atoms are near each other, the motions of the electrons in the two atoms affect each other so that the instantaneous dipole moments are aligned head-to-tail. We say that the motion of the electrons is correlated. This effect leads to an instantaneous dipole-dipole attraction between the two atoms.

TABLE 15.4 Relation between number of electrons in some atoms and nonpolar molecules and their enthalpies of vaporization

Substance	Number of electrons	ΔH_{vap}/kJ·mol^{-1}
He	2	0.08
Ne	10	1.71
Ar	18	6.43
Kr	36	9.08
Xe	54	12.57
F_2	18	6.62
Cl_2	34	20.04
Br_2	70	29.96
I_2	106	41.57

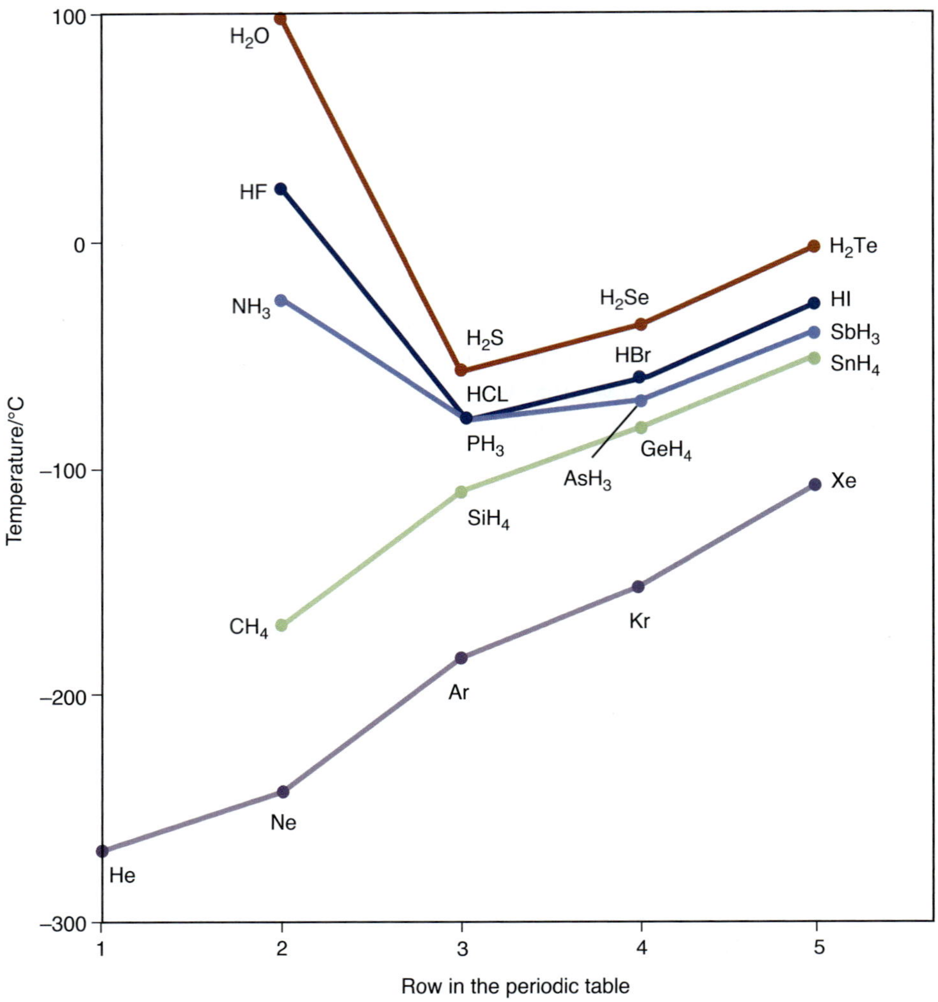

Figure 15.10 Boiling points of the noble gases and hydrides of the nonmetallic elements. Note the abnormally high values for hydrogen fluoride, water, and ammonia, which are the result of hydrogen bonding.

lar and have no dipole-dipole forces contributing to their total intermolecular forces.

The attractive forces between molecules, be they dipole-dipole forces or London forces, are collectively called **van der Waals forces.** Table 15.5 compares the various intermolecular forces that we have discussed in this section.

TABLE 15.5 Various types of attractive forces between ions and molecules

Type	Examples	Typical value of ΔH_{vap}/kJ·mol^{-1}
ion-ion	NaCl, KBr	~100
hydrogen bonding	H_2O, NH_3	20–40
London	Ar, CH_4	5–40
dipole-dipole	H_2CO, HCl	10–20

EXAMPLE 15-4: Without referring to any of the tables in the book, rank the following liquids in order of increasing molar enthalpies of vaporization and boiling points:

$$NaCl(l) \qquad H_2(l) \qquad C_2H_4(l) \qquad CH_3OH(l)$$

Solution: The only ionic compound listed is $NaCl(l)$. Thus, we predict that $NaCl(l)$ has the largest value of ΔH_{vap} and the highest boiling point. The Lewis formula for a methanol molecule, CH_3OH (margin), shows that there is a hydrogen atom bonded to an oxygen atom. Thus, we predict that in $CH_3OH(l)$ there will be hydrogen bonding between the molecules. The Lewis formula for an ethylene molecule, C_2H_4 (margin), indicates that C_2H_4 is a nonpolar molecule. H_2 is also a nonpolar molecule. However, there are many more electrons in a C_2H_4 molecule, so we predict that

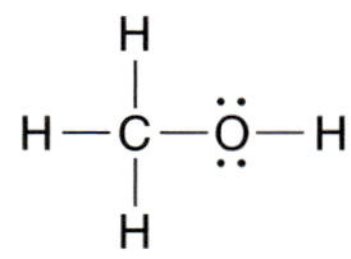

methanol

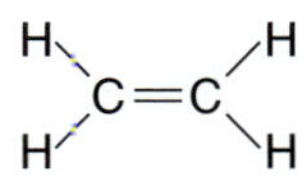

ethene (ethylene)

$$\Delta H_{vap}[NaCl(l)] > \Delta H_{vap}[CH_3OH(l)] > \Delta H_{vap}[C_2H_4(l)] > \Delta H_{vap}[H_2(l)]$$

The boiling points are in the same order. The actual values of ΔH_{vap} are 170 $kJ\cdot mol^{-1}$, 35.2 $kJ\cdot mol^{-1}$, 13.5 $kJ\cdot mol^{-1}$, and 0.90 $kJ\cdot mol^{-1}$, respectively; and the boiling points are 1738 K, 338 K, 169 K, and 20.3 K, respectively.

PRACTICE PROBLEM 15-4: Which of the following molecules (common names given in parentheses) would you predict to exhibit hydrogen bonding? Explain why.

hydrogen | 2-propanone (acetone) | hydrogen fluoride | ethanoic acid (acetic acid)

Answer: A hydrogen molecule contains no nitrogen, oxygen, or fluorine atoms, and so does not exhibit hydrogen bonding. Although an acetone molecule contains hydrogen and oxygen atoms, the hydrogen atoms are not directly bonded to the oxygen atom, and so hydrogen bonding is not observed in acetone. Hydrogen fluoride and acetic acid both exhibit hydrogen bonding because in each molecule there is a hydrogen atom directly bonded to a highly electronegative atom.

15-5. Viscosity, Surface Tension, and Capillary Action Are Properties of Liquids

Several properties of liquids that depend on intermolecular forces are of practical importance to chemists. The **viscosity** of a liquid is a measure of its resistance to flow. Substances that flow very slowly, such as molasses, are said to be viscous. Viscosity can be measured experimentally by timing how long it takes a certain volume of a liquid to flow through a thin tube. Many engineering processes involve the flow of liquids through pipes, so engineering design is

TABLE 15.6 Physical properties of a variety of liquids at 25°C

Compound	Formula	Viscosity (relative to water)	Surface tension/mJ·m^{-2}	Polarity
benzene	C_6H_6	0.68	28	nonpolar
bromine	Br_2	1.1	41	nonpolar
ethanol	C_2H_5OH	1.2	22	slightly polar
glycerol	$C_3H_5(OH)_3$	1050	63	slightly polar
hexane	C_6H_{14}	0.34	18	nonpolar
methanol	CH_3OH	0.89	22	slightly polar
octane	C_8H_{18}	0.57	19	nonpolar
2-propanone (acetone)	$(CH_3)_2CO$	0.34	23	slightly polar
sulfuric acid	H_2SO_4	27.6	—	polar
water	H_2O	1.00	72	polar

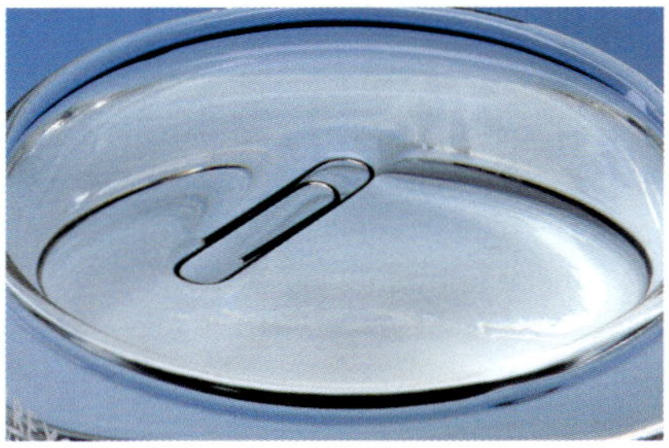

Figure 15.11 A metal paper clip floats on water because of the distribution of its mass and the high surface tension of the water. The surface of water is like an elastic skin that resists penetration.

often concerned with the viscosities of liquids and liquid mixtures. Viscosity decreases with increasing temperature because, as the temperature increases, molecules have greater kinetic energy and therefore can more readily overcome the intermolecular interactions. The viscosities of a variety of compounds are given in Table 15.6.

If a paper clip is carefully placed on a water surface, it floats even though the density of the paper clip is greater than that of water (Figure 15.11). The clip is held up by **surface tension**. Water striders and some other insects can walk on water, being supported by the surface tension of the water, which resists penetration of the surface (Figure 15.12).

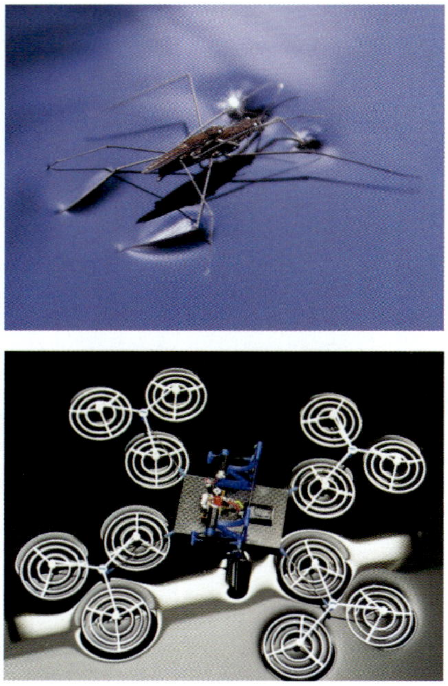

Figure 15.12 (*top*) Some insects like this water strider use surface tension to walk on water. (*bottom*) Researchers at Carnegie Mellon University are working to develop a microrobot that uses surface tension to walk on water.

What is the cause of the surface tension of a liquid? A molecule in the body of a liquid is subject to attractive forces in all directions, but a molecule at the surface experiences a net attractive force toward the interior of the liquid (Figure 15.13). Thus, molecules at the surface of a liquid experience a net inward force. This force tends to minimize the number of molecules at the surface and so minimize the surface area of the liquid. This force is the surface tension. Any liquid whose molecules attract one another strongly has a high surface tension; water is a good example. The surface tension of a liquid tends to hold a drop of liquid in a spherical shape because a sphere is the shape that has the smallest surface area for a given volume (Figure 15.14). The higher the surface tension is, the more nearly spherical the drop is (Figure 15.15). Table 15.6 lists the surface tension for a variety of liquids. Note that surface tension has units of energy per unit area. Surface tension can be thought of as the energy that it takes to create an area of surface.

Certain compounds, such as sodium dodecylsulfate, $NaC_{12}H_{25}SO_3(s)$, lower the surface tension of a liquid by concentrating at the liquid surface. Such molecules are called **surfactants** (surface active agents). A 0.1% sodium dodecylsulfate solution has a surface tension of 20 mJ·m^{-2}, whereas pure water has a surface tension of 72 mJ·m^{-2}. Reduction of the surface tension of water by surfactants is the basis of detergent action. In a detergent solution the surface tension of the water is comparable to that of the grime (oils) on clothing, for example, thereby allowing the solution to wet the grime.

Closely related to surface tension is **capillary action,** the rise of a liquid in a thin tube. Capillary action occurs when the adhesive forces between the molecules of the surface of a capillary wall and the molecules of the liquid are sufficiently great that the liquid adheres to, or wets, the solid surface. The adhesive force pulls the liquid up into the capillary. The liquid column rises until the upward adhesive force is balanced by the downward gravitational force. Capillary action plays a major role in the movement of water in plants, animals, and soil. The water is pulled by the capillary action up into and through living structures (Figure 15.16).

One important consequence of surface tension and capillary action is the formation of a **meniscus,** which is the shape formed by the surface of a liquid in a capillary (Figure 15.17). For a liquid that adheres to glass (for example, water), the liquid rises in a glass capillary and the meniscus is concave. For a liquid that does not adhere to glass (for example, mercury), the liquid is lower where it contacts the glass and the meniscus is convex.

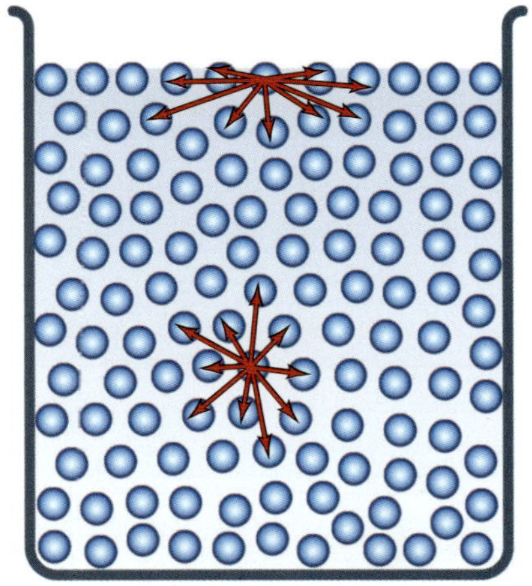

Figure 15.13 Molecules in the interior of a liquid are attracted in all directions, but the molecules at the surface experience a net inward attraction that minimizes the surface area of the liquid and results in surface tension.

Figure 15.14 Surface tension causes drops of liquid to assume a spherical shape because a sphere is the shape with the smallest surface area for a given volume.

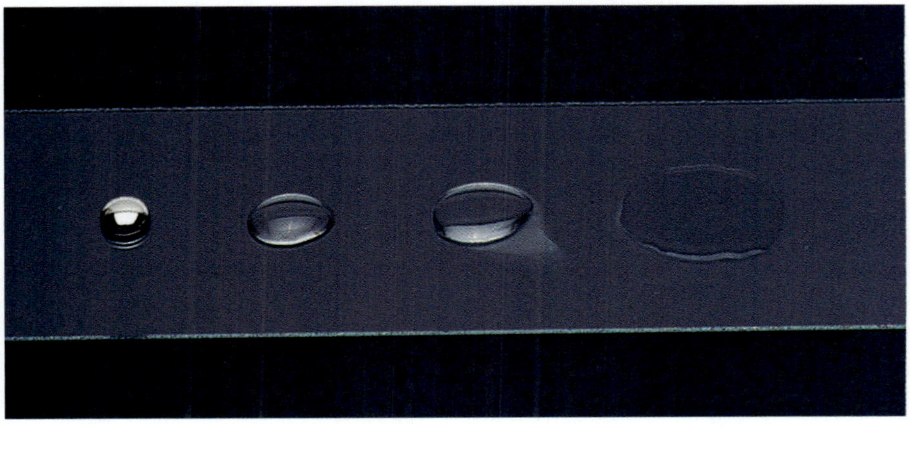

Figure 15.15 Shapes of equal volumes of $Hg(l)$, $H_2O(l)$, $(CH_3)_2SO_3(l)$ (dimethylsulfoxide), and $(CH_3)_2CO(l)$ (acetone), from left to right. Surface tension holds the drops in a spherical shape, and gravity flattens them. The effect of gravity is the same for all the drops; thus, the higher the surface tension, the more nearly spherical is the drop. Surface tension has units of energy per unit area, usually expressed as millijoules per square meter (mJ·m^{-2}).

Figure 15.16 Capillary action is shown here as colored water rises in a celery stalk and in these glass tubes. The cells in a celery stalk form a capillary structure.

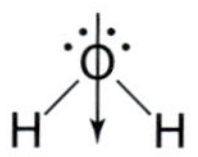

water is polar

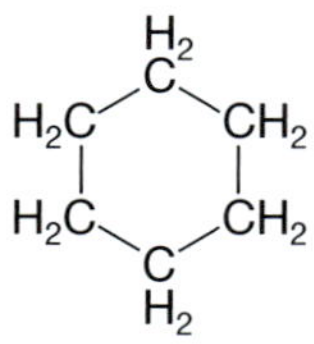

cyclohexane is nonpolar

One final property of liquids that we shall discuss is called **polarity.** Generally speaking, liquids comprised of molecules that have large dipole moments form polar liquids. For example, water is a very polar solvent because individual water molecules have a large dipole moment. On the other hand, cyclohexane, $C_6H_{12}(l)$, is a nonpolar liquid because individual cyclohexane molecules have no dipole moment (margin).

Water and other polar liquids tend to be good solvents for polar species such as ionic compounds. In contrast, cyclohexane and other nonpolar liquids are poor solvents for ionic compounds, but tend to be good solvents for nonpolar species, such as oils, which do not dissolve readily in water or other polar solvents. A general rule of thumb to remember is, "like dissolves like."

Table 15.6 lists the relative viscosities, the surface tensions, and polarity of a variety of compounds. The most important thing to notice is the relative values of these quantities.

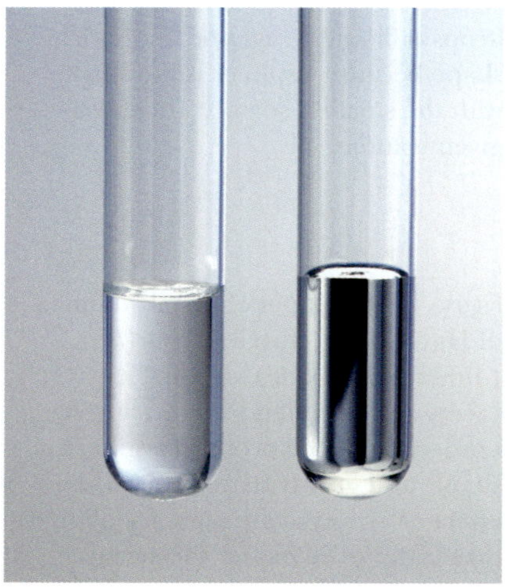

Figure 15.17 A meniscus formed by water (*left*) in a glass tube is concave, whereas the meniscus formed by mercury (*right*) is convex.

15-6. A Liquid Has a Unique Equilibrium Vapor Pressure at Each Temperature

Let's look more closely at the process of vaporization. Suppose that we cover a beaker containing a liquid with a bell jar, as shown in Figure 15.18, and maintain a constant temperature. Now suppose that we evacuate all the gas from the bell jar, so that the space above the beaker and its contents is a vacuum. The molecules in the liquid are in constant motion; some at the surface will break away from the liquid and form a vapor phase above it. The pressure of the vapor is observed to increase rapidly at first and then progressively more slowly until a constant pressure is reached. Let's see why.

In order to break free of the liquid, a molecule at the surface must have enough kinetic energy to overcome the attractive force of its neighbors and be moving in the right direction. The number of molecules that leave the surface is proportional to the surface area of the liquid. Because the surface area is constant, the rate of evaporation is constant (Figure 15.19). There are no molecules in the vapor phase initially, so there is no condensation from the vapor phase to the liquid phase. As the concentration of molecules in the vapor phase increases, the pressure of the vapor increases and the number of vapor phase molecules that collide with the liquid surface increases. As a result, the rate of condensation of the vapor increases. Eventually, a state is reached where the rate of evaporation from the liquid surface is equal to the rate of condensation from the vapor phase. The pressure of the vapor no longer increases but takes on a constant value. The evaporation-condensation process appears to have stopped, and we say that the system is at **equilibrium,** meaning that no change appears to be taking place (Figure 15.20). The pressure of the vapor is now a constant value.

The equilibrium between the liquid and the vapor is a **dynamic equilibrium;** that is, the liquid continues to evaporate and the vapor continues to condense,

Figure 15.18 When a liquid is placed in a closed container that has been evacuated, the pressure of the vapor above the liquid eventually reaches a constant value that depends upon the particular liquid and the temperature.

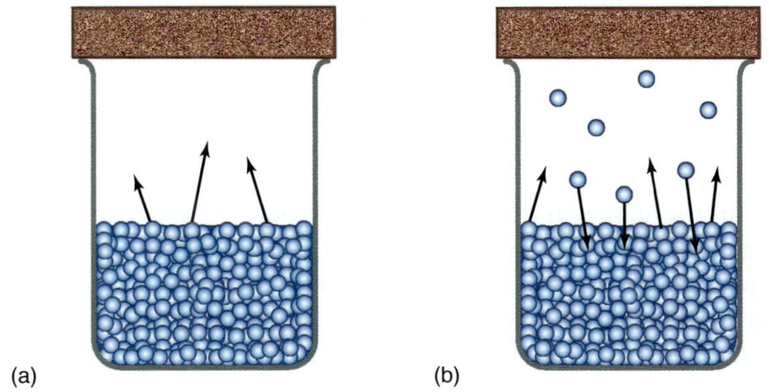

Figure 15.19 When a liquid is placed in a closed container, (a) the rate at which molecules escape from the surface is constant, but the rate at which molecules enter the liquid from the vapor is proportional to the number of molecules in the vapor. (b) When the number of molecules in the vapor is such that the rate of escape from the surface is equal to the rate of condensation from the vapor, the liquid and vapor are in equilibrium with each other.

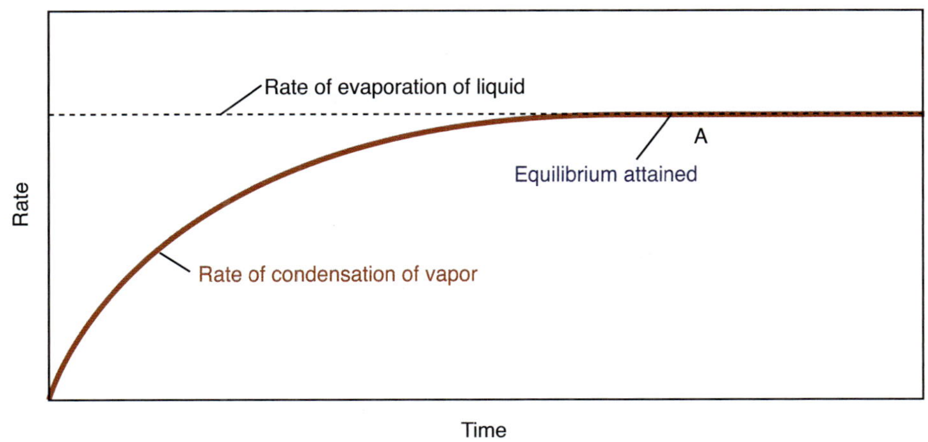

Figure 15.20 Equilibrium is attained (at A) when the rate of evaporation of the liquid equals the rate of condensation of the vapor. At equilibrium the pressure of the vapor is constant and is called the equilibrium vapor pressure.

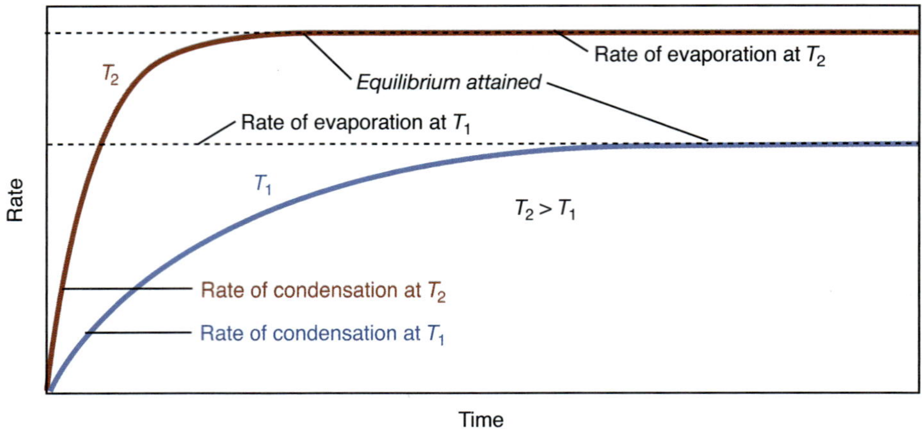

Figure 15.21 The change in the rate of condensation with time for the vapor over a liquid as it approaches equilibrium. Because the rate of evaporation increases with increasing temperature, the equilibrium vapor pressure increases with increasing temperature ($T_2 > T_1$).

Figure 15.22 Solid iodine in equilibrium with gaseous iodine at 40°C (lighter-colored vapor) and at 90°C (darker-colored vapor). The equilibrium vapor pressure increases with increasing temperature, as shown by the more intense color.

but the rate of evaporation is exactly equal to the rate of condensation and thus there is no *net* change. In such a case, we have the equilibrium condition

$$\text{rate of evaporation} = \text{rate of condensation}$$

The pressure of the vapor at equilibrium is called the **equilibrium vapor pressure.** We shall show that the value of the equilibrium vapor pressure depends on the particular liquid and the temperature.

Let's consider the approach to a dynamic liquid-vapor equilibrium at two different temperatures. The higher the temperature, the more rapidly the molecules in the liquid phase move and the higher the rate of evaporation is. Figure 15.21 shows that, because the rate of evaporation at T_2 is greater than the rate of evaporation at T_1 (given that $T_2 > T_1$), the equilibrium vapor pressure at T_2 is greater than that at T_1. Thus, we see that the value of the equilibrium vapor pressure of a liquid increases with increasing temperature (Figure 15.22).

At each temperature, a liquid has a definite equilibrium vapor pressure. The equilibrium vapor pressures of water at various temperatures are given in Table 15.7. The equilibrium vapor pressure of water plotted as a function of the temperature is called a **vapor pressure curve**. Figure 15.23 shows the equilibrium vapor pressure curve for water and for ethanol, $CH_3CH_2OH(l)$. In Chapter 23 we discuss a mathematical expression that describes the equilibrium vapor pressure as a function of temperature.

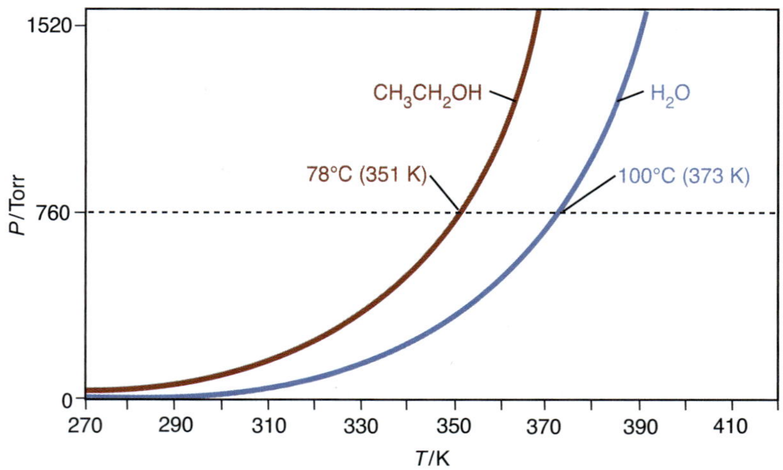

Figure 15.23 The equilibrium vapor pressure curves for water, $H_2O(l)$, and ethanol, $CH_3CH_2OH(l)$, over the temperature range 270 to 420 K (–3° to 147°C). Observe the rapid increase in vapor pressure with increasing temperature. The equilibrium vapor pressure curve for ethanol lies above that for water because ethanol has a higher equilibrium vapor pressure than water at the same temperature.

TABLE 15.7 Equilibrium vapor pressure of water as a function of temperature

t/°C	P/bar	P/atm	P/Torr	t/°C	P/bar	P/atm	P/Torr
0	0.00611	0.00603	4.59	60	0.199	0.197	149.5
5	0.00873	0.00861	6.55	65	0.250	0.247	187.7
10	0.0123	0.0121	9.21	70	0.312	0.308	233.8
15	0.0171	0.0168	12.8	75	0.386	0.381	289.2
20	0.0234	0.0231	17.5	80	0.474	0.467	355.3
25	0.0317	0.0313	23.8	85	0.578	0.571	433.6
30	0.0425	0.0419	31.8	90	0.701	0.692	525.9
35	0.0563	0.0555	42.2	95	0.845	0.834	634.0
40	0.0738	0.0728	55.4	100	1.013	1.000	760.0
45	0.0959	0.0946	71.9	105	1.208	1.192	906.0
50	0.123	0.122	92.6	110	1.432	1.414	1074.4
55	0.158	0.155	118.0	120	1.985	1.959	1488.7

EXAMPLE 15-5: A 0.0896-gram sample of water is placed in a 250.0-mL sealed container. Is there any liquid present when the temperature is held at 70.0°C?

Solution: First, we calculate the pressure assuming that all the water is gaseous. If this pressure is greater than the equilibrium vapor pressure of $H_2O(l)$ at 70.0°C, then $H_2O(g)$ will condense until the pressure is equal to the equilibrium vapor pressure. If the pressure is less than the equilibrium vapor pressure, then no liquid will be present. The pressure is given by

$$P = \frac{nRT}{V} = \frac{(0.0896\text{ g})\left(\dfrac{1\text{ mol}}{18.02\text{ g}}\right)(0.08206\text{ L}\cdot\text{atm}\cdot\text{mol}^{-1}\cdot\text{K}^{-1})(343.2\text{ K})}{0.250\text{ L}}$$

$$= 0.560\text{ atm}$$

According to Table 15.7, the equilibrium vapor pressure of water is 0.308 atm at 70.0°C, so some of the water vapor will condense until its vapor pressure is 0.308 atm.

PRACTICE PROBLEM 15-5: What mass of water will condense in Example 15-5?

Answer: 0.0403 g

The vapor pressure of a liquid depends on the attractive forces between its constituent molecules. Substances with relatively strong intermolecular attractions will have relatively low vapor pressures. At a given temperature, relatively few of the molecules of such a substance will have sufficient kinetic energy to overcome the attractive forces of the other molecules and enter the vapor phase.

The normal boiling point of a liquid is the temperature at which its vapor pressure is equal to exactly one atmosphere. The normal boiling point of water, for example, is 100.0°C. We saw in Chapter 13, however, that the pressure unit of the bar is replacing the atmosphere, so some people prefer to define the normal boiling point of a liquid as the temperature at which its vapor pressure is equal to 1.01325 bar (which is equal to one atmosphere).

The boiling point of a liquid is the temperature at which its vapor pressure equals the external pressure. The **normal boiling point**, the boiling point at exactly one atm, of water is 100.0°C. If the external pressure is less than the one atm, then the temperature at which the vapor pressure of liquid water equals this pressure is less than 100.0°C. For example, the elevation at Vail, Colorado, is about 2500 meters (8200 feet). The atmospheric pressure at this elevation is about 0.75 atm, so water boils at 92°C. In a pressure cooker, on the other hand, when the pressure is 2.0 atm, water boils at 120°C. Because atmospheric pressure decreases with increasing elevation and because the rate at which food cooks depends on the temperature, it requires a significantly longer time to cook food by boiling in an open container at high elevations than at sea level. An egg must be boiled for almost 5 minutes at Vail, Colorado, in order to be cooked to the same extent as one boiled for 3 minutes at sea level.

15-7. Relative Humidity Is Based on the Vapor Pressure of Water

The vapor pressure of water in the atmosphere is expressed in terms of relative humidity. **Relative humidity** is the ratio of the partial pressure of the water vapor in the atmosphere to the equilibrium vapor pressure of water at the same temperature expressed as a percentage, as given by

$$\text{relative humidity} = \left(\frac{P_{\mathrm{H_2O}}}{P^{\circ}_{\mathrm{H_2O}}}\right) \times 100 \tag{15.4}$$

where $P_{\mathrm{H_2O}}$ is the partial pressure of the water vapor in the air and $P^{\circ}_{\mathrm{H_2O}}$ is the equilibrium vapor pressure of water at the same temperature. At 20.0°C, the equilibrium vapor pressure of water is 17.5 Torr. If the partial pressure of the water vapor in the air is 11.2 Torr, then the relative humidity is

$$\text{relative humidity} = \left(\frac{11.2\ \text{Torr}}{17.5\ \text{Torr}}\right) \times 100 = 64.0\%$$

If the temperature of the air is lowered to 13.0°C, where the equilibrium vapor pressure of water is 11.2 Torr, then the relative humidity is

$$\text{relative humidity} = \left(\frac{11.2\ \text{Torr}}{11.2\ \text{Torr}}\right) \times 100 = 100\%$$

At 13.0°C, air that contains water vapor at a partial pressure of 11.2 Torr is saturated with water vapor. At this temperature, the water vapor begins to condense as dew or fog, which consists of small droplets of water. The air temperature at which the relative humidity reaches 100% is called the **dew point.** Most people begin to feel uncomfortable when the dew point rises above 20°C, and air with a dew point above 24°C is generally regarded as extremely humid or muggy.

EXAMPLE 15-6: Calculate the relative humidity and the dew point for a day when the partial pressure of water vapor in the air is 22.2 Torr and the temperature of the air is 30.0°C. The equilibrium vapor pressure of water at 30.0°C is 31.8 Torr (Table 15.7).

Solution: The relative humidity, given by Equation 15.4, is

$$\text{relative humidity} = \left(\frac{P_{H_2O}}{P^{\circ}_{H_2O}}\right) \times 100 = \left(\frac{22.2\ \text{Torr}}{31.8\ \text{Torr}}\right) \times 100 = 69.8\%$$

The dew point is the temperature at which the equilibrium vapor pressure of water is equal to 22.2 Torr. According to Table 15.7, this is about 24°C. Such a day would be considered very uncomfortable.

PRACTICE PROBLEM 15-6: Calculate the dew point for a day when the relative humidity is 78% at 20.0°C.

Answer: approximately 17°C

15-8. A Phase Diagram Displays the Regions of All the Phases of a Pure Substance Simultaneously

The vapor pressure curve of a pure substance can be combined with two other useful quantities, the sublimation pressure curve and the melting point curve, into a single diagram called a **phase diagram.** The phase diagram of water is shown in Figure 15.24.

Along the vapor pressure curve (the red curve in Figure 15.24), liquid and vapor exist together at equilibrium. To the left of this curve, at lower temperatures, the water exists as a liquid. To the right of this curve, at higher temperatures, the water exists as a vapor. The equilibrium vapor pressure of a liquid increases with temperature up to the **critical point** (Figure 15.24), where the vapor pressure curve terminates abruptly. Above the **critical temperature,** the gas and the liquid phases become indistinguishable. A gas above its critical temperature cannot be liquefied no matter how high a pressure is applied. The critical point for water occurs at 218 atm (221 bar) and 647 K. Water vapor above 647 K cannot be liquefied by the application of pressure.

Along the **sublimation pressure curve** (gray curve in Figure 15.24), solid and vapor exist together at equilibrium. To the left of this curve, at lower temperatures, water exists as a solid (ice). To the right of this curve, at higher temperatures, water exists as a vapor.

Along the **melting point curve** (blue curve in Figure 15.24), solid and liquid exist together in equilibrium. To the left of this curve, water exists as a solid, and to the right of this curve, water exists as a liquid. The melting point at a pressure of exactly one atm is called the **normal melting point** or the **normal freezing point**. However, melting points are only weakly dependent on pressure, so a melting point curve is an almost vertical line (Figure 15.24). For almost all

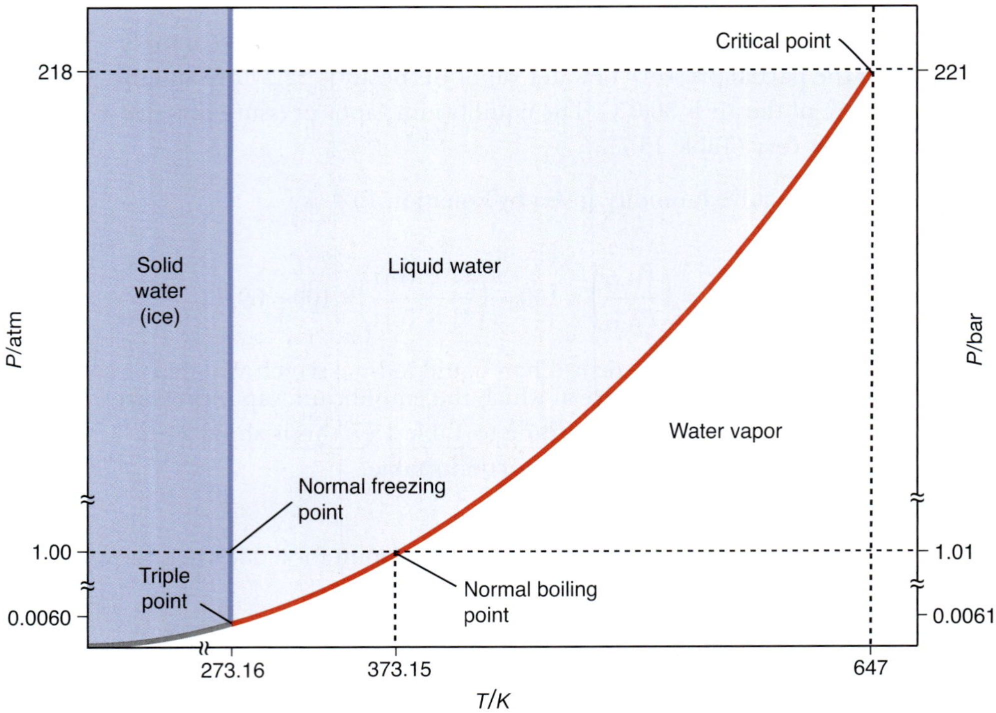

Figure 15.24 The phase diagram of water (not to scale; see the breaks in both the horizontal and vertical axes), displays simultaneously the sublimation pressure curve (gray), the vapor pressure curve (red), and the melting point curve (blue). The triple point, the critical point, the normal boiling point, and the normal freezing point are indicated in the figure. The phase equilibrium lines are the boundaries between the regions of stability of the solid, liquid, and vapor phases, which are labeled solid water (ice), liquid water, and water vapor.

substances, the melting point increases with increasing pressure at a rate of 0.01 to 0.03 $K\cdot atm^{-1}$. Water is anomalous because its melting point decreases with increasing pressure. The melting point of ice decreases by about 0.01 $K\cdot atm^{-1}$ of applied pressure. Consequently, unlike most other solids, ice can be melted by the application of pressure. We discuss why this is so in Section 19-7.

Notice that the three curves in Figure 15.24 separate regions in which water exists as a solid, a liquid, or a vapor. Let's use Figure 15.24 to follow the behavior of water as it is heated from −50.0°C to 200.0°C at a constant pressure of one atm. At −50.0°C (223.2 K) and one atm, water exists as ice. As we heat the ice at one atm, we move horizontally from left to right along the dashed line in Figure 15.24. At 0°C (273.2 K), we cross the melting point curve and pass from the solid region into the liquid region. At 100.0°C (373.2 K), we cross the equilibrium vapor pressure curve and pass from the liquid region into the vapor region.

The three curves in Figure 15.24 intersect at a point called the **triple point.** At the triple point, and only at the triple point, all three phases—solid, liquid, and gas—coexist in equilibrium. The triple point for water occurs at 4.58 Torr (0.0060 atm) and 273.16 K. Notice that if we heat ice at a constant pressure less than 4.58 Torr, then the ice sublimes rather than melts. The sublimation of water is used in the process of "freeze-drying."

EXAMPLE 15-7: Use the phase diagram of water given in Figure 15.24 to predict the result of increasing the pressure of water vapor initially at one atm and 500 K, keeping the temperature constant.

Solution: At one atm and 500 K, water exists as a vapor. As the pressure is increased, we cross the liquid-vapor curve below the critical point at a pressure of about 150 atm, and the vapor condenses to a liquid.

PRACTICE PROBLEM 15-7: Given the following data for iodine and the fact that solid iodine is more dense than liquid iodine, sketch the phase diagram for iodine.

triple point	113°C	0.12 atm
critical point	512°C	118 atm
normal melting point	114°C	1 atm
normal boiling point	184°C	1 atm

Solution: Your sketch should be similar in shape to that in Figure 15.25 (page 544) for carbon dioxide, but with the values for the various points taken from the table above.

EXAMPLE 15-8: The vapor pressures (in Torr) of solid and liquid argon are given by

$$\ln P_s = 17.283 - \frac{919.0\text{ K}}{T} \qquad \text{(solid)}$$

$$\ln P_l = 15.236 - \frac{700.5\text{ K}}{T} \qquad \text{(liquid)}$$

where T is the Kelvin temperature. Calculate the temperature and pressure of the triple point of argon.

Solution: The vapor pressures of the solid and the liquid are equal at the triple point, so we have

$$17.283 - \frac{919.0\text{ K}}{T} = 15.236 - \frac{700.5\text{ K}}{T}$$

Solving for T gives $T = 107$ K. Substituting this value into either of the vapor pressure equations gives

$$\ln P = 17.283 - \frac{919.0\text{ K}}{107\text{ K}} = 8.69$$

or

$$P = 6.0 \times 10^3 \text{ Torr} = 8.0 \text{ bar}$$

(When we take the inverse logarithm of a number, we get as many significant figures as the logarithm of the number had decimal places; see Appendix A.2.).

PRACTICE PROBLEM 15-8: The vapor pressures (in Torr) of solid and liquid bromine are given by

$$\ln P_s = 22.383 - \frac{4699 \text{ K}}{T} \qquad \text{(solid)}$$

$$\ln P_l = 15.400 - \frac{2579 \text{ K}}{T} \qquad \text{(liquid)}$$

where T is the Kelvin temperature. Calculate the temperature and pressure at the triple point of bromine.

Answer: $T = 303.6$ K; $P = 998$ Torr

The phase diagram of carbon dioxide is shown in Figure 15.25. Although it looks similar to that of water, there are several important differences. The melting point curve of carbon dioxide goes up and to the right, a direction indicating that the melting point of carbon dioxide increases with increasing pressure. Recall that the melting point curve of water points up and slightly to

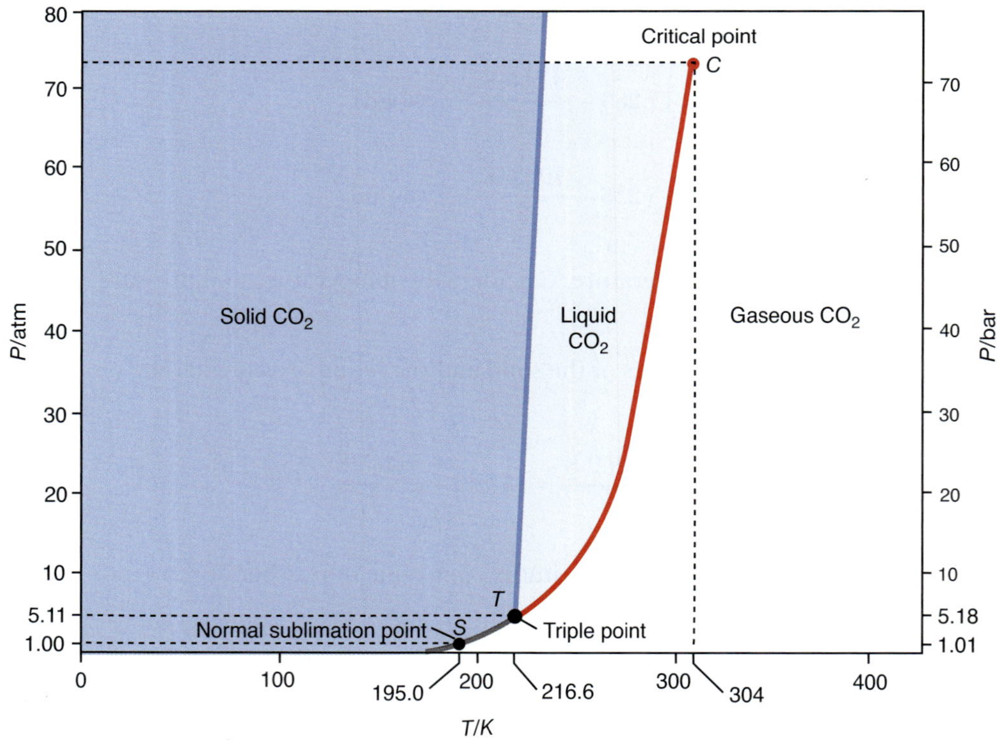

Figure 15.25 The phase diagram of carbon dioxide. The point C is the critical point, and the point T is the triple point. We see that the triple point lies above one atm and thus $CO_2(s)$ at one atm does not melt—it sublimes. The point S is the normal sublimation point of carbon dioxide.

the left, a direction indicating that the melting point of $H_2O(s)$ decreases with increasing pressure.

Another difference between Figure 15.24 and Figure 15.25 is the position of the triple points. The triple point for CO_2 occurs at 5.11 atm (5.18 bar) and 216.6 K. Because the pressure at the triple point is greater than one atm, $CO_2(s)$ does not melt when it is heated at one atm as water does. Instead, $CO_2(s)$ sublimes. The **normal sublimation point** of CO_2 is 195 K (−78°C), which is the temperature of solid $CO_2(s)$ at one atm. Liquid $CO_2(l)$ can be obtained by compressing $CO_2(g)$ at a temperature below its critical point of 304 K (31°C). A pressure of about 60 atm is required to liquefy $CO_2(g)$ at 25°C. A carbon dioxide–filled fire extinguisher at 25°C contains $CO_2(l)$ at a pressure of about 60 atm.

15-9. X-Ray Diffraction Patterns Yield Information About the Structures of Crystalline Solids

In the remaining sections of this chapter, we shall discuss solids. Solid-state chemistry has profoundly affected the world we live in through the creation and use of semiconductors, transistors, computer chips, and many other solid-state devices. Solids can be classified into crystalline and noncrystalline solids. We discuss crystalline solids first and then briefly discuss amorphous solids and liquid crystals.

A distinguishing characteristic of crystalline solids is the ordered nature of the molecules or ions in the solid state, an arrangement we refer to as a **crystal lattice.** We can actually obtain a "picture" of a crystal lattice by passing X-rays through the crystal. The presence of a definite ordered array of atoms in the crystal produces a characteristic **X-ray diffraction pattern** that can be recorded as an array of spots (Figure 15.26).

Figure 15.26 The X-ray diffraction pattern produced by a crystal of sodium chloride. The symmetry and spacing of the dots carry detailed information regarding the arrangement of atoms in the crystals.

To get a feel for the origin of X-ray diffraction patterns from crystals, let's examine the **optical diffraction patterns** formed by light passing through tiny holes in an opaque sheet (Figure 15.27). The size and arrangement of the holes yield a particular diffraction pattern that can be used to determine the arrangement of the holes that produced it. Just as the optical diffraction patterns carry information regarding the relative positions and the spacings of the holes, X-ray diffraction patterns provide information about the arrangement of the atoms in crystals (sidebox).

The smallest subunit of a crystal lattice that can be used to generate the entire lattice is called a **unit cell.** A crystal lattice is thus a repeating pattern of unit cells. Figure 15.28 illustrates in two dimensions how a unit cell can generate a crystal lattice. There are a variety of three-dimensional unit cells that differ in their shapes, but, for simplicity, we shall discuss only cubic unit cells because most metallic elements have one of three kinds of cubic unit cells (Figure 15.29).

The **simple cubic unit cell** has atoms only at the vertices of the cube, and each one of these atoms is shared by a total of eight unit cells. Because each of the eight atoms of a simple cubic unit cell is shared by a total of eight unit cells, we can assign one atom to a simple cubic unit cell. Only one metal, polonium, occurs as a simple cubic lattice. A **body-centered cubic unit cell** is similar to a simple cubic unit cell, except that there is an atom at the center of the unit cell. This atom is shared by no other unit cells, so we can assign a total of two atoms to a body-centered unit cell. Some metals that occur as a body-centered cubic lattice are barium, cesium, potassium, lithium, molybdenum, tantalum,

X-RAY DIFFRACTION X-ray diffraction is a powerful analytical method for determining molecular structure. X-rays are a penetrating form of radiation, providing information about the bulk structure of solids, not just the surface. For example, X-rays can be used to check the integrity of steel beams inside of walls following an earthquake or to look for stress fractures in airplane wings. X-rays are electromagnetic radiation with wavelengths from 10^{-8} to 10^{-11} meters. Because the wavelength of X-rays is comparable to the size of atoms (about 10^{-10} meters), they are ideally suited for probing the structural arrangement of atoms and molecules in various materials. The presence of a definite ordered array of atoms in a crystal produces a characteristic X-ray diffraction pattern. The electron density contour map of benzoic acid is shown below.

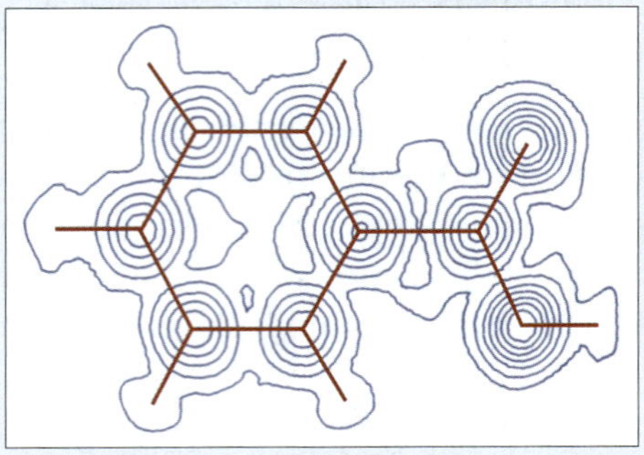

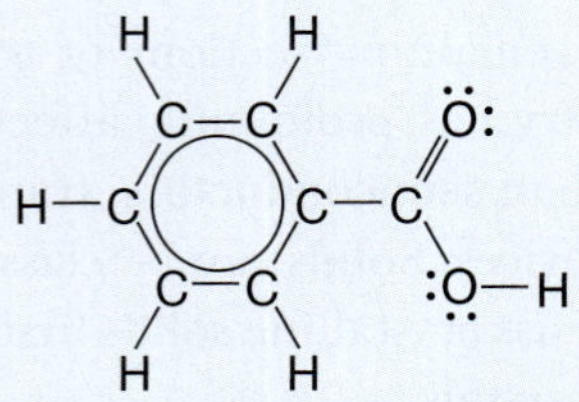

Lewis formula of benzoic acid

The general outline of the planar molecule is clearly discernible. The characteristic shape of a benzene ring is also evident. Such information is typical of that obtainable with X-ray diffraction techniques. X-ray diffraction patterns provide information about the spatial arrangement of atoms in crystals and allow us to determine the structure of various compounds, metals, and even large biomolecules. The first person to apply X-ray diffraction to proteins was Nobel laureate Dorothy Crowfoot Hodgkin (Frontispiece). Since her pioneering work in the 1930s, the structures of hundreds of proteins have been determined by X-ray diffraction methods. The image below shows the results obtained for a crystal of insulin. The famous double-helical structure of DNA was determined by Watson and Crick using X-ray diffraction data obtained by Rosalind Franklin. X-ray diffraction is one of the most powerful methods available for the determination of molecular structure and is used extensively by chemists and biologists.

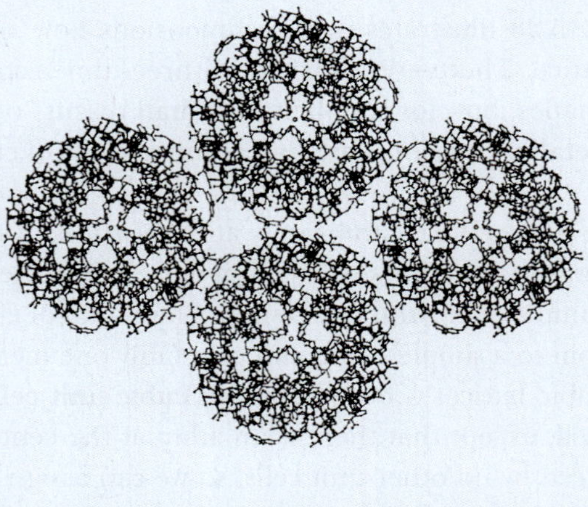

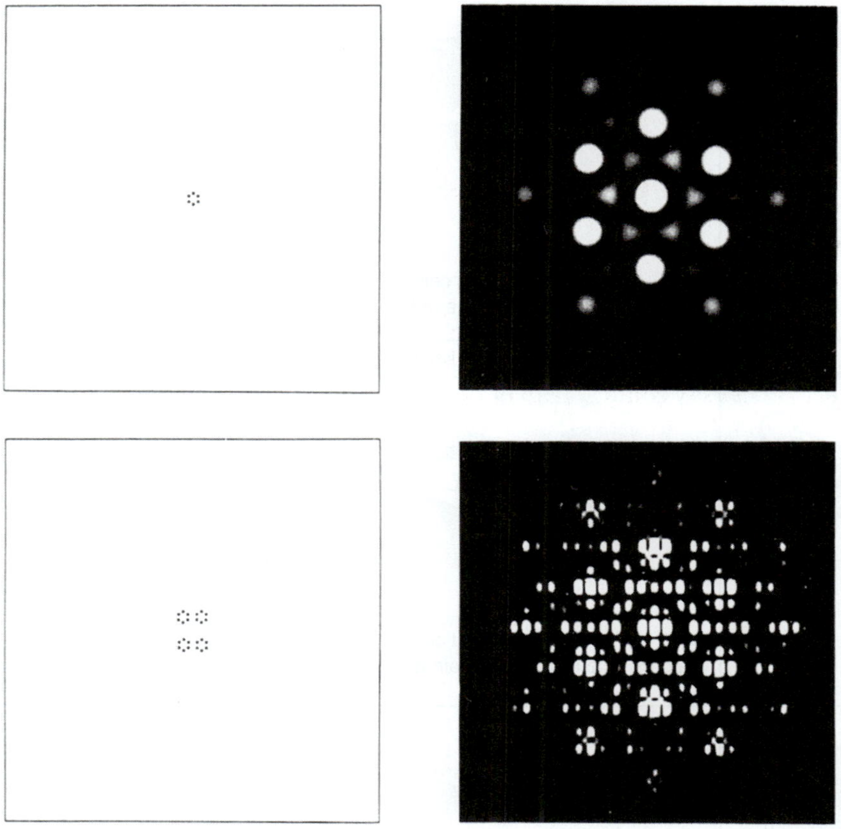

Figure 15.27 These optical diffraction patterns were produced by a light beam passing through holes in opaque sheets. Each arrangement of holes (*left*) yields a characteristic diffraction pattern (*right*). It is possible to work backward to determine the arrangement of holes that gives a particular diffraction pattern.

uranium, and vanadium. A **face-centered cubic unit cell** has an atom at each vertex, and also one at each face of the unit cube. An atom at one of the faces is shared by two unit cells. Because there are six faces on a cube and each of these atoms is shared by two unit cells, we assign a total of three atoms to the unit cell from the faces. In addition, as in the case of a simple cubic unit cell and a body-centered unit cell, we assign one atom to this unit cell from those at the eight vertices, giving a total of four atoms. Face-centered cubic lattices are found in silver, aluminum, gold, copper, nickel, lead, strontium, platinum, and the noble gases.

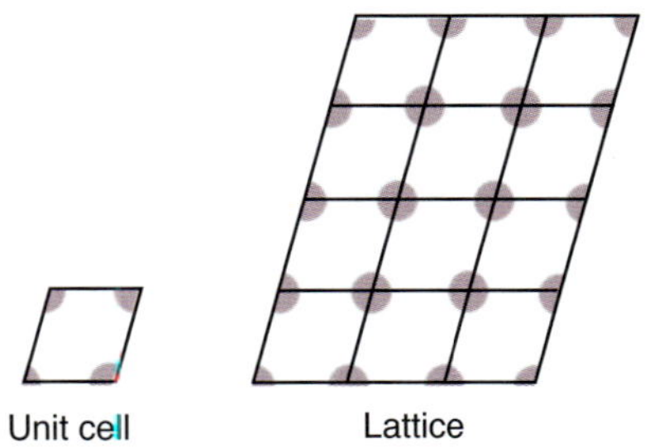

Figure 15.28 A two-dimensional illustration of the generation of a crystal lattice by a unit cell. Only a portion of each dot lies within the unit cell because dots are shared with adjoining cells.

EXAMPLE 15-9: Copper exists as a face-centered cubic lattice. How many copper atoms are there in a unit cell?

Solution: Reference to Figure 15.29 shows that each of the eight copper atoms at the corners of the unit cell are shared by eight unit cells, so we assign one copper atom $(8 \times \frac{1}{8} = 1)$ to each unit cell. Each of the six atoms at the faces is shared by two unit cells, so we assign three more copper atoms $(6 \times \frac{1}{2} = 3)$ to each unit cell, thereby giving a total of four copper atoms in a unit cell. Figure 15.29 also illustrates the counting process for the simple cubic and body-centered cubic unit cells.

PRACTICE PROBLEM 15-9: Cesium exists as a body-centered cubic lattice. How many cesium atoms are there in a unit cell?

Answer: 2

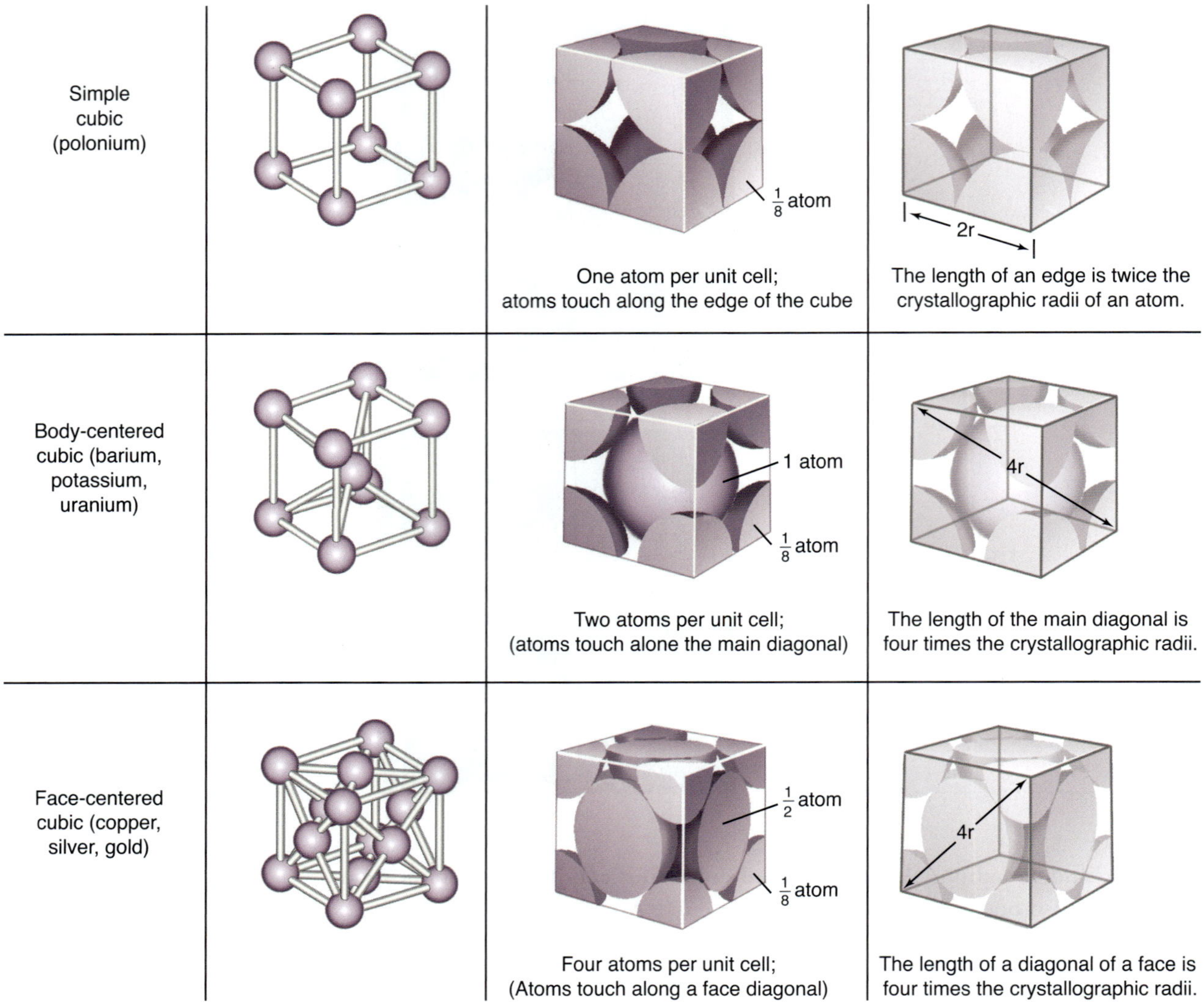

Figure 15.29 The three cubic unit cells: simple cubic, body-centered cubic, and face-centered cubic. (*left*) An open perspective of the unit cells. Note that the body-centered cubic unit cell has an atom at the center of the unit cell and that a face-centered unit cell has an atom at the center of each face. (*center*) The sharing of atoms by adjacent unit cells. (*right*) The crystallographic radii of atoms may be determined by measuring the length of the edge or a diagonal of a unit cell.

Figure 15.29 (center) implies that the atoms in the unit cell touch one another. We noted in Chapter 5 that atoms do not have well-defined radii, but we can use the assumption that the atoms do touch to calculate effective radii, or **crystallographic radii,** from the dimensions of atomic crystals as illustrated in Figure 15.29 (right). Referring to Figure 15.29 (right), we see that the length of a simple cubic unit cell is equal to twice the crystallographic radii of the constituent atoms. The length of the main diagonal (the line that goes from one vertex to another and passes through the center of the cube) of a body-centered unit cell is four times the crystallographic radii of the atoms; and the length of the diagonal of a face in a face-centered cubic lattice is also four times the crystallographic radii.

If we know the volume of the unit cell, we can calculate the length of an edge or a diagonal, and hence the crystallographic radius of the constituent atoms. We can determine the volume of a unit cell of an atomic crystalline solid if we know its density and its crystal structure. Let V_{mol} be the molar volume (volume per mole) of the substance and let $V_{\text{unit cell}}$ be the volume per unit cell. If n is the number of atoms in a unit cell (1 for a simple cubic lattice, 2 for a body-centered cubic lattice, and 4 for a face-centered cubic lattice), then $V_{\text{unit cell}}/n$ is the cubic volume element *around each atom* in the crystal. If we multiply this quantity by Avogadro's number, N_A, then we have the molar volume:

$$V_{\text{mol}} = \left(\frac{V_{\text{unit cell}}}{n}\right) N_A$$

Next we solve this equation for $V_{\text{unit cell}}$:

$$V_{\text{unit cell}} = \left(\frac{n}{N_A}\right) V_{\text{mol}} \tag{15.5}$$

Finally, we can calculate the molar volume in terms of the density by using the formula for the density,

$$d = \frac{m}{V} \tag{15.6}$$

If we replace m by the molar mass, M, then V in Equation 15.6 is the molar volume, V_{mol}, and so we have

$$d = \frac{M}{V_{\text{mol}}}$$

or

$$V_{\text{mol}} = \frac{M}{d} \tag{15.7}$$

We get the equation for $V_{\text{unit cell}}$ by substituting Equation 15.7 into Equation 15.5:

$$V_{\text{unit cell}} = \frac{nM}{dN_A} \tag{15.8}$$

The following Example illustrates the use of Equation 15.8.

EXAMPLE 15-10: Copper, which crystallizes as a face-centered cubic lattice, has a density of 8.96 g·cm^{-3} at 20°C. Calculate the crystallographic radius of a copper atom.

Solution: From Equation 15.8 we can calculate the volume of the unit cell. A face-centered cubic lattice has four atoms per unit cell, so $n = 4$ in Equation 15.8. Thus, we write

$$V_{\text{unit cell}} = \frac{(4\ \text{atom}\cdot\text{unit cell}^{-1})\,(63.55\ \text{g}\cdot\text{mol}^{-1})}{(8.96\ \text{g}\cdot\text{cm}^{-3})\,(6.022\times 10^{23}\ \text{atom}\cdot\text{mol}^{-1})}$$

$$= 4.71 \times 10^{-23}\ \text{cm}^3\cdot\text{unit cell}^{-1}$$

Because the unit cell is cubic in shape, the length of an edge, l, is given by the cube root of $V_{\text{unit cell}}$:

$$l = (V_{\text{unit cell}})^{1/3} = (4.71 \times 10^{-23}\ \text{cm}^3)^{1/3} = 3.61 \times 10^{-8}\ \text{cm} = 361\ \text{pm}$$

Figure 15.29, right, shows that the effective radius of an atom in a face-centered cubic lattice is given by one-fourth of the length of the diagonal of a face. The length of a diagonal is given by (margin)

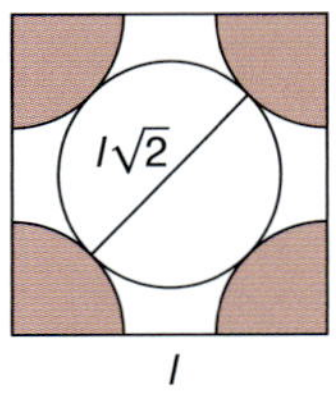

$$\text{diagonal} = l\sqrt{2} = (361\ \text{pm})\sqrt{2} = 511\ \text{pm}$$

so the crystallographic radius of a copper atom is

$$\text{radius} = \frac{511\ \text{pm}}{4} = 128\ \text{pm}$$

PRACTICE PROBLEM 15-10: Europium, which crystallizes in a body-centered cubic lattice, has a density of 5.244 g·cm^{-3} at 25°C. Calculate the crystallographic radius of a europium atom, given that $4r = l\sqrt{3}$ for a body-centered cubic unit cell (see Problem 15-55).

Answer: 198.4 pm

If the density of an atomic solid and the dimensions of its unit cell are known from X-ray diffraction, then we can use Equation 15.8 to determine Avogadro's number. The following Example illustrates this type of calculation.

EXAMPLE 15-11: Potassium crystallizes in a body-centered cubic lattice, and the edge length of a unit cell is determined from X-ray analysis to be 532.1 pm. Given that the density of potassium is 0.862 g·cm^{-3}, calculate the value of Avogadro's number.

Solution: We solve Equation 15.8 for N_A, giving

$$N_A = \frac{nM}{dV_{\text{unit cell}}}$$

Because the potassium unit cell is cubic in shape, its volume is

$$V_{\text{unit cell}} = l^3 = (532.1 \times 10^{-12}\ \text{m})^3 = (532.1 \times 10^{-10}\ \text{cm})^3 = 1.507 \times 10^{-22}\ \text{cm}^3$$

The unit cell is body-centered cubic, so $n = 2$. Thus, N_A is given by

$$N_A = \frac{(2\ \text{atom}\cdot\text{unit cell}^{-1})(39.0983\ \text{g}\cdot\text{mol}^{-1})}{(0.862\ \text{g}\cdot\text{cm}^{-3})(1.507\times 10^{-22}\ \text{cm}^3)} = 6.02\times 10^{23}\ \text{atom}\cdot\text{mol}^{-1}$$

PRACTICE PROBLEM 15-11: At 25°C we find from X-ray diffraction that cerium crystallizes in a face-centered cubic lattice, and the edge length of the unit cell is 516.10 pm. Given that the density of cerium is 6.770 $\text{g}\cdot\text{cm}^{-3}$, calculate Avogadro's number.

Answer: $6.022 \times 10^{23}\ \text{atom}\cdot\text{mol}^{-1}$

In 2002, the committee on data for science and technology recommended the value $6.022\,1415 \pm 10 \times 10^{23}$ for Avogadro's number. The ±10 indicates the uncertainty in the last two digits. One of the most precise measurements of Avogadro's number was made by scientists at the National Institute of Standards and Technology using X-ray measurements on ultra-pure silicon.

15-10. Crystals Can Be Classified According to the Forces Between the Constituent Particles

Crystal structures are determined by the size of the atoms, ions, or molecules making up the lattice and by the nature of the forces that act between these particles. The examples that we discussed in the previous section consisted of particles of the same size. Such crystals are called **atomic crystals,** a good example being crystals of the noble gases, which crystallize as face-centered cubic crystals. In this section we discuss the properties of some other kinds of crystals.

As discussed in Chapter 6, **ionic crystals** are held together by the electrostatic attraction between ions of opposite charge. The crystalline structure of ionic crystals depends on how the cations and anions can be packed together to form a lattice. Consequently, the difference in the sizes of the cation and the anion plays a key role. For example, the unit cells of the salts sodium chloride and cesium chloride are shown in Figure 15.30. Notice that the chloride ions in sodium chloride form a face-centered cubic lattice, while the chloride ions in cesium chloride form a simple cubic lattice. Also notice that each sodium ion in $NaCl(s)$ is surrounded by an octahedral arrangement of six larger chloride ions, while each cesium ion in $CsCl(s)$ is surrounded by a cubic arrangement of four larger chloride ions. The different packing arrangements for $NaCl(s)$ and $CsCl(s)$ are a direct consequence of the fact that cesium ions are larger than sodium ions.

Each ion in an ionic crystal is surrounded by ions of opposite charge. The total electrostatic interaction energy of the lattice accounts for much of the lattice energy (Section 6-7). Because electrostatic interactions are relatively strong, ionic solids usually have high melting points and low vapor pressures.

Crystals composed of neutral molecules are called **molecular crystals.** We discussed the forces that hold the molecules together in molecular crystals in Section 15-4. These forces include both dipole-dipole and London forces. Because these forces tend to be weaker than those between ions, molecular crystals generally have lower melting points and higher vapor pressures than ionic crystals.

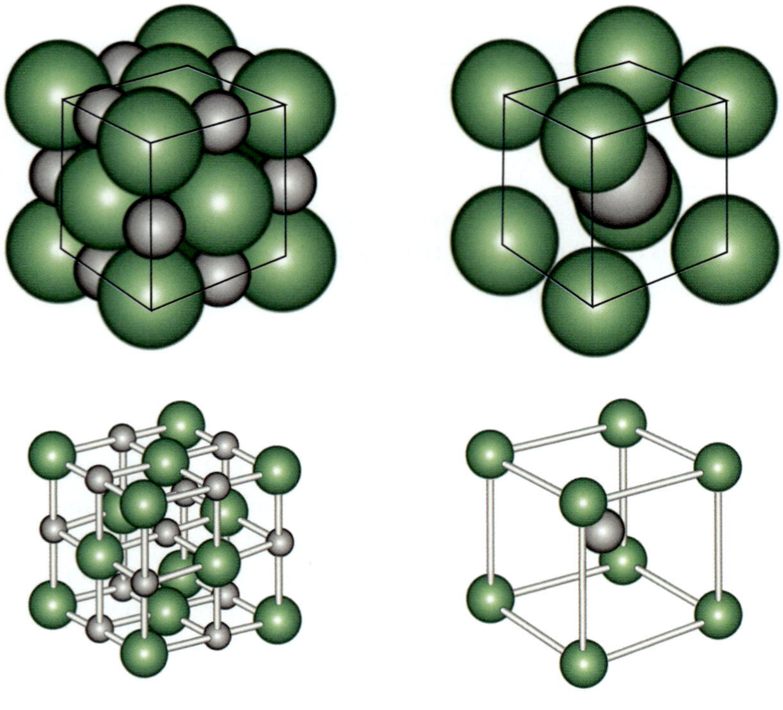

Figure 15.30 Space-filling and ball-and-stick representations of the unit cells of (a) NaCl(*s*) and (b) CsCl(*s*). The different crystalline structures in the two cases are a direct consequence of the relative sizes of the cations and the anions. Recall that cations are positively charged ions and that anions are negatively charged ions.

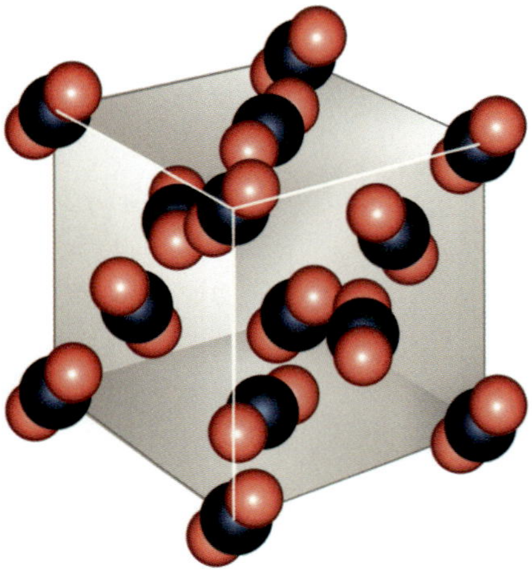

Figure 15.31 The unit cell of crystalline $CO_2(s)$. The molecules have been reduced in size for clarity.

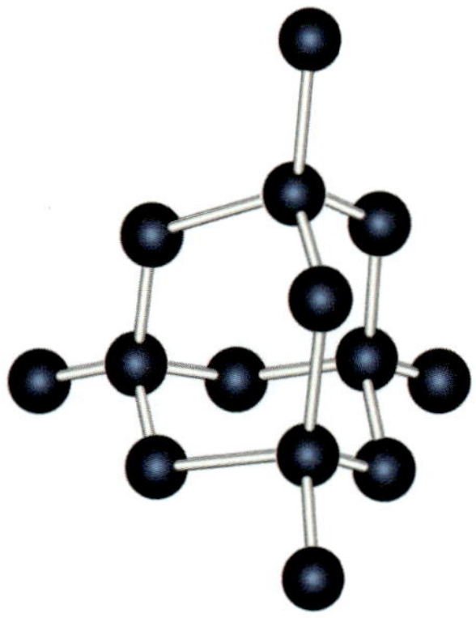

Figure 15.32 The crystalline structure of diamond. Each carbon atom is covalently bonded to four other carbon atoms, forming a tetrahedral network. A diamond crystal is essentially one gigantic molecule.

Molecular crystals come in a great variety of types. For example, methane crystallizes in a face-centered cubic structure; carbon dioxide crystals have the unit cell shown in Figure 15.31. Large biological molecules such as proteins and DNA also form molecular crystals (sidebox, page 546).

A few substances form **covalent network crystals,** in which the constituent particles are held together by covalent bonds. Diamond and graphite, two forms of carbon, are good examples of such substances (see also Interchapter M). Diamond has an extended, covalently bonded tetrahedral structure. Each carbon atom lies at the center of a tetrahedron formed by four other carbon atoms (Figure 15.32). The carbon-carbon bond distance is 154 pm, which is the same as the carbon-carbon bond distance in ethane. The diamond crystal is, in effect, a gigantic molecule. The hardness of diamond is due to the fact that each carbon atom throughout the crystal is covalently bonded to four others; thus, many strong covalent bonds must be broken in order to cleave a diamond.

Graphite has the unusual layered structure shown in Figure 15.33. The carbon-carbon bond distance within a layer is 139 pm, which is close to the carbon-carbon bond distance in benzene. The distance between layers is about 340 pm. The bonding within a layer is covalent, but the interaction between layers is weak. Therefore, the layers easily slip past one another, thereby producing the molecular basis for the lubricating action of graphite. The "lead" in lead pencils is actually graphite. Layers of the graphite slide from the pencil onto the paper. Graphite is also used as a solid lubricant in many high-temperature applications where oils would be unsuitable.

Not all solids are crystalline. When liquids of high viscosity are cooled rapidly, a rigid structure forms before the molecules have time to orient properly to form a crystalline lattice. The molecules of these materials lack the high degree of spatial order of crystalline substances and have more of a "frozen-in" liquid structure. Such materials are called **amorphous solids.** A wide variety of materi-

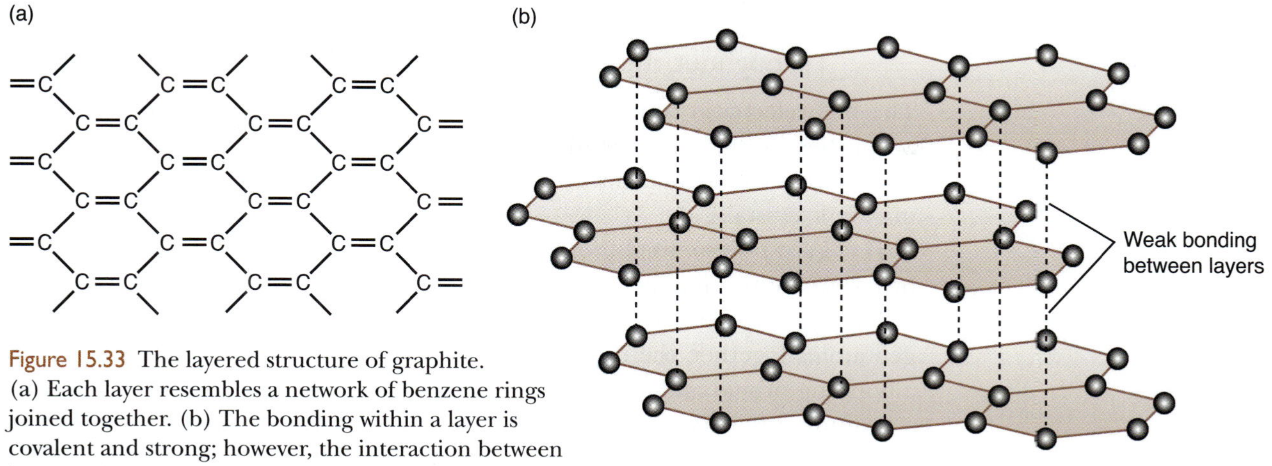

Figure 15.33 The layered structure of graphite. (a) Each layer resembles a network of benzene rings joined together. (b) The bonding within a layer is covalent and strong; however, the interaction between layers is due only to London forces, so is relatively weak. Consequently, the layers, which are separated by 340 pm, easily slip past each other, giving graphite its slippery feel and its use as a lubricant.

als form amorphous solids under the proper conditions, but the most familiar examples are materials with large molecules, such as plastics and rubber (see Interchapter S), as well as glasses. Amorphous solids are distinguished by the lack of a sharp melting point. Crystalline solids melt at a precise temperature, but amorphous solids gradually soften over a wide temperature range.

Amorphous solids are an active area of research in chemistry, physics, and engineering. Research in fiber optics has enabled millions of bits of information to be transmitted through glass fibers with diameters less than that of a hair (Figure 15.34). The laying of fiber optic cable below the ocean is what made the global Internet revolution possible. Another recent development is amorphous silicon, which has significantly reduced the cost of converting sunlight to electricity. Table 15.8 summarizes the various types of solids discussed in this chapter.

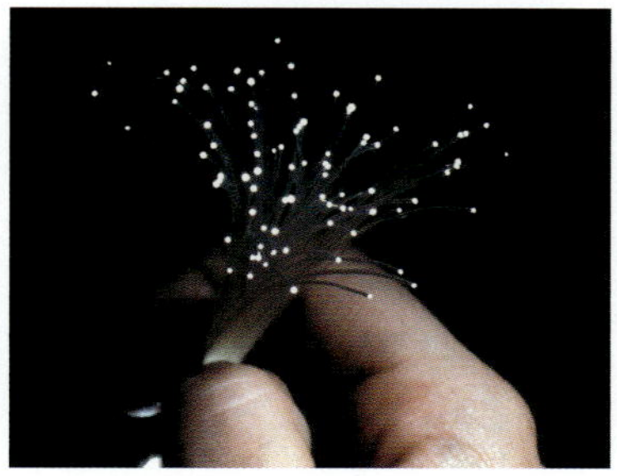

Figure 15.34 Fiber optics. Light is transmitted through thin strands of silica glass that form an optical-waveguide fiber bundle that acts as an optical pathway for the light.

TABLE 15.8 Types of solids

Type of solid	Constituent particles	Type of attractive forces	General properties	Examples
ionic crystal	cations and anions	coulombic charge-charge attractions between ions	hard and brittle high melting point poor electrical conductor	$NaCl(s)$, $KCl(s)$, $CaO(s)$
molecular crystal	molecules	van der Waals	soft low melting point poor electrical conductor	$CO_2(s)$, $I_2(s)$, $C_{12}H_{22}O_{11}(s)$
network crystal	atoms	extended network of covalent bonds	very hard high melting point	$C(s)$, $SiC(s)$, $SiO_2(s)$
amorphous	groups of molecules	various	lack of sharp melting point	glass, plastics
metallic crystal	cations at lattice points and delocalized electrons	delocalized molecular orbitals	good electrical conductor	$Na(s)$, $Ag(s)$, $Cu(s)$

15-11. The Electrons in Metals Are Delocalized Throughout the Crystal

The high electrical conductivity of metals is one of their most characteristic properties. To see why a metal has a high electrical conductivity and an insulator has a low electrical conductivity, we must discuss the electronic energy levels in atomic crystals.

Let's see how we might form a crystal of metallic sodium. Imagine that we bring Avogadro's number of widely separated sodium atoms to their positions in the crystalline lattice. We learned in Chapter 9 that when we bring two hydrogen atoms together, their 1*s* orbitals combine to form two molecular orbitals, a bonding orbital and an antibonding orbital (Figure 15.35a). When we form a sodium crystal from its constituent atoms, we bring Avogadro's number of sodium atoms together. Because a sodium atom has a neonlike electronic core and an outer 3*s* orbital, it's the 3*s* orbitals of the sodium atoms that combine to form molecular orbitals. In this case, however, the resultant molecular orbitals are delocalized over the entire crystalline lattice. Furthermore, because we are bringing Avogadro's number of sodium atoms together, we get Avogadro's number of these delocalized molecular orbitals. The energies of these orbitals are so close to one another that they essentially form a continuum of energy levels, from the bonding orbital of lowest energy to the antibonding orbital of highest energy, as depicted in Figure 15.35b. Thus, we picture a **metallic crystal** as a lattice occupied by ions that result from having the valence electrons of its atoms occupy orbitals that are delocalized over the entire crystal lattice.

Because a sodium atom has a neonlike core with a single unpaired outer 3*s* electron, only one-half of the delocalized orbitals of the crystalline lattice are occupied by the valence electrons of the sodium atoms. Of the remaining unfilled orbitals, many are so close in energy to the filled orbitals (valence orbitals) that relatively little energy is required to excite some electrons into the unfilled orbitals. Because these excited electrons are unpaired and the orbitals they occupy are only partially filled, these electrons are easily displaced by an applied electric field, a property accounting for the high electrical conductivity of the metal.

Now let's consider an insulator, such as diamond (carbon). Its extended covalent network can be described in terms of sp^3 hybrid orbitals on each carbon atom. Again, imagine bringing Avogadro's number of carbon atoms together.

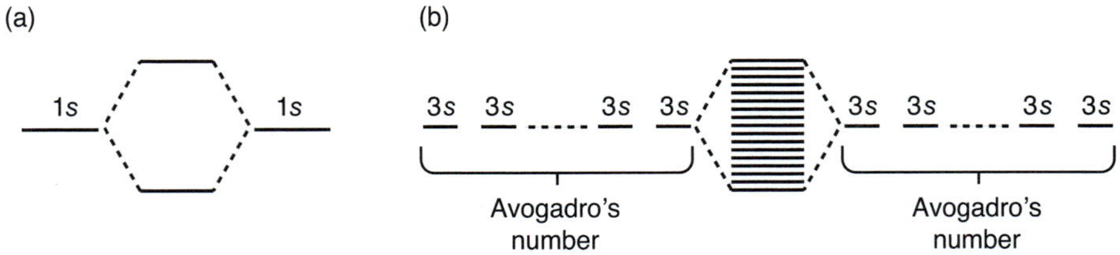

Figure 15.35 (a) When two hydrogen atoms are brought together, their 1*s* orbitals combine to form a bonding and an antibonding molecular orbital. (b) When Avogadro's number of sodium atoms are brought together to form a sodium crystal, the 3*s* orbitals (the valence orbitals) of the sodium atoms combine to form Avogadro's number of molecular orbitals that are delocalized over the entire crystal lattice. The energies of these orbitals are so close together that they form essentially a continuum of energy levels, as illustrated schematically here.

In this case, however, there is a relatively large gap in energy between the band of delocalized bonding orbitals (called the **valence band**) and the band of delocalized antibonding orbitals (called the **conduction band**) (Figure 15.36). The valence electrons of the carbon atoms occupy only the valence band, and the conduction band is empty. Furthermore, the energy of the gap between the two bands (called the **band gap**) is so large that the electrons in the valence band are not easily promoted into the conduction band. Consequently, no electrons are available to conduct an electric current, so diamond is an insulator. In contrast, in a good conductor, such as sodium metal, there is no gap between the energies of the bonding orbitals and the antibonding orbitals; in other words, the band gap is zero in a good conductor.

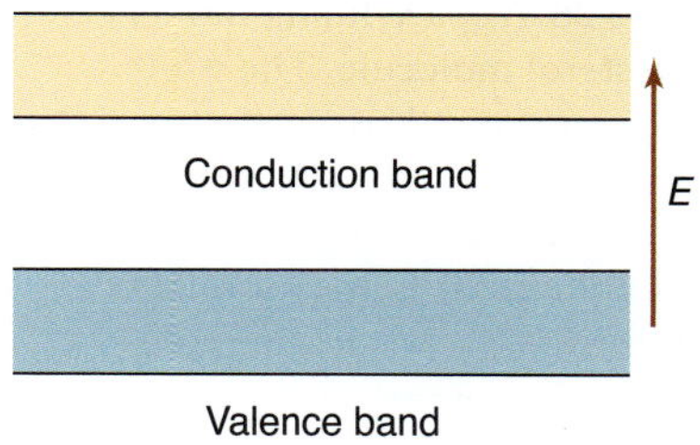

Figure 15.36 When the atoms of a crystal are brought together to form the crystal lattice, the valence orbitals of the atoms combine to form two sets of energy levels called the valence band and the conduction band.

Some substances, such as silicon and germanium, have relatively small band gaps. These substances have properties that are intermediate between conductors and insulators; therefore, they are called **semiconductors.** Semiconductors are the basis for transistors, computer chips, and many other electronic devices (see Interchapter M). Figure 15.37 summarizes the band structure of metals, semiconductors, and insulators. As the temperature of a semiconductor is increased, more electrons acquire sufficient energy to be promoted from the valence band to the conduction band, so the electrical conductivity of semiconductors increases with increasing temperature. This temperature dependence is the opposite of that of metals, whose electrical conductivity decreases with increasing temperature.

WWW See Interchapter M at www.McQuarrieGeneralChemistry.com

15-12. Liquid Crystals Are Semifluid Arrangements of Molecules

In the late 1800s scientists discovered that making some simple chemical modifications to cholesterol (Figure 15.38) produced substances with unusual properties. For example, some of these had two distinct freezing points upon cooling and two corresponding melting points upon heating with colors that were quite temperature sensitive. Subsequent analysis found that these substances consist

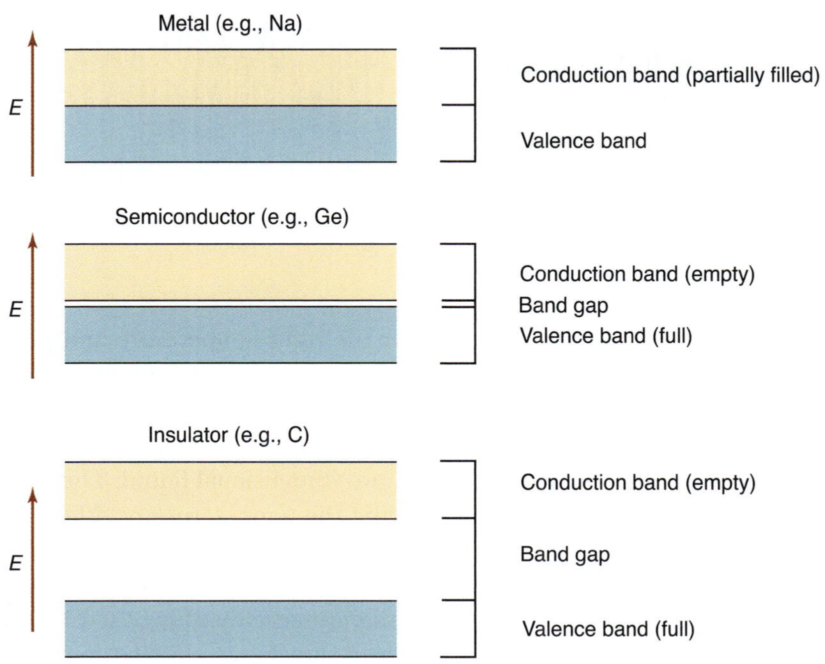

Figure 15.37 A comparison of the energy separations between the valence bands and conduction bands of metals, semiconductors, and insulators. Metals have no band gap, semiconductors have a small band gap, and insulators have a large band gap.

Figure 15.38 A space-filling model of a cholesterol molecule. The white spheres represent hydrogen atoms, the black ones represent carbon atoms, and the red one an oxygen atom. The (condensed) Lewis formula of cholesterol is

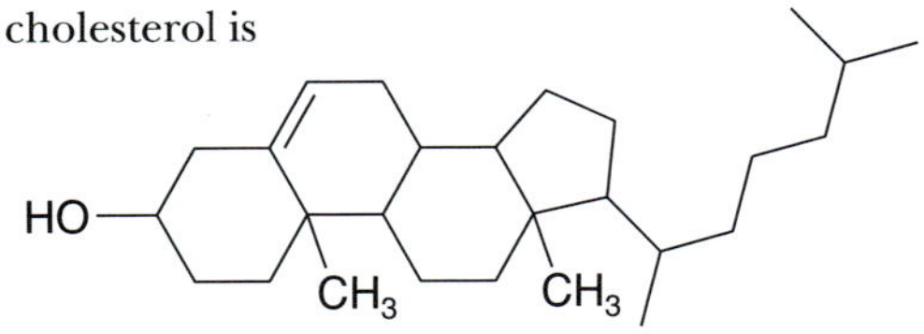

Cholesterol itself does not form liquid crystals, but molecules in which the –OH group has been altered do form liquid crystals.

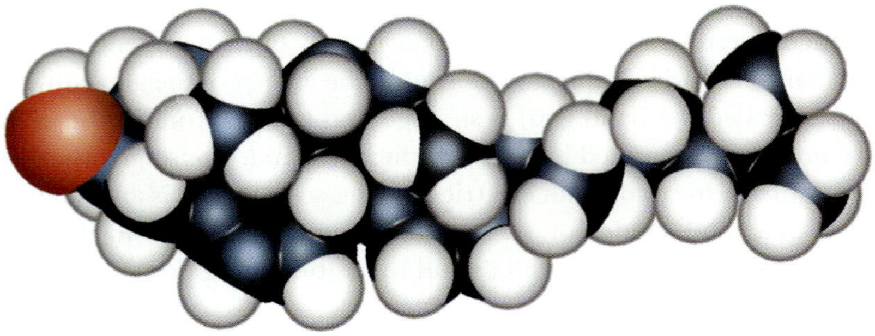

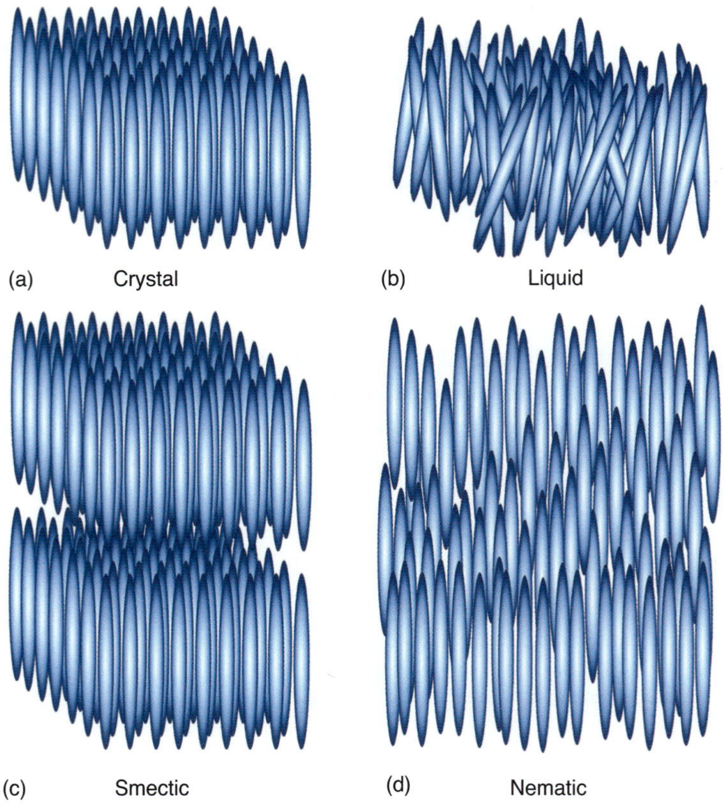

Figure 15.39 An illustration of (a) a solid crystal and (b) a liquid consisting of rodlike molecules, and two common arrangements of the molecules in a liquid crystal. In (c), called a smectic phase, the molecules are aligned in layers. In (d), called a nematic phase, the molecules are still aligned, but are not arranged in layers.

of long rod-shaped molecules that under the right temperature and pressure conditions can become aligned. Figure 15.39 compares the alignment of rod-shaped molecules in several different arrangements with that of a liquid. In one arrangement called a **smectic phase** (Figure 15.39c), the molecules are aligned in layers, where each layer is essentially a two-dimensional liquid. The spacings between the layers can be varied by varying the temperature or other parameters. In another arrangement, called a **nematic phase** (Figure 15.39d), the molecules are still aligned, but do not form layers. In both arrangements, van der Waals forces cause the molecules to align themselves along their long axes. These two phases are neither rigid as in a crystal (Figure 15.39a), nor random as in a liquid (Figure 15.39b), giving rise to the term **liquid crystal**. Liquid crystals have properties that are intermediate to those of solids and liquids.

Liquid crystals remained a laboratory curiosity for about 80 years until the discovery that they could be used to make temperature-sensing devices. Certain liquid crystals form repeating layers of molecules aligned along different axes in a helical structure, as illustrated in Figure 15.40. Liquid crystals with this structure are called **cholesteric**. These oriented layers in the liquid crystal strongly diffract light with a wavelength comparable to the distance between repeating layers. Because this distance is very sensitive to temperature, the wavelength of the diffracted light, and consequentially the perceived "color" of the crystal, changes with temperature. This effect is used to make devices such as liquid crystal thermometers and temperature-sensitive paints.

Because the orientation of the molecules in some liquid crystals is sensitive to electric fields, these crystals can be used to produce **liquid crystal displays**, or **LCDs**, for handheld electronics, computer monitors, and televisions. When an electric field is applied, the liquid crystal is aligned in such a way as to alter the ability of light to pass through the device.

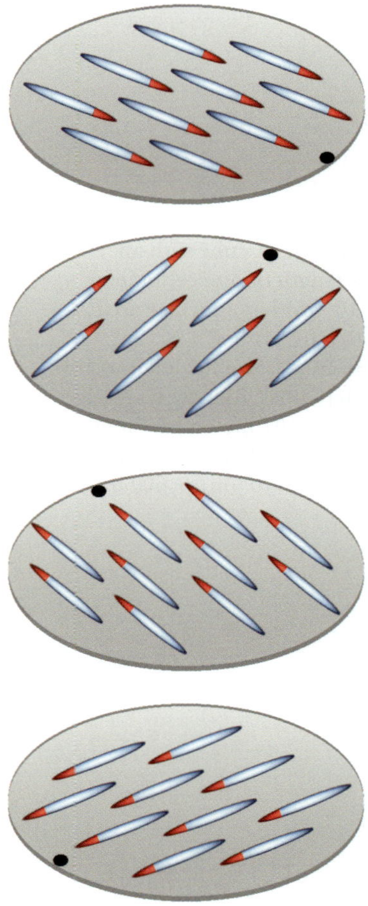

Figure 15.40 In one type of liquid crystal, called cholesteric, the molecules are arranged in layers with different orientations, forming a helical structure that repeats over a fixed distance. The distance between layers is sensitive to temperature.

15-13. Colloidal Dispersions Do Not Separate upon Standing

As we saw in Chapter 12, in a solution, the solute particles are dispersed as individual ions or molecules. A solution is homogeneous at the molecular level. At the other extreme we have what is called a **suspension**. In a suspension the dispersed particles are large enough to be seen with the naked eye, or at least under a microscope. A solution will remain homogeneous indefinitely, but the dispersed particles of a suspension will eventually settle out upon standing. There is an interesting and important case between these two extremes in which the particles are much larger than individual small molecules but are too small to be seen even with a microscope. These systems are called **colloidal dispersions** or sometimes just **colloids**. Although not an "official" state of matter like a solid or a liquid, colloidal dispersions have special properties that depend upon the size of the constituent particles.

The distinction between a solution, a colloidal dispersion, and a suspension is not rigid because it depends upon the size of the dispersed particles, but it is generally accepted that particles of colloidal dispersions are about 1 nm to 1000 nm in size. This figure is actually comparable to the sizes of macromolecules such as starches and proteins, and solutions of these molecules exhibit colloidal behavior even though the dispersed particles are individual molecules.

A common type of colloidal dispersion consists of solid colloidal particles dispersed in a liquid. Figure 15.41 shows a colloidal dispersion of gold in water. The dispersion is red because the size of the gold particles is such that they scatter red light, which is captured by the eye. This is but one example of how the properties of colloidal systems differ from a true solution.

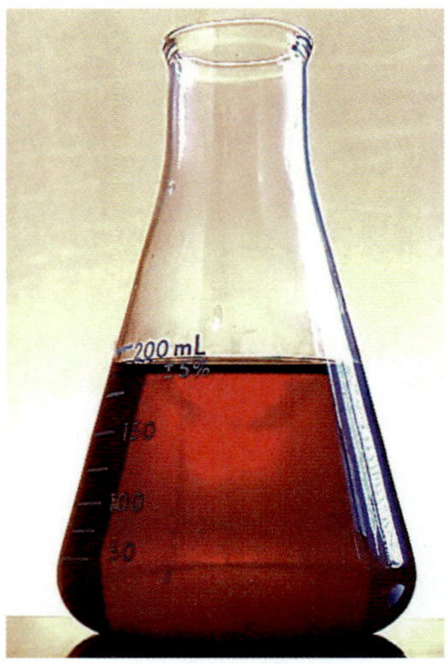

Figure 15.41 A colloidal dispersion of gold in water.

The properties of colloidal dispersions are due to the fact that the dispersed particles have a very large surface area. For example, the surface area of a cube that is one centimeter on a side is 6×10^{-4} m^2, but if the cube is subdivided into cubes that are 10^{-8} meters on a side (which is a typical size for colloids), then the surface area of all the little cubes is 600 m^2; the area is increased by a millionfold (Figure 15.42).

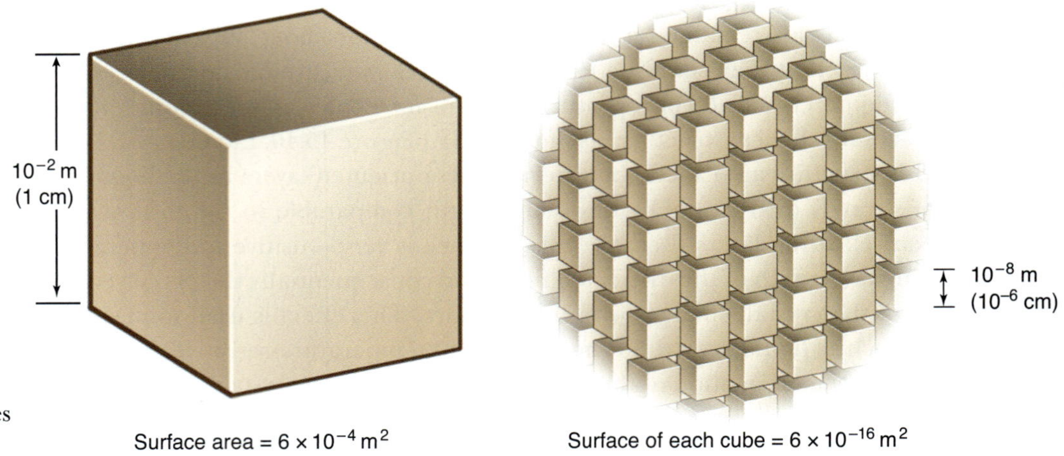

Figure 15.42 A cube with an edge length of one centimeter (or 10^{-2} m) has a surface area of $(6)(10^{-2}\text{ m})^2 = 6 \times 10^{-4}\text{ m}^2$. If the cube is subdivided into 10^{18} smaller cubes, each with an edge length of 10^{-8} meters, the total surface area of all the smaller cubes is $(10^{18})(6)(10^{-8}\text{ m})^2 = 600\text{ m}^2$.

Before we discuss other types of colloidal dispersions, we must introduce two terms. The particles that are dispersed are called the **dispersed phase** and the medium in which they are dispersed is called the **dispersion medium**. Thus, in the case of the colloidal gold-water system in Figure 15.41, gold is the dispersed phase and water is the dispersion medium. This type of colloidal dispersion is called a **sol**. Milk is a colloidal dispersion of a liquid in a liquid and is called an **emulsion**. Smoke is a colloidal dispersion of a solid in a gas, and is called an **aerosol**. Aerosols play an important role in atmospheric chemistry and meteorology. Various other types of colloidal dispersions are given in Table 15.9. Notice that all the combinations of a solid, a liquid, and a gas can form a colloidal dispersion except for a gas and a gas. A mixture of two gases always forms a true solution. In this section we shall be mainly concerned with sols and emulsions.

Colloidal dispersions can be prepared in a number of ways. Some dispersions are easy to prepare. For example, a colloidal dispersion of starch in water can be prepared by simply adding starch to water. The colloidal gold sol shown in Figure 15.41, on the other hand, is more difficult to prepare. The reaction of gold(III) chloride with formaldehyde, $CH_2O(l)$, and water produces colloidal gold according to

$$2\,AuCl_3(aq) + 3\,H_2O(l) + 3\,CH_2O(aq) \rightarrow 2\,Au(colloidal) + 3\,HCOOH(aq) + 6\,HCl(aq)$$

The concentrations of the reactants and the temperature must be carefully controlled to avoid the growth of the gold particles and their subsequent precipitation. Other sols, for example, those of various metallic oxides and hydroxides, can be prepared in a similar manner. For instance, pouring a solution of $FeCl_3(aq)$ into a beaker of boiling water produces a deep red sol of $Fe(OH)_3(colloidal)$ according to

$$FeCl_3(aq) + 3\,H_2O(l) \rightarrow Fe(OH)_3(colloidal) + 3\,HCl(aq)$$

TABLE 15.9 Types of colloidal dispersions

Dispersion medium	Dispersed phase	Name of system	Examples
gas	liquid	liquid aerosol	fog, mist, clouds
gas	solid	solid aerosol	dust, smoke
liquid	gas	foam	whipped cream, shaving cream
liquid	liquid	emulsion	water-based paints, milk, mayonnaise, detergents
liquid	solid	sol	gold in water, oil-based paint pigments, ink, blood, clay
solid	gas	solid foam	pumice stone, styrofoam, aerogel
solid	liquid	gel	opal, cheese, gelatin, jelly
solid	solid	solid sol	ruby glass, some colored gemstones

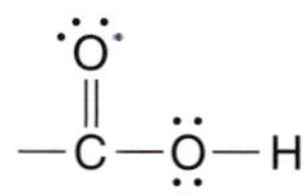
carboxylic acid group

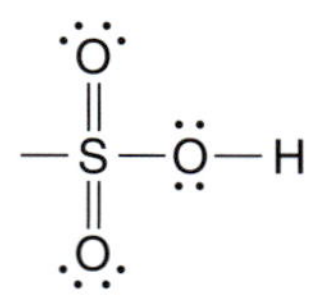
sulfonic acid group

Colloidal dispersions can also be prepared by physical grinding (using a colloid mill) or by vigorous mixing, as in the case of the formation of whipped cream (a foam) or mayonnaise (an emulsion).

An emulsion consists of droplets of one liquid dispersed in another. Emulsions are generally unstable unless a third substance, called an emulsifying agent, is added. Soaps and detergents act as emulsifying agents for oil-water emulsions. The molecules of soaps and detergents consist of a long hydrocarbon chain with a polar group, such as a carboxylic acid or sulfonic acid group (margin). These molecules can be represented schematically as

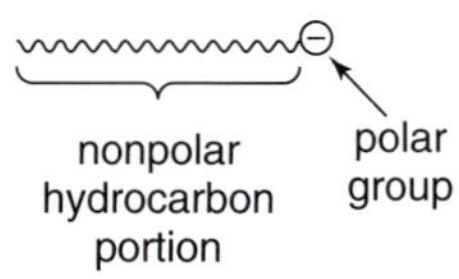

The hydrocarbon ends of these molecules dissolve in the oil droplets (remember that "like dissolves like") and the polar ends stick out of the droplets and dissolve in the water. The resulting particles are called **micelles** (Figure 15.43). Micelles stabilize the oil-water emulsion because the acid groups sticking out of the surfaces of the micelles dissociate, giving the surface a net negative charge, which repels other micelles.

One characterization of a colloidal solution is that it will scatter a beam of light. This effect was first characterized by the Irish natural philosopher John Tyndall and is known as the **Tyndall effect** (Figure 15.44). This is the same ef-

A completely formed micelle

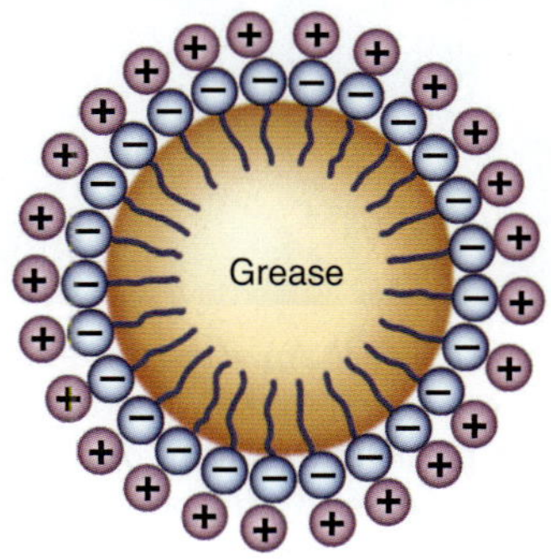

Figure 15.43 A schematic illustration of a micelle. Micelles contain emulsifying agents such as organic acids with long hydrocarbon chains. A micelle is formed when the hydrocarbon ends of these organic acids dissolve in a droplet of an oil-like substance, leaving the polar, acidic ends of the molecules sticking out into the solvent (usually water). The acidic groups dissociate and impart a net negative charge on the surface of the micelle. This charge helps to stabilize the micelles by hindering them from coming together and forming aggregates.

Figure 15.44 An illustration of the Tyndall effect. A colloidal dispersion (red) scatters a beam of light, while a true solution (blue) does not.

Figure 15.45 (*top*) An array containing blocks of aerogel that was used to capture particles from the comet Wild 2 and subsequently transported back to earth. (*bottom*) A sample of the mineral olivine collected in aerogel from the comet.

fect observed when automobile headlights are scattered by a thick fog. A nice, simple demonstration of the Tyndall effect is to mix a single drop of milk into a liter of water. Shining a flashlight or a laser pointer through the liquid dramatically shows the scattering due to the resulting colloid.

An interesting example of a colloid is aerogel, that was recently used to capture dust from the comet Wild 2 in 2004 in NASA's Stardust mission. Aerogel is a silicon-based colloidal solid with a porous spongelike structure that is 99.8% empty space and 1000 times less dense than glass. Particles of comet dust the size of a grain of sand and traveling at speeds up to six times that of a rifle bullet were slowed and captured in aerogel, which was then transported back to earth for study. The first samples arrived successfully in early 2006, still intact in the blue-tinged aerogel they were collected in (Figure 15.45).

SUMMARY

All molecules attract one another, but the molecular interactions are much greater in solids and liquids than in gases. The quantities ΔH_{vap} and ΔH_{fus} are measures of the strength of these interactions in liquids and in solids, respectively. The attractions between covalently bonded molecules are collectively called van der Waals forces. Two nonpolar molecules attract each other because their electrons redistribute such that the instantaneous dipole moments on each molecule are head-to-tail; the resulting attraction is called a London force. If the molecules are polar, then in addition to the London forces, they also attract each other by dipole-dipole forces. The polar molecules orient themselves so that their dipole moments are head-to-tail; this orientation minimizes their energy. When a hydrogen atom is covalently bonded to a highly electronegative species such as an oxygen, nitrogen, or fluorine atom within a molecule, a particularly strong intermolecular force called hydrogen bonding results. Hydrogen bonding is responsible for many of the unique properties of water.

Pure liquids have a unique equilibrium vapor pressure at each temperature. This vapor pressure increases with increasing temperature until it equals the external pressure, at which point the liquid boils. The normal boiling point of a liquid is the temperature at which the vapor pressure of the liquid is equal to one atmosphere.

The various phases of a pure substance can be displayed simultaneously on a phase diagram. The normal melting point is the temperature at which the pure solid and liquid phases are in equilibrium at one atmosphere. The normal boiling point is the temperature at which the pure liquid and gas phases are in equilibrium at one atmosphere. The triple point is the only point at which the solid, liquid, and gas phases are simultaneously in equilibrium. The critical point is the temperature above which a gas cannot be liquefied by the application of pressure.

The structure of a crystalline solid can be determined from the diffraction pattern of X-rays passed through the crystal. A unit cell is the smallest subunit of a crystal lattice that can be used to generate the entire lattice. There are a variety of different unit cells observed in crystalline solids. The three cubic ones are the simple cubic, body-centered cubic, and face-centered cubic unit cells. Crystals can be classified according to the type of forces between the constituent particles. Important types of crystals are atomic crystals, ionic crystals, molecular crystals, covalent network crystals, and metallic crystals. Metals are good conductors of electricity because the valence electrons are delocalized over the entire crystal and because there is no band gap between the valence band and conduction band. Amorphous solids have no distinct crystal structure.

Liquid crystals are a semifluid arrangement of molecules in which rodlike molecules are aligned in various orientations. The properties of liquid crystals allow them to be used to construct devices that change color with temperature and to make the displays used on various electronic devices. Colloidal dispersions are heterogeneous mixtures that, unlike a suspension, do not separate into two distinct components upon standing. The particles in a colloidal dispersion are typically 1 nm to 1000 nm in size and can scatter light. All combinations of a solid, a liquid, and a gas can form a colloidal dispersion except for a gas and a gas.

TERMS YOU SHOULD KNOW

translational motion *519*
heating curve *521*
molar enthalpy of fusion, ΔH_{fus} *522*
molar enthalpy of vaporization, ΔH_{vap} *522*
sublimation *526*
molar enthalpy of sublimation, ΔH_{sub} *526*
dipole-dipole attraction *528*
intermolecular forces *528*
intramolecular forces *528*
hydrogen bond *528*
London force *531*
van der Waals forces *532*
viscosity *533*
surface tension *534*
surfactant *534*
capillary action *535*
meniscus *535*
polarity *536*
equilibrium *537*
dynamic equilibrium *537*
equilibrium vapor pressure *538*
vapor pressure curve *538*
normal boiling point *540*
relative humidity *540*
dew point *540*
phase diagram *541*
critical point *541*
critical temperature *541*
sublimation pressure curve *541*
melting point curve *541*
normal melting point *541*
normal freezing point *541*
triple point *542*
normal sublimation point *545*
crystal lattice *545*
X-ray diffraction pattern *545*
optical diffraction pattern *545*
unit cell *545*
simple cubic unit cell *545*
body-centered cubic unit cell *545*
face-centered cubic unit cell *547*
crystallographic radius *548*
atomic crystal *551*
ionic crystal *551*
molecular crystal *551*
covalent network crystal *552*
amorphous solid *552*
metallic crystal *554*
valence band *555*
conduction band *555*
band gap *555*
semiconductor *555*
smectic phase *556*
nematic phase *556*
liquid crystal *557*
cholesteric *557*
liquid crystal display (LCD) *557*
suspension *557*
colloidal dispersion *557*
colloid *557*
dispersed phase *558*
dispersion medium *558*
sol *558*
emulsion *558*
aerosol *558*
micelle *559*
Tyndall effect *559*

EQUATIONS YOU SHOULD KNOW HOW TO USE

$q_{\text{P}} = nC_{\text{P}}(T_2 - T_1)$ (15.1) (energy as heat required to change the temperature of n moles of a substance from T_1 to T_2)

$q_{\text{fus}} = n\Delta H_{\text{fus}}$ (15.2) (energy as heat required to melt n moles of a substance)

$q_{\text{vap}} = n\Delta H_{\text{vap}}$ (15.3) (energy as heat required to vaporize n moles of a substance)

$$\text{relative humidity} = \left(\frac{P_{H_2O}}{P^{\circ}_{H_2O}}\right) \times 100 \quad (15.5) \quad \text{(definition of relative humidity)}$$

$$V_{\text{unit cell}} = \frac{nM}{dN_A} \quad (15.8) \quad \text{(volume of a unit cell)}$$

PROBLEMS

HEATS OF VAPORIZATION, FUSION, AND SUBLIMATION

15-1. Ammonia is used as a refrigerant in some industrial refrigeration units. The molar enthalpy of vaporization of liquid ammonia is 23.33 kJ·mol^{-1}. Calculate the amount of heat absorbed in the vaporization of 5.00 kilograms of $NH_3(l)$.

15-2. Given that 23.6 kilojoules of heat are required to completely vaporize 60.0 grams of benzene, $C_6H_6(l)$, at 80.1°C, calculate the molar enthalpy of vaporization, ΔH_{vap}, of benzene.

15-3. Calculate the energy as heat released when 20.1 grams of liquid mercury at 25.0°C are converted to solid mercury at its melting point. The heat capacity of $Hg(l)$ is 28.0 J·mol^{-1}·K^{-1}.

15-4. The chlorofluorocarbon refrigerant Freon-12, CCl_2F_2, was banned in 1995 to help protect the ozone layer. Most new air conditioning units now use the refrigerant tetrafluoroethane, CF_3CH_2F. The enthalpy of vaporization of Freon-12 is 155 J·g^{-1} and that of tetrafluoroethane is 215.9 J·g^{-1}. Estimate the number of grams of Freon-12 that must be vaporized to freeze a tray of 16 one-ounce (1 oz = 28 g) ice cubes with the water initially at 18°C. How many grams of tetrafluoroethane are required to perform the same task?

15-5. The metal gallium melts when held in the hand; its melting point is 29.76°C. How much energy as heat is removed from the hand when 5.00 grams of gallium initially at 20.0°C melts? The value of ΔH_{fus} is 5.576 kJ·mol^{-1} and the specific heat of gallium is 0.374 J·g^{-1}·K^{-1}. Take the final temperature to be 29.76°C.

15-6. The enthalpy of vaporization of einsteinium was determined to be 128 kJ·mol^{-1} using only a 100-μg sample. How much heat is required to vaporize 100 μg of einsteinium?

15-7. Calculate the amount of energy as heat absorbed by the sublimation of 100.0 grams of dry ice (solid carbon dioxide). The value of ΔH_{sub} for CO_2 is 25.2 kJ·mol^{-1}.

15-8. Calculate the number of moles of water at 0°C that can be frozen by one mole of dry ice (solid carbon dioxide). See Table 15.3 for the necessary data. The value of ΔH_{sub} for CO_2 is 25.2 kJ·mol^{-1}.

HEATING CURVES

15-9. Sketch a heating curve for 7.50 grams of mercury from 200 K to 800 K using a heat input rate of 100 J·min^{-1}. Refer to Table 15.3 for some of the necessary data for mercury. The molar heat capacities of solid, liquid, and gaseous mercury are 28.3 J·mol^{-1}·K^{-1}, 28.0 J·mol^{-1}·K^{-1}, and 20.8 J·mol^{-1}·K^{-1}, respectively.

15-10. What would take longer, heating 10.0 grams of water at 50.0°C to 100.0°C or vaporizing the 10.0 grams at 100.0°C if the rate of heating in both cases is 5 J·s^{-1}?

15-11. Heat was added to 25.0 grams of solid sodium chloride, $NaCl(s)$, at the rate of 3.00 kJ·min^{-1}. The temperature remained constant at 800.7°C, the normal melting point of $NaCl(s)$, for 241 seconds. Calculate the molar enthalpy of fusion of $NaCl(s)$.

15-12. Heat was added to a 45.0-gram sample of liquid propane, $C_3H_8(l)$, at the rate of 500.0 J·min^{-1}. The temperature remained constant at −42.1°C, the normal boiling point of propane, for 38.9 minutes. Calculate the molar enthalpy of vaporization of propane.

VAN DER WAALS FORCES

15-13. Which of the following molecules have polar interactions?

Cl_2 $\quad$ ClF $\quad$ NF_3 $\quad$ F_2

15-14. Which of the following exhibit primarily only London forces?

H_2O He Cl_2 HCl

15-15. Which of the following molecules can form hydrogen bonds?

H_2 HF CH_4 CH_3OH

15-16. Which of the following molecules do you predict to have unusually high boiling points due to hydrogen bonding?

HI H_2S CH_3OCH_3 CH_3CH_2OH

15-17. Arrange the following compounds in order of increasing boiling point:

KBr CH_3CH_2OH C_2H_6 Ne

15-18. Arrange the following compounds in order of increasing boiling point:

MgO NH_3 PH_3 KCl

15-19. Arrange the following molecules in order of increasing molar enthalpy of vaporization:

CH_4 C_2H_6 CH_3OH CH_3CH_2OH

15-20. Arrange the following molecules in order of increasing molar enthalpy of vaporization:

CCl_4 $SiCl_4$ CH_4 $SiBr_4$

VAPOR PRESSURE

15-21. A 0.75-gram sample of ethanol is placed in a sealed 400-milliliter container. Is there any liquid present when the temperature is held at 60°C?

15-22. Mexico City lies at an elevation of 2300 meters (7400 feet). If water boils at 93°C in Mexico City, what is the atmospheric pressure there?

15-23. A sample of ethanol vapor in a vessel of constant volume exerts a pressure of 300 Torr at 75.0°C. Use the ideal-gas law to plot pressure versus temperature of the vapor between 80.0°C and 40.0°C. Assume no condensation. Compare your result with the vapor pressure curve for ethanol shown in Figure 15.23. Estimate the temperature at which condensation occurs upon cooling from 80.0°C.

15-24. Atmospheric pressure decreases with altitude. Plot the following data (you may use either units of meters or feet):

Altitude/m	Altitude/ft	Atmospheric pressure/bar
1500	5000	0.83
3000	10000	0.70
4500	15000	0.58
6000	20000	0.47

Using your plot and the vapor pressure curve of water (Figure 15.23), estimate the boiling point of water at the following locations:

Location	Altitude/m	Altitude/ft
Denver	1610	5280
Mount Kilimanjaro	5895	19340
Mount Washington	1917	6290
the Matterhorn	4478	14690

15-25. Compare the dew points of two days with the same relative humidity of 70% but with temperatures of 20°C and 30°C, respectively.

15-26. The relative humidity in a greenhouse at 40°C is 92%. Calculate the vapor pressure of water vapor in the greenhouse.

PROPERTIES OF LIQUIDS

15-27. The surface tension of water is 72 $mJ\cdot m^{-2}$. What is the energy required to change a spherical drop of water with a diameter of 2 mm to two smaller spherical drops of equal size? The surface area of a sphere of radius r is $4\pi r^2$ and the volume is $4\pi r^3/3$.

15-28. The surface tension of water is 72 $mJ\cdot m^{-2}$. Calculate the amount of energy required to disperse one spherical drop of radius 3.0 mm into spherical drops of radius 3.0×10^{-3} mm. The surface area of a sphere of radius r is $4\pi r^2$ and the volume is $4\pi r^3/3$.

15-29. The simple hydrocarbon n-heptane has a structural formula of $CH_3(CH_2)_5CH_3$. Would you expect water or cyclohexane to be a better solvent for heptane?

15-30. Although it is only slightly polar, ethanol, $CH_3CH_2OH(l)$, is completely miscible in water. Explain this apparent contradiction to the "like dissolves like" adage for determining solubility. (*Hint*: What special property do water and ethanol have in common?)

PHASE DIAGRAMS

15-31. Determine whether water is a solid, liquid, or gas at the following pressure and temperature combinations (use Figure 15.24):

(a) 373 K, 0.70 atm (b) -100°C, 0.006 atm

(c) 400 K, 200 atm (d) 0°C, 300 atm

15-32. Referring to Figure 15.25, state the phase of carbon dioxide under the following conditions:

(a) 127°C, 8 atm (b) -60°C, 40 atm

(c) 50°C, 1 atm (d) -80°C, 5 atm

15-33. Sketch the phase diagram for oxygen using the following data:

	Triple point	Critical point
temperature/K	54.3	154.6
pressure/Torr	1.14	37826

The normal melting point and normal boiling point of oxygen are -218.8°C and -183.0°C. Does oxygen melt under an applied pressure as water does?

15-34. Sketch the phase diagram for nitrogen given the following data:

triple point, 63.15 K and 139 Torr

normal melting point, 63.15 K

normal boiling point, 77.35 K

critical point, 126.21 K and 33.9 bar

CRYSTAL STRUCTURES

15-35. Potassium exists as a body-centered cubic lattice. How many potassium atoms are there per unit cell?

15-36. Crystalline potassium fluoride has the NaCl-type structure shown in Figure 15.30. How many potassium ions and fluoride ions are there per unit cell?

15-37. The density of silver is 10.50 $g\cdot cm^{-3}$ at 20°C. Given that the unit cell of silver is face-centered cubic, calculate the length of an edge of a unit cell.

15-38. The density of tantalum is 16.654 $g\cdot cm^{-3}$ at 20°C. Given that the unit cell of tantalum is body-centered cubic, calculate the length of an edge of a unit cell.

15-39. Copper crystallizes in a face-centered cubic lattice with a density of 8.96 $g\cdot cm^{-3}$. Given that the length of an edge of a unit cell is 361.5 pm, calculate Avogadro's number.

15-40. Chromium crystallizes in a body-centered cubic lattice with a density of 7.20 $g\cdot cm^{-3}$. Given that the length of an edge of a unit cell is 288.4 pm, calculate Avogadro's number.

15-41. Crystalline potassium fluoride has the NaCl-type structure shown in Figure 15.30. Given that the density of $KF(s)$ is 2.481 $g\cdot cm^{-3}$ at 20°C, calculate the unit cell length and the nearest-neighbor distance in $KF(s)$. (The nearest-neighbor distance is the shortest distance between the centers of any two adjacent ions in the lattice.)

15-42. Crystalline cesium bromide has the CsCl-type structure shown in Figure 15.30. Given that the density of $CsBr(s)$ is 4.43 $g\cdot cm^{-3}$ at 25°C, calculate the unit cell length and the nearest-neighbor distance (see the previous Problem) in $CsBr(s)$.

15-43. Given that the density of $KBr(s)$ is 2.75 $g\cdot cm^{-3}$ and that the length of an edge of a unit cell is 654 pm, determine how many formula units of KBr there are in a unit cell. Does the unit cell have a $NaCl(s)$ or a $CsCl(s)$ structure? (See Figure 15.30.)

15-44. Given that the density of $CaO(s)$ is 3.34 $g\cdot cm^{-3}$ and that the length of an edge of a unit cell is 481.08 pm, determine how many formula units of CaO there are in a unit cell. Does the unit cell have a $NaCl(s)$ or a $CsCl(s)$ structure? (See Figure 15.30.)

ADDITIONAL PROBLEMS

15-45. Why do packages of "minute rice" and instant noodles often contain instructions to "boil longer at altitude"?

15-46. Moisture often forms on the outside of a glass containing a mixture of ice and water. Use the principles developed in this chapter to explain this phenomenon.

15-47. As illustrated in Figure 15.11, it is possible to float a paper clip on the surface of water. However, if a small amount of detergent is added to the water the clip will no longer float. Why is this?

15-48. In describing vapor pressure we described the equilibrium between the gas and liquid phases as being *dynamic*. Why is the equilibrium dynamic? Can you give an example of another system that is in a state of "dynamic equilibrium"?

15-49. Is it possible to boil water at a temperature lower than 100°C in the laboratory? If so, explain how this might be achieved.

15-50. (a) What is the difference between evaporation and boiling? (b) Can a solid evaporate?

15-51. Two students are arguing whether an equal mass of steam or hot water at 100°C would be more scalding if it came in contact with your skin. Is there a difference?

15-52. A pressure cooker is a sealed container that uses water to cook foods faster than by boiling on a stovetop. Why do foods cook faster in a pressure cooker?

15-53. The stained glass windows in many old churches are thicker at the bottom than the top. At one point scientists thought that this was evidence that the glass had "flowed" downward over the years. However, it has since been shown that this effect is due to the manufacturing process that was used and that glass does not flow appreciably. Why had scientists first thought it might be possible for glass to flow?

15-54. Suppose you are stranded in a mountain cabin by a snowstorm. You have plenty of wood, but only a limited supply of food and wish to conserve it as long as possible. You remember reading that you should melt snow to get water to drink and not eat the snow directly because the body expends energy to melt the snow. How many nutritional Calories are expended by your body in melting enough snow to make a liter of water (1 nutritional Calorie = 1 kcal = 4184 J)?

15-55. Prove that $4r = l\sqrt{3}$ for a body-centered cubic unit cell, where l is the length of an edge of the unit cell and r is the radius of the atoms.

15-56. Trouton's rule states that the molar enthalpy of vaporization of a liquid that does not have strong molecular interactions such as hydrogen bonding or ion-ion attractions is given by

$$\Delta H_{\text{vap}} = (85\ \text{J·K}^{-1}\text{·mol}^{-1})T_{\text{b}}$$

where T_{b} is the normal boiling point of the liquid in kelvins. Use Trouton's rule to estimate the value of ΔH_{vap} for each of the noble gases listed in Table 15.3.

15-57. Use Trouton's rule, given in the previous Problem, to estimate the value of ΔH_{vap} for each of the halogens listed in Table 15.3.

15-58. Apply Trouton's rule, given in Problem 15-56, to estimate the value of ΔH_{vap} for chloromethane, water, and hydrogen sulfide, listed in Table 15.3. What is the percentage error in each case? Suggest a molecular explanation for any discrepancy with the values of ΔH_{vap}.

15-59. What is the relationship between ΔH_{fus}, ΔH_{vap}, and ΔH_{sub} in the vicinity of the triple point?

15-60. Why is H_2S a gas at −10°C, whereas H_2O is a solid at this temperature?

15-61. Sulfur melts at 119°C to a thin, pale-yellow liquid consisting of S_8 rings:

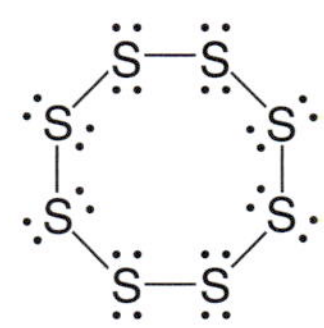

As sulfur is heated to 150°C and higher, it becomes so viscous that it hardly pours. Explain these observations in terms of the breaking of the S_8 rings.

15-62. Both silicon carbide, SiC(s), and boron nitride, BN(s), are about as hard as diamond. What does this suggest about their crystal structures?

15-63. In 2002 researchers at Lawrence Livermore National Laboratory demonstrated that the metal osmium, which does not form a covalent network, is less compressible than diamond, the former record holder. The researchers measured the compressibility under pressures up to 600 000 bars using X-ray diffraction. Explain how X-ray diffraction may be used to determine the compressibility of a solid.

15-64. Although the temperature may not exceed 0°C, the amount of ice on a sidewalk decreases owing to sublimation. A source of heat for the sublimation is solar radiation. The average daily solar radiation in February for Boston is 8.1 $MJ \cdot m^{-2}$. Calculate how much ice will disappear from a 1.0-m^2 area in one day assuming that all the radiation is used to sublime the ice. Take the density of ice to be 0.917 $g \cdot cm^{-3}$ and the molar enthalpy of sublimation to be 50.9 $kJ \cdot mol^{-1}$ at 0°C.

15-65. Commercial refrigeration units in the United States are rated in tons. During 24 hours of operation a one-ton unit is capable of removing an amount of heat equal to that released when 1.00 ton of water at 0°C is converted to ice. Calculate the number of kilojoules of heat per hour that can be removed by a four-ton home air conditioner (1 ton = 2000 pounds).

15-66. In the 1800s, British surveyors were prevented from extending their survey of India into the Himalayas because entry into Tibet was banned. In 1865, the Indian Nain Singh secretly entered Lhasa, the capital city of Tibet, and determined its correct location for map placement. Singh was not able to bring instruments for measuring altitude with him, but he did have a thermometer. He estimated that Lhasa was 3420 meters above sea level. (Its true elevation is 3540 meters or 11 600 feet) Describe how he was able to estimate the altitude from a measurement of the boiling point of water.

15-67. The vapor pressures (in Torr) of solid and liquid chlorine are given by

$$\ln P_s = 24.320 - \frac{3777\text{ K}}{T} \qquad \text{(solid)}$$

$$\ln P_l = 17.892 - \frac{2669\text{ K}}{T} \qquad \text{(liquid)}$$

where T is the absolute temperature. Calculate the temperature and pressure at the triple point of chlorine.

15-68. The relative humidity is 65% on a certain day on which the temperature is 30°C. As the air cools during the night, what will be the dew point?

15-69. Sodium chloride has the crystal structure shown in Figure 15.30. By X-ray diffraction, it is determined that the shortest distance between a sodium ion and a chloride ion is 282 pm. Using the fact that the density of sodium chloride is 2.163 $g \cdot cm^{-3}$, calculate Avogadro's number.

15-70. Cesium chloride has the crystal structure shown in Figure 15.30. The length of a side of a unit cell is determined by X-ray diffraction to be 412.1 pm. What is the density of cesium chloride?

15-71. The unit cell of lithium is body-centered cubic, and the length of an edge of a unit cell is 351 pm at 20°C. Calculate the density of lithium at 20°C.

15-72. Calculate the concentration in moles per liter of water vapor in air saturated with water vapor at 25°C.

15-73. Arrange the following substances in order of increasing polarity:

$$CCl_4 \quad CHCl_3 \quad CH_2Cl_2 \quad CH_3Cl$$

15-74. The phase diagram for sulfur is shown below.

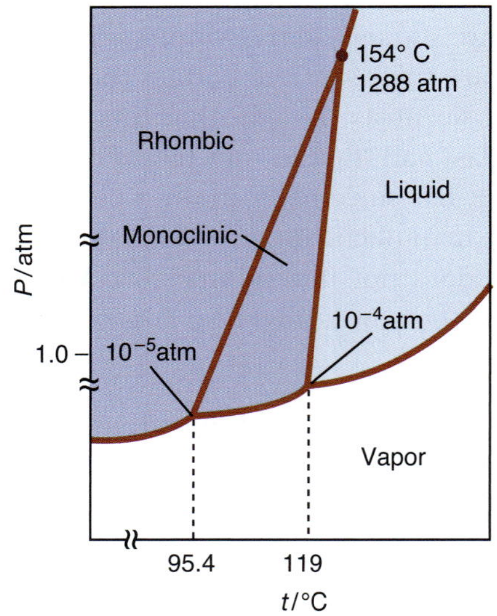

The regions labeled Rhombic and Monoclinic indicate two different crystalline forms of sulfur. How many triple points are there? Describe what happens if sulfur is heated from 40°C at one atm to 200°C at one atm. Below what pressure will sublimation occur?

15-75. Use hybrid orbitals to describe the bonding in diamond.

15-76. Use hybrid orbitals to describe the bonding within a layer and between the molecular layers in graphite.

15-77. (*) Two 20.0-gram ice cubes at −21.0°C are placed into 250 milliliters of water at 25.0°C. Calculate the final temperature of the water in the glass after all the ice melts. Assume no energy is transferred as heat to or from the surroundings. Take the density of water to be 1.00 $g\cdot mL^{-1}$ and the heat capacity of ice and water to be 37.7 $J\cdot mol^{-1}\cdot K^{-1}$ and 75.3 $J\cdot mol^{-1}\cdot K^{-1}$, respectively.

15-78. The vapor pressures (in Torr) of solid and liquid uranium hexafluoride are given by

$$\ln P_s = 24.513 - \frac{5892.5\text{ K}}{T} \qquad \text{(solid)}$$

$$\ln P_l = 17.357 - \frac{3479\text{ K}}{T} \qquad \text{(liquid)}$$

where T is the absolute temperature. Calculate the temperature and pressure at the triple point of UF_6.

15-79. It has been suggested that the reason why ice skates slide on ice is that the pressure exerted by the skater melts a small area of water below the blade. Determine the pressure on ice created by a 75-kilogram skater standing on a skate with a blade that measures 3 mm by 20 cm (recall that $P = F/A$, and $F = mg$ where $g = 9.8\ m\cdot s^{-2}$). The melting point of ice decreases by about 0.01°C per atm of applied pressure (Figure 15.24). Show that pressure alone cannot explain why a skater glides over ice at temperatures of 5 to 10°C below freezing. It turns out that the heat generated by friction is also insufficient to explain this phenomenon. Recent work by Gabor Somorjai at Lawrence Berkeley National Laboratory suggests that the surface of ice has a disordered, quasi-fluid layer that makes ice "slippery," even at temperatures below 144 K (−129°C).

15-80. A 0.677-gram sample of zinc reacts completely with sulfuric acid according to

$$Zn(s) + H_2SO_4(aq) \rightarrow ZnSO_4(aq) + H_2(g)$$

A volume of 263 milliliters of hydrogen gas is collected over water; the water level in the collecting vessel is the same as that surrounding the vessel. The atmospheric pressure is 756.0 Torr and the temperature is 25°C. Calculate the atomic mass of zinc.

15-81. (*) Steel can be made by inserting atoms of carbon into the spaces between the atoms in metallic iron. Iron has a body-centered crystal structure with a density of 7.86 $g\cdot cm^{-3}$. Carbon has an atomic radius of 77 pm. Show that there is enough space between the iron atoms to hold a carbon atom.

15-82. The Clapeyron–Clausisus equation (which we will derive in Chapter 23) is

$$\ln\left(\frac{P_2}{P_1}\right) = \frac{\Delta H_{vap}}{R}\left[\frac{1}{T_1} - \frac{1}{T_2}\right]$$

This equation, which assumes that ΔH_{vap} does not vary with temperature, relates the change in vapor pressure and temperature to a substance's enthalpy of vaporization (or sublimation), where R is the molar gas constant. Use this relationship and the fact that ΔH_{vap} of water at 25°C is 43.99 $kJ\cdot mol^{-1}$ to calculate the vapor pressure of water at 5°C, 25°C, 50°C, and 95°C. Compare your results to the values in Table 15.7. What might account for the slight discrepancy between the values in Table 15.7 and your results?

15-83. (*) Use the Clapeyron–Clausisus equation (see the previous Problem) to determine the molar enthalpy of vaporization and the normal boiling point of chlorine given the equations in Problem 15-67.

Jacobus Henricus van't Hoff (1852–1911) was born in Rotterdam, the Netherlands, where his father was a physician. He had hoped to pursue a career in chemistry, but his parents persuaded him to study engineering because they believed that the prospects for pure research in chemistry were poor. Nonetheless, he later studied chemistry at several universities and finally completed a doctorate at the University of Utrecht in 1874. In the same year, he published a paper using a tetrahedral model of the carbon atom to explain the structural properties of organic molecules and optical isomerization. However, he did not present this as his thesis for his doctorate because he feared that it was too controversial to be accepted. This work was strongly attacked by some prominent organic chemists, but eventually the model of the tetrahedral carbon atom became the foundation of modern organic structural chemistry. Because it was difficult to find a university position, van't Hoff first gave private lessons and then found a position in 1876 as a physics lecturer at the Royal Veterinary School in Utrecht. The next year he received an appointment at the University of Amersterdam, where he remained for 18 years. He eventually left for a research professorship at the University of Berlin, where they built a new laboratory for him. Working with Svante Arrhenius (Chapter 6 Frontispiece), he studied the dissociation of ionic substances in solution. He later turned his attention to investigating the rates of chemical reactions and chemical equilibrium. In 1887, Wilhelm Ostwald and van't Hoff founded *Zeitschrift fur Physikalische Chemie*, the first journal devoted to physical chemistry. Van't Hoff was awarded the first-ever Nobel Prize for Chemistry in 1901 "for his discovery of the laws of chemical dynamics and osmotic pressure in solutions."

16. Colligative Properties of Solutions

Why can some salts melt ice? Why does the same antifreeze added to the coolant water in a car radiator prevent both overheating in summer and freezing in winter? What factors operate to keep biological cells from collapsing? Why do divers get the bends when they ascend too quickly to the water's surface? These and many other questions can be answered when the properties of solutions are understood from a molecular point of view. The major emphasis of this chapter is on the colligative properties of solutions. Colligative properties depend primarily on the ratio of the number of solute particles to the number of solvent particles in the solution rather than on the chemical nature of the solute. The colligative properties of solutions are vapor pressure lowering, boiling point elevation, freezing point depression, and osmotic pressure.

We showed in Chapter 15 that pure water has a unique equilibrium vapor pressure at each temperature. We show in this chapter that the equilibrium vapor pressure of pure water, or any other solvent, decreases when a nonvolatile substance is dissolved in it. The vapor pressure–lowering effect is the key to understanding the colligative properties of solutions.

16-1. Solutes Affect the Properties of the Solvent

Recall from Chapter 12 that when a solute is dissolved in a solvent, the solution formed usually is homogeneous throughout, right down to the molecular level. We also noted there that solute molecules or ions interact with solvent molecules. The electrostatic interactions between dissolved ions and polar solvent molecules are especially strong. As a result, ions (especially cations) become **solvated** in the solution, that is, attached to solvent molecules (see Figure 12.2). Because water molecules are polar, water is a good solvent for a wide variety of electrolytes (e.g., sodium chloride and magnesium sulfate) and polar nonelectrolytes (e.g., sucrose and ethanol).

The strong interactions between solute particles (molecules or ions) and solvent molecules dramatically affect many of the properties of the solvent. Such properties, called **colligative properties,** depend primarily on the *ratio* of the number of solute particles to the number of solvent molecules in a solution. Consequently, it is important to understand the different ways in which this ratio is described. The four commonly used measures of solute concentration are molarity (denoted by M), percentage by mass (denoted by % mass), mole fraction (denoted by x), and molality (denoted by m). We shall define each of these measures in turn below.

We are already familiar with the concentration unit molarity (Section 12-2). Molarity is defined as the number of moles of solute per liter of solution. Molarity is the most widely used measure of solute concentration because it is the easiest to use. We need only weigh out the appropriate amount of solute, place it in the appropriate volumetric flask, add solvent to dissolve the solute, and dilute up to the volume mark; making certain by swirling that the solution is homogeneous (Figure 12.4). However, molarity has a disadvantage as a measure of concentration in that the molarity of a given solution changes as the temperature changes because the volume of the solution changes as the temperature changes. Usually, the volume of the solution increases as the temperature increases; thus, the molarity of the solution usually decreases as the temperature increases.

In contrast to molarity, solute concentrations expressed as percentage mass, mole fraction, and molality are all temperature independent because they are defined in terms of mass and mass ratios, which, unlike ratios involving volume, do not depend on the temperature.

Solute concentrations are most commonly expressed as **% mass** (percentage by mass) when either we do not know the chemical formula of the solute (e.g., with an unknown) or the substance cannot be described by a single chemical formula because it is a mixture of substances (e.g., a plant extract, an environmental sample, or a beverage). For these cases, we can characterize the solute concentration by simply stating the ratio of the number of grams of dissolved solute to the total mass of the solution (solute plus solvent); we then multiply the ratio by 100 to get the percentage by mass of solute, that is,

$$\% \text{ mass} = \left(\frac{\text{mass of solute}}{\text{mass of solute} + \text{mass of solvent}} \right) \times 100 \tag{16.1}$$

For example, if we dissolve 5.85 grams of an unknown solute in 100.0 grams of water, then the % mass of solute in the solution is

$$\% \text{ mass} = \left(\frac{5.85 \text{ g}}{5.85 \text{ g} + 100.0 \text{ g}} \right) \times 100 = 5.53\%$$

Recall from Section 13-8 that the **mole fraction** of component 1 in a solution, x_1, is defined as the ratio of the number of moles of component 1 to the total number of moles of all components in the solution:

$$x_1 = \frac{n_1}{n_{\text{tot}}} \tag{16.2}$$

If there are only two components (call them 1 and 2 to denote solvent and solute, respectively) in the solution, then the mole fraction of component 1 is given by

$$x_1 = \frac{n_1}{n_1 + n_2} \tag{16.3}$$

Similarly, the mole fraction of component 2 in the solution is

$$x_2 = \frac{n_2}{n_1 + n_2} \tag{16.4}$$

Because mole fractions are defined in terms of mole ratios, they are dimensionless quantities. Also, being fractions, the sum of the mole fractions is equal to one, as you can see from Equations 16.3 and 16.4. In general, the sum of the mole fractions of all the components in a solution must equal one (unity).

The numerical value of the mole fraction of the solvent in a dilute solution is close to one because in such a case n_1 is much greater than n_2, or $n_1 >> n_2$, making $n_1 + n_2 \approx n_1$, and so

$$x_1 = \frac{n_1}{n_1 + n_2} \approx \frac{n_1}{n_1} = 1$$

Consequently, mole fraction is not a convenient measure of concentration for dilute solutions. In dilute solutions, we usually find it more convenient to use a concentration unit called molality that is directly proportional to the mole fraction of solute in a dilute solution. We define the **molality,** m, of a solute as the number of moles of solute per 1000 grams or one kilogram of solvent:

$$\text{molality} = \frac{\text{moles of solute}}{\text{1000 g of solvent}} = \frac{\text{moles of solute}}{\text{1 kilogram of solvent}} \tag{16.5}$$

For example, a solution prepared by dissolving 0.100 moles of sodium chloride (5.84 grams) in 1.00 kilograms of water is 0.100 molal (0.100 m) in NaCl(aq). The unit of molality is the molal designated by m.

EXAMPLE 16-1: Calculate (a) the molality and (b) the mole fractions of a solution prepared by dissolving 20.0 grams of sucrose, $C_{12}H_{22}O_{11}(s)$, in 0.500 kilograms of water.

Solution: (a) The number of moles of sucrose in 20.0 grams is

$$\text{moles of sucrose} = (20.0\text{ g sucrose})\left(\frac{1\text{ mol sucrose}}{342.3\text{ g sucrose}}\right) = 0.0584\text{ mol}$$

When 0.0584 moles of sucrose are dissolved in 0.500 kilograms of water, the molality of sucrose in the resulting solution is given by Equation 16.5:

$$m = \frac{0.0584\text{ mol}}{0.500\text{ kg}} = 0.117\text{ mol·kg}^{-1} = 0.117\text{ m}$$

(b) To calculate the mole fractions of sucrose and water in the solution, we need to know both the number of moles of sucrose and the number of moles of water. From part (a), we know that there are 0.0584 moles of sucrose in the solution. The number of moles of water in the solution is

$$\text{moles of H}_2\text{O} = (0.500\text{ kg H}_2\text{O})\left(\frac{1000\text{ g}}{1\text{ kg}}\right)\left(\frac{1\text{ mol H}_2\text{O}}{18.02\text{ g H}_2\text{O}}\right) = 27.7\text{ mol}$$

The mole fractions of sucrose, x_s, and of water, x_w, in the solution are

$$x_s = \frac{0.0584\text{ mol}}{(27.7\text{ mol} + 0.0584\text{ mol})} = 2.10 \times 10^{-3}$$

$$x_w = \frac{27.7\text{ mol}}{(27.7\text{ mol} + 0.0584\text{ mol})} = 0.998$$

Note that $x_s + x_w = 0.00210 + 0.998 = 1.000$ and also that $x_w \approx 1$.

PRACTICE PROBLEM 16-1: Calculate the molality of a solution prepared by dissolving 5.25 grams of potassium permanganate, $KMnO_4(s)$, in 250.0 grams of water.

Answer: 0.133 m

The molality of a solute, *m*, is not the same as the molarity, *M*. To prepare 500.0 mL of a 0.117-M solution of sucrose in water, we dissolve 0.0585 moles of sucrose in less than 500.0 mL of water and dilute the resulting solution with enough water to yield exactly 500.0 mL of *solution*. Compare this procedure with the procedure described in Example 16-1 for the preparation of a 0.117-m sucrose solution.

In the following sections of this chapter, we discuss the colligative properties of solutions, namely, vapor pressure lowering, boiling point elevation, freezing point depression, and osmotic pressure. It is essential for the understanding of colligative properties to realize that it is the ratio of the number of solute particles to the number of solvent particles that determines the magnitude of a colligative effect. A 0.10-m aqueous sodium chloride solution has twice as many solute particles per mole of water as a 0.10-m aqueous sucrose solution because $NaCl(aq)$ is a strong electrolyte and dissociates completely in water to form $Na^+(aq)$ and $Cl^-(aq)$ ions, whereas sucrose exists in solution as intact $C_{12}H_{22}O_{11}(aq)$ molecules. We can express this result as an equation by writing

$$m_c = im \tag{16.6}$$

where *i* is the number of solute particles produced per formula unit when the solute is dissolved in the solvent. We thus distinguish between the molality, denoted by *m*, and the **colligative molality,** denoted by m_c. The distinction is illustrated numerically in Table 16.1.

TABLE 16.1 Comparison of molality and colligative molality of some aqueous solutions

Solute	Solute molality/m	Solute particles per formula unit, /*i*	Colligative molality/m_c
$C_{12}H_{22}O_{11}$	0.10	1	0.10
NaCl	0.10	2	0.20
$CaCl_2$	0.10	3	0.30

EXAMPLE 16-2: What is the colligative molality of a 0.20-m potassium sulfate, $K_2SO_4(aq)$, solution?

Solution: We first note that $K_2SO_4(s)$ is a strong electrolyte (see Section 12-3). Because one formula unit of K_2SO_4 produces one $SO_4^{2-}(aq)$ ion and two $K^+(aq)$ ions in aqueous solution, the colligative molality is three times the molality, that is, $m_c = im = (3)(0.20\text{ m}) = 0.60\text{ m}_c$.

PRACTICE PROBLEM 16-2: A 1.00-mole sample of each of the following substances is dissolved in 0.500 kilograms of water: (a) $CH_3OH(l)$, methanol, an organic alcohol; (b) $AgNO_3(s)$; (c) $Ca(ClO_4)_2(s)$. Determine the colligative molality of each of the resulting solutions.

Answer: (a) 2.00 m_c; (b) 4.00 m_c; (c) 6.00 m_c

16-2. The Equilibrium Partial Pressure of a Pure Liquid Always Decreases When a Substance Is Dissolved in the Liquid

Consider a solution of a nonvolatile solute such as sucrose dissolved in a volatile solvent such as water. As Figure 16.1 suggests, the vapor pressure of the solvent over a solution will be less than the vapor pressure of pure solvent at the same temperature.

The equilibrium vapor pressure results when the rate of evaporation of the solvent from the solution is equal to the rate of condensation of the solvent from the vapor (Section 15-6). The rate of evaporation of the solvent from a solution is less than that of the pure solvent because the presence of solute molecules at the surface of the solution decreases the number of solvent molecules per unit area of surface (Figure 16.1). The rate of condensation is directly proportional to the number of molecules per unit volume in the vapor, which in turn is proportional to the vapor pressure. The lower rate of evaporation is balanced by a lower rate of condensation; the result is a lower equilibrium vapor pressure (Figure 16.2).

As Figure 16.1 implies, the vapor pressure of the solvent over a solution is directly proportional to the fraction of solvent particles at the surface. Because

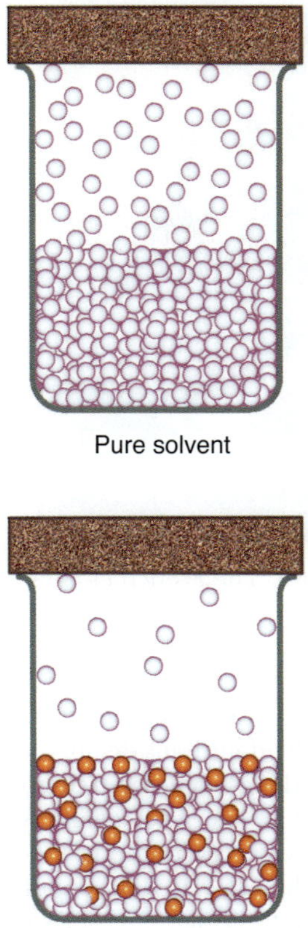

Figure 16.1 The effect of a nonvolatile solute on the equilibrium vapor pressure of a solvent at a fixed temperature. The solute molecules lower the equilibrium vapor pressure of the solvent relative to that of the pure solvent by partially blocking the escape of solvent molecules from the surface of the solution.

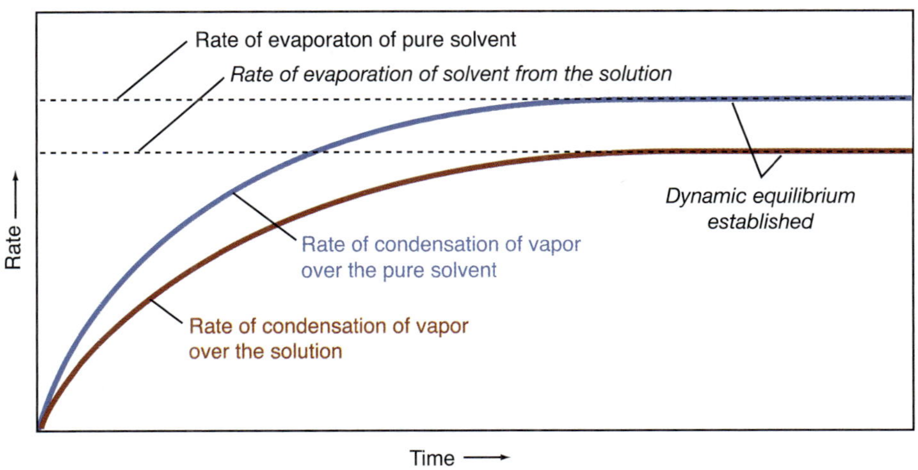

Figure 16.2 The effect of a nonvolatile solute on the solvent evaporation rate. The solute molecules lead to a lower rate of evaporation of the solvent, which in turn leads to a lower equilibrium vapor pressure of the solvent.

Figure 16.3 The French chemist François-Marie Raoult (1830–1901). Raoult taught at the University of Grenoble situated in the French Alps and later became department chair, a post he held until his death. Raoult's early work was concerned with voltaic cells, but he later became interested in the properties of solutions. He was the first to characterize how solutes could lower the freezing point of a solution and later showed how this effect could be used to determine the molecular mass of the solute. He is perhaps most famous for his discovery of the relation between solute concentration and vapor pressure lowering that now bears his name.

Generally, we shall denote the properties of the solvent by a subscript 1 and those of the solute by a subscript 2; or in the case of an aqueous solution, by *w* (for water) and *s* (for solute).

the number of moles is proportional to the number of particles, the mole fraction of the solvent (Equation 16.3) can be considered a particle fraction. Therefore, we can write the relation between the equilibrium vapor pressure of the solvent P_1 and the mole fraction of solvent x_1 as

$$P_1 \propto x_1$$

By introducing a proportionality constant, k, we can write

$$P_1 = kx_1 \tag{16.7}$$

The value of the proportionality constant is determined as follows. When the mole fraction of the solvent is unity, that is, when we have pure solvent, then P_1 is equal to the vapor pressure of the pure solvent, P_1°. Thus, when $x_1 = 1$ in Equation 16.7, we have $P_1 = P_1^\circ$. If we substitute $x_1 = 1$ and $P_1 = P_1^\circ$ into Equation 16.7, then we see that $k = P_1^\circ$. Using this determination of k, we can rewrite Equation 16.7 as

$$P_1 = x_1 P_1^\circ \tag{16.8}$$

Equation 16.8 was discovered by the French chemist F. M. Raoult and is known as **Raoult's law** (Figure 16.3). Raoult's law tells us that the equilibrium vapor pressure of the solvent over a solution is directly proportional to the mole fraction of the solvent in the solution. A solvent always obeys Raoult's law if the solution is sufficiently dilute.

The amount by which the vapor pressure of a solution is less than the vapor pressure of the pure solvent, that is, $P_1^\circ - P_1$, is called the **vapor pressure lowering.** Using Equation 16.8, we can express the vapor pressure lowering, ΔP_1, as

$$\Delta P_1 = P_1^\circ - P_1 = P_1^\circ - x_1 P_1^\circ = (1 - x_1)P_1^\circ \tag{16.9}$$

or

$$\Delta P_1 = x_2 P_1^\circ \tag{16.10}$$

where we have used the fact that $x_1 + x_2 = 1$ and where x_2 is the mole fraction of the solute.

EXAMPLE 16-3: The vapor pressure of water at 80°C is 355 Torr. (a) Use Raoult's law to calculate the vapor pressure of a solution made by dissolving 60.8 grams of the nonvolatile solid sucrose, $C_{12}H_{22}O_{11}(s)$, in 100.0 grams water. (b) Calculate the vapor pressure lowering of the water in the solution.

Solution: (a) The mole fraction of water in the solution, x_w, is

$$x_w = \frac{n_w}{n_w + n_s} = \frac{(100.0\text{ g})\left(\dfrac{1\text{ mol water}}{18.02\text{ g water}}\right)}{(100.0\text{ g})\left(\dfrac{1\text{ mol water}}{18.02\text{ g water}}\right) + (60.8\text{ g})\left(\dfrac{1\text{ mol sucrose}}{342.3\text{ g sucrose}}\right)}$$
$$= 0.969$$

The equilibrium vapor pressure of water over the solution is given by Raoult's law (Equation 16.8):

$$P_w = x_w P_w^\circ = (0.969)(355 \text{ Torr}) = 344 \text{ Torr}$$

(b) The vapor pressure lowering of water produced by the dissolved sucrose is

$$\Delta P_w = P_w^\circ - P_w = 355 \text{ Torr} - 344 \text{ Torr} = 11 \text{ Torr}$$

which corresponds to a vapor pressure of water decrease of 3.1%.

Note that we could also have obtained the same result for the vapor pressure lowering using Equation 16.9,

$$\Delta P_w = (1 - x_w)P_w^\circ = (1 - 0.969)(355 \text{ Torr}) = (0.031)(355 \text{ Torr}) = 11 \text{ Torr}$$

PRACTICE PROBLEM 16-3: A 41.3-gram sample of a solution of naphthalene, $C_{10}H_8(s)$, in benzene, $C_6H_6(l)$, has a vapor pressure at 20.0°C of 63.9 Torr. The vapor pressure of pure benzene at 20.0°C is 75.0 Torr. Assuming that naphthalene has a negligible vapor pressure over the solution, use the vapor pressure–lowering expression given by Equation 16.10 to calculate the mole fraction of naphthalene and the number of grams of naphthalene in the solution.

Answer: $x_{\text{naphth}} = 0.148$; 9.16 grams

The above examples of vapor pressure–lowering calculations involved only nonionic solutions. When the solute is an ionic substance, we must be sure to take into account that the solute dissociates into ions in solution when we calculate the mole fraction of solute to be used in Equation 16.10.

EXAMPLE 16-4: Calculate the vapor pressure lowering at 30°C of a 3.00-m sodium acetate, $NaCH_3COO(aq)$, solution.

Solution: We use Equation 16.10. The mole fraction of $NaCH_3COO(aq)$ is given by

$$x_{NaCH_3COO} = \frac{n_{NaCH_3COO}}{n_{H_2O} + n_{NaCH_3COO}}$$

Because $NaCH_3COO(aq)$ is a strong electrolyte, it dissociates completely into $Na^+(aq)$ and $CH_3COO^-(aq)$ ions. Assume a one-kilogram solution. The number of moles of ions in one kilogram of a 3.00-m $NaCH_3COO(aq)$ solution is

$$\text{moles of ions} = (2)(3.00 \text{ mol·kg}^{-1})(1.00 \text{ kg}) = 6.00 \text{ mol}$$

and the number of moles of water is 1000 g/18.02 g·mol^{-1} = 55.49 moles. Therefore,

$$x_{NaCH_3COO} = \frac{6.00 \text{ mol}}{55.49 \text{ mol} + 6.00 \text{ mol}} = 0.0976$$

The vapor pressure of water at 30°C is 31.8 Torr (Table 15.7), and so Equation 16.10 tells us that

$$\Delta P_{H_2O} = x_{NaCH_3COO} P^{\circ}_{H_2O} = (0.0976)(31.8 \text{ Torr}) = 3.10 \text{ Torr}$$

PRACTICE PROBLEM 16-4: Calculate the vapor pressure lowering at 40°C of a 0.70-m ammonium sulfate, $(NH_4)_2SO_4(aq)$, solution.

Answer: 2.0 Torr

16-3. Nonvolatile Solutes Increase the Boiling Point of a Liquid

Recall that the boiling point is the temperature at which the equilibrium vapor pressure over the liquid phase equals the atmospheric pressure. We know from Section 16-2 that the equilibrium vapor pressure of the solvent over a solution containing a nonvolatile solute is less than that for the pure solvent at the same temperature. Therefore, the temperature at which the equilibrium vapor pressure reaches atmospheric pressure is higher for the solution than for the pure solvent (Figure 16.4). In other words, the boiling point of the solution, T_b, is higher than the boiling point of the pure solvent, T_b°. The amount by which the boiling point of the solution exceeds the boiling point of the pure liquid, that is, $T_b - T_b^{\circ}$, is called the **boiling point elevation** (Figure 16.5).

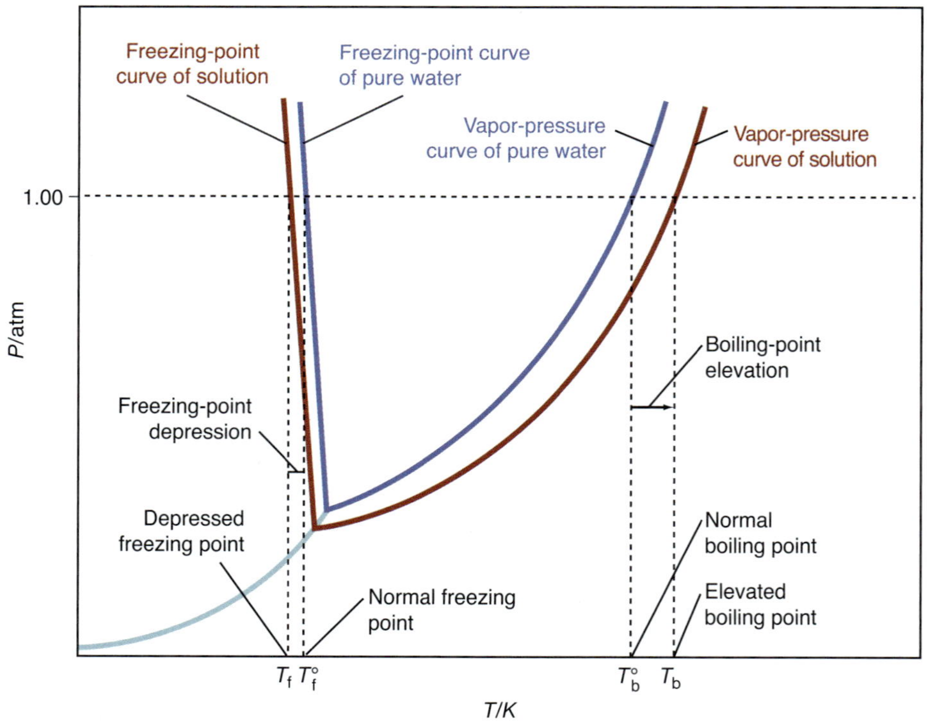

Figure 16.4 Phase diagrams for pure water (blue lines) and for water containing a nonvolatile solute (red lines). The presence of the solute lowers the vapor pressure of the solvent. The reduced vapor pressure of the solvent results in an *increase* in the boiling point of the solution relative to that of the pure solvent and a *decrease* in the freezing point of the solution relative to that of the pure solvent.

In solutions with colligative molalities less than about 1.0 m_c, the mole fraction of the solute is directly proportional to the colligative molality of the solute, that is, $x_2 \propto m_c$ (see Problem 16-86). Because the vapor pressure lowering of the solvent, $P_1^\circ - P_1$, is directly proportional to the mole fraction of the solute, x_2 (Equation 16.10), and $x_2 \propto m_c$, the vapor pressure lowering is also proportional to the colligative molality.

$$P_1^\circ - P_1 \propto m_c \tag{16.11}$$

Figure 16.5 Pure water open to the atmosphere boils at 100°C, whereas a 50-50 mixture by volume of water and automobile antifreeze (ethylene glycol) boils at about 109°C. The pure water and the antifreeze mixture in this photo are at the same temperature (100°C), but only the pure water is boiling.

Figure 16.6 shows an enlarged version of the region of Figure 16.4 where several vapor pressure curves intersect the horizontal line at $P_1^\circ = 1.00$ atm. Because the vapor pressure lowering is small for most solutions, the vapor pressure curves are essentially straight lines over a small temperature range. Consequently, the vapor pressure lowering $P_1^\circ - P_1$ is proportional to the boiling point elevation, $T_b - T_b^\circ$. Combination of the result, $(P_1^\circ - P_1) \propto (T_b - T_b^\circ)$, with Equation 16.11 yields the result that the boiling-point elevation, $T_b - T_b^\circ$, is directly proportional to the colligative molality, m_c:

$$T_b - T_b^\circ \propto m_c \tag{16.12}$$

Equation 16.12 can be written as an equality by introducing the proportionality constant K_b

$$T_b - T_b^\circ = K_b m_c \tag{16.13}$$

Equation 16.13 tells us that the increase in the boiling point of a solution containing a nonvolatile solute is directly proportional to the colligative molality.

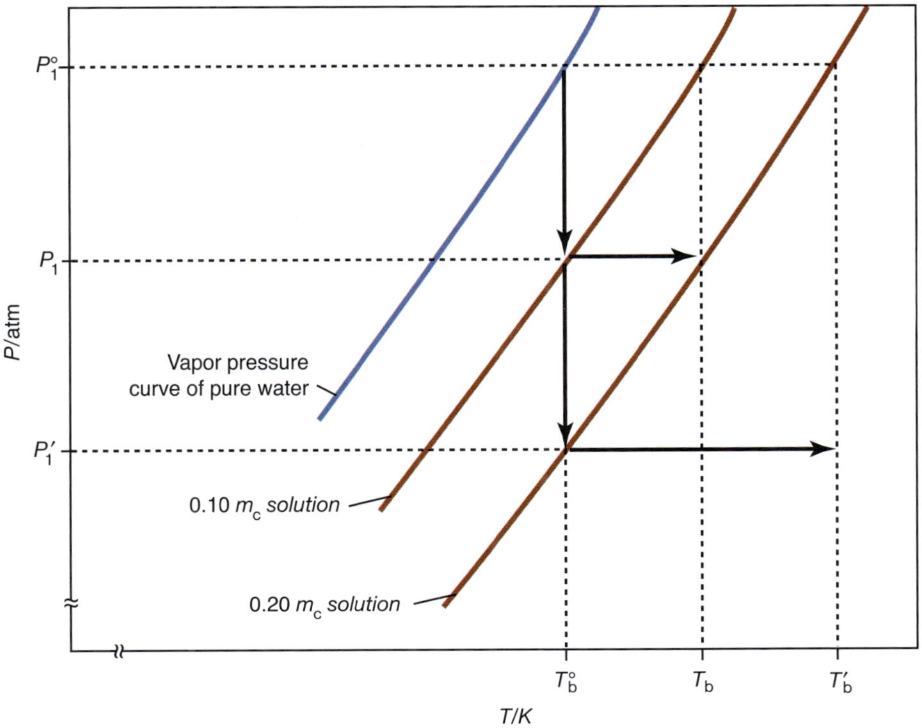

Figure 16.6 Expanded version of Figure 16.4 in the region where the vapor pressure curves intersect the $P = 1.00$ atm line (denoted P_1° here). Because the vapor pressure lowering is small unless m_c is large, the vapor pressure curves are essentially parallel straight lines over the temperature range T_b° to T_b, and therefore, $(P_1^\circ - P_1) \propto (T_b - T_b^\circ)$. The blue line is a portion of the vapor pressure curve for pure water, and the red lines are the vapor pressure curves for two different concentrations of a nonvolatile solute in water.

TABLE 16.2 Boiling point elevation constants (K_b) and freezing point depression constants (K_f) for various solvents

Solvent	Boiling point/°C	K_b/K·m_c^{-1}	Freezing point/°C	K_f/K·m_c^{-1}
benzene	80.09	2.64	5.49	5.07
camphor	(207.4)*	5.95	176	37.8
cyclohexane	80.73	2.92	6.59	20.8
ethanoic acid (acetic acid)	117.9	3.22	16.64	3.63
nitrobenzene	210.8	5.2	5.7	6.87
trichloromethane	61.17	3.80	−63.41	4.68
water	100.00	0.513	0.00	1.86

*sublimation point

That is, the boiling point elevation is directly proportional to the solute (particle) concentration. The value of K_b depends only on the solvent and is called the **boiling point elevation constant.** Values of K_b for several solvents are given in Table 16.2. Strictly speaking, Equation 16.13 is valid for dilute solutions only.

The magnitude of the boiling point elevation is usually small. For example, we see from Equation 16.13 and Table 16.2 that for a 1.00-m_c solution of glucose in water

$$T_b - T_b^\circ = (0.513\ \text{K}\cdot\text{m}_c^{-1})(1.00\ \text{m}_c) = 0.513\ \text{K}$$

EXAMPLE 16-5: Calculate the boiling point of a benzene-naphthalene, C_6H_6-$C_{10}H_8$, solution that is 20.0% by mass naphthalene. Assume that the vapor pressure due to naphthalene is negligible.

Solution: Taking a 100.0-gram sample for convenience, we have

$$\text{molality of naphthalene} = (20.0\ \text{g})\left(\frac{1\ \text{mol}}{128.17\ \text{g C}_{10}\text{H}_8}\right)\left(\frac{1}{0.0800\ \text{kg C}_6\text{H}_6}\right)$$

$$= 1.95\ \text{m} = 1.95\ \text{m}_c$$

We use Equation 16.13 along with the information in Table 16.2 to write

$$\Delta T_b = T_b - T_b^\circ = K_b m_c = (2.64\ \text{K}\cdot\text{m}_c^{-1})(1.95\ \text{m}_c) = 5.15\ \text{K} = 5.15^\circ\text{C}$$

where we have used the fact that the value of ΔT_b in kelvins is equal to the value of Δt_b in degrees Celsius.

$$t_b = 80.09^\circ\text{C} + 5.15^\circ\text{C} = 85.24^\circ\text{C}$$

PRACTICE PROBLEM 16-5: Seawater contains about 3.5% by mass dissolved solids. As a rough approximation, assume that the dissolved solids are predominantly NaCl(*aq*) [that is, 3.5 grams of NaCl(*aq*) in 96.5 grams of $H_2O(l)$] and estimate the normal boiling point of seawater.

Answer: 100.64°C

16-4. Solutes Decrease the Freezing Point of a Liquid

When an aqueous solution begins to freeze, the solid that separates out is usually pure ice. For example, ice formed from seawater is free of salt, and freezing is one method of preparing fresh water from seawater to provide pure drinking water (Figure 16.7). When an aqueous solution such as NaCl(*aq*) or seawater begins to freeze, we have pure ice in equilibrium with the solution.

Pure ice and a salt solution can coexist in equilibrium only if their vapor pressures are equal. But the vapor pressure of water over a NaCl (*aq*) solution is less than that of pure water, so the vapor pressure of the ice must be less than if it were in equilibrium with pure water. Consequently, the temperature of the ice must be less than 0°C, and the freezing point of the solution must be less than that of pure water (Figure 16.4). Thus, we see that the lowering of the vapor pressure of a solvent by a solute leads to a lowering of the freezing point of the solution relative to that of the pure solvent. This effect is called the **freezing point depression.**

Figure 16.7 Even though they are formed directly from seawater, icebergs that break off from the arctic ice shelves are composed of pure water. Because the density of ice is only about 91% of the density of liquid water, about 91% of the total mass of an iceberg is below the surface of the water.

By arguments analogous to those used to obtain the boiling point elevation equation, the magnitude of the freezing point depression produced by a solute is found to be proportional to its colligative molality m_c:

$$T_f^\circ - T_f = K_f m_c \qquad (16.14)$$

where T_f° is the freezing point of the pure solvent, T_f is the freezing point of the solution ($T_f^\circ > T_f$), and K_f is the **freezing point depression constant.** The value of K_f depends only on the solvent. Values of K_f for various solvents are given in Table 16.2. Once again we note that these results are valid only for dilute solutions of nonvolatile solutes.

The value of the freezing point depression constant of water is 1.86 $\text{K}\cdot\text{m}_c^{-1}$ (Table 16.2). Thus, we predict that an aqueous solution with a colligative molality of 0.50 m_c has a freezing point depression of

$$T_f^\circ - T_f = K_f m_c = (1.86\ \text{K}\cdot\text{m}_c^{-1})(0.50\ \text{m}_c) = 0.93\ \text{K}$$

and the freezing point is $T_f = 273.15\ \text{K} - 0.93\ \text{K} = 272.22\ \text{K}$ or −0.93°C.

Figure 16.8 A 30% by volume ethylene glycol (automobile antifreeze) and water mixture at −13.3°C. Note that the mixture remains as a liquid at −13.3°C, whereas pure water would be frozen solid.

The freezing point depression due to a dissolved substance is the basis of the action of **antifreezes.** The most commonly used antifreeze is ethylene glycol (systematic name: ethane-1,2-diol), which has a boiling point of 197°C and a freezing point of −17.4°C. The addition of ethylene glycol to water depresses the freezing point and elevates the boiling point of the solution relative to that of pure water. An ethylene glycol–water solution in which ethylene glycol is 30% by volume has a freezing point of about −14°C (Figure 16.8).

EXAMPLE 16-6: Estimate the freezing point of an aqueous solution of ethylene glycol that is 50.0% by mass (about 45% by volume) ethylene glycol.

Solution: As usual, we take a 100-gram sample. The molality of the ethylene glycol is

$$\text{molality of ethylene glycol} = (50.0\ \text{g})\left(\frac{1\ \text{mol}}{62.07\ \text{g}}\right)\left(\frac{1}{0.0500\ \text{kg H}_2\text{O}}\right) = 16.1\ \text{m}$$

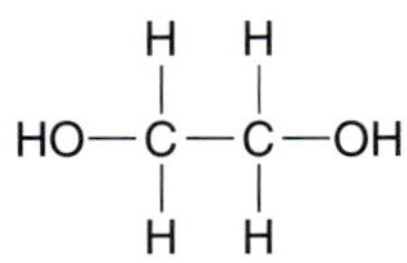

ethylene glycol

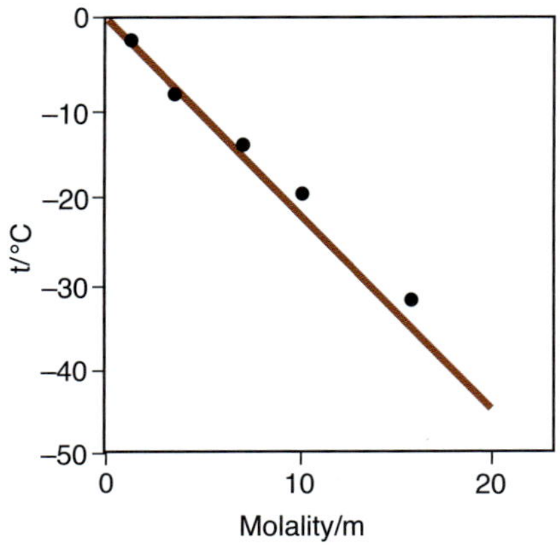

Figure 16.9 The freezing point of an aqueous ethylene glycol solution as a function of concentration calculated from Equation 16.14 (red line) compared to experimental data (black points). Observe how the two agree at low concentrations, but deviate at higher molalities. We say that the solution behaves ideally at low concentrations (where Raoult's law is valid) and nonideally at high concentrations (where Raoult's law is not valid).

because ethylene glycol is a nonionic solute, 16.1 m = 16.1 m_c. The freezing point depression of the solution is computed from Equation 16.14:

$$T_f^\circ - T_f = K_f m_c = (1.86\ K{\cdot}m_c^{-1})(16.1\ m_c) = 30.0\ K = 30.0°C$$

The freezing point of the solution is −30.0°C.

PRACTICE PROBLEM 16-6: Compute the molality (m) of a $CaCl_2(aq)$ solution that freezes at −5.25°C.

Answer: 0.941 m

Equation 16.14 (which is based on Raoult's law) is valid only for dilute solutions, or at most a few molal. In Example 16-6, the molality is 16.1 m, which is beyond the range of validity of Equation 16.14. Figure 16.9 shows the prediction of Equation 16.14 with actual experimental data for ethylene glycol–water solutions. Notice that the curve and data agree at low concentrations, but they deviate from each other at higher concentrations. For a 50.0% by mass ethylene glycol solution, the experimental value of ΔT_f is about 40°C instead of the 30°C given by Equation 16.14. Nevertheless, Equation 16.14 still gives us a rough estimate. The effectiveness of ethylene glycol as an antifreeze is a result of its high boiling point, its chemical stability, and the tendency of the ice that freezes out of the solution to form a slushy mass rather than a solid block. In the absence of a sufficient amount of antifreeze, the 9% volume expansion of water on freezing can generate a force of 200 000 kPa (30 000 lb·in^{-2}). If there is no room available to accommodate the increase in volume, such a force is more than sufficient to rupture a radiator or even a metal engine block.

The freezing-point depression Equation 16.14 can be combined with Equation 16.6 to obtain the equation

$$T_f^\circ - T_f = K_f i m \tag{16.15}$$

The i in Equation 16.15 is called the **van't Hoff *i*-factor**. When Equation 16.15 is applied to dilute solutions of strong electrolytes, i comes out to be equal to the number of ions produced per formula unit. For example, $i = 2$ for $NaCl(aq)$, $i = 3$ for $CaCl_2(aq)$, and so on. When Equation 16.15 is applied to solutions of weak electrolytes, however, the van't Hoff i-factor gives us a measure of the percentage of dissociation of the electrolyte in an aqueous solution around 0°C. For example, consider the weak electrolyte acetic acid, $CH_3COOH(aq)$, which partially dissociates in aqueous solution, as shown by the equation

$$CH_3COOH(aq) + H_2O(l) \rightleftharpoons CH_3COO^-(aq) + H_3O^+(aq)$$

Only a small percentage (< 5%) of the $CH_3COOH(aq)$ dissociates in solution. The partial dissociation gives rise to a van't Hoff i-factor in the range $1.00 < i < 2.00$. Suppose that the freezing point of a 0.0500-m aqueous solution of acetic acid is

NEW ANTIFREEZE MOLECULES FROM FISH BLOOD?

The freezing point of seawater is about −1.85°C; therefore, this temperature is also that of the seawater surrounding the polar ice shelves. Several species of fishes can live in the cold waters of the Ross Sea of Antarctica near the sea ice (Figure 16.10). On the basis of the total concentration of solutes dissolved in the blood serum of these fishes, the freezing point of the serum should be −1.46°C. So the blood of these fishes should freeze in the −1.85°C water, but it does not because the fishes are protected from freezing by "antifreeze" proteins in their blood. These proteins have an enhanced capacity to lower the freezing point of water. In fact, the freezing point depression of solutions containing the antifreeze proteins is much greater than that predicted using Equation 16.14. The freezing point of the fish blood serum, after removal of all the salts (but not the antifreeze proteins), is −0.60°C. The measured concentration of antifreeze proteins in the fish blood is about 3×10^{-4} m_c. According to Equation 16.14, the predicted freezing point depression of a 3×10^{-4} m_c solution of the antifreeze protein is only

$$T_f^\circ - T_f = (1.86\ \mathrm{K{\cdot}m_c^{-1}})(3 \times 10^{-4}\ \mathrm{m_c}) = 0.0006\ \mathrm{K}$$

Such a solution should freeze at −0.0006°C, but it actually freezes at −0.60°C. The observed freezing point of the antifreeze protein solution is thus 1000 times greater than that predicted by Equation 16.14. How the antifreeze proteins work is not entirely clear. The currently accepted hypothesis is that the proteins are adsorbed onto the surfaces of ice crystal nuclei, thereby stopping further growth of the crystals. Recently synthesized analogs of these proteins may be useful in making ice-resistant coatings for aircraft, roads, or crops and perhaps as additives to prevent ice crystal formation in frozen foods.

Figure 16.10 Antarctic mackerel icefish, *Champeocephalus gunnari*, one of the species of fish that have "antifreeze" proteins in their blood.

found to be −0.095°C; what is the value of i? Rearranging Equation 16.15, we find that

$$i = \frac{T_f^\circ - T_f}{K_f m} = \frac{0.095\ \mathrm{K}}{(1.86\ \mathrm{K{\cdot}m_c^{-1}})(0.0500\ \mathrm{m})} = 1.02$$

Because each acetic acid molecule that dissociates yields two solute particles, a value of $i = 1.02$ means that about 2% [i.e., $(1.02 - 1.00) \times 100 = 2\%$] of the acetic acid molecules are dissociated in an aqueous solution that is 0.0500 m in acetic acid. The *i*-factor was first noted by the Dutch chemist Jacobus Henricus van't Hoff (Frontispiece). He and Swedish chemist Svante August Arrhenius (Chapter 6 Frontispiece) later showed i to be related to the percent dissociation of a solute in solution. Both chemists were awarded Nobel prizes for their contributions to our understanding of solutions.

EXAMPLE 16-7: The freezing point depression of a 0.050-m HF(aq) solution is −0.100°C. Calculate the percentage of the HF(aq) molecules in the solution that are dissociated into ions according to

$$HF(aq) + H_2O(l) \rightarrow H_3O^+(aq) + F^-(aq)$$

Solution: Rearranging Equation 16.15, we find that

$$i = \frac{T_f^\circ - T_f}{K_f m} = \frac{0.100\ \text{K}}{(1.86\ \text{K}\cdot\text{m}_c^{-1})(0.050\ \text{m})} = 1.08$$

Thus, about 8% of the HF molecules are dissociated.

PRACTICE PROBLEM 16-7: A 1.00-m solution of acetic acid, CH_3COOH, in benzene has a freezing point of 2.96°C. Use the data in Table 16.2 to calculate the value of i and suggest an explanation for the unusual result. (Hint: If i is less than 1.0, each formula unit that dissolves yields less than one solute particle, an outcome suggesting aggregation of solute particles.)

Answer: $i = 0.50$; formation of dimers of composition $(CH_3COOH)_2$

One method used to melt ice on streets and sidewalks is to spread rock salt, NaCl(s), crystals on the ice. The solubility of NaCl(s) in liquid water around 0°C is 4.8 m. In water sodium chloride completely dissociates into $Na^+(aq)$ and $Cl^-(aq)$, and so the colligative molality is twice the molality. Thus the freezing-point depression of a saturated aqueous NaCl(aq) solution is

$$T_f^\circ - T_f = K_f m_c = (1.86\ \text{K}\cdot\text{m}_c^{-1})(9.6\ \text{m}_c) = 18\ \text{K}$$

The freezing point of the solution, therefore, is −18°C. On contact with ice, salt dissolves to form a saturated aqueous solution with a freezing point of −18°C. Above −18°C spreading salt on ice causes the ice to melt as it forms a concentrated aqueous salt solution. When the ice temperature is below −18°C, spreading salt on it will not melt the ice because ice freezes out from the saturated NaCl(aq) solution. This effect is also used in the preparation of homemade ice cream to obtain a low enough temperature to solidify cream. The temperature of 0°F on the Fahrenheit temperature scale was chosen as the freezing point of a saturated solution of sodium chloride in water and thus 0°F corresponds to −18°C, whereas the freezing point of pure water on the Fahrenheit temperature scale is 32°F.

One problem with the use of rock salt as a roadway deicer is its impact on the environment, including salt contamination of local water sheds, and accelerated corrosion of automobiles, bridges, and other structures. Some less environmentally damaging but significantly more costly alternatives are calcium magnesium acetate and potassium acetate, the latter currently being used in Yosemite National Park.

16-5. Osmotic Pressure Requires a Semipermeable Membrane

Suppose we place pure water in one beaker and an equal volume of seawater in another beaker and then place both beakers under a bell jar, as shown in Figure 16.11a. We observe that as time passes the volume of pure water decreases and the volume of seawater increases (Figure 16.11b). The pure water has a higher equilibrium vapor pressure than the seawater; thus, the rate of condensation of the water into the seawater is greater than the rate of evaporation of the water from the seawater. The net effect is the transfer of water, via the vapor phase, from the beaker with pure water to the beaker with seawater. This transfer continues until no pure liquid water remains, and the seawater ends up diluted.

If pure water and seawater are separated by a membrane that is permeable to water but not to the ions in seawater, then the water passes directly through the membrane from the pure water side of the membrane to the seawater side (Figure 16.12). Such a membrane can be composed of polymer molecules inter-

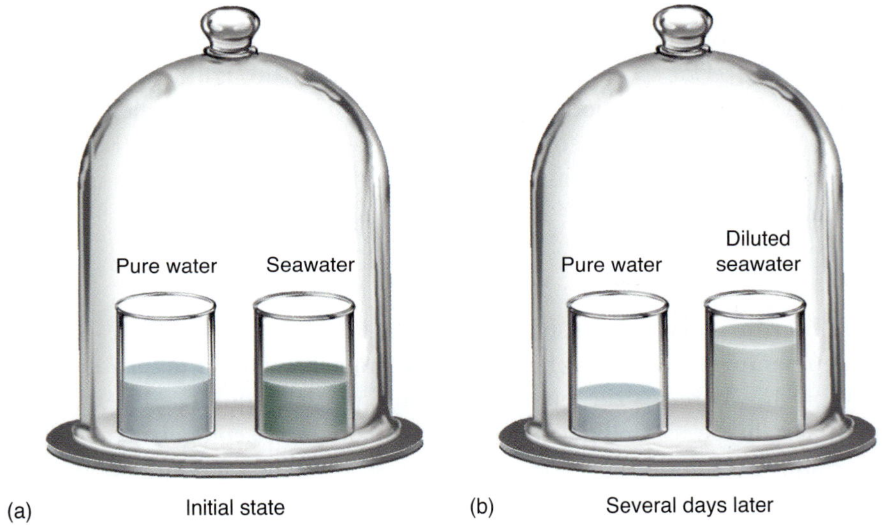

Figure 16.11 One beaker contains pure water, and the other contains seawater. (a) The equilibrium vapor pressure over the pure water is greater than that over the seawater solution. (b) As time passes, pure water is transferred via the vapor phase from the beaker containing pure water to the beaker containing seawater, thereby diluting the seawater. If we wait long enough, all the pure liquid water will transfer to the seawater beaker.

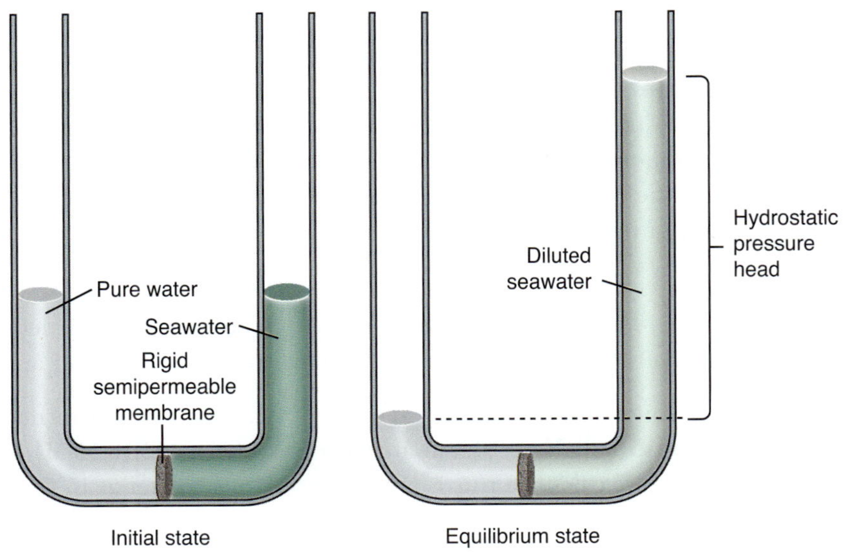

Figure 16.12 Passage of water through a rigid semipermeable membrane separating pure water from seawater. The water passes through the membrane until the escaping tendency of the water from the seawater equals the escaping tendency of the pure water. The escaping tendency of water from the seawater increases as the seawater is diluted and also as the hydrostatic pressure due to the seawater column increases. At equilibrium the osmotic pressure, Π, is equal to the hydrostatic pressure.

twined like a mass of cooked spaghetti. The small neutral water molecules can move between cavities in the strands, whereas the solvated ions such as $Na^{+}(aq)$ and $Cl^{-}(aq)$ are much too large to move through the membrane channels. This kind of membrane is called a **semipermeable membrane** (Figure 16.13). The tendency of the water to pass through the membrane is called the **escaping tendency** and is directly proportional to the vapor pressure of water over the solution. The escaping tendency of water from pure water is greater than the escaping tendency of water from seawater because pure water has a higher vapor pressure than seawater. As water passes through the membrane to the seawater side, the escaping tendency of the water in the seawater increases, not only because the seawater is being diluted, but also because of the increased pressure on the seawater side of the membrane. This pressure increase arises from the pressure due to the height of the water column that develops. Recall from Chapter 13 that a column of liquid exerts a pressure proportional to the height of the column of liquid. The column of seawater rises until the escaping tendency of the water in the seawater is equal to the escaping tendency of the pure water. When this condition is attained, an equilibrium exists and the column of seawater no longer rises. The pressure of the liquid column produced in this process is called the **osmotic pressure** (Figure 16.12). The spontaneous passage of solvent through a semipermeable membrane from one solution to a more concentrated solution is called **osmosis.**

Figure 16.13 A molecular view of a semipermeable membrane. The membrane is composed of intertwined polymer molecules (see Interchapter S). The cavities and chains that occur throughout the membrane are large enough to allow water molecules to pass through, but are too small to allow hydrated ions to pass through.

www

See Interchapter S at www.McQuarrieGeneralChemistry.com.

The relationship between osmotic pressure, Π, solute concentration, and temperature for a solution was also discovered by van't Hoff (Frontispiece) and is given by the equation

$$\Pi = RTM_c \tag{16.16}$$

where R is the molar gas constant ($R = 0.083145\ \text{L·bar·mol}^{-1}\text{·K}^{-1}$), T is the absolute temperature, and M_c is the **colligative molarity.** The value of M_c is given by $M_c = iM$, where i is the number of solute particles produced per formula unit that is dissolved. Colligative molarity is analogous to colligative molality. In dilute aqueous solutions $M_c \approx m_c$. If we express M_c as the total number of moles of dissolved solute divided by the volume of the solution, n/V, then Equation 16.16 has the same form as the ideal gas equation, namely, $\Pi V = nRT$.

We have shown that we can increase the escaping tendency of water from a solution either by decreasing the solute concentration of the solution or by increasing the temperature of the solution. The escaping tendency of water from a solution also can be increased by applying pressure to the solution. The increase in pressure increases the average energy of the solvent molecules and thereby increases their escaping tendency. This effect is the basis of osmotic pressure. The osmotic pressure is the pressure that must be applied to the solution to increase the vapor pressure (escaping tendency) of the solvent to a value equal to the vapor pressure of the pure solvent at that temperature.

EXAMPLE 16-8: As a rough approximation, seawater can be regarded as a 0.55 M $NaCl(aq)$ solution. Estimate the osmotic pressure of seawater at 15°C.

Solution: Because NaCl dissociates in water to yield two ions per formula unit, the colligative molarity of seawater is approximately (2)(0.55 M). From Equation 16.16, we calculate the osmotic pressure of seawater at 15°C as

$$\Pi = RTM_c = (0.083145 \text{ L·bar·K}^{-1}\text{·mol}^{-1})(288 \text{ K})(2)(0.55 \text{ mol·L}^{-1}) = 26 \text{ bar}$$

The osmotic pressure of seawater is 26 times higher than atmospheric pressure (one bar). The osmotic pressure effect is by far the largest in magnitude of all the colligative property effects.

PRACTICE PROBLEM 16-8: Determine the osmotic pressure at 25°C of a 0.10 M $CaCl_2(aq)$ solution.

Answer: 7.4 bar

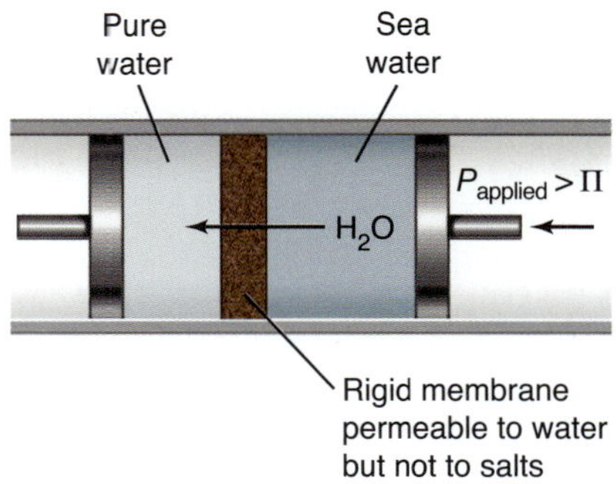

Figure 16.14 Reverse osmosis. A rigid semipermeable membrane separates pure water from seawater. A pressure in excess of the osmotic pressure of seawater (26 bar at 15°C) is applied to the seawater, and this increases the escaping tendency of water from the seawater to a value above that of pure water. Under these conditions, the net flow of water is from the seawater side through the semipermeable membrane to the pure water side, thus producing fresh water from seawater.

If a pressure in excess of 26 bar is applied to seawater at 15°C (Example 16.8), then the escaping tendency of the water in the seawater will exceed that of pure water. Consequently, pure water can be obtained from seawater by using a rigid semipermeable membrane and an applied pressure in excess of the osmotic pressure of 26 bar. This process is known as **reverse osmosis** (Figure 16.14). Reverse osmosis units are commercially available and are used to obtain fresh water from salt water using a variety of semipermeable membranes, the most common of which is cellulose acetate (Figures 16.15 and 16.16).

The magnitude of the osmotic pressure effect makes osmotic pressure measurements an especially powerful method for the determination of the molecular masses of proteins. Proteins have large molecular masses and thus yield a relatively small number of solute particles for a given dissolved mass. In fact, osmotic pressure is the only colligative effect sufficiently sensitive to provide useful molecular information about proteins.

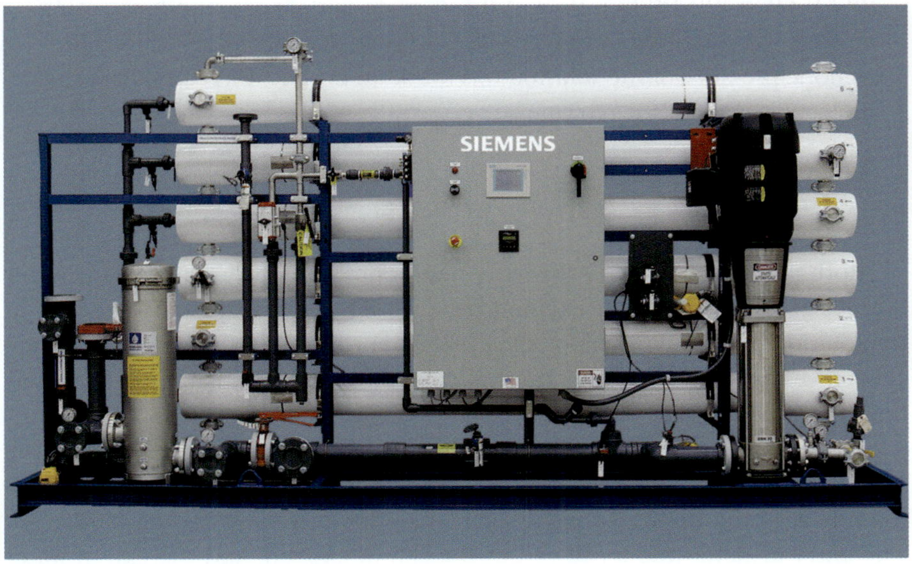

Figure 16.15 A commercial reverse osmosis unit. Suspended solids, including bacteria, are blocked by mechanical exclusion, and dissolved salts are removed using semipermeable membranes (Figure 16.14). The unit shown reduces the salt concentration to less than 5% of the initial concentration.

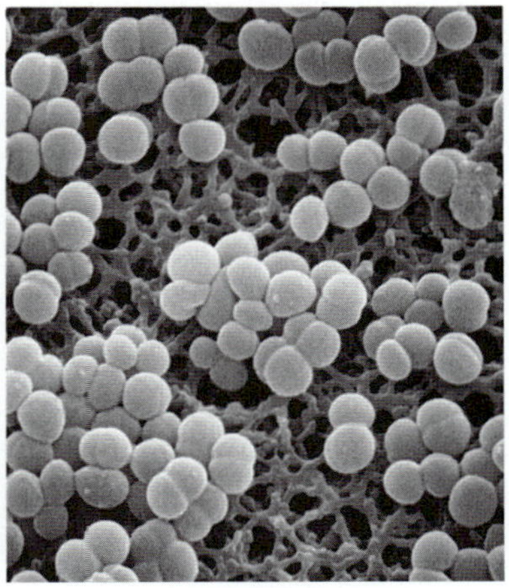

Figure 16.16 Bacterial cells on a membrane filter. A microporous membrane filter has pores so small that bacteria are trapped and thereby removed from the water. Similar membranes are used to produce "cold-filtered" bacteria-free beverages such as beer, thereby eliminating the need for pasteurization (heating) to kill bacteria.

EXAMPLE 16-9: A 4.00-gram sample of human hemoglobin (nondissociating in solution) was dissolved in water to make 0.100 L of solution. The osmotic pressure of the solution at 7.4°C was found to be 13.4 mbar. Calculate the molecular mass of the hemoglobin.

Solution: The concentration of the hemoglobin in the aqueous solution is (Equation 16.16)

$$M_c = \frac{\Pi}{RT} = \frac{13.4 \times 10^{-3}\ \text{bar}}{(0.083145\ \text{L}\cdot\text{bar}\cdot\text{K}^{-1}\cdot\text{mol}^{-1})(280.6\ \text{K})} = 5.74 \times 10^{-4}\ \text{mol}\cdot\text{L}^{-1}$$

The molecular mass can be calculated from the concentration of the protein because the dissolved mass is known. We have the correspondence

$$5.74 \times 10^{-4}\ \text{mol} \cdot \text{L}^{-1} \Bumpeq \frac{4.00\ \text{g}}{0.100\ \text{L}} = 40.0\ \text{g}\cdot\text{L}^{-1}$$

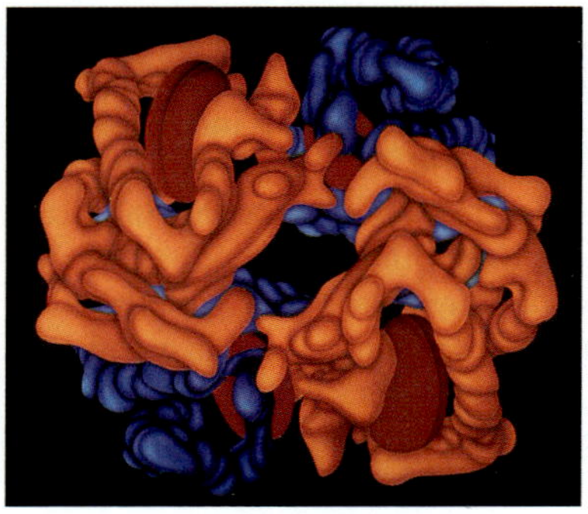

Figure 16.17 A computer-generated model of human hemoglobin with a molecular mass of about 70 000.

and therefore

$$5.74 \times 10^{-4}\ \text{mol} \Bumpeq 40.0\ \text{g}$$

By dividing both sides of this stoichiometric correspondence by 5.74×10^{-4}, we find that

$$1\ \text{mol} \Bumpeq 69\ 700\ \text{g}$$

The molecular mass of the hemoglobin is 69 700 (Figure 16.17). Protein molecular masses can be as large as 1 000 000.

PRACTICE PROBLEM 16-9: A 0.550-gram sample of a nondissociating enzyme was dissolved in water at 25°C to yield 50.0 mL of solution. The osmotic pressure of the solution was found to be 19.0 Torr. Calculate the molecular mass of the enzyme.

Answer: 1.08×10^4

Both plant and animal cells have membranes that are permeable to water but not to some solutes, for example, sucrose. The colligative concentration of the solution inside a typical biological cell is approximately 0.3 M_c. Most biological cells have about the same internal colligative molarity as the extracellular fluid in which the cells reside.

Water passes spontaneously through a biological cell membrane from the side with the lower colligative molarity (higher water-escaping tendency) to the side with the higher colligative molarity (lower water-escaping tendency). The entry of water into a cell causes the cell to expand, and the exit of water from the cell causes the cell to contract. The cell assumes its normal volume when it is placed in a solution with a colligative molarity of 0.3 M_c (Figure 16.18).

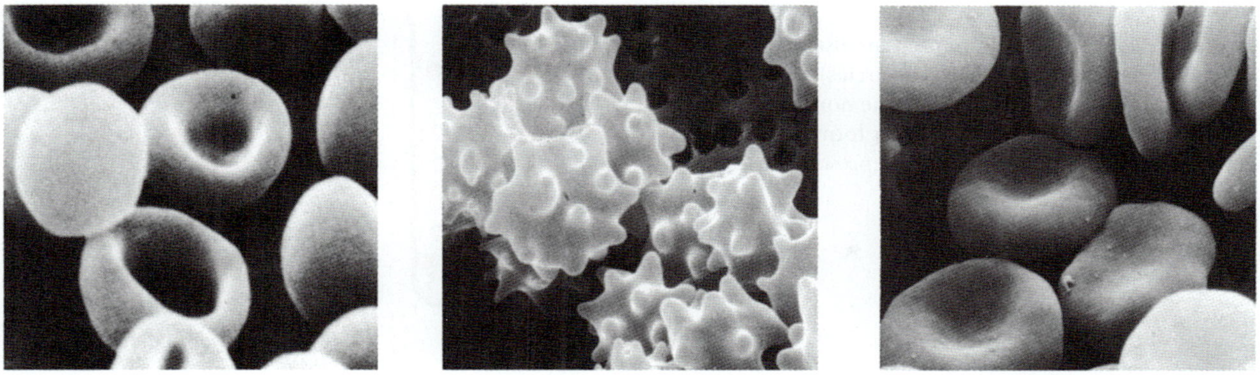

(a) (b) (c)

Figure 16.18 Osmosis in blood cells. (a) Blood cells neither contract nor expand when placed in a solution whose colligative molarity is equal to that of the solution inside the cell (isotonic solution). (b) Blood cells contract as a result of water loss when placed in a solution with a colligative molarity greater than 0.3 M_c (hypertonic solution). (c) Blood cells expand when placed in a solution with a colligative molarity less than 0.3 M_c (hypotonic solution).

More concentrated solutions cause a cell to contract, and less concentrated solutions cause it to expand. When cells are placed in distilled water at 27°C, equilibrium should be achieved at an internal cell pressure equal to the osmotic pressure of a 0.30-M_c solution, that is,

$$\Pi = RTM_c = (0.083145\ \text{L·bar·K}^{-1}\text{·mol}^{-1})(300\ \text{K})(0.30\ \text{mol·L}^{-1}) = 7.5\ \text{bar}$$

Because a pressure of 7.5 bar cannot be sustained by most animal cell membranes, the cells burst (lysis). Plant cell walls, in contrast, are rigid and can tolerate a pressure of 7.5 bar. In fact, the entry of water into plant cells gives green plants the rigidity required to stand erect, and causes them to wilt when dry.

16-6. The Components of an Ideal Solution Obey Raoult's Law

In the preceding sections of this chapter, we have discussed solutions of nonvolatile solutes in volatile solvents. In this section we shall discuss solutions consisting of two volatile liquids. We define an **ideal solution** as follows: a solution of two components, A and B, is said to be ideal if the interactions between an A molecule and a B molecule are the same as those between two A molecules or two B molecules. In an ideal solution the A and B molecules are randomly distributed throughout the solution, including the region near the surface (Figure 16.19). When the molecules of the two components are very similar, the solution is essentially ideal. For example, benzene and toluene, which have similar shapes and charge distributions (see margin), form ideal solutions.

benzene, C_6H_6

toluene, $C_6H_5CH_3$

In an ideal solution of two volatile liquids, the vapor pressure of each component is given by Raoult's law, so

$$P_A = x_A P_A^\circ \quad \text{and} \quad P_B = x_B P_B^\circ \tag{16.17}$$

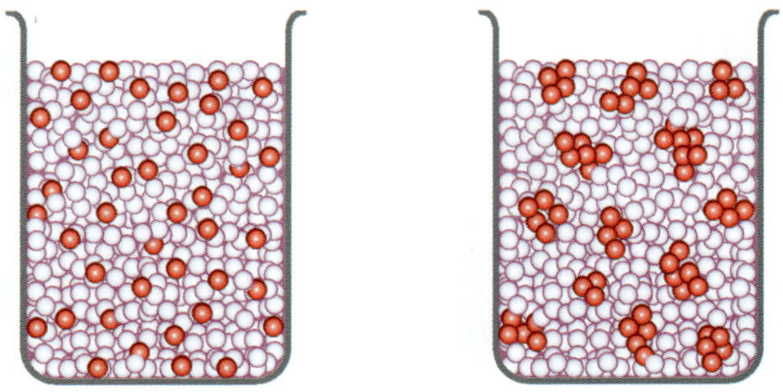

Figure 16.19 Random and nonrandom distribution of two types of molecules, A (open circles) and B (solid circles), in solution. Note in the nonrandom case that AAA ... and BBB ... clusters form in the solution. The random distribution represents an ideal solution.

In fact, Equations 16.17 serve as another definition of an ideal solution. The vapor over such a solution consists of both components, so the total vapor pressure is

$$P_{\text{total}} = P_{\text{A}} + P_{\text{B}} \tag{16.18}$$

Substituting Equation 16.17 into Equation 16.18 gives

$$P_{\text{total}} = x_{\text{A}} P_{\text{A}}^{\circ} + x_{\text{B}} P_{\text{B}}^{\circ} \tag{16.19}$$

Because $x_{\text{A}} + x_{\text{B}} = 1$, we can set $x_{\text{B}} = 1 - x_{\text{A}}$ and write Equation 16.19 as

$$P_{\text{total}} = P_{\text{B}}^{\circ} + x_{\text{A}} (P_{\text{A}}^{\circ} - P_{\text{B}}^{\circ}) \tag{16.20}$$

Notice that for pure B, $x_{\text{A}} = 0$, so $P_{\text{total}} = P_{\text{B}}^{\circ}$, and that when $x_{\text{A}} = 1$ (pure A), $P_{\text{total}} = P_{\text{A}}^{\circ}$.

If we plot P_{total} versus x_{A} for an ideal solution, then we obtain a straight line with an intercept of P_{B}° and a slope of $P_{\text{A}}^{\circ} - P_{\text{B}}^{\circ}$ (Appendix A.6). The total equilibrium vapor pressure of a benzene-toluene solution is plotted against the mole fraction of benzene in Figure 16.20. These data confirm that a benzene-toluene solution is an ideal solution. Note also from Figure 16.20 that because the total vapor pressure over an ideal solution is always between the vapor pressures of the two pure components, the solution will boil at a temperature between the boiling points of the two pure components.

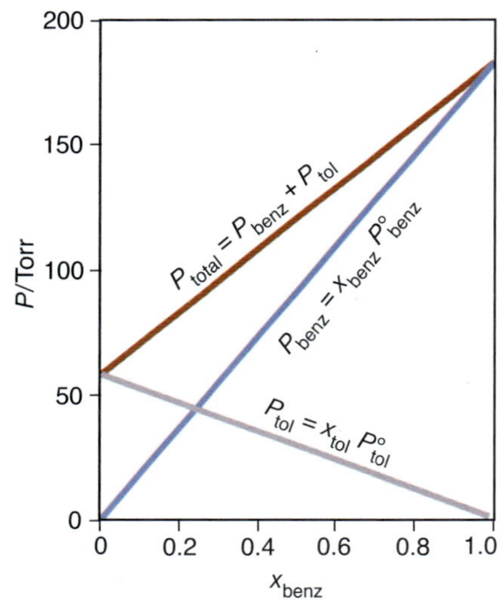

Figure 16.20 Equilibrium vapor pressure at 40°C versus the mole fraction of benzene in solutions of benzene (benz) and toluene (tol). The solutions are essentially ideal, and the equilibrium vapor pressures of benzene and toluene are given by Raoult's law (Equation 16.17). The total pressure is linearly related to x_{benz} when Raoult's law holds for both components, that is, when the solution is ideal (Equation 16.20).

EXAMPLE 16-10: Given that the equilibrium vapor pressures of pure benzene and pure toluene are 183 Torr and 59.2 Torr, respectively, calculate (a) the total vapor pressure over a $x_{\text{benz}} = x_{\text{tol}} = 0.500$ solution and (b) the mole fraction of benzene in the vapor.

Solution: (a) The partial pressures of benzene and toluene over the solution are given by Raoult's law:

$$P_{\text{benz}} = x_{\text{benz}} P_{\text{benz}}^{\circ} = (0.500)(183\ \text{Torr}) = 91.5\ \text{Torr}$$

$$P_{\text{tol}} = x_{\text{tol}} P_{\text{tol}}^{\circ} = (0.500)(59.2\ \text{Torr}) = 29.6\ \text{Torr}$$

The total vapor pressure is the sum of the partial pressures:

$$P_{\text{total}} = P_{\text{benz}} + P_{\text{tol}} = 91.5\ \text{Torr} + 29.6\ \text{Torr} = 121.1\ \text{Torr}$$

(b) The pressure of a gas is directly proportional to the number of moles of the gas, so the mole fraction of benzene in the vapor, y_{benz}, is given by

$$y_{benz} = \frac{n_{benz}}{n_{benz} + n_{tol}} = \frac{P_{benz}}{P_{benz} + P_{tol}} = \frac{91.5 \text{ Torr}}{121.1 \text{ Torr}} = 0.756$$

We have used y for the mole fraction in the vapor to distinguish it from the mole fraction, x, in the solution.

PRACTICE PROBLEM 16-10: Given that the total pressure over a solution of benzene and toluene is 100.0 Torr, calculate the mole fraction of benzene in the liquid and the vapor phases. (Take $P^{\circ}_{tol} = 59.2$ Torr and $P^{\circ}_{benz} = 183$ Torr.)

Answer: $x_{benz} = 0.329$; $y_{benz} = 0.602$

From Example 16-10 we see that the vapor over a benzene-toluene solution is richer in benzene, the more volatile component, than is the solution. If this vapor is condensed and then reevaporated, the resulting vapor will be even richer in benzene. If this condensation-evaporation process is repeated many times, a separation of the benzene and toluene is achieved. Such a process is called **fractional distillation** and is carried out automatically in a distillation column of the type shown in Figure 16.21. A fractional distillation column differs from an ordinary distillation column (Section 2-3) in that the former is packed with glass beads, glass rings, or glass wool. The packing material provides a large surface area for the repeated condensation-evaporation process.

A plot of the mole fraction of benzene in a benzene-tuoluene mixture during successive stages of fraction distilation is shown in Figure 16.22. Looking at Figure 16.22, if we start with a benzene-toluene mixture that has a mole fraction of benzene of $x_{benz} = 0.20$, then in the first stage of boiling, the vapor will be richer in benzene (the more volatile component) with a mole fraction $x_{benz} \approx 0.39$. This results in a condensate on the first stage of the column that is similarly richer in benzene. When this condensate again boils and recondenses (stage two), the resulting condensate has a new mole fraction of benzene of $x_{benz} \approx 0.59$, about three times richer in benzene than the original mixture. If this process is allowed to continue up the column, the condensate coming out of the top of the column will be essentially pure benzene.

Remarkable separations can be achieved with elaborate fractional distillation units. Only 0.015% of the hydrogen atoms in regular water are the deuterium isotope. Fractional distillation is one method that can be used to separate heavy water, $D_2O(l)$, from regular water. Heavy water is used on a large scale as a coolant in heavy water nuclear power plants and in innumerable chemical investigations ranging from spectroscopy to thermodynamics to chemical synthesis. Regular water has a normal boiling point of 100.00°C, whereas heavy water has a normal boiling point of 101.42°C. Heavy water distillation plants produce almost pure $D_2O(l)$ from regular water at a total cost of around \$500 per kilogram of $D_2O(l)$. Such distillation plants have over 300 successive distillation stages and require an input of over one metric ton of water per gram

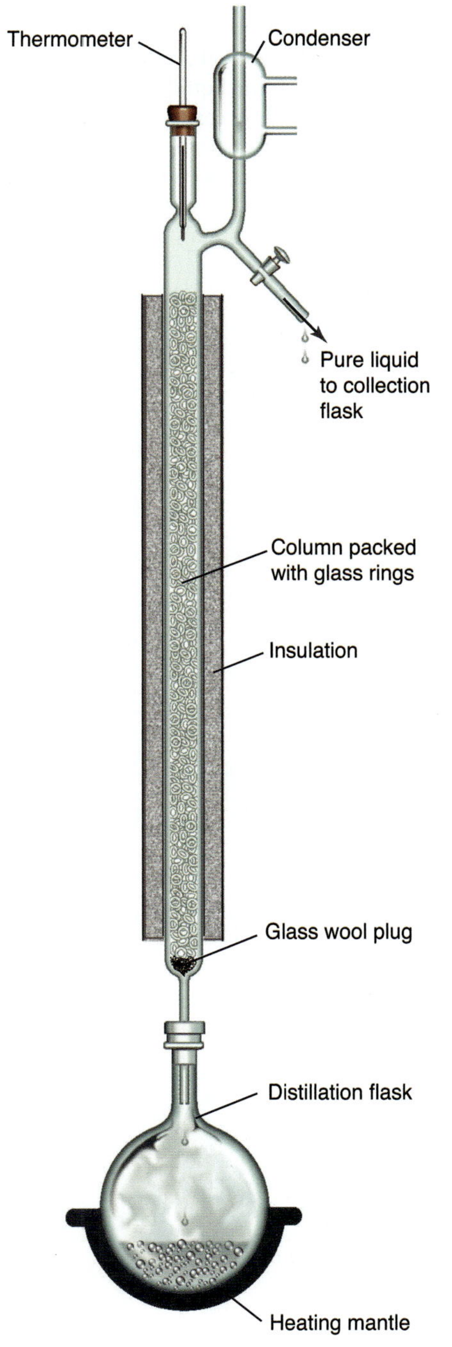

Figure 16.21 A simple fractional distillation column. Because repeated condensation and reevaporation occur along the entire column, the vapor becomes progressively richer in the more volatile component as it moves up the column.

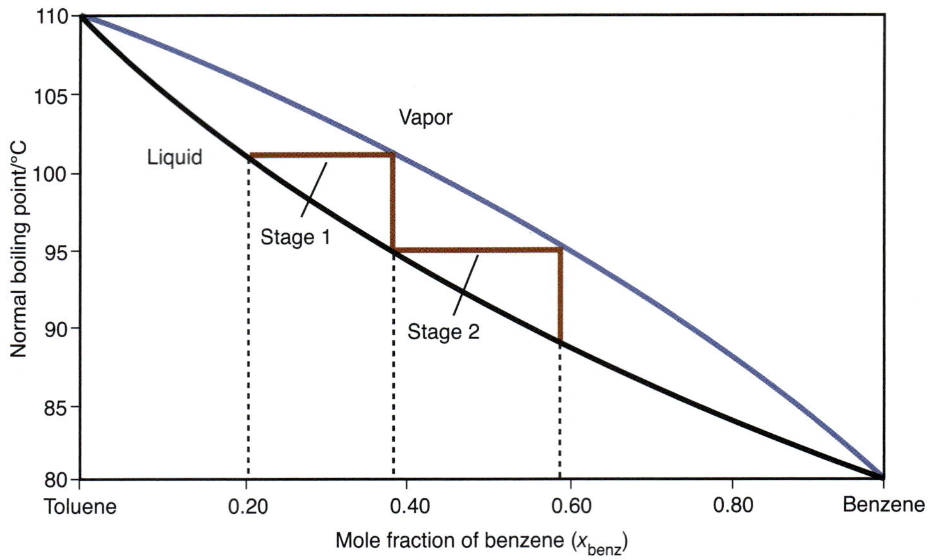

Figure 16.22 Vapor and liquid phase compositions as a function of boiling point at one bar for a mixture of benzene and toluene during fractional distillation. After each successive stage of boiling and condensation in fractional distillation, the condensate is richer in benzene than the previous mixture. If the liquid is boiled and condensed successively, then after many such stages the resulting condensate will be essentially pure benzene.

of $D_2O(l)$ produced. Following the closure of the Bruce Heavy Water Plant in Ontario, India has become the world's largest producer of heavy water.

Not all solutions can be separated completely using distillation. For instance, if we attempt to separate ethanol (drinking alcohol) from water by distillation, we find that no matter how effective a distillation column we use, the distillate issuing from the column has a maximum alcohol content of 95.6%. A 95.6% ethanol-plus-water solution distills as if it were a pure liquid. A solution that distills without change in composition is called an **azeotrope**. Azeotropes are solutions that do not behave ideally because of either attractive or repulsive forces between the various components of the solution, resulting in deviations from Raoult's law. Although simple distillation can generate up to 95.6% ethanol, 100% (pure) ethanol can be prepared from a 95.6% solution ethanol by dehydration with calcium oxide, which removes the water from the solution according to the equation

$$\mathrm{CaO}(s) + \mathrm{H_2O}(soln) \rightarrow \mathrm{Ca(OH)_2}(s)$$

The $Ca(OH)_2(s)$ is filtered off to leave 100% ethanol.

16-7. The Solubility of a Gas in a Liquid Is Directly Proportional to the Pressure of the Gas over the Liquid

The solubility of a gas in a liquid is directly proportional to the partial pressure of the gas in contact with the liquid. If we express the solubility as the molarity of the dissolved gas, M_{gas}, and the partial pressure of the gas as P_{gas}, then we can write

$$P_{gas} = k_h M_{gas} \tag{16.21}$$

Equation 16.21 is called **Henry's law.** The value of the proportionality constant, k_h, depends upon the gas, the solvent, and the temperature, and is called the **Henry's law constant.** Henry's law tells us that if we double the pressure of oxygen gas over liquid water, then the concentration of oxygen dissolved in the water also doubles. A doubling of the pressure of a gas over a solution doubles the concentration of the gas and thus doubles the rate at which the gas molecules enter the solution. This in turn doubles the concentration of the gas in the solution so that the rate of escape of the dissolved molecules from the solution balances the rate of entry of the gas molecules to the solution.

EXAMPLE 16-11: Calculate the concentration of $O_2(g)$ in water that is in equilibrium with air at 25°C. The Henry's law constant for $O_2(g)$ in water at 25°C is 790 bar·M^{-1}. Air is 21% oxygen by volume.

Solution: The partial pressure of $O_2(g)$ in the atmosphere is 0.21 bar; thus, from Equation 16.22,

$$M_{O_2} = \frac{P_{O_2}}{k_h} = \frac{0.21\text{ bar}}{790\text{ bar}\cdot\text{M}^{-1}} = 2.7 \times 10^{-4}\text{ M}$$

PRACTICE PROBLEM 16-11: The Henry's law constant for $H_2S(g)$ dissolving in water at 25°C is 10 bar·M^{-1}. Calculate the concentration of dissolved $H_2S(g)$ in an aqueous solution in equilibrium with $H_2S(g)$ gas at a pressure of 1.00 bar.

Answer: 0.10 M

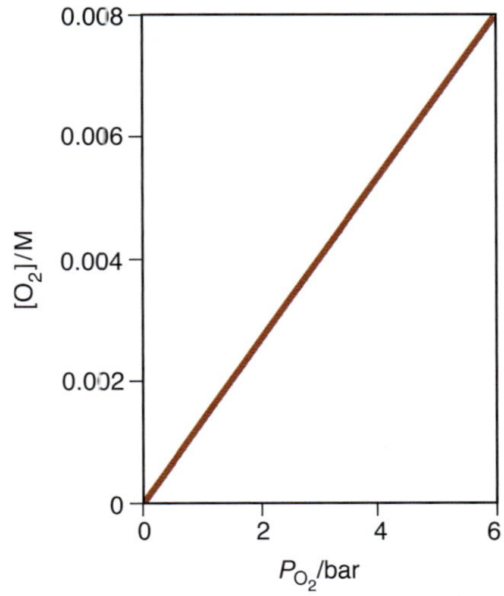

Figure 16.23 The solubility of oxygen in water at 25°C plotted against the pressure of the oxygen in contact with the water. The linear plot confirms that Henry's law holds for oxygen in water over the range of 0 to 6 bar $O_2(g)$.

Figure 16.23 shows the solubility of oxygen in water as a function of the pressure of the oxygen in contact with the water. The resulting straight line is in accord with Henry's law. Henry's law constants for several common gases are given in Table 16.3. The smaller the value of this constant for a gas, the greater the solubility of the gas, because $M_{gas} = P_{gas}/k_h$.

Carbonated beverages are pressurized with $CO_2(g)$ at a pressure above one bar; sodas are pressurized with $CO_2(g)$ at about two bar, and champagne is pressurized at four to five bar. The $CO_2(g)$ pressure is responsible for the rush of escaping gas that causes the "pop" when the carbonated drink container is opened (sidebox, page 592). The loss of $CO_2(g)$ from the solution occurs because the average atmospheric partial pressure of $CO_2(g)$ is only 3×10^{-4} bar. The bubbles that form in the liquid are mostly $CO_2(g)$ plus some water vapor at about one bar total pressure.

The air breathed by a diver under water is significantly above atmospheric pressure because the diver must exhale the air into an environment that has a pressure greater than atmospheric pressure. For example, at a depth of 30 meters (100 feet), the pressure is about three bar, so the diver must breathe air at three bar. At this pressure the solubilities of $N_2(g)$ and $O_2(g)$ in the blood are three times greater than they are at sea level. If a diver ascends too rapidly, then the sudden pressure drop causes the dissolved nitrogen to form numerous small gas bubbles in the blood. This phenomenon, which is extremely painful and can result in death, is called the bends because it causes the diver to bend over in pain. Oxygen, which is readily metabolized, does not accumulate as bubbles in the blood. However, pure oxygen cannot be used for breathing for extended periods because at high oxygen pressures, the need to breathe is greatly reduced. Thus, $CO_2(g)$ accumulates in the bloodstream and causes $CO_2(g)$ asphyxia-

TABLE 16.3 Henry's law constants for gases in water at 25°C

Gas	k_h/bar·M^{-1}	k_h/atm·M^{-1}
He	2.7×10^3	2.7×10^3
N_2	1.6×10^3	1.6×10^3
O_2	7.9×10^2	7.8×10^2
CO_2	29	29
H_2S	10	9.9

WHY DO SODAS "EXPLODE"? Every kid knows that if you shake up a can of soda and then open it, the soda "explodes" (sprays out of the bottle). Why does this happen? Recall from Chapter 15 that water has a significant surface tension. Even though the carbon dioxide dissolved in a soda is at a pressure of about two bar, the surface tension of water resists the formation of bubbles within the solution, causing the dissolved $CO_2(g)$ to escape slowly. When you shake up a soda you mechanically generate bubbles in the solution that help the $CO_2(g)$ to escape when the soda is opened. If a shaken soda is left to stand, these bubbles will eventually float to the top returning the soda to its original (unshaken) state. When an ice cube is dropped in a glass of soda, the solution fizzes because of small sites on the surface of the ice that assist in the formation of bubbles. One way to create a dramatic explosion with soda is to quickly drop a roll of mint candies, such as Wint-O-Green Lifesavers® or Mentos®, into an open soda bottle. These candies have many more such sites on their surfaces than ice and catalyze the formation of bubbles so rapidly that a "fountain" of soda pop results! (This experiment is quite messy and should not be done indoors.)

tion. The solution to these problems was proposed by the American chemist Joel Hildebrand (Chapter 12 Frontispiece). Hildebrand's solution was to substitute helium for nitrogen. Helium is only about half as soluble in blood as nitrogen; thus, the magnitude of the problem is cut in half. Divers' "air" tanks contain a mixture of $He(g)$ and $O_2(g)$ adjusted so that the pressure of $O_2(g)$ is about 0.20 bar at maximum dive depth, thereby also avoiding the $CO_2(g)$ asphyxiation problem.

The solubility of gases in liquids decreases with increasing temperature, because of the increase in escaping tendency of the solute gas with increasing temperature. Cold water in equilibrium with air has a higher concentration of dissolved oxygen than warm water. The decreased solubility of oxygen with increased temperature is one reason why more active fishes, such as trout and tuna, tend to live in cooler water.

SUMMARY

The concentration of a solute can be expressed in various concentration units, such as mass percent, molarity, molality, and mole fraction.

Vapor pressure lowering, boiling point elevation, freezing point depression, and osmotic pressure are colligative properties. The key point in understanding the colligative properties of solutions is realizing that the equilibrium vapor pressure of a solvent is reduced when a solute is dissolved in it. If a solution is sufficiently dilute, then the equilibrium vapor pressure of the solvent is given by Raoult's law. Colligative properties of solutions depend only on the solute particle concentration and are independent of the chemical nature of the solute.

Osmotic pressure is the largest of the colligative effects and can be used to determine the molecular mass of proteins. Osmotic pressure effects are important in biological systems because osmotic pressure keeps biological cells inflated.

Raoult's law also applies to ideal solutions consisting of two volatile liquids. Ideal solutions result when the interactions between unlike molecules in the solution are the same as the interactions between like molecules. If this is not the case, then the solution is not ideal and deviations from Raoult's law are observed. Fractional distillation can be used to separate a mixture of two volatile liquids.

Henry's law states that the solubility of a gas in a liquid is directly proportional to the pressure of the gas over the solution. The solubility of gases in liquids decreases with increasing temperature.

TERMS YOU SHOULD KNOW

solvated *569*
colligative property *569*
% mass (percentage by mass) *570*
mole fraction, x *570*
molality, m *571*
colligative molality, m_c *572*
Raoult's law *574*
vapor pressure lowering *574*
boiling point elevation *576*
boiling point elevation constant, K_b *578*
freezing point depression *579*
freezing point depression constant, K_f *579*
antifreeze *579*
van't Hoff *i*-factor *580*
semipermeable membrane *584*
escaping tendency *584*
osmotic pressure, Π *584*
osmosis *584*
colligative molarity, M_c *584*
reverse osmosis *585*
ideal solution *587*
fractional distillation *589*
azeotrope *590*
Henry's law *590*
Henry's law constant, k_h *590*

EQUATIONS YOU SHOULD KNOW HOW TO USE

$x_1 = \dfrac{n_1}{n_{tot}}$	(16.2)	(mole fraction)
$m = \dfrac{\text{moles of solute}}{\text{1 kilogram of solvent}}$	(16.5)	(definition of molality)
$m_c = im$	(16.6)	(colligative molality)
$P_1 = x_1 P_1^\circ$	(16.8)	(Raoult's law)
$\Delta P_1 = x_2 P_1^\circ$	(16.10)	(vapor pressure lowering)
$T_b - T_b^\circ = K_b m_c$	(16.13)	(boiling point elevation)
$T_f^\circ - T_f = K_f m_c$	(16.14)	(freezing point depression)
$T_f^\circ - T_f = K_f im$	(16.15)	(van't Hoff *i*-factor equation)
$\Pi = RTM_c$	(16.16)	(osmotic pressure)
$P_{total} = x_A P_A^\circ + x_B P_B^\circ$	(16.19)	(Raoult's law for a two-component solution)
$P_{total} = P_B^\circ + x_A (P_A^\circ - P_B^\circ)$	(16.20)	
$P_{gas} = k_h M_{gas}$	(16.21)	(Henry's law)

PROBLEMS

MOLE FRACTION

16-1. Calculate the mole fractions of ethanol and water in a solution that is made up of 20.0 grams of ethanol, CH_3CH_2OH, and 80.0 grams of water.

16-2. A solution of 40% formaldehyde, H_2CO, 10% methanol, CH_3OH, and 50% water by mass is called formalin. Calculate the mole fractions of formaldehyde, methanol, and water in formalin. Formalin is used to disinfect dwellings, ships, storage houses, and so forth.

16-3. Describe how you would prepare 1.00 kilogram of an aqueous solution of acetone, $(CH_3)_2CO$, in which the mole fraction of acetone is 0.19.

16-4. Describe how you would prepare 500.0 grams of a solution of sucrose, $C_{12}H_{22}O_{11}$, in water in which the mole fraction of sucrose is 0.125.

16-5. Calculate the mole fraction of isopropanol, $(CH_3)_2CHOH$, in a solution that is 70.0% isopropanol and 30.0% water by volume. Take the density of water as 1.00 $g{\cdot}cm^{-3}$ and the density of isopropanol as 0.785 $g{\cdot}cm^{-3}$.

16-6. Some forms of so-called maintenance-free car batteries use a lead-calcium alloy (solid solution). Given that the alloy is 95.0% lead and 5.0% calcium by mass, calculate the mole fraction of calcium in the solid solution.

MOLALITY

16-7. Describe how you would prepare a solution of formic acid, HCOOH, in acetone, $(CH_3)_2CO$, that is 2.50 m in formic acid.

16-8. Describe how you would prepare an aqueous solution that is 1.75 m in $Ba(NO_3)_2$.

16-9. The solubility of iodine, I_2, in carbon tetrachloride, CCl_4, is 2.603 grams per 100.0 grams of carbon tetrachloride at 35°C. Calculate the molality of iodine in a saturated solution.

16-10. How many kilograms of water would have to be added to 18.0 grams of oxalic acid, $H_2C_2O_4(s)$, to prepare a 0.050-m solution?

16-11. A 1.0-mole sample of each of the following substances is dissolved in 1000 grams of water. Determine the colligative molality of the substance in each case.

(a) $MgSO_4$ (b) $Cu(NO_3)_2$

(c) CH_3CH_2OH (d) $Al_2(SO_4)_3$

16-12. A 1.0-mole sample of each of the following substances is dissolved in 1.0 kilogram of water. Determine the colligative molality of the substance in each case.

(a) methanol, CH_3OH (b) aluminum nitrate

(c) iron(II) nitrate (d) ammonium dichromate

VAN'T HOFF FACTOR

16-13. Predict the van't Hoff *i*-factor for each of the following salts dissolved in water:

(a) silver nitrate (b) magnesium chloride

(c) potassium sulfate

16-14. Predict the van't Hoff *i*-factor for each of the following salts dissolved in water:

(a) $Na_2CO_3(s)$ (b) $(NH_4)_2SO_4(s)$ (c) $Pb(NO_3)_2(s)$

16-15. Which of the following compounds would you expect to have the largest van't Hoff *i*-factor?

(a) nitric acid (b) acetic acid (c) sulfurous acid

16-16. Which of these compounds would you expect to have the largest van't Hoff *i*-factor?

(a) $HCl(aq)$ (b) $HClO_2(aq)$ (c) $HClO(aq)$

RAOULT'S LAW AND VAPOR PRESSURE LOWERING

16-17. The vapor pressure of pure water at 37°C is 47.1 Torr. Use Raoult's law to estimate the vapor pressure of an aqueous solution at 37°C containing 20.0 grams of glucose, $C_6H_{12}O_6(s)$, dissolved in 500.0 grams of water. Also compute the vapor pressure lowering.

16-18. Water at 37°C has a vapor pressure of 6.27 kPa. Calculate the vapor pressure of water if 50.0 grams of glycerin, $C_3H_8O_3(l)$, is added to 100.0 milliliters of water. The density of water at 37°C is 0.993 $g{\cdot}mL^{-1}$. Also calculate the vapor pressure lowering.

16-19. The vapor pressure of pure water at 25°C is 23.76 Torr. Use Raoult's law to estimate the vapor pressure of an aqueous solution at 25°C containing 20.00 grams of sucrose, $C_{12}H_{22}O_{11}(s)$, dissolved in 195.0 grams of water. Also calculate the vapor pressure lowering.

16-20. The vapor pressure of pure water at 100°C is 1.00 atm. Use Raoult's law to estimate the vapor pressure of water over an aqueous solution at 100°C containing 50.0 grams of ethylene glycol, $C_2H_6O_2\,(l)$, dissolved in 100.0 grams of water. Also calculate the vapor pressure lowering for water.

16-21. Calculate the vapor pressure in kPa of an ethanol solution at 25°C containing 20.0 grams of the nonvolatile solute urea, $(NH_2)_2CO(s)$, dissolved in 100.0 grams of ethanol, $CH_3CH_2OH(l)$. The vapor pressure of ethanol at 25°C is 7.89 kPa. Also calculate the vapor pressure lowering.

16-22. Calculate the vapor pressure of ethanol, CH_3CH_2OH, over 80 proof (40.0% ethanol by volume) vodka at 19°C. The vapor pressure of pure ethanol at 19°C is 40.0 Torr. The density of ethanol is 0.79 $g\cdot mL^{-1}$ and the density of water is 1.00 $g\cdot mL^{-1}$.

16-23. Given that the vapor pressure of water is 17.54 Torr at 20°C, calculate the vapor pressure lowering of aqueous solutions that are 0.25 m in

(a) sodium chloride (b) calcium chloride

(c) sucrose, $C_{12}H_{22}O_{11}$ (d) aluminum perchlorate

16-24. The colligative molality of the contents of a typical human cell is about 0.30 m_c. Compute the equilibrium vapor pressure of water at 37°C for the cell solution. Take $P^{\circ}_{H_2O} = 62.8$ mbar at 37°C.

16-25. Calculate the vapor pressure lowering of the following aqueous solutions at 25°C ($P^{\circ}_{H_2O} = 0.0317$ bar):

(a) 2.00 m sucrose, $C_{12}H_{22}O_{11}(aq)$

(b) 2.00 m $NaCl(aq)$

(c) 2.00 m $CaCl_2(aq)$

16-26. Calculate the vapor pressure lowering of the following solutions at 25°C ($P^{\circ}_{H_2O} = 31.7$ mbar):

(a) 1.50 m ethanol, $CH_3CH_2OH(aq)$

(b) 0.50 m thallium(III) chloride

(c) 0.25 m potassium sulfate

16-27. The observed vapor pressure lowering of an aqueous sucrose, $C_{12}H_{22}O_{11}(aq)$, solution is 0.475 kPa at 25.0°C. Calculate the concentration of sucrose in the solution.

16-28. The observed vapor pressure lowering of a $CaCl_2(aq)$ solution is 5.00 Torr at 20.0°C. Calculate the molality of the $CaCl_2(aq)$ in the solution.

16-29. Given that an aqueous solution of ethylene glycol in water has a vapor pressure lowering of 16.0 mbar at 90.0°C, calculate the mole fraction of ethylene glycol in the solution.

16-30. Calculate the molalities of the following electrolytes that will produce a vapor pressure lowering of 2.00 Torr in water at 20°C:

(a) potassium iodide (b) strontium chloride

(c) ammonium sulfate

BOILING-POINT ELEVATION

16-31. Calculate the boiling point of a 2.0-m aqueous solution of $Sc(ClO_4)_3(aq)$.

16-32. How much $NaCl(s)$ would have to be dissolved in 1000.0 grams of water in order to raise the boiling point by 1.0°C?

16-33. Calculate the boiling point of a solution of 10.0 grams of picric acid, $C_6H_2(OH)(NO_2)_3(s)$, dissolved in 100.0 grams of cyclohexane, $C_6H_{12}(s)$. Assume that the colligative molality and the molality are the same for picric acid in cyclohexane.

16-34. The colligative molality of seawater is approximately 1.10 m_c. Calculate the boiling point of seawater at 1.00 atm and its vapor pressure at 15°C. The vapor pressure of pure water at 15°C is 12.79 Torr.

16-35. Calculate the boiling point of a solution containing 25.0 grams of urea, $(H_2N)_2CO(s)$, dissolved in 1.50 kilograms of nitrobenzene, $C_6H_5NO_2(l)$.

16-36. Calculate the boiling point of a solution of 25.0 grams of urea, $(H_2N)_2CO(s)$, plus 25.0 grams of thiourea, $(H_2N)_2CS(s)$, in 0.500 kilograms of trichloromethane (chloroform), $CHCl_3(l)$.

FREEZING POINT DEPRESSION

16-37. Calculate the freezing point of an aqueous solution of 60.0 grams of glucose, $C_6H_{12}O_6(s)$, dissolved in 200.0 grams of water.

16-38. Calculate the freezing point of an aqueous solution of 20.0 grams of $Ca(NO_3)_2(s)$ dissolved in 0.500 kilograms of water.

16-39. Calculate the freezing point of a solution of 5.00 grams of diphenyl, $C_{12}H_{10}(s)$, and 7.50 grams of naphthalene, $C_{10}H_8(s)$, dissolved in 200.0 grams of benzene.

16-40. Quinine is a natural product extracted from the bark of the cinchona tree, which is native to South America. Quinine is used as an antimalarial agent. When 1.00 gram of quinine is dissolved in 10.0 grams of cyclohexane, the freezing point is lowered by 6.42 K. Calculate the molecular mass of quinine.

16-41. Vitamin K is involved in normal blood clotting. When 0.500 grams of vitamin K is dissolved in 10.0 grams of camphor, the freezing point of the solution is lowered by 4.20 K. Calculate the molecular mass of vitamin K.

16-42. Don Juan Pond in the Wright Valley of Antarctica freezes at −57°C. The major solute in the pond is $CaCl_2$. Estimate the concentration of $CaCl_2$ in the pond water.

16-43. Menthol is a crystalline substance with a peppermint taste and odor. A solution of 6.54 grams of menthol per 100.0 grams of cyclohexane freezes at −2.13°C. Determine the molecular mass of menthol.

16-44. When 2.67 grams of an organic compound that is known to be 39.12% carbon, 8.76% hydrogen, and 52.12% oxygen are dissolved in 65.3 grams of camphor, the freezing point of the solution is 159.2°C. Determine the molecular formula of the compound.

16-45. An aqueous solution of mercury(II) chloride, $HgCl_2(aq)$, is a poor conductor of electricity. A 4.50-gram sample of $HgCl_2(s)$ is dissolved in 100.0 grams of water, and the freezing point of the solution is found to be −0.314°C. Explain why $HgCl_2$ in solution is a poor conductor of electricity.

16-46. Mayer's reagent, $K_2HgI_4(s)$, is used in analytical chemistry. In order to determine its extent of dissociation in water, its effect on the freezing point of water is investigated. A 0.25-m aqueous solution is prepared, and its freezing point is found to be −1.41°C. Suggest a possible dissociation reaction that takes place when $K_2HgI_4(s)$ is dissolved in water.

OSMOTIC PRESSURE

16-47. Calculate the osmotic pressure of a 0.25-M aqueous solution of sucrose, $C_{12}H_{22}O_{11}(aq)$, at 37°C.

16-48. Calculate the osmotic pressure of seawater at 37°C. Take $M_c = 1.10\ mol\cdot L^{-1}$ for seawater.

16-49. Insulin is a small protein hormone that regulates carbohydrate metabolism by decreasing blood glucose levels. A deficiency of insulin leads to diabetes. A 20.0-mg sample of insulin is dissolved in enough water to make 10.0 mL of solution, and the osmotic pressure of the solution at 25°C is found to be 6.48 Torr. Calculate the molecular mass of insulin.

16-50. Pepsin is the principal digestive enzyme of gastric juice. A 3.00-mg sample of pepsin is dissolved in enough water to make the 10.0 mL of solution, and the osmotic pressure of the solution at 25°C is found to be 0.162 Torr. Calculate the molecular mass of pepsin.

16-51. What is the minimum pressure in kPa that must be applied at 25°C to obtain pure water by reverse osmosis from water that is 0.15 M in sodium chloride and 0.015 M in magnesium sulfate?

16-52. In reverse osmosis, water flows out of a salt solution until the osmotic pressure of the solution equals the applied pressure. If a pressure of 75 bar is applied to seawater, what will be the final concentration of the seawater at 20°C when reverse osmosis stops? Assuming that seawater is a 1.1-M_c solution of $NaCl(aq)$, calculate how many liters of seawater are required to produce 15 liters of freshwater at 20°C with an applied pressure of 75 bar.

RAOULT'S LAW FOR TWO COMPONENTS

16-53. Given that the equilibrium vapor pressures of benzene and toluene at 81°C are 768 Torr and 293 Torr, respectively, (a) calculate the total vapor pressure at 81°C over a benzene-toluene solution with $x_{benz} = 0.250$, and (b) calculate the mole fraction of benzene in the vapor phase over the solution.

16-54. 1-propanol (*n*-propanol), $CH_3CH_2CH_2OH$, and 2-propanol (isopropanol), $CH_3CHOHCH_3$, form ideal solutions in all proportions. Calculate the partial pressure of each component in equilibrium at 25°C with a solution of compositions $x_{prop} = 0.25$, 0.50, and 0.75, given that $P^{\circ}_{prop} = 20.9$ Torr and $P^{\circ}_{iso} = 45.2$ Torr at 25°C. Calculate the composition of the vapor phase as well.

HENRY'S LAW

16-55. Calculate the concentration of nitrogen in water at a nitrogen gas pressure of 0.79 bar and a temperature of 25°C.

16-56. Of the gases $N_2(g)$, $O_2(g)$, and $CO_2(g)$, which has the highest concentration in water at 25°C when each gas has a pressure of 1.0 bar?

16-57. The Henry's law constant for $CO_2(g)$ in water at 25°C is 29 bar·M^{-1}. Estimate the concentration of dissolved $CO_2(g)$ in a carbonated soft drink pressurized with 2.0 bar of $CO_2(g)$.

16-58. Calculate the masses of oxygen and nitrogen that are dissolved in 1.00 liter of aqueous solution in equilibrium with air at 25°C and 760 Torr. Assume that air is 21% oxygen and 78% nitrogen by volume.

ADDITIONAL PROBLEMS

16-59. Explain why the unit molality, *m*, is generally preferable to the unit molarity, *M*, when calculating the colligative properties of solutions.

16-60. Describe how you would prepare a 0.100-m aqueous solution of sucrose, $C_{12}H_{22}O_{11}(aq)$. Would this solution also be 0.100 M in sucrose? Why or why not?

16-61. Is a 1.00-m_c aqueous solution of sucrose the same as a 1.00-m $C_{12}H_{22}O_{11}(aq)$ solution (see previous problem)? Is a 1.00-m_c aqueous solution of sodium chloride the same as a 1.00-m $NaCl(aq)$ solution?

16-62. It is often observed when cooking that adding a pinch of salt to boiling water causes the water to stop boiling for a while. Why does this occur?

16-63. What is meant by an "ideal" solution?

16-64. Most home ice-cream makers work by mixing milk, cream, eggs, sugar, and other ingredients inside a bath of ice water and rock salt. Why can't ice alone be used to make ice cream?

16-65. Kimchi is traditional Korean food made by fermenting cabbage or other vegetables in a mixture of spices and salt. Kimchi originally developed as a method of preserving vegetables before the advent of modern refrigeration. Explain why the use of a salt solution in kimchi acts as a preservative against bacterial decay.

16-66. During the age of exploration many seafarers lost their lives to thirst. Why can't sailors simply drink seawater to quench their thirst?

16-67. One method of determining the purity of an organic substance is to measure the melting point of some crystals. Impure crystals tend to melt at a temperature that is lower than that of the pure substance. Explain this phenomenon in terms of colligative properties of mixtures.

16-68. Which of the following solutions will have the highest boiling point? Why?

(a) 0.1 mole of glucose, $C_6H_{12}O_6$, dissolved in 1.0 kilogram of water

(b) 0.1 mole of magnesium chloride dissolved in 1.0 kilogram of water

(c) 0.1 mole of sodium hydrogen carbonate dissolved in 1.0 kilogram of water

(d) 0.1 mole of ammonium nitrate dissolved in 1.0 kilogram of water

16-69. Immunoglobulin G, formerly called gamma globulin, is a principal antibody in blood serum. A 0.500-gram sample of immunoglobulin G is dissolved in enough water to make 0.100 liters of solution, and the osmotic pressure of the solution at 25°C is found to be 0.825 mbar. Calculate the molecular mass of immunoglobulin G.

16-70. Most wines are about 12% ethanol, CH_3CH_2OH, by volume, and many hard liquors are about 80 proof (40% ethanol by volume). Assuming that the only major nonaqueous constituent of wine and vodka (a hard liquor) is ethanol, calculate the freezing points of wine and vodka. Take the density of ethanol to be 0.79 g·mL^{-1} and the density of water to be 1.00 g·mL^{-1}.

16-71. The boiling point of ethylene glycol is 197°C, and the boiling point of ethanol is 78°C. Ethylene glycol is called a "permanent" antifreeze and ethanol a "temporary" one. Explain the difference between "permanent" and "temporary" antifreezes.

16-72. Your friend is driving to Alaska and wants to winterize his car by adding antifreeze. Although the label on the container recommends using a 50-50 ethylene glycol water mixture under "normal conditions," he figures that by putting pure ethylene glycol into his cooling system, the car will be even better protected from freezing under the "extreme conditions" found in Alaska. Explain to your friend why this is a bad idea. (*Hint:* Look up the physical properties of ethylene glycol using the Internet).

16-73. Scientists have discovered that some insects produce an antifreeze in cold weather; the antifreeze is glycerol, $HOCH_2CHOHCH_2OH$. How much glycerol must an insect produce per gram of body fluid (taken to be water) to survive at −5.0°C?

16-74. A semipermeable membrane separates two aqueous solutions at 20°C. For each of the following cases, name the solution into which a net flow of water (if any) will occur:

(a) 0.10 M $NaCl(aq)$ and 0.10 M $KBr(aq)$

(b) 0.10 M $Al(NO_3)_3(aq)$ and 0.20 M $NaNO_3(aq)$

(c) 0.10 M $CaCl_2(aq)$ and 0.50 M $CaCl_2(aq)$

16-75. The density of a glycerol-water solution that is 40.0% glycerol by mass is 1.101 g·mL^{-1} at 20°C. Calculate the molality and the molarity of glycerol in the solution at 20°C. What is the molality at 0°C? The formula of glycerol is $C_3H_8O_3$.

16-76. Calculate the molality, the colligative molality, the freezing point, and the boiling point for each of the following solutions:

(a) 5.00 grams of $K_2SO_4(s)$ in 0.250 kilogram of water

(b) 5.00 grams of ethanol, $CH_3CH_2OH\ (l)$, in 0.250 kilogram of water

16-77. Calculate the vapor pressures of carbon tetrachloride, CCl_4, and ethyl acetate, $CH_3COOC_2H_5$, in a solution at 50°C containing 25.0 grams of carbon tetrachloride dissolved in 100.0 grams of ethyl acetate. The vapor pressures of pure $CCl_4(l)$ and $CH_3COOC_2H_5(l)$ at 50°C are 306 Torr and 280 Torr, respectively.

16-78. When 2.74 grams of phosphorus are dissolved in 100.0 milliliters of carbon disulfide, the boiling point is 46.71°C. Given that the normal boiling point of pure carbon disulfide is 46.30°C, that its density is 1.261 g·mL^{-1}, and that its boiling-point elevation constant is $K_b = 2.34$ K·m$_c^{-1}$, determine the molecular formula of phosphorus.

16-79. A 2.0-gram sample of the polymer polyisobutylene, $[CH_2C(CH_3)_2]_n$, is dissolved in enough cyclohexane to make 10.0 milliliters of solution at 20°C and produces an osmotic pressure of 2.0×10^{-2} bar. Determine the formula mass and the number of units (n) in the polymer.

16-80. It is possible to convert from molality to molarity if the density of a solution is known. The density of a 2.00-m aqueous sodium hydroxide solution is 1.22 g·mL^{-1}. Calculate the molarity of this solution.

16-81. In many fields outside chemistry, solution concentrations are expressed in mass percent. Calculate the molality of an aqueous solution that is 24.0% potassium chromate, K_2CrO_4, by mass. Given that the density of the solution is 1.21 g·mL^{-1}, calculate the molarity.

16-82. Radiator antifreeze also provides "antiboiling" protection for automobile cooling systems. Using Equation 16.13, estimate the boiling point of a solution composed of 50.0 grams of water and 50.0 grams of ethylene glycol. Assume that the vapor pressure of ethylene glycol is negligible at 100°C. The formula of ethylene glycol is $HOCH_2CH_2OH$.

16-83. Using Equation 16.14, estimate the molality of ethylene glycol, $HOCH_2CH_2OH$, in water that is necessary to give antifreeze protection down to −40°C.

16-84. The equations for boiling point elevation and freezing point depression given in this chapter are valid for dilute solutions only. Some actual freezing point data for aqueous sodium chloride solutions are given below. Calculate the percentage error in the values of the freezing point predicted using Equation 16.14. At approximately what value of the colligative molality, m_c, does Equation 16.14 yield a 2% error for the predicted value of the freezing point of a sodium chloride solution?

Mass percent NaCl	Freezing point/°C
0.50	−0.30
1.0	−0.59
5.0	−3.05
10.0	−6.56

16-85. (*)Show that the relation between the mole fraction of solute and the molality of the solution is given by

$$x_2 = \frac{\left(\dfrac{M_1 m}{1000}\right)}{\left(1 + \dfrac{M_1 m}{1000}\right)}$$

where M_1 is the molar mass of the solvent and m is the molality. Now argue that

$$x_2 \approx \frac{M_1 m}{1000}$$

if the solution is dilute.

16-86. Show that in an aqueous solution the mole fraction of the solute is given by

$$x_2 = \frac{m_c}{m_c + 55.5\ \text{mol} \cdot \text{kg}^{-1}} \approx \frac{m_c}{55.5\ \text{mol} \cdot \text{kg}^{-1}}$$

where the approximation holds to within about 2% when $m_c \leq 1.00\ \text{m}_c$.

16-87. Using the properties of similar triangles and the curves in Figure 16.6, show that

$$(P_1^\circ - P_1) \propto (T_b - T_b^\circ)$$

16-88. Given that the freezing point depression of a 1.00-m solution of $H_2SO_4(aq)$ is 3.74°C, determine the number of solute particles per formula unit of $H_2SO_4(aq)$.

16-89. (*) A mixture is found to contain both table salt, NaCl(*s*), and sugar, $C_{12}H_{22}O_{11}(s)$. When 2.00 grams of the mixture are dissolved in 500.0 milliliters of water, the freezing point is observed to be depressed by 0.101 K. What is the mass percentage of each component in the mixture?

16-90. (*) A 2.00-gram sample of an ionic compound commonly used as a fertilizer is dissolved in 250.0 milliliters of water and the freezing point of the solution found to be −0.372°C. Chemical analysis shows that the compound contains 35.00% nitrogen, 59.96% oxygen, and 5.04% hydrogen by mass. What is the chemical name of the compound? (Take the density of water as 1.00 g·mL^{-1}.)

16-91. Given the following freezing point depression data, determine the number of ions produced per formula unit when the indicated substance is dissolved in aqueous solution to produce a 1.00-m solution.

Formula	**ΔT/K**
$PtCl_2 \cdot 4NH_3$	5.58
$PtCl_2 \cdot 3NH_3$	3.72
$PtCl_2 \cdot 2NH_3$	1.86
$KPtCl_3 \cdot NH_3$	3.72
K_2PtCl_4	5.58

16-92. Consider the two beakers in Figure 16.11. Suppose instead of one beaker of pure water and one of seawater, that the first beaker contains 200.0 milliliters of a 0.100 M NaCl solution and the second beaker contains 300.0 milliliters of a 0.200 M NaCl solution. If we wait long enough, what will be the final volume and concentration of each of the solutions?

16-93. (*) The reason most samples may be purified using fractional distillation is that the vapor phase is richer in the more volatile component than the liquid phase. Let the mole fraction of component A in the vapor over a two-component ideal solution be y_A. Show that y_A is given by

$$y_A = \frac{x_A P_A^\circ}{x_A(P_A^\circ - P_B^\circ) + P_B^\circ}$$

Show that $y_A > x_A$ if $P_A^\circ > P_B^\circ$; in other words, show that the vapor phase is richer than the liquid phase in the more volatile component.

Dudley Herschbach (*right*), Yuan Tseh Lee (*top left*), and John Polanyi (*bottom left*) were awarded the Nobel Prize in Chemistry in 1986 "for their contributions to the understanding of the dynamics of chemical elementary processes."

Dudley Herschbach (1932–) was born in San Jose, California. He received his Ph.D. in chemical physics from Harvard University in 1958. After teaching several years at the University of California at Berkeley, he returned to Harvard in 1963, where he remains today. He pioneered the use of crossing beams of molecules in the gas phase to study chemical kinetics and reaction mechanisms in much the same way that modern atom smashers unravel how nuclear processes occur. Herschbach has demonstrated a special concern for undergraduates by serving as comaster with his wife at one of the residence halls at Harvard, a job that involves 40 hours a week outside teaching and research.

Yuan Tseh Lee (1936–) was born in Hsinchu, Taiwan. He received his Ph.D. from the University of California at Berkeley in three years and later was a postdoctoral fellow with Dudley Herschbach at Harvard in 1965. After spending six years at the University of Chicago, he returned to the University of California at Berkeley as a professor of chemistry. Lee advanced Herschbach's work and applied these techniques to study the reactions of larger molecules. He returned to Taiwan in 1994 as president of Academia Sinica, a post similar to being president of the National Academy of Sciences in the United States. In addition to these duties, Lee continues to study reaction dynamics, including reactions of significance for combustion chemistry and atmospheric chemistry.

John C. Polanyi (1929–) was born in Berlin, Germany, but grew up in Manchester, England. He received his Ph.D. from the University of Manchester in 1952, and in 1956 joined the faculty of the University of Toronto, where he remains today. Polanyi developed fast spectroscopic techniques for studying reaction dynamics. These techniques allowed him to determine how molecules change and interact during the course of a chemical reaction. His research has yielded great insight into how reactions occur on the molecular level and paved the way for modern techniques using ultrafast lasers to probe reaction dynamics. In addition to Polanyi's scientific papers, he has published almost 100 articles on science policy, on the control of armaments, and on the impact of science on society.

17. Chemical Kinetics: Rate Laws

Different chemical reactions take place at different rates. Some reactions, such as the reaction between $AgNO_3(aq)$ and $KCl(aq)$ to produce a precipitate of $AgCl(s)$, seem to occur almost instantaneously, whereas other reactions, such as the reaction between $H_2(g)$ and $N_2(g)$ to produce $NH_3(g)$, occur very slowly. Chemists study the rates of chemical reactions to determine the conditions under which reactions can be made to proceed at favorable rates. Certainly, if a chemist desires to produce a particular product, then the reaction used must take place at an appreciable rate. This condition is particularly important for reactions used for the commercial production of chemicals. As we shall see, reaction kinetics can also be applied to the rate of decay of radioactive isotopes, which are important in nuclear chemistry, medicine, archaeology, and geology.

A principal goal of a chemical kinetics experiment is to deduce how the rate of a reaction depends upon the concentrations of the species involved. This result is called the rate law of the reaction. The reaction rate law can be determined only by experiment. In the next chapter, we shall see how the reaction rate law provides the most important clue to the reaction mechanism, which is the sequence of steps by which the reactants are converted to products. An understanding of reaction mechanisms may enable us to adjust reaction conditions in order to produce a desired rate of reaction and may increase our understanding of how chemical reactions occur at the molecular level. Such an understanding enables us to control chemical reactions better and to manufacture chemical products in the most economical way.

17-1. A Rate of Concentration Change Tells Us How Fast a Quantity of Reactant or Product Is Changing with Time

Let's consider the decomposition of azomethane, $CH_3N_2CH_3(g)$, to ethane, $C_2H_6(g)$, and nitrogen, $N_2(g)$, which can be described by the chemical equation

$$CH_3{-}N{=}N{-}CH_3(g) \rightarrow CH_3CH_3(g) + N_2(g) \tag{17.1}$$

The rate of this reaction can be determined spectroscopically by measuring how the spectrum of any one of these three species changes in intensity as a function of time. Because two moles of gaseous products are formed for each mole of reactant that is consumed in Equation 17.1, we could also follow the

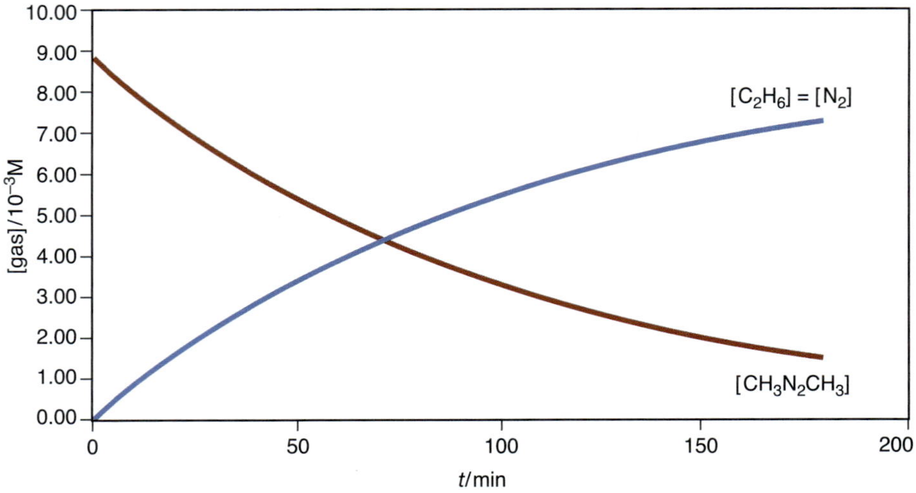

Figure 17.1 The concentrations of $CH_3N_2CH_3(g)$, $C_2H_6(g)$, and $N_2(g)$ as a function of time for the reaction described by

$$CH_3{-}N{=}N{-}CH_3(g) \rightarrow CH_3CH_3(g) + N_2(g)$$

at 576K. The initial concentrations are $[CH_3N_2CH_3]_0 = 8.70 \times 10^{-3}$M and $[C_2H_6]_0 = [N_2]_0 = 0$M. The numerical data are given in Table 17.1.

course of this reaction by measuring the increase in pressure as a function of time.

Figure 17.1 shows the concentrations of $CH_3N_2CH_3(g)$, $C_2H_6(g)$, and $N_2(g)$ plotted against time, and Table 17.1 lists numerical values of the concentrations of each species at 30-minute time intervals for the reaction described by Equation 17.1. We shall denote the concentration of a species by enclosing its chemical formula in square brackets. Figure 17.1 shows that $[CH_3N_2CH_3]$ decreases with time, while $[C_2H_6]$ and $[N_2]$ increase with time. Note also that $[C_2H_6]$ and $[N_2]$ increase at the same rate because they are produced on a one-to-one mole basis.

Let's see how we can use the change in concentration of a species in a reaction to determine the rate of a reaction. Suppose that we know the concentration of the ethane gas at some time, t_1, to be $[C_2H_6]_1$, and suppose we find that at some later time, t_2, it is $[C_2H_6]_2$. We can write the change in its concentration as

$$\Delta[C_2H_6] = [C_2H_6]_2 - [C_2H_6]_1$$

Here the Δ (delta) stands for change, or later value minus earlier value. Because the concentration of the product increases with time, we see that $\Delta[C_2H_6]$ will be a positive quantity. Similarly, the change in time is $\Delta t = t_2 - t_1$, which is also a positive quantity. We can now define the **average rate of concentration change** as the average rate at which ethane is produced, or the change in concentration of ethane divided by the elapsed time:

$$\begin{pmatrix}\text{average concentration}\\ \text{change of } C_2H_6(g)\end{pmatrix} = \frac{\text{change in } [C_2H_6]}{\text{elapsed time}} = \frac{\Delta[C_2H_6]}{\Delta t} \qquad (17.2)$$

TABLE 17.1 The concentrations of $CH_3N_2CH_3(g)$, $C_2H_6(g)$, and $N_2(g)$ as a function of time for the reaction described by

$$CH_3{-}N{=}N{-}CH_3(g) \rightarrow CH_3CH_3(g) + N_2(g)$$

at 576K. A plot of these concentrations versus time is shown in Figure 17.1.

t/min	$[CH_3N_2CH_3]/10^{-3}$ M	$[C_2H_6]/10^{-3}$ M	$[N_2]/10^{-3}$ M
0	8.70	0.00	0.00
30	6.52	2.18	2.18
60	4.89	3.81	3.81
90	3.67	5.03	5.03
120	2.75	5.95	5.95
150	2.06	6.64	6.64
180	1.55	7.15	7.15

This is a positive quantity because both $\Delta[C_2H_6]$ and Δt are positive. The units of the rate of concentration change are often expressed as moles per liter per second ($mol\cdot L^{-1}\cdot s^{-1}$ or $M\cdot s^{-1}$), or, more generally, moles per liter per unit time.

The average rate of concentration change as we have defined it in Equation 17.2 depends upon the time interval that we choose. What we are really interested in is the rate of concentration change at some particular instant of time, or what we call the **instantaneous rate of concentration change**. Suppose we wish to determine the instantaneous rate of concentration change of $C_2H_6(g)$ for the reaction described by Equation 17.1 at $t = 90.0$ minutes. We can use the data in Table 17.1 to obtain an estimate of the instantaneous rate of concentration change. As a first, and not especially good, estimate, we use the average rate over the time interval 0 to 180.0 minutes.

$$\begin{pmatrix}\text{average rate of}\\ \text{concentration change}\\ \text{0 to 180.0 min}\end{pmatrix} = \frac{(7.15\times10^{-3}-0)\ \text{M}}{(180.0-0)\ \text{min}} = 3.97\times10^{-5}\ \text{M}\cdot\text{min}^{-1}$$

Note that 90.0 minutes lies in the middle of this time interval. We can obtain a better estimate of the instantaneous rate of concentration change at 90.0 minutes by using a tighter time interval surrounding 90.0 minutes. If we choose 30.0 to 150.0 minutes, then we have

$$\begin{pmatrix}\text{average rate of}\\ \text{concentration change}\\ \text{30.0 to 150.0 min}\end{pmatrix} = \frac{(6.64\times10^{-3}-2.18\times10^{-3})\ \text{M}}{(150.0-30.0)\ \text{min}} = 3.72\times10^{-5}\ \text{M}\cdot\text{min}^{-1}$$

The best that we can do with the data in Table 17.1 is to use the time interval 60.0 to 120.0 minutes to obtain

$$\begin{pmatrix}\text{average rate of}\\ \text{concentration change}\\ \text{60.0 to 120.0 min}\end{pmatrix} = \frac{(5.95\times10^{-3}-3.81\times10^{-3})\ \text{M}}{(120.0-60.0)\ \text{min}} = 3.57\times10^{-5}\ \text{M}\cdot\text{min}^{-1}$$

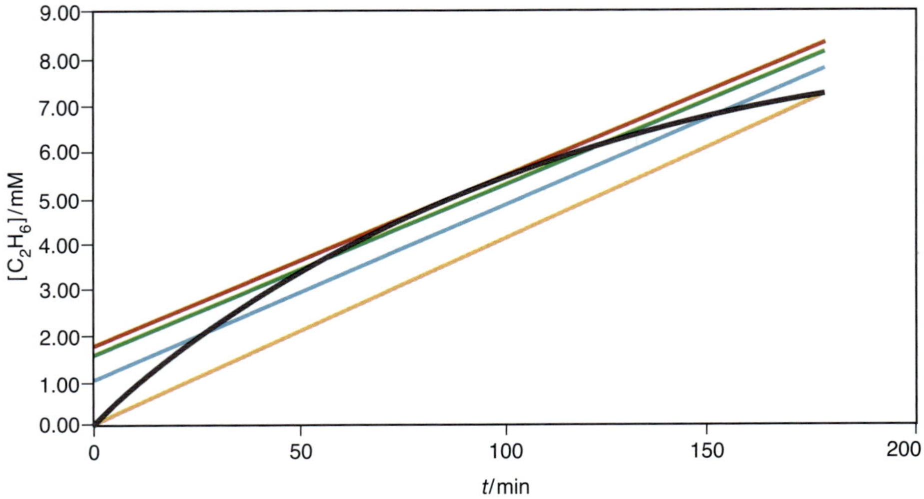

Figure 17.2 The instantaneous rate of reaction at 90.0 minutes is equal to the slope of the straight line that is tangent to the $[C_2H_6]$ curve (black) at 90.0 minutes (red). The slopes of the other straight lines in the figure are average rates of reaction calculated over a 60.0 to 120.0 minute time interval (green), a 30.0 to 150.0 minute time interval (blue), and a 0 to 180.0 minute time interval (yellow). Note that the average rates of reaction tend toward the instantaneous rate of reaction as the time interval gets shorter.

The instantaneous rate at 90.0 minutes is equal to the slope of the straight line that is tangent to the $[C_2H_6]$ versus time curve at 90.0 minutes, as shown in Figure 17.2. The slope of the tangent line in Figure 17.2 is 3.60×10^{-5} M·min^{-1}, which is close to the value of 3.57×10^{-5} M·min^{-1} that we calculated above.

We shall define the rate of concentration change by using the delta notation that we have introduced, but we shall always assume that the time interval we use is as small as possible.

Because the stoichiometric coefficient of $C_2H_6(g)$ in Equation 17.1 is equal to one, we can use the rate of concentration change of $C_2H_6(g)$ as a direct measure of the rate of the reaction. In an equation, we have

$$\text{rate of reaction} = \text{rate of concentration change of } C_2H_6(g) = \frac{\Delta[C_2H_6]}{\Delta t}$$

We can also express the rate of a reaction in terms of the rate at which a reactant is consumed. In this case we are interested in the number of moles per liter *reacted* per second. Let's look at the rate of concentration change of $CH_3N_2CH_3(g)$ for the reaction described by Equation 17.1. The rate of concentration change of $CH_3N_2CH_3(g)$ is negative in this case because $\Delta[CH_3N_2CH_3] = [CH_3N_2CH_3]_2 - [CH_3N_2CH_3]_1$ is negative and $\Delta t = t_2 - t_1$ is positive, yielding

$$\text{rate of concentration change of } CH_3N_2CH_3(g) = \frac{\Delta[CH_3N_2CH_3]}{\Delta t} < 0$$

We can use the rate of concentration change of $CH_3N_2CH_3(g)$ as a measure of the rate of the reaction if we write

$$\text{rate of reaction} = -\text{rate of concentration change of } CH_3N_2CH_3(g)$$

$$= -\frac{\Delta[CH_3N_2CH_3]}{\Delta t}$$

The inclusion of the negative sign here makes the rate of reaction a positive quantity.

The decomposition of $CH_3N_2CH_3(g)$ is a particularly simple case because all the stoichiometric coefficients of the reaction are equal to one, and so we are able to express the rate of reaction as

$$\text{rate of reaction} = \frac{\Delta[C_2H_6]}{\Delta t} = \frac{\Delta[N_2]}{\Delta t} = -\frac{\Delta[CH_3N_2CH_3]}{\Delta t}$$

All these expressions are stoichiometrically equivalent.

Let's now consider the case of the decomposition of nitrogen dioxide to form oxygen and nitrogen oxide as described by the equation

$$2\,NO_2(g) \rightarrow O_2(g) + 2\,NO(g)$$

We can express the rate of this reaction in terms of the rate of production of $O_2(g)$ as

$$\text{rate of reaction} = \frac{\Delta[O_2]}{\Delta t} \tag{17.3}$$

We can also express the rate in terms of $\Delta[NO_2]/\Delta t$ and $\Delta[NO]/\Delta t$ as

$$\text{rate of reaction} \propto -\frac{\Delta[NO_2]}{\Delta t} \quad \text{and} \quad \text{rate of reaction} \propto \frac{\Delta[NO]}{\Delta t}$$

We included a negative sign in front of $\Delta[NO_2]/\Delta t$ because $NO_2(g)$ is a reactant and so is consumed in the reaction. We intentionally did not use equal signs in the above expressions because according to the reaction stoichiometry, two molecules of $NO_2(g)$ react to produce two molecules of $NO(g)$ and one molecule of $O_2(g)$. Therefore, the rate of formation of $O_2(g)$ is only one-half the rate of decomposition of $NO_2(g)$ and one-half the rate of formation of $NO(g)$. If we define the rate of reaction in terms of the change in concentration of $O_2(g)$ as stated in Equation 17.3, then we should write

$$\text{rate of reaction} = \frac{\Delta[O_2]}{\Delta t} = \frac{1}{2}\left(\frac{\Delta[NO]}{\Delta t}\right) = -\frac{1}{2}\left(\frac{\Delta[NO_2]}{\Delta t}\right)$$

The numerical factors here assure us that all the rates are expressed on the same basis, and the signs assure us that the rate of reaction is always a positive quantity. All three expressions are stoichiometrically equivalent.

The rate of reaction is always a positive quantity.

EXAMPLE 17-1: Let the rate of the reaction described by the equation

$$N_2O_5(g) \rightarrow 2\,NO_2(g) + \tfrac{1}{2}O_2(g) \tag{17.4}$$

be expressed as $-\Delta[N_2O_5]/\Delta t$. Express the rate of the reaction in terms of the rate of concentration change of each product.

Solution: The rate of loss of $N_2O_5(g)$ is twice as great as the rate of production of $O_2(g)$ because two $N_2O_5(g)$ molecules are consumed for each $O_2(g)$ molecule produced. Thus, we have

$$\text{rate of reaction} = -\frac{\Delta[N_2O_5]}{\Delta t} = \frac{2\Delta[O_2]}{\Delta t}$$

The rate of loss of $N_2O_5(g)$ is one-half the rate of production of $NO_2(g)$ because two $NO_2(g)$ molecules are formed for each $N_2O_5(g)$ molecule that decomposes. Thus,

$$\text{rate of reaction} = -\frac{\Delta[N_2O_5]}{\Delta t} = \frac{1}{2}\left(\frac{\Delta[NO_2]}{\Delta t}\right)$$

or

$$\text{rate of reaction} = -\frac{\Delta[N_2O_5]}{\Delta t} = \frac{1}{2}\left(\frac{\Delta[NO_2]}{\Delta t}\right) = \frac{2\Delta[O_2]}{\Delta t}$$

All three expressions are stoichiometrically equivalent.

PRACTICE PROBLEM 17-1: Let the rate of the reaction described by the equation

$$H_2(g) + Br_2(g) \rightarrow 2\,HBr(g)$$

be expressed as $-\Delta[H_2]/\Delta t$. Express the rate of the reaction in terms of the rate of concentration change of the other reactants or products.

Answer:

$$\text{rate of reaction} = -\frac{\Delta[H_2]}{\Delta t} = -\frac{\Delta[Br_2]}{\Delta t} = \frac{1}{2}\left(\frac{\Delta[HBr]}{\Delta t}\right)$$

In all cases that we have discussed so far, the equivalent expressions for the rate of reaction depend upon the stoichiometric coefficients in the chemical equation. For example, the rate of the reaction for the equation describing the decomposition of $N_2O_5(g)$ in Example 17-1 can be written as

$$\text{rate of reaction} = -\frac{1}{1}\left(\frac{\Delta[N_2O_5]}{\Delta t}\right) = \frac{1}{2}\left(\frac{\Delta[NO_2]}{\Delta t}\right) = \frac{1}{\left(\frac{1}{2}\right)}\left(\frac{\Delta[O_2]}{\Delta t}\right) \quad (17.5)$$

The numerical factor in the denominator of each expression is the same as the stoichiometric coefficient of the species in the chemical equation (Equation 17.4). For the general reaction equation,

$$aA + bB \rightarrow cC + dD$$

the expression for the rate of reaction is given as

$$\text{rate of reaction} = -\frac{1}{a}\left(\frac{\Delta[\text{A}]}{\Delta t}\right) = -\frac{1}{b}\left(\frac{\Delta[\text{B}]}{\Delta t}\right) = \frac{1}{c}\left(\frac{\Delta[\text{C}]}{\Delta t}\right) = \frac{1}{d}\left(\frac{\Delta[\text{D}]}{\Delta t}\right) \quad (17.6)$$

Equation 17.6 expresses the relation between the rate of a reaction and the rate of concentration change of the various species in the reaction.

17-2. The Rate of Reaction Varies with Time

Figure 17.3 shows the concentrations of $N_2O_5(g)$, $NO_2(g)$, and $O_2(g)$ as a function of time for the reaction described by

$$N_2O_5(g) \rightarrow 2\,NO_2(g) + \tfrac{1}{2}O_2(g) \quad (17.7)$$

at 45°C. The corresponding numerical data are given in Table 17.2. From these data we can see that the rate of reaction decreases with time. For example, the average rate of reaction of $N_2O_5(g)$ over the first 10.0 minutes of the reaction is

$$\text{average rate of reaction} = -\frac{\Delta[N_2O_5]}{\Delta t} = -\frac{(0.92\times10^{-2} - 1.24\times10^{-2})\ \text{M}}{(10.0 - 0)\ \text{min}}$$
$$= 3.2\times10^{-4}\ \text{M·min}^{-1}$$

whereas the average rate of reaction over the period 10.0 minutes to 20.0 minutes is

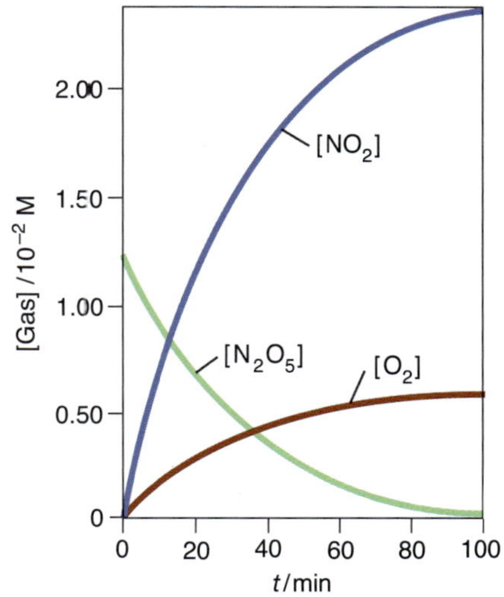

Figure 17.3 The concentrations of $N_2O_5(g)$, $NO_2(g)$, and $O_2(g)$ as a function of time for the reaction described by the equation

$$N_2O_5(g) \rightarrow 2\,NO_2(g) + \tfrac{1}{2}O_2(g)$$

at 45°C. The data are given in Table 17.2.

TABLE 17.2 Concentrations of $N_2O_5(g)$, $NO_2(g)$, and $O_2(g)$ as a function of time at 45°C for the reaction

$$N_2O_5(g) \rightarrow 2\,NO_2(g) + \tfrac{1}{2}O_2(g)$$

t/min	$[N_2O_5]/10^{-2}$ M	$[NO_2]/10^{-2}$ M	$[O_2]/10^{-2}$ M
0	1.24	0	0
10.0	0.92	0.64	0.16
20.0	0.68	1.12	0.28
30.0	0.50	1.48	0.37
40.0	0.37	1.74	0.44
50.0	0.28	1.92	0.48
60.0	0.21	2.06	0.52
70.0	0.14	2.18	0.55
80.0	0.11	2.26	0.57
90.0	0.08	2.32	0.58
100.0	0.06	2.36	0.59

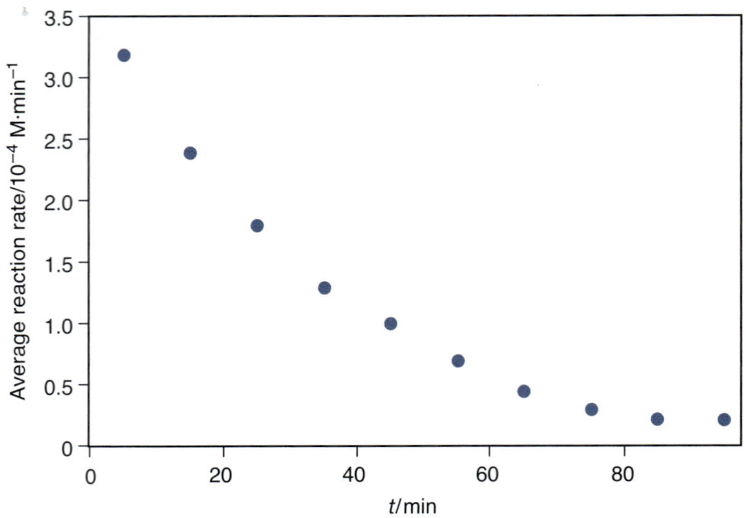

Figure 17.4 A plot of the average rate of reaction of the decomposition of $N_2O_5(g)$ according to

$$N_2O_5(g) \rightarrow 2\,NO_2(g) + \tfrac{1}{2}O_2(g)$$

at 45°C against time for consecutive 10-minute time intervals.

$$\text{average rate of reaction} = -\frac{(0.68\times10^{-2} - 0.92\times10^{-2})\ \text{M}}{(20.0 - 10.0)\ \text{min}}$$

$$= 2.4\times10^{-4}\ \text{M}\cdot\text{min}^{-1}$$

Figure 17.4 shows the average rates over the other 10-minute intervals in Table 17.2. Note that these rates decrease with increasing time. The average rate of reaction decreases with time because $[N_2O_5]$ decreases with time.

EXAMPLE 17-2: Use the data in Table 17.2 to calculate the average rate of production of $NO_2(g)$ over the first 10.0-minute time interval for the decomposition reaction of $N_2O_5(g)$ as described by Equation 17.4.

Solution: Because $NO_2(g)$ is a product, the average rate of production of $NO_2(g)$ is given by $\Delta[NO_2]/\Delta t$. For the first 10.0 minutes of the reaction

$$\text{average rate of concentration change} = \frac{\Delta[NO_2]}{\Delta t} = \frac{(0.64\times10^{-2} - 0)\ \text{M}}{(10.0 - 0)\ \text{min}}$$

$$= 6.4\times10^{-4}\ \text{M}\cdot\text{min}^{-1}$$

Notice that the average rate of production of $NO_2(g)$ over the first 10 minutes is twice as great as the average rate of consumption of $N_2O_5(g)$ over the same time period (Figure 17.4). This result is true for all the other 10-minute time intervals in Table 17.2 because, as we have shown from the reaction stoichiometry in Example 17-1,

$$\text{rate of reaction} = -\frac{\Delta[N_2O_5]}{\Delta t} = \frac{1}{2}\left(\frac{\Delta[NO_2]}{\Delta t}\right)$$

PRACTICE PROBLEM 17-2: Use the data in Table 17.2 to calculate the average rate of production of $O_2(g)$ for the reaction equation given in Equation 17.4 over the time interval 10.0 to 20.0 minutes.

Answer: 1.2×10^{-4} M·min^{-1}

It is important to realize that, given the initial concentrations in Table 17.2, the concentrations of $NO_2(g)$ and $O_2(g)$ can be calculated from the concentration of $N_2O_5(g)$ at any time. The following Example illustrates such a calculation.

EXAMPLE 17-3: Given that the initial concentration of $N_2O_5(g)$ is $[N_2O_5]_0 = 1.24 \times 10^{-2}$ M, if $[N_2O_5] = 0.23 \times 10^{-2}$ M at $t = 55.0$ min, calculate $[NO_2]$ and $[O_2]$ at $t = 55.0$ min.

Solution: The chemical equation is

$$N_2O_5(g) \rightarrow 2\,NO_2(g) + \tfrac{1}{2}O_2(g)$$

The number of moles per liter of $N_2O_5(g)$ that have reacted in 55.0 minutes is the difference between the initial concentration and the concentration at 55.0 minutes.

$$\begin{aligned}\text{mol·L}^{-1}\ N_2O_5(g)\ \text{reacted} &= [N_2O_5]_0 - [N_2O_5] \\ &= 1.24 \times 10^{-2}\ \text{M} - 0.23 \times 10^{-2}\ \text{M} = 1.01 \times 10^{-2}\ \text{M}\end{aligned}$$

According to the chemical equation, two moles of $NO_2(g)$ are produced for each mole of $N_2O_5(g)$ that reacts; so, per liter, we have

$$\begin{pmatrix}\text{mol·L}^{-1}\ \text{of}\ NO_2(g) \\ \text{produced}\end{pmatrix} = (1.01 \times 10^{-2}\ \text{M}\ N_2O_5)\left(\frac{2\ \text{M}\ NO_2}{1\ \text{M}\ N_2O_5}\right) = 2.02 \times 10^{-2}\ \text{M}$$

Similarly, for $O_2(g)$ we have

$$\begin{pmatrix}\text{mol·L}^{-1}\ \text{of}\ O_2(g) \\ \text{produced}\end{pmatrix} = (1.01 \times 10^{-2}\ \text{M}\ N_2O_5)\left(\frac{\frac{1}{2}\ \text{M}\ O_2}{1\ \text{M}\ N_2O_5}\right) = 0.505 \times 10^{-2}\ \text{M}$$

PRACTICE PROBLEM 17-3: Using the initial concentrations and the fact that $[O_2] = 0.55 \times 10^{-2}$ M at $t = 70.0$ minutes in Table 17.2, calculate $[N_2O_5]$ and $[NO_2]$ at $t = 70.0$ minutes.

Answer: $[N_2O_5] = 1.4 \times 10^{-3}$ M; $[NO_2] = 0.022$ M

17-3. The Rate Law of a Reaction Can Be Determined by the Method of Initial Rates

We have noted that the rate of a reaction changes as the reactants are consumed and the products are formed. This change occurs because the rates of chemical reactions often depend on the concentrations of one or more of the reactants. The mathematical equation that relates the concentrations of the reactants to the rate of reaction is called the **rate law** of the reaction. Many experiments show that the rate of a reaction is typically proportional to the concentrations of the reactants raised to small integer powers. Thus, for the case of the thermal decomposition of $N_2O_5(g)$ described by Equation 17.4, we shall assume that

$$\text{rate of reaction} \propto [N_2O_5]^x \tag{17.8}$$

We can write this proportionality as an equation by inserting a proportionality constant:

$$\text{rate of reaction} = k[N_2O_5]^x \tag{17.9}$$

The proportionality constant, k, in a rate law is called the **rate constant** of the reaction.

The value of x in Equation 17.9 must be determined *experimentally;* it is not necessarily the same as the balancing coefficient of $N_2O_5(g)$ in the chemical equation. *There is no relation between the balancing coefficients in a chemical equation and the reaction rate law.* For example, the rate law for the reaction described by the equation

$$N_2O_3(g) \rightarrow NO(g) + NO_2(g)$$

is

$$\text{rate of reaction} = k[N_2O_3]$$

whereas the rate law for the reaction described by the equation

$$NOBr(g) \rightarrow NO(g) + \tfrac{1}{2}Br_2(g)$$

is

$$\text{rate of reaction} = k[NOBr]^2$$

Reaction rate laws can often be determined from data involving reaction rates at the early stages of reactions. Using the **method of initial rates**, we measure the rate of a reaction over an initial time interval short enough that the concentrations of the reactants do not vary appreciably from their initial values. If we use zeros as subscripts to denote the initial values of the rate of reaction and of the various concentrations, then for any reaction described by an equation of the form

$$aA + bB + cC \rightarrow \text{products}$$

we can write

$$(\text{rate of reaction})_0 = k[\text{A}]_0^x[\text{B}]_0^y[\text{C}]_0^z \tag{17.10}$$

We can then determine the values of x, y, and z by varying the initial concentration of each reactant in turn, while keeping the other reactant concentrations constant. As a simple example, let's consider the $N_2O_5(g)$ decomposition reaction, in which there is only one reactant, as described by the equation

$$N_2O_5(g) \rightarrow 2NO_2(g) + \tfrac{1}{2}O_2(g)$$

Suppose that we run this reaction using successively doubled concentrations of $N_2O_5(g)$ and obtain the following initial-rate data at 45°C:

Run	$[N_2O_5]_0$/mol·L^{-1}	(rate of reaction)$_0$/mol·L^{-1}·h^{-1}
1	0.010	0.018
2	0.020	0.036
3	0.040	0.072

Notice that when the initial concentration, $[N_2O_5]_0$, is increased by a factor of two, the initial rate of reaction, (rate of reaction)$_0$, also increases by a factor of two. These results show that (rate of reaction)$_0$ is directly proportional to $[N_2O_5]_0$, so we can write

$$(\text{rate of reaction})_0 = k[N_2O_5]_0 \tag{17.11}$$

showing that $x = 1$ in Equation 17.10.

Extensive experimental studies show that in many cases the reaction rate law determined from initial-rate data describes the rate of reaction as the reaction proceeds. Thus, we can drop the subscript zero in Equation 17.11 to obtain the rate law:

$$\text{rate of reaction} = k[N_2O_5] \tag{17.12}$$

Because $[N_2O_5]$ is raised to the first power in Equation 17.12, the rate law is said to be a **first-order rate law,** and the reaction is said to be a **first-order reaction.** We say that the rate law is first order in $[N_2O_5]$.

We can determine the value of the rate constant k in Equation 17.12 by using the tabulated initial-rate data. If we substitute the fact that (rate of reaction)$_0$ = 0.072 mol·L^{-1}·h^{-1} when $[N_2O_5]_0$ = 0.040 mol·L^{-1}, then we obtain

$$k = \frac{(\text{rate of reaction})_0}{[N_2O_5]_0} = \frac{0.072\ \text{mol·L}^{-1}\text{·h}^{-1}}{0.040\ \text{mol·L}^{-1}} = 1.8\ \text{h}^{-1}$$

Notice that the units of the rate constant for a first-order reaction are reciprocal time (time^{-1}).

The following Example shows that the initial concentrations in the various runs do not have to be simple whole multiples of one another.

EXAMPLE 17-4: For the reaction described by the equation

$$SO_2Cl_2(g) \rightarrow SO_2(g) + Cl_2(g)$$

the following initial rate data were obtained:

Run	$[SO_2Cl_2]_0/\text{mol}\cdot\text{L}^{-1}$	$(\text{rate of reaction})_0/\text{mol}\cdot\text{L}^{-1}\cdot\text{s}^{-1}$
1	0.74	1.63×10^{-5}
2	1.25	2.75×10^{-5}
3	1.86	4.09×10^{-5}

Determine the order of the reaction and the value of the rate constant.

Solution: We first write the rate law as

$$(\text{rate of reaction})_0 = k[SO_2Cl_2]_0^x$$

where the value of x is to be determined. We now take the ratios of the results for any two runs, say, runs 2 and 3:

$$\frac{(\text{rate of reaction})_0 \text{ for run 2}}{(\text{rate of reaction})_0 \text{ for run 3}} = \frac{k[SO_2Cl_2]_0^x \text{ for run 2}}{k[SO_2Cl_2]_0^x \text{ for run 3}}$$

If we substitute the given data into these ratios, we obtain

$$\frac{2.75 \times 10^{-5}}{4.09 \times 10^{-5}} = \frac{\cancel{k}(1.25)^x}{\cancel{k}(1.86)^x}$$

$$0.672 = \left(\frac{1.25}{1.86}\right)^x$$

$$0.672 = (0.672)^x$$

Therefore, the value of x is 1, so the reaction is first order, and

$$\text{rate of reaction} = k[SO_2Cl_2]$$

The value of the rate constant, k, can be obtained from any run; if we arbitrarily choose run 2, then we have

$$k = \frac{(\text{rate of reaction})_0}{[SO_2Cl_2]_0} = \frac{2.75 \times 10^{-5}\ \text{mol}\cdot\text{L}^{-1}\cdot\text{s}^{-1}}{1.25\ \text{mol}\cdot\text{L}^{-1}} = 2.20 \times 10^{-5}\ \text{s}^{-1}$$

Assuming that the form of the rate law remains the same with time, we have

$$\text{rate of reaction} = (2.20 \times 10^{-5}\ \text{s}^{-1})\ [SO_2Cl_2]$$

PRACTICE PROBLEM 17-4: The decomposition of azomethane, $CH_3N_2CH_3(g)$, which can be described by the equation

$$CH_3N_2CH_3(g) \rightarrow CH_3CH_3(g) + N_2(g)$$

was studied at 300°C and the following initial rate data were obtained:

Run	$[CH_3N_2CH_3]_0$/mol·L^{-1}	(rate of reaction)$_0$/mol·L^{-1}·s^{-1}
1	0.604	2.42×10^{-4}
2	0.913	3.65×10^{-4}
3	1.701	6.80×10^{-4}

Determine the order of the reaction and the value of the rate constant from these data.

Answer: first order; $k = 4.00 \times 10^{-4}\ s^{-1}$

So far we have encountered only first-order reactions. Initial-rate data are given below for the reaction described by

$$2\,NO_2(g) \rightarrow 2\,NO(g) + O_2(g) \qquad (17.13)$$

Run	$[NO_2]_0$/mol·L^{-1}	(rate of reaction)$_0$/mol·L^{-1}·s^{-1}
1	0.85	0.39
2	1.10	0.65
3	1.60	1.38

We assume that the rate law is of the form

$$(\text{rate of reaction})_0 = k[NO_2]_0^x$$

and arbitrarily use runs 1 and 3

$$\frac{(\text{rate of reaction})_0 \text{ for run 3}}{(\text{rate of reaction})_0 \text{ for run 1}} = \frac{k[NO_2]_0^x \text{ for run 3}}{k[NO_2]_0^x \text{ for run 1}}$$

or

$$\frac{1.38}{0.39} = \frac{k(1.60)^x}{k(0.85)^x} = \left(\frac{1.60}{0.85}\right)^x$$

$$3.54 = (1.88)^x$$

By simple trial and error, we find that $x = 2$, so the rate law (assuming that the initial-rate law does not change with time) is

$$\text{rate of reaction} = k[NO_2]^2 \qquad (17.14)$$

Here we could also have determined the value of x by taking logarithms of both sides. (For a review of logarithms, see Appendix A.) Using the property of logarithms, $\log a^x = x \log a$,

$$3.54 = (1.88)^x$$

$$\log(3.54) = x \log(1.88)$$

$$x = \frac{\log(3.54)}{\log(1.88)} = \frac{0.549}{0.274} = 2$$

However, because most rate laws involve simple integer powers (or at least simple fractions), we can usually determine the power by direct inspection of the data.

Because the exponent of $[NO_2]$ in the rate law in Equation 17.14 is 2, we say that the reaction rate law is **second order** with respect to NO_2. Because $[NO_2]$ is the only concentration that appears in the rate law expression, we say that the decomposition of $NO_2(g)$ to $NO(g)$ and $O_2(g)$ is a **second-order reaction.**

We can also determine rate laws for reactions involving more than one reactant. For example, consider the following initial-rate data for the reaction described by

$$2\,NO_2(g) + F_2(g) \rightarrow 2\,NO_2F(g) \tag{17.15}$$

Run	$[NO_2]_0$/M	$[F_2]_0$/M	(rate of reaction)$_0$/M·s^{-1}
1	1.00	1.00	1.00×10^{-4}
2	2.00	1.00	2.00×10^{-4}
3	1.00	2.00	2.00×10^{-4}
4	2.00	2.00	4.00×10^{-4}

A comparison of the data for runs 1 and 2 shows that a twofold increase in the initial concentration of $NO_2(g)$ with $[F_2]_0$ held fixed increases the initial rate by a factor of two; thus, the rate of reaction must be proportional to the first power of $[NO_2]_0$,

$$(\text{rate of reaction})_0 \propto [NO_2]_0 \qquad \text{at constant } [F_2]_0$$

A comparison of the data for runs 1 and 3 shows that a twofold increase in the initial concentration of $F_2(g)$ with $[NO_2]_0$ held fixed increases the initial rate of reaction by a factor of two; thus, the rate must also be proportional to the first power of $[F_2]_0$, that is,

$$(\text{rate of reaction})_0 \propto [F_2]_0 \qquad \text{at constant } [NO_2]_0$$

We can combine these two results to get the reaction-rate law,

$$(\text{rate of reaction})_0 = k[NO_2]_0[F_2]_0$$

Comparison of runs 1 and 4 shows that a simultaneous twofold increase in $[NO_2]_0$ and $[F_2]_0$ increases the initial rate of reaction by a factor of four, a result consistent with the preceding rate law.

The value of the reaction rate constant is (from run 1)

$$k = \frac{(\text{rate of reaction})_0}{[NO_2]_0[F_2]_0} = \frac{1.00 \times 10^{-4}\ \text{M}\cdot\text{s}^{-1}}{(1.00\ \text{M})(1.00\ \text{M})} = 1.00 \times 10^{-4}\ \text{M}^{-1}\cdot\text{s}^{-1}$$

Thus, assuming that the rate law does not change with time, we take the rate law for the reaction to be

$$\text{rate of reaction} = (1.00 \times 10^{-4}\ \text{M}^{-1}\cdot\text{s}^{-1})[NO_2][F_2]$$

This rate law is first order in both $[NO_2]$ and $[F_2]$; and we say that the **overall order** of the reaction is second order because the sum of the exponents is 2. Note that the units of k for a second-order reaction are reciprocal concentration per time ($\text{M}^{-1}\cdot\text{s}^{-1}$).

More generally, if a reaction-rate law is of the form

$$\text{rate of reaction} = k[\text{A}]^x[\text{B}]^y[\text{C}]^z$$

then the rate law is x order in [A], y order in [B], and z order in [C] with an overall order of $x + y + z$. For the case we are considering, the sum of $x(= 1)$ and $y(= 1)$ is 2, so the rate law is first order in $[NO_2]$, first order in $[F_2]$, and second order overall. Both reactions described by Equations 17.13 and 17.15 represent second-order reactions.

EXAMPLE 17-5: At 325°C, $NO_2(g)$ reacts with $CO(g)$ to yield $NO(g)$ and $CO_2(g)$, as described by the reaction equation

$$NO_2(g) + CO(g) \rightarrow NO(g) + CO_2(g)$$

Using the following initial-rate data for the reaction,

Run	$[NO_2]_0/\text{mol}\cdot\text{L}^{-1}$	$[CO]_0/\text{mol}\cdot\text{L}^{-1}$	$(\text{rate of reaction})_0/\text{mol}\cdot\text{L}^{-1}\cdot\text{s}^{-1}$
1	0.15	0.15	0.011
2	0.30	0.15	0.045
3	0.60	0.30	0.18
4	0.60	0.60	0.18

determine the rate law and the value of the rate constant.

Solution: We see from runs 3 and 4 that the initial rate of reaction does not depend on the concentration of $CO(g)$. When we double the concentration of $[CO]_0$ from 0.30 M to 0.60 M while keeping $[NO_2]_0$ constant, the rate remains unchanged. Comparing the rates for runs 3 and 4, we find

$$\frac{(\text{rate of reaction})_0 \text{ for run 3}}{(\text{rate of reaction})_0 \text{ for run 4}} = \frac{k[\text{NO}_2]_0^x[\text{CO}]_0^y \text{ for run 3}}{k[\text{NO}_2]_0^x[\text{CO}]_0^y \text{ for run 4}}$$

$$\frac{0.18}{0.18} = \frac{\cancel{k(0.60)^x}(0.30)^y}{\cancel{k(0.60)^x}(0.60)^y}$$

$$1 = \left(\frac{0.30}{0.60}\right)^y = \left(\frac{1}{2}\right)^y$$

The only value of y that satisfies the relationship $1 = (\frac{1}{2})^y$ is $y = 0$ because any quantity raised to the zero power is equal to one. Thus, we say that the **reaction is zero order** in [CO]. In other words, the rate of reaction is *independent* of the concentration of [CO]. This is a surprising result! Who would think from the stated reaction equation that the rate of the reaction would not depend on the concentration of carbon monoxide present? We shall explain why this is the case for this reaction in the next chapter. This result does not mean that this reaction will occur when no CO(*g*) is present (that is clearly impossible); just that under the conditions of this experiment, the rate of the reaction does not depend on how much CO(*g*) is available, and that an increase or decrease in [CO] does not affect the rate of reaction.

Although the rate does not depend on $[\text{CO}]_0$, the rate does depend on $[\text{NO}_2]_0$. Here simple inspection of the data suffices to determine the rate law: As we double $[\text{NO}_2]_0$ in going from 0.15 M to 0.30 M while keeping $[\text{CO}]_0$ constant, we quadruple the rate of reaction. In going from 0.15 M to 0.60 M, a fourfold increase in concentration, we increase the rate of reaction by a factor of 16 (in this case the change in the concentration of $[\text{CO}]_0$ does not matter, because we have already determined the reaction to be zero order in [CO], or independent of CO(*g*) concentration). Thus, we see that the rate of reaction is proportional to the square of $[\text{NO}_2]_0$, so we can write the initial-rate law as

$$(\text{rate of reaction})_0 = k[\text{NO}_2]_0^2$$

Assuming that the rate law expression does not change with time, we take the rate law to be

$$\text{rate of reaction} = k[\text{NO}_2]^2$$

Because the exponent on the concentration term is 2, the rate law is a second-order rate law. We can evaluate k by using (from run 2)

$$k = \frac{(\text{rate of reaction})_0}{[\text{NO}_2]_0^2} = \frac{0.045\ \text{M}\cdot\text{s}^{-1}}{(0.30\ \text{M}\cdot\text{s}^{-1})^2} = 0.50\ \text{M}^{-1}\cdot\text{s}^{-1}$$

Notice, once again, that the units of k for a second-order rate law are $\text{M}^{-1}\cdot\text{s}^{-1}$, or liters per mole per second ($\text{L}\cdot\text{mol}^{-1}\cdot\text{s}^{-1}$).

PRACTICE PROBLEM 17-5: The following initial-rate data were obtained for the reaction described by the equation

$$2\,NO(g) + Br_2(g) \rightarrow 2\,NOBr(g)$$

Run	$[NO]_0$/M	$[Br_2]_0$/M	(rate of reaction)$_0$/M·min^{-1}
1	1.00	1.00	1.30×10^{-3}
2	1.50	1.00	2.93×10^{-3}
3	1.50	3.00	8.78×10^{-3}

Determine the reaction-rate law and the value of the rate constant.

Answer: rate of reaction = $(1.30 \times 10^{-3}\ M^{-2}\cdot min^{-1})[NO]^2[Br_2]$

17-4. A Plot of ln[A] Versus Time Gives a Straight Line for a First-Order Reaction

The rate law for a first-order reaction, say,

$$A \rightarrow B$$

is

$$\text{rate of reaction} = -\frac{\Delta[A]}{\Delta t} = k[A] \tag{17.16}$$

The spontaneous radioactive decay of certain unstable isotopes is an example of a first-order process. By using elementary calculus (see Problem 17-78), we can convert the expression given in Equation 17.16 into the form

$$\ln[A] = \ln[A]_0 - kt \tag{17.17}$$

where $[A]_0$ is the concentration of A at $t = 0$ and ln is the natural logarithm.

The general properties of natural logarithms are the same as those of common logarithms. (For a review of working with logarithms see Appendix A; the rules for significant figures and logarithms are given in the sidebox on the next page.) For example, using the relation $\ln(a/b) = \ln a - \ln b$, we can rewrite Equation 17.17 in another form that is often useful:

$$\ln\frac{[A]}{[A]_0} = -kt \tag{17.18}$$

Alternative forms of Equation 17.18 are

$$\frac{[A]}{[A]_0} = e^{-kt} \quad \text{or} \quad [A] = [A]_0 e^{-kt} \tag{17.19}$$

The ratio $[A]/[A]_0$ in these equations is the fraction of A that remains at time t.

SIGNIFICANT FIGURES AND LOGARITHMS The rules for significant figures of common and natural logarithms differ from those for multiplication and division. The number of significant figures in a number determines the number of significant digits listed after the decimal place in that number's logarithm. For example,

$$\log(\underbrace{6.022}_{\text{4 significant figures}} \times 10^{23}) = 23.\underbrace{7797}_{\text{4 significant digits following the decimal place}}$$

Because 6.022×10^{23} has four significant figures, its logarithm has four significant digits following the decimal place. This rule is reversed when taking the antilogarithm of a number. For example, the base *e* antilogarithm (or the inverse natural logarithm) of −10.55 is

$$e^{-10.\overbrace{55}^{\text{2 significant digits}}} = \underbrace{2.6}_{\text{2 significant figures}} \times 10^{-5}$$

Because the number −10.55 has two significant digits following the decimal point, its antilogarithm is given to two significant figures.

We showed in Section 17-3 that the reaction described by

$$N_2O_5(g) \rightarrow 2\,NO_2(g) + \tfrac{1}{2}O_2(g)$$

is first order with a rate constant $k = 1.8\ \text{h}^{-1}$ at 45°C. Using Equation 17.18, we can write

$$\ln\frac{[N_2O_5]}{[N_2O_5]_0} = -(1.8\ \text{h}^{-1})t$$

Taking the initial concentration of $N_2O_5(g)$ to be $[N_2O_5]_0 = 1.24 \times 10^{-2}$ M from Table 17.2, we can use this expression to calculate some of the values of $[N_2O_5]$ in Table 17.2. At $t = 50.0$ minutes, we have

$$\ln\frac{[N_2O_5]}{1.24 \times 10^{-2}\ \text{M}} = -(1.8\ \text{h}^{-1})(50.0\ \text{min})\left(\frac{1\ \text{h}}{60\ \text{min}}\right) = -1.5$$

Taking the antilogarithm of this expression yields

$$\frac{[N_2O_5]}{1.24 \times 10^{-2}\ \text{M}} = e^{-1.5} = 0.2$$

Thus,

$$[N_2O_5] = (0.2)(1.24 \times 10^{-2}\ \text{M}) = 0.3 \times 10^{-2}\ \text{M}$$

a result in agreement with the entry in Table 17.2.

EXAMPLE 17-6: The rate law for the decomposition of aqueous hydrogen peroxide at 70°C as described by the equation

$$2\,H_2O_2(aq) \rightarrow 2\,H_2O(l) + O_2(g)$$

is first order in $[H_2O_2]$ with a rate constant $k = 0.0347\ \text{min}^{-1}$. Given that the initial concentration is $[H_2O_2]_0 = 0.10$ M, calculate the value of $[H_2O_2]$ 60.0 minutes after the solution is prepared.

Solution: We use Equation 17.18,

$$\ln\frac{[H_2O_2]}{[H_2O_2]_0} = -kt$$

Substituting in the given values of $[H_2O_2]_0$, k, and t, we have

$$\ln\frac{[H_2O_2]}{0.10\ \text{M}} = -(0.0347\ \text{min}^{-1})(60.0\ \text{min}) = -2.08$$

or

$$\frac{[H_2O_2]}{0.10\ \text{M}} = e^{-2.08} = 0.13$$

Thus,

$$[H_2O_2] = (0.13)\,(0.10\ \text{M}) = 0.013\ \text{M}$$

PRACTICE PROBLEM 17-6: Cyclobutane decomposes to ethene at elevated temperatures according to

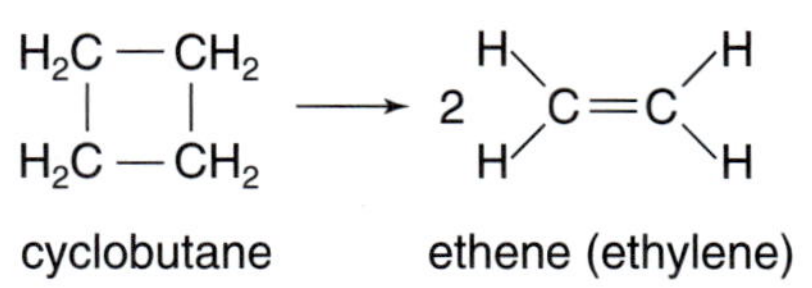

The reaction is first order in [cyclobutane], and at 700 K the value of the rate constant is $0.015\ \text{min}^{-1}$. Calculate the fraction of cyclobutane that will remain after 1.0 hour.

Answer: 0.41

It is convenient to plot Equation 17.17, $\ln[A] = \ln[A]_0 - kt$, as a straight line. Equation 17.17 may not look like it is of the form $y = mx + b$, but if we let $y = \ln[A]$ and $x = t$, then we have

$$\ln[A] = mt + b$$

By comparing this equation with Equation 17.17, we can see that the intercept, b, is

$$b = \ln[A]_0$$

Notice that we divide $[N_2O_5]$ by M here so that $[N_2O_5]/M$ is unitless; you can take logarithms only of unitless quantities.

TABLE 17.3 $[N_2O_5]$ and $\ln([N_2O_5]/M)$ as a function of time for the reaction described by

$$N_2O_5(g) \rightarrow 2NO_2(g) + \frac{1}{2}O_2(g)$$

t/min	$[N_2O_5]/10^{-2}$ M	$\ln([N_2O_5]/M)$
0	1.24	–4.39
10.0	0.92	–4.69
20.0	0.68	–4.99
30.0	0.50	–5.30
40.0	0.37	–5.60
50.0	0.28	–5.88
60.0	0.20	–6.21
70.0	0.15	–6.50
80.0	0.11	–6.81
90.0	0.08	–7.13
100.0	0.06	–7.42

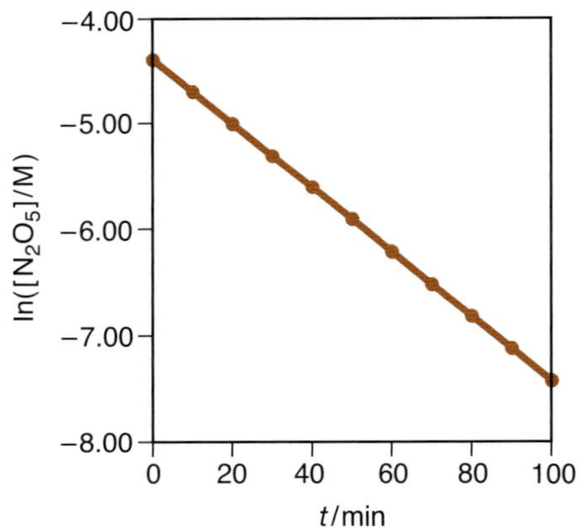

Figure 17.5 Plot of $\ln([N_2O_5]/M)$ versus time (t) for the reaction described by

$$N_2O_5(g) \rightarrow 2NO_2(g) + \frac{1}{2}O_2(g)$$

The plot is a straight line, confirming that the reaction is first order. The numerical values are given in Table 17.3.

and that the slope, m, is

$$m = -k$$

Therefore, for a first-order reaction, a plot of $\ln[A]$ versus t is a straight line, and we can determine the rate constant from the slope of the line. If the rate law is not first order in [A], then a plot of $\ln[A]$ versus t will not be linear. This result provides a simple test to determine whether a particular reaction is first order in [A].

Let's apply this procedure to our reaction involving $N_2O_5(g)$. First we must convert the $[N_2O_5]$ data to $\ln([N_2O_5]/M)$ data. The results are given in Table 17.3 and are plotted in the form $\ln([N_2O_5]/M)$ versus t in Figure 17.5. The plot is linear, confirming our earlier determination that the rate law is first order in $[N_2O_5]$. (We show how to find the equation of this line in Appendix A.6).

17-5. The Half-Life for a First-Order Reaction Is Independent of the Initial Concentration

The numerical value of a rate constant is a measure of the rate of a reaction. Other things being equal, a large rate constant indicates a fast reaction, whereas a small rate constant indicates a slow reaction. Another useful measure of the rate of a reaction is its **half-life**, $t_{1/2}$, the time it takes for one-half of a quantity of a reactant to react.

Let's return to the data in Table 17.2 on the $N_2O_5(g)$ decomposition reaction. Figure 17.6 shows a plot of $[N_2O_5]$ versus time on a larger scale than that shown in Figure 17.3. The initial concentration of $N_2O_5(g)$ is $[N_2O_5]_0 = 1.24 \times 10^{-2}$ M. The time required for $[N_2O_5]$ to decrease from 1.24×10^{-2} M to

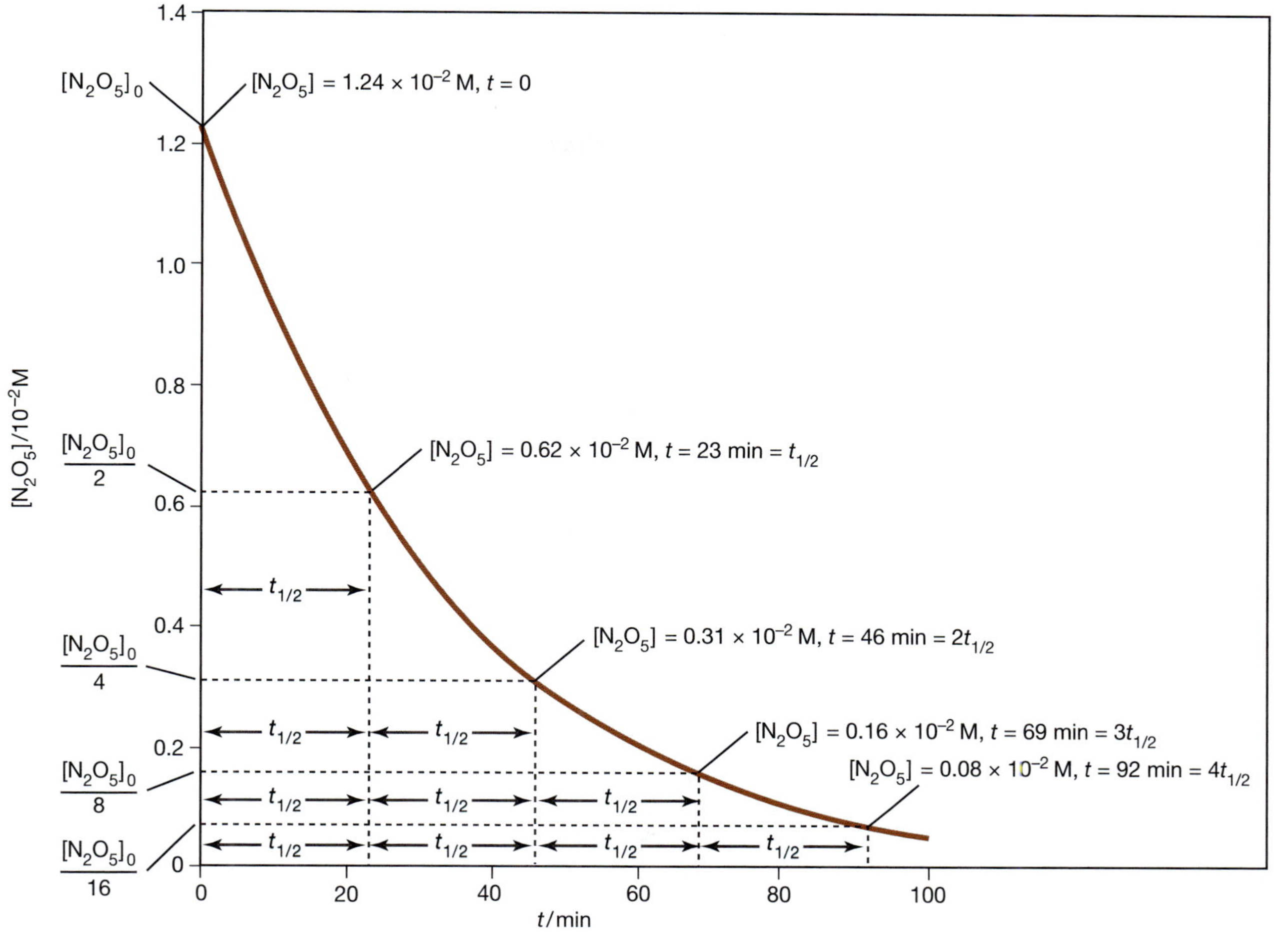

Figure 17.6 Plot of $[N_2O_5]$ versus time, illustrating the half-life ($t_{1/2}$) for $[N_2O_5]$ in the reaction equation

$$N_2O_5(g) \rightarrow 2\,NO_2(g) + \tfrac{1}{2}O_2(g)$$

at 45°C.

$\frac{1}{2}$ (1.24 × 10^{-2} M) = 0.62 × 10^{-2} M can be read off the plot in Figure 17.6 as $t_{1/2}$ = 23 minutes. Note also in Figure 17.6 that another 23 minutes are required for $[N_2O_5]$ to decrease from 0.62 × 10^{-2} M to $\frac{1}{2}$(0.62 × 10^{-2} M) = 0.31 × 10^{-2} M. The time required for the concentration of $N_2O_5(g)$ to decrease by a factor of two, the half-life of the reaction, is *independent* of the concentration of unreacted $N_2O_5(g)$. This result is shown in Table 17.4.

Notice that the initial concentration decreases by a factor of two for each half-life as the reaction proceeds. After the first 23 minutes,

$$[N_2O_5] = \tfrac{1}{2}[N_2O_5]_0 = 0.62 \times 10^{-2}\ \text{M} \qquad (t = t_{1/2} = 23\ \text{minutes})$$

After the second 23-minute interval (23 minutes to 46 minutes)

$$[N_2O_5] = \left(\tfrac{1}{2}\right)\left(\tfrac{1}{2}\right)[N_2O_5]_0 = \left(\tfrac{1}{2}\right)^2[N_2O_5]_0 = 0.31 \times 10^{-2}\ \text{M}$$
$$(t = 2t_{1/2} = 46\ \text{minutes})$$

TABLE 17.4 Half-life values for the N_2O_5 decomposition reaction

Twofold decrease in $[N_2O_5]$ /10^{-2} M	$t_{1/2}$/min	Total time/min
$1.24 \rightarrow 0.62$	23	$23 = (1 \times 23)$
$0.62 \rightarrow 0.31$	23	$46 = (2 \times 23)$
$0.31 \rightarrow 0.16$	23	$69 = (3 \times 23)$
$0.16 \rightarrow 0.08$	23	$92 = (4 \times 23)$

and in the third 23-minute interval (46 minutes to 69 minutes)

$$[N_2O_5] = \left(\tfrac{1}{2}\right)\left(\tfrac{1}{2}\right)\left(\tfrac{1}{2}\right)[N_2O_5]_0 = \left(\tfrac{1}{2}\right)^3[N_2O_5]_0 = 0.16 \times 10^{-2}\text{ M}$$
$$(t = 2t_{1/2} = 69\text{ minutes})$$

In general, after n half-lives, we have

$$[N_2O_5] = \left(\tfrac{1}{2}\right)^n[N_2O_5]_0 \qquad (t = nt_{1/2})$$

We can show that the rate law is consistent with the data in Figure 17.6 and Table 17.4. Because the half-life, $t_{1/2}$, of a reactant is defined as the time it takes for the concentration of the reactant to decrease by a factor of two, we substitute $t = t_{1/2}$ and $[A] = [A]_0/2$ in Equation 17.18. This substitution gives

$$\ln\left(\tfrac{1}{2}\right) = -0.693 = -kt_{1/2}$$

Solving this equation for $t_{1/2}$ yields

$$t_{1/2} = \frac{\ln(2)}{k} = \frac{0.693}{k} \tag{17.20}$$

where we have used the fact that $-\ln\left(\tfrac{1}{2}\right) = \ln(2)$. Equation 17.20 confirms that the half-life for a first-order reaction is independent of the initial concentration of the reactant. Conversely, if it is observed that $t_{1/2}$ for a reactant species A is independent of $[A]_0$, then the rate law must be first order.

Equation 17.20 also shows that there is an inverse relation between the half-life and the rate constant of a reaction. The larger the rate constant, the smaller the half-life, indicating a rapid reaction. The smaller the rate constant, the longer the half-life, indicating a slow reaction.

EXAMPLE 17-7: The rate law for the decomposition of aqueous hydrogen peroxide at 70°C as described by

$$2\,H_2O_2(aq) \rightarrow 2\,H_2O(l) + O_2(g)$$

is first order in $[H_2O_2]$. The half-life for the $H_2O_2(aq)$ decomposition at 70°C is $t_{1/2} = 20$ minutes. Given that the initial concentration is $[H_2O_2]_0 = 0.10$ M, calculate the value of $[H_2O_2]$ 60 minutes after the 0.10 M $H_2O_2(aq)$ solution is prepared.

Solution: A reaction time of 60 minutes corresponds to three half-lives:

$$\frac{60 \text{ min}}{20 \text{ min}\cdot\text{half-life}^{-1}} = 3 \text{ half-lives}$$

The concentration of a reactant A that remains unreacted after n half-lives is given by

$$[A] = \left(\frac{1}{2}\right)^n [A]_0 \tag{17.21}$$

where $[A]_0$ is the initial reactant concentration. Using Equation 17.21, we calculate

$$[H_2O_2] = \left(\frac{1}{2}\right)^3 [H_2O_2]_0 = \left(\frac{1}{2}\right)^3 (0.10 \text{ M}) = \left(\frac{1}{8}\right) 0.10 \text{ M} = 0.013 \text{ M}$$

Note that this problem is the same as Example 17-6, but worked in a different way.

PRACTICE PROBLEM 17-7: For the reaction equation given in Example 17-4, calculate $t_{1/2}$ and use the method of half-lives to find the fraction of $SO_2Cl_2(g)$ remaining after 35 hours.

Answer: $t_{1/2} = 3.15 \times 10^4 \text{s} = 8.75$ h; 0.0625 (35 hours is four half-lives)

17-6. The Rate of Decay of a Radioactive Isotope Is a First-Order Process

Recall from Chapter 2 that atomic nuclei consist of protons and neutrons and that a particular isotope of an element is described by the notation ^A_ZX, where X is the symbol of the element, Z is its atomic number, and A is its mass number. We also saw in Section 2-9 that certain nuclei are unstable and can spontaneously transform to more stable nuclei by emitting small particles such as α-particles (He-4 nuclei), β-particles (electrons), or γ-rays (high-energy radiation similar to X-rays). We call this process **radioactive decay**, and we say that the decaying nuclei are **radioactive**. For example, samples containing uranium-238 nuclei emit α-particles, which we can write as a helium-4 nuclei using the notation $^4_2$He. When a $^{238}_{92}$U nucleus emits a helium nucleus, the mass number of the uranium decreases by 4 and the atomic number decreases by 2. The change in atomic number tells us that a different element has been produced. We can describe this process by a **nuclear equation** as

$$^{238}_{92}\text{U} \rightarrow {}^{234}_{90}\text{Th} + {}^4_2\text{He}$$

The product here is thorium because the atomic number of the resulting nucleus is 90 (= 92 − 2). This nuclear equation is balanced: the total nuclear charge (given by the numbers of protons) and the total of the mass numbers (the total number of protons and neutrons) are the same on both sides.

We can write similar nuclear equations for β-particle and γ-ray emission. We use the notation ${}^{0}_{-1}e$ to represent the high-energy electrons formed during beta decay. The superscript 0 refers to the small mass of an electron relative to that of a proton or neutron, and the subscript −1 refers to the negative charge on an electron. Similarly, we use the notation ${}^{0}_{0}\gamma$ (or simply γ) to represent a gamma ray or high-energy photon. Table 17.5 summarizes the three forms of radioactive decay we have considered here. Although there are other forms of radioactive decay, only these three are important for purposes of our discussion here. Some examples of other chemically important nuclear processes are discussed in Interchapter O.

See Interchapter O at www.McQuarrieGeneralChemistry.com.

TABLE 17.5 Various particles emitted in radioactive processes

		Change in nucleus		
Emission	Symbol	Mass number	Atomic number	Example
α	${}^{4}_{2}He$	decreases by 4	decreases by 2	${}^{238}_{92}U \rightarrow {}^{234}_{90}Th + {}^{4}_{2}He$
β	${}^{0}_{-1}e$	no change	increases by 1	${}^{14}_{6}C \rightarrow {}^{14}_{7}N + {}^{0}_{-1}e$
γ	${}^{0}_{0}\gamma$	no change	no change	${}^{16}_{7}N \rightarrow {}^{16}_{8}O + {}^{0}_{-1}e + \gamma$

EXAMPLE 17-8: Fill in the missing symbols in the following nuclear equations:

(a) ${}^{214}_{82}Pb \rightarrow {}^{214}_{83}Bi + ?$ (b) ${}^{222}_{86}Rn \rightarrow ? + {}^{4}_{2}He$ (c) $? \rightarrow {}^{0}_{-1}e + {}^{97}_{41}Nb$

Solution: (a) The missing particle has a charge of −1 (82 = 83 − 1), and A does not change; thus, the missing particle is a β-particle, ${}^{0}_{-1}e$. (b) The missing product has an atomic number, $Z = 86 - 2 = 84$, and a mass number, $A = 222 - 4 = 218$. The element that has $Z = 84$ is polonium, so the missing product is ${}^{218}_{84}Po$. (c) The missing nucleus has $Z = 40$ and $A = 97$. The element that has $Z = 40$ is zirconium, so the nucleus that decays is ${}^{97}_{40}Zr$.

PRACTICE PROBLEM 17-8: Phosphorus-32 (a β-emitter) is used extensively as a **radiotracer,** that is, a radioactive isotope used to assist in mapping chemical and biochemical reaction pathways. Write a balanced nuclear equation to describe the decay of phosphorus-32.

Answer: ${}^{32}_{15}P \rightarrow {}^{0}_{-1}e + {}^{32}_{16}S$

How can we explain the emission of an electron when there are no electrons in atomic nuclei? The emission of a β-particle can be viewed as the result of the conversion of a neutron to a proton within the nucleus. The process can be represented as ${}^{1}_{0}\mathrm{n} \rightarrow {}^{1}_{1}\mathrm{p} + {}^{0}_{-1}\mathrm{e}$, where ${}^{1}_{0}\mathrm{n}$ and ${}^{1}_{1}\mathrm{p}$ denote a neutron and a proton, respectively.

When a radioactive nucleus emits a particle and transforms to another nucleus, we say that it *decays* to that nucleus. Thus, the expression radioactive decay refers to a process in which one nucleus is spontaneously converted into another.

Not all radioactive nuclei decay at the same rate. Some radioactive samples decay in a few millionths of a second; the same amount of another isotope may take billions of years to decay. Radioactive decay is an excellent example of a first-order kinetic process. An important aspect of radioactive decay is that its rate is independent of external factors such as temperature, pressure, and chemical bonding, at least under terrestrial conditions. It is customary to express the first-order rate law for radioactive decay in terms of half-life. If we solve Equation 17.20 for k and substitute the result into Equation 17.18, then we obtain

$$\ln \frac{[\mathrm{A}]}{[\mathrm{A}]_0} = -\left(\frac{0.693}{t_{1/2}}\right) t \qquad (17.22)$$

Table 17.6 lists the half-lives and decay modes of various common radioisotopes.

The number of radioactive nuclei in a sample is proportional to the concentration of the radioactive species, or $[\mathrm{A}]/[\mathrm{A}]_0 = N/N_0$, and so Equation 17.22 can be written as

TABLE 17.6 Half-lives of various radioisotopes

Isotope	Half-Life	Mode of Decay
${}^{3}_{1}\mathrm{H}$	12.33 years	β
${}^{14}_{6}\mathrm{C}$	5730 years	β
${}^{25}_{11}\mathrm{Na}$	1.0 minutes	β + γ
${}^{32}_{15}\mathrm{P}$	14.28 days	β
${}^{60}_{27}\mathrm{Co}$	5.271 years	β + γ
${}^{87}_{37}\mathrm{Rb}$	4.88×10^{10} years	β
${}^{90}_{38}\mathrm{Sr}$	29.1 years	β
${}^{131}_{53}\mathrm{I}$	8.02 days	β + γ
${}^{214}_{84}\mathrm{Po}$	163.7 microseconds	α
${}^{226}_{88}\mathrm{Ra}$	1600 years	α + γ
${}^{238}_{92}\mathrm{U}$	4.47×10^{9} years	α + γ
${}^{239}_{93}\mathrm{Pu}$	2.41×10^{4} years	α + γ

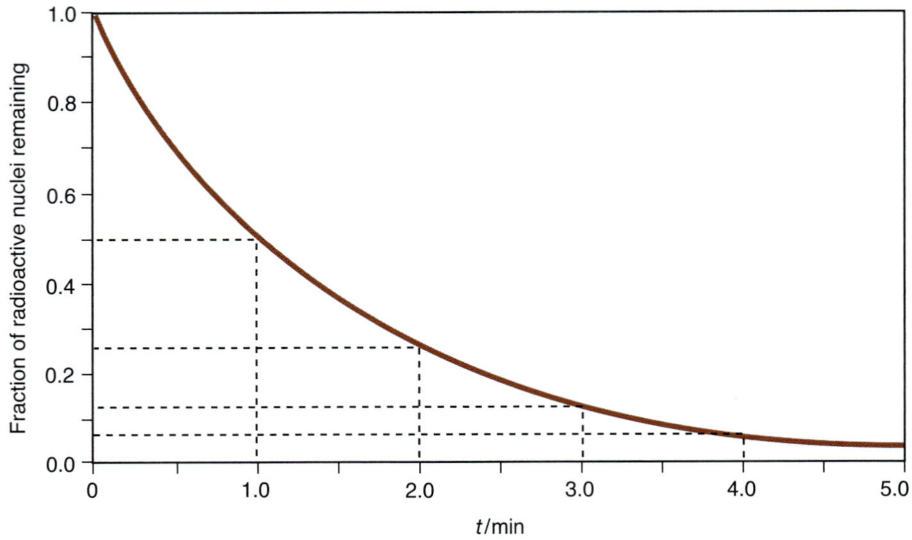

Figure 17.7 A graph of the decay of sodium-25 showing the fraction of sodium-25 nuclei present in a sample over time. This curve is representative of a first-order kinetic process: the rate of decay is proportional to the number of sodium-25 nuclei in the sample. The half-life ($t_{1/2}$) of sodium-25 is 1.0 minutes. One-half of the sample decays in the first minute, one-half of what remains decays in the next minute, and so on. The fraction remaining, N/N_0, after n half-lives is given by $N/N_0 = \left(\frac{1}{2}\right)^n$, where n need not be an integer.

$$\ln\frac{N}{N_0} = -\left(\frac{0.693}{t_{1/2}}\right)t \tag{17.23}$$

Using Equation 17.23, we can determine the number of radioactive nuclei, N, remaining in a sample at time t, given N_0, the number of nuclei present initially, and $t_{1/2}$, the half-life of the radioactive substance.

Figure 17.7 shows a plot of the fraction of radioactive nuclei, N/N_0, remaining in a sample of sodium-25, versus time. Sodium-25 emits a β-particle when it decays, which we can express as

$$^{25}_{11}\text{Na} \rightarrow {}^{25}_{12}\text{Mg} + {}^{0}_{-1}e \qquad t_{1/2} = 1.0\text{ min}$$

The following Examples illustrate the use of half-lives in radioactive decay.

EXAMPLE 17-9: The radioisotope cobalt-60 is used to destroy cancerous cells by directing its emitted γ-rays into the cancerous cell tissue. Calculate the fraction of a cobalt-60 sample left after 20.0 years.

Solution: From Table 17.6, we see that the half-life of cobalt-60 is 5.271 years. If we use Equation 17.23, then we have

$$\ln\frac{N}{N_0} = -\left(\frac{0.693}{t_{1/2}}\right)t = -\frac{(0.693)(20.0\text{ years})}{5.271\text{ years}} = -2.63$$

Therefore, $N/N_0 = e^{-2.63} = 0.072$. In other words, 7.2% of the original sample remains after 20.0 years.

PRACTICE PROBLEM 17-9: Calculate the number of years required for the amount of radioactivity in a sample of plutonium-239 to decrease to 1.0% of its original value. The half-life for plutonium-239 is 2.41×10^4 years. (*Hint:* A decrease to 1.0% of its original value means that $N = 0.010N_0$.)

Answer: 1.6×10^5 years

EXAMPLE 17-10: A 2.000-mg sample of the radioisotope phosphorus-32 is found to contain 0.400 mg of phosphorus-32 after 33.3 days. Calculate the half-life of this isotope.

Solution: Solving Equation 17.23 for $t_{1/2}$, we obtain

$$t_{1/2} = -\frac{0.693t}{\ln(N/N_0)} = \frac{0.693t}{\ln(N_0/N)} \qquad (17.24)$$

where we have used the fact that $\ln(N/N_0) = -\ln(N_0/N)$. The value of N_0/N after 33.3 days is given by

$$\frac{N_0}{N} = \frac{(2.000 \times 10^{-3}\text{ g})\left(\frac{1\text{ mol}}{32.0\text{ g}}\right)(6.022 \times 10^{23}\text{ atom}\cdot\text{mol}^{-1})}{(0.400 \times 10^{-3}\text{ g})\left(\frac{1\text{ mol}}{32.0\text{ g}}\right)(6.022 \times 10^{23}\text{ atom}\cdot\text{mol}^{-1})}$$

$$= \frac{2.000}{0.400} = 5.00$$

Notice that we have just shown that

$$\frac{N_0}{N} = \frac{\text{initial mass of phosphorus-32}}{\text{present mass of phosphorus-32}}$$

Using the value of 5.00 for N_0/N, we calculate the half-life of phosphorous-32:

$$t_{1/2} = \frac{(0.693)(33.3\text{ days})}{\ln(5.00)} = 14.3\text{ days}$$

This result agrees with the value given in Table 17.6.

PRACTICE PROBLEM 17-10: Calculate the number of half-lives required for the amount of radioactivity in a sample to decrease to 0.010% of its original value.

Answer: 13.3 half-lives

17-7. Carbon-14 Can Be Used to Date Certain Archaeological Objects

Naturally occurring uranium-238, which is radioactive, decays through a series of processes to lead-206. The *overall* process is

$$^{238}_{92}\text{U} \rightarrow {}^{206}_{82}\text{Pb} + 8\,{}^{4}_{2}\text{He} + 6\,{}^{0}_{-1}\text{e}$$

Figure 17.8 Willard F. Libby (1908–1980) was born in Grand Valley, Colorado, but grew up in Sebastopol, California. He was a tall and powerfully built young man and played tackle on his high school football team, where he earned the nickname "Wild Bill." He received his B.S. and Ph.D. in chemistry from the University of California at Berkeley, where he remained on the faculty until the outbreak of World War II. He worked on the Manhattan Project at Columbia University during the war years. In 1945 he became a professor at the University of Chicago, where he and his students developed the carbon-14 dating method that proved so useful to archaeologists. Libby was the first chemist to be appointed to the Atomic Energy Commission. He was a strong advocate for the industrial use of radioisotopes and nuclear energy for nonmilitary purposes. In 1959 he returned to the academic world, accepting a faculty position at the University of California at Los Angeles, where he remained until his retirement in 1976. Libby was awarded the Nobel Prize in Chemistry in 1960 for the development of radiocarbon dating.

The first step of this decay series is

$$^{238}_{92}\text{U} \rightarrow {}^{234}_{90}\text{Th} + {}^{4}_{2}\text{He}$$

with a half-life of 4.47×10^9 years. The half-life for the overall decay process is 4.51×10^9 years, and so once the U-238 decays to Th-234, the subsequent decay to Pb-206 occurs relatively quickly. This means that we can treat the decay of U-238 to Pb-206 as if it occurs in a single step. For example, suppose that a sample of uranium ore is found to contain equal molar quantities of U-238 and Pb-206. Because half of the U-238 nuclei initially present have decayed since the rock was formed, we conclude that the rock is 4.51×10^9 years old. This method is called **uranium-lead dating.** The oldest rocks analyzed by this method are 3.6 billion years old, which is about one billion years less than the present estimate of the age of the earth. Rocks that were obtained from the moon by the Apollo program indicate that the moon is about the same age as the earth.

Radiodating methods used to determine the age of rocks and ores are not useful for materials less than about a million years old, because the half-lives of the radioactive minerals used in this method are more than one billion years. Consequently, not enough decay occurs in less than one million years to be measured accurately. In the 1940s, Willard Libby (Figure 17.8) developed a method of using carbon-14 to date carbon-containing objects derived from formerly living materials. His method is useful for dating objects less than about 30 000 years old. It has found wide use in archaeology, and Libby was awarded the 1960 Nobel Prize in Chemistry for his work.

The idea behind Libby's **carbon-14 dating** method, also called **radiocarbon dating,** is as follows. The earth's upper atmosphere is being bombarded constantly by radiation from the sun and other parts of the universe. As a result, a small but fairly constant amount of carbon-14 is produced in the reaction between cosmic ray neutrons and atmospheric nitrogen:

$$^{14}_{7}\text{N} + {}^{1}_{0}\text{n} \rightarrow {}^{14}_{6}\text{C} + {}^{1}_{1}\text{H}$$

Carbon-14 is radioactive and decays by the reaction

$$^{14}_{6}\text{C} \rightarrow {}^{14}_{7}\text{N} + {}^{0}_{-1}\text{e}$$

The half-life of carbon-14 is 5730 years. The carbon-14 produced eventually reacts with oxygen and appears in the atmosphere as $^{14}CO_2(g)$. Because the $^{14}CO_2(g)$ diffuses throughout the earth's atmosphere, a small but fairly constant fraction of atmospheric $CO_2(g)$ contains carbon-14.

Living plants absorb $CO_2(g)$ to build carbohydrates through photosynthesis, so the atmospheric carbon-14 is incorporated into plants and into the food chain of animals. As a result, all living plants and animals (including you) contain a fraction of carbon-14 atoms that is the same as that in atmospheric $CO_2(g)$. Although the carbon-14 is continually decaying, as long as the organism continues to incorporate carbon-containing materials derived from atmospheric $CO_2(g)$, the amount of carbon-14 per gram of carbon in the organism remains constant. The number of the carbon-14 disintegrations per unit time in all living organisms is a constant: the measured value is 15.3 disintegrations per minute per gram of total carbon. When the organism dies, it no longer incorporates carbon-14, so the quantity of carbon-14 decreases with time according to Equation 17.23, with $t_{1/2}$ equal to 5730 years.

To determine the age of an object, we first solve Equation 17.23 for t, to obtain

$$t = \left(\frac{t_{1/2}}{0.693}\right)\ln\frac{N_0}{N} \tag{17.25}$$

If we substitute the value of the half-life of carbon-14 (5730 years) into Equation 17.25, we obtain

$$t = (8.27 \times 10^3 \text{ years})\ln\frac{N_0}{N} \tag{17.26}$$

We can convert the ratio N_0/N to a more convenient form by realizing that because radioactive decay is a first-order process, the 15.3 disintegrations per minute per gram of carbon in living things is proportional to N_0, the initial number of carbon-14 nuclei per gram of carbon present in the sample. We can write this as $N_0 \propto 15.3$. Similarly, if R is the *present* disintegration rate of carbon-14 per minute per gram of carbon in the nonliving sample, then $R \propto N$. Thus, the ratio N_0/N is given by

$$\frac{N_0}{N} = \frac{15.3}{R}$$

If we substitute this equation into Equation 17.26, we get

$$t = (8.27 \times 10^3 \text{ years})\ln\left(\frac{15.3 \text{ disintegration}\cdot\text{min}^{-1}\cdot\text{g}^{-1}}{R}\right) \tag{17.27}$$

where, once again, R is the present disintegration rate of carbon-14 in units of disintegrations per minute per gram of carbon. If we assume that the atmospheric level of carbon-14 is the same now as when the artifact to be analyzed was living matter, then the age of the artifact can be determined by measuring R.

The assumptions of carbon-14 dating have been tested extensively against other archaeological dating techniques, such as tree-ring dating methods. These comparisons show that the carbon-14 level in the atmosphere fluctuates over time, but these fluctuations are not large.

The following Example illustrates the use of Equation 17.27.

Figure 17.9 Stonehenge, an ancient megalithic site in southern England that presumably was used to make astronomical observations.

EXAMPLE 17-11: Stonehenge is an ancient megalithic site in southern England that apparently was designed to make astronomical observations (Figure 17.9). Charcoal samples taken from a series of holes at Stonehenge have a disintegration rate of $R = 9.65$ disintegrations per minute per gram of carbon. Calculate the age of the charcoal sample.

Solution: Using Equation 17.27, we have

$$t = (8.27 \times 10^3 \text{ years})\ln\left(\frac{15.3 \text{ disintegration}\cdot\text{min}^{-1}\cdot\text{g}^{-1}}{9.65 \text{ disintegration}\cdot\text{min}^{-1}\cdot\text{g}^{-1}}\right)$$

$$= (8.27 \times 10^3 \text{ years})(0.461) = 3.81 \times 10^3 \text{ years}$$

Thus, we estimate that the charcoal pits at Stonehenge are at least 3800 years old (about 1800 B.C.). This result is in agreement with evidence based on other archaeological data.

PRACTICE PROBLEM 17-11: In 1988 measurements on samples from the Shroud of Turin gave a carbon-14 content that was 0.928 times that for a contemporary carbon sample of biological origin. Estimate the age of the Shroud of Turin.

Answer: $t = 618$ years (in 1988), or about 1370 A.D.

Radiocarbon dating has been used to date many archaeological objects. Figure 17.10 shows a plot of ln R versus t and the ages of artifacts that have been determined by the carbon-14 method.

17-8. A Plot of 1/[A] Versus Time Is Linear for a Second-Order Reaction

We can also use a plot to help us understand a second-order reaction. For the second-order decomposition of a single reactant A, the rate expression is

$$\text{rate of reaction} = -\frac{\Delta[\text{A}]}{\Delta t} = k[\text{A}]^2 \qquad (17.28)$$

By using elementary calculus (see Problem 17-79), we can find the concentration of A at any time. The result is

$$\frac{1}{[\text{A}]} = \frac{1}{[\text{A}]_0} + kt \qquad (17.29)$$

By comparing Equation 17.29 to the equation $y = mx + b$, we find that a plot of 1/[A] versus time, t, is a straight line whose slope is the rate constant, k.

Let's do a numerical example. We saw in Example 17-5 that the rate law for the reaction described by

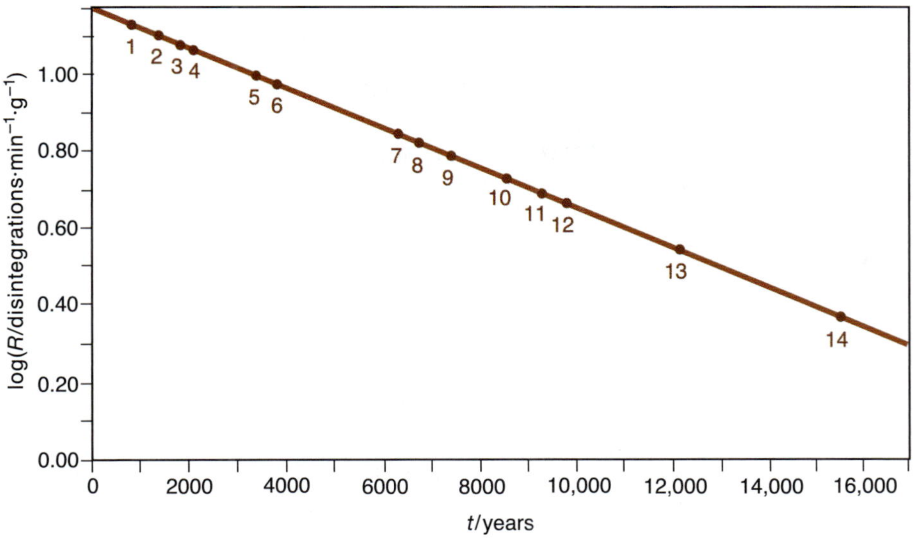

Figure 17.10 Plot of the logarithm of R, the number of disintegrations per minute per gram of carbon, versus time for various samples dated using the carbon-14 method. As Equation 17.27 indicates, this plot is a straight line. The numbers on the curve designate archaeological objects that have been dated by the carbon-14 method: (1) Charcoal from earliest Polynesian culture in Hawaii (946 ± 180 years). (2) Wooden lintels from a Mayan site in Tikal, Guatemala (1503 ± 110 years). (3) Linen wrappings from the Dead Sea Scrolls (1917 ± 200 years). (4) Wood from a coffin from the Egyptian Ptolemaic period (2190 ± 450 years). (5) Samples of oak from an ancient cooking place at Killeens, County Cork (3506 ± 230 years). (6) Charcoal sample from Stonehenge (3798 ± 275 years). (7) Charcoal from a tree destroyed by the explosion of Mount Mazama, the explosion that formed Crater Lake in Oregon (6453 ± 250 years). (8) Land-snail shells found at Jarmo, Iraq (6707 ± 320 years). (9) Charcoal from an archaeological site near Beer-Sheba, Israel (7240 ± 520 years). (10) Burned animal bones found near a site inhabited by humans in Palli Aike Cave in southern Chile (8639 ± 450 years). (11) Woven rope sandals found in Fork Rock Cave, Oregon (9053 ± 350 years). (12) Buried bison bone from Folsom Man site near Lubbock, Texas (9883 ± 350 years). (13) Glacial wood found near Skunk River, Iowa, (12 200 ± 500 years). (14) Charcoal from the Lascaux cave in France, which contains many cave paintings (15 516 ± 900 years).

$$NO_2(g) + CO(g) \rightarrow NO(g) + CO_2(g)$$

is

$$\text{rate of reaction} = (0.50\ M^{-1}\cdot s^{-1})\ [NO_2]^2 \qquad (17.30)$$

at 325°C. According to Equation 17.29, we can write the dependence of the concentration of $NO_2(g)$ on time as

$$\frac{1}{[NO_2]} = \frac{1}{[NO_2]_0} + kt \qquad (17.31)$$

Table 17.7 gives values of $[NO_2]$ and $1/[NO_2]$ for various times, and Figure 17.11 shows that a plot of $1/[NO_2]$ versus t is indeed a straight line. The slope of the line, $0.50\ M^{-1}\cdot s^{-1}$, is the rate constant, k.

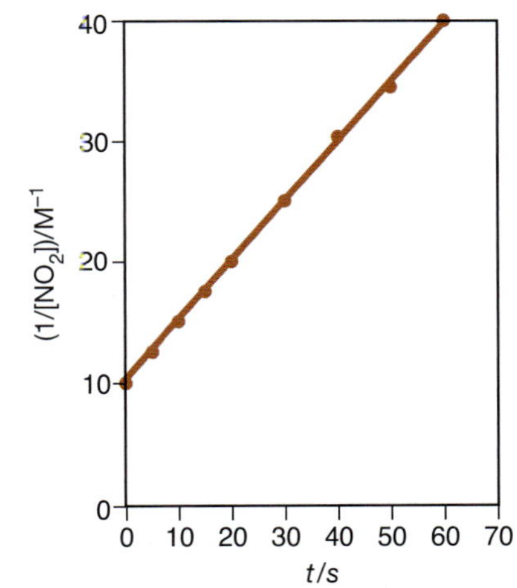

Figure 17.11 Plot of $1/[NO_2]$ versus time for the reaction described by the equation

$$NO_2(g) + CO(g) \rightarrow NO(g) + CO_2(g)$$

The numerical values are given in Table 17.7. The plot is linear, confirming that the reaction is second order.

TABLE 17.7 Kinetic data for the reaction described by the equation

$$NO_2(g) + CO(g) \rightarrow NO(g) + CO_2(g) \quad \text{at } 325°C$$

t/s	$[NO_2]$/M	$(1/[NO_2])/M^{-1}$
0	0.100	10.0
5.0	0.080	12.5
10.0	0.067	15.0
15.0	0.057	17.5
20.0	0.050	20.0
30.0	0.040	25.0
40.0	0.033	30.3
50.0	0.029	34.5
60.0	0.025	40.0

TABLE 17.8 Kinetic data for the reaction

$$NOBr(g) \rightarrow NO(g) + \tfrac{1}{2}Br_2(g)$$

t/s	[NOBr]/M
0	0.0250
6.2	0.0191
10.8	0.0162
14.7	0.0144
20.0	0.0125
24.6	0.0112

EXAMPLE 17-12: Table 17.8 gives values of [NOBr] versus time at a fixed temperature for the reaction described by

$$NOBr(g) \rightarrow NO(g) + \tfrac{1}{2}Br_2(g)$$

Prove that the reaction is second order and evaluate the rate constant.

Solution: We must show that a plot of 1/[NOBr] versus t is a straight line, so we shall make a table of 1/[NOBr] versus time (Table 17.9). These data are plotted in Figure 17.12, which confirms that the reaction is second order in [NOBr]. The slope of the line, 2.00 $M^{-1}\cdot s^{-1}$, is the rate constant.

PRACTICE PROBLEM 17-12: The data in Table 17.10 were obtained for the thermal decomposition of $NO_2(g)$ as described by the equation

$$NO_2(g) \rightarrow NO(g) + \tfrac{1}{2}O_2(g)$$

Determine the order of the reaction and the value of the rate constant.

Answer: second order; $k = 0.54\ M^{-1}\cdot s^{-1}$

In Example 17-12 and Practice Problem 17-12, we were given the concentrations of the reactant at various times and were asked to determine the order of the reaction. One method of following reactions like these is to monitor the total pressure. Note that we may express the concentration of a gaseous species in units of either molarity or pressure. For an ideal gas the pressure and concentration are related by the equation $PV = nRT$ as

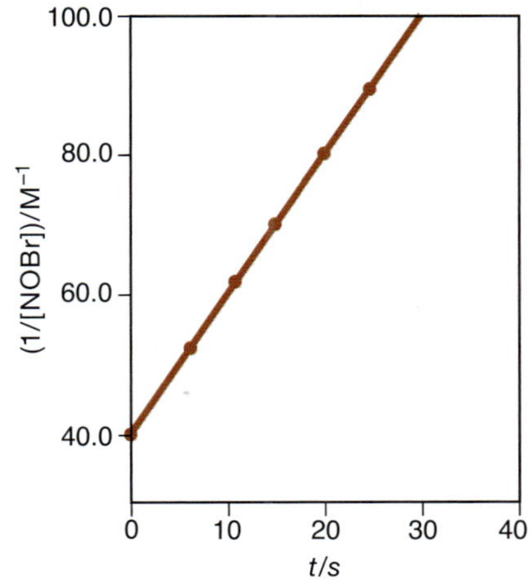

Figure 17.12 Plot of 1/[NOBr] versus time for the reaction described by the equation

$$\text{NOBr}(g) \rightarrow \text{NO}(g) + \tfrac{1}{2}\text{Br}_2(g)$$

The numerical values are given in Table 17.9. The fact that the plot is linear confirms that the reaction is second order.

TABLE 17.9 Data for the thermal decomposition of NOBr(g)

t/s	(1/[NOBr])/M^{-1}
0	40.0
6.2	52.3
10.8	61.7
14.7	69.9
20.0	80.0
24.6	89.3

TABLE 17.10 Data for the thermal decomposition of $NO_2(g)$

t/s	[NO_2]/M
0	0.20
10.0	0.096
20.0	0.063
30.0	0.047
40.0	0.038
50.0	0.032

$$M = \frac{\text{number of moles}}{\text{liter}} = \frac{n}{V} = \frac{P}{RT} \quad \text{or} \quad P = M(RT) \tag{17.32}$$

Let's consider the reaction in Example 17-12 for which the stated equation is

$$\text{NOBr}(g) \rightarrow \text{NO}(g) + \tfrac{1}{2}\text{Br}_2(g)$$

Because all the species are gaseous, the concentration of the reactant can be determined from the total pressure. The total pressure at any time is given by Dalton's law of partial pressure (Equation 13.14),

$$P_{\text{total}} = P_{\text{NOBr}} + P_{\text{NO}} + P_{\text{Br}_2}$$

The quantity of NOBr(g) that decomposes is proportional to $P^{\circ}_{\text{NOBr}} - P_{\text{NOBr}}$, where P°_{NOBr} is the initial pressure of NOBr(g). Assuming that only NOBr(g) was present initially, from the chemical reaction we have

$$P_{\text{NO}} = P^{\circ}_{\text{NOBr}} - P_{\text{NOBr}}$$

because one mole of NO(g) is produced for each mole of NOBr(g) that decomposes. Similarly, because half a mole of $Br_2(g)$ is produced for each mole of NOBr(g) that decomposes, we have

$$P_{\text{Br}_2} = \tfrac{1}{2}\,(P^{\circ}_{\text{NOBr}} - P_{\text{NOBr}})$$

If we substitute each of these two equations into the equation for P_{total}, we obtain

$$P_{\text{total}} = P_{\text{NOBr}} + (P^{\circ}_{\text{NOBr}} - P_{\text{NOBr}}) + \tfrac{1}{2}\,(P^{\circ}_{\text{NOBr}} - P_{\text{NOBr}}) = \tfrac{3}{2}\,P^{\circ}_{\text{NOBr}} - \tfrac{1}{2}P_{\text{NOBr}}$$

Multiplying through by two and solving for P_{NOBr} gives

$$P_{\text{NOBr}} = 3\,P^{\circ}_{\text{NOBr}} - 2\,P_{\text{total}}$$

Thus, we can calculate the pressure of NOBr(g) at any time in terms of the total pressure and its initial pressure. The following Example illustrates this type of calculation.

TABLE 17.11 Total pressure data for the dimerization of $C_5H_6(g)$

t/s	P_{total}/Torr
0	500
10.0	434
20.0	396
30.0	370
40.0	353

EXAMPLE 17-13: For the dimerization reaction described by the equation

$$2\,C_5H_6(g) \rightarrow C_{10}H_{12}(g)$$

the total pressure as a function of time is tabulated in Table 17.11. Assuming that only $C_5H_6(g)$ was present initially, determine the rate of reaction and the value of the rate constant.

Solution: First, we must determine the pressure of $C_5H_6(g)$ as a function of time. We use the fact that the total pressure is given by

$$P_{total} = P_{C_5H_6} + P_{C_{10}H_{12}} \tag{17.33}$$

Because one mole of $C_{10}H_{12}(g)$ is produced from every two moles of $C_5H_6(g)$ that react, we can write

$$P_{C_{10}H_{12}} = \frac{1}{2}\,(P^{\circ}_{C_5H_6} - P_{C_5H_6}) \tag{17.34}$$

In this equation, $P^{\circ}_{C_5H_6}$ is the initial pressure of $C_5H_6(g)$. If we substitute this result into Equation 17.33, we obtain

$$P_{total} - P_{C_5H_6} = P_{C_{10}H_{12}} = \frac{1}{2}(P^{\circ}_{C_5H_6} - P_{C_5H_6})$$

Solving for $P_{C_5H_6}$ gives

$$P_{C_5H_6} = 2\,P_{total} - P^{\circ}_{C_5H_6}$$

TABLE 17.12 Pressure of $C_5H_6(g)$ for the dimerization of $C_5H_6(g)$

t/s	$P_{C_5H_6}$/Torr
0	2(500) − 500 = 500
10.0	2(434) − 500 = 368
20.0	2(396) − 500 = 292
30.0	2(370) − 500 = 240
40.0	2(353) − 500 = 206

This equation gives the pressure of $C_5H_6(g)$ at any time t in terms of its initial pressure and the total pressure. The value of $P_{C_5H_6}$ at the various times are calculated in Table 17.12 using the data given in Table 17.11. As Figure 17.13 shows, a plot of ln ($P_{C_5H_6}$/Torr) versus t is *not* linear, as is required for a first-order reaction, but a plot of $1/P_{C_5H_6}$ is linear, so the reaction is second order.

The value of the rate constant can be determined from the slope of the straight line, or from the equation

$$\frac{1}{P_{C_5H_6}} = \frac{1}{P^{\circ}_{C_5H_6}} + kt$$

for any set of the pairs of data. Using the slope of the line, the value of the rate constant is: $k = 7.16 \times 10^{-5}$ Torr^{-1}·s^{-1} for the chemical equation as written.

PRACTICE PROBLEM 17-13: The data in Table 17.13 were obtained for the thermal decomposition of dinitrogen trioxide, $N_2O_3(g)$, as described by the equation

$$N_2O_3(g) \rightarrow NO(g) + NO_2(g)$$

Assuming that only $N_2O_3(g)$ was present initially, determine the reaction-rate law and the value of the rate constant.

Answer: first order; $k = 0.060\ s^{-1}$

TABLE 17.13 Data for the thermal decomposition of $N_2O_3(g)$

t/s	P_{total}/Torr
0	400
5.0	504
10.0	581
15.0	638
20.0	680

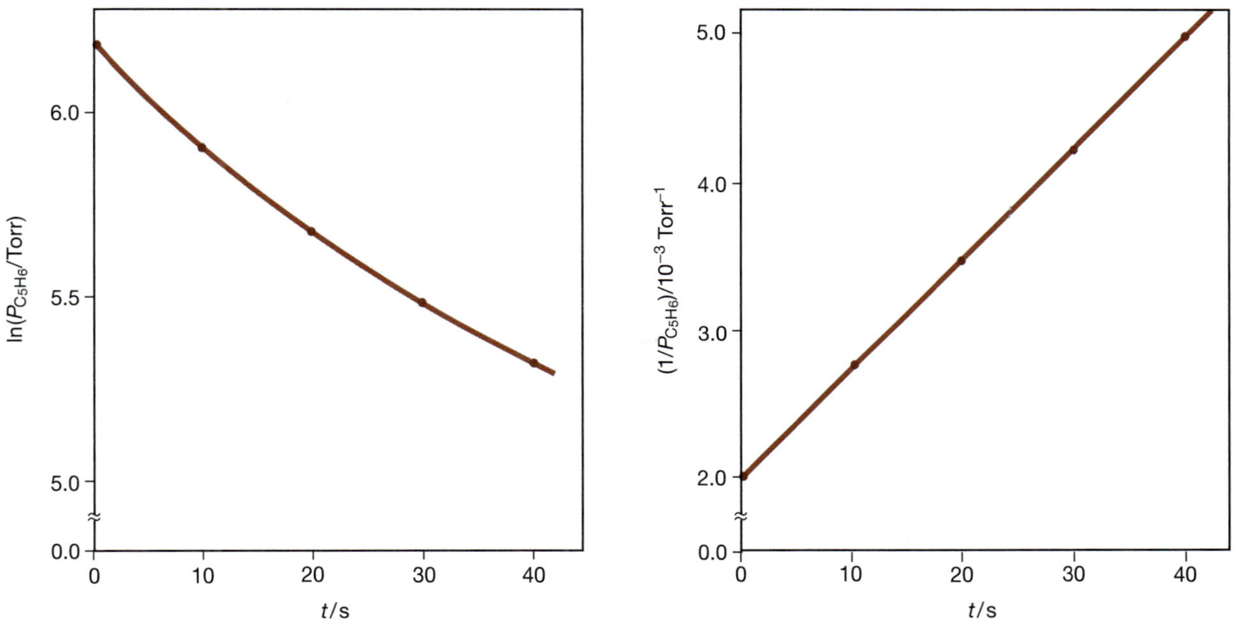

Figure 17.13 (*left*) Plot of ln ($P_{C_5H_6}$/Torr) versus time for the reaction described by the equation

$$2\,C_5H_6(g) \rightarrow C_{10}H_{12}(g)$$

(*right*) Plot of the reciprocal of $P_{C_5H_6}$ versus time for the same reaction.

17-9. The Half-Life of a Second-Order Reaction Depends on the Initial Concentration

As we saw in Example 17-5, the rate law for the reaction described by

$$NO_2(g) + CO(g) \rightarrow NO(g) + CO_2(g)$$

is

$$\text{rate of reaction} = (0.50\ M^{-1})\ [NO_2]^2$$

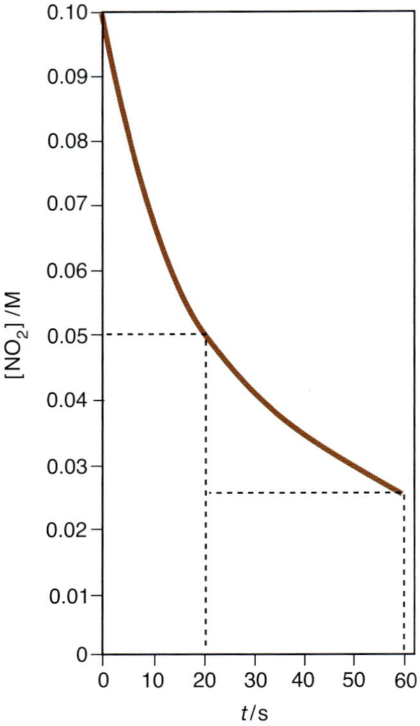

Figure 17.14 Plot of $[NO_2]$ versus time for the reaction described by the equation

$$NO_2(g) + CO(g) \rightarrow NO(g) + CO_2(g)$$

showing that the half-life is not independent of the initial concentration for a second-order reaction.

Figure 17.14 shows $[NO_2]$ plotted against time for this second-order reaction (see also Table 17.7). Notice that it takes 20.0 seconds for the concentration to decrease from 0.100 M to 0.050 M, but then it takes 40.0 seconds for it to decrease from 0.050 M to 0.025 M. In contrast to a first-order reaction, the half-life of a second-order reaction is *not* independent of the initial concentration. In Figure 17.14, $t_{½} = 20.0$ seconds when $[NO_2]_0 = 0.100$ M and $t_{½} = 40.0$ seconds when $[NO_2]_0 = 0.050$ M.

We can use Equation 17.29 to derive an equation for the half-life of a second-order reaction. If we substitute $t = t_{½}$ and $[A] = [A]_0/2$ into Equation 17.29, we obtain

$$\frac{2}{[A]_0} = \frac{1}{[A]_0} + kt_{1/2}$$

Solving this equation for $t_{½}$ gives

$$t_{1/2} = \frac{1}{k[A]_0} \qquad (17.35)$$

Note that $t_{½}$ is inversely proportional to $[A]_0$ for a second-order reaction. If we decrease $[A]_0$ by a factor of 2, then it takes twice as long for [A] to decrease by a factor of 2.

EXAMPLE 17-14: Calculate the initial half-life of the reaction system described in Example 17-12.

Solution: We use Equation 17.35 with $[NOBr]_0 = 0.0250$ M and $k = 2.00\ M^{-1}\cdot s^{-1}$:

$$t_{1/2} = \frac{1}{k[NOBr]_0} = \frac{1}{(2.00\ M^{-1}\cdot s^{-1})(0.0250\ M)} = 20.0\ s$$

PRACTICE PROBLEM 17-14: Calculate the half-life of the reaction system described in Example 17-13.

Answer: 27.9 seconds when $P^{\circ}_{C_5H_6} = 500$ Torr

Figure 17.15 Sample test plots for confirming reaction order. If the indicated test plot is linear, then the postulated reaction order is confirmed.

(a) If ln [A] versus t is linear, the reaction is first order in [A] and the slope $= -k$.

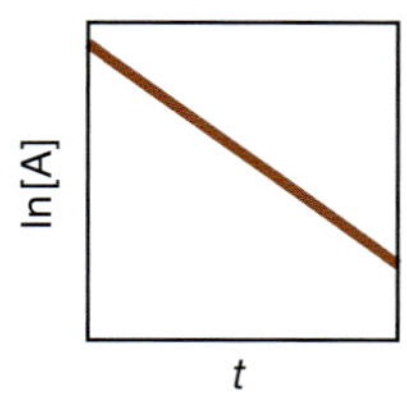

(b) If 1/[A] versus t is linear, the reaction is second order in [A] and the slope $= k$.

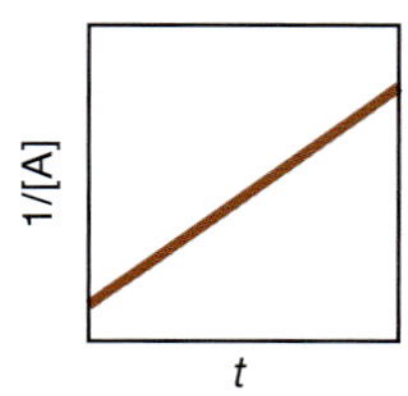

Table 17.14 summarizes the properties of the first-order and second-order reactions that we have studied so far, and Figure 17.15 shows sample test plots for each.

TABLE 17.14 Kinetic properties of first-order and second-order reaction-rate laws. Sample test plots are shown in Figure 17.15.

Order	Rate law	Units of k	Dependence of [A] on time	Half-life	Test plot
1	rate = k[A]	s^{-1}	$\ln[A] = \ln[A]_0 - kt$	$t_{1/2} = \frac{0.693}{k}$	ln[A] vs. t
2	rate = $k[A]^2$	$M^{-1}\cdot s^{-1}$	$\frac{1}{[A]} = \frac{1}{[A]_0} + kt$	$t_{1/2} = \frac{1}{k[A]_0}$	$\frac{1}{[A]}$ vs. t

SUMMARY

The rate of reaction tells us how fast a reactant is consumed or how fast a product is formed in a chemical reaction. The reaction-rate law describes the dependence of the rate of reaction on the concentrations of the various species involved. A first-order rate law is proportional to the first power of the reactant concentration; a second-order rate law is proportional to the second power of the reactant concentration; and a zero-order reaction is independent of reactant concentration.

The decay of a radioactive substance is a first-order kinetic process. An important property of a first-order kinetic process is that its rate of reaction is characterized by a half-life, which is the time required for one-half of the amount of a sample to react. The half-life is characteristic of a given radioisotope and is a direct measure of how rapidly it decays. The half-lives of radioisotopes range from less than a picosecond to billions of years. Rates of radioactive decay are used to determine the age of rocks and archaeological objects.

A test plot can be used to determine if a reaction is first or second order in [A]. If a reaction is first order in [A], a plot of ln[A] versus t will be linear. If a reaction is second order in [A], a plot of 1/[A] versus t will be linear. The half-life of a second-order reaction depends upon the initial concentration $[A]_0$.

TERMS YOU SHOULD KNOW

average rate of concentration change *602*
instantaneous rate of concentration change *603*
rate law *610*
rate constant *610*
method of initial rates *610*
first-order rate law *611*
first-order reaction *611*
second-order rate law *614*
second-order reaction *614*
overall order *615*
zero-order reaction *616*
half-life, $t_{1/2}$ *620*
radioactive decay *623*
radioactive *623*
nuclear equation *623*
radiotracer *624*
uranium-lead dating *628*
carbon-14 dating *628*
radiocarbon dating *628*

EQUATIONS YOU SHOULD KNOW HOW TO USE

$$\text{rate of reaction} = -\frac{1}{a}\left(\frac{\Delta[A]}{\Delta t}\right) = -\frac{1}{b}\left(\frac{\Delta[B]}{\Delta t}\right) = \frac{1}{c}\left(\frac{\Delta[C]}{\Delta t}\right) = \frac{1}{d}\left(\frac{\Delta[D]}{\Delta t}\right)$$ (17.6) (definition of rate of reaction for a reaction equation of the form $aA + bB \rightarrow cC + dD$)

$$(\text{rate of reaction})_0 = k[A]_0^x[B]_0^y[C]_0^z$$ (17.10) ($x + y + z$ = overall order of reaction)

$$\text{rate of reaction} = k[A]$$ (17.16) (first-order rate law)

$$\ln[A] = \ln[A]_0 - kt$$ (17.17) (an expression of the first-order rate law)

$$\ln\frac{[A]}{[A]_0} = -kt \quad \text{or} \quad [A] = [A]_0 e^{-kt}$$ (17.18 and 17.19) (alternative expressions of the first-order rate law)

$$t_{1/2} = \frac{\ln(2)}{k} = \frac{0.693}{k}$$ (17.20) (half-life for a first-order reaction)

$$[A] = \left(\frac{1}{2}\right)^n[A]_0$$ (17.21) (an alternative expression of the first-order rate law)

$$\ln \frac{N}{N_0} = -\left(\frac{0.693}{t_{1/2}}\right) t$$ (17.23) (relationship between the half-life and number of particles in first-order nuclear decay)

$$t = (8.25 \times 10^3 \text{ years}) \ln \left(\frac{15.3 \text{ disintegration}\cdot\text{min}^{-1}\cdot\text{g}^{-1}}{R}\right)$$ (17.27) (carbon-14 dating)

$$\text{rate of reaction} = k[\text{A}]^2$$ (17.28) (second-order rate law)

$$\frac{1}{[\text{A}]} = \frac{1}{[\text{A}]_0} + kt$$ (17.29) (an expression of the second-order rate law)

$$P = M(RT)$$ (17.32) (relationship between concentration and pressure for an ideal gas)

$$t_{1/2} = \frac{1}{k[\text{A}]_0}$$ (17.35) (half-life of a second-order reaction)

PROBLEMS

REACTION RATES

17-1. Express the rate of the following reaction equation in terms of the rate of concentration change for each of the three species involved:

$$2\,\text{NOCl}(g) \rightarrow 2\,\text{NO}(g) + \text{Cl}_2(g)$$

17-2. Express the rate of the following reaction equation in terms of the rate of concentration change for each of the three species involved:

$$2\,\text{SO}_2(g) + \text{O}_2(g) \rightarrow 2\,\text{SO}_3(g)$$

17-3. Suggest an experimental method for measuring the rate of the reactions described by the following chemical equations:

(a) $\underset{\text{yellow}}{\text{I}_2(aq)} + \underset{\text{colorless}}{2\,\text{S}_2\text{O}_3^{2-}(aq)} \rightarrow \underset{\text{colorless}}{\text{S}_4\text{O}_6^{2-}(aq)} + \underset{\text{colorless}}{2\,\text{I}^-(aq)}$

(b) $4\,\text{PH}_3(g) \rightarrow \text{P}_4(g) + 6\,\text{H}_2(g)$

17-4. Suggest an experimental method for measuring the rate of the reactions described by the following chemical equations:

(a) $2\,\text{H}_2\text{O}_2(aq) \rightarrow 2\,\text{H}_2\text{O}(l) + \text{O}_2(g)$

(b) $\underset{\text{colorless}}{2\,\text{HBr}(g)} \rightarrow \underset{\text{colorless}}{\text{H}_2(g)} + \underset{\text{reddish-brown}}{\text{Br}_2(g)}$

17-5. The rate law for the reaction described by the equation

$$\text{O}_3(g) + \text{NO}(g) \rightarrow \text{O}_2(g) + \text{NO}_2(g)$$

at 310 K is

$$\text{rate of reaction} = (3.0 \times 10^6\ \text{M}^{-1}\cdot\text{s}^{-1})[\text{O}_3][\text{NO}]$$

Given that $[\text{O}_3] = 6.0 \times 10^{-4}$ M and $[\text{NO}] = 4.0 \times 10^{-5}$ M at $t = 0$, calculate the rate of the reaction at $t = 0$. What is the overall order of this reaction?

17-6. The rate law for the reaction described by the equation

$$2\,\text{NO}(g) + \text{Br}_2(g) \rightarrow 2\,\text{NOBr}(g)$$

is

$$\text{rate of reaction} = (1.3 \times 10^{-3}\ \text{M}^{-2}\cdot\text{min}^{-1})[\text{NO}]^2[\text{Br}_2]$$

Calculate the rate of reaction when $[\text{NO}]_0 = [\text{Br}_2]_0 = 3.0 \times 10^{-4}$ M. What is the overall order of this reaction?

17-7. For the reaction described by

$$2\,\text{O}_3(g) \rightarrow 3\,\text{O}_2(g)$$

$-\Delta P_{\text{O}_3}/\Delta t$ was found to be 5.0×10^{-4} bar·s^{-1}. Determine the value of $\Delta P_{\text{O}_2}/\Delta t$ in bar·s^{-1} during this period of time.

17-8. Using the data plotted in Figure 17.3 graphically estimate (a) the average rate for the formation of $\text{NO}_2(g)$ from $t = 0$ minutes to $t = 100$ minutes, (b) the instantaneous rate of formation at $t = 50$ minutes, and (c) the initial rate of formation.

17-9. The decomposition of N_2O_5 in carbon tetrachloride solution can be described by the equation

$$2\,\text{N}_2\text{O}_5(soln) \rightarrow 4\,\text{NO}_2(soln) + \text{O}_2(g)$$

Given the following data for the reaction at 45°C, calculate the average rate of reaction for each successive time interval.

t/s	$[N_2O_5]$/M
0	1.48
175	1.32
506	1.07
845	0.87

17-10. The value of $[H^+]$ in the reaction described by the equation

$$CH_3OH(aq) + H^+(aq) + Cl^-(aq) \rightarrow CH_3Cl(aq) + H_2O(l)$$

was measured over a period of time:

t/s	$[H^+]$/M
0	2.12
31	1.90
61	1.78
121	1.61

Find the average rate of disappearance of $H^+(aq)$ for the time interval between each measurement. What is the average rate of disappearance of $CH_3OH(aq)$ and what is the average rate of appearance of $CH_3Cl(aq)$ for the same time intervals?

INITIAL RATES

17-11. Sulfuryl chloride decomposes according to the equation

$$SO_2Cl_2(g) \rightarrow SO_2(g) + Cl_2(g)$$

Using the following initial-rate data, determine the order of the reaction with respect to SO_2Cl_2:

$[SO_2Cl_2]_0$/mol·L^{-1}	Initial rate of reaction of $SO_2Cl_2(g)$/mol·L^{-1}·s^{-1}
0.10	2.2×10^{-6}
0.20	4.4×10^{-6}
0.30	6.6×10^{-6}
0.40	8.8×10^{-6}

Calculate the value of the rate constant.

17-12. Nitrosyl bromide decomposes according to the equation

$$2\,NOBr(g) \rightarrow 2\,NO(g) + Br_2(g)$$

Using the following initial-rate data, determine the order of the reaction with respect to $NOBr(g)$:

$[NOBr]_0$/mol·L^{-1}	Initial rate of formation of $Br_2(g)$/mol·L^{-1}·s^{-1}
0.20	0.80
0.40	3.20
0.60	7.20
0.80	12.80

Calculate the value of the rate constant.

17-13. The reaction described by the equation

$$C_2H_5Cl(g) \rightarrow C_2H_4(g) + HCl(g)$$

was studied at 300 K, and the following initial-rate data were collected:

Run	$[C_2H_5Cl]_0$/M	Initial rate of formation of $C_2H_4(g)$/M·s^{-1}
1	0.33	2.40×10^{-30}
2	0.55	4.00×10^{-30}
3	0.90	6.54×10^{-30}

Determine the rate law and the value of the rate constant for the reaction.

17-14. The reaction described by the equation

$$2\,C_5H_6(g) \rightarrow C_{10}H_{12}(g)$$

was studied at 373 K, and the following initial-rate data were collected:

Run	$P_{C_5H_6}$/Torr	Initial rate of formation of $C_{10}H_{12}(g)$/Torr·s^{-1}
1	200	5.76
2	314	14.2
3	576	47.8

Determine the rate law in terms of pressure rather than concentration, and calculate the value of the rate constant for the reaction.

17-15. The reaction described by the equation

$$2\,NOCl(g) \rightarrow 2\,NO(g) + Cl_2(g)$$

was studied at 400 K, and the following data were collected:

Run	$[NOCl]_0/M$	Initial rate of formation of $Cl_2(g)/M\cdot s^{-1}$
1	0.25	1.75×10^{-6}
2	0.42	4.94×10^{-6}
3	0.65	1.18×10^{-5}

Determine the rate law and the value of the rate constant for the reaction.

17-16. The following initial-rate data were obtained for the decomposition of $N_2O_3(g)$ as described by the equation

$$N_2O_3(g) \rightarrow NO(g) + NO_2(g)$$

Run	Initial pressure of $P_{N_2O_3}$/Torr	Initial rate of formation of $NO_2(g)$/Torr·s^{-1}
1	0.91	5.5
2	1.4	8.4
3	2.1	13

Determine the rate law for the reaction, expressed in terms of $P_{N_2O_3}$ rather than $[N_2O_3]$. Calculate the value of the rate constant for the reaction.

17-17. Consider the reaction described by the equation

$$Cr(H_2O)_6^{3+}(aq) + SCN^-(aq) \rightarrow Cr(H_2O)_5SCN^{2+}(aq) + H_2O(l)$$

for which the following initial-rate data were obtained at 25°C:

$[Cr(H_2O)_6^{3+}]_0/M$	$[SCN^-]_0/M$	Initial rate of formation of $Cr(H_2O)_5SCN^{2+}(aq)$ $/M\cdot s^{-1}$
1.0×10^{-4}	0.10	2.0×10^{-11}
1.0×10^{-3}	0.10	2.0×10^{-10}
1.5×10^{-3}	0.20	6.0×10^{-10}
1.5×10^{-3}	0.50	1.5×10^{-9}

Determine the rate law and the value of the rate constant for the reaction.

17-18. The reaction described by the equation

$$CoBr(NH_3)_5^{2+}(aq) + OH^-(aq) \rightarrow Co(NH_3)_5OH^{2+}(aq) + Br^-(aq)$$

was studied at 25°C, and the following initial-rate data were collected:

$[CoBr(NH_3)_5^{2+}]_0/M$	$[OH^-]_0/M$	Initial rate of formation of $Co(NH_3)_5OH^{2+}(aq)$ $/M\cdot s^{-1}$
0.030	0.030	1.37×10^{-3}
0.060	0.030	2.74×10^{-3}
0.030	0.090	4.11×10^{-3}
0.090	0.090	1.23×10^{-2}

Determine the rate law, the overall order of the rate law, and the value of the rate constant for the reaction.

17-19. Given the following initial-rate data at 300 K for the reaction described by the chemical equation

$$2\,NO_2(g) + O_3(g) \rightarrow N_2O_5(g) + O_2(g)$$

$[NO_2]_0/M$	$[O_3]_0/M$	Initial rate of formation of $O_2(g)/M\cdot s^{-1}$
0.65	0.80	2.61×10^{4}
1.10	0.80	4.40×10^{4}
1.70	1.55	1.32×10^{5}

determine the reaction rate law and the value of the rate constant.

17-20. Given the following initial-rate data for the reaction described by the chemical equation

$$CH_3COCH_3(aq) + Br_2(aq) \xrightarrow{H^+(aq)} CH_3COCH_2Br(aq) + H^+(aq) + Br^-(aq)$$

$[CH_3COCH_3]_0$ /M	$[Br_2]_0$ /M	$[H^+]_0$ /M	Initial rate of formation of $H^+(aq)$/M·s^{-1}
1.00	1.00	1.00	4.0×10^{-3}
1.75	1.00	1.00	7.0×10^{-3}
1.75	1.40	1.00	9.8×10^{-3}
1.00	1.40	2.00	11.3×10^{-3}

determine the reaction rate law and the value of the rate constant.

FIRST-ORDER REACTIONS

17-21. The reaction described by the equation

$$SO_2Cl_2(g) \rightarrow SO_2(g) + Cl_2(g)$$

is first order with a rate constant of 2.2×10^{-5} s^{-1} at 320°C. What fraction of a sample of $SO_2Cl_2(g)$ will remain if it is heated for 5.0 hours at 320°C?

17-22. The rate constant for the first-order reaction described by the equation

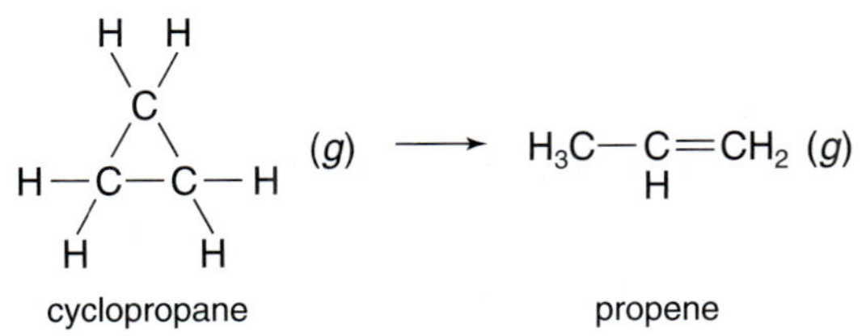

at 500°C is 5.5×10^{-4} s^{-1}. Calculate the half-life of cyclopropane at 500°C. Given an initial cyclopropane concentration of 1.00×10^{-3} M at 500°C, calculate the concentration of cyclopropane that remains after 2.0 hours.

17-23. Azomethane, $CH_3N_2CH_3(g)$, decomposes according to the equation

$$CH_3N_2CH_3(g) \rightarrow CH_3CH_3(g) + N_2(g)$$

Given that the decomposition is a first-order process with $k = 4.0 \times 10^{-4}$ s^{-1} at 300°C, calculate the fraction of azomethane that remains after 1.0 hour.

17-24. Iodomethane, $CH_3I(g)$, decomposes according to the equation

$$2\,CH_3I(g) \rightarrow C_2H_6(g) + I_2(g)$$

Given that the decomposition is a first-order process with $k = 1.5 \times 10^{-4}$ s^{-1} at 300°C, calculate the fraction of iodomethane that remains after one minute.

17-25. The table below gives the concentrations of $C_2H_4O(g)$ as a function of time at 690 K for the following reaction equation:

$$C_2H_4O(g) \rightarrow CH_4(g) + CO(g)$$

$[C_2H_4O]$/M	t/min
0.0860	0
0.0465	50
0.0355	72
0.0274	93
0.0174	130

Verify that this is a first-order reaction by plotting $\ln([C_2H_4O]/M)$ versus time and determine the value of the rate constant.

17-26. The table below gives the concentration of $SO_2Cl_2(g)$ as a function of time for the reaction described by the equation

$$SO_2Cl_2(g) \rightarrow SO_2(g) + Cl_2(g)$$

$[SO_2Cl_2]$/M	t/min
0.0345	0
0.0245	3.8
0.0212	5.6
0.0154	9.3
0.0103	14.0

Verify that this reaction is a first-order reaction by plotting $\ln([SO_2Cl_2] = M)$ versus time and determine the value of the rate constant.

17-27. Peroxydisulfate ion, $S_2O_8^{2-}(aq)$ decomposes in aqueous solution according to the equation

$$S_2O_8^{2-}(aq) + H_2O(l) \rightarrow 2\,SO_4^{2-}(aq) + \tfrac{1}{2}O_2(g) + 2\,H^+(aq)$$

Given the following data from an experiment with $[S_2O_8^{2-}]_0 = 0.100$ M in a solution with $[H^+]$ fixed at 0.100 M, determine the reaction rate law and calculate the value of the rate constant:

t/min	$[S_2O_8^{2-}]$/M
0	0.100
17	0.050
34	0.025
51	0.012

17-28. At 400 K oxalic acid decomposes according to

$$H_2C_2O_4(g) \rightarrow CO_2(g) + HCHO_2(g)$$

The rate of this reaction can be studied by measurement of the total pressure. Determine the rate law and the value of the rate constant of this reaction from the following measurements that give the total pressure reached after 2.00×10^4 seconds from the given starting pressure of oxalic acid:

$P^\circ_{H_2C_2O_4}$/Torr (at $t = 0$)	P_{total}/Torr (at $t = 2.00 \times 10^4$ s)
5.0	7.2
7.0	10
8.4	12

RATES OF NUCLEAR DECAY

17-29. The radioisotope argon-41 is used to measure the rate of the flow of gases from smokestacks. It is a γ-emitter with a half-life of 109.2 minutes. Calculate the fraction of an argon-41 sample that remains after one day.

17-30. The radioisotope bromine-82 is used as a tracer for organic materials in environmental studies. Its half-life is 35.3 hours. Calculate the fraction of a sample of bromine-82 that remains after one day.

17-31. A sample of sodium-24 chloride containing 0.055 milligrams of sodium-24 is injected into an animal to study sodium balance. How much sodium-24 remains 6.0 hours later? The half-life of sodium-24 is 14.96 hours.

17-32. Cesium-137 is produced in nuclear reactors. If this isotope has a half-life of 30.2 years, how many years will it take for it to decay to one tenth of a percent of its initial amount?

17-33. You order a sample of Na_3PO_4 containing the radioisotope phosphorus-32 ($t_{1/2} = 14.28$ days). If the shipment is delayed in transit for two weeks, how much of the original activity will remain when you receive the sample?

17-34. Strontium-90 is a radioactive isotope that is produced in nuclear explosions. It decays by β-emission with a half-life of 29.1 years. Suppose that an infant ingests strontium-90 in mother's milk. Calculate the fraction of the ingested strontium-90 that remains in the body when the infant reaches 74 years of age, assuming no loss of strontium-90 except by radioactive decay.

DATING

17-35. Burned animal bones found near a site inhabited by humans in Palli Aike Cave in southern Chile have a disintegration rate of 5.37 disintegrations per minute per gram of carbon. Estimate the age of the site.

17-36. In 1955 the French explorer Fernand Navarra discovered a log on Mt. Ararat in eastern Turkey, the legendary resting spot of Noah's ark. Navarra claims that the log is a beam from the ark. Samples of the wood have a disintegration rate of 13.19 disintegrations per minute per gram of carbon. Calculate the age of the log.

17-37. A sample of $CaCO_3(s)$ from the shell of a preserved ancient egg has a disintegration rate of 498 disintegrations per hour per gram of carbon. Estimate the age of the shell.

17-38. Samples of oak from an ancient Irish cooking site at Killeens, County Cork, have a carbon-14 content equal to 65 percent that in living matter. Estimate the age of the wood.

17-39. A sample of ocean sediment is found to contain 1.50 milligrams of uranium-238 and 0.460 milligrams of lead-206. Estimate the age of the sediment. The half-life for the conversion of uranium-238 to lead-206 is 4.51×10^9 years.

17-40. A sample of uranite is found to have a ${}^{206}_{82}Pb/{}^{238}_{92}U$ mass ratio of 0.395. Estimate the age of the uranite. The half-life of the conversion of uranium-238 to lead-206 is 4.51×10^9 years.

SECOND-ORDER REACTIONS

17-41. The gas-phase decomposition of $CH_3CHO(g)$ occurs according to the equation

$$CH_3CHO(g) \rightarrow CH_4(g) + CO(g)$$

and is second order. The value of the rate constant is 0.105 $M^{-1} \cdot s^{-1}$ at 490°C. If the concentration of $CH_3CHO(g)$ is 0.012 M initially, what will be its concentration 5.0 minutes later?

17-42. The reaction described by the chemical equation

$$3\,BrO^-(aq) \rightarrow BrO_3^-(aq) + 2\,Br^-(aq)$$

is second order in $BrO^-(aq)$ in basic solution with a rate constant equal to 0.056 $M^{-1} \cdot s^{-1}$ at 80°C. If $[BrO^-]_0 = 0.212$ M, what will $[BrO^-]$ be 1.00 minute later?

17-43. The following table gives $[NO_2]$ as a function of time for the reaction described by

$$NO_2(g) \rightarrow NO(g) + \tfrac{1}{2}O_2(g)$$

$[NO_2]$/M	t/s
0.0831	0
0.0666	4.2
0.0567	7.9
0.0497	11.4
0.0441	15.0

Show that this reaction is second order by plotting $1/[NO_2]$ versus time and determine the value of the rate constant.

17-44. The following table gives [HBr] as a function of time for the thermal decomposition of $HBr(g)$ according to the equation

$$2\,HBr(g) \rightarrow H_2(g) + Br_2(g)$$

[HBr]/M	t/s
0.0714	0.00
0.0520	1.02
0.0430	1.81
0.0371	2.53
0.0332	3.16

Show that this reaction is second order by plotting 1/[HBr] versus time and determine the value of the rate constant.

17-45. The rate law for the reaction described by the equation

$$2\,N_2O(g) \rightarrow 2\,N_2(g) + O_2(g)$$

is second order in $[N_2O]$. The reaction was carried out at 900 K with an initial concentration of $N_2O(g)$ of 2.0×10^{-2} M. It took 4500 seconds for $[N_2O]$ to fall to half its initial value. Determine the value of the rate constant for this reaction.

17-46. The decomposition of $NO_2(g)$ as described by the equation

$$2\,NO_2(g) \rightarrow 2\,NO(g) + O_2(g)$$

is second order in $NO_2(g)$. The rate constant for the reaction at 300°C is 0.54 $M^{-1} \cdot s^{-1}$. If $[NO_2]_0 = 1.25$ M, what is the value of $[NO_2]$ after the reaction has run for 2.0 minutes?

ADDITIONAL PROBLEMS

17-47. Discuss the difference between the average rate of reaction and the instantaneous rate of reaction.

17-48. Based on the plot shown in Figure 17.3, over approximately what time interval can this reaction be said to be running under "initial conditions"? Is this time interval the same for all reactions?

17-49. How does doubling the concentration of a reactant change the rate of a reaction that is first order in that reactant? How does it change the rate of a reaction that is second order in that reactant?

17-50. What are the units of the rate constant for a zero-order, first-order, second-order, and third-order reaction?

17-51. The National Institute of Science and Technology (NIST) kinetics database lists the rate constant of a reaction as 1.2×10^{-10} $cm^3 \cdot molecule^{-1} \cdot s^{-1}$ at 298 K. Determine the rate constant in units of $M^{-1} \cdot s^{-1}$ and $Torr^{-1} \cdot s^{-1}$. What is the order of this reaction?

17-52. The rate of the reaction described by the equation

$$2\,CO(g) \rightarrow CO_2(g) + C(s)$$

was studied by injecting some $CO(g)$ into a reaction vessel and measuring the total pressure while maintaining a constant reaction volume:

P_{total}/Torr	t/s
250	0
238	398
224	1002
210	1801
196	3000

Assuming that only $CO(g)$ is present initially, determine the value of the reaction rate constant.

17-53. Given the following data for the reaction described by the chemical equation

$$ClO_3^-(aq) + 9I^-(aq) + 6H^+(aq) \rightarrow 3I_3^-(aq) + Cl^-(aq) + 3H_2O(l)$$

$[I^-]_0$ /M	$[ClO_3^-]_0$ /M	$[H^+]_0$ /M	Initial rate of reaction of $ClO_3^-(aq)$/M·s^{-1}
0.10	0.10	0.10	1.00×10^{-3}
0.10	0.17	0.10	1.70×10^{-3}
0.26	0.17	0.10	4.42×10^{-3}
0.26	0.17	0.16	1.13×10^{-2}

determine the reaction rate law.

17-54. A reaction of importance in the formation of smog is that between ozone and nitrogen monoxide described by

$$O_3(g) + NO(g) \rightarrow O_2(g) + NO_2(g)$$

The rate law for this reaction is

$$\text{rate of reaction} = k[O_3][NO]$$

Given that $k = 2.99 \times 10^6$ M^{-1}·s^{-1} at 310 K, calculate the initial reaction rate when $[O_3]$ and [NO] remain essentially constant at the values $[O_3]_0 = 2.0 \times 10^{-6}$ M and $[NO]_0 = 6.0 \times 10^{-5}$ M, owing to continuous production from separate sources. Calculate the number of moles of $NO_2(g)$ produced per hour per liter of air.

17-55. Suppose that you place 100 bacteria into a flask containing nutrients for the bacteria and that you find the following data at 37°C:

t/min	Number of bacteria
0	100
15	200
30	400
45	800
60	1600

What is the order of the rate of production of the bacteria? How many bacteria do you predict there will be after two hours? What is the rate constant for the process?

17-56. Show that for a first-order reaction, the time required for 99.9% of the reaction to take place is about 10 times that required for 50% of the reaction to take place.

17-57. Fill in the missing symbols in the following nuclear equations:

(a) ${}^{72}_{30}\text{Zn} \rightarrow {}^{0}_{-1}\text{e} + ?$ (b) ${}^{230}_{92}\text{U} \rightarrow {}^{4}_{2}\text{He} + ?$

(c) ${}^{136}_{57}\text{La} \rightarrow {}^{0}_{+1}\text{e} + ?$ (d) ${}^{14}_{7}\text{N} + ? \rightarrow {}^{1}_{1}\text{H} + {}^{14}_{6}\text{C}$

17-58. One of the waste products of a uranium-powered nuclear reactor is plutonium-239, which has a half-life of 2.41×10^9 years. How long will it take for 10% of the plutonium-239 waste from a nuclear power plant to decay? How long will it take for 99% of the same waste to decay? Why is the storage of nuclear waste such a difficult problem to solve?

17-59. The radioisotope hydrogen-3 (tritium) is used in fusion reactors. It is a β-emitter with a half-life of 12.33 years. Calculate the fraction of a hydrogen-3 sample that will remain after 50.0 years.

17-60. A sample of radioactive $Na^{128}I$ is injected into a patient as part of radioiodine treatment of a thyroid condition. If the sample has an activity of 10 000 disintegrations·min^{-1} at 8 A.M., the time of injection, what is the activity at 2 P.M. the same day? The half-life of iodine-128 is 25.00 minutes.

17-61. No stable isotope of the halogen astatine exists in nature. The radioisotope astatine-211 is made by bombarding bismuth-209 with α-particles. Plot the following data and determine the half-life of astatine-211.

Fraction of ^{211}At remaining	Time/h
0.909	1.0
0.825	2.0
0.681	4.0
0.464	8.0
0.215	16.0

17-62. A 1.00-mL sample of blood is withdrawn from an animal, and the red blood cells are labeled with phosphorus-32 ($t_{1/2}$ = 14.28 days). The activity of this sample is 50 000 disintegrations·min^{-1}. The sample is then reinjected into the animal. A few hours later, another 1.00-mL sample is withdrawn, and its activity is 10.0 disintegrations·min^{-1}. Determine the volume of blood in the animal. Assume that phosphorus-32 is uniformly distributed throughout the blood and that the activity due to phosphorus-32 remains constant during the experiment. By using similar methods, it has been found that the human body contains about 75 mL of blood per kilogram of body weight.

17-63. Identify in each of the following cases the order of the reaction rate law with respect to the reactant A, where A → Products:

(a) The half-life of A is independent of the initial concentration of A.

(b) The rate of decrease of A is a constant.

(c) A twofold increase in the initial concentration of A leads to a 1.41-fold increase in the initial rate.

(d) A twofold increase in the initial concentration of A leads to a fourfold increase in the initial rate.

(e) The time required for $[A]_0$ to decrease to $[A]_0/2$ is equal to the time required for [A] to decrease from $[A]_0/2$ to $[A]_0/4$.

17-64. Given the following initial-rate data for the reaction described by the equation

$$BrO_3^-(aq) + 9I^-(aq) + 6H^+(aq) \rightarrow 3I_3^-(aq) + Br^-(aq) + 3H_2O(l)$$

$[I^-]_0$ /M	$[BrO_3^-]_0$ /M	$[H^+]_0$ /M	Initial rate of reaction of $BrO_3^-(aq)$/M·s^{-1}
0.10	0.10	0.10	3.00×10^{-4}
0.14	0.18	0.10	7.56×10^{-4}
0.10	0.18	0.10	5.40×10^{-4}
0.31	0.18	0.20	1.67×10^{-3}

determine the reaction rate law.

17-65. The rate law for the reaction described by the equation

$$C_2H_4Br_2(aq) + 3I^-(aq) \rightarrow C_2H_4(g) + 2Br^-(aq) + I_3^-(aq)$$

at 300 K is

$$\text{rate of reaction} = (5.0 \times 10^{-3}\ M^{-1}\cdot s^{-1})[C_2H_4Br_2][I^-]$$

Fill in the missing entries in the following table:

Run	$[C_2H_4Br_2]_0$ /M	$[I^-]_0$ /M	Initial rate of formation of $C_2H_4(g)$/M·s^{-1}
1	0.20	0.20	
2	0.20		4.0×10^{-4}
3		0.20	8.0×10^{-4}

17-66. The U.S. Public Health Service requires that milk fresh from a pasteurizer may contain no more than 20 000 bacteria per milliliter. It has been reported that the number of bacteria in milk stored in a refrigerator at 4°C (40°F) may double in 39 hours. If a milk sample had 20 000 bacteria per milliliter after pasteurization, what is the bacteria count per milliliter after 10 days at this temperature?

17-67. Calculate the time required for the concentration to decrease by 10.0% of its initial value for a first-order reaction with k = 10.0 s^{-1}.

17-68. Show that for a first-order reaction, the time required for 99.99% of the reaction to take place is twice as long as the time required for 99.0% of the reaction to take place.

17-69. Phosphine, $PH_3(g)$ decomposes according to the equation

$$4PH_3(g) \rightarrow P_4(g) + 6H_2(g)$$

The kinetics of the decomposition of phosphine at 950 K was followed by measuring the total pressure in the system as a function of time. The following data were obtained in a run where the reaction chamber contained only pure phosphine at the start of the reaction:

t/min	P_{total}/Torr
0	100
40	150
80	167
100	172

Determine the reaction rate law and calculate the value of the rate constant.

17-70. The reaction described by

$$NO_2(g) \rightarrow NO(g) + \frac{1}{2}O_2(g)$$

is a second-order reaction with a rate constant $k = 0.54\ M^{-1}\cdot s^{-1}$. How long will it take for $[NO_2]$ to be 10.0% of its initial value of 2.00 M?

17-71. Acetaldehyde, $CH_3CHO(g)$, decomposes in the gas phase according to

$$CH_3CHO(g) \rightarrow CH_4(g) + CO(g)$$

Determine the order of the reaction and the value of the rate constant from the following data where the reaction chamber contained only pure acetaldehyde at the start of the reaction:

t/s	P_{total}/Torr
0	363
73	417
190	477
310	517
480	587

17-72. Hydrogen peroxide is catalytically decomposed by $I^-(aq)$ according to the equation

$$2H_2O_2(aq) \xrightarrow{I^-(aq)} 2H_2O(l) + O_2(g)$$

A reaction flask containing 50.0 milliliters of reaction mixture is connected to a gas buret, and the reaction is followed by measuring the volume of $O_2(g)$ collected over water at a barometric pressure of 730.0 Torr. The following data were obtained at 20.0°C with $[H_2O_2]_0 = 0.250$ M.

t/min	V_{O_2}/mL
0	0
10	16.0
20	29.5
35	47.8
50	63.9
65	77.4

Determine the order of the reaction and the value of the rate constant.

17-73. The reaction between carbon disulfide and ozone described by the equation

$$CS_2(g) + 2O_3(g) \rightarrow CO_2(g) + 2SO_2(g)$$

was studied using a large excess of $CS_2(g)$. The pressure of ozone as a function of time is given in the following table. Is the reaction first order or second order with respect to ozone?

t/s	P_{O_3}/Torr
0	1.76
30	1.04
60	0.79
120	0.52
180	0.37
240	0.29

17-74. Determine the rate law for the reaction described by

$$NO(g) + H_2(g) \rightarrow \text{products}$$

from the initial-rate data tabulated below:

Initial pressure of P_{H_2}/Torr	Initial pressure of P_{NO}/Torr	Initial rate of reaction/Torr·s^{-1}
400	159	34
400	300	125
289	400	160
205	400	110
147	400	79

Calculate the value of the rate constant for this reaction.

17-75. Consider the reaction described by

$$\text{OCl}^-(aq) + \text{I}^-(aq) \xrightarrow{\text{OH}^-(aq)} \text{OI}^-(aq) + \text{Cl}^-(aq)$$

Use the following initial-rate data to determine the rate law and the corresponding value of the rate constant for the reaction.

$[\text{OCl}^-]_0$ /M	$[\text{I}^-]_0$ /M	$[\text{OH}^-]_0$ /M	Initial rate of reaction/M·s^{-1}
1.62×10^{-3}	1.62×10^{-3}	0.52	3.06×10^{-4}
1.62×10^{-3}	2.88×10^{-3}	0.52	5.44×10^{-4}
2.71×10^{-3}	1.62×10^{-3}	0.84	3.16×10^{-4}
1.62×10^{-3}	2.88×10^{-3}	0.91	3.11×10^{-4}

17-76. (*) Potassium-40 decays by the two different paths, shown below.

$$^{40}_{19}\text{K} \rightarrow {}^{40}_{20}\text{Ca} + {}^{0}_{-1}\text{e} \qquad (89.3\%)$$

$$^{40}_{19}\text{K} \xrightarrow{\text{EC}} {}^{40}_{18}\text{Ar} \quad (10.7\%)$$

where EC stands for electron capture. The overall half-life for the decay of $^{40}_{19}\text{K}$ is 1.248×10^9 years. The potassium-40 to argon-40 reaction is thought to be the source of argon in the earth's atmosphere. Potassium-argon dating is used in geology and archaeology to date sedimentary rocks. Estimate the age of sedimentary rocks with a ^{40}Ar-to-^{40}K ratio of 0.0102.

17-77. Uranyl nitrate decomposes according to

$$\text{UO}_2(\text{NO}_3)_2(aq) \rightarrow \text{UO}_3(s) + 2\,\text{NO}_2(g) + \tfrac{1}{2}\text{O}_2(g)$$

The rate law is first order in the concentration of uranyl nitrate. The following data were recorded for the reaction at 25.0°C.

t/min	$[\text{UO}_2(\text{NO}_3)_2]$/M
0	0.01413
20.0	0.01096
60.0	0.00758
180.0	0.00302
360.0	0.00055

Calculate the value of the rate constant for this reaction at 25.0°C.

17-78. (*) *This problem is appropriate only if you have taken calculus.* Derive Equation 17.17 from the first-order rate expression $-d[\text{A}]/dt = k[\text{A}]$.

17-79. (*) *This problem is appropriate only if you have taken calculus.* Derive Equation 17.29 from the second-order rate expression $-d[\text{A}]/dt = k[\text{A}]^2$.

17-80. (*) *This problem is appropriate only if you have taken calculus.* Show that for a zero-order reaction in a single reactant, A, that the integrated rate expression is given by the expression $[\text{A}] = [\text{A}]_0 - kt$, and the half-life is given by $t_{1/2} = [\text{A}]_0 / 2k$. What is the test plot for a zero-order reaction?

Leonor Michaelis (1875–1949) was born in Berlin, Germany. He studied medicine at the University of Berlin, where he received his doctorate in 1897. He was an assistant to several prominent medical scientists from 1898 to 1906. Because no university and research positions were available at the time, he accepted the directorship of the bacteriology laboratory in the Charité Hospital in Berlin. In spite of limited funds and research space, he did significant medical and biochemical research, attracting many students from around the world. In 1922, he was invited to spend three years at the medical school of the University of Nagoya in Japan, where he continued his research on the theory of pH. After three years at Johns Hopkins University, he received a permanent position at the Rockefeller Institute, where he remained until his death. In addition to his expertise in medicine and chemistry, he was proficient in mathematics, pioneering the use of mathematics in the biological sciences. Michaelis had a talent for languages and became fluent in Japanese as well as several other languages. He was also a talented musician with a lifelong interest in the theory of music.

Maud Menten (1879–1960) was born in Port Lambton, Ontario, Canada. She graduated from the University of Toronto in 1904 and remained there to earn her M.D. in 1911, one of the first women in Canada to receive a medical doctorate. Because women were not allowed to do research in Canada at the time, she left to pursue her career in Germany and then the United States. In 1912, she spent a year with Leonor Michaelis in Berlin working on enzyme kinetics, resulting in the famous Michaelis-Menten equation. She later received her Ph.D. in biochemistry from the University of Chicago in 1916. Unable to find a suitable position in Canada, she joined the department of pathology in the medical school at the University of Pittsburgh in 1919. In spite of her productivity, she was not promoted to full professor until 1949, when she was 70 years old. After retiring in 1950 she returned to Canada, where she died in 1960. Despite suffering from arthritis, she was an accomplished musician and painter; many of her paintings were exhibited in Pittsburgh art galleries and museums.

18. Chemical Kinetics: Mechanisms

In the previous chapter we learned about rate laws for chemical reactions. We often describe the reaction rate by specifying its order with respect to each of the reactants. Common reaction orders are first order and second order. We also saw that it is not possible to deduce the rate law of a reaction from its chemical equation. In this chapter, we shall introduce the idea of a reaction mechanism; that is, a sequence of so-called elementary reactions that add up to give the overall reaction. Each of the elementary reactions is easy to deal with, allowing us to visualize a reaction in terms of simple (elementary) steps. We then go on to study the conditions necessary for a chemical reaction to occur and learn the relationship between reaction rates and temperature. An understanding of these conditions along with a reaction mechanism allows us to optimize the conditions under which a reaction takes place and provides insight into how chemical processes occur at the molecular level. We then learn about catalysts. A catalyst is a substance that increases the rate of a chemical reaction, but is not consumed in the reaction. We shall see that a catalyst functions by supplying an alternative, more favorable, chemical pathway for a reaction. Catalysts are of great importance in chemistry, as they allow us to produce reaction products faster with less energy expenditure. In the final section we shall look at the catalysts used by living systems, called enzymes, and learn how enzymes affect the rates of some important biological processes.

18-1. Many Reactions Involve More Than One Step

As we emphasized in the previous chapter, a balanced chemical equation shows the stoichiometric relationship between the reactants and the products. However, it gives no indication of the rate law or how the reactants are converted into products. For example, a reaction equation that we discussed in the previous chapter is

$$NO_2(g) + CO(g) \rightarrow NO(g) + CO_2(g)$$

You might guess from this equation that a $NO_2(g)$ molecule collides with a $CO(g)$ molecule and that during the collision an oxygen atom is transferred from the $NO_2(g)$ molecule to the $CO(g)$ molecule. It would seem that increasing the concentration of either would make a collision more likely to occur, so that the rate of the reaction should depend on both $[NO_2]$ and $[CO]$. But as we saw in Example 17-5, the rate law for this reaction is rate of reaction = $k[NO_2]^2$, which is *independent* of $[CO]$.

It has been determined experimentally that the reaction is more involved than a single collision between a $NO_2(g)$ molecule and a $CO(g)$ molecule, and actually occurs in two steps. In the first step, two $NO_2(g)$ molecules collide and an oxygen atom is transferred from one $NO_2(g)$ molecule to the other according to

$$\text{(1)} \qquad NO_2(g) + NO_2(g) \rightarrow NO_3(g) + NO(g)$$

Now the $NO_3(g)$ species collides with a $CO(g)$ molecule, and an oxygen atom is transferred from $NO_3(g)$ to the $CO(g)$ molecule as described by the equation

$$\text{(2)} \qquad NO_3(g) + CO(g) \rightarrow NO_2(g) + CO_2(g)$$

Each of the two steps described by equations (1) and (2) is called an **elementary reaction**. Elementary reactions are reactions that occur in a single step. The sum of these two elementary reaction equations gives the equation for the overall reaction:

$$\begin{array}{rl} \text{(1)} & NO_2(g) + NO_2(g) \rightarrow NO_3(g) + NO(g) \\ \text{(2)} & NO_3(g) + CO(g) \rightarrow NO_2(g) + CO_2(g) \\ \hline \text{overall:} & NO_2(g) + CO(g) \rightarrow NO(g) + CO_2(g) \end{array}$$

A series of elementary reaction equations that add up to give a reaction equation is called a **reaction mechanism**. A reaction mechanism is a step-by-step description of the molecular pathway of a chemical reaction. Chemists are interested in not only what the products of a reaction are but also just how the products are formed from the reactants. One of the goals of chemical kinetics is the elucidation of reaction mechanisms. If we know the mechanism of a reaction, then we can often manipulate the reaction conditions to produce greater yields of a desired product or to control its rate of production. The task of determining a reaction mechanism often involves a great deal of experimental detective work. For example, the innocent-looking reaction between hydrogen and oxygen to produce water that we express as

$$2H_2(g) + O_2(g) \rightarrow 2H_2O(l)$$

is known to involve over ten steps, and no universally accepted mechanism has yet been proposed.

Note that the species $NO_3(g)$ in the two-step mechanism of the $NO_2(g)$ + $CO(g)$ reaction cancels out when the two elementary reaction equations are added to give the equation for the overall reaction. Such a species, called an **intermediate species** or simply an **intermediate**, is produced in one step of a reaction and consumed in a subsequent step. An intermediate, therefore, does not appear in the overall equation for the reaction. An intermediate is usually an unstable, reactive species. However, it does exist in the reaction mixture, and its presence can often be detected experimentally. Using modern spectroscopic techniques, it is possible to detect species with lifetimes of only microseconds or even nanoseconds. It is the possible involvement of one or more intermediates that makes it impossible to deduce a reaction mechanism solely from the overall reaction stoichiometry.

Because an elementary reaction takes place in a single step, we can write its rate law directly from the stoichiometry of the elementary reaction equation. This, in fact, is just another way of defining an elementary reaction. This observation suggests that the rate law for the elementary reaction

$$\text{(1)} \qquad NO_2(g) + NO_2(g) \xrightarrow{k_1} NO_3(g) + NO(g) \qquad (18.1)$$

is

$$\text{rate of reaction 1} = k_1[NO_2][NO_2] = k_1[NO_2]^2$$

and that for

$$\text{(2)} \qquad NO_3(g) + CO(g) \xrightarrow{k_2} NO_2(g) + CO_2(g) \qquad (18.2)$$

is

$$\text{rate of reaction 2} = k_2[NO_3][CO]$$

Elementary reactions that involve the reaction of two molecules (which may or may not be chemically distinct) are referred to as **bimolecular reactions**. Equations 18.1 and 18.2 describe bimolecular reactions. In contrast, reactions involving only a single species are called **unimolecular reactions**. The spontaneous decay of radioactive nuclei is an example of a unimolecular reaction.

We stress that you can write the reaction-rate law from the stoichiometry of a chemical equation of that reaction *only* for an elementary reaction equation. You cannot deduce the reaction-rate law from the stoichiometry of a general (nonelementary) chemical equation; it must always be determined experimentally. Once you have experimentally determined a reaction-rate law, you can then propose a mechanism consisting of one or more elementary steps that is consistent with this rate law.

EXAMPLE 18-1: The reaction described by

$$2NO_2(g) + F_2(g) \rightarrow 2NO_2F(g)$$

is thought to proceed via the following two-step mechanism:

$$\text{(1)} \qquad NO_2(g) + F_2(g) \xrightarrow{k_1} NO_2F(g) + F(g)$$

$$\text{(2)} \qquad F(g) + NO_2(g) \xrightarrow{k_2} NO_2F(g)$$

where k_1 and k_2 are rate constants. Identify any intermediate species in the reaction mechanism and deduce the rate law for each step.

Solution: The fluorine atom (also called a fluorine radical because it has an unpaired electron) is an intermediate in the reaction mechanism. Atomic fluorine is produced in the first step and consumed in the second step, and the sum of these two steps gives the overall reaction stoichiometry.

Because a proposed reaction mechanism consists of a series of postulated elementary reactions, we can derive the rate law for each step directly from its stoichiometry. Therefore, the rate laws for the above two elementary reaction equations are

$$\text{rate of reaction 1} = k_1[\text{NO}_2][\text{F}_2]$$

$$\text{rate of reaction 2} = k_2[\text{F}][\text{NO}_2]$$

PRACTICE PROBLEM 18-1: A possible mechanism for the reaction described by

$$\text{CO}(g) + \text{Cl}_2(g) \rightarrow \text{COCl}_2(g)$$

is the following four-step mechanism:

(1) $\text{Cl}_2(g) \xrightarrow{k_1} 2\,\text{Cl}(g)$

(2) $2\,\text{Cl}(g) \xrightarrow{k_2} \text{Cl}_2(g)$

(3) $\text{Cl}(g) + \text{CO}(g) \xrightarrow{k_3} \text{COCl}(g)$

(4) $\text{COCl}(g) + \text{Cl}_2(g) \xrightarrow{k_4} \text{COCl}_2(g) + \text{Cl}(g)$

Identify any intermediate species and write the rate law for each step.

Answer: $\text{Cl}(g)$ and $\text{COCl}(g)$ are intermediate species. The rate laws are $k_1[\text{Cl}_2]$, $k_2[\text{Cl}]^2$, $k_3[\text{Cl}][\text{CO}]$, and $k_4[\text{COCl}][\text{Cl}_2]$, respectively.

18-2. Reactants Must Surmount an Energy Barrier to React

Because an elementary reaction occurs in a single step, it is the simplest type of reaction to treat theoretically. One approach is called the **collision theory of reaction rates.** Its basic postulate is that two species must collide with each other in order to react. In one liter of a mixture of the gases $\text{A}(g)$ and $\text{B}(g)$ at one bar and 25°C, the collision frequency between $\text{A}(g)$ and $\text{B}(g)$ molecules is about $10^{31}\ \text{s}^{-1}$. If every collision led to a reaction, then the initial reaction rate would be about

$$\text{rate of reaction} = \left(\frac{10^{31}\ \text{s}^{-1}}{1\ \text{L}}\right)\left(\frac{1\ \text{mol}}{6.022 \times 10^{23}}\right) \approx 10^{7}\ \text{mol·L}^{-1}\text{·s}^{-1}$$

In fact, very few reactions occur at this extremely high rate; most occur at a much lower rate. The inescapable conclusion is that most collisions do not lead to a chemical reaction. Unless the following two conditions are met, the colliding molecules simply bounce off each other unchanged.

The first and more obvious condition is that the molecules must collide with sufficient energy either to break or to rearrange bonds. The other condition is that the molecules must collide in some preferred relative orientation. Figure 18.1 illustrates these conditions for the collision of a $\text{F}_2(g)$ molecule with a $\text{NO}_2(g)$ molecule (see Example 18-1). A reactive collision will occur only if the two molecules have sufficient energy and if, in addition, the $\text{F}_2(g)$ molecule strikes the nitrogen atom of the $\text{NO}_2(g)$ molecule. Even if the two molecules collide very energetically, collisions with any other relative orientation will not lead to a reaction.

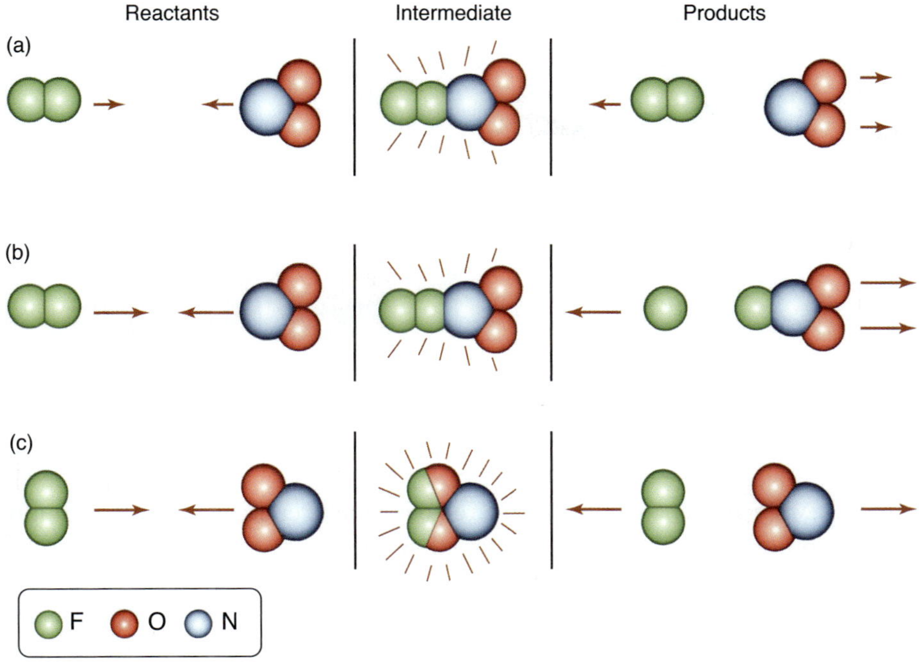

Figure 18.1 Molecular view of the reaction described by the equation

$$F_2(g) + NO_2(g) \rightarrow F(g) + FNO_2(g)$$

(a) Nonreactive collision. Molecules bounce off one another without reacting because the energy of the colliding particles is less than the energy required to break the F_2 bond and form the N–F bond. (b) Reactive collision. Molecules collide with sufficient energy and react because they also have the correct orientation for reaction. (c) Nonreactive collision. Molecules collide with sufficient energy but do not react because they do not have the correct orientation for reaction.

These two requirements are the basis of the collision theory of reaction rates. To incorporate them, we write the reaction rate for the elementary reaction equation

$$A(g) + B(g) \rightarrow C(g) + D(g)$$

as

$$\text{rate of reaction} = \begin{pmatrix}\text{collision}\\ \text{frequency}\end{pmatrix}\begin{pmatrix}\text{fraction of}\\ \text{collisions with}\\ \text{the required}\\ \text{energy}\end{pmatrix}\begin{pmatrix}\text{fraction of collisions}\\ \text{in which molecules}\\ \text{have the required}\\ \text{relative orientation}\end{pmatrix} \tag{18.3}$$

Let's look at each of the terms in parentheses in turn. The two major factors that affect the collision frequency are concentration and temperature. The greater the concentration of the reactant molecules, the greater the frequency of collisions. For two reactant species, A(g) and B(g), the collision frequency is proportional to the product of their concentrations, or

$$\text{collision frequency} \propto [A][B] \tag{18.4}$$

The collision frequency also increases with temperature because the root-mean-square molecular speed increases with temperature (see Equation 13.29). The faster the molecules are moving, the more frequently they collide.

We showed in Sections 13-8 and 13-9 that the molecules in a gas do not all have the same speed but have a distribution of speeds. Consequently, they have a distribution of energies as shown in Figure 13.24 and Figure 18.2.

Now let's consider the second term in parentheses in Equation 18.3. Figure 18.2 shows a plot of the number of collisions per second of the A(g) and B(g) molecules versus the kinetic energy of the colliding molecules. The fraction of collisions with a kinetic energy in excess of some fixed energy, such as E_a in the figure, increases with increasing temperature. The quantity E_a in Figure 18.2 represents the minimum energy necessary to cause a reaction between the colliding molecules and is called the **activation energy** of the

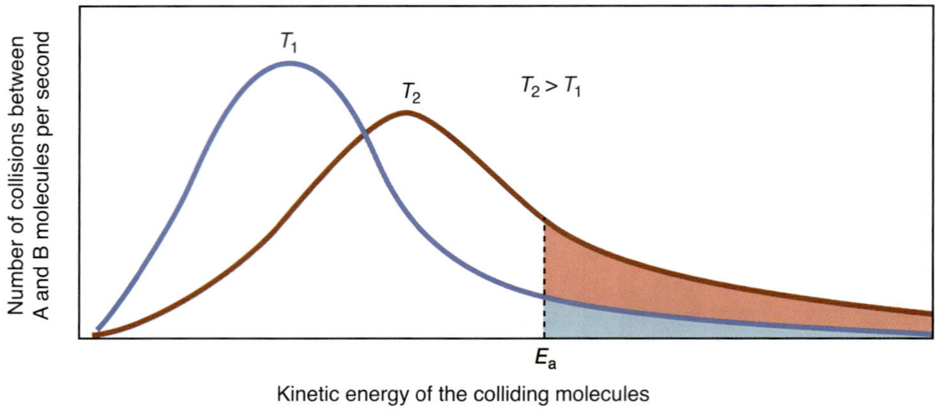

Figure 18.2 Plots of the number of collisions between A(*g*) and B(*g*) molecules per unit time versus the kinetic energy of the colliding molecules at two temperatures, T_1 and T_2 ($T_2 > T_1$). Observe that the area under the curve beyond E_a is greater for the T_2 curve than for the T_1 curve. The number of collisions with a kinetic energy E_a or greater increases with increasing temperature.

process. Molecules that collide with a relative kinetic energy less than E_a simply bounce off each other. A colliding pair that does have the required energy E_a can react, provided they have the required relative orientation.

Now we consider the final term in Equation 18.3. As we have seen, two molecules can react only if they have the correct relative orientation. As we saw in Figure 18.1, this depends on the shapes of the colliding molecules.

In Equation 18.3, then, only the collision frequency depends on the concentrations of A(*g*) and B(*g*). If we substitute Equation 18.4 into 18.3, we see that for an elementary reaction involving two species A and B

$$\text{rate of reaction} = k[\text{A}][\text{B}] \tag{18.5}$$

where the rate constant k depends on the other factors present in Equation 18.3; that is, the rate constant depends on temperature and the shapes of the molecules. Equation 18.5 gives us the rate law for an elementary reaction directly from its chemical equation. As we stated in the previous section, this is true, of course, only for an elementary reaction. Equation 18.5 can also be applied to reactions that occur in solution.

Figure 18.3 illustrates the activation energy for the elementary reaction described by A + B → C + D. The vertical axis in Figure 18.3 represents the energy, E, of the reactants and products, and the horizontal axis is a schematic representation of the reaction pathway, beginning with the reactants A and B and

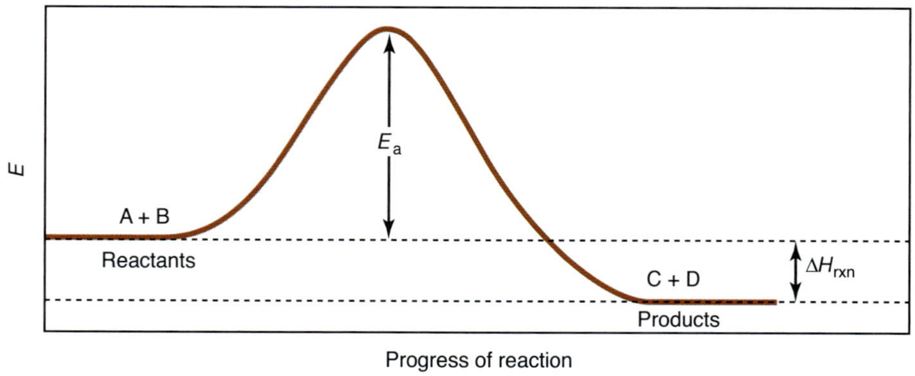

Figure 18.3 The energy of the molecules as the reaction described by A + B → C + D proceeds. The value of E_a has no relationship to the value of the overall energy change, ΔU_{rxn}, or the enthalpy change, ΔH_{rxn}. (Recall from Chapter 14 that $\Delta U_{rxn} \approx \Delta H_{rxn}$, for most reactions.)

ending up with the products C and D. The horizontal axis is labeled "progress of reaction" and represents how far the reaction has proceeded from reactants to products. Figure 18.3 is called an **activation energy diagram**. For the reaction to take place, the molecules must overcome an energy barrier of height E_a; they must collide with sufficient energy to go over the activation energy "hump."

For the reaction depicted in Figure 18.3, the products have a lower energy than the reactants, so $\Delta H_{rxn} < 0$; therefore, Figure 18.3 represents an exothermic reaction (see Section 14-1). In addition, the activation energy for the reverse reaction (going from products to reactants) is as follows: $E_a + |\Delta H_{rxn}|$, where $|\Delta H_{rxn}|$ denotes the absolute magnitude of ΔH_{rxn}, that is, the value of ΔH_{rxn} without its negative sign (remember that ΔH_{rxn} is a negative quantity in Figure 18.3).

EXAMPLE 18-2: Sketch an activation energy diagram for an endothermic reaction. Determine the activation energy in both the forward and reverse directions.

Solution: In an endothermic reaction, the energy of the reactants is lower than that of the products, so $\Delta H_{rxn} > 0$. Therefore, the activation energy diagram looks like the curve in Figure 18.4. The activation energy in the forward direction is E_a, but the activation energy in the reverse reaction is $E_a - |\Delta H_{rxn}|$.

PRACTICE PROBLEM 18-2: If the value of ΔH_{rxn} for a reaction is $-80\ \mathrm{kJ\cdot mol^{-1}}$ and its activation energy is $50\ \mathrm{kJ\cdot mol^{-1}}$, what is the value of the activation energy in the reverse direction?

Answer: $130\ \mathrm{kJ\cdot mol^{-1}}$

The maximum energy (the top of the energy hump) in an activation energy diagram like that in Figures 18.3 and 18.4 has a useful physical interpretation.

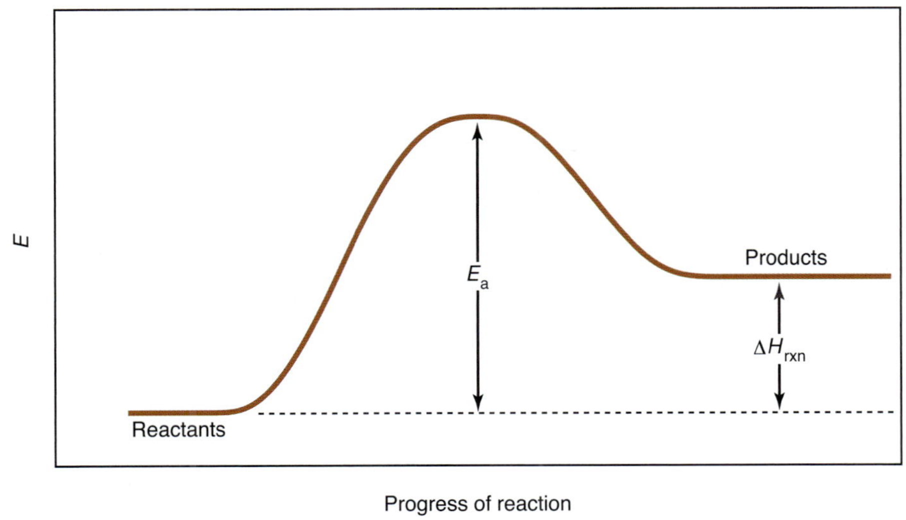

Figure 18.4 Activation energy diagram for an endothermic reaction.

To develop this idea, let's consider the reaction between chloromethane, $CH_3Cl(aq)$, and a hydroxide ion, $OH^-(aq)$, to produce methanol, $CH_3OH(aq)$, and a chloride ion, $Cl^-(aq)$, as described by

$$CH_3Cl(aq) + OH^-(aq) \rightarrow CH_3OH(aq) + Cl^-(aq)$$

This reaction has been determined to take place in one step, and the rate law has been determined experimentally to be

$$\text{rate of reaction} = k[CH_3Cl][OH^-]$$

We can view this reaction as occurring by the following pathway:

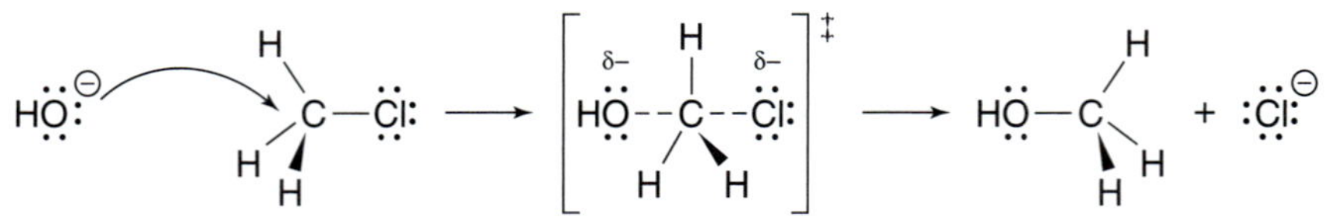

As the hydroxide ion approaches the chloromethane molecule, a carbon-oxygen bond begins to form, while the carbon-chlorine bond begins to break. The species shown in square brackets with a ‡ superscript is called an **activated complex**. An activated complex represents the state with the least amount of additional energy necessary to pass from reactants to products. It is an intrinsically unstable species because it sits at the maximum in the activation energy diagram, as shown in Figure 18.5. An activated complex cannot be isolated or detected by ordinary means, but can be *inferred* from fast spectroscopic data (Figure 18.6). In this regard, an activated complex is completely different from an intermediate species, which may have a lifetime millions of times longer than an activated complex and can often be detected experimentally, or occasionally even isolated. Although activated complexes cannot be isolated, they are useful hypothetical species in describing reaction mechanisms, as we shall see in the next few sections.

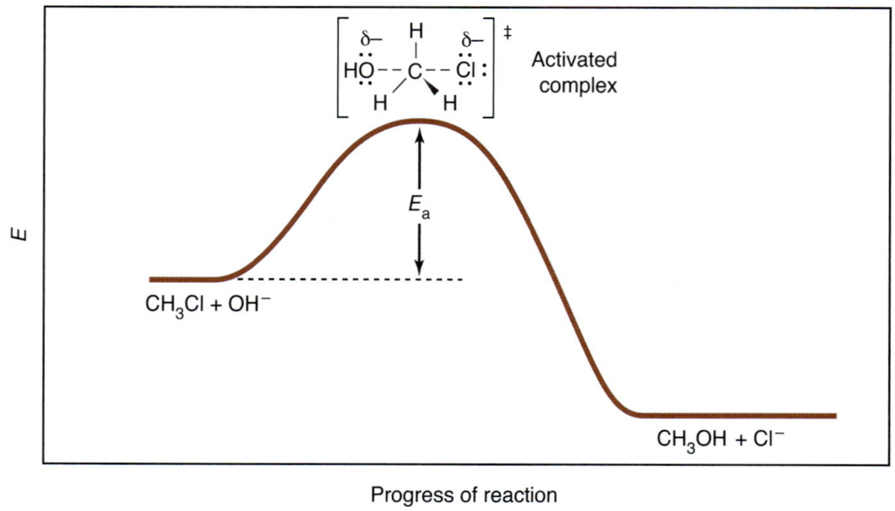

Figure 18.5 Activation energy diagram for the replacement of a chlorine atom by a hydroxyl group in a molecule of chloromethane. The activated complex, which involves both reactant molecules, lies at the top of the energy "hump" and so is an intrinsically unstable species.

Figure 18.6 One of the pioneers in the use of ultrafast spectroscopy to elucidate reaction mechanisms is Ahmed Zewail (1946–). Zeweil was born in Egypt and obtained his BS degree in chemistry and his MS degree in spectroscopy at Alexandria University and his Ph.D. at the University of Pennsylvania in 1974. Upon graduation, he continued his studies at the University of California at Berkeley for two years, where he developed an interest in ultrafast spectroscopy. He then took a position as a professor at Caltech, where he has held the Linus Pauling chair of Chemical Physics since 1990. Dr. Zewail's ground-breaking research has focused on the use of ultrafast lasers to take "stop action" images of reactions in progress on time scales on the order of one femtosecond (10^{-15} s). Because the interaction of two molecules in a chemical reaction occurs on timescales of tens to hundreds of femtoseconds, this work has enabled scientists to "watch" individual reactions occur on a molecular level. In 1999, he was awarded the Nobel Prize in Chemistry for, "showing that it is possible with rapid laser technique to see how atoms in a molecule move during a chemical reaction."

18-3. The Arrhenius Equation Describes the Temperature Dependence of a Reaction-Rate Constant

In almost all cases, an increase in temperature will increase the rate of a reaction. In this section we study the effect of temperature on reaction-rate constants.

Let's go back to Equation 18.3, which expresses the rate of a reaction as the product of three factors: the collision frequency, the fraction of collisions with sufficient kinetic energy, and the fraction of collisions in which the colliding pair of molecules has the appropriate relative orientation. We have seen that two of these terms depend on the temperature. We can use the kinetic theory of gases to evaluate each of the terms. The final result for the reaction-rate constant, k, can be expressed as

$$k = pze^{-E_a/RT} \tag{18.6}$$

In this equation, p, called a **steric factor,** is the fraction of collisions with the correct relative orientations; z is the collision frequency; E_a is the activation energy in units of $J\cdot mol^{-1}$; R is the molar gas constant ($8.3145\ J\cdot mol^{-1}\cdot K^{-1}$); and T is the Kelvin temperature.

The factor $e^{-E_a/RT}$ is the fraction of collisions with energy greater than E_a. This factor varies strongly with temperature, but the quantities p and z do not vary appreciably. We shall denote their product by A and consider it to be a constant. Thus, Equation 18.6 can be written as

$$k = Ae^{-E_a/RT} \tag{18.7}$$

Equation 18.7, which was first proposed in 1889 by the Swedish chemist Svante Arrhenius (Chapter 6 Frontispiece), is called the **Arrhenius equation**. It expresses the temperature dependence of a reaction-rate constant.

If we take the natural logarithm of both sides of Equation 18.7, we obtain

$$\ln k = -\frac{E_a}{RT} + \ln A \tag{18.8}$$

Equation 18.8 enables us to compare the values of rate constants at different temperatures. We can write Equation 18.8 for k_2 at temperature T_2 and for k_1 at temperature T_1.

$$\ln k_2 = -\left(\frac{E_a}{R}\right)\left(\frac{1}{T_2}\right) + \ln A$$

$$\ln k_1 = -\left(\frac{E_a}{R}\right)\left(\frac{1}{T_1}\right) + \ln A$$

Now subtract the second equation from the first:

$$\ln k_2 - \ln k_1 = -\frac{E_a}{R}\left(\frac{1}{T_2} - \frac{1}{T_1}\right) = -\frac{E_a}{R}\left(\frac{T_1 - T_2}{T_1 T_2}\right)$$

Using the property of logarithms that $\ln k_2 - \ln k_1 = \ln(k_2/k_1)$, and rearranging the right-hand side to get rid of the negative sign, we obtain

$$\ln\frac{k_2}{k_1} = \frac{E_a}{R}\left(\frac{T_2 - T_1}{T_1 T_2}\right) \tag{18.9}$$

We can use this equation to calculate an activation energy if the rate constant is known at two different temperatures. For example, consider the reaction described by the equation

$$2\,NO_2(g) \rightarrow 2\,NO(g) + O_2(g)$$

We can determine the activation energy, given that $k = 0.71\ M^{-1}\cdot s^{-1}$ at 604 K and $1.81\ M^{-1}\cdot s^{-1}$ at 627 K. We let $k_1 = 0.71\ M^{-1}\cdot s^{-1}$, $T_1 = 604$ K, $k_2 = 1.81\ M^{-1}\cdot s^{-1}$, and $T_2 =$ 627 K and substitute these values into Equation 18.9 to obtain

$$\ln\left(\frac{1.81\ M^{-1}\cdot s^{-1}}{0.71\ M^{-1}\cdot s^{-1}}\right) = \left(\frac{E_a}{8.3145\ J\cdot mol^{-1}\cdot K^{-1}}\right)\left(\frac{627\ K - 604\ K}{(604\ K)(627\ K)}\right)$$

$$0.94 = \frac{E_a}{1.4 \times 10^5\ J\cdot mol^{-1}}$$

Solving for E_a, we have that

$$E_a = (1.4 \times 10^5\ J\cdot mol^{-1})(0.94) = 1.3 \times 10^5\ J\cdot mol^{-1} = 130\ kJ\cdot mol^{-1}$$

Equation 18.9 also can be used to calculate the value of a rate constant at one temperature if its value is known at some other temperature and if the value of E_a is known.

EXAMPLE 18-3: The activation energy for the reaction described by the equation

$$2\,NO_2(g) + F_2(g) \rightarrow 2\,NO_2F(g)$$

is $E_a = 43.5\ kJ\cdot mol^{-1}$. Estimate the increase in the rate constant of the reaction for a 10-degree increase in temperature from 298 K to 308 K.

Solution: We use Equation 18.9:

$$\ln\frac{k_2}{k_1} = \frac{E_a}{R}\left(\frac{T_2 - T_1}{T_1T_2}\right)$$

Inserting the quantities $E_a = 43.5\ kJ\cdot mol^{-1}$, $T_1 = 298$ K, and $T_2 = 308$ K into this equation yields

$$\ln\frac{k_2}{k_1} = \left(\frac{43.5\times10^3\ J\cdot mol^{-1}}{8.3145\ J\cdot mol^{-1}\cdot K^{-1}}\right)\left(\frac{308\ K - 298\ K}{(298\ K)(308\ K)}\right)$$

Thus,

$$\frac{k_2}{k_1} = e^{0.57} = 1.8$$

Thus, the reaction-rate constant increases by about a factor of two for the 10-degree temperature increase.

PRACTICE PROBLEM 18-3: The major reaction that occurs when an egg is boiled in water is the denaturation of the egg protein. Given that $E_a \approx 42\ kJ\cdot mol^{-1}$ for this process, calculate how much time is required to cook an egg at 92.2°C [the boiling point of water in Vail, Colorado, at about 2400 meters (or 8000 feet) elevation] to the same extent as an egg cooked for 3.0 minutes at 100.0°C (the boiling point of pure water at sea level).

Answer: 4.0 min

18-4. Some Reaction Mechanisms Have a Rate-Determining Step

Let's consider once again the $NO_2(g) + CO(g)$ reaction described by the equation

$$NO_2(g) + CO(g) \rightarrow NO(g) + CO_2(g)$$

We proposed in Section 18-1 that this reaction has a two-step mechanism:

(1) $NO_2(g) + NO_2(g) \xrightarrow{k_1} NO_3(g) + NO(g)$ rate of reaction 1 = $k_1[NO_2]^2$

(2) $NO_3(g) + CO(g) \xrightarrow{k_2} NO_2(g) + CO_2(g)$ rate of reaction 2 = $k_2[NO_3][CO]$

It turns out that the first step is much slower than the second, and because the reaction proceeds through both steps, the slow step acts as a bottleneck. In general, the rate of the overall reaction can be no faster than the rate of the slowest step. In this case, the rate of the overall reaction will be given by the rate of the first step alone. The overall reaction rate for the $NO_2(g) + CO(g)$ reaction will thus be

$$\text{rate of reaction} = k_1[NO_2]^2$$

which, in fact, is the experimentally observed rate law. In effect, the $CO(g)$ molecules have to wait around for $NO_3(g)$ molecules to be produced. Once formed, these are consumed very rapidly by reaction with $CO(g)$. This is why, as we saw in Example 17-5, this reaction is zero order in $CO(g)$.

If one step in a reaction mechanism is much slower than any of the other steps, then that step effectively controls the overall reaction rate and is called the **rate-determining step**. Not all reaction mechanisms have a rate-determining step, but when one does occur, the overall reaction rate is limited by the rate of the rate-determining step.

EXAMPLE 18-4: In Example 18-1 we claimed that the reaction described by

$$2\,NO_2(g) + F_2(g) \rightarrow 2\,NO_2F(g)$$

proceeds by a two-step mechanism. If the rate of the first step is much slower than that of the second step, then deduce the rate law of the overall reaction.

Solution: The first step is a rate-determining step; hence the rate law for the overall reaction is the rate law of the first step,

$$\text{rate of reaction} = k_1[NO_2][F_2]$$

PRACTICE PROBLEM 18-4: A possible mechanism for the reaction described by

$$Fe^{2+}(aq) + HNO_2(aq) + H^+(aq) \rightarrow Fe^{3+}(aq) + NO(g) + H_2O(l)$$

is the two-step mechanism

(1) $Fe^{2+}(aq) + HNO_2(aq) \xrightarrow{k_1} Fe^{3+}(aq) + OH^-(aq) + NO(g)$ (slow)

(2) $H^+(aq) + OH^-(aq) \xrightarrow{k_2} H_2O(l)$ (fast)

If the first step occurs much more slowly than the second step, what is the rate law of the overall reaction?

Answer: rate of reaction = $k_1[Fe^{2+}][HNO_2]$
The reaction is first order in $[Fe^{2+}]$, first order in $[HNO_2]$, and second order overall.

18-5. Some Reaction Mechanisms Have a Fast Equilibrium Step

Until this point we have discussed reactions occurring in one direction only. However, in many cases both the forward and reverse reactions are significant. For example, the elementary reaction described by the equation

$$N_2O_4(g) \underset{k_r}{\overset{k_f}{\rightleftharpoons}} 2\,NO_2(g) \qquad (18.10)$$

proceeds at an appreciable rate in both the forward direction and the reverse direction. We indicate this by writing two arrows in Equation 18.10. Because Equation 18.10 represents an elementary reaction, the forward and reverse reaction-rate laws are:

$$\text{rate of forward reaction} = k_f\,[N_2O_2]$$

and

$$\text{rate of reverse reaction} = k_r[NO_2]^2$$

where k_f and k_r are the forward and reverse reaction-rate constants, respectively.

If the reaction is allowed to proceed, it will eventually reach a state of **equilibrium** where the rate of the forward reaction is equal to the rate of the reverse reaction. (A discussion of equilibrium is the subject of our next chapter.) Under these conditions we can write

$$\text{rate of forward reaction} = \text{rate of reverse reaction} \quad \text{(at equilibrium)} \qquad (18.11)$$

Thus, for the reaction at equilibrium, we have

$$k_f\,[N_2O_2] = k_r[NO_2]^2 \quad \text{(at equilibrium)} \qquad (18.12)$$

Many mechanisms have a step involving an equilibrium and particularly an equilibrium that is assumed to be established very rapidly. For example, let's consider the reaction between nitrogen oxide and oxygen gas described by the equation

$$2\,NO(g) + O_2(g) \rightarrow 2\,NO_2(g) \qquad (18.13)$$

The rate law for this reaction has been determined experimentally to be

$$\text{rate of reaction} = k[NO]^2[O_2]$$

One proposed mechanism for this reaction is

(1) $$2\,NO(g) \underset{k_{-1}}{\overset{k_1}{\rightleftharpoons}} N_2O_2(g) \quad \text{(fast equilibrium)} \qquad (18.14)$$

(2) $$N_2O_2(g) + O_2(g) \xrightarrow{k_2} 2\,NO_2(g) \quad \text{(slow)} \qquad (18.15)$$

where k_1 and k_{-1} are the forward and reverse rate constants for the first step in the mechanism and k_2 is the rate constant for the second step. We can see that this mechanism is consistent with the stoichiometry of the overall chemical reaction by adding steps (1) and (2) and canceling the $N_2O_2(g)$ intermediate to obtain

$$2\,NO(g) + O_2(g) \rightarrow 2\,NO_2(g)$$

which is Equation 18.13.

What we mean by "fast equilibrium" in the first step above is that both the forward and reverse rates are fast, so that the equilibrium is established very rapidly. In particular, for our purposes, when an $N_2O_2(g)$ molecule reacts with an $O_2(g)$ molecule in the second step, the equilibrium in the first step is re-established essentially instantaneously. Thus, during the course of the reaction, we always have the equilibrium condition (Equation 18.11)

$$k_1[\mathrm{NO}]^2 = k_{-1}[\mathrm{N_2O_2}] \tag{18.16}$$

Because the second step (Equation 18.15) is slow, we may assume that it is rate determining and write for the overall reaction

$$\text{rate of reaction} = k_2[\mathrm{N_2O_2}][\mathrm{O_2}] \tag{18.17}$$

This rate law depends on the concentration of the intermediate species, $N_2O_2(g)$. In general, we do not include intermediates in the rate law because these species tend to be short-lived and their concentrations are often difficult to measure. Thus, we want to eliminate $[N_2O_2]$ from Equation 18.17. To do so, we solve Equation 18.16 for $[N_2O_2]$, the concentration of the intermediate species we wish to eliminate,

$$[\mathrm{N_2O_2}] = \left(\frac{k_1}{k_{-1}}\right)[\mathrm{NO}]^2$$

and substitute this result into Equation 18.17 to obtain

$$\text{rate of reaction} = k_2\left(\frac{k_1}{k_{-1}}\right)[\mathrm{NO}]^2[\mathrm{O_2}] = k[\mathrm{NO}]^2[\mathrm{O_2}]$$

where $k = k_2k_1/k_{-1}$. We have combined the three rate constants into one rate constant that we can measure. This rate law agrees with the observed experimental rate law for the reaction.

EXAMPLE 18-5: The reaction described by

$$\mathrm{Hg_2^{2+}}(aq) + \mathrm{Tl^{3+}}(aq) \rightarrow 2\,\mathrm{Hg^{2+}}(aq) + \mathrm{Tl^{+}}(aq) \tag{18.18}$$

has the observed rate law

$$\text{rate of reaction} = k[\mathrm{Hg_2^{2+}}][\mathrm{Tl^{3+}}][\mathrm{Hg^{2+}}]^{-1}$$

One proposed mechanism for this reaction is

(1) $\mathrm{Hg_2^{2+}}(aq) + \mathrm{Tl^{3+}}(aq) \underset{k_{-1}}{\overset{k_1}{\rightleftharpoons}} \mathrm{HgTl^{3+}}(aq) + \mathrm{Hg^{2+}}(aq)$ (fast equilibrium)

(2) $\mathrm{HgTl^{3+}}(aq) \xrightarrow{k_2} \mathrm{Hg^{2+}}(aq) + \mathrm{Tl^{+}}(aq)$ (slow)

Show that this mechanism is consistent with Equation 18.18 and with the observed rate law.

Solution: If we add steps 1 and 2, the intermediate species $HgTl^{3+}(aq)$ cancels and we obtain Equation 18.18. The proposed mechanism says that step 1 is rapidly established, and so using Equation 18.11, we can assume that

$$k_1[\mathrm{Hg_2^{2+}}][\mathrm{Tl^{3+}}] = k_{-1}[\mathrm{HgTl^{3+}}][\mathrm{Hg^{2+}}] \tag{18.19}$$

throughout the course of the reaction. The overall rate of the reaction is controlled by step 2, and so we write

$$\text{rate of reaction} = k_2[\text{HgTl}^{3+}] \tag{18.20}$$

Because $HgTl^{3+}(aq)$ is an intermediate species, we solve Equation 18.19 for $[HgTl^{3+}]$ and substitute the result into Equation 18.20, to give

$$\text{rate of reaction} = \frac{k_1k_2}{k_{-1}}\left(\frac{[\text{Hg}_2^{2+}][\text{Tl}^{3+}]}{[\text{Hg}^{2+}]}\right) = k[\text{Hg}_2^{2+}][\text{Tl}^{3+}][\text{Hg}^{2+}]^{-1}$$

in agreement with the observed rate law. The rate constant for the overall reaction, k, is a composite of the rate constants from the elementary reactions that constitute the mechanism. In an equation, we have $k = k_1k_2/k_{-1}$.

PRACTICE PROBLEM 18-5: An alternative mechanism proposed for the reaction described in Equation 18.13 is

(1) $NO(g) + O_2(g) \underset{k_{-1}}{\overset{k_1}{\rightleftharpoons}} NO_3(g)$ (fast equilibrium)

(2) $NO(g) + NO_3(g) \xrightarrow{k_2} 2NO_2(g)$ (slow)

Determine whether this mechanism is consistent with Equation 18.13 and with the experimentally determined rate law, rate of reaction = $k[NO]^2[O_2]$.

Answer: Yes, this mechanism is consistent with the chemical equation and the experimentally determined rate law.

We see from the preceding Practice Problem that two different mechanisms can be used to explain the reaction described by Equation 18.13. In fact, we can often write several mechanisms that are consistent with any given rate law. For this reason, all we can say is that a given mechanism is *plausible* if it is shown to be consistent with the experimentally determined rate law and other available data.

The determination of reaction mechanisms is an important area of chemical research. Chemists use reaction mechanisms to model complex processes such as those occurring in the atmosphere. For example, a study of the reaction dynamics of ozone led to the discovery that chlorofluorocarbons were destroying ozone in the stratosphere, resulting in a global ban on such compounds. Today scientists use reaction mechanisms to model processes such as the formation of smog over cities and global climate change.

18-6. A Catalyst Is a Substance That Increases the Reaction Rate but Is Not Consumed in the Reaction

The rates of many reactions are increased by catalysts. A catalyst increases the rate of a reaction by providing a different and faster reaction mechanism than that in the absence of the catalyst. For example, the reaction-rate law for the reaction described by

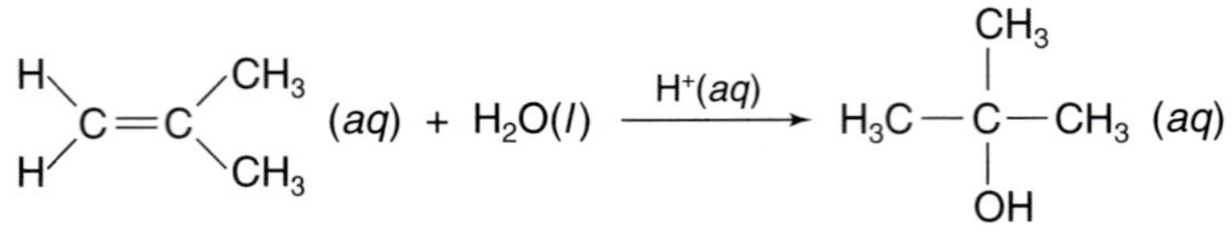

an alkene an alcohol

is

$$\text{rate of reaction} = k[\text{alkene}][\text{H}^+]$$

The solvated hydrogen ion, $H^+(aq)$, does not appear as a reactant in this reaction, but nevertheless the rate of reaction is proportional to $[H^+]$. The $H^+(aq)$ ion presumably facilitates the reaction by attaching to the carbon atom that is bonded to the two hydrogen atoms. Thus, a plausible mechanism for this reaction is as follows:

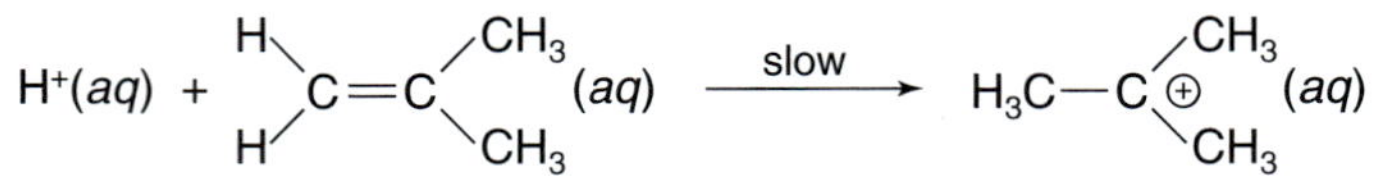

intermediate

The intermediate reacts rapidly with water to form the alcohol:

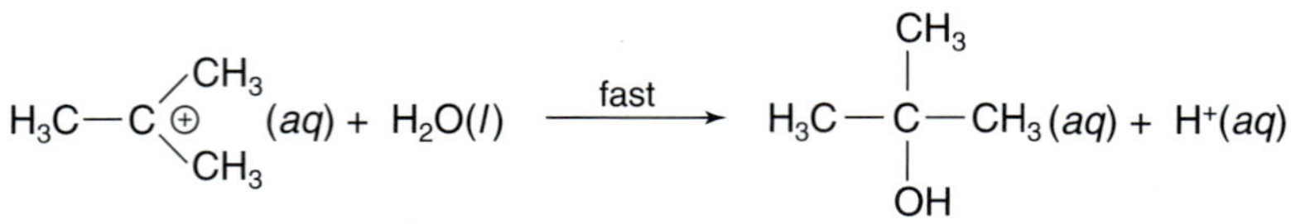

Note that $H^+(aq)$ is regenerated in the second step, so it is not consumed in the reaction.

Because the presence of $H^+(aq)$ increases the rate of reaction, but $H^+(aq)$ is not consumed in the reaction, we say that the reaction is catalyzed by $H^+(aq)$. The catalyst $H^+(aq)$ acts by providing a new reaction pathway with a lower activation energy and thus a larger rate constant. Because the rate of a reaction is proportional to the rate constant, the fact that the rate constant is larger means that the reaction goes faster.

The role of a catalyst is illustrated in Figure 18.7. Note that the catalyzed reaction pathway has a lower activation energy than that of the uncatalyzed reaction. This lower activation energy implies not only that it is easier to go from reactants to products but also that it is easier to go from products to reactants.

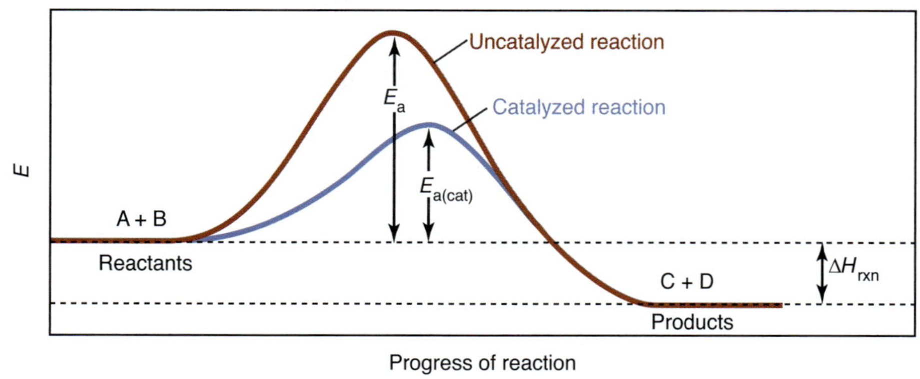

Figure 18.7 A comparison of the activation energies for the uncatalyzed, E_a, and catalyzed, $E_{a(cat)}$, reaction as described by $A + B \rightarrow C + D$. The catalyst lowers the activation energy barrier to the reaction and thereby increases the reaction rate.

TABLE 18.1 Some properties of catalysts

Provide a lower energy pathway for the reaction
Not consumed by the reaction
Increase the rates of both the forward and reverse reactions
Do not change the final amounts of reactants and products produced
Bring a reaction to completion (or equilibrium) faster
Often listed above or below the reaction arrow in the overall chemical equation

Because a catalyst increases the rates of both the forward and reverse reactions, a catalyst does not affect the final amounts of reactants and products. In effect, a catalyst helps to get the job done faster, but the final result is the same. A summary of the properties of catalysts is given in Table 18.1.

Most catalysts act by providing a new reaction pathway (or mechanism) with a lower activation energy than that of the uncatalyzed reaction. Consider the reaction of aqueous cerium(IV) ions, $Ce^{4+}(aq)$, with aqueous thallium(I) ions, $Tl^{+}(aq)$, described by the equation

$$2\,Ce^{4+}(aq) + Tl^{+}(aq) \rightarrow 2\,Ce^{3+}(aq) + Tl^{3+}(aq)$$

This reaction occurs very slowly and its rate law is

$$\text{rate of reaction} = k[Tl^{+}][Ce^{4+}]^2$$

The low reaction rate is thought to be a consequence of the requirement that the reactive event, that is, the simultaneous transfer of two electrons from a $Tl^{+}(aq)$ ion to two different $Ce^{4+}(aq)$ ions, requires that two $Ce^{4+}(aq)$ ions be present simultaneously near a $Tl^{+}(aq)$ ion. Three-body encounters, or **termolecular reactions**, are much less likely to occur than two-body encounters (bimolecular reactions); for this reason, the reaction rate is low. The $2\,Ce^{4+}(aq) + Tl^{+}(aq)$ reaction is catalyzed by $Mn^{2+}(aq)$. The catalytic action of $Mn^{2+}(aq)$ has been attributed to the availability of the Mn^{3+} and Mn^{4+} ionic states, which provide a new reaction pathway involving a three-step sequence of bimolecular elementary reactions as described by the equations

(1) $$Ce^{4+}(aq) + Mn^{2+}(aq) \rightarrow Mn^{3+}(aq) + Ce^{3+}(aq) \quad \text{(slow)}$$

(2) $$Ce^{4+}(aq) + Mn^{3+}(aq) \rightarrow Mn^{4+}(aq) + Ce^{3+}(aq) \quad \text{(fast)}$$

(3) $$Tl^{+}(aq) + Mn^{4+}(aq) \rightarrow Mn^{2+}(aq) + Tl^{3+}(aq) \quad \text{(fast)}$$

The sum of these three equations corresponds to the overall reaction stoichiometry and the $Mn^{2+}(aq)$ is not consumed in the process. The rate law for the $Mn^{2+}(aq)$-catalyzed reaction is determined by the slowest step in the mechanism; thus, the rate law is

$$\text{rate of reaction} = k_{cat}[Ce^{4+}][Mn^{2+}]$$

where k_{cat} is the rate constant for the catalyzed reaction. Note that the rate law for the catalyzed reaction is different from that for the uncatalyzed reaction. Different mechanisms usually (but not always) give rise to different rate laws.

(a)

(b)

Figure 18.8 Platinum metal acts as a heterogeneous catalyst for many reactions. (a) When a cool stream of hydrogen gas flows through a platinum gauze, it reacts with oxygen without the need of a spark or any other initiator because the platinum surface acts as a catalyst. (b) When the catalytic effect of platinum metal on the combustion of hydrogen was discovered in the early 1800s, the process was used to produce cigar lighters, which became very fashionable.

EXAMPLE 18-6: Suppose that a reaction is catalyzed by two different catalysts and that the activation energies of the two catalyzed reactions at 25°C are 30.0 $kJ\cdot mol^{-1}$ and 50.0 $kJ\cdot mol^{-1}$, respectively. Use Equation 18.8 to calculate the ratio of the rate constants for the two catalyzed reactions, assuming that the value of A in Equation 18.8 is the same for both reactions.

Solution: We write Equation 18.8 for one of the catalyzed reactions as

$$\ln k_1 = -\frac{E_{a_1}}{RT} + \ln A_1$$
$$= -\frac{30.0 \times 10^3\ \text{J}\cdot\text{mol}^{-1}}{(8.3145\ \text{J}\cdot\text{mol}^{-1}\cdot\text{K}^{-1})(298\ \text{K})} + \ln A_1 = -12.1 + \ln A_1$$

and for the other catalyzed reaction

$$\ln k_2 = -\frac{50.0 \times 10^3\ \text{J}\cdot\text{mol}^{-1}}{(8.3145\ \text{J}\cdot\text{mol}^{-1}\cdot\text{K}^{-1})(298\ \text{K})} + \ln A_2 = -20.2 + \ln A_2$$

If we subtract the second equation from the first, and assume $\ln A_2$ is equal to $\ln A_1$, then we have

$$\ln k_1 - \ln k_2 = \ln\frac{k_1}{k_2} = 8.1$$

or

$$\frac{k_1}{k_2} = e^{8.1} = 3 \times 10^3$$

Thus, the rate of the reaction is about 3000 times larger with one catalyst than with the other.

PRACTICE PROBLEM 18-6: Calculate the ratio of the rate constants for a catalyzed reaction and for an uncatalyzed reaction at 25.0°C if the activation energy is lowered by 5.00 $kJ\cdot mol^{-1}$ by the introduction of a catalyst (assume that the Arrhenius equation A factors are the same).

Answer: 7.5

In each of the two catalyzed reactions that we have discussed so far, the catalyst is in the same phase as the reaction mixture. We call such a catalyst a **homogeneous catalyst**. In contrast, **heterogeneous catalysts** are in a different phase than the reaction mixture. Many industrial chemical reactions are catalyzed by metal surfaces, which are heterogeneous catalysts (Figure 18.8). For example, platinum and palladium are used as surface catalysts for a variety of reactions, such as the hydrogenation of double bonds:

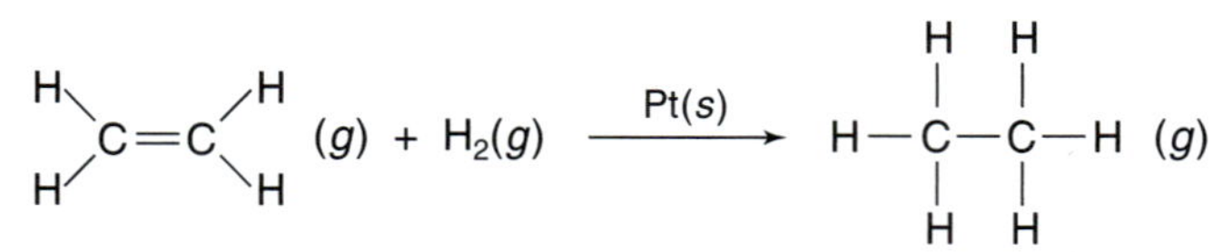

ethene (ethylene) ethane

The first step in this hydrogenation involves the adsorption of hydrogen molecules onto the platinum surface; this step is followed by dissociation of the adsorbed hydrogen molecules into adsorbed hydrogen atoms:

$$H_2(\text{surface}) \rightarrow 2\,H(\text{surface})$$

The adsorbed hydrogen atoms can move around on the platinum surface and eventually react stepwise with adsorbed ethylene molecules, $H_2C{=}CH_2$, to form ethane molecules, H_3CCH_3 according to

$$(1)\quad H_2C{=}CH_2(\text{surface}) + H(\text{surface}) \rightarrow H_2\dot{C}{-}CH_3(\text{surface})$$

$$(2)\quad H_2\dot{C}{-}CH_3(\text{surface}) + H(\text{surface}) \rightarrow H_3C{-}CH_3(g)$$

The ethane produced does not interact strongly with the platinum surface and thus leaves the surface immediately after it is formed.

Platinum metal can also be used to catalyze a wide variety of reactions, including the conversion of $SO_2(g)$ to $SO_3(g)$ during the contact process used to manufacture sulfuric acid (although in practice a vanadium pentoxide catalyst is often used because it is less costly, see Interchapter J). Sulfuric acid is the industrial chemical with the greatest production worldwide (Appendix H). The first step in this process is the adsorption of $O_2(g)$ onto the catalyst's surface. The adsorbed oxygen molecules dissociate into oxygen atoms to form a surface layer of reactive oxygen atoms. The final step involves the rapid reaction of $SO_2(g)$ with surface oxygen atoms (Figure 18.9) as described by

See Interchapter J at www.McQuarrieGeneralChemistry.com.

$$SO_2(g) + O(\text{surface}) \rightarrow SO_3(g)$$

Heterogeneous metal catalysts are used in catalytic converters that facilitate the conversion of $NO(g)$ and $NO_2(g)$ to $N_2(g)$ and the conversion of unburned or partially oxidized hydrocarbons to $CO_2(g)$ and $H_2O(l)$. Platinum metal is also used as a catalyst in fuel cells. Fuel cells are predicted by some to replace gasoline engines in future automobiles (see Chapter 25).

Because catalysts can significantly lower the energy needed to run a reaction and can facilitate reactions that would otherwise be impossible to perform under normal conditions, the development of new catalytic materials represents significant savings for industrial processes. Consequently, the development of new catalytic materials and better understanding of the mechanisms of catalysis are important active areas of chemical and biochemical research.

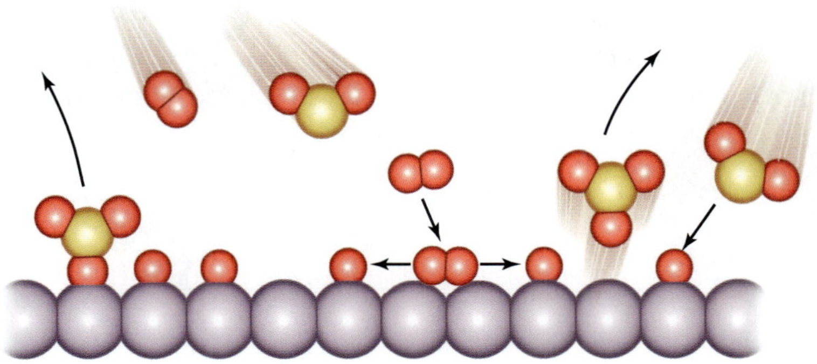

Figure 18.9 Heterogeneous (contact) catalysis by a platinum metal surface. The catalyzed reaction equation is

$$2\,SO_2(g) + O_2(g) \xrightarrow{Pt(s)} 2\,SO_3(g)$$

The platinum surface catalyzes the reaction by causing the dissociation of adsorbed oxygen molecules into oxygen atoms. The surface-bound oxygen atoms then react with $SO_2(g)$ to give $SO_3(g)$.

18-7. The Michaelis-Menten Equation Describes the Rates of Many Enzyme-Catalyzed Reactions

One of the most important classes of catalyzed reactions consists of the biological processes that involve enzymes. **Enzymes** are protein molecules that catalyze specific biochemical reactions. Without enzymes, many of the reactions necessary to sustain life would occur at negligible rates and life as we know it would not exist. The reactant molecule acted upon by an enzyme is called the **substrate**. The region of the enzyme where the substrate reacts is called the **active site**. The active site is only a small part of the enzyme molecule. For example, consider the enzyme hexokinase, which catalyzes the reaction of glucose to glucose 6-phosphate according to the following equation:

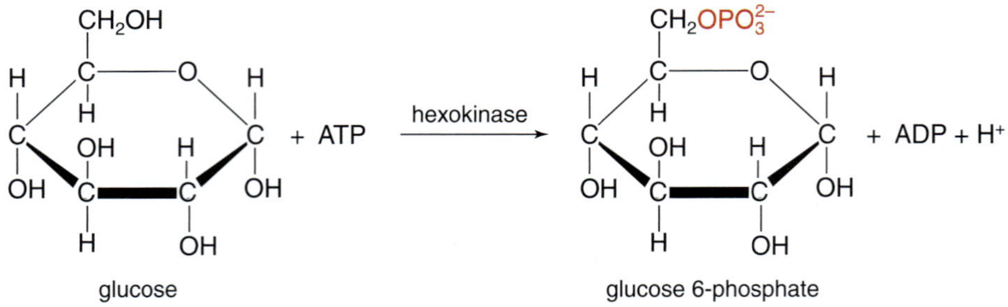

The essence of this reaction is the addition of a phosphate group (red) to the glucose substrate, a process called phosphorylation in biochemistry. ATP (adenosine triphosphate) and ADP (adenosine diphosphate) are molecules that the body uses to store and transfer energy in the form of phosphate bonds.

Figure 18.10a shows a space-filling model of hexokinase. Typical of proteins, it is a very large molecule with a molecular mass of about 50 000. We see that the protein has a cleft at its active site. The glucose molecule enters this cleft, and the protein closes around it (Figure 18.10b).

Experimental studies reveal that the rate law for many enzyme-catalyzed reactions has the form

$$\text{rate of reaction} = R = \frac{\Delta[\mathrm{P}]}{\Delta t} = \frac{k[\mathrm{S}]}{[\mathrm{S}] + K_\mathrm{M}} \tag{18.21}$$

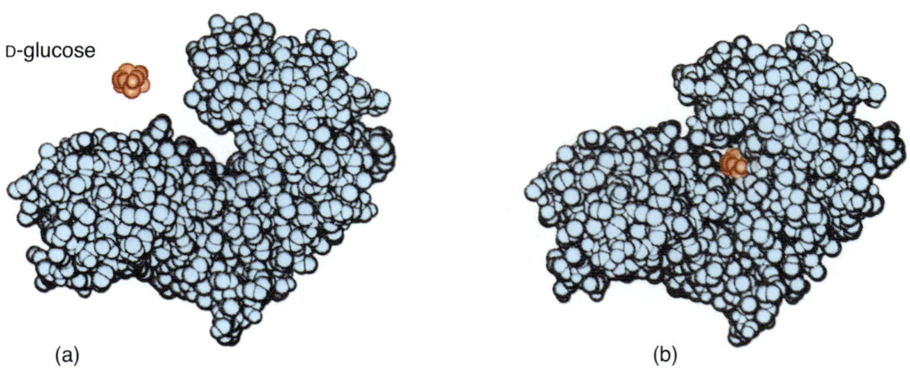

Figure 18.10 Space-filling model of the two conformations of hexokinase. (a) The active site is not occupied. There is a cleft in the enzyme structure that allows the substrate molecule (glucose) to access the active site. (b) The active site is occupied. The enzyme has closed around the substrate.

where [S] is the concentration of the substrate, [P] is the concentration of the product, and k and K_M are constants. Figure 18.11 shows the dependence of the rate, R, of an enzyme-catalyzed reaction on the substrate concentration. At low substrate concentrations, the rate is proportional to [S] and is thus first order with respect to the substrate concentration. As the concentration of substrate is increased, the reaction rate levels off and attains a maximum value, R_{max}. At these concentrations, the rate is independent of [S] and is thus zero order with respect to the substrate concentration. Physically, this means that there is so much substrate relative to the amount of enzyme present that the substrate must wait around for an active site to become available. In this range of substrate concentration, the enzyme is said to be saturated. The reaction rate is controlled by the (constant) amount of enzyme and is independent of the amount of substrate present.

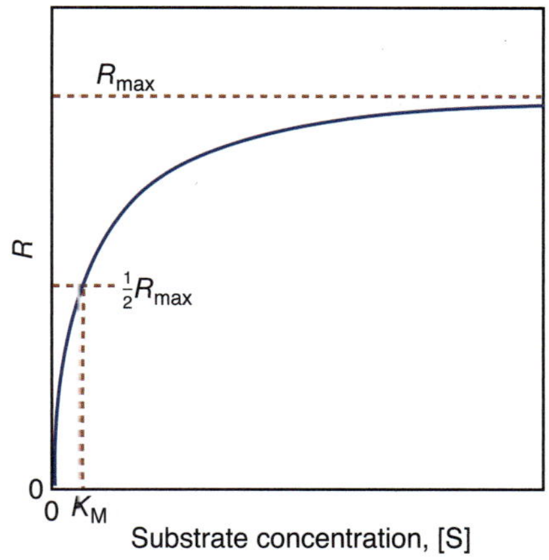

Figure 18.11 The effect of substrate concentration on the rate of an enzyme-catalyzed reaction. The reaction reaches half its maximum rate, $\frac{1}{2}R_{max}$ when [S] = K_M.

Let's see how Equation 18.21 is consistent with Figure 18.11. For small values of [S], we can neglect [S] with respect to K_M in the denominator of Equation 18.21, in which case we have $R = k[S]/K_M$, or that the reaction-rate law is first order in [S]. For large values of [S], we can neglect K_M with respect to [S] to obtain $R = k = R_{max}$. The reaction-rate law is zero order in [S] because the rate of the reaction is equal to a constant. Thus, we see that Equation 18.21 is in accord with Figure 18.11.

A simple mechanism that accounts for this rate law was first proposed by Leonor Michaelis and Maude Menten (Frontispiece) in 1913. The **Michaelis-Menten mechanism** is a two-step process that involves the formation of an intermediate complex between the enzyme and the substrate, denoted by ES (such as that depicted in Figure 18.10b) and can be represented by

(1) $$E + S \underset{k_{-1}}{\overset{k_1}{\rightleftharpoons}} ES$$

(2) $$ES \xrightarrow{k_2} E + P$$

Note that the overall reaction is S → P. Also note that ES, the enzyme-substrate complex, is an intermediate species in this mechanism. Unlike the reactions discussed in Section 18-4 or 18-5, there is no rate-determining step, nor is there a fast equilibrium step indicated in the mechanism. Because ES is an intermediate species and is very reactive, its concentration never builds up and so is small throughout the course of the reaction. This is especially true if the rate of step 2 is greater than that of step 1. Consequentially, the concentration of ES is small and essentially constant during the course of the reaction. Figure 18.12 shows how the concentration of free enzyme [E], substrate [S], enzyme-substrate complex [ES], and product [P] vary with time. You can see that [S] decreases and that [P] increases with time, but the concentration [ES] is essentially constant after a very brief initial period. At this point, the concentration [ES] has reached a steady state, in which ES is consumed as rapidly as it is formed. The amounts of E and ES in Figure 18.12 are greatly exaggerated for clarity. The concentration of enzyme and consequently that of the enzyme-substrate complex are generally very small in enzyme-catalyzed reactions.

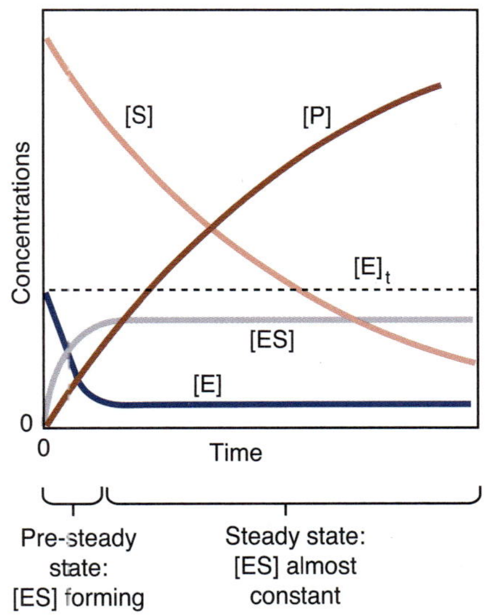

Figure 18.12 The steady state in enzyme kinetics. For a simple enzyme-catalyzed reaction E + S ⇌ ES → E + P, the graph shows how the concentration of free enzyme [E], substrate [S], enzyme-substrate complex [ES], and product [P] vary with time. After a very brief initial period, the concentration of ES reaches a steady state, in which ES is consumed as rapidly as it is formed. Enzyme concentrations are typically very small; the amounts of E and ES in this figure are greatly exaggerated for clarity.

The rate of the overall reaction can be described by the rate of formation of product, which we write as

$$R = \frac{\Delta[P]}{\Delta t} = k_2[ES] \qquad (18.22)$$

Because ES is an intermediate species, we want to express [ES] in this equation in terms of the concentration of the other species.

The rate of formation of the ES complex is given by

$$\text{rate of formation of ES} = k_1[\text{E}][\text{S}]$$

and the rate of breakdown of the ES complex by

$$\text{rate of breakdown of ES} = k_{-1}[\text{ES}] + k_2[\text{ES}] = (k_{-1} + k_2)[\text{ES}]$$

Notice that both the reverse of the reaction in step 1 and the reaction in step 2 contribute to the breakdown of the ES complex in the stated mechanism.

As we have argued above, the concentration of the ES complex is essentially constant during the course of the reaction. For this to be so, the rate of formation of ES must equal its rate of breakdown, and so we write

$$k_1[\text{E}][\text{S}] = (k_{-1} + k_2)[\text{ES}] \quad \text{(steady-state approximation)} \tag{18.23}$$

Equation 18.23 is called the **steady-state approximation** because it is based on the assumption that the rate at which ES is formed is equal to the rate at which it is broken down and thus [ES] maintains a steady state value during the course of the reaction (Figure 18.12). Solving Equation 18.23 for [ES] gives

$$[\text{ES}] = \frac{k_1[\text{E}][\text{S}]}{k_{-1} + k_2} \tag{18.24}$$

We can simplify Equation 18.24 by defining a new constant, called the **Michaelis-Menten constant**,

$$K_\text{M} = \frac{k_{-1} + k_2}{k_1} \tag{18.25}$$

so that Equation 18.24 becomes

$$[\text{ES}] = \frac{[\text{E}][\text{S}]}{K_\text{M}} \tag{18.26}$$

The concentration of free enzyme, [E], in Equation 18.26 is not known; only the total concentration of enzyme, $[\text{E}]_\text{t}$, is known. The total concentration of enzyme is the sum of the concentration of free enzyme, [E], and that of the enzyme-substrate complex [ES] (Figure 18.12). In an equation, we have

$$[\text{E}]_\text{t} = [\text{E}] + [\text{ES}] \tag{18.27}$$

If we solve Equation 18.27 for [E] and substitute the result into Equation 18.26, then we obtain

$$[\text{ES}] = \frac{([\text{E}]_\text{t} - [\text{ES}])[\text{S}]}{K_\text{M}}$$

Solving for [ES] gives

$$[\text{ES}] = \frac{[\text{E}]_\text{t}[\text{S}]}{[\text{S}] + K_\text{M}}$$

Substituting this into Equation 18.22 gives us our final result

$$R = \frac{k_2[\mathrm{E}]_t[\mathrm{S}]}{[\mathrm{S}] + K_\mathrm{M}} \tag{18.28}$$

Observe that Equation 18.28 has the same form as the experimentally observed rate law, Equation 18.21, where $k = k_2[\mathrm{E}]_t$.

When $[\mathrm{S}] >> K_\mathrm{M}$, $[\mathrm{S}] + K_\mathrm{M} \approx [\mathrm{S}]$ and Equation 18.28 becomes

$$R = \frac{k_2[\mathrm{E}]_t[\mathrm{S}]}{[\mathrm{S}]} = k_2[\mathrm{E}]_t \tag{18.29}$$

Because the substrate concentration is so high under these conditions, essentially all the enzyme in the system is present as the ES complex, and so the rate of the reaction is a maximum (in accord with Figure 18.11). Thus, we find that

$$R_\mathrm{max} = k_2[\mathrm{E}]_t \tag{18.30}$$

where R_max is the maximum rate of the reaction. We point out that this expression is zero order in [S], in that the rate of the reaction is equal to a constant. Using this result, Equation 18.28 can be written in the form

$$R = \frac{R_\mathrm{max}[\mathrm{S}]}{[\mathrm{S}] + K_\mathrm{M}} \tag{18.31}$$

Notice that $R = \frac{1}{2}R_\mathrm{max}$ when $K_\mathrm{M} = [\mathrm{S}]$ as shown in Figure 18.11. Thus, we see that the Michaelis-Menten constant, K_M, is the value of [S] for which $R = \frac{1}{2}R_\mathrm{max}$.

When working with experimental data, it is always convenient to deal with straight lines. Equation 18.31 can be cast as a straight line by taking its reciprocal and writing it as

$$\frac{1}{R} = \left(\frac{K_\mathrm{M}}{R_\mathrm{max}}\right)\left(\frac{1}{[\mathrm{S}]}\right) + \frac{1}{R_\mathrm{max}} \tag{18.32}$$

Equation 18.32 is called the **Lineweaver-Burk equation**. If $1/R$ is plotted against $1/[\mathrm{S}]$, we obtain a straight line with an intercept of $1/R_\mathrm{max}$ and a slope of $K_\mathrm{M}/R_\mathrm{max}$. Such a plot, called a **Lineweaver-Burk plot**, is shown in Figure 18.13. The value of R_max depends upon the total enzyme concentration, but the value of K_M is characteristic of the enzyme itself. One of the many uses of the Michaelis-Menten constant is for the quantitative assay of enzyme activity in tissues and in enzyme purification. It also provides a useful index for the analyses of some enzyme regulatory mechanisms. Table 18.2 contains values of K_M for a few enzymes.

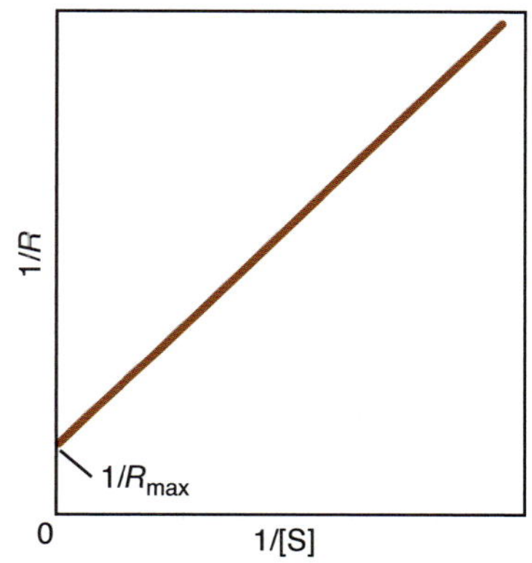

Figure 18.13 A Lineweaver-Burk plot. Here $1/R$ is graphed versus $1/[\mathrm{S}]$ according to Equation 18.32. The slope of this plot is $K_\mathrm{M}/R_\mathrm{max}$ and the intercept is $1/R_\mathrm{max}$.

TABLE 18.2 Values of the Michaelis-Menten constant, K_M, for some enzymes

Enzyme	$K_\mathrm{M}/\mu\mathrm{M}$
acetylcholinesterase	95
catalase	25 000
carbonic anhydrase	12 000
urease	25 000
chymotrypsin	5000
penicillinase	4.9

EXAMPLE 18-7: Penicillin is rendered inactive by an enzyme called penicillinase, which is present in some penicillin-resistant bacteria. The amount of penicillin rendered inactive in one minute in a one-liter solution containing 0.10 micrograms of purified penicillinase was measured as a function of the concentration of penicillin. Determine the value of R_max and K_M for this reaction from the following data.

[penicillin]/μM	[inactivated]/μM·min⁻¹
1.0	0.011
3.0	0.025
5.0	0.034
10.0	0.045
30.0	0.058
50.0	0.061

Solution: The first column of data represents the substrate concentration, [S], and the second column of data represents the corresponding rate of reaction, R, in Equation 18.31. To construct a Lineweaver-Burk plot, we express the data in terms of 1/[S] and $1/R$:

(1/[S])/μM⁻¹	(1/R)/min·μM⁻¹
1.000	90.9
0.333	40.0
0.200	29.4
0.100	22.2
0.033	17.2
0.020	16.4

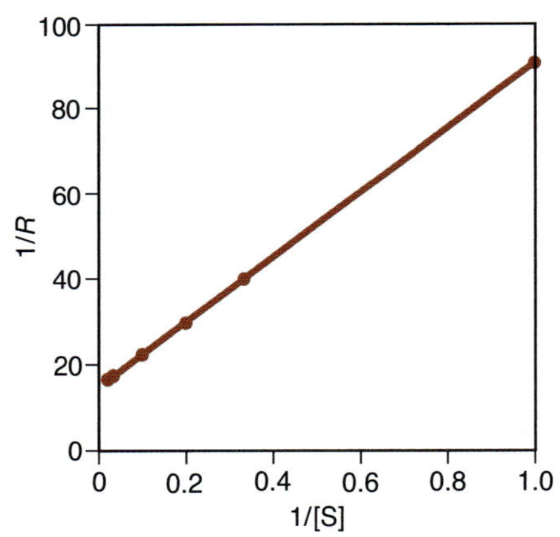

Figure 18.14 A plot of $1/R$ versus 1/[S] for the data in Example 18-7.

Figure 18.14 shows a plot of $1/R$ against 1/[S]. Because the points are quite linear, we can select any two data points to estimate the slope of the line. Choosing the second and third points (arbitrarily), we get

$$\text{slope} = \frac{40.0\ \text{min}\cdot\mu\text{M}^{-1} - 29.4\ \text{min}\cdot\mu\text{M}^{-1}}{0.333\ \mu\text{M}^{-1} - 0.200\ \mu\text{M}^{-1}} = 79.7\ \text{min}$$

We can visually estimate the intercept to be 16 min·μM⁻¹. Using the fact that the intercept is $1/R_{max}$, we find that $R_{max} = 0.063\ \mu\text{M}\cdot\text{min}^{-1}$. The slope is equal to K_M/R_{max}, and so we find that

$$K_M = (79.7\ \text{min})(0.063\ \mu\text{M}\cdot\text{min}^{-1}) = 5.0\ \mu\text{M}$$

PRACTICE PROBLEM 18-7: Carboxypeptidase, an enzyme which is present in the pancreas, produces the amino acid tryptophan (see Interchapter T). Use the following data to evaluate the value of R_{max} and the Michaelis-Menten constant, K_M, for this reaction.

[S]/mM	2.5	5.0	10.0	15.0	20.0
R/mM·s⁻¹	0.024	0.036	0.053	0.060	0.064

Answer: $R_{max} = 0.085\ \text{mM}\cdot\text{s}^{-1}$, $K_M = 6.4$ mM

WWW

See Interchapter T at www.McQuarrieGeneralChemistry.com.

SUMMARY

An elementary reaction is a chemical reaction that occurs in a single step. In this case, the rate law can be deduced from the reaction stoichiometry. Most chemical reactions are not elementary reactions, so their rate laws cannot be deduced from the reaction stoichiometry. The reaction mechanism is the sequence of elementary reactions by which reactants are converted to products. A reaction mechanism that involves more than one elementary reaction necessarily involves reaction intermediates. An intermediate is a species that is formed from the reactants and is involved in the conversion of reactants to products but does not appear in the overall reaction equation.

In order to react, molecules have to collide. However, only those collisions in which the molecules have the correct orientation and have a combined energy equal to or greater than the activation energy for the reaction lead to products. The temperature dependence of a rate constant is given by the Arrhenius equation (Equation 18.7).

Two devices that help us write the rate law for an overall reaction are the possibility of a rate-determining step and that of a fast equilibrium step. If either of these occurs in a mechanism, then it is generally straight forward to write the rate law for the overall reaction.

A catalyst is a species that increases the rate of a reaction but is not consumed as a reactant. A catalyst provides a different and faster reaction pathway than is possible in its absence. A catalyst lowers the activation energy of a reaction but has no effect on the equilibrium concentration of a species.

Enzymes are proteins that catalyze chemical reactions in living systems. Many enzyme-catalyzed reactions can be described by the Michaelis-Menten mechanism. This mechanism invokes the steady-state approximation, where we assume that the concentration of an enzyme-substrate complex remains essentially constant throughout the course of the reaction. Enzyme properties can be obtained from kinetic data using a Lineweaver-Burk plot.

TERMS YOU SHOULD KNOW

elementary reaction *650*
reaction mechanism *650*
intermediate species *650*
bimolecular reaction *651*
unimolecular reaction *651*
collision theory of reaction rates *652*
activation energy, E_a *653*
activation energy diagram *655*
activated complex *656*
steric factor *657*
Arrhenius equation *657*
rate-determining step *660*
equilibrium *661*
termolecular reaction *665*
homogeneous catalyst *666*
heterogeneous catalyst *666*
enzyme *668*
substrate *668*
active site *668*
Michaelis-Menten mechanism *669*
steady-state approximation *670*
Michaelis-Menten constant, K_M *670*
Lineweaver-Burk equation *671*
Lineweaver-Burk plot *671*

EQUATIONS YOU SHOULD KNOW HOW TO USE

$k = Ae^{-E_a/RT}$	(18.7)	(Arrhenius equation)
$\ln k = -\frac{E_a}{RT} + \ln A$	(18.8)	(alternative form of the Arrhenius equation)
$\ln\frac{k_2}{k_1} = \frac{E_a}{R}\left(\frac{T_2 - T_1}{T_1 T_2}\right)$	(18.9)	(alternative form of the Arrhenius equation)
rate of forward reaction = rate of reverse reaction	(18.11)	(definition of equilibrium)
rate of reaction $= R = \frac{\Delta[\text{P}]}{\Delta t} = \frac{k[\text{S}]}{[\text{S}] + K_\text{M}}$	(18.21)	(rate of an enzyme-catalyzed reaction)
$R = \frac{R_\text{max}[\text{S}]}{[\text{S}] + K_\text{M}}$	(18.31)	(steady state rate of an enzyme-catalyzed reaction)
$\frac{1}{R} = \left(\frac{K_\text{M}}{R_\text{max}}\right)\left(\frac{1}{[\text{S}]}\right) + \frac{1}{R_\text{max}}$	(18.32)	(Lineweaver-Burk equation)

PROBLEMS

ARRHENIUS EQUATION

18-1. The decomposition of N_2O_5 takes place in carbon tetrachloride at room temperature according to the equation

$$2\,N_2O_5(soln) \rightarrow 2\,N_2O_4(soln) + O_2(g)$$

The rate constant is $2.35 \times 10^{-4}\ \text{s}^{-1}$ at 293.0 K and $9.15 \times 10^{-4}\text{s}^{-1}$ at 303.0 K. Calculate E_a for this reaction.

18-2. The rate constant for the reaction described by

$$H_2(g) + I_2(g) \rightarrow 2HI(g)$$

was determined to be $0.0234\ \text{M}^{-1}\cdot\text{s}^{-1}$ at 400°C and $0.750\ \text{M}^{-1}\cdot\text{s}^{-1}$ at 500°C. Calculate E_a for this reaction.

18-3. The activation energy for the reaction described by

$$C_4H_8(g) \rightarrow 2\,C_2H_4(g)$$

is 262 kJ·mol^{-1}. At 600.0 K the rate constant is $6.1 \times 10^{-8}\text{s}^{-1}$. What is the value of the rate constant at 800.0 K?

18-4. The activation energy for the decomposition of $N_2O_5(g)$ according to the equation

$$2\,N_2O_5(g) \rightarrow 4\,NO_2(g) + O_2(g)$$

is 102 kJ·mol^{-1}. At 45.0°C the rate constant is $5.0 \times 10^{-4}\text{s}^{-1}$. What is the rate constant at 65.0°C?

18-5. The denaturation of a certain virus is a first-order process with an activation energy of 586 kJ·mol^{-1}. The half-life of the reaction at 29.6°C is 4.5 hours. Calculate the half-life at 37.0°C.

18-6. Cryosurgical procedures involve lowering the body temperature of the patient prior to surgery. Given that the activation energy for the beating of the heart muscle is about 30 kJ, estimate the pulse rate at 22°C. Assume the pulse rate at 37°C (body temperature) to be 75 beats·min^{-1}.

REACTION MECHANISMS

18-7. Write the rate law for each of the following elementary reaction equations:

(a) $N_2O(g) + O(g) \rightarrow 2NO(g)$

(b) $2O_2(g) \rightarrow O(g) + O_3(g)$

(c) $ClCO(g) + Cl_2(g) \rightarrow Cl_2CO(g) + Cl(g)$

18-8. Write the rate law for each of the following elementary reaction equations:

(a) $K(g) + HCl(g) \rightarrow KCl(g) + H(g)$

(b) $H_2O_2(g) \rightarrow H_2O(g) + O(g)$

(c) $2O_2(g) + Cl(g) \rightarrow ClO(g) + O_3(g)$

(d) $NO_3(g) + CO(g) \rightarrow NO_2(g) + CO_2(g)$

18-9. For each of the elementary reaction equations in Problem 18-7, state the overall order of the reaction and the order with respect to each reactant. Classify the reaction as unimolecular, bimolecular, or termolecular.

18-10. For each of the elementary reaction equations in Problem 18-8, state the overall order of the reaction and the order with respect to each reactant. Classify the reaction as unimolecular, bimolecular, or termolecular.

18-11. The reaction of carbon dioxide with hydroxide ion in aqueous solution described by

$$CO_2(aq) + 2OH^-(aq) \rightarrow CO_3^{2-}(aq) + H_2O(l)$$

is postulated to occur according to the mechanism

(1) $CO_2(aq) + OH^-(aq) \rightarrow HCO_3^-$ (slow)

(2) $HCO_3^-(aq) + OH^-(aq) \rightarrow CO_3^{2-}(aq) + H_2O(l)$ (fast)

The rate law for the disappearance of $CO_2(aq)$ was found experimentally to be

$$\text{rate of reaction} = k[CO_2][OH^-]$$

Is this mechanism consistent with the observed rate law? Justify your answer.

18-12. The reaction described by

$$2NO_2(g) \rightarrow 2NO(g) + O_2(g)$$

is postulated to occur via the mechanism

(1) $NO_2(g) + NO_2(g) \rightarrow NO(g) + NO_3(g)$ (slow)

(2) $NO_3(g) \rightarrow NO(g) + O_2(g)$ (fast)

The rate law for the reaction is

$$\text{rate of reaction} = k[NO_2]^2$$

Is this mechanism consistent with the observed rate law? Justify your answer.

18-13. For the reaction mechanism given in Problem 18-11, classify each step as unimolecular, bimolecular, or termolecular and identify the reaction intermediate(s).

18-14. For the reaction mechanism given in Problem 18-12, classify each step as unimolecular, bimolecular, or termolecular and identify the reaction intermediate(s).

18-15. A proposed mechanism for the reaction described by

$$2NO(g) + H_2(g) \rightarrow N_2O(g) + H_2O(g)$$

is

(1) $2NO(g) \underset{k_{-1}}{\overset{k_1}{\rightleftharpoons}} N_2O_2(g)$ (fast equilibrium)

(2) $N_2O_2(g) + H_2(g) \xrightarrow{k_2} N_2O(g) + H_2O(g)$ (slow)

Show that this mechanism is consistent with the observed rate law,

$$\text{rate of reaction} = k[NO]^2[H_2]$$

18-16. The experimentally determined rate law for the reaction described by the equation

$$2NO_2(g) + Br_2(g) \rightarrow 2NO_2Br(g)$$

is

$$\text{rate of reaction} = k[NO_2][Br_2]$$

One proposed mechanism for this reaction is

(1) $NO_2(g) + Br_2(g) \underset{k_{-1}}{\overset{k_1}{\rightleftharpoons}} NO_2Br_2(g)$ (fast equilibrium)

(2) $NO_2Br_2(g) \xrightarrow{k_2} NO_2Br(g) + Br(g)$ (slow)

(3) $Br(g) + NO_2(g) \xrightarrow{k_3} NO_2Br(g)$ (fast)

Is this mechanism consistent with the experimentally determined rate law?

18-17. The reaction described by

$$2NO(g) + H_2(g) \rightarrow N_2O(g) + H_2O(g)$$

has an experimentally determined rate law of

$$\text{rate of reaction} = k[NO]^2[H_2]$$

A proposed mechanism for this reaction is

$$\text{(1)}\ 2\,NO(g) \underset{k_{-1}}{\overset{k_1}{\rightleftharpoons}} N_2O_2(g) \quad \text{(fast equilibrium)}$$

$$\text{(2)}\ N_2O_2(g) + H_2(g) \xrightarrow{k_2} N_2O(g) + H_2O(g) \quad \text{(slow)}$$

Show that this mechanism is consistent with the observed rate law.

18-18. The reaction described by

$$H_2(g) + I_2(g) \rightarrow 2\,HI(g)$$

has an experimentally determined rate law of

$$\text{rate of reaction} = k[H_2][I_2]$$

Some proposed mechanisms for this reaction are given below:

Mechanism a.

$$\text{(1)}\ H_2(g) + I_2(g) \xrightarrow{k_1} 2\,HI(g) \quad \text{(one-step reaction)}$$

Mechanism b.

$$\text{(1)}\ I_2(g) \underset{k_{-1}}{\overset{k_1}{\rightleftharpoons}} 2\,I(g) \quad \text{(fast equilibrium)}$$

$$\text{(2)}\ H_2(g) + 2\,I(g) \xrightarrow{k_2} 2\,HI(g) \quad \text{(slow)}$$

Mechanism c.

$$\text{(1)}\ I_2(g) \underset{k_{-1}}{\overset{k_1}{\rightleftharpoons}} 2\,I(g) \quad \text{(fast equilibrium)}$$

$$\text{(2)}\ I(g) + H_2(g) \underset{k_{-2}}{\overset{k_2}{\rightleftharpoons}} H_2I(g) \quad \text{(fast equilibrium)}$$

$$\text{(3)}\ H_2I(g) \xrightarrow{k_3} HI(g) + I(g) \quad \text{(slow)}$$

(a) Which of these mechanisms are consistent with the observed rate law?

(b) In 1967 J. H. Sullivan showed that this reaction was dramatically catalyzed by light when the energy of the light was sufficient to break the I–I bond in an I_2 molecule. Which mechanism(s) are consistent with both the rate law and this additional observation? Justify your answer.

18-19. The rate law for the reaction between $CO(g)$ and $Cl_2(g)$ to form phosgene, $Cl_2CO(g)$, described by

$$Cl_2(g) + CO(g) \rightarrow Cl_2CO(g)$$

is

$$\text{rate of reaction} = k[Cl_2]^{3/2}[CO]$$

Show that the following mechanism is consistent with this rate law

$$\text{(1)}\ Cl_2(g) \underset{k_{-1}}{\overset{k_1}{\rightleftharpoons}} 2\,Cl(g) \quad \text{(fast equilibrium)}$$

$$\text{(2)}\ Cl(g) + CO(g) \underset{k_{-2}}{\overset{k_2}{\rightleftharpoons}} ClCO(g) \quad \text{(fast equilibrium)}$$

$$\text{(3)}\ ClCO(g) + Cl_2(g) \xrightarrow{k_3} Cl_2CO(g) + Cl(g) \quad \text{(slow)}$$

Express k in terms of the rate constants for the individual steps of the reaction mechanism.

18-20. Nitramide, O_2NNH_2, decomposes in water according to the chemical equation

$$O_2NNH_2(aq) \rightarrow N_2O(g) + H_2O(l)$$

The experimentally determined rate law for this reaction is

$$\text{rate of reaction} = k\frac{[O_2NNH_2]}{[H^+]}$$

A proposed mechanism for this reaction is

$$\text{(1)}\ O_2NNH_2(aq) \underset{k_{-1}}{\overset{k_1}{\rightleftharpoons}} O_2NNH^-(aq) + H^+(aq) \quad \text{(fast equilibrium)}$$

$$\text{(2)}\ O_2NNH^-(aq) \xrightarrow{k_2} N_2O(g) + OH^-(aq) \quad \text{(slow)}$$

$$\text{(3)}\ H^+(aq) + OH^-(aq) \xrightarrow{k_3} H_2O(l) \quad \text{(fast)}$$

Is this mechanism consistent with the observed rate law? If so, what is the relationship between the observed value of k and the rate constants for the individual steps of the mechanism?

18-21. The decomposition of perbenzoic acid in water as described by

$$2\,C_6H_5CO_3H(aq) \rightarrow 2\,C_6H_5CO_2H(aq) + O_2(g)$$

is proposed to occur by the following mechanism:

$$\text{(1)}\ C_6H_5CO_3H(aq) \underset{k_{-1}}{\overset{k_1}{\rightleftharpoons}} C_6H_5CO_3^-(aq) + H^+(aq) \quad \text{(fast equilibrium)}$$

$$\text{(2)}\ C_6H_5CO_3H(aq) + C_6H_5CO_3^-(aq) \xrightarrow{k_2} C_6H_5CO_2H(aq) + C_6H_5CO_2^-(aq) + O_2(g) \quad \text{(slow)}$$

$$\text{(3)}\ C_6H_5CO_2^-(aq) + H^+(aq) \xrightarrow{k_3} C_6H_5CO_2H(aq) \quad \text{(fast)}$$

Derive an expression for the rate of formation of $O_2(g)$ in terms of the reactant concentration and $[H^+]$.

18-22. An alternative mechanism for the chemical reaction described by

$$Cl_2(g) + CO(g) \rightarrow Cl_2CO(g)$$

(see Problem 18-19) is

(1) $Cl_2(g) \underset{k_{-1}}{\overset{k_1}{\rightleftharpoons}} 2\,Cl(g)$ (fast equilibrium)

(2) $Cl(g) + Cl_2(g) \underset{k_{-2}}{\overset{k_2}{\rightleftharpoons}} Cl_3(g)$ (fast equilibrium)

(3) $Cl_3(g) + CO(g) \xrightarrow{k_3} Cl_2CO(g) + Cl(g)$ (slow)

Show that this mechanism also gives the observed rate law. How would you go about determining whether this mechanism or the one in Problem 18-19 is correct?

CATALYSIS

18-23. The aqueous decomposition of hydrogen peroxide in the presence of $Br^-(aq)$ and $H^+(aq)$ occurs according to the equation

$$2\,H_2O_2(aq) \rightarrow 2\,H_2O(l) + O_2(g)$$

and has the rate law

$$\text{rate of reaction} = k[H_2O_2][H^+][Br^-]$$

(a) Identify the catalyst(s) for the reaction.

(b) What is the overall order of the reaction?

(c) Suppose $[H_2O_2]_0 = 0.10$ M, $[H^+]_0 = 1.00 \times 10^{-3}$ M, and $[Br^-]_0 = 1.00 \times 10^{-3}$ M are the initial concentrations of these species in the reaction. Sketch the concentrations of these three species as a function of time given that $k = 1.0 \times 10^3\ M^{-2}{\cdot}s^{-1}$.

18-24. From the following initial-rate data for the decomposition of $H_2O_2(aq)$ according to the equation

$$2\,H_2O_2(aq) \rightarrow 2\,H_2O(l) + O_2(g)$$

(a) determine the rate law.

(b) Pick out the catalyst(s), if any.

$[H_2O_2]_0$/M	$[I^-]_0$/M	$[H^+]_0$/M	Initial rate of formation of O_2/M·s^{-1}
0.20	0.010	0.010	2.0×10^{-3}
0.40	0.010	0.010	4.0×10^{-3}
0.20	0.020	0.010	8.0×10^{-3}
0.20	0.020	0.020	1.6×10^{-2}

18-25. Figure 18.9 outlines the platinum-catalyzed mechanism for the reaction described by

$$2\,SO_2(g) + O_2(g) \rightarrow 2\,SO_3(g)$$

It is observed that, except for very low pressures, the rate of the catalyzed reaction is independent of the pressures of $SO_2(g)$ and $O_2(g)$. That is, the rate law is zero-order in both reactants:

$$\text{rate of reaction} = k$$

Explain how the mechanism outlined in Figure 18.9 leads to this rate law.

18-26. The rate of decomposition of gases on hot metal surfaces often is found to be independent of the concentration of the available gas; that is, the rate law is zero-order in the reactant gas, or rate of reaction = k. Such a situation is found for the catalytic decomposition of ammonia on tungsten,

$$2\,NH_3(g) \xrightarrow{W(s)} N_2(g) + 3\,H_2(g)$$

How do you explain these observations in mechanistic terms?

ENZYME KINETICS

18-27. The following data have been collected for a certain enzyme-catalyzed reaction:

[S]/µmol·L^{-1}	5.0	10.0	20.0	50.0	100.0	200.0
R/µmol·L^{-1}·min^{-1}	22	39	65	102	120	135

Use these data to determine the values of R_{max} and K_M, the Michaelis-Menten constant of the enzyme.

18-28. The following data have been collected for a certain enzyme-catalyzed reaction:

[S]/µmol·L^{-1}	3.0	5.0	10.0	30.0	90.0
R/µmol·L^{-1}·min^{-1}	10.4	14.5	22.5	33.8	40.5

Use these data to determine the values of R_{max} and K_M, the Michaelis-Menten constant of the enzyme.

18-29. An enzyme called carbonic anhydrase catalyzes both the forward and reverse reactions for the hydration of $CO_2(aq)$ according to

$$H_2O(l) + CO_2(aq) \rightleftharpoons HCO_3^-(aq) + H^+(aq)$$

Carbon dioxide is produced in tissue as one of the final products of respiration. It then diffuses into the blood system, where it is converted to the hydrogen carbonate ion by carbonic anhydrase. The reverse reactions occur in the lungs, where $CO_2(g)$ is expelled. Use the following data to determine the values of R_{max} and K_M, the Michaelis-Menten constant.

(*Hint:* Notice that each row of data is given in different units of concentration.)

$[CO_2]/\text{mmol}\cdot\text{L}^{-1}$	1.25	2.50	5.00	20.00
$R/\mu\text{mol}\cdot\text{L}^{-1}\cdot\text{s}^{-1}$	27.8	50.0	83.3	166

18-30. The hydrolysis of sucrose to glucose and fructose is catalyzed by an enzyme called invertase. Use the following data to determine the value of R_{max} and K_M, the Michaelis-Menten constant. (*Hint:* Notice that each row of data is given in different units of concentration).

$[S]/10^{-2}\ \text{mol}\cdot\text{L}^{-2}$	3.57	5.88	8.33	15.4	22.2
$R/\mu\text{mol}\cdot\text{L}^{-1}\cdot\text{min}^{-1}$	7.04	8.93	10.5	12.3	13.0

18-31. The enzyme fumarase catalyzes the conversion of fumarate to malate according to

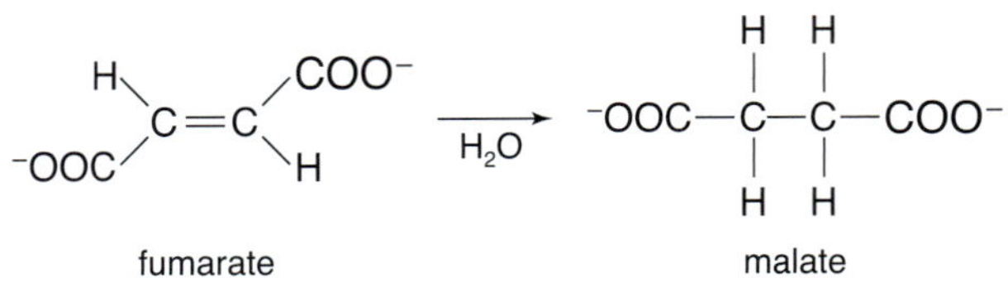

Using the following data, determine the value of R_{max} and K_M, the Michaelis-Menten constant.

$[S]/\mu\text{mol}\cdot\text{L}^{-1}$	1.0	2.0	5.0	10.0	20.0
$R/10^2\ \mu\text{mol}\cdot\text{L}^{-1}\cdot\text{s}^{-1}$	2.6	4.3	7.2	9.3	10.8

18-32. The enzyme catalase catalyzes the decomposition of $H_2O_2(aq)$ to $H_2O(l)$ and $O_2(g)$. Using the following data, determine the value of R_{max} and K_M, the Michaelis-Menten constant.

$[H_2O_2]_0/\text{mmol}\cdot\text{L}^{-1}$	1.0	2.0	5.0
$R/\text{mmol}\cdot\text{L}^{-1}\cdot\text{s}^{-1}$	1.38	2.67	6.00

18-33. We define a quantity called the *turnover number* to be the maximum number of substrate molecules that can be converted into product molecules per unit time by an enzyme molecule. The concentration of enzyme active sites is not necessarily equal to the concentration of enzyme molecules because some enzyme molecules have more than one active site. If the enzyme molecule has one active site, the turnover number is given by

$$\text{turnover number} = \frac{R_{max}}{[E]_t} = k_2$$

(see Equation 18.30). If the enzyme molecule has more than one active site, then we multiply $[E]_t$ by the number of active sites to determine its effective concentration.

Determine the value of the turnover number of carbonic anhydrase using the data given in Problem 18-29 and given that $[E]_t = 2.32\ \text{nmol}\cdot\text{L}^{-1}$. Carbonic anhydrase has a single active site.

18-34. The turnover number is defined in the previous problem. Using the data given in Problem 18-30 and $[E]_t = 1.1\ \mu\text{mol}\cdot\text{L}^{-1}$, determine the value of the turnover number of the enzyme molecule invertase, which has a single active site.

18-35. The turnover number is defined in Problem 18-33. Using the data given in Problem 18-31 and $[E]_t = 1.0\ \mu\text{mol}\cdot\text{L}^{-1}$, determine the value of the turnover number of the enzyme molecule fumarase, which has a single active site.

18-36. The turnover number is defined in Problem 18-33. Using the data given in Problem 18-32 and $[E]_t = 4.0\ \text{nmol}\cdot\text{L}^{-1}$, determine the value of the turnover number of the enzyme molecule catalase, which has a single active site.

ADDITIONAL PROBLEMS

18-37. What is meant by an "elementary reaction equation"? What information can we gain from an elementary reaction equation that we cannot gain from the overall reaction equation? Can an elementary reaction equation and the overall reaction equation ever be the same?

18-38. Do all collisions with energies greater than the activation energy result in a reaction?

18-39. An experimenter writes a mechanism that is consistent with the experimentally determined rate law for a reaction and predicts the existence of several intermediate species that are later confirmed experimentally. Is this evidence sufficient to prove that the experimenter's proposed mechanism is correct?

18-40. Discuss the difference between a reaction intermediate and an activated complex.

18-41. Although the combustion of gasoline is a highly exothermic reaction, gasoline may be stored indefinitely in the presence of oxygen at room temperature. Why is this?

18-42. Correct the mistakes in the following student's definition of a catalyst: "A catalyst increases the yield of a reaction by lowering the activation energy, but it is not used up because it does not participate in the reaction."

18-43. Can a catalyst lower the enthalpy change for a reaction?

18-44. What is a zero-order reaction?

18-45. Under what conditions does an enzyme-catalyzed reaction become zero-order?

18-46. A student wishes to study the kinetics of fruit ripening for a class project. She chooses bananas for her study and defines "ripening" as the day on which 50% or more of the skin turns brown. How might she conduct an experiment to determine the activation energy for this process?

18-47. Some reactions have no activation energy, that is, $E_a = 0$. What would the rate of such a reaction depend on?

18-48. Many recombination reactions between two free radicals, such as

$$2\,CH_3(g) \rightarrow C_2H_6(g)$$

proceed at the diffusion-controlled limit; that is, essentially every collision leads to a reactive event. Why do you think this is so? (Explain in terms of chemical bonds.)

18-49. The rapid reaction between nitrogen oxide and oxygen gases described by

$$2\,NO(g) + O_2(g) \rightarrow 2\,NO_2(g)$$

has an experimentally determined rate law:

$$\text{rate of reaction} = k[NO]^2[O_2]$$

Although the reaction stoichiometry is consistent with the observed rate law, why is it unlikely that this reaction occurs in a single step?

18-50. For the reaction rate law described in the previous problem, indicate whether each of the following would *increase, decrease,* or *have no effect* on the value of the rate constant, k, and the overall reaction rate.

(a) Increasing the concentration of $NO(g)$.

(b) Adding a catalyst.

(c) Decreasing the concentration of $O_2(g)$.

(d) Decreasing the temperature.

18-51. The activation energy for the reaction,

$$N_2(g) + 3H_2(g) \rightarrow 2\,NH_3(g)$$

is $E_a = 940\ \text{kJ}\cdot\text{mol}^{-1}$.

(a) Use this and the data from Table 14.3 to calculate the activation energy of the reverse reaction.

(b) Make a sketch of the activation energy diagram for this reaction. Label the reactants, products, E_a(forward), E_a(reverse), and ΔH_{rxn} on your sketch. What does the top of the hump in your diagram represent? Is this an exothermic or endothermic reaction?

(c) This reaction is catalyzed in the presence of an iron surface, which lowers the forward activation energy to about 80 kJ·mol^{-1}. Add a dashed line representing the overall pathway for the catalyzed reaction to your diagram. What is the magnitude of the activation energy for the catalyzed reaction in the reverse direction (going from products to reactants)?

18-52. A surface-catalyzed decomposition reaction that is known to be zero-order has a rate constant $k = 7.4 \times 10^{-1}\ \text{Torr}\cdot\text{s}^{-1}$. If the pressure of the reactant at the start of the reaction is 275 Torr, how much time will it take for half of the reactant to decompose? The integrated zero-order rate law for the reaction, $A \rightarrow$ Products, is $P_A = P_A^\circ - kt$.

18-53. Equation 18.8 says that a plot of ln k versus $1/T$ is a straight line. If we compare our results with the algebraic equation for a straight line, $y = mx + b$, then we see that the slope, m, and the intercept, b, are given by

$$m = -\frac{E_a}{R} \qquad b = \ln A$$

Thus, from a plot of ln k versus $1/T$, we can determine the activation energy of the reaction from the slope of the line. The rate constants at several different temperatures for the reaction

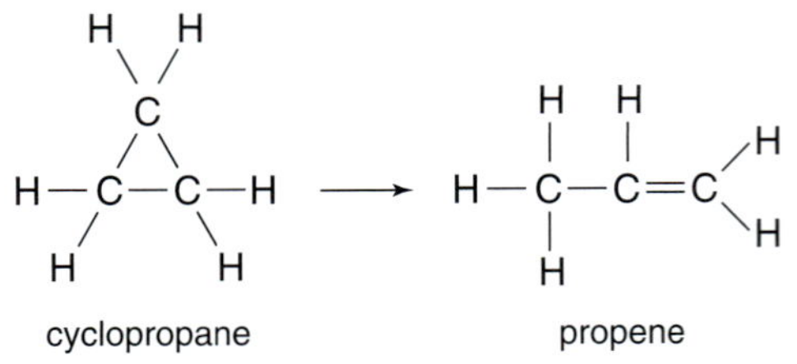

are given below.

t/°C	k/10^{-4} s^{-1}
470	1.10
485	2.61
500	5.70
510	10.2
519	16.4
530	28.6

Determine the value of the activation energy for this reaction by plotting ln k versus $1/T$.

18-54. Given the following rate-constant data on the decomposition of $NO_2(g)$ according to the chemical equation

$$2\,NO_2(g) \rightarrow 2\,NO(g) + O_2(g)$$

Plot ln k versus $1/T$ (see previous problem).

T/K	k/$M^{-1}\cdot s^{-1}$
600	0.70
625	1.83
650	4.46
700	21.8

Use your plot to estimate the value of k at 500 K and calculate the value of E_a for the reaction.

18-55. The decomposition of dinitrogen pentoxide is a first-order process. The temperature dependence of the observed rate constant is given in the following table:

t/°C	k/10^{-5} s^{-1}
0	0.0787
25	3.46
45	49.8
65	487

Plot ln k versus $1/T$ and calculate the value of E_a for the reaction. Use your plot to estimate the value of k at 50°C (see Problem 18-53).

18-56. The rate constant for the low pressure decomposition of $N_2O_4(g)$ is given in a table of reference data as

$$k = 3.3 \times 10^{-7}\ \text{cm}^3\cdot\text{molecule}^{-1}\cdot\text{s}^{-1}\ e^{-46.6\ \text{kJ}\cdot\text{mol}^{-1}/RT}$$

What is the order of this reaction? What is the activation energy for this decomposition? If the initial pressure of $N_2O_4(g)$ is 50.0 mbar at 250 K, calculate the pressure of $N_2O_4(g)$ remaining after 10.0 milliseconds. Assume that any reverse reactions are insignificant under the experimental conditions.

18-57. The rate of decomposition of acetaldehyde, $CH_3CHO(g)$, into $CH_4(g)$ and $CO(g)$ in the presence of $I_2(g)$ at 800 K follows the rate law

$$\text{rate of reaction} = k[CH_3CHO][I_2]$$

The decomposition is believed to occur by the following two-step mechanism:

(1) $CH_3CHO(g) + I_2(g) \rightarrow CH_3I(g) + HI(g) + CO(g)$

(2) $CH_3I(g) + HI(g) \rightarrow CH_4(g) + I_2(g)$

(a) What is the catalyst for the reaction?

(b) Which step in the proposed mechanism is most likely the rate-limiting step?

18-58. The reaction described by the equation

$$H_2(g) + Br_2(g) \rightarrow 2\,HBr(g)$$

has a fractional dependence on bromine. The observed rate law is

$$\text{rate of reaction} = k[H_2][Br_2]^{1/2}$$

One proposed mechanism for this reaction is

(1) $Br_2(g) \underset{k_{-1}}{\overset{k_1}{\rightleftharpoons}} 2\,Br(g)$ (fast equilibrium)

(2) $Br(g) + H_2(g) \xrightarrow{k_2} HBr(g) + H(g)$ (slow)

(3) $H(g) + Br_2(g) \xrightarrow{k_3} HBr(g) + Br(g)$ (fast)

(a) Show that the observed rate law is consistent with the proposed mechanism.

(b) What are the units of the rate constant for this reaction?

(c) Is it possible that this reaction occurs in a single-step mechanism?

18-59. Use the data in Tables 9.1 and 9.2 to predict which of the following dissociation reactions will have the greater activation energy:

(1) $F_2(g) \rightarrow 2F(g)$

(2) $H_2(g) \rightarrow 2H(g)$

18-60. As a rule of thumb, the rate of many reactions doubles around room temperature for a 10°C increase in the temperature. Calculate the value of the activation energy if a temperature increase from 20°C to 30°C doubles the rate constant.

18-61. Hydrogen peroxide decomposes spontaneously to yield water and oxygen gas according to the reaction equation,

$$2H_2O_2(aq) \rightarrow 2H_2O(l) + O_2(g)$$

The activation energy for this reaction is about 75 kJ·mol^{-1}. In the presence of a metal catalyst the activation energy is lowered to about 49 kJ·mol^{-1}.

(a) At what temperature would the non-catalyzed reaction need to be run to have a rate equal to that of the metal-catalyzed reaction at 25°C?

(b) The enzyme catalase (found in blood) lowers the activation energy even further to about 8 kJ·mol^{-1}. At what temperature would the non-catalyzed reaction have to be run to have a rate equal to the enzyme-catalyzed reaction?

18-62. Some values of the rate constant versus temperature for the reaction

$$N_2O_5(g) \rightarrow 2NO_2(g) + \tfrac{1}{2}O_2(g)$$

are

t/°C	k/s^{-1}
24.5	3.47×10^{-5}
34.5	1.35×10^{-4}
45.3	4.17×10^{-4}
54.7	1.51×10^{-3}
64.7	4.90×10^{-3}

Show that a plot of ln k versus $1/T$ is linear and determine the value of the activation energy from the slope of the plot (see Problem 18-53).

18-63. (*) A suggested mechanism for the reaction described by

$$NO_2(g) + CO(g) \rightarrow CO_2(g) + NO(g)$$

is

(1) $2NO_2(g) \xrightarrow{k_1} NO_3(g) + NO(g)$

(2) $NO_3(g) + CO(g) \xrightarrow{k_2} NO_2(g) + CO_2(g)$

Assuming that $[NO_3]$ is governed by steady-state conditions, derive the rate law for the production of $CO_2(g)$.

18-64. (*) For the case of a species A reacting on a solid surface, it is often assumed that the rate of reaction is proportional to the fraction of the surface covered by A. If θ_A denotes this fraction, then

$$\text{rate of reaction} = k\theta_A$$

We can derive a simple expression for θ_A in the following way. Consider a gas at a pressure P_A in equilibrium with the gas adsorbed on a solid surface. Assume that the rate of adsorption is proportional to the pressure and to the fraction of surface with no adsorbed molecules $(1 - \theta_A)$, so that

$$\text{rate of absorption} = k_A P_A (1 - \theta_A)$$

Now assume that the rate of desorption is proportional to the fraction of the surface covered, so that

$$\text{rate of desorption} = k_2 \theta_A$$

These two rates will be equal at equilibrium. Solve this equality and then solve the resulting expression for θ_A and substitute this result into the equation

$$\text{rate of reaction} = k\theta_A$$

to obtain

$$\text{rate of reaction} = \frac{kk_1 P_A}{k_1 P_A + k_2} = \frac{kbP_A}{bP_A + 1}$$

where $b = k_1/k_2$. Compare this equation to the equation for the Michaelis-Menten mechanism. Show that the rate of the surface-catalyzed reaction is first-order at low pressures and zero-order at high pressures. Give a physical interpretation of this result.

18-65. This problem is an application of the result for the previous problem. Germanium hydride, GeH_4, decomposes on a germanium surface according to

$$GeH_4(ad) \rightarrow Ge(s) + 2H_2(ad)$$

where in this case *ad* stands for adsorbed. This is a zero-order reaction at 55 K. Given that an initial pressure of $GeH_4(g)$ of 41.0 kPa was reduced to 11.6 kPa after 20.0 minutes, calculate the value of the rate constant and the half-life for this reaction. (*Hint*: The integrated rate law for this zero-order reaction is $P_A = P_A^\circ - kt$.)

18-66. Determine the value of K_M, the Michaelis-Menten constant, from the following data for the myosin-catalyzed hydrolysis of ATP.

[ATP]/µmol·L^{-1}	7.5	12.5	20.0	43.5	62.5
R/pmol·L^{-1}·s^{-1}	67	95	119	155	166

18-67. Given that the concentration of myosin for the data in the previous problem was 21 pmol·L^{-1}, determine the turnover number of the enzyme molecule myosin, which has a single active site. (See Problem 18-33 for a discussion of turnover number).

18-68. The following data were obtained for the bovine carbonic anhydrase-catalyzed reaction described by

$$CO_2(aq) + H_2O(l) \rightarrow H^+(aq) + HCO_3^-(aq)$$

[CO_2]/mmol·L^{-1}	1.25	2.50	5.00	20.00
R/µmol·L^{-1}·s^{-1}	28	48	80	155

Determine the value of K_M, the Michaelis-Menten constant, for these data.

18-69. Given that the concentration of bovine carbonic anhydrase for the data in the previous problem was 2.8 pmol·L^{-1}, determine the turnover number of the enzyme molecule bovine carbonic anhydrase, which has a single active site. (See Problem 18-33 for a discussion of turnover number.)

18-70. The protein catalase catalyzes the reaction described by

$$2\,H_2O_2(aq) \rightarrow 2\,H_2O(l) + O_2(g)$$

and has a Michaelis-Menten constant of $K_M = 25$ mM and a turnover number (see Problem 18-33) of 4.0×10^7 s^{-1}. Calculate the initial rate of this reaction if the total enzyme concentration is 0.016 µM and the initial substrate concentration is 4.32 µM. Calculate the value of R_{max} for this enzyme. Catalase has a single active site.

18-71. Antibiotic-resistant bacteria have an enzyme called penicillinase that catalyzes the decomposition of penicillin and some penicillin-derived antibiotics. The molecular mass of penicillinase is 30 000. The turnover number (see Problem 18-33) of the enzyme at 28°C is 2000 s^{-1}. If 640 µg of penicillinase catalyzes the destruction of 3.11 mg of amoxicillin, an antibiotic with a molecular mass of 364, in 20.0 seconds at 28°C, how many active sites does the enzyme have?

18-72. (*) A proposed mechanism for the reaction described by the equation

$$2\,N_2O_5(g) \rightarrow 4\,NO_2(g) + O_2(g)$$

is

$$(1)\ N_2O_5(g) \underset{k_{-1}}{\overset{k_1}{\rightleftharpoons}} NO_2(g) + NO_3(g)$$

$$(2)\ NO_2(g) + NO_3(g) \xrightarrow{k_2} NO_2(g) + O_2(g) + NO(g)$$

$$(3)\ NO(g) + N_2O_5(g) \xrightarrow{k_3} 3\,NO_2(g)$$

Assuming a steady-state condition for [NO_3], show that the predicted rate law for the overall reaction is

$$\text{rate of reaction} = \frac{\Delta[O_2]}{\Delta t} = \frac{k_1 k_2 [N_2O_5]}{k_2 + k_{-1}}$$

18-73. (*) At high concentrations of $CH_3NC(g)$, the reaction described by

$$CH_3NC(g) \rightarrow CH_3CN(g)$$

obeys the first-order rate law

$$\text{rate of reaction} = -\frac{\Delta[CH_3NC]}{\Delta t} = k[CH_3NC]$$

and at low concentrations, the second-order rate law

$$\text{rate of reaction} = -\frac{\Delta[CH_3NC]}{\Delta t} = k'[CH_3NC]^2$$

A mechanism that predicts a first-order rate law at high gas concentration and a second-order at low gas concentration for reactions such as this was proposed independently by the British chemists J.A. Christiansen in 1921 and F.A. Lindemann in 1922. The mechanism is generally referred to as the *Lindemann mechanism*.

Lindemann proposed that the energy source for a reaction such as this results from bimolecular collisions. He further postulated that there must be a time lag between the collision (or energizing step) and the ensuing reaction. Depending on the collision rate in the gas and the lag time between a collision and the subsequent reaction, the molecule could possibly undergo a deactivating bimolecular collision before it has a chance to react. In terms of chemical equations, the Lindemann mechanism for a reaction of the type $A(g) \rightarrow B(g)$ is described by the equations

$$\text{(1)}\quad A(g) + M(g) \underset{k_{-1}}{\overset{k_1}{\rightleftharpoons}} A^*(g) + M(g)$$

$$\text{(2)}\quad A^*(g) \xrightarrow{k_2} B(g)$$

The symbol $A^*(g)$ in equations (1) and (2) represents an energized reactant molecule, and $M(g)$ is the collision partner whose collision energizes molecule $A(g)$. The molecule $M(g)$ can be a second reactant molecule, a product molecule, or a nonreactive buffer gas such as $N_2(g)$ or $Ar(g)$.

Using the steady-state approximation for $[A^*]$, show that the rate law for the overall reaction is

$$\text{rate of reaction} = -\frac{\Delta[A]}{\Delta t} = \frac{k_1 k_2 [A][M]}{k_{-1}[M] + k_2}$$

Show that at sufficiently high pressure, the rate law becomes

$$\text{rate of reaction} = k[A]$$

where the reaction rate is first-order in A in this high pressure limit. Now show that at sufficiently low pressures, the rate law for the overall reaction becomes

$$\text{rate of reaction} = k'[A][M]$$

where in this low-pressure limit, the rate law is first-order in both A and M and has an overall reaction order of two. Notice that if $M(g)$ is a second reactant molecule, then we have

$$\text{rate of reaction} = k'[A]^2$$

as observed for the reaction of $CH_3NC(g)$ described above. One of the great successes of the Lindemann mechanism was its ability to predict the experimentally observed change from first-order kinetics to second-order kinetics with decreasing pressure.

18-74. (*) This problem explores the connection between the observed activation parameters for a chemical reaction and the activation parameters of the individual steps of the reaction mechanism. Specifically, suppose that the rate constant for each step of the Lindemann mechanism in the previous problem shows Arrhenius behavior. How are the overall measured values of A and E_a for the reaction related to the values of A and E_a for the individual steps of the mechanism for the high-concentration reaction?

Cato Guldberg (1836–1902), *left*, was born in Christiania (now Oslo), Norway, the son of a minister. Peter Waage (1833–1900), *right*, was born on the island of Hitter (now Hidra) near Flekkefjord, Norway, the son of a shipmaster. Guldberg and Waage both entered the University of Christiania in 1854, forming a lifelong friendship. While undergraduates, they helped form a club to discuss problems of physics and chemistry. Guldberg majored in mathematics with a minor in physics and chemistry and later became a professor of applied mathematics at the University of Christiania. Waage studied medicine, but then switched to mineralogy and chemistry in his third year, eventually becoming chair of the chemistry department at Christiania. Guldberg and Waage were related through two marriages. Guldberg married his cousin Bodil Riddervold and Waage married her sister Johanne Riddervold, making the two friends brothers-in-law. After Johanne's death in 1869, Waage married Guldberg's sister, Mathilde, making them brothers-in-law for a second time. Together they published a series of papers that became the foundation of our modern theory of equilibrium. Their recognition that it is the concentrations of reacting substances that determine the equilibrium resulting from the forward and reverse reactions is known today as the Guldberg and Waage law.

19. Chemical Equilibrium

In the 1800s, Norwegian scientists Cato Guldberg and Peter Waage (Frontispiece) developed a quantitative theory to explain the direction of chemical change based on concentration. They showed that a chemical reaction takes place in both the forward and reverse directions simultaneously. As a reaction proceeds (from reactants to products), the rate of the forward reaction slows down. Meanwhile, as the concentration of products builds up, the rate of the reverse reaction (from products to reactants) increases, until the rate of the reverse reaction equals that of the forward reaction. When that happens, the reaction appears to have stopped. No more reactant disappears, and no more product appears. Nevertheless, on a molecular scale reactant molecules continue to react to produce product molecules, and product molecules continue to react to produce reactant molecules, but the concentrations of reactants and products no longer change. The system has reached a state of equilibrium.

So far we have usually assumed that chemical reactions go in one direction, forward from reactants to products. However, as we shall see, this is really a special case in which the equilibrium state lies so far to the product side that it appears as if the reaction runs in only one direction.

If we prepare a mixture of reactants and products, in what direction does the reaction proceed toward equilibrium and what are the final concentrations? If conditions then change, disturbing the equilibrium, in which direction does the reaction shift? In this chapter, we show how to answer these questions. We show that a chemical reaction at a given temperature is characterized by a quantity called an equilibrium constant. By studying reaction systems at equilibrium, we determine how to maximize the amount of a desired product.

19-1. A Chemical Equilibrium Is a Dynamic Equilibrium

Consider the chemical reaction described by the equation

$$\underset{\text{colorless}}{N_2O_4(g)} \leftrightharpoons \underset{\text{brown}}{2NO_2(g)} \tag{19.1}$$

in which the colorless gas dinitrogen tetroxide, $N_2O_4(g)$, dissociates into the reddish-brown gas nitrogen dioxide, $NO_2(g)$. As we discussed briefly in Section 18-5, the reaction described by Equation 19.1, like all chemical reactions, is really two opposing reactions. The **forward reaction** is the dissociation of $N_2O_4(g)$ molecules into $NO_2(g)$ molecules, and the **reverse reaction** is the association of $NO_2(g)$ molecules into $N_2O_4(g)$ molecules. We recognize the existence of both a forward reaction and reverse reaction by separating the two sides of Equation 19.1 by *two* arrows pointing in opposite directions.

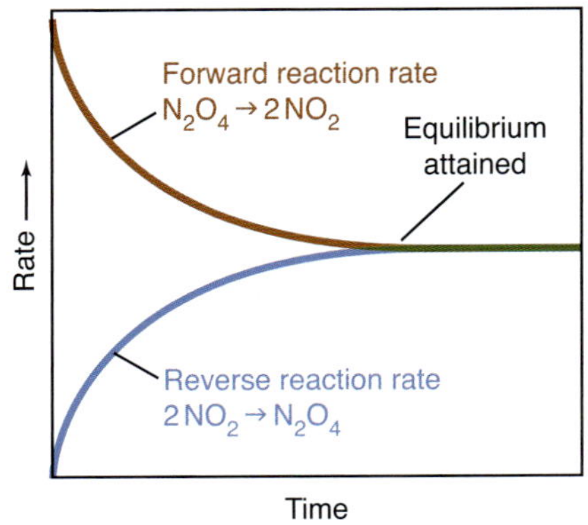

Figure 19.1 Dinitrogen tetroxide, $N_2O_4(g)$, injected into an empty vessel will dissociate according to

$$N_2O_4(g) \leftrightharpoons 2NO_2(g)$$

As the $N_2O_4(g)$ dissociates, its concentration decreases, so the rate of the forward reaction decreases (red curve). As the concentration of $NO_2(g)$ builds up, the rate of the reverse reaction increases (blue curve). Eventually, the two rates become equal, and a state of equilibrium results. The concentrations of $N_2O_4(g)$ and $NO_2(g)$ no longer change with time.

Suppose that we start with only $N_2O_4(g)$. Initially, the reaction mixture is colorless. As $N_2O_4(g)$ molecules dissociate into $NO_2(g)$ molecules, the mixture becomes reddish-brown. As the concentration of $NO_2(g)$ increases, more and more $NO_2(g)$ molecules collide and associate back into $N_2O_4(g)$ molecules. Thus, the reverse rate of the process described by Equation 19.1 increases with time. As $N_2O_4(g)$ molecules dissociate, there are fewer $N_2O_4(g)$ molecules remaining, so the forward rate decreases. Eventually, the forward rate and the reverse rate become equal and a state of **chemical equilibrium** exists (Figure 19.1). At equilibrium, the concentrations of $N_2O_4(g)$ and $NO_2(g)$ no longer change with time. A chemical equilibrium is a **dynamic equilibrium** because $N_2O_4(g)$ molecules are still dissociating into $NO_2(g)$ molecules and $NO_2(g)$ molecules are still associating into $N_2O_4(g)$ molecules. The rates of these two processes are exactly the same, however, so there is no net change in the concentrations of $N_2O_4(g)$ and $NO_2(g)$.

We can perform a simple demonstration to prove that a chemical equilibrium is dynamic. When calcium carbonate, $CaCO_3(s)$, is heated in a closed container, it decomposes to calcium oxide, $CaO(s)$, and carbon dioxide, $CO_2(g)$, according to

$$CaCO_3(s) \leftrightharpoons CaO(s) + CO_2(g) \tag{19.2}$$

Eventually, the rate of the reverse reaction will equal the rate of the forward reaction, and a state of equilibrium results. At equilibrium, the pressure of the carbon dioxide will be constant (Figure 19.2). Suppose now that we add a small amount of calcium carbonate containing radioactive carbon to the equilibrium mixture. We denote this calcium carbonate by $Ca^*CO_3(s)$ to emphasize that the carbon atoms are radioactive. After we add the $Ca^*CO_3(s)$, we observe that the pressure of the $CO_2(g)$ does not change, an observation indicating that the **equilibrium state**, the concentration of each of the species present at equilibrium, has not been disturbed. Yet the radioactivity, which can be measured with an instrument such as a Geiger counter, does *not* remain solely in the calcium carbonate. In time, some of the radioactivity is observed in the carbon dioxide (Figure 19.3). If a chemical equilibrium were a static equilibrium, then the radioactivity would remain in the calcium carbonate phase. It gets distributed between the calcium carbonate and the carbon dioxide, however, because the equilibrium is a dynamic equilibrium. Both the forward and the reverse reactions in Equation 19.2 occur constantly, but at exactly the same rate, so there is no *net* change.

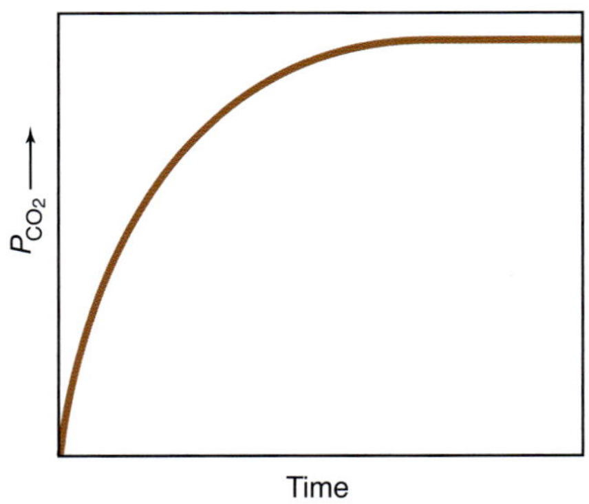

Figure 19.2 If $CaCO_3(s)$ is placed in a closed container and heated, it will decompose into $CaO(s)$ and $CO_2(g)$. The pressure of the $CO_2(g)$ initially rises and then levels off as the state of equilibrium is reached.

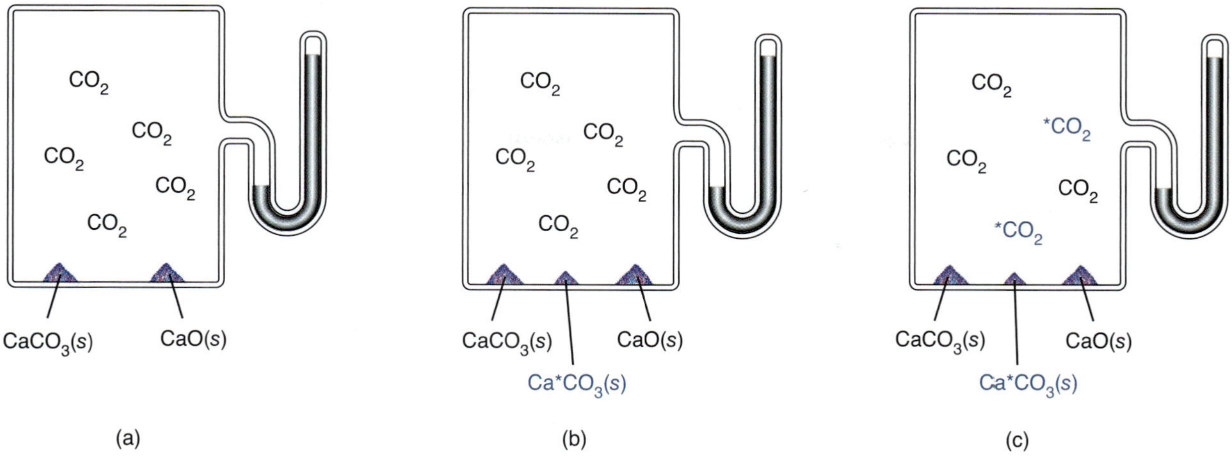

Figure 19.3 Chemical equilibrium is a dynamic state. If a small amount of calcium carbonate containing the radioactive isotope carbon-14, $Ca^*CO_3(s)$, is added to an equilibrium mixture of $CaCO_3(s)$, $CaO(s)$, and $CO_2(g)$, the equilibrium described by

$$CaCO_3(s) \leftrightharpoons CaO(s) + CO_2(g)$$

is not disturbed. We see this in (a) and (b) where the pressure of the $CO_2(g)$ remains constant. The radioactive carbon atoms, however, do not remain in the solid calcium carbonate phase. Instead, as shown in (c), they become distributed between the calcium carbonate and the carbon dioxide.

19-2. A Chemical Equilibrium Can Be Attained from Either Direction

We can study the approach to equilibrium of the reaction described by Equation 19.1 quantitatively. If we determine the equilibrium concentration of $NO_2(g)$ and if we know the initial concentrations of $N_2O_4(g)$ and $NO_2(g)$, then it is a simple matter to determine the equilibrium concentration of $N_2O_4(g)$. For example, suppose that we start with 1.00 $mol{\cdot}L^{-1}$ of $N_2O_4(g)$ and no $NO_2(g)$. We can denote these initial concentrations by $[N_2O_4]_0 = 1.00$ M and $[NO_2]_0 = 0$ M, where the subscript 0 indicates an initial concentration. Suppose also that at equilibrium we find that $[NO_2] = 0.40$ M. Then from the reaction stoichiometry given in Equation 19.1, we have

$$\begin{pmatrix}\text{change in moles}\\ \text{per liter of } N_2O_4\end{pmatrix} = \begin{pmatrix}\text{moles per liter}\\ \text{of } NO_2 \text{ produced}\end{pmatrix}\left(\frac{1\text{ mol }N_2O_4}{2\text{ mol }NO_2}\right)$$

$$= \left(\frac{0.40\text{ mol }NO_2}{1\text{ L}}\right)\left(\frac{1\text{ mol }N_2O_4}{2\text{ mol }NO_2}\right) = 0.20\text{ mol}{\cdot}\text{L}^{-1}$$

Therefore, the value of $[N_2O_4]$ when $[NO_2] = 0.40$ M is

$$[N_2O_4] = \underset{\text{initial concentration}}{1.00\text{ M}} - \underset{\text{change in concentration}}{0.20\text{ M}} = \underset{\text{equilibrium concentration}}{0.80\text{ M}}$$

The following Example illustrates the calculation of equilibrium concentrations in reaction mixtures.

EXAMPLE 19-1: Suppose that the methanol synthesis reaction described by the equation

$$CO(g) + 2H_2(g) \leftrightharpoons CH_3OH(g)$$

is carried out with the initial concentrations

$$[CO]_0 = 2.00\text{ M} \qquad [H_2]_0 = 0.50\text{ M} \qquad [CH_3OH]_0 = 0\text{ M}$$

and that at equilibrium we find that $[CH_3OH] = 0.20$ M. Calculate the equilibrium values of $[CO]$ and $[H_2]$.

Solution: From the stoichiometry of the chemical equation, we have

$$\left(\begin{array}{c}\text{change in moles}\\ \text{per liter of CO}\end{array}\right) = \left(\begin{array}{c}\text{moles per liter of}\\ CH_3OH\text{ produced}\end{array}\right)\left(\frac{1\text{ mol CO}}{1\text{ mol }CH_3OH}\right)$$

$$= \left(\frac{0.20\text{ mol }CH_3OH}{1\text{ L}}\right)\left(\frac{1\text{ mol CO}}{1\text{ mol }CH_3OH}\right) = 0.20\text{ M}$$

Therefore, the value of [CO] when $[CH_3OH] = 0.20$ M is

$$\underset{\text{initial concentration}}{[CO] = 2.00\text{ M}} - \underset{\text{change in concentration}}{0.20\text{ M}} = \underset{\text{equilibrium concentration}}{1.80\text{ M}}$$

Similarly, we calculate for $[H_2]$ at equilibrium:

$$\left(\begin{array}{c}\text{change in moles}\\ \text{per liter of }H_2\end{array}\right) = \left(\frac{0.20\text{ mol }CH_3OH}{1\text{ L}}\right)\left(\frac{2\text{ mol }H_2}{1\text{ mol }CH_3OH}\right) = 0.40\text{ M}$$

and so

$$\underset{\text{initial concentration}}{[H_2] = 0.50\text{ M}} - \underset{\text{change in concentration}}{0.40\text{ M}} = \underset{\text{equilibrium concentration}}{0.10\text{ M}}$$

Note that twice as much $H_2(g)$ as $CO(g)$ is consumed in the reaction, as required by the stoichiometry of the reaction.

PRACTICE PROBLEM 19-1: A reaction that is used in the production of sulfuric acid can be described by the chemical equation

$$2\,SO_2(g) + O_2(g) \leftrightarrows 2\,SO_3(g)$$

If initially $[SO_2]_0 = 8.500$ M, $[O_2]_0 = 3.500$ M, and $[SO_3]_0 = 5.575$ M, and if $[O_2] = 0.075$ M at equilibrium, then calculate the equilibrium values of $[SO_2]$ and $[SO_3]$.

Answer: $[SO_2] = 1.650$ M and $[SO_3] = 12.425$ M

Let's continue our discussion of the $N_2O_4(g)$-$NO_2(g)$ reaction (Equation 19.1). Equilibrium values of $[N_2O_4]$ and $[NO_2]$ for several sets of initial conditions are given in Table 19.1. The most remarkable feature of the data in Table 19.1 is shown in the last column. The value of the ratio $[NO_2]^2/[N_2O_4]$ at equilibrium for the reaction described by

$$N_2O_4(g) \leftrightarrows 2\,NO_2(g)$$

is equal to a constant. From the data in Table 19.1, we see that

$$\frac{[NO_2]^2}{[N_2O_4]} = 0.20\text{ M} \qquad \text{(at equilibrium at 100°C)} \qquad (19.3)$$

TABLE 19.1 Initial and equilibrium values at 100°C of $[N_2O_4]$ and $[NO_2]$ for the reaction $N_2O_4(g) \leftrightharpoons 2NO_2(g)$

Initial concentration		Equilibrium concentration		Value at equilibrium of the quantity
$[N_2O_4]_0/M$	$[NO_2]_0/M$	$[N_2O_4]/M$	$[NO_2]/M$	$([NO_2]^2/[N_2O_4])/M$
1.00	0	0.80	0.40	0.20
2.00	0	1.71	0.58	0.20
0	2.00	0.80	0.40	0.20
0	1.00	0.36	0.27	0.20
1.00	1.00	1.25	0.50	0.20

For example, if we know that $[NO_2] = 0.175$ M at equilibrium, then $[N_2O_4]$ would have to be 0.15 M so that $[NO_2]^2/[N_2O_4]$ equals 0.20 M. Table 19.1 shows that the constant value of $[NO_2]^2/[N_2O_4]$ at equilibrium is independent of the initial values, $[N_2O_4]_0$ and $[NO_2]_0$. It does not matter whether we start our reaction with only $N_2O_4(g)$, with only $NO_2(g)$, or with a mixture of both; in all cases, the value of the ratio $[NO_2]^2/[N_2O_4]$ at equilibrium is the same. These data show that a chemical equilibrium can be attained from either direction.

19-3. The Equilibrium Constant for a Chemical Equation Is Equal to the Ratio of Product Concentration Terms to Reactant Concentration Terms

For the $N_2O_4(g)$-$NO_2(g)$ reaction (Equation 19.1), we found that regardless of the initial concentrations of $N_2O_4(g)$ and $NO_2(g)$, Equation 19.3 is satisfied at equilibrium. Equation 19.3 says that the ratio of equilibrium product concentration to equilibrium reactant concentration, each raised to a power equal to its balancing coefficient in the chemical equation for the reaction, is a constant value at a given temperature, regardless of the initial concentrations of the reactants and products. A similar relation holds for *any* reaction equilibrium. For example, for the reaction described by

$$2H_2(g) + CO(g) \leftrightharpoons CH_3OH(g)$$

we find experimentally that at equilibrium

$$\frac{[CH_3OH]}{[H_2]^2[CO]} = 1.3 \times 10^7\ M^{-2} \qquad \text{(at 25°C)}$$

for any initial concentrations. Similarly, for

$$C(s) + CO_2(g) \leftrightharpoons 2CO(g)$$

we find experimentally that at equilibrium

$$\frac{[CO]^2}{[CO_2]} = 0.023\ \text{M} \qquad \text{(at 1000 K)} \tag{19.4}$$

for any initial concentrations of $CO(g)$ and $CO_2(g)$. Notice that the concentration of $C(s)$ does not appear in Equation 19.4. The concentration of a pure liquid or a pure solid can be expressed by its density, but the densities of pure liquids and pure solids do not vary much with temperature or pressure and so are essentially constant. Consequently, they contribute only a constant factor to the equilibrium constant. By convention, we set this factor equal to one so that [C] does not appear in Equation 19.4.

The general pattern was first deduced by two Norwegian chemists, Cato Guldberg and Peter Waage (Frontispiece). On the basis of their experimental observations of a number of chemical reactions, they postulated that for the balanced general chemical equation

$$a\,A(g) + b\,B(soln) + c\,C(s) \leftrightharpoons x\,X(g) + y\,Y(soln) + z\,Z(l)$$

the ratio

$$\frac{[X]^x[Y]^y}{[A]^a[B]^b} \tag{19.5}$$

is a constant at a given temperature. Once again, notice that the concentrations of the solid and liquid species do not appear in Equation 19.5. We call the constant ratio in Equation 19.5 an **equilibrium constant,** K_c, and write

$$K_c = \frac{[X]^x[Y]^y}{[A]^a[B]^b} \tag{19.6}$$

Equation 19.6 is the mathematical form of the **equilibrium-constant expression**. The subscript c indicates that the equilibrium constant is expressed in terms of concentrations. We shall soon show that we can also express equilibrium constants involving gaseous species in terms of pressures. In that case, we write K_p instead of K_c. Equation 19.6 was originally deduced experimentally, but later the same type of expression was derived theoretically using thermodynamic concepts that we shall discuss in Chapter 23. Application of Equation 19.6 to the equilibrium equation

$$N_2O_4(g) \leftrightharpoons 2\,NO_2(g) \tag{19.7}$$

yields

$$K_c = \frac{[NO_2]^2}{[N_2O_4]} \tag{19.8}$$

From the results in the preceding section, we determine for the $N_2O_4(g)$ dissociation reaction that $K_c = 0.20$ M at 100°C.

It is important to realize that an equilibrium-constant expression and its corresponding value depend upon how we choose to represent the reaction by a chemical equation. If, for instance, we had chosen to describe the reaction of $N_2O_4(g)$ as

$$\tfrac{1}{2}N_2O_4(g) \leftrightharpoons NO_2(g) \qquad (19.9)$$

instead of by Equation 19.7, then we would have

$$K_c' = \frac{[NO_2]}{[N_2O_4]^{1/2}} \qquad (19.10)$$

Note that Equations 19.8 and 19.10 differ by a square root, so that the value of K_c' is

$$K_c' = K_c^{1/2} = (0.20\ \text{M})^{1/2} = 0.45\ \text{M}^{1/2} \qquad (19.11)$$

We have shown that there is no such thing as an equilibrium constant for a reaction. There is only an equilibrium constant for the chemical equation that is used to represent a reaction. We shall see more examples of this in Section 19-6. We learned in Chapter 14 that the values of thermochemical quantities also depend upon how we choose to write the chemical equation that represents a reaction. In Chapter 23 we shall see a nice relation between thermochemical quantities and equilibrium constants for a given chemical equation.

Note that the equilibrium constants in Equations 19.3, 19.4, and 19.11 clearly have units, in accord with the recommendation of the International Union of Pure and Applied Chemistry. When we study thermodynamics in Chapter 23, we shall define a more fundamental equilibrium constant, K (without a subscript), called a thermodynamic equilibrium constant, that is related to K_c and K_p but is unitless. In this chapter, however, all our equilibrium constants will have units because their defining expressions have units.

The important points to know about equilibrium-constant expressions are:

1. The value of K_c or K_p for a stated chemical equation is equal to a constant at a given temperature.
2. The equilibrium product concentrations appear in the numerator, and the equilibrium reactant concentrations appear in the denominator.
3. The equilibrium concentration of each species is raised to a power equal to the stoichiometric coefficient of that species in the balanced chemical equation.
4. Reactants and products that appear as pure solids or pure liquids in the chemical equation do not appear in the equilibrium-constant expression, or by convention are assigned a numeric value of one.
5. Equilibrium is a dynamic state; at equilibrium the rates of the forward and reverse reactions are equal so that concentrations of the products and reactants are constant.

EXAMPLE 19-2: Write the equilibrium-constant expressions for the following chemical equations:

(a) $N_2(g) + 3H_2(g) \leftrightharpoons 2NH_3(g)$

(b) $\tfrac{1}{2}N_2(g) + \tfrac{3}{2}H_2(g) \leftrightharpoons NH_3(g)$

(c) $PCl_3(l) + Cl_2(g) \leftrightharpoons PCl_5(s)$

Solution: (a) All the reactants and products are gases, so

$$K_c = \frac{[NH_3]^2}{[N_2][H_2]^3}$$

(b) The appropriate expression is

$$K_c = \frac{[NH_3]}{[N_2]^{1/2}[H_2]^{3/2}}$$

This expression is the square root of that given in (a).

(c) The appropriate expression is simply

$$K_c = \frac{1}{[Cl_2]}$$

Neither $PCl_3(l)$ nor $PCl_5(s)$ appear in this K_c expression because they are a liquid and a solid.

PRACTICE PROBLEM 19-2: Write the equilibrium-constant expressions for the following chemical equations:

(a) $3\,O_2(g) \leftrightharpoons 2\,O_3(g)$

(b) $SO_2(g) + 2\,H_2S(g) \leftrightharpoons 3\,S(l) + 2\,H_2O(g)$

(c) $2\,Hg(l) + O_2(g) \leftrightharpoons 2\,HgO(s)$

Answer: (a) $K_c = \dfrac{[O_3]^2}{[O_2]^3}$; (b) $K_c = \dfrac{[H_2O]^2}{[SO_2][H_2S]^2}$; (c) $K_c = \dfrac{1}{[O_2]}$

We can calculate the value of K_c from the initial values of the concentrations of all the species if we know the equilibrium value of the concentration of one of them. Here is an example of such a calculation.

EXAMPLE 19-3: Suppose that the methanol synthesis reaction, represented by

$$CO(g) + 2\,H_2(g) \leftrightharpoons CH_3OH(g) \qquad (19.12)$$

is carried out with the following initial concentrations:

$$[CO]_0 = 1.75\text{ M} \qquad [H_2]_0 = 0.80\text{ M} \qquad [CH_3OH]_0 = 0.65\text{ M}$$

At equilibrium, we find that $[CO] = 1.60$ M. Calculate the value of the equilibrium constant for this equation.

Solution: The equilibrium-constant expression associated with Equation 19.12 is

$$K_c = \frac{[CH_3OH]}{[CO][H_2]^2}$$

It is often helpful in working equilibrium calculations to set the problem up in tabular form with the initial concentrations, change in concentrations, and equilibrium concentrations of each species placed in separate rows directly below the species in the chemical equation. From the data given, we can construct the following table:

Concentration	$CO(g)$	+	$2H_2(g)$	$\leftrightharpoons$	$CH_3OH(g)$
initial	1.75 M		0.80 M		0.65 M
change					
equilibrium	1.60 M				

To determine the value of K_c, we must determine each of the equilibrium concentrations in the table above. Because the initial concentration of $CO(g)$ is 1.75 M and its equilibrium value is 1.60 M, we see that the change in the concentration of $CO(g)$ is simply the difference between these two values, or 0.15 M. Thus, we have the following:

Concentration	$CO(g)$	+	$2H_2(g)$	$\leftrightharpoons$	$CH_3OH(g)$
initial	1.75 M		0.80 M		0.65 M
change	−0.15 M				
equilibrium	1.60 M				

where the change in concentration of $CO(g)$ is negative because the concentration of $CO(g)$ decreased. From the reaction stoichiometry, we have

$$\left(\begin{array}{c}\text{change in moles}\\ \text{per liter of } H_2\end{array}\right) = \left(\begin{array}{c}\text{change in moles}\\ \text{per liter of CO}\end{array}\right)\left(\frac{2 \text{ mol } H_2}{1 \text{ mol CO}}\right) = -(0.15 \text{ M})(2) = -0.30 \text{ M}$$

and

$$\left(\begin{array}{c}\text{change in moles}\\ \text{per liter of } CH_3OH\end{array}\right) = \left(\begin{array}{c}\text{change in moles}\\ \text{per liter of CO}\end{array}\right)\left(\frac{1 \text{ mol } CH_3OH}{1 \text{ mol CO}}\right) = 0.15 \text{ M}$$

Putting these values into our table yields:

Concentration	$CO(g)$	+	$2H_2(g)$	$\leftrightharpoons$	$CH_3OH(g)$
initial	1.75 M		0.80 M		0.65 M
change	−0.15 M		−0.30 M		+0.15 M
equilibrium	1.60 M				

Here the change in concentration of $H_2(g)$ is negative because, as we have already observed from the change in concentration of $CO(g)$, the concentrations of the reactants are decreasing as the reaction approaches equilibrium. Similarly, the change in concentration of the $CH_3OH(g)$ is positive because the concentration of the $CH_3OH(g)$ product increases as the reaction approaches equilibrium. Notice that we could also have determined these values from that of the $CO(g)$ simply by using the stoichiometric coefficients given in the header of the table, which is one of the advantages of listing the data in tabular form.

Using these values, our completed **concentration table** is

Concentration	$CO(g)$	+	$2H_2(g)$	$\leftrightharpoons$	$CH_3OH(g)$
initial	1.75 M		0.80 M		0.65 M
change	−0.15 M		−0.30 M		+0.15 M
equilibrium	1.60 M		0.50 M		0.80 M

The value of the equilibrium constant is given by

$$K_c = \frac{[CH_3OH]}{[CO][H_2]^2} = \frac{0.80\text{ M}}{(1.60\text{ M})(0.50\text{ M})^2} = 2.0\text{ M}^{-2}$$

PRACTICE PROBLEM 19-3: Initially, $[SO_2]_0 = 3.00$ M, $[O_2]_0 = 6.00$ M, and $[SO_3]_0 = 1.00$ M for the reaction whose equation is

$$2SO_2(g) + O_2(g) \leftrightharpoons 2SO_3(g)$$

If $[SO_3] = 1.40$ M at equilibrium, then calculate the value of the equilibrium constant.

Answer: 0.0500 M^{-1}

An equilibrium-constant expression involves the equilibrium product concentrations in the numerator and the equilibrium reactant concentrations in the denominator. Therefore, a reaction equilibrium in which there is much more product than reactant will typically have a large value of K_c. For example, at 1200°C the value of K_c for the reaction described by

$$H_2(g) + Cl_2(g) \leftrightharpoons 2HCl(g)$$

is 2.5×10^4. For this reaction, we say that the equilibrium lies far to the right. Although there is some $H_2(g)$ and $Cl_2(g)$ present at equilibrium, the amount is so small that the reaction goes essentially to completion.

The equilibrium constant for this reaction is unitless because the sum of the stoichiometric coefficients of the gaseous reactants and the gaseous products are the same in the chemical equation.

$$K_c = \frac{[HCl]^2}{[H_2][Cl_2]} = \frac{M^2}{M \cdot M}$$

is unitless.

For a reaction equilibrium in which there is little product compared with reactants, the value of K_c will typically be small. For example, at 1700°C the value of K_c for the reaction described by

$$N_2(g) + O_2(g) \leftrightharpoons 2NO(g)$$

is 4.1×10^{-4}. For this reaction, we say that the equilibrium lies far to the left. In general, reactions with large values of K_c give a large amount of product at equilibrium, whereas those with small values of K_c give little product at equilibrium. Thus, the yield of a chemical reaction depends on the value of K_c. Notice that the magnitude of K_c tells us nothing about how long a reaction takes to reach equilibrium. Reaction rates depend on the specific reaction kinetics and *not* on the value of the equilibrium constant.

19-4. Equilibrium Constants Can Be Expressed in Terms of Partial Pressures

For reactions involving only gases, it is often convenient to express the equilibrium constant in terms of partial pressures rather than concentrations. An equilibrium-constant expression written in terms of partial pressures is denoted by K_p. For example, the K_p expression for the equation

$$\mathrm{CaCO_3}(s) \leftrightharpoons \mathrm{CaO}(s) + \mathrm{CO_2}(g)$$

is $K_p = P_{CO_2}$. Note once again that solids and liquids do not appear in the equilibrium constant expression.

Because K_p is a constant (at a given temperature), the equilibrium pressure of $CO_2(g)$ is the same, regardless of the amount of $CaCO_3(s)$ or $CaO(s)$ present. This effect is illustrated in Figure 19.4.

For most chemical equations, $K_p \neq K_c$. To see the connection between K_p and K_c, we use the ideal-gas law, $PV = nRT$. The concentration of the gas is $[\text{gas}] = n/V$, so

$$P = \left(\frac{n}{V}\right)RT = [\text{gas}]RT \tag{19.13}$$

Thus, the pressure of a gas is directly proportional to the concentration of the gas at constant temperature. Let's apply this result to the equation

$$\mathrm{C}(s) + \mathrm{CO_2}(g) \leftrightharpoons 2\,\mathrm{CO}(g) \tag{19.14}$$

We start with

$$K_p = \frac{P^2_{CO}}{P_{CO_2}} = \frac{([\mathrm{CO}]RT)^2}{[\mathrm{CO_2}]RT} = \frac{[\mathrm{CO}]^2RT}{[\mathrm{CO_2}]}$$

But

$$K_c = \frac{[\mathrm{CO}]^2}{[\mathrm{CO_2}]}$$

so for Equation 19.14

$$K_p = K_cRT \tag{19.15}$$

The only case in which $K_p = K_c$ occurs is when the sum of the stoichiometric coefficients of the gaseous reactants is equal to that of the gaseous products. For example, consider the high-temperature reaction

$$\mathrm{H_2}(g) + \mathrm{I_2}(g) \leftrightharpoons 2\,\mathrm{HI}(g) \tag{19.16}$$

In this case

$$K_c = \frac{[\mathrm{HI}]^2}{[\mathrm{H_2}][\mathrm{I_2}]} = \frac{\left(\frac{P_{HI}}{RT}\right)^2}{\left(\frac{P_{H_2}}{RT}\right)\left(\frac{P_{I_2}}{RT}\right)} = \frac{P^2_{HI}}{P_{H_2}P_{I_2}} = K_p \tag{19.17}$$

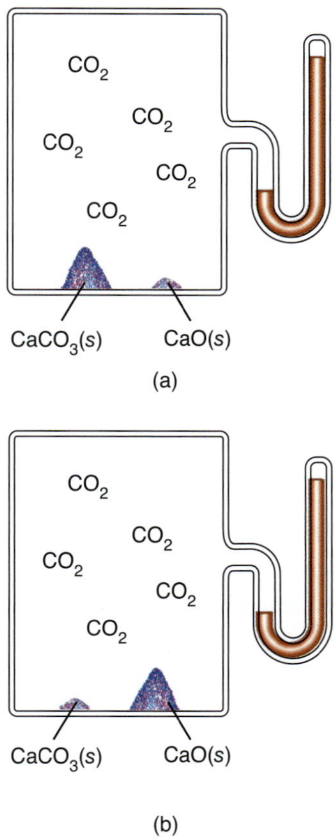

Figure 19.4 The equilibrium constant for the reaction described by the equation

$$\mathrm{CaCO_3}(s) \leftrightharpoons \mathrm{CaO}(s) + \mathrm{CO_2}(g)$$

does not depend on how much $CaCO_3(s)$ and $CaO(s)$ are present (as long as there is some of each). It is given by $K_p = P_{CO_2}$. Thus, the pressure of $CO_2(g)$ at equilibrium does not depend on the relative amounts of $CaCO_3(s)$ and $CaO(s)$, as shown in the two cases (a) and (b).

The factors of RT cancel because the total exponent in the denominator is equal to the total exponent in the numerator. Also note again that K_c and K_p are unitless in cases like this.

We can derive a general expression for the relationship between K_p and K_c. For the general gas phase reaction described by

$$aA(g) + bB(g) \leftrightharpoons cC(g) + dD(g)$$

K_p is given by

$$K_p = \frac{P_C^c P_D^d}{P_A^a P_B^b} = \frac{([C]RT)^c([D]RT)^d}{([A]RT)^a([B]RT)^b} = \frac{[C]^c[D]^d}{[A]^a[B]^b} \times \frac{(RT)^c(RT)^d}{(RT)^a(RT)^b}$$
$$= K_c(RT)^{(c+d)-(a+b)}$$

If we let $\Delta\nu_{gas} = (c + d) - (a + b)$, then we have

$$K_p = K_c(RT)^{\Delta\nu_{gas}} \qquad (19.18)$$

Recall from Section 14-3 that $\Delta\nu_{gas}$ is the difference between the (unitless) stoichiometric coefficients of the products and those of the reactants for the gas phase species in a chemical equation.

The value of K_p and its units depend upon which value of R, the gas constant, is used in Equation 19.18.

For the reaction described by Equation 19.14, $C(s) + CO_2(g) \leftrightharpoons 2CO(g)$, we have $\Delta\nu_{gas} = 2 - 1 = 1$. Substituting this value into Equation 19.18, we find

$$K_p = K_c(RT)^1 = K_cRT$$

which is the same result we obtained in Equation 19.15.

Similarly, applying Equation 19.18 to the reaction given in Equation 19.16, $H_2(g) + I_2(g) \leftrightharpoons 2HI(g)$, we have $\Delta\nu_{gas} = 2 - (1 + 1) = 0$, so

$$K_p = K_c(RT)^0 = K_c$$

which is the same result we obtained in Equation 19.17.

EXAMPLE 19-4: (a) Find the relation between K_p and K_c for

$$N_2(g) + 3H_2(g) \leftrightharpoons 2NH_3(g) \qquad (19.19)$$

(b) At 225°C the value of K_c is 170 M^{-2}. Find the value of K_p at this temperature.

Solution: (a) We see from Equation 19.19 that

$$\Delta\nu_{gas} = 2 - (1 + 3) = -2$$

Substituting this result into Equation 19.18, we obtain

$$K_p = K_c(RT)^{-2} \qquad (19.20)$$

(b) If we substitute a value of the gas constant, 0.083145 L·bar·mol^{-1}·K^{-1} (Table 13.2) in this case, and the temperature in kelvin, 498 K, into Equation 19.20, then we obtain

$$K_p = K_c(RT)^{-2} = \frac{K_c}{(RT)^2} = \frac{170\text{ M}^{-2}}{[(0.083145\text{ L·bar·mol}^{-1}\text{·K}^{-1})(498\text{ K})]^2}$$

$$= 0.099\text{ bar}^{-2}$$

PRACTICE PROBLEM 19-4: For the equilibrium described by

$$\text{CaO}(s) + \text{CO}_2(g) \leftrightharpoons \text{CaCO}_3(s) \qquad K_p = 25\text{ atm}^{-1}\text{ at 1020 K}$$

(a) Find the relation between K_p and K_c.
(b) Find the value of K_c at this temperature.

Answer: (a) $K_p = K_c(RT)^{-1}$; (b) $K_c = 2.1 \times 10^3$ M^{-1}

19-5. Equilibrium Constants Are Used in a Variety of Calculations

Most of the reactions that we have considered in the previous chapters of this text have been ones for which the value of the equilibrium constant is quite large and so the reaction goes essentially in one direction, toward the formation of products. It is under these conditions that the method of limiting reactants we learned in Chapter 11 applies. Here we shall learn how to perform similar calculations for reactions where the equilibrium state does not necessarily lie entirely on the products side. Such equilibrium calculations are best illustrated by example.

EXAMPLE 19-5: Suppose that 2.00 moles of $N_2O_4(g)$ are injected into a 1.00-liter reaction vessel held at 100°C and that equilibrium is attained:

$$\text{N}_2\text{O}_4(g) \leftrightharpoons 2\,\text{NO}_2(g) \qquad K_c = 0.20\text{ M} \tag{19.21}$$

Calculate the equilibrium values of $[N_2O_4]$ and $[NO_2]$.

Solution: The equilibrium-constant expression for this equation is

$$K_c = \frac{[\text{NO}_2]^2}{[\text{N}_2\text{O}_4]} = 0.20\text{ M}$$

Once again, it is helpful to set up a table. From the data given, we have that

Concentration	$N_2O_4(g)$	$\leftrightharpoons$	$2NO_2(g)$
initial	2.00 M		0 M
change			
equilibrium			

A review of the quadratic formula is given in Appendix A.

We now use the reaction stoichiometry to express the change in the initial concentrations of $N_2O_4(g)$ and $NO_2(g)$ needed to reach equilibrium. Let the number of moles per liter of $N_2O_4(g)$ that dissociates into $NO_2(g)$ be x. Then the change in concentration of $N_2O_4(g)$ is $-x$ and the equilibrium concentration of $N_2O_4(g)$ is $[N_2O_4] = 2.00\ M - x$. According to Equation 19.21, each mole of $N_2O_4(g)$ that dissociates produces two moles of $NO_2(g)$, so the change in the initial concentration of $NO_2(g)$ is $+2x$ and the equilibrium concentration of $NO_2(g)$ expressed in terms of x is $[NO_2] = 2x$.

Our completed concentration table is now

Concentration	$N_2O_4(g)$	$\leftrightharpoons$	$2NO_2(g)$
initial	2.00 M		0 M
change	$-x$		$+2x$
equilibrium	$2.00\ M - x$		$2x$

Substituting the equilibrium concentration data from this table into the K_c expression yields

$$K_c = \frac{[NO_2]^2}{[N_2O_4]} = \frac{(2x)^2}{2.00\ M - x} = 0.20\ M$$

or

$$4x^2 = (0.20\ M)(2.00\ M - x) = 0.40\ M^2 - (0.20\ M)x$$

We can rearrange the terms in this equation to get the standard form of a **quadratic equation**,

$$ax^2 + bx + c = 0$$

as

$$4x^2 + (0.20\ M)x - 0.40\ M^2 = 0$$

If we let $a = 4$, $b = 0.20$ M, and $c = -0.40\ M^2$, then the two roots of this equation are given by the **quadratic formula**, as

$$x = \frac{-b \pm \sqrt{b^2 - 4ac}}{2a} = \frac{-0.20\ M \pm \sqrt{(0.20\ M)^2 - (4)(4)(-0.40\ M^2)}}{(2)(4)}$$

$$= 0.29\ M \text{ and } -0.34\ M$$

We reject the negative root because an equilibrium concentration of $[NO_2] = 2x = -0.68$ M is physically unacceptable; only positive values of concentrations have physical meaning. From the value $x = 0.29$ M, we calculate that at equilibrium

$$[NO_2] = 2x = (2)(0.29\ M) = 0.58\ M$$

$$[N_2O_4] = 2.00\ M - x = 2.00\ M - 0.29\ M = 1.71\ M$$

As a final check, we note that $[NO_2]^2/[N_2O_4]$ is equal to 0.20 M, the value of K_c.

PRACTICE PROBLEM 19-5: Carbon disulfide, which is liquid at room temperature and is used extensively as a solvent, can be prepared by heating sulfur and charcoal at 900 K via the reaction described by

$$2\,S(g) + C(s) \leftrightharpoons CS_2(g) \qquad K_c = 9.40\ M^{-1}$$

What concentration of $CS_2(g)$ can be produced by heating a reaction system containing 0.700 M $S(g)$ with excess charcoal at 900 K until equilibrium is established?

Answer: 0.266 M

EXAMPLE 19-6: At a certain elevated temperature, the reaction between carbon dioxide and excess hot graphite can be described by

$$C(s) + CO_2(g) \leftrightharpoons 2\,CO(g) \qquad K_p = 1.90\ \text{atm} \qquad (19.22)$$

Given that at equilibrium the total pressure in the reaction vessel is 2.00 atm, calculate the values of P_{CO_2} and P_{CO} at equilibrium. Note that this problem involves K_p rather than K_c.

Solution: Here we are given the total pressure at equilibrium, rather than the initial pressures of the two gases, so we must use a different method from the previous Examples to find the gas pressures at equilibrium. Because the total pressure is equal to the sum of the partial pressures of $CO(g)$ and $CO_2(g)$, we have

$$P_{total} = P_{CO} + P_{CO_2} = 2.00\ \text{atm}$$

or $P_{CO_2} = 2.00\ \text{atm} - P_{CO}$. The K_p expression is given by

$$K_p = \frac{P_{CO}^2}{P_{CO_2}} = \frac{P_{CO}^2}{2.00\ \text{atm} - P_{CO}} = 1.90\ \text{atm}$$

Again, we rearrange the terms to get the standard form of a quadratic equation:

$$P_{CO}^2 + (1.90\ \text{atm})P_{CO} - 3.80\ \text{atm}^2 = 0$$

The two roots are 1.22 atm and −3.12 atm. Because pressure must be a positive quantity, we reject −3.12 atm, so $P_{CO} = 1.22$ atm. The pressure of $CO_2(g)$ at equilibrium is thus

$$P_{CO_2} = 2.00\ \text{atm} - P_{CO} = 2.00\ \text{atm} - 1.22\ \text{atm} = 0.78\ \text{atm}$$

As a final check, note that $P_{CO}^2/P_{CO_2} = 1.90\ \text{atm} = K_p$.

PRACTICE PROBLEM 19-6: For the reaction in Example 19-6, suppose that initially only $CO_2(g)$ is injected into the reaction vessel at a pressure of 0.22 atm. What will be the total pressure at equilibrium?

Answer: 0.39 atm

EXAMPLE 19-7: At a certain temperature, the value of the equilibrium constant for the reaction

$$CO(g) + Cl_2(g) \leftrightarrows COCl_2(g) \tag{19.23}$$

is $K_c = 4.0\ M^{-1}$. Calculate the equilibrium concentrations of $CO(g)$, $Cl_2(g)$, and $COCl_2(g)$ when 0.33 moles of $CO(g)$ and 0.33 moles of $Cl_2(g)$ are added to a 1.5-liter reaction vessel.

Solution: The equilibrium-constant expression for the chemical equation is

$$K_c = \frac{[COCl_2]}{[CO][Cl_2]}$$

The initial value of [CO] is 0.33 mol/1.5 L = 0.22 M and that of $[Cl_2]$ is 0.33 mol/1.5 L = 0.22 M. We set up a concentration table, letting x be the change in the concentration of each species needed to reach equilibrium.

Concentration	$CO(g)$	+	$Cl_2(g)$	$\leftrightarrows$	$COCl_2(g)$
initial	0.22 M		0.22 M		0
change	$-x$		$-x$		$+x$
equilibrium	$0.22\ M - x$		$0.22\ M - x$		x

Notice that the changes in concentration of $CO(g)$ and $Cl_2(g)$ are negative and that of $COCl_2(g)$ is positive because in this case the reaction proceeds from reactants toward products in order to reach equilibrium. Also, notice that the coefficient in front of each x is one, in accordance with the reaction stoichiometry.

Substituting the equilibrium concentrations into the K_c expression yields

$$\frac{x}{(0.22\ M - x)^2} = 4.0\ M^{-1}$$

If we multiply both sides of this equation by $(0.22\ M - x)^2$, we obtain

$$x = 0.1936\ M - 1.76x + (4.0\ M^{-1})x^2$$

Rearranging into the standard form of a quadratic formula yields

$$(4.0\ M^{-1})x^2 - 2.76x + 0.1936\ M = 0$$

The two roots of the quadratic equation are given by

$$x = \frac{2.76 \pm \sqrt{7.6176 - 3.098}}{8.0\ M^{-1}} = 0.61\ M \text{ and } 0.08\ M$$

As often happens, we have more than one choice for x, which in this case is $[COCl_2]$. Because we started with 0.22 M $CO(g)$ and 0.22 M $Cl_2(g)$, the *greatest* that $[COCl_2]$ could be is 0.22 M. Thus, we reject the spurious root 0.61 M, and so we have the following concentration table:

Concentration	$CO(g)$	+	$Cl_2(g)$	$\leftrightharpoons$	$COCl_2(g)$
initial	0.22 M		0.22 M		0
change	−0.08 M		−0.08 M		+ 0.08 M
equilibrium	0.14 M		0.14 M		0.08 M

As a final check, we note that

$$K_c = \frac{[COCl_2]}{[CO][Cl_2]} = \frac{0.08\ \text{M}}{(0.14\ \text{M})(0.14\ \text{M})} = 4\ \text{M}^{-1}$$

which agrees (within round-off error) with the stated value of K_c.

PRACTICE PROBLEM 19-7: At elevated temperatures, ammonium iodide decomposes according to

$$NH_4I(s) \leftrightharpoons NH_3(g) + HI(g)$$

(a) If $K_p = 18\ \text{bar}^2$, calculate the total pressure, assuming that only $NH_4I(s)$ was present initially.
(b) What would be the total pressure at equilibrium if the initial pressure of $NH_3(g)$ was 5.0 bar, with $NH_4I(s)$ but no $HI(g)$ present?

Answer: (a) 8.5 bar; (b) 9.8 bar

EXAMPLE 19-8: At 1200°C the value of K_c for the reaction described by

$$H_2(g) + Cl_2(g) \leftrightharpoons 2HCl(g)$$

is 2.5×10^{-4}. Calculate the equilibrium concentrations of $H_2(g)$, $Cl_2(g)$, and $HCl(g)$ when 4.00 moles of each species are added to a 2.0-liter reaction vessel.

Solution: The equilibrium-constant expression is

$$K_c = \frac{[HCl]^2}{[H_2][Cl_2]}$$

Because the same number of moles of each species were added to the reaction vessel, the initial concentration of each species is 4.00 mol/2.00 L = 2.00 M. We set up a concentration table, letting x be the change in the number of moles per liter of $H_2(g)$ and $Cl_2(g)$, and $2x$ be the corresponding change in the number of moles per liter of $HCl(g)$, in accordance with the reaction stoichiometry.

Concentration	$H_2(g)$	+	$Cl_2(g)$	$\leftrightharpoons$	$2HCl(g)$
initial	2.00 M		2.00 M		2.00 M
change	$-x$		$-x$		$+2x$
equilibrium	2.00 M $- x$		2.00 M $- x$		2.00 M $+ 2x$

Substituting the equilibrium concentrations into the K_c expression yields

$$\frac{(2.00\text{ M} + 2x)^2}{(2.00\text{ M} - x)^2} = 2.5 \times 10^{-4}$$

In this case the left-hand side of this expression is a perfect square, so we do not have to solve a quadratic equation. Taking the square root of both sides, we find that

$$\frac{2.00\text{ M} + 2x}{2.00\text{ M} - x} = \pm\sqrt{2.5 \times 10^{-4}}$$

which yields $x = -0.976$ M and -1.024 M. We must think about these two roots. The concentration of HCl(g) must be a positive quantity. If we were to use the $x = -1.024$ M root, then the equilibrium concentration of HCl(g), which is 2.00 M + $2x$, will be a negative quantity. Thus, we must reject this root as spurious. What about the $x = -0.976$ M root? All the concentrations come out to be positive if we use this root. The fact that x is negative here means that we *lose* HCl(g) and *produce* H_2(g) and Cl_2(g) as the reaction comes to equilibrium. In writing our concentration table, we implied that the reaction shifts toward the product side by *subtracting* x from each of the reactants and by *adding* $2x$ to the products; in fact, we should have *added* x to the initial concentrations of H_2(g) and Cl_2(g) and *subtracted* $2x$ from the initial HCl(g) concentration. Because the value of K_c is small, the equilibrium state lies toward the reactants. The negative sign tells us that our initial assumption about the direction the reaction shifts to attain equilibrium was wrong. Nevertheless, we can use the negative value of x we obtained to determine the equilibrium concentrations of each of the species:

Concentration	$H_2(g)$	+	$Cl_2(g)$	⇆	$2HCl(g)$
initial	2.00 M		2.00 M		2.00 M
change	+0.98 M		+0.98 M		−1.95 M
equilibrium	2.98 M		2.98 M		0.05 M

Note that this case is different from that in Example 19-5, where the negative root produced physically unacceptable results.

As a final check, we note that

$$K_c = \frac{[HCl]^2}{[H_2][Cl_2]} = \frac{(0.05\text{ M})^2}{(2.98\text{ M})(2.98\text{ M})} = 3 \times 10^{-4}$$

which agrees (within round-off error) with the given value of K_c.

PRACTICE PROBLEM 19-8: (a) For the reaction given in Example 19-8, calculate the equilibrium concentrations when 1.00 mole of each species are initially present in the 2.00-liter reaction vessel. (b) Are the equilibrium concentrations the same as in Example 19-8? Explain.

Answer: (a) $[H_2] = 0.74$ M; $[Cl_2] = 0.74$ M; $[HCl] = 0.012$ M. (b) They are not the same because the total number of moles has changed. Although K_c is a constant, the values of the specific concentrations at equilibrium depend upon the initial conditions.

19-6. Equilibrium Constants for Chemical Equations Can Be Combined to Obtain Equilibrium Constants for Other Equations

It is sometimes useful to be able to calculate the equilibrium constant for a chemical equation from the equilibrium constants for other algebraically related chemical equations. As an example, consider the pair of equations

$$CO(g) + 2H_2(g) \rightleftharpoons CH_3OH(g) \qquad K_f = \frac{[CH_3OH]}{[CO][H_2]^2} \tag{19.24}$$

$$CH_3OH(g) \rightleftharpoons CO(g) + 2H_2(g) \qquad K_r = \frac{[CO][H_2]^2}{[CH_3OH]} \tag{19.25}$$

The reaction described by Equation 19.25 is simply the reverse of the reaction described by Equation 19.24. Comparison of the equilibrium-constant expressions for these two chemical equations leads to the conclusion that the equilibrium constant of the equation for the reverse reaction is equal to the reciprocal of the equilibrium constant of the equation for the forward reaction:

$$K_r = \frac{1}{K_f} \tag{19.26}$$

Equation 19.26 is a general result that is easily verified for any particular case.

Next, consider the two reactions described by the following equations and their corresponding equilibrium expressions

$$C(s) + H_2O(g) \rightleftharpoons CO(g) + H_2(g) \qquad K_1 = \frac{[CO][H_2]}{[H_2O]} \tag{19.27}$$

$$CO_2(g) + 2H_2(g) \rightleftharpoons 2H_2O(g) + C(s) \qquad K_2 = \frac{[H_2O]^2}{[CO_2][H_2]^2} \tag{19.28}$$

If we add Equation 19.28 to Equation 19.27 and cancel like terms on the two sides of the new equation, then we obtain

$$CO_2(g) + H_2(g) \leftrightharpoons CO(g) + H_2O(g) \tag{19.29}$$

Multiplication of the equilibrium-constant expressions for Equations 19.27 and 19.28 yields

$$K_1K_2 = \frac{[CO][H_2]}{[H_2O]} \times \frac{[H_2O]^2}{[CO_2][H_2]^2} = \frac{[CO][H_2O]}{[CO_2][H_2]}$$

which is the equilibrium-constant expression for Equation 19.29. Thus, we see that

$$K_{(1+2)} = K_1 K_2 \tag{19.30}$$

Equation 19.30 is a general result. If we add two chemical equations to obtain a new equation, then the equilibrium constant of the new equation is equal to the product of the equilibrium constants of the two equations that are added together.

Suppose that we add two identical chemical equations, such as

$$\tfrac{1}{2}\mathrm{H_2}(g) + \tfrac{1}{2}\mathrm{I_2}(g) \leftrightharpoons \mathrm{HI}(g)$$

$$\tfrac{1}{2}\mathrm{H_2}(g) + \tfrac{1}{2}\mathrm{I_2}(g) \leftrightharpoons \mathrm{HI}(g)$$

to get

$$\mathrm{H_2}(g) + \mathrm{I_2}(g) \leftrightharpoons 2\,\mathrm{HI}(g)$$

If the equilibrium constant for each of the first two equations is K_1, then Equation 19.30 gives $K_{(1+1)} = K_1^2$ as the equilibrium constant for their sum. Thus, if we double the stoichiometric coefficients in a chemical equation, then the equilibrium constant for the resulting equation is the square of the equilibrium constant for the original equation. For example, if we have

$$\mathrm{NO_2}(g) \leftrightharpoons \tfrac{1}{2}\mathrm{N_2O_4}(g) \qquad K_\mathrm{p} = 1.22\ \mathrm{atm}^{-1/2}$$

then the value of K_p for

$$2\,\mathrm{NO_2}(g) \leftrightharpoons \mathrm{N_2O_4}(g)$$

is $K_\mathrm{p} = (1.22\ \mathrm{atm}^{-1/2})^2 = 1.49\ \mathrm{atm}^{-1}$.

More generally, if we multiply a chemical equation with a corresponding equilibrium constant, K_p, by some number n, then the equilibrium constant of the resulting equation will be K_p^n. (The same relation holds true for expressions described by K_c.)

EXAMPLE 19-9: At 240°C,

$$\mathrm{NO_2}(g) \leftrightharpoons \mathrm{NO}(g) + \tfrac{1}{2}\mathrm{O_2}(g) \qquad K_\mathrm{p} = 0.0386\ \mathrm{bar}^{1/2}$$

What is the value of K_p for

$$2\,\mathrm{NO}(g) + \mathrm{O_2}(g) \leftrightharpoons 2\,\mathrm{NO_2}(g)$$

Solution: To obtain the second equation from the first, we multiply the first equation by 2 and then reverse it. Consequently, we square the given equilibrium constant and take its reciprocal to obtain $K_\mathrm{p} = 671\ \mathrm{bar}^{-1}$.

PRACTICE PROBLEM 19-9: Given

$$2\,SO_3(g) \leftrightharpoons 2\,SO_2(g) + O_2(g) \qquad K_p = 0.29\text{ atm}$$

determine the value of K_p for

$$SO_2(g) + \frac{1}{2}O_2(g) \leftrightharpoons SO_3(g)$$

Answer: $K_p = 1.9\text{ atm}^{-1/2}$

Notice that several of the examples in this section once again show that the equilibrium-constant expression and its associated value depend upon how we choose to represent a chemical reaction by a chemical equation.

19-7. Le Châtelier's Principle Is Used to Predict the Direction of the Shift in a Chemical Reaction Displaced from Equilibrium

Consider a chemical reaction that is initially at equilibrium. Often a change in conditions will displace the reaction from equilibrium. The reaction then shifts toward one direction or the other (left to right or right to left as the equation is written) as it proceeds to a new equilibrium state. Henri Le Châtelier (pronounced luh shat′elyay), a French physical chemist, was the first to explain how to predict the direction of the shift (Figure 19.5). We do not need to know the numerical value of the equilibrium constant or perform calculations to apply **Le Châtelier's principle**, that can be stated as follows:

> *If a chemical reaction at equilibrium is subjected to a change in conditions that displaces it from equilibrium, then the reaction adjusts toward a new equilibrium state. The reaction proceeds in the direction that—at least partially—offsets the change in conditions.*

Figure 19.5 Henri Le Châtelier (1850–1936), a French chemist, received his training in chemistry and mathematics from his father, an accomplished engineer who helped to establish France's aluminum industry. His mother was a strong disciplinarian and a devout Catholic from Italy who fostered in him a strong sense of order and discipline. Despite his training in engineering, Le Châtelier chose to pursue an academic career in chemistry, investigating the nature of ceramics, cements, and alloys. He is most famous for postulating the principle that now bears his name. In 1907 Le Châtelier was elected to the French Academy of Sciences.

The quantities whose change may affect a reaction equilibrium are

1. concentration of a reactant or product
2. reaction volume or applied pressure
3. temperature

Although Le Châtelier's principle sounds imposing, it is actually simple to apply. A key point to recognize is that the value of the equilibrium constant depends only on the temperature; it does not change when the reactant or product concentrations, the reaction volume, or the applied pressure is changed.

Consider the reaction equilibrium described by the equation

$$C(s) + CO_2(g) \leftrightharpoons 2\,CO(g)$$

If we disturb the equilibrium by injecting some additional $CO_2(g)$ into the reaction vessel, then the concentration of $CO_2(g)$ is increased. In response to the change in conditions, the reaction equilibrium shifts from left to right, so more $CO_2(g)$ is consumed; this shift helps reduce the increased $CO_2(g)$ concentration. In the new equilibrium state, the concentration of $CO_2(g)$ and the concentration of $CO(g)$ are both greater than in the original equilibrium state, but the concentration of $CO_2(g)$ in the new equilibrium state is less than it was immediately after the additional $CO_2(g)$ was injected (Figure 19.6).

If we disturb the reaction equilibrium by injecting some additional $CO(g)$ into the reaction vessel, then the concentration of $CO(g)$ is increased. This time the reaction equilibrium shifts from right to left, because a shift in this direction decreases the $CO(g)$ concentration (Figure 19.7).

If we add some additional $C(s)$ to the reaction vessel, while keeping the total pressure constant, then there is no shift in the reaction equilibrium because the concentration of a solid is independent of the amount present (Figure 19.8). In other words, the addition or removal of some $C(s)$ does not displace the reaction from equilibrium. Typically, the further addition or partial removal of any pure solid or liquid reactant or product does not shift the reaction equilibrium.

In general, a decrease in volume shifts a reaction equilibrium toward the side with the smaller number of moles of gas. If the number of moles of gaseous products is greater than the number of moles of gaseous reactants, as in $N_2O_4(g) \leftrightharpoons 2NO_2(g)$, then a decrease in volume shifts the reaction equilib-

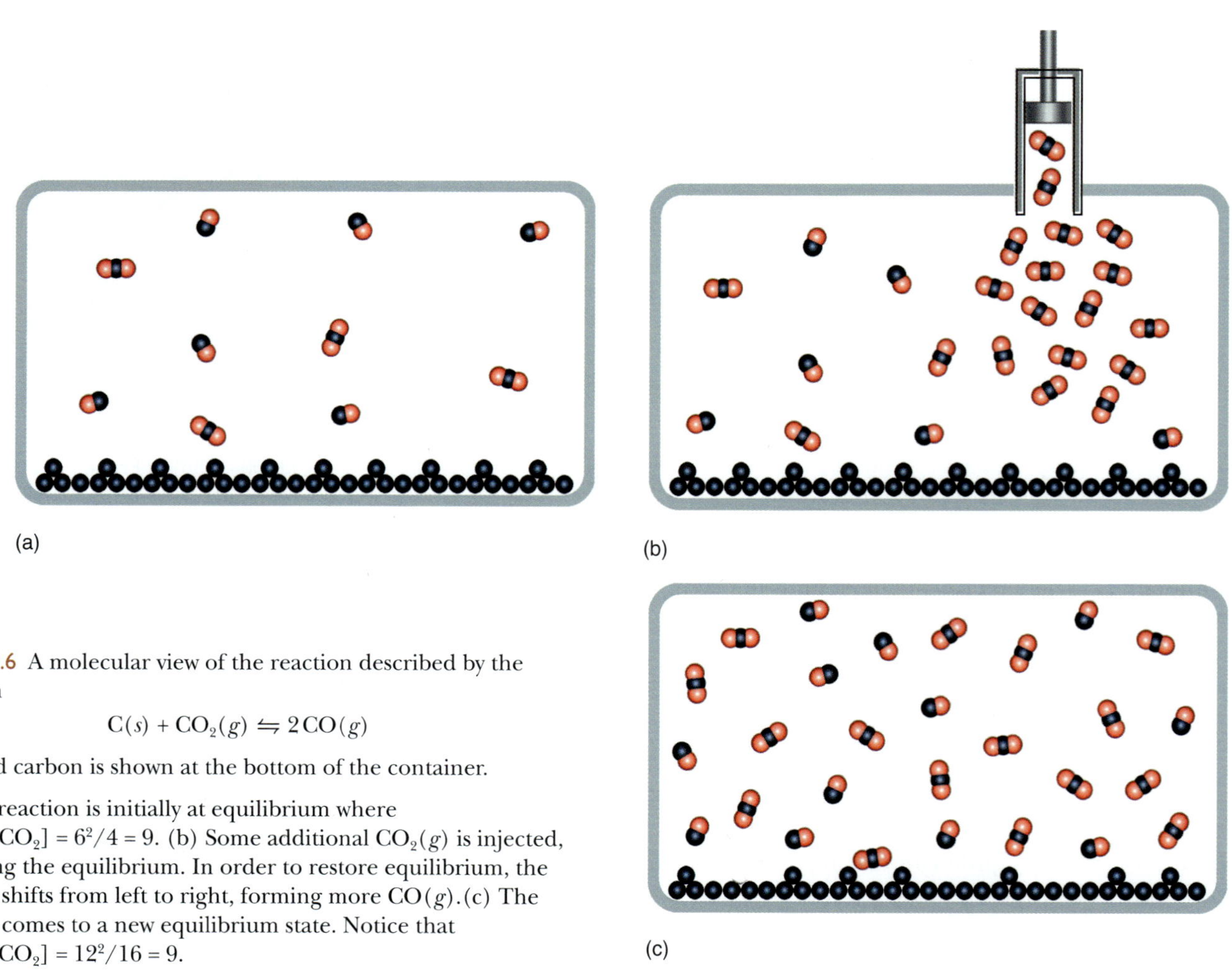

Figure 19.6 A molecular view of the reaction described by the equation

$$C(s) + CO_2(g) \leftrightharpoons 2CO(g)$$

The solid carbon is shown at the bottom of the container.

(a) The reaction is initially at equilibrium where $[CO]^2/[CO_2] = 6^2/4 = 9$. (b) Some additional $CO_2(g)$ is injected, disturbing the equilibrium. In order to restore equilibrium, the reaction shifts from left to right, forming more $CO(g)$. (c) The reaction comes to a new equilibrium state. Notice that $[CO]^2/[CO_2] = 12^2/16 = 9$.

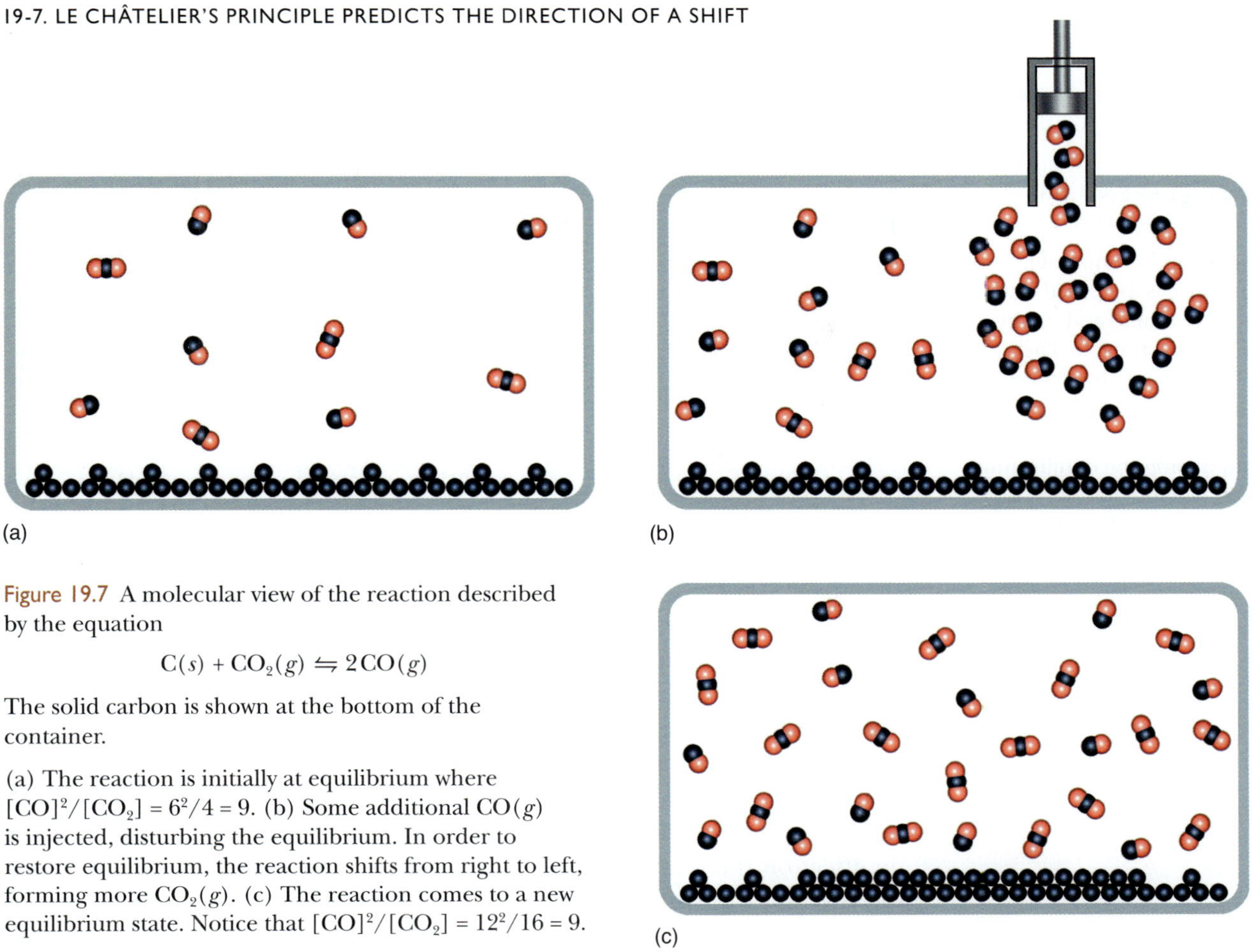

Figure 19.7 A molecular view of the reaction described by the equation

$$C(s) + CO_2(g) \leftrightharpoons 2CO(g)$$

The solid carbon is shown at the bottom of the container.

(a) The reaction is initially at equilibrium where $[CO]^2/[CO_2] = 6^2/4 = 9$. (b) Some additional $CO(g)$ is injected, disturbing the equilibrium. In order to restore equilibrium, the reaction shifts from right to left, forming more $CO_2(g)$. (c) The reaction comes to a new equilibrium state. Notice that $[CO]^2/[CO_2] = 12^2/16 = 9$.

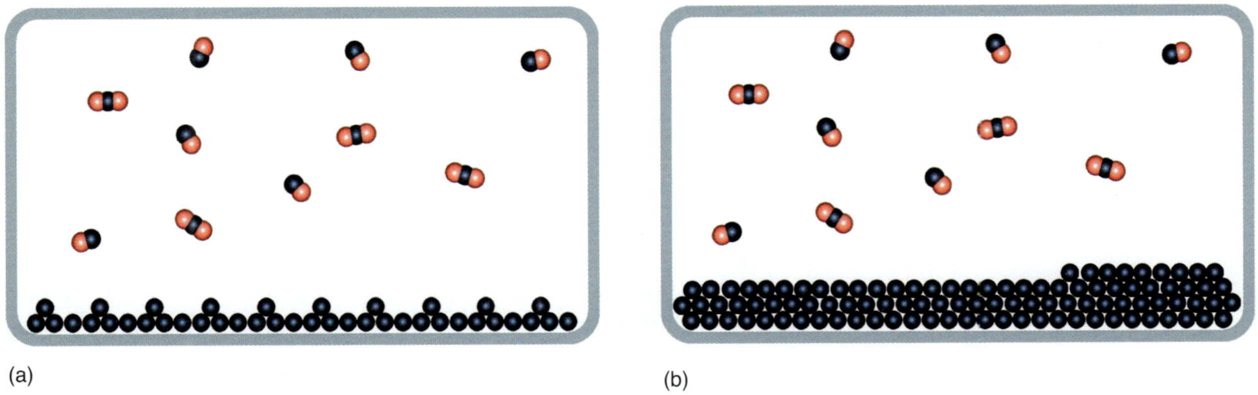

Figure 19.8 A molecular view of the reaction described by the equation

$$C(s) + CO_2(g) \leftrightharpoons 2CO(g)$$

The solid carbon is shown at the bottom of the container.

(a) The reaction is initially at equilibrium where $[CO]^2/[CO_2] = 6^2/4 = 9$. (b) Some additional $C(s)$ is added. Because the carbon is a solid, it does not appear in the equilibrium expression. The equilibrium is undisturbed and there is no change to the equilibrium position.

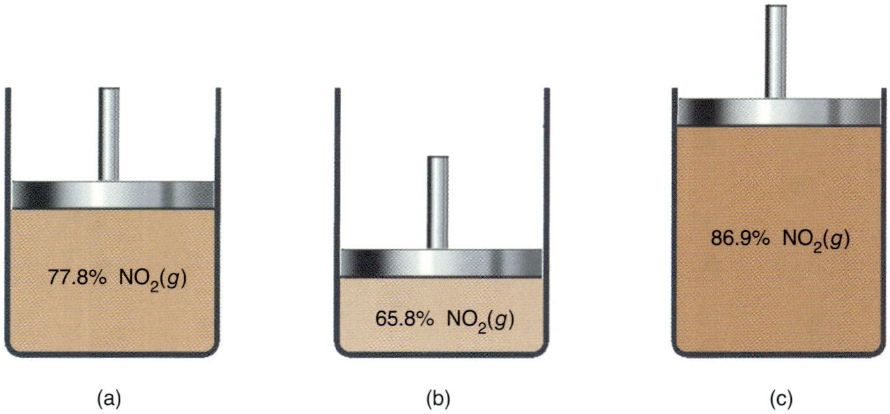

Figure 19.9 The effect of pressure (at a fixed temperature) on the equilibrium

$$\underset{\text{colorless}}{N_2O_4(g)} \leftrightharpoons \underset{\text{brown}}{2NO_2(g)}$$

(a) An initial equilibrium state. (b) The pressure is increased, thus decreasing the volume and driving the equilibrium to the left. The fraction of $NO_2(g)$ in (b) is therefore less than that in (a). (c) The pressure is decreased, thus increasing the volume and driving the equilibrium to the right. The fraction of $NO_2(g)$ in (c) is greater than that in (a) or (b).

rium from right to left (Figure 19.9). If the number of moles of gaseous products is less than that of gaseous reactants, as in $N_2(g) + 3H_2(g) \leftrightharpoons 2NH_3(g)$, then a decrease in volume shifts the reaction equilibrium from left to right. In both cases the observed shift partially offsets the increased pressure that results when the volume is decreased because a shift to the side of the reaction with the smaller number of moles of gas decreases the total number of gas molecules in the reaction system. If the number of moles of gas is the same on both sides, as in the equation $H_2(g) + I_2(g) \leftrightharpoons 2HI(g)$, then a change in volume has no effect on the reaction equilibrium because there is no direction in which the reaction can shift to change the total number of gas molecules per unit volume.

EXAMPLE 19-10: Should the total volume of the equilibrium reaction mixture described by the equation

$$C(s) + CO_2(g) \leftrightharpoons 2CO(g)$$

be increased or decreased in order to increase the extent of conversion of $C(s)$ and $CO_2(g)$ to $CO(g)$?

Solution: An increase in total volume favors the side of the equation with the greater number of moles of gas. Thus, an increase in volume would shift this equilibrium from left to right, thereby increasing the production of $CO(g)$ from $C(s)$ and $CO_2(g)$. In assessing the effect of a volume change on a reaction equilibrium, we do not have to consider pure liquid and solid phases because the change in total gas volume does not affect the concentration of species in solid and liquid phases.

PRACTICE PROBLEM 19-10: For the equilibrium discussed in Example 19-10, use Le Châtelier's principle to predict the effect on the equilibrium concentration of $CO_2(g)$ by (a) decreasing the concentration of $CO(g)$, (b) decreasing the amount of solid carbon in the system.

Answer: (a) $[CO_2]$ decreases, (b) no effect

From Boyle's law, we know that the volume of an ideal gas is inversely proportional to the applied pressure. Thus, a decrease in volume has the same effect as an increase in applied pressure and vice versa. For instance, in Example 19-10, Le Châtelier's principle says that we should decrease the pressure on the reaction system in order to produce more $CO(g)$. A decrease in the applied pressure is equivalent to an increase in the reaction volume.

Changes in external pressure can significantly affect even equilibria involving only solids and liquids. The conversion between diamond and graphite, the two allotropic forms of carbon (Figure 19.10), illustrates this application of Le Châtelier's principle. Graphite is the stable form at ordinary pressures, whereas diamond is the stable form at high pressures, such as thousands of bars. The molar volume of graphite is 5.5 $cm^3 \cdot mol^{-1}$, and the molar volume of diamond is 3.4 $cm^3 \cdot mol^{-1}$. When a pressure in excess of about 15000 bars is applied to graphite, some of the effect of the pressure can be relieved by converting graphite to diamond, which has a smaller molar volume. This effect also explains why ice floats on water (Figure 19.11) and melts under pressure: liquid water has a smaller molar volume than ice.

Figure 19.10 Two allotropes of carbon are graphite and diamond.

Let's now consider the effect of a change in temperature upon a reaction at equilibrium. The value of an equilibrium constant varies with temperature. For some reactions, the value of the equilibrium constant increases with increasing temperature, whereas for others it decreases. Let's first consider the endothermic reaction described by

$$\underset{\text{colorless}}{N_2O_4(g)} \leftrightharpoons \underset{\text{brown}}{2\,NO_2(g)} \qquad \Delta H^\circ_{rxn} = +55.3\ \text{kJ}\cdot\text{mol}^{-1}$$

This reaction requires the input of energy as heat, so we can write the equation for the reaction as

$$N_2O_4(g) + 55.3\ \text{kJ}\cdot\text{mol}^{-1} \leftrightharpoons 2\,NO_2(g)$$

or simply

$$N_2O_4(g) + \text{heat} \leftrightharpoons 2\,NO_2(g)$$

Suppose now that we increase the temperature, making more energy as heat available. Le Châtelier's principle tells us that the system will absorb heat to

Figure 19.11 Water is unusual in that its solid form (ice) is less dense than its liquid form. The beaker on the right contains ice and water; the beaker on the left contains solid and liquid benzene. Observe that the ice floats on the water, but solid benzene, being more dense than liquid benzene, sits at the bottom of the beaker.

Figure 19.12 Effect of temperature on the reaction equilibrium

$$\underset{\text{colorless}}{N_2O_4(g)} \leftrightharpoons \underset{\text{brown}}{2NO_2(g)}$$

An increase in temperature from 0°C (ice water) to 25°C converts some of the $N_2O_4(g)$ to $NO_2(g)$ and results in a darker color for the reaction mixture.

offset this change, so the equilibrium shifts from left to right. The left-to-right shift produces an increase in the equilibrium concentration of $NO_2(g)$ in the system. A decrease in temperature produces a shift in the opposite direction (Figure 19.12).

The value of ΔH°_{rxn} for the chemical equation

$$2Cu(s) + O_2(g) \leftrightharpoons Cu_2O(s)$$

is $-168.6\ kJ\cdot mol^{-1}$. In this case energy as heat is evolved by the reaction, and we write

$$2Cu(s) + O_2(g) \leftrightharpoons Cu_2O(s) + \text{heat}$$

An increase in temperature increases the availability of energy as heat to the reaction. Thus, Le Châtelier's principle tells us that the equilibrium shifts from right to left, thereby producing an increase in the pressure of $O_2(g)$. Again, the result of a decrease in temperature is a shift in the opposite direction.

We can summarize these results as follows:

Type of reaction	Temperature change	Direction of change
endothermic	increase	→
($\Delta H^\circ_{rxn} > 0$)	decrease	←
exothermic	increase	←
($\Delta H^\circ_{rxn} < 0$)	decrease	→

EXAMPLE 19-11: For the reaction described by the equation

$$N_2(g) + 3H_2(g) \leftrightharpoons 2NH_3(g)$$

$\Delta H^\circ_{rxn} = -91.8\ kJ\cdot mol^{-1}$. Will an increase in the temperature increase or decrease the extent of conversion of $N_2(g)$ and $H_2(g)$ to $NH_3(g)$?

Solution: The value of ΔH°_{rxn} is negative, so the reaction evolves energy as heat:

$$N_2(g) + 3H_2(g) \leftrightharpoons 2NH_3(g) + \text{heat}$$

An increase in temperature favors the absorption of energy as heat and thus shifts the equilibrium to the left, thereby decreasing the yield of ammonia.

PRACTICE PROBLEM 19-11: Use the data in Table 14.3 to determine whether an increase or decrease in temperature will favor the production of $SO_3(g)$ by the process

$$SO_2(g) + \tfrac{1}{2}O_2(g) \leftrightharpoons SO_3(g)$$

Answer: decrease in temperature

Unlike the effect of a change in the concentrations of any of the reactants or products, or a change in the reaction volume or applied pressure, a change in temperature changes the value of the equilibrium constant. For an exothermic reaction, the equilibrium shifts to the left with increasing temperature, so the value of the equilibrium constant decreases. For an endothermic reaction, the equlibrium shifts to the right with increasing temperature, so the value of the equilibrium constant increases. For example, for the exothermic reaction described by

$$H_2(g) + I_2(g) \leftrightharpoons 2HI(g) \qquad \Delta H^\circ_{rxn} = -9.4\ \text{kJ}\cdot\text{mol}^{-1}$$

the value of the equilibrium constant decreases with increasing temperature. Some experimental values of K_c as a function of temperature for this reaction are given in Table 19.2. For the endothermic reaction described by

$$N_2(g) + O_2(g) \leftrightharpoons 2NO(g) \qquad \Delta H^\circ_{rxn} = +182.6\ \text{kJ}\cdot\text{mol}^{-1}$$

the value of the equilibrium constant increases with increasing temperature. Some experimental values of K_c as a function of temperature for this reaction are given in Table 19.3. A quantitative discussion of the temperature dependence of K_c will be given in Chapter 23 when we study thermodynamics.

An important application of the ideas in this section is the **Haber process**, developed by German Nobel laureate Fritz Haber. In the Haber process, ammonia is produced commercially by the direct reaction of nitrogen and hydrogen according to

$$N_2(g) + 3H_2(g) \leftrightharpoons 2NH_3(g)$$

Because of the tremendous scale of ammonia production (10 762 metric tons were produced in the United States in 2005, Appendix H), it is important that the reaction be run under the most favorable conditions (Figure 19.13).

Figure 19.13 An ammonia plant. This plant produces 750 tons of ammonia per day from hydrogen and nitrogen. The nitrogen comes from the air and the hydrogen is obtained from the reaction between methane and steam. Ammonia is used in the production of fertilizers, explosives, polymers, and household cleansers.

Because there are more moles of gaseous species on the left of the above equation, a decrease in the total reaction volume, or equivalently, an increase in the total reaction pressure favors the conversion of reactants to products. Thus, the percentage conversion of $N_2(g)$ and $H_2(g)$ to $NH_3(g)$ (the yield) increases as the total pressure increases. The value of ΔH°_{rxn} for the reaction is -92 kJ·mol^{-1}, so we write

$$N_2(g) + 3H_2(g) \leftrightharpoons 2NH_3(g) + \text{heat}$$

TABLE 19.2 The variation of K_c with temperature for the exothermic reaction $H_2(g) + I_2(g) \leftrightharpoons 2HI(g)$

t/°C	K_c
340	70.8
380	61.9
420	53.7
460	46.8

TABLE 19.3 The variation of K_c with temperature for the endothermic reaction $N_2(g) + O_2(g) \leftrightharpoons 2NO(g)$

T/K	$K_c/10^{-4}$
2000	4.08
2200	11.0
2400	25.1
2600	50.3

TABLE 19.4 Relation between the equilibrium constant and increasing temperature for the ammonia synthesis reaction described by $N_2(g) + 3H_2(g) \leftrightharpoons 2NH_3(g)$ where $K_p = P^2_{NH_3}/P_{N_2}P^3_{H_2}$

t/°C	K_p/bar^{-2}
25	5.8×10^5
227	0.097
500	1.5×10^{-5}

TABLE 19.5 Percentage of conversion of reactants to ammonia at 500°C

P_{total}/bar	P_{NH_3}/bar	Conversion of 3:1 ratio of H_2:N_2 to NH_3/%
1.00	1.26×10^{-3}	0.25
300	152	30

Thus, an increase in temperature favors the conversion of products back to reactants; that is, the equilibrium shifts from right to left when the temperature is increased. The equilibrium constant decreases with increasing temperature, as shown in Table 19.4. Thus, at a given total pressure the percentage conversion of nitrogen and hydrogen to ammonia *decreases* as the temperature increases.

The application of Le Châtelier's principle to the ammonia synthesis reaction leads to the prediction that *at equilibrium* the yield of ammonia is greater the higher the total pressure and the lower the temperature. However, the *rate* of the reaction at 25°C is negligibly slow. A high yield is of no commercial value if it takes forever to achieve conversion. Because the rate of most reactions increases with increasing temperature, the ammonia production reaction is run at an elevated temperature (500°C), even though the equilibrium yield is not as favorable as at lower temperatures. In order to make the process economically feasible, the low value of K_p at 500°C is offset by using a very high pressure (300 bar), as shown in Table 19.5. The Haber process is thus based on a compromise between equilibrium (yield) and rate (speed of reaction) considerations.

19-8. Le Châtelier's Principle Has a Quantitative Basis

We can discuss the changes predicted by Le Châtelier's principle quantitatively. For example, consider the high-temperature reaction given by

$$CO_2(g) + H_2(g) \leftrightharpoons CO(g) + H_2O(g)$$

Suppose that at equilibrium the concentrations are $[CO_2] = 0.075$ M, $[H_2] = 0.060$ M, $[CO] = 0.100$ M, and $[H_2O] = 0.025$ M. Now suppose that additional $CO_2(g)$ is injected into the mixture and its concentration suddenly jumps to 0.125 M. What will be the concentration of each gas when equilibrium is reestablished?

We can calculate the value of the equilibrium constant from the first set of equilibrium concentrations:

$$K_c = \frac{[CO][H_2O]}{[CO_2][H_2]} = \frac{(0.100\text{ M})(0.025\text{ M})}{(0.075\text{ M})(0.060\text{ M})} = 0.56$$

To determine the new equilibrium concentrations, we must set up a concentration table. From Le Châtelier's principle, we know that the position of the equilibrium will shift to the right in order to offset the change caused by the addition of the $CO_2(g)$. Thus, some $CO_2(g)$ and $H_2(g)$ will react to produce equivalent amounts of $CO(g)$ and $H_2O(g)$. If we let x be the change in the number of moles per liter of each species required to reestablish the equilibrium, then the concentration table will be

Concentration	$CO_2(g)$	+	$H_2(g)$	$\leftrightharpoons$	$CO(g)$	+	$H_2O(g)$
initial	0.125 M		0.060 M		0.100 M		0.025 M
change	$-x$		$-x$		$+x$		$+x$
equilibrium	$0.125\text{ M} - x$		$0.060\text{ M} - x$		$0.100\text{ M} + x$		$0.025\text{ M} + x$

We substitute the new equilibrium concentrations into the equilibrium-constant expression,

$$\frac{(0.100\text{ M} + x)(0.025\text{ M} + x)}{(0.125\text{ M} - x)(0.060\text{ M} - x)} = 0.56$$

We now perform the multiplication steps and then collect like terms to obtain

$$0.44x^2 + (0.2286\text{ M})x - 0.0017\text{ M}^2 = 0$$

The two roots are given by

$$x = \frac{-0.2286\text{ M} \pm 0.235\text{ M}}{0.88} = 0.007\text{ M and } -0.527\text{ M}$$

We reject the $x = -0.527$ M root because this value would result in negative equilibrium concentrations for [CO] and $[H_2O]$. Thus, $x = 0.007$ M and the new equilibrium concentrations are given in the following table:

Concentration	$CO_2(g)$	+	$H_2(g)$	$\leftrightharpoons$	$CO(g)$	+	$H_2O(g)$
initial	0.125 M		0.060 M		0.100 M		0.025 M
change	−0.007 M		−0.007 M		+0.007 M		+0.007 M
equilibrium	0.118 M		0.053 M		0.107 M		0.032 M

Notice that $[H_2]$ has dropped from 0.060 M to 0.053 M as a result of the added $CO_2(g)$. Both [CO] and $[H_2O]$ have increased, and $[CO_2]$ is less than it was immediately after the addition, but greater than it was before.

As a final check, we show that

$$\frac{[CO][H_2O]}{[CO_2][H_2]} = \frac{(0.107\text{ M})(0.032\text{ M})}{(0.118\text{ M})(0.053\text{ M})} = 0.55$$

which agrees (within round-off error) with $K_c = 0.56$ calculated above.

EXAMPLE 19-12: At 200°C, for an equilibrium mixture of $PCl_3(g)$, $Cl_2(g)$, and $PCl_5(g)$, the partial pressures are $P_{PCl_3} = 0.84$ atm, $P_{Cl_2} = 1.32$ atm, and $P_{PCl_5} = 5.89$ atm. More $PCl_5(g)$ is injected into the mixture, and the total pressure of the mixture jumps to 9.29 atm. What are the final partial pressures once equilibrium is reestablished? The equation for the reaction between these gases is

$$PCl_3(g) + Cl_2(g) \leftrightharpoons PCl_5(g)$$

Solution: The value of the equilibrium constant is given by the first set of equilibrium pressures:

$$K_p = \frac{P_{PCl_5}}{P_{PCl_3}P_{Cl_2}} = \frac{5.89\text{ atm}}{(0.84\text{ atm})(1.32\text{ atm})} = 5.31\text{ atm}^{-1}$$

The pressure of $PCl_5(g)$ immediately after the addition of $PCl_5(g)$ can be calculated from the relation for the total pressure

$$P_{total} = P_{PCl_3} + P_{Cl_2} + P_{PCl_5}$$

Substituting the given values into this relation yields

$$9.29\text{ atm} = 0.84\text{ atm} + 1.32\text{ atm} + P_{PCl_5}$$

or

$$P_{PCl_5} = 7.13\text{ atm}$$

To determine the new equilibrium conditions, we set up a pressure table. From Le Châtelier's principle, we know that the position of the equilibrium will shift from right to left in response to the addition of $PCl_5(g)$. Thus, some $PCl_5(g)$ will decompose to produce equivalent amounts of $PCl_3(g)$ and $Cl_2(g)$. If we let x be the number of atmospheres of $PCl_5(g)$ that decomposes, then the pressure table will be

Pressure	**$PCl_3(g)$**	+	**$Cl_2(g)$**	⇋	**$PCl_5(g)$**
initial	0.84 atm		1.32 atm		7.13 atm
change	$+x$		$+x$		$-x$
equilibrium	0.84 atm + x		1.32 atm + x		7.13 atm − x

We substitute these equilibrium pressures into the equilibrium-constant expression to get

$$\frac{7.13\text{ atm} - x}{(0.84\text{ atm} + x)(1.32\text{ atm} + x)} = 5.31\text{ atm}^{-1}$$

This expression can be written out as

$$(5.31\text{ atm}^{-1})x^2 + 12.47x - 1.24\text{ atm} = 0$$

The two roots of this equation are

$$x = 0.10\text{ atm and } -2.44\text{ atm}$$

Because x must be positive in this case, we reject the negative root to obtain the following table:

Pressure	**$PCl_3(g)$**	+	**$Cl_2(g)$**	⇋	**$PCl_5(g)$**
initial	0.84 atm		1.32 atm		7.13 atm
change	+0.10 atm		+0.10 atm		−0.10 atm
equilibrium	0.94 atm		1.42 atm		7.03 atm

We see that not all the added $PCl_5(g)$ has reacted (1.24 atm were added, but only 0.10 atm reacted). The net result is that the pressure of each reactant and product is greater at the new equilibrium than at the initial equilibrium. As a final check, we show that

$$\frac{P_{PCl_5}}{P_{PCl_3}P_{Cl_2}} = \frac{7.03 \text{ atm}}{(0.94 \text{ atm})(1.42 \text{ atm})} = 5.3 \text{ atm}^{-1}$$

which agrees to within round-off error with $K_p = 5.31 \text{ atm}^{-1}$.

PRACTICE PROBLEM 19-12: At 250°C an equilibrium mixture of $SbCl_3(g)$, $Cl_2(g)$, and $SbCl_5(g)$ has the partial pressures 0.670 bar, 0.438 bar, and 0.228 bar, respectively. Calculate the new equilibrium pressures if the volume of the reaction vessel is doubled.

Answer: $P_{Cl_2} = 0.258$ bar; $P_{SbCl_3} = 0.374$ bar; $P_{SbCl_5} = 0.075$ bar

19-9. Chemical Reactions Always Proceed Toward Equilibrium

At 100°C the equilibrium constant for the chemical reaction described by the equation

$$N_2O_4(g) \leftrightharpoons 2NO_2(g)$$

is

$$K_c = \frac{[NO_2]^2}{[N_2O_4]} = 0.20 \text{ M} \tag{19.31}$$

We now define the **reaction quotient** Q_c as the quantity that has exactly the same algebraic form as the equilibrium-constant expression but involves arbitrary (not necessarily equilibrium) concentrations. Thus, for the $N_2O_4(g)$-$NO_2(g)$ reaction above, we have the Q_c expression

$$Q_c = \frac{[NO_2]_0^2}{[N_2O_4]_0} \tag{19.32}$$

where the zero subscripts denote arbitrary concentrations. Remember that only equilibrium values of the concentrations can be used in the K_c expression because Equation 19.31 applies only at equilibrium. We stress that *any* values of the concentrations can be used in Equation 19.32.

The reaction quotient Q_c is not a constant; instead, its value depends on how we prepare the reaction system. For example, suppose we mix 2.00 moles of $N_2O_4(g)$ with 2.00 moles of $NO_2(g)$ in a 1.00-liter reaction vessel at 100°C. Then

$$Q_c = \frac{[NO_2]_0^2}{[N_2O_4]_0} = \frac{(2.00 \text{ M})^2}{2.00 \text{ M}} = 2.00 \text{ M}$$

The value of K_c at 100°C for this reaction is 0.20 M. Because $Q_c \neq K_c$, we know that the reaction mixture is not at equilibrium. At equilibrium, the ratio of $[NO_2]^2$ to $[N_2O_4]$ must equal 0.20 M (Equation 19.31). For the reaction system to attain equilibrium, Q_c must decrease from 2.00 M to 0.20 M. Therefore,

TABLE 19.6 Relation between the value of Q_c/K_c and the direction in which a reaction proceeds to attain equilibrium (the same relations apply to the value of Q_p/K_p)

Value of Q_c/K_c	Direction of reaction to attain equilibrium
$Q_c/K_c < 1$ or $Q_c < K_c$	$\rightarrow$
$Q_c/K_c > 1$ or $Q_c > K_c$	$\leftarrow$
$Q_c/K_c = 1$ or $Q_c = K_c$	no net change (equilibrium state)

$[NO_2]_0$ must decrease and $[N_2O_4]_0$ must increase; consequently, the reaction proceeds from right to left toward its equilibrium concentrations. When a reaction system reaches equilibrium, the value of Q_c is equal to the value of K_c; likewise, at equilibrium $Q_p = K_p$ for pressure units. For any reaction at equilibrium,

$$\frac{Q_c}{K_c} = 1 \quad \text{or} \quad Q_c = K_c \quad \text{(at equilibrium)} \tag{19.33}$$

with a similar relation between Q_p and K_p.

The **direction of reaction spontaneity** is always toward equilibrium. The numerical value of the ratio Q_c/K_c tells us the direction (left to right or right to left) in which a reaction system not at equilibrium spontaneously proceeds. The various possibilities are given in Table 19.6. In other words, a system that is not in equilibrium proceeds toward equilibrium in the direction in which Q_c approaches K_c in magnitude, or Q_p approaches K_p in magnitude.

EXAMPLE 19-13: Suppose that $CO_2(g)$ and $CO(g)$ are brought into contact with $C(s)$ at 1000 K, where $P_{CO_2} = 2.00$ bar and $P_{CO} = 0.50$ bar. Is the reaction described by the equation

$$C(s) + CO_2(g) \leftrightharpoons 2CO(g) \qquad K_p = 1.93 \text{ bar}$$

at equilibrium? If not, in what direction will the reaction proceed to attain equilibrium?

Solution: The value of Q_p for the reaction system as prepared is

$$Q_p = \frac{P_{CO}^2}{P_{CO_2}} = \frac{(0.50 \text{ bar})^2}{2.00 \text{ bar}} = 0.13 \text{ bar}$$

The value of K_p is given as 1.93, so

$$\frac{Q_p}{K_p} = \frac{0.13 \text{ bar}}{1.93 \text{ bar}} < 1$$

Therefore, the reaction system is not at equilibrium. Because $Q_p/K_p < 1$, the reaction proceeds from left to right, with P_{CO} increasing and P_{CO_2} decreasing until $P_{CO}^2/P_{CO_2} = 1.93$ bar, that is, until equilibrium is attained.

PRACTICE PROBLEM 19-13: For the reaction equation in Example 19-13, suppose that $P_{CO_2} = 150$ Torr and $P_{CO} = 485$ Torr. Is the reaction at equilibrium? If not, in what direction will the reaction proceed to attain equilibrium?

Answer: no; right to left

SUMMARY

A chemical equilibrium is a dynamic equilibrium. At equilibrium the rates of the forward and reverse reactions are equal and there is no net change in the system. A chemical equilibrium is characterized quantitatively by the equilibrium-constant expression for the reaction. The important points to remember about equilibrium-constant expressions are that:

1. The value of K_c or K_p is equal to a constant at a given temperature.
2. The equilibrium product concentrations appear in the numerator and the equilibrium reactant concentrations appear in the denominator.
3. Each equilibrium concentration is raised to a power equal to its stoichiometric coefficient in the balanced chemical equation.
4. Reactants and products that appear as pure solids or pure liquids in the chemical equation do not appear in the equilibrium-constant expression or by convention are assigned a numeric value of one.
5. The equilibrium constant of the equation for the reverse reaction is equal to the reciprocal of the equilibrium constant of the equation for the forward reaction.
6. The equilibrium constant for a chemical equation obtained by algebraically adding two chemical equations is equal to the product of the equilibrium constants for the two equations that are added together.
7. Equilibrium is a dynamic state; at equilibrium the rates of the forward and reverse reactions are equal.
8. The value of an equilibrium constant depends on how the chemical equation that describes the reaction is written.

Le Châtelier's principle predicts the direction in which a reaction mixture in equilibrium shifts in response to a change in conditions, such as a change in reactant or product concentration, or in temperature, or in volume. The direction in which a reaction mixture not at equilibrium proceeds to attain equilibrium can also be predicted from the value of the ratio of the reaction quotient to the reaction equilibrium constant, Q_c/K_c or Q_p/K_p.

TERMS YOU SHOULD KNOW

forward reaction *686*
reverse reaction *686*
chemical equilibrium *686*
dynamic equilibrium *686*
equilibrium state *686*
equilibrium constants, K_c and K_p *690*
equilibrium-constant expression *690*
concentration table *694*
quadratic equation *698*
quadratic formula *698*
Le Châtelier's principle *705*
Haber process *711*
reaction quotients, Q_c and Q_p *715–716*
direction of reaction spontaneity *716*

EQUATIONS YOU SHOULD KNOW HOW TO USE

$K_c = \dfrac{[X]^x[Y]^y}{[A]^a[B]^b}$	(19.6)	(definition of equilibrium constant for $aA(g) + bB(soln) + cC(s) \leftrightharpoons xX(g) + yY(soln) + zZ(l)$)
$K_p = K_c(RT)^{\Delta\nu_{gas}}$	(19.18)	(relationship between K_p and K_c)
$K_r = \dfrac{1}{K_f}$	(19.26)	(relation between equilibrium constants of forward and reverse reactions)
$K_{(1+2)} = K_1K_2$	(19.30)	(equilibrium constant of the sum of two chemical equations)
Value of Q_c/K_c or Q_p/K_p compared with unity	(Table 19.6)	(direction in which a reaction proceeds to attain equilibrium)

PROBLEMS

CALCULATION OF EQUILIBRIUM CONCENTRATION

19-1. Antimony pentachloride decomposes according to the equation

$$SbCl_5(g) \leftrightharpoons SbCl_3(g) + Cl_2(g)$$

Suppose that the initial concentrations are $[SbCl_5]_0 = 0.165$ M, $[SbCl_3]_0 = 0.0955$ M, and $[Cl_2]_0 = 0.210$ M. If it is determined that $[SbCl_5] = 0.135$ M at equilibrium, calculate the equilibrium values of $[SbCl_3]$ and $[Cl_2]$.

19-2. Sulfur trioxide decomposes according to the equation

$$2SO_3(g) \leftrightharpoons 2SO_2(g) + O_2(g)$$

Suppose that the initial concentrations are $[SO_3]_0 = 0.176$ M, $[SO_2]_0 = 0.625$ M, and $[O_2]_0 = 0.436$ M. If it is determined that $[O_2] = 0.387$ M at equilibrium, calculate the equilibrium values of $[SO_3]$ and $[SO_2]$.

EQUILIBRIUM-CONSTANT EXPRESSION

19-3. Write the equilibrium-constant expression (K_c) for the following equations:

(a) $ZnO(s) + CO(g) \leftrightharpoons Zn(l) + CO_2(g)$

(b) $2C_5H_6(g) \leftrightharpoons C_{10}H_{12}(g)$

(c) $2N_2O_5(soln) \leftrightharpoons 4NO_2(soln) + O_2(g)$

What are the units of K_c in each case?

19-4. Write the equilibrium-constant expression (K_c) for the following equations:

(a) $2SO_2(g) + O_2(g) \leftrightharpoons 2SO_3(g)$

(b) $2NaHCO_3(s) \leftrightharpoons Na_2CO_3(s) + CO_2(g) + H_2O(g)$

(c) $C(s) + 2H_2(g) \leftrightharpoons CH_4(g)$

What are the units of K_c in each case?

19-5. Write the equilibrium-constant expression (K_c) for each of the following equations:

(a) $SO_2Cl_2(g) \leftrightharpoons SO_2(g) + Cl_2(g)$

(b) $2H_2O_2(g) \leftrightharpoons 2H_2O(l) + O_2(g)$

(c) $2CaSO_4{\cdot}H_2O(s) + 2H_2O(g) \leftrightharpoons 2CaSO_4{\cdot}2H_2O(s)$

What are the units of K_c in each case?

19-6. Write the equilibrium-constant expression (K_c) for each of the following equations:

(a) $NH_2COONH_4(s) \leftrightharpoons 2NH_3(g) + CO_2(g)$

(b) $2HgO(s) \leftrightharpoons 2Hg(l) + O_2(g)$

(c) $N_2(g) + 2O_2(g) \leftrightharpoons N_2O_4(g)$

What are the units of K_c in each case?

19-7. Write K_p expressions for the chemical equations in Problem 19-5. What are the units of K_p in each case? Express pressures in bar.

19-8. Write K_p expressions for the chemical equations in Problem 19-6. What are the units of K_p in each case? Express pressures in bar.

CALCULATION OF EQUILIBRIUM CONSTANTS

19-9. Consider the chemical equation

$$\mathrm{CuSO_4{\cdot}4NH_3}(s) \leftrightharpoons \mathrm{CuSO_4{\cdot}2NH_3}(s) + 2\,\mathrm{NH_3}(g)$$

At 20°C, the equilibrium pressure of $NH_3(g)$ is 62 Torr. Calculate the value of K_p for this equation and include the corresponding units.

19-10. At 1000°C, methane and water react according to the chemical equation

$$\mathrm{CH_4}(g) + \mathrm{H_2O}(g) \leftrightharpoons \mathrm{CO}(g) + 3\,\mathrm{H_2}(g)$$

At equilibrium, it was found that $P_{CH_4} = 0.31$ bar, $P_{H_2O} = 0.84$ bar, $P_{CO} = 0.58$ bar, and $P_{H_2} = 2.29$ bar. Calculate the value of K_p for this equation and state the corresponding units.

19-11. Phosgene, $COCl_2(g)$, a toxic gas used in the synthesis of a variety of organic compounds, decomposes according to

$$\mathrm{COCl_2}(g) \leftrightharpoons \mathrm{CO}(g) + \mathrm{Cl_2}(g)$$

A sample of phosgene gas at an initial concentration of 0.500 M is heated at 527°C in a reaction vessel. At equilibrium, the concentration of $CO(g)$ was found to be 0.046 M. Calculate the equilibrium constant for the reaction equation at 527°C.

19-12. The decomposition of phosphorus pentachloride is described by

$$\mathrm{PCl_5}(g) \leftrightharpoons \mathrm{PCl_3}(g) + \mathrm{Cl_2}(g)$$

A sample of $PCl_5(g)$ at an initial concentration of 1.10 M is placed in a reaction vessel held at 250°C. When equilibrium is attained, the concentration of $PCl_5(g)$ is 0.33 M. Calculate the value of K_c for the reaction equation.

19-13. A mixture of 1.00 mole of $H_2(g)$ and 1.00 mole of $I_2(g)$ is placed in a 2.00-liter container held at a constant temperature. After equilibrium is attained, 1.56 moles of $HI(g)$ are found. Calculate the value of K_c for the chemical equation

$$\mathrm{H_2}(g) + \mathrm{I_2}(g) \leftrightharpoons 2\,\mathrm{HI}(g)$$

and include the corresponding units.

19-14. Nitrogen dioxide decomposes at high temperatures according to the equation

$$2\,\mathrm{NO_2}(g) \leftrightharpoons 2\,\mathrm{NO}(g) + \mathrm{O_2}(g)$$

Suppose initially we have pure $NO_2(g)$ at 1000 K and 0.500 bar. If the total pressure is 0.732 bar when equilibrium is reached, what is the value of K_p? What are its units?

19-15. Nitrosyl chloride decomposes according to the chemical equation

$$2\,\mathrm{NOCl}(g) \leftrightharpoons 2\,\mathrm{NO}(g) + \mathrm{Cl_2}(g)$$

Suppose initially we have pure $NOCl(g)$ at 400 K and 2.75 bar. If the total pressure is 3.58 bar when equilibrium is reached, what is the value of K_p? What are the corresponding units?

19-16. Hydrogen sulfide decomposes at 1400 K according to the chemical equation

$$2\,\mathrm{H_2S}(g) \leftrightharpoons 2\,\mathrm{H_2}(g) + \mathrm{S_2}(g)$$

Suppose initially we have pure $H_2S(g)$ at a pressure of 0.956 bar. If the total pressure is 1.26 bar when equilibrium is reached, what is the value of K_p and its corresponding units?

EQUILIBRIUM CALCULATIONS

19-17. Given that $[\mathrm{Ni(CO)_4}] = 0.85$ M at equilibrium for the equation

$$\mathrm{Ni}(s) + 4\,\mathrm{CO}(g) \leftrightharpoons \mathrm{Ni(CO)_4}(g) \qquad K_c = 5.0 \times 10^4\ \mathrm{M^{-3}}$$

calculate the concentration of $CO(g)$ at equilibrium.

19-18. The equilibrium constant for the chemical equation

$$\mathrm{C}(s) + \mathrm{CO_2}(g) \leftrightharpoons 2\,\mathrm{CO}(g)$$

at 1000 K is 1.90 bar. If the equilibrium pressure of $CO(g)$ is 1.50 bar, what is the equilibrium pressure of $CO_2(g)$?

19-19. Phosphorus pentachloride decomposes according to the chemical equation

$$\mathrm{PCl_5}(g) \leftrightharpoons \mathrm{PCl_3}(g) + \mathrm{Cl_2}(g) \qquad K_c = 1.8\ \mathrm{M\ at\ 250°C}$$

A 0.50-mole sample of $PCl_5(g)$ is injected into an empty 2.0-liter reaction vessel held at 250°C. Calculate the concentrations of $PCl_5(g)$ and $PCl_3(g)$ at equilibrium.

19-20. Carbon disulfide is prepared by heating sulfur and charcoal. The chemical equation is

$$\mathrm{S_2}(g) + \mathrm{C}(s) \leftrightharpoons \mathrm{CS_2}(g) \qquad K_c = 9.40\ \mathrm{at\ 900\ K}$$

How many grams of $CS_2(g)$ can be prepared by heating 10.0 moles of $S_2(g)$ with excess carbon in a 5.00-liter reaction vessel held at 900 K until equilibrium is attained?

19-21. At 1200°C, $K_c = 2.5 \times 10^4$ for the equation

$$H_2(g) + Cl_2(g) \leftrightharpoons 2HCl(g)$$

If 0.50 moles of $H_2(g)$ and 0.50 moles of $Cl_2(g)$ are introduced initially into a reaction vessel, how many moles of $HCl(g)$ are there at equilibrium?

19-22. At 1000°C, $K_p = 0.263\ \text{bar}^{-1}$ for the equation

$$C(s) + 2H_2(g) \leftrightharpoons CH_4(g)$$

Calculate the equilibrium pressure of $CH_4(g)$ in bars if 0.250 moles of $CH_4(g)$ is placed in a 4.00-liter container at 1000°C.

19-23. Ammonium hydrogen sulfide decomposes according to the chemical equation

$$NH_4HS(s) \leftrightharpoons NH_3(g) + H_2S(g)$$

The equilibrium constant, K_c, is $1.81 \times 10^{-4}\ \text{M}^2$ at 25°C. If $NH_4HS(s)$ is placed in an evacuated reaction vessel at 25°C, what is the total gas pressure (in bars) in the vessel when equilibrium is attained?

19-24. Sodium hydrogen carbonate, commonly called sodium bicarbonate, is used in baking soda and in fire extinguishers as a source of $CO_2(g)$. It decomposes according to the equation

$$2NaHCO_3(s) \leftrightharpoons Na_2CO_3(s) + CO_2(g) + H_2O(g)$$

Given that $K_p = 0.26\ \text{bar}^2$ at 125°C, calculate the partial pressures of $CO_2(g)$ and $H_2O(g)$ in units of bars at equilibrium when $NaHCO_3(s)$ is heated to 125°C in a closed vessel.

19-25. The equilibrium constant for the equation

$$2ICl(g) \leftrightharpoons I_2(g) + Cl_2(g)$$

is $K_c = 0.11$. Calculate the equilibrium concentrations of $ICl(g)$, $I_2(g)$, and $Cl_2(g)$ when 0.65 moles of $I_2(g)$ and 0.33 moles of $Cl_2(g)$ are mixed in a 1.5-liter reaction vessel.

19-26. Suppose that 5.00 moles of $CO(g)$ are mixed with 2.50 moles of $Cl_2(g)$ in a 10.0-liter reaction vessel and the reaction attains equilibrium according to the equation

$$CO(g) + Cl_2(g) \leftrightharpoons COCl_2(g)$$

Given that $K_c = 4.00\ \text{M}^{-1}$, calculate the equilibrium values of $[CO]$, $[Cl_2]$, and $[COCl_2]$.

19-27. Suppose that $N_2O_4(g)$ and $NO_2(g)$ are mixed together in a reaction vessel and that the total pressure at equilibrium is 1.45 atm. Calculate $P_{N_2O_4}$ and P_{NO_2} at equilibrium when the value of K_p is 4.90 atm for the equation

$$N_2O_4(g) \leftrightharpoons 2NO_2(g)$$

19-28. Given that $H_2(g)$ reacts with $I_2(s)$ according to the chemical equation

$$H_2(g) + I_2(s) \leftrightharpoons 2HI(g) \quad K_p = 8.6\ \text{bar}$$

and that at equilibrium the total pressure in the reaction vessel is 4.5 bar, calculate P_{HI} and P_{H_2} at equilibrium. Neglect the vapor pressure of $I_2(s)$.

19-29. Zinc metal is produced by the reaction of its oxide with carbon monoxide at high temperature. The equilibrium constant for the equation

$$ZnO(s) + CO(g) \leftrightharpoons Zn(s) + CO_2(g)$$

is $K_p = 6.00 \times 10^2$. At equilibrium the total pressure in the reaction vessel is 1.80 bar. Calculate P_{CO_2} and P_{CO} at equilibrium.

19-30. At equilibrium, the total pressure in the reaction vessel for the reaction between carbon and hydrogen described by the chemical equation,

$$C(s) + 2H_2(g) \leftrightharpoons CH_4(g)$$

is 2.11 bar. Given that $K_p = 0.263\ \text{bar}^{-1}$ at 1000°C, calculate P_{H_2} and P_{CH_4} at equilibrium.

EQUILIBRIUM CONSTANTS OF EQUATIONS FOR COMBINATIONS OF REACTIONS

19-31. Given that

$$(1)\ CO(g) + H_2O(g) \leftrightharpoons CO_2(g) + H_2(g) \qquad K_{p_1} = 1.44$$

$$(2)\ CH_4(g) + H_2O(g) \leftrightharpoons CO(g) + 3H_2(g) \qquad K_{p_2} = 25.6\ \text{atm}^2$$

calculate the value of K_p for the equation

$$(3)\ CH_4(g) + 2H_2O(g) \leftrightharpoons CO_2(g) + 4H_2(g)$$

19-32. Given that

(1) $C(s) + 2H_2O(g) \leftrightharpoons CO_2(g) + 2H_2(g)$ $\quad K_{P_1} = 3.85$ bar

(2) $H_2(g) + CO_2(g) \leftrightharpoons H_2O(g) + CO(g)$ $\quad K_{P_2} = 0.71$

calculate the value of K_p for the equation

(3) $C(s) + CO_2(g) \leftrightharpoons 2CO(g)$

19-33. Given that at 973 K

(1) $2MgCl_2(s) + O_2(g) \leftrightharpoons 2MgO(s) + 2Cl_2(g)$ $\quad K_{P_1} = 8.70$ bar

(2) $MgCl_2(s) + H_2O(g) \leftrightharpoons MgO(s) + 2HCl(g)$ $\quad K_{P_2} = 8.40$ bar

determine the equilibrium constant at 973 K for the equation

(3) $2Cl_2(g) + 2H_2O(g) \leftrightharpoons 4HCl(g) + O_2(g)$

19-34. Given the equilibrium constants at 1000 K for the following equations

(1) $CaCO_3(s) \leftrightharpoons CaO(s) + CO_2(g)$ $\quad K_{P_1} = 0.040$ bar

(2) $C(s) + CO_2(g) \leftrightharpoons 2CO(g)$ $\quad K_{P_2} = 1.9$ bar

determine the equilibrium constant, K_p, at 1000 K for the equation

(3) $CaCO_3(s) + C(s) \leftrightharpoons CaO(s) + 2CO(g)$

QUALITATIVE APPLICATION OF LE CHÂTELIER'S PRINCIPLE

19-35. Consider the chemical equilibrium described by the equation

$$H_2(g) + CO_2(g) \leftrightharpoons H_2O(g) + CO(g)$$

Use Le Châtelier's principle to predict the effect on the equilibrium pressure of $CO_2(g)$ and of $CO(g)$ resulting from

(a) an increase in the pressure of $H_2O(g)$

(b) an increase in the reaction volume

19-36. Consider the chemical equilibrium described by the equation

$$2NO(g) + Br_2(g) \leftrightharpoons 2NOBr(g)$$

Use Le Châtelier's principle to predict the effect on the equilibrium concentration of $NOBr(g)$ and of $NO(g)$ resulting from

(a) an increase in the concentration of $Br_2(g)$

(b) a twofold decrease in the reaction volume

19-37. Consider the chemical equilibrium described by the equation

$$C(s) + 2H_2(g) \leftrightharpoons CH_4(g) \quad \Delta H^\circ_{rxn} = -74.6\ \text{kJ}\cdot\text{mol}^{-1}$$

Predict the way in which the equilibrium will shift in response to each of the following changes in conditions (if the equilibrium is unaffected by the change, then write *no change*):

(a) decrease in temperature

(b) decrease in reaction volume

(c) decrease in P_{H_2}

(d) increase in P_{CH_4}

(e) addition of $C(s)$

19-38. For the chemical equilibrium described by the equation

$$N_2(aq) \leftrightharpoons N_2(g) \quad \Delta H^\circ_{rxn} > 0$$

in which direction will the equilibrium shift in response to the following changes in conditions?

(a) increase in temperature

(b) increase in volume over the solution

(c) addition of $H_2O(l)$

(d) addition of $N_2(g)$

19-39. For the exothermic dissolution reaction of hydrofluoric acid described by

$$HF(aq) \leftrightharpoons H^+(aq) + F^-(aq)$$

in which direction will the equilibrium shift in response to the following changes?

(a) decrease in temperature

(b) addition of $H_2O(l)$

(c) addition of $KF(s)$

(d) addition of $NaOH(s)$ (*Hint*: What will this strong base react with in solution?)

19-40. For the chemical equilibrium described by the equation

$$PbCl_2(s) \leftrightharpoons Pb^{2+}(aq) + 2\,Cl^-(aq)$$

in which direction will the equilibrium shift in response to the following changes?

(a) addition of $PbCl_2(s)$

(b) addition of $NaCl(s)$

(c) addition of $H_2O(l)$

(d) addition of $AgNO_3(s)$ (*Hint*: What is the solubility of silver chloride?)

19-41. Several key reactions in coal gasification are (1) the synthesis gas reaction, (2) the water-gas-shift reaction, and (3) the catalytic methanation reaction as described, respectively, by the following three chemical equations:

(1) $C(s) + H_2O(g) \leftrightharpoons CO(g) + H_2(g)$ $\quad \Delta H^\circ_{rxn} = +131.3\ kJ\cdot mol^{-1}$

(2) $CO(g) + H_2O(g) \leftrightharpoons CO_2(g) + H_2(g)$ $\quad \Delta H^\circ_{rxn} = -41.2\ kJ\cdot mol^{-1}$

(3) $CO(g) + 3\,H_2(g) \leftrightharpoons H_2O(g) + CH_4(g)$ $\quad \Delta H^\circ_{rxn} = -205.9\ kJ\cdot mol^{-1}$

(a) Write the equilibrium-constant expressions in terms of concentrations, K_c, for each of these three equations.

(b) Predict the direction in which each equilibrium shifts in response to an increase in temperature or a decrease in reaction volume.

19-42. An important modern chemical problem is the liquefaction of coal because coal is still relatively abundant whereas oil is a dwindling resource (Interchapter L). The first step is heating the coal with steam to produce synthesis gas, as described by the equation

(1) $C(s) + H_2O(g) \leftrightharpoons CO(g) + H_2(g)$ $\quad \Delta H^\circ_{rxn} = +131.3\ kJ\cdot mol^{-1}$

Carbon monoxide can be hydrogenated to form the important chemical methanol, $CH_3OH(g)$, according to

(2) $CO(g) + 2\,H_2(g) \leftrightharpoons CH_3OH(g)$ $\quad \Delta H^\circ_{rxn} = -90.5\ kJ\cdot mol^{-1}$

Use Le Châtelier's principle to suggest conditions that maximize the yield of $CH_3OH(g)$ from $CO(g)$ and $H_2(g)$.

QUANTITATIVE APPLICATION OF LE CHÂTELIER'S PRINCIPLE

19-43. At 320 K an equilibrium mixture of $N_2O_4(g)$ and $NO_2(g)$ has partial pressures of 292 Torr and 393 Torr, respectively. A quantity of $NO_2(g)$ is injected into the mixture, and the total pressure jumps to 812 Torr. Calculate the new partial pressures after equilibrium is reestablished. The appropriate chemical equation is

$$N_2O_4(g) \leftrightharpoons 2\,NO_2(g)$$

19-44. An equilibrium mixture of $PCl_5(g)$, $PCl_3(g)$, and $Cl_2(g)$ has partial pressures of 217 Torr, 13.2 Torr, and 13.2 Torr, respectively. A quantity of $Cl_2(g)$ is injected into the mixture, and the total pressure jumps to 263 Torr. Calculate the new partial pressures after equilibrium is reestablished. The appropriate chemical equation is

$$PCl_3(g) + Cl_2(g) \leftrightharpoons PCl_5(g)$$

19-45. Ammonium hydrogen sulfide decomposes according to

$$NH_4HS(s) \leftrightharpoons NH_3(g) + H_2S(g)$$

In a certain experiment, $NH_4HS(s)$ is placed in a sealed 1.00-liter container at 25°C and the total equilibrium pressure is observed to be 0.664 bar, with a small amount of $NH_4HS(s)$ remaining. A quantity of $NH_3(g)$ is injected into the container and the total pressure jumps to 0.906 bar. Calculate the total pressure after equilibrium is reestablished.

19-46. Dinitrogen tetroxide decomposes according to

$$N_2O_4(g) \leftrightharpoons 2\,NO_2(g)$$

In a certain experiment, $N_2O_4(g)$ at an initial pressure of 0.554 bar is introduced into an empty reaction container; after equilibrium is established, the total pressure is 0.770 bar. A quantity of $NO_2(g)$ is injected into the container and the total pressure jumps to 0.906 bar. Calculate the total pressure after equilibrium is reestablished.

19-47. Consider the reaction described by

$$CO_2(g) + H_2(g) \leftrightharpoons CO(g) + H_2O(g)$$

An equilibrium mixture of these gases has the partial pressures $P_{CO} = 512$ Torr, $P_{H_2O} = 77$ Torr, $P_{H_2} = 192$ Torr, and $P_{CO_2} = 384$ Torr. If the volume of the reaction container is doubled, what will the new values of the partial pressures be?

19-48. Ammonium bromide decomposes according to

$$NH_4Br(s) \leftrightharpoons NH_3(g) + HBr(g)$$

Some $NH_4Br(s)$ is introduced into an empty reaction container; after equilibrium is established, the total pressure is 26.4 Torr. Calculate the total pressure if the volume of the reaction container is halved.

REACTION QUOTIENT CALCULATIONS

19-49. At 900 K the equilibrium constant for the equation

$$2SO_2(g) + O_2(g) \leftrightharpoons 2SO_3(g)$$

is 13 M^{-1}. If we mix the following concentrations of the three gases, predict in which direction the reaction will proceed toward equilibrium:

Mixture	$[SO_2]/M$	$[O_2]/M$	$[SO_3]/M$
(a)	0.40	0.20	0.10
(b)	0.05	0.10	0.30

19-50. Suppose that $H_2(g)$ and $CH_4(g)$ are brought into contact with C(s) at 500°C with $P_{H_2} = 0.20$ bar and $P_{CH_4} = 3.0$ bar. Is the reaction described by the equation

$$C(s) + 2H_2(g) \leftrightharpoons CH_4(g) \quad K_p = 2.69\times10^3 \text{ bar}^{-1}$$

at equilibrium under these conditions? If not, in what direction will the reaction proceed to attain equilibrium?

19-51. Suppose we have a mixture of the gases $H_2(g)$, $CO_2(g)$, $CO(g)$, and $H_2O(g)$ at 1260 K, with $P_{H_2} = 0.55$ bar, $P_{CO_2} = 0.20$ bar, $P_{CO} = 1.25$ bar, and $P_{H_2O} = 0.10$ bar. Is the reaction described by the equation

$$H_2(g) + CO_2(g) \leftrightharpoons CO(g) + H_2O(g) \quad K_p = 1.59$$

at equilibrium under these conditions? If not, in what direction will the reaction proceed to attain equilibrium?

19-52. Given that $K_p = 2.25 \times 10^4$ atm^{-2} at 25°C for the equation

$$2H_2(g) + CO(g) \leftrightharpoons CH_3OH(g)$$

predict the direction in which a reaction mixture for which $P_{CH_3OH} = 10.0$ atm, $P_{H_2} = 0.010$ atm, and $P_{CO} = 0.0050$ atm proceeds to attain equilibrium.

ADDITIONAL PROBLEMS

19-53. If the concentrations of the products and reactants are constant at equilibrium, why is an equilibrium state said to be dynamic?

19-54. Explain the difference between an equilibrium constant and an equilibrium state. For a specific reaction equation at a given temperature, is it possible to have more than one equilibrium constant? Is it possible to have more than one equilibrium state?

19-55. What does the magnitude of the equilibrium constant tell you about a reaction?

19-56. A problem gives the equilibrium constant for the reaction equation

$$2ICl(g) \leftrightharpoons I_2(g) + Cl_2(g)$$

as $K = 0.11$. Is this the value of K_p or K_c? Explain.

19-57. Prior to learning about equilibrium states, we solved stoichiometric problems using the concept of "limiting reactants." Under what conditions does the method of limiting reactants apply?

19-58. What is the difference between the equilibrium constant, K_c, and the reaction quotient, Q_c?

19-59. How is the value of the ratio of the reaction quotient to the equilibrium constant, Q_c/K_c, related to Le Châtelier's principle?

19-60. Iodine crystals have a significant vapor pressure at room temperature and form a rich purple vapor when allowed to stand in a sealed container (Figure 15.22). A sealed container is filled with 10.0 grams of solid iodine crystals and allowed to reach equilibrium between its solid and vapor phases at 25°C. How will the vapor color change if

(a) the container is heated to 90°C?

(b) the container is cooled to −5°C?

(c) an additional 10.0 grams of iodine crystals are added to the container?

19-61. Is a reaction with a large value of K_c faster than a different reaction with a smaller value of K_c?

19-62. A student answering a question about equilibrium writes, "As a reaction approaches equilibrium from the left, the concentration of reactants decreases and the concentration of products increases until the two are equal. At this point a balance is reached and the reaction stops in a state of equilibrium." Rewrite the response, correcting the student's mistakes.

19-63. Does the addition of a catalyst affect the value of the equilibrium constant? Explain your answer in terms of the forward and reverse reaction kinetics.

19-64. (a) When the volume of a gas-phase reaction is increased (at constant pressure), in what direction does the equilibrium shift? (b) Under what conditions does increasing the volume of a gas-phase reaction have no effect on the equilibrium state? (c) What is the equivalent to increasing the volume of a gas-phase reaction for a reaction that occurs in solution?

19-65. The value of K_p for the chemical equation

$$CuSO_4{\cdot}4NH_3(s) \leftrightharpoons CuSO_4{\cdot}2NH_3(s) + 2NH_3(g)$$

is 6.66×10^{-3} atm^2 at 20°C. Calculate the equilibrium pressure of ammonia at 20°C.

19-66. The equilibrium constant for the chemical equation

$$N_2(g) + 3H_2(g) \leftrightharpoons 2NH_3(g)$$

is $K_p = 0.099$ bar^{-2} at 227°C. Calculate the value of K_c for the reaction at 227°C.

19-67. At 500°C hydrogen iodide decomposes according to

$$2HI(g) \leftrightharpoons H_2(g) + I_2(g)$$

For HI(g) heated to 500°C in a 1.00-liter reaction vessel, chemical analysis gave the following concentrations at equilibrium: $[H_2] = 0.42$ M, $[I_2] = 0.42$ M, and $[HI] = 3.52$ M. If an additional mole of HI(g) is introduced into the reaction vessel, what are the equilibrium concentrations after the new equilibrium has been reached?

19-68. The equilibrium constant for the methanol synthesis equation

$$2H_2(g) + CO(g) \leftrightharpoons CH_3OH(g)$$

is $K_p = 2.19 \times 10^4$ bar^{-2} at 25°C. (a) Calculate the value of P_{CH_3OH} at equilibrium when $P_{H_2} = 0.020$ bar and $P_{CO} = 0.010$ bar. (b) Given that at equilibrium $P_{total} = 10.0$ bar and $P_{H_2} = 0.020$ bar, calculate P_{CO} and P_{CH_3OH}.

19-69. Given the reaction equation

$$SO_2(g) + NO_2(g) \leftrightharpoons SO_3(g) + NO(g)$$
$$\Delta H^\circ_{rxn} = -40.8 \text{ kJ}\cdot\text{mol}^{-1}$$

Predict the effect of the following changes on the item listed:

Change	Effect on
(a) decrease in total volume	$[SO_2]$
(b) increase in temperature	[NO]
(c) increase in partial pressure of $NO_2(g)$	K_c
(d) decrease in partial pressure of products	$[SO_2]$

19-70. According to Table 19.1, $K_c = 0.20$ M at 100°C for the chemical equation

$$N_2O_4(g) \leftrightharpoons 2NO_2(g)$$

Calculate K_p at the same temperature in units of bars.

19-71. Tin can be prepared by heating $SnO_2(s)$ ore with hydrogen gas, according to:

$$SnO_2(s) + 2H_2(g) \leftrightharpoons Sn(s) + 2H_2O(g)$$

When the reactants are heated to 500°C in a closed vessel, $[H_2O] = [H_2] = 0.25$ M at equilibrium. If more hydrogen is added so that its new initial concentration is 0.50 M, what are the concentrations of $H_2(g)$ and $H_2O(g)$ when equilibrium is restored?

19-72. The equilibrium constant for the chemical equation

$$SO_2(g) + NO_2(g) \leftrightharpoons SO_3(g) + NO(g)$$

is 3.0. Calculate the number of moles of $NO_2(g)$ that must be added to 2.4 moles of $SO_2(g)$ in order to form 1.2 moles of $SO_3(g)$ at equilibrium.

19-73. Diatomic chlorine dissociates to chlorine atoms at elevated temperatures according to

$$Cl_2(g) \leftrightharpoons 2Cl(g)$$

For example, $K_p = 0.578$ bar at 2000°C. Calculate the fraction of chlorine molecules that are dissociated at 2000°C if the initial concentration of $Cl_2(g)$ is 0.050 M.

19-74. The value of the equilibrium constant for the equation

$$H_2(g) + I_2(g) \leftrightharpoons 2HI(g)$$

is $K_p = 85$ at 553 K. (a) Is it possible at 553 K to have an equilibrium reaction mixture for which $P_{HI} = P_{H_2} = P_{I_2}$? (b) Suppose a 5.0-gram sample of HI(*g*) is heated to 553 K in a 2.00-liter vessel. Calculate the composition of the equilibrium reaction mixture.

19-75. The value of the equilibrium constant for the chemical equation

$$H_2(g) + I_2(g) \leftrightharpoons 2HI(g)$$

is $K_p = 85$ at 280°C. Calculate the value of K_p at 280°C for the equation

$$HI(g) \leftrightharpoons \tfrac{1}{2}H_2(g) + \tfrac{1}{2}I_2(g)$$

19-76. The equilibrium constant at 823 K for the chemical equation

$$MgCl_2(s) + \tfrac{1}{2}O_2(g) \leftrightharpoons MgO(s) + Cl_2(g)$$

is $K_p = 1.75\ \text{atm}^{1/2}$. Suppose that 50.0 grams of $MgCl_2(s)$ is placed in a reaction vessel with 2.00 liters of oxygen at 25°C and 1.00 atm and that the reaction vessel is sealed and heated to 823 K until equilibrium is attained. Calculate P_{Cl_2} and P_{O_2} at equilibrium in atm.

19-77. Consider the reaction equilibrium described by the equation

$$COCl_2(g) \leftrightharpoons CO(g) + Cl_2(g)$$

If 2.00 moles of $COCl_2(g)$ is introduced into a 10.0-liter flask at 1000°C, calculate the equilibrium concentrations of all species at this temperature. At 1000°C, $K_c = 0.329$ M.

19-78. Osmium dioxide occurs either as a black powder or as brown crystals. The density of the black powder form is 7.7 $g\cdot cm^{-3}$ and the density of the brown crystalline form is 11.4 $g\cdot cm^{-3}$. Which is the more stable form at high pressure?

19-79. Ammonia, $NH_3(g)$, and hydrogen chloride, HCl(*g*), with ammonia in excess, are injected into a reaction vessel maintained at 300°C. A white powder of ammonium chloride, $NH_4Cl(s)$, is observed to form according to

$$NH_3(g) + HCl(g) \leftrightharpoons NH_4Cl(s)$$

with $K_p = 17.1\ \text{bar}^{-2}$ at 300°C. When the system comes to equilibrium, the total pressure is 2.776 bar. Calculate the partial pressure of each gas at equilibrium.

19-80. Phosphorus pentachloride decomposes according to the chemical equation

$$PCl_5(g) \leftrightharpoons PCl_3(g) + Cl_2(g)$$

with $K_p = 1.78$ atm at 250°C. How many moles of $PCl_5(g)$ must be added to a 1.00-liter container at 250°C to obtain a concentration of 1.00 M of $PCl_3(g)$?

19-81. An equilibrium mixture of CO(*g*), $Cl_2(g)$, and $COCl_2(g)$ has partial pressures $P_{CO} = P_{Cl_2} = 1.09$ bar and $P_{COCl_2} = 0.144$ bar. A quantity of CO(*g*) is suddenly injected into the reaction vessel and the total pressure jumps to 3.31 bar. Calculate the total pressure after equilibrium is reestablished. The relevant chemical equation is

$$CO(g) + Cl_2(g) \leftrightharpoons COCl_2(g)$$

19-82. For the reaction described by

$$H_2(g) + I_2(g) \leftrightharpoons 2HI(g)$$

the value of K_p is 54.4 at 355°C. What percentage of $I_2(g)$ will be converted to HI(*g*) if 0.200 moles each of $H_2(g)$ and $I_2(g)$ are mixed and allowed to come to equilibrium at 355°C in a 1.00-liter container?

19-83. Antimony pentachloride, $SbCl_5(s)$, decomposes to antimony trichloride, $SbCl_3(s)$, and chlorine, $Cl_2(g)$, according to

$$SbCl_5(s) \leftrightharpoons SbCl_3(s) + Cl_2(g)$$

A 0.50-mole sample of $SbCl_5(s)$ is put into a closed 1.00-liter container and heated to 250°C. At equilibrium the mole fraction of $Cl_2(g)$ is found to be 0.428. Calculate the value of K_p.

19-84. Calculate the partial pressures in the equilibrium gas mixture that results when 26.1 Torr of $CO_2(g)$ and 26.1 Torr of $H_2(g)$ are mixed at 1000°C. The relevant chemical equation is

$$CO_2(g) + H_2(g) \leftrightharpoons CO(g) + H_2O(g) \qquad K_P = 0.719$$

19-85. Calculate the equilibrium partial pressures in the previous problem if the volume of the reaction container is doubled.

19-86. Ammonium chloride decomposes according to the equation

$$NH_4Cl(s) \leftrightharpoons NH_3(g) + HCl(g)$$

with $K_P = 5.82 \times 10^{-2}$ bar^2 at 300°C. Calculate the equilibrium partial pressure of each gas and the number of grams of $NH_4Cl(s)$ produced if equal molar quantities of $NH_3(g)$ and $HCl(g)$ at an initial total pressure of 8.87 bar are injected into a 2.00-liter container at 300°C.

19-87. Sulfuryl chloride, $SO_2Cl_2(g)$, decomposes to $SO_2(g)$ and $Cl_2(g)$ at 100°C according to

$$SO_2Cl_2(g) \leftrightharpoons SO_2(g) + Cl_2(g)$$

A 6.175-gram sample of $SO_2Cl_2(g)$ is placed in an evacuated 1.00-liter container at 100°C, and the total pressure at equilibrium is found to be 2.41 bar. Calculate the partial pressures of $SO_2Cl_2(g)$, $SO_2(g)$, and $Cl_2(g)$ and the value of K_P.

19-88. An equilibrium mixture contains 0.20 moles of hydrogen gas, 0.80 moles of carbon dioxide, 0.10 moles of carbon monoxide, and 0.40 moles of water vapor in a 1.00-liter container. How many moles of carbon dioxide would have to be added at constant temperature and volume to increase the amount of carbon monoxide to 0.20 moles? The equation for the reaction is

$$CO(g) + H_2O(g) \leftrightharpoons CO_2(g) + H_2(g)$$

19-89. The decomposition of ammonium hydrogen sulfide is an endothermic reaction. The equation for the reaction is

$$NH_4HS(s) \leftrightharpoons NH_3(g) + H_2S(g)$$

A 5.260-gram sample of solid ammonium hydrogen sulfide is placed in an evacuated 3.00-liter container at 25°C. After equilibrium is established, the total pressure inside the vessel is 0.659 atm. Some solid ammonium hydrogen sulfide remains in the flask. (a) What is the value of the equilibrium constant, K_P? (b) What percentage of the solid placed in the flask has reacted?

19-90. Prove that the relationship between K_P referenced to units of bars and K_P referenced to units of atmospheres is,

$$K_{P_{bar}} = K_{P_{atm}}(1.01325 \text{ bar·atm}^{-1})^{\Delta\nu_{gas}}$$

where $\Delta\nu_{gas}$ is the difference between the sum of the stoichiometric coefficients of the products and that of the reactants for the gas-phase species in a given reaction equation.

19-91. Given:

(1) $2BrCl(g) \leftrightharpoons Cl_2(g) + Br_2(g) \qquad K_{P_1} = 0.45$

(2) $2IBr(g) \leftrightharpoons Br_2(g) + I_2(g) \qquad K_{P_2} = 21.0$

Determine the value of K_P for the reaction equation

$$BrCl(g) + \tfrac{1}{2}I_2(g) \leftrightharpoons IBr(g) + \tfrac{1}{2}Cl_2(g)$$

19-92. Acetic acid in the vapor phase is in equilibrium with its dimer as described by

$$2CH_3COOH(g) \leftrightharpoons (CH_3COOH)_2(g) \qquad K_P = 3.67 \text{ bar}^{-1} \text{ at } 100°C$$

If the total pressure is 1.50 bar, calculate the partial pressure of the dimer.

19-93. Dinitrogen tetraoxide decomposes according to

$$N_2O_4(g) \leftrightharpoons 2NO_2(g)$$

Pure $N_2O_4(g)$ was placed in an empty container at 127°C at a pressure of 0.0438 atm. After the system reached equilibrium, the total pressure was 0.0743 atm. Calculate the value of K_P for this equation.

19-94. The decomposition of ammonium carbamate, $NH_2COONH_4(s)$, takes place according to the chemical equation

$$NH_2COONH_4(s) \leftrightharpoons 2NH_3(g) + CO_2(g)$$

Show that if all the $NH_3(g)$ and $CO_2(g)$ result from the decomposition of ammonium carbamate, then $K_p = (4/27)P^3$, where P is the total pressure at equilibrium.

19-95. (*) The equilibrium constant for the chemical equation

$$N_2(g) + 3H_2(g) \leftrightharpoons 2NH_3(g)$$

is $K_p = 0.097\ \text{bar}^{-2}$ at 227°C. (a) Given that at equilibrium $P_{N_2} = 1.00$ bar and $P_{H_2} = 3.00$ bar, calculate P_{NH_3} at equilibrium. (b) Given that at equilibrium the total pressure is 2.00 bar and also that the mole fraction of $H_2(g)$, x_{H_2}, is 0.20, calculate x_{NH_3}.

19-96. (*) Consider the methanation reaction described by the chemical equation

$$CO(g) + 3H_2(g) \leftrightharpoons CH_4(g) + H_2O(g)$$

It was found that 0.613 moles of $CO(g)$, 0.387 moles of $H_2(g)$, 0.387 moles of $CH_4(g)$, and 0.387 moles of $H_2O(g)$ were present in an equilibrium mixture in a 1.00-liter container. All the water vapor was removed and the system allowed to come to equilibrium again. Calculate the concentration of all gases in the new equilibrium system. (You must solve the resulting equation by trial and error.)

19-97. (*) Referring to Problem 19-73, show that the fraction of chlorine molecules (x) that dissociate is given by

$$\frac{x^2}{1-x} = \frac{K_c}{4[Cl_2]_0}$$

for any initial concentration of $Cl_2(g)$.

Søren Peter Lauritz Sørensen (1868–1939) was born in Havrebjerg, Denmark, into a farming family. He was a high-strung boy who found refuge in his schoolwork. He first studied medicine at the University of Copenhagen, but soon changed to chemistry, graduating in 1881. He received his Ph.D. doing research in inorganic chemistry at the Technical University of Denmark in Copenhagen. In 1901, he was invited to be the head of the chemical department of the Carlsberg Laboratory, which had been established by J. C. Jacobsen, the founder of the Carlsberg Brewery in Copenhagen in 1876. The laboratory was a renowned center of research in biochemistry and later supported the work of Niels Bohr and his institute in the development of the quantum theory.

When Sørensen joined the Carlsberg Laboratory, he continued his studies in protein chemistry and was later assisted by his second wife, Margrethe Hoyrup Sørensen. He soon realized that the aqueous hydrogen ion concentration was critical to the function of most proteins, especially enzymes. At that time, there was no convenient way to determine and express the concentration of the hydrogen ion. In addition, the wide range of its concentration, from about 1 M to 10^{-14} M, was numerically inconvenient. He devised the seemingly simple method of expressing the hydrogen ion scale on a logarithmic scale, which he called the PH scale—now called pH. The letters pH are an abbreviation for "pondus hydrogenii" (potential hydrogen). His pH scale was quickly adopted by chemists because of its convenience. The use of pH became routine in laboratories when Arnold Beckman introduced the portable pH meter in 1935.

The concept of pH is so ingrained in us that it is difficult to appreciate the importance of Sørensen's contribution. At the time, there was no way to conveniently express the acidity of a solution, although it had become increasingly clear that some sort of expression was desperately needed.

20. The Properties of Acids and Bases

A quantitative understanding of the chemistry of acids and bases is essential to an understanding of many chemical reactions and most biochemical reactions. In this chapter, we first present a definition of acids and bases incorporating the key role that water plays in acid-base chemistry. Then we introduce the concept of pH that is a convenient measure of the acidity of a solution. We shall show that acids and bases can be classified as either strong or weak. A central, quantitative theme of this chapter is the calculation of the acidity, or pH, of an acidic or a basic solution as a function of the concentration of the acid or base. We shall learn that solutions of salts can be acidic, basic, or neutral.

20-1. An Acid Is a Proton Donor and a Base Is a Proton Acceptor

In Chapter 10 we defined an acid as a substance that produces $H^+(aq)$ ions in aqueous solution and a base as a substance that produces $OH^-(aq)$ ions in aqueous solution. For example, we write the equations

$$HCl(aq) \rightarrow H^+(aq) + Cl^-(aq)$$

and

$$NaOH(aq) \rightarrow Na^+(aq) + OH^-(aq)$$

as expressions of the acidic nature of hydrochloric acid, $HCl(aq)$, and the basic nature of sodium hydroxide, $NaOH(aq)$. We noted that acidic solutions taste sour (for example, vinegar and lemon juice) and that basic solutions feel slippery and taste bitter (for example, soap and ammonia).

The definitions of acids and bases that we have used up to now are due to Arrhenius (Chapter 6 Frontispiece), and substances like $HCl(aq)$ and $NaOH(aq)$

are called **Arrhenius acids and bases**. The Arrhenius definition of acids and bases is limited to aqueous solutions, however. Although most of the reactions in solution that we study in general chemistry take place in aqueous solution, many important reactions take place in other solvents, such as ethanol, $CH_3CH_2OH(l)$, acetone, $(CH_3)_2CO(l)$, or ammonia $NH_3(l)$. In this chapter we use a more general definition of acids and bases, based on the classification scheme proposed independently by the Danish chemist Johannes Brønsted and the English chemist Thomas Lowry. Their definition emphasizes the role of the solvent in acid-base systems.

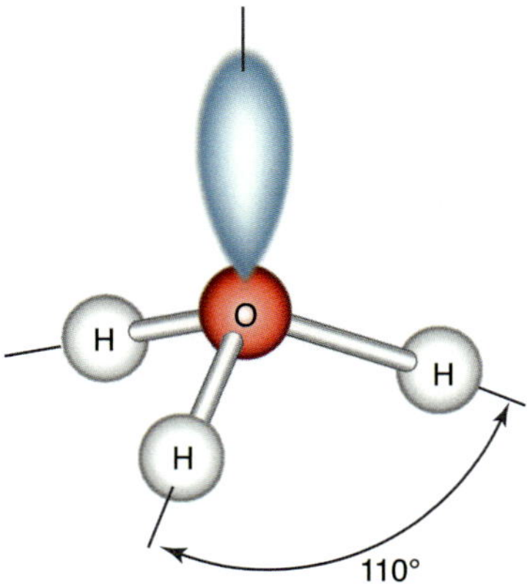

Figure 20.1 The hydronium ion has a trigonal pyramidal shape. All the oxygen-hydrogen bond lengths are identical (106 pm) and all the H–O–H bond angles are identical (110°). The ion has one lone pair of electrons; the Lewis formula is

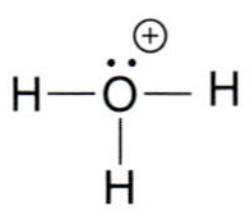

Up to now we have generally used $H^+(aq)$ to represent the species formed when an acid is dissolved in water. Experiments show, however, that the proton is bonded to a water molecule (recall that H^+ is often referred to as a proton), so a more accurate representation of the species present is $H_3O^+(aq)$. This species is called the **hydronium ion** and has the trigonal pyramidal structure shown in Figure 20.1. The $H_3O^+(aq)$ notation also tells us that the hydronium ion is further solvated by other water molecules that are represented by the symbol (aq). The equation that represents the acidic nature of $HCl(aq)$ in the Brønsted-Lowry sense is

$$HCl(aq) + H_2O(l) \rightarrow H_3O^+(aq) + Cl^-(aq)$$

Note that the role of the solvent is explicitly included.

In this system, we define **Brønsted-Lowry acids** as **proton donors**, and **Brønsted-Lowry bases** as **proton acceptors**. In our example, $HCl(aq)$ is said to be a Brønsted-Lowry acid because it donates a proton to the solvent $H_2O(l)$ to produce a hydronium ion. The $H_2O(l)$ functions as a Brønsted-Lowry base in accepting a proton from $HCl(aq)$.

A reaction involving the transfer of a proton from one molecule to another is called a **proton-transfer reaction** or a **protonation reaction**. **Acid-base reactions** are proton transfer reactions.

An even more general theory of acids and bases was put forth by G. N. Lewis in 1923 (Chapter 7 Frontispiece). In this system, a **Lewis acid** is defined as an **electron-pair acceptor** and a **Lewis base** is defined as an **electron-pair donor**. These more general definitions of an acid and a base include both the Arrhenius and Brønsted-Lowry definitions. For example, the reaction of HCl with water,

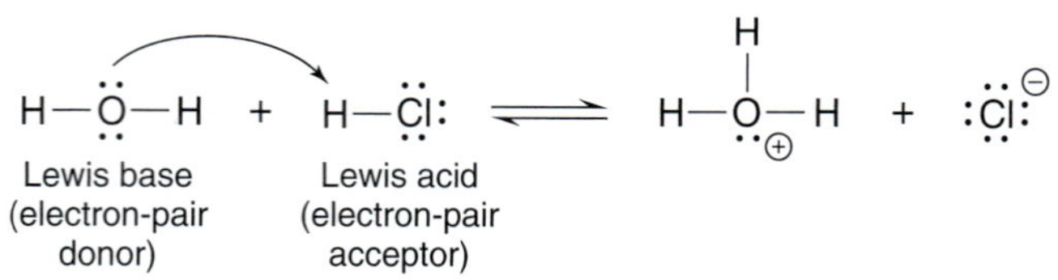

may also be viewed as the donation of one of the lone pairs of electrons in the H_2O molecule to the hydrogen atom in the HCl molecule, which in turn accepts the lone pair to form the products. Thus, in this reaction $H_2O(l)$ acts as a Lewis base and $HCl(aq)$ acts as a Lewis acid.

Because the Lewis definition of acids and bases is more general than that of Arrhenius or Brønsted-Lowry, there are many reactions that may be considered to be acid-base reactions in the Lewis sense that would not be in, say, the Brønsted-Lowry sense. For example, consider the reaction between ammonia and boron trifluoride:

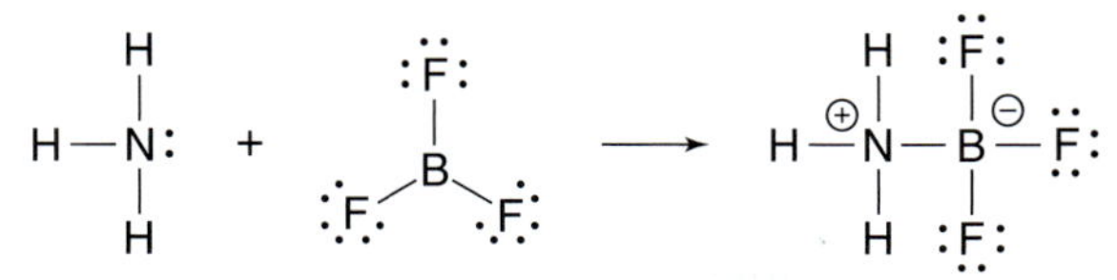

Lewis base Lewis acid donor-acceptor complex

A BF_3 molecule reacts with an NH_3 molecule by accepting the lone pair of electrons on the nitrogen atom to form a nitrogen-boron bond. Note that BF_3 is not an acid in the Brønsted-Lowry sense because it does not donate a proton, but it is a Lewis acid because it acts as an electron-pair acceptor. In general, an electron-deficient species (Section 7-7) can act as a Lewis acid, and a species with a lone pair of electrons can act as a Lewis base.

The Lewis acid-base classification scheme is especially useful in understanding the mechanisms of organic chemical reactions. However, the Brønsted-Lowry classification system is more convenient in most of the cases that we shall consider here.

20-2. In an Aqueous Solution the Product of the Ion Concentrations [H_3O^+] and [OH^-] Is a Constant

Pure water contains a small number of hydronium ions, $H_3O^+(aq)$, and hydroxide ions, $OH^-(aq)$, that arise from the equilibrium

$$H_2O(l) + H_2O(l) \leftrightharpoons H_3O^+(aq) + OH^-(aq) \tag{20.1}$$

In terms of Lewis formulas, this reaction can be represented as follows:

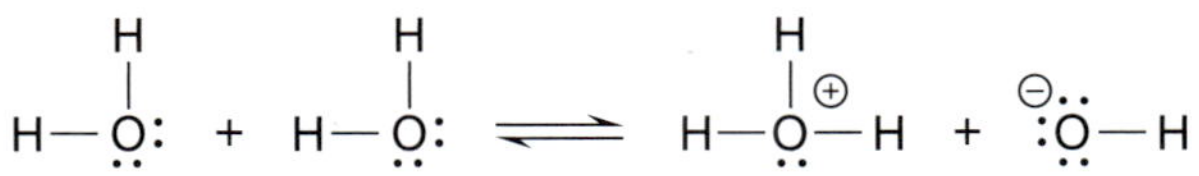

In this reaction, a water molecule transfers a proton to another water molecule. Thus, water acts both as an acid (proton donor) and as a base (proton acceptor) in the Brønsted-Lowry system. The reaction described by Equation 20.1 is called an **autoprotonation reaction**.

The equilibrium-constant expression for Equation 20.1 is

$$K_w = [H_3O^+][OH^-] \tag{20.2}$$

where the subscript w refers to water. Note that the concentration of water does not appear in the K_w expression because the water is a liquid and the value of [H_2O] is effectively constant in aqueous solutions (Section 19-3). The equilibrium constant K_w is called the **ion-product constant of water**. At 25°C the experimental value of K_w is

$$K_w = [H_3O^+][OH^-] = 1.0 \times 10^{-14}\ M^2 \tag{20.3}$$

This small value of K_w means that in pure water the concentrations of $H_3O^+(aq)$ and $OH^-(aq)$ are low; that is, the equilibrium represented by Equation 20.1 lies far to the left. From the stoichiometry of Equation 20.1 we note that if we start with pure water, then $H_3O^+(aq)$ and $OH^-(aq)$ are produced on a one-for-one basis. Therefore, *in pure water* we have the equality

$$[H_3O^+] = [OH^-]$$

Using this equation to eliminate $[OH^-]$ in Equation 20.3 yields

$$[H_3O^+]^2 = 1.0 \times 10^{-14}\ M^2$$

Taking the square root of both sides yields

$$[H_3O^+] = 1.0 \times 10^{-7}\ M$$

Because $[H_3O^+] = [OH^-]$, we conclude that

$$[OH^-] = 1.0 \times 10^{-7}\ M$$

Thus, both $[H_3O^+]$ and $[OH^-]$ are equal to 1.0×10^{-7} M in pure water at 25°C. Although $[H_3O^+] = [OH^-]$ for pure water, this is not necessarily true when substances are dissolved in the water. We have the following classification:

A **neutral** aqueous solution is defined as one in which

$$[H_3O^+] = [OH^-] \qquad \text{(neutral solution)}$$

An **acidic** aqueous solution is defined as one in which

$$[H_3O^+] > [OH^-] \qquad \text{(acidic solution)}$$

A **basic** aqueous solution is defined as one in which

$$[OH^-] > [H_3O^+] \qquad \text{(basic solution)}$$

20-3. Strong Acids and Strong Bases Are Completely Dissociated in Aqueous Solutions

We noted in Section 12-3 that aqueous solutions of electrolytes conduct an electric current. The electrical conductance is proportional to the number of ions that are available to conduct a current. Conductivity measurements on dilute $HCl(aq)$ solutions show that the HCl molecules in the solution are completely dissociated, forming $H_3O^+(aq)$ and $Cl^-(aq)$ ions. There are essentially no undissociated HCl molecules in aqueous solution. Acids that are completely dissociated are referred to as **strong acids**. The term *strong* refers to the ability of such acids to donate protons to water molecules. Strong acids transfer essentially all their dissociable protons to water molecules.

EXAMPLE 20-1: Calculate $[H_3O^+]$, $[Cl^-]$, and $[OH^-]$ in a 0.15-M aqueous solution of $HCl(aq)$ at 25°C.

Solution: Because $HCl(aq)$ is a strong acid, it is completely dissociated, so $[H_3O^+] = 0.15$ M and $[Cl^-] = 0.15$ M. The corresponding value of $[OH^-]$ in this solution can be computed from the K_w expression:

$$K_w = [H_3O^+][OH^-] = 1.0 \times 10^{-14}\ M^2$$

Solving for $[OH^-]$, we have that

$$[OH^-] = \frac{1.0 \times 10^{-14}\ M^2}{[H_3O^+]} = \frac{1.0 \times 10^{-14}\ M^2}{0.15\ M} = 6.7 \times 10^{-14}\ M$$

Because $[H_3O^+] >> [OH^-]$, the solution is strongly acidic. Here we have ignored the contribution to $[H_3O^+]$ arising from the autoprotonation of water, which is very small compared with 0.15 M.

PRACTICE PROBLEM 20-1: Calculate $[H_3O^+]$, $[NO_3^-]$, and $[OH^-]$ at 25°C in a 0.60-M aqueous solution of $HNO_3(aq)$, a strong acid.

Answer: $[H_3O^+] = [NO_3^-] = 0.60\ M$; $[OH^-] = 1.7 \times 10^{-14}\ M$

The equilibrium expression $K_w = [H_3O^+][OH^-]$ is *always* true for water, regardless of what else is dissolved in the water; but $[H_3O^+]$ equals $[OH^-]$ only for a neutral solution.

Conductivity measurements also show that sodium hydroxide in water is completely dissociated; that is, it exists as $Na^+(aq)$ and $OH^-(aq)$ ions:

$$NaOH(s) \xrightarrow[H_2O(l)]{} Na^+(aq) + OH^-(aq)$$

There are essentially no undissociated NaOH molecules present in aqueous solution. Sodium hydroxide is a base because $OH^-(aq)$ is a proton acceptor:

$$H_3O^+(aq) + OH^-(aq) \rightarrow 2H_2O(l)$$

Completely dissociated bases such as $NaOH(aq)$ are referred to as **strong bases**.

EXAMPLE 20-2: Calculate $[OH^-]$, $[Na^+]$, and $[H_3O^+]$ at 25°C in a 0.15-M aqueous solution of $NaOH(aq)$.

Solution: Because $NaOH(aq)$ is a strong base, it is completely dissociated, so $[OH^-] = 0.15\ M$ and $[Na^+] = 0.15\ M$. The value of $[H_3O^+]$ can be computed from the K_w expression:

$$[H_3O^+] = \frac{1.0 \times 10^{-14}\ M^2}{[OH^-]} = \frac{1.0 \times 10^{-14}\ M^2}{0.15\ M} = 6.7 \times 10^{-14}\ M$$

Because $[OH^-] >> [H_3O^+]$, the solution is strongly basic.

PRACTICE PROBLEM 20-2: Calculate $[Ba^{2+}]$, $[OH^-]$, and $[H_3O^+]$ at 25°C in a 0.25-M aqueous solution of barium hydroxide, a strong base.

Answer: $[Ba^{2+}] = 0.25\ M$; $[OH^-] = 0.50\ M$; $[H_3O^+] = 2.0 \times 10^{-14}\ M$

There are only a few strong acids and bases in water. Most acids and bases when dissolved in water are only partly dissociated into their constituent ions. Acids that are incompletely dissociated are called **weak acids**, and bases that are incompletely dissociated are called **weak bases**.

TABLE 20.1 Strong acids and strong bases in water

Strong acids		Strong bases	
Formula	**Name**	**Formula**	**Name**
$HClO_4$	perchloric acid	LiOH	lithium hydroxide
HNO_3	nitric acid	NaOH	sodium hydroxide
H_2SO_4	sulfuric acid*	KOH	potassium hydroxide
HCl	hydrochloric acid	RbOH	rubidium hydroxide
HBr	hydrobromic acid	CsOH	cesium hydroxide
HI	hydroiodic acid	TlOH	thallium(I) hydroxide
		$Ca(OH)_2$	calcium hydroxide
		$Sr(OH)_2$	strontium hydroxide
		$Ba(OH)_2$	barium hydroxide

*First proton only

TABLE 20.2 Hydrogen halide molar bond enthalpies

HX	Molar bond enthalpy/kJ·mol^{-1}
HF	570
HCl	432
HBr	366
HI	298

Table 20.1 lists common strong acids and bases. (Note that these are the first two classes of strong electrolytes listed in Section 12-3.) You should memorize their formulas because this information is essential in working problems in acid-base chemistry. Three of the six strong acids are halogen acids: $HCl(aq)$, $HBr(aq)$, and $HI(aq)$. Five of the nine strong bases are alkali metal hydroxides: $LiOH(aq)$, $NaOH(aq)$, $KOH(aq)$, $RbOH(aq)$, and $CsOH(aq)$. Another three of the strong bases are alkaline-earth metal hydroxides: $Ca(OH)_2(aq)$, $Sr(OH)_2(aq)$, and $Ba(OH)_2(aq)$. Unlike the other halogen acids, $HF(aq)$ is a weak acid because it has a much stronger bond than the H–X bonds in the other halogen acids (Table 20.2).

We can use the data in Table 20.1 to understand why certain salts form neutral solutions. For example, when $NaCl(s)$ is dissolved in water, the result is a neutral solution. That is, $[H_3O^+] = [OH^-]$, even though both $Na^+(aq)$ and $Cl^-(aq)$ ions are present in the water. Let's first examine the $Cl^-(aq)$ anion. This is the anion present in the strong acid $HCl(aq)$. Because $HCl(aq)$ is a strong acid, it dissociates completely in water according to

$$HCl(aq) + H_2O(l) \rightarrow H_3O^+(aq) + Cl^-(aq)$$

where we have used a right-hand arrow to show that there are essentially no undissociated $HCl(aq)$ molecules present in aqueous solution. Thus, $Cl^-(aq)$ has no tendency to undergo a proton transfer reaction with water. That is, the reaction described by

$$Cl^-(aq) + H_2O(l) \rightarrow HCl(aq) + OH^-(aq) \quad \text{(does not occur)}$$

does not occur to any significant extent because the products of this reaction immediately react to re-form $Cl^-(aq)$ and water. Thus, we say that $Cl^-(aq)$ ions are neutral in water (in the acid-base sense, not the electrical sense).

Now let's turn our attention to the $Na^+(aq)$ cation. $Na^+(aq)$ is the cation present in the strong base $NaOH(aq)$. As we have already seen, like strong acids, strong bases dissociate completely in water. Therefore, a $Na^+(aq)$ cation has no tendency to form $NaOH(aq)$ by undergoing a proton transfer reaction with water, and so $Na^+(aq)$ is a neutral cation in water.

Therefore, we can conclude that the cations of the nine strong bases and the anions of five of the strong acids listed in Table 20.1 form **neutral ions** in aqueous solutions. Sulfuric acid, H_2SO_4, is a strong acid for only the dissociation of the first proton. The anion, $HSO_4^-(aq)$, is a weak acid as we shall discuss in Section 20-12. These neutral ions are listed in Table 20.3. Because $NaCl(s)$ dissociates into neutral cations $Na^+(aq)$ and neutral anions $Cl^-(aq)$, an $NaCl(aq)$ solution is neutral. In contrast, because $NaOH(s)$ dissociates into neutral cations $Na^+(aq)$ and strongly basic anions $OH^-(aq)$, a $NaOH(aq)$ solution is basic. This illustrates a very important point: in order to determine whether a given solution is acidic, basic, or neutral, we must look at *all* the ions present in the solution. We shall return to the question of predicting whether various salts form acidic, basic, or neutral solutions in Section 20-11.

TABLE 20.3 Neutral cations and anions

Neutral cations		Neutral anions
$Li^+(aq)$	$Tl^+(aq)$	$Cl^-(aq)$
$Na^+(aq)$	$Mg^{2+}(aq)$	$Br^-(aq)$
$K^+(aq)$	$Ca^{2+}(aq)$	$I^-(aq)$
$Rb^+(aq)$	$Sr^{2+}(aq)$	$ClO_4^-(aq)$
$Cs^+(aq)$	$Ba^{2+}(aq)$	$NO_3^-(aq)$

20-4. Almost All Organic Acids Are Weak Acids

The most common organic acids are the **carboxylic acids**, that have the general formula RCOOH, where R represents a hydrogen atom, an **alkyl group** such as methyl (CH_3–) or ethyl (CH_3CH_2–), or some other organic group (see Interchapter F). The –COOH group is called the **carboxyl group**.

See Interchapter F at www.McQuarrieGeneralChemistry.com.

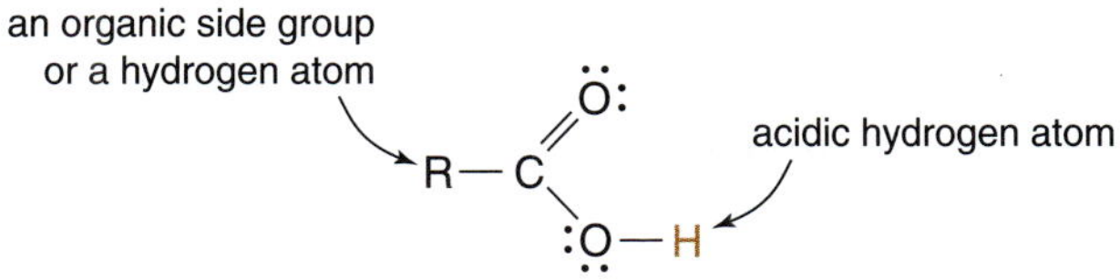

a carboxylic acid

The two simplest carboxylic acids are methanoic acid and ethanoic acid, where R = H and R = CH_3, respectively. Although these are the IUPAC or systematic names of these compounds, they are almost always referred to by their common names, formic acid and acetic acid (as we shall do here).

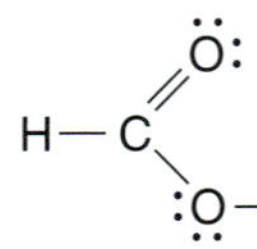

methanoic acid (formic acid, HCOOH)

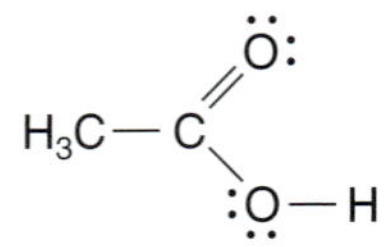

ethanoic acid (acetic acid, CH_3COOH)

Formic acid is the major irritant in the bite of ants and was first isolated in the 1600s from the distillation of ants. Acetic acid is familiar as vinegar, which is a 5% aqueous solution of acetic acid.

A carboxyl group produces hydronium ions in water, for example,

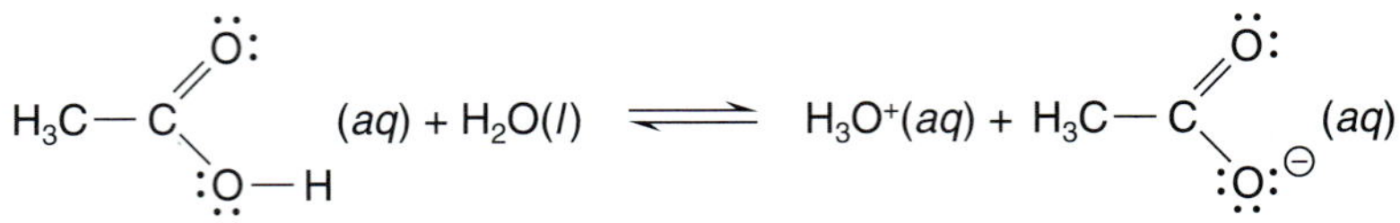

acetic acid acetate ion

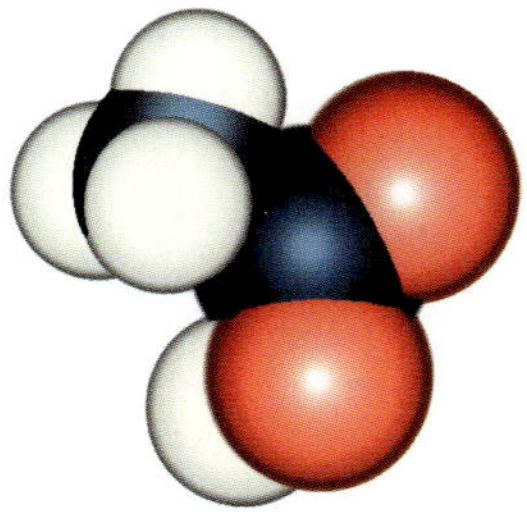

acetic acid

As noted previously, the hydrogen atom in the carboxyl group is the acidic hydrogen atom. Organic acids also react with bases such as sodium hydroxide to produce salts and water:

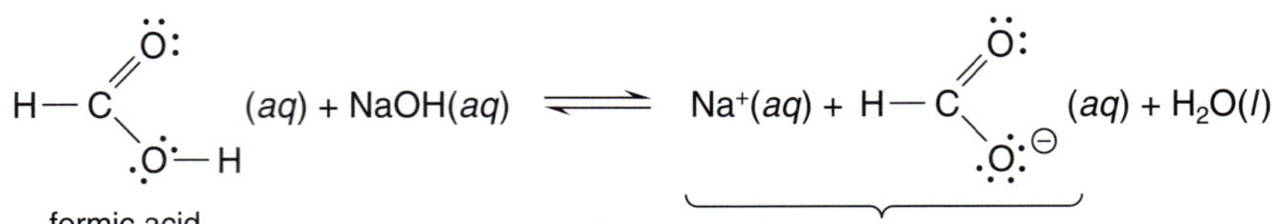

formic acid sodium formate, a salt

formic acid

The formulas of acetic acid and the acetate anion may also be written as $HC_2H_3O_2(aq)$ and $C_2H_3O_2^-(aq)$, respectively. However, in order to emphasize their structures, we shall usually write these formulas as $CH_3COOH(aq)$ and $CH_3COO^-(aq)$. Similarly, we shall write the formulas of formic acid and its anion as $HCOOH(aq)$ and $HCOO^-(aq)$, rather than as $HCHO_2(aq)$ and $CHO_2^-(aq)$.

EXAMPLE 20-3: Complete and balance the chemical equation

$$CH_3COOH(aq) + Ca(OH)_2(aq) \rightarrow$$

and name the salt produced in the reaction.

Solution: Each acetic acid molecule contributes one hydrogen ion, so two moles of acetic acid are required to neutralize completely one mole of calcium hydroxide. The balanced chemical equation is

$$2\,CH_3COOH(aq) + Ca(OH)_2(aq) \rightarrow Ca(CH_3COO)_2(aq) + 2\,H_2O(l)$$

The product is a salt of calcium hydroxide and acetic acid. To name the anion of this calcium salt, we change the *-ic* ending of the acid to *-ate* and drop the word *acid*. Thus, we have, in this case, calcium acetate.

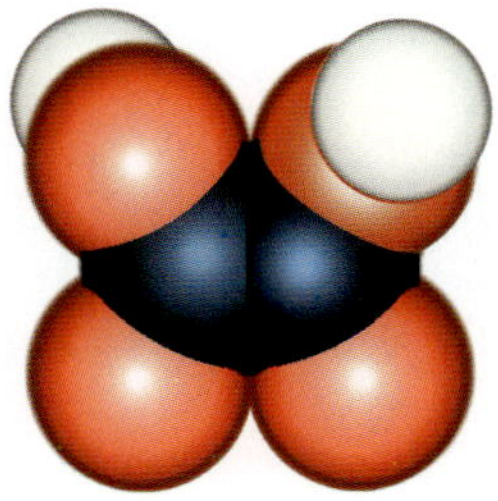

oxalic acid

PRACTICE PROBLEM 20-3: Oxalic acid, HOOCCOOH(aq), occurs naturally in many plants, such as tea, spinach, and rhubarb. The Lewis formula of oxalic acid is

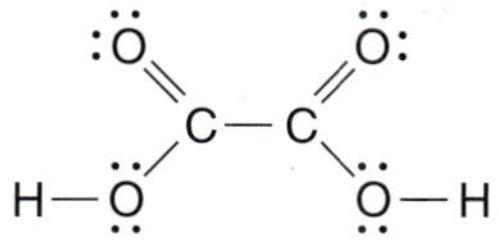

Because an oxalic acid molecule has two carboxylic groups, it is called a dicarboxylic acid. Complete and balance the equation

$$\mathrm{HOOCCOOH}(aq) + \mathrm{KOH}(aq) \rightarrow$$

Answer: $\mathrm{HOOCCOOH}(aq) + 2\,\mathrm{KOH}(aq) \rightarrow \mathrm{K_2(OOCCOO)}(aq) + 2\,\mathrm{H_2O}(l)$

The anion formed by a carboxylic acid is called a **carboxylate ion** and has the general formula $RCOO^-$. The carboxylate ion may be described by the two resonance formulas

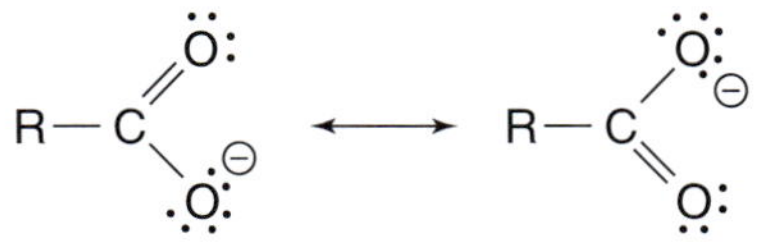

whose resonance hybrid is

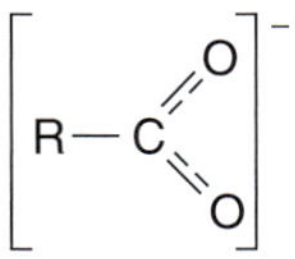

The Lewis formula of the hybrid shows that the negative charge is distributed equally between the two oxygen atoms. The two carbon–oxygen bonds in formic acid, for example, have different lengths, but in the formate ion the two carbon–oxygen bond lengths are identical and are intermediate between those of single and double carbon–oxygen bonds:

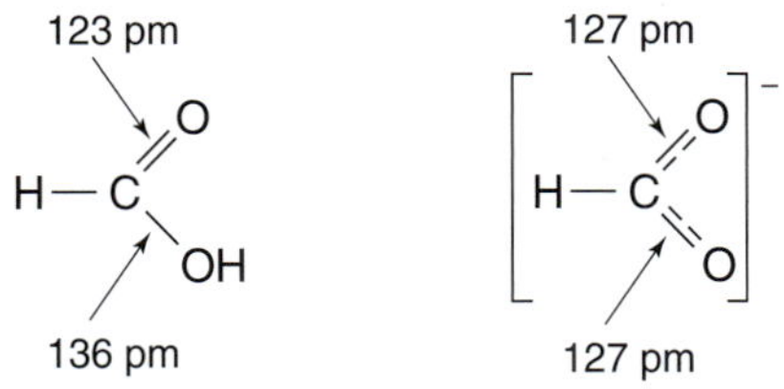

formic acid   formate ion

The delocalization of the negative charge over the two oxygen atoms confers an added degree of stability to a carboxylate ion. Recall from Chapters 7 and 9 that resonance structures are particularly stable. Because the loss of a hydrogen atom from a carboxyl group leaves behind a resonance-stabilized carboxylate ion, carboxyl groups tend to give up their hydrogen atoms, forming acidic solutions. In contrast, other groups such as methyl, $-CH_3$, and the $-OH$ group found in various alcohols, such as ethanol $CH_3CH_2OH(l)$ and methanol $CH_3OH(l)$, do not form resonance-stabilized anions and so are generally not acidic in solution.

Alcohols and other organic molecules containing –OH groups should not be confused with bases. Although the hydroxide group in an ionic compound, such as NaOH(*s*), readily dissociates to form a basic solution of $OH^-(aq)$ ions, the –OH group in an alcohol molecule, such as CH_3OH, is *covalently* bonded to a carbon atom, and so does not tend to dissociate in solution.

20-5. pH Is a Measure of the Acidity of an Aqueous Solution

You will find throughout your study of chemistry that the rates of many chemical reactions depend on the concentration of $H_3O^+(aq)$ in the reaction mixture. This may occur even if $H_3O^+(aq)$ is not one of the reactants or products. Thus, the addition of a small amount of $H_3O^+(aq)$ can dramatically alter the rates of many reactions (Section 18-6).

As we shall see, concentrations of $H_3O^+(aq)$ often lie in the range from 1 M to 1×10^{-14} M. Such a wide range of concentrations of $H_3O^+(aq)$ made plotting rates of reactions versus $[H_3O^+]$ very difficult. Søren Sørenson (Frontispiece) proposed expressing the hydronium ion scale on a logarithmic scale, now called the pH scale. The quantity called **pH** is defined as

$$\text{pH} = -\log([H_3O^+]/\text{M}) \tag{20.4}$$

We divide $[H_3O^+]$ by the unit M in Equation 20.4 so that $[H_3O^+]/\text{M}$ is unitless; you can take logarithms only of unitless quantities. Chemists quickly adopted the pH scale because of its convenience in plotting chemical properties versus pH instead of $[H_3O^+]$.

We use a common logarithm, that is, a logarithm to the base 10 in Equation 20.4. (The properties of logarithms are reviewed in Appendix A.) Equation 20.4 enables us to calculate the pH corresponding to various values of $[H_3O^+]$. For example, the pH of a 5.0×10^{-10} M solution of $H_3O^+(aq)$ is given by

$$\text{pH} = -\log([H_3O^+]/\text{M}) = -\log(5.0 \times 10^{-10}) = 9.30$$

Recall that the number of significant figures in a number determines the number of significant digits following the decimal place in that number's logarithm (see sidebox in Section 17-4, page 618).

EXAMPLE 20-4: Calculate the pH of an aqueous solution at 25°C that has been prepared by dissolving 0.26 grams of calcium hydroxide, $Ca(OH)_2(s)$, in water and diluting the solution to a final volume of 0.500 liters.

Solution: We first calculate the number of moles of $Ca(OH)_2(s)$. The formula mass of $Ca(OH)_2(s)$ is 74.09, so the number of moles is given by

$$\text{moles of Ca(OH)}_2 = (0.26\text{ g Ca(OH)}_2)\left(\frac{1\text{ mol Ca(OH)}_2}{74.09\text{ g Ca(OH)}_2}\right)$$
$$= 3.5 \times 10^{-3}\text{ mol}$$

The molarity of the solution is the number of moles per liter of solution:

$$\text{molarity} = \frac{\text{moles of solute}}{\text{liters of solution}} = \frac{3.5 \times 10^{-3}\ \text{mol Ca(OH)}_2}{0.500\ \text{L}} = 7.0 \times 10^{-3}\ \text{M}$$

Calcium hydroxide is a strong base and yields two moles of $OH^-(aq)$ ions per mole of $Ca(OH)_2(aq)$. Therefore, the molarity of the $OH^-(aq)$ is

$$[\text{OH}^-] = [7.0 \times 10^{-3}\ \text{M Ca(OH)}_2]\left(\frac{2\ \text{mol OH}^-}{1\ \text{mol Ca(OH)}_2}\right) = 1.4 \times 10^{-2}\ \text{M}$$

The value of $[H_3O^+]$ is calculated by using the ion-product constant of water:

$$[\text{H}_3\text{O}^+] = \frac{1.0 \times 10^{-14}\ \text{M}^2}{[\text{OH}^-]} = \frac{1.0 \times 10^{-14}\ \text{M}^2}{1.4 \times 10^{-2}\ \text{M}} = 7.1 \times 10^{-13}\ \text{M}$$

The pH of the solution is

$$\text{pH} = -\log([\text{H}_3\text{O}^+]/\text{M}) = -\log(7.1 \times 10^{-13}) = 12.15$$

PRACTICE PROBLEM 20-4: Calculate the pH of a solution that results when 20.0 mL of 6.0 M $NaOH(aq)$ is diluted with water to a final volume of 75.0 mL.

Answer: pH = 14.20

We can solve problems like Example 20-4 a little more quickly if we first introduce a new quantity called pOH. We take the logarithm of Equation 20.3 divided by M^2 to obtain

$$\log([\text{H}_3\text{O}^+][\text{OH}^-]/\text{M}^2) = \log(1.0 \times 10^{-14}) = -14.00$$

We use the fact that

$$\log([\text{H}_3\text{O}^+][\text{OH}^-]/\text{M}^2) = \log([\text{H}_3\text{O}^+]/\text{M}) + \log([\text{OH}^-]/\text{M}) = -14.00$$

Recall that $\log ab = \log a + \log b$

and then multiply by –1 to obtain

$$[-\log([\text{H}_3\text{O}^+]/\text{M})] + [-\log([\text{OH}^-]/\text{M})] = 14.00$$

Notice that the quantity $-\log([H_3O^+]/M)$ on the left side of this equation is the quantity pH. If we define a new quantity called **pOH** as

$$\text{pOH} = -\log([\text{OH}^-]/\text{M}) \qquad (20.5)$$

then we can say that, at 25°C,

$$\text{pH} + \text{pOH} = 14.00 \qquad (20.6)$$

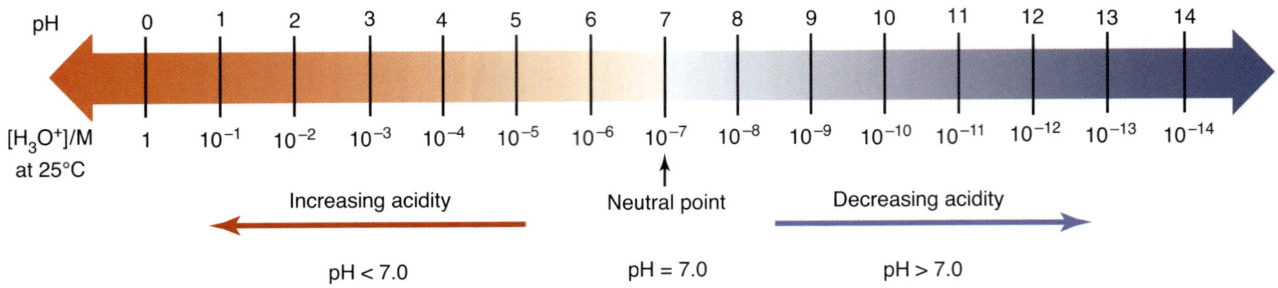

Equations 20.5 and 20.6 are often useful in solving problems involving basic solutions. In Example 20-4 above, we found that $[OH^-] = 1.4 \times 10^{-2}$ M. We substitute this value in Equation 20-5 to get pOH = 1.85, so the pH is 14.00 – 1.85 = 12.15.

At 25°C, pure water has a hydronium ion concentration of $[H_3O^+] = 1.0 \times 10^{-7}$ M; therefore, the pH of a neutral aqueous solution at 25°C is

$$pH = -\log([H_3O^+]/M) = -\log(1.0 \times 10^{-7}) = 7.00$$

At 25°C acidic solutions have $[H_3O^+]$ values greater than 1.0×10^{-7} M. Thus, acidic solutions have pH values less than 7.00. Basic solutions have $[H_3O^+]$ value less than 1.0×10^{-7} M and so pH values greater than 7.00. The pH scale is shown schematically above. Notice that a change in pH of one unit corresponds to a 10-fold change in $[H_3O^+]$.

The pH values of various common aqueous solutions are given in Figure 20.2.

Figure 20.2 The range of pH values for common aqueous solutions.

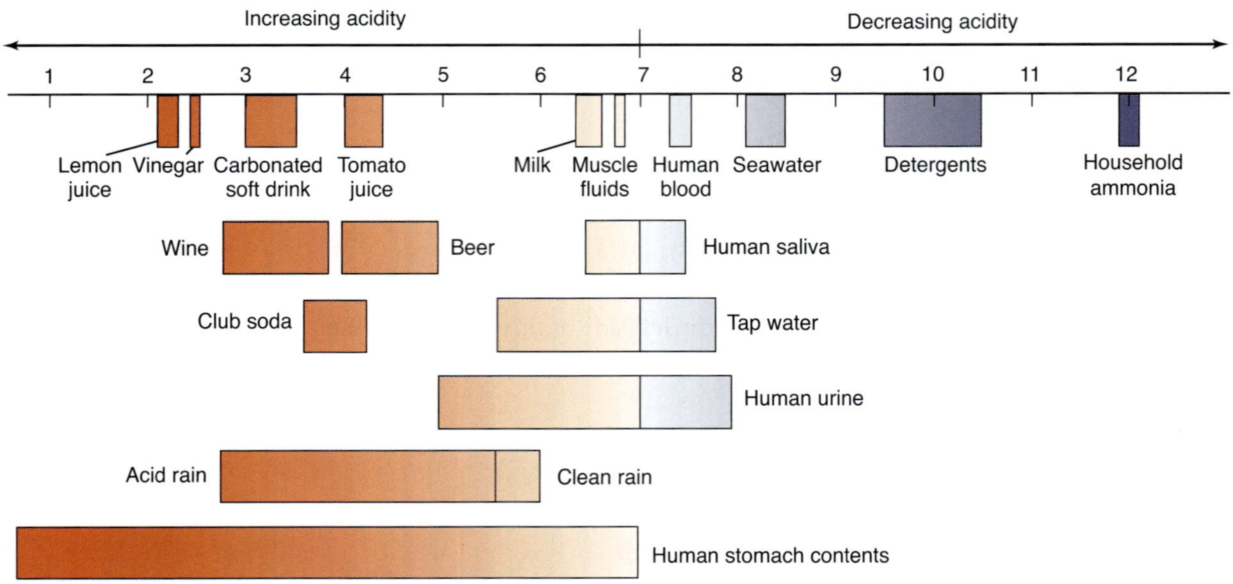

EXAMPLE 20-5: Compare the values of the pH of two solutions for which $[H_3O^+] = 5.5 \times 10^{-5}$ M and $[H_3O^+] = 5.5 \times 10^{-3}$ M.

Solution: The two solutions have pH values of $pH = -\log(5.5 \times 10^{-5}) = 4.26$ and $pH = -\log(5.5 \times 10^{-3}) = 2.26$. The pH of the two solutions differ by 2 pH units, whereas the hydronium ion concentrations differ by a factor of 100. The pH scale is a logarithmic scale, just like the Richter scale of earthquake intensities. One unit on each scale corresponds to a factor of 10.

PRACTICE PROBLEM 20-5: Calculate the difference in pH values of two solutions for which the $OH^-(aq)$ concentrations differ by a factor of 10 000.

Answer: 4 pH units

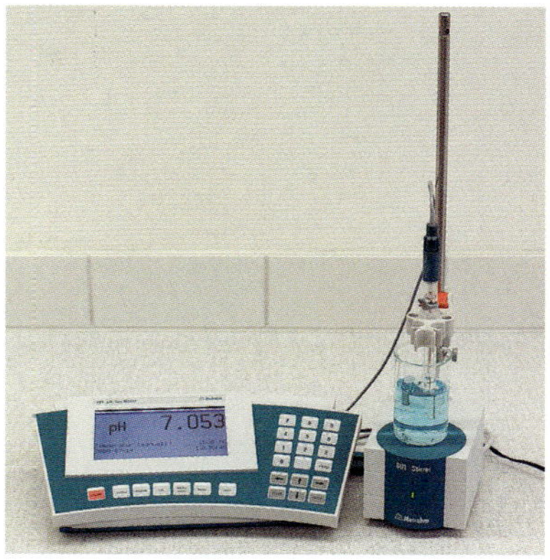

Figure 20.3 A pH meter. A glass electrode sensitive to the concentration of $H_3O^+(aq)$ ions is placed in the solution to measure the pH. Such a meter contains a reference solution (which must be changed periodically). Because the concentration of this reference solution changes slowly with time, pH meters must be calibrated periodically using standardized solutions of known pH.

The pH of a solution is conveniently measured in the laboratory with a **pH meter**, an electronic device that measures the concentration of $H_3O^+(aq)$ ions in a solution (Figure 20.3).

We have already seen how to calculate pH from $[H_3O^+]$. It is often necessary to do the inverse calculation, that is, to calculate $[H_3O^+]$ from the measured pH of the solution. The following Example illustrates such a calculation.

EXAMPLE 20-6: The pH of milk is about 6.5. Calculate the value of $[H_3O^+]$ for milk.

Solution: From the definition of pH, we know that

$$pH = -\log([H_3O^+]/M)$$

From the definition of logarithms, we know that if $y = \log x$, then $x = 10^y$. Thus,

$$[H_3O^+]/M = 10^{-pH}$$

The pH of milk is 6.5, so

$$[H_3O^+]/M = 10^{-6.5}$$

or

$$[H_3O^+] = 10^{-6.5}\ M = 3 \times 10^{-7}\ M$$

PRACTICE PROBLEM 20-6: The pH of a household ammonia solution is about 12. Calculate the concentration of $OH^-(aq)$ in household ammonia.

Answer: 1×10^{-2} M

20-6. Weak Acids and Weak Bases Are Dissociated Only Partially in Water

If 0.10 moles of HCl(g) are dissolved in enough water to make 1.00 liter of aqueous solution, then the observed pH of the resulting solution is 1.00. However, if 0.10 moles of HF(g) are dissolved in enough water to make 1.00 liter of aqueous solution, then the observed pH of the resulting solution is 2.10. In each case we calculate the value of $[H_3O^+]$ from the pH by using the relation (see Example 20-6)

$$[H_3O^+] = 10^{-pH}\ M \tag{20.7}$$

The values that we obtain using Equation 20.7 are as follows:

	Measured pH	$[H_3O^+]$/M
0.10 M HCl(aq)	1.00	0.10
0.10 M HF(aq)	2.10	7.9×10^{-3}

Comparison of these two values of $[H_3O^+]$ shows that, unlike HCl(aq), HF(aq) is only partially dissociated in water. The **percentage of dissociation** (% dissociation) of hydrofluoric acid in the 0.10 M HF(aq) solution is only

$$\%\ \text{dissociation} = \frac{[H_3O^+]}{[HF]_0} \times 100 = \frac{0.0079\ M}{0.10\ M} \times 100 = 7.9\%$$

where the subscript 0 denotes the stoichiometric concentration. In contrast, hydrochloric acid is completely dissociated:

$$\%\ \text{dissociation} = \frac{[H_3O^+]}{[HCl]_0} \times 100 = \frac{0.10\ M}{0.10\ M} \times 100 = 100\%$$

By **stoichiometric concentration**, we mean the concentration at which a solution is prepared. For example, a HF(aq) solution of a stoichiometric concentration of 0.10 M means that the solution was prepared by adding 0.10 moles of HF(g) to enough water to make exactly one liter of solution. The label on a bottle of such a solution would read "0.10 M hydrofluoric acid" or "0.10 M HF(aq)." Because hydrofluoric acid only partially dissociates in water, however, the actual concentration of undissociated HF(aq) molecules at equilibrium is less than 0.10 M.

Figure 20.4 The reactions of a strong acid (*left*) and of a weak acid (*right*) with magnesium. The strong acid produces a much more vigorous reaction.

At the same stoichiometric concentrations, a weak electrolyte such as HF(aq) will be a poorer conductor than a strong electrolyte such as HCl(aq) because fewer ions are available in HF(aq) to conduct a current. The ratio of the electrical conductivity of a 1.0 M HF(aq) solution to that of a 1.0 M HCl(aq) solution is about 0.08, in accord with their relative amounts of dissociation. Similarly, strong acids tend to react more vigorously than weak acids. Figure 20.4 illustrates the difference between the reactions of a strong acid and a weak acid with magnesium.

We now consider a weak base in aqueous solution. Ammonia is a base in aqueous solution because it accepts a proton from water, according to

$$NH_3(aq) + H_2O(l) \leftrightharpoons NH_4^+(aq) + OH^-(aq)$$

Ammonia is said to be a weak base because the position of the above equilibrium does not lie far to the right; in other words, not many of the ammonia molecules are protonated.

EXAMPLE 20-7: The pH of a 0.20 M $NH_3(aq)$ solution is 11.27. Calculate the percentage of ammonia molecules that are protonated in this solution.

Solution: The equation for the reaction is

$$NH_3(aq) + H_2O(l) \leftrightharpoons NH_4^+(aq) + OH^-(aq)$$

The percentage of ammonia molecules that are protonated is given by

$$\% \text{ protonated} = \frac{[NH_4^+]}{[NH_3]_0} \times 100$$

According to the stoichiometry of the chemical equation, $[NH_4^+] = [OH^-]$, and so we have

$$\% \text{ protonated} = \frac{[OH^-]}{[NH_3]_0} \times 100$$

We need to determine $[OH^-]$. The pH is 11.27, so

$$pOH = 14.00 - 11.27 = 2.73$$

The value of $[OH^-]$ can be calculated from the pOH:

$$[OH^-] = 10^{-pOH}\text{ M} = 10^{-2.73}\text{ M} = 1.9 \times 10^{-3}\text{ M}$$

The percentage of ammonia molecules that are protonated in a 0.20-M $NH_3(aq)$ solution is

$$\% \text{ protonated} = \frac{1.9 \times 10^{-3}\text{ M}}{0.20\text{ M}} \times 100 = 0.95\%$$

Thus, fewer than 1% of the ammonia molecules are protonated.

PRACTICE PROBLEM 20-7: The pH of a 0.40-M solution of formic acid, $HCOOH(aq)$, is 2.08 and that of a 0.40-M solution of hydrocyanic acid, $HCN(aq)$, is 4.80. Compare the percentages of acid molecules that are dissociated in the two solutions.

Answer: HCOOH, 2.1%; HCN, 0.0040%

20-7. Acids with Large Values of K_a Are Stronger than Acids with Smaller Values of K_a

The equilibrium-constant expression for an **acid dissociation reaction** is written according to Equation 19.6. Let's consider acetic acid, $CH_3COOH(aq)$, a weak acid that produces $H_3O^+(aq)$ by the protonation of water according to

$$CH_3COOH(aq) + H_2O(l) \leftrightharpoons H_3O^+(aq) + CH_3COO^-(aq) \qquad (20.8)$$

The equilibrium-constant expression for this equation, as given by Equation 19.6, is

$$K_a = \frac{[H_3O^+][CH_3COO^-]}{[CH_3COOH]} \tag{20.9}$$

where the subscript a on K reminds us that K_a is an **acid dissociation constant.** Notice that the $H_2O(l)$ concentration does not appear in the K_a expression. The experimental value of K_a at 25°C for acetic acid is 1.8×10^{-5} M.

The small value of K_a for acetic acid means that only a small fraction of the acetic acid molecules dissociate. Most of them remain in the undissociated form as $CH_3COOH(aq)$, so we expect the equilibrium concentration of undissociated acetic acid not to differ significantly from the stoichiometric concentration. Consider a 0.050-M acetic acid solution. We know that all the acetic acid either remains as $CH_3COOH(aq)$ or dissociates to $CH_3COO^-(aq)$, so

$$0.050\text{ M} = [CH_3COOH] + [CH_3COO^-] \tag{20.10}$$

This equation expresses the fact that all the CH_3COOH units in the solution are in the form of either undissociated acetic acid, $CH_3COOH(aq)$, or acetate ions, $CH_3COO^-(aq)$. A condition such as Equation 20.10, which accounts for all of a certain type of unit, is called a **material balance condition.**

How do we calculate the pH of a 0.050 M $CH_3COOH(aq)$ solution? The dissociation of acetic acid is one source of $H_3O^+(aq)$. Another source of $H_3O^+(aq)$ is the reaction represented by Equation 20.1,

$$H_2O(l) + H_2O(l) \leftrightharpoons H_3O^+(aq) + OH^-(aq) \tag{20.11}$$

for which

$$K_w = [H_3O^+][OH^-] = 1.0 \times 10^{-14}\text{ M}^2 \tag{20.12}$$

at 25°C. It would appear at first sight that we must consider both Equations 20.8 and 20.11, along with their associated equilibrium-constant expressions. However, because K_a is so much larger than K_w, we can neglect the contribution that Equation 20.11 makes to the overall concentration of $H_3O^+(aq)$. We use just Equation 20.8 to calculate the pH of the acetic acid solution and write

$$x = [H_3O^+] = [CH_3COO^-] \tag{20.13}$$

Next, we set up a concentration table for the dissociation reaction as we did in Chapter 19:

Concentration	$CH_3COOH(aq)$	+	$H_2O(l)$	$\leftrightharpoons$	$H_3O^+(aq)$	+	$CH_3COO^-(aq)$
initial	0.050 M		—		≈ 0 M		0 M
change	$-x$		—		$+x$		$+x$
equilibrium	$0.050\text{ M} - x$		—		x		x

Note that we have set the initial concentration of $H_3O^+(aq)$ essentially equal to 0 because we are neglecting the proton contribution of the reaction described by Equation 20.11. We substitute the values from the concentration table into Equation 20.9 to get

$$1.8 \times 10^{-5}\ \mathrm{M} = \frac{x^2}{0.050\ \mathrm{M} - x} \tag{20.14}$$

Equation 20.14 can be written in the standard form of a quadratic equation:

$$x^2 + (1.8 \times 10^{-5}\ \mathrm{M})x - 9.0 \times 10^{-7}\ \mathrm{M}^2 = 0$$

The two solutions to this equation are

$$x = 9.4 \times 10^{-4}\ \mathrm{M} \text{ and } -9.6 \times 10^{-4}\ \mathrm{M}$$

We reject the physically unacceptable negative concentration, so the final concentration table is:

Concentration	$CH_3COOH(aq)$ +	$H_2O(l)$ ⇋	$H_3O^+(aq)$	+ $CH_3COO^-(aq)$
initial	0.050 M	—	≈ 0 M	0 M
change	-9.4×10^{-4} M	—	$+9.4 \times 10^{-4}$ M	$+9.4 \times 10^{-4}$ M
equilibrium	0.049 M	—	9.4×10^{-4} M	9.4×10^{-4} M

Thus, the pH of a 0.050-M acetic acid solution is

$$\mathrm{pH} = -\log([H_3O^+]/\mathrm{M}) = -\log(9.4 \times 10^{-4}) = 3.03$$

Note that the concentration of $H_3O^+(aq)$ from the dissociated acetic acid, 9.4×10^{-4} M, is much larger than the concentration of $H_3O^+(aq)$ due to the autoprotonation of water at 25°C, 1.0×10^{-7} M, so our initial assumption that we could neglect the contribution from Equation 20.11 is justified. However, in cases where the resulting concentration of $H_3O^+(aq)$ is on the order of 10^{-6} M or less, we must include the contribution from Equation 20.11 in our calculations. In general this occurs only when the value of K_a is close to that of K_w or in the case of extremely dilute solutions (see Problem 20-94).

We can see that the equilibrium concentration of undissociated acetic acid, 0.049 M, is indeed very close to its stoichiometric concentration of 0.050 M. The percentage of dissociation of acetic acid in a 0.050-M solution of $CH_3COOH(aq)$ is

$$\%\ \text{dissociation} = \frac{[H_3O^+]}{[CH_3COOH]_0} \times 100 = \frac{9.4 \times 10^{-4}\ \mathrm{M}}{0.050\ \mathrm{M}} \times 100 = 1.9\%$$

Thus, we see that over 98% of the acetic acid in a 0.050-M solution remains undissociated.

Figure 20.5 shows how the percentage of dissociation of acetic acid increases as the acetic acid concentration decreases. The equation for the dissociation of acetic acid in water is

$$CH_3COOH(aq) + H_2O(l) \leftrightharpoons H_3O^+(aq) + CH_3COO^-(aq)$$

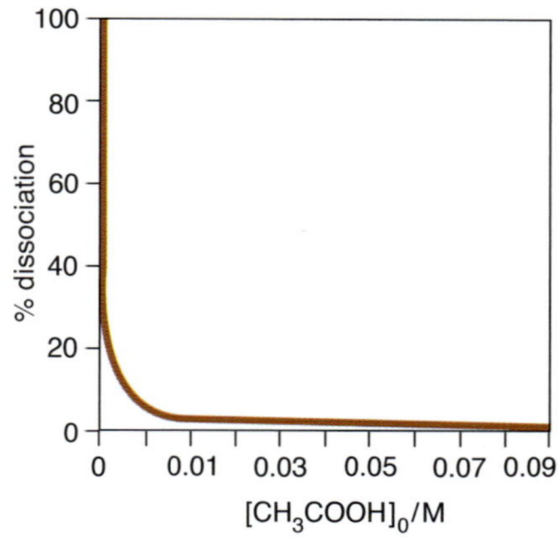

Figure 20.5 The percentage dissociation of acetic acid as a function of the stoichiometric (initial) concentration of acid, $[CH_3COOH]_0$. The calculations are carried out by solving the quadratic equation

$$K_a = \frac{[H_3O^+]^2}{[CH_3COOH]_0 - [H_3O^+]} = 1.8 \times 10^{-5}\ \mathrm{M}$$

for $[H_3O^+]$, for various values of $[CH_3COOH]_0$. The percentage dissociation is then computed by using

$$\%\ \text{dissociation} = \frac{[H_3O^+]}{[CH_3COOH]_0} \times 100$$

The procedure works down to $[CH_3COOH]_0 = 10^{-5}$ M; below this concentration the contribution to $[H_3O^+]$ from the autoprotonation reaction of water (Equation 20.1) must be taken into account.

This effect can be understood in terms of Le Châtelier's principle (Section 19-7). The addition of water increases the volume available to the reaction. Thus, the reaction equilibrium shifts to the side with the greater number of moles of aqueous solute; that is, the equilibrium shifts toward the formation of more $H_3O^+(aq)$ and $CH_3COO^-(aq)$ ions.

EXAMPLE 20-8: Calculate the pH and the concentrations of all the species in a 0.050-M aqueous solution of chloroacetic acid, $ClCH_2COOH(aq)$. The value of K_a for $ClCH_2COOH(aq)$ is 1.4×10^{-3} M at 25°C.

Solution: The relevant equation for the dissociation reaction is

$$ClCH_2COOH(aq) + H_2O(l) \rightleftharpoons H_3O^+(aq) + ClCH_2COO^-(aq) \qquad K_a = 1.4 \times 10^{-3}\ \text{M}$$

The only other source of $H_3O^+(aq)$ is

$$H_2O(l) + H_2O(l) \rightleftharpoons H_3O^+(aq) + OH^-(aq) \qquad K_w = 1.0 \times 10^{-14}\ \text{M}^2$$

Because $K_a >> K_w$, we can neglect the autoprotonation source of $H_3O^+(aq)$ in setting up a concentration table.

Concentration	$ClCH_2COOH(aq)$	+	$H_2O(l)$	$\rightleftharpoons$	$H_3O^+(aq)$	+	$ClCH_2COO^-(aq)$
initial	0.050 M		—		≈ 0 M		0 M
change	$-x$		—		$+x$		$+x$
equilibrium	0.050 M $- x$		—		x		x

The equilibrium-constant expression is

$$K_a = \frac{[H_3O^+][ClCH_2COO^-]}{[ClCH_2COOH]} = 1.4 \times 10^{-3}\ \text{M}$$

Substituting the appropriate entries from the table gives

$$\frac{x^2}{0.050\ \text{M} - x} = 1.4 \times 10^{-3}\ \text{M}$$

We then write this equation in the standard form of a quadratic equation:

$$x^2 + (1.4 \times 10^{-3}\ \text{M})x - 7.0 \times 10^{-5}\ \text{M}^2 = 0$$

The two solutions to this equation are $x = 7.7 \times 10^{-3}$ M and -9.1×10^{-3} M. We reject the physically unacceptable negative concentration, and so our final concentration table is

Concentration	$ClCH_2COOH(aq)$	+	$H_2O(l)$	$\rightleftharpoons$	$H_3O^+(aq)$	+	$ClCH_3COO^-(aq)$
initial	0.050 M		—		≈ 0 M		0 M
change	-7.7×10^{-3} M		—		$+7.7 \times 10^{-3}$ M		$+7.7 \times 10^{-3}$ M
equilibrium	0.042 M		—		7.7×10^{-3} M		7.7×10^{-3} M

The pH of the solution is given by

$$pH = -\log([H_3O^+]/M) = -\log(7.7 \times 10^{-3}) = 2.11$$

and the concentration of $OH^-(aq)$ is given by Equation 20.12

$$[OH^-] = \frac{1.0 \times 10^{-14}\ M^2}{[H_3O^+]} = \frac{1.0 \times 10^{-14}\ M^2}{7.7 \times 10^{-3}\ M} = 1.3 \times 10^{-12}\ M$$

Observe that the pH of the 0.050 M $ClCH_2COOH(aq)$ solution is almost an entire pH unit less than that of the 0.050 M $CH_3COOH(aq)$ solution calculated above. In other words, 0.050 M $ClCH_2COOH(aq)$ is about 10 times more acidic than 0.050 M $CH_3COOH(aq)$. The percentage of dissociation of 0.050 M $ClCH_2COOH(aq)$ is

$$\%\ \text{dissociation} = \frac{[H_3O^+]}{[ClCH_2COOH]_0} \times 100 = \frac{7.7 \times 10^{-3}\ M}{0.050\ M} \times 100 = 15\%$$

compared with only 1.9% for 0.050 M $CH_3COOH(aq)$.

PRACTICE PROBLEM 20-8: Calculate the pH and the concentrations of all the species in a 0.250-M aqueous solution of benzoic acid, $C_6H_5COOH(aq)$. The value of K_a for benzoic acid is 6.3×10^{-5} M.

Answer: $[H_3O^+] = [C_6H_5COO^-] = 4.0 \times 10^{-3}$ M; $[C_6H_5COOH] = 0.246$ M; $[OH^-] = 2.5 \times 10^{-12}$ M; pH = 2.40

For a given stoichiometric concentration, the percentage of dissociation of an acid depends on the value of K_a. The larger the value of K_a, the stronger the acid. The small value of K_a of acetic acid reflects the fact that it is only slightly dissociated in aqueous solution. Table 20.4 gives the K_a values for a number of weak acids.

Because K_a values for aqueous acids range over many powers of 10 (Table 20.4), it is convenient to define a quantity $\mathbf{p}K_a$ as

$$pK_a = -\log(K_a/M) \tag{20.15}$$

Note the similarity in the definitions of pH (Equation 20.4) and pK_a. The pK_a values at 25°C for some weak acids in water are given in Table 20.4. From these pK_a values, we see that the stronger the acid, the smaller the value of pK_a.

EXAMPLE 20-9: Given that $K_a = 1.8 \times 10^{-5}$ M for $CH_3COOH(aq)$ at 25°C, calculate pK_a.

Solution: Using Equation 20.15, we have

$$pK_a = -\log(K_a/M) = -\log(1.8 \times 10^{-5}) = 4.74$$

in agreement with the value in Table 20.4. Note that like pH and pOH, pK_a values are also unitless.

PRACTICE PROBLEM 20-9: Given that the value of pK_a for iodic acid, $HIO_3(aq)$, is 0.78 at 25°C, calculate the corresponding value of K_a for the acid.

Answer: 0.17 M

TABLE 20.4 Values of K_a and pK_a for weak acids in water at 25°C*

Acid	Formula	K_a/M	pK_a
acetic	CH_3COOH	1.8×10^{-5}	4.74
benzoic	C_6H_5COOH	6.3×10^{-5}	4.20
chloroacetic	$ClCH_2COOH$	1.4×10^{-3}	2.87
chlorous	$HClO_2$	1.2×10^{-2}	1.94
cyanic	HCNO	3.5×10^{-4}	3.46
dichloroacetic	$Cl_2CHCOOH$	4.5×10^{-2}	1.35
fluoroacetic	FCH_2COOH	2.6×10^{-3}	2.59
formic	HCOOH	1.8×10^{-4}	3.75
hydrazoic	HN_3	3×10^{-5}	4.6
hydrocyanic	HCN	6.2×10^{-10}	9.21
hydrofluoric	HF	6.3×10^{-4}	3.20
hypochlorous	HClO	4.0×10^{-8}	7.40
iodic	HIO_3	0.17	0.78
lactic	$CH_3CHOHCOOH$	1.4×10^{-4}	3.86
nitrous	HNO_2	5.6×10^{-4}	3.25
propanoic	CH_3CH_2COOH	1.4×10^{-5}	4.87

*pK_a values from the *CRC Handbook of Chemistry and Physics,* 87th Ed., 2006–2007

20-8. The Method of Successive Approximations Is Often Used in Solving Acid-Base Equilibrium Problems

The use of the quadratic formula to solve equations like Equation 20.14 is effective but tedious. An alternate, and often faster, method of solution of a quadratic equation is the **method of successive approximations**. Let's reconsider Equation 20.14:

$$1.8 \times 10^{-5}\ \text{M} = \frac{x^2}{0.050\ \text{M} - x}$$

Because acetic acid is a weak acid (as indicated by the small value of K_a), we expect that the percentage of dissociation of the acid is small. Furthermore, because essentially all the $H_3O^+(aq)$ arises from dissociation of the acid, we expect that $[H_3O^+]$ will be small relative to the initial concentration of the acid. We can express this algebraically as

$$0.050\ \text{M} - x \approx 0.050\ \text{M}$$

so that Equation 20.14 becomes

$$1.8 \times 10^{-5}\ \text{M} \approx \frac{x^2}{0.050\ \text{M}}$$

Solving for x^2, we find

$$x^2 \approx (0.050\ \text{M})(1.8 \times 10^{-5}\ \text{M})$$

and so

$$x \approx \sqrt{(0.050\ \text{M})(1.8 \times 10^{-5}\ \text{M})} = 9.5 \times 10^{-4}\ \text{M}$$

This first approximation is close to the value of 9.4×10^{-4} M obtained by the solution of the full quadratic equation using the quadratic formula in the previous section.

We can now use this approximate value of x in the denominator of the right-hand side of Equation 20.14 to obtain a second, more accurate, approximation; that is, we take

$$1.8 \times 10^{-5}\ \text{M} \approx \frac{x^2}{0.050\ \text{M} - 9.5 \times 10^{-4}\ \text{M}}$$

Thus, we calculate that

$$x \approx \sqrt{(0.049\ \text{M})(1.8 \times 10^{-5}\ \text{M})} = 9.4 \times 10^{-4}\ \text{M}$$

which is the same as the exact solution to two significant figures. If we repeat the approximation procedure, using 9.4×10^{-4} M in the denominator, then we find no further change in the value of x to two significant figures. Therefore, we have found that

$$1.8 \times 10^{-5}\ \text{M} = \frac{(9.4 \times 10^{-4}\ \text{M})^2}{0.050\ \text{M} - 9.4 \times 10^{-4}\ \text{M}}$$

Identical values of x in the numerator and the denominator means that the original equation (Equation 20.14) is satisfied. In other words, an unchanged value of x on successive approximations means that we have found the correct solution.

The method of successive approximations is usually much faster and easier than the use of the full quadratic equation. We shall use the method of successive approximations often to solve equilibrium problems. More examples can be found in Appendix A5.

EXAMPLE 20-10: The value of K_a for an aqueous solution of hypochlorous acid, HClO(*aq*), is 4.0×10^{-8} M. Calculate the pH of a 0.050-M HClO(*aq*) solution at 25°.

Solution: The equilibrium equation is

$$\mathrm{HClO}(aq) + \mathrm{H_2O}(l) \leftrightharpoons \mathrm{H_3O^+}(aq) + \mathrm{ClO^-}(aq)$$

The only other source of $H_3O^+(aq)$ is from the reaction described by Equation 20.1; but, because $K_a >> K_w$, we can ignore that source of $H_3O^+(aq)$ and set up the following concentration table for initial and equilibrium concentrations of the species in solution:

Concentration	HClO(*aq*)	+	$H_2O(l)$	⇋	$H_3O^+(aq)$	+	$ClO^-(aq)$
initial	0.050 M		—		≈ 0 M		0 M
change	$-x$		—		$+x$		$+x$
equilibrium	0.050 M $- x$		—		x		x

If we combine the entries in this table with the expression for K_a, then we have

$$K_a = \frac{[\mathrm{H_3O^+}][\mathrm{ClO^-}]}{[\mathrm{HClO}]} = \frac{x^2}{0.050\ \mathrm{M} - x} = 4.0 \times 10^{-8}\ \mathrm{M}$$

We use the method of successive approximations to solve this equation. Neglecting x relative to 0.050 M, we obtain the first approximate solution:

$$x \approx \sqrt{(0.050\ \mathrm{M})(4.0 \times 10^{-8}\ \mathrm{M})} = 4.5 \times 10^{-5}\ \mathrm{M}$$

Because x is very small relative to 0.050 M, the second approximation yields the same value of x as the first approximation; namely,

$$\frac{x^2}{0.050\ \mathrm{M} - 4.5 \times 10^{-5}\ \mathrm{M}} \approx 4.0 \times 10^{-8}\ \mathrm{M}$$

and

$$x \approx \sqrt{(0.050\ \mathrm{M} - 4.5 \times 10^{-5}\ \mathrm{M})(4.0 \times 10^{-8}\ \mathrm{M})} = 4.5 \times 10^{-5}\ \mathrm{M}$$

Because $[H_3O^+] = x$, the pH of the solution is

$$\mathrm{pH} = -\log([\mathrm{H_3O^+}]/\mathrm{M}) = -\log(4.5 \times 10^{-5}) = 4.35$$

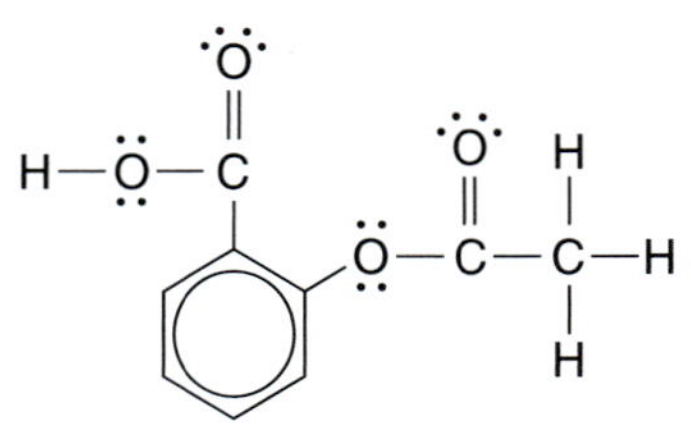

acetylsalicylic acid
(aspirin)

PRACTICE PROBLEM 20-10: The chemical name for aspirin is acetylsalicylic acid, $CH_3COOC_6H_4COOH$ (its Lewis formula is shown in the margin). Given that $pK_a = 3.48$ for aspirin, calculate the pH at 25°C of a solution that contains 324 mg (a standard 5-grain aspirin tablet) in 100.0 mL of aqueous solution.

Answer: pH = 2.64

20-9. Bases with Large Values of K_b Are Stronger than Bases with Smaller Values of K_b

Ammonia is a base in aqueous solution because it reacts with water to accept a proton according to the chemical equation

$$NH_3(aq) + H_2O(l) \leftrightharpoons NH_4^+(aq) + OH^-(aq)$$

The equilibrium constant for this equation is 1.8×10^{-5} M, a value indicating that the reaction equilibrium lies far to the left. In terms of Lewis formulas, the protonation of ammonia can be described by

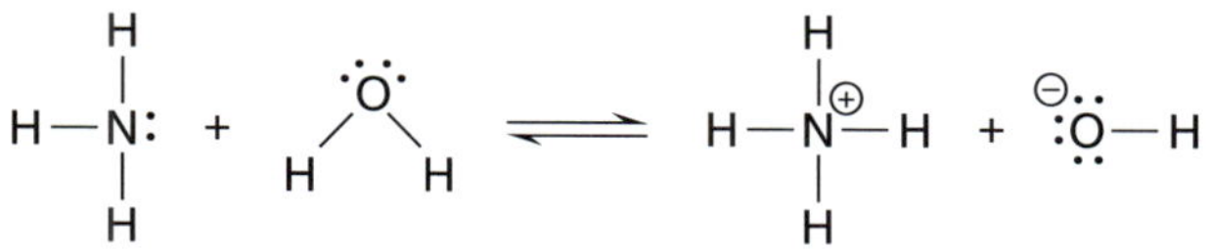

The lone pair of electrons on the nitrogen atom in the ammonia molecule is able to bond to the proton donated by the water molecule.

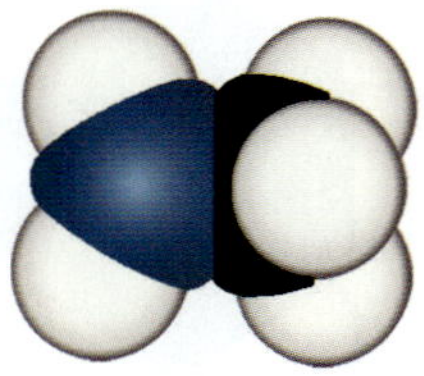

methylamine

Other common weak bases include organic compounds called **amines**. Amines are derivatives of ammonia in which one or more of the hydrogen atoms in an NH_3 molecule are replaced by alkyl groups. The common names of alkyl amines are often formed by combining the names of the attached alkyl groups in alphabetical order with the word *amine*. Some simple examples of amines are

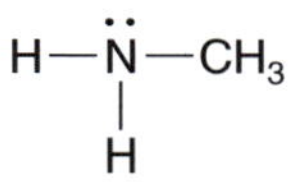

methylamine

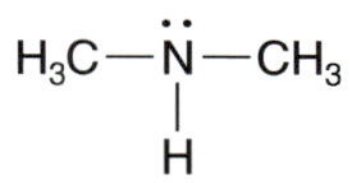

dimethylamine

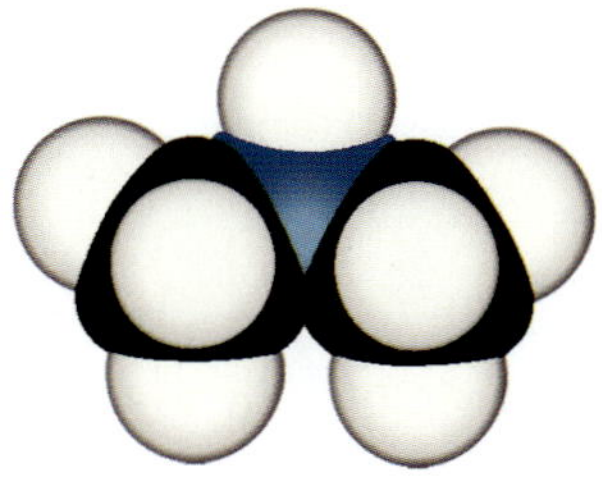

dimethylamine

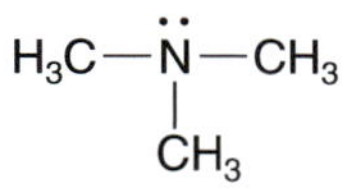

trimethylamine

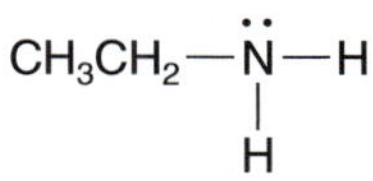

ethylamine

Ammonia and most amines react with acids to form nonvolatile salts. The cation formed by the protonation of an amine is named by changing the amine suffix in the name of the amine to ammonium and adding the word ion as shown below. For example,

$$\underset{\text{dimethylamine}}{(CH_3)_2NH(aq)} + HBr(aq) \rightarrow \underbrace{\underset{\text{dimethylammonium ion}}{(CH_3)_2NH_2^+(aq)} + \underset{\text{bromide ion}}{Br^-(aq)}}_{\text{a salt}}$$

trimethylamine

The salts of amines are generally more water-soluble than the amines themselves.

Many alkyl amines have strong odors, like that of decaying fish. The common names of some diamines (i.e., molecules containing two amino groups) suggest even worse odors:

ethylamine

$H_2NCH_2CH_2CH_2CH_2NH_2$ 1,4-butanediamine (putrescine)

$H_2NCH_2CH_2CH_2CH_2CH_2NH_2$ 1,5-pentanediamine (cadaverine)

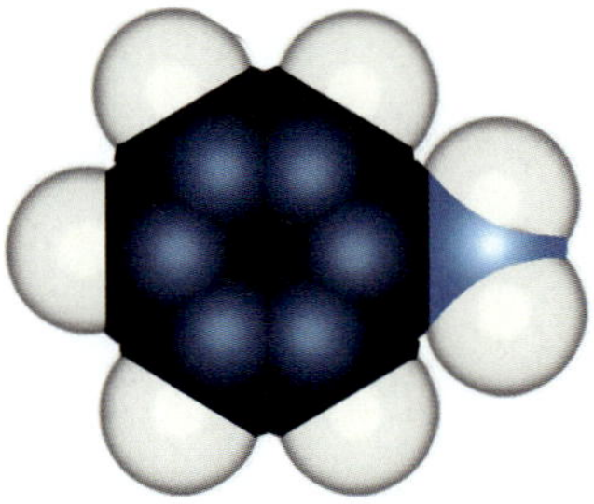
aniline

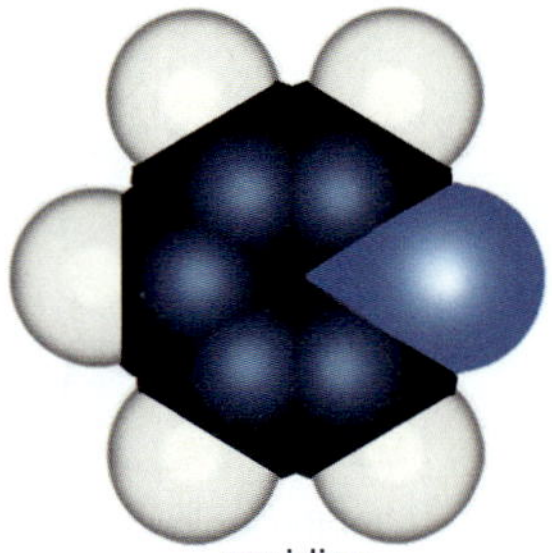
pyridine

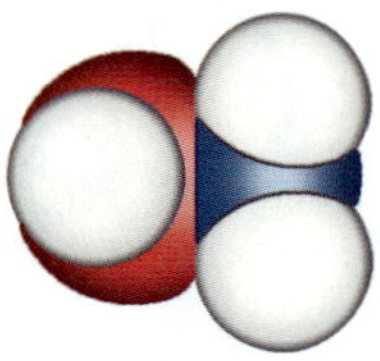
hydroxylamine

Some other organic bases are shown below.

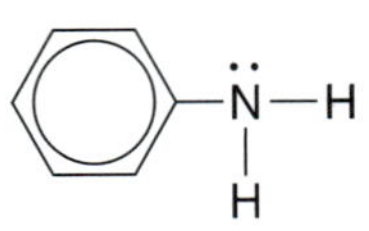
aniline

pyridine

Notice that the nitrogen atom in each of these bases has a lone pair of electrons. Pyridine, for example, is basic in aqueous solution because of the reaction described by the following equation:

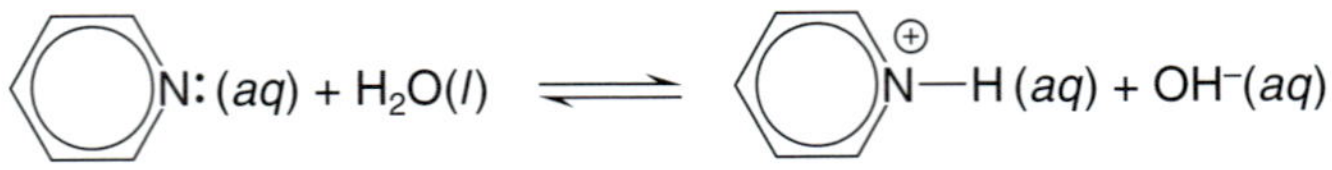

pyridine pyridinium ion

If we abbreviate pyridine by Py and the pyridinium ion by PyH^+, then the equilibrium-constant expression for this equation is

$$K_b = \frac{[PyH^+][OH^-]}{[Py]}$$

The subscript b on K indicates that the reaction is that of a weak base with water, that is, a base protonation reaction. Thus, K_b is called the **base protonation constant**. By analogy to pK_a, $\mathbf{p}\boldsymbol{K}_\mathbf{b}$ is defined as

$$pK_b = -\log(K_b/M) \qquad (20.16)$$

The smaller the value of pK_b, the stronger the base. The values of K_b and pK_b for various weak bases are given in Table 20.5.

EXAMPLE 20-11: Calculate the pH and the concentrations of all the species in a 0.75-M hydroxylamine, $HONH_2(aq)$, solution at 25°C.

Solution: The equation for the base protonation equilibrium is

$$HONH_2(aq) + H_2O(l) \leftrightharpoons HONH_3^+(aq) + OH^-(aq) \qquad K_b = 8.7 \times 10^{-9}\ M$$

where the value of K_b is from Table 20.5. The only other source of $OH^-(aq)$ is from the reaction described by Equation 20.1; but, because $K_b >> K_w$, we can ignore this source of $OH^-(aq)$ and set up the following concentration table.

Concentration	$HONH_2(aq)$	+	$H_2O(l)$	$\leftrightharpoons$	$HONH_3^+(aq)$	+	$OH^-(aq)$
initial	0.75 M		—		0 M		≈ 0 M
change	$-x$		—		$+x$		$+x$
equilibrium	$0.75\ M - x$		—		x		x

The equilibrium-constant expression for this equation is

TABLE 20.5 Values of K_b and pK_b for weak bases in water at 25°C*

Base	Formula	K_b/M	pK_b
ammonia	$H-\ddot{N}(H)-H$	1.8×10^{-5}	4.75
methylamine	$H-\ddot{N}(H)-CH_3$	4.6×10^{-4}	3.34
dimethylamine	$H_3C-\ddot{N}(H)-CH_3$	5.4×10^{-4}	3.27
trimethylamine	$H_3C-\ddot{N}(CH_3)-CH_3$	6.3×10^{-5}	4.20
ethylamine	$CH_3CH_2-\ddot{N}(H)-H$	4.5×10^{-4}	3.35
hydroxylamine	$H-\ddot{N}(H)-OH$	8.7×10^{-9}	8.06
aniline	$C_6H_5-\ddot{N}(H)-H$	7.4×10^{-10}	9.13
pyridine	$C_5H_5N:$	1.7×10^{-9}	8.77

*pK_b values derived from data in the *CRC Handbook of Chemistry and Physics*, 87th Ed., 2006–2007

$$K_b = \frac{[HONH_3^+][OH^-]}{[HONH_2]} = \frac{x^2}{0.75\text{ M} - x} = 8.7 \times 10^{-9}\text{ M}$$

The method of successive approximations gives $x = 8.1 \times 10^{-5}$ M, and so the concentration table is now:

Concentration	$HONH_2(aq)$	+ $H_2O(l)$ ⇋	$HONH_3^+(aq)$	+ $OH^-(aq)$
initial	0.75 M	—	0 M	≈ 0 M
change	-8.1×10^{-5} M	—	$+8.1 \times 10^{-5}$ M	$+8.1 \times 10^{-5}$ M
equilibrium	0.75 M	—	8.1×10^{-5} M	8.1×10^{-5} M

Because $[OH^-] = 8.1 \times 10^{-5}$ M, the pOH is

$$\text{pOH} = -\log([OH^-]/\text{M}) = -\log(8.1 \times 10^{-5}) = 4.09$$

and the pH is

$$\text{pH} = 14.00 - \text{pOH} = 14.00 - 4.09 = 9.91$$

The concentration of $H_3O^+(aq)$ is given by

$$[H_3O^+] = \frac{1.0 \times 10^{-14}\ M^2}{8.1 \times 10^{-5}\ M} = 1.2 \times 10^{-10}\ M$$

We could also have found $[H_3O^+]$ from the pH by using Equation 20.7, that is,

$$[H_3O^+] = 10^{-\text{pH}}\ M = 10^{-9.91}\ M = 1.2 \times 10^{-10}\ M$$

PRACTICE PROBLEM 20-11: Aniline, $C_6H_5NH_2(l)$, is used in the manufacture of dyes and various pharmaceuticals. The solubility of aniline in water at 25°C is 3.50 grams per 100.0 mL of solution. Calculate the pH and the concentrations of all the species in a saturated aqueous solution of aniline at 25°C.

Answer: $[OH^-] = [C_6H_5NH_3^+] = 1.7 \times 10^{-5}$ M; $[C_6H_5NH_2] = 0.376$ M; $[H_3O^+] = 5.9 \times 10^{-10}$ M; pH = 9.23

20-10. The Pair of Species HA(*aq*), A□(*aq*) Is Called a Conjugate Acid-Base Pair

When an acid such as $CH_3COOH(aq)$ reacts with water, water acts as a base by accepting a proton from the acid according to the equation

$$\underset{\text{acid}}{CH_3COOH(aq)} + \underset{\text{base}}{H_2O(l)} \leftrightharpoons H_3O^+(aq) + CH_3COO^-(aq)$$

If we look at the reverse reaction equation,

$$\underset{\text{acid}}{H_3O^+(aq)} + \underset{\text{base}}{CH_3COO^-(aq)} \leftrightharpoons CH_3COOH(aq) + H_2O(l)$$

we see that $H_3O^+(aq)$ donates a proton to $CH_3COO^-(aq)$. In other words, $H_3O^+(aq)$ acts as an acid by donating a proton and $CH_3COO^-(aq)$ acts as a base by accepting a proton. The base $CH_3COO^-(aq)$ is called the **conjugate base** of the acid $CH_3COOH(aq)$. Acetic acid and the acetate anion are called a **conjugate acid-base pair**. Similarly, the hydronium ion, $H_3O^+(aq)$, is the **conjugate acid** of the base $H_2O(l)$. The conjugate acid-base pairs for the dissociation of acetic acid in water are

$$CH_3COOH(aq) + H_2O(l) \rightleftharpoons H_3O^+(aq) + CH_3COO^-(aq)$$

conjugate acid-base pair ($H_2O(l)$ and $H_3O^+(aq)$)

conjugate acid-base pair ($CH_3COOH(aq)$ and $CH_3COO^-(aq)$)

Note that a conjugate base has one less proton than its corresponding conjugate acid.

EXAMPLE 20-12: Determine the conjugate base for each of the following acids: (a) $HClO(aq)$; (b) $NH_4^+(aq)$; (c) $HSO_4^-(aq)$.

Solution: A conjugate base is the ion that is produced when its conjugate acid donates a proton to water. The chemical equation that represents this process in each case is

(a) $$HClO(aq) + H_2O(l) \leftrightharpoons H_3O^+(aq) + ClO^-(aq)$$

Thus, the conjugate base of $HClO(aq)$ is $ClO^-(aq)$. $HClO(aq)$ and $ClO^-(aq)$ are a conjugate acid-base pair.

(b) $$NH_4^+(aq) + H_2O(l) \leftrightharpoons H_3O^+(aq) + NH_3(aq)$$

The conjugate base of $NH_4^+(aq)$ is $NH_3(aq)$; $NH_4^+(aq)$ and $NH_3(aq)$ are a conjugate acid-base pair.

(c) $$HSO_4^-(aq) + H_2O(l) \leftrightharpoons H_3O^+(aq) + SO_4^{2-}(aq)$$

The conjugate base of $HSO_4^-(aq)$ is $SO_4^{2-}(aq)$; $HSO_4^-(aq)$ and $SO_4^{2-}(aq)$ are a conjugate acid-base pair.

PRACTICE PROBLEM 20-12: What are the conjugate acids of (a) $CN^-(aq)$; (b) $HSO_3^-(aq)$; (c) $SO_3^{2-}(aq)$?

Answer: (a) $HCN(aq)$; (b) $H_2SO_3(aq)$; (c) $HSO_3^-(aq)$

An aqueous solution containing acetate ions may be basic because the acetate ion is the conjugate base of a weak acid, acetic acid. To see more explicitly that the $CH_3COO^-(aq)$ anion is a weak base, consider the equation for the reverse of the acid dissociation:

(1) $H_3O^+(aq) + CH_3COO^-(aq) \leftrightharpoons CH_3COOH(aq) + H_2O(l)$

Because this reaction equation is the reverse of the equation associated with the equilibrium constant K_a of the acetic acid, we see that the equilibrium constant of equation (1) is $K_1 = 1/K_a$.

The equation for the dissociation of water is

(2) $H_2O(l) + H_2O(l) \leftrightharpoons H_3O^+(aq) + OH^-(aq)$

(Equation 20.1), and its equilibrium constant is $K_2 = K_w$. If we add equations (1) and (2) above, we obtain

(3) $CH_3COO^-(aq) + H_2O(l) \leftrightharpoons CH_3COOH(aq) + OH^-(aq)$ (20.17)

The equilibrium constant for equation (3) is the product of the equilibrium constants for equations (1) and (2) above (Equation 19.30), and so we have

$$K_3 = K_1 K_2 = \frac{K_w}{K_a}$$

Equation 20.17 shows explicitly that the $CH_3COO^-(aq)$ anion accepts a proton from water to yield hydroxide ions. It is the same type of equation as that for the reaction between ammonia and water. Because the acetate ion acts as a base in solution, we denote the equilibrium constant for Equation 20.17 by K_b instead of K_3:

$$CH_3COO^-(aq) + H_2O(l) \rightleftharpoons CH_3COOH(aq) + OH^-(aq) \qquad K_b = \frac{K_w}{K_a} \qquad (20.18)$$

The value of K_a for $CH_3COOH(aq)$ at 25°C is 1.8×10^{-5} M, so at 25°C the value of K_b for the acetate anion is

$$K_b = \frac{1.0 \times 10^{-14}\ \text{M}^2}{1.8 \times 10^{-5}\ \text{M}} = 5.6 \times 10^{-10}\ \text{M}$$

The small value of K_b means that the $CH_3COO^-(aq)$ anion is a weak base in water.

Once again we emphasize that you must consider *all* the species present in a solution in order to determine if the solution is acidic, basic, or neutral. Just because $CH_3COO^-(aq)$ ions are basic, their presence in a solution does not guarantee that the solution containing these ions will be basic. For example, even though $CH_3COOH(aq)$ dissociates to form $CH_3COO^-(aq)$ anions according to

$$CH_3COOH(aq) + H_2O(l) \leftrightharpoons H_3O^+(aq) + CH_3COO^-(aq)$$

the resulting solution is acidic because of the presence of essentially equal quantities of $H_3O^+(aq)$ and $CH_3COO^-(aq)$ ions and the fact that $H_3O^+(aq)$ ions are strongly acidic, while $CH_3COO^-(aq)$ ions are weakly basic. In contrast, a solution of $NaCH_3COO(aq)$ will be basic. The $Na^+(aq)$ ions are neutral (Section 20-3), but the $CH_3COO^-(aq)$ ions react with water according to Equation 20.18 to form equal quantities of $CH_3COOH(aq)$ and $OH^-(aq)$ ions. The resulting solution is basic because $OH^-(aq)$ ions are strongly basic, while $CH_3COOH(aq)$ is weakly acidic. Therefore, we must be careful to examine each species present in a solution before concluding whether the solution is acidic, basic, or neutral.

Generally, the conjugate base of any weak acid is itself a weak base. In the case of a weak acid comprised of a proton and an anion, the anion is the conjugate base of the weak acid. If we denote a weak acid by $HA(aq)$ and its conjugate base, by $A^-(aq)$, then we have the equations

$$HA(aq) + H_2O(l) \rightleftharpoons H_3O^+(aq) + A^-(aq) \qquad K_a = \frac{[H_3O^+][A^-]}{[HA]} \qquad (20.19)$$

and

$$A^-(aq) + H_2O(l) \rightleftharpoons HA(aq) + OH^-(aq) \qquad K_b = \frac{[HA][OH^-]}{[A^-]} \qquad (20.20)$$

The first equation explicitly shows $HA(aq)$ is an acid, and the second equation explicitly shows that its conjugate base $A^-(aq)$ is a base. As can be seen by multiplying the equilibrium constants of Equations 20.19 and 20.20 together,

$$K_aK_b = \frac{[H_3O^+][A^-]}{[HA]} \times \frac{[HA][OH^-]}{[A^-]} = [H_3O^+][OH^-]$$

the relation between K_a and K_b for a conjugate acid-base pair is

$$K_aK_b = K_w \qquad (20.21)$$

Table 20.6 lists the values of K_a, pK_a, K_b, and pK_b for a number of conjugate acid-base pairs.

TABLE 20.6 Values of K_a, pK_a, K_b, and pK_b at 25°C for conjugate acid-base pairs (see also Appendix E)*

Name of acid	Acid†	K_a/M	pK_a	Base‡	K_b/M	pK_b
sulfurous acid	H_2SO_3	1.4×10^{-2}	1.85	HSO_3^-	7.1×10^{-13}	12.15
hydrogen sulfate ion	HSO_4^-	1.0×10^{-2}	1.99	SO_4^{2-}	9.8×10^{-13}	12.01
phosphoric acid	H_3PO_4	6.9×10^{-3}	2.16	$H_2PO_4^-$	1.5×10^{-12}	11.84
chloroacetic acid	$ClCH_2COOH$	1.4×10^{-3}	2.87	$ClCH_2COO^-$	7.4×10^{-12}	11.13
hydrofluoric acid	HF	6.3×10^{-4}	3.20	F^-	1.6×10^{-11}	10.80
nitrous acid	HNO_2	5.6×10^{-4}	3.25	NO_2^-	1.8×10^{-11}	10.75
cyanic acid	HCNO	3.5×10^{-4}	3.46	CNO^-	2.9×10^{-11}	10.54
formic acid	HCOOH	1.8×10^{-4}	3.75	$HCOO^-$	5.6×10^{-11}	10.25
benzoic acid	C_6H_5COOH	6.3×10^{-5}	4.20	$C_6H_5COO^-$	1.6×10^{-10}	9.79
acetic acid	CH_3COOH	1.8×10^{-5}	4.74	CH_3COO^-	5.6×10^{-10}	9.25
pyridinium ion	$C_5H_5NH^+$	5.9×10^{-6}	5.23	C_5H_5N	1.7×10^{-9}	8.77
carbonic acid	H_2CO_3	4.5×10^{-7}	6.35	HCO_3^-	2.2×10^{-8}	7.65
hydrosulfuric acid	H_2S	8.9×10^{-8}	7.05	HS^-	1.1×10^{-7}	6.95
hydrogen sulfite ion	HSO_3^-	6.7×10^{-8}	7.17	SO_3^{2-}	1.5×10^{-7}	6.82
dihydrogen phosphate ion	$H_2PO_4^-$	6.2×10^{-8}	7.21	HPO_4^{2-}	1.6×10^{-7}	6.79
hypochlorous acid	HClO	4.0×10^{-8}	7.40	ClO^-	2.5×10^{-7}	6.60
boric acid	$B(OH)_3$	5.4×10^{-10}	9.27	$B(OH)_4^-$	1.9×10^{-5}	4.73
ammonium ion	NH_4^+	5.6×10^{-10}	9.25	NH_3	1.8×10^{-5}	4.75
hydrocyanic acid	HCN	6.2×10^{-10}	9.21	CN^-	1.6×10^{-5}	4.79
hydrogen carbonate ion	HCO_3^-	4.7×10^{-11}	10.33	CO_3^{2-}	2.2×10^{-4}	3.67
hydrogen phosphate ion	HPO_4^{2-}	4.8×10^{-13}	12.32	PO_4^{3-}	2.1×10^{-2}	1.68
hydrogen sulfide ion	HS^-	1.2×10^{-13}	12.91	S^{2-}	8.1×10^{-2}	1.09

← increasing acid strength (upward) | increasing base strength (downward) →

*pK_a and pK_b values from the *CRC Handbook of Chemistry and Physics,* 87th Ed., 2006–2007

†For the acids: $HA(aq) + H_2O(l) \rightleftharpoons H_3O^+(aq) + A^-(aq) \quad K_a = \frac{[H_3O^+][A^-]}{[HA]}$

‡For the bases: $A^-(aq) + H_2O(l) \rightleftharpoons HA(aq) + OH^-(aq) \quad K_b = \frac{[HA][OH^-]}{[A^-]}$

By taking the logarithm of both sides of Equation 20.21 and then multiplying through by –1, we can write (at 25°C)

$$pK_a + pK_b = 14.00 \qquad (20.22)$$

which is analogous to Equation 20.6.

EXAMPLE 20-13: Given that $K_a = 6.3 \times 10^{-4}$ M at 25°C for the acid HF(*aq*), calculate K_b and pK_b for $F^-(aq)$.

Solution: The fluoride ion is the conjugate base of the weak acid HF(*aq*). The equilibrium equation associated with K_b is

$$F^-(aq) + H_2O(l) \leftrightharpoons HF(aq) + OH^-(aq)$$

where

$$K_b = \frac{[HF][OH^-]}{[F^-]}$$

Thus, from Equation 20.21, we have at 25°C

$$K_b = \frac{K_w}{K_a} = \frac{1.0 \times 10^{-14}\ M^2}{6.3 \times 10^{-4}\ M} = 1.6 \times 10^{-11}\ M$$

The value of pK_b for $F^-(aq)$ is obtained by using Equation 20.16:

$$pK_b = -\log(K_b/M) = -\log(1.6 \times 10^{-11}) = 10.80$$

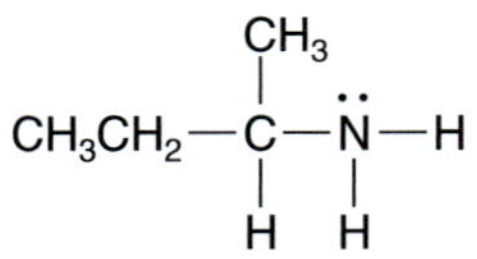

2-aminobutane

PRACTICE PROBLEM 20-13: 2-aminobutane, $CH_3CH_2CH(CH_3)NH_2(aq)$ (margin), is a weak base that is sometimes used as a fungicide. Given that $pK_b = 3.40$ at 25°C for 2-aminobutane, write the chemical formula for its conjugate acid and calculate the values of pK_a and K_a for the acid.

Answer: $CH_3CH_2CH(CH_3)NH_3^+(aq)$; $pK_a = 10.60$; $K_a = 2.5 \times 10^{-11}$ M

20-11. Aqueous Solutions of Many Salts Are Either Acidic or Basic

Suppose we dissolve sodium fluoride, NaF(*s*), in water. The resulting solution contains sodium ions, $Na^+(aq)$, and fluoride ions, $F^-(aq)$. Because $F^-(aq)$ is the conjugate base of the weak acid HF(*aq*), a fraction of the $F^-(aq)$ ions react with water according to the equation

$$F^-(aq) + H_2O(l) \leftrightharpoons HF(aq) + OH^-(aq) \qquad (20.23)$$

to form hydrofluoric acid and hydroxide ions, species that are weakly acidic and strongly basic, respectively. The sodium ions, on the other hand, do not

TABLE 20.7 Acid-base properties of some common cations and anions in water

Acidic cations	Neutral cations		Basic cations	
$NH_4^+(aq)$	$Li^+(aq)$	$Mg^{2+}(aq)$	none	
$Al^{3+}(aq)$	$Na^+(aq)$	$Ca^{2+}(aq)$		
$Pb^{2+}(aq)$	$K^+(aq)$	$Sr^{2+}(aq)$		
$Sn^{2+}(aq)$	$Rb^+(aq)$	$Ba^{2+}(aq)$		
transition metal ions	$Cs^+(aq)$			
	$Tl^+(aq)$			
Acidic anions	**Neutral anions**		**Basic anions**	
$HSO_4^-(aq)$	$Cl^-(aq)$	$ClO_4^-(aq)$	$F^-(aq)$	$CO_3^{2-}(aq)$
$H_2PO_4^-(aq)$	$Br^-(aq)$	$NO_3^-(aq)$	$CH_3COO^-(aq)$	$S^{2-}(aq)$
	$I^-(aq)$		$NO_2^-(aq)$	$SO_4^{2-}(aq)$
			$HCO_3^-(aq)$	$HPO_4^{2-}(aq)$
			$CN^-(aq)$	$PO_4^{3-}(aq)$
				(many others)

react with water to produce hydronium ions because $Na^+(aq)$ is the cation of the strong base $NaOH(aq)$, and the cations of strong bases form neutral ions in aqueous solution (Section 20-3). The net result is that the sodium fluoride solution is basic because it consists of a neutral cation and a basic anion.

Various ions react with water to produce hydronium ions or hydroxide ions. The acidic, neutral, or basic properties of a number of ions are given in Table 20.7, from which you should note the following:

1. The **neutral cations** are the cations of the strong bases listed in Table 20.1, and the **neutral anions** are the anions of the strong acids listed in Table 20.1, except for sulfuric acid. Being neutral, these ions do not undergo proton-transfer reactions with water; for example,

$$Na^+(aq) + H_2O(l) \rightarrow \text{no reaction}$$
$$ClO_4^-(aq) + H_2O(l) \rightarrow \text{no reaction}$$

2. The **basic anions** are the conjugate bases of weak acids. Because most acids are weak, most anions are basic. These ions react with water in aqueous solution to produce $OH^-(aq)$; for example,

$$CN^-(aq) + H_2O(l) \leftrightharpoons HCN(aq) + OH^-(aq)$$

3. There are no common basic cations.

4. The **acidic cations** consist of two types:

   a. The conjugate acids of weak bases are acidic because of reactions like that described by

$$NH_4^+(aq) + H_2O(l) \leftrightharpoons NH_3(aq) + H_3O^+(aq)$$

Ions such as $[Fe(H_2O)_6]^{3+}(aq)$ that involve a transition metal and one or more attached molecules are called complex ions and are denoted by enclosing the formula in square brackets. We discuss the chemistry of complex ions in Chapter 26.

Although only $NH_4^+(aq)$ is listed in Table 20.7, substituted ammonium ions, such as a methylammonium ion, $CH_3NH_3^+(aq)$, are also acidic.

b. Many metal ions are acidic in aqueous solution. These ions exist in aqueous solution with a certain number of water molecules bonded to them and are said to be solvated. An example of an acidic cation is $[Fe(H_2O)_6]^{3+}(aq)$. The equation for the acid dissociation reaction of this ion is

$$[Fe(H_2O)_6]^{3+}(aq) + H_2O(l) \leftrightharpoons [Fe(OH)(H_2O)_5]^{2+}(aq) + H_3O^+(aq) \quad (20.24)$$

The value of K_a for this reaction is 6.7×10^{-3} M at 25°C. The acidic character of the $[Fe(H_2O)_6]^{3+}(aq)$ ion is attributable to the loss of a proton from an attached water molecule. Table 20.8 lists values of K_a and pK_a for various metal cations in aqueous solution at 25°C.

TABLE 20.8 Values of K_a and pK_a for various metal cations in aqueous solution at 25°C

Cation*	K_a/M	pK_a
$[Al(H_2O)_6]^{3+}(aq)$	1.2×10^{-5}	4.91
$[Cd(H_2O)_6]^{2+}(aq)$	4.1×10^{-10}	9.39
$[Fe(H_2O)_6]^{3+}(aq)$	6.7×10^{-3}	2.17
$[Ga(H_2O)_6]^{3+}(aq)$	1.7×10^{-3}	2.77
$[In(H_2O)_6]^{3+}(aq)$	1.3×10^{-4}	3.87
$[Sc(H_2O)_6]^{3+}(aq)$	1.1×10^{-4}	3.94
$[Tl(H_2O)_6]^{3+}(aq)$	7.0×10^{-2}	1.15
$[V(H_2O)_6]^{3+}(aq)$	1.2×10^{-3}	2.92
$[Zn(H_2O)_6]^{2+}(aq)$	3.3×10^{-10}	9.48

*In each case the cation is acidic because of a reaction of the type described by

$$[M(H_2O)_6]^{n+}(aq) + H_2O(l) \leftrightharpoons [M(OH)(H_2O)_5]^{(n-1)+}(aq) + H_3O^+(aq)$$

as in Equation 20.24.

As we discussed in Section 20-3, the salts of neutral cations and neutral anions dissolved in water produce neutral solutions (pH = 7.0). For example,

$$KBr(s) \xrightarrow[H_2O(l)]{} K^+(aq) + Br^-(aq)$$

$K^+(aq)$ is the cation of the strong base $KOH(aq)$ and $Br^-(aq)$ the anion of the strong acid $HBr(aq)$, so the solution is neutral. Here again we stress the need to memorize the list of strong bases and acids in order to recognize their respective cations and anions.

Salts of neutral cations and basic anions dissolved in water produce basic solutions. For example, when sodium nitrite, $NaNO_2(s)$, is dissolved in water, according to

$$NaNO_2(s) \xrightarrow[H_2O(l)]{} Na^+(aq) + NO_2^-(aq)$$

The $Na^+(aq)$ ion does not react with $H_2O(l)$, but the protonation reaction of the weak base $NO_2^-(aq)$, described by

$$NO_2^-(aq) + H_2O(l) \leftrightharpoons HNO_2(aq) + OH^-(aq)$$

produces $OH^-(aq)$, and thus yields a basic solution.

Salts of acidic cations and neutral anions produce acidic solutions when dissolved in water. For example, when aluminum nitrate is dissolved in water,

$$Al(NO_3)_3(s) + 6\,H_2O(l) \xrightarrow[H_2O(l)]{} [Al(H_2O)_6]^{3+}(aq) + 3\,NO_3^-(aq)$$

Because $NO_3^-(aq)$ is the anion of the strong acid $HNO_3(aq)$, it does not react with $H_2O(l)$; but the acid dissociation reaction described by

$$[Al(H_2O)_6]^{3+}(aq) + H_2O(l) \leftrightharpoons [Al(OH)(H_2O)_5]^{2+}(aq) + H_3O^+(aq)$$

shows explicitly the production of $H_3O^+(aq)$.

If we have a salt with an acidic cation and a basic anion, for example, $NH_4CN(aq)$, then we need to know the K_a value of the cation and the K_b value of the anion to predict whether the salt solution is acidic, basic, or neutral. If $K_a > K_b$, then the solution will be acidic; and if $K_b > K_a$, the solution will be basic.

EXAMPLE 20-14: Predict whether the following salts produce acidic, basic, or neutral solutions when dissolved in water: $CaI_2(s)$, $Na_2SO_4(s)$, and $NH_4Br(s)$.

Solution:

Salt	Cation	Anion	Resulting solution
$CaI_2(s)$	neutral	neutral	neutral solution (pH = 7.0)
$Na_2SO_4(s)$	neutral	basic	basic solution
$NH_4Br(s)$	acidic	neutral	acidic solution

PRACTICE PROBLEM 20-14: Predict whether the following salts produce acidic, basic, or neutral solutions when dissolved in water: (a) $HCN(s)$, (b) $KClO_4(s)$, and (c) $FeCl_3(s)$.

Answer: (a) acidic; (b) neutral; (c) acidic

EXAMPLE 20-15: Using the data in Table 20.6 where necessary, predict whether the following substances produce acidic, basic, or neutral solutions when dissolved in water: (a) pyridine, $C_5H_5N(l)$; (b) ammonium fluoride, $NH_4F(s)$; and (c) methanol, $CH_3OH(l)$.

Solution: (a) $C_5H_5N(l)$ forms a basic solution because the lone-pair electrons on the nitrogen atom can accept a proton from water to form pyridinium, $C_5H_5NH^+(aq)$, ions (page 752).

(b) $NH_4F(s)$ forms a basic solution because the value of K_a for $NH_4^+(aq)$ (5.6×10^{-10} M) is greater than that of K_b for F^- (1.6×10^{-11} M).

(c) $CH_3OH(l)$ forms a neutral solution because the –OH group is covalently bonded to the carbon atom and so does not dissociate in water to form $OH^-(aq)$ ions.

PRACTICE PROBLEM 20-15: Using the data in Table 20.6 and Table 20.7, predict whether the following solutions are acidic, basic, or neutral: (a) $NH_4CN(aq)$, (b) $CH_3NH_2(aq)$, (c) $Pb(NO_3)_2(aq)$.

Answer: (a) basic; (b) basic; (c) acidic

In addition to predicting whether a salt solution is acidic, basic, or neutral, we also can calculate its pH by using the methods already developed. Let's calculate the pH of a 0.050 M $NH_4Cl(aq)$ solution at 25°C. $NH_4Cl(s)$ is a strong electrolyte (Section 12-3), and so when $NH_4Cl(s)$ is dissolved in water, solvated ammonium ions and chloride ions are produced according to

$$NH_4Cl(s) \xrightarrow[H_2O(l)]{} NH_4^+(aq) + Cl^-(aq)$$

Chloride ions are neutral and do not react with $H_2O(l)$, but ammonium ions are acidic because of the acid dissociation equilibrium described by

$$NH_4^+(aq) + H_2O(l) \leftrightharpoons H_3O^+(aq) + NH_3(aq)$$

We find from Table 20.6 that $K_a = 5.6 \times 10^{-10}$ M for this reaction. The only other equilibrium to consider is

$$2H_2O(l) \leftrightharpoons H_3O^+(aq) + OH^-(aq) \quad K_w = 1.0 \times 10^{-14}\ M^2$$

Because $K_a >> K_w$, we shall ignore the $H_3O^+(aq)$ from the autoprotonation reaction of water (Equation 20.1). The concentration table is shown below.

Concentration	$NH_4^+(aq)$	+ $H_2O(l)$	$\leftrightharpoons$ $H_3O^+(aq)$	+ $NH_3(aq)$
initial	0.050 M	—	≈ 0 M	0 M
change	$-x$	—	$+x$	$+x$
equilibrium	$0.050\ M - x$	—	x	x

The expression for K_a becomes

$$K_a = \frac{[H_3O^+][NH_3]}{[NH_4^+]} = \frac{x^2}{0.050\ M - x} = 5.6 \times 10^{-10}\ M$$

We first neglect x relative to 0.050 M in the denominator and obtain

$$x \approx \sqrt{(0.050\ M)(5.6 \times 10^{-10}\ M)} = 5.3 \times 10^{-6}\ M$$

The method of successive approximations confirms this result for x. Because $[H_3O^+] = x$, the pH of the solution is

$$pH = -\log([H_3O^+]/M) = -\log(5.3 \times 10^{-6}) = 5.28$$

Notice that the solution is acidic (pH < 7.0); NH_4Cl consists of an acidic cation and a neutral anion.

EXAMPLE 20-16: $NaCH_3COO(s)$ is a strong electrolyte. Calculate the pH of a 0.050 M $NaCH_3COO(aq)$ solution at 25°C.

Solution: Because it is a strong electrolyte, when $NaCH_3COO(s)$ is dissolved in water, solvated sodium ions and acetate ions are produced according to

$$NaCH_3COO(s) \xrightarrow[H_2O(l)]{} Na^+(aq) + CH_3COO^-(aq)$$

The sodium ions are neutral, so they do not react with $H_2O(l)$; but the acetate ions, being the conjugate base of a weak acid, are basic and undergo the proton-transfer reaction with water described by the equation

$$CH_3COO^-(aq) + H_2O(l) \leftrightharpoons CH_3COOH(aq) + OH^-(aq)$$

From Table 20.6 we see that $K_b = 5.6 \times 10^{-10}$ M for this equation at 25°C. Because $K_a >> K_w$, we ignore the $OH^-(aq)$ from the autoprotonation reaction of water and write the following concentration table:

Concentration	$CH_3COO^-(aq)$	+	$H_2O(l)$	$\leftrightharpoons$	$CH_3COOH(aq)$	+	$OH^-(aq)$
initial	0.050 M		—		≈ 0 M		0 M
change	0.050 M − x		—		+ x		+ x
equilibrium	0.050 M − x		—		x		x

The corresponding expression for K_b is

$$K_b = \frac{[CH_3COOH][OH^-]}{[CH_3COO^-]} = \frac{x^2}{0.050\ M - x} = 5.6 \times 10^{-10}\ M$$

Neglecting x relative to 0.050 M, we obtain

$$x \approx \sqrt{(0.050\ M)(5.6 \times 10^{-10}\ M)} = 5.3 \times 10^{-6}\ M$$

which is negligible compared with 0.050 M, as confirmed by successive approximations. Therefore, $[OH^-] = x = 5.3 \times 10^{-6}$ M. The value of pOH is given by

$$pOH = -\log([OH^-]/M) = \log(5.3 \times 10^{-6}) = 5.28$$

The pH of the solution is

$$pH = 14.00 - pOH = 8.72$$

which shows that the solution is basic (pH > 7.0).

PRACTICE PROBLEM 20-16: Using the data in Table 20.8, calculate the pH at 25°C of a 0.25-M aqueous solution of thallium(III) chloride, a strong electrolyte.

Answer: pH = 1.00

The salt $NH_4Cl(aq)$ has an acidic cation and a neutral anion, whereas $NaCH_3COO(aq)$ consists of a neutral cation and a basic anion. For a salt in which neither ion is neutral, such as $NH_4CH_3COO(aq)$, the calculation of the pH is more involved and we shall not consider it here.

20-12. A Polyprotic Acid Can Donate More Than One Proton in Solution

We have already seen that some substances can donate more than one proton in solution; for example, sulfuric acid, $H_2SO_4(aq)$, can donate two protons in solution. We call an acid that can donate two acidic protons a **diprotic acid**. Similarly, phosphoric acid, $H_3PO_4(aq)$, can donate up to three acidic protons in solution and so is called a **triprotic acid**. Acids that donate more than one acidic proton are collectively known as **polyprotic acids**.

Let's look at the dissociation of oxalic acid, a diprotic acid. Here we shall write its chemical formula as HOOCCOOH to emphasize the two carboxylic acid groups (see Practice Problem 20-3). Oxalic acid can undergo two sequential acid dissociation reactions as described by

$$(1)\quad HOOCCOOH(aq) + H_2O(l) \leftrightharpoons H_3O^+(aq) + HOOCCOO^-(aq) \qquad K_{a_1} = 5.6 \times 10^{-2}\ M$$

$$(2)\quad HOOCCOO^-(aq) + H_2O(l) \leftrightharpoons H_3O^+(aq) + {}^-OOCCOO^-(aq) \qquad K_{a_2} = 1.5 \times 10^{-4}\ M$$

We often label the acid dissociation constants of a polyprotic species as K_{a_1}, K_{a_2}, and so forth, to indicate the donation of sequential protons from the same species.

Table 20.9 lists values of the acid-dissociation constants of some polyprotic acids. A quick inspection of Table 20.9 shows that the values of successive pK_a's for any given species increase by about 3 to 6 pK_a units for each proton donated. That is, each sequential value of K_a is about 10^{-3} to 10^{-6} times the value of the

TABLE 20.9 Successive acid dissociation constants of various polyprotic acids in water at 25°C*

Acid	Formula	pK_{a_1}	pK_{a_2}	pK_{a_3}
Diprotic				
carbonic	H_2CO_3	6.35	10.33	
chromic	H_2CrO_4	0.74	6.49	
hydroselenic	H_2Se	3.89	11.0	
hydrosulfuric	H_2S	7.05	12.91	
malonic	$HOOCCH_2COOH$	2.85	5.70	
oxalic	HOOCCOOH	1.25	3.81	
phosphorous acid	H_3PO_3	1.3	6.70	
sulfuric	H_2SO_4	strong	1.99	
sulfurous	H_2SO_3	1.85	7.2	
Triprotic				
arsenic	H_3AsO_4	2.26	6.76	11.29
citric	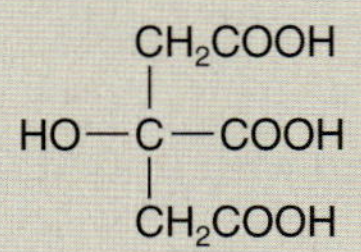	3.13	4.76	6.40
phosphoric	H_3PO_4	2.16	7.21	12.32

*pK_a values from the *CRC Handbook of Chemistry and Physics*, 87th Ed., 2006–2007

preceding one. In other words, each succeeding acidic proton is more difficult to remove, primarily because of the extra energy required to separate a positively charged proton from a negatively charged ion. When sequential values of K_a differ by a large magnitude, we can treat each sequential acidic dissociation as a separate equilibrium problem and so solve them independently, as illustrated by the following Example.

EXAMPLE 20-17: Calculate the pH of a 0.10 M $H_2SO_4(aq)$ solution at 25°C.

Solution: The fact that the first dissociation of $H_2SO_4(aq)$ is strong (Table 20.9) means that 0.10 M $H_2SO_4(aq)$ dissociates to form stoichiometric concentrations of ions, and so is 0.10 M in $H_3O^+(aq)$ and 0.10 M in $HSO_4^-(aq)$:

$$\underset{0.10\ \mathrm{M}}{H_2SO_4(aq)} + H_2O(l) \rightarrow \underset{0.10\ \mathrm{M}}{H_3O^+(aq)} + \underset{0.10\ \mathrm{M}}{HSO_4^-(aq)}$$

Because of the second dissociation, however, we also have the equilibrium described by

$$HSO_4^-(aq) + H_2O(l) \leftrightharpoons H_3O^+(aq) + SO_4^{2-}(aq)$$

with

$$K_{a_2} = \frac{[H_3O^+][SO_4^{2-}]}{[HSO_4^-]} = 0.010\ \text{M}$$

where we have used Equation 20.15 to obtain the value of K_{a_2} from the pK_{a_2} value of 1.99 listed in Table 20.9.

Because of this second dissociation reaction, both $[H_3O^+]_0$ and $[HSO_4^-]_0$ will be slightly different from 0.10 M. If we let x be the additional $[H_3O^+]$ due to the dissociation of $HSO_4^-(aq)$, then we get the following concentration table:

Concentration	$HSO_4^-(aq)$	+	$H_2O(l)$	$\leftrightharpoons$	$H_3O^+(aq)$	+	$SO_4^{2-}(aq)$
initial	0.10 M		—		0.10 M		0 M
change	$-x$		—		$+x$		$+x$
equilibrium	0.10 M $- x$		—		0.10 M $+ x$		x

Substituting these values into the equilibrium-constant expression gives

$$K_{a_2} = \frac{(0.10\ \text{M} + x)x}{0.10\ \text{M} - x} = 0.010\ \text{M} \tag{20.25}$$

If we assume that x is small compared with 0.10 M, then as a first approximation

$$x_1 \approx \frac{(0.10\ \text{M})(0.010\ \text{M})}{0.10\ \text{M}} = 0.010\ \text{M}$$

To obtain a better approximation, we substitute $x = 0.010$ M into the 0.010 M $+ x$ and 0.010 M $- x$ terms in Equation 20.25 to obtain a second approximation.

$$\frac{(0.10\ \text{M} + 0.010\ \text{M})x_2}{0.10\ \text{M} - 0.010\ \text{M}} = 0.010\ \text{M}$$

or

$$x_2 = \frac{(0.10\ \text{M} - 0.010\ \text{M})(0.010\ \text{M})}{0.10\ \text{M} + 0.010\ \text{M}} = 0.0082\ \text{M}$$

If we use the method of successive approximations twice more, we obtain successively 0.0085 M and 0.0085 M. (We could also have obtained this result directly by solving Equation 20.25 using the quadratic formula.) Thus, we have the values shown below.

Concentration	$HSO_4^-(aq)$	+	$H_2O(l)$	$\leftrightharpoons$	$H_3O^+(aq)$	+	$SO_4^{2-}(aq)$
initial	0.10 M		—		0.10 M		0 M
change	$-$ 0.0085 M		—		+ 0.0085 M		+ 0.0085 M
equilibrium	0.09 M		—		0.11 M		0.0085 M

Because $[H_3O^+] = 0.11$ M, the pH is 0.96.

PRACTICE PROBLEM 20-17: Use the data in Table 20.9 to calculate the pH and the concentrations of all the species in a 0.250-M aqueous solution of oxalic acid at 25°C. (*Hint:* Use the fact that $K_{a_1} >> K_{a_2}$ to separate the problem into two sequential steps, creating a new concentration table for each step.)

Answer: pH = 1.03; $[H_3O^+] = [HC_2O_4^-] = 0.094$ M; $[H_2C_2O_4] = 0.156$ M; $[C_2O_4^{2-}] = 1.5 \times 10^{-4}$ M

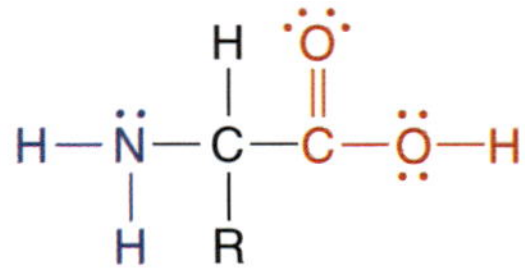

Figure 20.6 Amino acids have both an acidic carboxylic acid group (red) and a basic amine group (blue), making them amphoteric in solution. The letter R here represents one of twenty specific organic side groups. The simplest amino acid, glycine, for example, has R = H. The structures of the other 19 amino acids are given in Interchapter T.

See Interchapter T at www.McQuarrieGeneralChemistry.com.

Notice that in both Example 20-17 and Practice Problem 20-17, the contribution of the first acidic proton to the pH of the solution is much more significant than that of the second acidic proton because $K_{a_1} >> K_{a_2}$.

The salt of a weak polyprotic acid can act as both an acid and a base in solution. Such species are said to be **amphoteric**. Amphoteric species are of particular importance in biology where many molecules have both an acidic and a basic functional group, such as the amino acids (Figure 20.6). An interesting example of an amphoteric species is sodium hydrogen carbonate (or sodium bicarbonate), $NaHCO_3(aq)$, which is the sodium salt of the weak diprotic acid $H_2CO_3(aq)$. Many laboratories use saturated $NaHCO_3(aq)$ solutions for treating *both* base and acid burns (Figure 20.7). Let's see how the same solution can be used to neutralize both a base and an acid. When $NaHCO_3(s)$ is dissolved in water, it dissociates to form $Na^+(aq)$ and $HCO_3^-(aq)$ ions. The sodium ion is neutral. However, if we write out the proton-transfer reactions associated with the $HCO_3^-(aq)$ ion, we find that it can act as an acid according to the equation

$$HCO_3^-(aq) + H_2O(l) \leftrightharpoons H_3O^+(aq) + CO_3^{2-}(aq)$$

or as a base according to the equation

$$HCO_3^-(aq) + H_2O(l) \leftrightharpoons H_2CO_3(aq) + OH^-(aq)$$

Thus, $HCO_3^-(aq)$ is an amphoteric species. Because $HCO_3^-(aq)$ can act as both an acid and a base, a safety solution containing $HCO_3^-(aq)$ ions can be used to treat both base and acid burns. The chemical equations for the neutralization reactions are

$$HCO_3^-(aq) + H_3O^+(aq) \leftrightharpoons H_2CO_3(aq) + H_2O(l)$$

and

$$HCO_3^-(aq) + OH^-(aq) \leftrightharpoons CO_3^{2-}(aq) + H_2O(l)$$

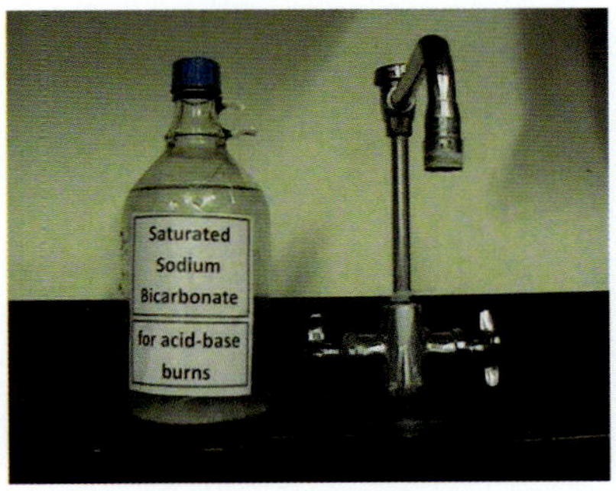

Figure 20.7 Bottles of saturated sodium hydrogen carbonate (or sodium bicarbonate), $NaHCO_3(aq)$, are often used as laboratory safety solutions for the treatment of acid and base burns. Sodium hydrogen carbonate is commonly known as baking soda.

SUMMARY

There are three major systems for classifying acids and bases in order of increasing generality: (1) An Arrhenius acid is a substance that produces $H^+(aq)$ ions in aqueous solution and an Arrhenius base is a substance that produces $OH^-(aq)$ ions in aqueous solution. (2) A Brønsted-Lowry acid is a species that donates protons and a Brønsted-Lowry base is a species that accepts protons. In this system, acid-base reactions are proton-transfer reactions. (3) A Lewis acid is defined as an electron-pair acceptor and a Lewis base is defined as an electron-pair donor.

Water plays a central role in acid-base reactions because it can act either as an acid or as a base. Water undergoes an autoprotonation reaction to produce $H_3O^+(aq)$ and $OH^-(aq)$ ions. The product of the equilibrium concentrations of $H_3O^+(aq)$ and $OH^-(aq)$ ions is fixed by the value of the ion-product constant of water, $K_w = [H_3O^+][OH^-]$, which equals $1.0 \times 10^{-14}\ M^2$ at 25°C.

For aqueous solutions, if $[H_3O^+] > [OH^-]$, the solution is acidic; if $[H_3O^+] < [OH^-]$, the solution is basic; and if $[H_3O^+] = [OH^-]$, the solution is neutral. The value of $[H_3O^+]$ is more conveniently presented on a pH scale. The pH of a solution is defined through the equation $pH = -\log([H_3O^+]/M)$. At 25°C, if $pH < 7$, the solution is acidic; if $pH > 7$, the solution is basic; and if $pH = 7$, the solution is neutral. The pOH of a solution is related to the pH of the solution by the expression $pOH = 14.00 - pH$ at 25°C.

Strong acids and strong bases are completely dissociated in solution; weak acids are only partially dissociated and weak bases are only partially protonated. The strength of a weak acid is governed by the value of K_a, its acid dissociation constant; the strength of a weak base is governed by the value of K_b, its base protonation constant. The larger the value of K_a or K_b, the stronger the acid or the base, respectively. The treatment of weak acids and weak bases can be unified by the introduction of the idea of a conjugate acid-base pair. The values of K_a and K_b for a conjugate acid-base pair obey the relation $K_w = K_aK_b$. The stronger an acid, the weaker its conjugate base; the stronger a base, the weaker its conjugate acid. The pK_a value of an acid is given by $pK_a = -\log(K_a/M)$; the smaller the value of pK_a, the stronger the acid. The pK_b value of a base is given by $pK_b = -\log(K_b/M)$; the smaller the value of pK_b, the stronger the base. For any conjugate acid-base pair, $pK_a + pK_b = 14.00$ at 25°C.

Many cations and anions have acidic or basic properties. Salts that consist of these cations or anions may form acidic or basic aqueous solutions. An acid that can donate more than one acidic proton in solution is called a polyprotic acid. An amphoteric species is one that can act as either an acid or a base in solution.

TERMS YOU SHOULD KNOW

Arrhenius acids and bases *730*
hydronium ion, $H_3O^+(aq)$ *730*
Brønsted-Lowry acids and bases *730*
proton donor *730*
proton acceptor *730*
proton-transfer reaction *730*
protonation reaction *730*
acid-base reaction *730*
Lewis acids and bases *730*
electron-pair acceptor *730*
electron-pair donor *730*
autoprotonation reaction *731*
ion-product constant of water, K_w *731*
neutral solution, $[H_3O^+] = [OH^-]$ *732*
acidic solution, $[H_3O^+] > [OH^-]$ *732*
basic solution, $[H_3O^+] < [OH^-]$ *732*
strong acid *732*
strong base *733*
weak acid *733*
weak base *733*
neutral ion *735*
carboxylic acid *735*
alkyl group *735*
carboxyl group *735*
carboxylate ion *737*
pH *738*
pOH *739*
pH meter *741*

percentage of dissociation, % dissociation *742*
stoichiometric concentration *742*
acid dissociation reaction *743*
acid dissociation constant, K_a *744*
material balance condition *744*
pK_a *747*
method of successive approximations *748*
amine *751*
base protonation constant, K_b *752*
pK_b *752*
conjugate base *754*
conjugate acid-base pair *754*
conjugate acid *754*
neutral cation *759*
neutral anion *759*
basic anion *759*
acidic anion *759*
acidic cation *759*
diprotic acid *764*
triprotic acid *764*
polyprotic acid *764*
amphoteric *767*

EQUATIONS YOU SHOULD KNOW HOW TO USE

(The values of the dissociation constants for most of the acids and bases listed in this chapter are compiled in Appendix E).

$K_w = [H_3O^+][OH^-] = 1.0 \times 10^{-14}\ M^2$	(20.3)	(ion-product constant for water at 25°C)
$pH = -\log([H_3O^+]/M)$	(20.4)	(definition of pH)
$pOH = -\log([OH^-]/M)$	(20.5)	(definition of pOH)
$pH + pOH = 14.00$	(20.6)	(relation between pH and pOH at 25°C)
$[H_3O^+] = 10^{-pH}\ M$	(20.7)	(calculation of $[H_3O^+]$ from pH)
$pK_a = -\log(K_a/M)$	(20.15)	(definition of pK_a)
$pK_b = -\log(K_b/M)$	(20.16)	(definition of pK_b)
$K_a = \dfrac{[H_3O^+][A^-]}{[HA]}$	(20.19)	(acid dissociation constant expression for an acid with formula HA)
$K_b = \dfrac{[HA][OH^-]}{[A^-]}$	(20.20)	(base protonation constant expression for a base with formula A^-)
$K_a K_b = K_w$	(20.21)	(relation between K_a and K_b for a conjugate acid-base pair)
$pK_a + pK_b = 14.00$	(20.22)	(relation between pK_a and pK_b for a conjugate acid-base pair at 25°C)

PROBLEMS

STRONG ACIDS AND BASES

20-1. Calculate $[H_3O^+]$, $[ClO_4^-]$, and $[OH^-]$ in an aqueous solution that is 0.150 M in $HClO_4(aq)$ at 25°C. Is the solution acidic or basic?

20-2. Calculate $[OH^-]$, $[K^+]$, and $[H_3O^+]$ in an aqueous solution that is 0.25 M in $KOH(aq)$ at 25°C. Is the solution acidic or basic?

20-3. Calculate $[Tl^+]$, $[OH^-]$, and $[H_3O^+]$ for a solution that is prepared by dissolving 2.00 grams of $TlOH(s)$ in enough water to make 500.0 milliliters of solution at 25°C.

20-4. Calculate $[Ca^{2+}]$, $[OH^-]$, and $[H_3O^+]$ for a solution that is prepared by dissolving 0.600 grams of $Ca(OH)_2(s)$ in enough water to make 1.00 liters of solution at 25°C.

pH CALCULATIONS

20-5. Calculate the pH of an aqueous solution that is 0.020 M in $HNO_3(aq)$ at 25°C. Is the solution acidic or basic?

20-6. Calculate the pH of an aqueous solution that is 0.20 M in $CsOH(aq)$ at 25°C. Is the solution acidic or basic?

20-7. Calculate the pH and the pOH of an aqueous solution that is 0.035 M in $HCl(aq)$ and 0.045 M in $HBr(aq)$ at 25°C.

20-8. Calculate the pH and the pOH of an aqueous solution that is 0.020 M in $Ba(OH)_2(aq)$ at 25°C.

20-9. Calculate the pH and the pOH of an aqueous solution prepared by dissolving 2.0 grams of $KOH(s)$ pellets in water and diluting to a final volume of 0.500 liters at 25°C.

20-10. A solution of $NaOH(aq)$ contains 6.3 grams of $NaOH(s)$ per 100.0 milliliters of solution. Calculate the pH and the pOH of the solution at 25°C.

CALCULATION OF $[H_3O^+]$ AND $[OH^-]$ FROM pH

20-11. The pH of human muscle fluids is 6.8. Calculate the value of $[H_3O^+]$ in muscle fluid at 25°C.

20-12. The pH of a soap solution at 25°C is about 11.0. Calculate the value of $[OH^-]$ for the soap solution.

20-13. The pH of the contents of the human stomach can be as low as 1.0. Calculate the value of $[H_3O^+]$ in the stomach when the pH = 1.0.

20-14. Normal rainwater has a pH of about 5.6, whereas what is called acid rain has been observed to have pH values as low as 3.0. Calculate the ratio of $[H_3O^+]$ in pH = 3.0 acid rain to that in normal rain.

20-15. The pH of human blood is fairly constant at 7.4. Calculate the hydronium ion concentration and the hydroxide ion concentration in human blood at 25°C.

20-16. The pH of the world's oceans is remarkably constant at 8.15. Calculate the hydronium ion and hydroxide ion concentrations in the ocean. Assume a temperature of 25°C.

CALCULATION OF K_a FROM pH

20-17. The pH of a 0.050-M aqueous solution of pyruvic acid, $CH_3COCOOH(aq)$, an intermediate in the metabolism of glucose, is found to be 1.91. Calculate K_a, the acid-dissociation constant, for pyruvic acid.

20-18. The pH of a 0.20-M aqueous solution of hexanoic acid, $CH_3(CH_2)_4COOH(aq)$, a fatty acid derived from various animal oils, is found to be 2.78. Calculate K_a, the acid-dissociation constant, for hexanoic acid.

20-19. The pH of a 1.00×10^{-2} M solution of an unknown acid is 4.79 at 25°C. Calculate the value of K_a for the acid at 25°C.

20-20. The pH of a 0.10-M aqueous solution of formic acid, HCOOH, is 2.38 at 25°C. Calculate the value of K_a for formic acid.

CALCULATION OF pH FROM K_a

20-21. The value of K_a in water at 25°C for benzoic acid, $C_6H_5COOH(aq)$, is 6.3×10^{-5} M. Calculate the pH and the concentration of the other species in a 0.0200-M aqueous solution of $C_6H_5COOH(aq)$.

20-22. The value of K_a in water at 25°C for hypochlorous acid, $HClO(aq)$, is 4.0×10^{-8} M. Calculate the pH and the concentrations of the other species in a 0.15-M aqueous solution of $HClO(aq)$.

20-23. The value of K_a in water at 25°C for trichloroacetic acid, $Cl_3CCOOH(aq)$, is 2.2×10^{-1} M. Calculate the pH and the concentrations of the other species in a 0.030-M aqueous solution of $Cl_3CCOOH(aq)$.

20-24. Calculate the pH and the concentrations of the other species in a 0.10-M aqueous solution of chloroacetic acid, $ClCH_2COOH(aq)$, given that $K_a = 1.4 \times 10^{-3}$ M at 25°C.

CALCULATIONS INVOLVING K_b

20-25. The measured pH of a 0.100-M solution of $NH_3(aq)$ at 25°C is 11.12. Calculate K_b for $NH_3(aq)$ at 25°C.

20-26. The pH of a 0.50-M solution of the weak base ethylamine, $CH_3CH_2NH_2(aq)$, is 12.17 at 25°C. Determine the value of K_b for ethylamine.

20-27. The organic solvent pyridine, $C_5H_5N(aq)$, is a base with a strong, irritating odor. Calculate the pH of a 0.300-M aqueous solution of pyridine at 25°C.

20-28. The base hydroxylamine, $HONH_2(aq)$, is used to synthesize a variety of organic compounds. Calculate the pH of a 0.125-M aqueous solution of $HONH_2(aq)$ at 25°C.

20-29. Calculate the pH of a 0.060-M aqueous solution of dimethylamine, $(CH_3)_2NH(aq)$ at 25°C.

20-30. Calculate the pH of a household ammonia cleaning solution prepared by dissolving $NH_3(g)$ in water to yield a 0.20 M $NH_3(aq)$ solution at 25°C.

LE CHÂTELIER'S PRINCIPLE

20-31. Use Le Châtelier's principle to predict the direction in which the acid-dissociation equilibrium described by the following equation shifts in response to the indicated change in conditions:

$$CO_2(aq) + 2H_2O(l) \leftrightharpoons H_3O^+(aq) + HCO_3^-(aq)$$

if

(a) $[CO_2]$ is decreased

(b) $[HCO_3^-]$ is decreased

(c) $[H_3O^+]$ is decreased

(d) the solution is diluted with water

20-32. Use Le Châtelier's principle to predict the direction in which the acid-dissociation equilibrium described by the following equation shifts in response to the indicated change in conditions:

$$HCOOH(aq) + H_2O(l) \leftrightharpoons H_3O^+(aq) + HCOO^-(aq)$$

(a) addition of $NaOH(s)$

(b) addition of $NaHCOO(s)$

(c) dilution of a 0.1-M solution to 0.01 M

(d) addition of $HCl(g)$

20-33. Use Le Châtelier's principle to predict the direction in which the equilibrium described by the following equation shifts in response to the indicated change in conditions:

$$C_6H_5COOH(aq) + H_2O(l) \leftrightharpoons H_3O^+(aq) + C_6H_5COO^-(aq) \quad \Delta H^\circ_{rxn} \approx 0$$

(a) evaporation of water from the solution at a fixed temperature

(b) decrease in the temperature of the solution

(c) addition of $KC_6H_5COO(s)$

(d) addition of $NH_3(g)$

20-34. Use Le Châtelier's principle to predict the direction in which the equilibrium described by the following equation shifts in response to the indicated change in conditions:

$$HNO_2(aq) + H_2O(l) \leftrightharpoons H_3O^+(aq) + NO_2^-(aq) \quad \Delta H^\circ_{rxn} < 0$$

(a) increase in the temperature of the solution

(b) dissolution of $NaNO_2(s)$

(c) dissolution of $NaOH(s)$

(d) removal of $NO_2^-(aq)$ as $AgNO_2(s)$ by addition of $AgNO_3(s)$

CONJUGATE ACIDS AND BASES

20-35. Give the conjugate base for each of the following acids:

(a) $HClO(aq)$ (b) $NH_4^+(aq)$

(c) $HN_3(aq)$ (d) $HS^-(aq)$

20-36. Give the conjugate base for each of the following acids:

(a) $HNO_3(aq)$ (b) $HCOOH(aq)$

(c) $C_6H_5COOH(aq)$ (d) $CH_3NH_3^+(aq)$

20-37. Identify which of the following species are Brønsted-Lowry acids and which are Brønsted-Lowry bases in water. In each case give the chemical formula for the conjugate member of the conjugate acid-base pair:

(a) $HCNO(aq)$ (b) $OBr^-(aq)$

(c) $HClO_3(aq)$ (d) $CH_3NH_3^+(aq)$

(e) $ClNH_2(aq)$ (f) $HONH_2(aq)$

20-38. Identify which of the following species are Brønsted-Lowry acids and which are Brønsted-Lowry bases in water. In each case give the chemical formula for the conjugate member of the conjugate acid-base pair.

(a) $ClCH_2COOH(aq)$ (b) $NH_3(aq)$
(c) $ClO^-(aq)$ (d) $HCOO^-(aq)$
(e) $HN_3(aq)$ (f) $NO_2^-(aq)$

20-39. Given the following acids and their dissociation constants at 25°C, calculate K_b for the conjugate bases:

Acid	K_a/M
(a) CH_3CH_2COOH, propanoic acid	1.4×10^{-5}
(b) NH_2SO_3H, sulfamic acid	8.9×10^{-2}
(c) NH_4^+, ammonium ion	5.6×10^{-10}
(d) $HTeO_4^-$, hydrogen tellurate ion	1.0×10^{-11}

20-40. Given the following bases and their values of K_b at 25°C, calculate K_a for the conjugate acids:

Base	K_b/M
(a) C_5H_5N, pyridine	1.5×10^{-9}
(b) CN^-, cyanide ion	2.1×10^{-5}
(c) NH_2OH, hydroxylamine	8.7×10^{-9}
(d) $(CH_3)_2NH$, dimethylamine	5.4×10^{-4}

ACID-BASE PROPERTIES OF SALTS

20-41. Predict whether the following salts, when dissolved in water, produce acidic, basic, or neutral solutions:

(a) $Al(NO_3)_3(s)$ (b) $NH_4Br(s)$
(c) $NaHCO_3(s)$ (d) $LiCNO(s)$

20-42. Predict whether the following salts, when dissolved in water, produce acidic, basic, or neutral solutions:

(a) $FeBr_3(s)$ (b) $NaNO_3(s)$
(c) $KHSO_4(s)$ (d) $NaF(s)$

20-43. Predict whether the following salts, when dissolved in water, produce acidic, basic, or neutral solutions:

(a) $Na_2CO_3(s)$ (b) $KClO_4(s)$
(c) $RbClO(s)$ (d) $Al(ClO_4)_3(s)$

20-44. Predict whether the following salts, when dissolved in water, produce acidic, basic, or neutral solutions:

(a) $KCN(s)$ (b) $Pb(NO_3)_2(s)$
(c) $NaHSO_4(s)$ (d) $CaCl_2(s)$

20-45. Some soaps are produced by reacting caustic soda, $NaOH(s)$, with animal fats, which contain a type of organic acid called a fatty acid. An example is stearic acid, whose chemical formula is $CH_3(CH_2)_{16}COOH$. Fatty acids, like most organic acids, are weak acids. Is a soap solution acidic, basic, or neutral?

20-46. Various aluminum salts, such as $Al_2(SO_4)_3(s)$ and $KAl(SO_4)_2(s)$, are used as additives to increase the acidity of soils for "acid-loving" plants such as azaleas and tomatoes. Explain how these salts increase soil acidity.

pH CALCULATIONS OF SALT SOLUTIONS

20-47. Sodium hypochlorite, $NaClO(s)$, is a bleaching agent. Calculate the pH and the concentration of $HClO(aq)$ in an aqueous solution that is 0.030 M in $NaClO(aq)$ at 25°C.

20-48. Sodium propanoate, $NaCH_3CH_2COO(s)$, is used as a food preservative. Calculate the pH at 25°C of a 0.20-M solution of $NaCH_3CH_2COO(aq)$, taking $K_a = 1.4 \times 10^{-5}$ M for propanoic acid.

20-49. A solution of sodium cyanate, $NaCNO(aq)$, is prepared at a concentration of 0.20 M. Calculate the equilibrium concentrations of $OH^-(aq)$, $HCNO(aq)$, $CNO^-(aq)$, and $H_3O^+(aq)$ and the pH of the solution at 25°C.

20-50. Calculate the equilibrium concentrations of $OH^-(aq)$, $HNO_2(aq)$, $NO_2^-(aq)$, and $H_3O^+(aq)$ and the pH at 25°C of a solution that is 0.25 M in $NaNO_2$.

20-51. A saturated aqueous solution of ammonium perchlorate contains 23.7 grams of $NH_4ClO_4(s)$ per 100.0 mL of solution at 25°C. Calculate the pH of this solution at 25°C.

20-52. Calculate the pH at 25°C of a solution that is prepared by dissolving 25.0 grams of barium acetate, $Ba(CH_3COO)_2(s)$, in enough water to make exactly 1.00 liter of solution.

20-53. The acid-dissociation constant at 25°C for the equilibrium described by

$$[Fe(H_2O)_6]^{3+}(aq) + H_2O(l) \leftrightharpoons H_3O^+(aq) + [Fe(OH)(H_2O)_5]^{2+}(aq)$$

is $K_a = 6.7 \times 10^{-3}$ M. Calculate the pH of a 0.20-M solution of $Fe(NO_3)_3(aq)$ at 25°C.

20-54. Calculate the pH at 25°C of a solution that is 0.10 M in $TlBr_3(aq)$. The acid-dissociation constant at 25°C for the equilibrium described by

$$[Tl(H_2O)_6]^{3+}(aq) + H_2O(l) \leftrightharpoons H_3O^+(aq) + [Tl(OH)(H_2O)_5]^{2+}(aq)$$

is $K_a = 7.0 \times 10^{-2}$ M.

POLYPROTIC ACIDS

20-55. Oxalic acid, HOOCCOOH(aq), is a dicarboxylic acid (see Practice Problem 20-3). Use the data from Table 20.9 to determine the pH of a 0.125-M oxalic acid solution at 25°C. What assumptions must you make in performing this calculation?

20-56. Use the data in Table 20.9 to calculate the pH of a 0.50-M sulfurous acid, $H_2SO_3(aq)$, solution at 25°C. What assumptions must you make in performing this calculation?

20-57. Phosphoric acid, $H_3PO_4(aq)$, is a triprotic acid. Using the pK_a values listed in Table 20.9, estimate the pH and concentrations of $H_3O^+(aq)$, $H_3PO_4(aq)$, $H_2PO_4^-(aq)$, $HPO_4^{2-}(aq)$, $PO_4^{3-}(aq)$, and $OH^-(aq)$ in a 0.100-M phosphoric acid solution at 25°C. What assumptions must you make in performing this calculation?

20-58. Citric acid, $HO(CH_2COOH)_2COOH(aq)$, is a triprotic acid found in fruits and vegetables and is central to the metabolic cycles of most higher organisms. Using the data in Table 20.9, calculate the concentration of $H_3O^+(aq)$ and the pH of a 0.050-M citric acid solution at 25°C. Assume that each consecutive acidic-proton may be treated as an independent equilibrium problem. Based on your results, what fraction of the total $H_3O^+(aq)$ concentration in the solution comes from the first, second, and third dissociated protons of citric acid?

ADDITIONAL PROBLEMS

20-59. What is the difference between a strong acid and a weak acid?

20-60. What is the relationship between: (a) the magnitude of K_a, pK_a, and the strength of an acid? (b) the magnitude of K_b, pK_b, and the strength of a base?

20-61. What is the relationship between the strength of a weak acid and that of its conjugate base?

20-62. Is it possible to have a negative value of pH?

20-63. Determine whether each of the following substances is an Arrhenius acid, a Brønsted-Lowry acid, or a Lewis acid (it is possible for each to be of more than one type):

(a) $HCl(aq)$ (b) $AlCl_3(aq)$ (c) $BCl_3(aq)$

20-64. Determine whether each of the following substances is an Arrhenius base, a Brønsted-Lowry base, or a Lewis base (it is possible for each to be of more than one type):

(a) $NH_3(aq)$ (b) $Br^-(aq)$ (c) $NaOH(aq)$

20-65. Explain why a 0.10-M aqueous solution of sodium hydroxide, NaOH(aq), is highly basic, whereas a 0.10-M aqueous solution of methanol, $CH_3OH(aq)$, is essentially neutral.

20-66. Explain why the hydrogen halides HCl(g) and HBr(g) form acidic aqueous solutions, whereas the metal hydrides NaH(s) and KH(s) form basic aqueous solutions.

20-67. Given that the neutralization reaction between a strong acid and a strong base is highly exothermic, would you predict that the pH of neutral water would be less than, greater than, or equal to 7.00 at temperatures greater than 25°C?

20-68. A student observes that when lithium metal, Li(s), is added to water, the solution bubbles and the pH becomes basic; whereas when lithium chloride, LiCl(s), is added to water, the solution remains neutral. Explain these observations.

20-69. Classify each of the following organic compounds as acidic, basic, or neutral in aqueous solution. In each case state the reason for your choice.

(a)

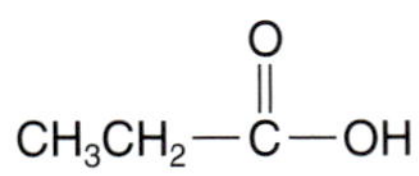

(b)

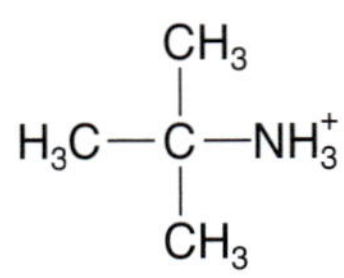

(c)

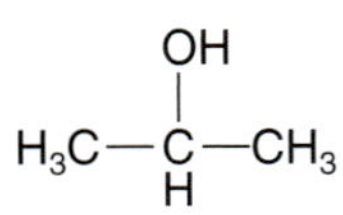

(d)

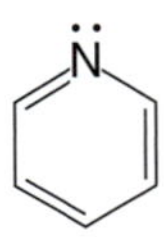

20-70. Nitrites, such as $NaNO_2$, are added to processed meats and hamburger both as a preservative and to give the meat a redder color by binding to hemoglobin in the red blood cells. When the nitrite ion is ingested, it reacts with stomach acid to form nitrous acid, $HNO_2(aq)$, in the stomach. Given that $K_a = 5.6 \times 10^{-4}$ M for nitrous acid at 25°C, compute the value of the ratio $[HNO_2]/[NO_2^-]$ in the stomach following ingestion of $NaNO_2(aq)$ when $[H_3O^+]$ is 0.10 M. Assume a temperature of 25°C.

20-71. Calculate the pH at 25°C of a 3.00-M acetic acid, $CH_3COOH(aq)$, solution. Calculate the pH of the resulting solution when 2.00 milliliters of the 3.00-M acetic acid is diluted to make a 250.0-milliliter solution. The value of K_a for acetic acid at 25°C is 1.8×10^{-5} M.

20-72. Given that K_w for water is 2.40×10^{-14} M^2 at 37°C, compute the pH of a neutral aqueous solution at 37°C, which is the normal human body temperature. Is a pH = 7.00 solution acidic or basic at 37°C?

20-73. Without referring to the tables in the text, rank 0.050-M aqueous solutions of the following compounds in order of increasing pH: $KOH(aq)$, $LiBr(aq)$, $Fe(ClO_4)_3(aq)$, $H_2SO_4(aq)$, $NaHCOO(aq)$.

20-74. A saturated solution of the strong base $Mg(OH)_2(aq)$ at 25°C has a pH of 10.52. Estimate the solubility of $Mg(OH)_2(s)$ in water at 25°C in grams per 100 milliliters.

20-75. Use the data in Table 20.6 to compute the equilibrium constants at 25°C for the following equations:

(a) $HCNO(aq) + NO_2^-(aq) \leftrightharpoons HNO_2(aq) + CNO^-(aq)$

(b) $NH_4^+(aq) + HCOO^-(aq) \leftrightharpoons NH_3(aq) + HCOOH(aq)$

20-76. Vinegar contains about 5.5% acetic acid by mass. Assuming that the density of vinegar is the same as that of water, ($1.0\ g\cdot mL^{-1}$), calculate the pH of a typical vinegar solution at 25°C.

20-77. A solution of household bleach contains 5.25% sodium hypochlorite, $NaOCl(s)$, by mass. Assuming that the density of bleach is the same as water, ($1.0\ g\cdot mL^{-1}$), calculate the volume of household bleach that should be diluted with water to make 500.0 mL of a pH = 9.80 solution at 25°C.

20-78. Sodium benzoate, $NaC_6H_5COO(s)$, is used as a food preservative because of its antimicrobial action. The K_a of benzoic acid at 25°C is 6.3×10^{-5} M. Calculate the ratio of the concentration of benzoic acid to the concentration of benzoate in a food with a pH of 3.00 at 25°C.

20-79. Calculate the pH and the concentrations of all the species in 0.0250 M $H_2Se(aq)$ at 25°C.

20-80. Ascorbic acid (vitamin C) has the structure shown below.

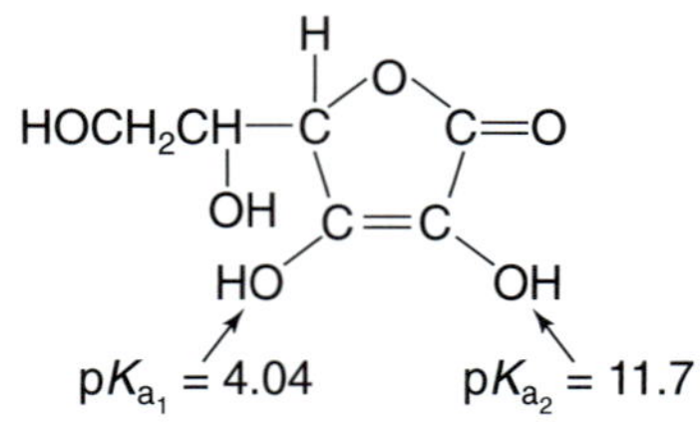

Calculate the pH of an aqueous vitamin C solution obtained by dissolving 500.0 milligrams of vitamin C in enough water to make 1.00 liter of solution at 25°C.

20-81. Autoprotonation occurs in solvents other than water. For liquid ammonia, the autoprotonation equilibrium is described by the equation

$$2NH_3(l) \leftrightharpoons NH_4^+(amm) + NH_2^-(amm)$$

with $K \approx 1 \times 10^{-30}$ M^2 at −50°C, where (*amm*) denotes a solute in liquid ammonia. Estimate the concentration of the ammonium ion in liquid ammonia at −50°C. How many molecules of ammonia are dissociated per mole of ammonia? Take the density of liquid ammonia as $0.77\ g\cdot mL^{-1}$ at −50°C.

20-82. The autoprotonation constant for the solvent ethanol, $CH_3CH_2OH(l)$, is 8×10^{-20} M^2 at 25°C. The autoprotonation equilibrium equation is

$$2CH_3CH_2OH(l) \leftrightharpoons CH_3CH_2OH_2^+(alc) + CH_3CH_2O^-(alc)$$

where (*alc*) denotes a solute in alcohol solution.

(a) We define a pH scale in alcohol by the equation

$$pH = -\log([CH_3CH_2OH_2^+]/M)$$

Calculate the pH of a neutral alcohol solution at 25°C.

(b) Calculate the pH of a 0.010-M solution of sodium ethoxide, $CH_3CH_2O^-Na^+(alc)$, in alcohol at 25°C. Assume that $CH_3CH_2O^-Na^+(alc)$ is completely dissociated in alcohol.

20-83. Uric acid is an end product of the metabolism of certain biological compounds. Gout is a disease of the joints that is due to the precipitation of sodium urate crystals. Given that under physiological conditions

$$\underset{\text{uric acid}}{HC_5H_3N_4O_3(aq)} + H_2O(l) \leftrightharpoons \underset{\text{urate ion}}{C_5H_3N_4O_3^-(aq)} + H_3O^+(aq)$$

with $K_a = 1.6 \times 10^{-6}$ M at 25°C. Determine the pH values for which [urate] > [uric acid]

20-84. The value of the ion-product constant for water, K_w, at 0°C is 0.12×10^{-14} M^2. Calculate the pH of a neutral aqueous solution at 0°C. Is an aqueous solution with a pH = 7.25 acidic or basic at 0°C?

20-85. A saturated solution of $Sr(OH)_2(aq)$ at 25°C has a measured pH of 13.50. Estimate the solubility of $Sr(OH)_2(s)$ in water at 25°C in grams per 100 milliliters of solution.

20-86. Calculate the pH of a 0.200 M $GaCl_3(aq)$ solution at 25°C.

20-87. Calculate the pH and the concentrations of all the species in 0.150 M formic acid, $HCOOH(aq)$ at 25°C.

20-88. Calculate the pH and the concentrations of all the species in 0.150 M dimethylamine, $(CH_3)_2NH(aq)$ at 25°C.

20-89. Experimental measurement shows that 0.200 M arsenious acid, $H_2AsO_3(aq)$, is 0.0051% dissociated at 25°C. Determine the value of the acid-dissociation constant of $H_2AsO_3(aq)$ at 25°C.

20-90. Calculate the percentage of dissociation of 0.100 M $HCNO(aq)$ at 25°C.

20-91. Calculate the pH of a 0.010-M sulfurous acid solution at 25°C.

20-92. (*) Although the pH of neutral water is said to be 7.00 at 25°C, the experimentally measured pH of a typical water sample is about 5.6. This is true both for tap water and for samples of store-bought distilled or deionized water that contain no dissolved metal ions. Explain this observation.

20-93. (*) The neutralization of an acid with a base yields a salt and water in aqueous solution.

(a) Calculate the resulting pH when 50.0 milliliters of 0.250-M hydrochloric acid solution is mixed with 50.0 milliliters of 0.250-M sodium hydroxide solution at 25°C.

(b) Calculate the resulting pH when 50.0 milliliters of 0.250-M acetic acid solution is mixed with 50.0 milliliters of 0.250-M sodium hydroxide solution at 25°C.

(c) Explain why the pH of the resulting solution in part (b) is not 7.00.

20-94. (*) What is the pH of a 2.60×10^{-8} M solution of $HCl(aq)$ at 25°C? (*Hint*: What must you also consider at this acid concentration?)

20-95. (*) The principal reaction that occurs when an amphoteric salt (one containing an anion that can act as either an acid or a base), such as $NaHCO_3(s)$, is dissolved in water is of the type

$$2\,HCO_3^-(aq) \leftrightharpoons CO_3^{2-}(aq) + H_2CO_3(aq)$$

Show that the $[H_3O^+]$ of the solution is given by

$$[H_3O^+] \approx (K_{a_1}K_{a_2})^{1/2}$$

and that the pH of the solution is given by

$$pH = \frac{pK_{a_1} + pK_{a_2}}{2}$$

where K_{a_1} and K_{a_2} are the first and second acid-dissociation constants of $H_2CO_3(aq)$. Note that the pH of the solution is independent of the salt concentration. Calculate the pH of a saturated $NaHCO_3(aq)$ solution at 25°C.

Lawrence Joseph Henderson (1878–1942) was born in Lynn, Massachusetts, and spent most of his professional life at Harvard University, where he received an M.D. in 1902. Much of his early career was devoted to the study of the role of blood in the transport of oxygen to various organs of the body. He was aware that blood could resist changes in acidity and basicity, but the relationship between the hydrogen ion concentration, the composition of a buffer, and its buffering capacity was not understood. In 1908, Henderson published two papers in the American Journal of Physiology, and in them put forward a simple formula linking the aqueous hydrogen ion concentration and the composition of a buffer. Henderson's equation represented a significant advance in the understanding of acid-base chemistry at the time. One year after Henderson's papers were published, the Danish biochemist, Søren Sørensen (Chapter 20 Frontispiece), created the pH scale. The pH scale found immediate acceptance among biochemists, who had investigated the ability of living organisms to buffer against excessive acidity or alkalinity. In 1916, Karl Albert Hasselbalch (1874–1962), not pictured, from the University of Copenhagen, merged Henderson's buffer formula with Sørensen's pH scale and wrote an expression now known as the Henderson-Hasselbalch equation. Hasselbalch's simple idea of expressing the Henderson equation in terms of logarithms made his and Henderson's names household names in chemistry. It is remarkable that it was physiologists and medical scientists working with biological fluids who introduced the concepts of buffers and pH into the field of chemistry.

21. Buffers and the Titration of Acids and Bases

This chapter continues our treatment of acid-base chemistry. We shall discuss two important topics: buffers and acid-base titrations. Buffers are solutions that are able to resist changes in pH and so are used to maintain the pH in chemical and biochemical systems. For instance, chemically formulated cosmetics, such as "pH balanced" hair care products, use buffers to maintain a pH for optimal lathering. Biochemical systems, such as our blood, employ natural buffers to control enzyme activity, which depends markedly upon pH. In addition, many natural systems, such as lakes and streams, contain natural buffers that maintain the pH of the water, which is critical to the survival of many aquatic species. In this chapter, we explain the chemical basis of buffer action in terms of Le Châtelier's principle, and we show how to calculate the pH of buffer solutions. We then discuss acid-base indicators that provide a measure of the acidity of a solution. We end the chapter with a discussion of acid-base titration curves that allow us to determine the total acidity or total basicity of a solution.

21-1. The Henderson-Hasselbalch Equation Often Can Be Used to Calculate the pH of a Buffer Solution

A solution that contains both a weak acid and its conjugate base can resist a change in pH by neutralizing either an added acid or an added base. Such a solution is called a **buffer.**

As an example, consider a solution that contains a mixture of $CH_3COOH(aq)$ and its conjugate base, $CH_3COO^-(aq)$. One way to prepare such a solution is by mixing a solution of acetic acid with a solution of sodium acetate. Notice that although $CH_3COOH(aq)$ is an acid and $CH_3COO^-(aq)$ is a base, the two can not neutralize each other because they share a common anion. That is, a proton transfer reaction between acetic acid and the acetate ion results in no net change to the overall concentrations of the species present in solution. In terms of an equation, we have

$$CH_3COOH(aq) + CH_3COO^-(aq) \leftrightharpoons CH_3COO^-(aq) + CH_3COOH(aq)$$
(no net reaction)

Thus, a buffer solution, such as the acetic acid–acetate solution we have described here, results from a mixture of an acid with its *own* conjugate base.

If an acid is added to this buffer solution, then essentially all the added $H_3O^+(aq)$ ions are removed from the solution by the reaction of $H_3O^+(aq)$ with $CH_3COO^-(aq)$ according to:

Here we assume sufficient buffer to neutralize essentially all the added acid or base.

$$H_3O^+(aq) + CH_3COO^-(aq) \leftrightharpoons CH_3COOH(aq) + H_2O(l) \qquad (21.1)$$

This equation is the reverse of the equation that represents the dissociation of $CH_3COOH(aq)$ (Equation 20.8), so the value of K_c for this equation is the reciprocal of K_a (Table 20.6) for $CH_3COOH(aq)$, or $K_c = 1/K_a = 5.6 \times 10^4\ M^{-1}$. This large value of K_c indicates that the reaction goes essentially to completion and converts almost all the added $H_3O^+(aq)$ into $CH_3COOH(aq)$.

Suppose instead that a base is added to the buffer solution. Essentially all the added $OH^-(aq)$ ions are removed from the solution by the reaction with acetic acid according to

$$CH_3COOH(aq) + OH^-(aq) \leftrightharpoons CH_3COO^-(aq) + H_2O(l) \qquad (21.2)$$

This equation is the reverse of the equation that represents the basic nature of $CH_3COO^-(aq)$, so the value of K_c for this equation is the reciprocal of K_b (Table 20.6) for $CH_3COO^-(aq)$, or $K_c = 1/K_b = 1.8 \times 10^9\ M^{-1}$. The large value of K_c indicates that essentially all the added $OH^-(aq)$ ions are removed from solution by this reaction. Thus, a solution containing both $CH_3COOH(aq)$ and $CH_3COO^-(aq)$ can react completely with added acid or added base thereby suppressing changes in pH when a small amount of either an acid or a base is added.

This buffer solution contains both $CH_3COOH(aq)$ and $CH_3COO^-(aq)$, and so we have to consider the two equilibrium equations:

Throughout this chapter, we shall take all solutions to be at 25°C.

$$CH_3COOH(aq) + H_2O(l) \leftrightharpoons H_3O^+(aq) + CH_3COO^-(aq) \qquad (21.3)$$
$$K_a = 1.8 \times 10^{-5}\ M$$

$$CH_3COO^-(aq) + H_2O(l) \leftrightharpoons OH^-(aq) + CH_3COOH(aq) \qquad (21.4)$$
$$K_b = 5.6 \times 10^{-10}\ M$$

Because the equilibrium constants for both of these equations are small, both equilibria lie far to the left and neither $[CH_3COOH]$ nor $[CH_3COO^-]$ will differ significantly from its stoichiometric concentration. Consequently, we can write

$$[CH_3COOH] \approx [CH_3COOH]_0$$

$$[CH_3COO^-] \approx [CH_3COO^-]_0$$

where the subscript zero emphasizes that these are stoichiometric concentrations.

Now let's calculate the pH of a buffer solution that is, say, 0.10 M in $CH_3COOH(aq)$ and 0.15 M in $CH_3COO^-(aq)$. The conjugate acid-base equilibrium equation involving these two species is given by Equation 21.3.

If we neglect the source of $H_3O^+(aq)$ from Equation 20.1,

$$H_2O(l) + H_2O(l) \leftrightharpoons H_3O^+(aq) + OH^-(aq) \qquad K_w = 1.0 \times 10^{-14}\ M$$

then we can write the concentration table associated with Equation 21.3 as follows:

Concentration	$CH_3COOH(aq)$	+ $H_2O(l)$ ⇋	$H_3O^+(aq)$	+ $CH_3COO^-(aq)$
initial	0.10 M	—	≈ 0 M	0.15 M
change	$-x$	—	$+x$	$+x$
equilibrium	$0.10\ M - x$	—	x	$0.15\ M + x$

Substituting these values into the equilibrium-constant expression gives

$$\frac{x(0.15\ M + x)}{0.10\ M - x} = 1.8 \times 10^{-5}\ M$$

We could solve this equation for x by using the quadratic equation, but because $CH_3COOH(aq)$ is a weak acid, we expect that the value of x will be small. Therefore, we neglect x with respect to both 0.10 M and 0.15 M, and write

$$\frac{x(0.15\ M)}{0.10\ M} \approx 1.8 \times 10^{-5}\ M$$

which yields

$$x \approx \frac{(0.10\ M)(1.8 \times 10^{-5}\ M)}{0.15\ M} = 1.2 \times 10^{-5}\ M$$

This small value for x justifies our approximation of neglecting x with respect to 0.10 M and 0.15 M. Because $[H_3O^+] = x$, the pH of the solution is given by

$$pH = -\log([H_3O^+]/M) = -\log(1.2 \times 10^{-5}) = 4.92$$

Generally, a buffer solution consists of a weak acid $HA(aq)$ and its conjugate base $A^-(aq)$. The two equilibria equations that we must consider are

$$HA(aq) + H_2O(l) \leftrightharpoons H_3O^+(aq) + A^-(aq)$$

and

$$A^-(aq) + H_2O(l) \leftrightharpoons OH^-(aq) + HA(aq)$$

If both K_a and K_b are small (less than 10^{-3} M, for example), then we can write

$$[HA] \approx [HA]_0 \quad \text{(stoichiometric amount of acid)} \qquad (21.5a)$$

$$[A^-] \approx [A^-]_0 \quad \text{(stoichiometric amount of base)} \qquad (21.5b)$$

The equilibrium-constant expression involving the acid dissociation is

$$K_a = \frac{[H_3O^+][A^-]}{[HA]} \qquad (21.6)$$

If we substitute Equations 21.5 into Equation 21.6, then we obtain

$$K_a \approx \frac{[H_3O^+][A^-]_0}{[HA]_0}$$

or

$$[H_3O^+] \approx \frac{K_a[HA]_0}{[A^-]_0}$$

If we divide both sides by M to cancel the units, take the logarithm of both sides, and use the fact that log ab = log a + log b, we obtain

$$\log([H_3O^+]/M) \approx \log(K_a/M) + \log\left(\frac{[HA]_0}{[A^-]_0}\right)$$

Finally, multiplying both sides by –1 and using pH = –log($[H_3O^+]$/M), pK_a = –log(K_a/M), and the fact that log a/b = –log b/a, we get

$$pH \approx pK_a + \log\frac{[base]_0}{[acid]_0} \qquad (21.7)$$

Equation 21.7 is known as the **Henderson-Hasselbalch equation** (Frontispiece). It enables us to use the stoichiometric concentrations directly without the need for a stepwise analysis of the equilibrium. The Henderson-Hasselbalch equation is used extensively in biochemistry. Whenever Equation 21.7 is applied, several general conditions must be met:

One must always be careful to check these conditions before applying the Henderson-Hasselbalch equation.

1. The value of K_a for the conjugate acid-base pair should be in the range of about 10^{-4} M to 10^{-11} M.
2. The ratio $[base]_0/[acid]_0$ must be between roughly 0.1 and 10.
3. The values of $[base]_0$ and $[acid]_0$ should be in the range of roughly 10^{-3} M to 1 M.

While these conditions may appear restrictive, in fact the majority of the acid-base systems of interest to us fall within these ranges.

Let's calculate the pH of a buffer that is 0.25 M in formic acid, HCOOH(*aq*), and 0.15 M in potassium formate, KHCOO(*aq*). Because potassium formate is a strong electrolyte consisting of potassium ions and formate ions, the solution will be 0.25 M in HCOOH(*aq*) and 0.15 M in $HCOO^-$(*aq*). We see from Table 20.6 that the value of K_a for formic acid is 1.8×10^{-4} M and the value of K_b for the formate anion is 5.6×10^{-11} M. The K^+(*aq*) ion is a neutral ion (Table 20.7) and so does not participate in the buffer activity. Because the value of K_a is between 10^{-4} M and 10^{-11} M, the ratio of initial concentrations of base to acid is 0.15 M/0.25 M = 0.6, and the molarities of the conjugate acid-base pair are within the range 10^{-3} M to 1 M, we can use the Henderson-Hasselbalch equation to obtain the pH of the buffer solution:

$$pH \approx pK_a + \log\frac{0.15\ M}{0.25\ M} = 3.75 - 0.22 = 3.53$$

EXAMPLE 21-1: Calculate the pH of a buffer solution that is 0.050 M in $NH_4Cl(aq)$ and 0.20 M in $NH_3(aq)$.

Solution: Because $NH_4Cl(aq)$ is a strong electrolyte, it dissociates completely to form $NH_4^+(aq)$ ions and $Cl^-(aq)$ ions. Therefore, the solution will be 0.050 M in $NH_4^+(aq)$ and 0.20 M in $NH_3(aq)$. Because the chloride ion is a neutral ion, it does not participate in the buffer activity. In this case $NH_4^+(aq)$ is the conjugate acid and $NH_3(aq)$ is the conjugate base. We see from Table 20.6 that the value of K_a for NH_4^+ is 5.6×10^{-10} M and that the value of K_b for $NH_3(aq)$ is 1.8×10^{-5} M. Because the value of K_a is between 10^{-4} M and 10^{-11} M, the ratio of concentrations is 4.0, and the concentrations of the conjugate acid-base pair are between 10^{-3} M and 1 M, we can use the Henderson-Hasselbalch equation to obtain the pH of this buffer solution:

$$\text{pH} \approx \text{p}K_a + \log\frac{0.20\text{ M}}{0.050\text{ M}} = 9.25 + 0.60 = 9.85$$

PRACTICE PROBLEM 21-1: Calculate the pH of a buffer solution that is 0.85 M in $CH_3COOH(aq)$ and 0.30 M in $NaCH_3COO(aq)$.

Answer: pH = 4.29

21-2. A Buffer Solution Suppresses a Change in pH When a Small Amount of Either an Acid or a Base Is Added

Now let's consider the response of a buffer solution to the addition of small amounts of an acid or a base. Let the buffer be composed of a solution that is 0.100 M in $CH_3COOH(aq)$ and 0.100 M in $CH_3COO^-(aq)$. Suppose that 10.0 mL of 0.100 M $HCl(aq)$ is added to 100.0 mL of the buffer solution. The number of millimoles of acid in 10.0 mL of 0.100 M $HCl(aq)$ is

As we saw in Chapter 12, using units of millimoles when working with dilute solutions is more convenient than using units of moles.

$$\text{mmol of H}_3\text{O}^+ \text{ added} = (10.0\text{ ml})(0.100\text{ M}) = 1.00\text{ mmol}$$

The $H_3O^+(aq)$ reacts with the $CH_3COO^-(aq)$ in the buffer solution according to

$$H_3O^+(aq) + CH_3COO^-(aq) \rightarrow CH_3COOH(aq) + H_2O(l)$$

Recall from Section 21.1 that the equilibrium constant for this equation is quite large (5.6×10^4 M), so the reaction goes essentially to completion. Before the $HCl(aq)$ is added, the number of millimoles of acetate ion in the buffer solution is

$$\begin{pmatrix}\text{mmol of CH}_3\text{COO}^- \\ \text{before HCl added}\end{pmatrix} = (100.0\text{ mL})(0.100\text{ M}) = 10.0\text{ mmol}$$

According to the chemical equation above, the number of millimoles of $CH_3COO^-(aq)$ after the addition of 10.0 mL of 0.10 M $HCl(aq)$ is

$$\begin{pmatrix}\text{mmol of CH}_3\text{COO}^- \\ \text{after HCl added}\end{pmatrix} = \begin{pmatrix}\text{mmol of CH}_3\text{COO}^- \\ \text{before HCl added}\end{pmatrix} - \begin{pmatrix}\text{mmol of} \\ \text{H}_3\text{O}^+ \text{ added}\end{pmatrix}$$

$$= 10.0\text{ mmol} - 1.00\text{ mmol} = 9.0\text{ mmol}$$

Similarly, before the HCl(*aq*) is added the number of millimoles of acetic acid in the buffer solution is

$$\left(\begin{matrix}\text{mmol of } CH_3COOH \\ \text{before HCl added}\end{matrix}\right) = (100.0\text{ ml})(0.100\text{ M}) = 10.0\text{ mmol}$$

From the reaction stoichiometry, the number of millimoles of $CH_3COOH(aq)$ after the addition of 1.00 mmol of $H_3O^+(aq)$ is

$$\left(\begin{matrix}\text{mmol of } CH_3COOH \\ \text{after HCl added}\end{matrix}\right) = \left(\begin{matrix}\text{mmol of } CH_3COOH \\ \text{before HCl added}\end{matrix}\right) + \left(\begin{matrix}\text{mmol of} \\ H_3O^+ \text{ added}\end{matrix}\right)$$

$$= 10.0\text{ mmol} + 1.00\text{ mmol} = 11.0\text{ mmol}$$

Thus, we see that the addition of 1.00 mmol of acid results in a decrease in the amount of $CH_3COO^-(aq)$ present from 10.00 mmol to 9.00 mmol, and a corresponding increase in the amount of $CH_3COOH(aq)$ present from 10.00 mmol to 11.00 mmol.

The pH of the buffer solution before the addition of the HCl(*aq*) is

$$\text{pH} \approx \text{p}K_a + \log\frac{[\text{base}]_0}{[\text{acid}]_0} = 4.74 + \log\frac{0.100\text{ M}}{0.100\text{ M}} = 4.74$$

and the pH of the buffer solution after the addition of the HCl(*aq*) is

$$\text{pH} \approx 4.74 + \log\frac{(9.0\text{ mmol}/110.0\text{ ml})}{(11.0\text{ mmol}/110.0\text{ ml})} = 4.65$$

Thus, the pH of the solution decreases by less than 0.10 pH unit. The pH decreased because we added an acid. Notice that the volumes (110.0 mL) cancel out in the logarithm term. Therefore, we can work directly with the number of millimoles of the conjugate acid and base without the need to convert to molarities before taking the logarithm of the ratio.

In contrast to the above result, if we add 10.0 mL of 0.100 M HCl(*aq*) to 100.0 mL of pure water, then the pH of the resulting solution is

$$[H_3O^+] = \frac{\text{mmol of } H_3O^+}{\text{mL of solution}} = \frac{(10.0\text{ mL})(0.100\text{ M})}{110.0\text{ ml}} = 9.09\times10^{-3}\text{ M}$$

or

$$\text{pH} = -\log([H_3O^+]/\text{M}) = -\log(9.09\times10^{-3}) = 2.04$$

The pH of pure water is 7.00, so we see that the pH changes by about 5 pH units. Because pH is a logarithmic scale, the value of $[H_3O^+]$ changes by a factor of 10^5, or 100000. In contrast, the pH of the buffer solution changes by less than 0.1 pH unit, which corresponds to the value of $[H_3O^+]$ changing by about a factor of 1.2. This demonstrates the amazing power of a buffer solution to resist changes in pH.

EXAMPLE 21-2: Calculate the change in pH when 10.0 mL of 0.100 M $NaOH(aq)$ is added to 100.0 mL of a buffer that is 0.100 M in $CH_3COOH(aq)$ and 0.100 M in $CH_3COO^-(aq)$. Take $pK_a = 4.74$ for $CH_3COOH(aq)$.

Solution: The number of millimoles of base in 10.0 mL of 0.100 M $NaOH(aq)$ is

$$\text{mmol of OH}^- \text{ added} = (10.0 \text{ ml})(0.100 \text{ M}) = 1.00 \text{ mmol}$$

The $OH^-(aq)$ reacts with $CH_3COOH(aq)$ in the buffer according to the equation

$$OH^-(aq) + CH_3COOH(aq) \rightarrow H_2O(l) + CH_3COO^-(aq)$$

The number of millimoles of $CH_3COOH(aq)$ in the buffer solution before the addition of $NaOH(aq)$ is

$$\begin{pmatrix}\text{mmol of CH}_3\text{COOH} \\ \text{before NaOH added}\end{pmatrix} = (100.0 \text{ ml})(0.100 \text{ M}) = 10.0 \text{ mmol}$$

According to the chemical equation, the number of millimoles of $CH_3COOH(aq)$ after the addition of 10.0 mL of 0.100 M $NaOH(aq)$ is

$$\begin{pmatrix}\text{mmol of CH}_3\text{COOH} \\ \text{after NaOH added}\end{pmatrix} = \begin{pmatrix}\text{mmol of CH}_3\text{COOH} \\ \text{before NaOH added}\end{pmatrix} - \begin{pmatrix}\text{mmol of} \\ \text{OH}^- \text{ added}\end{pmatrix}$$

$$= 10.0 \text{ mmol} - 1.00 \text{ mmol} = 9.0 \text{ mmol}$$

Similarly, the initial number of millimoles of $CH_3COO^-(aq)$ in the buffer solution is 10.0 mmol. The number of millimoles of $CH_3COO^-(aq)$ present following the addition of 1.00 mmol of $NaOH(aq)$ is

$$\begin{pmatrix}\text{mmol of CH}_3\text{COO}^- \\ \text{after NaOH added}\end{pmatrix} = \begin{pmatrix}\text{mmol of CH}_3\text{COO}^- \\ \text{before NaOH added}\end{pmatrix} + \begin{pmatrix}\text{mmol of} \\ \text{OH}^- \text{ added}\end{pmatrix}$$

$$= 10.0 \text{ mmol} + 1.00 \text{ mmol} = 11.0 \text{ mmol}$$

Because K_a is between 10^{-4} M and 10^{-11} M, the ratio $[\text{base}]_0/[\text{acid}]_0$ is $11.0/9.0 = 1.2$, and the concentrations of the conjugate acid-base pair are 9.0 mmol/110 mL = 8.2×10^{-2} M and 11.0 mmol/110 mL = 0.100 M, we can use the Henderson-Hasselbalch equation. The pH is

$$\text{pH} \approx \text{p}K_a + \log\frac{[\text{base}]_0}{[\text{acid}]_0} = 4.74 + \log\frac{11.0 \text{ mmol}}{9.0 \text{ mmol}} = 4.83$$

Once again, notice that the pH changes by less than 0.1 pH unit.

PRACTICE PROBLEM 21-2: Calculate the change in pH when (a) 10.0 mL of 0.250 M $HCl(aq)$ or (b) 10.0 mL of 0.250 M $KOH(aq)$ are added to 100.0 mL of a buffer that is 0.500 M in $NH_4Cl(aq)$ and 0.500 M in $NH_3(aq)$.

Answer: (a) −0.04 pH unit; (b) +0.04 pH unit

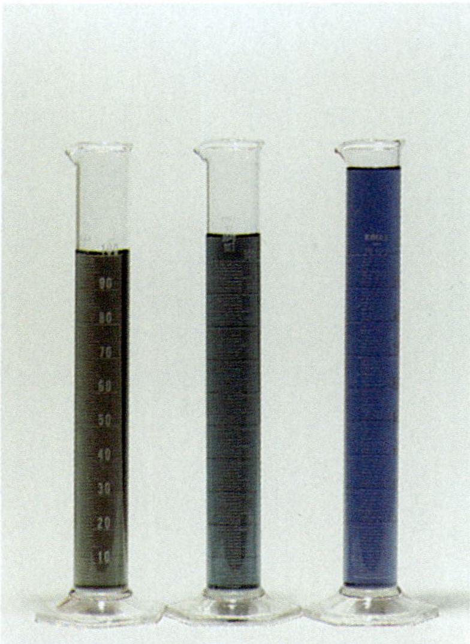

Figure 21.1 The left cylinder contains 100 mL of a sodium acetate and acetic acid buffer solution with bromcresol green indicator that indicates that the pH of the solution is about 5 (see Figure 21.7). The middle cylinder contains 100 mL of the $CH_3COO^-(aq)$–$CH_3COOH(aq)$ buffer solution plus 7.0 mL of 0.10 M $NaOH(aq)$. The fact that the solution remains green indicates that the pH of the solution is still about 5. The right cylinder contains 100.0 mL of the $CH_3COO^-(aq)$–$CH_3COOH(aq)$ buffer solution plus 25.0 mL of 0.10 M $NaOH(aq)$. The change in the indicator color from green to blue shows that the pH has increased significantly and that the buffer has been overwhelmed by the addition of a large amount $NaOH(aq)$.

The capacity of a buffer to resist changes in pH is not unlimited. If sufficient acid (or base) is added to neutralize all the conjugate base (or all the conjugate acid), then the pH of the solution will change significantly. In such a case the buffer is simply overwhelmed (Figure 21.1). The number of millimoles of acid or base that may be added to a buffer before it is overwhelmed is called the **buffer capacity** (see Problems 21-11 and 21-12).

Another property of buffers is their ability to resist changes in pH when diluted with solvent. This property is readily understood from the Henderson-Hasselbalch equation. If we dilute a buffer solution by, say, a factor of 2, then the stoichiometric concentration of the base, $[\text{base}]_0$, and the stoichiometric concentration of the acid, $[\text{acid}]_0$, both decrease by a factor of 2, but the ratio of the stoichiometric concentrations does not change:

$$\frac{\frac{1}{2}[\text{base}]_0}{\frac{1}{2}[\text{acid}]_0} = \frac{[\text{base}]_0}{[\text{acid}]_0}$$

Thus the pH of the buffer solution does not change with dilution. The effects of the addition of $HCl(aq)$ and $NaOH(aq)$ and of dilution on the pH of the buffer solutions $CH_3COOH(aq)$–$CH_3COO^-(aq)$ and $NH_4^+(aq)$–$NH_3(aq)$ are summarized in Table 21.1.

The activity of many enzymes depends on pH, and numerous biological systems have natural buffers to control enzyme activities. Essentially constant pH values are required for maintaining the delicate balances in the complex sequences of biochemical reactions essential to the existence of life. For example, blood is buffered by a mixture of carbonates, phosphates, and proteins and exhibits a remarkably constant pH value of 7.4. At pH values lower than 7.3, the blood cannot efficiently remove carbon dioxide from cells, and at pH values greater than 7.7, blood cannot efficiently release carbon dioxide to the lungs. Blood pH values outside the range 7.0 to 7.8 cannot sustain human life.

The following Example and Practice Problem illustrate how to prepare a buffer solution with a specified pH.

TABLE 21.1 Resistance of two buffer solutions to change in pH upon addition of acid or base and upon dilution

Buffer	Initial pH	pH after addition of 10.0 mL of 0.100 M $HCl(aq)$ to 100.0 mL of buffer	pH after addition of 10.0 mL of 0.100 M $NaOH(aq)$ to 100.0 mL of buffer	pH after two-fold dilution with water
0.100 M $CH_3COOH(aq)$ and 0.100 M $CH_3COO^-(aq)$	4.74	4.65	4.83	4.74
0.100 M $NH_4^+(aq)$ and 0.100 M $NH_3(aq)$	9.25	9.16	9.34	9.25

EXAMPLE 21-3: To study a certain chemical reaction, the pH of the solution must be fixed and maintained throughout the experiment. In the experiment, the pH must be 4.40. You have available a solution of 0.100 M $CH_3COOH(aq)$ and another solution of 0.100 M $NaCH_3COO(aq)$. Describe how to prepare 100.0 mL of a buffer solution with a pH of 4.40.

Solution: We see from Table 20.6 that $K_a = 1.8 \times 10^{-5}$ M for $CH_3COOH(aq)$. Because K_a is within the range 10^{-3} M to 10^{-11} M and the concentrations of both solutions are 0.100 M, we can use the Henderson-Hasselbalch equation. Substituting the values pH = 4.40 and pK_a = 4.74 into Equation 21.7, we find

$$4.40 \approx 4.74 + \log \frac{[CH_3COO^-]_0}{[CH_3COOH]_0}$$

Taking the antilogarithm of both sides and solving for $[CH_3COO^-]_0/[CH_3COOH]_0$, we find

$$\frac{[CH_3COO^-]_0}{[CH_3COOH]_0} = 10^{(4.40-4.74)} = 10^{-0.34} = 0.457$$

Now let x equal the number of milliliters of the 0.100 M $NaCH_3COO(aq)$ solution and y equal the number of milliliters of the 0.100 M $CH_3COOH(aq)$ solution that we should use. This gives us

$$0.457 = \frac{[NaCH_3COO]_0}{[CH_3COOH]_0} = \frac{(x \text{ mL } NaCH_3COO)(0.100 \text{ M } NaCH_3COO)}{(y \text{ mL } CH_3COOH)(0.100 \text{ M } CH_3COOH)}$$

or simply

$$0.457 = \frac{x}{y}$$

We need another relation between x and y. Our second condition is that we want to prepare 100.0 mL of the buffer using our two 0.100 M solutions. This condition gives us

$$x + y = 100.0 \text{ mL}$$

Substituting this second equation into our first equation, we find that

$$x = 31.4 \text{ mL} \quad \text{and} \quad y = 68.6 \text{ mL}$$

Thus, to prepare the buffer we mix 31.4 mL of the 0.100 M $NaCH_3OO(aq)$ solution with 68.6 mL of the 0.100 M $CH_3COOH(aq)$ solution.

As a check, we can substitute these values back into Equation 21.7 to yield

$$\text{pH} \approx 4.74 + \log \frac{[CH_3COO^-]_0}{[CH_3COOH]_0} \approx 4.74 + \log \frac{(31.4 \text{ ml})(0.100 \text{ M})}{(68.6 \text{ ml})(0.100 \text{ M})} \approx 4.40$$

which is indeed our desired pH.

PRACTICE PROBLEM 21-3: Which of the following conjugate acid-base pairs would be best for preparing a pH = 4.00 buffer? Justify your selection.

(a) $ClCH_2COOH(aq)–ClCH_2COO^-(aq)$

(b) $NH_4^+(aq)–NH_3(aq)$

(c) $C_6H_5COOH(aq)–C_6H_5COO^-(aq)$

Answer: (c) The value of pK_a of C_6H_5COOH (4.74) is closest to the desired pH of the buffer (Table 20.6). For choices (a) and (b), the ratio $[base]_0/[acid]_0$ needed to achieve the desired pH will not be between 0.1 and 10 and so the solution will not be an effective buffer.

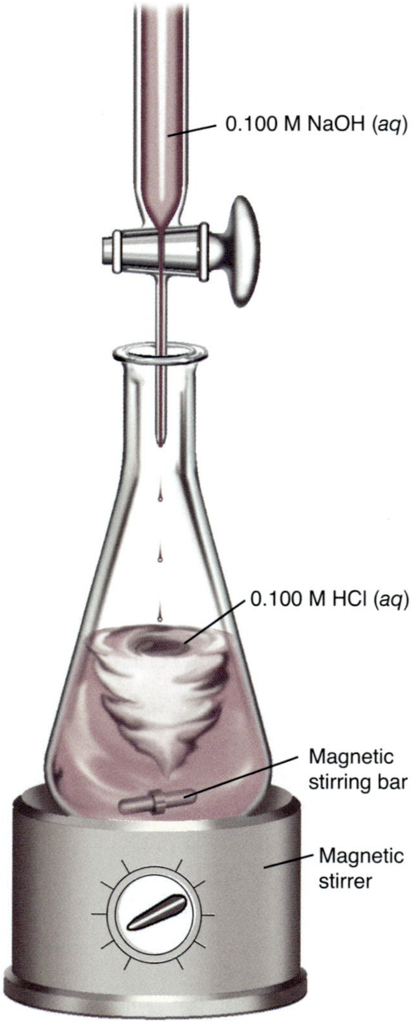

Figure 21.2 A diagrammatic representation of the setup used for titration. The solution is stirred constantly and the NaOH(*aq*) is added drop by drop until the end point is approached. The end point is signaled by a change in color of the indicator (see Figure 21.3).

21-3. An Indicator Is Used to Signal the End Point of a Titration

Chemists often need to determine the amount of a substance present in a solution. The determination of the concentration of a solution of an acid or a base serves as a good example. A commonly used technique is **titration,** the addition of a precisely measured volume of a solution of known concentration to a precisely measured volume of the solution of unknown concentration (Figure 21.2). The solutions are chosen so that they react completely, and the completion of the reaction is signaled by the sudden change in some physical property, such as the color of a compound that has been added in small quantity to the reaction mixture (Figure 21.3).

In the titration of an acid by a base, a plot of the pH of the resulting solution as a function of the volume of the added base is called a **titration curve.** The titration curve of a 50.0-mL sample of 0.100 M HCl(*aq*) titrated with 0.100 M NaOH(*aq*) is shown in Figure 21.4. The net ionic equation for the titration of HCl(*aq*) with NaOH(*aq*) is

$$H_3O^+(aq) + OH^-(aq) \rightarrow 2\,H_2O(l) \tag{21.8}$$

The value of the equilibrium constant for the reaction represented by Equation 21.8 is

$$K = \frac{1}{K_w} = 1.0 \times 10^{14}\ M^{-2}$$

The large value of K indicates that essentially all the $H_3O^+(aq)$ and $OH^-(aq)$ react to form $H_2O(l)$; the reaction goes to completion.

At the **equivalence point,** stoichiometrically equivalent amounts of acid and base have reacted, and the reaction is complete. This occurs when the number of millimoles of $OH^-(aq)$ added from the NaOH(*aq*) is equal to the number of millimoles of $H_3O^+(aq)$ present initially from the HCl(*aq*). Therefore, we have

$$\begin{pmatrix}\text{mmol of } OH^- \\ \text{required}\end{pmatrix} = \begin{pmatrix}\text{mmol of } H_3O^+ \\ \text{present initially}\end{pmatrix} = (50.0\text{ ml})(0.100\text{ M}) = 5.00\text{ mmol}$$

which corresponds to a volume of 0.100 M NaOH(*aq*) solution equal to

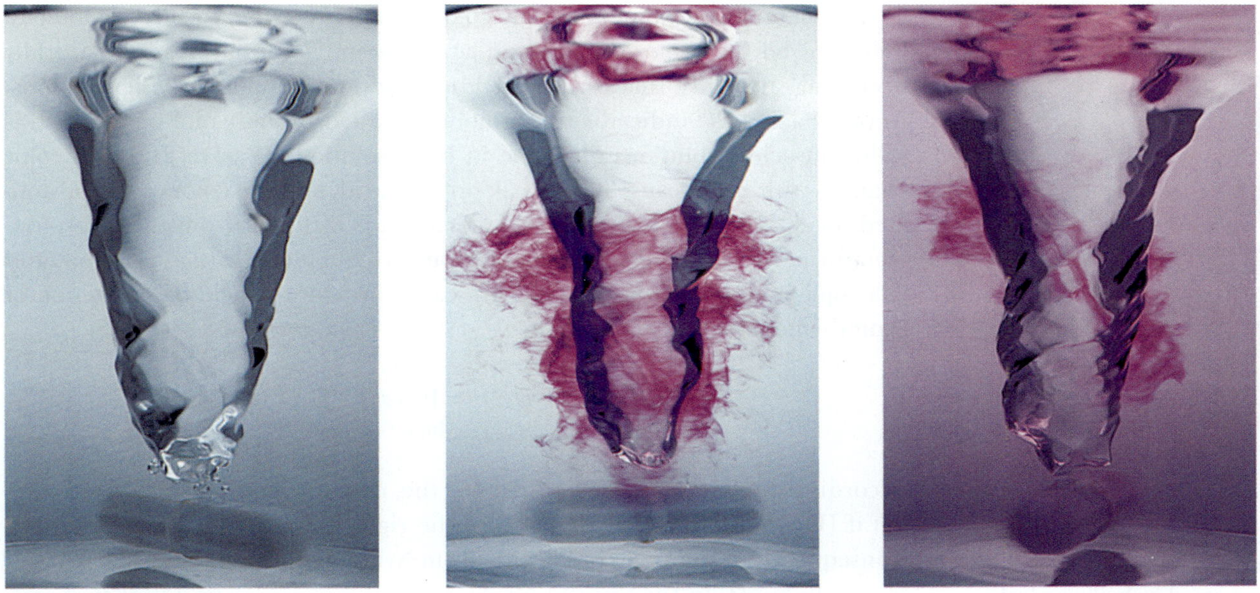

Figure 21.3 Titration of $HCl(aq)$ with $NaOH(aq)$. A 50.0-mL sample of $HCl(aq)$ is placed in the reaction flask together with a magnetic bar for stirring the reaction mixture during the titration. (a) Two or three drops of the acid-base indicator phenolphthalein is added to the $HCl(aq)$, and then (b) $NaOH(aq)$ is added from a buret. As the end point is approached, the base is added dropwise until (c) a single drop turns the entire reaction mixture a pale pink color that does not revert to colorless with stirring.

$$V = \frac{n}{M} = \frac{5.00 \text{ mmol}}{0.100 \text{ M}} = 50.0 \text{ mL}$$

Thus, we see that the equivalence point in Figure 21.4 occurs when 50.0 mL of 0.100 M $NaOH(aq)$ has been added.

At the equivalence point, the solution is simply an aqueous solution of $NaCl(aq)$. Because $NaCl(aq)$ is a neutral salt, the pH is 7.0 at the equivalence point (Figure 21.4). When we titrate an acid of unknown concentration with a base of known concentration, or a base of unknown concentration with an acid of known concentration, we need some way to tell when the equivalence point has been reached. Let's consider the case of titrating $HCl(aq)$ with $KOH(aq)$. Because $KCl(aq)$ is a neutral salt, the pH at the equivalence point is 7.0. Before the equivalence point is reached, not all the $HCl(aq)$ has been neutralized, so the solution is acidic (pH < 7). Beyond the equivalence point, where all the $HCl(aq)$ has been neutralized and excess $KOH(aq)$ has been added, the solution is basic (pH > 7). If there were a substance that was one color (e.g., yellow) at pH < 7 and another color (e.g., blue) at pH > 7, then we could add a few drops of this substance to the $HCl(aq)$ being titrated and it would signal the occurrence of the equivalence point as the point when the solution just changes from yellow to blue. We call such a substance an **indicator,** because it indicates the pH of the solution by its color.

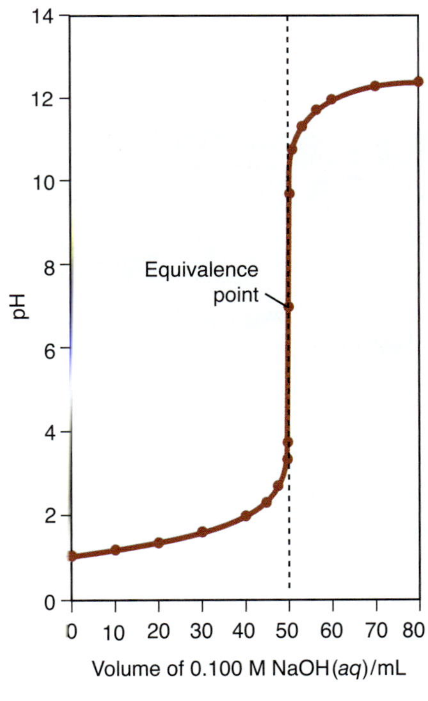

Figure 21.4 Titration curve for the titration of 50.0 mL of 0.100 M $HCl(aq)$ with 0.100 M $NaOH(aq)$. The pH of the solution changes slowly except in the vicinity of the equivalence point where it changes very rapidly.

Figure 21.5 Red cabbage juice is a natural indicator. Cabbage juice extract (prepared by boiling a leaf from a red cabbage in water) turns green in basic solutions, reddish-purple in neutral solutions, and pink in acids.

In fact, many substances change color with pH (Figure 21.5). In Chapter 10 we discussed the use of litmus as an acid-base indicator. Litmus is a naturally occurring substance that is red in acidic solutions and blue in basic solutions. Another acid-base indicator is a compound called bromthymol blue, which is yellow in acidic solutions and blue in basic solutions (Figure 21.6). Its color changes with pH because it is a weak organic acid, and the color of the protonated form, HIn(*aq*) (In stands for indicator), is different from the unprotonated form, $In^-(aq)$. The acid form of bromthymol blue, HIn(*aq*), is yellow, and its conjugate base, $In^-(aq)$, is blue. We can represent the acid-base reaction of bromthymol blue by the equation

$$\underset{\text{yellow}}{\mathrm{HIn}(aq)} + \mathrm{H_2O}(l) \leftrightharpoons \underset{\text{blue}}{\mathrm{In^-}(aq)} + \mathrm{H_3O^+}(aq) \qquad (21.9)$$

According to Le Châtelier's principle, the reaction equilibrium lies to the left if $[H_3O^+]$ is large (low pH) and to the right if $[H_3O^+]$ is small (high pH). Consequently, a solution containing bromthymol blue is yellow at low pH and blue at high pH. In the transition region (about pH = 7 for bromthymol blue), the solution is green, which is a mixture of yellow and blue.

We can make this discussion quantitative by considering the acid-dissociation constant of bromthymol blue. The acid-dissociation constant, K_{ai}, for a general indicator (Equation 21.9) is

$$K_{ai} = \frac{[H_3O^+][In^-]}{[HIn]} \qquad (21.10)$$

Figure 21.6 Bromthymol blue is an acid-base indicator that is yellow if the pH is less than 7 and blue if the pH is greater than 7. In the transition region, around a pH of 7, it appears green (yellow + blue).

As we have noted, a solution containing bromthymol blue is green in the transition region because both the HIn(*aq*) and $In^-(aq)$ forms are present simultaneously in comparable concentrations; that is,

$$[HIn] \approx [In^-]$$

Substitution of the condition $[HIn] \approx [In^-]$ into Equation 21.10 yields

$$[H_3O^+] \approx K_{ai}$$

By taking the negative of the logarithms of the terms in this expression and using the definitions of pH and pK_a, we obtain

$$pH \approx pK_{ai} \qquad (21.11)$$

In other words, if bromthymol blue is green in a solution, then the pH of the solution is approximately equal to the pK_{ai} value for bromthymol blue. The value of pK_{ai} for bromthymol blue is 7.3, so the pH of the green solution is about 7. Generally, the pH range over which an indicator changes color is approximately equal to $pK_{ai} \pm 1$. Even though the color change might occur across two pH units, it occurs suddenly during a titration because most titration curves are very steep around the equivalence point (Figure 21.4). Table 21.2 lists the values of the pK_{ai}'s and the color changes of a number of indicators.

Because of the intense color of indicators, only a very small concentration is necessary to produce a visible color. Thus, its contribution to the total acidity of

TABLE 21.2 The pK_{ai} values and color changes of various indicators

Indicator	pK_{ai}	pH range of color change	Color change
thymol blue	1.65	1.2–2.8	red to yellow
2,4-dinitrophenol	3.96	2.0–4.7	colorless to yellow
methyl orange	3.46	3.2–4.4	red to orange
methyl red	5.00	4.8–6.0	red to yellow
bromcresol purple	6.40	5.2–6.8	yellow to purple
bromthymol blue	7.30	6.0–7.6	yellow to blue
cresol red	8.46	7.0–8.8	yellow to red
thymol blue	9.20	8.0–9.6	yellow to blue
phenolphthalein	9.5	8.2–10.0	colorless to red
thymolphthalein	10.0	9.4–10.6	colorless to blue
alizarin yellow R	11.2	10.1–12.0	yellow to violet

the solution is negligible. Several indicators, together with their colors at various pH values, are shown in Figure 21.7. Note the relation between Table 21.2 and Figure 2.7. Using the indicators in Figure 21.7, we can estimate the pH of an aqueous solution to within about 0.5 pH unit. Of course, the solution must be

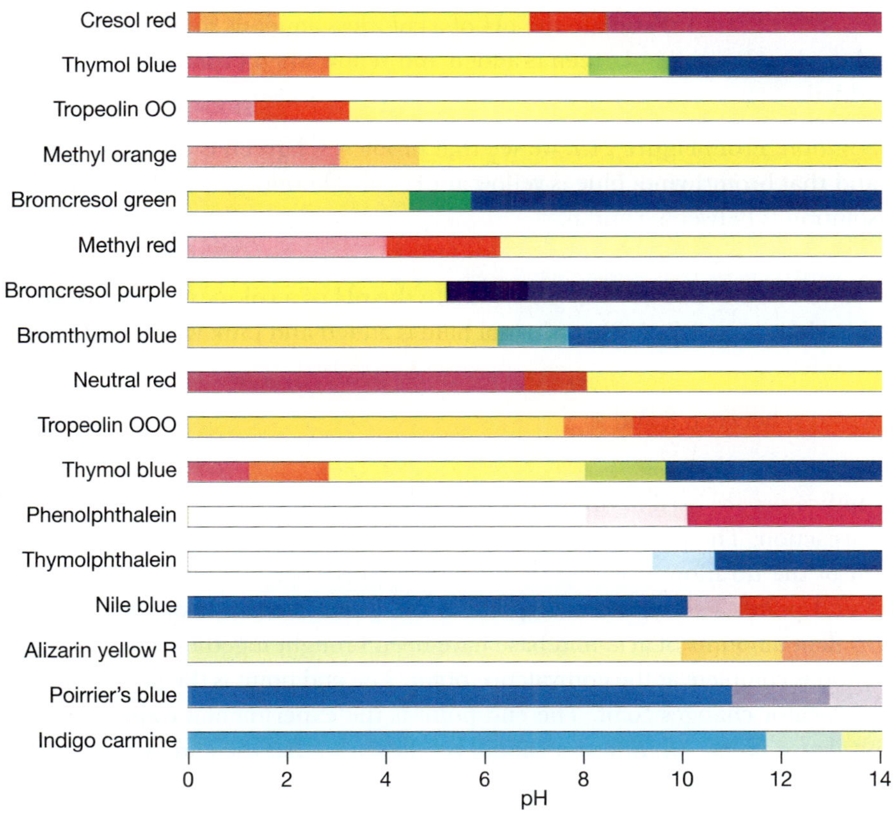

Figure 21.7 The colors of various indicators at different pH values.

Figure 21.8 Aqueous solutions of the acid-base indicator thymol blue at pH values of 1 through 14 (left to right). Observe that the indicator undergoes two color changes.

colorless initially, otherwise the color change of the indicator may be obscured. Some indicators can undergo multiple color changes. Figure 21.8 shows the color range of thymol blue, an indicator that changes color from red to yellow at around pH 1.2–2.8 and again from yellow to blue at around pH 8.0–9.6.

EXAMPLE 21-4: Estimate the pH of a colorless aqueous solution that turns blue when bromcresol green is added and yellow when bromthymol blue is added.

Solution: From Figure 21.7, we see that bromcresol green is blue at $pH > 5$ and that bromthymol blue is yellow at $pH < 6$. Therefore, the pH of the solution is between 5 and 6.

PRACTICE PROBLEM 21-4: Estimate the pH of a colorless aqueous solution that turns yellow when thymol blue is added and pink when methyl red is added.

Answer: between 3 and 4

Indicators are used in titration solutions to signal the completion of the acid-base reaction. The point at which the indicator changes color is called the **end point** of the titration. It is important to distinguish between equivalence point and end point. The equivalence point is the point at which stoichiometrically equivalent amounts of acid and base have been brought together. The acid-base reaction is complete at the equivalence point. The end point is the point at which the indicator changes color. The end point is the experimental estimate of the equivalence point. An indicator should be chosen so that the end point signaled by the indicator corresponds as closely as possible with the equivalence point.

EXAMPLE 21-5: By referring to Figures 21.4 and 21.7, choose an indicator to signal the equivalence point shown in Figure 21.4.

Solution: From Figure 21.4, we note that the equivalence point occurs at pH = 7.0. The titration curve is very steep in the vicinity of the equivalence point, however, so an indicator with a color transition range lying between pH = 5 and pH = 9 would be suitable in this case. Referring to Figure 21.7, we see that there are several possible choices, for example, bromthymol blue or phenolphthalein. We choose our indicator such that the end point and the equivalence point are the same within the required accuracy of the titration.

PRACTICE PROBLEM 21-5: Choose an indicator from Figure 21.7 (other than methyl red) that can be used to signal the equivalence point for the titration shown in Figure 21.11 (page 802).

Answer: bromcresol green or bromcresol purple

21-4. The pH Changes Abruptly at the Equivalence Point of the Titration of a Strong Acid with a Strong Base

Titration curves are obtained experimentally by measuring the pH as a function of the volume of added acid or base. In order to understand the origin of experimental titration curves and to interpret them, in this section we shall calculate some of the points on the titration curve of a strong acid with a strong base (Figure 21.4). In particular, we shall titrate 50.0 mL of 0.100 M HCl(*aq*) with 0.100 M NaOH(*aq*).

The concentration of $H_3O^+(aq)$ in the 0.100 M HCl(*aq*) solution before any base is added is $[H_3O^+] = 0.100$ M, so the pH of the solution initially is 1.00. After 10.0 mL of the 0.100 M NaOH(*aq*) is added, the total volume of the resulting solution is 50.0 mL + 10.0 mL = 60.0 mL. The $OH^-(aq)$ from the NaOH(*aq*) reacts with the $H_3O^+(aq)$ from the HCl(*aq*) to produce water according to:

$$H_3O^+(aq) + OH^-(aq) \rightarrow 2\,H_2O(l)$$

Thus, when 10.0 mL of 0.100 M NaOH(*aq*) is added to 50.0 mL of 0.100 M HCl(*aq*), the number of millimoles of $H_3O^+(aq)$ that reacts is equal to the number of millimoles of $OH^-(aq)$ added:

$$\begin{pmatrix}\text{mmol of OH}^-\\ \text{added}\end{pmatrix} = (10.0\text{ ml})(0.100\text{ M}) = 1.00\text{ mmol} = \begin{pmatrix}\text{mmol of H}_3\text{O}^+\\ \text{reacted}\end{pmatrix}$$

The total number of millimoles of $H_3O^+(aq)$ initially present in 50.0 mL of 0.100 M HCl(*aq*) is

$$\begin{pmatrix}\text{mmol of H}_3\text{O}^+\\ \text{initially present}\end{pmatrix} = (50.0\text{ ml})(0.100\text{ M}) = 5.00\text{ mmol}$$

The concentration of $H_3O^+(aq)$ that remains after the addition of 10.0 mL of 0.100 M NaOH(*aq*) is equal to the number of millimoles of $H_3O^+(aq)$ that remains unreacted divided by the total volume in milliliters of the resulting solution. The number of millimoles of $H_3O^+(aq)$ unreacted is

$$\begin{pmatrix}\text{mmol of } H_3O^+ \\ \text{unreacted}\end{pmatrix} = \begin{pmatrix}\text{mmol of } H_3O^+ \\ \text{initially present}\end{pmatrix} - \begin{pmatrix}\text{mmol of } H_3O^+ \\ \text{reacted}\end{pmatrix}$$

$$= 5.00 \text{ mmol} - 1.00 \text{ mmol} = 4.00 \text{ mmol}$$

and the molarity of unreacted $H_3O^+(aq)$ is

$$[H_3O^+] = \frac{4.00 \text{ mmol}}{60.0 \text{ mL}} = 6.67 \times 10^{-2} \text{ M}$$

Therefore, the pH of the solution after the addition of 10.0 mL of $NaOH(aq)$ is

$$\text{pH} = -\log([H_3O^+]/\text{M}) = -\log(6.67 \times 10^{-2}) = 1.18$$

Proceeding in an analogous fashion, we can calculate the pH of the resulting solution after the incremental additions of 10.0-mL of $NaOH(aq)$ (Table 21.3). Beyond the equivalence point, essentially all the $H_3O^+(aq)$ has reacted, and we simply have a diluted solution of $NaOH(aq)$.

EXAMPLE 21-6: Calculate the pH of a solution obtained by adding 60.0 mL of 0.100 M $NaOH(aq)$ to 50.0 mL of 0.100 M $HCl(aq)$.

Solution: The total number of millimoles of $H_3O^+(aq)$ in 50.0 mL of 0.100 M $HCl(aq)$ is

$$\begin{pmatrix}\text{mmol of } H_3O^+ \\ \text{initially present}\end{pmatrix} = (50.0 \text{ mL})(0.100 \text{ M}) = 5.00 \text{ mmol}$$

The total number of millimoles of $OH^-(aq)$ in 60.0 mL of 0.100 M $NaOH(aq)$ is

$$\begin{pmatrix}\text{mmol of } OH^- \\ \text{added}\end{pmatrix} = (60.0 \text{ mL})(0.100 \text{ M}) = 6.00 \text{ mmol}$$

Note that the number of millimoles of $NaOH(aq)$ added exceeds the number of millimoles of $H_3O^+(aq)$ present initially. The number of millimoles of $OH^-(aq)$ that remain unreacted after the $NaOH(aq)$ addition is

$$\begin{pmatrix}\text{mmol of } OH^- \\ \text{unreacted}\end{pmatrix} = 6.00 \text{ mmol} - 5.00 \text{ mmol} = 1.00 \text{ mmol}$$

The total volume of the solution is 50.0 mL + 60.0 mL = 110.0 mL, so the concentration of unreacted $OH^-(aq)$ is

$$[OH^-] = \frac{1.00 \text{ mmol}}{110.0 \text{ mL}} = 9.09 \times 10^{-3} \text{ M}$$

The pOH of the solution is

TABLE 21.3 Calculation of various points for the titration of 50.0 mL of 0.100 M HCl(*aq*) with 0.100 M NaOH(*aq*)

Up to the Equivalence Point:

Volume of 0.100 M NaOH(*aq*) added/mL	OH^-(*aq*) added/mmol	Unreacted H_3O^+(*aq*)/mmol	Total volume of solution/mL	Concentration of unreacted H_3O^+(*aq*)/M	pH
10.0	1.00	4.00	60.0	6.67×10^{-2}	1.18
20.0	2.00	3.00	70.0	4.29×10^{-2}	1.37
30.0	3.00	2.00	80.0	2.50×10^{-2}	1.60
40.0	4.00	1.00	90.0	1.11×10^{-2}	1.96
45.0	4.50	0.50	95.0	5.26×10^{-3}	2.28
49.0	4.90	0.10	99.0	1.01×10^{-3}	3.00
49.5	4.95	0.05	99.5	5.03×10^{-4}	3.30
50.0 (equivalence point)	5.00	1.00×10^{-5} (from H_2O dissociation)	100.0	1.00×10^{-7}	7.00

Beyond the Equivalence Point:

Volume of 0.100 M NaOH(*aq*) added/mL	OH^-(*aq*) added/mmol	Unreacted OH^-(*aq*)/mmol	Total volume of solution/mL	Concentration of unreacted OH^-(*aq*)/M	pH
50.5	5.05	0.05	100.5	4.98×10^{-4}	10.70
51.0	5.10	0.10	101.0	9.90×10^{-4}	11.00
55.0	5.50	0.50	105.0	4.76×10^{-3}	11.68
60.0	6.00	1.00	110.0	9.09×10^{-3}	11.96
70.0	7.00	2.00	120.0	1.67×10^{-2}	12.22
80.0	8.00	3.00	130.0	2.31×10^{-2}	12.36
90.0	9.00	4.00	140.0	2.86×10^{-2}	12.46

$$\text{pOH} = -\log([\text{OH}^-]/\text{M}) = -\log(9.09 \times 10^{-3}) = 2.04$$

so the pH of the solution is

$$\text{pH} = 14.00 - \text{pOH} = 14.00 - 2.04 = 11.96$$

which is the result shown in Figure 21.4 and Table 21.3 for 60.0 mL of added base. You should reproduce some of the other values given in Table 21.3.

PRACTICE PROBLEM 21-6: Construct a table similar to Table 21.3 for the titration of 30.0 mL of 0.200 M $Ba(OH)_2(aq)$ with 0.300 M $HNO_3(aq)$.

Answer:

Volume HNO_3/mL	0	10.0	20.0	30.0	39.5	40.0	40.5	50.0	60.0
pH	13.60	13.35	13.08	12.70	11.33	7.00	2.67	1.43	1.18

The equivalence point is at pH 7.00.

21-5. Weak Acids Can Be Titrated with Strong Bases

The titration curve of a weak acid with a strong base looks somewhat different from the titration curve of a strong acid with a strong base shown in Figure 21.4. The titration curve for 50.0 mL of 0.100-M acetic acid, $CH_3COOH(aq)$, titrated with 0.100 M $NaOH(aq)$, is shown in Figure 21.9. We see that the pH of the acetic acid solution is about 2.9 initially and increases slowly until the equivalence point is reached. Near the equivalence point, the pH changes from 6 to 11. Note also that the equivalence point occurs at pH 9, *not* at pH 7.

We can calculate the pH of the 0.100 M $CH_3COOH(aq)$ solution before any $NaOH(aq)$ is added by the method developed in Section 20-7. The equilibrium equation is

$$CH_3COOH(aq) + H_2O(l) \leftrightharpoons H_3O^+(aq) + CH_3COO^-(aq) \qquad (21.12)$$

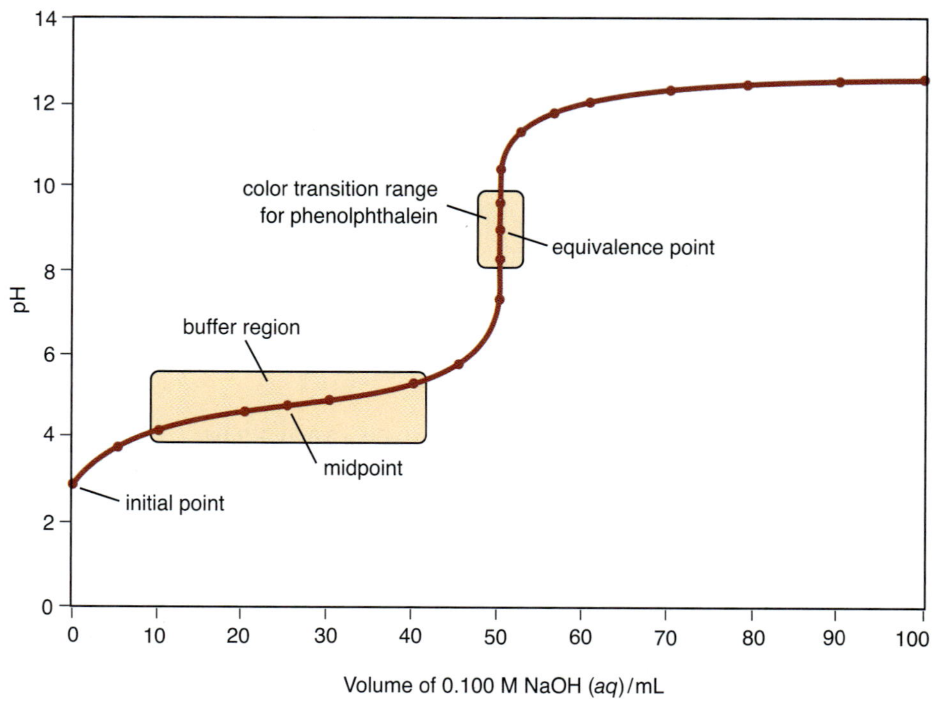

Figure 21.9 Titration curve for the titration of 50.0 mL of 0.100 M $CH_3COOH(aq)$ with 0.100 M $NaOH(aq)$. The indicator used in this case is phenolphthalein, which changes from colorless to pink around pH = 9. The relatively flat portion of the curve between 10 mL and 40 mL of added base illustrates graphically the $CH_3COOH(aq)$–$CH_3COO^-(aq)$ buffer action and is called the buffer region.

with

$$K_a = \frac{[H_3O^+][CH_3COO^-]}{[CH_3COOH]} = 1.8 \times 10^{-5}\ M \tag{21.13}$$

As usual, we set up a concentration table:

Concentration	$CH_3COOH(aq)$	+ $H_2O(l)$	$\leftrightharpoons$ $H_3O^+(aq)$	+ $CH_3COO^-(aq)$
initial	0.100 M	—	≈ 0 M	0 M
change	$-x$		$+x$	$+x$
equilibrium	$0.100\ M - x$	—	x	x

where we have neglected the contribution to x from the autoprotonation of water because $K_a >> K_w$. Thus, Equation 21.13 becomes

$$\frac{x^2}{0.100\ M - x} = 1.8 \times 10^{-5}\ M \tag{21.14}$$

Solving Equation 21.14 by the method of successive approximations or by using the quadratic formula yields $x = 1.3 \times 10^{-3}$ M. Because $[H_3O^+] = x$, we have

$$pH = -\log([H_3O^+]/M) = -\log(1.3 \times 10^{-3}) = 2.89$$

in agreement with Figure 21.9.

Now let's calculate the pH at the equivalence point. To do so, we must consider the principal species in the solution at the equivalence point, where the number of moles of $NaOH(aq)$ added is equal to the number of moles of $CH_3COOH(aq)$ initially present. The equation for the reaction is

$$CH_3COOH(aq) + NaOH(aq) \rightarrow NaCH_3COO(aq) + H_2O(l) \tag{21.15}$$

It is important to realize that the reaction described by Equation 21.15 goes essentially to completion even though $CH_3COOH(aq)$ is a weak acid. We can see this quantitatively by calculating the equilibrium constant of Equation 21.15. The net ionic equation for the titration of $CH_3COOH(aq)$ with $NaOH(aq)$ is

$$CH_3COOH(aq) + OH^-(aq) \leftrightharpoons CH_3COO^-(aq) + H_2O(l) \tag{21.16}$$

Equation 21.16 is simply the reverse of the base protonation reaction equation for $CH_3COO^-(aq)$, so $K_c = 1/K_b = 1.8 \times 10^9\ M^{-1}$ (see Equation 21.4). This large value of K_c indicates that the equilibrium described by Equation 21.16 lies far to the right and that $CH_3COOH(aq)$ reacts essentially completely when a strong base is added to the solution.

Because the reaction described by Equation 21.16 goes essentially to completion, the solution at the equivalence point of the titration is essentially just a $NaCH_3COO(aq)$ solution. To calculate the concentration of this $NaCH_3COO(aq)$ solution, we use the fact that we started with 50.0 mL of 0.100 M $CH_3COOH(aq)$

and that $[NaCH_3COO] = [CH_3COO^-]$ from Equation 21.15. The initial number of millimoles of $CH_3COOH(aq)$ is

$$\begin{pmatrix}\text{mmol of } CH_3COOH\\ \text{initially present}\end{pmatrix} = (50.0\text{ mL})(0.100\text{ M}) = 5.00\text{ mmol}$$

According to Equation 21.16

$$\begin{pmatrix}\text{mmol of } CH_3COO^-\\ \text{at the equivalence point}\end{pmatrix} = \begin{pmatrix}\text{mmol of } CH_3COOH\\ \text{initially present}\end{pmatrix} = 5.00\text{ mmol}$$

The total volume of the solution at the equivalence point is 100.0 mL, that is, 50.0 mL of $CH_3COOH(aq)$ plus 50.0 mL of added $NaOH(aq)$, so the stoichiometric concentration of $CH_3COO^-(aq)$ at the equivalence point is

$$[CH_3COO^-]_0 = \frac{5.00\text{ mmol}}{100.0\text{ mL}} = 0.0500\text{ M}$$

We emphasize here that 0.0500 M is the *stoichiometric* concentration of $CH_3COO^-(aq)$. The actual concentration of $CH_3COO^-(aq)$ will be slightly less than 0.0500 M because of the reaction described by

$$CH_3COO^-(aq) + H_2O(l) \leftrightharpoons OH^-(aq) + CH_3COOH(aq) \qquad K_b = 5.6 \times 10^{-10}\text{ M} \tag{21.17}$$

We calculated the pH of a solution of $NaCH_3COO(aq)$ in Section 20-11. In fact, Example 20-16 involves the calculation of the pH of a 0.050 M $NaCH_3COO(aq)$ solution, where we found a value of pH = 8.72. Thus, we see that the pH at the equivalence point in Figure 21.9 is 8.72.

EXAMPLE 21-7: Compare the use of bromthymol blue and phenolphthalein as indicators for the titration of 50.0 mL of 0.100 M $CH_3COOH(aq)$ with 0.100 M $NaOH(aq)$. The pH at the equivalence point is 8.72.

Solution: Referring to Figure 21.7, we see that phenolphthalein is a suitable indicator for the titration. Bromthymol blue is not because it changes color from yellow to blue at too low a pH.

PRACTICE PROBLEM 21-7: Using the data from Appendix E, calculate the pH at the equivalence point when 250.0 mL of 0.500-M hypobromous acid, $HOBr(aq)$, is titrated with 0.275 M $Sr(OH)_2(aq)$.

Answer: pH = 10.99

21-6. pH = pK_a at the Midpoint in the Titration of a Weak Acid with a Strong Base

We have calculated the pH at two points (the initial point and the equivalence point) on the titration curve of a weak acid with a strong base. We can use the Henderson-Hasselbalch equation to calculate the pH at the midpoint and at points around the midpoint of the titration. The **midpoint** is the point at which one-half of the acid has reacted with the strong base. Let's now calculate the pH at the midpoint, which for the titration of 50.0 mL of 0.100 M $CH_3COOH(aq)$ with 0.100 M $NaOH(aq)$ occurs at the point indicated on the titration curve in Figure 21.9. We can calculate the stoichiometric concentrations of $CH_3COOH(aq)$ and $CH_3COO^-(aq)$ at the midpoint. The total number of millimoles of $CH_3COOH(aq)$ present initially is

$$\begin{pmatrix}\text{mmol of } CH_3COOH \\ \text{initially present}\end{pmatrix} = (50.0\text{ mL})(0.100\text{ M}) = 5.00\text{ mmol}$$

The number of millimoles of $NaOH(aq)$ added is one half of 5.00 mmol, or 2.50 mmol, which corresponds to a volume of 0.100 M $NaOH(aq)$ solution equal to

$$V = \frac{n}{M} = \frac{2.50\text{ mmol}}{0.100\text{ M}} = 25.0\text{ mL}$$

Because the neutralization reaction (Equation 21.16) goes essentially to completion, all the $OH^-(aq)$ reacts with $CH_3COOH(aq)$. Furthermore, the number of millimoles of $CH_3COO^-(aq)$ produced is equal to the number of millimoles of $OH^-(aq)$ added, so at the midpoint we have 2.50 millimoles of $CH_3COO^-(aq)$. The number of millimoles of $CH_3COOH(aq)$ unreacted is

$$\begin{pmatrix}\text{mmol of } CH_3COOH \\ \text{unreacted}\end{pmatrix} = \begin{pmatrix}\text{mmol of } CH_3COOH \\ \text{initially present}\end{pmatrix} - \begin{pmatrix}\text{mmol of } CH_3COO^- \\ \text{produced}\end{pmatrix}$$

$$= 5.00\text{ mmol} - 2.50\text{ mmol} = 2.50\text{ mmol}$$

Thus, we find that the corresponding stoichiometric concentrations of both $CH_3COOH(aq)$ and $CH_3COO^-(aq)$ are

$$[CH_3COOH]_0 = [CH_3COO^-]_0 = \frac{2.50\text{ mmol}}{75.0\text{ mL}} = 0.0333\text{ M} \qquad (21.18)$$

at the midpoint.

The actual concentrations of $CH_3COOH(aq)$ and $CH_3COO^-(aq)$ will differ slightly from their stoichiometric concentrations because of the reactions described by Equations 21.12 and 21.17. Because the concentrations of both $CH_3COOH(aq)$ and $CH_3COO^-(aq)$ are large compared with those of the other species in solution, they are called the **principal species.** Notice that for the principal species at the midpoint of a titration, we have $[\text{base}]_0 = [\text{acid}]_0$, and so the solution is a buffer solution. Thus, we can also use a titration to prepare a buffer solution. Because K_a for $CH_3COOH(aq)$ is between 10^{-3} M and 10^{-11} M and because the concentrations of $[\text{base}]_0$ and $[\text{acid}]_0$ are both 0.0333 M, we can apply the Henderson-Hasselbalch equation to determine the pH. At the midpoint of a titration, therefore, we see that

$$\text{pH} \approx \text{p}K_a + \log\frac{[\text{base}]_0}{[\text{acid}]_0} = \text{p}K_a + \log\frac{0.0333\text{ M}}{0.0333\text{ M}} = \text{p}K_a$$

and so pH = pK_a = 4.74.

Note that pH at the midpoint of the titration of a weak acid by a strong base is equal to the value pK_a of the acid. Although we have shown this only for the titration of acetic acid with sodium hydroxide, this is a general result if $[H_3O^+]$ is negligible compared with $[\text{acid}]_0$ and $[\text{base}]_0$, or equivalently, when K_a and K_b are much smaller than $[\text{acid}]_0$ and $[\text{base}]_0$. In such cases (which are very common), we can write

$$\text{pH} = \text{p}K_a \qquad \text{(midpoint)} \tag{21.19}$$

We can also apply the Henderson-Hasselbalch equation to points around the midpoint of a titration curve where, as previously discussed, the ratio $[\text{base}]_0/[\text{acid}]_0$ is between roughly 0.1 and 10, K_a is between 10^{-3} M and 10^{-11} M, and the concentrations of the conjugate acid-base pair are between about 10^{-3} M and 1 M. The points along a titration curve described by this condition are collectively called the **buffer region**. In Figure 21.9, the buffer region lies between 10 mL and 40 mL of NaOH(*aq*) added. Example 21-8 illustrates the use of the Henderson-Hasselbalch equation to determine the pH within the buffer region of the CH_3COOH(*aq*)–NaOH(*aq*) titration curve.

EXAMPLE 21-8: Calculate the pH of the solution that results when 20.0 mL of 0.100 M NaOH(*aq*) is added to 50.0 mL of 0.100 M CH_3COOH(*aq*).

Solution: The number of millimoles of NaOH(*aq*) added is

$$\begin{pmatrix}\text{mmol of NaOH}\\ \text{added}\end{pmatrix} = (20.0\text{ mL})(0.100\text{ M}) = 2.00\text{ mmol}$$

The NaOH(*aq*) reacts with CH_3COOH(*aq*) initially present according to the net ionic equation,

$$\text{OH}^-(aq) + \text{CH}_3\text{COOH}(aq) \rightarrow \text{H}_2\text{O}(l) + \text{CH}_3\text{COO}^-(aq)$$

The number of millimoles of CH_3COOH(*aq*) in the solution before the addition of NaOH(*aq*) is

$$\begin{pmatrix}\text{mmol of CH}_3\text{COOH}\\ \text{before NaOH added}\end{pmatrix} = (50.0\text{ mL})(0.100\text{ M}) = 5.00\text{ mmol}$$

According to the chemical equation, the number of millimoles of CH_3COOH(*aq*) after the addition of 20.0 mL of 0.100 M NaOH(*aq*) is

$$\begin{pmatrix}\text{mmol of CH}_3\text{COOH}\\ \text{after NaOH added}\end{pmatrix} = \begin{pmatrix}\text{mmol of CH}_3\text{COOH}\\ \text{before NaOH added}\end{pmatrix} - \begin{pmatrix}\text{mmol of NaOH}\\ \text{added}\end{pmatrix}$$
$$= 5.00\text{ mmol} - 2.00\text{ mmol} = 3.00\text{ mmol}$$

Additionally, from the net ionic equation we see that all the NaOH(*aq*) added goes to produce CH_3COO^-(*aq*), so that

$$\begin{pmatrix}\text{mmol of CH}_3\text{COO}^-\\ \text{produced}\end{pmatrix} = \begin{pmatrix}\text{mmol of NaOH}\\ \text{added}\end{pmatrix} = 2.00\text{ mmol}$$

Because K_a is between 10^{-3} M and 10^{-11} M, the ratio $[\text{base}]_0/[\text{acid}]_0$ is 2.00/3.00, and the molarity of the acid-base pair is between 10^{-3} M and 1 M, we can use the Henderson-Hasselbalch equation. The pH is

$$\text{pH} \approx \text{p}K_a + \log\frac{[\text{base}]_0}{[\text{acid}]_0} = 4.74 + \log\frac{2.00 \text{ mmol}}{3.00 \text{ mmol}} = 4.56$$

in agreement with the point at 20.0 mL NaOH(*aq*) in Figure 21.9.

PRACTICE PROBLEM 21-8: We just calculated the pH a little below the midpoint in Example 21-7. Now calculate the pH on the other side of the midpoint when 30.0 mL of 0.100 M NaOH(*aq*) is added to 50.0 mL of CH_3COOH(*aq*).

Answer: pH = 4.92

From Figure 21.9 we see that the buffer region extends from a little beyond the initial point to a little before the equivalence point of the titration curve. In most cases, we can use the Henderson-Hasselbalch equation to determine the pH of a solution being titrated at points within the buffer region. We emphasize here that the Henderson-Hasselbalch equation does *not* apply to points outside the buffer region, specifically, at points close to and including the initial point and the equivalence point of a titration where the ratio of $[\text{base}]_0/[\text{acid}]_0$ is outside the range of 0.1 to 10.

The calculation of the entire titration curve of a weak acid with a strong base is fairly lengthy and for the most part is left to the problems. The calculations of the pH at the initial point, the midpoint (pH = pK_a), and the equivalence point are almost sufficient to sketch the titration curve in Figure 21.9. All we really need is the pH at a point a little beyond the equivalence point. The following Example illustrates such a calculation.

EXAMPLE 21-9: Calculate the pH of the solution that results when 60.0 mL of 0.100 M NaOH(*aq*) is added to 50.0 mL of 0.100 M CH_3COOH(*aq*).

Solution: The number of millimoles of NaOH(*aq*) added is

$$\left(\begin{array}{c}\text{mmol of NaOH}\\ \text{added}\end{array}\right) = (60.0 \text{ mL})(0.100 \text{ M}) = 6.00 \text{ mmol}$$

There are only 5.00 mmol of CH_3COOH(*aq*) in the initial solution, so we have added an excess of NaOH(*aq*). The number of millimoles of unreacted NaOH(*aq*) is

$$\left(\begin{array}{c}\text{mmol of NaOH}\\ \text{unreacted}\end{array}\right) = \left(\begin{array}{c}\text{mmol of NaOH}\\ \text{added}\end{array}\right) - \left(\begin{array}{c}\text{mmol of CH}_3\text{COOH}\\ \text{available}\end{array}\right)$$

$$= 6.00 \text{ mmol} - 5.00 \text{ mmol} = 1.00 \text{ mmol}$$

The stoichiometric unreacted (excess) concentration of $OH^-(aq)$ in the final solution is

$$[OH^-]_0 = \frac{1.00 \text{ mmol}}{110.0 \text{ mL}} = 9.09 \times 10^{-3} \text{ M}$$

The stoichiometric concentration of $CH_3COO^-(aq)$ in the final solution is

$$[CH_3COO^-]_0 = \frac{5.00 \text{ mmol}}{110.0 \text{ mL}} = 4.55 \times 10^{-2} \text{ M}$$

The titration solution beyond the equivalence point consists of a mixture of $NaCH_3COO(aq)$ and $NaOH(aq)$. The principal species are $CH_3COO^-(aq)$ and $OH^-(aq)$, so we shall use a chemical equation and its corresponding equilibrium-constant expression that involves both of these species. A chemical equation that contains both species is Equation 21.17,

$$CH_3COO^-(aq) + H_2O(l) \leftrightharpoons OH^-(aq) + CH_3COOH(aq)$$

for which the corresponding value of K_b is 5.6×10^{-10} M. We set up a concentration table, letting x denote the concentration of $OH^-(aq)$ formed in the reaction described by Equation 21.17; in other words, x is the concentration of $[OH^-]$ formed *in addition* to the stoichiometric excess of added $NaOH(aq)$.

Concentration	$CH_3COO^-(aq)$ +	$H_2O(l)$ $\leftrightharpoons$	$CH_3COOH(aq)$ +	$OH^-(aq)$
initial	0.0455 M	—	0 M	0.00909 M
change	$-x$		$+x$	$+x$
equilibrium	$0.0455 \text{ M} - x$	—	x	$0.00909 \text{ M} + x$

Substituting these values into the equilibrium-constant expression for Equation 21.17 gives

$$\frac{x(0.00909 \text{ M} + x)}{0.0455 \text{ M} - x} = 5.6 \times 10^{-10} \text{ M}$$

If we assume that x is small compared with both 0.00909 M and 0.0455 M, then we obtain

$$x \approx \frac{(0.0455 \text{ M})(5.6 \times 10^{-10} \text{ M})}{0.00909 \text{ M}} = 2.8 \times 10^{-9} \text{ M}$$

which indeed *is* small compared with both 0.00909 M and 0.0455 M. Thus, $[OH^-] = 0.00909 \text{ M} + 2.8 \times 10^{-9} \text{ M} = 0.00909 \text{ M}$.

We can obtain this result more easily by the following argument. Because K_b is very small, the equilibrium described by Equation 21.17 lies far to the left. The equilibrium is driven even further to the left (Le Châtelier's principle) by the presence of excess $OH^-(aq)$ in the solution at a stoichiometric concentration of $[OH^-]_0 = 9.09 \times 10^{-3}$ M. Thus, the reaction described by Equation 21.17 makes a negligible contribution to the total concentration of $OH^-(aq)$, and we have

$$[OH^-] \approx [OH^-]_0 = 9.09 \times 10^{-3} \text{ M} \tag{21.20}$$

The pOH of the solution is

$$\text{pOH} = -\log([OH^-]/\text{M}) = -\log(9.09 \times 10^{-3}) = 2.04$$

and the pH of the solution is

$$\text{pH} = 14.00 - \text{pOH} = 14.00 - 2.04 = 11.96$$

in agreement with the point at 60.0 mL NaOH(*aq*) in Figure 21.9.

PRACTICE PROBLEM 21-9: Calculate the pH of the solution that results when 55.0 mL of 0.100 M NaOH(*aq*) is added to 25.0 mL of 0.200 M $CH_3COOH(aq)$.

Answer: pH = 11.80

As a final note, Figure 21.10 shows the concentrations of undissociated acetic acid and acetate ion plotted against the volume of added base. Note that $[CH_3COOH]$ decreases steadily, becoming essentially 0 at and beyond the equivalence point. Also, $[CH_3COO^-]$ increases steadily up to the equivalence point and then decreases steadily beyond it. No more $CH_3COO^-(aq)$ is produced once the equivalence point is reached, and it becomes diluted as base is added.

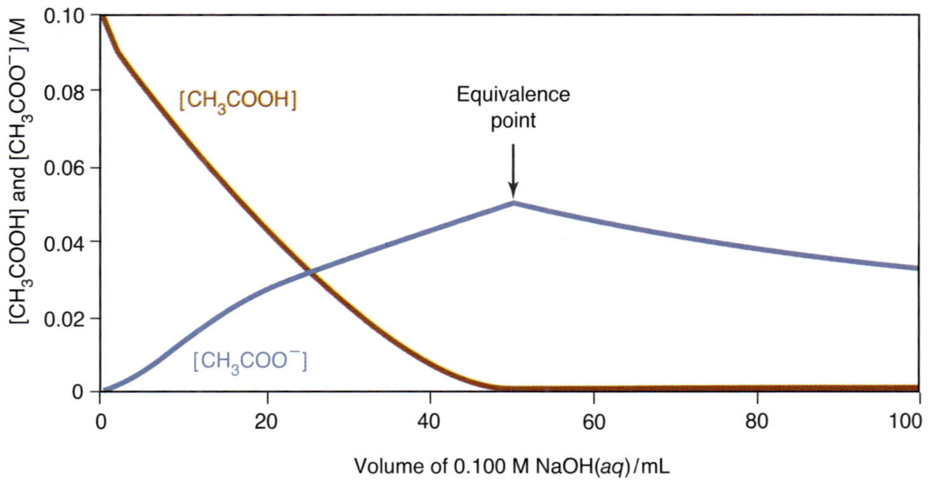

Figure 21.10 The concentration of undissociated acetic acid, $[CH_3COOH]$, and acetate ion, $[CH_3COO^-]$, versus the volume of 0.100 M NaOH(*aq*) used in the titration of 50.0 mL of 0.100 M $CH_3COOH(aq)$.

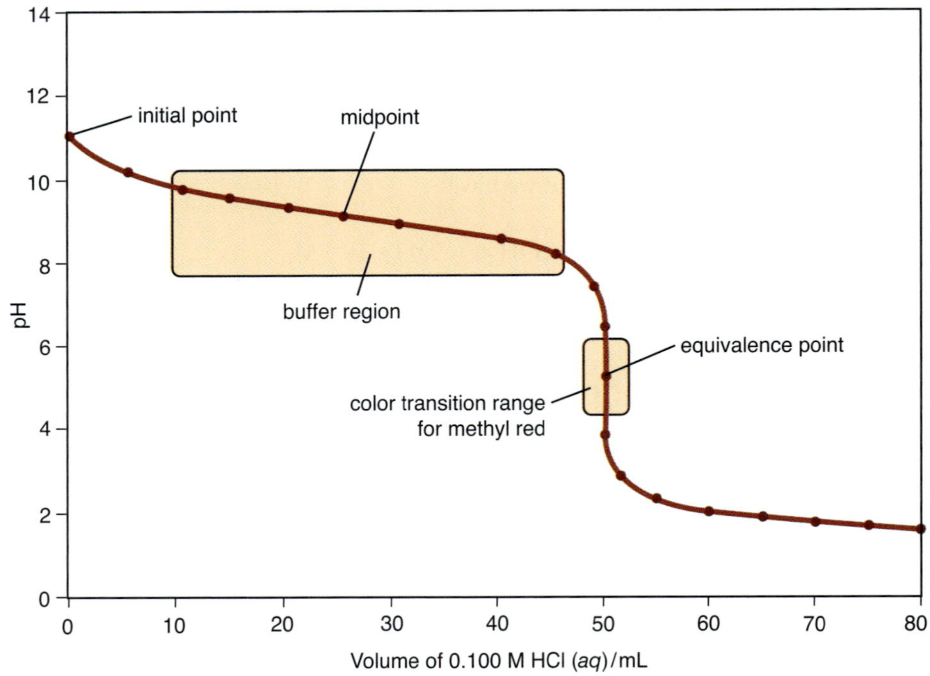

Figure 21.11 Titration curve for the titration of 50.0 mL of 0.100 M $NH_3(aq)$ with 0.100 M $HCl(aq)$. The pH at the equivalence point is acidic because at that point we have a 0.050-M solution of $NH_4Cl(aq)$, and $NH_4^+(aq)$ is an acidic ion while $Cl^-(aq)$ is a neutral ion.

21-7. Weak Bases Can Be Titrated with Strong Acids

We have considered the titration of a strong acid with a strong base and a weak acid with a strong base. Let's now discuss the titration of a weak base with a strong acid. The titration curve for 50.0 mL of 0.100 M $NH_3(aq)$ with 0.100 M $HCl(aq)$ is shown in Figure 21.11. The net ionic equation for the reaction is

$$NH_3(aq) + H_3O^+(aq) \rightarrow NH_4^+(aq) + H_2O(l) \qquad (21.21)$$

$Cl^-(aq)$ is a spectator ion and so does not appear in the net ionic equation. Like the reaction of a weak acid with a strong base, this reaction goes essentially to completion.

EXAMPLE 21-10: Calculate the value of K_c for Equation 21.21.

Solution: The reaction represented by Equation 21.21 is simply the reverse of the acid-dissociation reaction of $NH_4^+(aq)$ described by

$$NH_4^+(aq) + H_2O(l) \leftrightharpoons NH_3(aq) + H_3O^+(aq) \qquad K_a = 5.6 \times 10^{-10}\ M$$

Thus, the equilibrium constant for Equation 21.21 is

$$K_c = \frac{1}{K_a} = \frac{1}{5.6 \times 10^{-10}\ M} = 1.8 \times 10^9\ M^{-1}$$

This large value of K_c indicates that the reaction goes essentially to completion.

PRACTICE PROBLEM 21-10: Calculate the value of the equilibrium constant for the reaction that occurs in the titration of pyridine, $C_5H_5N(aq)$, with a strong acid.

Answer: $1.7 \times 10^5\ M^{-1}$

Figure 21.12 Aqueous solutions containing the acid-base indicator methyl red at pH values of 2 (red), 5 (orange), and 10 (yellow).

Note the general features of the titration curve in Figure 21.11: The initial pH is around 11, and it decreases slowly until near the equivalence point, where the pH is around 5. After the equivalence point, the pH levels off at values around 2 or less. A suitable indicator for the titration is methyl red, which has a color transition range centered around pH 5 (Figure 21.12).

As in the case of the titration of a weak acid with a strong base, we can sketch the main features of the titration curve by calculating the initial pH, the pH at the midpoint ($pH = pK_a$), the pH at the equivalence point, and a point after the equivalence point. The following Example and Practice Problem illustrate the calculations involved in the titration of a weak base with a strong acid.

EXAMPLE 21-11: Calculate the pH for each of the following cases in the titration of 50.0 mL of 0.100 M $NH_3(aq)$ solution with 0.100 M $HCl(aq)$:

(a) before addition of any $HCl(aq)$ (the initial point)
(b) after addition of 20.0 mL of $HCl(aq)$ (5.0 mL before the midpoint)
(c) after addition of 25.0 mL of $HCl(aq)$ (the midpoint)
(d) after addition of 30.0 mL of $HCl(aq)$ (5.0 mL after the midpoint)
(e) after addition of 50.0 mL of $HCl(aq)$ (the equivalence point)
(f) after addition of 55.0 mL of $HCl(aq)$ (5.0 mL after the equivalence point)

Solution:
(a) Before the addition of any $HCl(aq)$ we have a solution containing 50.0 mL of 0.100 M $NH_3(aq)$ which reacts with water according to,

$$NH_3(aq) + H_2O(l) \leftrightharpoons NH_4^+(aq) + OH^-(aq) \qquad (21.22)$$

We begin by making a concentration table,

Concentration	$NH_3(aq)$	+ $H_2O(l)$ ⇋	$NH_4^+(aq)$	+ $OH^-(aq)$
initial	0.100 M	—	0 M	≈ 0 M
change	$-x$		$+x$	$+x$
equilibrium	$0.100\ M - x$	—	x	x

The equilibrium-constant expression for Equation 21.22 (Table 20.6) is

$$K_b = \frac{x^2}{0.100\ M - x} = 1.8 \times 10^{-5}\ M$$

The method of successive approximations gives $x = [OH^-] = 1.3 \times 10^{-3}$ M, so the pOH is 2.88 and the pH = 11.12, in agreement with Figure 21.11.

(b) 20.0 mL of 0.100 M $HCl(aq)$ is equivalent to adding 2.00 mmol of $HCl(aq)$. The original solution contained (50.0 mL)(0.100 M) = 5.00 mmol $NH_3(aq)$. Because K_c for the reaction described in Equation 21.21 is large (see Example 21-10), essentially all the $HCl(aq)$ added goes to form $NH_4^+(aq)$ according to $NH_3(aq) + HCl(aq) \rightarrow NH_4^+(aq) + Cl^-(aq)$. Thus, after the addition of the $HCl(aq)$, we have 2.00 mmol of $NH_4^+(aq)$ and (5.00 mmol − 2.00 mmol) = 3.00 mmol $NH_3(aq)$ as the principal species in solution. The resulting solution is a buffer. Because K_a is between 10^{-4} M and 10^{-11} M, the ratio $[\text{base}]_0/[\text{acid}]_0 = 3.00/2.00$, and both $[\text{base}]_0$ and $[\text{acid}]_0$ are within the range 10^{-3} M to 1 M, we can apply the Henderson-Hasselbalch equation. The pH is given by

$$\text{pH} \approx \text{p}K_a + \log\frac{[\text{base}]_0}{[\text{acid}]_0} = 9.25 + \log\frac{[NH_3]_0}{[NH_4^+]_0} = 9.25 + \log\frac{3.00\ \text{mmol}}{2.00\ \text{mmol}} = 9.43$$

in agreement with the point at 20.0 mL $HCl(aq)$ in Figure 21.11. Notice that regardless of whether we are titrating a weak acid or a weak base, we use pK_a, *not* pK_b, in the Henderson-Hasselbalch equation.

(c) At this point we have added (25.0 mL)(0.100 M) = 2.50 mmol $HCl(aq)$. Because the original amount of $NH_3(aq)$ was 5.00 mmol [see part (b) above], we are at the midpoint of the titration and $[NH_3]_0 = [NH_4^+]_0$. Because both K_a and K_b are small, we have

$$\text{pH} \approx \text{p}K_a = 9.25$$

in agreement with the midpoint plotted in Figure 21.11.

(d) At this point we have added (30.0 mL)(0.100 M) = 3.00 mmol of $HCl(aq)$, and so we are slightly past the midpoint along the titration curve. Because the original amount of $NH_3(aq)$ was 5.00 mmol [see part (b) above], we have 5.00 mmol − 3.00 mmol = 2.00 mmol of $NH_3(aq)$ and 3.00 mmol of $NH_4^+(aq)$. Once again, we apply the Henderson-Hasselbalch equation to obtain

$$\text{pH} \approx \text{p}K_a + \log\frac{[\text{base}]_0}{[\text{acid}]_0} = 9.25 + \log\frac{[NH_3]_0}{[NH_4^+]_0} = 9.25 + \log\frac{2.00\ \text{mmol}}{3.00\ \text{mmol}} = 9.07$$

in agreement with the data plotted in Figure 21.11.

(e) Here we have added (50.0 mL)(0.100 M) = 5.00 mmol $HCl(aq)$. Since this is equal to the number of millimoles of $NH_3(aq)$ originally present, we are at the equivalence point of the titration and all the $NH_3(aq)$ is converted to $NH_4^+(aq)$ according to Equation 21.21. Thus, we now have a problem involving the solution of the weak acid $NH_4^+(aq)$, which dissociates according to the equation

$$NH_4^+(aq) + H_2O(l) \leftrightharpoons NH_3(aq) + H_3O^+(aq)$$

Because the total volume of solution is (50.0 mL + 50.0 mL) = 100.0 mL, we have that

$$[NH_4^+] = \frac{5.00 \text{ mmol}}{100.0 \text{ mL}} = 0.0500 \text{ M}$$

Ignoring the contribution to $[H_3O^+]$ from the autoprotonation reaction of water, the concentration table for this equation is as follows:

Concentration	$NH_4^+(aq)$	+ $H_2O(l)$	⇋ $NH_3(aq)$	+ $H_3O^+(aq)$
initial	0.0500 M	—	0 M	≈ 0 M
change	$-x$		$+x$	$+x$
equilibrium	0.0500 M − x	—	x	x

The expression for K_a becomes

$$\frac{x^2}{0.0500 \text{ M} - x} = 5.6 \times 10^{-10} \text{ M}$$

Using the method of successive approximations, we find $x = [H_3O^+] = 5.3 \times 10^{-6}$ M, which gives a pH of 5.28, in agreement with Figure 21.11.

(f) 55.0 mL of 0.100 M $HCl(aq)$ is equivalent to adding 5.50 mmol of $HCl(aq)$. Because the original solution contained 5.00 mmol of $NH_3(aq)$, we are now past the equivalence point, and have added an excess of (5.50 mmol − 5.00 mmol) = 0.50 mmol of $HCl(aq)$. Because the value of K_a for the ammonium ion is small, we can neglect the $[H_3O^+]$ coming from the dissociation of the $NH_4^+(aq)$ and write,

$$[H_3O^+] \approx \frac{0.50 \text{ mmol}}{55.0 \text{ mL} + 50.0 \text{ mL}} = 4.8 \times 10^{-3} \text{ M}$$

and the pH = 2.32, in accord with the curve shown in Figure 21.11. Notice how quickly the pH drops after going beyond the equivalence point.

PRACTICE PROBLEM 21-11: Methylamine, $CH_3NH_2(aq)$, is a weak base for which $K_b = 4.6 \times 10^{-4}$ M (Table 20.5). Calculate the pH for each of the following cases in the titration of 100.0 mL of 0.250 M $CH_3NH_2(aq)$ solution with 0.500 M $HBr(aq)$:

(a) before addition of any $HBr(aq)$ (the initial point)
(b) after addition of 15.0 mL of $HBr(aq)$ (10.0 mL before the midpoint)
(c) after addition of 25.0 mL of $HBr(aq)$ (the midpoint)
(d) after addition of 35.0 mL of $HBr(aq)$ (10.0 mL after the midpoint)
(e) after addition of 50.0 mL of $HBr(aq)$ (the equivalence point)
(f) after addition of 60.0 mL of $HBr(aq)$ (10.0 mL after the equivalence point)

(g) Sketch a titration curve for the reaction, label the initial point, midpoint, equivalence point, and buffer region.

(h) Suggest a suitable indicator for this titration.

Answer: (a) 12.02; (b) 11.03; (c) 10.66; (d) 10.29; (e) 5.72; (f) 1.51; (g) See the solutions acompanying this text; (h) methyl red or bromcresol purple

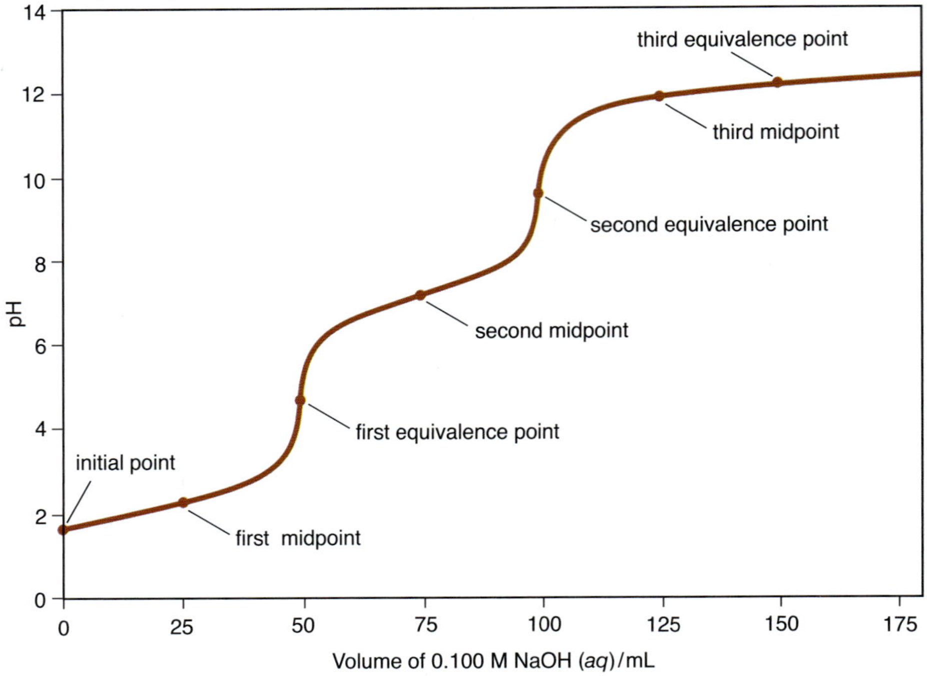

Figure 21.13 Titration curve for the titration of 50.0 mL of 0.100 M $H_3PO_4(aq)$ with 0.100 M $NaOH(aq)$. Because phosphoric acid is a tripotic acid, the curve has the three equivalence points. The third equivalence point is not sharply defined because $HPO_4^{2-}(aq)$ is a very weak acid, so $PO_4^{3-}(aq)$ can compete with $H_2O(l)$ for protons. Notice that the curve has three distinct sections: 0 to 50 mL, 50 mL to 100 mL, and 100 mL to 150 mL. In each section the behavior of pH versus volume of base is analogous to a titration curve for a monoprotic acid.

Many of the acids found in nature, particularly those of biological significance, are polyprotic species. Let's examine the titration of phosphoric acid, $H_3PO_4(aq)$, a triprotic acid, with a strong base. The titration curve for 50.0 mL of 0.100 M $H_3PO_4(aq)$ titrated with 0.100 M $NaOH(aq)$ is shown in Figure 21.13. The titration curve has three equivalence points because $H_3PO_4(aq)$ has three acidic protons. The calculation of a titration curve like that shown in Figure 21.13 is fairly long, and we shall not consider it here. It is important to understand the general shape of the curve and to see that for a polyprotic acid there generally is a distinct section and equivalence point for each acidic proton released.

So far we have considered three types of acid-base titrations. The common feature of the three types of titrations is that they all involve an acid-base reaction with a large equilibrium constant. For an acid-base titration to be useful in chemical analysis, the equation for the titration reaction must have a large value of K_c. For this reason, weak acid-weak base titrations are not usually employed.

SUMMARY

A buffer is a mixture of a weak acid and its conjugate base. A buffer suppresses the change in pH that would otherwise result from the addition of an acid or a base to a solution. The pH of a buffer solution can often be calculated by using the Henderson-Hasselbalch equation. The validity of this equation requires that the equilibrium constants for the weak acid and its conjugate base are in the range of 10^{-4} M to 10^{-11} M, that their concentration ratio is between 0.1 and 10, and that the stoichiometric concentrations of the acid and its conjugate base are greater than about 10^{-3} M but less than 1 M.

A weak acid whose conjugate base has a different color can be used as a pH indicator. Indicators can be used to signal the end point in an acid-base titration. This point is chosen to be as close as possible to the equivalence point, the point on the titration curve at which stoichiometrically equivalent amounts of acid and base have reacted. The pH of a titration curve usually undergoes an especially rapid change in the vicinity of the equivalence point. Most of a titration curve of a weak acid or a weak base between the starting point and the equivalence point is within the buffer region. In many cases we can apply the Henderson-Hasselbalch equation to determine the pH of the solution within this region. We can also use a titration to prepare a buffer solution of a desired pH.

TERMS YOU SHOULD KNOW

buffer *777*
Henderson-Hasselbalch equation *780*
buffer capacity *784*
titration *786*
titration curve *786*
equivalence point *786*
indicator *787*
end point *790*
midpoint *797*
principal species *797*
buffer region *798*

EQUATIONS YOU SHOULD KNOW HOW TO USE

$$pH \approx pK_a + \log\frac{[\text{base}]_0}{[\text{acid}]_0} \qquad (21.7) \qquad \text{(Henderson-Hasselbalch equation}^*\text{)}$$

$$pH \approx pK_{ai} \qquad (21.11) \qquad \text{(pH region in which an indicator changes color)}$$

*This equation only applies for buffers where K_a is between 10^{-4} M and 10^{-11} M, the ratio $[\text{base}]_0/[\text{acid}]_0$ is within the range 0.1 to 10, and $[\text{base}]_0$ and $[\text{acid}]_0$ are within the range 10^{-3} M to 1 M.

PROBLEMS

Assume all solutions to be at 25°C unless stated otherwise.

BUFFER CALCULATIONS

21-1. Calculate the pH of a solution that is 0.050 M in $CH_3COOH(aq)$ and 0.050 M in $NaCH_3COO(aq)$.

21-2. Calculate the pH of a solution that is 0.10 M in $CH_3COOH(aq)$ and 0.20 M in $NaCH_3COO(aq)$.

21-3. Calculate the pH of a solution that is 0.15 M in $HCOOH(aq)$ and 0.25 M in $NaHCOO(aq)$.

21-4. Calculate the pH of a solution that is 0.20 M in $HNO_2(aq)$ and 0.15 M in $NaNO_2(aq)$.

21-5. Calculate the pH of an aqueous solution that is 0.200 M in pyridine, $C_5H_5N(aq)$, and 0.250 M in pyridinium chloride, $C_5H_5NH^+Cl^-(aq)$.

21-6. Calculate the pH of a $NH_4^+(aq)$–$NH_3(aq)$ buffer solution that is 0.40 M in $NH_4Cl(aq)$ and 0.20 M in $NH_3(aq)$.

ADDITION OF ACIDS AND BASES TO BUFFERS

21-7. The higher the concentration of acid and conjugate base in a buffer, the smaller is the pH change when acid or base is added. Calculate the pH change in the following two buffers when 1.00 grams of $KOH(s)$ is added:

(a) 500.0 mL of a 0.10 M $NH_4Cl(aq)$–0.10 M $NH_3(aq)$ buffer

(b) 500.0 mL of a 1.00 M $NH_4Cl(aq)$–1.00 M $NH_3(aq)$ buffer

21-8. (a) Calculate the change in pH when 5.00 mL of 0.100 M $HCl(aq)$ is added to 100.0 mL of a buffer solution that is 0.100 M in $NH_3(aq)$ and 0.100 M in $NH_4Cl(aq)$. (b) Calculate the change in pH when 5.00 mL of 0.100 M $NaOH(aq)$ is added to the original buffer solution.

21-9. Calculate the pH of a solution that is prepared by mixing 2.16 grams of propanoic acid, $CH_3CH_2COOH(s)$, and 0.56 grams of $NaOH(s)$ in enough water to make exactly 100.0 mL of solution. Take $pK_a = 4.87$ for propanoic acid.

21-10. If 6.52 grams of pyridine, $C_5H_5N(l)$, is added to 30.0 mL of 0.950 M $HCl(aq)$, what will be the pH of the resulting solution? Take the final volume of the solution to be 36.0 mL.

21-11. A buffer is prepared such that $[CH_3COOH]_0 = [CH_3COO^-]_0 = 0.050$ M. Calculate the volume of 0.10 M NaOH that can be added to 100.0 mL of the solution before its buffering capacity is lost. Assume the buffer capacity is lost when the ratio $[\text{base}]_0/[\text{acid}]_0$ is less than 0.1 or greater than 10.

21-12. A buffer solution is prepared such that $[H_3PO_4]_0 = [H_2PO_4^-]_0 = 0.20$ M. Calculate the volume of 0.10 M HCl that can be added to 200.0 mL of the solution before its buffering capacity is lost. Assume the buffer capacity is lost when the ratio $[\text{base}]_0/[\text{acid}]_0$ is less than 0.1 or greater than 10.

PREPARING BUFFER SOLUTIONS

21-13. Suppose you are performing an experiment during which the pH must be maintained at 3.70. What would be an appropriate buffer to use? (See Table 20.6.)

21-14. Suppose you are performing an experiment during which the pH must be maintained at 5.20. What would be an appropriate buffer to use? (See Table 20.6.)

21-15. Consider the following mixtures. In each case either calculate the pH if the resulting solution is a buffer solution or state why the solution does not form a buffer.

(a) 25.0 mL of 0.100 M $CH_3COOH(aq)$ and 25.0 mL of 0.200 M $NaCH_3COO(aq)$

(b) 25.0 mL of 0.200 M $CH_3COOH(aq)$ and 25.0 mL of 0.100 M $Ba(OH)_2(aq)$

(c) 25.0 mL of 0.200 M $CH_3COOH(aq)$ and 25.0 mL of 0.100 M $HCl(aq)$

(d) 25.0 mL of 0.200 M $CH_3COOH(aq)$ and 25.0 mL of 0.100 M $NaOH(aq)$

21-16. Consider the following mixtures. In each case either calculate the pH if the resulting solution is a buffer solution or state why the solution does not form a buffer.

(a) 25.0 mL of 0.100 M $HCl(aq)$ and 25.0 mL of 0.100 M $NaOH(aq)$

(b) 25.0 mL of 0.100 M $NH_4Cl(aq)$ and 15.0 mL of 0.100 M $NH_3(aq)$

(c) 25.0 mL of 0.100 M $NH_4Cl(aq)$ and 25.0 mL of 0.100 M $NaOH(aq)$

(d) 25.0 mL of 0.100 M $NH_4Cl(aq)$ and 15.0 mL of 0.100 M $NaOH(aq)$

21-17. Describe how you would prepare 100.0 mL of a pH = 4.00 buffer solution from a solution of 0.100-M benzoic acid, $C_6H_5COOH(aq)$, and a solution of 0.150-M sodium benzoate, $NaC_6H_5COO(aq)$. Take the pK_a of benzoic acid to be 4.20.

21-18. Describe how you would prepare 50.0 mL of a pH = 7.00 buffer solution from a solution of 0.200-M potassium hydrogen sulfite, $KHSO_3(aq)$, and a solution of 0.100-M potassium sulfite, $K_2SO_3(aq)$. Take the pK_a of the hydrogen sulfite ion to be 7.17.

21-19. Describe how you would prepare 150.0 mL of a pH = 5.00 buffer solution from a solution of 0.200-M aniline, $C_6H_5NH_2(aq)$, a weak base, and a solution of 0.100-M hydrochloric acid. Take the pK_b of aniline to be 9.13.

21-20. Describe how you would prepare 200.0 mL of a pH = 4.00 buffer solution from a solution of 0.150-M lactic acid, $CH_3CHOHCOOH(aq)$, and a solution of 0.100-M barium hydroxide, $Ba(OH)_2(aq)$. Take the pK_a of lactic acid to be 3.86.

INDICATORS

21-21. Estimate the pH of a colorless aqueous solution that turns blue when Nile blue is added and turns blue when thymol blue is added.

21-22. Estimate the pH of a colorless aqueous solution that turns yellow when methyl orange is added and yellow when bromcresol purple is added.

21-23. A colorless aqueous solution was obtained from a rain gauge. The following indicators were added to separate samples of the solution; the colors observed were

bromcresol purple: yellow
bromcresol green: green

Estimate the pH of the solution.

21-24. We wish to estimate the pH of a colorless aqueous solution. Litmus paper indicates that the solution is basic. The following indicators are added to separate samples of the solution, and the colors observed are

Nile blue:	red
alizarin yellow R:	deep yellow

Estimate the pH of the solution.

21-25. A certain bacterium grows best in an acidic medium. In an experiment with this bacterium, the pH must be maintained at around 5. What indicator should be added to indicate a pH of 5 and to monitor changes in the pH of the medium?

21-26. The pH of the nutrient broth used to maintain cultures of tissue samples must be greater than 7 but should not rise above 8. What indicator can be added to the nutrient broth to monitor the pH?

21-27. The pH indicator bromcresol green changes from yellow to blue over the pH range 4 to 5. Estimate the value of K_{ai} for bromcresol green.

21-28. The pH indicator Nile blue changes from blue to pink over the pH range 10 to 11. Estimate the value of K_{ai} of Nile blue.

TITRATIONS INVOLVING STRONG ACIDS AND STRONG BASES

21-29. Calculate the pH of the resulting solution if 20.0 mL of 0.20 M $HCl(aq)$ is added to

(a) 25.0 mL of 0.20 M $NaOH(aq)$

(b) 30.0 mL of 0.25 M $NaOH(aq)$

21-30. Calculate the pH of the resulting solution if 20.0 mL of 0.20 M $HCl(aq)$ is added to

(a) 40.0 mL of 0.10 M $NaOH(aq)$

(b) 20.0 mL of 0.15 M $NaOH(aq)$

21-31. Determine the volume at the equivalence point if a 0.100 M $NaOH(aq)$ solution is used to titrate the following acidic solutions:

(a) 50.0 mL of 0.200 M $HBr(aq)$

(b) 30.0 mL of 0.150 M $HNO_3(aq)$

21-32. Determine the volume at the equivalence point if a 0.100 M $NaOH(aq)$ solution is used to titrate the following acidic solutions:

(a) 30.0 mL of 0.600 M $HClO_4(aq)$

(b) 50.0 mL of 0.400 M $HI(aq)$

21-33. Sketch, but do not calculate, the titration curve for the titration of 50.0 mL of 0.100 M $NaOH(aq)$ with 0.100 M $HCl(aq)$.

21-34. Sketch, but do not calculate, the titration curve for the titration of 50.0 mL of 0.25 M $HNO_3(aq)$ with 0.50 M $KOH(aq)$.

21-35. Calculate the molarity of an aqueous nitric acid solution if it requires 35.6 mL of 0.165 M $NaOH(aq)$ to neutralize 25.0 mL of the $HNO_3(aq)$ solution.

21-36. Calculate the molarity of an aqueous sodium hydroxide solution if it requires 34.7 mL of 0.125 M $HCl(aq)$ to neutralize 15.0 mL of the $NaOH(aq)$ solution.

TITRATION INVOLVING WEAK ACIDS OR WEAK BASES

21-37. Calculate the pH of the solution obtained by titrating 25.0 mL of a 0.100-M solution of the herbicide cacodylic acid, $HC_2H_6AsO_2(aq)$, with 0.0950 M $NaOH(aq)$ to the equivalence point. Take $K_a = 5.4 \times 10^{-7}$ M for cacodylic acid. What indicator should be used to signal the equivalence point?

21-38. Calculate the pH of the solution obtained by titrating 50.0 mL of 0.100 M $HNO_2(aq)$ with 0.150 M $NaOH(aq)$ to the equivalence point. Take $K_a = 5.6 \times 10^{-4}$ M for $HNO_2(aq)$. What indicator should be used to signal the equivalence point?

21-39. Calculate the pH for each of the following cases in the titration of 50.0 mL of 0.150-M hypochlorous acid, $HClO(aq)$, with 0.150 M $KOH(aq)$:

(a) before addition of any $KOH(aq)$

(b) after addition of 25.0 mL of $KOH(aq)$

(c) after addition of 35.0 mL of $KOH(aq)$

(d) after addition of 50.0 mL of $KOH(aq)$

(e) after addition of 60.0 mL of $KOH(aq)$

21-40. Calculate the pH for the following cases in the titration of 25.00 mL of 0.200-M acetic acid, $CH_3COOH(aq)$, with 0.200 M $NaOH(aq)$:

(a) before addition of any $NaOH(aq)$

(b) after addition of 5.00 mL of $NaOH(aq)$

(c) after addition of 12.50 mL of $NaOH(aq)$

(d) after addition of 25.00 mL of $NaOH(aq)$

(e) after addition of 26.00 mL of $NaOH(aq)$

21-41. Calculate the pH for each of the following cases in the titration of 25.0 mL of 0.150-M pyridine, $C_5H_5N(aq)$, with 0.150 M $HBr(aq)$:

(a) before addition of any $HBr(aq)$

(b) after addition of 10.0 mL of $HBr(aq)$

(c) after addition of 24.0 mL of $HBr(aq)$

(d) after addition of 25.0 mL of $HBr(aq)$

(e) after addition of 26.0 mL of $HBr(aq)$

21-42. Calculate the pH for each of the following cases in the titration of 35.0 mL of 0.200-M methylamine, $CH_3NH_2(aq)$, with 0.200 M $HCl(aq)$:

(a) before addition of any $HCl(aq)$

(b) after addition of 17.5 mL of $HCl(aq)$

(c) after addition of 34.9 mL of $HCl(aq)$

(d) after addition of 35.0 mL of $HCl(aq)$

(e) after addition of 35.1 mL of $HCl(aq)$

21-43. Sketch the titration curve for Problem 21-39. Label the initial point, midpoint, buffer region, and equivalence point on your curve.

21-44. Sketch the titration curve for Problem 21-42. Label the initial point, midpoint, buffer region, and equivalence point on your curve.

21-45. A 1.50-gram sample of ascorbic acid, vitamin C, is dissolved in 100.0 mL of water and titrated with 0.250 M $NaOH(aq)$ to the equivalence point. The volume of base consumed is 34.1 mL. Calculate the molecular mass of vitamin C, assuming one acidic proton per molecule.

21-46. A 0.772-gram sample of benzoic acid, a monoprotic acid found in most berries, is dissolved in 50.0 mL of water and titrated to the equivalence point with 0.250 M $NaOH(aq)$. The volume of base consumed is 25.3 mL. Calculate the molecular mass of benzoic acid.

ADDITIONAL PROBLEMS

21-47. What is required to form a buffer solution? Can a buffer solution resist a change in pH due to the addition of acid or base indefinitely?

21-48. Will a 50-50 mixture of equal moles of a strong acid and a strong base form a buffer solution?

21-49. Under what conditions can we apply the Henderson-Hasselbalch equation to determine the pH of a buffer solution?

21-50. What is the difference between the end point and the equivalence point of a titration? Are these two points ever the same?

21-51. State whether the pH at the equivalence point is *acidic, basic, neutral,* or *can't tell* for

(a) The titration of a weak acid with a strong base.

(b) The titration of a strong acid with a strong base.

(c) The titration of a weak base with a strong acid.

Assume all species are monoprotic or monobasic.

21-52. You are asked to make up an acetic acid–acetate buffer solution. The stockroom has available 0.100-M solutions of acetic acid, sodium acetate, hydrochloric acid, and sodium hydroxide. Without performing any calculations, describe in words as many different procedures as you can think of for making the desired solution.

21-53. Although indicators are weak acids, why does the addition of an indicator generally not affect the pH of a solution being titrated?

21-54. Why is the pH at the equivalence point for the titration of a strong acid with a strong base equal to 7.00, but generally not equal to 7.00 for the titration of a weak acid with a strong base?

21-55. State the sort of calculation (for example, weak acid dissociation-reaction calculation, Henderson-Hasselbalch equation) that can be used to determine the pH at the initial point, midpoint, close to the midpoint, and at the equivalence point for the titration curves plotted in Figures 21.4 and Figure 21.11.

21-56. Use Le Châtelier's principle to explain why the addition of a strong acid to a solution containing a weak acid *reduces* the percentage dissociation of the weak acid.

21-57. You are given a sample of a solid unknown monoprotic weak acid. Describe how you could use a titration curve to determine both the molecular mass and the value of pK_a of the unknown acid. Would this be sufficient information to positively identify the unknown acid?

21-58. A solution of sodium hydrogen carbonate, $NaHCO_3(aq)$, commonly known as baking soda, can act like a buffer and resist changes in pH, even though only one principle species is dissolved in solution. Explain the buffering property of a $NaHCO_3(aq)$ solution.

21-59. Calculate the pH of a buffer formed by titrating 45.0 mL of a 0.020-M acetic acid solution with 25.0 mL of a 0.015-M sodium hydroxide solution using units of millimoles for all your calculations. Repeat your calculation using units of moles. Which is the more convenient unit?

21-60. A student is preparing for the titration of 25.0 mL of an approximately 0.5-M solution of a weak base using hydrochloric acid. If she has a 50.0-mL buret, which of the following solutions would be best for use in her titration? Explain your choice.

(a) 6 M concentrated $HCl(aq)$

(b) 2.00 M $HCl(aq)$

(c) 0.500 M $HCl(aq)$

(d) 0.250 M $HCl(aq)$

21-61. Which of the following procedures may be used to make a pH = 4.74 buffer solution:

(a) Combine 50.0 mL of 0.10 M $CH_3COOH(aq)$ with 50.0 mL of 0.10 M $NH_3(aq)$.

(b) Combine 50.0 mL of 0.20 M $CH_3COOH(aq)$ with 50.0 mL of 0.10 M $NaOH(aq)$.

(c) Combine 50.0 mL of 0.20 M $KCH_3COO(aq)$ with 50.0 mL of 0.20 M $HCl(aq)$.

(d) Combine 50.0 mL of 0.10 M $CH_3COOH(aq)$ with 50.0 mL of 0.10 M $KCH_3COO(aq)$.

(e) Titrate 50.0 mL of 0.20 M $HNO_3(aq)$ with NaOH until the pH equals 4.74.

21-62. Various antacid tablets contain water-insoluble metal hydroxides, such as $Mg(OH)_2(s)$. Given that stomach acid is about 0.10 M in $HCl(aq)$, calculate the number of milliliters of stomach acid that can be neutralized by 1.00 gram of $Mg(OH)_2(s)$.

21-63. An unknown sample is thought to be either benzoic acid, $C_6H_5COOH(s)$, or chlorobenzoic acid, $C_6H_4ClCOOH(s)$. When 1.89 grams is dissolved in water, 15.49 mL of 1.00 M $NaOH(aq)$ are required to reach the equivalence point. Which acid is the unknown sample?

21-64. The electronic meters used to measure pH are calibrated using standard buffer solutions of known pH. For example, the directions for preparing a certain buffer at 25°C are as follows. Dissolve 3.40 grams of $KH_2PO_4(s)$ and 3.55 grams of $Na_2HPO_4(s)$ in sufficient water to make 1.00 L of solution. What pH do you calculate for this buffer solution?

21-65. Suppose that you wish to determine whether a solution of unknown composition is buffered. Explain how you could do this with only two pH measurements.

21-66. It is important always to remember that the Henderson-Hasselbalch equation is based on the assumption that $[\text{acid}] \approx [\text{acid}]_0$ and $[\text{base}] \approx [\text{base}]_0$. Calculate $[CH_3COOH]$, $[CH_3COOH]_0$, $[CH_3COO^-]$, and $[CH_3COO^-]_0$ for a solution that is 0.100 M in both $CH_3COOH(aq)$ and $NaCH_3COO(aq)$. Is $[CH_3COOH] \approx [CH_3COOH]_0$ and is $[CH_3COO^-] \approx [CH_3COO^-]_0$? Can you see from your calculation why this is so?

21-67. A 2.500-gram sample of oxalic acid, a diprotic acid, is dissolved in 250.0 mL of water and titrated with 1.000 M $NaOH(aq)$ to the first equivalence point. The volume of base required is 27.77 mL. Calculate the molecular mass of oxalic acid.

21-68. Alka-Seltzer contains sodium hydrogen carbonate, $NaHCO_3(s)$, and the triprotic acid citric acid, $H_3C_6H_5O_7(s)$, in addition to 324 mg of aspirin. Write the acid-base reaction that gives rise to the fizz (CO_2 gas evolution) when an Alka-Seltzer tablet is dissolved in water.

21-69. Suppose that 80.0 mL of an aqueous 0.200-M methylamine solution, $CH_3NH_2(aq)$, is titrated with a 0.400 M $HCl(aq)$ solution. Calculate the pH at 10.0-mL intervals of added $HCl(aq)$, up to 50.0 mL added, and plot the titration curve. Take $pK_b = 3.34$ for methylamine.

21-70. Several types of commercial antacid tablets contain $Al(OH)_3(s)$ as the active ingredient. Given that stomach acid is about 0.10 M in $HCl(aq)$, calculate the number of milliliters of stomach acid that can be neutralized by 500.0 mg of $Al(OH)_3(s)$.

21-71. Suppose that you are titrating 25.0 mL of a 0.250 M $HBr(aq)$ solution with a 0.500 M $NaOH(aq)$ solution. The equivalence point occurs when 12.5 mL of the $NaOH(aq)$ solution has been added. The pH at this point is 7.0. Calculate the pH change if one more drop (0.02 mL) of the $NaOH(aq)$ solution is added.

21-72. A 0.550-gram sample of butyric acid is dissolved in 100.0 mL of water and titrated with 0.100 M $NaOH(aq)$ to the equivalence point. The volume of base consumed is 62.4 mL. Calculate the molecular mass of butyric acid, which has one acidic proton per molecule.

21-73. Indicate for which of the following solutions the Henderson-Hasselbalch equation cannot be used to calculate the pH:

(a) 0.15 M $HNO_2(aq)$ plus 0.20 M $NaNO_2(aq)$

(b) 0.15 M $HNO_2(aq)$

(c) 0.20 M $NaNO_2(aq)$

(d) 0.10 M $Na_2HPO_4(aq)$ plus 0.20 M $KH_2PO_4(aq)$

In each case that the Henderson-Hasselbalch equation cannot be used, state why.

21-74. Calculate the volume of 0.10 M $NaOH(aq)$ required to neutralize completely 25.0 mL of a 0.10-M aqueous solution of oxalic acid, $H_2C_2O_4(aq)$, a diprotic acid.

21-75. Calculate the number of grams of $NaHCO_3(s)$ required to neutralize 2.00 grams of arsenic acid, $H_3AsO_4(s)$, in water. Arsenic acid is a triprotic acid.

21-76. Determine how many of the hydrogen atoms in citric acid, $C_6H_8O_7(s)$, are acidic if it requires 147.2 mL of a 0.135 M $NaOH(aq)$ solution to completely titrate a 1.270-gram sample of citric acid.

21-77. Calculate the pH of the solution that results when 2.00 grams of $Mg(OH)_2(s)$ is dissolved in 850.0 mL of 0.160 M $HCl(aq)$.

21-78. Vinegar is an aqueous solution of acetic acid. A 21.0-mL sample of vinegar requires 38.5 mL of 0.400 M $NaOH(aq)$ to neutralize the $CH_3COOH(aq)$. Given that the density of the vinegar is 1.060 $g\cdot mL^{-1}$, calculate the mass percentage of acetic acid in the vinegar.

21-79. A 2.00-gram sample of acetylsalicylic acid, better known as aspirin, is dissolved in 100.0 mL of water and titrated with 0.200 M $NaOH(aq)$ to the equivalence point. The volume of base required is 55.5 mL. Calculate the molecular mass of the acetylsalicylic acid, which has one acidic proton per molecule.

21-80. A commonly used buffer in biological experiments is a phosphate buffer containing $NaH_2PO_4(aq)$ and $Na_2HPO_4(aq)$. Estimate the pH of an aqueous solution that contains

(a) 0.050 M $NaH_2PO_4(aq)$ and 0.050 M $Na_2HPO_4(aq)$

(b) 0.050 M $NaH_2PO_4(aq)$ and 0.10 M $Na_2HPO_4(aq)$

(c) 0.10 M $NaH_2PO_4(aq)$ and 0.050 M $Na_2HPO_4(aq)$

The relevant equation is

$$H_2PO_4^-(aq) + H_2O(l) \leftrightharpoons H_3O^+(aq) + HPO_4^{2-}(aq) \qquad K_a = 6.2 \times 10^{-8}\ M$$

21-81. Calculate the pH of a buffer solution obtained by dissolving 10.0 grams of $KH_2PO_4(s)$ and 20.0 grams of $Na_2HPO_4(s)$ in water and then diluting to 1.00 liter. The relevant equation is

$$H_2PO_4^-(aq) + H_2O(l) \leftrightharpoons H_3O^+(aq) + HPO_4^{2-}(aq) \qquad pK_a = 7.21$$

21-82. It is found that it takes 37.5 mL of 0.200 M $NaOH(aq)$ to titrate a 50.0-mL sample of an unknown monoprotic acid to its equivalence point. The pH after adding 15.0 mL of $NaOH(aq)$ was 4.67. Determine the value of pK_a of the unknown acid.

21-83. Calculate the pH at the equivalence point in the titration of 50.0 mL of 0.125-M trimethylamine with 0.175 M $HCl(aq)$.

21-84. Calculate the pH at the equivalence point in the titration of 17.5 mL of 0.098-M pyridine with 0.117 M $HI(aq)$.

21-85. Calculate the mass of $NH_4Cl(s)$ that must be added to 1.00 L of a 0.200 M $NH_3(aq)$ solution to obtain a solution of pH 9.50. Assume no change in volume.

21-86. Calculate the mass of $NaOH(s)$ that must be added to 500.0 mL of 0.120 M $CH_3COOH(aq)$ to yield a solution of pH = 4.52. Assume no change in volume.

21-87. Describe how you would prepare a buffer solution of pH 9.20 starting with 0.10 M $NH_3(aq)$ and 0.10 M $HCl(aq)$.

21-88. How many grams of $NaNO_2(s)$ must be added to 300.0 mL of 0.200 M $HNO_2(aq)$ to give a pH of 3.70? Assume no volume change when the salt is added.

21-89. A 25.0-gram sample of $NH_4Cl(s)$ is added to 300.0 mL of 0.500 M $NH_3(aq)$. What is the pH? Assume no volume change when the salt is added.

21-90. What is the pH of the solution that results from the addition of 25.0 mL of 0.200 M $KOH(aq)$ to 50.0 mL of 0.150 M $HNO_2(aq)$?

21-91. A buffer solution consists of 1.00 M each of $HNO_2(aq)$ and $NaNO_2(aq)$. Calculate the change in pH if 50.0 mL of 0.650 M $HCl(aq)$ is added to a liter of the buffer solution.

21-92. (*) Phosphorous acid, $H_3PO_3(aq)$, is a diprotic oxyacid that is an important compound in industry and agriculture. Its formula is better expressed as

$HP(O)(OH)_2$ (shown at right). The values of pK_{a_1} and pK_{a_2} for phosphorous acid are 1.3 and 6.70, respectively. Calculate the pH for each of the following points in the titration of 50.0 mL of a 3.0-M aqueous solution of $H_3PO_3(aq)$ with 3.0 M $KOH(aq)$:

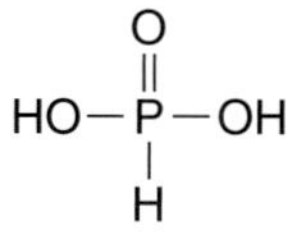

(a) before addition of any $KOH(aq)$

(b) after addition of 25.0 mL of $KOH(aq)$

(c) after addition of 50.0 mL of $KOH(aq)$ (*Hint*: See Problem 20-95)

(d) after addition of 75.0 mL of $KOH(aq)$

(e) after addition of 100.0 mL of $KOH(aq)$

Sketch the titration curve and label the various points and regions.

21-93. (*) The principal reaction when a salt composed of an acidic cation and a basic anion, for example, $NH_4CH_3COO(s)$, is dissolved in water is of the type described by

$$NH_4^+(aq) + CH_3COO^-(aq) \leftrightarrows NH_3(aq) + CH_3COOH(aq)$$

(a) Show that the value of the equilibrium constant for this reaction is given by

$$K_c \approx \frac{K_{\mathrm{a,NH_4^+}}}{K_{\mathrm{a,CH_3COCH}}}$$

(b) Given the above stoichiometry, show that the $[H_3O^+]$ of the solution is equal to

$$[H_3O^+] \approx \sqrt{(K_{\mathrm{a,NH_4^+}})(K_{\varepsilon,\mathrm{CH_3COOH}})}$$

Note that $[H_3O^+]$ and thus the pH of the solution are independent of the concentration of the salt.

(c) Is an $NH_4CH_3COO(aq)$ solution a buffer? Explain why or why not.

George Charles de Hevesy (1885–1966) was born in Budapest, Hungary. After earning his doctorate degree at the University of Freiburg in 1908, he worked as a research assistant at various laboratories throughout Europe. During this time, de Hevesy got the idea of using radioisotopes as "tracers" in reactions, where they would be chemically indistinguishable from nonradioactive species of the same element, but could be readily measured by their radioactivity. He measured the solubility of highly insoluble lead compounds by monitoring the radioactivity of their solutions. This work ended in 1915 when he was drafted into the Austro-Hungarian army. After the war, he applied his tracer technique to biology and undertook the first clinical studies using radioactive isotopes. This work, however, was limited to the study of the absorption and elimination of compounds involving the heavy metals lead, bismuth, and thallium (elements with naturally occurring radioisotopes). The discovery of methods for producing synthetic radioisotopes in the 1930s allowed de Hevesy's tracer technique to be applied to other, less toxic elements, opening the door for the use of tracer elements in medical studies. During World War II, de Hevesy helped hide the valuable gold in the Nobel Prize medals of several Nobel laureates from the Nazis by dissolving them in aqua regia and storing the solutions on the shelf. There, despite occupation of the laboratory and a thorough search, they safely waited out the war. Later, the Nobel Society recast the medals from their original gold. In 1943, de Hevesy was awarded the Nobel Prize in Chemistry, "for his work on the use of isotopes as tracers in the study of chemical processes." His method of using radioactive tracers is still widely used today in chemistry, biology, and medicine.

22. Solubility and Precipitation Reactions

Chemists often are faced with the problem of determining which species are present in a sample of an unknown material. Analysis of this kind is called qualitative analysis, and many general chemistry laboratory courses include experiments dealing with the techniques of qualitative analysis. Many of the procedures of qualitative analysis involve the formation and the separation of precipitates. We learned in Chapter 10 that many substances are insoluble or have a low solubility in water. In this chapter we treat solubility within the framework of chemical equilibria; this approach enables us to calculate the solubility of a solid not only in pure water, but also in solutions of acids, bases, and salts. The types of calculations that we discuss in this chapter have many practical applications. For example, many metals are obtained from their ores by a series of reactions involving precipitates; the geological formation of minerals and rocks is governed by the relative solubilities of various substances; and the ultrapurification of materials used in computers and electronic devices often involves precipitation reactions.

22-1. The Solubility of an Ionic Solid Can Be Determined Using the Equilibrium-Constant Expression

In Chapter 10, we presented a few rules that enable us to determine which ionic solids (salts) are soluble in water and which are not. We defined the **solubility** of a substance as the quantity of that substance that is dissolved in a saturated solution. We chose to call any substance whose solubility is less than 0.01 M insoluble. In this section we shall discuss solubility more quantitatively. In particular, we shall cast solubility in terms of chemical equilibria, using the idea of an equilibrium constant.

Consider the equilibrium between solid silver bromate and its constituent ions in water as described by the chemical equation

$$AgBrO_3(s) \leftrightharpoons Ag^+(aq) + BrO_3^-(aq) \qquad (22.1)$$

We can write the equilibrium-constant expression for Equation 22.1 as

$$K_{sp} = [Ag^+][BrO_3^-]$$

The subscript sp stands for solubility product, and K_{sp} is called the **solubility-product constant**. Note that $AgBrO_3(s)$ does not appear in the K_{sp} expression. As we discussed in Chapter 19, a pure solid does not appear in an equilibrium-constant expression because its concentration does not vary.

The experimental value of K_{sp} at 25°C for Equation 22.1 is

$$K_{sp} = [Ag^+][BrO_3^-] = 5.4 \times 10^{-5}\ M^2 \tag{22.2}$$

Equation 22.2 states that if $AgBrO_3(s)$ is in equilibrium with an aqueous solution of $AgBrO_3(aq)$ at 25°C, then the product of the concentrations of $Ag^+(aq)$ and $BrO_3^-(aq)$ at equilibrium must equal $5.4 \times 10^{-5}\ M^2$.

A K_{sp} expression can be used to calculate the solubility of a solid. For example, suppose that excess $AgBrO_3(s)$ is in contact with water at 25°C. Then at equilibrium

$$5.4 \times 10^{-5}\ M^2 = [Ag^+][BrO_3^-]$$

From the reaction stoichiometry of Equation 22.1, we have that $[Ag^+] = [BrO_3^-]$ because each formula unit of $AgBrO_3$ that dissolves produces one $Ag^+(aq)$ ion and one $BrO_3^-(aq)$ ion, and because $AgBrO_3(s)$ is the only source of $Ag^+(aq)$ and $BrO_3^-(aq)$ ions. If we denote the solubility of $AgBrO_3(s)$ in units of molarity by s, then

$$s = \text{solubility of } AgBrO_3(s) \text{ in water} = [Ag^+] = [BrO_3^-]$$

because the concentration of either $Ag^+(aq)$ or $BrO_3^-(aq)$ is equal to the number of moles of dissolved salt per liter of solution. Thus, from the K_{sp} expression, we have

$$K_{sp} = 5.4 \times 10^{-5}\ M^2 = [Ag^+][BrO_3^-] = s^2$$

and so

$$s = (5.4 \times 10^{-5}\ M^2)^{1/2} = 7.3 \times 10^{-3}\ M$$

The formula mass of $AgBrO_3(s)$ is 235.8, so the number of grams of $AgBrO_3(s)$ that dissolves in one liter of solution at 25°C is

$$\text{solubility in grams per liter} = (7.3 \times 10^{-3}\ \text{mol}\cdot\text{L}^{-1})\left(\frac{235.8\ \text{g AgBrO}_3}{1\ \text{mol AgBrO}_3}\right) = 1.7\ \text{g}\cdot\text{L}^{-1}$$

EXAMPLE 22-1: The value of K_{sp} for $BaCrO_4(s)$ in equilibrium with an aqueous solution of its constituent ions at 25°C is $1.2 \times 10^{-10}\ M^2$. Write the chemical equation that represents the solubility equilibrium for $BaCrO_4(s)$ and calculate its solubility in water in grams per liter at 25°C.

Solution: The chemical equation that describes the solubility equilibrium is

$$BaCrO_4(s) \leftrightharpoons Ba^{2+}(aq) + CrO_4^{2-}(aq)$$

The K_{sp} expression for this equation is

$$K_{sp} = [Ba^{2+}][CrO_4^{2-}] = 1.2 \times 10^{-10}\ M^2$$

If $BaCrO_4(s)$ is equilibrated with pure water, then, from the reaction stoichiometry, we have at equilibrium

$$[Ba^{2+}] = [CrO_4^{2-}] = s$$

where s is the solubility of $BaCrO_4(s)$ in pure water. Thus,

$$K_{sp} = s^2 = 1.2 \times 10^{-10}\ M^2$$

and

$$s = (1.2 \times 10^{-10}\ M^2)^{1/2} = 1.1 \times 10^{-5}\ M$$

The solubility in grams per liter is given by

$$s = (1.1 \times 10^{-5}\ mol{\cdot}L^{-1})\left(\frac{253.3\ g\ BaCrO_4}{1\ mol\ BaCrO_4}\right) = 2.8 \times 10^{-3}\ g{\cdot}L^{-1}$$

PRACTICE PROBLEM 22-1: Calculate the solubility of $TlBrO_3(s)$ in water in grams per liter at 25°C given that $K_{sp} = 1.1 \times 10^{-4}\ M^2$.

Answer: $3.5\ g{\cdot}L^{-1}$

We saw that when excess $AgBrO_3(s)$ is in equilibrium with pure water, we have that $[Ag^+] = [BrO_3^-]$ because each $AgBrO_3(s)$ unit that dissolves yields one $Ag^+(aq)$ ion and one $BrO_3^-(aq)$ ion and there is no other source of $Ag^+(aq)$ or $BrO_3^-(aq)$. Now consider the problem of calculating the solubility in water of copper(II) iodate, $Cu(IO_3)_2(s)$, which yields *two* $IO_3^-(aq)$ ions and one $Cu^{2+}(aq)$ ion for each $Cu(IO_3)_2(s)$ formula unit that dissolves.

The solubility equilibrium of $Cu(IO_3)_2(s)$ in water can be described by the equation

$$Cu(IO_3)_2(s) \leftrightharpoons Cu^{2+}(aq) + 2\ IO_3^-(aq)$$

The K_{sp} expression for this equation is

$$K_{sp} = [Cu^{2+}][IO_3^-]^2$$

The experimental value of K_{sp} at 25°C is $7.4 \times 10^{-8}\ M^3$, so we have at 25°C

$$K_{sp} = [Cu^{2+}][IO_3^-]^2 = 7.4 \times 10^{-8}\ M^3 \qquad (22.3)$$

Note that it is the *square* of the concentration of $IO_3^-(aq)$ that appears in the K_{sp} expression for $Cu(IO_3)_2(s)$ because each formula unit of $Cu(IO_3)_2(s)$ that dissolves produces two iodate ions. Thus, when $Cu(IO_3)_2(s)$ is in equilibrium with its constituent ions in water, the concentration of iodate ions is twice as great as the concentration of copper(II) ions if there is no other source of $Cu^{2+}(aq)$ and $IO_3^-(aq)$ ions, and so we have $[IO_3^-] = 2[Cu^{2+}]$. The solubility of $Cu(IO_3)_2(s)$ in pure water is equal to $[Cu^{2+}]$ because each mole of $Cu(IO_3)_2(s)$ that dissolves yields one mole of $Cu^{2+}(aq)$. If we denote the solubility of $Cu(IO_3)_2(s)$ in pure water by s, then $[Cu^{2+}] = s$ and $[IO_3^-] = 2s$. Combining these results with the K_{sp} expression, Equation 22.3, yields

$$7.4 \times 10^{-8}\ \mathrm{M}^3 = [Cu^{2+}][IO_3^-]^2 = (s)(2s)^2 = 4s^3$$

and

$$s = \left(\frac{7.4 \times 10^{-8}\ \mathrm{M}^3}{4}\right)^{1/3} = 2.6 \times 10^{-3}\ \mathrm{M}$$

Therefore $[Cu^{2+}] = s = 2.6 \times 10^{-3}$ M and $[IO_3^-] = 2s = 5.3 \times 10^{-3}$ M. The solubility of $Cu(IO_3)_2(s)$ in grams per liter is

$$s = (2.6 \times 10^{-3}\ \mathrm{mol{\cdot}L^{-1}})\left(\frac{413.4\ \mathrm{g}\ Cu(IO_3)_2}{1\ \mathrm{mol}\ Cu(IO_3)_2}\right) = 1.1\ \mathrm{g{\cdot}L^{-1}}$$

The values of solubility-product constants for a number of compounds are given in Table 22.1.

EXAMPLE 22-2: The solubility-product constant for silver chromate in equilibrium with its constituent ions in water at 25°C is 1.1×10^{-12} M^3. Calculate the value of $[Ag^+]$ that results when pure water is saturated with $Ag_2CrO_4(s)$.

Solution: The $Ag_2CrO_4(s)$ solubility equilibrium can be described by

$$Ag_2CrO_4(s) \leftrightharpoons 2Ag^+(aq) + CrO_4^{2-}(aq)$$

and the corresponding solubility-product expression is

$$K_{sp} = [Ag^+]^2[CrO_4^{2-}] = 1.1 \times 10^{-12}\ \mathrm{M}^3$$

Each $Ag_2CrO_4(s)$ formula unit that dissolves yields two $Ag^+(aq)$ ions and one $CrO_4^{2-}(aq)$ ion; thus, $[Ag^+] = 2s$ and $[CrO_4^{2-}] = s$. Substitution of this result into the K_{sp} expression for $Ag_2CrO_4(s)$ yields

$$K_{sp} = (2s)^2(s) = 1.1 \times 10^{-12}\ \mathrm{M}^3$$

Solving for s yields

$$s = 6.5 \times 10^{-5}\ \mathrm{M}$$

The value of $[Ag^+]$ is

$$[Ag^+] = 2s = 1.3 \times 10^{-4}\ \mathrm{M}$$

TABLE 22.1 Solubility-product constants for various salts in water at 25°C

Bromates	K_{sp}	**Cyanides**	K_{sp}	**Oxalates**	K_{sp}
$AgBrO_3$	5.4×10^{-5} M^2	AgCN	6.0×10^{-17} M^2	$Ag_2C_2O_4$	5.4×10^{-12} M^3
$Ba(BrO_3)_2$	2.4×10^{-4} M^3	CuCN	3.5×10^{-20} M^2	CaC_2O_4	4×10^{-9} M^2
$Pb(BrO_3)_2$	7.9×10^{-6} M^3	$Hg_2(CN)_2$*	5×10^{-40} M^3	MgC_2O_4	7×10^{-7} M^2
$TlBrO_3$	1.1×10^{-4} M^2	$Zn(CN)_2$	3×10^{-16} M^3	SrC_2O_4	4×10^{-7} M^2
Bromides	K_{sp}	**Fluorides**	K_{sp}	**Sulfates**	K_{sp}
AgBr	5.4×10^{-13} M^2	BaF_2	1.8×10^{-7} M^3	Ag_2SO_4	1.2×10^{-5} M^3
CuBr	6.3×10^{-9} M^2	CaF_2	3.5×10^{-11} M^3	$BaSO_4$	1.1×10^{-10} M^2
Hg_2Br_2*	6.4×10^{-23} M^3	LiF	1.8×10^{-3} M^2	$CaSO_4$	4.9×10^{-5} M^2
$HgBr_2$	6.2×10^{-20} M^3	MgF_2	5.2×10^{-11} M^3	Hg_2SO_4	6.5×10^{-7} M^2
$PbBr_2$	6.6×10^{-6} M^3	PbF_2	3.3×10^{-8} M^3	$PbSO_4$	2.5×10^{-8} M^2
TlBr	3.7×10^{-6} M^2	SrF_2	4.3×10^{-9} M^3	$SrSO_4$	3.4×10^{-7} M^2
Carbonates	K_{sp}	**Hydroxides**	K_{sp}	**Sulfides**	K_{sp}
Ag_2CO_3	8.5×10^{-12} M^3	$Al(OH)_3$	1.3×10^{-33} M^4	Ag_2S	8×10^{-51} M^3
$BaCO_3$	2.6×10^{-9} M^2	$Ca(OH)_2$	5.0×10^{-6} M^3	CdS	8.0×10^{-27} M^2
$CaCO_3$	3.4×10^{-9} M^2	$Cd(OH)_2$	7.2×10^{-15} M^3	CoS	5×10^{-22} M^2
$CdCO_3$	1.0×10^{-12} M^2	$Co(OH)_2$	5.9×10^{-15} M^3	CuS	6.3×10^{-36} M^2
$CoCO_3$	1.0×10^{-10} M^2	$Cr(OH)_3$	6.3×10^{-31} M^4	FeS	6.3×10^{-18} M^2
$CuCO_3$	1.4×10^{-10} M^2	$Cu(OH)_2$	2.2×10^{-20} M^3	HgS	4×10^{-53} M^2
$FeCO_3$	3.1×10^{-11} M^2	$Fe(OH)_2$	4.9×10^{-17} M^3	MnS	2.5×10^{-13} M^2
$MgCO_3$	6.8×10^{-6} M^2	$Fe(OH)_3$	2.8×10^{-39} M^4	NiS	1.3×10^{-25} M^2
$MnCO_3$	2.2×10^{-11} M^2	$Mg(OH)_2$	5.6×10^{-12} M^3	PbS	8.0×10^{-28} M^2
$NiCO_3$	1.4×10^{-7} M^2	$Ni(OH)_2$	5.5×10^{-16} M^3	SnS	1.0×10^{-25} M^2
$PbCO_3$	7.4×10^{-14} M^2	$Pb(OH)_2$	1.4×10^{-20} M^3	Tl_2S	6×10^{-22} M^3
$SrCO_3$	5.6×10^{-10} M^2	$Sn(OH)_2$	5.5×10^{-27} M^3	ZnS	1.6×10^{-24} M^2
$ZnCO_3$	1.5×10^{-10} M^2	$Zn(OH)_2$	1.0×10^{-15} M^3		
Chlorides	K_{sp}	**Iodates**	K_{sp}	**Thiocyanates**	K_{sp}
AgCl	1.8×10^{-10} M^2	$AgIO_3$	3.2×10^{-8} M^2	AgSCN	1.0×10^{-12} M^2
CuCl	1.7×10^{-7} M^2	$Ba(IO_3)_2$	4.0×10^{-9} M^3	CuSCN	1.8×10^{-13} M^2
Hg_2Cl_2*	1.4×10^{-18} M^3	$Ca(IO_3)_2$	6.5×10^{-6} M^3	$Cu(SCN)_2$	4.0×10^{-14} M^3
$PbCl_2$	1.5×10^{-5} M^3	$Cd(IO_3)_2$	2.5×10^{-8} M^3	$Hg_2(SCN)_2$*	3.2×10^{-20} M^3
TlCl	1.9×10^{-4} M^2	$Cu(IO_3)_2$	7.4×10^{-8} M^3	$Hg(SCN)_2$	2.8×10^{-20} M^3
		$Pb(IO_3)_2$	3.7×10^{-13} M^3	TlSCN	1.6×10^{-4} M^2
		$TlIO_3$	3.1×10^{-6} M^2		
		$Zn(IO_3)_2$	3.9×10^{-6} M^3		
Chromates	K_{sp}	**Iodides**	K_{sp}		
Ag_2CrO_4	1.1×10^{-12} M^3	AgI	8.5×10^{-17} M^2		
$BaCrO_4$	1.2×10^{-10} M^2	CuI	1.3×10^{-12} M^2		
$CuCrO_4$	3.6×10^{-6} M^2	Hg_2I_2*	5.2×10^{-29} M^3		
Hg_2CrO_4*	2.0×10^{-9} M^2	HgI_2	2.9×10^{-29} M^3		
$PbCrO_4$	2.8×10^{-13} M^2	PbI_2	9.8×10^{-9} M^3		
Tl_2CrO_4	8.7×10^{-13} M^3	TlI	5.5×10^{-8} M^2		

*Hg(I) exists as $Hg_2^{2+}(aq)$ in aqueous solution.

PRACTICE PROBLEM 22-2: Calculate the solubility in grams per liter of mercury(I) chloride, $Hg_2Cl_2(s)$, in water at 25°C given that $K_{sp} = 1.4 \times 10^{-18}$ M^3.

Answer: 3.3×10^{-4} g·L^{-1}

EXAMPLE 22-3: The solubility of $PbI_2(s)$ in water at 25°C is 0.59 g·L^{-1}. Calculate the value of K_{sp} for $PbI_2(s)$.

Solution: The solubility of $PbI_2(s)$ in moles per liter is

$$s = (0.59\ \text{g}\cdot\text{L}^{-1})\left(\frac{1\ \text{mol PbI}_2}{461.0\ \text{g PbI}_2}\right) = 0.0013\ \text{M}$$

The chemical equation for the solubility equilibrium is

$$PbI_2(s) \leftrightharpoons Pb^{2+}(aq) + 2I^-(aq)$$

The solubility s is related to $[Pb^{2+}]$ and $[I^-]$ by

$$[Pb^{2+}] = s \qquad \text{and} \qquad [I^-] = 2s$$

Therefore,

$$K_{sp} = [Pb^{2+}]\,[I^-]^2 = (s)(2s)^2 = 4s^3 = 4(0.0013\ \text{M})^3 = 8.8 \times 10^{-9}\ \text{M}^3$$

The value of K_{sp} calculated here differs somewhat from that in Table 22.1 because the solubility-product constant for a species assumes the formation of simple ions from that species, such as $Pb^{2+}(aq)$ and $I^-(aq)$ in the case of $PbI_2(s)$. In actuality the chemistry that occurs in solution can be more complex than this and so the actual experimentally determined solubility of a species can differ somewhat from the value determined from its solubility-product constant. Nevertheless, calculations involving solubility and K_{sp} are often reasonably accurate and allow us to make useful predictions about the solubility of ionic species.

PRACTICE PROBLEM 22-3: Calculate the pH of a saturated aqueous solution of $Mg(OH)_2(s)$ at 25°C. The value of K_{sp} is 5.6×10^{-12} M^3.

Answer: pH = 10.35

You may be wondering how scientists determine the solubility of sparingly soluble salts for which less than a milligram of the salt can be dissolved in each liter of water. Prior to the 1900s such work was painstakingly difficult, requiring the careful evaporation of large volumes of water from saturated solutions followed by precise weighing of the remaining residue. In the early 1900s, however, George de Hevesy (Frontispiece) pioneered the technique of using radioisotopes as "tracers" to measure the solubility of sparingly soluble salts, such as those of lead and other heavy metals for which stable radioisotopes were then available. First he would saturate a solution with a salt formed from a radioisotope, such as $^*PbSO_4(s)$, where we denote the lead atoms with an asterisk to emphasize that they are radioactive. He then separated out any undissolved solid and finally measured the radioactivity of the resulting solution. From this, he was able to accurately calculate the concentration of $^{*}Pb^{2+}(aq)$ ions dissolved in the solution, and thus determine the solubility of the salt and its corresponding value of K_{sp}. The use of radioactive tracers in chemistry and medicine is further discussed in Interchapter O.

See Interchapter O at www.McQuarrieGeneralChemistry.com.

22-2. The Solubility of an Ionic Solid Decreases When a Common Ion Is Present in the Solution

In Section 22-1 we discussed the solubility of an ionic solid in pure water. In this section we discuss the solubility of ionic solids in solutions that already contain one of the ionic constituents of the salt.

Consider the problem of calculating the solubility of silver bromate, $AgBrO_3(s)$, in an aqueous solution at 25°C that is 0.10 M in sodium bromate, $NaBrO_3(aq)$, which is a strong electrolyte. As we learned in the previous section, the solubility equilibrium of $AgBrO_3(s)$ can be described by

$$AgBrO_3(s) \leftrightharpoons Ag^+(aq) + BrO_3^-(aq)$$

and the corresponding solubility-product constant expression is

$$K_{sp} = [Ag^+][BrO_3^-] = 5.4 \times 10^{-5}\ M^2 \qquad (22.4)$$

It is important to realize that the ionic concentrations in Equation 22.4 are the *total* ionic concentrations, *regardless of the source of each ionic species*. In the case of $AgBrO_3(s)$ dissolved in a 0.10 M $NaBrO_3(aq)$ solution, the $Ag^+(aq)$ ions come only from the $AgBrO_3(s)$ that dissolves. The $BrO_3^-(aq)$ ions, on the other hand, come from two sources: from the 0.10 M $NaBrO_3(aq)$ initially present, which is completely dissociated into $Na^+(aq)$ ions and $BrO_3^-(aq)$ ions, and from the $AgBrO_3(s)$ that dissolves. The $Na^+(aq)$ from the $NaBrO_3(aq)$ is simply a spectator ion and does not enter into any of our calculations here.

If we let s be the solubility of $AgBrO_3(s)$ in 0.10 M $NaBrO_3(aq)$, then we can set up a concentration table as shown below.

Concentration	$AgBrO_3(s)$	$\leftrightharpoons$	$Ag^+(aq)$	+	$BrO_3^-(aq)$
initial	—		0 M		0.10 M
change	—		$+s$		$+s$
equilibrium	—		s		0.10 M + s

Substituting the equilibrium values into Equation 22.4, we obtain

$$s(0.10\ M + s) = 5.4 \times 10^{-5}\ M^2 \qquad (22.5)$$

Because $AgBrO_3(s)$ is only sparingly soluble in pure water, we expect the value of s in Equation 22.5 to be small compared to 0.10 M. In fact, the largest that s can be is the solubility of $AgBrO_3(s)$ in pure water, which we calculated in Section 22-1 to be 7.3×10^{-3} M. Le Châtelier's principle shows that the solubility of $AgBrO_3(s)$ in a solution that already contains $BrO_3^-(aq)$ will be even smaller than this. Therefore, we can neglect s relative to 0.10 M and write

$$s(0.10 \text{ M}) \approx 5.4 \times 10^{-5} \text{ M}^2$$
$$s \approx 5.4 \times 10^{-4} \text{ M}$$

The small value of $[Ag^+]$ justifies the assumption that $s << 0.10$ M. As a check, if we substitute $s = 5.4 \times 10^{-4}$ M into Equation 22.5, we find that it is satisfied. Thus, we see that the **common ion**, $BrO_3^-(aq)$, decreases the solubility of $AgBrO_3(s)$. The solubility of $AgBrO_3(s)$ in pure water is 7.3×10^{-3} M, about 13 times greater than its solubility in a 0.10 M $NaBrO_3(aq)$ solution. The lowering of the solubility of a salt by the presence of a common ion in the solution is called the **common ion effect**.

EXAMPLE 22-4: Calculate the solubility at 25°C in moles per liter of copper(II) iodate, $Cu(IO_3)_2(s)$, ($K_{sp} = 7.4 \times 10^{-8}$ M^3) in an aqueous solution that is 0.20 M in copper(II) perchlorate, $Cu(ClO_4)_2(s)$, which is a strong electrolyte.

Solution: The equilibrium equation that describes the solubility of $Cu(IO_3)_2(s)$ in water is

$$Cu(IO_3)_2(s) \leftrightarrows Cu^{2+}(aq) + 2IO_3^-(aq)$$

and the corresponding solubility-product expression is

$$K_{sp} = [Cu^{2+}][IO_3^-]^2 = 7.4 \times 10^{-8} \text{ M}^3 \quad (22.6)$$

The $ClO_4^-(aq)$ from the $Cu(ClO_4)_2(aq)$ is simply a spectator ion and does not enter into any of our calculations here. The only source of $IO_3^-(aq)$ is from the $Cu(IO_3)_2(s)$ that dissolves. If we let s be the solubility of $Cu(IO_3)_2(s)$ in 0.20 M $Cu(ClO_4)_2(aq)$, then $[IO_3^-] = 2s$ because two $IO_3^-(aq)$ ions are produced for each formula unit of $Cu(IO_3)_2(s)$ that dissolves. The $Cu^{2+}(aq)$ ions comes from both the 0.20 M $Cu(ClO_4)_2(aq)$ and the $Cu(IO_3)_2(s)$ that dissolves; thus, $[Cu^{2+}] = 0.20 \text{ M} + s$. If we substitute these expressions for $[Cu^{2+}]$ and $[IO_3^-]$ into Equation 22.6, then we obtain

$$(0.20 \text{ M} + s)(2s)^2 = 7.4 \times 10^{-8} \text{ M}^3 \quad (22.7)$$

Because $Cu(IO_3)_2(s)$ is a slightly soluble salt, we expect the value of s to be small. Therefore, we can neglect s relative to 0.20 M in Equation 22.7 and write

$$(0.20 \text{ M})(2s)^2 \approx 7.4 \times 10^{-8} \text{ M}^3$$

Solving for s yields

$$s \approx \left(\frac{7.4 \times 10^{-8} \text{ M}^3}{(0.20 \text{ M})(4)}\right)^{1/2} = 3.0 \times 10^{-4} \text{ M}$$

As a check, we note that the value of s that we obtain is indeed negligible relative to 0.20 M. The presence of the common ion $Cu^{2+}(aq)$ in the solution lowers the solubility of $Cu(IO_3)_2(s)$. The solubility of $Cu(IO_3)_2(s)$ in pure water is 2.6×10^{-3} M, about nine times greater than its solubility in a solution that is 0.20 M in $Cu^{2+}(aq)$.

PRACTICE PROBLEM 22-4: Using the data in Table 22.1, calculate the solubility in grams per liter at 25°C of $Ag_2CrO_4(s)$ in an aqueous solution that is 0.65 M in $AgNO_3(aq)$, which is a strong electrolyte.

Answer: 8.6×10^{-10} g·L^{-1}

The common ion effect is readily understood in terms of Le Châtelier's principle. Consider the equation for the solubility equilibrium for silver bromate,

$$AgBrO_3(s) \leftrightharpoons Ag^+(aq) + BrO_3^-(aq)$$

for which

$$K_{sp} = [Ag^+][BrO_3^-] = 5.4 \times 10^{-5}\ M^2$$

An increase in the concentration of $BrO_3^-(aq)$—for example, by adding $NaBrO_3(s)$, a strong electrolyte, to the solution—shifts the solubility equilibrium from right to left and thereby decreases the solubility of $AgBrO_3(s)$. The larger the value of $[BrO_3^-]$ at equilibrium, the smaller the value of $[Ag^+]$ because the product $[Ag^+][BrO_3^-]$ must equal $5.4 \times 10^{-5}\ M^2$ at equilibrium at 25°C. Similarly, an increase in the concentration of $Ag^+(aq)$—for example, by adding $AgNO_3(s)$, which is soluble in water—also shifts the solubility equilibrium from right to left and thereby decreases the solubility of $AgBrO_3(s)$. The common ion effect for $AgBrO_3(s)$ is illustrated in Figure 22.1. The common ion effect is often used in separations, such as those performed in qualitative analysis (Section 22-8).

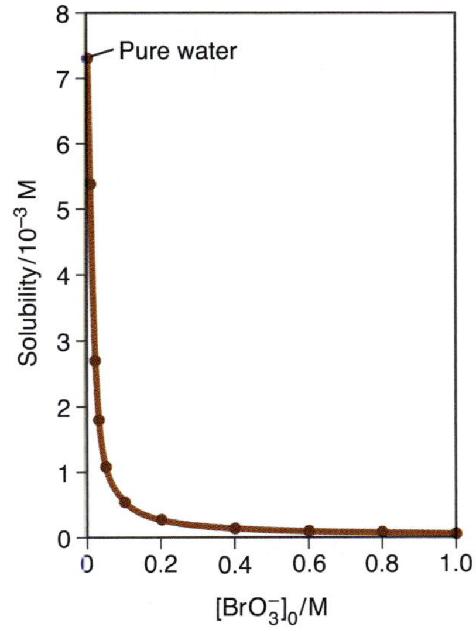

Figure 22.1 The solubility of $AgBrO_3(s)$ in water as a function of the bromate-ion concentration. The bromate-ion concentration can be controlled by adding a soluble salt such as $NaBrO_3(s)$ to an aqueous solution of $AgBrO_3(aq)$. The plot illustrates the common ion effect, whereby the solubility of $AgBrO_3(s)$ is decreased by the addition of $BrO_3^-(aq)$.

22-3. The Solubility of an Ionic Solid Is Increased by the Formation of a Soluble Complex Ion

When working with solubilities and K_{sp}, you must be aware of the possibility of other important equilibria involving the ions. For example, $Ag^+(aq)$ reacts with $NH_3(aq)$ according to

$$Ag^+(aq) + 2NH_3(aq) \leftrightharpoons [Ag(NH_3)_2]^+(aq) \qquad K_f = 2.0 \times 10^7\ M^{-2} \quad (22.8)$$

The product here is called a **complex ion**, that is a metal ion with small molecules or ions attached to it. Equation 22.8 represents a **complexation reaction**. We often denote a complex ion by enclosing its formula in square brackets as we have done here. Another example of a complexation reaction is shown in Figure 22.2. (We shall study complex ions and complexation reactions in Chapter 26.) The subscript f on K denotes an equilibrium involving the *formation* of a complex ion.

To see the consequence of the reaction described by Equation 22.8, consider the addition of 0.50 M $NH_3(aq)$ to an aqueous solution in equilibrium

Figure 22.2 Copper(II) hydroxide, $Cu(OH)_2(s)$, is essentially insoluble in pure water, but is soluble in an aqueous ammonia solution, $NH_3(aq)$, because of the formation of the complex ion $[Cu(NH_3)_4]^{2+}(aq)$. The equation for the reaction is

$Cu^{2+}(aq) + 4NH_3(aq) \leftrightharpoons [Cu(NH_3)_4]^{2+}(aq)$

with a precipitate of $AgBrO_3(s)$. If we add Equation 22.8 to the equation representing the solubility equilibrium of $AgBrO_3(s)$,

$$AgBrO_3(s) \leftrightharpoons Ag^+(aq) + BrO_3^-(aq) \qquad K_{sp} = 5.4 \times 10^{-5}\ M^2 \qquad (22.9)$$

we obtain

$$AgBrO_3(s) + 2NH_3(aq) \leftrightharpoons [Ag(NH_3)_2]^+(aq) + BrO_3^-(aq) \qquad (22.10)$$

Equation 22.10 is the sum of Equations 22.8 and 22.9, and so the equilibrium constant for Equation 22.10 is the product of the equilibrium constants for Equations 22.8 and 22.9, or

$$K_c = K_{sp}K_f = (5.4 \times 10^{-5}\ M^2)(2.0 \times 10^7\ M^{-2}) = 1.1 \times 10^3 \qquad (22.11)$$

Notice that the value of K_c for this equation is much larger than the value of K_{sp} for $AgBrO_3(s)$. This relation indicates that $AgBrO_3(s)$ is much more soluble in $NH_3(aq)$ than in pure water.

Because $K_c >> K_{sp}$, we expect that essentially all the silver ions in solution are in the form of the $[Ag(NH_3)_2]^+(aq)$ complex ion. Thus, we assume that $[Ag(NH_3)_2]^+(aq)$, along with $BrO_3^-(aq)$, are the principal species in the solution, in the sense that their concentrations are large compared with the other species in the solution. Because Equation 22.10 involves these species and because $K_c >> K_{sp}$, we can neglect Equation 22.9 and use Equation 22.10 to calculate the solubility of $AgBrO_3(s)$ in 0.50 M $NH_3(aq)$. First we set up a concentration table:

Concentration	$AgBrO_3(s)$	+	$2NH_3(aq)$	$\leftrightharpoons$	$[Ag(NH_3)_2]^+(aq)$	+	$BrO_3^-(aq)$
initial	—		0.50 M		0 M		≈ 0 M
change	—		$-2x$		$+x$		$+x$
equilibrium	—		0.50 M $- 2x$		x		x

We use the square bracket notation for a complex ion when we are discussing the complex ion. However, we drop the square brackets when we are discussing the concentration of the complex ion in order to avoid the awkward use of double brackets. Notice that the charge on the complex ion is denoted inside the square brackets when referring to its concentration.

We say that $[BrO_3^-] \approx 0$ initially because we are neglecting the small quantity of $BrO_3^-(aq)$ due to the reaction described by Equation 22.9.

The equilibrium-constant expression for Equation 22.10 is

$$K_c = \frac{x^2}{(0.50 - 2x)^2} = 1.1 \times 10^3$$

Taking the square root of both sides yields

$$\frac{x}{0.50 - 2x} = \pm 33$$

The negative root leads to a negative value for $[NH_3]$, so we shall ignore it. Using the positive root, we find $x = [Ag(NH_3)_2^+] = [BrO_3^-] = 0.246\ M \approx 0.25\ M$. Note that the contribution to $[BrO_3^-]$ of 7.3 x 10^{-3} M from Equation 22.9 is, indeed, negligible. Thus, if $NH_3(aq)$ is added to an aqueous solution in equilibrium with $AgBrO_3(s)$, then the solubility of $AgBrO_3(s)$ is enhanced as a result of the formation of the complex ion $[Ag(NH_3)_2]^+(aq)$. The shift in the solubility equilibrium from left to right in Equation 22.10 leads to an increased amount of dissolved $AgBrO_3(s)$.

We also can calculate the concentrations of the other species, $Ag^+(aq)$ and $NH_3(aq)$. The concentration of $Ag^+(aq)$ is given by

$$K_{sp} = [Ag^+][BrO_3^-] = [Ag^+](0.25\text{ M}) = 5.4 \times 10^{-5}\text{ M}^2$$

so that

$$[Ag^+] = 2.2 \times 10^{-4}\text{ M}$$

Because all the dissolved silver from the $AgBrO_3(s)$ ends up as either $Ag^+(aq)$ ions or $[Ag(NH_3)_2]^+(aq)$ complex ions in the solution, the total solubility of $AgBrO_3(s)$ is given by

$$s = [Ag^+] + [Ag(NH_3)_2^+] = 2.2 \times 10^{-4}\text{ M} + 0.25\text{ M} = 0.25\text{ M}$$

to two significant figures. The formation of the $[Ag(NH_3)_2]^+(aq)$ complex ion increases the solubility of $AgBrO_3(s)$ in 0.50 M $NH_3(aq)$ about 35-fold relative to its solubility in pure water (0.25 M versus 0.0073 M). Figure 22.3 shows the increase in the solubility of $AgBrO_3(s)$ with the addition of $NH_3(aq)$.

If we try to calculate $[NH_3]$ by using $0.05\text{ M} - 2x$ from the concentration table, we find that $[NH_3] = 0.50\text{ M} - (2)(0.25\text{ M}) = 0\text{ M}$. This result does not imply that $[NH_3]$ is actually 0 M, only that it is equal to 0 M within the numerical accuracy given. This result means that the solubility of the $AgBrO_3(s)$ is limited by the 0.50 M $NH_3(aq)$. With higher $[NH_3]$, the solubility would be even greater. The actual concentration of $NH_3(aq)$ is most easily calculated using the equilibrium-constant expression of Equation 22.8:

$$K_f = \frac{[Ag(NH_3)_2^+]}{[Ag^+][NH_3]^2} = 2.0 \times 10^7\text{ M}^{-2}$$

Using the values of $[Ag(NH_3)_2^+]$ and $[Ag^+]$ that we have already calculated, we find

$$[NH_3]^2 = \frac{[Ag(NH_3)_2^+]}{[Ag^+]K_f} = \frac{0.25\text{ M}}{(2.2 \times 10^{-4}\text{ M})(2.0 \times 10^7\text{ M}^{-2})} = 5.7 \times 10^{-5}\text{ M}^2$$

or

$$[NH_3] = (5.7 \times 10^{-5}\text{ M}^2)^{1/2} = 7.5 \times 10^{-3}\text{ M}$$

Figure 22.3 Silver bromate is only slightly soluble in water at 25°C. If $NH_3(aq)$ is added to a saturated solution of $AgBrO_3(aq)$, the solubility of $AgBrO_3(s)$ is increased as a result of the formation of $[Ag(NH_3)_2]^+(aq)$ complex ions. (*left*) Test tube with 1.0 gram of $AgBrO_3(s)$. (*center*) Test tube with 1.0 gram of $AgBrO_3(s)$ and 20 mL of water. (*right*) Test tube with 1.0 gram of $AgBrO_3(s)$ and 20 mL of 6 M $NH_3(aq)$.

EXAMPLE 22-5: Calculate the solubility of $AgCl(s)$ in 1.0 M $NH_3(aq)$.

Solution: The two relevant equations are

$$AgCl(s) \leftrightharpoons Ag^+(aq) + Cl^-(aq) \qquad K_{sp} = 1.8 \times 10^{-10}\text{ M}^2 \qquad (22.12)$$

and

$$Ag^+(aq) + 2NH_3(aq) \leftrightharpoons [Ag(NH_3)_2]^+(aq) \qquad K_f = 2.0 \times 10^7\text{ M}^{-2} \qquad (22.13)$$

By adding these two equations, we obtain

$$AgCl(s) + 2NH_3(aq) \leftrightharpoons [Ag(NH_3)_2]^+(aq) + Cl^-(aq) \qquad (22.14)$$

The corresponding equilibrium-constant expression is given by the product of K_{sp} and K_f, or

$$K_c = K_{sp}K_f = (1.8 \times 10^{-10}\ M^2)(2.0 \times 10^7\ M^{-2}) = 3.6 \times 10^{-3}$$

As in the previous calculation, because $K_c >> K_{sp}$, we shall ignore Equation 22.12 and use only Equation 22.14 to set up a concentration table.

Concentration	$AgCl(s)$	+	$2NH_3(aq)$	$\leftrightharpoons$	$[Ag(NH_3)_2]^+(aq)$	+	$Cl^-(aq)$
initial	—		1.0 M		0 M		≈ 0 M
change	—		$-2x$		$+x$		$+x$
equilibrium	—		$1.0\ M - 2x$		x		x

The equilibrium-constant expression is

$$K_c = \frac{x^2}{(1.0\ M - 2x)^2} = 3.6 \times 10^{-3}$$

Taking the square root of both sides gives

$$\frac{x}{1.0\ M - 2x} = \pm 0.060$$

The negative root leads to a negative value of x, and the positive root gives $x = 0.054$ M. Thus, the equilibrium values are as follows:

Concentration	$AgCl(s)$	+	$2NH_3(aq)$	$\leftrightharpoons$	$[Ag(NH_3)_2]^+(aq)$	+	$Cl^-(aq)$
initial	—		1.0 M		0 M		≈ 0 M
change	—		− 2(0.054 M)		+ 0.054 M		+ 0.054 M
equilibrium	—		0.89 M		0.054 M		0.054 M

We can calculate $[Ag^+]$ by using the equilibrium-constant expression for Equation 22.13:

$$K_f = \frac{[Ag(NH_3)_2{}^+]}{[Ag^+][NH_3]^2} = 2.0 \times 10^7\ M^{-2}$$

or

$$[Ag^+] = \frac{[Ag(NH_3)_2{}^+]}{(2.0 \times 10^7\ M^{-2})[NH_3]^2} = \frac{0.054\ M}{(2.0 \times 10^7\ M^{-2})(0.89\ M)^2} = 3.4 \times 10^{-9}\ M$$

The total solubility of $AgCl(s)$ is given by

$$s = [Ag^+] + [Ag(NH_3)_2{}^+] = 3.4 \times 10^{-9}\ M + 0.054\ M \approx 0.054\ M$$

Notice that the solubility of $AgCl(s)$ in 1.0 M $NH_3(aq)$ is about 4000 times greater than it is in pure water (0.054 M versus 1.3×10^{-5} M).

PRACTICE PROBLEM 22-5: Calculate the solubility of $AgCl(s)$ at 25°C in an aqueous solution that is 0.100 M in $NH_3(aq)$ and 0.100 M in $NaCl(aq)$, given the following equilibrium data:

$$AgCl(s) \leftrightharpoons Ag^+(aq) + Cl^-(aq) \qquad K_{sp} = 1.8 \times 10^{-10}\ M^2$$

$$Ag^+(aq) + 2NH_3(aq) \leftrightharpoons [Ag(NH_3)_2]^+(aq) \qquad K_f = 2.0 \times 10^7\ M^{-2}$$

Answer: 3.6×10^{-4} M; 0.052 g·L^{-1}

The formation of complex ions plays an important role in the separation of mixtures. For example, a mixture containing $Ag^+(aq)$ and $Hg_2^{2+}(aq)$ cations can be separated by first adding $Cl^-(aq)$ to precipitate the two cations as $AgCl(s)$ and $Hg_2Cl_2(s)$, which are quite insoluble in water. If we then add $NH_3(aq)$ to this mixture, the $AgCl(s)$ reacts with the $NH_3(aq)$ according to Equation 22.14 to form soluble $[Ag(NH_3)_2]^+(aq)$ complex ions. Because the $Hg_2Cl_2(s)$ reacts with $NH_3(aq)$ to form insoluble products, it remains as a precipitate. We can now centrifuge the solution, resulting in a clear solution containing the dissolved silver ions and a precipitate containing the mercury(I) ions. We then either decant the solution or filter it to completely separate the silver from the mercury (Figure 22.4). Processes such as these are often used in reclamation of metals from chemical waste such as that generated during laboratory experiments, and in qualitative analysis (see Section 22-8).

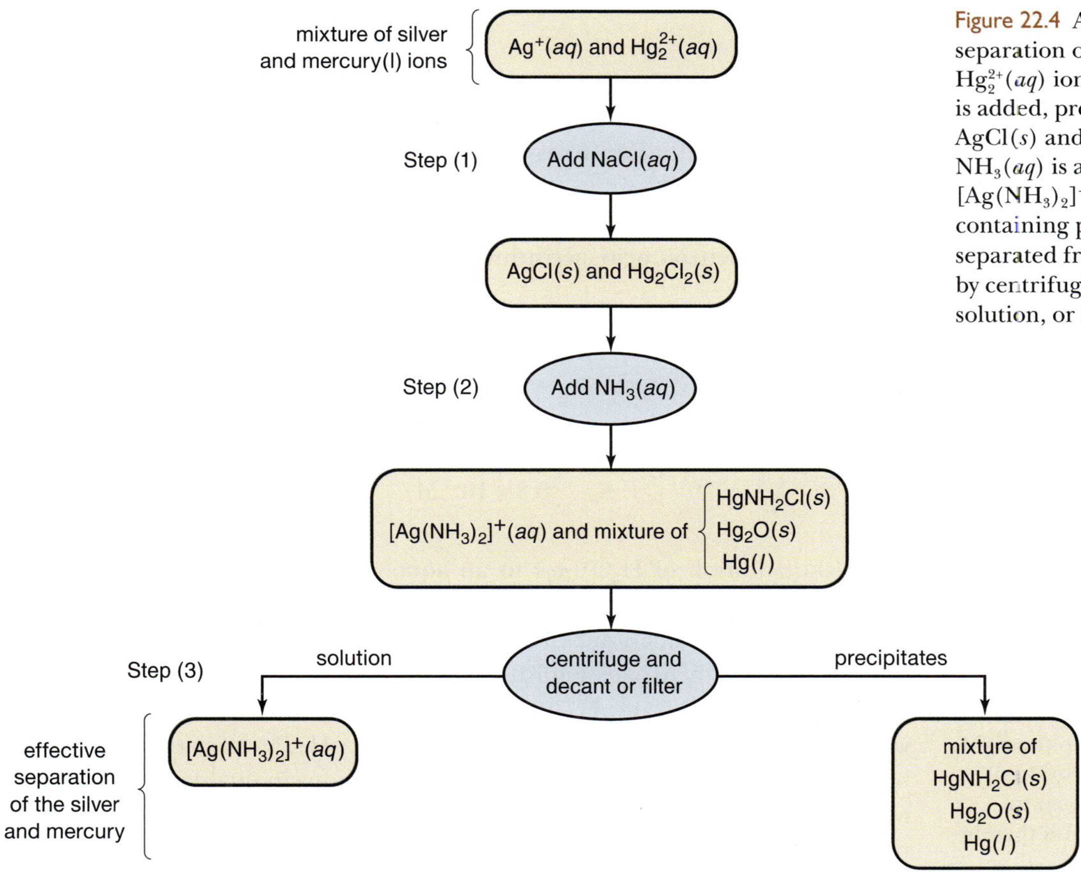

Figure 22.4 A flowchart showing the separation of $Ag^+(aq)$ ions from $Hg_2^{2+}(aq)$ ions. In step (1) $NaCl(aq)$ is added, precipitating the ions as $AgCl(s)$ and $Hg_2Cl_2(s)$. In step (2) $NH_3(aq)$ is added, forming soluble $[Ag(NH_3)_2]^+(aq)$ ions. The mercury containing precipitate can now be separated from the dissolved silver by centrifuging and decanting the solution, or by filtration.

22-4. Salts of Weak Acids Are More Soluble in Acidic Solutions Than in Neutral or Basic Solutions

Sodium benzoate, $NaC_6H_5COO(s)$, is the salt of the strong base sodium hydroxide and the weak acid benzoic acid, $C_6H_5COOH(s)$. It has the Lewis formula

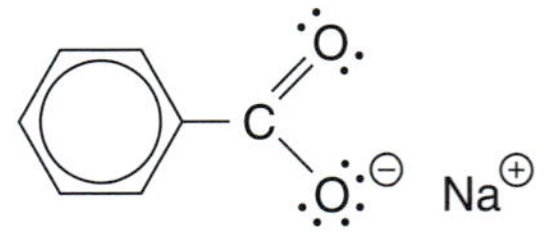

Sodium benzoate is a water-soluble food additive that functions as an antimicrobial agent in foods with pH values lower than about 4. Numerous acidic beverages, syrups, jams, jellies, and processed fruits contain about 0.1% sodium benzoate to prevent the growth of yeasts and harmful bacteria. Many benzoate salts are insoluble above pH = 4, and the formation of insoluble benzoate salts at these higher pH values removes benzoate ions from solution, thus rendering the benzoate ion ineffective as an antimicrobial agent. At pH ≤ 4, an appreciable fraction of the benzoate ion exists as benzoic acid ($K_a = 6.3 \times 10^{-5}$ M at 25°C), which is the biologically active form.

We shall use the benzoate salt, silver benzoate, $AgC_6H_5COO(s)$, to illustrate the effect of pH on the solubility of a salt of a weak acid. The equation for the solubility equilibrium of silver benzoate is

$$AgC_6H_5COO(s) \leftrightharpoons Ag^+(aq) + C_6H_5COO^-(aq) \qquad (22.15)$$

with

$$K_{sp} = [Ag^+][C_6H_5COO^-] = 2.5 \times 10^{-5}\ M^2$$

The solubility equilibrium fixes the value of the product $[Ag^+][C_6H_5COO^-]$. We must recognize, however, that the benzoate ion is the conjugate base of a weak acid and so in the presence of acid some of it reacts with $H_3O^+(aq)$ to form undissociated benzoic acid, according to the equation

$$C_6H_5COO^-(aq) + H_3O^+(aq) \leftrightharpoons C_6H_5COOH(aq) + H_2O(l) \qquad (22.16)$$

with

$$K_c = \frac{1}{K_a} = \frac{1}{6.3 \times 10^{-5}\ M} = 1.6 \times 10^4\ M^{-1}$$

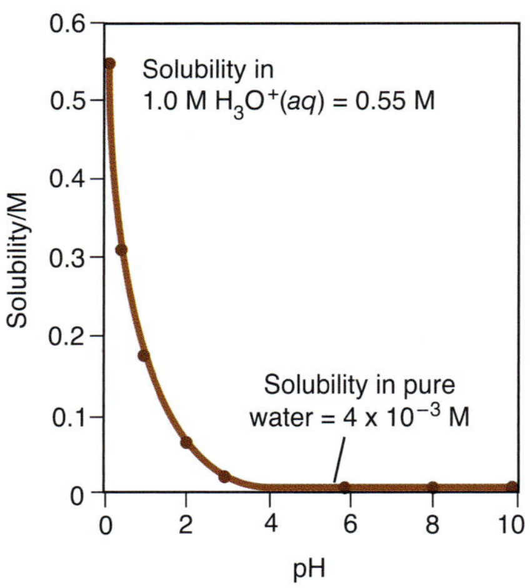

Figure 22.5 Solubility of silver benzoate in water as a function of pH. The addition of $H_3O^+(aq)$ shifts the solubility equilibrium to the right (increase in solubility) as $H_3O^+(aq)$ combines with $C_6H_5COO^-(aq)$ to produce $C_6H_5COOH(aq)$ and thereby to decrease $[C_6H_5COO^-]$.

Thus, the addition of $H_3O^+(aq)$ to an aqueous solution in equilibrium with $AgC_6H_5COO(s)$ decreases $[C_6H_5COO^-]$ and thus shifts the solubility equilibrium described by Equation 22.15 from left to right, thereby increasing the solubility of the silver benzoate. Figure 22.5 shows the solubility of $AgC_6H_5COO(s)$ in water as a function of pH. It also illustrates that the solubility of a salt of a weak acid increases dramatically with decreasing pH. The variation in the solubility of such salts with pH is utilized to separate ions in qualitative analysis, as we shall see in Section 22-8. This phenomenon is also responsible for the formation of limestone caves (sidebox).

pH, SOLUBILITY, AND THE GEOLOGY OF LIMESTONE CAVE FORMATION

One of the most dramatic examples of the solubility of ionic compounds as a function of pH can be seen in the formation of limestone caves. Limestone is a mineral that is predominantly composed of calcium carbonate, $CaCO_3(s)$. Limestone has a low solubility in pure water,

$$\text{(1)} \qquad CaCO_3(s) \leftrightharpoons Ca^{2+}(aq) + CO_3^{2-}(aq) \qquad K_{sp} = 3.4 \times 10^{-9}\ M^2$$

However, the pH of natural waters tends to be slightly acidic due to the dissolution of $CO_2(g)$ according to the acid-forming reaction described by

$$\text{(2)} \qquad CO_2(g) + 2H_2O(l) \leftrightharpoons H_3O^+(aq) + HCO_3^-(aq)$$

Because $CO_3^{2-}(aq)$ is a weak base, it reacts with $H_3O^+(aq)$ according to

$$\text{(3)} \qquad CO_3^{2-}(aq) + H_3O^+(aq) \rightarrow HCO_3^-(aq) + H_2O(l)$$

The removal of $CO_3^-(aq)$ by way of equation (3) increases the solubility of $CaCO_3(s)$ in equation (1). Thus, the presence of $H_3O^+(aq)$ increases the solubility of $CaCO_3(s)$, helping groundwater to carve out large limestone caves over millions of years. The overall reaction for the dissolution of limestone in groundwater is obtained by adding equations (1), (2), and (3) to give

$$CaCO_3(s) + CO_2(aq) + H_2O(l) \leftrightharpoons Ca^{2+}(aq) + 2HCO_3^-(aq)$$

In addition to carving out the caves, groundwater containing dissolved $Ca(HCO_3)_2(aq)$ that drips from the ceilings of these caves can release $CO_2(g)$. This results in the gradual formation of limestone stalactites (hanging from the ceiling of the cave) and stalagmites (built up from the cave floor). Such formations are a magnificent display of the chemistry of solubility!

EXAMPLE 22-6: Calculate the solubility of silver benzoate in an aqueous solution buffered at pH = 2.0 at 25°C. Compare your result with Figure 22.5.

Solution: Adding Equations 22.15 and 22.16 we have the overall reaction equation,

$$AgC_6H_5COO(s) + H_3O^+(aq) \leftrightharpoons Ag^+(aq) + C_6H_5COOH(aq) + H_2O(l)$$

with

$$K_c = K_{sp}\left(\frac{1}{K_a}\right) = (2.5 \times 10^{-5}\ M^2)(1.6 \times 10^4\ M^{-1}) = 0.40\ M$$

For every formula unit of $AgC_6H_5COO(s)$ that dissolves, one $Ag^+(aq)$ ion occurs in solution. Therefore, the solubility can be expressed as $s = [Ag^+]$. Because the value of K_c for this reaction is much greater than the value of K_{sp} for Equation 22.15 we can ignore the additional contribution to s from

Equation 22.15. The pH is buffered at 2.0, so $[H_3O^+]$ is essentially constant and may be taken as equal to 1.0×10^{-2} M throughout the reaction. In addition, according to the overall reaction stoichiometry,

$$[Ag^+] = [C_6H_5COOH]$$

So at equilibrium $s = [Ag^+] = [C_6H_5COOH]$. Constructing a concentration table, we have the following:

Concentration	$AgC_6H_5COO(s)$ +	$H_3O^+(aq)$ ⇋	$Ag^+(aq)$ +	$C_6H_5COOH(aq)$ +	$H_2O(l)$
initial	—	1.0×10^{-2} M	0 M	0 M	—
change	—	(buffered, no change)	$+s$	$+s$	—
equilibrium	—	1.0×10^{-2} M	s	s	—

Substituting these values into the equilibrium expression, we get

$$0.40\text{ M} = \frac{[Ag^+][C_6H_5COOH]}{[H_3O^+]} = \frac{s^2}{1.0 \times 10^{-2}\text{ M}}$$

Solving for s and taking the positive root, we find $s = 0.063$ M, in agreement with Figure 22.5.

PRACTICE PROBLEM 22-6: Calculate the solubility of silver acetate, $AgCH_3COO(s)$, in an aqueous solution buffered at pH = 4.00 at 25°C. Take the value of K_{sp} for silver acetate to be 1.9×10^{-3} M^2.

Answer: 0.10 M, or 17 g·L^{-1}

22-5. The Relative Magnitudes of Q_{sp} and K_{sp} Can Be Used to Predict Whether an Ionic Solid Can Precipitate

In Chapter 19, where we first studied chemical equilibria, we introduced a quantity Q_c called the reaction quotient. The reaction quotient has the same form as an equilibrium constant, but is expressed in terms of *arbitrary* concentrations. We showed that if $Q_c < K_c$, then the reaction proceeds from left to right until equilibrium is established, at which point $Q_c = K_c$. Conversely, if $Q_c > K_c$, then the reaction proceeds from right to left to equilibrium. We can use this same idea here to determine whether a precipitate can form when two solutions are added to each other by defining Q_{sp} as the **concentration quotient** for chemical equations involving solubility equilibria.

Consider the silver bromate solubility equilibrium described by the equation

$$AgBrO_3(s) \leftrightharpoons Ag^+(aq) + BrO_3^-(aq) \qquad (22.17)$$

for which

$$K_{sp} = [Ag^+][BrO_3^-] = 5.4 \times 10^{-5}\text{ M}^2$$

and

$$Q_{sp} = [Ag^+]_0[BrO_3^-]_0$$

If we prepare a solution with arbitrary values of $[Ag^+]_0$ and $[BrO_3^-]_0$, then the

TABLE 22.2 Criteria for the formation of a precipitate from a solution prepared with the constituent ions

For any arbitrary ion concentrations:	
$Q_{sp} > K_{sp}$	precipitate forms
$Q_{sp} < K_{sp}$	no precipitate forms
When equilibrium is disturbed:	
$Q_{sp} > K_{sp}$	more precipitate forms until $Q_{sp} = K_{sp}$
$Q_{sp} < K_{sp}$	precipitate dissolves either until $Q_{sp} = K_{sp}$ or until solid phase disappears completely

criterion for whether or not $AgBrO_3(s)$ will precipitate depends on the relative values of Q_{sp} and K_{sp}. If $Q_{sp} > K_{sp}$, then precipitation will occur until $Q_{sp} = K_{sp}$, that is, until equilibrium is established. If, on the other hand, $Q_{sp} < K_{sp}$, then precipitation will not occur and it is possible to dissolve additional $AgBrO_3(s)$ until $Q_{sp} = K_{sp}$ and equilibrium is established.

If we already have a saturated solution of $AgBrO_3(aq)$ that is at equilibrium and the equilibrium is disturbed in such a way that Q_{sp} becomes less than K_{sp}, then additional $AgBrO_3(s)$ dissolves until the solution becomes saturated or all the $AgBrO_3(s)$ dissolves. These conditions are summarized in Table 22.2.

For example, suppose we mix 50.0 mL of 1.0 M $AgNO_3(aq)$ with 50.0 mL of 0.010 M $NaBrO_3(aq)$ at 25°C. Does $AgBrO_3(s)$ precipitate? The value of K_{sp} for $AgBrO_3(s)$ at 25°C is 5.4×10^{-5} M^2. The initial concentrations of $Ag^+(aq)$ and $BrO_3^-(aq)$ in the 100.0-mL mixture of the two solutions are

$$[Ag^+]_0 = \frac{(50.0\text{ mL})(1.0\text{ M})}{100.0\text{ mL}} = 0.50\text{ M}$$

and

$$[BrO_3^-]_0 = \frac{(50.0\text{ mL})(0.010\text{ M})}{100.0\text{ mL}} = 0.0050\text{ M}$$

The initial value of Q_{sp} for the mixture is

$$Q_{sp} = [Ag^+]_0[BrO_3^-]_0 = (0.50\text{ M})(0.0050\text{ M}) = 2.5 \times 10^{-3}\text{ M}^2$$

The fact that $Q_{sp} > K_{sp}$ means that $AgBrO_3(s)$ can precipitate. Once started, precipitation continues until $Q_{sp} = K_{sp}$, that is, until equilibrium is attained.

Although the relative values of Q_{sp} and K_{sp} for a stated equation tell us whether precipitation is possible, they do not guarantee that precipitation will actually occur. A situation may arise where the initial formation of crystals (nucleation) is very slow, leading to a **supersaturated solution** (Figure 22.6). Such situations are not common, however, and so we shall assume that precipitation is rapid in the cases considered in this chapter.

(a) (b) (c)

Figure 22.6 A supersaturated solution of sodium acetate, $NaCH_3OO(aq)$. When a single seed crystal of $NaCH_3OO(s)$ is added to the solution rapid precipitation occurs.

EXAMPLE 22-7: A 1.0×10^{-3} M $NaIO_3(aq)$ solution is made 0.010 M in $Cu^{2+}(aq)$ by dissolving the soluble salt $Cu(ClO_4)_2(s)$. Does $Cu(IO_3)_2(s)$ ($K_{sp} = 7.4 \times 10^{-8}\ M^3$) precipitate from the solution at 25°C?

Solution: The chemical equation for the solubility equilibrium is

$$Cu(IO_3)_2(s) \leftrightharpoons Cu^{2+}(aq) + 2\ IO_3^-(aq)$$

The value of Q_{sp} is

$$Q_{sp} = [Cu^{2+}]_0[IO_3^-]_0^2 = (0.010\ M)(1.0 \times 10^{-3}\ M)^2 = 1.0 \times 10^{-8}\ M^3$$

Because $Q_{sp} < K_{sp}$, no precipitate of $Cu(IO_3)_2(s)$ forms.

PRACTICE PROBLEM 22-7: Will a precipitate of PbI_2 form if we mix 30.0 mL of 0.050 M $Pb(NO_3)_2(aq)$ with 50.0 mL of 0.015 M $KI(aq)$ at 25°C?

Answer: yes; $Q_{sp} = 1.7 \times 10^{-6}\ M^3$ and $K_{sp} = 9.8 \times 10^{-9}\ M^3$

EXAMPLE 22-8: Suppose we mix 50.0 mL of 1.00×10^{-2} M $AgNO_3(aq)$ with 100.0 mL of 2.00×10^{-4} M $K_2CrO_4(aq)$ at 25°C. Does $Ag_2CrO_4(s)$ precipitate from the solution? If yes, then calculate how many millimoles of $Ag_2CrO_4(s)$ precipitate.

Solution: The chemical equation for the precipitation reaction is

$$2Ag^+(aq) + CrO_4^{2-}(aq) \leftrightharpoons Ag_2CrO_4(s) \qquad (22.18)$$

and from Table 22.1, we have for the solubility equilibrium defined by the reverse of Equation 22.18

$$[Ag^+]^2[CrO_4^{2-}] = K_{sp} = 1.1 \times 10^{-12}\ M^3$$

The concentration of $Ag^+(aq)$ immediately after mixing is

$$[Ag^+]_0 = \frac{(50.0\ mL)(1.00 \times 10^{-2}\ M)}{150.0\ mL} = 3.33 \times 10^{-3}\ M$$

and the concentration of $CrO_4^{2-}(aq)$ immediately after mixing is

$$[CrO_4^{2-}]_0 = \frac{(100.0\ mL)(2.00 \times 10^{-4}\ M)}{150.0\ mL} = 1.33 \times 10^{-4}\ M$$

Therefore, the value of Q_{sp} is

$$Q_{sp} = [Ag^+]_0^2[CrO_4^{2-}]_0 = (3.33 \times 10^{-3}\ M)^2(1.33 \times 10^{-4}\ M) = 1.47 \times 10^{-9}\ M^3$$

Because $Q_{sp} > K_{sp}$, precipitation of $Ag_2CrO_4(s)$ results.

To determine how much $Ag_2CrO_4(s)$ precipitates, we must determine which, if either, reactant in Equation 22.18 is a limiting reactant. The amount of $Ag^+(aq)$ added is given by $(50.0\ mL)(1.00 \times 10^{-2}\ M) = 0.500$ mmol, and the amount of $CrO_4^{2-}(aq)$ added is given by $(100.0\ mL)(2.00 \times 10^{-4}\ M) =$

0.0200 mmol. Because the 0.0200 mmol of $CrO_4^{2-}(aq)$ requires only 0.0400 mmol of $Ag^+(aq)$ to react completely, we see that $Ag^+(aq)$ is in great excess and that $CrO_4^{2-}(aq)$ is a limiting reactant. Therefore, essentially 0.0200 mmol of $Ag_2CrO_4(s)$ precipitates.

Because the value of K_{sp} for the dissolution of $Ag_2CrO_4(s)$ is extremely small (1.1×10^{-12} M^3), we can neglect the amount of product that remains in solution compared to the amount that precipitates.

PRACTICE PROBLEM 22-8: Suppose we mix 30.0 mL of 0.025 M $Hg_2(CH_3COO)_2(aq)$ with 20.0 mL of 0.0065 M $K_2SO_4(aq)$ at 25°C. Does a precipitate form? If so, what is the mass of the precipitate?

Answer: yes; 0.065 grams

22-6. It Is Often Possible to Separate One Compound from Another by Selective Precipitation

Consider a solution that is 0.010 M in $Pb^{2+}(aq)$ and 0.010 M in $Hg_2^{2+}(aq)$. The iodide salt of each of these ions is insoluble in pure water:

$$PbI_2(s) \leftrightharpoons Pb^{2+}(aq) + 2I^-(aq) \qquad K_{sp} = 9.8 \times 10^{-9}\ \text{M}^3$$

$$Hg_2I_2(s) \leftrightharpoons Hg_2^{2+}(aq) + 2I^-(aq) \qquad K_{sp} = 5.2 \times 10^{-29}\ \text{M}^3$$

Is it possible to separate the $Pb^{2+}(aq)$ and $Hg_2^{2+}(aq)$ ions completely by selectively precipitating with $I^-(aq)$? To answer this question, we must first agree on what we mean by "completely." Let's say that we want to precipitate 99.99% of the $Hg_2^{2+}(aq)$ without precipitating $Pb^{2+}(aq)$. The concentration of $Hg_2^{2+}(aq)$ remaining in solution would be 0.01% of 0.010 M, or 1.0×10^{-6} M. This quantity is clearly very small in comparison with the initial amount of $Hg_2^{2+}(aq)$, so we shall consider it a negligible residual. You can never reduce the residual $[Hg_2^{2+}]$ to 0 M, because $[Hg_2^{2+}][I^-]^2$ must equal 5.2×10^{-29} M^3.

The concentration of $I^-(aq)$ required to give a value of $[Hg_2^{2+}]$ of 1.0×10^{-6} M is given by the solubility-product expression of $Hg_2I_2(s)$:

$$K_{sp} = [Hg_2^{2+}][I^-]^2 = 5.2 \times 10^{-29}\ \text{M}^3$$

Because we wish the value of $[Hg_2^{2+}]$ to be 1.0×10^{-6} M, the required value of $[I^-]$ is

$$[I^-] = \left(\frac{K_{sp}}{[Hg_2^{2+}]}\right)^{1/2} = \left(\frac{5.2 \times 10^{-29}\ \text{M}^3}{1.0 \times 10^{-6}\ \text{M}}\right)^{1/2} = 7.2 \times 10^{-12}\ \text{M}$$

Although we want to precipitate the $Hg_2^{2+}(aq)$ ions as $Hg_2I_2(s)$, we do not want to precipitate the $Pb^{2+}(aq)$ ions. In order to achieve an effective separation, we must ensure that the $Pb^{2+}(aq)$ ions remain in solution. Therefore, we must determine whether this concentration of $[I^-]$ will also precipitate $PbI_2(s)$. To decide this, we calculate Q_{sp} for $PbI_2(s)$:

$$Q_{sp} = [Pb^{2+}]_0[I^-]_0^2 = (0.010\ \text{M})(7.2 \times 10^{-12}\ \text{M})^2 = 5.2 \times 10^{-25}\ \text{M}^3$$

Thus, $Q_{sp} < K_{sp}$ for $PbI_2(s)$, so no $Pb^{2+}(aq)$ will precipitate. We conclude that "complete" separation of $Hg_2^{2+}(aq)$ and $Pb^{2+}(aq)$ can be achieved. We say that the $Pb^{2+}(aq)$ and the $Hg_2^{2+}(aq)$ have been separated by **selective precipitation**.

EXAMPLE 22-9: A solution is 0.100 M in $Ag^+(aq)$ and 0.100 M in $Pb^{2+}(aq)$. Can these two ions be separated by precipitation with $Cl^-(aq)$? Assume that "separation" means that 99.0% of either ion is precipitated without precipitating the other ion.

Solution: The two solubility equilibrium equations and their respective solubility-product constants are

$$AgCl(s) \leftrightharpoons Ag^+(aq) + Cl^-(aq) \qquad K_{sp} = 1.8 \times 10^{-10}\ M^2$$
$$PbCl_2(s) \leftrightharpoons Pb^{2+}(aq) + 2\,Cl^-(aq) \qquad K_{sp} = 1.5 \times 10^{-5}\ M^3$$

The minimum values of $[Cl^-]$ required to initiate precipitation of $Ag^+(aq)$ and $Pb^{2+}(aq)$ are given by solving each solubility-product expression for $[Cl^-]$. For $Ag^+(aq)$, we have

$$K_{sp} = [Ag^+][Cl^-]$$

so that

$$[Cl^-] = \frac{K_{sp}}{[Ag^+]} = \frac{1.8 \times 10^{-10}\ M^2}{0.100\ M} = 1.8 \times 10^{-9}\ M$$

For $Pb^{2+}(aq)$, we have

$$K_{sp} = [Pb^{2+}][Cl^-]^2$$

so

$$[Cl^-] = \left(\frac{K_{sp}}{[Pb^{2+}]}\right)^{1/2} = \left(\frac{1.5 \times 10^{-5}\ M^3}{0.100\ M}\right)^{1/2} = 1.2 \times 10^{-2}\ M$$

Because a much smaller concentration of $Cl^-(aq)$ is required to precipitate the $Ag^+(aq)$ ions than to precipitate the $Pb^{2+}(aq)$ ions, the $Ag^+(aq)$ will precipitate before the $Pb^{2+}(aq)$.

If we precipitate 99.0% of the $Ag^+(aq)$ according to the statement of the problem, then the concentration of $Ag^+(aq)$ that would remain in solution is 1.0% of 0.100 M, or 1.0×10^{-3} M. The concentration of $Cl^-(aq)$ that will give this concentration of $Ag^+(aq)$ is obtained from

$$[Cl^-] = \frac{K_{sp}}{[Ag^+]} = \frac{1.8 \times 10^{-10}\ M^2}{1.0 \times 10^{-3}\ M} = 1.8 \times 10^{-7}\ M$$

To determine whether $Pb^{2+}(aq)$ will or will not precipitate with this concentration of $Cl^-(aq)$, we calculate Q_{sp}:

$$Q_{sp} = [Pb^{2+}]_0\,[Cl^-]_0^2 = (0.100\ M)(1.8 \times 10^{-7}\ M)^2 = 3.2 \times 10^{-15}\ M^3$$

We see that $Q_{sp} < K_{sp}$ for $PbCl_2(s)$. Therefore, $PbCl_2(s)$ will not precipitate, and the separation of the $Ag^+(aq)$ and $Pb^{2+}(aq)$ can be accomplished under these conditions.

PRACTICE PROBLEM 22-9: A solution is 0.225 M in $KCl(aq)$ and 0.175 M in $KI(aq)$. Can $Cl^-(aq)$ be separated from $I^-(aq)$ by selective precipitation with $Ag^+(aq)$? Assume that "separation" means that 99.0% of either ion is precipitated without precipitating the other.

Answer: yes, by adding $Ag^+(aq)$ at a concentration between 5×10^{-14} M and 8×10^{-8} M

Aqua regia (Latin for royal water) is a corrosive, fuming yellow or red solution of concentrated nitric acid and concentrated hydrochloric acid in the ratio 1:3 by volume. It was so named because it can dissolve the "royal metals," gold and platinum. Aqua regia was known and used by alchemists since the early 800s.

Selective precipitation is another important technique used for separating ions in solution during qualitative analysis and the cleaning up of chemical waste. For example, selective precipitation of gold from an aqua regia solution (which dissolves most metals) is one of the techniques used to recover gold from discarded electronic circuit boards.

22-7. Amphoteric Metal Hydroxides Dissolve in Both Highly Acidic and Highly Basic Solutions

Many metal oxides and hydroxides that are insoluble in neutral aqueous solutions dissolve in *both* acidic and basic solutions. Such hydroxides are called **amphoteric metal hydroxides**. Aluminum hydroxide, $Al(OH)_3(s)$, is an example (Figure 22.7). In acidic solutions, $Al(OH)_3(s)$ dissolves because of a reaction that is similar to an acid-base neutralization reaction, as described by

$$Al(OH)_3(s) + 3H_3O^+(aq) \leftrightharpoons Al^{3+}(aq) + 6H_2O(l) \qquad (22.19)$$

and in basic solutions, it dissolves because of the formation of a soluble hydroxy complex ion, $[Al(OH)_4]^-(aq)$, according to

$$Al(OH)_3(s) + OH^-(aq) \leftrightharpoons [Al(OH)_4]^-(aq) \qquad (22.20)$$

Figure 22.7 Aluminum hydroxide, $Al(OH)_3(s)$, occurs as a white, flocculent precipitate that is used to clarify water.

The total solubility of $Al(OH)_3(s)$ at any pH is given by

$$s = [Al^{3+}] + [Al(OH)_4^-] \qquad (22.21)$$

The value of $[Al^{3+}]$ can be obtained from the solubility-product constant expression of $Al(OH)_3(s)$,

$$K_{sp} = [Al^{3+}][OH^-]^3 = 1.3 \times 10^{-33}\ M^4 \qquad (22.22)$$

and that of $[Al(OH)_4^-]$ from the equilibrium-constant expression for the reaction described by Equation 22.20,

$$K_f = \frac{[Al(OH)_4^-]}{[OH^-]} = 40 \qquad (22.23)$$

where the value of K_f is given in Table 22.3.

Let's determine the solubility at 25°C of $Al(OH)_3(s)$ in a base at pH = 12.00. When the value of the pH is 12.00, the value of pOH is 2.00 and that of $[OH^-]$ is 1.0×10^{-2} M, so Equation 22.22 gives

$$[Al^{3+}] = \frac{1.3 \times 10^{-33}\ M^4}{[OH^-]^3} = \frac{1.3 \times 10^{-33}\ M^4}{(1.0 \times 10^{-2}\ M)^3} = 1.3 \times 10^{-27}\ M$$

TABLE 22.3 Equilibrium constants for the formation of metal hydroxy complex ions in water at 25°C

Reaction equation	K_f
$Al(OH)_3(s) + OH^-(aq) \leftrightharpoons [Al(OH)_4]^-(aq)$	40
$Pb(OH)_2(s) + OH^-(aq) \leftrightharpoons [Pb(OH)_3]^-(aq)$	0.08
$Zn(OH)_2(s) + 2OH^-(aq) \leftrightharpoons [Zn(OH)_4]^{2-}(aq)$	0.05 M^{-1}
$Cr(OH)_3(s) + OH^-(aq) \leftrightharpoons [Cr(OH)_4]^-(aq)$	0.04
$Sn(OH)_2(s) + OH^-(aq) \leftrightharpoons [Sn(OH)_3]^-(aq)$	0.01

and Equation 22.23 gives

$$[Al(OH)_4^-] = 40[OH^-] = (40)(1.0 \times 10^{-2}\ M) = 0.4\ M$$

Therefore, the solubility of $Al(OH)_3(s)$ at pH = 12.00 is

$$s = [Al^{3+}] + [Al(OH)_4^-] = 1.3 \times 10^{-27}\ M + 0.4\ M \approx 0.4\ M$$

If we computed the solubility of $Al(OH)_3(s)$ at a pH of 12.00 without considering the reaction described by Equation 22.20, then our result would be in error by a factor of about 10^{26}!

At lower values of pH (from 4.00 to 10.00), the reactions described by Equations 22.19 and 22.20 yield very little $Al^{3+}(aq)$ or $[Al(OH)_4]^-(aq)$. For example, at pH = 9.00, $[OH^-] = 1.0 \times 10^{-5}$ M, so Equation 22.22 gives

$$[Ag^{3+}] = \frac{1.3 \times 10^{-33}\ M^4}{[OH^-]^3} = \frac{1.3 \times 10^{-33}\ M^4}{(1.0 \times 10^{-5}\ M)^3} = 1.3 \times 10^{-18}\ M$$

and Equation 22.23 gives

$$[Al(OH)_4^-] = 40[OH^-] = (40)(1.0 \times 10^{-5}\ M) = 4 \times 10^{-4}\ M$$

so that

$$s = 1.3 \times 10^{-18}\ M + 4 \times 10^{-4}\ M \approx 4 \times 10^{-4}\ M$$

For lower values of pH, however, Equation 22.19 gives significant values of $Al^{3+}(aq)$. For example, at pH = 3.00, we have from Equation 22.22,

$$[Al^{3+}] = \frac{1.3 \times 10^{-33}\ M^4}{(1.0 \times 10^{-11}\ M)^3} = 1.3\ M$$

whereas Equation 22.23 gives

$$[Al(OH)_4^-] = 40[OH^-] = (40)(1.0 \times 10^{-11}\ M) = 4 \times 10^{-10}\ M$$

so that

$$s = 1.3\ M + 4 \times 10^{-10}\ M \approx 1.3\ M$$

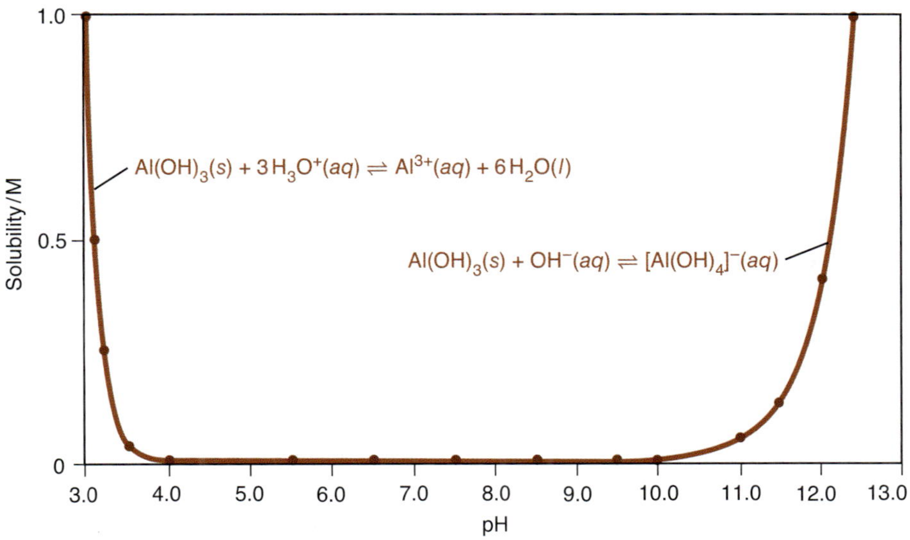

Figure 22.8 The solubility of $Al(OH)_3(s)$ as a function of pH. The amphoteric nature of $Al(OH)_3(s)$ is clearly shown by its solubility in both highly acidic and highly basic solutions. Note that $Al(OH)_3(s)$ is essentially insoluble over the pH range 4 to 10.

Thus, we see that $Al(OH)_3(s)$ dissolves in acidic solutions and in basic solutions, but not in neutral solutions. The amphoteric behavior of $Al(OH)_3(s)$ is illustrated in Figure 22.8. Some other examples of amphoteric metal hydroxides are given in Table 22.3. The solubility of amphoteric metal salts as a function of pH can sometimes result in the redissolution of a metal hydroxide precipitate upon the addition of excess hydroxide ions (Figure 22.9).

EXAMPLE 22-10: Use the equilibrium-constant data for zinc hydroxide, $Zn(OH)_2(s)$, in Table 22.3 to calculate its solubility in a solution buffered at pH = 7.0, 10.0, and 14.0 at 25°C.

Solution: The two relevant equilibrium expressions are

$$Zn(OH)_2(s) \leftrightharpoons Zn^{2+}(aq) + 2\,OH^-(aq)$$

and

$$Zn(OH)_2(s) + 2\,OH^-(aq) \leftrightharpoons [Zn(OH)_4]^{2-}(aq)$$

and the corresponding equilibrium-constant expressions are

$$K_{sp} = [Zn^{2+}][OH^-]^2 = 1.0 \times 10^{-15}\ M^3$$

and

$$K_f = \frac{[Zn(OH)_4^{2-}]}{[OH^-]^2} = 0.05\ M^{-1}$$

The solubility of $Zn(OH)_2(s)$ at any pH is given by

$$s = [Zn^{2+}] + [Zn(OH)_4^{2-}]$$

At pH = 7.0, $[OH^-] = 1.0 \times 10^{-7}$ M, so

$$[Zn^{2+}] = \frac{1.0 \times 10^{-15}\ M^3}{(1.0 \times 10^{-7}\ M)^2} = 0.10\ M$$

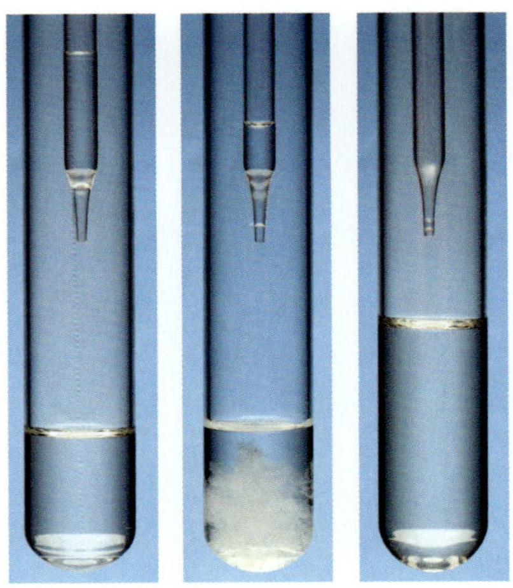

Figure 22.9 (a) A solution containing $Al(NO_3)_3(aq)$, a highly soluble salt, is clear. (b) Upon the addition of a few drops of dilute $NaOH(aq)$, a white precipitate of $Al(OH)_3(s)$ forms. (c) When excess $NaOH(aq)$ is added, the solution becomes highly basic (pH > 10) and the precipitate redissolves, forming soluble $[Al(OH)_4]^-(aq)$ complex ions.

and

$$[Zn(OH)_4^{2-}] = (0.05\ M^{-1})(1.0 \times 10^{-7}\ M)^2 = 5 \times 10^{-16}\ M$$

The total solubility of $Zn(OH)_2(s)$ is given by

$$s = [Zn^{2+}] + [Zn(OH)_4^{2-}] = 0.10\ M + 5 \times 10^{-16}\ M = 0.10\ M$$

At pH = 10.0, $[OH^-] = 1.0 \times 10^{-4}$ M, so

$$[Zn^{2+}] = \frac{1.0 \times 10^{-15}\ M^3}{(1.0 \times 10^{-4}\ M)^2} = 1.0 \times 10^{-7}\ M$$

and

$$[Zn(OH)_4^{2-}] = (0.05\ M^{-1})(1.0 \times 10^{-4}\ M)^2 = 5 \times 10^{-10}\ M$$

Thus,

$$s = [Zn^{2+}] + [Zn(OH)_4^{2-}] = 1.0 \times 10^{-7}\ M + 5 \times 10^{-10}\ M = 1.0 \times 10^{-7}\ M$$

At pH = 14.0, $[OH^-] = 1.0$ M, so

$$[Zn^{2+}] = \frac{1.0 \times 10^{-15}\ M^3}{(1.0\ M)^2} = 1.0 \times 10^{-15}\ M$$

and

$$[Zn(OH)_4^{2-}] = (0.05\ M^{-1})(1.0\ M)^2 = 0.05\ M$$

Thus,

$$s = [Zn^{2+}] + [Zn(OH)_4^{2-}] = 1.0 \times 10^{-15}\ M + 0.05\ M = 0.05\ M$$

Notice that the solubility of $Zn(OH)_2(s)$ increases with increasing pH, but not as dramatically as in the case of $Al(OH)_3(s)$, because the value of K_f for the complexation reaction (Table 22.3) is not as large as that of $Al(OH)_3(s)$.

PRACTICE PROBLEM 22-10: Calculate the solubility of $Pb(OH)_2(s)$ in water at 25°C as a function of pH over the range pH = 4.0 to pH = 14.0. Use increments of two pH units.

Answer:

pH	4.0	6.0	8.0	10.0	12.0	14.0
***s*/M**	1.4	1.4×10^{-4}	9×10^{-8}	8×10^{-6}	8×10^{-4}	0.08

22-8. Qualitative Analysis Is the Identification of the Species Present in a Sample

Some of the laboratory work in your introductory chemistry courses is likely to involve **qualitative analysis**. The objective of qualitative analysis is to identify various cations and anions in an unknown solution or in a mixture of solids. In qualitative analysis, we seek only to identify the ions, not to determine their concentrations. The determination of the amounts present or the percentage compositions is called **quantitative analysis.**

The basic approach used in the qualitative analysis of a mixture of ions is to add a reactant that precipitates certain ions but not others. We do so using techniques detailed in the previous sections of this chapter, such as the common ion effect, selective precipitation, and adjustment of pH. Separating some of the ions simplifies the analytical problem because each of the two resulting samples (precipitates and remaining solution) has fewer constituents than the original sample.

An essential feature of a qualitative analysis scheme for a large group of ions is the *successive* removal of subgroups of the ions by precipitation reactions. It is essential to carry out the separation steps in a *systematic* fashion; otherwise, ions that are presumed to have been removed may interfere with subsequent steps in the analytical scheme. One such qualitative analysis scheme for separating out major groups of common cations for subsequent analysis is shown in Figure 22.10. Schemes like this provide a systematic method of separating various groups of ions.

As a simple example, consider the separation and analysis of a solution that may contain either $KNO_3(aq)$ and $AgNO_3(aq)$ or both. Because $AgCl(s)$ is

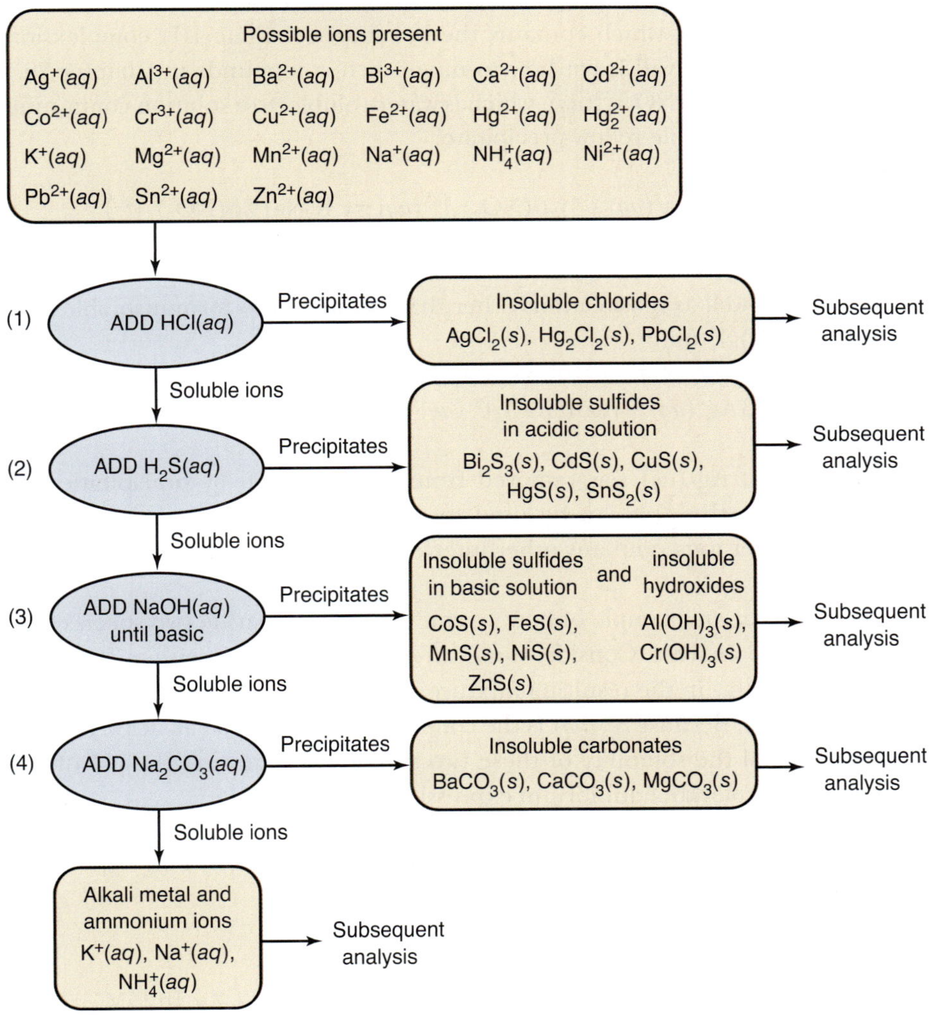

Figure 22.10 A qualitative analysis scheme for the separation of various groups of common cations by selective precipitation.

insoluble in pure water but KCl(s) is soluble, the formation of a white precipitate upon addition of 6 M HCl(aq) (the first step in Figure 22.10) suggests that Ag^+(aq) is present:

$$Ag^+(aq) + Cl^-(aq) \leftrightharpoons AgCl(s)$$
$$K^+(aq) + Cl^-(aq) \leftrightharpoons \text{no reaction}$$

Figure 22.11 Precipitation of Ag^+(aq) as AgCl(s) by addition of HCl(aq) to a solution containing Ag^+(aq).

Addition of excess HCl(aq) precipitates essentially all the Ag^+(aq) as AgCl(s) (Figure 22.11). The settling of the AgCl(s) is hastened by using a centrifuge, which spins the sample and thereby accelerates separation of the two phases. The resulting **supernatant** solution is then decanted (poured off) and tested separately for the presence of K^+(aq). Water-insoluble silver chloride is soluble in 6 M NH_3(aq) because of the formation of the $[Ag(NH_3)_2]^+$(aq) complex ion:

$$AgCl(s) + 2NH_3(aq) \leftrightharpoons [Ag(NH_3)_2]^+(aq) + Cl^-(aq)$$

The solubilization of the white AgCl(s) in NH_3(aq) is a confirmatory test for Ag^+(aq).

Notice that potassium is one of the few ions that remains in solution at the end of the qualitative analysis scheme shown in Figure 22.10. This is because almost all potassium salts are water-soluble. An exception is the insoluble salt $K_2Na[Co(NO_2)_6]$(s), which contains the hexanitrocobaltate(III) complex ion $[Co(NO_2)_6]^{3-}$. (You will learn how to name such compounds in Chapter 26.) Addition of $Na_3[Co(NO_2)_6]$(aq), which is water-soluble, to a solution containing K^+(aq) produces a pale yellow precipitate:

$$2K^+(aq) + Na^+(aq) + [Co(NO_2)_6]^{3-}(aq) \leftrightharpoons \underset{\text{pale yellow}}{K_2Na[Co(NO_2)_6](s)}$$

Silver ions, as well as most cations other than sodium, also form insoluble salts with $[Co(NO_2)_6]^{3-}$(aq):

$$3Ag^+(aq) + [Co(NO_2)_6]^{3-}(aq) \leftrightharpoons Ag_3[Co(NO_2)_6](s)$$

But recall that Ag^+(aq) was removed from the unknown by precipitation as AgCl(s). Thus, this test can be used to confirm the presence of potassium ions, once silver ions (and any other interfering ions) have been systematically removed.

Let's look at an example of how we can use pH to separate two species during qualitative analysis. Consider a mixture that may contain either ZnS(s) or FeS(s) or both, as in the resulting mixture of precipitates at the end of step (3) in Figure 22.10. Because S^{2-}(aq) is the conjugate base of the weak acid HS^-(aq), we can control the solubility of these two sulfides by adjusting the pH of the solution. The relevant equilibrium expressions are

$$H_2S(aq) + H_2O(l) \leftrightharpoons H_3O^+(aq) + HS^-(aq) \qquad K_{a_1} = 8.9 \times 10^{-8}\ \text{M}$$

and

$$HS^-(aq) + H_2O(l) \leftrightharpoons H_3O^+(aq) + S^{2-}(aq) \qquad K_{a_2} = 1.2 \times 10^{-13}\ \text{M}$$

The sum of these two equations is

$$H_2S(aq) + 2\,H_2O(l) \leftrightharpoons 2\,H_3O^+(aq) + S^{2-}(aq)$$

for which

$$K_c = K_{a_1}K_{a_2} = (8.9 \times 10^{-8}\text{ M})(1.2 \times 10^{-13}\text{ M}) = 1.1 \times 10^{-20}\text{ M}^2$$

The equilibrium-constant expression for this equation,

$$K_c = \frac{[H_3O^+]^2[S^{2-}]}{[H_2S]}$$

can be rearranged to

$$[S^{2-}] = \frac{K_c[H_2S]}{[H_3O^+]^2} = \frac{(1.1 \times 10^{-20}\text{ M}^2)[H_2S]}{[H_3O^+]^2} \qquad (22.24)$$

If we saturate the solution with hydrogen sulfide, $H_2S(aq)$, then it turns out that $[H_2S] = 0.10$ M at 25°C. If we substitute this value into Equation 22.24, then we obtain

$$[S^{2-}] = \frac{(1.1 \times 10^{-20}\text{ M}^2)(0.10\text{ M})}{[H_3O^+]^2} = \frac{1.1 \times 10^{-21}\text{ M}^3}{[H_3O^+]^2} \qquad [\text{saturated } H_2S(aq) \text{ at } 25°\text{C}] \qquad (22.25)$$

Equation 22.25 shows that the sulfide ion concentration in a saturated $H_2S(aq)$ solution can be controlled by the hydronium ion concentration or, equivalently, by the pH.

Let's see how we can use this pH dependence to separate our $ZnS(s)$ and $FeS(s)$ mixture. Suppose we add a buffer to our mixture, so that the value of the pH is 2.00. Let's calculate the solubility of $ZnS(s)$ and $FeS(s)$ at pH = 2.00. The hydronium ion concentration at pH 2.00 is 0.010 M, so for a saturated $H_2S(aq)$ solution at 25°C, we have

$$[S^{2-}] = \frac{1.1 \times 10^{-21}\text{ M}^3}{(0.010\text{ M})^2} = 1.1 \times 10^{-17}\text{ M}$$

Using the K_{sp} values for $ZnS(s)$ and $FeS(s)$ from Table 22.1, we have

$$[Zn^{2+}][S^{2-}] = 1.6 \times 10^{-24}\text{ M}^2$$

and

$$[Fe^{2+}][S^{2-}] = 6.3 \times 10^{-18}\text{ M}^2$$

Therefore, the concentrations of $Zn^{2+}(aq)$ and $Fe^{2+}(aq)$ in equilibrium with excess $ZnS(s)$ and $FeS(s)$ in a saturated $H_2S(aq)$ solution at pH = 2.0 at 25°C, are

$$[Zn^{2+}] = \frac{1.6 \times 10^{-24}\text{ M}^2}{[S^{2-}]} = \frac{1.6 \times 10^{-24}\text{ M}^2}{1.1 \times 10^{-17}\text{ M}} = 1.5 \times 10^{-7}\text{ M}$$

Because $H_2S(g)$ is poisonous at high concentrations, $H_2S(aq)$ is often prepared directly in solution by the decomposition of thioacetamide on gentle heating:

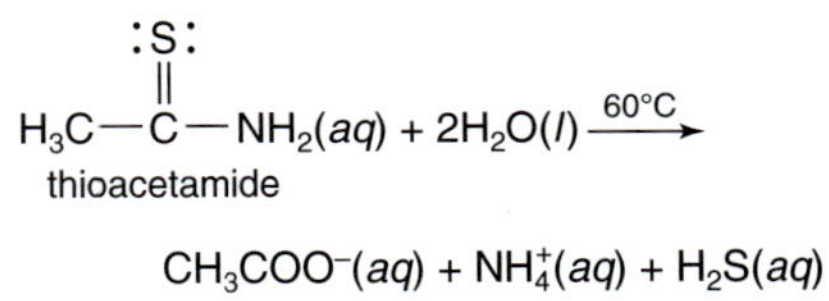

The use of thioacetamide as a source of $H_2S(aq)$ and the use of laboratory ventilation hoods make it possible to saturate the solution and yet keep the $H_2S(g)$ concentration in the laboratory air at a safe level.

and

$$[Fe^{2+}] = \frac{6.3 \times 10^{-18}\ M^2}{[S^{2-}]} = \frac{6.3 \times 10^{-18}\ M^2}{1.1 \times 10^{-17}\ M} = 0.57\ M$$

Because one formula unit of $Zn^{2+}(aq)$ occurs in solution for each formula unit of $ZnS(s)$ that dissolves, $[Zn^{2+}]$ is equal to the solubility of $ZnS(s)$. Similarly, $[Fe^{2+}]$ is equal to the solubility of $FeS(s)$. We see, then, that the solubility of $FeS(s)$ is large (0.57 M), whereas the solubility of $ZnS(s)$ is small (1.5×10^{-7} M). Therefore, in a saturated aqueous solution of $H_2S(aq)$ at 25°C and at a pH of 2.0, $FeS(s)$ is soluble and $ZnS(s)$ is insoluble (Figure 22.12). We can take advantage of the pH dependence of the solubility of species containing acidic or basic cations or anions to separate such species during qualitative analysis.

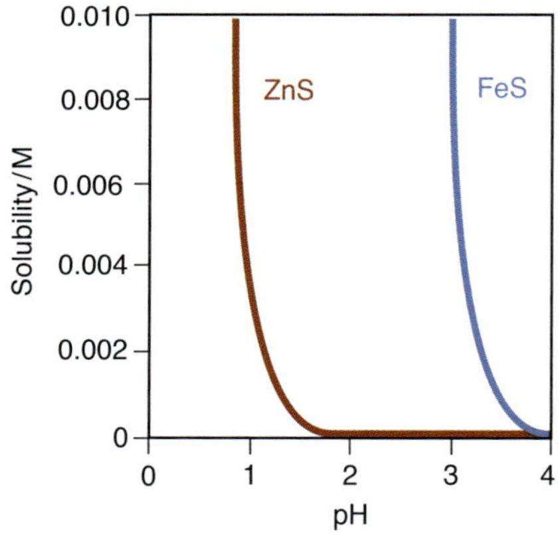

Figure 22.12 Solubility of $ZnS(s)$ and $FeS(s)$ in water saturated with $H_2S(aq)$ at one bar and 25°C at various pH values. We can separate $Zn^{2+}(aq)$ from $Fe^{2+}(aq)$ by saturating a solution buffered at about pH = 2.0 with $H_2S(aq)$. The $Zn^{2+}(aq)$ precipitates as $ZnS(s)$, but the $Fe^{2+}(aq)$ remains in solution.

EXAMPLE 22-11: Consider a mixture of $Cd(OH)_2(s)$ and $Cu(OH)_2(s)$, for which the values of K_{sp} at 25°C are $7.2 \times 10^{-15}\ M^3$ and $2.2 \times 10^{-20}\ M^3$, respectively. Can this mixture be separated by adjusting the pH of the solution?

Solution: Let's begin by considering the $Cd(OH)_2(s)$. The dissolution equation and corresponding K_{sp} expression for $Cd(OH)_2(s)$ are

$$Cd(OH)_2(s) \leftrightharpoons Cd^{2+}(aq) + 2OH^-(aq)$$

and

$$K_{sp} = [Cd^{2+}][OH^-]^2 = 7.2 \times 10^{-15}\ M^3$$

The solubility of $Cd(OH)_2(s)$ in water can be calculated from the K_{sp} expression:

$$s = [Cd^{2+}] = \frac{7.2 \times 10^{-15}\ M^3}{[OH^-]^2} \qquad (22.26)$$

The concentration of $OH^-(aq)$ can be related to $[H_3O^+]$ by using the ion product constant expression for water:

$$[OH^-] = \frac{K_w}{[H_3O^+]} = \frac{1.0 \times 10^{-14}\ M^2}{[H_3O^+]}$$

Substitution of this equation into Equation 22.26 yields

$$s = [Cd^{2+}] = \frac{7.2 \times 10^{-15}\ M^3\ [H_3O^+]^2}{(1.0 \times 10^{-14}\ M^2)^2} \qquad (22.27)$$

Now let's consider the $Cu(OH)_2(s)$, for which we have

$$Cu(OH)_2(s) \leftrightharpoons Cd^{2+}(aq) + 2OH^-(aq)$$

and

$$K_{sp} = [Cu^{2+}][OH^-]^2 = 2.2 \times 10^{-20}\ M^3$$

We can find the solubility of $Cu(OH)_2(s)$ in a manner analogous to that for $Cd(OH)_2(s)$ by using the K_{sp} expression for $Cu(OH)_2(s)$ and ion product concentration expression of water to obtain

$$s = [Cu^{2+}] = \frac{2.2 \times 10^{-20}\ M^3}{[OH^-]^2} = \frac{2.2 \times 10^{-20}\ M^3\ [H_3O^+]^2}{(1.0 \times 10^{-14}\ M^2)^2} \qquad (22.28)$$

From Equations 22.27 and 22.28, we can calculate the solubility of $Cd(OH)_2(s)$ and $Cu(OH)_2(s)$ at various pH values, as shown in Table 22.4 and Table 22.5. These results are plotted in Figure 22.13. From Figure 22.13 we see that $Cd(OH)_2(s)$ can be separated from $Cu(OH)_2(s)$ by adjusting the pH of the solution to about 6.5 using a buffer solution. At pH ≈ 6.5, the $Cd^{2+}(aq)$ dissolves, but the $Cu(OH)_2(s)$ remains in solution.

PRACTICE PROBLEM 22-11: Using the data in Table 22.1, calculate the solubilities at 25°C of $MnS(s)$ and $PbS(s)$ at pH values of 4.0 and 7.0 in a saturated solution of $H_2S(aq)$ at 25°C. Can we achieve an effective separation of these solids at either value of the pH?

Answer: At pH = 4.0, $[Mn^{2+}] = 2.3$ M and $[Pb^{2+}] = 7.3 \times 10^{-15}$ M. At pH = 7.0, $[Mn^{2+}] = 2.3 \times 10^{-6}$ M and $[Pb^{2+}] = 7.3 \times 10^{-21}$ M. Reasonable separation can be achieved at pH = 4.0, but not at pH = 7.0.

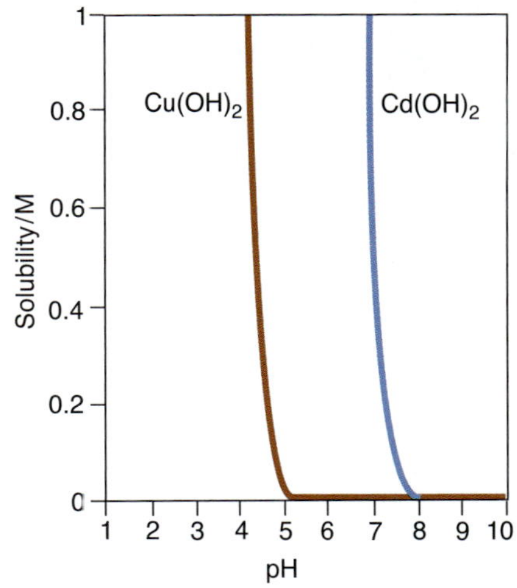

Figure 22.13 Solubilities of $Cd(OH)_2(s)$ and $Cu(OH)_2(s)$ as a function of pH. Note that a lower pH is required to dissolve $Cu(OH)_2(s)$ than to dissolve $Cd(OH)_2(s)$. Therefore, at pH = 6.5, for example, $Cu(OH)_2(s)$ precipitates and $Cd^{2+}(aq)$ remains in solution. The $Cu(OH)_2(s)$ can be filtered off, thereby separating copper from cadmium.

The methods of qualitative analysis developed in general chemistry laboratories are not, in most cases, the same as those used by professional analytical chemists to identify the constituents of an unknown sample. Analytical chemists working in, say, a criminal investigations laboratory face much more difficult challenges to their chemical ingenuity than those encountered in a general chemistry laboratory. The major differences are that in such cases the possible constituents of an unknown sample are essentially unlimited and often truly unknown. Furthermore, in many cases the available sample is very small and irreplaceable. Consequently, practicing analytical chemists tend to rely on spec-

TABLE 22.4 Solubility of $Cd(OH)_2(s)$ in water at 25°C at various pH values

pH	$[H_3O^+]$/M	$[H_3O^+]^2/M^2$	$[Cd^{2+}]$/M
7.00	1.0×10^{-7}	1.0×10^{-14}	0.72
7.20	6.3×10^{-8}	4.0×10^{-15}	0.29
7.40	4.0×10^{-8}	1.6×10^{-15}	0.11
7.60	2.5×10^{-8}	6.3×10^{-16}	0.045
7.80	1.6×10^{-8}	2.6×10^{-16}	0.018
8.00	1.0×10^{-8}	1.0×10^{-16}	0.0072

TABLE 22.5 Solubility of $Cu(OH)_2(s)$ in water at 25°C at various pH values

pH	$[H_3O^+]$/M	$[H_3O^+]^2/M^2$	$[Cu^{2+}]$/M
4.20	6.3×10^{-5}	4.0×10^{-9}	0.88
4.40	4.0×10^{-5}	1.6×10^{-9}	0.35
4.60	2.5×10^{-5}	6.3×10^{-10}	0.14
5.00	1.0×10^{-5}	1.0×10^{-10}	0.022
5.20	6.3×10^{-6}	4.0×10^{-11}	0.0088
5.40	4.0×10^{-6}	1.6×10^{-11}	0.0035

Figure 22.14 Although chemists now often use a variety of sophisticated electronic instruments in performing chemical analyses, they still employ "wet lab" chemical techniques in certain cases, such as shown here at the Smithsonian Institution.

troscopic techniques, rather than on those outlined here. Nonetheless, all the chemical principles used in qualitative analysis are used by analytical chemists (Figure 22.14).

SUMMARY

The solubility of an ionic solid is a measure of its concentration in a saturated solution. The equilibrium between the salt and the saturated solution is characterized by a solubility-product constant, K_{sp}.

The solubility of a salt is less in a solution that contains one of its constituent ions than it is in pure water. For example, the solubility of $AgCl(s)$ is less in 0.10 M $KCl(aq)$ than it is in pure water. The lowering of the solubility of a salt in a solution containing one of its constituent ions is called the common ion effect.

Many metal ions in aqueous solution react with small molecules or ions such as $NH_3(aq)$ or $OH^-(aq)$ to form complex ions. Complex ion formation can enhance the solubility of compounds containing these metal ions. For example, the solubility of $AgBrO_3(s)$ is much greater in $NH_3(aq)$ than in pure water because of the formation of a soluble $[Ag(NH_3)_2]^+(aq)$ complex ion.

Slightly soluble salts containing an anion that is the conjugate base of a weak acid become more soluble as the pH of the solution is lowered. The increase in solubility is a consequence of the protonation of the anion to form the weak acid, which shifts the solubility equilibrium.

The relative values of the solubility-product constant, K_{sp}, and the solubility-product quotient, Q_{sp}, can be used to predict whether a salt dissolves in or precipitates out of a solution. If $Q_{sp} > K_{sp}$, a precipitate will form; and if $Q_{sp} < K_{sp}$, no precipitate will form.

Selective precipitation can often be used to separate ions in solution. Reasonable separations are achieved when each ion forms an insoluble salt with the same reactant and the values of K_{sp} of the two salts differ significantly.

Amphoteric metal hydroxides are soluble in both acidic and basic solutions; the solubility of amphoteric metal hydroxides in strong bases is a consequence of the formation of a soluble hydroxy ion complex of the metal ion.

Qualitative analysis schemes are based on the sequential separation of ions in an unknown sample. The separated ions can be identified by their characteristic chemical reactions. The pH dependence of the solubility of salts containing the conjugate base of a weak acid, especially hydroxides and sulfides, can be used to achieve separations of various metal ions in qualitative analysis by successive adjustments of pH and addition of precipitating agents.

TERMS YOU SHOULD KNOW

solubility *815*
solubility-product constant, K_{sp} *816*
common ion *822*
common ion effect *822*
complex ion *823*
complexation reaction *823*
concentration quotient, Q_{sp} *830*
supersaturated solution *831*
selective precipitation *834*
amphoteric metal hydroxide *835*
qualitative analysis *838*
quantitative analysis *839*
supernatant *840*

EQUATIONS YOU SHOULD KNOW HOW TO USE

For any arbitrary ion concentrations:

$Q_{sp} > K_{sp}$ precipitate forms
$Q_{sp} < K_{sp}$ no precipitate forms

When equilibrium is disturbed:

$Q_{sp} > K_{sp}$ more precipitate forms until $Q_{sp} = K_{sp}$
$Q_{sp} < K_{sp}$ precipitate dissolves either until $Q_{sp} = K_{sp}$ or until solid phase disappears

PROBLEMS

REVIEW OF SOLUBILITY RULES

22-1. Use the solubility rules from Chapter 10 to predict whether the following compounds are soluble or insoluble in water:

(a) $AgI(s)$ (b) $Pb(ClO_4)_2(s)$
(c) $NH_4Br(s)$ (d) $SrCO_3(s)$

22-2. Use the solubility rules from Chapter 10 to predict whether the following compounds are soluble or insoluble in water:

(a) $Al_2O_3(s)$ (b) $CuCl_2(s)$
(c) $KNO_3(s)$ (d) $Hg_2Br_2(s)$

22-3. Use the solubility rules from Chapter 10 to predict whether the following barium salts are soluble or insoluble in water:

(a) $BaCO_3(s)$ (b) $Ba(ClO_4)_2(s)$
(c) $BaCl_2(s)$ (d) $BaSO_4(s)$

22-4. Use the solubility rules from Chapter 10 to predict whether the following silver salts are soluble or insoluble in water:

(a) $AgBr(s)$ (b) $AgNO_3(s)$
(c) $Ag_2S(s)$ (d) $AgClO_4(s)$

22-5. In each of the following cases, the two solutions indicated are mixed. In cases for which a precipitate forms on mixing, write the complete chemical equation and the net ionic equation. If no precipitate forms, then write "no reaction." Use the solubility rules (Chapter 10) and assume that all solutions before mixing are 0.20 M and that equal volumes of the two solutions are mixed.

(a) $CuCl_2(aq) + Na_2S(aq) \rightarrow$
(b) $MgBr_2(aq) + K_2CO_3(aq) \rightarrow$
(c) $BaCl_2(aq) + K_2SO_4(aq) \rightarrow$
(d) $Hg_2(NO_3)_2(aq) + KCl(aq) \rightarrow$

22-6. In each of the following cases, the two solutions indicated are mixed. In cases for which a precipitate forms on mixing, write the complete chemical equation and the net ionic equation. If no precipitate forms, then write "no reaction." Use the solubility rules (Chapter 10) and assume that all solutions before mixing are 0.20 M and that equal volumes of the two solutions are mixed.

(a) $H_2SO_4(aq) + Ca(ClO_4)_2(aq) \rightarrow$

(b) $AgNO_3(aq) + NaClO_4(aq) \rightarrow$

(c) $Hg_2(NO_3)_2(aq) + NaC_6H_5COO(aq) \rightarrow$

(d) $Na_2SO_4(aq) + AgF(aq) \rightarrow$

K_{sp} CALCULATIONS

22-7. The value of K_{sp} for $PbCrO_4(s)$ in equilibrium with water at 25°C is 2.8×10^{-13} M^2. Write the chemical equation that represents the solubility equilibrium for $PbCrO_4(s)$ and calculate its solubility in grams per liter in water at 25°C.

22-8. The value of K_{sp} for $TlCl(s)$ in equilibrium with water at 25°C is 1.9×10^{-4} M^2. Write the chemical equation that represents the solubility equilibrium for $TlCl(s)$ and calculate its solubility in grams per liter in water at 25°C.

22-9. The solubility-product constant for $Mg(OH)_2(s)$ in equilibrium with water at 25°C is 5.6×10^{-12} M^3. Calculate the solubility in grams per liter of $Mg(OH)_2(s)$ in water at 25°C.

22-10. The solubility-product constant for $PbBr_2(s)$ in equilibrium with water at 25°C is 6.6×10^{-6} M^3. Calculate the solubility in grams per liter of $PbBr_2(s)$ in water at 25°C.

22-11. Potassium perchlorate, $KClO_4(s)$, is soluble in water to the extent of 0.70 grams per 100.0 mL at 0°C. Calculate the value of K_{sp} of $KClO_4(s)$ at 0°C.

22-12. Lithium fluoride, $LiF(s)$, dissolves in water to the extent of 0.13 grams per 100.0 mL at 20°C. Calculate the value of K_{sp} of $LiF(s)$ at 20°C.

22-13. The solubility of silver bromide, $AgBr(s)$, in pure water at 18°C is 1.3×10^{-4} g·L^{-1}. Calculate the value of K_{sp} for silver bromide at 18°C. Is the value of K_{sp} you calculated the same as that listed in Table 22.1? Suggest a possible reason for any discrepancy.

22-14. The solubility of lead(II) iodate in pure water is 0.76 g·L^{-1} at 25°C. Calculate the value of K_{sp} for lead(II) iodate at 25°C. Is the value of K_{sp} you calculated the same as that listed in Table 22.1? Suggest a possible reason for any discrepancy.

COMMON-ION EFFECT

22-15. Calculate the solubility in grams per liter of silver sulfate, $Ag_2SO_4(s)$, in a 0.55-M silver nitrate, $AgNO_3(aq)$, solution at 25°C.

22-16. Calculate the solubility in grams per liter of barium chromate, $BaCrO_4(s)$, in a 0.0553-M ammonium chromate, $(NH_4)_2CrO_4(aq)$, solution at 25°C.

22-17. Calculate the solubility in grams per liter of $AgI(s)$ in 0.20 M $CaI_2(aq)$ at 25°C.

22-18. Calculate the solubility in grams per liter of $CaSO_4(s)$ in 0.25 M $Na_2SO_4(aq)$ at 25°C.

22-19. Calculate the solubility in grams per liter of $Pb(IO_3)_2(s)$ in 0.10 M $KIO_3(aq)$ at 25°C.

22-20. Calculate the solubility in grams per liter of $Ag_2SO_4(s)$ in 0.15 M $K_2SO_4(aq)$ at 25°C.

FORMATION OF SOLUBLE COMPLEX IONS

22-21. When sodium hydroxide is added to a solution containing lead(II) nitrate, a solid precipitate forms. However, when additional sodium hydroxide is added, the precipitate redissolves, forming a soluble $[Pb(OH)_4]^{2-}(aq)$ complex ion. Write the balanced chemical equation for each of these reactions.

22-22. When aqueous ammonia is added to a solution containing copper(II) nitrate, a solid precipitate forms. However, when additional aqueous ammonia is added, the precipitate redissolves, forming a soluble $[Cu(NH_3)_4]^{2+}(aq)$ complex ion. Write the balanced chemical equation for each of these reactions.

22-23. Calculate the solubility in grams per liter of $AgBr(s)$ in 0.35 M $NH_3(aq)$. Take $K_f = 2.0 \times 10^7$ M^{-2} for the complexation reaction of $Ag^+(aq)$ to produce $[Ag(NH_3)_2]^+(aq)$. Assume a temperature of 25°C.

22-24. Calculate the solubility in grams per liter of $AgI(s)$ in 0.60 M $NH_3(aq)$. Take $K_f = 2.0 \times 10^7$ M^{-2} for the complexation reaction of $Ag^+(aq)$ to produce $[Ag(NH_3)_2]^+(aq)$. Assume a temperature of 25°C.

22-25. Copper(I) ions in aqueous solution react with $NH_3(aq)$ according to

$$Cu^+(aq) + 2NH_3(aq) \leftrightharpoons [Cu(NH_3)_2]^+(aq)$$

with $K_f = 6.3 \times 10^{10}$ M^{-2}. Calculate the solubility in grams per liter of $CuBr(s)$ in 0.50 M $NH_3(aq)$ at 25°C.

22-26. Calculate the solubility in grams per liter of $AgBr(s)$ in a solution that is 0.200 M in $KBr(aq)$ and 0.200 M in $NH_3(aq)$. Take $K_f = 2.0 \times 10^7$ M^{-2} for the complexation reaction of $Ag^+(aq)$ to produce $[Ag(NH_3)_2]^+(aq)$. Assume a temperature of 25°C.

SOLUBILITY AND pH

22-27. Lead(II) fluoride, $PbF_2(s)$, is slightly soluble in water. Predict the effect on its solubility when

(a) the pH of the solution is decreased to 3

(b) $Pb(NO_3)_2(s)$ is added to the solution

(c) $NaF(s)$ is added to the solution

22-28. Magnesium oxalate, $MgC_2O_4(s)$, is sparingly soluble in water. Predict the effect on its solubility when

(a) the solution is made more acidic

(b) the solution is made more basic

(c) $Mg(NO_3)_2(s)$ is added to the solution

22-29. Indicate for which of the following compounds the solubility increases as the pH of the solution is lowered:

(a) $CaCO_3(s)$ (b) $KClO_4(s)$

(c) $Fe(OH)_3(s)$ (d) $ZnS(s)$

22-30. Indicate for which of the following compounds the solubility increases as the pH of the solution is lowered:

(a) $PbCrO_4(s)$ (b) $Ag_2C_2O_4(s)$

(c) $Hg_2I_2(s)$ (d) $Ag_2O(s)$

22-31. Calculate the solubility of $Mg(OH)_2(s)$ in grams per liter in an aqueous solution buffered at pH = 8.50 at 25°C.

22-32. Calculate the solubility of $AgC_6H_5COO(s)$ in grams per liter in an aqueous solution buffered at pH = 4.00 at 25°C. Given $K_a = 6.3 \times 10^{-5}$ M for $C_6H_5COOH(aq)$ and that $K_{sp} = 2.5 \times 10^{-5}$ M^2 for $AgC_6H_5COO(s)$.

Q_{sp} CALCULATIONS

22-33. A 100.0-mL sample of water from a salt lake has a chloride ion concentration of 0.25 M. Does $AgCl(s)$ precipitate from solution if 5.0 mL of 0.10 M $AgNO_3(aq)$ is added to the sample?

22-34. Suppose we mix 50.0 mL of 0.20 M $AgNO_3(aq)$ with 150.0 mL of 0.10 M $H_2SO_4(aq)$. Does $Ag_2SO_4(s)$ precipitate from the solution?

22-35. If we mix 40.0 mL of 3.00 M $Pb(NO_3)_2(aq)$ with 20.0 mL of 2.00×10^{-3} M $NaI(aq)$, does $PbI_2(s)$ precipitate from the solution? If yes, then calculate how many moles of $PbI_2(s)$ precipitate and the values of $[Pb^{2+}]$, $[I^-]$, $[NO_3^-]$, and $[Na^+]$ at 25°C at equilibrium.

22-36. Suppose we mix 50.0 mL of 0.50 M $AgNO_3(aq)$ with 50.0 mL of 1.00×10^{-4} M $NaBr(aq)$. Does $AgBr(s)$ precipitate from the solution? If yes, then calculate how many moles of $AgBr(s)$ precipitate and the values of $[Ag^+]$, $[Br^-]$, $[Na^+]$, and $[NO_3^-]$ at 25°C at equilibrium.

22-37. Suppose we mix 100.0 mL of a 2.00 M $NaCl(aq)$ solution with 100.0 mL of a 0.020 M $AgNO_3(aq)$ solution at 25°C. Determine

(a) the number of grams of $AgCl(s)$ that precipitate from the solution

(b) the concentration of $Ag^+(aq)$ at equilibrium following the precipitation of $AgCl(s)$

22-38. Suppose that 10.0 mL of a 0.30 M $Zn(NO_3)_2(aq)$ solution is added to 10.0 mL of a 2.00×10^{-4} M $Na_2S(aq)$ solution at 25°C. Calculate

(a) the number of milligrams of $ZnS(s)$ that precipitate

(b) the concentrations of $Zn^{2+}(aq)$ and $S^{2-}(aq)$ at equilibrium

SELECTIVE PRECIPITATION

22-39. Consider a solution containing $Pb^{2+}(aq)$ and $Hg_2^{2+}(aq)$, each at a concentration of 0.010 M. Is it possible to separate the $Pb^{2+}(aq)$ and the $Hg_2^{2+}(aq)$ by selectively precipitating the $Hg_2^{2+}(aq)$ with $Cl^-(aq)$ at 25°C? Use a criterion of precipitating 99.99% of the $Hg_2^{2+}(aq)$ without causing $Pb^{2+}(aq)$ to precipitate.

22-40. Consider a solution containing $Ca^{2+}(aq)$ and $Ba^{2+}(aq)$, each at a concentration of 0.10 M. Is it possible to separate the $Ca^{2+}(aq)$ and the $Ba^{2+}(aq)$ by selectively precipitating the $Ba^{2+}(aq)$ with $SO_4^{2-}(aq)$ at 25°C? Use a criterion of precipitating 99.99% of the $Ba^{2+}(aq)$ without causing $Ca^{2+}(aq)$ to precipitate.

22-41. A solution contains 0.050 M $Ca^{2+}(aq)$ and 0.025 M $Ag^+(aq)$. Can 99% of either ion be precipitated by adding $SO_4^{2-}(aq)$, without precipitating the other metal ion at 25°C?

22-42. A solution contains 0.0100 M $Pb^{2+}(aq)$ and 0.0100 M $Sr^{2+}(aq)$. Can 99% of either ion be precipitated by adding $SO_4^{2-}(aq)$, without precipitating the other metal ion at 25°C?

SEPARATION OF CATIONS AS HYDROXIDES AND SULFIDES

22-43. Calculate the solubility of $CuS(s)$ in a solution buffered at pH = 2.00 and saturated with hydrogen sulfide so that $[H_2S]$ = 0.10 M. Assume a temperature of 25°C.

22-44. Calculate the solubility of $SnS(s)$ in a solution buffered at pH = 2.00 and saturated with hydrogen sulfide so that $[H_2S]$ = 0.10 M. Assume a temperature of 25°C.

22-45. Calculate the solubility of $Cr(OH)_3(s)$ and $Ni(OH)_2(s)$ in an aqueous solution buffered at pH = 5.00 at 25°C. Can $Cr(OH)_3$ be separated from $Ni(OH)_2$ at this pH?

22-46. Calculate the solubility of $Cu(OH)_2(s)$ and $Zn(OH)_2(s)$ in an aqueous solution buffered at pH = 4.00 at 25°C. Can $Cu(OH)_2$ be separated from $Zn(OH)_2$ at this pH?

22-47. What must the pH of a buffered solution saturated with hydrogen sulfide ($[H_2S]$ = 0.10 M) be at 25°C in order to precipitate $PbS(s)$ leaving $[Pb^{2+}]$ = 1.0×10^{-6} M, without precipitating any $MnS(s)$? The original solution is 0.025 M in both $Pb^{2+}(aq)$ and $Mn^{2+}(aq)$.

22-48. Iron(II) sulfide, $FeS(s)$, is used as the pigment in black paint. A sample of $FeS(s)$ is suspected of containing lead(II) sulfide, $PbS(s)$, which can cause lead poisoning if ingested. Suggest a scheme based on pH for separating $FeS(s)$ from $PbS(s)$.

22-49. Calculate the solubilities of $Mg(OH)_2(s)$ and $Ca(OH)_2(s)$ in an aqueous solution at 25°C as a function of pH. In what pH range can the two hydroxides be separated?

22-50. Calculate the solubilities of $Cd(OH)_2(s)$ and $Fe(OH)_3(s)$ in an aqueous solution at 25°C as a function of pH. In what pH range can the two hydroxides be separated?

AMPHOTERIC METAL HYDROXIDES

22-51. Use the equilibrium-constant data in Table 22.3 to estimate the solubility of tin(II) hydroxide, $Sn(OH)_2(s)$, in a solution buffered at pH = 13.00 at 25°C.

22-52. Use the equilibrium-constant data in Table 22.3 to estimate the solubility of lead(II) hydroxide, $Pb(OH)_2(s)$, in a solution buffered at pH = 13.00 at 25°C.

22-53. Use the equilibrium-constant data in Table 22.3 to estimate the solubility at 25°C of zinc hydroxide, $Zn(OH)_2(s)$, in a solution buffered at pH = 7.0, 10.0, and 14.0.

22-54. Use the equilibrium-constant data in Table 22.3 to calculate the solubility of $Pb(OH)_2(s)$ in water at 25°C as a function of pH over the range pH = 4.0 to pH = 16.0. Use increments of two pH units.

ADDITIONAL PROBLEMS

22-55. Define precipitation. Why in meteorology are both snow and rain called precipitation?

22-56. What is meant by a common ion? Give an example of adding a species with a common ion to a solution.

22-57. Does the presence of a common ion always decrease the solubility of a salt containing that ion?

22-58. Describe an experimental procedure to determine the value of K_{sp} for an unknown solid.

22-59. Why is it possible to separate a mixture of $Pb^{2+}(aq)$ and $Hg_2^{2+}(aq)$ by selectively precipitating out the mercury as $Hg_2I_2(s)$; but not possible to separate the same mixture by selectively precipitating out the lead as $PbI_2(s)$?

22-60. Can we ever achieve total separation of two substances in solution using selective precipitation?

22-61. What is the difference between qualitative analysis and quantitative analysis?

22-62. Which is more soluble in pure water, silver chloride, $AgCl(s)$, or silver chromate, $Ag_2CrO_4(s)$?

22-63. $BaSO_4(s)$ is a good absorber of *x*-rays and often used as a contrasting agent for imaging the soft tissue of the intestines. Although $Ba^{2+}(aq)$ is poisonous, why is it safe for patients to consume a suspension of $BaSO_4(s)$?

22-64. An unknown sample is known to contain one or more of the following ions: $Ba^{2+}(aq)$, $Ag^{+}(aq)$, and $Cd^{2+}(aq)$. Using only the solubility rules that we learned in Chapter 10, devise a flowchart for determining which of these ions are present in the unknown sample.

22-65. One treatment for poisoning by soluble lead compounds is to give $MgSO_4(aq)$ or $Na_2SO_4(aq)$ as soon as possible. Explain in chemical terms why this procedure is effective.

22-66. Calculate the solubility of $FeS(s)$ and $CdS(s)$ at pH = 2.00 and 4.00 for aqueous solutions that are saturated with hydrogen sulfide at $[H_2S] = 0.10$ M at 25°C. Can $FeS(s)$ and $CdS(s)$ be separated at either pH?

22-67. It is observed that a precipitate forms when a 2.0 M $NaOH(aq)$ solution is added dropwise to a 0.10 M $Cr(NO_3)_3(aq)$ solution and that, on further addition of $NaOH(aq)$, the precipitate dissolves. Explain these observations using balanced chemical equations.

22-68. $Pb(OH)_2(s)$ and $Sn(OH)_2(s)$ are formed when sodium hydroxide is added to a solution containing $Pb^{2+}(aq)$ and $Sn^{2+}(aq)$. At what pH can $Pb(OH)_2(s)$ be separated from $Sn(OH)_2(s)$ at 25°C? Assume that an effective separation requires a maximum concentration of the less soluble hydroxide of 1×10^{-6} M.

22-69. A solution 0.30 M in $H_3O^{+}(aq)$ containing $Mn^{2+}(aq)$, $Cd^{2+}(aq)$, and $Fe^{2+}(aq)$, all at 0.010 M, was saturated with $H_2S(g)$ at 25°C. Calculate the equilibrium concentrations of $Mn^{2+}(aq)$, $Cd^{2+}(aq)$, and $Fe^{2+}(aq)$. Assume that the solution is continuously saturated with $H_2S(aq)$ and that the pH remains constant.

22-70. It is observed that a precipitate forms when a 2.0 M $KOH(aq)$ solution is added dropwise to a 0.20 M $Zn(ClO_4)_2(aq)$ solution and that, similarly, a precipitate forms when a 2.0 M $KOH(aq)$ solution is added dropwise to a 0.20 M $Mg(ClO_4)_2(aq)$ solution. However, on further addition of $KOH(aq)$, the precipitate from the $Zn(ClO_4)_2(aq)$ solution dissolves, whereas the precipitate from the $Mg(ClO_4)_2(aq)$ solution does not. Explain these observations using balanced chemical equations. What can we conclude about the properties of these respective metal ions?

22-71. A deposit of limestone is analyzed for its calcium and magnesium content. A sample is dissolved, and then the calcium and magnesium are precipitated as $Ca(OH)_2(s)$ and $Mg(OH)_2(s)$. At what pH can $Ca(OH)_2(s)$ be separated from $Mg(OH)_2(s)$ at 25°C? Assume that an effective separation requires a maximum concentration of the less soluble hydroxide of 1×10^{-6} M.

22-72. Given the equation

$$Ag^{+}(aq) + 2NH_3(aq) \rightleftharpoons [Ag(NH_3)_2]^{+}(aq) \qquad K_f = 2.0 \times 10^{7}\ M^{-2}$$

determine the concentration of $NH_3(aq)$ that is required to dissolve 250 mg of $AgCl(s)$ in 100.0 mL of solution.

22-73. The image on black-and-white film is created by the exposure of $AgBr(s)$ to light, forming metallic silver. During the development of black-and-white film, unexposed $AgBr(s)$ is removed by the use of "hypo," a solution of $Na_2S_2O_3(aq)$, which forms a soluble $[Ag(S_2O_3)_2]^{3-}(aq)$ complex with silver ions. Compare the solubility of $AgBr(s)$ in (a) water and in (b) "hypo," given that the value of K_f for the formation of the $[Ag(S_2O_3)_2]^{3-}(aq)$ complex ion is $5 \times 10^{13}\ M^{-2}$ at 25°C.

22-74. Use the K_{sp} data in Table 22.1 to calculate the equilibrium constants for the following chemical equations at 25°C:

(1) $Ag_2CrO_4(s) + 2Br^{-}(aq) \rightleftharpoons 2AgBr(s) + CrO_4^{2-}(aq)$

(2) $PbCO_3(s) + Ca^{2+}(aq) \rightleftharpoons CaCO_3(s) + Pb^{2+}(aq)$

22-75. Calculate the pH at which $Ca(OH)_2(s)$ will begin to precipitate from an aqueous solution that is 2.0×10^{-2} M in $Ca^{2+}(aq)$ at 25°C.

22-76. Excess $HgI_2(s)$ was equilibrated with a solution that is 0.10 M in $KI(aq)$. Calculate the solubility of $HgI_2(s)$ in this solution at 25°C given

$$HgI_2(s) \rightleftharpoons Hg^{2+}(aq) + 2I^{-}(aq) \qquad K_{sp} = 2.9 \times 10^{-29}\ M^3$$

$$HgI_2(s) + 2I^{-}(aq) \rightleftharpoons [HgI_4]^{2-}(aq) \qquad K_f = 0.79\ M^{-1}$$

22-77. Given that $K_f = 0.05\ M^{-1}$ for

$$Zn(OH)_2(s) + 2OH^{-}(aq) \rightleftharpoons [Zn(OH)_4]^{2-}(aq)$$

calculate the solubility of $Zn(OH)_2(s)$ in a solution buffered at pH = 12.00 at 25°C.

22-78. Using the K_{sp} data in Table 22.1, calculate the value of the equilibrium constant at 25°C for the reaction described by

$$Ag_2SO_4(s) + Ca^{2+}(aq) \rightleftharpoons CaSO_4(s) + 2Ag^{+}(aq)$$

Calculate $[Ag^+]$ and $[Ca^{2+}]$ when excess $CaSO_4(s)$ is equilibrated with 0.100 M $AgNO_3(aq)$ at 25°C.

22-79. The equilibrium constant for the equation

$$AgCl(s) + 2S_2O_3^{2-}(aq) \leftrightharpoons [Ag(S_2O_3)_2]^{3-}(aq) + Cl^-(aq)$$

is 5.2×10^3 at 25°C. Calculate the solubility of $AgCl(s)$ in a solution whose *equilibrium* concentration of $S_2O_3^{2-}(aq)$ is 0.010 M.

22-80. Copper(I) ions in aqueous solution react with $NH_3(aq)$ at 25°C according to

$$Cu^+(aq) + 2NH_3(aq) \leftrightharpoons [Cu(NH_3)_2]^+(aq) \qquad K_f = 6.3 \times 10^{10}\ M^{-2}$$

Calculate the solubility of $CuBr(s)$ in a solution in which the equilibrium concentration of $NH_3(aq)$ is 0.15 M at 25°C.

22-81. The equilibrium constant for the equation

$$Al(OH)_3(s) + OH^-(aq) \leftrightharpoons [Al(OH)_4]^-(aq)$$

is $K_f = 40$ at 25°C. Calculate the solubility of $Al(OH)_3(s)$ in a solution buffered at pH = 12.00 at 25°C.

22-82. The equilibrium constant for the equation

$$Zn(OH)_2(s) + 2OH^-(aq) \leftrightharpoons [Zn(OH)_4]^{2-}(aq)$$

is $K_f = 0.050\ M^{-1}$ at 25°C. Calculate the solubility of $Zn(OH)_2(s)$ in a 0.10 M $NaOH(aq)$ solution at 25°C.

22-83. Calculate the pH of a saturated $Zn(OH)_2(aq)$ solution at 25°C.

22-84. Given that the pH of a saturated $Ca(OH)_2(aq)$ solution is 12.45, calculate the solubility of $Ca(OH)_2(s)$ in water at 25°C.

22-85. Calculate the solubility of $Cd(OH)_2(s)$ in an aqueous solution buffered at pH = 9.00 at 25°C.

22-86. Use Le Châtelier's principle to predict the effect on the solubility of

(a) $ZnS(s)$ when $HNO_3(aq)$ is added to a saturated $ZnS(aq)$ solution

(b) $AgI(s)$ when $NH_3(g)$ is added to a saturated $AgI(aq)$ solution

22-87. Silver chromate is sparingly soluble in aqueous solutions. What is the solubility in moles per liter of silver chromate at 25°C

(a) in a 1.5-M potassium chromate aqueous solution?

(b) in a 1.5-M silver nitrate aqueous solution?

(c) in pure water?

22-88. Calculate the equilibrium chloride ion concentration in a solution made by mixing 50.0 mL of 1.00-M sodium chloride with 50.0 mL of 1.00-M mercury(I) nitrate at 25°C.

22-89. Calculate the equilibrium chromate ion concentration in a solution made by mixing 200.0 mL of 0.200-M silver nitrate with 200.0 mL of 0.100-M potassium chromate at 25°C.

22-90. Calculate the equilibrium mercury(I) ion concentration in a solution made by mixing 100.0 mL of 0.200-M mercury(I) nitrate with 150.0 mL of 0.100-M aluminum chloride at 25°C? What fraction of the mercury(I) ion is not precipitated?

22-91. Calculate the equilibrium silver ion concentration in a solution made by mixing 500.0 mL of 0.200-M silver nitrate with 1200.0 mL of 0.100-M potassium chloride at 25°C? What fraction of the silver ion is not precipitated?

22-92. Determine the molar solubility of silver iodide in 14.0-M aqueous ammonia at 25°C. Use the criterion that soluble means at least 0.10 moles of the salt dissolve per liter of solution to determine whether silver iodide is soluble in aqueous ammonia. Take the value of K_f to be $2.0 \times 10^7\ M^{-2}$.

22-93. Some oil-well brines contain iodide ion. One particular brine sample had 6.5 mg of iodide ion per liter at 25°C. If equal volumes of this brine sample are mixed with a solution that is 0.0010 M each in lead(II) ion and silver(I) ion, which metal ion will precipitate as an iodide?

22-94. (*) Using a computer, generate a table of solubility as a function of pH for chromium(III) hydroxide and tin(II) hydroxide at 25°C. Have the computer graph your results by plotting the solubility as a function of pH for each species to generate a plot similar to that shown in Figure 22.13. Over what pH range is the best separation achieved for a mixture of these two species?

22-95. (*) Calculate the solubility of silver acetate, $AgCH_3COO(s)$, in solutions buffered at pH = 2.00, 4.00, 6.00, 8.00, and 10.00 at 25°C. Take the value of K_{sp} for $AgCH_3COO(s)$ to be 1.9×10^{-3} M^2 at 25°C.

22-96. (*) Using a computer, generate a table of solubility as a function of pH for silver acetate at 25°C (see the previous problem). Have the computer graph your results by plotting the solubility as a function of pH to generate a plot similar to that shown in Figure 22.13.

22-97. (*) Marble is predominantly composed of calcium carbonate, $CaCO_3(s)$. Calculate the solubility of marble in normal rainwater at pH = 5.60, and in acidic rainwater at pH = 4.40. Explain why acid rain is so damaging to buildings and statues composed of marble and why you should never use an acidic cleanser on marble tile. The relevant chemical equations are

(1) $CaCO_3(s) \leftrightharpoons Ca^{2+}(aq) + CO_3^{2-}(aq)$

(2) $CO_3^{2-}(aq) + H_3O^+(aq) \leftrightharpoons HCO_3^-(aq) + H_2O(l)$

(3) $HCO_3^-(aq) + H_3O^+(aq) \leftrightharpoons H_2CO_3(aq) + H_2O(l)$

22-98. (*) Given the following data at 25°C:

solubility of $I_2(s)$ in $H_2O(l)$:	0.0013 M
solubility of $I_2(s)$ in 0.100 M $KI(aq)$:	0.051 M

calculate the equilibrium constants for the following set of equations:

(1) $I_2(s) \leftrightharpoons I_2(aq)$

(2) $I_2(s) + I^-(aq) \leftrightharpoons I_3^-(aq)$

(3) $I_2(aq) + I^-(aq) \leftrightharpoons I_3^-(aq)$

Josiah Willard Gibbs (1839–1903) was born in New Haven, Connecticut. His father was a professor of sacred literature at Yale University. Gibbs studied at Yale and received his Ph.D. in engineering in 1863, the second doctorate in science and the first in engineering awarded in the United States. He continued his engineering studies and in 1866 received a patent for an improved railway brake. From 1866 to 1868, he studied in Paris, Berlin, and Heidelberg, and upon returning to New Haven, he began his studies of thermodynamics. He stayed on at Yale for years without salary, and remained there for the rest of his life. In 1878, Gibbs published a long, original treatise on thermodynamics entitled, "On the Equilibrium of Heterogeneous Substances" in the *Transactions of the Connecticut Academy of Sciences.* Between its austere writing style and the obscurity of the journal in which it was published, this important work was not as widely appreciated as it deserved to be. Fortunately, Gibbs sent copies to a number of prominent European scientists, who immediately appreciated the significance of the work and made it known in Europe. Eventually, Gibbs received the recognition that was his due, and Yale finally offered him a salaried position in 1880. He is known as the founder of thermodynamics, which is presented in courses today much as he formulated it over 150 years ago. His father died in 1855 and his mother died in 1861. Afterwards, Gibbs, an unassuming, modest person, lived in the family home with his sisters in New Haven, Connecticut, for most of his life.

23. Chemical Thermodynamics

It is natural to ask why some substances react with each other and others do not. Why do some reactions occur spontaneously and others require a continuous input of energy from an external source? In this chapter, we discover the condition that must be met for a reaction to be spontaneous—a condition called the *criterion for reaction spontaneity*.

The observation that all highly exothermic reactions are spontaneous led the French chemist Marcelin Berthelot to put forth, in the 1860s, the hypothesis that all spontaneous reactions are exothermic. Berthelot's criterion of spontaneity for a chemical reaction was based on the sign of the enthalpy change for the reaction, ΔH_{rxn}. According to Berthelot, if ΔH_{rxn} is negative, then a reaction is spontaneous.

Berthelot's criterion of reaction spontaneity was shown to be incorrect, however; and it was superseded by the criterion of reaction spontaneity developed by the American thermodynamicist J. Willard Gibbs (Frontispiece). Gibbs showed that reaction spontaneity is not just a matter of enthalpy changes. Another property, called entropy, also must be considered when determining whether a reaction is spontaneous. In this chapter, we shall discuss how the entropy concept is based on a fundamental principle called the second law of thermodynamics. We shall also show that the Gibbs criterion of reaction spontaneity is given by the sign of a new quantity called the Gibbs energy change that depends on both enthalpy and entropy changes for the reaction. In fact, as we shall see, the Gibbs energy change is equal to the maximum amount of energy in the form of work that can be extracted from a reaction system to perform tasks, such as propelling a rocket or powering an electric motor.

23-1. Not All Spontaneous Reactions Evolve Energy

In Chapter 14, we discussed enthalpy changes for chemical equations. Recall that an exothermic chemical reaction is one in which energy is evolved as

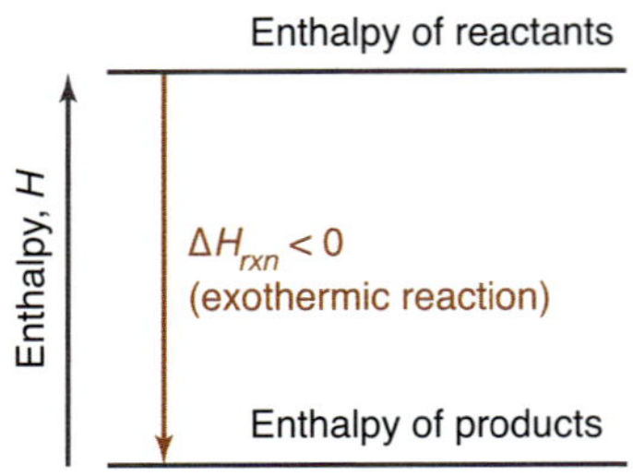

Figure 23.1 In an exothermic reaction, the total enthalpy of the products is less than the total enthalpy of the reactants.

heat. Exothermic reactions go energetically downhill, in the sense that the total enthalpy of the products is less than the total enthalpy of the reactants (Figure 23.1). For an exothermic reaction, the enthalpy change, ΔH_{rxn}, is negative. For an exothermic reaction run at standard conditions (all species at one bar pressure, all solutes at one molar concentration), we have

$$\Delta H^\circ_{rxn} = H^\circ[\text{products}] - H^\circ[\text{reactants}] < 0 \qquad \text{(exothermic)}$$

where the superscript degree signs denote *standard* values for the thermodynamic quantities. Thus, ΔH°_{rxn} is the standard enthalpy change for the equation.

In Chapter 14, we learned how to calculate ΔH°_{rxn} from the values in a table of enthalpies of formation (Table 14.3). Recall that the value of ΔH°_{rxn} is approximately equal to the value of the standard energy change, ΔU°_{rxn}, for a reaction. Also, recall that $\Delta H^\circ_{rxn} \approx \Delta H_{rxn}$, because enthalpy is relatively independent of pressure.

A process that takes place without the input of energy from an external source is said to be **spontaneous**. The natural tendency of simple mechanical systems is to undergo processes that lead to a decrease in the energy of the system. For example, water flows spontaneously downhill. Water at the bottom of a waterfall has a lower potential energy than water at the top. To get water back to the top of the waterfall, we have to use a pump, which requires energy input. Thus, the flow of water uphill is not a spontaneous process. As another example, if we release the ball at the top of the hill as shown in Figure 23.2, then it spontaneously rolls down the hill and eventually comes to rest at the bottom of the valley. The ball at the lowest point of the valley has the lowest possible potential energy for this system.

We have all observed a wide variety of spontaneous chemical processes. For example, methane (natural gas), $CH_4(g)$, once ignited, burns spontaneously in air to yield carbon dioxide and water (Figure 23.3) according to the chemical equation

$$CH_4(g) + 2\,O_2(g) \rightarrow CO_2(g) + 2\,H_2O(g) \qquad \Delta H^\circ_{rxn} = -802.5\ \text{kJ}\cdot\text{mol}^{-1}$$

Recall from Chapter 14 that this equation and the corresponding value of ΔH°_{rxn} mean that when one mole of $CH_4(g)$ and two moles of $O_2(g)$ react to produce one mole of $CO_2(g)$ and two moles of $H_2O(g)$, that −802.5 kilojoules of energy as heat is released at standard conditions.

Two other examples of spontaneous processes are that Fe(*s*), on exposure to air and moisture, spontaneously rusts, according to

$$4\,Fe(s) + 3\,O_2(g) \xrightarrow[H_2O(l)]{} 2\,Fe_2O_3(s) \qquad \Delta H^\circ_{rxn} = -1648.4\ \text{kJ}\cdot\text{mol}^{-1}$$

and Zn(*s*) reacts spontaneously with 1.0 M HCl(*aq*) to yield $H_2(g)$ and aqueous $ZnCl_2(aq)$, according to

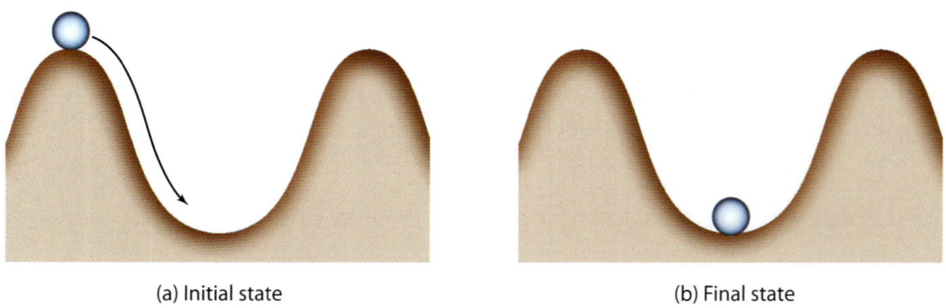

Figure 23.2 A ball at the top of a hill will roll down the hill and eventually come to rest. The potential energy of the ball at the bottom of the hill is less than that at the top. The ball spontaneously goes from a state of high potential energy to a state of low potential energy.

$$\mathrm{Zn}(s) + 2\,\mathrm{HCl}(aq) \rightarrow \mathrm{H_2}(g) + \mathrm{ZnCl_2}(aq) \qquad \Delta H^\circ_{\mathrm{rxn}} = -153.9\ \mathrm{kJ{\cdot}mol^{-1}}$$

Furthermore, these reactions as described do not occur spontaneously in the reverse direction.

The three chemical equations we have considered so far have been highly exothermic, and, by analogy with mechanical processes that move spontaneously to lower energy states (such as a ball rolling downhill), we predict (correctly) that they occur spontaneously. Many spontaneous processes, however, are not exothermic. Let's consider two gases, such as $N_2(g)$ and $I_2(g)$, occupying two separate containers, as shown in Figure 23.4. We know that, if allowed to, the gases will mix spontaneously. Furthermore, simple gaseous mixtures do not separate spontaneously. It would be a potentially disastrous occurrence if the air in a room suddenly separated so that part of the room contained pure oxygen and the rest contained nitrogen. Although the mixing of two gases is a spontaneous process, it turns out that $\Delta H^\circ \approx 0$ for such a process. Surely it is not the enthalpy change that drives the gas-mixing process.

Not only do some spontaneous processes have $\Delta H^\circ \approx 0$, but others are even endothermic. The enthalpy of the products is *greater* than that of the reactants. For example, ordinary ice at any temperature greater than 0°C spontaneously melts:

$$\mathrm{H_2O}(s) \rightarrow \mathrm{H_2O}(l) \qquad \Delta H^\circ_{\mathrm{fus}} = +6.0\ \mathrm{kJ{\cdot}mol^{-1}}$$

Another example is that table salt, $NaCl(s)$, spontaneously dissolves in water at 25°C:

$$\mathrm{NaCl}(s) \xrightarrow[\mathrm{H_2O}(l)]{} \mathrm{Na^+}(aq) + \mathrm{Cl^-}(aq) \qquad \Delta H^\circ_{\mathrm{solution}} = +3.9\ \mathrm{kJ{\cdot}mol^{-1}}$$

An especially interesting example of a spontaneous endothermic reaction is the reaction of solid barium hydroxide, $Ba(OH)_2(s)$, and solid ammonium nitrate, $NH_4NO_3(s)$, described by

$$\mathrm{Ba(OH)_2}(s) + 2\,\mathrm{NH_4NO_3}(s) \rightarrow \mathrm{Ba(NO_3)_2}(s) + 2\,\mathrm{H_2O}(l) + 2\,\mathrm{NH_3}(aq)$$

The energy absorbed from the reaction system is sufficient to drop the temperature of the reaction system from 20°C (room temperature) to well below the freezing point of water, as evidenced by the freezing of the water placed between the reaction flask and a block of wood (Figure 23.5). Mixtures of chemicals such as these are used in commercially available cold packs for the first-aid treatment of bruises and strains.

Figure 23.3 The combustion of methane, $CH_4(g)$, in air to yield gaseous carbon dioxide, $CO_2(g)$, and water vapor, $H_2O(g)$, is an example of a spontaneous, exothermic process.

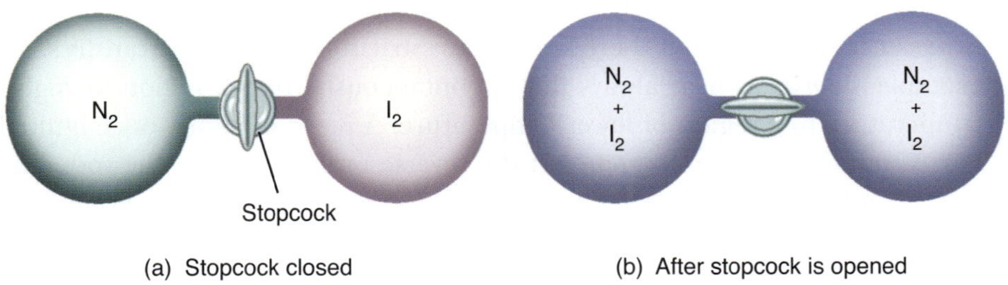

Figure 23.4 A simple example of a spontaneous process with $\Delta H^\circ \approx 0$ is the mixing of two gases.

Figure 23.5 (a) Samples of ammonium nitrate, $NH_4NO_3(s)$, and barium hydroxide, $Ba(OH)_2(s)$. The water in the eyedropper is placed on the wood block, the chemicals are mixed in the flask, and the flask is then placed on the wood block over the water. (b) The endothermic reaction between $Ba(OH)_2(s)$ and $NH_4NO_3(s)$ produces a large temperature drop that causes the water to freeze under the flask. The ice bonds the flask to the wood block.

(a)

(b)

That spontaneous processes can occur with negative, zero, or positive values of ΔH clearly indicates that the spontaneity of a process is not determined solely by the sign of the enthalpy change. There is an additional factor involved—the entropy change for the process.

23-2. The Second Law of Thermodynamics Places an Additional Restriction on Energy Transfers

The only restriction placed on energy transfers by the first law of thermodynamics (Chapter 14) is the conservation of energy. We know from everyday experience, however, that there are other restrictions on energy transfers. For example, energy as heat always flows spontaneously from a region of higher temperature to a region of lower temperature. A piece of paper, once ignited, burns spontaneously in oxygen, but the reverse process, that is, the spontaneous recombination of the combustion products to the piece of paper and oxygen gas, has never been observed in nature. Nor will any incantation cause a scrambled egg to reassemble itself. Innumerable other naturally occurring processes are also unidirectional. Spontaneous, unidirectional processes are often referred to as **irreversible processes**.

In thermodynamics, we always distinguish between the **system**, that part of the universe where the change of interest occurs, and its **surroundings**, which are the rest of the universe. Thermodynamics states that when a process occurs spontaneously, the system and its surroundings cannot both be restored exactly to their original states. If, for example, we restore the system to its initial state by doing work on the system, then it is impossible to restore the surroundings to its original state; if we restore the surroundings to its original state, then it is impossible to restore the system to its original state. All naturally occurring processes are in this sense irreversible. Nature is never exactly the same today as it was yesterday. When a process occurs in nature, the universe is irreversibly changed.

The second law of thermodynamics and the concept of entropy, which is embodied in the second law, arose from an analysis of the operation of heat engines (especially steam engines) by the French engineer Sadi Carnot in the early 19th century. Carnot's work led directly to the discovery of a new thermodynamic state function (Sections 14-1 and 14-2) called **entropy** and given the symbol S that is closely associated with the transfer of energy as heat. The key result is most compactly expressed in the equation

$$\Delta S_{\text{sys}} \geq \frac{q_{\text{sys}}}{T_{\text{sys}}} \tag{23.1}$$

Equation 23.1 is a mathematical statement of the **second law of thermodynamics**; it tells us that if energy is transferred either to or from a system as heat, q_{sys}, at the Kelvin temperature, T_{sys}, then the **entropy change** of the system, ΔS_{sys}, is greater than or equal to $q_{\text{sys}}/T_{\text{sys}}$. The equality sign applies only if the process is reversible. A **reversible process** is a process where the direction can be reversed at any point by an exceedingly small change in some parameter. For example, a temperature difference serves as the driving force for the flow of energy as heat from one object to another.

Suppose we have two metals in contact. If the temperature of metal A is greater than that of metal B, then energy as heat will flow from metal A to metal B. If the temperature of metal B is greater than that of metal A, then energy as heat will flow from metal B to metal A. If the temperature difference is exceedingly small, then we can change the direction of the flow of energy by changing the temperature difference slightly. In this case, we say that the process is reversible.

Another example of a reversible process is the vaporization of a liquid at a given pressure. If we increase the external pressure just slightly, then the liquid will condense, and if we decrease the external pressure just slightly, then the liquid will vaporize. In this case, then, we can reverse the process of vaporization-condensation by a slight change in the external pressure.

Strictly speaking, a reversible process is an idealization because the driving force (the temperature difference in the case of the two metals in contact or the pressure change in the case of vaporization-condensation of a liquid) has to be infinitesimally small, which cannot be realized in practice. Consequently, all naturally occurring processes are irreversible to some extent. The inequality sign in Equation 23.1 applies to irreversible processes, or essentially to all pro-

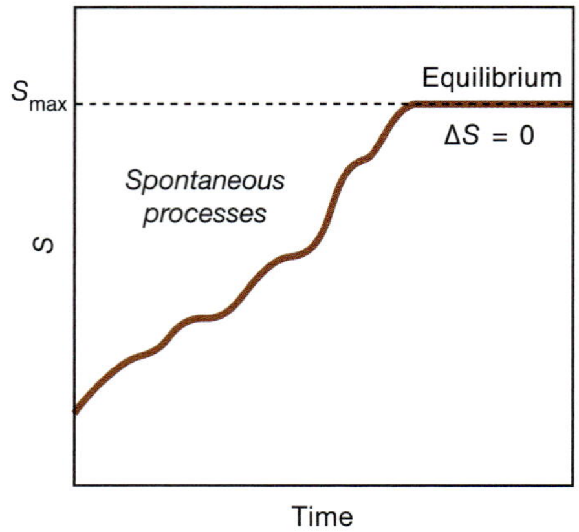

Figure 23.6 As a system spontaneously evolves toward equilibrium, it continues to generate entropy until equilibrium is reached, at which point the entropy is a maximum.

cesses that occur naturally. Note from Equation 23.1 that the units of entropy are joules per kelvin ($J \cdot K^{-1}$).

Equation 23.1 takes on a simpler form for an **isolated system**, that is, a system that cannot exchange energy with its surroundings, so that it can neither do work nor absorb energy as heat. We can picture an isolated system as one with rigid, thermally insulating, impermeable walls. For example, a liquid in a stoppered Dewar flask is an example of an isolated system because it is enclosed in a rigid, thermally insulating, impermeable container. For such a system, $q_{sys} = 0$ and Equation 23.1 becomes

$$\Delta S_{sys} = 0 \quad \text{(equilibrium in an isolated system)} \qquad (23.2a)$$

or

$$\Delta S_{sys} > 0 \quad \text{(a spontaneous process in an isolated system)} \qquad (23.2b)$$

Consider an isolated system that is not in equilibrium. As the system undergoes spontaneous changes and inevitably evolves toward equilibrium, it continually generates entropy until equilibrium is attained, at which point the entropy will achieve its maximum value (Figure 23.6). Unlike energy, entropy is not conserved, and is generated by any spontaneously occurring process.

We can give another interpretation to Equations 23.2. Consider a system in contact with its surroundings, as illustrated in Figure 23.7. We can isolate the system and its surroundings so that we can view the two as one isolated system and write Equations 23.2 as

$$\Delta S_{sys} + \Delta S_{surr} = 0 \quad \text{(equilibrium)} \qquad (23.3a)$$

or

$$\Delta S_{sys} + \Delta S_{surr} > 0 \quad \text{(spontaneous process)} \qquad (23.3b)$$

Because the most general example of a system and its surroundings includes the entire universe, and because all naturally occurring processes are spontaneous, Equation 23.3b may also be expressed as $\Delta S_{universe} > 0$ for a spontaneous process. Thus, the second law of thermodynamics tells us that the entropy of the universe always increases when a natural process occurs. In fact, the first and second laws of thermodynamics can be summarized in this pithy statement: The energy of the universe is constant and its entropy is continuously increasing.

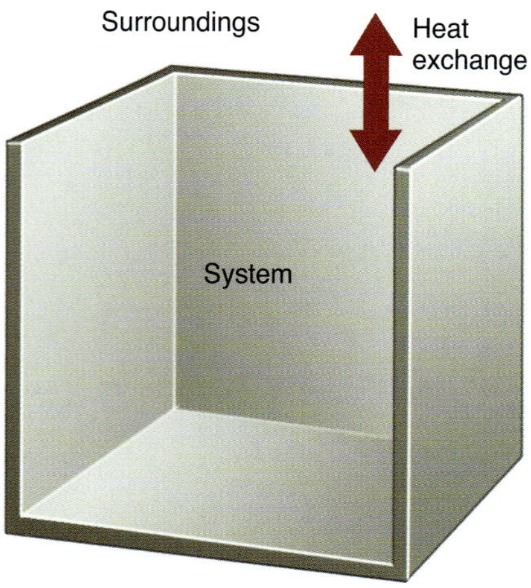

Figure 23.7 A system in contact with its surroundings.

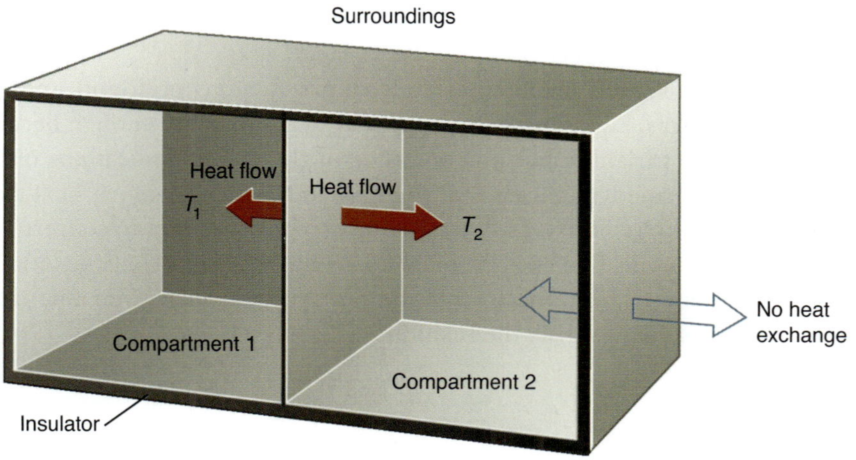

Figure 23.8 Two compartments in contact but isolated from their surroundings.

Neither Equation 23.1 nor Equation 23.3 says that ΔS_{sys} must be positive. In fact, ΔS_{sys} may indeed be negative. If it is negative, Equation 23.3 says that ΔS_{surr} must be positive to the extent that it at least cancels the negative contribution of ΔS_{sys} so that $\Delta S_{sys} + \Delta S_{surr} \geq 0$. This misunderstanding of the second law occasionally leads to bizarre interpretations, especially outside of the physical sciences.

Let's use Equation 23.1 to show that energy as heat will flow naturally from a region of high temperature to a region of low temperature. Consider the situation in Figure 23.8 where two compartments at different temperatures, T_1 and T_2, are in thermal contact, but are separated from their surroundings by a rigid, thermally insulating, impermeable wall, so that this two-compartment system is effectively isolated. The only flow of energy as heat will be from one compartment to the other. The entropy change of compartment 1 satisfies $\Delta S_1 \geq q_1/T_1$ and that of compartment 2 satisfies $\Delta S_2 \geq q_2/T_2$. Therefore, the entropy change of the two-compartment system satisfies

$$\Delta S = \Delta S_1 + \Delta S_2 \geq \frac{q_1}{T_1} + \frac{q_2}{T_2} \tag{23.4}$$

Because $q_2 = -q_1$, that is, the energy gained as heat by one compartment is equal to the energy lost as heat by the other, we can write Equation 23.4 as

$$\Delta S \geq \frac{q_1}{T_1} - \frac{q_1}{T_2} \geq q_1\left(\frac{1}{T_1} - \frac{1}{T_2}\right) \tag{23.5}$$

The two-compartment system is isolated, so we can use Equations 23.2 to write

$$\Delta S \geq q_1\left(\frac{1}{T_1} - \frac{1}{T_2}\right) \geq 0 \tag{23.6}$$

Equation 23.6 says that if the system is at equilibrium, then $\Delta S = 0$, and, therefore, $T_1 = T_2$. If the system is not at equilibrium, then $\Delta S > 0$. Let's assume first that $q_1 > 0$, meaning that energy as heat flows from compartment 2 to compartment 1. In this case, the condition $\Delta S > 0$ says that $1/T_1 > 1/T_2$, or that $T_2 > T_1$, meaning that energy as heat will flow spontaneously from compartment 2 to compartment 1 only if compartment 2 is at a higher temperature than compartment 1. If, on the other hand, $q_1 < 0$, meaning that energy as heat flows from compartment 1 to compartment 2, then the condition $\Delta S > 0$ requires that $T_1 > T_2$, as you should expect.

Actually, the second law of thermodynamics is quite profound and has many consequences. Equation 23.1 may appear to be innocent enough because it has only three variables, but it is not. One of the many implications of the second law of thermodynamics is that you have to transfer energy as heat from a high-temperature region to a low-temperature region in order to convert heat energy into mechanical energy. Problem 23-88 has you show that the amount of mechanical energy (or work, w) that you can extract from a given amount of energy as heat (q) must satisfy the relation

$$\frac{w}{q} < \frac{T_2 - T_1}{T_2} \tag{23.7}$$

where T_2 is the temperature of the high-temperature region and T_1 is that of the low-temperature region. Notice that if the two temperatures are the same, then no mechanical energy is produced, that is, $w/q = 0$. Also notice that Equation 23.7 says that it is not possible to convert all the energy as heat into mechanical energy ($w/q = 1$), unless $T_1 = 0$ K, which is essentially an unattainable temperature. Thus, we see that the second law of thermodynamics imposes severe restrictions on the conversion of energy as heat into mechanical energy. One sad consequence of this restriction is that even though your car is surrounded by the atmosphere, containing a vast amount of thermal energy, it is of no use for powering your car because your car and the atmosphere are at the same temperature. Nevertheless, the allure of getting something for nothing is so irresistible that patent offices around the world are beset with proposals of such devices, many of which are wonderfully ingenious. But alas, they are ultimately shown to violate the second law of thermodynamics.

The concept of entropy also can be developed in molecular terms. In the next few sections, we shall see how the entropy concept gives us insight into chemical reactivity.

23-3. Entropy May Be Considered as a Measure of the Amount of Disorder or Randomness in a System

On a molecular level, it is sometimes convenient to consider the entropy of a substance to be a quantitative measure of the amount of disorder. Disorder is of two types: **positional disorder** that refers to the distribution of the particles in space, and **thermal disorder** that refers to the distribution of the available energy among the particles. Any process that produces a more random distribution of the particles in space gives rise to an increase in the total entropy of the substance. Any constant-pressure process that increases the temperature of the particles also gives rise to an increase in the total entropy of the substance.

As a simple example of an increase in positional disorder, consider the **isothermal** (constant temperature) expansion of a gas into a vacuum, as illustrated in Figure 23.9. When the stopcock between the bulb containing the gas on the left and the evacuated bulb on the right is opened, the gas expands to fill both bulbs. When equilibrium is attained, the gas occupies both bulbs instead

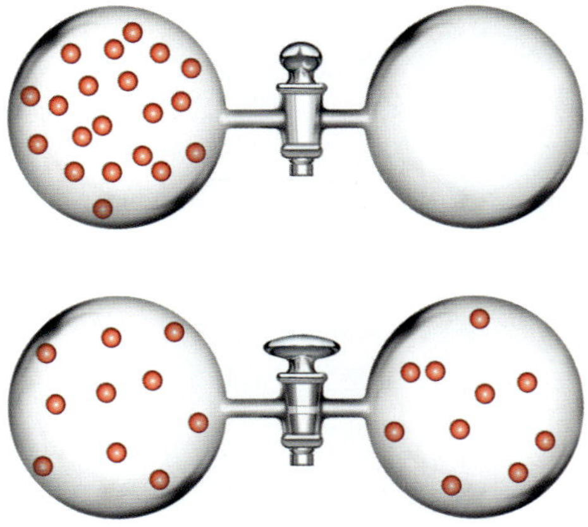

Figure 23.9 An illustration of a process driven by positional disorder. (a) The left side of the container contains a gas, the right side is evacuated. (b) When the stopcock is open, the gas on the left expands to fill both chambers, leading to an increase in positional disorder.

of just one and so has greater positional disorder. This increase in positional disorder is accompanied by an increase in entropy, which is the driving force for this process.

Now let's consider a process that involves an increase in both positional disorder and thermal disorder. Figure 23.10 shows the molar entropy of oxygen plotted against temperature from 0 K to 300 K. The entropy is zero at 0 K. At this temperature (absolute zero), there is no positional disorder because the O_2 molecules in the crystal are perfectly arrayed in the lattice. Nor is there any thermal disorder because all the molecules are in the lowest possible energy state. Entropy is associated with disorder—no disorder, no entropy. The statement that the entropy of a perfect crystalline substance is zero at absolute zero is known as the **third law of thermodynamics**. Absolute zero (0 K) is a theoretical state that can never be reached in practice, although temperatures as low as 100 picokelvin have been achieved by some experimenters (Figure 23.11).

As the temperature is increased from 0 K, the oxygen molecules begin to vibrate more freely about their lattice sites. The increase in temperature causes

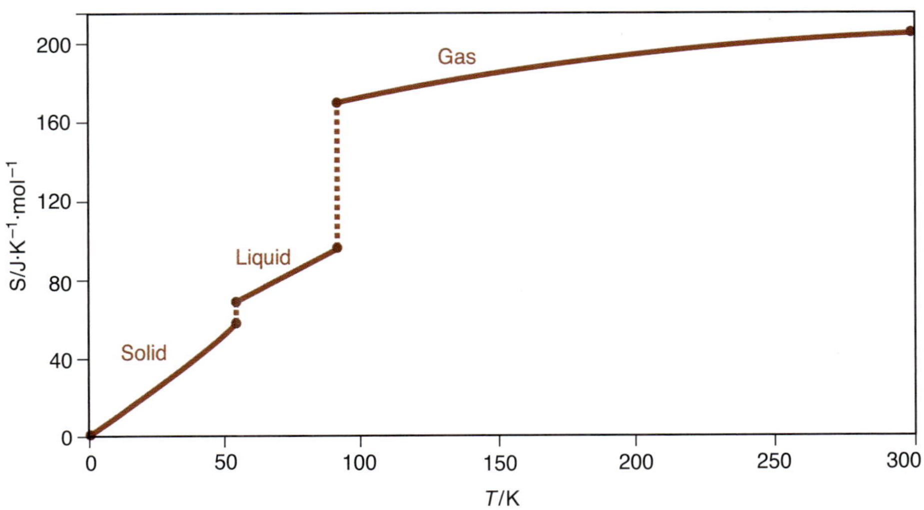

Figure 23.10 The molar entropy of oxygen as a function of temperature at one bar. Note that the entropy of each phase increases smoothly with increasing temperature. This gradual increase in entropy with an increase in temperature is a result of the increase in thermal disorder. The jumps in entropy that occur when the solid melts and the liquid vaporizes are the result of an increase in positional disorder of the oxygen molecules.

Figure 23.11 Although absolute zero (0 K) cannot be achieved in practice, temperatures as low as 100 picokelvin have been achieved in the laboratory using light and magnetic fields to trap and slow the rotation and vibration of molecules, such as in this apparatus at the Massachusetts Institute of Technology used to cool sodium atoms down to 500 picokelvin. The atoms were magnetically confined by the coil in the center, which is one centimeter in diameter. Such ultracold atoms may one day be the basis for a new generation of quantum computers and atomic lasers.

an increase in thermal disorder that leads to an increase in entropy (Figure 23.10). This increase in entropy is associated with the greater amount of energy that must be distributed among the molecules. We learned in Chapter 9 that molecules, like atoms, are restricted to discrete, quantized energy levels. The greater the amount of energy stored in a substance, the greater the number of ways in which the energy can be distributed among these levels, and thus the greater the thermal disorder and the greater the entropy.

Figure 23.10 also shows that at fixed pressure, the entropy of liquid and gaseous oxygen, like that for solid oxygen, increases as the temperature increases. As the temperature increases, the motions of the liquid and the gaseous oxygen molecules become more vigorous. The average molecular speed increases with temperature (Section 13-10), and this increased molecular speed leads to an increase in thermal disorder, which is the molecular basis for the increase in entropy.

Although the entropy of a particular phase increases smoothly with increasing temperature due to its increasing thermal disorder with increasing temperature, at a phase transition there is a jump in entropy. Let's investigate the origin of the two entropy jumps in Figure 23.10. When solid oxygen melts at 54.36 K, the solid in an ordered lattice breaks down into a liquid, where the molecules no longer are confined to lattice sites. Each molecule moves throughout the liquid volume, so there is an increase in the positional disorder of the oxygen (Figure 23.12). This increase in positional disorder leads to an increase in entropy. Because all the entropy increase occurs at the melting point, which is a fixed temperature, ΔS of melting (fusion) appears as a vertical jump in Figure 23.10. A jump also occurs at the boiling point because the molecules in the gas phase can move throughout a much larger volume than the molecules in the liquid phase. Therefore, there is a sudden increase in positional disorder as a substance goes from the liquid phase to the gas phase (Figure 23.13 and Figure 23.14).

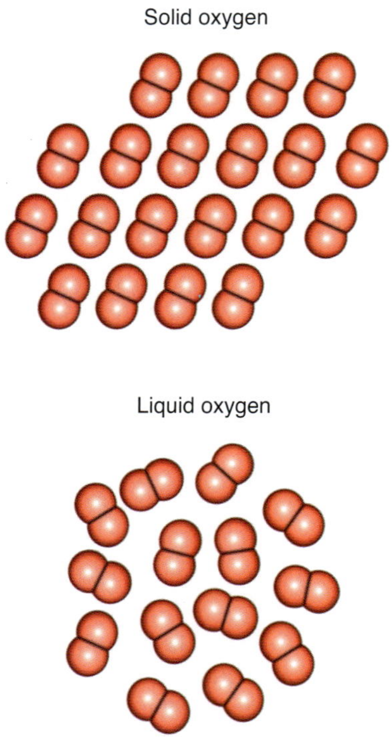

Figure 23.12 When a solid melts, the molecules become free to move throughout the volume of the liquid. This increased positional disorder in the liquid implies that, under the same conditions, the entropy of a liquid is greater than the entropy of the solid.

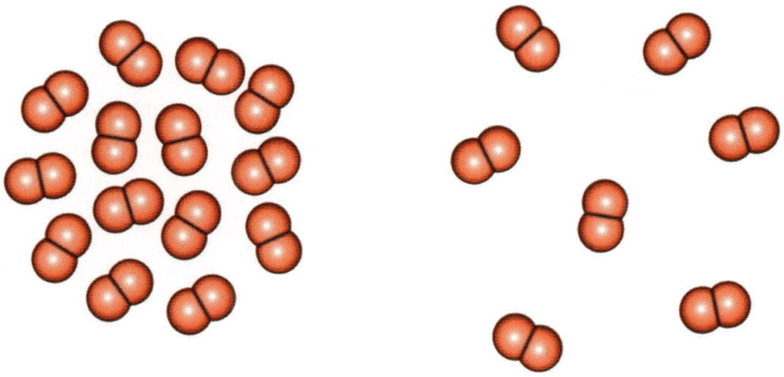

Figure 23.13 When a liquid vaporizes, there is a large increase in positional disorder and hence a large increase in entropy.

EXAMPLE 23-1: Predict the sign of the entropy change, ΔS_{sys}, for the following processes:

(a) $H_2O(g)$ (1 bar, 25°C) $\rightarrow$ $H_2O(g)$ (0.01 bar, 25°C)

(b) The dissolution of sodium chloride in water

(c) $NaCl(aq)$ (4.8 M, 20°C) $\rightarrow$ $NaCl(aq)$ (1.0 M, 20°C)

(d) $C_6H_6(g)$ (10 Torr, 25°C) $\rightarrow$ C_6H_6 (adsorbed on a metal surface at 25°C)

Solution: (a) A gas at a lower pressure, and thus with greater volume, is more disordered than a gas at a higher pressure at the same temperature; therefore,

$$\Delta S_{gas} \text{ (constant-temperature expansion)} > 0$$

(b) Upon dissolving, the sodium and chloride ions are dispersed throughout the solvent and are much more disordered than in the crystal, so

$$\Delta S_{solid} \text{ (dissolution)} > 0$$

(c) Dilution with solvent increases the positional disorder of the solute, so

$$\Delta S_{solute} \text{ (dilution)} > 0$$

(d) The adsorption of a gas onto a surface produces a decrease in the entropy of adsorbed molecules because of the decreased freedom of movement (see Figure 23.15), so

$$\Delta S_{benzene} \text{ (adsorption)} < 0$$

PRACTICE PROBLEM 23-1: Predict the sign of the entropy change, ΔS_{sys}, for the following processes:

(a) $H_2O(l) \rightarrow H_2O(s)$ (0°C, 1 bar)

(b) $Cu(s)$ (1 bar, 500°C) $\rightarrow$ $Cu(s)$ (1 bar, 25°C)

(c) $4Al(s) + 3O_2(g) \rightarrow 2Al_2O_3(s)$ (25°C, 1 bar)

Answer: (a) $\Delta S_{sys} < 0$; (b) $\Delta S_{sys} < 0$; (c) $\Delta S_{sys} < 0$

Figure 23.14 Liquid nitrogen has a normal boiling point of 77 K. This photograph shows the liquid boiling at atmospheric pressure. The heat energy necessary to boil the liquid comes from the surroundings at 20°C. Because the boiling point is so low, no other heat source (such as a flame) is needed to boil the liquid.

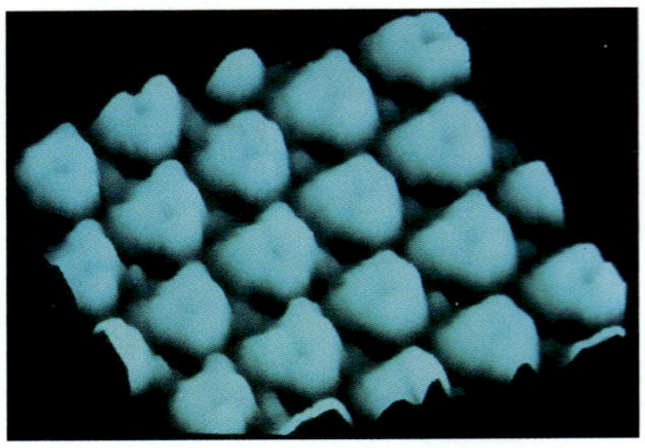

Figure 23.15 Scanning tunneling micrograph of benzene molecules adsorbed on the surface of a rhodium crystal. The hexagonal shape of the benzene molecules is revealed in this remarkable micrograph.

We can use Equation 23.1 to calculate the change in the entropy of one mole of a substance upon fusion, called the **molar entropy of fusion** and written as ΔS_{fus}. Because melting (as well as vaporization) can be carried out under essentially reversible conditions (recall that we can make the substance freeze or melt by changing the external pressure by a very small amount), we use the equality sign in Equation 23.1, together with the relation $q_P = \Delta H_{fus}$, to obtain

$$\Delta S_{fus} = \frac{\Delta H_{fus}}{T_m} \tag{23.8}$$

In Equation 23.8, ΔH_{fus} is the molar enthalpy of fusion (Chapter 15) and T_m is the melting point in kelvin. From Equation 23.8 and Table 15.3, we find that the entropy increase of one mole of oxygen upon melting at 54.36 K (the normal melting point of oxygen) is

$$\Delta S_{fus} = \frac{\Delta H_{fus}}{T_m} = \frac{440\ \text{J}\cdot\text{mol}^{-1}}{54.36\ \text{K}} = 8.1\ \text{J}\cdot\text{K}^{-1}\cdot\text{mol}^{-1}$$

Note that ΔS_{fus} is positive. As expected, the molar entropy of liquid oxygen is greater than the molar entropy of solid oxygen. We expected it to be greater, but we didn't predict by how much.

We can calculate the change in the entropy of a substance upon vaporization by using a formula that is analogous to Equation 23.8. If we let ΔS_{vap} be the entropy change of one mole of a substance on vaporization, called the **molar entropy of vaporization**, then using Equation 23.1 and the fact that $q_P = \Delta H_{vap}$, we have

$$\Delta S_{vap} = \frac{\Delta H_{vap}}{T_b} \tag{23.9}$$

where ΔH_{vap} is the molar enthalpy of vaporization (Chapter 15) and T_b is the boiling point in kelvin. Using Equation 23.9 and Table 15.3, we find that the change in the entropy of one mole of liquid oxygen upon being vaporized at 90.20 K (the normal boiling point of oxygen) is

$$\Delta S_{vap} = \frac{\Delta H_{vap}}{T_b} = \frac{6820\ \text{J}\cdot\text{mol}^{-1}}{90.20\ \text{K}} = 75.6\ \text{J}\cdot\text{K}^{-1}\cdot\text{mol}^{-1}$$

Observe that ΔS_{vap} is positive. As expected, the molar entropy of gaseous oxygen is greater than the molar entropy of liquid oxygen. Notice also that $\Delta S_{vap} > \Delta S_{fus}$ because the molecules in the gas phase are much more positionally disordered than the molecules in the liquid or the solid phase. This order of magnitude is a general result.

EXAMPLE 23-2: The molar enthalpy of fusion of water is ΔH_{fus} = 6.01 kJ·mol^{-1} at 0°C, and the molar enthalpy of vaporization of water is ΔH_{vap} = 40.65 kJ·mol^{-1} at 100°C. Calculate the values of ΔS_{fus} and ΔS_{vap} for water at these two phase transitions at a constant pressure of one atm.

Solution: The value of ΔS_{fus} is calculated using Equation 23.8. The melting point of water is 0°C, or 273.15 K; so we have

$$\Delta S_{fus} = \frac{\Delta H_{fus}}{T_m} = \frac{6.01 \times 10^3 \text{ J·mol}^{-1}}{273.15 \text{ K}} = 22.0 \text{ J·K}^{-1}\text{·mol}^{-1}$$

The value of ΔS_{vap} is calculated using Equation 23.9, with $T_b = 373.15$ K:

$$\Delta S_{vap} = \frac{\Delta H_{vap}}{T_b} = \frac{40.65 \times 10^3 \text{ J·mol}^{-1}}{373.15 \text{ K}} = 108.9 \text{ J·K}^{-1}\text{·mol}^{-1}$$

Notice that both ΔS_{fus} and ΔS_{vap} are positive, and $\Delta S_{vap} > \Delta S_{fus}$.

PRACTICE PROBLEM 23-2: Recall from Chapter 15 that the process where a solid is converted directly to a gas without passing through the liquid phase is called sublimation. Why is the molar entropy of sublimation always greater than the molar entropy of vaporization at the same temperature and pressure?

Answer: $\Delta S_{sub} = \Delta S_{fus} + \Delta S_{vap} > \Delta S_{vap}$

23-4. Molar Entropy Depends upon Molar Mass and Molecular Structure

So far we have calculated only entropy changes. Because the molar entropy is zero at 0 K—by the third law of thermodynamics—we can establish a scale of **absolute molar entropy** values. Recall that the superscript degree sign on a thermodynamic quantity tells us that it is the value at standard conditions—that is, at exactly one bar pressure and, for solutes, at a concentration of exactly one molar. Table 23.1 gives the **standard molar entropies** of substances at 25°C, denoted by $S°$. Table 23.1 also contains values of some other thermodynamic quantities (ΔH_f° and ΔG_f°) that cannot be specified absolutely. We shall learn about ΔG_f° values in Section 23-9.

A more complete listing of thermodynamic data appears in Appendix D.

The molar entropy of a gas or solute depends strongly on concentration because of the change in positional disorder with concentration. This property of entropy differs sharply from enthalpy, which is relatively insensitive to pressure or concentration.

Let's look at the standard molar entropy values in Table 23.1 and try to determine some trends. First notice that the values of the molar entropies of the gaseous substances are the largest, and the values of the molar entropies of the solid substances are the smallest. We have already discussed the reason for this.

Now consider the standard molar entropies of the noble gases. The values at standard conditions and 25°C are given in Table 23.2. The increase in molar entropy of the noble gases is a consequence of their increasing mass as we move down the periodic table. It is known from quantum theory that the greater the molecular mass, the more closely spaced are the energy levels. Thus, an increase in mass leads to an increase in thermal disorder (more quantum levels

TABLE 23.1 Standard molar entropies ($S°$), enthalpies of formation (ΔH_f°), and Gibbs energies of formation (ΔG_f°) of various substances at 25°C and one bar (see also Appendix D)*

Substance	$S°$/ J·K^{-1}·mol^{-1}	ΔH_f°/ kJ·mol^{-1}	ΔG_f°/ kJ·mol^{-1}	Substance	$S°$/ J·K^{-1}·mol^{-1}	ΔH_f°/ kJ·mol^{-1}	ΔG_f°/ kJ·mol^{-1}
Ag(*s*)	42.6	0	0	H_2O_2(*l*)	109.6	−187.8	−120.4
AgCl(*s*)	96.3	−127.0	−109.8	H_2S(*g*)	205.8	−20.6	−33.4
C(*s*, diamond)	2.4	1.9	2.9	N(*g*)	153.3	472.7	455.5
C(*s*, graphite)	5.7	0	0	N_2(*g*)	191.6	0	0
CH_4(*g*)	186.3	−74.6	−50.5	NH_3(*g*)	192.8	−45.9	−16.4
C_2H_2(*g*)	200.9	227.4	209.9	N_2H_4(*l*)	121.2	50.6	149.3
C_2H_4(*g*)	219.3	52.4	68.4	NO(*g*)	210.8	91.3	87.6
C_6H_6(*l*)	173.4	49.1	124.5	NO_2(*g*)	240.1	33.2	51.3
CH_3OH(*l*)	126.8	−239.2	−166.6	N_2O(*g*)	220.0	81.6	103.7
CH_3Cl(*g*)	234.6	−81.9	−58.4	N_2O_4(*g*)	304.4	11.1	99.8
CH_3Cl(*l*)	145.3	−102	−51.5	N_2O_5(*s*)	178.2	−43.1	113.9
CH_2Cl_2(*g*)	270.2	−95.4	−68.8	Na(*g*)	153.7	107.5	77.0
CH_2Cl_2(*l*)	177.8	−124.2	−70.0	Na(*s*)	51.3	0	0
$CHCl_3$(*g*)	295.7	−102.7	6.0	O(*g*)	161.1	249.2	231.7
$CHCl_3$(*l*)	201.7	−134.1	−73.7	O_2(*g*)	205.2	0	0
CO(*g*)	197.7	−110.5	−137.2	P(*s*, white)	41.1	0	0
CO_2(*g*)	213.8	−393.5	−394.4	P(*s*, red)	22.8	−17.6	−12.1
Cl(*g*)	165.2	121.3	105.3	PCl_3(*g*)	311.8	−287.0	−267.8
Cl_2(*g*)	223.1	0	0	PCl_5(*g*)	364.6	−374.9	−305.0
H(*g*)	114.7	218.0	203.3	S(*s*, rhombic)	28.5	0	0
H_2(*g*)	130.7	0	0	S(*s*, monoclinic)	32.6	0.3	0.1
H_2O(*g*)	188.8	−241.8	−228.6	SO_2(*g*)	248.2	−296.8	−300.1
H_2O(*l*)	70.0	−285.8	−237.1	SO_3(*g*)	256.8	−395.7	−371.1

*Most data from *CRC Handbook of Chemistry and Physics*, 87th Online Edition, 2006–2007.

TABLE 23.2 Standard molar entropies ($S°$) for the noble gases at 25°C and one bar

Noble gas	$S°$/J·K^{-1}·mol^{-1}
He(*g*)	126.2
Ne(*g*)	146.3
Ar(*g*)	154.8
Kr(*g*)	164.1
Xe(*g*)	169.7

TABLE 23.3 Standard molar entropies ($S°$) for the halogen gases at 25°C and one bar

Halogen	$S°$/J·K^{-1}·mol^{-1}
$F_2(g)$	202.8
$Cl_2(g)$	223.1
$Br_2(g)$	245.5
$I_2(g)$	260.7

populated) and a greater entropy. The same trend can be seen by comparing the standard molar entropies at 25°C of the gaseous halogens (Table 23.3). For molecules with the same number of atoms, the standard molar entropy increases with increasing molecular mass.

Generally speaking, the more atoms of a given type there are in a molecule, the greater the capacity of the molecule to take up energy and thus the greater the entropy (the greater the number of atoms, the more different ways in which the molecules can vibrate). This trend is illustrated by the series $C_2H_2(g)$, $C_2H_4(g)$, and $C_2H_6(g)$. Their Lewis formulas and standard molar entropies are shown in the margin.

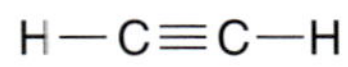

ethyne (acetylene)
$S° = 200.9$ J·K^{-1}·mol^{-1}

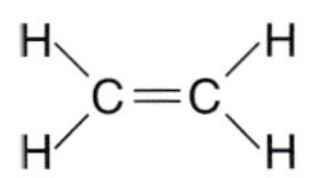

ethene (ethylene)
$S° = 219.3$ J·K^{-1}·mol^{-1}

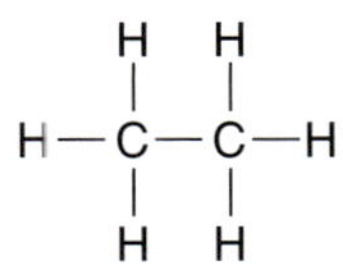

ethane
$S° = 229.2$ J·K^{-1}·mol^{-1}

EXAMPLE 23-3: Without referring to Table 23.1, arrange the following molecules in order of increasing standard molar entropy: $CH_2Cl_2(g)$; $CHCl_3(g)$; $CH_3Cl(g)$.

Solution: The number of atoms is the same in each case, but chlorine has a greater mass than hydrogen. Thus, we predict that

$$S°(CH_3Cl) < S°(CH_2Cl_2) < S°(CHCl_3)$$

The values of the standard molar entropies at 25°C are 234.6 J·K^{-1}·mol^{-1}, 270.2 J·K^{-1}·mol^{-1}, and 295.7 J·K^{-1}·mol^{-1}, respectively, in agreement with our predicted order.

PRACTICE PROBLEM 23-3: Arrange the following species in order of increasing standard molar entropy: $N_2O_4(g)$; $NO_2(g)$; $NO(g)$; $N_2O(g)$; and $N_2O_5(g)$.

Answer: $NO(g) < N_2O(g) < NO_2(g) < N_2O_4(g) < N_2O_5(g)$

An interesting comparison is that of the isomers 2-propanone, commonly called acetone, and oxetane, commonly called trimethylene oxide. Their Lewis formulas and standard molar entropies at 25°C are shown in the margin. The

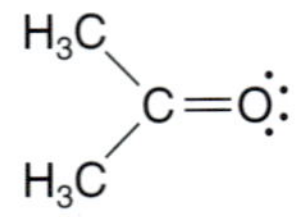

2-propanone (acetone)
$S^\circ = 295.3\ \text{J·K}^{-1}\text{·mol}^{-1}$

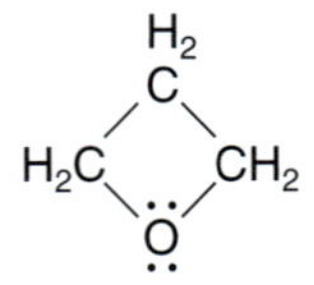

oxetane (trimethylene oxide)
$S^\circ = 274\ \text{J·K}^{-1}\text{·mol}^{-1}$

entropy of acetone is higher than that of trimethylene oxide because of the free rotation of the methyl groups about the carbon-carbon bonds in the acetone molecule. The relatively rigid ring structure of the trimethylene oxide molecule restricts the movement of the atoms in the ring. This restriction gives rise to a lower entropy because the rigid trimethylene oxide isomer has fewer ways to take up energy than the more flexible acetone molecule, which has more possibilities for intermolecular motion. For molecules with approximately the same molecular masses, the more compact the molecule, the smaller its entropy.

23-5. $\Delta S^\circ_{\text{rxn}}$ Equals the Standard Entropy of the Products Minus the Standard Entropy of the Reactants

Consider the general chemical equation

$$aA + bB \rightarrow yY + zZ \tag{23.10}$$

We define the **standard entropy change** for a chemical equation, $\Delta S^\circ_{\text{rxn}}$, as

$$\Delta S^\circ_{\text{rxn}} = S^\circ[\text{products}] - S^\circ[\text{reactants}] \tag{23.11}$$

Application of Equation 23.11 to Equation 23.10 yields

$$\Delta S^\circ_{\text{rxn}} = yS^\circ[\text{Y}] + zS^\circ[\text{Z}] - aS^\circ[\text{A}] - bS^\circ[\text{B}] \tag{23.12}$$

where the brackets serve to separate the formula of the compound from the symbol S°.

We can use the values of the standard molar entropies from Table 23.1 to calculate the value of the standard entropy change of a reaction, $\Delta S^\circ_{\text{rxn}}$. Let's use Equation 23.12 to calculate the value of $\Delta S^\circ_{\text{rxn}}$ at 25°C for the equation

$$CO(g) + 3H_2(g) \rightarrow CH_4(g) + H_2O(g)$$

From Equation 23.12 we have

$$\Delta S^\circ_{\text{rxn}} = S^\circ[CH_4(g)] + S^\circ[H_2O(g)] - S^\circ[CO(g)] - 3S^\circ[H_2(g)]$$

From Table 23.1 we obtain

$$\begin{aligned}\Delta S^\circ_{\text{rxn}} &= (1)(186.3\ \text{J·K}^{-1}\text{·mol}^{-1}) + (1)(188.8\ \text{J·K}^{-1}\text{·mol}^{-1}) \\ &\quad - (1)(197.7\ \text{J·K}^{-1}\text{·mol}^{-1}) - (3)(130.7\ \text{J·K}^{-1}\text{·mol}^{-1}) \\ &= -214.7\ \text{J·K}^{-1}\text{·mol}^{-1}\end{aligned}$$

There is a large negative change in entropy for this reaction because there are only two moles of gaseous products but four moles of gaseous reactants. At the same temperature and pressure, the volume of the products will be only about one-half the volume of the reactants. The reactant state has more positional disorder than the product state, so the entropy of the products is less than that of the reactants. Because the molar entropies of gases are generally much larger than those for solids or liquids, the value of ΔS°_{rxn} for reactions involving gases is generally dominated by the gaseous species in the equation.

By the same argument, we predict that the standard entropy change for the equation

$$2\,H_2O(l) \rightarrow 2\,H_2(g) + O_2(g)$$

is positive. The volume, and hence the positional disorder of the products, is much greater than the positional disorder of the reactants at the same temperature and pressure. Using data from Table 23.1, we obtain

$$\begin{aligned}\Delta S^\circ_{rxn} &= 2\,S^\circ[H_2(g)] + S^\circ[O_2(g)] - 2\,S^\circ[H_2O(l)]\\ &= (2)(130.7\,J\cdot K^{-1}\cdot mol^{-1}) + (1)(205.2\,J\cdot K^{-1}\cdot mol^{-1}) - (2)(70.0\,J\cdot K^{-1}\cdot mol^{-1})\\ &= +326.6\,J\cdot K^{-1}\cdot mol^{-1}\end{aligned}$$

We see that ΔS°_{rxn} is positive, just as we predicted. In general, the greater the difference between the total number of moles of gaseous products and the total number of moles of gaseous reactants, the greater is the value of ΔS°_{rxn}.

23-6. The Sign of ΔG_{rxn} Determines Reaction Spontaneity

The ill-fated principle of Berthelot that we discussed in the introduction to the chapter stated that a reaction has to be exothermic to be spontaneous. Berthelot reasoned that reactions evolving energy are spontaneous because the products are energetically downhill from the reactants. However, we have seen that there are spontaneous reactions in which the products either are energetically uphill from the reactants (such as the spontaneous melting of ice) or have essentially the same energy as the reactants (such as the spontaneous mixing of gases). Such reactions are said to be **entropy-driven reactions**. A reaction for which the total entropy of the products is greater than the total entropy of the reactants ($\Delta S^\circ_{rxn} > 0$) is an **entropy-favored reaction**.

When the energy of the products is less than the energy of the reactants ($\Delta H_{rxn} \approx \Delta U_{rxn} < 0$), the reaction is said to be **energy favored**. If there is little or no change in entropy ($\Delta S^\circ_{rxn} \approx 0$), then a system will change in such a way that its energy is decreased. Once the energy of the system is minimized, the system no longer changes and is at equilibrium. The simple mechanical examples of energy minimization for spontaneous processes that we discussed in Section 23-1 (for example, water or a ball moving downhill) are spontaneous processes for which there is little or no change in entropy.

In contrast, if there is little or no change in energy ($\Delta H_{rxn} \approx \Delta U_{rxn} \approx 0$), then a system will change in such a way that its entropy increases. When the entropy of the system is maximized, the system no longer changes and is at equilibrium. The mixing of two gases is a good example of this case. The gases mix spontaneously because the entropy of the final state (the mixture) is greater than the entropy of the initial state (separate gases). You may have had some idea that systems try to minimize energy, but the idea of entropy maximization may be new to you. Both ideas are equally important, however.

All processes that are both energy favored *and* entropy favored are spontaneous. That is, if $\Delta H_{\text{rxn}} < 0$ and $\Delta S_{\text{rxn}} > 0$, the process is spontaneous. Conversely, if $\Delta H_{\text{rxn}} > 0$ and $\Delta S_{\text{rxn}} < 0$, the process does not occur spontaneously. But what about the many processes for which the energy (enthalpy) and entropy factors oppose each other? In fact, many reactions have values of ΔH_{rxn} and ΔS_{rxn} that oppose each other. For such cases, we have either

(1) $\Delta H_{\text{rxn}} > 0$ and $\Delta S_{\text{rxn}} > 0$

or

(2) $\Delta H_{\text{rxn}} < 0$ and $\Delta S_{\text{rxn}} < 0$

Case (1) is entropy favored but energy disfavored; case (2) is energy favored but entropy disfavored. Such reactions may or may not be spontaneous, depending on the temperature and on the relative magnitudes of ΔH_{rxn} and ΔS_{rxn}.

The spontaneity of such reactions was explained over 100 years ago by J. Willard Gibbs, one of America's greatest scientists (Frontispiece). Gibbs introduced a quantity now called the **Gibbs energy**, G, which, as we shall show, serves as a compromise between the enthalpy change and the entropy change of a reaction. For a reaction run at a constant temperature, the **Gibbs energy change**, ΔG_{rxn}, is given by

$$\Delta G_{\text{rxn}} = \Delta H_{\text{rxn}} - T\Delta S_{\text{rxn}} \tag{23.13}$$

where T is the Kelvin temperature.

Chemical reactions are forced by nature to seek a compromise between energy minimization and entropy maximization. For a reaction that occurs at a constant temperature and pressure, the nature of the compromise is given by the sign and magnitude of ΔG_{rxn}. The **Gibbs criteria for reaction spontaneity** are as follows:

1. If $\Delta G_{\text{rxn}} < 0$, the reaction is spontaneous and additional products can form.
2. If $\Delta G_{\text{rxn}} > 0$, the reaction is not spontaneous and no additional products can form without the input of energy from an external source. Without such an input of energy, the reverse reaction will be spontaneous.
3. If $\Delta G_{\text{rxn}} = 0$, the reaction is at equilibrium and no further net change occurs.

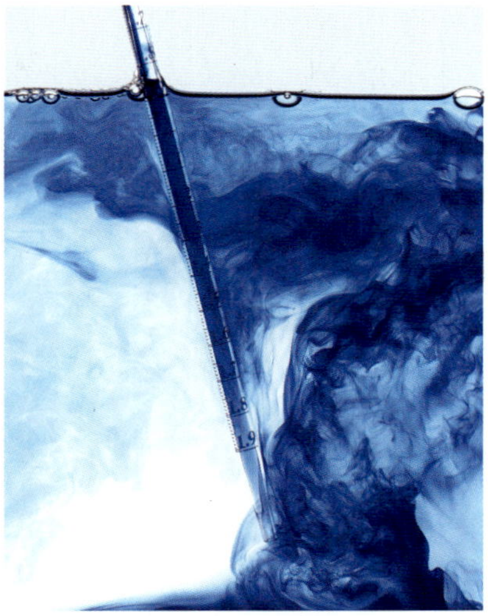

Figure 23.16 The dispersal of ink throughout the entire volume of water is an example of a spontaneous process.

When we say *the reaction is spontaneous*, we are referring to the reaction *as written* and are using the word spontaneous to mean *progressing naturally in the forward direction* (that is, from reactants toward products, or from left to right) without the input of energy from an external source.

The spontaneous mixing of two gases (Figure 23.4) and the spreading of a drop of ink throughout a volume of water (Figure 23.16) are examples of processes for which $\Delta H \approx 0$ and $\Delta S > 0$. For these processes,

$$\Delta G = \Delta H - T\Delta S \approx 0 - T\Delta S < 0$$

so the Gibbs criteria predict that two gases mix spontaneously and that a drop of ink becomes uniformly dispersed throughout a volume of water. Furthermore, note that $\Delta G > 0$ for the reverse processes because $\Delta S < 0$ for the reverse processes. Therefore, the Gibbs criteria predict that we shall never see the reverse processes occur spontaneously; we shall never see two gases separate spontaneously, nor a dispersal of ink in water revert spontaneously to its undispersed state.

Equation 23.8 says that $\Delta S_{fus} = \Delta H_{fus}/T_m$. When we substitute this result into Equation 23.13, then we see that $\Delta G_{fus} = 0$. At the standard melting point, the solid and liquid phases are in equilibrium with each other, and thus $\Delta G_{fus} = 0$, as the third Gibbs criteria states. It should be clear that this should also be the case at the boiling point, and so $\Delta G_{vap} = 0$ (see Equation 23.9).

Let's consider the melting point of a substance once again, which we describe by

$$A(s) \leftrightharpoons A(l) \tag{23.14}$$

For this process, Equation 23.13 can be written as

$$\Delta G_{fus} = \Delta H_{fus} - T_m \Delta S_{fus} = 0 \tag{23.15}$$

If the temperature is raised slightly above T_m, then $\Delta G_{fus} < 0$ (assuming ΔH_{fus} and ΔS_{fus} do not vary much with temperature). Thus, according to the first of the Gibbs spontaneity criteria, the process described by Equation 23.14 is spontaneous, and the substance melts, as we should expect since we raised the temperature. If we lower the temperature slightly below T_m, then Equation 23.15 says that $\Delta G_{fus} > 0$. According to the second of the Gibbs spontaneity criteria, the reverse of Equation 23.14 occurs spontaneously, and the substance freezes, as we should expect since we lowered the temperature.

In the case of fusion or vaporization, both ΔH_{rxn} and ΔS_{rxn} are positive, and so we see that the reaction will proceed spontaneously when the temperature is sufficiently high so that the value of the $-T\Delta S_{rxn}$ term in $\Delta G_{rxn} = \Delta H_{rxn} - T\Delta S_{rxn}$ causes the value of ΔG_{rxn} to be less than zero. Table 23.4 summarizes the different cases for determining reaction spontaneity that we have discussed here.

Highly exothermic processes, like most combustion reactions, have large negative values of ΔH_{rxn}. In such cases the $\Delta H_{rxn} < 0$ term in Equation 23.13 tends to dominate the $-T\Delta S_{rxn}$ term, even when $\Delta S_{rxn} < 0$. Thus, for combustion reactions or other highly exothermic reactions, ΔH_{rxn} is generally an adequate criterion of reaction spontaneity. It was in his generalization of this observation to all reactions that Berthelot erred.

TABLE 23.4 Spontaneity criteria for the equation $\Delta G_{rxn} = \Delta H_{rxn} - T\Delta S_{rxn}$ at constant pressure.

ΔH_{rxn}	ΔS_{rxn}	Reaction spontaneity*
< 0	> 0	always spontaneous ($\Delta G_{rxn} < 0$)
< 0	< 0	spontaneous ($\Delta G_{rxn} < 0$) when $\lvert\Delta H_{rxn}\rvert > \lvert T\Delta S_{rxn}\rvert$
> 0	> 0	spontaneous ($\Delta G_{rxn} < 0$) when $\lvert T\Delta S_{rxn}\rvert > \lvert\Delta H_{rxn}\rvert$
> 0	< 0	never spontaneous ($\Delta G_{rxn} > 0$)

*Vertical lines (e.g. $\lvert\Delta H_{rxn}\rvert$) denote magnitude or absolute value.

EXAMPLE 23-4: Predict under what conditions the following processes will be spontaneous:

(a) $Br_2(l) \rightarrow Br_2(g)$

(b) $CH_4(g) + 2O_2(g) \rightarrow CO_2(g) + 2H_2O(l)$

(c) $Ar(g, 1L) \rightarrow Ar(g, 2L)$ (free expansion at constant T)

(d) $3O_2(g) \rightarrow 2O_3(g)$ (endothermic)

Solution: (a) The vaporization of bromine requires an input of energy as heat, so that $\Delta H_{vap} > 0$. However, $Br_2(g)$ has a greater positional disorder than $Br_2(l)$, so $\Delta S_{vap} > 0$. Thus, the reaction will be spontaneous when $T\Delta S_{vap} > \Delta H_{vap}$ because the value of ΔG_{vap} will be less than zero.

(b) This is a combustion reaction and is highly exothermic. Therefore, the magnitude of the $\Delta H_{rxn} < 0$ term is much greater than that of the $-T\Delta S_{rxn}$ term in Equation 23.13 and the reaction is spontaneous under most conditions.

(c) For the free expansion of argon gas at constant temperature, $\Delta H_{rxn} \approx 0$. However, $\Delta S_{rxn} > 0$ because the entropy of a gas increases with increasing volume. Thus, the reaction is always spontaneous.

(d) This reaction is endothermic, so $\Delta H_{rxn} > 0$. Furthermore, two moles of a gas have a lower entropy than three moles of a gas, so that $\Delta S_{rxn} < 0$. Thus, the reaction is not spontaneous.

PRACTICE PROBLEM 23-4: The production of glucose in plants can be described by the chemical equation

$$6CO_2(g) + 6H_2O(l) \rightarrow \underset{\text{glucose}}{C_6H_{12}O_6(s)} + 6O_2(g)$$

(This is a simplification of the overall process that actually occurs). For this equation $\Delta H^\circ_{rxn} = 2826\ \text{kJ}\cdot\text{mol}^{-1}$ and $\Delta S^\circ_{rxn} = -260.0\ \text{kJ}\cdot\text{mol}^{-1}$. Under what conditions is this reaction spontaneous? How are plants able to use the reaction described by this equation to produce glucose?

Answer: This reaction is not spontaneous under any conditions. Plants use energy from the sun (photosynthesis) to drive the reaction and produce glucose. A nonspontaneous reaction will only form products upon the input from an external source of the energy needed to drive the reaction.

In general, a system will change spontaneously in such a way that its Gibbs energy is minimized. The Gibbs energy of a substance consists of an enthalpy term and an entropy term. The enthalpy term is fairly independent of pressure (for a gaseous species) or concentration (for a species in solution), but as we showed in Section 23-3, the entropy term depends strongly on these quantities. In any spontaneous process, the composition of a system will change through the consumption of some species and the formation of others, thereby changing their pressures or concentrations so that the Gibbs energy of the system is minimized.

23-7. The Reaction Quotient, Equilibrium Constant, and ΔG°_{rxn} Are Related

In Chapter 19 we presented another set of criteria that govern the direction in which a chemical reaction proceeds. These criteria involve the ratio of the reaction quotient Q_c to the equilibrium constant K_c. Recall that for a given reaction equation we have the following:

1. If $Q_c/K_c > 1$ or $Q_c > K_c$, the reaction proceeds spontaneously from right to left.
2. If $Q_c/K_c < 1$ or $Q_c < K_c$, the reaction proceeds spontaneously from left to right.
3. If $Q_c/K_c = 1$ or $Q_c = K_c$, the reaction is at equilibrium.

Similar relations apply to Q_p and K_p.

Because the sign of ΔG_{rxn} also determines whether a reaction is spontaneous, we should expect a relation between ΔG_{rxn} and Q_c/K_c or Q_p/K_p; the relation is

$$\Delta G_{rxn} = RT \ln\left(\frac{Q_c}{K_c}\right) \quad \text{or} \quad \Delta G_{rxn} = RT \ln\left(\frac{Q_p}{K_p}\right) \tag{23.16}$$

Because the units of Q_c and K_c are the same, the units cancel in the ratio Q_c/K_c and similarly for Q_p/K_p. Also note the following from Equation 23.16:

1. If $Q_c/K_c > 1$, then $\Delta G_{rxn} = RT\ln(Q_c/K_c) > 0$ and so ΔG_{rxn} is positive.
2. If $Q_c/K_c < 1$, then $\Delta G_{rxn} = RT\ln(Q_c/K_c) < 0$ and so ΔG_{rxn} is negative.
3. If $Q_c/K_c = 1$, then $\Delta G_{rxn} = RT\ln(Q_c/K_c) = 0$ and so ΔG_{rxn} is zero.

Similar relations are true for Q_p and K_p.

In applying the Gibbs criteria to a reaction, it is worth remembering that even when a reaction is spontaneous, it may not occur at a detectable rate. *Spontaneous is not synonymous with immediate.* On the other hand, if $\Delta G_{rxn} > 0$, then the reaction will *not* occur in the absence of an input of energy from an external source under the prevailing conditions. The *no* of thermodynamics is emphatic; the *yes* of thermodynamics is actually a *maybe*. For a reaction to occur spontaneously, it is *absolutely necessary* that $\Delta G_{rxn} < 0$; however, a negative value of ΔG_{rxn} is not sufficient to guarantee that the reaction will occur at a detectable rate. In order to determine the rate of a reaction, we must apply the concepts discussed in Chapters 17 and 18 on chemical kinetics. For example, a sizeable energy barrier may effectively prevent a reaction from occurring (see Section 18-2), even though ΔG_{rxn} is negative and the reaction is thermodynamically spontaneous.

A spontaneous reaction moves towards equilibrium in the forward direction as written; a nonspontaneous reaction moves toward equilibrium in the reverse direction as written. Spontaneous is not synonymous with instantaneous and does not indicate the rate of reaction.

The following Example illustrates the use of Equation 23.16.

EXAMPLE 23-5: The reaction system described by the chemical equation

$$2HI(g) \leftrightharpoons H_2(g) + I_2(g)$$

is prepared at 598 K with the initial pressures $P_{HI} = 0.500$ bar, $P_{H_2} = 0.150$ bar, and $P_{I_2} = 4.25 \times 10^{-2}$ bar. Given that $K_p = 1.08 \times 10^{-2}$ at 598 K, calculate the value of ΔG_{rxn} for this reaction system and indicate in which direction the reaction as written will proceed spontaneously.

Solution: The value of Q_p for this equation is given by

$$Q_p = \frac{P_{H_2}P_{I_2}}{(P_{HI})^2} = \frac{(0.150\text{ bar})(4.25 \times 10^{-2}\text{ bar})}{(0.500\text{ bar})^2} = 0.0255$$

Thus,

$$\Delta G_{rxn} = RT\ln\left(\frac{Q_p}{K_p}\right) = (8.3145\text{ J}\cdot\text{K}^{-1}\cdot\text{mol}^{-1})(598\text{ K})\ln\left(\frac{0.0255}{1.08 \times 10^{-2}}\right)$$
$$= 4.27\text{ kJ}\cdot\text{mol}^{-1}$$

where, like the other thermodynamic variables we encountered in Chapter 14, the subscript rxn and the unit of mol^{-1} here refer to one mole of the chemical equation as written. Also recall from Chapter 19 that the values of Q_p and K_p (or Q_c and K_c) depend upon how we choose to write the chemical equation (see Problem 23-62).

The positive value of ΔG_{rxn} (which results from the fact that $Q_p > K_p$) implies that the reaction system will evolve in such a way that the concentrations of $H_2(g)$ and $I_2(g)$ will decrease and that of $HI(g)$ will increase. That is, the reaction described by

$$2\,HI(g,\ 0.500\text{ bar}) \leftrightharpoons H_2(g,\ 0.150\text{ bar}) + I_2(g,\ 0.0425\text{ bar})$$

proceeds spontaneously from right to left as written.

PRACTICE PROBLEM 23-5: Suppose we add 5.0 mL of 0.10 M $AgNO_3(aq)$ to 100.0 mL of water that is in equilibrium with $AgCl(s)$ at 25°C. Calculate the value of ΔG_{rxn} for the reaction described by

$$AgCl(s) \leftrightharpoons Ag^+(aq) + Cl^-(aq)$$

immediately after adding the $AgNO_3(aq)$. Take $K_{sp} = 1.8 \times 10^{-10}\text{ M}^2$ for $AgCl(s)$. Is your result consistent with the common ion effect discussed in Chapter 22?

Answer: $\Delta G_{rxn} = +14\text{ kJ}\cdot\text{mol}^{-1}$; yes

At this point, you may think that the introduction of ΔG_{rxn} has not given us anything new because we can determine the direction in which a reaction will proceed spontaneously by simply using the value of Q_c/K_c. But what if we are not given the value of the equilibrium constant? We shall show later in this chapter that we can use Equation 23.16 to calculate the value of the equilibrium constant for a reaction described by a given chemical equation and thus determine reaction spontaneity. Another advantage of introducing ΔG_{rxn} is that the value of ΔG_{rxn} is related to the energy that can be obtained as work from a process. If ΔG_{rxn} is negative (spontaneous reaction), then its magnitude is equal to the *maximum* energy that can be obtained from the reaction to perform work, such as moving a piston. For example, the value of ΔG_{rxn} at 25°C for the reaction described by

$$2\,H_2(g,\,1\text{ bar}) + O_2(g,\,1\text{ bar}) \rightarrow 2\,H_2O(l)$$

is −474.2 kJ·mol^{-1}. This reaction can provide a maximum of 474.2 kJ·mol^{-1} of energy to an external device, such as an electric motor. This maximum amount of energy is an ideal value. In practice we always obtain less than the maximum amount.

If ΔG_{rxn} is positive, then its value is the *minimum* energy that must be *supplied* to the reaction in order to make it occur. For instance, Example 23-5 shows that ΔG_{rxn} = +4.27 kJ·mol^{-1} at 598 K for the reaction described by the equation

$$2\,HI(g,\,0.500\text{ bar}) \leftrightharpoons H_2(g,\,0.150\text{ bar}) + I_2(g,\,0.0425\text{ bar})$$

Thus, we must supply at least 4.27 kJ·mol^{-1} to make the reaction occur under the stated conditions.

23-8. The Values of ΔG_{rxn} and ΔG°_{rxn} Are Related

We can use the first expression in Equation 23.16 to define a **thermodynamic equilibrium constant**, K (no subscript), by

$$K = \frac{K_c}{Q^{\circ}_c} \qquad (23.17)$$

where Q°_c is a **standard reaction quotient**. The numerical value of Q°_c is unity, and its units, by convention, are expressed using concentrations in molarity. Therefore, K has the same numerical value as K_c, but is unitless because the units of K_c and the units of Q°_c cancel.

The International Union of Pure and Applied Chemistry (IUPAC) recommends a standard state of one molar for solutions and one bar for gases.

We can also use the second expression in Equation 23.16 to define a thermodynamic equilibrium constant in terms of K_p instead of K_c. By analogy with Equation 23.17, we write

$$K = \frac{K_p}{Q^{\circ}_p} \qquad (23.18)$$

where Q°_p has a numerical value of unity, and its units are expressed using bars. Thus, K and K_p have the same numerical value, but K is unitless.

EXAMPLE 23-6: The value of K_c for the equation

$$CH_3COO^-(aq) + H_2O(l) \leftrightharpoons CH_3COOH(aq) + OH^-(aq)$$

is 5.6×10^{-10} M at 25°C. Determine the value of the thermodynamic equilibrium constant, K, for this equation.

Solution: In this case, Q°_c = 1 M, and so the value of K is 5.6×10^{-10} (no units).

PRACTICE PROBLEM 23-6: Given that $K_p = 1.90$ bar at a certain temperature for the equation

$$C(s) + CO_2(g) \leftrightarrows 2CO(g)$$

determine the value of K for this equation.

Answer: $K = 1.90$ (unitless)

Notice that Q_c° and Q_p° reflect what we have called standard conditions (concentrations are 1 M and pressures are 1 bar). Under these conditions $\Delta G_{rxn} = \Delta G^\circ_{rxn}$, $K = K_c/Q_c^\circ$ or $K = K_p/Q_p^\circ$, and Equation 23.16 becomes

$$\Delta G^\circ_{rxn} = -RT \ln K \qquad (23.19)$$

where we have used the fact that $\ln(1/a) = -\ln a$ (Appendix A).

Equation 23.19 shows that the magnitude of ΔG°_{rxn}, the **standard Gibbs energy change** for a reaction, is determined by the magnitude of the thermodynamic equilibrium constant for the reaction and its corresponding equation at a given temperature. If K is greater than unity, then $\ln K$ is positive and the value of ΔG°_{rxn} is negative. In other words, if $K > 1$, then the reaction is spontaneous *under standard conditions,* that is, for solutes at exactly one molar concentration and all species at exactly one bar pressure. If K is less than unity, then $\ln K$ is negative, the value of ΔG°_{rxn} is positive, and the reaction is not spontaneous *under standard conditions.*

A thermodynamic equilibrium constant does not have units and so it is legitimate to take its logarithm in Equation 23.19. However, because thermodynamic equilibrium constants do not have units, it may not be clear to what standard states they are referred. For this reason, it is the responsibility of the person that reports a value of K to specify just what standard states are used. For the rest of this chapter, we'll continue to refer to reactions involving solutions to a standard state of exactly one molar concentration and those involving gases to a standard state of exactly one bar pressure. As we have already seen, these are conventional standard states in chemistry and are the ones recommended by the International Union of Pure and Applied Chemistry.

EXAMPLE 23-7: The thermodynamic equilibrium constant at 25°C for the chemical equation

$$CH_3COOH(aq) + H_2O(l) \leftrightarrows H_3O^+(aq) + CH_3COO^-(aq)$$

is $K = 1.8 \times 10^{-5}$. Calculate the value of ΔG°_{rxn} at 25°C.

Solution: The value of ΔG°_{rxn} can be calculated from the value of K by using Equation 23.19:

$$\begin{aligned}\Delta G^\circ_{rxn} &= -RT\ln K = -(8.3145\ \text{J·mol}^{-1}\text{·K}^{-1})(298\ \text{K})\ln(1.8 \times 10^{-5}) \\ &= +2.71 \times 10^4\ \text{J·mol}^{-1} = +27.1\ \text{kJ·mol}^{-1}\end{aligned}$$

The positive value of ΔG°_{rxn} means that the reaction described by

$$CH_3COOH(aq,\ 1\ M) + H_2O(l) \leftrightharpoons H_3O^+(aq,\ 1\ M) + CH_3COO^-(aq,\ 1\ M)$$

will proceed spontaneously from right to left under standard conditions. We say that the reaction as written is not spontaneous from left to right *under standard conditions*. The units of $kJ\cdot mol^{-1}$ mean that this is the value of ΔG°_{rxn} for one mole of the reaction as written.

PRACTICE PROBLEM 23-7: Acetylene gas can react with itself to form liquid benzene. Calculate the value of ΔG°_{rxn} at 25°C for the reaction described by the equation

$$3\,C_2H_2(g) \rightarrow C_6H_6(l)$$

given that $K = 1.54 \times 10^{88}$ for this reaction as written.

Answer: $\Delta G^{\circ}_{rxn} = -503\ kJ\cdot mol^{-1}$ and the reaction as written is spontaneous at 25°C under standard conditions.

It is important to keep in mind the difference between ΔG_{rxn} and ΔG°_{rxn}. We can use Equation 23.19 to write Equation 23.16 in another useful form. Let's start with the first expression in Equation 23.16 and write it as

$$\Delta G_{rxn} = RT\ln\left(\frac{Q_c}{K_c}\right) = RT\ln\left(\frac{Q_c}{K_c}\cdot\frac{Q^{\circ}_c}{Q^{\circ}_c}\right) = RT\ln\left(\frac{Q^{\circ}_c}{K_c}\cdot\frac{Q_c}{Q^{\circ}_c}\right)$$

Now use the fact that $\ln ab = \ln a + \ln b$ to write this equation as

$$\Delta G_{rxn} = RT\ln\frac{Q^{\circ}_c}{K_c} + RT\ln\frac{Q_c}{Q^{\circ}_c} = -RT\ln\frac{K_c}{Q^{\circ}_c} + RT\ln\frac{Q_c}{Q^{\circ}_c}$$

Finally, we use Equations 23.17 and 23.19 to write

$$\Delta G_{rxn} = \Delta G^{\circ}_{rxn} + RT\ln Q \qquad (23.20)$$

where

$$Q = \frac{Q_c}{Q^{\circ}_c} \qquad (23.21)$$

is the **thermodynamic reaction quotient**. Like the thermodynamic equilibrium constant, Q (no subscript) has the same numerical value as Q_c, but is unitless.

Using Equation 23.19, we can also express Equation 23.20 as

$$\Delta G_{rxn} = RT\ln\left(\frac{Q}{K}\right) \qquad (23.22)$$

This equation has the same form as Equation 23.16, but with Q and K instead of Q_c/K_c or Q_p/K_p. The results are the same because either

$$\frac{Q}{K} = \frac{Q_c}{Q_c^\circ} \cdot \frac{Q_c^\circ}{K_c} \quad \text{or} \quad \frac{Q}{K} = \frac{Q_P}{Q_P^\circ} \cdot \frac{Q_P^\circ}{K_P}$$

If all the reactants and products are at standard conditions, that is, if all concentrations are exactly one molar and all the pressures are exactly one bar, then $Q = 1$ in Equation 23.20 and

$$\Delta G_{rxn} = \Delta G^\circ_{rxn} \qquad \text{(standard conditions)}$$

We emphasize standard conditions here to point out that ΔG°_{rxn} is equal to ΔG_{rxn} *only* if $Q = Q^\circ = 1$.

To see clearly the distinction between ΔG_{rxn} and ΔG°_{rxn}, consider the familiar reaction of the dissociation of the weak acid acetic acid, $CH_3COOH(aq)$, at 25°C:

$$CH_3COOH(aq) + H_2O(l) \leftrightharpoons H_3O^+(aq) + CH_3COO^-(aq) \qquad K = 1.8 \times 10^{-5}$$

In Example 23-7 we calculated a value of $\Delta G^\circ_{rxn} = +27.1\ \text{kJ}\cdot\text{mol}^{-1}$ for this equation. The fact that ΔG°_{rxn} is positive does *not* mean that no $CH_3COOH(aq)$ dissociates when we dissolve acetic acid in water at 25°C. Some $CH_3COOH(aq)$ does dissociate because it is ΔG_{rxn}, *not* ΔG°_{rxn}, that determines whether a reaction occurs. The value of ΔG_{rxn} at 25°C for the dissociation of $CH_3COOH(aq)$ in water is given by Equation 23.20 as

$$\Delta G_{rxn} = \Delta G^\circ_{rxn} + RT \ln Q = 27.1\ \text{kJ}\cdot\text{mol}^{-1} + (2.48\ \text{kJ}\cdot\text{mol}^{-1}) \ln Q$$

where Q is the thermodynamic reaction quotient.

Let's consider a 0.050 M $CH_3COOH(aq)$ solution. Initially (at the instant of mixing), $[CH_3COOH]_0 = 0.050$ M, $[H_3O^+]_0 \approx 1.0 \times 10^{-7}$ M, and $[CH_3COO^-]_0 \approx 0$ M. Thus, if we let Q_0 be the initial value of Q,

$$Q_0 = \frac{([H_3O^+]_0/M)([CH_3COO^-]_0/M)}{[CH_3COOH]_0/M} \approx 0$$

Because the logarithm of a very small number is a large negative number, the value of ΔG_{rxn} obtained from Equation 23.20 for this initial value of Q is very large and negative. Therefore, the dissociation of $CH_3COOH(aq)$ initially takes place spontaneously. The concentration of $CH_3COOH(aq)$ decreases and the concentrations of $H_3O^+(aq)$ and $CH_3COO^-(aq)$ increase until equilibrium is reached. The equilibrium state is determined by the condition $\Delta G_{rxn} = 0$. If we set $\Delta G_{rxn} = 0$ in Equation 23.20, then we obtain

$$\Delta G^\circ_{rxn} = -RT \ln Q_{eq} \qquad \text{(at equilibrium)}$$

which, when compared with Equation 23.19, gives

$$Q_{eq} = K \qquad \text{(at equilibrium)}$$

We can show this behavior graphically. Figure 23.17 shows ΔG_{rxn} plotted

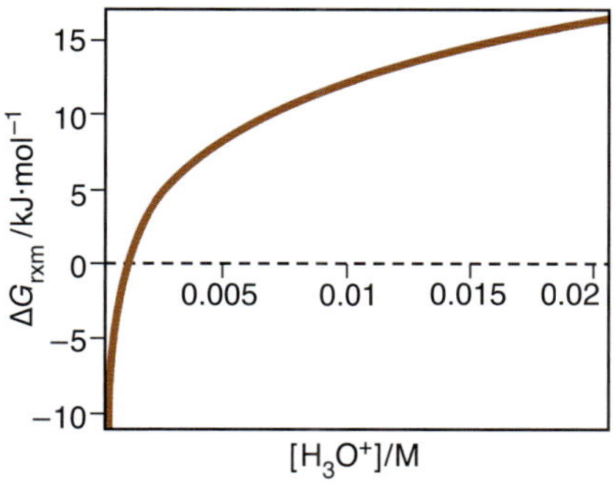

Figure 23.17 A plot of ΔG_{rxn} against $[H_3O^+]/M$ for the dissociation of 0.050 M $CH_3COOH(aq)$ described by

$$CH_3COOH(aq) + H_2O(l) \leftrightharpoons H_3O^+(aq) + CH_3COO^-(aq)$$

The value of ΔG_{rxn} is initially large and negative and continually increases with increasing $[H_3O^+]$, passing through $\Delta G_{rxn} = 0$, where $[H_3O^+] = 9.4 \times 10^{-4}$ M. This is the equilibrium value of $[H_3O^+]$ at 25°C.

against $[H_3O^+]/M$ for the dissociation of 0.050 M $CH_3COOH(aq)$ at 25°C. You can see from the figure that ΔG_{rxn} is large and negative initially and then increases with increasing $[H_3O^+]$ until $\Delta G_{rxn} = 0$, where the reaction reaches equilibrium. As we saw in Section 20-7, for a 0.050 M $CH_3COOH(aq)$ solution at 25°C, the equilibrium concentrations of the various species in solution are $[H_3O^+] = [CH_3COO^-] = 9.4 \times 10^{-4}$ M and $[CH_3COOH] = 0.050\text{ M} - 9.4 \times 10^{-4}$ M $= 0.049$ M. Substituting these values into the expression for Q_{eq} gives

$$Q_{eq} = K = \frac{([H_3O^+]/M)([CH_3COO^-]/M)}{[CH_3COOH]/M} = \frac{(9.4 \times 10^{-4})^2}{0.049} = 1.8 \times 10^{-5}$$

Figure 23.18 shows Q plotted against $[H_3O^+]/M$ for this same reaction, showing that Q increases with increasing $[H_3O^+]$ until $Q = K = 1.8 \times 10^{-5}$ when $[H_3O^+] = 9.4 \times 10^{-4}$ M. Beyond this point, $Q > K$ and the reaction is no longer spontaneous. Figure 23.17 and Figure 23.18 say the very same thing, but Figure 23.17 is expressed in terms of reaction criteria involving the sign of ΔG_{rxn} and Figure 23.18 is expressed in terms of reaction criteria involving the relative values of K and Q. The value of $[H_3O^+]$ that we obtain here is the very same value that we obtained in Section 20-7 for the dissociation of 0.050 M $CH_3COOH(aq)$ at 25°C. All three calculations are superficially different, but they are essentially the same.

The above calculations emphasize the fact that the value of ΔG_{rxn} depends on the concentrations of the reactants and products through the quantity Q in

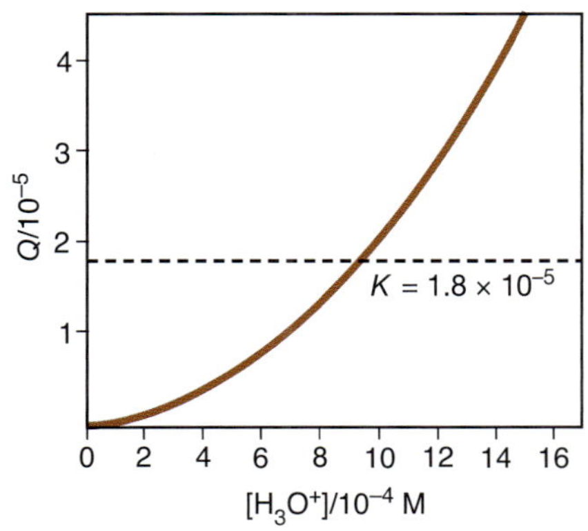

Figure 23.18 A plot of Q against $[H_3O^+]/M$ for the dissociation of 0.050 M $CH_3COOH(aq)$ described by

$$CH_3COOH(aq) + H_2O(l) \leftrightharpoons H_3O^+(aq) + CH_3COO^-(aq)$$

The value of Q is initially zero, and increases with increasing $[H_3O^+]$ until it intersects the horizontal line given by $K = 1.8 \times 10^{-5}$. At this point $[H_3O^+] = 9.4 \times 10^{-4}$ M, which is the equilibrium value of $[H_3O^+]$, in agreement with the value found in Figure 23.17.

Equation 23.22. The sign of ΔG_{rxn} depends on the ratio Q/K. The value of ΔG°_{rxn}, on the other hand, is fixed at any given temperature and requires that all reactants and products be at standard conditions. At equilibrium, $\Delta G_{rxn} = 0$, but, in general, $\Delta G^\circ_{rxn} \neq 0$ at equilibrium.

EXAMPLE 23-8: The value of K for the equation

$$\text{AgCl}(s) \leftrightharpoons \text{Ag}^+(aq) + \text{Cl}^-(aq) \qquad (23.23)$$

is 1.8×10^{-10} at 25°C. Calculate the value of ΔG_{rxn} at 25°C for the equation

$$\text{Ag}^+(aq,\ 0.10\ \text{M}) + \text{Cl}^-(aq,\ 0.30\ \text{M}) \rightarrow \text{AgCl}(s) \qquad (23.24)$$

Solution: Equation 23.24 is the reverse of Equation 23.23, so the thermodynamic equilibrium constant is given by

$$K = \frac{1}{1.8 \times 10^{-10}} = 5.6 \times 10^9$$

Reaction spontaneity is governed by ΔG_{rxn}. From Equation 23.22, we have

$$\Delta G_{rxn} = RT \ln\left(\frac{Q}{K}\right)$$

The value of Q is given by Equation 23.21 as

$$Q = \frac{Q_c}{Q_c^\circ} = \frac{1}{([\text{Ag}^+]/\text{M})([\text{Cl}^-]/\text{M})} = \frac{1}{(0.10)(0.30)} = 33$$

and so

$$\Delta G_{rxn} = (8.3145\ \text{J}\cdot\text{K}^{-1}\cdot\text{mol}^{-1})(298\ \text{K})\ln\left(\frac{33}{5.6 \times 10^9}\right) = -47.0\ \text{kJ}\cdot\text{mol}^{-1}$$

Thus, a precipitate of AgCl(s) forms if 0.10 M $Ag^+(aq)$ is added to 0.30 M $Cl^-(aq)$ at 25°C.

PRACTICE PROBLEM 23-8: Redo the calculation in Example 23-8 with 2.0×10^{-6} M $Ag^+(aq)$ and 2.0×10^{-6} M $Cl^-(aq)$. Will any AgCl(s) form?

Answer: $\Delta G_{rxn} = +9.41\ \text{kJ}\cdot\text{mol}^{-1}$; no AgCl($s$) forms

23-9. ΔG°_{rxn} Values and Equilibrium Constants Can Be Calculated from Tabulated ΔG°_f Values

In Chapter 14, we described the procedure for setting up a table of standard molar enthalpies of formation of compounds from their constituent elements. The procedure for setting up a table of **standard molar Gibbs energies of**

formation of compounds from their constituent elements is analogous to that used to set up the table of ΔH°_f values described in Section 14-5. For the general chemical equation

$$aA + bB \rightarrow yY + zZ$$

we have

$$\Delta G^\circ_{\text{rxn}} = \Delta G^\circ_f[\text{products}] - \Delta G^\circ_f[\text{reactants}] \quad (23.25)$$

and

$$\Delta G^\circ_{\text{rxn}} = y\Delta G^\circ_f[Y] + z\Delta G^\circ_f[Z] - a\Delta G^\circ_f[A] - b\Delta G^\circ_f[B] \quad (23.26)$$

Table 23.1 lists ΔG°_f values for a variety of compounds. As with enthalpy, our ΔG°_f values are relative quantities. By convention, we take ΔG°_f of an element in its most stable physical state at 25°C and one bar to be zero. This convention is analogous to the one we introduced for ΔH°_f in Chapter 14. Let's consider the equation

$$2H_2(g) + O_2(g) \rightarrow 2H_2O(l)$$

The ΔG°_f data in Table 23.1 yields

$$\Delta G^\circ_{\text{rxn}} = 2\Delta G^\circ_f[H_2O(l)] - 2\Delta G^\circ_f[H_2(g)] - \Delta G^\circ_f[O_2(g)]$$
$$= (2)(-237.1\ \text{kJ}\cdot\text{mol}^{-1}) - (2)(0\ \text{kJ}\cdot\text{mol}^{-1}) - (1)(0\ \text{kJ}\cdot\text{mol}^{-1}) = -474.2\ \text{kJ}\cdot\text{mol}^{-1}$$

for the equation above.

At standard conditions, $P_{O_2} = P_{H_2} = 1$ bar, $Q = 1$, and $\Delta G_{\text{rxn}} = \Delta G^\circ_{\text{rxn}}$. The large negative value of ΔG_{rxn} tells us that the process described by the equation

$$2H_2(g, 1\ \text{bar}) + O_2(g, 1\ \text{bar}) \rightarrow 2H_2O(l)$$

is highly spontaneous, as we know from the explosive nature of hydrogen and oxygen mixtures. Still, it is important to keep in mind that spontaneous does not mean instantaneous—a mixture of hydrogen and oxygen gas can be stored indefinitely under these conditions and will not react in the absence of a spark or other reaction initiators.

EXAMPLE 23-9: Calculate the values of $\Delta G^\circ_{\text{rxn}}$ and K at 25°C for the equation

$$PCl_3(g) + Cl_2(g) \leftrightharpoons PCl_5(g)$$

Solution: Application of Equation 23.26 to this equation yields

$$\Delta G^\circ_{\text{rxn}} = \Delta G^\circ_f[PCl_5(g)] - \Delta G^\circ_f[PCl_3(g)] - \Delta G^\circ_f[Cl_2(g)]$$

The data from Table 23.1 yield

$$\Delta G^\circ_{rxn} = (1)(-305.0 \text{ kJ·mol}^{-1}) - (1)(-267.8 \text{ kJ·mol}^{-1}) - (1)(0 \text{ kJ·mol}^{-1})$$
$$= -37.2 \text{ kJ·mol}^{-1}$$

The negative value of ΔG°_{rxn} tells us that the reaction between $PCl_3(g)$ and $Cl_2(g)$ to produce $PCl_5(g)$ is a spontaneous process under standard conditions at 25°C. The value of the equilibrium constant, *K*, for the reaction as described by the above equation is calculated using Equation 23.19:

$$\ln K = -\frac{\Delta G^\circ_{rxn}}{RT} = -\frac{-37.2 \times 10^3 \text{ J·mol}^{-1}}{(8.3145 \text{ J·K}^{-1}\text{·mol}^{-1})(298 \text{ K})} = 15.0$$

Recall that the number of decimal places in the logarithm of a number determines the number of significant figures when we take its antilogarithm.

Thus,

$$K = e^{15.0} = 3 \times 10^6$$

PRACTICE PROBLEM 23-9: Calculate the value of *K* at 25°C for the reaction described by the equation

$$COCl_2(g) \leftrightharpoons CO(g) + Cl_2(g)$$

Take the value of ΔG°_f for $COCl_2(g)$ to be -205.9 kJ·mol^{-1}.

Answer: $\Delta G^\circ_{rxn} = 68.7$ kJ·mol^{-1}; $K = 9 \times 10^{-13}$

Tables of ΔG°_f, ΔH°_f, and S° values are especially useful for the thermodynamic analysis of chemical reactions. In particular, when combined with Equation 23.20, they enable us to predict whether a reaction is spontaneous under the stated conditions. Appendix D gives a more extensive list of ΔG°_f, ΔH°_f, and S° values.

EXAMPLE 23-10: Will the reaction described by the equation

$$CO(g, 0.010 \text{ bar}) + 2H_2(g, 0.010 \text{ bar}) \rightarrow CH_3OH(l)$$

occur spontaneously at 25°C?

Solution: We first use Table 23.1 to calculate ΔG°_{rxn}:

$$\Delta G^\circ_{rxn} = \Delta G^\circ_f[CH_3OH(l)] - \Delta G^\circ_f[CO(g)] - 2\Delta G^\circ_f[H_2(g)]$$
$$= (1)(-166.6 \text{ kJ·mol}^{-1}) - (1)(-137.2 \text{ kJ·mol}^{-1}) - (2)(0 \text{ kJ·mol}^{-1})$$
$$= -29.4 \text{ kJ·mol}^{-1}$$

The negative value of ΔG°_{rxn} means that the reaction described by the equation

$$CO(g, 1 \text{ bar}) + 2H_2(g, 1 \text{ bar}) \rightarrow CH_3OH(l)$$

proceeds spontaneously at standard conditions. However, it does not tell us whether the reaction will occur when the pressures of $CO(g)$ and

$H_2(g)$ both are 0.010 bar, as originally stated. To find out, we must use Equation 23.20 to calculate ΔG_{rxn}:

$$\begin{aligned}\Delta G_{rxn} &= \Delta G^\circ_{rxn} + RT \ln Q \\ &= -29.4 \text{ kJ}\cdot\text{mol}^{-1} + (8.3145 \text{ J}\cdot\text{K}^{-1}\cdot\text{mol}^{-1})(298 \text{ K}) \ln\left(\frac{1}{(0.010)(0.010)^2}\right) \\ &= -29.4 \text{ kJ}\cdot\text{mol}^{-1} + 34.2 \text{ kJ}\cdot\text{mol}^{-1} = +4.8 \text{ kJ}\cdot\text{mol}^{-1}\end{aligned}$$

Because $\Delta G_{rxn} > 0$, the reaction will not occur spontaneously if the pressures of $CO(g)$ and $H_2(g)$ are 0.010 bar, even though the reaction is spontaneous at standard conditions.

PRACTICE PROBLEM 23-10: Calculate the minimum operating pressure for which the reaction described in Example 23-10 will be spontaneous at 25°C, where $P_{CO} = P_{H_2}$.

Answer: 0.019 bar

The reaction described in the previous Example and Practice Problem is used on a commercial scale to produce methanol, but the pressures used in the process are well above one bar, due to both thermodynamic and kinetic considerations.

23-10. The van't Hoff Equation Governs the Temperature Dependence of Equilibrium Constants

Now we derive an equation that tells us how the value of a thermodynamic equilibrium constant varies with temperature. Equation 23.13 at standard conditions is

$$\Delta G^\circ_{rxn} = \Delta H^\circ_{rxn} - T\Delta S^\circ_{rxn}$$

The value of ΔG°_{rxn} is also given by Equation 23.19 as:

$$\Delta G^\circ_{rxn} = -RT \ln K$$

Equating these two expressions for ΔG°_{rxn} yields

$$-RT \ln K = \Delta H^\circ_{rxn} - T\Delta S^\circ_{rxn}$$

Solving this equation for ln K yields

$$\ln K = -\frac{\Delta H^\circ_{rxn}}{RT} + \frac{\Delta S^\circ_{rxn}}{R} \tag{23.27}$$

The first term on the right-hand side of Equation 23.27, $\Delta H^\circ_{rxn}/RT$, is inversely proportional to the absolute temperature, whereas the second term, $\Delta S^\circ_{rxn}/R$, does not involve explicitly the absolute temperature. Thus, as T gets increasingly large, the $\Delta H^\circ_{rxn}/RT$ term becomes relatively small, and the $\Delta S^\circ_{rxn}/R$ term becomes more important in determining the magnitude of K. The entropy change for a chemical reaction is the dominant factor in determining the equilibrium distribution of species at high temperatures, whereas the enthalpy change (approximately equal to the energy change) for a chemical reaction is the dominant factor in determining the equilibrium distribution of species at low temperatures. In general, at intermediate temperatures, both terms affect the equilibrium distribution of species. In other words, attainment of equilibrium in chemical systems involves a compromise between minimization of energy and maximization of entropy.

Assuming that ΔH°_{rxn} and ΔS°_{rxn} do not vary appreciably with temperature, we see from Equation 23.27 that a plot of ln K against $1/T$ will yield a straight line with a slope of $-\Delta H^\circ_{rxn}/R$ and an intercept of $\Delta S^\circ_{rxn}/R$. Thus, from measurements of the equilibrium constant as a function of temperature, we can determine the values of ΔH°_{rxn} and ΔS°_{rxn}.

We can express Equation 23.27 in another useful form by considering a reaction that is run at two different temperatures, T_1 and T_2, for which we have

$$\ln K_2 = -\frac{\Delta H^\circ_{rxn}}{RT_2} + \frac{\Delta S^\circ_{rxn}}{R}$$

and

$$\ln K_1 = -\frac{\Delta H^\circ_{rxn}}{RT_1} + \frac{\Delta S^\circ_{rxn}}{R}$$

In writing these two equations, we have assumed that the values of ΔH°_{rxn} and ΔS°_{rxn} are the same at the two temperatures T_1 and T_2. The assumption that ΔH°_{rxn} and ΔS°_{rxn} are independent of temperature is usually an adequate approximation, provided the temperature change is not large. Subtracting the second equation from the first yields

$$\ln K_2 - \ln K_1 = -\frac{\Delta H^\circ_{rxn}}{RT_2} + \frac{\Delta H^\circ_{rxn}}{RT_1}$$

Using the fact that $\ln a - \ln b = \ln(a/b)$, we get

$$\ln\left(\frac{K_2}{K_1}\right) = \frac{\Delta H^\circ_{rxn}}{R}\left(\frac{1}{T_1} - \frac{1}{T_2}\right)$$

or

$$\ln\left(\frac{K_2}{K_1}\right) = \frac{\Delta H^\circ_{rxn}}{R}\left(\frac{T_2 - T_1}{T_1 T_2}\right) \qquad (23.28)$$

Equation 23.28, which describes the dependence of an equilibrium constant on the temperature, is called the **van't Hoff equation**. The van't Hoff equation is used to calculate the value of the equilibrium constant for a chemical equation at, say, temperature T_2, given the value of K at temperature T_1, together with the value of ΔH°_{rxn} for the chemical equation. Conversely, if we know the values of K at two temperatures, T_1 and T_2, then Equation 23.28 can be used to calculate the value of ΔH°_{rxn} for the chemical equation.

If $\Delta H^\circ_{rxn} > 0$ (endothermic reaction) and $T_2 > T_1$ in Equation 23.28, then $\ln(K_2/K_1) > 0$, or $K_2 > K_1$. Thus, we see that the value of the equilibrium constant for an endothermic reaction increases with increasing temperature, in accord with Le Châtelier's principle (Chapter 19). Conversely, if $\Delta H^\circ_{rxn} < 0$ (exothermic reaction) and $T_2 > T_1$, then $\ln(K_2/K_1) < 0$, or $K_2 < K_1$. The value of K for an exothermic reaction decreases with increasing temperature.

EXAMPLE 23-11: One step in the conversion of coal to liquid fuels involves the reaction between $H_2(g)$ and $CO(g)$ to produce methanol, $CH_3OH(g)$.

$$CO(g) + 2H_2(g) \leftrightharpoons CH_3OH(g) \qquad \Delta H^\circ_{rxn} = -90.5\ \text{kJ}\cdot\text{mol}^{-1} \qquad (23.29)$$

The equilibrium constant at 25°C for this chemical equation is $K = 2.5 \times 10^4$. Calculate the value of K at 325°C.

Solution: We can apply the van't Hoff equation (Equation 23.28) to the reaction described by Equation 23.29, using the value of ΔH°_{rxn} given. At $T_1 = 298$ K, $K_1 = 2.5 \times 10^4$. Thus, at $T_2 = 573$ K, we have for K_2

$$\ln\left(\frac{K_2}{2.5 \times 10^4}\right) = \left(\frac{-90.2 \times 10^3\ \text{J}\cdot\text{mol}^{-1}}{8.3145\ \text{J}^{-1}\cdot\text{K}\cdot\text{mol}^{-1}}\right)\left(\frac{598\ \text{K} - 298\ \text{K}}{(598\ \text{K})(298\ \text{K})}\right)$$

and

$$K_2 = (2.5 \times 10^4)(e^{-18.3}) = 3 \times 10^{-4}$$

The large decrease in K with increasing T for this chemical reaction is a consequence of the large negative value of ΔH°_{rxn} for the equation.

PRACTICE PROBLEM 23-11: The value of ΔH°_{rxn} for the reaction described by the equation

$$2H_2O(l) \leftrightharpoons H_3O^+(aq) + OH^-(aq)$$

is $+55.8\ \text{kJ}\cdot\text{mol}^{-1}$. Given that $K = 1.0 \times 10^{-14}$ at 25°C, calculate the value of K at 37°C. What is the pH of water at 37°C? Is this result consistent with Le Châtelier's principle?

Answer: $K = 2.4 \times 10^{-14}$ at 37°C; pH = 6.81; yes

The temperature dependence of an equilibrium constant can be presented in graphical form using the form of the van't Hoff equation given in Equation 23.27. Figure 23.19 shows a plot of $-\ln K$ versus $1/T$ for the dissociation reaction of water.

In Chapter 15, we studied the equilibrium between a liquid and its vapor and between a solid and its vapor. For example, recall that liquid water has a unique equilibrium vapor pressure at each temperature:

$$H_2O(l) \leftrightharpoons H_2O(g) \qquad \Delta H^\circ_{vap} = 40.65\ \text{kJ}\cdot\text{mol}^{-1} \qquad (23.30)$$

The thermodynamic equilibrium constant expression for Equation 23.30 is given simply by

$$K = P_{H_2O}/\text{bar}$$

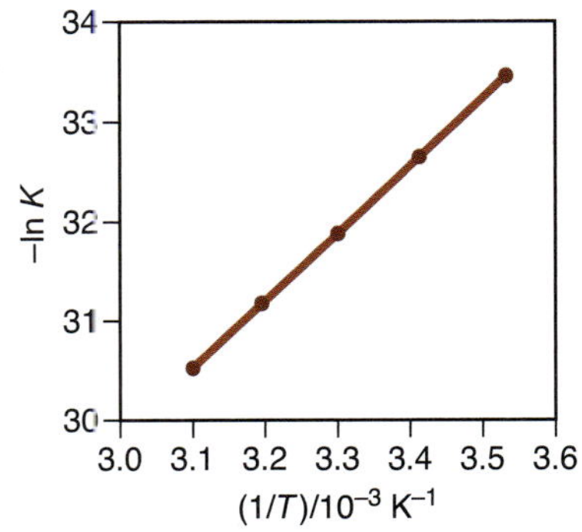

Figure 23.19 A plot of $-\ln K$ versus $1/T$ for the equation

$$2H_2O(l) \leftrightharpoons H_3O^+(aq) + OH^-(aq)$$

Note that the plot is linear, in accord with Equation 23.27.

where P_{H_2O} denotes the equilibrium vapor pressure at the prevailing temperature. Application of the van't Hoff equation (Equation 23.28) to a liquid-vapor equilibrium, with $K_2 = P_2/\text{bar}$ and $K_1 = P_1/\text{bar}$, gives us an equation that describes the dependence of equilibrium vapor pressure on temperature. This equation is called the **Clapeyron-Clausius equation**:

$$\ln\left(\frac{P_2}{P_1}\right) = \frac{\Delta H^\circ_{vap}}{R}\left(\frac{T_2 - T_1}{T_1 T_2}\right) \tag{23.31}$$

In this equation P_2 is the equilibrium vapor pressure at the kelvin temperature T_2, P_1 is the equilibrium vapor pressure at the kelvin temperature T_1, R is the molar gas constant, $8.3145\ \text{J}\cdot\text{K}^{-1}\cdot\text{mol}^{-1}$, and ΔH°_{vap} is the molar enthalpy of vaporization.

Given the value of the molar enthalpy of vaporization of a liquid, we can use the Clapeyron-Clausius equation to calculate the vapor pressure at one temperature if we know the vapor pressure at some other temperature. Let's calculate the vapor pressure of water at 110.0°C, given that the boiling point of water is 99.6°C at 1.00 bar. Taking $P_1 = 1.00$ bar at $T_1 = 372.8$ K, we let the vapor pressure at $T_2 = 383.2$ K be P_2. Given that $\Delta H_{vap} = 40.65\ \text{kJ}\cdot\text{mol}^{-1}$, we write

$$\ln\left(\frac{P_2}{1.00\ \text{bar}}\right) = \left(\frac{40.65\times 10^3\ \text{J}\cdot\text{mol}^{-1}}{8.3145\ \text{J}^{-1}\cdot\text{K}\cdot\text{mol}^{-1}}\right)\left(\frac{383.2\ \text{K} - 372.8\ \text{K}}{(372.8\ \text{K})(383.2\ \text{K})}\right) = 0.356$$

Therefore,

$$\frac{P_2}{1.00\ \text{bar}} = e^{0.356} = 1.43$$

or

$$P_2 = 1.43\ \text{bar}$$

in excellent agreement with the entry at 110°C in Table 15.7.

SUMMARY

Not all chemical reactions that evolve energy are spontaneous. Reaction spontaneity is determined by both energy and entropy changes. In general, a chemical equilibrium involves a compromise between minimization of the energy and maximization of the entropy. In any naturally occurring process, the second law of thermodynamics requires that the total entropy of the universe (system + surroundings) increase.

It is often convenient to consider entropy to be a measure of the disorder, or randomness, of a system. The entropy of a compound increases with increasing temperature. Both melting and vaporization processes lead to an increase in the entropy of a compound because molecules in a liquid are more disordered than the same molecules in the solid phase, and molecules in a gas are more disordered than the same molecules in the liquid phase.

The entropy of a perfect crystalline substance is zero at 0 K (third law of thermodynamics). This property of crystals is the basis of the absolute entropy scale. The entropy of a compound at $T > 0$ K is always positive and increases with increasing temperature. The standard entropy change for a chemical reaction, ΔS°_{rxn}, is expressed in terms of the standard entropies of the products minus the standard entropies of the reactants.

The Gibbs criterion for reaction spontaneity is that the Gibbs energy change for the reaction, ΔG_{rxn}, must be less than zero. The value of ΔG_{rxn} depends on the values of ΔH_{rxn} and ΔS_{rxn}. For a reaction with $\Delta G_{rxn} < 0$, the value of ΔG_{rxn} equals the maximum amount of work that can be obtained from the reaction under the stated conditions. The value of ΔG_{rxn} is related to the value of Q_c/K_c by

$$\Delta G_{rxn} = RT \ln\left(\frac{Q_c}{K_c}\right)$$

The criteria for reaction spontaneity are summarized in the table below, with corresponding relationships for Q_p/K_p.

Reaction type	**Value of (Q_c/K_c)**	**Gibbs energy change**
spontaneous reaction	< 1	$\Delta G_{rxn} < 0$
nonspontaneous reaction	> 1	$\Delta G_{rxn} > 0$
reaction at equilibrium	$= 1$	$\Delta G_{rxn} = 0$

"Spontaneous" in thermodynamics is defined as being able to proceed in the forward direction, from reactants toward products for a stated chemical equation. Spontaneous is not synonymous with immediate. The fact that $\Delta G_{rxn} < 0$ is not sufficient to ensure that a reaction proceeds toward equilibrium at a detectable rate.

The standard Gibbs energy change, ΔG°_{rxn}, is equal to the value of ΔG_{rxn} when all products and reactants are at standard conditions (all species at exactly one bar pressure, all solutes at exactly one molar concentration). The value of ΔG°_{rxn} can be calculated from a table of standard molar Gibbs energies of formation ΔG°_f (Table 23.1 and Appendix D). The standard Gibbs energy change associated with a chemical equation can be used to define a thermodynamic equilibrium constant. A thermodynamic equilibrium constant has the same numerical value as K_c or K_p, but is unitless. The value of ΔG°_{rxn} can be used to calculate the value of the thermodynamic equilibrium constant using Equation 23.19.

The van't Hoff equation describes the change in the value of an equilibrium constant with temperature, and the Clapeyron-Clausius equation describes the change in equilibrium vapor pressure with temperature.

TERMS YOU SHOULD KNOW

spontaneous process *854*
irreversible process *857*
system *857*
surroundings *857*
entropy, S *857*
second law of thermodynamics *857*
entropy change, ΔS *857*
reversible process *857*
isolated system *858*
positional disorder *860*
thermal disorder *860*
isothermal *860*
third law of thermodynamics *861*
molar entropy of fusion, ΔS_{fus} *864*
molar entropy of vaporization, ΔS_{vap} *864*
absolute molar entropy *865*
standard molar entropy, S° *865*
standard entropy change for a reaction, ΔS°_{rxn} *868*
entropy-driven reaction *869*
entropy-favored reaction *869*
energy-favored reaction *869*
Gibbs energy, G *870*
Gibbs energy change, ΔG_{rxn} *870*
Gibbs criteria for reaction spontaneity *870*
thermodynamic equilibrium constant, K *875*
standard reaction quotient, Q°_c *875*
standard Gibbs energy change, ΔG°_{rxn} *876*
thermodynamic reaction quotient, Q *877*
standard molar Gibbs energy of formation, ΔG°_f *880–881*
van't Hoff equation *884*
Clapeyron-Clausius equation *886*

EQUATIONS YOU SHOULD KNOW HOW TO USE

$\Delta S_{sys} + \Delta S_{surr} \geq 0$	(23.3)	(second law of thermodynamics; the entropy change of the universe is never negative)
$\Delta S_{fus} = \dfrac{\Delta H_{fus}}{T_m}$	(23.8)	(entropy change on fusion)
$\Delta S_{vap} = \dfrac{\Delta H_{vap}}{T_b}$	(23.9)	(entropy change on vaporization)
$\Delta S^\circ_{rxn} = S^\circ[\text{products}] - S^\circ[\text{reactants}]$	(23.11)	(standard entropy change for a reaction in terms of tabulated standard molar entropies)
$\Delta G_{rxn} = \Delta H_{rxn} - T\Delta S_{rxn}$	(23.13)	(Gibbs energy change for a reaction in terms of ΔH_{rxn} and ΔS_{rxn})
$\Delta G_{rxn} = RT\ln\left(\dfrac{Q_c}{K_c}\right)$ or $\Delta G_{rxn} = RT\ln\left(\dfrac{Q_p}{K_p}\right)$	(23.16)	(relation between the Gibbs energy change for a reaction and Q_c/K_c or Q_p/K_p)
$K = \dfrac{K_c}{Q^\circ_c}$ or $K = \dfrac{K_p}{Q^\circ_p}$	(23.17) (23.18)	(definition of thermodynamic equilibrium constant)
$\Delta G^\circ_{rxn} = -RT\ln K$	(23.19)	(relation between the standard Gibbs energy change for a reaction and the equilibrium constant)
$\Delta G_{rxn} = \Delta G^\circ_{rxn} + RT\ln Q$	(23.20)	(relation between the Gibbs energy change for a reaction and the standard Gibbs energy change for the reaction)
$\Delta G_{rxn} = RT\ln\left(\dfrac{Q}{K}\right)$	(23.22)	(relation between the Gibbs energy change for a reaction and Q/K)
$\Delta G^\circ_{rxn} = \Delta G^\circ_f[\text{products}] - \Delta G^\circ_f[\text{reactants}]$	(23.25)	(relation between the standard Gibbs energy change for a reaction and the standard Gibbs energies of formation of the products and the reactants)
$\ln K = -\dfrac{\Delta H^\circ_{rxn}}{R}\left(\dfrac{1}{T}\right) + \dfrac{\Delta S^\circ_{rxn}}{R}$	(23.27)	(The relation between ΔH°_{rxn}, ΔS°_{rxn}, and K; a plot of $\ln K$ versus $1/T$ is linear with slope $-\Delta H^\circ_{rxn}/R$ and intercept $\Delta S^\circ_{rxn}/R$)
$\ln\left(\dfrac{K_2}{K_1}\right) = \dfrac{\Delta H^\circ_{rxn}}{R}\left(\dfrac{T_2 - T_1}{T_1 T_2}\right)$	(23.28)	(the van't Hoff equation)
$\ln\left(\dfrac{P_2}{P_1}\right) = \dfrac{\Delta H^\circ_{vap}}{R}\left(\dfrac{T_2 - T_1}{T_1 T_2}\right)$	(23.31)	(the Clapeyron-Clausius equation)

PROBLEMS

ENTROPIES OF FUSION AND VAPORIZATION

23-1. From the following data, calculate ΔS_{fus} and ΔS_{vap} for the compounds methane, CH_4; ethane, C_2H_6; and propane, C_3H_8.

Compound	t_m/°C	ΔH_{fus}/ kJ·mol^{-1}	t_b/°C	ΔH_{vap}/ kJ·mol^{-1}
CH_4	−182.5	0.9370	−161.5	8.907
C_2H_6	−182.8	2.859	−88.6	15.65
C_3H_8	−187.6	3.525	−42.1	20.13

23-2. From the following data, calculate ΔS_{fus} and ΔS_{vap} for hydrogen fluoride, HF; hydrogen chloride, HCl; hydrogen bromide, HBr; and hydrogen iodide, HI.

Compound	t_m/°C	ΔH_{fus}/ kJ·mol^{-1}	t_b/°C	ΔH_{vap}/ kJ·mol^{-1}
HF	−83.11	4.577	19.54	25.18
HCl	−114.3	1.991	−84.9	17.53
HBr	−86.96	2.406	−67.0	19.27
HI	−50.91	2.871	−35.38	21.16

23-3. From the following data, calculate ΔS_{fus} and ΔS_{vap} for hydrogen sulfide, H_2S.

Compound	t_m/°C	ΔH_{fus}/ kJ·mol^{-1}	t_b/°C	ΔH_{vap}/ kJ·mol^{-1}
H_2S	−85.6	2.38	−60.7	18.7

Note the differences between these values and those for water (Example 23-2). Give a simple molecular interpretation for the differences.

23-4. Arrange the compounds $NH_3(l)$, $CH_4(l)$, and $H_2O(l)$ in order of increasing ΔS_{vap} values. Describe the reasoning that you used to reach your conclusions.

ENTROPY: MOLECULAR MASS, STRUCTURE, AND PHYSICAL STATE

23-5. In each case, predict which molecule of the pair has the greater molar entropy under the same conditions (assume gaseous species):

(a) H_2O water — D_2O heavy water

(b) CH_3CH_2OH ethanol — 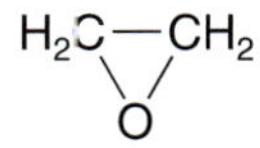 epoxyethane (ethylene oxide)

(c) $CH_3CH_2CH_2CH_2NH_2$ butylamine — 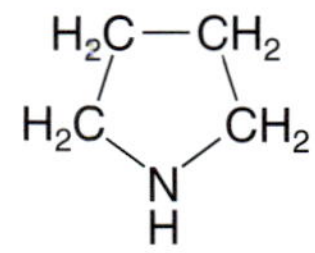 pyrrolidine

23-6. In each case, predict which molecule of the pair has the greater molar entropy under the same conditions (assume gaseous species):

(a) CO carbon monoxide — CO_2 carbon dioxide

(b) $CH_3CH_2CH_3$ propane — 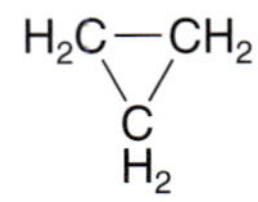 cyclopropane

(c) $CH_3CH_2CH_2CH_2CH_3$ pentane — 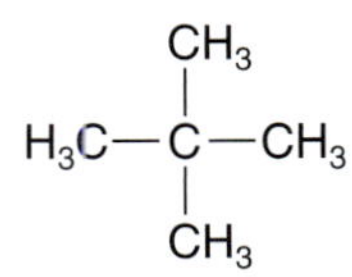 1,2-dimethylpropane (neopentane)

23-7. Without referring to Table 23.1, rank the following compounds in order of increasing molar entropy at one bar:

$CH_3Cl(g)$ $CH_4(g)$ $CH_3OH(g)$

23-8. Without referring to Table 23.1, rank the following compounds in order of increasing molar entropy at one bar:

$CH_4(g)$ $H_2O(g)$ $NH_3(g)$ $CH_3OH(g)$ $CH_3OD(g)$

23-9. Explain why the total molar entropy of $PCl_3(g)$ is less than that of $PCl_5(g)$ at 25°C and one bar pressure.

23-10. Explain why the molar entropy of $H_2O(g)$ is greater than that of $H_2O(l)$ when both are at 100°C and one bar pressure.

23-11. Predict whether the entropy of the substance increases, decreases, or remains the same in the following processes:

(a) $Ar(l) \rightarrow Ar(g)$

(b) $O_2(g, 200\text{ kPa}, 300\text{ K}) \rightarrow O_2(g, 100\text{ kPa}, 300\text{ K})$

(c) $Cu(s, 300\text{ K}) \rightarrow Cu(s, 800\text{ K})$

(d) $CO_2(g) \rightarrow CO_2(s)$

23-12. Predict whether the entropy of the substance increases, decreases, or remains the same in the following processes:

(a) $H_2O(g, 75\text{ Torr}, 300\text{ K}) \rightarrow H_2O(g, 150\text{ Torr}, 300\text{ K})$

(b) $Br_2(l, 1\text{ bar}, 25°C) \rightarrow Br_2(g, 1\text{ bar}, 25°C)$

(c) $I_2(g, 1\text{ bar}, 125°C) \rightarrow I_2(g, 1\text{ bar}, 200°C)$

(d) $Fe(s, 250°C, 1\text{ bar}) \rightarrow Fe(s, 25°C, 1\text{ bar})$

VALUES OF ΔS°_{rxn}

23-13. Arrange the following reaction equations according to increasing ΔS°_{rxn} values (do not consult any references):

(a) $S(s) + O_2(g) \rightarrow SO_2(g)$

(b) $H_2(g) + O_2(g) \rightarrow H_2O_2(l)$

(c) $CO(g) + 3H_2(g) \rightarrow CH_4(g) + H_2O(g)$

(d) $C(s) + H_2O(g) \rightarrow CO(g) + H_2(g)$

23-14. Arrange the following reaction equations according to increasing ΔS°_{rxn} values (do not consult any references):

(a) $2H_2(g) + O_2(g) \rightarrow 2H_2O(l)$

(b) $NH_3(g) + HCl(g) \rightarrow NH_4Cl(s)$

(c) $K(s) + O_2(g) \rightarrow KO_2(s)$

(d) $N_2(g) + 3H_2(g) \rightarrow 2NH_3(g)$

23-15. Use the data in Appendix D to calculate the value of ΔS°_{rxn} for each of the following reaction equations at 25°C:

(a) $4NH_3(g) + 7O_2(g) \rightarrow 4NO_2(g) + 6H_2O(g)$

(b) $CO(g) + 2H_2(g) \rightarrow CH_3OH(l)$

(c) $C(s, \text{graphite}) + H_2O(g) \rightarrow CO(g) + H_2(g)$

(d) $2CO(g) + O_2(g) \rightarrow 2CO_2(g)$

23-16. Use the data in Appendix D to calculate the value of ΔS°_{rxn} for each of the following reaction equations at 25°C:

(a) $2H_2O_2(l) + N_2H_4(l) \rightarrow N_2(g) + 4H_2O(g)$

(b) $N_2(g) + O_2(g) \rightarrow 2NO(g)$

(c) $2CH_4(g) + O_2(g) \rightarrow 2CH_3OH(l)$

(d) $C_2H_4(g) + H_2(g) \rightarrow C_2H_6(g)$

SPONTANEITY AND ΔG_{rxn}

23-17. Water slowly evaporates at 25°C and one bar. Is the process described by the equation

$$H_2O(l) \rightarrow H_2O(g)$$

spontaneous? What are the signs of ΔG_{rxn}, ΔH_{rxn}, and ΔS_{rxn} at 25°C? What drives the reaction?

23-18. Naphthalene, the active component of one variety of mothballs, sublimes at room temperature. Is the process described by the equation

$$\text{naphthalene}(s) \rightarrow \text{naphthalene}(g)$$

spontaneous? What are the signs of ΔG_{rxn}, ΔH_{rxn}, and ΔS_{rxn} at 25°C and one bar? What drives the reaction?

23-19. For the reaction described by the chemical equation

$$3C_2H_2(g) \leftrightharpoons C_6H_6(l)$$
$$\Delta H^\circ_{rxn} = -633.1\text{ kJ}\cdot\text{mol}^{-1}$$

use the data in Appendix D to calculate the value of ΔS°_{rxn} at 25.0°C. Combine your calculated value of ΔS°_{rxn} with the value of ΔH°_{rxn} and calculate that of ΔG°_{rxn}. Indicate the direction in which the reaction is spontaneous at 25.0°C and one bar pressure.

23-20. For the reaction described by the chemical equation

$$C_2H_4(g) + H_2O(l) \leftrightharpoons C_2H_5OH(l)$$
$$\Delta H^\circ_{rxn} = -44.2\text{ kJ}\cdot\text{mol}^{-1}$$

use the data in Appendix D to calculate the value of ΔS°_{rxn} at 25.0°C. Combine your calculated value of ΔS°_{rxn} with the value of ΔH°_{rxn} and calculate that of ΔG°_{rxn}. Indicate the direction in which the reaction is spontaneous at 25.0°C and one bar pressure.

23-21. For the reaction described by the chemical equation

$$C_2H_4(g) + 3O_2(g) \leftrightharpoons 2CO_2(g) + 2H_2O(g)$$
$$\Delta H^\circ_{rxn} = -1323.0\text{ kJ}\cdot\text{mol}^{-1}$$

use the data in Appendix D to calculate the value of ΔS°_{rxn} at 25.0°C. Combine your calculated value of ΔS°_{rxn} with the value of ΔH°_{rxn} and calculate that of ΔG°_{rxn}. Indicate the direction in which the reaction is spontaneous at 25.0°C when the pressures are as follows: $P_{C_2H_4} = 0.010$ bar; $P_{O_2} = 0.020$ bar; $P_{CO_2} = 20.0$ bar; and $P_{H_2O} = 0.010$ bar.

23-22. For the reaction described by the chemical equation

$$C(s, \text{graphite}) + CO_2(g) \rightarrow 2\,CO(g)$$
$$\Delta H^\circ_{rxn} = +172.5 \text{ kJ·mol}^{-1}$$

use the data in Appendix D to calculate the value of ΔS°_{rxn} at 25.0°C. Combine your calculated value of ΔS°_{rxn} with the value of ΔH°_{rxn} and calculate that of ΔG°_{rxn}. Indicate the direction in which the reaction is spontaneous at 25.0°C when the pressures are as follows: $P_{CO} = 5.0 \times 10^{-4}$ bar; $P_{CO_2} = 20.0$ bar.

23-23. A critical reaction in the production of energy to do work or drive chemical reactions in biological systems is the hydrolysis of adenosine triphosphate, ATP, to adenosine diphosphate, ADP, as described by

$$ATP(aq) + H_2O(l) \leftrightharpoons ADP(aq) + HPO_4^{2-}(aq)$$

for which $\Delta G^\circ_{rxn} = -30.5$ kJ·mol^{-1} at 37.0°C and pH 7.0. Calculate the value of ΔG_{rxn} in a biological cell in which [ATP] = 5.0 mM, [ADP] = 0.50 mM, and $[HPO_4^{2-}]$ = 5.0 mM. Is the hydrolysis of ATP spontaneous under these conditions?

23-24. The thermodynamic equilibrium constant for the equation

$$2\,SO_2(g) + O_2(g) \leftrightharpoons 2\,SO_3(g)$$

is $K = 14$ at 900 K. Calculate the value of ΔG_{rxn} for this process at 900 K when it takes place with the indicated gas pressures:

$$2\,SO_2(1.0 \times 10^{-3} \text{ bar}) + O_2(0.20 \text{ bar}) \rightarrow 2\,SO_3(1.0 \times 10^{-4} \text{ bar})$$

EQUILIBRIUM CONSTANTS AND ΔG°_{rxn}

23-25. The equilibrium constant at 250°C for the equation

$$PCl_5(g) \leftrightharpoons PCl_3(g) + Cl_2(g)$$

is $K_p = 4.5 \times 10^3$ bar. Calculate the value of ΔG°_{rxn} at 250°C. In which direction is the reaction spontaneous when $PCl_3(g)$, $Cl_2(g)$, and $PCl_5(g)$ are at standard conditions? Calculate the value of ΔG_{rxn} when P_{PCl_3} = 0.20 bar; $P_{Cl_2} = 0.80$ bar; and $P_{PCl_5} = 1.0 \times 10^{-6}$ bar. In which direction is the reaction spontaneous under these conditions?

23-26. The equilibrium constant at 527°C for the equation

$$COCl_2(g) \leftrightharpoons CO(g) + Cl_2(g)$$

is $K_c = 4.6 \times 10^{-3}$ M. Calculate the value of ΔG°_{rxn} at 527°C. In which direction is the reaction spontaneous at 527°C when the concentrations of $CO(g)$, $Cl_2(g)$, and $COCl_2(g)$ are 1.00 M? Calculate the value of ΔG_{rxn} when [CO] = 0.010 M, $[Cl_2]$ = 0.010 M, and $[COCl_2]$ = 1.00 M. In which direction is the reaction spontaneous?

23-27. The equilibrium constant for the equation

$$HNO_2(aq) + H_2O(l) \leftrightharpoons H_3O^+(aq) + NO_2^-(aq)$$

is $K_a = 5.6 \times 10^{-4}$ M at 25.0°C. Calculate the value of ΔG°_{rxn} at 25.0°C. Will nitrous acid spontaneously dissociate when $[NO_2^-] = [H_3O^+] = [HNO_2]$ = 1.00 M? When $[NO_2^-] = [H_3O^+] = 1.0 \times 10^{-5}$ M and $[HNO_2]$ = 1.0 M?

23-28. The equilibrium constant for the equation

$$HClO(aq) + H_2O(l) \leftrightharpoons H_3O^+(aq) + ClO^-(aq)$$

is $K_a = 4.0 \times 10^{-8}$ M at 25.0°C. Calculate the value of ΔG°_{rxn} at 25.0°C. Will hypochlorous acid spontaneously dissociate when $[ClO^-] = [H_3O^+] = [HClO]$ = 1.0 M? When $[ClO^-] = [H_3O^+] = 1.0 \times 10^{-6}$ M and [HClO] = 0.10 M?

23-29. The equilibrium constant for the equation

$$ClCH_2COOH(aq) + H_2O(aq) \leftrightharpoons H_3O^+(aq) + ClCH_2COO^-(aq)$$

is $K_a = 1.4 \times 10^{-3}$ M at 25.0°C. Calculate the value of ΔG°_{rxn} at 25.0°C. Will chloroacetic acid spontaneously dissociate when $[ClCH_2COO^-] = [H_3O^+] = [ClCH_2COOH]$ = 1.0 M? When $[ClCH_2COO^-]$ = 0.0010 M, $[H_3O^+] = 1.0 \times 10^{-5}$ M, and $[ClCH_2COOH]$ = 0.10 M?

23-30. The equilibrium constant for the equation

$$NH_3(aq) + H_2O(l) \leftrightharpoons NH_4^+(aq) + OH^-(aq)$$

is $K_b = 1.8 \times 10^{-5}$ M at 25.0°C. Calculate the value of ΔG°_{rxn} at 25.0°C. In which direction is the reaction spontaneous when $NH_3(aq)$, $NH_4^+(aq)$, and $OH^-(aq)$ are at standard conditions? Will ammonia react with water when both $[NH_4^+]$ and $[OH^-] = 1.0 \times 10^{-6}$ M and $[NH_3] = 0.050$ M?

23-31. The equilibrium constant for the equation

$$AgCl(s) \xrightarrow[H_2O(l)]{} Ag^+(aq) + Cl^-(aq)$$

is the solubility-product constant, $K_{sp} = 1.8 \times 10^{-10}$ M^2 at 25.0°C. Calculate the value of ΔG°_{rxn} at 25.0°C. Is it possible to prepare a solution that is 1.0 M in both $Ag^+(aq)$ and $Cl^-(aq)$?

23-32. The equilibrium constant for the equation

$$CaCO_3(s) \xrightarrow[H_2O(l)]{} Ca^{2+}(aq) + CO_3^{2-}(aq)$$

is the solubility-product constant, $K_{sp} = 3.4 \times 10^{-9}$ M^2 at 25.0°C. Calculate the value of ΔG°_{rxn} at 25.0°C. What happens when a solution is prepared in which $[Ca^{2+}] = [CO_3^{2-}] = 1.0$ M?

23-33. The equilibrium constant at 25.0°C for the equation

$$Ag^+(aq) + 2NH_3(aq) \leftrightharpoons [Ag(NH_3)_2]^+(aq)$$

is $K_f = 2.5 \times 10^3$ M^{-2}. Calculate the value of ΔG°_{rxn} at 25.0°C. In which direction is the reaction spontaneous when $Ag^+(aq)$, $NH_3(aq)$, and $[Ag(NH_3)_2]^+(aq)$ are at standard conditions? Calculate the value of ΔG_{rxn} when $[Ag^+] = 1.0 \times 10^{-3}$ M, $[NH_3] = 0.10$ M, and $[Ag(NH_3)_2^+] = 1.0 \times 10^{-3}$ M. In which direction is the reaction spontaneous under these conditions?

23-34. The equilibrium constant at 25.0°C for the equation

$$Co^{3+}(aq) + 6NH_3(aq) \leftrightharpoons [Co(NH_3)_6]^{3+}(aq)$$

is $K_f = 2.0 \times 10^7$ M^{-6}. Calculate the value of ΔG°_{rxn} at 25.0°C. In which direction is the reaction spontaneous when $Co^{3+}(aq)$, $NH_3(aq)$, and $[Co(NH_3)_6]^{3+}(aq)$ are at standard conditions? Calculate the value of ΔG_{rxn} when $[Co^{3+}] = 0.0050$ M, $[NH_3] = 0.10$ M, and $[Co(NH_3)_6^{3+}] = 1.00$ M. In which direction is the reaction spontaneous under these conditions?

CALCULATION OF ΔG°_{rxn} FROM TABULATED DATA

23-35. Use the data in Appendix D to calculate the value of ΔG°_{rxn} and K at 25°C for the following equations:

(a) $CO(g) + 2H_2(g) \rightarrow CH_3OH(l)$

(b) $C(s, \text{graphite}) + H_2O(g) \rightarrow CO(g) + H_2(g)$

(c) $CO(g) + 3H_2(g) \rightarrow CH_4(g) + H_2O(g)$

23-36. Use the data in Appendix D to calculate the value of ΔG°_{rxn} and K at 25°C for the following equations:

(a) $2H_2O_2(l) + N_2H_4(l) \rightarrow N_2(g) + 4H_2O(g)$

(b) $N_2(g) + O_2(g) \rightarrow 2NO(g)$

(c) $2CH_4(g) + O_2(g) \rightarrow 2CH_3OH(l)$

23-37. Use the data in Appendix D to calculate the values of ΔG°_{rxn} and ΔH°_{rxn} at 25.0°C for the equation

$$2HCl(g) + F_2(g) \leftrightharpoons 2HF(g) + Cl_2(g)$$

Calculate the value of the equilibrium constant at 25.0°C for the equation.

23-38. Use the data in Appendix D to calculate the values of ΔG°_{rxn} and ΔH°_{rxn} at 25.0°C for the equation

$$Fe_3O_4(s) + 2C(s, \text{graphite}) \rightarrow 3Fe(s) + 2CO_2(g)$$

Calculate the value of the equilibrium constant at 25.0°C for the equation.

23-39. Use the data in Appendix D to calculate the values of ΔG°_{rxn}, ΔH°_{rxn}, and ΔS°_{rxn} at 25°C for the reaction described by the equation

$$H_2(g) + CO_2(g) \leftrightharpoons H_2O(g) + CO(g)$$

What drives the reaction as written and in what direction at standard conditions?

23-40. Use the data in Appendix D for $CH_4(g)$, $Cl_2(g)$, and $HCl(g)$ to calculate the values of ΔG°_f and ΔH°_f at 25°C for $CCl_4(l)$, given that $\Delta G^\circ_{rxn} = -396.0$ kJ·mol^{-1} and $\Delta H^\circ_{rxn} = -422.8$ kJ·mol^{-1} for the equation

$$CH_4(g) + 4Cl_2(g) \leftrightharpoons CCl_4(l) + 4HCl(g)$$

Compare your results with the values of ΔG°_f and ΔH°_f for $CCl_4(l)$ given in Appendix D.

23-41. Calculate the maximum amount of work that can be obtained from the combustion of 1.00 mole of ethane, $C_2H_6(g)$, at 25°C and standard conditions.

23-42. Calculate the maximum amount of work that can be obtained from the combustion of 1.00 mole of methane, $CH_4(g)$, at 25°C and standard conditions.

TEMPERATURE DEPENDENCE OF EQUILIBRIUM CONSTANTS

23-43. For the equation $N_2(g) + O_2(g) \leftrightharpoons 2NO(g)$, use the following data to calculate the value of ΔH°_{rxn}:

T/K	$K_p/10^{-4}$
2000	4.08
2100	6.86
2200	11.0
2300	16.9
2400	25.1

Compare your answer to the value calculated from the data in Appendix D.

23-44. For the dissociation of $Br_2(g)$ into $2Br(g)$, use the following data to calculate the value of ΔH°_{rxn}:

t/°C	$K_p/10^{-3}$ bar
850	0.608
900	1.47
950	3.30
1000	6.97

Compare your answer to the value calculated from the data in Appendix D.

23-45. Use the data in Appendix D to calculate the value of ΔH°_{rxn} for the equation

$$PCl_3(g) + Cl_2(g) \leftrightharpoons PCl_5(g)$$

Given that $K_p = 0.555\ \text{bar}^{-1}$ at 250.0°C, calculate the value of K_p at 400.0°C.

23-46. Use the data in Appendix D to calculate the value of ΔH°_{rxn} for the equation

$$H_2(g) + I_2(g) \leftrightharpoons 2HI(g)$$

Given that $K = 58.0$ at 400.0°C, calculate the value of K at 500.0°C.

CLAPEYRON-CLAUSIUS EQUATION

23-47. Acetone, a widely used solvent (as nail-polish remover, for example), has a normal boiling point of 56.05°C and a molar enthalpy of vaporization of 29.10 $kJ\cdot mol^{-1}$. Calculate the equilibrium vapor pressure of acetone at 20.00°C in torr.

23-48. Diethyl ether is a volatile liquid whose vapor is highly combustible. The equilibrium vapor pressure over ether at 20.0°C is 455 Torr. Calculate the vapor pressure over ether when it is stored in the refrigerator at 4.0°C ($\Delta H_{vap} = 26.52\ kJ\cdot mol^{-1}$).

23-49. The heat of vaporization of benzene, $C_6H_6(l)$, is 30.72 $kJ\cdot mol^{-1}$. Given that the vapor pressure of benzene is 404.5 Torr at 60.0°C, calculate the normal boiling point of benzene.

23-50. Mercury is an ideal substance to use in manometers and in studying the effect of pressure on the volume of gases. Its surface is fairly inert and few gases are soluble in mercury. We now consider whether the partial pressure of mercury vapor contributes significantly to the pressure of the gas above mercury. Using the data in Table 15.3, calculate the vapor pressure of mercury at 25°C and 100°C in torr.

23-51. Carbon tetrachloride, $CCl_4(l)$, has a vapor pressure of 92.7 Torr at 23.5°C and 221.6 Torr at 45.0°C. Calculate the value of ΔH°_{vap} for $CCl_4(l)$.

23-52. The vapor pressure of bromine is 133.0 Torr at 20.0°C and 48.1 Torr at 0.0°C. Calculate the value of ΔH°_{vap} for bromine.

23-53. Using the data from Problem 23-51, calculate the normal boiling point of carbon tetrachloride.

23-54. Using the data from Problem 23-52, calculate the normal boiling point of bromine.

23-55. The molar heat of vaporization of lead is 179.5 $kJ\cdot mol^{-1}$. Calculate the ratio of the vapor pressure of lead at 1300°C to that at 550°C.

23-56. The molar heat of vaporization of sodium chloride is 180 $kJ\cdot mol^{-1}$. Calculate the ratio of the vapor pressure of sodium chloride at 1100°C to that at 900°C.

ADDITIONAL PROBLEMS

23-57. Define the word spontaneous as used in thermodynamics and in chemistry.

23-58. Combustion reactions are generally spontaneous ($\Delta G_{rxn} < 0$). Why then is it possible to store gasoline and other combustible fuels indefinitely in air?

23-59. Is it possible to make a nonspontaneous process occur?

23-60. Some sources give an example of an ordered room becoming spontaneously messy over time to illustrate that the entropy of a system naturally increases. What is wrong with this example?

23-61. Suppose that you see an advertisement for a catalyst that decomposes water into hydrogen and oxygen at room temperature. Would you be skeptical of this claim?

23-62. (a) Show that in Example 23-5, had we chosen to arbitrarily represent the reaction equation as $HI(g) \leftrightharpoons \frac{1}{2}H_2(g) + \frac{1}{2}I_2(g)$ rather than as $2HI(g) \leftrightharpoons H_2(g) + I_2(g)$, that the corresponding value of ΔG_{rxn} would likewise be halved. (b) Would this change have any effect on the physical reaction described by the equation? (c) Is it possible for the sign of ΔG_{rxn} to change as a consequence of how you choose to write the chemical equation that represents the reaction?

23-63. Arrange the following isomers of pentane, $C_5H_{12}(g)$, in order of increasing molar entropy

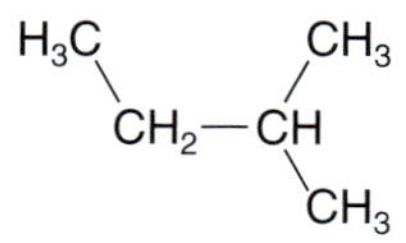

2-methylbutane
(isopentane)

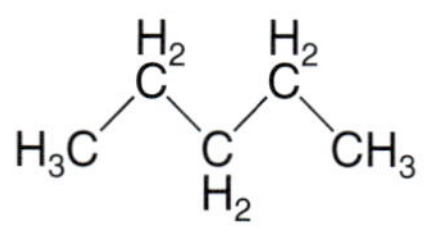

n-pentane

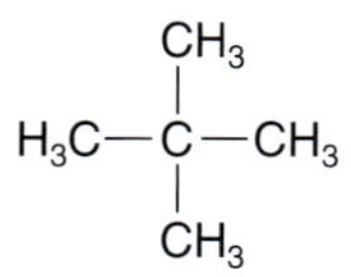

2,2-dimethylpropane
(neopentane)

23-64. Use the data in Appendix D to calculate the value of ΔS°_{rxn} for each of the following reaction equations at 25°C:

(a) $I_2(s) \rightarrow I_2(g)$

(b) $BaCO_3(s) \rightarrow BaO(s) + CO_2(g)$

(c) $CH_4(g) + Cl_2(g) \rightarrow CH_3Cl(g) + HCl(g)$

(d) $2NaBr(s) + Cl_2(g) \rightarrow 2NaCl(s) + Br_2(l)$

23-65. For each of the following processes, give the signs of ΔH°_{rxn} and ΔS°_{rxn}:

(a) $2C_2H_2(g) + 5O_2(g) \rightarrow 4CO_2(g) + 2H_2O(g)$

(b) $CO_2(s) \rightarrow CO_2(g)$

(c) one liter of He(g) and one liter of Ar(g) are allowed to mix, forming two liters of gas at constant temperature

23-66. Table 15.3 lists ΔH_{vap} for dichloromethane, CH_2Cl_2, as 28.06 kJ·mol^{-1}. However, using the data from Table 23.1, we find that

$$\begin{aligned}\Delta H_{vap} &= \Delta H^\circ_f[CH_2Cl_2\,(g)] - \Delta H^\circ_f[CH_2Cl_2\,(l)] \\ &= (-95.4\ \text{kJ·mol}^{-1}) - (-124.2\ \text{kJ·mol}^{-1}) \\ &= 28.8\ \text{kJ·mol}^{-1}\end{aligned}$$

What is the reason for this slight difference in the two values?

23-67. Using the following data, calculate ΔS_{fus} and ΔS_{vap} for the alkali metals:

Metal	T_m/K	ΔH_{fus}/ kJ·mol^{-1}	T_b/K	ΔH_{vap}/ kJ·mol^{-1}
Li	454	2.99	1615	134.7
Na	371	2.60	1156	89.6
K	336	2.33	1033	77.1
Rb	312	2.34	956	69
Cs	302	2.10	942	66

23-68. Given the following possibilities for ΔG°_{rxn}, what can you say in each case about the value of the equilibrium constant for the reaction as written?

(a) $\Delta G^\circ_{rxn} > 0$ (b) $\Delta G^\circ_{rxn} = 0$ (c) $\Delta G^\circ_{rxn} < 0$

23-69. Hydrogen peroxide can be prepared in several ways. One method is the reaction between hydrogen and oxygen, as described by the equation

$$H_2(g) + O_2(g) \leftrightharpoons H_2O_2(l)$$

Another method is the reaction between water and oxygen, as described by

$$H_2O(l) + \tfrac{1}{2}O_2(g) \leftrightharpoons H_2O_2(l)$$

Calculate the value of ΔG°_{rxn} for both reaction equations. Which method requires less energy under standard conditions?

23-70. For the equation

$$H_2(g) + CO_2(g) \leftrightharpoons CO(g) + H_2O(g)$$

use the following data to calculate the value of ΔH°_{rxn}:

t/°C	K
600	0.39
700	0.64
800	0.96
900	1.34
1000	1.77

Compare your answer to the value calculated from the data in Appendix D.

23-71.(*) The solubility of $PbI_2(s)$ is measured as a function of temperature and the following data are collected.

Temp/°C	10.0	25.0	40.0	60.0	80.0
Solubility/g·L^{-1}	0.058	0.088	0.133	0.207	0.366

Use these data to determine the value of ΔH°_{rxn} for the dissolution of $PbI_2(s)$ in water. The actual value of ΔH°_{rxn} is 63.4 kJ·mol^{-1}.

23-72. Given the following Gibbs energies at 25°C, calculate the solubility-product constant of (a) $AgCl(s)$ and (b) $AgBr(s)$.

Substance	G_f°/kJ·mol^{-1}
$Ag^+(aq)$	77.1
$Cl^-(aq)$	−131.2
$AgCl(s)$	−109.8
$Br^-(aq)$	−104.0
$AgBr(s)$	−96.9

23-73. Calculate the value (in kilojoules per mole) of the change in ΔG°_{rxn} that corresponds to a ten-fold change in K at 25.0°C.

23-74. The value of ΔH°_{rxn} is 34.78 kJ·mol^{-1} at 1000 K for the reaction described by

$$H_2(g) + CO_2(g) \leftrightharpoons CO(g) + H_2O(g)$$

Given that the value of K_p is 0.236 at 800 K, estimate the value of K_p at 1200 K, assuming that ΔH°_{rxn} is independent of temperature.

23-75. Given that the standard molar Gibbs energy of formation of $Cl(g)$ is 5.081 kJ·mol^{-1} at 2000 K and −56.297 kJ·mol^{-1} at 3000 K, determine the value of K (the thermodynamic equilibrium constant) at each temperature for the reaction described by

$$\tfrac{1}{2}Cl_2(g) \leftrightharpoons Cl(g)$$

Assuming that ΔH°_{rxn} is independent of temperature, determine the value of ΔH°_{rxn} from these data.

23-76. Glucose, $C_6H_{12}O_6(s)$, is a primary fuel in the production of energy in biological systems. Given that $\Delta G_f^\circ = -916$ kJ·mol^{-1} for glucose, calculate the maximum amount of work that can be obtained from the complete combustion of 1.00 moles of glucose under standard conditions as described by the chemical equation

$$C_6H_{12}O_6(s) + 6O_2(g) \leftrightharpoons 6CO_2(g) + 6H_2O(l)$$

23-77. The solubility of gases in water decreases with increasing temperature. What does this tell you about the enthalpy of solution of gases?

23-78. Estimate the value of the molar enthalpy of solution of $AgCl(s)$ in water from the following data:

t/°C	K_{sp}/M^2
50.0	13.2×10^{-10}
100.0	2.15×10^{-8}

23-79.(*) The variation of the Henry's law constant with temperature for the dissolution of $CO_2(g)$ in water is shown below.

t/°C	k_h/bar·M^{-1}
0	13.4
25	29.8

Calculate the value of ΔH°_{rxn} for the process described by

$$CO_2(aq) \leftrightharpoons CO_2(g)$$

23-80. The vapor pressure of water above equilibrium mixtures of $CuCl_2 \cdot H_2O(s)$ and $CuCl_2 \cdot 2H_2O(s)$ is 3.72 Torr at 18.0°C and 91.2 Torr at 60.0°C, respectively. Calculate the value of ΔH°_{rxn} for the equilibrium described by the chemical equation

$$CuCl_2 \cdot 2H_2O(s) \leftrightharpoons CuCl_2 \cdot H_2O(s) + H_2O(g)$$

23-81. Given the following data, calculate the values of ΔH°_{rxn}, ΔS°_{rxn}, and ΔG°_{rxn} at 298 K for the equilibrium described by the chemical equation

$$Mg(s) + 2HCl(aq) \leftrightharpoons H_2(g) + MgCl_2(aq)$$

Species	ΔH°_f/kJ·mol^{-1}	S°_f/J·K^{-1}·mol^{-1}
$H_2(g)$	0	130.7
$HCl(aq)$	−167.2	56.5
$Mg(s)$	0	32.7
$MgCl_2(aq)$	−801.3	−24.9

23-82. Given that ΔG°_f = 3.142 kJ·mol^{-1} for $Br_2(g)$ at 25.0°C, calculate the vapor pressure of bromine at 25.0°C.

23-83. Use the following equations to calculate the value of ΔG°_f for $HBr(g)$ at 25.0°C.

	Equation	ΔG°_{rxn}/kJ·mol^{-1}
(1)	$Br_2(l) \leftrightharpoons Br_2(g)$	3.14
(2)	$HBr(g) \leftrightharpoons H(g) + Br(g)$	339.12
(3)	$H_2(g) \leftrightharpoons 2H(g)$	406.53
(4)	$Br_2(g) \leftrightharpoons 2Br(g)$	161.71

23-84. The Clapeyron-Clausius equation can be used to calculate the equilibrium vapor pressure of a solid by writing ΔH_{sub} in place of ΔH_{vap}. Given that ΔH_{sub} = 25.2 kJ·mol^{-1} for $CO_2(s)$ and that $P_{CO_2} = 1.0 \times 10^2$ kPa at −78.6°C, calculate the equilibrium vapor pressure of $CO_2(s)$ at −100.0°C. Compare your results to the experimental value of 14 kPa.

23-85. The thermal decomposition of ammonium chloride can be described by the equation:

$$NH_4Cl(s) \leftrightharpoons NH_3(g) + HCl(g)$$

Given the ΔG°_f values $NH_4Cl(s)$ = −202.9 kJ·mol^{-1}, $NH_3(g)$ = −16.4 kJ·mol^{-1}, and $HCl(g)$ = −95.3 kJ·mol^{-1} at 25.0°C

(a) Calculate the value of ΔG°_{rxn} for the equation above at 25.0°C.

(b) Calculate the value of the thermodynamic-equilibrium constant for the equation at 25.0°C.

(c) Determine the equilibrium partial pressure of $HCl(g)$ above a one-gram sample of $NH_4Cl(s)$ in a 1.0-L container at 25.0°C.

23-86. The value of ΔG°_f for $HCl(g)$ is −95.3 kJ·mol^{-1} at 25.0°C.

(a) What is the value of ΔG_{rxn} for the formation of $HCl(g)$ at 298.0 K if the partial pressures are P_{H_2} = 3.5 bar, P_{Cl_2} =1.5 bar, and P_{HCl} = 0.31 bar?

(b) Is the process *more* or *less* favorable under these conditions than under standard state conditions?

23-87. (*) At 2000 K and one bar, water vapor is 0.53% dissociated according to the chemical equation

$$H_2O(g) \leftrightharpoons H_2(g) + \tfrac{1}{2}O_2(g)$$

At 2100 K and one bar, it is 0.88% dissociated. Calculate the value of ΔH°_{rxn} for the dissociation of water at one bar, assuming that the enthalpy of the reaction is constant over the range from 2000 K to 2100 K.

23-88. (*) The concept of entropy and the second law of thermodynamics was first developed by a French engineer named Sadi Carnot in the 1820s in a study of the efficiency of the newly developed steam engines and other types of heat engines. Basically, a steam engine works in a cyclic manner; in each cycle, it withdraws energy as heat from some high-temperature thermal reservoir, uses some of this energy to do work, and then discharges the rest of the energy as heat to a lower-temperature thermal reservoir. The maximum amount of work will be obtained if the cyclic process is carried out reversibly. Of course, the maximum amount of work cannot be achieved in practice because the reversible process is idealized, but the results give us a measure of the maximum efficiency that can be expected. Because the process is cyclic and reversible, we have

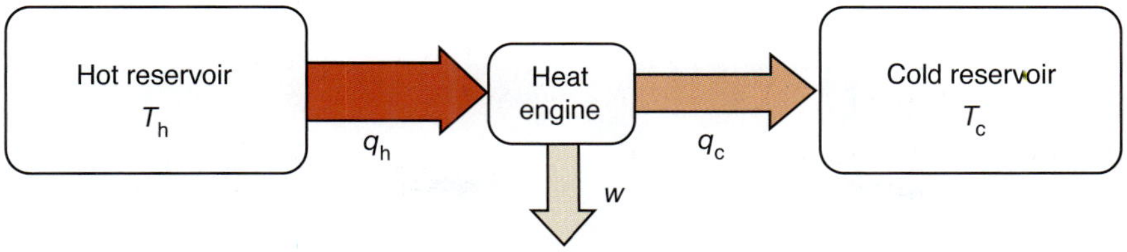

$$\Delta U_{engine} = q_{rev,h} - q_{rev,c} - w = 0$$

where $q_{rev,h}$ is the energy withdrawn reversibly as heat from the high-temperature reservoir at temperature T_h, $q_{rev,c}$ is the energy discharged reversibly as heat to the lower-temperature reservoir at temperature T_c, and w is the work performed. Defining the efficiency of such an engine as the ratio of the work performed by the engine to the amount of heat extracted from the high-temperature reservoir, w/q_h, use the above equation and Equation 23.1 to show that the efficiency of such an engine is given by

$$\text{efficiency} = \frac{w}{q_h} < \frac{T_h - T_c}{T_h}$$

Calculate the maximum efficiency of a steam engine that extracts energy in the form of heat from boiling water at 100°C and releases it into its surroundings at 20°C. Under what conditions would this engine achieve 100% efficiency?

Rudolph "Rudy" Arthur Marcus (1923–) was born in Montreal, Quebec. In 1943, he earned a B.S. in chemistry and in 1946 a Ph.D. from McGill University, doing experimental work on chemical kinetics. Upon graduation, Marcus obtained a postdoctoral fellowship at the newly formed National Research Council of Canada (NRC) studying the rate of photochemical reactions. At this time, there was little to no theoretical chemistry being done in Canada, and, together with a colleague, he began studying theoretical research papers. He had always excelled in mathematics and had taken extra mathematics courses at McGill. He realized that he needed to leave Canada to continue in theoretical chemistry, and so took a second postdoctoral position at the University of North Carolina in 1949, where he began to develop theories on the rates of chemical reactions. While at North Carolina, he met his future wife, who was a graduate student in sociology. In 1951, he took a position at the Polytechnic Institute of Brooklyn, where he developed the theory now known as "Marcus theory," that describes the rate of electron-transfer (oxidation-reduction) reactions. In 1964, he joined the faculty of the University of Illinois in Urbana-Champaign, concentrating on theoretical work. In 1978, he accepted an offer from the California Institute of Technology as the Arthur Amos Noyes Professor of Chemistry, where he remains today. Over his long and distinguished career, he has received numerous honors, including the National Medal of Science in 1989 and election to the National Academy of Sciences in 1970, the Royal Society of London in 1978, and the Royal Society of Canada in 1993. In 1992, he was awarded the Nobel Prize in Chemistry "for his contributions to the theory of electron-transfer reactions in chemical systems." Marcus believes that his experimental work always enriched his theoretical work.

24. Oxidation-Reduction Reactions

All chemical reactions can be assigned to one of two classes: reactions in which electrons are transferred from one reactant to another and reactions in which electrons are not transferred. We learned in Chapter 10 that reactions in which electrons are transferred from one reactant to another are called oxidation-reduction reactions or electron-transfer reactions. Oxidation-reduction reactions are important in many technological applications such as corrosion prevention, electroplating, batteries, photography, and combustion reactions. Many reactions of biological significance such as photosynthesis, respiration, the transport of energy within the body, the prevention of food spoilage, and even the aging of wine involve oxidation-reduction reactions (Figure 24.1).

Figure 24.1 Port wine undergoes a color change from ruby to tawny as it ages—an oxidation-reduction process involving the anaerobic oxidation of the wine.

We shall use the idea of oxidation states to determine whether a reaction involves electron transfer and, if so, which species is oxidized and which species is reduced. Much of this chapter is devoted to the use of oxidation states to balance oxidation-reduction reaction equations. The final two sections of the chapter discuss oxidation-reduction reactions used in chemical analyses and the corrosion of metals by oxidation-reduction reactions. A thorough understanding of oxidation-reduction reactions is an essential prerequisite to our study of electrochemistry in the next chapter.

24-1. Oxidation-Reduction Reactions Involve the Transfer of Electrons Between Species

We learned in Chapter 10 that reactions in which electrons are transferred from one species to another are called **oxidation-reduction reactions**. Consider the reaction of Na(s) with S(s) to produce the ionic compound sodium sulfide, $Na_2S(s)$, which we discussed in Section 10-11 (Figure 24.2). The equation that describes this reaction is

$$\underset{2(0)}{2\,Na(s)} + \underset{(0)}{S(s)} \rightarrow \overset{2(+1)}{\underset{(-2)}{Na_2S(s)}}$$

Each Na atom donates one electron. Each S atom accepts two electrons.

Figure 24.2 Sodium reacts with sulfur to produce sodium sulfide, $Na_2S(s)$.

Note that two sodium atoms react with one sulfur atom. The ionic charge of each of the two sodium atoms goes from 0 to +1 and that of the sulfur atom

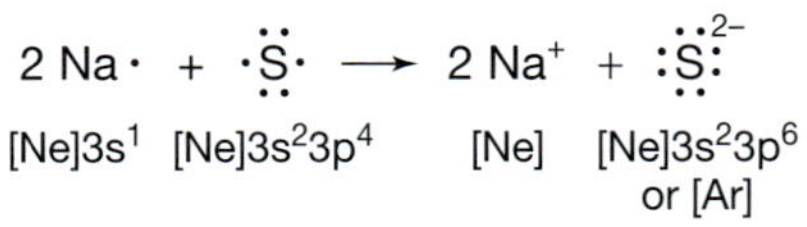

goes from 0 to –2. The two sodium atoms give up two electrons and the sulfur atom acquires the two electrons. As this equation shows, the electron transfer between species in a chemical equation must be balanced; that is, in any oxidation-reduction reaction the total number of electrons lost always equals the total number of electrons gained. The ionic products in this equation have a noble-gas configuration (margin). The stability of these ions accounts for the driving force of this reaction; we go from relatively reactive reactants to relatively stable products.

We can use the idea of ionic charge and the conservation of electrons to determine the balancing coefficients in chemical equations involving oxidation-reduction reactions. The concept of an ionic charge on an element is so convenient in balancing oxidation-reduction equations that it has been generalized. This type of ionic charge is called an **oxidation state.**

We can assign an oxidation state to each atom in a chemical species on the basis of a set of rules. The rules originate from a consideration of the number of electrons in a neutral atom of an element relative to the number of electrons that we assign to that element when it is incorporated in a molecule or ion. In the case of monatomic ions (e.g., Na^+, Ca^{2+}, O^{2-}, and Cl^-), the assigned oxidation state is simply equal to the ionic charge on the atomic ion. For chemical species involving two or more atoms, the assigned oxidation states are often not equal to the actual charges on the atoms. In such species, the assignment of oxidation states is, in essence, a bookkeeping device that is useful for identifying and balancing oxidation-reduction equations.

Rules for Assigning Oxidation States

The general procedure for assigning oxidation states to elements in chemical species containing two or more atoms is given by the following set of rules *that take priority in the order given*:

1. Free elements are assigned an oxidation state of 0.
2. The sum of the oxidation states of all the atoms in a species must be equal to the net charge on the species.
3. The alkali metal atoms (Li, Na, K, Rb, and Cs) in compounds are *always* assigned an oxidation state of +1.
4. Fluorine atoms in compounds are *always* assigned an oxidation state of –1.
5. The alkaline-earth atoms (Be, Mg, Ca, Sr, Ba, and Ra) and Zn and Cd in compounds are *always* assigned an oxidation state of +2.
6. Hydrogen atoms in compounds are assigned an oxidation state of +1.
7. Oxygen atoms in compounds are assigned an oxidation state of –2.
8. The preceding seven rules can be used to determine oxidation states in many but not all cases. For those cases where these rules are not sufficient, we can always use Lewis formulas to determine oxidation states. This method is more cumbersome to apply, but it is very general and covers all situations. We assign oxidation states by the following additional four steps:
   a. Write the Lewis formula for the molecule or ion.
   b. Assign all the electrons in each bond to the more electronegative atom in the bond (Table 24.1). If the two atoms have identical electronegativities, then divide the bonding electrons equally between them.

c. Add up the total number of valence electrons assigned to each atom in step 8b.

d. Assign an oxidation state to each *main-group element* in the species according to the formula

$$\begin{pmatrix}\text{oxidation}\\ \text{state}\end{pmatrix} = \begin{pmatrix}\text{number of valence}\\ \text{electrons in the free}\\ \text{(nonbonded) atom}\end{pmatrix} - \begin{pmatrix}\text{number of valence}\\ \text{electrons assigned to the}\\ \text{atom in the molecule}\end{pmatrix} \quad (24.1)$$

The application of these rules to the assignment of oxidation states is best illustrated by example.

TABLE 24.1 Electronegativities of selected elements*

H	C	N	O	F
2.1	2.55	3.04	3.44	3.98
	Si	P	S	Cl
	1.90	2.19	2.58	3.16
		As	Se	Br
		2.18	2.55	2.96
				I
				2.66

* A full table of electronegativity values is given in Figure 7.13.

EXAMPLE 24-1: Assign an oxidation state to each atom in the following compounds:

(a) cesium chloride, CsCl (b) manganese heptoxide, Mn_2O_7
(c) hydrogen peroxide, H_2O_2 (d) potassium superoxide KO_2
(e) perchloric acid, $HClO_4$

Solution: (a) Applying Rule 3, we assign the cesium atom an oxidation state of +1. Because CsCl is a neutral species, we see from Rule 2 that $(+1) + x = 0$, where x is the oxidation state of the chlorine atom. Thus, the chlorine atom in CsCl is assigned an oxidation state of –1.

(b) Applying Rule 7, we assign each oxygen atom an oxidation state of –2. Because Mn_2O_7 is a neutral species, we see from Rule 2 that $2x + 7(-2) = 0$, where x is the oxidation state of the manganese atom. Thus, each manganese atom in Mn_2O_7 is assigned an oxidation state of +7.

(c) Applying Rule 6, we assign each hydrogen atom an oxidation state of +1. Because H_2O_2 is a neutral species, we see from Rule 2 that $2(+1) + 2x = 0$, where x is the oxidation state of each oxygen atom. Thus, each oxygen atom in H_2O_2 is assigned an oxidation state of –1. Remember that Rule 6 takes precedence over Rule 7. The –1 oxidation state of each oxygen atom in H_2O_2 is characteristic of peroxides.

(d) Applying Rule 3, we assign an oxidation state of +1 to the potassium atom. Because KO_2 is a neutral species, we have $1 + 2x = 0$, where x is the oxidation state of each oxygen atom. Thus, the oxidation state of each oxygen atom in KO_2 is $-\frac{1}{2}$. Keep in mind that Rule 3 takes precedence over Rule 7 in this case. The $-\frac{1}{2}$ oxidation state of the oxygen atoms in KO_2 is characteristic of superoxides.

(e) Applying Rules 6 and 7, we assign an oxidation state of +1 to the hydrogen atom and –2 to each oxygen atom. Because $HClO_4$ is a neutral species, we have $+1 + x + 4(-2) = 0$, where x is the oxidation state of the chlorine atom. Thus, the chlorine atom is assigned an oxidation state of +7 in $HClO_4$.

Example 24-1 involves only neutral compounds whose net charge must be 0. For ionic species, the sum of the oxidation states of each atom must equal the net charge on the ion.

PRACTICE PROBLEM 24-1: Assign an oxidation state to each atom in the following polyatomic ions (ions that consist of more than one atom):

(a) dichromate ion, $Cr_2O_7^{2-}$ (b) hydrogen difluroide ion, HF_2^-

(c) ammonium ion, NH_4^+ (d) triiodide ion, I_3^-

Answer: (a) Cr(+6), O(−2); (b) H(+1), F(−1); (c) H(+1), N(−3); (d) I($-\frac{1}{3}$)

All the examples so far could be answered without invoking Rule 8, where you must use Lewis formulas to determine the value of the oxidation states. The following Example and Practice Problem illustrate the use of Lewis formulas and Rule 8.

EXAMPLE 24-2: Use Rule 8 to determine the oxidation states of all the atoms in the molecules

(a) thionyl chloride, $SOCl_2$ (b) hydrogen cyanide, HCN
(c) nitrosyl chloride, NOCl (d) phosphorous pentachloride, PCl_5

Solution: (a) We first write the Lewis formula of a $SOCl_2$ molecule:

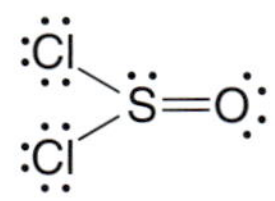

According to Table 24.1, both the oxygen atom and the chlorine atoms are more electronegative than the sulfur atom, and so according to Rule 8b we write

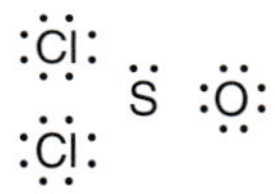

A neutral sulfur atom has six valence electrons, and the sulfur atom has been assigned two valence electrons in the formula. Applying Equation 24.1, Rule 8d, we have

$$\text{oxidation state of S in SOCl}_2 = 6 - 2 = +4$$

A neutral chlorine atom has seven valence electrons, and each chlorine atom has been assigned eight valence electrons in the formula. Thus,

$$\text{oxidation state of Cl in SOCl}_2 = 7 - 8 = -1$$

A neutral oxygen atom has six valence electrons, and each oxygen atom has been assigned eight valence electrons in the formula. Thus,

$$\text{oxidation state of O in SOCl}_2 = 6 - 8 = -2$$

Notice that the sum of the oxidation states of all the atoms is zero, as it must be according to Rule 2. Also notice that we determined the oxidation state of each atom independently using Rule 8 rather than determining two of them and then using Rule 2. In this way, we have a nice check that the sum of the oxidation states is equal to the net charge on the species.

(b) We first write the Lewis formula for a HCN molecule,

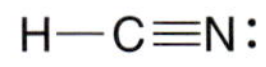

According to Table 24.1, a nitrogen atom is more electronegative than a carbon atom, and a carbon atom is more electronegative than a hydrogen atom. Therefore, we write

Notice that Rules 1 through 7 cannot be applied to HCN, and so we must use Rule 8 to assign the oxidation states.

A neutral nitrogen atom has five valence electrons, and the nitrogen atom in the formula has been assigned eight valence electrons. Thus,

$$\text{oxidation state of N in HCN} = 5 - 8 = -3$$

A neutral carbon atom has four valence electrons, and the carbon atom in the formula has been assigned two valence electrons. Thus,

$$\text{oxidation state of C in HCN} = 4 - 2 = +2$$

A neutral hydrogen atom has one valence electron, and the hydrogen atom in the formula has been assigned no valence electrons. Thus,

$$\text{oxidation state of H in HCN} = 1 - 0 = +1$$

Note that the sum of the oxidation states of all the atoms is zero, as it must be according to Rule 2.

(c) The Lewis formula of a NOCl molecule is

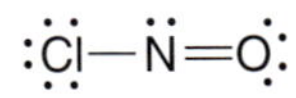

Assigning the electrons to the more electronegative atoms according to Table 24.1 gives

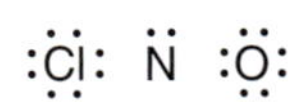

A neutral chlorine atom has seven valence electrons, and the chlorine atom in the formula has been assigned eight valence electrons. Thus,

$$\text{oxidation state of Cl in NOCl} = 7 - 8 = -1$$

A neutral oxygen atom has six valence electrons, and the oxygen atom in the formula has been assigned eight valence electrons. Thus,

$$\text{oxidation state of O in NOCl} = 6 - 8 = -2$$

A neutral nitrogen atom has five valence electrons, and the nitrogen atom in the formula has been assigned two valence electrons. Thus,

$$\text{oxidation state of N in NOCl} = 5 - 2 = +3$$

Once again, we see that the sum of the oxidation states of all the atoms is zero, as it must be according to Rule 2.

(d) The Lewis formula of a PCl_5 molecule is

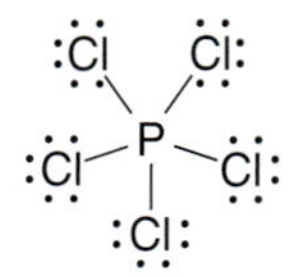

Assigning the electrons to the more electronegative atoms according to Table 24.1 gives

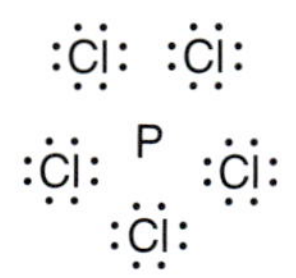

A neutral phosphorous atom has five valence electrons, and the phosphorus atom in the formula has been assigned no valence electrons. Thus,

$$\text{oxidation state of P in PCl}_5 = 5 - 0 = +5$$

A neutral chlorine atom has seven valence electrons, and each chlorine atom in the formula has been assigned eight valence electrons. Thus,

$$\text{oxidation state of Cl in PCl}_5 = 7 - 8 = -1$$

Once again, the sum of the oxidation states is zero, as it must be for a neutral species.

PRACTICE PROBLEM 24-2: Use Rule 8 to determine the oxidation states of all the atoms in the interhalogen species:

(a) bromine chloride, BrCl (b) iodine bromide, IBr

(c) iodine dichloride ion, ICl_2^- (d) bromine trichloride, $BrCl_3$

Answer: (a) Br(+1), Cl(−1); (b) I(+1), Br(−1); (c) I(+1), Cl(−1); (d) Br(+3), Cl(−1)

Problems 24-1 through 24-8 at the end of the chapter provide some additional practice in assigning oxidation states.

The previous Example and Practice Problem show how to use Lewis formulas to deduce the value of oxidation states. In many cases we can also use oxidation states to help write Lewis formulas. This is particularly so for species involving transition metals. Although many compounds of transition metals are ionic compounds, many others are covalent. To write Lewis formulas of covalent compounds or polyatomic ions involving transition metals, we must first determine the number of valence electrons. For the main-group elements, the number of valence electrons in the free atom is simply determined by its position in the periodic table. For the transition metals, however, the situation is not so simple. The difficulty involves the inverted order of filling of the two outermost occupied orbitals (recall, for example, that the 4*s* orbital fills before the 3*d* orbital in a neutral first row transition metal atom) as well as the similarity of the energies of these orbitals. We define the number of valence electrons in a transition metal to be equal to the oxidation state of the transition metal in the compound being considered. Thus, for example, the titanium atom in a TiF_4 molecule has four valence electron because the oxidation state of titanium in TiF_4 is +4 (Rules 2 and 4). We assign a total of $4 + (4 \times 7) = 32$ valence electrons in TiF_4, so its Lewis formula is

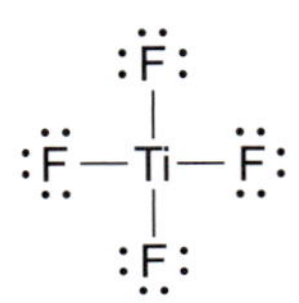

We should point out that this molecule is an AX_4 species in the notation of VSEPR theory (Chapter 8) and so is tetrahedral.

The use of oxidation states to write Lewis formulas of species involving transition metals is particularly important for oxyanions such as the manganate, MnO_4^{2-}, and vanadate, VO_3^-, ions, for example. Let's consider the manganate ion. Using Rules 2 and 7, we see that the oxidation state of the manganese atom in a MnO_4^{2-} ion is +6. Therefore, we assign the manganese atom six valence electrons for a total of $6 + (4 \times 6) + 2 = 32$ valence electrons in the MnO_4^{2-} ion. The Lewis formula with the least separation of formal charge is

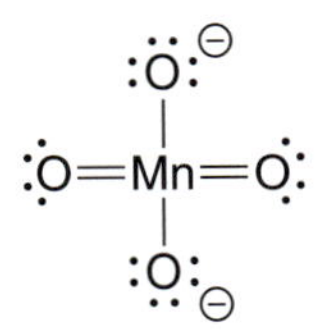

with other resonance forms. This ion is an AX_4 species in the notation of VSEPR theory and so is tetrahedral.

EXAMPLE 24-3: Determine the number of valence electrons for the vanadium atom in the vanadate ion, VO_3^-, and write its Lewis formula. Use VSEPR theory to predict its shape.

Solution: The oxidation state of the vanadium atom in the vanadate ion is +5 (Rules 2 and 7), and so we assign five valence electrons to the vanadium atom. Therefore, there is a total of $5 + (3 \times 6) + 1 = 24$ valence electrons in a VO_3^- ion. We can write the Lewis formula as

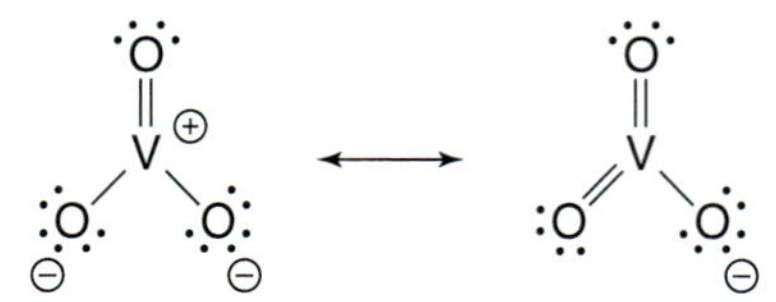

three of these three of these

or simply

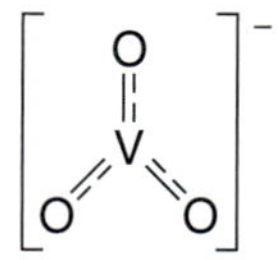

This is an AX_3 ion, and so we predict that it is trigonal planar.

PRACTICE PROBLEM 24-3: Determine the number of valence electrons in a Mn_2O_7 molecule and write its Lewis formula. (*Hint*: There is a Mn-O-Mn linkage in Mn_2O_7.)

Answer: Mn_2O_7 has 56 valence electrons and its Lewis formula is

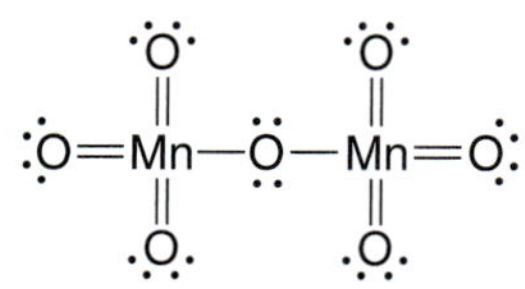

and other resonance forms with formal charge separation.

24-2. Oxidation-Reduction Reactions Involve the Transfer of Electrons from One Reactant to Another

Electrons are transferred from one reactant to another in an oxidation-reduction reaction. In fact, oxidation-reduction reactions are often called **electron-transfer reactions**. The reactant that gains or accepts electrons is called the **oxidizing agent**, and the reactant that loses or donates electrons is called the **reducing agent**. Usually, one particular type of atom in the oxidizing agent accepts the electrons, and hence its oxidation state decreases. Similarly, often one particular type of atom in the reducing agent donates the electrons, and hence its oxidation state increases. Thus, the oxidizing agent contains atoms that are reduced, and the reducing agent contains atoms that are oxidized. Table 24.2 summarizes these results.

Consider the reaction described by the equation

$$Fe_2O_3(s) + 3\,CO(g) \rightarrow 2\,Fe(s) + 3\,CO_2(g)$$

TABLE 24.2 Summary of oxidation-reduction reactions

The reducing agent:
is the electron donor
contains the atom that is oxidized
contains the atom whose oxidation state increases
The oxidizing agent:
is the electron acceptor
contains the atom that is reduced
contains the atom whose oxidation state decreases

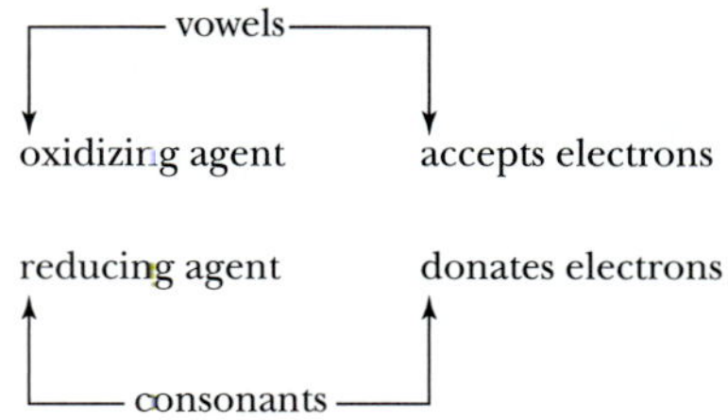

oxidizing agent ~ atoms reduced
reducing agent ~ atoms oxidized

The oxidation state of each iron atom decreases from +3 in $Fe_2O_3(s)$ to 0 in $Fe(s)$, and the oxidation state of each carbon atom increases from +2 in $CO(g)$ to +4 in $CO_2(g)$. Therefore, $Fe_2O_3(s)$ is the oxidizing agent because it contains the atoms that are reduced, and $CO(g)$ is the reducing agent because it contains the atoms that are oxidized.

EXAMPLE 24-4: In the following chemical equation, identify the atom that is oxidized, the atom that is reduced, the oxidizing agent, and the reducing agent:

$$MnO_2(s) + 4\,HCl(aq) \rightarrow MnCl_2(aq) + Cl_2(g) + 2\,H_2O(l)$$

Solution: The oxidation state of the manganese atom is +4 in $MnO_2(s)$ and +2 in $MnCl_2(s)$. The oxidation state of the chlorine atom is −1 in $HCl(aq)$ and 0 in $Cl_2(g)$. Therefore, the chlorine atom is oxidized and the manganese atom is reduced in this reaction. Because $MnO_2(s)$ contains the atom that is reduced, it is the oxidizing agent. Because the $HCl(aq)$ contains the atom that is oxidized, it is the reducing agent. Notice that two electrons are transferred in this reaction: one manganese atom accepts two electrons and each of the two chlorine atoms donates one electron. This reaction is used on a laboratory scale to prepare chlorine gas for use in chemical reactions.

An *agent* is something that performs a task. An *oxidizing agent* causes another species to be oxidized; a *reducing agent* causes another species to be reduced.

PRACTICE PROBLEM 24-4: In the following chemical equation, identify the reducing agent, the oxidizing agent, the species reduced, and the species oxidized:

$$O_2(aq) + 6\,I^-(aq) + 2\,H_2O(l) \rightarrow 2\,I_3^-(aq) + 4\,OH^-(aq)$$

Answer: $O_2(aq)$ is the oxidizing agent and the species that is reduced; $I^-(aq)$ is the reducing agent and the species that is oxidized.

(a)

(b)

Figure 24.3 (a) The oxidation-reduction reaction between the purple-black oxidizing agent $KMnO_4(s)$ and a few drops of glycerin, $HOCH_2(CHOH)CH_2OH(l)$, is very vigorous, as shown in (b). The primary reaction products are $MnO_2(s)$, $CO_2(g)$, $H_2O(g)$, and $K_2O(s)$. Organic compounds should not be brought into contact with strong oxidizing agents, except under carefully controlled conditions—otherwise, serious injury could result.

The oxidation-reduction reactions between organic compounds and strong oxidizing agents such as $KMnO_4(s)$ can be very vigorous, as illustrated in Figure 24.3.

24-3. Electron-Transfer Reactions Can Be Separated into Two Half Reactions

The electron-transfer reaction described by the equation

$$\mathrm{Zn}(s) + \mathrm{Cu}^{2+}(aq) \rightarrow \mathrm{Zn}^{2+}(aq) + \mathrm{Cu}(s) \tag{24.2}$$

can be separated into two **half reaction** equations, one representing the oxidation process and the other representing the reduction process:

$$\mathrm{Zn}(s) \rightarrow \mathrm{Zn}^{2+}(aq) + 2\,\mathrm{e}^- \qquad \text{(oxidation)} \tag{24.3}$$

$$\mathrm{Cu}^{2+}(aq) + 2\,\mathrm{e}^- \rightarrow \mathrm{Cu}(s) \qquad \text{(reduction)} \tag{24.4}$$

If we add Equations 24.3 and 24.4, then we obtain Equation 24.2. The half reaction equation in which electrons appear on the right side (Equation 24.3) is called the **oxidation half reaction** (recall that oxidation is a *loss* of electrons). The half reaction equation in which electrons appear on the left side (Equation 24.4) is called the **reduction half reaction** (recall that reduction is a *gain* of electrons). The oxidation half reaction supplies electrons to the reduction half reaction (Table 24.2).

EXAMPLE 24-5: Write the oxidation and reduction half reactions for the reaction described by the equation

$$\mathrm{Tl}^{+}(aq) + 2\,\mathrm{Ce}^{4+}(aq) \rightarrow 2\,\mathrm{Ce}^{3+}(aq) + \mathrm{Tl}^{3+}(aq)$$

This reaction is used in analytical chemistry for determining the concentration of thallium(I).

Solution: The oxidation state of the thallium atom increases from +1 in Tl^{+} to +3 in Tl^{3+} and so it is oxidized. The oxidation state of the cerium atom decreases from +4 in Ce^{4+} to +3 in Ce^{3+} and so it is reduced.

We identify the two half reactions by writing the equations for the oxidation and the reduction reactions separately:

$$\mathrm{Tl}^{+}(aq) \rightarrow \mathrm{Tl}^{3+}(aq) + 2\,\mathrm{e}^- \qquad \text{(oxidation)}$$

$$\mathrm{Ce}^{4+}(aq) + \mathrm{e}^- \rightarrow \mathrm{Ce}^{3+}(aq) \qquad \text{(reduction)}$$

Because the Tl^{+} ion is a two-electron reducing agent, whereas the Ce^{4+} ion is a one-electron oxidizing agent, two moles of Ce^{4+} are required to oxidize one mole of Tl^{+}. Thus, the complete balanced equation is obtained by multiplying the cerium half reaction equation by two and adding the result to the thallium half reaction equation:

$$Tl^{+}(aq) + 2\,Ce^{4+}(aq) \rightarrow Tl^{3+}(aq) + 2\,Ce^{3+}(aq)$$

PRACTICE PROBLEM 24-5: Write the oxidation and the reduction half reactions for the reaction described by the equation

$$2\,Cu^{+}(aq) \rightarrow Cu(s) + Cu^{2+}(aq)$$

Answer:

$$Cu^{+}(aq) \rightarrow Cu^{2+}(aq) + e^{-} \qquad \text{(oxidation)}$$

$$Cu^{+}(aq) + e^{-} \rightarrow Cu(s) \qquad \text{(reduction)}$$

$Cu^{+}(aq)$ is both the reducing agent and the oxidizing agent in this reaction.

24-4. Equations for Oxidation-Reduction Reactions Can Be Balanced by Balancing Each Half Reaction Separately

Consider the equation for the reaction between iron and aqueous chlorine (Figure 24.4):

$$Fe(s) + Cl_2(aq) \rightarrow Fe^{3+}(aq) + Cl^{-}(aq) \qquad \text{(not balanced)}$$

This equation as it stands is not balanced. If we write

$$Fe(s) + Cl_2(aq) \rightarrow Fe^{3+}(aq) + 2\,Cl^{-}(aq) \qquad \text{(not balanced)}$$

then the equation is balanced with respect to the elements but *not* with respect to charge. The net charge on the left side is zero, whereas the net charge on the right side is $(+3) + 2(-1) = +1$. The balanced equation,

$$2\,Fe(s) + 3\,Cl_2(aq) \rightarrow 2\,Fe^{3+}(aq) + 6\,Cl^{-}(aq)$$

has the same number of atoms of each type on both sides *and* the same net charge (zero in this case) on both sides.

Although fairly simple oxidation-reduction reaction equations such as this one can be balanced by inspection, attempting to balance more complex oxidation-reduction reaction equations by guessing the balancing coefficients can be time consuming and frustrating. Therefore, we shall now develop a more general method of balancing oxidation-reduction reactions: the **method of half reactions**. This method can be used to balance even the most complicated reaction equation in a straightforward and systematic way. We first apply the method of half reactions to equations for reactions that take place in acidic solution, and then we discuss its application to equations for reactions that take place in basic solution.

(a)

(b)

Figure 24.4 Chlorine in aqueous solution oxidizes iron metal to $Fe^{3+}(aq)$, which is orange in the presence of $Cl^{-}(aq)$ as a result of the formation of the orange complex ion $[FeCl]^{2+}(aq)$. (a) Before addition of iron powder to $Cl_2(aq)$. (b) After addition of iron powder and occurrence of the oxidation-reduction reaction.

Figure 24.5 Addition of a 0.10-M solution of the oxidizing agent potassium dichromate, $K_2Cr_2O_7(aq)$ (orange solution), to a solution containing 0.10-M iron(II) sulfate, $FeSO_4(aq)$ (pale green), and 0.10 M $H_2SO_4(aq)$. The iron atom is oxidized from $Fe^{2+}(aq)$ to $Fe^{3+}(aq)$, and the chromium atom is reduced from Cr(VI) in $Cr_2O_7^{2-}(aq)$ to Cr(III) in the green $Cr^{3+}(aq)$ ion.

Let's illustrate the method of half reactions by balancing the equation involving the oxidation of $Fe^{2+}(aq)$ by the dichromate ion, $Cr_2O_7^{2-}(aq)$, a reaction used in the determination of the iron content of ores (Figure 24.5):

$$Fe^{2+}(aq) + Cr_2O_7^{2-}(aq) \xrightarrow{H^+(aq)} Fe^{3+}(aq) + Cr^{3+}(aq) \qquad (24.5)$$

The $H^+(aq)$ over the arrow in the chemical equation indicates that the reaction occurs in acidic solution.

To balance oxidation-reduction equations in acidic solution by the method of half reactions, we use the following sequence of steps:

1. *Identify the species oxidized and reduced and then separate the equation into two equations representing the oxidation half reaction and the reduction half reaction.*

The oxidation state of the iron atom increases from +2 to +3, and the oxidation state of each chromium atom decreases from +6 in the $Cr_2O_7^{2-}$ ion to +3 in the Cr^{3+} ion. Thus, the equations for the two half reactions are

$$Fe^{2+} \rightarrow Fe^{3+} \qquad \text{(oxidation)}$$

$$Cr_2O_7^{2-} \rightarrow Cr^{3+} \qquad \text{(reduction)}$$

2. *Balance the equation for each half reaction with respect to all atoms other than oxygen and hydrogen.*

The equation for the iron half reaction is already balanced with respect to iron atoms. We balance the equation for the chromium half reaction with respect to chromium atoms by placing a 2 in front of Cr^{3+}:

$$Fe^{2+} \rightarrow Fe^{3+}$$

$$Cr_2O_7^{2-} \rightarrow 2\,Cr^{3+}$$

3. *Balance each half reaction equation with respect to oxygen atoms. To accomplish this step, add the appropriate number of H_2O molecules to the side deficient in oxygen atoms.*

Only the chromium half reaction equation involves oxygen atoms. There are seven oxygen atoms on the left and none on the right. Therefore, we balance the oxygen atoms by adding seven H_2O molecules to the right side of the equation for the chromium half reaction:

$$Fe^{2+} \rightarrow Fe^{3+}$$

$$Cr_2O_7^{2-} \rightarrow 2\,Cr^{3+} + 7\,H_2O$$

4. *Balance each half reaction equation with respect to hydrogen atoms by adding the appropriate number of H^+ ions to the side deficient in hydrogen atoms.*

Only the chromium half reaction equation involves hydrogen atoms. There are 14 hydrogen atoms on the right and none on the left. Therefore, we balance the hydrogen atoms by adding 14 H^+ ions to the left side of the equation for the chromium half reaction:

$$Fe^{2+} \rightarrow Fe^{3+}$$

$$14\,H^+ + Cr_2O_7^{2-} \rightarrow 2\,Cr^{3+} + 7\,H_2O$$

The two half reaction equations are now balanced with respect to atoms, but they are not balanced with respect to charge.

5. *Balance each half reaction equation with respect to charge by adding the appropriate number of electrons to the side with the excess positive charge.*

The equation for the iron half reaction has a charge of +2 on the left and +3 on the right. Thus, we balance the charge by adding one electron to the right side:

$$Fe^{2+} \rightarrow Fe^{3+} + e^- \qquad \text{(oxidation)}$$

The equation for the chromium half reaction has a net charge of $14(+1) + (-2) = +12$ on the left and $2(+3) + 7(0) = +6$ on the right. Thus, we balance the charge by adding six electrons to the left side:

$$14\,H^+ + Cr_2O_7^{2-} + 6\,e^- \rightarrow 2\,Cr^{3+} + 7\,H_2O \qquad \text{(reduction)}$$

The two half reaction equations are now balanced. Note that the iron half reaction donates electrons and the chromium half reaction accepts electrons.

6. *Make the number of electrons supplied by the oxidation half reaction equation equal to the number of electrons consumed by the reduction half reaction equation.*

The iron half reaction supplies one electron for each Fe^{2+} ion that is oxidized to Fe^{3+}, and the chromium half reaction consumes six electrons for each $Cr_2O_7^{2-}$ ion that is reduced to Cr^{3+}. Therefore, we multiply the equation for the iron half reaction equation by 6:

$$6\,Fe^{2+} \rightarrow 6\,Fe^{3+} + 6\,e^-$$

$$14\,H^+ + Cr_2O_7^{2-} + 6\,e^- \rightarrow 2\,Cr^{3+} + 7\,H_2O$$

Now the number of electrons supplied and the number consumed are equal.

7. *Obtain the complete balanced equation by adding the two balanced half reaction equations and canceling or combining any like terms.*

Adding the equations for the two half reactions and canceling the $6\,e^-$ terms that appear on both sides yields

$$\begin{array}{rl} & 6\,Fe^{2+} \rightarrow 6\,Fe^{3+} + \cancel{6e^-} \\ + & \cancel{6e^-} + 14\,H^+ + Cr_2O_7^{2-} \rightarrow 2\,Cr^{3+} + 7\,H_2O \\ \hline & 6\,Fe^{2+} + 14\,H^+ + Cr_2O_7^{2-} \rightarrow 6\,Fe^{3+} + 2\,Cr^{3+} + 7\,H_2O \end{array}$$

No electrons should appear in the complete balanced equation because electrons are conserved. This fact serves as a check on your results. You should also check that the equation is balanced with respect to each element and with respect to charge. As a final step, we rewrite the balanced equation with phases indicated:

$$6\,Fe^{2+}(aq) + 14\,H^+(aq) + Cr_2O_7^{2-}(aq) \rightarrow 6\,Fe^{3+}(aq) + 2\,Cr^{3+}(aq) + 7\,H_2O(l)$$

Although the method of half reactions involves numerous steps, it is actually simple to use and, with a little practice, becomes straightforward. Even though it may seem arbitrary, the use of $H_2O(l)$ and $H^+(aq)$ to balance the half reactions (steps 3 and 4) is logical because in acidic aqueous solution these species are always present in appreciable concentrations and therefore are readily available to participate in chemical reactions.

EXAMPLE 24-6: Balance the following oxidation-reduction reaction equation:

$$Fe^{2+}(aq) + O_2(g) \xrightarrow{H^+(aq)} Fe^{3+}(aq) + H_2O(l)$$

The reaction described by this equation occurs in the air oxidation of aqueous solutions containing $Fe^{2+}(aq)$, such as $FeSO_4(aq)$.

Solution: The $H^+(aq)$ over the arrow indicates that the reaction takes place in an acidic aqueous solution. The oxidation state of the iron atom changes from +2 in $Fe^{2+}(aq)$ to +3 in $Fe^{3+}(aq)$, and that of the oxygen atom changes from 0 in $O_2(g)$ to −2 in $H_2O(l)$. Thus, the equations for the two half reactions are

$$Fe^{2+} \rightarrow Fe^{3+} \qquad \text{(oxidation)}$$

$$O_2 \rightarrow H_2O \qquad \text{(reduction)}$$

Let's balance the equation for each half reaction in turn. The oxidation half reaction equation is balanced with respect to iron atoms. To balance it with respect to charge, we add one electron to the right side:

$$Fe^{2+} \rightarrow Fe^{3+} + e^- \quad \text{(oxidation)}$$

To balance the reduction half-reaction equation with respect to oxygen atoms, we place a 2 in front of the H_2O molecule on the right side:

$$O_2 \rightarrow 2\,H_2O$$

To balance it with respect to hydrogen atoms, we add 4 H^+ ions to the left side:

$$4H^+ + O_2 \rightarrow 2H_2O$$

Now we add $4e^-$ to the left side to balance it with respect to charge:

$$4H^+ + O_2 + 4e^- \rightarrow 2H_2O \qquad \text{(reduction)}$$

The oxidation half reaction as written supplies one electron, and the reduction half reaction as written consumes four electrons. If we multiply the oxidation half reaction equation by 4, then both half reaction equations will involve four electrons:

$$4Fe^{2+} \rightarrow 4Fe^{3+} + 4e^- \qquad \text{(oxidation)}$$

$$4H^+ + O_2 + 4e^- \rightarrow 2H_2O \qquad \text{(reduction)}$$

Addition of these two half reaction equations yields

$$4Fe^{2+} + 4H^+ + O_2 \rightarrow 4Fe^{3+} + 2H_2O$$

Finally, we indicate the phases and write

$$4Fe^{2+}(aq) + 4H^+(aq) + O_2(g) \rightarrow 4Fe^{3+}(aq) + 2H_2O(l)$$

This equation is balanced with respect to each atom and with respect to charge.

PRACTICE PROBLEM 24-6: Balance the oxidation-reduction equation

$$Br^-(aq) + MnO_4^-(aq) \xrightarrow{H^+(aq)} BrO_3^-(aq) + Mn^{2+}(aq)$$

which is used to prepare bromate ions, $BrO_3^-(aq)$, from bromide ions, $Br^-(aq)$.

Answer:
$5Br^-(aq) + 18H^+(aq) + 6MnO_4^-(aq) \rightarrow 5BrO_3^-(aq) + 6Mn^{2+}(aq) + 9H_2O(l)$

The study of the oxidation-reduction chemistry of a species is facilitated by considering the possible half reactions it can undergo. Thus, it is often necessary to balance the equations for half reactions without reference to a complete balanced equation. The following Example illustrates the balancing of an equation for a single half reaction in acidic solution.

EXAMPLE 24-7: Hydrogen peroxide is used extensively as an oxidizing agent in acidic aqueous solution. For example, $H_2O_2(aq)$ is used as a bleaching agent for human hair and to kill bacteria (antiseptic agent). Write a balanced half reaction equation for $H_2O_2(aq)$ acting as an oxidizing agent in acidic aqueous solution.

Solution: The oxidation state of each oxygen atom in H_2O_2 is –1 (see Example 24-1c). Because $H_2O_2(aq)$ is acting as an oxidizing agent, the oxygen atom is reduced to an oxidation state of –2. This is the lowest possible oxidation state for an oxygen atom because the addition of two electrons yields a noble-gas electron configuration. The oxidation state of the oxygen atom in a water molecule is –2, and so we write

$$H_2O_2 \rightarrow H_2O \qquad \text{(reduction)}$$

Balancing the oxygen atoms by adding another H_2O molecule to the right side yields

$$H_2O_2 \rightarrow 2\,H_2O$$

and balancing the hydrogen atoms by adding 2 H^+ ions to the left side yields

$$2\,H^+ + H_2O_2 \rightarrow 2\,H_2O$$

We now balance the charge by adding $2e^-$ to the left side and indicate phases to obtain

$$2\,H^+(aq) + H_2O_2(aq) + 2\,e^- \rightarrow 2\,H_2O(l)$$

PRACTICE PROBLEM 24-7: Write the balanced half reaction equation for $H_2O_2(aq)$ acting as a reducing agent in acidic aqueous solution.

Answer: $H_2O_2(aq) \rightarrow O_2(g) + 2\,H^+(aq) + 2\,e^-$

24-5. Chemical Equations for Oxidation-Reduction Reactions Occurring in Basic Solution Are Balanced Using OH^- and H_2O

The reactions considered up to this stage have all taken place in acidic aqueous solution, where $H^+(aq)$ and $H_2O(l)$ are readily available and thus can be used in balancing the equations for the half reactions. In basic solution, however, $H^+(aq)$ is not available at significant concentrations to use in balancing the equations for the half reactions with respect to hydrogen. Therefore, for reactions that take place in basic solution, we must change step 4 by using OH^- ions instead of H^+ ions. If we simply add OH^- ions to balance an equation with respect to hydrogen atoms, however, then each OH^- ion adds an oxygen atom in addition to a hydrogen atom and the oxygen-atom balance that was attained in step 3 will be lost. To overcome this problem, we must modify step 3 as well, to preserve the oxygen-atom balance. The idea is again to add one H_2O molecule for each excess oxygen atom, but this time to the side with *excess* oxygen atoms. Then we add *two* OH^- ions to the other side of the equation for each H_2O molecule added. These two steps replace steps 3 and 4 that we previously used to balance half reactions in acidic solution. Thus, steps 3 and 4 for balancing oxidation-reduction reaction equations in basic solutions are as follows:

3. *Add a number of H_2O molecules equal to the number of excess oxygen atoms to the side with the excess oxygen atoms.*
4. *Balance each half reaction with respect to hydrogen and oxygen atoms by adding a number of OH^- ions twice as great as the number of H_2O molecules added in step 3, but to the side opposite the added H_2O molecules.*

The procedure outlined above is best illustrated by example. Let's balance the equation for the reduction half reaction

$$ClO^-(aq) \xrightarrow{OH^-(aq)} Cl^-(aq) \qquad \text{(reduction)}$$

in basic solution as indicated by $OH^-(aq)$ over the arrow in the equation. The half reaction equation is already balanced with respect to chlorine atoms. Because the left side of this equation has one extra oxygen atom, to balance the equation with respect to oxygen and hydrogen atoms, we add one H_2O molecule to the left side (Step 3) and two OH^- ions to the right side (Step 4):

$$H_2O + ClO^- \rightarrow Cl^- + 2\,OH^-$$

This procedure simultaneously balances the reaction equation with respect to oxygen and hydrogen atoms. We now balance the equation with respect to charge by adding $2\,e^-$ to the left side:

$$H_2O + ClO^- + 2\,e^- \rightarrow Cl^- + 2\,OH^-$$

The balanced equation for the half reaction with phases indicated is

$$H_2O(l) + ClO^-(aq) + 2\,e^- \rightarrow Cl^-(aq) + 2\,OH^-(aq)$$

Finally, in cases involving metal hydroxides in basic solution, we add OH^- ions directly, skipping steps 3 and 4. This is illustrated in balancing the following half reaction equation:

$$Cu(OH)_2(s) \rightarrow Cu(s)$$

$$Cu(OH)_2(s) \rightarrow Cu(s) + 2\,OH^-(aq)$$

$$Cu(OH)_2(s) + 2\,e^- \rightarrow Cu(s) + 2\,OH^-(aq)$$

The following Example and Practice Problem provide additional examples of balancing oxidation-reduction equations in basic solution.

EXAMPLE 24-8: Given that the oxidation-reduction reaction described by the following equation takes place in basic aqueous solution, balance the equation

$$BrO_3^-(aq) + F_2(g) \xrightarrow{OH^-(aq)} BrO_4^-(aq) + F^-(aq)$$

Solution: The oxidation state of the bromine atom changes from +5 in $BrO_3^-(aq)$ to +7 in $BrO_4^-(aq)$ and that of the fluorine atoms changes from 0 in $F_2(g)$ to −1 in $F^-(aq)$. Thus, we have the two half reaction equations

$$BrO_3^- \rightarrow BrO_4^- \qquad \text{(oxidation)}$$

$$F_2 \rightarrow 2F^- \qquad \text{(reduction)}$$

where we have added a 2 to balance the fluorine atoms in the reduction half reaction equation. Both half reaction equations are now balanced with respect to atoms other than oxygen and hydrogen. Let's balance each one in turn. The oxidation half reaction equation involves one excess oxygen atom on the right side, so we add one H_2O molecule to that side and two OH^- ions to the left side:

$$2OH^- + BrO_3^- \rightarrow BrO_4^- + H_2O$$

To balance with respect to charge, we add $2e^-$ to the right side:

$$2OH^- + BrO_3^- \rightarrow BrO_4^- + H_2O + 2e^- \qquad \text{(oxidation)}$$

To balance the reduction half reaction equation with respect to charge, we add $2e^-$ to the left side:

$$F_2 + 2e^- \rightarrow 2F^- \qquad \text{(reduction)}$$

Because the number of electrons is the same in both half reaction equations, we simply add the two half reaction equations and include the phase designations to obtain a complete balanced equation:

$$2OH^-(aq) + BrO_3^-(aq) + F_2(g) \rightarrow BrO_4^-(aq) + 2F^-(aq) + H_2O(l)$$

PRACTICE PROBLEM 24-8: Balance the following oxidation-reduction equation in basic aqueous solution:

$$Fe(OH)_2(s) + O_2(g) \xrightarrow{OH^-(aq)} Fe(OH)_3(s)$$

Answer: $4Fe(OH)_2(s) + 2H_2O(l) + O_2(g) \rightarrow 4Fe(OH)_3(s)$

TABLE 24.3 Procedure for balancing the equations for oxidation-reduction reactions in acidic and basic aqueous solutions

Step	Procedure	
1.	Separate the equation into an oxidation half reaction equation and a reduction half reaction equation.	
2.	Balance each half reaction equation with respect to all atoms other than oxygen and hydrogen.	
	In Acidic Solution:	*In Basic Solution:* (For metal hydroxides, directly add OH^- ions and skip Steps 3 and 4.)
3.	Balance each half reaction equation with respect to oxygen atoms by adding the appropriate number of H_2O molecules to the side deficient in oxygen atoms.	3. Add a number of H_2O molecules equal to the number of excess oxygen atoms to the side with the excess oxygen atoms.
4.	Balance each half reaction equation with respect to hydrogen atoms by adding the appropriate number of H^+ ions to the side deficient in hydrogen atoms.	4. Balance each half reaction equation with respect to hydrogen and oxygen atoms by adding a number of OH^- ions twice as great as the number of H_2O molecules added in Step 3 to the side opposite to the added H_2O molecules.
5.	Balance each half reaction equation with respect to charge by adding the appropriate number of electrons to the side with the excess positive charge.	
6.	Multiply each half reaction equation by an integer that makes the number of electrons supplied by the oxidation half reaction equation equal to the number of electrons accepted by the reduction half reaction equation.	
7.	Obtain the complete balanced equation by adding the two half reaction equations and canceling or combining any like terms.	

If you remember to use OH^- ions and H_2O molecules, or, in cases involving metal hydroxides, just OH^- ions, then balancing equations in basic aqueous solution is straightforward. The procedure for balancing oxidation-reduction equations using the method of half reactions for acidic and basic aqueous solutions is summarized in Table 24.3.

24-6. Oxidation-Reduction Reactions Are Used in Chemical Analysis

Oxidation-reduction reactions are used extensively in chemical analysis. To illustrate this application of oxidation-reduction reactions, let's look at the analytical determination of iron(II) in a sample of iron ore. Suppose that we dissolve a 3.532-gram sample of iron ore in $H_2SO_4(aq)$ and that we reduce any iron(III) present to iron(II) by adding powdered zinc to the solution. We now titrate the resulting filtered $Fe^{2+}(aq)$ solution with the oxidizing agent potassium permanganate, $KMnO_4(aq)$ (Figure 24.6). If 34.58 mL of 0.1108 M $KMnO_4(aq)$ are required to oxidize all the $Fe^{2+}(aq)$, what is the mass percentage of iron in the ore? The balanced oxidation-reduction reaction equation is

Figure 24.6 Titration of a solution containing $Fe^{2+}(aq)$ with a solution containing the purple-colored oxidizing agent permanganate, $MnO_4^-(aq)$. The MnO_4^- oxidizes $Fe^{2+}(aq)$ to $Fe^{3+}(aq)$ and is itself reduced to $Mn^{2+}(aq)$.

$$5\,Fe^{2+}(aq) + MnO_4^-(aq) + 8\,H^+(aq) \rightarrow 5\,Fe^{3+}(aq) + Mn^{2+}(aq) + 4\,H_2O(l)$$

The equilibrium constant of this equation at 25°C is very large ($K_c = 3 \times 10^{62}$), so essentially all the added $KMnO_4(aq)$ oxidizes the $Fe^{2+}(aq)$ to $Fe^{3+}(aq)$. Such a reaction is said to be a **quantitative reaction**, because it goes essentially to completion, that is, essentially all the $Fe^{2+}(aq)$ is converted to $Fe^{3+}(aq)$ by the permanganate ion.

The number of millimoles of $KMnO_4(aq)$ required to oxidize the $Fe^{2+}(aq)$ to $Fe^{3+}(aq)$ is given by

$$\text{millimoles of } KMnO_4 = MV = \left(\frac{0.1108 \text{ mmol}}{1 \text{ mL}}\right)(34.58 \text{ mL}) = 3.831 \text{ mmol}$$

From the balanced equation, we see that five millimoles of $Fe^{2+}(aq)$ are oxidized for each millimole of $KMnO_4(aq)$ added, so

$$\text{millimoles of } Fe^{2+} = (3.831 \text{ mmol of } KMnO_4)\left(\frac{5 \text{ mmol } Fe^{2+}}{1 \text{ mmol } KMnO_4}\right)$$
$$= 19.16 \text{ mmol} = 0.01916 \text{ mol}$$

The atomic mass of iron is 55.845, so the mass of iron in the sample is

$$\text{mass of Fe} = \text{mass of } Fe^{2+} = (0.01916 \text{ mol})\left(\frac{55.845 \text{ g Fe}}{1 \text{ mol Fe}}\right) = 1.070 \text{ g}$$

The mass percentage of iron in the ore sample is

$$\text{mass \% Fe} = \left(\frac{1.070 \text{ g Fe}}{3.523 \text{ g ore}}\right) \times 100 = 30.37\%$$

EXAMPLE 24-9: The concentration of ozone in a sample of air can be determined by reacting the sample with a buffered solution of potassium iodide, $KI(aq)$. The $O_3(g)$ oxidizes the $I^-(aq)$ ion to $I_3^-(aq)$ according to

$$O_3(g) + 3\,I^-(aq) + 2\,H^+(aq) \rightarrow O_2(g) + I_3^-(aq) + H_2O(l)$$

The concentration of the $I_3^-(aq)$ ion formed in the reaction is determined by titration with a sodium thiosulfate solution, $Na_2S_2O_3(aq)$, of known concentration according to

$$\underset{\text{thiosulfate}}{2\,S_2O_3^{2-}(aq)} + I_3^-(aq) \rightarrow \underset{\text{tetrathionate}}{S_4O_6^{2-}(aq)} + 3\,I^-(aq)$$

Suppose that a 50.00-mL sample of KI(*aq*) has reacted with a 43.15-gram sample of air. If 34.56 mL of 0.002475 M $Na_2S_2O_3(aq)$ are required to titrate the $I_3^-(aq)$ produced, calculate the mass percentage of $O_3(g)$ in the sample.

Solution: The number of millimoles of $Na_2S_2O_3(aq)$ required is

$$\text{millimoles of } S_2O_3^{2-} = (0.002475 \text{ M})(34.56 \text{ mL}) = 8.554 \times 10^{-2} \text{ mmol}$$

The number of millimoles of $I_3^-(aq)$ reduced by the $S_2O_3^{2-}(aq)$ is given by

$$\text{millimoles of } I_3^-(aq) = (8.554 \times 10^{-2} \text{ mmol } S_2O_3^{2-})\left(\frac{1 \text{ mmol } I_3^-}{2 \text{ mmol } S_2O_3^{2-}}\right)$$
$$= 4.277 \times 10^{-2} \text{ mmol}$$

According to the stoichiometry of the equation for the ozone-plus-iodide reaction,

$$\text{millimoles of } O_3 = \text{millimoles of } I_3^- = 4.277 \times 10^{-2} \text{ mmol}$$

The mass of ozone is given by

$$\text{mass of } O_3 = (4.277 \times 10^{-2} \text{ mmol})\left(\frac{48.00 \text{ mg } O_3}{1 \text{ mmol } O_3}\right) = 2.053 \text{ mg}$$

and the mass percentage of ozone in the air sample is

$$\text{mass \% } O_3 = \left(\frac{\text{mass of ozone}}{\text{mass of sample}}\right) \times 100 = \left(\frac{2.053 \times 10^{-3} \text{ g}}{43.15 \text{ g}}\right) \times 100$$
$$= 4.758 \times 10^{-3} \text{ \%}$$

Solutions containing the thiosulfate ion, $S_2O_3^{2-}(aq)$, are used extensively in analytical chemistry to determine the concentration of $I_3^-(aq)$ or $I_2(aq)$ in a solution (Figure 24.7).

Figure 24.7 The reaction of sodium thiosulfate, $Na_2S_2O_3(aq)$, in the presence of starch, with $I_2(aq)$. Iodine, $I_2(aq)$, combines with starch to form a blue starch-iodine complex. When the $I_3^-(aq)$ ions are reduced by thiosulfate, $Na_2S_2O_3(aq)$, to iodine, the blue color appears.

PRACTICE PROBLEM 24-9: The amount of ethanol in air exhaled from the lungs can be determined using a breath analyzer or "breathalyzer." This procedure is used by law enforcement agencies to determine the level of intoxication of drivers of motor vehicles suspected of driving under the influence. The exhaled air is bubbled through a solution containing the orange-yellow compound potassium dichromate, $K_2Cr_2O_7(aq)$, dissolved in aqueous sulfuric acid. The ethanol, $CH_3CH_2OH(aq)$, is oxidized by the $Cr_2O_7^{2-}(aq)$ to acetic acid, $CH_3COOH(aq)$, and the orange-yellow dichromate is reduced to $Cr^{3+}(aq)$, which is green. The amount of ethanol present is directly proportional to the decrease in the concentration of $Cr_2O_7^{2-}(aq)$. (a) Balance the oxidation-reduction equation

$$CH_3CH_2OH(aq) + \underset{\text{orange-yellow}}{Cr_2O_7^{2-}(aq)} \xrightarrow{H^+(aq)} CH_3COOH(aq) + \underset{\text{green}}{Cr^{3+}(aq)}$$

(b) Calculate the number of milligrams of ethanol required to decrease the concentration of 10.0 mL of a 0.0100 M $K_2Cr_2O_7(aq)$ solution by 10%.

Answer:

(a) $3\,CH_3CH_2OH(aq) + 2\,Cr_2O_7^{2-}(aq) + 16\,H^+(aq) \rightarrow 3\,CH_3COOH(aq) + 4\,Cr^{3+}(aq) + 11\,H_2O(l)$

(b) 0.69 mg

Oxidation-reduction reactions are used extensively in analytical chemistry to determine unknown concentrations of reducing and oxidizing agents. These reactions are especially useful in analytical chemistry because we can choose a reactant that makes the reaction quantitative—that is, we choose a reactant that gives the resulting oxidation-reduction reaction a very large equilibrium constant. We can determine the equilibrium constant for an oxidation-reduction reaction based on the principles of electrochemistry, as described in Chapter 25.

Figure 24.8 (*top*) There are a wide variety of iron oxides and iron oxide hydrides giving rise to the diverse colors of rust and iron deposits found in nature. (*bottom*) The red color of Martian soil is due primarily to $Fe_2O_3(s)$ (other iron oxides are also found on Mars).

24-7. Billions of Dollars Are Spent Each Year to Protect Metals from Corrosion

We are all familiar with corrosion, the best-known example of which is the rusting of iron and steel. Brown rust is iron(III) oxide, and the corrosion of iron proceeds by air oxidation of the iron:

$$4\,Fe(s) + 3\,O_2(g) \xrightarrow{H_2O(l)} \underset{\text{brown rust}}{2\,Fe_2O_3(s)}$$

Iron(III) oxide (which can also be red in color) is the pigment used in rouge, and purified $Fe_2O_3(s)$ is used to coat magnetic tapes and hard disks. There are a wide variety of iron oxides in addition to $Fe_2O_3(s)$, some of which include $Fe_3O_4(s)$ (black rust), $Fe(OH)_2(s)$ (green rust), and $FeO(s)$ (a black powder). In addition, many of these oxides can form various hydrides—giving rise to a diverse range of colors and chemistry (Figures 24.8 and 24.9).

Most metals when exposed to air develop an oxide film. In some cases this film is very thin and protects the metal, helping to maintain its luster. Examples are nickel and chromium. However, depending on the metal, the humidity, the acidity, and the presence of certain anions, corrosion can completely destroy a metal. For example, the $Cl^-(aq)$ anion, which is present in sea spray and on roads treated with rock salt, promotes corrosion through the formation of chloro complexes. Other anions act by increasing the electrical conductivity of the solution causing the corrosion. Certain gaseous species, such as the oxides of sulfur and nitrogen, combine with water to form acids that attack metals.

Corrosion is a major problem costing billions of dollars annually for replacements of corroded parts. There is a great deal of research being done to

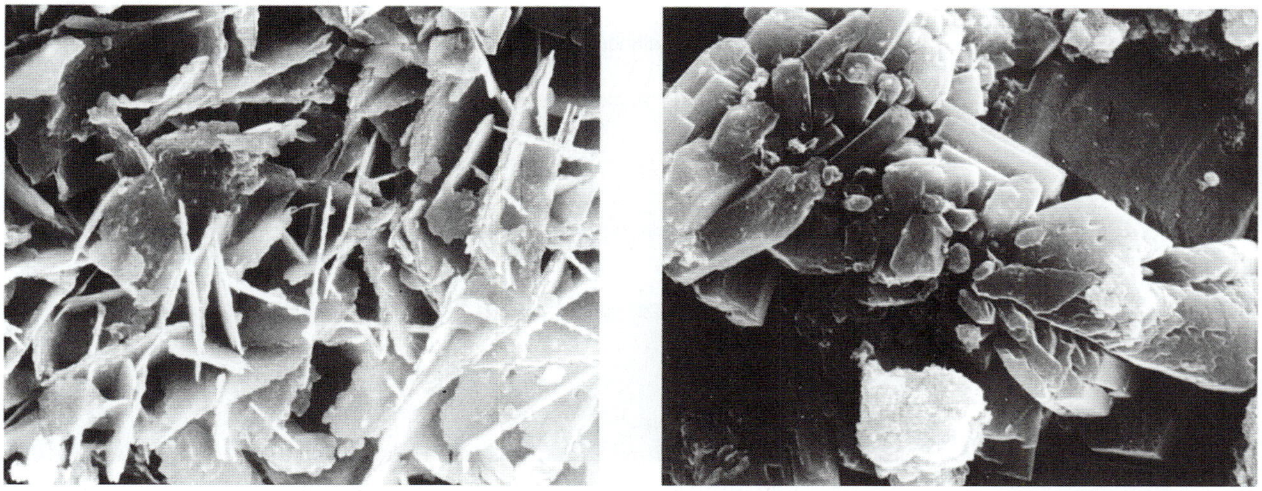

Figure 24.9 Scanning electron photomicrographs of iron oxides. (*left*) The angular crystals are green rust, Fe(II) oxides. (*right*) The column-shaped and rounded crystals are black rust, $Fe_3O_4(s)$, and brown rust, $Fe_2O_3(s)$, respectively. Because this type of microscope does not use optical wavelengths, the different rust colors are not actually seen in such pictures.

understand corrosion mechanisms because a detailed understanding of these mechanisms can provide important clues to methods for preventing the process.

The corrosion of metals involves oxidation-reduction reactions between different sections of the same piece of metal or between two dissimilar metals in electrical contact with each other. One metal piece acts as the reducing agent, and the other provides the conducting surface on which reduction occurs. For example, iron in contact with air and moisture corrodes according to the mechanism sketched in Figure 24.10. The oxidation step is described by the equation

$$2\,Fe(s) + 4\,OH^-(aq) \rightarrow 2\,Fe(OH)_2(s) + 4\,e^- \quad \text{(oxidation)}$$

and the reduction step is

$$2\,H_2O(l) + O_2(g) + 4\,e^- \rightarrow 4\,OH^-(aq) \qquad \text{(reduction)}$$

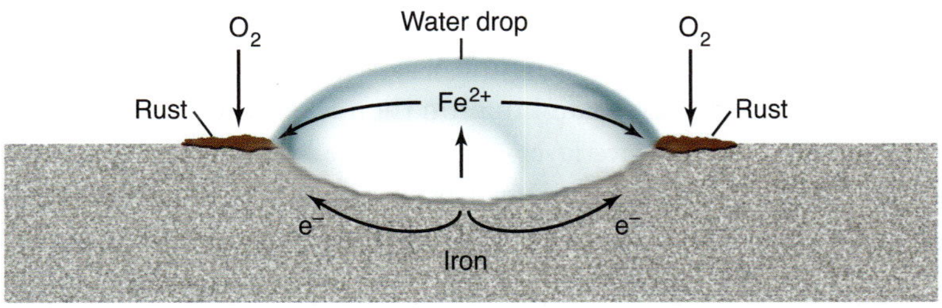

Figure 24.10 Corrosion of iron. A drop of water on an iron surface can act as a corrosion center. The iron is oxidized by oxygen from the air. Moisture is necessary for corrosion because the mechanism involves the formation of dissolved Fe(II) ions. Salts promote the corrosion by enabling a larger current flow between the active regions.

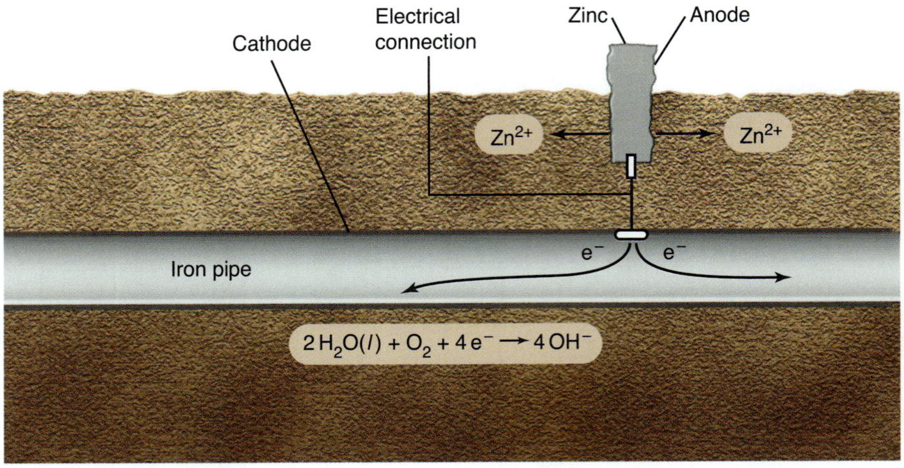

Figure 24.11 Protection of an iron pipe from corrosion with sacrificial zinc metal. Zinc is a stronger reducing agent than iron and thus is preferentially oxidized. The electrons produced in the oxidation of the zinc flow to the iron pipe, on the surface of which $O_2(aq)$ is reduced to hydroxide ion. The net process is

$$2\,Zn(s) + O_2(aq) + 2\,H_2O(l) \rightarrow 2\,Zn(OH)_2(s)$$

and the iron remains intact.

where the $O_2(g)$ comes from air. The $Fe(OH)_2(s)$ formed is rapidly air-oxidized in the presence of water to $Fe(OH)_3(s)$ as described by the equation

$$4\,Fe(OH)_2(s) + O_2(g) + 2\,H_2O(l) \rightarrow 4\,Fe(OH)_3(s)$$

This in turn converts spontaneously to iron(III) oxide trihydrate, $Fe_2O_3{\cdot}3\,H_2O(s)$, one of the most common forms of rust, according to:

$$2\,Fe(OH)_3(s) \rightarrow Fe_2O_3{\cdot}3\,H_2O(s)$$

The corrosion of aluminum in air is not so pronounced as that of iron because the $Al_2O_3(s)$ film that forms is tough, adherent, and impervious to oxygen. The same is true for chromium and nickel.

The simplest method of corrosion prevention is to provide a protective layer of paint or of a corrosion-resistant metal, such as chromium or nickel. The weakness of such methods is that any scratch or crack in the protective layer exposes the metal surface. The exposed surface, even though small in area, can act as an electron donor in conjunction with other exposed metal parts, that act as electron acceptors. This combination then leads to corrosion of the metal under the no-longer-protective layer.

Another anticorrosion technique uses a replaceable sacrificial metal that is a piece of metal electrically connected to a less active metal (Figure 24.11). The more active metal is the stronger reducing agent and is thus preferentially oxidized while oxygen is reduced on the surface of the less active metal. Sacrificial metals are used to protect nails, automobile body parts, water pipes, and ship propellers. The process of coating or impregnating iron or steel with zinc is called **galvanization.** A crack in a zinc coating does not affect the corrosion protection provided because a zinc coating functions like the zinc metal shown in Figure 24.11. The less active metal (iron) promotes the corrosion of the more active metal (zinc) by providing a metal surface for the reduction of oxygen.

SUMMARY

To determine whether a reaction involves electron transfer, we first assign oxidation states to each element on both sides of the equation. Oxidation states are assigned using the rules in Section 24-1. In an electron-transfer reaction, the oxidation state of one atom increases and the oxidation state of another atom decreases.

Electron-transfer reaction equations can be separated into two half reaction equations: the oxidation half reaction equation (electrons on the right) and the reduction half reaction equation (electrons on the left). The oxidation half reaction supplies electrons to the reduction half reaction. The equations for electron-transfer reactions can be balanced by a systematic procedure. The procedure for balancing equations for oxidation-reduction reactions in acidic and basic aqueous solutions is summarized in Table 24.3.

Oxidation-reduction reactions are often used in chemical analysis. Corrosion of metals and metal alloys usually involves air oxidation of the metal. Anticorrosion methods involve either protective coatings or the use of a more reactive metal that is preferentially oxidized (corroded).

TERMS YOU SHOULD KNOW

oxidation-reduction reaction *899*
oxidation state *900*
electron-transfer reaction *906*
oxidizing agent *906*
reducing agent *906*
half reaction *908*
oxidation half reaction *908*
reduction half reaction *908*
method of half reactions *909*
quantitative reaction *918*
galvanization *923*

EQUATIONS YOU SHOULD KNOW HOW TO USE

$$\begin{pmatrix}\text{oxidation}\\\text{state}\end{pmatrix} = \begin{pmatrix}\text{number of valence}\\\text{electrons in the free}\\\text{(nonbonded) atom}\end{pmatrix} - \begin{pmatrix}\text{number of valence}\\\text{electrons assigned to the}\\\text{atom in the molecule}\end{pmatrix} \quad (24.1)$$

(Lewis formula method of assigning oxidation states)

PROBLEMS

OXIDATION STATES

24-1. Assign oxidation states to the oxygen atoms in

(a) $O_2(g)$ (b) $KO_2(s)$

(c) $Na_2O_2(s)$ (d) $OF_2(g)$

24-2. Assign oxidation states to the oxygen atoms in

(a) $HO_2(g)$ (b) $S_2O_8^{2-}(aq)$ $[O_3S{-}O{-}O{-}SO_3]^{2-}$

(c) $O_3^-(aq)$ (d) $H_3COOCH_3(l)$

24-3. Assign oxidation states to the atoms in

(a) $ClO^-(aq)$ (b) $ClO_2^-(aq)$

(c) $ClO_3^-(aq)$ (d) $ClO_4^-(aq)$

24-4. Assign oxidation states to the atoms in

(a) $Cl_2O(g)$ (b) $ClO_2(g)$

(c) $Cl_2O_7(l)$ (d) $Cl_2O_5(l)$

24-5. Assign oxidation states to the atoms in

(a) $KMnO_4(s)$ (b) $MnO_4^{2-}(aq)$

(c) $MnO_2(s)$ (d) $Mn(ClO_4)_3(s)$

24-6. Assign oxidation states to the atoms in

(a) $Fe_2O_3{\cdot}3H_2O(s)$ (b) $Fe_3O_4(s)$

(c) $KCN(s)$ (d) $KCNO(s)$

24-7. Using Lewis formulas (Rule 8 in Section 24-1), assign oxidation states to the atoms in

(a) $H_2CO(g)$

(b) $CH_3OH(l)$

(c) 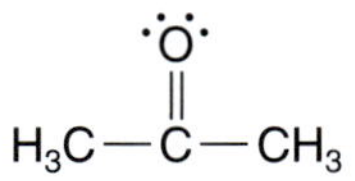

(d) $CH_3COOH(l)$

24-8. Using Lewis formulas (Rule 8 in Section 24-1), assign oxidation states to the atoms in

(a) $CS_2(l)$

(b) $CH_3S\text{–}SCH_3(l)$

(c) $HCONH_2(l)$

(d) 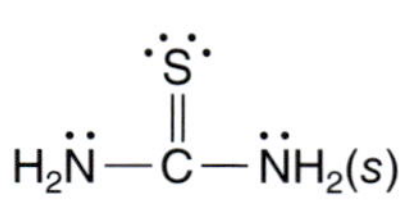

TRANSITION METAL COMPOUNDS

24-9. Write Lewis formulas for the covalent ions

(a) $VO_2^+(aq)$ (b) $VO_3^{2-}(aq)$

24-10. Write Lewis formulas for the covalent titanium ions

(a) $TiF_6^{2-}(aq)$ (b) $TiBr_5^-(aq)$

24-11. Using VSEPR theory (Chapter 8), predict the shapes of the following species involving transition metals:

(a) $TiF_6^{2-}(aq)$ (b) $VO_2^+(aq)$

(c) $VOCl_3(g)$ (d) $CrO_4^{2-}(aq)$

24-12. Using VSEPR theory (Chapter 8), predict the shapes of the following species involving mercury:

(a) $HgCl_2(g)$ (b) $HgCl_4^{2-}(aq)$ (c) $HgCl_3^-(aq)$

OXIDIZING AGENTS AND REDUCING AGENTS

24-13. Identify the oxidizing and reducing agents in the equation

$$I_2(s) + 2\,Na_2S_2O_3(aq) \rightarrow 2\,NaI(aq) + Na_2S_4O_6(aq)$$

24-14. Sodium sulfide is manufactured by reacting sodium sulfate with carbon in the form of coke according to the equation

$$Na_2SO_4(s) + 4\,C(s) \rightarrow Na_2S(s) + 4\,CO(g)$$

Identify the oxidizing and reducing agents in this equation.

24-15. Sodium nitrite, an important chemical in the dye industry, is manufactured by the reaction between sodium nitrate and lead according to

$$NaNO_3(aq) + Pb(s) \rightarrow NaNO_2(aq) + PbO(s)$$

Identify the oxidizing and reducing agents in this equation.

24-16. Sodium chlorite, an industrial bleaching agent, is prepared as shown by the equation

$$4\,NaOH(aq) + Ca(OH)_2(aq) + C(s) + 4\,ClO_2(g) \rightarrow 4\,NaClO_2(aq) + CaCO_3(s) + 3\,H_2O(l)$$

Identify the oxidizing and reducing agents in this equation.

24-17. Identify the oxidizing and reducing agents and write the oxidation and reduction half-reaction equations for the following chemical equations:

(a) $2\,Fe^{3+}(aq) + 2\,I^-(aq) \rightarrow 2\,Fe^{2+}(aq) + I_2(s)$

(b) $2\,Ti^{2+}(aq) + Co^{2+}(aq) \rightarrow 2\,Ti^{3+}(aq) + Co(s)$

24-18. Identify the oxidizing and reducing agents and write the oxidation and reduction half-reaction equations for the following chemical equations:

(a) $In^+(aq) + 2\,Fe^{3+}(aq) \rightarrow 2\,Fe^{2+}(aq) + In^{3+}(aq)$

(b) $H_2S(aq) + ClO^-(aq) \rightarrow S(s) + Cl^-(aq) + H_2O(l)$

24-19. Which of the following are oxidation-reduction reactions?

(a) $2\,H_2(g) + O_2(g) \rightarrow 2\,H_2O(l)$

(b) $2\,C_8H_{18}(l) + 25\,O_2(g) \rightarrow 16\,CO_2(g) + 18\,H_2O(g)$

(c) $Pb(NO_3)_2(aq) + Na_2SO_4(aq) \rightarrow 2\,NaNO_3(aq) + PbSO_4(s)$

(d) $NaOH(aq) + CH_3COOH(aq) \rightarrow CH_3COONa(aq) + H_2O(l)$

(e) $Zn(s) + 2HCl(aq) \rightarrow ZnCl_2(aq) + H_2(g)$

24-20. Which of the following are oxidation-reduction reactions?

(a) $6Fe^{2+}(aq) + 14H^+(aq) + Cr_2O_7^{2-}(aq) \rightarrow 6Fe^{3+}(aq) + 2Cr^{3+}(aq) + 7H_2O(l)$

(b) $Ag^+(aq) + Cl^-(aq) \rightarrow AgCl(s)$

(c) $CH_4(g) + 2O_2(g) \rightarrow CO_2(g) + 2H_2O(g)$

(d) $H_2SO_4(aq) + 2KOH(aq) \rightarrow K_2SO_4(aq) + 2H_2O(l)$

(e) $Cl_2(g) + NaBr(aq) \rightarrow Br_2(l) + NaCl(aq)$

24-21. For each of the reactions in Problem 24-19 that is an oxidation-reduction reaction, identify the following: the species oxidized, the species reduced, the oxidizing agent, and the reducing agent.

24-22. For each of the reactions in Problem 24-20 that is an oxidation-reduction reaction, identify the following: the species oxidized, the species reduced, the oxidizing agent, and the reducing agent.

24-23. Potassium superoxide, $KO_2(s)$, is a strong oxidizing agent. Explain why.

24-24. Lithium aluminum hydride, $LiAlH_4(s)$, is a strong reducing agent. Explain why.

BALANCING HALF REACTIONS

24-25. Balance the following equations for half reactions that occur in acid solution:

(a) $Mo^{3+}(aq) \rightarrow MoO_2^{2+}(aq)$

(b) $P_4(s) \rightarrow H_3PO_4(aq)$

(c) $S_2O_8^{2-}(aq) \rightarrow HSO_4^-(aq)$

24-26. Balance the following equations for half reactions that occur in acidic solution:

(a) $H_2BO_3^-(aq) \rightarrow BH_4^-(aq)$

(b) $ClO_3^-(aq) \rightarrow Cl_2(g)$

(c) $Cl_2(g) \rightarrow HClO(aq)$

24-27. Balance the following half-reaction equations:

(a) $WO_3(s) \rightarrow W_2O_5(s)$ (acidic)

(b) $U^{4+}(aq) \rightarrow UO_2^+(aq)$ (acidic)

(c) $Zn(s) \rightarrow [Zn(OH)_4]^{2-}(aq)$ (basic)

24-28. Balance the following half-reaction equations:

(a) $OsO_4(s) \rightarrow Os(s)$ (acidic)

(b) $S(s) \rightarrow SO_3^{2-}(aq)$ (basic)

(c) $Sn(s) \rightarrow HSnO_2^-(aq)$ (acidic)

24-29. Balance the following equations for half reactions that occur in basic solution:

(a) $SO_3^{2-}(aq) \rightarrow S_2O_4^{2-}(aq)$

(b) $Cu(OH)_2(s) \rightarrow Cu_2O(s)$

(c) $AgO(s) \rightarrow Ag_2O(s)$

24-30. Balance the following equations for half reactions:

(a) $[Au(CN)_2]^-(aq) \rightarrow Au(s) + CN^-(aq)$ (acidic)

(b) $MnO_4^-(aq) \rightarrow MnO_2(s)$ (acidic)

(c) $Cr(OH)_3(s) \rightarrow CrO_4^{2-}(aq)$ (basic)

BALANCING OXIDATION-REDUCTION EQUATIONS

24-31. Balance the following equations for reactions that occur in acidic solution:

(a) $MnO(s) + PbO_2(s) \rightarrow MnO_4^-(aq) + Pb^{2+}(aq)$

(b) $As_2S_5(s) + NO_3^-(aq) \rightarrow H_3AsO_4(aq) + HSO_4^-(aq) + NO_2(g)$

For each of these reactions, identify the species oxidized, species reduced, oxidizing agent, and reducing agent.

24-32. Balance the following equations for reactions that occur in acidic solution:

(a) $ZnS(s) + NO_3^-(aq) \rightarrow Zn^{2+}(aq) + S(s) + NO(g)$

(b) $MnO_4^-(aq) + HNO_2(aq) \rightarrow NO_3^-(aq) + Mn^{2+}(aq)$

For each of these reactions, identify the species oxidized, species reduced, oxidizing agent, and reducing agent.

24-33. Complete and balance the following equations:

(a) $NH_4^+(aq) + NO_3^-(aq) \rightarrow N_2O(g)$ (acidic)

(b) $Fe(s) + O_2(g) \rightarrow Fe_2O_3 \cdot 3H_2O(s)$ (basic)

24-34. Complete and balance the following equations:

(a) $CoCl_2(s) + Na_2O_2(aq) \rightarrow Co(OH)_3(s) + Cl^-(aq) + Na^+(aq)$ (basic)

(b) $C_2O_4^{2-}(aq) + MnO_2(s) \rightarrow Mn^{2+}(aq) + CO_2(g)$ (acidic)

24-35. Complete and balance the following equations:

(a) $Fe(OH)_2(s) + O_2(g) \rightarrow Fe(OH)_3(s)$ (basic)

(b) $Cu(s) + NO_3^-(aq) \rightarrow Cu^{2+}(aq) + NO(g)$ (acidic)

24-36. Complete and balance the following equations:

(a) $Cr_2O_7^{2-}(aq) + I^-(aq) \rightarrow Cr^{3+}(aq) + I_2(s)$ (acidic)

(b) $CuS(s) + NO_3^-(aq) \rightarrow Cu^{2+}(aq) + S(s) + NO(g)$ (acidic)

24-37. Use the method of half reactions to balance the following equations:

(a) $IO_4^-(aq) + I^-(aq) \rightarrow IO_3^-(aq) + I_3^-(aq)$ (basic)

(b) $H_2MoO_4(aq) + Cr^{2+}(aq) \rightarrow Mo(s) + Cr^{3+}(aq)$ (acidic)

24-38. Use the method of half reactions to balance the following equations:

(a) $N_2H_4(aq) + Cu(OH)_2(s) \rightarrow N_2(g) + Cu(s)$ (basic)

(b) $H_3AsO_3(aq) + I_2(aq) \rightarrow H_3AsO_4(aq) + I^-(aq)$ (acidic)

24-39. Use the method of half reactions to balance the following equations:

(a) $CrO_4^{2-}(aq) + Cl^-(aq) \rightarrow Cr^{3+}(aq) + ClO_2^-(aq)$ (acidic)

(b) $Cu^{2+}(aq) + S_2O_3^{2-}(aq) \rightarrow Cu^+(aq) + S_4O_6^{2-}(aq)$ (acidic)

24-40. Use the method of half reactions to balance the following equations:

(a) $Co(OH)_2(s) + SO_3^{2-}(aq) \rightarrow SO_4^{2-}(aq) + Co(s)$ (basic)

(b) $IO_3^-(aq) + I^-(aq) \rightarrow I_3^-(aq)$ (acidic)

CALCULATIONS INVOLVING OXIDATION-REDUCTION EQUATIONS

24-41. The amount of $I_3^-(aq)$ in a solution can be determined by titration with a solution containing a known concentration of thiosulfate ion, $S_2O_3^{2-}(aq)$. The determination is based on the equation (see Example 24-9)

$$I_3^-(aq) + 2S_2O_3^{2-}(aq) \rightarrow 3I^-(aq) + S_4O_6^{2-}(aq)$$

Given that it requires 36.4 mL of 0.330 M $Na_2S_2O_3(aq)$ to titrate the $I_3^-(aq)$ in a 15.0-mL sample, calculate the molarity of $I_3^-(aq)$ in the solution.

24-42. The amount of $Fe^{2+}(aq)$ in an $FeSO_4(aq)$ solution can be determined by titration with a solution containing a known concentration of $Ce^{4+}(aq)$. The determination is based on the reaction described by

$$Fe^{2+}(aq) + Ce^{4+}(aq) \rightarrow Fe^{3+}(aq) + Ce^{3+}(aq)$$

Given that it requires 37.5 mL of 0.0965 M $Ce^{4+}(aq)$ to completely oxidize the $Fe^{2+}(aq)$ in a 35.0-mL sample to $Fe^{3+}(aq)$, calculate the molarity of $Fe^{2+}(aq)$ and the number of milligrams of iron in the sample.

24-43. The quantity of antimony in a sample can be determined by an oxidation-reduction titration with an oxidizing agent. A 9.62-gram sample of stibnite, an ore of antimony, is dissolved in hot, concentrated $HCl(aq)$ and passed over a reducing agent so that all the antimony is in the form $Sb^{3+}(aq)$. The $Sb^{3+}(aq)$ is completely oxidized by 43.7 mL of a 0.125-M aqueous solution of $KBrO_3(aq)$. The unbalanced equation for the reaction is

$$BrO_3^-(aq) + Sb^{3+}(aq) \rightarrow Br^-(aq) + Sb^{5+}(aq)$$

Calculate the amount of antimony in the sample and its percentage in the ore.

24-44. An ore is to be analyzed for its iron content by an oxidation-reduction titration with permanganate ion. A 4.24-gram sample of the ore is dissolved in hydrochloric acid and passed over a reducing agent so that all the iron is in the form $Fe^{2+}(aq)$. The $Fe^{2+}(aq)$ is completely oxidized by 31.6 mL of a 0.0512-M aqueous solution of $KMnO_4(aq)$. The unbalanced equation for the reaction is

$$KMnO_4(aq) + HCl(aq) + FeCl_2(aq) \rightarrow MnCl_2(aq) + FeCl_3(aq) + H_2O(l) + KCl(aq)$$

Calculate the amount of iron in the sample and its mass percentage in the ore.

24-45. A rock sample is to be assayed for its tin content by an oxidation-reduction titration with $I_3^-(aq)$. A 10.0-gram sample of the rock is crushed, dissolved in sulfuric acid, and passed over a reducing agent so that all the tin is in the form $Sn^{2+}(aq)$. The $Sn^{2+}(aq)$ is completely oxidized by 34.6 mL of a 0.556-M solution of $NaI_3(aq)$. The unbalanced equation for the reaction is

$$I_3^-(aq) + Sn^{2+}(aq) \rightarrow Sn^{4+}(aq) + I^-(aq)$$

Calculate the amount of tin in the sample and its mass percentage in the rock.

24-46. Sodium chlorite, $NaClO_2(s)$, is a powerful but stable oxidizing agent used in the paper industry, especially for the final whitening of paper. Sodium chlorite is capable of bleaching materials containing cellulose without oxidizing the cellulose. Sodium chlorite is made by the reaction described by

$$NaOH(aq) + Ca(OH)_2(s) + C(s) + ClO_2(g) \rightarrow NaClO_2(aq) + CaCO_3(s)$$

Balance this equation and calculate the number of kilograms of $ClO_2(g)$ required to make 1.00 metric ton of $NaClO_2$.

24-47. Solid phosphorus, $P_4(s)$, reacts with $BaSO_4(s)$ under oxygen-free, anhydrous conditions to produce $P_4O_{10}(s)$ and $BaS(s)$; write a balanced equation for this process. How much phosphorus is required to react completely with 2.16 grams of $BaSO_4(s)$?

24-48. A solution of $I_3^-(aq)$ can be standardized by using it to titrate $As_4O_6(aq)$. The titration of 0.1021 grams of $As_4O_6(s)$ dissolved in 30.00 mL of water requires 36.55 mL of $I_3^-(aq)$. Calculate the molarity of the $I_3^-(aq)$ solution. The (unbalanced) equation is

$$As_4O_6(s) + I_3^-(aq) \rightarrow As_4O_{10}(s) + I^-(aq)$$

ADDITIONAL PROBLEMS

24-49. What is an oxidation-reduction reaction?

24-50. Explain the difference between a species that is oxidized and an oxidizing agent.

24-51. Correct the mistake(s) in the following definition: "An oxidation-reduction reaction is a reaction in which the charge of one or more species in the reaction changes."

24-52. Could sodium metal be used as a sacrificial anode to protect a ship's propeller made predominantly out of iron?

24-53. Nitrogen atoms in molecules generally have oxidation states ranging from −3 to +5. Why is this the case?

24-54. Assign oxidation states to the nitrogen atoms in each of the following species:

(a) $NH_3(g)$ (b) $N_2(g)$ (c) $N_2H_4(g)$
(d) $N_2O(g)$ (e) $NO(g)$

24-55. Assign oxidation states to the nitrogen atoms in each of the following species:

(a) $NO_2^-(aq)$ (b) $NO_3^-(aq)$
(c) $NH_2OH(s)$ (d) $N_2O_4(g)$

24-56. Assign oxidation states to the atoms in the following species. In cases where none of the rules for assigning oxidation states apply, assign the states by analogy with similar elements in the periodic table.

(a) $MoSe_2(s)$ (b) $SiS_2(s)$
(c) $GaAs(s)$ (d) $K_2S_2O_3(s)$

24-57. Aqueous solutions of potassium permanganate decompose according to

$$MnO_4^-(aq) \rightarrow MnO_2(s) + O_2(g)$$

Balance the equation for this reaction under basic conditions.

24-58. Iodate, $IO_3^-(aq)$, can be used to titrate $Tl^+(aq)$ in a concentrated solution of $HCl(aq)$. Balance the equation

$$IO_3^-(aq) + Tl^+(aq) + Cl^-(aq) \rightarrow ICl_2^-(aq) + Tl^{3+}(aq)$$

24-59. Balance the following equations, which represent oxidation of $I^-(aq)$ to $I_3^-(aq)$:

(a) $Cr_2O_7^{2-}(aq) + I^-(aq) \rightarrow Cr^{3+}(aq) + I_3^-(aq)$ (acidic)
(b) $IO_4^-(aq) + I^-(aq) \rightarrow I_3^-(aq)$ (acidic)

24-60. Peroxydisulfate, $S_2O_8^{2-}(aq)$, is a strong oxidizing agent that can oxidize $Mn^{2+}(aq)$ to $MnO_4^-(aq)$; $Cr^{3+}(aq)$ to $Cr_2O_7^{2-}(aq)$; and $V^{4+}(aq)$ to $V^{5+}(aq)$. Excess reagent can be destroyed by boiling the solution after oxidation is complete. Balance the equation

$$S_2O_8^{2-}(aq) \rightarrow SO_4^{2-}(aq) + O_2(g) \quad \text{(acidic)}$$

24-61. Silver(II) oxide dissolves in concentrated inorganic acids to produce $Ag^{2+}(aq)$, which is a powerful oxidizing agent. Excess $Ag^{2+}(aq)$ can be removed by boiling the solution. Balance the equation

$$Ag^{2+}(aq) \rightarrow Ag^+(aq) + O_2(g) \quad \text{(acidic)}$$

24-62. Potassium permanganate solutions can be standardized by titration with sodium oxalate, $Na_2C_2O_4(aq)$. Balance the equation

$$KMnO_4(aq) + Na_2C_2O_4(aq) \rightarrow Mn^{2+}(aq) + CO_2(g) \quad \text{(acidic)}$$

24-63. Chromium(II), one of the most commonly used reducing agents, can be prepared by reducing $K_2Cr_2O_7(aq)$ with $H_2O_2(aq)$, followed by reduction of $Cr^{3+}(aq)$ to $Cr^{2+}(aq)$ by zinc. Balance the equations

(a) $Cr_2O_7^{2-}(aq) + H_2O_2(aq) \rightarrow Cr^{3+}(aq) + O_2(g)$ (acidic)

(b) $Cr^{3+}(aq) + Zn(s) \rightarrow Cr^{2+}(aq) + Zn^{2+}(aq)$ (acidic)

24-64. A 32.15-mL sample of a solution of $MoO_4^{2-}(aq)$ was passed through a Jones reductor (a column of zinc powder) in order to convert all the $MoO_4^{2-}(aq)$ to $Mo^{3+}(aq)$. The filtrate required 20.85 mL of 0.0955 M $KMnO_4(aq)$ for the reaction given by

$$MnO_4^-(aq) + Mo^{3+}(aq) \rightarrow Mn^{2+}(aq) + MoO_2^{2+}(aq)$$

Balance this equation in acidic solution and then calculate the concentration of the original $MoO_4^{2-}(aq)$ solution.

24-65. Hydrogen peroxide can act as either an oxidizing agent or a reducing agent, depending on the species present in solution. Write balanced half-reaction equations for each of the following:

(a) $H_2O_2(aq)$ acting as an oxidizing agent in an acidic solution

(b) $H_2O_2(aq)$ acting as a reducing agent in an acidic solution

A disproportionation reaction is one in which a single species oxidizes and reduces itself. Write a balanced equation for the disproportionation reaction of $H_2O_2(aq)$.

24-66. A 3.651-gram sample of a lanthanum sulfate ore is dissolved in nitric acid, and the lanthanum precipitated as $La(IO_3)_3(s)$ by the addition of 40.00 mL of 0.1105 M $KIO_3(aq)$. When an excess of $KI(aq)$ is added to the acidified filtrate, $I_2(s)$ results from the reaction of the remaining $IO_3^-(aq)$, as given by

$$IO_3^-(aq) + 5\,I^-(aq) + 6\,H^+(aq) \rightarrow 3\,I_2(s) + 3\,H_2O(l)$$

If it requires 12.65 mL of 0.0650 M $Na_2S_2O_3(aq)$ to react with this $I_2(s)$, according to

$$I_2(s) + 2\,Na_2S_2O_3(aq) \rightarrow Na_2S_4O_6(aq) + 2\,NaI(aq)$$

calculate the mass percentage of $La_2(SO_4)_3(s)$ in the sample.

24-67. We described in Example 24-9 the procedure to find the amount of ozone, $O_3(g)$, in an air sample by reaction with $I^-(aq)$ followed by titration with $S_2O_3(aq)$. An 8.65-gram sample of a mixture of oxygen-ozone was reacted with 50.0 milliliters of $KI(aq)$. Given that 22.50 mL of 0.0100 M $Na_2S_2O_3(aq)$ are required to titrate the $I_3^-(aq)$ produced, calculate the mass percentage of $O_3(g)$ in the sample.

24-68. The quantity of atmospheric $SO_2(g)$ can be determined by reaction with $H_2O_2(aq)$, according to

$$H_2O_2(aq) + SO_2(g) \rightarrow H_2SO_4(aq)$$

The amount of sulfuric acid produced can be determined by titration with sodium hydroxide of known concentration. An 812.1-gram sample of an air that is known to contain sulfur dioxide was reacted with 50.0 mL of hydrogen peroxide. Given that 18.50 mL of a 0.00250 M $NaOH(aq)$ were required to neutralize the $H_2SO_4(aq)$ produced in the 50.0-mL solution, calculate the mass percentage of $SO_2(g)$ in the air sample.

24-69. Iodine pentoxide is a reagent for the quantitative determination of carbon monoxide. The equation for the reaction is

$$5\,CO(g) + I_2O_5(s) \rightarrow I_2(s) + 5\,CO_2(g)$$

The iodine produced is dissolved in $KI(aq)$ to form $I_3^-(aq)$ and then determined by reaction with $Na_2S_2O_3(aq)$ according to

$$2\,S_2O_3^{2-}(aq) + I_3^-(aq) \rightarrow 3\,I^-(aq) + S_4O_6^{2-}(aq)$$

Calculate the mass percentage of $CO(g)$ in a 56.04-gram sample of air if it requires 10.0 mL of 0.0350 M $Na_2S_2O_3(aq)$ to react completely with the $I_3^-(aq)$ produced.

24-70. The concentration of $H_2S(g)$ in air can be determined by the following method. A sample of air is passed through a $Cd^{2+}(aq)$ solution, where the sulfur is precipitated as $CdS(s)$. This precipitate is then treated with an excess of $I_2(aq)$, which oxidizes the sulfide ion to elemental sulfur, $S(s)$. The amount of excess $I_2(aq)$ is determined by titration with $Na_2S_2O_3(aq)$ as described in the previous problem. Suppose that a 10.75-gram sample of air is passed through a $Cd^{2+}(aq)$ solution. A 30.00-mL sample of 0.0115 M $I_2(aq)$ is then added to the $CdS(s)$. The unreacted $I_2(aq)$ required 7.65 mL of 0.0750 M $Na_2S_2O_3(aq)$ for reduction to $I^-(aq)$. Calculate the mass percentage of $H_2S(s)$ in the air.

24-71. (*) Use the method of half reactions to balance the following equations:

(a) $CrI_3(s) + Cl_2(g) \rightarrow CrO_4^{2-}(aq) + IO_4^-(aq) + Cl^-(aq)$ (basic)

(b) $C_2H_5OH(aq) + I_3^-(aq) \rightarrow CO_2(g) + CHO_2^-(aq) + CHI_3(aq) + I^-(aq)$ (acidic)

24-72. (*) You are given an unknown salt that contains a mixture of $FeCl_2(s)$ and $FeCl_3(s)$. Describe a procedure for determining the mass percentage of each compound in the salt.

Michael Faraday (1791–1867) was born in London, England, into a rather poor family. He received little formal education and at age 14 was apprenticed to a bookbinder and bookseller. He educated himself by reading books from his workplace, which led to his interest in science. In 1812, he began attending public lectures on various scientific topics by Humphry Davy, a prominent chemist. He presented a 300-page book that he had written based on the lectures to Davy, who was so impressed that he soon appointed Faraday as his assistant at the Royal Institution. Davy seems to have become increasingly jealous of Faraday's successful experiments and had him work on a long series of useless experiments. Upon Davy's death in 1829, Faraday was able to return to his early successful work on electricity and magnetism. He received a lifetime appointment at the Royal Institution as the Fullerian Professor of Chemistry. His most important work in chemistry are his laws of electrolysis, but his greatest work was in the physics of electricity and magnetism. He is considered one of the greatest experimentalists of all time.

Faraday was a lay preacher in the obscure Sandemanian sect, an offshoot of the Scottish Presbyterian Church, and met his wife through the sect. He was a very religious man, and due to his religious beliefs, he refused knighthood and the presidency of the Royal Society. He preferred burial in a Sandemanian plot and refused burial at Westminster Abbey. He was a hugely popular lecturer on science and gave an annual Christmas reading for children of his book *The Chemical History of a Candle*. He received many honors during his life, and in the 1990s his picture was on the British 20-pound banknote, a rare honor for a scientist.

As a result of his being self-taught and not having had a formal education, Faraday was not well trained in mathematics, something that he regretted his entire life. In spite of his receiving countless honors and awards and enjoying international fame as an experimentalist, he said in his later life, "If I could live my life over again, I would study mathematics; it is a great mistake not to do so, but it is too late now."

25. Electrochemistry

In this chapter, we shall study electrochemistry. Electrochemical principles explain the operation of the batteries that power our cell phones and handheld calculators. They allow physicians to guide the rhythm of a heartbeat and enable our brains and senses to function. Chemists rely on electrochemical sensors to measure pH, to perform potentiometric titrations, and to monitor a wide variety of ionic species and enzymes, even in the presence of numerous other species. Electrochemistry is also used to combat corrosion and to produce many important industrial chemicals through electrolysis.

In Chapter 23 we used the Gibbs energy to predict whether a reaction is spontaneous. We showed that if the Gibbs energy change is greater than zero, then energy must be supplied to drive the reaction. For an oxidation-reduction reaction, that energy can be provided by an electric current. Conversely, when an oxidation-reduction reaction is spontaneous, we can use it to obtain electricity directly from chemicals, without the need for heat engines driven by combustion reactions. These two processes are the subject of this chapter on electrochemistry.

In this chapter we show how to calculate the current needed to produce a given amount of reaction product. We also show how to predict the voltage produced in an electrochemical reaction and the work that can be obtained. We shall apply these results to a variety of commercial processes.

25-1. Chemical Reactions Can Occur as a Result of the Passage of an Electric Current Through a Solution

The science of electrochemistry began at the end of the 18th century when Luigi Galvani, an Italian scientist, showed that the contraction of a frog's leg produces an electric current. While other scientists of that time were invoking mysterious forces to explain Galvani's "animal electricity," his countryman, Alessandro Volta, was using experiments to show that the phenomenon could be explained in purely physical and chemical terms. Volta constructed a device

You might try your own hand at making a voltaic pile by alternating pieces of dissimilar metals found in everyday objects, such as aluminum from a soda can, pennies, nickels, or other coins, separated by a paper towel damped with a salt solution.

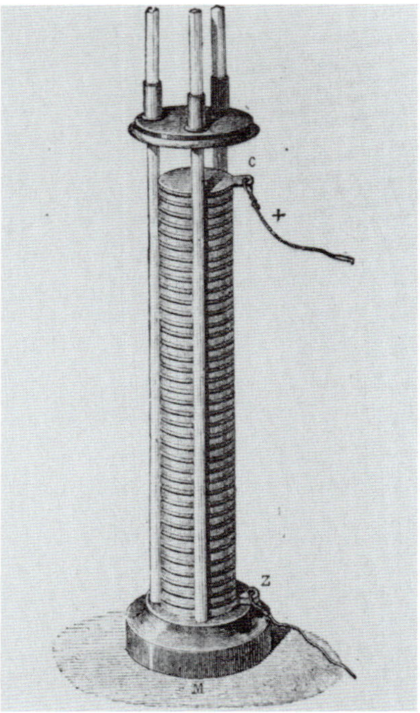

Figure 25.1 Diagram of Volta's apparatus for the generation of an electric current from a chemical reaction. This drawing is from Volta's publication of his work in the *Philosophical Transactions of the Royal Society* in 1800.

that produced an electric current of considerable power. In this **voltaic pile** (Figure 25.1), alternate disks of dissimilar metals, such as zinc and copper, are separated by damp cloths soaked in salt water. The electric current is produced by a chemical reaction between the substances in the pile. The reaction occurs when an electrical connection is made between the top and the bottom of the pile. The voltage developed by a voltaic pile depends on the number and nature of the pair of metal disks used to construct the pile.

To study the chemical reaction in a voltaic pile, we need a measure of how strongly the electric current that is produced is driven through the wire connecting the top and bottom of the pile. **Voltage**, named after Volta, is this driving force. Voltage plays a role in electrical systems analogous to force in mechanical systems. The SI unit of voltage is the **volt**, also named in honor of Volta, and denoted by V.

In mechanical systems, energy is equal to force times distance. In electrical systems, the electrical energy, U, is equal to the voltage, V, times charge, Z:

$$U = VZ \tag{25.1}$$

Recall that the SI unit of electrical charge is the **coulomb**, C. In terms of SI units, we see from Equation 25.1 that

$$1 \text{ joule} = 1 \text{ volt} \times 1 \text{ coulomb}$$

or that one volt is equivalent to one joule per coulomb or $1\text{ V} = 1\text{ J}\cdot\text{C}^{-1}$.

Electric currents are measured in **amperes**, A. One ampere is the flow of one coulomb, C, of charge per second, or $1\text{ A} = 1\text{ C}\cdot\text{s}^{-1}$.

The total charge that flows in a given time is equal to the current times the time. Representing charge (in coulombs) by Z, current (in amperes) by I, and time (in seconds) by t, we have

$$Z = It \tag{25.2}$$

Using SI units, we have coulombs = amperes × seconds or $1\text{ C} = 1\text{ A}\cdot\text{s}$.

VOLTS AND AMPERES

A good analogy for remembering the difference between voltage and amperage is that of balls rolling down a hill. In this analogy the height of the hill represents the electrical voltage. The balls in this analogy represent electrons, each with a uniform charge in coulombs. The current, measured in amperes, is proportional to the number of balls rolling down the hill per second.

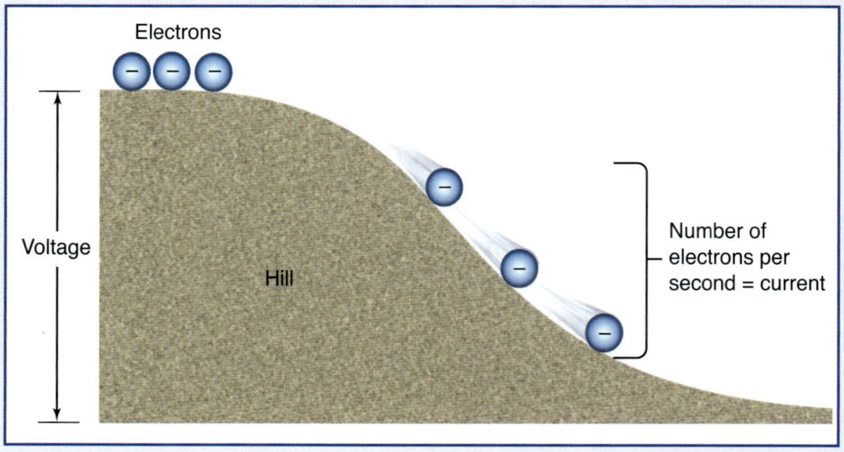

TABLE 25.1 Terminology and symbols used in electrical notation

Quantity	Symbol	Equation	SI Unit
charge	Z	—	coulomb (C)
voltage	V	$V = \frac{U}{Z}$	volt (V); $1\ \mathrm{V} = 1\ \mathrm{J \cdot C^{-1}}$
current	I	$I = \frac{Z}{t}$	ampere (A); $1\ \mathrm{A} = 1\ \mathrm{C \cdot s^{-1}}$

The symbols and units used in electrical notation are summarized in Table 25.1.

EXAMPLE 25-1: If a current of 1.5 amperes flows for 5.0 minutes, what quantity of charge has flowed?

Solution: The total charge is equal to the current multiplied by time, or $Z = It$. Thus,

$$Z = (1.5\ \mathrm{A})(5.0\ \mathrm{min}) = (1.5\ \mathrm{C \cdot s^{-1}})(5.0\ \mathrm{min})(60\ \mathrm{s \cdot min^{-1}}) = 450\ \mathrm{C}$$

PRACTICE PROBLEM 25-1: The charge of an electron has a magnitude of 1.602×10^{-19} C. Calculate the number of electrons that move through a cross section of a metal wire in 1.0 minute when the current is 0.50 amperes.

Answer: 1.9×10^{20} electrons

As soon as Volta introduced the voltaic pile to generate an electric current, chemists applied the apparatus to study chemical reactions. One of the first applications was to decompose water into hydrogen and oxygen by passing an electric current through water, as shown in Figure 25.2. One of the most

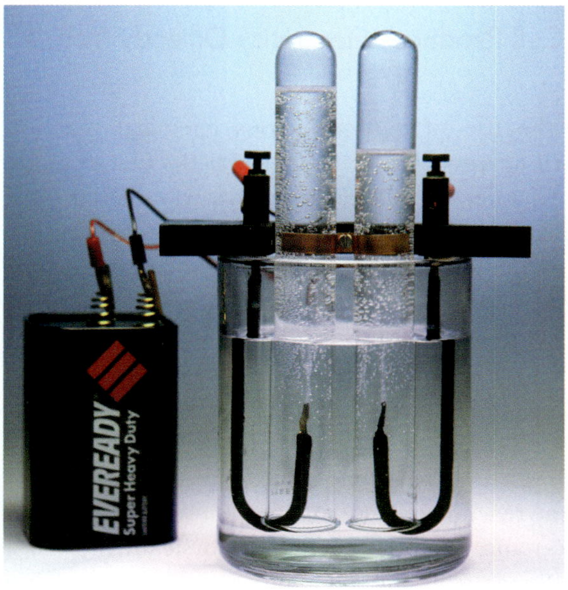

Figure 25.2 Preparation of hydrogen and oxygen by electrolysis of an aqueous solution. Hydrogen gas is liberated at the negative terminal, and oxygen gas is liberated at the positive terminal. As is required by the reaction stoichiometry, the volume of $H_2(g)$ liberated is twice as great as that of $O_2(g)$:

$$2H_2O(l) \xrightarrow{\text{electrolysis}} 2H_2(g) + O_2(g)$$

Figure 25.3 Humphry Davy (1778–1829) was one of the most famous chemists of the early 19th century. Although best known for having isolated pure sodium and potassium metals by the electrolysis of their molten salts, his earlier work was an investigation of the medical effects of inhaling various gases. He discovered the intoxicating effect of nitrous oxide, $N_2O(g)$, called "laughing gas," which led to a craze of nitrous oxide parties. Davy was knighted in 1812 and was created a baronet in 1818.

Figure 25.4 Humphry Davy's lectures at the Royal Institution were very popular and became a fashionable social event. In this famous caricature he is demonstrating the intoxicating effect of nitrous oxide.

spectacular applications was made by Sir Humphry Davy, who was professor of chemistry at the Royal Institution of London (Figures 25.3 and 25.4). Using a voltaic pile of over 250 plates, one of the most powerful then available, Davy passed an electric current through the molten salts of sodium chloride and potassium chloride and was able to isolate pure sodium and potassium metals. He then went on to isolate barium, magnesium, calcium, and strontium by similar methods. The highly reactive nature of these metals provided chemists with many new reactions to explore.

25-2. An Electrochemical Cell Produces Electricity Directly from a Chemical Reaction

In this section we consider the use of a spontaneous chemical reaction ($\Delta G_{rxn} < 0$) to produce an electric current. The voltaic pile was the first example of such a device. Recall from Chapter 24 that electrons are transferred from one substance to another in oxidation-reduction (electron-transfer) reactions. Figure 25.5 shows that when a zinc rod is immersed in an aqueous solution of copper sulfate, the oxidation-reduction reaction described by the following equation occurs spontaneously:

Here we use a single right-hand arrow to emphasize the fact that the reaction described by the equation proceeds spontaneously from left to right.

$$Zn(s) + Cu^{2+}(aq) \rightarrow Cu(s) + Zn^{2+}(aq)$$

It is possible to use oxidation-reduction reactions to produce an electric current. The basic idea is to keep the reactants [$Zn(s)$ and $Cu^{2+}(aq)$ in the equation above] and the products [$Cu(s)$ and $Zn^{2+}(aq)$] separated physically in such a way that the electrons are transferred from the reducing agent to the oxidizing agent after passing through an external circuit. A setup in which an electric

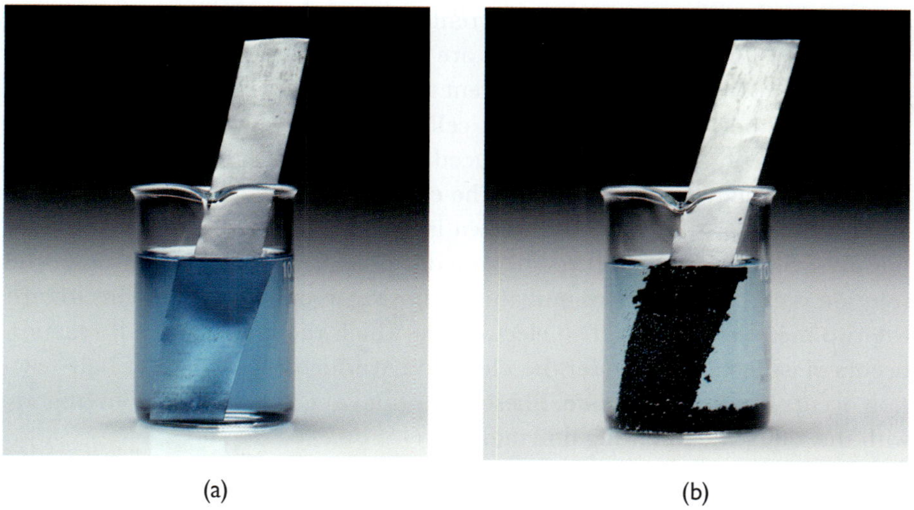

Figure 25.5 (a) When a zinc rod is placed in a copper sulfate solution, zinc replaces the copper in solution, and (b) elemental copper forms.

current is obtained from a chemical reaction is called an **electrochemical cell**. The electrochemical cell shown in Figure 25.6 has the remarkable property of being able to supply an electric current large enough to power a lightbulb or an electric motor. Such cells provided the power for the telegraph system in the United States for decades. This cell consists of two different metals, Zn(*s*) and Cu(*s*), each immersed in an electrolyte solution containing their respective metal ions, $Zn^{2+}(aq)$ and $Cu^{2+}(aq)$. The zinc and copper metals also act as the cell's electrodes. In general, an **electrode** is any solid on the surface of which oxidation-reduction reactions occur. The two solutions in the cell are connected electrically through a **salt bridge** that consists of a saturated KCl(*aq*) solution mixed with agar (a substance that forms a gel similar to gelatin). The purpose of the gel is to hold the salt solution in the tube, thus preventing the solutions from mixing while permitting the passage of current carried by ions. The salt

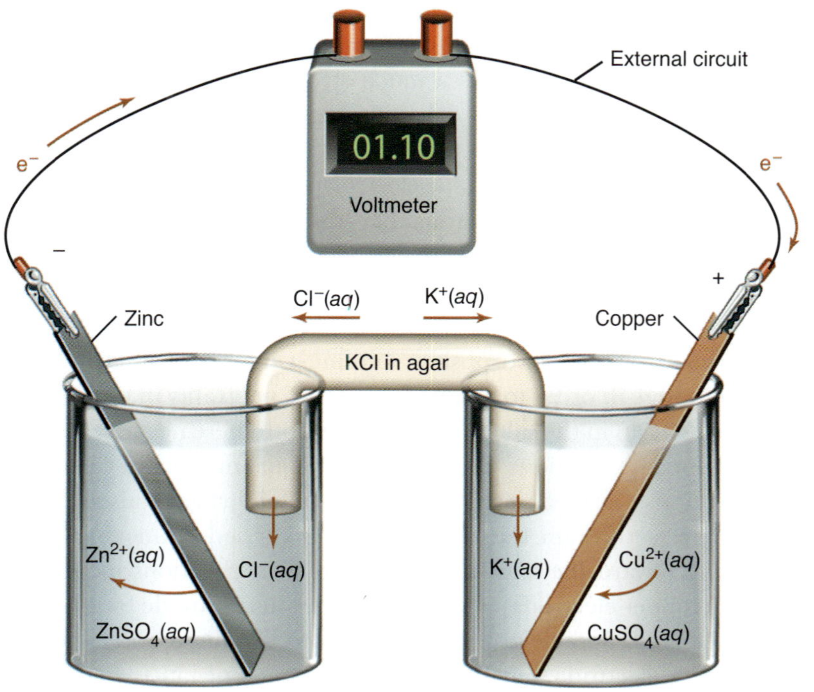

Figure 25.6 A zinc-copper electrochemical cell. When the cell operates, electrons flow through the external circuit from the zinc electrode to the copper electrode.

bridge therefore provides an ionic current path between the $ZnSO_4(aq)$ and the $CuSO_4(aq)$ solutions. The electrodes are attached to metal leads (wires) that enable the cell to deliver electric current to the **external circuit**.

Let's see how an electrochemical cell produces electricity directly from a chemical reaction. The voltage produced by the cell can be measured using a **voltmeter**, as shown in Figure 25.6. The current through the metal sections of an electrochemical cell like that shown in Figure 25.6 is carried by electrons, but electrons do not pass through aqueous phases. The current through the solution is carried by cations and anions moving in opposite directions toward the two metal electrodes. The electrode at which the reduction half reaction occurs is called the **cathode**; the electrode at which the oxidation half reaction occurs is called the **anode**. These definitions are remembered conveniently with the aid of the following mnemonic:

$$\text{consonants}\begin{cases}\textit{c}\text{athode}\\ \textit{r}\text{eduction}\end{cases}\qquad \text{vowels}\begin{cases}\textit{a}\text{node}\\ \textit{o}\text{xidation}\end{cases}$$

When current flows through an electrochemical cell, the cations in the solution move toward the cathode and the anions in the solution move toward the anode:

$$\textit{cat}\text{ions} \rightarrow \textit{cat}\text{hode} \qquad \textit{an}\text{ions} \rightarrow \textit{an}\text{ode}$$

When the cell shown in Figure 25.6 is used as a source of electricity, the reactions that occur at the electrodes are given by

$$Zn(s) \rightarrow Zn^{2+}(aq) + 2\,e^- \qquad \text{(oxidation of Zn at anode)}$$

$$Cu^{2+}(aq) + 2\,e^- \rightarrow Cu(s) \qquad \text{(reduction of } Cu^{2+} \text{ at cathode)}$$

The zinc-copper cell is based on the coupling of two half reactions. When the external circuit is completed by an electrical connection between the $Zn(s)$ and $Cu(s)$ electrodes, electrons can flow in the external circuit, that is, through the wire leads and the voltmeter (Figure 25.6). The electrons produced at the zinc electrode by the oxidation of $Zn(s)$ travel through the external circuit to the $Cu(s)$ electrode, where they are consumed in the reduction of $Cu^{2+}(aq)$ to $Cu(s)$. The fact that zinc metal is a much stronger reducing agent (electron donor) than copper metal is what drives the electron flow from the zinc to the copper in the external circuit.

As noted earlier, the current in the electrolyte solutions and in the salt bridge is carried by ions and not by electrons. Because $Zn^{2+}(aq)$ ions are produced in the solution containing the $Zn(s)$ electrode, negative ions must enter the $ZnSO_4(aq)$ solution from the salt bridge in order to maintain electrical neutrality in the solution. Because $Cu^{2+}(aq)$ ions are removed from the solution containing the $Cu(s)$ electrode, positive ions must enter the $CuSO_4(aq)$ solution from the salt bridge in order to maintain electrical neutrality in that solution. In this case, the $KCl(aq)$ in the salt bridge provides both these negative and positive ions to the respective solutions. By providing an equal number of $Cl^-(aq)$ anions and $K^+(aq)$ cations, the electrical neutrality of the salt bridge is also maintained. Thus, the current through the cell solutions is carried by moving ions, with anions moving toward the anode (zinc electrode) and cations moving toward the cathode (copper electrode). Oxidation occurs at the anode by the oxidation of $Zn(s)$ to $Zn^{2+}(aq)$ and electrons. The electrons produced enter the external circuit from this electrode, making the zinc electrode negative. Because reduction occurs at the cathode, electrons that arrive from the external circuit at the copper electrode are consumed in the reduction of $Cu^{2+}(aq)$ to $Cu(s)$, making the copper electrode positive. The total current through the cell electrolyte, which must equal the current in the external circuit, is carried by anions moving toward the zinc electrode, where $Zn^{2+}(aq)$ ions are produced, and by cations moving toward the copper electrode, where $Cu^{2+}(aq)$ ions are consumed.

The components of an electrochemical cell must be set up in such a manner that the transfer of electrons from the reducing agent to the oxidizing agent proceeds via the external circuit. If the two electrolyte solutions were simply mixed together, the cell would not produce a current because, as shown in Figure 25.5, $Cu^{2+}(aq)$ would be spontaneously deposited as copper metal on the zinc electrode. In other words, the cell would be internally short-circuited. For this reason, the reducing agent, $Zn(s)$, must be separated physically from the oxidizing agent, $Cu^{2+}(aq)$; the separation is achieved in this case with a salt bridge.

EXAMPLE 25-2: Consider the electrochemical cell shown in Figure 25.7. Indicate the current flow in the external circuit and in the cell electrolyte solutions.

Solution: From the figure we see that the $Pb(s)$ electrode is the negative electrode and the $Ag(s)$ electrode is the positive electrode. Therefore, electrons flow in the external circuit from the $Pb(s)$ electrode to the $Ag(s)$ electrode. The electrons are produced at the $Pb(s)$ electrode according to

$$Pb(s) \rightarrow Pb^{2+}(aq) + 2e^-$$

and are consumed at the $Ag(s)$ electrode according to

$$2Ag^+(aq) + 2e^- \rightarrow 2Ag(s)$$

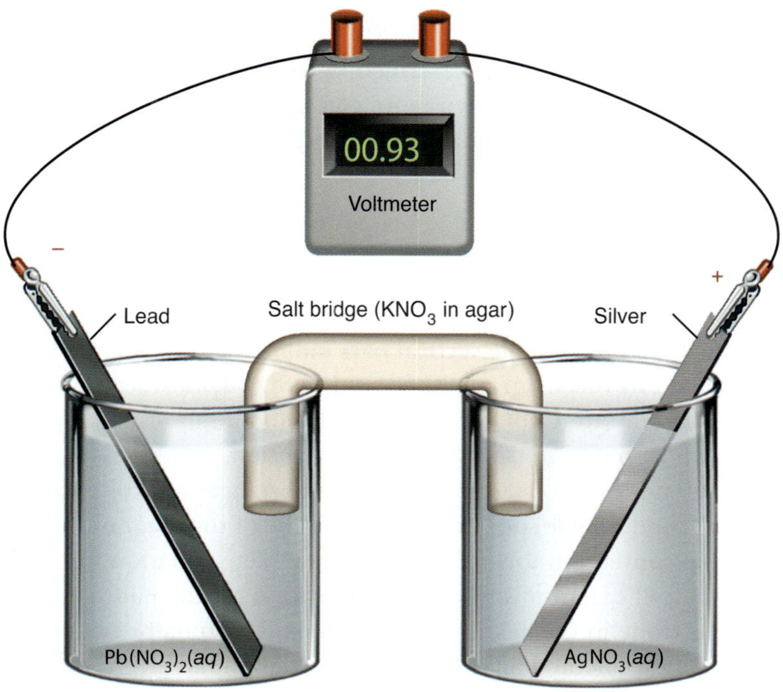

Figure 25.7
A lead-silver electrochemical cell.

Note that two silver ions are reduced for each lead atom oxidized in order to maintain electrical neutrality.

Positive $Pb^{2+}(aq)$ ions are produced at the lead electrode. For each $Pb^{2+}(aq)$ ion produced, two $NO_3^-(aq)$ ions flow from the salt bridge into the $Pb(NO_3)_2(aq)$ solution to maintain electrical neutrality. Two positive $Ag^+(aq)$ ions are consumed at the silver electrode for each $Pb^{2+}(aq)$ ion produced; thus, two $K^+(aq)$ ions flow simultaneously from the salt bridge into the $AgNO_3(aq)$ solution. This flow of ions maintains electrical neutrality both in the two cell solutions and in the salt bridge solution. The net ionic equation that describes the overall cell reaction is

$$Pb(s) + 2Ag^+(aq) \rightarrow Pb^{2+}(aq) + 2Ag(s)$$

If we simply place a lead rod in $AgNO_3(aq)$, then the spontaneous reduction of $Ag^+(aq)$ by $Pb(s)$ occurs directly on the surface of the lead, where silver metal crystals form. Separating the two half reactions in an electrochemical cell forces the reaction to proceed through the external circuit and thereby prevents the direct reaction of lead metal with silver ions.

PRACTICE PROBLEM 25-2: Using Example 25-2 as a guide, sketch a cell in which the reaction described by the following equation occurs:

$$3Cd(s) + 2In^{3+}(aq) \rightarrow 2In(s) + 3Cd^{2+}(aq)$$

Answer: Replace the lead electrode in Figure 25.7 by $Cd(s)$ and replace $Pb(NO_3)_2(aq)$ by $Cd(NO_3)_2(aq)$; replace $Ag(s)$ by $In(s)$ and replace $AgNO_3(aq)$ by $In(NO_3)_3(aq)$.

25-3. A Cell Diagram Is Used to Represent an Electrochemical Cell

Electrochemical cells are often described by means of a **cell diagram**, a kind of shorthand notation that is often more convenient than sketching the cell or writing out all the chemical equations. For example, the cell diagram for the cell shown in Figure 25.6 is

By convention the oxidation half reaction is written on the left-hand side of a cell diagram and the reduction half reaction on the right. This convention is easy to remember because *reduction* and *right* both begin with *r*.

$$Zn(s)|ZnSO_4(aq)||CuSO_4(aq)|Cu(s)$$

The single vertical bars in the cell diagram indicate boundaries of phases that are in contact, and the double vertical bars indicate a salt bridge. Thus, in the cell represented by this diagram, $Zn(s)$ and $ZnSO_4(aq)$ are separate phases in physical contact, as are $CuSO_4(aq)$ and $Cu(s)$, and a salt bridge separates the $ZnSO_4(aq)$ and $CuSO_4(aq)$ solutions. By convention, we write the metal electrodes at the ends of the diagram, insoluble substances and/or gases adjacent to the metals, and soluble species near the middle of the diagram.

The basic convention for obtaining the complete equation for a cell reaction from the cell diagram is to write the equation for the half reaction of the left-hand electrode in the cell diagram as an oxidation half-reaction equation and the equation for the half reaction of the right-hand electrode in the cell diagram as a reduction half-reaction equation. This convention enables us to write the equation for the complete cell reaction unambiguously. For the above cell diagram, then, we have, with oxidation taken to occur at the left-hand electrode,

$$Zn(s) \rightarrow Zn^{2+}(aq) + 2\,e^- \qquad \text{(oxidation at left-hand electrode)}$$

If oxidation takes place at the left-hand electrode, then reduction must take place at the right-hand electrode:

$$Cu^{2+}(aq) + 2\,e^- \rightarrow Cu(s) \qquad \text{(reduction at right-hand electrode)}$$

The equation for the net cell reaction is given by the sum of the equations for the two electrode half reactions:

$$Zn(s) + Cu^{2+}(aq) \rightarrow Cu(s) + Zn^{2+}(aq)$$

Consider the oxidation-reduction reaction described by the equation

$$Zn(s) + 2\,H^+(aq) \rightarrow H_2(g) + Zn^{2+}(aq)$$

Let's construct an electrochemical cell that uses this reaction. This example also illustrates how we deal with cell reactions that involve a gas. We first note that $Zn(s)$ is oxidized to $Zn^{2+}(aq)$; thus, the equation for the oxidation half reaction of the cell is

$$Zn(s) \rightarrow Zn^{2+}(aq) + 2\,e^- \qquad \text{(oxidation, left-hand electrode)}$$

We next note that $H^+(aq)$ is reduced to hydrogen gas; thus, the equation for the reduction half reaction of the cell is

$$2\,H^+(aq) + 2\,e^- \rightarrow H_2(g) \qquad \text{(reduction, right-hand electrode)}$$

This reduction half reaction requires a nonreactive metal electrode that can provide a pathway for the electrons coming through the external circuit from the zinc electrode to enter the cell compartment containing the $H^+(aq)$ solution. The reduction of the $H^+(aq)$ to $H_2(g)$ takes place on the surface of this metal electrode, which acts as a source of electrons. In this case, the necessary nonreactive electrode is provided by inserting a platinum coil or strip into the cell compartment containing $H^+(aq)$ and $H_2(g)$. Platinum is a relatively unreactive metal that simply provides a metallic surface on which the reduction half reaction occurs. Other relatively unreactive metals commonly used as electrodes are gold and carbon (graphite). An electrode involving a gaseous species and nonreactive metal is called a **gas electrode**. The cell diagram is

$$Zn(s)|Zn^{2+}(aq)||H^+(aq)|H_2(g)|Pt(s)$$

An electrochemical cell that incorporates the zinc oxidation and $H^+(aq)$ reduction half reactions is shown in Figure 25.8. Hydrogen gas is bubbled continuously through the right-hand compartment in order to provide $H_2(g)$ at a known pressure. Because the cell is open to the atmosphere, the pressure of $H_2(g)$ in the cell is essentially equal to the atmospheric pressure.

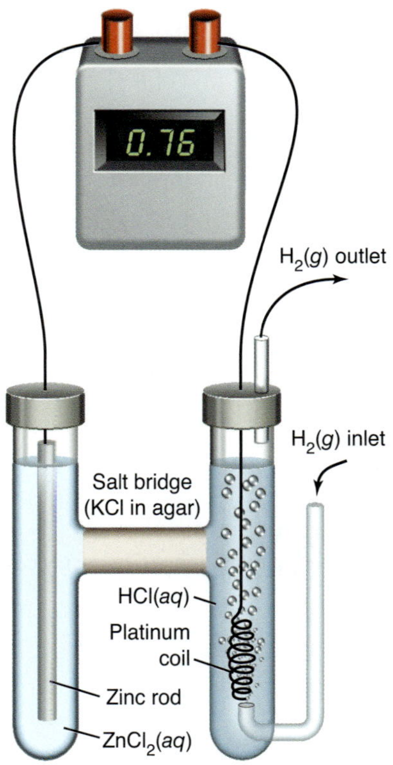

Figure 25.8 A zinc-hydrogen electrochemical cell. Note the H-type geometry of the cell, which holds the salt bridge and separates the two electrolyte solutions.

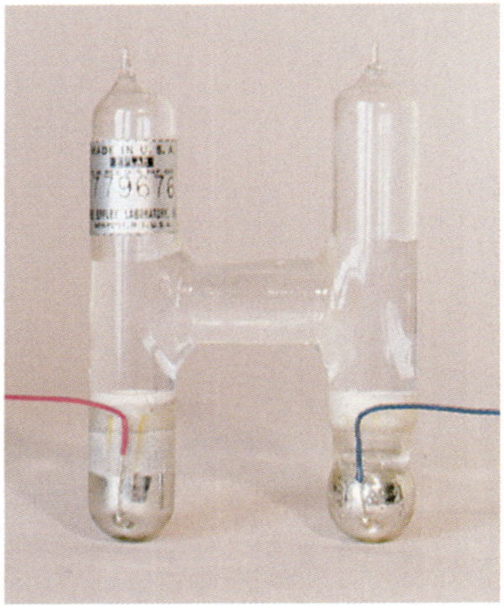

Figure 25.9 A Weston standard cell. The photograph shows the cell in the reverse of the conventional orientation, that is

$$Hg(l)|Hg_2SO_4(s)|CdSO_4(aq)|Cd(Hg)$$

with the anode on the right. The cell shown has two porcelain spacers with center holes that prevent the mixing of the cadmium amalgam, Cd(*Hg*), and the mercury. The grayish powder on the left is $Hg_2SO_4(s)$, which sits on top of the $Hg(l)$.

EXAMPLE 25-3: Write the equations for the electrode half reactions and the net cell reaction for the electrochemical cell

$$Cd(s)|CdSO_4(aq)|Hg_2SO_4(s)|Hg(l)$$

This cell does not have a salt bridge because it has only one electrolyte solution, $CdSO_4(aq)$ (all the other species are either liquids or solids).

Solution: By our convention, oxidation is assumed to take place at the left-hand electrode in the cell diagram. The oxidation of $Cd(s)$ yields $Cd^{2+}(aq)$, so we write

$$Cd(s) \rightarrow Cd^{2+}(aq) + 2e^- \qquad \text{(oxidation)}$$

Because oxidation occurs at the left-hand electrode, reduction must occur at the right-hand electrode. The only element besides cadmium that appears in two different oxidation states in the cell is mercury, which has an oxidation state of 0 in $Hg(l)$ and +1 in water-insoluble $Hg_2SO_4(s)$. For the reduction at the mercury electrode, therefore, we write

$$Hg_2SO_4(s) \rightarrow 2Hg(l) \qquad \text{(not balanced)}$$

The balanced equation for the electrode half reaction is

$$Hg_2SO_4(s) + 2e^- \rightarrow 2Hg(l) + SO_4^{2-}(aq) \qquad \text{(reduction)}$$

The sum of the equations for the oxidation and reduction half reactions gives the equation for the net cell reaction:

$$Cd(s) + Hg_2SO_4(s) \rightarrow 2Hg(l) + Cd^{2+}(aq) + SO_4^{2-}(aq)$$

The cell described here is similar to the Weston standard cell (Figure 25.9), which is widely used as a source of accurately known voltage. The voltage of the cell is 1.018 V and is essentially independent of temperature, a desirable feature for a voltage reference. As we shall show in Section 25-4, in most cases the voltage of a cell changes significantly with the temperature.

PRACTICE PROBLEM 25-3: Prior to the development of the more environmentally friendly silver-zinc (button) battery (see Interchapter U), mercury cells were used in devices such as hearing aids, photographic light meters, and watches where a current source with a stable voltage was required. A simplified cell diagram for the mercury cell is

$$Zn(s)|ZnO(s)|NaOH(aq)|HgO(s)|Hg(l)$$

Write the equations for the electrode half reactions and the net cell reaction.

Answer:

$$2OH^-(aq) + Zn(s) \rightarrow ZnO(s) + H_2O(l) + 2e^- \qquad \text{(oxidation)}$$
$$H_2O(l) + HgO(s) + 2e^- \rightarrow Hg(l) + 2OH^-(aq) \qquad \text{(reduction)}$$
$$Zn(s) + HgO(s) \rightarrow ZnO(s) + Hg(l) \qquad \text{(overall)}$$

WWW

See Interchapter U at www.McQuarrieGeneralChemistry.com.

25-4. The Nernst Equation Can Be Used to Determine the Cell Voltage for a Stated Reaction

Walther Nernst (Figure 25.10), who received a Nobel Prize in Chemistry in 1920, performed pioneering work in electrochemistry, investigating how the cell voltage depends on the concentration of the electrolytes, the size of the electrodes, and other factors. He found, for example, that the voltage of the cell,

$$Zn(s)|ZnSO_4(aq)||CuSO_4(aq)|Cu(s)$$

is independent of the size of the cell, the size of the electrodes, and the volume of the $ZnSO_4(aq)$ and $CuSO_4(aq)$ solutions. He also found that the voltage of the cell increases when the concentration of $Cu^{2+}(aq)$ is increased and decreases when the concentration of $Zn^{2+}(aq)$ is increased. The effect of a change in reactant or product concentration on cell voltage is easily understood in qualitative terms by applying Le Châtelier's principle to the cell reaction, given that the voltage of the cell is a measure of the driving force of the cell reaction.

The reaction of the zinc-copper cell (Figure 25.6) can be described by the equilibrium equation

$$Zn(s) + Cu^{2+}(aq) \leftrightharpoons Cu(s) + Zn^{2+}(aq)$$

An increase in the value of $[Cu^{2+}]$ increases the driving force of the reaction from left to right, thereby increasing the cell voltage; whereas an increase in the value of $[Zn^{2+}]$ increases the driving force of the reaction from right to left, thereby decreasing the cell voltage.

Figure 25.10 In addition to developing the Nernst equation from his studies of electrochemistry, German physicist Walther Nernst (1864–1941) also made contributions to the fields of thermodynamics, solid state chemistry, solution chemistry, and photochemistry. His "heat theorem," which later became known as the third law of thermodynamics, earned Nernst the 1920 Nobel Prize in Chemistry.

EXAMPLE 25-4: Consider an electrochemical cell in which the cell reaction described by the equation

$$H_2(g) + 2AgCl(s) \leftrightharpoons 2Ag(s) + 2H^+(aq) + 2Cl^-(aq)$$

takes place. In this reaction, $H_2(g)$ reduces $AgCl(s)$ to silver metal. Predict the effect of the following changes on the observed cell voltage:

(a) increase in the $H_2(g)$ pressure

(b) increase in the concentration of $H^+(aq)$

(c) increase in the amount of $AgCl(s)$

Solution: Using Le Châtelier's principle, we predict the following:

(a) An increase in P_{H_2} leads to an increase in the concentration of $H_2(g)$, and thereby increases the reaction driving force from left to right. The cell voltage thus increases.

(b) An increase in $[H^+]$ increases the reaction driving force from right to left and thus decreases the cell voltage.

(c) An increase in the amount of $AgCl(s)$ has no effect on the cell voltage because a change in the amount of a solid reactant has no effect on the driving force of a reaction and thus no effect on the voltage.

PRACTICE PROBLEM 25-4: Although the size of the electrodes and the volume of the electrolytes affect the physical dimensions of an electrochemical cell, these factors do not affect the voltage of the cell (compare 1.5 volt AAA and D-size batteries). However, they do affect the total amount of current that can be drawn from the cell. Why is this so?

Answer: The larger the cell (all other factors being equal), the more reactants available and thus the greater the amount of current that can be drawn from the cell reaction.

Strictly speaking, the terms voltage and electromotive force are not synonomous. The electromotive force of a cell is the measured voltage when the current drawn from the cell is extremely small (essentially zero).

Equation 25.1 says that charged particles gain or lose energy when subjected to a voltage difference. If the voltage difference is due to a chemical reaction that constitutes an electrochemical cell operating at a constant temperature and pressure, then the voltage difference is called the **electromotive force** (**emf**) of the cell, which we denote by E_{cell}. The corresponding energy change is equivalent to the Gibbs energy change of the reaction. Therefore, we can write Equation 25.1 in the notation

$$\Delta G_{\text{rxn}} = ZE_{\text{cell}} \tag{25.3}$$

Equation 25.3 is essentially Equation 25.1 written in a notation appropriate for the study of electrochemical cells. In this equation, Z represents the charge transferred in the cell reaction as written. For example, consider the $\text{Cd}(s)|\text{CdSO}_4(aq)|\text{Hg}_2\text{SO}_4(s)|\text{Hg}(l)$ cell presented in Example 25-3. The oxidation-reduction equation describing the cell reaction is

$$\text{Cd}(s) + \text{Hg}_2\text{SO}_4(s) \rightarrow 2\,\text{Hg}(l) + \text{Cd}^{2+}(aq) + \text{SO}_4^{2-}(aq) \tag{25.4}$$

and the two half reactions are

$$\text{Cd}(s) \rightarrow \text{Cd}^{2+}(aq) + 2\,\text{e}^- \qquad \text{(oxidation)}$$

$$\text{Hg}_2\text{SO}_4(s) + 2\,\text{e}^- \rightarrow 2\,\text{Hg}(l) + \text{SO}_4^{2-}(aq) \qquad \text{(reduction)}$$

Thus, when one mole of this reaction takes place, two moles of electrons are transferred and the quantity Z in Equation 25.3 is equal to the charge transferred by two moles of electrons. Because we will be dealing with the charges transferred by a number of moles of electrons, it is convenient to introduce a quantity called **Faraday's constant**. Faraday's constant, which is named after the English scientist Michael Faraday (Frontispiece), is defined as the charge on one mole (Avogadro's number) of protons. Thus, the value of Faraday's constant, which we denote by F, is equal to

$$\begin{aligned} F &= \text{(charge on a proton)(Avogadro's number)} \\ &= (1.60218 \times 10^{-19}\ \text{C})(6.02214 \times 10^{23}\ \text{mol}^{-1}) = 96\,485\ \text{C}\cdot\text{mol}^{-1} \end{aligned}$$

Because electrons are negatively charged, the charge transferred when one mole of the reaction described by Equation 25.4 takes place is $-2F = -192\,970\ \text{C}\cdot\text{mol}^{-1}$, where the unit mol^{-1} here refers to one mole of the reaction equation as written. We can generalize this result to any oxidation-reduction reaction that constitutes an electrochemical cell by writing Z as $-\nu_e F$, where ν_e is the (unitless) stoichiometric coefficient of the electrons in the two half reaction equations that make up the overall reaction equation. Using this notation, Equation 25.3 becomes

$$\Delta G_{\text{rxn}} = -\nu_e F E_{\text{cell}} \tag{25.5}$$

The units of ΔG_{rxn} are joules per mole (J·mol^{-1}) and the units on the right side of Equation 25.5 are $(\text{C·mol}^{-1})(\text{V}) = \text{C·V·mol}^{-1} = \text{J·mol}^{-1}$, as they must be.

In Chapter 23 we found that ΔG_{rxn} is related to the value of Q/K, the ratio of the thermodynamic reaction quotient to the thermodynamic equilibrium constant, by Equation 23.22,

$$\Delta G_{\text{rxn}} = RT \ln\left(\frac{Q}{K}\right)$$

Recall that the thermodynamic equilibrium constant and the thermodynamic reaction quotient are unitless quantities; $K = K_c/Q_c^\circ$ and $Q = Q_c/Q_c^\circ$, where Q_c° is what we called the standard reaction quotient in Section 23-8, its numerical value is unity, and its units cancel those of both K_c and Q_c.

Combining the above equation with Equation 25.5 gives us

$$-\nu_e F E_{\text{cell}} = RT \ln\left(\frac{Q}{K}\right)$$

Solving this equation for E_{cell}, we obtain

$$E_{\text{cell}} = -\left(\frac{RT}{\nu_e F}\right)\ln\left(\frac{Q}{K}\right) \tag{25.6}$$

Equation 25.6 is called the **Nernst equation**. The Nernst equation tells us how the cell voltage depends on the concentrations of the products and the reactants via the thermodynamic reaction quotient, Q, the thermodynamic equilibrium constant, K, and the kelvin temperature, T. As we have seen, the value of ν_e is that of the (unitless) stoichiometric coefficient of the number of electrons in the two half-reaction equations for the cell.

We often apply the Nernst equation at 25.0°C. Evaluating RT/F at 25.0°C, we have

$$\frac{RT}{F} = \frac{(8.3145\ \text{J·mol}^{-1}\text{·K}^{-1})(298.2\ \text{K})}{96\,485\ \text{C·mol}^{-1}} = 0.02570\ \text{J·C}^{-1} = 0.02570\ \text{V}$$

Thus, at 25.0°C the Nernst equation (Equation 25.6) becomes

$$E_{\text{cell}} = -\frac{0.02570\ \text{V}}{\nu_e}\ln\left(\frac{Q}{K}\right) \qquad \text{(at 25.0°C)} \tag{25.7}$$

Recall from Chapter 19 that, if $Q/K < 1$, then the reaction is spontaneous from left to right as written. In this case, $\ln(Q/K)$ is negative because it is the loga-

TABLE 25.2 Sign of various parameters and the direction of spontaneity for electrochemical reactions

Q/K	$\ln(Q/K)$	E_{cell}	ΔG_{rxn}	Spontaneity
$\frac{Q}{K} < 1$	$\ln\frac{Q}{K} < 0$	$E_{cell} > 0$	$\Delta G_{rxn} < 0$	cell reaction is spontaneous from left to right as written
$\frac{Q}{K} = 1$	$\ln\frac{Q}{K} = 0$	$E_{cell} = 0$	$\Delta G_{rxn} = 0$	cell reaction is at chemical equilibrium (no net driving force)
$\frac{Q}{K} > 1$	$\ln\frac{Q}{K} > 0$	$E_{cell} < 0$	$\Delta G_{rxn} > 0$	cell reaction is spontaneous from right to left as written

rithm of a number less than 1. Thus, if $Q/K < 1$, then Equation 25.7 shows that the cell voltage is positive and, from Equation 25.5, ΔG_{rxn} is negative. The other possible values of Q/K, the corresponding voltages, and ΔG_{rxn} values can be analyzed in the same way. We conclude that the sign of the cell voltage tells us the direction in which the cell reaction as written is spontaneous. A summary of the results for the various possibilities is given in Table 25.2.

The value of the cell voltage is a quantitative measure of the driving force of the cell reaction toward equilibrium. The larger the voltage, the greater is the reaction driving force. If the reaction is at chemical equilibrium, then the corresponding cell voltage is zero, and there is no driving force for the reaction; the cell is completely discharged.

Equation 25.6 can be written in a form that is often more convenient for calculations. If we write the relation $\ln(Q/K) = \ln Q - \ln K$ and combine this equation with Equation 25.6, then we obtain

$$E_{cell} = \left(\frac{RT}{\nu_e F}\right)\ln K - \left(\frac{RT}{\nu_e F}\right)\ln Q \qquad (25.8)$$

If all solution species involved in a reaction are at a concentration of exactly one molar and the pressure of all gaseous species involved are at exactly one bar, then $Q = 1$ for the cell equation. In this case all species are in their standard states, and the resulting cell voltage is called the **standard cell voltage**, denoted by E°_{cell}. Substituting $Q = 1$ into Equation 25.8 and noting that $\ln(1) = 0$, we obtain

$$E^\circ_{cell} = \left(\frac{RT}{\nu_e F}\right)\ln K \qquad (25.9)$$

Solving for $\ln K$ and using $RT/F = 0.02570$ V at 25.0°C gives

$$\ln K = \frac{\nu_e E^\circ_{cell}}{0.02570\ \text{V}} \qquad (\text{at } 25.0^\circ\text{C}) \qquad (25.10)$$

Equation 25.10 enables us to calculate the value of the thermodynamic equilibrium constant, K, for the equation of the cell reaction if the value of E°_{cell} is known, or to calculate the value of E°_{cell} if the value of K is known.

EXAMPLE 25-5: The value of E°_{cell} at 25.0°C for the equation

$$\text{Zn}(s) + \text{Cu}^{2+}(aq) \leftrightharpoons \text{Cu}(s) + \text{Zn}^{2+}(aq) \qquad (25.11)$$

is 1.10 V. Calculate the equilibrium constant at 25.0°C for this equation.

Solution: From Equation 25.10, we have

$$\ln K = \frac{\nu_e E^\circ_{\text{cell}}}{0.02570\ \text{V}}$$

For each mole of Zn(s) in the oxidation-reduction reaction described by Equation 25.11, two moles of electrons are transferred to a mole of $Cu^{2+}(aq)$. Thus, the value of ν_e is 2. Therefore,

$$\ln K = \frac{(2)(1.10\ \text{V})}{0.02570\ \text{V}} = 85.6$$

and

$$K = \frac{[\text{Zn}^{2+}]_{\text{eq}}}{[\text{Cu}^{2+}]_{\text{eq}}} = e^{85.6} = 1.50 \times 10^{37} \approx 2 \times 10^{37}$$

The very large value of K means that at equilibrium, the ratio of $[Zn^{2+}]$ to $[Cu^{2+}]$ is very large or, in other words, that the value of $[Cu^{2+}]$ at equilibrium is very small. (The subscript eq on a concentration term denotes a chemical equilibrium value of that concentration.)

PRACTICE PROBLEM 25-5: Suppose the electrical connections to the electrodes in the cell

$$\text{Zn}(s)|\text{ZnSO}_4(aq,\ 0.100\ \text{M})||\text{CuSO}_4(aq,\ 0.100\ \text{M})|\text{Cu}(s)$$

are connected and the cell is allowed to discharge completely at 25.0°C. What is the resulting voltage of the cell? What are the final concentrations of $Zn^{2+}(aq)$ and $Cu^{2+}(aq)$? (See Example 25-5 for any data necessary to carry out the calculations.)

Answer: $E_{\text{final}} = 0$ V; $[Zn^{2+}] = 0.200$ M and $[Cu^{2+}] = 1 \times 10^{-38}$ M ≈ 0 M

Substituting $E^\circ_{\text{cell}} = (RT/\nu_e F)\ln K$ from Equation 25.9 into Equation 25.8, we obtain an especially useful form of the Nernst equation:

$$E_{\text{cell}} = E^\circ_{\text{cell}} - \left(\frac{RT}{\nu_e F}\right)\ln Q \qquad (25.12)$$

Using $RT/F = 0.02570$ V at 25.0°C yields

$$E_{\text{cell}} = E^\circ_{\text{cell}} - \left(\frac{0.02570\ \text{V}}{\nu_e}\right)\ln Q \qquad (\text{at } 25.0^\circ\text{C}) \qquad (25.13)$$

where the units of E_{cell} and E°_{cell} are volts and all values are at 25.0°C. Equations 25.12 and 25.13 tell us that the cell voltage differs from the standard cell voltage when the reaction quotient, Q, is not equal to 1. If $Q < 1$, then $E_{cell} > E^{\circ}_{cell}$; if $Q > 1$, then $E_{cell} < E^{\circ}_{cell}$.

EXAMPLE 25-6: The measured voltage at 25.0°C of a cell in which the reaction described by the equation

$$\mathrm{Zn}(s) + \mathrm{Cu}^{2+}(aq,\ 1.00\ \mathrm{M}) \leftrightharpoons \mathrm{Cu}(s) + \mathrm{Zn}^{2+}(aq,\ 0.100\ \mathrm{M})$$

occurring at the concentrations given is 1.13 V. Calculate the value of E°_{cell} for the cell.

Solution: Using Equation 25.13 with $\nu_e = 2$, we obtain

$$E_{cell} = E^{\circ}_{cell} - \left(\frac{0.02570\ \mathrm{V}}{2}\right)\ln\frac{[\mathrm{Zn}^{2+}]/\mathrm{M}}{[\mathrm{Cu}^{2+}]/\mathrm{M}}$$

($\mathrm{Zn}(s)$ and $\mathrm{Cu}(s)$ do not appear in the Q expression because they are both solids.) Therefore,

$$1.13\ \mathrm{V} = E^{\circ}_{cell} - \left(\frac{0.02570\ \mathrm{V}}{2}\right)\ln\left(\frac{0.100}{1.00}\right)$$

from which we calculate

$$E^{\circ}_{cell} = 1.13\ \mathrm{V} + (0.01285\ \mathrm{V})\ln(0.100) = 1.10\ \mathrm{V}$$

Note that $E_{cell} > E^{\circ}_{cell}$ because $Q < 1$.

PRACTICE PROBLEM 25-6: Use the Nernst equation and the result of Example 25-6 to calculate the voltage of the zinc-copper cell at 25°C when $[\mathrm{Zn}^{2+}] = 1.00$ M and $[\mathrm{Cu}^{2+}] = 0.010$ M.

Answer: $E_{cell} = 1.04$ V

25-5. E°_{red} Values Can Be Assigned to Half-Reaction Equations

It is not possible to measure the voltage of a single electrode; only the difference in voltage between two electrodes can be measured. However, if we agree to choose a numerical value for the standard voltage of a particular electrode, then we can assign voltages to half-reaction equations.

In this section, we see how the voltages assigned to half reactions are used to calculate values of E°_{cell} for complete cell-reaction equations. Our convention is to set the voltage of the **standard hydrogen electrode** (abbreviated SHE) that consists of a nonreactive metal (such as platinum) in a one-molar $\mathrm{H}^{+}(aq)$ solution through which hydrogen gas is bubbled at a pressure of one bar equal to zero. That is, we assign the value $E^{\circ} = 0$ V (at all temperatures) to the electrode half reaction

$$2\,\mathrm{H}^{+}(aq,\ 1\ \mathrm{M}) + 2\,\mathrm{e}^{-} \rightarrow \mathrm{H}_2(g,\ 1\ \mathrm{bar}) \qquad E^{\circ}_{red} = 0\ \mathrm{V} \qquad \text{(by convention)} \qquad (25.14)$$

Equation 25.14 assigns a **standard reduction voltage**, denoted E°_{red}, to a half reaction written as a reduction reaction, that is, written with electrons on the left. A complete cell involves two half reactions: a reduction half reaction, with a standard reduction voltage, E°_{red}, and an oxidation half reaction, with a standard oxidation voltage, E°_{ox}. The voltage of a complete cell, which involves a reduction half reaction and an oxidation half reaction, is

$$E^\circ_{\text{cell}} = E^\circ_{\text{red}} + E^\circ_{\text{ox}} \tag{25.15}$$

Equation 25.15 is used together with Equation 25.14 to set up a table of standard reduction voltages. Because, for a particular half reaction, oxidation is the reverse of reduction, we have

$$E^\circ_{\text{ox}} = -E^\circ_{\text{red}} \qquad \text{(same half reaction)} \tag{25.16}$$

Thus, given that for the reduction half-reaction equation

$$\text{Cu}^{2+}(aq, 1\text{ M}) + 2\text{e}^- \rightarrow \text{Cu}(s) \qquad E^\circ_{\text{red}} = +0.34\text{ V}$$

we have for the corresponding oxidation half-reaction equation

$$\text{Cu}(s) \rightarrow \text{Cu}^{2+}(aq, 1\text{ M}) + 2\text{e}^- \qquad E^\circ_{\text{ox}} = -0.34\text{ V}$$

Note that if we double the coefficients in, say, the reduction half-reaction equation,

$$2\text{Cu}^{2+}(aq, 1\text{ M}) + 4\text{e}^- \rightarrow 2\text{Cu}(s) \qquad E^\circ_{\text{red}} = +0.34\text{ V}$$

the value of E°_{red} is unchanged because the voltage difference (or height of the hill in our analogy from the sidebox in Section 25-1) remains the same. Recall that current, not voltage, increases with cell size (see Practice Problem 25-4). Moreover, because E°_{cell} is a physically measurable quantity, the magnitude of E°_{cell}, and consequentially E°_{red} for a half-reaction equation, cannot depend on how we (arbitrarily) choose to write the equations that describe a cell.

To see how standard reduction voltages are obtained, consider the cell in Figure 25.8:

$$\text{Zn}(s)|\text{ZnCl}_2(aq)||\text{HCl}(aq)|\text{H}_2(g)|\text{Pt}(s)$$

The experimental value for the standard cell voltage at 25.0°C is

$$E^\circ_{\text{cell}} = E^\circ_{\text{red}} + E^\circ_{\text{ox}} = 0.76\text{ V}$$

The right-hand electrode (cathode) involves the pair $H^+|H_2$, and the left-hand electrode (anode) involves the pair $Zn|Zn^{2+}$. Thus,

$$E^\circ_{\text{cell}} = E^\circ_{\text{red}}[\text{H}^+|\text{H}_2] + E^\circ_{\text{ox}}[\text{Zn}|\text{Zn}^{2+}] = 0.76\text{ V} \tag{25.17}$$

where the brackets enclose the oxidized and reduced species for each electrode. We indicate the process as a reduction reaction in $E^\circ_{\text{red}}[\text{H}^+|\text{H}_2]$ and as an oxidation reaction in $E^\circ_{ox}[Zn|Zn^{2+}]$. Substitution of $E^\circ_{red}[H^+|H_2] = 0$ V into Equation 25.17 yields

$$E^\circ_{cell} = 0\text{ V} + E^\circ_{ox}[Zn|Zn^{2+}] = 0.76\text{ V}$$

or

$$E^\circ_{ox}[Zn|Zn^{2+}] = 0.76\text{ V}$$

Using Equation 25.16, we find that the standard reduction voltage for the corresponding reduction half reaction, $Zn^{2+}(aq, 1\text{ M}) + 2e^- \rightarrow Zn(s)$, at 25.0°C is

$$E^\circ_{red}[Zn^{2+}|Zn] = -E^\circ_{ox}[Zn^{2+}|Zn] = -0.76\text{ V}$$

A more complete listing of standard reduction voltages appears in Appendix G.

Once we know the standard reduction voltage of an electrode relative to the hydrogen electrode, we can use that value to determine the standard reduction voltage of other electrodes. Tables of E°_{red} values, such as Table 25.3, are used to calculate E°_{cell} values for cells. These E°_{cell} values can, in turn, be used to calculate equilibrium constants for chemical equations (Equation 25.10). The procedure for using standard reduction voltages to calculate a standard cell voltage is illustrated in the following Example.

EXAMPLE 25-7: The standard voltage of the cell

$$Zn(s)|ZnSO_4(aq)||CuSO_4(aq)|Cu(s)$$

at 25.0°C is $E^\circ_{cell} = 1.10$ V. Given that $E^\circ_{red} = -0.76$ V for the electrode half reaction described by

$$Zn^{2+}(aq, 1\text{ M}) + 2e^- \rightarrow Zn(s)$$

calculate E°_{red} at 25.0°C for the electrode half reaction given by

$$Cu^{2+}(aq, 1\text{ M}) + 2e^- \rightarrow Cu(s)$$

Solution: The oxidation takes place at the left-hand electrode, so the half reaction equations are

$$Zn(s) \rightarrow Zn^{2+}(aq) + 2e^- \qquad \text{(oxidation)}$$

$$Cu^{2+}(aq) + 2e^- \rightarrow Cu(s) \qquad \text{(reduction)}$$

Because

$$E^\circ_{ox}[Zn|Zn^{2+}] = -E^\circ_{red}[Zn|Zn^{2+}] = +0.76\text{ V}$$

we have

$$E^\circ_{cell} = E^\circ_{red}[Cu^{2+}|Cu] + E^\circ_{ox}[Zn|Zn^{2+}] = E^\circ_{red}[Cu^{2+}|Cu] + 0.76\text{ V} = 1.10\text{ V}$$

and so

$$E^\circ_{red}[Cu^{2+}|Cu] = 1.10\text{ V} - 0.76\text{ V} = +0.34\text{ V}$$

TABLE 25.3 Standard reduction voltages at 25.0°C for aqueous solutions (see also Appendix G)*

Electrode half reaction	E°_{red}/V
Acidic solutions	
$F_2(g) + 2e^- \rightarrow 2F^-(aq)$	+2.866
$O_3(g) + 2H^+(aq) + 2e^- \rightarrow O_2(g) + H_2O(l)$	+2.076
$Co^{3+}(aq) + e^- \rightarrow Co^{2+}(aq)$	+1.92
$Cl_2(g) + 2e^- \rightarrow 2Cl^-(aq)$	+1.358
$O_2(g) + 4H^+(aq) + 4e^- \rightarrow 2H_2O(l)$	+1.229
$Pt^{2+}(aq) + 2e^- \rightarrow Pt(s)$	+1.18
$NO_3^-(aq) + 4H^+(aq) + 3e^- \rightarrow NO(g) + 2H_2O(l)$	+0.957
$Ag^+(aq) + e^- \rightarrow Ag(s)$	+0.7996
$Cu^+(aq) + e^- \rightarrow Cu(s)$	+0.521
$Cu^{2+}(aq) + 2e^- \rightarrow Cu(s)$	+0.342
$Hg_2Cl_2(s) + 2e^- \rightarrow 2Hg(l) + 2Cl^-(aq)$	+0.268
$AgCl(s) + e^- \rightarrow Ag(s) + Cl^-(aq)$	+0.2223
$Cu^{2+}(aq) + e^- \rightarrow Cu^+(aq)$	+0.153
$2H^+(aq) + 2e^- \rightarrow H_2(g)$	+0.0
$Pb^{2+}(aq) + 2e^- \rightarrow Pb(s)$	−0.126
$V^{3+}(aq) + e^- \rightarrow V^{2+}(aq)$	−0.255
$Fe^{2+}(aq) + 2e^- \rightarrow Fe(s)$	−0.447
$Zn^{2+}(aq) + 2e^- \rightarrow Zn(s)$	−0.762
$Mn^{2+}(aq) + 2e^- \rightarrow Mn(s)$	−1.185
$Al^{3+}(aq) + 3e^- \rightarrow Al(s)$	−1.662
$H_2(g) + 2e^- \rightarrow 2H^-(aq)$	−2.23
$Mg^{2+}(aq) + 2e^- \rightarrow Mg(s)$	−2.372
$Na^+(aq) + e^- \rightarrow Na(s)$	−2.71
$Ca^{2+}(aq) + 2e^- \rightarrow Ca(s)$	−2.868
$K^+(aq) + e^- \rightarrow K(s)$	−2.931
$Li^+(aq) + e^- \rightarrow Li(s)$	−3.0401
Basic solutions	
$O_2(g) + 2H_2O(l) + 4e^- \rightarrow 4OH^-(aq)$	+0.401
$Cu(OH)_2(s) + 2e^- \rightarrow Cu(s) + 2OH^-(aq)$	−0.222
$Fe(OH)_3(s) + e^- \rightarrow Fe(OH)_2(s) + OH^-(aq)$	−0.56
$2H_2O(l) + 2e^- \rightarrow H_2(g) + 2OH^-(aq)$	−0.8277
$2SO_3^{2-}(aq) + 2H_2O(l) + 2e^- \rightarrow S_2O_4^{2-}(aq) + 4OH^-(aq)$	−1.12

(Left margin arrow, pointing up: increasing strength of oxidizing agents; right margin arrow, pointing down: increasing strength of reducing agents)

*Data from *CRC Handbook of Chemistry and Physics,* 87th ed., ed. David R. Lide, CRC Press, 2006–2007

Thus, the standard reduction voltage of the $Cu^{2+}(aq)|Cu(s)$ electrode is $E^{\circ}_{red} = +0.34$ V at 25.0°C, in agreement with the value listed in Table 25.3. Therefore, the standard reduction voltage of an electrode can be obtained from the standard cell voltage of a cell for which the standard reduction voltage of the other electrode is known.

PRACTICE PROBLEM 25-7: The standard voltage of the cell

$$Sn(s)|SnCl_2(aq)||AgNO_3(aq)|Ag(s)$$

at 25.0°C is $E^{\circ}_{cell} = 0.938$ V. Given that $E^{\circ}_{red} = 0.800$ V for the electrode half reaction occurring at the cathode described by

$$Ag^{+}(aq, 1\ M) + 1e^{-} \rightarrow Ag(s)$$

calculate the value of E°_{red} at 25.0°C for the electrode half reaction that occurs at the anode.

Answer: $E^{\circ}_{red} = -0.138$ V

The stronger a reducing agent a metal is, the greater its activity.

In the arrangement of E°_{red} values used in Table 25.3, more positive E°_{red} values for half-reaction equations indicate more powerful oxidizing agents, as indicated by the arrows alongside Table 25.3. Therefore, the more positive the value of E°_{red} for a half reaction, the stronger is the oxidizing agent (electron acceptor) in the half reaction. The more negative the value of E°_{red} for a half reaction, the stronger is the reducing agent (electron donor) in the half reaction. Thus, fluorine is the strongest oxidizing agent (most positive value of E°_{red}) and lithium is the strongest reducing agent (most negative value of E°_{red}) in Table 25.3. The arrangement of half reactions in order of the standard reduction voltages is the basis for the activity series of metals given in Section 10-7. Compare the order of the metals listed in Table 10.8 to those in Table 25.3 (some of the metals listed in Table 10.8 are not given in Table 25.3, but may be found in Appendix G).

Looking at Table 25.3 we see that platinum is a weaker reducing agent than any of the other metals listed in the table. Thus, platinum metal is not easily oxidized and so is less active than most metals. This and its high electrical conductivity is why platinum makes an excellent nonreactive metal electrode.

We can also use Table 25.3 to see why some metals react with acids to produce $H_2(g)$ and others do not under standard conditions at 25.0°C. Let's consider the reaction of $Zn(s)$ with $HCl(aq)$, which we know occurs readily. The equation for the reaction is

$$Zn(s) + 2\,HCl(aq) \rightarrow H_2(g) + ZnCl_2(aq)$$

and the equations for the two half reactions are

$$Zn(s) \rightarrow Zn^{2+}(aq) + 2\,e^{-} \qquad \text{(oxidation)}$$

and

$$2\,H^{+}(aq) + 2\,e^{-} \rightarrow H_2(g)^{-} \qquad \text{(reduction)}$$

The corresponding value of E°_{cell} is given by

$$E^\circ_{cell} = E^\circ_{red} + E^\circ_{ox} = E^\circ_{red}[H^+|H_2] + E^\circ_{ox}[Zn|Zn^{2+}] = 0\ V - E^\circ_{red}[Zn^{2+}|Zn] = +0.76\ V$$

Because $E^\circ_{cell} > 0$, the reaction is spontaneous and $Zn(s)$ will produce $H_2(g)$ from its reaction with $HCl(aq)$ under standard conditions at 25.0°C. The reason that $E^\circ_{cell} > 0$ is that $E^\circ_{red}[Zn^{2+}|Zn] < 0$. In general, metals for which the value of the standard reduction voltage is less than zero will react with acids to produce $H_2(g)$ under standard conditions. Conversely, metals for which this value is greater than zero will not.

Table 25.3 and Appendix G contain a tremendous amount of information on the chemistry of species in aqueous solutions. The E°_{red} values are used to predict many of the reactions that can occur between oxidizing and reducing agents.

EXAMPLE 25-8: Use the data in Table 25.3 to determine whether $Co^{3+}(aq)$, a fairly strong oxidizing agent, is capable of oxidizing water to $O_2(g)$ in acidic aqueous solution under standard conditions at 25.0°C, according to the equation

$$4\,Co^{3+}(aq) + 2\,H_2O(l) \rightarrow 4\,Co^{2+}(aq) + O_2(g) + 4\,H^+(aq)$$

Solution: The oxidation of $H_2O(l)$ to $O_2(g)$ by $Co^{3+}(aq)$ under standard conditions ($Q = 1$) will be a spontaneous process if $E^\circ_{cell} = E^\circ_{red} + E^\circ_{ox}$ is greater than zero. The two half reaction equations are

$$4\,Co^{3+}(aq) + 4\,e^- \rightarrow 4\,Co^{2+}(aq) \qquad E^\circ_{red} = +1.92\ V$$

$$2\,H_2O(l) \rightarrow O_2(g) + 4\,H^+(aq) + 4\,e^- \qquad E^\circ_{ox} = -E^\circ_{red} = -1.23\ V$$

The E°_{red} values for the two half reactions were obtained from Table 25.3. The value of E°_{cell} is

$$E^\circ_{cell} = E^\circ_{red}[Co^{3+}|Co^{2+}] + E^\circ_{ox}[H_2O|O_2] = (1.92\ V) + (-1.23\ V) = +0.69\ V$$

Again note that we do not multiply the value of $E^\circ_{red}[Co^{3+}|Co^{2+}]$ by 4 because the magnitude of a cell voltage or half-cell voltage is independent of the quantity of material involved or how we choose to (arbitrarily) write the equations that describe the cell reaction. The positive value of E°_{cell} means that $Co^{3+}(aq)$ is capable of oxidizing water at 25.0°C under standard conditions. The rate of oxidation of water by $Co^{3+}(aq)$ is fairly rapid at 25.0°C, and so $Co^{3+}(aq)$ does not persist at appreciable concentrations in water.

PRACTICE PROBLEM 25-8: The $V^{2+}(aq)$ ion is a moderately strong reducing agent, $V^{2+}(aq) \rightarrow V^{3+}(aq) + e^-$, in acidic aqueous solution. Is $V^{2+}(aq)$ capable of liberating $H_2(g)$ from an acidic aqueous solution under standard conditions at 25.0°C?

Answer: Yes. The value of E°_{cell} is +0.255 V, and so $V^{2+}(aq)$ is capable of liberating $H_2(g)$ from an acidic, aqueous solution at standard conditions ($Q = 1$) at 25.0°C.

EXAMPLE 25-9: Write out the cell equation for the cell

$$Fe(s)|Fe^{2+}(aq, 0.100\ M)||Fe^{2+}(aq, 2.00\ M)|Fe(s)$$

Calculate the voltage generated by this "concentration cell" at 25.0°C. Why do you think this cell is called a concentration cell?

Solution: The two half reactions and the corresponding values of E°_{red} from Table 25.3 are

$$Fe^{2+}(aq, 2.00\ M) + 2e^- \rightarrow Fe(s) \qquad E^\circ_{red} = -0.45\ V$$

$$Fe(s) \rightarrow Fe^{2+}(aq, 0.100\ M) + 2e^- \qquad E^\circ_{ox} = -E^\circ_{red} = +0.45\ V$$

Because the oxidation half reaction is the opposite of the reduction half reaction for this cell, $E^\circ_{ox} = -E^\circ_{red}$, and so

$$E^\circ_{cell} = E^\circ_{red} + E^\circ_{ox} = (-0.45\ V) + (0.45\ V) = 0\ V$$

Thus, the cell generates no voltage at standard conditions. However, the concentrations of the various species given are *not* at standard conditions and so we must apply the Nernst equation (Equation 25.13),

$$E_{cell} = E^\circ_{cell} - \left(\frac{0.02570\ V}{\nu_e}\right) \ln Q$$

$$= E^\circ_{cell} - \left(\frac{0.02570\ V}{2}\right) \ln \frac{[Fe^{2+}(aq, 0.100\ M)]/M}{[Fe^{2+}(aq, 2.00\ M)]/M}$$

$$= 0\ V - (0.01285\ V) \ln \left(\frac{0.100}{2.00}\right) = 0.03850\ V$$

The overall chemical equation for the concentration cell in Example 25-9 is

$$Fe^{2+}(aq, 2.00\ M) \rightarrow Fe^{2+}(aq, 0.100\ M)$$

and so the thermodynamic reaction quotient, Q, is

$$Q = \frac{[Fe^{2+}(aq, 0.100\ M)]/M}{[Fe^{2+}(aq, 2.00\ M)]/M}$$

Thus, the cell generates 38.50 mV of electricity at 25.0°C. This sort of cell is called a **concentration cell** because it uses the difference in concentrations between the two solutions to generate a voltage. This is the electrical equivalent of the osmotic pressure discussed in Section 16-5.

PRACTICE PROBLEM 25-9: Some biological cells afford an example of a concentration cell. These cell walls are much more permeable to $K^+(aq)$ ions than other ions, and the concentration of $K^+(aq)$ ions inside the cell is maintained to be about 20 times greater than that on the outside by a biochemical pumping mechanism. Use the Nernst equation (Equation 25.12) for a concentration cell to estimate the magnitude of the voltage difference across the cell wall when the cell is at 37°C.

Answer: 80 mV, in good agreement with the measured value.

25-6. Electrochemical Cells Can Be Used to Determine the Concentration of Ions

We have shown in Section 25-4 how the Nernst equation can be used to calculate the voltage of an electrochemical cell when the concentrations of the species involved in the cell reaction are known. Conversely, a measured cell voltage can be used to determine the concentration of a species in solution.

Consider the cell reaction described by

$$Zn(s) + Cu^{2+}(aq) \leftrightharpoons Cu(s) + Zn^{2+}(aq) \tag{25.18}$$

Application of the Nernst equation (Equation 25.13) to Equation 25.18 at 25.0°C yields

$$E_{cell} = E^{\circ}_{cell} - \left(\frac{0.02570\text{ V}}{2}\right)\ln\frac{[Zn^{2+}]/M}{[Cu^{2+}]/M}$$

We know from Example 25-7 that $E^{\circ}_{cell} = 1.10$ V at 25.0°C for Equation 25.18, so

$$E_{cell} = 1.10\text{ V} - (0.01285\text{ V})\ln\frac{[Zn^{2+}]/M}{[Cu^{2+}]/M} \tag{25.19}$$

If we measure E°_{cell} at a known value of, say, $[Cu^{2+}]$, then we can use Equation 25.19 to calculate the value of $[Zn^{2+}]$ in a solution containing an unknown concentration of $Zn^{2+}(aq)$. For example, suppose that when $[Cu^{2+}] = 0.10$ M, we find that $E_{cell} = 1.20$ V at 25.0°C. Substitution of these values for E_{cell} and $[Cu^{2+}]$ into Equation 25.19 yields

$$1.20\text{ V} = 1.10\text{ V} - (0.01285\text{ V})\ln\frac{[Zn^{2+}]/M}{0.10}$$

from which we calculate

$$\ln\frac{[Zn^{2+}]/M}{0.10} = \frac{1.10\text{ V} - 1.20\text{ V}}{0.01285\text{ V}} = -7.8$$

or

$$[Zn^{2+}] = (0.10\text{ M})(e^{-7.8}) = 4 \times 10^{-5}\text{ M}$$

Electrochemical cells are used extensively in biology and analytical chemistry to determine the concentrations of ions in solution. For example, microelectrodes can be inserted into biological cells, making possible the monitoring of ion concentrations across cell membranes or within individual cells. Electrochemical cells are also used routinely in electroanalytical devices such as pH meters.

Let's see how we can measure pH with an electrochemical cell. Consider the cell described by the cell diagram

$$Pt(s)|H_2(g, 1.00\text{ bar})|H^+(aq, \textit{unknown}\text{ M})||Cl^-(aq, 1.00\text{ M})|AgCl(s)|Ag(s)$$

The two half reactions associated with the cell diagram are described by

$$H_2(g) \rightarrow 2H^+(aq) + 2e^- \qquad \text{(oxidation)}$$

$$2AgCl(s) + 2e^- \rightarrow 2Ag(s) + 2Cl^-(aq) \qquad \text{(reduction)}$$

and the equation for the overall reaction is

$$H_2(g) + 2\,AgCl(s) \rightarrow 2\,H^+(aq) + 2\,Cl^-(aq) + 2\,Ag(s) \tag{25.20}$$

The standard cell voltage is given by (Table 25.3)

$$E^\circ_{cell} = E^\circ_{red}[AgCl|Ag] + E^\circ_{ox}[H_2|H^+] = (0.222\text{ V}) + (0\text{ V}) = +0.222\text{ V}$$

and the Nernst equation (Equation 25.13) in this case reads

$$E_{cell} = 0.222\text{ V} - \left(\frac{0.02570\text{ V}}{2}\right)\ln Q \tag{25.21}$$

The thermodynamic reaction quotient for Equation 25.20 is

$$Q = \frac{([H^+]/M)^2([Cl^-]/M)^2}{P_{H_2}/\text{bar}} = \frac{([H^+]/M)^2(1.00)^2}{1.00} = ([H^+]/M)^2$$

Substituting this result into Equation 25.21 gives

$$E_{cell} = 0.222\text{ V} - (0.02570\text{ V})\ln([H^+]/M) \tag{25.22}$$

where we have used the relation $\ln x^2 = 2 \ln x$.

We normally work with natural logarithms in the Nernst equation, but because we want to use Equation 25.22 to measure the pH, we shall express it in terms of base-10 logarithms as $pH = -\log([H^+]/M)$. This is easy to do using the relation $\ln x = 2.303 \log x$, and so Equation 25.22 becomes

A review of the properties of logarithms is given in Appendix A.

$$\begin{aligned} E_{cell} &= 0.222\text{ V} - (0.02570\text{ V})(2.303)\log([H^+]/M) \\ &= 0.222\text{ V} - (0.0592\text{ V})\log([H^+]/M) \\ &= 0.222\text{ V} + (0.0592\text{ V})\text{ pH} \end{aligned}$$

Solving this equation for pH gives

$$pH = \frac{E_{cell} - 0.222\text{ V}}{0.0592\text{ V}} \tag{25.23}$$

Thus, you can see that the pH is given directly by the measured cell voltage. For example, if the measured cell voltage is 0.340 V, then the pH is

$$pH = \frac{0.340\text{ V} - 0.222\text{ V}}{0.0592\text{ V}} = 1.99$$

EXAMPLE 25-10: Another commonly used electrode that is used to measure pH is described by the cell diagram $Cl^-(aq)|Hg_2Cl_2(s)|Hg(l)$ and the corresponding (reduction) half reaction

$$Hg_2Cl_2(s) + 2\,e^- \rightarrow 2\,Hg(l) + 2\,Cl^-(aq) \qquad E^\circ_{red} = 0.268\text{ V}$$

This electrode is called a **calomel electrode** because $Hg_2Cl_2(s)$ used to be called calomel. The cell diagram of the complete cell is

$$Pt(s)|H_2(g,\ 1.00\text{ bar})|H^+(aq,\ \textit{unknown}\text{ M})||Cl^-(aq,\ 1.00\text{ M})|Hg_2Cl_2(s)|Hg(l)$$

If the cell voltage is measured to be 0.432 V at 25.0°C, calculate the corresponding pH.

Solution: The equation for the overall cell reaction is

$$H_2(g) + Hg_2Cl_2(s) \rightarrow 2\,H^+(aq) + 2\,Cl^-(aq) + 2\,Hg(l)$$

and the standard cell voltage is given by

$$E^\circ_{cell} = E^\circ_{red}[Hg_2Cl_2|Hg] + E^\circ_{ox}[H_2|H^+] = +0.268\text{ V}$$

The thermodynamic reaction quotient for the above equation under the stated conditions is

$$Q = \frac{([H^+]/M)^2([Cl^-]/M)^2}{P_{H_2}/\text{bar}} = ([H^+]/M)^2$$

Substituting this into the Nernst equation (Equation 25.13) gives

$$E_{cell} = 0.268\text{ V} - (0.02570\text{ V})\ln([H^+]/M)$$

Again, using the conversion $\ln x = 2.303 \log x$, we have

$$E_{cell} = 0.268\text{ V} - (0.0592\text{ V})\log([H^+]/M)$$

or

$$\text{pH} = -\log([H^+]/M) = \frac{E_{cell} - 0.268\text{ V}}{0.0592\text{ V}} \qquad (25.24)$$

Given that $E_{cell} = 0.432$ V, we see that the pH is

$$\text{pH} = \frac{0.432\text{ V} - 0.268\text{ V}}{0.0592\text{ V}} = 2.77$$

PRACTICE PROBLEM 25-10: Some of the first pH meters used the concentration cell

$$\mathrm{Pt}(s)|\mathrm{H_2}(g,\ 1\ \mathrm{bar})|\mathrm{H^+}(aq,\ unknown\ \mathrm{M})||\mathrm{H^+}(aq,\ 1\ \mathrm{M})|\mathrm{H_2}(g,\ 1\ \mathrm{bar})|\mathrm{Pt}(s)$$

to measure the pH of an unknown solution. Derive a general expression for the pH of the unknown solution using this cell. Express your results in terms of R, T, F, and E_{cell}.

Answer: $$\mathrm{pH} = \left(\frac{F}{2.303\ RT}\right)E_{\mathrm{cell}} \qquad (25.25)$$

Example 25-10 and Practice Problem 25-10 show how cells can be used to measure pH. Both Equations 25.23 and 25.24 may be rewritten as

$$\mathrm{pH} = \frac{E_{\mathrm{cell}} - E^\circ_{\mathrm{ref}}}{0.0592\ \mathrm{V}} \qquad (\text{at } 25.0°\mathrm{C}) \qquad (25.26)$$

where E°_{ref} is the known standard half-reaction voltage of a reference cell at standard conditions, which is +0.222 V in Equation 25.23 and +0.268 V in Equation 25.24.

If we solve Equation 25.26 for E°_{cell}, then we obtain

$$E_{\mathrm{cell}} = E^\circ_{\mathrm{ref}} + (0.0592\ \mathrm{V})\mathrm{pH} = E^\circ_{\mathrm{ref}} - (0.0592\ \mathrm{V})\log([\mathrm{H^+}]/\mathrm{M})$$

We can see from this equation that a change in the hydrogen-ion concentration of a factor of 10 (one pH unit) changes the cell voltage by 0.0592 V = 59.2 mV, a result well known to people who frequently use pH meters in their work.

The standard hydrogen electrode, $\mathrm{Pt}(s)|\mathrm{H_2}(g)|\mathrm{H^+}(aq)$, is not suitable for routine measurements because, among other reasons, a hydrogen electrode requires a continuous supply of hydrogen gas and the exiting hydrogen gas poses an explosion hazard. Most modern pH meters now use special glass electrodes. One such meter uses a probe containing a silver–silver chloride electrode inside a bulb consisting of a thin glass membrane connected to a reference cell (often also contained within the probe). An abbreviated cell diagram for the complete cell is

$$\text{glass electrode} \mid \mathrm{H^+}(aq,\ unknown\ \mathrm{M}) \parallel \text{reference electrode}$$

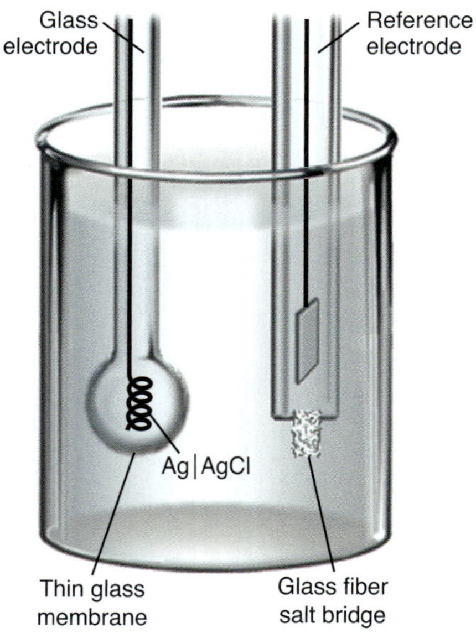

Figure 25.11 A schematic diagram of a hydrogen-ion–sensitive glass electrode and a reference electrode.

Figure 25.11 shows a schematic diagram of a glass electrode and accompanying reference electrode that can be used to measure pH. It is found experimentally that these glass electrodes respond to pH quantitatively the same as a standard hydrogen electrode. The development of glass electrodes has made it routine to measure pH to ±0.01 pH units with an easily portable device. Glass electrodes that respond to the concentrations of ions other than $\mathrm{H^+}(aq)$ have also been developed. These ion-selective electrodes have led to significant advances in the chemical, biochemical, medical, and environmental sciences.

25-7. The Electrical Energy Released from an Electrochemical Cell Can Do Useful Work

We have shown that the voltage of an electrochemical cell is a measure of the driving force of the cell reaction toward equilibrium. A spontaneous cell reaction creates a voltage difference between the electrodes that can be used to produce an electric current. Electrochemical cells therefore provide a mechanism for converting chemical energy into electrical energy. By calculating the Gibbs energy change for a reaction, we can determine how much electrical energy can be supplied to an external device, such as an electric motor. Equation 25.5, $\Delta G_{rxn} = -\nu_e F E_{cell}$, relates the value of ΔG_{rxn} directly to the cell voltage. From Chapter 23, we know that ΔG_{rxn} is equal to the maximum amount of work, here in the form of electrical energy, that can be obtained from the reaction.

One of the most important examples of an electrochemical cell is a fuel cell. A **fuel cell** is an electrochemical cell in which the substance oxidized is capable of being used as a fuel in a heat engine, that is, capable of being burned. A fuel cell, then, is an electrochemical cell that carries out the oxidation of a fuel. The reaction of the fuel with the oxidizer is the same as that for the combustion of the fuel, but the cell generates electricity directly, without a flame. The reactants are continuously fed into the system so that it can derive power without interruption. The simplest type of fuel cell uses hydrogen and oxygen as fuel. Fuel cells are used as a source of electrical power in the Space Shuttle (Figure 25.12) and in some newer electric-powered buses and automobiles (Figure 25.13). Fuel cells are also being developed to power personal electronic devices and are used as a method of storing electrical energy in some solar-powered homes. The development of fuel cells has received a great deal of attention in recent years, especially in view of our present energy shortfalls.

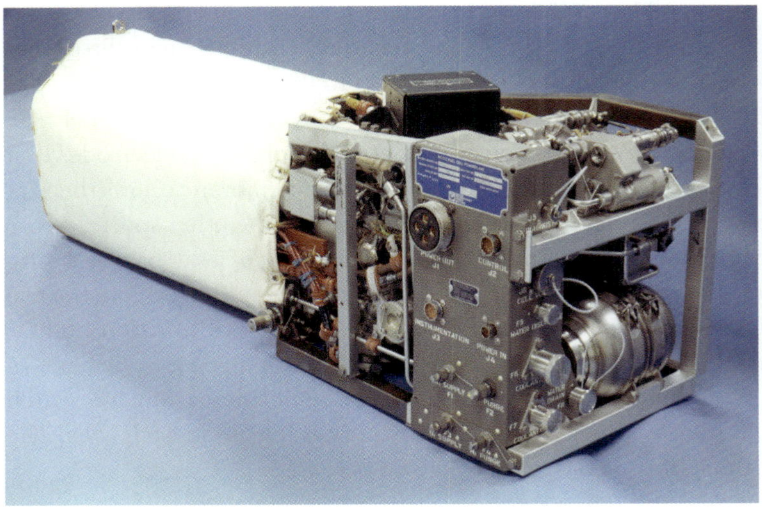

Figure 25.12 The power plant used in the Space Shuttle. Each Space Shuttle is equipped with three hydrogen-oxygen fuel cell power plants. Each power plant is capable of supplying 12 kilowatts at peak output and 7 kilowatts average power to the spacecraft's electrical system.

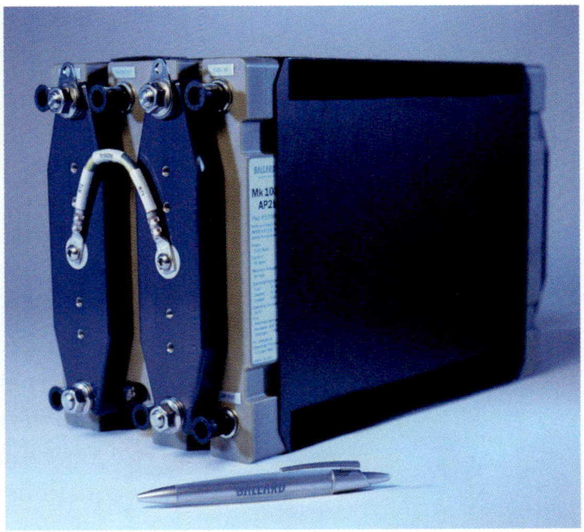

Figure 25.13 This fuel cell stack produced by Ballard Power Systems in British Columbia can deliver up to 1030 watts of electrical power from hydrogen gas. About 15 of these stacks can generate enough energy to power a medium-sized four-passenger electric car with only pure water as a waste product.

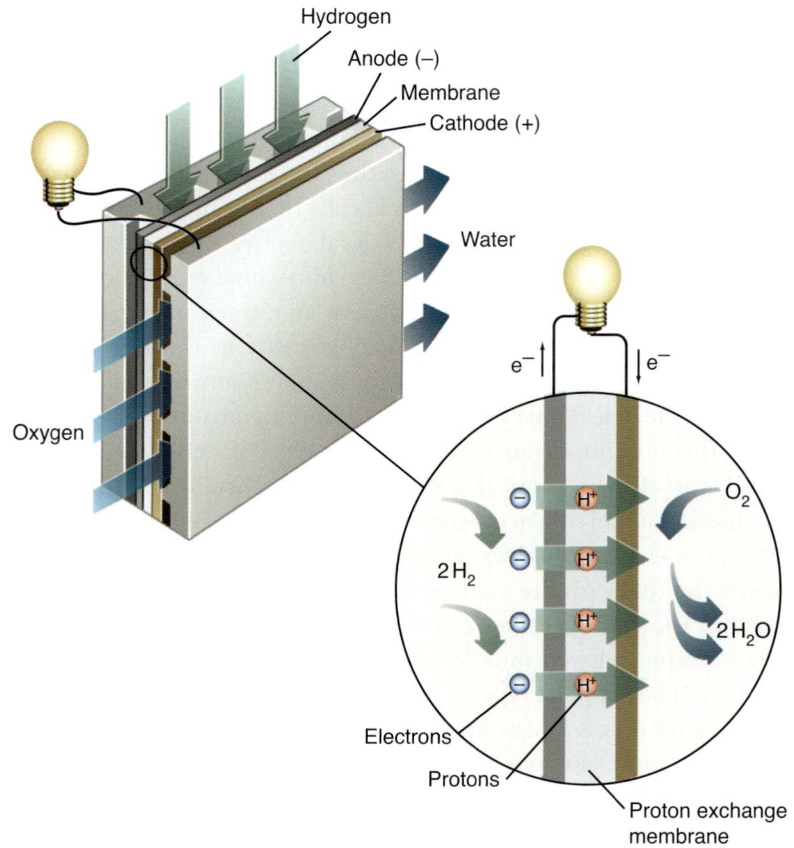

Figure 25.14 A schematic diagram of a hydrogen-oxygen fuel cell. Hydrogen flows over an inert metal anode, where it is oxidized to $H^+(aq)$ according to

$$2H_2(g) \rightarrow 4H^+(aq) + 4e^- \quad \text{(oxidation)}$$

Oxygen flows over an inert metal cathode, where it is reduced according to

$$O_2(g) + 4H^+(aq) + 4e^- \rightarrow 2H_2O(l) \quad \text{(reduction)}$$

The two half reactions are separated by a proton exchange membrane that permits the flow of protons between the two electrodes.

Fuel cells offer the possibility of achieving high thermodynamic efficiency in the conversion of Gibbs energy into mechanical work. Equation 23.7 shows that internal combustion engines can convert at best only the fraction $(T_2 - T_1)/T_2$ of the energy as heat of combustion into mechanical work. In this relation, which comes from the second law of thermodynamics, T_2 is the temperature of the gas in the combustion region and T_1 is its exhaust temperature. Fuel cells do not operate like a heat engine and so are not subject to this limitation.

Figure 25.14 shows a schematic diagram of a hydrogen-oxygen fuel cell. Hydrogen and oxygen are continuously pumped over platinum electrodes that are separated by a proton exchange membrane. The membrane allows the flow of protons between the two electrodes, while excluding the other species in the reaction. The function of the membrane is similar to that of a salt bridge; it provides a path for the ion current while preventing the reactants at each of the two electrodes from mixing. The platinum electrodes in the cell function both as an inert electrode and a catalyst for the oxidation of hydrogen gas. Because of the high cost of platinum, which limits commercial applications, much research is now aimed at finding lower-cost catalytic materials for use in such cells. The equations for the half-cell reactions are shown below.

Anode:	$2H_2(g) \rightarrow 4H^+(aq) + 4e^-$	$E^\circ_{ox} = 0$ V
Cathode:	$O_2(g) + 4H^+(aq) + 4e^- \rightarrow 2H_2O(l)$	$E^\circ_{red} = 1.229$ V
Overall:	$2H_2(g) + O_2(g) \rightarrow 2H_2O(l)$	$E^\circ_{cell} = 1.229$ V

For a hydrogen-oxygen fuel cell operating reversibly at constant temperature and pressure under standard conditions, the electrical work available is given by

$$w_{\text{elec}} = \Delta G^{\circ}_{\text{cell}} = -\nu_e F E^{\circ}_{\text{cell}}$$
$$= -(4)(96\,485\ \text{C·mol}^{-1})(1.229\ \text{V}) = -474.3\ \text{kJ·mol}^{-1} \qquad (25.27)$$

In contrast, for the reaction run in a heat engine at 700°C with a low-temperature reservoir at 27°C, Equation 23.7 shows that the maximum work is only

Recall from Chapter 14 that the sign of the work is negative when energy in the form of work is transferred from the system to the surroundings.

$$w_{\text{engine}} = (-474.3\ \text{kJ·mol}^{-1})\left(\frac{973\ \text{K} - 300\ \text{K}}{973\ \text{K}}\right) = -328\ \text{kJ·mol}^{-1} \qquad (25.28)$$

Because of the hazards of using hydrogen as a fuel and the cost of preparing it from other sources, much research has also been directed to the development of other types of fuel cells. Other promising fuel cells use methanol, $CH_3OH(l)$, and methane, $CH_4(g)$ (natural gas), as fuels. These types of fuel cells have several advantages over hydrogen fuel cells. For instance, methanol is an easily stored liquid and natural gas pipelines are already present in most major cities. The major disadvantage of such cells is that they produce $CO_2(g)$ as a byproduct.

EXAMPLE 25-11: A schematic diagram for a methane-oxygen fuel cell is shown in Figure 25.15. The equations for the half-cell reactions are given below.

Anode: $CH_4(g) + 2H_2O(l) \rightarrow CO_2(g) + 8H^+(aq) + 8e^-$

Cathode: $2O_2(g) + 8H^+(aq) + 8e^- \rightarrow 4H_2O(l)$

Overall: $CH_4(g) + 2O_2(g) \rightarrow CO_2(g) + 2H_2O(l)$

Use the thermodynamic data in Appendix D to calculate the standard cell voltage produced by this methane-oxygen fuel cell.

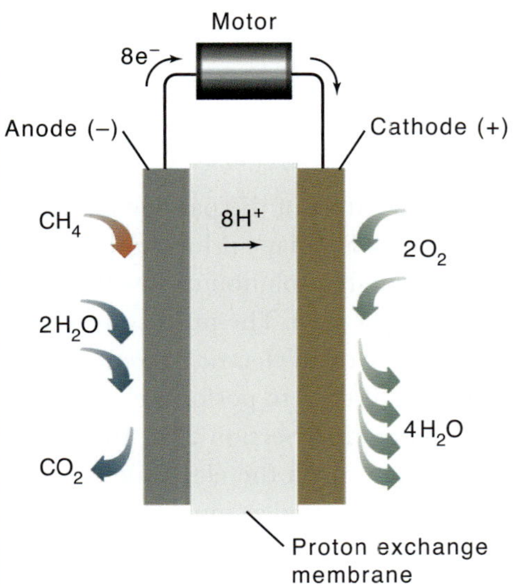

Figure 25.15 A (simplified) schematic diagram of a methane-oxygen fuel cell. Methane, $CH_4(g)$, flows over an inert metal anode, where it is oxidized according to

$$CH_4(g) + 2H_2O(l) \rightarrow CO_2(g) + 8H^+(aq) + 8e^-$$

Oxygen flows over an inert metal cathode, where it is reduced according to

$$2O_2(g) + 8H^+(aq) + 8e^- \rightarrow 4H_2O(l)$$

The two half reactions are separated by a proton exchange membrane.

Solution:

$$\Delta G^\circ_{rxn} = \Delta G^\circ_f[CO_2(g)] + 2\Delta G^\circ_f[H_2O(l)] - \Delta G^\circ_f[CH_4(g)] - 2\Delta G^\circ_f[O_2(g)]$$
$$= (-394.4 \text{ kJ}\cdot\text{mol}^{-1}) + (2)(-237.1 \text{ kJ}\cdot\text{mol}^{-1}) - (-50.5 \text{ kJ}\cdot\text{mol}^{-1}) - 0$$
$$= -818.1 \text{ kJ}\cdot\text{mol}^{-1}$$

Using Equation 25.5 at standard conditions gives us

$$E^\circ_{cell} = \frac{-\Delta G^\circ_{rxn}}{\nu_e F} = \frac{818.1 \times 10^3 \text{ J}\cdot\text{mol}^{-1}}{(8)(96\,485 \text{ C}\cdot\text{mol}^{-1})} = 1.060 \text{ V}$$

The key information that we obtain from the equations for the half reactions is the value of ν_e. You really don't need the details of the half reactions, however, if you realize that the oxidation state of the carbon atom goes from −4 in $CH_4(g)$ to +4 in $CO_2(g)$, for a total change of 8. The following Practice Problem utilizes this approach.

PRACTICE PROBLEM 25-11: Use the thermodynamic data in Appendix D to calculate the standard cell voltage produced by a propane-oxygen fuel cell. The molecular formula of propane gas is $C_3H_8(g)$.

Answer: $\nu_e = 20$, $E^\circ_{cell} = 1.093$ V

The energies that we calculated in Example 25-11 and Practice Problem 25-11 are ideal (maximum) values. In practice, we would obtain less than these amounts because some of the available energy is lost as heat. This heat loss is a consequence of the second law of thermodynamics.

As we learned in Chapter 23, if ΔG°_{rxn} is positive, then its value is the minimum energy that must be *supplied* in order to make the reaction occur. Equation 25.27 shows that $\Delta G^\circ_{rxn} = -474.3 \text{ kJ}\cdot\text{mol}^{-1}$ for the reaction that occurs in a hydrogen-oxygen fuel cell. Thus, for the reverse reaction, that is, the decomposition of water into $H_2(g)$ and $O_2(g)$ at one bar and 25.0°C, we must supply at least 474.3 kJ to decompose two moles (36.04 grams) of water. This energy could be supplied by an electric current at an appropriate voltage, as is done in electrolysis, which we discuss in the next section.

25-8. Electrolysis Is Described Quantitatively by Faraday's Laws

Michael Faraday (Frontispiece) investigated the effect of the passage of an electric current through various electrolyte solutions. Faraday's primary observation was that the passage of an electric current through a solution causes the occurrence of chemical reactions *that would not otherwise occur.* The process by which a chemical reaction is made to occur by the passage of an electric current through the solution is called **electrolysis**, and a cell that is used to perform electrolysis is called an **electrolytic cell** (Figure 25.16). As noted in Section 25-2, the current enters and leaves the solution in an electrolytic cell via the electrodes. Faraday discovered that the metal ions of many salts are deposited as pure metal when an electric current is passed through aqueous solutions of their salts. In his studies of electrolysis, Faraday carried out an extensive series of experiments to determine

Figure 25.16 Electrodeposition of silver from $AgNO_3(aq)$ (*left*), and copper from $Cu(NO_3)_2(aq)$ (*right*). The same quantity of electricity flows through the two solutions because they are placed in series. The number of moles of silver deposited after a given time is twice as great as the number of moles of copper deposited because the reduction of one mole of $Ag^+(aq)$ requires one mole of electrons, whereas the reduction of one mole of $Cu^{2+}(aq)$ requires two moles of electrons.

the amount of electricity required to deposit measured quantities of metals on electrodes as a result of the passage of current through his electrolysis apparatus. The electrodes in Faraday's experiments were electrochemically inert, in the sense that the electrode itself [e.g., Pt(s)] was unchanged by the passage of current.

Faraday found that the quantity of a metal deposited at an electrode when an electric current is passed through a solution of its salt depends on both the current and the stoichiometry of the half reaction that occurs. For example, when pure silver is deposited from a solution of $AgNO_3(aq)$ and pure copper is deposited from a separate solution of $Cu(NO_3)_2(aq)$, the reactions that take place at the cathodes can be described by the half-reaction equations

$$Ag^+(aq) + e^- \rightarrow Ag(s) \qquad \text{(electrolysis of silver nitrate)} \qquad (25.29)$$

and

$$Cu^{2+}(aq) + 2\,e^- \rightarrow Cu(s) \qquad \text{(electrolysis of copper nitrate)} \qquad (25.30)$$

Supplying one mole of electrons deposits one mole of silver from $Ag^+(aq)$, but two moles of electrons are needed to deposit one mole of copper from $Cu^{2+}(aq)$ (Figure 25.16). The number of electrons supplied to drive the reaction can be controlled by regulating the electric current through the solution, which, along with the stoichiometry, determines the quantity of metal deposited.

Suppose that a current of 0.850 amperes flows through each of the two solutions described above for 20.0 minutes. The total charge, Z, that passes through each solution is (Equation 25.2)

$$Z = It = (0.850\ \mathrm{C{\cdot}s^{-1}})(20.0\ \mathrm{min})(60\ \mathrm{s{\cdot}min^{-1}}) = 1020\ \mathrm{C}$$

The number of moles of electrons that corresponds to a given amount of charge that flows through a solution, Z, can be determined using Faraday's constant, F. Therefore,

$$\text{moles of electrons} = (1020\ \text{C})\left(\frac{1\ \text{mol e}^-}{96\,485\ \text{C}}\right) = 1.06 \times 10^{-2}\ \text{mol}$$

From the stoichiometry of Equation 25.29 we see that

$$\text{moles of Ag deposited} \approx \text{moles of electrons}$$

Thus, the number of grams of silver (atomic mass = 107.9) deposited by the passage of 0.850 amperes through a $AgNO_3(aq)$ solution for 20.0 minutes is

$$\text{mass of Ag deposited} = (1.06 \times 10^{-2}\ \text{mol e}^-)\left(\frac{1\ \text{mol Ag}}{1\ \text{mol e}^-}\right)\left(\frac{107.9\ \text{g Ag}}{1\ \text{mol Ag}}\right) = 1.14\ \text{g}$$

Note that the stoichiometric ratio, 1 mol Ag/1 mol e^-, in this expression is simply equal to the magnitude of the change in the oxidation state of silver during the reaction described by Equation 25.29.

These results for silver deposition are an illustration of **Faraday's laws of electrolysis**:

First law: The extent to which an electrochemical reaction occurs depends solely on the quantity of electricity that is passed through a solution.

Second law: The mass of a substance that is produced as a result of the passage of a given quantity of electricity is directly proportional to the molar mass of the substance divided by the change in magnitude of the oxidation state of the substance.

We can write Faraday's second law as an equation. First, we denote the molar mass of the metal or gas produced by M and divide by the (unitless) stoichiometric coefficient of the electrons, ν_e, in the half-reaction equation required to produce one formula unit of the substance. Then we have

$$\left(\begin{array}{c}\text{mass produced per mole of}\\ \text{electrons used in the electrolysis}\end{array}\right) = \frac{M}{\nu_e}$$

Some examples of M and ν_e per formula unit of product are shown below.

Process	M/g·mol^{-1}	ν_e	M/ν_e
$Cu^{2+}(aq) + 2e^- \rightarrow Cu(s)$	63.55	2	31.78
$Ag^+(aq) + e^- \rightarrow Ag(s)$	107.9	1	107.9
$2H^+(aq) + 2e^- \rightarrow H_2(g)$	2.016	2	1.008

The number of moles of electrons used in the electrolysis is equal to the total charge, Z, that is passed through the solution divided by Faraday's constant, F. Thus,

$$\left(\begin{array}{c}\text{moles of electrons}\\ \text{used in the electrolysis}\end{array}\right) = \frac{Z}{F} = \frac{It}{F}$$

where we have used Equation 25.2, that is, $Z = It$. Therefore, we can express Faraday's laws as

$$m = \begin{pmatrix}\text{mass deposited}\\\text{as metal or}\\\text{evolved as gas}\end{pmatrix} = \begin{pmatrix}\text{moles of}\\\text{electrons used in}\\\text{the electrolysis}\end{pmatrix}\begin{pmatrix}\text{mass deposited as}\\\text{metal or evolved as gas}\\\text{per mole of electrons}\\\text{used in the electrolysis}\end{pmatrix}$$

and so we can write the equation

$$m = \left(\frac{It}{F}\right)\left(\frac{M}{\nu_e}\right) \qquad (25.31)$$

Notice that the first term in Equation 25.31, It/F, expresses Faraday's first law, and the second term, M/ν_e, expresses Faraday's second law.

Applying Equation 25.31 to the deposition of copper (atomic mass = 63.55) when 0.850 amperes of current flows through the cell described by Equation 25.30 for 20.0 minutes (1200 seconds), and taking $\nu_e = 2$ for the stoichiometric coefficient of the electrons in the half-reaction equation $Cu^{2+}(aq) + 2e^- \rightarrow Cu(s)$, we find that

$$m = \left(\frac{It}{F}\right)\left(\frac{M}{\nu_e}\right) = \left[\frac{(0.850\ \text{C·s}^{-1})(1200\ \text{s})}{96\,485\ \text{C·mol}^{-1}}\right]\left(\frac{63.55\ \text{g·mol}^{-1}}{2}\right) = 0.336\ \text{g}$$

Thus, 0.336 grams of copper metal are produced at the cathode.

EXAMPLE 25-12: An aqueous solution containing aqueous gold(III) ions was electrolyzed with a current of 0.0381 amperes until 1.47 grams of gold was deposited at the cathode. Determine the duration of the experiment.

Solution: The equation for the reaction at the cathode is

$$Au^{3+}(aq) + 3\,e^- \rightarrow Au(s)$$

Thus, one mole of electrons (96485 C) deposits one-third of a mole of Au(s). The number of coulombs required to deposit 1.47 grams of Au(s) is given by

$$Z = (1.47\ \text{g Au})\left(\frac{1\ \text{mol Au}}{197.0\ \text{g Au}}\right)\left(\frac{3\ \text{mol e}^-}{1\ \text{mol Au}}\right)\left(\frac{96\,485\ \text{C}}{1\ \text{mol e}^-}\right) = 2160\ \text{C}$$

The time required at a current of 0.0381 amperes is given by (Equation 25.2)

$$t = \frac{Z}{I} = \frac{2160\ \text{C}}{0.0381\ \text{C·s}^{-1}} = 5.67 \times 10^4\ \text{s} = 15.7\ \text{h}$$

PRACTICE PROBLEM 25-12: Given that the equation for the half reaction at the anode in Example 25-12 is

$$2\,H_2O(l) \rightarrow 4\,H^+(aq) + O_2(g) + 4\,e^-$$

calculate the volume of $O_2(g)$ produced at 0°C and 1.0 bar.

Answer: 0.13 liters

Figure 25.17 A chlor-alkali plant in which chlorine gas and an aqueous solution of sodium hydroxide are obtained by electrolysis from an aqueous solution of sodium chloride.

25-9. Many Chemicals Are Prepared on an Industrial Scale by Electrolysis

The alkali and alkaline-earth metals are prepared on a commercial scale by electrolysis. All the sodium hydroxide (about 10 million metric tons annually) and most of the chlorine (about 12 million metric tons annually) produced in the United States are made by the **chlor-alkali process**, which involves the electrolysis of concentrated aqueous solutions of sodium chloride (Figures 25.17 and 25.18). The overall electrolysis reaction is described by the equation

$$2\,\text{NaCl}(aq) + 2\,\text{H}_2\text{O}(l) \xrightarrow{\text{electrolysis}} 2\,\text{NaOH}(aq) + \text{Cl}_2(g) + \text{H}_2(g)$$

The corresponding net ionic equation is

$$2\,\text{Cl}^-(aq) + 2\,\text{H}_2\text{O}(l) \rightarrow 2\,\text{OH}^-(aq) + \text{Cl}_2(g) + \text{H}_2(g)$$

The two half-reaction equations are the reduction of water at the cathode,

$$2\,\text{H}_2\text{O}(l) + 2\,\text{e}^- \rightarrow 2\,\text{OH}^-(aq) + \text{H}_2(g) \qquad \text{(cathode)}$$

(a)

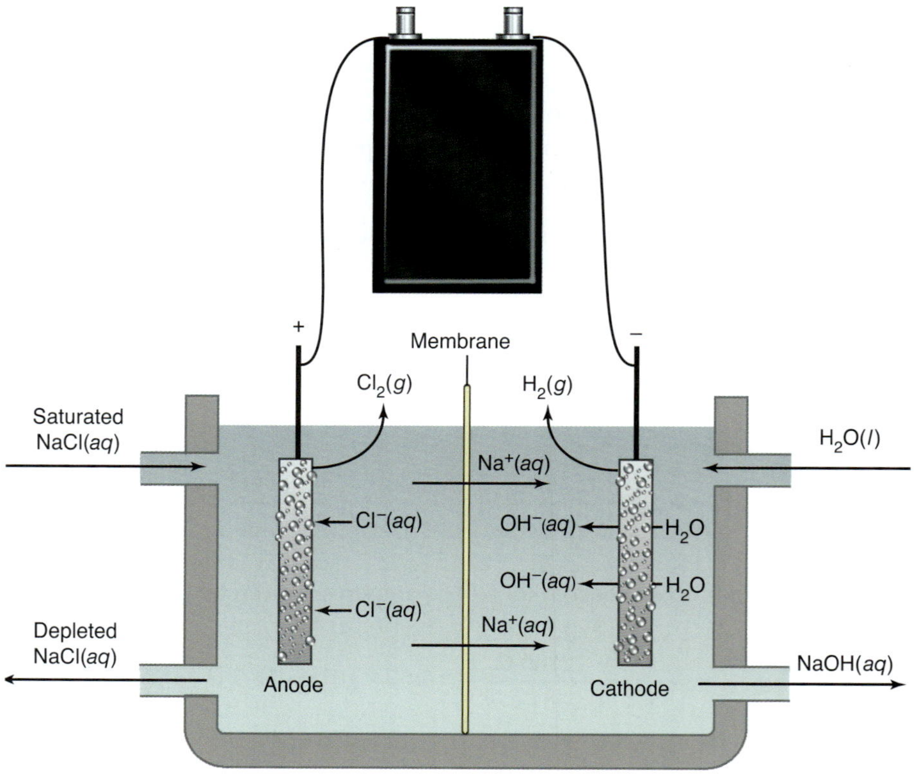

(b)

Figure 25.18 A Nafion membrane and a chlor-alkali membrane cell. (a) A Nafion separator membrane. The membrane is a polymeric material with negatively charged groups ($-SO_3^-$) that permit the passage of $Na^+(aq)$ from $-SO_3^-$ group to $-SO_3^-$ group in the membrane. The membrane is supported by a Teflon grid for additional mechanical strength. (b) A schematic view of the cell. Hydrogen gas and aqueous sodium hydroxide solution are produced at the cathode (−) and chlorine gas is produced at the anode (+). The migration of $Na^+(aq)$ through the membrane maintains equal numbers of positive and negative charges in the separate cell solutions and also carries the current through the cell solutions.

and the oxidation of chloride ions at the anode,

$$2\,Cl^-(aq) \rightarrow Cl_2(g) + 2\,e^- \qquad \text{(anode)}$$

The electrolysis must be carried out in an apparatus in which the sodium hydroxide and hydrogen produced at the cathode and the chlorine produced at the anode are separated by a special membrane. If they were not separated, the sodium hydroxide and chlorine would react to form sodium hypochlorite, $NaClO(aq)$, according to

$$2\,NaOH(aq) + Cl_2(g) \rightarrow NaClO(aq) + NaCl(aq) + H_2O(l)$$

In fact, sodium hypochlorite, which is a commonly used bleaching agent, is prepared in this manner. Household bleach is an aqueous solution that is about 5% by mass of $NaClO(aq)$.

Figure 25.18 shows a schematic drawing of a chlor-alkali membrane cell. The anode and cathode compartments are separated by a Nafion membrane, which has a high internal negative charge. The Nafion membrane excludes the negatively charged $Cl^-(aq)$ ions but allows the positively charged $Na^+(aq)$ ions to pass through. A saturated sodium chloride solution enters the anode compartment, where the $Cl^-(aq)$ is oxidized to $Cl_2(g)$ according to the anode half-reaction equation. The excess $Na^+(aq)$ migrates through the membrane to the cathode, where water is reduced to $H_2(g)$ and $OH^-(aq)$ according to the cathode half-reaction equation.

Many pure metals are produced from their ores using electrolysis (Figure 25.19). An interesting example is the production of aluminum. Over two million metric tons of aluminum metal are produced by electrolysis each

Figure 25.19 An electrochemical plant used to produce high-purity copper metal from impure copper ores. Gleaming sheets of pure copper, in the process of being removed from one of the production units, are visible in the center of the photograph.

Figure 25.20 Charles Hall developed the electrolytic process for the production of aluminum while an undergraduate student at Oberlin College. His sister, Julia, who was also a chemistry major at Oberlin, was of great assistance to him, helping with experiments, keeping laboratory notes, and most important, giving him sound business advice. Upon forming the Pittsburgh Reduction Company for the large-scale production of aluminum, the price of aluminum went from over 12 dollars per pound to 30 cents per pound. In 1907, the company was renamed the Aluminum Company of America, which was later shortened to Alcoa.

Figure 25.21 Paul Héroult first produced aluminum electrolytically after a number of unsuccessful attempts when he was 23 years old. Both Hall and Héroult filed for patents in 1886, and after a prolonged patent litigation, a settlement was reached between them. Héroult later invented the electric steel furnace, which was universally adopted in the United States. Héroult was a free spirit. He loved games, the company of women, travel, and good food and drink. He spent his last years living on a 100-foot yacht in the Mediterranean. Both he and Hall were born in the same year (1863) and died in the same year (1914).

year in the United States. Most aluminum metal is produced by the electrolytic **Hall-Héroult process**, which was independently discovered around the same time by Charles Hall (Figure 25.20) in America and Paul Héroult (Figure 25.21) in France. Prior to the development of this process, aluminum was considered a semiprecious metal that was produced by a process using potassium and later sodium at a cost per gram greater than that of gold! The process for the electrochemical production of aluminum was patented independently by Hall and Héroult in 1886, when they were each 23 years old. After prolonged litigation, they reached an agreement on the patent rights. Hall conceived his process while an undergraduate student at Oberlin College in Ohio, performing all his experiments in a shed in his backyard using a blacksmith's forge to heat the ore. Héroult did all his work in old tannery buildings left to him by his late father.

They faced two major challenges in devising this process. The first was that aluminum metal cannot be produced by the electrolysis of an aqueous solution containing $Al^{3+}(aq)$, because $H^{+}(aq)$ is reduced at an applied voltage much lower than that required to reduce $Al^{3+}(aq)$ at the cathode. The second challenge was that aluminum ores melt at temperatures above 2000°C, which exceeds the boiling point of aluminum metal, thus ruling out electrolysis of the molten ore. The solution to these problems was to use a molten salt electrolyte consisting of powdered aluminum oxide, $Al_2O_3(s)$, dissolved in the mineral cryolite, $Na_3AlF_6(s)$. Aluminum oxide dissolves in cryolite to form a conductive solution with a melting point low enough to allow the operation of the cell at temperatures below 1000°C.

The electrolysis apparatus used today is essentially the same as that used in Hall and Héroult's original process. Figure 25.22 shows a portion of an electrochemical aluminum production facility in Massena, New York. The electrolysis is carried out at about 980°C, at which temperature aluminum is a liquid and so can be siphoned off from the cathode compartment. Electrical contact to the molten aluminum cathode is made through a carbon shell that constitutes the bottom of the electrode compartment. The consumable anodes are composed of a petroleum coke that is obtained by heating to dryness the heavy petroleum fraction remaining from petroleum refining. The equation for the overall electrochemical reaction is

$$2\,Al_2O_3(soln) + 3\,C(s) \rightarrow 4\,Al(l) + 3\,CO_2(g)$$

The production of aluminum is highly energy intensive, consuming about 5% of all the electricity produced in the United States. In recent years the energy cost of aluminum production has dropped significantly, in large part due to recycling. The energy cost of recovering aluminum from recycled materials is only 6% that of manufacturing new aluminum from its ore. Prior to 1960 less than 18% of all aluminum produced was made from recycled materials; today that figure has climbed to about 48%.

Aluminum, which is a fairly reactive metal, protects itself from corrosion by spontaneously forming a thin, tough, adherent layer of $Al_2O_3(s)$. The oxide coating is what gives aluminum metal its dull cast. Electrolysis can be used to protect other reactive metals that do not self-protect as aluminum does. Using electrolysis, a thin layer of a relatively nonreactive metal such as nickel, chromium, tin, silver, or gold is deposited on the surface of the reactive metal. The production of a layer of protective metal by electrolysis is called **electroplating**. For example, electroplating can be used to coat base metals with a thin layer of gold that functions both as a decorative and a protective coating. Electroplating has innumerable industrial applications, ranging from the manufacture of heavy machinery to the production of microcircuits.

Figure 25.22 The Hall-Héroult process on an industrial scale. Shown here is an Alcoa potline (198 pots) at the Massena, New York, plant.

SUMMARY

An electrochemical cell provides the means for obtaining electricity from an electron-transfer reaction. The cell consists of a pair of metal electrodes in contact with an electrolyte solution. The dependence of the cell voltage on the concentrations of the reactants and products of the cell reaction is described quantitatively by the Nernst equation.

The standard cell voltage, E°_{cell}, is the voltage of the cell when the value of the reaction quotient is unity ($Q = 1$). The equilibrium constant of the cell reaction can be calculated from the value of E°_{cell}. Electrode reactions can be arranged in a series of increasing electrode reduction voltages. The assignment of standard reduction voltages to electrode reactions is achieved by setting $E^\circ_{red} = 0$ for the standard hydrogen electrode (SHE) reaction. The E°_{red} values for electrode reactions given in Table 25.3 and Appendix G can be used to calculate E°_{cell} values for reactions and to predict the thermodynamic stability of oxidizing agents and reducing agents. Electrochemical cells can also be used to determine the concentration of various ions in a solution and to measure pH.

The maximum possible amount of energy that can be obtained from an electrochemical cell for the performance of useful work is given by $\Delta G_{rxn} = -\nu_e F E_{cell}$, where E_{cell} is the cell voltage, F is the Faraday constant (96 485 coulombs per mole of protons), and ν_e is the unitless stoichiometric coefficient of the number of electrons transferred in the stated chemical equation describing the cell reaction.

Electrolysis is the process by which a chemical reaction is driven by the application of a voltage across electrodes placed in a solution. The extent to which the electrochemical reaction occurs is proportional to the current that flows through the solution, according to Faraday's laws of electrolysis.

TERMS YOU SHOULD KNOW

EQUATIONS YOU SHOULD KNOW HOW TO USE

$U = VZ$ (25.1) (energy of a charge Z at a voltage V)

$Z = It$ (25.2) (charge that flows in a given time)

$$\Delta G_{rxn} = -\nu_e F E_{cell}$$
$$\Delta G^\circ_{rxn} = -\nu_e F E^\circ_{cell}$$
(25.5) (relation between Gibbs energy and cell voltage)

$$\ln K = \frac{\nu_e E^\circ_{cell}}{0.02570\ \text{V}}$$
(25.10) (relation between the equilibrium constant and the standard cell voltage at 25.0°C)

$$E_{cell} = E^\circ_{cell} - \left(\frac{RT}{\nu_e F}\right)\ln Q$$
(25.12) (the Nernst equation)

$$E_{cell} = E^\circ_{cell} - \left(\frac{0.02570\ \text{V}}{\nu_e}\right)\ln Q$$
(25.13) (the Nernst equation at 25.0°C)

$$E^\circ_{cell} = E^\circ_{red} + E^\circ_{ox}$$
(25.15) (relation between E°_{cell} for a cell and the values of E° for the oxidation and reduction half reactions)

$$E^\circ_{ox} = -E^\circ_{red}$$
(25.16) (relation between the standard reduction voltage and standard oxidation voltage of a half reaction)

$$\text{pH} = \frac{E_{cell} - E^\circ_{ref}}{0.0592\ \text{V}}$$
(25.26) (the pH in terms of a cell voltage and a standard reference voltage at 25.0°C)

$$m = \left(\frac{It}{F}\right)\left(\frac{M}{\nu_e}\right)$$
(25.31) (Faraday's law of electrolysis)

PROBLEMS

CELL SETUPS

25-1. Consider a zinc-silver electrochemical cell. The negative electrode is a zinc rod immersed in a $ZnCl_2(aq)$ solution, and the positive electrode is a silver rod in a $AgNO_3(aq)$ solution. The two solutions are connected by a KNO_3-agar salt bridge. Sketch a diagram of the cell, indicating the flow of electrons. Write the equation for the reaction that occurs at each electrode and write the cell diagram.

25-2. Consider a manganese-chromium electrochemical cell. The negative electrode is a manganese rod immersed in a $MnSO_4(aq)$ solution, and the positive electrode is a chromium rod in a $CrSO_4(aq)$ solution. The two solutions are connected by a salt bridge. Sketch a diagram of the cell, indicating the flow of electrons. Write the equation for the reaction that occurs at each electrode and write the cell diagram.

25-3. Consider a vanadium-copper electrochemical cell. The negative electrode is a vanadium rod immersed in a $VI_2(aq)$ solution, and the positive electrode is a copper rod in a $CuSO_4(aq)$ solution. The two solutions are connected by a salt bridge. Sketch a diagram of the cell, indicating the flow of electrons. Write the equation for the reaction that occurs at each electrode and write the cell diagram.

25-4. Consider a cobalt-lead electrochemical cell. The negative electrode is a cobalt rod immersed in a $Co(NO_3)_2(aq)$ solution, and the positive electrode is a lead rod in a $Pb(NO_3)_2(aq)$ solution. The two solutions are connected by a salt bridge. Sketch a diagram of the cell, indicating the flow of electrons. Write the equation for the reaction that occurs at each electrode and write the cell diagram.

25-5. Consider a hydrogen-cadmium electrochemical cell. The negative electrode is a platinum wire immersed in an $HCl(aq)$ solution over which hydrogen gas is bubbled, and the positive electrode is a strip of cadmium metal in a $Cd(NO_3)_2(aq)$ solution. The two solutions are connected by a salt bridge. Sketch a diagram of the cell, indicating the flow of electrons. Write the equation for the reaction that occurs at each electrode and write the cell diagram.

25-6. Consider a zinc-mercury electrochemical cell. The negative electrode is a zinc rod immersed in a $ZnCl_2(aq)$ solution, and the positive electrode is a platinum wire in electrical contact with a mixture of liquid mercury metal and $Hg_2Cl_2(s)$, which are also in the $ZnCl_2(aq)$ solution. The $Hg(l)$ and $Hg_2Cl_2(s)$ are separated from the zinc rod by means of an H-shaped cell. Sketch a diagram of the cell, indicating the flow of electrons. Write the equation for the reaction that occurs at each electrode and write the cell diagram.

CELL DIAGRAMS

25-7. The cell diagram for an electrochemical cell is given as

$$In(s)|In(ClO_4)_3(aq)||CdCl_2(aq)|Cd(s)$$

Write the equations for the half reactions that occur at the two electrodes and the net cell reaction.

25-8. The cell diagram for an electrochemical cell is given as

$$Sn(s)|SnCl_2(aq)||AgNO_3(aq)|Ag(s)$$

Write the equations for the half reactions that occur at the two electrodes and the net cell reaction.

25-9. Write the equations describing the electrode reactions and the net cell reaction for the electrochemical cell

$$Pb(s)|PbI_2(s)|HI(aq)|H_2(g)|Pt(s)$$

25-10. Write the equations describing the electrode reactions and the net cell reaction for the electrochemical cell

$$Cu(s)|Cu(ClO_4)_2(aq)||AgClO_4(aq)|Ag(s)$$

25-11. Write the net cell reaction equation for the electrochemical cell

$$Pt(s)|H_2(g)|HCl(aq)|Hg_2Cl_2(s)|Hg(l)$$

25-12. Write the equation for the net cell reaction for the electrochemical cell

$$Pb(s)|PbSO_4(s)|K_2SO_4(aq)|Hg_2SO_4(s)|Hg(l)$$

ELECTROCHEMICAL CELLS AND LE CHÂTELIER'S PRINCIPLE

25-13. Consider an electrochemical cell in which the reaction is described by the equation

$$2\,HCl(aq) + Ca(s) \leftrightharpoons CaCl_2(aq) + H_2(g)$$

Predict the effect of the following changes on the cell voltage:

(a) decrease in amount of $Ca(s)$

(b) increase in pressure of $H_2(g)$

(c) increase in [HCl]

(d) dissolution of $Ca(NO_3)_2(s)$ in the $CaCl_2(aq)$ solution

25-14. Consider an electrochemical cell in which the reaction is described by the equation

$$Pb(s) + 2\,Ag^+(aq) + SO_4^{2-}(aq) \leftrightharpoons PbSO_4(s) + 2\,Ag(s)$$

Predict the effect of the following changes on the observed cell voltage:

(a) increase in $[Ag^+]$

(b) increase in amount of $PbSO_4(s)$

(c) increase in $[SO_4^{2-}]$

(d) doubling the size of the cell (keeping the concentrations constant)

25-15. Given the following equation for an electrochemical cell reaction

$$H_2(g) + PbCl_2(s) \leftrightharpoons Pb(s) + 2\,HCl(aq) \qquad \Delta H^\circ_{rxn} > 0$$

Predict the effect of the following changes on the observed cell voltage:

(a) increase in the amount of $PbCl_2(s)$

(b) dilution of the cell solution with $H_2O(l)$

(c) dissolution of $NaOH(s)$ in the cell solution

(d) increase in the temperature

25-16. Given the following equation for an electrochemical cell reaction

$$H_2(g) + PbSO_4(s) \leftrightharpoons 2\,H^+(aq) + SO_4^{2-}(aq) + Pb(s)$$

Predict the effect of the following changes on the observed cell voltage:

(a) increase in size of $Pb(s)$ electrode

(b) decrease in pH of cell electrolyte

(c) dilution of cell electrolyte with water

(d) dissolution of a small amount of $NaOH(s)$ in the cell electrolyte

25-17. Consider the following equation for a reaction taking place in an electrochemical cell:

$$2\,Cr^{2+}(aq) + HClO(aq) + H^+(aq) \leftrightharpoons 2\,Cr^{3+}(aq) + Cl^-(aq) + H_2O(l)$$

Predict the effect of the following changes on the observed cell voltage:

(a) increase in [HClO]

(b) increase in size of the inert electrodes

(c) increase in pH of cell solution

(d) dissolution of $KCl(s)$ in the cell solution containing $Cl^-(aq)$

25-18. Given that the equation for the reaction in an electrochemical cell is

$$Fe^{2+}(aq) + Ag^+(aq) \leftrightharpoons Ag(s) + Fe^{3+}(aq)$$

Predict the effect of the following changes on the observed cell voltage:

(a) increase in $[Ag^+]$

(b) increase in $[Fe^{3+}]$

(c) twofold decrease in both $[Fe^{3+}]$ and $[Fe^{2+}]$

(d) decrease in amount of $Ag(s)$

(e) decrease in $[Fe^{2+}]$

(f) addition of $NaCl(aq)$ to $Ag^+(aq)$ solution

NERNST EQUATION

25-19. Determine the value of ν_e in the Nernst equation for the following equations:

(a) $CH_4(g) + 2\,O_2(g) \rightarrow CO_2(g) + 2\,H_2O(l)$

(b) $2\,Zn(s) + Ag_2O_2(s) + 2\,H_2O(l) + 4\,OH^-(aq) \rightarrow 2\,Ag(s) + 2[Zn(OH)_4]^{2-}(aq)$

25-20. Determine the value of ν_e in the Nernst equation for the following equations:

(a) $Cu(s) + Mg^{2+}(aq) \rightarrow Cu^{2+}(aq) + Mg(s)$

(b) $2\,H_2O(l) + 2\,Na(s) \rightarrow 2\,Na^+(aq) + 2\,OH^-(aq) + H_2(g)$

25-21. The value of E°_{cell} at 25°C for the equation

$$Pb(s) + Cu^{2+}(aq) \leftrightharpoons Pb^{2+}(aq) + Cu(s)$$

is 0.47 V. Calculate the value of the equilibrium constant at 25°C for this equation.

25-22. The value of E°_{cell} at 25°C for the equation

$$H_2(g) + 2AgCl(s) \leftrightharpoons 2Ag(s) + 2HCl(aq)$$

is 0.22 V. Calculate the value of the equilibrium constant at 25°C for this equation.

25-23. The measured voltage at 25°C of a cell in which the reaction described by the equation

$$Cd(s) + Pb^{2+}(aq, 0.150\text{ M}) \leftrightharpoons Pb(s) + Cd^{2+}(aq, 0.0250\text{ M})$$

takes place at the concentrations shown is 0.293 V. Calculate the values of E°_{cell} and K, the equilibrium constant, for the cell equation.

25-24. The measured voltage at 25°C of a cell in which the reaction described by the equation

$$Co(s) + Sn^{2+}(aq, 0.18\text{ M}) \leftrightharpoons Sn(s) + Co^{2+}(aq, 0.020\text{ M})$$

takes place at the concentrations shown is 0.168 V. Calculate the values of E°_{cell} and K, the equilibrium constant, for the cell equation.

25-25. The measured voltage at 25°C of a cell in which the reaction described by the equation

$$Al(s) + Fe^{3+}(aq, 0.0050\text{ M}) \leftrightharpoons Al^{3+}(aq, 0.250\text{ M}) + Fe(s)$$

takes place at the concentrations shown is 1.591 V. Calculate the values of E°_{cell} and K, the equilibrium constant, for the cell equation.

25-26. The measured voltage at 25°C of a cell in which the reaction described by the equation

$$Zn(s) + Hg_2^{2+}(aq, 0.30\text{ M}) \leftrightharpoons 2Hg(l) + Zn^{2+}(aq, 0.50\text{ M})$$

takes place at the concentrations shown is 1.553 V. Calculate the values of E°_{cell} and K, the equilibrium constant, for the cell equation.

25-27. Consider the electrochemical cell described by

$$Zn(s)|ZnCl_2(aq)|Hg_2Cl_2(s)|Hg(l)$$

(a) Write a balanced oxidation-reduction equation describing the cell reaction.

(b) The standard cell voltage for the cell at 25°C is E°_{cell} = 1.030 V. Use the Nernst equation to calculate the cell voltage when the concentration of $ZnCl_2(aq)$ is 0.040 M.

25-28. The standard voltage of the cell described by

$$Pt(s)|H_2(g)|H^+(aq)||Cd^{2+}(aq)|Cd(s)$$

is −0.403 V. Write the cell equation and calculate the cell voltage when $[H^+]$ = 0.10 M, P_{H_2} = 0.10 bar, and $[Cd^{2+}] = 2.5 \times 10^{-3}$ M at 25°C.

25-29. Calculate the voltage generated at 25°C by the aluminum concentration cell described by

$$Al(s)|Al^{3+}(aq, 0.010\text{ M})||Al^{3+}(aq, 0.500\text{ M})|Al(s)$$

25-30. Calculate the voltage generated at 25°C by the hydrogen concentration cell described by

$$Pt(s)|H_2(g, 1.00\text{ bar})|H^+(aq, 0.030\text{ M})|| H^+(aq, 0.250\text{ M})|H_2(g, 1.00\text{ bar})|Pt(s)$$

CONCENTRATIONS FROM CELL MEASUREMENTS

25-31. The reaction described by the equation

$$Zn(s) + Hg_2Cl_2(s) \leftrightharpoons 2Hg(l) + Zn^{2+}(aq) + 2Cl^-(aq)$$

is run in an electrochemical cell. The measured voltage of the cell at 25°C is 1.03 V when Q = 1.00. Suppose that the measured voltage of the cell at 25°C is 1.21 V when $[Cl^-]$ = 0.10 M in the cell solution and $[Zn^{2+}]$ is unknown. Calculate the value of $[Zn^{2+}]$ in the cell solution.

25-32. The voltage generated by the zinc concentration cell described by

$$Zn(s)|Zn^{2+}(aq, 0.100\text{ M})||Zn^{2+}(aq)|Zn(s)$$

is 20.0 mV at 25°C. Calculate the concentration of the $Zn^{2+}(aq)$ ion at the cathode.

25-33. The measured voltage of the cell described by

$$Pt(s)|H_2(g, 1.00\text{ bar})|H^+(aq)||Ag^+(aq, 1.00\text{ M})|Ag(s)$$

is 0.915 V at 25°C. Given $E^\circ_{cell} = 0.7996$ V, calculate the pH of the solution.

25-34. The measured voltage of the cell

$$Zn(s)|Zn^{2+}(aq, M)||Ag^{+}(aq, 0.10\text{ M})|Ag(s)$$

is 1.502 V at 25°C. Given $E^\circ_{cell} = 1.561$ V, calculate the value of $[Zn^{2+}]$.

USE OF TABULATED $E°$ VALUES

25-35. Using the data in Appendix G, calculate the values of E°_{cell} for the following equations:

(a) $Cu(s) + [Fe(CN)_6]^{3-}(aq) \leftrightharpoons Cu^{+}(aq) + [Fe(CN)_6]^{4-}(aq)$

(b) $Fe^{3+}(aq) + Ag(s) \leftrightharpoons Fe^{2+}(aq) + Ag^{+}(aq)$

(c) $Zn(s) + F_2(g) \leftrightharpoons Zn^{2+}(aq) + 2F^{-}(aq)$

25-36. Using the data in Appendix G, calculate the values of E°_{cell} for the following equations:

(a) $2Na(s) + 2H_2O(l) \leftrightharpoons 2Na^{+}(aq) + 2OH^{-}(aq) + H_2(g)$

(b) $2H^{+}(aq) + Pd(s) \leftrightharpoons Pd^{2+}(aq) + H_2(g)$

(c) $4Fe(OH)_2(s) + O_2(g) + 2H_2O(l) \leftrightharpoons 4Fe(OH)_3(s)$

25-37. The standard voltage for the equation

$$HClO(aq) + H^{+}(aq) + 2Cr^{2+}(aq) \leftrightharpoons 2Cr^{3+}(aq) + Cl^{-}(aq) + H_2O(l)$$

is $E^\circ_{cell} = 1.889$ V. Use data from Appendix G to calculate the value of E°_{red} for the half-reaction equation

$$HClO(aq) + H^{+}(aq) + 2e^{-} \leftrightharpoons Cl^{-}(aq) + H_2O(l)$$

25-38. The standard voltage for the equation

$$NO_3^{-}(aq) + 2H^{+}(aq) + Cu^{+}(aq) \leftrightharpoons NO_2(g) + H_2O(l) + Cu^{2+}(aq)$$

is $E^\circ_{cell} = 0.65$ V. Use data from Appendix G to calculate the value of E°_{red} for the half-reaction equation

$$NO_3^{-}(aq) + 2H^{+}(aq) + e^{-} \leftrightharpoons NO_2(g) + H_2O(l)$$

25-39. Use data from Appendix G to calculate the value of E°_{cell} at 25°C for the equation

$$Cd^{2+}(aq) + Zn(s) \leftrightharpoons Zn^{2+}(aq) + Cd(s)$$

Will zinc displace cadmium from the compound $Cd(NO_3)_2(aq)$ if $[Zn^{2+}] = [Cd^{2+}] = 1.00$ M? Is the reaction spontaneous if $[Cd^{2+}] = 0.0010$ M and $[Zn^{2+}] = 1.00$ M?

25-40. Use data from Appendix G to calculate the value of E°_{cell} at 25°C for the equation

$$S_2O_8^{2-}(aq) + 2H_2O(l) \leftrightharpoons H_2O_2(aq) + 2SO_4^{2-}(aq) + 2H^{+}(aq)$$

Is an aqueous solution of potassium peroxodisulfate, $K_2S_2O_8(aq)$, stable over a long period of time?

25-41. Calculate the value of E°_{cell} and the equilibrium constant at 25°C for the cell equation

$$V^{2+}(aq) + H^{+}(aq) \leftrightharpoons V^{3+}(aq) + \tfrac{1}{2}H_2(g)$$

Can $V^{2+}(aq)$ at 1.0 M liberate $H_2(g)$ at 1.00 bar from a solution with $[H^{+}] = 1.0$ M and $[V^{3+}] = 1.00 \times 10^{-4}$ M? See Appendix G for the necessary $E°$ data.

25-42. Calculate the value of E°_{cell} and the equilibrium constant at 25°C for the cell reaction described by

$$Cu(s) + 2Ag^{+}(aq) \leftrightharpoons 2Ag(s) + Cu^{2+}(aq)$$

Predict whether the reaction is spontaneous when $[Ag^{+}] = 0.10$ M and $[Cu^{2+}] = 1.00 \times 10^{-4}$ M. See Appendix G for the necessary $E°$ data.

25-43. Calculate the voltage at 25°C of an electrochemical cell for the reaction described by the equation

$$2Zn(s) + O_2(g, 0.20\text{ bar}) + 4H^{+}(aq, 0.20\text{ M}) \leftrightharpoons 2Zn^{2+}(aq, 0.0010\text{ M}) + 2H_2O(l)$$

See Appendix G for the necessary $E°$ data.

25-44. Calculate the voltage at 25°C of an electrochemical cell for the reaction described by the equation

$$Cd(s) + Pb(NO_3)_2(aq, 0.10\text{ M}) \leftrightharpoons Cd(NO_3)_2(aq, 0.010\text{ M}) + Pb(s)$$

See Appendix G for the necessary $E°$ data.

25-45. The standard voltage for the equation

$$S_2O_3^{2-}(aq) + 2OH^{-}(aq) + O_2(g) \leftrightharpoons 2SO_3^{2-}(aq) + H_2O(l)$$

is $E^\circ_{cell} = 0.972$ V at 25°C. Write the two half-reaction equations and use data from Appendix G to determine the value of E°_{red} for the $SO_3^{2-}(aq)|S_2O_3^{2-}(aq)$ half reaction in basic solution at 25°C.

25-46. The standard voltage for the equation

$$BH_4^-(aq) + 8\,OH^-(aq) + 8\,O_2(g) \leftrightharpoons H_2BO_3^-(aq) + 5\,H_2O(l) + 8\,O_2^-(aq)$$

at 25°C is $E^\circ_{cell} = 0.68$ V, and the value of E°_{red} for the $O_2(g)|O_2^-(aq)$ half reaction in basic solution is −0.56 V. Write the two half-reaction equations and use these data to determine the value of E°_{red} for the $H_2BO_3^-(aq)|BH_4^-(aq)$ half reaction in basic solution at 25°C.

E_{cell} AND SPONTANEITY

25-47. Suppose we have an aqueous solution at 25°C with $[Co^{3+}] = 0.20$ M, $[Co^{2+}] = 1.0 \times 10^{-4}$ M, and $[H^+] = 0.30$ M, which is exposed to air ($P_{O_2} = 0.20$ bar). Use the Nernst equation to determine whether the oxidation of water by $Co^{3+}(aq)$ is spontaneous under the given conditions.

25-48. Using the data in Appendix G, determine whether air ($P_{O_2} = 0.20$ bar) is capable of oxidizing $Fe^{2+}(aq)$ to $Fe^{3+}(aq)$ in a solution with $[Fe^{2+}] = [Fe^{3+}] = 0.10$ M and pH = 2.0 at 25°C.

25-49. Using the data in Appendix G, determine whether $Co^{3+}(aq, 0.010\text{ M})$ is capable of liberating $O_2(g)$ at 25°C and an air pressure of 1.0 bar from water at pH = 1.0.

25-50. Use the data in Appendix G to calculate the value of E°_{cell} for the equation

$$O_3(g) + H_2O(l) + Cu(s) \leftrightharpoons Cu(OH)_2(s) + O_2(g)$$

Is the oxidation of copper by ozone a spontaneous process at 25°C under the conditions $P_{O_3} = 1.00 \times 10^{-4}$ bar and $P_{O_2} = 0.20$ bar?

CELLS AND ΔG_{rxn}

25-51. An electrochemical cell is set up so that the reaction described by the equation

$$Zn(s) + Cu^{2+}(aq) \leftrightharpoons Zn^{2+}(aq) + Cu(s)$$

occurs. At 25°C the measured cell voltage is 1.05 V. Calculate the value of ΔG_{rxn} for the reaction equation.

25-52. An electrochemical cell is set up so that the reaction described by the equation

$$H_2O_2(aq) + Fe(s) + 2\,H^+(aq) \leftrightharpoons Fe^{2+}(aq) + 2\,H_2O(l)$$

occurs. At 25°C the measured cell voltage is 2.03 V. Calculate the value of ΔG_{rxn}.

25-53. An electrochemical cell is set up so that the reaction described by the equation

$$2\,NO_3^-(aq) + 4\,H^+(aq) + Cu(s) \leftrightharpoons 2\,NO_2(g) + 2\,H_2O(l) + Cu^{2+}(aq)$$

occurs. At 25°C, the standard cell voltage is 0.65 V. Calculate the value of ΔG°_{rxn}.

25-54. An electrochemical cell is set up so that the reaction described by the equation

$$Cr_2O_7^{2-}(aq) + 14\,H^+(aq) + 6\,Fe^{2+}(aq) \leftrightharpoons 2\,Cr^{3+}(aq) + 6\,Fe^{3+}(aq) + 7\,H_2O(l)$$

occurs. At 25°C, the standard cell voltage is 0.461 V. Calculate the value of ΔG°_{rxn}.

25-55. Use the data in Appendix G to calculate the value of ΔG°_{rxn} for each of the following equations:

(a) $2\,Ag(s) + F_2(g) \leftrightharpoons 2\,Ag^+(aq) + 2\,F^-(aq)$

(b) $\frac{1}{2}H_2(g) + Fe^{3+}(aq) \leftrightharpoons Fe^{2+}(aq) + H^+(aq)$

25-56. Use the data in Appendix G to calculate the value of ΔG°_{rxn} for each of the following equations:

(a) $Zn(s) + Cu^{2+}(aq) \leftrightharpoons Zn^{2+}(aq) + Cu(s)$

(b) $Ag(s) + Fe^{3+}(aq) \leftrightharpoons Fe^{2+}(aq) + Ag^+(aq)$

25-57. For the electrochemical cell described by

$$Zn(s)|Zn^{2+}(aq, 0.010\text{ M})||Cd^{2+}(aq, 0.050\text{ M})|Cd(s)$$

write the cell equation. Use the data in Appendix G to calculate the values of E°_{cell}, ΔG°_{rxn}, ΔG_{rxn}, and E_{cell} at 25°C for the cell equation.

25-58. For the electrochemical cell

$$Co(s)|Co^{2+}(aq, 0.0155\ M)||Ag^{+}(aq, 1.50\ M)|Ag(s)$$

write the cell equation. Use the data in Appendix G to calculate the values of E°_{cell}, ΔG°_{rxn}, ΔG_{rxn}, and E_{cell} at 25°C for the cell equation.

FARADAY'S LAWS

25-59. Cesium metal is produced by the electrolysis of molten cesium cyanide. Calculate the maximum amount of $Cs(s)$ that can be deposited from $CsCN(l)$ in 30.0 minutes by a current of 500.0 mA.

25-60. Beryllium occurs naturally in the form of beryl. The metal is produced from its ore by electrolysis after the ore has been converted to the oxide and then to the chloride. Calculate the maximum amount of $Be(s)$ that can be deposited from a $BeCl_2(l)$ melt by a current of 5.0 amperes that flows for 1.0 hour.

25-61. Fluorine is manufactured by the electrolysis of hydrogen fluoride dissolved in molten potassium fluoride. The equation is

$$2\,HF(KF) \rightarrow H_2(g) + F_2(g)$$

The potassium fluoride acts as a solvent for hydrogen fluoride and as the conductor of electricity. A commercial cell for producing fluorine operates at a current of 1500 amperes. What is the maximum amount of $F_2(g)$ that can be produced per 24 hours? Why isn't the electrolysis of $HF(l)$ alone used?

25-62. Suppose that it is planned to electrodeposit 200.0 mg of gold onto the surface of a steel object via the process

$$[Au(CN)_2]^{-}(aq) + e^{-} \rightarrow Au(s) + 2\,CN^{-}(aq)$$

If the electric current in the circuit is set at 30.0 mA, for how long should the current be passed?

25-63. Gallium is produced by the electrolysis of a solution obtained by dissolving gallium oxide in concentrated $NaOH(aq)$. Calculate the amount of $Ga(s)$ that can be deposited from a Ga(III) solution by a current of 0.50 amperes that flows for 30 minutes.

25-64. Hydrogen and oxygen can be produced by the electrolysis of water according to

$$2H_2O(l) \xrightarrow{\text{electrolysis}} 2H_2(g) + O_2(g)$$

Calculate the volume of $O_2(g)$ produced at 25°C and 1.00 atm when a current of 30.35 amperes is passed through a $K_2SO_4(aq)$ solution for 2.00 hours.

ADDITIONAL PROBLEMS

25-65. The current in a wire is carried by moving electrons. What carries the current through the solution in an electrochemical cell?

25-66. What is the function of a salt bridge in an electrochemical cell? What would happen if the salt bridge were removed?

25-67. What is meant by "a nonreactive metal"? Why must nonreactive metals be used as electrodes in certain cells? Why can't we simply use *any* metal, such as, say, copper, as an electrode in a cell?

25-68. What is the difference between an electrochemical cell and an electrolytic cell?

25-69. How does doubling the size of an electrochemical cell (while keeping the concentration of all species the same) affect the current produced and the cell voltage?

25-70. Using the data in Appendix G, determine the placement of cerium metal in the metals activity series given in Table 10.8.

25-71. Write the balanced chemical equation for the cell reaction in the following electrochemical cell:

$$Pt(s)|MnO_4^{-}(aq), Mn^{2+}(aq), H^{+}(aq)|| \\ IO_3^{-}(aq), I^{-}(aq), H^{+}(aq)|Pt(s)$$

25-72. An oxide cell, involving a $Ag_2O(s)|Ag(s)$ cathode is used to power a wristwatch. The cell is estimated to last 1000 hours while drawing a current of only 0.10 mA. Calculate the mass of silver metal that will be produced over the lifetime of the cell.

25-73. A hydrogen gas electrode is connected to a $Br_2(l)|Br^-(aq, 0.500\text{ M})$ reference cell. Hydrogen gas is bubbled through a solution of unknown acid concentration at an atmospheric pressure of 756.5 Torr. If the cell voltage is 1.205 V at 25°C, calculate the pH of the acidic solution.

25-74. Bauxite, the principal source of aluminum oxide, contains about 55% $Al_2O_3(s)$ by mass.

(a) How much bauxite is required to produce the 2.0 million metric tons of aluminum metal produced each year by electrolysis?

(b) Estimate the amount of energy needed to produce this much aluminum per year using Hall-Héroult process if the operating voltage of a typical aluminum generating cell is 5.0 V.

(c) What fraction of the 13×10^{15} kJ of electrical energy generated annually in the United States does this represent?

25-75. Many metals can be refined electrolytically. The impure metal is used as the anode, and the cathode is made of the pure metal. The electrodes are placed in an electrolyte containing a salt of the metal being refined. When an electric current is passed between these electrodes, the metal leaves the impure anode and is deposited in a pure form on the cathode. If it requires two moles of electrons for every mole of copper metal refined electrolytically, how many ampere-hours of electricity are required to refine 1.0 metric ton of copper? (An ampere-hour is an ampere times an hour.)

25-76. Why can a solid copper penny be dissolved in nitric acid, but not in hydrochloric acid? Modern pennies are made from zinc metal plated with copper. What happens to a modern penny when it is placed inside a bath of hydrochloric acid after the penny is scratched, exposing some of the zinc?

25-77. Two electrolytic cells are placed in series in a manner similar to the cells in Figure 25.16. One cell contains a solution of $AgClO_4(aq)$ and the other cell contains a solution of $Cd(ClO_4)_2(aq)$. An electric current is passed through the two cells until 0.876 grams of $Ag(s)$ is deposited. How many grams of $Cd(s)$ will be deposited in the same time?

25-78. Given that $E^\circ_{cell} = 0.728$ V at 25°C for the cell

$$Ag(s)|Br^-(aq)|AgBr(s)||Ag^+(aq)|Ag(s)$$

write the cell equation and determine the solubility-product constant of $AgBr(s)$ in water at 25°C.

25-79. The deteriorating iron framework inside the Statue of Liberty was replaced with stainless steel as part of a major restoration project. The work was finished in 1986, exactly one hundred years after the statue was first completed. To avoid any electrochemical contact between the metals, the new stainless steel frame and the external copper plates covering the statue were separated using Teflon spacers. The original statue was constructed using asbestos pads as insulating spacers. Apparently, the pads were still able to act as a conductor (in conjunction with moisture and gases from the atmosphere). Why was the iron framework on the interior of the statue most in need of repair and not the copper plating exposed to the atmosphere on the exterior of the statue?

25-80. Consider a fuel cell that uses the combustion of ethanol to produce electricity,

$$CH_3CH_2OH(l) + 3O_2(g) \rightarrow 2CO_2(g) + 3H_2O(l)$$

Use the thermodynamic data in Appendix D to determine the value of E°_{cell} for this cell at 25°C.

25-81. The Weston standard cell is given by

$$Cd(Hg, 12.5\%\ Cd)|CdSO_4(aq, satd)|Hg_2SO_4(s)|Hg(l)$$

Write the reaction equation that occurs in the cell. Ten Weston standard cells that use a saturated $CdSO_4(aq)$ solution are maintained at the U.S. Bureau of Standards as the official unit of voltage. The voltage of each cell is virtually constant at 1.01857 V at 25°C. Explain why the voltage remains constant. Here Cd(Hg, 12.5% Cd) stands for a cadmium mercury amalgam containing 12.5% cadmium in mercury by mass.

25-82. Suppose a zinc rod is dipped into a 1.0 M $CuSO_4(aq)$ solution containing a copper rod and the system is allowed to stand for several hours. What do you predict for the voltage measured between the $Zn(s)$ and $Cu(s)$ rods?

25-83. A battery that operates at −50°C was developed for the exploration of the moon and Mars. The electrodes are magnesium metal-magnesium chloride and silver chloride-silver. The electrolyte is potassium thiocyanate, KSCN, in liquid ammonia. Write the equation for the cell reaction and the cell diagram for the cell.

25-84. Suppose the leads of an electrochemical cell are connected together external to the cell and the cell is allowed to come to equilibrium. What will be the value of the cell voltage at equilibrium?

25-85. Electrolysis can be used to determine atomic masses. A current of 0.600 amperes deposits 2.42 grams of a certain metal from its salt solution in 1.00 hour. A second experiment shows that the dissolved salt of the metal is in a +1 oxidation state. What is the metal?

25-86. (*) A solution containing a nickel salt undergoes electrolysis, forming 39.12 grams of nickel metal at the cathode for every 16.00 grams of $O_2(g)$ evolved at the anode. What is the oxidation state of nickel in the solution? (The oxygen formed at the anode comes from the water that the salt is dissolved in.)

25-87. (*) Write a balanced equation for the cell reaction in the following electrochemical cell and calculate the cell voltage, E_{cell}, at 25°C:

$$Pt(s)|H_2(g, 0.50\text{ bar})|H_2SO_4(aq, 1.00\text{ M})|PbSO_4(s)|Pb(s)$$

25-88. (*) The cell diagram for the lead-acid cell that is used in automobile and truck batteries is

$$Pb(s)|PbSO_4(s)|H_2SO_4(aq)|PbO_2(s), PbSO_4(s)|Pb(s)$$

where the comma between $PbO_2(s)$ and $PbSO_4(s)$ denotes a heterogeneous mixture of the two solids and the right-hand lead electrode is nonreactive.

(a) Determine the equation for the net cell reaction.

(b) Use the data in Appendix G to calculate the value of E°_{cell}.

(c) Calculate the value of ΔG°_{rxn}.

(d) Calculate the value of E_{cell} at 25°C if $[H_2SO_4]$ = 10.0 M

(e) How many lead-acid cells are in a 12 V car battery?

25-89. (*) A battery that operates at 500°C was developed for the exploration of Venus. The electrodes are a magnesium metal anode and a mixture of copper(I) and copper(II) oxides in contact with an inert steel cathode. The electrolyte is a mixture of LiCl and KCl, which is melted to activate the cell. The MgO that is produced is sparingly soluble in the molten salt mixture and precipitates. Write the equation for the cell reaction and the cell diagram for the cell.

25-90. (*) Given the standard reduction voltages at 25°C for the half-reaction equations

$$Cr^{3+}(aq) + e^- \rightarrow Cr^{2+}(aq) \qquad E^\circ_{red} = -0.407\text{ V}$$

$$Cr^{2+}(aq) + 2\,e^- \rightarrow Cr(s) \qquad E^\circ_{red} = -0.913\text{ V}$$

use these data to determine the standard reduction voltage at 25°C for the half-reaction equation

$$Cr^{3+}(aq) + 3\,e^- \rightarrow Cr(s)$$

Compare your answer to the value listed in Appendix G. Why can't we simply add the reduction voltages of the two half-reaction equations above to get this value?

25-91. (*) From 1882 to 1895 home electricity was provided as direct current rather than as alternating current, as is now the case. Thomas Edison invented a meter to measure the amount of electricity used by a consumer. A small amount of current was diverted to an electrolysis cell that consisted of zinc electrodes in a zinc sulfate solution. Once a month the cathode was removed, washed, dried, and weighed. The bill was figured in ampere-hours (see Problem 25-75). In 1888, Boston Edison Company had 800 chemical meters in service. In one case, in one 30-day period, 65 grams of zinc was deposited on the cathode. The meter used 11% of the current into the house. How many coulombs were used in the month? Calculate the current used in ampere-hours.

25-92. (*) It is well known that you can produce electricity by sticking a copper metal electrode and a zinc metal electrode (such as a penny and a galvanized nail) into a lemon. Explain how this works, given that there are no dissolved copper or zinc ions inside the lemon.

Alfred Werner (1866–1919) was born in Mulhouse, Alsace (then part of France but later annexed by Germany in 1871) to a working family. From 1885 to 1886, he did his military service in Karlsruhe. He then moved to Zurich and attended lectures at the Federal Technical High School, where he became interested in the spatial arrangement of atoms in molecules. He received his doctorate from the University of Zurich in 1892, during which time he laid the foundation of his work on coordination chemistry. Werner was appointed professor of chemistry at the University of Zurich at the age of 29, where he first gave lectures in organic chemistry and later in inorganic chemistry as well. The structure of the compound $CoCl_3{\cdot}6NH_3(s)$, which was an unsolved problem at the time, came to him in a dream, where he pictured a central Co^{3+} ion bonded to six NH_3 molecules situated at the vertices of a regular octahedron to produce a $[Co(NH_3)_6]^{3+}$ ion. At this time, he began the experiments that confirmed his theory of coordination, thus opening up the field of inorganic chemistry. He remained at the University of Zurich until his retirement, although he received many other offers of positions in Europe. He became a Swiss citizen in 1895. Werner was awarded the Nobel Prize in Chemistry in 1913 for his work on coordination chemistry. He was the only inorganic chemist to receive the Nobel Prize before 1973.

Werner was a sociable man who was fond of billiards, chess, and the Swiss card game, Jass. He was an enthusiastic lecturer with a talent for making difficult concepts clear to his students. He developed arteriosclerosis, which forced him to give up his lectures in 1915 and then to retire in 1919. He died at the early age of 53.

26. The Chemistry of the Transition Metals

Transition metals and their compounds are the structural backbone of modern civilization. Human development itself was marked by progress to the Bronze Age and then to the Iron Age. The Industrial Revolution was powered by steam engines made from steels. Today we rely on countless exotic transition metal compounds developed to meet specialized requirements as diverse as those of space probes, personal computers, and ordinary house paint.

We shall not discuss the chemistry of all the transition metals in this chapter but shall focus mainly on the first transition metal series. Just as the first member in the main-group families differs significantly from the others, the chemistry of the first transition metal series also differs appreciably from that of the other series. In particular, the aqueous solution chemistry of the first transition metal series is simpler than that of the heavier transition metals.

The transition metals have a rich and fascinating chemistry. In the first half of the chapter we shall study the physical properties, uses, and important compounds of some selected transition metals. Then, in the second half of the chapter, we shall discuss the coordination chemistry of the transition metals and the formation of transition metal complexes. Transition metal complexes are species in which several anions or neutral molecules, called ligands, bond to a transition metal atom or ion. The chemistry of the transition metals is especially rich and interesting because of these complexes, which occur in a variety of geometries and oxidation states. Much of the chemistry of transition metal complexes can be understood in terms of the electron occupancy of the *d* orbitals of the metal ion. We use *d*-orbital electron configurations to explain many of the spectral, magnetic, and structural properties of transition metal complexes.

26-1. The Maximum Oxidation States of Scandium Through Manganese Are Equal to the Total Number of 4*s* and 3*d* Electrons

Both the websites Wikipedia and Periodic Table Live! contain a wealth of additional information on each of the metals listed here. See links to these sites at www.McQuarrieGeneralChemistry.com.

The ten members of each ***d*-block** transition metal series correspond to the ability of a *d* subshell to hold a maximum of ten electrons. Figure 26.1 gives the ground state outer electron configurations of the 3*d* series (Sc to Zn), 4*d* series (Y to Cd), and 5*d* series (Lu to Hg). Note that the filling of the *d* orbitals is not perfectly regular in all cases. For example, the ground state outer electron configuration of chromium is $4s^13d^5$ rather than $4s^23d^4$ and that of copper is $4s^13d^{10}$ rather than $4s^23d^9$.

Table 26.1 lists the properties and uses of the 3*d* transition metals (Figure 26.2). The chemistry of even this series is especially varied because of the several oxidation states available to many of the metals. We shall briefly discuss the chemical properties of the metals in this series in the next few sections.

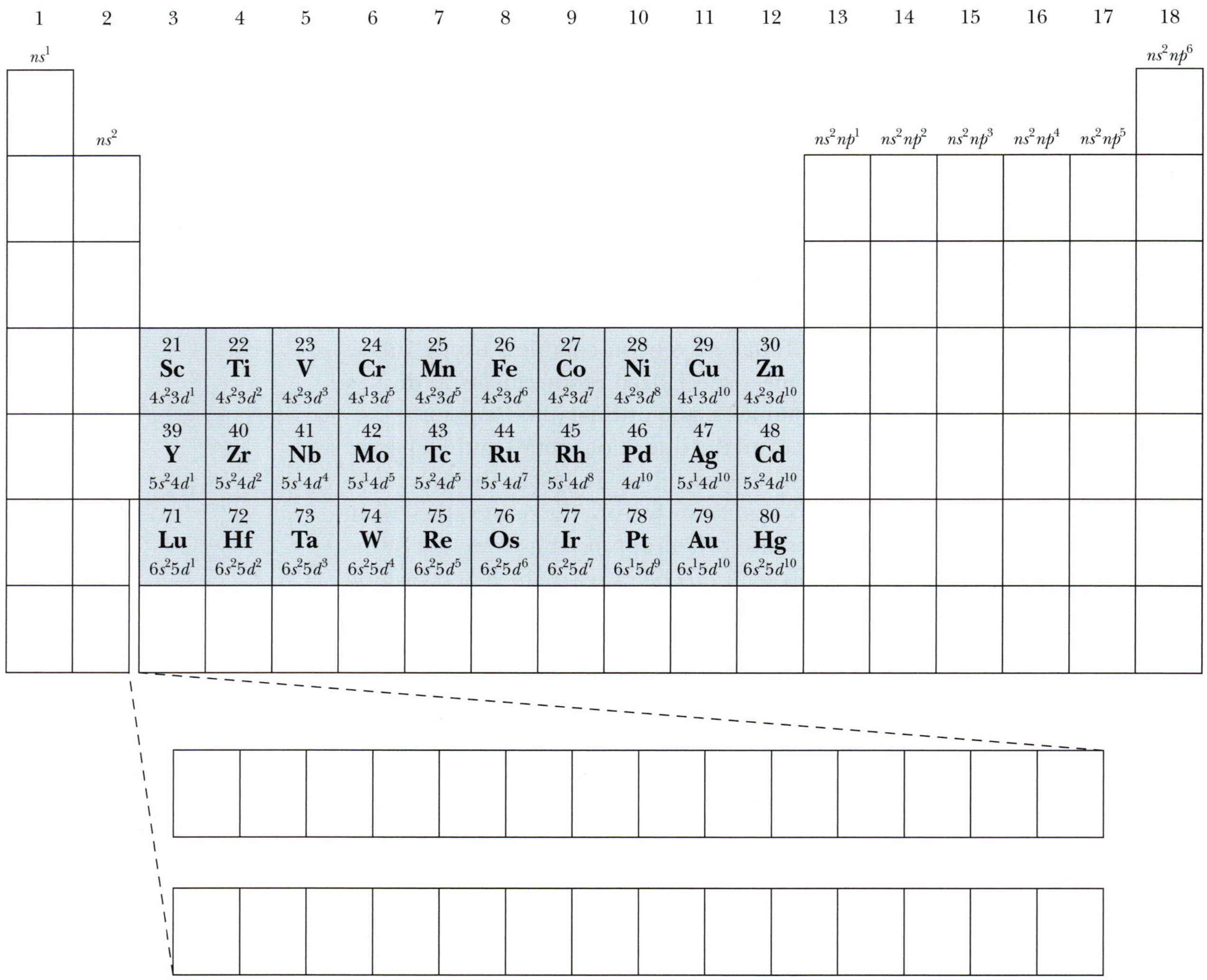

Figure 26.1 The ground state outer electron configurations of the 3*d*, 4*d*, and 5*d* transition metal series.

TABLE 26.1 Properties and uses of the 3*d* transition metals

Element	Density/ $g\cdot cm^{-3}$	Melting point/°C	Principal sources	Uses
scandium	2.99	1541	thortveitite, $(Sc,Y)_2Si_2O_7(s)$	no major industrial uses
titanium	4.506	1668	rutile, $TiO_2(s)$	high-temperature, lightweight steel alloys; $TiO_2(s)$ in white paints
vanadium	6.0	1910	vanadinite, $(PbO)_9(V_2O_5)_3PbCl_2(s)$	vanadium steels (rust-resistant)
chromium	7.15	1907	chromite, $FeCr_2O_4(s)$	stainless steels; chrome plating; catalysis
manganese	7.3	1246	pyrolusite, $MnO_2(s)$ manganosite, $MnO(s)$ nodules on ocean floor	alloys
iron	7.87	1538	hematite, $Fe_2O_3(s)$ magnetite, $Fe_3O_4(s)$	steels; magnetic storage media
cobalt	8.86	1495	cobaltite, $CoS_2\cdot CoAs_2(s)$	linnaeite, $Co_3S_4(s)$ alloys; cobalt-60 radiology; magnets; stainless steels; hard disc media; blue coloring in glasses and glazes
nickel	8.90	1455	pentlandite, $(Fe, Ni)_9S_8(s)$ limonite, $(Fe, Ni)O(OH)(s)$ garnierite, $(Ni, Mg)_3Si_2O_5(OH)(s)$	nickel plating; coins; magnets; catalysts; nickel steels for armor plating; batteries; green coloring in glasses and glazes
copper	8.96	1085	chalcopyrite, $CuFeS_2(s)$	chalcocite, $Cu_2S(s)$ malachite, $Cu_2(CO_3)(OH)_2(s)$ bronzes; brass; coins; electric conductors; algaecide
zinc	7.134	420	zinc blende, $ZnS(s)$ smithsonite, $ZnCO_3(s)$	galvanizing; bronze; brass; dry cells

Figure 26.2 The 3*d* transition metal series. Top row from left to right: scandium, titanium, vanadium, chromium, and manganese. Bottom row: iron, cobalt, nickel, copper, and zinc.

TABLE 26.2 The common oxidation states of the 3*d* transition metals

Element	Common oxidation states
scandium	+3
titanium	+4
vanadium	+2, +3, +4, +5
chromium	+2, +3, +6
manganese	+2, +4, +7
iron	+2, +3
cobalt	+2, +3
nickel	+2
copper	+1, +2
zinc	+2

Scandium

The ground-state electron configuration of scandium is $[Ar]4s^2 3d^1$. Scandium is somewhat similar to aluminum in its chemical properties. Scandium has a +3 oxidation state in almost all its compounds. The common oxidation states of the 3*d* transition metals are given in Table 26.2. It forms a very stable oxide, $Sc_2O_3(s)$, and forms halides with the formula ScX_3. The addition of a base to $Sc^{3+}(aq)$ produces a white, gelatinous precipitate with the formula $Sc_2O_3 \cdot nH_2O(s)$. Like $Al(OH)_3(s)$, hydrated $Sc_2O_3(s)$ is amphoteric (see Section 22-7). Scandium and its compounds have little technological importance at present.

Titanium

The ground state electron configuration of titanium is $[Ar]4s^2 3d^2$. Its most common and stable oxidation state by far is +4, as in the compounds $TiO_2(s)$ and $TiCl_4(l)$, which are covalently bonded.

Pure titanium is a lustrous white metal (Figure 26.3) and is the second most abundant transition metal. It is used to make lightweight alloys that are stable at high temperatures and are used in missiles, high-performance aircraft, bicycles, machine tool coatings, and dental work. Titanium is as strong as most steels but 50% lighter. It is 60% denser than aluminum but twice as strong. In addition, it has excellent resistance to corrosion.

Pure titanium is difficult to prepare because the metal is very reactive at high temperatures. The most important ore of titanium is rutile, which is primarily $TiO_2(s)$ (Figure 26.4). Pure titanium metal is produced by first converting $TiO_2(s)$ to $TiCl_4(l)$ by heating $TiO_2(s)$ to red heat in the presence of carbon and chlorine, as described by

$$TiO_2(s) + C(s) + 2\,Cl_2(g) \xrightarrow{\text{high temperature}} TiCl_4(l) + CO_2(g)$$

The $TiCl_4(l)$, which is a colorless, fuming covalent liquid, is reduced to the metal by reacting it with magnesium in an inert atmosphere of argon, according to

$$TiCl_4(l) + 2\,Mg(s) \rightarrow Ti(s) + 2\,MgCl_2(s)$$

Figure 26.3 Titanium has a relatively low density, high strength, and excellent corrosion resistance.

Figure 26.4 Rutile, an ore of titanium.

TABLE 26.3 Important compounds of titanium, vanadium, chromium, and manganese

Compound	Uses
titanium dioxide, $TiO_2(s)$	ceramic colorant; white paints and lacquers; inks and plastics; gemstones
titanium tetrachloride, $TiCl_4(l)$	smoke screens; iridescent glass; artificial pearls
vanadium (*V*) oxide, $V_2O_5(s)$	production of sulfuric acid; manufacture of yellow glass
vanadyl sulfate, $VOSO_4(s)$	blue- and green-colored glasses and glazes on pottery
chromium(IV) oxide, $CrO_2(s)$	constituent of high-fidelity magnetic recording tapes
chromium(III) oxide, $Cr_2O_3(s)$	constituent of abrasives, refractory materials, and semiconductors; green pigment, especially for coloring glass
sodium dichromate, $Na_2Cr_2O_7(s)$	leather tanning; textile manufacture; metal corrosion inhibitor
manganese(IV) oxide, $MnO_2(s)$	manufacture of manganese steel; alkaline batteries; printing and dyeing textiles; pigment in brick industry
manganese(II) sulfate, $MnSO_4(s)$	dyeing; red glazes on porcelain; fertilizers for vines, tobacco
potassium permanganate, $KMnO_4(s)$	oxidizing agent; medical disinfectant (bladder infection); water and air purification

Most titanium is used in the production of titanium steels, but $TiO_2(s)$, which is white when pure, is used as the white pigment in many paints and plastics. Titanium tetrachloride is used to make smoke screens; when it is sprayed into the air, it reacts with moisture to produce a dense and persistent white cloud of $TiO_2(s)$. The chemical equation for the reaction of $TiCl_4(l)$ with water vapor is

$$TiCl_4(l) + 2H_2O(g) \rightarrow TiO_2(s) + 4HCl(g)$$

Some commercially important titanium compounds are given in Table 26.3.

Vanadium

The ground state electron configuration of vanadium is $[Ar]4s^23d^3$. Its maximum oxidation state is +5, although the +2, +3, and +4 oxidation states are common, with the +2 state being the least common. Vanadium(V) oxide is obtained when vanadium is burned in excess oxygen. It is used as a catalyst in several industrial processes, including the oxidation of $SO_2(g)$ to $SO_3(g)$ that is one step in the production of sulfuric acid by the contact process (Interchapter J).

Vanadium(V) oxide is amphoteric. It dissolves in concentrated bases such as $NaOH(aq)$ to produce a colorless solution in which the principal species above pH = 13 is believed to be $VO_4^{3-}(aq)$. As the pH is lowered, the solution turns orange, and at pH = 2, a precipitate occurs that redissolves at a lower pH to give a pale yellow solution, in which the principal species is believed to be $VO_2^+(aq)$ (Figure 26.5).

WWW

See Interchapter J at www.McQuarrieGeneralChemistry.com.

Figure 26.5 Solutions of $V_2O_5(s)$ at various pH values. Left, $V_2O_5(s)$ dissolved in $NaOH(aq)$ at pH = 13, where the principal species is $VO_4^{3-}(aq)$. Center left, $V_2O_5(s)$ dissolved in hydrochloric acid at pH = 0 where the principal species is the vanadyl ion $VO_2^+(aq)$. Center right, $V_2O_5(s)$ dissolved in hydrochloric acid at pH = 2 with the formation of a precipitate. Right, $V_2O_5(s)$ dissolved in hydrochloric acid at pH = 4.

Except for $V_2O_5(s)$, the compounds of vanadium have limited commercial importance (Table 26.3), but vanadium itself is used in alloy steels, particularly ferrovanadium steel.

26-2. The +6 Oxidation State of Chromium and the +7 Oxidation State of Manganese Are Strongly Oxidizing

Chromium

Chromium, with the ground state electron configuration $[Ar]4s^1 3d^5$, has a maximum oxidation state of +6, although the +2 and +3 states are common (Figure 26.6). Whereas the +4 oxidation state for titanium and the +5 state of vanadium are only mildly oxidizing, the +6 oxidation state of chromium is strongly oxidizing. Indeed, the dichromate ion, $Cr_2O_7^{2-}(aq)$ in acidic solution is a strong oxidizing agent:

$$14\,H^+(aq) + Cr_2O_7^{2-}(aq) + 6\,e^- \rightarrow 2\,Cr^{3+}(aq) + 7\,H_2O(l) \qquad E^\circ_{red} = 1.232\text{ V}$$

Recall from our discussion in Section 25-5 that the larger a positive value of E°_{red}, the stronger the agent's oxidizing power. In contrast to chromium(VI) compounds, the chromium(II) ion is a fairly strong reducing agent:

Figure 26.6 Left, $Na_2CrO_4(aq)$ in $NaOH(aq)$ at pH = 8, where $CrO_4^{2-}(aq)$ is the principal species. Center, the $Na_2CrO_4(aq)$ solution with pH adjusted to 4, where the principal species is $Cr_2O_7^{2-}(aq)$. Right, the $Na_2CrO_4(aq)$ solution with pH adjusted to 0, where the principal species is $H_2CrO_4(aq)$.

$$Cr^{3+}(aq) + e^- \rightarrow Cr^{2+}(aq) \qquad E^\circ_{red} = -0.407\ V$$

A **chromous bubbler** is a freshly prepared solution of $Cr^{2+}(aq)$ that is used to remove traces of oxygen from gases (Figure 26.7). The chromous bubbler solution is prepared by reacting zinc from a zinc-mercury amalgam, Zn(Hg)(*s*), with an aqueous chromium(III) solution to form $Cr^{2+}(aq)$ as described by

$$Zn(Hg)(s) + 2\,Cr^{3+}(aq) \rightarrow 2\,Cr^{2+}(aq) + Zn^{2+}(aq)$$

The resulting chromium(II) nitrate, $Cr(NO_3)_2(aq)$, solution reduces oxygen to water as oxygen is bubbled through the solution according to

$$4\,Cr^{2+}(aq) + O_2(g) + 4\,H^+(aq) \rightarrow 4\,Cr^{3+}(aq) + 2\,H_2O(l)$$

The equilibrium constant for this reaction equation is very large, making it essentially a quantitative reaction.

Chromium metal is used in catalysis, to harden steel, to manufacture stainless steel, and to form various alloys. Chromium metal is also used in electroplating to form corrosion-resistant "chrome" surfaces. Sources and uses of chromium are summarized in Table 26.1, and some important chromium compounds are given in Table 26.3.

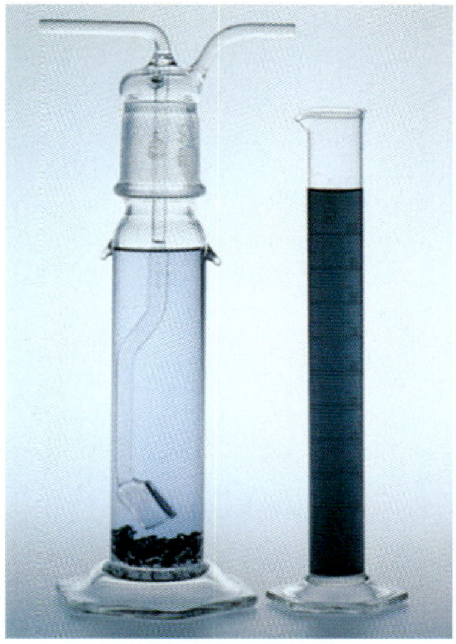

Figure 26.7 Left, a chromous bubbler, which is used to remove traces of oxygen from unreactive gases such as nitrogen or argon. The blue color of the solution is due to $Cr^{2+}(aq)$, and the solid at the bottom of the tube is a zinc amalgam, Zn(Hg)(*s*). The gas to be deoxygenated enters the bubbler at the left inlet tube, bubbles up through the solution of $Cr^{2+}(aq)$, and exits from the right tube. A glass membrane at the bottom of the inlet tube breaks up the entering gas stream into small bubbles. Right, the green solution in the graduated cylinder is $Cr_2(NO_3)_3(aq)$, which is used to prepare the chromous bubbler. The green $Cr^{3+}(aq)$ is reduced to the blue $Cr^{2+}(aq)$ by the zinc amalgam in the bubbler.

Manganese

The ground state electron configuration of manganese is $[Ar]4s^23d^5$. The highest oxidation state of manganese is +7, which is best known in the strongly oxidizing permanganate ion, $MnO_4^-(aq)$, which reacts in both acidic and basic solutions according to

Acidic solution

$$MnO_4^-(aq) + 8\,H^+(aq) + 5\,e^- \rightarrow Mn^{2+}(aq) + 4\,H_2O(l) \qquad E^\circ_{red} = +1.507\ V$$

Basic solution

$$MnO_4^-(aq) + 2\,H_2O(l) + 3\,e^- \rightarrow MnO_2(s) + 4\,OH^-(aq) \qquad E^\circ_{red} = +0.595\ V$$

The most important permanganate salt is potassium permanganate, $KMnO_4(s)$, that is used as an oxidizing agent in industry and medicine, as well as in many general chemistry laboratories. Freshly prepared solutions of $KMnO_4(aq)$ have a deep purple color but turn brown on long standing because $MnO_4^-(aq)$ oxidizes water to oxygen and is thereby reduced to $MnO_2(s)$ (Figure 26.8). The net equation is

$$\underset{\text{purple}}{4\,MnO_4^-(aq)} + 2\,H_2O(l) \rightarrow \underset{\text{brown}}{4\,MnO_2(s)} + 3\,O_2(g) + 4\,OH^-(aq)$$

The reaction is catalyzed by $MnO_2(s)$ and thus is **autocatalytic**.

Manganese(II) forms soluble salts with most common anions. For the +3 and +4 oxidation states, the most important compounds are the oxides $Mn_2O_3(s)$ and $MnO_2(s)$. Manganese(IV) oxide as the mineral pyrolusite is an important ore of manganese and is the source of most manganese compounds. Commercially important manganese compounds are given in Table 26.3.

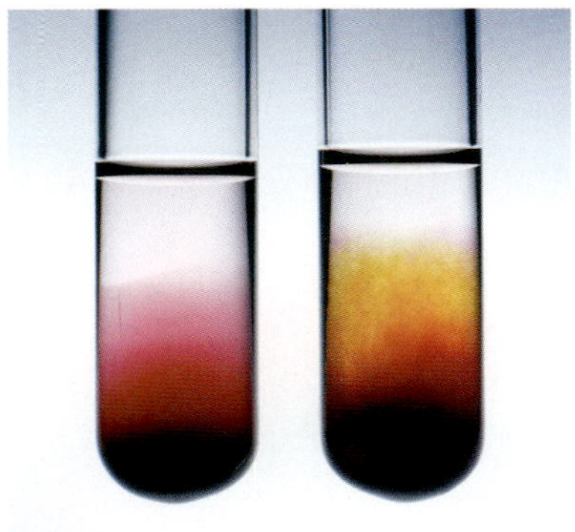

Figure 26.8 An aqueous solution of potassium permanganate, $KMnO_4(aq)$, is a strong oxidizing agent. Freshly prepared $KMnO_4(aq)$ is purple (*left*). As the solution stands, brown $MnO_2(s)$ precipitates as a result of the reduction of $KMnO_4(aq)$ by $H_2O(l)$ (*right*).

26-3. Iron Is Produced in a Blast Furnace

Although the ground state electron configuration of iron is [Ar]$4s^2 3d^6$, starting with iron and continuing across the row of $3d$ transition metals through zinc, we no longer associate the highest oxidation state with the total number of $4s$ and $3d$ electrons. The highest known oxidation state of iron is +6, but it is very rare. Only the +2 and +3 oxidation states of iron are common.

Iron is found in nature as hematite, $Fe_2O_3(s)$; magnetite, $Fe_3O_4(s)$; siderite, $FeCO_3(s)$; and iron pyrite, $FeS_2(s)$, called fool's gold (Figure 26.9). It is the most abundant transition metal, constituting 4.7% by mass of the earth's crust. It is the cheapest metal, and, in the form of steel, the most useful. Pure iron is a silvery white, soft metal that rusts rapidly in moist air. We have already discussed this familiar example of corrosion in Section 24-7. Iron is little used as the pure element but is strengthened greatly by the addition of small amounts of carbon and of various other transition metals. Its principal use is in the production of steel, but some other important compounds and their uses are given in Table 26.4.

Figure 26.9 Iron ores. Clockwise from the left: magnetite, $Fe_3O_4(s)$; siderite, $FeCO_3(s)$; iron pyrite, $FeS_2(s)$; and hematite, $Fe_2O_3(s)$.

TABLE 26.4 Important iron compounds

Compound	Uses
iron(III) chloride, $FeCl_3(s)$	treatment of wastewater; etching for engraving copper for printed circuitry; feed additive
iron(III) oxide, $Fe_2O_3(s)$	metallurgy; paint and rubber pigment; memory cores for computers; magnetic tapes; polishing agent for glass, precious metals, and diamonds
iron(II) sulfate, $FeSO_4(s)$	flour enrichment; wood preservative; water and sewage treatment; manufacture of ink
iron(III) sulfate, $Fe_2(SO_4)_3(s)$	soil conditioner; disinfectant; etching aluminum; wastewater treatment

Iron is also a vital constituent of mammalian cells. It occurs in a number of proteins, such as hemoglobin, which transports oxygen from the lungs to muscles and other tissues, where the oxygen is used in the cellular oxidation reactions. It has been estimated that an adult human being contains about five grams of iron.

Millions of tons of iron are produced annually in the United States by the reaction of $Fe_2O_3(s)$ with **coke** (coal from which the volatile components have been driven off). This reaction is carried out in a **blast furnace**. About 30 meters high and 8 meters wide, the modern blast furnace produces about 5000 metric tons of iron daily (Figure 26.10). A mixture of iron ore, coke, and limestone, $CaCO_3(s)$, is loaded into the top, and preheated compressed air and oxygen are blown in near the bottom. The reaction of the coke and the oxygen to produce carbon dioxide gives off a great deal of heat, and the temperature in the lower region of a blast furnace is around 1900°C. As the $CO_2(g)$ rises, it reacts with more coke to produce hot carbon monoxide, which reduces the iron ore to iron according to

$$Fe_2O_3(l) + 3\,CO(g) \xrightarrow{\text{high temp}} 2\,Fe(l) + 3\,CO_2(g)$$

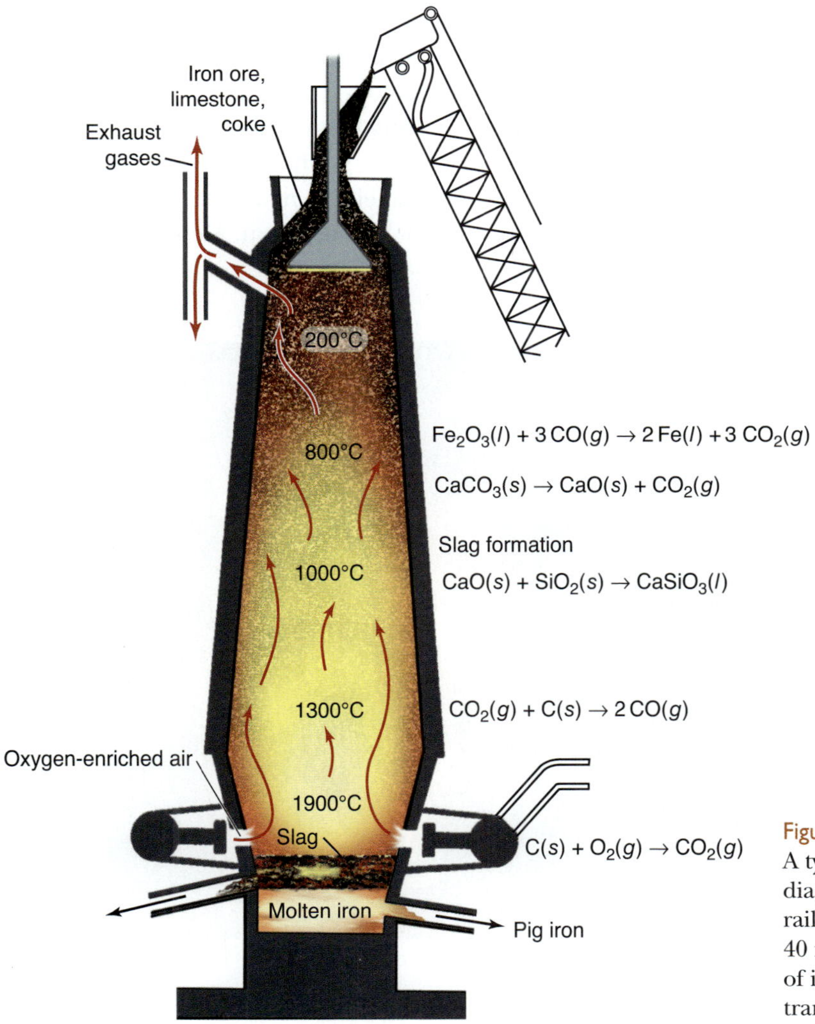

Figure 26.10 Iron is produced in a blast furnace. A typical blast furnace like the one shown in this diagram runs continuously and consumes about 120 railroad cars of iron ore, 50 railroad cars of coke, and 40 railroad cars of limestone per day. The 5000 tons of iron produced requires about 75 railroad cars to transport it.

The molten iron metal is denser than the other substances and drops to the bottom, where it can be drained off to form ingots of what is called pig iron.

The function of the limestone is to remove the sand and gravel that normally contaminate the iron ore. The intense heat decomposes the limestone to CaO(*s*) and $CO_2(g)$. The CaO(*s*) combines with the sand and gravel (both of which are primarily silicon dioxide) to form molten calcium silicate, $CaSiO_3(l)$.

$$\mathrm{CaO}(s) + \mathrm{SiO_2}(s) \rightarrow \underset{\text{slag}}{\mathrm{CaSiO_3}(l)}$$

The molten calcium silicate, called slag, floats on top of the molten iron and is drained off periodically. It is used in building materials, such as cement and concrete aggregate, rock-wool insulation, cinder blocks, and as railroad ballast.

Pig iron contains about 4 or 5% carbon together with lesser amounts of silicon, manganese, phosphorus, and sulfur. It is brittle, difficult to weld, and not strong enough for structural applications. To be useful, pig iron must be converted to steel, which is an alloy of iron with small but controlled amounts of other metals and between 0.1 and 1.5% carbon. Steel is made from pig iron in several different processes, all of which use oxygen to oxidize most of the impurities. One such process is the **basic oxygen process**, in which hot, pure $O_2(g)$ is blown through molten pig iron (Figure 26.11). The oxidation of carbon and phosphorus is complete in less than one hour. The desired carbon content of the steel is then achieved by adding high-carbon steel alloy.

There are two types of steels: carbon steels and alloy steels. Both types contain carbon, but carbon steels contain essentially no other metals besides iron. About 90% of all steel produced is carbon steel. Carbon steel that contains less than 0.2% carbon is called **mild steel**. Mild steels are malleable and ductile and are used where load-bearing ability is not a consideration. **Medium steels** that

Figure 26.11 Molten iron being charged into a basic oxygen furnace. Most steel is produced by a process called the basic oxygen process. A typical basic oxygen furnace is charged with about 200 tons of molten pig iron, 100 tons of scrap iron, and 20 tons of limestone (to form a slag). A stream of hot oxygen is blown through the molten mixture, and the oxidized impurities are blown out of the iron. High-quality steel is produced in an hour or less.

contain 0.2 to 0.6% carbon are used for such structural materials as beams and girders and for railroad equipment. **High-carbon steels** contain 0.8 to 1.5% carbon and are used to make drill bits, knives, and other tools in which hardness is important.

Alloy steels contain other metals in small amounts. Different metals give different properties to steels. The alloy steels called **stainless steels** contain high percentages of chromium and nickel. Stainless steels resist corrosion and are used for cutlery and hospital equipment. The most common stainless steel contains 18% chromium and 8% nickel.

26-4. The +2 Oxidation State Is the Most Important Oxidation State for Cobalt, Nickel, Copper, and Zinc

As we go from iron to zinc, there is an increasing prominence of the +2 oxidation state. Most compounds of cobalt and almost all compounds of nickel and zinc involve the metal in the +2 oxidation state. Only copper, which has an important +1 oxidation state, and cobalt, which has an important +3 oxidation state, have extensive chemistries not involving the +2 state.

Cobalt

Cobalt is a fairly rare element and is usually found associated with nickel in nature. It is a hard, bluish-white metal that is used in the manufacture of high-temperature alloys such as alnico, an aluminum-nickel-cobalt alloy used in permanent magnets. The pure metal is relatively unreactive and dissolves only slowly in dilute mineral acids. When cobalt is burned in oxygen, a mixture of $CoO(s)$ and $Co_3O_4(s)$ is obtained. Blue "cobalt" glass contains a small amount of $CoO(s)$ (Figure 26.12). Most simple cobalt salts involve Co(II). The species $Co^{3+}(aq)$ is a strong oxidizing agent, $E^{\circ}_{red}[Co^{3+}|Co^{2+}] = 1.92$ V, and can oxidize water. Cobalt is an important element in the computer industry; various alloys of cobalt are used as the media (magnetic surface material) of most hard disk drives. Other commercially important cobalt compounds are given in Table 26.5.

Figure 26.12 Cobalt glass. The characteristic blue color is due to the presence of small amounts of $CoO(s)$.

TABLE 26.5 Important compounds of cobalt and nickel

Compound	Uses
cobalt(II) oxide, $CoO(s)$	glass and ceramic coloring and decolorization
cobalt(II) phosphate, $Co_3(PO_4)_2(s)$	lavender pigment in paints and ceramics
cobalt(II) sulfate, $CoSO_4(s)$	storage batteries; agent in ceramics, enamels, glazes to prevent discoloring
nickel(II) cyanide, $Ni(CN)_2(s)$	metallurgy; electroplating
nickel(II) chloride, $NiCl_2(s)$	nickel plating; absorbent for $NH_3(g)$ in gas masks

Nickel

Nickel occurs in a variety of sulfide ores, the most important deposit being found in the Sudbury basin of Ontario, Canada. The metal is obtained by roasting the ore to obtain $NiO(s)$ and then reducing with hydrogen or carbon. Very pure nickel is obtained by electrolysis. Nickel is a silvery metal that takes a beautiful high polish, which is protected by a spontaneously formed transparent layer of the oxide $NiO(s)$. It is used in a number of magnetic alloys and in the alloy Monel, which is used to handle fluorine and other reactive fluorine compounds.

Nickel is more reactive than cobalt and dissolves readily in dilute acids. The aqueous solution chemistry of nickel involves primarily the species $Ni^{2+}(aq)$. Important nickel compounds are given in Table 26.5.

Copper

Copper is only slightly less abundant than nickel. Deposits of the free metal are very rare; it generally occurs as various sulfides. Most copper-containing deposits have a copper content of less than 1%, although some richer deposits have up to 4% copper (Figure 26.13). Copper ores contain other metals and semimetals such as selenium and tellurium, which are important by-products of copper production. Important copper minerals are chalcocite, $Cu_2S(s)$; chalcopyrite, $CuFeS_2(s)$; and malachite, $CuCO_3{\cdot}Cu(OH)_2(s)$. The total world output of copper metal is about 10 million metric tons, with U.S. production accounting for over 10% of the total. The extraction of copper from its ores is a multistep process, the final step being purification by electrolysis (Section 25-9).

Copper is a reddish, soft, ductile metal that takes on a bright metallic luster. Its important use is as an electrical conductor. The only metal that is a better conductor is silver, but the price of silver precludes its widespread use. Although copper is fairly unreactive, its surface turns green after long exposure to the atmosphere. The green patina (Figure 26.14) is due to the surface formation of copper hydroxocarbonate, $Cu_2(OH)_2CO_3(s)$ and copper hydroxosulfate, $Cu_3(OH)_4SO_4(s)$.

Figure 26.13 An open pit copper mine at Bingham Canyon, located about 25 miles southwest of Salt Lake City, Utah. Since opening in 1903, this mine has produced more than 18 million tons of copper ore.

Brass, an alloy of copper with zinc, and **bronze**, an alloy of copper with tin, are among the earliest known alloys. Bronze usually contains from 5 to 10% tin and is very resistant to corrosion. It is used for casting, marine equipment, fine arts work, and spark-resistant tools. Yellow brasses contain about 35% zinc and have good ductility and high strength. Brass is used for piping, valves, hose nozzles, marine equipment, and jewelry and in the fine arts.

Copper does not replace hydrogen from dilute acids, but it does react with oxidizing acids such as concentrated nitric acid or hot concentrated sulfuric acid according to

$$3\,Cu(s) + 8\,HNO_3(aq) \rightarrow 3\,Cu(NO_3)_2(aq) + 2\,NO(g) + 4\,H_2O(l)$$

$$Cu(s) + 2\,H_2SO_4(conc) \rightarrow CuSO_4(aq) + 2\,H_2O(l) + SO_2(g)$$

Commercially important copper compounds are given in Table 26.6.

Most compounds of copper involve copper(II), many of which are blue or bluish-green. The most common copper(II) salt is copper(II) sulfate pentahydrate, $CuSO_4 \cdot 5H_2O(s)$, which occurs as beautiful blue crystals (Figure 26.15). When the crystals are heated gently, the water of hydration is driven off to produce anhydrous copper(II) sulfate, $CuSO_4(s)$, which is a white powder (Figure 26.15). When $NH_3(aq)$ is added to aqueous solutions containing $Cu^{2+}(aq)$, an intense dark blue color occurs as a result of the formation of a copper-ammonia complex ion according to

$$[Cu(H_2O)_6]^{2+}(aq) + 4\,NH_3(aq) \rightarrow [Cu(NH_3)_4(H_2O)_2]^{2+}(aq) + 4\,H_2O(l)$$

Figure 26.14 The green patina of the Statue of Liberty is due to the surface formation of copper hydroxocarbonate and hydroxosulfate compounds. A major restoration of the statue's surface ended in 1986.

We discuss the chemistry of transition metal complexes in Sections 26-7 through 26-13.

TABLE 26.6 Important compounds of copper and zinc

Compound	Uses
copper(II) arsenite, $Cu(AsO_2)_2(s)$ (Scheele's green)	pigment; wood preservative; insecticide, fungicide, rodenticide, mosquito control
copper(II) nitrate, $Cu(NO_3)_2(s)$	light-sensitive paper; insecticide for vines; wood preservative
copper(I) oxide, $Cu_2O(s)$	antifouling paints for wood and steel exposed to seawater; fungicide; porcelain red glaze; red glass
copper(II) sulfate pentahydrate, $CuSO_4 \cdot 5H_2O(s)$	soil and feed additive; germicide; petroleum and rubber industries; laundry and metal-marking inks
zinc carbonate, $ZnCO_3(s)$	white pigment; porcelain, pottery, and rubber manufacture; astringent and antiseptic
zinc chloride, $ZnCl_2(s)$	deodorants; disinfectants; fireproofing and preserving wood; adhesives; dental cements; taxidermist fluid; artificial silk; parchment paper
zinc oxide, $ZnO(s)$	ointment; pigment and mold inhibitor in paints; floor tile; cosmetics; color photography; dental cements; automobile tires

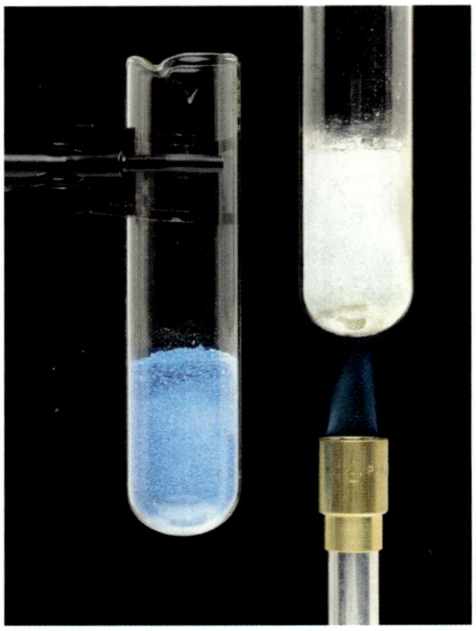

Figure 26.15 Copper(II) sulfate pentahydrate, $CuSO_4 \cdot 5H_2O(s)$, crystals are blue, and copper(II) sulfate, $CuSO_4(s)$, is a white powder.

Figure 26.16 A comparison of the color of $[Cu(H_2O)_6]^{2+}(aq)$ and $[Cu(NH_3)_4(H_2O)_2]^{2+}(aq)$ (top layer) after the addition of NH_3 (aq).

A comparison of the colors of $[Cu(H_2O)_6]^{2+}(aq)$ and $[Cu(NH_3)_4(H_2O)_2]^{2+}(aq)$ is shown in Figure 26.16.

Copper(I) salts are often colorless and only slightly soluble in water. The $Cu^+(aq)$ ion is unstable and undergoes an oxidation-reduction reaction with itself of the type described by: $2Cu^+(aq) \rightarrow Cu(s) + Cu^{2+}(aq)$.

Zinc

Zinc, which is widely distributed in nature, is about as abundant as copper. Its principal ores are sphalerite (or zinc blende), $ZnS(s)$, and smithsonite, $ZnCO_3(s)$, from which zinc is obtained by roasting and reduction of the resultant $ZnO(s)$ with carbon. Zinc is a shiny white metal with a bluish-gray luster.

The $3d$ subshell of zinc is completely filled, so zinc behaves more like a Group 2 metal than like a transition metal. Metallic zinc is a strong reducing agent. It dissolves readily in dilute acids and combines with oxygen, sulfur, phosphorus, and the halogens upon being heated. The only important oxidation state of zinc is +2, and zinc(II) salts are colorless, unless color is imparted by the anion. Zinc is found in nearly every cell in the body, particularly in the skin, hair, nails, eyes, and testes. It is required for the proper healing of wounds, healthy skin and hair, and male virility. The human body contains about two grams of zinc. Commercially important zinc(II) salts are given in Table 26.6.

26-5. Gold, Silver, and Mercury Have Been Known Since Ancient Times

Gold has been mined since at least 4000 BC. Many gold artifacts have been found in Egyptian tombs dating as early as 2500 BC; a beautiful example is the gold mask found in the tomb of King Tutankhamun. Silver is believed to have been mined as early as gold. The ancient Greeks and Romans were familiar with silver and had developed techniques to mine it. Mercury is thought to have been discovered later than gold and silver and has been found in Egyptian tombs dating from 1500 BC.

Figure 26.17 Gold often occurs in the free state in nature. Here we see native gold on a quartz crystal.

Gold

Gold is a very dense, soft, yellow metal with a high luster (Figure 26.17). It is found in nature as the free element and in tellurides. It occurs in veins and alluvial deposits and is often separated from rocks and other minerals by sluicing or panning. Over two-thirds of the gold produced by the Western world comes from South Africa. In many mining operations, about five grams of gold is recovered from one ton of rock.

Pure gold is soft and often alloyed to make it harder. The amount of gold in an alloy is expressed in karats: pure gold is 24 karat; coinage gold is 22 karat, or $(22/24) \times 100 = 92\%$. White gold, which is used in jewelry, is usually an alloy of gold and nickel. Gold is very unreactive, so it has a remarkable resistance to corrosion. It is also an excellent conductor of electricity. In addition to its use in jewelry and as a world monetary standard, gold is used in microelectronic devices (Figure 26.18). It is also used extensively in dentistry for tooth crowns.

Figure 26.18 Gold is used in the production of printed circuits.

Gold is extracted from ores by reaction with sodium cyanide, $NaCN(aq)$, and oxygen according to

$$4\,Au(s) + 8\,CN^-(aq) + O_2(g) + 2\,H_2O(l) \rightarrow 4[Au(CN)_2]^-(aq) + 4\,OH^-(aq)$$

This oxidation-reduction reaction is driven by the formation of the very stable

TABLE 26.7 Important compounds of gold, silver, and mercury

Compound	Uses
gold(I) stannate, $Au_2SnO_2(s)$	manufacture of ruby glass, colored enamels, and porcelain
tetrachloroauric(III) acid, $HAuCl_4(s)$	photography; gold plating; gilding glass and porcelain
silver iodide, $AgI(s)$	dispersed in clouds to induce rain; fast film photography
silver nitrate, $AgNO_3(s)$	manufacture of mirrors; silver plating; hair-darkening agent; eyedrops for newborn infants
silver oxide, $Ag_2O(s)$	Anode material in silver-zinc button batteries
silver bromide, $AgBr(s)$	photography; photosensitive lenses
mercury(I) chloride, $Hg_2Cl_2(s)$ (calomel)	calomel electrodes; formerly used as a fungicide and for control of root maggots on cabbage and onions
mercury(II) chloride, $HgCl_2(s)$	formerly used as a preservative for wood and anatomical specimens; embalming agent; photographic intensifier
mercury(II) oxide, $HgO(s)$	formerly used as marine paints and porcelain pigments

$[Au(CN)_2]^-(aq)$ complex ion. The gold is recovered from the $[Au(CN)_2]^-(aq)$ complex ion either by the replacement reaction described by

$$2[Au(CN)_2]^-(aq) + Zn(dust) \rightarrow [Zn(CN)_4]^{2-}(aq) + 2Au(s)$$

or by electrolysis. Important gold compounds are listed in Table 26.7.

Silver

Silver is a lustrous, white metal whose ductility and malleability are exceeded only by those of gold and palladium. Pure silver also has the highest electrical conductivity of all metals. Most of the silver produced today is a by-product of the production of other metals such as copper, lead, and zinc. Its uses include jewelry, silverware, wires, coinage, and in silver-zinc button batteries (see Interchapter U). Other important compounds appear in Table 26.7.

WWW

See Interchapter U at www.McQuarrieGeneralChemistry.com.

The silver halides illustrate the phenomenon that the first member of a family in the periodic table may differ some from the others. Although $AgCl(s)$, $AgBr(s)$, and $AgI(s)$ are insoluble in water, $AgF(s)$ is very soluble, having a solubility in water of $1800\ g\cdot L^{-1}$.

Mercury

Figure 26.19 Cinnabar, $HgS(s)$, and mercury, $Hg(l)$.

Like gold and silver, mercury has been known for thousands of years; it used to be called quicksilver because it is the only metal that is a liquid at 25°C and often forms small balls that move around rapidly on uneven surfaces. The principal ore of mercury is cinnabar, $HgS(s)$, which was widely used in the ancient world as a vermilion pigment (Figure 26.19). The most extensive and richest deposits of cinnabar occur in the Almaden region of Spain, the world's largest producer of mercury. The metal is easily recovered from its ore by roasting as described by the chemical equation

$$HgS(s) + O_2(g) \rightarrow Hg(l) + SO_2(g)$$

The mercury is then purified by distillation.

Mercury is not very reactive. On being heated, it reacts with oxygen, sulfur, and the halogens, but not with nitrogen, phosphorus, hydrogen, or carbon. When mercury is heated in air to around 300°C, it reacts with oxygen to produce the bright orange–red mercury(II) oxide, $HgO(s)$. When $HgO(s)$ is heated to about 400°C, it decomposes according to

$$2HgO(s) \xrightarrow{400^\circ C} 2Hg(l) + O_2(g)$$

Like copper, mercury does not replace hydrogen from acids, but it does react with oxidizing acids such as concentrated nitric acid or hot concentrated sulfuric acid according to

$$3Hg(l) + 8HNO_3(aq) \rightarrow 3Hg(NO_3)_2(aq) + 2NO(g) + 4H_2O(l)$$

$$Hg(l) + 2H_2SO_4(aq) \rightarrow HgSO_4(aq) + SO_2(aq) + 2H_2O(l)$$

Mercury compounds occur as Hg(I) or Hg(II), with Hg(II) being the more common oxidation state. Except for the nitrate, acetate, and perchlorate salts,

Hg(I) salts are insoluble. A notable feature of Hg(I) salts is that they consist of the diatomic Hg_2^{2+} ion. Many of the salts of Hg(I) and Hg(II) are covalently bonded. Some mercury compounds are listed in Table 26.7.

Mercury compounds are poisonous and were once used in insecticides, fungicides, rodenticides, and disinfectants. Mercury was believed by alchemists to be an essential ingredient in longevity potions, often to deadly effect. In the last century, mercury compounds were used in the production of felt for hats; the felt workers suffered from a nervous disorder called "hatter's shakes," which led to the expression "mad as a hatter." The discharge of mercury-containing industrial wastes into rivers, lakes, and oceans caused serious environmental problems, culminating in the Minamata disaster in Japan in 1952, when over 50 people died of mercury poisoning. The mercury effluent was converted to the organomercury compounds by certain sedimentary bacteria, entered the marine food chain, and became concentrated in fish, which was the main diet of the fishing village of Minamata. Since then, the disposal of mercury wastes has been regulated and mercury levels in food are constantly monitored. In 1972, more than 90 nations agreed on an international ban on the dumping of mercury wastes. In addition, many countries have now banned or have placed limits on the use of mercury and mercury-containing compounds.

26-6. Each *d*-Block Transition Metal Ion Has a Characteristic Number of *d* Electrons

Much of the chemistry of the transition metals is determined by the shape of the *d* orbitals. The shapes and relative spatial orientations of the five *d* orbitals are shown in Figure 26.20. (see also Section 5-11). These orbitals are distinguished by *x*, *y*, and *z* subscripts that define the orientation of the orbitals with respect to the *x*, *y*, and *z* coordinate axes. Thus, the *d* orbitals are named d_{xy}, d_{xz}, d_{yz}, $d_{x^2-y^2}$, and d_{z^2} **orbitals**. In the absence of any external electric or magnetic field, the energies of the five *d* orbitals for a given value of the principal quantum number are equal.

Now consider transition metal ions. Recall from Chapters 5 and 6 that the order of filling of the orbitals in these ions is regular in the sense that it follows the arithmetic sequence: the $n = 1$ shell is filled first, then $n = 2$, then $n = 3$, and so on, until all the electrons are accommodated. For example, the ten ions of the 3*d* series with oxidation state +2 are

Just as we observed that not all atoms follow the same orbital filling order, so too there are exceptions to the orbital filling order of certain transition metal ions. However, for simplicity, we shall take the orbital filling order of all such ions to be regular.

d^1	d^2	d^3	d^4	d^5	d^6	d^7	d^8	d^9	d^{10}
Sc(II)	Ti(II)	V(II)	Cr(II)	Mn(II)	Fe(II)	Co(II)	Ni(II)	Cu(II)	Zn(II)
21	22	23	24	25	26	27	28	29	30

We write the *d*-orbital electrons as d^x, where the *d* indicates the set of orbitals and the *x* indicates the number of electrons in the *d* orbitals. In general, a transition-metal ion with *x d* electrons is called a $\boldsymbol{d^x}$ **ion**. Below the designation d^x, we have given the symbol for the ion, with its oxidation state in parentheses. The row of numbers below the ion symbols shows the corresponding atomic number of each element. Note that the second digit in the atomic number for the M(II) ions is the same as the number of *d* electrons, except for zinc where 0 is interpreted as 10. For example, scandium is element 21 and Sc(II) is a d^1 ion; manganese is element 25 and Mn(II) is a d^5 ion. We can use this rule to quickly find the number of *d* electrons in any 3*d*-series transition-metal ion.

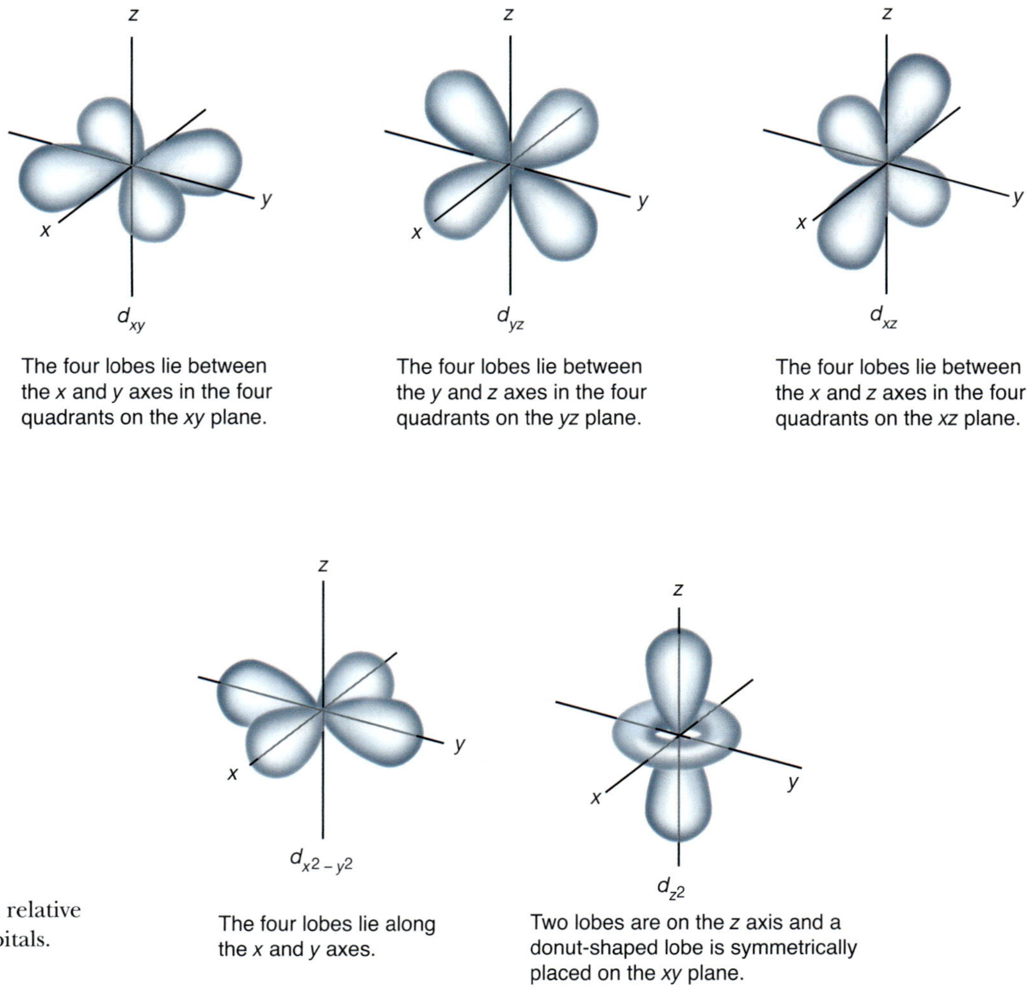

Figure 26.20 The shapes and relative orientations of the five *d* orbitals.

The M(III) ions of the 3*d* series have one fewer electron than the M(II) ions. Thus, for example, Fe(II) and Co(III) both have six *d* electrons; for this reason they both are called d^6 ions. Removal of an electron from Fe(II) yields Fe(III), which is a d^5 ion.

For ions in the 4*d* series, we can quickly determine the number of *d* electrons by noting the position of the element relative to those of the 3*d* series. For example, rhodium is directly below cobalt (element 27); thus, Rh(II) is a d^7 ion. For ions in the 5*d* series, the second digit of the atomic number again correlates with the number of *d* electrons in the M(II) ion. Thus, Pt(II) ($Z = 78$) is a d^8 ion.

EXAMPLE 26-1: Determine the number of outer-shell *d* electrons in Ir(III), Pt(IV), and Mo(III).

Solution: Iridium(II) ($Z = 77$) is a d^7 ion; so Ir(III), which has one fewer electron, is a d^6 ion. For platinum, $Z = 78$; so Pt(II) is a d^8 ion. Platinum(IV) has two fewer electrons than Pt(II); so Pt(IV) is a d^6 ion. Molybdenum ($Z = 42$) is below chromium ($Z = 24$); so Mo(II) is a d^4 ion and Mo(III), which has one fewer electron than Mo(II), is a d^3 ion.

PRACTICE PROBLEM 26-1: Give three examples of d^6 ions with an oxidation state of +3.

Answer: Co(III), Rh(III), and Ir(III)

The wide variety of possible oxidation states of transition metals is the primary reason for their extensive and interesting chemistry. Nowhere can we find a clearer illustration of this fact than that afforded by an examination of their ability to form numerous complex ions.

26-7. Transition Metal Complexes Consist of Central Metal Atoms or Ions That Are Bonded to Ligands

Simple cyanide salts, such as sodium cyanide, NaCN(*s*), are deadly poisons. An aqueous solution of NaCN(*aq*) contains the ions $Na^+(aq)$ and $CN^-(aq)$, and the solution is colorless. The toxicity is due to the $CN^-(aq)$ ions, which bind to iron in the enzyme cytochrome c oxidase found in our cells' mitochondria, blocking the binding of oxygen to this protein. This effectively shuts down our cells' ability to produce the energy needed for tissue such as our heart and nervous system. If excess iron(II) nitrate, $Fe(NO_3)_2(aq)$, is added to NaCN(*aq*), then the solution turns yellow. Chemical tests show that $CN^-(aq)$ is no longer present and that the solution, although still somewhat poisonous, no longer causes essentially instantaneous death. What happens is that $CN^-(aq)$ reacts with $Fe^{2+}(aq)$ to form the ion $[Fe(CN)_6]^{4-}(aq)$, in which six cyanide ions are bonded directly to the iron ion in an octahedral structure. The $[Fe(CN)_6]^{4-}(aq)$ ions exist in solution as integral units called complex ions (Figure 26.21). The charge on the complex ion $[Fe(CN)_6]^{4-}(aq)$ is −4, and its sodium salt has the formula $Na_4[Fe(CN)_6](s)$. If $Na_4[Fe(CN)_6](s)$ is dissolved in water, then the resulting solution contains the ions $Na^+(aq)$ and $[Fe(CN)_6]^{4-}(aq)$

$$Na_4[Fe(CN)_6](s) \xrightarrow[H_2O(l)]{} 4Na^+(aq) + [Fe(CN)_6]^{4-}(aq)$$

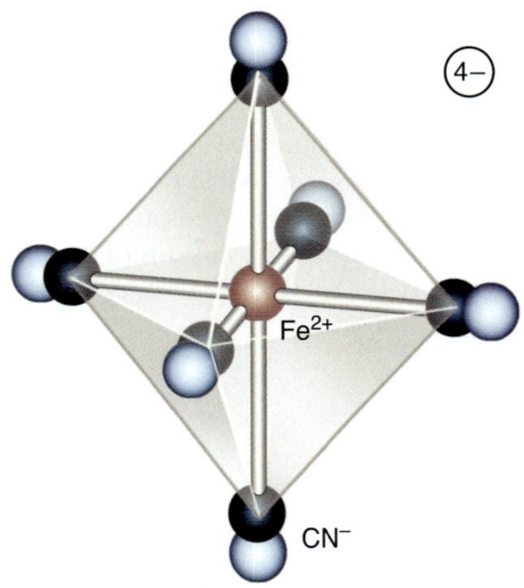

Figure 26.21 The complex ion $[Fe(CN)_6]^{4-}$ is octahedral, with the six cyanide ions bonded to the central iron atom. Aqueous solutions of $[Fe(CN)_6]^{4-}(aq)$ are yellow in color.

Generally, a transition metal **complex ion** contains a central metal ion to which are attached anions, neutral molecules, or some of each. As our example of $[Fe(CN)_6]^{4-}(aq)$ formation shows, a complex ion is a distinct chemical species, with properties different from those of its constituents. The anions or neutral molecules attached directly to the metal ion or atom are called **ligands**, and the number of ligands that are attached to the central metal ion or atom is called the **coordination number**. Transition metal complexes often are called **coordination complexes** to emphasize the bonding (i.e., the coordinations) of the ligands to the central metal ion or atom.

The sodium ions in the salt, $Na_4[Fe(CN)_6](s)$, are called **counter ions** because the positively charged sodium atoms counter (or balance) the negative charge of the transition metal complex. Note that counter ions are not ligands. For a soluble salt containing a transition metal complex, we can distinguish between the ligands and the counter ions by observing what ions form upon dissolution in water. For example, a soluble salt with the empirical formula $PtCl_4(NH_3)_6(s)$ is found to dissociate in water according to the equation

$$PtCl_4(NH_3)_6(s) \xrightarrow[H_2O(l)]{} [Pt(NH_3)_6]^{4+}(aq) + 4Cl^-(aq)$$

Thus, we see that $[Pt(NH_3)_6]^{4+}$ is the formula of the transition metal complex ion and the four chloride ions in the empirical formula are counter ions. The chemical formula for the salt is written correctly as $[Pt(NH_3)_6]Cl_4(s)$, where we enclose the formula of the transition metal complex in brackets. As with other salts, we list cations first and anions second.

Transition metal ions are capable of bonding to a wide variety of anions and neutral molecules to form complex ions and **neutral complexes** (i.e., complexes with zero net charge). For example, some transition metal ions in aqueous solution form complex ions by bonding to water molecules. A $Ni^{2+}(aq)$ ion, for instance, forms the octahedral complex ion $[Ni(H_2O)_6]^{2+}(aq)$ in aqueous solution (Figure 26.22). A solution of nickel(II) perchlorate, $Ni(ClO_4)_2(aq)$, is brilliant green because of the presence of the $[Ni(H_2O)_6]^{2+}(aq)$ complex ions. If $NH_3(aq)$ is added to the solution, the color changes from green to blue-violet (Figure 26.23). The chemical reaction responsible for this color change involves a change in the ligands attached to the nickel(II) ion, as described by

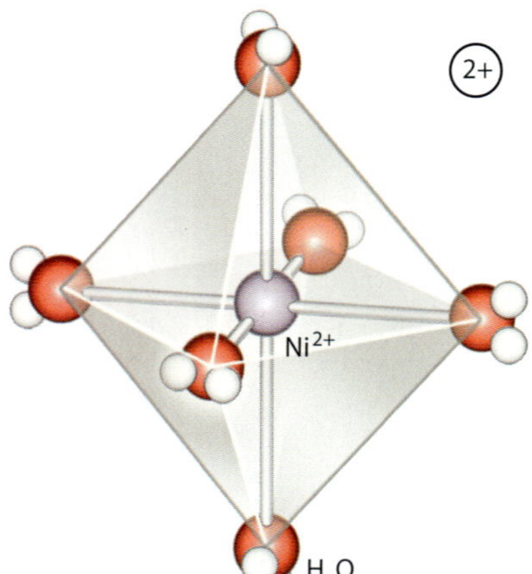

Figure 26.22 Octahedral structure of the complex ion $[Ni(H_2O)_6]^{2+}$. All six nickel–oxygen bonds are equivalent. Aqueous solutions of $[Ni(H_2O)_6]^{2+}(aq)$ are green in color.

$$\underset{\text{green}}{[Ni(H_2O)_6]^{2+}(aq)} + \underset{\text{colorless}}{6NH_3(aq)} \leftrightharpoons \underset{\text{blue-violet}}{[Ni(NH_3)_6]^{2+}(aq)} + \underset{\text{colorless}}{6H_2O(l)}$$

The H_2O ligands attached to the nickel(II) ion are displaced by the NH_3 molecules which then become the new ligands bonded to the nickel(II) ion. The complex ion $[Ni(NH_3)_6]^{2+}$ is octahedral like $[Ni(H_2O)_6]^{2+}$, with NH_3 molecules instead of H_2O molecules surrounding the central nickel ion. A reaction involving a change in the ligands bonded to the central metal ion in a complex ion is called a **ligand-substitution reaction**.

The change in ligands from H_2O to NH_3 around nickel(II) produces a modified electrical environment around the nickel(II) ion that changes the energies of the eight *d* electrons in the nickel(II) ion. This in turn causes a change in the wavelength of the light absorbed by the ion and thus results in a color change. In general, light-induced electronic transitions of transition metal species give rise to their color (Figure 26.23).

Not all complex ions are octahedral. For example, if we add $NaCN(aq)$ to a solution containing the blue-violet $[Ni(NH_3)_6]^{2+}(aq)$ complex ion, then cyanide ions displace the NH_3 ligands from the nickel(II) ion to form the orange-yellow $[Ni(CN)_4]^{2-}(aq)$ complex ion (Figure 26.23). The reaction is described by

Figure 26.23 Aqueous solutions of Ni(II) complex ions: $[Ni(H_2O)_6]^{2+}(aq)$ (green), $[Ni(NH_3)_6]^{2+}(aq)$ (blue-violet), and $[Ni(CN)_4]^{2-}(aq)$ (orange-yellow).

$$\underset{\text{blue-violet}}{[Ni(NH_3)_6]^{2+}(aq)} + \underset{\text{colorless}}{4CN^-(aq)} \leftrightharpoons \underset{\text{orange-yellow}}{[Ni(CN)_4]^{2-}(aq)} + \underset{\text{colorless}}{6NH_3(aq)}$$

The structure of the complex ion $[Ni(CN)_4]^{2-}$ is square planar (Figure 26.24); the four CN^- ligands are arranged in a plane at the four corners of an imaginary square around the nickel(II) ion, and the cyanide ions are bonded to the Ni(II) ion through the carbon atoms of the cyanide ions.

The overall charge on the $[Ni(CN)_4]^{2-}$ complex ion is −2, because the nickel(II) ion contributes a charge of +2 and the four CN^- ligands each contribute a charge of −1.

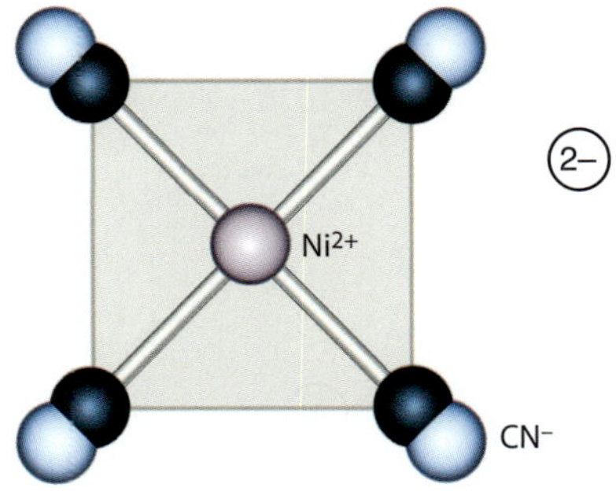

Figure 26.24 The $[Ni(CN)_4]^{2-}$ complex ion is square planar. Aqueous solutions of $[Ni(CN)_4]^{2-}(aq)$ are orange-yellow in color.

EXAMPLE 26-2: Determine the oxidation state of the platinum atom in the $[Pt(NH_3)_4Cl_2]^{2+}$ complex ion.

Solution: The complex ion $[Pt(NH_3)_4Cl_2]^{2+}$ has two kinds of ligands around the central metal ion: four NH_3 ligands and two Cl^- ligands. The charge on each NH_3 ligand is 0; the charge on each Cl^- ligand is −1; and the overall charge of the complex ion is +2. Denoting the charge on the platinum ion as x, we have

$$\overset{\text{charge on Pt}}{x} + \underset{4NH_3}{4(0)} + \overset{2Cl^-}{2(-1)} = \underset{\text{overall charge on complex ion}}{+2}$$

or

$$x + 2(-1) = +2$$
$$x = +4$$

Thus, the oxidation state of the platinum atom in the complex ion $[Pt(NH_3)_4Cl_2]^{2+}$ is +4.

PRACTICE PROBLEM 26-2: Determine the oxidation state of the transition metal ion in each of the following complex ions: (a) $[Fe(CN)_6]^{3-}$; (b) $[PtCl_6]^{2-}$; (c) $[Pt(NH_3)_3Cl_3]^+$.

Answer: (a) +3; (b) +4; (c) +4

Devices that indicate humidity levels by changing color are based on ligand-substitution reactions. For example, consider the ligand substitution described by the equation

$$\underset{\text{pink}}{2[Co(H_2O)_6]Cl_2(s)} \leftrightharpoons \underset{\text{blue}}{Co[CoCl_4](s)} + 12H_2O(g)$$

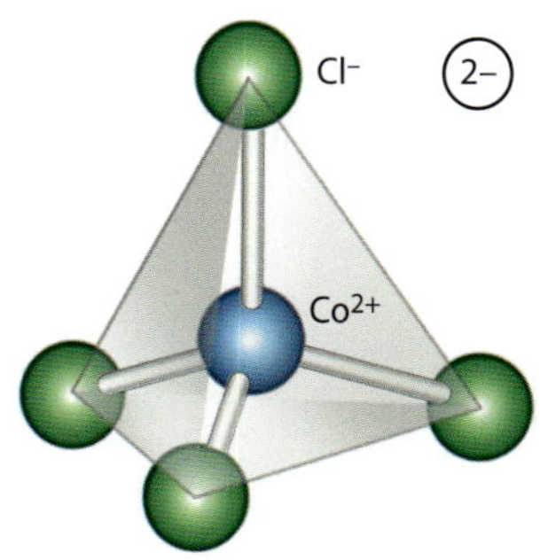

Figure 26.25 The $[CoCl_4]^{2-}$ complex ion is tetrahedral. Aqueous solutions of $[CoCl_4]^{2-}(aq)$ are blue in color.

Compounds containing the $[Co(H_2O)_6]^{2+}$complex ion are pink, and those containing the $[CoCl_4]^{2-}$ complex ion are blue. When the humidity is high, $Co[CoCl_4](s)$ reacts with the water vapor in the air to form pink $[Co(H_2O)_6]Cl_2(s)$. When the humidity is low, the equilibrium shifts from left to right, forming blue $Co[CoCl_4](s)$. The $[CoCl_4]^{2-}$ complex ion is tetrahedral, with the cobalt(II) ion at the center of a tetrahedron formed by the four chloride ligands (Figure 26.25).

In many qualitative analysis schemes (Section 22-8), $AgCl(s)$ is separated from other insoluble chlorides by the addition of $NH_3(aq)$ to form the soluble salt $[Ag(NH_3)_2]Cl(aq)$, which contains the complex ion $[Ag(NH_3)_2]^+$. The reaction equation is

$$AgCl(s) + 2NH_3(aq) \rightarrow [Ag(NH_3)_2]^+(aq) + Cl^-(aq)$$

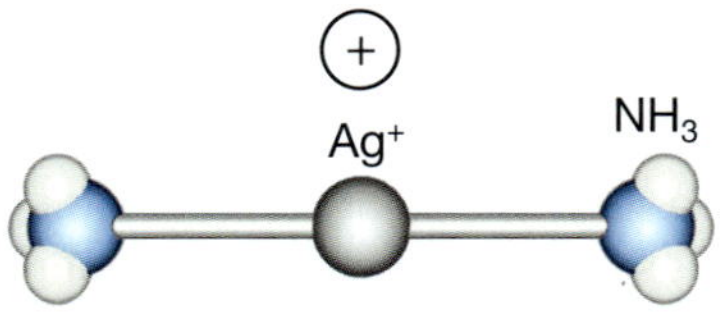

Figure 26.26 The two nitrogen atoms and the silver atom in the complex ion $[Ag(NH_3)_2]^+$ are in a straight line.

In the complex ion $[Ag(NH_3)_2]^+$, the N–Ag–N atoms are arranged in a straight line; for this reason, the complex ion is usually referred to as linear (Figure 26.26).

The most common geometries for transition metal complexes are octahedral, tetrahedral, square planar, and linear. Of these four geometries, octahedral is by far the most common. Examples of transition metal complexes with these four geometries are given in Table 26.8.

TABLE 26.8 Examples of transition metal complexes of various geometries

Octahedral	Tetrahedral	Square planar	Linear
$[Fe(H_2O)_6]^{2+}$	$[Zn(NH_3)_4]^{2+}$	$[Pt(CN)_4]^{2-}$	$[AgCl_2]^-$
$[Co(NO_2)_6]^{3-}$	$[Ni(CO)_4]$	$[AuCl_4]^-$	$[CuI_2]^-$
$[Cr(CO)_6]$	$[HgI_4]^{2-}$	$[Cu(NH_3)_4]^{2+}$	$[AuCl_2]^-$

26-8. Transition Metal Complexes Have a Systematic Nomenclature

The wide variety and large number of possible transition metal complexes make a systematic procedure for naming them essential. An example of a systematic name for a compound that contains a transition metal complex ion is

$[Ni(NH_3)_6](NO_3)_2(s)$ hexaamminenickel(II) nitrate

Let's analyze the name of this compound. As with any salt, the cation is named first. Thus, this name tells us that the compound consists of a hexaamminenickel(II) cation and nitrate anions. The Greek prefix *hexa-* denotes six, and *ammine-* denotes the ligand NH_3. Thus, the *hexaammine-* part of the name tells us that there are six NH_3 ligands in the cation. The Roman numeral II tells us that the nickel atom is in the +2 oxidation state. Because ammonia is a neutral molecule and the nickel atom is in a +2 oxidation state, the charge on the complex cation is +2. Its formula is $[Ni(NH_3)_6]^{2+}$. This complex ion combines with two NO_3^- ions to form a neutral salt, resulting in the formula given.

A simplified set of nomenclature rules for naming transition metal complexes is as follows:

1. *State the name of the cation first and then the anion:* for example, potassium tetracyanonickelate(II), $K_2[Ni(CN)_4]$.
2. *In any complex ion or neutral complex, name the ligands first and then the metal to which they are attached:* for example, hexaamminenickel(II) or

tetracyanonickelate(II). If there is more than one type of ligand in the complex, then name them in alphabetical order: for example, diammine dichloroplatinum(II), $[Pt(NH_3)_2Cl_2]$.

3. *End the names of negative ligands in the letter o, but give neutral ligands the name of the ligand molecule.* Some common neutral ligands have special names, such as *aqua* for H_2O, *ammine* for NH_3, and *carbonyl* for CO. Table 26.9 lists the names of a number of ligands.
4. *Denote the number of ligands of a particular type by a Greek prefix, such as di-, tri-, tetra-, penta-, or hexa-. (Mono is not used.)*
5. *If the complex ion is a cation or is neutral, then use the ordinary name for the metal; if the complex ion is an anion, then end the name of the metal in -ate:* for example, tetrachlorocobaltate(II), $[CoCl_4]^{2-}$, where the suffix *-ate* on the metal name tells us that the complex ion is an anion. Table 26.10 lists a few exceptions to this rule for anions; in these cases the Latin name of the metal is used.
6. *Denote the oxidation state of the metal by a Roman numeral or zero in parentheses following the name of the metal.*

TABLE 26.9 Names for common ligands

Ligand*	Name as ligand
Anions:	
F^-	fluoro
Cl^-	chloro
Br^-	bromo
I^-	iodo
$\underline{C}N^-$	cyano
$\underline{O}H^-$	hydroxo
$\underline{N}O_2^-$	nitro
$\underline{O}NO^-$	nitrito
$\underline{S}CN^-$	thiocyanato
$\underline{N}CS^-$	isothiocyanato
$\underline{O}CO_2^{2-}$	carbonato
O^{2-}	oxo
Neutral ligands:	
$\underline{N}H_3$	ammine
$H_2\underline{O}$	aqua
$\underline{C}O$	carbonyl
$\underline{N}O$	nitrosyl

*For ligands with two or more different atoms, the underlined atom is the one bonded to the metal.

TABLE 26.10 Exceptions to rule 5 for naming metal complexes in anions

Metal	Name in complex anion	Complex anion	Name of complex anion
silver	argentate	$[AgCl_2]^-$	dichloroargentate(I)
gold	aurate	$[Au(CN)_4]^-$	tetracyanoaurate(III)
copper	cuprate	$[CuCl_4]^{2-}$	tetrachlorocuprate(II)
iron	ferrate	$[Fe(CN)_6]^{3-}$	hexacyanoferrate(III)

The application of these rules is illustrated in the following Example.

EXAMPLE 26-3: The water-soluble yellow-orange compound $Na_3[Co(NO_2)_6](s)$ is used in some qualitative analysis schemes to test for $K^+(aq)$. Almost all potassium salts are water-soluble, but $K_2Na[Co(NO_2)_6](s)$ is only slightly soluble in water. The net ionic equation is

$$Na^+(aq) + 2K^+(aq) + [Co(NO_2)_6]^{3-}(aq) \rightarrow K_2Na[Co(NO_2)_6](s)$$

Name the $[Co(NO_2)_6]^{3-}(aq)$ complex ion.

Solution: The oxidation state of cobalt in the complex ion is determined as follows: the overall charge on the complex ion is –3, and there are six nitrite ions in the complex, each with a charge of –1. Denoting the oxidation state of cobalt as x, we have

$$\underset{\text{Co}}{x} + \underset{6\,NO_2^-}{6(-1)} = \underset{\text{net charge on complex ion}}{-3}$$

or $x = +3$. The complex ion is called hexanitrocobaltate(III), where the *-ate* ending tells us that the complex is an anion.

PRACTICE PROBLEM 26-3: Name the salt $[Ag(NH_3)_2]_3[Fe(CN)_6](s)$.

Answer: diamminesilver(I) hexacyanoferrate(III)

Note that in Practice Problem 26-3 we do not give silver the special name *argentate* listed in Table 26.10 because the $[Ag(NH_3)_2]^+$ transition metal complex is a cation.

Other examples of the nomenclature of transition metal complexes are

$K_2[Ni(CN)_4](s)$ — potassium tetracyanonickelate(II)

- potassium: cation
- tetracyanonickelate(II): complex anion
- tetracyano: $4CN^-$ ligands
- (II): Ni in +2 oxidation state

and

$[Co(H_2O)_4Cl_2]Cl(s)$ — tetraaquadichlorocobalt(III) chloride

- tetraaquadichlorocobalt(III): complex cation
- chloride: anion
- tetraaqua: $4\,H_2O$ ligands
- dichloro: $2\,Cl^-$ ligands
- (III): Co in +3 oxidation state

The rules for writing a chemical formula from the name of a transition metal complex follow from the nomenclature rules. For example, the formula for the compound named potassium hexacyanoferrate(II) is determined as follows:

1. The cation is potassium, K^+.
2. The complex anion contains six CN^- (hexacyano) ions and an iron atom. The oxidation state of the iron atom is +2, as indicated by the Roman numeral. The ending *-ate* tells us that the complex is an anion.
3. The charge on the complex anion is calculated by adding up the charges on the metal ion and the ligands:

$$(+2) + 6(-1) = -4$$

(+2): Fe(II); 6(−1): $6\,CN^-$; −4: net charge on complex

4. The formulas for complex ions and neutral complexes are enclosed in brackets, so we write the formula for the complex anion as $[Fe(CN)_6]^{4-}$. The formula for the salt is $K_4[Fe(CN)_6](s)$ because four K^+ ions are required to balance the −4 charge on the complex anion.

EXAMPLE 26-4: Give the formula for the compound hexamminecobalt(III) hexachlorocobaltate(III).

Solution: In this case both the cation and the anion are complex ions. The cation has six NH_3 ligands and one cobalt atom in a +3 oxidation state. Therefore, the formula for the cation is

$$[Co(NH_3)_6]^{3+}$$

The anion has six Cl^- ligands, each with a charge of −1 plus one cobalt atom in a +3 oxidation state. Therefore, the net charge on the complex anion is $6(-1) + (+3) = -3$ and its formula is

$$[CoCl_6]^{3-}$$

The magnitudes of the charges on the complex cation and the complex anion are equal; thus, the complex ions appear in the formula for the salt on a one-to-one basis:

$$[Co(NH_3)_6][CoCl_6](s)$$

PRACTICE PROBLEM 26-4: Give the chemical formula for the compound hexaaquanickel(II) diaquatetrabromochromate(III).

Answer: $[Ni(H_2O)_6][Cr(H_2O)_2Br_4]_2(s)$

26-9. Polydentate Ligands Bind to More Than One Coordination Position Around the Metal Ion

Certain ligands can bond to a metal ion or atom at more than one point of attachment, or **coordination position**. Examples of ligands that bond to two coordination positions are an oxalate ion (abbreviated ox) and an ethylenediamine molecule (abbreviated en):

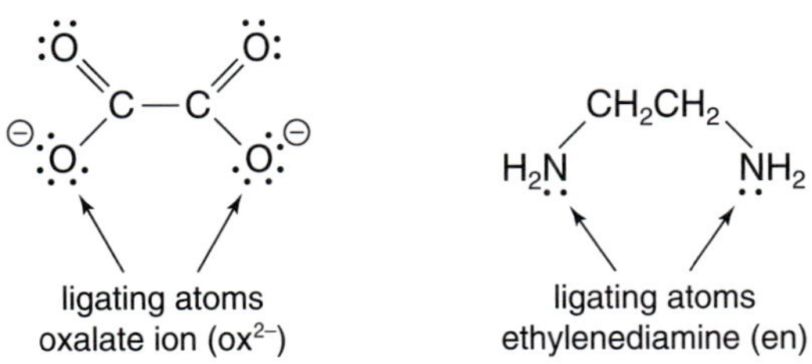

The atoms of the ligand that attach to the metal ion are called **ligating atoms**. Two complexes involving these two ligands are shown in Figure 26.27.

Ligands that attach to a metal ion at more than one coordination position are called **polydentate ligands** or **chelating ligands**. The resulting complex is called a **chelate**, which comes from the Greek word meaning *claw*. We can visualize the attachment of a chelating ligand as a species grasping a metal ion with molecular claws. A chelating ligand that attaches to two metal coordination positions is called **bidentate** (two teeth); one that attaches to three positions is **tridentate** (three teeth); and so on.

Ethylenediaminetetraacetate ion (abbreviated EDTA), is the best-known example of a hexadentate (i.e., six-coordinate) ligand:

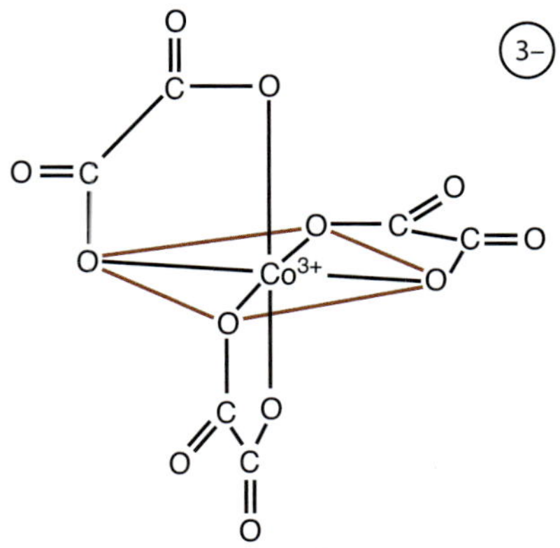

$[Co(ox)_3]^{3-}$

(a)

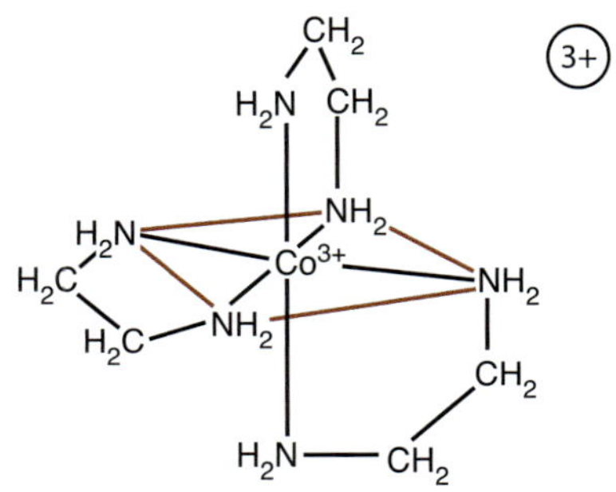

$[Co(en)_3]^{3+}$

(b)

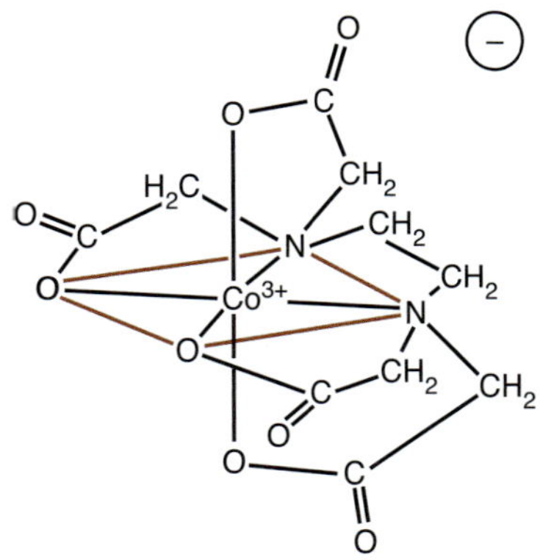

[Co(EDTA)]

(c)

Figure 26.27 Complexes of cobalt(III), involving the ligands (a) oxalate (ox^{2-}), $C_2O_4^{2-}$; (b) ethylenediamine (en), $C_2H_4(NH_2)_2$; and (c) ethylenediaminetetraacetate ($EDTA^{4-}$), $(CH_2N(CH_2CO_2)_2)_2^{4-}$.

EDTA^{4-}

The six ligating atoms are shown in red. Figure 26.27c shows the structure of the ethylenediaminetetraacetocobaltate(III) complex ion, $[Co(EDTA)]^-$. The EDTA^{4-} hexadentate ligand binds strongly to a number of metal ions and has a great variety of uses. It is used as an antidote for poisoning by heavy metals such as lead and mercury because it complexes with these metals, effectively rendering them inert and allowing them to pass harmlessly through the body. It is also used as a food preservative because it complexes with and renders inactive metal ions that catalyze the reactions involved in the spoiling process. In addition, EDTA is used as an analytical reagent in the analysis of the hardness of water, to reduce soap residue in detergents, soaps, and shampoos, and to decontaminate radioactive surfaces.

The nomenclature for complex ions and molecules that contain polydentate ligands follows the rules listed in Section 26-8, with one additional rule:

7. *If the ligand attached to the metal ion is a polydentate ligand, then enclose the ligand name in parentheses and use the prefix bis- for two ligands and tris- for three ligands:* for example, tris(ethylenediamine)cobalt(III), $[Co(H_2NCH_2CH_2NH_2)_3]^{3+}$, abbreviated as $[Co(en)_3]^{3+}$. The parentheses are not used if the complex contains only one of the polydentate ligands of a particular type.

EXAMPLE 26-5: Give the chemical formula for ammonium tris(oxalato) ferrate(III).

Solution: The cation is the ammonium ion, NH_4^+. The anion is a complex ion with an iron atom in the +3 oxidation state and three (tris) oxalate ions, $C_2O_4^{2-}$. (Recall from rule 3 that we end the names of negative ligands in the letter o, thus oxalate becomes oxalato.) The net charge on the complex anion is

$$\underbrace{(+3)}_{\text{Fe(III)}} + \underbrace{3(-2)}_{3\,C_2O_4^{2-}} = \underbrace{-3}_{\text{net charge}}$$

Therefore, the formula for the complex anion is $[Fe(C_2O_4)_3]^{3-}$ and the formula for ammonium tris(oxalato)ferrate(III) is $(NH_4)_3[Fe(C_2O_4)_3](s)$. Notice that we do not say triammonium, because the number of ammonium ions is unambiguously fixed by the net charge of −3 on the complex anion. There are four ions per formula unit in $(NH_4)_3[Fe(C_2O_4)_3]$, three NH_4^+ ions, and one $[Fe(C_2O_4)_3]^{3-}$ complex ion.

PRACTICE PROBLEM 26-5: Name the following compounds: (a) $K_2[Fe(EDTA)](s)$; (b) $Na[Co(C_2O_4)_2(en)](s)$.

Answer: (a) potassium ethylenediaminetetraacetoferrate(II); (b) sodium ethylenediaminebis(oxalato)cobaltate(III)

The nomenclature of transition metal complexes may appear cumbersome at first because of the length of the names, but with a little practice you will find it straightforward. Table 26.11 gives several additional examples of names of transition metal complexes. You should try to name each one from the formula and write the formula from each name.

TABLE 26.11 Examples of nomenclature for transition metal compounds

Compound	Name
$[Co(NH_3)_6]Cl_3$	hexaamminecobalt(III) chloride
$K[AuCl_4]$	potassium tetrachloroaurate(III)
$Cu_2[Fe(CN)_6]$	copper(II) hexacyanoferrate(II)
$[Pt(NH_3)_6]Cl_4$	hexaammineplatinum(IV) chloride
$[Cu(NH_3)_4(H_2O)_2]Cl_2$	tetraamminediaquacopper(II) chloride
$K[V(CO)_6]$	potassium hexacarbonylvanadate(–I)
$K_3[CoF_6]$	potassium hexafluorocobaltate(III)
$Cr(CO)_6$	hexacarbonylchromium(0)
$[CoCl_2(en)_2]NO_3$	dichlorobis(ethylenediamine)cobalt(III) nitrate

26-10. Some Octahedral and Square Planar Transition Metal Complexes Can Exist in Isomeric Forms

Consider the compound

$$[Pt(NH_3)_2Cl_2](s) \qquad \text{diamminedichloroplatinum(II)}$$

Platinum(II) complexes are invariably square planar. As shown in Figure 26.28, there are two possible arrangements of the four ligands around the central platinum(II) ion; these two complexes are ***cis* and *trans* isomers** (Section 9-11). *Cis* and *trans* compounds are **geometric isomers** because they differ in the spatial arrangements of the constituent atoms. The designation *cis* ("on the same side") tells us that identical ligands are placed adjacent to each other (on the same side) in the structure. The designation *trans* ("opposite") tells us that the identical ligands are placed directly opposite each other in the structure. The *cis* and *trans* isomers of $[Pt(NH_3)_2Cl_2](s)$ are different compounds with different physical and chemical properties. For example, the *cis*-diamminedichloroplatinum(II) isomer is manufactured as the potent anticancer drug cisplatin, whereas the *trans* isomer

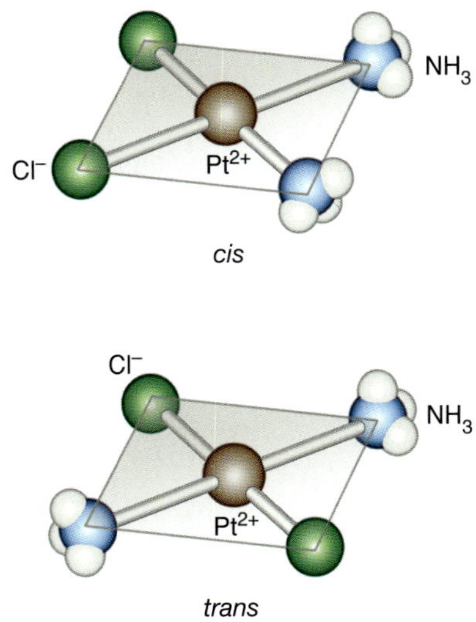

Figure 26.28 *Cis* and *trans* isomers of the square planar complexes of diamminedichloroplatinum(II), $[Pt(NH_3)_2Cl_2]$.

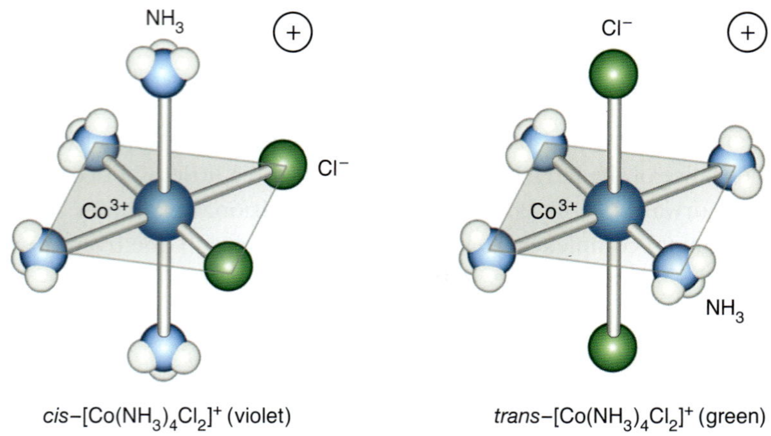

Figure 26.29 *Cis* and *trans* isomers of the octahedral complex ion tetraamminedichlorocobalt(III), $[Co(NH_3)_4Cl_2]^+$.

See Interchapter T at www.McQuarrieGeneralChemistry.com.

does not exhibit anticancer activity. Cisplatin interferes with the cell duplication process by inserting into the DNA double helix (Interchapter T).

Cis and *trans* isomers are also found in certain octahedral complexes. Consider the octahedral complex ion tetraamminedichlorocobalt(III), $[Co(NH_3)_4Cl_2]^+$. The two Cl^- ligands can be placed in adjacent (*cis*) or opposite (*trans*) positions around the central cobalt(III) ion, as shown in Figure 26.29. Because the six coordination positions in an octahedral complex are equivalent, any other *cis* placement of the two Cl^- ligands around the cobalt(III) ion yields a structure identical to the *cis* structure shown in the figure, except for its orientation in space. This equivalency is also true for the *trans* placement of the two Cl^- ligands. Thus, there are only two geometric isomers of a tetraamminedichlorocobalt(III) complex ion.

EXAMPLE 26-6: The compound $[Co(NH_3)_3Cl_3](s)$ exists in two isomeric forms. Draw the structures of the two isomers of this neutral complex.

Solution: The structures of the two isomers are shown in Figure 26.30. One compound is denoted *cis-cis* because each Cl^- ligand is adjacent to the other two. The other compound is denoted *cis-trans* because one Cl^- ligand is adjacent to the third chloride ligand and one Cl^- ligand is opposite it.

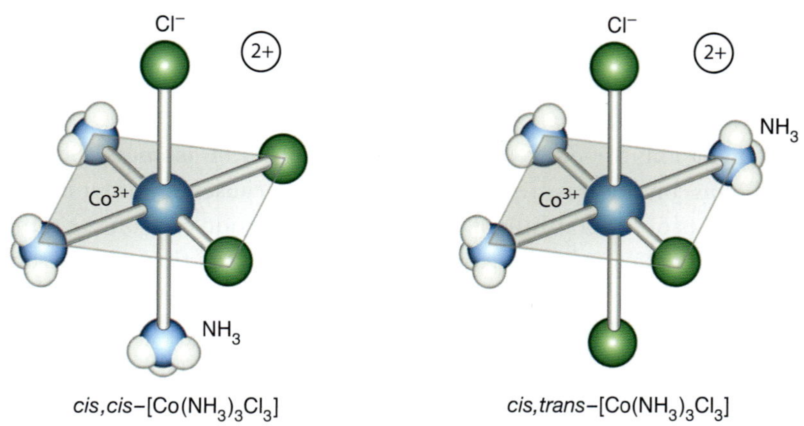

Figure 26.30 *Cis-cis* and *cis-trans* geometrical isomers of the octahedral complex triamminetrichlorocobalt(III), $[Co(NH_3)_3Cl_3]$.

PRACTICE PROBLEM 26-6: Draw the structures of the various possible geometric isomers of the $[CoCl_4BrI]^{3-}$ complex ion.

Answer: There are two geometric isomers, one with the Br^- and I^- ligands *trans* to each other and one with the Br^- and I^- ligands *cis* to each other. Because of the symmetry of an octahedron, there are no other isomers.

Many transition metal complexes also exhibit **optical isomerism** or **chirality**. Recall from Section 8-10 that **optical isomers** are nonsuperimposable isomers that are mirror images of each other (Figure 26.31).

If you have trouble visualizing optical isomers, it may help to construct models of these compounds.

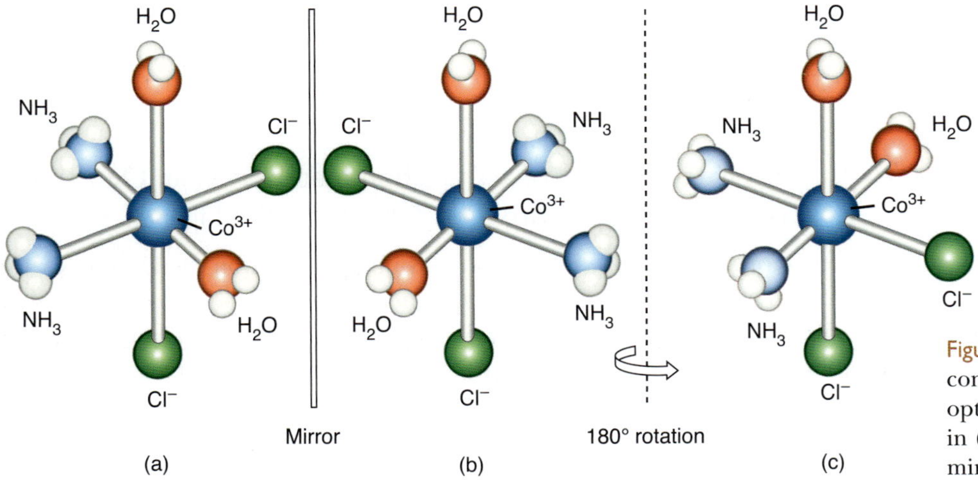

Figure 26.31 The all-*cis* form of the complex ion $[Co(NH_3)_2(H_2O)_2Cl_2]^+$ is optically active. The structures shown in (a) and (b) are nonsuperimposable mirror images of each other. The structure shown in (c) is the same as that in (b) but rotated 180 degrees. Notice that structure (c) cannot be superimposed on (a).

EXAMPLE 26-7: How many geometric and optical isomers can exist for the complex ion $[CoCl_2(en)_2]^+$?

Solution: The $[CoCl_2(en)_2]^+$ complex ion has three isomers: two *cis-trans* isomers and one optical isomer (the *cis* isomer is optically active), as shown in Figure 26.32.

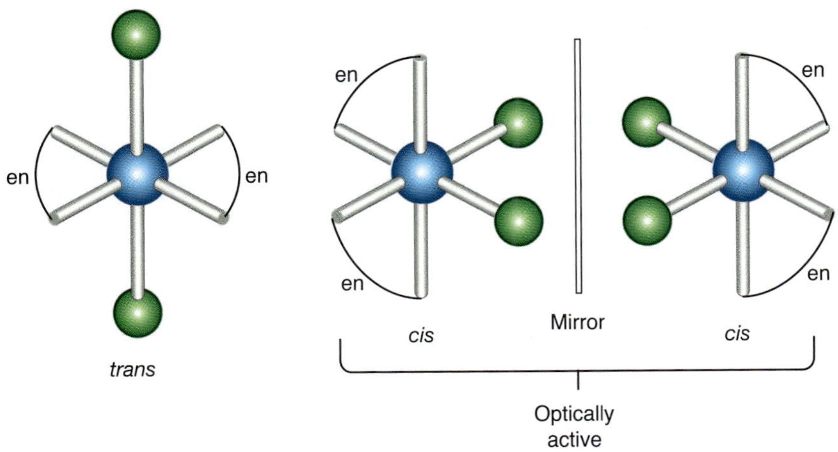

Figure 26.32 The complex ion $[CoCl_2(en)_2]^+$ can form a *trans* and a *cis* isomer. The *cis* form is chiral.

PRACTICE PROBLEM 26-7: Are either of the isomers of $[Co(NH_3)_3Cl_3]$ shown in Figure 26.30 optically active?

Answer: No.

The structures of transition metal complexes were worked out by the Swiss chemist Alfred Werner (Frontispiece) in the late nineteenth and early twentieth centuries, without the aid of modern X-ray structure determination methods. In 1893, at the age of 26, he proposed a correct structural theory based on the number of different types of complexes, including isomers, that could be prepared for platinum(II), platinum(IV), and cobalt(III) amminechloro complexes. In 1913 Werner was awarded the Nobel Prize for his research in transition metal chemistry.

This section completes our introduction to the structure and nomenclature of transition metal complexes. In the next section we turn to a consideration of interactions between the central metal ion *d* orbitals and the coordinated ligands. We shall show how the ligands perturb the various *d* orbitals in different ways, depending on the type and the spatial arrangement of the coordinated ligands.

26-11. The Five *d* Orbitals of a Transition Metal Ion in an Octahedral Complex Are Split into Two Groups by the Ligands

The five *d* orbitals (Figure 26.20) in a gas-phase transition metal atom or ion without any attached ligands all have the same energy. However, when six identical ligands are attached to the transition metal ion to form an octahedral complex, the *d* orbitals on the metal ion are split into two sets, each set with a different energy (Figure 26.33). This outcome is called ***d*-orbital splitting**.

The difference in the energies of the two sets of *d* orbitals means that a *d* electron can make a transition from one set of orbitals to the other by absorbing light. We shall show that this absorption of light by electrons in the *d* orbitals accounts for the colors of many coordination compounds. The lower-energy set of orbitals, called **t_{2g} orbitals**, consists of three orbitals (d_{xy}, d_{xz}, and d_{yz}), and the higher-energy set of orbitals, called **e_g orbitals**, consists of two orbitals ($d_{x^2-y^2}$ and d_{z^2}). (The designation t_{2g} and e_g may seem a bit mysterious to you. The origin of the notation lies in the mathematical theory of the symmetry of these species, and is really of no concern to us here.) The magnitude of the splitting of the *d* orbitals depends on both the central metal ion and the ligands. We shall let Δ_o (where the subscript o stands for octahedral) be the energy difference between the t_{2g} and e_g orbitals. The t_{2g} orbitals can accommodate up to six electrons, and the e_g orbitals can accommodate up to four electrons.

The splitting of the *d* orbitals in an octahedral complex can be explained using an electrostatic model called **crystal field theory**. Consider the octahedral

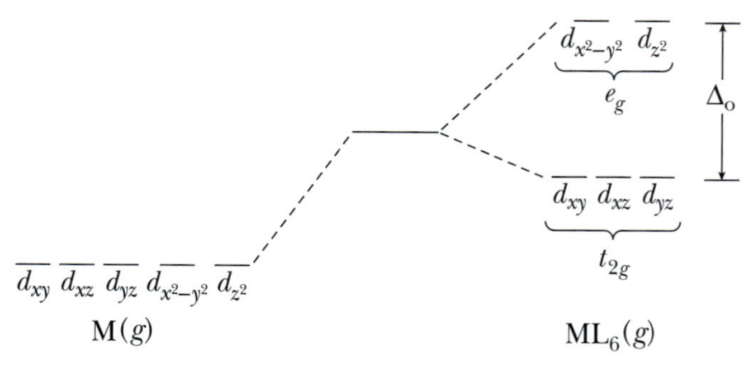

Figure 26.33 *d*-orbital splitting pattern for a regular octahedral complex, where M stands for the metal atom and L for the ligand.

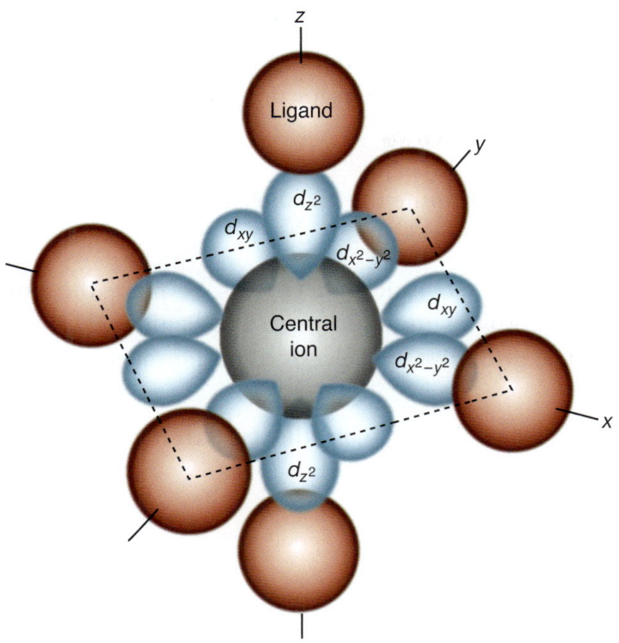

Figure 26.34 A regular octahedral complex, showing the orientation of the d_{z^2}, $d_{x^2-z^2}$, and d_{xy} orbitals relative to the ligands, which are brought in along the *x*, *y*, and *z* axes toward the central metal ion. For simplicity, the d_{xz} and d_{yz} orbitals are not shown. The d_{xy}, d_{xz}, and d_{yz} lobes point toward positions between the ligands, as shown here for d_{xy}.

complex shown in Figure 26.34, where the ligands are located along the positive and negative *x*, *y*, and *z* axes. Each ligand possesses a lone pair of electrons that for the moment can be considered to be localized on its ligating atom and directed toward the transition metal atom. Note that the d_{z^2} and $d_{x^2-y^2}$ orbitals (the e_g set) point *directly* toward these ligand lone pairs, whereas the d_{xy}, d_{xz}, and d_{yz} orbitals (the t_{2g} set) point *between* the ligand lone pairs. An electron placed in a d_{z^2} or a $d_{x^2-y^2}$ orbital will, therefore, experience a greater electrostatic repulsion (like charges repel) than one placed in a d_{xy}, d_{xz}, or d_{yz} orbital because the e_g electron will have a higher probability of being close to the ligand lone-pair electrons. Thus, electrons in the e_g set of orbitals will have a higher energy than those in the t_{2g} set. The *d*-orbital splitting pattern in an octahedral complex is thus seen to be a consequence of the positions of the ligands relative to the *d* orbitals.

To determine the ground state *d*-orbital configuration of a transition metal in a complex, we place electrons in the t_{2g} and e_g orbitals in a manner similar to that used in Chapter 5 for the electrons in atoms. Here also we must observe the restrictions of the Pauli exclusion principle (Section 5-6) and Hund's rule (Section 5-8). Thus, the orbitals of lower energy (the t_{2g} set) are occupied first, and as these orbitals are all of the same energy, each one must contain one electron before any one can contain two electrons. For example, an octahedral Cr(III) complex has three *d* electrons; and the *d* electron configuration would be $t_{2g}^3e_g^0$ (Figure 26.35). Note that Hund's rule requires that each t_{2g} orbital contain one electron and that all have the same spin.

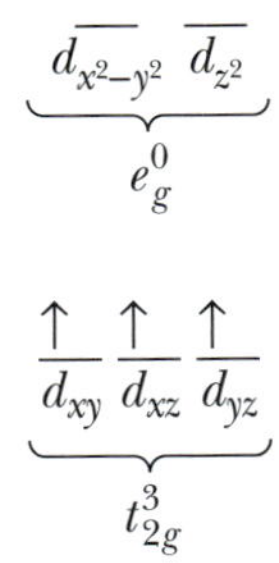

Figure 26.35 Ground state *d*-electron configuration of an octahedral d^3complex.

For an octahedral Ni(II) complex with eight *d* electrons, we can accommodate six of these eight electrons in the t_{2g} orbitals (spins paired). The remaining two electrons must occupy the higher-energy set (the e_g orbitals), giving the configuration $t_{2g}^6e_g^2$ (Figure 26.36). By Hund's rule the two e_g orbitals have one electron each with parallel spins. This procedure can be used to obtain the ground state d-electron configurations for d^1, d^2, and d^3 and d^8, d^9, and d^{10} ions. For d^4, d^5, d^6, and d^7 ions, on the other hand, two different d electron configurations are found to be possible. The reasons for this are explained in the next section.

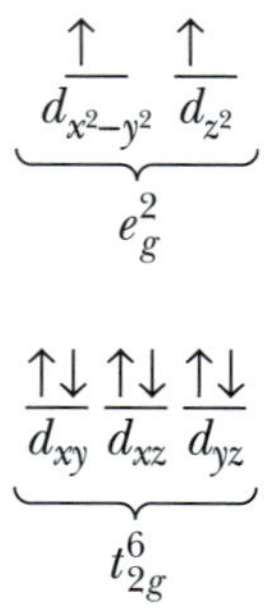

Figure 26.36 Ground state d-electron configuration of an octahedral d^8 complex.

EXAMPLE 26-8: Determine the ground state d electron configuration of copper(II) in an octahedral complex.

Solution: Copper(II) has nine d electrons. We place the nine d electrons in the t_{2g} and e_g orbitals in accord with the Pauli exclusion principle and Hund's rule. The ground state d electron configuration of copper(II) is $t_{2g}^6e_g^3$.

PRACTICE PROBLEM 26-8: Give the d electron configuration of V(III) in an octahedral complex.

Answer: $t_{2g}^2e_g^0$

The magnitude of the energy difference between the t_{2g} and e_g orbitals, denoted by Δ_o, depends on the central metal ion and the ligands. For most cases, the value of the energy difference, Δ_o (also called the **octahedral splitting energy**), corresponds to the values of photon energies in the visible region of the electromagnetic spectrum. In other words, the frequency of the radiation that is absorbed, which obeys the relation

$$E = h\nu = \Delta_o$$

is often in the visible region of the spectrum. Thus, many transition metal complexes absorb light in the visible region as a result of d-electron transitions and are colored (Figure 26.37).

It is possible to understand the variety of colors of octahedral complex ions in terms of the magnitudes of Δ_o values and the electron occupancy of the t_{2g} and e_g orbitals. For example, consider the purple $[Ti(H_2O)_6]^{3+}(aq)$ complex ion (Figure 26.38). Titanium is the second member of the $3d$ transition series, and

Figure 26.37 Crystals of the chloro complexes of the transition metals: (*top row*) scandium through manganese; (*bottom row*) iron through zinc. Most compounds of the transition metals are colored.

so Ti(II) has two *d* electrons. The $[Ti(H_2O)_6]^{3+}(aq)$ complex ion, which contains Ti(III), has one fewer *d* electron than Ti(II). Therefore, the titanium atom in $[Ti(H_2O)_6]^{3+}(aq)$ is a d^1 ion. The ground state and excited state *d*-electron configurations of this d^1 ion are

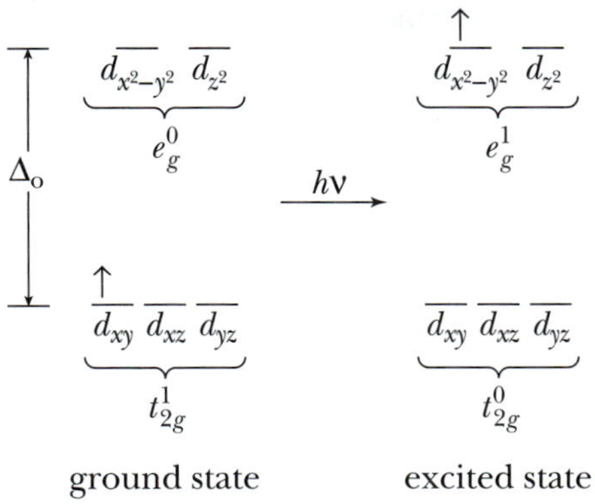

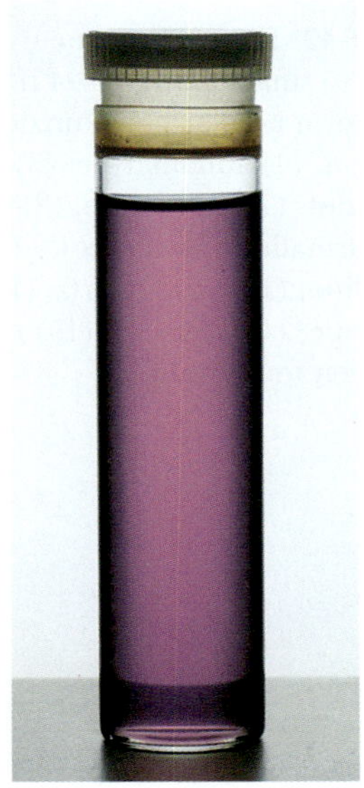

Figure 26.38 An aqueous solution of titanium(III) chloride. The d^1 complex ion $[Ti(H_2O)_6]^{3+}$ has a purple color.

The absorption spectrum $[Ti(H_2O)_6]^{3+}(aq)$ in the visible region is shown in Figure 26.39. The absorption of a photon in the blue through orange region excites the *d* electron from the lower-energy (t_{2g}) set of *d* orbitals to the higher-energy (e_g) set.

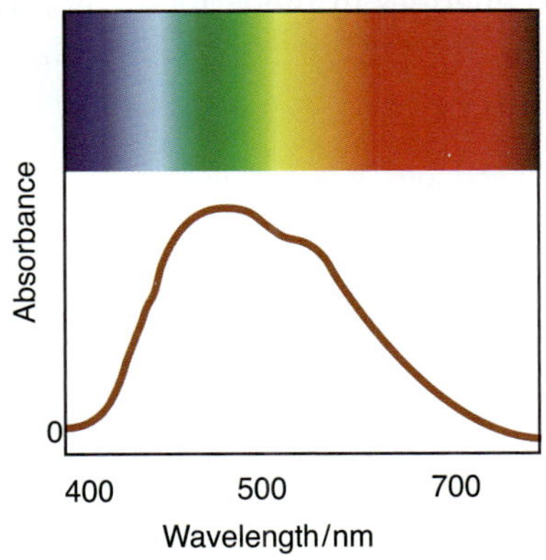

Figure 26.39 Absorption spectrum of the red-purple complex ion $[Ti(H_2O)_6]^{3+}(aq)$. The complex absorbs light in the blue, green, yellow, and orange regions, whereas most of the light in the red and violet regions passes through the sample and therefore is detected by the eye. Thus, the $[Ti(H_2O)_6]^{3+}(aq)$ complex ion appears purple in color.

Transition metal complexes are not the only species that derive their color from transition metals. The colors of many familiar gemstones (Figure 26.40) are due to the presence of transition metal ions. For example, the mineral corundum, $Al_2O_3(s)$, is colorless when pure. However, when certain transition metal ions are present in trace amounts, various gemstones result (Table 26.12). All the corundum and beryl gemstones in Table 26.12 can be made in the laboratory. Very large ruby rods are used in ruby lasers (Figure 26.41). The quality of synthetic rubies is at least as good as that of any natural ruby. Synthetic emeralds differ slightly in water content from natural emeralds, but the difference cannot be detected by the eye. Synthetic emeralds can withstand much higher temperatures than natural emeralds without shattering. The monetary values of many types of gemstones are artificially maintained by limiting production of the synthetic ones.

Figure 26.40 The colors of many gemstones are due to small quantities of transition metal ions. Shown here are (1) kunzite, (2) garnet, (3) zircon, (4) aquamarine, (5) amethyst, (6) peridot, (7) morganite, (8) topaz, (9) ruby, (10) tourmaline (indicolite), (11) chrome tourmaline, (12) rose quartz, (13) rubellite tourmaline, (14) kyanite, (15) citrine, and (16) green tourmaline.

Figure 26.41 Synthetic ruby rod prepared by melting $Al_2O_3(s)$ with about 3% $Cr_2O_3(s)$ and recrystallizing.

TABLE 26.12 Colors originating from transition metal ions in gemstones

Metal ion	Color	Gemstone	Mineral	Formula
Cr^{3+}	red	ruby	corundum	Al_2O_3
	green	emerald	beryl	$Be_3Al_2(Si_6O_{18})$
	pink	topaz	topaz	$Al_2SiO_4(F,OH)_2$
Fe^{3+}	yellow	citrine	quartz	SiO_2
	yellow	sapphire	corundum	Al_2O_3
	yellow	heliodor	beryl	$Be_3Al_2(Si_6O_{18})$
V^{3+}	green	green beryl ("emerald")	beryl	$Be_3Al_2(Si_6O_{18})$
Ni^{2+}	green	chrysoprase	quartz	SiO_2
Mn^{2+}	pink	morganite	beryl	$Be_3Al_2(Si_6O_{18})$
Cu^{2+}	blue	chrysocolla	chrysocolla	$Cu_2H_2Si_2O_5(OH)_4$
Fe^{2+}–Ti^{4+}	blue	sapphire	corundum	Al_2O_3
Fe^{2+}–Fe^{3+}	blue	aquamarine	beryl	$Be_3Al_2(Si_6O_{18})$

26-12. *d*-Orbital Electron Configuration Is the Key to Understanding Many Properties of the *d*-Block Transition Metal Ions

Not only the spectral properties but also the magnetic properties of transition metal complexes can be understood on the basis of the *d*-electron configuration of the metal ion in the complex. Unpaired electrons act like tiny magnets as a result of their intrinsic spin. Molecules with unpaired electrons can be magnetized by an external field and are called **paramagnetic**. As we showed for oxygen in Section 9-4, paramagnetic molecules line up in an external magnetic field with the electron spins parallel to the applied field. Thus, a paramagnetic substance behaves like a collection of magnets and is drawn into an externally applied magnetic field (Figure 26.42). In contrast, the magnetic fields from paired electrons, which have opposite spins, cancel out. Therefore, molecules with no unpaired electrons cannot be magnetized (in fact, they are repelled) by an external magnetic field and are called **diamagnetic**. Because a diamagnetic substance is not drawn into an applied magnetic field, it is easily distinguished from a paramagnetic substance. In some cases it is possible to determine the number of unpaired electrons by measuring the force with which the paramagnetic substance is drawn into the magnetic field.

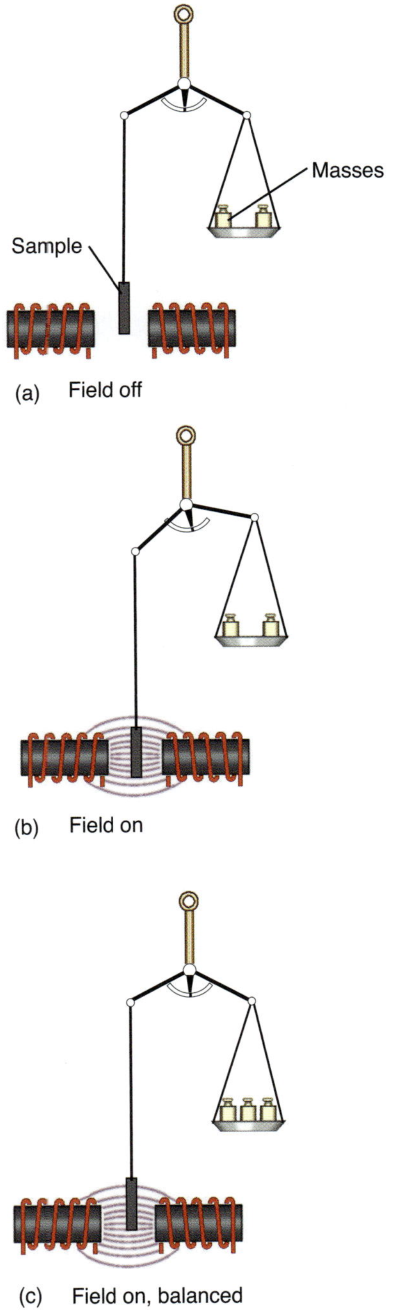

Figure 26.42 Attraction of a paramagnetic substance into a magnetic field. The magnetic attractive force on the sample makes it appear heavier. The number of unpaired electrons in the sample can be calculated from the apparent mass gain.

Magnetic experiments have shown that the compound $K_4[Fe(CN)_6](s)$ is diamagnetic, whereas the compound $K_4[FeF_6](s)$ is paramagnetic. Furthermore, there are four unpaired electrons in the $[FeF_6]^{4-}$ complex ion. We can explain these observations in terms of *d*-orbital electron configurations. Let's consider the $[FeF_6]^{4-}$ complex ion. Iron(II) has six *d* electrons, and, as noted in the previous section, d^6 systems have two possible *d*-electron configurations. The two possibilities for the *d*-electron configuration of an octahedral iron(II) complex ion are shown below:

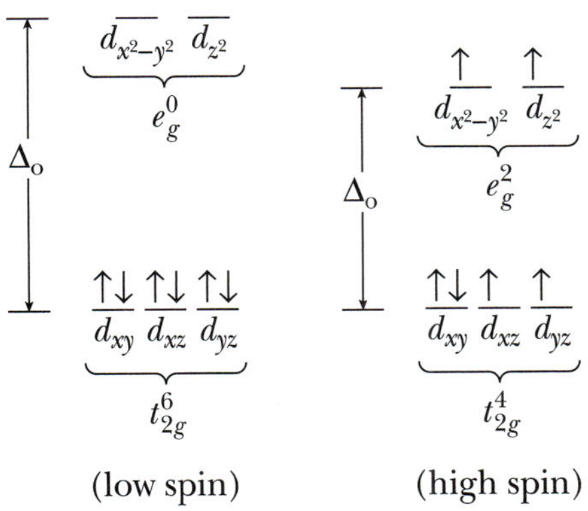

In one case the *d*-electron configuration is $t_{2g}^6e_g^0$, and in the other case it is $t_{2g}^4e_g^2$. In the $t_{2g}^6e_g^0$ configuration, the spins of all the electrons are paired; in the $t_{2g}^4e_g^2$ configuration, the electrons are in different orbitals with four unpaired electrons, in accord with Hund's rule. The $t_{2g}^6e_g^0$ configuration is said to be a **low-spin configuration**, and the $t_{2g}^4e_g^2$ configuration is said to be a **high-spin configuration**.

The value of Δ_o determines whether a *d*-electron configuration is low spin or high spin. If Δ_o is large, then the *d* electrons fill the t_{2g} orbitals completely before occupying the higher-energy e_g orbitals. If Δ_o is small, then the *d* electrons occupy the e_g orbitals before they pair up in the t_{2g} orbitals. For example, the d^6 complex ion, $[Fe(CN)_6]^4$, is low spin and has no unpaired electrons (Figure 26.43a). This configuration occurs because Δ_o is large relative to the **electron-pairing energy**, the energy required to pair up the electrons in the t_{2g} orbitals. In other words, it

Figure 26.43 (a) A low-spin octahedral complex results when Δ_o is greater than the pairing energy. (b) A high-spin octahedral complex results when Δ_o is less than the pairing energy.

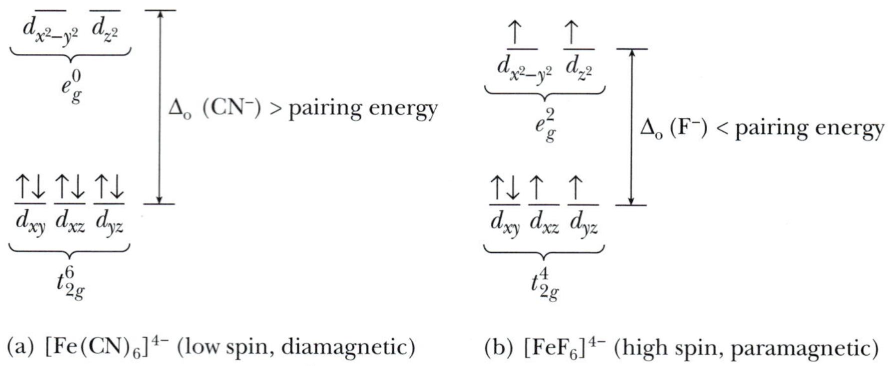

requires less energy to pair up two more electrons in the t_{2g} orbitals in $[Fe(CN)_6]^{4-}$ than to place the two electrons in the high-energy (large Δ_o) e_g orbitals. The d^6 complex ion, $[FeF_6]^{4-}$, on the other hand, is high spin and has four unpaired electrons (Figure 26.43b). This configuration occurs because Δ_o is small relative to the energy required to pair up the electrons in the t_{2g} orbitals. Thus, we see that $[Fe(CN)_6]^{4-}$ is diamagnetic, with no unpaired electrons, and that $[FeF_6]^{4-}$ is paramagnetic, with four unpaired electrons. The Δ_o values obtained from the spectra of the complexes show that the CN^- ligands interact much more strongly with iron(II) than do the F^- ligands; that is, $\Delta_o(CN^-) >> \Delta_o(F^-)$. The increased *d*-orbital splitting energy for CN^- ligand is sufficiently great to overcome the additional electron-electron repulsions that result from pairing up the electrons. A low-spin complex results whenever the energy difference between the t_{2g} and e_g orbitals is greater than the electron-pairing energy.

As we saw in the previous section, for d^1, d^2, and d^3 and d^8, d^9, and d^{10} octahedral complexes there is only one possible arrangement of the electrons in the ground state configuration. However, as we have just seen, for d^4, d^5, d^6, and d^7 octahedral complexes there are two possible *d*-electron configurations: a high-spin configuration and a low-spin configuration. The high-spin and low-spin *d*-electron configurations of these ions are shown in Figure 26.44. The high-spin configuration has the maximum possible number of unpaired *d* electrons and the low-spin configuration has the minimum possible number of unpaired *d* electrons.

EXAMPLE 26-9: Give the *d* electron configuration of the low-spin complex $[Pt(NH_3)_6]^{4+}(aq)$.

Solution: Referring to Figure 26.1, we see that platinum is the eighth member of the 5*d* transition series. Therefore, platinum(II) is a d^8 ion. The platinum ion in the $[Pt(NH_3)_6]^{4+}(aq)$ complex ion is platinum(IV). Platinum(IV) is a d^6 ion because it has two fewer *d* electrons than platinum(II). The *d*-electron configuration of a low-spin d^6 ion is $t_{2g}^6e_g^0$ (Figure 26.44). The complex ion is diamagnetic because it has no unpaired electrons.

PRACTICE PROBLEM 26-9: Predict which of the following ions can form both low-spin and high-spin complexes depending on the ligands: (a) Cr(III); (b) Mn(II); (c) Cu(III).

Answer: (b) Mn(II)

26-13. Ligands Can Be Ordered According to Their Ability to Split the Transition Metal *d* Orbitals

The analysis of many spectroscopic experiments with transition metal complexes demonstrates that the value of Δ_o depends primarily on the type of ligands and is largely independent of the metal ion for octahedral complexes with the same metal oxidation state in the same transition metal series. The arrangement of ligands in order of increasing ability to split *d* orbitals on the central transition metal atom is shown in Figure 26.45. This list is called the **Fajans-Tsuchida spectrochemical series** in honor of two pioneers in the spectroscopic studies of transition metal complexes. The most important thing to observe from the series is that metal-halide complexes are always high spin and that metal-cyanide complexes and most metal-carbonyl complexes are low spin. The splitting due to the other ligands can be either low spin or high spin, depending on the oxidation state of the transition metal and whether it is a 3*d*, 4*d*, or 5*d* transition metal.

$$\underbrace{I^- < Br^- < Cl^- < F^-}_{\text{low }\Delta\text{, high spin}} < OH^- < H_2O < NH_3 < \underbrace{NO_2^- < CN^- < CO}_{\text{high }\Delta\text{, low spin}}$$

$$\xrightarrow[\text{increasing }\Delta]{}$$

Figure 26.45 The Fajans-Tsuchida spectrochemical series for some common ligands.

> **EXAMPLE 26-10:** Predict the number of unpaired electrons in (a) the hexachlorochromate(II) complex ion and (b) the hexacyanocobaltate(III) complex ion.
>
> Solution: (a) The chemical formula of the complex ion is $[CrCl_6]^{4-}$. Chromium(II) is a d^4 ion, and according to Figure 26.45 the value of Δ_o for a Cl^- ligand is relatively small. Thus, we predict that $[CrCl_6]^{4-}$ is a high-spin complex ($t_{2g}^3e_g^1$) with four unpaired electrons. (b) The chemical formula of the complex ion is $[Co(CN)_6]^{3-}$. Cobalt(III) is a d^6 ion, and according to Figure 26.45 the value of Δ_o is relatively large. Thus, we predict that $[Co(CN)_6]^{3-}$ is a low-spin complex ($t_{2g}^6e_g^0$) with no unpaired electrons.
>
> **PRACTICE PROBLEM 26-10:** For the following two salts, predict the ground state *d*-electron configurations and whether each is paramagnetic or diamagnetic: (a) $K_3[Fe(CN)_6](s)$; (b) $K_2Na[Co(NO_2)_6](s)$.
>
> Answer: (a) $t_{2g}^5e_g^0$, paramagnetic; (b) $t_{2g}^6e_g^0$, diamagnetic

The ligands OH^-, H_2O, and NH_3 give rise to Δ_o values for M(II) 3*d* ions that are roughly equal to the pairing energy; thus, some complexes containing these ligands are high spin and some are low spin. For a given metal and ligand, the value of Δ_o increases as the oxidation state of the metal increases because the higher ionic charge leads to a greater electrostatic interaction between the metal and the ligand. Spectroscopic measurements on solutions of complex

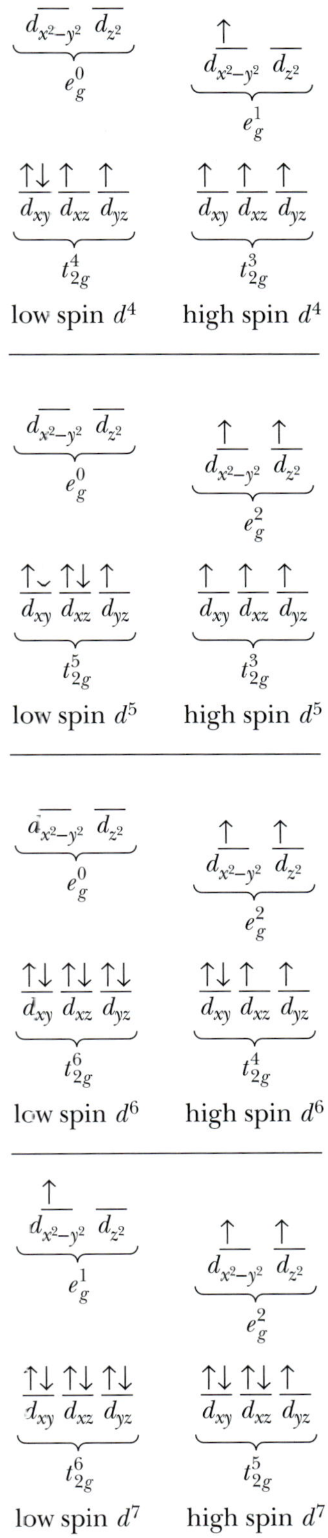

Figure 26.44 Possible ground state *d*-electron configurations for transition metal ions in octahedral complexes. For each of the d^4, d^5, d^6, and d^7 ions, there are two possibilities: a high-spin and a low-spin configuration.

TABLE 26.13 Octahedral splitting energy (Δ_o) values for various transition metal complexes

Complex	Δ_o/kJ·mol^{-1}	Observations
$[Co(NH_3)_6]^{2+}$	121	Δ_o increases with increasing oxidation states of cobalt
$[Co(NH_3)_6]^{3+}$	274	
$[Rh(NH_3)_6]^{3+}$	408	Δ_o increases as we move down a column in the periodic table for a given oxidation state
$[Ir(NH_3)_6]^{3+}$	478	

ions show that the value of Δ_o increases as we go from the $3d$ to the $4d$ to the $5d$ transition series for a particular number of d electrons. These trends are illustrated by the data given in Table 26.13. The magnitude of Δ_o values is such that, with the exception of metal-halide (F^-, Cl^-, Br^-, I^-) complexes, all ions of the $4d$ and $5d$ transition metal series in M(III) and higher oxidation states are low spin. Also, as shown in Figure 26.45, all NO_2^-, CN^-, and CO complexes are low spin. The orbitals of square planar and tetrahedral complexes are also split in the presence of ligands. If you pursue your studies in chemistry, you will learn about these splittings and many other aspects of the rich and fascinating chemistry of the transition metals in a course on inorganic chemistry.

SUMMARY

There are three d-block transition metal series ($3d$, $4d$, and $5d$), each having 10 members. A transition metal ion with x d electrons is called a d^x ion. The keys to understanding the chemistry of the d-block transition metal series are the electron occupancy of the five d orbitals of the metal ion and the influence of the ligands on the relative energies and splitting patterns of the orbitals. Ligands are anions or neutral molecules that bind to metal ions or neutral metal atoms to form distinct chemical species called complexes, complex ions, or neutral complexes. Chelating or polydentate ligands are ligands that attach to two or more coordination positions on the metal.

The geometry of a complex with identical ligands may be octahedral (the most common), tetrahedral, square planar, or linear. The d-orbital splitting pattern is different for each geometry. The determining factor of the splitting of the d orbitals is the placement of the ligands in the complex relative to the positions of the d orbitals. The splitting of the d orbitals gives rise to the possibility of low-spin and high-spin complexes for octahedral d^4, d^5, d^6, and d^7 ions. The d-orbital splittings are denoted as Δ_o. A paramagnetic complex ion has unpaired electrons. Magnetic measurements can be used to detect the presence of unpaired electrons in a complex. Certain octahedral and square planar complexes exist as *cis* and *trans* isomers. Some complexes also exhibit optical isomerization.

TERMS YOU SHOULD KNOW

d-block elements *980*
chromous bubbler *985*
autocatalytic *985*
coke *987*
blast furnace *987*
basic oxygen process *988*
mild steel *988*
medium steel *988*
high-carbon steel *989*
stainless steel *989*
brass *991*
bronze *991*
d_{xy}, d_{xz}, d_{yz}, $d_{x^2-y^2}$, d_{z^2} orbitals *995*
d^x ion *995*
complex ion *997*
ligand *997*
coordination number *997*
coordination complex *997*
counter ion *997*
neutral complex *998*
ligand-substitution reaction *998*
coordination position *1003*
ligating atoms *1003*
polydentate ligand *1003*
chelating ligand *1003*
chelate *1003*
bidentate *1003*
tridentate *1003*
cis and *trans* isomers *1005*
geometric isomers *1005*
optical isomerism *1007*
chirality *1007*
optical isomers *1007*
d-orbital splitting *1008*
t_{2g} and e_g orbitals *1008*
crystal field theory *1008*
octahedral splitting energy, Δ_o *1010*
paramagnetic *1013*
diamagnetic *1013*
low-spin configuration *1013*
high-spin configuration *1013*
electron-pairing energy *1013*
Fajans-Tsuchida spectrochemical series *1015*

PROBLEMS

TRANSITION METAL CHEMISTRY

26-1. Which is the most abundant transition metal?

26-2. Describe how titanium metal is produced from its ore.

26-3. Give the highest oxidation states for Sc, Ti, V, Cr, and Mn.

26-4. Name a catalyst in the production of sulfuric acid by the contact process.

26-5. A small quantity of $Na_2Cr_2O_7(s)$ is often added to water stored in steel drums used in some passive home solar heating systems. The dichromate acts as a corrosion inhibitor by forming an impervious layer of $Cr_2O_3(s)$ on the iron surface. Write a balanced chemical equation for the process in which $Cr_2O_3(s)$ is formed.

26-6. A chromous bubbler is used to remove traces of oxygen from various gases, for example, from tank nitrogen. The bubbler solution is prepared by reducing $Cr^{3+}(aq)$ to $Cr^{2+}(aq)$ with excess zinc metal. Chromium(II) rapidly reduces $O_2(g)$ to water and forms chromium(III), which is then reduced back to Cr(II) by the zinc. Write balanced chemical equations for the various chemical reactions involved in the operation of the bubbler.

26-7. Why are solutions of potassium permanganate stored in dark bottles?

26-8. Write the chemical equation that describes the reaction of titanium tetrachloride with water to produce a titanium dioxide smokescreen.

26-9. Describe the principal reactions that take place in a blast furnace.

26-10. Describe the basic oxygen process.

26-11. Why must solutions of $Co^{3+}(aq)$ be prepared fresh just before use?

26-12. Which two elements are important by-products of copper production?

26-13. When copper reacts with nitric acid, $NO(g)$ is evolved. Although $NO(g)$ is colorless, it appears as though a brown-red gas is evolved. Why?

26-14. What is the percentage of gold in 14-karat gold?

26-15. Alloys are usually mixtures of metals. Give the elemental compositions of the common alloys: brass, bronze, common solder, and white gold. You may have to do an Internet search or consult a handbook of chemistry in the library for the necessary information.

26-16. The chief ore of chromium is chromite,

$FeCr_2O_4(s)$. Chromium of high purity can be obtained from chromite by oxidizing chromium(III) to chromium(VI) in the form of sodium dichromate, $Na_2Cr_2O_7(s)$, and then reducing it with carbon according to

$$Na_2Cr_2O_7(s) + 2C(s) \rightarrow Cr_2O_3(s) + Na_2CO_3(s) + CO(g)$$

The oxide is then reduced with aluminum by the thermite reaction

$$Cr_2O_3(s) + 2Al(s) \rightarrow Al_2O_3(s) + 2Cr(s)$$

If an ore is 65.0% chromite, how many grams of pure chromium can be obtained from 100.0 grams of ore?

26-17. A copper ore consists of 2.65% chalcopyrite, $CuFeS_2(s)$. How many tons of ore must be processed to obtain one metric ton of copper?

26-18. Suppose that an iron ore consists of 50% $Fe_2O_3(s)$ and 50% $SiO_2(s)$. How many metric tons of iron and slag will be produced from 10 000 metric tons of ore?

26-19. Gold is recovered by leaching crushed ores with aqueous cyanide solutions. The resultant complex, $[Au(CN)_2]^-(aq)$, is reduced with zinc metal to produce gold. Write the chemical equations for the reactions that take place.

26-20. Ores containing as little as 0.25% copper are sometimes used to obtain copper metal. What mass of such an ore is needed to produce 91 metric tons of copper, the amount used in the Statue of Liberty?

ELECTRON CONFIGURATIONS AND OXIDATION STATES

26-21. Write the ground state electron configuration for

(a) Mn(II) (b) V(III)
(c) Ru(II) (d) Pt(IV)

26-22. Write the ground state electron configuration for

(a) Co(III) (b) Ti(IV)
(c) Au(III) (d) Cu(I)

26-23. How many outer-shell d electrons are there in each of the following transition metal ions?

(a) Ag(I) (b) Pd(IV)
(c) Ir(III) (d) Co(II)

26-24. How many outer-shell d electrons are there in each of the following transition metal ions?

(a) Re(III) (b) Sc(III)
(c) Ru(IV) (d) Hg(II)

26-25. Give three examples of
(a) M(III) d^6 ions (b) M(IV) d^4 ions
(c) M(I) d^{10} ions

26-26. Give three examples of

(a) M(II) d^3 ions (b) M(I) d^8 ions
(c) M(IV) d^0 ions

26-27. Give the oxidation state of the metal in

(a) $[Os(NH_3)_4Cl_2]^+$ (b) $[CoCl_6]^{3-}$
(c) $[Fe(CN)_6]^{4-}$ (d) $[Nb(NO_2)_6]^{3-}$

26-28. Give the oxidation state of the metal in

(a) $[Ir(H_2O)_6]^{3+}$ (b) $[Co(NH_3)_3(CO)_3]^{3+}$
(c) $[CuCl_4]^{2-}$ (d) $[Ni(CN)_4]^{2-}$

26-29. Give the oxidation state of the metal in

(a) $[Cd(CN)_4]^{2-}$ (b) $[Pt(NH_3)_6]^{2+}$
(c) $[Pt(NH_3)_4Cl_2]$ (d) $[RhBr_6]^{3-}$

26-30. Give the oxidation state of the metal in

(a) $[Mo(CO)_4Cl_2]^+$ (b) $[Ta(NO_2)_3Cl_3]^{3-}$
(c) $[Co(CN)_6]^{3-}$ (d) $[Ni(CO)_4]$

IONS FROM COMPLEX SALTS

26-31. Name the complex ion present when the following compounds are dissolved in water and determine the number of moles of each species present if one mole of the compound is dissolved:

(a) $K_3[Fe(CN)_6](s)$
(b) $[Ir(NH_3)_6](NO_3)_3(s)$
(c) $[Pt(NH_3)_4Cl_2]Cl_2(s)$
(d) $[Ru(NH_3)_6]Br_3(s)$

26-32. Name the complex ion present when the following compounds are dissolved in water and determine the number of moles of each species present if one mole of the compound is dissolved:

(a) $[Cr(NH_3)_6]Br_3(s)$
(b) $[Pt(NH_3)_3Cl_3]Cl(s)$
(c) $[Mo(H_2O)_6]Br_3(s)$
(d) $K_4[Cr(CN)_6](s)$

26-33. Some of the first complexes discovered by Werner in the 1890s had the empirical formulas given below. Also given are the number of chloride ions per formula unit precipitated by the addition of $Ag^{+}(aq)$. Explain these observations.

Empirical formula	Number of Cl^{-} per formula unit precipitated by $Ag^{+}(aq)$
$PtCl_4 \cdot 6NH_3$	4
$PtCl_4 \cdot 5NH_3$	3
$PtCl_4 \cdot 4NH_3$	2
$PtCl_4 \cdot 3NH_3$	1
$PtCl_4 \cdot 2NH_3$	0

26-34. Some of the first complexes discovered by Werner in the 1890s had the empirical formulas given below. Also given are the number of chloride ions per formula unit precipitated by the addition of $Ag^{+}(aq)$. Explain these observations.

Empirical formula	Number of Cl^{-} per formula unit precipitated by $Ag^{+}(aq)$
$PtCl_2 \cdot 4NH_3$	2
$PtCl_2 \cdot 3NH_3$	1
$PtCl_2 \cdot 2NH_3$	0

CHEMICAL FORMULAS AND NAMES

26-35. Give the systematic name for

(a) $K_3[Cr(CN)_6]$

(b) $[Cr(H_2O)_5Cl](ClO_4)_2$

(c) $[Co(CO)_4Cl_2]ClO_4$

(d) $[Pt(NH_3)_4Br_2]Cl_2$

26-36. Give the systematic name for

(a) $K_3[Fe(CN)_6]$

(b) $[Ni(CO)_4]$

(c) $[Ru(H_2O)_6]Cl_3$

(d) $Na[Al(OH)_4]$

26-37. Give the systematic name for

(a) $(NH_4)_3[Co(NO_2)_6]$

(b) $[Ir(NH_3)_4Br_2]Br$

(c) $K_2[CuCl_4]$

(d) $[Ru(CO)_5]$

26-38. Give the systematic name for

(a) $Na[Au(CN)_4]$

(b) $[Cr(H_2O)_6]Cl_3$

(c) $[V(en)_3]Cl_3$

(d) $[Cu(NH_3)_6]Cl_2$

26-39. Give the chemical formula for

(a) sodium pentacyanocarbonylferrate(II)

(b) ammonium *trans*-dichlorodiiodoaurate(III)

(c) potassium hexacyanocobaltate(III)

(d) calcium hexanitritocobaltate(III)

26-40. Give the chemical formula for

(a) sodium bromochlorodicyanonickelate(II)

(b) rubidium tetranitritocobaltate(II)

(c) potassium hexachlorovanadate(III)

(d) pentaamminechlorochromium(III) acetate

26-41. Give the chemical formula for

(a) triamminechloroplatinum(II) nitrate

(b) sodium tetrafluorocuprate(II)

(c) lithium hexanitrocobaltate(II)

(d) bis(ethylenediamine)oxalatocadmium(II)

26-42. Give the chemical formula for

(a) barium hexacyanoferrate(II)

(b) chlorohydroxobis(ethylenediamine)cobalt(III) nitrate

(c) lithium dinitrobis(oxalato)platinate(IV)

(d) bis(ethylenediamine)oxalatovanadium(III) acetate

ISOMERS

26-43. Draw the structure for

(a) *trans*-dichlorodibromoplatinum(IV) (square planar)

(b) potassium *trans*-dichlorodiodoaurate(III) (square planar)

(c) *cis,cis*-triamminetrichlorocobalt(III)

(d) *cis,trans*-triamminetrichloroplatinum(IV) chloride

26-44. Indicate whether each of the following complex ions has geometric isomers:

(a) $[Cr(NH_3)_4Cl_2]^{+}$

(b) $[Cr(NH_3)_5Cl]^{2+}$

(c) $[Co(NH_3)_2Cl_2]^{2-}$ (tetrahedral)

(d) $[Pt(NH_3)_2Cl_2]$ (square planar)

26-45. Draw all the geometric and any optical isomers for

(a) $[Co(en)_2Br_2]$ (b) $[RuCl_2Br_2(NO_2)_2]^{3-}$

26-46. Draw all the geometric and any optical isomers for

(a) $[Pd(C_2O_4)_2I_2]^{2-}$ (b) $[PtCl_3Br_3]^{2-}$

HIGH-SPIN AND LOW-SPIN COMPLEXES

26-47. Write the *d*-orbital electron configurations for the following octahedral complex ions:

(a) a Nb(III) complex

(b) a Mo(II) complex if Δ_o is greater than the electron-pairing energy

(c) a Mn(II) complex if Δ_o is less than the electron-pairing energy

(d) a Au(I) complex

(e) an Ir(III) complex if Δ_o is greater than the electron-pairing energy

26-48. Write the *d*-orbital electron configurations for the following octahedral complex ions:

(a) a high-spin Ni(II) complex

(b) a high-spin Mn(II) complex

(c) a low-spin Fe(III) complex

(d) a Ti(IV) complex

(e) a Ni(II) complex

26-49. Classify the following complex ions as high spin or low spin:

(a) $[Fe(CN)_6]^{4-}$ (no unpaired electrons)

(b) $[Fe(CN)_6]^{3-}$ (one unpaired electron)

(c) $[Co(NH_3)_6]^{2+}$ (three unpaired electrons)

(d) $[CoF_6]^{3-}$ (four unpaired electrons)

(e) $[Mn(H_2O)_6]^{2+}$ (five unpaired electrons)

26-50. Classify the following complex ions as high spin or low spin:

(a) $[Mn(NH_3)_6]^{3+}$ (two unpaired electrons)

(b) $[Rh(CN)_6]^{3-}$ (no unpaired electrons)

(c) $[Co(C_2O_4)_3]^{4-}$ (three unpaired electrons)

(d) $[IrBr_6]^{4-}$ (three unpaired electrons)

(e) $[Ru(NH_3)_6]^{3+}$ (one unpaired electron)

PARAMAGNETISM IN COMPLEX IONS

26-51. Predict the number of unpaired electrons in

(a) $[VCl_6]^{3-}$

(b) $[Cr(CN)_6]^{4-}$

(c) $[Cr(CO)_6]$

26-52. Predict the number of unpaired electrons in

(a) $[Rh(NH_3)_6]^{3+}$ (low spin)

(b) $[FeF_6]^{3-}$

(c) $[Ir(H_2O)_6]^{3+}$ (low spin)

26-53. Indicate whether each of the following complexes is paramagnetic:

(a) $[Cu(NH_3)_6]^{2+}$ (b) $[Co(en)_3]^{3+}$ (low spin)

(c) $[CrF_6]^{3-}$ (d) $[Zn(H_2O)_6]^{2+}$

26-54. The complex $[Fe(H_2O)_6]^{2+}$ is paramagnetic, whereas $[Fe(CN)_6]^{4-}$ is diamagnetic. Explain the difference.

ADDITIONAL PROBLEMS

26-55. Give the systematic name for the following complex ions:

(a) $[Pt(en)_3](NO_3)_4$

(b) $K_4[Co(ox)_3]$

(c) $[Ni(H_2O)_4(OH)_2]$

(d) $Na_2[Fe(EDTA)]$

26-56. Give the chemical formula for the following complex ions:

(a) hexanitrocobaltate(III)

(b) *trans*-dichlorobis(ethylenediamine)platinum(IV)

(c) pentacyanocarbonylferrate(II)

(d) *trans*-dichlorodiiodoaurate(III)

26-57. Write the chemical formulas for the following compounds:

(a) hexaaquanickel(II) perchlorate

(b) triamminetrichloroplatinum(IV) bromide

(c) potassium chloropentacyanoferrate(III)

(d) strontium hexacyanoferrate(II)

26-58. Draw all the geometric and optical isomers for

(a) tetraamminedibromoiron(III) ion

(b) diamminebromochloroplatinum(II) (square-planar)

(c) $[Pt(NH_3)_2Cl_2F_2]$

(d) [CoBrCl(en)] (tetrahedral)

26-59. Arrange the following complexes in order of increasing values of Δ_o:

$$[Cr(H_2O)_6]^{3+} \quad [Co(NH_3)_6]^{3+} \quad [CrF_6]^{3-} \quad [Cr(CN)_6]^{3-} \quad [Ru(CN)_6]^{3-}$$

26-60. Use VSEPR theory to predict the shapes of

(a) $TiCl_4$ (b) VF_5

(c) CrO_4^{2-} (d) MnO_4^-

26-61. Comparing $[Co(CN)_6]^{3-}$ with $[CoCl_6]^{4-}$, indicate whether each of the following statements is true or false:

(a) $[Co(CN)_6]^{3-}$ has more *d* electrons than $[CoCl_6]^{4-}$.

(b) $[Co(CN)_6]^{3-}$ has the same number of *d* electrons as $[CoCl_6]^{4-}$.

(c) $[Co(CN)_6]^{3-}$ is paramagnetic, whereas $[CoCl_6]^{4-}$ is diamagnetic.

(d) $[Co(CN)_6]^{3-}$ is diamagnetic, whereas $[CoCl_6]^{4-}$ is paramagnetic.

26-62. A scientist wishes to determine if a newly discovered neutral ligand, *X*, produces high-spin or low-spin octahedral complexes. He has available samples of the compounds $[Fe(X)_6]Cl_2(s)$ and $[Ni(X)_6]Cl_2(s)$. Which, if either, of these compounds is better suited for this analysis? Why?

26-63. Copper metal reacts with concentrated nitric acid but not with concentrated hydrochloric acid. Explain these observations by using standard reduction voltages. Write balanced chemical equations for the two reactions that occur between copper and dilute and concentrated nitric acid.

26-64. Predict whether the glycinate ion

$$H_2N–CH_2–COO^-$$

is a monodentate, bidentate, or tridentate ligand. What experiment could you perform to test this prediction?

26-65. How many unpaired electrons would you predict for the following complex ions? Indicate if the complex ion is paramagnetic or diamagnetic.

(a) $[Co(CO)_6]^{3+}$ (b) $[Fe(CN)_6]^{3-}$

26-66. (*) Give the systematic name for:

$$[Pt(NH_3)_4I_2][PtI_4].$$

26-67. (*) Explain why most zinc and silver complexes are colorless.

26-68. (*) How many geometric and optical isomers are possible for the $[Pt(en)_3]^{4+}$ complex ion?

26-69. (*) Silver nitrate was added to solutions of the following octahedral complexes and AgCl(*s*) was precipitated immediately in the mole ratios indicated:

Formula of the complex	(mol AgCl/ mol complex)
$CoCl_3(NH_3)_6$	3
$CoCl_3(NH_3)_5$	2
$CoCl_3(NH_3)_4$ (purple)	1
$CoCl_3(NH_3)_4$ (green)	1

(a) Draw the structures expected for each of these complexes.

(b) Explain the fact that $CoCl_3(NH_3)_4$ can be purple or green but that both forms give one mole of AgCl(*s*) per mole complex.

26-70. (*) Excess $Pb_2[Fe(CN)_6](s)$ was equilibrated at 25°C with an aqueous solution of NaI(*aq*). The equilibrium concentrations of $I^-(aq)$ and $[Fe(CN)_6]^{4-}(aq)$ were found by chemical analysis to be 0.57 M and 0.11 M, respectively. Estimate the value of K_{sp} of $Pb_2[Fe(CN)_6](s)$. See Appendix F for the value of K_{sp} of $PbI_2(s)$.

APPENDIX A

A Mathematical Review

A1. Scientific Notation and Exponents

The numbers encountered in chemistry are often extremely large (such as 8 180 000 000) or extremely small (such as 0.000 004 613). When working with such numbers, it is convenient to express them in **scientific notation**, where we write the number as a number between 1 and 10 multiplied by 10 raised to the appropriate power. For example, the number 171.3 is $1.713 \times 100 = 1.713 \times 10^2$ in scientific notation. Some other examples are

$$7320 = 7.32 \times 10^3$$
$$1\,623\,000 = 1.623 \times 10^6$$

Because the zeros in these numbers serve as placeholders only, they are not regarded as significant figures and are dropped in scientific notation. Notice that in each case the power of 10 is the number of places that the decimal point has been moved to the left:

$$\underset{\text{3 places}}{7320.} \qquad \underset{\text{6 places}}{1\,623\,000.}$$

When numbers that are smaller than 1 are expressed in scientific notation, the 10 is raised to a negative power. For example, 0.614 becomes 6.14×10^{-1}. Recall that a negative exponent is governed by the relation

$$10^{-n} = \frac{1}{10^n} \qquad \text{(A1.1)}$$

It is good practice always to include a zero before a leading decimal point so that the decimal point does not get overlooked. For example, you should write 0.345 instead of just .345. You can see here that the zero nicely alerts you to the presence of the following decimal point.

Some other examples are

$$0.0005 = 5 \times 10^{-4}$$
$$0.000\,000\,000\,446 = 4.46 \times 10^{-10}$$

Notice that the power of 10 in each case is the number of places that the decimal point has been moved to the right:

$$\underset{\text{4 places}}{0.0005} \qquad \underset{\text{10 places}}{0.000\,000\,000\,446}$$

One of the best time investments that you can make is to become proficient in using your calculator, not only for general chemistry but for other courses that you will take. Reworking the examples in this Appendix using your calculator is a good way to test your proficiency in working with numbers in scientific notation.

When performing chemical calculations, it is necessary to be able to work with numbers in scientific notation.

To add or subtract two or more numbers expressed in scientific notation, the power of 10 must be the same in both. For example, consider the sum

$$5.127 \times 10^4 + 1.073 \times 10^3$$

We rewrite the first number as

$$5.127 \times 10^4 = 51.27 \times 10^3$$

Note that we have changed the 10^4 factor to 10^3, so we must make the factor in front of 10^3 one power of 10 larger. Thus, we have

$$\begin{aligned} 5.127 \times 10^4 + 1.073 \times 10^3 &= (51.27 + 1.073) \times 10^3 \\ &= 52.34 \times 10^3 = 5.234 \times 10^4 \end{aligned}$$

To change a number such as 52.34×10^3 to 5.234×10^4, we make the number in front one factor of 10 smaller; so we must make 10^3 one factor of 10 larger. Similarly, we have

$$\begin{aligned} (4.708 \times 10^{-6}) - (2.1 \times 10^{-8}) &= (4.708 - 0.021) \times 10^{-6} \\ &= 4.687 \times 10^{-6} \end{aligned}$$

In changing 2.1×10^{-8} to 0.021×10^{-6}, we make 2.1 two factors of 10 (or $10 \times 10 = 100$ times) smaller and 10^{-8} two factors of 10 (or 100 times) larger.

When multiplying two numbers, we add the powers of 10 because of the relation

$$(10^x)(10^y) = 10^{x+y} \tag{A1.2}$$

For example,

$$\begin{aligned} (5.00 \times 10^2)(4.00 \times 10^3) &= (5.00)(4.00) \times 10^{2+3} \\ &= 20.0 \times 10^5 = 2.00 \times 10^6 \end{aligned}$$

and

$$\begin{aligned} (3.014 \times 10^3)(8.217 \times 10^{-6}) &= (3.014)(8.217) \times 10^{3-6} \\ &= 24.77 \times 10^{-3} = 2.477 \times 10^{-2} \end{aligned}$$

To divide, we subtract the power of 10 of the number in the denominator from the power of 10 of the number in the numerator because of the relation

$$\frac{10^x}{10^y} = 10^{x-y} \tag{A1.3}$$

For example,

$$\begin{aligned} \frac{4.0 \times 10^{12}}{8.0 \times 10^{23}} &= \left(\frac{4.0}{8.0}\right) \times 10^{12-23} \\ &= 0.50 \times 10^{-11} = 5.0 \times 10^{-12} \end{aligned}$$

and

$$\begin{aligned} \frac{2.80 \times 10^{-4}}{4.73 \times 10^{-5}} &= \left(\frac{2.80}{4.73}\right) \times 10^{-4+5} \\ &= 0.592 \times 10^1 = 5.92 \end{aligned}$$

To raise a number to a power, we use the fact that

$$(10^x)^n = 10^{nx} \tag{A1.4}$$

For example,

$$\begin{aligned}(2.187 \times 10^2)^3 &= (2.187)^3 \times 10^{3\times 2}\\ &= 10.46 \times 10^6 = 1.046 \times 10^7\end{aligned}$$

To take a root of a number, we use the relation

$$\sqrt[n]{10^x} = (10^x)^{1/n} = 10^{x/n} \tag{A1.5}$$

Thus, the power of 10 must be expressed such that it is divisible by the root. For example,

$$\begin{aligned}\sqrt[3]{2.70 \times 10^{10}} &= (2.70 \times 10^{10})^{1/3} = (27.0 \times 10^9)^{1/3}\\ &= (27.0)^{1/3} \times 10^{9/3} = 3.00 \times 10^3\end{aligned}$$

and

$$\begin{aligned}\sqrt{6.40 \times 10^5} &= (6.40 \times 10^5)^{1/2} = (64.0 \times 10^4)^{1/2}\\ &= (64.0)^{1/2} \times 10^{4/2} = 8.00 \times 10^2\end{aligned}$$

Today you can easily do all such operations using a handheld calculator or a computer. However, you should also be proficient at carrying out such calculations by hand because a quick estimate of the "order of magnitude" of a desired result often serves as a good check that the operations were carried out correctly.

A2. Common Logarithms

You know that $100 = 10^2$, $1000 = 10^3$, and so on. You also might know that

$$\sqrt{10} = 10^{1/2} = 10^{0.500} = 3.16$$

(where we have rounded off the answer to two decimal places).

By taking the square root of both sides of

$$10^{0.500} = 3.16$$

we find that

$$\sqrt{10^{0.500}} = 10^{(1/2)0.500} = 10^{0.250} = \sqrt{3.16} = 1.78$$

Furthermore, because

$$(10^x)(10^y) = 10^{x+y}$$

we can write

$$10^{0.250} \times 10^{0.500} = 10^{0.750} = (3.16)(1.78) = 5.62$$

By continuing this process, we can express any number y as

$$y = 10^x \tag{A2.1}$$

When taking a logarithm, $x = \log y$, the number of decimal places in x is equal to the number of significant figures in y. So, for example,

$23.\underline{780} = \log \underline{6.02} \times 10^{23}$
3 decimal places — 3 significant figures

The number x to which 10 must be raised to get y is called the **logarithm** of y and is written as

$$x = \log y \tag{A2.2}$$

We say that Equations A2.1 and A2.2 are the inverse of each other. Because $10^0 = 1$, we have $\log 1 = 0$. Thus far, then, we've shown that

$$\begin{aligned} \log 1.00 &= 0.000 \\ \log 1.78 &= 0.250 \\ \log 3.16 &= 0.500 \\ \log 5.62 &= 0.750 \\ \log 10.00 &= 1.0000 \end{aligned}$$

The last entry here follows from the fact that $10 = 10^1$. We could go on like this and build up a complete table of common logarithms, but this has already been done for us. Not only are there extensive tables of logarithms, but any scientific calculator has a log key.

Because logarithms are exponents ($y = 10^x$), they have certain special properties, such as

$$\log xy = \log x + \log y \tag{A2.3}$$

You should try to verify these equations for yourself.

$$\log \frac{x}{y} = \log x - \log y \tag{A2.4}$$

$$\log x^n = n \log x \tag{A2.5}$$

$$\log \sqrt[n]{x} = \log x^{1/n} = \frac{1}{n} \log x \tag{A2.6}$$

If we let $x = 1$ in Equation A2.4, then we have

$$\log \frac{1}{y} = \log(1) - \log y = -\log y$$

or

$$\log \frac{1}{y} = -\log y \tag{A2.7}$$

Thus, we change the sign of a logarithm when we take the reciprocal of its argument.

Let's go back to Equation A2.1, $y = 10^x$. If $x = 0$, then $y = 1$. Therefore, if $x \geq 0$, then $y \geq 1$; and if $x \leq 0$, then $y \leq 1$, or

$$\begin{aligned} \log y &\geq 0 \qquad \text{if } y \geq 1 \\ \log y &\leq 0 \qquad \text{if } y \leq 1 \end{aligned} \tag{A2.8}$$

Figure A.1 shows a plot of log y against y illustrating these relationships.

One property of logarithms that you may not have encountered before is that you cannot take the logarithm of a quantity that has units. You can take the logarithms only of numbers. It makes no sense to ask what the logarithm of 2.43 grams is. You can take the logarithm of 2.43, but not of 2.43 grams. Neither x nor y can have units in Equation A2.1.

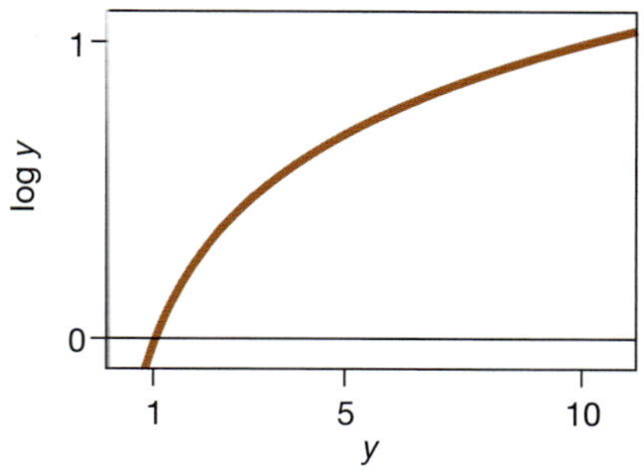

Figure A.1 The common logarithm of y plotted against y. Notice that log y takes on positive values for y greater than one and negative values for y less than one. The curve increases rather slowly for increasing y and decreases rather quickly for small values of y. It passes through the horizontal axis at $y = 1$ because log 1 = 0.

Up to this point we have found the value of x in $y = 10^x$ when y is given. It is often necessary to find the value of y when x is given. Because x is called the logarithm of y, y is called the **antilogarithm**, or **inverse logarithm**, of x. For example, the antilogarithm of $x = 2$ is $y = 100$. Less obvious is that the antilogarithm of $x = 6.0969$ is 1.250×10^6. You can get this last result using your hand calculator. Different calculators have different ways of finding antilogarithms, so you should practice this on your own calculator.

When taking an antilogarithm, $y = 10^x$, the number of significant figures in y is equal to the number of decimal places in x. So, for example,

$\underline{8.79} \times 10^{-18} = 10^{-17.\underline{056}}$

3 significant figures — 3 decimal places

A3. Natural Logarithms

The logarithms that we have just discussed in the previous section are called **common logarithms**, or **logarithms to the base 10** because we start with Equation A2.1, $y = 10^x$, to define $x = \log y$. In fact, some authors use the notation $x = \log_{10} y$ to emphasize the base 10. The definition of pH, the Richter earthquake scale, and the decibel sound scale, among others, are expressed in terms of common logarithms. Logarithms to another base arise naturally in calculus. The base is a number called e, where

$$e = 2.718\,281\,828\,46\ldots \tag{A3.1}$$

The logarithms to this base, called **natural logarithms**, are designated by ln instead of log. Thus, we have

$$x = \ln y \tag{A3.2}$$

The rules for determining the number of significant figures for natural logarithms are the same as those for common logarithms.

and its inverse

$$y = e^x \tag{A3.3}$$

Even if you have not taken a course in calculus, you needn't worry: the functions ln y and e^x are on all scientific calculators. For example, you can use your calculator to show that

$$e^2 = 7.389\,056\ldots$$

and

$$e^{-2} = 0.135\,335\ldots$$

Note that $e^{-2} = 1/e^2$, as you might expect. In fact, the mathematical properties of e^x and natural logarithms are similar to those of 10^x and $\log y$. For example,

$$\ln xy = \ln x + \ln y \tag{A3.4}$$

$$\ln \frac{x}{y} = \ln x - \ln y \tag{A3.5}$$

$$\ln x^n = n \ln x \tag{A3.6}$$

$$\ln \sqrt[n]{x} = \ln x^{1/n} = \frac{1}{n} \ln x \tag{A3.7}$$

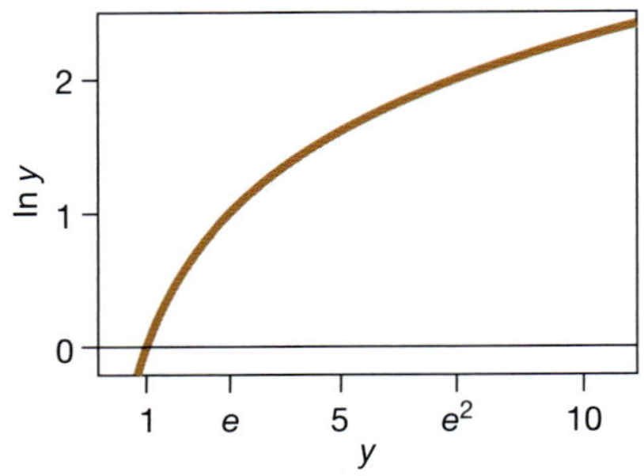

Figure A.2 The natural logarithm of y plotted against y. Note that $\ln y$ takes on positive values for y greater than one and negative values for y less than one. The curve increases rather slowly for increasing y and decreases rather quickly for small values of y. It passes through the horizontal axis at $y = 1$ because $\ln 1 = 0$.

Because $e = e^1$, we see that $\ln e = 1$, just as $\log_{10} 10 = 1$. Because $e^0 = 1$, we have $\ln 1 = 0$ and

$$\begin{aligned} \ln y &\geq 0 \quad \text{if } y \geq 1 \\ \ln y &\leq 0 \quad \text{if } y \leq 1 \end{aligned} \tag{A3.8}$$

just as in the case of common logarithms (Equations A2.8). Figure A.2 shows a plot of $\ln y$ against y illustrating these relationships.

EXERCISE: Using your hand calculator, determine the following quantities:
(a) $e^{0.37}$ (b) $\ln(4.07)$
(c) $e^{-6.02}$ (d) $\ln(0.00965)$

Solution: (a) 1.4; (b) 1.404; (c) 2.4×10^{-3}; (d) −4.641

EXERCISE: Determine the value of y for (a) $\ln y = 3.065$ and (b) $\ln y = -0.605$.

Solution: (a) $y = e^{3.065} = 21.4$; (b) $y = e^{-0.605} = 0.546$

It is sometimes necessary to convert from natural logarithms to base-10 logarithms. We will take a few lines to derive the relation between $\ln y$ and $\log y$. We start with $y = 10^x$ (Equation A2.1) and write $10 = e^a$ (we'll determine the value of a shortly), so that we have

$$y = 10^x = (e^a)^x = e^{ax}$$

Now take log and ln of this equation to write

$$\log y = x \log 10 = x$$

and

$$\ln y = ax \ln e = ax$$

Substituting $x = \log y$ into this equation, we have

$$\ln y = a \log y$$

We can determine the value of a by taking ln of $10 = e^a$ to get

$$a = \ln 10$$

Using a calculator, we find that

$$a = 2.302585\ldots$$

Therefore, the relation between $\ln y$ and $\log y$ is given by

$$\ln y = 2.303 \log y \tag{A3.9}$$

to four significant figures.

EXERCISE: Using your hand calculator, verify that $\ln 120.6 = 2.303 \log 120.6$ to four significant figures.

Solution: $\ln 120.6 = 4.7925$ and $2.303 \log 120.6 = 4.7933$. The two numbers are the same to four significant figures.

EXERCISE: Generalize Equation A3.9 to

$$(\log_b a)(\log_a y) = \log_b y$$

where $y = a^x$. Show that this equation is consistent with Equation A3.9.

Solution: Let $a = b^c$, where c is a constant to be determined, so that

$$y = a^x = b^{cx}$$

Therefore,

$$\log_a y = x \log_a a = x$$

and

$$\log_b y = cx \log_b b = cx$$

where we have used the fact that $\log_a a = 1$ and $\log_b b = 1$. Combining these two expressions yields

$$\log_b y = cx = c \log_a y$$

Substituting $y = a^x$ into the left side of this expression gives

$$x \log_b a = cx$$

Dividing this through by x we get

$$\log_b a = c$$

Finally, we have

$$\log_b y = c \log_a y = (\log_b a)(\log_a y).$$

To get Equation A3.9, simply let $b = e$ and $a = 10$.

A4. The Quadratic Formula

The standard form for a quadratic equation in x is

$$ax^2 + bx + c = 0 \qquad \text{(A4.1)}$$

where a, b, and c are constants. The two solutions to the quadratic equation are

Do you remember how to derive this equation? Try completing the square of Equation A4.1.

$$x = \frac{-b \pm \sqrt{b^2 - 4ac}}{2a} \qquad \text{(A4.2)}$$

Equation A4.2 is called the **quadratic formula** and is used to obtain the solutions to a quadratic equation expressed in the standard form. For example, let's find the solutions to the quadratic equation

$$2x^2 - 3x - 1 = 0$$

In this case, $a = 2$, $b = -3$, and $c = -1$ and Equation A4.2 gives

$$\begin{aligned} x &= \frac{3 \pm \sqrt{(-3)^2 - (4)(2)(-1)}}{2(2)} \\ &= \frac{3 \pm 4.123}{4} \\ &= 1.781 \quad \text{and} \quad -0.281 \end{aligned}$$

Note that you get two solutions to a quadratic equation. To use the quadratic formula, it is first necessary to put the quadratic equation in the standard form so that we know the values of the constants a, b, and c. For example, consider the problem of solving for x in the quadratic equation

$$\frac{x^2}{0.50 - x} = 0.040$$

To identify the constants a, b, and c, we must write this equation in the standard quadratic form. Multiplying both sides by $0.50 - x$ yields

$$x^2 = (0.50 - x)(0.040) = 0.020 - 0.040x$$

Rearrangement to the standard quadratic form yields

$$x^2 + 0.040x - 0.020 = 0$$

Thus, $a = 1$, $b = 0.040$, and $c = -0.020$. Using Equation A4.2, we have

$$x = \frac{-0.040 \pm \sqrt{(0.040)^2 - (4)(1)(-0.020)}}{(2)(1)}$$

from which we compute

$$\begin{aligned} x &= \frac{-0.040 \pm \sqrt{0.0816}}{2} \\ &= \frac{-0.040 \pm 0.286}{2} \\ &= 0.123 \quad \text{and} \quad -0.163 \end{aligned}$$

If x represents, say, a concentration of a solution or the pressure of a gas, then the only physically possible value is $x = 0.123$ because concentrations and pressures cannot have negative values.

EXERCISE: Solve the equation

$$\frac{(x + 0.235)x}{x - 0.514} = 2x + 0.174$$

Solution: 1.17 and −0.0765

A5. Successive Approximations

Many problems involving chemical equilibria lead to a quadratic equation of the form

$$\frac{x^2}{M_0 - x} = K \qquad \text{(A5.1)}$$

where x is the concentration of a particular species, M_0 is an initial concentration, and K is often a small constant. For example, the equation

$$\frac{[\text{A}]^2}{0.100\ \text{M} - [\text{A}]} = 6.25 \times 10^{-5}\ \text{M} \qquad \text{(A5.2)}$$

is the equation for the concentration of species A in a certain solution (the square brackets around A denote concentration). If the value of K is small (which it is in this case), then it is more convenient to solve an equation like Equation A5.1 by the **method of successive approximations** than by using the quadratic equation.

The first step in the method of successive approximations is to neglect the unknown in the denominator on the left-hand side of Equation A5.2. This approximation allows the unknown to be found by simply multiplying through by the initial concentration and taking the square root of both sides. Making this approximation in Equation A5.2 yields

$$[A]_1 \approx [(0.100 \text{ M})(6.25 \times 10^{-5} \text{ M})]^{1/2} = 2.50 \times 10^{-3} \text{ M} \tag{A5.3}$$

We have subscripted [A] with a 1 in this result because it represents a first approximation to the value of [A]. To obtain a second approximation, we use this value of $[A]_1$ in the denominator of the left-hand side of Equation A5.2, multiply both sides by the result in the denominator, and then take the square root:

$$[A]_2 \approx [(0.100 \text{ M} - 2.50 \times 10^{-3} \text{ M})(6.25 \times 10^{-5} \text{ M})]^{1/2} = 2.47 \times 10^{-3} \text{ M}$$

We now carry out the cycle, called an **iteration**, again to obtain a third approximation:

$$[A]_3 \approx [(0.100 \text{ M} - 2.47 \times 10^{-3} \text{ M})(6.25 \times 10^{-5} \text{ M})]^{1/2} = 2.47 \times 10^{-3} \text{ M}$$

Observe that $[A]_3 \approx [A]_2$. When this occurs, we say that the procedure has **converged**. After convergence is achieved, the same result will occur in any subsequent iteration, and the value obtained is the solution to the original equation because the equation is satisfied by the same value of [A] in the numerator as in the denominator.

The method of successive approximations is particularly convenient for rapidly solving a quadratic equation using a hand calculator. While it is usually necessary to carry out several iterations to obtain the solution, each cycle is easy to perform on a calculator, and the total effort involved usually is less than that of using the quadratic formula.

Usually, you will obtain convergence after only a few iterations. If you don't see the successive iterations approaching some value after just a few iterations, then probably it is better to use the quadratic formula.

Here are some examples to practice with:

1. $\dfrac{x^2}{0.500 - x} = 1.07 \times 10^{-3}$ $(x_1 = 2.31 \times 10^{-2}, x_2 = 2.26 \times 10^{-2}, x_3 = 2.26 \times 10^{-2})$

2. $\dfrac{x^2}{0.0100 - x} = 6.80 \times 10^{-4}$ $(x_1 = 2.61 \times 10^{-3}, x_2 = 2.24 \times 10^{-3}, x_3 = 2.30 \times 10^{-3}, x_4 = 2.29 \times 10^{-3}, x_5 = 2.29 \times 10^{-3})$

3. $\dfrac{x^2}{0.150 - x} = 0.0360$ $(x_1 = 7.35 \times 10^{-2}, x_2 = 5.25 \times 10^{-2}, x_3 = 5.92 \times 10^{-2}, x_4 = 5.72 \times 10^{-2}, x_5 = 5.78 \times 10^{-2}, x_6 = 5.76 \times 10^{-2}, x_7 = 5.77 \times 10^{-2}, x_8 = 5.77 \times 10^{-2})$

Even in this last case, which requires eight iterations, the method of successive approximations is easier than using the quadratic formula.

A6. Plotting Data

The human eye and brain are quite sensitive to recognizing straight lines, so it is always desirable to plot equations or experimental data such that a straight line is obtained. The mathematical equation for a straight line is of the form

$$y = mx + b \qquad \text{(A6.1)}$$

In this equation, m and b are constants: m is the **slope** of the line and b is its **intercept** with the y axis. The slope of a straight line is a measure of its steepness; it is defined as the ratio of any vertical height to its corresponding horizontal distance (sometimes called rise over run).

Let's plot the straight line

$$y = 2x - 2$$

We first make a table of values of x and y:

x	−3	−2	−1	0	1	2	3	4	5
y	−8	−6	−4	−2	0	2	4	6	8

The result is plotted in Figure A.3. The line intersects the y axis at $y = -2$ and so $b = -2$. The line has a slope (rise over run) of 2 and so $m = 2$.

Often the equation to be plotted will not appear to be of the form of Equation A6.1 at first. For example, consider the Boyle's law (Chapter 13) relation between the volume of a gas and its pressure. For a 0.29-gram sample of air at 25°C, Boyle's law says that

$$V = \frac{0.244\ \text{L}\cdot\text{atm}}{P} \qquad \text{(constant temperature)} \qquad \text{(A6.2)}$$

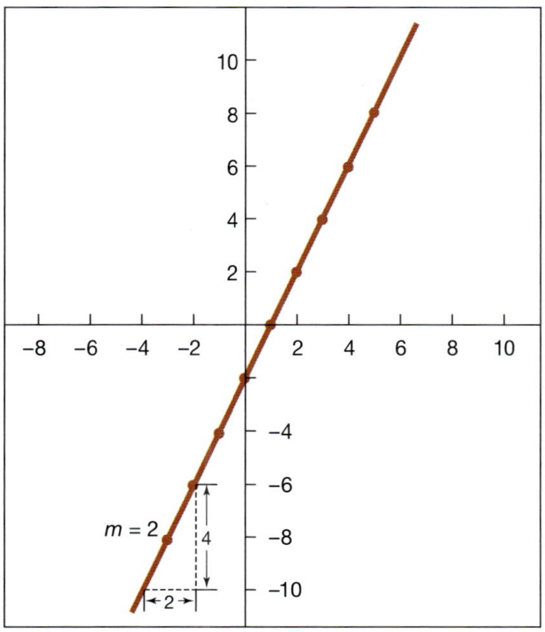

Figure A.3 Plot of the equation $y = 2x - 2$.

Figure A.4 shows that a plot of V against P is certainly not a straight line. The data for this plot are given in Table A.1. We can plot Equation A6.2 as a straight line, however, if we let $V = y$ and $1/P = x$, because then Equation A6.2 becomes

$$y = cx$$

Thus, if we plot V versus $1/P$, instead of V versus P, we get a straight line. The data in Table A.1 are plotted as V versus $1/P$ in Figure A.5.

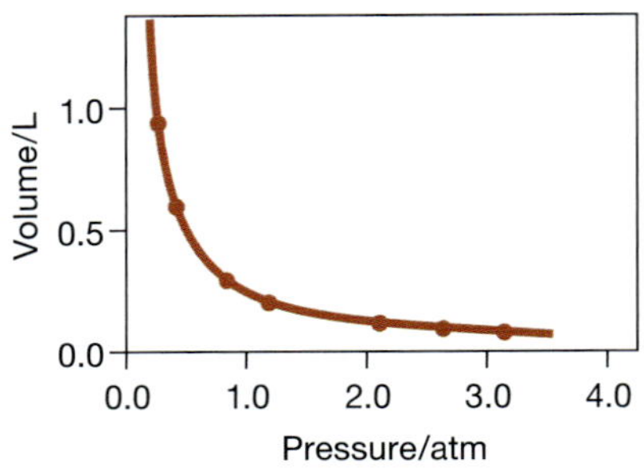

Figure A.4 The volume of 0.29 grams of air plotted against pressure at 25°C. The data are given in Table A.1.

TABLE A.1 Pressure-volume data for 0.29 grams of air at 25°C

P/atm	V/L	$(1/P)$/atm^{-1}
0.26	0.938	3.85
0.41	0.595	2.44
0.83	0.294	1.20
1.20	0.203	0.83
2.10	0.116	0.48
2.63	0.093	0.38
3.14	0.078	0.32

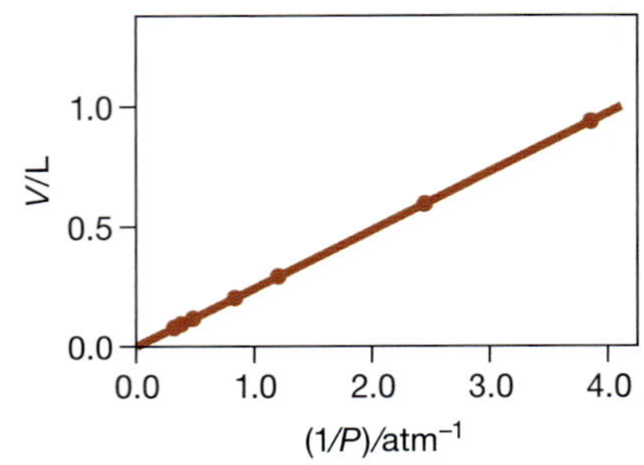

Figure A.5 The volume of 0.29 grams of air plotted against the reciprocal of the pressure, $1/P$, in units of atm^{-1} at 25°C. Straight lines are much easier to work with than other curves, so it is usually desirable to plot equations and data in the form of a straight line.

EXAMPLE: Plot the equation

$$\ln(P/\text{Torr}) = -\frac{1640\ \text{K}}{T} + 10.560 \tag{A6.3}$$

as a straight line. The quantity T is the Kelvin temperature. We take the natural logarithm of P/Torr here, which is unitless because the /Torr serves to eliminate the units. For example, 123 Torr/Torr = 123.

Solution: Comparing Equation A6.3 to Equation A6.1, we see that we can let

$$y = \ln(P/\text{Torr})$$

and

$$x = \frac{1}{T}$$

which suggests that a straight line will result if we plot $\ln(P/\text{Torr})$ versus $1/T$. Table A.2 shows the numerical results and Figure A.6 shows $\ln(P/\text{Torr})$ plotted against $1/T$.

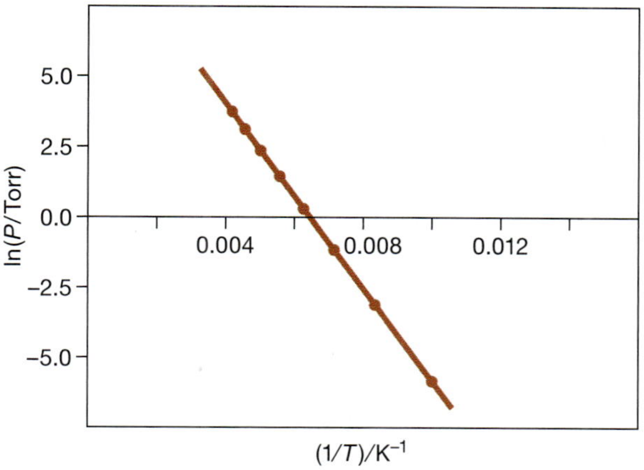

Figure A.6 Plot of $\ln(P/\text{Torr})$ as a function of $1/T$ (in units of K^{-1}) for Equation A6.3.

TABLE A.2 Numerical results for plotting Equation A6.3 as a straight line

T/K	$(1/T)$/K^{-1}	ln (P/Torr)
100	0.0100	−5.84
120	0.00833	−3.11
140	0.00714	−1.15
160	0.00625	0.31
180	0.00556	1.45
200	0.00500	2.36
220	0.00455	3.11
240	0.00417	3.73

EXAMPLE: $N_2O_5(g)$ undergoes spontaneous decomposition to $NO_2(g)$ and $O_2(g)$.
(a) Plot the following data for the natural logarithm of the concentration of $N_2O_5(g)$ in a reaction vessel as a function of time.

t/min	$\ln([N_2O_5]/M)$
0	−4.39
20.0	−4.99
40.0	−5.60
60.0	−6.21
80.0	−6.81
100.0	−7.42

(b) Find the equation of the straight line describing the time dependence of the reaction.

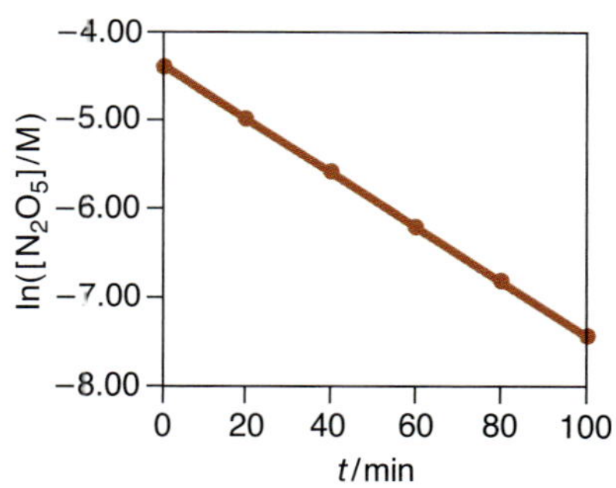

Figure A.7 Plot of $\ln([N_2O_5]/M)$ versus time, t, in minutes for the decomposition of $N_2O_5(g)$. The plot is a straight line.

Solution: (a) Figure A.7 shows that a plot of $\ln([N_2O_5]/M)$ against t is a straight line.
(b) The equation of the straight line is

$$\ln([N_2O_5]/M) = mt + b \qquad \text{(A6.4)}$$

To specify this equation, we must determine the values of the slope, m, and the intercept, b. To determine the value of m, we select two points at the extreme ends of the best-fit line to the data. The **best-fit line** is a line drawn such that it passes as close to the most points on the graph as possible. In this case the graph is fairly straight (Figure A.7), so we shall choose the first and last points given in the data set to find the slope, m, which is given by

$$m = \frac{(-7.42) - (-4.39)}{(100 - 0)\ \text{min}} = -0.0303\ \text{min}^{-1}$$

Some computer programs and some calculators can find the equation of the best-fit line automatically using a mathematical technique known as linear regression.

Substituting this value of the slope into Equation A6.4 yields

$$\ln([N_2O_5]/M) = (-0.0303\ \text{min}^{-1})t + b$$

To find the value of the intercept, b, we select one of the given data points, say, $t = 0$ and $\ln([N_2O_5]/M) = -4.39$, and substitute these into this equation to obtain

$$-4.39 = (-0.0303\ \text{min}^{-1})(0\ \text{min}) + b$$

from which we find that

$$b = -4.39$$

Thus, the equation of the straight line that describes the time dependence of the reaction is

$$\ln([N_2O_5]/M) = (-0.303\ \text{min}^{-1})t - 4.39$$

APPENDIX B

SI Units and Conversion Factors

Measurements and physical quantities in the sciences are expressed in the **metric system,** which is a system of units that was formalized by the French National Academy in 1790. There are several subsystems of units in the metric system; and in an international effort to achieve uniformity, the International System of Units (abbreviated SI from the French *Système International d'Unités)* was adopted by the General Conference of Weights and Measures in 1960 as *the* recommended units for science and technology. The SI system of units is constructed from a set of basic units. The six units that are used frequently in general chemistry are given in Table B.1. Each has a technical definition that serves to define the unit in an unambiguous, reproducible way. Here we give both the technical definition and the relation between the corresponding English system of measures for some SI units commonly used in chemistry:

1. Length: In 1983, one meter was redefined as the distance that light travels through space in 1/299 792 458 seconds. One meter is equivalent to 1.0936 yards, or to 39.370 inches. Thus, a meter stick is about 3 inches longer than a yardstick.
2. Mass: The kilogram is the only SI unit still based on an artifact (Figure B.1). Although a variety of definitions for the kilogram that are not based on artifacts have been proposed, none has yet to be formally adopted. One kilogram is equivalent to 2.2046 pounds. The mass of a substance is determined by balancing it against a set of standard masses using a balance.
3. Temperature: The kelvin, unit of thermodynamic temperature, is the fraction 1/273.15 of the thermodynamic temperature of the triple point of water. On the Celsius scale, the freezing point of water at 760 Torr is 0°C

Figure B.1 The U.S. National Prototype Kilogram is currently the primary standard for measuring mass in the United States. It is periodically recertified against the primary international standard Kilogram held at the *Bureau International des Poids et Mesures* in Sèvres, France (pictured here).

TABLE B.1 Some basic SI units

Physical quantity	Name of unit	Symbol
length	meter	m
mass	kilogram	kg
time	second	s
temperature	kelvin	K
amount of substance	mole	mol
electric current	ampere	A

and its boiling point at 760 Torr is 100°C. The Kelvin and Celsius scales are related by the equation (Chapter 1)

$$T/\mathrm{K} = t/^\circ\mathrm{C} + 273.15 \tag{B.1}$$

Recall that the freezing point of water is 32°F and its boiling point (at sea level) is 212°F on the Fahrenheit scale. The relation between the Celsius and Fahrenheit scales is given by

$$t/^\circ\mathrm{C} = \frac{5}{9}(t/^\circ\mathrm{F} - 32) \tag{B.2}$$

Thus, for example, 50°F corresponds to 10°C and 86°F corresponds to 30°C. Note that the symbol for kelvin is K and not °K.

4. Amount of substance: One mole is the amount of substance that contains as many elementary entities as there are atoms in exactly 0.012 kg of carbon-12 (Chapter 11).

An important feature of the SI system of units is the use of prefixes to designate multiples of the basic units (Table B.2).

The units of all quantities not listed in Table B.1 involve combinations of the basic SI units and are called **derived units.** The derived units frequently used in general chemistry are given in Table B.3. Many of these units may not be familiar

TABLE B.2 Prefixes used for multiples and fractions of SI units

Prefix	Symbol	Multiple	Example
peta-	P	10^{15}	petajoule, 1PJ = 10^{15} J
tera-	T	10^{12}	terawatt, 1 TW = 10^{12} W
giga-	G	10^{9}	gigavolt, 1 GV = 10^{9} V
mega-	M	10^{6}	megawatt, 1 MW = 10^{6} W
kilo-	k	10^{3}	kilometer, 1 km = 10^{3} m
hecto-	h	10^{2}	hectometer, 1 hm = 10^{2} m
deca-	da	10^{1}	decagram, 1 dag = 10^{1} g
deci-	d	10^{-1}	decimeter, 1 dm = 10^{-1} m
centi-	c	10^{-2}	centimeter, 1 cm = 10^{-2} m
milli-	m	10^{-3}	millimole, 1 mmol = 10^{-3} mol
micro-	μ*	10^{-6}	microampere, 1 μA = 10^{-6} A
nano-	n	10^{-9}	nanosecond, 1 ns = 10^{-9} s
pico-	p	10^{-12}	picometer, 1 pm = 10^{-12} m
femto-	f	10^{-15}	femtosecond, 1 fs = 10^{-15} s
atto-	a	10^{-18}	attojoule, 1 aJ = 10^{-18} J

*This is the Greek letter mu, pronounced "mew."

TABLE B.3 Names and symbols for SI-derived units

Quantity	Unit	Symbol	Definition
area	square meter	m^2	
volume	cubic meter	m^3	
mass	metric ton	t	10^3 kg
density	kilogram per cubic meter	$kg{\cdot}m^{-3}$	
speed	meter per second	$m{\cdot}s^{-1}$	
frequency	hertz	Hz	s^{-1} (cycles per second)
force	newton	N	$kg{\cdot}m{\cdot}s^{-2}$
pressure	pascal	Pa	$N{\cdot}m^{-2} = kg{\cdot}m^{-1}{\cdot}s^{-2}$
energy	joule	J	$kg{\cdot}m^2{\cdot}s^{-2} = N{\cdot}m$
electric charge	coulomb	C	$A{\cdot}s$
electric potential difference	volt	V	$J{\cdot}A^{-1}{\cdot}s^{-1} = kg{\cdot}m^2{\cdot}s^{-3}{\cdot}A^{-1}$

to you unless you have had a course in physics. For example, the SI unit of force is a **newton** (N) that is defined as the force required to give a 1-kg body an acceleration of 1 $m{\cdot}s^{-2}$. The SI unit of pressure is the **pascal** (Pa). Pressure is force per area, and a pascal is defined as the pressure produced by a force of 1 N acting on an area of 1 m^2. The SI unit of energy is the **joule** (J). A joule is the energy that a mass gains when it is acted upon by a force of 1 N through a distance of 1 m. Thus, we have $J = N{\cdot}m$, or $J = kg{\cdot}m^2{\cdot}s^{-2}$.

Although the SI is gradually becoming the universally accepted system of units, a number of older units are frequently used (Table B.4). For example, volume is usually expressed in **liters** (L). A liter is defined as a cubic decimeter and is slightly larger than a quart, being equivalent to 1.0567 qt. The glassware

TABLE B.4 Commonly used non-SI units

Quantity	Unit	Symbol	SI definition
length	ångström	Å	10^{-10} m
length	micron	μ	10^{-6} m = 1 μm
volume	liter	L	10^{-3} m^3
energy	calorie	cal	4.184 J
energy	nutritional Calorie	Cal	4.184 kJ
pressure	atmosphere	atm	101.325 kPa
pressure	Torr	Torr	133.322 Pa
pressure	bar	bar	10^5 Pa

in your laboratory is measured in milliliters (mL). One milliliter is equivalent to one cubic centimeter (cm^3).

The SI unit of pressure, the pascal, is rarely used in the United States. The most commonly used units of pressure are the **atmosphere** (atm), the **bar**, and the **Torr**—which is equivalent to **millimeters of mercury** (mmHg). The English unit of pressure is **pounds per square inch** (psi). Definitions of these units are given in Table 13.1.

EXAMPLE: Use the relation between atmospheres and pascals to derive a relation between liter-atmospheres and joules. Use this relationship to find the value of the gas constant, $R = 0.082058\ \text{L·atm·mol}^{-1}\text{·K}^{-1}$ in units of $\text{J·mol}^{-1}\text{·K}^{-1}$.

Answer: We start with

$$1\ \text{atm} = 101.325\ \text{kPa} = 1.01325 \times 10^{5}\ \text{Pa}$$

and multiply both sides by 1 L:

$$1\ \text{L·atm} = (1.031325 \times 10^{5}\ \text{Pa})(1\ \text{L})$$

Using the relations

$$\text{Pa} = \text{N·m}^{-2} \qquad \text{J} = \text{N·m} \qquad \text{L} = \text{dm}^{3} = 10^{-3}\ \text{m}^{3}$$

we obtain

$$1\ \text{L·atm} = (1.01325 \times 10^{5}\ \text{N·m}^{-2})(10^{-3}\ \text{m}^{3}) = 101.325\ \text{N·m} = 101.325\ \text{J}$$

or writing this result as a unit conversion factor, we have

$$101.325\ \text{J} = 1\ \text{L·atm}$$

From this conversion factor, we find

$$\begin{aligned} 0.082058\ \text{L·atm·mol}^{-1}\text{·K}^{-1} &= (0.082058\ \text{L·atm·mol}^{-1}\text{·K}^{-1})(101.325\ \text{J·L}^{-1}\text{·atm}^{-1}) \\ &= 8.3145\ \text{J·mol}^{-1}\text{·K}^{-1} \end{aligned}$$

The SI units and their conversion factors are given inside the back cover of this book.

APPENDIX C

Summary of IUPAC Nomenclature Rules

C1. Naming of Ionic Compounds

Binary Ionic Compounds with Single Oxidation States (Section 2-7):

When the two elements that make up a binary compound are a metal and a nonmetal that combine in only one fixed ratio, we name the compound by first writing the name of the metal and then that of the nonmetal, with the ending of the name of the nonmetal changed to *-ide*. For example,

$BaO(s)$	barium oxide
$ZnCl_2(s)$	zinc chloride
$Na_2S(s)$	sodium sulfide

Binary Ionic Compounds with Multiple Oxidation States (Section 6-4):

When the two elements that make up a binary compound are a metal and a nonmetal that can combine in more than one ratio because the metal forms two or more possible types of ions (Table C.1), the oxidation state of the metal in the compound is indicated by a Roman numeral in parentheses following the

TABLE C.1 Common oxidation states of selected metals

Metals with one common oxidation state	
Group 1 metals:	all +1; e.g. Na^+
Group 2 metals:	all +2; e.g. Mg^{2+}
Ag^+	Ni^{2+}
Cd^{2+}	Sc^{3+}
Al^{3+}	N^{3+}
Metals with two common oxidation states	
Au^+ and Au^{3+}	Co^{2+} and Co^{3+}
Cu^+ and Cu^{2+}	Fe^{2+} and Fe^{3+}
Hg_2^{2+} (†) and Hg^{2+}	Pb^{2+} and Pb^{4+}
$Sb^{3+}$and Sb^{5+}	Sn^{2+} and Sn^{4+}
Ti^{3+} and Ti^{4+}	Tl^+ and Tl^{3+}
Metals with three common oxidation states	
Cr^{2+}, Cr^{3+}, Cr^{6+}	Mn^{2+}, Mn^{4+}, Mn^{7+}

† Mercury(I) ion is dimeric (Hg_2^{2+}); that is, it involves a molecular ion composed of two Hg(I) ions bonded together.

name of the metal. The ending of the name of the nonmetal is changed to *-ide.* For example,

$PbCl_2(s)$	lead(II) chloride
$PbO_2(s)$	lead(IV) oxide
$Hg_2Cl_2(s)$	mercury(I) chloride

Note that dimeric mercury, Hg_2^{2+}, is called mercury(I).

Naming of Ionic Compounds Containing Polyatomic Ions (Section 10-2): Compounds containing polyatomic ions are named by writing the name of the cation and then that of the anion, indicating the oxidation state of metals as necessary according to the rules above. The ending of the names of polyatomic anions are not changed, but the names of nonmetal anions that combine with polyatomic cations are changed to *-ide.* For example,

$NH_4Cl(s)$	ammonium chloride
$NH_4NO_3(s)$	ammonium nitrate
$Pb(CH_3COO)_2(s)$	lead(II) acetate
$K_2Cr_2O_7(s)$	potassium dichromate

A list of common polyatomic anions is given in Table C.2.

TABLE C.2 Common polyatomic ions*

OH^-	hydroxide	O_2^{2-}	peroxide
CN^-	cyanide	CO_3^{2-}	carbonate
SCN^-	thiocyanate	SO_3^{2-}	sulfite
HCO_3^-	hydrogen carbonate (bicarbonate)	SO_4^{2-}	sulfate
HSO_3^-	hydrogen sulfite (bisulfite)	$S_2O_3^{2-}$	thiosulfate
HSO_4^-	hydrogen sulfate (bisulfate)	$C_2O_4^{2-}$	oxalate
$C_2H_3O_2^-$	acetate (also written CH_3COO^-)	CrO_4^{2-}	chromate
NO_2^-	nitrite	$Cr_2O_7^{2-}$	dichromate
NO_3^-	nitrate		
MnO_4^-	permanganate	PO_3^{3-}	phosphite
ClO^-	hypochlorite	PO_4^{3-}	phosphate
ClO_2^-	chlorite		
ClO_3^-	chlorate	NH_4^+	ammonium
ClO_4^-	perchlorate	Hg_2^{2+}	mercury(I)

*Often encountered common names are given in parenthesis.

Hydrogen (Section 2-7):
Hydrogen can act as a metal or a nonmetal. In ionic compounds hydrogen is listed first when it acts as a metal and second when it acts as a nonmetal, for example

$HCl(g)$	hydrogen chloride
$NaH(s)$	sodium hydride

Hydrates (Section 10-5):
Hydrates are ionic salts that combine with water in specific ratios. In a hydrate, the chemical formula of the salt is written followed by a dot and then the number of waters of hydration. To name a hydrate, we write the name of the anhydrous compound followed by the appropriate Greek prefix (Table C.3) to indicate the number of waters of hydration and then add the word *hydrate.* For example,

$MgSO_4{\cdot}7H_2O(s)$	magnesium sulfate heptahydrate
$CuSO_4{\cdot}5H_2O(s)$	copper(II) sulfate pentahydrate
$Na_2CO_3{\cdot}H_2O(s)$	sodium carbonate monohydrate

TABLE C.3 Greek prefixes

Number	Greek prefix*
1	mono-
2	di-
3	tri-
4	tetra-
5	penta-
6	hexa-
7	hepta-
8	octa-
9	nona-
10	deca-

*The final *a* or *o* is dropped from the prefix when it is combined with a name beginning with a vowel.

C2. Naming of Covalent Compounds

Binary Covalent Compounds (Section 2-7):

1. When the two elements that make up a binary compound are both nonmetals other than hydrogen, list the names of the elements in order of their group number in the periodic table, except for compounds formed between oxygen and a halogen, in which case the name of the halogen comes first. Hydrogen is always listed second in binary covalent compounds, except for water. In cases where both compounds are from the same group, we write the one with the greater atomic number first.
2. Use Greek prefixes (Table C.3) to specify the number of each atom in the formula of the compound with the ending of the name of the second element listed changed to *-ide.*
3. The prefix *mono-* is never used for naming the first element and is generally dropped from the second. Notable exceptions are *carbon monoxide* and sometimes *nitrogen monoxide.* The final *a* or *o* is dropped from the name of the prefix when it is combined with a name beginning with a vowel.
4. Notable exceptions are $CH_4(g)$, $NH_3(g)$, and $H_2O(l)$, which are named, *methane, ammonia,* and *water,* respectively.

Some examples of naming binary covalent compounds are,

$N_2O_5(g)$	dinitrogen pentoxide
$BN(s)$	boron nitride
$ClO_2(g)$	chlorine dioxide
$NO(g)$	nitrogen oxide (or nitrogen monoxide)
$AsH_3(g)$	arsenic trihydride

C3. Naming of Inorganic Acids

Binary Acids (Section 10-3):

Binary acids consist of two elements, one of which must be hydrogen. Binary acids are named by adding the prefix *hydro-* to the root of the anion and the ending *-ic* plus the word *acid.*

Some examples of naming binary acids are given below,

$HF(aq)$	hydrofluoric acid
$H_2S(aq)$	hydrosulfuric acid

Oxyacids (Section 10-3):

Oxyacids contain hydrogen, oxygen, and another element. The nomenclature of oxyacids is based on the name of the anion from which the acid is derived. If the name of the anion ends in *-ite* then the corresponding acid name ends in *-ous acid.* If the name of the anion ends in *-ate* then the corresponding acid name ends in *-ic acid.*

Some examples of naming oxyacids are shown below,

$HNO_2(aq)$	nitrous acid	
$HNO_3(aq)$	nitric acid	
$HClO_4(aq)$	perchloric acid	
$HC_2H_3O_2(aq)$	acetic acid	[usually written as $CH_3COOH(aq)$]

Most inorganic acids are only named as acids when present in solution. For example, $HCl(g)$ is *hydrogen chloride,* but $HCl(aq)$ is *hydrochloric acid.*

APPENDIX D
Thermodynamic Data

Standard molar entropies, enthalpies of formation, Gibbs energies of formation, and heat capacities of various substances at 25°C and 1 bar*

Substance	$S°/J·K^{-1}·mol^{-1}$†	$\Delta H_f°/kJ·mol^{-1}$	$\Delta G_f°/kJ·mol^{-1}$	$C_p/J·K^{-1}·mol^{-1}$
aluminum				
$Al(s)$	28.3	0	0	24.4
$Al_2O_3(s)$	50.9	−1675.7	−1582.3	79.0
argon				
$Ar(g)$	154.8	0	0	20.8
barium				
$Ba(s)$	62.5	0	0	28.1
$BaCO_3(s)$	112.1	−1213.0	−1134.4	86.0
$BaO(s)$	72.1	−548.0	−520.3	47.3
$Ba^{2+}(aq)$	9.6	−537.6	−560.8	—
bromine				
$Br(g)$	175.0	111.9	82.4	20.8
$Br_2(g)$	245.5	30.9	3.1	36.0
$Br_2(l)$	152.2	0	0	75.7
$Br^-(aq)$	82.4	−121.6	−104.0	—
calcium				
$Ca(s)$	41.6	0	0	25.9
$CaC_2(s)$	70.0	−59.8	−64.9	62.7
$CaCO_3(s)$	91.7	−1207.6	−1129.1	83.5
$CaO(s)$	38.1	−634.9	−603.3	42.0
$CaSO_4(s)$	106.5	−1434.5	−1322.0	99.7
$Ca^{2+}(aq)$	−53.1	−542.8	−553.6	—

Substance	$S°/J\cdot K^{-1}\cdot mol^{-1}$†	$\Delta H_f°/kJ\cdot mol^{-1}$	$\Delta G_f°/kJ\cdot mol^{-1}$	$C_p/J\cdot K^{-1}\cdot mol^{-1}$
carbon				
$C(s$, diamond)	2.4	1.9	2.9	6.1
$C(s$, graphite)	5.7	0	0	8.5
$C_{60}(s$, fullerene)	426.0	2327.0	2302.0	520.0
$CH_4(g)$	186.3	−74.6	−50.5	35.7
$C_2H_2(g)$	200.9	227.4	209.9	44.0
$C_2H_4(g)$	219.3	52.4	68.4	42.9
$C_2H_6(g)$	229.2	−84.0	−32.0	52.5
$C_3H_8(g)$	270.3	−103.8	−23.4	73.6
$C_6H_6(l)$	173.4	49.1	124.5	136.0
$CH_3OH(g)$	239.9	−201.0	−162.3	44.1
$CH_3OH(l)$	126.8	−239.2	−166.6	81.1
$C_2H_5OH(g)$	281.6	−234.8	−167.9	65.6
$C_2H_5OH(l)$	160.7	−277.6	−174.8	112.3
$CH_3Cl(g)$	234.6	−81.9	−58.4	40.8
$CH_3Cl(l)$	145.3	−102	−51.5	—
$CH_2Cl_2(g)$	270.2	−95.4	−68.8	51.0
$CH_2Cl_2(l)$	177.8	−124.2	−70.0	101.2
$CHCl_3(g)$	295.7	−102.7	6.0	65.7
$CHCl_3(l)$	201.7	−134.1	−73.7	114.2
$CCl_4(g)$	308.7	−95.7	−60.6	83.3
$CCl_4(l)$	215.4	−128.2	−65.3	130.7
$CO(g)$	197.7	−110.5	−137.2	29.1
$CO_2(g)$	213.8	−393.5	−394.4	37.1
$CO_3^{2-}(aq)$	−56.9	−677.1	−527.9	—
chlorine				
$Cl(g)$	165.2	121.3	105.3	21.8
$Cl_2(g)$	223.1	0	0	33.9
$Cl^-(aq)$	56.5	−167.2	−131.2	—
copper				
$Cu(s)$	33.2	0	0	24.4
$CuO(s)$	42.6	−157.3	−129.7	42.3
$Cu_2O(s)$	93.1	−168.6	−146.0	63.6
$Cu^{2+}(aq)$	−99.6	64.8	65.5	—

Substance	$S°/J·K^{-1}·mol^{-1}$†	$ΔH_f°/kJ·mol^{-1}$	$ΔG_f°/kJ·mol^{-1}$	$C_p/J·K^{-1}·mol^{-1}$
fluorine				
$F(g)$	158.8	79.4	62.3	22.7
$F_2(g)$	202.8	0	0	31.3
$F^-(aq)$	–13.8	–332.6	–278.8	—
helium				
$He(g)$	126.2	0	0	20.8
hydrogen				
$H(g)$	114.7	218.0	203.3	20.8
$H_2(g)$	130.7	0	0	28.8
$H_2O(g)$	188.8	–241.8	–228.6	33.6
$H_2O(l)$	70.0	–285.8	–237.1	75.3
$H_2O_2(l)$	109.6	–187.8	–120.4	89.1
$HF(g)$	173.8	–273.3	–275.4	29.1
$HCl(g)$	186.9	–92.3	–95.3	29.1
$HBr(g)$	198.7	–36.3	–53.4	29.1
$HI(g)$	206.6	26.5	1.7	29.2
$H_2S(g)$	205.8	–20.6	–33.4	34.2
$H^+(aq)$	0	0	0	—
$H_3O^+(aq)$	70.0	–285.8	–237.1	—
iodine				
$I(g)$	180.8	106.8	70.2	20.8
$I_2(g)$	260.7	62.4	19.3	36.9
$I_2(s)$	116.1	0	0	54.4
$I^-(aq)$	111.3	–55.2	–51.6	—
iron				
$Fe(s)$	27.3	0	0	25.1
$FeO(s)$	60.75	–272.0	–251.4	49.92
$Fe_2O_3(s)$	87.4	–824.2	–742.2	103.9
$Fe_3O_4(s)$	146.4	–1118.4	–1015.4	143.3
$Fe^{2+}(aq)$	–137.7	–89.1	–78.9	—
$Fe^{3+}(aq)$	–315.9	–48.5	–4.7	—
krypton				
$Kr(g)$	164.1	0	0	20.8

Substance	$S°$/J·K^{-1}·mol^{-1}†	$\Delta H_f°$/kJ·mol^{-1}	$\Delta G_f°$/kJ·mol^{-1}	C_p/J·K^{-1}·mol^{-1}
lead				
$Pb(s)$	64.8	0	0	26.4
$PbCl_2(s)$	136.0	−359.4	−314.1	—
$PbO(s$, massicot)	68.7	−217.3	−187.9	45.8
$PbO(s$, litharge)	66.5	−218.0	−188.9	45.8
$PbSO_4(s)$	148.5	−920.0	−813.0	103.2
$Pb^{2+}(aq)$	10.5	−1.7	−24.4	—
magnesium				
$Mg(s)$	32.7	0	0	24.9
$MgO(s)$	27.0	−601.6	−569.3	37.2
$MgCO_3(s)$	65.7	−1095.8	−1012.1	75.5
$Mg^{2+}(aq)$	−138.1	−466.9	−454.8	—
neon				
$Ne(g)$	146.3	0	0	20.8
nitrogen				
$N(g)$	153.3	472.7	455.5	20.8
$N_2(g)$	191.6	0	0	29.1
$NH_3(g)$	192.8	−45.9	−16.4	35.1
$N_2H_4(l)$	121.2	50.6	149.3	98.9
$NO(g)$	210.8	91.3	87.6	29.9
$NO_2(g)$	240.1	33.2	51.3	37.2
$N_2O(g)$	220.0	81.6	103.7	38.6
$N_2O_4(g)$	304.4	11.1	99.8	79.2
$N_2O_4(l)$	209.2	−19.5	97.5	142.7
$N_2O_5(s)$	178.2	−43.1	113.9	143.1
$NOCl(g)$	261.7	51.7	66.1	44.7
$NH_3(aq)$	111.3	−80.3	−26.5	—
$NH_4^+(aq)$	113.4	−132.5	−79.3	—
$NO_3^-(aq)$	146.4	−207.4	−111.3	—
oxygen				
$O(g)$	161.1	249.2	231.7	21.9
$O_2(g)$	205.2	0	0	29.4
$O_3(g)$	238.9	142.7	163.2	39.2
$OH^-(aq)$	−10.8	−230.0	−157.2	—

Substance	$S°$/J·K^{-1}·mol^{-1}†	ΔH_f°/kJ·mol^{-1}	ΔG_f°/kJ·mol^{-1}	C_p/J·K^{-1}·mol^{-1}
phosphorus				
P(*s*, white)	41.1	0	0	23.8
P(*s*, red)	22.8	−17.6	−12.1	21.2
$P_4O_{10}(s)$	228.9	−2984	−2698	211.7
$POCl_3(g)$	325.5	−558.5	−512.9	84.9
$POCl_3(l)$	222.5	−597.1	−520.8	138.8
$PCl_3(g)$	311.8	−287.0	−267.8	71.8
$PCl_5(g)$	364.6	−374.9	−305.0	112.8
$PH_3(g)$	210.2	5.4	13.5	37.1
potassium				
KOH(*s*)	81.2	−424.6	−379.4	68.9
KCl(*s*)	82.6	−436.5	−408.5	51.3
$KClO_3(s)$	143.1	−397.7	−296.3	100.3
$K^+(aq)$	102.5	−252.4	−283.3	—
silver				
Ag(*s*)	42.6	0	0	25.4
AgBr(*s*)	107.1	−100.4	−96.9	52.4
AgCl(*s*)	96.3	−127.0	−109.8	50.8
$Ag_2SO_4(s)$	200.4	−715.9	−618.4	131.4
$Ag^+(aq)$	72.7	105.6	77.1	—
sodium				
Na(*g*)	153.7	107.5	77.0	20.8
Na(*s*)	51.3	0	0	28.2
$NaHCO_3(s)$	101.7	−950.8	−851.0	87.6
$Na_2CO_3(s)$	135.0	−1130.7	−1044.4	112.3
$Na_2O(s)$	75.1	−414.2	−375.5	69.1
NaOH(*s*)	64.4	−425.8	−379.7	59.5
NaF(*s*)	51.1	−576.6	−546.3	46.9
NaCl(*s*)	72.1	−411.2	−384.1	50.5
NaBr(*s*)	86.8	−361.1	−349.0	51.4
NaI(*s*)	98.5	−287.8	−286.1	52.1
$Na^+(aq)$	59.0	−240.1	−261.9	—

Substance	$S°$/J·K^{-1}·mol^{-1}†	$\Delta H_f°$/kJ·mol^{-1}	$\Delta G_f°$/kJ·mol^{-1}	C_p/J·K^{-1}·mol^{-1}
sulfur				
S(*s*, rhombic)	28.5	0	0	22.6
S(*s*, monoclinic)	32.6	0.3	0.1	—
$SO_2(g)$	248.2	−296.8	−300.1	39.9
$SO_3(g)$	256.8	−395.7	−371.1	50.7
$SF_6(g)$	291.5	−1220.5	−1116.5	97.0
$SO_4^{2-}(aq)$	20.1	−909.3	−744.5	—
tin				
Sn(*s*, white)	51.2	0	0	27.0
Sn(*s*, gray)	44.1	−2.1	0.1	25.8
SnO(*s*)	57.2	−280.7	−251.9	44.3
$SnO_2(s)$	49.0	−577.6	−515.8	52.6
xenon				
Xe(*g*)	169.7	0	0	20.8
zinc				
Zn(*s*)	41.6	0	0	25.4
ZnO(*s*)	43.7	−350.5	−320.5	40.3
ZnS(*s*)	57.7	−206.0	−201.3	46.0
$Zn^{2+}(aq)$	−112.1	−153.9	−147.1	46.0

*Data from *CRC Handbook of Chemistry and Physics*, 87th online edition, 2006–2007, except where noted in blue.

†Solution-phase entropies are measured relative to a defined standard of $S°$ for $H^+(aq) \equiv 0$; thus, the given entropies of some aqueous compounds may be negative.

APPENDIX E

Data for Selected Acids and Bases

Acid dissociation constants of weak acids in water at 25°C[*]

Name	Formula[†]	K_a/M	pK_a
acetic acid	CH_3COOH	1.8×10^{-5}	4.74
arsenic acid	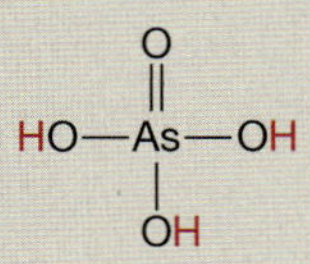	5.5×10^{-3} 1.7×10^{-7} 5.1×10^{-12}	2.26 6.76 11.29
benzene-1,2,3-tricarboxylic acid (trimellitic acid)	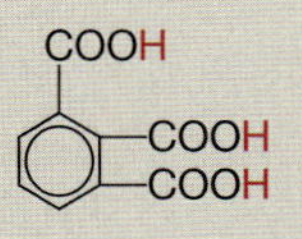	1.3×10^{-3} 1.8×10^{-5} 7.4×10^{-8}	2.88 4.75 7.13
benzoic acid	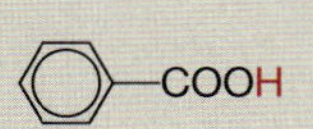	6.3×10^{-5}	4.20
boric acid	$HOB(OH)_2$	5.4×10^{-10}	9.27
bromoacetic acid	$BrCH_2COOH$	1.3×10^{-3}	2.90
butanoic acid	$CH_3CH_2CH_2COOH$	1.5×10^{-5}	4.83
carbonic acid	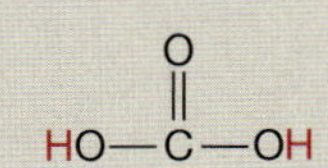	4.5×10^{-7} 4.7×10^{-11}	6.35 10.33
chloroacetic acid	$ClCH_2COOH$	1.4×10^{-3}	2.87
chlorous acid	O=Cl—OH	1.2×10^{-2}	1.94
cyanic acid	HCNO	3.5×10^{-4}	3.46
dichloroacetic	$Cl_2CHCOOH$	4.5×10^{-2}	1.35
fluoroacetic	FCH_2COOH	2.6×10^{-3}	2.59
formic acid	HCOOH	1.8×10^{-4}	3.75
hydrazoic (hydrogen azide)	HN_3	3×10^{-5}	4.6
hydrocyanic acid	HCN	6.2×10^{-10}	9.21
hydrofluoric acid	HF	6.3×10^{-4}	3.20

Name	Formula[†]	K_a/M	pK_a
hydrosulfuric acid	H_2S	8.9×10^{-8} 1.2×10^{-13}	7.05 12.91
thiocyanic acid	HSCN	63.1	–1.8
hypobromous acid	HOBr	2.8×10^{-9}	8.55
hypochlorous acid	HOCl	4.0×10^{-8}	7.40
hypoiodous acid	HOI	3×10^{-11}	10.5
iodic acid	HIO_3	0.17	0.78
lactic acid	$CH_3CHOHCOOH$	1.4×10^{-4}	3.86
nitrous acid	HNO_2	5.6×10^{-4}	3.25
oxalic acid	HOOC–COOH	0.056 1.5×10^{-4}	1.25 3.81
phosphoric acid	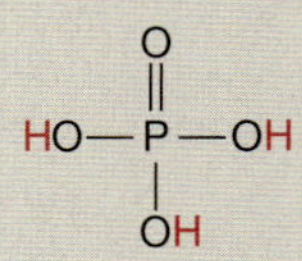	6.9×10^{-3} 6.2×10^{-8} 4.8×10^{-13}	2.16 7.21 12.32
phosphorous acid	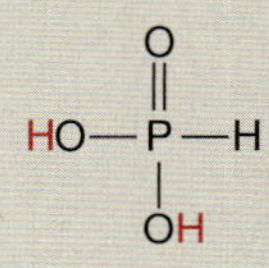	5×10^{-2} 2.0×10^{-7}	1.3 6.70
propanoic acid	CH_3CH_2COOH	1.4×10^{-5}	4.87
sulfuric acid	H_2SO_4	strong 1.0×10^{-2}	strong 1.99
sulfurous acid	H_2SO_3	1.4×10^{-2} 6×10^{-8}	1.85 7.2
thiosulfuric acid	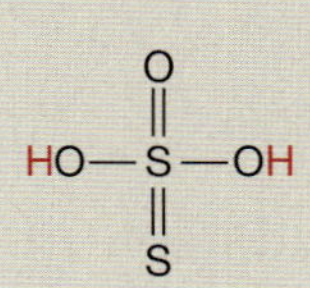	0.30 3.0×10^{-2}	0.52 1.52

*pK_a values from the *CRC Handbook of Chemistry and Physics,* 87th ed., 2006–2007, except where noted in blue.

[†]Acidic protons are shown in red.

Base protonation constants of weak bases in water at 25°C*

Name	Formula	Protonated form	K_b/M	pK_b*
ammonia	NH_3	NH_4^+	1.8×10^{-5}	4.75
aniline	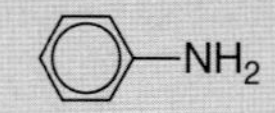	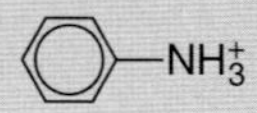	7.4×10^{-10}	9.13
benzylamine	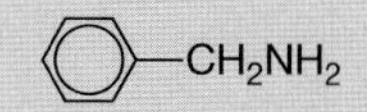	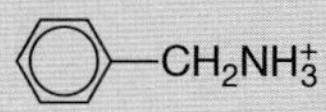	2.2×10^{-5}	4.66
butylamine	$CH_3CH_2CH_2CH_2NH_2$	$CH_3CH_2CH_2CH_2NH_3^+$	4.0×10^{-4}	3.40
cyclohexylamine	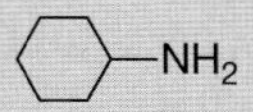	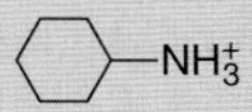	4.4×10^{-4}	3.36
diethylamine	$(C_2H_5)_2NH$	$(C_2H_5)_2NH_2^+$	6.9×10^{-4}	3.16
dimethylamine	$(CH_3)_2NH$	$(CH_3)_2NH_2^+$	5.4×10^{-4}	3.27
ethanolamine	$HOCH_2CH_2NH_2$	$HOCH_2CH_2NH_3^+$	3.2×10^{-5}	4.50
ethylamine	$CH_3CH_2NH_2$	$CH_3CH_2NH_3^+$	4.5×10^{-4}	3.35
hydroxylamine	$HONH_2$	$HONH_3^+$	8.7×10^{-9}	8.06
methylamine	CH_3NH_2	$CH_3NH_3^+$	4.6×10^{-4}	3.34
piperdine	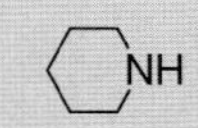	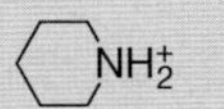	1.3×10^{-3}	2.88
propylamine	$CH_3CH_2CH_2NH_2$	$CH_3CH_2CH_2NH_3^+$	3.5×10^{-4}	3.46
pyridine	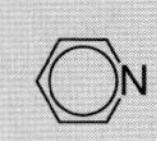	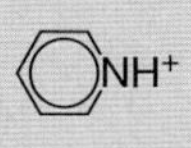	1.7×10^{-9}	8.77
trimethylamine	$(CH_3)_3N$	$(CH_3)_3NH^+$	6.3×10^{-5}	4.20

*pK_b values derived from data in the *CRC Handbook of Chemistry and Physics,* 87th ed., 2006–2007.

APPENDIX F

Solubility of Ionic Compounds

General Solubility Rules for Ionic Compounds (Section 10-9), *applied in the order given*:

1. Most alkali metal salts and ammonium salts are soluble.
2. Most nitrates, acetates, and perchlorates are soluble.
3. Most silver, lead, and mercury(I) salts are insoluble.
4. Most chlorides, bromides, and iodides are soluble.
5. Most carbonates, chromates, sulfides, oxides, phosphates, and hydroxides are insoluble; *except* for hydroxides of Ba^{2+}, Ca^{2+}, and Sr^{2+} which are slightly soluble.
6. Most sulfates are soluble; *except* for calcium sulfate and barium sulfate which are insoluble.

Solubility-product constants for various ionic compounds in water at 25°C†

Bromates	K_{sp}	**Cyanides**	K_{sp}	**Oxalates**	K_{sp}
$AgBrO_3$	5.4×10^{-5} M^2	AgCN	6.0×10^{-17} M^2	$Ag_2C_2O_4$	5.4×10^{-12} M^3
$Ba(BrO_3)_2$	2.4×10^{-4} M^3	CuCN	3.5×10^{-20} M^2	CaC_2O_4	4×10^{-9} M^2
$Pb(BrO_3)_2$	7.9×10^{-6} M^3	$Hg_2(CN)_2$*	5×10^{-40} M^3	MgC_2O_4	7×10^{-7} M^2
$TlBrO_3$	1.1×10^{-4} M^2	$Zn(CN)_2$	3×10^{-16} M^3	SrC_2O_4	4×10^{-7} M^2
Bromides	K_{sp}	**Fluorides**	K_{sp}	**Sulfates**	K_{sp}
AgBr	5.4×10^{-13} M^2	BaF_2	1.8×10^{-7} M^3	Ag_2SO_4	1.2×10^{-5} M^3
CuBr	6.3×10^{-9} M^2	CaF_2	3.5×10^{-11} M^3	$BaSO_4$	1.1×10^{-10} M^2
Hg_2Br_2*	6.4×10^{-23} M^3	LiF	1.8×10^{-3} M^2	$CaSO_4$	4.9×10^{-5} M^2
$HgBr_2$	6.2×10^{-20} M^3	MgF_2	5.2×10^{-11} M^3	Hg_2SO_4	6.5×10^{-7} M^2
$PbBr_2$	6.6×10^{-6} M^3	PbF_2	3.3×10^{-8} M^3	$PbSO_4$	2.5×10^{-8} M^2
TlBr	3.7×10^{-6} M^2	SrF_2	4.3×10^{-9} M^3	$SrSO_4$	3.4×10^{-7} M^2

Carbonates	K_{sp}	Hydroxides	K_{sp}	Sulfides	K_{sp}
Ag_2CO_3	8.5×10^{-12} M^3	$Al(OH)_3$	1.3×10^{-33} M^4	Ag_2S	8×10^{-51} M^3
$BaCO_3$	2.6×10^{-9} M^2	$Ca(OH)_2$	5.0×10^{-6} M^3	CdS	8.0×10^{-27} M^2
$CaCO_3$	3.4×10^{-9} M^2	$Cd(OH)_2$	7.2×10^{-15} M^3	CoS	5×10^{-22} M^2
$CdCO_3$	1.0×10^{-12} M^2	$Co(OH)_2$	5.9×10^{-15} M^3	CuS	6.3×10^{-36} M^2
$CoCO_3$	1.0×10^{-10} M^2	$Cr(OH)_3$	6.3×10^{-31} M^4	FeS	6.3×10^{-18} M^2
$CuCO_3$	1.4×10^{-10} M^2	$Cu(OH)_2$	2.2×10^{-20} M^3	HgS	4×10^{-53} M^2
$FeCO_3$	3.1×10^{-11} M^2	$Fe(OH)_2$	4.9×10^{-17} M^3	MnS	2.5×10^{-13} M^2
$MgCO_3$	6.8×10^{-6} M^2	$Fe(OH)_3$	2.8×10^{-39} M^4	NiS	1.3×10^{-25} M^2
$MnCO_3$	2.2×10^{-11} M^2	$Mg(OH)_2$	5.6×10^{-12} M^3	PbS	8.0×10^{-28} M^2
$NiCO_3$	1.4×10^{-7} M^2	$Ni(OH)_2$	5.5×10^{-16} M^3	SnS	1.0×10^{-25} M^2
$PbCO_3$	7.4×10^{-14} M^2	$Pb(OH)_2$	1.4×10^{-20} M^3	Tl_2S	6×10^{-22} M^3
$SrCO_3$	5.6×10^{-10} M^2	$Sn(OH)_2$	5.5×10^{-27} M^3	ZnS	1.6×10^{-24} M^2
$ZnCO_3$	1.5×10^{-10} M^2	$Zn(OH)_2$	1.0×10^{-15} M^3		

Chlorides	K_{sp}	Iodates	K_{sp}	Thiocyanates	K_{sp}
AgCl	1.8×10^{-10} M^2	$AgIO_3$	3.2×10^{-8} M^2	AgSCN	1.0×10^{-12} M^2
CuCl	1.7×10^{-7} M^2	$Ba(IO_3)_2$	4.0×10^{-9} M^3	CuSCN	1.8×10^{-13} M^2
Hg_2Cl_2*	1.4×10^{-18} M^3	$Ca(IO_3)_2$	6.5×10^{-6} M^3	$Cu(SCN)_2$	4.0×10^{-14} M^3
$PbCl_2$	1.5×10^{-5} M^3	$Cd(IO_3)_2$	2.5×10^{-8} M^3	$Hg_2(SCN)_2$*	3.2×10^{-20} M^3
TlCl	1.9×10^{-4} M^2	$Cu(IO_3)_2$	7.4×10^{-8} M^3	$Hg(SCN)_2$	2.8×10^{-20} M^3
		$Pb(IO_3)_2$	3.7×10^{-13} M^3	TlSCN	1.6×10^{-4} M^2
		$TlIO_3$	3.1×10^{-6} M^2		
		$Zn(IO_3)_2$	3.9×10^{-6} M^3		

Chromates	K_{sp}	Iodides	K_{sp}
Ag_2CrO_4	1.1×10^{-12} M^3	AgI	8.5×10^{-17} M^2
$BaCrO_4$	1.2×10^{-10} M^2	CuI	1.3×10^{-12} M^2
$CuCrO_4$	3.6×10^{-6} M^2	Hg_2I_2*	5.2×10^{-29} M^3
Hg_2CrO_4*	2.0×10^{-9} M^2	HgI_2	2.9×10^{-29} M^3
$PbCrO_4$	2.8×10^{-13} M^2	PbI_2	9.8×10^{-9} M^3
Tl_2CrO_4	8.7×10^{-13} M^3	TlI	5.5×10^{-8} M^2

*Recall that Hg(I) exists as $Hg_2^{2+}(aq)$ in aqueous solution.

†K_{sp} values from the *CRC Handbook of Chemistry and Physics*, 87th ed., 2006–2007, except where noted in blue.

APPENDIX G

Standard Reduction Voltages for Aqueous Solutions at 25.0°C*

Elements	Half reaction	E°_{red}/V
aluminum	$Al^{3+}(aq) + 3e^- \leftrightharpoons Al(s)$	–1.662
	$Al(OH)_4^-(aq) + 3e^- \leftrightharpoons Al(s) + 4OH^-(aq)$	–2.328
barium	$Ba^{2+}(aq) + 2e^- \leftrightharpoons Ba(s)$	–2.912
beryllium	$Be^{2+}(aq) + 2e^- \leftrightharpoons Be(s)$	–1.847
bromine	$BrO_3^-(aq) + 6H^+(aq) + 5e^- \leftrightharpoons \frac{1}{2}Br_2(l) + 3H_2O(l)$	1.482
	$Br_2(l) + 2e^- \leftrightharpoons 2Br^-(aq)$	1.066
	$BrO^-(aq) + H_2O(l) + 2e^- \leftrightharpoons Br^-(aq) + 2OH^-(aq)$	0.761
	$BrO_3^-(aq) + 3H_2O(l) + 6e^- \leftrightharpoons Br^-(aq) + 6OH^-(aq)$	0.61
cadmium	$Cd^{2+}(aq) + 2e^- \leftrightharpoons Cd(s)$	–0.403
calcium	$Ca^{2+}(aq) + 2e^- \leftrightharpoons Ca(s)$	–2.868
	$CaSO_4(s) + 2e^- \leftrightharpoons Ca(s) + SO_4^{2-}(aq)$	–2.936
cerium	$Ce^{4+}(aq) + e^- \leftrightharpoons Ce^{3+}(aq)$	1.72
	$Ce^{3+}(aq) + 3e^- \leftrightharpoons Ce(s)$	–2.336
cesium	$Cs^+(aq) + e^- \leftrightharpoons Cs(s)$	–3.026
chlorine	$HClO(aq) + H^+(aq) + e^- \leftrightharpoons \frac{1}{2}Cl_2(g) + H_2O(l)$	1.611
	$ClO_3^-(aq) + 6H^+(aq) + 5e^- \leftrightharpoons \frac{1}{2}Cl_2(g) + 3H_2O(l)$	1.47
	$Cl_2(g) + 2e^- \leftrightharpoons 2Cl^-(aq)$	1.358
	$ClO_4^-(aq) + 2H^+(aq) + 2e^- \leftrightharpoons ClO_3^-(aq) + H_2O(l)$	1.189
	$ClO^-(aq) + H_2O(l) + 2e^- \leftrightharpoons Cl^-(aq) + 2OH^-(aq)$	0.81
chromium	$Cr_2O_7^{2-}(aq) + 14H^+(aq) + 6e^- \leftrightharpoons 2Cr^{3+}(aq) + 7H_2O(l)$	1.232
	$CrO_4^{2-}(aq) + 4H_2O(l) + 3e^- \leftrightharpoons Cr(OH)_3(aq) + 5OH^-(aq)$	–0.13
	$Cr^{3+}(aq) + e^- \leftrightharpoons Cr^{2+}(aq)$	–0.407
	$Cr^{3+}(aq) + 3e^- \leftrightharpoons Cr(s)$	–0.744
	$Cr^{2+}(aq) + 2e^- \leftrightharpoons Cr(s)$	–0.913

*Data from *CRC Handbook of Chemistry and Physics,* 87th ed., Ed. David R. Lide, CRC Press, 2006–2007.

Elements	Half reaction	E°_{red}/V
cobalt	$Co^{3+}(aq) + e^- \leftrightharpoons Co^{2+}(aq)$	1.92
	$[Co(NH_3)_6]^{3+}(aq) + e^- \leftrightharpoons [Co(NH_3)_6]^{2+}(aq)$	0.108
	$Co^{2+}(aq) + 2e^- \leftrightharpoons Co(s)$	−0.28
	$Co(OH)_2(s) + 2e^- \leftrightharpoons Co(s) + 2OH^-(aq)$	−0.73
copper	$Cu^+(aq) + e^- \leftrightharpoons Cu(s)$	0.521
	$Cu^{2+}(aq) + 2e^- \leftrightharpoons Cu(s)$	0.342
	$Cu^{2+}(aq) + e^- \leftrightharpoons Cu^+(aq)$	0.153
	$Cu(OH)_2(s) + 2e^- \leftrightharpoons Cu(s) + 2OH^-(aq)$	−0.222
fluorine	$F_2(g) + 2e^- \leftrightharpoons 2F^-(aq)$	2.866
gadolinium	$Gd^{3+}(aq) + 3e^- \leftrightharpoons Gd(s)$	−2.279
gallium	$Ga^{3+}(aq) + 3e^- \leftrightharpoons Ga(s)$	−0.549
gold	$Au^+(aq) + e^- \leftrightharpoons Au(s)$	1.692
	$Au^{3+}(aq) + 2e^- \leftrightharpoons Au^+(aq)$	1.401
	$AuCl_2^-(aq) + e^- \leftrightharpoons Au(s) + 2Cl^-(aq)$	1.154
	$AuCl_4^-(aq) + 2e^- \leftrightharpoons AuCl_2^-(aq) + 2Cl^-(aq)$	0.926
hydrogen	$H_2O_2(aq) + 2H^+(aq) + 2e^- \leftrightharpoons 2H_2O(l)$	1.776
	$2H^+(aq) + 2e^- \leftrightharpoons H_2(g)$	0.0000
	$H_2O(l) + e^- \leftrightharpoons \frac{1}{2}H_2(g) + OH^-(aq)$	−0.828
	$H_2(g) + 2e^- \leftrightharpoons H^-(aq)$	−2.23
indium	$In(OH)_3(s) + 3e^- \leftrightharpoons In(s) + 3OH^-(aq)$	−0.99
iodine	$IO_3^-(aq) + 6H^+(aq) + 5e^- \leftrightharpoons \frac{1}{2}I_2(l) + 3H_2O(l)$	1.195
	$I_2(s) + 2e^- \leftrightharpoons 2I^-(aq)$	0.536
	$I_3^-(aq) + 2e^- \leftrightharpoons 3I^-(aq)$	0.536
iron	$Fe^{3+}(aq) + e^- \leftrightharpoons Fe^{2+}(aq)$	0.771
	$[Fe(CN)_6]^{3-}(aq) + e^- \leftrightharpoons [Fe(CN)_6]^{4-}(aq)$	0.358
	$Fe^{3+}(aq) + 3e^- \leftrightharpoons Fe(s)$	−0.037
	$Fe^{2+}(aq) + 2e^- \leftrightharpoons Fe(s)$	−0.447
	$Fe(OH)_3(s) + e^- \leftrightharpoons Fe(OH)_2(s) + OH^-(aq)$	−0.56
lanthanum	$La^{3+}(aq) + 3e^- \leftrightharpoons La(s)$	−2.379
lead	$PbO_2(s) + 4H^+(aq) + SO_4^{2-}(aq) + 2e^- \leftrightharpoons PbSO_4(s) + 2H_2O(l)$	1.691
	$PbO_2(s) + 4H^+(aq) + 2e^- \leftrightharpoons Pb^{2+}(s) + 2H_2O(l)$	1.455
	$Pb^{2+}(aq) + 2e^- \leftrightharpoons Pb(s)$	−0.126
	$PbF_2(s) + 2e^- \leftrightharpoons Pb(s) + 2F^-(aq)$	−0.344
	$PbSO_4(s) + 2e^- \leftrightharpoons Pb(s) + SO_4^{2-}(aq)$	−0.359

Elements	Half reaction	E°_{red}/V
lithium	$Li^+(aq) + e^- \leftrightharpoons Li(s)$	−3.040
magnesium	$Mg^{2+}(aq) + 2e^- \leftrightharpoons Mg(s)$	−2.372
manganese	$MnO_4^-(aq) + 4H^+(aq) + 3e^- \leftrightharpoons MnO_2(s) + 2H_2O(l)$	1.679
	$Mn^{3+}(aq) + e^- \leftrightharpoons Mn^{2+}(aq)$	1.542
	$MnO_4^-(aq) + 8H^+(aq) + 5e^- \leftrightharpoons Mn^{2+}(aq) + 4H_2O(l)$	1.507
	$MnO_2(s) + 4H^+(aq) + 2e^- \leftrightharpoons Mn^{2+}(aq) + 2H_2O(l)$	1.224
	$MnO_4^-(aq) + e^- \leftrightharpoons MnO_4^{2-}(aq)$	0.558
	$2MnO_2(s) + H_2O(l) + e^- \leftrightharpoons Mn_2O_3(s) + 2OH^-(aq)$	0.118
	$Mn^{2+}(aq) + 2e^- \leftrightharpoons Mn(s)$	−1.185
	$Mn(OH)_2(s) + 2e^- \leftrightharpoons Mn(s) + 2OH^-(aq)$	−1.56
mercury	$2Hg^{2+}(aq) + 2e^- \leftrightharpoons Hg_2^{2+}(aq)$	0.920
	$Hg^{2+}(aq) + 2e^- \leftrightharpoons Hg(l)$	0.851
	$Hg_2^{2+}(aq) + 2e^- \leftrightharpoons 2Hg(l)$	0.797
	$Hg_2SO_4(s) + 2e^- \leftrightharpoons 2Hg(l) + SO_4^{2-}(aq)$	0.613
	$Hg_2Cl_2(s) + 2e^- \leftrightharpoons 2Hg(l) + 2Cl^-(aq)$	0.268
	$Hg_2Br_2(s) + 2e^- \leftrightharpoons 2Hg(l) + 2Br^-(aq)$	0.139
	$HgO(s) + H_2O(l) + 2e^- \leftrightharpoons Hg(l) + 2OH^-(aq)$	0.0977
nickel	$Ni^{2+}(aq) + 2e^- \leftrightharpoons Ni(s)$	−0.257
	$Ni(OH)_2(s) + 2e^- \leftrightharpoons Ni(s) + 2OH^-(aq)$	−0.72
nitrogen	$N_2O(g) + 2H^+(aq) + 2e^- \leftrightharpoons N_2(g) + H_2O(l)$	1.766
	$2NO(aq) + 2H^+(aq) + 2e^- \leftrightharpoons N_2O(g) + H_2O(l)$	1.591
	$HNO_2(aq) + H^+(aq) + e^- \leftrightharpoons NO(g) + H_2O(l)$	0.983
	$NO_3^-(aq) + 4H^+(aq) + 3e^- \leftrightharpoons NO(g) + 2H_2O(l)$	0.957
	$NO_3^-(aq) + 3H^+(aq) + 2e^- \leftrightharpoons HNO_2(aq) + H_2O(l)$	0.934
	$NO_3^-(aq) + 2H^+(aq) + e^- \leftrightharpoons \frac{1}{2}N_2O_4(g) + H_2O(l)$	0.803
oxygen	$O_3(g) + 2H^+(aq) + 2e^- \leftrightharpoons O_2(g) + H_2O(l)$	2.076
	$O_3(g) + H_2O(l) + 2e^- \leftrightharpoons O_2(g) + 2OH^-(aq)$	1.24
	$O_2(g) + 4H^+(aq) + 4e^- \leftrightharpoons 2H_2O(l)$	1.229
	$O_2(g) + 2H^+(aq) + 2e^- \leftrightharpoons H_2O_2(l)$	0.695
	$O_2(g) + 2H_2O(l) + 4e^- \leftrightharpoons 4OH^-(aq)$	0.401
palladium	$Pd^{2+}(aq) + 2e^- \leftrightharpoons Pd(s)$	0.951
phosphorous	$H_3PO_4(aq) + 2H^+(aq) + 2e^- \leftrightharpoons H_3PO_3(aq) + H_2O(l)$	−0.276
	$H_3PO_3(aq) + 2H^+(aq) + 2e^- \leftrightharpoons H_3PO_2(aq) + H_2O(l)$	−0.499
platinum	$Pt^{2+}(aq) + 2e^- \leftrightharpoons Pt(s)$	1.18

Elements	**Half reaction**	E°_{red}/V
potassium	$K^+(aq) + e^- \leftrightharpoons K(s)$	−2.931
rubidium	$Rb^+(aq) + e^- \leftrightharpoons Rb(s)$	−2.98
scandium	$Sc^{3+}(aq) + 3e^- \leftrightharpoons Sc(s)$	−2.077
silver	$Ag^+(aq) + e^- \leftrightharpoons Ag(s)$	0.7996
	$Ag_2O(s) + H_2O(l) + 2e^- \leftrightharpoons 2Ag(s) + 2OH^-(aq)$	0.342
	$AgCl(s) + e^- \leftrightharpoons Ag(s) + Cl^-(aq)$	0.2223
	$AgBr(s) + e^- \leftrightharpoons Ag(s) + Br^-(aq)$	0.0713
	$AgI(s) + e^- \leftrightharpoons Ag(s) + I^-(aq)$	−0.152
	$Ag_2S(s) + 2e^- \leftrightharpoons 2Ag(s) + S^{2-}(aq)$	−0.691
sodium	$Na^+(aq) + e^- \leftrightharpoons Na(s)$	−2.71
strontium	$Sr^{2+}(aq) + 2e^- \leftrightharpoons Sr(s)$	−2.899
sulfur	$S_2O_8^{2-}(aq) + 2e^- \leftrightharpoons 2SO_4^{2-}(aq)$	2.010
	$S_2O_6^{2-}(aq) + 4H^+(aq) + 2e^- \leftrightharpoons 2H_2SO_3(aq)$	0.564
	$SO_4^{2-}(aq) + 4H^+(aq) + 2e^- \leftrightharpoons H_2SO_3(aq) + H_2O(l)$	0.172
	$S_4O_6^{2-}(aq) + 2e^- \leftrightharpoons 2S_2O_3^{2-}(aq)$	0.08
	$2SO_4^{2-}(aq) + 4H^+(aq) + 2e^- \leftrightharpoons S_2O_6^{2-}(aq) + H_2O(l)$	−0.22
	$SO_4^{2-}(aq) + H_2O(l) + 2e^- \leftrightharpoons SO_3^{2-}(aq) + 2OH^-(aq)$	−0.93
	$2SO_3^{2-}(aq) + 2H_2O(l) + 2e^- \leftrightharpoons S_2O_4^{2-}(aq) + 4OH^-(aq)$	−1.12
thallium	$Tl^{3+}(aq) + 2e^- \leftrightharpoons Tl^+(aq)$	1.252
	$Tl^+(aq) + e^- \leftrightharpoons Tl(s)$	−0.336
	$TlCl(s) + e^- \leftrightharpoons Tl(s) + Cl^-(aq)$	−0.557
tin	$Sn^{4+}(aq) + 2e^- \leftrightharpoons Sn^{2+}(aq)$	0.151
	$Sn^{2+}(aq) + 2e^- \leftrightharpoons Sn(s)$	−0.138
vanadium	$VO_2^+(aq) + 2H^+(aq) + e^- \leftrightharpoons VO^{2+}(aq) + H_2O(l)$	0.991
	$VO^{2+}(aq) + 2H^+(aq) + e^- \leftrightharpoons V^{3+}(aq) + H_2O(l)$	0.337
	$V^{3+}(aq) + e^- \leftrightharpoons V^{2+}(aq)$	−0.255
	$V^{2+}(aq) + 2e^- \leftrightharpoons V(s)$	−1.175
zinc	$Zn^{2+}(aq) + 2e^- \leftrightharpoons Zn(s)$	−0.762
	$Zn(OH)_4^{2-}(aq) + 2e^- \leftrightharpoons Zn(s) + 4OH^-(aq)$	−1.199
	$Zn(OH)_2(s) + 2e^- \leftrightharpoons Zn(s) + 2OH^-(aq)$	−1.249
	$ZnO(s) + H_2O(l) + 2e^- \leftrightharpoons Zn(s) + 2OH^-(aq)$	−1.260

APPENDIX H
World Chemical Production

World Chemical Production—*Output in thousands of metric tons (unless otherwise specified)*

Inorganic chemicals	Formula	U.S.	Europe	Canada	China	Japan
Aluminum sulfate	$Al_2(SO_4)_3$	922		167		
Ammonia	NH_3	10 762		4 996		1 340
Ammonium nitrate	NH_4NO_3	6 021		1 096		
Ammonium sulfate	$(NH_4)_2SO_4$	2 643				1 526
Carbon black	C		738	223		804
Chlorine	Cl_2	12 166	9 856	1 057		619
Hydrochloric acid	HCl	5 012	2 457	149	6 007	2 324
Hydrogen	H_2	17 698	4 511			
Hydrogen peroxide	H_2O_2		229	244		196
Nitric acid	HNO_3	6 703		1 219		
Nitrogen	N_2	26 675	12 407			11 281
Oxygen	O_2	19 539	13 672			11 278
Phosphoric acid	H_3PO_4	11 463	305			
Sodium carbonate	Na_2CO_3		1 439		12 668	
Sodium chlorate	$NaClO_3$	658		1 183		
Sodium hydroxide	NaOH	9 508	5 622	1 146	10 603	4 493
Sodium silicate	Na_2SiO_3					577
Sodium sulfate	Na_2SO_4	473	873			
Sulfuric acid	H_2SO_4	37 515	4 816	3 933	38 249	6 444
Titanium oxide	TiO_2		439			253

Organic chemicals		U.S.	Europe	Canada	China	Japan	S. Korea	Taiwan
Acetic acid	CH_3COOH		876			589		
Acetone	$(CH_3)_2CO$		564			539		
Acrylonitrile	$CH_2{=}CHCN$	1598	n/a			711		379
Aniline	$C_6H_5NH_2$	813						
Benzene (liters)	C_6H_6	8781	4265	915	2556	4758	3462	1088
Butadiene	$CH_2{=}CHCH{=}CH_2$	2204	2222	289		1041	917	412
Butanol	$CH_3CH_2CH_2CH_2OH$		535			506		
Caprolactam	$C_6H_{11}NO$				228	503		216
Cumene	$C_6H_5CH(CH_3)_2$	3736						
Cyclohexane	C_6H_{12}					676		
Ethylbenzene	$C_6H_5C_2H_5$	5779	858					
Ethene (ethylene)	C_2H_4	25682	21408	5095	6266	7570	5945	2864
Ethylene dichloride	$ClCH_2CH_2Cl$	12163	3276			3594		
Ethylene glycol	$HOCH_2CH_2OH$		277			786		1459
Ethylene oxide	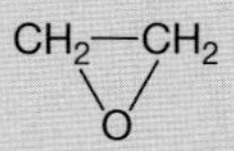	3772	874			941		
Formaldehyde	HCHO		1461	269				
Methanol	CH_3OH		1823		4406			
Octanol	$C_8H_{17}OH$					307		
Phenol	C_6H_5OH		827			966		
Phthalic anhydride	$C_8H_4O_3$		259			257		
Propylene	$CH_3CH{=}CH_2$	15345	15123	939		5767	3892	1995
Propylene glycol	$CH_3CHOHCH_2OH$		348			346		
Propylene oxide	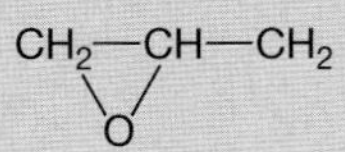		1048					
Styrene	$C_6H_5CH{=}CH_2$	5394	1666			3345		1247
Toluene	$C_6H_5CH_3$		853	n/a		1634		140
Urea	$(NH_2)_2CO$	5755		3654				
Vinyl acetate	$CH_3COOCH{=}CH_2$	1431	154					
Vinyl chloride	$CH_2{=}CHCl$						1498	1763
Xylenes	$C_6H_4(CH_3)_2$		594	351		5395		

From "Facts and Figures for the Chemical Industry," *Chemical & Engineering News*, July 11, 2005, Volume 83, Number 28.
(Blank spaces indicate no data published for the specific compound or country in this source)

APPENDIX I

Answers to Selected Even-Numbered Problems

Chapter 1

1-6. (a) 299 800 000 (b) 0.000 548 580 (c) 0.000 000 000 052 92 (d) 155 000 000 000 000 0000

1-8. 556.0×10^3

1-10. (a) 10^{-12} (b) 10^9 (c) 10^{-9} (d) 10^3 (e) 10^{-18} (f) 10^{-15}

1-12. $f < d < b < e < a < c$

1-14. 8×10^{-30} M^3

1-16. 1.61 cm^3

1-18. *Celsius* 233 or *Kelvin* 506

1-20. 86m·s^{-1}

1-22. 31 m

1-24. $55

1-26. 2%

1-28. 0.6%

1-30. (a) three (b) three (c) exact (d) nine (e) two

1-32. (a) 33209 (b) 254 (c) 0.0143 877 (d) -1.26×10^{-13}

1-34. 2.35×10^6 pm^3

1-36. (a) 1.259 kJ (b) 2.18 aJ (c) 5.5×10^3 kJ (d) 7.5×10^3 fs (e) 2.0×10^6 mL

1-38. (a) 99.1 m (b) 154 pm and 0.154 nm (c) 79.4 kg

1-40. 7.44 L

1-42. 2.8 min

1-44. (a) 10^6 mL (b) 10^3 L

1-46. 0.650 g·mL^{-1}

1-48. Use Time/s and Distance/ft as table headings

1-50. The table is

Height/cm	Time/s
27	6.09
42.7	11.65
60.7	18.11
129	30.41

1-60. class average = 11.31 g·cm^{-3}; class percentage error = 0.00%

1-62. 88.2°C

1-64. 63 mL

1-66. 430 kJ

1-68. area = 56m^2; pay for themselves in 3750 days

1-70. 9.0×10^{10} kJ

1-72. 1.4 g

1-74. 1500 K and 2200°F

1-76. 1.60 g·mL^{-1}

1-78. 2.472 acre

1-80. 0.0200 mm

1-84. 0.12 mm

1-86. 2.0%

Chapter 2

2-14. 85.4% La and 14.6% O

2-16. 75.74% Sn and 24.26% F

2-18. 52.2% C; 13% H; 34.6% O

2-28. (a) 287.92 (b) 231.53 (c) 209.94 (d) 537.50 (e) 222.84

2-30. (a) 286.45 (b) 300.81 (c) 376.36 (d) 793.30 (e) 176.12

2-32. (a) 153.18 (b) 162.23

2-34. 63.6483% N and 36.3517% O

2-36. 32.85171% Na; 12.85194% Al; 54.2963% F

2-38. 40.050% S and 2.237 grams S

2-40. 68.420% Cr in Cr_2O_3; 29.2 g Cr in Cr_2O_3; 29.2% Cr in the ore

2-42. (a) 15 protons, 15 electrons, 15 neutrons
(b) 43 protons, 43 electrons, 54 neutrons
(c) 26 protons, 26 electrons, 29 neutrons
(d) 95 protons, 95 electrons, 145 neutrons

2-44. The completed table is

Symbol	Atomic number	Number of electrons	Mass number
$^{48}_{20}Ca$	20	28	48
$^{90}_{40}Zr$	40	50	90
$^{131}_{53}I$	53	78	131
$^{99}_{42}Mo$	42	57	99

2-46. The completed table is

Symbol	Atomic number	Number of electrons	Mass number
$^{39}_{19}K$	19	20	39
$^{56}_{26}Fe$	26	30	56
$^{84}_{36}Kr$	36	48	84
$^{120}_{50}Sn$	50	70	120

2-48. 24.31

2-50. 28.0854

2-52. 68.925

2-54. 19.9% boron-100 and 80.1% boron-11

2-56. 47.8% europium-151 and 52.2% europium-153

2-58. (a) 36 (b) 18 (c) 46 (d) 18

2-60. (a) 54 (b) 54 (c) 2 (d) 28

2-66. (a) 18.038 (b) 33.007 (c) 178.77 (d) 243.69

2-72. 0.023% D in DHO and 0.00013% D inD_2O

2-74. 35 100 if there is one Co atom per protein molecule

2-76. 183.84

2-78. ratio = 2.28/1.14 = 2/1

2-80. 1.333

Chapter 3

There are no numerical problems in Chapter 3.

Chapter 4

4-4. 11.45 aJ

4-6. 7.624 MJ

4-10. $\cdot\ddot{\underset{\cdot\cdot}{O}}\cdot$ $\cdot\ddot{\underset{\cdot\cdot}{S}}\cdot$ $\cdot\ddot{\underset{\cdot\cdot}{Se}}\cdot$ $\cdot\ddot{\underset{\cdot\cdot}{Te}}\cdot$

4-12. B^{3+} $:\ddot{\underset{\cdot\cdot}{N}}:^{3-}$ $:\ddot{\underset{\cdot\cdot}{F}}:^{-}$ Na^{+}

4-16. $5.088 \times 10^{14}\ s^{-1}$

4-18. 8.52 mm to 8.77 mm

4-20. No, $E = 2.5$ aJ

4-22. 5.70×10^{8} J

4-24. 26 mJ

4-26. 7.77×10^{8} J

4-28. Will not eject an electron; $\nu = 3.57 \times 10^{14}\ s^{-1}$

4-30. $5.96 \times 10^{14}\ s^{-1}$

4-32. 2.37×10^{-34} m

4-34. $7.58 \times 10^{6}\ m\cdot s^{-1}$

4-36. 6.6×10^{-29} kg

4-40. $2.9244 \times 10^{15}\ s^{-1}$, Lyman series in the ultraviolet region

4-42. Infrared region.

n_f	4	5	6	7	8
$\lambda/\mu m$	1.875	1.282	1.094	1.005	0.9544

4-44. $n = 6$

4-46. $n = 2$

4-50.

n_f	5	6	7	8	9
E/aJ	0.1962	0.3028	0.3670	0.4087	0.4373
λ/μm	1.012	0.65560	0.5413	0.4860	0.4542

The infrared-visible region corresponding to visible emission lines in stars.

4-62. $h = 6.63 \times 10^{-34}$ J·s; $\nu_0 = 1.1 \times 10^{15}$ s^{-1}

4-64. $n = 6$

4-66. slope = 3.29×10^{15} s^{-1}

4-68. 2.6×10^{17} photons·s^{-1}·cm^{-2}

Chapter 5

5-2. 4 m·s^{-1}

5-4. 2.8×10^{-38} m·s^{-1}; No

5-8. (a) $3p$ orbital (b) $5s$ orbital (c) $2p$ orbital (d) $4f$ orbital

5-12. The possible sets of quantum numbers are

n	l	m_l	m_s
4	3	−3	+1/2 or −1/2
4	3	−2	+1/2 or −1/2
4	3	−1	+1/2 or −1/2
4	3	0	+1/2 or −1/2
4	3	1	+1/2 or −1/2
4	3	2	+1/2 or −1/2
4	3	3	+1/2 or −1/2

5-14. 2 electrons; 8 electrons; 18 electrons; 32 electrons

5-20. We have the following:

(a) $1s^22s^22p^63s^23p^1$	13 electrons, aluminum
(b) $1s^22s^22p^63s^23p^64s^23d^3$	23 electrons, vanadium
(c) $1s^22s^22p^5$	9 electrons, fluorine
(d) $1s^22s^22p^63s^23p^64s^23d^{10}4p^1$	31 electrons, gallium
(e) $1s^22s^22p^4$	8 electrons, oxygen

5-22.
(a) Si: $[Ne]3s^23p^2$ (b) Ni: $[Ar]4s^23d^8$
(c) Se: $[Ar]4s^23d^{10}4p^4$ (d) Cd: $[Kr]5s^24d^{10}$
(e) Mg: $[Ne]3s^2$

5-24.
(a) Ba: $[Xe]6s^2$ (b) Ag: $[Kr]5s^14d^{10}$
(c) Gd: $[Xe]4f^75d^16s^2$ (d) Pd: $[Kr]4d^{10}$
(e) Sn: $[Kr]5s^24d^{10}5p^2$

5-26. The electron configuration for sulfur is

$1s$	$2s$	$2p$	$3s$	$3p$
⇅	⇅	⇅ ⇅ ⇅	⇅	⇅ ↑ ↑

The electron configuration for chromium is

$1s$	$2s$	$2p$	$3s$	$3p$	$4s$	$3d$
⇅	⇅	⇅ ⇅ ⇅	⇅	⇅ ⇅ ⇅	↑	↑ ↑ ↑ ↑ ↑

5-28. The ground state electron configurations are as follows:
(a) Ca: $[Ar]4s^2$ (b) Br: $[Ar]4s^23d^{10}4p^5$
(c) B: $[He]2s^22p^1$ (d) Zn: $[Ar]4s^23d^{10}$
(e) W: $[Xe]6s^24f^{14}5d^4$

5-30. (a) cerium, Ce (b) vanadium, V
(c) copper, Cu (d) sulfur, S

5-32. (a) one (b) six (c) two (d) none

5-34. (a) 2 electrons; Ca^{2+}, argon
(b) 1 electron, Li^+, helium
(c) 1 electron, Na^+, neon
(d) 2 electrons, Mg^{2+}, neon

5-36. (a) $1s^22s^22p^3$; isoelectronic with nitrogen
(b) $1s^22s^22p^3$; isoelectronic with nitrogen
(c) $1s^22s^22p^4$; isoelectronic with oxygen
(d) $1s^22s^22p^2$; isoelectronic with carbon

5-38. (a) none (b) one (c) one (d) none (e) none

5-42. (a) $I(g) + e^- \rightarrow I^-(g)$
$[Kr]5s^24d^{10}5p^5 + e^- \rightarrow [Kr]5s^24d^{10}5p^6$ or [Xe]
(b) $K(g) + F(g) \rightarrow K^+(g) + F^-(g)$
$[Ar]4s^1 + [He]2s^22p^5 \rightarrow [Ar] + [He]2s^22p^6$ or [Ne]

5-44. (a), (b), and (c)

5-46. (a) 0 valence electrons He
(b) 8 valence electrons $:\ddot{\underset{\cdot\cdot}{N}}:^{3-}$
(c) 7 valence electrons $\cdot\ddot{\underset{\cdot\cdot}{F}}\cdot^{+}$
(d) 1 valence electron Na·
(e) 0 valence electrons K^+

5-48. (a) O > F (b) Xe > Kr (c) Cl > F (d) C > Mg

5-50. (a) Li < Na < Rb < Cs (b) P < Al < Mg < Na
(c) Mg < Ca < Sr < Ba

5-52. $Mg^{2+} < Na^+ < F^- < O^{2-} < N^{3-}$

5-56. (a) B < O < F < Ne (b) Sn < Te < I < Xe (c) Cs < Rb < K < Ca (d) Na < Al < S < Ar

5-60. (a) nickel (b) silver (c) sulfur (d) lead

5-62. (a) aluminum (b) oxygen (c) zinc (d) krypton

5-64. (b), (c), and (e)

5-66. (a) 2; O (b) 0; Cd (c) 0; Hg (d) 1; Cl

5-68. $2s^1$, $2s^12p_x^1$, $2s^12p_x^12p_y^1$, $2s^12p_x^12p_y^12p_z^1$, $2s^22p_x^12p_y^12p_z^1$, $2s^22p_x^22p_y^12p_z^1$, $2s^22p_x^22p_y^22p_z^1$, $2s^22p_x^22p_y^22p_z^2$

Chapter 6

6-4. (a) Li^+ and O^{2-}; lithium oxide
(b) Ca^{2+} and S^{2-}; calcium sulfide
(c) Mg^{2+} and N^{3-}; magnesium nitride
(d) Al^{3+} and S^{2-}; aluminum sulfide

6-8. (a) Al_2S_3 (b) Na_2O (c) BaF_2 (d) LiH

6-10. (a) Cs_2O (b) Na_2Se (c) Li_2S (d) CaI_2

6-16. (a) Ru_2S_3 (b) ScF_3 (c) OsO_4 (d) MnS (e) $PtCl_4$

6-18. (a) $TlCl_3$ (b) CdI_2 (c) Zn_3As_2 (d) $AlBr_3$

6-24. (a) $2\,CO(g) + O_2(g) \rightarrow 2\,CO_2(g)$
(b) $2\,Cs(s) + Br_2(l) \rightarrow CsBr_2(s)$
(c) $2\,NO(g) + O_2(g) \rightarrow 2\,NO_2(g)$
(d) $4\,NH_3(g) + 5\,O_2(g) \rightarrow 4\,NO(g) + 6\,H_2O(l)$

6-26. (a) $Ga([Ar]4s^23d^{10}4p^1) + 3\,F([He]2s^22p^5) \rightarrow Ga^{3+}([Ar]3d^{10}) + 3\,F^-([Ne]) \rightarrow GaF_3(g)$
(b) $Ag([Kr]5s^14d^{10}) + Cl([Ne]3s^23p^5) \rightarrow Ag^+([Kr]4d^{10}) + Cl^-([Ar]) \rightarrow AgCl(g)$
(c) $3\,Li([He]2s^1) + N([He]2s^22p^3) \rightarrow 3\,Li^+([He]) + N^{3-}([Ne]) \rightarrow Li_3N(g)$

6-28. (a) $3\,Li\cdot + \cdot\dot{\underset{\cdot\cdot}{N}}\cdot \rightarrow 3\,Li^+ + :\ddot{\underset{\cdot\cdot}{N}}:^{3-}$
(b) $\cdot\dot{Al}\cdot + 3\,:\ddot{\underset{\cdot\cdot}{Cl}}\cdot \rightarrow Al^{3+} + 3\,:\ddot{\underset{\cdot\cdot}{Cl}}:^-$
(c) $2\,Li\cdot + \cdot\ddot{\underset{\cdot\cdot}{O}}\cdot \rightarrow 2\,Li^+ + :\ddot{\underset{\cdot\cdot}{O}}:^{2-}$

6-30. (a) $Ru^{2+}([Kr]4d^6)$ (b) $W^{3+}([Xe]4f^{14}5d^3)$ (c) $Pd^{2+}([Kr]4d^8)$ (d) $Ti^+([Ar]3d^3)$

6-32. (a) six (b) ten (c) three (d) eight

6-36. (d) and (e)

6-38. (a) H (b) Fe^{2+} (c) S^{2-} (d) O^{2-}

6-40. (a) Cl^- (b) Au^+ (c) Cr^+ (d) P^{3-}

6-42. $Mo^{+6} < Y^{3+} < Rb^+ < Br^- < Se^{2-}$

6-44. K > Na > B > H > He

6-48. 2.56 J

6-50. −0.983 aJ

6-52. 0.35 aJ

6-54. 1.45 aJ

6-66. (a) Zn^{2+}, Cd^{2+}, and Hg^{2+} (b) Ti^{4+}, Zr^{4+}, Hf^{4+}

6-70. chloride ion

6-76. −2.55 aJ

Chapter 7

There are no numerical problems in Chapter 7.

Chapter 8

8-2. PF_5 and AlF_6^{3-}

8-4. SeF_6 and BrF_2^-

8-6. (a) bent (b) bent (c) bent (d) linear

8-8. NH_2Cl, PF_3, BF_3

8-10. (a) square planar (b) tetrahedral (c) tetrahedral (d) seesaw shaped

We have for the given fluorides

(a) AX_6	octahedral	90°
(b) AX_4	tetrahedral	109.5°
(c) AX_3E_2	T-shaped	90°
(d) AX_5E	square pyramidal	< 90°

8-14. We have for the given ions

(a) AX_4	tetrahedral	109.5°
(b) AX_6	octahedral	90°
(c) AX_3	trigonal planar	120°
(d) AX_6	octahedral	90°

8-16. We have for the given molecules

(a) AX_4	tetrahedral	109.5°
(b) AX_3E	trigonal pyramidal	< 120°
(c) AX_6	octahedral	90°
(d) AX_4E	seesaw-shaped	< 120° and < 180°

8-18. We have for the given molecules

(a) AX_2E_2	bent	109.5°
(b) AX_6	octahedral	90°
(c) AX_4E	seesaw-shaped	< 120°, < 180°
(d) AX_4E_2	square planar	90°

8-20. (a) square pyramidal < 90° (b) octahedral 90° (c) tetrahedral ~ 109.5° (d) tetrahedral ~ 109.5°

8-22. (a) seesaw-shaped < 90°, < 120° (b) bent < 109.5° (c) bent < 120° (d) trigonal planar < 120°

8-24. (a) tetrahedral ~ 109.5° (b) linear 180° (c) tetrahedral < 109.5° (d) bent < 109.5°

8-26. (a) tetrahedral 109.5° (b) bent < 109.5° (c) trigonal planar 120° (d) trigonal pyramidal < 109.5°

8-28. (a) a bent ion, < 120° (b) a trigonal planar ion, 120° (c) a linear ion, 180° (d) a tetrahedral ion, 109.5°

8-36. (a) tetrahedral (no dipole moment) (b) bent (dipole moment) (c) octahedral (no dipole moment) (d) T-shaped (dipole moment)

8-38. (a) octahedral (no dipole moment) (b) square pyramidal (dipole moment) (c) tetrahedral (no dipole moment) (d) AX_2E_2 bent (dipole moment)

8-40. (a) seesaw-shaped, polar (b) trigonal planar, nonpolar (c) octahedral, polar (d) trigonal bipyramidal, polar

8-48. (a) and (b)

8-56. (a) trigonal pyramidal (b) tetrahedral (c) seesaw shaped (d) octahedral

8-58. (a) bent (b) trigonal pyramidal (c) tetrahedral

8-60. The PCl_4^+ ions are tetrahedral and the Pl_6^- ions are octahedral.

8-62. SOF_2

8-64. $BeCl_2$

8-66. (a) linear (b) square planar (c) bent (d) seesaw shaped

8-68. (a) 1 and 2 (b) 3 (c) 2 and 9 (d) 1 and 2

8-70. (a) 90° (b) 90°, 120° (c) 90° (d) 120°

Chapter 9

9-2. one

9-10. We have

	Ground-state electron configuration	Bond order
NO	$(\sigma_{1s})^2(\sigma_{1s}^*)^2(\sigma_{2s})^2(\sigma_{2s}^*)^2(\pi_{2p})^4(\sigma_{2p})^2(\pi_{2p}^*)^1$	$2\frac{1}{2}$
NO^+	$(\sigma_{1s})^2(\sigma_{1s}^*)^2(\sigma_{2s})^2(\sigma_{2s}^*)^2(\pi_{2p})^4(\sigma_{2p})^2$	3
NO^-	$(\sigma_{1s})^2(\sigma_{1s}^*)^2(\sigma_{2s})^2(\sigma_{2s}^*)^2(\pi_{2p})^4(\sigma_{2p})^2(\pi_{2p}^*)^2$	2

9-12. (a) 13 electrons; $(\sigma_{1s})^2(\sigma_{1s}^*)^2(\sigma_{2s})^2(\sigma_{2s}^*)^2(\pi_{2p})^4(\sigma_{2p})^1$; bond order = $2\frac{1}{2}$
(b) 9 electrons; $(\sigma_{1s})^2(\sigma_{1s}^*)^2(\sigma_{2s})^2(\sigma_{2s})^2(\pi_{2p})^1$; bond order = $\frac{1}{2}$
(c) 3 electrons; $(\sigma_{1s})^2(\sigma_{1s}^{*1})$; bond order = $\frac{1}{2}$
(d) 20 electrons; $(\sigma_{1s})^2(\sigma_{1s}^*)^2(\sigma_{2s})^2(\sigma_{2s}^*)^2(\pi_{2p})^2(\pi_{2p})^4(\pi_{2p}^*)^4(\sigma_{2p}^*)^2$; bond order = 0

9-14. (a) $\frac{1}{2}$; stable (b) 0, stable (c) $1\frac{1}{2}$; stable (d) 0, stable

9-18. $(\sigma_{1s})^2(\sigma_{1s}^*)^2(\sigma_{2s})^2(\sigma_{2s}^*)^2(\pi_{2p})^4(\sigma_{2p})^2(\pi_{2p}^*)^4(\sigma_{2p}^*)^2(\sigma_{3s})^2(\sigma_{3s}^*)^2(\pi_{3p})^4(\sigma_{3p})^2$; bond order is 3; :P≡P:

9-20. 16 valence electrons; Use *sp* hybrid orbitals.

9-22. 20 valence electrons; Use sp^2 hybrid orbitals on S atom.

9-26. Use sp^2 hybrid orbitals on C atom.

9-28. 26 valence electrons; Use sp^3 hybrid orbitals on C atom.

9-30. 22 valence electrons; Use sp^3d hybrid orbitals on Xe atom.

9-32. 42 valence electrons; Use sp^3d^2 hybrid orbitals on Te atom.

9-34. Use sp^3 hybrid orbitals on the two C atoms and O atom. ~105°

9-36. Use sp^3 hybrid orbitals on the two C atoms and N atom. 9 σ-bonds, 1 lone pair; tetrahedral; trigonal pyramidal

9-38. Use sp^3 hybrid orbitals on the three C atoms and O atom. 11 localized bond orbitals

9-40. (a) five σ bonds and one π bond (b) seven σ bonds and two π bonds (c) six σ bonds and two π bonds (d) eight σ bonds and two π bonds

9-42. five σ bonds and two π bonds; Bond order of the C–C bond is 1; the bond order of the C≡N bond is 3

9-50. 26 σ bonds and seven delocalized π bonds

9-60. $2\frac{1}{2}$

9-62. $(\sigma_{1s})^2(\sigma^*_{1s})^2(\sigma_{2s})^2(\sigma^*_{2s})^2(\pi_{2p})^4(\sigma_{2p})^2$; 3;$N_2$ and CO

9-68. Use sp^2 hybrid orbitals on the central C atom and O atom. Trigonal planar around central C atom

9-70. Use sp^3 hybrid orbitals on the C atom bonded to H atoms and and sp hybrid orbitals on cyanide C atom and N atom. Eight σ bonds and two π bonds

9-72. (a) eight σ bonds and one π bond (b) six σ bonds and one π bond (c) five σ bonds and two π bonds (d) eight σ bonds

9-78. $1\frac{1}{3}$

Chapter 10

10-2. (a) sodium acetate (b) calcium chlorate (c) ammonium carbonate (d) barium nitrate

10-4. (a) ammonium thiosulfate (b) sodium sulfite (c) potassium carbonate (d) sodium thiosulfate

10-6. (a) chromium(II) sulfate (b) cobalt(II) cyanide (c) tin(II) nitrate (d) copper(I) carbonate

10-8. (a) $HC_2H_3O_2$ (b) $HClO_3$ (c) H_2CO_3 (d) $HClO_4$

10-10. (a) $NaClO_4$ (b) $KMnO_4$ (c) $CaSO_3$ (d) LiCN

10-12. (a) $Hg_2(C_2H_3O_2)_2$ (b) $Hg(CN)_2$ (c) $Fe(ClO_4)_2$ (d) $CrSO_3$

10-14. (a) $NaClO(s)$ (b) $H_2O_2(l)$ (c) $KOH(s)$ (d) $CH_3COOH(aq)$

10-16. (a) basic (b) acidic (c) acidic (d) basic (e) basic

10-18. (a) and (b)

10-20. (a) organic acid (b) oxyacid (c) organic acid (d) oxyacid

10-22. (a) nitrous acid (b) hyposulfurous acid (c) chlorous acid (d) iodic acid

10-24. (a) copper(II) hypochlorite (b) scandium(III) iodate (c) iron(III) bromate (d) ruthenium(III) periodate

10-30. (a) barium hydroxide octahydrate (b) lead(II) chloride dihydrate (c) lithium hydroxide monohydrate (d) lithium chromate dihydrate

10-32. (a) $H_2C_2O_4\cdot 2H_2O$ (b) $Al_2(SO_4)_3\cdot 8H_2O$ (c) $NdI_3\cdot 9H_2O$ (d) $Na_2HPO_4\cdot 7H_2O$

10-34. (a) decomposition (b) single replacement (c) double-replacement (d) single replacement

10-42. (a) $H^+(aq) + OH^-(aq) \rightarrow H_2O(l)$
(b) $Pb^{2+}(aq) + CO_3^{2-}(aq) \rightarrow PbCO_3(s)$
(c) $2Ag^+(aq) + SO_4^{2-}(aq) \rightarrow Ag_2SO_4(s)$
(d) $S^{2-}(aq) + Zn^{2+}(aq) \rightarrow ZnS(s)$

10-44. We have
(a) $2AgNO_3(aq) + Na_2S(aq) \rightarrow Ag_2S(s) + 2NaNO_3(aq)$
$2Ag^+(aq) + S^{2-}(aq) \rightarrow Ag_2S(s)$
(b) $H_2SO_4(aq) + Pb(NO_3)_2(aq) \rightarrow PbSO_4(s) + 2HNO_3(aq)$
$SO_4^{2-}(aq) + Pb^{2+}(aq) \rightarrow PbSO_4(s)$
(c) $Hg(NO_3)_2(aq) + 2NaI(aq) \rightarrow HgI_2(s) + 2NaNO_3(aq)$
$Hg^{2+}(aq) + 2I^-(aq) \rightarrow HgI_2(s)$
(d) $CdCl_2(aq) + 2AgClO_4(aq) \rightarrow 2AgCl(s) + Cd(ClO_4)_2(aq)$
$Cl^-(aq) + Ag^+(aq) \rightarrow AgCl(s)$

10-46. (a) insoluble, Rule 5 (b) soluble, Rule 4 (c) soluble, Rule 1 (d) insoluble, Rule 3 (e) insoluble, Rule 3

10-48. (a) insoluble, Rule 3 (b) soluble, Rule 2 (c) insoluble, Rule 3 (d) soluble, Rule 2 (e) insoluble, Rule 3

10-50. (a) soluble; $FeBr_3(s) \xrightarrow[H_2O(l)]{} Fe^{3+}(aq) + 3Br^-(aq)$
(b) insoluble (c) soluble;
$(NH_4)_2CO_3(s) \xrightarrow[H_2O(l)]{} 2NH_4^+(aq) + CO_3^{2-}(aq)$
(d) soluble; $K_2S(s) \xrightarrow[H_2O(l)]{} 2K^+(aq) + S^{2-}(aq)$

10-52. (a) $CaSO_4$ is insoluble by Rule 6.
$H_2SO_4(aq) + Ca(ClO_4)_2(aq) \rightarrow CaSO_4(s) + 2HClO_4(aq)$
$SO_4^{2-}(aq) + Ca^{2+}(aq) \rightarrow CaSO_4(s)$
(b) All compounds are soluble; thus there is no reaction. (c) $Hg_2(C_7H_5O_2)_2$ is insoluble by Rule 3.
$Hg_2(NO_3)_2(aq) + 2NaC_7H_5O_2(aq) \rightarrow Hg_2(C_7H_5O_2)_2(s) + 2NaNO_3(aq)$
$Hg^{2+}(aq) + 2C_7H_5O_2^-(aq) \rightarrow Hg_2(C_7H_5O_2)_2(s)$

(d) $PbBr_2$ is insoluble by Rule 3.

$Pb(CH_3COO)_2(aq) + 2KBr(aq) \rightarrow PbBr_2(s) + 2KCH_3COO(aq)$

$Pb^{2+}(aq) + 2Br^-(aq) \rightarrow PbBr_2(s)$

10-54. (a) $NH_4NO_3(aq) + NaOH(aq) \rightarrow NaNO_3(aq) + H_2O(l) + NH_3(aq)$

$NH_4^+(aq) + OH^-(aq) \rightarrow H_2O(l) + NH_3(aq)$

(b) $2HNO_3(aq) + BaCO_3(s) \rightarrow Ba(NO_3)_2(aq) + H_2O(l) + CO_2(g)$

$2H^+(aq) + BaCO_3(s) \rightarrow Ba^{2+}(aq) + H_2O(l) + CO_2(g)$

(c) $2H_2O_2(aq) \rightarrow 2H_2O(l) + O_2(g)$

10-56. (a) $2HClO_4(aq) + Ca(OH)_2(aq) \rightarrow \underbrace{Ca(ClO_4)_2(aq)}_{\text{calcium perchlorate}} + 2H_2O(l)$

$H^+(aq) + OH^-(aq) \rightarrow H_2O(l)$

(b) $2HCl(aq) + CaCO_3(s) \rightarrow \underbrace{CaCl_2(aq)}_{\text{calcium chloride}} + \underbrace{CO_2(g)}_{\text{carbon dioxide}} + H_2O(l)$

$2H^+(aq) + CaCO_3(s) \rightarrow Ca^{2+}(aq) + CO_2(g) + H_2O(l)$

(c) $6HNO_3(aq) + Al_2O_3(s) \rightarrow \underbrace{2Al(NO_3)_3(aq)}_{\text{aluminum nitrate}} + 3H_2O(l)$

$6H^+(aq) + Al_2O_3(s) \rightarrow 2Al^{3+}(aq) + 3H_2O(l)$

(d) $H_2SO_4(aq) + Cu(OH)_2(s) \rightarrow \underbrace{CuSO_4(aq)}_{\text{copper(II) sulfate}} + 2H_2O(l)$

$2H^+(aq) + Cu(OH)_2(s) \rightarrow Cu^{2+}(aq) + 2H_2O(l)$

10-58. (a) $K_2CrO_4(aq) + Pb(NO_3)_2(aq) \rightarrow 2KNO_3(aq) + PbCrO_4(s)$

(b) $2HCl(aq) + Na_2S(aq) \rightarrow 2NaCl(aq) + H_2S(g)$

(c) $Ba(OH)_2(aq) + ZnSO_4(aq) \rightarrow Zn(OH)_2(s) + BaSO_4(s)$

(d) $2HNO_3(aq) + CaO(s) \rightarrow Ca(NO_3)_2(aq) + H_2O(l)$

10-60. (a) precipitation reaction (b) gas forming reaction (c) precipitation reaction (d) acid-base reaction

10-62. (a) $Li(s)$ is the reducing agent and $Se(s)$ is the oxidizing agent.

(b) $Sc(s)$ is the reducing agent and $I_2(g)$is the oxidizing agent.

(c) $Ga(s)$ is the reducing agent and $P_4(s)$ is the oxidizing agent.

(d) $K(s)$ is the reducing agent and $F_2(g)$ is the oxidizing agent.

10-64. (a) 2 (b) 12 (c) 12 (d) 2

10-66. (a) $CH_4(g)$ is the reducing agent and $O_2(g)$ is the oxidizing agent. 8

10-72. (a) $2Na(s) + H_2(g) \rightarrow 2NaH(s)$

(b) $2Al(s) + 3S(s) \rightarrow Al_2S_3(s)$

(c) $H_2O(g) + C(s) \rightarrow CO(g) + H_2(g)$

(e) $PCl_3(l) + Cl_2(g) \rightarrow PCl_5(s)$

10-74. $HCHO_2(aq) + NH_3(aq) \rightarrow NH_4CHO_2(aq)$

10-76. (a) $Cl_2(g) + 2NaI(aq) \rightarrow 2NaCl(aq) + I_2(s)$

(b) $Br_2(l) + 2NaI(aq) \rightarrow 2NaBr(aq) + I_2(s)$

(c) no reaction (d) no reaction

10-78. (a) $ZnS(s) + 2HCl(aq) \rightarrow ZnCl_2(aq) + H_2S(g)$

(b) $2PbO_2(s) \rightarrow 2PbO(s) + O_2(g)$

(c) $3CaCl_2(aq) + 2H_3PO_4(aq) \rightarrow Ca_3(PO_4)_2(s) + 6HCl(aq)$

10-80. (a) $C_{12}H_{22}O_{11}(s) \rightarrow 12C(s) + 11H_2O(l)$

(b) $Cl_2(g) + 2NaBr(aq) \rightarrow 2NaCl(aq) + Br_2(l)$

(c) $Li_2O(s) + H_2O(l) \rightarrow 2LiOH(aq)$

10-82. (a) $Na_2CO_3 \cdot 10H_2O(s) \rightarrow Na_2CO_3(s) + 10H_2O(g)$

(b) $Pb(NO_3)_2(aq) + Na_2SO_4(aq) \rightarrow PbSO_4(s) + 2NaNO_3(aq)$

(c) $2Fe(s) + 3Pb(NO_3)_2(aq) \rightarrow 2Fe(NO_3)_3(aq) + 3Pb(s)$

10-86. $2Pb(l) + O_2(g) \rightarrow 2PbO(s)$

$Ag(l) + O_2(g) \rightarrow$ no reaction

10-88. $2HgS(s) + 2O_2(g) \xrightarrow{\text{heat}} HgO(s) + SO_2(g)$

$HgO(s) + HgS(s) \xrightarrow{\text{heat}} Hg(g) + SO_2(g)$

$Hg(g) \xrightarrow{\text{cold}} Hg(l)$

10-92. $3CH_3CH_2OH(aq) + 2K_2Cr_2O_7(aq) + 8H_2SO_4(aq) \rightarrow 3CH_3COOH(aq) + 2Cr_2(SO_4)_3(aq) + 2K_2SO_4(aq) + 11H_2O(l)$

Chapter 11

11-2. (a) 602 g (b) 332.2 g (c) 18 g (d) 56 g

11-4. 75.35% C; 8.959% H; 7.323% N; 8.365% O

11-6. (a) 37.48% C; 49.93% O; 12.58% H (b) 88.81% O; 11.19% H (c) 94.07% O; 5.926% H (d) 9.861% Mg; 13.01% S; 71.40% O; 5.725% H

11-8. (a) 5.26 g (b) 0.081 g (c) 3.69 g (d) 1.67 g

11-10. Fe_2O_3

11-12. FeS

11-14. Al_2O_3

11-16. (a) TlBr, thallium bromide (b) $PbCl_2$, lead(II) chloride (c) NH_3, ammonia (d) Mg_3N_2, magnesium nitride

11-18. 47.7; titanium, Ti

11-20. 35.4; chlorine, Cl

11-22. $C_6H_{12}O_6$

11-24. 59 700

11-26. $C_{10}H_{10}Fe$

11-28. $C_{15}H_{24}O$

11-30. $C_8H_{20}Pb$

11-32. 59.1 g

11-34. 3.92 g

11-36. 58.5 metric tons

11-38. 2.46×10^3 kg

11-40. 9.41 metric tons

11-42. 17.0 g

11-44. 78.0%

11-46. 43.3% K_2SO_4 and 56.7% $MnSO_4$

11-48. 62.3% Al and 37.7% Mg

11-50. 5.49 g

11-52. 19.4 g

11-54. (a) $CaCO_3(s) + 2HCl(aq) \rightarrow CaCl_2(aq) + H_2O(l) + CO_2(g)$ (b) 4.50 g $CaCO_3$ remaining

11-56. (a) $CdCl_2(aq) + 2AgClO_4(aq) \rightarrow Cd(ClO_4)_2(aq) + 2AgCl(s)$ (b) excess of 1.8 g $CdCl_2$

11-58. $Hg(NO_3)_2(aq) + 2NaBr(aq) \rightarrow HgBr_2(s) + 2NaNO_3(aq)$; excess of 1.3 g $Hg(NO_3)_2$

11-60. 97.8%

11-62. 82.3%

11-64. (a) $Fe(s) + H_2SO_4(aq) \rightarrow FeSO_4(aq) + H_2(g)$ (b) 29.2 kg

11-70. 6.95×10^{25}

11-72. 1280 g Na_2CS_3; 442 g Na_2CO_3; 225 g H_2O

11-74. (a) 1.02 g (b) 1.09 g (c) 0.825 g (d) 0.744 g

11-76. 29.8 g O_2; 40.9 g CO_2

11-78. 23% NaCl and 77% $CdCl_2$

11-80. 6 water molecules

11-82. B_5H_9

11-84. 48.2% Na_2SO_4 and 51.8% $NaHSO_4$

11-86. 1.58×10^5 kg

11-88. $C_{17}H_{21}NO_4$, it may be cocaine

11-90. 4.5 ton as the ad claims

11-92. $C_6H_{12}SO_2$

Chapter 12

12-2. 14.3 M

12-4. 0.00643 M

12-6. (a) 1.00×10^{-5} mol (b) 4.00×10^{-6} mol

12-8. 20.8 mL

12-10. Dissolve 2.50 grams in a 50 mL volumetric flask and add water to the 50 mL mark

12-12. 1.7 $g \cdot mL^{-1}$

12-16. 0.050 M in $Ni^{3+}(aq)$ and 0.150 M in $Cl^-(aq)$

12-18. 27.7 mL

12-20. 4.33 g

12-22. 0.456 g

12-24. 320 g Br_2 and 142 g Cl_2

12-26. 32.2 mL

12-28. 17.5 g AgI; 0.319 M Ca^{2+}, 0.496 M NO_3^-, and 0.141 M I^-

12-30. 0.738 M

12-32. 170 mL by $Mg(OH)_2$ and 190 mL by $Al(OH)_3$

12-34. 0.11 M NaOH and 0.0286 M NaBr

12-36. 92.8%; we assumed that any impurities do not react with HCl(*aq*).

12-38. 60.1

12-44. 30.0 M

12-46. 13.2 M

12-48. 70.8 g

12-50. 4.2 mL

12-52. 3.75 M; Add 3.75 mL to a 250 mL volumetric flask and fill to the 250 mL mark.

12-54. 9.5×10^{-5} M

12-56. 23.2 g

12-58. 1.00 kg

12-60. 184 g

12-62. 177.8 mL

12-64. 2 acidic protons

12-66. 2.57%

12-68. 2.10 g

12-70. 5.15 g

12-72. 0.2300 M

12-74. 6.89%

12-76. 0.0310 M

12-78. 63.6%

12-80. 190 $g \cdot L^{-1}$

12-82. 2.77 g

12-84. 14.1 mL

Chapter 13

13-2. 739 Torr and 0.972 atm

13-4. 12.6 mL

13-6. two cylinders

13-8. 1.0 L

13-10. –78°C

13-14. 0.28 L O_2; 0.55 L H_2O

13-16. 676 g

13-18. 1.93×10^5 Pa

13-20. 56.2 μg

13-22. 3.3×10^{13} molecules

13-24. 5.40 $g \cdot L^{-1}$

13-26. C_4H_6

13-28. C_2H_4

13-30. C_3H_5O

13-32. 0.4818 H_2; 0.4277 N_2; 0.0906 Ar

13-34. 24.8% O_2; 75.2% N_2. The discrepancy suggests that air consists of more than just nitrogen and oxygen.

13-36. 103 kPa

13-38. 466 Toor N_2; 318 Torr O_2; 784 Torr total

13-40. 52.3 L of each

13-42. 34.9 L; 50.9 L

13-44. 7.43 metric tons

13-46. 986 L; 56 atm

13-48. zero

13-50. 408 $m \cdot s^{-1}$; 518 $m \cdot s^{-1}$; 1140 $m \cdot s^{-1}$

13-52. 11.22

13-54. 7.01 mL

13-56. 46.4%

13-58. 0.123 atm; 1.23×10^{-4} atm; 1.23×10^{-7} atm

13-60. 1.0×10^{-4} $collisions \cdot s^{-1}$

13-64. 20.9 bar; 22.1 bar from the ideal gas equation

13-68. 29.5 in

13-70. 3.0 m

13-74. 8300 gallons. Information is approximately correct.

13-76. 21.2 mL

13-78. 0.58 bar

13-82. 287 L CO_2; 144 L O_2

13-84. 238 mL

13-86. 3.0 L

13-88. 43% NaH and 57% CaH_2

13-92. 62.5%

Chapter 14

14-2. 3.2 kJ

14-4. +87.50 J

14-6. 200 J

14-8. –860 $kJ \cdot mol^{-1}$

14-10. –601 $kJ \cdot mol^{-1}$

14-12. –73.8 kJ

14-14. –1154 $kJ \cdot mol^{-1}$

14-16. 333 J; –73.5 kJ

14-18. 0.879 $kJ \cdot g^{-1}$

14-20. 82.9 $kJ \cdot mol^{-1}$

14-22. –521 $kJ \cdot mol^{-1}$

14-24. 1.0 $kJ \cdot mol^{-1}$

14-26. –130 $kJ \cdot mol^{-1}$

14-28. –11.3 $kJ \cdot mol^{-1}$

14-30. (a) –196.0 $kJ \cdot mol^{-1}$; exothermic
(b) –100.7 $kJ \cdot mol^{-1}$; exothermic
(c) –902.0 $kJ \cdot mol^{-1}$; exothermic

14-32. (a) –725.9 $kJ \cdot mol^{-1}$ (b) –622.2 $kJ \cdot mol^{-1}$

14-34. –2226 $kJ \cdot mol^{-1}$

14-36. (a) 40 $kJ \cdot mol^{-1}$ (b) 32.5 $kJ \cdot mol^{-1}$

14-38. –2808.7 $kJ \cdot mol^{-1}$; 7.015×10^6 kJ

14-40. 192 $kJ \cdot mol^{-1}$

14-42. –435$kJ \cdot mol^{-1}$

14-44. 943 $kJ \cdot mol^{-1}$

14-46. 324.40 $kJ \cdot mol^{-1}$

14-48. 195 $J \cdot K^{-1} \cdot mol^{-1}$

14-50. 7.46×10^{-4} g

14-52. 3.2°

14-54. 2.6 $J \cdot K^{-1} \cdot g^{-1}$

14-56. –26.0 $kJ \cdot mol^{-1}$

14-58. –66.3 $kJ \cdot mol^{-1}$

14-60. 33.3 $kJ \cdot K^{-1}$

14-62. –2820 $kJ \cdot mol^{-1}$

14-64. –236 kJ; –56.4 Calories

14-66. –1343 $kJ \cdot mol^{-1}$; –695 $kJ \cdot mol^{-1}$

14-74. 1210 Cal; 5060 kJ

14-76. 48 g

14-78. –110 J

14-80. TlCl

14-82. (a) –49.94 $kJ \cdot g^{-1}$ (b) –50.30 $kJ \cdot g^{-1}$
(c) –51.89 $kJ \cdot g^{-1}$

14-84. –352.7 $kJ \cdot mol^{-1}$

14-86. –0.557 kJ; 0.666°C

14-88. 5.35 $Cal \cdot g^{-1}$

14-90. 1.62×10^5 $kJ \cdot day^{-1}$; 7 $barrel \cdot yr^{-1}$

Chapter 15

15-2. 30.7 $kJ \cdot mol^{-1}$

15-4. 1180 g; 848 g

15-6. 5.08×10^{-2} J

15-8. 4.19 mol

15-10. Takes longer to vaporize the water.

15-12. 19.1 $kJ \cdot mol^{-1}$

15-14. Cl_2

15-16. CH_3CH_2OH

15-22. 580 Torr

15-24. 95°C; 80°C; 95°C; 85°C

15-26. 51 Torr

15-28. 8.1 mJ

15-36. Four

15-38. 330.51 pm

15-40. 6.02×10^{23} atom·mol^{-1}

15-42. 4.31×10^{-8} cm; 373 pm

15-44. Four; NaCl type

15-54. 117 Cal

15-56. 0.36 kJ·mol^{-1}; 2.3 kJ·mol^{-1}; 7.4 kJ·mol^{-1}; 10 kJ·mol^{-1}; 14 kJ·mol^{-1}

15-58. 21 kJ·mol^{-1}, 2%; 32 kJ·mol^{-1}, 20%; 18 kJ·mol^{-1}, 4%

15-64. 0.313 cm

15-68. 23°C

15-70. 3.994 g·cm^{-3}

15-72. 1.28 mol·L^{-1}

15-78. 1100 Torr

15-80. 65.3 g·mol^{-1}

15-82. 5.92 Torr; 21.2 Torr; 83.9 Torr; 622 Torr

Chapter 16

16-2. 0.30 H_2CO; 0.071 CH_3OH; 0.63 H_2O

16-4. 365 grams of sucrose in 135 grams of water

16-6. 0.21

16-8. 457 grams of $Ba(NO_3)_2(s)$ in 1000 grams of water

16-10. 4.0 kg

16-12. (a) 1.0 m_c (b) 4.0 m_c (c) 3.0 m_c (d) 3.0 m_c

16-14. (a) 3 (b) 3 (c) 3

16-16. HCl

16-18. 5.71 kPa; 0.56 kPa

16-20. 0.873 atm; 0.13 atm

16-22. 6.8 Torr

16-24. 62.5 mbar

16-26. (a) 0.834 mbar (b) 1.10 mbar (c) 0.423 mbar

16-28. 7.37 m

16-30. (a) 3.58 m (b) 2.38 m (c) 2.38 m

16-32. 57 g

16-34. 100.6°C; 12.5 Torr

16-36. 66.83°C

16-38. -1.36°C

16-40. 324

16-42. 10m

16-44. $C_3H_8O_3$

16-46. $K_2HgI_4 \xrightarrow[H_2O]{} 2K^+(aq) + HgI_4^{2-}(aq)$

16-48. 28.0 atm

16-50. 34 400

16-52. 23 L

16-54. P_{prop} = 5.2 Torr; 10 Torr; 16 Torr and P_{iso} = 34 Torr; 23 Torr; 11 Torr; in the vapor, x_{prop} = 0.13; 0.30; 0.59 and x_{iso} = 0.87; 0.70; 0.41

16-56. $CO_2(g)$

16-58. 8.6 mg O_2 and 14 mg N_2

16-68. $MgCl_2$ solution

16-70. –4.4°C wine and –21°C vodka

16-76. (a) 0.115 m; 0.345 m_c; –9.64°C; 100.18°C (b) 0.434 m; 0.434 m_c; –81°C; 100.22°C

16-78. P_4

16-80. 2.26 M

16-82. 108.26°C

16-88. 2.01

16-90. ammonium nitrate, NH_4NO_3

16-92. 125 mL and 375 mL

Chapter 17

17-2. $-\frac{1}{2}\frac{\Delta[SO_2]}{\Delta t}; -\frac{\Delta[O_2]}{\Delta t}; \frac{1}{2}\frac{\Delta[SO_3]}{\Delta t}$

17-6. 3.5×10^{-14} M·min^{-1}; third order

17-8. (a) 2.36×10^{-4} M·min^{-1} (b) 1.7×10^{-4} M·min^{-1} (c) 6.4×10^{-4} M·min^{-1}

17-12. (20 mol^{-1}·L·s^{-1})[NOBr]^2

17-14. (1.44×10^{-4} Torr^{-1}·s^{-1})$P^2_{C_5H_6}$

17-16. (6.2 s^{-1})$P_{N_2O_3}$

17-18. (1.5 M^{-1}·s^{-1})[$CoBr(NH_3)_5^{2+}$][OH^-]

17-20. (4.0×10^{-3} M^{-1}·s^{-1})[CH_3COCH_3][H^+][B_2]

17-22. 1.9×10^{-5} M

17-24. 0.99

17-26. 0.0864 min^{-1}

17-28. (2.8×10^{-5} s^{-1})$P_{H_2C_2O_4}$

17-30. 0.624

17-32. 300 years

17-34. 0.17

17-36. 1230 years old

17-38. 3600 years old

17-40. 2.44×10^9 years

17-42. 0.12 M

17-44. 5.11 M^{-1}·s^{-1}

17-46. 0.015 M

17-52. (1.01×10^{-6} Torr^{-1}·s^{-1})P^2_{CO}

17-54. 3.6×10^{-4} M·s^{-1}; 1.3 mol·L^{-1}

17-58. 1.6×10^8 years; 1.6×10^{10} years

17-60. 0.464 disintegrations·min^{-1}

17-62. 5.00 L

17-64. (0.030 M^{-1}·s^{-1})[BrO_3^-][I^-]

17-66. 1.4×10^6 per milliliter

17-70. 8.3 s

17-72. 0.0102 min^{-1}

17-74. 3.40×10^{-6} Torr^{-2}·s^{-1}

17-76. 170 million years old

Chapter 18

18-2. 1.50×10^2 kJ·mol^{-1}

18-4. 4.9×10^{-3} s^{-1}

18-6. 41 beats·min^{-1}

18-8. (a) rate of reaction = k[K][HCl] (b) rate of reaction = k[H_2O_2] (c) rate of reaction = k[O_2]^2[Cl] (d) rate of reaction = k[NO_3][CO]

18-12. Yes

18-16. Yes

18-18. First two mechanisms; mechanism b

18-20. Yes; $\frac{k_2 k_1}{k_{-1}}$

18-24. k[H_2O_2][I^-]^2[H^+]; $H^+(aq)$ and $I^-(aq)$

18-28. 50 μmol·L^{-1}·min^{-1}; 11 μmol·L^{-1}

18-30. 15 mmol·L^{-1}·min^{-1}; 38 mmol·L^{-1}

18-32. 40 mmol·L^{-1}·s^{-1}; 28 mmol·L^{-1}·s^{-1}

18-34. 1.4×10^4 min^{-1}

18-36. 1.0×10^8 s^{-1}

18-50. (a) increase (b) increase (c) decrease (d) decrease

18-52. 3.1 min

18-54. 4.99×10^{-3} M^{-1}·s^{-1}; 120 kJ·mol^{-1}

18-56. second order; 46.06 kJ·mol^{-1}; 27 mbar

18-60. 51 kJ·mol^{-1}

18-62. 101 kJ·mol^{-1}

18-66. 15.0 μmol·L^{-1}

18-68. 9.20 mmol·L^{-1}

18-70. 110 μmol·s^{-1}; R_{max} =0.64 M·s^{-1}

Chapter 19

19-2. 0.527 M SO_2 and 0.274 M SO_3

19-4. (a) $K_c = \frac{[SO_3]^2}{[SO_2]^2[O_2]}$ M^{-1}

(b) K_c = [CO_2][H_2O] M^2 (c) $K_c = \frac{[CH_4]}{[H_2]^2}$ M^{-1}

19-6. (a) $K_c = [O_2]$ M^3 (b) $K_c = \frac{[N_2O_4]}{[N_2][O_2]^2}$ M (c) $K_c = \frac{[N_2O_4]}{[N_2][O_2]^2}$ M^{-2}

19-8. (a) $K_p = P^2_{NH_3}P_{CO_2}$ bar^3 (b) $K_p = P_{O_2}$ bar (c) $K_p = \frac{P_{N_2O_4}}{P_{N_2}P^2_{O_2}}$ bar^{-2}

19-10. 27 bar^2

19-12. 1.8 M

19-14. 39 bar

19-16. 0.83 bar

19-18. 1.18 bar

19-20. 688 g

19-22. 4.54 bar

19-24. both are 0.51 bar

19-26. $[COCl_2] = 0.146$ M; $[Cl_2] = 0.104$ M; $[CO] = 0.354$ M

19-28. $P_{H_2} = 1.2$ bar and $P_{HI} = 3.3$ bar

19-30. $P_{H_2} = 1.51$ bar and $P_{CH_4} = 0.60$ bar

19-32. 1.9 bar

19-34. 0.076 bar^2

19-38. (a) to the right (b) to the right (c) to the left (d) to the left

19-40. (a) no change (b) to the left (c) to the right (d) to the right

19-44. $P_{PCl_3} = 6.7$ Torr; $P_{PCl_5} = 224$ Torr; $P_{Cl_3} = 26.7$ Torr

19-46. $P_{N_2O_4} = 0.390$ bar; $P_{NO_2} = 0.464$ bar; $P_{tot} = 0.854$ bar

19-48. 26.4 Torr

19-50. No; from left to right

19-52. From right to left

19-60. (a) more intense (b) less intense (c) no change

19-66. 170 M^{-2}

19-68. (a) 0.088 bar (b) $P_{CH_3OH} = 8.95$ bar and $P_{CO} = 1.03$ bar

19-70. 6.2 bar

19-72. 1.60 moles

19-74. (a) No (b) $[H_2] = [I_2] = 1.7 \times 10^{-3}$ M; $[HI] = 0.016$ M

19-76. $P_{Cl_2} = 2.24$ atm; $P_{O_2} = 1.64$ atm

19-80. 25.1 moles

19-82. 78.5%

19-84. $P_{CO} = P_{H_2O} = 12.0$; Torr $P_{CO_2} = P_{H_2} = 14.1$ Torr

19-86. $P_{NH_3} = P_{HCl} = 0.24$ bar; 9.41 g

19-88. 3.3 moles

19-92. 0.982 bar

19-96. $[H_2O] = 0.0559$ M; $[CH_4] = 0.443$ M; $[CO] = 0.557$ M; $[H_2] = 0.219$ M

Chapter 20

20-2. $[OH^-] = 0.25$ M; $[K^+] = 0.25$ M; $[H_3O^+] = 4.0 \times 10^{-14}$ M; basic

20-4. $[OH^-] = 0.0162$ M; $[Ca^{2+}] = 8.10 \times 10^{-3}$ M; $[H_3O^+] = 6.17 \times 10^{-13}$ M

20-6. 13.30; basic

20-8. pOH = 1.40 and pH = 12.60

20-10. pOH = −0.20 and pH = 14.20

20-12. 1×10^{-3} M

20-14. 300

20-16. $[H_3O^+] = 7.1 \times 10^{-9}$ M and $[OH^-] = 1.4 \times 10^{-6}$ M

20-18. 1.4×10^{-5} M

20-20. 1.8×10^{-4} M

20-22. pH = 4.11; $[ClO^-] = [H_3O^+] = 7.7 \times 10^{-5}$ M; $[HClO] \approx 0.150$ M; $[OH^-] = 1.3 \times 10^{-10}$ M

20-24. pH = 1.96; $[ClCH_2COO^-] = [H_3O^+] = 0.011$ M; $[ClCH_2COOH] = 0.089$ M; $[OH^-] = 9.1 \times 10^{-13}$ M

20-26. 4.6×10^{-4} M

20-28. 9.52

20-30. 11.28

20-40. (a) $K_a = 6.7 \times 10^{-6}$ M for $C_6H_5NH^+$
(b) $K_a = 4.8 \times 10^{-10}$ M for HCN
(c) $K_a = 1.1 \times 10^{-6}$ M for NH_3OH^+
(d) $K_a = 1.9 \times 10^{-11}$ M for $(CH_3)_2NH_2^+$

20-48. 9.08

20-50. $[OH^-] = 2.1 \times 10^{-6}$ M; $[HNO_2] = 2.1 \times 10^{-6}$ M; $[NO_2^-] \approx 0.25$ M; $[H_3O^+] = 4.8 \times 10^{-9}$ M; pH = 8.32

20-52. 9.00

20-54. 1.25

20-56. 1.11

20-58. $[H_3O^+] = 5.7 \times 10^{-3}$ M; $f_1 = 1.0$; $f_2 = 3.0 \times 10^{-3}$; $f_3 = 7.0 \times 10^{-5}$

20-70. 180

20-72. 6.81

20-74. 9.62×10^{-4} g per 100 mL of solution

20-76. 2.39

20-78. 16

20-80. 3.34

20-82. (a) 9.5 (b) 17.1

20-84. 7.46; acidic

20-86. 1.74

20-88. $[(CH_3)_2NH_2^+] = [OH^-] = 8.7 \times 10^{-3}$ M; $[(CH_3)_2NH] = 0.141$ M; $[H_3O^+] = 1.1 \times 10^{-12}$ M; pH = 11.96

20-90. 5.7%

20-94. 6.943

Chapter 21

21-2. 5.04

21-4. 3.13

21-6. 8.95

21-8. (a) −0.04 (b) 0.04

21-10. 5.51

21-12. 330 mL

21-14. almost equal concentrations of pyridinium chloride and pyridine

21-16. (b) 9.03 (d) 9.43

21-18. Mix 28.7 mL of 0.100 M $K_2SO_3(aq)$ with 21.3 mL of 0.200 M $KHSO_3(aq)$.

21-20. React 139 mL of the lactic acid solution with 61.0 mL of the $Ba(OH)_2(aq)$ solution.

21-30. (a) 7.00 (b) 1.60

21-32. (a) 210 mL (b) 250 mL

21-36. 0.289 M

21-38. 8.01; phenolphthalein

21-40. (a) 2.72 (b) 4.14 (c) 4.74 (d) 8.88 (e) 11.59

21-42. (a) 11.97 (b) 10.66 (c) 8.11 (d) 5.82 (e) 3.55

21-46. 122

21-62. 340 mL

21-64. 7.21

21-66. $[CH_3COOH] \approx [CH_3COOH]_0$ and $[CH_3COO^-] \approx [CH_3COO^-]_0$

21-70. 190 mL

21-72. 88.1

21-74. 50.0 mL

21-76. three acidic protons

21-78. 4.15%

21-80. (a) 7.21 (b) 7.51 (c) 6.91

21-82. 4.85

21-84. 3.25

21-86. 0.900 grams

21-88. 11.7 grams

21-90. 3.55

21-92. (a) 0.4 (b) 1.3 (c) 4.0 (d) 6.70 (e) 10.34

Chapter 22

22-6. (a) $H_2SO_4(aq) + Ca(ClO_4)_2(aq) \rightarrow CaSO_4(s) + 2\,HClO_4(aq)$

$Ca^{2+}(aq) + SO_4^{2-}(aq) \rightarrow CaSO_4(s)$

(b) no reaction

(c) $Hg_2(NO_3)_2(aq) + 2\,NaC_6H_5COO(aq) \rightarrow Hg_2(C_6H_5COO)_2(s) + 2\,NaNO_3(aq) Hg_2^{2+}(aq) + 2\,C_6H_5COO^-(aq) \rightarrow Hg_2(C_6H_5COO)_2(s)$

(d) $Na_2SO_4(aq) + 2\,AgF(aq) \rightarrow Ag_2SO_4(s) + 2\,NaF(aq) 2\,Ag^+(aq) + SO_4^{2-}(aq) \rightarrow Ag_2SO_4(s)$

22-8. 3.3 $g \cdot L^{-1}$

22-10. $4.3 \times 10^{-3}\ g \cdot L^{-1}$

22-12. $2.5 \times 10^{-3}\ M^2$

22-14. $1.1 \times 10^{-8}\ M^2$

22-16. $5.5 \times 10^{-7}\ g \cdot L^{-1}$

22-18. 0.027 $g \cdot L^{-1}$

22-20. 1.4 $g \cdot L^{-1}$

22-22. $Cu(NO_3)_2(aq) + 2\,NH_3(aq) + 2\,H_2O(l) \rightarrow Cu(OH)_2(s) + 2\,NH_4NO_3(aq) Cu(OH)_2(s) + 4\,NH_3(aq) \rightarrow [Cu(NH_3)_4]^{2+}(aq) + 2\,OH^-(aq)$

22-24. $5.8 \times 10^{-3}\ g \cdot L^{-1}$

22-26. $4.1 \times 10^{-4}\ g \cdot L^{-1}$

22-30. (a) $PbCrO_4$ (b) $Ag_2C_2O_4$ (d) Ag_2O

22-32. 0.14 $g \cdot L^{-1}$

22-34. Yes

22-36. Yes; 5.4×10^{-6} mol; $[Ag^+] = [NO_3^-] = 0.25$ M; $[Br^-] = 2.2 \times 10^{-12}$ M; $[Na^+] = 5.0 \times 10^{-5}$ M

22-38. (a) 0.19 mg (b) 1.1×10^{-23} M

22-42. No

22-44. 9.1×10^{-9} M

22-46. 2.2 M $Cu(OH)_2$ and 1.0×10^5 M $Zn(OH)_2$; No

22-50. For $Cd(OH)_2$, $s = (7.2 \times 10^{13}\ M^{-1})[H_3O^+]^2$; for $Fe(OH)_3$, $s = (2.8 \times 10^3\ M^{-2}[H_3O^+]^3$; pH from 2 to 8

22-52. 0.008 M

22-62. Ag_2CrO_4

22-66. For FeS, $s = 0.57$M at pH 2.00 and 5.7×10^{-5} M at pH 4.00; 0.0061 M

22-78. For CdS, $s = 7.3 \times 10^{-10}$ M at pH 2.00 and 7.3×10^{-14} M at pH 4.00; pH of 2.00

22-68. 3.0

22-72. 0.29 M

22-74. (a) $K_3 = 3.8 \times 10^{12}\ M^{-1}$ (b) $K_3 = 2.2 \times 10^{-5}$

22-76. 0.0061 M

22-78. $K_3 = \dfrac{[Ag^+]^2}{[Ca^{2+}]} = 0.24$; $[Ca^{2+}] = 0.018$ M and $[Ag^+] = 0.064$ M

22-80. 3.0 M

22-80. 5.0×10^{-4} M

22-84. 1.4×10^{-2} M; 1.0 $g \cdot L^{-1}$

22-88. 2.4×10^{-9} M

22-90. 3.5×10^{-15} M; 4.4×10^{-14}

22-92. 5.8×10^{-4} M; not soluble

22-94. For chromium,

$$s = (6.3 \times 10^{11}\ M^{-1})[H_3O^+]^3 + \frac{4.0 \times 10^{-16}\ M^2}{[H_3O^+]};$$

For tin, $s = (55\ M^{-1})[H_3O^+]^2 + \dfrac{1 \times 10^{-16}\ M^2}{[H_3O^+]}$; between pH 4 and 7

22-98. (1) $K_1 = 0.013$ M (2) $K_2 = \dfrac{I_3^-}{[I^-]} = 0.61$ (3) $K_3 = 47\ M^{-1}$

Chapter 23

23-2. For HF, $\Delta S_{fus} = 24.08\ J \cdot K^{-1} \cdot mol^{-1}$; $\Delta S_{vap} = 86.03\ J \cdot K^{-1} \cdot mol^{-1}$
For HCl, $\Delta S_{fus} = 12.53\ J \cdot K^{-1} \cdot mol^{-1}$; $\Delta S_{vap} = 93.10\ J \cdot K^{-1} \cdot mol^{-1}$
For HBr, $\Delta S_{fus} = 12.92\ J \cdot K^{-1} \cdot mol^{-1}$; $\Delta S_{vap} = 93.45\ J \cdot K^{-1} \cdot mol^{-1}$
For HI, $\Delta S_{fus} = 12.92\ J \cdot K^{-1} \cdot mol^{-1}$; $\Delta S_{vap} = 88.99\ J \cdot K^{-1} \cdot mol^{-1}$

23-16. (a) $606.4\ J \cdot K^{-1} \cdot mol^{-1}$; (b) $24.8\ J \cdot K^{-1} \cdot mol^{-1}$ (c) $-324.2\ J \cdot K^{-1} \cdot mol^{-1}$ (d) $-120.8\ J \cdot K^{-1} \cdot mol^{-1}$

23-20. $\Delta S^\circ_{rxn} = -128.6\ J \cdot K^{-1} \cdot mol^{-1}$; $\Delta G^\circ_{rxn} = -5.9\ kJ \cdot mol^{-1}$; left to right

23-22. $\Delta S^\circ_{rxn} = 175.9\ J \cdot K^{-1} \cdot mol^{-1}$; $\Delta G^\circ_{rxn} = 120.0\ kJ \cdot mol^{-1}$; $\Delta G_{rxn} = 75.0\ kJ \cdot mol^{-1}$; right to left

23-24. $-42.2\ kJ \cdot mol^{-1}$; left to right

23-26. $\Delta G^\circ_{rxn} = 35.8\ kJ \cdot mol^{-1}$; right to left; $\Delta G_{rxn} = -25.5\ kJ \cdot mol^{-1}$; left to right

23-28. $\Delta G^\circ_{rxn} = 42.2\ kJ\cdot mol^{-1}$; No; $\Delta G_{rxn} = -20.6\ kJ\cdot mol^{-1}$; Yes

23-30. $\Delta G^\circ_{rxn} = 27.1\ kJ\cdot mol^{-1}$ right to left; $\Delta G_{rxn} = -34.0\ kJ\cdot mol^{-1}$; Yes

23-32. $48.3\ kJ\cdot mol^{-1}$; Insoluble $CaCO_3(s)$ will precipitate.

23-34. $\Delta G^\circ_{rxn} = -41.7\ kJ\cdot mol^{-1}$ left to right; $\Delta G_{rxn} = 5.71\ kJ\cdot mol^{-1}$; right to left

23-36. (a) $-822\ kJ\cdot mol^{-1}$; 1×10^{144} (b) $175\ kJ\cdot mol^{-1}$; 2×10^{-31} (c) $-232.2\ kJ\cdot mol^{-1}$; 4.7×10^{40}

23-38. $\Delta G^\circ_{rxn} = 226.6\ kJ\cdot mol^{-1}$; $\Delta H^\circ_{rxn} = 331.4\ kJ\cdot mol^{-1}$; 2.0×10^{-40}

23-40. $\Delta G^\circ_f[CCl_4] = -65.3\ kJ\cdot mol^{-1}$; $\Delta H^\circ_f[CCl_4] = -128.2\ kJ\cdot mol^{-1}$

23-42. 818.1 kJ

23-44. $193\ kJ\cdot mol^{-1}$

23-46. $\Delta H^\circ_{rxn} = -9.4\ kJ\cdot mol^{-1}$; 46

23-48. 243 Torr

23-50. 0.323 Torr

23-52. $33.87\ kJ\cdot mol^{-1}$

23-54. 62.0°C

23-56. 15

23-64. (a) $144.6\ J\cdot K^{-1}\cdot mol^{-1}$ (b) $173.8\ J\cdot K^{-1}\cdot mol^{-1}$ (c) $12.1\ J\cdot K^{-1}\cdot mol^{-1}$ (d) $-100.3\ J\cdot K^{-1}\cdot mol^{-1}$

23-70. $39.5\ kJ\cdot mol^{-1}$; The discrepancy is due to the large temperature difference between 25°C and the data.

23-72. (a) 2×10^{-10} (b) 6×10^{-13}

23-74. 1.35

23-76. 2873 kJ

23-78. $55.9\ kJ\cdot mol^{-1}$

23-80. $61.4\ kJ\cdot mol^{-1}$

23-82. 0.282 bar

23-84. 15 kPa

23-86. (a) $-100\ kJ\cdot mol^{-1}$

23-88. 21%

Chapter 24

24-2. (a) $-1/2$ (b) -1 (c) $-1/3$ (d) -1

24-4. Each O atom is -2 (a) $+1$ (b) $+4$ (c) $+7$ (d) $+5$

24-6. H is +1, O is −2, Fe is +3 (b) Fe is +8/3 (c) K is +1, C is +4, N is −3 (d) C is 0, N is +1, O is −2

24-8. S is −2, C is +4 (b) H is +1, S is −1, C is −2 (c) H is +1, C is +2, O is −2, N is −3 (d) H is +1, C is +4, N is −3, S is −2

24-12. (a) linear (b) tetrahedral (c) trigonal planar

24-18. (a) $In^+(aq)$ is the reducing agent and $Fe^{3+}(aq)$ is the oxidizing agent.
$In^+(aq) \rightarrow In^{3+}(aq) + 2e^-$ (oxidation half reaction)
$2Fe^{3+}(aq) + 2e^- \rightarrow 2Fe^{2+}(aq)$ (reduction half reaction)
(b) $H_2S(aq) \rightarrow S(s) + 2H^+(aq) + 2e^-$ (oxidation half reaction)
$ClO^-(aq) + 2H^+(aq) + 2e^- \rightarrow Cl^-(aq) + H_2O(l)$ (reduction half reaction)

24-26. (a) $H_2BO_3^-(aq) + 8H^+(aq) + 8e^- \rightarrow BH_4^-(aq) + 3H_2O(l)$
(b) $2ClO_3^-(aq) + 12H^+(aq) + 10e^- \rightarrow Cl_2(g) + 6H_2O(l)$
(c) $Cl_2(g) + 2H_2O(l) \rightarrow 2HClO(aq) + 2H^+(aq) + 2e^-$

24-28. (a) $OsO_4(s) + 8H^+(aq) + 8e^- \rightarrow Os(s) + 4H_2O(l)$
(b) $S(s) + 6OH^-(aq) \rightarrow SO_3^{2-}(aq) + 3H_2O(l) + 4e^-$
(c) $Sn(s) + 2H_2O(l) \rightarrow HSnO_2^-(aq) + 3H^+(aq) + 2e^-$

24-30. (a) $[Au(CN)_2]^-(aq) + e^- \rightarrow Au(s) + 2CN^-(aq)$
(b) $MnO_4^-(aq) + 4H^+(aq) + 3e^- \rightarrow MnO_2(s) + 2H_2O(l)$
(c) $Cr(OH)_3(s) + 5OH^-(aq) \rightarrow CrO_4^{2-}(aq) + 4H_2O(l) + 3e^-$

24-32. (a) $3ZnS(s) + 2NO_3^-(aq) + 8H^+(aq) \rightarrow 3S(s) + 3Zn^{2+}(aq) + 2NO(g) + 4H_2O(l)$
oxidizing agent, $NO_3^-(aq)$; reducing agent, $ZnS(s)$; species oxidized, S; species reduced, N
(b) $2MnO_4^-(aq) + 5HNO_2(aq) + H^+(aq) \rightarrow 5NO_3^-(aq) + 2Mn^{2+}(aq) + 3H_2O(l)$
oxidizing agent, $MnO_4^-(aq)$; reducing agent, $HNO_3(aq)$; species oxidized, N; species reduced, Mn

24-34. (a) $2CoCl_2(s) + Na_2O_2(aq) + 2H_2O(l) + 2OH^-(aq) \rightarrow 2Co(OH)_3(s) + 4Cl^-(aq) + 2Na^+(aq)$
(b) $C_2O_4^{2-}(aq) + MnO_2(s) + 4H^+(aq) \rightarrow Mn^{2+}(aq) + 2CO_2(g) + 2H_2O(l)$

24-36. (a) $Cr_2O_7^{2-}(aq) + 6I^-(aq) + 14H^+(aq) \rightarrow 2Cr^{3+}(aq) + 3I_2(s) + 7H_2O(l)$
(b) $3CuS(s) + 2NO_3^-(aq) + 8H^+(aq) \rightarrow 3S(s) + 3Cu^{2+}(aq) + 2NO(g) + 4H_2O(l)$

24-38. (a) $N_2H_4(aq) + 2Cu(OH)_2(s) \rightarrow N_2(g) + 2Cu(s) + 4H_2O(l)$
(b) $H_3AsO_3(aq) + I_2(aq) + H_2O(l) \rightarrow H_3AsO_4(aq) + 2I^-(aq) + 2H^+(aq)$

24-40. (a) $Co(OH)_2(s) + SO_3^{2-}(aq) \rightarrow SO_4^{2-}(aq) + Co(s) + H_2O(l)$
(b) $3IO_3^-(aq) + 24I^-(aq) + 18H^+(aq) \rightarrow 9I_3^-(aq) + 9H_2O(l)$

24-42. 0.103 M; 202 mg

24-44. 0.452 g; 10.7%

24-46. $4NaOH(aq) + Ca(OH)_2(aq) + C(s) + 4ClO_2(g) \rightarrow 4NaClO_2(aq) + CaCO_3(aq) + 3H_2O(l)$; 0.746 metric tons

24-48. 0.02824 M

24-54. (a) H is +1, N is −3 (b) N is +1
(c) H is +1, N is −2 (d) N is +1, O is −2
(e) N is +2, O is −2

24-56. (a) Se is −2, Mo is +4 (b) S is −2, Si is +4
(c) Ga is +3, As is −3 (d) K is +1, O is −2, S is +2

24-58. $2Tl^+(aq) + IO_3^-(aq) + 2Cl^-(aq) + 6H^+(aq) \rightarrow 2Tl^{3+}(aq) + ICl_2^-(aq) + 3H_2O(l)$

24-60. $2S_2O_8^{2-}(aq) + 2H_2O(l) \rightarrow 4SO_4^{2-}(aq) + O_2(g) + 4H^+(aq)$

24-62. $2KMnO_4(aq) + 16H^+(aq) + 5Na_2C_2O_4(aq) \rightarrow 2Mn^{2+}(aq) + 2K^+(aq) + 8H_2O(l) + 10CO_2(g) + 10Na^+(aq)$

24-64. $3MnO_4^-(aq) + 4H^+(aq) + 5Mo^{3+}(aq) \rightarrow 3Mn^{2+}(aq) + 5MoO_2^{2+}(aq) + 2H_2O(l)$; 0.103 M

24-66. 11.07%

24-68. 1.825×10^{-4}%

24-70. 0.018%

Chapter 25

25-2. At negative electrode: $Mn(s) \rightarrow Mn^{2+}(aq) + 2e^-$, at positive electrode: $Cr^{2+}(aq) + 2e^- \rightarrow 2Cr(s)$; $Mn(s)|MnSO_4(aq)||CrSO_4(aq)|Cr(s)$

25-4. At negative electrode: $Co(s) \rightarrow Co^{2+}(aq) + 2e^-$, at positive electrode: $Pb^{2+}(aq) + 2e^- \rightarrow Pb(s)$; $Co(s)|Co(NO_3)_2(aq)||Pb(NO_3)_2(aq)|Pb(s)$

25-6. At negative electrode: $Zn(s) \rightarrow Zn^{2+}(aq) + 2e^-$, at positive electrode: $Hg_2Cl_2(s) + 2e^- \rightarrow 2Hg(l) + 2Cl^-(aq)$; $Zn(s)|ZnCl_2(aq)|Hg_2Cl_2(aq)|Hg(l)|Pt(s)$

25-8. At left electrode: $Sn(s) \rightarrow Sn^{2+}(aq) + 2e^-$
At right electrode: $Ag^+(aq) + e^- \rightarrow Ag(s)$ $Sn(s) + 2Ag^+(aq) \rightarrow Sn^{2+}(aq) + 2Ag(s)$

25-10. At left electrode: $Cu(s) \rightarrow Cu^{2+}(aq) + 2e^-$
At right electrode: $Ag^+(aq) + e^- \rightarrow Ag(s)$ $Cu(s) + 2Ag^+(aq) \rightarrow Cu^{2+}(aq) + 2Ag(s)$

25-10. $Pb(s) + Hg_2SO_4(s) \rightarrow 2Hg(l) + PbSO_4(s)$

25-22. 2.7×10^7

25-24. 0.140 V; 5.4×10^4

25-26. 1.560 V; 5.0×10^{52}

25-28. $H_2(g) + Cd^{2+}(aq) \rightarrow 2H^+(aq) + Cd(s)$; −0.450 V

25-30. 0.0272 V

25-32. 0.47M

25-34. 0.99 M

25-36. (a) 1.88 V (b) −0.954 V (c) 0.96 V

25-38. +0.80 V

25-40. +0.234 V; No

25-42. +0.458 V; 3.0×10^{15}; spontaneous

25-44. 0.307 V

25-46. $BH_4^-(aq) + 8OH^-(aq) \rightarrow H_2BO_3^-(aq) + 5H_2O(l) + 8e^-$ oxidation; $8O_2(g) + 8e^- \rightarrow 8O_2^-(aq)$; $E^\circ_{red}[O_2|O_2^-] = -0.56$ V reduction; −1.24 V

25-48. $E_{cell} = 0.329$ V; spontaneous

25-50. E_{cell} = 1.36 V; spontaneous

25-52. –392 kJ·mol^{-1}

25-54. –267 kJ·mol^{-1}

25-56. –213.0 kJ·mol^{-1}

25-58. E°_{cell} = 1.08 V; ΔG°_{rxn} = –208 kJ·mol^{-1}; ΔG°_{rxn} = –220 kJ·mol^{-1}; E_{cell} = 1.14 V

25-60. 0.84 g

25-62. 54 min

25-64. 13.9 L

25-70. Between Mg and Al

25-72. 0.402 g

25-74. (a) 6.9 × 10^6 metric tons (b) 1.1 × 10^{14} kJ (c) 0.85%

25-80. 1.1447 V

25-86. +3

25-88. (a) $Pb(s) + PbO_2(s) + 4H^+(aq) + 2SO_4^{2-}(aq) \leftrightharpoons 2PbSO_4(s) + 2H_2O(l)$
(b) 2.050 V (c) –396.6 kJ·mol^{-1}
(d) 2.05 V (e) six cells

25-90. –0.744 V

Chapter 26

26-14. 58%

26-16. 30.2 g

26-18. 3500 metric tons of iron and 9670 metric tons of slag

26-20. 36 400 metric tons

26-22. (a) [Ar]3d^6 (b) [Ar] (c) [Xe]4f^{14}5d^8 (d) [Ar]3d^{10}

26-24. (a) 4 (b) none (c) 4 (d) 10

26-28. (a) +3 (b) +3 (c) +2 (d) +2

26-30. (a) +3 (b) +3 (c) +3 (d) 0

26-32. (a) hexaamminochromium(III)
(b) triamminetrichloroplatinum(IV)
(c) hexaaquamolybdenum(III)
(d) hexacyanochromate(II)

26-36. (a) potassium hexacyanoferrate(III)
(b) tetracarbonylnickel(0)
(c) hexaaquaruthenium(III) chloride
(d) sodium tetrahydroxoaluminate(III)

26-38. (a) sodium tetracyanoaurate(III)
(b) hexaaquachromium(III) chloride
(c) tris(ethylenediamine) vanadium(III) chloride
(d) hexaamminecopper(II) chloride

26-40. (a) $Na_2[Ni(CN)_2BrCl]$ (b) $Rb_2[Co(NO_2)_4]$
(c) $K_3[VCl_6]$ (d) $[Cr(NH_3)_5Cl](CH_3COO)_2$

26-42. (a) $Ba_2[Fe(CN)_6]$ (b) $[CoCl(OH)(en)_2]NO_3$
(c) $Li_2[Pt(NO_2)_2(ox)_2]$ (d) $[V(en)_2ox]CH_3COO$

26-48. (a) $t_{2g}^6e_g^2$ (b) $t_{2g}^3e_g^2$ (c) $t_{2g}^5e_g^0$ (d) $t_{2g}^0e_g^0$ (e) $t_{2g}^6e_g^2$

26-50. $[Fe(CN)_6]^{4-}$ is low spin; $[Fe(CN)_6]^{3-}$ is low spin; $[Co(NH_3)_6]^{2+}$ is high spin; $[CoF_6]^{3-}$ is high spin; and $[Mn(H_2O)_6]^{2+}$ is high spin.

26-52. (a) none (b) 5 (c) none

26-56. (a) $[Co(NO_2)_6]^{3-}$ (b) *trans*-$[PtCl_2(en)_2]^{2+}$
(c) $K_3[FeCl(CN)_5]$ (d) $Sr_2[Fe(CN)_6]$

26-66. tetraamminediiodoplatinum(IV); tetraiodoplatinate(II)

26-68. two optical isomers and no geometric isomers

26-70. 1.0 × 10^{-16} M^3

PHOTO CREDITS

All photos © Chip Clark except where listed below.

Chapter 1

Frontispiece, *Antoine-Laurent Lavoisier (1743–1749) and His Wife, Marie-Anne-Pierrette Paulze (1758–1836),* 1788. Oil on canvas, The Metropolitan Museum of Art, New York, NY, USA, image copyright © The Metropolitan Museum of Art/Art Resource, NY; 1.1, © Joel Gordon Photography; 1.2, courtesy of Dow Chemical; 1.5, SPL/Photo Researchers, Inc.; 1.7, © Joel Gordon Photography; 1.9 (*top*), © 2005 Richard Megna, Fundamental Photographs, NYC; 1.9 (*middle*), © 1990 Richard Megna, Fundamental Photographs, NYC; 1.9 (*bottom*), © 2002 Richard Megna, Fundamental Photographs, NYC; 1.12, permission granted from istockphoto, www.istockphoto.com; 1.15, courtesy of CRC Press; 1.16, Maximilian Stock Ltd./Photo Researchers, Inc.; page 38 (*top*), © 2002 Richard Megna, Fundamental Photographs, NYC; page 38 (*bottom*), © 2006 Warren Rosenberg, Fundamental Photographs, NYC.

Chapter 2

Frontispiece, Science Photo Library Photo/Photo Researchers, Inc.; 2.9, courtesy of Texasgulf; 2.12, © 2000 Larry Stepanowicz, Fundamental Photographs, NYC; 2.13, Omikron/Photo Researchers, Inc.; 2.14, Photo Researchers, Inc.; 2.17, reproduced with permission from The Cavendish Laboratory, University of Cambridge; 2.19, Library of Congress/Photo Researchers, Inc.; 2.22, photo by J. Ellis and by courtesy of Drew University.

Chapter 3

Frontispiece, courtesy of www.ilyarepin.org; 3.2, courtesy of New York Public Library; 3.6, with permission from Alexander Boden, Boden Books Pty Ltd., Fitzroy & Chapel Sts., Marrickville NsN 2204 Australia; 3.7, © Joel Gordon Photography; 3.13, Kenneth Edward/Photo Researchers, Inc.

Chapter 4

Frontispiece, AIP Emilio Segre Visual Archives, Margrethe Bohr Collection; 4.11, photo provided courtesy of Dr. Richard Zare, Stanford University, all rights reserved; 4.16, courtesy of Wabash Instrument Co.; 4.19, AIP Emilio Segre Visual Archives, W. F. Meggers Gallery of Nobel Laureates; 4.20, SPL/Photo Researchers, Inc.; 4.21, David Scharf/Photo Researchers, Inc.; 4.22, Education Development Center, Newton, MA; 4.27 (*all*), © Joel Gordon Photography.

Chapter 5

Frontispiece, photo by Wolfgang Pfaundler, Innsbruck, Austria, courtesy AIP Emilio Segre Visual Archives; 5.1, AIP Emilio Segre Visual Archives; page 147, Niels Bohr Archive P006; 5.13, photograph by Francis Simon, courtesy AIP Emilio Segre Visual Archives.

Chapter 6

Frontispiece, Elliot and Fry, courtesy AIP Emilio Segre Visual Archives; 6.10, AIP Emilio Segre Visual Archives, E. Scott Barr Collection.

Chapter 7

Frontispiece, Lawrence Berkeley National Laboratory/Photo Researchers, Inc.

Chapter 8

Frontispiece (*left*), with permission from Ronald Gillespie; frontispiece (*right*), © Godfrey Argent Studio; 8.3, with permission from Museum Boerhaave Leiden; 8.22b, courtesy of Argonne National Laboratory.

Chapter 9

Frontispiece, Tom Hollyman/Photo Researchers, Inc.; 9.12, Charles D. Winters/Photo Researchers, Inc.

Chapter 10

Frontispiece, with permission from DePauw Archives; 10.1, Andrew Lambert Photography/Photo Researchers, Inc.; 10.2, © Joel Gordon Photography; 10.3, © Joel Gordon Photography; 10.9, © 2002 Richard Megna, Fundamental Photographs, NYC; 10.10, © 1994 Richard Megna, Fundamental Photographs, NYC; 10.11, Cheryl Power/Photo Researchers, Inc.; 10.13, CC Studio/Photo Researchers, Inc.; 10.14, © 1997 Richard Megna, Fundamental Photographs, NYC; 10.15, © 2001 Jeff J. Daly, Fundamental Photographs, NYC; 10.16, © Joel Gordon Photography; 10.21, © 1995 Richard Megna, Fundamental Photographs, NYC.

Chapter 11

Frontispiece, with permission from Science Museum/SSPL; 11.4, Bibliotheque Nationale de France; 11.8, Dennis Harding, Chevron Corp.; 11.10, © 1992 Richard Megna, Fundamental Photographs, NYC; 11.11, courtesy of JPL/NASA.

Chapter 12

Frontispiece, with permission from Bancroft Library, University of California, Berkeley; page 396, © 1995 Richard Megna, Fundamental Photographs, NYC; 12.9, © 1998 Richard Megna, Fundamental Photographs, NYC; 12.10, © Joel Gordon Photography; 12.13, © 1998 Richard Megna, Fundamental Photographs, NYC.

Chapter 13

Frontispiece (*left*), SPL/Photo Researchers, Inc.; frontispiece (*right*), permission granted from University of Pennsylvania Libraries; 13.4, National Maritime Museum, Greenwich, England; 13.5 (*top*), permission granted from istockphoto, www.istockphoto.com; 13.5 (*second from top*), © 1994 Richard Megna, Fundamental Photographs, NYC; 13.5 (*second from bottom and bottom*), courtesy of Kurt J. Lesker Company; 13.6 (*left*), Science Source/Photo Researchers, Inc.; 13.6 (*right*), © Photo Deutsches Museum; 13.7, The Royal Society; 13.10 (*left*), Sheila Terry/Photo Researchers, Inc.; 13.10 (*right*), Library of Congress/SPL/Photo Researchers, Inc.; 13.15, Science Source/Photo Researchers, Inc.; 13.18, © 1997 Richard Megna, Fundamental Photographs, NYC; 13.19, Travis Amos; 13.25, Trinity College Library, Cambridge University, courtesy AIP Emilio Segre Visual Archives; 13.26, AIP Emilio Segre Visual Archives, Segre Collection; 13.27, permission granted from istockphoto, www.istockphoto.com; 13.33, SPL/Photo Researchers, Inc.

Chapter 14

Frontispiece, The Royal Society; 14.1, permission granted from istockphoto, www.istockphoto.com; 14.5, © Andrew Lambert Photography/Photo Researchers, Inc.; 14.7, © 1992 Richard Megna, Fundamental Photographs, NYC; 14.9, General Motors; 14.13, Charles D. Winters/Photo Researchers, Inc.; 14.14, Department of the Navy; 14.18a, Charles D. Winters/Photo Researchers, Inc.

Chapter 15

Frontispiece, AIP Emilio Segre Visual Archives, Physics Today Collection; 15.11, © 2002 Richard Megna, Fundamental Photographs, NYC; 15.12 (*top*), permission granted from istockphoto, www.istockphoto.com; 15.12 (*bottom*), courtesy of Metin Sitti, Carnegie Mellon University; 15.14, John Shaw/Tom Stack & Associates; 15.16 (*left*), © 2009 Jeff J. Daly, Fundamental Photographs, NYC; 15.16 (*right*), © 2005 Richard Megna, Fundamental Photographs, NYC; 15.17, © 1987 Richard Megna, Fundamental Photographs, NYC; 15.26, John Shaw/Tom Stack & Associates; page 546 (*upper left*), courtesy of Professor H. Hope, University of California, Davis; 15.34, © 1988 Paul Silverman, Fundamental Photographs, NYC; 15.41, Permission granted from Christina Bauer, Ph.D.; 15.44, © 1987 Richard Megna, Fundamental Photographs, NYC; 15.45, NASA/Photo Researchers, Inc.

Chapter 16

Frontispiece, AIP Emilio Segre Visual Archives, W. Cady Collection; 16.3, Wikimedia Commons; 16.5, John Lythgoe/Planet Earth Pictures; 16.7, permission granted from istockphoto, www.istockphoto.com; 16.10, Philip Saters, Planet Earth Pictures; 16.15, courtesy of Siemens Water Technologies Corp.; 16.16, Millipore; 16.17, Francis Leroy, Biocosmos/Photo Researchers, Inc.; 16.18, M. Sheetz, R. Painter, and S. Singer, *J. Cell Biol.*, 70: 193 (1976); page 592, Ted Kinsman/Photo Researchers, Inc.

Chapter 17

Frontispiece (*top left*), courtesy of University of California, Berkeley/LBNL; frontispiece (*right*), courtesy of Dudley Herschbach; frontispiece (*bottom left*), courtesy of AIP Meggers Gallery of Nobel Laureates; 17.8, Digital Photo Archive, Department of Energy (DOE), courtesy AIP Emilio Segre Visual Archives; 17.9, permission granted from istockphoto, www.istockphoto.com.

Chapter 18

Frontispiece (*left*) photo by Kaiden-Kazanjian, from Corner, George, *A History of the Rockefeller Institute, 1901–1953,* New York City: The Rockefeller Institute Press, 1964; frontispiece (*right*), University Archives, University of Pittsburgh; 18.6, courtesy of the Archives, California Institute of Technology; 18.8b, Science Museum, London.

Chapter 19

Frontispiece, permission granted from University of Pennsylvania Libraries; 19.5, SPL/Photo Researchers, Inc.; 19.10, © 1989 Paul Silverman, Fundamental Photographs, NYC; 19.13, © Joel Gordon Photography.

Chapter 20

Frontispiece, courtesy of Carlsberg Archives; 20.3, Metrohm/Wikimedia Commons; 20.7, courtesy of Randall Ishimaru.

Chapter 21

Frontispiece, courtesy of David Bruce Dill, Harvard Fatigue Laboratory Collections, MSS 517, Mandeville Special Collections Library, University of California, San Diego; 21.5, © 2003 Richard Megna, Fundamental Photographs, NYC; 21.6, © 1995 Richard Megna, Fundamental Photographs, NYC.

Chapter 22

Frontispiece, Austrian Academy of Sciences; page 829 Michael Szoenyi/Photo Researchers, Inc; 22.6, © 1990 Richard Megna, Fundamental Photographs, NYC; 22.9, © 1990, 1995 Richard Megna, Fundamental Photographs, NYC; 22.11, © 1995 Richard Megna, Fundamental Photographs, NYC; 22.14, California Water Service Company.

Chapter 23

Frontispiece, AIP Emilio Segre Visual Archives, Brittle Books Collection; 23.3, Pacific Gas and Electric; 23.5, © 2006 Richard Megna, Fundamental Photographs, NYC; 23.11, courtesy of W. Ketterle, MIT; 23.14, Charles D. Winters/Photo Researchers, Inc.; 23.15, reprinted with permission from Shirley Chiang, University of California, Davis/reprinted with permission from *Phys. Rev. Lett.* 60: 2398 (1988), image originally created by IBM Corporation; 23.16, © 1995 Paul Silverman, Fundamental Photographs, NYC.

Chapter 24

Frontispiece, IP Emilio Segre Visual Archives; 24.1, courtesy of Ethan Gallogly; 24.8 (*top*), © 1988 Kip Peticolas, Fundamental Photographs, NYC; 24.8 (*bottom*), NASA/Science Source/Photo Researchers, Inc.; 24.9, Rodney Cotterill and Flemming Kragh.

Chapter 25

Frontispiece, Royal Astronomical Society/Photo Researchers, Inc.; 25.1, SPL/Photo Researchers, Inc.; 25.2, Charles D. Winters/Photo Researchers, Inc.; 25.3, Science Museum/SSPL; 25.4, Jean-Loup Charmet/Photo Researchers, Inc.; 25.5, © 1987 Richard Megna, Fundamental Photographs, NYC; 25.9, Travis Amos; 25.10, with permission from Museum Boerhaave Leiden; 25.12, image used with permission from UTC Power Corporation; 25.13, courtesy of Ballard Power Systems; 25.17, Vulcan Materials Company; 25.18a, E.I. Du Pont de Nemours & Co.; 25.19, Manley Prim Photography, Inc.; 25.20, AIP Emilio Segre Visual Archives, Brittle Books Collection; 25.21, SPL/Photo Researchers, Inc.; 25.22, Alcoa.

Chapter 26

Frontispiece, SPL/Photo Researchers, Inc.; 26.3, Martin Marietta; 26.8, © 2003 Richard Megna, Fundamental Photographs, NYC; 26.11, Bethlehem Steel; 26.12, © Joel Gordon Photography; 26.13, David R. Frazier Photolibrary, Inc./Photo Researchers, Inc.; 26.14, permission granted from istockphoto, www.istockphoto.com; 26.15, © 1994 Paul Silverman, Fundamental Photographs, NYC; 26.18, permission granted from istockphoto, www.istockphoto.com.

Appendix B

B.1, © BIPM, reproduced with permission from BIPM Library.

INDEX

T

Physical Constants

Constant	Symbol	Value
atomic mass unit	u (formerly amu)	$1.660\,538\,782 \times 10^{-27}$ kg
Avogadro's number	N_A	$6.022\,141\,79 \times 10^{23}$ mol^{-1}
Bohr radius	a_0	$5.291\,7721 \times 10^{-11}$ m
Boltzmann constant	k	$1.380\,6504 \times 10^{-23}$ J·K^{-1}
charge of a proton	e	$1.602\,1765 \times 10^{-19}$ C
Faraday constant	F	96 485.34 C·mol^{-1}
gas constant	R	8.314 472 J·K^{-1}·mol^{-1} 0.083 1447 L·bar·mol^{-1}·K^{-1} 0.082 0575 L·atm·mol^{-1}·K^{-1}
mass of an electron	m_e	$9.109\,382\,15 \times 10^{-31}$ kg $5.485\,779\,094 \times 10^{-4}$ u
mass of a neutron	m_n	$1.674\,9272 \times 10^{-27}$ kg 1.008 664 916 u
mass of a proton	m_p	$1.672\,621\,637 \times 10^{-27}$ kg 1.007 276 467 u
Planck's constant	h	$6.626\,069 \times 10^{-34}$ J·s
speed of light	c	$2.997\,924\,58 \times 10^{8}$ m·s^{-1}

SI Prefixes

Prefix	Symbol	Multiple	Prefix	Symbol	Multiple
peta-	P	10^{15}	deci-	d	10^{-1}
tera-	T	10^{12}	centi-	c	10^{-2}
giga-	G	10^{9}	milli-	m	10^{-3}
mega-	M	10^{6}	micro-	μ	10^{-6}
kilo-	k	10^{3}	nano-	n	10^{-9}
hecto-	h	10^{2}	pico-	p	10^{-12}
deca-	da	10^{1}	femto-	f	10^{-15}
			atto-	a	10^{-18}